D1146342

WITHDRAWN

PHILIP'S

SCIENCE & TECHNOLOGY ENCYCLOPEDIA

PHILIP'S

SCIENCE & TECHNOLOGY ENCYCLOPEDIA

Philip's Science and Technology Encyclopedia

First published in Great Britain in 1998
by George Philip Limited, Michelin House,
81 Fulham Road, London SW3 6RB
a division of Octopus Publishing Group Ltd

EDITOR
Steve Luck

TEXT EDITORS
Chris Humphries
Frances Adlington

ART EDITOR
Mike Brown

PICTURE RESEARCH
Sarah Moule

PRODUCTION
Gudrun Hughes

Reproduction by Colourpath Ltd, London

Printed and bound in Spain by Cayfosa

A CIP catalogue record for this book is available from the British Library.

ISBN 0-540-07574-4

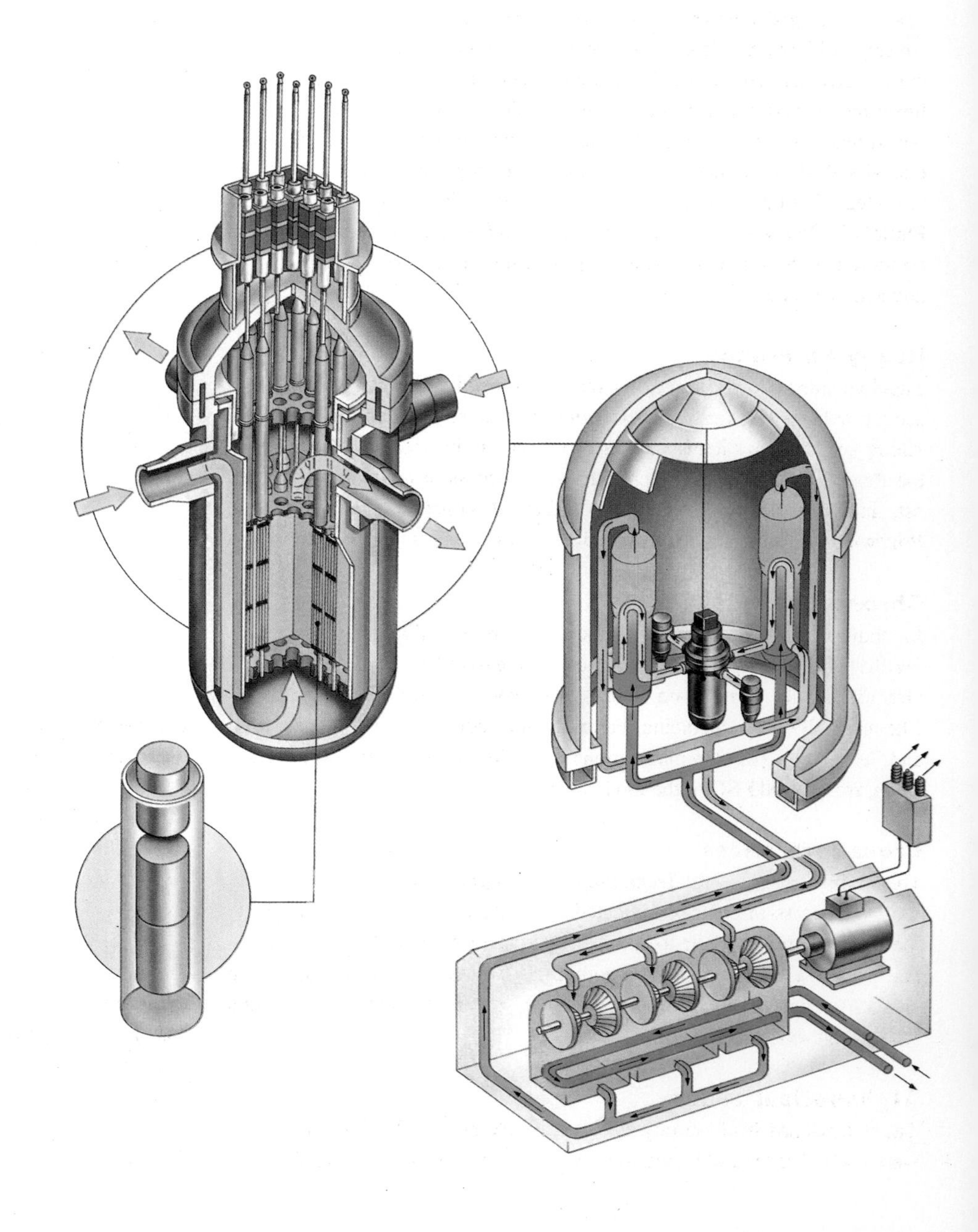

PREFACE

The Philip's *Science and Technology Encyclopedia* has been created for both students of science and for those who are simply enthusiasts – as a study aid for secondary school and college students, a first stop for general enquiries, and a fascinating mine of knowledge for browsers. The 6,500 alphabetically organized entries provide clear, essential information on a vast variety of scientific and technological subjects, from earth science to life science, and from astronomy to particle physics. The extraordinary wealth of colour illustration, including photographs, technological "cutaways" and artworks conveys visual information far beyond the descriptive power of many thousands of words.

When choosing a single-volume encyclopedia, the most important consideration for the user is the criteria by which the articles have been selected. Philip's *Science and Technology Encyclopedia* has been created with secondary school and college students particularly in mind. Key areas of science and technology, including maths, physics, chemistry, biology and information technology have been given the greatest attention. The articles, compatible with and complementary to what is learned in the classroom, are up-to-date and clearly written.

An encyclopedia, however, must contain more than just a comprehensive coverage of key areas if it is capture and maintain an interest. Equally full in their treatment are the articles that cover cutting-edge technology, such as linear motors and virtual reality, many of which are illustrated and placed in boxed features to help clarify the accompanying article, making this encyclopedia ideal for home reference as well as an important resource at school or college. For this reason also, unlike most encyclopedias of this type, the Philip's *Science and Technology Encyclopedia* features biographical articles on more than 850 famous scientists, providing information on their careers and achievements.

Ready Reference

Supplementing the main body of the Encyclopedia, the Ready Reference section collates the kind of information best presented in tables and lists. Along with standard items such as conversion units and a list of elements, the Ready Reference section provides rapid access to many areas of interest, including taxonomic groups, a list of constellations and Nobel Prizewinners. The contents are listed in full on the first page of the section.

Chronology

In addition to the Ready Reference section, the Philip's *Science and Technology Encyclopedia* also features an extensive Chronology of Science. Organized into nine main areas – Astronomy and Space, Physics, Chemistry, Biology, Medicine, Farming and Food, Transport, Engineering and Technology and Communications – the Chronology spans thousands of years, from 10,000 BC to the modern-day.

Cross-references

The Philip's *Science and Technology Encyclopedia* has more than 20,000 individual cross-references, indicated by SMALL CAPITAL letters, which take the reader from one article to other articles that provide useful, related information. For example, contained within the "satellite television" article are cross-references to TELEVISION, SATELLITES, GEOSTATIONARY ORBIT, TRANSMITTER and ANTENNA.

Alphabetical order

The order of articles is strictly alphabetical, except that Mc is treated as if it were spelt Mac, and abbreviations as if spelt out in full. Articles that have more than one word in the heading, such as "saturated solution", are ordered as if there were no space between the words.

Nomenclature

In keeping with the methods taught in schools and colleges, modern scientific names have been used. For example information on "acetaldehyde" will be found under "ethanal". Where less well-known modern names are used, cross-references from the old names will take the reader to the new headings.

Metric units have been used throughout with corresponding imperial measurements following in brackets.

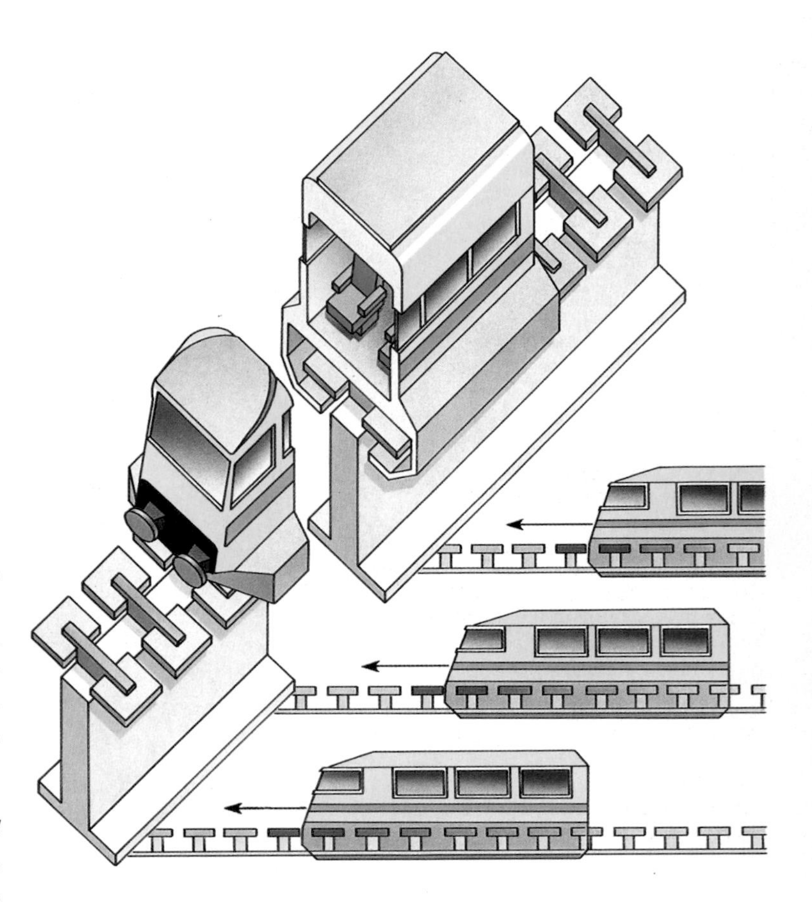

Correspondence

In compiling the first edition of the Philip's *Science and Technology Encyclopedia*, Philip's editorial staff, along with all the contributors and consultants, have set rigorous standards to ensure that the information provided is accurate and well informed. However, in our continual striving for excellence and the maximum usefulness to our readers, we welcome all suggestions and comments that will help to improve and develop future editions.

Please address correspondence to:

The Editor,
Philip's Science and Technology Encyclopedia,
George Philip Limited,
Michelin House,
81 Fulham Road,
London
SW3 6RB

CONSULTANTS

Anatomy/ Dentistry
Prof. A.E. Walsby *University of Bristol*

Animals and Agriculture
Prof. K. Simkiss *University of Reading*

Aviation
Dr I. Hall *Victoria University of Manchester*

Biology
Prof. T. Halliday *Open University*

Botany
Prof. A.E. Walsby *University of Bristol*

Chemistry
Dr A.S. Bailey *University of Oxford*

Earth Sciences/ Atmospheric Sciences/ Geology
Prof. Michael Tooley *University of St Andrews*

Mathematics
Dr Jan Brandts *University of Bristol*
Dr Kevin Thompson *University of Durham*

Medicine
Prof. S.L. Lightman *University of Bristol*

Philosophy/ Computing
Prof. Ron Chrisley *University of Sussex*

Physics
Prof. John Gribbin *University of Sussex*

Technology and Industrial Processes/ Engineering/ Material Science
Dr P.L. Domone *University College London*

EDITORIAL CREDITS

Peter Astley
Jill Bailey
John Bailie
Richard Brzezinski
Ian Chilvers
Roy Carr
John O.E. Clark
Sean Connolly
Chris Cooper
Mike Darton
Stephanie Driver
Roger Few
Keith Lye
Eddie Mizzi
Paulette Pratt
A.T.H. Rowland-Entwistle
Tom Ruppel
Clint Twist
Keith Wicks
Richard Widdows
John D. Wright

Curriculum consultants
Duncan Hawley
David J. McHugh
Silvia Newton
Brian Speed
Jane Wheatley

PICTURE CREDITS

NASA 315.
Oxford Scientific Films/E Bartov 286, /S Dalton 290, /J Frazier 313, /H Taylor 13.
Planet Earth Pictures/L Collier 316, /R Coomber 57, /P David 39, /T Dressler 206, /R Hessler 2, /F Schulke 93.
Rex Features Ltd. 75, 174, 340 /**Sipa** 22, 94, 363, /**Today** 208.
Science Photo Library 28, 44, 51 top, 119, 120,137 top, 200, 215, 218, 231, 253, 257, 261, 352,/Aeroservice 12,/A Bartel 243, /M Bond 388,/B Brake 146,/Dr J Burgess 294,/T Buxton 33,/T Craddock 348 below,/Cern, P. Loiez 51 below,/CNRI 302, /Custom Medical Stock Photo 270,/Deep Light Productions 128,/Eye of Science 29, /R Folwell 112,/Prof P Fowler 311,/R de Gugliemo 296,/J Heseltine 147,/M Land 260,/Los Alamos National Laboratory 137 below,/T Malyon & P Biddle 5,/P Menzel 159,/NASA 168,182, 334, /NASA/Space Telescope Science Institute 282,/NOAA 266, /D Nunuk 166, 235,/D Parker 98, 280, /P Plailly 179,/C Pouedras 256, /Dr M Read 300, /Restec, Japan 207,/Royal Observatory, Edinburgh 348 top,/G Sams 151, /Sandia National Laboratories 249, /Science Source 323,/G Tompkinson 269, /US Geological Survey 229, /US Library of Congress 389.
Technology Illustrations Ted McCausland

aa Type of LAVA with a block-like structure. *See also* BLOCKY LAVA

abacus Counting device used in the Middle and Far East for addition and subtraction. The most common form of abacus consists of beads strung on wires and arranged in columns which represent units, tens, hundreds and so on. Various types have been used for thousands of years.

abdomen In VERTEBRATES, that portion of the body between the chest and the pelvis containing the abdominal cavity and the abdominal viscera, including most of the digestive organs. In ARTHROPODS it is the posterior part of the body, containing the reproductive organs and part of the digestive system.

Abel, John Jacob (1857–1938) US biochemist, best known for the first identification (1898) of a hormone, ADRENALINE (epinephrine). Abel was professor of pharmacology (1893–1932) at John Hopkins. He made his key discovery after many years of studying the chemical composition of body tissues. Later, Abel isolated amino acids from blood by the process of DIALYSIS, and discovered insulin in crystalline form.

aberration In astronomy, the apparent slight change of position of a star due to the effect of the Earth's orbital motion and the finite velocity of light. A telescope must be inclined by an angle of up to about 20° to compensate for it. Aberration of light was discovered in 1729 and used to prove that the Earth orbits the Sun.

aberration In physics, a defect in lens or mirror images. **Spherical** aberration occurs when rays falling on the periphery of a lens or curved mirror are not brought to the same focus as light at the centre; the image is blurred. **Chromatic** aberration occurs because the varying wavelengths of coloured light are not brought to the same focus; the image is falsely coloured.

ABERRATION (ASTRONOMY)

(A) A ball is falling vertically and the moving tube must be tilted in order to catch it cleanly. This phenomenon is called aberration. Similarly, light falling from a star (S_1) is aberrated so that the telescope (T), which is mounted on a moving Earth, must be tilted like the tube. (C) As the motion of the Earth is circular, the light falling from a star (S_2) lying in the plane (e) is aberrated through a maximum of 20" of arc (1) when the motion is perpendicular to the light-path; and then through 0" (2) when it is parallel to it 3 months later; (3) is a minimum of −20" and (4) a further 0". In this way the star appears to wobble in the sky during the course of 1 year.

abiotic factors Various non-living factors – favourable or harmful – that contribute to the environment of living things. They include, for example, the atmosphere, climate, geology, amount of light and temperature of the ecosystem. *See also* ECOSYSTEM; ENVIRONMENT

ablation In geology, a measure of glacial loss through melting, evaporation, wind erosion, or calving (formation of icebergs). Sometimes used in geomorphology to mean the loss of surface soil or rock by wind or water action. *See also* ACCUMULATION; EROSION

abortion Loss of a fetus before it is sufficiently advanced to survive outside the UTERUS; commonly called a miscarriage. A medical abortion is the termination of pregnancy by drugs or surgery. A legal abortion is one sanctioned by law. In Britain, legal abortion is only permissible until the 24th week of pregnancy, although a late termination can be carried out if the mother's life would be endangered by allowing the pregnancy to continue.

abrasion In geology, mechanical wearing down of rock surface by the dragging of rock particles across it. Common agents of abrasion are the bed load of streams, rock debris at the base of glaciers, and sand transported by wind or waves. *See also* EROSION

abrasive Hard and rough substance used to grind and polish surfaces. Some abrasives are used as fine powders, others in larger fragments with sharp cutting edges. Most natural abrasives are minerals, such as diamond, garnet, emery, corundum, pumice, flint and quartz. Synthetic abrasives include silicon carbide, aluminium oxide, synthetic diamond and boron carbide.

abscissa In mathematics, the distance of a point from the *y*-axis in a CARTESIAN COORDINATE SYSTEM. That is the *x*-coordinate of the pair (*x*,*y*), which locates a point in the plane.

abscission Deliberate shedding of a part of a plant. It commonly occurs with fruit and leaves, and is controlled by abscisic acid and other growth substances. An abscission layer of cells forms in an abscission zone at the base of the part. As these cells break down, the part falls off the plant, usually dislodged by the wind or rain.

absolute alcohol Pure ETHANOL (ethyl alcohol). It is made from distilled ethanol (a constant-boiling mixture containing 95.6% ethanol) by the addition of sodium metal or another drying agent.

absolute zero Temperature at which all parts of a

ABERRATION (PHYSICS)

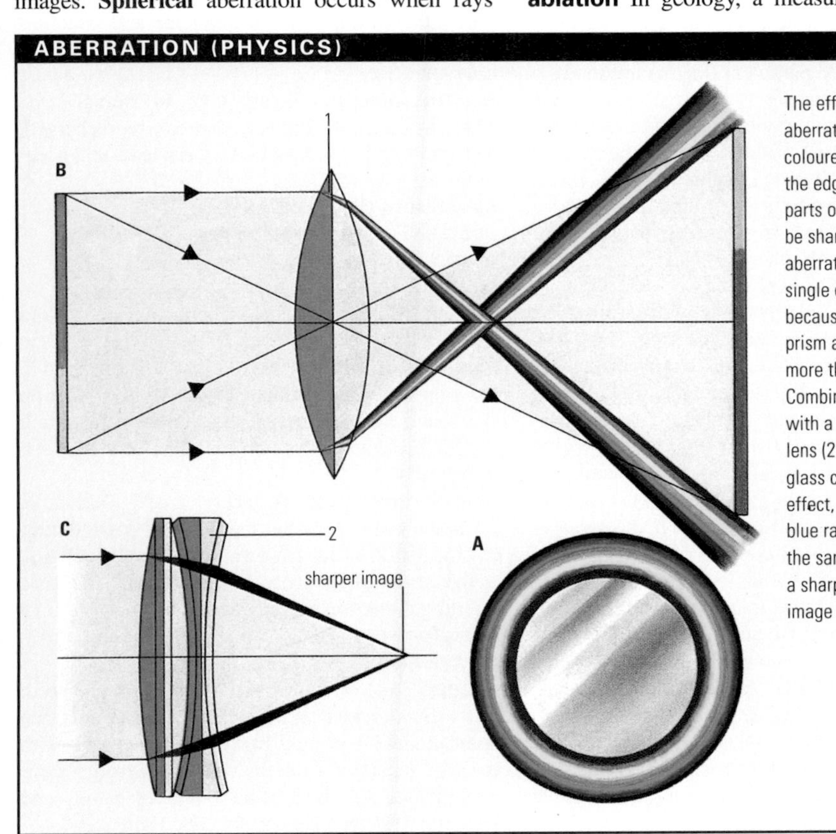

The effects of chromatic aberration result in coloured fringes around the edges of the lens, and parts of the image may not be sharp (A). This aberration occurs with a single convex lens (1) because it behaves like a prism and bends blue light more than red light (B). Combining the convex lens with a weaker concave lens (2) made of a different glass cancels out this effect, and both red and blue rays are brought into the same focus to produce a sharper, more distinct image (C).

► **abyssal zone** The deepest parts of the ocean floor are referred to as the abyssal zone. Shown here is the Mariana deep-sea trench, near Guam in the Pacific Ocean. It is the site of the Challenger Deep, which, at 11,033m (36,198ft), is the greatest known depth in the Pacific. Challenger Deep is deeper than Mount Everest is high.

system are at the lowest energy permitted by the laws of QUANTUM MECHANICS; zero on the Kelvin temperature scale, which is −273.15°C (−459.67°F). At this temperature the system's ENTROPY, its energy available for useful work, is also zero, although the total energy of the system may not be zero.

absorbent Substance with the power to absorb large quantities of other substances. Absorbents are usually porous materials: examples are activated charcoal and zeolites. They have many uses, including separating, purifying, decolorizing and deodorizing.

absorption Taking up chemically or physically of one substance into another. The absorbed matter permeates all of the absorber. This includes a gas taken in by a liquid, such as carbon dioxide by sodium hydroxide, and a liquid or gas absorbed by a solid, such as water by a GEL. The process is often utilized commercially, an example being the purification of natural gas by the absorption of hydrogen sulphide in aqueous ethanolamine. Absorption occurs in all scientific disciplines; in nuclear physics, neutrons produced by fission are absorbed by elements such as boron. *See also* ADSORPTION

absorption spectrum *See* SPECTRUM

abyssal Term to describe oceanic features occurring at great depths, usually more than *c*.3,000m (10,000ft) below sea level. Abyssal **plains** cover *c*.30% of the Atlantic and nearly 75% of the Pacific ocean floors. They are covered by deposits of biogenic oozes formed by the remains of microscopic plankton, and nonbiogenic sediments (red clays). The gradient is less than 1:1,000, except for the occasional low, oval-shaped abyssal **hills**. The plains are characterized by stable temperatures from −1°C to 5°C (30–41°F) and the relative absence of water currents. The abyssal **zone** is the deepest area of the ocean. It receives no sunlight, so there are no seasons and no plants, but there are many forms of life, such as glass sponges, crinoids (sea lilies), and brachiopods (lamp shells).

Acanthocephala Phylum of spiny-headed parasitic worms once thought to be NEMATODES. They are identified by a retractable spiny proboscis and an elongated, cylindrical body. The young are parasitic in arthropods and the adults are parasitic in vertebrates, attaching themselves to the intestinal lining. Average length: 1–2cm (0.4–0.8in). *Echinorhyncus* is the main genus.

acceleration Amount by which the VELOCITY of an object increases in a certain time. For example, a stone dropped over a cliff accelerates from zero velocity at a rate of 9.81m (32.2ft) per second per second, this acceleration being due to the pull of the Earth's gravity. The rate of acceleration can be found by applying the equation acceleration = (change in velocity)/(time taken for change). For example, if an object increases in velocity from 5 metres per second (ms^{-1} or m/s) to $7ms^{-1}$ in one second, its velocity has increased by $2ms^{-1}$ in one second; the acceleration is thus 2 metres per second per second, written as $2ms^{-2}$ or $2m/s^2$. When an object rotates about an axis, its angular acceleration is given as the change in angular velocity divided by elapsed time, expressed in radians per second per second (written rad s^{-2} or rad/s^2). A RADIAN is a measure of angle: 2π radians = 360°.

acceleration of free fall (acceleration due to gravity, symbol g) Acceleration experienced by bodies falling freely in the Earth's gravitational field. It varies from place to place around the globe, but is assigned a standard value of $9.80665ms^{-2}$ ($32.1740fts^{-2}$). Ignoring air resistance, the acceleration does not vary with the size or shape of the falling body. Pilots can lose consciousness at vertical accelerations in excess of $3g$.

accelerator (particle accelerator) In PARTICLE PHYSICS, machine for increasing the energy of charged particles by increasing their speed by use of alternating electric fields in an evacuated chamber. The particles must enter the high-frequency field as it begins increasing (negative particles) or decreasing (positive particles) for maximum energy increase. Magnetic fields are used to focus the particles into a narrow stable beam and to maintain the required curvature of the beam. As the particle velocity rises a relativistic increase in mass occurs. Accelerators are used mostly in particle physics experiments, in which high-energy particles are forced to collide with other particles. The way the fragments of particles produced behave following the collision provides physicists with information on the forces found within atoms. In a LINEAR accelerator, the particles travel in a straight line, usually accelerated by an electric field. In a CYCLOTRON, particles are accelerated in a spiral path between pairs of D-shaped magnets with an alternating voltage between them. In a SYNCHROCYCLOTRON, the accelerating voltage is synchronized with the time it takes the particles to make one revolution. A SYNCHROTRON consists of an (often very large) circular tube with magnets to deflect the particles in a curve and radio-frequency fields to accelerate them. The most advanced modern accelerators are **colliders**, in which beams of particles moving in opposite directions are allowed to collide with each other, thus achieving higher energy of interaction. *See also* BUBBLE CHAMBER

accelerator, electrostatic Early type of particle ACCELERATOR in which a constant high voltage is applied between a pair of electrodes in an evacuated tube. A large static charge builds up, producing an electric field along the tube. Charged particles are accelerated in the electric field between the electrodes, gaining an energy eV, where e is the electron charge and V the applied voltage. The VAN DE GRAAFF GENERATOR (developed 1931) and the COCKCROFT-WALTON GENERATOR (1932) can be used as high-voltage sources.

accelerometer Device used to measure acceleration. A simple example is a plumb bob suspended from the accelerating object, whose angle with the vertical is proportional to the acceleration. A more sophisticated version, used in ballistic missiles, is an electromechanical device that translates the acceleration into electric current.

acclimatization Temporary adjustment of an organism to a new environment, climate or circumstance. It involves a gradual, natural change in physiology allowing an organism to exploit new regions, but unlike EVOLUTION, does not involve any genetic change.

accommodation Process by which the EYE focuses on objects at various distances. In the human eye, focusing is achieved when the muscles of the CILIARY BODY contract or relax to change the shape (curvature) of the lens and bring light rays into focus on the light-sensitive RETINA found at the back of the eye. Presbyopia, an eye condition that occurs after middle age in which people have difficulty in focusing on distant or very near objects, is caused by impaired accommodation.

accretion Continually growing or building up; a term frequently used to describe certain modes of geological deposition. The term is also used in astronomy to describe the gradual building up of larger celestial bodies from smaller ones by gravitational attraction; also the accumulation of matter by a star or other celestial object. Accretion is an important factor in the evolution of stars, planets and comets.

accumulation In geology, the addition of snow and ice to a GLACIER. It usually occurs near the head of the glacier where the air temperature is well below freezing. Snow from avalanches and precipitation falls onto the glacier and becomes compressed under its own weight, forming ice. *See also* ABLATION

accumulator (secondary CELL or storage BATTERY) Voltaic cell (battery) that can be recharged. The commonly used car battery is a lead-acid accumulator.

acetaldehyde *See* ETHANAL

acetate Former name for ETHANOATE

acetic acid *See* ETHANOIC ACID

acetone Former name for PROPANONE

acetylcholine (ACh) Chemical compound released by certain nerve cells that serves as a transmitter in the NERVOUS SYSTEM. It is involved in the transmission of impulses across the junction (SYNAPSE) between nerve and muscle cells which trigger the contraction of muscles, and between nerve cells.

acetyl coenzyme A (acetyl CoA) Derivative formed when PYRUVIC ACID is broken down during the formation of fat from carbohydrates. CoA acts as the carrier of the acetyl group, which is the basic building block for the fatty acids.

acetylene *See* ETHYNE

acetylsalicylic acid *See* ASPIRIN

achene Type of fruit that contains only one seed. It is a dry, indehiscent fruit (that is, it does not open spontaneously) formed from a single carpel, with the seed separated from the wall of the fruit. Examples include members of the buttercup family and clematis. Winged achenes are called samaras.

Achilles tendon Strong band of elastic connective tissue at the back of the ankle. One of the largest tendons in the body, it connects the calf muscles to the heel bone. The spring provided by this tendon is very important in walking, running and jumping.
achromatic lens Lens designed to reduce or eliminate chromatic ABERRATION. Chromatic aberration occurs in a simple lens because the different component wavelengths (colours) of white light are brought to a different focus, resulting in a distorted image surrounded by coloured fringes. An achromatic lens brings all colours approximately to the same focus. It is usually constructed by combining two lenses made from different kinds of glass (and therefore having different refractive indexes), so that the errors introduced by one lens are cancelled out by the other. The addition of a third lens forms an **apochromatic** lens.
acid Chemical compound containing hydrogen that can be replaced by a metal or other positive ION to form a SALT. Acids dissociate in water to yield aqueous hydrogen ions (H^+), thus acting as proton donors; the solutions are corrosive, have a sour taste, and give a red colour to indicators, and have a pH below 7. **Strong** acids, such as sulphuric acid, are fully dissociated into ions whereas **weak** acids, such as ethanoic acid, are only partly dissociated. *See also* BASE
acid number (acid value) Number of milligrams of potassium hydroxide necessary to neutralize the free fatty acids in one gram of a specified substance. The acid number is used to measure the fatty acid content of fats, oils, etc.
acid rain Rain that is highly acidic because of sulphur oxides, nitrogen oxides, and other air pollutants dissolved in it. Normal rain is slightly acidic, with a pH of 6. (The pH scale ranges from 1 for extremely acidic to 14 for extremely basic or alkaline, with 7 being neutral.) Acid rain may have a pH value as low as 2.8. Acid rain can severely damage both plant and animal life. Certain lakes, for example, have lost all fish and plant life because of acid rain. *See diagram on page 4*
acid rock Any IGNEOUS ROCK that contains a high proportion of SILICA (more than 66%). The main silica mineral is QUARTZ, but some FELDSPARS are also quite rich in silica. Acid igneous rocks are usually light in colour. The most common is GRANITE. When they are weathered to form soil, they are generally infertile.
acid salt Salt of an acid that has two or more replaceable hydrogen atoms in which not all of the hydrogens are replaced by a metal or its equivalent. For example, the dibasic acid carbonic acid (H_2CO_3) forms acid salts such as sodium hydrogencarbonate (sodium bicarbonate, $NaHCO_3$). The tribasic acid phosphoric(V) acid (orthophosphoric acid, H_3PO_4) forms two series of acid salts, such as sodium dihydrogen phosphate (NaH_2PO_4) and disodium hydrogen phosphate (Na_2HPO_4).
acoustics Study of sound, especially the behaviour of sound waves. Architects apply acoustics in the design of public rooms such as concert and lecture halls to acquire a clearer sound. Engineers use it in their designs of sound detectors such as microphones and of sound producers such as loudspeakers and devices using ULTRASONICS. Audiologists use acoustics to assess degrees of abnormality in

ACCELERATOR

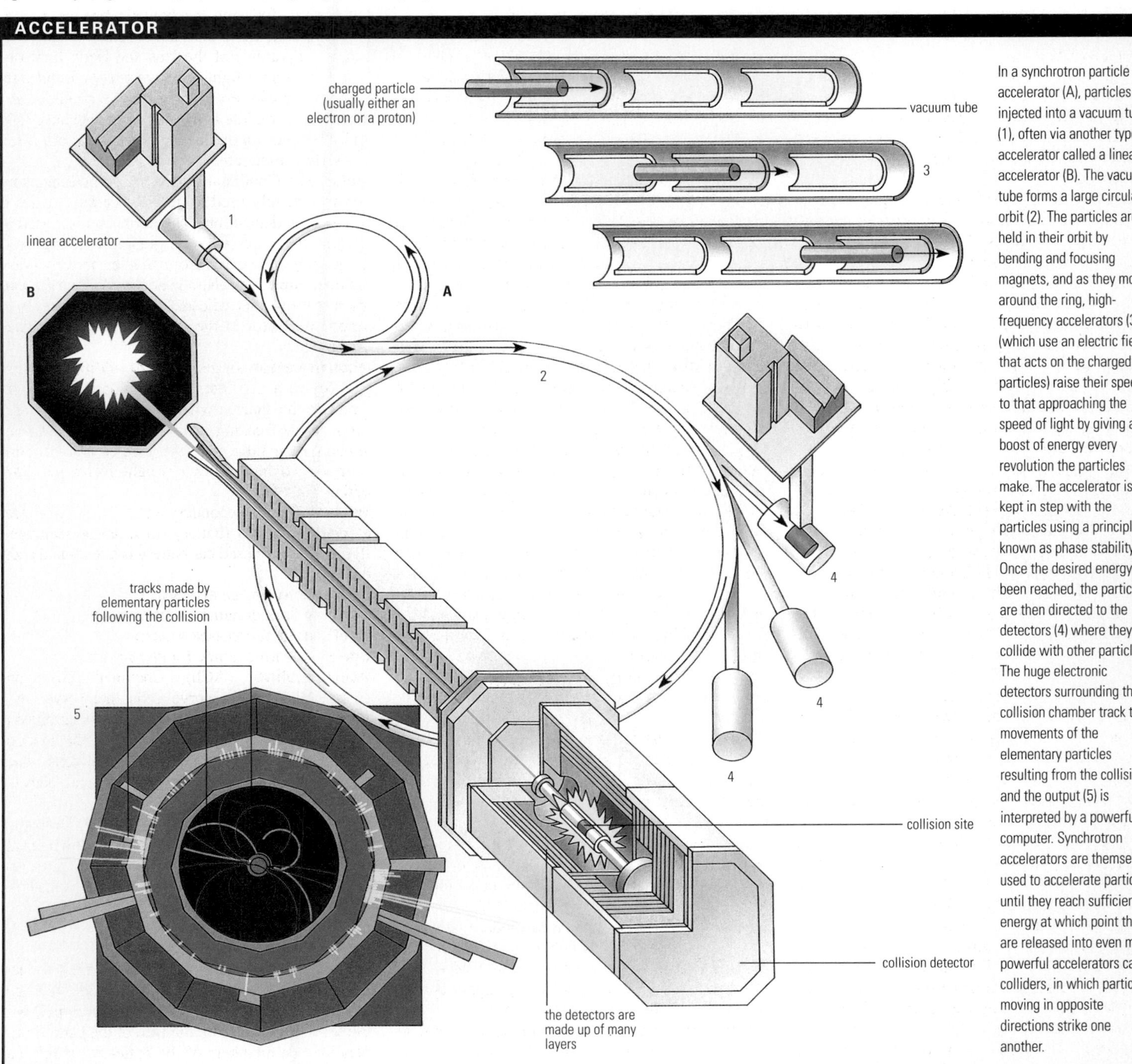

In a synchrotron particle accelerator (A), particles are injected into a vacuum tube (1), often via another type of accelerator called a linear accelerator (B). The vacuum tube forms a large circular orbit (2). The particles are held in their orbit by bending and focusing magnets, and as they move around the ring, high-frequency accelerators (3) (which use an electric field that acts on the charged particles) raise their speed to that approaching the speed of light by giving a boost of energy every revolution the particles make. The accelerator is kept in step with the particles using a principle known as phase stability. Once the desired energy has been reached, the particles are then directed to the detectors (4) where they collide with other particles. The huge electronic detectors surrounding the collision chamber track the movements of the elementary particles resulting from the collision and the output (5) is interpreted by a powerful computer. Synchrotron accelerators are themselves used to accelerate particles until they reach sufficient energy at which point they are released into even more powerful accelerators called colliders, in which particles moving in opposite directions strike one another.

A

the hearing of their patients. Craftsmen of musical instruments use it to improve their designs. *See also* ANECHOIC CHAMBER

acoustic shielding Deadening of unwanted sounds, usually by introducing sound-absorbing materials in the path of the sound waves. Sound enters a room in three ways: through the air, through the structure, or by the diaphragm action of floors, walls and ceilings. Airborne sounds are absorbed by using thick and absorbent walls in which there are no cracks or ducts; doors and windows should fit tightly and be seated in rubber or felt liners. Windows should be double-glazed. Structure-born vibrations and diaphragm effects are reduced by using double walls with insulating material but as few ties as possible between them.

acquired characteristic Feature that develops during the lifetime of an organism. The enlarged muscles of a manual worker are an example of an acquired characteristic. Because they are not genetically controlled, acquired characteristics cannot be passed on to offspring – that is, they cannot be inherited (although the French naturalist Jean Baptiste LAMARCK incorrectly thought that they could).

acquired immune deficiency syndrome (AIDS) Fatal disease caused by a retrovirus, called Human Immunodeficiency Virus (HIV), that main-

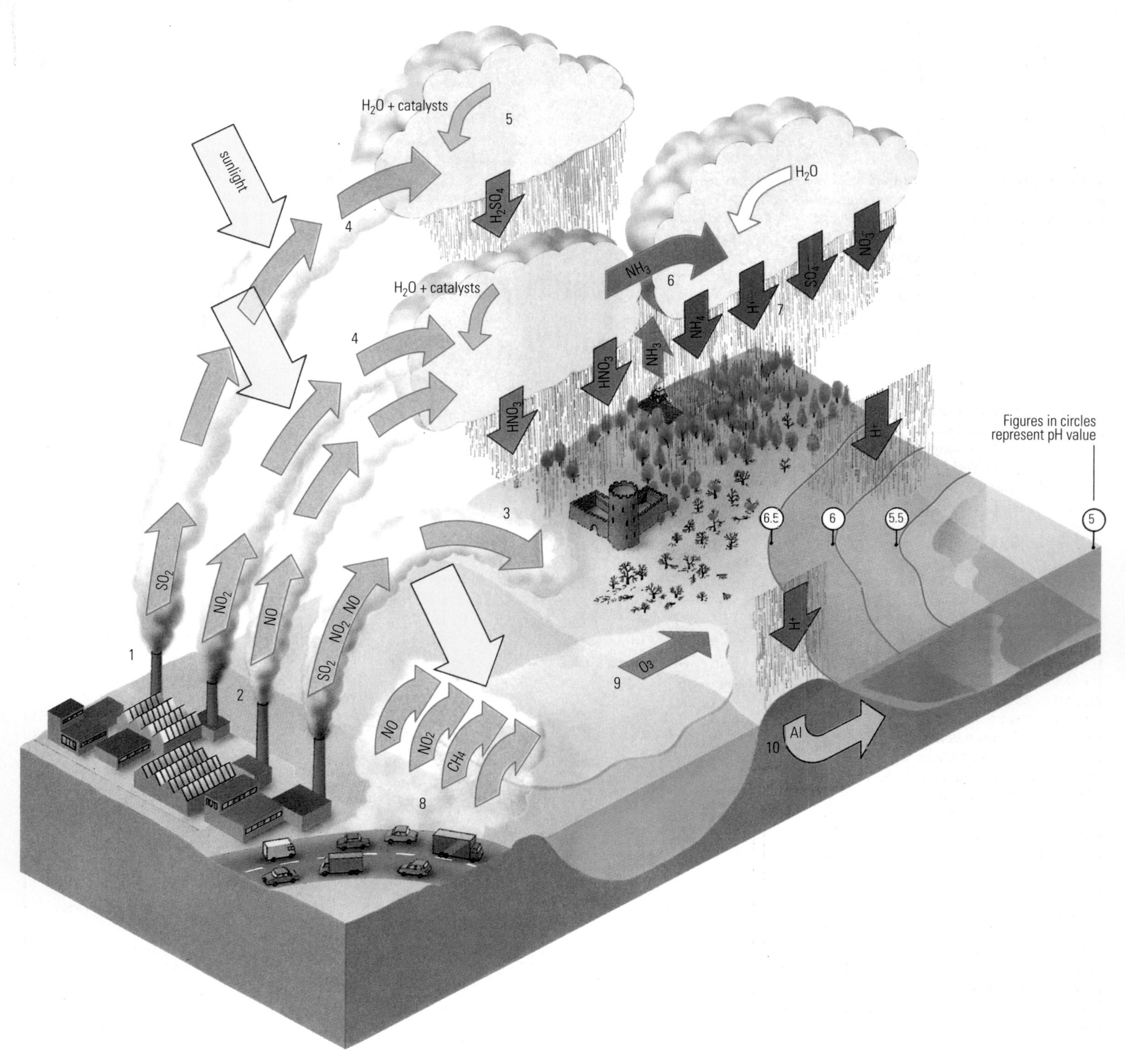

▲ **acid rain** Most sulphur (1) leaves factory chimneys as the gaseous sulphur dioxide (SO_2) and most nitrogen (2) is also emitted as one of the nitrogen oxides (NO or NO_2), both of which are gases. The gases may be dry deposited – absorbed directly by the land, by lakes or by the surface vegetation (3). If they are in the atmosphere for any time, the gases will oxidize (gain an oxygen atom) and go into solution as acids (4). Sulphuric acid (H_2SO_4) and the nitrogen oxides will become nitric acid (HNO_3). The acids usually dissolve in cloud droplets and may travel great distances before being precipitated as acid rain. Catalysts such as hydrogen peroxide, ozone and ammonium help promote the formation of acids in clouds (5). More ammonium (NH_4) can be formed when some of the acids are partially neutralized (6) by airborne ammonia (NH_3). Acidification increases with the number of active hydrogen (H^+) ions dissolved in an acid (7). Hydrocarbons emitted by, for example, car exhausts (8) will react in sunlight with nitrogen oxides to produce ozone (9). Although it is invaluable in the atmosphere, low-level ozone causes respiratory problems and also hastens the formation of acid rain. When acid rain falls on the ground it dissolves and liberates heavy metals and aluminium (Al) (10). When it is washed into lakes, aluminium irritates the outer surfaces of many fish. As acid rain falls or drains into the lake the pH of the lake falls. Forests suffer the effects of acid rain through damage to leaves, through the loss of vital nutrients, and through the increased amounts of toxic metals liberated by acids, which damage roots and soil microorganisms.

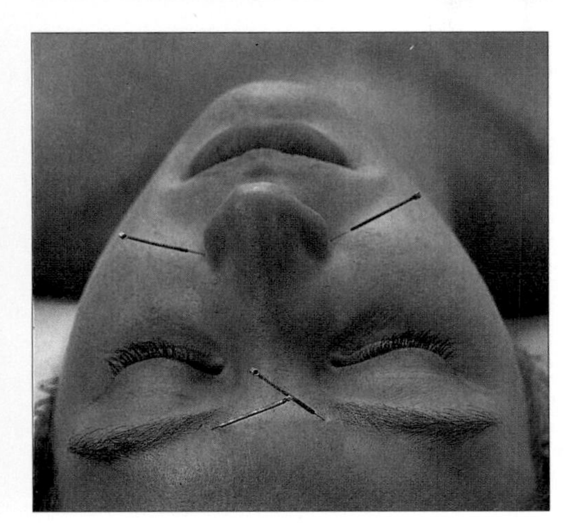

▲ **acupuncture** Patient undergoing acupuncture treatment for hay fever. Needles are inserted into the skin of the face at points on the bladder and large intestine meridians. In traditional Chinese medicine, illness is considered to be due to an imbalance of "vital energy" or Ch'i, which flows through the body in twelve pathways or meridians, each meridian corresponding to one of the vital organs. The acupuncturist corrects the imbalance by inserting needles at specific points on the meridians and rotating the free ends to stimulate the energy flow.

ly attacks T-4 cells (which help the production of ANTIBODIES) and renders the body's IMMUNE SYSTEM incapable of resisting infection. The first diagnosis was in New York in 1979. In 1983–84 scientists at the Pasteur Institute in France and the National Cancer Institute in the USA isolated HIV as the cause of the disease. The virus can remain dormant in infected T-cells for up to 10 years. Early symptoms include severe weight loss and fatigue. The condition progresses to full-blown AIDS, characterized by secondary infections, neurological damage and cancers (such as Kaposi's sarcoma and B-cell lymphomas). AIDS is transmitted only by a direct exchange of body fluids. Methods of transmission include: sexual intercourse, sharing of contaminated needles by intravenous drug users, and infected mothers to their babies in the UTERUS. Before effective screening procedures were introduced many haemophiliacs were infected through transfusions of contaminated blood. Development of a vaccine has been hampered by the virus' constantly shifting genetic composition. Research has centred on treatment, and on delaying the onset of the full-blown syndrome. In 1986–87 the first chemical was produced which was partially successful in controlling symptoms, zidovudine or azidothymidine (AZT). More recently the emphasis has been on reducing viral load—the amount of virus within the body—using a combination of zidovudine and newer drugs known as protease inhibitors. By late 1997 some 11.7 million people had died from AIDS and more than 30 million were infected. AIDS raises many legal and moral issues of disclosure and discrimination.

acre Unit of area measurement in English-speaking countries, equal to 4,840sq yd (0.405ha).

acridine Organic chemical compound, the molecule of which is a PYRIDINE ring held between two BENZENE rings. It is used in making drugs and dyes.

acromegaly Condition in which overproduction of PITUITARY growth hormone after adulthood causes enlarged hands, feet and facial features.

acrylic Type of plastic, one of a group of synthetic, short-chain unsaturated CARBOXYLIC ACID derivatives. Variation in the reagents and the method of formation yields either hard and transparent, soft and resilient, or liquid products. Their transparency, toughness and dimensional stability make acrylics useful for moulded structural parts, lenses, adhesives and paints.

ACTH Abbreviation of ADRENOCORTICOTROPHIC HORMONE

actin Protein involved in cellular contractile processes. Found mainly in muscle cells, it reacts with MYOSIN to form ACTOMYOSIN.

actinide series Group of radioactive elements with atomic numbers from 89 to 103; ACTINIUM (89), THORIUM, PROTACTINIUM, URANIUM, NEPTUNIUM, PLUTONIUM, AMERICIUM, CURIUM, BERKELIUM, CALIFORNIUM, EINSTEINIUM, FERMIUM, MENDELEVIUM, NOBELIUM and LAWRENCIUM (103). Each element is analogous to the corresponding LANTHANIDE SERIES (rare-earth) group. The most important of the group is uranium. Those having atomic numbers greater than 92 are called **transuranium elements**.

actinium (symbol Ac) Radioactive metallic element, the first of the ACTINIDE SERIES, first discovered in 1899. It is found associated with uranium ores. Actinium-227, a decay product of Uranium-235, emits beta particles (electrons) during disintegration. Properties: at.no.89; r.d.10.07 (calc.); m.p.1,100°C (1,900°F); b.p. 3,200°C (5,800°F); most stable isotope ^{227}Ac (half-life 21.8 yr).

action potential Change that occurs in the electrical potential between the outside and the inside of a nerve fibre or muscle fibre when stimulated by the transmission of a nerve impulse. At rest the fibre is electrically negative inside and positive outside; this is called **resting potential**. When the nerve or muscle is stimulated, the charges are momentarily reversed.

activation energy Smallest amount of energy necessary to make a chemical reaction take place. As chemical bonds are broken and formed during a reaction, the energy of the system increases from that of the reactants, reaches a maximum, and then decreases to that of the products. The difference between the energy of the reactants and the maximum is the activation energy. Often this energy has to be supplied to the reaction mixture in the form of heat, although some chemical reactions take place spontaneously merely by bringing the reactants together.

activator Agent that accelerates or augments chemical activity, for example, an impurity that increases luminescence.

active transport Energy-requiring process by which molecules or ions are transported across the membranes of living CELLS against a concentration gradient. It is particularly important in the uptake of food across the gut lining, in the reabsorption of water and salts from the urine in the kidney before excretion, in secretion of substances from gland cells, in the transmission of nerve impulses, and in the uptake of minerals by the plant root. The energy is supplied by the chemical ADENOSINE TRIPHOSPHATE (ATP), which is broken down by an enzyme at the site of active transport to form ADENOSINE DIPHOSPHATE (ADP), with the release of energy. Active transport enables cells to maintain an internal chemical environment which is of a different composition from that of their surroundings.

actomyosin Complex of the proteins ACTIN and MYOSIN. This forms the basic unit of contraction in muscle cells.

acupuncture System of medical treatment in which long needles are inserted into the body to assist healing, relieve pain or for anaesthetic purposes. Of ancient Chinese origin, it enables surgery to be done with the patient conscious and free of pain. It has so far defied scientific explanation.

ACV Abbreviation of AIR-CUSHION VEHICLE

acyclic In organic chemistry, term used to describe any compound that has no rings in its molecular make-up. *See also* CYCLIC

acyl group In organic chemistry, group consisting of an ALKYL GROUP joined to CO–. An example is the ethanoyl (acetyl) group, CH_3CO–.

Adams, John Couch (1819–92) English astronomer. While an undergraduate at Cambridge, he set out to analyse the motion of URANUS. The planet's observed path was not in agreement with its calculated orbit, and Adams believed that the discrepancies could be accounted for by the gravitational influence of an undiscovered planet. He calculated an orbit for the new planet, but no search was mounted. It was Johann Galle who located NEPTUNE, as it was subsequently called, near a position predicted by the French astronomer Urbain LEVERRIER.

adaptation Adjustment by a living organism to its surroundings. Animals and plants adapt to changes in their environment through variations in

ADAPTATION

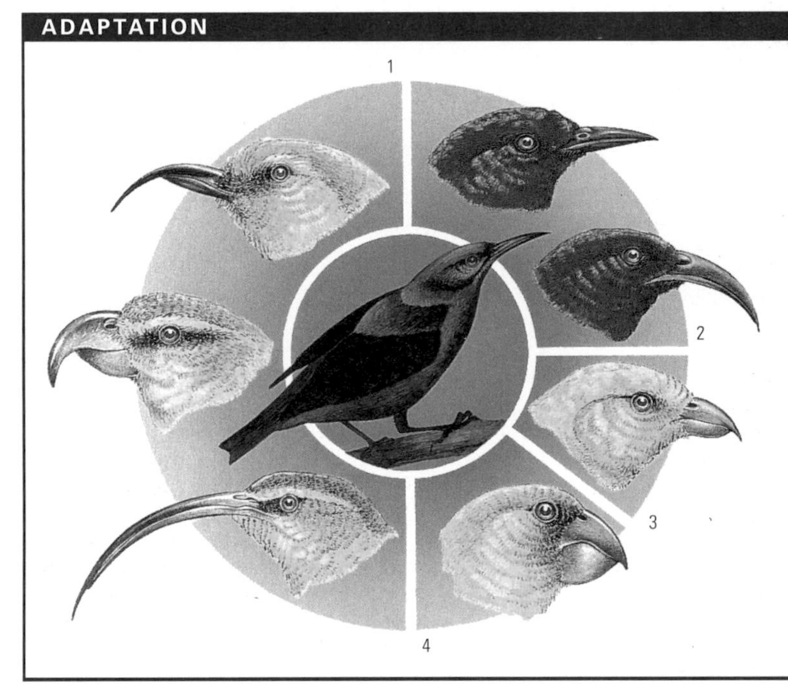

The various honeycreepers of Hawaii in the Pacific evolved from one species of bird now long extinct (centre). Over millions of years the honeycreepers evolved different methods of feeding. This ensured the island's various habitat niches could be exploited, resulting in less competition among the birds and therefore allowing more to survive. The main adaptation was the dramatic change in the shape of the beaks. A few species evolved beaks best suited to feed on nectar (1), others feed purely on insects (2), while some feed on fruit (3) or seeds (4).

ADRENAL GLAND

Located just above the kidneys (1), the two adrenal glands (2) are well supplied with blood entering from the aorta (3), and from the tributaries of the renal arteries (4). Each gland consists of an outer layer (the cortex) and a central medulla. The cortex produces steroid hormones, and hormones involved in maintaining water-balance, and small quantities of sex hormones. The medulla produces adrenaline and noradrenaline, both of which prepare the body for an emergency or stress situation.

structure, reproduction or organization within communities. Some such changes are temporary (ACCLIMATIZATION), while others may involve changes in the genetic material (DNA) and be inherited by offspring (EVOLUTION). The word is also used to describe a particular characteristic, of body size, shape, colour, physiology or behaviour, that fits an organism to survive in its environment.

adaptive radiation In biology, the EVOLUTION of different forms of living organisms from a common ancestral stock, as different populations adapt to different environmental conditions or modes of life. Eventually the populations may become so different that they constitute separate species. If, for instance, a species in time becomes distributed over several types of surroundings, the populations of each area may develop specialized features suited to the new environment. Examples are the many different kinds of finches in the Galápagos Islands, which diversified to specialize in different kinds of food, feeding methods and HABITATS. *See also* ADAPTATION

addiction Inability to control the ingestion or inhalation of a drug or other substance, resulting in physiological or psychological dependence. Symptoms of addiction include craving and habitual or compulsive behaviour. It is most frequently associated with drug addiction. In a medical context, addiction requires a **physical** dependence. When the dose of a drug is reduced or withdrawn from the user, the addict experiences withdrawal symptoms. **Psychological** dependence on activities such as gambling or exercise are difficult to distinguish from a personality disorder or mania. The wide variety of theories on addiction, such as the psychological theory of addictive personality or the physiological concentration on the brain, produce a range of therapeutic treatments.

Addison, Thomas (1793–1860) British physician who identified the adrenal hormone deficiency known as Addison's disease (1855) and Addison's anaemia. After 1837 he worked at Guy's Hospital, London, and was the joint author of *Elements of the Practice of Medicine* (1839). *See also* ADRENAL GLANDS

addition Arithmetical operation signified by + (plus sign). This is said to be a BINARY operation, since at least two numbers (or elements) are required for the operation to make sense.

addition reaction Chemical reaction in which two substances combine to form a third substance, with no other substance being produced. Addition reactions are most commonly used in organic CHEMISTRY, particularly by adding a simple molecule across a carbon-carbon double bond in an UNSATURATED COMPOUND. For example, the addition reaction between chlorine (Cl_2) and ethene ($CH_2{=}CH_2$) produces 1,2-dichloroethane ($ClCH_2(CH_2)Cl$). Some plastics (such as polystyrene) are made by addition polymerization, usually initiated by the action of a free radical. *See also* SUBSTITUTION

adenine Nitrogen-containing, organic BASE of the PURINE group found combined in the NUCLEIC ACIDS.

adenoids Masses of lymphatic tissue in the upper part of the PHARYNX (throat) behind the NOSE; part of a child's defences against disease, they normally disappear by the age of ten. When infected, the adenoids become swollen, and may need to be removed surgically.

adenosine diphosphate (ADP) Nucleotide chemical involved in energy-generating reactions during cell METABOLISM. ADP consists of the PURINE base ADENINE linked to the sugar D-ribose, which in turn carries two phosphate groups. It is formed by the HYDROLYSIS of ADENOSINE TRIPHOSPHATE (ATP) – with the release of energy – or the phosphorylation of ADENOSINE MONOPHOSPHATE (AMP) – which requires the input of energy; both reactions are catalysed by ENZYMES.

adenosine monophosphate (AMP) Nucleotide chemical involved in energy-generating reactions during cell METABOLISM. AMP consists of the PURINE base ADENINE linked to the sugar D-ribose, which in turn carries a phosphate group. It is formed by the HYDROLYSIS of ADENOSINE TRIPHOSPHATE (ATP) – with the release of energy – catalysed by ENZYMES.

adenosine triphosphate (ATP) Nucleotide chemical consisting of ADENINE, D-ribose, and three phosphate groups. It is found in all plant and animal cells, and is fundamental in the biochemical reactions required to support life. In animals, during RESPIRATION ATP can be broken down through the process of HYDROLYSIS, coupled with PHOSPHORYLATION, to form ADENOSINE DIPHOSPHATE (ADP) and phosphate, or ADENOSINE MONOPHOSPHATE (AMP) and pyrophosphate. In either case, the reactions, which are catalysed by ENZYMES, yield large amounts of energy. The energy is used either to turn simple molecules into more complex ones required by individual cells, or to control an activity such as muscle contraction. Conversely energy is used up when ATP is indirectly synthesized from ADP and AMP by complex biochemical reactions known as the ELECTRON CHAIN SYSTEM during the KREBS CYCLE. In plants, ATP is synthesized during PHOTOSYNTHESIS.

adenovirus Type of VIRUS that contains DNA. Adenoviruses infect cattle, chickens, monkeys, rodents and human beings, in whom they cause infections of the respiratory tract, similar to the common cold and gastrointestinal infections. They may also be the cause of a type of cancerous tumour.

Ader, Clément (1841–1926) French pioneer of flight. A self-taught engineer and inventor, he was the first to demonstrate that a heavier-than-air machine could become airborne.

ADH Abbreviation of ANTIDIURETIC HORMONE

adhesion Attraction of molecules of one substance to the molecules of another. GUM, GLUE and paste use the property of adhesion to join various substances together. *See also* COHESION

adhesion In medicine, fibrous band of connective tissue developing at a site of inflammation or damage; it may bind together adjacent tissues, such as loops of intestine, occasionally causing obstruction. Most adhesions result from inflammation or surgery.

adhesive Substance, or two substances, that have the ability to stick items together. Natural adhesives, usually called GLUES, are made by boiling down animal skin, bones, horns and hooves. It consists of a jelly of hydrolysed collagen (fibrous protein) – mainly gelatin – mixed with many other animal substances. It dries to form a hard, tough skin or film. There are also vegetable glues made from starch (flour and water), rubber, soybeans and other sources; synthetic adhesives include EPOXY RESIN and a hardner that react together, and THERMOSETTING and THERMOPLASTIC RESINS.

adiabatic process Thermodynamic change without any gain or loss of heat or mass into or out of a system during expansion or compression of the gas or fluid composing the system. Truly adiabatic changes must take place in short time intervals so that the heat content of the system remains unchanged, or else the system must be perfectly insulated (a practical impossibility). *See also* ISOTHERM

AEROFOIL

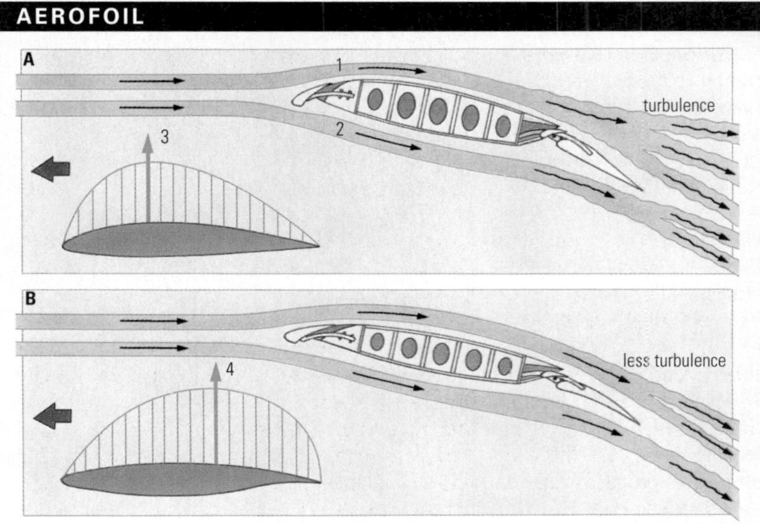

(A) As air passes over the top edge of an aerofoil (1) has to travel further than the air flowing beneath it (2), an area of low pressure forms above the wing that generates lift (3). In modern high-lift aerofoils (B) the centre of pressure is further towards the rear of the wing (4). By moving the point of maximum lift backwards the aerofoil has a more even distribution of lift allowing a plane to fly slower without stalling.

adipose tissue (fatty tissue) Connective tissue made up of body cells which store large globules of fat.

adolescence Period in human development from PUBERTY to adulthood.

ADP Abbreviation of ADENOSINE DIPHOSPHATE

adrenal gland One of a pair of small endocrine glands situated on top of the KIDNEYS. They produce many STEROIDS that regulate the blood's salt and water balance and are concerned with the METABOLISM of carbohydrates, proteins and fats, and the HORMONES ADRENALINE and noradrenaline, which are closely involved in preparing the body to meet conditions of stress, such as pain, shock, intense cold or physical danger. *See also* ENDOCRINE SYSTEM

adrenaline (US epinephrine) HORMONE secreted by the medulla of the ADRENAL GLANDS, important in preparing the body's response to stress (often referred to as "fight or flight"). It has widespread effects in the body, increasing the strength and rate of heart beat and the rate and depth of breathing, diverting blood from the skin and digestive system to the heart and muscles, and stimulating the release of GLUCOSE from the liver to increase energy supply by promoting increased RESPIRATION. Synthetic adrenaline is used medicinally in some situations, especially in the resuscitation of patients in shock or following cardiac arrest.

adrenocorticotrophic hormone (ACTH) Protein hormone secreted by the anterior lobe of the PITUITARY GLAND at the base of the brain. It stimulates the cortex of the ADRENAL GLANDS to release steroid hormones.

adsorption Attraction of a gas or liquid to the surface of a solid or liquid, unlike ABSORPTION which implies incorporation (the absorbed substance spreading into the absorbing one, for example, a sponge soaking up water). The amounts adsorbed and the rate of adsorption depend on the structure exposed, the chemical identities and concentrations of the substances involved, and the temperature.

adventitious root Root that grows in an unusual place on a plant. Normally roots grow underground at the base of a plant. But some plants produce roots from their stems and branches. With ivy, for example, adventitious roots provide nourishment and enable the plant to cling onto other plants or buildings for support.

Aeolian formation In geology, structure created by wind-transported material. It may be a dune on a riverbank or ripple marks in sand on a beach or desert, or the growth phase of DUNE building. It also can be used to describe shapes carved in rock by the wearing away of softer materials. *See also* BUTTE

aeration Bringing air into contact with a fluid by bubbling through or by agitation. Compressed air, providing oxygen to promote bacterial action, is blown into a reagent tank in the treatment of sewage. Aeration is also used in the fermentation and soft drinks industries, as well as in the manufacture of penicillin and other antibiotics.

aerial *See* ANTENNA

aerobe Minute organism that usually grows only in the presence of free atmospheric oxygen. Some aerobes can, however, remain alive even in the absence of oxygen and are called **facultative** ANAEROBES.

aerobic Connected with or dependent on the presence of free oxygen or air. An aerobic organism can only function normally in the presence of oxygen and depends on it for breaking down GLUCOSE and other foods to release energy. This process is called **aerobic respiration**. *See also* ANAEROBIC

AERODYNAMICS

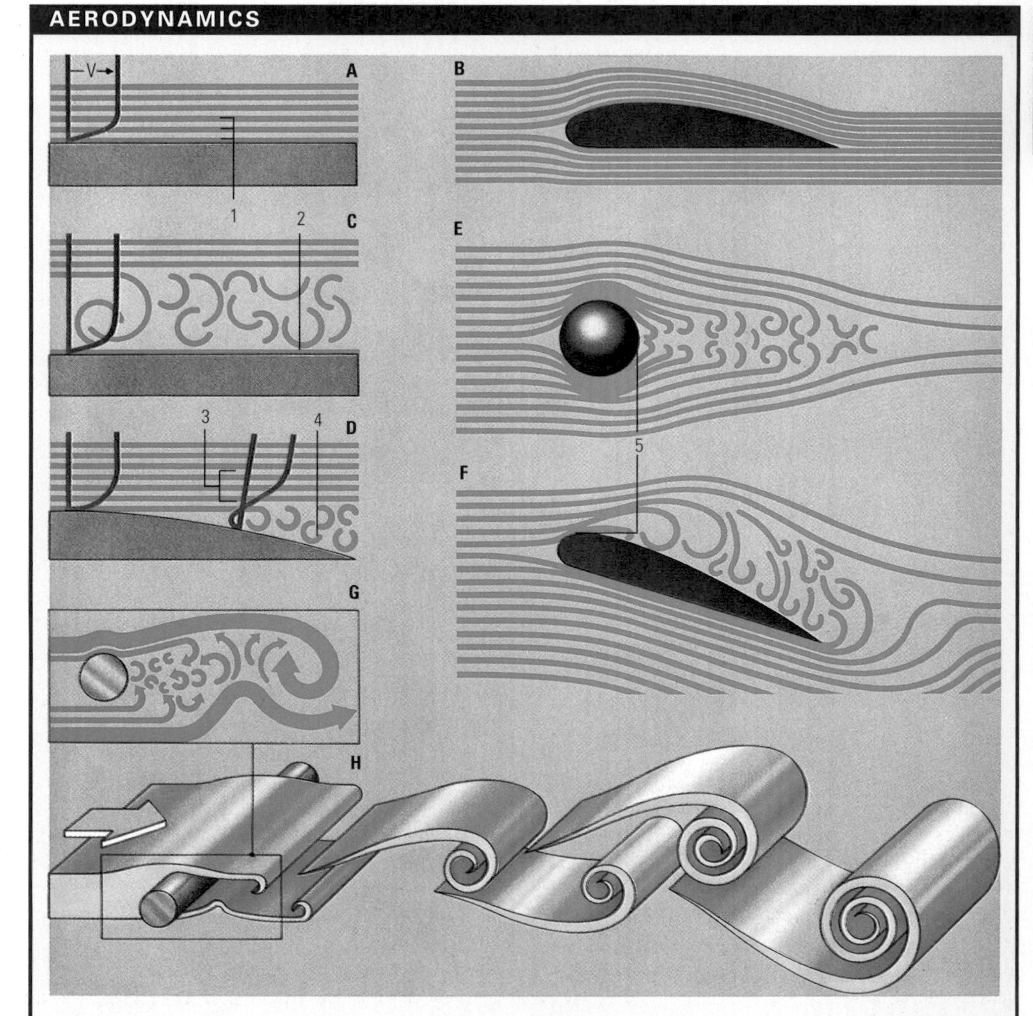

Any fluid, that is a gas or liquid, has a viscosity which resists flow across a solid surface. When flow is streamlined (A), viscosity causes the flow speed (V) to vary with distance from the surface in the manner shown. The region of varying speed (1) is called a laminar boundary layer. A perfectly streamlined aircraft wing (B) has a continuous boundary layer of this kind. A boundary layer may also, for example if the flow is very rapid, become turbulent (C). Usually there is still a laminar sub-layer (2); the speed tends to be nearly uniform in the turbulent layer due to mixing of fast and slow-moving fluid. Where a surface curves (D), the boundary layer (3) often leaves the surface, and turbulence sets in between them (4). "Boundary layer separation" (5) occurs when air flows rapidly over a sphere (E) or an inclined wing (F). Turbulence is characterized by rotational motion. In the simple case of rapid gas motion past a cylinder (such as wind through telegraph-wires), the small rotational motions may become organized (G and H) into large regularly spaced, vortices which produce an audible note.

aerodynamics Science of gases in motion and the forces acting on objects, such as aircraft, in motion through the air. An aircraft designer must consider four main factors and their interrelationships: weight of the aircraft and the load it will carry; lift to overcome the pull of gravity; drag, or the forces that retard motion; and thrust, the driving force. An aeroplane must have sufficient thrust to propel itself at speeds high enough for its wings to generate enough lift to overcome the pull of gravity. Air resistance increases as the square of an object's speed and must be minimized by streamlining to limit turbulence (which increases drag). Engineers use a WIND TUNNEL and computer systems to predict aerodynamic performance.

aerofoil Any shape or surface, such as a wing, tail or propeller blade on an aircraft, the major function of which is the deflection of airflow to produce a pressure differential or lift. A typical aerofoil has a leading and trailing edge, and an upper and lower camber.

aerolite STONY METEORITE, as opposed to an iron (SIDERITE) one. Aerolites are composed of silicate materials, with less iron and nickel than siderites.

aeronautics Study of flight and the control of AIRCRAFT involving AERODYNAMICS, aircraft structures and methods of propulsion. Aeronautics started with the study of the BALLOON, which mainly concerned the raising of a load by means of BUOYANCY. It later included the heavier-than-air flight of gliders, planes, helicopters and ROCKETS. A HELICOPTER utilizes lift provided by a rotor (rotating wing). Gliders and planes use wings to provide LIFT, but a minimum forward speed is essential to maintain height. A plane is pulled forward by a propeller, or is pushed by the reaction forces of hot expanding gases from one or more JET or rocket engines. The increased speeds of modern aircraft to supersonic (speeds in excess of that of sound; *c.*1,225km/h or 760mph) and the accompanying shock waves this produces has brought changes in wing and fuselage designs to improve streamlining. While with hypersonic speeds (in excess of five times the speed of sound or Mach 5), the forces involved again change fundamentally requiring further design adjustments.

aeronomy Study of the Earth's upper ATMOSPHERE, including its composition, density, temperature and chemical changes as recorded by satellites.

aeroplane *See* AIRCRAFT

A

AEROSOL

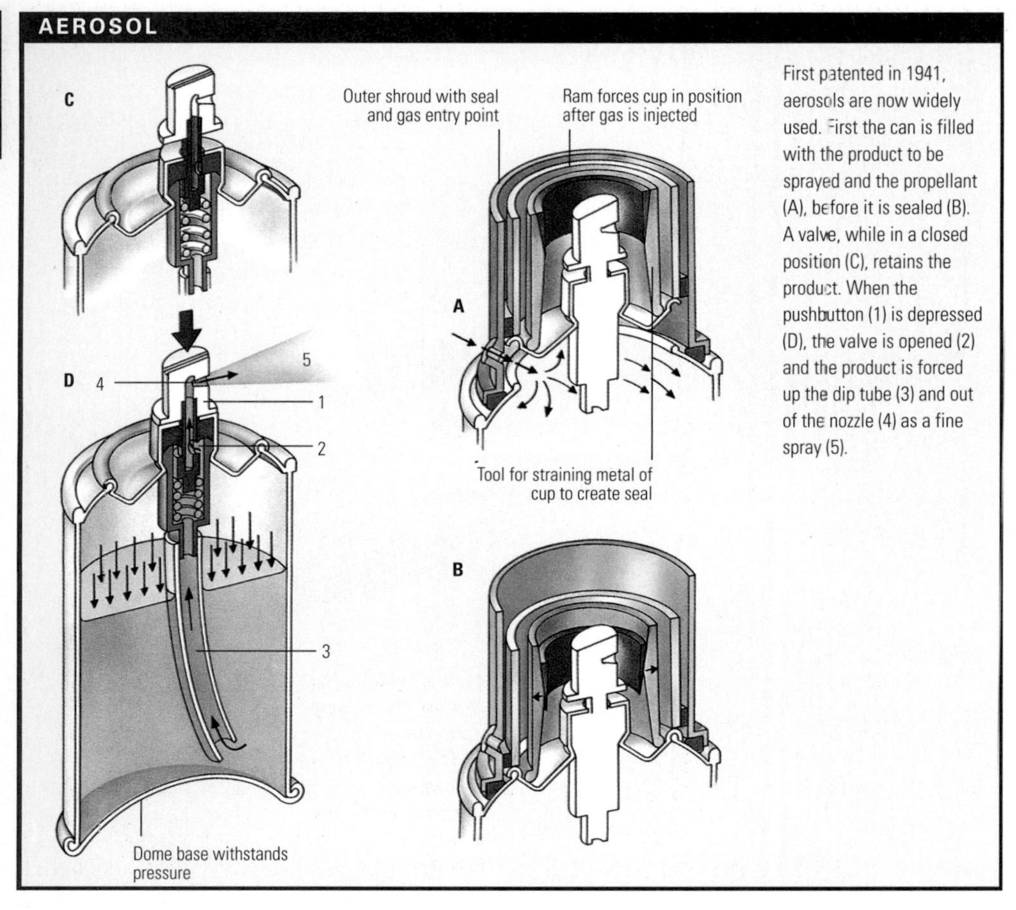

First patented in 1941, aerosols are now widely used. First the can is filled with the product to be sprayed and the propellant (A), before it is sealed (B). A valve, while in a closed position (C), retains the product. When the pushbutton (1) is depressed (D), the valve is opened (2) and the product is forced up the dip tube (3) and out of the nozzle (4) as a fine spray (5).

aerosol Suspension of liquid or solid particles in a gas. Fog – millions of tiny water droplets suspended in air – is a liquid-based example; airborne dust or smoke is a solid-based equivalent. Manufactured aerosols are used in products such as deodorants, cosmetics, paints and household sprays to name just a few. They use propellants to force the product through a fine nozzle. Chlorofluorocarbons (CFCs), once commonly used as propellants, are being phased out because they damage the ozone layer.

aestivation In biology, a state of prolonged inactivity in some animals that occurs during a period of hot weather or drought. Bodily activities such as feeding, movement and respiration slow down, and the animal expends very little energy; in this respect, it is similar to DORMANCY. It occurs in some insects, snails and lungfish, for example. In botany, the term aestivation describes the way petals and sepals are arranged in a flower bud. *See also* HIBERNATION

aetiology In medicine, the science of the origins and causes of diseases.

afforestation Planting of trees in non-wooded areas. Trees may be planted for several reasons: to reduce erosion and hold the soil together, to act as windbreaks to shelter arable crops, and to produce timber, particularly softwoods for wood pulp. Usually trees are planted where they have not grown before (at least in the recent past), such as moorland and steppes.

afterbirth PLACENTA, UMBILICAL CORD and fetal membranes expelled from the womb after childbirth. The expulsion is brought about naturally by contraction of the womb.

after-image Image retained by the brain and seen even after looking away from an object. The colour of the after-image is the complement of that of the object viewed. Television and motion pictures make use of this phenomenon.

agar Complex substance extracted from seaweed; its powder forms a "solid" gel in solution. It is used as a thickening agent or emulsifier in foods; as an adhesive; and as a medium for growing bacteria, mould, YEAST and other microorganisms; as a medium for TISSUE CULTURE; and as a gel for ELECTROPHORESIS.

Agassiz, Alexander (1835–1910) US marine zoologist, b. Switzerland. He was influential in the development of modern systematic zoology, and made important studies of the seafloor. In 1873 he succeeded his father, Louis, as curator of the Harvard Museum of Natural History, to which he also made major financial contributions.

agate Microscrystalline form of quartz with parallel bands of colour. Extracted from rock cavities, it is regarded as a semi-precious stone and is used for making ornaments and jewellery. Hardness *c.*6.5; s.g. *c.*2.6.

agglomerate In geology, coarse volcanic rock that includes both rounded and angular fragments in a finer matrix, thus combining the characteristics of CONGLOMERATES and BRECCIAS.

agglutination Clumping of BACTERIA or red blood cells by ANTIBODIES that react with ANTIGENS on the cell surface.

aggregation In mathematics, treatment of a number of terms in a mathematical expression as a single unit, denoted by enclosing the terms in parentheses. Thus $7(x^2 = 2)$ is $7x^2 = 14$.

Agnatha Class or subphylum of jawless FISH. Only the lamprey and the hagfish live today, but many fossil species date from as long ago as 450 million years.

agronomy Science of soil management and improvement in the interests of agriculture. It includes the studies of particular plants and soils and their interrelationships. Agronomy involves disease-resistant plants, selective breeding and the development of chemical fertilizers.

AIDS Acronym for ACQUIRED IMMUNE DEFICIENCY SYNDROME

Aiken, Howard (1900–73) US mathematician

AIR CONDITIONING

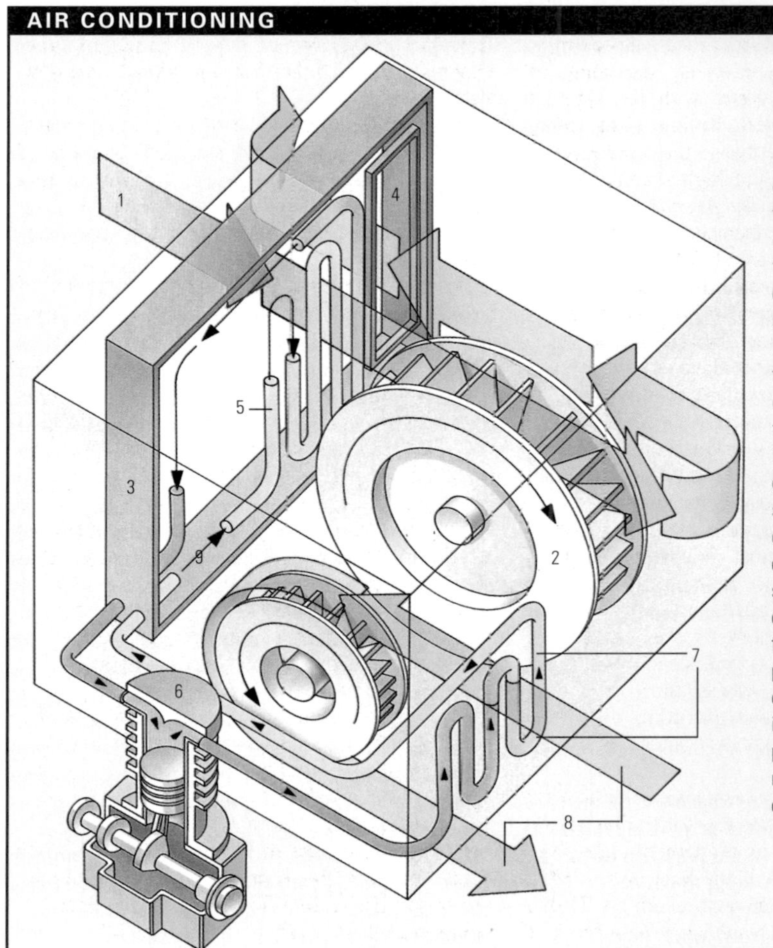

Air conditioning is used to cool the air in a sealed room. Warm air from the room (1) is drawn by a centrifugal fan (2) across the coils of a heat exchanger (3) and returned to the room through a filter (4). The cold liquid refrigerant at low pressure in the coils (5) is heated by the air from the room, evaporates and passes to the compressor (6) where it is compressed. The hot, compressed gas enters a set of coils on the exterior of the unit (7) where it is cooled and condensed by a stream of outside air (8) drawn over it by a second fan. The refrigerant then passes through an expansion valve (9) into the interior coil, as the pressure drops the refrigerant cools and the cycle begins again.

who did pioneering work in the development of computers. When Aiken was working for IBM in the USA in 1939 he designed an electromechanical automatic calculating machine. By 1944 he and his associates had built the Mark 1 Automatic Sequence Controlled Calculator, one of the first programmable computers. The input medium was punched cards or punched paper tape.

aileron Hinged control surface on the outer trailing edge of each wing of an AIRCRAFT. By moving a control, the pilot deflects the ailerons down or up in opposite directions to increase or decrease lift, causing the aeroplane to roll, or bank.

air Gases above the Earth's surface. *See* ATMOSPHERE

air conditioning Process of controlling the temperature, humidity, flow and sometimes odour and dust content of air in any enclosed space. Air is cooled by refrigeration or heated by steam or hot water. Odours and dust are removed by filters and the moisture content is adjusted by humidifiers before the air is circulated by fans which also remove stale air.

aircraft Any vehicle capable of travelling in the Earth's atmosphere. By far the most common aircraft is the airplane, aeroplane or plane. This is a heavier-than-air flying machine that depends upon fixed wings for LIFT in the air, as it moves under the THRUST of its engines. This thrust may be provided by an airscrew (propeller) turned by a piston or turbine engine, or by the exhaust gases of a JET engine or rocket motor. GLIDERS differ from planes only in their complete dependence upon air currents to keep them airborne. The main body of a plane is the fuselage, to which are attached the wings and tail assembly. Engines may be incorporated into or slung below the wings, but are sometimes mounted on the fuselage towards the tail or, as in some fighter aircraft, built into the fuselage near the wings. The landing gear or undercarriage, with its heavy wheels and stout shock absorbers, is usually completely retractable into the wings or fuselage. Wing design varies with the type of plane; high-speed fighters having slim, often swept-back or adjustable wings that create minimal air resistance (drag) at high speeds. At the other extreme heavy air freighters need broader wings in order to achieve the necessary lift at take-off. A delta wing is a broad wing, or fuselage extension, that is aerodynamically suited for both large and small high-speed planes. A plane is steered by the pilot moving flaps and AILERONS on the wings, and rudder and elevators on the tail assembly. This deflects the pressure of air on various AEROFOIL surfaces, causing the plane to rise or descend, to bank (tilt) or swing and turn in the air. A pilot needs to refer to many control instruments, including an altimeter showing height, an artificial horizon showing tilt, and indicators for air speed, ground speed, rate of climb and rate of banking and turning. Engine and airframe systems are monitored by numerous additional gauges, meters and warning lights. RADAR systems aid navigation, an AUTOPILOT keeps the aircraft steady on a fixed course, and pressurized cabins allow passenger planes to fly at heights exceeding 10,000m (32,800ft). *See also* AIRSHIP; AERODYNAMICS; BALLOON; HELICOPTER

aircraft carrier Military vessel with a wide open deck which serves as a runway for the launching and landing of aircraft. A modern nuclear-powered carrier may have a flight deck about 300m (1,000ft) long, a displacement of about 75,000 tonnes, a 4,000-man crew and carry 90 aircraft of various types – launched one a minute with each of four steam catapults. Some carriers have large, angled decks to permit launching and landing simultaneously.

AIR-CUSHION VEHICLE (ACV)

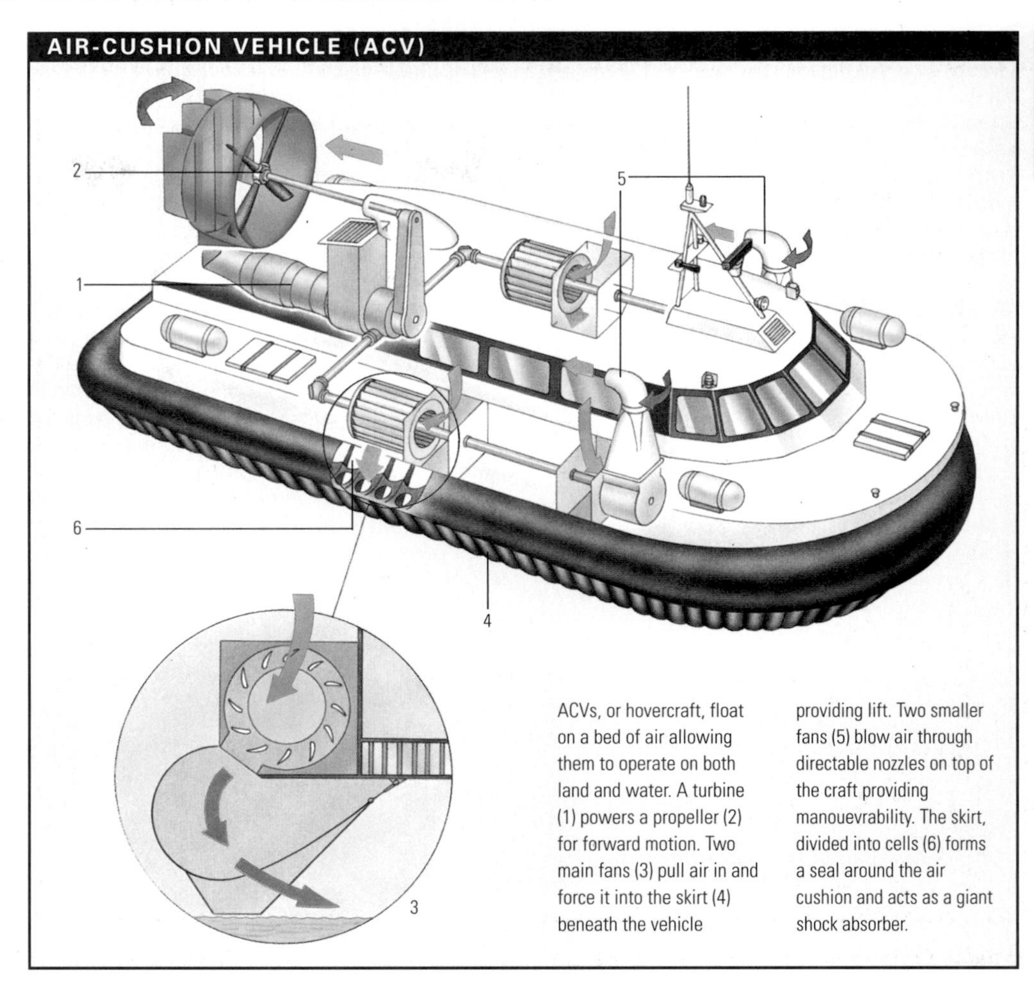

ACVs, or hovercraft, float on a bed of air allowing them to operate on both land and water. A turbine (1) powers a propeller (2) for forward motion. Two main fans (3) pull air in and force it into the skirt (4) beneath the vehicle providing lift. Two smaller fans (5) blow air through directable nozzles on top of the craft providing manouevrability. The skirt, divided into cells (6) forms a seal around the air cushion and acts as a giant shock absorber.

air-cushion vehicle (ACV) Vehicle that is lifted from the ground by air as it is forced out from under the craft. The best-known example is a HOVERCRAFT.

airglow Faint permanent glow of the Earth's IONOSPHERE resulting from the recombination of molecules, such as oxygen and nitrogen, that have been ionized by ultraviolet radiation from the Sun and probably by cosmic-ray and solar-wind particles.

airlock Hermetically sealed passageway between two leakproof doors connecting an airless environment, such as the sea depths or outer space, with a pressurized cabin or work area. Entry is made through the outer door, which is then closed; air is pumped into the airlock until its pressure is equal to that inside and then the inner door opens.

air mass Large body of the atmosphere nearly uniform horizontally in temperature, pressure, and vapour content and therefore in weather effects. High and low pressure zones on weather maps correspond to air masses, covering large areas horizontally although only a few miles vertically. Fronts, or frontal zones, occur between air masses, the movement of which brings the weather. Air masses are often classified in terms of origin as polar (cold) or tropical (warm), maritime (wet, over oceans) or continental (dry, over land). *See also* CIRCULATION, ATMOSPHERIC; FRONT

air pollution *See* POLLUTION

air pressure *See* ATMOSPHERIC PRESSURE

air sac In birds, one of the many thin-walled sacs connected to the lungs or to cavities in the bones. Air sacs improve the efficiency of air ventilation in the lungs and contribute to the lightness of the bones. In insects, air sacs connected to the tracheae (breathing tubes) improve gas exchange by increasing the surface area available for respiration.

airship (dirigible) Powered, lighter-than-air craft able to control its direction of motion. A rigid airship, or Zeppelin, maintains its form with a framework of girders covered by fabric or aluminium alloy. A gas that is less dense than air, nowadays helium, provides lift. Non-rigid airships, or blimps, have no internal structure. They rely on the pressure of the contained gas to maintain the shape. A semi-rigid airship has a rigid or jointed keel to stiffen the envelope. *See illustration on page 10*

air speed indicator Instrument that measures the speed of an aircraft relative to the air through which it passes. It measures the difference in pressures of air rushing through a **Pitot tube** (generally mounted on the wing) and static air from a vent on the fuselage; air-speed is usually indicated in miles per hour or knots.

Airy, Sir George Biddell (1801–92) British astronomer who also made discoveries in optics, such as a cylindrical lens for correcting the eye defect ASTIGMATISM. He was Astronomer Royal from 1853 to 1881.

alabaster Fine-grained, massive variety of GYPSUM (calcium sulphate), snow-white and translucent in its natural form. It can be dyed or made opaque by heating. For centuries alabaster has been used for making statues and other ornaments.

alanine ($CH_3C(NH_2)COOH$) Colourless, soluble AMINO ACID that occurs widely in PROTEINS, for example those derived from silk.

albedo Fraction of light or other radiation that is reflected from a surface. An ideal reflector has an albedo of 1; those of real reflectors are less. The albedo of snow varies from 0.45 to 0.90; that of the Earth, viewed from satellites, is 0.35. Much has been deduced about the surfaces of planets by comparing their albedos with those of known substances.

A

albino Person or animal with a rare hereditary absence of pigment from the skin, hair and eyes. The hair is white and the skin and eyes are pink because, in the absence of pigment, the blood vessels are visible. The eyes are abnormally sensitive to light and vision is often poor.

albumin (albumen) Type of water-soluble PROTEIN occurring in animal tissues and fluids. It coagulates if heated. Principal forms are EGG albumin (egg white), milk albumin and blood albumin.

albuminuria Presence of PROTEIN (serum albumin or globulin) in the urine. It may result from eating certain foods, or from certain diseases, notably those of the kidney.

alchemy Primitive form of chemistry practised in w Europe from early Christian times until the 17th century, popularly supposed to involve a search for the **Philosopher's Stone** – capable of transmuting base metals into gold – and the elixir of life. In fact, alchemy involved a combination of practical chemistry, astrology, philosophy and mysticism. Similar movements existed in China and India.

alcohol Organic compound having a hydroxyl (–OH) group bound to a carbon atom, for example ETHANOL (C_2H_5OH). Alcohols are used to make dyes and perfumes and as SOLVENTS in lacquers and varnishes.

aldehyde Member of a class of organic compounds, characterized by the group –CHO. The simplest example is METHANAL (HCOH), used as a preservative when dissolved in water, and also used in the manufacture of certain plastics. Other examples include ETHANAL (CH_3CHO) and BENZALDEHYDE (C_6H_5CHO). Aldehydes can be made by oxidizing primary ALCOHOLS. They are generally reducing agents, being oxidized to CARBOXYLIC ACIDS. Their systematic names are formed using the suffix –al. *See also* OXIDATION

Alfvén, Hannes (1908–95) Swedish physicist who shared the 1970 Nobel Prize for physics with Louis Néel of France. He laid the foundation of what has become known as magnetohydrodynamics (mhd); he advanced the theory of frozen-in flux, which states that if a magnetic field passes through a PLASMA, the movement of the plasma is constrained by the magnetic lines of flux.

algae Large group of essentially aquatic photosynthetic organisms, belonging to the kingdom PROTOCTISTA. Algae are found in salt and freshwater throughout the world, and are a primary source of food for molluscs, fish and other aquatic animals. Algae are directly important to humans as food (especially in Japan) and as FERTILIZERS. They range in size from unicellular microscopic organisms such as those that form green pond scum to huge brown seaweeds more than 45m (150ft) long. *See also* CYANOBACTERIA; CHLOROPHYTE; PHOTOSYNTHESIS; RHODOPHYTE

algal bloom Sudden increase in the amount of algae growing in fresh water. The common species involved are CYANOBACTERIA (blue-green algae), although other types of phytoplankton may also rapidly grow in numbers. It is usually caused by an increase in nitrates and other nutrients in the water, often from artificial fertilizers washed off nearby fields or from effluent containing sewage. Algal blooms often occur in lakes, and may be prolific enough to prevent light from reaching the lower depths of the water. They are a factor in EUTROPHICATION.

algebra Branch of MATHEMATICS dealing with the study of equations that are written using numbers and alphabetic symbols, which themselves represent quantities to be determined. An algebraic equation may be thought of as a constraint on the possible values of the alphabetic symbols. For example, $y + x = 8$ is an algebraic equation involving the variables x and y. Given any value of x the value of y may be determined, and vice-versa. The term algebra literally means "to find the unknown" from the Arabic *al-jabr*. *See also* ALGEBRAIC OPERATION; BOOLEAN ALGEBRA; SIMULTANEOUS EQUATIONS

algebraic function (algebraic equation) Function that is expressible using rational powers of the variable. For instance,

$$f(x) = \pi x^3 + x^{1/4} - 2/x$$

is an algebraic function. In contrast log x, is a transcendental function expressed only by an infinite series. An algebraic equation is any expression that can be written $f(x) = 0$, where f is an algebraic function. All polynomials are algebraic. *See also* NUMBER, TRANSCENDENTAL

algebraic operation Operation of ordinary algebra: that is, the arithmetical operations addition, subtraction, multiplication and division. Operations that involve infinite series and functions such as log x are not algebraic – they depend on the use of limits. The term algebraic is also applied to arithmetical operations carried out with due regard to sign. Thus, the algebraic sum of a and $-b$ is $a - b$, the algebraic product of $-a$ and $-b$ is $+ab$, etc. *See also* ARITHMETIC

algorithm Step-by-step set of instructions or procedures followed to obtain a result from an initial set of input data. The term is used in COMPUTER PROGRAMMING to refer to a set of instructions in a computer-readable format which incorporates an established series of steps to obtain the solution to a problem.

alicyclic In organic chemistry, term used to describe an ALIPHATIC COMPOUND that has one or more rings in its molecular make-up. An example is cyclohexane (C_6H_{12}). *See also* CYCLIC

alimentary canal Digestive tract of an animal that begins with the MOUTH, continues through the OESOPHAGUS to the STOMACH and INTESTINES, and ends at the anus. It is about 9m (30ft) long in humans. *See also* DIGESTIVE SYSTEM

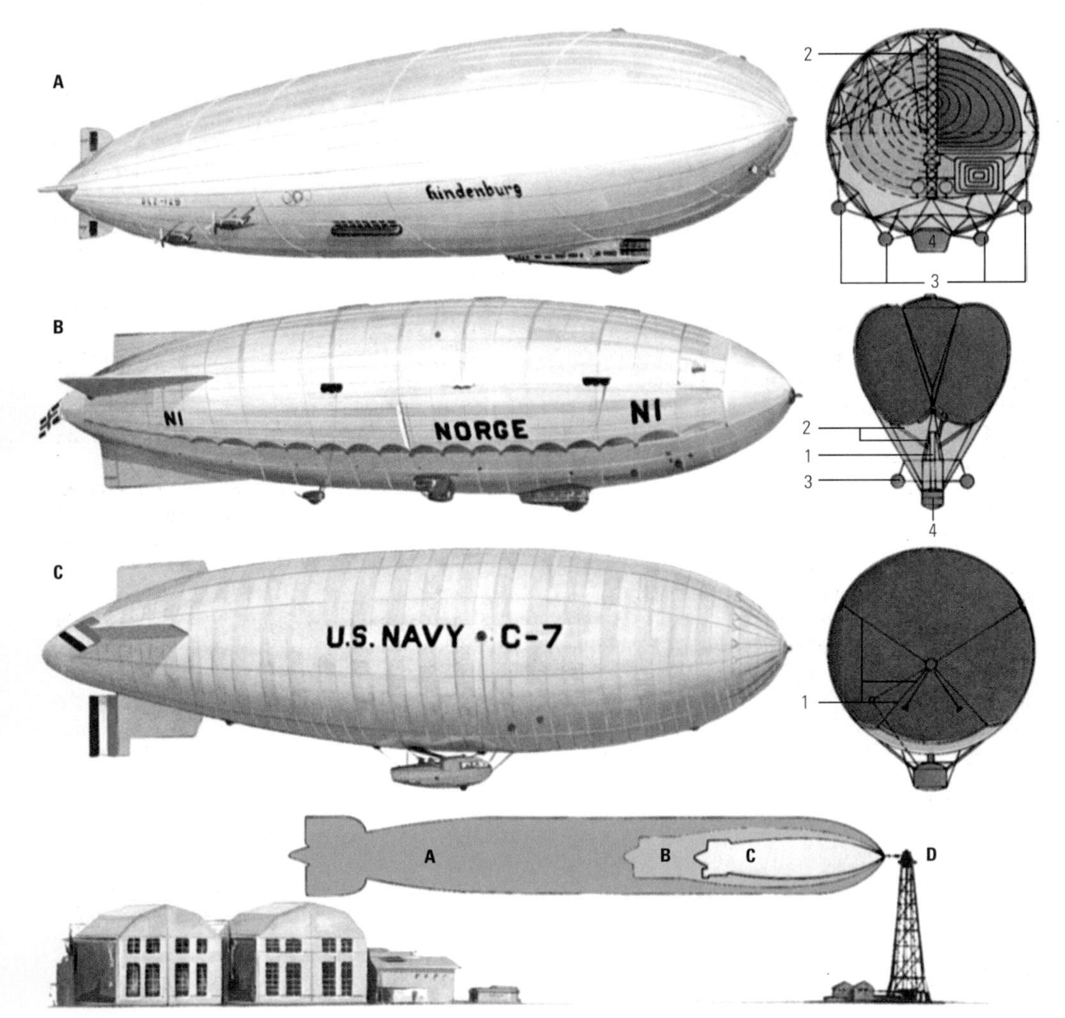

▲ **airship** There are three main types of airship. Rigid airships (A), such as the *Hindenburg* (1936), have a skeleton framework that contains the lifting gas in a series of bags (1). The total capacity of all the bags amounted to 200,000m³ (7 million cu ft) of hydrogen, giving a lift of about 232 tonnes. The bags were housed inside the aluminium framework (2) from which were hung the four 1,050hp diesel engines (3) and the payload (4), comprising 50 passengers and their baggage and 12 tonnes of freight and mail. In 1937, it exploded, killing 36 people. Semi-rigid airships (B), such as the *Norge* (1924), were uncommon. Although there was no rigid structure inside the gas envelop, a rigid keel (1) ran from bow to stern, and served as a structure to which everything could be attached. Cords and cables (2) extended from the keel to secure the envelop, preserve its shape and enable it to lift the whole ship. Below the keel were braced frames carrying the control car (3) and engines (4). Non-rigid airships, or blimps, (C) have flexible envelopes that are stabilized by being inflated to a pressure slightly higher than that of the surrounding atmosphere. The load is suspended by a system of ropes or wires (1) that distributes the weight around the fabric envelope. The blimp shown dates to from 1913, but closely resembles many used in World War 2. Similar airships are still used albeit infrequently. They are filled with nonflammable helium. The comparative sizes of the three different types (D) show just how much larger rigid airships could be.

ALBINO

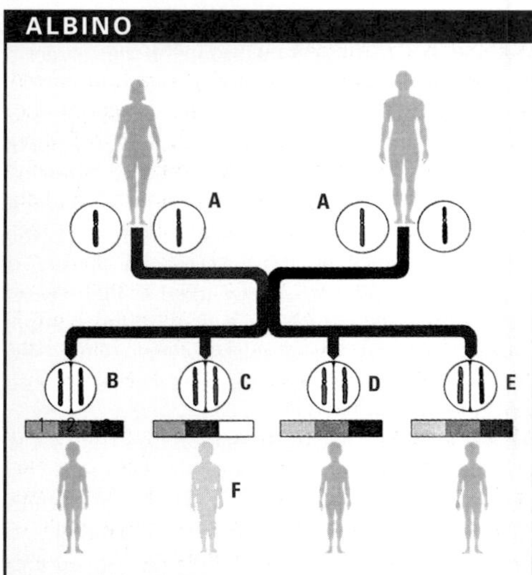

A male and female both carrying a recessive gene (green) for albinism (A) will have three normally pigmented children (B, D, and E) to every one albino (C). The corresponding normal gene (orange or purple) in both produces normal skin colour. This is due to an amino acid, phenyl-alanine (1), which has been converted to tyrosine (2) and then to the pigment melanin (3). However, a recessive gene in double quantity only allows the conversion of phenyl-alanine to tyrosine (F) resulting in albinism.

aliphatic compound Any organic chemical compound the carbon atoms of which are linked in straight chains, not closed rings. They include the ALKANES (paraffins), ALKENES (olefines) and ALKYNES (acetylenes).

alkali Soluble BASE that reacts with an ACID to form a SALT and water. A solution of an alkali has a pH greater than 7 and turns litmus dye blue. Alkali solutions are used as cleaning materials. Strong alkalis include the hydroxides of the ALKALI METALS (lithium, sodium, potassium, rubidium and caesium) and ammonium hydroxide. The carbonates (salts of carbonic acid) of these metals, and ammonium carbonate, are weak alkalis.

alkali metals Univalent metals forming Group I of the periodic table: LITHIUM, SODIUM, POTASSIUM, RUBIDIUM, CAESIUM and FRANCIUM. They are soft, silvery-white metals that tarnish rapidly in air and react violently with water to form hydroxides, which make strongly basic solutions.

alkaline-earth metals Bivalent metals forming Group II of the periodic table: BERYLLIUM, MAGNESIUM, CALCIUM, STRONTIUM, BARIUM and RADIUM. They are all light, soft and highly reactive. All, except beryllium and magnesium, react with cold water to form hydroxides (magnesium reacts with hot water). Radium is important for its radioactivity.

alkaloid Member of a class of complex nitrogen-containing organic compounds found in certain plants. They are sometimes bitter and highly poisonous substances, used as DRUGS. Examples include codeine, morphine, nicotine and quinine.

alkane (PARAFFIN) HYDROCARBON compound with the general formula C_nH_{2n+2}. Alkanes have a single carbon-carbon bond and form an HOMOLOGOUS SERIES whose first members are METHANE (CH_4), ETHANE (C_2H_6), PROPANE (C_3H_8) and BUTANE (C_4H_{10}). Because alkanes are SATURATED COMPOUNDS they are relatively unreactive. Alkanes are used as fuels.

alkene (olefin) Unsaturated HYDROCARBON compound with the general formula C_nH_{2n}. Alkenes have a carbon-carbon double bond and form an homologous series whose first members are ETHENE (ethylene, $CH_2{=}CH_2$) and PROPENE (propylene, $CH_3CH{=}CH_2$). They are reactive, particularly in ADDITION REACTIONS. Alkenes are made by the dehydration of alcohols, and are used as fuels and to make POLYMERS.

Al-Khwarizmi (active 820) Persian mathematician. He is credited as being the first to solve the quadratic equation a x^2 + bx + c = 0. In his book *Calculation with Hindu Numerals* he described a number notation which was used at the time. Following the translation of this work (in the 13th century) the Hindu number system was adopted in Europe. It is the system that we use today. Consisting of ten digits, including zero, this number system is usually (in fact falsely) referred to as the Arabic system and the digits 0,1,..,9 are called the Arabic numerals.

alkyd Synthetic resin, generally made from phthalic acid and glycerol. Alkyds can be moulded at high speed under low pressure and cured quickly. They are widely used as paint resins and in products where good insulation, high strength and good temperature, voltage and humidity stability are important, such as in car ignition systems. *See also* POLYESTER

alkyl group In organic chemistry, a group consisting of an ALKANE with one hydrogen atom removed. For example, the alkanes methane (CH_4) and ethane (C_2H_6) give rise to the alkyl groups methyl (CH_3 –) and ethyl (C_2H_5 –).

alkyl halide *See* HALOALKANE

alkyne (acetylene) Unsaturated HYDROCARBON compound with the general formula C_nH_{2n-2}. Alkynes have a carbon-carbon triple bond and form an homologous series whose first members are ETHYNE (acetylene, $CH{\equiv}CH$) and propyne (methyl acetylene, $CH_3C{\equiv}CH$). Ethyne is very reactive and prepared by adding water to calcium dicarbide. It is used as a fuel gas, particularly in oxyacetylene welding. It produces a range of POLYMERS when heated to high temperatures.

allantois Sac-like membrane that grows from the hindgut in the embryos of birds, mammals and reptiles. It is involved in respiration and the absorption of nutrients (in mammals), and in birds and reptiles acts as a bladder for the storage of waste products within the egg. *See also* PLACENTA

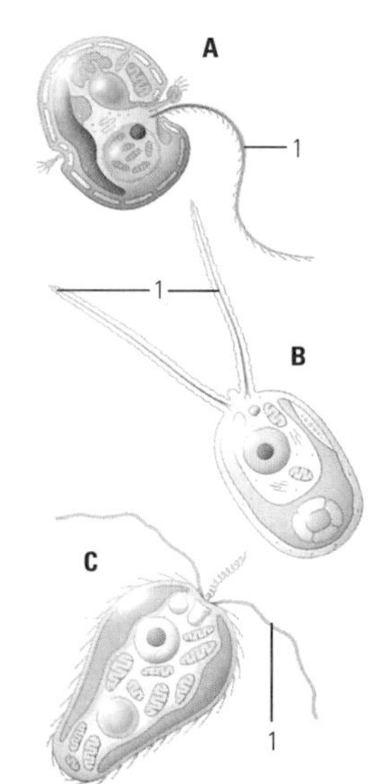

◀ **algae** Many unicellular (single-celled) algae are said to be motile, that is they move in response to changes in their environment, in particular to light. This is achieved with tail-like flagella (1) which propel them through the water. The illustration shows three types of algae. (A) *Gonyaulax tamarensis*, (B) *Chlamydomonas* and (C) *Prymnesium parvum*. Algae have been found in rocks over 2,700 million years old. They are a vital source of oxygen.

allele One of two or more alternative forms of a particular GENE. Different alleles may give rise to different forms of the characteristic for which the gene codes; for example, different flower colour in peas is due to the presence of different alleles of a single gene. *See* MENDEL, GREGOR JOHANN

allergy Disorder in which the body mounts a hypersensitive reaction to one or more substances (allergens) not normally considered harmful. Common allergens include pollen, animal hair, fungi, dust, occasionally foods and some drugs, such as PENICILLIN. Typical allergic reactions are sneezing (hay fever), "wheezing" and difficulty in breathing (ASTHMA) and skin eruptions and itching (ECZEMA). A tendency to allergic reactions is often hereditary. Treatment usually requires the identification of the allergen and its avoidance if possible, or a course of desensitization. Drugs can be useful in relieving acute symptoms.

allopathy Treatment of diseases by methods and remedies that produce effects contrasting with those caused by the disease. The term is usually applied to conventional medicine, as opposed to HOMEOPATHY.

allopatric speciation Formation of one or more new species through geographical isolation. One of

◀ **algal bloom** Water of a lake discoloured by *Cyanobacteria* (blue-green algae). This phenomenon is also known as water bloom, an abnormal growth of algae. It may have natural causes, such as an unusual amount of sunshine, or it may be due to a surplus of phosphorus or nitrogen derived from human activities. Some species of *Cyanobacteria* liberate a harmful toxin for zooplankton, fishes, aquatic birds and human beings. Water bloom may also lead to eutrophication, the progressive consumption of oxygen in the water, resulting in the death of all animal and vegetable species.

A

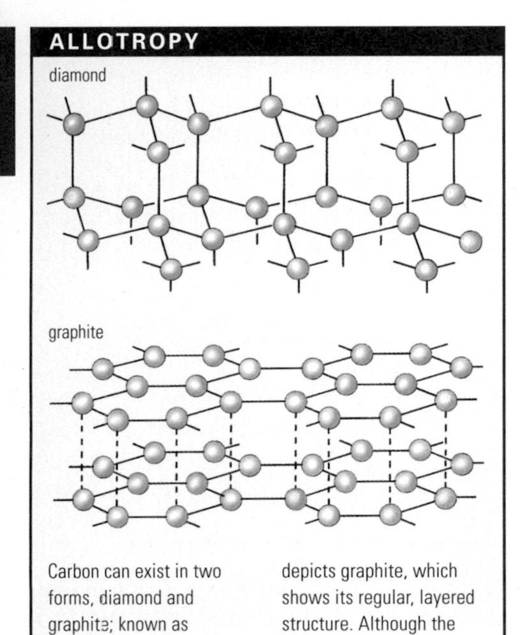

Carbon can exist in two forms, diamond and graphite; known as allotropes of carbon. The top diagram shows the atomic structure of diamond, revealing its sharp, angular structure. The bottom diagram depicts graphite, which shows its regular, layered structure. Although the carbon atoms bond strongly in sheets, the bonds between the layers are weak, enabling us to utilize graphite as a lubricant.

the criteria that defines a SPECIES is that its members are capable of interbreeding. But if members of a species become separated by a geographical barrier, such as an ocean or a mountain range, they can no longer interbreed (the separated groups are allopatric) and in time may diverge so much from each other that they are no longer even capable of interbreeding. They have therefore become different species.

allotropy Property of some chemical elements that enables them to exist in two or more distinct physical forms. Each form (called an **allotrope**) can have different chemical properties but can be changed into another allotrope – given suitable conditions. Examples of allotropes are molecular oxygen and ozone; white and yellow phosphorous; and graphite and diamond (carbon).

alloy Combination of two or more metals. An alloy's properties are different from those of its constituent elements. Alloys are generally harder and stronger, and have lower melting points. Combinations with the lowest melting points are called **eutectic** mixtures. Most alloys are prepared by mixing when molten. Some mixtures that combine a metal with a nonmetal, such as STEEL, are also referred to as alloys. The proportions in which the component metals are mixed greatly affects the properties of an alloy.

alluvial fan Generally fan-shaped area of ALLUVIUM (water-borne sediment) deposited by a river when the stream reaches a plain on lower ground, and the water velocity is abruptly reduced. Sand, silt and gravel are the main constituents of the alluvium, with the coarser material deposited near the head of the fan. Organic matter is also transported, making the soil highly fertile. Valuable minerals such as CASSITERITE (tin ore, SnO_2), diamonds, gold and platinum are often found in alluvial fans.

alluvium Fine mud or silt deposited by rivers. It accumulates on river beds but can be spread over large areas of a river valley at times of flood. All rivers have some alluvium deposits, but large rivers such as the Mississippi or Amazon have alluvial plains which are several kilometres wide. Alluvial deposits at the foot of steep mountains often accumulate in a triangular or cone shape, rather like an inland delta.

alp Area of grassland vegetation high up on the side of any mountain range. The alpine pasture is used by grazing animals in the summer but is snow-covered in winter.

Alpha Centauri Brightest star in the constellation Centaurus, and the third-brightest star in the sky. It is a visual BINARY STAR, meaning that both it and its partner star, the fainter Proxima Centauri, are visible through a telescope. It is located at a distance of 4.3 light years.

alpha particle (alpha ray) Stable, positively charged particle emitted spontaneously from the nuclei of certain radioactive ISOTOPES when undergoing RADIOACTIVE DECAY called **alpha decay**. They were discovered and investigated by Ernest RUTHERFORD and later identified as the nuclei of helium atoms consisting of two PROTONS and two NEUTRONS bound together (Helium-4). Their penetrating power is low compared with that of BETA PARTICLES (electrons) but they cause intense IONIZATION along their track. This ionization is used to detect them. Alpha particles can be stopped by a sheet of paper.

alternating current (AC) *See* ELECTRIC CURRENT

alternation of generations Two-generation cycle by which plants and some algae reproduce. The asexual diploid SPOROPHYTE form produces haploid SPORES that, in turn, grow into the sexual (GAMETOPHYTE) form. The gametophyte produces the egg cell that is fertilized by a male gamete to produce a diploid zygote that grows into another sporophyte.

alternative energy *See* RENEWABLE ENERGY

alternator Electrical GENERATOR that produces an alternating current.

altimeter Instrument for measuring altitude. Pilots, scientists, surveyors and mountain climbers use altimeters. The simplest type is a form of aneroid BAROMETER. As height increases, air pressure decreases, so the barometer scale can be calibrated to show altitude. Some aircraft have a radar altimeter. This instrument measures the delay between sending a radio signal to the ground and receiving its reflection. From this delay, the instrument calculates the distance to the ground.

altitude In astronomy, the angular distance of a celestial body above the observer's horizon. It is measured in degrees from 0 (on the horizon) to 90 (at the zenith) along the GREAT CIRCLE passing through the body and the zenith. If the object is below the horizon, the altitude is negative.

altruism Principle of acting for the welfare or interests of someone else. The word was coined by the French philosopher Auguste Comte from the Italian *altrui* (others). It is used in biology to describe the behaviour of an animal that acts in a way which enhances the prospects of survival or reproduction of another individual at the expense of its own interests. Most cases of altruism among animals involve close relatives or young, with whom the animal shares many GENES. So it is, in effect, perpetuating its own genes in the process.

alum Double sulphate of aluminium (or another trivalent metal) and another, univalent metal. The most important is potash alum, potassium aluminium sulphate; it is used as a mordant in dyeing.

alumina (aluminium oxide, Al_2O_3) Mineral used as an abrasive, electrical insulator and furnace lining. The impure hydrated form of alumina, BAUXITE, is the chief ore of aluminium. Other forms of alumina include corundum, two impure varieties of which are the gemstones SAPPHIRE and RUBY.

aluminium (symbol Al) Metallic, silvery white element of group III of the periodic table, first obtained in pure form in 1827. It is the most common metal in the Earth's crust; the chief ore is BAUXITE. Bauxite undergoes two processes to yield aluminium. It is first refined to obtain pure alumina (Al_2O_3), which is then smelted to produce aluminium. The Bayer process, involving digestion, clarification, precipitation and calcination, is the most common refining process yielding half the original amount of ore. Smelting involves dissolving alumina in melted cryolite and passing a current through the mixture (ELECTROLYSIS) to yield pure aluminum and carbon dioxide (the Hall-Heroult process). Alloyed with other metals, it is used extensively in machined and moulded articles, particularly where lightness is important, as in aircraft. It is protected from oxidation (corrosion) by a thin, natural layer of oxide. Properties: at.no. 13; r.a.m. 26.98; r.d. 2.69; m.p. 660.2°C (1,220.38°F); b.p. 1,800°C (3,272°F); most common isotope ^{27}Al. *See also* ANODIZING

Alvarez, Luis Walter (1911–88) US physicist in the field of subatomic particles who won the 1968 Nobel Prize for physics for developing the liquid-hydrogen BUBBLE CHAMBER. He used it to identify many "resonances", very short-lived par-

◀ **alluvial fan** A coloured photograph of the Colorado River delta, in the Gulf of California, USA. The river is the dark hemispherical shape at the bottom of the picture. Its waters branch out (dark) like the boughs of a tree through sandbars (the alluvium), which are seen as the peach coloured areas.

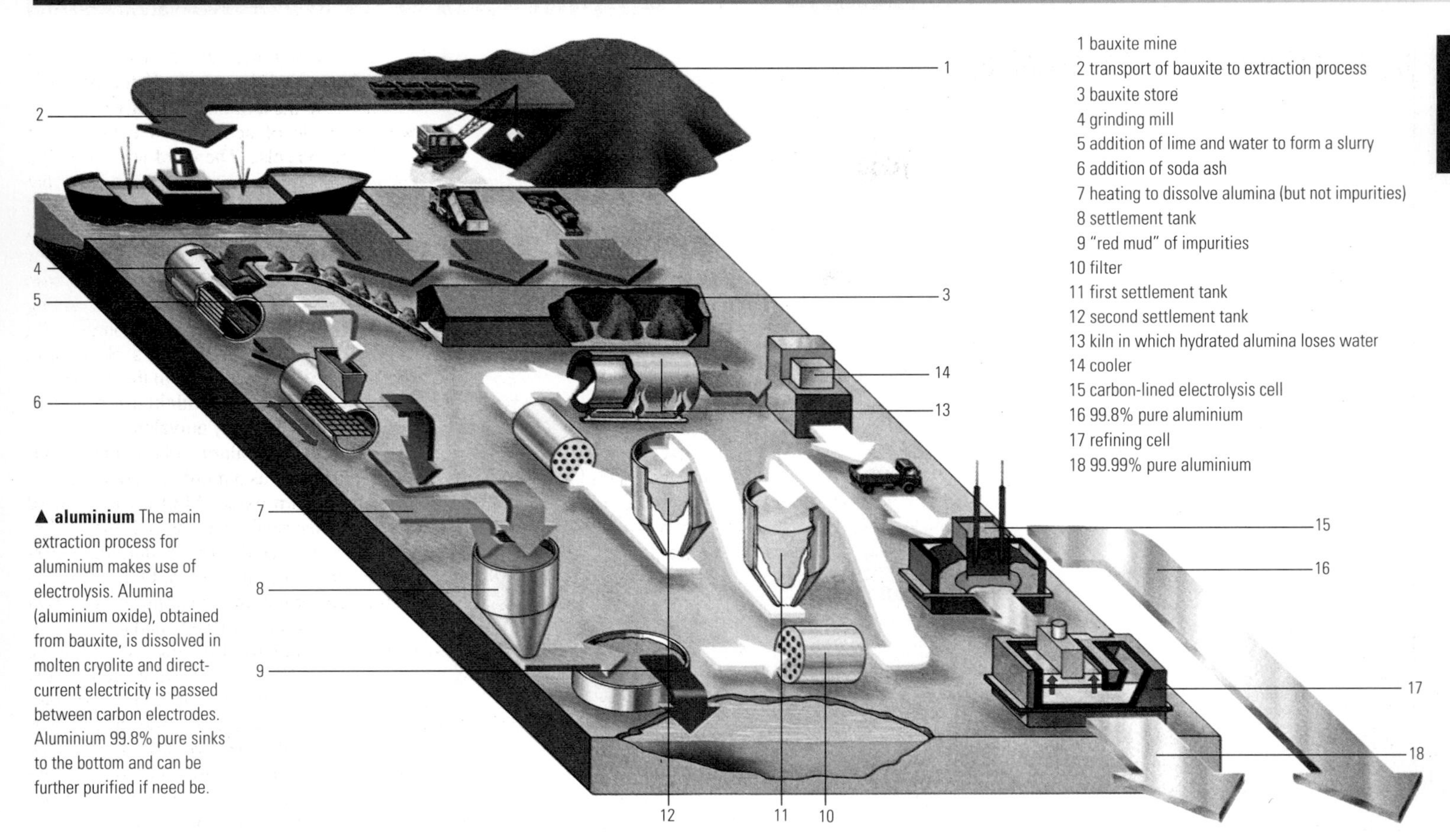

▲ **aluminium** The main extraction process for aluminium makes use of electrolysis. Alumina (aluminium oxide), obtained from bauxite, is dissolved in molten cryolite and direct-current electricity is passed between carbon electrodes. Aluminium 99.8% pure sinks to the bottom and can be further purified if need be.

ticles. He also helped construct the first proton linear ACCELERATOR. Alvarez worked on the MANHATTAN PROJECT to develop the atom bomb and invented the radar guidance system for aircraft landings. He and his son, the geologist Walter Alvarez, were in the group who formed the theory that dinosaurs became extinct as a result of changes to the atmosphere after a massive meteorite crashed into the Earth.

alveolus (pl. alveoli) One of a cluster of microscopic air sacs that open out from the alveolar ducts at the far end of each bronchiole in the LUNGS. The alveolus is the site for the exchange of gases between the air and the bloodstream, and its walls contain a network of CAPILLARY blood vessels. *See also* GAS EXCHANGE; RESPIRATORY SYSTEM

Alzheimer's disease Degenerative condition characterized by memory loss and progressive mental impairment; it is the commonest form of DEMENTIA. Sometimes seen in the middle years, Alzheimer's becomes increasingly common with advancing age. Many factors have been implicated, but the precise cause is unknown.

AM Abbreviation of AMPLITUDE MODULATION

amalgam Solid or liquid alloy of mercury with other metals. Dentists once commonly filled teeth with amalgams usually containing copper and zinc. Most metals dissolve in mercury, although iron and platinum are exceptions.

amatol High explosive made from ammonium nitrate and TNT.

amber Hard yellow or brown, translucent fossil resin, mainly from pine trees. Amber is most often found in alluvial soils, in lignite beds or around seashores, especially near the Baltic Sea. The resin occurs as rods or as irregular nodules, sometimes with embedded fossil insects or plants. Amber can be polished to a high degree and is used to make necklaces and other items of jewellery.

americium (symbol Am) Radioactive metallic element of the ACTINIDE SERIES, first made in 1944 by neutron bombardment of plutonium. It is used in home smoke detectors, and ^{241}Am is a source of gamma rays. Properties: at.no. 95; r.a.m. 243.13; r.d. 13.67; m.p. 995°C (1,821°F); b.p. unknown; most stable isotope ^{243}Am (half-life 7,650 yr).

amethyst Transparent, violet variety of crystallized QUARTZ, containing more iron oxide than other varieties. It is found mainly in Brazil, Uruguay, Ontario, Canada and North Carolina, USA. Amethyst is valued as a semi-precious gem and may be made synthetically.

amide Organic chemical compound that contains the amide group $-CONH_2$. Amides are generally made by heating the ammonium salt of a CARBOXYLIC ACID. For example, heating ammonium ethanoate (acetate) produces ethanamide (acetamide, CH_3CONH_2), which is a volatile solid.

amine Any of a group of organic compounds derived from ammonia by replacing hydrogen atoms with alkyl groups. Methylamine (CH_3NH_2), has one hydrogen replaced. Replacement of two hydrogens gives a secondary amine and of three hydrogens, a tertiary amine. Amines are produced in the putrefaction of organic matter and are weakly basic. ANILINE is an aromatic amine compound used in dyeing.

amino acid Organic acid containing at least one carboxyl group (COOH) and at least one amino group (NH_2). Amino acids are of great biological importance because they combine together to form PROTEINS. Amino acids form PEPTIDES by the reaction of adjacent amino (NH_2) and carboxyl (COOH) groups. Proteins are polypeptide chains consisting of hundreds of amino acids. About 20 amino acids occur in proteins; not all organisms are able to synthesize all of them. **Essential amino acids** are those that an organism has to obtain ready-made from its environment. There are ten essential amino acids for humans: ARGININE, HISTIDINE, ISOLEUCINE, LEUCINE, LYSINE, METHIONINE, PHENYLALANINE, THREONINE, TRYPTOPHAN and VALINE.

ammeter Instrument for measuring electric current in amperes (amps). An ammeter is connected in series in a circuit. In the moving-coil type for direct current (DC), the current to be measured passes through a coil suspended in a magnetic field and deflects a needle attached to the coil. In the moving-iron type for both direct and alternating current (AC), current through a fixed coil magnetizes two pieces of soft iron that repel each other and deflect the needle. Digital ammeters are now also commonly used. *See also* ELECTRIC CURRENT

ammonia (NH_3) Colourless, nonflammable pungent gas, a compound of hydrogen and nitrogen. It is used to make nitrogenous fertilizers. The gas is extremely soluble in water, forming an alkaline solution of ammonium hydroxide (NH_4OH) that can give rise to ammonium salts containing the ion NH_4^+. Ammonia solutions are used in cleaning and

▲ **amber** Small fly embedded in Baltic amber. Amber is fossilized resin, and in some deposits, insects and spiders are preserved complete. Amber occurs worldwide but is most common in the Baltic region. IIt has been highly valued since prehistoric times.

A

bleaching. Chief properties: r.d. 0.59; m.p. −77.7°C (−107.9°F); b.p. −33.4°C (−28.1 °F).

ammonite Any member of the order Ammonitida, an extinct group of shelled cephalopod MOLLUSCS. Most ammonites had a spiral shell, and they are believed to be related to the nautiloids – the only surviving form being the pearly nautilus. They are common as fossils in marine rocks.

ammonium chloride (NH_4Cl) Ammonium salt of hydrochloric acid. Formerly called sal ammoniac, ammonium chloride is a volatile, highly soluble white solid, made by reacting ammonia with hydrogen chloride. It is used in dry batteries and as a mordant in the dyeing and printing of cloth.

amnion Membrane or sac that encloses the EMBRYO of a reptile, bird or mammal. The embryo floats in the amniotic fluid within the sac. *See also* UTERUS

amniotic fluid Liquid filling the amniotic cavity which protects the fetus from injury. This fluid is released at delivery when the membranes rupture.

amoeba (genus *Amoeba*) Microscopic, almost transparent, single-celled PROTOZOA of the phylum Rhizopoda. Amoebas have a constantly changing, irregular shape. Found in ponds, damp soil and animal intestines, they consist of a thin outer cell membrane, a large nucleus, food and contractile VACUOLES and fat globules. They reproduce by binary FISSION. Length: up to 3mm (0.1in). Species include the common *Amoeba proteus* and *Entamoeba histolytica*, which causes amoebic DYSENTERY.

amorphous substance Non-crystalline solid; its atoms or molecules have no regular order. Supercooled liquids such as glass, rubber and some plastics are amorphous. Many powders appear amorphous but are microcrystalline in structure. *See also* CRYSTAL

AMP Abbreviation of ADENOSINE MONOPHOSPHATE

ampere (Symbol A) SI unit of ELECTRIC CURRENT. It is defined as the current in a pair of straight, parallel conductors of infinite length and 1m (39in) apart in a vacuum that produces a force of 2×10^{-7} newton per metre (N/m) in their length. This force may be measured on a current balance instrument, the standard against which current meters, such as an AMMETER, are calibrated.

Ampère, André Marie (1775–1836) French physicist and mathematician. He was professor of chemistry and physics at Bourg and later of mathematics at the École Polytechnique in Paris. He founded electrodynamics (now called ELECTROMAGNETISM) and performed numerous experiments to investigate the magnetic effects of ELECTRIC CURRENTS. He was the first to devise techniques for detecting and measuring currents, and constructed an early type of galvanometer. Ampère's law – proposed by him – is a mathematical description of the magnetic force between two electric currents. His name is also commemorated in the fundamental unit of current, the AMPERE (A), used in the SI system of units.

AMMONITE

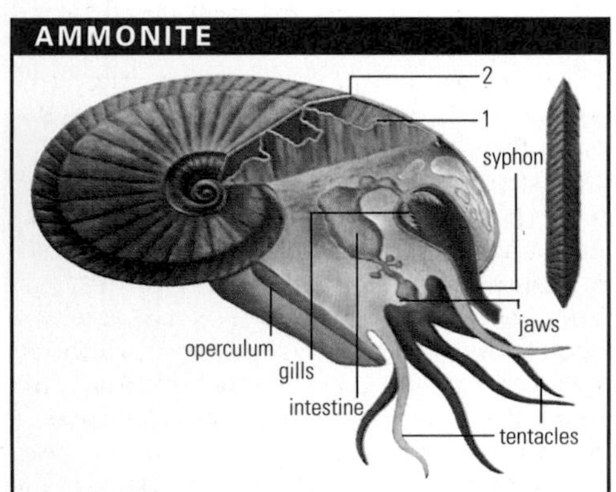

Ammonites had a soft anatomy, similar to that of the modern nautilus which lives in the open end of its shell. As the animal grew it secreted more and more shell and moved forward into the new part, walling off the old section with a septum (1). The walled-off chambers were used for buoyancy, being supplied with air from a tissue filament or siphuncle (2) that connected them all. The septa met the shell wall in suture lines that had identifiable patterns for each species and became increasingly complex as the group advanced.

amphetamine Name for a number of DRUGS (1-phenyl-2-aminopropane and its derivatives) that stimulate the CENTRAL NERVOUS SYSTEM, inhibit sleep and suppress appetite. These drugs (known as "pep pills" or "speed"), if used long-term, can lead to drug abuse and ADDICTION. After the immediate sense of well-being wears off, the user can experience fatigue and depression.

amphibian Member of a class (Amphibia) of egg-laying VERTEBRATES, whose larval stages (tadpoles) are usually spent in water but whose adult life is normally spent on land. Amphibians have smooth, moist skin and are cold-blooded. Larvae breathe through gills; adults usually have lungs. All adults are carnivorous but larvae are frequently herbivorous. There are three living orders: Urodela (NEWTS and SALAMANDERS); Anura (FROGS and TOADS); and Apoda or CAECILIANS.

amphibole Any of a large group of complex rock-forming minerals characterized by a double-chain silicate structure (Si_4O_{11}). They all contain water as OH^- ions and usually calcium, magnesium and iron. Found in IGNEOUS and METAMORPHIC ROCKS, they form wedge-shaped fragments on cleavage. Their orthorhombic or monoclinic crystals are often needle-like or fibrous. Common varieties are hornblende, tremolite, actinolite and anthophyllite. Some varieties are used in commercial ASBESTOS.

Amphineura Class of MOLLUSCS that includes the CHITONS, small dwellers on rock surfaces. It also includes *Neopilina*, a small marine mollusc recently dredged from great depths, which is so primitive as to show some similarities to the ANNELID worms.

amphoteric In chemistry, term used to describe a substance that has both acidic and alkaline properties. Aluminium hydroxide is a typical amphoteric compound. It behaves like the base $Al(OH)_3$ when it reacts with acids (to form aluminium salts); and it behaves like the acid H_3AlO_3 when it reacts with alkalis (to form aluminates). Zinc hydroxide is also amphoteric. In organic chemistry, AMINO ACIDS can also be regarded as amphoteric because they contain both an acid −COOH group and a basic $-NH_2$ group.

amplifier Device for changing the magnitude of a signal such as voltage, current or mechanical motion but not the way it varies. Amplifiers are used in radio and television transmitters and receivers, and in audio equipment. The first electronic amplifiers used thermionic valves activated by heat, such as the triode (having three electrodes) invented in 1906 by the American physicist Lee De Forest. Valve amplifiers were once widely used but are now almost completely replaced by amplifiers that employ TRANSISTORS. *See also* THERMIONICS

amplitude modulation (AM) Form of RADIO transmission. Broadcasts on the short-, medium- and long-wave bands are transmitted by amplitude modulation. The sound signals to be transmitted are superimposed on a constant-amplitude radio signal called the **carrier**. The resulting modulated radio signal varies in amplitude according to the strength of the sound signal. *See also* FREQUENCY MODULATION (FM)

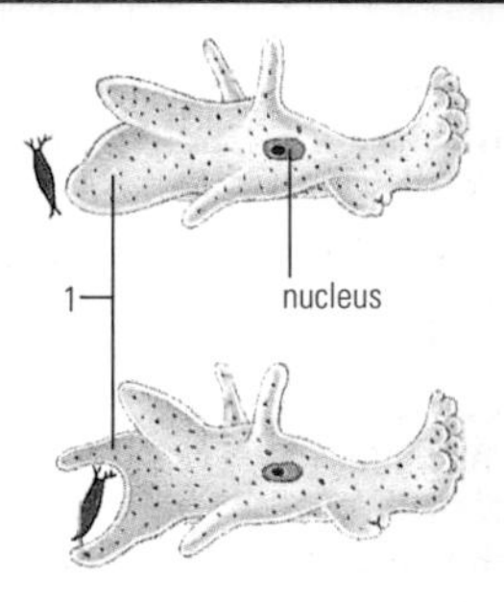

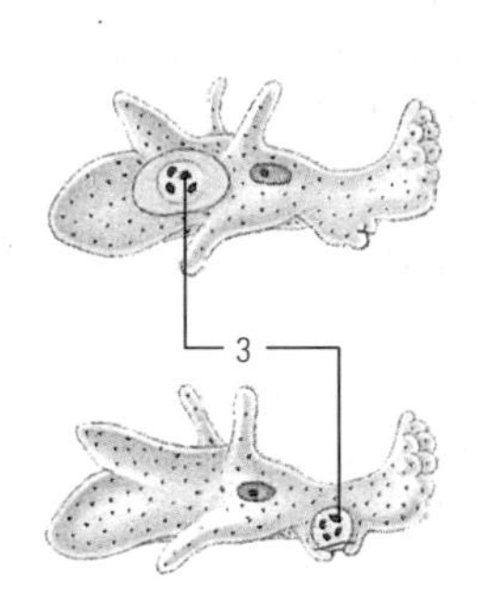

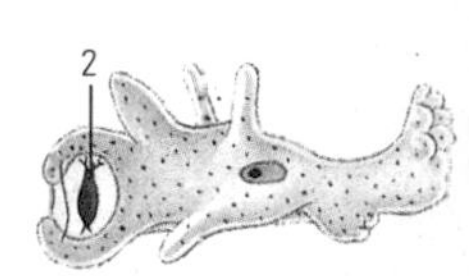

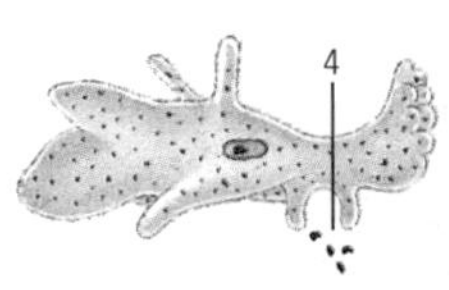

◀ **amoeba** In order to move, an amoeba pushes out projections called pseudopods (lit. fake foot) from its body. Cytoplasm – the fluid content of the cell – flows into the pseudopod, constantly enlarging it until all the cytoplasm has entered and the amoeba as a whole has moved. Pseudopods are also used in feeding: they move out to engulf a food particle (1), which then becomes enclosed in a membrane-bound food vacuole (2). Digestive enzymes enter the vacuole, which gradually shrinks as the food is broken down (3). Undigested material is discharged by the vacuole and left behind as the amoeba moves on (4).

ampulla Dilated end of a body duct or canal.

amu Abbreviation of ATOMIC MASS UNIT

amygdala Almond-shaped mass of nerve cells deep in the cerebral hemispheres of the BRAIN. It is concerned with emotion.

amyl alcohol ($C_5H_{11}OH$) ALCOHOL which has eight isomers. Commercial amyl alcohol, a mixture of these, has a sharp odour and is used as a solvent.

amylase Digestive enzyme secreted by the SALIVARY GLANDS (salivary amylase) and the PANCREAS (pancreatic amylase). It aids digestion by breaking down starch into MALTOSE (a disaccharide) and then GLUCOSE (a monosaccharide).

anabolic steroid Any of a group of HORMONES that stimulate the growth of tissue. Synthetic versions are used in medicine to treat OSTEOPOROSIS in older women and some types of ANAEMIA; they may also be prescribed to aid weight gain in the elderly, sick or in severely ill patients. These drugs are associated with a number of side-effects, including acne, fluid retention, liver damage and masculinization in women. Some athletes, such as sprinters and weight-lifters, have been known to abuse anabolic steroids in order to increase muscle bulk.

anabolism *See* METABOLISM

anacoustic zone (zone of silence) Region above 160km (100mi) altitude where distance between air molecules becomes greater than the wavelength of sound, so that no sound waves can be propagated. With increasing height, high-frequency (short-wavelength) sounds disappear first, and only lower tones can be heard.

anaemia Condition in which there is a shortage of HAEMOGLOBIN, the oxygen-carrying pigment contained in red blood cells. Symptoms include fatigue, weakness, pallor, breathlessness, faintness, palpitations and lowered resistance to infection. It may be due to a decrease in the production of haemoglobin or red blood cells or excessive destruction of red blood cells or blood loss. There are several types of anaemia. Worldwide, iron deficiency is the commonest cause of anaemia. It usually results from poor nutrition, intestinal parasites, heavy periods or chronic bleeding from an ulcer or haemorrhoids (piles). Pernicious anaemia is an impairment of the body's ability to absorb vitamin B_{12}, a VITAMIN necessary for the normal development of red blood cells. Deficiencies in other substances, such as folic acid or other B vitamins, may also reduce red blood cell production. In haemolytic anaemia there is excessive destruction of these cells. It may be due to hereditary abnormalities (such as SICKLE-CELL ANAEMIA), adverse reaction to drugs, or the production of antibodies to one's own red blood cells. Red blood cells are produced in the bone marrow and damage to it because of radiation, drugs or chemicals, cancer and so on, may also cause anaemia.

anaerobe Minute organism that grows only in the absence of free atmospheric oxygen. Anaerobic BACTERIA can be a hazard in food canning because they can multiply in foods even under vacuum. *See also* AEROBE.

anaerobic Connected with the absence of oxygen or air or not dependent on oxygen or air for survival. An anaerobic organism, or ANAEROBE, is a microorganism that can survive by releasing energy from GLUCOSE and other foods in the absence of oxygen. The process by which it does so is called **anaerobic respiration.** Most anaerobes can survive in oxygen but do not need it for RESPIRATION. *See also* AEROBIC

anaesthesia State of insensibility or loss of sensation produced by disease or by various anaesthetic drugs used during surgical procedures. During general, or total, anaesthesia the entire body becomes insensible and the patient sleeps; in local anaesthesia only a specific part of the body is rendered insensible and the patient remains conscious. A general anaesthetic may be either an injected drug, such as the barbiturate thiopentone, used to induce unconsciousness, or an inhalation agent such as halothane, which is used to maintain anaesthesia for surgery. Local anaesthetics such as lignocaine numb the relevant part of the body by blocking the transmission of impulses through the sensory nerves that supply it.

anaesthetics Branch of medicine involving the use of anaesthetic agents to prevent pain during surgical and other procedures and care of the anaesthetized patient.

analgesic DRUG that relieves or prevents pain without causing loss of consciousness. It does not cure the cause of the pain, but helps to deaden the sensation. Some analgesics are also NARCOTICS, and many also have valuable anti-inflammatory properties. Common analgesics include aspirin, codeine, and morphine. *See also* ANAESTHESIA

analogue computer Machine that processes continuously variable information. Information is usually first converted into proportional electrical quantities. These are manipulated by amplifiers and other circuits that perform various mathematical operations. In other words, the COMPUTER solves problems by dealing with quantities (voltages) that are analogous (similar) to the quantities in the problem. Analogue computers are time consuming to set up and operate. Most work once done on analogue computers is now carried out on digital computers, which are simpler and quicker to use.

analogue signal In telecommunications and electronics, transmission of information by means of variation in a continuous waveform. An analogue signal varies (usually in amplitude or FREQUENCY) in direct proportion to the information content of the signal.

analysis In chemistry, any method of determining the composition of a substance. There are two main types of chemical analysis. **Qualitative** analysis is concerned with finding out what elements or compounds are present in a sample. It makes use of various specific chemical tests. **Quantitative** analysis is used to determine how much of a known substance is present in a sample. There are various quantitative techniques, including volumetric analysis (which depends on measuring volumes of reactants), gravimetric analysis (which relies on weighing), SPECTROSCOPY and spectrometry (which identifies elements by their spectra) and CHROMATOGRAPHY.

analysis In mathematics, field that is concerned with the concepts of CONVERGENCE, CONTINUITY, DIFFERENTIATION and INTEGRATION. **Real** analysis is the branch of mathematics where such ideas are applied to real NUMBERS or functions of real numbers. Similarly, **complex** analysis applies these concepts to complex numbers.

analytic geometry *See* COORDINATE GEOMETRY

anaphase Stage or stages in cell division. In MITOSIS, anaphase occurs when chromatids (from the CHROMOSOMES) separate and go to each end of the spindle. In MEIOSIS, there are two anaphases. In the first, paired chromosome separate and move apart. In the second, as in mitosis the chromatids separate and move to the poles of the spindle.

anaphylaxis (anaphylactic shock) Acute, serious allergic reaction that occurs after a person eats, inhales, or is injected with a substance (ANTIGEN), such as penicillin, insect venom, or certain foods, to which they have been previously sensitized and against which they have developed ANTIBODIES. Second exposure to the offending substance can lead to paleness, faintness, palpitations, hives, swelling, difficulty in breathing, and, if untreated, SHOCK.

anatomy Branch of biological science concerned with the structure of an organism. The study of anatomy can be divided in several ways. On the basis of size, there is **gross** anatomy, which is studying structures with the naked eye; **microscopic** anatomy, studying finer detail with a light microscope; **submicroscopic** anatomy, studying even finer structural detail with an electron MICROSCOPE; and **molecular** anatomy, studying with sophisticated instruments the molecular make-up of an organism. Microscopic and submicroscopic anatomy involve two closely related sciences: HISTOLOGY, the study of tissue that makes up a body organ, and CYTOLOGY, the study of cells that make up tissue. Anatomy can also be classified according to the type of organism studied: plant, invertebrate, vertebrate or human anatomy. *See also* PHYSIOLOGY

andalusite One of many crystalline forms of aluminium silicate, occurring in contact metamorphic rock and in other deposits. It is mined commercially in the USA, Kazakstan and South Africa to make temperature-resistant and insulating porcelains.

Anderson, Carl David (1905–91) US physicist who shared the 1936 Nobel Prize for physics with Victor HESS. In 1932 he discovered the first known particle of antimatter, the POSITRON or anti-electron, while studying cosmic rays. He later also helped discover the MUON, an ELEMENTARY PARTICLE.

andesite Fine-grained, intermediate volcanic rock, found most frequently in recent or ancient continental margin areas. It is largely composed of finely crystalline FELDSPARS, with occasional larger crystals called **phenocrysts**. It is chemically and mineralogically similar to DIORITE.

androecium Male part of a flower. It comprises the STAMENS, each consisting of a two-lobed, pollen-carrying ANTHER on a thin stalk called a **filament**. *See also* GYNOECIUM

androgen General name for male sex HORMONES, such as TESTOSTERONE. They are produced mainly by the TESTES following the release of LUTEINIZING HORMONE, although the ADRENAL GLANDS and OVARIES produce much smaller amounts.

Andromeda Large constellation of the N sky, adjoining the Square of Pegasus. The main stars lie in a line leading away from Pegasus, and the star Alpha Andromedae actually forms one corner of the Square. In mythology, Andromeda was the princess rescued by Perseus. The most famous object in the constellation is the ANDROMEDA GALAXY.

Andromeda Galaxy Spiral GALAXY 2.2 million light-years away in the constellation Andromeda, the most distant object visible to the naked eye. The Andromeda Galaxy has a mass of over 300,000 million stars. To the naked eye, it appears as a smudge; through binoculars and small telescopes it is seen to have an elongated structure. Its true diameter is *c*.150,000 light-years, somewhat larger than our own Galaxy.

anechoic chamber (dead room) Room designed to be echo-free so that it can be used in acoustic laboratories to measure sound reflection and transmission, and to test audio equipment. The walls, floor and ceiling must be insulated and all surfaces covered with an absorbent material such as asbestos fibre or rubber, often over inward-pointing pyramid shapes to reduce reflection. The room is usually asymmetrical to reduce stationary waves. *See also* ACOUSTICS

anemometer Instrument for measuring wind speed. It consists of three cups afixed to an upright length of metal, which in turn drives a mechanism that adjusts a dial. The cups are blown round by the wind, and the speed of the wind can read from the dial. More sophisticated instruments keep a record of wind speed and can record the direction as well. An anemometer should ideally be located in an open space, well away from buildings and trees, and about 10m (30ft) above ground level.

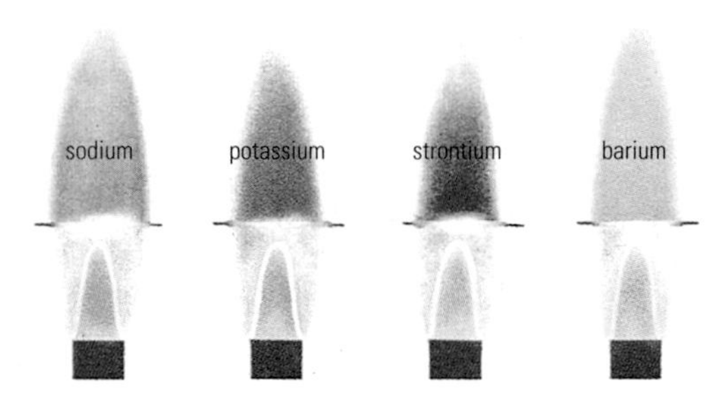

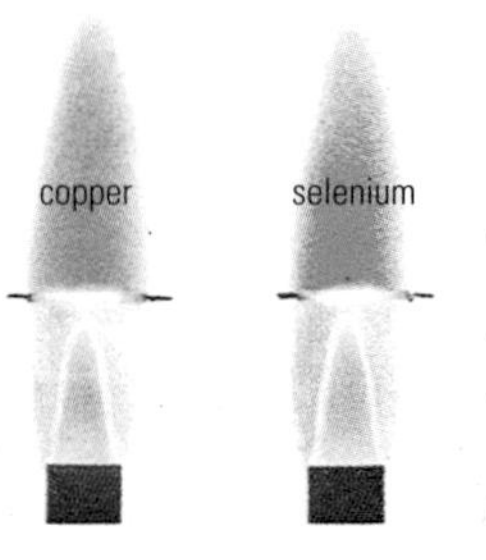

◄ **analysis** Most elements will burn, and this fact is used in flame spectral analysis. The diagram shows simple flame tests and the colour imparted to the flame by various elements.

A

ANILINE

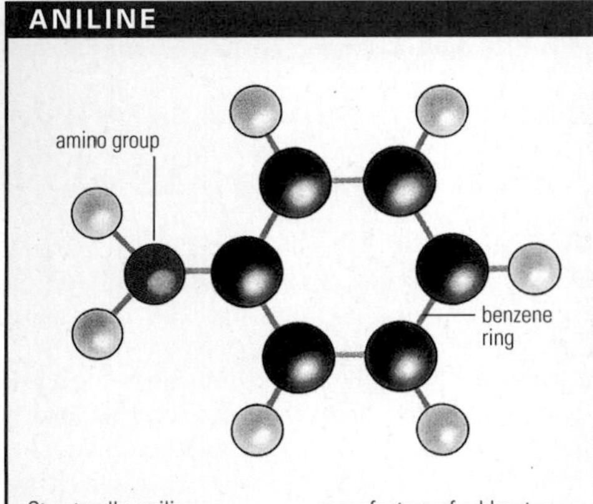

Structurally, aniline ($C_6H_5NH_2$) consists of a planer benzene ring, with one amino group attached, giving the molecule basic properties. As an organic base, aniline is used in the manufacture of rubber tyres, where it acts as an accelerator in the process of vulcanization by catalysing the chemical reactions involved – shortening the heating time needed.

anemophily Type of POLLINATION that relies on the wind to carry the pollen. Most grasses and many trees have anemophilous flowers, which lack scent and nectar, and have small or nonexistent petals. Male and female flowers are usually separate, as with the catkins of many trees, which bloom before the trees come into leaf.

aneroid barometer *See* BAROMETER

Anfinsen, Christian Boehmer (1916–) US biochemist who in 1972 shared the Nobel Prize for chemistry with Stanford MOORE and William STEIN. Anfinsen carried out research into the relationships between the biological function and molecular structure of PROTEINS.

angiography X-RAY examination of major blood vessels. It is most commonly used to investigate CORONARY ARTERY DISEASE or any disruption of the blood supply to the brain. The technique requires the injection of a radiopaque dye into the bloodstream so that the diseased vessel appears in clear silhouette on the X-ray screen.

angioplasty Surgical repair of diseased or damaged blood vessels. It is a MINIMAL ACCESS SURGERY (keyhole) procedure, in which a balloon-tipped catheter is passed into a main ARTERY, often in the groin, and advanced until it comes to rest in the narrowed vessel; the balloon is then inflated to widen the artery. One of the best-known interventions of this kind, used to treat coronary heart disease, is percutaneous transluminal coronary angioplasty (PTCA), a method of improving blood flow in either of the two arteries supplying the heart muscle.

angiosperm Any of about 250,000 species of plant of the phylum Angiospermophyta, which produce FLOWERS, FRUITS and SEEDS. Angiosperms include most herbs, shrubs, many trees, fruits, vegetables and cereals. Their seeds are protected by an outer covering – angiosperm means "plant with enclosed seeds". Angiosperms are further subdivided into two classes, MONOCOTYLEDONS (Monocotyledonae) and DICOTYLEDONS (Dicotyledonae). Many flowering plants are used for food and timber and in medicine.

angiotensin PEPTIDE in the blood that increases blood pressure by inducing contraction of narrow blood vessels. *See also* RENIN.

angle Measure of the inclination of two straight lines or planes to one another; also a means of describing the amount of rotation. One complete revolution is divided into 360° (degrees) or 2π radians. A right-angle measures 90° or $\pi/2$ radians. One degree may be subdivided into 60' (minutes), and one minute into 60" (seconds).

angle of incidence In optics, the angle at which a light ray strikes a mirror or other surface. The angle is measured between the ray and the normal – the line at right angles to the surface where the ray meets it. For a mirror, it is equal to the ANGLE OF REFLECTION.

angle of reflection In optics, the angle at which a light ray leaves a reflecting surface. The angle is measured between the ray and the normal – the line at right angles to the surface where the ray leaves it. For a mirror, it is equal to the ANGLE OF INCIDENCE.

angle of refraction In optics, the angle at which a light ray is refracted at the surface between two different transparent media. The angle is measured between the refracted ray and the normal – the line at right angles to the surface where the ray is refracted.

anglesite Sulphate mineral, a form of lead sulphate ($PbSO_4$). It is usually found together with cerussite in hydrothermal veins as an alternation product of galena. Its crystals are usually orthorhombic, and can be tabular or prismatic. It can be colourless, white or grey. Hardness 2.5–3; s.g. 6.4.

angstrom (angstrom unit, symbol Å) Obsolete unit of length, equal to 10^{-10} m or 0.1nm (NANOMETRE). It was used to express the wavelength of light and ultraviolet radiation, and interatomic and intermolecular distances. The nanometre is now the preferred unit.

angular acceleration Rate of change of angular velocity. The average angular acceleration of an object whose angular velocity changes from ω_1 to ω_2 over a time t is $(\omega_2 - \omega_1)/t$. Instantaneous angular acceleration is the value approached as t becomes small. The direction of the angular acceleration VECTOR is perpendicular to the plane of motion. The tangential acceleration a_T of a particle at a distance r from a fixed point is $r\alpha$, where α is its angular acceleration.

angular diameter In astronomy, the apparent diameter of a celestial body expressed in angular measure (usually in arc minutes and seconds). It is the angle subtended at the observer by the true diameter of the body under observation. If the distance of the celestial body from the observer is known, its true diameter can be calculated.

angular distance In astronomy, the distance on the CELESTIAL SPHERE between two celestial bodies measured along an arc of a great circle passing through them with the observer at the centre. For example, the angular distance of the Pointers of Ursa Major (The Plough) is 5°.

angular momentum (symbol L) Product of the moment of inertia I and the angular velocity ω of an object or the centre of gravity of a system of particles. Angular momentum is a VECTOR quantity that is conserved at all times.

angular velocity Rate of change of an object's angular position relative to a fixed point. Average angular VELOCITY ω of an object moving from angle θ_1 to θ_2 in time t is $(\theta_2 - \theta_1)/t$. Instantaneous angular velocity is the value approached by ω as t tends to zero. The direction associated with the angular velocity of an object is perpendicular to the plane of its motion. The speed v of an object at a distance r from a fixed point is directly proportional to the magnitude ω of its angular velocity: $v=r\omega$.

anhydride Chemical compound derived from another compound by removing water. Thus, sulphur trioxide (SO_3) is the anhydride of sulphuric acid (H_2SO_4).

anhydrite Mineral form of calcium sulphate ($CaSO_4$), usually found in sedimentary rocks in salt beds. Its crystals are orthorhombic, and usually occur in large deposits. It has a glassy or pearly lustre and is colourless when pure. Hardness: 3–3.5; s.g. 3. *See also* GYPSUM

aniline (phenylamine, $C_6H_5NH_2$) Highly poisonous, colourless, oily liquid made by the reduction of nitrobenzene. Also called aminobenzene, it is an important starting material for making organic compounds such as drugs, explosives and particularly dyestuffs. Properties: r.d. 1.02; m.p. −6.2°C (20.8°F); b.p. 184.1°C (363.4°F). *See also* AMINE

animal Living organism of the kingdom Animalia, usually distinguishable from members of the plant kingdom (Plantae) by its power of locomotion (at least during some stage of its existence); a well-defined body shape; limited growth; its feeding exclusively on organic matter; the production of two different kinds of sex cells; and the formation of an embryo or larva during the developmental stage. Higher animals, such as the VERTEBRATES, are easily distinguishable from plants, but the distinction becomes blurred with the lower forms. Some one-celled organisms could easily be assigned to either category. Scientists have classified about a million different kinds of animals in more than twenty phyla. The simplest (least highly evolved) animals include the PROTOZOA, SPONGES, jellyfish and worms. Other invertebrate phyla include ARTHROPODS (arachnids, crustaceans and insects), MOLLUSCS (shellfish, octopus and squid) and ECHINODERMS (sea urchins and starfish). Vertebrates belong to the CHORDATA phylum, which includes fish, amphibians, reptiles, birds and mammals.

animal behaviour *See* ETHOLOGY

animal classification Systematic grouping of animals into categories based on shared characteristics. The first major classification was drawn up by Aristotle. The method now used was devised by Carolus LINNAEUS, a Swedish botanist, in the 1750s. Each animal is given a two-part Latin name (*see* BINOMIAL NOMENCLATURE), the first part indicating its GENUS, the second its SPECIES. A species is composed of animals capable of interbreeding in nature. A genus includes all similar and related species. The FAMILY takes in all related genera, and an ORDER is made up of all related families. Similar orders are grouped in a CLASS, and related classes make up a PHYLUM. More than twenty separate phyla comprise the animal kingdom. For example, the dog is classified: phylum Chordata; class Mammalia; order Carnivora; family Canidae; genus *Canis* and species *familiaris*.

anion Negative ION which is attracted to the ANODE during electrolysis.

annealing Slow heating and cooling of a metal, alloy or glass to relieve internal stresses and make up crystal dislocations or vacancies that may have been introduced during mechanical shaping, such as rolling or extruding (ejection). Annealing increases the material's workability and durability. Machine tools, wire and sheet metal are annealed during manufacture. *See also* EXTRUSION; TEMPERING

annelid Member of the Annelida phylum of segmented worms. All have encircling grooves usually corresponding to internal partitions of the body. A digestive tube, nerves and blood vessels run through the entire body, but each segment has its own set of internal organs. Annelids form an important part of the diets of many animals. The three main classes are: Polychaeta, marine worms; Oligochaeta, freshwater or terrestrial worms; and Hirudinea, leeches.

annihilation In particle physics, the complete

conversion of matter into energy. An ELEMENTARY PARTICLE and its antiparticle are converted, on collision, into gamma radiation. An electron and positron annihilate to produce two gamma-ray photons, travelling in opposite directions to conserve momentum. Each has an energy of 0.511MeV, which is equivalent to the rest mass of the electron or positron.

annual Plant that completes its life cycle in one growing season, such as sweet pea, sunflower and wheat. Annual plants overwinter as seeds. *See also* BIENNIAL; PERENNIAL

annual paralax Angle subtended at a celestial object by the radius of the Earth's orbit, which is 1 astronomical unit.

annual ring (growth ring) Concentric circles visible in cross-sections of woody stems or trunks. Each year the CAMBIUM layer produces a layer of XYLEM, the vessels of which are large and thin-walled in the spring and smaller and thick-walled in the summer, creating the contrast between the rings. Used to determine the approximate age of trees, the thickness of these rings reveals the environmental conditions during a tree's lifetime.

anode Positive ELECTRODE of an electrolytic cell which attracts ANIONS during ELECTROLYSIS.

anodizing Electrolytic process to coat ALUMINIUM or MAGNESIUM with a thin layer of oxide to help prevent corrosion. The process makes the metal the ANODE in an acid solution. The protective coating is insoluble and a good insulator. It can be dyed bright colours, many of which are resistant to sunlight.

anomalistic period Time taken for a celestial body to make one complete revolution around another, starting and finishing at the same point, such as PERIGEE or PERIHELION. It is slightly longer than the SIDEREAL PERIOD – 27.55455 days for the Moon and 365.25964 days for the Earth.

anorexia nervosa Abnormal loss of the desire to eat. A pathological condition, it is seen mainly in young women anxious to lose weight. It can result in severe emaciation and in rare cases become life-threatening.

anorthosite Coarse-grained, basic IGNEOUS ROCK that forms deep underground and in DYKES and INTRUSIONS. It is composed mainly of FELDSPAR, with some olivine and pyroxine. Usually light in colour, it may have bands of darker minerals.

ANS Abbreviation of AUTONOMIC NERVOUS SYSTEM

antagonist In biology, a muscle that acts in opposition to another muscle. For example, the biceps and triceps are an **antagonistic pair**. In medicine, a drug or hormone that acts to oppose the action of another drug or hormone. In each case, the opposing muscle, drug or hormone is called the agonist. The antagonist to a drug may also be termed an ANTIDOTE.

antenna (aerial) Part of a radio system from which the signal is radiated into space (transmitting antenna) or by which it is received from space (receiving antenna). The shape and structure of the antenna depend on the frequency of the radiation and the directional requirements of the system. For example, a beam antenna is required to transmit a radar beam while a simple wire may suffice for an AMPLITUDE-MODULATED (AM) radio receiver.

anthelion Phenomenon in which there appears to be a second Sun in the sky opposite the real Sun. The "ghost" Sun is at the same altitude as the real Sun and is caused by the refraction of sunlight by small crystals of ice in the atmosphere. In polar regions, the faint halo sometimes seen fringing the shadow of an object cast onto fog or a bank of cloud is also called an anthelion.

anther Element of a FLOWER that produces, contains and distributes POLLEN (which contains the male GAMETES, or sex cells). The pollen is formed within two chambers called **pollen sacs**, and the chambers themselves are located in two lobes. Together with its connecting **filament**, the anther forms a STAMEN.

antheridium Male sex organ of fungi and various primitive plants. As well as in fungi, antheridia are found in algae, bryophytes (liverworts and mosses), pteridophytes (club mosses, ferns and horsetails) and cycads. They produce male gametes (sex cells) called **antherozoids**. These move independently in a film of water to the female gametes to achieve fertilization.

anthracene ($C_6H_4(CH)_2C_6H_4$) Organic chemical compound, the molecular structure of which consists of three BENZENE rings linked together. It is produced in the distillation of coal, and is used for the manufacture of dyes.

anthracite Form of COAL consisting of more than 90% CARBON, which is relatively hard, black and with a metallic lustre and sub-conchoidal fracture. It burns with the hot non-luminous flame of complete combustion. It is the final form in the series of fuels: PEAT, lignite, bituminous coal and black coal.

anthrax Contagious disease, chiefly of livestock, caused by the microbe *Bacillus anthracis*. The earliest reports of the disease date back to about the first century BC, although it is likely to have existed before then. Human beings can catch anthrax from contact with infected animals or their hides. Symptoms include dark pustules on the skin and severe pneumonia. Treatment is with antibiotics. Since World War 2, the disease has been harnessed as a potential biological weapon.

Anthropoidea Suborder of primates including monkeys, apes and human beings. Anthropoids have flatter, more human like faces, larger brains and are larger in size than prosimian PRIMATES.

antibiotic Substance that is capable of stopping the growth of, or destroying, BACTERIA and other microorganisms. Many antibiotics are themselves produced by microorganisms (bacteria and MOULDS). Antibiotics are germicides that are safe enough to be swallowed or injected into the body. The introduction of antibiotics, from about the time of World War 2, has revolutionized medical science, making possible the virtual elimination of once widespread and often fatal diseases, including typhoid fever, plague and cholera. Some antibiotics are selective – that is, effective against specific microorganisms; those effective against a large number of microorganisms are known as **broad-spectrum** antibiotics. Some important antibiotics are: PENICILLIN, the first widely used antibiotic, streptomycin and the tetracyclines and cephalosporins. Because some bacteria, once sensitive to certain antibiotics, have now become resistant to them, there is a constant search for new antibiotics. *See also* ANTISEPTIC

ANTENNA

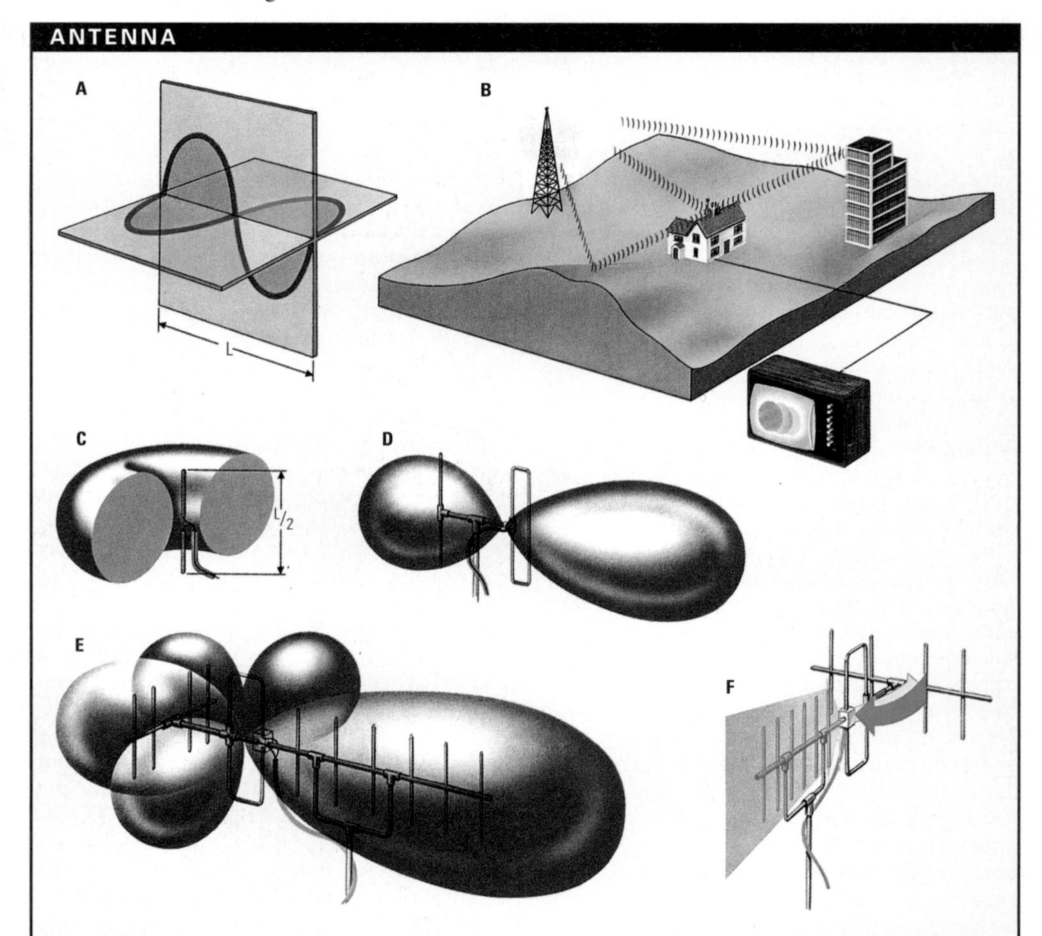

The signal emitted by a transmitting antenna (A) can be visualized as a magnetic and an electrical sine wave perpendicular to one another. The electrical wave, which carries television signals, is shown to be vertical and this situation is known as vertical polarization. The most favourable length of antenna to receive this signal is $^1/_2$L (or $^L/_2$). An antenna may receive a direct signal as well as a number of reflected signals. The effect of this is to produce a multiple image called ghosting (B). This is because a plain antenna is equally receptive in all directions and the shape representing this is shown in C. This defect can be partially overcome by the addition of a reflector (D), or a system of reflectors and directors (E). The effect is to make the antenna sensitive in the forward direction. The mechanism of the directors is to concentrate the incoming wave and the reflector behaves in the same way as a mirror (F).

A

ANTIBODY

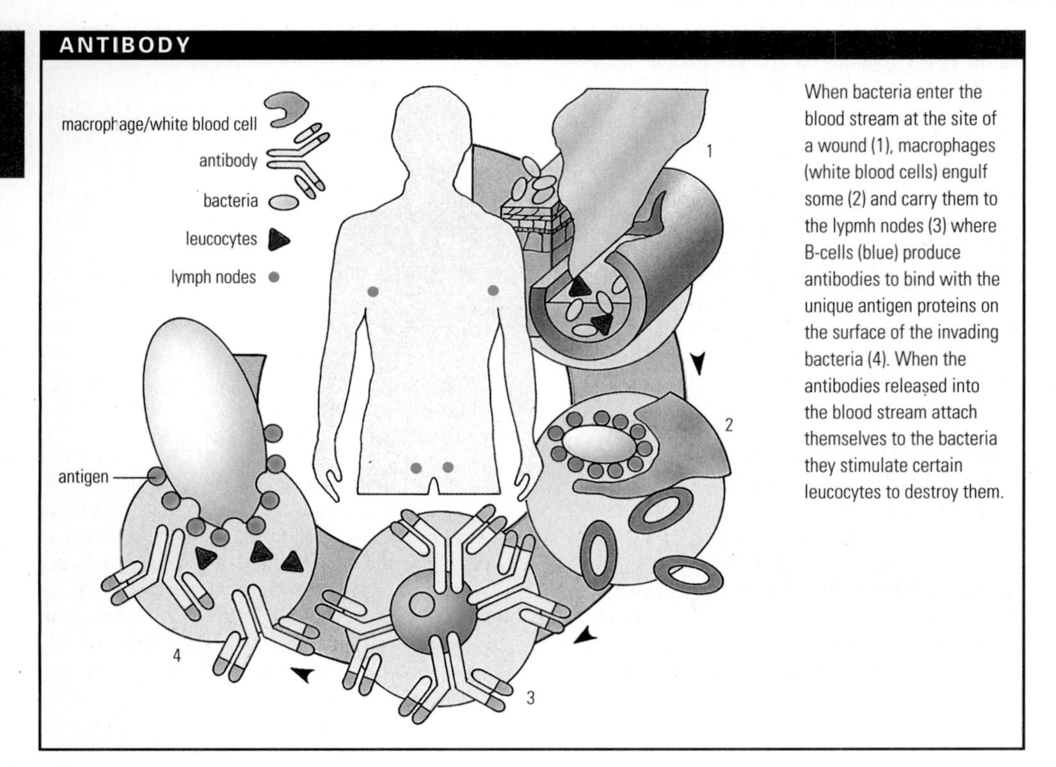

When bacteria enter the blood stream at the site of a wound (1), macrophages (white blood cells) engulf some (2) and carry them to the lypmh nodes (3) where B-cells (blue) produce antibodies to bind with the unique antigen proteins on the surface of the invading bacteria (4). When the antibodies released into the blood stream attach themselves to the bacteria they stimulate certain leucocytes to destroy them.

antibody PROTEIN synthesized in the blood by LYMPHOCYTES in response to the entry of "foreign" substances or organisms (ANTIGENS) into the body. Each episode of bacterial or viral infection prompts the production of a specific antibody to fight the antigens in question. After the infection has cleared, the antibody remains in the blood to fight off any future invasion. This process is a significant part of the bodies IMMUNE SYSTEM.

anticline Arch-shaped fold in rock strata, closing upwards. Unless the formation has been overturned, the oldest rocks are found in the centre with younger rocks symmetrically on each side of it.

anticoagulant Substance that prevents or counteracts coagulation, or clotting. Anticoagulants are used to treat and prevent diseases caused by blood clots, such as THROMBOSIS. Heparin is a commonly used blood anticoagulant.

anticyclone (high) An area of high pressure where air is subsiding, causing dry weather. Winds are normally light and the weather is settled. Major high-pressure regions are found in the HORSE LATITUDES, about 20° to 30° N and S of the Equator. There are also large anticyclones in the Arctic and Antarctic. High pressure over Britain in summer will bring hot and sunny weather because the air will have come from the Sahara and Mediterranean regions, but in winter the high pressure may have come from the Arctic and thus results in cold weather. An anticyclone is characterized by clockwise wind movement in the Northern Hemisphere and anticlockwise movement in the Southern Hemisphere. *See also* CYCLONE

antidepressant *See* DRUG

antidiuretic hormone (ADH) (vasopressin) HORMONE that controls the concentration of fluid in the body by stimulating the kidneys to absorb water and thereby produce less urine. ADH is secreted by the posterior part of the PITUITARY GLAND, although it is actually produced in the nearby HYPOTHALAMUS of the BRAIN. A deficiency of ADH causes the disorder diabetes insipidus, which is characterized by the production of large quantities of dilute urine.

antidote Term generally used to describe a drug or other agent used to counter the effects of a POISON. Antidotes may be either specific or general in their effect.

antifreeze Substance dissolved in a liquid to lower its freezing point. Ethane diol (ethylene glycol, HOC_2H_4OH) is commonly used in car radiators.

antigen Any substance or organism that is recognized as "foreign" by the IMMUNE SYSTEM. The presence of an antigen triggers the production of an ANTIBODY, part of the body's defence mechanism against disease. The antibody reacts specifically with the antigen and either neutralizes it, causes it to destroy itself or attracts LEUCOCYTES to carry out the destruction.

antihistamine Any one of certain drugs that counteract or otherwise prevent the effects of histamine, a natural substance released by the body in response to injury or more often as part of an allergic reaction. Histamine can produce symptoms such as sneezing, running nose and burning eyes. *See also* HAY FEVER

antilogarithm *See* LOGARITHM

antimatter Matter made up of antiparticles, which are identical to ordinary particles in every way except that the ELECTRIC CHARGE, SPIN and MAGNETIC MOMENT are reversed. When an antiparticle, such as a positron (anti-electron), antiproton or antineutron meets its respective particle, both are annihilated. It is possible, though unlikely, that there are stars or galaxies composed entirely of antimatter. *See also* SUBATOMIC PARTICLES

antimony (symbol Sb) Toxic, semimetallic element of group V of the periodic table. Stibnite (a sulphide) is its commonest ore. It is used in some alloys, particularly in hardening lead for batteries and type metal, and in semiconductors. The element has two allotropes: a silvery white metallic form and an amorphous grey form. Properties: at.no. 51; r.a.m. 121.75; r.d. 6.68; m.p. 630.5°C (1,166.9°F); b.p. 1,750 °C (3,182°F); most common isotope ^{121}Sb (57.25%).

antineutron In particle physics, the antiparticle of a NEUTRON. It has no electric charge, and a magnetic moment equal but opposite to that of the neutron. An antineutron is sometimes formed in a bubble chamber when a PROTON collides with an antiproton. *See also* ANTIMATTER

antinode Point of maximum displacement on a standing WAVE. A standing, or stationary, wave is one whose shape remains constant and that does not appear to move through a medium. It occurs when a travelling wave is reflected along its own path. Points of minimum displacement are called nodes. Antinodes and nodes are separated by a quarter of a wavelength.

antioxidant Chemical additive designed to reduce oxidation. Antioxidants are used to prevent fatty foods becoming rancid, deterioration of rubber, formation of gums in petrol, etc. Most are organic amines and phenols, functioning by terminating the free-radical chain reaction causing oxidation. Sulphur dioxide and ascorbic acid are often used as antioxidants for foodstuffs.

antiparticle *See* ANTIMATTER

antiseptic Chemical that destroys or stops the growth of many microorganisms. Antiseptics are weak germicides that can be used on the skin. The surgeon Joseph LISTER pioneered the use of antiseptics in 1863. One commonly used is ALCOHOL. *See also* ANTIBIOTIC

antitoxin ANTIBODY produced by the body in response to a TOXIN. It is specific in action and neutralizes the toxin. Antitoxin sera are used to treat and prevent bacterial diseases such as TETANUS and DIPHTHERIA.

Antlia (Air Pump) Small constellation of the S sky. It has no brighter star than magnitude 4.3.

anus Opening at the end of the alimentary canal in most animals, through which waste material and undigested food are passed from the body in the form of faeces. *See also* DIGESTIVE SYSTEM

aorta Principal ARTERY in the body of higher vertebrates. In humans, carrying freshly oxygenated blood, the aorta leaves the left ventricle of the HEART and descends the length of the trunk, finally dividing to form the two femoral arteries that serve the legs. *See also* CIRCULATORY SYSTEM

apatite Phosphate mineral, usually found as calcium phosphate associated with hydroxyl, chloride or fluoride ions. It occurs in igneous rocks and sedimentary deposits, as prismatic or tabular hexagonal crystals, as granular aggregates or in massive crusts. It is usually too soft for cutting and polishing, but there are two gem varieties. Hardness: 5; s.g. 3.1–3.4.

aperture In photography, a hole that allows light to pass through the lens onto the film. Modern cameras usually have a diaphragm aperture which works like the iris of a human eye. The photographer can widen or narrow the diaphragm according to a series of points on the lens dial called "f-numbers" or "f-stops". The individual f-numbers represent the focal length of the lens divided by the diameter of the aperture. As the aperture narrows (larger f-number), it gives a longer depth of field.

apex In astronomy, point on the CELESTIAL SPHERE, located in the constellation of Hercules, towards which the Sun appears to be moving. As the Sun slowly orbits the galactic centre, nearby stars (as seen from Earth) appear to move away from it because of the Sun's relative velocity.

aphelion Point at which the orbit of a celestial body that travels round the Sun is farthest from the Sun. All non-circular solar orbits, such as those of planets, asteroids, comets and some meteors, have an aphelion, as do the orbits of space probes that orbit the Sun. The point at which they are closest to the Sun is called PERIHELION.

aphotic Describing a region in which there is no light. In the seas and oceans, the aphotic zone exists at depths below which light does not penetrate, generally below 1,500m (5,000ft).

apocrine gland Type of sweat-producing gland in human skin, restricted mainly to the armpits and groin. In response to sex and stress stimuli, it secretes a liquid which is readily decomposed by bacteria giving rise to odour. *See also* ECCRINE GLAND

apogee Point in the Moon's or an artificial satellite's orbit at which it is farthest from Earth.

Apollonius of Perga (262–190 BC) Greek mathematician and astronomer. Apollonius was known as "The Great Geometer" and much of his work built on the foundations laid by EUCLID. In his only surviving book, *On Conic Sections*, he showed that an ELLIPSE, a PARABOLA and a HYPERBOLA can be obtained by taking plane sections at different angles through a cone. In astronomy, he described the motion of the planets in terms of epicycles. (An epicycle is a path drawn by a point moving on a circle which, in turn, moves so that its centre is on the circumference of a larger circle.) This was the basis of the Ptolemaic system used until the time of COPERNICUS.

Apollo project US space programme that was set up with the specific aim of landing men on the Moon and returning them safely to Earth. Initiated in May 1961 by President John F. Kennedy, it achieved its objective on 21 July 1969, when Neil ARMSTRONG was the first man to set foot on the lunar surface during the Apollo 11 mission. There was a total of 17 Apollo missions, of which six comprised lunar landings. The programme terminated with the successful Apollo- SOYUZ link-up in space during July 1975, having placed more than 30 astronauts in space and 12 on the Moon.

appendix In some mammals, finger-shaped organ, *c.*10cm (4in) long, located near the junction of the small and large intestines, usually in the lower right part of the abdomen. It has no known function in humans but can become inflamed or infected (appendicitis).

Appleton, Sir Edward Victor (1892–1965) British physicist. He was awarded the 1947 Nobel Prize for physics for his discovery of the Appleton layer of the IONOSPHERE. This layer reflects radio waves, and its discovery spurred the development of RADAR.

applied mathematics Application of mathematics with the specific aim of describing or modelling some real, idealized or hypothetical phenomena. It is a body of knowledge and theory with a mathematical structure which uses the formal rules of pure MATHEMATICS together with physical measurable quantities as input from the natural world. Mathematics has been used to represent, and make predictions about, almost every area of scientific interest over the last 3,000 years. Mathematics is practically applied in computing, electronics, engineering, meteorology, aeronautics, space flight, and so on, as well as in more theoretical areas such as MATHEMATICAL PHYSICS and MATHEMATICAL BIOLOGY.

apsis (pl. apsides) Either of two points in an object's orbit that are respectively closest to and farthest from the primary body, that which it is orbiting. The closest point is known as the **periapsis** and the farthest is known as the **apapsis**. The apsides of the Earth's orbit are its PERIHELION and APHELION; in the Moon's orbit they are its PERIGEE and APOGEE. The line of apsides is the line connecting these points.

Apus (Bird of Paradise) Far-S constellation. Its brightest star is magnitude 3.8.

aqualung *See* SCUBA DIVING

aquamarine *See* BERYL

aqua regia (Lat. royal water) Mixture of concentrated nitric and hydrochloric acids. It is so called because the acid dissolves the "royal" metal gold.

Aquarius (Water-bearer) Constellation of the S sky; it is situated between PISCES and CAPRICORNUS. Alpha and Beta are jointly the constellations brightest stars, magnitude 2.9. It also contains two important planetary nebulae, Saturn Nebula and Helix Nebula.

aqueduct Artificial channel for conducting water from its source to its distribution point. While the ancient Romans were not the first to build these conduits, their aqueducts are the most famous because they not only conveyed water over large areas but were also graceful architectural structures in their own right. One of their most extensive water systems, which served Rome itself, consisted of 11 aqueducts and took 500 years to complete. California has the world's largest conduit system: it carries water into its cities over a distance of more than 800km (500mi).

aqueous humour Watery fluid that fills the eyeball between the lens and the CORNEA. It serves a double purpose: to nourish the eyeball and to refract, or bend, light rays, so helping them to focus on the RETINA. *See also* VITREOUS HUMOUR

aqueous solution Solution in which water is the solvent. Water dissolves many polar substances, that is, IONIC COMPOUNDS and COVALENT compounds with molecular dipoles, because the water molecules themselves are polar, with a negative charge on the oxygen atoms. They tend to cluster around positive ions (solvation), making the solution energetically favourable.

aquifer Rock, often sandstone or limestone, which is capable of both storing and transmitting water owing to its porosity and permeability. Much of the world's human population depends on aquifers for its water supply. These may be directly exploited by sinking wells.

Aquilia (Eagle) Distinctive constellation of the N hemisphere led by the star Altair of first magnitude. The Milky Way is very rich in this region.

Ara (Altar) Constellation of the S sky, between Scorpius and Triangulum Australe. Its brightest star is Beta, magnitude 2.9.

Arabic numeral Any of the common notations 0 to 9. The system is strictly Hindu. *See* NUMBER

arachnid ARTHROPOD (animal with an exoskeleton and jointed limbs) of the class Arachnida, which includes the SPIDER, tick, mite, scorpion, and harvestman. Arachnids have four pairs of jointed legs, two distinct body segments (cephalothorax and abdomen), and chelicerate jaws (consisting of clawed pincers). They lack antennae and wings.

aragonite Carbonate mineral, calcium carbonate ($CaCO_3$). It is formed under special conditions, generally in caverns and hot springs. Present-day mollusc shells are formed of aragonite. It readily converts to calcite (another mineral of $CaCO_3$). It occurs as orthorhombic system groups of needle-like crystals or massive deposits. Its appearance is glassy white. Hardness 3.5–4; s.g. 2.9.

arc Portion of a curve. For a circle, of radius r, the length (s) of an arc covering θ degrees on that circle is given by

$$s = 2 \times r \times \pi \times \theta/360.$$

If the angle is measured in RADIANS the length of the arc is given by the product of the radius and the angle, that is $s = r\theta$. The arc is said to subtend the angle θ at the centre of the circle.

arch Upward-pointing or curving arrangement of masonry blocks or other load-bearing materials, also used in architectural decoration. The ancient Romans invented traditional masonry arches but later cultures extended their repertoire to include many different and sometimes quite elaborate shapes. The basic structure of a masonry arch consists of wedge-shaped blocks ("voussoirs") placed on top of each other and a central **keystone** which holds them together at the top. Modern materials, such as steel and reinforced concrete, are strong and flexible enough to stand on their own and they can also stretch across much wider spans. The form of an arch is useful for historians who want to date a building.

Archaean Sub-division of Precambrian geological time, from about 4,000 to 2,500 million years ago.

Archaebacteria Subkingdom of the kingdom PROKARYOTAE, comprising BACTERIA which on the basis of both RNA and DNA composition and biochemistry differ significantly from other bacteria. They are thought to resemble the ancient bacteria that first arose in extreme environments such as hot springs, boiling muds, sulphur-rich deep-sea vents, volcanoes and highly saline seashores, habitats in which they are still found today. Archaebacteria have unique protein-like cell walls, unique cell membrane chemistry, and RIBOSOMES (microscopic structures that synthesize proteins) of a distinctive shape. They include methane-producing bacteria, which use simple organic compounds such as methanol and ethanoate as food, combining them with carbon dioxide and hydrogen gas from the air, and releasing methane as a by-product. The bacteria of hot springs and saline areas have a variety of ways of obtaining food and energy, including the use of minerals instead of organic compounds. They include both AEROBIC and ANAEROBIC bacteria. Some hot spring bacteria can tolerate temperatures of up to 88°C (190°F) and acidities as low as pH 0.9. One species, *Thermoplasma*, may be related to the ancestor of the nucleus and cytoplasm of the more advanced EUKARYOTE cells. Some taxonomists consider the Archaebacteria to be so different from other living organisms that they belong not even in their own separate kingdom, but in a higher grouping called a DOMAIN. *See also* TAXONOMY

archegonium Flask-shaped female sex organ of all plants except algae. Bryophytes (liverworts and mosses), pteridophytes (club mosses, ferns and horsetails) and many GYMNOSPERMS (seed-bearing plants) have archegonia. Female gametes (ova) develop in the base of the "flask". At fertilization the cells forming the "neck" of the flask disintegrate to allow male gametes (**antherozoids**) to enter.

Archimedes (*c.*287–212 BC) Greek mathematician and engineer. He developed a method for expressing large numbers and made outstanding discoveries about the determination of areas and volumes, which led to a new accurate method of calculating π (pi). *See also* ARCHIMEDES' PRINCIPLE; ARCHIMEDES' SCREW

Archimedes' principle Observation by ARCHIMEDES that a body immersed in a FLUID is

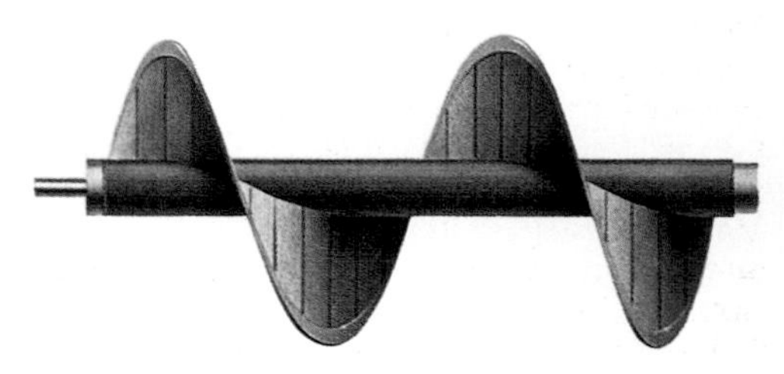

▲ **Archimedes** Thought to have been invented by Archimedes in the 3rd century BC, the Archimedes' screw was primarily used for raising water. However, in the 19th century the same principle was applied to propelling ships. During trials, however, this particular design of screw was broken in half, and the result was an immediate increase in the speed of the vessel. In this way, by happy accident, the science of marine propulsion advanced by 50 years.

A

pushed up by a force equal to the weight of the displaced fluid. He supposedly formulated this principle after stepping into a bath and watching it overflow. According to legend, he became so excited that he ran out naked shouting "Eureka! Eureka!" ("I have found it! I have found it!").

Archimedes' screw Machine used for raising water, thought to have been invented by Archimedes in the 3rd century BC. The most common form of the machine is a cylindrical pipe enclosing a helix, inclined at a 45° angle to the horizontal with its lower end in the water. When the machine rotates, water moves up through the pipe.

arc-lamp Device in which light is produced by an electric arc between two electrodes. Many modern arc-lamps, used as sources of intense light, have metal and oxide electrodes immersed in a gas which becomes luminous during arcing.

area Two-dimensional measurement of a plane, figure or body (such as this page) given in square units, such as cm^2 (sq cm) or m^2 (sq m). The area of a rectangle of sides a and b is ab; the areas of triangles and other polygons can be determined using TRIGONOMETRY. Areas of curved figures and surfaces can be determined by integral CALCULUS.

arenaceous Describing something that is associated with sand; most commonly used to describe types of SEDIMENTARY ROCK. In botany, an arenaceous plant is one that thrives in sandy soil; while in zoology, arenaceous animals live in sand. The shells of some microscopic animals, which consist mainly of sand particles, are also described as being arenaceous.

arête Sharp ridge formed by erosion where the heads of two GLACIERS meet. An example is the Matterhorn on the Swiss-Italian border.

argentite (silver glance) Mineral that is an important ore of silver. Argentite is a dark grey mineral that consists of silver sulphide (Ag_2S). It occurs in veins, often together with native silver, in parts of Germany, Mexico and the USA.

argillaceous Describing something that consists of very fine grains, such as those associated with clay, shale or silt. Argillaceous rocks, such as kaolinite, are sedimentary rocks the particles of which are similar in size to those of clay or silt.

arginine ($H_2NC(NH)NH(CH_2)_3CH(NH_2)COOH$) One of the essential AMINO ACIDS. It is a colourless, crystalline compound that occurs in decomposing animal and vegetable proteins, and in the muscles of invertebrates. It can also be made synthetically.

argon (symbol Ar) Monatomic (single-atom), colourless and odourless gaseous element that is the most abundant NOBLE GAS (inert gas). Argon was discovered in 1894 in air by Lord RAYLEIGH and Sir William RAMSAY. It makes up 0.93% of the atmosphere by volume, and 99.6% of this is ^{40}Ar, with the remainder being ^{36}Ar (0.34%) and ^{38}Ar (0.06%). Obtained commercially by the fractionation of liquid air, it is used in electric LIGHT BULBS, fluorescent tubes and argon LASERS, and in arc welding of aluminium and semiconductor production. The element has no known true compounds. Properties: at.no. 18; r.a.m. 39.948; r.d. 0.0017837gcm^{-3}; m.p. −189.4 °C (−308.9°F); b.p. −185.9 °C (−302.6°F).

argument In mathematics, the name given to the dependent variable. For instance, in the function $f(x) = x^2 + 3$, x is the argument.

Aries (Ram) Constellation in the N sky; it is situated between PISCES and TAURUS. Its brightest stars are Alpha, magnitude 2.0, and Beta, magnitude 2.6.

Aristarchus of Samos (310–230 BC) Greek mathematician and astronomer. He tried to calculate the distances of the Sun and Moon from the Earth, as well as their sizes, but although his method was sound the results were inaccurate. He is thought to be the first to propose that the Sun is the centre of the Universe (**heliocentric theory**); the idea was not taken up because it did not seem to make the calculation of planetary positions any easier.

Aristotle (384–322 BC) Greek philosopher and encyclopedist. He developed a view of the Universe based on the four "elements" (earth, air, fire and water) and the system of concentric spheres proposed by Eudoxus of Cnidus upon which various celestial bodies were carried. Aristotle added more spheres so as to account for the motion of all celestial bodies. The celestial bodies were, he said, made of a substance called aether, and are perfect and incorruptible, unlike the Earth. He attempted to estimate the size of the Earth, which he maintained was spherical, but unmoving – the centre of the Universe. This "Aristotelian" view of the world remained almost unchallenged until COPERNICUS.

arithmetic Calculations using the operations of addition, subtraction, multiplication and division. The procedures of arithmetic were put on a formal axiomatic basis by Guiseppe PEANO in the late 19th century. Using certain postulates, including that there is a unique natural number, 1, it is possible to give formal definition of the set of natural numbers and the arithmetical operations. Multiplication can be thought of as repeated additions: subtraction and division are the inverse operations of addition and multiplication. The operations can be extended to negative, rational and irrational NUMBERS. *See* AXIOMATIC METHOD

arithmetic progression Sequence of numbers in which each term is produced by adding a constant (the common difference d) to the preceding term. It has the general form $a, a + d, a + 2d,...$ and so on. The sum of such a progression

$a + (a + d) + (a + 2d) + ...$

is an arithmetic series. For n terms, the sum has a value $an + dn(n - 1)/2$. An example of a finite arithmetic progression with 10 terms is the sequence 5, 7, 9, 11, ..., 21, 23. In this case $a = 5$, $d = 2$ and $n = 10$; the sum is 140.

arkose Comparatively young, medium-grained SEDIMENTARY ROCK resembling SANDSTONE, consisting of QUARTZ with up to 30% FELDSPAR. It forms in marine and freshwater sediments and continental deposits, usually from the breakdown of granite. It is pale grey to pinkish in colour.

Arkwright, Sir Richard (1732–92) British inventor and industrialist. He introduced powered machinery to the textile industry with his water-driven frame for spinning; he started work on this machine in 1764 and completed it four years later, although it was not patented until the following year. He opened textile factories in Nottingham.

armature Central part of an ELECTRIC MOTOR or GENERATOR. In most machines the armature consists of several coils of wire on a spindle and rotates in a magnetic field. In an electric motor the armature rotates when an electric current passes through it, thus providing the driving force of the motor.

armour Protection against weaponry. Soldiers' body armour (usually made of metal or thick leather) has been employed since antiquity to prevent injury. Except for helmets (and occasionally bulletproof vests) body armour ceased to be used in European battle in the 17th century. Warships were first armoured in the mid-19th century, using metal plating to deflect small-arms fire. Vehicles acquired armour-plating during World War 1. Armoured vehicles include TANKS, armoured personnel carriers and mechanized ARTILLERY.

Armstrong, Neil Alden (1930–) US astronaut. He was chosen as a NATIONAL AERONAUTICS AND SPACE ADMINISTRATION (NASA) astronaut in 1962 and was the command pilot for the GEMINI 8 orbital flight in 1966. In July 1969, during the APOLLO 11 mission, he became the first man to step onto the Moon, remarking that it was "one small step for man, one giant leap for mankind."

aromatic compound Organic chemical compound that contains atoms of carbon (often six) joined to form a stable ring-shaped molecule, as seen in BENZENE. Some of these compounds have rings also containing nitrogen, oxygen or sulphur.

Arrhenius, Svante August (1859–1927) Swedish chemist and physicist. Arrhenius won the Nobel Prize for chemistry in 1903 for his theory of electrolytic dissociation. His later work was concerned with reaction rates, biochemistry, the structure of the Universe and the revelation that light exerts pressure.

arsenic (symbol As) Toxic, semimetallic element of group V of the periodic table, probably obtained in 1250. Compounds containing arsenic are used as a poison for rodents, insects and weeds, and to harden lead and make semiconductors. Three allotropes are known: white arsenic, black arsenic and a yellow, nonmetallic form. Properties: at.no. 33; r.a.m. 74.9216; r.d. 5.7; sublimes 613°C (1,135°F); most common isotope ^{75}As (100%).

arsenopyrite (iron arsenide-sulphide, FeAsS) Sulphide mineral, the major ore of ARSENIC. It is found with precious metal ores in high-temperature veins. It occurs as monoclinic system prismatic crystals or granular masses. It is a metallic white-grey. Hardness 5.5–6; s.g. 6.

arteriography Imaging technique for examining blood vessels. A dye opaque to X-rays

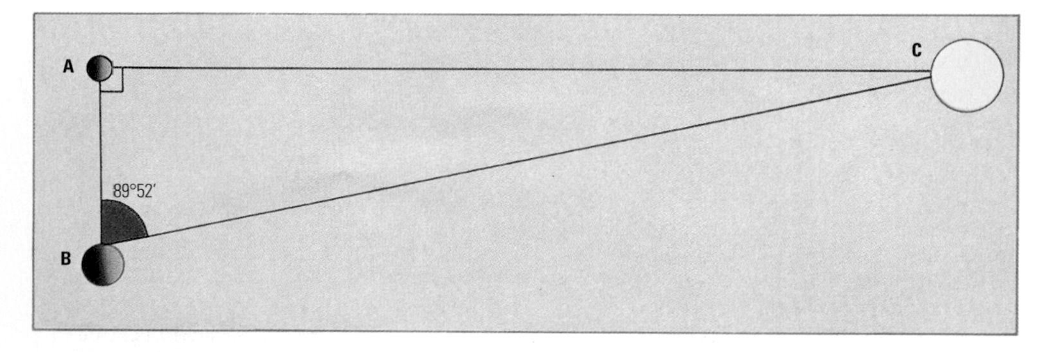

▲ **Aristarchus of Samos** The Greek astronomer Aristarchus measured the relative distances of the Sun and Moon. When the Moon is at its first quarter (A), the angle it makes with the Sun (C) is near 90°. By measuring the angle of the Earth (B) Aristarchus could determine from the triangle the relative distances from Earth. He found the angle to be 87°, instead of the true value of 89°52′; but a small error leads to a large discrepancy in the ratio of the distances of the Sun and Moon respectively from Earth. Aristarchus' ratio was 19:1; the true ratio is 370:1.

ARTHRITIS

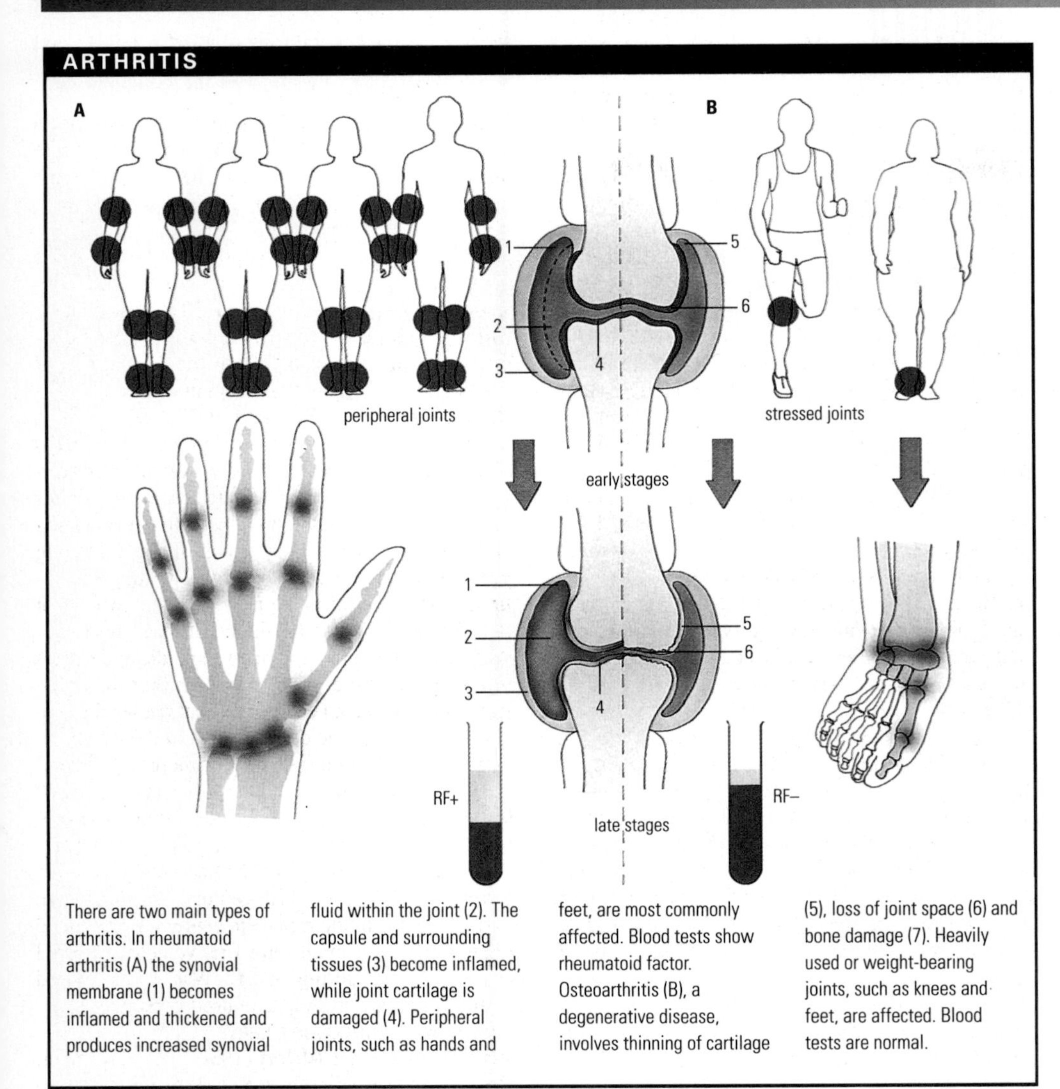

There are two main types of arthritis. In rheumatoid arthritis (A) the synovial membrane (1) becomes inflamed and thickened and produces increased synovial fluid within the joint (2). The capsule and surrounding tissues (3) become inflamed, while joint cartilage is damaged (4). Peripheral joints, such as hands and feet, are most commonly affected. Blood tests show rheumatoid factor. Osteoarthritis (B), a degenerative disease, involves thinning of cartilage (5), loss of joint space (6) and bone damage (7). Heavily used or weight-bearing joints, such as knees and feet, are affected. Blood tests are normal.

(radiopaque) is injected into the patient's bloodstream and the arteries are revealed in an X-ray photograph. The technique is particularly useful in detecting ARTERIOSCLEROSIS.

arteriosclerosis Blanket term for degenerative diseases of the arteries, in particular **atherosclerosis** (hardening of the arteries). It is caused by deposits of fatty materials and scar tissue on the ARTERY walls which narrow the channel and restrict blood flow. There is usually an absence of overt symptoms until the disease is well advanced, when there is an increased risk of heart disease, stroke or gangrene. Evidence suggests that predisposition to the disease is hereditary and risks factors include cigarette smoking, inactivity, obesity and a diet rich in animal fats and refined sugar. Treatment is by drugs and, in some cases, surgery to replace a diseased length of artery.

artery One of the BLOOD VESSELS that carry BLOOD away from the HEART. The pulmonary artery carries deoxygenated blood to the lungs, but all other arteries carry oxygenated blood to the body's tissues. An artery is usually protected and embedded in muscle; its walls are thick, elastic and muscular and pulsate with the heartbeat. A severed artery causes major HAEMORRHAGE.

artesian well Well from which water is forced out naturally under pressure. Artesian wells are bored where water in a layer of porous rock is sandwiched between two layers of impervious rock. The water-filled layer is called an AQUIFER. Water flows up to the surface because distant parts of the aquifer are higher than the well-head.

arthritis Inflammation of the joints, with pain and restricted mobility. The most common forms are osteoarthritis and rheumatoid arthritis. **Osteoarthritis**, common among the elderly, occurs with erosion of joint cartilage and degenerative changes in the underlying bone. It is treated with analgesics and anti-inflammatories and, in some cases (especially a diseased hip), by joint replacement surgery. **Rheumatoid arthritis**, more common in women, is generally more disabling. It is an AUTOIMMUNE DISEASE which may disappear of its own accord but is usually slowly progressive. Treatment includes analgesics to relieve pain. The severest cases may need to be treated with CORTISONE injections, drugs to suppress immune activity or joint replacement surgery. *See also* RHEUMATISM

arthropod Any member of the largest animal phylum, Arthropoda. Living forms include CRUSTACEA, ARACHNIDS, centipedes, millipedes and INSECTS. Fossil forms include the extinct TRILOBITE. The species (numbering well over 1 million) are thought to have evolved from ANNELIDS. All have a hard outer skin of CHITIN that is attached to the muscular system on the inside, and is periodically shed during growth. The body is divided into segments, modified among different groups, with each segment originally carrying a pair of walking or swimming jointed legs. In some animals some of the legs have evolved into jaws, sucking organs or weapons. Arthropods have well-developed digestive, circulatory and nervous systems. Land forms use tracheae for respiration.

artificial insemination Method of inducing PREGNANCY without sexual intercourse by injecting SPERM into the female genital tract. Artificial insemination of livestock allows proven sires to impregnate many females at low cost.

artificial intelligence (AI) Science concerned with developing COMPUTERS and COMPUTER PROGRAMS that model human intelligence. The most common form of AI involves a computer being programmed to answer questions on a specialized subject. Such "expert systems" are said to display the human ability to perform expert analytical tasks. A similar system in a WORD PROCESSOR may highlight incorrect spellings, and be "taught" new words. A closely related science, sometimes known as "artificial life", is concerned with more low-level intelligence. For example, a ROBOT may be programmed to find its way around a maze, displaying the basic ability to physically interact with its surroundings.

artificial selection Breeding of plants, animals or other organisms in which the parents are individually selected by humans in order to perpetuate certain desired traits and eliminate others from the captive population. By this means, most of our domestic crops, livestock and pets have arisen. The many breeds of dog have been developed by artificial selection. Crops such as wheat have been produced by artificial selection over thousands of years. Artificial selection can be accelerated today by techniques such as plant TISSUE CULTURE, in which selected plants can be cloned and grown directly into new plants without having to wait for seed production and germination; and ARTIFICIAL INSEMINATION of livestock. In the near future the cloning of livestock embryos may also be possible. GENETIC ENGINEERING is the ultimate advance in artificial selection, allowing humans to combine specific sequences of DNA to combine desired GENES. *See also* CLONE

artillery Projectile-firing weapons with a carriage or mount. An artillery piece is generally one of four types: gun, howitzer, mortar or missile launcher. Modern artillery is classified according to calibre; ranging from under 105mm for light artillery to more than 155mm for heavy. The exact origin of artillery using gunpowder is unknown, although such weapons appeared in Europe and the Near East in the 14th century. Used by the Turks to take Constantinople in 1453, artillery changed the whole strategy and tactics of siege warfare. Advances in the 19th century included smokeless powder, elongated shells, rifling and rapid-fire breach loading and made artillery indispensable in battle. *See also* CANNON

Artiodactyla Order of mammals characterized by hoofs with an even number (2 or 4) of toes. All herbivores, the order includes giraffes, hippopotamus-

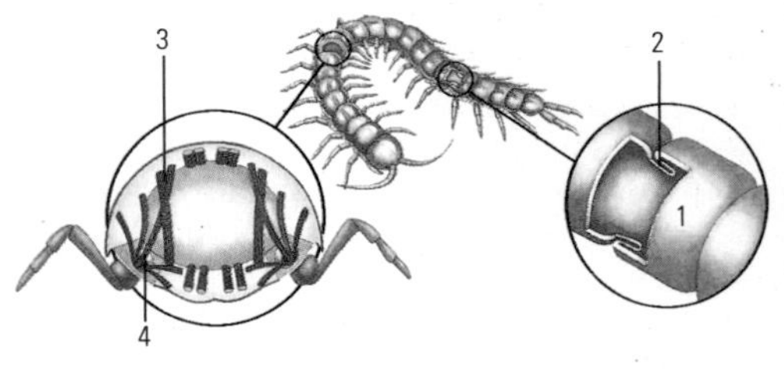

▲ **arthropod** The most numerous invertebrates (animals without a backbone) are the arthropods (joint-legged animals), such as the centipede shown here. They owe their success to the exoskeleton that covers their bodies and allows the development of jointed limbs. Body segments are encased in a rigid protein cuticle (1) and body flexibility is permitted by an overlapping membrane (2). The strength of the exoskeleton ensures that muscles (3) can be anchored to the inside of the cuticle. Groups of muscles (4) are used to move the legs. Centipedes differ from millipedes in that they have one pair of legs attached to each segment rather than the millipede's two.

► **ash** Plymouth, Monsterrat's capital, covered by volcanic ash after the eruption of Soufrière on 3 August 1997. Soufrière erupted so violently that by 5 August it became obvious that the island's 350-year old capital, Plymouth, abandoned because of earlier volcanic activity, would have to be scrapped for good, leaving its ancient churches buried by the ash.

es, deer, cattle, pigs, sheep, goats and camels. They are mostly of Old World origin and range in size from the 3.6kg (8lb) mouse deer to the 4.5 tonne hippopotamus. *See also* PERISSODACTYLA

aryl group Organic chemical group consisting of an AROMATIC COMPOUND with one hydrogen atom removed. For example, BENZENE (C_6H_6) gives rise to the phenyl group (C_6H_5 –).

asbestos Group of fibrous, naturally occurring silicate minerals used in insulating, fireproofing, brake linings and astronaut suits. Asbestos is used in the form of wool, fabric and various asbestos-cement compounds. Several types exist, the most important being white asbestos. Many countries have banned the use of asbestos, since it can cause lung cancer and asbestosis, a lung disease.

Ascomycetes Largest group of the four main classes of FUNGUS, characterized by the formation of **asci** (enlarged, commonly elongated, cells in which usually eight spores are formed). Examples are yeasts, truffles, and blue and green moulds.

ascorbic acid *See* VITAMIN

asdic *See* SONAR

asepsis Absence of disease-causing organisms. Aseptic technique is practised to exclude bacteria during surgery, wound dressing or other medical procedures. In an operating theatre, clothing, gloves, equipment and even the air are sterilized.

asexual reproduction Type of reproduction in organisms that does not involve the union of male and female reproductive cells. It occurs in several forms: FISSION – simple division of a single individual, as in bacteria and protozoa; BUDDING – growing out and eventual splitting off of a new individual, as in hydra; spore formation, as in fungi; VEGETATIVE REPRODUCTION – in which a plant sends out runners that take root to form new plants, as in strawberries, or new plants grow from organs such as bulbs. *See also* CLONE; SEXUAL REPRODUCTION

ash, volcanic Fine particles of LAVA thrown up by a volcanic explosion. The cone of compound volcanoes consists of built-up layers of ash and lava.

asparagine ($NH_2COCH_2CH(NH_2)COOH$) Crystalline AMINO ACID that occurs in proteins. It was the first amino acid to be discovered (in 1806 in asparagus).

aspartame Artificial sweetener. Consisting chemically of a compound that has two AMINO ACIDS (aspartic acid and phenylalanine) joined by a methylene group, aspartame is non-fattening and much used in "diet" foods and drinks, and in foods prepared for diabetics. It is 200 times as sweet as sugar and has none of saccharin's disadvantages. It breaks down, however, at high temperatures and is therefore unsuitable for prolonged cooking.

aspartic acid ($HOOCCH_2CH(NH_2)COOH$) AMINO ACID, it is formed by the hydrolysis of another amino acid ASPARAGINE.

asphalt *See* BITUMEN

aspirin (acetylsalicylic acid) DRUG widely used to reduce fever, and as an ANALGESIC to relieve minor pain. It is effective against inflammation, as caused by arthritis, and low-intensity pain such as headache or muscular aches. Recent evidence indicates aspirin can inhibit the formation of blood clots and in low doses can reduce the danger of heart attack and stroke. Aspirin, however, can irritate the stomach and in overdose is toxic and can cause death. It should not be given to children under the age of 12 years.

assay Test to determine the amount of a metal present in a sample of material such as ores and alloys. The term is normally reserved for finding the proportion of gold, silver or platinum present.

assembly language COMPUTER LANGUAGE for writing COMPUTER PROGRAMS in a form that is closely related to the form that computers can understand directly. Assembly language is a **low-level language**. Each instruction to be carried out by the computer is represented by a simple code. Programs written using these codes are translated by an "assembler" into a form the computer can understand.

assimilation Process by which an organism uses substances taken in from its surroundings to make new living protoplasm or to provide energy for metabolic processes. It includes the incorporation of the products of food digestion into living tissues in animals, and the synthesis of new organic material by plants during photosynthesis.

associative law Rule of combination in mathematics in which the result of two or more operations does not depend on the order in which the operations are carried out. Thus, normal addition and multiplication of numbers are operations which follow the associative law, since

$a + (b + c) = (a + b) + c$

and

$a \times (b \times c) = (a \times b) \times c$

astatine (symbol At) Semimetallic, radioactive element that is one of the HALOGENS (group VII of the periodic table). It is rare in nature, and is found in radioactive decay. Astatine-211 will collect in the THYROID GLAND and is used in medicine as a radioactive tracer. Properties: at.no. 85; r.a.m. 211; m.p. 302°C (575.6°F); b.p. 377°C (710.6°F); most stable isotope ^{210}At (half-life 8.3hr).

asteroid Small body in an independent orbit around the Sun. The majority move between the orbits of Mars and Jupiter, in the main asteroid belt. The largest asteroid (and the first to be discovered) was CERES. The smallest measured are the so-called, near-Earth asteroids. There are thought to be a million asteroids with a diameter greater than 1km (0.6mi); below this, they decrease in size to dust particles. Some very small objects find their way to Earth as METEORITES. So far nearly 6,000 asteroids have been catalogued and have had their orbits calculated, and this figure is increasing by several hundred a year. At least 10,000 more have been observed, but not often enough for an orbit to be calculated. Some of the larger asteroids are spherical, but most are irregularly shaped, and a wide variety of compositional types have been identified. Asteroids almost certainly originate from the time of the formation of the SOLAR SYSTEM and are not remnants of a large planet that disintegrated, as was once thought.

asthenosphere Weak zone within the Earth's upper MANTLE, extending from 100 to 700km (60–450mi) below the surface. It is detected by a slowing of seismic waves passing through the mantle.

asthma Disorder of the respiratory system in which the bronchi (air passages) of the LUNGS go into spasm, making breathing difficult. It can be triggered by infection, air pollution, allergy, certain drugs, exertion or emotional stress. The chest feels tight and breathing is difficult, with coughing and a characteristic "wheeze". **Allergic** asthma may be treated by injections aimed at lessening sensitivity to specific allergens. Otherwise treatment is with bronchodilators to relax the bronchial muscles and ease breathing; in severe asthma inhaled steroids may be given. Children often outgrow asthma, while some people suddenly acquire the disease in middle age. In the acute form, the attacks are usually severe and follow longish periods of otherwise good health. The chronic condition is marked less by severe attacks than by a general breathing impairment. *See also* BRONCHITIS; EMPHYSEMA

astigmatism Defect of vision in which the curvature of the lens differs from one perpendicular plane to another. It can be compensated for by use of corrective lenses.

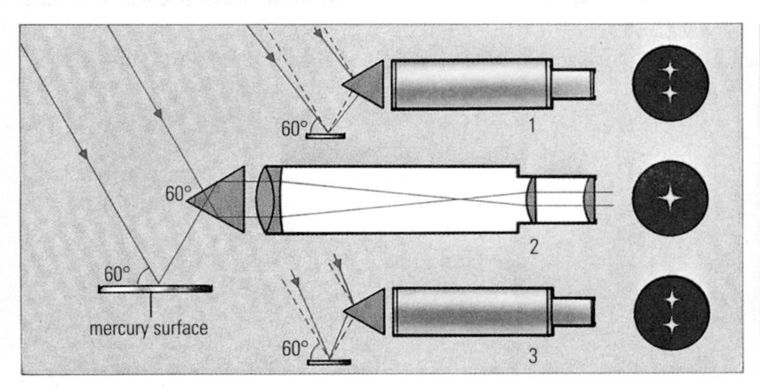

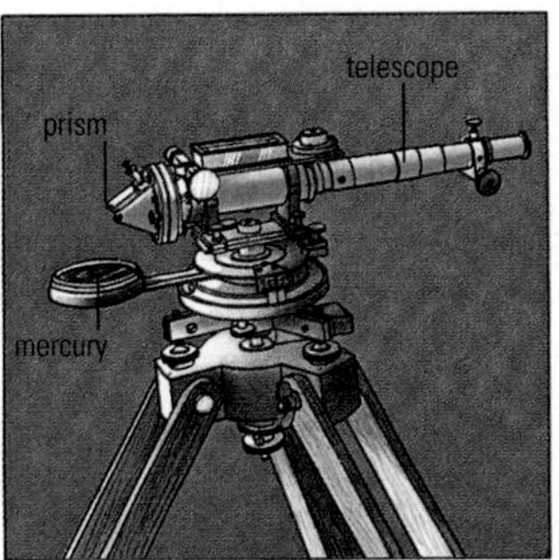

▲ **astrolabe** The prism astrolabe was used in conjunction with a chronometer to measure the exact moment at which a given star passes through a set altitude above the horizon (60°). This information is useful in fixing one's position on the Earth's surface. Essentially it was a highly accurate sighting device. 1) Before the star reaches 60° altitude, two images are seen through the telescope section of the instrument. 2) At the correct altitude a ray of star-light reflected off the mercury surface and into the prism forms a coincident image with that of the other ray, which is reflected from the lower face of the prism. The mercury was used because a liquid surface defines a horizontal plane. Only one image can therefore be seen. 3) When the star's altitude is too great, and has therefore passed beyond the 60° limit, two images are seen again.

Aston, Francis William (1877–1945) British physicist awarded the 1922 Nobel Prize for chemistry for his work on ISOTOPES. He developed the MASS SPECTROGRAPH that separates isotopes, and he used it to identify 212 naturally occurring isotopes.

astringent Drug or lotion which contracts the tissues or restricts fluid secretion. Astringents are used in a number of preparations to protect the skin and promote healing.

astrobiology *See* EXOBIOLOGY

astrodynamics Application of CELESTIAL MECHANICS, BALLISTICS, mathematical PERTURBATION theory, and the principle of observation reduction to determine, predict and correct orbits and trajectories in space.

astrogeology Study of rocks, craters and other surface features of the Moon, Mars, and other planets. It includes on site and laboratory studies of Moon rocks, seismological data (Moon only), photographic mappings, magnetic field measurements, and micrometeorite data.

astrolabe Early astronomical instrument for showing the appearance of the CELESTIAL SPHERE at a given moment and for determining the altitude of celestial bodies. The basic form consisted of two concentric disks one with a star map and one with a scale of angles around its rim joined and pivoted at their centres (rather like a modern planisphere), with a sighting device attached. Astrolabes were used from the time of the ancient Greeks through the 17th century for navigation, measuring time, and terrestrial measurement of height and angles.

astronaut (Rus. *cosmonaut*) Person who navigates or rides in a space vehicle, or is trained to do so. The first man to orbit the Earth was Yuri GAGARIN in 1961. The first man to walk on the Moon was Neil ARMSTRONG in 1969. The first woman in space was Valentina Tereshkova of Russia, in 1963.

astronautics (astronautical engineering) Practical study of the principles of space flight. Astronautics includes ASTRODYNAMICS, propulsion theory, ASTROGEOLOGY, EXOBIOLOGY, communications principles, and the design, materials analysis, guidance and control of spacecraft.

astronomical telescope *See* TELESCOPE

astronomical unit (symbol AU) Mean distance between the Earth and the Sun, used as a fundamental unit of distance, particularly for distances in the Solar System. 1AU is equal to 149,598,000km (92,956,000mi).

astronomy Branch of science studied since ancient times and concerned with the Universe and its components in terms of the relative motions of celestial bodies, their positions on the celestial sphere, physical and chemical structure, evolution and the phenomena occurring on them. It includes celestial mechanics, ASTROPHYSICS, COSMOLOGY and astrometry. Waves in all regions of the ELECTROMAGNETIC SPECTRUM can now be studied either with ground-based instruments or, where no atmospheric window exists, by observations and measurements made from satellites, space probes and rockets. **History** Astronomy was first used practically to develop a calendar, the units of which were determined by observing the heavens. The Chinese had a calendar in the 14th century BC. The Greeks developed astronomy significantly between 600 BC and AD 200. THALES introduced geometrical ideas and PYTHAGORAS saw the Universe as a series of concentric spheres. ARISTOTLE believed the Earth to be stationary but he explained lunar eclipses correctly. ARISTARCHUS put forward a heliocentric theory. HIPPARCHUS used trigonometry to determine astronomical distances. The system devised by PTOLEMY was a geometrical representation of the SOLAR SYSTEM that predicted the motions of the planets with great accuracy. From then on astronomy remained dormant until the scientific revolution of the 16th and 17th centuries, when COPERNICUS stated his theory that the Earth rotates on its axis and, with all the other planets, revolves round the Sun, which had a profound effect upon the religion and philosophy of the day. KEPLER and his laws of planetary motion refined the theory of heliocentric motion, and his contemporary, GALILEO, made use of the TELESCOPE and discovered the moons of Jupiter. Isaac NEWTON combined the sciences of astronomy and physics. His laws of motion and universal theory of GRAVITATION provided a physical basis for Kepler's laws and the work of many astronomers from then on – such as, in the prediction of HALLEY'S COMET and the discovery of the planets Uranus, Neptune and Pluto. By the early 19th century the science of celestial mechanics, the study of the motions of bodies in space as they move under the influence of their mutual gravitation, had become highly advanced and new mathematical techniques permitted the solution of the remaining problems of classical gravitation theory as applied to the Solar System. In the second half of the 19th century astronomy was revolutionized by the introduction of techniques based on photography and SPECTROSCOPY. These encouraged investigation into the physical composition of stars, rather than their position. Ejnar HERTZSPRUNG and Henry RUSSELL studied the relationship between the colour of a star and its luminosity. By this time also

ASTRONAUT

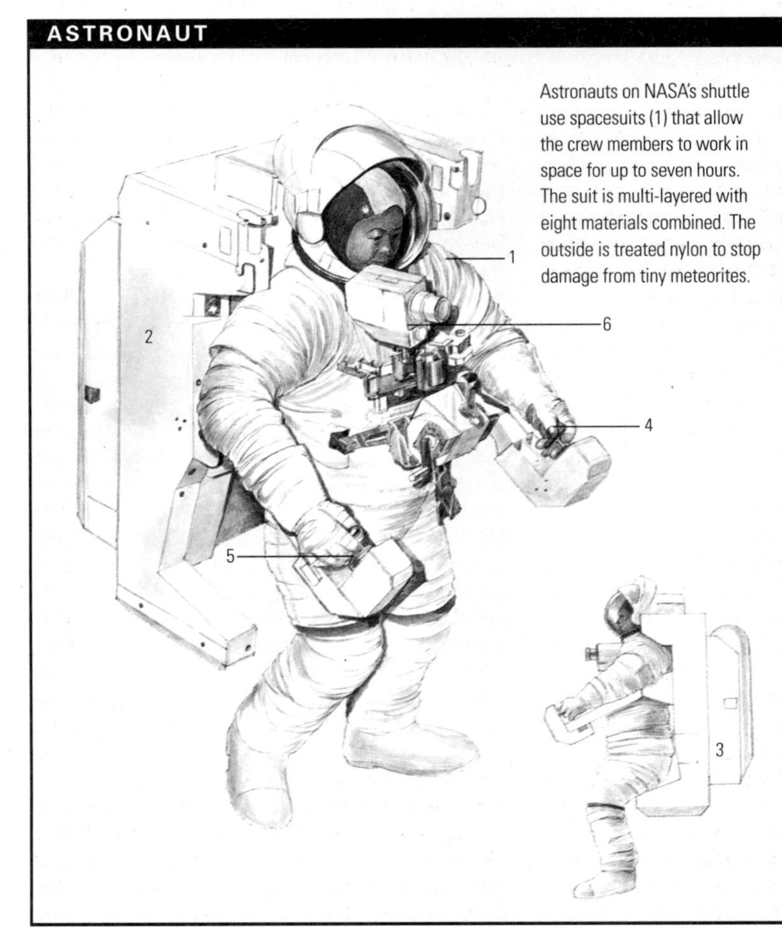

Astronauts on NASA's shuttle use spacesuits (1) that allow the crew members to work in space for up to seven hours. The suit is multi-layered with eight materials combined. The outside is treated nylon to stop damage from tiny meteorites. Four layers of aluminium material then provide a heat shield from solar radiation backed by a fire and tear-resistant layer. The astronaut is protected from the vacuum of space by a pressure suit of nylon coated with polyeurathane and is kept comfortable in extremes of heat and cold by water pumped through a network of tubes in a nylon chiffon undergarment. (2) The manned manouevring unit (MMU) allows an astronaut to move away from the shuttle. Power comes from 24 thrusters arranged at the corners of the MMU. By releasing pressurized nitrogen from two tanks (3) through nozzles the astronaut can propel himself/herself through the vacuum. The hand controllers regulate rotation (4) and speed (5). A video camera (6) sends pictures to the shuttle and records the work carried out.

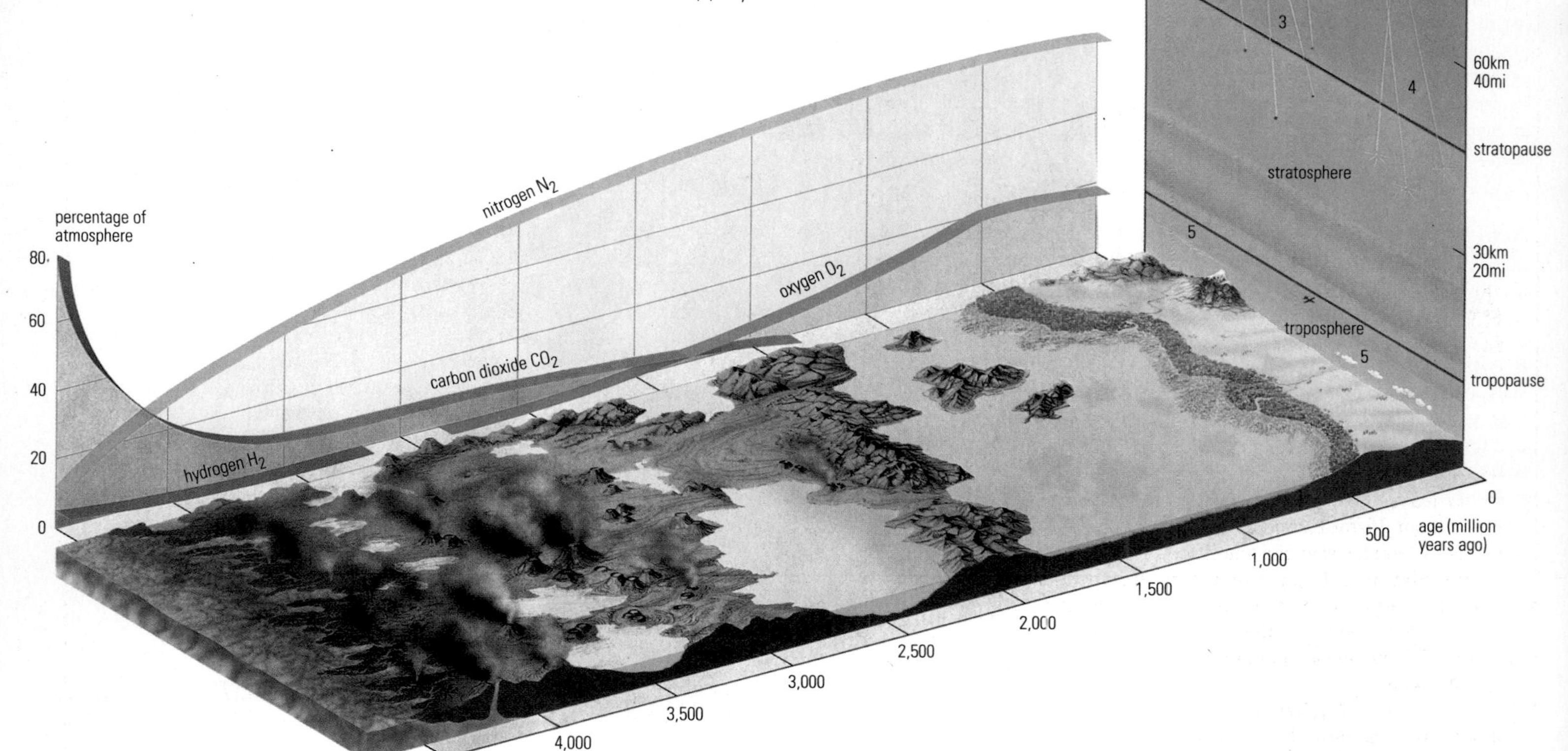

▼ **atmosphere** The evolution of the Earth's atmosphere reveals how oxygen levels began to increase 2,000 million years ago, as shown in the formation of extensive "red bed" sediments – sands coloured with oxidized (ferric) iron. By 4,500 million years ago, the carbon dioxide in the atmosphere was beginning to be lost in sediments. The vast amounts of carbon deposited in limestone, coal and oil indicate that carbon dioxide concentration must once have been many times greater than today, when it stands at only 0.04%. The first carbonate deposits appeared about 1,700 million years ago, the first sulphate deposits about 1,000 million years ago. The decreasing carbon dioxide was balanced by an increase in the nitrogen content of the air. The forms of "respiration" practiced advanced from fermentation 4,000 million years ago, to anaerobic photosynthesis 3,000 million years ago to aerobic photosynthesis 1,500 million years ago. The aerobic respiration that is so familiar today only began to appear about 500 million years ago. Today's atmosphere comprises some interesting phenomena. Aurorae (1) can be seen in the thermosphere, which extends from 80km (50mi) to 400km (250mi) up. Noctilucent clouds (2) only occur around the mesopause – the line between the thermosphere and the mesosphere. Some meteors (3) reach the surface of the Earth, but most burn up in the mesosphere. Cosmic rays (4) penetrate to the stratosphere. Most human activities, and the weather that directly affects us (5), occur in the troposphere.

larger telescopes were being constructed, which extended the limits of the known Universe. Harlow SHAPLEY determined the shape and size of our galaxy and Edwin HUBBLE's study of distant galaxies led to his theory of an expanding Universe. The "BIG BANG" and STEADY-STATE THEORY of the origins of the Universe were formulated. In recent years space exploration and observation in different parts of the electromagnetic spectrum have contributed to the discovery and postulation of such phenomena as the QUASAR, PULSAR and BLACK HOLE. There are various branches of modern astronomy: **Gamma-ray** astronomy is concerned with the study of gamma rays, which cannot penetrate the Earth's atmosphere and must be studied from satellites. The Sun is the principal source. **Infrared** astronomy is concerned with detecting infrared IR waves, determining their source and studying their spectra. Most IR radiation is absorbed by the Earth's atmosphere. The Sun and the centre of the galaxy are sources of IR radiation. **Optical** astronomy is the oldest branch which studies sources of light in space. Light rays can penetrate the atmosphere but, because of disturbances, many observations are now made from above the atmosphere, as in the case of the HUBBLE SPACE TELESCOPE. **Radar** astronomy is used to determine distance, orbital motion and surface features of objects in the Solar System. Radar pulses are bounced back to Earth from the object. **Radio** astronomy uses radio telescopes to detect radio waves from space and determine their source and energy spectrum. The most powerful sources include the Sun, interstellar clouds of hot hydrogen, such as the Orion Nebula, supernova remnants and pulsars, like the Crab Nebula, quasars and radio galaxies. **Ultraviolet** astronomy studies ultraviolet (UV) waves from space, their source and spectra. Higher wavelengths can be studied from the ground but lower wavelengths must be studied from satellites and balloons. Sources include the Sun, the Orion Nebula, and Wolf-Rayet stars. **X-ray** astronomy is concerned with studying X-ray sources. X-rays are absorbed by the Earth's atmosphere so detecting instruments must be carried in satellites. Sources include the Sun, the Crab Nebula, Cygnus A and the galaxy M87.

astrophysics Branch of ASTRONOMY that studies the physical and chemical nature of celestial bodies and their evolution. Many branches of physics, including nuclear physics, plasma physics, relativity and spectroscopy, are used to predict properties of stars, planets and other celestial bodies. Astrophysicists also interpret the information obtained from astronomical studies of the electromagnetic spectrum, including light, X-rays and radio waves.

asynchronous transmission Fastest method of program execution in computers; the beginning of each operation is signalled by the stop bit at the end of the previous one.

atavism Reversion by an organism to a characteristic of its ancestors after an interval of at least one generation in which the trait was absent. The term is no longer in scientific use since the reappearance of ancestral traits is now understood to be the expression of RECESSIVE genes.

atmolysis Process of separating mixed gases by allowing them to pass through a porous material.

atmosphere Envelope of gases surrounding the Earth. It shields the planet from the harsh environment of space, while the gases it contains are vital to life. About 95% by weight of the Earth's atmosphere lies below 25km (15mi) altitude; the mixture of gases in the lower atmosphere is commonly called air. The atmosphere's composition by weight is: nitrogen 78.09%, oxygen 20.9%, argon 0.93%, 0.03% of carbon dioxide, plus 0.05% of hydrogen, other gases and varying amounts of water vapour. The atmosphere can be seen as concentric shells; the innermost is the **troposphere**, in which dust and water vapour create the clouds and weather. The **stratosphere** extends from 10–55km (8–36mi) and is cooler and clearer and contains ozone. Above, to a height of 70km (43mi), is the **mesosphere** in which chemical reactions occur, powered by sunlight. The temperature climbs steadily in the thermosphere, which gives way to the **exosphere** at about 400km (250mi), where helium and hydrogen may be lost into space. The **ionosphere** ranges from about 50km (30mi) out into the VAN ALLEN RADIATION BELTS.

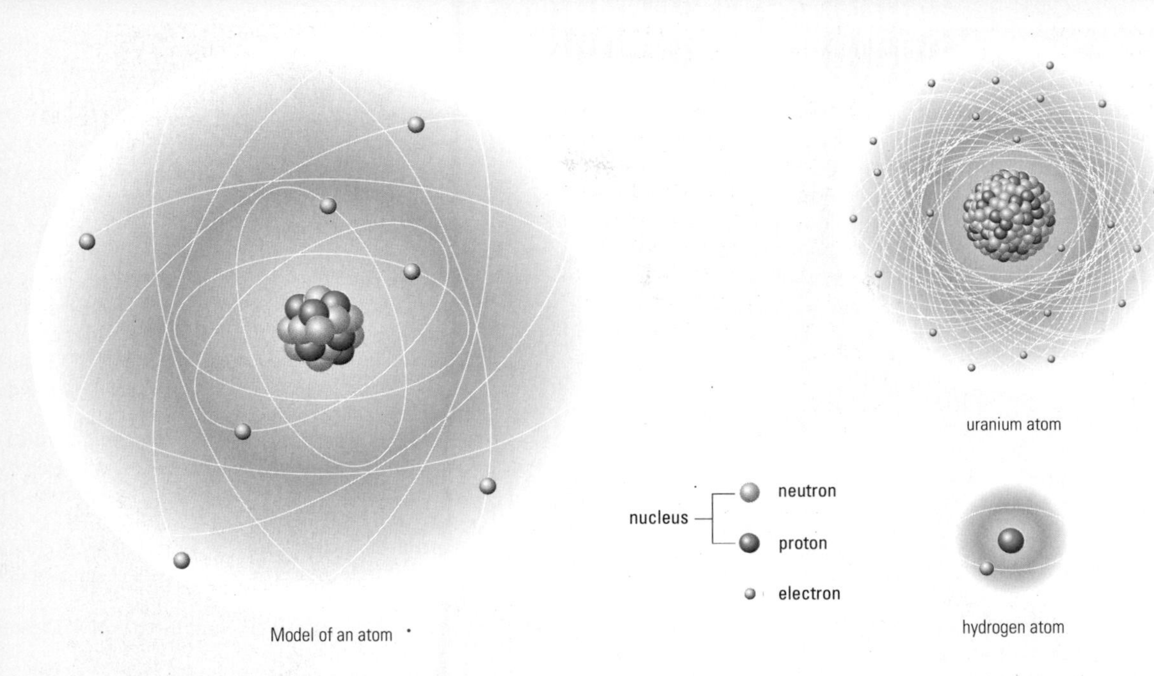

Model of an atom

◀ **The mass** of an atom depends on the size of the nucleus. The nucleus makes up the very great majority of the atom's mass as electrons weigh nothing. A uranium atom, for example, is the heaviest naturally occurring atom. It has 146 neutrons, 92 protons, and 92 electrons. A hydrogen atom on the other hand is the lightest atom. It has only 1 proton and 1 electron. However, while a uranium atom weighs 230 times as much as a hydrogen atom it is only 3 times bigger. The weight of an atom is measured in atomic mass units (amu), symbol u.

Atoms are made of even smaller particles known as SUBATOMIC PARTICLES. The three basic subatomic particles are PROTONS (positive electronic charge), NEUTRONS (electrically neutral) and ELECTRONS (negatively charged). Clusters of protons and neutrons form a NUCLEUS at the centre of atoms of all the elements (except hydrogen, which has just one proton). The electrons "orbit" the nucleus some distance away in terms of the size of the atom. If, for example, the nucleus of a helium atom were the size of a tennis ball, the electrons would orbit at a distance of around 6km (4mi). There are 112 different types of atom, corresponding to the elements in the PERIODIC TABLE. The atoms of elements vary both in ATOMIC NUMBER and ATOMIC MASS.

The nucleus

The mass of an atom lies mostly within the comparatively dense nucleus. Both protons and neutrons have a mass *c.*1,840 times greater than that of an electron. Because protons are positively charged and neutrons are neutral, the nucleus of an atom is always positively charged. As opposite charges attract, the nucleus keeps in orbit the negatively charged electrons. Protons and neutrons are made up of smaller subatomic particles, such as QUARKS.

Electrons

The number of electrons in an atom determines its chemical properties. Unlike the planets of the Solar System, electrons orbit haphazardly, but also at fixed distances from the nucleus, forming "shells". The more energy an electron has, the further it can escape the pull of the positively charged nucleus. In a neutral atom, the positive charge of the electrons balances that of the protons in the nucleus. Therefore, removal or addition of an atomic electron produces a charged ion. Electrons are arranged in shells at fixed distances from the nucleus, depending on their energy. Each electron shell is labelled with a number, the shell closest to the nucleus is given the number 1. The most shells an atom can have is 7, and each shell can only support a certain number of electrons. Given sufficient energy, an electron can jump from one shell to a higher shell. When it falls back to a lower shell, it emits radiation in the form of a PHOTON. The electron belongs to the LEPTON class of particles, its antiparticle is the POSITRON.

A NUCLEAR CHAIN REACTION

In a nuclear fission explosion, such as that of an atomic bomb, a neutron hits a uranium-235 nucleus (that is a nucleus with a total of 235 protons and neutrons). The neutron is absorbed to make uranium-236. This is very unstable, and splits into two smaller nuclei giving off massive amounts of energy and several neutrons. Each neutron released can, in turn, smash apart another uranium nucleus. If the initial conditions are critical (the quantity of uranium-235 exceeds the critical mass) then enough neutron collisions will occur to multiply the reaction at lightning speed to give a chain reaction. In a nuclear reactor, the heat from the reaction is harnessed to create steam which in turn drives a turbine generator that creates electricity.

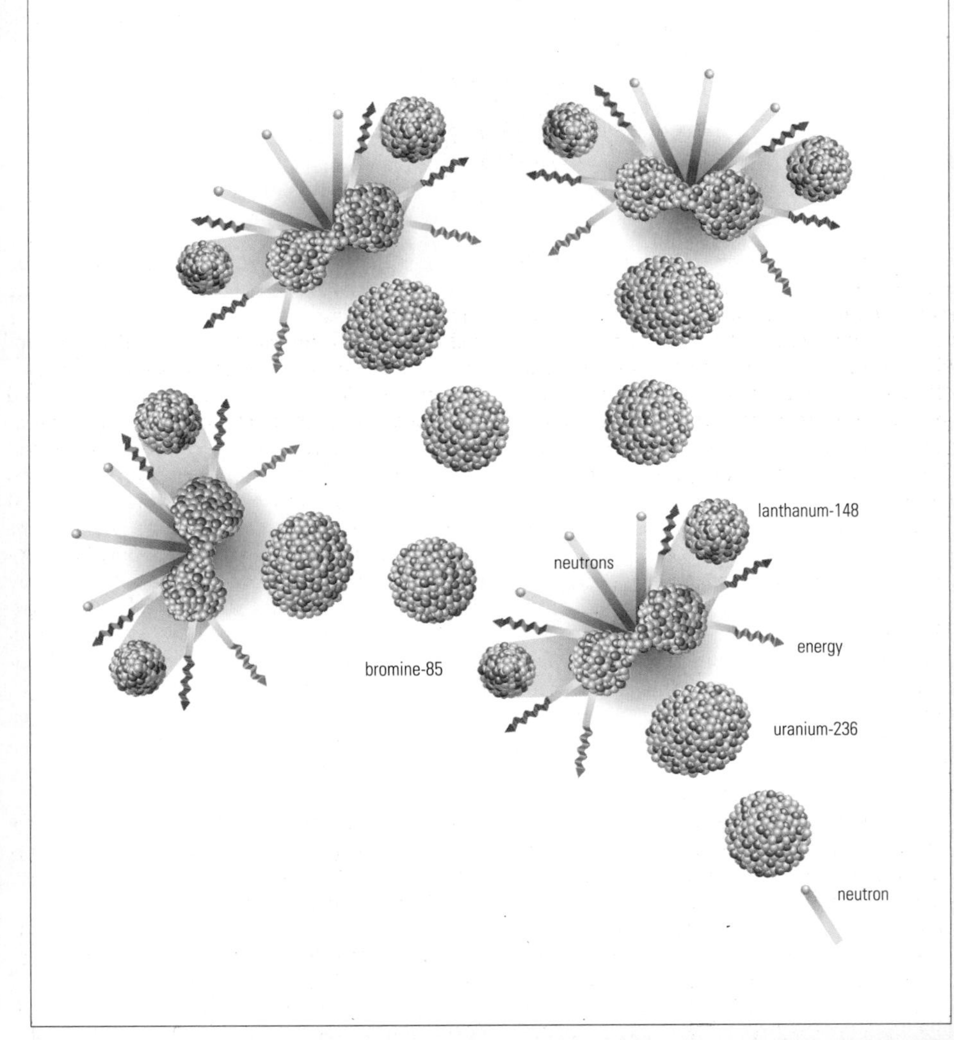

AURORA

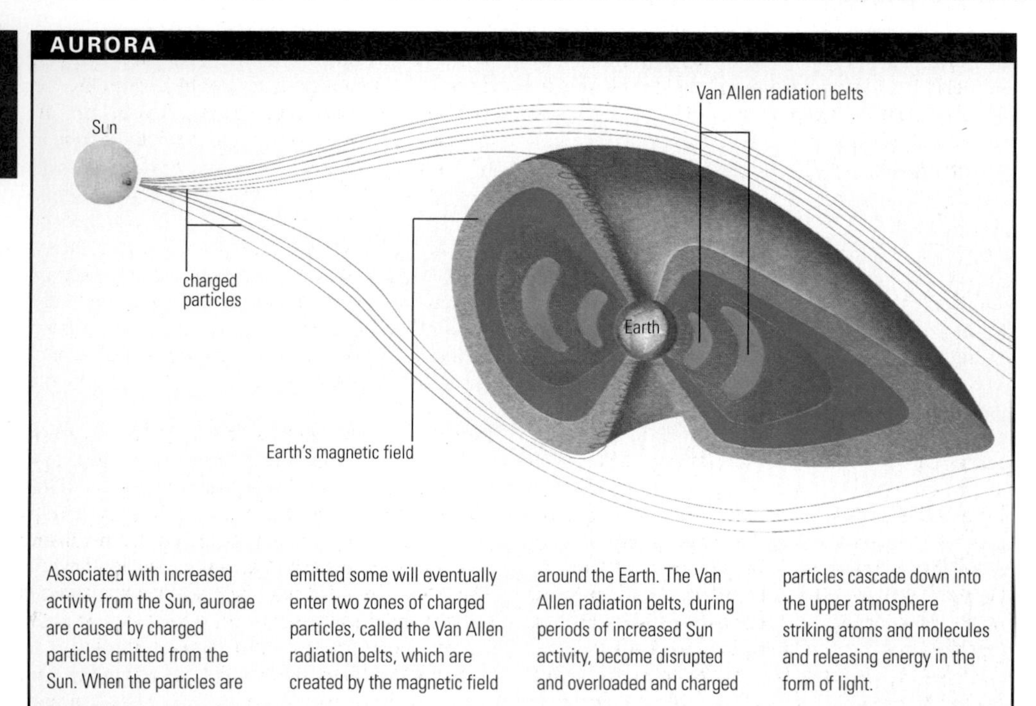

Associated with increased activity from the Sun, aurorae are caused by charged particles emitted from the Sun. When the particles are emitted some will eventually enter two zones of charged particles, called the Van Allen radiation belts, which are created by the magnetic field around the Earth. The Van Allen radiation belts, during periods of increased Sun activity, become disrupted and overloaded and charged particles cascade down into the upper atmosphere striking atoms and molecules and releasing energy in the form of light.

atmospheric pressure Downward force exerted by the atmosphere because of its weight (gravitational attraction to the Earth or other body), measured by barometers and usually expressed in units of millibars (mb). Standard atmospheric pressure at sea level is 1013.25mb. The column of air above $1cm^2$ ($0.15in^2$) of Earth's surface weighs *c.*1kg (2.2lbs).

atoll Ring-shaped REEF of CORAL enclosing a shallow LAGOON. An atoll begins as a fringing reef surrounding a slowly subsiding island, usually volcanic. As the island sinks the coral continues to grow upwards until eventually the island is below sea level and only a ring of coral is left at the surface.

atom Smallest particle of matter that can take part in a chemical reaction, every element having its own characteristic atoms. The atom, once thought indivisible, consists of a central, positively charged NUCLEUS orbited by negatively charged ELECTRONS. The nucleus (identified in 1911 by Ernest RUTHERFORD) is composed of tightly packed protons and neutrons. It occupies a small fraction of the atomic space but accounts for almost all of the mass of the atom. In 1913 Niels BOHR suggested that electrons moved in fixed orbits. The study of QUANTUM MECHANICS has since modified the concept of orbits: the Heisenberg UNCERTAINTY PRINCIPLE says it is impossible to know the exact position and MOMENTUM of a subatomic particle. The number of electrons in an atom and their configuration determine its chemical properties. Adding or removing one or more electrons produces an ION. *See feature on page 25*

atomic battery (nuclear battery) Device that converts energy from radioactive particles into electric current. Common sources include tritium (T), Hydrogen–3 and Krypton–85, both emitters of beta particles.

atomic bomb *See* NUCLEAR WEAPON

atomic clock Most accurate type of CLOCK. It is an electric clock regulated by such natural periodic phenomena as emitted radiation or atomic vibration, with the atoms of the metal caesium most commonly used. Clocks that run on the radiation from hydrogen atoms lose only one second in about 1,700,000 years.

atomic energy *See* NUCLEAR ENERGY

atomic force microscope (AFM) Type of microscope that depends on the forces between atoms to produce an image of a surface. A spring-loaded arm holds a tiny piece of diamond in contact with the surface while it scans across it. The probe is lowered or raised to keep the tracking force constant, and the necessary movements are analysed by a computer to produce a "contour map" of the surface displayed on the computer screen. It can scan non-conducting biological specimens, unlike the SCANNING TUNNELLING MICROSCOPE which works only with metallic samples.

atomicity Number of atoms of an element in a particular molecule. For example, neon (Ne) has an atomicity of 1, nitrogen (N_2) has an atomicity of 2, ozone (O_3) has an atomicity of 3 and certain forms of sulphur (S_8) have an atomicity of 8.

atomic mass number *See* MASS NUMBER

atomic mass unit (symbol u) Unit of mass used to compare RELATIVE ATOMIC MASS (R.A.M.), defined since 1961 as 1/12th the mass of the most abundant isotope of carbon, carbon-12 (which has 6 electrons, 6 protons and 6 neutrons). It is equal to 1.66033×10^{-27} kg.

atomic number (proton number, symbol *Z*) Number of protons in the nucleus of an atom of an element, which is equal to the number of electrons moving around that nucleus. It is abbreviated to at.no. It is signified by a subscript number before the symbol, for example, the atomic number for carbon is written $_6C$. The atomic number determines the chemical properties of an element and its position in the periodic table. ISOTOPES of an element all have the same atomic number but a different atomic MASS NUMBER because they have different numbers of neutrons.

atomic spectrum Spectrum of sharp lines characteristic of the element involved and produced by the radiation emitted when electrons jump between energy levels of the ATOM.

atomic volume Quantity arrived at by dividing the RELATIVE ATOMIC MASS (R.A.M.) of an element by its density.

atomic weight Former term for RELATIVE ATOMIC MASS (R.A.M.)

ATP Abbreviation of ADENOSINE TRIPHOSPHATE

atrium (auricle) Either of the two upper chambers of the four-chambered HEART. They are comparatively thin-walled since they only pump blood down into the muscular VENTRICLES of the heart. The term is also used for various other chambers in animals.

atropine ($C_{17}H_{23}NO_3N$) Poisonous ALKALOID drug obtained from certain plants such as *Atropa belladonna* (deadly nightshade). Atropine is used medicinally as an antispasmodic, to dry secretions and regularize the heartbeat during anaesthesia, to dilate the pupil of the eye and to treat motion sickness.

Attenborough, Sir David Frederick (1926–) British naturalist and broadcaster; brother of Sir Richard Attenborough. He was controller of BBC2 television (1965–68). Since 1954, he has travelled on zoological and ethnographical filming expeditions, which have formed the basis of such landmark natural history series as *Life on Earth* (1979), *The Living Planet* (1984), *The Trials of Life* (1990), and *The Private Life of Plants* (1995). He was knighted in 1985.

AUTOMATION

Employed in a wide variety of manufacturing and distribution processes, the diagram shows how with sophisticated automated machines one controller using a computer console can store, select and distribute goods.

attitude Name given to the orientation, usually of ships, aircraft and spacecraft, with respect to a given set of axes. Three variables specify attitude: **pitch**, the angle with the horizontal made by the long axis of the craft; **yaw**, the angle through which the bow turns left or right of the "forward" direction; and **roll**, the angle of rotation about the long axis.

attrition In geology, a type of EROSION in which the particles produced by the erosive process go on to erode each other. The process may take place in water, where sand particles in the river's load erode each other, or where pebbles on a beach tumble against each other. Eventually particles affected by attrition become completely rounded.

audiometry Technique to evaluate hearing ability. An audiometer is an instrument that measures the sensitivity of the ear to sounds; it is used in the assessment of hearing loss.

audio-visual aid Use of sound and pictures to assist learning. The use of film, television and computers, together with the ability to record, select and replay, has made it possible to develop understanding and knowledge quicker than was possible using only the spoken and written word.

auditory canal Tube leading from the outer EAR to the eardrum. It is about 2.5cm (1in) long.

auditory nerve Bundle of nerves that carry impulses related to hearing and balance from the inner EAR to the brain.

auditory ossicles Three small bones in the middle EAR that transmit vibrations from the eardrum to the cochlea in the inner ear.

Auger, Pierre Victor (1899–) French physicist who discovered the **Auger effect**, a change in an atom from an excited to a lower energy state with the emission of an (Auger) electron but without radiation. He became director general of the European Space Research Organization (ESRO) in 1962.

augite *See* PYROXENES

Auriga (Charioteer) Large constellation of the N sky, containing the first-magnitude star Capella. It also contains the eclipsing binary Epsilon Aurigae.

aurora Sporadic display of coloured light in the night sky, usually green, caused by charged particles from SOLAR FLARES interacting with atoms and air molecules in the Earth's upper atmosphere. The charged particles from the Sun are attracted by the Earth's magnetic field into zones called VAN ALLEN RADIATION BELTS. Auroras occur in polar regions and are known as aurora borealis, or "northern lights", in the N, and aurora australis in the S.

auscultation Practice of listening to sounds within the body as an aid to diagnosis. The invention of the stethoscope in 1819 replaced the earlier method of placing the ear against the body.

Australopithecus *See* HUMAN EVOLUTION

autoimmune disease Any one of a group of disorders caused by the body's production of ANTIBODIES which attack the body's own tissues. One example of such an autoimmune disease is systemic lupus erythematosus (SLE), an inflammation of the connective tissue occurring most often in young women. The occasional presence of so-called auto-antibodies in an individual does not necessarily indicate autoimmune disease. Treatment depends on the specific disease process, but often includes drugs such as CORTISONE which damp down the immune response.

autolysis In biology, process in which a tissue, cell or part of a cell self-destructs. The process is brought about by ENZYMES that act on cells, produced by LYSOSOMES within the cell itself, normally after the cell has died.

automation Use of self-governing machines to carry out manufacturing, distribution and other processes automatically. By using FEEDBACK, sensors check a system's operations and send signals to a computer that automatically regulates the process. *See also* MASS PRODUCTION; ROBOT

automobile (car) Road vehicle that first appeared in the 19th century. The first cars were propelled by steam, but were not a success. The age of the motor car really dates from the introduction (1885–86) of the petrol-driven carriages of Gottlieb DAIMLER and Karl BENZ. The INTERNAL COMBUSTION ENGINE for these cars had been developed earlier by several engineers (most notably Nikolaus Otto in 1876). The main components of a motor car remain unchanged. A body (**chassis**) to which are attached all other parts including: an **engine** or power plant; a **transmission** system for transferring the drive to the wheels, and steering, braking and suspension for guiding, stopping and supporting the car. Early cars were assembled by a few experts, but modern mass-production began in the early 1900s by Henry FORD and R.E. Olds in the USA. In most modern motor factories, component parts are put together on assembly lines. Each worker has a specific task (such as fitting doors or crankshafts). Bodies and engines are made on separate assembly lines which converge when the engine is installed. Overhead rail conveyors move heavy components along the assembly lines, lowering them into position. The final stages of assembly include the fitting of items such as lamps and paint spraying. Electrical, braking and control systems are checked. The assembled car is tested before sale. Recent technology has seen the introduction of robots (properly, robotic arms secured to the workshop floor) on the assembly line. They are usually used for welding and painting. Increasing concern over the environmental impact of the car (such as congestion, pollution and energy consumption) has encouraged governments to examine alternative forms of mass transport, oil companies to produce cleaner fuels and car manufacturers to look at alternative power plants (such as electric- or gas-powered motors).

autonomic nervous system (ANS) In mammals, part of the body's NERVOUS SYSTEM that regulates the body's involuntary functions. In conjunction with the PERIPHERAL NERVOUS SYSTEM it helps prepare the body for action or rest by regulating functions such as heart beat, digestion and sweating. The ANS is divided into the SYMPATHETIC NERVOUS SYSTEM and the PARASYMPATHETIC NERVOUS SYSTEM. *See also* HOMEOSTASIS; INVOLUNTARY MUSCLE

autopilot (automatic pilot) Electronic and mechanical control system that ensures an aircraft follows a pre-programmed flight plan. It monitors the course and speed of the aircraft and corrects any deviations from the flight plan. Systems range from

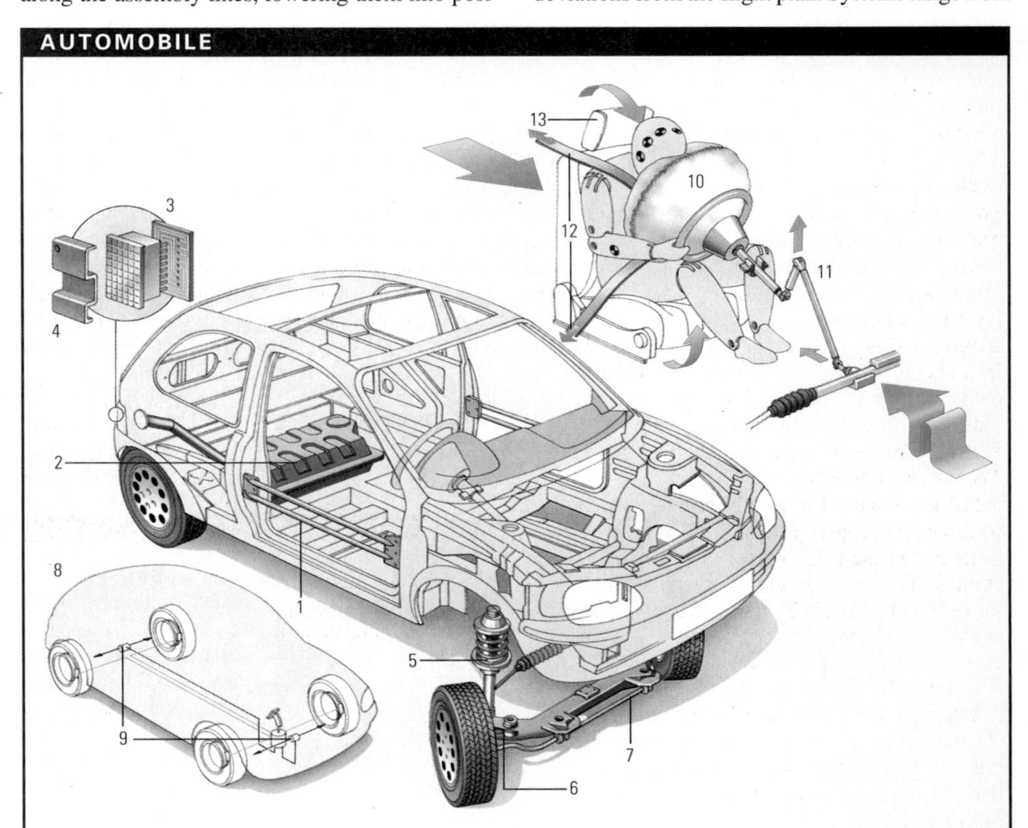

A modern car is designed with crumple zones at the front and rear to absorb the energy of a crash and protect the car's passengers. Side-impact protection bars (1) give strength to the side of the vehicle and spread energy to either side of the passenger cell. Fuel tanks (2) are situated in front of the rear axle to protect the tank if the car is hit from behind. Some manufacturers have replaced the traditional rear brake lights with LEDs (3) which light more quickly. The cover is stepped (4) to prevent the light being obscured by dirt. The suspension, a MacPherson strut (5) system, allows vertical movement through the spring (5) while the wishbone (6) and anti-roll bar (7) keep the wheels in position and stop excessive roll respectively. Anti-skid braking systems (ABS) (8) prevent the wheels locking under heavy braking or in poor weather. Sensors (9) detect when a wheel is about to lock and release the brake pads for a fraction of a second. An explosive charge inflates the air bag (10) which prevents the driver or passenger from hitting the steering wheel or dashboard. The steering column (11) is designed to collapse so the driver is not impaled. Seat belt tensioners use the impact to pull the belt tight (12) holding the passenger in place. The headrest (13) helps stop whiplash injuries when heads snap back in the aftermath of the impact.

AUTOPILOT

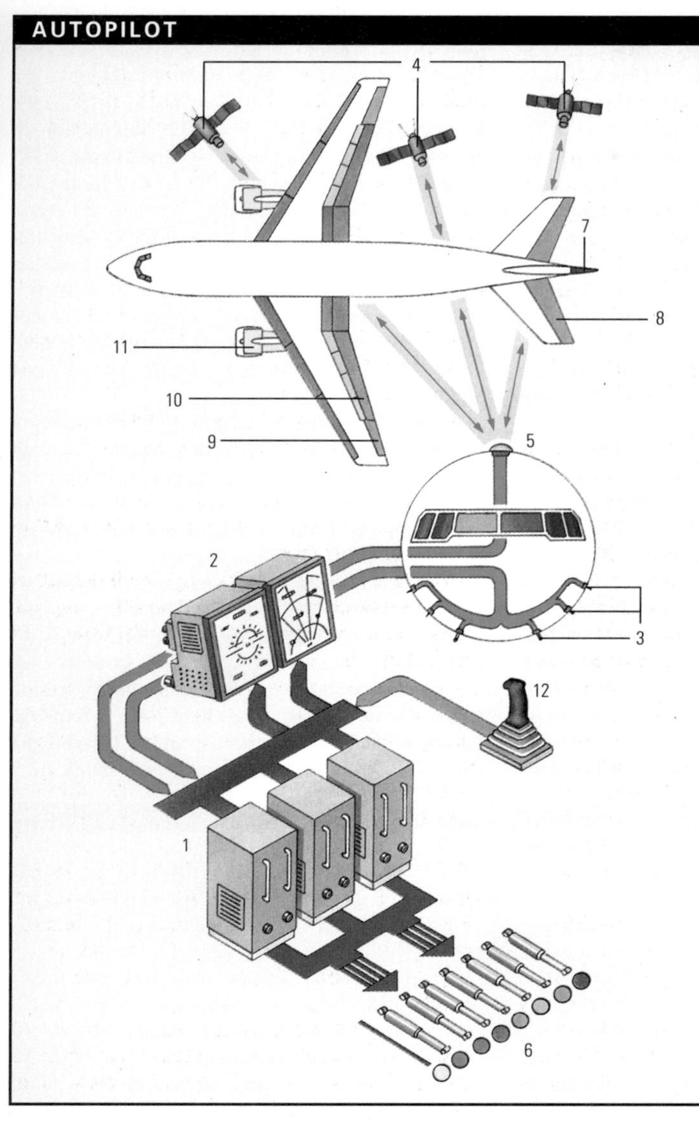

The diagram shows how a typical autopilot system works. A pre-programmed flight plan is loaded into the aircraft's computers (1). After take-off the autopilot is engaged. Two visual display units (2) show the aircraft's position, its intended route and its attitude. The change in movement of small vanes (3) on the outside of the aircraft alert the computers to any change in the aircraft's orientation. (4) The aircraft uses a Global Positioning System (GPS) to determine its position. The receiver is located on top of the aircraft (5). The computers track the aircraft's route and automatically make any adjustments via servos (6) which control the rudder (7), elevators (8), aerlirons (9), flaps (10) and throttle settings on the engines (11). The pilots can override the automatic system at anytime and revert to manual controls (12).

simple wing-levellers in light aircraft to computer-operated units consisting of: a GYROSCOPE; an electric SERVOMECHANISM unit; and an ACCELEROMETER, which measures the acceleration of the aircraft.

autotroph Organism that manufactures its own organic food from simple inorganic chemicals in a process known as **autotrophic nutrition**. Green plants are typical autotrophs; they make sugars by PHOTOSYNTHESIS from carbon dioxide and water using the energy of sunlight. In FOOD CHAINS, autotrophs make up the primary producers, providing energy (as food) for organisms higher up the chain. *See also* HETEROTROPH

auxin Plant hormone produced mainly in the growing tips of plant stems. Auxins accelerate plant growth by stimulating cell division and enlargement, and by interacting with other hormones. Actions include the elongation of cells (by increasing the elasticity of cell walls, allowing the cells to take up more water) in GEOTROPISM and PHOTOTROPISM, and fruit drop and leaf fall. Synthetic auxins are the basis of rooting powders and selective weedkillers.

average In statistics, the one score that most typifies an entire set of scores. It is the arithmetic MEAN of the scores. Other calculations that are also used to express what is typical in a set of scores are the mode (the one score that occurs most often), and the median (the middle score in a range which thus divides the set of scores into upper and lower halves).

Aves *See* BIRD

Avogadro, Amedeo, Conte di Quaregna (1776–1856) Italian physicist and chemist. In 1811 Avogadro put forward the hypothesis (now known as **Avogadro's law**) that equal volumes of gases at the same pressure and temperature contain an equal number of molecules. This led later physicists to determine that the number of molecules in one gram molecule (the relative molecular mass expressed in grams) is constant for all gases. This number, called the AVOGADRO CONSTANT. It is both the ratio of the universal gas constant to BOLTZMANN CONSTANT and of Faraday's constant to the charge of the electron.

Avogadro constant (formerly Avogadro's number, symbol *L*) Constant equal to 6.022×10^{23}, giving the number of atoms or molecules present in one MOLE of a substance.

Axelrod, Julius (1912–) US biochemist. Axelrod shared the 1970 Nobel Prize for physiology or medicine with Ulf Svante von Euler and Sir Bernard Katz for their work on the chemistry of nerve impulse transmission. He discovered how the NEUROTRANSMITTER NORADRENALINE is inactivated after it has performed its function.

axil Angular space between a growing leaf, bract or branch and the stem it grows from. Buds or shoots that grow in axils are known as axillary buds or shoots.

axiom Assumption used in mathematics or logic as a basis for deductive reasoning. *See* AXIOMATIC METHOD

axiomatic method Method of mathematical reasoning based on logical deduction from assumptions (axioms). The method is fundamental to the philosophy of mathematics: it was used by the Greeks and formalized early in the 20th century by David HILBERT. In an axiomatic system, certain undefined entities (terms) are taken and described by a set of axioms. Other, often surprising, relationships (theorems) are then deduced by logical reasoning. *See also* GÖDEL, KURT

axis Imaginary straight line about which a body rotates, or about which rotation is conceived. In mechanics an axis runs longitudinally through the centre of an axle or rotating shaft. In geography and astronomy, it is a line through the centre of a planet or star, about which the planet or star rotates. The Earth's axis between the north and south geographic poles is 12,700km (7,900mi) long and is inclined at an angle of 66.5° to the plane in which the Earth orbits the Sun. A mathematical axis is a fixed line, such as the *x*-, *y*- or *z*-axis, chosen for reference.

axon Part of a nerve cell, or NEURON, that carries a nerve impulse beyond and away from the cell body, such as an impulse for movement to a muscle. There is typically only one axon per neuron, and it is generally long and unbranched. It is encased by a fatty, pearly MYELIN SHEATH in all peripheral nerves, and in all central nerves except those of the brain and spinal cord. Axons in peripheral nerves are covered by an additional delicate sheath, a neurilemma, which helps to regenerate damaged nerves.

azimuth Angle between the vertical plane through a celestial body and the N-S direction. Astronomers measure the angle E from the N point of the observer's horizon. Navigators and surveyors measure it W from the S point. Altitude and azimuth form an astronomical coordinate system for defining position.

azo and diazo compounds Nitrogenous organic chemical compounds used widely for organic synthesis, particularly in the manufacture of dyes. Diazo compounds include reactive salts having the general formula Ar-N_2^+X, where X is the anion.

azurite Basic copper carbonate mineral found in the oxidized parts of copper ore veins, often as earthy material with MALACHITE. The crystals are blue, brilliant and transparent. In the past the gemstones were used as pigments in wall paintings. Hardness 3.5–4; s.g. 3.77–3.89

▲ **Avogadro** Although originally trained in law, Avogadro turned to science and held professorships in physics for much of his life. He worked on thermodynamics, electricity and the properties of liquids. His major contribution to science was his explanation of Gay-Lussac's law of combining volumes in gases. His hypothesis, later known as Avogadro's Law, was not widely accepted until after his death.

Baade, Walter (1893–1960) US astronomer, b. Germany. From Mount Wilson Observatory in the 1943 wartime blackout, he was able to observe individual stars in the Andromeda Galaxy and distinguish the younger, bluer Population I stars from the older, redder Population II stars. He went on to improve the use of CEPHEID VARIABLE stars as distance indicators, and showed that the Universe was older and larger than had been thought.

Babbage, Charles (1791–1871) British mathematician. A professor of mathematics, working on the theory of FUNCTIONS and ALGEBRA, Babbage's main interest was the development of a mechanical calculating machine that could evaluate mathematical tables without error. A small prototype of the "difference engine", which could calculate tables of LOGARITHMS, was made, but a full-working model was never completed due to insufficient financial support. He then planned another machine (the "analytical engine") that could carry out any set of arithmetic operations using punched cards (a forerunner of the modern computer). But again lack of money, and the fact that it was too ambitious and intricate for the mechanical devices of the day, ensured it was never completed.

Babbitt metal Any alloy with a high tin content as well as copper and antimony – specifically an alloy invented by Isaac Babbitt (1799–1862) in 1839 as bearing material for steam engines.

Babcock, Harold Delos (1882–1968) US physicist and spectroscopist who developed a modern theory to account for the formation of SUNSPOTS.

bacillus Genus of rod-like BACTERIA present everywhere in the air and soil. One example of a species that is pathogenic in people is *Bacillus anthracis*, which causes ANTHRAX.

backcross Offspring of a first generation HYBRID and either of its parents. Backcrosses are used by biologists as a means of testing the GENOTYPE of the hybrid.

background radiation RADIATION that is normally present in an environment. Such radiation must be taken into account when measuring radiation from a particular source. On Earth, background radiation is caused by the decay of naturally occurring radioactive substances in surface rocks. In space, the so-called "microwave background" is attributed to the BIG BANG.

backwash Movement of seawater down a beach after a wave has broken on it. An incoming wave is taken up the beach by the **swash**; backwash takes it back again. On a hard beach the backwash is on the surface; or it may be within the shingle or sand on a softer beach.

bacteria Simple unicellular microscopic organisms belonging to the kingdom Prokaryotae. They lack a clearly defined nucleus and most are without CHLOROPHYLL. Many are motile, swimming about by means of whip-like FLAGELLA. Most multiply by FISSION. Under adverse conditions many can remain dormant inside highly resistant SPORES with thick, protective coverings. Bacteria may be AEROBIC or ANAEROBIC. Although pathogenic bacteria are a major cause of human disease, many bacteria are harmless or even beneficial to humans by providing an important link in FOOD CHAINS, such as in decomposing plant and animal tissue, and in converting free nitrogen and sulphur into AMINO ACIDS and other compounds that plants and animals can use. Some contain a form of chlorophyll and carry out PHOTOSYNTHESIS. *See also* ARCHAEBACTERIA; EUBACTERIA; PROKARYOTE; *see feature on page 30*

bacteriology Scientific study of BACTERIA. These single-celled organisms were first observed in the 17th century by the amateur microscopist Anton van LEEUWENHOEK, but it was not until the researches of Louis PASTEUR and later Robert KOCH that bacteriology was established as a scientific discipline.

bacteriophage VIRUS that lives on and infects BACTERIA. It has a protein head containing a core of DNA and a protein tail. Since its discovery in 1915 it has been important in the study of GENETICS.

badlands Eroded, barren plateau in an arid or semi-arid area characterized by steep gullies and ravines. Because of the lack of adequate vegetation (due to climate or human intervention), the rainwater runs off very quickly and erodes soft and exposed rock. The best-known examples are the badlands of SW South Dakota and NW Nebraska, USA.

Baekeland, Leo Hendrik (1863–1944) US chemist, b. Belgium. Baekeland invented a type of photographic paper, Velox, capable of being exposed under artificial light. He also invented (1909) the first thermosetting PLASTIC, BAKELITE, a substance that led to the development of the plastics industry.

Baeyer, Johann Friedrich Wilhelm Adolph von (1835–1917) German chemist. Baeyer specialized in organic chemistry. He was awarded the 1905 Nobel Prize for chemistry for his synthesis of INDIGO. He also investigated uric acid derivatives, discovered the phthalein dyes and devised a "strain" theory to account for the stability and conformation of carbon compounds with five-membered and six-membered rings.

Bailey, Sir Donald Coleman (1901–85) British civil engineer who designed the Bailey bridge for military use. The steel lattice girders can be assembled by groups of six men from easily manhandled component parts and cantilevered across a gap to provide a bridge to carry 70 tonnes over spans of up to 45m (150ft).

Bailey, Liberty Hyde (1858–1954) US botanist. He helped to establish HORTICULTURE as an applied science through the systematic study of cultivated plants. His work had an important influence on the development of GENETICS and plant pathology. He founded and directed the Bailey Hortorium at Cornell University.

Baird, John Logie (1888–1946) Scottish electrical engineer and inventor of TELEVISION. In 1926 he demonstrated the first working television to members of the Royal Institution, London. In 1928 he transmitted television to a ship at sea, and in 1929 was granted experimental broadcasting facilities by the British Broadcasting Corporation (BBC). His 240-line, part-mechanical, television system was used for the world's first public television service by the BBC in 1936. In 1937 it was superseded by a fully electronic model.

Bakelite Trade name (coined by Leo BAEKELAND) for a thermosetting PLASTIC used for insulating purposes and in making paint. It was the first plastic made (1909) by CONDENSATION REACTION, in which many molecules of two chemicals, in this case PHENOL and METHANAL (formaldehyde), are joined together to form large polymer molecules, by splitting off water molecules.

Baker, Sir Benjamin (1840–1907) British civil engineer. Baker was the principal designer of the Forth railway bridge in Scotland, completed in 1890, which at 518m (1,700ft) had the longest span in the world until 1918. He was also responsible for the construction of much of London's underground railways and the first Aswan dam.

baking powder Mixture used in cooking as a substitute for yeast. It contains SODIUM HYDROGENCARBONATE (sodium carbonate) mixed with an acid component, such as tartaric acid or cream of tartar. During cooking, the acid reacts with the sodium hydrogencarbonate to generate carbon dioxide gas, causing the food to rise without the fermentation effects of yeast.

baking soda Common name for SODIUM HYDROGENCARBONATE, so-called because it is a constituent of BAKING POWDER.

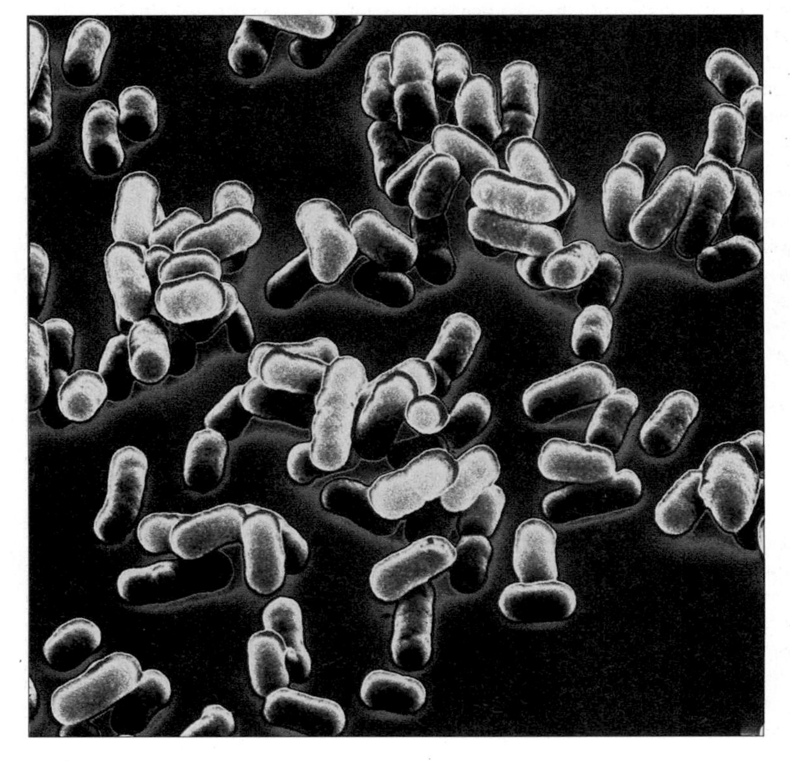

◄ **bacillus** Coloured scanning electron micrograph (SEM) of clusters of the bacteria *Yersinia pestis*, cause of bubonic plague. The bacteria are rod-shaped, gram-negative, non-motile bacilli. *Yersinia pestis* is primarily carried by the fleas of rats. Transfer to humans occurs when such fleas feed on human blood. Infection is rapid, causing swollen lymph nodes, and leading to septicaemia and pulmonary infection. Extensive control measures against rats and their fleas have eliminated plague from Europe, but it still occurs in other regions of the world.

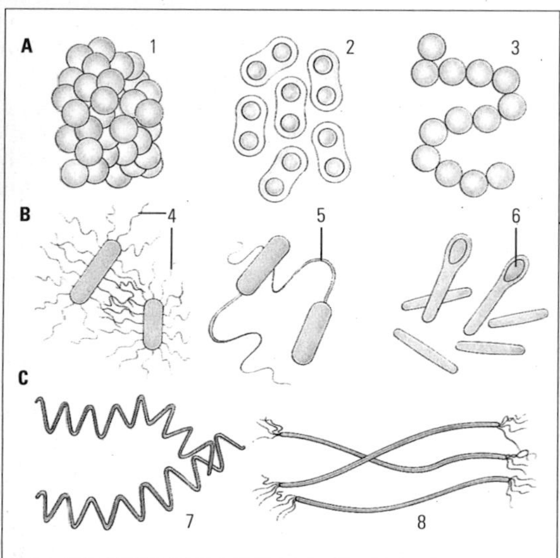

◄ **Bacteria occur** in three basic shapes and forms: spherical forms called cocci (A), rod-like bacilli (B), and spiral spirilla (C). Cocci can occur in clumps known as staphylococci (1), groups of two called diplococci (2) or chains called streptococci (3). Unlike cocci, which do not move, bacilli are freely mobile; some are termed peritrichous and use many flagellae (4) to swim about, while other monotrichous forms (as seen below) use a single flagellum (5). Bacilli can also form spores (6) to survive unfavourable conditions. Spirilla may be either cork-screw-shaped spirochaetes like Leptospira (7), or less coiled and flagellated, such as Spirillum (8) (magnification x 5,000).

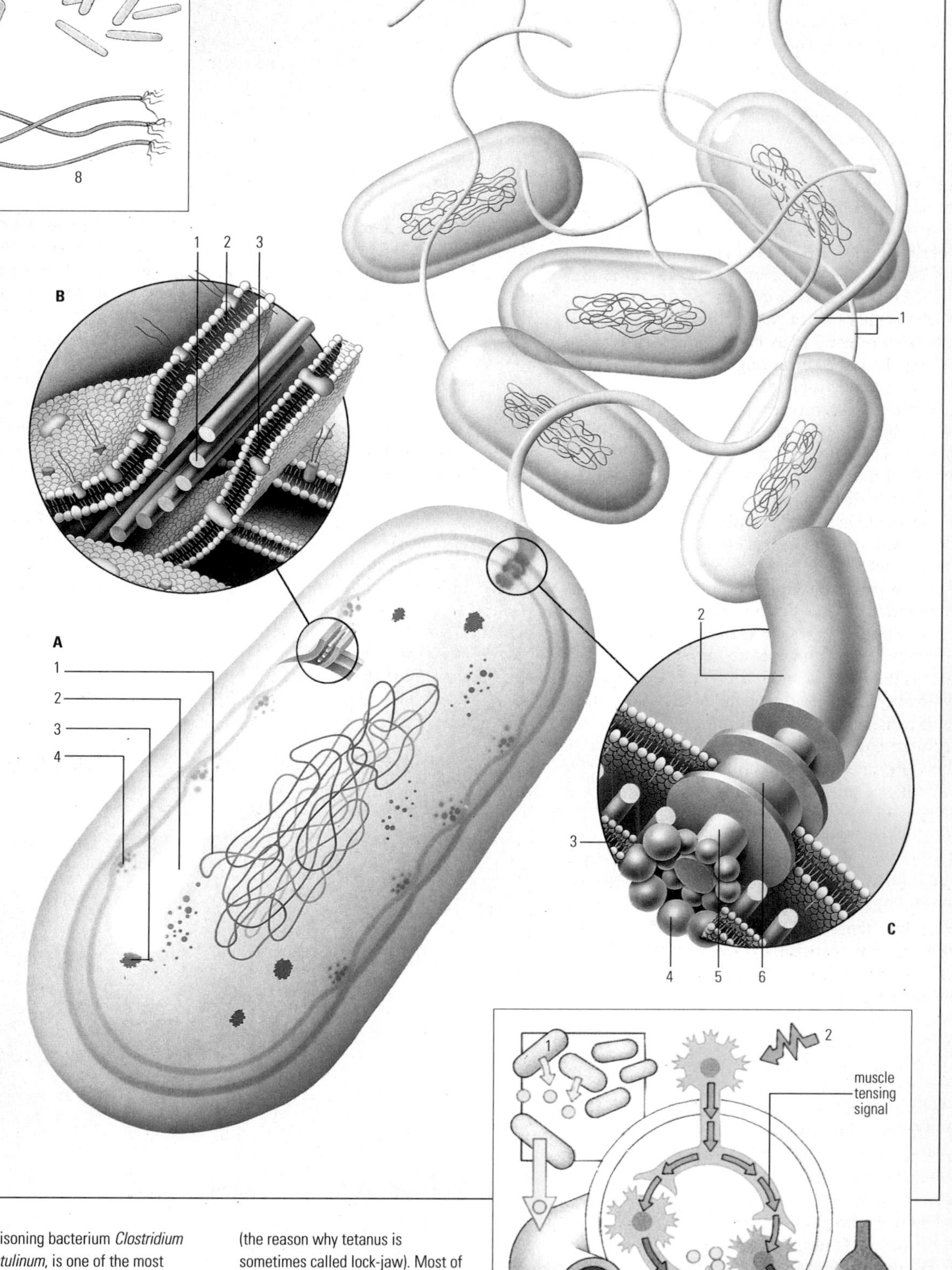

► **Bacteria** (A) have no nucleus; instead, they have a nucleoid (1), a single loop of DNA. This carries the genes, chemically coded instructions that define the bacterium. The average bacterium has about 3,000 genes, compared to a human's 100,000. The cytoplasm (2) also contains glycogen (food) granules (3); and ribosomes (4), which give the cytoplasm a grainy appearance and are the site of protein production. In many bacteria it also contains minute genetic elements called plasmids. Most, but not all, bacteria have rigid, protective cell walls (B). There are two main types. One has a single thick (10–50nm) layer. Bacteria with this type of cell wall are called Gram-positive, because they stain bright purple with the Gram stain. Gram-negative bacteria, as shown, have a thinner wall (1) with an extra layer of proteins and lipids on the outside (2). This type of cell does not stain purple, a basic distinction useful in medicine. The defensive cells of the body recognize bacteria by their cell walls. A cell membrane (3) surrounds the cytoplasm. It is a few molecules thick, made of proteins and lipids, and is the barrier at which a living cell controls what enters and leaves it. Some bacteria move (C) using flagella (1) whirled about by a hook (2). The motion is powered by a flow of protons across the cell membrane (3), which rotates a disc of protein molecules (4) in the membrane. A rod (5) connects this protein "rotor" to the hook via another disc (6) which seals the cell wall.

► **Before the advent** of effective sanitation and the discovery of antibiotics, recurrent epidemics of serious bacterial diseases swept Europe. The symptoms of many bacterial diseases are caused by toxic proteins (called toxins) produced by the bacteria. Botulinum toxin, produced by the food-poisoning bacterium *Clostridium botulinum*, is one of the most powerful poisons known. Tetanus toxin, produced by the related *Clostridium tetani* (1), can infect deep, dirty wounds. When a nerve impulse (2) tenses a muscle cell, the toxin blocks the relaxing part of the signal so the muscle stays tensed (the reason why tetanus is sometimes called lock-jaw). Most of the real killers among bacteria are now under control in the developed world, where tuberculosis is rare and diphtheria seldom a problem. In the developing world, however, bacterial diseases still take a dreadful toll.

balance Piece of apparatus for making accurate weighings. In a **beam** balance, two pans are suspended at the ends of a beam pivoted at its centre. The mass of the object to be weighed, located in one pan, is counterbalanced by known weights added to the other pan. In a **substitution** balance, there is only one pan suspended from the end of an arm with a fixed counterweight at the other end. The object to be weighed is placed in the pan, and calibrated weights are removed from the arm until it is counterbalanced. An **electronic** balance uses an electromagnetic force to restore the balance of a lever arm, the current flowing in the magnetic coil being a measure of the mass involved. Rough weighings may be made with the aid of a **spring** balance (or newton meter), in which the amount of extension of a vertical spring supporting a pan is a measure of mass.

ball-and-socket joint Type of joint that allows one part to rotate at almost any angle with respect to another. Such joints are used in machinery and occur naturally, as in the human hip joint.

ball bearing Type of bearing for rotating machinery. It consists of two concentric rings of steel, which together form a "race" for a number of steel balls (also referred to as ball bearings) positioned between them. The inner ring is fixed to a shaft and the outer one to a support.

ballistics Science of projectiles, including bullets, shells, bombs, rockets and guided MISSILES. **Interior** ballistics examines the propulsion and motion of the projectile within the firing device. **Exterior** ballistics investigates the trajectory of the projectile in flight. **Terminal** ballistics is concerned with the impact and effect of the projectile at the target. At each stage, scientists try to maximize the performance of the gun and projectile by improving their design. Ballistic technology has developed alongside ARTILLERY, and with the invention of instruments to monitor variables, such as the ignition and burning of the propellant explosive, the stress on a gun barrel, or the effect of air resistance and gravity on the trajectory.

balloon Unsteerable, lighter-than-air craft. Balloons are used for recreation and for scientific and military purposes. Either hot air or a gas that is lighter than air lifts the balloon from the ground. The first balloons to fly were of the open-necked hot-air type. Unmanned military, meteorological or other scientific balloons are usually filled with hydrogen, least dense of all gases but dangerously flammable. Manned balloons are now generally filled with the safer gas helium, or with hot air. *See also* AIRSHIP

ballpoint pen Pen that employs as its writing point a small ball bearing that rolls against and picks up semi-solid ink from a reservoir. It was invented by John Loud in 1888 but Lazlo Biro was the first to make a reliable model in 1938, which was patented in 1943.

Baltimore, David (1938–) US microbiologist. He shared the 1975 Nobel Prize for physiology or medicine for his work on the "interaction between tumour viruses and the genetic material of the cell". Working independently of co-laureate Howard TEMIN, he proved the existence of reverse transcriptase, the enzyme necessary for viral genetic information to be incorporated into an animal cell. *See also* REVERSE TRANSCRIPTION

bandwidth Range of frequencies spanned by a radio signal of a particular nominal frequency. If a transmitted radio signal is modulated, the bandwidth is the range of frequencies employed on either side of the CARRIER WAVE signal. It is therefore also the range of frequencies used for a particular radio transmission within a particular waveband. The term is also used to describe the frequency range over which a device, such as an amplifier or radio receiver, should not significantly differ from its maximum value. In communications and computing, bandwidth describes that rate at which data is transmitted (for example, by a modem), usually measured in bits per second.

Banks, Sir Joseph (1743–1820) English botanist. He was the senior scientist of the group who sailed to Tahiti with Captain James Cook aboard HMS *Endeavour* in 1768. At Botany Bay, Australia, in 1770, Banks collected examples of plants hitherto unknown in Europe, including the shrub banksia, named in his honour. Upon his return, he helped set up the Royal Botanic Gardens at Kew, W London, and financed international plant-collecting expeditions. In 1778 Banks became president of the Royal Society.

Banting, Sir Frederick Grant (1891–1941) Canadian physician. He shared, with J.J.R. MACLEOD, the 1923 Nobel prize for physiology or medicine for his work in extracting the hormone INSULIN from the PANCREAS. This made possible the effective treatment of DIABETES.

bar Unit of pressure; the pressure created by a column of mercury 75.006cm high at 0°C (32°F). It is equal to 10^5 PASCALS. Standard atmospheric pressure (at sea level) is 1.01325 bars, or 1,013.25 millibars.

Bárány, Robert (1876–1936) Swedish physician, b. Austria. He was awarded the 1914 Nobel Prize for physiology or medicine for his work on the physiology and pathology of the vestibular apparatus of the EAR. His studies of the normal and abnormal functioning of this important balancing mechanism greatly advanced understanding of the ear.

barbiturate DRUG used as a sedative or to induce sleep or sedation. Highly addictive and dangerous in high doses or when combined with other drugs such as alcohol or tranquillizers, most barbiturates are no longer prescribed. Short-acting barbiturates are used in surgery to induce general anaesthesia; long-acting formulations are sometimes prescribed for epilepsy.

barchan Crescent-shaped DUNE found in sandy DESERTS throughout the world where the wind is constant in speed and direction. Barchans also occur fairly frequently in coastal regions; they are quickly shifted by the wind, particularly when small.

bar code Coded information consisting of thick and thin black lines, and designed for computer recognition. A laser beam scans the bar code and a light-sensitive detector picks up the reflected signal, which consists of a pattern of pulses. Bar codes are used on many products for sale in shops and supermarkets. The store's computer translates the bar code into information, including the product's name, weight or size. The computer refers to a price list data file to see how much to charge the customer and records the sale so that the shop can monitor its stockholding levels.

Bardeen, John (1908–91) US physicist known for his research into SEMICONDUCTORS. He worked with the Bell Telephone Laboratories (1945–51) and was then professor of physics at Illinois University (1951–75). He was the first person to win the Nobel prize twice in the same field, physics: in 1956 he shared it with William SHOCKLEY and Walter BRATTAIN, for their joint invention of the TRANSISTOR, and in 1972 with Leon COOPER and John SCHRIEFFER, for their theory of SUPERCONDUCTIVITY.

barite *See* BARYTE

barium (symbol Ba) Silver-white, metallic element of the ALKALINE-EARTH METALS, discovered in 1808 by Sir Humphry DAVY. It is a soft metal the chief sources of which are baryte (barium sulphate) and witherite (barium carbonate). Barium compounds are used as rodent poison, pigments for paints and as drying agents. Barium sulphate ($BaSO_4$) is swallowed to allow X-ray examination of the stomach and intestines because barium atoms are opaque to X-rays; this is called a "barium meal". Properties: at.no. 56; r.a.m. 137.34;

BAR CODE

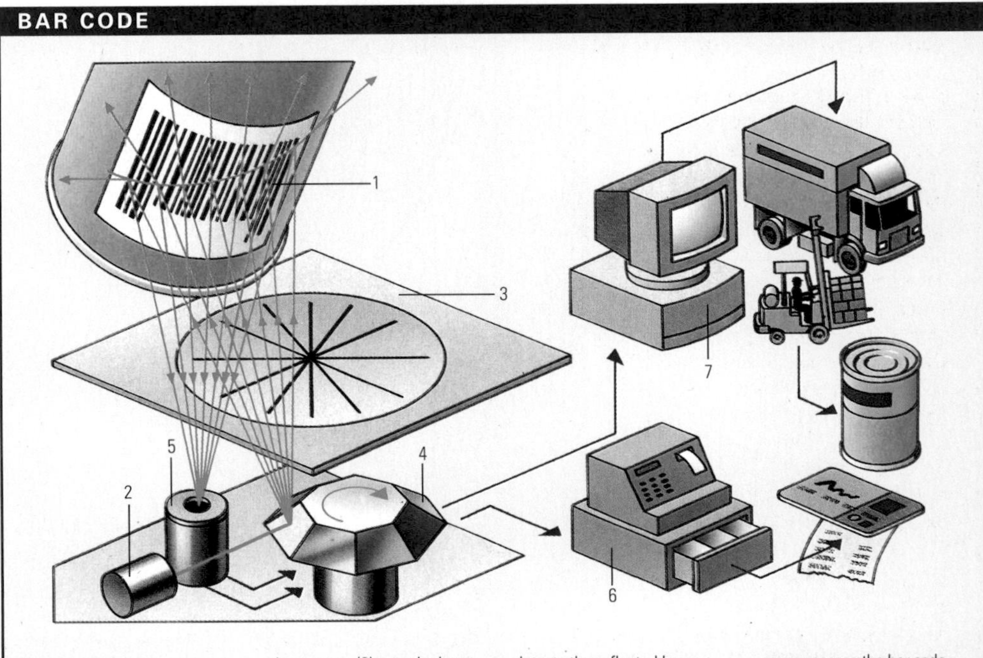

Bar codes represent information concerning a product and its manufacturer in a series of thick and thin black and white lines (1). A laser (2) is reflected through a glass screen (3) onto the bar code by a rotating multi-faceted mirror (4). The laser light is scattered by the white lines and absorbed by the black lines. A sensor (5) detects the reflected laser light and compares the relative width of the lines. Because the relative widths are compared, the bar code does not have to be on a flat surface. The sensor passes the bar code information to the till (6) for billing the customer, and a central store computer (7) monitors stock levels and order supplies of goods.

r.d. 3.51; m.p. 725°C (1,337°F); b.p. 1,640°C (2,984°F); most common isotope ^{138}Ba (71.66%).

bark Outer protective covering of a woody plant stem. It is made up of several layers. The CORK layer, waxy and waterproof, is the thickest and hardens into the tough, fissured outer covering. Lenticels (pores) in the bark allow GAS EXCHANGE between the stem and the atmosphere. *See also* CAMBIUM

Barkla, Charles Glover (1877–1944) British physicist who was awarded the 1917 Nobel Prize for physics for his discovery that elements have a characteristic X-RAY spectrum, which enables them to be identified. He also formulated laws governing the scattering of X-rays and their transmission through solids.

barn (symbol *b*) Scientific unit of area used in nuclear physics to measure the cross-sections in interactions of particles. A barn equals 10^{-24} cm^2. This area is a measure of the probability that FISSION will occur when a neutron passes near a heavy nucleus.

Barnard, Christiaan (1922–) South African surgeon. He was the first to perform a human heart transplant (3 December 1967). In 1974 he was the first to implant a second heart in a patient and to link the circulations of the two hearts so that they worked together as one.

Barnard's Star RED DWARF star 6 light-years away, in the constellation Ophiuchus. It is the closest star to the Sun after the ALPHA CENTAURI system and was discovered in 1916 by the US astronomer Edward E. Barnard (1857–1923).

barograph Recording BAROMETER. It consists of an aneroid barometer that has, instead of a pointer, a pen that continuously records ATMOSPHERIC PRESSURE on a paper chart wrapped round a revolving drum, usually driven by clockwork.

barometer Instrument for measuring ATMOSPHERIC PRESSURE. There are two main types: the **mercury** barometer, and the **aneroid** barometer, which is the less accurate of the two but is the type normally used in BAROGRAPHS and found in the home. The mercury barometer consists of a glass tube containing mercury, which is inverted over a small reservoir of mercury; as the pressure rises, the weight of the atmosphere forces mercury out of the reservoir and up the tube. The tube has a scale on it, so that the pressure can be read off at the top of the column of mercury. As the pressure falls, the column of mercury gradually moves down. A column of mercury 760mm (30in) in length represents the average pressure at sea level.

barred spiral galaxy Type of GALAXY as classified by Edwin HUBBLE.

barrel Unit of volume used to measure liquids, particularly petroleum (crude oil). For oil and petroleum products, a barrel equals 158.98 litres or 35 gallons (42 US gallons). For beer, a barrel equals 163.66 litres or 36 gallons (43.24 US gallons). For dry goods (such as grain), a barrel equals $0.1156m^3$ or 3.180 bushels (3.283 US bushels).

barrier island Any long, low island of sand parallel to a shore and permanently separated from it. It may be composed of dunes, swamps and areas of vegetation.

barrier reef Long, narrow CORAL REEF lying some distance from and roughly parallel to the shore, but separated from it by a deep LAGOON. The Great Barrier Reef off the coast of Queensland, NE Australia, is the most famous.

Barton, Sir Derek Harold Richard (1918–98) British chemist. He did innovatory research into **conformational analysis** (study of the geometric structure of complex molecules), discovering that the reactivity of a molecule was related to its preferred shape. For this work he shared the 1969 Nobel Prize for chemistry with Odd HASSEL. One of Barton's many advances in organic chemistry was his use of FREE RADICALS, which enabled him to invent new reactions. Barton was professor (1957–70), Hofmann Professor (1970–78) and then Emeritus Professor (1978–98) of organic chemistry at Imperial College, London University. He was director (1977–85) of the Institut de Chimie des Substances Naturelles (CNRS).

baryon Any of several types of ELEMENTARY PARTICLE affected by the STRONG NUCLEAR FORCE. A baryon consists of three QUARKS, which are indivisible elementary particles. Baryons and MESONS (made up of two quarks) are subclasses of HADRONS. The only stable baryons are the PROTON and (provided it is inside a nucleus) the NEUTRON. Heavier baryons are called HYPERONS. *See also* LEPTON

baryte (barite or heavy spar) Barium-containing mineral, the chief ore of the metal. Baryte consists of barium sulphate ($BaSO_4$). It usually occurs as white crystals, although it may be brown, grey or yellow, and is often found in association with lead or zinc ores. Because of its high density, it is used to make muds for lubricating and cooling oil-well drilling equipment. It is also used in the chemical industry for paper-making, rubber manufacture and high-quality paints. Hardness 3–3.5; r.d. 4.5

basal metabolic rate (BMR) Minimum amount of energy required by the body to sustain basic life processes, including breathing, circulation and tissue repair. It is calculated by measuring oxygen consumption. Metabolic rate increases well above basal metabolic rate (BMR) during vigorous physical activity, fever or under the influence of some DRUGS (including CAFFEINE). It falls below BMR during sleep, general ANAESTHESIA or starvation. BMR is highest in childhood and decreases with age.

basalt Hard, fine-grained, basic IGNEOUS ROCK, which may be an INTRUSIVE or EXTRUSIVE ROCK. Its colour can be dark green, brown, dark grey or black. If it originally solidified quickly it can have a glassy appearance. There are many types of basalt with different proportions of elements. It may be compact or vesicular (porous) because of gas bubbles contained in the lava while it was cooling. If the vesicles are subsequently filled with secondary minerals, such as quartz or calcite, it is called **amygdaloidal** basalt. Basalts are the main rocks of ocean floors, and on continental areas form the world's major lava flows, such as the Deccan Trap in India.

base In chemistry, any compound that accepts protons. A base will neutralise an ACID to form a SALT and water. Most are oxides or hydroxides of metals; others, such as ammonia are compounds that yield hydroxide IONS in water. Soluble bases are called ALKALIS. **Strong** bases are fully dissociated into ions; **weak** bases are partly dissociated in solution. *See also* NEUTRALIZATION

base In geometry, the side opposite the vertex from which an altitude is drawn in a TRIANGLE.

base In mathematics, the number of units in a number system that is equivalent to one unit in the next higher counting place in that system. 10 is the base of the decimal system. In base 5 only the digits 0 to 4 can be used in each place. Hence, the counting numbers in base 5 are 1, 2, 3, 4, 10, 11, 12, 13, 14, 20, 21... etc.. The number 45 in base 7 is equal to 33 in base 10 since

$$(4 \times 7) + (5 \times 1) = 33$$

BAROMETER

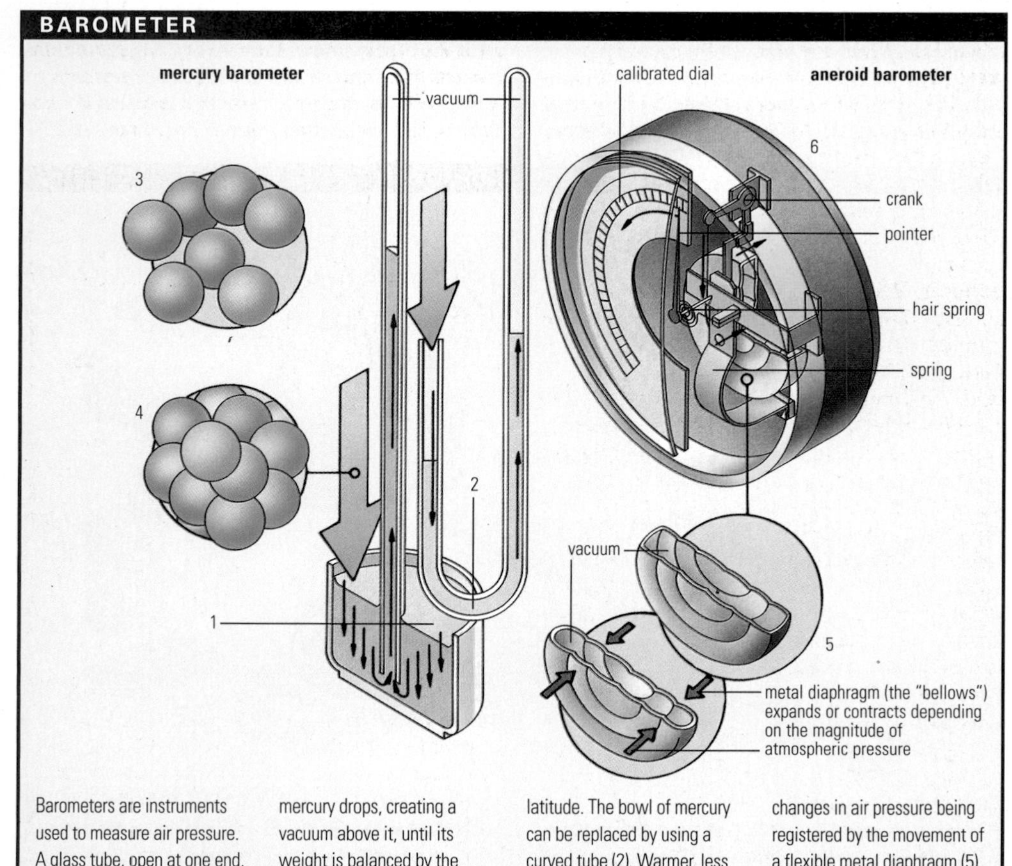

Barometers are instruments used to measure air pressure. A glass tube, open at one end, is filled with mercury (which is used because of its density) and inverted in a bowl of mercury. The column of mercury drops, creating a vacuum above it, until its weight is balanced by the weight of the air pressing on the mercury surface (1). The column of mercury is 760mm tall at sea level and 45° latitude. The bowl of mercury can be replaced by using a curved tube (2). Warmer, less dense air (3) will exert less pressure than cold dense air (4). The aneroid barometer dispenses with the mercury, changes in air pressure being registered by the movement of a flexible metal diaphragm (5) containing a vacuum, which is connected by a series of levers to a sprung indicator pointer and scale (6).

◀ **barrier reef** Edge of the Great Barrier Reef, Australia; an aerial view taken off the Queensland coast. The shallows (foreground) reveal submerged coral, and beyond the wave zone the Australian continental shelf drops to deep water. A reef is a ridge or island chain that projects above the sea during tidal cycles. A barrier reef is one which develops offshore so that a deep lagoon is formed between the reef and the mainland coastline. At 2,000km (1,200mi) long and 50–150km (30–90mi) offshore, the Great Barrier Reef is the longest of its kind and the largest accumulation of coral known.

Similarly, the number 2142 in base 5 is equal to 297 in base 10 since

$$(2 \times 125) + (1 \times 25) + (4 \times 5) + (2 \times 1) = 297$$

The BINARY SYSTEM is used in computing since any number can be represented by a string of just two digits 0 and 1, which match states of gates in computer circuitry that can be either open or closed.

base metal Common, cheap metal. Typical examples are lead and iron which, like other base metals, tarnish or oxidize on exposure to air or moisture. Such properties distinguish base metals from gold, platinum, silver and other so-called precious or NOBLE METALS. In chemical terms, base metals are those at the negative end of the ELECTROMOTIVE SERIES. In ancient times, it was an unfulfilled ambition of alchemists to convert base metals into gold. *See also* ALCHEMY

base pair Combination of two nitrogenous BASES in a molecule of DNA. There are four bases in DNA: ADENINE, CYTOSINE, GUANINE and THYMINE. Using HYDROGEN BONDS, adenine always pairs with thymine, and cytosine always pairs with guanine. The pairs form the "rungs" of the double-helix structure of DNA and form the basis of the GENETIC CODE. RNA also has base pairs, but URACIL replaces thymine.

BASIC (**B**eginners' **A**ll-purpose **S**ymbolic **I**nstruction **C**ode) Computer programming language that is easy to learn and uses many everyday words. It is commonly used by both amateur and professional programmers. To run, BASIC COMPUTER PROGRAMS usually require a separate program called an **interpreter**, which converts the BASIC code into the machine code required by the computer's processor.

basic oxygen process Widely used steelmaking process in which high-pressure oxygen is blown over the surface of the molten charge to help remove impurities. It has superseded the BESSEMER PROCESS and OPEN-HEARTH PROCESS.

basic salt Salt of a BASE that has had one or more, but not all, of its oxide or hydroxide ions replaced by other negative ions. Lead nitrate hydroxide ($Pb(OH)NO_3$) is a basic salt.

Basidiomycota Phylum of FUNGI (formerly the class Basidiomycetes), including the jelly fungus, the bird's nest fungus, the ear fungus and mushrooms. They reproduce through sexual spores (**basidiospores**) produced by a specialized cell, the BASIDIUM.

basidium Cell that produces spores (**basidiospores**) used in sexual reproduction by certain types of fungi of the phylum BASIDIOMYCOTA. Basidia are cylindrical or club-shaped, often grouped to form the external fruiting bodies (gills) of bracket fungi, mushrooms and puffballs. Each basidium produces four spores.

basilar membrane Part of the COCHLEA, a structure in the inner EAR.

Basov, Nikolai Gennadiyevich (1922–) Soviet physicist who developed the MASER which amplifies microwaves and the LASER which amplifies light. For these contributions to science, Basov and his co-worker Alexander PROKHOROV shared the 1964 Nobel Prize for physics with Charles TOWNES (who had made similar discoveries independently).

bast Type of vegetable fibre or matting made from the inner bark (PHLOEM) of plants. Examples include jute, hemp and flax.

Bates, Henry Walter (1825–92) British naturalist. His work on NATURAL SELECTION in animal MIMICRY lent support to Charles DARWIN's theory of EVOLUTION. Bates collected over 8,000 previously unrecorded species of insects during 11 years in the Amazon (1848–59) and recounted his experiences in *The Naturalist on the River Amazon* (1863).

Bateson, William (1861–1926) British biologist. He founded and named the science of GENETICS. Bateson translated much of Gregor MENDEL's pioneering work on inheritance in plants, so bringing it recognition. By his own experiments he also extended Mendel's theories to animals, which provided a foundation for the modern understanding of HEREDITY.

batholith Huge mass of IGNEOUS ROCK at the Earth's surface that has an exposed surface of more than 100sq km (40sq mi). It may have originated as an INTRUSIVE igneous structure that was gradually eroded and which became surface material. Most batholiths consist of GRANITE rock types, and are associated with the OROGENESIS (mountain-building phases) of PLATE TECTONICS.

bathyal zone Region of the ocean floor from the edge of the CONTINENTAL SHELF, *c.*133m (436ft) in depth, to about 2,000m (6,560ft), where the ABYSSAL PLAIN begins. Only a feeble light penetrates to the upper layers of this zone.

bathysphere and bathyscaphe Manned vehicles for deep-sea exploration. The **bathysphere**, invented by Otis Barton and William BEEBE in the USA, had steel walls and thick, toughened glass windows through which underwater observations could be made. Used mainly during the 1930s, it was released from a surface vessel and was lowered on a steel cable to depths of more than 900m (3,000ft). The **bathyscaphe** was invented by August PICCARD of Switzerland, and was first used in 1948. It consists of a bathysphere slung below a tank called a float. The whole device can be sunk or made buoyant in a controlled manner, while propellers move the craft horizontally. In January 1960, Piccard and Don Walsh of the US Navy, made a descent in the bathyscaphe *Trieste* to a depth of 10,916m (35,810ft) (more or less equivalent to the altitude of commercial jets) in the Pacific Ocean's Mariana Trench off the island of Guam. This record still stands.

battery Collection of voltaic CELLS that convert chemical energy into direct current (DC) electricity. The term is also commonly used for a single cell, particularly a dry cell as used in portable electronic equipment. Most primary cell batteries are not rechargeable; some types of primary cell – such as nickel-cadmium (Nicad) batteries – and all accumulators (storage batteries) can be recharged when a current passed through them in the reverse direction restores the original chemical state.

baud Unit for measuring the speed at which a digital communications device or system carries information. One baud is equal to one BIT per second. Although the term baud rate is still widely used, the speed of modern equipment is often expressed in kilobits per second.

Baudot, Jean-Maurice-Emile (1845–1903) French electrical engineer who devised a telegraph code that used the presence or absence of an electrical pulse, rather than the short and long pulses of the MORSE CODE. He also invented a distributor system which permitted several messages to be transmitted along the same wire.

bauxite Rock from which nearly all aluminium is extracted. Bauxite is a mixture of several minerals, such as diaspore, gibbsite, boehmite and iron. It is formed by prolonged weathering and

BATTERY

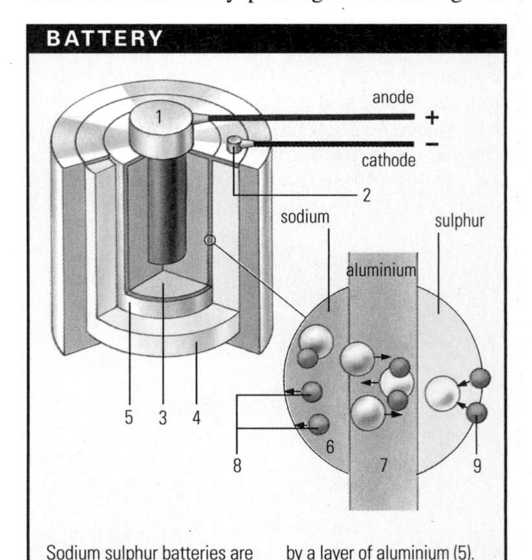

Sodium sulphur batteries are the newest type of battery and are much lighter than nickel cadmium varieties. They have a carbon anode (1) and a metal cathode (2). The reactants are arranged in rings around the central anode. An inner core of sodium (3) is separated from an outer ring of sulphur (4) by a layer of aluminium (5). The sodium (6) reacts with the aluminium layer (7) giving up electrons (8) which stream to the anode. The cathode gives electrons to the sulphur atoms (9), which bond with the sodium ions to form sodium sulphide. This process creates a voltage.

B

leaching of rocks containing aluminium silicates. Large deposits occur in France, Hungary, USA, Guyana, Jamaica, Surinam, Italy, Greece, Russia, Azerbaijan and Kazakstan.

bay bar Ridge of sediment that is deposited across the mouth of a bay. It is usually attached to the shore at each end. The deposit may be carried to the bar by LONGSHORE DRIFT, or it may be carried there by a river. When it grows large enough it may completely block the bay from the sea, so that a DELTA forms on the landward side as further deposits pile up.

Bayliss, Sir William Maddock (1860–1924) British physiologist. He studied blood pressure, circulation and digestion, and with Professor Ernest STARLING discovered (1902) SECRETIN, a hormone secreted by the DUODENUM.

BCG (Bacille Calmette-Guérin) Vaccine against tuberculosis. It was named after its discoverers, the French bacteriologists Albert Calmette (1863–1933) and Camille Guérin (1872–1964).

BCS theory Theory that explains SUPERCONDUCTIVITY. The theory is named after three American physicists, John BARDEEN, Leon COOPER and John SCHRIEFFER. It states that, in a superconductor, electrostatic forces between positive sites on the metal lattice and a conducting electron distort the lattice. A second electron is then affected and, as a result, the current is carried not by individual electrons but by bound pairs of electrons called **Cooper pairs**. Such a pair is not affected by the lattice and so carries current indefinitely.

beach Sloping zone of the shore of the sea or a lake, covered by sediment, sand or pebbles, which extends from the low-water line to the limit of the highest storm waves. The sediment is derived from coastal erosion or river ALLUVIUM. Waves breaking on the beach move the sediment back and forth so the heavier pebbles remain on the upper beach and the sand moves downwards towards low water. *See also* LONGSHORE DRIFT

beacon, radio Navigational aid for ships and aircraft. It consists of a transmitter that broadcasts radio signals of a fixed frequency, peculiar to one beacon only. A navigator or pilot, by referring to navigation charts, can find his or her position from the bearings of two or more beacons, using a DIRECTIONAL ANTENNA to receive the signals.

Beadle, George Wells (1903–89) US geneticist. During his study of MUTATIONS in the bread mould *Neurospora crassa*, he and Edward TATUM found that GENES are responsible for the synthesis of ENZYMES and that these enzymes control each step of all biochemical reactions occurring in an organism. For this discovery they shared with Joshua LEDERBERG the 1958 Nobel prize for physiology or medicine.

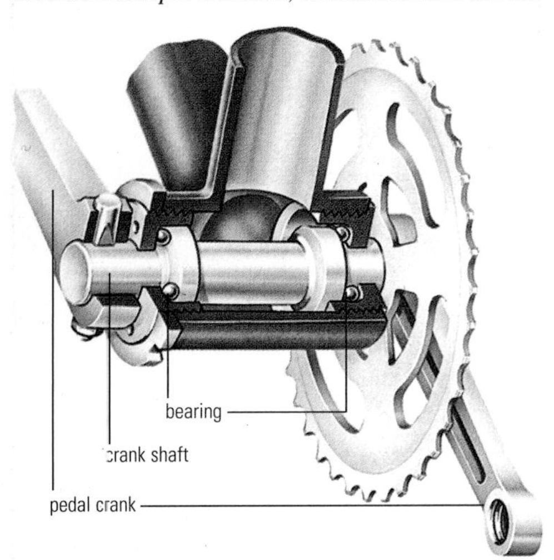

▲ **bearing** The bottom bracket assembly of a typical bicycle (shown here) must be very strong to stand up to the driving force of the pedals. The bearings are of the cup and cone type, the right-hand cup having a left-hand thread and the left-hand cup being adjustable.

Beagle, HMS British survey ship that carried Charles DARWIN as ship's naturalist. The *Beagle* left England in December 1831 and for five years explored parts of South America and the Pacific islands. Darwin's observations formed the basis for his theory of EVOLUTION of species by NATURAL SELECTION.

Bear, Great and Little Popular names for URSA MAJOR and URSA MINOR, respectively

bearing Supporting component, usually of a resistant metal alloy, used to minimize friction between moving parts of machines. Some bearings are cylindrical shells which separate the surfaces of moving parts by trapping a thin film of lubricant between them. BALL BEARINGS use balls confined to a race to reduce the areas in contact; ROLLER BEARINGS use rollers.

Beaufort wind scale Range of numbers from 0 to 17 representing the force of winds, together with descriptions of the corresponding land or sea effects. The Beaufort number 0 means calm wind less than 1km/h (0.62mph), with smoke rising vertically. Beaufort 3 means light breeze, 12–19km/h (8–12mph), with leaves in constant motion. Beaufort 11 is a violent storm, 103–117km/h (64–72mph) and Beaufort 12 is a hurricane, 118+km/h (73+mph), with devastation. Beaufort 13–17 indicates the force of the hurricane. The scale is named after its inventor, Admiral Sir Francis Beaufort (1774–1857). *For complete scale see ready reference section.*

Beaumont, William (1785–1853) US surgeon. He gained valuable knowledge of the functioning of the human stomach through a series of experiments he performed on the stomach of Alexis St. Martin, who, as the result of a gunshot wound, had a small opening into his body over his stomach. It provided Beaumont with a unique opportunity to study stomach action.

Becquerel, (Antoine) Henri (1852–1908) French physicist. He was professor of physics at the Paris Museum of Natural History, and later at the École Polytechnique. In 1896 he discovered RADIOACTIVITY in uranium salts, for which he shared the 1903 Nobel prize for physics with Pierre and Marie CURIE. He also studied the rotation of the plane of polarized light in a magnetic field. The BECQUEREL was named after him. *See also* BETA PARTICLE

becquerel (symbol Bq) SI unit of radioactivity. It is defined as the activity of a radioactive material that decays at an average rate of one transition per second. It is named after the French physicist Henri BECQUEREL. It replaces the former unit, the CURIE, which equals 3.7×10^{10} Bq.

bed In geology, a layer of SEDIMENTARY ROCK greater than 1cm (0.4in) in thickness, representing a single episode of sediment deposition. The original sediment of a bed is usually deposited in a horizontal sheet.

bedding plane In geology, the upper or lower surface of a BED. The lower surface is known as the **sole** of the bed.

bedrock Solid rock that occurs below the soil and any REGOLITH (rock fragments). It is the C or D horizon in a soil profile (*see* SOIL HORIZON). On hillsides the bedrock may be only a few centimetres below the surface, whereas on prairies and plains it may be several metres down.

Beebe, Charles William (1877–1962) US naturalist and explorer, Curator of Ornithology (1899–1919) at the New York Zoological Gardens and Director of Tropical Research (1919–52) at the New York Zoological Society. He led explorations in central and s America, the West Indies and the Orient. He also made undersea descents of 1,000m (3,300ft) in his invention, the BATHYSPHERE. His numerous books include *Galápagos* (1923) and *Beneath Tropic Seas* (1928).

behavioural ecology Study of the complex relationship between environment and animal behaviour. This involves drawing on natural history, to study the adaptive features of an organism within its habitat. Human behaviour is similarly studied. *See also* ADAPTATION; ECOLOGY; ETHOLOGY

Behring, Emil Adolph von (1854–1917) German bacteriologist and pioneer immunologist. In 1901 he was awarded the first Nobel Prize for physiology or medicine for his work on serum therapy, especially for developing immunization against DIPHTHERIA (1890) and TETANUS (1892) by injections of ANTITOXINS, a word he introduced. His discoveries led to the treatment of many childhood diseases.

Békésy, Georg von (1899–1972) US physicist, b. Hungary. As director of the Hungarian Telephone System Research Laboratory (1923–46), Békésy worked on problems of communication and the mechanics of human hearing. At Harvard University, he carried on his research (1949–66) on the COCHLEA of the EAR. For this research Békésy received the 1961 Nobel Prize for physiology or medicine. Extending his studies to other senses, he developed a unified picture of the role of sensory and nervous processes in perception.

belemnite Extinct group (order) of CEPHALOPODS which appeared in the Jurassic period (213–144 million years ago) and lived through the Cretaceous period (144–65 million years ago), with a few persisting into the early part of the Tertiary period (65–2 million years ago) up to about 45 million years ago. Entirely marine creatures, belemnites were characterized by a bullet-shaped, cylindrical shell.

Bell, Alexander Graham (1847–1922) Scottish-born scientist, inventor of the TELEPHONE. He first worked with his father, the inventor of a system for educating the deaf. The family moved to Canada in 1870, and Bell taught speech at Boston University (1873–77). His work on the transmission of sound by electricity led to the first demonstration of the telephone in 1876 and the founding of the Bell Telephone Co. (USA) in 1877.

Bell, Sir Charles (1774–1842) Scottish anatomist and surgeon, the first man to distinguish between sensory and motor nerves in the brain and to describe the facial paralysis known as Bell's palsy. He was professor of anatomy and surgery at the Royal College of Surgeons, London, and from 1836 was professor of surgery at Edinburgh University.

Benioff zone Area of deep-focus earthquakes that dips from the surface to a depth of about 700km (450mi). The zone is thought to indicate areas of active SUBDUCTION of the Earth's tectonic plates. *See also* PLATE TECTONICS

benitoite Glassy, blue to violet mineral, barium titanium silicate ($BaTi(SiO_3)_3$), found in San Benito, USA. It forms hexagonal system tabular, triangular crystals and is a valuable gem when transparent and without flaws. Hardness 6-6.5; r.d. 3.6.

Bentham, George (1800–84) British botanist. His classification of GYMNOSPERMS (seed-plants) provided a foundation for modern systems. With

BENZENE

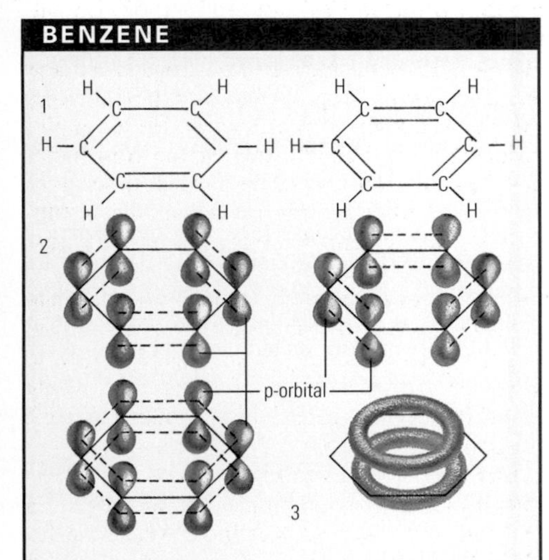

The Kekulé structures (named after the German chemist Friedrich Kekulé) for benzene, once generally accepted, usually represent the molecule as a six-membered, hexagonal carbon ring with alternating single and double bonds. However, the carbon-to-carbon bond lengths are equal, intermediate between single and double bonds, and the double bonds do not belong to any particular bond. The six p-electrons are considered to be delocalized and free to move within the overlapping p-orbitals of all six carbon atoms.
Key 1) The Kekulé structures representation for benzene. 2) The double bonds in the Kekulé formulae may be represented by the overlapping of p-orbitals in two different ways. 3) The p-orbitals can overlap equally in both directions, forming a "molecular orbital" which may be represented as a pair of thick rings, or ring-shaped electron clouds.

William Hooker (1785–1865) he wrote the multi-volume *Genera Plantarum* (1862–83), a definitive work in its time. Bentham also wrote *Handbook of British Flora* (1858).

benthos Flora and fauna of the seabed. The fauna includes sedentary forms such as sponges, creeping creatures such as crabs and snails, and burrowing animals such as worms, and countless bacteria.

Benz, Karl (1844–1929) German pioneer of the INTERNAL COMBUSTION ENGINE. After some success with an earlier TWO-STROKE ENGINE, he built a FOUR-STROKE ENGINE in 1885 which was first applied to a tricycle. Benz achieved great success when he installed the new engine in a four-wheel vehicle in 1893. Benz was the first to make and sell light, self-propelled vehicles built to a standardized pattern. Hundreds had been built by 1900.

benzaldehyde *See* BENZENECARBALDEHYDE

benzene (C_6H_6) Colourless, volatile, sweet-smelling, flammable liquid HYDROCARBON, a product of petroleum refining. A benzene molecule is a hexagonal ring of six unsaturated carbon atoms (benzene ring), and is the simplest AROMATIC COMPOUND. It is a raw material for manufacturing many organic chemicals and plastics, drugs and dyes. Benzene is carcinogenic and should be handled with caution. Properties: r.d. 0.88; m.p. 5.5°C (41.9°F); b.p. 80.1°C (176.2°F).

benzenecarbaldehyde (benzaldehyde, C_6H_5CHO) Simplest of the AROMATIC compounds in which an aldehyde group, –CHO, is attached directly to a benzene ring. It is a colourless liquid having a smell of bitter almonds; it is used as a chemical reagent, flavouring material and for making perfumes and dyes. Properties: r.d. 1.04; m.p. −26°C (−14.8°F); b.p. 178.1°C (352.6°F)

benzenecarboxylic acid (benzoic acid, C_6H_5COOH) White, crystalline weak CARBOXYLIC ACID made from toluene. It is used to make dyestuffs and for preserving fruit juices. Properties: r.d. 1.266 (15°C); m.p. 122.4°C (252.3°F); b.p. 249°C (480.2°F).

benzodiazepine Any of a group of mood-altering DRUGS, such as Librium and Valium, that are used primarily to treat severe anxiety or insomnia. These drugs, which intervene in the transmission of nerve signals in the CENTRAL NERVOUS SYSTEM, were originally developed as muscle relaxants. Today they are widely prescribed as tranquillizers. Benzodiazepines may produce a number of side-effects, including drowsiness, unsteadiness and confusion. Also, they are known to cause dependence. For these reasons, they are recommended only for short-term use.

benzoic acid *See* BENZENECARBOXYLIC ACID

benzoin Fragrant, resinous POLYMER once obtained from the balsam resin found in the trees of the genus *Styrax* in tropical SE Asia, but now made synthetically. It is used in perfumes and in decongestant cough linctuses.

benzyne (C_6H_4) Unstable organic compound derived from BENZENE. Benzyne molecules consist of a hexagonal ring of carbon atoms joined by two double bonds and one triple bond. It probably exists only briefly in various reactions of AROMATIC COMPOUNDS.

Bergius, Friedrich (1884–1949) German chemist who developed a method of treating coal with hydrogen, under high pressure, to produce oil. It became known as the **Bergius process**, and in 1931 Bergius shared the Nobel Prize for chemistry with Carl BOSCH. He also developed a process for converting wood into sugar by HYDROLYSIS.

bergschrund Deep, wide crevasse or a series of parallel narrow crevasses in a GLACIER, produced by tension within the ice, often at the point where the moving ice pulls away from the rock slope at the head of the glacier CIRQUE.

berkelium (symbol Bk) Synthetic radioactive, metallic element of the ACTINIDE SERIES. It was first made in 1949 by alpha-particle bombardment of americium-241 at the University of California at Berkeley (after which it is named). Nine isotopes are known. Properties: at.no. 97; r.d. (calculated) 14; m.p. 986°C (1,807°F); most stable isotope ^{247}Bk (half-life 1.4×10^3 yr).

berm Narrow ridge of debris above the foreshore. During storms, extra large waves carry debris up the beach so that it forms a narrow horizontal shelf or ridge above the normal high-water mark.

Bernal, John Desmond (1901–71) Irish physicist, remembered for his work in the field of X-RAY CRYSTALLOGRAPHY; he used the technique to study atomic structures. He was professor of physics, and then of crystallography, at London University (1938–68).

Bernard, Claude (1813–78) French physiologist. He defined the role of the PANCREAS in digestion, the role of the LIVER in regulating blood sugar levels and carbohydrate stores, and the regulation of blood supply by vasomotor nerves.

Bernoulli, Daniel (1700–82) Swiss mathematician and physicist, member of a famous family of mathematicians. His work on HYDRODYNAMICS demonstrated that pressure in a FLUID decreases as the velocity of fluid flow increases. This fact, which explains the LIFT of an aircraft wing, has become known as BERNOULLI'S LAW.

Bernoulli's law For a steadily flowing FLUID (liquid or gas), the sum of the pressure, kinetic energy per unit volume and potential energy per unit volume is constant at any point in the fluid. Using this relationship, which was formulated by Daniel BERNOULLI, it is possible to measure the velocity of a fluid by measuring its pressure at two points, as with a MANOMETER or PITOT TUBE.

Berthelot, Pierre Eugène Marcelin (1827–1907) French chemist. By making some organic compounds in the laboratory, including ethyne (acetylene), ethanol, methanol, benzene and methane, he helped to prove that the distinction between organic compounds (as substances formed only in living things) and inorganic compounds was wrong. In the 1860s he did important work in THERMOCHEMISTRY. He created the terms ENDOTHERMIC and EXOTHERMIC reactions.

beryl Mineral, beryllium aluminium silicate. Its crystals are usually hexagonal prisms of the hexagonal system. Gemstone varieties are aquamarine (pale blue-green) from Brazil; emerald (deep green) from Colombia; and morganite (pink) from Madagascar. Cut stones have little brilliance, but are valued for their intense colour. Hardness 8; r.d. 2.6–2.8.

beryllium (symbol Be) Strong, light, silver-grey member of the ALKALINE-EARTH METALS, first isolated in 1828. Beryllium occurs in many minerals including aquamarine, emerald and morganite (all forms of BERYL), and is used in alloys that combine lightness with rigidity. Properties: at.no. 4; r.a.m. 9.012; r.d. 1.85; m.p. 1,285°C (2,345°F); b.p. 2,970°C (5,378°F); most common isotope ^{9}Be (100%).

Berzelius, Jöns Jakob, Baron (1779–1848) Swedish chemist, one of the founders of modern chemistry. Berzelius' accomplishments include the discovery of cerium, selenium and thorium; the isolation of the elements silicon, zirconium and titanium; the determination of RELATIVE ATOMIC MASSES; and the devising of a modern system of chemical symbols. He prepared the first table of relative atomic masses and contributed to the founding of the theory of atomic RADICALS. *See also* PERIODIC TABLE

Bessel, Friedrich Wilhelm (1784–1846) German astronomer and mathematician. He was the first astronomer to use a technique called PARALLAX to measure how distant a star is from the

BERNOULLI'S LAW

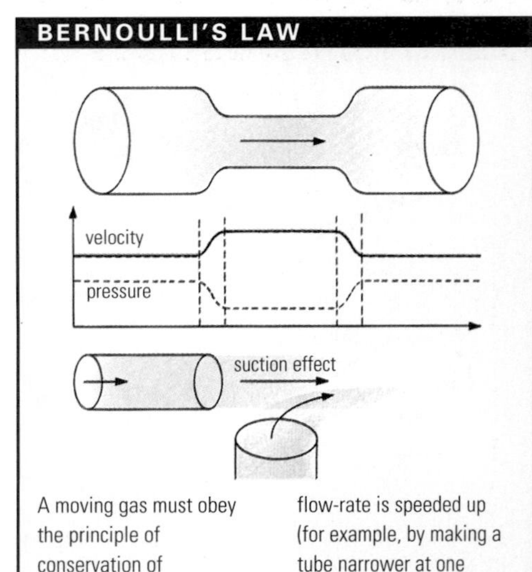

A moving gas must obey the principle of conservation of momentum. If viscosity is ignored, and if flow is streamlined, this principle predicts the Bernoulli effect (often known as Bernoulli's law): if the flow-rate is speeded up (for example, by making a tube narrower at one place) the pressure falls. This effect is responsible for the familiar suction of gas from an orifice (such as a chimney) by a rapid stream of gas across it.

Earth using the apparent motion of the star relative to the Earth as the Earth moves in its orbit. The parallax of the binary star Cygni was found to be 0.3 seconds of ARC which indicated that the star was just over 10 light-years away from Earth. He devised **Bessel functions,** a type of mathematical function used to described the observed perturbations of the planets and stars.

Bessemer process First method for the mass production of steel. The process was patented in 1856 by the British engineer and inventor Sir Henry Bessemer (1813–98) – although a US steelmaker, William KELLY had used a similar process in *c.*1850. In a **Bessemer converter**, pig iron is converted into steel by blowing air through the molten iron to remove impurities. Precise amounts of carbon and metals are then added to give the desired properties to the steel.

Best, Charles Herbert (1899–1978) Canadian physiologist. He and Frederick BANTING discovered INSULIN in 1921. He was head of the department of physiology at the University of Toronto (1929–65) and chief of the Banting-Best department of medical research there after Banting's death.

beta-blocker Any of a class of DRUGS that block impulses to certain nerve receptors (beta receptors) in various tissues throughout the body, including the heart, airways and peripheral arteries. These drugs are mainly prescribed to regulate the heartbeat, reduce blood pressure, relieve ANGINA and improve survival following a heart attack. However, they are also being used in an increasingly wide range of other conditions, including GLAUCOMA, liver disease, thyrotoxicosis, MIGRAINE and anxiety states. Beta-blockers are not suitable for all heart patients and are not used for patients with asthma or severe lung disease.

beta particle (beta ray) Particle emitted spontaneously by certain radioactive ISOTOPES undergoing RADIOACTIVE DECAY called **beta decay**. Beta particles were discovered in 1876 by Henri BECQUEREL, and are now known to be energetic ELECTRONS. Beta decay results when a NEUTRON (in the nucleus) is converted into a PROTON, with the emission of an electron (beta particle) and an antineutrino or NEUTRINO. They cause IONIZATION along their path and have fairly high penetrating power. Their energy, ranging from 0.003–13 MeV, is characteristic of the emitting isotope.

betatron Type of particle ACCELERATOR consisting of a hollow, evacuated circular ring in which ELECTRONS are confined and accelerated by a rapidly changing magnetic field. When they reach very high energies, of the order of 350MeV, they are focused magnetically to strike a target. The output can be either an electron beam or secondarily produced X-rays.

Betelgeuse (Alpha Orionis) Red SUPERGIANT star and the second-brightest in the constellation of Orion. It is a pulsating variable whose diameter fluctuates between 300–400 times that of the Sun. Characteristics: apparent mag. 0.85 (mean); absolute mag. −5.5 (mean); spectral type M2; distance 500 light-years.

Bethe, Hans Albrecht (1906–) US nuclear physicist, b. Germany. He left Germany when Hitler came to power, going first to Britain and then to the USA to become professor of theoretical physics at Cornell University (1935–75) until his retirement. He worked on stellar energy processes and helped to develop the atomic bomb. He is noted for his theories on atomic and nuclear properties and was awarded the 1967 Nobel Prize for physics for his work on the origin of solar and stellar energy. *See also* GAMOW, GEORGE

BeV Abbreviation formerly used for one billion electron volts (eV), a unit of energy equal to a thousand million (10^9) electron volts; now symbolized by GeV.

bicarbonate of soda Popular name for SODIUM HYDROGENCARBONATE

biceps Large muscle on the front of the upper arm, easily felt when the arm is bent. It contracts to raise the forearm towards the upper arm, and also contracts to turn the inturned hand outwards (flexion).

bicycle Two-wheeled vehicle propelled by the rider. The earliest design dates from about 1790. Karl von Drais of Germany developed an improved version *c.*1816. An Englishman, James Starley, demonstrated the first successful chain drive in 1871. Bicycles have been a popular means of transport and recreation in many countries since the late 1800s. Some modern bicycles feature recent innovations such as disc brakes, front fork suspension and the use of carbon fibre to add strength while reducing overall weight.

biennial Plant that completes its life cycle in two years, producing flowers and seed during the second year; an example is the onion. This distinguishes it from an ANNUAL, which germinates, flowers, fruits and dies in one season, and a PERENNIAL, which is a plant that lives for three years or more.

Big Bang Theory advanced to explain the origin of the Universe. It was developed in the 1940s by George GAMOW from the ideas of Georges LEMAÎTRE. According to the Big Bang theory, a giant explosion 10 to 20 thousand million years ago began the expansion of the Universe, which still continues. Everything in the Universe once constituted an exceedingly hot and compressed gas with a temperature exceeding 10,000 million degrees. When the Universe was only a few minutes old, its temperature would have been 1,000 million degrees. As it cooled, nuclear reactions would have taken place that would have led to the material emerging from the fireball consisting of about 75% hydrogen and 25% helium by mass, the composition of the Universe as we observe it today. There were local fluctuations in the density or expansion rate. Slightly denser regions of gas, whose expansion rate lagged behind the mean value, collapsed to form galaxies when the Universe was perhaps 10% of its present age. The cosmic microwave background radiation detected in 1965 is considered to be the residual radiation of the Big Bang explosion.

Big Dipper *See* URSA MAJOR

bile Bitter yellow, brown or green alkaline fluid, secreted by the LIVER and stored in the GALL BLADDER. Important in digestion, it enters the duodenum via the bile duct. The bile salts it contains emulsify fats (allowing easier digestion and absorption) and neutralize stomach acids. Its

BICYCLE

Modern designs of mountain bike have reduced weight without sacrificing strength and have added suspension to both front (1) and rear wheels (2) to allow greater speed over rough terrain. The front suspension has twin pistons in the forks with elastomer cores (3) that allow travel and help to absorb vibration. Oil and air can also be used in the pistons. The rear suspension has a single, oil-filled piston (4) which damps the spring (5). Rear suspension units are designed in a variety of different forms (6). New, lightweight but strong frame materials include carbon fibre, titanium and aluminium. A V-shaped frame (7) allows bike designers to use the more exotic substances which are difficult to use in a traditional tubular frame. Some bikes have softer compounds of rubber for the back wheels to give greater grip on steep slopes.

BIMETALLIC STRIP THERMOMETER

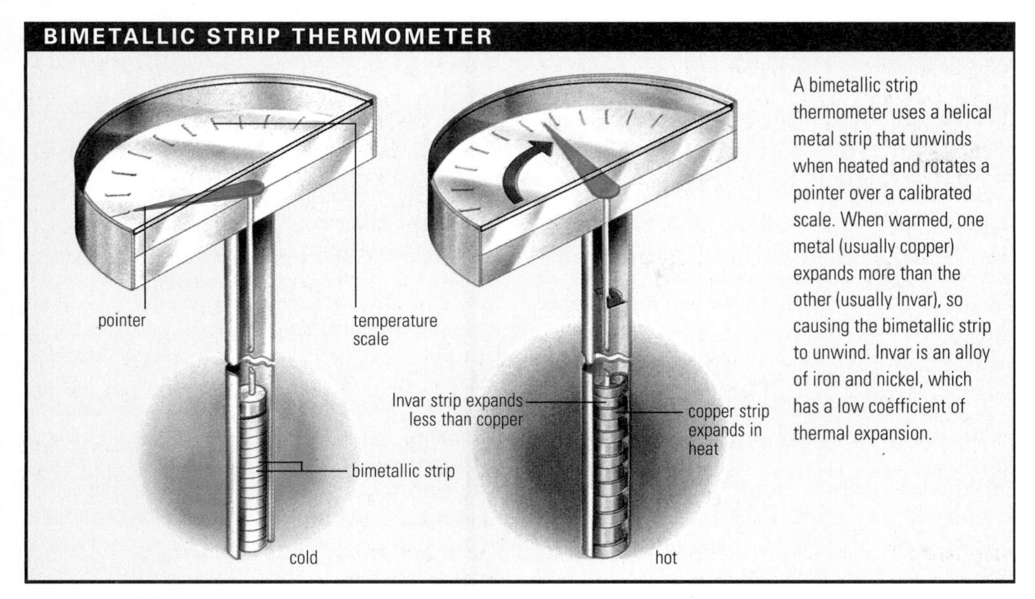

A bimetallic strip thermometer uses a helical metal strip that unwinds when heated and rotates a pointer over a calibrated scale. When warmed, one metal (usually copper) expands more than the other (usually Invar), so causing the bimetallic strip to unwind. Invar is an alloy of iron and nickel, which has a low coefficient of thermal expansion.

colour comes from the breakdown products of ERYTHROCYTES (red blood cells).

bile acids Group of steroid acids present in BILE. In humans, the commonest is cholic acid ($C_{24}H_{40}O_5$), which is conjugated by its carboxyl group to the amino groups of the AMINO ACIDS glycine and taurine. The bile acids are emulsifiers for fats and fat-soluble vitamins, thus promoting their absorption by the intestine. *See also* CARBOXYLIC ACIDS

bilirubin Pigment derived from blood. When ERYTHROCYTES (red blood cells) have reached the end of their useful lives, they are broken down, mainly by the LIVER. Useful materials, such as the haem of HAEMOGLOBIN, are recycled in the body. Among the waste materials is the green-orange pigment bilirubin, which is converted into stercobilin, the colouring matter of faeces.

bimetallic strip Device used in THERMOSTATS and mechanical thermometers. It consists of bonded strips of two metals with dissimilar coefficients of thermal expansion. When heated the bimetallic strip bends because one metal expands more than the other. This reversible distortion is used to move a dial pointer or to open or close a switch in an electrical circuit.

binary fission *See* FISSION

binary star Two stars in orbit around a common centre of mass. **Visual** binaries are those whose components can be seen as separate stars with the naked eye or through a telescope. In an **eclipsing** binary, one star periodically passes in front of the other, so that the total light output appears to fluctuate. Most eclipsing binaries are also **spectroscopic** binaries, systems in which the components are too close for their separation to be measured visually and must be measured spectroscopically. *See also* SPECTROSCOPY

binary system In mathematics, number system having a BASE of 2 (the decimal system has a base of 10). It is most appropriate to computers since it is simple and corresponds to the open (0) and the closed (1) states of a switch, or logic GATE, on which computers are based.

binding energy Energy that must be supplied to an atomic nucleus in order to split it into its constituent nucleons (neutrons and protons). A nucleus must be supplied with its binding energy before it will undergo FISSION (except in the case of radioactive decay). The mass of a nucleus is slightly less than the mass of its constituent particles. According to EINSTEIN's law $E = mc^2$, this difference in mass is equivalent to the energy released when the nucleons bind together. This is the energy source of the hydrogen bomb and the FUSION reaction. *See also* NUCLEAR ENERGY

binoculars Optical instrument, used with both eyes simultaneously, that produces a magnified image of a distant object or scene. It consists of a pair of identical telescopes, one for each eye, both containing an objective lens, an eyepiece lens and an optical system (usually prisms), to form an upright image.

binomial nomenclature System of categorizing organisms by giving them a two-part Latin name. The first part of the name is the GENUS and the second part the SPECIES. For example, *Homo sapiens* is the binomial name for humans. The system was developed by the Swedish botanist Carolus LINNAEUS in the 18th century. *See also* CLASSIFICATION

binomial theorem Mathematical rule for the expansion of the algebraic expression $(x + y)^n$ as a series in powers of the numerical quantities x and y (n is a positive integer). For $n = 2$, its expansion is given by

$$(x + y)^2 = x^2 + 2xy + y^2$$

biochemical oxygen demand (BOD) Chemical test for determining the level of pollution of water by organic matter. Two equal samples of water are taken, and the first is treated chemically to "mop up" any dissolved oxygen. Both samples are then incubated in the dark for some days, after which the second sample is given a similar chemical treatment. This allows the amount of "mopped-up" oxygen to be estimated. The difference between the two estimations represents the amount of oxygen that has been used up by living and dead organic matter in the water during the period of storage.

biochemistry Science of the CHEMISTRY of life. It attempts to use the methods and concepts of organic and physical chemistry to investigate living matter and systems. Biochemists study both the structure and properties of all the constituents of living matter (such as FATS, PROTEINS, ENZYMES, HORMONES, VITAMINS, DNA, CELLS, MEMBRANES and ORGANS) together with the complex reactions and pathways of these in METABOLISM.

biodegradable Property of a substance that enables it to be decomposed by microorganisms. The end result of decay is stable, simple compounds (such as water and carbon dioxide). This property has been designed into products such as many plastics to aid refuse disposal and reduce pollution.

bioelectricity Electricity generated in plants and animals. In animals bioelectricity is associated with nerve impulses and muscle contractions. Different electric POTENTIALS are built up within the organism by the process of ionic separation across a membrane. Some fish have electric organs that create more powerful external currents used for sensing or to stun prey.

bioengineering Application of engineering techniques to medical and biological problems. It includes the engineering of devices to aid or

BINARY STAR

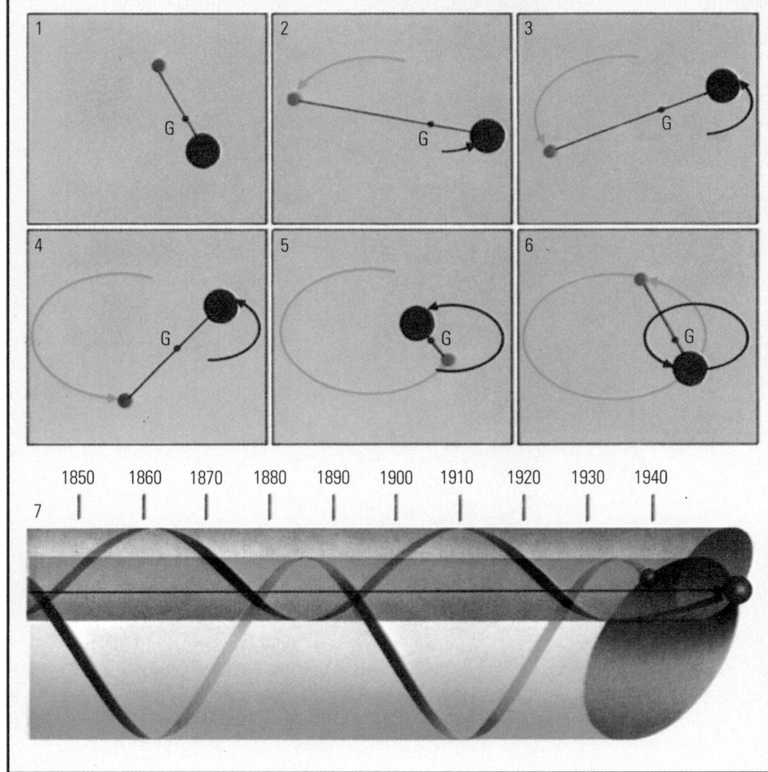

Binary stars, such as Sirius A and B, are double stars that revolve around each other. In fact, they revolve about their common centre of gravity (G), which is not halfway between them but nearer to the more massive component. Moreover, they do not rotate in circles around G but in co-planar ellipses so that the distance between them varies. The rate of rotation, too, is not uniform. In (2) they are greatly separated and moving slowly. In (6) they are closer and moving more rapidly. (7) shows the proper motion of Sirius A and B over a century. The paths of the two components in space are drawn out to elliptical helices and the diagram also shows that the direction of the proper motion is not perpendicular to the plane of the ellipses.

replace defective or inadequate body parts, as in the production of artificial limbs and hearing aids.

biofeedback In alternative medicine, the use of monitoring systems providing changing information about body processes to enable them to be controlled voluntarily. By observing data on events which are normally involuntary, such as breathing and the heart beat, many people learn to gain control over them to some extent in order to improve well-being. The technique has proved helpful in a number of conditions, including headaches and hypertension.

biogas Fuel gas derived from the decomposition of biological material. It is a mixture consisting mainly (up to 60%) of methane, which is combustible, with some carbon dioxide. It is formed by the ANAEROBIC breakdown – termed FERMENTATION – of waste matter such as domestic and agricultural sewage. The waste is contained in a digester where it is acted on by methanogenic bacteria.

biogenesis Biological principle maintaining that all living organisms derive from parent(s) generally similar to themselves. This long-held principle was originally established in opposition to the idea of SPONTANEOUS GENERATION of life. On the whole, it still holds good, despite variations in individuals caused by mutations, hybridization and other genetic effects. *See also* GENETICS

biogenetic law (recapitulation theory) Principle that the stages that an organism goes through during embryonic development reflect the stages of that organism's evolutionary development.

biological clock Internal system in organisms that relates behaviour to certain rhythms. Functions, such as growth, feeding or reproduction, coincide with certain external events, including day and night, tides and seasons. This innate sense of timing makes some animals feed during the day when food is available and they can see best. These "clocks" seem to be set by environmental conditions, but if organisms are isolated from these conditions, they still function according to the usual rhythm. If conditions change gradually, the organisms adjust their behaviour gradually.

biological control Use of biological methods to control pests. Usually the term describes the deliberate introduction of a parasite or other natural enemy of the pest, and is often preferred to the use of chemical pesticides. Examples include the introduction of ladybirds to kill scale insects on US citrus crops, and the use of cactus moth caterpillars to control prickly pear cactus on Australian farmland. In another method, successfully employed to control screw worm flies that parasitize cattle, male flies sterilized by radiation are released into the natural population to produce a reduction in the fly population (because of the preponderance of sterile matings). Parasitic wasps are also commercially available to gardeners to control white fly and aphids in greenhouses. Care has to be taken with biological control so as not to upset the natural ecological balance.

biological shield Thick wall containing such materials as concrete, steel, magnetite and lead, used to protect workers in nuclear power stations and radiochemical laboratories from the possible harmful effects of radiation. In remote-handling "caves" the biological shield contains a viewing window *c.*30cm (12in) thick, also with radiation absorbers.

biological warfare Use of disease microbes and their toxins in warfare. The extensive use of mustard gas during World War 1 prompted the prohibition of biological warfare by the Geneva Convention (1925). However, many nations have maintained costly research programmes for the production of harmful microorganisms and discovery of more effective antidotes to their pathogenic effects. These microbes include plant pathogens for the destruction of food crops. None has yet been used, although US forces employed a variety of biological warfare, such as the use of the defoliant Agent Orange during the Vietnam War. Saddam Hussein used gas to kill thousands of Kurds after the Iran-Iraq War. Allied troops were inoculated and heavily protected against biological weapons during the Iran-Iraq War. Worries have recently surfaced about its terrorist uses, such as the Sarin gas attack on the Tokyo underground in 1996.

biology Science of life and living organisms. Its branches include BOTANY, ZOOLOGY, ECOLOGY, PHYSIOLOGY, CYTOLOGY, GENETICS, TAXONOMY, EMBRYOLOGY and MICROBIOLOGY. These sciences deal with the origin, history, structure, development and function of living organisms, their relationships to each other and their environment, and the differences between living and nonliving organisms.

bioluminescence Production of light, with very little heat, by some living organisms. Its biological function is varied: in some species, such as fireflies, it is a recognition signal in mating; in others, such as squids, it is a method of diverting predators for protection; and in deep-sea anglerfish it is used as a lure to attract prey. The light-emitting substance (luciferin) in most species is an organic molecule that emits light when it is oxidized by molecular oxygen in the presence of an enzyme (luciferase). Each species has different forms of luciferin and luciferase. *See also* FLUORESCENCE

biomass Total mass (excluding water content) of the plants and/or animals in a particular place. The term is often used to refer to the totality of living things on Earth; or those occupying a part of the Earth, such as the oceans. It may also refer to plant material that can be exploited, either as fuel or as raw material for an industrial or chemical process.

biome Natural and extensive community of animals and plants whose make-up is determined by the type of soil and the climate. There is generally distinctive, dominant vegetation, and characteristic climate and animal life. Ecologists divide the earth (including the seas, lakes and rivers) into ten biomes.

biophysics Study of biological phenomena in terms of the laws and techniques of physics.

BIOLOGICAL CLOCK

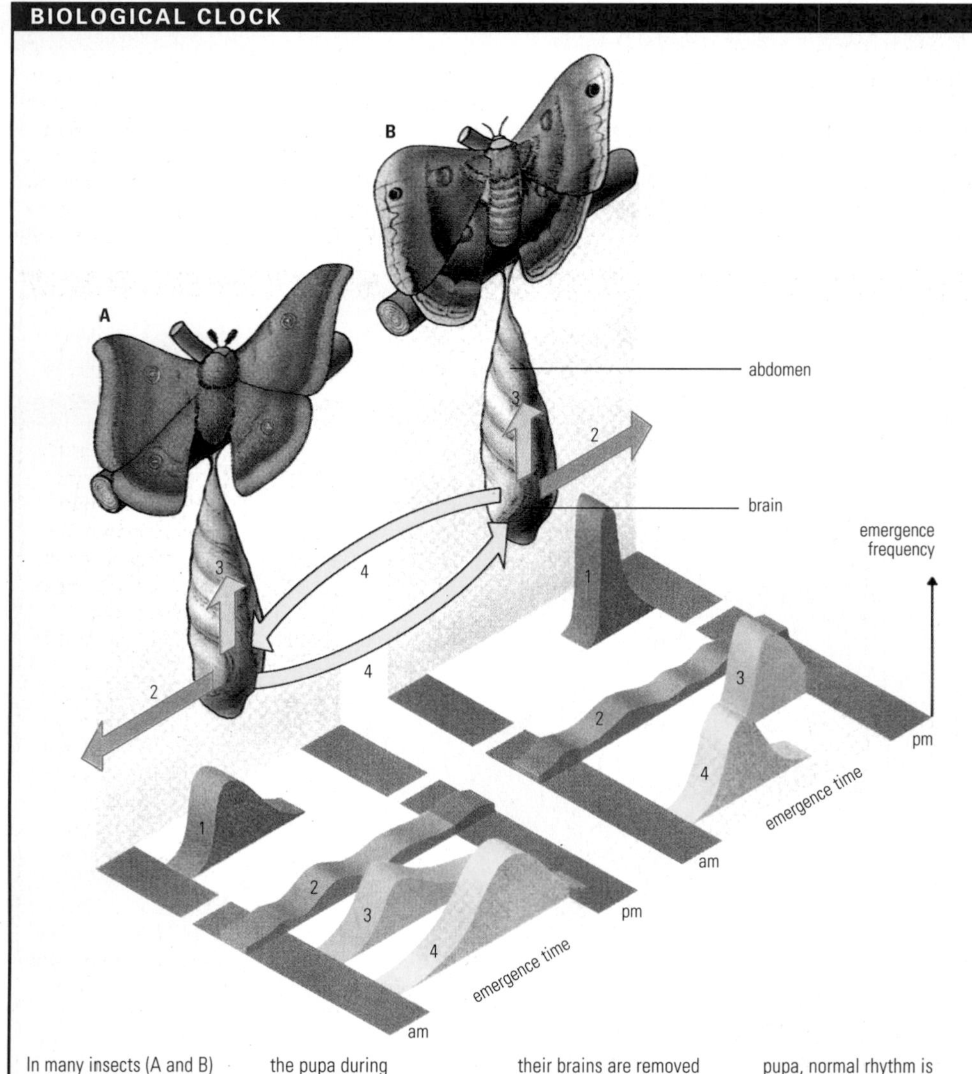

In many insects (A and B) the master clock lies in the brain. It uses hormones to exert its influence on many processes, including the emergence of the adult from the pupa during metamorphosis. The graphs show (1) normal emergence times of the Chinese oak silkmoth (A) and the robin moth (B). When parts of their brains are removed hormonal control is lost and the moths emerge at random times (2). If the brain is replaced in the abdomen of the brainless pupa, normal rhythm is restored (3). If the brains of the two species are exchanged, the moths can be made to swap emergence rhythms (4).

▲ **bioluminescence** The deep-sea anglerfish inhabits water that it so deep that virtually no light penetrates. The anglerfish emits light in order to lure its prey. The photograph shows its head, esca and barbel.

Techniques include those of X-RAY DIFFRACTION and SPECTROSCOPY. Subjects studied include the structure and function of biological molecules, the conduction of electricity by nerves, the visual mechanism, the transport of molecules across cell membranes, muscle contraction (using electron microscopy) and energy transformations in living organisms.

biopsy Removal of a small piece of tissue from a patient so that it can be examined in the laboratory for evidence of disease. A common example is the cervical biopsy ("smear test"), performed in order to screen for the pre-cancerous changes that may lead to cervical cancer.

biorhythm Any regular pattern of changes in METABOLISM or activity in living things. These are usually synchronized with daily, monthly, seasonal or annual changes in the environment. Examples of daily, or circadian, rhythms are the opening and closing of flowers, feeding cycles of animals during the day or the night, the response of marine organisms to the tides and, in humans, changes in body temperature and blood pressure. Monthly changes include the MENSTRUAL CYCLE in women. Annual rhythms include hibernation, migration and reproductive activity. The internal mechanism of these rhythms (sometimes referred to as the BIOLOGICAL CLOCK) is not yet fully understood.

biosphere (zone of life) Portion of the Earth from its CRUST to the surrounding ATMOSPHERE, encompassing and including all living organisms, animal and vegetable. It is self-sufficient except for energy and extends a few kilometres above and below sea-level.

biosynthesis Process in living cells by which complex chemical substances, such as PROTEINS, are made from simpler substances. A GENE "orders" a molecule of RNA to be made, which carries the genetic instructions from the DNA. On the RIBOSOMES of the CELL, the protein is built up from molecules of AMINO ACIDS, in the order determined by the genetic instructions carried by the RNA.

Biot, Jean Baptiste (1774–1862) French astronomer, mathematician and physicist who made fundamental discoveries about POLARIZED LIGHT. He developed the technique of estimating the concentration of certain organic substances (particularly sugars) by measuring the angle through which they rotate the plane of polarization of light.

biotechnology Use of biological processes for medical, industrial or manufacturing purposes. Humans have long used yeast for brewing and bacteria for products such as cheese and yoghurt. Biotechnology now enjoys a wider application. By growing microorganisms in the laboratory, new drugs and chemicals are produced. GENETIC ENGINEERING techniques of cloning, splicing and mixing genes facilitate, for example, the growing of crops outside their normal environment, and the production of vaccines to fight specific diseases. Hormones are also produced, such as INSULIN for treating diabetes.

biotin ($C_{10}H_{16}O_3N_2S$) Member of the VITAMIN B complex. It is vital to the body, but is needed in only minute quantities, which are synthesized by intestinal bacteria.

biotite Common mineral of the MICA group. It is a silicate of aluminium, iron, potassium and magnesium. Its colour ranges from greenish-brown to black. Its lustrous, monoclinic crystals are opaque to translucent, and cleave to form flexible sheets. It is found in IGNEOUS ROCK (such as granite), METAMORPHIC ROCK (such as schist and gneiss) and SEDIMENTARY ROCK.

bird Any one of about 8,600 species of feathered VERTEBRATES of the class Aves. They occupy most natural habitats from deserts and tropics to polar wastes. Birds are warm-blooded and have forelimbs modified as wings, hind-limbs for walking and jaws elongated into a toothless beak. They lay eggs (usually in nests), incubate the eggs and care for young. As a group they feed on seeds, nectar, fruit and carrion, and hunt live prey ranging from insects to small mammals, although individual species may be very specialized in their diet. Sight is the dominant sense, smell the poorest. Size ranges from the bee hummingbird, 6.4cm (2.5in) to the wandering albatross, whose wingspread reaches 3.5m (11.5ft). The 2.5m (8ft) tall ostrich is the largest of living birds, but several extinct flightless birds were even bigger. Of the 27 orders of birds, the perching birds (Passeriformes) include more species than all others combined. A bird's body is adapted primarily for flight, with all its parts modified accordingly. There are several groups of large flightless land birds, including the ostrich, rhea, emu, cassowary, kiwi and penguin. Birds are descended from Theocodonts (reptiles), and the first fossil bird, ARCHAEOPTERYX, dates from late Jurassic times. *See feature on pages 40–41*

birth (parturition) Bringing forth of live, partly or fully formed offspring. All mammals (except the echidna and the platypus), some reptiles and sharks, and various insects and other invertebrate animals give birth to live young (*see* VIVIPARITY). All birds, most reptiles, amphibians and fishes, and the majority of insects lay eggs from which the live young later emerge, this process being called hatching (*see* OVIPARITY). In humans, birth involves the delivery of the FETUS at the end of pregnancy. There are three stages in this process. First, contractions of the UTERUS begin and the CERVIX dilates in readiness; the sac containing the amniotic fluid ruptures. In the second stage the contractions strengthen and the baby is propelled (normally head first) through the birth canal. The

BIOTECHNOLOGY

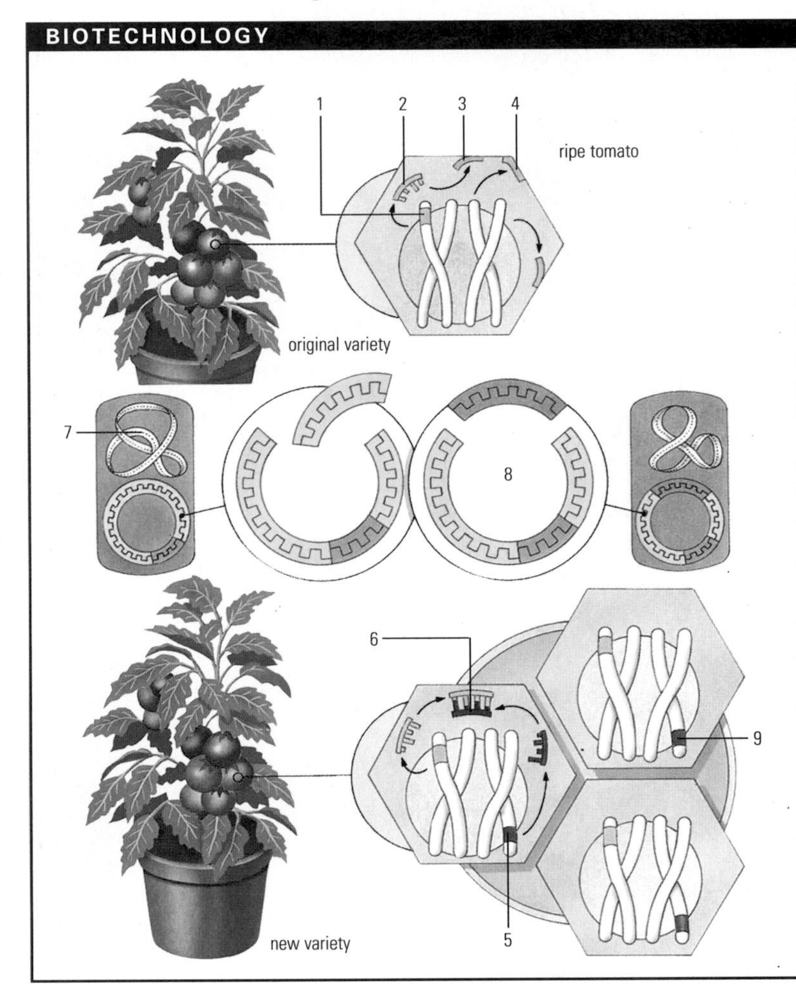

In a ripe tomato, rotting is caused by an enzyme formed by the copying of a gene in the plant DNA (1) in a messenger molecule mRNA (2). The mRNA is changed into the enzyme (3) which damages the cell wall (4). In a genetically altered tomato, a mirror duplicate of the gene that starts the process is present (5). The result is that two mirror-image mRNA molecules are released (6) and they bind together preventing the creation of the rotting enzyme. The result is longer-lasting tomatoes. Introducing the necessary DNA through the rigid cell wall is accomplished by using a bacteria (7) which naturally copies its own DNA onto that of a plant. It is easy to introduce the mirror DNA (8) into the bacteria and once the bacteria has infected the cell the DNA is transferred (9). All cells then replicated have the new DNA in their chromosomes and can be grown to create the new variety of plant.

Birds inhabit diverse regions of the world – the Antarctic ice-sheets, tropical rainforests, arid deserts and the open oceans, are all home to birds of one kind or another. There are an estimated 9,300 species of bird alive today, belonging to 28 orders, all of which are members of the class Aves. All birds are remarkably similar in basic structure in that they have a body-plan which evolved primarily as an adaptation for flight. Even the flightless species evolved from ancestors that could fly, and therefore they share many typical features. Birds' bones are light and strong, and the skeleton has the form of a rigid box, with a large breastbone, or sternum. The really unique feature of birds, however, is their covering of feathers. No other vertebrates have these extraordinary outgrowths, which, in all their modifications, provide birds with many attributes.

▼ The archaeopteryx is the earliest known recognizable bird. It dates from the upper Jurassic period. The presence of wings and feathers define it as a bird, but the skeleton is quite reptilian. The wings, instead of being specialized flying limbs, were really elongated forelimbs, complete with claws. The tail resembles a lizard's and the skull had teeth. The small breastbone shows it was a poor flyer.

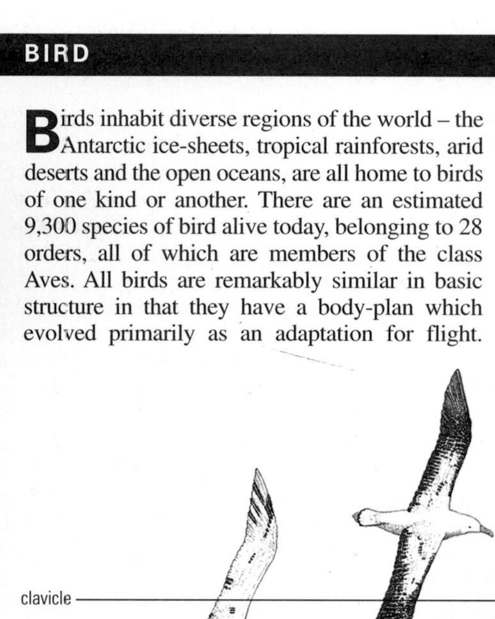

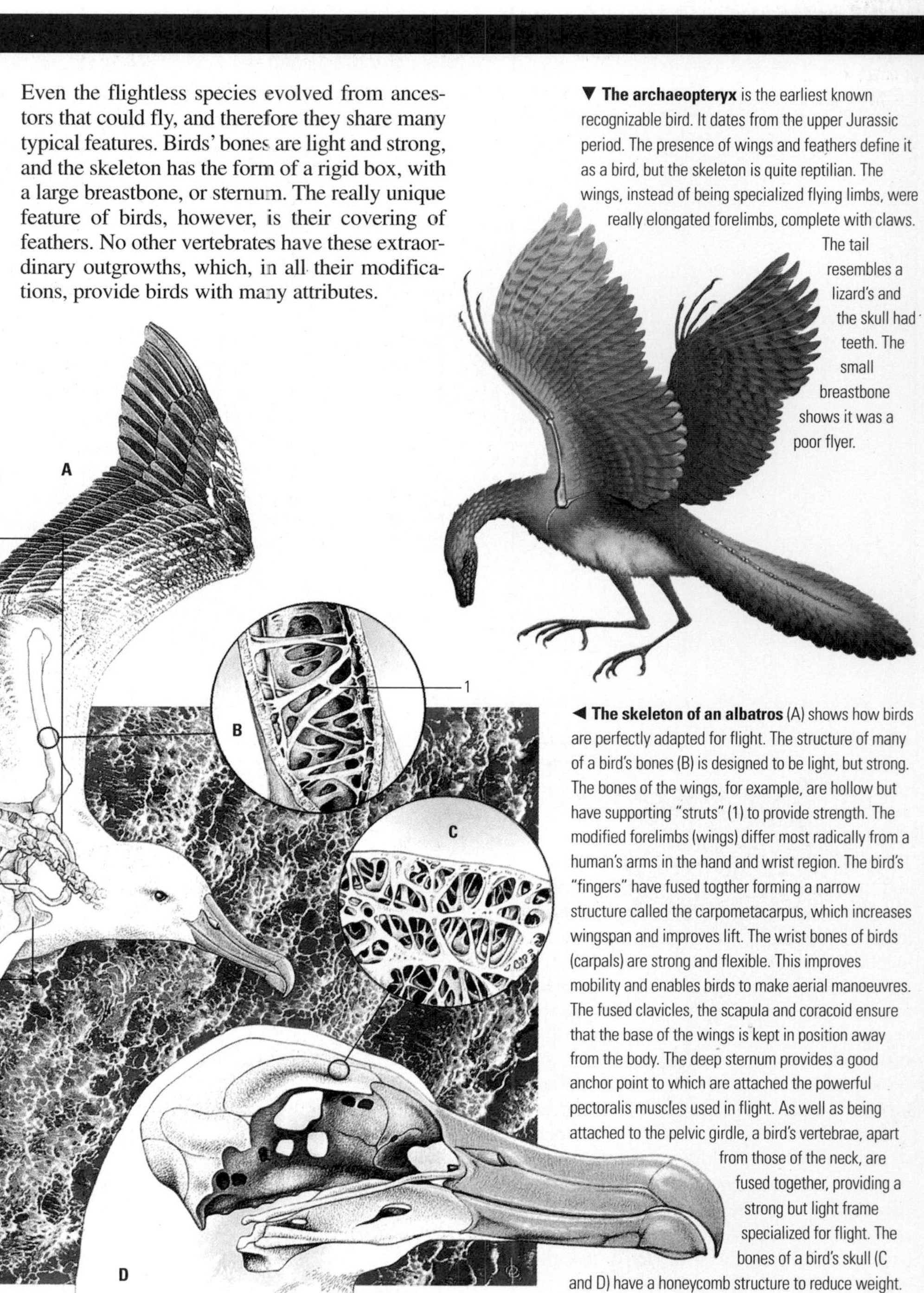

◄ The skeleton of an albatros (A) shows how birds are perfectly adapted for flight. The structure of many of a bird's bones (B) is designed to be light, but strong. The bones of the wings, for example, are hollow but have supporting "struts" (1) to provide strength. The modified forelimbs (wings) differ most radically from a human's arms in the hand and wrist region. The bird's "fingers" have fused togther forming a narrow structure called the carpometacarpus, which increases wingspan and improves lift. The wrist bones of birds (carpals) are strong and flexible. This improves mobility and enables birds to make aerial manoeuvres. The fused clavicles, the scapula and coracoid ensure that the base of the wings is kept in position away from the body. The deep sternum provides a good anchor point to which are attached the powerful pectoralis muscles used in flight. As well as being attached to the pelvic girdle, a bird's vertebrae, apart from those of the neck, are fused together, providing a strong but light frame specialized for flight. The bones of a bird's skull (C and D) have a honeycomb structure to reduce weight.

FILTER FEEDING

Flamingoes feed by lowering their heads into the water so that their bills are upside down (1). Its crooked shape allows the front half of the bill to lie horizontally in the water (2). Tiny hook-like lamellae (3) strain food as the water is pumped through them by a backward and forward motion of the tongue. Protuberances on the tongue (4) scrape the particles of food off the lamellae for ingestion as the tongue moves back and forth.

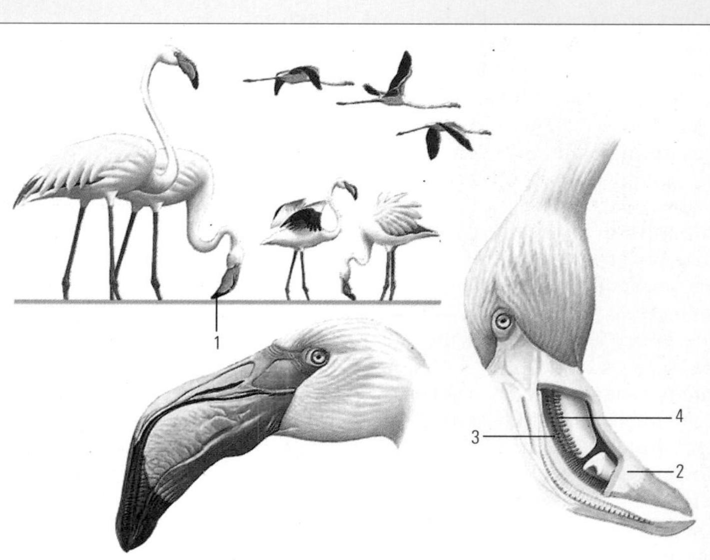

EGGS

A hen's fertilized ovum moves from the ovary (1) into the enlarged head of the oviduct (2). First it passes along the main section of oviduct, or magnum (3), the walls of which add layers of egg white (albumen). After about 3 or 4 hours, it enters another section called the isthmus (4), where the egg and shell membranes are deposited. The shell itself forms over the egg in a wider section of the oviduct just beyond the isthmus. This is called the uterus (5), or shell gland, and the process of shell formation takes about 20 hours. The egg is laid thought the cloaca (6), by contractions of the vagina.

Eggs vary enourmously in shape and size:
1) wood warbler
2) dunnock
3) blackbird
4) tawny owl
5) Egyptian vulture
6) emu
7) ostrich

◄ ► **Birds' beaks** show great variety in shape and size; usually this is an adaptation due to their feeding habits. The sword-billed humming bird (A), for example, uses its long, thin bill to extract nectar from flowers. Alternatively, a bird's bill is important in courtship, as in the case of the great hornbill (B).

▼ **Feathers** are not just for flying. The male superb lyrebird of Australia uses his impressive tail feathers for extravagant courtship displays.

◄ **Flightless birds**, as well as those that fly, have adapted to survive in a wide variety of environments. The rhea (A), for example, lives in the open pampas regions of South America. Its long, strong legs make it an excellent runner. While penguins, such as the royal penquin (B), survive in cold Southern Hemisphere regions. Their wings, which have adapted to flippers, and their webbed feet ensure that they are powerful swimmers.

NESTS

Weaver birds live in warm regions of Europe, W Asia and N Africa. The weaver birds' nest is built by the male. Using no adhesive, he loops, twists and knots leaf strips to make an enclosed hanging structure. Starting with a ring attached to a forked twig (1), he gradually adds a roof and entrance (2). When the finished nest is accepted by the female, she inserts soft, feathery grass tops or feathers to make a thick, soft lining around the egg chamber base. "Weaving" the nest is a little misleading, since the green leaves used are rarely over three times the bird's length. Young male birds' first nests tend to be untidy; gradually, however, they learn how to make neater and better nests, using good vision and well-coordinated head movements to direct claws and beak to manipulate the nest material. The leaf fastenings the bird uses include half-hitch knots and slip knots.

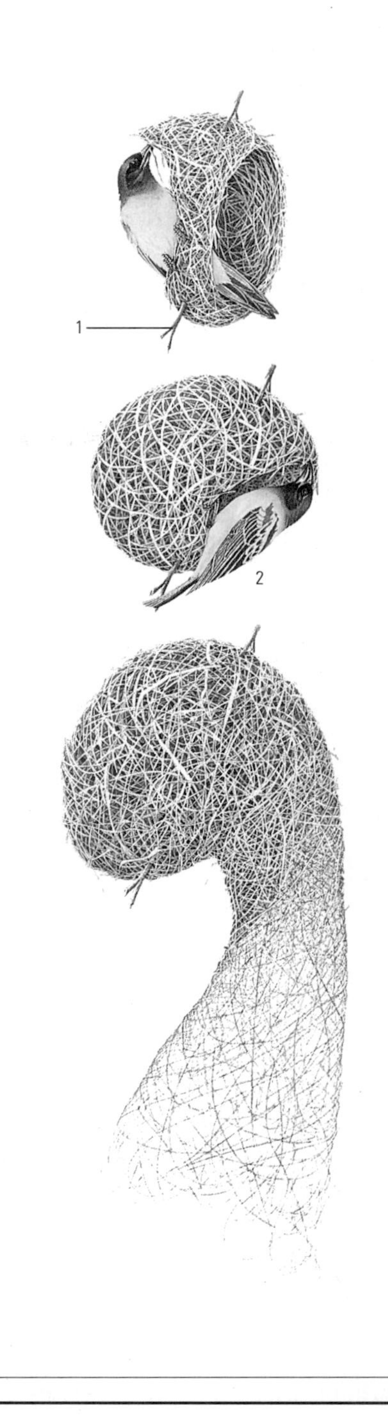

third and final stage, once the baby has been born, is the expulsion of the placenta and fetal membranes, together known as the afterbirth.

birth control Alternative term for CONTRACEPTION

bismuth (symbol Bi) Metallic, silver-white, element of group V of the periodic table, first identified as a separate element in 1753. The chief ores are bismite (Bi_2O_3) and bismuthinite (Bi_2S_3). A poor heat conductor, it is put into low-melting alloys used in automatic sprinkler systems. Bismuth is also used in insoluble compounds to treat gastric ulcers and skin injuries. It expands when it solidifies, a property exploited in several bismuth alloys used in making castings. Properties: at.no. 83; r.a.m. 208.98; r.d. 9.75; m.p. 271.3°C (520.3°F); b.p. 1,560°C (2,840°F); most common isotope ^{209}Bi (100%).

bit In computing, abbreviation for binary digit, a 1 or 0 as used in the BINARY SYSTEM. A bit is the smallest element of storage. Groups of bits form a BYTE of binary code representing letters and other characters. Binary code is used in computing because it is easy to represent each 1 or 0 using electrical components that can be switched between two states (such as on and off). The code is also easy to store on disk as a magnetic or optical pattern.

bitumen (asphalt) Material used for roadmaking and for proofing timber against rot. It consists of a mixture of hydrocarbons and other organic chemical compounds. Some bitumen occurs naturally in pitch lakes, notably in Trinidad and Venezuela. The material is also made by distilling tar from coal or wood, and a little is obtained during the refining of petroleum.

bivalent In biology, describing a pair of HOMOLOGOUS CHROMOSOMES formed during MEIOSIS. In chemistry, the term is used to describe an atom or group that has a VALENCE of two.

bivalve Any animal of the class Bivalvia, which has a shell with two halves or parts hinged together. The term most usually applies to a class of MOLLUSCS – Pelecypoda or Lamellibranchiata – with left and right shells, such as clams, cockles, mussels and oysters. It also refers to animals of the phylum Brachiopoda (BRACHIOPOD) with dorsal and ventral shells. Length: 2mm–1.2m (0.17in–4ft).

A

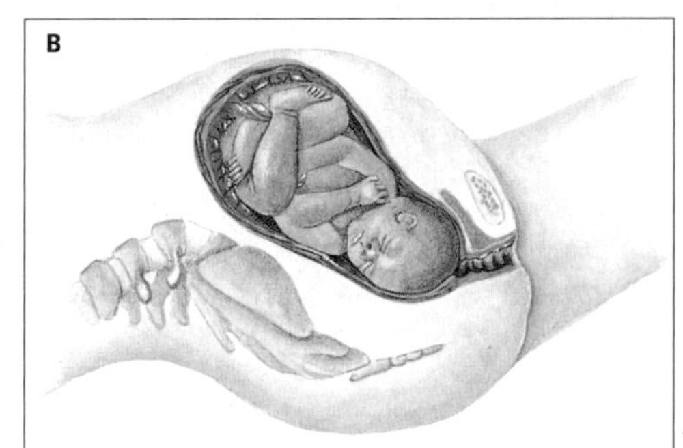

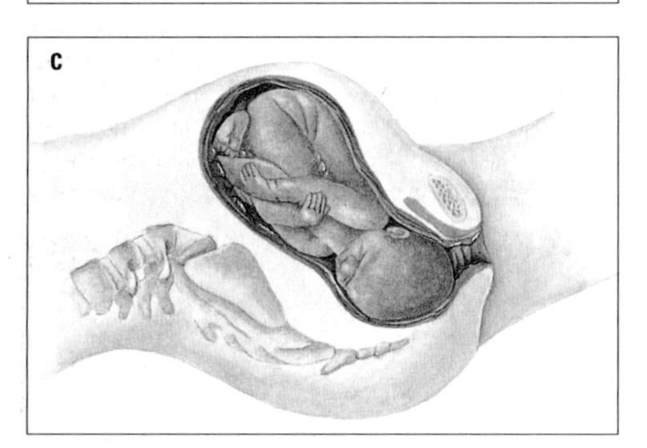

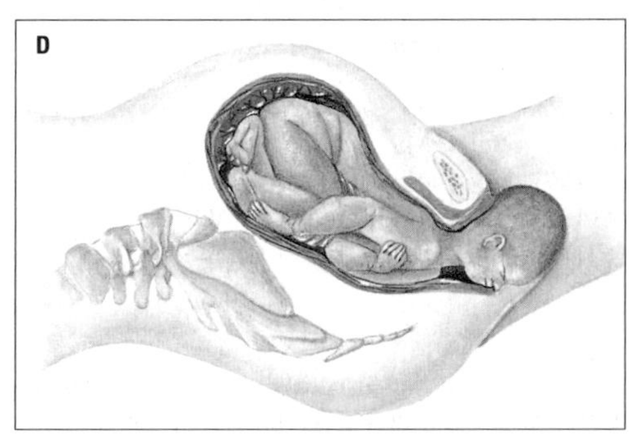

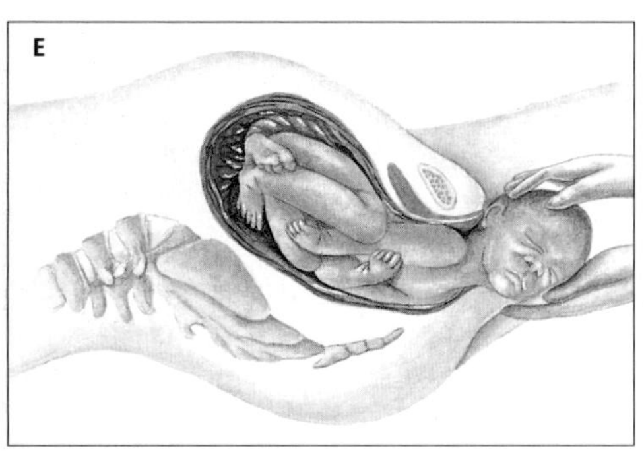

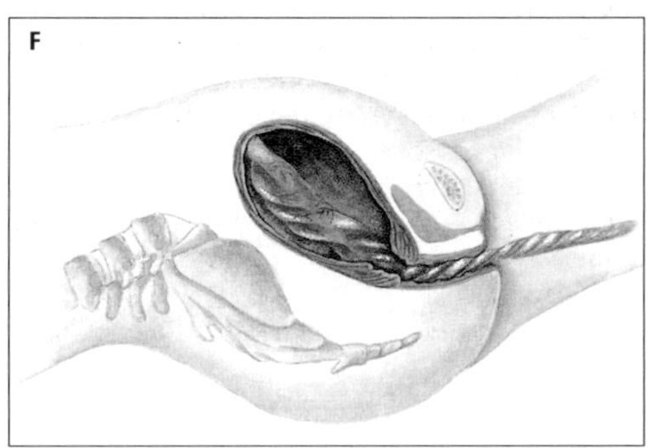

▲ **birth** Sequence of normal childbirth. During the last few weeks of the pregnancy uterine contractions become stronger. They help to keep the fetal head engaged in the pelvis (A). The first stage of labour starts with backache and regular uterine contractions that become gradually stronger and by the end occur about every two minutes. As the head is forced deeper into the pelvis it turns sideways (B) and the plug of cervical mucus is dislodged causing a "show" of blood. The cervix is gradually stretched, thinned and dilated until it is wide enough (C) for the head to pass into the vagina. The second stage starts with the cervix fully dilated (C). The head moves deeper into the pelvis and again turns. The mother has a "bearing down" sensation and can help the powerful uterine contractions to force the fetus through the vagina. The vulva distend around the head and finally it is "crowned" when the greatest circumference is reached. The head extends (D) round the symphysis as it is pushed out. The baby's first breath is taken and is followed by a cry. This expands the lungs and starts respiration, as well as obliterating the fetal circulation. The baby can now be delivered (E) with the next contraction as the smaller shoulders and body easily follow the large head. The third stage follows 10–20 minutes later when the placenta has separated from the uterus and is expelled (F) along with the umbilical cord and membranes.

Bjerknes, Vilhelm Friman Koren (1862–1951) Norwegian physicist who laid the foundation of a revolution in weather forecasting by applying theories of fluid forces and motions to the circulation of the atmosphere with the development of air masses, fronts and cyclones.

Black, Joseph (1728–99) British chemist and physicist. Rediscovering "fixed air" (carbon dioxide), he found that this gas is produced by RESPIRATION, burning of charcoal and FERMENTATION, that it behaves as an ACID, and that it is probably found in the atmosphere. He also discovered hydrogencarbonates (bicarbonates) and investigated LATENT HEAT and the concept of specific heat but was unable to reconcile it with the PHLOGISTON THEORY that said burning materials lose a substance to the atmosphere.

blackband iron ore Thin seam of dark-coloured iron ore that occurs in layers of coal. Historically it was important to the development of iron and steel industries and often accounted for the location of smelters near coal mines. Once the blackband ore was exhausted, iron ore had to be imported and it became more economic to locate steel works near ports.

black body In physics, an ideal body that absorbs all incident radiation and reflects none. Such a body would look "perfectly" black – hence the name. The study of black bodies has been important in the history of physics. WIEN's law, Stefan's law (*see* BOLTZMANN) and PLANCK's law of black body radiation grew out of this study, as did Planck's discoveries in quantum mechanics.

black earth *See* CHERNOZEM

Blackett, Patrick Maynard Stuart, Baron (1897–1974) English physicist who was awarded the 1948 Nobel Prize for physics for his research on COSMIC RADIATION. He spent 10 years at the Cavendish Laboratory, Cambridge, developing the Wilson cloud chamber into an instrument to study cosmic radiation. He also used it to prove the existence of the POSITRON, the antiparticle of the ELECTRON. Blackett was professor of physics at Manchester University (1937–53) and then professor of physics at Imperial College, London (1953–65). *See also* ANTIMATTER

black hole Localized region of space from which neither matter nor radiation can escape – in other words, the escape velocity exceeds the velocity of light. The boundary of this region is called the EVENT HORIZON. Its radius, the **Schwarzschild radius**, depends on the amount of matter that has fallen into the region and it increases linearly as the mass increases. A black hole of stellar mass is thought to form when a massive star undergoes total gravitational collapse. For stars up to about 1.4 solar masses, gravitational collapse can be physically halted to produce a WHITE DWARF. A slightly more massive object will collapse to form a NEUTRON STAR. If, however, the mass exceeds about 3 solar masses, even after a supernova explosion has blasted away the outer layers of the star, the collapse continues beyond even the neutron star stage. As it contracts below its Schwarzschild radius, the object becomes a black hole and effectively disappears. A star of 3 solar masses has a Schwarzschild radius of about 9km (5.5mi). Inside the event horizon of the black hole, space and time are highly distorted and the stellar matter is increasingly compressed until it forms an infinitely dense singularity. Black holes can have an immense range

BLAST FURNACE

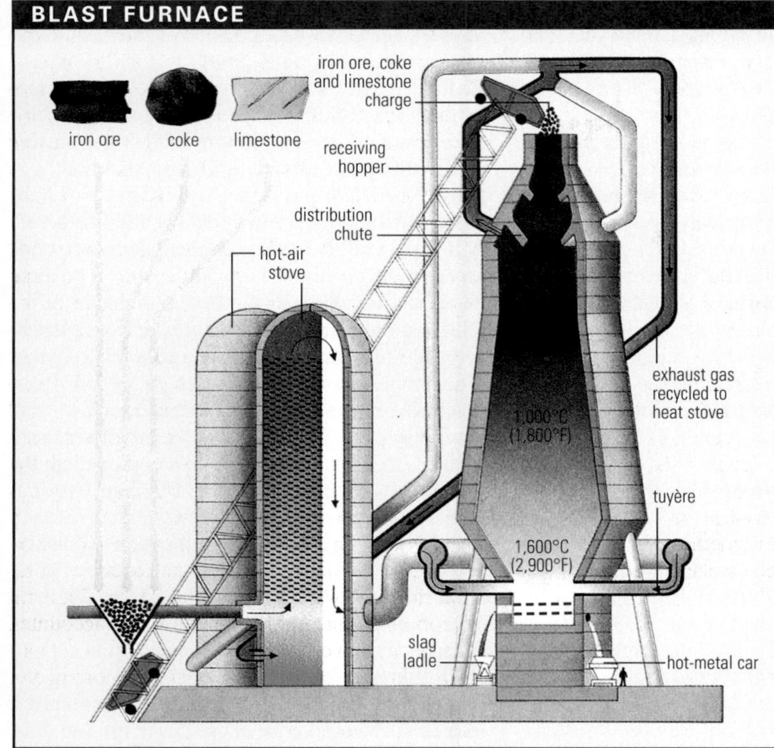

The blast furnace gains its name from the blasts of superheated air that are blown into the furnace. The furnace is charged from the top with coke, iron ore (collectively called sinter) and limestone. Air, heated by hot cowper stoves, is then blown through the blast pipes (tuyères) into the lower part of the furnace. The ore is converted to the liquid pig iron and trickles down into the hearth from where it is discharged every three or six hours. The limestone combines with impurities in the ore, primarily silica, to form a slag. The molten slag floats on the pig iron and is discharged through the slagnotch.

of sizes. Supermassive black holes with up to a billion solar masses could be the source of energy in quasars and other types of active galaxy. At the other end of the scale, some primordial black holes (see below) could be truly microscopic. Since no light or other radiation can escape from black holes, they are extremely difficult to detect. Any matter encountered by the black hole will most likely go into orbit first rather than being drawn directly into it. A rapidly spinning disk of matter, known as an ACCRETION disk, forms around the object and heats up through friction to such high temperatures that it emits X-rays. Black holes may therefore appear as X-ray sources in BINARY STARS. The most famous candidate is Cygnus X-1. Not all black holes result from stellar collapse. During the Big Bang, some regions of space might have become so compressed that they formed so-called **primordial black holes**. Such black holes would not be completely black, because radiation could still "tunnel out" of the event horizon at a steady rate, leading to the evaporation of the hole. Primordial black holes could thus be very hot. *See also* STELLAR EVOLUTION

Blackwell, Elizabeth (1821–1910) US physician, b. England. In 1847 she began to study medicine at the Geneva Medical College in New York. She graduated in 1849, becoming the first woman doctor in the USA. She established the New York Infirmary (1857), which combined health services and medical training.

bladder Large, elastic-walled organ in the lower abdomen in which URINE is stored. Urine passes from each KIDNEY by way of two narrow tubes (URETERS) to the bladder, where it is stored until it can be voided. When pressure in the bladder becomes too great, nervous impulses signal the need for emptying. Urine leaves the bladder through a tube called the URETHRA.

blast furnace Cylindrical smelting furnace. It is used in the extraction of metals, mainly iron and copper, from their ores. The ore is mixed with coke and a FLUX (limestone in the case of iron ore). A blast of hot compressed air is piped in at the bottom of the furnace to force up temperatures to where the reduction of the oxide ore to impure metal occurs. The molten metal sinks to the bottom and is tapped off. Waste "slag" floats to the top of the metal and is piped off. *See also* OXIDATION-REDUCTION

blastula Stage in the development of the EMBRYO in animals.The blastula consists of a hollow cavity (**blastocoel**) surrounded by one or more spherical layers of cells. Commonly called the hollow ball of cells stage, it occurs at or near the end of cleavage (*see* MORULA) and precedes the GASTRULA stage. The blastula stage in mammals is known as a **blastocyst** or **germinal vessel**.

bleach Substance used to remove the colour from materials such textiles and paper. Most bleaches act by oxidizing the pigment. Chlorine, or compounds such as calcium and sodium hypochlorites, are commonly used. Other oxidizing bleaches are hydrogen peroxide, used for hair, and sodium perborate. For some applications reducing agents, such as sulphur dioxide, are employed. *See also* OXIDATION

bleaching powder White powder of imprecise composition used for bleaching. It is made by bubbling chlorine gas through calcium hydroxide (slaked lime) solution, and contains calcium chlorate and calcium chloride. It releases chlorine when treated with dilute acid and is used for purifying water and bleaching cloth and wood pulp for paper-making.

bleeding Loss of blood from the circulation. *See* HAEMORRHAGE

Blériot, Louis (1872–1936) French aircraft designer and aviator. In 1909, he became the first man to fly an aircraft across the English Channel. The flight from Calais to Dover took 37 minutes. As a designer, Blériot was responsible for various innovations, including a system by which the pilot could operate ailerons by remote control.

blindness Severe impairment of, or absence of, vision. It may be due to heredity, accident, disease or old age. Worldwide, the commonest cause of blindness is TRACHOMA. In developed countries it is most often due to severe DIABETES, GLAUCOMA, CATARACT or degenerative changes associated with ageing.

blind spot Small area on the RETINA of the EYE where no visual image can be formed because of the absence of light-sensitive cells, the rods and cones. It is the area where the OPTIC NERVE leaves the eye.

BL Lacertae object (BL Lac object) Highly luminous object located in the nuclei of some galaxies, the prototype of which, BL Lacertae, was originally classified as a variable star. BL Lac objects are variable at all wavelengths from radio to X-rays, sometimes over a timescale of just a few hours, and their optical spectra are unusual in being completely featureless, containing neither absorption nor emission lines. In this respect they differ from QUASARS. BL Lac objects are thought to lie in the centres of gas-free galaxies such as giant ellipticals.

Bloch, Felix (1905–83) US nuclear physicist, b. Switzerland. From 1934 he was professor of theoretical physics at Stanford University in California. He shared the 1952 Nobel Prize for physics with Edward PURCELL for their separate development of the technique of NUCLEAR MAGNETIC RESONANCE (NMR) used to study the interactions between atomic nuclei. Bloch was the first director (1954–55) of the Conseil Européen pour la Recherche Nucléaire (CERN), the European centre in Geneva for research into high-energy PARTICLE PHYSICS.

Bloch, Konrad Emil (1912–) US biochemist, b. Germany. He studied METABOLISM of FATTY ACIDS, and in particular the way CHOLESTEROL is synthesized in the body. For this work he shared the 1964 Nobel Prize for physiology or medicine with Feodor LYNEN.

block mountain Uplift that is the result of block faulting. This is one type of FAULT in which the crustal portions of the Earth are broken into structural blocks of different elevations and positions.

blocky lava Surface of hot molten lava covered with a "skin" of large angular lava blocks formed by the break up of a solidified surface of a lava flow. Blocky lavas are characteristic of high viscosity lavas with relatively high silica content.

blood Fluid circulating in the body that trans-

BLOOD

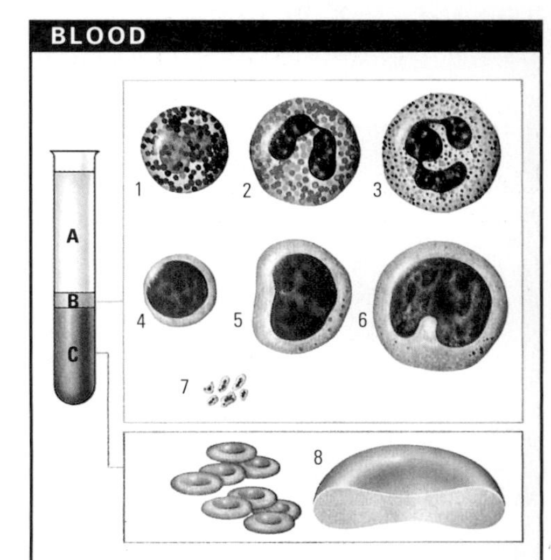

Spun in a high-speed centrifuge, blood separates out into plasma (A), layers of white cells and platelets (B) and red cells (C). Fluid plasma, almost 90% water, contains salts and proteins. Three main types of white cells shown in magnification, are polymorphonuclearcytes (1-3), responsible for the destruction of invading bacteria and removal of dead or damaged tissue, small and large lymphocytes (4, 5), which are involved in the body's immunity, and monocytes (6), which form a further line of the body's defence. Platelets (7) are vital clotting agents. Red cells (8) are the most numerous, and are concerned with the transport of oxygen around the body.

BLOOD VESSELS

Arteries (A) and veins (B) are both blood vessels that conduct blood around the body. They have a common structure consisting of an inner avascular coat and vascular middle and outer coats. The inner coat is composed of (1) a cellular lining with (2) associated basement membrane surrounded by (3) an elastic tissue/layer with longitudinal fibres. The middle coat contains smooth muscle and elastic fibres and collagen connective tissue. Arteries have much thicker walls and a smaller channel diameter. The middle coat is very thick and gives arteries their elastic, muscular properties. In veins, inner coat layers are often indistinguishable. A comparison of the two vessels is shown in half sections of actual sizes of arteries and veins found in the body. Very small vessels have a similar structure.

ports oxygen and nutrients to all the cells and removes wastes such as carbon dioxide. In a healthy human, it constitutes about 5% of the body's total weight; by volume, it comprises about 5.5l (9.7 pints). It is composed of a colourless, transparent fluid called plasma in which are suspended microscopic ERYTHROCYTES (red blood cells), which make up nearly half the blood volume; LEUCOCYTES (white blood cells); and PLATELETS.

blood-brain barrier Mechanism that prevents harmful substances from reaching the brain via the bloodstream. The mechanism involves various defensive strategies involving the permeability of cells which prevent "foreign" particles from entering the CEREBROSPINAL FLUID. A semi-permeable lipid membrane allows the passage of solutions (containing essential chemicals) but blocks large molecules such as harmful microorganisms and most drugs.

blood clotting Protective mechanism that prevents excessive blood loss after injury. A mesh of tight fibres (of insoluble FIBRIN) coagulates at the site of injury through a complex series of chemical reactions. This mesh traps blood cells to form a clot which dries to form a scab. This prevents further loss of blood, and also prevents bacteria getting into the wound. Normal clotting takes place within five minutes. The clotting mechanism is impaired in some diseases such as HAEMOPHILIA.

blood group Any of the types into which blood is classified according to which ANTIGENS are present on the surface of its red cells. There are four major types: A, B, AB and O. Each group in the ABO system may also contain the rhesus factor (Rh), in which case it is Rh-positive; otherwise it is Rh-negative. Such typing is essential before BLOOD TRANSFUSION since using blood of the wrong group may produce a dangerous or even fatal reaction. *See also* LANDSTEINER, KARL

blood plasma *See* PLASMA

blood pressure Force exerted by circulating BLOOD on the walls of blood vessels due to the pumping action of the HEART. Blood pressure is measured, using a gauge known as a SPHYGMOMANOMETER, in millimetres of mercury. It is greatest when the heart contracts and lowest when it relaxes. High blood pressure (HYPERTENSION) is associated with an increased risk of heart attacks and strokes; abnormally low blood pressure (HYPOTENSION) is mostly seen in people in shock or following excessive loss of fluid or blood.

blood test Analysis of a sample of blood as an aid to diagnosis. There are a great many tests available, from blood typing, in case a transfusion is required, to counts of red or white cells or elaborate assays to trace minute amounts of hormones. Most are performed to evaluate the condition of the blood itself, to detect infection or to probe chemical changes taking place in the body.

blood transfusion Transfer of blood or a component of blood (such as plasma or red cells) from one body to another in order to make up for some deficiency. It is often done to counteract life-threatening SHOCK following excessive blood loss. The BLOOD GROUP of the donor must be compatible with that of the recipient to avoid the donated blood coagulating in the recipient's blood vessels. Donated blood is scrutinized for readily transmissible organisms such as the HEPATITIS B virus and HIV, the virus responsible for AIDS.

blood vessel Closed channel that carries blood throughout the body. An ARTERY carries oxygenated blood away from the heart; these give way to smaller arterioles and finally to tiny capillaries deep in the tissues, where oxygen and nutrients are exchanged for cellular wastes. The deoxygenated blood is returned to the heart by way of the VEINS.

blowhole In geology, hole in a cliff or cave through which seawater spurts at high tide. Blowholes are formed by erosion. Initially, waves acting on a joint or other line of weakness in a cliff face carve out a cave. Continued HYDRAULIC ACTION by pounding waves on the roof of the cave forms a sort of chimney. Eventually this breaks through the surface of the ground near the edge of the cliff. At high tide, incoming waves force water out of the top of the blowhole. In biology, blowhole refers to the nostrils (single or double) of whales, situated towards the back of the skull.

blubber Layer of subcutaneous fat that surrounds the bodies of many aquatic mammals. The fat, a feature of whales and seals, acts as an energy store and insulating layer to prevent the loss of heat to cold water. During the days of whaling, blubber was rendered down by boiling and used as a source of oil for many applications, all of which can now be supplied by synthetic substances.

blue-green algae *See* CYANOBACTERIA

blueprint Photographic image on paper with white lines against a blue background, frequently used for engineering drawings. The paper is coated with a solution of ammonium ferric citrate and potassium ferricyanide and exposed to intense light under the sheet of drawings to be copied. The blueprint is "developed" in water.

blue shift Effect in which the lines in the SPECTRUM of a celestial object are displaced towards the blue end of the spectrum. It results from the DOPPLER EFFECT because the object and the observer are moving towards each other. The closing speed can be calculated from the extent of the shift. *See also* RED SHIFT

bluff Steep slope at the side of a bend in a river. A bluff forms on the outside of a MEANDER by the erosive action of the slow-moving river water. Some bluffs are as much as 100m (330ft) tall.

boat Vehicle for passenger and freight transport by water. Today the term "boat" is often reserved for a small craft that can easily be taken out of the water; a larger vessel is called a SHIP. The first boats, made in prehistoric times, included rafts, hollowed-out logs and vessels made from plaited reeds. Among the first maritime peoples were the Phoenicians. They built fleets of galleys for their extensive trading in the Mediterranean and adjoining areas. These galleys, and the later ones of the Greeks and Romans, were propelled by sails, supplemented by oars. The later Viking long-boats, also square-sailed, were slimmer and speedier. Lateen (triangular) sails were probably imported from the Persian Gulf and introduced to the West by the empire-building Arabs. Modern boats include sailing vessels, used mainly for pleasure, motorboats and launches.

BOD Abbreviation of BIOLOGICAL OXYGEN DEMAND

Bode's law Empirical, numerical relationship for the mean distances of the planets from the Sun. It is named after the German astronomer Johann Bode (1747–1826), who popularized it in the late 18th century. If the number 4 is added to 0, 3, 6, 12, 24, 48, 96 and 192, and each sum is divided by 10, the figure arrived at is the mean distance in astronomical units of the planets from the Sun, from Mercury to Uranus, including the asteroid belt. The law does not work for Neptune and Pluto.

bogie Small four- or six-wheeled undercarriage pivoted near the end of a rigid vehicle. Commonly used in pairs on long railway wagons or passenger coaches, so that they can better negotiate

▲ **Bohr** Niels Bohr was educated at the University of Copenhagen, receiving his doctorate in 1911. Whilst at Cambridge in 1912–16, he proposed the idea of electron shells and energy levels. He explained the absorption and emission of radiation by atoms as a change in electron energy, further demonstrating that electrons occupied discrete energy levels. For this work he was awarded the 1922 Nobel Prize for physics.

B

BONE

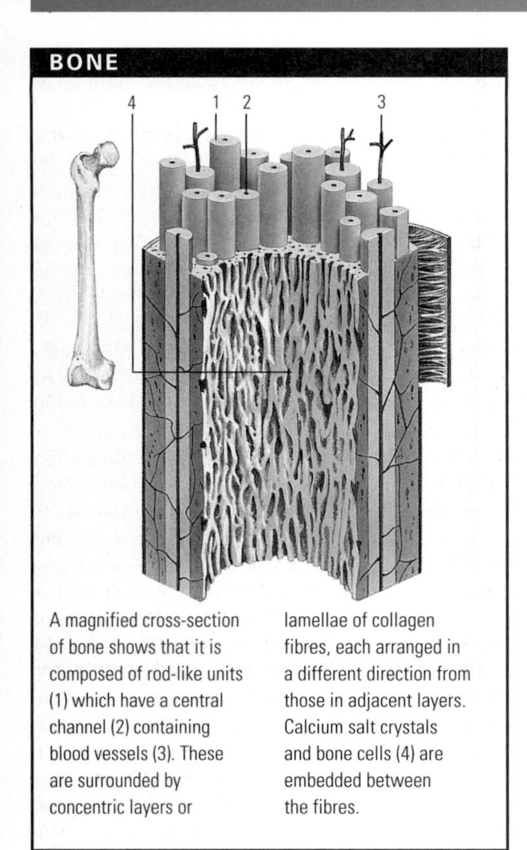

A magnified cross-section of bone shows that it is composed of rod-like units (1) which have a central channel (2) containing blood vessels (3). These are surrounded by concentric layers or lamellae of collagen fibres, each arranged in a different direction from those in adjacent layers. Calcium salt crystals and bone cells (4) are embedded between the fibres.

curves, bogies are also used at the front of large locomotives to "lead" the vehicle into curves. The term bogie is also used for a short wheel-base railway truck used for carrying minerals.

Bohr, Aage Niels (1922–) Danish physicist. He was educated at the University of Copenhagen and the University of London. From 1956 he was professor of physics at Copenhagen University. He was the director (1963–70) of the Institute of Theoretical Physics, founded by his father, Niels BOHR, and then directed (1975–81) the Nordic Institute for Theoretical Nuclear Physics (Nordita). With Ben MOTTELSON (also of Nordita) and James RAINWATER he shared the 1975 Nobel Prize for physics for devising a "collective model" of the atomic nucleus.

Bohr, Niels Henrik David (1885–1962) Danish physicist who was a major developer of the QUANTUM THEORY. He was the first person to apply it with success to the problem of atomic structure. Bohr worked with J.J. THOMSON and Ernest RUTHERFORD in Britain before teaching theoretical physics at the University of Copenhagen. He escaped from German-occupied Denmark during World War 2 and worked briefly on developing the atom bomb in the USA. He later returned to Copenhagen and worked for international cooperation. Bohr used the quantum theory to explain the spectrum of hydrogen and in the 1920s helped develop the "standard model" of the quantum theory, known as the Copenhagen Interpretation. He was awarded the 1922 Nobel Prize for physics for his work on atomic structure and in 1957 received the first Atoms for Peace Award. He is the father of Aage BOHR.

boiler Vessel for heating and converting water to steam. The boiler is an essential part of STEAM-ENGINES and TURBINES. It consists of a furnace for burning fuel and a container where water is evaporated into steam. The term is also applied to devices used to provide hot, though not boiling, water in central heating systems.

boiling point Temperature at which a substance changes phase (state) from a liquid to a vapour or gas. The boiling point increases as the external pressure increases and falls as it decreases. It is usually measured at a standard pressure of one atmosphere (760mm of mercury, 101,325 pascals). The boiling point of pure water at standard pressure is 100°C (212°F).

bolometer Instrument for detecting and measuring ELECTROMAGNETIC RADIATION such as light and heat. One type uses a thin blackened metal strip, the electrical resistance of which varies according to the amount of radiation falling on it. Bolometers are often mounted in telescopes to measure the energy given off by stars.

Boltwood, Bertram Borden (1870–1927) US chemist and physicist who did important research in radioactivity which led to the development of the theory of ISOTOPES. Boltwood discovered (1907) ionium – a radioactive isotope of thorium. He was professor of radiochemistry (1910–27) at Yale University.

Boltzmann, Ludwig (1844–1906) Austrian physicist who helped to establish the foundations of statistical mechanics and who was acclaimed for his major contribution to the kinetic theory of gases. His statistical research into the velocity distribution of gas molecules extended the ideas of Scottish physicist James MAXWELL. Boltzmann's general law said a system will approach a state of thermodynamic equilibrium because this is the most probable state. He introduced the "Boltzmann equation" in 1877 relating the kinetic energy of a gas atom or molecule to temperature. The formula involves the gas constant per molecule called the " BOLTZMANN CONSTANT ". In 1884 he derived a law, often termed the "Stefan-Boltzmann law", for BLACK BODY radiation discovered by his Viennese teacher, Josef Stefan (1835–93). After being attacked for his belief in the atomic theory of matter, Boltzmann committed suicide.

Boltzmann constant (symbol k) Ratio of the universal GAS CONSTANT to the AVOGADRO CONSTANT, equal to 1.381×10^{-23} joules per kelvin. It shows the relationship between the kinetic energy of a gas particle (atom or molecule) and its absolute temperature. It is named after Ludwig BOLTZMANN.

bomb, volcanic Piece of solid material, usually cooled lava, ejected by a volcano. Volcanic bombs are usually greater than 32mm (1.3in) in diameter, and may be very large.

bond *See* CHEMICAL BOND

bone CONNECTIVE TISSUE that forms the skeleton of most vertebrates, protects its internal organs, serves as a lever during locomotion and when lifting objects, and stores calcium and phosphorus. Bone is composed of a strong, compact layer of COLLAGEN and calcium phosphate and a lighter, porous, inner spongy layer containing MARROW, in which ERYTHROCYTES and some LEUCOCYTES are produced.

bone china Hard-paste PORCELAIN. It consists of kaolin, china stone and bone ash. Josiah SPODE perfected the manufacture of bone china in the 19th century and was largely responsible for its popularity. Like true porcelain, it is translucent and lends itself well to underglaze decoration.

bony fish *See* FISH

Boole, George (1815–64) English mathematician. Largely self-taught, he was appointed professor of mathematics at Cork University in 1849. He is remembered for his invention of a set of symbols to represent logical operations. *See also* BOOLEAN ALGEBRA

Boolean algebra Branch of mathematics that deals with rules that can be applied to sets and to logical "and" or "or" propositions. For example, in Boolean algebra, xy means x **and** y, while $x + y$ means x **or** y. It is applied widely to computer design, where the BINARY SYSTEM (0 and 1) correspond to the logical propositions a computer uses to function. It was named after George BOOLE

booster engine ROCKET engine that powers a missile or space vehicle in the early stages of its flight and then drops off, reducing deadweight. Several booster engines may be attached to the first stage of a launch vehicle for extra power.

boot (bootstrap, boot-up) In computing, the

BOOLEAN ALGEBRA

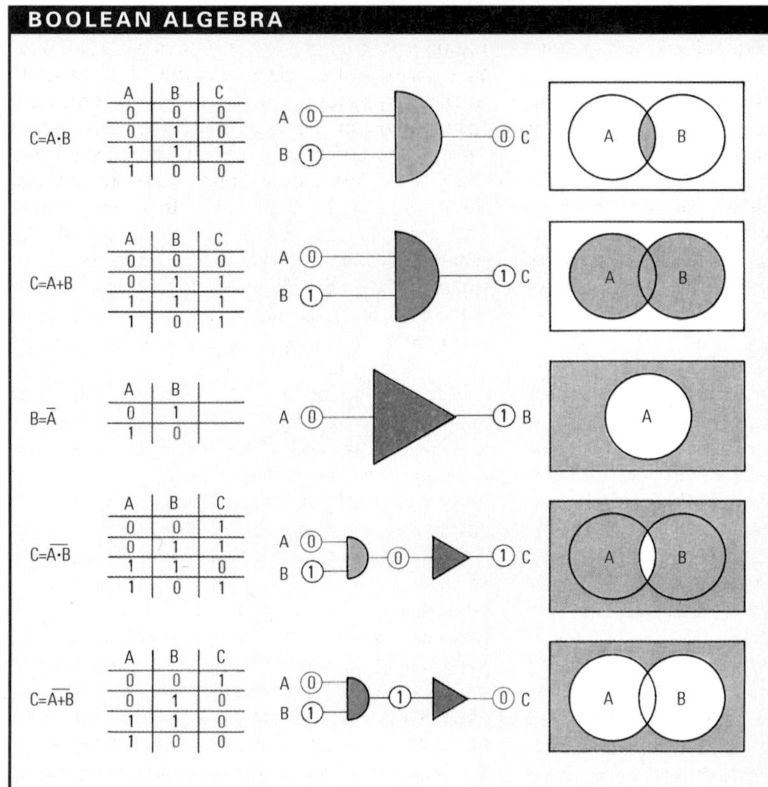

The algebra of logic. Five basic logic statements are illustrated here. In any of them, if A is true the table will show a "1", if A is false there will be a "0". In the statement AND, C is true (i.e. there is a "1" in the truth table) when A and B are true, but it is false if either of A and B is false. With OR, C is true if either A or B is true and it is only false if both A and B are false. NOT has one input and one output, its function is to reverse "true" and "false"; when applied to AND and OR, their inverses, NAND and NOR, result. The Boolean statements shown here are also illustrated by symbolic circuit elements (inputs on the left and outputs on the right) and by Venn diagrams as in set theory (the outcome is shown by the shaded areas).

loading of a short program that allows the machine to run other programs – most importantly, a DISK OPERATING SYSTEM. With personal computers, the booting program is stored on the hard disk and runs automatically when the computer is switched on. The terms boot and bootstrap may also refer to the booting program itself.

Boötes (herdsman) Prominent constellation of the N sky; it contains the bright orange star Arcturus.

borane HYDRIDE of BORON. The simplest borane is diborane (B_2H_6), made by treating magnesium boride with acid. The higher boranes have up to 10 boron atoms. They are all reactive and readily oxidize, sometimes explosively.

borate Any salt of BORIC ACID or of more complex oxyacids of BORON. Borates are inorganic compounds; many exist as minerals, the most important of which are BORAX, colemanite and kernite – all used as sources of boron compounds.

borax (hydrated sodium borate, $Na_2B_4O_7.10H_2O$) Most common borate mineral. It is found in large deposits in dried-up alkaline lakes in arid regions as crusts or masses of crystals. It may be colourless or white, transparent or opaque. It is used to make heat-resistant glass, pottery glaze, water softeners in washing powders, fertilizers and pharmaceuticals.

boric acid (boracic acid, H_3BO_3) Soft white crystalline solid, that occurs naturally in certain volcanic hot springs. It is used as a metallurgical FLUX, preservative, antiseptic, and an insecticide for ants and cockroaches.

boride Any of a number of chemical compounds formed between a metal and the nonmetallic element BORON. Borides are made by combining the elements at a very high temperature. All are extremely hard and are used as abrasives and refractories.

boring machine Device for making accurate and smoothly finished holes in metal objects. Drilled holes are enlarged with a cutting bore, generally tipped with steel, carbide or diamond, and gripped in an adjustable boring head attached to a rotating spindle. The work-holding table can usually be moved in two perpendicular directions so that holes can be spaced accurately.

Borlaug, Norman Ernest (1914–) US agronomist. As a director of the Rockefeller Foundation in Mexico, he led a team of scientists experimenting with the improvement of cereal crops. He was awarded the Nobel Peace Prize in 1970 for his accomplishments in the "GREEN REVOLUTION", developing improved wheat seed, a higher-yielding rice and better ways of using fertilizer and water.

Born, Max (1882–1970) German-British physicist. He was professor of physics at Göttingen University from 1921 but left Germany in 1933 and went to Britain, teaching physics at Cambridge (1933–6) and the University of Edinburgh (1936–53). He returned to Germany in 1954. For his work in QUANTUM MECHANICS, the basis of atomic and nuclear physics, he was awarded the 1954 Nobel Prize for physics, which he shared with Walther BOTHE.

bornite (peacock ore, Cu_5FeS_4) Common copper mineral, a copper iron sulphide. It generally occurs in masses, and sometimes as crystals (cubic system) in INTRUSIVE IGNEOUS rocks and METAMORPHIC rocks. It is opaque and bronze with an iridescent purple tarnish. Hardness 3; r.d. 5.0.

boron (symbol B) Nonmetallic element of group III of the periodic table, first isolated in 1808 by Sir Humphry DAVY. It occurs in several minerals, notably KERNITE (its chief ore) and BORAX. It has two allotropes (different forms): **amorphous** boron, an impure brown powder, is made by reducing the oxide with magnesium; **metallic** boron, a black to silver-grey hard crystalline material, is obtained by decomposing boron tribromide vapour on a hot metal filament. The element is used in semiconductor devices and the stable isotope ^{10}B is a good neutron absorber, used in NUCLEAR REACTORS and particle counters. Properties: at.no. 5; r.a.m. 10.81; r.d. 2.34 (cryst.), 2.37 (amorph.); m.p. 2,079°C (3,774°F); sublimes 2,550°C (4,622°F); most common isotope ^{11}B (80.22%).

BOYLE'S LAW

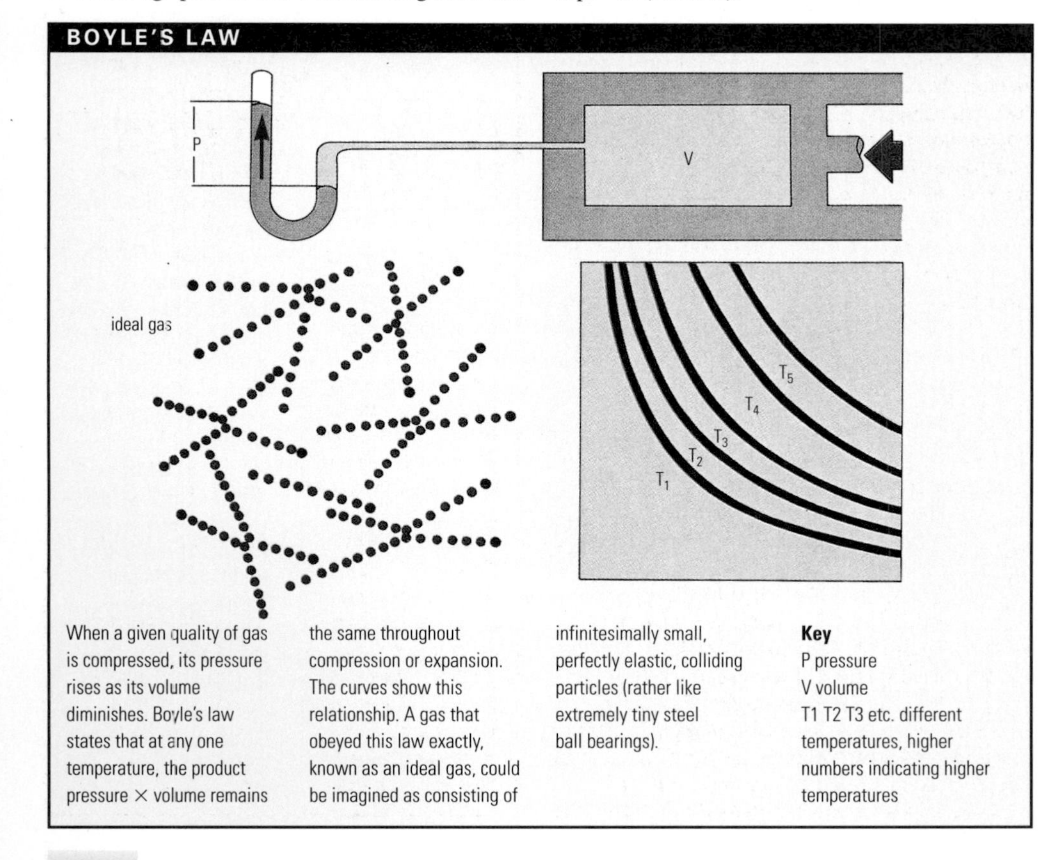

When a given quality of gas is compressed, its pressure rises as its volume diminishes. Boyle's law states that at any one temperature, the product pressure × volume remains the same throughout compression or expansion. The curves show this relationship. A gas that obeyed this law exactly, known as an ideal gas, could be imagined as consisting of infinitesimally small, perfectly elastic, colliding particles (rather like extremely tiny steel ball bearings).

Key
P pressure
V volume
T1 T2 T3 etc. different temperatures, higher numbers indicating higher temperatures

Bosch, Carl (1874–1940) German industrial chemist who adapted Fritz HABER's method of synthesizing ammonia to an industrial scale. Bosch's work involved finding metallic catalysts that could promote this high-pressure synthesis. His invention of the **Bosch process** (in which water gas and steam at high temperatures are passed over a catalyst) aided the large-scale preparation of hydrogen. Bosch was awarded the 1931 Nobel Prize for chemistry, with Friedrich BERGIUS, for his high-pressure techniques.

Bose, Sir Jagadis Chandra (1858–1937) Indian plant physiologist and physicist. He invented highly sensitive instruments capable of detecting tissue responses of plants to external stimuli.

Bose, Satyendranath (1894–1974) Indian physicist and mathematician who contributed significantly to the theory of QUANTUM MECHANICS and STATISTICAL MECHANICS. Bose made the initial advances to describe the statistical properties of certain ELEMENTARY PARTICLES (now called BOSONS). These particles, which include all those that mediate force (like the PHOTON which carries the ELECTROMAGNETIC FORCE), all have the property that any number of them can occupy the same quantum state: that is they do not obey Enrico FERMI's EXCLUSION PRINCIPLE. Bose's work was developed further by EINSTEIN and the statistics which such particles obey is now called **Bose-Einstein statistics**.

boson ELEMENTARY PARTICLE which has an integer SPIN. Named after the physicist Satyendranath BOSE, bosons are those particles that are not covered by the EXCLUSION PRINCIPLE. This means that the number of bosons occupying the same quantum state is not restricted. Bosons are force-transmitting particles, such as PHOTONS and GLUONS (the particles that hold QUARKS together). *See also* FERMION

botanical garden Large garden preserve for display, research and teaching purposes. Wild and cultivated plants from all climates are maintained outdoors and in greenhouses. Although organized gardens date from ancient Rome, the first botanical gardens were established during the Middle Ages. In the 16th century, gardens existed in Pisa, Bologna, Padua and Leiden. Aromatic and medicinal herbs were arranged in rows and still exist in the Botanical Garden of Padua. The first US botanical garden was established by John Bartram in Philadelphia in 1728. Famous botanical gardens include the Royal Botanical Gardens in Kew, near London (established 1759); Botanical Gardens of Berlin-Dahlem (1646); and Botanical Gardens in Schönbrunn, Vienna (1753).

botany Study of plants and algae, including their classification, structure, physiology, reproduction and evolution. The discipline used to be studied in two halves: **lower** (non-flowering) plants, which included the ALGAE (now in the kingdom PROTOCTISTA), MOSS and FERNS; and **higher** (seed-bearing) plants, including most flowers, trees and shrubs. Botany also studies the importance of plants to people.

Bothe, Walther Wilhelm Georg Franz (1891–1957) German physicist. From 1934 he was director of the Max Planck Institute, Heidelberg, and in 1946 was professor of physics at the university there. During World War 2 he worked on Germany's nuclear energy project and built Germany's first CYCLOTRON. He shared the 1954 Nobel Prize for physics with Max BORN for his development of the **coincidence method** which can detect two particles emitted simultaneously from the same nucleus during RADIOACTIVE DECAY.

B

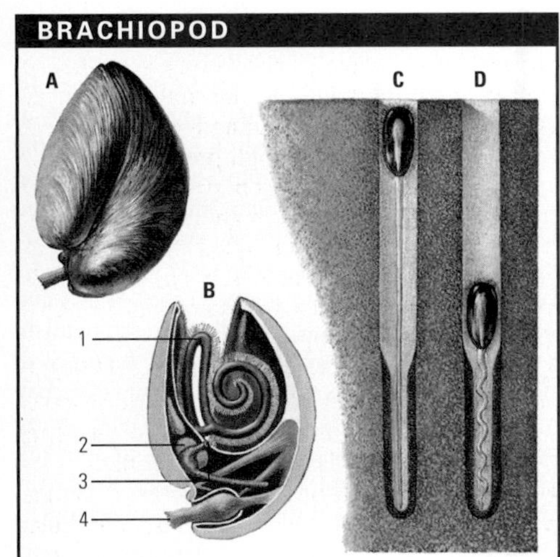

The marine animal known as a brachiopod, or lampshell, lives in holes in mud flats. It comprises (A) a hinged shell and a stalk with which it grips the rocks. The cross-section (B) shows; lophophore (1) which bears ciliated tenatacles used for feeding, digestive gland (2), mouth (3) and stalks (4). When feeding *Lingula*, (C), which resembles fossil forms of 500 million years ago rests at the surface of its burrow using feathery cilia to filter water for food particles. When disturbed, its stalk contracts, drawing the animal into the burrow (D).

botulism Rare but potentially lethal form of food poisoning caused by a toxin produced by the bacterium *Clostridium botulinum*. The toxin attacks the nervous system, causing paralysis and cessation of breathing. The most likely source of botulism is imperfectly canned meat. Botulinum toxin is used medicinally as a treatment for some neuromuscular disorders.

Bouguer anomaly Observation that gravity measured on a great rock mass, such as a mountain range, is higher than average. This is due to the gravitational force exerted by the rock mass itself. It is named after its discoverer, the French mathematician Pierre Bouguer (1698–1758).

boulder clay *See* TILL

boundary-layer flow Behaviour of a FLUID in the region immediately next to a solid body immersed in the fluid. In this region, layers of the fluid "slide" over one another, and the measure of this internal friction, viscosity, becomes the determining characteristic of the flow. *See also* AERODYNAMICS; FLUID MECHANICS

Bovet, Daniele (1907–92) Italian pharmacologist. He was awarded the 1957 Nobel Prize for physiology or medicine for his pioneering work on the development of ANTIHISTAMINES and the muscle relaxants used in surgery. He also studied the effects of mental illness on the chemistry of the brain.

bovine spongiform encephalopathy (BSE) ("mad cow disease") In cattle, degenerative disease of the brain caused by infectious particles or PRIONS. The consumption of infected beef or beef products can transmit the disease to other animals and human beings. *See also* CREUTZFELDT-JAKOB DISEASE (CJD); SCRAPIE

bowden cable Device for the flexible transmission of small mechanical movements, generally to the controls of a machine. It consists of a strong steel cable (usually multi-stranded and lubricated) in a flexible outer tube, often made in the form of a long coiled spring. The outer tube is anchored at the ends so that the inner cable can move independently. Bowden cables are commonly used for throttle (accelerator), clutch, choke and sometimes brake controls on cars and motorcycles.

Bowen, Norman Levi (1887–1956) Canadian petrologist and mineralogist. His contributions to the study of the origins, structure and chemistry of rocks were outstanding. His most important book is *The Evolution of Igneous Rocks* (1928).

Bowen's reaction series Explanation of the order of crystallization of minerals within a cooling magma, derived from laboratory experiments on a rock melt carried out by Norman BOWEN. He concluded that two independent lines of mineral crystallization occur; (1) the discontinuous reaction series, consisting of ferromagnesian minerals, where each crystallization "step" produces a different mineral; (2) the continuous reaction series, consisting largely of plagioclase FELDSPARS, where there is a gradual change in feldspar composition as cooling and crystallization take place.

Bowman's capsule (renal capsule) Funnel-shaped end of a NEPHRON in a KIDNEY. It has special epithelial cells (**podocytes**) that allow fluid filtered from the blood to pass into the nephron, from where it eventually passes into the pelvis of the kidney as urine. There are about a million Bowman's capsules in each human kidney.

Boyle, Robert (1627–91) British scientist, b. Ireland. He is often regarded as the father of modern chemistry. At his laboratories in Oxford and London, Boyle conducted research into air, vacuum, metals, combustion and sound. He was a founding member of the ROYAL SOCIETY. Boyle's *Sceptical Chymist* (1661) proposed an early atomic theory of matter. He made an efficient vacuum pump, which he used to establish (1662) BOYLE'S LAW. Boyle formulated the first chemical definitions of an element and a reaction.

Boyle's law Law that states that the volume of a gas at constant temperature is inversely proportional to the pressure. This means that as pressure increases, the volume of a gas decreases. First stated (1662) by Robert BOYLE, Boyle's law is a special case of the IDEAL GAS LAW (involving a hypothetical gas that perfectly obeys the gas laws).

brachiopod Any of about 260 species of small, bottom-dwelling, marine invertebrates of the phylum Brachiopoda. They are similar in outward appearance to BIVALVE molluscs, having a shell composed of two valves; however, unlike bivalves, there is a line of symmetry running through the valves. They live attached to rocks by a **pedicle** (stalk), or buried in mud or sand. Most modern brachiopods are less than 5cm (2in) across.

bract Modified leaf found on a flower stalk or the flower base. Bracts are usually small and scalelike. In some species they are large and brightly coloured, such as dogwood and poinsettia.

Bragg, Sir (William) Lawrence (1890–1971) English physicist, b. Australia. He was director (1938–53) of the Cavendish Laboratory at Cambridge. With his father, Sir William Henry Bragg, he determined the mathematics involved in X-ray DIFFRACTION, showed how to compute X-ray wavelengths and studied CRYSTAL structure by X-ray diffraction. For these advances, they were jointly awarded the 1915 Nobel Prize for physics.

Brahe, Tycho (1546–1601) Danish astronomer. Under the patronage of King Frederick II of Denmark he became the most skilled observer of the pre-telescope era, expert in making accurate naked-eye measurements of the positions of stars and planets. He built an observatory on the island of Hven (1576) and calculated the orbit of the comet seen in 1577. This, together with his study of the supernova, showed that ARISTOTLE was wrong in picturing an unchanging heavens. Brahe could not, however, accept the world system put forward by COPERNICUS. In his own planetary theory (the **Tychonian system**), the planets move around the Sun, and the Sun itself, like the Moon, moves round the stationary Earth. In 1597 he left Denmark and settled in Prague, where Johann KEPLER became his assistant. The accurate series of planetary observations made by Tycho Brahe were used by Kepler in deriving his laws of planetary motion, which ironically demonstrated the validity of the Copernican theory.

braided stream Network of small, shallow interlacing streams. It forms when a river deposits sediment, causing it to divide up into new channels, which then redivide and rejoin, resembling a braided cord.

Braille System of reading and writing for the blind. It was invented by Louis Braille (1809–52), who himself lost his sight at the age of three. Braille was a scholar, and later a teacher, at the National Institute of Blind Youth in Paris. He developed a system of embossed dots to enable blind people to read by touch. This was first published in 1829, and a more complete form appeared in 1837. There are also Braille codes for music and mathematics..

brain Mass of nerve tissue which regulates all physical and mental activity; it is continuous with the SPINAL CORD. Weighing about 1.5kg (3.3lb) in an adult human (about 2% of body weight), the human brain has three parts – the HINDBRAIN, MIDBRAIN and FOREBRAIN. The hindbrain is where basic physiological processes such as breathing and the heartbeat are coordinated. The midbrain links the hindbrain and the forebrain. The forebrain is the seat of all higher functions and attributes (personal-

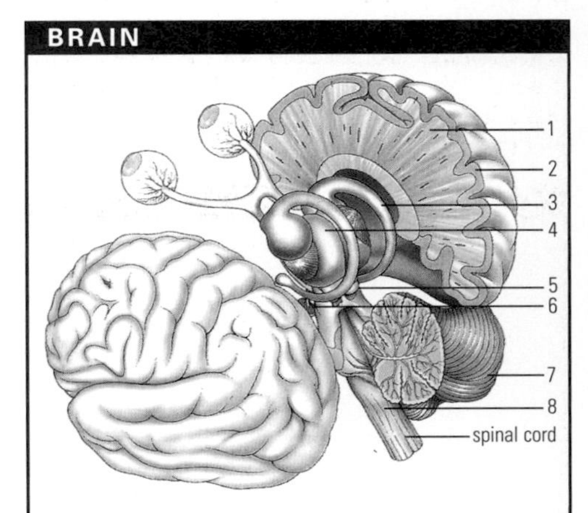

The human brain has three major structural and functional regions – the forebrain, the midbrain and the hindbrain. Part of the forebrain, the cerebrum (1), has developed into a deeply fissured structure. It comprises large regions concerned with association, reasoning and judgement. Its outer layer, the cortex (2), contains areas that coordinate movement and sensory information. The limbic system (3) controls emotional responses. The thalamus (4) coordinates sensory and motor signals, and relays them to the cerebrum: the hypothalamus (5) along with the pituitary glands control the body's hormonal system. Visual, tactile and auditory inputs are coordinated by the tectum (6). In the hindbrain, the cerebellum (7) controls muscle activity needed for refined limb movements. The medulla (8) contains reflex centres that are involved in respiration, heartbeat regulation and gastric function.

ity, intellect, memory, emotion), as well as being involved in sensation and initiating voluntary movement. *See also* CENTRAL NERVOUS SYSTEM

brain damage Result of any harm done to brain tissue causing the death of nerve cells. It may arise from a number of causes, including oxygen deprivation, brain or other disease or head injury. The nature and extent of damage varies. Sudden failure of the oxygen supply to the brain may result in widespread (global) damage, whereas a blow to the head may affect only one part of the brain (local damage). Common effects of brain damage include weakness of one or more limbs, impaired balance, memory loss and personality change; epilepsy may develop. Although brain cells do not regenerate, there is some hope of improvement following mild to moderate brain damage, especially with skilled rehabilitation. The survivor of major brain injury is likely to remain severely disabled, possibly even permanently unconscious.

brain disorder Disturbances of physical or mental function due to abnormality or disease of the brain. Some such disorders are hereditary; some are congenital (present at birth); and some are acquired during life. Brain disorders should be distinguished from psychological (psychogenic) mental disturbances in which the functioning of the brain itself is not impaired. Brain disorders are associated with impairment of memory, orientation, comprehension and judgment, and also by shallowness of emotional expression. Secondary personality changes may occur, depending upon such factors as the strength and type of personality and the amount of psychological and social stress present. Brain disorders are divided into acute and chronic types. **Acute** disorders are temporary, and are generally due to disruption of brain function rather than destruction of brain tissue. They may be caused by such things as infection, drug or alcohol intoxication, and brain trauma. **Chronic** brain disorders are irreversible, and include such things as CONGENITAL DISORDER, hereditary disease, senility and BRAIN DAMAGE.

brainstem Stalk-like portion of the BRAIN in vertebrates that includes everything except the CEREBELLUM and the CEREBRAL HEMISPHERES. It provides a channel for all signals passing between the spinal cord and the higher parts of the brain. It also controls automatic functions such as breathing and heartbeat.

brake Device for slowing the speed of a vehicle or machine. Braking can be accomplished by a mechanical, hydraulic (liquid) or pneumatic (air) system that presses a non-rotating part into contact with a rotating part, so that friction stops the motion. In a car, the non-rotating part is called a shoe or pad, and the rotating part is a drum or disc attached to a wheel. Some vehicles use electromagnetic effects to oppose the motion and cause braking. A "power" brake utilizes a vacuum system.

braking, atmospheric Drag exerted on a space vehicle as it re-enters the Earth's atmosphere and encounters increased air density. The friction slows down the vehicle, which must be built to withstand the intense heat created. *See also* ABLATION

branching Growth extension of vascular plants. A branch develops from the stem and consists of the growth from the previous year (branchlet) and new growth (twig). New twigs are produced during the next season of growth, both from the terminal bud at the end of the twig and from lateral buds in leaf axils along it.

Brandt, Georg (1694–1768) Swedish chemist who discovered, isolated and named COBALT in 1730. Brandt was a research director for the Council of Mines, Stockholm, and later became the warden of the Swedish Mint. He rejected alchemy and discredited many spurious processes for the production of gold.

brass Alloy of mainly copper (55%–95%) and zinc (5%–45%). Brass is yellowish or reddish, malleable and ductile, and can be hammered, machined or cast. Its properties can be altered by varying the amounts of copper and zinc, or by adding other metals, such as tin, lead and nickel. Brass is widely used for pipe and electrical fittings, screws, musical instruments and ornamental metalwork.

Brattain, Walter Houser (1902–87) US physicist. In 1929 he joined the Bell Telephone Laboratories as a research physicist and concentrated on research into SOLID-STATE PHYSICS. In 1956 he shared the Nobel Prize for physics with John BARDEEN and William SHOCKLEY for their development of the TRANSISTOR and research into SEMICONDUCTORS. In 1967 Brattain became professor at Whitman College in Walla Walla, Washington, USA.

Braun, (Karl) Ferdinand (1850–1918) German physicist. From 1895 he was professor of physics and director of the Physical Institute, Strasbourg. In 1897 he invented the oscilloscope – using a CATHODE-RAY TUBE known as the **Braun tube**, the forerunner of the television tube. In 1909, with Guglielmo MARCONI, he shared the Nobel Prize for physics for the development of wireless telegraphy.

BRAINSTEM

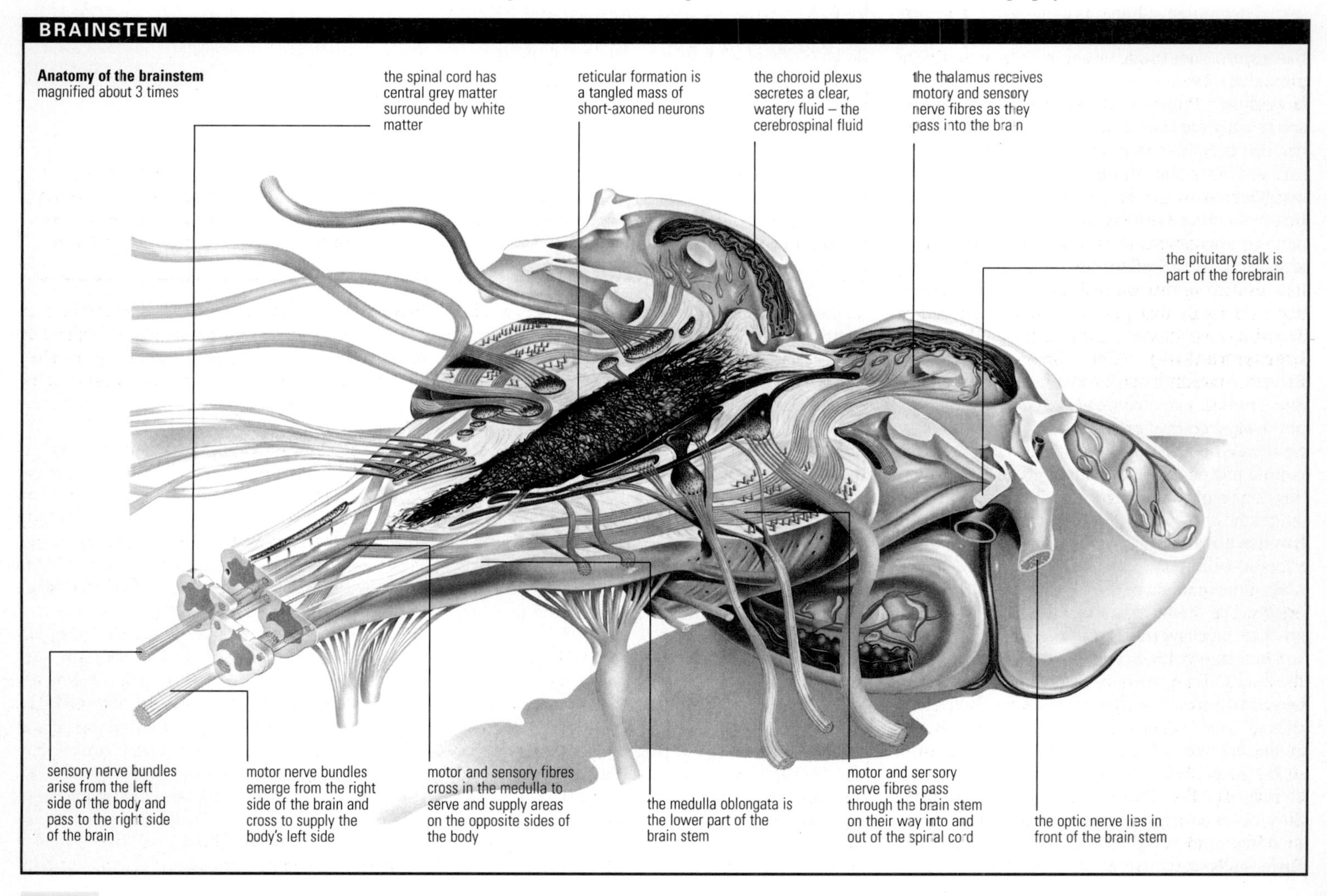

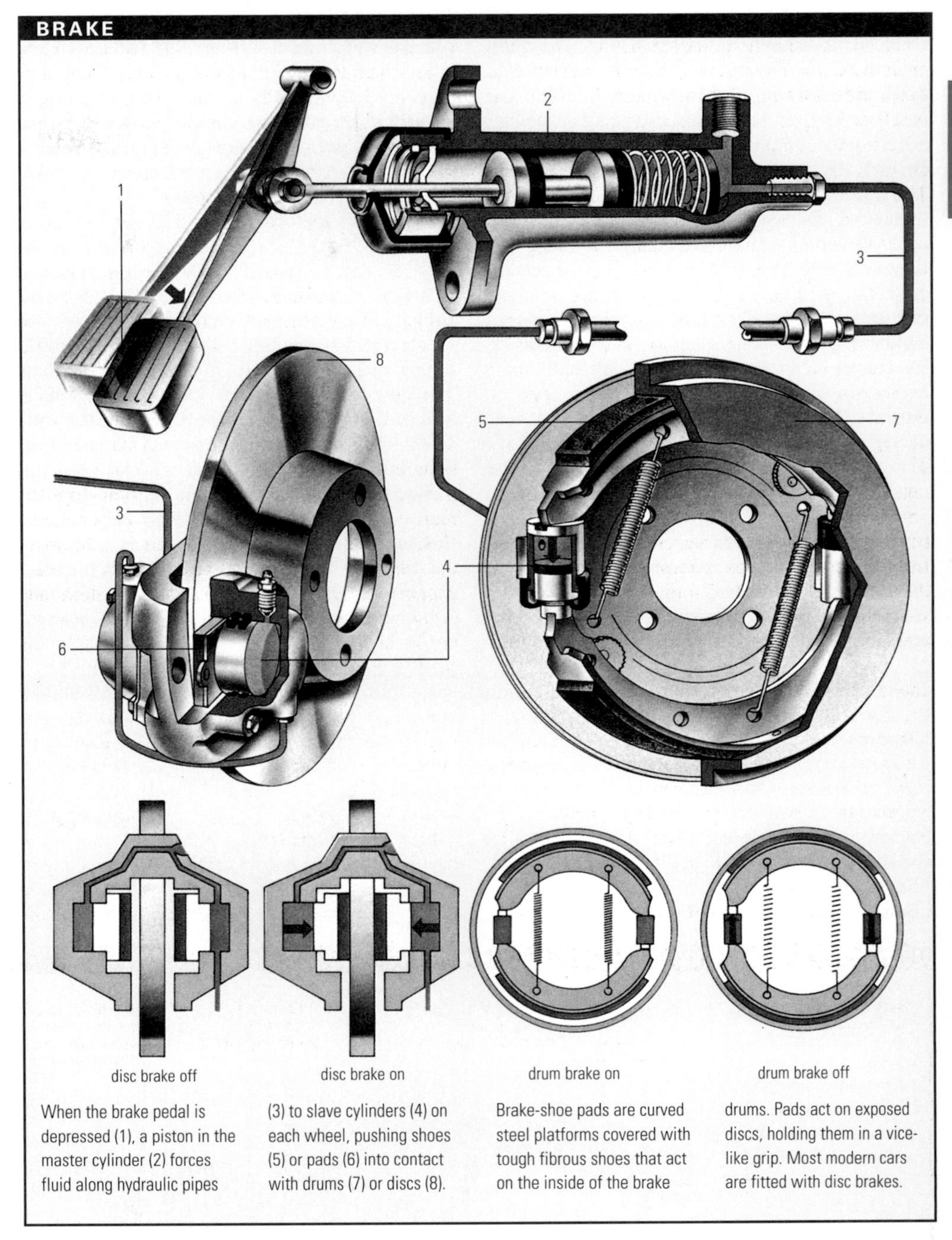

When the brake pedal is depressed (1), a piston in the master cylinder (2) forces fluid along hydraulic pipes (3) to slave cylinders (4) on each wheel, pushing shoes (5) or pads (6) into contact with drums (7) or discs (8). Brake-shoe pads are curved steel platforms covered with tough fibrous shoes that act on the inside of the brake drums. Pads act on exposed discs, holding them in a vice-like grip. Most modern cars are fitted with disc brakes.

Braun, Wernher von (1912–77) US rocket engineer, b. Germany. By 1937 he was director of a rocket research station at Peenemünde, Germany, where he perfected V-2 rocket missiles in the early 1940s. In 1945 he went to the USA, becoming a US citizen in 1955. In 1958 von Braun was largely responsible for launching the first US satellite, *Explorer 1*. He later worked on the development of the *Saturn* rocket (for the APOLLO PROJECT) and was deputy associate administrator of NATIONAL AERONAUTICS AND SPACE ADMINISTRATION (NASA) (1970–72).

brazing Process in which metallic parts are joined by the fusion of alloys that have lower melting points than the parts themselves. The bonding alloy is either preplaced or fed into the joint as the parts are heated. Brazed joints are very reliable and are used extensively in the aerospace industry.

breast *See* MAMMARY GLAND

breastbone *See* STERNUM

breathing Process by which air is taken into and expelled from the lungs for the purpose of GAS EXCHANGE. When breathing in (inhaling), the intercostal muscles raise the ribs, increasing the volume of the THORAX and drawing air into the LUNGS. During breathing out (exhalation), the ribs are lowered, and air is forced out through the nose, and sometimes also the mouth.

breccia SEDIMENTARY ROCK formed by the cementation of sharp-angled fragments in a finer matrix of the same or different material. It can also refer to rock formed by movements of the crust (**fault** breccia) and to volcanic rock from a vent (**volcanic** breccia). *See also* CONGLOMERATE

breeder reactor NUCLEAR REACTOR in which more fissile (capable of undergoing FISSION) material is produced than is consumed. In it, surplus high-energy neutrons transmute a "breeding blanket" of nonfissile uranium-238 into fissile plutonium-239.

breeding Process of producing offspring, specifically the science of changing or promoting certain genetic characteristics in animals and plants. This is done through careful selection and combination of the parent stock. Breeding may involve outbreeding (mating of unrelated individuals) or INBREEDING to produce the desired characteristics in the offspring. Scientific breeding has resulted in disease-resistant strains of crops, and in animals that give improved food yields. *See also* GENE; GENETIC ENGINEERING

bremsstrahlung (Ger. braking radiation) ELECTROMAGNETIC RADIATION in the form of X-rays emitted when charged particles slow down or change course rapidly. This happens when high-speed electrons enter the electric field of an atomic nucleus. Such radiation covers a continuous range of wavelengths within the ELECTROMAGNETIC SPECTRUM.

brewing Preparation of beer and stout by using yeast as a catalyst in the alcoholic FERMENTATION of liquors containing malt and hops. In beer brewing, a malt liquor (wort) is made from crushed germinated barley grains. Hops are added to the boiling wort both to impart a bitter flavour, and also to help to clarify the beer and keep it free from spoilage by microbes. The clear, filtered wort is cooled and inoculated with brewer's yeast, which ferments part of the sugar from malt into alcohol.

Brewster, Sir David (1781–1868) Scottish physicist who performed significant experiments in OPTICS and POLARIZED LIGHT. He discovered **Brewster's law**, which relates the polarizing angle (the angle at which light strikes a substance) to the REFRACTIVE INDEX of the substance.

brick Hardened block of clay used for building and paving. Usually rectangular, bricks are made in standard sizes by machines that either mould bricks or cut off extruded sections of stiff clay which are conveyed into a continuously operating kiln where they are baked at temperatures of up to 1,300°C (2,372°F). The first, sun-dried bricks were used in the Tigris-Euphrates basin about 5,000 years ago. Mechanized processes replaced the old hand-moulding methods during the 19th century.

bridge In civil engineering, structure providing a continuous passage over a body of water, roadway or valley. Bridges are built for people, vehicles, pipelines or power transmission lines. Prehistoric in origin, the first bridges were probably merely logs over rivers or chasms. Modern bridges take a great variety of forms including beams, ARCHES, CANTILEVERS, SUSPENSION BRIDGES and CABLE-stayed bridges. They can also be movable or floating pontoons. They can be made from a variety of materials, including brick or stone (for arches), steel and/or concrete.

bridge In dentistry, a partial denture held in place by anchorage to adjacent teeth. Depending on the situation it may be permanently installed or removable; anchorages are of various sorts. Usually, only one or a few teeth in a series are artificially replaced in this manner.

Bridges, Calvin Blackman (1889–1938) US geneticist. He helped to prove the chromosomal basis of HEREDITY and sex. His work with Thomas Hunt MORGAN on the fruit fly (*Drosophila*) proved that inheritable variations could be traced to observable changes in the CHROMOSOMES. These experiments resulted in the construction of "gene maps".

Bridgman, Percy Williams (1882–1961) US physicist. From 1919 he was a professor at Harvard University. He was awarded the 1946 Nobel Prize for physics for his work in high-pressure physics. He is known also for his work on crystal properties and for his writings on the philosophy of science. His books include *The Nature of Modern Physics* (1927); his collected papers were published in 1964.

Bright, Sir Charles Tilston (1832–88) British engineer who helped to develop submarine telegraph cables. In 1856 he became engineer-in-chief of the Atlantic Telegraph Co. and in 1858 supervised the laying of the first transatlantic cable from Ireland to Newfoundland. With the engineer Joseph Clark he also invented an asphalt covering for submarine cables.

Bright, Richard (1789–1858) British physician who was the first to describe the kidney disorder (NEPHRITIS) that used to be known as Bright's disease. He graduated in Edinburgh in 1813 and became a physician at Guy's Hospital, London, in 1820.

Britannia metal Tin alloy used for tableware, similar to pewter but containing *c*.7% antimony, 2% copper but no lead. It is a lustrous, hard, malleable alloy that is easy to cast.

British thermal unit (BTU) Amount of heat energy necessary to raise the temperature of 1lb of water from 59.5°F to 60.5°F. One BTU is the equivalent of 778.3 foot-pounds, 252 CALORIES or 1,055 JOULES. It has been largely superseded by the metric unit of energy, the joule.

broad Small, shallow lake that occurs in East Anglia, England. Many of the broads were formed in hollows made by people digging out peat in the Middle Ages. The Norfolk Broads are linked by a network of rivers and have proved popular with holiday-makers, but leisure use has caused pollution and silting up. They have now been given National Park status.

broadcasting Transmission of sound or images by radio waves to a widely dispersed audience through RADIO or TELEVISION receivers. The first US commercial radio company, KDKA, began broadcasting in Pittsburgh in 1920. By 1992 there were over 6,000 FM stations operating in the USA. In the UK, the British Broadcasting Company began radio transmission in 1922 and, as the British Broadcasting Corporation (BBC), was incorporated as a public body in 1927. Until 1973 it enjoyed a monopoly in radio broadcasting. There are five national BBC radio stations. Television is a huge market in the USA. By the early 1990s, there were 1,100 commercial television stations and 215 million TV sets. The size of this market has led to an expansion in choice via alternative technologies. In 1962 *Telstar* delivered the first transatlantic, satellite television broadcast. By the 1990s, 60% of all US households had CABLE TELEVISION. Cable systems can deliver more than 100 channels. UK public television broadcasting began in 1936 from Alexandra Palace, London. The BBC transmitted on one channel, using 405 lines to build up an image. A second channel, *ITV*, run by Independent Television (formerly the Independent Broadcasting Authority (IBA)), was set up in 1955. *BBC2* started broadcasting in 1964, and *Channel 4* commenced transmission in 1982. *BBC1* and *BBC2* do not carry commercial advertising, and receive their funding from a licence fee. All four terrestrial channels may be received on sets which use 625 lines. Rupert Murdoch's *Sky Television* satellite service began broadcasting in 1989, since when cable television as well as SATELLITE TELEVISION has become increasingly popular. Further developments included a fifth channel (1997) and the introduction of digital television.

Broglie, Louis Victor de (1892–87) French physicist who put forward the theory that all ELEMENTARY PARTICLES have an associated wave. He devised the formula that predicts this wavelength, and its existence was proven in 1927 by others. De Broglie developed this form of QUANTUM MECHANICS, called WAVE MECHANICS, for which he was awarded the 1929 Nobel Prize for physics. Erwin SCHRÖDINGER advanced de Broglie's ideas with his equation that describes the wave function of a particle.

BRIDGE

The simplest form of bridge is a solid beam spanning a gap. The weight of a load crossing the bridge is resisted by bending in the material. The top of the beam will be in compression and the bottom held in tension. For a given weight of metal or reinforced concrete used to build a bridge, it is better to make a deep I-shaped beam or box than have a solid structure. Stone and concrete resist compression much better than tension and are used for arches, which transfer the weight of the load to the piers. Suspension bridges support a steel or concrete deck from continuous steel cables anchored at the bank. Cantilever (and cable stay bridges) balance forces on either side of a fulcrum or pier and can be built out from a central support. Short sections of suspended beam bridges built from a network of girders (making a box) are used to join the cantilevers.

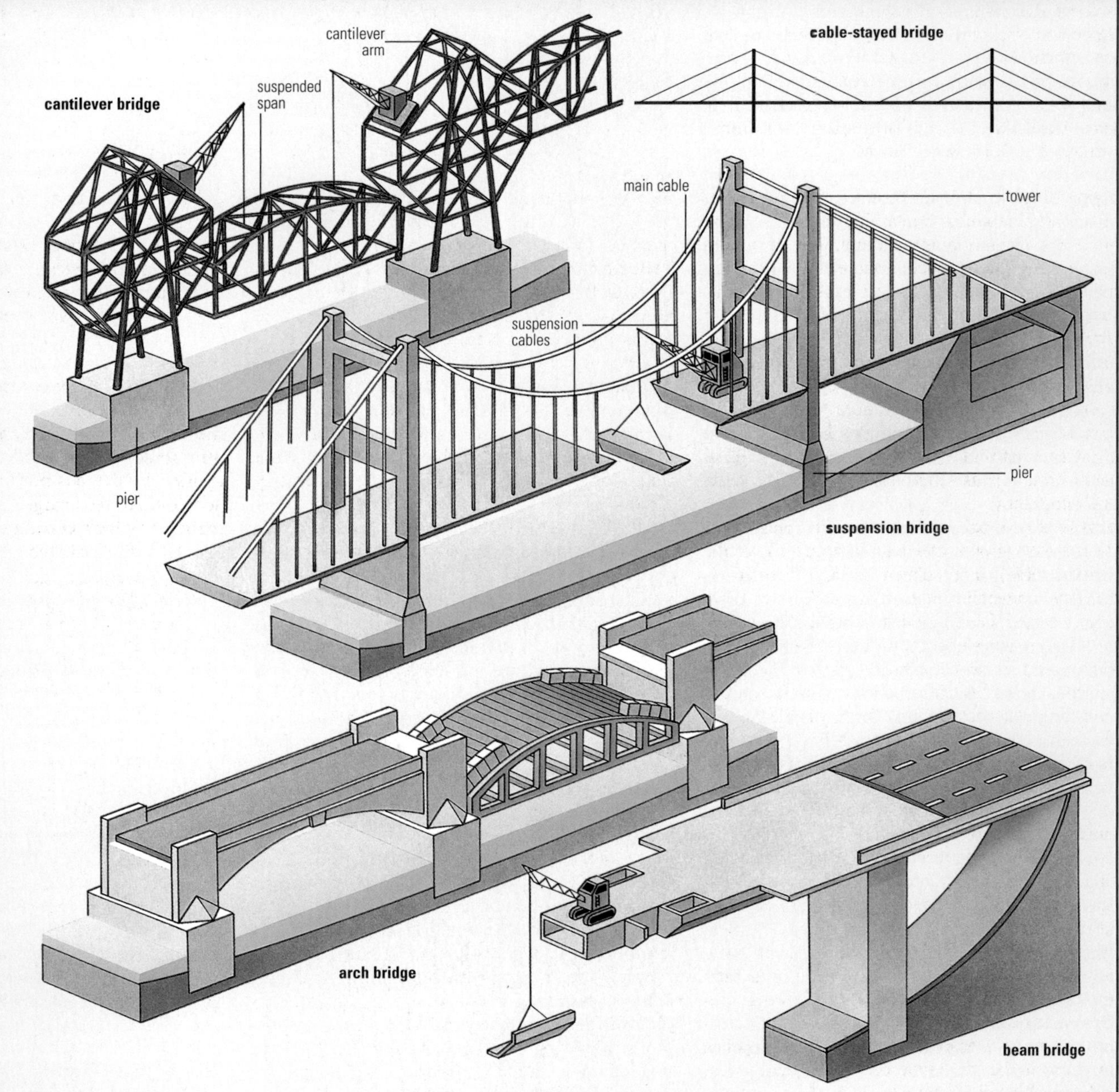

▲ **Brunel** Isambard Kingdom Brunel showed an early talent for drawing and geometry. In 1825 he helped his father, Marc Isambard Brunel, to construct the first tunnel under the Thames in London, now used by London Underground. In 1829 he won the competition for a design for the Clifton Suspension Bridge at Bristol, his first independent work. In 1833 he was appointed engineer of the Great Western Railway. He later designed huge steam ships capable of crossing the Atlantic.

bromide Salt of hydrobromic acid or certain organic compounds containing BROMINE. The bromides of ammonium, sodium, potassium and certain other metals were once extensively used medically as sedatives. Silver bromide is light-sensitive and is used in photography.

bromination Introduction of a bromine atom into a chemical compound, either by reaction with an unsaturated molecule or by replacing hydrogen, as in the preparation of bromobenzene from BENZENE. Organic bromides are widely used as intermediates in dye and drug syntheses, alkali metal bromides as sedatives, and silver bromide in photography.

bromine (symbol Br) Volatile, liquid element of the HALOGEN group (elements in group VII of the periodic table), first isolated in 1826. Bromine is the only nonmetallic element that is liquid at room temperature. It is extracted from soluble bromides, especially magnesium bromide, found in sea-water. The bromine is displaced by the more active halogen chlorine and is extracted as a by-product from the saline concentrates left after crystallization of the main salt products. A reddish-brown, fuming liquid having an unpleasant odour, bromine is used in commercially useful compounds, such as those used to manufacture photographic film and anti-knock additives for petrol. Chemically, it resembles CHLORINE but is less reactive. Properties: at.no. 35; r.a.m 79.904; r.d. 3.12; m.p. −7.2°C; (19.04°F); b.p. 58.8°C (137.8°F); most common isotope ^{79}Br (50.54%).

bronchiectasis Enlargement of one or more bronchial tubes usually resulting from bronchitis or tuberculosis. Symptoms are coughing and copious expectoration.

bronchus (pl. bronchi) One of two branches into which the TRACHEA or windpipe divides, with one branch leading to each of the LUNGS. The bronchus divides into smaller and smaller branches, called **bronchioles**, which extend throughout the lung, opening into the air sacs or ALVEOLI. The bronchi are supported and kept open by rings of CARTILAGE.

Bronowski, Jacob (1908–75) British biologist and man of letters. He lectured at many US universities and was made a fellow of Jesus College, Cambridge, in 1967. Among his books are *Science and Human Values* (1958), *Nature and Knowledge* (1969) and three studies of William Blake. His celebrated BBC television series on man's intellectual history was published as *The Ascent of Man* (1973).

bronze Traditionally an ALLOY of copper and no more than 33% tin. It is hard and resistant to corrosion, but easy to work. It has long been used in sculpture and in bell-casting. Other metals are often added for specific properties and uses, such as aluminium in aircraft parts and tubing, silicon in marine hardware and chemical equipment, and phosphorus in springs, gunmetal and electrical parts.

brown coal *See* LIGNITE

brown dwarf Star with mass less than 0.08 solar masses, in which the core temperature does not rise high enough to initiate thermonuclear reactions. Such a star is, however, luminous, for as it slowly shrinks in size it radiates away its gravitational energy. The surface temperature of a brown dwarf is below the 2,500 K lower limit of a RED DWARF.

Brownian movement Random, zigzag movement of particles suspended in a FLUID (liquid or gas). It is caused by the unequal bombardment of the larger particles, from different sides, by the smaller molecules of the fluid. The movement is named after the Scottish botanist Robert Brown (1773–1858) who in 1827 observed the movement of plant spores floating in water.

brucite Mineral form of magnesium hydroxide ($Mg_3(OH)_6$), derived from periclase. The crystals are either glassy or waxy, can be white, pale green, grey or blue, and occur in fibrous masses or plate aggregates. It is found in SERPENTINE, SCHISTS and other METAMORPHIC ROCKS. It is used to make magnesia refractory materials.

bruise (contusion) Area of discolouration on the SKIN due to escape of blood from damaged underlying vessels. It is caused by injury, usually a blow.

Brunel, Isambard Kingdom (1806–59) British marine and railway engineer. A man of remarkable foresight, imagination and daring, he revolutionized engineering in Britain. In 1829 he designed the Clifton Suspension Bridge (completed 1864) and in 1833 he became chief engineer for the Great Western Railway, for whom he designed the Royal Albert Bridge across the River Tamar. He is also famous for his ships: *Great Western* (designed 1837), the first trans-Atlantic wooden steamship, *Great Britain* (1843), the first iron-hulled screw-driven steamship, and *Great Eastern* (1858), a steamship powered by screws and paddles which was the largest vessel of its time. His father was Sir Marc Isambard BRUNEL.

Brunel, Sir Marc Isambard (1769–1849) Architect and engineer. A refugee from the French Revolution, he went to the USA in 1793. He was chief engineer of New York, before moving to England in 1799, where he invented machinery for making ships' pulley blocks. He was responsible for the construction of the first tunnel under the River Thames (1825–43). He is the father of Isambard Kingdom BRUNEL.

bryophyte Any member of the phylum Bryophyta – small, green, rootless, non-VASCULAR PLANTS, including MOSSES and liverworts. Bryophytes grow on damp surfaces exposed to light, including rocks and tree bark, almost everywhere from the Arctic to the Antarctic. There are about 24,000 species. *See also* ALTERNATION OF GENERATIONS

bryozoan (moss animal) Any member of the phylum Ectoprocta – small invertebrate, mainly marine, animals. Growing to about 1mm (0.05in) in length, they live in colonies attached to rocks, seaweed or large shells.

BSE Abbreviation for BOVINE SPONGIFORM ENCEPHALOPATHY

bubble chamber Device for detecting and identifying ELEMENTARY PARTICLES. It consists of a sealed chamber filled with a liquefied gas, usually liquid hydrogen, kept just below its boiling point by high pressure in the chamber. When the pressure is released, the boiling point is lowered and a charged particle passing through the superheated liquid leaves a trail of tiny gas bubbles that can be illuminated and photographed before the pressure is restored. If a magnetic field is applied to the chamber, the tracks are curved according to the charge, mass and velocity of the particles, which can thus be identified. Donald GLASER received the 1960 Nobel Prize for

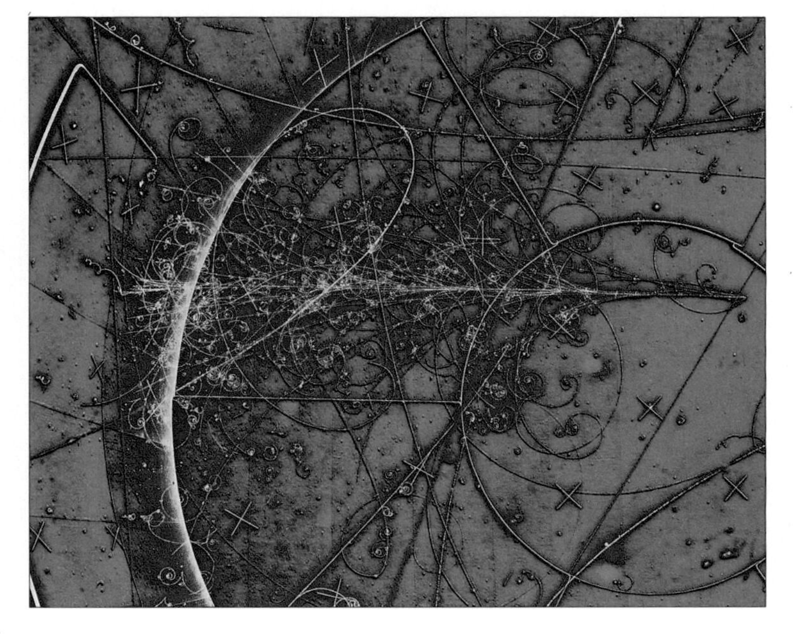

◀ **bubble chamber** Coloured image showing a collection of tracks left by subatomic particles in a bubble chamber. A charged particle leaves behind a trail of tiny bubbles as the liquid hydrogen boils in its wake. The tracks are curved due to an intense applied magnetic field. The tightly wound spiral tracks are due to electrons and positrons.

physics for inventing the bubble chamber, and it was developed by Luis ALVAREZ.

Büchner, Eduard (1860–1917) German biochemist. In 1896 Büchner discovered that it was not the actual yeast cells, but rather the ENZYMES contained within them, that actively caused the alcoholic FERMENTATION of sugar. In 1903 he isolated zymase, the enzyme mixture responsible for fermentation. He was awarded the 1903 Nobel Prize for chemistry.

Buckminsterfullerene (buckyball) Allotrope of CARBON that consists of many carbon atoms bonded together in the shape of a hollow sphere. The simplest Buckminsterfullerene has 60 carbon atoms arranged as 12 regular pentagons and 20 hexagons (like the panels on a modern football). It can be made by exposing graphite to a laser beam or electric arc in an inert atmosphere. It is a yellow crystalline solid that dissolves in benzene. It gets its name from the US architect Richard Buckminster FULLER, who invented the GEODESIC DOME structure that the substance's molecules resemble. Derivatives containing metal atoms trapped inside the carbon-atom sphere are being assessed as semiconductors. Stable polymers have also been produced. *See also* ALLOTROPY

buckyball *See* BUCKMINSTERFULLERENE

bud In botany, a small swelling or projection consisting of a short stem with overlapping, immature leaves covered by scales. Leaf buds develop into leafy twigs, and flower buds develop into blossoms. A bud at the tip of a twig is a **terminal** bud and contains the growing point; **lateral** buds develop in leaf AXILS along a twig.

budding Method of ASEXUAL REPRODUCTION that produces a new organism from an outgrowth of the parent. Hydras (small, freshwater polyps), for example, often bud in spring and summer. A small bulge appears on the parent and grows until it breaks away as a new individual.

buffer solution Solution to which a moderate quantity of a strong acid or a strong base can be added without making a significant change to its pH value (acidity or alkalinity). Buffer solutions usually consist of either a mixture of a weak acid and one of its salts, a mixture of an acid salt and its normal salt or a mixture of two acid salts.

bulb In botany, a food storage organ consisting of a short stem and swollen scale leaves. Food is stored in the scales, which are either layered in a series of rings, as in the onion, or loosely attached to the stem, as in some lilies. Small buds between the scale leaves give rise to new shoots each year. New bulbs are produced in the AXILS of the outer scale leaves. *See also* ASEXUAL REPRODUCTION

bulbil Small BULB-like structure that grows on a plant. Bulbils may develop above the ground at the base of a plant or on stems from an axillary BUD, or they form instead of flowers (as with some onions). Bulbils that become detached and fall to the ground can develop into new plants, and so function as a method of ASEXUAL REPRODUCTION.

bulk modulus Physical constant of solids and fluids that indicates their elastic properties when they are under pressure over their entire surfaces. The bulk modulus is also called the **incompressibility:** if a solid or a fluid (liquid or gas) has a high bulk modulus, then it is difficult to compress.

Bullard, Sir Edward Crisp (1907–80) English geophysicist. He is noted for his work in GEOMAGNETISM, especially his theory of the geomagnetic dynamo, based on convective motion in the Earth's core. He was an early supporter of the

BURGESS SHALE

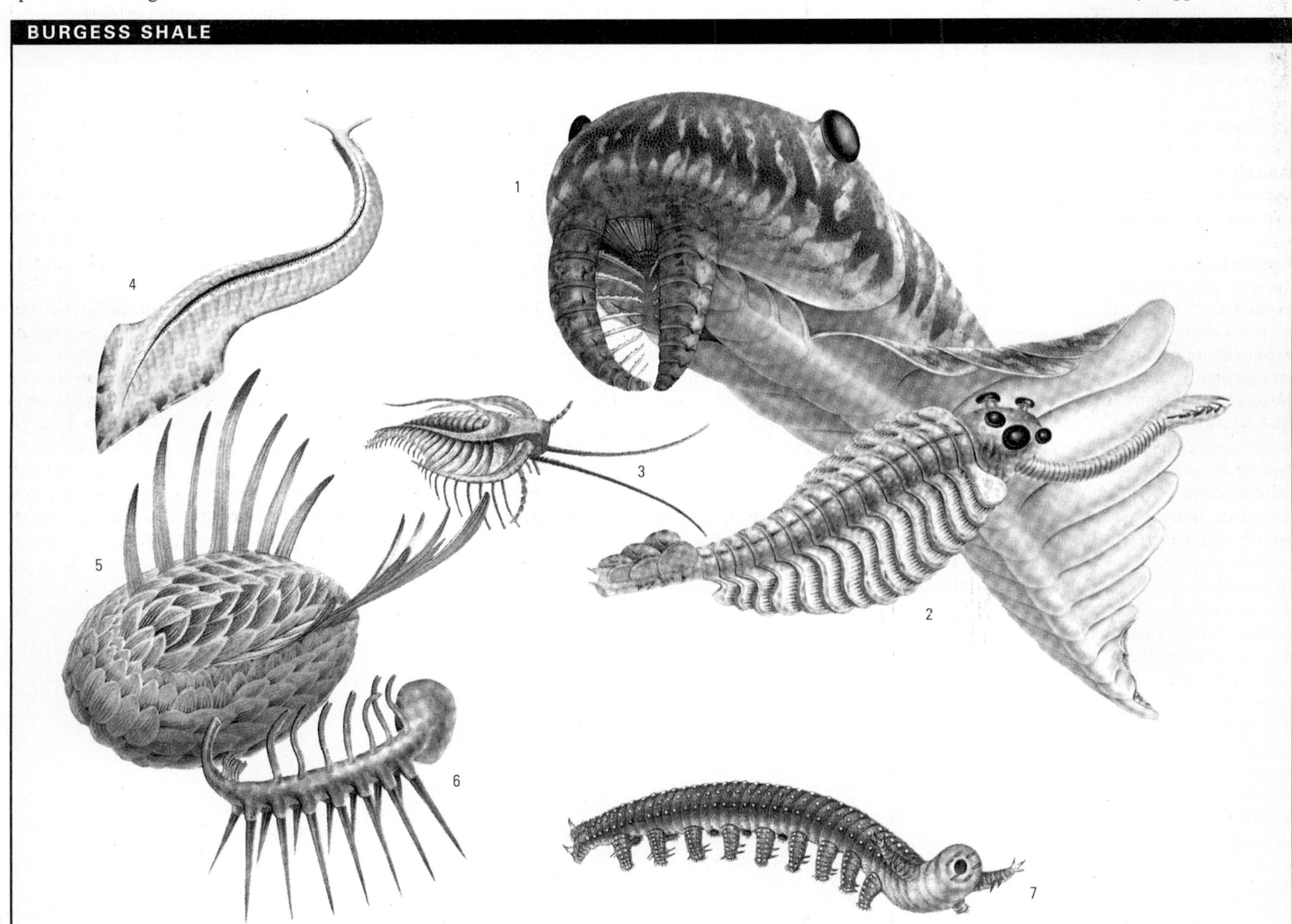

The Burgess shale organisms provide a fascinating glimpse of marine life 570 million years ago. These creatures lived during the Cambrian explosion, a period of intense evolutionary diversification when the ancestors of probably all the modern animal groups we know today came about. The majority (perhaps 90%) of those organisms became extinct, and with them many experiments in animal design. *Anomalocaris* (1) was the largest of the Burgess shale creatures, at over 2ft (0.6m) long, with powerful circular jaws that crunched up trilobites. *Opabinia* (2) was a strange animal with a bizarre vacuum-like frontal nozzle. *Marella* (3), an arthropod, was the first – and most abundant – creature found in the Burgess shale. *Pikaia* (4), a worm-like animal, is significant among all the Burgess organisms because it is the first known chordate. *Wiwaxia* (5) was covered in plates and spines, which presumably acted as protection against predators. If *Hallucigenia* (6) existed as an independent organism, it was probably a bottom dweller, supported by its peculiar struts and feeding with its many tentacles. It has been suggested, though, that it may simply be part of a larger, undiscovered creature. *Aysheaia* (7) was probably a parasite, living and feeding on ancient sponges on the seabed.

theory of PLATE TECTONICS, producing a computer-generated map matching the continental shelves of Africa and South America.

Bunsen, Robert Wilhelm (1811–99) German chemist. Bunsen did important work with organo-arsenic compounds, and discovered an arsenic poisoning antidote. He later evolved a method of gas ANALYSIS. With his assistant, Gustav KIRCHHOFF, Bunsen used SPECTROSCOPY to discover two new elements (caesium and rubidium). He invented various kinds of laboratory equipment, such as the Bunsen cell, a carbon-zinc electric cell that was used in arc lamps. Bunsen also improved a gas burner that was later named after him.

Bunsen burner Gas burner widely used in science laboratories. It is named after the German chemist Robert BUNSEN. The burner is a 5-in (13-cm) upright tube, usually of brass, attached to a gas source. It has a variable air inlet at its base to control the intensity of its flame.

buoyancy Upward pressure exerted on an object by the FLUID in which it is immersed. The object is subjected to pressure from all sides, but the pressure on its lower part is greater because of the increasing depth of the fluid. The result of all these pressures is a force acting upwards that is equal to the weight of the fluid displaced. *See also* ARCHIMEDES' PRINCIPLE

Burgess shale Layer of siltstone in a quarry in Yoho National Park in E British Columbia, Canada. Discovered in 1909 by a US scientist, Charles D. Walcott, it contains a large number of fossils of animals that lived in the CAMBRIAN period, more than 500 million years ago. The fossils and the silt that covered them accumulated at the foot of an undersea cliff. The silt has preserved traces of many of the soft bodies of sea creatures, which normally do not survive. The fossils include the oldest known chordate, a forerunner of all animals with backbones. They also include a number of kinds of animals that have completely vanished, and apparently do not belong to any of the 32 or so phyla (major groups) of animals we know today.

burglar alarm Device that gives warning of an actual or attempted illegal entry to premises. In a typical system electrified metal tape is placed at all doors and other entrances to a building. The tape is in circuit with an electromechanical relay, which trips and so energizes an alarm if any tape is disturbed or broken. Other alarm systems employ photoelectric cells (magic eyes), ultrasonic sensors or laser beams.

burnout, rocket Point on a rocket's trajectory at which the fuel supply is exhausted or cut off. In solid fuel rockets, controlled burnout is achieved by lowering the pressure in the combustion chamber or by opening vents to drop the temperature below the combustion point. In liquid fuel rockets the fuel supply is simply turned off.

Bush, Vannevar (1890–1974) US electrical engineer. From 1932–38 he was vice president and dean of engineering at Massachusetts Institute of Technology. He headed the World War 2 Office of Scientific Research and Development, overseeing the initial stages of the atom bomb project. From 1938–55 he was president of the Carnegie Institute of Washington.

bushel Unit of dry measure equal in volume to eight imperial gallons of water (36.5l), and to four pecks. It is traditionally used to measure grain and fruit. The name comes from a word in Old French for a much smaller measure, "the amount one can hold in the hand".

butadiene (buta-1,3-diene, $CH_2{=}CHCH{=}CH_2$) Flammable hydrocarbon made from butenes or by "cracking" NAPHTHA. It is copolymerized with STYRENE to produce synthetic rubbers. Properties: r.d. 0.62; m.p. −108.9°C (−164.0°F); b.p. −4.4°C (24.1°F).

butane (C_4H_{10}) Colourless, flammable gas, the fourth member of the ALKANE series of HYDROCARBONS. It has two ISOMERS: n-butane is obtained from natural gas; isobutane (2-methylpropane) is a by-product of PETROLEUM refining. Butane can be liquefied under pressure at normal temperatures and is used in the manufacture of fuel gas and synthetic rubber. Properties: b.p. (n-butane) −0.3°C (31.5°F) and (isobutane) −10.3°C (13.46°F).

Butenandt, Adolf Friedrich Johann (1903–95) German biochemist who did important research into sex hormones. In 1931 Butenandt isolated the male sex hormone ANDROSTERONE, and investigated (1934) the chemical structure and properties of the female sex hormone PROGESTERONE. For this work he was awarded the 1939 Nobel Prize for chemistry, but had to decline the award because of a Nazi decree prohibiting acceptance.

BULB

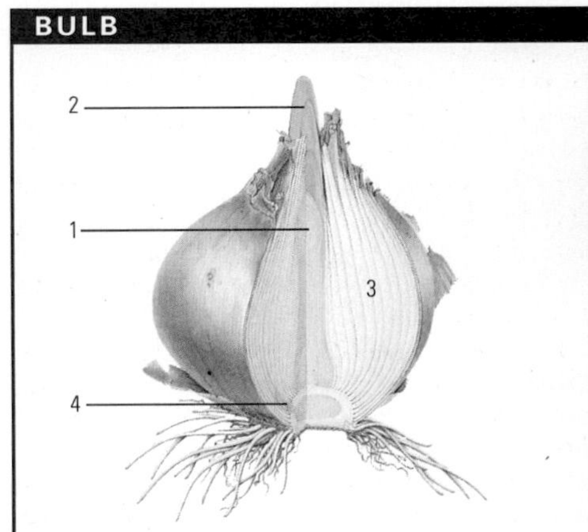

As well as serving as underground storage organs, bulbs may also provide flowering plants with a means of vegetative reproduction. In spring, the flower bud (1) and young foliage leaves (2) will develop into a flowering plant using the food and water stored in the bulb's fleshy scale leaves (3). When the flower has died, the leaves live on, and continue to make food which is transported downwards to the leaf bases. These swell and develop into new bulbs. Axillary buds (4) may develop into daughter bulbs which break off to form new plants.

butene (butylene) Any of three hydrocarbon gases having the general formula C_4H_8: butene-1 ($CH_3CH_2CH{=}CH_2$), butene-2 ($CH_3CH{=}CHCH_3$) and isobutene ($[CH_3]_2C{=}CH_2$) or 2-methylpropene). Butene is made from petroleum and used as a starting material for other organic compounds.

butte Isolated, flat-topped, steep-sided hill. It is formed when a remnant of hard rock overlies and protects softer rock underneath from being worn down, while the surrounding areas continue to be eroded.

byte Binary number used to represent the letters, numbers and other characters in a computer system. Each byte consists of the same number of BITS. Byte is a contraction of the words "by eight", and originally meant an eight-bit byte, such as 01101010 (representing j on most systems). Many computers now use 16-, 32-, or 64-bit bytes. *See also* GIGABYTE; MEGABYTE

cable Wire for mechanical support, for conducting electricity or for carrying signals. In civil and mechanical engineering, a cable is a rope made of twisted strands of steel wire. They range in size from small BOWDEN CABLES to massive supporting cables on the decks of SUSPENSION BRIDGES. In electrical engineering, a cable is a CONDUCTOR consisting of one or more insulated wires, which may be either single or multistranded. They also range greatly in size, from the small cables used for domestic wiring to the large, armoured underwater cables. These are used for telephone, radio, television and data signals. In a COAXIAL CABLE one conductor is cylindrical and surrounds the other. Fibre optic cables consist of OPTICAL FIBRES and carry signals in the form of coded pulses of light.

cable car Passenger vehicle for ascending and descending mountains, often used as a tourist facility. A car is slung from a trolley pulley running on a steel cable. It is moved along by a steel traction cable, continuously looped around terminal drums which are turned by an electric motor. In the USA a tram "powered" by a moving cable beneath the tracks is called a cable car.

cable television Generally refers to community ANTENNA television. CATV does not broadcast, but picks up signals at a central antenna and delivers them to individual subscribers via COAXIAL CABLES. Originally designed for areas with poor reception and no local station, cable television, run by private franchise, now serves to increase the variety of local viewing by transmitting channels brought by microwave relay. In the USA, there are currently over 11,000 cable operating systems, serving approximately 60% of all US households. In both the USA and the UK, channels are required to carry a proportion of community access programming. Viewers pay a monthly subscription for basic cable services such as the Home Box Office (HBO). For an additional monthly charge, they can also receive network services, known as Pay TV.

cadmium (symbol Cd) Silver-white, metallic element in group II of the periodic table, first isolated in 1817. Cadmium is found in greenockite (a sulphide) but is mainly obtained as a by-product in the extraction of zinc and lead. Malleable and ductile, its main uses are as a protective electroplated coating, an absorber of neutrons in NUCLEAR REACTORS and in nickel-cadmium BATTERIES. Chemically it resembles zinc. Properties: at.no. 48; r.a.m. 112.4; r.d. 8.65; m.p. 320.9°C (609.6°F); b.p. 765°C (1,409°F); most common isotope ^{114}Cd (28.86%). *See also* ELECTROPLATING

caecum Dilated pouch at the junction of the small and large INTESTINES, terminating in the APPENDIX. It has no known function in humans. In rabbits and horses, the caecum contains large numbers of microorganisms, which help to break down the cellulose cell walls of the plants they eat, thus making more nutrients available to their hosts.

Caelum Constellation of the s sky, adjoining Columba. It has no star above magnitude 4.5. One variable, R Caeli, ranges from magnitude 6.7 to 13.7 with a period of 391 days.

caesium (symbol Cs) Rare, silver-white, metallic element in group I of the periodic table; the most alkaline and electropositive element. Discovered in 1860, caesium is ductile and used commercially in photoelectric cells. The isotope ^{137}Cs is used in cancer treatments. Properties: at.no. 55; r.a.m. 132.9055; r.d. 1.87; m.p.28.4°C (83.1°F); b.p. 678°C (1,252.4°F); most common isotope ^{133}Cs (100%).

caffeine ($C_8H_{10}N_4O_2$) White, bitter ALKALOID that occurs in coffee, tea and other substances such as cocoa and ilex (holly) plants. In beverages it acts as a mild, harmless stimulant and DIURETIC, although an excessive dose can cause insomnia and delirium.

Cainozoic Alternative name for the CENOZOIC era of geological time.

caisson Watertight structure or chamber used in underwater excavation or construction. It allows people to work underwater and makes it easier to remove excavated material. Caissons are frequently used in building bridges, where they keep out water during the construction of a foundation. They may be open, with "walls" that project above the surface of the water, or closed at the top and submerged (**pneumatic** caissons), using air pressure to exclude water.

calamine Pinkish, odourless powder of zinc carbonate and some ferric oxide, dissolved in mineral oils and used in skin lotions to alleviate such disorders as chickenpox, hives and other skin rashes.

calcareous Describing a rock that is rich in calcium carbonate ($CaCO_3$). CHALK is a common calcareous rock, as are other LIMESTONES, although these may also contain many other minerals. Soils that develop on such rocks are also termed calcareous.

calciferol (vitamin D_2) Fat-soluble VITAMIN formed by the action of sunlight on the skin and further converted into compounds necessary for the intestinal uptake of dietary calcium for bone formation.

calcination Process of heating solids to high temperatures (but not to their fusion point) to remove volatile substances, partially to oxidize them, or to render them friable (easily crumbled). Lead, zinc, calcium, copper and iron ores calcine to sintered oxides, which are used as pigments or as intermediates in metal extraction.

calcite Mineral, calcium carbonate ($CaCO_3$). Calcite is a major constituent of CALCAREOUS SEDIMENTARY ROCK, especially LIMESTONE. The crystals are in the hexagonal system and vary in form from tabular (rare) to prismatic or needle-like. Calcite is usually glassy white but may be red, pink or yellow. It reacts with dilute hydrochloric acid. Hardness 3; r.d. 2.7.

calcitonin HORMONE secreted by the THYROID GLAND in mammals. It lowers the concentration of calcium in the blood. It opposes the action of the PARATHYROID HORMONE.

calcium (symbol Ca) Common, silver-white, metallic element of the ALKALINE-EARTH METALS; first isolated in 1808. Calcium occurs in many rocks and minerals, notably limestone and gypsum, and in bones. In the body it helps to regulate the heartbeat and is essential for strong bones and teeth. The metal, which is soft and malleable, has few commercial applications but its compounds are widely used. Chemically it is a reactive element, combining readily with oxygen, nitrogen and other nonmetals. Properties: at.no. 20; r.a.m. 40.08; r.d. 1.55; m.p. 839°C (1,542°F); b.p. 1,484°C (2,703°F); most common isotope ^{40}Ca (96.95%).

calcium bicarbonate Former name of CALCIUM HYDROGENCARBONATE

calcium carbide (calcium acetylide, CaC_2) Chemical made commercially by heating coke and calcium oxide (lime, CaO) in an ELECTRIC FURNACE. It reacts with water to yield ETHYNE (acetylene, C_2H_2), a gas that burns with a luminous flame; for this reason it was once much used in carriage and locomotive lamps. Calcium carbide is also used to manufacture ETHANOIC ACID (acetic acid, CH_3COOH) and ETHANAL (acetaldehyde, CH_3CHO).

calcium carbonate ($CaCO_3$) White compound, insoluble in water, that occurs naturally in such forms as MARBLE, CHALK, LIMESTONE and CALCITE. It also forms the shells of molluscs. Crystals are in the hexagonal system and vary in form from tabular (rare) to prismatic or needle-

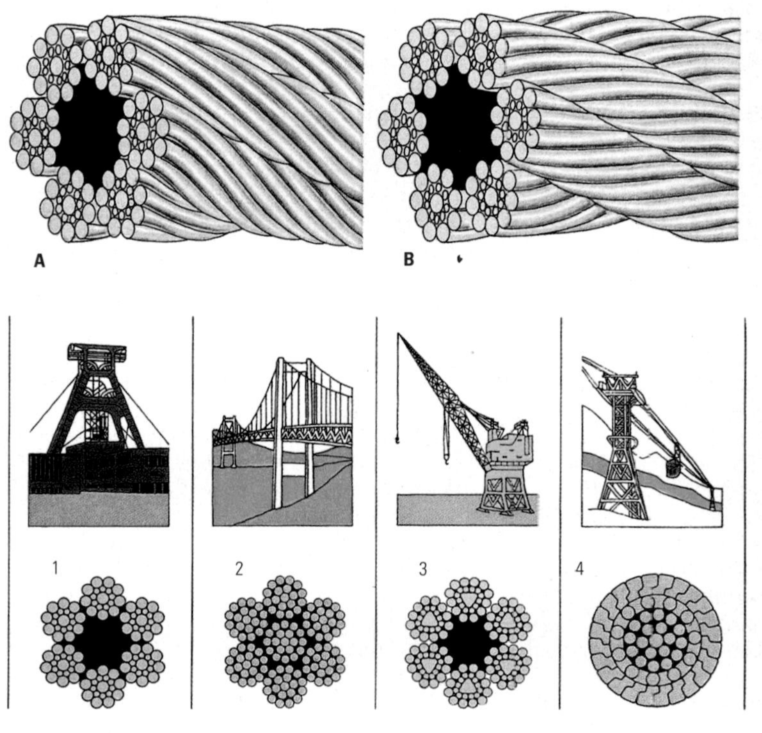

◀ **cable** Wire cables are made from preformed wires that are wound into strands and twisted together. The two common ways of winding wire rope are indicated by the lay of the rope. In lang lay (A) the wires in the strand and the strands themselves are laid in the same direction. Regular lay (B) has the wires in the strand laid in one direction and the completed strands laid perpendicular to the wires.
(C) Cross sections of typical cables:
1 fibre core cable for haulage and driven conveyors
2 wire rope for bridges and structural engineering
3 steel core triangular strand cable for earth-moving machinery
4 locked coil cable for aerial ropeways

CALORIMETER

Calorimeters measure the amount of heat absorbed or let out during a chemical reaction. In a high-pressure flow calorimeter, the apparatus is contained in a vacuum (1) for insulation. A constant flow of liquid or gas enters the calorimeter (2). A platinum resistance thermometer (3) measures the temperature of the substance on entry. A heater (4) puts a known amount of energy into the liquid or gas inside a radiation shield (5) which further lessens any dispersion of energy. The change in temperature is measured by a second thermometer (6) again shielded (7).

like. Calcium carbonate is used in the manufacture of cement, iron, steel and lime, to neutralize soil acidity and as a constituent of antacids.

calcium hydrogencarbonate ($Ca(HCO_3)_2$) Salt that is responsible for temporary HARDNESS OF WATER. The hardness is removed when the water is heated and insoluble calcium carbonate is precipitated from the water, forming a "fur" inside pipes and kettles. This also happens when water drips inside caves and forms STALAGMITES and STALACTITES.

calcium hydroxide (slaked lime, $Ca(OH)_2$) White solid obtained by the action of water on calcium oxide. It is used in mortar, plaster, cements and in water softening and agriculture. *See also* LIME WATER

calcium oxide (quicklime, CaO) White solid made by heating calcium carbonate ($CaCO_3$) at high temperatures. It is used industrially to treat acidic soil and make porcelain and glass, bleaching powder, caustic soda, mortar and cement, and in the recovery of AMMONIA from the ammonia-soda process. Calcium oxide reacts with water to form CALCIUM HYDROXIDE (slaked lime, $Ca(OH)_2$), which dissolves in water to give lime water, used in laboratories as a test for carbon dioxide gas (CO_2). Calcium oxide was used in the 19th century for producing limelight illumination, an incandescence that occurs when coal-gas is burned in air against a block of quicklime.

calcium sulphate ($CaSO_4$) Chemical compound that occurs naturally as the mineral anhydrite. The hydrated form ($CaSO_4.2H_2O$) is the mineral GYPSUM, which loses water when heated to form plaster of Paris (calcium sulphate, $(CaSO_4)_2.H_2O$). This, when mixed with water, quickly sets hard to a crystalline form of the hydrate.

calculus Branch in mathematics involving the techniques of DIFFERENTIATION and INTEGRATION. **Differential calculus** deals with differentiation: the process of finding the instantaneous rate of change of a function at any time. Limiting values of the increment of the function are taken as the increment of one of its variables tends to zero: the ratio of these increments is the DERIVATIVE. Differential calculus is used to find slopes of curves. **Integral calculus** deals with the operation of integration: the finding of a function, one or more derivatives of which are given. To take a simple example, the INTEGRAL of x with respect to x is $^1/_2x^2 + c$, where c is a constant. Integral calculus is used to determine the areas or volumes enclosed by a given boundary. For more than 200 years it was believed that calculus was discovered (independently) by Gottfried LEIBNIZ and Isaac NEWTON. However in 1934 a note written by Newton was discovered in which he acknowledged that his formulation of calculus was based on a method of drawing tangents first developed by Pierre de FERMAT.

caldera Large CRATER formed when a VOLCANO collapses due to the migration of the MAGMA under the Earth's crust. The caldera of an extinct volcano, if fed by floodwater, rain, or springs, can become a CRATER LAKE.

calibration Testing of scientific instruments for the purpose of affixing measuring scales. For example, thermometers are calibrated in degrees Fahrenheit or Celsius, or in kelvins, and pressure gauges are calibrated in Nm^{-2} or lb/sq ft. All calibrations are based on such fundamental standard units as length and temperature.

californium (symbol Cf) Radioactive, metallic element of the ACTINIDE SERIES, first made in 1950 at the University of California, Berkeley, by alpha-particle bombardment of the curium isotope curium-242. Californium presents biological dangers because one microgram releases 170 million neutrons a minute. Properties: at.no. 98; most stable isotope ^{251}Cf (half-life 800 yr). *See also* TRANSURANIC ELEMENTS

Callisto Second-largest of Jupiter's GALILEAN SATELLITES, with a diameter of 4,800km (2,980mi), and the outermost of them. The surface of the satellite is dark. It is the most heavily cratered object known. Since the main period of cratering was early in the Solar System's history, this indicates that no subsequent geological activity has disturbed Callisto's surface. As well as the dense craters, there are several large, multi-ringed impact features, the largest of which is Valhalla, with an overall diameter of 4,000km (2,500mi).

callus In botany, a protective mass of undifferentiated plant cells formed at the site of a wound in a woody plant. Callus tissue is also formed at the base of cuttings as they start to take root. Callus tissue is important as the starting point for TISSUE CULTURE of plants.

caloric theory of heat *See* HEAT

calorie Unit of HEAT. A calorie is the amount of heat required to raise 1g of water 1°C between the temperatures of 14.5 and 15.5°C (58.1 and 59.9°F). The SI system of units uses the JOULE (1 calorie = 4.184 joules) instead of the calorie. A dietitian's "calorie" is the kilocalorie, which is 1,000 times larger than a calorie.

calorimeter Apparatus used for experiments involving heat measurements. It is usually a conducting container, such as a copper vessel, that is thermally insulated. There are many types designed for special purposes, such as measuring calories in a food, specific heat capacities, specific latent heats, heats of reaction and heats of formation. A **bomb** calorimeter ignites sealed oxygen to measure heats of combustion, like calories in food.

Calvin, Melvin (1911–97) US chemist. Calvin conducted experiments using radioactive carbon-14 as a trace to label carbon dioxide and track the process by which plants turned it into glucose by means of PHOTOSYNTHESIS. The series of reactions that take place during photosynthesis are known as the CALVIN CYCLE. Calvin received the Nobel Prize for chemistry in 1961.

Calvin cycle Sequence of reactions that take place during the dark stage of PHOTOSYNTHESIS. The overall effect of the cycle is the REDUCTION of carbon dioxide to form carbohydrate. First, an ENZYME (ribulose biphosphate carboxylase) causes carbon dioxide to combine with ribulose biphosphate to form a six-carbon compound that quickly decomposes to give two molecules of glycerate 3-phosphate (a three-carbon compound). Then, after being changed to glyceraldehyde 3-phosphate, this re-forms ribulose biphosphate with the release of the sugars fructose and glucose. The whole cycle takes place in the CHLOROPLASTS in leaves or other green parts of the plant. It was worked out, using radioactive tracers, by Melvin CALVIN.

calyx SEPALS of a flower, taken as a group. Usually green in colour, the calyx protects the flower when it is a bud. When the flower opens, it surrounds the CARPELS, PETALS and STAMENS, and forms the outer whorl of the PERIANTH.

cam Mechanical device consisting of an eccentric projection on a rotating shaft, shaped so as to give some desired reciprocating linear motion to another component. Cams are used in many different kinds of machinery.

cambium Layer of cells parallel to the surface of stems and roots of plants that divides to produce new cells to allow for growth in diameter of the stem and roots. Once a plant cell has differentiated, it is unable to divide again. To allow for growth, certain bands of cells remain undifferentiated, retaining their ability to divide. The cambium is made up of such cells. There are two main types of cambium, the **vascular** cambium and the **cork** cambium. The vascular cambium produces new PHLOEM on the outside and XYLEM on the inside, leaving narrow bands of thin-walled cells that allow nutrients and gases to diffuse to the centre of the plant. The cork cambium forms a cylinder just below the epidermis, and produces cork cells to replace the epidermis, which ruptures as the stem and root expand, forming the bark and the corky outer layer of the older root. *See also* MERISTEM

Cambrian Earliest period of the PALAEOZOIC era, lasting from *c*.590 million to 505 million years ago. The rocks of this period are the earliest to preserve the hard parts of animals as fossils. Cambrian rocks

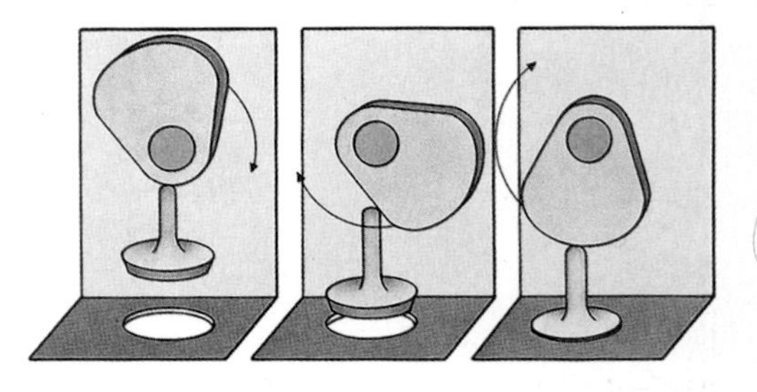

▲ **cam** A disc cam can be regarded as a lever of variable length that changes rotary motion into an up-and-down or side-to-side (reciprocating) motion. Disc cams are often used to operate the valves in a car engine, as above.

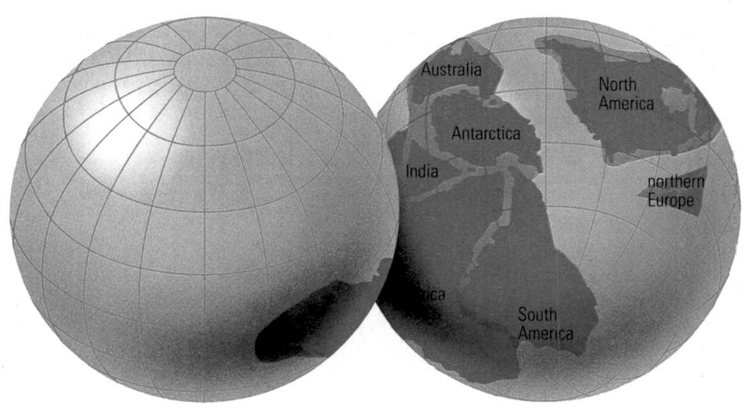

► **Cambrian** The palaeography of the early Palaeozoic era is highly conjectural but the continental plates had formed and were moving in relation to one another. The palaeographic maps only show the relative positions of the tectonic plates; they do not show such features as continental seas.

contain a large variety of fossils, including all the animal phyla with the exception of the vertebrates. At that time the animals lived in the seas, while the land was barren. The commonest animal forms were trilobites, brachiopods, sponges and snails. Plant life consisted mainly of seaweeds.

Camelopardalis (Giraffe) Barren constellation of the N sky. Its brightest star, Beta, is of magnitude 4.0. NGC 1502 is an open cluster visible with binoculars.

camera Apparatus for taking photographs, consisting essentially of a lightproof box containing photographic film. When a shutter is opened, usually briefly, light from the scene outside is focused by a lens system onto the film. The amount of light falling on the film is controlled by the shutter speed and by the diameter of the lens APERTURE, which can also be varied using an adjustable iris diaphragm. Most cameras also have a rangefinder, enabling a focused image to be produced for a given object distance, and a built-in exposure meter to determine the correct combination of shutter speed and aperture for the prevailing light conditions. Both of these functions are automated in a camera with a computerized program (to set the correct combination of shutter speed and aperture) and an autofocus lens system. Recently digital cameras have been developed, in which a CHARGE-COUPLED DEVICE (CCD) "records" the photograph; which can then be downloaded onto a computer and a print made.

camera lucida Apparatus for drawing and copying in perspective, developed in 1812 by William WOLLASTON. A prism is set between the draughtsman's eye and the paper in such a way that light is reflected from the object he is copying to form an image on the paper.

camera obscura Optical device consisting of a darkened room into which an inverted image is thrown through a convex lens. A portable version of this was used by 17th- and 18th-century artists to trace scenes from nature. By the 19th century it had become a box, fitted with a lens and mirror and placed on a tripod. It eventually developed into the modern CAMERA.

camouflage In zoology, means by which some animals blend in with their natural environment, usually by resembling their surroundings in colour. Camouflage enables animals to conceal themselves from their prey or to protect themselves from predators. The leaf insect is a notable example. It is flat and green and the female has large leathery forewings with markings that resemble a pattern of leaf veins.

camphor ($C_{10}H_{16}O$) Organic compound, the molecule of which has a complex ring structure. It has a strong odour, which also occurs in its original source, the wood and leaves of the camphor tree, *Cinammonum camphora*, native to Taiwan. Camphor is used in medicine for liniments, in the manufacture of celluloid, lacquers and explosives, and as an ingredient of mothballs.

canal Artificial waterway for irrigation, drainage, navigation or in conjunction with hydroelectric DAMS. The first canals were built 4,000 years ago in ancient Mesopotamia. The longest canal able to accommodate large ships today connects the Baltic and White seas in N Europe and is 227km (141mi) long. The heyday of canal building in England was from the late 18th–early 19th century.

Cancer (Crab) Constellation of the N sky, between Gemini and Leo. It contains two open clusters: M44, the Praesepe or Beehive Nebula (NGC 2632), and M67 (NGC 2692). The brightest star is Beta Cancri, magnitude 3.5.

cancer Group of diseases featuring the uncontrolled proliferation of cells (tumour formation). Malignant (cancerous) cells spread (metastasize) from their original sites to other parts of the body. There are many different cancers. Known causative agents (carcinogens) include smoking, certain industrial chemicals, asbestos dust and radioactivity. Viruses are implicated in the causation of some cancers. Some people have a genetic predisposition towards particular types of cancer. Treatments include surgery, chemotherapy with cell-destroying drugs and radiotherapy (or sometimes a combination of all three). Early diagnosis holds out the best chance of successful treatment.

Cancer, Tropic of Line of latitude, *c*.23.5° N of the Equator, which marks the N boundary of the tropics. It indicates the farthest N position at which the Sun appears directly overhead at noon. The Sun is vertical over the Tropic of Cancer on about 21 June, the summer SOLSTICE in the Northern Hemisphere.

candela (symbol cd) SI unit of luminous intensity. It is defined as the luminous intensity of a source of monochromatic light of frequency 540×10^{12} Hz of radiant intensity 1/683 watt per steradian in a given direction.

Canes Venatici (Hunting Dogs) Constellation

CAMERA

silicon dioxide
pixel
electrode
electron
silicon atom
light photons from subject
masking strip

The digital camera takes pictures like a conventional camera (1) but stores the images on an internal memory, or a PCMCIA card (2) so that they can be quickly downloaded to a computer (3), at which point the image can be manipulated and/or printed. Instead of film, the camera uses a (charge-coupled device) CCD chip. Light striking silicon atoms on the surface of the chip knocks off electrons (4), which are attracted to the positive electrode (5). Periodically the levels of charge in the electrodes are measured in vertical columns (6) and horizontal rows (7) to build up a picture (and clear the pixels of residual charge) (8).

CANCER

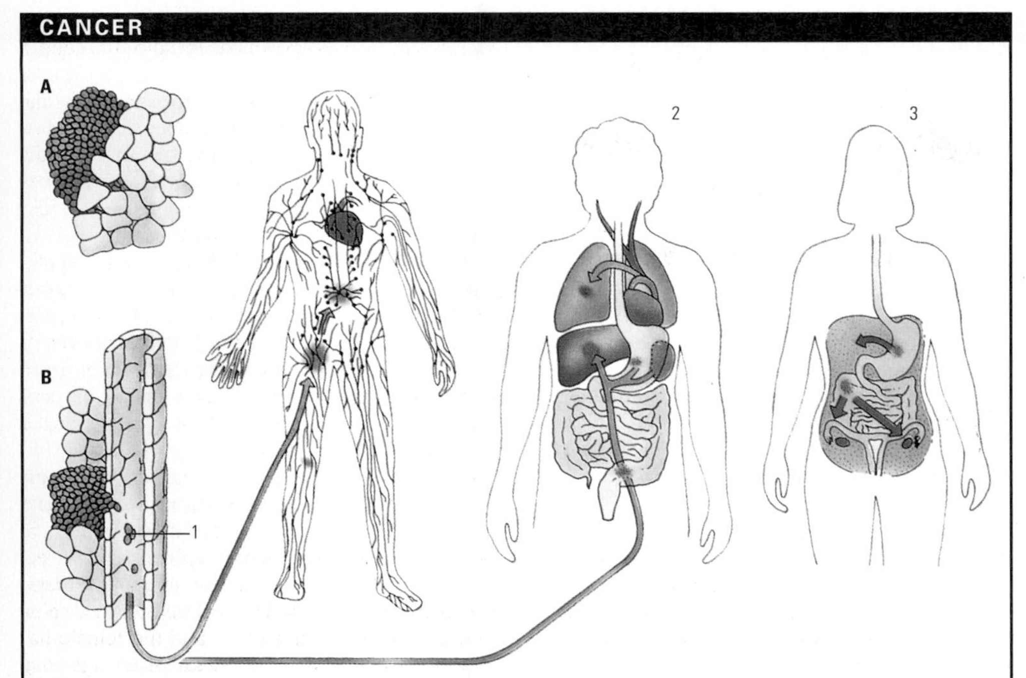

Cancer can spread in two ways. First, by direct growth into adjacent tissues, called "direct extension" (A), when cancer cells penetrate into bone, soft connective tissue and the walls of veins and lymphatic vessels. Alternatively, a cancer cell separates from its tumour and is transported to another part of the body. This spread of cancer is called "metastasis" (B). In metastasis after the tumour has grown to some size, cancer cells or small groups of cells enter a blood or lymph vessel through the vessel wall (1). They travel through the vessel until they are stopped by a barrier, such as a lymph node, where additional tumours may develop, before releasing more cells which may develop on other lymph nodes. Such cancers, usually carcinomas, may also invade the blood stream and establish more distant secondary growths. Another type of cancer, sarcomas, tend to spread via venous blood vessels frequently establishing tumours in the lungs, gastrointestinal tract or the genito-urinary tract (2). In abdominal cancers, metastases may also arise as a result of travel across body cavities, such as the peritoneal, oral or pleural cavities (3).

of the N sky. The only star above magnitude 4.0 is Cor Caroli, magnitude 2.9. It does contain several galaxies including the famous Whirlpool Galaxy, M51.

canine Any one of four sharp "stabbing" TEETH in the frontal dentition of most mammals. In humans they are also called eye teeth.

Canis Major (Great Dog) Constellation of the S sky situated S of Monoceros. It contains the bright open cluster M41 (NGC 2287). The brightest star is Alpha Canis Majoris or SIRIUS (Dog Star), the brightest star in the sky, with a magnitude of –1.47

Canis Minor (Little Dog) Constellation of the N sky. It is notable only for containing the eighth-brightest star in the sky, Procyon, which has a magnitude of 0.38.

canning Method of FOOD PRESERVATION by sealing it in cans. Cans are made from tin-plated steel sheet, or from aluminium. Any bacteria present in the food are killed, either by dry heat or by steam heat. A temperature of about 100°C (212°F) is necessary for fruit, and somewhat higher temperatures for meat and vegetables. Once a can is sealed it must remain airtight to prevent infection by bacteria, particularly those that cause FOOD POISONING. *See illustration on page 58*

Cannon, Walter B. (Bradford) (1871–1945) US physiologist. He studied the regulation of hunger and thirst in animals, introduced the term HOMEOSTASIS to describe the ability of an organism to maintain its internal environment, and proposed a theory that placed the seat of emotions in the THALAMUS of the brain.

cannon ARTILLERY piece consisting of a metal tube, used to aim and fire missiles propelled by the explosion of gunpowder in the closed end of the cylinder. Cannon, first used in the 14th century, were originally made of bronze or iron. Major improvements in the 19th century included the introduction of steel barrels, more powerful chemical propellants, breech-loading mechanisms and standardized parts. By World War 1 recoil devices were employed to absorb the shock of firing.

◀ **canyon** The N rim of the Grand Canyon, Arizona, USA, photographed just after sunrise. Carved by the Colorado River, the Grand Canyon is 450km (280mi) long and varies from 6km (4mi) to 18km (11mi) in width. Its magnificent, multicoloured rock formations reveal hundreds of millions of years of geological history. The Grand Canyon is considered one of the great natural wonders of the world.

Canopus (Alpha Carinae) Second-brightest star observed from the Earth; it has a magnitude of –0.72. Its luminosity and distance are not accurately known, but one estimate classifies it as a bright giant, 800 times as luminous as the Sun, and 74 light-years away.

cantilever Either a projecting beam that is rigidly supported at one end with force applied at the free end (such as a diving board), or a BRIDGE supported by two projecting beams, joined in the centre by a connecting member and supported on piers and anchored by counterbalancing members.

Cantor, Georg (1845–1918) German mathematician, b. Russia. Cantor was professor of mathematics (1869–1913) at the University of Halle, Germany. His work on the concept of INFINITY challenged the existing deductive processes of mathematics. Cantor developed SET THEORY and provided a new definition of irrational numbers (*see* NUMBER, IRRATIONAL). **Cantor's paradox** arises with the attempt to construct a universal set that includes all other sets because every set has more subsets than members and yet the universal set should contain the subsets as well.

canyon Deep, narrow depression in the Earth's crust. Land canyons are the result of erosion by rivers of comparatively recent origin flowing through arid terrain. Marine canyons may be formed when a river-bed and the surrounding terrain is submerged, or by turbulence produced by deep-water currents.

capacitance (symbol *C*) Ability of an ELECTRIC CIRCUIT or component to store charge. Capacitance is measured in farads (1 farad is a capacitance needing a charge of 1 coulomb to raise its potential by 1 volt), but most capacitances are small enough to be measured in microfarads (one millionth of a farad) or picofarads (one million-millionth of a farad).

capacitor (condenser) ELECTRIC CIRCUIT component that has CAPACITANCE. Having at least two metal plates, it is used principally in alternating-current (AC) circuits. The various types include parallel-plate condensers and electrolytic capacitors. *See also* ELECTRIC CURRENT

CANNING

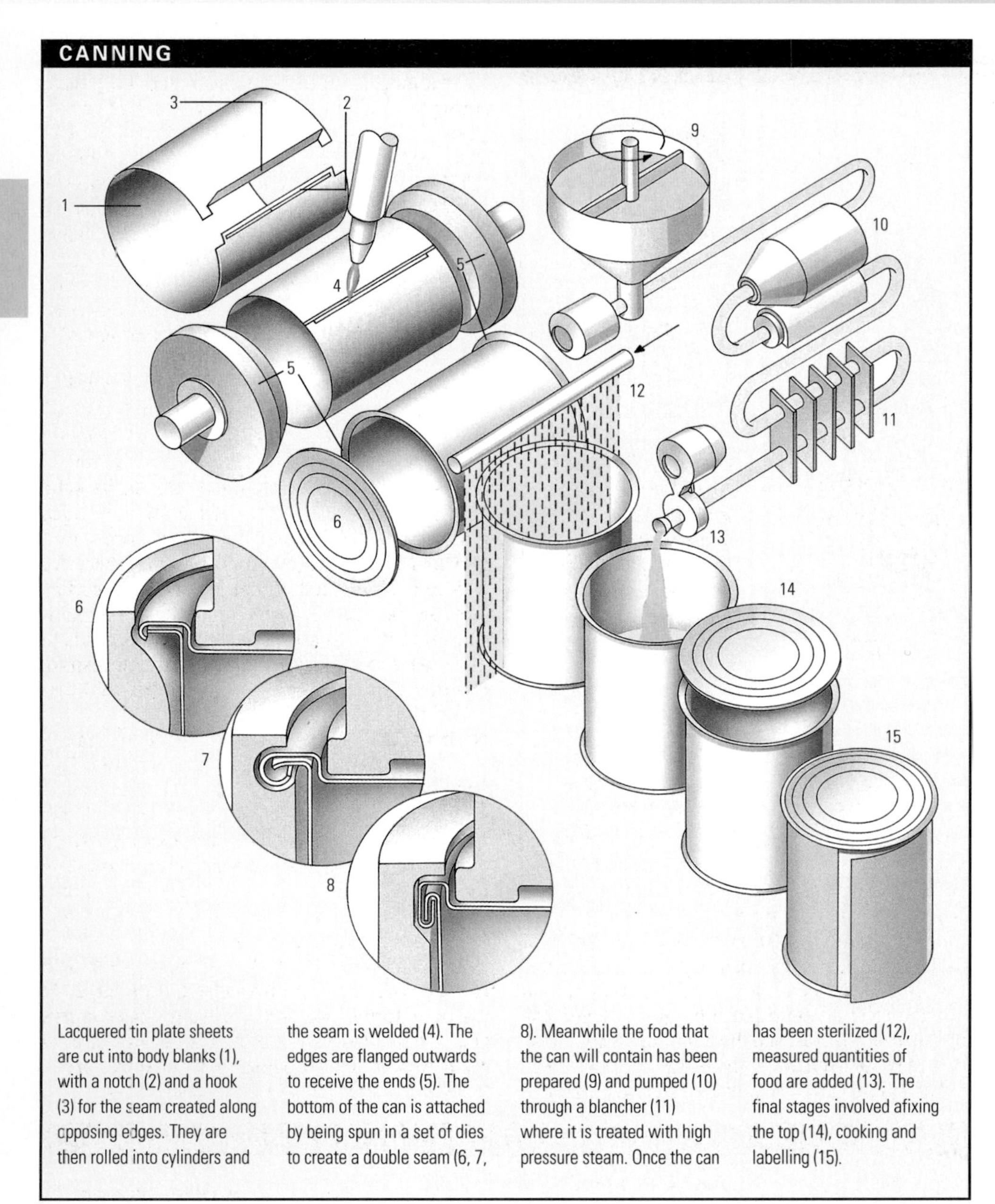

Lacquered tin plate sheets are cut into body blanks (1), with a notch (2) and a hook (3) for the seam created along opposing edges. They are then rolled into cylinders and the seam is welded (4). The edges are flanged outwards to receive the ends (5). The bottom of the can is attached by being spun in a set of dies to create a double seam (6, 7, 8). Meanwhile the food that the can will contain has been prepared (9) and pumped (10) through a blancher (11) where it is treated with high pressure steam. Once the can has been sterilized (12), measured quantities of food are added (13). The final stages involved afixing the top (14), cooking and labelling (15).

capillarity Movement of a liquid in a narrow opening caused by the surface tension between the liquid and the surrounding material. This is most often seen in a vertical, narrow glass capillary tube, but capillarity also occurs in various directions – as when a sponge or blotting paper soaks up water.

capillary Smallest of BLOOD VESSELS, connecting arteries and veins. Capillary walls consist of only a single layer of cells, so that fluid containing dissolved oxygen and other nutrients (as well as carbon dioxide and other wastes) can pass easily between the blood and surrounding tissues.

capillary constant Measure of surface tension defined, for two immiscible fluids in contact, by $2T/g(D-d)$, where T is the surface tension for the interface, g is the acceleration due to gravity and D and d are the densities of the fluids. Usually one of these is air. The capillary constant is equal to the rise of a liquid in a capillary tube multiplied by the radius of the tube.

Capricorn, Tropic of Line of latitude, *c.*23.5°s of the Equator, which marks the s boundary of the tropics. It indicates the farthest s position at which the Sun appears directly overhead at noon. The Sun is vertical over the Tropic of Capricorn on about 22 December, which is the summer SOLSTICE in the Southern Hemisphere.

Capricornus Constellation of the s sky, situated on the ecliptic between Sagittarius and Aquarius. Usually referred to as Capricorn (Ram) only for astrological purposes, this constellation contains the faint globular cluster M30 (NGC 7099). Its brightest star is Delta Capricorni (Deneb Algiedi), an eclipsing BINARY STAR with magnitude 2.9.

capsule In botany, a dry type of fruit that releases its seeds when it ripens. The capsule is formed by the fusion of several CARPELS, which may split apart to allow the seeds to scatter. Other capsules have a lid that opens, or holes through which the seeds escape. The spore-containing structures of mosses and liverworts are also called capsules. In zoology, the sticky layer that surrounds the cell walls of some bacteria is called a capsule, as are various other surrounding structures such as the sheath of connective tissue surrounding the bones in a joint and the membranous envelope around the kidneys and spleen.

capsule, space That part of a spacecraft that carries the payload, which may be a crew of astronauts or a variety of scientific instruments. A space capsule is designed to withstand extremes of temperature and pressure, shocks and vibration during acceleration, spinning or tumbling, radiation and meteoric impacts, while maintaining a stable environment for the payload. Some of the features that help to achieve these ends are an ABLATION heat shield for re-entering the atmosphere; slow, continuous rotation to reduce exposure to the Sun; and highly reflecting exteriors. The capsule of a manned spacecraft contains the flight controls, and television cameras and instruments for monitoring the activities of the crew and controlling their environment.

car *See* AUTOMOBILE

carapace In zoology, the back of the shell of certain animals. The shield-like part of the skeleton that covers and protects the head and thorax of crabs and other crustaceans is called a carapace, as is the dorsal (upper) part of the shell of a tortoise or turtle. In a tortoise, it is made up of bony plates joined to the backbone and ribs, covered by an outer horny layer.

carat Unit of weight used for gemstones. Traditionally it varied from country to country, perhaps because of its method of origin: the name carat derives from that of the seeds of the Mediterranean locust tree, which were once used to weigh gemstones. By 1913 the carat was fixed internationally at 200 milligrams. The carat measure of gold alloys (usually gold-silver or gold-copper) indicates the number of parts of gold in 24 parts. Thus 22-carat gold contains 22 parts of gold in 24 parts of the alloy.

carbide (acetylide) Inorganic compound of carbon with metals or other more ELECTROPOSITIVE elements. BORON and SILICON both form extremely hard carbides which are used as abrasives. Many TRANSITION ELEMENTS also form carbides, in which carbon atoms occupy spaces between adjacent atoms in the metal lattice. Some electropositive metals form ionic carbon compounds; the best known is CALCIUM CARBIDE (CaC_2), which reacts with water to give ETHYNE (acetylene, C_2H_2).

carbohydrate Organic compound of carbon, hydrogen and oxygen that is a constituent of many foodstuffs. The hydrogen and oxygen atoms are in the ratio of 2:1, as they are in water (H_2O). The simplest carbohydrates are the sugars, usually with five or six carbon atoms in each molecule. Glucose and fructose are MONOSACCHARIDES, naturally occurring sugars; they have the same formula ($C_6H_{12}O_6$) but different structures. One molecule of each combines with the loss of water to make sucrose ($C_{12}H_{22}O_{11}$), a DISACCHARIDE which also occurs naturally in sugar cane and sugar beet. Starch and cellulose are POLYSACCHARIDES, carbo-

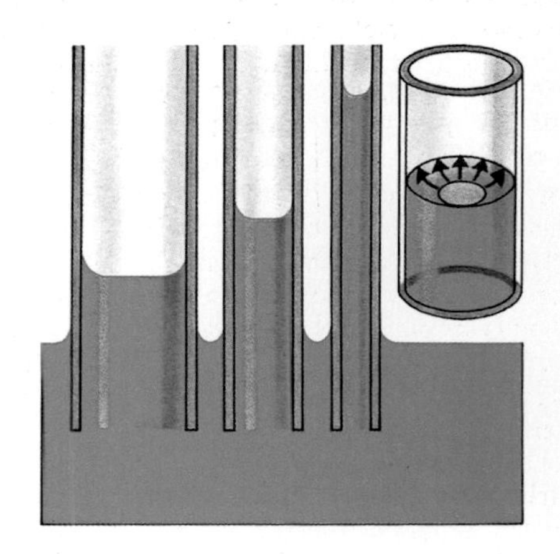

▲ **capillarity** The cohesion between a liquid and a solid results in the liquid surface curving near the solid, to meet it at a definite angle. Water curves upwards against glass, and the force of cohesion is exerted along the water surface, tending to lift it. The lifting force is proportional to the circumference of the water surface; in a narrow tube this force becomes powerful enough to lift a tall column of water.

CARBON CYCLE

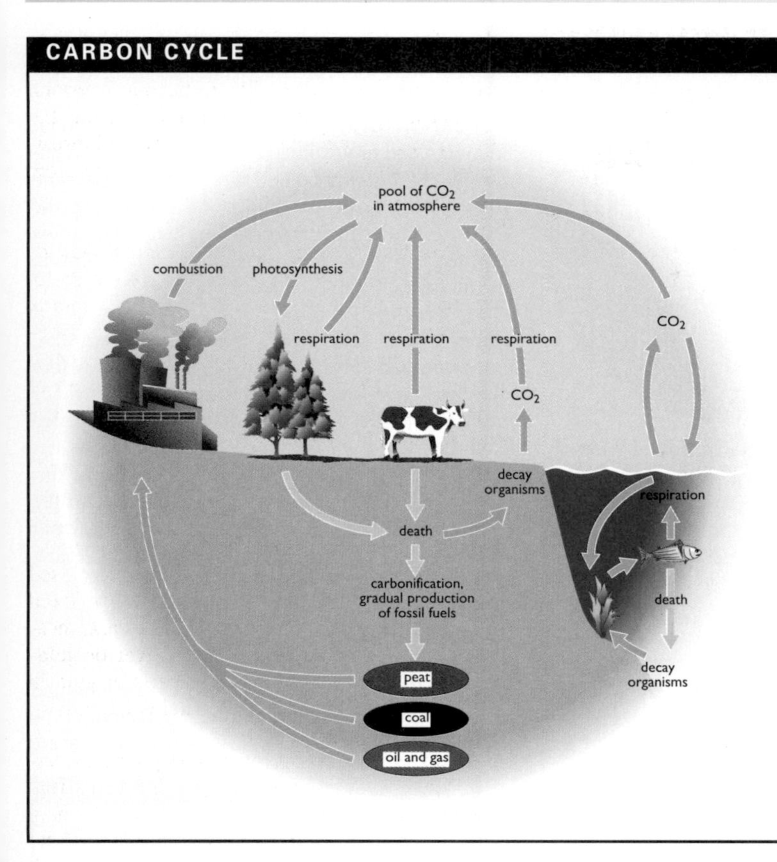

Elemental carbon is in constant flux. Gaseous carbon dioxide (CO_2) is first incorporated into simple sugars by photosynthesis in green plants. These may be broken down (respired) to provide energy, a process that releases CO_2 back into the atmosphere. Alternatively, animals which eat the plants also metabolize the sugars and release CO_2 in the process. Geological processes also affect the Earth's carbon balance, with carbon being removed from the cycle when it is accumulated within fossil fuels such as coal, oil and gas. Conversely, large amounts of carbon dioxide are released into the atmosphere when such fuels are burned.

hydrates consisting of hundreds of glucose molecules linked together. *See also* SACCHARIDE

carbolic acid *See* PHENOL

carbon (symbol C) Common, nonmetallic element of group IV of the periodic table. Carbon forms a vast number of compounds, which, with hydrogen (hydrocarbons) and other nonmetals, form the basis of organic CHEMISTRY. Until recently, it was believed there were two crystalline carbon ALLOTROPES: GRAPHITE and DIAMOND. In 1996 a third type, BUCKMINSTERFULLERENE ("buckyballs"), was discovered, the molecules of which are shaped like geodesic domes. Various amorphous (noncrystalline) forms of carbon also exist, such as coal, coke, charcoal, soot and lampblack. Amorphous carbon has many uses, for example as pigment for inks and a filler for rubber for tyres. A recently made synthetic form of carbon is CARBON FIBRE. The radioactive isotope ^{14}C is used for CARBON DATING of archaeological specimens. Properties: at.no. 6; r.a.m. 12.011; r.d. 1.9–2.3 (graphite), 3.15–3.53 (diamond); m.p. *c*.3,550°C (6,422°F); sublimes at 3,367°C (6,093°F); b.p. *c*.4,200°C (7,592°F); most common isotope ^{12}C (98.89%).

carbonate Salt of CARBONIC ACID (H_2CO_3), which is formed when carbon dioxide (CO_2) dissolves in water, as when rain falls through the air. Carbonic acid is an extremely weak acid and both it and many of its salts are unstable, decomposing readily to release carbon dioxide. Nevertheless, large parts of the Earth's crust are made up of carbonates, such as CALCIUM CARBONATE and DOLOMITE.

carbonation In geology, the process of chemical EROSION of rocks by rainwater containing carbon dioxide. When carbon dioxide gas in the atmosphere dissolves in rainwater, it forms CARBONIC ACID. This will dissolve CALCAREOUS rocks such as chalk and limestone. The dissolved chemicals can drip into caves where they may form STALACTITES and STALAGMITES.

carbon black Form of the element carbon made by heating or burning hydrocarbon gases in a restricted supply of air. The product contains a little hydrogen, oxygen and sulphur as well as carbon. Carbon black is used to reinforce rubber (for vehicle tyres) and other materials, and in various dark pigments for paint and inks.

carbon cycle Circulation of carbon in the biosphere. It is a complex chain of events. The most important components are the taking up of carbon dioxide from the atmosphere by green plants during PHOTOSYNTHESIS, and the return of carbon dioxide to the atmosphere by the respiration and eventual decomposition of animals which eat the plants.

carbon dating (radiocarbon dating) Method of determining the age of organic materials by measuring the amount of radioactive decay of an ISOTOPE of carbon, carbon-14 (^{14}C). This radio-isotope decays to form nitrogen, with a half-life of 5,730 years. When a living organism dies, it ceases to take carbon dioxide into its body, so that the amount of ^{14}C it contains is fixed relative to its total weight. Over the centuries, this quantity steadily diminishes. Refined chemical and physical analysis is used to determine the exact amount remaining, and from this the age of a specimen is deduced.

carbon dioxide (CO_2) Colourless, odourless gas that occurs in the ATMOSPHERE (0.03%) and as a product of the combustion of fossil fuels and respiration in plants and animals. It is taken up by green plants in PHOTOSYNTHESIS. In its solid form (dry ice) it is used in refrigeration; as a gas it is used in carbonated beverages, fire extinguishers and provides an inert atmosphere for welding. If inhaled in large amounts, carbon dioxide can cause suffocation. Research indicates that its increase in the atmosphere leads to the GREENHOUSE EFFECT and GLOBAL WARMING. Properties: m.p. −56.6°C (−69.9°F); sublimes −78.5°C (−109.3°F).

carbon fibre Form of carbon made by heating textile fibres to very high temperatures. The result is fibres (typically 0.001cm in diameter) which are, weight-for-weight, some of the strongest of all fibres. They are too short to be woven into a super-strong yarn. Instead they are incorporated into plastics, ceramics and glass, which give the materials great strength and resistance to breakage.

carbonic acid (H_2CO_3) Extremely weak acid formed when carbon dioxide (CO_2) dissolves in water. It forms two series of salts, CARBONATES and HYDROGENCARBONATES. Carbonates are of great importance in nature both in rock formations and in the shells of various animals.

CARBON FIBRE

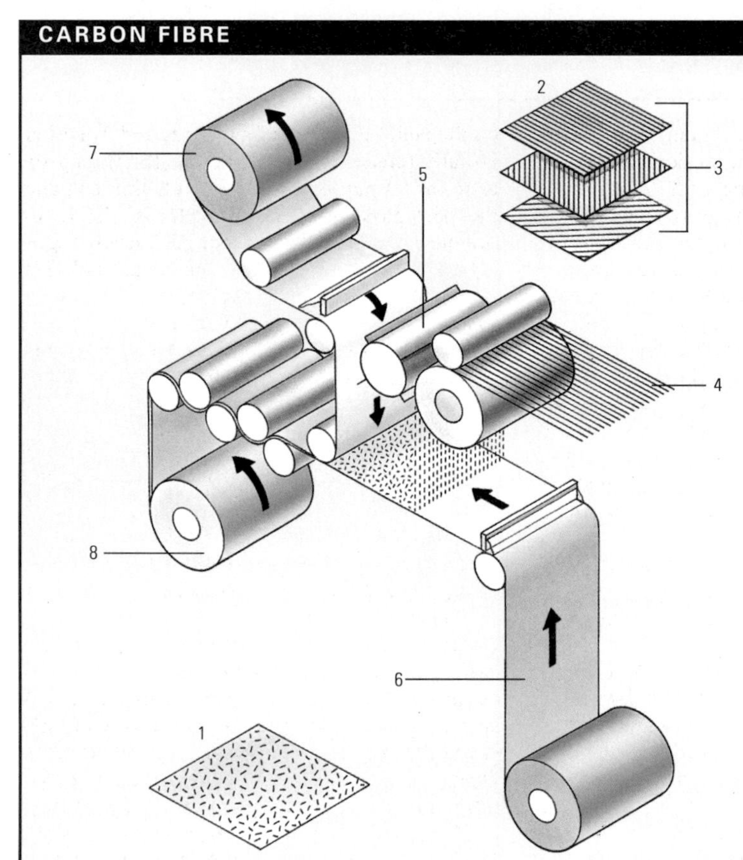

Carbon fibre composites combine twisted carbon fibres with a resin to produce a material up to four times stronger than steel for its weight. Two main forms exist. The carbon fibre is either chopped into short lengths and attached to the resin backing at random (1) or laid down in strips (2). The former has strength in all directions but the latter has great strength against the grain in different directions to give overall strength in all directions (3). Strands of twisted carbon (4) pass through a chopper (5) and fall onto a backing material (6) coated with the bonding resin (7). Another layer of resin and backing material is then laid on top and the composite is rolled for storage (8).

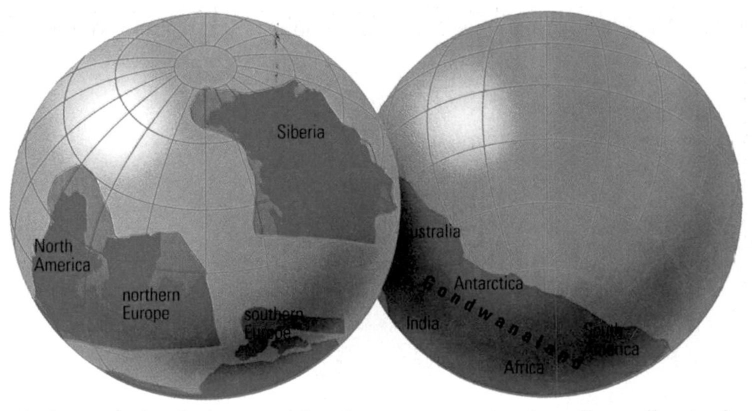

► Carboniferous Gondwana and was still whole during the Carboniferous period. North America and N Europe combined were moving towards S Europe.

Carboniferous Fifth geologic period of the PALAEOZOIC era, lasting from 360 to 286 million years ago. It is divided into two series; the Lower Carboniferous is characterized by marine limestones with a coral-rich fauna; the Upper Carboniferous is dominated by river and deltaic sediments containing coal seams formed from extensive swampy forests of conifers and tree ferns. In North America, the Lower Carboniferous is known as the Mississippian period and the Upper Carboniferous as the Pennsylvanian period.

carbon monoxide (CO) Colourless, odourless, poisonous gas formed during the incomplete combustion of fossil fuels, occurring for example in coal gas and the exhaust fumes of cars. Carbon monoxide poisons by combining with the HAEMOGLOBIN in ERYTHROCYTES (red blood cells) thus preventing them from carrying oxygen round the body. (This happens if the inhaled air contains only 0.1% of carbon monoxide by volume.) It is used as a reducing agent in metallurgy. Properties: density 0.968 (air = 1); m.p. −205°C (−337°F); b.p. −191.5°C (−312.7°F).

carbon tetrachloride *See* TETRACHLOROMETHANE

carbonyl compound Compound containing the carbonyl group (–CO). Organic carbonyls include KETONES, ALDEHYDES and CARBOXYLIC ACIDS. Inorganic carbonyls contain carbonyl groups bound directly to metal atoms.

carborundum Originally a tradename for silicon carbide (SiC) abrasives and refractories prepared by heating silica (SiO_2) with carbon in an electric furnace. It is used in grinding wheels, grinding papers, abrasive grains and powders, valve-grinding compounds, and in refractory bricks and blocks. Nearly as hard as diamond, it oxidizes slowly at temperatures above 1,000°C (2,000°F).

carboxylic acid Member of a class of organic chemical compounds containing the group –COOH. The commonest example is ETHANOIC ACID (acetic acid, CH_3COOH), which is present in vinegar. These acids are weakly acidic, forming salts with bases and esters with alcohols. Esters of high-molecular weight, carboxylic acids, such as stearic, lauric and oleic acids, are present in animal and vegetable fats; for this reason carboxylic acids are often called FATTY ACIDS. The systematic names are formed using the suffix -oic.

carburation Mixing of air and fuel vapour in the correct proportions for an INTERNAL COMBUSTION ENGINE. In a petrol engine, air sucked into the cylinders passes through the CARBURETTOR, creating a partial vacuum which draws petrol in through a jet. The resulting air/petrol mixture rapidly burns in the cylinders and the expansion of the gases produced provides the power.

carburettor Component of some petrol-powered INTERNAL COMBUSTION ENGINES, used to vaporize and mix fuel with air in the correct proportion for proper combustion. Generally steady speed requires a ratio of 15:1 air to fuel. Richer ratios of 10:1 air to fuel are necessary for starting cold engines. Efficient CARBURATION is essential for smooth running and efficient engine performance. No carburettor is needed in engines that use FUEL INJECTION.

carcinogen External substance or agent that causes CANCER, including chemicals, radiation and some viruses. Many different chemical carcinogens have been found to cause cancer in animals, but more research is needed to confirm their ability to cause cancer in humans. It is thought that tobacco smoke causes lung cancer because of carcinogenic hydrocarbons.

Cardano, Girolamo (1501–76) Italian mathematician and physician. He wrote several influential texts on mathematics. In *De Subtilitate Rerum* (1550) he attempted to summarize the existing scientific beliefs of his time. His most notable contribution is thought to be the solution of the general cubic equation

$ax^3 + bx^2 + cx + d = 0$

often called **Cardano's solution**. TARTAGLIA, however, had solved the problem 10 years before Cardano announced the solution.

cardiac cycle Sequence of events in between each heartbeat. Blood enters the heart while it is relaxed, filling the ATRIA and VENTRICLES. Contraction of the ventricles forces blood out of the heart and, at the end of the contraction, the ventricles again relax and the heart starts to fill again, ready for the next cycle.

cardiac muscle *See* MUSCLE

cardinal number *See* NUMBER, CARDINAL

cardiology Branch of medicine that deals with the diagnosis and treatment of the diseases and disorders of the HEART and vascular system.

caries Decay and disintegration of teeth or BONE

CARBURETTOR

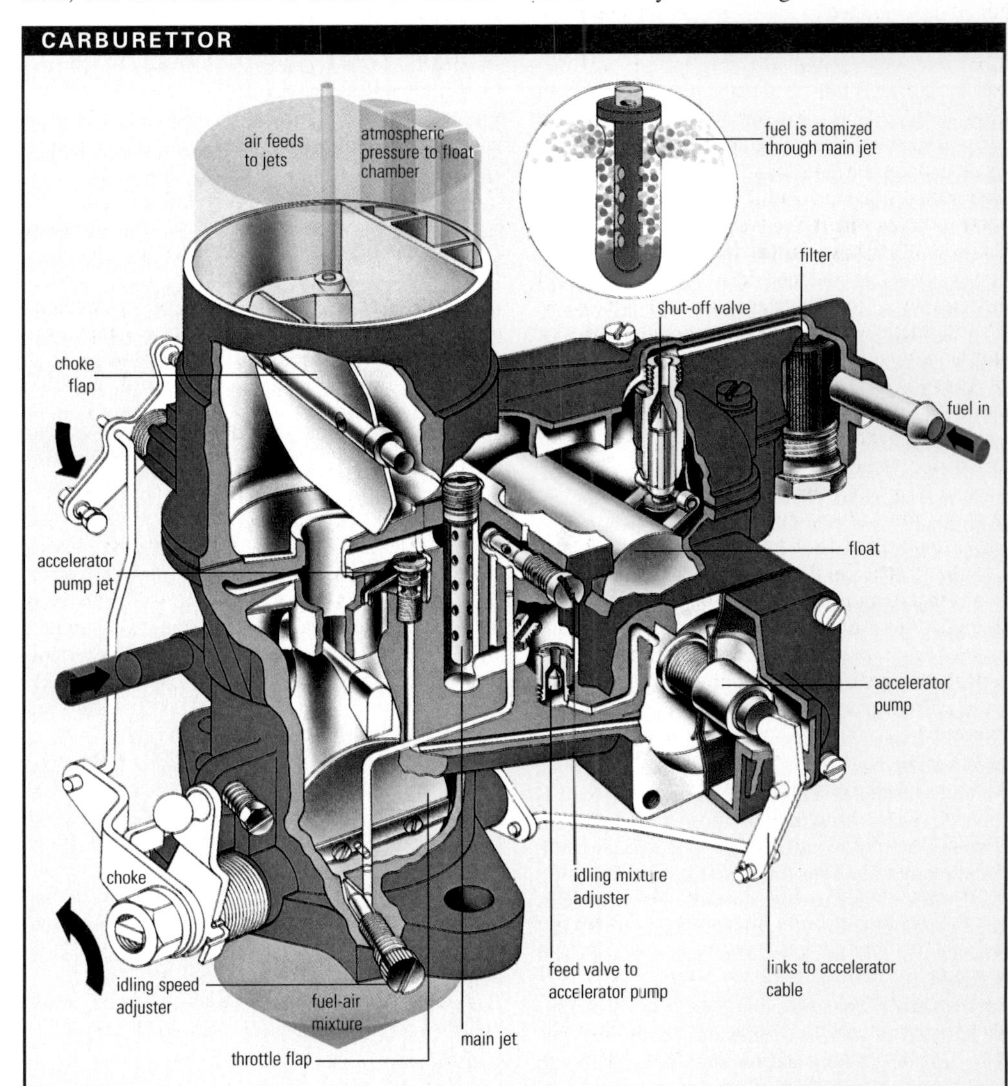

The purpose of the carburettor is to provide the required amount of fuel and air at the correct mixture to the combustion chambers of an internal combustion engine. Air is drawn through the carburettor into the combustion chamber when the piston moves down the cylinder with the inlet valve open. The amount of air is controlled by the throttle flap, which is operated by the accelerator pedal. When the throttle is closed the idling system, which by-passes the throttle, keeps the engine "ticking over". Another jet, the accelerator pump jet, forces more fuel into the engine when required for acceleration.

substance. Tooth decay is caused mainly by acids produced when bacteria present in the mouth break down sugars in food. Regular brushing, a reduced sugar intake and fluoride prevent caries. Eating between meals encourages tooth decay, because it maintains a supply of food for the bacteria and keeps acidity levels high.

Carina (Keel) Part of the dismembered constellation Argo Navis, the ship Argo of the S sky. It is the brightest and richest part of Argo, and contains CANOPUS, the second-brightest star in the sky. Apart from Canopus, its brightest stars are Beta, magnitude 1.7, and Epsiolon, magnitude 1.9.

Carlson, Chester (1906–68) US physicist, inventor of XEROGRAPHY (1938). He patented it in 1940, but had great difficulty in convincing anyone of its commercial value. Finally, in 1947, he signed an agreement with the Haloid Company (now the Xerox Corporation). His royalties made him a multi-millionaire.

carnivore Any member of the order of flesh-eating mammals (Carnivora). Mustelids – weasels, martens, minks and the wolverine – make up the largest family. The cats are the most specialized killers among the carnivores; dogs, bears and raccoons are much less exclusively meat eaters; and civets, mongooses and their relatives also have a mixed diet. Related to the civets, but in a separate family, are the hyenas, large dog-like scavengers. More distantly related to living land carnivores are the seals, sea lions and walruses; they evolved from ancient land carnivores who gave rise to early weasel- or civet-like forms. Extinct carnivores include the sabretooth cats, which died out during the Pliocene epoch 2 million years ago.

carnivorous plant *See* INSECTIVOROUS PLANT

Carnot, (Nicolas Léonard) Sadi (1796–1832) French engineer and physicist whose work laid the foundation for the science of THERMODYNAMICS. His major work, *Réflexions sur la puissance motrice du feu* (1824), provided the first theoretical background for the STEAM ENGINE and introduced the concept of the second law of thermodynamics (involving ENTROPY), which was formulated later by Rudolf CLAUSIUS. Carnot's work was ignored until quoted and extended in 1834 by the railway engineer Emile Clapeyron and recognized in 1848 by William KELVIN.

Carnot cycle In thermodynamics, a cycle of events that demonstrates the impossibility of total efficiency in HEAT ENGINES. Named after Sadi CARNOT, it shows how an engine can never convert all the heat energy supplied to it from its burning fuel into mechanical energy. Some heat energy always remains unused in a "cold sink". In an INTERNAL COMBUSTION ENGINE, this can be thought of as the engine itself. More properly, the cold sink is the entire Universe.

carnotite ($K_2(UO_2)_2(VO_4)_2.nH_2O$) Secondary vanadate mineral, potassium uranium vanadate, an ore of uranium and radium, important for nuclear energy. It occurs in the Colorado Plateau, Australia and Congo as yellow-green crusts or cavity fillings in sandstone and in fossilized wood. It is finely crystalline (monoclinic), dull or earthy; r.d. 3–5.

carotene Plant pigment that is converted to vitamin A by the liver. It occurs in various fruits and vegetables (such as carrots).

carotenoid One of a group of fat-soluble plant pigments ranging in colour from yellow to red. Carotenoids also occur in some animal fats. They include isomers of CAROTENE, a pigment that is converted in the liver into vitamin A, necessary for normal vision and healthy skin.

Carothers, Wallace Hume (1896–1937) US chemist who discovered the synthetic polyamide fibre now called NYLON.

CASTING

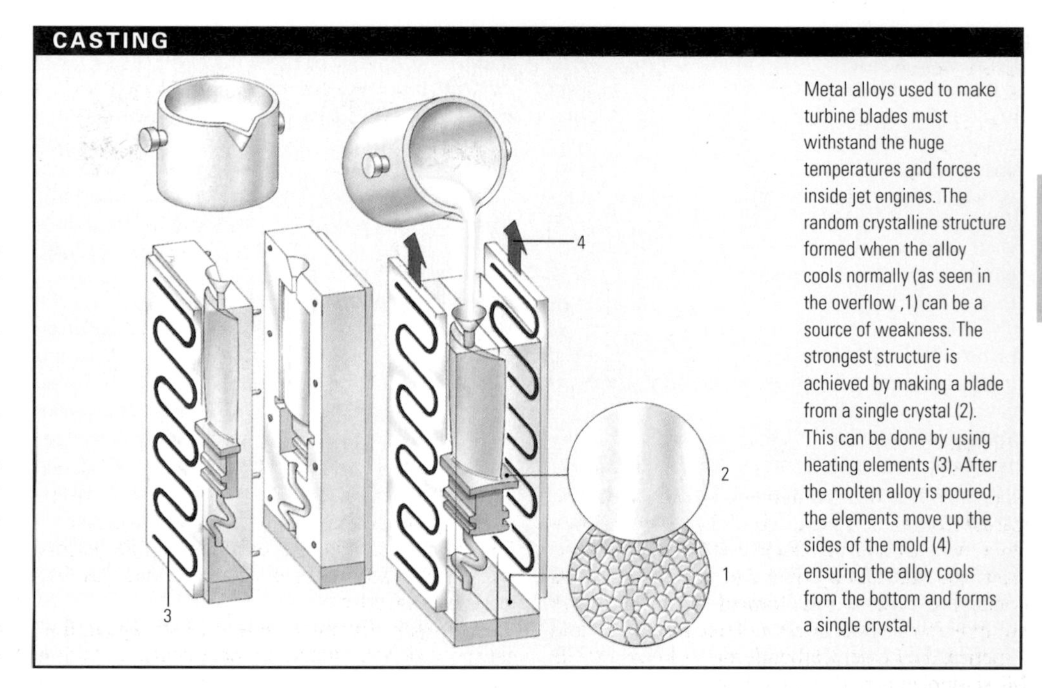

Metal alloys used to make turbine blades must withstand the huge temperatures and forces inside jet engines. The random crystalline structure formed when the alloy cools normally (as seen in the overflow ,1) can be a source of weakness. The strongest structure is achieved by making a blade from a single crystal (2). This can be done by using heating elements (3). After the molten alloy is poured, the elements move up the sides of the mold (4) ensuring the alloy cools from the bottom and forms a single crystal.

carpal One of the small bones in the wrist.

carpel Female reproductive element of a FLOWER. A carpel consists of a STIGMA (which receives pollen), a STYLE and an OVARY (containing ovules). A group of carpels make up the GYNOECIUM, the complete female reproductive structure within a flower.

Carrel, Alexis (1873–1944) US surgeon and experimental biologist, b. France. Carrel was a member (1912–39) of the Rockefeller Institute, New York City. He was awarded the 1912 Nobel Prize for physiology or medicine for his development of **anastomosis**, a surgical technique for stitching together blood vessels end-to-end. Carrel and Charles LINDBERGH invented an artificial heart.

carrier In medicine, apparently healthy person who harbours a disease which, while it does not affect him or her, can be transmitted to other people. Carriers are difficult to detect and are often unaware of the infection they carry.

carrier wave High-frequency, ELECTROMAGNETIC RADIATION that carries information signals from a radio or television transmitting ANTENNA. The fixed wavelength and frequency of the transmitter's wave is the signal, or channel, to which a receiving antenna must be tuned in order to receive its broadcast. To prevent interference, the frequencies of transmitter carrier waves differ by 10kHz.

Carson, Rachel Louise (1907–64) US writer and marine biologist. Carson is best known for her popular books on marine ecology. *The Sea Around Us* (1951) won a National Book Award. *Silent Spring* (1962) directed public attention to the dangers of agricultural pesticides and was a pioneering work in the development of the environmental movement. Other works include *Under the Sea Wind* (1941) and *The Edge of the Sea* (1955).

Cartesian coordinate system System introduced by René DESCARTES in which the position of a point is specified by its distances from intersecting lines (the axes). In the simplest Cartesian system the axes are perpendicular (called y and x axes). The position of a point is then given by a pair of numbers (x, y). The **abscissa**, x, is the point's distance from the y-axis, measured in the direction of the x-axis, and the **ordinate**, y, is the distance from the x-axis. The axes in such a system need not be at right angles but should not be parallel to each other. Three axes are needed to describe the location of points in three dimensions.

cartilage (gristle) Flexible supporting tissue made up of the tough protein COLLAGEN. In the vertebrate embryo, the greater part of the SKELETON consists of cartilage, which is gradually replaced by BONE during development, except in areas of wear such as the ends of bones and the INTERVERTEBRAL DISCS, where caps of cartilage help to protect the bone below. In humans, cartilage is also present in the larynx, nose and external ear, and rings of cartilage help to support the windpipe (TRACHEA) and BRONCHI.

cartilaginous fish *See* FISH

Cartwright, Edmund (1743–1823) British inventor. Cartwright invented the power LOOM (patented 1785) after visiting Sir Richard ARKWRIGHT's cotton spinning mill, although it was not used commercially until the early 19th-century. He also invented a wool-combing machine (patented 1789) and an alcohol engine (1797).

casein Principal protein in milk, containing 0.71% phosphorus, 0.72% sulphur and about 15 AMINO ACIDS. It constitutes about 80% of the proteins in cow's milk and about 40% in human milk. Obtained from milk by the addition of either acid or the enzyme rennet, casein is used to make plastics, cosmetics, paper coatings, adhesives, paints, glues, textile sizing, cheeses and animal feed.

Cassini, Giovanni Domenico (1625–1712) French astronomer. He was the first to measure accurately the dimensions of the SOLAR SYSTEM. He discovered the division in the rings of SATURN that now bears his name, and also four of Saturn's satellites. He measured Jupiter's rotation period.

Cassiopeia Distinctive constellation of the N sky, representing in mythology the mother of Andromeda. The five leading stars make up a "W" or "M" pattern, on the opposite side of the Pole Star from URSA MAJOR. Alpha (Schedar) is its brightest star, magnitude 2.2.

cassiterite Translucent black or brown mineral, tin oxide (SnO_2); the major ore of tin. It occurs in PLACER DEPOSITS, chiefly in the Malay peninsula, and in pegmatites and other INTRUSIVE IGNEOUS ROCKS. It takes the form of short tetragonal prismatic crystals, or masses and radiating fibres. Hardness 6–7; r.d. 7.

casting Forming objects by pouring molten metal into moulds and allowing it to cool and solidify. Many castings are made by pouring metal into sand or clay moulds. Specialized processes, such as plastic moulding, composite moulding, CIRE PERDUE casting and die casting, give greater dimensional accuracy, smoother surfaces and finer detail.

C

cast iron General term applied to various grades of iron, especially **grey** iron and **pig** iron (directly out of a BLAST FURNACE). It includes a wide range of iron-carbon-silicon alloys containing from 1.7–4.5% carbon with varying amounts of other elements. Grey iron (so-called because its fracture looks greyish) is the most widely used for casting vehicle engines, machinery parts and many other products.

castration Removal of the sexual glands from an animal or human – the TESTES in a male, the OVARIES in a female. In human beings it has been a form of punishment, a way of sexually incapacitating slaves to produce eunuchs or artificially creating soprano voices (castrati) and as a way of stopping the spread of cancerous growths. It can make livestock animals tamer and improve the quality of their meat; it prevents breeding.

catabolism *See* METABOLISM

catalysis Modification of the rate of a chemical reaction, by the addition of a CATALYST, which is not consumed in the reaction. Catalytic action can reveal the reaction mechanism; many industrial processes rely on catalysis to accelerate reactions and occasionally to inhibit undesirable ones. Catalysis plays a key part in most biochemical reactions. *See also* BIOCHEMISTRY

catalyst Substance that speeds up the rate of a chemical reaction without itself being consumed. Many industrial processes rely on catalysts, for example the HABER PROCESS for manufacturing AMMONIA uses iron as a catalyst. Metals or their compounds catalyse by adsorbing gases to their surface, forming intermediates that then readily react to form the desired product while regenerating the original catalytic surface. The METABOLISM of all living organisms depends on biological catalysts called ENZYMES. Without these catalysts most reactions would happen so slowly that life would not be possible.

catalytic converter Anti-pollution device used in internal combustion engines. It consists of a bed of catalytic agents through which flow the gaseous exhaust of fuel combustion. Converters located in silencers reduce harmful unburned hydrocarbons and carbon monoxide. These converters are adversely affected by tetraethyl lead found in some types of petrol.

catalytic cracking *See* CRACKING

cataract In medicine, opacity in the lens of an eye, causing blurring of vision. Most cases are due to degenerative changes in old age but it can also be congenital, the result of damage to the lens, or some metabolic disorder such as diabetes. Treatment is by surgical removal of the cataract followed by implantation of an artificial lens.

cataract In physical geography, term usually applied to that section of a rapidly flowing river where the running water falls suddenly in a sheer drop. When the drop is less steep, the fall is known as a cascade.

catastrophe theory Mathematical technique published in 1972 by the French mathematician René Thom (1923–). The theory concerns processes that do not always take place through continuously changing conditions but, instead, may at some point undergo a catastrophe – an event in which certain mathematical quantities change by a sudden, discontinuous leap. The technique is particularly useful for describing situations in which small changes in initial conditions or input can cause sudden or drastic changes in a system's behaviour or output.

catechol (1,2-dihydroxybenzene, $C_6H_4(OH)_2$) Organic chemical compound. It forms clear aqueous solutions which soon turn brown because, under the influence of light, they react strongly with oxygen. For these reasons catechol is used to protect stored materials against oxidation, and in photographic developers. It also has uses in making dyes.

catecholamine Any of a group of AMINES with important biological activity. Catecholamines are derived from DOPA (dihydroxyphenylalanine) and include nerve transmitters and hormones such as dopamine, adrenaline (epinephrine) and noradrenaline (norepinephrine).

catena Variation on the vertical profile of soils derived from the same parent material, usually resulting from topographic changes, such as a hillslope, which controls other soil-forming factors such as microclimate and drainage. *See also* SOIL PROFILE

catheter Fine tube introduced into the body to deliver or remove fluids. The most common is the **urinary** catheter, fed into the bladder by way of the URETHRA.

cathode Negative ELECTRODE of an electrolytic cell or ELECTRON TUBE. It attracts positive ions (CATIONS) during ELECTROLYSIS, the process that uses electrical energy to bring about chemical changes.

cathode ray Radiation emitted by the CATHODE of a thermionic (heated) electron valve containing a gas at low pressure. The rays were identified in 1897 by J. J. THOMSON as streams of charged particles having extremely low mass, later called ELECTRONS. Some electrons are emitted because the cathode is heated but most because CATIONS (positive ions) formed in the valve collide with the cathode. *See also* CATHODE-RAY TUBE; THERMIONICS

cathode-ray tube Evacuated ELECTRON tube used for TELEVISION picture tubes, OSCILLOSCOPES and display screens in radar sets and computers. An electron gun shoots a beam of electrons, focused by anodes. The electrons strike a fluorescent screen and produce a spot of light. In a television tube, an electrostatic or magnetic field deflects the beam so that it scans a number of lines on the screen, controlled by the incoming picture signals.

cation Positive ION which is attracted to the CATHODE during ELECTROLYSIS.

catkin Drooping, scaly spike of unisexual flowers without petals, such as those of the hazel or poplar. This deciduous flower cluster is also typical of birches and some beeches.

CAT scan Abbreviation of COMPUTERIZED AXIAL TOMOGRAPHY

cat's eye *See* CHRYSOBERYL

Cauchy, Augustin Louis, Baron (1789–1857) French mathematician who created complex analysis and, following EULER, formalized many of the ideas of CALCULUS. Cauchy defined the notion of a limit and of a continuous function and formalized the definitions of derivative and integral that are still used today. He produced more than a thousand papers and it is said that more theorems and concepts are named after Cauchy than after any other mathematician.

caustic potash *See* POTASSIUM HYDROXIDE

caustic soda *See* SODIUM HYDROXIDE

cave Natural underground cavity. There are several kinds of caves, including **coastal** caves, formed by wave erosion, **ice** caves, formed in glaciers, and **lava** caves. By far the largest caves are formed in carbonate rocks such as limestone. Such rocks are impervious, but dissolve in underground streams or ground water formed by rain.

Cavendish, Henry (1731–1810) British chemist and physicist. Cavendish discovered hydrogen and the compositions of water and air; he estimated the Earth's mass and density, and calculated the gravitational constant by a method now known as the "Cavendish experiment". He also measured the specific gravity of carbon

CATALYTIC CONVERTER

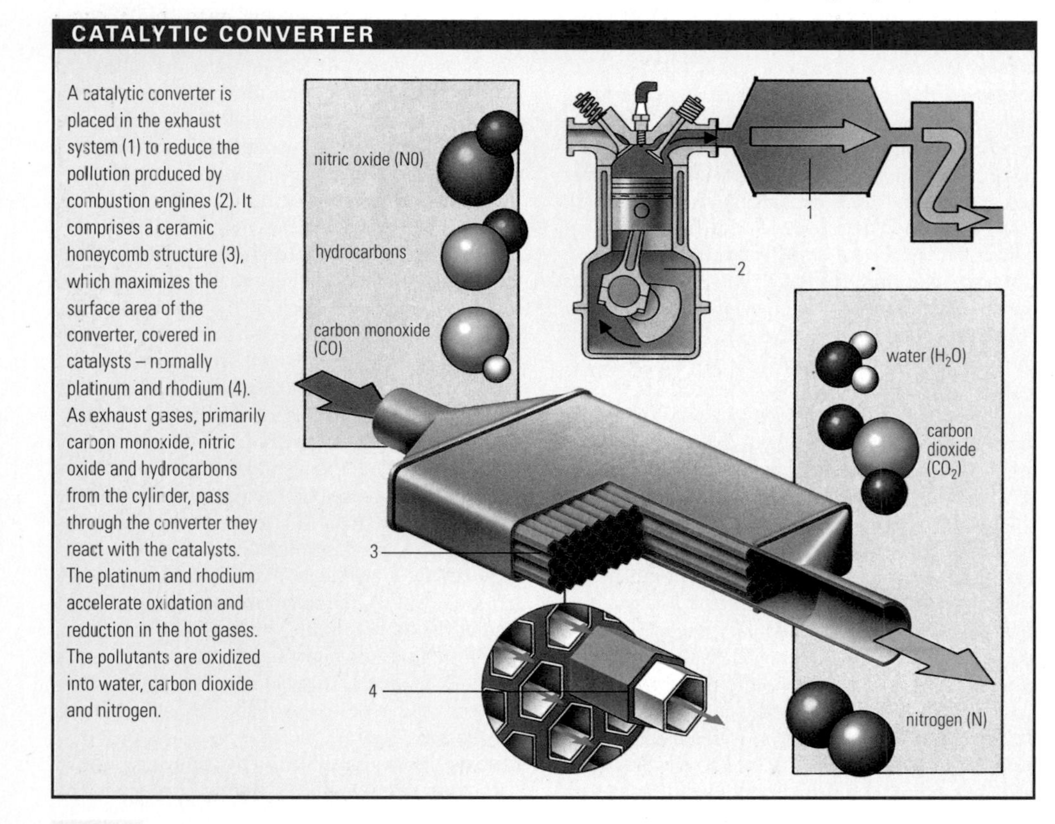

A catalytic converter is placed in the exhaust system (1) to reduce the pollution produced by combustion engines (2). It comprises a ceramic honeycomb structure (3), which maximizes the surface area of the converter, covered in catalysts – normally platinum and rhodium (4). As exhaust gases, primarily carbon monoxide, nitric oxide and hydrocarbons from the cylinder, pass through the converter they react with the catalysts. The platinum and rhodium accelerate oxidation and reduction in the hot gases. The pollutants are oxidized into water, carbon dioxide and nitrogen.

CATHODE-RAY TUBE

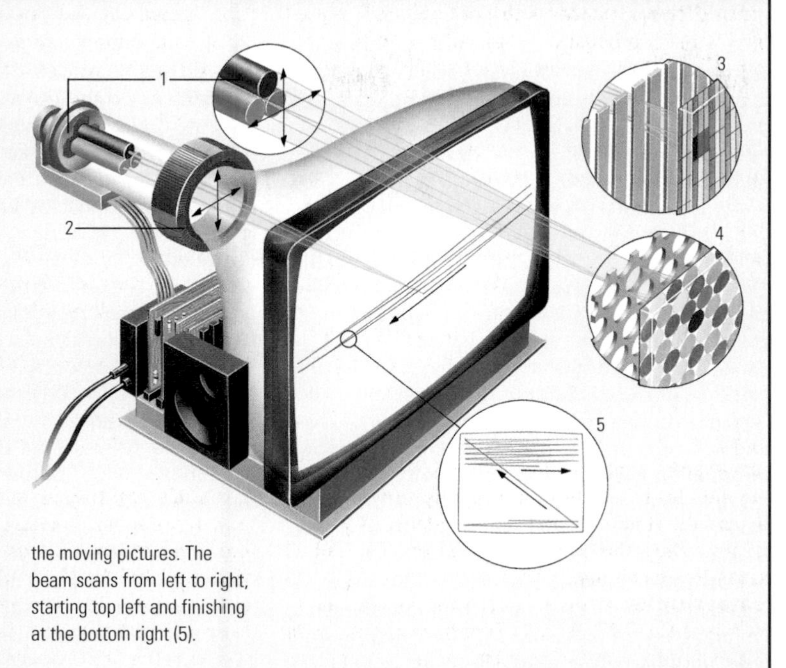

Television receivers are a type of cathode-ray tube. Three electron guns (1) receive colour signals from a colour decoder which splits the colour signal into red, green and blue. The guns fire three beams of electrons through vertical and horizontal deflection coils (2) onto the screen of a "shadow mask tube" (3). This is made up of about a million dots (4), a third of which glow red when bombarded, a third blue and the remaining third, green. The dots compose the colour picture received by the television. The beam of electrons scans hundreds of lines on the screen (525 in the USA, 625 in Europe) making up the moving pictures. The beam scans from left to right, starting top left and finishing at the bottom right (5).

dioxide and hydrogen, and stated the inverse square law for the interaction of charged particles. The Cavendish Laboratory at Cambridge University, England, is named after him.

Caventou, Joseph Bienaimé (1795–1877) French chemist who collaborated in 1817 with Pierre Joseph Pelletier (1788–1842) to isolate and name CHLOROPHYLL (choosing the Greek words for "green leaf"). They are considered the founders of alkaloid chemistry, having isolated veratrine, strychnine, brucine, quinine and cinchonine. Caventou also extracted caffeine from coffee beans in 1822. The son of an apothecary, Caventou studied in Paris where he was professor of toxicology (1835–60) at the École de Pharmacie. He and Pelletier concentrated on research into plant chemistry in order to find the component that renewed air. They discovered that chlorophyll was the compound which turned plants green, and it was later shown that the molecules absorbed the Sun's light energy for PHOTOSYNTHESIS.

Caxton, William (1422–91) First English printer. Following a period in Cologne (1470–72), where he learned printing, he set up his own printing press in 1476 at Westminster. He published more than 100 items on a wide variety of subjects, many of them his own translations from French, Latin and Dutch.

Cayley, Arthur (1821–95) British mathematician. He published more than 900 papers in pure mathematics and geometry, but is best know for the invention of matrices. *See also* MATRIX

Cayley, Sir George (1773–1857) British inventor who founded the science of AERODYNAMICS. He built the first glider to carry a person successfully and developed the basic form of the early aeroplane. He also invented a caterpillar tractor and founded London's Regent Polytechnic.

CD-I (COMPACT DISC interactive) System for storing and reproducing sounds and still and moving pictures on a disc resembling an ordinary CD. The system is described as interactive because, when playing a game, for example, actions by the user control what appears on the screen. CD-i players have either remote or wired control units, and use a television set for reproducing the sound and vision.

CD-ROM (COMPACT DISC read-only memory) Optical storage device for computer data and programs. It resembles a COMPACT DISC (CD) used in hi-fi systems. A CD-ROM can store much more data than a comparably priced portable MAGNETIC DISK. Computer games, encyclopedias and other software are now available in this form. To use a CD-ROM, the disc is placed in a specialized player connected, or built in, to a computer.

celestial equator Great circle on the CELESTIAL SPHERE, lying midway between the CELESTIAL POLES in the same plane as the Earth's Equator.

celestial mechanics Branch of ASTRONOMY concerned with the relative motions of stars and planets that are associated in systems, such as the Solar System or a BINARY STAR system, by gravitational fields. Introduced by Isaac NEWTON in the 17th century, celestial mechanics, rather than general RELATIVITY, is usually sufficient to calculate the various factors determining the motion of planets, satellites, comets, stars and galaxies around a centre of gravitational attraction.

celestial meridian Circle on the CELESTIAL SPHERE passing through the CELESTIAL POLES, the ZENITH and the NADIR and crossing the horizon at the N and S points.

celestial poles Two diametrically opposite points at which the extension of the Earth's axis meets the CELESTIAL SPHERE. The celestial sphere rotates about a line through the celestial poles.

celestial sphere Imaginary sphere of infinite radius used to define the positions of celestial bodies as seen from Earth, the centre of the sphere. The sphere rotates, once in 24 hours,

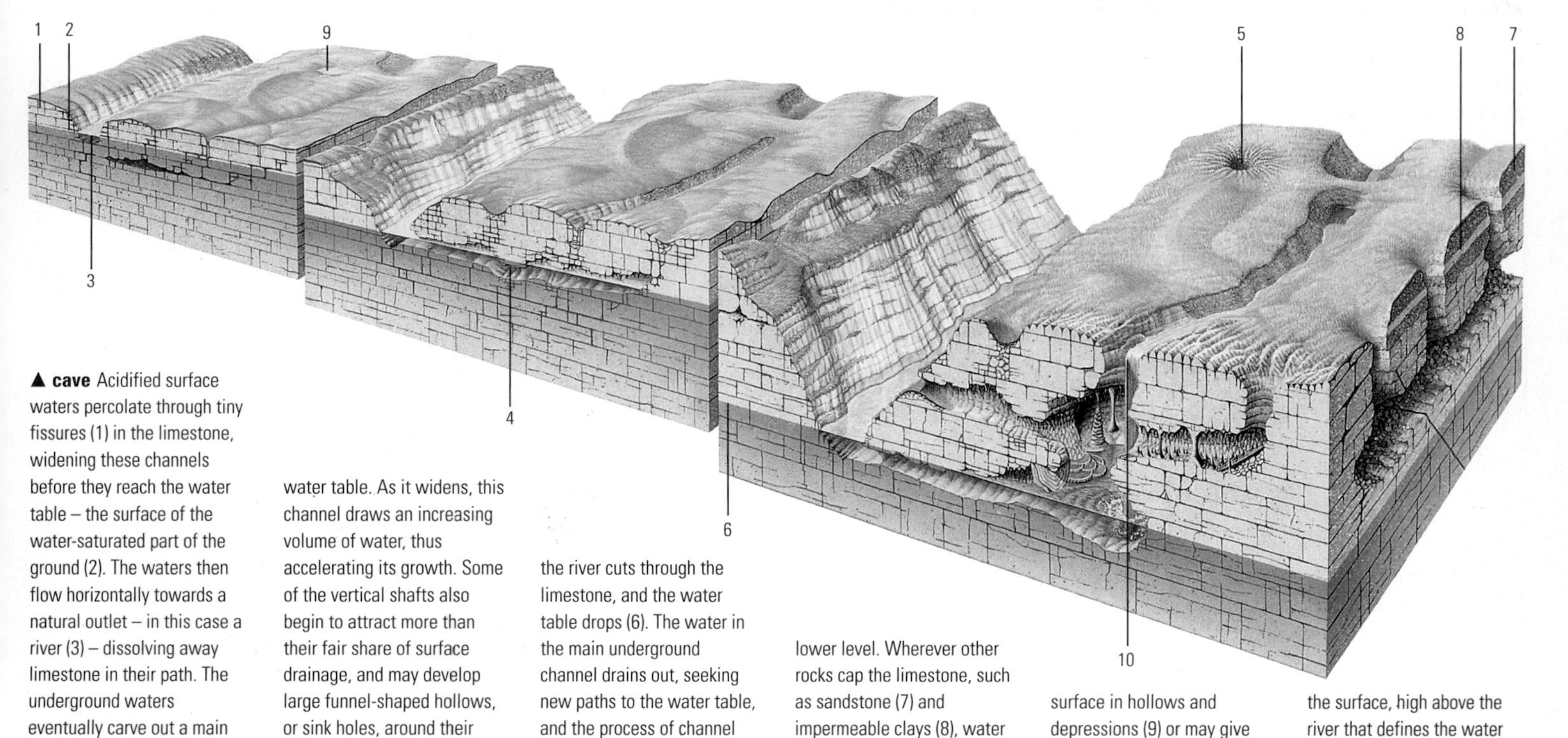

▲ **cave** Acidified surface waters percolate through tiny fissures (1) in the limestone, widening these channels before they reach the water table – the surface of the water-saturated part of the ground (2). The waters then flow horizontally towards a natural outlet – in this case a river (3) – dissolving away limestone in their path. The underground waters eventually carve out a main channel (4) at the depth of the water table. As it widens, this channel draws an increasing volume of water, thus accelerating its growth. Some of the vertical shafts also begin to attract more than their fair share of surface drainage, and may develop large funnel-shaped hollows, or sink holes, around their mouths (5). On the surface, the river cuts through the limestone, and the water table drops (6). The water in the main underground channel drains out, seeking new paths to the water table, and the process of channel carving begins again at the lower level. Wherever other rocks cap the limestone, such as sandstone (7) and impermeable clays (8), water may be trapped on the surface in hollows and depressions (9) or may give rise to rivers that run along the surface, high above the river that defines the water table (10).

about a line that is an extension of the Earth's axis. The position of a celestial body is the point at which a radial line through it meets the surface of the sphere. The position is defined in terms of coordinates, such as declination and right ascension or altitude and azimuth, which refer to great circles on the sphere, such as the CELESTIAL EQUATOR or the ecliptic.

celestine Mineral, strontium sulphate ($SrSO_4$), with distinctive pale blue or white, glassy, orthorhombic crystals, sometimes occurring in fibrous masses. It is found chiefly in SEDIMENTARY ROCK and also as gangue material in ore veins. There are deposits in Britain, Sicily and the USA. It is an important source of strontium and some of its compounds.

cell In biology, basic unit of which all plant and animal tissues are composed. The cell is the smallest unit of life that can exist independently, with its own self-regulating chemical system. Most cells consist of a MEMBRANE surrounding jelly-like CYTOPLASM with a central NUCLEUS. The nucleus is the main structure in which DNA is stored in CHROMOSOMES. Animal cells vary widely in shape. An ERYTHROCYTE (red blood cell), for instance, is a biconcave disc, whereas a NEURON has a long fibre. The cells of plants and algae are enclosed in a cell wall, which gives them a more rigid shape. Bacterial cells also have a cell wall, but do not have nuclei or chromosomes; instead, they have a loop of DNA floating in the cytoplasm. More advanced cells (those that have nuclei), often have other membrane-bounded structures inside the cell, such as MITOCHONDRIA and CHLOROPLASTS. *See also* EUKARYOTE; PROKARYOTE *See also feature on pages 66 and 67*

cell In physics, device from which electricity is obtained due to a chemical reaction. A cell consists of two ELECTRODES (a positive ANODE and a negative CATHODE) immersed in a solution (ELECTROLYTE). A chemical reaction takes place between the electrolyte and one of the electrodes. In a primary cell, current is produced from an irreversible chemical reaction, and the chemicals must be renewed at intervals. In a secondary cell (BATTERY), the chemical reaction is reversible, and the cell can be recharged by passing a current through it. *See also* ELECTROLYSIS

cell differentiation Way in which an embryo's cells become specialized for specific functions. Initially all embryonic cells are similar but soon begin to differentiate into specific kinds such as bone cells, brain cells, heart cells, muscle cells, and so on. Differentiation also occurs in mature plants and animals during regeneration of lost or damaged tissue.

CENTRIFUGE

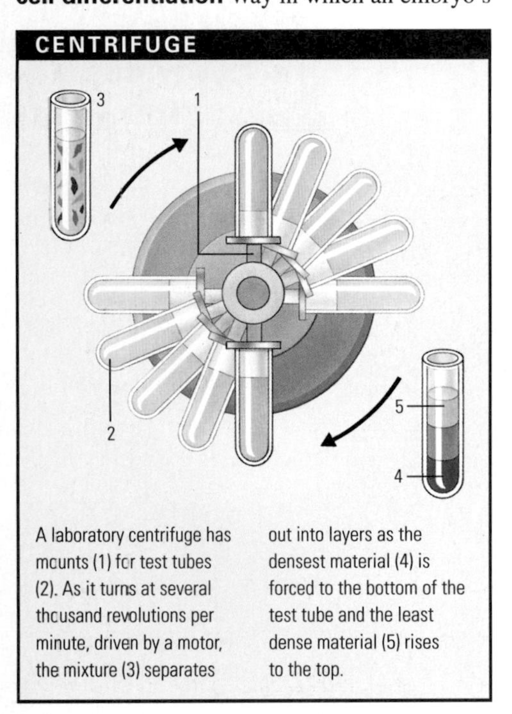

A laboratory centrifuge has mounts (1) for test tubes (2). As it turns at several thousand revolutions per minute, driven by a motor, the mixture (3) separates out into layers as the densest material (4) is forced to the bottom of the test tube and the least dense material (5) rises to the top.

cell division Process by which living cells reproduce and thereby allow an organism to grow. In EUKARYOTE cells, a single cell splits in two, first by division of the NUCLEUS (occurring by MITOSIS or MEIOSIS), then by fission of the CYTOPLASM. For growth and asexual reproduction, where the daughter cells are required to be genetically identical to their parents, mitosis is used. Meiosis results in daughter cells having half the number of chromosomes (HAPLOID). This type of division results in the production of haploid sex cells or GAMETES, which allows genetic information from two parents to be combined at FERTILIZATION, when the DIPLOID number of chromosomes is restored. *See also* ALTERNATION OF GENERATIONS

cellophane Flexible, transparent film made of regenerated CELLULOSE and used mostly as a wrapping material. It is made by dissolving wood pulp or other plant material in an ALKALI, to which carbon disulphide is added to form viscose. This is forced through a narrow slit into a dilute acid, where it forms a film of cellulose.

cell potential Potential difference between the electrodes of an electric CELL. It depends on the internal resistance (r) of the cell and the external resistance (R) through which the current flows. It is given by the formula $ER/R(r + R)$, where E is the ELECTROMOTIVE FORCE on open circuit.

cellular telephone *See* MOBILE TELEPHONE

celluloid Hard plastic invented by US scientist John Hyatt (1837–1920) in 1869. Hyatt made the plastic by mixing cellulose nitrate with pigments and fillers in a solution of camphor and alcohol. When heated, it can be moulded into a variety of shapes, and hardens on cooling. The first major plastic, it was used for early motion picture film. It is highly flammable.

cellulose ($[C_6H_{10}O_5]_n$) POLYSACCHARIDE CARBOHYDRATE that is the structural constituent of the cell walls of plants and algae. Consisting of parallel unbranched chains of GLUCOSE units cross-linked together into a stable structure, it forms the basic material of the paper and textile industries.

cellulose nitrate (nitrocellulose or guncotton) Organic compound made by reacting the natural polymer CELLULOSE with nitric acid. It was chiefly used in the 19th and 20th centuries (together with camphor) to make CELLULOID – one of the first plastics – or as a propellant explosive (called cordite in Britain).

cellulose plastic THERMOPLASTIC derivative of CELLULOSE. Cellulose plastics include: CELLULOSE NITRATE (or nitrocellulose), used in explosives and propellants; various cellulose acetates for making textile fibres and packaging films; and ethyl cellulose, used in the manufacture of shock-resistant materials. *See also* RAYON

Celsius, Anders (1701–44) Swedish astronomer who invented the CELSIUS, or centigrade, scale in 1742. He was a strong supporter of the introduction of the Gregorian calendar. In 1733 he published a collection of 316 observations of the AURORA Borealis made by himself and other astronomers.

Celsius Temperature scale based on the freezing point of water (0°C) and the boiling point of water (100°C). The interval between these points is divided into 100 degrees. The name "Celsius" officially replaced "centigrade" in 1948. Degrees Celsius are converted to degrees FAHRENHEIT by multiplying by 1.8 and then adding 32. The scale was devised by Anders CELSIUS.

cement In geology, a mineral material, such as CALCITE, that fills open pore space in sediments and binds grains to form a SEDIMENTARY ROCK. In construction, the term usually refers to Portland cement. This is made by heating a mixture of limestone and clay, grinding it, and adding GYPSUM.

cementite (Fe_3C) Hard, iron-carbon compound present as a major component of steel. The first of all steel-making processes, practised in the Hittite civilization of the 2nd millennium BC, is the cementite process; the ancient smiths beat carbon from hot wood ashes into wrought iron, so forming cementite.

Cenozoic Most recent era of geological time, beginning about 65 million years ago and extending up to the present. It is subdivided into the TERTIARY and QUATERNARY periods. It is the era during which the modern world with its present geographical features, and plants and animals developed.

Centaurus (Centaur) Brilliant constellation of the S sky. It has 13 stars above magnitude 3.5, the brightest of which is ALPHA CENTAURI. The third-brightest star in the sky, it has a magnitude of –0.27. Centaurus also contains Omega Centauri, the brightest globular cluster galaxy in the sky.

centigrade *See* CELSIUS

centimetre Unit of length defined as one hundredth part of a metre. *See* WEIGHTS AND MEASURES.

central nervous system (CNS) In some advanced invertebrates, a neural pathway along which are located clusters of NEURONS called GANGLIA; these are involved in the movement of limbs, wings, and so on. In vertebrates, part of the NERVOUS SYSTEM that comprises the brain and spinal cord, and is connected to the PERIPHERAL NERVOUS SYSTEM, a branching network of sensory and motor nerves. In humans, the CNS coordinates all neural activity including that producing movement, thought, emotion and REFLEX ACTIONS. *See also* AUTONOMIC NERVOUS SYSTEM (ANS)

central processing unit (CPU) Part of a digital computer that controls all operations. In most modern computers, the CPU consists of one complex INTEGRATED CIRCUIT (IC), a chip called a MICROPROCESSOR. A CPU contains temporary storage circuits that hold data and instructions; an arithmetic and logic unit (ALU) that performs calculations; and a control unit that organizes operations.

centre of gravity Point at which the weight of a body can be considered to be concentrated and around which its weight is evenly balanced. An object in free flight rotates around its centre of gravity, which in turn moves along the trajectory that would be followed by a small object of the same mass as the object. In a uniform gravitational field, the centre of gravity is the same as the CENTRE OF MASS.

centre of mass Point at which the whole mass of an object or group of objects is considered to be concentrated. The gravitational forces between spherical celestial bodies can be correctly calculated by regarding their masses as being located at their centres.

centrifugal force *See* CENTRIPETAL FORCE

centrifuge Rotating device used for separating substances. In laboratories, centrifuges separate particles from suspensions, and erythrocytes (red blood cells) from plasma. In the food industry, centrifuges separate cream from milk and sugar from syrup. In each case, the denser substance is forced to the outside of a rotating container. A spin dryer uses the same principle to remove water from clothes.

centriole Dense body consisting of microtubules near the nucleus of a cell. It occurs in all cells except those of ANGIOSPERMS and the sperm cells of ferns and GYMNOSPERMS. During CELL DIVISION centrioles reproduce before the rest of the cell and move to each pole to form the spindle.

centripetal force In circular or curved motion, the force acting on an object that keeps it moving in a circular path. For example, if an object attached to a rope is swung in a circular motion above a person's head, the centripetal force acting on the object is the tension in the rope. Similarly, the centripetal force acting on the Earth as it orbits the Sun is gravity. In accordance with Newton's laws, the reaction to this can be regarded as a **centrifugal force**, equal in magnitude and opposite in direction.

centromere Part of a CHROMOSOME that appears only during CELL DIVISION. When chromosomes shrink during MEIOSIS or MITOSIS, the centromere appears as a narrowing that contains no genes. It connects the chromosomes to the spindle fibres.

centrosome Region in a cell where microtubules are broken down and assembled during CELL DIVISION. Located alongside the nucleus, the centrosome contains two CENTRIOLES. At METAPHASE, the centrioles separate and the two regions of the centrosome containing them move to opposite sides of the cell as the microtubules form a spindle between them. The spindle eventually divides the chromosomes into the two daughter cells. *See also* HAPLOID

cephalochordates (lancelets or amphioxus) Small, marine animals that look and swim rather like fish but have no head or paired fins. They are related to vertebrates, having a NOTOCHORD throughout their lives. Their method of filter-feeding is, however, much more like that of many animals without backbones. Cephalochordates inhabit shallow, temperate and tropical seawater.

cephalopod Any of more than 600 species of predatory marine MOLLUSC of the class Cephalopoda, including squid, nautilus, octopus and cuttlefish. Each has eight or more arms surrounding the mouth, which typically has a beak. The nervous system is well developed, permitting great speed and alertness; the large eyes have an image-forming ability equal to that of vertebrates. Most squirt an inky fluid to alarm attackers. Cephalopods move by squirting water from their mantle edge. Their heavily yolked eggs develop into larval young that resemble the adults. Members of this class vary dramatically in size from 4cm (1.5in) to the giant squid, which may reach 20m (65ft).

cephalosporin Class of semi-synthetic ANTIBIOTIC drugs derived from fungi of the genus *Cephalosporium*. Similar to PENICILLIN, they are effective against a wide spectrum of BACTERIA, including some that have become resistant to penicillin.

Cepheid variable One of an important class of VARIABLE STARS that pulsate in a regular manner, accompanied by changes in LUMINOSITY. Cepheids periodically expand and contract, changing in size by as much as 30% in each cycle. The average luminosity is 10,000 times that of the Sun. Cepheids became important in cosmology in 1912, when Henrietta LEAVITT discovered a simple relationship between the period of light variation and the absolute magnitude of a cepheid. This relationship, the **period-luminosity law**, enables the distances of stars to be ascertained, not only in our Galaxy but in other galaxies too.

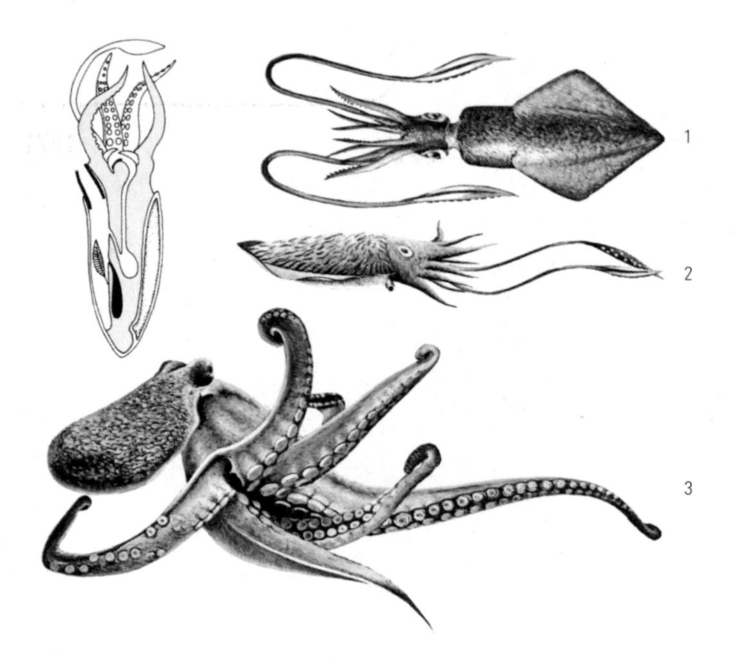

◀ **cephalopod** The squid (1), cuttlefish (2) and octopus (3) are all swimming molluscs of the Cephalopoda group. They have advanced, powerful eyes, and tentacles lined with sucker pads which are used to catch fish and small crustaceans. The horny jawed mouth is powerful enough to break up their prey before it is digested in the gut.

Cepheus Indistinctive constellation of the N sky, it adjoins URSA MAJOR. It contains the star Delta, important as a prototype CEPHEID VARIABLE.

ceramic In art and technology, article made from inorganic compounds formed in a plastic condition and hardened by heating in a furnace. **Earthenware** is a porous ceramic made from kaolin, ball clay and crushed flint. **Porcelain** is made from kaolin and feldspar, and heated to a higher temperature. It is nonporous and translucent. Special ceramics are made from pure aluminium oxide, silicon carbide, titranates and other compounds.

cerebellum Part of the BRAIN, often known as the "little brain", located at the base of the CEREBRUM. It is involved in maintaining muscle tone, balance and coordinated movement.

cerebral cortex Deeply fissured outer layer of the CEREBRUM. The cortex, or "grey matter", is the most sophisticated part of the BRAIN, responsible for the appreciation of sensation, initiating voluntary movement and for all higher functions, such as the emotions and intellect.

cerebral hemispheres Lateral halves of the CEREBRUM, the largest parts of the BRAIN and the sites of higher thought. Due to the crossing-over of nerve fibres from one cerebral hemisphere to the other, the right side controls most of the movements and sensation on the left side of the body, and vice versa. Damage to the cerebral hemispheres often produces personality changes.

cerebral palsy Disorder mainly of movement and coordination caused by damage to the BRAIN during or soon after birth. It may feature muscular spasm and weakness, lack of coordination and impaired movement or paralysis and deformities of the limbs; intelligence is not necessarily affected. The condition may result from any one of a number of causes, such as faulty development, oxygen deprivation, birth injury, haemorrhage or infection.

cerebrospinal fluid Clear fluid that cushions the BRAIN and SPINAL CORD, giving some protection against shock. It is found between the two innermost MENINGES (membranes) in the four ventricles of the brain and in the central canal of the spinal cord. A small quantity of the fluid can be withdrawn by lumbar puncture to aid diagnosis of some brain disorders.

cerebrum Largest and most highly developed part of the BRAIN, consisting of the CEREBRAL HEMISPHERES separated by a central fissure. It is covered by the CEREBRAL CORTEX. It coordinates all higher functions and voluntary activity.

Cerenkov, Pavel Alekseevich (1904–90) Russian physicist. Working at the Institute of Physics of the Soviet Academy of Science, he discovered that light (CERENKOV RADIATION) is emitted by charged particles travelling at very high speeds, a phenomenon known as the **Cerenkov effect**. He was awarded the 1958 Nobel Prize for physics with his co-workers, the Russian physicists I.M. Frank and I.Y. TAMM.

Cerenkov radiation Light emitted when energetic particles travel through a transparent medium, such as water, at a speed higher than the velocity of light in that medium. This action is called the **Cerenkov effect**. It is analogous to the SONIC BOOM, and a cone of light is emitted trailing the path of the particle. It is named after Pavel CERENKOV, who discovered the radiation in 1934. Cerenkov radiation is used in a Cerenkov counter or detector of energetic particles.

Ceres Largest ASTEROID and the first to be discovered, by Giuseppe PIAZZI on 1 January 1801. Its diameter is 913km (567mi). It orbits in the main asteroid belt, at an average distance from the Sun of 414 million km (257 million mi), the distance of the "missing" planet predicted by BODE'S LAW.

cerium (symbol Ce) Soft, ductile, iron-grey metallic element, the most abundant of the LANTHANIDE SERIES. It was first isolated in 1803. The chief ore is monazite. It is used in alloys (for use in lighter flints), in catalysts, nuclear fuels, in the making of glass and as the core of carbon electrodes in arc lamps. Properties: at.no. 58; r.a.m. 140.12; r.d. 6.77; m.p. 798°C (1,468°F); b.p. 3,257°C (5,895°F). The most common isotope is ^{140}Ce (88.48%).

cermet One of a group of hard and brittle heat-resistant materials. Cermet is a combination of ceramic material and metal. Applications include the manufacture of drilling tools, heat shields and turbine blades.

CERN (European Laboratory for Particle Physics) Nuclear research centre sited on the Franco-Swiss border W of Geneva. It was founded in 1954 as an intergovernmental organization,

All animal cells are remarkably similar in structure. Of the millions of different species of animal, from the simple sponge to the complex mammal, the cells of which they are made share much the same basic internal organization. The human body is made of 10 million million cells, and while each cell has its own specific function to perform, they have to cooperate and communicate to ensure the body survives.

Modern microscopy has revealed the complexity of the internal structure of animal cells. Some structures are responsible for maintaining the shape of the cell, others assemble and transport complex molecules, and yet more are involved with the essential processes of CELL DIVISION. Different cell processes occur within different types of compartment, called ORGANELLES. Many organelles are common to both plant and animal cells, but the most significant difference is that animal cells do not contain CHLOROPLASTS and are therefore unable to perform PHOTOSYNTHESIS, obtaining their energy rather from digested food.

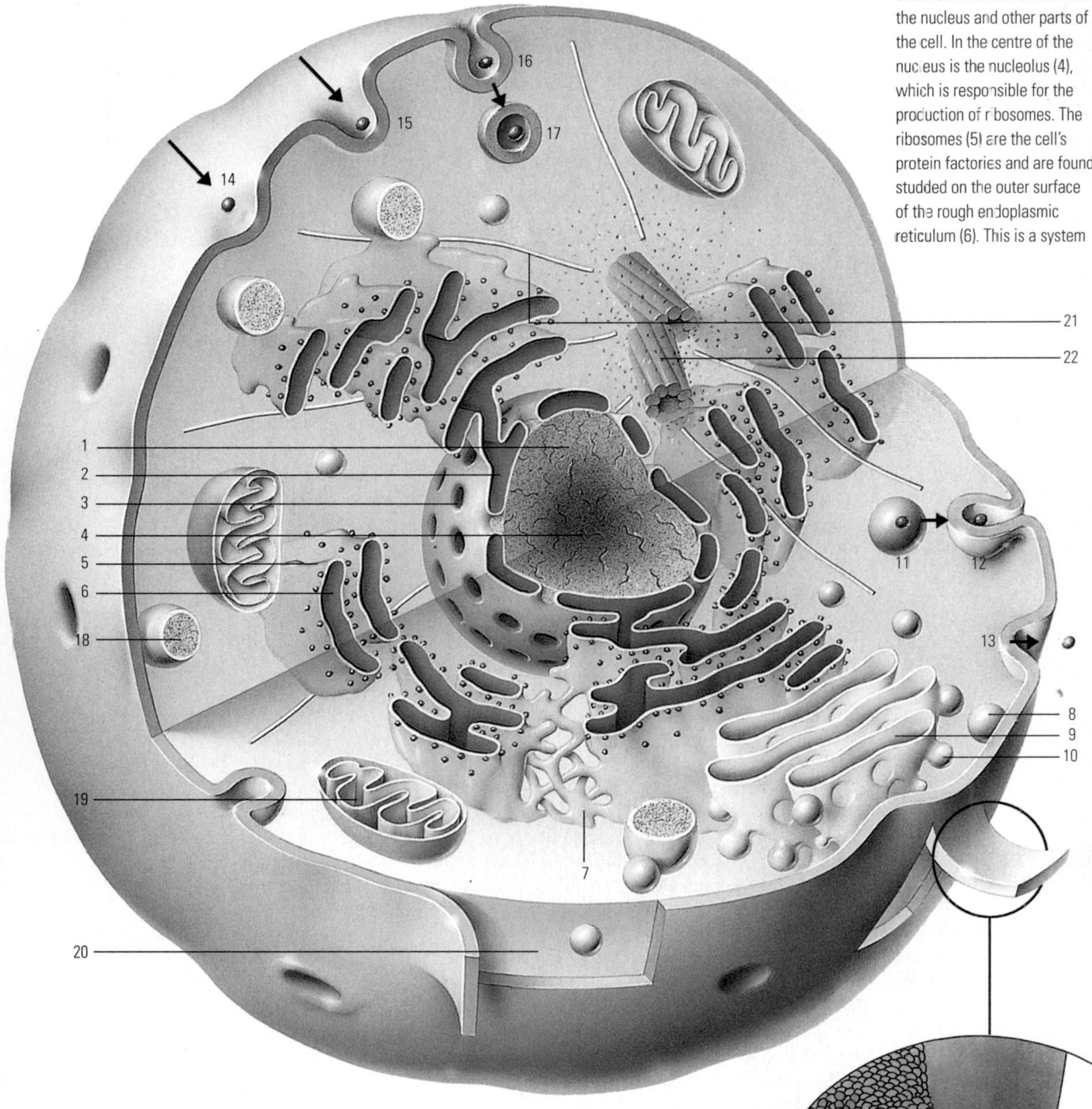

▼ Animal cells are compartmentalized into various organelles. The most prominent of these is the nucleus (1), the information centre of the cell, which contains the genetic material in the form of long thread-like chromosomes. It is bounded by the nuclear membrane (2), which has many pores (3) to allow communication between the nucleus and other parts of the cell. In the centre of the nucleus is the nucleolus (4), which is responsible for the production of ribosomes. The ribosomes (5) are the cell's protein factories and are found studded on the outer surface of the rough endoplasmic reticulum (6). This is a system of flattened sacs and tubes of membrane connected to the nuclear membrane. It brings the messenger RNA molecules – which direct protein synthesis – to the ribosomes. Lipids are also produced here which form part of the cell membrane. The smooth endoplasmic reticulum (7) – connected to the rough – produces small membranous spheres called vesicles (8) These transport proteins to the Golgi apparatus (9), which modifies, sorts and packs many large molecules into other vesicles, which bud off the apparatus (10). They are then sent to other organelles or secreted from the cell. The fusion of such vesicles with the cell membrane allows particles to be transported out of the cell (exocytosis) (11–13). Similarly, particles can be brought into the cell (14–17) in vesicles (endocytosis). Molecules entering the cell may be broken down by enzymes found in special vesicles, called lysosomes (18). The mitochondria (19) are the powerhouses of the cell, using oxygen and food to generate energy (as ATP), which is then used in many metabolic processes. The majority of these are chemical reactions which take place in the aqueous medium of the cytoplasm (20). Running through the cytoplasm there is a matrix of protein filaments (microtubules, 21) known as the cytoskeleton, which acts like scaffolding, giving the cell shape, and also providing a system for transport and movement. The cytoskeleton originates at the centrioles (22), which also help the chromosomes line up during cell division.

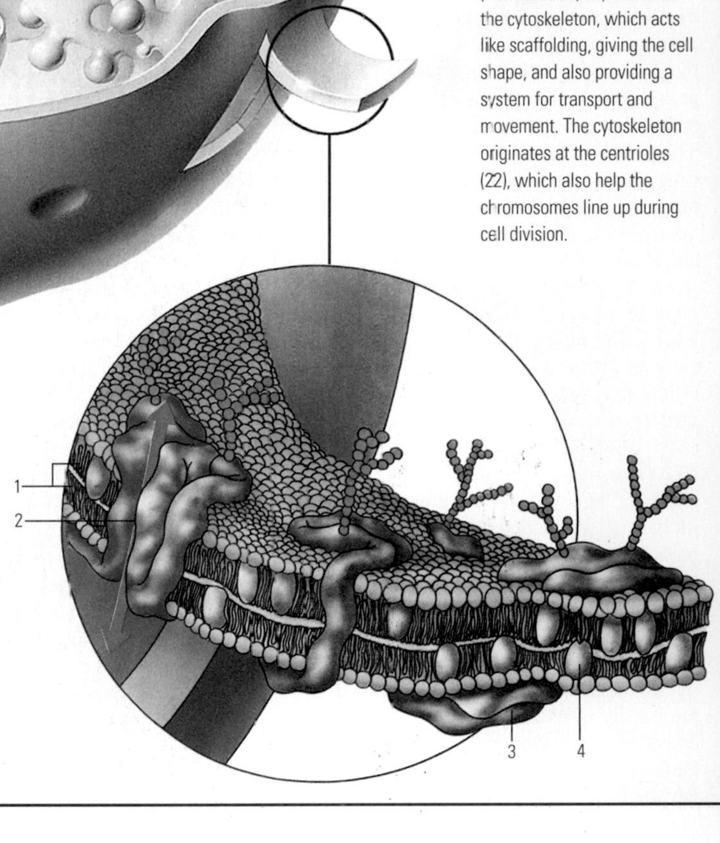

► The cell membrane is a thin, two-fold layer of lipid molecules (1) which surrounds the cytoplasm of all cells. Very few molecules can pass through the cell membrane unaided. Special transport proteins and protein-lined channels (2) in the membrane allow through sugars, amino acids and essential ions like sodium and calcium. Other proteins (3) act as receptors for chemical signals, and provide a chemical signature that allows recognition by other cells, particularly of the immune system. Cholesterol molecules (4) are important for the membrane's stability, though too many can cause the membrane to seize up.

PROTON POWERHOUSE

Cells are powered primarily by energy released from ATP (adenosine triphosphate) as it becomes ADP (adenosine diphosphate). ATP is made in the mitochondria (1) by recycling ADP. The first step is to split pyruvate (2) – a fuel molecule derived from glucose in the cytoplasm – into carbon dioxide, hydrogen and high-energy electrons. These electrons pass along a line of proteins in the inner membrane (3), giving them energy to pump out protons (4) into the intermembrane space (5). As more protons are pumped out, a pressure builds up in the space, forcing protons back across the membrane. But the protons can only flow back into the matrix via the ATP generator (6) – the enzyme ATP synthetase – and as they do so they drive round the blades of this turbine, producing ATP (7).

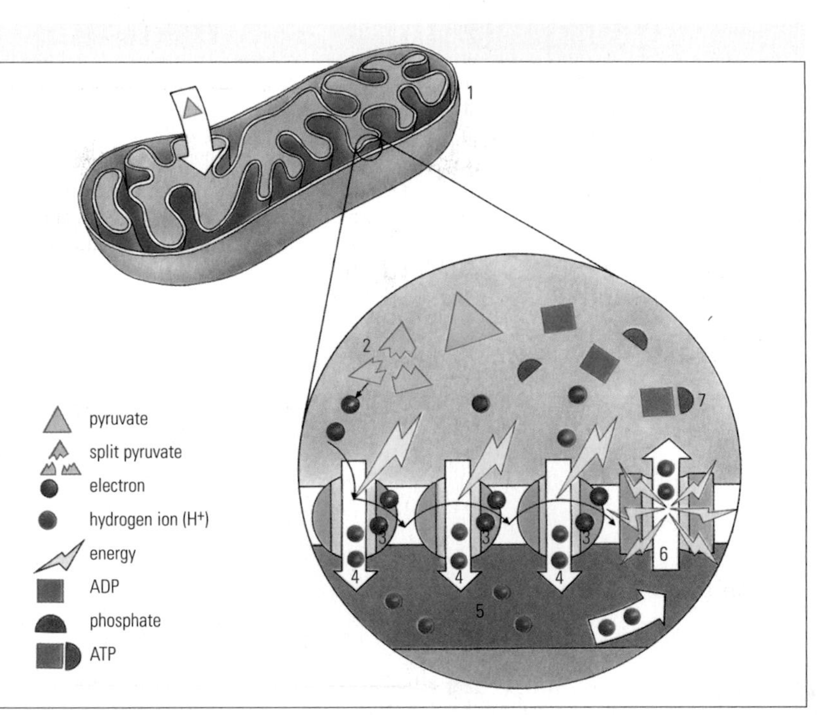

PLANT CELLS

Plant cells come in a great variety of shapes and sizes, and not all cells contain all the features in the "typical" cell below. However, they call have un inflexible cellulose cell wall on the outside of the membrane.

The earliest plant cells are thought to have formed more than 1,000 million years ago, when cells that fed on the nutrients of the primaeval seas were colonized by bacteria capable of PHOTOSYNTHESIS. Over time the bacteria lost their independence and developed into CHLOROPLASTS. The sugars resulting from photosynthesis can be broken down by MITOCHONDRIA releasing energy to fuel the cell's activity, or used as a source of carbon for larger molecules from which new plant material is made. The presence of structures that produce and store food is another feature that distinguishes plant cells from animal cells.

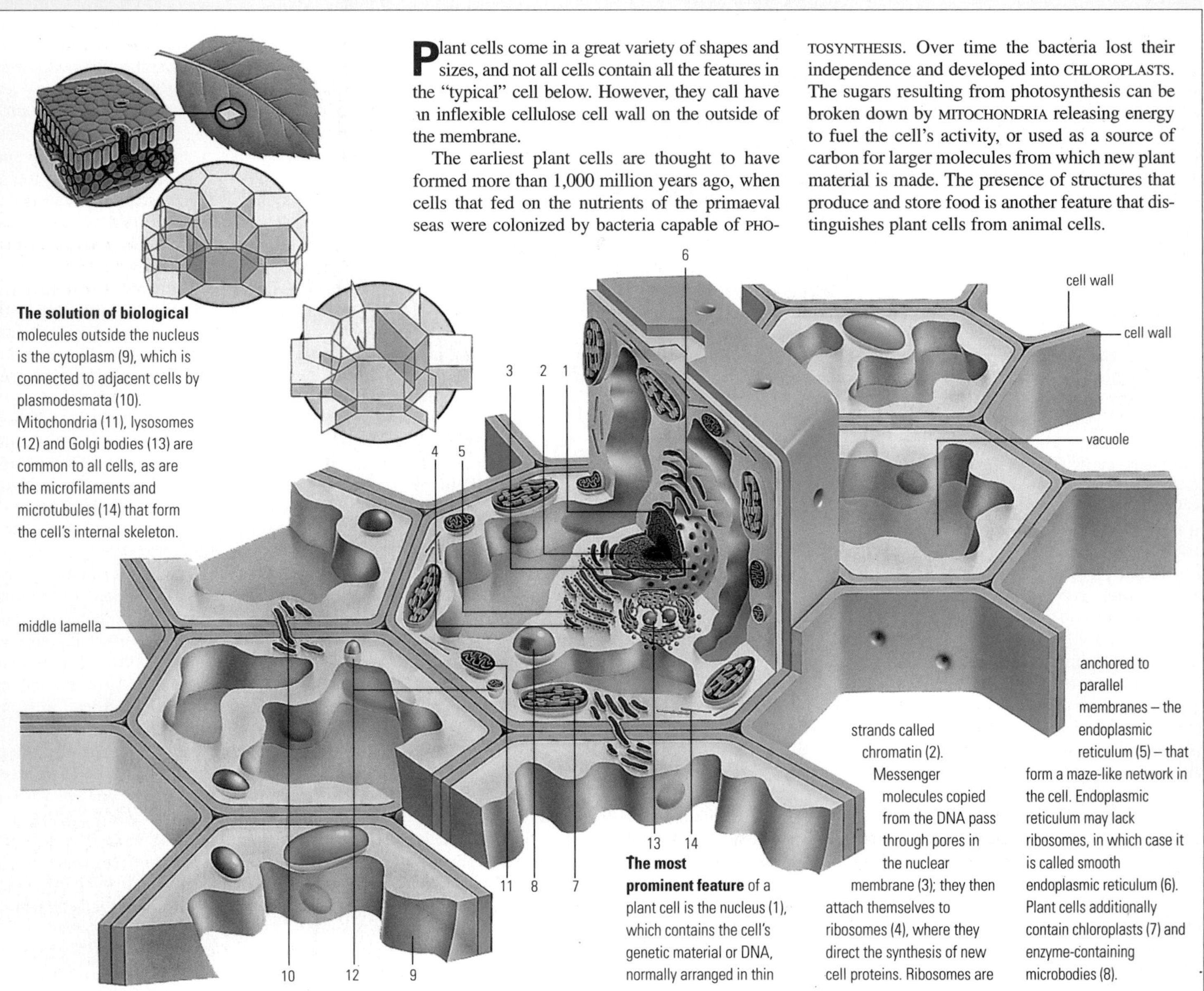

The solution of biological molecules outside the nucleus is the cytoplasm (9), which is connected to adjacent cells by plasmodesmata (10). Mitochondria (11), lysosomes (12) and Golgi bodies (13) are common to all cells, as are the microfilaments and microtubules (14) that form the cell's internal skeleton.

The most prominent feature of a plant cell is the nucleus (1), which contains the cell's genetic material or DNA, normally arranged in thin strands called chromatin (2). Messenger molecules copied from the DNA pass through pores in the nuclear membrane (3); they then attach themselves to ribosomes (4), where they direct the synthesis of new cell proteins. Ribosomes are anchored to parallel membranes – the endoplasmic reticulum (5) – that form a maze-like network in the cell. Endoplasmic reticulum may lack ribosomes, in which case it is called smooth endoplasmic reticulum (6). Plant cells additionally contain chloroplasts (7) and enzyme-containing microbodies (8).

C

CHAIN REACTION

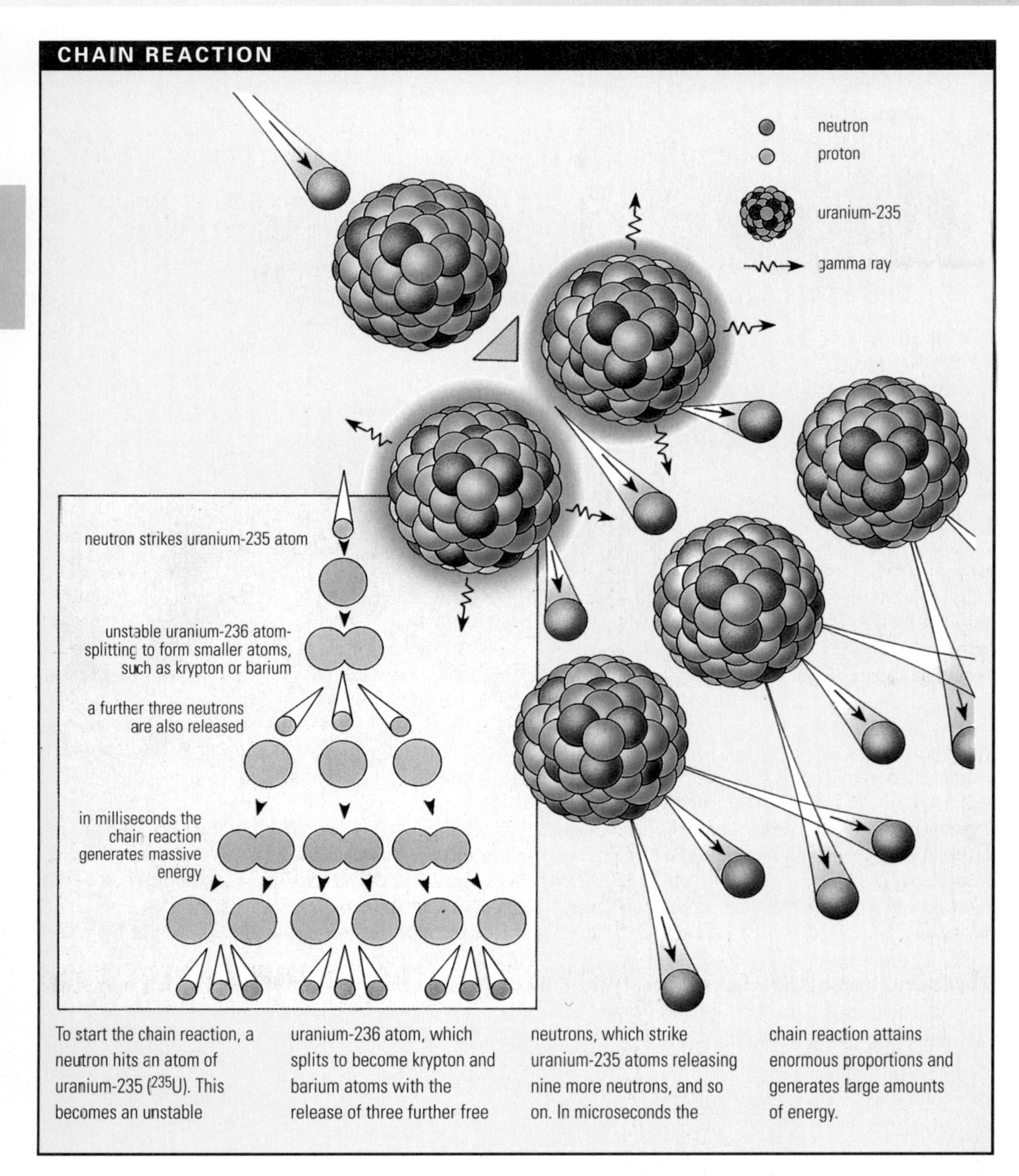

To start the chain reaction, a neutron hits an atom of uranium-235 (^{235}U). This becomes an unstable uranium-236 atom, which splits to become krypton and barium atoms with the release of three further free neutrons, which strike uranium-235 atoms releasing nine more neutrons, and so on. In microseconds the chain reaction attains enormous proportions and generates large amounts of energy.

when it was called Conseil Européen pour la Recherche Nucléaire. It is the principal European centre for research into PARTICLE PHYSICS.

cerussite (white lead) Mineral form of lead carbonate ($PbCO_3$). It forms in oxidized regions of mineral veins, particularly of GALENA. It has prismatic or needle-shaped crystals (orthorhombic) that are usually colourless or white, but may be green or grey with a resinous or vitreous lustre. Cerussite is mined as lead ore in Mexico; it is also used as a white pigment. Hardness: 3–3.5; r.d. 6.6.

cervical smear (pap test) Test for CANCER of the CERVIX, established by the US physiologist George Papanicolau (1883–1962). In this diagnostic procedure, a small sample of tissue is removed from the cervix and is examined under a MICROSCOPE for the presence of abnormal, pre-cancerous cells. Treatment in the early stages of cervical cancer can prevent the disease from developing.

cervix Neck of the UTERUS projecting downwards into the VAGINA. It dilates (expands) widely to allow the passage of a baby during BIRTH.

cestode Any member of the class Cestoda, which contains the endoparasitic TAPEWORMS. Cestodes usually live in the intestines, holding on to its lining by terminal hooks. They have no mouths and absorb their food through the body wall from their hosts. They are HERMAPHRODITE, one segment fertilizing another. Ripe segments drop off and are voided by the host after which, to survive, they usually need to be eaten by a second host.

cetacean Any member of the order Cetacea of aquatic mammals found in all oceans and some rivers. The larger cetaceans are WHALES, the smaller ones are PORPOISES or DOLPHINS. Cetaceans are streamlined, with a pair of front flippers and horizontal tail flukes. Their bodies are insulated by a thick layer of BLUBBER. Length: from 1.5m (4.9ft) to more than 30m (100ft); weight: from 36kg (80lb) to 100 tonnes.

Cetus (Whale) Fourth-largest constellation, but inconspicuous. It lies across the Equator. Its brightest star is Beta, magnitude 2.0

CFC Abbreviation of CHLOROFLUOROCARBON

CGS system System of metric units based on the centimetre, gram and second. The DYNE is the unit of force, the ERG the unit of energy. It has been largely superseded by the SI UNITS system.

Chadwick, Sir James (1891–1974) British physicist who discovered and named the NEUTRON and worked on the development of the atomic bomb. Chadwick worked on radioactivity with Ernest RUTHERFORD at the Cavendish Laboratory, Cambridge. In 1920 Rutherford had predicted a particle without electric charge in the nucleus of an atom, and in 1932 Chadwick proved the neutron's existence and calculated its mass. For this, he received the 1935 Nobel Prize for physics. He established a school of nuclear physics in Liverpool, constructing Britain's first CYCLOTRON there in 1935. During World War 2, he moved to the USA to head British research for the MANHATTAN PROJECT to develop the atom bomb.

chaetognath Any member of the phylum Chaetognatha, which includes arrow worms – small marine animals that are not at all closely related to any other group. They are between 3mm and 10cm (0.125–4in) long, and have narrow bodies, with bristly jaws for seizing their prey (PLANKTON). Although invertebrate animals, they swim like fish, by means of fins. They are HERMAPHRODITE. Some species are useful indicator organisms for marine biologists because they are sensitive to the temperature, salinity and depth of the water in which they live.

Chain, Sir Ernst Boris (1906–79) British biochemist, b. Germany. Chain shared the 1945 Nobel Prize for physiology or medicine with Howard FLOREY and Alexander FLEMING for the isolation and development of PENICILLIN as an ANTIBIOTIC. He also studied spreading factor, an enzyme that aids the dispersal of fluids in tissue.

chain reaction Self-sustaining nuclear FISSION reaction in which one reaction is the cause of a second, the second of a third and so on. The initial conditions are often critical, in that the quantity of fissionable material must exceed the CRITICAL MASS. Such a reaction produces energy for all commercial NUCLEAR REACTORS. *See also* NUCLEAR WEAPON

chalcanthite Mineral which consists mainly of hydrated copper sulphate ($CuSO_4.5H_2O$), although it is rarely used as a source of copper. It occurs as greenish-blue triclinic crystals or as fibrous veins or stalactites. It is soluble in water and has a nauseating taste. Hardness 2.5; r.d. 2.25.

chalcedony Microcrystalline form of QUARTZ. When cut and polished, it is used by gem engravers. It is waxy, lustrous and there are white, grey, blue and brown varieties. It is often coloured by artificial methods. Some varieties contain impurities giving a distinctive appearance, such as AGATE (coloured bands), ONYX (striped) and bloodstone (dark green with red flecks).

chalcocite Dark grey, metallic, soft mineral, copper sulphide (Cu_2S); one of the chalcocite group. It is a major ore of copper, found mainly in sulphur deposits. The crystals occur in orthorhombic granular masses, or rarely in prismatic form. Hardness 2.5 to 3.0; r.d. 5.7.

chalcopyrite (copper pyrites) Opaque, brass-coloured, copper iron sulphide ($CuFeS_2$); the most important copper ore. It is found in sulphide veins and in IGNEOUS and certain METAMORPHIC ROCKS (*see* CONTACT METAMORPHISM). The crystals are tetragonal but often occur in masses. Hardness 3.5–4; r.d. 4.2.

chalk Porous, fine-grained rock, mainly composed of CALCAREOUS skeletons of marine microorganisms, especially COCCOLITHS and FORAMINIFERA. It varies in properties and appearance; pure forms, such as CALCITE, contain up to 99% calcium carbonate. It is used in making putty, plaster and cement, and harder forms are occasionally used for building.

Challenger expedition (1872–76) British expedition in oceanographic research. The *Challenger* ship had a staff of six naturalists headed by Charles Wyville THOMSON. It sailed nearly 128,000km (69,000 nautical mi) making studies of the life, water and seabed in the three main oceans.

Chamaeleon Small constellation of the s sky, near the South Pole. Its brightest stars are only of magnitude 4.1.

Chamberlain, Owen (1920–) US physicist. After working on the development of the atom bomb between 1942 and 1946 he became professor of physics at the University of California,

Berkeley. With Emilio SEGRÈ, using the BEVATRON particle ACCELERATOR, he confirmed the existence of the antiproton and was awarded the Nobel Prize for physics in 1959. *See also* ANTIMATTER

Chandrasekhar, Subrahmanyan (1910–95) US astrophysicist, b. India. He formulated theories about the creation, life and death of stars, and calculated the maximum mass of a WHITE DWARF star before it becomes a NEUTRON STAR; the **Chandrasekhar limit**. It equals 1.4 times the mass of the Sun. He shared the 1983 Nobel Prize for physics with William FOWLER.

change of state (change of phase) In physics, change that takes place when matter turns from one physical PHASE (gas, liquid or solid) into another. Typical changes of state are EVAPORATION and CONDENSATION, SUBLIMATION, boiling, and melting and solidification. Change of state always involves the absorption or release of heat energy. But because of the phenomenon of LATENT HEAT, a change in temperature does not necessarily accompany a change of state.

Channel Tunnel (Chunnel) Railway tunnel under the English Channel, 49km (31mi) long. The first Channel tunnel was proposed in 1802 by a French engineer. A start was made in 1882 but soon abandoned for defence reasons. Another false start was made in the 1970s. In 1985 Eurotunnel, a joint French-English private company, was granted a 55-year concession to finance and operate the tunnel. The French and English sections were linked in 1990 and the tunnel became operational in 1994. It consists of two railway tunnels and one service tunnel, and links Folkestone, S England, with Calais, N France. A tunnel fire in November 1996 raised safety fears.

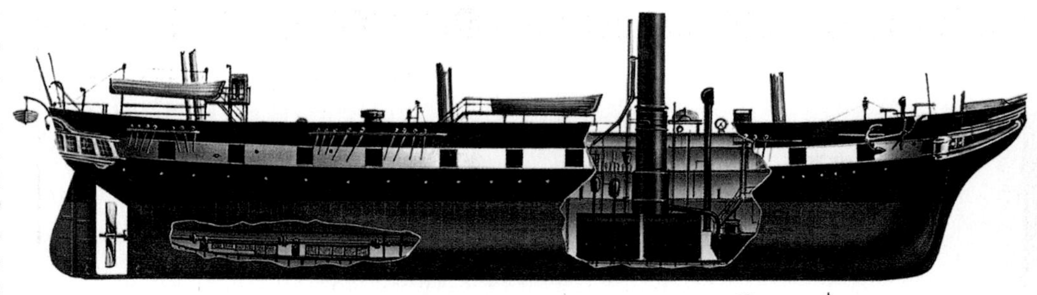

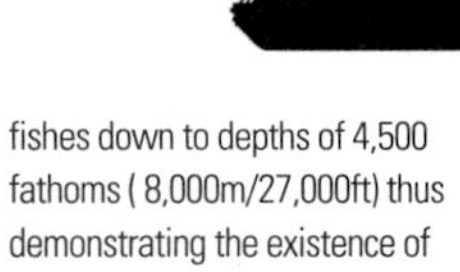

▲ ***Challenger* expedition** In 1872 the Royal Navy made HMS *Challenger*, a full-rigged, spar-decked corvette, available to the Royal Society for a three-and-a-half year oceanographic survey – the first of its kind. One of the most important scientific voyages ever made, the expedition laid the foundations for the modern science of oceanography. Two biologist, W.B. Carpenter and Wyville Thomson, persuaded the British Government to equip the expedition to study deep-sea circulation and the distribution of life in the seas. The voyage set a pattern for similar oceanographic cruises during the late 19th and early 20th centuries. The voyage was the first to discover manganese nodules on the deep ocean floor, which were found at all sites in the ocean basins. The *Challenger* expedition also sampled the benthic fauna and fishes down to depths of 4,500 fathoms (8,000m/27,000ft) thus demonstrating the existence of life in the abyssal depths. Hundreds of dredge samples were examined initially on board and the samples were preserved for analysis on return to England.

chaos theory Theory that attempts to describe and explain the highly complex behaviour of apparently chaotic or unpredictable systems which show an underlying order. The behaviour of some physical systems is impossible to describe using the standard laws of physics. This is because the mathematics needed to describe these systems is too difficult for even the largest supercomputers. Such systems are sometimes known as "nonlinear" or "chaotic" systems, and they include complex machines, electric circuits, and natural phenomena such as the weather. Non-chaotic systems can become chaotic, as when smoothly flowing water hits a rock and becomes turbulent. The lack of an adequate description means that a standard prediction of their behaviour is also impossible.

CHANNEL TUNNEL

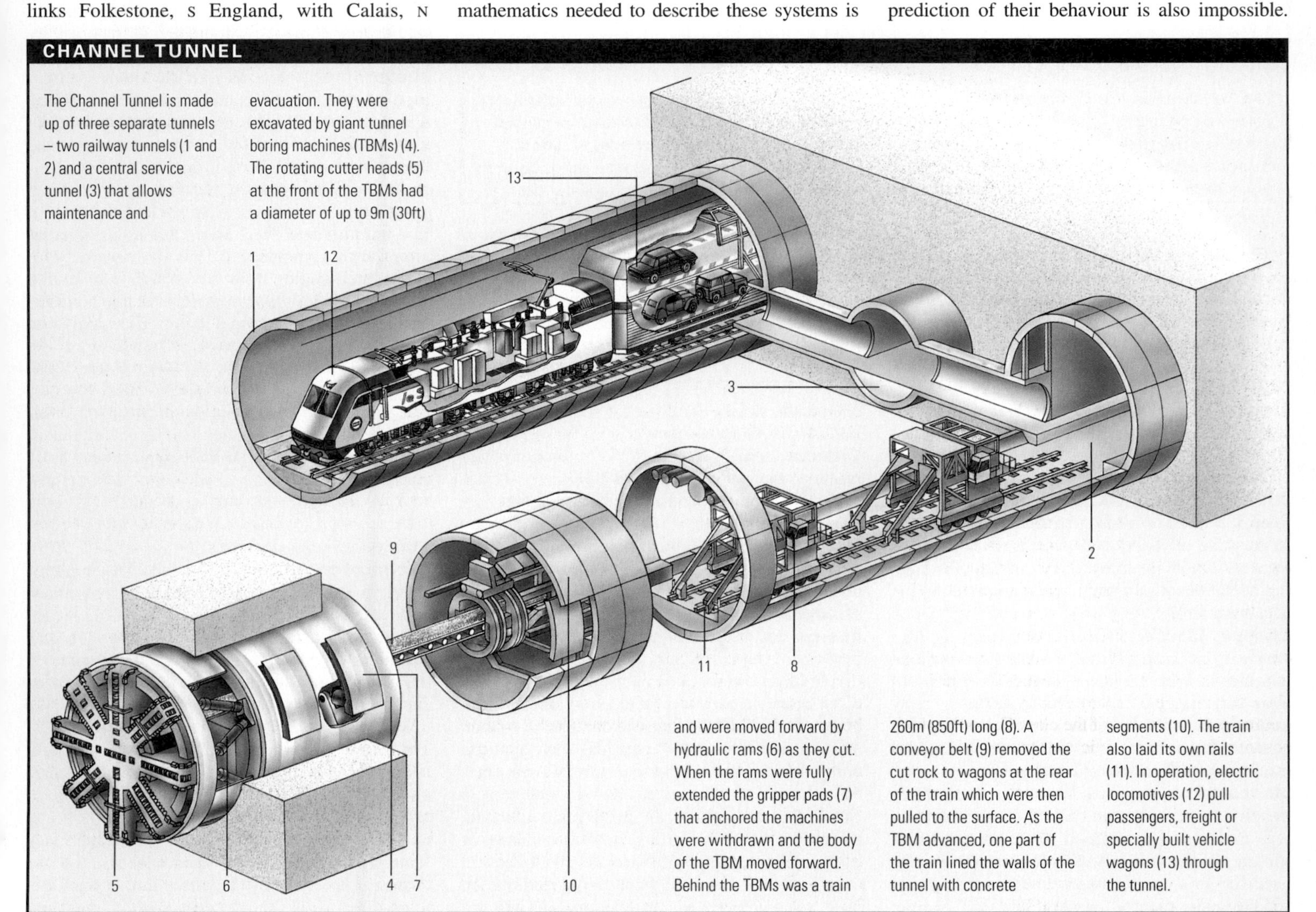

The Channel Tunnel is made up of three separate tunnels – two railway tunnels (1 and 2) and a central service tunnel (3) that allows maintenance and evacuation. They were excavated by giant tunnel boring machines (TBMs) (4). The rotating cutter heads (5) at the front of the TBMs had a diameter of up to 9m (30ft) and were moved forward by hydraulic rams (6) as they cut. When the rams were fully extended the gripper pads (7) that anchored the machines were withdrawn and the body of the TBM moved forward. Behind the TBMs was a train 260m (850ft) long (8). A conveyor belt (9) removed the cut rock to wagons at the rear of the train which were then pulled to the surface. As the TBM advanced, one part of the train lined the walls of the tunnel with concrete segments (10). The train also laid its own rails (11). In operation, electric locomotives (12) pull passengers, freight or specially built vehicle wagons (13) through the tunnel.

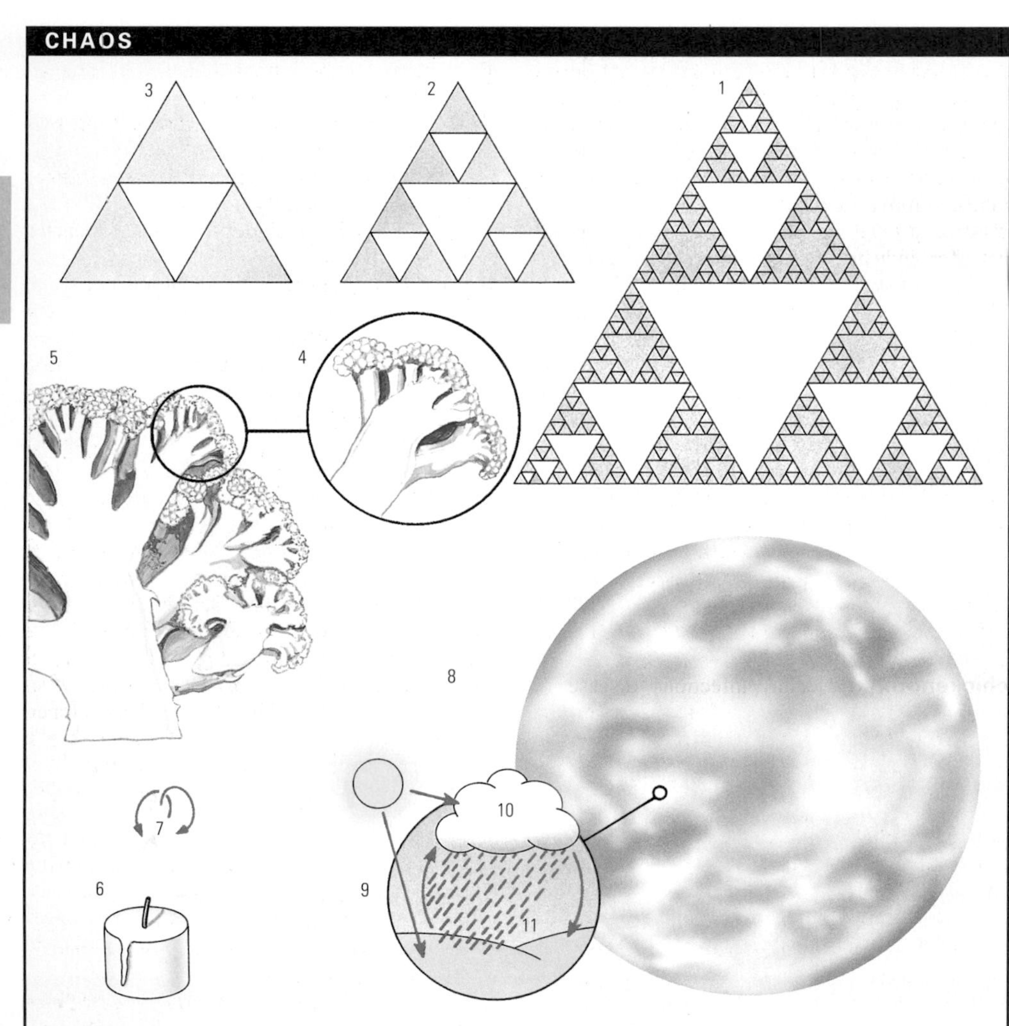

Chaos theory is used to explain seemingly complex phenomena, which can be mathematically modelled by applying simple numerical formula a great many times. Some chaotic systems are fractal, that is they exhibit self-similar geometries or structures – in other words a small section of the structure will resemble the whole, and if it is possible to describe the part mathematically the whole can be described also. The Sierpinsky gasket (1) is "constructed" by the repeated division of equilateral triangles (2–3), and is a fractal structure. The apparently complex structure of living things, such as the cauliflower, are also self-similar, a single floret (4) representing a whole head (5). The smoke from an extinguished candle (6) moves in a complex pattern that is difficult to understand, unless the effects of turbulence (7) on the laminar flow of the smoke are modelled. The climate of the Earth is extremely complex, but is governed by simple rules (8). Sunlight falling on the sea causes water to evaporate (9) and form clouds (10) which reflect the sunlight and prevent it reaching the sea or land. Temperature drops and rain may fall (11). If we could measure weather conditions on a large enough scale and make extremely detailed mathematical models, using computers, we could predict the weather perfectly.

Chaos theory provides mathematical methods needed to describe chaotic systems, and even allows some general prediction of a system's likely behaviour. Chaos theory also shows, however, that even the tiniest variation in the starting conditions of a system can lead to enormous differences in the state of the system some time later. Thus because it is impossible to know the precise starting conditions of a system, accurate prediction is also impossible.

Chappe, Claude (1763–1805) French engineer who, with his brother, Ignace, invented a SEMAPHORE telegraph in 1791. During the French Revolutionary Wars they used their telegraphs on towers to relay information about the state of the war between France and Austria, this method being much faster than other existing forms of communication.

charcoal Porous form of CARBON, made traditionally by heating wood in the absence of air, and used in W Europe until late medieval times for smelting iron ore. Today, charcoal is chiefly used for its absorptive properties, to decolorize food liquids such as syrups, and to separate chemicals. Artists use charcoal sticks for sketching. *See also* ABSORPTION

Charcot, Jean Martin (1825–93) French physician and founder of NEUROLOGY. He made classic studies of hypnosis and hysteria, and taught Sigmund FREUD and the French psychologist Pierre Janet (1869–1947). Charcot's work centred on discovering how behavioural symptoms of patients relate to neurological disorders.

charge *See* ELECTRIC CHARGE

charge-coupled device (CCD) Type of SILICON CHIP designed to capture images. The CCD is divided into a number of microscopic areas (pixels), arranged in rows. When a PHOTON hits a pixel, it knocks off an electron from a silicon atom, which becomes charged. An opposite charge in a layer on the base of the CCD confines this charged silicon atom, and in this way a charge builds up in each pixel relative to the number of photons hitting it. The contents of each pixel are read off 50 times a second, a row at a time, forming an electrical signal that can be used to create television pictures. CCDs are found in VIDEO cameras, FAX machines and digital CAMERAS. CCDs are much more sensitive to light than photographic film, and are used in most large telescopes to detect light from distant stars.

Charles, Jacques Alexandre César (1746–1823) French physicist, inventor and mathematician who was the first to use a hydrogen balloon. He discovered the law relating the expansion of a gas to its temperature rise. Gay-Lussac published this work some 15 years after Charles' discovery and it is alternately known as CHARLES' LAW or Gay-Lussac's law. He is credited with inventing a thermometric HYDROMETER and improving Fahrenheit's aerometer and Gravesande's hydrometer.

Charles' law Volume of a gas at constant pressure is directly proportional to its absolute temperature. The relationship was discovered by Jacques CHARLES in 1787. The law is a special case of the IDEAL GAS LAW. It is sometimes called Gay-Lussac's law, because Joseph GAY-LUSSAC established it more accurately in 1802.

Charon Satellite of PLUTO, discovered in 1978. Its diameter of 1,270km (790mi) makes it the largest satellite in relation to its primary (Pluto) in the Solar System. Estimates of Charon's mass vary from 8% to 16% of Pluto's. It is unique among planetary satellites in having a synchronous orbit with a period (6.4 days) that matches the rotation period of its primary; Charon therefore keeps the same face turned towards Pluto, and hangs motionless in Pluto's sky. Charon seems to have a greyish surface, probably of water-ice. There is almost certainly no permanent atmosphere, but some of the nitrogen outgassed by Pluto when at PERIHELION could be captured and retained by Charon as a temporary atmosphere.

chelation Chemical reaction in which a certain type of organic compound, termed a **chelating agent**, combines with a metal ION by forming coordinate bonds with two or more atoms of the organic compound. Tartaric acid $(CHOHCOOH)_2$ and ethylenediamine $(CH_2NH_2)_2$ are chelating agents. *See also* COORDINATION COMPOUND

chemical bond Mechanism that holds together atoms to form molecules. There are several types which arise either from the attraction of unlike charges or from the formation of stable configurations through electron-sharing. The number of bonds an atom can form depends upon its VALENCE. The main types are IONIC BOND, COVALENT BOND, METALLIC BOND and HYDROGEN BOND.

chemical engineering Application of engineering principles to the making of chemical products on an industrial scale. Unit processes of chemical engineering include OXIDATION-REDUCTION, HYDROGENATION, NITRATION and SULPHONATION, ELECTROLYSIS, POLYMERIZATION, ION EXCHANGE and FERMENTATION.

chemical equation Set of symbols used to represent a CHEMICAL REACTION. Equations show how atoms are rearranged as a result of a reaction, with **reactants** on the left-hand side and **products** on the right-hand side. For example, the formation of magnesium oxide when magnesium burns in oxygen is represented by

$$2Mg + O_2 \rightarrow 2MgO$$

The number of atoms of an element on the left-hand side of an equation must equal the number on the right.

chemical equilibrium Balance in a REVERSIBLE REACTION, when two opposing reactions proceed at constant equal rates with no net change in the system. The initial rate of the reactions falls off as the concentrations of reactants decrease and the build-up of products causes the rate of the reverse reaction to increase.

chemical reaction Change or process in which chemical substances convert into other substances. This usually involves the breaking and formation of CHEMICAL BONDS. Reaction mechanisms include ENDOTHERMIC, EXOTHERMIC, ADDITION, CONDENSATION, combination (formation of a COMPOUND), DECOMPOSITION and OXIDATION-REDUCTION reactions.

chemical warfare Employment of chemical weapons such as poison and nerve gases, defoliants and herbicides. Together with BIOLOGICAL WARFARE, chemical warfare was banned by the Geneva Convention (1925) after the use of mustard gas during World War 1. Despite this, many nations have developed chemical weapons. Mustard gas was used by Saddam Hussein to kill thousands of Kurds after the Iran–Iraq War. Herbicides, such as the defoliant Agent Orange, were used by US forces during the Vietnam War. During the Gulf War, Allied troops were inoculated and heavily protected against possible chemical warfare. Worries have recently surfaced about its terrorist uses, such as the Sarin nerve gas attack on the Tokyo underground system in 1995.

chemistry Branch of science concerned with the properties, structure and composition of substances and their reactions with one another. Today, chemistry forms a vast body of knowledge with a number of subdivisions: the major division is between organic and inorganic. **Inorganic** chemistry studies the preparation, properties and reactions of all chemical elements and their COMPOUNDS, except those of CARBON. However, such simple carbon compounds as CARBONATES, CARBIDES and carbon oxides come within the province of inorganic chemistry. The historic separation from organic chemistry is, in any case, a false one, because many "inorganic" compounds are found in living organisms, such as common salt (NaCl) in human blood. In education and industry, however, the distinction is frequently still made. **Organic** chemistry studies the reactions of carbon compounds. Because of carbon's unique ability to bond repeatedly with itself to form chains and rings of atoms (silicon is the only other element that has this property to any degree), carbon is able to form compounds of extreme complexity. For this reason it is predominantly the element of life, forming the "backbone" structures of such compounds as PROTEINS, FATS and CARBOHYDRATES. Organic compounds are about 100 times more numerous than nonorganic ones. Organic chemistry also studies an immense variety of molecules, including those of industrial compounds such as plastics, rubbers, dyes, drugs and solvents. **Analytical** chemistry deals with the composition of substances. PHYSICAL CHEMISTRY deals with the physical properties of substances, such as their boiling and melting points. Its subdivisions include ELECTROCHEMISTRY, THERMOCHEMISTRY and chemical KINETICS.

chemoreceptor Tiny region on the outer membrane of some biological cells that is sensitive to chemical stimuli. The chemoreceptor transforms a stimulus from an external molecule into a sensation, such as smell or taste.

chemotherapy Treatment of a disease (usually cancer) by a combination of chemical substances, or DRUGS, that kill or impair tumours or disease-producing organisms in the body. Specific drug treatment was first introduced in the early 1900s by Paul EHRLICH.

chemotropism Growth or movement of a plant or plant part in response to a chemical stimulus. In **positive** chemotropism, the movement is towards the chemical; in **negative** chemotropism movement is away from the chemical. An example occurs during pollination. The ovary releases sugars into the style of the flower, and these act positively to cause pollen to produce a pollen tube that moves down the style.

chemurgy Development of new chemical products for industry from organic raw materials, especially those of agricultural origin.

chernozem (black earth) Humus-rich type of dark soil typical of the grasslands of the steppes of Eurasia and the prairies of North America. It is prized for its agricultural qualities such as good structure and high nutrient content.

chert Impure, brittle type of flint. A cryptocrystalline variety of SILICA, its colour can be white, yellow, grey or brown. It occurs mainly in limestone and dolomite although its origin is unknown.

chiastolite Variety of andalusite, aluminium silicate (Al_2OSiO_4), found in METAMORPHIC ROCKS. It has elongated prismatic crystals, which in cross-section show a black cross on a grey ground. Hardness 7.5; r.d. 3.1–3.2.

chickenpox (varicella) Infectious disease of childhood caused by a virus of the HERPES group. After an incubation period of two to three weeks, a fever develops and red spots (which later develop into blisters) appear on the trunk, face and limbs. Recovery is usual within a week, although the possibility of contagion remains until the last scab has been shed.

childbirth *See* BIRTH

chilopod (centipede) Any member of the order Chilopoda, ARTHROPOD animals that have long, flat, segmented bodies, each bearing one pair of legs (the total number of legs being much less than 100, despite the name). The head has a large pair of claws with which the centipede seizes its prey and injects it with venom.

china clay *See* KAOLIN

china stone Partially decomposed granite, frequently used as a FLUX to produce VITRIFICATION and translucency, or mixed with silica and lime to form a glaze in the manufacture of PORCELAIN.

chinook Warm, dry foehn wind experienced on the E side of the Rocky Mountains in Canada and the USA, and in the European Alps. It commonly blows during winter and spring.

chip General term used to describe an INTEGRATED CIRCUIT.

chiropody Diagnosis and treatment of minor disorders of the foot. Typical conditions treated by a chiropodist include ingrowing toenails, veruccas and corns.

chiropractic Alternative medical practice based on the theory that the NERVOUS SYSTEM integrates all of the body's functions, including defence against disease. Chiropractors aim to remove nerve interference by manipulations of the affected musculo-skeletal parts, particularly in the spinal region.

chiropteran Any member of the order Chiroptera, which contains the 178 genera of bats. Bats are the only MAMMAL to have true flight (although a few others can glide). A bat's wing is formed by a sheet of skin stretched over a frame of greatly elongated bones. Bats are able to navigate in complete darkness by means of a kind of SONAR, which uses echoes of the bat's own supersonic squeaks to locate obstacles and prey. Bats are nocturnal and found in all tropical and temperate regions. Most are small, although they range in wingspan from 25cm to 147cm (10–58in).

chitin Hard, tough substance that occurs widely in nature, particularly in the hard shells (EXOSKELETONS) of ARTHROPODS such as crabs, insects, spiders and their relatives. The walls of HYPHAE – the microscopic tubes of fungi – are composed of slightly different chitin. Chemically chitin is, like CELLULOSE, a polysaccharide, derived from glucose.

Chladni, Ernst Florens Friedrich (1756–1827) German physicist who made studies of sound, particularly the way in which it vibrates metal plates and diaphragms. Patterns produced in fine powders on sound-vibrated plates are called **Chladni's figures**.

chlamydia Small, virus-like BACTERIA that live as PARASITES in human beings and animals and cause disease. One strain, *C. trachomatis*, is responsible for TRACHOMA, the leading cause of

CHEMICAL REACTION

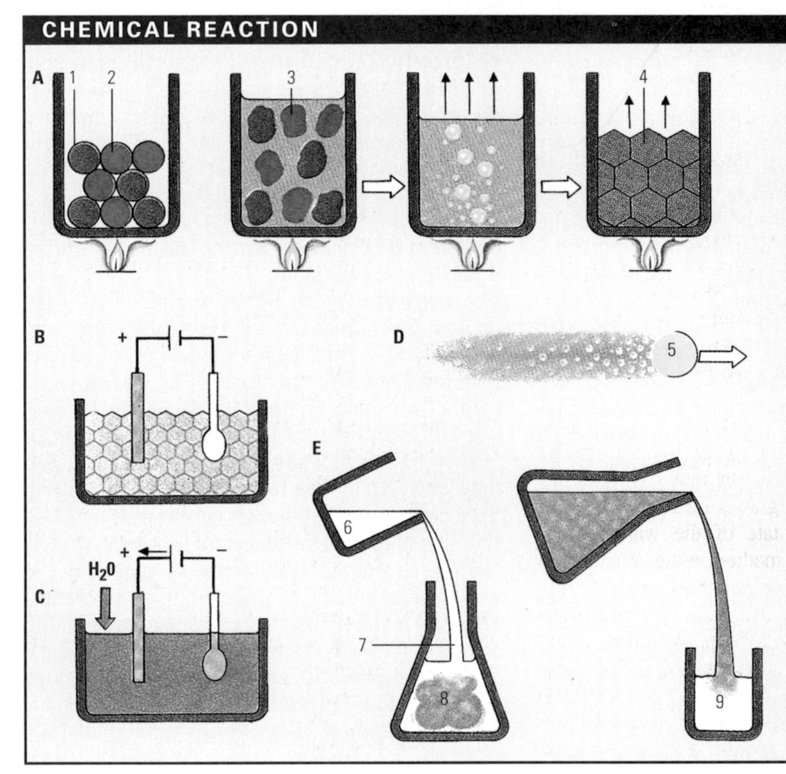

The application of heat (A) alone to aluminium sulphate (1) and potassium sulphate (2) produces no reaction, but they combine when dissolved in water (3). If heating is continued to evaporation, alum (4) forms. Dry copper sulphate crystals (B) do not conduct current; but dissolve them in water (C) and electrolysis can proceed. Metals may react with a liquid (D): a grain of sodium (5) dropped into a water bath melts, generating hydrogen. Solutions and other liquids react readily (E). Phenolphthalein (6) added to a solution of alkali (7) produces a pink solution (8). When this is added to an acid solution (9), the pink disappears.

CHLORINE

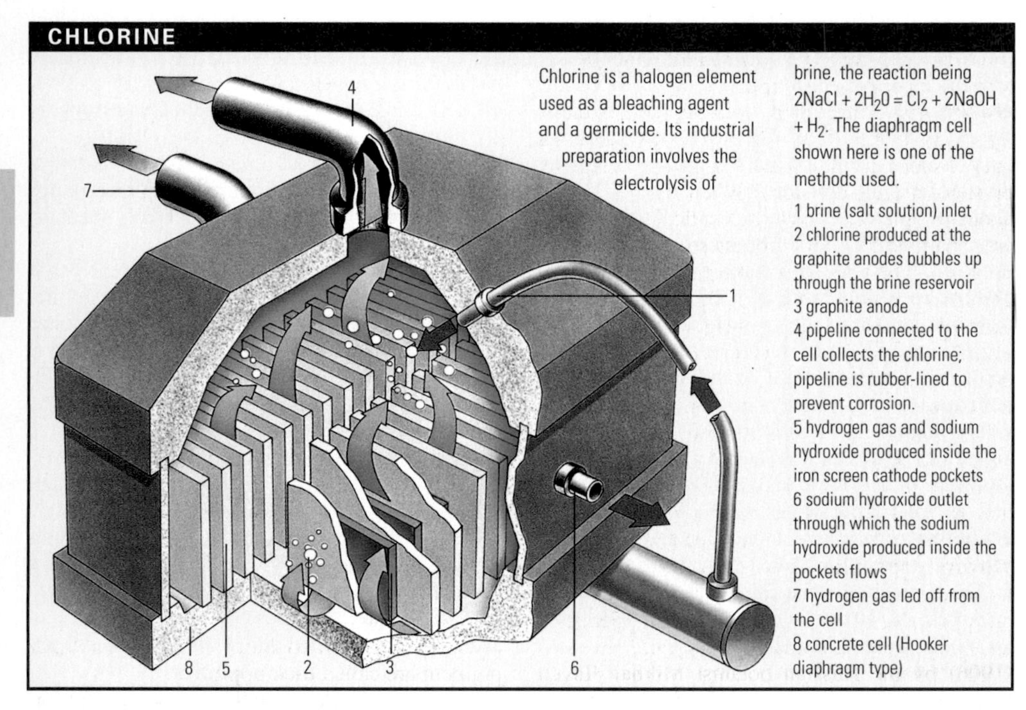

Chlorine is a halogen element used as a bleaching agent and a germicide. Its industrial preparation involves the electrolysis of brine, the reaction being $2NaCl + 2H_2O = Cl_2 + 2NaOH + H_2$. The diaphragm cell shown here is one of the methods used.
1 brine (salt solution) inlet
2 chlorine produced at the graphite anodes bubbles up through the brine reservoir
3 graphite anode
4 pipeline connected to the cell collects the chlorine; pipeline is rubber-lined to prevent corrosion.
5 hydrogen gas and sodium hydroxide produced inside the iron screen cathode pockets
6 sodium hydroxide outlet through which the sodium hydroxide produced inside the pockets flows
7 hydrogen gas led off from the cell
8 concrete cell (Hooker diaphragm type)

BLINDNESS in the developing world; it is also a major cause of pelvic inflammatory disease (PID) in women. *C. psittaci* causes PSITTACOSIS, a disease of birds that can be transmitted to human beings. Chlamydial infection is the most common SEXUALLY TRANSMITTED DISEASE in many developed countries.

chlorate Salt of chloric acid, containing the ion ClO_3^-. Chlorate salts are good oxidizing agents and are a useful source of oxygen, which they give off when heated. Sodium and potassium chlorates are used in explosives and also as weedkillers. They are manufactured by ELECTROLYSIS.

chloride Salt of HYDROCHLORIC ACID or some organic compounds containing CHLORINE, especially those with the negative ion Cl^-. The best-known example is common table salt, sodium chloride (NaCl). Most chlorides are soluble in water, except mercury(I) (mercurous) and silver chlorides.

chlorine (symbol Cl) Common, nonmetallic element, one of the HALOGENS (elements in group VII of the periodic table), first discovered in 1774. It occurs in common salt (NaCl). It is a greenish-yellow poisonous gas extracted by the electrolysis of brine (salt water). It is widely used, in a process called **chlorination**, to disinfect water used for drinking and in swimming pools. Chlorine is also used to bleach wood pulp, and in the manufacture of plastics, chloroform, pesticides and other compounds. Chemically, it is a reactive element and combines with most metals. Properties: at.no. 17; r.a.m. 35.453; m.p. −101°C (−149.8°F); b.p. −34.6°C (−30.28°F). The most common isotope is ^{35}Cl (75.53%).

chlorite Complex silicate mineral of iron, magnesium and aluminium, formed in low-grade METAMORPHIC ROCKS. It is similar to MICA and may be formed from it. Its scaly crystals are usually green, but may also be yellow, white, brown or even black. Hardness: 2–3; r.d. 2.6–3.3, depending on iron content.

chloroethene (vinyl chloride, CH_2CHCl) Gas with an ether-like odour manufactured by the chlorination of ETHENE (ethylene). It polymerizes to form POLYVINYL CHLORIDE (PVC) and is widely used in this form. Properties: m.p. −153.8°C (−244.8°F); b.p. −13.4°C (7.9°F).

chlorofluorocarbon (CFC) Chemical compound in which hydrogen atoms of a HYDROCARBON, such as an ALKANE, are replaced by atoms of fluorine, chlorine, and sometimes bromine. CFCs are inert, stable at high temperatures and are odourless, colourless, nontoxic, noncorrosive and nonflammable. Under the trade name of Freons, CFCs were widely used in aerosols, fire-extinguishers, refrigerators, and in the manufacture of foam plastics. The two most common are Freon 11 (trichlorofluoromethane, $CFCl_3$) and Freon 12 (dichlorodifluoromethane, CF_2Cl_2). When CFCs are used they slowly drift into the STRATOSPHERE and are broken down by the Sun's ultraviolet radiation into chlorine atoms that destroy the OZONE LAYER. It often takes more than 100 years for CFCs to disappear from the atmosphere. Growing environmental efforts led to a 1990 international agreement by governments to reduce and eventually phase out the use of CFCs and other chemicals harming the ozonosphere, and to develop safe substitutes.

chloroform *See* TRICHLOROMETHANE

chlorophyll Group of green pigments present in the CHLOROPLASTS of plants and ALGAE that absorb light for PHOTOSYNTHESIS. There are five types: chlorophyll *a* is present in all photosynthetic organisms except bacteria; chlorophyll *b* in plants and CHLOROPHYTES; and chlorophylls *c*, *d* and *e* present in some algae. It is similar in structure to HAEMOGLOBIN, with a magnesium atom replacing the iron atom.

chlorophyte (green algae) Any member of the phylum Chlorophyta, a large group of marine and freshwater ALGAE. Chlorophytes have several features that make them more like typical green plants than other types of algae: they have cup-shaped CHLOROPLASTS that contain CHLOROPHYLL *b*; they have cell walls made of CELLULOSE; and they store food in the form of STARCH. Some produce cells with flagella (*see* FLAGELLUM) at some stage in their lives. Green algae range in size from microscopic, single-cell types (some of which have flagella) to large, complex SEAWEEDS.

chloroplast Microscopic green structure within a plant cell in which PHOTOSYNTHESIS takes place. The chloroplast is enclosed in an envelope formed from two membranes and contains internal membranes to increase the surface area for reactions. Molecules of the light-absorbing pigment CHLOROPHYLL are embedded in these internal membranes.

chloroquine Synthetic drug used mainly to treat MALARIA and also sometimes HEPATITIS, rheumatoid ARTHRITIS and some skin conditions. It can have marked side-effects, such as blurred vision.

chlorosis In plant pathology, a yellowing or blanching of leaves of green plants caused by deficiencies of minerals, especially magnesium or potassium, or by plant parasites.

chlorpromazine Synthetic drug first developed to counter nausea and vomiting but subsequently found to have a powerful calming effect on the CENTRAL NERVOUS SYSTEM. It is used to treat psychotic conditions and also in terminal illness and occasionally in ANAESTHESIA. Its long-term use is associated with a number of unpleasant side-effects, including tremor, movement disorders and HYPOTENSION.

choke In electronics, coil of low-resistance wire that acts as an INDUCTOR. Chokes are used to limit the flow of high-frequency alternating current (AC) in radio circuits or to smooth the waveform of the direct current (DC) from a RECTIFIER. A familiar choke in domestic use is the "starter" of a fluorescent lighting unit. The term is also used to describe a mechanical valve-like device in many INTERNAL COMBUSTION ENGINES that reduces the amount of air in the air-fuel mixture in order to provide a rich enough mixture for the engine to start.

cholera Infectious disease caused by the bacterium *Vibrio cholerae*, which is transmitted in contaminated water. Cholera, which is prevalent in many tropical regions, produces almost continuous, watery diarrhoea often accompanied by vomiting and muscle cramps, and leads to severe dehydration. Untreated it can be fatal, but proper treatment, including fluid replacement and antibiotics, can result in a high recovery rate. There is a vaccine available.

cholesterol White, fatty STEROID. It occurs in large concentrations in the brain, spinal cord and liver. It is synthesized in the liver, intestines and skin and is an intermediate in the synthesis of vitamin D and many hormones. GALLSTONES are composed mainly of cholesterol. A diet rich in animal products may produce an excess of cholesterol deposits in blood vessels, and this can lead to ARTERIOSCLEROSIS.

choline Organic compound, sometimes classified as a B vitamin. It is necessary for the synthesis of ACETYLCHOLINE and FATTY ACIDS by the liver. Its dietary sources are egg yolk and some vegetable oils. It can be synthesized in the body from the AMINO ACID serine.

cholinesterase Enzyme that breaks down and inactivates the NEUROTRANSMITTER ACETYLCHOLINE at the nerve SYNAPSE.

Chondrichthyes Class of cartilaginous FISH

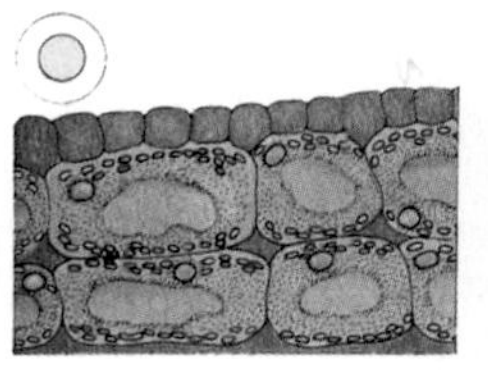

◀ **chloroplast** Found mostly in the cells of plant leaves, chloroplasts absorb sunlight and use it to manufacture special types of sugar. They are able to move about in order to receive the maximum amount of light possible. A section through a leaf reveals that during the day (top) chloroplasts have moved to the outer and inner walls in the direct line of light. During the night (bottom) they move to the inner and side walls only.

including Elasmobranchii (shark, ray and skate) and Holocephali (chimaera). These marine fish have cartilaginous (composed of CARTILAGE) skeletons, a well-developed lower jaw, paired fins, separate gill openings, no air bladder, bony teeth and plate-like (placoid) scales. Chondrichthyes evolved during the late DEVONIAN period.

chordate Any member of the phylum Chordata, a large group of VERTEBRATES and some marine INVERTEBRATES, which, at some stage in their lives, have rod-like, cartilaginous supporting structures (NOTOCHORDS). Invertebrate chordates are divided into three subphyla: Tunicata (seasquirts); Cephalochordata (amphioxus); and Hemichordata (acorn worms).

chorion Outermost of the protective membranes of the embryonic system of birds, reptiles or mammals, or of the insect egg. In birds it is the moist lining between the shell and the ALLANTOIS (the organ for GASEOUS EXCHANGE). In placental mammals the chorion contributes to PLACENTA formation, and through it the EMBRYO receives nourishment, oxygen and water.

chorionic gonadotrophin (CG) HORMONE produced by the PLACENTA in the UTERUS of most higher mammals in order to maintain the CORPUS LUTEUM. Present in urine, the detection of this hormone is the basis of most methods of pregnancy testing.

chorionic villus sampling Method of checking for fetal abnormalities during early pregnancy. The CHORION is a membrane that surrounds the fetus in the UTERUS. Small pieces of VILLI (outgrowths) of the membrane are removed via the woman's abdomen or cervix. These can be examined to study the chromosomes and, if necessary, DNA of the fetus to detect any abnormalities. The procedure can be carried out much earlier than AMNIOCENTESIS so that if an abnormality is found the parents can be offered a termination at an early stage of pregnancy.

chromatid Either of the two duplicate strands into which each CHROMOSOME in a biological CELL nucleus divides in the first phase of MITOSIS or MEIOSIS (cell division). When the nucleus is about to split, the pairs of identical chromatids are separated by a long fibrous structure made of proteins, called a **mitotic spindle**. The separated chromatids become identical "daughter" chromosomes of the same kind as those of the parent cell on opposite sides of the nucleus. *See also* HOMOLOGOUS CHROMOSOME

chromatin Substance that makes up CHROMOSOMES in the nucleus of a cell. It consists of DNA and some RNA, as well as HISTONES and other proteins. In a metabolically active cell nucleus, the chromatin expands to create a region in which MESSENGER RNA (MRNA) may be formed.

chromatography Name given to a number of techniques of chemical ANALYSIS by which substances are separated from one another, identified and measured. Chromatography was invented (1906) by the Russian botanist Mikhail Tsvett (1872–1920). There are several types, all involving a moving phase consisting of a liquid or gaseous mixture of the substances to be separated, and a stationary phase consisting of a material that differentially absorbs the substances in the mixture. *See also* GAS CHROMATOGRAPHY; PAPER CHROMATOGRAPHY

chromatophore Pigment-containing cell that occurs in the skin of some lower animals. Accumulation or dispersion of granules of pigment in chromatophores under nervous or hormonal stimulation allows some animals, such as chameleons, to change colour to suit their surroundings. Chromatophores that contain black pigment are called **melanophores**.

CHROMATOGRAPHY

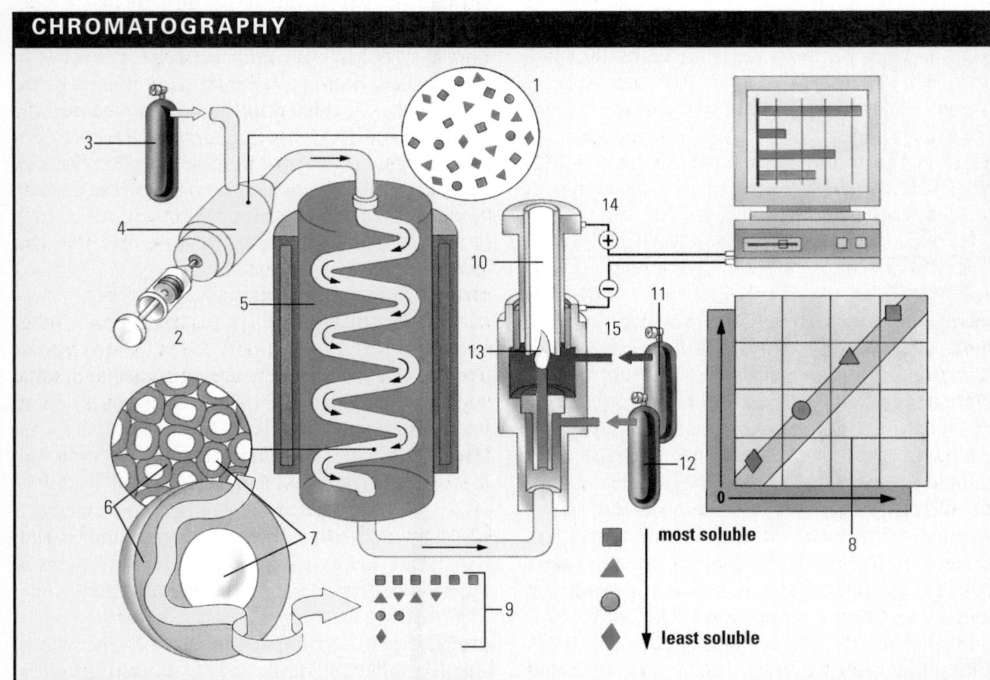

Gas-liquid chromatographs can separate the components of tiny amounts of an unknown mixture. A sample of the mixture (1) is injected (2) into a stream of helium (3), or another inert gas. Heating ensures the vaporized gas mixes fully with the helium. After impurities are removed (4) the gas mixture passes into a tube (5) packed with coated granules of silicon (6). A liquid with a very high boiling point (7) covers the 4mm (0.15in) granules. The components of the vaporized mixture have different solubilities (8) and so pass through the liquid around the silicon, and the whole tube, at different speeds. The whole tube is kept at a high temperature to prevent the vaporized gas condensing. As the now-separated parts of the mixture exit the tube (9) they enter a detector (10). Hydrogen (11) and oxygen (12) are added and the gas stream is then burnt (13). During burning, each compound produces ions that pass a charge between an anode (14) and a cathode (15). This charge is measured and can be compared to known results to determine the composition of the initial mixture.

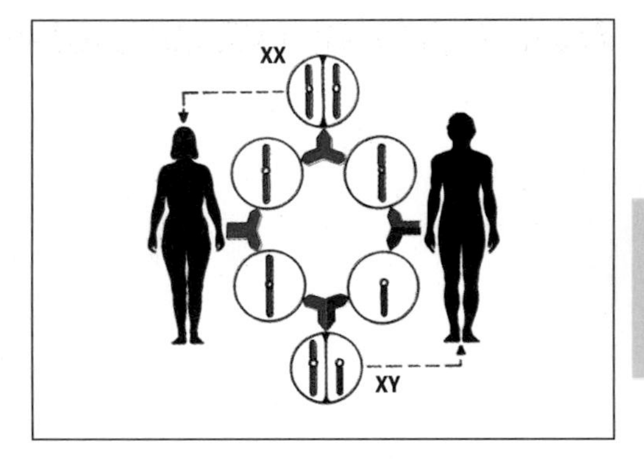

▲ **chromosome** The 46 chromosomes in somatic (non-reproductive) cells contain a single sex-determining pair, which consists of an X and Y chromosome in males, or an XX pair in females. Ova contain only the X chromosome, while spermatazoa contain X or Y chromosomes in equal proportions. At fertilization, therefore, there is a 50% chance of an XX or XY pair being formed.

chromite Black metallic mineral, ferrous chromic oxide ($FeOCr_2O_3$), the only important ore of chromium. It separates from magma when IGNEOUS ROCKS first form, occurring as octahedral crystals and as granular masses. It is weakly magnetic and opaque. Hardness 5.5; r.d. 4.6.

chromium (symbol Cr) Grey-white TRANSITION ELEMENT, first isolated in 1797. Its chief ore is CHROMITE. Chromium is a dull grey metal but takes a high polish and is extensively used as an electroplated coating. It is also an ingredient of many special steels, and chromium compounds are used in tanning and dyeing. It forms two series of salts, termed chromium(I) (chromous), and chromium(III) (chromic). It also forms chromates(VI), containing the ion CrO_4^{2-}. Properties: at.no. 24; r.a.m. 51.996; r.d. 7.19; m.p. 1,890°C (3,434°F); b.p. 2,672°C (4,842°F); most common isotope ^{52}Cr (83.76%).

chromosome Structure carrying the genetic information of an organism, found only in the cell nucleus of EUKARYOTES. Thread-like and composed of DNA, chromosomes carry a specific set of GENES. Each species usually has a characteristic number of chromosomes; these occur in pairs, members of which carry identical genes, so that most cells have a DIPLOID number of chromosomes. GAMETES (sex cells) carry a HAPLOID number of chromosomes. *See also* HEREDITY

chromosphere Layer of the Sun's atmosphere between the PHOTOSPHERE and the CORONA. The chromosphere is about 10,000km (6,000mi) thick and is normally invisible because of the glare of the photosphere shining through it. It is briefly visible near the beginning and end of a total solar ECLIPSE as a spiky red rim around the Moon's disk and at other times it can be studied by SPECTROSCOPY or with the use of a special filter. At its base the temperature of the chromosphere is about 4,000K, rising to 100,000K at the top, where it gives way to the corona. Powerful magnetic fields are believed to cause this rise in temperature.

chronometer Instrument for accurately measuring time. The chronometer was an important navigational aid for ships before the advent of radio time signals. Knowing the precise time at which a star reaches a certain position enables a navigator to work out a ship's longitude.

chrysalis Intermediate or pupal stage in the life cycle of all insects that undergo complete METAMOR-

CIRCULATORY SYSTEM

The blood is carried in a 96,500km (58,000mi)-network of blood vessels. The aorta leaves the aortic valve where its first two branches, the coronary arteries, subdivide into brachial and carotid arteries. The aorta then runs in front of the spine and behind the esophagus to the abdomen, with small and large branches to the vertebrae, diaphragm and intercostal muscles. In the abdomen it divides just above the pelvis into the common iliac arches to the legs. There are four main branches: gastric to the stomach; splanchnic to the intestine; renal to the kidneys; and splenic to the spleen. The venous blood returns to the superior and inferior venae cavae and thus to the right atrium.

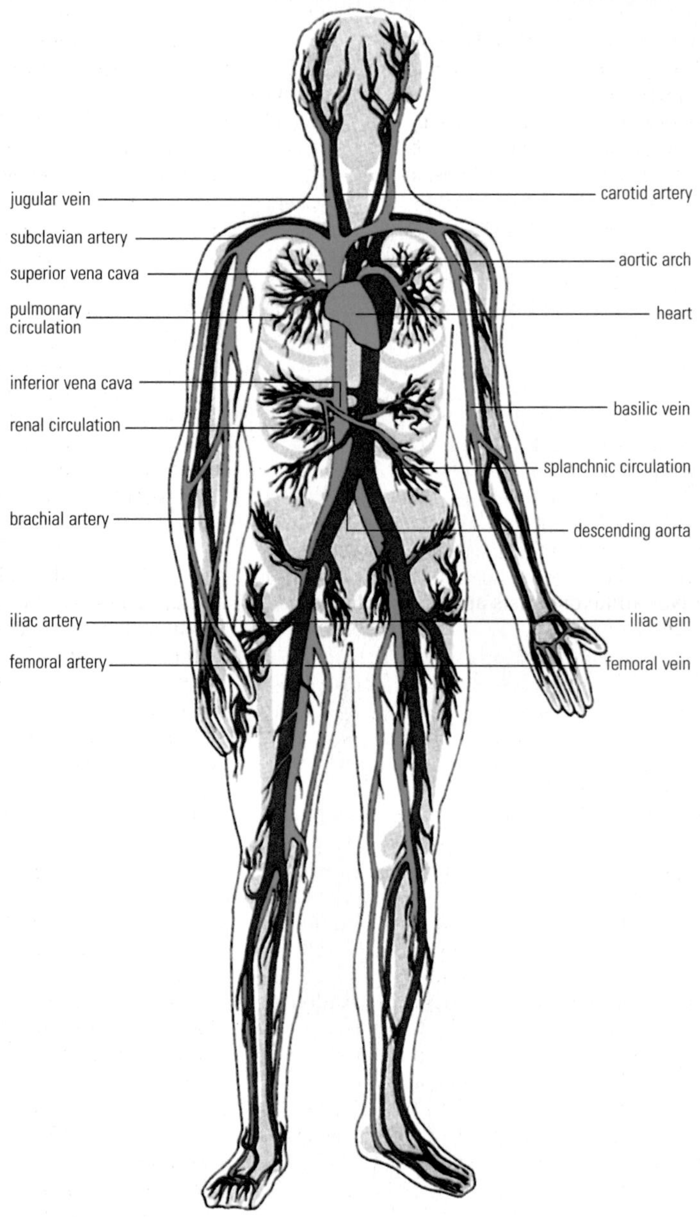

PHOSIS. The chrysalis is usually covered with a hard case, but some pupae, such as those of the silk moth, spin a silk COCOON around themselves. Within the chrysalis, feeding and locomotion stop and the final stages of the development take place. *See also* PUPA

chrysoberyl Oxide mineral, beryllium aluminum oxide ($BeAl_2O_4$). It is found in beryllium-rich pegmatite dykes. Its hexagonal system crystals are prismatic or tabular. It can be transparent green, yellow or brown, but bright yellow-green is most highly valued. Gem varieties include cat's eye and alexandrite. Hardness 8.5; r.d. 3.69.

chrysotile Fibrous serpentine mineral, from which comes most of the world's supply of ASBESTOS. Serpentines are hydrated magnesium silicates ($3MgO2SiO_2.2H_2O$). The variety called chrysotile has crystalline, tubular fibres that are particularly suitable for spinning and weaving into heat-resistant fabrics.

chyle Milky fluid found in the lymphatic vessels in the intestine and consisting of LYMPH containing minute fat droplets absorbed during DIGESTION. It travels through the LYMPHATIC SYSTEM and drains into the bloodstream from the thoracic duct in the neck.

chyme Stomach contents, a mixture of partly digested food, gastric acid and digestive enzymes. It passes from the stomach into the small intestine in semi-liquid form. *See also* DIGESTION; DIGESTIVE SYSTEM

chymotrypsin Substance produced in the body that aids in the digestion of food. Chymotrypsin is an ENZYME that breaks down proteins. It is made in the SMALL INTESTINE, by complex chemical reactions, from chymotrypsinogen. This, together with other digestive substances, is secreted by the PANCREAS through the pancreatic duct into the DUODENUM, which lies between the stomach and the small intestine. The secretion is controlled partly by the vagus nerve (the AUTONOMIC NERVOUS SYSTEM) and partly by HORMONES, which are released automatically when food enters the duodenum. *See also* TRYPSIN

cilia (sing. cilium) Small, hair-like structure on cell walls used for propulsion and feeding. Cilia are present in great quantities on some lining cells of the body, such as those along the respiratory tract. The wafting movements of the cilia help to propel foreign particles towards the exterior. They are also found on single-celled PROTOZOA, some of which, such as the trypanosome of sleeping sickness, are known as CILIATES.

ciliary body Band of tissue around the lens of the EYE. The ciliary body supports the lens and contains the ciliary muscles, which contract to change the shape of the lens and focus the eye (*see* ACCOMMODATION). The AQUEOUS HUMOUR, the liquid within the eyeball in front of the lens, is produced by the ciliary body.

ciliate Any one of the *c*.8,000 species of the phylum Ciliophora, characterized by hair-like CILIA used for locomotion and food collecting. Ciliates are the largest and the most complex of the PROTOZOA. They are found in both aquatic and terrestrial habitats and many are carnivorous. Ciliates have two nuclei **(macronucleus** and **micronucleus)** and a variety of organelles, such as a cystome (mouth). Subclasses include the Holotrichs (*Paramecium*), Spirotrichs (*Stentor*) and Peritrichs (*Vorticella*).

cinder cone Conical hill or mountain composed largely of unconsolidated material, mostly ash ejected by a volcano. It is characteristic of volcanoes that produce large amounts of gas and ash rather than lava. The coarser the ejected material, the steeper the sides of the cone.

cinnabar Deep red mineral, mercury(II) (mercuric) sulphide (HgS), the major ore of mercury. Its crystal system is trigonal. It occurs as rhombohedral crystals, often twinned, and as granular masses. It is found in hydrothermal veins and volcanic deposits. The ore is reduced to mercury by roasting. Hardness 2–2.5; r.d. 8.1.

cipher *See* CRYPTOGRAPHY

circadian rhythm Internal "clock" mechanism that normally corresponds roughly with the 24-hour day. It relates most obviously to the cycle of waking and sleeping, but is also involved in other cyclic variations, such as body temperature, hormone levels, metabolism and mental performance. It can be disrupted by irregular events such as shift-working and jet lag. *See also* BIOLOGICAL CLOCK

Circinus (Compasses) Small constellation of the s sky in the region of Alpha and Beta Centauri. Its brightest star is Alpha, magnitude 3.2.

circle Plane geometric figure that is the locus of points equidistant from a fixed point (the centre). This distance is the radius (*r*). The area of a circle is πr^2 and its perimeter (circumference) is $2\pi r$. *See also* CONIC

circuit *See* ELECTRIC CIRCUIT

circuit breaker Automatic switch that disconnects the electricity supply if dangerous conditions occur. A common type of circuit breaker switches off the current if it exceeds a certain value for more than a specified time. The circuit breaker performs a similar function to a FUSE, but is not destroyed in operation. Instead it can be reset when the fault condition has been cleared.

circular motion In physics, any type of movement that can be related to or analysed in terms of the motion of an object around a CIRCLE.

circulation *See* CIRCULATORY SYSTEM

circulation, atmospheric Flow of the ATMOSPHERE around the Earth. It is caused by temperature differences in the atmosphere, and the rotation of the Earth, which transfers heat from warm zones (the tropics) to cooler zones (towards the poles). The poleward circulation due to CONVECTION gives rise to large-scale eddies such as a CYCLONE and ANTICYCLONE, low-pressure troughs and high-pressure ridges. The eddies also take part in the longitudinal atmospheric circulation around the Earth, with the Earth's rotation maintaining easterly winds towards the Equator and westerly winds towards the poles. Narrow JET STREAMS

club moss Any of *c.*200 species of small, evergreen, spore-bearing plants of the phylum LYCOPODOPHYTA. Unlike the more primitive true MOSSES, club mosses have specialized tissues for transporting water, food and minerals. They are related to ferns and horsetails. The small leaves are arranged in tight whorls around the aerial stems. Millions of years ago their ancestors formed the large trees that dominated CARBONIFEROUS forests.

cluster, galaxy *See* GALAXY CLUSTER

cluster, globular *See* GLOBULAR CLUSTER

cluster, open *See* OPEN CLUSTER

cluster, stellar *See* STELLAR CLUSTER

clutch Any device placed between the rotating parts of an ENGINE or motor and the drive-shaft to facilitate their quick connection or disconnection. In a car, for example, temporary disengagement of the engine is essential during gear-changes. The clutch usually consists of a pair of friction plates, although there are fluid clutches as well.

clutch, electromagnetic Device that uses MAGNETISM to connect two rotating shafts. Some forms are disc clutches with energized coils and magnetic clutch plates. **Eddy current** clutches induce rotational movement in the shaft to be engaged and rotated. **Hysteresis** clutches also transmit rotation without slip. Other electromagnetic clutches employ magnetic metal particles, either dry-flowing or suspended in a liquid, to induce torque.

cnidarian Any one of the *c.* 9,000 species of the phylum Cnidaria (commonly called the **coelenterates**). Marine invertebrates, cnidarians include corals, jellyfish and sea anemones. Characterized by a digestive cavity that forms the main body cavity, they may have been the first animal group to reach the tissue level of organization. The cnidarians are radially symmetrical, jelly-like, and have a nerve net and one body opening. Reproduction is sexual and asexual; REGENERATION also occurs.

CNS Abbreviation of CENTRAL NERVOUS SYSTEM

coagulation Clumping together of colloidal particles (*see* COLLOID), as in the clotting of blood. The particles in a colloidal solution are completely dispersed and will not settle out. But the solution can be destabilized by adding ions, particularly those with a high charge. This process occurs when river water containing colloids of silt reaches the sea, where the ions in sea water coagulate the silt and it precipitates to form a delta. Heat is also used to coagulate colloids, particularly those containing proteins.

coal Blackish, carbon-rich deposit formed from the remains of fossil plants. In the Carboniferous and Tertiary periods, swamp vegetation subsided to form PEAT bogs. Sedimentary deposits buried the bogs, and the increasing pressure and heat produced LIGNITE (brown coal), then bituminous coal, and finally ANTHRACITE if temperature increased sufficiently. This is termed the **coal rank series**; each rank of coal represents an increase in carbon content and a reduction in the proportion of natural gases and moisture. Lignite, which has a low carbon content, is a poorer fuel than anthracite. Most coal seams are inter-stratified with shale, clay, sandstone and sometimes limestone.

coal tar By-product from the manufacture of COKE. The tar comes from the bituminous coal

COAL

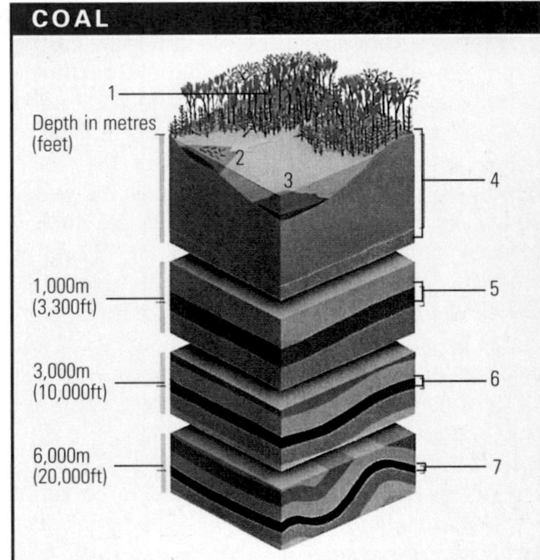

The process of making coal begins with plant debris (1). Dead vegetation lies in a swampy environment and forms peat (4), the first stage of coal formation. Underwater bacteria remove some oxygen, nitrogen and hydrogen from the organic material. Debris carried elsewhere and deposited by water forms a product called cannel coal (2). Algal material collected underwater forms boghead coal (3). If the dead organic material is buried by sediment, the weight on top of the peat and the higher temperature will turn the peat into lignite (5). With more heat and pressure at increasing depths, lignite becomes bituminous coal (6) and then anthracite (7).

▲ **cloud** The major cloud types have characteristic shapes and occur within broad altitudinal boundaries. Stratus (1) is a low-level cloud, usually featureless and grey. Its base may obscure hilltops or occasionally extend right down to the ground, and because of its low altitude it appears to move very rapidly on breezy days. Stratus can produce drizzle or snow, particularly over hills, and may occur in huge sheets covering thousands of square kilometres. Cumulus (2) clouds also seem to scuttle across the sky, reflecting their low altitude. These small "fleecy" clouds are short-lived, lasting no more than 15 minutes before dispersing. They are typically formed on sunny days, when localized convection currents are set up: these currents can form over power stations or even stubble fires, which may produce their own clouds. Cumulus may expand into stratocumulus (3) or into giant cumulonimbus (4), which are up to 10km (7mi) in diameter. These clouds typically form on summer afternoons: their high, flattened tops contain ice, which may fall to the ground in the form of heavy showers of rain or hail. Rising to middle altitudes, stratus and cumulus cloud (then termed altostratus (5) and alto cumulus (6)), appear to move more slowly because of their greater distance from the observer. Cirrus clouds (7) are named after the Latin for "tuft of hair", which they resemble. Relatively common over N Europe, they sometimes become associated with a jetstream and then appear to move rapidly despite their great height. Cirrocumulus (8) is often present with cirrus cloud in small amounts, while the presence of cirrostratus (9) is given away by haloes surrounding the Sun or Moon.

used in this destructive DISTILLATION process. Coal tar is a volatile substance, important for its organic chemical constituents. These are extracted by further distillation. Such chemicals, called **coal-tar crudes**, include xylene, toluene, naphthalene and phenanthrene. They are the basic ingredients from which many products, including explosives, dyes, drugs and perfumes, are made.

coaxial cable CABLE consisting of a central conductor with a surrounding insulator and tubular shield. Most television sets have an aerial connected by coaxial cable. Multiple coaxial cables, used for long-distance communications systems, contain many separate coaxial cables. They are used to transmit high-frequency signals.

cobalt (symbol Co) Grey TRANSITION ELEMENT, first discovered in 1737. It is found in cobaltite and smaltite, but most is obtained as a by-product during the processing of other ores. Cobalt is a constituent of vitamin B_{12}. It is used in high-temperature steels, artists' colours (especially cobalt blue), the manufacture of jet engines, cutting tools and (as an alloy) in magnets. Cobalt-60 (half-life 5.26 yr) is an artificial isotope used as a source of GAMMA RADIATION in radiotherapy and tracer studies. Properties: at.no. 27; r.a.m. 58.9332; r.d. 8.9; m.p. 1,495°C (2,723°F); b.p. 2,870°C (5,198°F); most common isotope ^{59}Co (100%).

COBOL (**C**ommon **B**usiness-**O**riented **L**anguage) COMPUTER LANGUAGE developed in 1959 for processing business data. Revised and improved since then, COBOL is now widely used.

cocaine White crystalline ALKALOID extracted from the leaves of the coca plant. Once used as a local anaesthetic, it is now primarily an illegal narcotic, with stimulant and hallucinatory effects. It is psychologically habit-forming, but increasing doses are not needed, as the body does not develop tolerance. Habitual use of cocaine results in physical and nervous deterioration. A smokeable derivative, crack, is a highly volatile, addictive and lethal substance.

coccolith Any microscopic, single-celled flagellate of the Coccolithophorida, a class of ALGAE of the phylum Chrysophyta. The cell is covered with round, chalky platelets only one or two thousandths of a millimetre in diameter. Many limestone and chalk cliffs are made up entirely of the remains of such platelets.

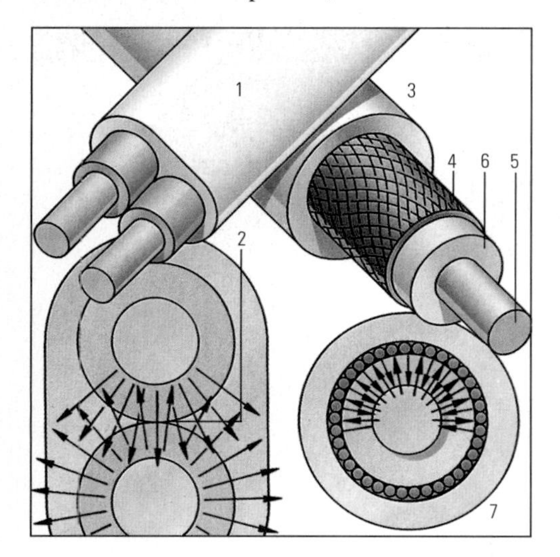

▲ coaxial cable The magnetic fields induced by the current flowing in a normal two-core cable (1) will interfere with each other and may induce unacceptable variations in the level of current flowing in the wire (2). This is particularly a problem when the current represents a communications signal, part of a computer network or the connection between a television and its aerial. Here coaxial cable is used, where one conductor (4) is wound round the other (5), separated by a layer of insulating material (6). The electromagnetic flux no longer causes such a damaging distortion of any transmitted signals (7).

COCHLEA

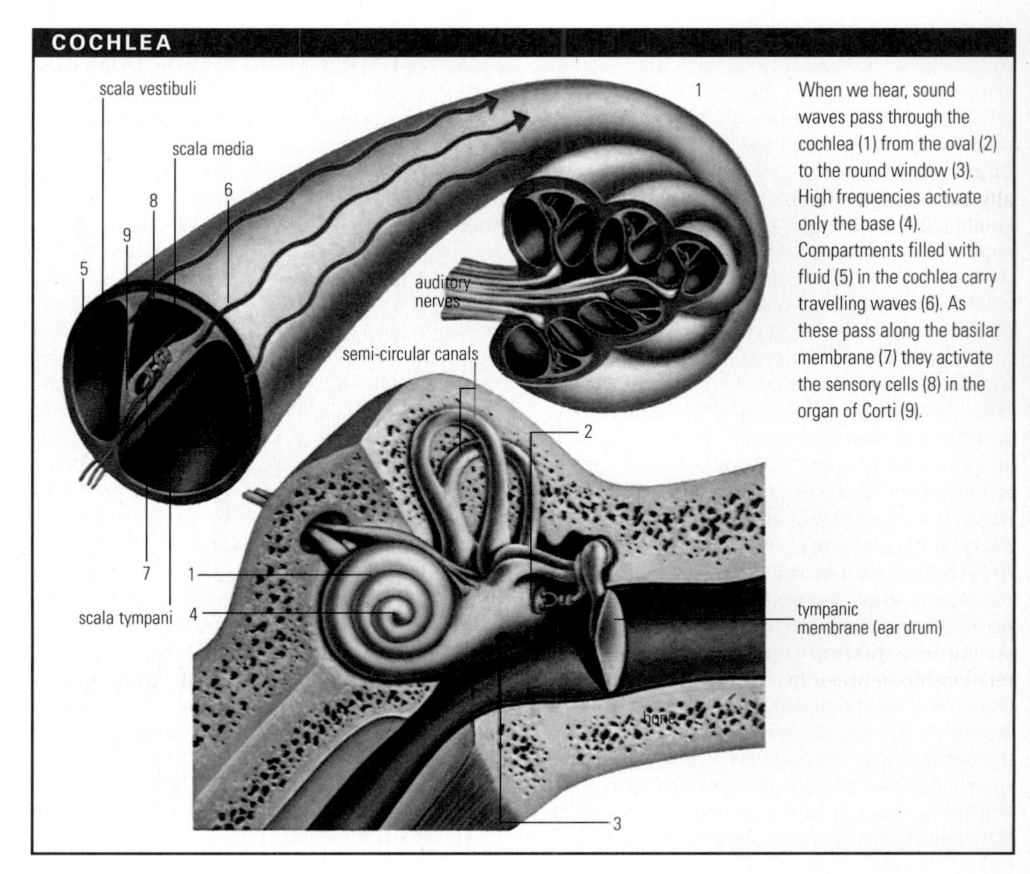

When we hear, sound waves pass through the cochlea (1) from the oval (2) to the round window (3). High frequencies activate only the base (4). Compartments filled with fluid (5) in the cochlea carry travelling waves (6). As these pass along the basilar membrane (7) they activate the sensory cells (8) in the organ of Corti (9).

coccus Spherical BACTERIUM with an average diameter of 0.5–1.25 micrometres. Some, such as *Streptococcus* and *Staphylococcus*, are a common cause of infection.

coccyx Triangular bone at the lower end of the vertebral column. It is formed by the fusion of three to five small VERTEBRAE. *See also* SPINE

cochlea Fluid-filled structure in the inner EAR which is essential to HEARING. It has a shape like a coiled shell, and is lined with hair cells which move in response to incoming sound waves, stimulating nerve cells to transmit impulses to the brain. Different groups of hair cells are stimulated by different pitches of sound, thus helping the brain to analyze the sound.

Cockcroft, Sir John Douglas (1897–1967) British physicist who, with his colleague Ernest WALTON, was the first person to split the atom. He and Walton constructed a particle ACCELERATOR and in 1932 created the first man-made nuclear reaction by bombarding lithium atoms with protons. They shared the 1951 Nobel Prize for physics for their use of particle accelerators to study atomic nuclei. Cockcroft also contributed to the development of NUCLEAR REACTORS and in 1946 became the first director of the UK's Atomic Energy Research Establishment at Harwell.

Cockcroft-Walton generator Device in which a high constant voltage is obtained by multiplying a low alternating voltage using a circuit of rectifiers and capacitors. *See also* ACCELERATOR, ELECTROSTATIC

Cockerell, Sir Christopher Sydney (1910–) British engineer who invented the HOVERCRAFT. In 1948 Cockerell started a boat-building business and, in the early 1950s, began research into the development of AIR-CUSHION VEHICLES. By 1954 he had developed the hovercraft. He filed his first patent in 1955. In 1959 the first SRN1 passenger-carrying hovercraft crossed the English Channel.

cocoon Case or wrapping produced by larval forms of animals (such as some moths, butterflies and wasps) for the resting or pupal stage in their life cycle. Some spiders spin a cocoon that protects their eggs. Most cocoons are made of silk, and those of the domestic silkworms provide most of the world's commercial silk. *See also* CHRYSALIS; PUPA

codeine White crystalline ALKALOID extracted from OPIUM by the methylation of MORPHINE, and with the properties of weak morphine. It is used in medicine as an ANALGESIC to treat mild to moderate pain, as a cough suppressant and to treat diarrhoea.

codon Unit of the genetic code in a molecule of DNA or MESSENGER RNA (mRNA). In messenger RNA it consists of a triplet of bases that usually specifies a particular AMINO ACID as part of a POLYPEPTIDE chain being built into a protein. Some codons carry instructions (such as "start" and "stop") involved in the process of protein building. *See also* GENE

coefficient Term multiplying a specified unknown quantity in an algebraic expression. In the expression $1 + 5x + 2x^2$, 5 and 2 are the coefficients of x and x^2 respectively. In physics, a coefficient is a number that specifies a given property of a substance such as the elasticity coefficient or coefficient of friction.

coelenterate Common name for a CNIDARIAN.

coeliac disease Disorder in which the SMALL INTESTINE fails to absorb food properly. It is caused by intolerance to GLUTEN, a protein in wheat and rye products. Symptoms include depression, diarrhoea and malnutrition.

coelom True body cavity, or space between an inner digestive tube and outer body wall, that provides room for additional organs. Indications of a coelom appear in roundworms but the first true coelom appears among earthworms and other segmented worms.

coenocyte Type of cell that has many nuclei. Coenocytes occur when only the nucleus of a cell divides repeatedly, without division of the whole cell or cell wall. The fibres of striated muscle, for example, are made up of such cells, giving a type of tissue called **syncytium**. Coenocytes are also found in some algae and fungi.

coenzyme Non-protein organic molecule, usually containing a vitamin and phosphorus. When combined with an **apoenzyme**, a protein molecule, it activates an ENZYME. A coenzyme always regains its original structure even though it may have been altered during a reaction.

coevolution Evolution of complementary features in two different species; a result of interaction between the two. Both species gain from the association, and evolve structures or behaviour of mutual benefit. The classical example occurs in plants that are pollinated by insects. The plant's flowers have a colour or smell that attracts insects, and a shape that makes it easy for insects to reach the nectar and in doing so remove pollen. The insects have evolved to be able to see and smell the flowers, and have mouthparts adapted to reach the nectar.

coherence In physics, the property of ELECTROMAGNETIC RADIATION that has a constant phase relationship between two or more sets of waves. If the peaks and troughs of one wave are always the same distance from those of another wave, the waves are coherent. It follows that the waves must also have exactly the same FREQUENCY. A pair of coherent waves produce INTERFERENCE. The light from a LASER is coherent.

cohesion Mutual attraction between the component atoms, ions or molecules of a substance. Weak cohesive forces permit the fluidity of liquids; those of solids are much stronger. Liquids form droplets because of surface tension caused by cohesion. *See also* ADHESION

coke Stiff, porous, grey mass of carbon mixed with small amounts of minerals, sulphur and residual volatiles. It is left behind as a residue in coking ovens after coal has been heated without air to drive off most of the volatile matter. It is used mainly in steelmaking for fuelling BLAST FURNACES. Its high calorific value is due to the high carbon content – more than 90%.

col In geology, a low point between two mountain peaks or on an upland ridge. In meteorology, a calm area between two DEPRESSIONS and two ANTICYCLONES.

colchicine ALKALOID obtained from the corm of the autumn crocus (*Colchicum autumnale*). It is employed in genetic and cellular research, as it inhibits MITOSIS (cell division). This property makes colchicine valuable in cancer research and as an IMMUNOSUPPRESSIVE DRUG.

cold-blooded *See* POIKILOTHERMIC

coleopteran Any member of the order Coleoptera, the largest order of the animal kingdom, generally the BEETLES. They are insects that undergo complete METAMORPHOSIS. Most possess two pairs of wings, the forward pair usually being hard covers that fold over the flying pair. More than 350,000 species are known, many of them pests called weevils. Beetles include some of the largest and many of the smallest of all insects. Their larvae are either maggot-like, or elongated with six legs.

coleoptile Sheath that covers and protects the shoot from an embryo of a grass seed. The sheath protects the shoot as it forces its way upwards through the soil and remains in place until it is split open by the development of the first true leaves.

collagen Protein substance that is the main constituent of bones, tendons, cartilage, connective tissue and skin. It is made up of inelastic fibres.

collar bone *See* CLAVICLE

collimator Lens system used to produce a parallel or near parallel beam of light, subatomic particles or other radiation. Light passing through a fine slit at the focus of a convex lens is rendered parallel by the lens. A collimated beam is essential for many optical purposes.

colloid Substance composed of fine particles that are dispersed throughout a second substance. A SOL is a solid dispersed in a liquid, an AEROSOL is a solid or a liquid in a gas, an EMULSION is a liquid in a liquid, and a FOAM is a gas in either a liquid or a solid.

colon Part of the large INTESTINE, the digestive tract that extends from the small intestine to the RECTUM. The colon absorbs water from digested food and allows bacterial action for the formation of faeces. *See also* DIGESTIVE SYSTEM

colony In biology, group of similar animals or plants living together for mutual benefit. Individuals perform like or varied functions and may be structurally separated or united.

colophony Alternative name for ROSIN

colorimetry Scientific methods in which an unknown colour is compared with known colours. It is used in chemical ANALYSIS, and for checking the colour of products such as paints, dyes and pharmaceuticals.

colostomy Operation to bring a part of the large intestine, called the COLON, out through the wall of the abdomen in order to bypass the lower section of the bowel. An artificial opening (**stoma**) is created so that faecal matter is passed into a collecting appliance worn on the exterior of the body. The site of the colostomy varies according to the location of the disease. A colostomy may be temporary, in which case it can be closed to restore continuity of the bowel; usually, however, it will be permanent, if the lower part of the bowel has been removed. *See also* DIGESTION

colostrum Milk-like fluid produced by the female MAMMARY GLAND soon after (or even before) the birth of young; it precedes true milk, which is produced about three days after birth. Colostrum is rich in protein and the antibodies it contains give the newborn some immunity from infection.

colour Sensation experienced when light of sufficient brightness and of a particular wavelength strikes the RETINA of the EYE. Normal daylight (white light) is made up of a spectrum of colours, each of which has a different wavelength. These colours can be arbitrarily assigned to seven bands – red, orange, yellow, green, blue, indigo and violet – of decreasing wavelength. A pure spectral colour is called a **hue**. Any actual colour is caused by light of a mixture of wavelengths. It can be matched by (that is indistinguishable from) a suitably chosen pure wavelength mixed with white light of appropriate brightness. If the colour is not pure but contains some white, it is "desaturated". **Saturation** is the degree to which a colour departs from white and approaches a pure hue. A colour is also defined by its luminosity, or brightness. Any colour of light can be matched by a mixture of three primary colours: red, green and blue-violet; red and green light mix to produce yellow; red and blue-violet mix to form purple, and green and blue-violet mix to form cyan. The production of colour by paints and pigments is more complicated. The colours of these result from the absorption of particular wavelengths and reflection of the rest. The "subtractive" primary colours used in printing inks are **magenta** (green-absorbing), **cyan** (red-absorbing) and **yellow** (blue-absorbing).

colour film Photographic film available either as positive film for projection or as negative film for making colour prints. It contains three layers of light-sensitive chemicals which respond, in turn, to the optical primary colours (roughly red, green and blue) present in the light passing through the lens of the camera. Polaroid colour film contains additionally a number of coloured dyes which are released by the light-sensitive chemicals as they react to the optical primaries, so developing a colour print directly.

colour vision Ability of the EYE to detect the different wavelengths (COLOURS) of light. This is achieved by three types of cone cell in the RETINA,

COLOUR VISION

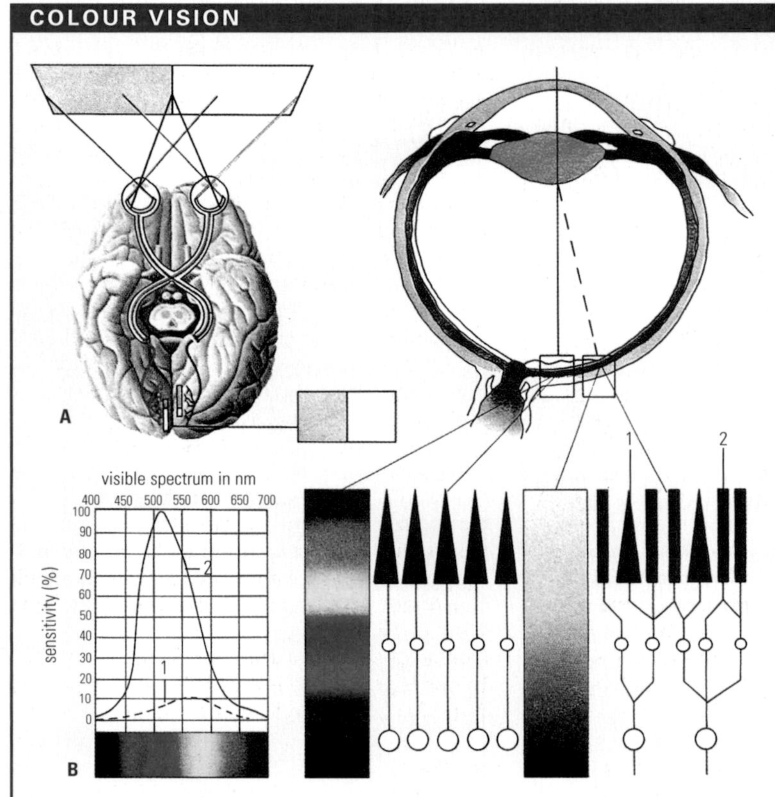

The surface of the retina contains light-sensitive rods and cones. These translate photons of light into nerve impulses, which travel to the brain, impulses from the right eye travelling to the left side of the brain and vice versa (A). The rods are sensitive to low levels of light. The cones, which are colour-sensitive, come into play when the light is strong. As the light dims, the cones lose their sensitivity and cease to respond. The response to colour also varies (B). Cones (1) are most responsive to the yellowish-green part of the spectrum, whereas rods (2), although still giving black and white vision, respond best to the blue and green wavelengths. The small area giving the most accurate vision in bright light is the fovea; this contains only cones.

C

COMET

There are three main types of comet. Short-period comets (A) often have their aphelia at approximately the distance of Jupiter's orbit (1). Their periods amount to a few years and all short-period comets are faint. Long-period comets (B) have aphelia near or beyond Neptune's orbit (2) – Halley's is the only conspicuous member of the class. Comets with very long periods (C) have such great orbital eccentricities that the paths are almost parabolic. Because only a short arc can be measured, it is impossible to calculate the periods of these comets accurately. All the really brilliant comets, apart from Halley's, are of this type.

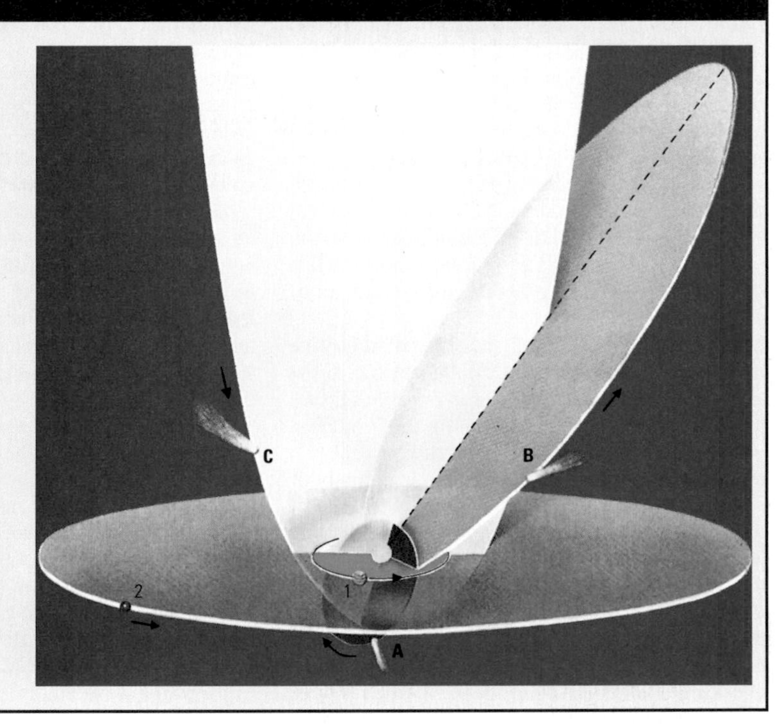

"red", "green" and "blue", which respond to light of those colours. The cone cells secrete one of three different pigments that break down and generate nerve impulses which the brain then interprets to give the range of colours we perceive.

Columba (Dove) Indistinctive constellation of the S sky, adjoining Canis Major and Carina. Its brightest star is Alpha, magnitude 2.6.

columbite Black oxide mineral of IRON, MANGANESE and NIOBIUM. Niobium is replaced by TANTALUM to form tantalite, with similar properties to columbite but denser. They are found in granite PEGMATITES in W Australia and Madagascar. Columbite crystallizes in the orthorhombic system, forming prismatic crystals with one distinct cleavage. Hardness: 6.0 to 6.5; r.d. 5.2 (columbite) to 8.0 (tantalite).

columbium *See* NIOBIUM

coma Spherical cloud of gas and dust surrounding the nucleus of an active COMET.

Coma Berenices (Berenice's Hair) Constellation of the N sky, adjoining Boötes. Its brightest stars are of 4th and 5th magnitudes.

combe Small hollow on a steep chalk hillside. Generally found on an ESCARPMENT, combes are formed when WEATHERING opens up and enlarges a joint in the rock. Weathered material that slips down to the bottom of the slope may form a mound known as a combe rock.

combustion Burning, usually in oxygen. The combustion of fuels is used to produce heat and light. An example is a fire. In solids and liquids, the reaction speed may be controlled both by the rate of oxygen flow to the surface, and by CATALYSTS. Industrial techniques harness the energy produced using combustion chambers and furnaces.

comet Small, icy Solar System body in an independent orbit around the Sun. The solid nucleus of a comet is small: for example, that of HALLEY'S COMET measures just 16 × 8km (10 × 5mi). The nucleus is made up of ices, largely water-ice, in which are embedded rock and dust particles, and is covered by a thin dark crust. As the comet approaches the Sun and gets warmer, evaporation begins, and jets of gas and dust escape through the crust to form the luminous, spherical COMA. The coma can be huge, a million kilometres or so across, but extremely tenuous. Later, radiation pressure from the Sun and the solar wind may send dust and gas streaming away as a tail, tens of millions of kilometres long. The gas, or ion, tail is straight and long, up to 10 million km, while the dust tail is shorter, broader, and curved. Comets are thought to originate outside the Solar System, from where they may be displaced from their orbit and sent towards the Sun. Nearly 800 comets have had their orbits calculated in detail, and several new comets are discovered every year. *See also* ASTEROID

commensalism Situation in nature in which two species live in close association but only one partner benefits. One of the species (the **commensal**) may gain from increased food supply, or by procuring shelter, support or means of locomotion, but the other (the **host**) neither gains nor loses from the relationship. *See also* MUTUALISM; PARASITE; SYMBIOSIS

communications Processes for sharing information and ideas. Facial expressions, hand signals, sirens, writing, speech, song and cinema are examples. The 15th-century invention of the printing press revolutionized communications. The 20th century has witnessed its own communications revolution, primarily in terms of increased access. Mass communication, principally through the medium of TELECOMMUNICATIONS, has facilitated rapid, international communications. The invention of the TELEPHONE, RADIO, TELEVISION and the COMPUTER NETWORK have enabled worldwide exchange of information, and in some cases have shaped 20th-century history. The INTERNET is the latest in a long line of technological innovations. *See also* SATELLITE, COMMUNICATIONS

community In ecology, naturally occurring group of plants and animals living within a restricted area. Communities are usually named after a prominent physical characteristic or a dominant species.

Community Antenna Television (CATV) System in which cables connect a central antenna to numerous television receivers, providing a wide range of commercial programs telecast by distant stations. Subscribers do not need separate antennas, but pay monthly fees for the cable service. CATV began about 1950. *See also* CABLE TELEVISION

commutative law Property of an operation and a set of elements in mathematics. Two elements a and b are said to commute under the operation * if $a * b = b * a$: that is if the result of the operation is the same regardless of the order of the elements. Addition and multiplication of numbers is commutative, since $a + b = b + a$ and $ab = ba$. Matrix multiplication and vector cross products are examples of operations that do not obey the commutative law.

commutator Cylinder made up from copper bars, part of a DC ELECTRIC MOTOR or GENERATOR. It makes contact via brushes between the electric terminals and the rotary ARMATURE winding. It also serves to reverse the current in these windings.

compact disc (CD) Disc used for high-quality DIGITAL sound reproduction. It is a plastic disc with a shiny metal layer and a transparent protective plastic coating. The sound signal consists of millions of minute pits, pressed into one side of the metal. On replay, a narrow laser beam is reflected

COMPACT DISC

A compact disc player reads digital information from a compact disc (1) using a focused laser (2). Music or other information is written on the underside of the disc in a spiral track of pits (3), representing a digital code of zeros and ones. The disc spins and the laser, mounted on a swing arm (4), moves as the disc plays. The laser passes through a semi-silvered mirror (5) and is focused on the disc (6). When the laser hits a flat area it is reflected back via the mirror to a sensor (7) and the information sent to a chip. When the laser hits a pit, it is scattered.

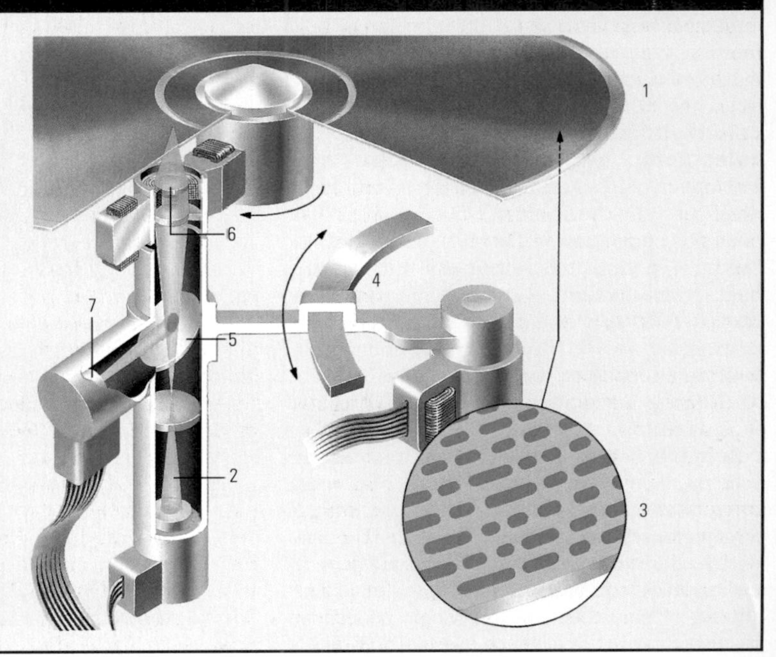

from the rotating disc's surface. A sensor detects changes in the beam, and forms an electrical signal of pulses. This is processed and decoded to form an ANALOGUE sound signal that can be amplified for reproduction on loudspeakers. *See also* CD-ROM

comparator Measuring device used to inspect a manufactured component for deviation from a specified value, normally by direct matching or by comparing it with a master part against pre-set tolerances. The decision to accept or reject a part may be done by an operator or the comparator makes the decision automatically.

compass Direction-finding instrument also used to show direction of a MAGNETIC FIELD. It consists of a horizontal magnetic needle pivoted so that its north-seeking end can turn to point towards magnetic north. Adjustments can be made to give true north. The first use of the magnetic compass was once attributed to the ancient Chinese, but this now seems unlikely. It was not until the 12th century that a compass was used in Europe, the "needle" being a piece of LODESTONE. In navigation today, the magnetic compass is often replaced by the motor-driven GYROCOMPASS.

competition In ecology, the state of affairs that exists between two or more organisms or groups of organisms when a shared resource is limited. For example, there may be direct confrontation over the amount of living space, food supply or the availability of mates. There are two main types: **intraspecific competition** between members of the same species, and **interspecific competition** between different species. The usual outcome is the classic "survival of the fittest", with at least a temporary reduction in the numbers of the less fit organism, and for this reason competition is a significant force in the process of EVOLUTION.

compiler COMPUTER PROGRAM that translates the symbols of a programming COMPUTER LANGUAGE into instructions readable directly by a computer. Most programs are written in HIGH-LEVEL LANGUAGES, such as "C", Pascal or BASIC, which are made up of words and symbols easily comprehended by humans. A compiler takes these programs and renders them into a form that is readable by the circuits of the computer itself.

complementary angles Two angles whose sum is a right-angle (90°). Supplementary angles have a sum of 180°.

complementary colours Any two COLOURS that combine to make white light. Any colour can be described in terms of the combination of red, green or blue (the "additive primary" colours) that has the same effect on the eyes. Since red, green and blue in the right proportions add to make white, it follows that the complementary colour of red is blue-green (**cyan**), of blue is **yellow** (which is produced by red and green combined) and of green is **magenta** (red-blue). In painting, combinations of colours get darker, not lighter. It remains true, however, that complementary colours tend to neutralize each other.

complex compound *See* COORDINATION COMPOUND

complex number *See* NUMBER, COMPLEX

composite Material such as concrete, fibreglass or plywood, made by combining two or more other materials. A composite usually has qualities superior to those of the materials from which it is made.

compound Substance formed by chemical combination of two or more ELEMENTS that cannot be separated by physical means. Compounds are produced by the rearrangement of VALENCE electrons (outer ELECTRONS of an atom) seeking to attain more stable configurations. They usually have properties quite different from those of their constituent elements. IONIC COMPOUNDS have IONIC BONDS – they are collections of oppositely charged ions. The ions are packed together in a regular arrangement called a CRYSTAL LATTICE. Ionic compounds, such as sodium chloride, are solids at room temperature and have high melting and boiling points, due to the strong electrostatic forces of attraction holding the lattice together. COVALENT BONDING occurs where nonmetal atoms share electrons. Such compounds can be classified as simple molecular structures (such as carbon dioxide) – with low melting and boiling points; or giant molecular structures (such as graphite and diamond). Their properties depend on the arrangement of the atoms in the macromolecule. *See also* FUNDAMENTAL FORCE; MOLECULE

compound eye Type of animal eye that consists of many individual **facets**, common in adult crustaceans and flying insects. Each facet is an individual visual unit called an **ommatidium**. It has a lens and several light-sensitive cells with nerve fibres that converge to form the optic nerve leading to the animal's brain. The ommatidia in the eyes of day-flying insects each produce a separate image to produce a mosaic of the whole visual field. In night-flying insects, each ommatidium "sees" a large part of the visual field so that individual images overlap to produce a bright view lacking in detail. Compound eyes are made up of from between 12 to more than 1,000 facets, depending on the species of animal. *See also* SIGHT

compressed air Air kept at pressures much higher than that of the atmosphere. It is prepared by pumping air into a reservoir from a pump or COMPRESSOR, and is used extensively as a source of power for machines such as PNEUMATIC hammers and other moving tools. It is easily transported in metal "bottles", and many factories have their own compressors.

compression-ignition engine *See* DIESEL ENGINE; ENGINE

compression of gas Reduction in the volume of gas achieved by applying external pressure. Some gases, including carbon dioxide, can be liquefied by pressure at room temperatures. Other gases must be cooled before they can be liquefied by pressure. The "critical temperature" of a gas is the highest temperature at which it can be liquefied by an increase of pressure.

compressor Machine that delivers air or gas at pressure. It is used for furnace blast systems, ventilation and refrigeration systems, pneumatic drills and for inflating vehicle tyres. **Reciprocating** and **rotary** compressors are the two basic types.

Compton, Arthur Holly (1892–1962) US physicist. Working in the field of X-RAYS, he discovered what is now known as the COMPTON EFFECT: wavelengths of X-rays increase when the rays collide with electrons. This helped to prove that X-rays could act as particles. He was awarded the 1927 Nobel Prize for physics jointly with the British physicist Charles WILSON. As head of the early phase of the MANHATTAN PROJECT to develop the atom bomb, Compton helped to create the first sustained nuclear CHAIN REACTION.

Compton effect Increase in the wavelength of X-ray or gamma ray PHOTONS as they collide with electrons. The effect was discovered (1923) by

COMPTON EFFECT

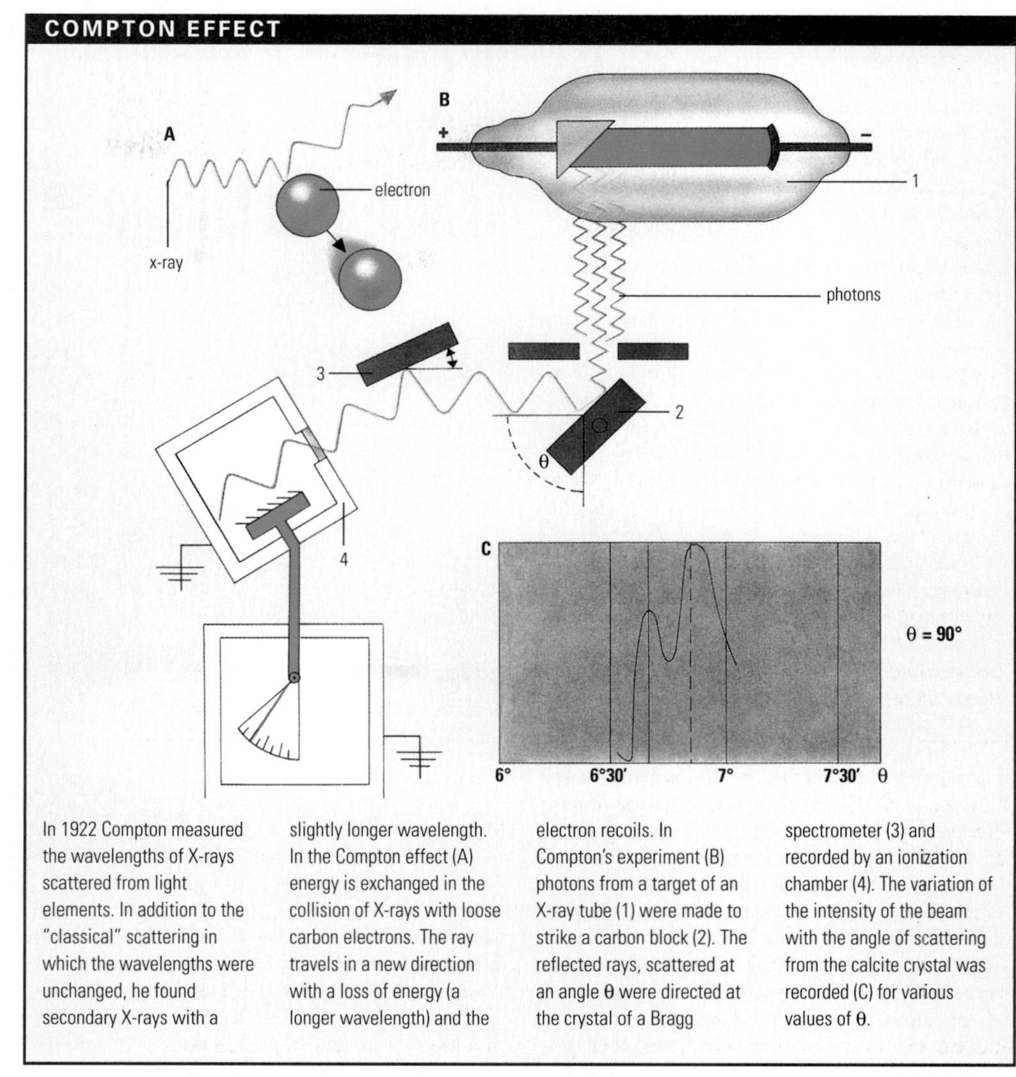

In 1922 Compton measured the wavelengths of X-rays scattered from light elements. In addition to the "classical" scattering in which the wavelengths were unchanged, he found secondary X-rays with a slightly longer wavelength. In the Compton effect (A) energy is exchanged in the collision of X-rays with loose carbon electrons. The ray travels in a new direction with a loss of energy (a longer wavelength) and the electron recoils. In Compton's experiment (B) photons from a target of an X-ray tube (1) were made to strike a carbon block (2). The reflected rays, scattered at an angle θ were directed at the crystal of a Bragg spectrometer (3) and recorded by an ionization chamber (4). The variation of the intensity of the beam with the angle of scattering from the calcite crystal was recorded (C) for various values of θ.

All digital computers work by manipulating data represented as numbers. The tallying principle of the ABACUS was mechanized in calculating machines, such as those devised by Charles BABBAGE, in which complicated calculations were processed by means of geared wheels or cogs. By the mid-1940s mechanical machines were replaced by electronic versions. Some used groups of electromagnetic switches, called relays, to register binary numbers. At any instant, each switch could be either "on" or "off", corresponding to the digits 1 or 0 in the BINARY SYSTEM. Stages in the long-term development of electronic digital computers are termed **computer generations**. A **first generation** computer was developed by engineers at the University of Pennsylvania in 1946. The 27-tonne machine, called ENIAC (Electronic Numerical Indicator and Computer), used electronic VALVES instead of relays. Programming ENIAC to do a particular task was a lengthy process that consisted of changing wired connections. John VON NEUMANN helped to develop techniques for storing programs in code to avoid this problem. In 1951, UNIVAC 1 became the first computer offered for general sale. This **second generation** computer used TRANSISTORS to perform the same role as the valves. As a result computers became smaller and more commonplace. In the 1960s a **third generation** of computers appeared with the invention of the INTEGRATED CIRCUIT, leading to a further reduction in size and greater processing power. **Fourth generation** computers, developed in the 1980s were even smaller, utilizing powerful MICROPROCESSORS. Microprocessors contain a complete CENTRAL PROCESSING UNIT (CPU), which controls operations. The latest microprocessors contain more than 1 million transistors and other components, all in a package little larger than a postage stamp.

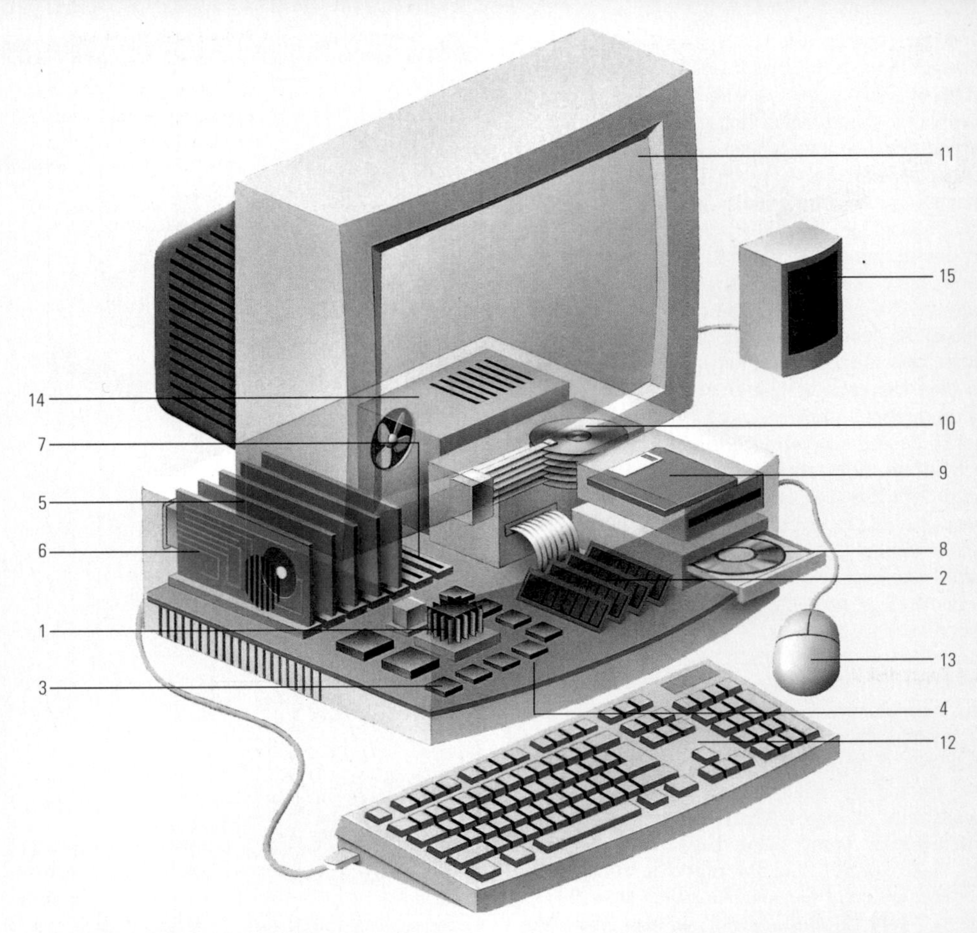

▲ **A typical personal computer** contains a central processing unit (CPU) (1), RAM (random access memory) (2), BIOS (basic input-output system) and ROM (read-only memory) chips (3), a mother board (4), expansion cards (5), a video card (6), expansion slots (7), optical disc drive (8), floppy disc drive (9), hard disc (10), monitor (11), keyboard (12), mouse (13), power supply (14), and loudspeaker (15).

▼ **A mouse** moves a cursor on a monitor. A ball (1) rotates as the mouse moves. As the ball rolls, spoked wheels (2) turn through horizontal and vertical axes. An LED (3) shines through the spokes (4). The rotation of both wheels is detected by sensors (5) which send information to the computer via a cable (6), moving the cursor. Buttons (7) are used to click on areas of the screen, or call up menus.

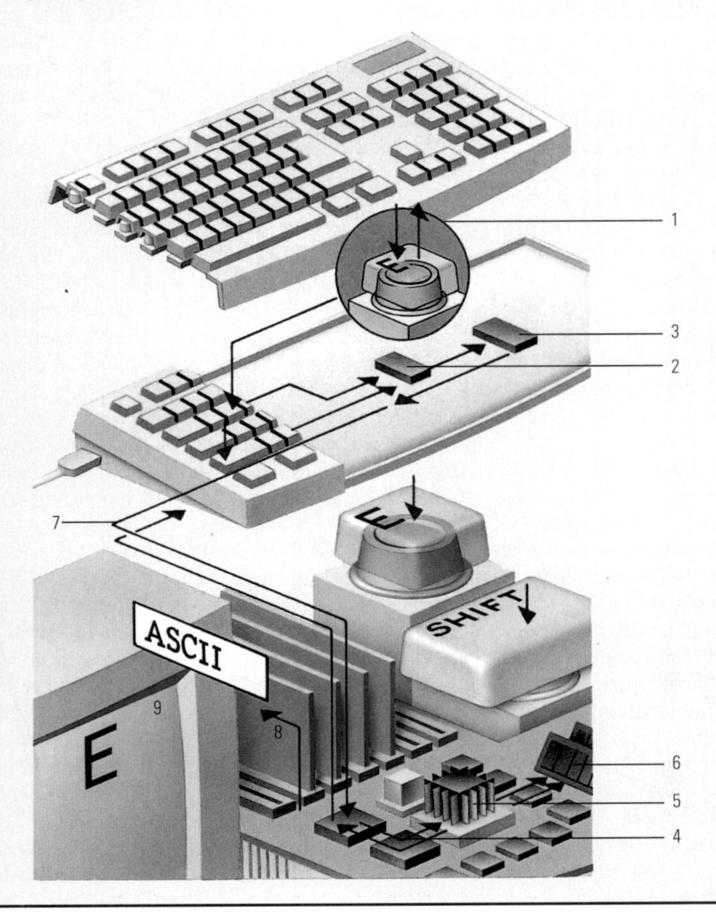

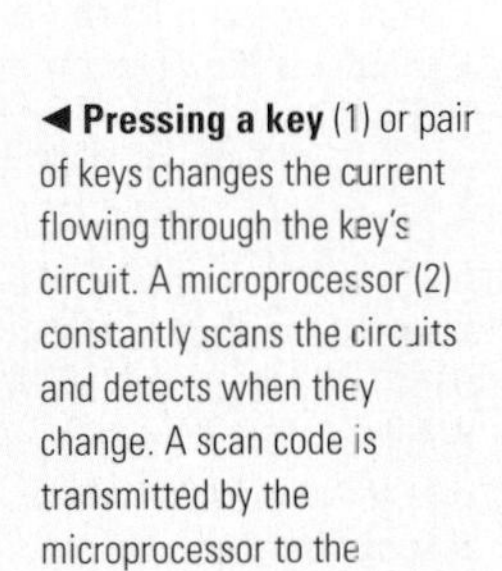

◄ **Pressing a key** (1) or pair of keys changes the current flowing through the key's circuit. A microprocessor (2) constantly scans the circuits and detects when they change. A scan code is transmitted by the microprocessor to the memory buffer in the keyboard (3). The scan code then travels through the cable connecting the keyboard to its controller chip (4) in the body of the computer. The controller chip informs the central processing unit (CPU) (5) which finds the keyboard program in read-only memory (ROM) (6) and cancels the scan code in the keyboard's memory buffer (7). The ROM converts the scan code into the PC's language, ASCII, and then instructs the monitor (8) to display the character, in this example, an uppercase E.

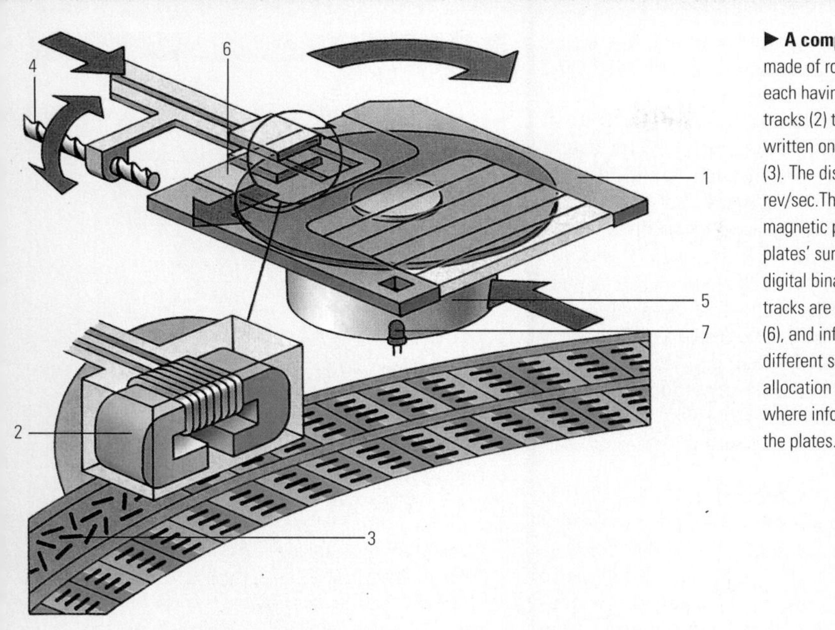

▶ **A computer hard disk** is made of rotating plates (1) each having circular magnetic tracks (2) that are read and written on by a magnetic head (3). The disk spins at about 100 rev/sec. The heads align magnetic particles on the plates' surfaces to represent digital binary code (5). The tracks are divided into sectors (6), and information is stored in different sectors (7). A file allocation table tells a chip (8) where information is held on the plates.

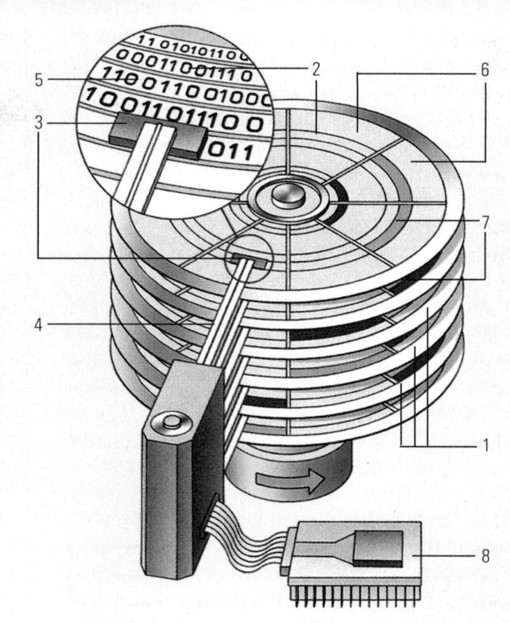

▲ **A floppy disk** (1) uses a magnetic head (2) to polarize magnetic particles on the surface of the disk (3). The polarization represents binary code. A screw (4) moves across the disk which is spun by a motor (5). When the disk is inserted, a lever moves aside the protective window over the disk (6). An LED (7) checks whether the disk is write protected, in which case information cannot be put onto the disk only read from it.

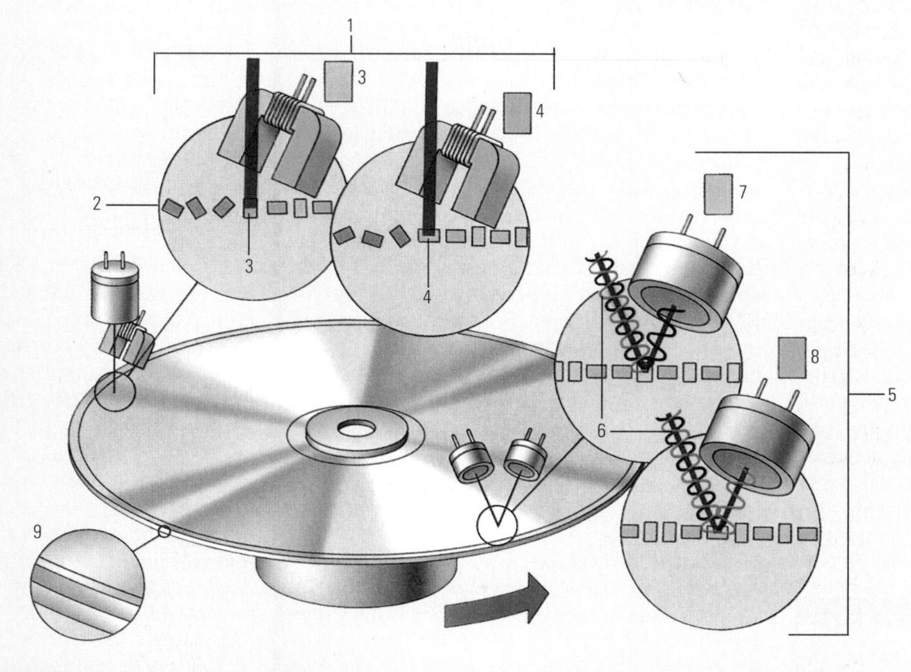

◀ **An optical disk** can store far more information than a magnetic disk, because information is read or written by a laser, make the disk much more accurate. The crystalline metal surface of the disk cannot be polarized by a magnet alone, unlike a floppy or hard disk. To write to the disk (1) a laser heats a thin strip of the metallic crystals (2), which then can be aligned to represent 0s (3) and 1s (4), as used in binary code. To read the disk a laser is again used (5). The alignment of the metallic crystals controls the way they reflect laser light (6). A sensor translates the reflected light into 0s (7) and 1s (8). The metallic alloy is protected by layers of plastic (9).

COMPUTER COMMUNICATION

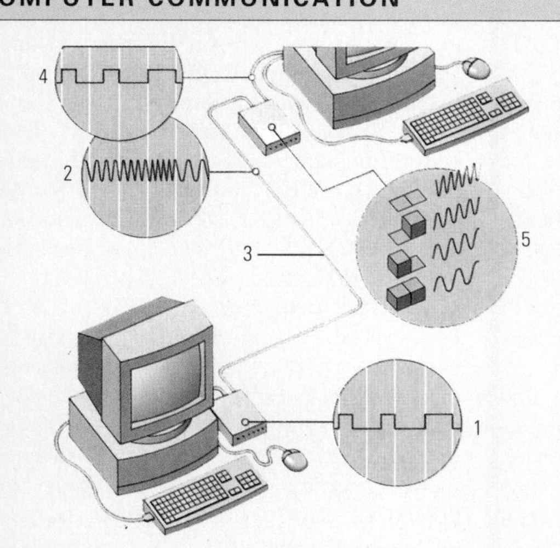

▲ **A modem** allows computers to communicate over telephone systems. It converts the digital signals (1) used by computers into an analogue signal (2) that travels over phone lines (3). The analogue signal is converted back into a digital signal (4) using binary code (5), which can then be read by the receiving computer.

▶ **The most common** local area network (LAN) provides a communication link for all office computers and printers. From a single cable (1), spurs (2) lead to individual machines. A machine sends (3) an address code at the start of each message. Receiving machines (4) return a message, consisting of an address code and confirmation. When two machines send messages simultaneously they collide (5). The electronic shock wave (6), produced when the messages hit, are picked up by all the machines on the LAN. The machines suspend message-sending for a random length of time (7), before repeating the message (8).

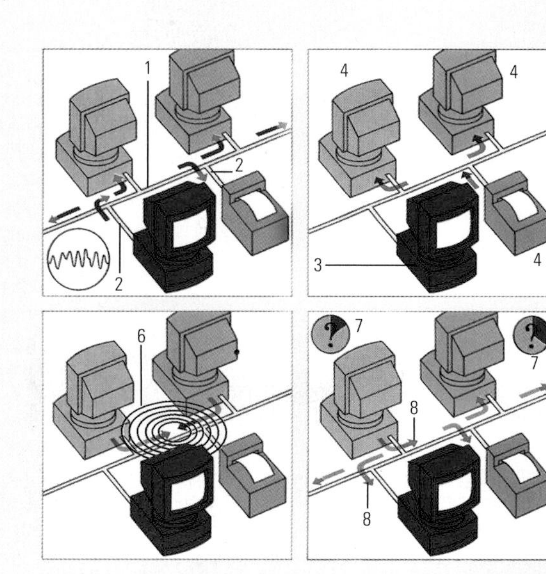

Arthur COMPTON and is evidence for the dual nature (wave and particle) of energy.

Compton, Karl Taylor (1887–1954) US physicist. An outstanding researcher (particularly in the fields of CRYSTALLOGRAPHY, IONIZATION, photoelectricity), he was a Princeton professor (1915–30) and president of the Massachusetts Institute of Technology (1930–48).

computer Device that processes data (information) by following a set of instructions called a COMPUTER PROGRAM. Read Only Memory (ROM) and Random Access Memory (RAM) chips act as permanent and temporary electronic memories for storing data. A typical desktop computer system consists of: a main unit containing a central processor together with memory chips and storage devices (usually MAGNETIC DISKS); a monitor containing a CATHODE-RAY TUBE; a keyboard and a mouse; and a printer. Computer programs are usually stored on disks and transferred to the machine's RAM when required. The keyboard and mouse are called **input devices**, since they allow the user to feed information into the computer. The **keyboard** enables the user to enter letters, numbers and other symbols. The **mouse** is a small device moved by hand, which enables the user to control the computer by positioning a pointer on the monitor screen, to select functions from a list. A magnetic disk drive, such as a HARD DISK, acts as both an input and output device. It can supply programs and data to the computer, and store its output. Many computers have CD-ROM drives. Many other **peripherals** are used, such as a scanner which converts images into an electronic signal so that they can be stored and displayed by the computer, or other HARDWARE for storing and manipulating sounds. The modern computer market is dominated by PCs – the generic term used to refer to machines based on the original IBM personal computer produced in the early 1980s. All these machines use an operating system (such as DOS (DISK OPERATING SYSTEM or Windows) produced by the giant SOFTWARE corporation, Microsoft. Other popular operating systems include Apple Macintosh (MacOS) and UNIX. *See feature on pages 82–83*

computer-aided design (CAD) Use of COMPUTER GRAPHICS to assist the design of, for example, fabrics, integrated circuits, buildings and vehicles. With CAD, designers can make alterations and analyse the effect of such changes. In fabric design, colours may be varied and the effect judged on the computer screen. In vehicle design, road conditions may be simulated and tested on the design before a prototype is actually built. In the design of integrated circuit boards and microprocessors, CAD programs can automatically find the most efficient way to route the highly complex interconnections between billions of components.

computer-aided instruction (CAI) Use of computers to assist teaching. It applies to education facilitated by computers, not to education in the use of computers. The area is in rapid growth, as technology advances. Two important developments have been CD-ROMs, enabling instruction to be accompanied by audio and video material, and the Internet, allowing students to access information from worldwide sources.

computer-aided manufacture (CAM) Use of computers to control industrial production. Its main applications involve the control of robots and automated machine tools in factories to achieve rapid and consistent manufacturing without the possibility of human error. Often CAM is linked to COMPUTER-AIDED DESIGN (CAD).

computer-aided software engineering (CASE) Use of computers to automate the writing of COMPUTER PROGRAMS.

computer graphics Illustrations produced on a COMPUTER. Simple diagrams and shapes may be produced by typing on the keyboard. Complex images require a mouse or an equivalent input device such as a graphics pad, painting or drawing SOFTWARE and sometimes special graphics HARDWARE.

computer interface Way that a COMPUTER PROGRAM or system interacts with its user. The simplest form of computer interface is the simple screen and keyboard, in which the user controls the computer by typing in commands. The most common type for personal computers is the GRAPHICAL USER INTERFACE (GUI).

computerized axial tomography (CAT) Method of taking X-rays that provides images of "slices" through the body. Inside a CAT scanner is an X-ray source, which produces a narrow beam of radiation. This passes through a patient's body and is detected by an electronic sensor. The X-ray source and detector are rotated around the patient's body so that views are taken from all angles. A computer analyses the output of the sensor from each of these angles, and uses the information to build up a picture of the slice of the body, showing whether it is damaged or diseased.

computer language System of words and rules used to program a COMPUTER. Most computers work using the binary system language (using 1s and 0s) called machine code. Rather than using a MACHINE CODE, a language consisting of words and symbols that relate more directly to normal language can be used to instruct a computer. A COMPILER, assembler or other such program then translates this into machine code. Several kinds of programming language have been designed for different purposes. Fortran is for scientific and mathematical use, COBOL for business programs, ALGOL for mathematical applications, and BASIC and Pascal were originally for use by learners. Today, the majority of applications for personal computers are written in a language called "C", or derivatives of it. *See also* ASSEMBLY LANGUAGE; COMPUTER PROGRAMMING

computer memory Part of a COMPUTER that stores information in "words" or their "BITS", each of which has an identification number (address) assigned to it for immediate use by the CENTRAL PROCESSING UNIT (CPU). It may consist of magnetic cores, tapes, drums or discs. Developments in INTEGRATED CIRCUITS have revolutionized computer memory technology, permitting greater retrieval speeds and miniaturization.

computer network Number of computers linked together and able to communicate with each other. A typical local area network (LAN) links computers within the same building, enabling staff to exchange data and share printers. Connections are usually through electric cables that link all the machines. A wide area network (WAN) covers longer distances and may link LANs. The interconnections are made through public TELEPHONE services via electronic units called MODEMS, or through the INTEGRATED SERVICES DIGITAL NETWORK (ISDN), a dedicated high-speed line that carries digital signals. *See also* INTERNET

computer program (SOFTWARE) Set of step-by-step instructions that enables a COMPUTER to carry out a task. A typical computer can carry out many tasks including word processing, calculating, drawing, communicating, as well as providing games and other forms of entertainment. Programs can be written in a variety of COMPUTER LANGUAGES. A simple program for carrying out a specific calculation might involve only a few lines of code and take five minutes to write. But a team of programmers might take a year, or more, to write and test the vast amounts of code needed for a complex program. It is usually stored on a magnetic DISK. To make a computer perform a particular task, a program is loaded into the computer's RAM.

COMPUTER GRAPHICS

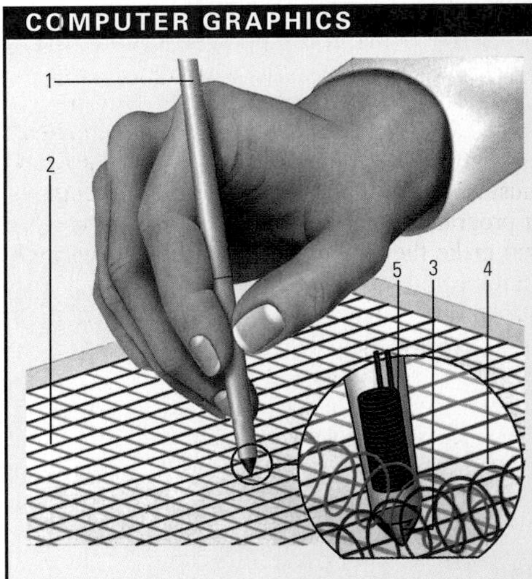

A graphics pad is a method for inputting information to a computer. It allows the operator to "draw" on the computer with a pen or stylus (1) via the pad (2). Just below the surface of the pad are current-carrying filaments (shown blue and green). The pen has a magnet in the tip (3). As the pen moves across the pad, the magnet interferes with the magnetic fields (4) created by the filaments (5) sending the location of the pen to the computer. The location of the interference is read hundreds of times a second by a chip, providing a constant stream of coordinates.

COMPUTER INTERFACE

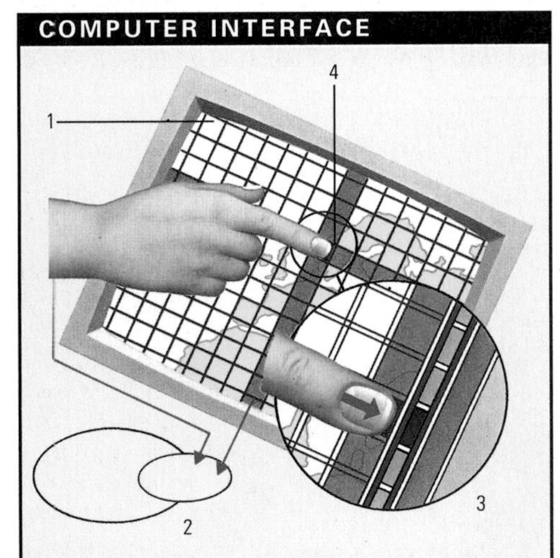

Computer touch-screens detect physical contact with the screen (1), such as a hand or finger. Most screens have two plastic layers with a thin transparent coating of a conducting material (2). The pressure of a finger brings the two sheets slightly closer together (3). The screen reads the position of the touch horizontally and vertically (4) to ascertain its position. In consumer touch-screens, such as automated teller machines, a small selection of options is presented to the user with each section of the screen relating to one of them.

computer programming Preparation of a COMPUTER so that it can perform a specific task. Before being given data, a computer must be given a set of instructions, or a COMPUTER PROGRAM, telling it how to process the data. Each instruction is a single step and all information must be in the BINARY SYSTEM form. For computer programmers, languages have been developed that make the task of programming increasingly intuitive.

computer virus Sequence of COMPUTER PROGRAM code that is able to copy itself from one computer to another and is usually designed to disrupt the normal operation of a computer. Some viruses attached to widely circulated free programs find their way into computers all over the world. A virus may remain undetected for months and then suddenly go into action. A "Friday the thirteenth" virus, for example, waits until this "unlucky" date before causing chaos. The most destructive viruses cause loss or alteration of data held on the computer. A more friendly virus may simply flash a message on the screen from time to time.

concentration (formerly molarity) Measure of the concentration of a SOLUTION, expressed as the number of MOLES. (It is different from **molality**, which is the number of moles in a litre of solvent.) For example, a solution containing 0.2 moles per litre of solution has a concentration of 0.2 molar (written 0.2M).

concentric circles Two or more circles lying in the same plane and having the same point as their centre.

conception Start of pregnancy. The term is used to describe the fertilization of an OVUM (egg cell) by the SPERM, or the implantation of the fertilized ovum in the wall of the UTERUS.

concrete Hard, strong building material made by mixing Portland CEMENT, sand, gravel and water. Concrete is an important building material for general, large-scale construction and for making smaller items including slabs and pipes. It can be strengthened by steel rods, which are embedded in it to produce **reinforced** concrete. **Prestressed** concrete contains tensioned wires instead of steel. Its modern use dates from the early 19th century, although the Romans made extensive use of concrete.

condensation Formation of a liquid from a gas or vapour, caused by cooling or an increase in pressure. More particularly, it is the changing of water vapour in the air into water droplets, forming mist, cloud, rain or drops on cold surfaces.

condensation nucleus Small liquid or solid particle, such as dust, on which water vapour in the atmosphere begins to condense in tiny water droplets or ice crystals resulting in CLOUD formation.

condensation reaction Type of CHEMICAL REACTION in which two simple molecules combine to form a third more complex molecule, with the elimination of a simple substance. The molecules of the substance produced are larger than those of the combining molecules, while the eliminated substance (typically ammonia or water) consists of very small molecules. Condensation reactions occur mainly in organic CHEMISTRY. An example is the reaction between a KETONE and hydroxylamine to produce an OXIME and water. A similar process involving MONOMERS can be used to produce a POLYMER in a type of reaction called condensation POLYMERIZATION. *See also* ADDITION REACTION

condenser *See* CAPACITOR

condenser, steam Device for changing steam to water by cooling. Condensers are used in steam power plants where the steam from a TURBINE or steam engine must be condensed to water, which may then be pumped back for re-use in the boiler.

conditioning In experimental psychology, learning in which human or animal subjects learn to respond in a certain way to a stimulus. Most of the procedures and terminology of **classical (or Pavlovian) conditioning** stem from the work of Ivan PAVLOV. A response elicited by a provided stimulus is called a **conditioned reflex**. **Operant (or instrumental) conditioning** was first described by Burrhus SKINNER in the 1930s. It involved punishing or rewarding certain actions to positively or negatively reinforce behaviour. It can give rise to new behavioural patterns, unlike classical conditioning which can modify only the existing behaviour of animals. Operant conditioning occurs naturally, and many emotional responses are learned on this principle.

Condon, Edward Uhler (1902–74) US physicist. A former director of the National Bureau of Standards, he also applied QUANTUM MECHANICS to investigations of the atom and its nucleus. He wrote a controversial report on UFOs.

conductance (Symbol G) Ability of a material to conduct electricity. In a direct current (DC) circuit, it is the reciprocal of electrical RESISTANCE. For example, a conductor of resistance R has a conductance of 1/R. In an alternating current (AC) circuit, it is the resistance divided by the square of IMPEDANCE (the opposition of a circuit to the passage of a current); $G = R/Z^2$. SI units of conductance are SIEMENS (symbol *S*). *See also* ELECTRIC CURRENT

conduction **Thermal** conduction is the transfer of heat from a hot region of a body to a cold region. If one end of a metal rod is placed in a flame, the heat energy received causes increased vibratory motion of the molecules in that end. These molecules bump into others farther along the rod, and the increased motion is passed along until finally the end not in the flame becomes hot. **Electrical** conduction is the progress of charged particles through a substance resulting in an electric charge. In metals it is the flow of free ELECTRONS. In gases it is the flow of IONS. *See also* SEMICONDUCTOR; SUPERCONDUCTIVITY

conductivity Measure of the ease with which a material allows electricity (**electrical** conduction) or heat (**thermal** conduction) to pass through it. For a solid substance, the electrical conductivity is the CONDUCTANCE (ability to conduct electricity) between the opposite faces of a unit cube at a specific temperature. It is the reciprocal of RESISTIVITY (a material's ability to oppose the flow of an ELECTRIC CURRENT) and is expressed in SIEMENS per metre. For an ELECTROLYTE, conductivity is the ratio of the current density to the field strength. Thermal conductivity of a metal rod is determined by $H = -k(A/l)(T_2 - T_1)$, where H equals the rate of conductivity, k is constant, A is the cross-section of the rod, l its length and $T_2 - T_1$, the difference in temperature.

conductor Substance or object that allows easy passage of free ELECTRONS, thereby allowing heat energy or charged particles to flow easily. Conductors have a low electrical RESISTANCE. Metals, the best conductors, have free electrons that become an ELECTRIC CURRENT when made to move. The resistance of a metallic conductor increases with temperature because the lattice vibrations of atoms increase and scatter the free conduction electrons.

condyle Ball-shaped end of a bone that fits into a socket in another bone to form a movable JOINT. In many mammals the joints between the ends of the mandible (lower jaw) and the skull have condyles, permitting the up-and-down and side-to-side movements in chewing.

cone In botany, conical, spheroidal or cylindrical fruit- or seed-bearing structure borne by CONIFER trees and comprising clusters of stiff, overlapping, woody scales which separate to release seeds from naked ovules developed at their base.

cone In geometry, a figure swept out by a line (the **generator**) that joins a point moving in a closed curve in a plane, to a fixed point (the **vertex**) outside the plane. In a right circular cone the vertex lies above the centre of a circle (the **base**), and the generator joins the vertex to points on the circle.

◀ **cone** An important family of mathematical curves results from sectioning a cone at various angles. A horizontal section (1) gives a circle; an inclined one (2) gives an ellipse. A section parallel to one side of the cone (3) gives a parabola and still greater inclination (4) a hyperbola. All these have the same general equation $ax^2 + by^2 + 2hxy + 2gx + 2fy = c$. With h^2 greater than ab, it is a hyperbola; with $h^2 = ab$, a parabola; h^2 less than ab gives an ellipse of which the circle (h = 0, a = b) is a special case. The terms a, b, c, h, g and f are chosen constants; with b = c = h = g = 0, a = 1 and 2f = 1, the equation becomes $y = -x^2$. As $h^2 = ab$ (both = 0) this is the equation of a parabola.

Such a cone has a volume $^1/_3\pi r^2 h$ and a curved surface area πrs, where h is the vertical height, s the slant height, and r the radius of the base.

configuration, electron Arrangement of ELECTRONS around an atom. It is usually written in a notation using 1, 2, 3, etc for the principal shells and s, p, d and f for the subshells. The number of electrons is written as a superscript. The electron configuration of the helium atom is $1s^2$; that of the sodium atom is $1s^2 2s^2 2p^6 3s^1$.

congenital disorder Abnormal condition that is present from birth. It may be due to faulty development, infection, or the effects of the mother's exposure to some drugs or other toxic substances during pregnancy. Important examples of congenital disorders include spina bifida, with incomplete development of the spine, and congenital herpes, acquired from the mother during the passage through the birth canal.

conglomerate In geology, a SEDIMENTARY ROCK made up of rounded pebbles embedded in a fine matrix of sand or silt, commonly formed along beaches or on river beds.

congruent Equivalence of shape and size. Two congruent geometric figures will coincide exactly when superimposed. In contrast, two figures are said to be **similar** if the figures cannot be superimposed without one being scaled or reflected.

conic (conic section) Curve formed by the intersection of a plane with a CONE. CIRCLES, ELLIPSES, PARABOLAS and HYPERBOLAS are conic sections. Alternatively a conic is the locus of a point that moves so that the ratio of its distance from a fixed point (the **focus**) to its distance to a fixed line (the **directrix**) is constant. This ratio is called the eccentricity (e): $e = 1$ gives a parabola, $e > 1$ a hyperbola, $e < 1$ an ellipse, and $e = 0$ a circle.

conifer Cone-bearing tree, generally evergreen, such as pines, firs and redwoods. Some are the Earth's largest plants, reaching heights of up to 99m (325ft). They are major natural resource of the Northern Hemisphere. *See also* GYMNOSPERM

conjugation SEXUAL REPRODUCTION by FUSION of GAMETES. It is characteristic of certain simple animals, lower plants and bacteria. In some algae, for example, a temporary conjugation tube forms a passageway for the contents of one cell to enter another.

conjunction In astronomy, celestial configuration characterized by a coincidence in the longitudes of two celestial objects, usually two planets or a planet and the Sun, as viewed from the Earth; this is called **planetary** conjunction. **Inferior** conjunction occurs when either Mercury or Venus lies between the Earth and the Sun; **superior** conjunction occurs when they lie on the opposite side of the Sun to the Earth.

conjunction Logical proposition produced by joining two simple propositions by "and". An example is the proposition "Bill is kind and Bill is honest"; it is true only if both parts are separately true. The conjunction of two simple propositions, P and Q, is written PQ (or, equivalently $P \wedge Q$), read "P and Q". *See also* DISJUNCTION

connective tissue Type of animal tissue with a supporting and packing function; it helps to maintain a body's shape and hold it together. Connective tissue comprises a small number of cells (such as FIBROBLASTS) and protein fibres within a noncellular substance known as the **extracellular matrix**. BONE, LIGAMENT, CARTILAGE, SKIN and ADIPOSE TISSUE are all types of connective tissue.

conservation Term that has come to mean a number of different, if associated, things in the preservation of nature and natural resources. Conservation requires planning and organization to make the best use of resources or to preserve the natural landscape and wildlife. It may involve a combination of: 1) placing restrictions on the use of materials or areas; 2) using alternative materials; 3) providing areas of protection; and 4) applying methods which limit or control the amount of change or damage to the natural environment. The term is also used to describe the preservation, and sometimes renovation, of ancient and historic man-made structures.

conservation, laws of Physical laws stating that some property of a closed system is unaltered by changes in the system. The most important are the laws of **conservation of matter** and of **energy**. The former states that matter can be neither created nor destroyed; the total mass remains constant when chemical changes occur. The total energy of a system also remains the same; energy is converted from one form into another. Both these laws are only approximately true. Mass and energy are interconvertible according to the equation $E = mc^2$. What is conserved is the total mass and its equivalent in energy. A further law of conservation involves charge, as electric charge can neither be created nor destroyed.

conservation law, nuclear In a nuclear interaction, the total charge, SPIN, or other specific QUANTUM NUMBERS of the interacting particles must equal that of the resulting particles. In STRONG NUCLEAR FORCE interactions all quantum numbers are conserved. In WEAK NUCLEAR FORCE interactions several of the laws break down, notably that for PARITY.

constant In mathematics and science, a quantity or factor that does not change. It may be universal, such as π: the ratio of the circumference of a circle to its diameter; it may apply only in a particular circumstance, as a symbol that has a fixed value in an algebraic equation; or it may be a defined physical value of a substance, such as the BOLTZMANN CONSTANT for gases or the speed of light (c) in a vacuum.

constantan High-melting point alloy containing 45% nickel and 55% copper. It is used in THERMOCOUPLES and in conjunction with iron and copper to form BIMETALLIC STRIPS – the sensing element in some THERMOSTATS.

constellation Grouping of stars, forming an imaginary figure traced on the sky. The stars in a constellation lie at very different distances from the Earth, so the groupings have no physical significance. There are 88 constellations that have been assigned boundaries on the CELESTIAL SPHERE by the International Astronomical Union in 1930. Constellations range from the vast Hydra down to the tiny but brilliant Crux. Many of the names are drawn from ancient mythology, while more recent mapping of the Southern-Hemisphere skies introduced some modern-sounding names such as the Octant and the Telescope. *See also individual constellations*

construction engineering Branch of CIVIL ENGINEERING responsible for preparing the site, directing the placement of materials and organizing personnel and equipment.

constructive margin In PLATE TECTONICS, the boundary between two lithospheric plates, which are moving apart and new crust is being formed. Constructive margins are associated with shallow-focus earthquakes, high heat flow and eruptions and injections of basalt. Characteristically they produce submarine MID-OCEAN RIDGES, but Iceland is an example where a mid-ocean ridge has surfaced above sea level. *See also* LITHOSPHERE

consumer *See* FOOD CHAIN

contact lens Lenses worn on the eye to aid or correct defective vision. The earliest contact lenses were made of glass, but since 1938 plastic has generally been used. In the early 1970s "soft" lenses were introduced. Lenses cover only the CORNEA and float on a layer of tears.

contact metamorphism Recrystallization of rocks surrounding on igneous INTRUSION in

CONIFER

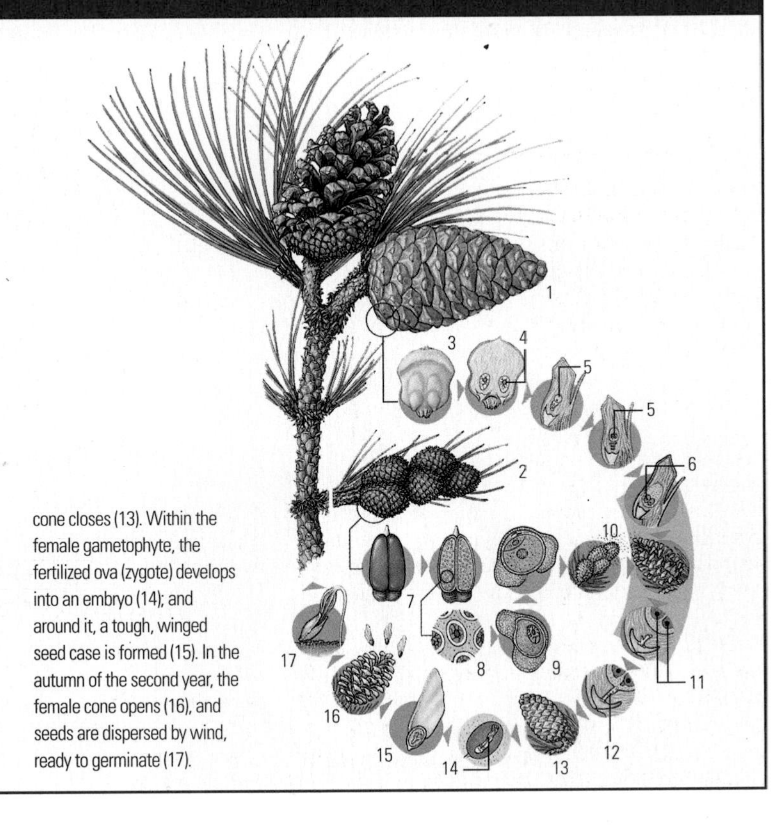

The reproductive cycle of the ponderosa pine is typical of many conifers. In summer, the mature tree bears both female cones (1) and male cones (2). A scale from the female cone (3) contains two ovules (4). Within each ovule, a spore cell (5) divides to develop into a female gametophyte (6). A scale from the male cone (7) contains many spores (8). Each of these develops into a male gametophyte within a winged pollen grain (9). This process lasts one year. Pollination occurs early the next summer, when female cones open so that airborne pollen grains enter an ovule. Inside the ovule, the female gametophyte develops two ova (11). Fertilization occurs during the spring of the following year, after the male gametophyte has matured and grown a pollen tube, and the cone closes (13). Within the female gametophyte, the fertilized ova (zygote) develops into an embryo (14); and around it, a tough, winged seed case is formed (15). In the autumn of the second year, the female cone opens (16), and seeds are dispersed by wind, ready to germinate (17).

C

CONSERVATION, LAWS OF

▼ **The law of conservation of energy** can be explained by considering a ball weighing 1kg released at the top of a 100m high drop. At first the total energy of the ball is its potential energy (PE). As the ball begins to descend, it progressively loses potential energy and gains kinetic energy (KE) to keep its total energy constant. Thus energy is conserved. A kinetic energy increases from zero to a maximum B potential energy decreases from the maximum value to zero C total energy, which is the sum of the kinetic and potential energies, remains constant

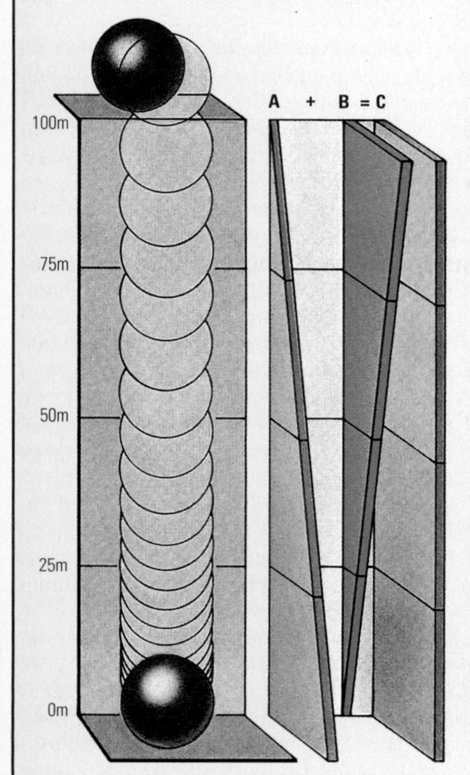

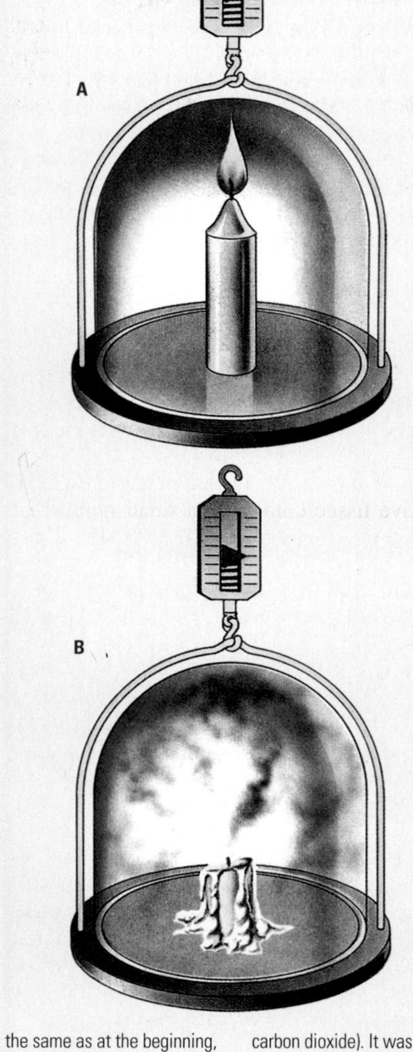

► **The law of conservation of matter** states that matter is neither created nor destroyed in a chemical reaction. This conservation of mass can be shown by a classic experiment in which a candle is burned inside a weighed bell jar (A). At the end of the experiment, the weight of the jar and its contents (B) are the same as at the beginning, although a part of the candle – made up largely of carbon and hydrogen – has "disappeared" as volatile reaction products (water and carbon dioxide). It was only after scientists accepted the principle of conservation of mass in the late 1700s that a quantitative approach to chemistry became possible.

response to the heat supplied by the intrusion. There is no significant increase in pressure and the affected rocks do not melt in the process.

contactor Device for the repeated switching on and off of ELECTRICAL CIRCUITS. Often, contactors contain a magnetic coil which diverts and extinguishes the electrical arc formed when a circuit is opened.

continent Any one of seven (or six) large land masses on the Earth's surface. The continents are Europe and Asia (or Eurasia), Africa, North America, South America, Australia and Antarctica. They are concentrated in the Northern Hemisphere. They cover about 30% of the Earth above sea level and extend below sea level forming CONTINENTAL SHELVES. All continents have four components which make up the continental crust: 1. **Shields** – areas of relatively level land within a few hundred metres height above sea level consisting of crystalline rocks that are generally very old, up to 3,800 million years in age; 2. **Stable platforms** – areas of the continent that have a thin covering of SEDIMENTARY ROCKS which are generally horizontal; 3. **Sedimentary basins** – broad, deep depressions filled with sedimentary rocks formed in shallow seas that sometimes covered parts of the ancient shields or their margins; 4. **Folded mountain belts** of younger sedimentary rocks, which typically occur along their margins and consist of long, linear zones of intensely folded and faulted rocks which have been metamorphosed and intruded by igneous and volcanic activity. The continental crust is composed of rocks that are less dense than the basaltic rocks in ocean basins and they appear to be riding on the asthenosphere of the underlying MANTLE, moving position over the surface of the Earth very slowly by CONTINENTAL DRIFT. Its thickness is mainly between 30–40km (20–25mi) except under large mountain chains where thickness can be 70km (45mi). Continental crust is thought to originate from the SUBDUCTION of early crust, causing partial melting which released lighter material that punched up through the crust. This material was too light to be subducted and the continents grew by the collision and fusion of these microcontinents.

continental climate Type of CLIMATE found in the interior of large continents, that is, hot summers with convectional rainfall and very cold, dry winters, with only light falls of snow. Summer temperatures are about 20°C (68°F), and winter temperatures are −10°C to −20°C (14°F to −4°F) in the coldest month. Total annual precipitation is about 500mm (20in). The natural vegetation consists of grassland. Areas of continental climate are found in Poland and Hungary, and on the steppes of Russia and the prairies of North America. In the Southern Hemisphere there are no areas of real continental climate because the land masses are relatively narrow, so that oceanic influences are present throughout. *See also* MARITIME CLIMATE; MEDITERRANEAN CLIMATE

continental drift Theory that the CONTINENTS change position very slowly, moving over the Earth's surface at a rate of a only a few centimetres per year, adding up to thousands of kilometres over geological time. Until the mid-20th century most scientists believed that the continents were in a fixed position. Early supporters of continental drift claimed that the jigsaw shapes of the present day continents could be pieced together to form an ancient landmass which, at sometime in the past, split and drifted apart. Evidence for the theory included matching the outlines of continents, rock types, geological structures and fossils. Fossil magnetism in rocks is used to calculate the ancient latitude of continents over time. Continental drift became accepted with the development of PLATE TECTONICS in the 1960s. In recent years continental movement has been confirmed by direct measurements made by global positioning satellites using laser beams.

continental margin Region of the ocean floor

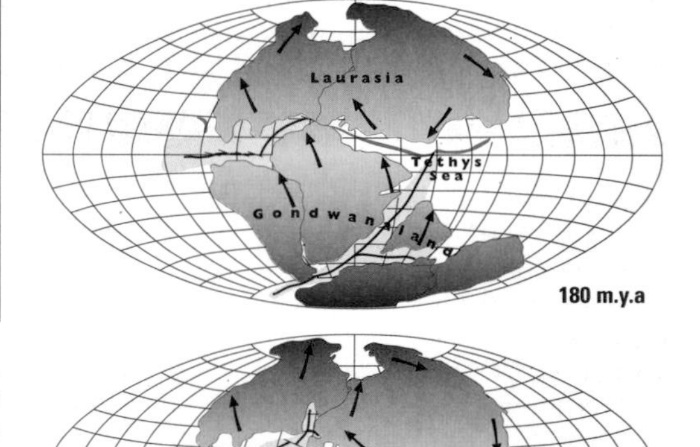

135 m.y.a

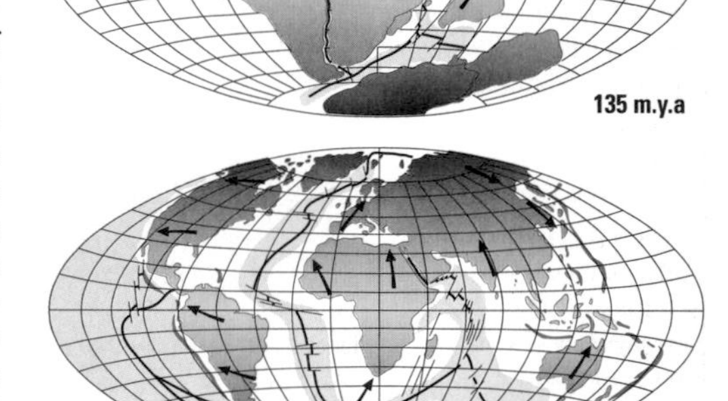

▲ **continental drift** About 200 million years ago, the original Pangaea land mass began to split into two continental groups, which further separated over time to produce the present-day configuration.

that lies between the shoreline and the deep-ocean floor. It includes the CONTINENTAL SHELF, the CONTINENTAL SLOPE and the CONTINENTAL RISE.

continental rise Gently sloping region of the CONTINENTAL MARGIN at the foot of the CONTINENTAL SLOPE. It is an area of thick deposits of sediments carried down by currents off the CONTINENTAL SHELF.

continental shelf Gently, seaward-sloping part of the CONTINENTAL MARGIN between the shoreline and the CONTINENTAL SLOPE, at a depth of *c.*150m (500ft). A continental shelf can be quite narrow, as, for example off the W coast of South America, but in places can be more than 150km (90mi) wide – for example, in the North Sea and around Britain. The shallow waters contain rich food for fish, and therefore some of the world's major fishing grounds are on continental shelves. Other areas of continental shelf, such as the North Sea and the Gulf of Mexico have been exploited for oil and natural gas.

continental slope Relatively steep slope in the seabed that lies between the CONTINENTAL SHELF and the CONTINENTAL RISE leading into the areas of much deeper water. In many places the continental slope is cut into by deep submarine canyons.

continuous spectrum Unbroken sequence of colours, merging one into the other, produced when light is decomposed by refraction through a prism. An incandescent solid, liquid, or dense gas emits a continuous spectrum. The spectrum of a star consists of a continuous spectrum crossed by **absorption lines**, which are produced when particular wavelengths of radiation from a hot source are absorbed by a cooler, intervening medium.

continuum *See* SPACE-TIME

contour In cartography, a line on a map joining places of equal elevation. Closely spaced contours indicate a steep slope, few or no contours mean flat or almost flat ground.

contraception (birth control) Use of devices or techniques to prevent PREGNANCY. **Oral contraceptives** include the female contraceptive pill, a hormone preparation containing synthesized OESTROGEN and PROGESTERONE, which prevents ovulation – the release of an OVUM (egg) – and thickens the cervical mucus thereby reducing the chances of fertilization should ovulation take place. The convenience and efficacy of the various versions of the "pill" make it the most reliable form of contraception available, although it has been associated with some serious and some minor side-effects and should be prescribed on an individual basis. Development of a male pill, containing TESTOSTERONE and progestin, is underway. Although initial tests have shown the male pill to be effective in reducing sperm count, it will not be available for some years. **Barrier contraceptives** include the male and female condom and the diaphragm. The male condom is a latex sheath which covers the penis and collects the ejaculated semen; the female condom lines the inside of the vagina, preventing any sperm entering the UTERUS. The use of condoms is widely advocated because they help protect against some sexually transmitted diseases, including ACQUIRED IMMUNE DEFICIENCY SYNDROME (AIDS). Devices such as diaphragms or caps cover the CERVIX thus preventing sperm entering the uterus. The intra-uterine device (IUD) is a small spring made from plastic or metal inserted into the uterus. It stops the fertilized egg embedding itself in the uterine lining. Emergency contraception, known as the morning-after pill, can be taken up to 72 hours after unprotected sexual intercourse; it prevents the fertilized ovum embedding itself in the uterus. It is not suitable to be used regularly. Techniques to prevent pregnancy include the less effective "rhythm method". This involves the avoidance of intercourse on days when conception is most likely (when the woman is ovulating). It is not a reliable method because ovulation cannot always be predicted accurately. Another technique is *coitus interruptus* – the withdrawing of the penis from the vagina before ejaculation. This is also very unreliable as small amounts of semen are often released from the penis before ejaculation. *See also* SEXUAL REPRODUCTION

contractile root Type of ADVENTITIOUS ROOT that develops from the base of a BULB or CORM. New bulbs or corms form higher in the soil at the tops of the contractile roots, which then shorten to pull the bulbs or corms down to a better level.

contractile vacuole *See* VACUOLE

control column Main steering component in an AIRCRAFT that controls roll and pitch. In both aeroplanes (joystick) and helicopters (collective pitch stick) it combines with foot controls which change direction above the vertical axis. Vertical movement in helicopters is controlled by the cyclic pitch stick.

control systems Means by which a process is

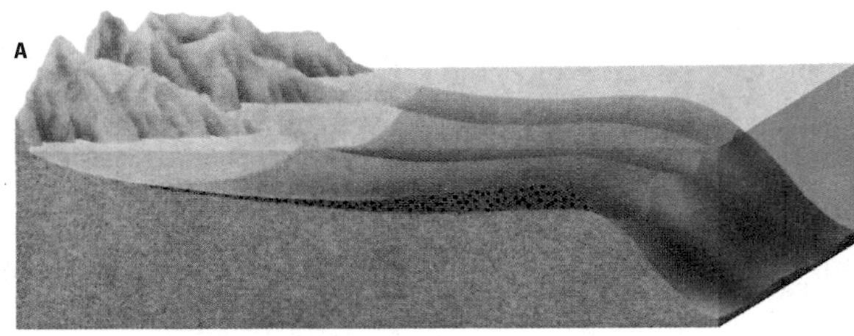

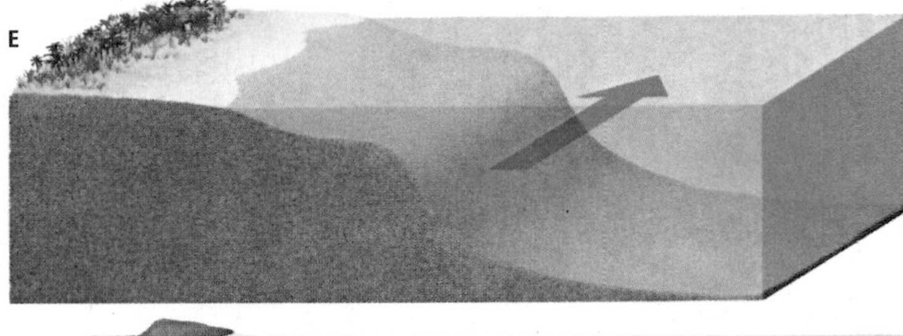

► **continental shelf** The regions immediately off the land masses are the continental shelves. There are several different sorts. Off areas of North America, such as California, the shelf (A) has deep gorges (1); they can be caused by river erosion before the land was submerged by the sea, or by turbidity currents. Mud and sediment-laden water often pour out of major estuaries scouring out gorges. Other regions of North America and Europe have shelves with a gentle relief (B), often with sandy ridges and barriers. In high latitudes, floating ice wears the shelf smooth (C) and in clear tropical seas a smooth shelf may be rimmed with a coral barrier like the Great Barrier Reef off eastern Australia (D), leaving an inner lagoonal area "damm the reef. Other types of contiental shelf often start with a gradual slope, but then drop suddenly. This can be caused by strong offshore currents washing sediment away (E). Continental shelves can also be affected by faulting (F) which disrupts the original shelf.

made to conform to prescribed instructions, either by maintaining the values of certain parameters at a constant level or by making them change according to a predetermined plan. Control systems may be mechanical, electromechanical, electronic, fluidic (operated by liquid or gas pressure), or a combination of any of these means. All systems depend on either feed-forward (such as a cutting tool that follows the shape of a model) or on FEEDBACK (for example a governor that reduces the input of fuel to an engine when the power exceeds a certain level). Many of the more complex systems used in industry are computer-controlled. *See also* ELECTRONIC CONTROL SYSTEM

convection Transfer of heat by flow of currents within fluids due to KINETIC THEORY. Warm fluids have a natural tendency to rise (because they are less dense), whereas cooler fluids tend to fall. This movement subsides when all areas of the fluid are at the same temperature. Convection in the ATMOSPHERE is the main method of heat transfer from one part of the Earth to another. Liquid convection is used in a car's cooling system and some domestic central heating systems.

convection cell Organized circular flow of FLUID, such as air or water, based on thermal changes in density and gravitational attraction, rising away from the heat source and sinking in cooler outer areas.

convection current In geology, heat generated from radioactivity deep within the Earth's MANTLE causing rock to flow towards the CRUST. At the top of the mantle the rising rock is deflected laterally below the crust before sinking. This mantle convection is thought to be the process driving PLATE TECTONICS. In meteorology, convection is the process by which air that is warmed close to the ground rises because it is less dense than the surrounding air. This can lead to condensation, cloud formation and rain, which is called **convectional rainfall**. It is frequently associated with cumulonimbus CLOUDS and thunderstorms.

convergence In mathematics, property of an infinite series (or sequence) having a unique and finite limiting value. Thus, for the series

$$1 + 1/2 + 1/2^2 + 1/2^3 + ...$$

the sum of the first two terms is 1.5, the first three 1.75, and the first four 1.875; as more and more terms are evaluated, the sum approaches the limiting value of 2. Such a series is said to converge. *See also* DIVERGENCE

convergent evolution Tendency of several different species to resemble each other and to develop similar characteristics in their attempt to adapt to similar environments. The most commonly cited example is the wings of bats and birds. *See also* EVOLUTION

convergent margin In PLATE TECTONICS, the boundary between two lithospheric plates, which are moving towards each other. It can form either a SUBDUCTION margin or a collision zone. *See also* LITHOSPHERE

convulsion Series of violent, involuntary contractions of the muscles, sometimes accompanied by loss of consciousness. Such a seizure may indicate EPILEPSY. However, isolated seizures can arise from many other causes, including intoxication, brain abscess or an abnormally low blood-sugar level (**hypoglycaemia**). Young children may suffer febrile seizures, brought on by high fever.

Cooke, Sir William Fothergill (1806–79) British electrical engineer. He collaborated with the physicist Charles WHEAT STONE to develop the first commercial electric telegraph in 1839. In 1845 they patented a single-needle apparatus.

coolant Liquid or gas used to transfer heat from a hot to a cool region or to remove excess heat. For example, in NUCLEAR REACTORS heat is transferred from the core to the generator by liquid sodium, pressurized water or a gas.

Coolidge, William David (1873–1975) US physicist. Coolidge developed (1908) a method for drawing tungsten into filaments for making LIGHT BULBS and radio valves. In 1916 he patented an X-ray tube (**Coolidge tube**) capable of producing accurate amounts of radiation. Coolidge also devised portable X-ray units and worked on construction techniques for industrial quality control. During World War 2 he devised a submarine-detection system with Irving LANGMUIR. Coolidge also worked on atom bomb research. *See also* ELECTRON TUBE

cooling tower Tall construction that uses atmospheric air to cool warm water in refrigeration and steam-power generation systems. There are three types of towers: **atmospheric** cooling towers, which depend on wind currents; **natural draught** towers, which depend on natural convection currents; and **mechanical draught** towers, which use fans to provide forced or induced convection.

Cooper, Leon Neil (1930–) US physicist. From 1974 he was professor at Brown University, Rhode Island, USA. With John BARDEEN and John Robert SCHRIEFFER, he shared the 1972 Nobel Prize for physics for research in SUPERCONDUCTIVITY. They demonstrated that electrons in a superconductor arrange themselves in so-called "Cooper pairs", rather than randomly dispersing energy and momentum.

coorbital satellite Satellite orbiting a planet at the same average distance as another satellite. Although the distances – and hence periods – are the same, the orbital inclinations and eccentricities may differ slightly.

coordinate geometry (analytical geometry) Branch of mathematics combining the methods of pure GEOMETRY with those of ALGEBRA. Any geometrical point can be given an algebraical value by relating it to coordinates, marked off from a frame of reference. Thus, if a point is marked on a square grid so that it is x_1 squares along the x-axis and y_1 squares along the y-axis, it has the coordinates (x_1, y_1). POLAR COORDINATES can also be used. It was introduced in the 17th century by René DESCARTES. *See also* CARTESIAN COORDINATE SYSTEM; CONIC

coordinate system Reference system used to locate a point in space. A point can be defined by numbers representing distances or angles measured from lines or points of reference. In a CARTESIAN COORDINATE SYSTEM a point is defined by distances from intersecting axes. In a POLAR COORDINATE system, a point is located by a distance from a fixed point (the origin) and an angle from a reference line. Two numbers are required to define a position in a plane: three numbers are required in three-dimensional space, and so on. A coordinate system enables curves or surfaces to be defined algebraically.

coordination compound (complex compound) Type of chemical compound in which one or more groups or molecules each form a coordinate bond (two atoms sharing two electrons from one of the atoms) with a central metal atom, usually a TRANSITION ELEMENT. The complex may be either a complex ion or a neutral molecule, as in nickel carbonyl ($Ni(CO)_4$). The coordinating species are known as LIGANDS. Some coordination compounds, such as haem and chlorophyll, have biochemical importance. *See also* CHELATION; ION

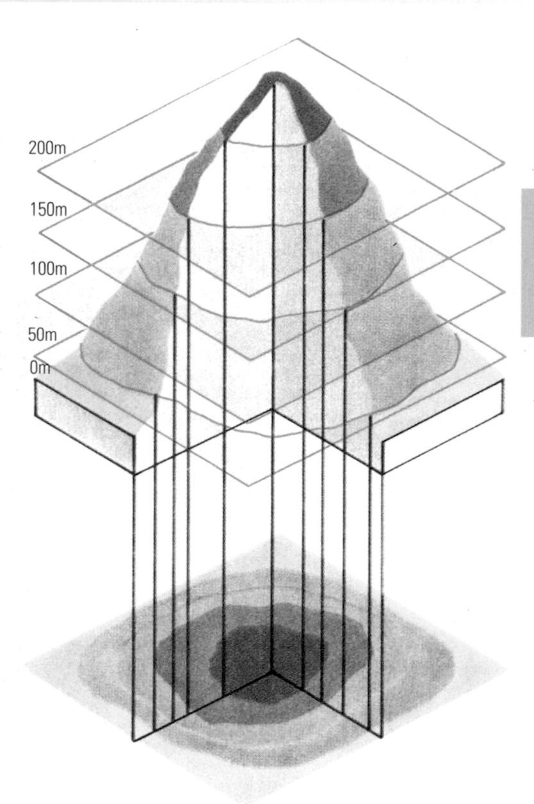

▲ **contour** Recording a three-dimensional shape on a flat surface is a problem that can be solved by the convention of contour scaling. This is shown when the cross-sections of a hill at 50-, 100-,150 and 200-m heights are projected onto a map of the hill. The hill can be envisaged fairly well from such a map, although the "coarseness" of the contour intervals loses some finer detail. The steepness of the sides can be judged by the contour lines on the map.

coordination number Number of chemical bonds made to a central atom in a specified complex salt, such as the number of ANIONS surrounding a CATION.

copepod Any marine or freshwater CRUSTACEAN of the subclass Copepoda. Copepods are possibly the most numerous type of animal in the world and are a major component of the marine FOOD CHAIN. Some are parasitic on aquatic animals, especially fish. Their segmented, cylindrical bodies have a single median eye and no carapace. Length: 0.5–2mm (0.02–0.08in); length of parasitic forms may be more than 30.5cm (1ft). There are 7,500 species.

Copernicus, Nicolas (1473–1543) (Mikotay Kopernik) Polish astronomer. The accepted view in his day was the geocentric (Earth-centred) Universe with its complicated system of orbits, as explained by PTOLEMY nearly 1,500 years before. Through his study of planetary motions, Copernicus developed a heliocentric (Sun-centred) theory of the Universe. In the **Copernican system** (as it is now called) the planets' motions in the sky were explained by orbiting the Sun. The motion of the sky was simply a result of the Earth turning on its axis. An account of his work, *De revolutionibus orbium coelestium*, was published in 1543. Most astronomers considered the new system as merely a means of calculating planetary positions, and continued to believe in ARISTOTLE's view of the world. *See also* GALILEO; KEPLER, JOHANNES

copper (symbol Cu) Red-pink TRANSITION ELEMENT. Reddish copper occurs native (free or uncombined) and in several ores including cuprite (an oxide) and chalcopyrite (a sulphide). The ores are separated from the surrounding rock and are then concentrated. They are smelted into a molten mass, called **matte**, and pure copper (**blister copper**) is extracted by oxidizing the impurities with air. Ores are often treated with acids and the copper recovered by ELECTROLYSIS. It is malleable, a good thermal and electrical CONDUCTOR, second only to silver, and is extensively used in boilers, pipes, electrical equipment and alloys, such as brass and bronze. Copper tarnishes in air, oxidizes at high temperatures and is attacked only by oxidizing acids. It forms two series of salts, termed copper(I) (cuprous) and copper(II) (cupric). Properties: at.no. 29; r.a.m. 63.546; r.d. 8.96; m.p. 1,083°C (1,981°F); b.p. 2,567°C (4,653°F); most common isotope ^{63}Cu (69.09%).

copper oxide Chemical compound of copper and oxygen which exists in two forms: copper(I) (cuprous) oxide (Cu_2O), a brilliant red powder found in nature as the mineral cuprite; and copper(II) (cupric) oxide (CuO), which is black and decomposes into copper(I) oxide and oxygen when heated. It is added to furnace melts as an oxidant.

copper plating Application of a coating of COPPER onto another material. Copper plating by ELECTRODEPOSITION is used to prepare steel for a top coating (also by ELECTROPLATING – a form of electrodeposition) of nickel, chromium or silver. Steel and brass are copper plated in the production of decorative oxidized copper finishes. Steel is sometimes copper plated to prevent surface hardening. The main copper salts used in electroplating solutions are copper sulphate and copper cyanide in the form of COORDINATION COMPOUNDS such as potassium cuprocyanide ($K_2Cu(CN)_3$).

copper sulphate Chemical compound which exists in two forms. Copper(I) (cuprous) sulphate (Cu_2SO_4) is a light grey powder that reacts instantly with atmospheric moisture to produce copper(II) (cupric) sulphate. This is usually seen as the bright blue crystals of the pentahydrate, blue vitriol ($CuSO_4.5H_2O$), used

COPPER

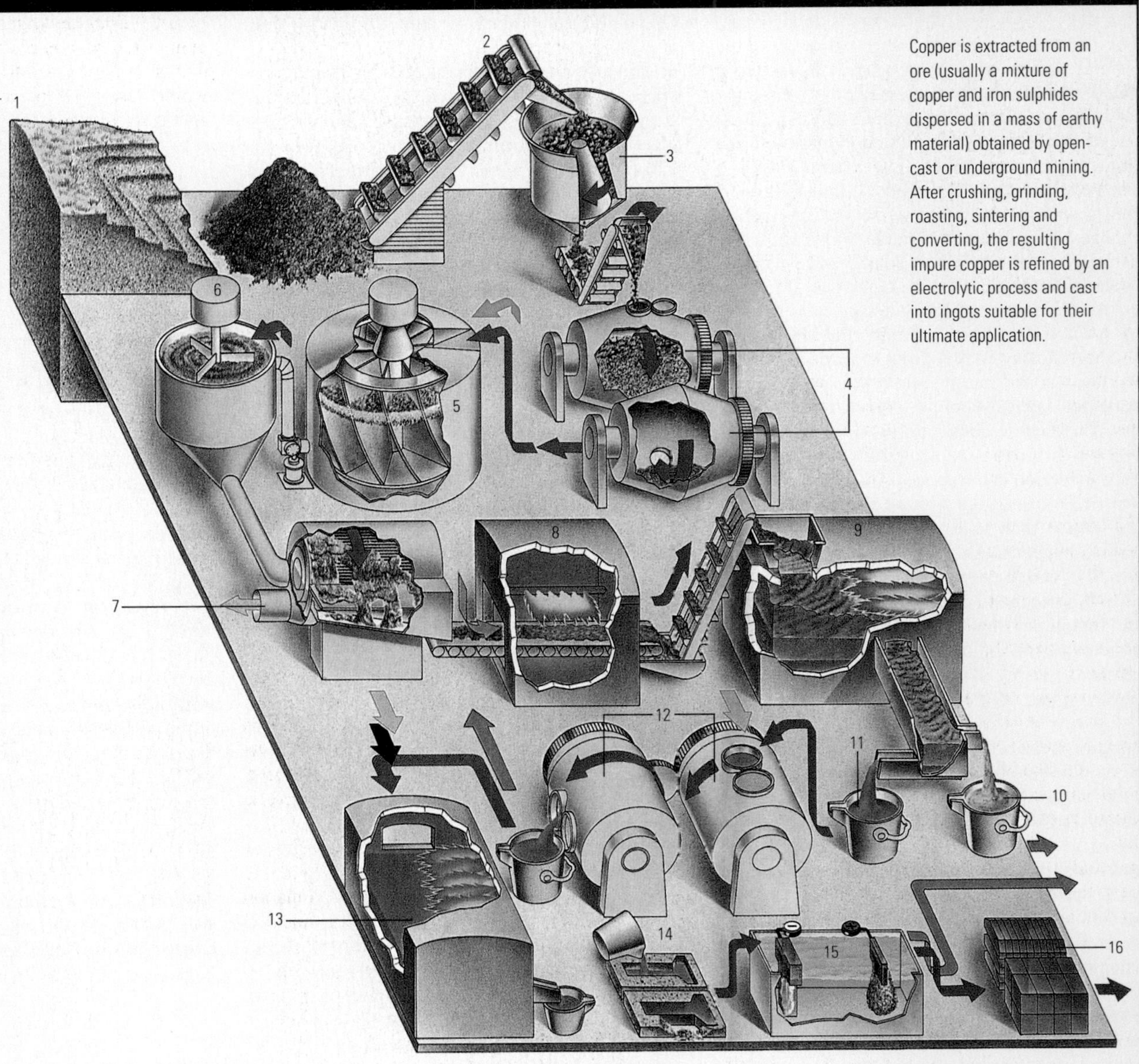

Copper is extracted from an ore (usually a mixture of copper and iron sulphides dispersed in a mass of earthy material) obtained by open-cast or underground mining. After crushing, grinding, roasting, sintering and converting, the resulting impure copper is refined by an electrolytic process and cast into ingots suitable for their ultimate application.

1) ore is mined by the open-cast method, still mixed with earthy material.
2) mixture of ore and earthy material is conveyed away from the mining area for processing.
3) ore is reduced to lumps of about 1.5cm (0.6in) in diameter by crushers. (A cone-type crusher is shown.) At this stage much of the earthy material is separated from the copper-bearing minerals.
4) crushed ore and other matter is mixed with water and ground in ball mills, to a powder capable of passing through a "100 mesh".
5) at this stage, the minerals are extracted by "flotation". More water and frothing agents are added, the mixture stirred and aerated, and activating agents introduced. Bubbles are formed around the copper-bearing minerals and float to form a surface-froth. The unwanted earthy material is precipitated as sediment.
6) froth is concentrated in a "thickener", which extracts much of the water.
7) any remaining water is removed in a drum filter.
8) mixed with fluxes, the dried ore is roasted, producing particles of optimum size and composition for the smelting process.
9) reverberatory furnace, where sintered ore is smelted. Any remaining earthy material sill attached to the minerals combines with the fluxes to form a slag. This floats to the surface of the melt and is run off – 10.
11) more fluxes are added to the remaining melt, which consists of molten copper and iron oxides called a "matte".
12) converter. Here, air is blown through the "matte" and combines with the iron to form oxides. The oxides dissolve in the fluxes to make a slag layer, which is then tapped off. Copper sulphide remains, and the air blown through this combines with sulphur in the melt to provide sulphur dioxide – a valuable by-product often used in the manufacture of sulphuric acid. The melt by now consists of "tough pitch copper", which is copper ready for refining and containing few impurities (mainly oxygen). This melt is tapped off.
13) reverberatory furnace. Here, air is blown through the molten copper. Carbon combines with oxygen to form carbon dioxide, and excess oxygen is removed by the injection of reducing gases. The copper is now in a reasonably pure state, and is poured off for casting into ingots.
14) casting ingots
15) electrorefining. Copper is further refined by electrolysis, using the impure fire-refined copper as anode, a thin sheet of very pure copper as cathode, and dipping both into a solution of acidified copper sulphate. On passing a carefully controlled current through the electrodes and solution, pure copper is taken from the anode and deposited on the cathode, forming an ingot. Noble metals, which may be present and which do not dissolve, are often collected in canvas bags draped around the anode.
16) very pure "cathode" ingot of copper is cast into ingots of various shapes, as required.

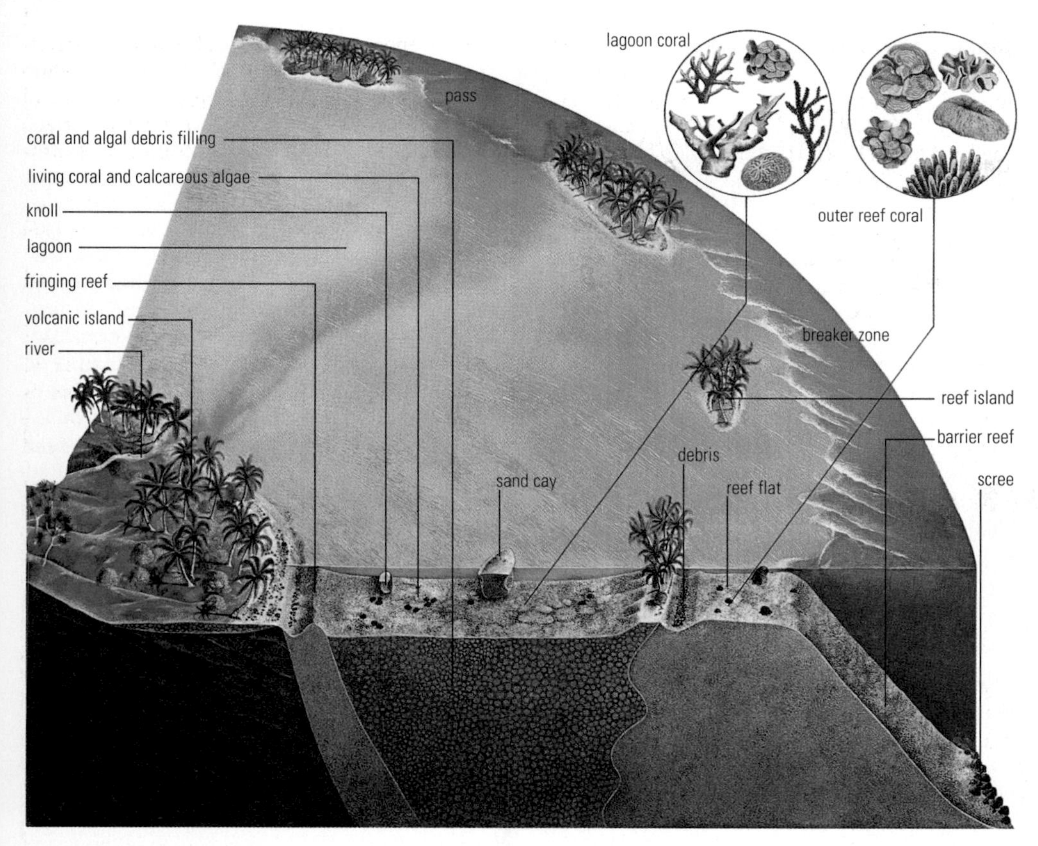

▲ **coral reef** Coral reefs are the largest structures constructed by living organisms. In reef island situations, typical characteristics arising from the growth of a barrier reef and a fringing reef include a water-covered coral platform known as a reef flat, sandy islands and a debris-filled lagoon supporting its own coral life. Beneath the living surface of the reef, dead coral and debris of ancient origin remain as the foundation for the present-day formation. Coral will only grow when it is continuously covered by sea water but at depths less than 50 metres where light can still penetrate. The water must be clear and contain plenty of food material for the coral. Only the surface layer of coral on any reef is living since it is in constant contact with the water, food, oxygen and light.

for copper plating, for preserving wood and for killing algae in ponds. It can also be dehydrated to the colourless anhydrous salt ($CuSO_4$), which, because it absorbs water and turns blue, is used as a desiccant and, more widely, as a moisture indicator.

coral Small coelenterate marine animal of class Anthozoa in the phylum Cnidaria, often found in colonies. The calcium carbonate skeletons secreted by each animal POLYP accumulate to form a CORAL REEF. Reef-building corals are found only in waters with temperatures in excess of 20°C (68°F).

coral reef Rock-like formation found in shallow tropical seas. Such reefs are formed from gradual build up of the calcium carbonate secreted by living CORAL organisms as protection against predators and wave action. When the individual coral dies, its skeleton remains and thus the reef expands. The way in which the coral, and therefore the reef, grows is strongly influenced by the depth, prevailing currents and the temperature of the surrounding sea-water. The main types are **fringing** reefs and BARRIER REEFS.

cordierite Silicate of aluminium, magnesium and iron, which occurs mostly in METAMORPHIC ROCKS. It forms crystals in the orthorhombic system. Hardness 7, r.d. 2.6.

cordillera Chain of mountains. The term is used particularly of parallel mountain ranges, such as those of the Rockies, Sierra Nevada and Coastal Ranges in W North America or of the Andes in South America.

cordite Smokeless, high-explosive shell and bullet propellant made by mixing CELLULOSE NITRATE (guncotton) – a highly nitrated form of CELLULOSE – and NITROGLYCERINE, both of which are themselves explosives. For stability, mineral jelly and acetone are added in small amounts. Cordite is generally manufactured in strands or cords, which give it its name.

core In geology, a cylindrical rock sample that has been gathered by drilling; on land usually by a rotary drill, from the sea often by a metal cylinder, with a cutter on the bottom, which is forced into the rock. The core is used to identify the various layers in the rock or sediment. The mechanism used for collecting the core is called a **corer.**

core Central area of the EARTH from a depth of 2,900km (1,800mi), which accounts for 16% of the Earth's volume and 31% of its mass. Information about the core is obtained from the measurement of SEISMIC WAVES. These indicate that the outer part of the core is liquid, because shear (S) waves will not travel through it, whereas the inner core from 5,150km (3,200mi) to the centre of the Earth is interpreted as solid because seismic velocities are lower. It is believed that the change from liquid to solid core occurs because of immense pressure conditions. The core is thought to be composed of iron-nickel alloy (90% iron, 10% nickel). Temperature estimates for the core vary from 4,000 to 7,000°C. CONVECTION in the iron liquid outer core is thought to be responsible for producing the Earth's magnetic field.

Cori, Carl Ferdinand (1896–1984) US biochemist, b. Czechoslovakia. Cori shared the 1947 Nobel Prize for physiology or medicine with his wife Gerty Theresa (1896–1957) – and Bernardo HOUSSAY – for their discovery of how the chemical energy of GLYCOGEN, a carbohydrate stored in the liver and muscle, is broken down into a form that can be used by the body.

Coriolis effect Apparent force on particles or objects due to the rotation of the Earth under them. It causes objects in motion, and ocean and atmospheric currents to be deflected towards the right in the Northern Hemisphere and towards the left in the Southern Hemisphere and therefore impacts greatly on the world's weather systems. The direction of water swirling round in a drain or whirlpool demonstrates this effect. It is named after the French mathematician Gaspard Gustave de Coriolis (1792–1843).

cork Outer dead, waterproof layer of the BARK of woody plants. The bark of the cork oak (*Quercus ruber*) is the chief source of commercial cork.

corm Fleshy, underground stem that produces a plant such as the gladiolus or crocus. In most plants, new corms form on top of old ones, which last for one season. *See also* ASEXUAL REPRODUCTION

▲ **Coriolis effect** A wind will tend to drag ocean surface water after it. However, as soon as the water starts to move, the Coriolis effect throws it off at an angle. Each successive thin sheet of water beneath the surface layer begins to move because it is linked by friction to the layer above. But each layer is thrown even farther off course by the Coriolis effect. So the surface current moves at an angle of 45° to the wind at about 2–3% of the wind speed. Lower layers move more slowly, at an increasingly eccentric angle. The resulting vertical profile of the motion is the Ekman spiral. At a depth called the Ekman depth, the water flows in the opposite direction to that of the surface current and at 0.043 times the speed of the surface current.

CORNEA

Focusing of light rays from distant objects (A) is mainly done by the cornea (1) with a little help from the lens (2). Ciliary muscles (3) encircling the lens relax and stretch ligaments (4), which pull the lens flat. Rays from a near object (B) are bent by a thick lens produced when the ligaments slacken as the ciliary muscles contract. This process, which is called accommodation, is essential for sharp focusing.

cornea Transparent membrane at the front of the EYE. It is curved and acts as a fixed lens, so that light entering the eye is to some extent focused before it reaches the eye LENS.

Cornu, Marie Alfred (1841–1902) French physicist who worked mainly in optics and astronomy. He devised a photographic method of measuring light intensities in DIFFRACTION patterns and calculated the velocity of light.

corolla PETALS of a flower. The corolla encloses the STAMENS and CARPELS. The petals, which may be free or fused, together form the inner whorl of the PERIANTH.

corollary In mathematics, a theorem that is incidentally proved in proving some other theorem or proposition. More generally, it is an immediate or natural consequence or an easily drawn conclusion.

corona Outermost layer of the Sun's atmosphere, extending for many millions of kilometres into space. It springs into view as a white halo during a total solar ECLIPSE; at other times it can be observed in visible light only by using a special instrument called a **coronagraph**. The corona emits strongly in the X-ray region, and has been studied by X-ray satellites. Its overall shape changes during the 11-year solar cycle, from regular and symmetrical at solar minimum to uneven with long streamers at solar maximum. The corona has a temperature of 1–2 million K.

Corona Australis (Southern Crown) Small constellation of the S sky. Although none of its stars are above magnitude 4, it is easy to identify by its arc of stars.

Corona Borealis (Northern Crown) Constellation of the N sky, adjoining Boötes. Its brightest star is Alpha, magnitude 2.2. The "bowl" of the crown contains the famous variable star R Coronae Borealis, the prototype of a group of stars that fade suddenly and unpredictably.

coronary artery disease Disease of the coronary blood vessels, the ARTERIES supplying blood to the HEART muscle. *See also* ARTERIOSCLEROSIS

coronary thrombosis Formation of a blood clot in one or other of the coronary ARTERIES supplying the HEART, preventing blood (and with it oxygen and nutrients) from reaching the heart. It leads to death of part of the heart muscle – a HEART ATTACK. A major attack, marked by severe chest pain, sweating and sometimes collapse, may be fatal.

corposant *See* ST ELMO'S FIRE

corpus luteum Mass of yellow tissue formed in the GRAAFIAN FOLLICLE in the OVARY of a mammal after the OVUM (egg cell) is released. If the ovum is fertilized, the corpus luteum secretes PROGESTERONE, a hormone needed to prepare the uterus for pregnancy. If the ovum is not fertilized, the corpus luteum becomes inactive.

corrasion Mechanical EROSION caused by loose material, such as sand or pebbles, during transportation. The material is carried by wind, water or ice, and it scrapes against the bed or sides of the river or glacial valley, or against the cliffs and the shore.

correlation In geology, method of linking rock strata of the same age. Geologists do this by comparing similar rocks in different (and sometimes widely separated) outcrops or by studying the FOSSILS they contain. *See also* PALAEONTOLOGY

Correns, Karl Erich (1864–1933) German botanist and geneticist. In 1900 he rediscovered, along with two others, the works of Gregor MENDEL outlining the principles of HEREDITY. Correns conducted experiments to determine the validity of Mendel's laws and helped to provide evidence that proved his theories.

correspondence In geometry, property of two geometric figures in which angles, lines and points in one bear a similar relationship to angles, lines and points in another. In set theory, two sets A and B are said to be in one-to-one correspondence if every element of A can be mapped to an element of B by a single function, with no two elements of A mapping to the same element in B.

corrie *See* CIRQUE

corrosion In chemistry, gradual tarnishing of the surface or major structural decomposition by chemical action on solids, especially metals and alloys. It commonly appears as a greenish deposit on copper and brass, a reddish-brown deposit (RUST) on iron or a grey deposit on aluminium, zinc and magnesium. Rust is the most important form of corrosion because of the extensive use of iron and steel, and their susceptibility to attack. Some metals, such as aluminium and magnesium, corrode readily forming an oxide, which then protects the undersurface from further corrosion.

corrosion In geology, chemical EROSION leading to the disintegration of rocks, such as by the action of running water or by SOLUTION. It is particularly common in limestone.

cortex In animal and plant anatomy, outer layer of a gland or tissue. Examples are the cortex of the ADRENAL GLANDS; the CEREBRAL CORTEX or outer layer of the BRAIN; and the cortical layers of tissue in plant roots and stems lying between the BARK or EPIDERMIS and the hard wood or conducting tissues.

corticosteroid Any HORMONE produced by the CORTEX (outer layer) of the ADRENAL GLANDS. There are two types of corticosteroids. **Mineralocorticoids**, such as aldersterone, regulate the balance of fluids and salts in the body. **Glucocorticoids**, such as CORTISONE and hydrocortisone, regulate the use of carbohydrates, fats and proteins. Some corticosteroids have been synthesized.

cortisone HORMONE produced by the CORTEX of the ADRENAL GLANDS and essential for carbohydrate, protein and fat metabolism, kidney function and disease resistance. Synthetic cortisone is used to treat adrenal insufficiency, rheumatoid arthritis and other inflammatory diseases, rheumatic fever, asthma, severe allergies and skin complaints. A potent, versatile drug, it can have unwanted side-effects such as body swelling.

corundum Translucent to transparent mineral in many hues, aluminium oxide (Al_2O_3). It is found in IGNEOUS and METAMORPHIC ROCKS, occurring as pyramidal or prismatic crystals in the rhombohedral class and as granular masses. It is the hardest natural substance after DIAMOND. Gemstone varieties are sapphire and ruby. It is used in watches and motors and is an important industrial abrasive. Hardness 9; r.d. 4.

Corvus (Crow) Small constellation of the N sky, adjoining Hydra and Crater. Its brightest stars are of the third magnitude.

cosecant In TRIGONOMETRY, ratio of the length of the HYPOTENUSE to the length of the side opposite an acute angle in a right-angled triangle. The cosecant of angle A is usually abbreviated cosec(A) and is equal to the reciprocal of sin(A).

cosine In TRIGONOMETRY, ratio of the length of the side adjacent to an acute angle to the length of the HYPOTENUSE in a right-angled triangle. The cosine of angle A is usually abbreviated cos(A).

cosmic dust Very fine particles of solid matter in any part of the Universe, including meteoric dust and interstellar material that can absorb starlight to form dark NEBULAE in galaxies. Spherical dust particles about 0.05mm (0.002in) in diameter, found in certain marine sediments, are thought to be the remains of some 5,000 tonnes of cosmic dust that fall to Earth each year.

cosmic microwave background Weak ELECTROMAGNETIC RADIATION from the Universe. It was first detected in 1965 by Arno PENZIAS and Robert WILSON. It is considered to be the remnant of the radiation from the hot, early Universe created by the BIG BANG. BACKGROUND RADIATION increased in wavelength as the Universe expanded, and today it peaks in the millimetre region of the ELECTROMAGNETIC SPECTRUM.

cosmic radiation (cosmic rays) Streams of SUBATOMIC PARTICLES from space that constantly bombard the Earth at velocities approaching the speed of light. **Primary** cosmic RADIATION is high-energy radiation that comes from the Sun and other sources in outer space. It consists mainly of atomic nuclei and PROTONS. When primary cosmic rays strike gas molecules in the upper atmosphere they yield showers of **secondary** cosmic radiation, which consists of energetic protons, NEUTRONS and PIONS. Further collisions yield MUONS, ALPHA PARTICLES, POSITRONS, ELECTRONS, GAMMA RADIATION and PHOTONS. Cosmic radiation contributes to BACKGROUND RADIATION.

cosmology Branch of scientific study that brings together the disciplines of astronomy, mathematics and physics in an effort to understand the make-up and evolution of the Universe. Once considered the province of theologians and philosophers, it is now an all-embracing science, which has made great strides in the 20th century. The discovery by Edwin HUBBLE in the 1920s that galaxies are receding from each other promoted the concept of an EXPANDING UNIVERSE, which perhaps started with an explosion – the BIG BANG theory. Associated with this is the OSCILLATING UNIVERSE THEORY. The other main theory is the STEADY-STATE THEORY, which says that matter is continually being created. Observations of the most remote regions of space, made in the mid-1990s, have shown astronomers what they consider to be a picture of the Universe at only a few hundred thousand years old. Such observations have been generally taken as confirmation of the Big Bang theory.

cosmonaut *See* ASTRONAUT

Cosmos (Gk. order) Universe considered as an ordered whole. PLATO and ARISTOTLE conceived of the Universe as ordered by an intelligent principle. The conviction of an ordered nature became the basis of modern natural science.

cotangent In TRIGONOMETRY, ratio of the length of the side adjacent to an acute angle to the length of the side opposite the angle in a right-angled triangle. The cotangent of angle A is usually abbreviated cot(A) and is equal to the reciprocal of tangent(A).

cotyledon First leaf or pair of leaves produced by the embryo of any ANGIOSPERM (flowering plant). Its function is to store and digest food for the embryo plant, and, if it emerges above ground, to PHOTOSYNTHESIZE for seedling growth. *See also* DICOTYLEDON; MONOCOTYLEDON

Coulomb, Charles Augustin de (1736–1806) French physicist. His invention of the TORSION BALANCE led to his experiments in electrostatics and the discovery of the inverse square law that bears his name. *See* COULOMB'S LAW

coulomb (symbol C) SI unit of ELECTRIC CHARGE. It is defined as the charge carried by a current of 1 AMPERE in 1 second. It is named after Charles COULOMB.

Coulomb's law One of the basic inverse square laws of physics. It states that the force between two point electric charges is proportional to the product of the charges and inversely proportional to the square of the distance between them. It can be stated mathematically as: F is proportional to Q_1Q_2/d^2, where F is the force, Q_1 and Q_2 are the charges and d is the distance between them. The constant of proportionality is called the PERMITTIVITY of the medium between the charges.

country rock In geology, rock which has been intruded into by a subsequent igneous INTRUSION.

couple Two equal and opposite parallel forces, which do not act in the same line. The forces produce a turning effect or TORQUE.

coupling Mechanical fastening connecting shafts together for power transmission. A **flexible** coupling is used to compensate for misalignment of shaft axes and mispositioning of shaft centre line; a **rigid** coupling is used for maximum power transfer with minimum misalignment.

Cousteau, Jacques Yves (1910–97) French oceanographer. Best known as the co-inventor (with Emile Gagnan) of the aqualung, Cousteau also invented a process of underwater television and conducted a series of undersea living experiments (Conshelf I–III, 1962–65). Many of the expeditions made by his research ship *Calypso* were filmed by him for television and cinema. Although immensely popular around the world with the public, academics accused him of showmanship and questioned the validity of his research. In his later life he became a prominent figure in ecological movements. *See also* SCUBA DIVING

covalent bond Chemical bond in which two atoms share a pair of ELECTRONS, one from each atom. Covalent bonds with one shared pair of electrons are called **single** bonds; **double** and **triple** bonds also exist. The molecules tend to have low melting and boiling points and to be soluble in nonpolar solvents. Covalent bonding is most common in organic compounds.

covellite Mineral form of copper sulphide (CuS). It occurs with other copper minerals and is mined as an ore. It forms hexagonal, platy crystals, which have a deep indigo blue colour, often tinged with purple. It cleaves into thin, flexible plates. Hardness: 1.5–2.0; r.d. 4.7.

crab nebula NEBULA located about 6,500 light-years away in the constellation Taurus. It is the remnant of a SUPERNOVA noted by Chinese astronomers in July 1054, when it shone as brightly as Venus, being visible even in daylight. The nebula was discovered in 1731 by the British astronomer John Bevis and independently by the French astronomer Charles Messier in 1758. It gained its popular name from a sketch by Lord Rosse.

cracking Stage in oil-refining during which the products of the first DISTILLATION are treated to break up large hydrocarbons into smaller molecules by the controlled use of heat, CATALYSTS and often pressure. The cracking of petroleum yields heavy oils, petrol, and gases such as ETHANE, ETHENE (ethylene) and PROPENE (propylene), which are used in the manufacture of plastics, textiles, detergents and agricultural chemicals. Cracking is therefore a means of yielding greater amounts of the lighter hydrocarbons, which are in greater demand, from heavier fractions such as lubricating oil.

crane Machine for lifting and placing heavy loads. The large tower cranes seen on building sites have a long jib, or arm, counterbalanced by a weight on a shorter arm. Other heavy cranes, such as those moving on wheels or rails in foundries, or those in railway goods yards, form a bridge over the load. Climbing and crawling cranes are used in the construction of high-rise buildings and large bridges. Small cranes, or hoists, are frequently mounted on road vehicles for breakdown work.

cranial nerves Twelve pairs of nerves that arise in the BRAIN and serve the muscles, glands and sensory apparatus of the head and neck as well as organs in the chest and abdomen. Traditionally in anatomy the cranial nerves are identified by Roman numerals in addition to their Latin names.

cranium Dome-shaped part of the SKULL that protects the BRAIN. It is composed of eight bones that are fused together.

Crater (Chalice) Small constellation of the N sky, adjoining Hydra and Corvus. Its brightest stars are of 4th magnitude.

crater Roughly circular depression in the surface of the Earth, the Moon, or some other planets, usually with steep sides. It is formed either by meteoric impact, when shock waves blast out a hole in the ground, or at the vent of a volcano, when lava is expelled explosively. Space probes have revealed such craters on Mars and Mercury, and some of the larger asteroids and satellites of other planets.

crater lake Accumulation of water, usually by precipitation of rain or snow but sometimes ground water, in a volcanic crater (CALDERA). Should an eruption occur, the resulting mud flow (**lahar**) is often more destructive than a lava flow, owing to its greater speed. A noted example is Crater Lake in Crater Lake Park, Oregon, USA. The waters of the lake were formed by precipitation and are maintained solely by rain and snow; there is no outlet.

craton Stable area of the Earth's CRUST that has not been affected by TECTONIC activity during the previous 1,000 million years. They form the SHIELD areas of the Earth's continents.

crazing (crackling) Method of decorative glazing of ceramics, in which fine cracks in the glaze result from different shrinkage rates of the clay and the glaze during firing. It was originated by the Chinese.

creatine AMINO ACID compound found mostly in the muscle tissue of vertebrate animals. Combined with phosphoric acid, it plays an important part in muscle contraction.

creep In geology, slow downward movement of material. This slow movement of soil is not associated with erosion. It may occur on all sorts of slopes including those that are covered with vegetation. The steady downflow of soil and rock carries the vegetation with it. Creep may be caused by the growth of plant roots, the burrowing of small animals or trampling of larger creatures, frost movement, or successive drying and wetting that leads to shrinkage and swelling of the soil.

creep In metallurgy, continuing deformation of a metal under stress so that is appears to flow like an extremely viscous liquid. Metal components can be tested for "creep rate" – the percentage elongation per 1,000 hours at a particular temperature. Engineers have to consider both the applied stress and the temperature of metal structures to avoid possible failure through creep.

Cretaceous Last period of the MESOZOIC era, lasting from 144 to *c*.65 million years ago. Dinosaurs flourished until the end of this period, when they died out in a mass extinction that

▲ **Cousteau** Jacques Cousteau became interested in underwater exploration as a naval officer in the 1930s. In the next decade he experimented with filming underwater, inventing a camera for the purpose. His films *The Silent World* (1956) and *The World Without Sun* (1964) won best documentary Academy Awards. He also had a successful television series.

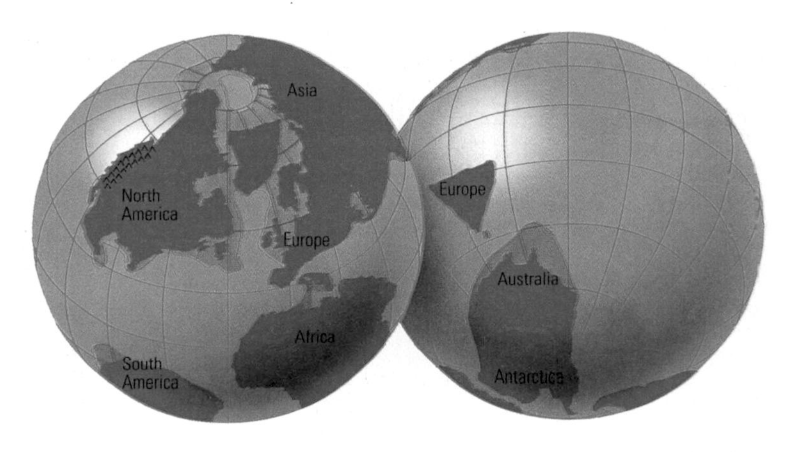

► **Cretaceous** The split-up of Gondwanaland continued during the Cretaceous period. North America broke away from Europe, leaving part of its original mass behind as part of Scotland and Norway. The movement of North America caused the Rocky Mountains to rise.

killed off a vast range of vertebrate and invertebrate life forms. The first true placental and MARSUPIAL mammals and flowering plants appeared. The chalk rocks of NW Europe were deposited during the Upper Cretaceous.

Creutzfeldt-Jakob disease (CJD) Rare, incurable disease of the brain that causes physical and mental deterioration, progressing to death usually within a year of onset. Now known to be caused by an abnormal protein called a PRION, it is one of a group of diseases which also includes SCRAPIE in sheep and BOVINE SPONGIFORM ENCEPHALOPATHY (BSE) – "mad cow disease". Normally CJD attacks older people, but in March 1996 came the first reports of young people in the UK dying of a variant of the disease. Eighteen months later scientists confirmed that the agent giving rise to the new variant (nvCJD) is identical to that of BSE, confirming the link between the human disease and the consumption of infected beef or beef products.

crevasse Deep crack in a GLACIER. It is the result of stress within the glacier or the movement of the glacier over uneven terrain.

Crick, Francis Harry Compton (1916–) British biophysicist. In the 1950s, in association with James WATSON and Maurice Wilkins (1916–), he established the double-helix molecular structure of deoxyribonucleic acid (DNA). The three were jointly awarded the Nobel Prize for physiology or medicine in 1962.

crinoid Any member of the class Crinoidea, a group of primitive ECHINODERMS, which includes the sea lily and the feather star. A crinoid's mouth, on a small disc with the other main organs, is surrounded by long feathery arms. It feeds on particles that fall through the water and respires by means of its tube-feet. They first appeared in the Ordovician period (510 million years ago) and fossil crinoids are an important constituent of Palaeozoic limestones.

critical angle Any angle at which a significant transition occurs. In optics, it is the ANGLE OF INCIDENCE with a medium at which total internal REFLECTION occurs. In telecommunications, it is the angle at which a radio wave just misses being reflected by the IONOSPHERE.

critical mass Minimum mass of fissionable (able to undergo FISSION) material required in a fission bomb or NUCLEAR REACTOR to sustain a CHAIN REACTION. The fissionable material of a fission bomb is divided into portions less than the critical mass; when brought together at the moment of detonation they exceed the critical mass and the high-energy chain reaction begins.

critical point Temperature and pressure above which a liquid and its vapour PHASE (a gas) can no longer coexist. If a gas is slowly compressed at temperatures above the critical temperature or cooled at pressures above the critical pressure, it changes gradually from gas to liquid rather than suddenly separating into two phases.

Crookes, Sir William (1832–1919) British chemist and physicist. He invented the RADIOMETER (which measures the intensity of ELECTROMAGNETIC RADIATION) and the CROOKES TUBE, which led to the discovery of the electron by J.J. THOMSON. He was the first to suggest that CATHODE RAYS consist of negatively charged particles. He also discovered the element THALLIUM. *See also* X-RAY

Crookes tube Highly evacuated tube developed by Sir William CROOKES for the study of electrical discharges at low pressure. It was used in 1897 by J.J. Thomson to demonstrate the existence of the ELECTRON, and was also used to generate X-rays.

cross-bedding In geology, a sedimentary structure comprising a series of inclined parallel or near parallel layers within a bed, found mainly in sandstones but also in some limestones and CONGLOMERATES. Cross-bedding forms where sediment is being moved as it is being deposited, building sand dunes, sand waves or ripples. It is useful for interpreting ancient environments as the direction and strength of former currents can be determined from the pattern and type of structure.

crossing over During CELL DIVISION, a mechanism by which pairs of HOMOLOGOUS CHROMOSOMES exchange strands of each chromosome, called CHROMATIDS. At the end of the first PROPHASE in MEIOSIS, the diverging chromosomes remain in contact at a number of places, called **chiasma**. Chromatids split and rejoin at each chiasma, with the result that sections of chromatids are exchanged. This alters the distribution of GENES along the chromosomes and so gives rise to genetic variation in the resulting GAMETES, essential in the process of EVOLUTION.

cross-pollination Transfer of POLLEN from the flower of one plant to another. The process may be used by plant breeders to hybridize different plant species the genetic make-up of which is compatible, thus producing "synthetic" varieties. The result of the process is called a HYBRID.

crucible process Method of making high-quality steel, invented in 1740 by Benjamin Huntsman (1704–76). Steel, charcoal and wrought-iron are melted in a fire-clay crucible. Coke fires were once used but were replaced after 1870 by the SIEMENS gas furnace, which could heat 100 crucibles to over 1,700°C (3,100°F). Electric furnaces have now replaced this process.

crude oil *See* PETROLEUM

cruise missile (aerodynamic MISSILE) Self-propelled, GUIDED MISSILE that travels, generally at low altitudes, using an advanced guidance system incorporating terrain contour matching (TERCOM). Modern cruise missiles can be extremely accurate and have the advantage of being able to fly low enough to avoid conventional radar defences. The siting of American cruise missiles on European soil during the 1980s led to large-scale public demonstrations. In the 1991 Gulf War, the US Navy used Tomahawk cruise missiles on ground targets in Iraq.

cruiser Warship smaller, lighter and faster than a BATTLESHIP, ranging in size from 7,500–21,000 tonnes. After World War 1, arms limitation treaties restricted its guns to 200mm (8in). Since World War 2, cruisers have replaced battleships as the major warships of a modern navy. The USS *Long Beach* (launched 1961) was the world's first nuclear-powered cruiser.

crust In geology, thin outermost solid layer of the EARTH. The crust represents less than 1% of the Earth's volume and varies in thickness from as little as 5km (3mi) beneath the oceans to *c*.70km (45mi) beneath mountain chains such as the Himalayas. Oceanic crust is generally 7km (4.5mi) thick and is basaltic in composition, whereas continental crust is mainly between 30–40km (20–25mi) thick and of granitic composition. The crust is defined by its seismicity (the velocity of EARTHQUAKE waves that pass through it). The lower boundary of the crust is defined by a marked increase in seismic velocity,

▲ **cruise missile** The earliest cruise missiles were developed during World War 2. Development continued after the war, especially in the 1970s, when the USA wanted to develop a relatively cheap means of delivering weapons accurately over relatively long distances. Cruise missiles fly at low altitudes, using an internal guidance system to follow the shape of the land, and thereby often evading radar systems. There are various cruise missile systems, including ground-launched missiles (GLCMs), air-launched missiles (ALCMs) and sea-launched missiles (which can be launched from either submarines or ships, SLCMs).

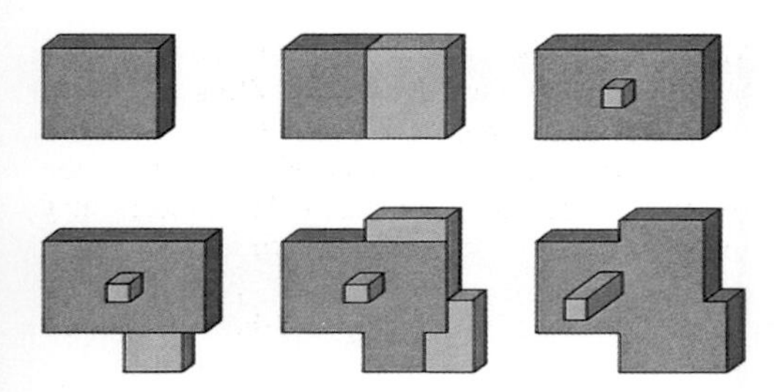

▲ **crystal structure** Crystals grow in a random fashion although the internal structures conform to six basic geometrical systems. The external shape of a crystal is dependent upon the internal arrangement of the atoms and molecules within the crystals. Crystals are built up by repeating the basic unit cell. The unit cell consists of a number of atoms or molecules in a specific spatial arrangement. These arrangements are responsible for the internal structure and external shape of the crystal.

known as the Mohorovičić discontinuity. *See also* MOHO; SIAL; SIMA

crustacean Any member of the class Crustacea, a class of *c*.30,000 species of ARTHROPOD. The class includes the DECAPODS (crabs, lobsters, shrimps and crayfish), isopods (pill millipedes and woodlice) and many varied forms, most of which have no common names. Most crustaceans are aquatic (marine or freshwater) and breathe through gills or the body surface. They are typically covered by a hard EXOSKELETON. They range in size from the Japanese spider crab, which grows to 3m (12ft) across, to the ocean plankton, as little as 1mm (0.04in) in diameter.

Crux (Cross) Smallest constellation in the s sky. Although small it is a distinctive constellation. Its brightest star is Alpha, magnitude 0.76.

cryogenics Branch of physics that studies materials and effects at low temperatures. Some materials exhibit highly unusual properties such as SUPERCONDUCTIVITY or SUPERFLUIDITY at such temperatures. Low temperatures can be reached in different ways, but among the most common is the JOULE-THOMSON EFFECT, when temperatures of 0.3K can be reached. To cool gases even further requires magnetic processes such as the ADIABATIC PROCESS. The applications of cryogenics vary from computer technology to medicine. Cryogenics has been used in a few cases to freeze human bodies in the hope that future technology may be able to revive the subjects. *See also* CRYOSURGERY

cryolite (kryolite) Brittle, icy-looking, red, brown or black halide mineral, sodium-aluminium fluoride (Na_3AlF_6) found in pegmatite DYKES (Greenland has the only large deposit) and used in ALUMINIUM processing. It occurs as crystals in the monoclinic system, occasionally the cubic system, sometimes as granular masses. The crystals are frequently twinned. Cryolite is also a source of aluminium salts and fluorides.

cryosurgery (cryotherapy) Surgery that utilizes CRYOGENICS, and is carried out using a freezing probe, cooled by liquid nitrogen, that destroys diseased tissue on contact. It is particularly effective in treating detached retinas and in the removal of tumours and of some skin blemishes.

crystal Solid with a regular geometrical form and with characteristic angles between its faces, having limited chemical composition. The structure of a crystal, such as common salt, is based upon a regular 3-D arrangement of atoms, ions or molecules (an ionic or CRYSTAL LATTICE). Crystals are produced when a substance passes from a gaseous or liquid PHASE to a solid state, or comes out of solution by evaporation or precipitation. The rate of CRYSTALLIZATION determines the size of crystal formed. Slow-cooling produces large crystals, whereas fast-cooling produces small crystals.

crystal index Set of three numbers that characterize the type of planar configuration of atoms in a lattice. The three integers give the ratio of the reciprocals of the intercepts of the planes on three axes.

crystal lattice Three-dimensional arrangement of atoms, ions, or molecules in a crystalline substance. The term is sometimes used in a more restricted sense to denote the diagrammatic abstraction of the pattern in which the atoms, ions or molecules are positioned. *See also* CRYSTAL

crystalline substance Solid substance in which the atoms or molecules form a regular ordered lattice. Most solids exist in the crystalline state, which is more stable, but it does not follow that they have crystals in the usual sense; metallic copper, for example, is crystalline only in that the copper atoms are arranged in a regular lattice. The BONDS in crystals can be any of the major types. *See also* AMORPHOUS SUBSTANCE; CHEMICAL BOND

crystallization Process of forming CRYSTALS by a substance passing from a gas or liquid to the solid state (SUBLIMATION or fusion) or coming out of solution (PRECIPITATION or EVAPORATION). In the fusion method a solid is melted by heating, and crystals form as the melt cools and solidifies. Ice crystals and monoclinic sulphur are formed in this way. Crystallization is an important laboratory and industrial technique for purifying and separating compounds.

crystallography Study of the formation and structure of crystalline substances. It includes the study of CRYSTAL formation, chemical bonding in crystals and the physical properties of solids. In particular, crystallography is concerned with the internal structure of crystals including substances that were not previously thought capable of forming crystals, such as DNA. *See also* X-RAY CRYSTALLOGRAPHY

crystal optics Optical properties of CRYSTALS. The transmission of light by crystals differs from that by glass since the REFRACTIVE INDEX may depend on the direction of incidence of the light (crystals are not, in general, ISOTROPIC). **Uniaxial** crystals (belonging to the tetragonal, hexagonal and trigonal systems) have two principal refractive indices and display double refraction. **Biaxial** crystals (orthorhombic, monoclinic, triclinic) have three principal refractive indices.

crystal packing Manner in which atoms, ions or molecules are packed together in a crystal. In simple terms, crystal structures can often be considered geometrically, as regularly arranged spheres of equal diameter. The most efficient ways of packing spheres are the arrangements known as hexagonal and cubic close-packing. In these packing configurations every sphere has 12 nearest neighbours and 73% of the available space is filled.

crystal structure Regular arrangement of atoms, linked together by CHEMICAL BONDS in repeating patterns. Because of this regularity, and sometimes a corresponding asymmetry, the physical properties of a crystal, such as strength and conductivity, vary in a regular manner depending on the crystal axis along which they are measured. Single crystals are distinguished from non-crystalline (amorphous) and from polycrystalline materials, the physical properties of which are either invariant (because of complete mixing) or vary in an irregular manner.

crystal systems Six types of crystal structure can be grouped from the 32 classes of crystal symmetry. The elements of symmetry found in crystals are axes, planes and centres. The six systems are: cubic, tetragonal, hexagonal (collectively called uniaxial), and orthorhombic, monoclinic and triclinic (called biaxial). A seventh system, the trigonal, is recognized by many mineralogists. The trigonal system has the same set of reference axes as the hexagonal system.

cube In mathematics, the result of multiplying a given number by itself twice. Thus the cube of a is $a \times a \times a$, written a^3. The cube is also described as the third power of a number. Cube is also the name of a regular six-sided solid figure (all its edges are equal in length and all its faces are squares).

cube root (symbol $\sqrt[3]{}$) Number that must be multiplied by itself twice over to give a specified number. For example, the cube root of 64 is 4, since $4 \times 4 \times 4 = 64$. In this case, the cube root is written $\sqrt[3]{64} = 4$. In algebraic terms, the cube root of x^3 is x and the cube root of x is $\sqrt[3]{x}$ or $x^{1/3}$.

cubic equation Algebraic equation of the third degree, that is one in which the highest power of the unknown variable is three. An example is

$$2x^3 + x^2 + 7 = 0.$$

Cubic equations have three ROOTS (solutions), two of which may be complex, or all three may be real. *See* TARTAGLIA, NICCOLÓ FONTANA

Cugnot, Nicolas Joseph (1725–1804) French engineer who, in 1769, invented the first self-propelled road vehicle. This three-wheeled, two-cylinder steam tractor was designed to pull guns, and travelled at nearly 5km/h (3mph).

cumulonimbus *See* CLOUD

cumulus *See* CLOUD

cuprite Reddish-brown, brittle, translucent oxide mineral, cuprous oxide (Cu_2O). It is formed by the OXIDATION of other ores such as copper sulphide and so is commonly found near the surface. Its octahedral, dodecahedral and cubic crystals occur in the cubic system, and it also occurs in grains. Hardness 3.5–4; r.d. 6.1.

cupronickel Alloy containing 75 parts of copper to 25 parts of nickel, widely used for coins. In

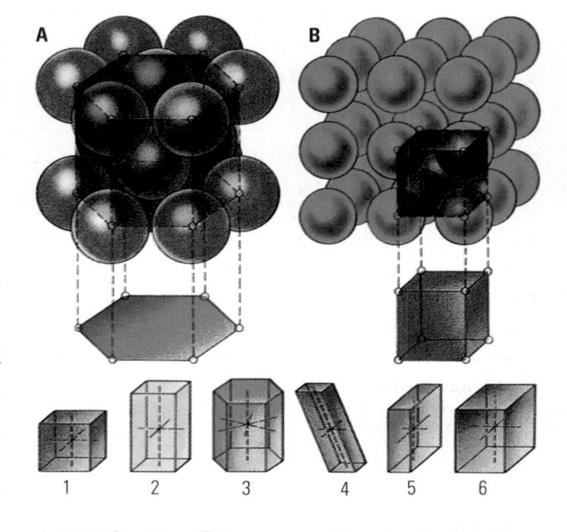

▲ **crystal systems** There are six main crystal structure systems, cubic (1), tetragonal (2), hexagonal (3), monoclinic (4), triclinic (5) and orthorhombic (6). Crystals were first studied by systematically examining their external shape and optical properties. The two examples in the diagram show a complex and simple structure. The molecular structure found in a quartz crystal, SiO_2, is shown in (A) and it can be seen how the hexagonal system is related to the stereochemical arrangement of the atoms. This crystal structure has three equal axes inclined at 120° to each other with a fourth unequal and perpendicular to the other three. The structure of the halite crystal (B), sodium chloride NaCl, was the first to be elucidated, having three axes equal and mutually perpendicular to each other.

JANUARY TEMPERATURE AND OCEAN CURRENTS
(Northern Hemisphere – Winter)

ACTUAL SURFACE TEMPERATURE
°C
30
20
10
0
–10
–20
–30
–40

OCEAN CURRENTS
Cold Warm Speed (knots)
Less than 0.5
0.5 – 1.0
Over 1.0

N. Pacific Current
Californian Current
Labrador C.
North Atlantic Drift
Oya Siwo
Kuro Siwo
Gulf Stream
N. Equatorial Current
Northern Equatorial Current
Equatorial Countercurrent
Southern Equatorial Current
Guinea C.
Benguela Current
Peru (Humboldt) Current
Brazil Current
N.E. Monsoon Drift
South Equatorial Current
Equatorial Countercurrent
Agulhas C.
Antarctic Circumpolar Current

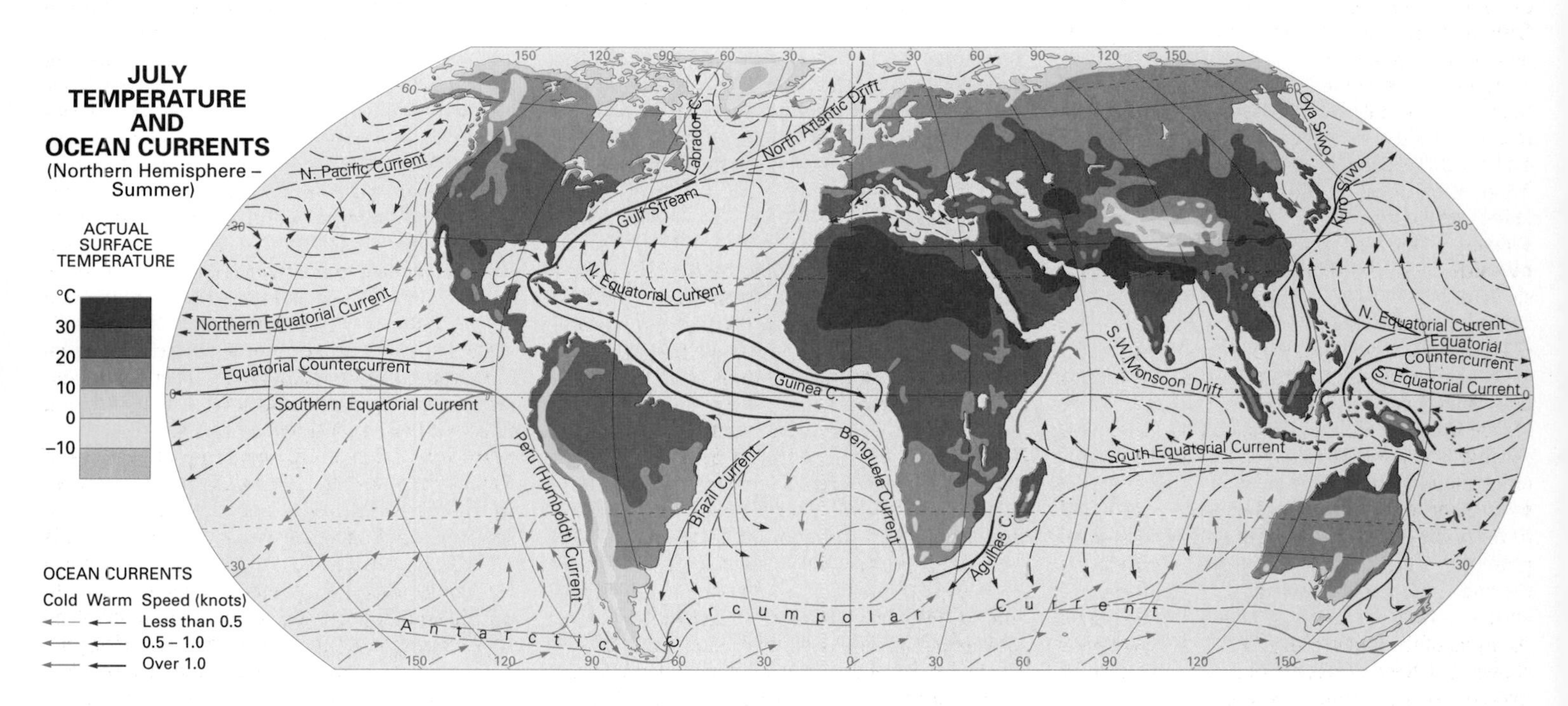

▲ **current** Moving immense quantities of energy as well as billions of tonnes of water every hour, the ocean currents are a vital part of the great "heat engine" that drive's the Earth's climatic conditions.

1947 it replaced the silver alloy used until that time for non-copper coins in the UK.

curare Poisonous, resinous extract obtained from various tropical South American plants of the genera *Chondodendron* and *Strychnos*. Most of its active elements are ALKALOIDS. Causing muscle paralysis, it is used on the poisoned arrows of Native South Americans when hunting. It has also been used as a muscle relaxant in surgery, but has been replaced by safer agents.

Curie, Marie (1867–1934) Polish scientist, who specialized in work on RADIOACTIVITY. Marie and her husband Pierre Curie (who specialized in the electrical and magnetic properties of crystals, and formulated a law relating magnetism to temperature) worked together on a series of radiation experiments in their Paris laboratory. In 1898 they discovered RADIUM and POLONIUM. In 1903 they shared the Nobel Prize for physics with Antoine BECQUEREL. In 1911 Marie Curie became the first person to be awarded a second Nobel Prize (this time for chemistry), for her work on radium and its compounds. She died of leukaemia caused by laboratory radiation. Their daughter, Irène JOLIOT-CURIE, and her husband, Frédéric Joliot-Curie, were awarded the 1935 Nobel Prize for chemistry for producing artificial radioactive substances by bombarding elements with ALPHA PARTICLES.

curie (symbol Ci) Unit formerly used to measure the activity of a radioactive substance. Named after Marie CURIE, it is defined as that quantity of an ISOTOPE that decays at the rate of 3.7×10^{10} disintegrations per second. The curie has been replaced by an SI unit, the much smaller BECQUEREL.

curium (symbol Cm) Synthetic, radioactive metallic element of the ACTINIDE SERIES. It was first made in 1944 by Glenn SEABORG and his colleagues by the alpha particle bombardment of plutonium-239 in a cyclotron ACCELERATOR. Silvery in colour, curium is chemically reactive, intensely radioactive and is toxic if absorbed by the body. A modern use is to provide power for orbiting satellites. Properties: at.no. 96; r.d. (calculated) 13.51; m.p. 1,340°C (2,444°F); 14 isotopes, most stable ^{247}Cm (half-life 1.6×10^{7} yr).

current Broad, slow drift of moving water, distinguishable from the water surrounding it by differences in temperature and/or salinity. A current can be caused by prevailing winds which sweep surface water along, forming drift currents. Currents are affected by the CORIOLIS EFFECT, by the shape of the ocean bed or by nearby land masses. Currents at a deeper level are the result of variation in density of the water which in turn varies according to temperature and salinity.

Curtiss, Glen Hammond (1878–1930) US aviation pioneer. He began, as did the WRIGHT BROTHERS, by building bicycles. By 1908 he had made the first US flight of more than 1km (0.6 mi). Curtiss built planes during World War 1 (1914–18), and his JN-4 ("Jenny") was a well-known trainer. His company and that of the Wrights merged to become the Curtiss-Wright Corporation.

curvature of the spine Exaggerated shaping of the SPINE. There are three major types. **Scoliosis**, or lateral curvature, can be due to bad posture or to abnormality. **Lordosis**, an accentuation of the inward curve of the neck region or more commonly of the lower back region, results in a sway-back appearance. **Kyphosis**, an accentuation of the outward curve behind the chest, can in severe form result in a hunchback appearance.

curve A continuous, one-dimensional set of points. In mathematical terms, a straight line is also a curve, and in analytic geometry, a curve is the locus of points that satisfy a mathematical equation.

Cushing, Harvey (1869–1939) US surgeon. His pioneering techniques for surgery on the brain and spinal cord helped advance NEUROSURGERY. He first described the condition produced by over-secretion of adrenal hormones that is now known as **Cushing's syndrome**. This syndrome is characterized by weight-gain in the face and trunk, high blood pressure, excessive growth of facial and body hair, demineralization of bone and diabetes-like effects.

cuticle Exposed outer layer of an animal. In humans this refers to the EPIDERMIS, especially the dead skin at the edge of fingers. In botany cuticle refers to the waxy layer on the outer surface of the epidermal cells of the leaves and stems of vascular plants. It helps to prevent excessive water loss.

Cuvier, Georges, Baron (1769–1832) French geologist and zoologist, a founder of comparative anatomy and palaeontology. His scheme of classification stressed the form of organs and their correlation within the body. He applied this system of classification to fossils, and came to reject the theory of the gradual development of the Earth and animals, favouring instead a theory of catastrophic changes. *See also* PUNCTUATED EQUILIBRIUM

cyanide Salt or ester of hydrocyanic acid (prussic acid, HCN). The most important cyanides are sodium cyanide (NaCN) and potassium cyanide (KCN), both of which are deadly poisonous and have a characteristic smell of almonds. Cyanides have many industrial uses – in ELECTROPLATING, for the heat treatment of metals, in the extraction of silver and gold, in photography, and in insecticides and pigments.

cyanobacteria (formerly blue-green algae) One of the major BACTERIA phyla, distinguished by the presence of the green pigment CHLOROPHYLL and the blue pigment **phycocyanin**. They perform PHOTOSYNTHESIS with the production of oxygen. Analysis of the genetic material of CHLOROPLASTS shows that they evolved from cyanobacteria by ENDOSYMBIOSIS. Many cyanobacteria perform NITROGEN FIXATION. They occur in soil, mud and deserts; they are most abundant in lakes, rivers and oceans. Some produce toxic BLOOMS.

cybernetics Study of communication and control systems in animals, organizations and machines. The term was first used in this sense in 1948 by Norbert WIENER. Cybernetics makes analogies between processes in the brain and nervous system and those in computers and other electronic systems, analysing, for example, the mechanisms of FEEDBACK and data processing in both. Thus a household thermostat might be compared with the body's mechanisms for temperature control and respiration. Cybernetics combines aspects of mathematics, neurophysiology, computer technology, INFORMATION THEORY and psychology.

cyberspace Popular term for the perceived "virtual" space within computer memory, especially if rendered graphically. The term is a product of science fiction, where it usually refers to situations involving direct interface between brain and computer. During the mid-1990s the term became widespread in reference to the INTERNET and the WORLD WIDE WEB.

cycad Any member of the phylum Cycadophyta, primitive palm-like shrubs and trees that grow in tropical and subtropical regions. They have feathery palm- or fern-like leaves (poisonous in most species) at the top of stout (usually unbranched) stems. In addition to their main roots, they also have special roots containing CYANOBACTERIA that carry out NITROGEN FIXATION. These plants first flourished about 225 million years ago. Most of the 100 or so surviving species are less than 6.1m (20ft) tall.

cyclamates White, odourless, soluble crystalline salts, calcium cyclamate ($[C_6H_{11}NHSO_3]_2Ca.2H_2O$) and sodium cyclamate ($CH_6H_{11}NHSO_3Na$). They have about 30 times the sweetening power of ordinary sugar (SUCROSE). Both were used as artificial sweeteners until they were shown to cause cancer in laboratory animals.

cycle In physics, series of changes through which any system passes which brings it back to its original state. For example, alternating current starts from zero voltage, rises to a maximum, declines through zero to a minimum and rises again to zero. In the INTERNAL COMBUSTION ENGINE, the TWO-STROKE ENGINE completes one cycle each downward plunge and return; the FOUR-STROKE ENGINE takes two such movements.

cyclic In organic chemistry, describing a substance the molecules of which contain one or more rings of atoms. In a **homocyclic** compound, all the atoms in the ring are the same; examples include benzene (C_6H_6) and cyclohexane (C_6H_{12}). In a **heterocyclic** compound, there are different atoms in the ring; examples include pyridine (C_5H_5N) and pyrrole (C_4H_5N).

cyclohexane (C_6H_{12}) Colourless, liquid hydrocarbon that occurs naturally in crude oil, but is made commercially by combining hydrogen with benzene, using a catalyst. It is one of a type of chemical compounds called **conformations**; they can exist in two forms differing only in the arrangement of their bonds, and may "flip" from one form to the other. Properties: m.p. 6°C (43°F); b.p. 81°C (178°F).

cyclone (low) System of atmospheric low pressure that occurs when a cold air mass moving S from the Arctic meets a warm air mass moving N from the tropics to form a circulating air mass. A cyclone is characterized by relatively low pressure at the centre, and by anticlockwise wind movement in the Northern Hemisphere and by clockwise motion in the Southern Hemisphere. There are two types of cyclone: the DEPRESSION, associated with temperate latitudes; and the tropical cyclone, which is much more violent but usually affects a smaller area. *See also* ANTICYCLONE

cyclotron Particle ACCELERATOR in which charged particles such as PROTONS or heavier ions are accelerated outwards in a spiral. It has two D-shaped ELECTRODES across which is applied an alternating voltage. The charged particles are attracted by electrostatic forces to the electrode of opposite charge, but since the polarity alternates, the particles are swung outwards at increasing speed. The cyclotron was invented by Ernest LAWRENCE in order to obtain high energy particles without having to use an extremely long LINEAR ACCELERATOR.

Cygnus One of the most distinctive constellations, often nicknamed the NORTHERN CROSS.

cylinder Solid figure or surface formed by rotating a rectangle about one side as axis. If the vertical height is h and the radius of the base r, then the volume is $\pi r^2 h$ and the curved surface area $2\pi rh$.

cylinder, engine One of the cylindrical chambers in an INTERNAL COMBUSTION ENGINE. Inside each, a tightly fitting piston is pushed down by the pressure of the expansive force of burning fuel. The movement of the piston is lubricated by oil from the sump. The cylinders become hot in use and can be cooled either by air or water cooling systems. Reciprocating STEAM ENGINES also have cylinders.

cylindrical polar coordinates System of COORDINATES used to represent the position of a point in space. There are three coordinates, r, θ and z. The point's projection on a reference plane is represented by the POLAR COORDINATES r and θ (where r is the distance of the projected point from the origin and θ its angle to some reference direction in that plane), and z is the point's height above (or below) the reference plane.

cyst Hollow cavity or sac in the body that contains liquid or semi-solid matter. It may occur in a glandular organ, such as the breast or prostate, or in the skin. It may be caused by infection or by a blocked duct. A cyst may be removed for medical or cosmetic reasons.

cysteine ($HSCH_2CH_2COOH$) Crystalline AMINO ACID that occurs in animal proteins, especially those in the hair, hooves and the keratin of the skin. The HS– group in it makes it the catalytic element in some enzymes.

cystic fibrosis Hereditary glandular disease in which the body produces abnormally thick mucus that obstructs the breathing passages, causing chronic lung disease. There is a deficiency of pancreatic enzymes, an abnormally high salt concentration in the sweat and a general failure to gain weight. The disease is treated with antibiotics, pancreatic enzymes and a high-protein diet; sufferers must undergo vigorous physiotherapy to keep the chest as clear as possible. Heart-lung transplantation is sometimes recommended where there is severe lung damage. There have been experiments using GENE THERAPY.

cytochromes Proteins containing haem, an iron group, as in HAEMOGLOBIN. They are fundamental to the process of RESPIRATION in all living cells that need atmospheric oxygen.

cytokinin (kinetin or kinin) Any of a group of plant hormones that stimulate cell division. Cytokinins work in conjunction with AUXINS to promote swelling and division in the plant cells producing lateral buds. They can also slow down the aging process in plants, encourage seeds to germinate and plants to flower, and are involved in plant responses to drought and waterlogging. They are used commercially to produce seedless grapes, to stimulate germination of barley in brewing, and to prolong the life of green-leaf vegetables. Cytokinins, such as zeatin, are derived from ADENINE.

cytology Study of living CELLS and their structure, behaviour and function. The discipline began with Robert HOOKE's microscopic studies of cork in 1665, and the various forms of MICROSCOPE are still the main tools of cytology. In the 19th century, a cell theory was developed which suggested that cells are the basic units of organisms. Recently cytological study has focused on the chemistry of cell components (cytochemistry).

cytoplasm Jelly-like matter inside a CELL and surrounding the NUCLEUS. Cytoplasm has a complex constituency and contains various bodies known as organelles, with specific metabolic functions. The proteins needed for cell growth and repair are produced in the cytoplasm. *See also* METABOLISM

cytosine Organic base, first isolated in 1894. Derivatives of cytosine made in the body are important in cellular METABOLISM and in the formation of RNA and DNA, thus being vital for the retention of genetic characteristics. *See also* GENE

D

Daimler, Gottlieb (1834–1900) German engineer and automobile manufacturer. In 1822, with Wilhelm MAYBACH, Daimler established a research laboratory where a year later they developed a lightweight petrol INTERNAL COMBUSTION ENGINE. Daimler used this to power a motorcycle and then his first car (1886). In 1890 he founded the Daimler Motor Company, which made Mercedes cars, and in 1926 became Daimler-BENZ and Company.

Dale, Sir Henry Hallett (1875–1968) British physiologist who shared the 1936 Nobel Prize for physiology or medicine with Otto LOEWI for discoveries relating to the chemical transmission of NERVE IMPULSES. Dale found that the chemical acetylcholine served to transmit nerve impulses across the tiny gap (SYNAPSE) from one NERVE CELL to another. His writings include *Adventures in Physiology* (1953) and *An Autumn Gleaning* (1954).

D'Alembert, Jean le Rond (1717–83) French philosopher and mathematician. He was a religious sceptic and physicist, and formulated d'Alembert's principle, used to solve certain problems in mechanics. He collaborated with Denis Diderot (1713–84) on the French *Encyclopédie* (1751–72), for which he contributed the *Preliminary Discourse* (1751). He wrote a systematic *Treatise on Dynamics* (1743) as well as several pieces on DIFFERENTIAL EQUATIONS.

Dalén, Niels Gustav (1869–1937) Swedish engineer responsible for the development of automatic lighting for lighthouses and railway signals by means of a "sunvalve", a gas valve responsive to the Sun's radiation. Dalén was awarded the Nobel Prize for physics in 1912, but was blinded in a laboratory explosion a year later. He also invented a substance that absorbs acetylene and prevents the risk of explosion in acetylene tanks.

Dalton, John (1766–1844) British chemist and physicist. His early interest in meteorology yielded important information on the trade winds, the cause of rain and the AURORA borealis. He described colour blindness (sometimes called Daltonism) based on his own experiences and those of his brother. His study of gases led to DALTON'S LAW of partial pressures. His atomic theory states that each element is made up of indestructible, identical, small particles; he also constructed a table of RELATIVE ATOMIC MASSES (r.a.m.).

Dalton's law Pressure exerted by each gas in a mixture of gases does not depend on the pressures of the other gases, provided no chemical reaction occurs. The total pressure of such a mixture is therefore the sum of the partial pressures exerted by each gas (as if it were alone in the same volume as the mixture occupies).

Dam, Carl Peter Henrik (1895–1976) Danish biologist. In 1934 he discovered vitamin K, the fat-soluble vitamin needed for blood clotting. He isolated it from hempseed and the seeds of other plants, and he also discovered it in liver. For this work, he received the 1943 Nobel Prize for physiology or medicine, which he shared with Edward DOISY. In addition, he examined the roles of other vitamins and LIPIDS.

dam Barrier built across a stream, river, estuary, or part of the sea. A dam confines water, or checks its flow, for irrigation, flood control or HYDROELECTRICITY generation. The first dams were probably constructed by the Egyptians at least 4,500 years ago. Common types are **gravity**, **arch** and **buttress** dams. Gravity dams are anchored by their own weight. Single-arch dams are curved (convex to the water they retain) and are supported at each end by the river banks. Multiple-arch and buttress dams are supported by buttresses rooted in the bedrock. The highest dam is Russia's Rogun, 335m (1,100ft) high. The Chapeton Dam in Argentina has a record structural volume of 296,000,000m^3 (387,568,000yd^3). The cheapest source of electricity that is used commercially comes from HYDROELECTRICITY schemes made possible by dams such as the Aswan High Dam in Egypt.

damping Gradual slowing down of the oscillations in a vibrating mechanical system. It occurs as the system loses energy to overcome friction or other resistance; this is why a swinging pendulum gradually slows down. A mechanism may be deliberately damped to facilitate reading an oscillating needle or to slow down an oscillating balance. Damping that stops the oscillation in the shortest possible time is called CRITICAL damping.

D and C operation (abbreviation for dilatation and curettage) Procedure in which the CERVIX (neck of the womb) is dilated so that the lining of the womb can be scraped away (curettage). It is done to treat disease or an incomplete miscarriage, to terminate pregnancy, or to secure a sample of tissue for testing (BIOPSY).

dark adaptation Slow change in sensitivity of the human EYE that takes place when someone passes into dim light from bright light. It is caused by a shift in functional dominance from cone cells to rod cells in the RETINA of the eye as the overall illumination is reduced. The complete process takes 35–40 minutes. *See also* LIGHT ADAPTATION

dark reaction One of the two distinct phases of PHOTOSYNTHESIS; the other being LIGHT REACTION. During the dark reaction, which can take place in light or in darkness, carbon dioxide is reduced to carbohydrate (sugars) by means of the CALVIN CYCLE.

Darwin, Charles Robert (1809–82) British naturalist, originator of a theory of EVOLUTION based on NATURAL SELECTION. In 1831 he joined a round-the-world expedition on HMS BEAGLE. Observations made of the FLORA and FAUNA of South America (especially the Galápagos Islands) formed the basis of his work on animal variation. The development of a similar theory by Alfred WALLACE led Darwin to present his ideas to the Linnean Society in 1858, and in 1859 he published *The Origin of Species*, one of the world's most influential science books. Darwin argued that organisms reproduce more than is necessary to replenish their population, creating competition for survival. Opposed to the ideas of LAMARCK, Darwin argued that each organism was a unique combination of genetic variations. The variations that prove helpful in the struggle to survive are passed down to the offspring of the survivors. He termed this process NATURAL SELECTION. Darwin did not distinguish between ACQUIRED CHARACTERISTICS and genetic variations. Instead, NEO-DARWINISM supplemented his ideas with modern research into HEREDITY, especially MUTATION.

Darwin, Sir Charles Galton (1887–1962) British physicist who trained under Ernest RUTHERFORD at Manchester, England. At Cambridge, Darwin developed the new methods of statistical mechanics that later served as a foundation for QUANTUM MECHANICS.

Darwinism *See* EVOLUTION

DAT Acronym for DIGITAL AUDIO TAPE

data Information, such as lists of words, quantities or measurements, or codes representing a picture. A COMPUTER PROGRAM works by processing data, which may be entered using a keyboard or other input device, stored as a data file on a COMPUTER DISK, downloaded from the INTERNET, or come from many other sources.

database Collection of DATA produced and

◀ **dam** The Hoover Dam on the Colorado River, USA. Built during 1931–36, this dam is 221m (725ft) high and 379m (1,240ft) wide at its crest. The reservoir behind it, Lake Mead, has an area of 694sq km (268sq mi). The dam of a hydroelectric power station, its turbines are capable of generating 1.5 million kilowatts of power. Electricity generated by hydroelectric stations is relatively cheap and has little impact on the environment other than on the river it is built on. The amount of hydroelectric power generated may be increased quickly simply by raising the flood gates. Hydroelectric power stations are commonly used during peak demand times.

retrieved by a computer. The data is usually stored on MAGNETIC DISK or DIGITAL AUDIOTAPE. A database program enables the computer to generate files of data and later search for and retrieve specific items or groups of items. For example, a library database system can list, on screen, all the books by a specified author, or all the books on a particular subject. The computer can then display further details of any selected book. *See also* INFORMATION STORAGE AND RETRIEVAL

data compression Technique in computing to reduce the amount of storage space occupied by data. Methods include representing common characters by fewer BITS than normal and storing frequently used words as shorter words (**tokenization**). Long sequences of repeated characters can be replaced by a single character and a count of how many there are, a technique called **run-length encoding**. By such means, text files can be reduced by up to 50% and digitized images by about 90%. Compression techniques can be divided into two main types: **lossy** and **non-lossy**. With non-lossy compression, there is no loss in the quality of the data. With lossy compression, the compression is greater but the quality of the data is reduced; when the data is uncompressed, it will be slightly different from before it was compressed. Lossy compression is used chiefly for images, video and music.

data processing Systematic sequence of operations performed on DATA, especially by a computer, in order to calculate new information or revise or update existing information stored on magnetic or optical DISK, magnetic tape and so on. The data may be in the form of numerical values representing measurements, scientific or technical facts, or lists of names, places or book titles. The main processing operations performed by a computer are arithmetical addition, subtraction, multiplication and division, and logical operations that involve decision-making based on comparison of data. For the latter, an instruction might read: "If condition *a* holds, then follow programmed instruction P; if *a* does not hold, then follow instruction Q".

data protection Measures taken to guard data against unauthorized access. Computer technology now makes it easy to store large amounts of data containing, for example, a person's medical or financial details. Many governments have passed data-protection legislation ensuring that such databases are registered and the information that they contain is used only for the purpose for which it was originally given.

data transmission Supply of information to or from a computer via hard wire (telephone lines) or electromagnetic (radio or microwave) broadcast from earth-stations or satellites. Computers or other electronic devices at each end of the transmission path regulate the information by encoding, decoding and synchronizing the transmitted signals.

dating, radioactive (radiometric dating) Any of several methods using the laws of RADIOACTIVE DECAY to assess the very considerable ages of archaeological remains, fossils, rocks and of the Earth itself. The specimens must contain a very long-lived radioisotope of known HALF-LIFE, which, with a measurement of the ratio of radioisotope to a stable isotope (usually the decay product), gives the age. In potassium-argon dating, the ratio of potassium-40 to its stable decay product argon-40 gives ages over ten million years. In rubidium-strontium dating, the ratio of rubidium-87 to its stable product strontium-87 gives ages up to several thousand million years. In CARBON DATING, the proportion of carbon-14 (half-life 5,730 years) to stable carbon-12 absorbed into once-living matter, such as wood or bone, gives ages up to several thousand years.

datolite Colourless or white orthosilicate mineral, basic calcium borosilicate ($CaBSiO_4[OH]$), found in cavities in trap-rock. It occurs as monoclinic system crystals or, rarely, as granular masses. Hardness 5–5.5; r.d. 2.9.

Davisson, Clinton Joseph (1881–1958) US physicist who worked on THERMIONICS, ELECTRON DIFFRACTION and the ELECTRON MICROSCOPE for Bell Telephone Laboratories. In 1927, working with Lester Germer (1896–1971), he was able to confirm (by means of DIFFRACTION by crystals) the wave nature of moving electrons, which had been hypothesized by Louis de BROGLIE. He shared the Nobel Prize for physics with Sir George THOMSON in 1937.

Davy, Sir Humphry (1778–1829) British chemist who discovered that ELECTROLYTIC CELLS produce electricity by chemical means. This led to the use of ELECTROLYSIS to isolate the elements sodium, potassium, barium, strontium, calcium and magnesium. He also proved that all acids contain hydrogen, which is responsible for their acidic properties. An investigation into the conditions under which firedamp (methane and other gases) and air explode, led to his invention of the miner's SAFETY LAMP.

Dawkins, Richard (1941–) English zoologist. He studied at Oxford before accepting a teaching post in California, where he remained until returning to Oxford in 1970. Dawkins' first popular science book, *The Selfish Gene* (1976), examines animal behaviour in the context of EVOLUTION, proposing that genes govern behaviour in order to survive. *The Blind Watchmaker* (1986) is a good introduction to NEO-DARWINISM.

DDT (dichlorodiphenyltrichloroethane) Colourless, crystalline, organic halogen compound, first used as an insecticide in 1939 against the Colorado potato beetle. It acts as a contact poison, disorganizing the nervous system. It was initially effective against most insect pests and had dramatic effects in reducing malaria. However, many insect species developed resistant populations, and it proved to have long-lasting toxic effects on birds and other animals. As a result, it is now banned in many countries, though still used for malaria control in others.

deafness Partial or total HEARING loss. **Conductive** deafness, where there is faulty transmission of sound to the sensory organs deep in the EAR, is usually due to infection or to inherited abnormalities of the tiny bones (OSSICLES) of the middle ear. **Perceptive** deafness may be inborn or acquired. It is due to injury or disease of the COCHLEA, auditory nerve or hearing centres in the BRAIN. The treatment of deafness depends on its cause and ranges from the removal of impacted wax to cochlear implant or delicate microsurgery to correct some congenital defect. Hearing aids, sign language and lip-reading are among techniques that help the deaf to communicate.

deamination Removal of an amino ($-NH_2$) group from a compound, particularly an AMINO ACID. It takes place in the liver of animals, where enzymes cause the amino group to be converted into ammonia, which is either excreted as it is or converted into urea or uric acid and then excreted in the urine.

death Cessation of life. Traditionally in medicine, death has been pronounced on cessation of the heartbeat. However, modern resuscitation and life-support techniques have led to the revival of some people whose hearts have stopped. In a tiny minority of such patients, while for a short time breathing and the heartbeat can be maintained artificially, the potential for life is extinct. In this context death may be pronounced when it is clear that the brain no longer controls vital functions.

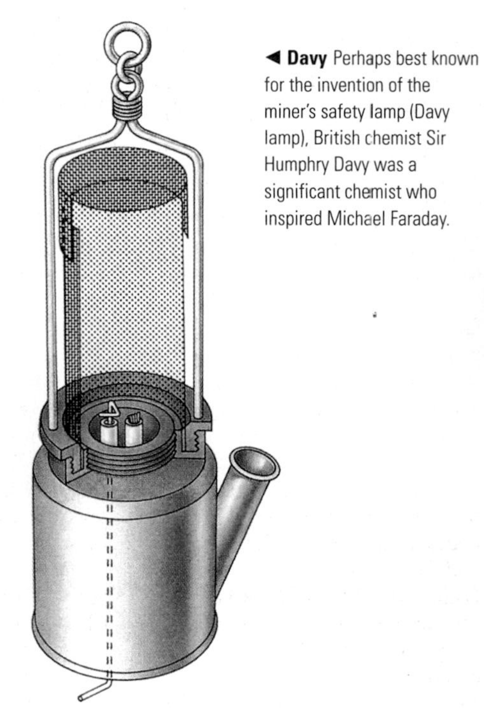

◀ **Davy** Perhaps best known for the invention of the miner's safety lamp (Davy lamp), British chemist Sir Humphry Davy was a significant chemist who inspired Michael Faraday.

DeBakey, Michael Ellis (1908–) US surgeon who pioneered surgical techniques for treatment of CIRCULATORY SYSTEM disorders. In 1966 DeBakey implanted the first mechanical device into a human chest to assist the heart. He had already devised the roller pump (1932), a part of the machinery used in open-heart surgery, and developed a method of grafting frozen blood vessels to replace those that were diseased as a means of correcting aortic aneurysms; later (1956), he replaced the grafts with plastic tubing.

de Broglie, Louis Victor *See* BROGLIE, LOUIS VICTOR DE

Debye, Peter Joseph Wilhelm (1884–1966) US physical chemist, b. Netherlands. He is best known for his work on molecular structure and dissociation in solution. He also pioneered X-RAY CRYSTALLOGRAPHY. He was awarded the 1936 Nobel Prize for chemistry.

decapod Any of about 8,500 species of the order Decapoda, mainly marine, higher crustaceans, including shrimps, lobsters and crabs. They have ten legs including the chelae (pincers), which are enlarged modified legs. The order Decapoda has been represented since the Permian or Triassic period.

decay (rot) Partial or complete deterioration of a substance caused by natural changes. Plant rot, caused by soil-borne bacteria and fungi, can affect any plant part, making it spongy, watery, hard or dry. *See also* DECOMPOSITION. The term is also used in physics to describe RADIOACTIVE DECAY.

decibel (symbol dB) Logarithmic unit, one tenth of a *bel*, used for comparing two power levels. It is frequently used for expressing the loudness of a sound in terms of a particular reference level. The faintest audible sound (corresponding to an excess air pressure of 2×10^{-5} pascal) is given an arbitrary value of 0dB. The human pain threshold is about 120dB. Ordinary conversations occur at about 50–60dB.

deciduous Term describing the annual or seasonal loss of all leaves from a TREE or shrub; it is the opposite of EVERGREEN.

decimal fraction Number in the DECIMAL SYSTEM written as a digit to the right of a decimal point. The number 52.437 represents an INTEGER (whole number, 52) added to a decimal fraction (0.437). It is composed of

$52 + (4 \times 10^{-1}) + (3 \times 10^{-2}) + (7 \times 10^{-3})$,

which may also be written

52 + (4/10 + 3/100 + 7/1000), 52 + 437/1000, or as the improper fraction 52437/1000. Decimal fractions are added, subtracted, multiplied and divided like integers, but the decimal point must be correctly positioned after each operation.

decimal place Number of places to the right of a decimal point required to specify a real number (*see* NUMBER, REAL) to certain accuracy. The number 4.893302 to three decimal places is written 4.893 (3d.p.). To five decimal places the same number is 4.89330 (5d.p.). The convention is to round up the last decimal place if the digit after that place is five or greater, and to round down if it is four or less. Thus the numbers 239.705 and 239.706 are both 239.71 (2d.p.) to two decimal places, whereas the number 239.704 to two decimal places is 239.70 (2d.p.). *See also* SIGNIFICANT FIGURES (S.F.)

decimal system Commonly used system of writing numbers using a base ten and the Arabic numerals 0 to 9. It is a positional number system in which each position to the left represents an extra power of ten. Thus 6,741.83 is $(6 \times 10^3) + (7 \times 10^2) + (4 \times 10^1) + (1 \times 10^0) + (8 \times 10^{-1}) + (3 \times 10^{-2})$. Numbers to the right of the decimal point are represented by a digit multiplied by ten raised to a negative power. Note that $10^0 = 1$.

declination (dec. symbol δ) Angular distance of a celestial object N or S of the CELESTIAL EQUATOR. It is calculated positively from 0 to 90 (from the Equator to the north celestial pole), and negatively from 0 to −90 (from the Equator to the south celestial pole). *See also* CELESTIAL SPHERE

declination, magnetic (deviation) Amount by which the direction indicated by a magnetic compass differs from the direction of true north. It arises because the magnetic north pole is not coincident with the true North Pole. As a result, declination varies from place to place on the Earth's surface. Also, because the north magnetic pole slowly moves over time, declination varies from year to year.

decoder In telecommunications and electronics, a device that converts the information content of a signal into a more usable form. In a satellite-television receiver, it decodes encrypted signals; in a colour TELEVISION, it separates the red, blue and green components; in stereo FM radio it separates the left and right sound channels.

DEEP SEA DRILLING PROJECT

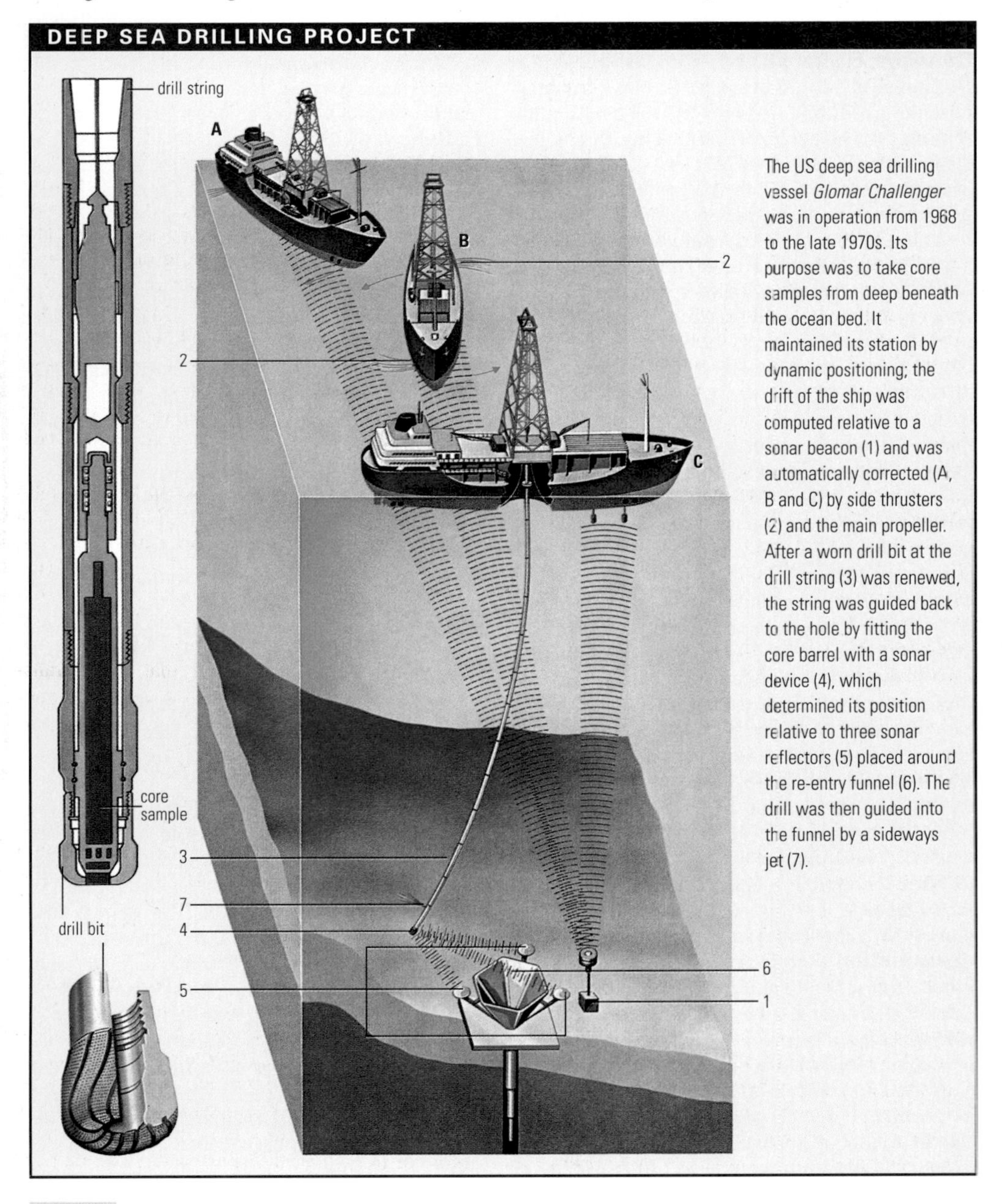

The US deep sea drilling vessel *Glomar Challenger* was in operation from 1968 to the late 1970s. Its purpose was to take core samples from deep beneath the ocean bed. It maintained its station by dynamic positioning; the drift of the ship was computed relative to a sonar beacon (1) and was automatically corrected (A, B and C) by side thrusters (2) and the main propeller. After a worn drill bit at the drill string (3) was renewed, the string was guided back to the hole by fitting the core barrel with a sonar device (4), which determined its position relative to three sonar reflectors (5) placed around the re-entry funnel (6). The drill was then guided into the funnel by a sideways jet (7).

decomposition Natural degradation of organic matter into simpler substances, such as carbon dioxide and water. Organisms of decay are usually bacteria and fungi. Decomposition recycles nutrients by releasing them back into the ecosystem.

decompression Rapid decrease in ATMOSPHERIC PRESSURE, experienced, for example, when the pressurized cabin of an aircraft is ruptured, or when a diver returns to the surface too quickly. When someone works under pressures greater than atmospheric pressure, the respiratory gases are compressed and abnormally large amounts are dissolved in the blood and tissues. Sudden release of this pressure causes the gases to bubble off and interrupt the body's oxygen supply.

decompression chamber Air-tight compartment capable of withstanding high pressures, used to house one or more divers or other personnel working in pressurized environments while they are gradually returned to normal atmospheric pressure. Decompression chambers are necessary to avoid **decompression** sickness (THE BENDS), in which nitrogen is released into the blood and tissue causing acute pain in the joints, dizzyness, nausea and paralysis. Deep-sea divers usually work in teams so that work progresses while divers who have been relieved are progressively decompressed in submerged chambers at various depths.

Dee, John (1527–1608) English mathematician and occultist. He was an associate of the alchemist Edward Kelley and was a favourite of Queen Elizabeth I. He also did preparatory work on the Gregorian CALENDAR.

deep freezing Method of FOOD PRESERVATION, usually at −5°C (23°F) or below. Some foods can last for years if properly frozen. Three methods are widely used for commercial deep freezing: in **blast** freezing, a flow of cold air is passed over the food; in contact freezing, the food is placed between refrigerated plates, or in a refrigerated alcohol bath; and in **vapour** freezing, liquid nitrogen or solid carbon dioxide (dry ice) is made to vapourize in the food compartment. In home freezing, various refrigerated cabinets are used.

Deep Sea Drilling Project (Ocean Drilling Program) US project to investigate the evolution of OCEANIC BASINS by studying the composition of the oceanic CRUST. The *Glomar Challenger*, equipped with well-drilling equipment and a satellite-controlled navigation system, went into operation in 1968 until the late 1970s. Measurements of the cores recovered, some of which were obtained at over 900m (3,000ft) beneath the seabed, have provided evidence for the theory of SEAFLOOR SPREADING and have shown that the present oceanic basins are relatively young. Seismic and magnetic studies are also carried out.

deficiency disease Illness caused by lack of an essential nutrient in the DIET. Scurvy is caused by a lack of vitamin C, rickets by a deficiency of vitamin D and pellagra by insufficient niacin (nicotinic acid) in the diet. Rare in the developed world, such disorders are treated by supplying the missing nutrient.

De Forest, Lee (1873–1961) US inventor of the audion triode valve (ELECTRON TUBE). This three-electrode valve, invented in 1907, was one of the most important inventions in electronics. It could amplify signals, and had numerous applications. Valves became essential in radio, television, radar, computer and other electronic systems, and remained so until the TRANSISTOR was invented in 1947.

deforestation Clearing away of forests and their ECOSYSTEMS, usually on a large scale. It may be done to create open areas for farming or building, or to make use of the timber from the trees. There is an immediate danger that the vital topsoil will be eroded by wind (as in the dust bowl in the USA in the 1930s) or, in hilly areas, by rain. Proposals to clear whole regions of the rainforests of Amazonia in Brazil, which play a key role in maintaining the oxygen balance of the Earth, could, if fully implemented, cause an environmental catastrophe. *See also* REFORESTATION

degaussing Demagnetization of a magnetized substance, performed by surrounding the substance with a coil carrying an alternating electric current of decreasing magnitude. The name is derived from the CGS unit of magnetic flux, the **gauss**.

degeneration In biology, term used to describe the effects disease may have on cells or groups of cells, resulting in loss of function or in extreme cases death. In evolution, the term refers to the gradual appearance of a VESTIGIAL STRUCTURE, such as the appendix in humans or the wings of emus, which serve little or no function in the evolved species.

degenerative disease Progressive wasting ailment in which a patient's condition deteriorates at a greater or lesser rate, depending on the treatment used and the response to it; often such a condition can be arrested, although it cannot be reversed.

degree In mathematics, unit of angular measure equal to one three-hundred-and-sixtieth (1/360) of a complete revolution (360°). One degree is written 1°, and can be divided into 60 parts called minutes or "minutes of arc" (one minute is denoted 1'), which may in turn be divided into 60 parts called seconds or "seconds of arc" (one second is denoted 1", so 3,600"= 60'= 1°). 360° are equal to 2π radians. In physics and engineering, a degree is one unit on any of various scales, such as the CELSIUS temperature scale or the **Baumé scale** of SPECIFIC GRAVITY.

dehiscent Describing a FRUIT or seed pod that opens spontaneously to release its seeds. Often violent, dehiscence disperses the seeds over a wide area. The explosive release of spores from a SPORANGIUM or pollen from an ANTHER is also described as dehiscence. *See also* INDEHISCENT

dehydration Removal or loss of water from a substance or tissue. Water molecules can be removed by heat, catalysts or a dehydrating agent such as concentrated sulphuric acid, which removes hydrogen and oxygen in the form of water from another chemical. Dehydration (drying) is used to preserve foodstuffs. In medicine, dehydration is excessive water loss from the body, often a symptom or result of disease or injury.

Delbrück, Max (1906–81) US biologist, b. Germany. He shared the 1969 Nobel Prize for physiology or medicine with Alfred HERSHEY and Salvador LURIA for work on reproduction in VIRUSES.

delocalization In nuclear physics, spreading of VALENCE electrons across more than one CHEMICAL BOND in a compound. Molecules with delocalized electrons tend to be more stable than molecules with electrons that are not delocalized.

Delphinus (Dolphin) Small but distinctive constellation of the N sky. All its leading stars are of fourth magnitude.

delta Fan-shaped body of sediment deposited at the mouth of a river. A delta is formed when a river deposits its sediment load as its speed decreases whilst entering the sea or a lake, and the waves, tides and currents are not sufficiently strong to carry the material away. The shape and size of a delta is controlled by a combination of factors, including climate, water discharge, sediment load, rate of subsidence of the sea or lake, and the nature of the river mouth processes. Most deltas are extremely fertile areas, but are subject to frequent flooding.

dementia Deterioration of personality and intellect that can result from disease of or damage to the brain. It is characterized by memory loss, impaired mental processes, personality changes, confusion, lack of inhibition and a deterioration in personal care and hygiene. Dementia can occur at any age, although the disease process is more common in the elderly. As well as being a degenerative disease in the brain, the causes of dementia include disease of the blood vessels, brain injury, chronic poisoning and infection. *See also* ALZHEIMER'S DISEASE

De Morgan, Augustus (1806–71) British mathematician. He contributed to the development of set theory and wrote a number of textbooks, including *Elements of Arithmetic* (1830). With his contemporary, George BOOLE, he created a mathematical theory of LOGIC.

dendrite Short branching projection from a nerve cell (NEURON). It carries impulses to the cell body and transmits impulses to other nerve cells over short gaps called SYNAPSES. There may be more than one dendrite per neuron.

dendritic drainage Tree-like branching pattern of streams and their tributaries. The pattern is common in regions where the land is essentially flat and the rock is homogeneous.

dendrochronology Means of estimating time by examination of the ANNUAL RINGS (growth rings) in trees. This data can be related to wood used in buildings, for instance. It may also indicate the HYDROLOGY of the region in which the tree grew, thus fixing points in the climatic history of that region. Chronology based on the bristle-cone pine extends back over 7,000 years.

dendron Major projection that protrudes from a nerve cell (NEURON) and is concerned with nerve transmission. Found mostly on MOTOR NEURONS, dendrons usually branch into smaller DENDRITES, which chemically transmit NERVE IMPULSES across a SYNAPSE to other neighbouring neurons.

Deneb (Alpha Cygni) Remote and very luminous white SUPERGIANT star in the constellation of Cygnus. It is 60,000 times more luminous than the Sun and located about 1,500 light-years away.

denitrification Process that chemically reduces ammonia, nitrites or nitrates to yield free nitrogen. In biology, under waterlogged conditions denitrifying BACTERIA change the nitrogen of ammonia into free nitrogen that enters the atmosphere or soil. *See also* NITROGEN CYCLE

denominator In mathematics, the part of a FRACTION below the line that serves as the divisor of the NUMERATOR. In the fraction $^7/_8$, the denominator is 8; 7 is the numerator.

densitometer Instrument for measuring the optical transmission or reflection (optical density) of a material such as a photographic film or plate. It is used in SPECTROSCOPY to determine the positions of spectral lines and bands and to measure their RELATIVE DENSITIES, and thus intensities.

density (symbol ρ) Ratio of mass to volume for a given substance, usually expressed in SI UNITS as kilograms per cubic metre (kgm^{-3}). It is an indication of the concentration of particles within a material. The density of a solid or liquid changes little over a wide range of temperatures and pressures. **Relative density** (symbol r.d.) or **specific gravity** (symbol s.g.) is the ratio of the density of one substance to that of a reference substance (usually water) at the same temperature and pressure. The density of a gas depends strongly on both pressure and temperature.

dental formula System for indicating the type and distribution of TEETH in an animal's jaw. Half of each jaw is considered and represented by two sets of four numbers, one set above and one set below a horizontal line. The numbers above the line stand for the upper jaw and the numbers below the line for the lower jaw. The first number in each set stands for the number of incisors, the second number for the canines, the third number for the premolars and the fourth number for the molars. An adult human with a full set of teeth has, in each half of each jaw, two incisors, one canine, two premolars and three MOLARS. The dental formula is written: $\frac{2\ 1\ 2\ 3}{2\ 1\ 2\ 3}$. Adding all the numbers and multiplying by 2 gives the total number of teeth (32). *See also* DENTITION

dentine Hard yellow matter of all TEETH, which consists of crystals of calcium and phosphate. Human beings and other higher animals have **tubular** dentine, so called because a line of dentine-producing cells (**odontoblasts**), surrounding the pulp, sends out tubules into the dentine; these transmit sensations to the nerve.

dentistry Profession concerned with the care and treatment of the mouth, the teeth and their supporting tissues. As well as general practice, dentistry includes such specialities as oral surgery, prosthodontics, periodontics, orthodontics and public health.

dentition Type, number and arrangement of TEETH (*see* DENTAL FORMULA). An adult human has 32. The incisors are used for cutting; the canines for gripping and tearing; and the molars and premolars for crushing and grinding food. A HERBIVORE has relatively unspecialized teeth that grow throughout life to compensate for wear (plant fibres are very tough) and are adapted for grinding. A CARNIVORE has a range of specialized teeth related to killing, gripping and crushing bones. In carnivores, unspecialized milk teeth are replaced by more specialized adult teeth, which have to last a lifetime.

denudation In geology, wearing away of land by WEATHERING and EROSION. Denudation is a broad term and includes all the natural agencies, such as Sun, rain, wind, rivers, frost, ice and sea, as well as heating and cooling, freezing and thawing, SOLUTION, ABRASION, CORRASION and CORROSION. In addition to weathering and erosion, the removal of material, that is, the transportation, is also part of denudation. Together with DEPOSITION, denudation is the major process which creates the Earth's landscape.

deoxyribonucleic acid *See* DNA

dependent variable *See* VARIABLE

depletion Special form of depreciation referring to the exhaustion of nonrenewable natural resources (such as oil, minerals or natural gas) as they are exploited for human use. Depletion is usually figured as the percentage of the estimated reserves of the resource that has been used.

deposition In geology, laying down of material which has been removed by DENUDATION. Most of the material will be sediment and therefore a possible source of SEDIMENTARY ROCKS when consolidated. After denudation has taken place, the material is transported, sometimes over quite considerable distances, before being deposited. Most of the material will eventually be dumped on the seabed by rivers. However, there will be some deposition on land – for example, silt on flood plains, TILL deposited by ice, LOESS deposited by wind. Together with denudation, deposition is the major process which creates the Earth's landscape.

depression In meteorology, a region of low ATMOSPHERIC PRESSURE with the lowest pressure at the centre. It usually brings unsettled or stormy weather. In the Northern Hemisphere, winds circulate anticlockwise in a depression; they circulate clockwise in the Southern Hemisphere. *See also* CYCLONE

depth perception Ability of the eyes to locate the position of objects in three-dimensional space. The RETINA is two-dimensional, so information about depth is created in the brain. The brain uses "depth clues", which include such factors as linear perspective, PARALLAX, relative size and the slightly different view each eye has of the object.

derivative Rate of change of the value of a mathematical function with respect to a change in the independent variable. The derivative is an expression of the instantaneous rate of change of the function's value. The derivative of a function $f(x)$, at the point x, is defined to be the value of the ratio $[f(x+h) - f(x)]/h$, in the limit where the number h approaches zero, and is denoted df/dx or $f'(x)$. Derivatives provide a means of describing many processes which can be viewed as changing continuously in time. For instance, if the position of an object at a time, t, is given by $p(t)$, then the VELOCITY (which is defined as the *instantaneous* rate of change of position), obtained by differentiation, is given by the derivative dp/dt. Derivatives tell us how fast a quantity changes at a given time. Geometrically, the derivative can be viewed as the gradient, or slope, of a curve at a particular point. *See also* CALCULUS; DIFFERENTIAL

dermapteran Any INSECT of the order Dermaptera, the earwigs. Slender, flattened, brownish-black insects, they are found in crevices and under tree bark. There are some 900 winged and wingless species worldwide. All have a pair of forceps at the hind end of the abdomen.

dermatitis Inflammation of the skin. In acute form it produces redness, itching and blisters or oozing. In chronic form it causes thickening, scaling and darkening of the skin. In **contact** dermatitis, a reaction occurs on contact with a particular substance, such as soap or nettles. In **atopic** dermatitis (often associated with HAY FEVER and ASTHMA), excessive dryness occurs, with redness at the neck, elbows and knees. In **stasis** dermatitis, heavy pigmentation and sometimes ulcers develop on the inner sides of the lower legs as a result of poor circulation. The term is sometimes used interchangeably with eczema.

dermatology Branch of medicine that deals with the diagnosis and treatment of skin diseases.

dermis Thick inner layer of the SKIN, which lies beneath the EPIDERMIS. Also known as the true skin, it consists mainly of loose CONNECTIVE TISSUE richly supplied with BLOOD and LYMPH vessels, nerve endings, sensory organs and sweat glands.

desalination Extraction of pure water (for drinking, industrial and chemical uses, or for irrigation) from water containing dissolved salts, usually seawater. The commonest and oldest method is to evaporate the water from a salt solution by DISTILLATION, and condense the vapour to form water. Another method is to freeze the salt solution; salt is excluded from the ice crystals which can then be melted. In the method of REVERSE OSMOSIS, pure water only passes through a semipermeable membrane against which salt water is pressurized. Other methods are based on the migration of ions between electrodes.

Descartes, René (1596–1650) French philosopher and mathematician. Descartes is often regarded as the father of modern philosophy. In 1619 he described an all-embracing science of the Universe. He evolved his own philosophical principles during the 1630s and 1640s. His works include *Discourse on Method* (1637), *Meditations on the First Philosophy* (1641) and *Principles of Philosophy* (1644). His methods of deduction and intuition inform modern metaphysics. By doubting all his ideas, he reached one indubitable proposition: "I am thinking", and from this that he existed: *cogito ergo sum* (I think, therefore I am). Descartes also founded analytic geometry, introduced the CARTESIAN COORDINATE SYSTEM and helped establish the science of OPTICS.

desensitization Medical treatment in which people who are allergic to certain substances (**allergens**) are injected with extracts of an allergen in a series of gradually increasing dosages. It is done in order to build up resistance to the allergen, thus decreasing sensitivity. The term desensitization is also used to describe a type of treatment in psychotherapy. *See also* ALLERGY

desert Area with a dry CLIMATE, which is sometimes defined as having a total annual rainfall of 250mm (10in) or less. The effectiveness of 250mm (10in) of rainfall can vary, however, depending on whether it falls within a short time or over a prolonged period. A desert is almost barren, although there are very few places with absolutely no vegetation – this occurs only on some patches of moving sand dunes or on bare rocky areas. Most deserts contain tufts of grass and scattered, usually thorny, bushes which have the ability to withstand long dry spells. In some deserts there are cactus plants and various species of flowering plant which have a very short life cycle. They spring up quickly after a shower of rain and go through their complete life cycle in a few days, before dying down and leaving seeds to lie dormant until the next rain. Some deserts contain large areas of sand DUNES, but most deserts are rocky. There are bare rock areas, but generally there will be a cover of loose stones with patches of moving sand. The stones and sand are gradually broken down by mechanical WEATHERING and wind action (AEOLIAN FORMATION). The world's major deserts, including the Australian, Kalahari and Sahara, are found in the HORSE LATITUDES, where the permanent high pressure causes drought throughout the year. Deserts occur in various W coast areas where the influence of cool CURRENTS offshore makes the land even drier, with less than 100mm (4in) rainfall in places; for example, the Atacama Desert in Peru (with the Humboldt Current offshore) and the Namib Desert in S Africa (with the Benguela Current offshore). There are deserts in the middle of the largest continents where no onshore winds can reach to bring any rainfall; for example, the Gobi Desert in Asia, and smaller examples in Nevada and New Mexico in the USA. Some cold regions are also regarded as deserts because they have less than 250mm (10in) annual precipitation. Antarctica and Greenland are ice deserts, and even the TUNDRA areas of N Canada, N Siberia and elsewhere have some similarities to deserts, although they have cool summers and very cold winters.

desertification Process by which a DESERT gradually spreads into neighbouring areas of semi-desert, transforming them into true desert. The change may result from a natural event, such as the destruction of the vegetation by fire or a slight climatic change, but it occurs most frequently as a result of human activity. In many semi-arid areas and dry grasslands, the vegetation becomes overgrazed by domestic animals, so that the land is left bare. Wind and rain then erode the soil, removing any residual fertility, so that no new vegetation can survive. Once the vegetation and soil have gone, the land becomes desert. An additional problem is the use of trees and bushes for firewood by the local people. With an increasing population there is an increasing need for fuel, and so landscapes become depleted of trees. Once it has been destroyed, restoring the land is a long and slow process. Extensive desertification

Existing deserts
Areas with high risk of desertification
Areas with moderate risk of desertification

◄ **desertification** The true causes of desertification are complex and still not entirely understood, but it is generally accepted that recent desertification is directly attributable to increased human intervention. On a large scale, the burning of fossil fuel is likely to shift climatic belts and increase areas of desert. More localized problems have occurred due to overgrazing of livestock and ill-planned irrigation processes.

DETERGENT

Detergents are needed for washing because water will not dissolve grease and oils. Detergents solve this problem. One end is water-attracted (hydrophilic); the other dissolves in oils (hydrophobic, or water repellent). Sodium dodecylbenzenesulphonate (1) is a synthetic detergent with a hydrophilic head (2) and hydrophobic tail. The detergent (3) is added to water and the dirty material (4). The hydrophobic tails stick into the grease (5), while the hydrophilic heads repel each other, forcing dirt and grease up into the water (6). Dirt particles (7) do not return to the cleaned material because both material and dirt now have the same electric charge and repel each other.

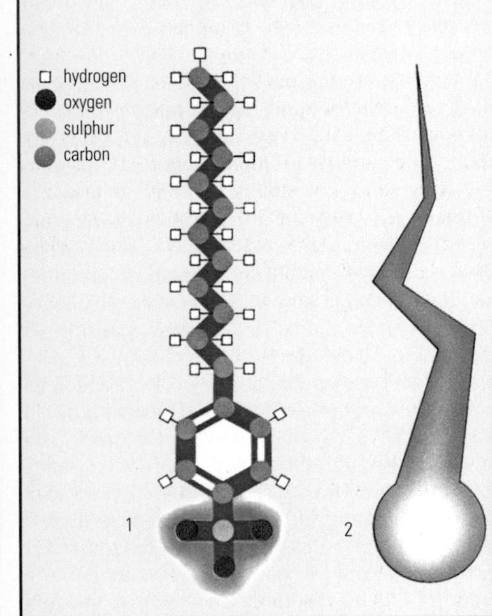

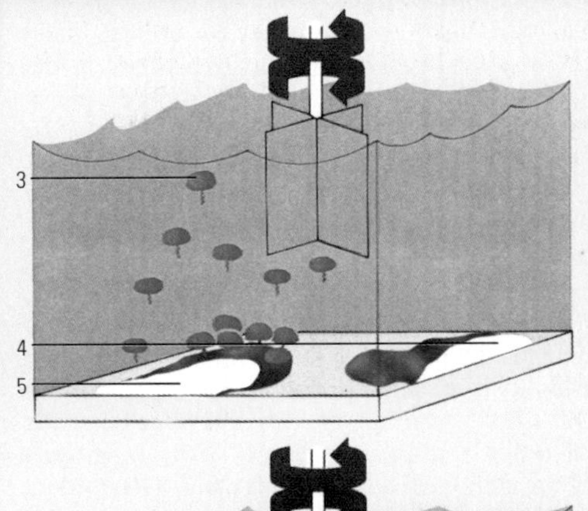

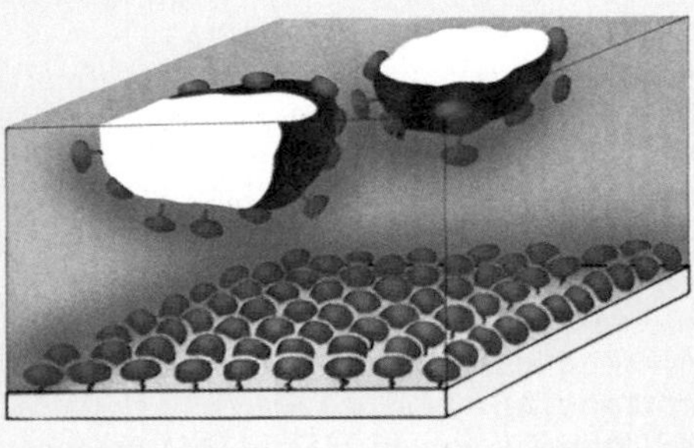

has been taking place in the Sahel, the S edge of the Sahara Desert, and there are also examples in S Africa, India and many other places where there are too many people subsisting on an inadequate landscape.

desiccator Chemical apparatus used for drying substances or for keeping them dry. It consists of a glass vessel with a tight-fitting lid. Specimens are supported on a perforated platform above a lower chamber containing a desiccant, such as anhydrous calcium chloride.

desk-top publishing (DTP) Use of a computer to prepare text and pictures for publication. The technique uses COMPUTER PROGRAMS that display documents on the computer screen. The operator can control the type font, size, line length and column organization of text, and incorporate scanned images if necessary. The result can be output in forms ready for conventional printing and publishing.

destructive distillation Chemical process of breaking down complex organic substances by heating them in the absence of air. Substances such as coal and wood yield such useful products as fuel gases (coal gas) and tar (coal tar), leaving carbon in the form of coke. Before the availability of crude oil (petroleum), this was the usual method of obtaining industrial organic compounds.

destructive margin In PLATE TECTONICS, the boundary between two lithospheric plates (*see* LITHOSPHERE), which are moving towards each other, and where oceanic crust is being recycled into the mantle by SUBDUCTION. Destructive margins can be detected by shallow- to deep-focus EARTHQUAKES (the BENIOFF ZONE), ocean TRENCHES and volcanic ISLAND ARCS. *See also* CONSTRUCTIVE MARGIN

detergent Synthetic chemical cleansing substance that, unlike soap, does not form scum in hard water. There are several types, the commonest being an alkyl sulphonate. Detergents have molecules that possess a long hydrocarbon chain (tail) attached to an ionized group (head). This chain attaches to grease and other non-polar substances, while the ionized group has an affinity for water (so the grease is washed away with the water). Domestic detergents often contain additives such as perfumes and optical brighteners.

determinant A single numerical value associated with a MATRIX. Matrices can be used to transform any geometric figure in the plane or higher dimensional spaces – causing rotations, reflections, scalings or stretches. In this geometric interpretation of a matrix, the determinant denotes the increase in area or volume of the figure resulting from the transformation. Matrices can also be used to solve sets of SIMULTANEOUS EQUATIONS. The equations can be solved provided the determinant is not zero. The determinant for the matrix $A=\begin{bmatrix}a&b\\c&d\end{bmatrix}$, is the number $\det(A) = ad-bc$.

detonator Device using sensitive chemicals (initiating explosives) to set off less sensitive main charges (high explosives). The detonating charge is generally housed in a thin-walled metal or plastic, waterproof capsule, such as the "cap" at the base of a rifle cartridge. Detonators may be exploded electrically, by mechanical shock or by igniting a fuse.

detritus In geology, sediment deposited by natural forces. It is classed by the size of the particles. *See also* ALLUVIUM

deuterium Isotope (D or ^{2}H) of hydrogen whose nuclei (DUETERONS) contain a neutron in addition to a proton. For every million hydrogen atoms in nature, there are about 156 deuterium atoms. Deuterium occurs in water as D_2O (HEAVY WATER), from which it is obtained by ELECTROLYSIS. Heavy water is used as a moderator in some FISSION reactors and it could become a fuel in FUSION NUCLEAR REACTORS. Properties: r.a.m. 2.0144.

deuteron Nucleus of a DEUTERIUM (heavy hydrogen) atom, consisting of one proton and one neutron. The nucleus of ordinary hydrogen contains one proton only.

Devonian Fourth-oldest period of the PALAEOZOIC era, lasting from 408 to 360 million years ago. It is sometimes called the Age of Fishes. Numerous marine and freshwater remains include jawless fish and forerunners of today's FISH. The first known land vertebrate, the amphibian *Ichthyostega*, appeared at this time. Land animals included scorpions, mites, spiders and the first insects. Land plants consisted of tall CLUB MOSSES, HORSETAILS and FERNS. In Devonian times much of the British Isles was desert mountain environment or semi-desert coastal plains, giving rise to the red rock known as Old Red Sandstone.

De Vries, Hugo (1848–1935) Dutch botanist. His experimental methods led to the rediscovery (1900) of Gregor MENDEL's laws of HEREDITY and the development of a theory of MUTATION. In his work *The Mutation Theory* (1901–03), De Vries argued that genetic mutation was the chief engine of EVOLUTION.

dew Water droplets formed, usually at night, by CONDENSATION on vegetation and other surfaces near the ground. Hoar frost is formed when temperatures are below freezing. Fog also deposits moisture on exposed surfaces.

Dewar, Sir James (1842–1923) Scottish chemist and physicist who carried out research

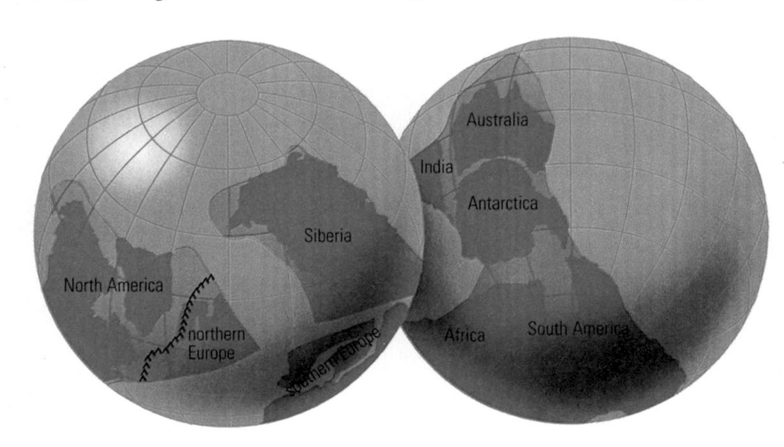

◀ **Devonian** By the Devonian period, North America had collided with N Europe and the sediments between them had been thrust up to form the Caledonian Mountains.

DIELS-ALDER REACTION

The Diels-Alder reaction has served an important role in many syntheses because it smoothly and stereospecifically unites two carbon compounds. A six-membered ring is formed by 1,4-addition of an alkene unit to a conjugated diene. The alkene unit (known as the dienophile) has electron-withdrawing groups attached which activate the diene. The illustration shows the Diels-Alder reaction using molecular orbital models. The addition reaction mechanism is shown by the rearrangement of the electron clouds.

Key
A) butadiene. This is a conjugated diene (two double bonds separated by a single bond). It is shown in the s-cis form and has the property of undergoing 1,4 addition very easily.
B) 1,2-dinitrile ethene. This is the dienophile, and the electron-withdrawing or accelerating groups are nitrile groups. It is these that activate the diene.
C) The diene and dienophile react to form 2,3-dinitrile benzene. The configurations of the diene and dienophile are retained in the product, this means that the reactants come together to give cis addition.

into materials at extremely low temperatures. In 1872 he invented the vacuum **Dewar** flask, the commercial adaptation of which is the Thermos flask. He built a device that could produce liquid oxygen and demonstrated that liquid oxygen and ozone are magnetic. Dewar also liquified hydrogen in 1898 and solidified it in 1899.

dew point Temperature to which air must be cooled for water vapour to begin to condense.

dextran Stable, water-soluble POLYSACCHARIDE sugar. It is used as a substitute or extender for blood plasma in transfusions.

diabetes Any metabolic disorder in which there is exaggerated thirst and copious output of urine. The term generally refers to **diabetes mellitus**, a disease characterized by lack of the pancreatic hormone INSULIN needed for sugar METABOLISM. Failure of insulin production leads to a build-up of sugar in the bloodstream (**hyperglycaemia**) and subsequently in the urine. Symptoms include abnormal thirst, over-production of urine and weight loss; degenerative changes occur in the blood vessels. Untreated, the condition progresses to diabetic coma and death. There are two forms of the disease. Type 1, or insulin-dependent, diabetes is the more severe form. It usually begins in childhood and is an AUTOIMMUNE DISEASE. People with type 1 diabetes owe their survival to insulin injections. In type 2 diabetes, which mostly begins after the age of 40, there is some insulin output but not enough for the body's needs. This milder form can usually be controlled with dietary restrictions, possibly together with oral drugs. Careful treatment of diabetes, including the control of high blood pressure (HYPERTENSION), can keep at bay serious complications such as blindness and kidney failure. Susceptibility to diabetes mellitus is inherited and the disease is slightly more common in males.

diagenesis Physical and chemical processes whereby sediments are transformed into solid rock, usually at low pressure and temperature. Pressure results in compaction, forcing grains together and eliminating air and water, and cementation binds individual particles together by the precipitation of a secondary mineral, commonly an iron oxide, calcite or silica.

dialysis Process for separating particles from a solution by virtue of differing rates of DIFFUSION through a SEMIPERMEABLE MEMBRANE. In the artificial kidney unwanted (smaller) molecules of waste products (urea and some salt) are separated out to purify the blood. **Electrodialysis** employs a direct electric current to accelerate the process, which is especially useful for isolating proteins.

diamagnetism Type of MAGNETISM possessed by all materials in which an applied **magnetic field** induces a weak magnetization in the opposite direction.

diamond Crystalline form of carbon (C); the hardest natural substance. Diamond is found in kimberlite pipes and in alluvial deposits as cubic system octahedral crystals. It is brilliant, transparent to translucent, colourless or of many hues including yellow, green, blue and brown, depending on the impurities contained. Bort, a variety of diamond inferior in crystal and colour, and carborondo, an opaque grey to black variety called black diamond, as well as all other non-gem varieties are used in industry. Industrial diamonds are used as abrasives, bearings in precision instruments such as watches, and in the cutting heads of drills for mining. Synthetic diamonds, made by subjecting graphite with a catalyst to high pressure and temperatures of about 3,000°C (5,400°F), have been made since 1955, but are fit only for industrial applications. Diamonds are weighed in carats (0.2gm) and in points (1/100 carat). The largest producer of diamonds is Australia. Hardness 10; r.d. 3.5.

diapause Time break in the development of an INSECT. During diapause, growth stops and METABOLISM slows right down. It occurs because of interruptions in an innate rhythm brought about by unfavourable changes in the environment, possibly because of seasonal changes in climate. In this way, the developing insect survives until conditions improve. *See also* BIOLOGICAL CLOCK

diaphragm Sheet of muscle that separates the ABDOMEN from the THORAX, the cavity of the chest, in mammals. During exhalation (breathing out) it relaxes and allows the chest to subside; on inhalation it contracts and flattens, causing the chest cavity to enlarge.

diastole Phase of the cardiac cycle in which the HEART muscle is relaxed and is filling with blood.

diathermy Medical treatment that uses heat generated by high-frequency (short-wave) alternating electric current. The heat stimulates local blood circulation and tissue repair. It is used to treat back pain, to aid muscle and tendon repair, and is often used in eye and neurosurgery.

diatom Any of a group of tiny, microscopic, single-celled ALGAE (phylum Bacillariophyta) characterized by a shell-like cell wall made of silica. The shell (**frustule**) consists of two halves that fit together like the two halves of a petri dish. Diatom frustules occur in a wide variety of highly symmetrical shapes. Diatoms live in nearly every environment that has water and is exposed to sunlight, including virtually all bodies of salt

DIESEL ENGINE

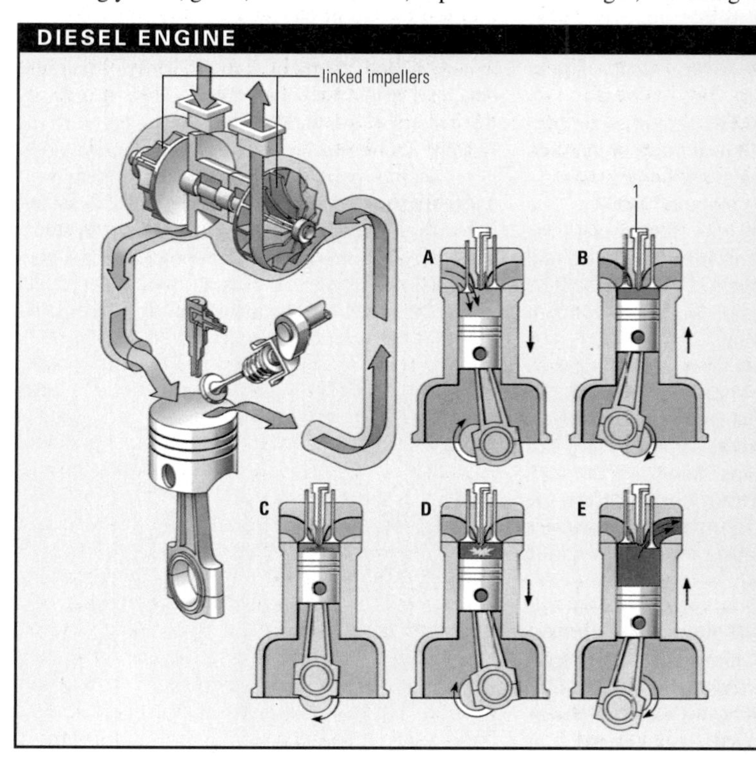

A diesel engine has no spark plugs, instead it works on the principle of compression-ignition. Air is introduced into the cylinder (A). An injector (1) squirts fuel into the cylinder where the upwards stroke of the cylinder (B and C) compresses the mixture. In these conditions the fuel ignites spontaneously (D) and the expansion of the combustion products forces the piston down. The rotation of the crankshaft pushes the piston up (E) and the exhaust gases are expelled. In the diesel engine illustrated, a turbocharger uses the energy of the exhaust gases to force air into the cylinder via linked impellers, achieving greater compression in the cylinder.

DIFFERENTIAL

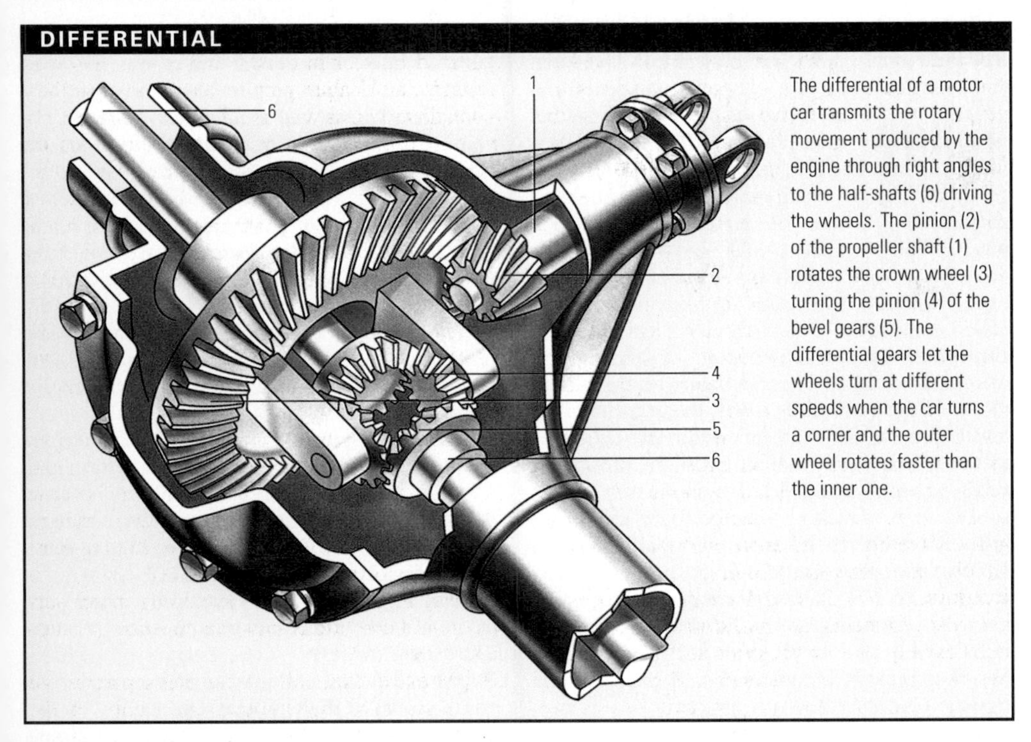

The differential of a motor car transmits the rotary movement produced by the engine through right angles to the half-shafts (6) driving the wheels. The pinion (2) of the propeller shaft (1) rotates the crown wheel (3) turning the pinion (4) of the bevel gears (5). The differential gears let the wheels turn at different speeds when the car turns a corner and the outer wheel rotates faster than the inner one.

and freshwater and even soil, damp rocks, tree bark and the undersides of icebergs. Light can pass through the transparent cell wall, and is used for PHOTOSYNTHESIS. The microscopic shells of long-dead diatoms accumulate in, for example, diatomite (kieselguhr), a mineral with abrasive, absorbent and refractory uses.

diazepam Tranquillizer in the benzodiazepine group that also acts as a skeletal muscle relaxant. It is prescribed for acute anxiety, delirium tremens and epilepsy and is also used in premedication for surgery. Side-effects include drowsiness, lethargy, vertigo and skin rashes.

diazo compound *See* AZO AND DIAZO COMPOUNDS

dicarboxylic acid In organic chemistry, a CARBOXYLIC ACID with two carboxyl groups (–COOH). Examples include ethanedioic acid (oxalic acid, $(COOH)_2$), hexanedioic acid (adipic acid, $(CH_2)_4(COOH)_2$), and phthalic acid ($C_6H_4(COOH)_2$).

dichroism Property of some CRYSTALS that makes them appear to have a different colour according to the direction of the incident light. The crystal allows light vibrations in one plane to pass through but absorbs light vibrations at right angles to this plane. Tourmaline is a natural dichroic crystal; polaroid is a synthetic material with this property. *See also* POLARIZED LIGHT

dicotyledon Any member of the class Dicotyledonae within the phylum Angiospermophyta – ANGIOSPERMS (flowering plants). Dicotyledonous plants are characterized by two seed leaves (COTYLEDONS) in the seed embryo. Other general features of dicotyledons include: broad leaves with branching veins; FLOWER parts (sepals, petals, stamens) arranged in whorls of fours or fives; VASCULAR BUNDLES arranged in a ring in the stem and ROOT; and having a main central root (TAPROOT). There are about 250 families of dicotyledons, including the rose, daisy and magnolia families. *See also* MONOCOTYLEDON

dielectric Nonconducting material such as an electrical insulator that separates two conductors in a CAPACITOR. The PERMITTIVITY of a dielectric is a measure of the extent to which it can resist the flow of charge. The dielectric strength is the maximum field that the dielectric can withstand without breaking down by becoming ionized. For many low-power applications such as tuning capacitors in radios, air is used as the dielectric.

Diels, Otto Paul Hermann (1876–1954) German chemist. Diels discovered (1906) the suboxide of carbon (tricarbon dioxide, C_3O_2), studied the structure and synthesis of cantharidine (a poison contained in the body fluid of blister beetles), and did research on STEROLS (solid, higher alcohols such as cholesterol). He was professor of chemistry (1916–48) at Kiel University. He shared the 1950 Nobel Prize for chemistry with the German chemist Kurt Alder (1902–58), with whom he discovered (1928) the synthesis of dienes (organic compounds containing two double bonds), known as the **Diels-Alder reaction**.

diesel engine INTERNAL COMBUSTION ENGINE in which the heat for igniting the fuel oil is produced by compressing air. The engine, invented by Rudolf DIESEL in the 1890s, is also referred to as a compression-ignition engine. The fuel-air mixture burns rapidly and expands to drive the pistons. Diesel engines usually have electric heaters called **glow plugs** for warming the cylinders during starting. Once the engine has fired, the heaters may be switched off.

Diesel, Rudolf (1858–1913) German inventor of the DIESEL ENGINE. Diesel developed, patented and built the engine in the 1890s. He later established an engine factory in Augsburg, Germany.

diesel fuel (diesel oil) Petroleum product heavier than kerosene but lighter than heating oil, used to power DIESEL ENGINES in industrial and agricultural vehicles such as lorries, buses, tractors, locomotives and ships. There are many grades of diesel fuels, which, unlike petrol, burn unevenly. They are graded against standardized mixtures of hexadecane and alpha methylnaphthalene to establish a cetane number. A cetane number of 30 to 45 is desirable.

diet Range of food and drink consumed by a person or animal. The human diet falls into five main groups of nutrient necessary for the maintenance of health: PROTEINS, CARBOHYDRATES, FATS, VITAMINS and MINERALS. An adult requires daily about 1g (0.035oz) of protein for 1kg (2.2lb) of body weight. Beans, peas, fish, eggs, milk and meat are important protein sources. The body breaks these down into AMINO ACIDS from which it synthesizes its own protein. Carbohydrates (starches and sugars), stored in the body as GLYCOGEN and fat, are the chief sources of energy and are found in cereals, root vegetables and sugars. Carbohydrates make up the bulk of most diets. Fats in the diet, such as butter and vegetable oils, are a concentrated source of energy and aid in the absorption of the fat-soluble vitamins (vitamins A, D, E and K). Vitamins function as co-enzymes in important body processes. Water and minerals such as iron, calcium, potassium and sodium are also essential. Specialized study of the human diet and nutrition is known as DIETETICS.

dietetics Specialized study of the human DIET and nutrition and its clinical management. A dietician plans what people should eat when they are ill, making sure they get what they need to aid recovery.

differential In mathematics, small change occurring in the value of a mathematical expression due to a small change in a variable. If $f(x)$ is a function of x, the differential of the function in response to a small change dx in x, is written df and is given by $f'(x)\,dx$, where $f'(x)$ is the DERIVATIVE of $f(x)$.

differential In a vehicle, set of circular GEARS that transmits power from the engine to the wheels. When a car is turning a corner, the differential allows the outside drive wheel to rotate faster than the inner one.

differential calculus *See* CALCULUS

differential equation Equation containing DERIVATIVES. Differential equations are used in almost all areas of APPLIED MATHEMATICS. *See also* CALCULUS

differential geometry Type of GEOMETRY that uses CALCULUS notation to analyse geometric concepts such as curves and surfaces. For example, a curve, such as the path of a projectile, or the orbit of a spaceship, can be described mathematically in terms of three mutually perpendicular VECTORS, which vary independently along the length of the curve.

differentiation In mathematics, name given to the method of evaluating the DERIVATIVE of some given function. The techniques of differentiation and INTEGRATION together constitute the subject of CALCULUS and are frequently required in almost all branches of APPLIED MATHEMATICS. *See also* DIFFERENTIAL EQUATION

diffraction Spreading of a wave, such as a light beam, on passing through a narrow opening or hitting the edge of an obstacle, such as sound being heard around corners. Diffraction provides information on the wavelength of light and the internal structure of the diffracting CRYSTALS. All waves are diffracted by obstacles. *See also* ELECTRON DIFFRACTION

diffraction grating Substrate consisting of par-

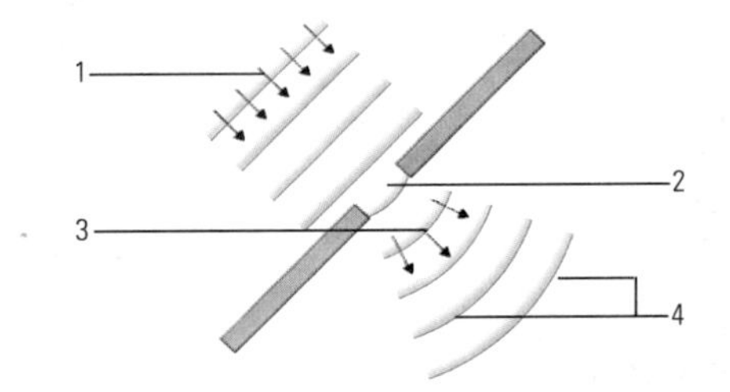

▲ **diffraction** When waves (1) pass through a narrow gap (2) they are diffracted (3) and spread out in concentric rings (4). An example would be ocean waves passing through the narrow entrance of a harbour.

D

DIFFUSION

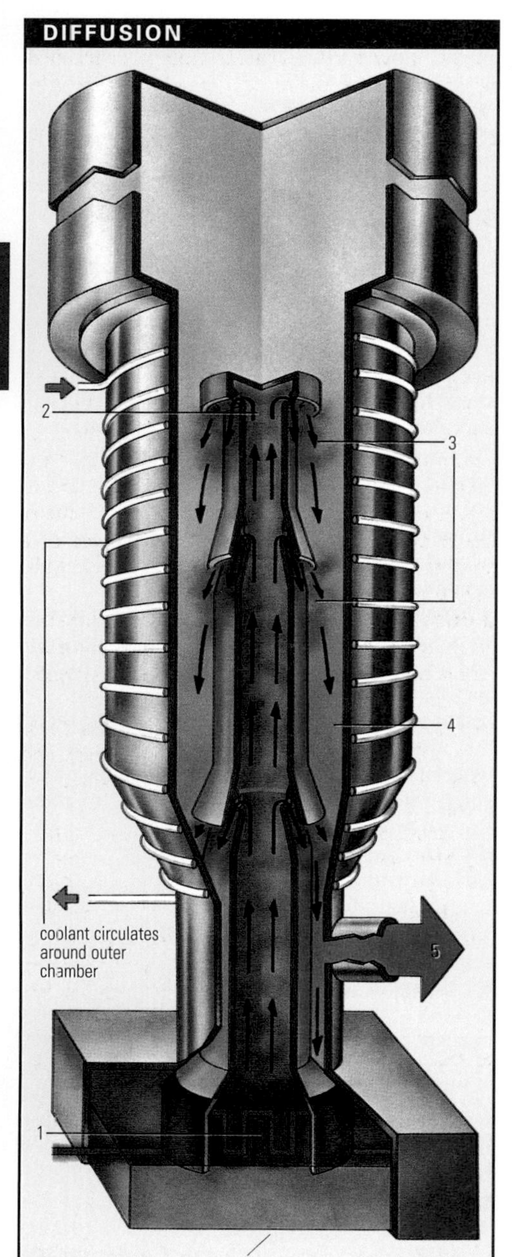

Diffusion pumps, a type of vacuum pump, are used in conjunction with mechanical pumps to achieve very low pressures (below 10^{-7} pa). Oil (or mercury) vaporizes from the heated reservoir (1) and rises in the inner chamber (2) where the evacuation occurs. The vapour passes by diffusion through small holes in the inner walls and is channelled downwards (3) by the circular shields towards the cooled walls of the outer chamber (4). As the molecules of the oil vapour move down, they carry with them the gas molecules in the vicinity – causing more gas molecules to flow into the outer chamber. Oil vapour condenses and runs to the reservoir while the gas molecules are extracted (5) by a force pump. The minimum pressure attainable is equal to the vapour pressure of oil at the temperature of the cooled walls.

allel equidistant lines (as many as 1,500 per mm) for producing spectra by DIFFRACTION of light. Transmission gratings are ruled on glass and are transparent; reflection gratings are opaque – they are ruled on metal films. Such gratings are much used in SPECTROSCOPY. *See also* DIFFRACTION; INTERFERENCE

diffusion Movement of a substance in a mixture from regions of high concentration to regions of low concentration, due to the random motion of the individual atoms or molecules. Diffusion ceases when there is no longer a concentration gradient. Its rate increases with temperature, since average molecular speed also increases with temperature. The process occurs quickly in gases and liquids, much more slowly in solids.

digestion Process of breaking down food mechanically and chemically into smaller molecules that can be readily absorbed. Digestion occurs mainly by means of chemical agents called ENZYMES. *See also* DIGESTIVE SYSTEM

digestive system (alimentary system) Group of organs of the body concerned with the DIGESTION of foodstuffs. In humans, the digestive system begins with the mouth where the action of teeth and enzymes in saliva begin the process of breaking down the food. It continues with the OESOPHAGUS, which carries food into the STOMACH. The stomach leads to the small intestine, which then opens into the COLON. After food is swallowed, it is pushed through the digestive tract by PERISTALSIS. On its journey, food is transformed into small molecules that can be absorbed into the bloodstream and carried to the tissues. CARBOHYDRATE is broken down to sugars, PROTEIN to AMINO ACIDS, and FAT to FATTY ACIDS and GLYCEROL. Indigestible matter, mainly cellulose, passes into the rectum, and is eventually eliminated from the body (as faeces) through the ANUS.

digit Any numeral from 0 to 9.

digital Describing information expressed in terms of numbers. DATA, such as words, pictures and music, are represented as a series of numbers (1s and 0s) in a BINARY SYSTEM, such as a COMPUTER. Digital also refers to any way of representing information as numbers as opposed to a continuously varying ANALOGUE fashion. For example, the figures of a digital clock are the counterpart to the ANALOGUE, or continuous, display of a traditional clock face. A COMPACT DISC stores information digitally, in the form of pits in its surface that represent 0s and 1s. A vinyl LP, by contrast, stores information in analogue form, as a "wavy" groove.

digital audio tape (DAT) Technology for recording data in DIGITAL form on MAGNETIC TAPE. DATs are smaller and longer than analog cassettes. They are used primarily for computer backups and studio-recording. In the latter, the original sound signal is encoded to form patterns of equal-strength pulses. These DIGITAL SIGNALS are recorded onto tape. On playback, the patterns are detected and decoded to produce an identical signal.

digital signal Group of electrical or other pulses in a COMPUTER or communications system. They may represent data, sounds or pictures. Pulses in a stream of digital signals are represented by 1s and 0s in the BINARY SYSTEM.

dimension In mathematics, the spatial dimension is the number specifying the extent of an object in different directions. A figure having length only is said to be one-dimensional; a figure having area but not volume, two-dimensional; and a figure having volume, three-dimensional. More generally the dimension of a figure equals the number of coordinates that are needed to specify all points of the figure. For example, in a plane

DIGESTIVE SYSTEM

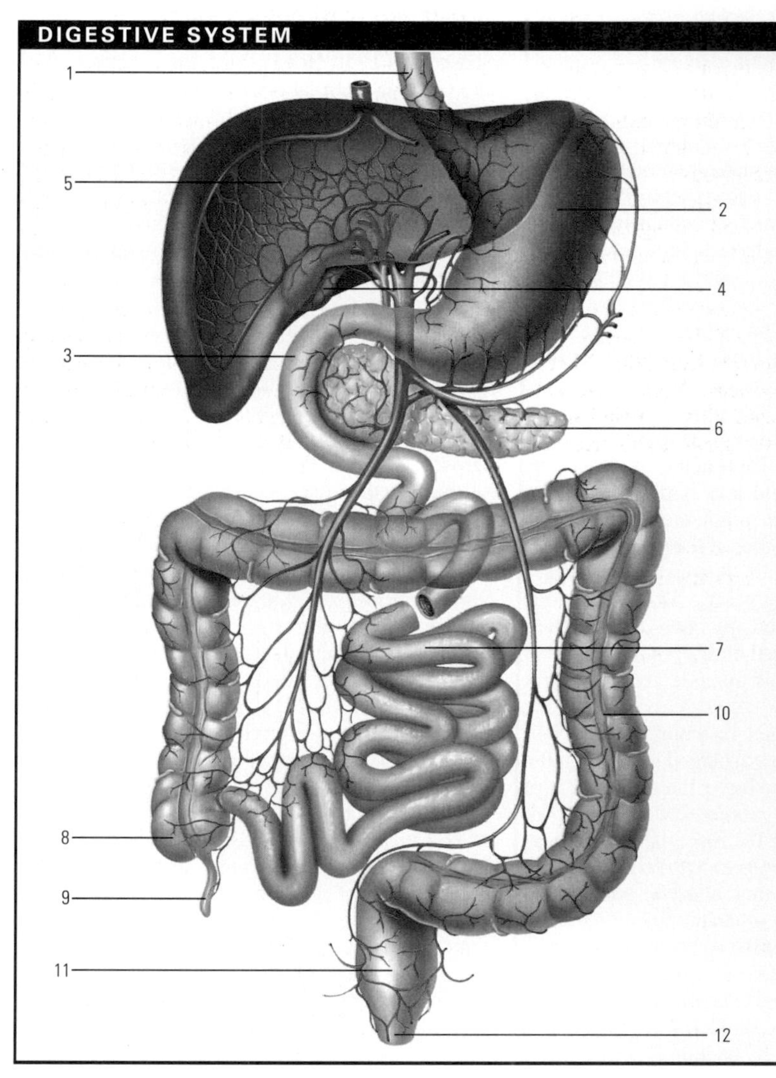

The digestion and absorption of food takes place within the digestive tract, a coiled tube some 10m (33ft) long which links mouth to anus. Food is passed down the oesophagus (1) to the stomach (2), where it is partially digested. Chyme is released into the duodenum (3), the first part of 7m (23ft) of small intestine. The duodenum receives bile secreted by the gall bladder (4) in the liver (5), and enzymes secreted by the pancreas (6). Most absorption occurs in the jejunum and ileum, the remaining parts of the small intestine (7). Any residue passes into the caecum (8), the pouch at the start of the large intestine. At one end of the caecum is the 10cm (4in) long vermiform appendix (9), which serves no useful purpose in man. Water is reabsorbed in the colon (10). Faeces form and collect in the rectum (11) before being expelled as waste through the anus (12).

DIGITAL AUDIO TAPE

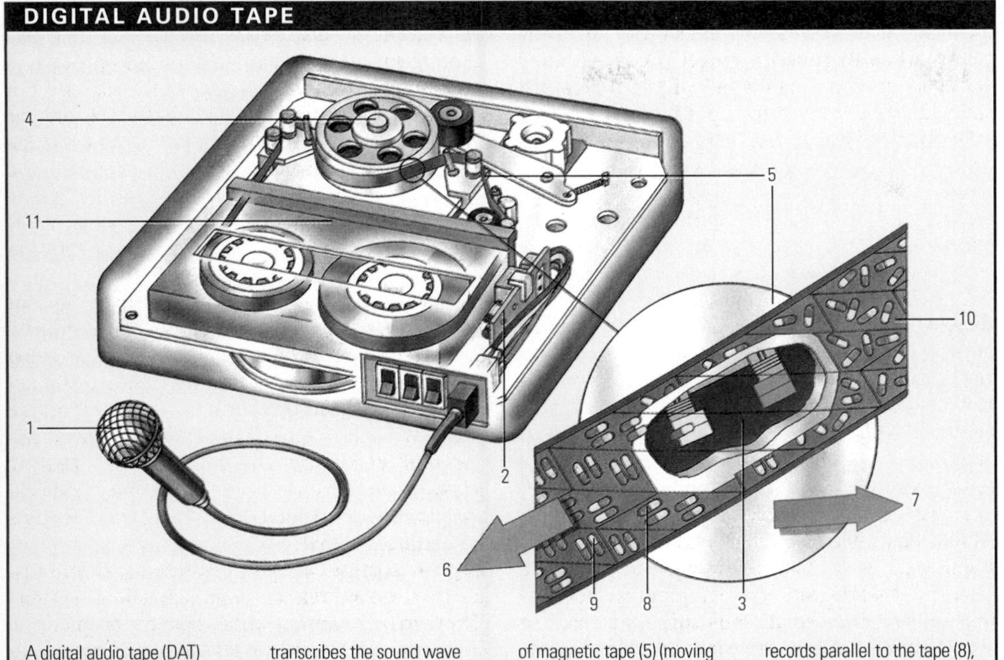

A digital audio tape (DAT) recorder records sound, an analogue signal, in digital form. The analogue signal enters via a microphone (1) and passes through a converter (2), which transcribes the sound wave into a series of 0s and 1s. Two magnetic heads, tiny electromagnets (3) in a rotating drum (4), receive the signal as electrical pulses which polarize diagonal strips of magnetic tape (5) (moving right to left) (6) as it is scanned diagonally (7). The heads align the magnetic elements of the tape representing a 0 or a 1. Each head records half of a stereo recording. One head records parallel to the tape (8), one perpendicular (9) to avoid interference. The units of an unrecorded tape are jumbled (10). When a tape (11) is played the head reads the polarization of the tape.

which is two-dimensional, two coordinates are required (x, y). *See also* COORDINATE SYSTEM

dinitrogen oxide (nitrous oxide, N_2O) Colourless gas that has a pleasant odour and is sometimes used as an anaesthetic or analgesic during surgical or dental operations. It is also known as "laughing gas" since it produces exhilaration, sometimes accompanied by laughter. It is also used in making pressurized foods.

dinosaur (Gr. terrible lizard) Any of a large number of REPTILES that lived during the MESOZOIC era, between 225 and 65 million years ago. They appeared during the TRIASSIC period, survived the JURASSIC and became extinct at the end of the CRETACEOUS. Dinosaurs were mostly egg-laying animals, ranging in size from 30in (91 cm) to the 90ft (27m) diplodocus. There were two orders: **Saurischia** ("lizard hips"), included the bipedal carnivores and the giant herbivores; the **Ornithiscia** ("bird hips") were smaller herbivores. Their posture, with limbs vertically beneath the body, distinguish them from other reptiles. There is evidence that some birds are the living descendants of ornithischians. Many theories are advanced to account for their extinction. It is possible that, as the climate changed, they were incapable of swift adaptation. A more catastrophic theory is that they died because of the devastating atmospheric effects from the impact of a large meteor.

diode Electronic component with two ELECTRODES, used mainly as a RECTIFIER to convert alternating current (AC) to direct current (DC). The SEMICONDUCTOR diode, which has largely replaced the ELECTRON TUBE (valve) type, usually has a single *p-n* junction. It allows electrons to flow freely in only one direction (*n*-type material to *p*-type), but only a small current flows in the reverse voltage direction. A **Zener** diode blocks current flow until a critical voltage is reached, when it breaks down and conducts. *See also* ELECTRIC CURRENT

dioecious plant Plant bearing either female, **pistillate** flowers (with CARPELS only) or male, **staminate** flowers (with STAMENS only); but not both.

Diophantus (active AD 250) Greek mathematician, believed to have produced one of the earliest works in ALGEBRA. He is best known for his investigation and description of a set of algebraic equations now called **Diophantine equations**. Work in this area is known as **Diophantine analysis**.

dioptre Optical unit of magnifying power of a lens system. The power of a lens is equal to one metre divided by the focal length in metres. An algebraic sign (+ or −) indicates whether the system is **convergent** or **divergent**; for example, a diverging lens of focal length 1/3 metre has a power of −3 dioptres. The power of a combination of lenses is the sum of their individual powers.

diorite Intermediate coarse-grained IGNEOUS ROCK, similar to GRANITE in its texture but made up mainly of plagioclase feldspar and hornblende, with BIOTITE or augite. It is usually dark grey.

dip In geology, the steepest angle of a tilted plane of a rock (such as bed, joint, cleavage and fault). It is measured from the horizontal, 90° being vertical, using a CLINOMETER.

dip, magnetic *See also* INCLINATION, MAGNETIC

diphtheria Acute, infectious disease characterized by the formation of a membrane in the throat which can cause asphyxiation; there is also release of a toxin which can damage the nerves and heart. It is caused by a bacterium, *Corynebacterium diphtheriae*, which often enters

▲ **digital signal** An analogue signal has a continuous wave form. A digital signal carries information in a binary form – 0s and 1s (A). This can be transmitted electrically with a voltage either on or off (B), or optically (C), with each pulse of light corresponding to a 1.

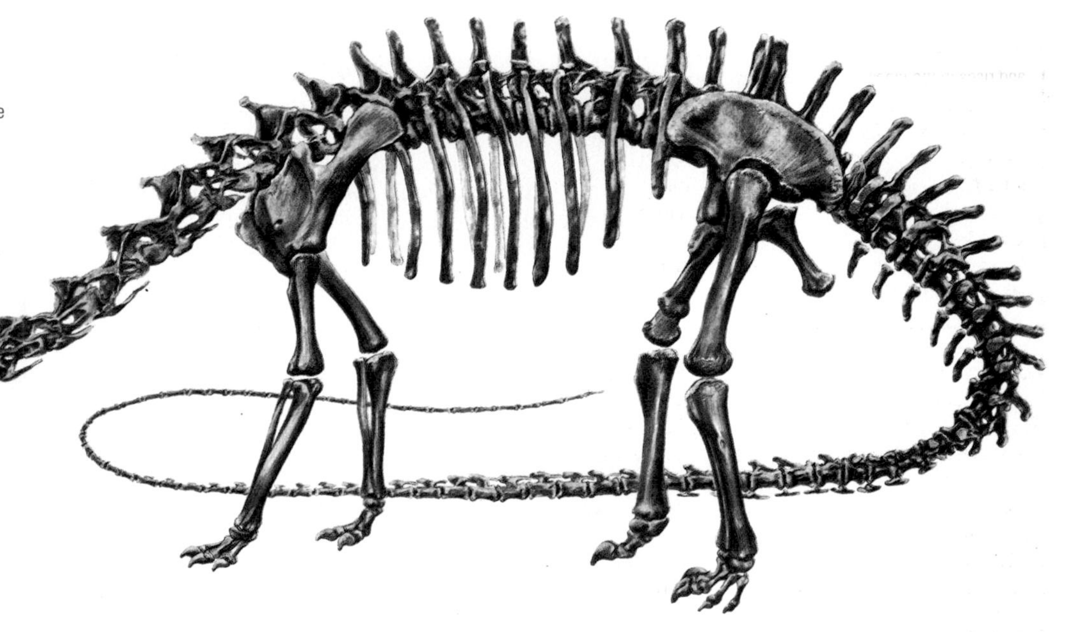

► **dinosaur** The diplodocus was the longest of the dinosaurs and the longest land animal ever to have lived, reaching a length of 25m (82ft). Its characteristically long neck and tail and small head are noticeable in the diplodocus skeleton pictured here. A member of the Saurischia order, the diplodocus was one of the giant herbivores. It flourished in the Jurassic period, dwelling in swamps.

D

DISTILLATION

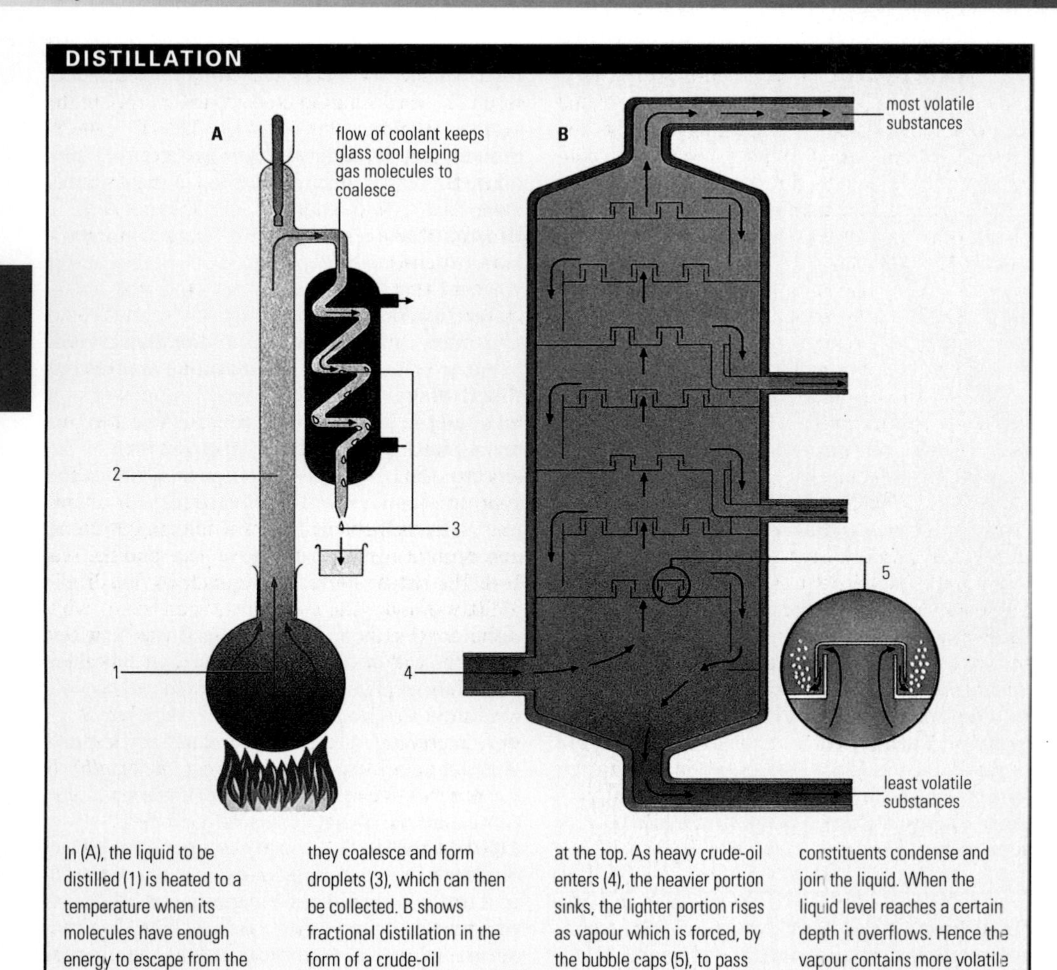

In (A), the liquid to be distilled (1) is heated to a temperature when its molecules have enough energy to escape from the liquid as a gas (2). When they encounter a colder surface they coalesce and form droplets (3), which can then be collected. B shows fractional distillation in the form of a crude-oil fractionating column, which is hotter at the bottom than at the top. As heavy crude-oil enters (4), the heavier portion sinks, the lighter portion rises as vapour which is forced, by the bubble caps (5), to pass through a layer of liquid fractions. The less volatile constituents condense and join the liquid. When the liquid level reaches a certain depth it overflows. Hence the vapour contains more volatile components as it travels up the column.

through the upper respiratory tract. It is treated with antitoxin and ANTIBIOTICS. Once common throughout the temperate regions, the disease is now rare in countries where there has been an active programme of immunization.

diplococcus Genus of BACTERIA characterized by gram-positive, spherical cells that grow in pairs. It includes the pneumococcus, which causes pneumonia.

diploid Term describing a nucleus or CELL that has its CHROMOSOMES in pairs. Almost all animal cells are diploid, except GAMETES (reproductive cells), which are HAPLOID (containing only a single set of chromosomes). The cells of ANGIOSPERMS (flowering plants) and GYMNOSPERMS are also diploid. Algae and lower plants, such as ferns and mosses, have two generations in their life cycle, one diploid and the other haploid. In a diploid cell, the chromosomes of each pair carry the same GENES. *See also* ALTERNATION OF GENERATIONS

diplopod Any member of the class Diplopoda, ARTHROPODS found worldwide. They are known as millipedes. Some attack plants, while others are predatory carnivores. They may have as many as 200 pairs of legs.

dipole Separation of electric charge in a molecule. In a COVALENT BOND between two atoms the electron pair is not necessarily equally shared between them. In hydrogen chloride (HCl), the electrons are attracted towards the more electronegative chlorine atom, giving it a partial negative charge and leaving an equal positive charge on the hydrogen atom. Such dipoles contribute to the chemical properties of molecules. Magnetic substances have dipoles consisting of north and south poles; they may be ordered or disordered to give strong or weak magnetic properties.

dip pole Either of two imaginary points on the Earth's surface where the direction of the Earth's magnetic field is vertical – downwards at the north magnetic pole and upwards at the south pole – and are offset in respect to the geographical poles, each by a different amount so that the S dip pole is not exactly opposite the N dip pole.

dipteran Any member of the order Diptera, the true flies. Adult flies have soft bodies and one pair of wings, the other being reduced to knob-like **halteres**; they have COMPOUND EYES and sucking mouthparts. They have a complete life cycle: the adult lays eggs that hatch into LARVAE (maggots); these pupate and become adults. Within the Diptera order are many important pests, such as mosquitoes, house flies and others that attack human beings, animals or crops or carry diseases. *See also* INSECT

Dirac, Paul Adrien Maurice (1902–84) British physicist who devised one version of QUANTUM MECHANICS. He extended this to combine RELATIVITY and quantum-mechanical descriptions of electron properties. He also predicted the existence of the POSITRON, which was later discovered by Carl ANDERSON. Dirac shared the 1933 Nobel Prize for physics with Erwin SCHRÖDINGER for their work in QUANTUM THEORY. He was professor of physics at Florida State University from 1971 until his retirement.

direct current (DC) *See* ELECTRIC CURRENT

directional antenna Radar or radio antenna for receiving or transmitting signals in one prime direction. A radio direction-finder for example, has a rectangular loop antenna which receives maximum signal when it is broadside to the station, and minimum signal when it faces the station end-on. The orientation of the antenna is indicated on a direction scale.

direction finding (df) In NAVIGATION, a method that determines position by the intersection of two projected bearing lines obtained from known landmarks. In communications, radio df instruments can determine a bearing by using a DIRECTIONAL ANTENNA which locates two radio signals from transmitters at known points.

direct motion Orbital motion of a planet, comet or other heavenly body around the Sun or of a satellite around its primary (the planet about which it orbits) in a west-to-east direction. Also the apparent west-to-east movement of a planet against a stellar background, as seen from the North Pole.

disaccharide Type of sugar (including common sugar) formed by the condensation of two MONOSACCHARIDES with the removal of water. Cane sugar (sucrose) is a disaccharide which, on HYDROLYSIS with dilute acid, yields both glucose and fructose (monosaccharides). Lactose (the sugar in milk) and maltose are other important disaccharides. *See also* POLYSACCHARIDE

disc In astronomy, circular appearance of the Sun, Moon or a planet, especially when viewed through a telescope. It contrasts markedly with the sharp point source image of a star.

discharge (symbol Q) Measure of waterflow at a particular point, measured in cubic metres per second (m^3s^{-1}). Discharge can be calculated with the formula $Q = V \times A$, where V is the velocity of the water and A the cross-sectional area of the river channel.

discharge tube Glass envelope containing gas at low pressure, which gives light due to the collision of electrons with gas molecules. Discharge lamps, filled with one or more of various gases, are used to give coloured light: mercury vapour gives purple, sodium vapour gives yellow and neon gives red. Early discharge tubes (called **Geissler tubes**) were used as toys. A FLUORESCENT LAMP is a discharge tube coated internally with a fluorescent substance.

discontinuity *See* MOHO

disease In medicine, any departure from health, with impaired functioning of the body as a whole or one or more of its parts. Disease may be: **acute**, producing severe symptoms for a short time; **chronic**, lasting a long time; or **recurrent**, returning periodically. Treatment depends on the cause and course of the disease but, in general, may be **symptomatic** (relieving symptoms, but not necessarily combating a cause) or **specific** (attempting to cure an underlying cause). Disease prevention involves eradication of harmful organisms, IMMUNIZATION, personal hygiene, public health measures, health education and routine medical checks.

disinfectant (germicide) Agent that kills, or inhibits the growth of, bacteria and other microorganisms on inanimate objects. ANTISEPTIC is used on contaminated living tissue. Joseph LISTER introduced carbolic acid (PHENOL) as a medical disinfectant in the 1870s. Today, chlorinated phenols are used in pharmaceutical products. Chlorine and chlorine compounds are commonly used to kill bacteria, especially in water. Iodine is used in food preparation. Alcohol is an effective disinfectant. Ammonium is the most common agent in household cleaners. It also acts as a DETERGENT.

disjunction Logical proposition produced by

joining two simple propositions by the word "or". An example is the proposition "John is intelligent or John is modest"; it is false if both parts are separately false, otherwise it is true. The disjunction is used in mathematics – in which the proposition is true if *either* or *both* components are true. The disjunction of two simple propositions, P or Q, is written $P \vee Q$, and read "P or Q". *See also* CONJUNCTION

disk Form of computer DATA storage. Disks come in many different forms, some using magnetic methods to store data (*see* MAGNETIC DISK, while others use optical systems such as the COMPACT DISC (CD-ROM).

disk operating system (DOS) Computer OPERATING SYSTEM, developed in the early 1980s by Bill GATES' Microsoft Corporation for use with early International Business Machines (IBM) personal computers. DOS is still used by many personal computers worldwide, although it is rapidly being replaced by windows-based operating systems (also developed by Microsoft).

dislocation In medicine, displacement of bones forming a joint, accompanied by damage to supporting ligaments and the enclosing capsule. It gives rise to swelling and loss of function; it is treated by manipulation, usually under a general anaesthetic. In physics, a dislocation is a fault in the CRYSTAL LATTICE.

dispersion, colloidal *See* COLLOID

dispersion, wave *See* WAVE DISPERSION

displacement In geology, relative movements on either side of a FAULT. It incorporates the direction of change and the specific amount of the movement. **Lateral** displacement is described as strike slip and strike separation, whereas **vertical** displacement is known as dip slip and dip separation.

disproportionation Simultaneous OXIDATION-REDUCTION of the same chemical substance. An example is the disproportionation of copper(I) chloride, involving oxidation to copper(II) chloride and reduction to metallic copper:
$2CuCl \rightarrow CuCl_2 + Cu$.

dissection Systematic cutting into plant or animal tissue, usually after death, to explore its anatomy or to discover abnormalities which may help in the understanding of the cause and effect of disease. It is the foundation of a proper understanding of how the healthy body works.

dissociation In chemistry, a reaction in which the molecules of a compound split into smaller components. In many dissociation reactions the smaller components are able to recombine under other conditions; such dissociation is called **reversible** dissociation. For example, heating causes the reversible dissociation of hydrogen iodide into hydrogen and iodine. The formation of IONS (electrically charged particles) when an acid dissolves in water may also be called dissociation. Known as **ionic** dissociation, this reaction explains many properties of electrolytic solutions. The extent to which dissociation proceeds is indicated by the **dissociation constant** (which is equal to the EQUILIBRIUM constant of the reaction).

distance ratio (velocity ratio) In a simple MACHINE, distance moved by the point of application of the effort (input force) divided by the distance moved by the point of application of the load (output force). *See also* FORCE RATIO

distillation Extraction of a liquid by boiling a solution in which it is contained and cooling the vapour so that it condenses and can be collected. The method is used, in the process of DESALINATION, to separate liquids that are in solution or liquid solvents from dissolved solids. Distillation is used to yield drinking water from sea water and to produce alcoholic beverages. **Fractional** distillation, which uses a vertical column for condensation, is the method used in oil refineries to separate the various fractions (portions) of crude oil. *See also* FRACTIONATION

distilled water Pure water prepared by boiling tap water and condensing the steam. It is essential for some chemical reactions, in which even the small traces of dissolved salts (such as the carbonates of calcium and magnesium) and other impurities present in tap water are intolerable. For many applications it is being replaced by deionized water. *See also* WATER SOFTENER

distilling Production of spiritous liquors by DISTILLATION, especially of alcohol (ethanol, ethyl alcohol). In such drinks as wines, FERMENTATION using yeast produces a maximum alcohol content of around 15%. Distillation concentrates the alcohol to a much higher degree in the distilled spirit: further alcohol may be later added as needed for the spirit's final desired proof content. Most spirits (whisky, brandy, gin, rum, vodka and so on) are around 40% proof.

distributive law Rule of combination in mathematics in which an operation applied to a combination of terms is equal to the combination of the operation applied to each individual term. Symbolically this is written $a * (x + y) = a * x + a * y$ where '*' is an operation such as multiplication and a, x and y are numbers. Thus, in arithmetic $3 \times (2 + 1) = (3 \times 2) + (3 \times 1)$: the multiplication is distributive over the addition.

distributor In a motor vehicle, an electrical device for distributing synchronized pulses of high-voltage secondary current from the induction coil, via a rotating contact (rotor arm), to the various sparking plugs of an INTERNAL COMBUSTION ENGINE in their proper firing order.

diuretic Drug used to increase the output of urine. It is used to treat raised blood pressure and oedema, the accumulation of fluid that occurs in some diseases.

diurnal rhythm *See* CIRCADIAN RHYTHM

divergence Mathematical property of an infinite series (or sequence) of not having a finite limiting value. Such a series is said to diverge. The harmonic series, $1 + 1/2 + 1/3 + 1/4 + ...$, is an example of a divergent series since the sum has no finite value. *See also* CONVERGENCE

diving bell Hollow structure providing a dry environment for underwater workers. Early diving-bells were bell-shaped, filled with compressed air, and open at the bottom to give access to the seabed. The BATHYSPHERE, which made its first dive in 1930, was spherical, made of steel, and could withstand considerable pressures, but it was restricted to depths within reach of supply and winch cables. It has been replaced by the BATHYSCAPHE.

division Arithmetical operation signified by ÷ or /, interpreted as the inverse of MULTIPLICATION. The quotient of two numbers, $a \div b$ (or a/b), is the number that must be multiplied by b to give a. a is called the **dividend** and b the **divisor**.

DNA (deoxyribonucleic acid) NUCLEIC ACID that is the major constituent of the CHROMOSOMES of EUKARYOTE cells and some VIRUSES. DNA is often referred to as the "building block" of life since it stores the GENETIC CODE that functions as the basis of HEREDITY. The molecular structure of

DNA FINGERPRINTING

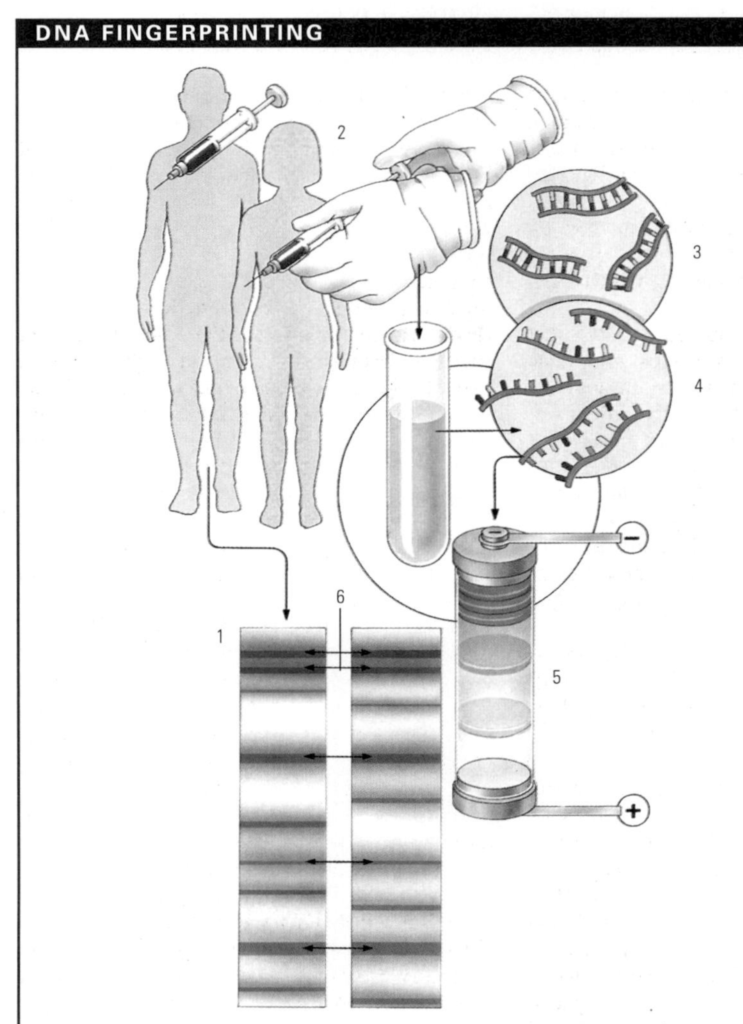

Using a technique known as DNA fingerprinting, a person can be accurately identified. The process allows a person's DNA to be represented in visual form (1). Each DNA pattern is unique (like a fingerprint) – with the exception of identical twins. In a case of disputed paternity, DNA fingerprinting allows the relationship to be settled beyond doubt. DNA is present in all cells, so a sample can be taken from blood (2), skin or even sweat. DNA is separated out (3) and an enzyme that divides DNA is added. The enzyme attacks the minisatellite region between the genes (4). The genes are then sorted by size by an electric field (5). Gel electrophoresis exploits the fact that snippets of DNA carry a charge to force them through a gel. The size of the snippets controls how far they travel, giving a pattern unique to each individual. A child combines DNA from both parents, so will have a partially similar pattern. Paternity is confirmed by the matching marks (6).

A single human cell contains 4m (13ft) of DNA (deoxyribonucleic acid), packed into a nucleus only 5,000ths of millimetre across. In this mass of tangled threads is contained all the information needed to make a human being. DNA directs development and maintains life of an organism by instructing cells to make proteins – the versatile molecules on which all life depends. The CELL's DNA is a vast library of coded commands; the long molecules are packaged into CHROMOSOMES on which GENES are arranged like beads on a string.

Each chromosome is believed to involve more than 100,000 different genes – shorter, functional units of DNA – each representing one of the instructions needed to make and maintain the organism from which it originated. The complete set of genes from a living organism is known as its GENOME, and every cell of that organism carries at least one copy of the basic set.

◄▼ **The structure of DNA** is crucial to its role as the cell's information store (A). The molecule is often called a double helix – a reference to its two spiral "backbones" (1, 2), which are made up of sugar and phosphate units. Linking the two backbones like the rungs of a ladder are the so-called bases (3) – adenine thymine, guanine and cytosine. Each backbone contributes one base to each rung, and the bases are paired according to strict rules – adenine (light blue) always pairs with thymine (dark blue), and cytosine (red) with guanine (yellow). The sequence of bases along one backbone therefore exactly mirrors, or complements, the sequence on the other; when DNA is replicated in cell division, this property makes base mispairing – which may constitute a damaging mutation – less likely. The bonds between paired bases are relatively weak, allowing the DNA molecule to be "unzipped" prior to the processes of replication or transcription.

► **DNA** is permanently locked into the nucleus. But the machinery for protein synthesis is situated in the cytoplasm – outside the cell membrane. DNA communicates with this machinery through a messenger molecule known as RNA. The messenger RNA (mRNA) is chemically similar to DNA itself, but has a single rather than double backbone, and the base uracil takes the place of the DNA's thymine. When a gene is active, the DNA base sequence corresponding to that gene is transcribed into m RNA. Enzymes in the nucleus "read" the base sequence and assemble a complementary strand of mRNA (4) from the base-sugar phosphate subunits (5). When the whole gene has been transcribed into mRNA, the messenger molecule (6) passes into the cytoplasm via pores in the nuclear membrane (7). The mRNA becomes attached to one or more ribosomes (8) – small cytoplasmic particles that are the sites of protein synthesis. A ribosome moves along the mRNA molecule sequentially passing each three-base "word" that specifies a particular amino acid. Another type of RNA, known as transfer RNA (tRNA) (9), then comes into play. This molecule acts as an adaptor between the three-letter words in mRNA and the amino acids that will be joined together to make a protein. At one end of each tRNA molecule is a sequence of three bases (10), complementary to a particular word on the mRNA: at the other end is the amino acid (11) specified by that word. The appropriate tRNAs plug into the mRNA, and the amino acids they carry are linked together by enzymes. As the ribosome passes along the mRNA strand, the protein chain gradually grows in length (12). A typical protein chain made this way may contain a sequence of between 100 and 500 amino acids linked by enzymes.

DNA

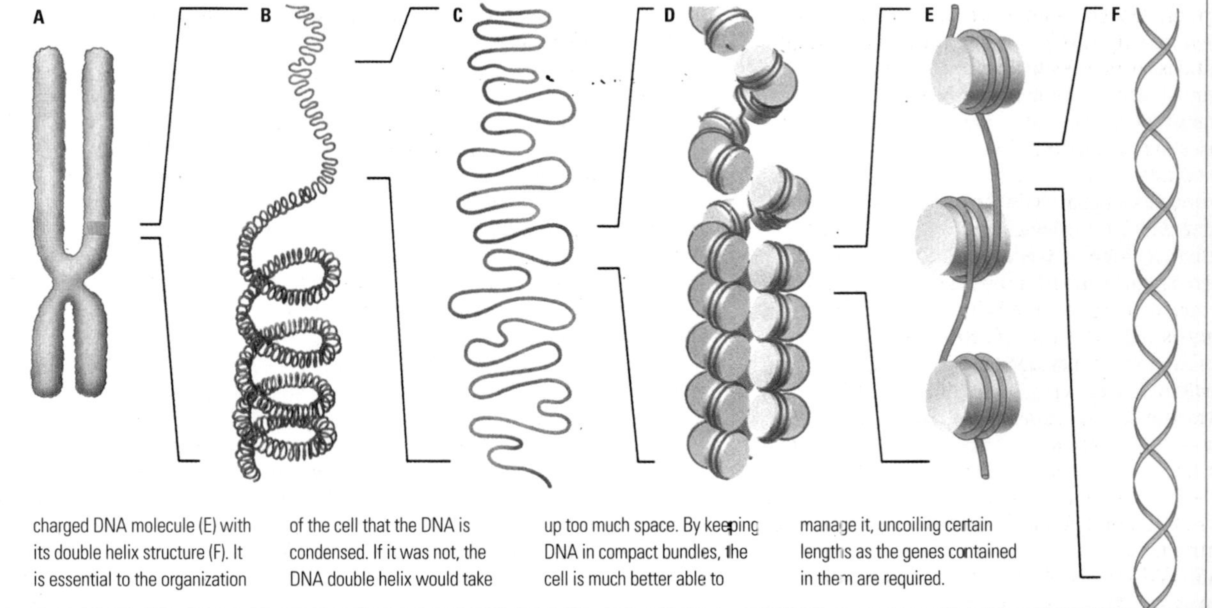

► **Viewed under** a microscope, a chromosome of a dividing cell has a simple cross-like shape (A) that belies the complex way in which it "packages" DNA. Magnifying a small section (B) reveals a tightly coiled strand of chromatin – DNA closely associated with protein. Further enlarging a segment of chromatin (C) shows it to be a tight coil of nucleosomes – bead-like subunits composed of a protein core wrapped by the DNA molecule (D). The protein core is positively charged, allowing it to bind to the negatively charged DNA molecule (E) with its double helix structure (F). It is essential to the organization of the cell that the DNA is condensed. If it was not, the DNA double helix would take up too much space. By keeping DNA in compact bundles, the cell is much better able to manage it, uncoiling certain lengths as the genes contained in them are required.

DNA was first proposed by James WATSON and Francis CRICK in 1953. It consists of a DOUBLE HELIX of two long strands of alternating sugar molecules and phosphate groups linked by nitrogenous bases. The whole molecule is shaped like a twisted rope ladder with the nitrogenous bases forming the rungs. The sugar is deoxyribose, and the four bases are ADENINE (A), CYTOSINE (C), GUANINE (G) and THYMINE (T). The bases are always paired in the same way: adenine always binds with thymine, guanine with cytosine. This regularity ensures accurate self-replication. During replication the two DNA strands separate, each providing a template for the synthesis of a new strand of RNA (MESSENGER RNA). This process of TRANSCRIPTION, mediated by ENZYMES, results in an identical copy of the original helix. The amount of DNA is constant for all cells of a species of animal or plant. In the process of replication the amount of DNA doubles as the chromosomes replicate themselves before MITOSIS; in the gametes, ovum and sperm (HAPLOID cells), the amount is half that of the body cells (*see* MEIOSIS). A base and its associated sugar and phosphate are known as a NUCLEOTIDE; the whole chain is a polynucleotide chain. The genetic code is stored in terms of the sequence of nucleotides: three nucleotides code for one specific AMINO ACID and a series of them constitute a GENE. *See also* BIOTECHNOLOGY; GENETIC ENGINEERING; MUTATION; RECOMBINANT DNA RESEARCH

DNA hybridization Method of comparing DNA from different organisms in order to discover genetic relationships between species. Single strands of DNA from each species are placed together and allowed to react. Some strands will "hybridize" to form double strands (the usual structure of DNA) and the extent to which they do so indicates how many base sequences are complementary. This in turn is a a measure of how similar the genes are.

Dobzhansky, Theodosius (1900–75) US geneticist, b. Russia. An authority on HUMAN EVOLUTION, he was influential in the development of population GENETICS as a separate study. His writings include *Genetics and the Origin of Species* (1937), *Mankind Evolving* (1962) and *Genetics of the Evolutionary Process* (1970).

dock In marine engineering, a wharf, pier, the waterway alongside or an artificial basin where ships are tended. A **wet** dock has watertight gates so that a ship inside does not rise and fall with the tides. A **dry** dock is an enclosure that can be pumped dry. It enables a ship or boat to be repaired below its water line. Floating docks can tend ships at sea and lift them out of the water for repairs.

Doisy, Edward Adelbert (1893–1986) US biochemist. He isolated the female sex hormones oestrone (1929) and oestradiol (1935). Doisy was awarded, with Henrik DAM, the 1943 Nobel Prize for physiology or medicine for his chemical analysis of vitamin K, which Dam discovered in 1934 and Doisy isolated in 1939.

doldrums Region of the ocean near the Equator, characterized by calms, and light and variable winds. It corresponds approximately to the belt of low pressure around the Equator.

dolerite Medium-grained, basic, dark-coloured INTRUSIVE ROCK found in DYKES, SILLS and VOLCANIC PLUGS.

Doll, William Richard (1912–) British physician who first demonstrated the correlation between cigarette smoking and lung CANCER. Working with Professor Bradford Hill (1897–1991), Doll announced his findings in 1950 based on an analysis of statistics showing the incidence of lung cancer in smokers and non-smokers. They later showed that the risks decrease when a smoker stops smoking.

dolomite Carbonate mineral, calcium-magnesium carbonate ($CaMg(CO_3)_2$), found in altered limestones. It is usually colourless or white. A calcite-like rhombohedral class prismatic crystal, dolomite is often found as a gangue mineral in hydrothermal veins, particularly associated with galena and sphalerite. Dolomite is also the name given to a pearly white or pink SEDIMENTARY rock, probably formed by the alteration of limestone by seawater, where 90% or more of the calcite has been replaced by calcium magnesium carbonate. Hardness 3.5–4; r.d. 2.8.

Domagk, Gerhard (1895–1964) German chemist. In 1927 he was made director of the research institute of the I.G. Farben industrial works. Domagk is known for his discovery (1932) of the antibacterial properties of the dye, prontosil. The first SULPHONAMIDE DRUG, sulphanilamide, was synthesized from the dye. Domagk was awarded the 1939 Nobel Prize for physiology or medicine, but was prevented by Nazi decree from accepting it until 1947.

domain In mathematics, a set of values that can be assigned to the independent variable in a FUNCTION or relation; the set of values of the dependent variable is called the **range**. For example, let the function be $y = x^2$, with x restricted to 0, 1, 2, 3, −3 and −5. Then y takes the values 0, 1, 4, 9 and 25 respectively. The domain is {0, 1, 2, 3, −3, −5} and the range is {0, 1, 4, 9, 25}.

domain In certain classification schemes, a category that is recognized as higher than KINGDOM. In these schemes, the two subkingdoms of PROKARYOTAE (ARCHAEBACTERIA and the EUBACTERIA) constitute two domains, called **Archaea** and **Bacteria**, while all other living organisms are included in a third domain, the EUKARYA.

domain *See* MAGNETISM

dominant In genetics, term that describes the ALLELE of a heterozygous pair of alleles that manifests itself over the other. For example, if an offspring has one allele each for brown eyes and blue eyes, the brown-eyed allele will dominate and manifest itself over the RECESSIVE, blue-eyed allele, although he or she will retain the ability to pass on recessive genes to his or her children. *See also* HETEROZYGOTE; HOMOZYGOTE

dongle In computing, device that prevents the illegal use of a commercial COMPUTER PROGRAM. It is usually attached between the computer and its printer (in the printer port). Without the dongle, the particular program – or a pirated copy of it – will not run.

dopamine Chemical normally found in the corpus striatum region of the human BRAIN, where insufficient levels are associated with PARKINSON'S DISEASE. Dopamine is a NEUROTRANSMITTER and a precursor in the production of ADRENALINE and NORADRENALINE. The low levels in Parkinson's disease can be treated by administering the drug L-DOPA.

Doppler, Christian Johann (1803–53) Austrian physicist and mathematician famous for his discovery of the DOPPLER EFFECT. He first described it in a paper on DOUBLE STARS published in 1842. Educated in Salzburg and Vienna, Doppler became professor of physics at Vienna in 1850.

Doppler effect Change in frequency of a wave (such as sound, light or radio) when there is relative motion between the wave source and the observer or detector. The amount of change depends on the velocities of the wave, source and observer. With a sound wave, the effect is demonstrated by the drop in pitch of a vehicle's siren as it passes the observer. With light, the velocity of the source or observer must be large for an appreciable effect to occur, as when light received from a rapidly receding galaxy has its spectral lines shifted towards the red end of the spectrum. *See also* NAVIGATION; RED SHIFT

Doppler radar Essentially a radar navigational aid that relies on the DOPPLER EFFECT – the change in frequency (or wavelength) due to motion of an object relative to an observer. Electronic accessories are used to measure the difference in frequency between the emitted pulse and the pulse received back. This shows the velocity of the target from which the radar beam has been reflected. This information is displayed on the radar screen.

Dorado (Swordfish) Small constellation of the S sky. It has no stars brighter than the third magnitude but it is notable for containing most of the Large Magellanic Cloud (Nubecula Major).

dormancy Temporary state of inaction or reduced METABOLISM. Animals may become dormant by hibernating during winter months when food resources are scarce; dormant plant seeds for a time cease to grow or develop. An organism can return to a fully active state when conditions, such as temperature, moisture or day length, change.

Dornier, Claude (1884–1969) German aircraft manufacturer who designed and manufactured the first all-metal aeroplane. While working with Ferdinand von Zeppelin, the AIRSHIP manufacturer, Dornier founded an aeroplane company and built wooden and metal planes which served in World War 1 (1914–18). In 1929 he designed the DO-X, a 12-engined passenger plane. From the late-1940s, Dornier's company built US-designed aircraft.

dorsal Upper surface of an organism – the surface that is turned away from the ground or other support. In an animal that walks on all fours, the dorsal surface is its back. By extension, in all VERTEBRATES it is the surface along which the spine runs. In animals that walk on two legs, such as humans, it is therefore the surface that faces backwards, or the posterior surface. On FISH, the dorsal fin is the main fin running along the top of the body.

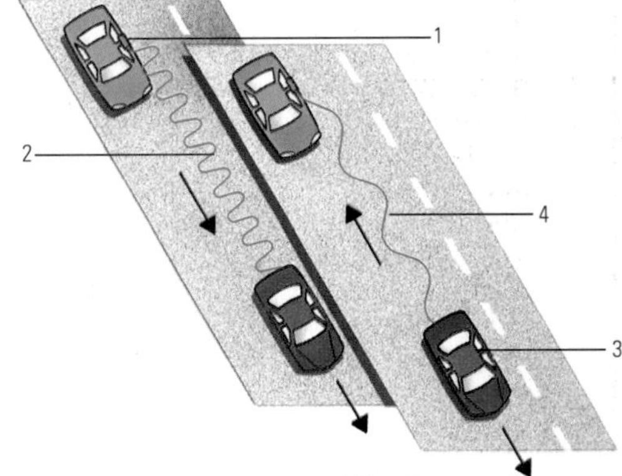

▲ **Doppler radar** The portable radar devices, speed guns (1), used by police to catch speeding motorists use the Doppler effect to measure speed. The "guns" send out a beam of radio waves (2), the frequency of which is known. When the wave strikes a moving car (3) the returning signal (4) has a different frequency. The equipment calculates the difference between the two frequencies and in this way the speed of the motorist is ascertained.

D

DOS Acronym for DISK OPERATING SYSTEM

dosimeter Instrument used for measuring RADIATION dose, usually a pocket ELECTROSCOPE. A quartz fibre, after being charged, is viewed against a scale; its deflection across the scale on discharge gives a measure of the dosage of radiation to which it was exposed.

double galaxy Two adjacent galaxies orbiting around a common centre of gravity, rather like a BINARY STAR. The components are often linked by tenuous bridges of intergalactic matter.

double helix Structure of DNA (deoxyribonucleic acid), the basic store of genetic information in the cells of each living organism. First demonstrated in 1953 by Francis CRICK and James WATSON, the knowledge of the structure of the DNA molecule has since revolutionized biological science.

double star Two stars that appear close together in the sky. There are two types of double star. If the stars are genuinely close together in space and are connected by gravity, they are known as a BINARY STAR. Two stars that are quite distant from each other, but appear close together as a result of chance alignment, are known as an **optical double**.

Down's syndrome (trisomy 21) Chromosomal abnormality giving rise to degrees of mental retardation ranging from mild to severe; the syndrome also brings a decreased life expectancy and may involve physical problems such as heart and respiratory disorders. The syndrome was first described by the British physician John Down (1828–96). It is caused by the presence of an extra copy of CHROMOSOME 21, so can be detected by counting chromosomes in the cells of the fetus during prenatal testing. The risk of having a child with the syndrome increases with maternal age. The condition was originally termed "mongolism" by Down, a reference to the characteristic flat facial appearance and slanting eyes of affected individuals.

Draco (Dragon) Long, winding constellation of the N sky; it is the eighth largest constellation, and extends between Ursa Major and Ursa Minor.

drag Force opposing the motion of a body through a FLUID (gas or liquid). Aircraft, for example, experience drag (or air resistance) as the friction of air over external surfaces and other parts of the aircraft, such as a fixed undercarriage, not associated with the development of lift. To combat drag and therefore make them more efficient, aircraft and cars are designed to be as streamlined as possible. *See also* AERODYNAMICS

drainage basin (catchment area) Region from which all precipitation such as rain and melted snow flows to a single stream or system of streams. By identifying such a basin, scientists can calculate DENUDATION rates and moisture balances from various hydrological measurements such as EVAPORATION rates. The boundary of a drainage basin is called a WATERSHED. The shape of a drainage basin may help in forecasting river flooding.

drainage network Relationship between the sizes, frequency, lengths and patterns of streams within a DRAINAGE BASIN. The drainage network can by analysed in any combination of these variables to help describe the characteristics of a drainage basin.

drainage system In geology, the system by which surface water is collected and removed by streams, rivers and lakes. The pattern of drainage is determined by the type of rock and the slope of the land over which the water flows.

Drew, Charles Richard (1904–50) US physician. His research at the Columbia Medical Center, USA in 1940 led to the discovery that BLOOD plasma could replace whole blood transfusions. During World War 2 he supervised the American Red Cross blood-donor project but was not allowed to donate blood himself because he was Afro-American; his continuing protest led to a change in policy.

drift In geology, gradual change in the land mass, related to soil CREEP. In oceanography it indicates slow, oceanic circulation. Sedimentary drift is a general description of surface debris carried by either a river or glacier. It may also indicate an accumulation such as a snowdrift or sand drift. CONTINENTAL DRIFT is the movement of continents.

drilling rig Platform carrying a derrick and all the associated equipment used for drilling for OIL or natural gas below the seafloor. Most rigs have long legs (usually three or four) located on the sea-bed, with a working platform up to 30m (100ft) above the water. They are generally sited in comparatively shallow water on the continental shelf up to 200km (125mi) offshore. The gas or oil is transported ashore either in tankers or through undersea pipelines. Drilling rigs are sited in the North Sea and off the coasts of Australia, California and the N of South America. *See also* OIL EXTRACTION; OIL WELL

◄ **drilling rig** Heather-A, shown here, is a fixed oil production platform in the North sea, owned by Union Oil. The arm extending out of the picture at the right is a flare boom which burns off excess gas from the well. The drilling tower where the well is made is just visible on the top deck, a grey tower on the right. The disc overhanging the top deck at the left is the helipad. A single crane stretches skyward behind the helipad. It is employed to move heavy goods about the massive platform.

drip-feeding Introduction of liquid nourishment directly into the veins of someone who is unable to eat or drink due to unconsciousness, weakness or illness. Glucose sugar solution is commonly drip-fed; it is held in a sterile container and passes under gravity through a tube which terminates in a vein – usually in the forearm.

drought Condition that occurs when EVAPORATION and TRANSPIRATION exceed PRECIPITATION for long periods. Four kinds are recognized: **permanent**, typical of desert and semi-arid regions; **seasonal**, in climates with well-defined dry and rainy seasons; **unpredictable**, an abnormal failure of expected rainfall; and **invisible**, when even frequent showers do not restore sufficient moisture.

drowned coast (submerged shoreline) Coastal strip that has been submerged under the sea, either because the sea level has risen, or because the land has sunk. Valleys become flooded and hills become islands.

drug In medicine, any substance used to diagnose, prevent or treat DISEASE or aid recovery from injury. Still today many drugs are obtained from natural sources: plants, animals or minerals. Many ANTIBIOTICS, for instance, have been developed from microbes present in the soil. However, since the time of Paul EHRLICH, the German chemist who coined the term "magic bullet", scientists have sought to develop synthetic drugs which will work only on target cells or microorganisms, leaving other tissues unaffected. There is a vast range of drugs in circulation. The most obvious magic bullet effect is that of the antibiotics, which destroy invading bacteria. Some drugs interfere in physiological processes – for instance, anti-coagulants render the blood less prone to clotting. Drugs may be given to make good some deficiency, such as hormone preparations that compensate for an under-active gland. There are a number of routes of drug delivery: orally, by injection, as a lotion or ointment, or by inhalation, pessary or transdermal patch. As well as direct effects, many drugs, too, are associated with side-effects. If these become too problematic, the dosage may have to be adjusted or a different drug substituted. Also, some people have an ALLERGY to particular drugs. Increasingly, new drugs are being developed and prescribed to combat the effects of certain psychological disorders. The use of some drugs on both psychological and physiological conditions can result in a dependance on that drug, a situation doctors are very careful to avoid where possible. Drugs are subject to extensive clinical trials before being licensed for use. The word is also used to describe illegal, or controlled substances, such as COCAINE and ECSTASY (MDMA).

drumlin Smooth, oval-shaped mound of glacial TILL, one end of which is blunt the other tapered. Drumlins usually occur in groups called a "drumlin field" or "drumlin swarm". They are believed to be formed beneath the outer zone of an advancing ice sheet, which deposits and streamlines material. The long axis of a drumlin

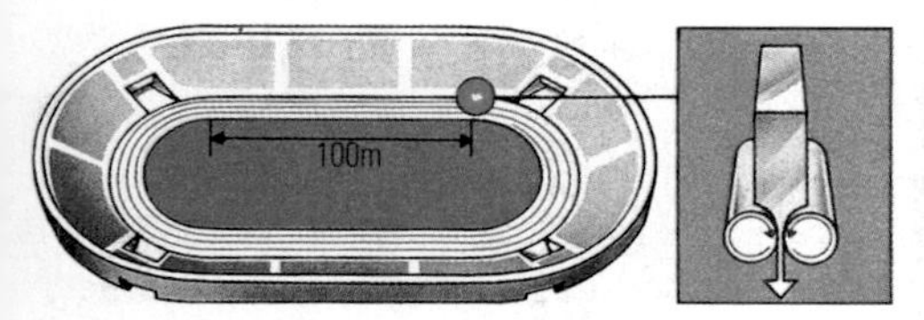

▲ **ductility** Aluminium is a very ductile metal. A 1kg (2.2lb) block of aluminium can be rolled to a foil 0.006mm (0.0002in) thick, 1,220m (4,000ft) long and 5cm (2in) wide. This is more than three times the length of the standard running track shown in the illustration.

lies parallel to the direction of the movement of the GLACIER.

drupe (stone fruit) Any FRUIT with a thin skin, fleshy pulp and hard stone or pip enclosing a single seed. Examples are plums, cherries, peaches, olives, almonds and coconuts.

dry cell Electric BATTERY, a type of small LECLANCHÉ CELL containing no free fluid. In the commonest type the ELECTROLYTE is ammonium chloride paste, the positive ELECTRODE (ANODE) a central carbon rod and the negative electrode (CATHODE) the zinc casing. Dry cells are widely used in torches and portable radios.

dry cleaning Method of cleaning fabrics using special solvent fluids and soaps without water. It was discovered by accident in 1849 by a French tailor, Jolly-Belin, when he noticed the cleaning effect of some spilt turpentine. Introduced into Britain in 1866, it now uses non-flammable solvents in special machines (washing by hand was originally used). Clothes are washed with solvent and dried in a warm-air tumbler. Special solvents are used for some stains.

dry ice Popular name for solid CARBON DIOXIDE.

drying oil Natural oil, such as linseed oil or cottonseed oil, that slowly hardens in air to form a tough film. Drying oils are extensively used in paints, varnishes, lacquers, printing inks and putty. They are UNSATURATED compounds that harden by OXIDATION or POLYMERIZATION.

dubnium (element 104, symbol Db) Synthetic, radioactive, metallic element, the first of the TRANSACTINIDE ELEMENTS. The longest-lived of its 10 ISOTOPES has a half-life of 70 seconds. Dubnium forms a volatile compound with chlorine. The University of California at Berkeley claimed to have synthesized the element in 1969, but the Joint Institute for Nuclear Research in Dubna, Russia, claimed production five years earlier. A name compromise of "unnilquodium" (symbol Unq) proved unappealing. "Dubnium" was adopted in 1995 over the proposed US name of "rutherfordium" (symbol Rf), honouring British physicist Ernest RUTHERFORD, and Russia's proposal of "kurchatovium" (symbol Ku) for Soviet physicist Igor Kurchatov (1903–60). RUTHERFORDIUM has now become element 106. Properties: at.no. 104.

ductility Ability of metals and some other materials to be stretched without being weakened. Copper is said to be a ductile metal as it is easily drawn out to form wire, and silver and gold are even more ductile than copper. The ductility of cast iron varies greatly according to the impurities present.

ductless gland *See* ENDOCRINE SYSTEM

Dulbecco, Renato (1914–) US virologist. He shared the 1975 Nobel Prize for physiology or medicine with David BALTIMORE and Howard TEMIN for his contributions to an understanding of the interaction between tumour viruses and the genetic material of the cell.

Dulong, Pierre Louis (1785–1838) French chemist and physicist. He discovered the explosive nitrogen trichloride. In 1819 he helped to formulate, along with Alexis PETIT, DULONG AND PETIT'S LAW of specific heats, which aided the determination of RELATIVE ATOMIC MASS (r.a.m.).

Dulong and Petit's law Physical rule stating that the product of SPECIFIC HEAT CAPACITY and RELATIVE ATOMIC MASS is approximately equal to 25 (when the specific heat capacity is measured in J mol−1 K−1) for the heavier solid elements. The specific heat capacity of an element was one of the easier parameters to measure during the early days of chemistry, so this law was used to obtain an approximation of the RELATIVE ATOMIC MASS (r.a.m.). The rule was formulated by Pierre DULONG and Alexis PETIT.

dumortierite Silicate mineral, hydrous aluminum borosilicate ($Al_8BSi_3O_{19}[OH]$). It is found scattered in METAMORPHIC ROCKS, and occurs usually in fibrous masses, rarely as orthorhombic crystals. It is glassy to pearly violet or blue. Properties: hardness 7; r.d. 3.3.

dump In computing, information copied from computer memory to an output or storage device. It may be the entire contents of a file copied to another disk, or the contents of the screen being printed out (a **screen** dump).

dune Hill or ridge of wind-blown particles, most often sand. Dunes are found wherever sandy particles carried by the wind are deposited. They occur in a variety of shapes depending on the direction of the wind, whether or not it is constant, and the surrounding landforms. *See also* BARCHAN; SEIF

dunite Coarse-grained IGNEOUS ROCK of colour ranging from light yellowish green to an emerald green, composed almost entirely of OLIVINE. It occurs at Dun Mountain, New Zealand, from which it takes its name.

Dunlop, John Boyd (1840–1921) Scottish inventor and pioneer of the pneumatic TYRE. He was a veterinary surgeon in Belfast until, in 1888, he re-invented the pneumatic tyre (previously patented in 1846 by Robert Thompson (1822–73) but undeveloped). His invention replaced solid tyres.

duodecimal system Number system with 12 as its BASE. It has advantages over the decimal system (base 10) inasmuch as, for example, one-third and two-thirds in duodecimal would be 0.4 and 0.8 respectively – much simpler than the recurring expressions 0.3333... and 0.6666... in the DECIMAL SYSTEM.

duodenum First section of the small INTESTINE, shaped like a horseshoe and part of the DIGESTIVE SYSTEM. The pyloric sphincter, a circular muscle, separates it from the STOMACH. Alkaline BILE and pancreatic juices are released into the duodenum to aid the DIGESTION of food.

Duralumin Alloy of aluminium with copper, magnesium and manganese composed within the following limits: aluminium – over 90%; copper – about 4%; magnesium – 0.5% to 1.5%; manganese – less than 1%.

Dutton, Clarence Edward (1841–1912) US geologist who developed the principle of isostasy, according to which the level of the Earth's crust is determined by its density. The rise of light materials forms continents and mountains; the sinking of heavy materials forms oceans and basins.

Duve, Christian René de (1917–78) Belgian CELL biologist who shared with Albert Claude and George PALADE the 1974 Nobel Prize for physiology or medicine for a detailed description of the structure and function of the cell and its parts. Analysis of biochemical activity in a cell led him to discover the LYSOSOME, an organelle that acts as the "stomach of the cell".

Du Vigneaud, Vincent *See* VIGNEAUD, VINCENT DU

dwarf star Most common type of star in our Galaxy, constituting 90% of its stars, including the Sun. They are also known as main-sequence stars, from their position on the HERTZSPRUNG-RUSSELL DIAGRAM. The term "dwarf" refers to LUMINOSITY rather than size, so dwarfs should be thought of as normal rather than diminutive.

dye Substance, natural or synthetic, used to impart colour to various substances, including textiles, leather, hair, wood and food. Natural dyes have been mostly replaced by synthetic dyes, most of which are derived from COAL TAR. For example alizarin comes from ANTHRACENE and indigo from ANILINE. Dyes are classified according to the way they are applied, which depends on their chemical composition and on the material being coloured. **Direct** dyes can be applied directly to a fabric because they bind to the fibres. **Indirect** dyes require a secondary process to fix the dye in the fabric. Substantive dyes, sulphur dyes and vat dyes act directly, ingrain dyes and mordant dyes act indirectly.

dyke In engineering, a barrier or embankment designed to confine or regulate the flow of water. Dykes are used in reclaiming land from the sea

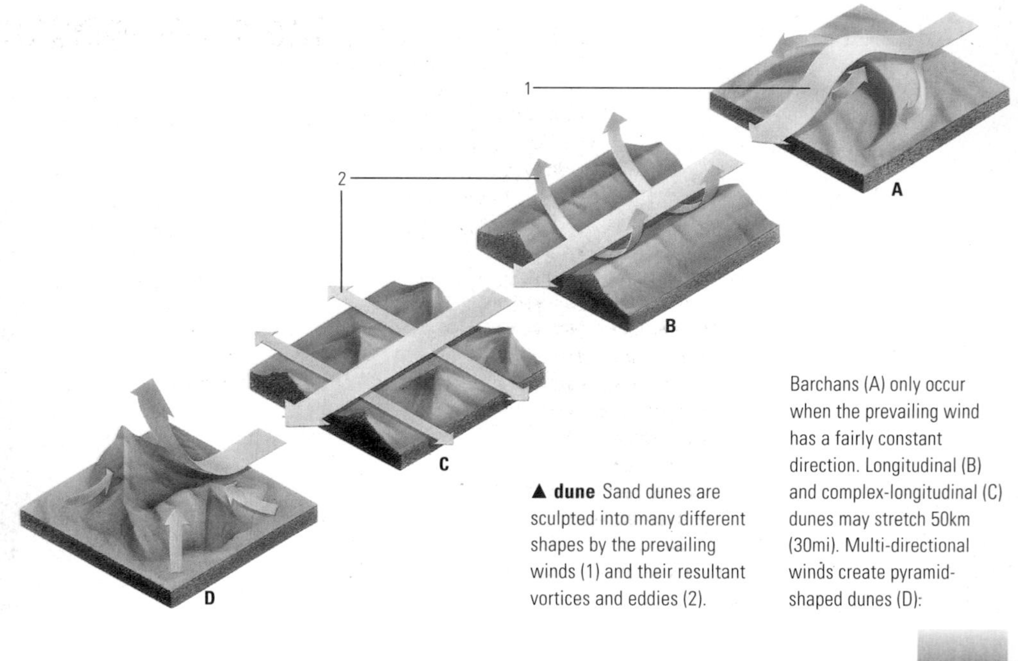

▲ **dune** Sand dunes are sculpted into many different shapes by the prevailing winds (1) and their resultant vortices and eddies (2). Barchans (A) only occur when the prevailing wind has a fairly constant direction. Longitudinal (B) and complex-longitudinal (C) dunes may stretch 50km (30mi). Multi-directional winds create pyramid-shaped dunes (D).

DYSENTERY

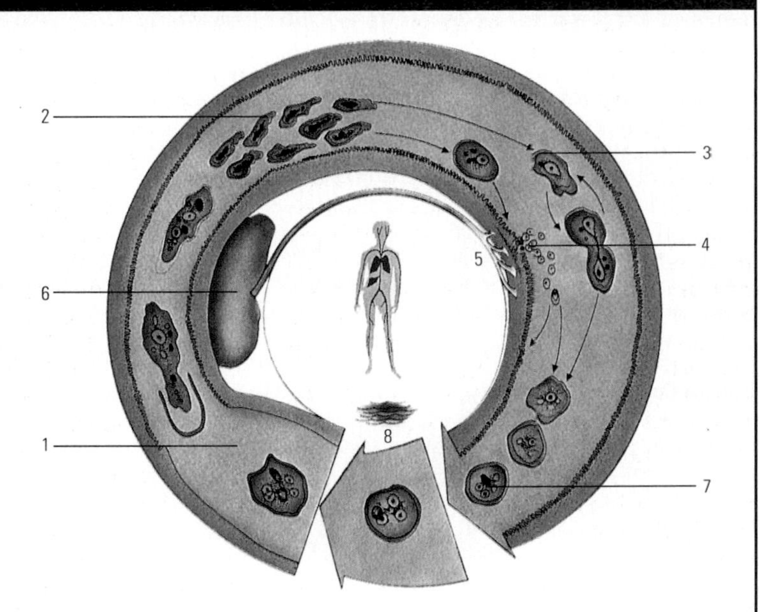

Amoebic dysentery is a widespread disease caused by a microscopic organism (Entamoeba histolytica), found in contaminated water and food. Entamoeba is a natural inhabitant of the gut, however, under certain conditions invasion of the gut wall occurs. Ingested Entamoeba cysts undergo division and multiplication in the large intestine (1). After division, eight trophozoites (feeding protozoa) are produced (2), non-infective trophozoites remain in the intestine (3) feeding on bacteria and food particles. Infective trophozoites invade the gut wall (4), multiply and dissolve away tissues by producing protein-digesting enzymes. If organisms enter the blood stream (5) they can be carried to the lungs (6), liver and brain where abscesses develop. Those released from gut abscesses reinvade tissues or form cysts (7) and are passed in the faeces. The disease is transmitted if flies carry cysts from faeces to food or, more commonly, by drinking contaminated water (8).

by sedimentation (as practised in The Netherlands), and also as controls against river flooding.

dyke (dike) In geology, an INTRUSION of IGNEOUS ROCK forming a seam which cuts across the structure of the surrounding rock, and is often vertical or near-vertical. The cross-cutting nature indicates that a dyke is younger than the rocks it has intruded. At the edge of a dyke the rock is finer-grained than at the centre, as a result of cooling and solidifying more quickly; and this is termed a **chilled margin**. The rock immediately next to the dyke may be altered by CONTACT METAMORPHISM, for example in intruded limestones there may be a thin zone of marble alongside each dyke.

dynamics Branch of MECHANICS that deals with objects in motion. Its two main branches are **kinematics**, which studies motion without regard to its cause, and KINETICS, which also takes into account forces that cause motion. *See also* INERTIA; MOMENTUM; MOTION, LAWS OF

dynamic testing Measurement of the forces exerted by moving bodies on resisting members. Dynamic testing is used to study such things as the effects of cars moving over a bridge and the landing shock on an aircraft's undercarriage.

dynamite Solid, blasting explosive. It contains NITROGLYCERINE incorporated in an absorbent base, such as charcoal or wood-pulp. Shock-resistant but easily detonated by heat or percussion, dynamite is used in mining, quarrying and engineering. Its properties can be varied by adding ammonium nitrate or sodium nitrate. It was invented by Alfred NOBEL in 1866.

dynamo (simple generator) Device that converts mechanical energy into electrical energy by the principle of ELECTROMAGNETIC INDUCTION. A dynamo has a CONDUCTOR, usually an open coil of wire called an armature, that is placed between the poles of a permanent magnet. This armature is rotated within the magnetic field, which causes a current to be induced within the coil. **Alternating-current** (AC) dynamos are usually called ALTERNATORS. *See also* ELECTRIC CURRENT

dynamometer Electrical measuring instrument in which a pointer is deflected as a result of the force exerted between fixed and moving coils, the pointer being attached to the latter. It has applications in AMMETERS, wattmeters and VOLTMETERS. The term is also used for a torque meter, which measures the power of rotating engines.

dyne (symbol dyn) Unit of force in the CGS (centimetre-gram-second) system of units. One dyne is the force that gives a mass of 1g an acceleration of 1cm s^{-2}. One NEWTON equals 100,000 dynes.

dysentery Infectious disease characterized by diarrhoea, with bleeding and abdominal cramps. It is spread in contaminated food and water, especially in the tropics. There are two types: **bacillary** dysentery, caused by BACTERIA of the genus *Shigella*; and **amoebic** dysentery, caused by a type of PROTOZOAN. Both forms are treated with antibacterials and fluid replacement.

dysprosium (symbol Dy) Silver-white, metallic element of the LANTHANIDE SERIES. Its chief ores are monazite and bastnaesite. Its capacity to absorb neutrons makes it important in NUCLEAR REACTOR control systems; dysprosium compounds are also used in lasers. Properties: at.no. 66; r.a.m. 162.5; r.d. 8.54; m.p. 1,409°C (2,568°F); b.p. 2,335°C (4,235°F); most common isotope ^{164}Dy (28.18%).

dystrophy, muscular Any of a group of inherited disorders featuring weakness or wasting of muscles; the muscle fibres degenerate, to be replaced by fatty tissue. The commonest form, **Duchenne** muscular dystrophy, is almost entirely confined to boys, developing usually before the age of four. There is no cure and death is usual before the age of 20.

e Irrational number that is the base of natural, or Napierian, LOGARITHMS. e is a real decimal number without end equal to 2.7182818284590.... e is the limit of the expression $(1 + 1/n)^n$ as the integer n tends to infinity. In fact the number is transcendental, since there is no ALGEBRAIC EQUATION with e as the solution. e also occurs in EXPONENTIAL functions such as e^x. *See also* NUMBER, IRRATIONAL; NUMBER, TRANSCENDENTAL

ear Organ of HEARING and BALANCE. It converts sound waves to nerve impulses which are carried to the BRAIN. In most mammals it consists of the outer, middle and inner ear. The **outer** ear carries sound to the eardrum. The **middle** ear is air-filled, and has three tiny bones (OSSICLES) that pass on and amplify sound vibrations to the fluid-filled, inner ear. The **inner** ear contains the COCHLEA. Vibrations stimulate tiny hairs in these organs which cause impulses to be sent via the AUDITORY NERVE to the brain. The inner ear also contains SEMICIRCULAR CANALS, which maintain orientation and balance.

Early Bird Satellite First commercial communications SATELLITE, launched on 6 April 1965. A stationary satellite, it was in a SYNCHRONOUS ORBIT so that it remained always over the same place, *c*.35,400km (22,000mi) above the mid-Atlantic. It could relay 240 telephone conversations simultaneously between North America and Europe.

Earth Third major planet from the Sun, and the largest of the four inner, or terrestrial, planets. From space, the Earth is predominantly the blue of the oceans, plus the browns and greens of its land masses, the white polar caps, and a continually changing pattern of white cloud. Some 70% of the surface is covered by water, and it is this and the Earth's average surface temperature of 13°C (55°F) that make it suitable for life. The continental land masses make up the other 30%. Our planet has one natural satellite, the MOON. Monitoring the propagation of SEISMIC WAVES from earthquakes has revealed the Earth's internal structure. Like all the terrestrial planets, there is a dense core rich in iron and nickel, surrounded by a mantle consisting of silicate rocks. The thin, outermost layer of lighter rock is called the CRUST. The inner core rotates at a different rate from the solid outer layers, and this, together with currents in the molten outer core, gives rise to the Earth's MAGNETIC FIELD. The Earth's ATMOSPHERE is divided into a number of layers according to the way in which its temperature varies with altitude. The TROPOSPHERE contains most of the atmosphere, and is where lifeforms are found and weather systems operate. The STRATOSPHERE contains the OZONE LAYER, which absorbs the high-energy, ULTRAVIOLET RADIATION from the Sun that is harmful to life. METEORS occur in the MESOSPHERE, and AURORAE in the THERMOSPHERE. The thermosphere is extremely rarefied, and its high temperature indicates the high kinetic energy of its molecules, rather than its heat content. Ionized atoms and molecules in the mesosphere and thermosphere constitute the IONOSPHERE. Above the thermosphere is the EXOSPHERE, which contains the Earth's MAGNETOSPHERE and VAN ALLEN RADIATION BELTS, and merges into interplanetary space. *See feature on page 116*

earthenware Vessels, other utensils and ornaments made of clay. The material is fired at relatively low temperatures, resulting in porous, opaque pieces. Such items can be made waterproof by glazing the surface.

earthquake Tremor below the surface of the Earth which causes shaking to occur in the CRUST. Shaking lasts for only a few seconds, but widespread devastation can result. According to PLATE TECTONICS, earthquakes are caused by the movement of crustal plates, which produces FAULT lines. The main earthquake regions are found along plate margins, especially on the edges of the Pacific, such as the San Andreas fault. When the shock takes place, three different SEISMIC WAVES are created: primary/push (P), secondary/shake (S) and longitudinal/surface (L). P and S waves originate from the seismic FOCUS (point of origin), up to 690km (430mi) deep. They travel to the surface and cause shaking. On the Earth's surface, they travel as L waves. The surface point directly above the seismic focus is the EPICENTRE, around which most damage is concentrated. A large earthquake is usually followed by smaller "aftershocks". An earthquake beneath the sea can often result in a TSUNAMI. Earthquake prediction is a branch of SEISMOLOGY. Present methods indicate only a probability of earthquake activity, and cannot be used to predict actual events. The world's largest recorded earthquake (1976) at Tangshan, China, killed over 250,000 people and measured 8.2 on the RICHTER SCALE. *See illustration on page 117*

earthquake engineering Term used to describe the designing of structures such as buildings, dams and roads to minimize damage to them during earthquakes. Constructions over FAULT lines are avoided and cushions of concrete or polymeric materials are used as foundations to help buildings float or slide during a earthquake without breaking in areas where strain build-up is evident.

earth sciences General term used to describe all the sciences concerned with the structure, age, composition and atmosphere of the Earth. It includes the basic subject of GEOLOGY, with its sub-classifications of GEOCHEMISTRY, GEOMORPHOLOGY, GEOPHYSICS; MINERALOGY and PETROLOGY; SEISMOLOGY and VOLCANISM; OCEANOGRAPHY; METEOROLOGY; and PALAEONTOLOGY.

Earthshine (Earthlight) Phenomenon most easily observed during the crescent phase of the Moon, when the darkened portion of the lunar disc is illuminated by an ashen light reflected onto it by the Earth.

Earth Summit United Nations Conference on Environment and Development, popularly known as the "Earth Summit", held in Rio de Janeiro, Brazil, in June 1992. Reportedly the largest gathering of world leaders ever held, it was the first serious global acknowledgement of the problems created by climate change and the impact of 20th-century industrial society on the environment. The Rio Declaration laid down principles of environmentally sound development, a "blueprint for action", and imposed limits on the emission of the "greenhouse gases" causing the GREENHOUSE EFFECT.

East Pacific Rise Part of the mid-oceanic ridge and a region of seismic activity in the E Pacific Ocean. It is also known as the Albatross Cordillera. Shallow-focus earthquakes occur along the crest of the rise, which is an area of SEAFLOOR SPREADING.

ebola Virus that causes a haemorrhagic fever. Thought to have been long present in animals, the Ebola virus was first identified in humans during an outbreak in Zaïre in 1976. Following an incubation period of between two and 21 days, the disease begins with a severe headache, muscle pains and high fever. A rash develops and there is severe diarrhoea and vomiting, with chest pain and abdominal cramps; haemorrhaging occurs from internal organs. Ebola is acquired through contact with contaminated body fluids. There is no specific treatment and the death rate can be as high as 90%.

Eccles, Sir John Carew (1903–1997) Australian physiologist. He made fundamental discoveries concerning the interaction of NEURONS (nerve cells) in the transmission of a nerve impulse. He shared the 1963 Nobel Prize for physiology or medicine with Alan HODGKIN and Andrew HUXLEY.

EAR

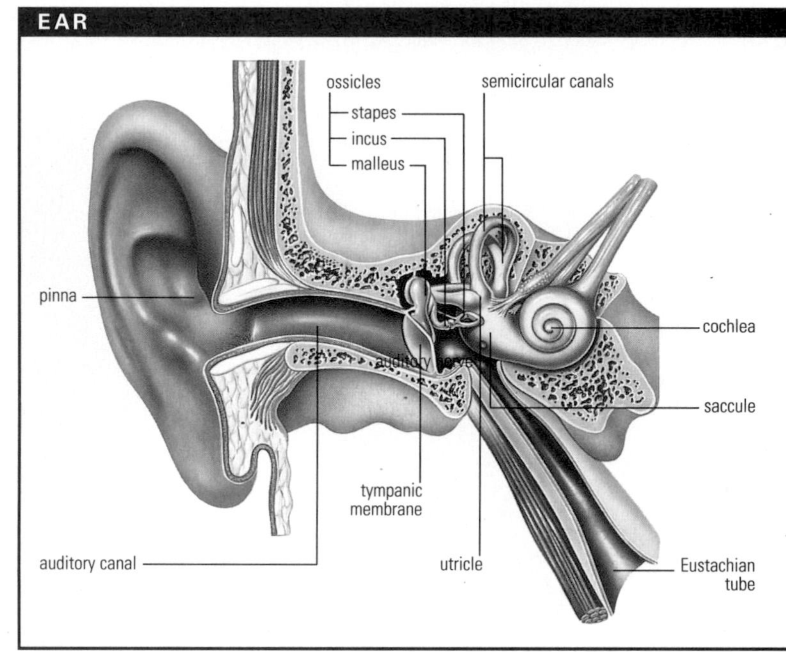

The ear is divided into three parts – the outer, middle and inner ear. The outer consists of the pinna and the auditory canal. The pinna funnels sound waves via the canal to the ear drum, tympanic membrane, of the middle ear. The sound waves are amplified and transmitted by tiny bones, the ossicles, which cause the oval window to vibrate. This sets the fluids of the inner ear in motion. Hair cells in structures of the inner ear, the cochlea and semicircular canals, are stimulated and generate impulses interpreted by the brain as sound.

The planet Earth probably began in the coming together (ACCRETION) of dust and gas particles around the newly formed Sun *c*.1,000 million years after the BIG BANG, which is thought to have occurred some 20,000 million years ago.

Extraterrestrial materials therefore provide a good way of investigating the Earth's internal structure. Meteorites are thought to be fragmented remains of planet-like bodies only a few tens of kilometres across. The different types of metorite would then represent different layers in the planetoid structure. Iron meteorites (siderites) are *c*.97% metal, mainly nickel-iron, and being so dense would probably represent the Earth's core. Chondrite meteorites are of silcate composition and are thought to represent the lower mantle. Finally, achondrites are also silica based, but with different composition. This type of meteorite may represent fragments of crust and upper mantle.

▼ The Earth's crust (A) may be up to 40km (25mi) thick under continents (B), or only 5km (3mi) thick under the sea (C). The crust and the very top of the mantle (A) form the lithosphere (1) which drifts on the plastic asthenosphere (2). The upper mantle (3) stretches down to overlie the lower mantle (4) at 700km (430mi) depth. From the surface the temperature inside the Earth increases by 30°C/km (85°F/mi) so that the asthenosphere is close to melting point. After 100km (60mi) depth the rate of temperature increase slows dramatically. Because we know so little about what happens deep within the Earth, there is a great uncertainty about the temperature of the core. Pressure also increases with depth: already in the asthenosphere the pressure is equal to 250,000 atmospheres. At the Earth's core it may go as high as 4 million atmospheres. The density near the surface of the Earth is only about 4g cm^{-3}, and this does not vary by much more than 30 per cent right down through the lower mantle. The increase in density that does occur is due to closer atomic packing under pressure. There is a leap in density to 10g cm^{-3} in passing to the outer core (5) and a further leap to between 13 and 16g cm^{-3} in the inner core (6). The heat at the core fuels vast convection currents (7) in the material of the mantle. Continental crust is made by many different processes, and is therefore difficult to generalize. A typical cross section (B) might well consist of deformed and metamorphosed sedimentary rocks at the surface (1), underlain by a granite intrusion (2). The remaining crust would consist of metamorphosed sedimentary rock and igneous rock (3) reaching to the mantle (4). In oceanic crust (C), ocean sediments (1) overlie a basalt which is pillowed and also chemically altered by sea water. Below this layer (2) the basalt is made up of dykes – small vertical intrusions that are packed so close together they intrude each other. Gabbros are coarse-grained igneous rocks with a very similar composition to basalt. They are sometimes layered.

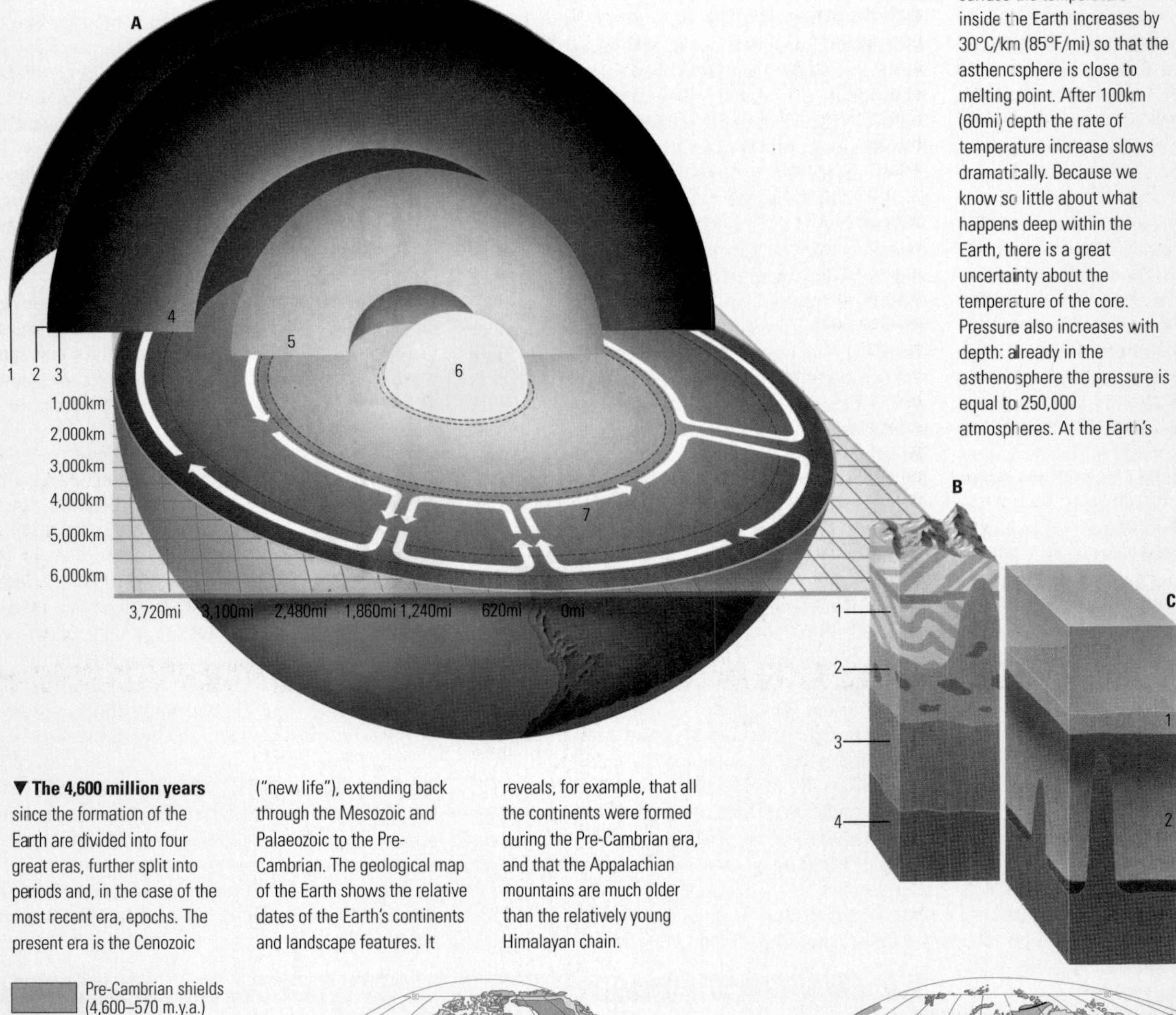

▼ The 4,600 million years since the formation of the Earth are divided into four great eras, further split into periods and, in the case of the most recent era, epochs. The present era is the Cenozoic ("new life"), extending back through the Mesozoic and Palaeozoic to the Pre-Cambrian. The geological map of the Earth shows the relative dates of the Earth's continents and landscape features. It reveals, for example, that all the continents were formed during the Pre-Cambrian era, and that the Appalachian mountains are much older than the relatively young Himalayan chain.

Pre-Cambrian shields (4,600–570 m.y.a.)
Sedimentary cover on Pre-Cambrian shields
Palaeozoic folding (570–225 m.y.a.)
Sedimentary cover on Palaeozoic folding
Mesozoic folding (225–65 m.y.a.)
Sedimentary cover on Mesozoic folding
Cenozoic folding (65–2 m.y.a.)
Sedimentary cover on Cenozoic folding
Intensive Mesozoic and Cenozoic volcanism
Principal faults
Oceanic marginal troughs
Midoceanic ridges
Overthrust faults

Canadian Shield
Rocky Mountains
Appalachians
Sierra Madre
East Pacific Ridge
Northern Mid-Atlantic Ridge
Guiana Shield
Amazonian Shield
Andes
Southern Mid-Atlantic Ridge
Pacific-Antarctic Ridge
Tropic of Cancer
Equator
Tropic of Capricorn
Baltic Shield
Urals
Alps
Atlas
Angara Shield
Altai
Tien Shan
Chinese Shield
Kunlun Shan
Hindu Kush
Himalayas
Zagros
Arabian Shield
Great Rift Valley
Ethiopian Shield
Indian Shield
Carlsberg Ridge
Atlantic - Indian Ridge
Mid-Indian Ridge
Australian Shield
Great Divide

eccentricity (symbol *e*) One of the elements of an ORBIT. It indicates how much an elliptical orbit departs from a circle. The eccentricity is found by dividing the distance between the two foci of the ellipse by the length of the major axis. A circle has an eccentricity of 0, and a PARABOLA has an eccentricity of 1.

eccrine gland One of two types of SWEAT GLAND in mammals, the other is the APOCRINE GLAND, both are found within layers of the SKIN. In humans and some apes, eccrine glands are the more numerous and are distributed generally over the body, not associated only with hair, as are apocrine glands.

ecdysone Hormone that stimulates MOULTING in some animals. It is a steroid hormone produced by CRUSTACEANS and INSECTS. By acting at particular sites on genes, it stimulates the production of the proteins that are involved in moulting (**ecdysis**) and METAMORPHOSIS.

echinoderm Any member of the phylum Echinodermata, a group of spiny-skinned, marine invertebrate animals, which includes sea urchins, sea cucumbers and starfish. Radially symmetrical with five axes, they have a skeleton of calcareous plates in their skin. Their hollow body cavity includes a complex, internal fluid-pumping system and tube feet. They reproduce sexually, producing bilaterally symmetrical larvae resembling that of CHORDATES; REGENERATION also occurs.

echo Portion of a SOUND or RADAR wave reflected from a surface so that it returns to the source and is heard or observed after a short interval. Sound echoes are often heard as the distorted repetition of the sound in a large empty room. High notes provide a better echo than low notes. Echoes are utilized in ECHOLOCATION, which is used by bats and whales and by navigational equipment. Auditoriums are designed to minimize echoes; they are eliminated by using absorbent material on the walls and by avoiding curved surfaces that can focus the echoes. *See also* ACOUSTICS; ANECHOIC CHAMBER

echocardiography Use of pulses of ultrasound to examine the functioning of the heart. This painless, non-intrusive technique produces a computerized image on a screen. It can detect any structural fault or malfunction that may be linked to heart disease, particularly congenital disorders.

echolocation In animals, system of navigation used principally by whales and bats. The animal emits a series of short, high-frequency sounds, and from the returning ECHO it gauges its environment. Bats and dolphins also use the system for hunting. *See also* SONAR

echo sounder Device that sends ULTRASONIC sound pulses through water and detects their reflection – from the sea bottom, shoals of fish or any submerged object. Continuous chart recordings can provide a profile of the sea bottom. *See also* SONAR

eclampsia Occurrence of convulsions not associated with any other disease, such as epilepsy, in a pregnant woman with abnormally high blood pressure and fluid retention (pre-eclampsia, toxaemia of pregnancy). The condition, which threatens the lives of mother and baby, must be treated immediately or coma and death will follow. Delivery may be hastened or Caesarean section performed. The cause of eclampsia is unknown.

eclipse Astronomical phenomenon that occurs when a celestial body temporarily obscures another, as seen from the Earth. The most familiar kinds are **solar** and **lunar** eclipses. A solar eclipse happens when the Moon passes between the Earth and the Sun so that the Sun's light is blocked from the part of the Earth on which the Moon's shadow, or **umbra**, falls. A lunar eclipse is caused by the Earth when it moves between Sun and Moon so that the Moon passes into the Earth's umbra, and cannot shine by reflected sun-

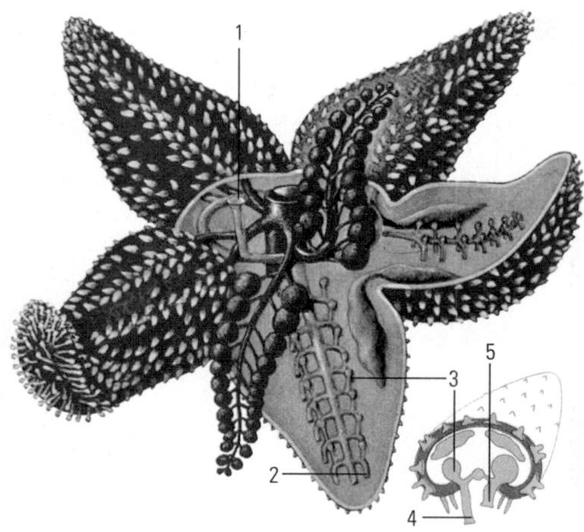

▲ **echinoderm** A starfish (an echinoderm) feeds by surrounding its prey, eventing its stomach through its mouth and partially digesting the food, which is then taken back into the stomach extensions (red). It moves by means of a water-vascular system (blue) unique to echinoderms. Water enters through the sieve plate (1) and is drawn by tiny hairs through the five radial canals into the many pairs of tube feet (2) armed with suckers. When the ampulla (3) of each tube foot contracts, water is forced into the foot (illustrated in cross-section), which extends (4) and allows attachment to the hard rock. Muscles in the foot then shorten it (5) forcing water back into the ampulla and drawing the animal forward.

EARTHQUAKE

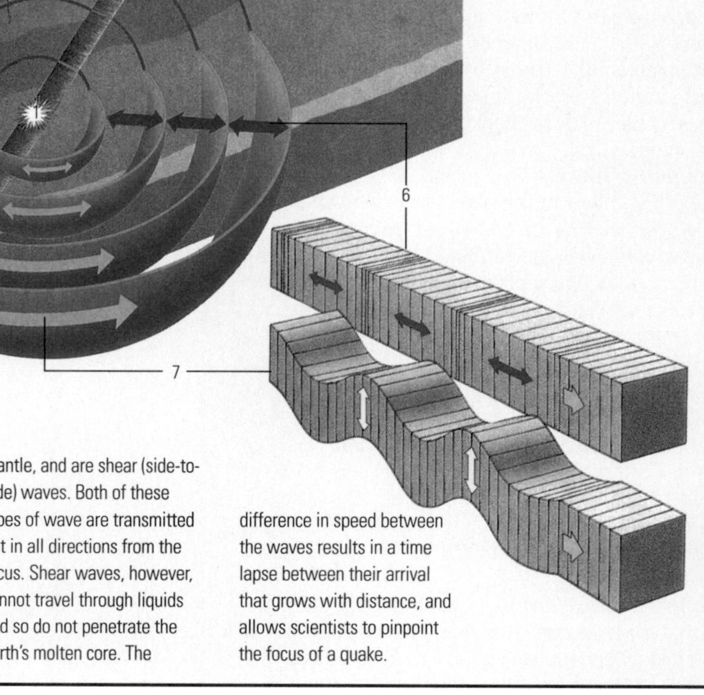

The focus of an earthquake – the point of rupture (1) – may be at the Earth's surface or up to 700km (450mi) below. The epicentre of the earthquake (2) is at ground level directly above. Most damage is done by earthquakes that occur at a depth of 10km (6mi) or less. Normally the longer the interval between movements on an active fault line, the greater the eventual shock. The San Andreas fault in California is a 1,200km (750mi) boundary between the Pacific (3) and the North American (4) tectonic plates. Although the Pacific plate is heading n–w at an average 6cm (2.5in) a year, most of its movement consists of sudden jumps. Earthquake science (seismology) is not advanced enough to predict such jumps accurately. In the 1989 Loma Prieta quake, the focus was 18km (11mi) below the surface. Although the Pacific plate slipped 2m (7ft) N-W and rode 1m (3.3ft) upwards on the North American plate, most of the energy of the quake was absorbed below ground, so there was only comparatively minor surface cracking (5). Primary, or P, waves (6), are compression-dilation (back-and-forth) waves, like sound waves. These travel through typical crustal rocks at approximately 5km/s (3mi/s). Secondary, or S, waves (7) travel more slowly, at approximately 3km/s (2mi/s), or 7km/s (4mi/s) deeper in the mantle, and are shear (side-to-side) waves. Both of these types of wave are transmitted out in all directions from the focus. Shear waves, however, cannot travel through liquids and so do not penetrate the Earth's molten core. The difference in speed between the waves results in a time lapse between their arrival that grows with distance, and allows scientists to pinpoint the focus of a quake.

ECLIPSE

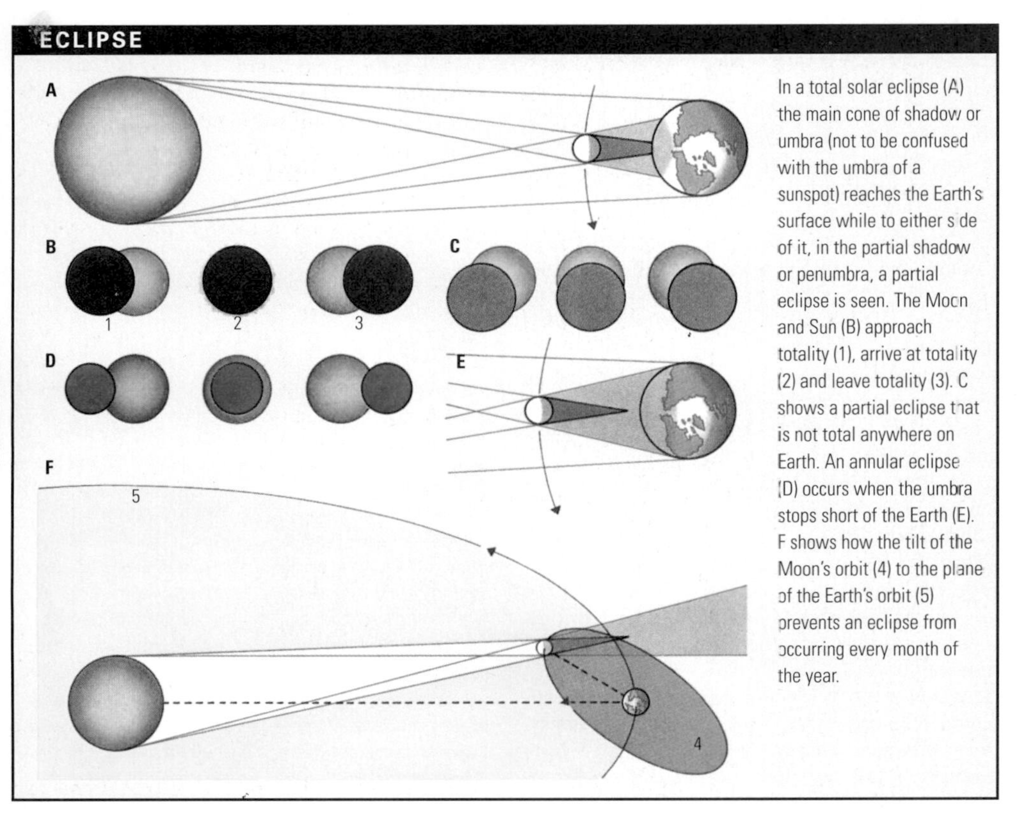

In a total solar eclipse (A) the main cone of shadow or umbra (not to be confused with the umbra of a sunspot) reaches the Earth's surface while to either side of it, in the partial shadow or penumbra, a partial eclipse is seen. The Moon and Sun (B) approach totality (1), arrive at totality (2) and leave totality (3). C shows a partial eclipse that is not total anywhere on Earth. An annular eclipse (D) occurs when the umbra stops short of the Earth (E). F shows how the tilt of the Moon's orbit (4) to the plane of the Earth's orbit (5) prevents an eclipse from occurring every month of the year.

light. Both kinds of eclipses can be **total**, when Earth or Moon pass completely into the other's umbra; or **partial**, when Earth or Moon remain partly in the other's partial shadow, or **penumbra**. If a solar eclipse happens when the Moon is at APOGEE (furthest point from Earth), its apparent size is less than the Sun's, and an **annular**, or ring, eclipse results, with the Sun appearing as a bright ring around the dark Moon. If Earth and Moon's orbits were in the same plane, solar and lunar eclipses would be monthly, occurring at new and full moons. But because the Moon's orbit is at an angle to Earth's, eclipses are possible only when the Moon crosses Earth's orbital plane and is in line with Earth and Sun. These conditions happen seven times a year at the most. To early peoples eclipses, especially of the Sun, were awe-inspiring events. It is not surprising that the knowledge that eclipses can be predicted added greatly to the authority of its discoverers, the astronomer-priests of Chaldea (Mesopotamia). Today solar eclipses enable astronomers to make studies that are not normally possible, such as checking Einstein's theory of RELATIVITY.

eclipsing binary Type of BINARY STAR in which the orbital plane of the two stars is viewed almost edge-on, leading to mutual eclipses and consequent variations in combined light output. Two eclipses may be expected during an orbital cycle, one usually causing a bigger drop in light than the other since the two stars are rarely equal in size or brightness.

ecliptic GREAT CIRCLE on the CELESTIAL SPHERE, inclined at 23.5° to the CELESTIAL EQUATOR. The ecliptic is the yearly path of the Sun as seen from Earth or the Earth's orbit as seen from the Sun. The plane of the ecliptic thus passes through the centre of both the Earth and the Sun. The planets (except PLUTO) are always located near the ecliptic.

ecliptic coordinate system Astronomical COORDINATE SYSTEM that refers to the plane of the ECLIPTIC. The coordinates used are celestial latitude and longitude. Celestial latitude is the angular distance between the object and the ecliptic, measured at right angles to the ecliptic. It is measured from 0° to 90°, positively towards the north ecliptic pole and negatively towards the south ecliptic pole. Celestial longitude is given by the angle between the object and the vernal EQUINOX, measured along the ecliptic from 0° to 360° E from the vernal equinox.

eclogite Coarse- to medium-grained METAMORPHIC ROCK consisting mainly of green pyroxine and red garnet. A very dense rock, it is formed deep in the Earth's crust under conditions of very high temperature and pressure. It occurs in SCHISTS (Norway) and in diamond-bearing pipes of kimberlite (South Africa).

ecology Biological study of relationships of organisms to their ENVIRONMENT and to one another. The term was coined (1866) by Ernst HAECKEL. Ecologists study **populations** (groups of individual organisms), **communities** (different organisms sharing the same environment), or ECOSYSTEMS (a community and its physical environment). The largest population that can be sustained by a particular environment's resources is called its **carrying capacity**. The role of a species within its community is termed its **ecological niche**. Within the BIOSPHERE, natural cycles (CARBON CYCLE, HYDROLOGICAL CYCLE, NITROGEN CYCLE and OXYGEN CYCLE) are assisted when the biological diversity of species fill these various ecological niches. This diversity produces stable or CLIMAX COMMUNITIES. If a climax community is extensive and well defined it is called a BIOME. **Applied ecology** is the practical management and preservation of natural resources and environments.

ecosphere Similar to BIOSPHERE, the total of all the ECOSYSTEMS of the Earth. The term BIOSPHERE is used solely to indicate the zone of life, but the term ecosphere includes the interaction of living organisms with their ENVIRONMENT.

ecosystem Basic unit in ECOLOGY, consisting of a community of organisms in a physical ENVIRONMENT. Study of these systems is based often on energy flow. The sun provides the energy that **primary producers**, such as green plants, convert into food by PHOTOSYNTHESIS. Plants also need ABIOTIC (non-living) substances from the water and soil to grow. In a typical FOOD CHAIN, herbivores or PRIMARY CONSUMERS, such as rabbits, eat the plants and in turn serve as food for SECONDARY CONSUMERS, such as foxes. **Decomposers**, such as bacteria, break down dead organic matter into simple nutrients completing the food chain. The chemicals necessary for life are recycled by the processes of the CARBON CYCLE, HYDROLOGICAL CYCLE, NITROGEN CYCLE and OXYGEN CYCLE. Interference with these finely balanced natural processes, such as POLLUTION, climate change or the loss of a species, can disrupt the entire ecosystem.

ECTOPIC PREGNANCY

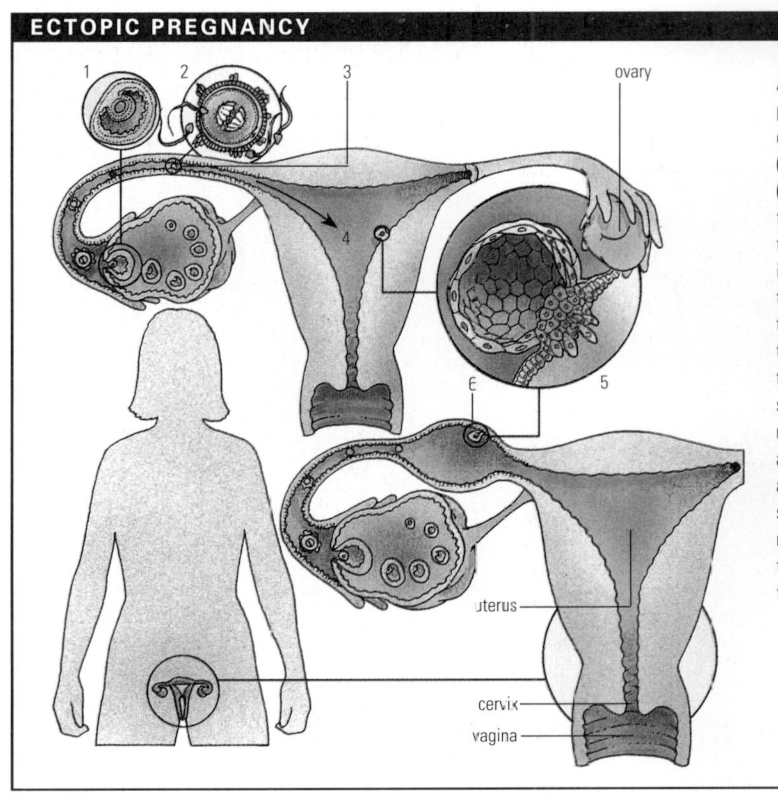

An ectopic or extrauterine pregnancy occurs when the egg released from the ovary (1) and fertilized by sperm (2) in the Fallopian tube (3) fails to pass down into the top of the uterus (4), where it would normally implant in the wall of the womb (5). If the zygote does implant in the wall of the Fallopian tube (6), there is not the space or the blood supply to nurture a fetus to full term, and the pregnancy usually aborts spontaneously. If not surgery is necessary to remove the fetus and save the mother from potentially fatal haemorrhaging.

▲ **Edison** This photograph shows the US inventor Thomas Edison at the age of 31. In 1869 Edison invented a paper tape "ticker" for sending stock prices across the country, and sold it for $30,000. He used this money to establish an industrial research laboratory, where he spent the rest of his life. His more notable inventions include the carbon granule microphone, the phonograph (seen here) and the electric light bulb. He also enriched physics by discovering that electricity flows from a heated filament to a nearby electrode, but not in reverse (the Edison effect).

ecotone Transitional zone between two neighbouring HABITATS.

ecotype In ecology, variety of a species that has special inherited characteristics that allow it to thrive in a particular habitat. In botany, variety of a plant species the distinguishing characteristics of which are mainly inherited rather than resulting from environmental pressure.

ecstasy (MDMA) (3,4-methylenedioxymethylamphetamine) An AMPHETAMINE-based drug, which raises body temperature and blood pressure by inducing the release of adrenaline and targeting the neurotransmitter, serotonin. Users experience short-term feelings of euphoria, rushes of energy and increased tactility. Withdrawal can involve bouts of DEPRESSION and INSOMNIA. In the 1980s taking of ecstasy in tablet form was widespread in club culture in Europe and North America. Some deaths have resulted from use of ecstasy.

ECT Abbreviation of ELECTROCONVULSIVE THERAPY

ectoderm One of three so-called GERM LAYERS of tissue formed in the early development of a fertilized OVUM (egg). It is the outer layer of the BLASTULA and later develops, in most animals, into a skin or shell, a nervous system, lining tissue, parts of some sense organs, and various miscellaneous tissues. Other germ layers are the ENDODERM and MESODERM.

ectoparasite *See* PARASITE

ectopic Occurrence of a PREGNANCY outside the UTERUS – in the FALLOPIAN TUBE or elsewhere. The embryo cannot develop normally and spontaneous ABORTION often occurs. If not, urgent surgery is necessary to save the mother from serious, possibly fatal haemorrhage.

ectoplasm In biology, outer layer of the CYTOPLASM of a CELL; it is usually semi-solid and transparent.

Eddington, Sir Arthur Stanley (1882–1944) English astronomer and physicist. Eddington pioneered the use of atomic theory to study the internal constitution of stars, and explained the role of radiation pressure in preventing stars from collapsing under gravity. Among his discoveries were the mass-luminosity relationship and the degeneration of matter by WHITE DWARFS. Eddington helped popularize Einstein's theory of RELATIVITY, and in 1919 obtained experimental proof of the general theory that gravity bends light by measuring stars close to the Sun during a solar eclipse.

eddy current Circulatory electric current induced in a CONDUCTOR when subjected to a varying magnetic field. Eddy currents cause a loss of energy in AC GENERATORS and MOTORS: the reaction between the eddy currents in a moving conductor and the field in which it moves retards the motion of the conductor.

Edelman, Gerald Maurice (1929–) US molecular biologist. He shared with Rodney PORTER the 1972 Nobel Prize for physiology or medicine for his work on unravelling the chemical structure of the ANTIBODY GAMMA GLOBULIN. The determination of the structure of gamma globulin was essential to research on how antibodies function in the IMMUNE system.

Edison, Thomas Alva (1847–1931) US inventor. With little formal education, he became the most prolific inventor of his generation. In 1876 Edison opened a laboratory in Menlo Park, New Jersey, USA. Here, he invented the carbon transmitter for TELEPHONES (1876), and the phonograph (1877). Using a carbon filament, Edison's invention of the first commercially viable ELECTRIC LIGHT (October 21, 1879) ensured his fame. In New York City, he built (1881–82) the world's first permanent electric power plant for distributing electric light. In 1887 Edison opened a larger research laboratory in West Orange, New Jersey, and in 1892 most of his companies were merged into the General Electric Company (GEC). In 1914 he developed an experimental talking motion picture. During World War 1 Edison worked for the US government on antisubmarine weapons. By the time of his death, he had patented more than 1,300 inventions.

EDTA Ethylenediaminetetraacetic acid (diaminoethane tetracarboxylic acid, $(HOOCCH_2)_2N(CH_2)_2N(CH_2COOH)_2$), a chelating agent (*see* CHELATION). The acid binds with metals such as calcium, magnesium and iron, and is used in chemical analysis, water softening and to counteract metallic poisons that have been swallowed. In such applications EDTA acts as a sequestering agent.

effector Term used for any agent or structure that brings about a physiological response. For example, a MOTOR NEURON that causes a muscle to contract or a hormone that stimulates a gland to function are both effectors. The target tissues are effector muscles and effector glands.

efficiency Work a MACHINE does (output) divided by the amount of work put in (input). It is usually expressed as a percentage, and a perfect machine would therefore have an efficiency of 100%. In mechanical systems there are invariably energy losses, such as those caused by FRICTION in the moving parts. The output never equals the input, and the efficiency is always less than 100% (although transformers and other electrical machines with no moving parts do have very high efficiencies). For very simple machines, efficiency can be defined as the FORCE RATIO (mechanical advantage) divided by the DISTANCE RATIO (velocity ratio).

egg For egg-laying animals, the fertilized female GAMETE (OVUM), once it has been laid or spawned by the mother. The developing EMBRYO is surrounded by ALBUMIN, shell, egg case or MEMBRANE, depending on the species. The egg provides a reserve of food for the embryo in the form of YOLK and protects it from a potential harmful external environment. For non-egg-laying animals, including almost all mammals, the term egg is synonymous with ovum.

Ehrenburg, Christian Gottfried (1795–1876) German biologist and founder of the science of micro-palaeontology. Ehrenburg studied at Berlin University and travelled widely as a naturalist, identifying and classifying many terrestrial and marine plants and microorganisms. He advanced the theory of "complete organisms", stating that all animals – even microscopic ones – have complete organ systems; it was later refuted.

Ehrlich, Paul (1854–1915) German bacteriologist. He shared with Élie METCHNIKOFF the 1908 Nobel Prize for physiology or medicine for his work on immunization, which included the development of basic standards and methods for studying toxins and antitoxins, with special reference to diphtheria antitoxins. His subsequent search for a "magic bullet" against disease and his discovery of salvarsan, a chemical effective against *Treponema pallidum*, the bacterium which causes syphilis, introduced the modern era of CHEMOTHERAPY (a term he coined).

Eichler, August Wilhelm (1839–87) German botanist who developed a system of PLANT CLASSIFICATION eventually accepted worldwide. It included division of plants into four major classes. He helped to edit some 15 volumes on the flora of Brazil and published a 2-volume study of the comparative structure of flowers.

Eigen, Manfred (1927–) German chemist who shared the 1967 Nobel Prize for chemistry with Ronald NORRISH and George PORTER for their

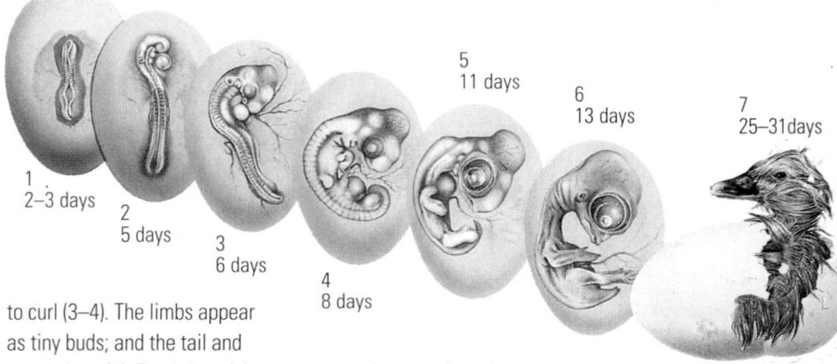

► **egg** A duck embryo grows from a patch of cells on the surface of the egg yolk. The yolk is its food store. First, a network of tiny blood vessels spreads over the yolk and a simple heart develops. The developing embryo (enlarged here for clarity) begins to elongate and develops a vertebral column (1). A head and bulging eye start to form, and the heart folds around into its final position (2). The gut forms, the brain begins to enlarge and the embryo starts to curl (3–4). The limbs appear as tiny buds; and the tail and mouth form (5). By 13 days (6) it is possible to identify the bird from its bill. Some species of bird hatch shortly after this stage, others, such as the mallard duck (7) continue to develop in the egg. Feathers grow, limbs become stronger and the bird hatches with its eyes open and able to see.

▲ **Einstein** The great physicist, Albert Einstein, was forced to flee Nazi persecution (1933), and worked at Princeton, USA. He formulated the special theory of relativity and advanced quantum theory (1905). Most of his work was on unified field theory.

research into very fast chemical reactions. Eigen devised techniques to disturb the balance of a system near equilibrium in order to make measurements of colours emitted as the system relaxed into a new equilibrium state. These techniques have been used in the study of RADIATION chemistry, and in reactions that use ENZYMES as CATALYSTS.

Eijkman, Christiaan (1858–1930) Dutch medical researcher and physician. In 1929 he shared with Frederick HOPKINS the Nobel Prize for physiology or medicine for his discovery of the anti-neuritic vitamin. Eijkman was the first to recognize a dietary deficiency disease, demonstrating that beriberi was produced by a lack of a certain dietary substance, later identified as vitamin B_1.

Einstein, Albert (1879–1955) US physicist, b. Germany, best known for his theories of RELATIVITY. In 1905 Einstein published four papers that revolutionized physical science. "The Electrodynamics of Moving Bodies" announced his special theory of relativity. Drawing on the work of Hendrik LORENTZ, Einstein's **special theory of relativity** discarded the notion of absolute motion in favour of the hypothesis that the speed of light is constant for all observers in uniform (unaccelerated) motion. Measurements in one uniformly moving system can be correlated with measurements in another uniform system, if their **relative** velocity is known. It demonstrated the relativity of time and asserted that the speed of light was the maximum velocity attainable in the Universe. A corollary of this special theory – the equivalence of mass and energy ($E = mc^2$) – was put forward in a second paper. A third paper, on BROWNIAN MOVEMENT, confirmed the atomic theory of MATTER. Lastly, Einstein explained the PHOTOELECTRIC EFFECT in terms of quanta or PHOTONS of light. For this insight, which forms the basis of modern QUANTUM THEORY, Einstein received the 1921 Nobel Prize for physics. In 1911 he asserted the equivalence of GRAVITATION and INERTIA. Einstein extended his special theory into a **general theory of relativity** (1916) that incorporated systems in non-uniform (accelerated) motion. This proved highly important in the field of COSMOLOGY. He asserted that matter in space causes curvature in the SPACE-TIME continuum, resulting in gravitational fields. This explained the peculiar motion of the planet Mercury and was confirmed (1919) by Eddington's study of starlight. Fearful of the rise of Nazism, Einstein accepted a post (1933–55) at the Institute of Advanced Study, Princeton, New Jersey. In 1934 he was deprived of his German citizenship because he was Jewish. In 1940 Einstein became a US citizen. He devoted the rest of his life to a UNIFIED FIELD THEORY that attempted to combine ELECTROMAGNETISM and GRAVITATION into a single theory.

einsteinium (symbol Es) Radioactive, synthetic metallic element of the ACTINIDE SERIES. The isotope ^{253}Es was first identified in 1952 at the University of California at Berkeley; this was after it was found as a decay product of ^{238}U produced by the first large hydrogen bomb explosion. Eleven isotopes have been identified. Properties: at.no. 99; most stable isotope ^{254}Es (half-life 276 days). *See also* TRANSURANIC ELEMENTS

Einthoven, Willem (1860–1927) Dutch physiologist. He became professor at Leiden University in 1886 and was awarded the 1924 Nobel Prize for physiology or medicine for his work on the string GALVANOMETER, the forerunner of the electrocardiograph, the apparatus used in an ELECTROCARDIOGRAM (ECG).

ejector seat Device for escaping from an aircraft in an emergency. In order to parachute successfully from high-speed aircraft it is necessary to blow the seat and the pilot clear of the fuselage and tail section with an explosive charge. In hypersonic craft the pilot can be protected by separating the entire cockpit from the fuselage and parachuting the capsule and occupant together to the ground.

elasticity Capability of a material to recover its size and shape after deformation by STRESS. When an external force (stress) is applied, the material develops STRAIN (a change in dimensions). If a material passes its elastic limit, it will not return to its original shape *See also* HOOKE'S LAW

elastomer Substance that regains its original size and shape after being deformed. Elastomers include natural RUBBER and various synthetic materials with similar properties. All are POLYMERS in which the molecular chains are folded; stretching the elastomer straightens the chains.

electrical engineering Branch of ENGINEERING that deals with the practical application of ELECTRICITY, especially as related to illumination, communication, COMPUTERS and automatic control of machinery. Electrical engineers are fully trained in the mathematics of electric circuits and

ELECTRIC CIRCUIT

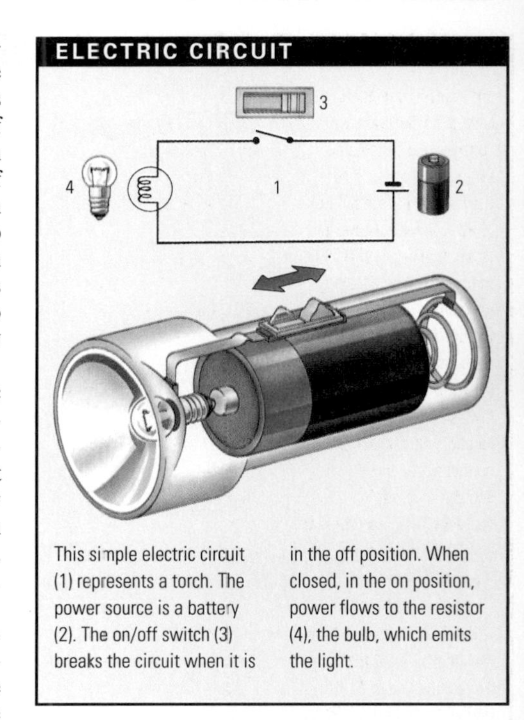

This simple electric circuit (1) represents a torch. The power source is a battery (2). The on/off switch (3) breaks the circuit when it is in the off position. When closed, in the on position, power flows to the resistor (4), the bulb, which emits the light.

ELECTRIC CAR

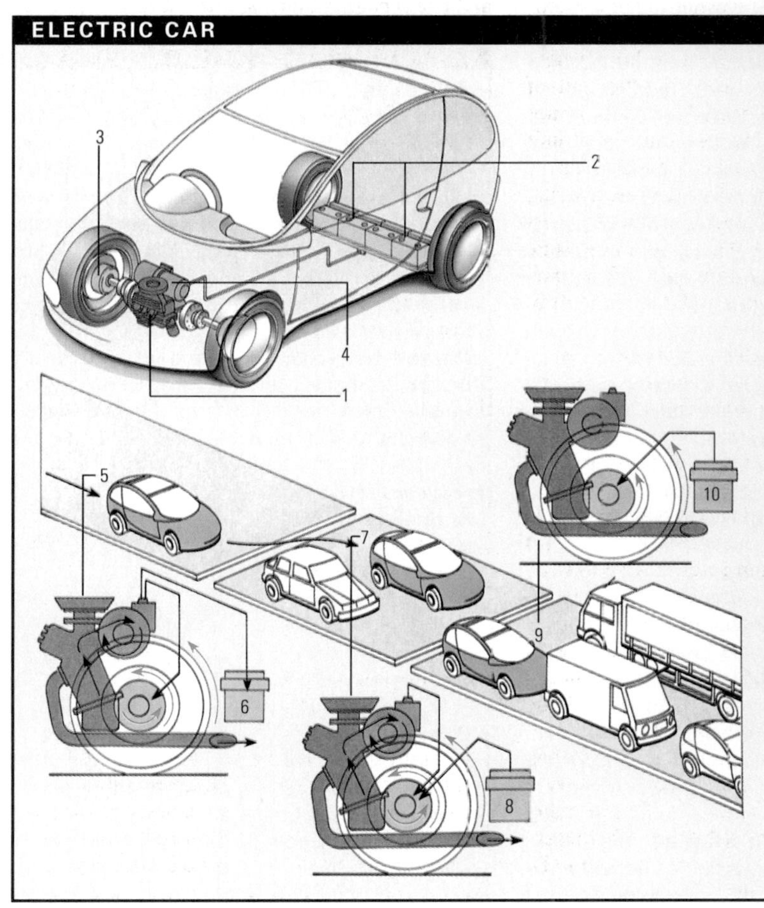

Hybrid car designs use lean-burn petrol engines (1) and batteries (2) in tandem to give low emissions as well as high speed and range. The petrol engine does not directly drive the wheels but powers electric motors mounted at the front wheels (3) through an electric generator (4). The car is powered in three ways. On the open road (5) the petrol engine powers the car and also charges the batteries (6). When extra power is required, for example during overtaking (7), the petrol engine is augmented by the batteries (8). In heavy traffic (9) the petrol engine shuts down and the car depends entirely on the batteries (10).

ELECTRIC CIRCUIT

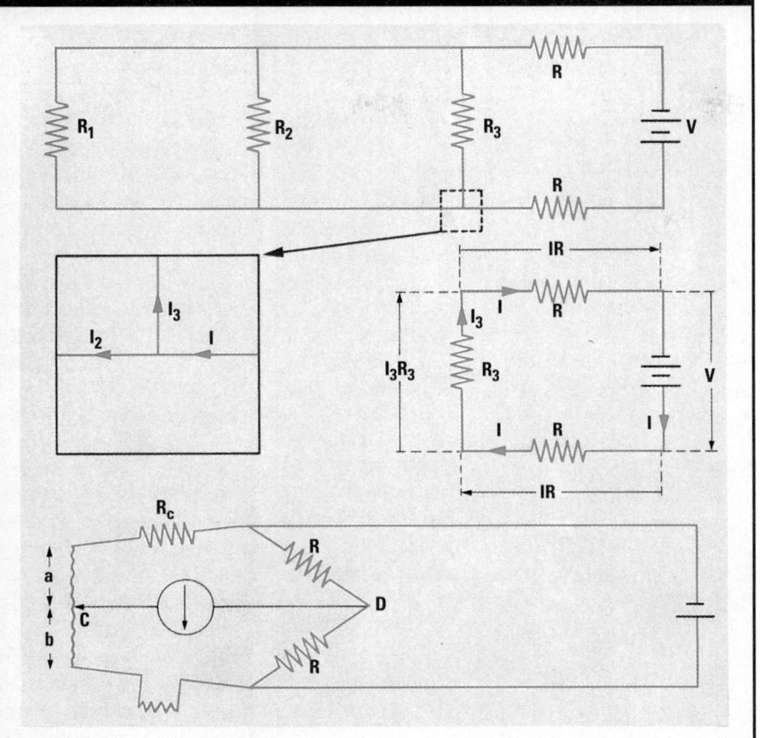

The analysis of electric circuits, to discover the voltages and currents at every point, employs two basic laws, illustrated here. The circuit shown supplies power from a battery (V) to three components (having resistance R_1, R_2, R_3) through power lines which each also have a resistance (R). First, Kirchhoff's "current law" states that at every junction, electric current neither appears nor vanishes: current is conserved, thus (centre left) $I = I_3 + I_2$; and so on around the circuit. Second, Kirchhoff's "voltage law" states that the total voltage around any constituent circuit is zero: thus (centre right, applying Ohm's law) $V + IR + I_3R_3 + IR = 0$; the same applies to the other two circuit loops. Given the battery voltage and the resistances, the currents can be calculated. The bottom diagram shows the circuit of an electrical resistance thermometer, in which changes in the probe resistance, r, are used to measure the temperature. Contact C is moved along a resistance-wire (ab) until the detector in CD shows zero current. If the two resistances R are equal, it is then true that $R_c + a = b + r$; r is read off from a pre-calibrated scale along the wire ab.

are involved in research, development, design, production, supervision and improvement of production methods.

electrical switch Any device used to stop and restart the flow of an ELECTRIC CURRENT within an ELECTRIC CIRCUIT. The basic principle always involves an interruption and restarting of the conducting medium, usually the wiring.

electrical unit Standard quantity in which electric energy is measured. The domestic unit of electricity is the kilowatt-hour. An electric heater rated at 1 kilowatt (1,000 watts) consumes 1 unit of electricity when it is switched on for 1 hour.

electric branch circuit Any ELECTRIC CIRCUIT that is an interconnected branch of the main circuit where KIRCHHOFF'S LAWS apply. These state that at any branch junction the sum of the currents flowing to that point is equal to the sum of the currents flowing away.

electric car AUTOMOBILE powered by a rechargeable BATTERY. The electrical energy stored in the battery is converted to mechanical power by ELECTRIC MOTOR. Electric cars are considered to be more environmetally-friendly than their gasoline-powered relatives. Firstly, they do not produce exhaust pollutants, thus reducing air pollution. Secondly, they do not consume scarce petroleum resources and lastly, they are quieter and would reduce noise pollution. The major disadvantage of electric cars is their limited range – they can only travel *c.*100mi (160km) before the battery needs recharging. Furthermore, the small size of the battery reduces performance. The maximum speed of an electric car is usually less than 60mph (100km/h). The first electric cars were produced in Europe in the late 1800s and many early cars were powered by battery. The INTERNAL COMBUSTION ENGINE, however, proved to be more powerful and cheaper.

electric charge (symbol q or Q) Property shown by some ELEMENTARY PARTICLES. Electric charges (measured in COULOMBS) are described as either **positive** or **negative**. If two particles have a positive charge, or both have a negative one, they repel one another; if one particle has a positive and one a negative charge, they attract. Electrical charges can be stored on insulated metal spheres (as in a VAN DE GRAAFF GENERATOR), on insulated plates (as in a CAPACITOR) and in chemical solutions (as in an electric BATTERY). An ELECTRIC CURRENT consists of electric charges in motion. *See also* COULOMB'S LAW; ELECTRON; PROTON

electric circuit System of electric CONDUCTORS, appliances or electronic components connected together so that they form a continuously conducting path for an ELECTRIC CURRENT. A circuit must include a source of ELECTROMOTIVE FORCE (EMF) to drive the current. Current flows according to several definite laws, the most important of which is OHM'S LAW. In modern electronic devices, circuits are often printed in copper on a plastic card (PRINTED CIRCUIT). *See also* CAPACITOR; INTEGRATED CIRCUIT (IC); TRANSISTOR

electric constant (symbol ε_0) Physical quantity relating the force between ELECTRIC CHARGES in a vacuum to the size of the charges and the distance between them. It was formerly called the PERMITTIVITY of free space. It appears in COULOMB'S LAW.

electric current (symbol I) Movement of ELECTRIC CHARGES, usually the flow of ELECTRONS along a CONDUCTOR or the movement of IONS through an ELECTROLYTE or gaseous discharge. It flows by convention from a positive (ANODE) to a negative terminal (CATHODE), although electrons actually flow along a wire in the opposite direction. It is measured in AMPERES. Direct current (DC) flows continuously in one direction, whereas alternating current (AC) regularly reverses direction. The frequency of AC current is measured in HERTZ (Hz). The frequency of the domestic supply is different in different parts of the world (in Europe, it is 50Hz, in North America, 60Hz). *See also* ELECTRICITY

electric displacement (symbol D) Density of ELECTRIC FLUX per unit area.

electric field (electrostatic field) Region around an ELECTRIC CHARGE in which any charged particle experiences a force. An object of opposite

ELECTRIC CURRENT (ALTERNATING)

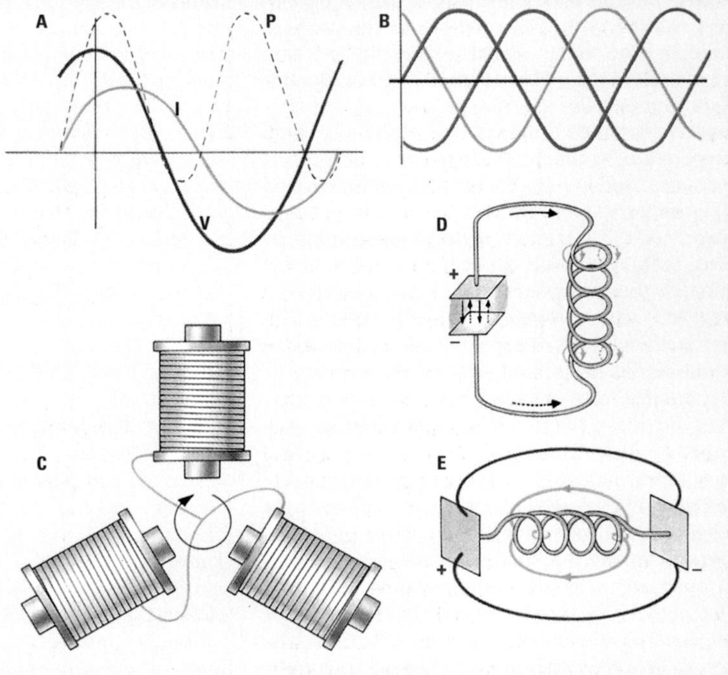

Electrical energy can be produced by induction in a generator, a primary voltage causing an alternating current to flow in an external circuit. The presence of an inductor or capacitor (or both) produces phase displacement (A) between the voltage, V, and the current, I. Here a capacitor has produced a current phase lead of 90°, resulting in zero average power, although the power curve is still a sine wave. The reduction of power, P, due to the phase displacement, is called the power factor. If three alternating currents have phase displacements of 120° relative to each other, the sum of their current or voltage is always zero (B). These three-phase currents are used in an electric motor with a "squirrel cage" rotor (C). This consists of three electromagnets which rotate in the magnetic field produced. Alternating currents are also produced in closed (D) and open (E) oscillating circuits. The high frequency electromagnetic waves used in some telecommunications originate from oscillating circuits.

ELECTRICITY SOURCES

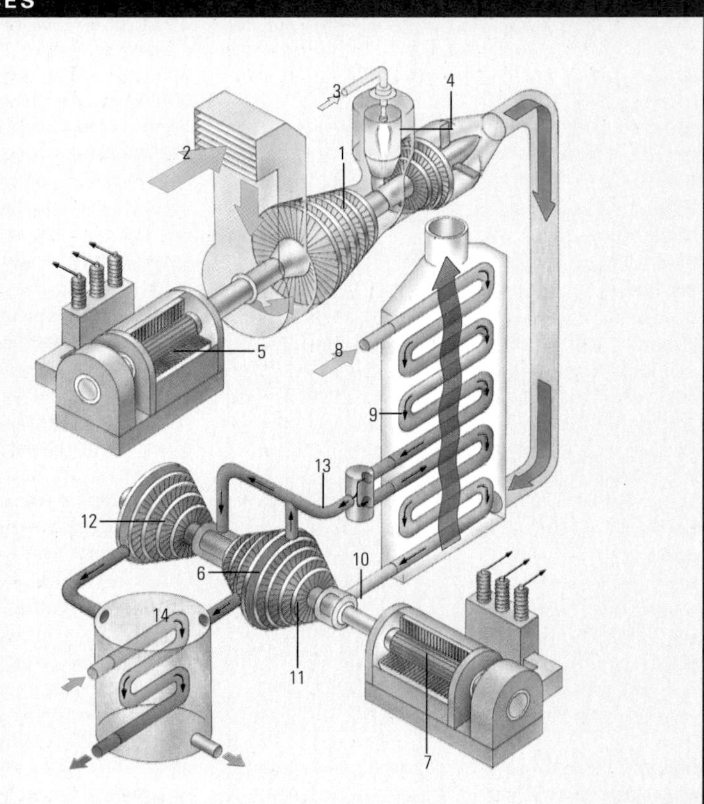

A combined cycle power station burns gas to generate electricity. It is considerably more efficient than traditional fossil fuel power stations. The first turbine (1) sucks in air (2) compressing it before mixing it with the fuel (3) and burning the mixture (4). Exhaust gases spin a second turbine, connected to the first turbine and a generator (5). The energy of the gases is harnessed to power a second multiple turbine (6) connected to another generator (7). Gases are used to superheat water (8) looping through a special vessel (9). To maximize power generation, superheated steam (10) turns a high-pressure turbine (11) before passing (at a slightly lower temperature) into a lower-pressure turbine (12). Steam is fed into the turbine directly from the heating loops (13), and is cooled (14) before going back into the circuit.

charge experiences an attractive force. An object of the same charge as that generating the electric field experiences a force of repulsion. The strength of the field upon unit charge at a distance r from a charge Q is equal to $Q/4\pi r^2\varepsilon$, where ε is the PERMITTIVITY of the medium surrounding the charge. A changing MAGNETIC FIELD can also create an electric field. *See also* ELECTROMAGNETISM

electric flux Quantity of ELECTRIC CHARGE that would be displaced across a CONDUCTOR imagined to be placed in an ELECTRIC FIELD. It is also the amount of charge that would cause the same electric field. It is proportional to the strength of the field and to the PERMITTIVITY of the medium through field passes through. The flux per unit area is called the electric flux density or **electric displacement**. *See also* GAUSS' LAW

electric furnace FURNACE heated to a very high temperature by an electric current. Electric furnaces are used in industry for melting metals and other materials, and for the production of high-grade steels. Three main methods of heating are used: striking an ARC between electrodes in the furnace; producing currents in the material by ELECTROMAGNETIC INDUCTION; and passing a high current through the material, heat being produced because of its RESISTANCE.

electricity Form of energy associated with STATIC or moving ELECTRIC CHARGES. Charge has two forms – positive and negative. Like charges repel each other and unlike attract; these forces are described in COULOMB'S LAW. Electric charges are acted upon by forces when they move in a MAGNETIC FIELD and this movement in turn generates an opposing magnetic field (FARADAY'S LAWS). Electricity and MAGNETISM are different aspects of the same phenomenon, ELECTROMAGNETISM. The flow of charges constitutes an ELECTRIC CURRENT, which in a conductor consists of negatively charged ELECTRONS. For a current to exist in a CONDUCTOR there must be an ELECTROMOTIVE FORCE (EMF) or POTENTIAL DIFFERENCE between the ends of the conductor. If the source of potential difference is a BATTERY, the current flows in one direction only, being direct current. If the source is the mains, the current changes direction twice every cycle, being alternating current (AC). The AMPERE is the unit of current, the COULOMB is the unit of charge, the OHM is the unit of RESISTANCE and the VOLT is the unit of electromotive force. OHM'S LAW and KIRCHHOFF'S LAWS (the summing of voltages and currents in a circuit) are the basic means of calculating circuit values. *See also* ELECTRIC CURRENT; ELECTRONICS

electricity sources Devices that convert other forms of energy into electricity. Most of our electricity is produced in power stations from the chemical energy of fossil fuels. The heat from burning coal, oil or natural gas turns water into steam. The steam drives a TURBINE, linked to an electricity GENERATOR. In a nuclear power station, heat comes from the FISSION of nuclei in a NUCLEAR REACTOR. A BATTERY and FUEL CELL convert chemical energy directly into electricity. SOLAR CELLS convert SOLAR ENERGY into electricity. Wind generators and water turbines produce electricity from the energy of movement in wind and water. *See also* ENERGY SOURCES; HYDROELECTRICITY; RENEWABLE ENERGY

electric light Artificial light produced by a flow of electricity in a wire or gas. In an ordinary LIGHT BULB, light is produced by INCANDESCENCE when a filament, such as tungsten-alloy wire, is heated by an electric current. In a fluorescent tube, the electric current passes through a gas. The gas atoms give off invisible ultraviolet rays. These strike a coating on the inside of the tube, causing it to emit light by FLUORESCENCE.

electric motor Machine that converts electrical energy into mechanical energy. In a simple form of electric motor, an ELECTRIC CURRENT powers a set of ELECTROMAGNETS on a rotor in the MAGNETIC FIELD of a permanent MAGNET. Magnetic forces set up between the permanent magnet and the electromagnet cause the rotor to turn. Electric motors are convenient and relatively quiet, and have replaced many other forms of motive power. They may use alternating current (AC) or direct current (DC), and vary in size from thumb-sized miniatures used in toys, through those in medium-sized domestic appliances (such as vacuum cleaners, food mixers and hand drills), to large motors such as those used in factories, lifts and diesel-electric locomotives.

electrocardiogram (ECG) Recording of the electrical activity of the heart traced on a moving strip of paper. It is made with recording apparatus called an **electrocardiograph**. Electrocardiography is used to diagnose heart disease.

electrochemical cell *See* CELL

electrochemical series *See* ELECTROMOTIVE SERIES

electrochemistry Branch of physical chemistry concerned with the relationship between electricity and chemical changes. It includes the properties and reactions of IONS in solution, the CONDUCTIVITY of ELECTROLYTES and the study of the processes occurring in electrochemical CELLS and in ELECTROLYSIS.

electroconvulsive therapy (ECT) Treatment of mental disturbance by means of an electric current passed through the brain to induce convulsions. Given under anaesthesia and with the use of a muscle relaxant, it is recommended mainly for severe depression that has failed to respond to other forms of treatment. Electrodes are attached to the scalp and current is applied to one or both sides of the brain in order to alter its electrical activity. The treatment can produce unpleasant side-effects, including confusion, memory loss and headaches, and there is continuing controversy about its use and effectiveness.

electrode Conductor, usually a wire or rod, through which an electric current flows into or leaves a medium. In ELECTROLYSIS, two electrodes – a positive (ANODE) and a negative (CATHODE) – are immersed in the ELECTROLYTE.

electrodeposition Deposition of an adhering

ELECTRIC MOTOR

Electric motors work using the interaction of a magnet (1) and a wire with a current passing through it (2). With the current flowing, the magnetic field produced by the loop interacts with the field of the magnet. A downward force acts on the right side, an upward force on the left side. When the loop reaches the vertical, the split ring (through which the current reaches the loop) (3) reverses the current and so the magnetic field. Electric motors use multiple coils (4) to ensure constant torque. In an electric drill, the turning shaft (5) emerges from the magnetic coils and is then geared (6) through to a chuck (7) to the drill bit (8).

► **electromagnetic spectrum** is ordered by either frequency or wavelength (or both). It ranges from low-frequency (high-wavelength) radio waves, through microwaves, infrared waves, light (the visible spectrum – red, orange, yellow, green, blue, indigo and violet), continues through ultraviolet waves and X-rays, to very high-frequency (short wavelength) gamma rays.

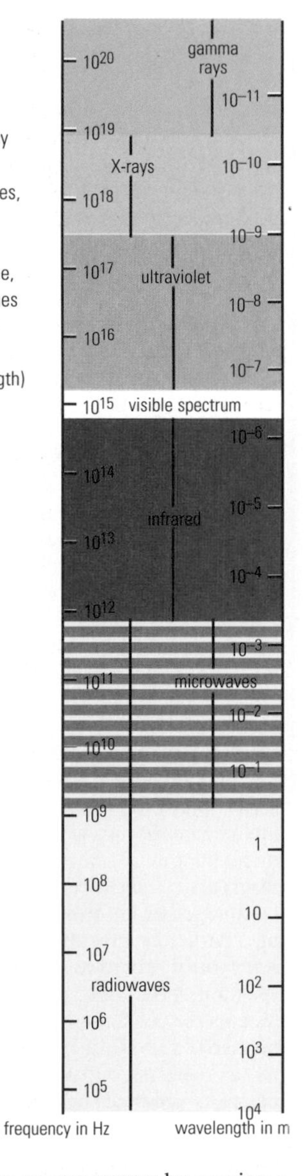

metallic coating onto an ELECTRODE by passing a low-voltage direct current (DC) through an ELECTROLYTE. Its most common application is ELECTROPLATING. *See also* ANODIZING

electrode potential Measure of the tendency of a reaction at an ELECTRODE. An electrode of an element (M) placed in a solution of its ions (M^+) constitutes a HALF-CELL. In general a POTENTIAL DIFFERENCE exists between the electrode and the solution, caused by reactions of the type $M \Leftrightarrow M^+ + e$. In practice it cannot be measured absolutely and standard electrode potentials are defined with reference to a hydrogen electrode under specified conditions of concentration, temperature and pressure. Such reactions are OXIDATIONS, and so the term oxidation potential is often also used. *See also* ELECTROMOTIVE SERIES

electrodialysis *See* DIALYSIS

electrodynamics In physics, study of interactions between electric and magnetic fields and charged bodies. It began in the 19th century with the theoretical work of James Clerk MAXWELL, and later became part of QUANTUM MECHANICS.

electroencephalogram (EEG) Technique for recording the electrical activity of the brain. Electrodes are attached to the scalp to pick up the tiny oscillating currents produced by brain activity. The brainwaves are recorded as a continuous trace on a paper strip. The technique is used mainly in the diagnosis and monitoring of EPILEPSY.

electroforming Making of metal articles by deposition of the metal onto a mould by ELECTROLYSIS. A notable application is the manufacture of metal negative replica moulds for vinyl records, where the accuracy of the deposition process provides faithful sound reproduction. *See also* ELECTRODEPOSITION

electroluminescence Production of light by certain substances, particularly phosphors, when they are placed in an alternating electric field. Part of the light from a FLUORESCENT LAMP is due to electroluminescence; the rest is due to PHOTOLUMINESCENCE.

electrolysis CHEMICAL REACTION caused by passing a direct current (DC) through an ELECTROLYTE. This results in positive IONS migrating to the negative ELECTRODE (CATHODE) and negative ions migrating to the positive electrode (ANODE). The type of electron transfer reaction occurring depends on the electrode potentials of the ions present, and the electrode material may also play a part in the reaction. For example, in the electrolysis of copper salts with a copper anode, atoms of the electrode ionize and enter into solution. Electrolysis is an important method of obtaining chemicals, particularly for extracting reactive elements such as sodium, magnesium, aluminium and chlorine. A commercial use is in ELECTROPLATING.

electrolyte Solution or molten salt that can conduct electricity, as in ELECTROLYSIS (which decomposes the electrolyte). In electrolytes, the current is carried by charged particles called IONS, rather than by electrons. In a lead-acid car battery, for example, the electrolyte is dilute sulphuric acid, which contains negative sulphate ions and positive hydrogen ions.

electromagnet Magnet constructed from a soft iron core around which is wound a coil of insulated wire. A MAGNETIC FIELD is set up when an electric current is passed through the wire, and it disappears when the current is switched off. This creates a magnet that can be turned on and off.

electromagnetic force One of the four FUNDAMENTAL FORCES in nature. Within an atom, the electromagnetic force binds the negatively charged electrons to the positively charged NUCLEUS. *See also* GRAND UNIFIED THEORY (GUT); UNIFIED FIELD THEORY

electromagnetic induction Use of MAGNETISM to produce ELECTRICITY. If a bar magnet is pushed through a wire coil, an electric current is set up, or "induced", in the coil, as long as the magnet is moving. This is the basic working principle of the DYNAMO, ELECTRIC MOTOR and TRANSFORMER. *See also* INDUCTANCE; INDUCTION

electromagnetic radiation Energy in the form of waves of a wide range of frequencies. Electromagnetic radiation travels through empty space at the speed of light, approximately 300,000km (186,000mi) per second, and through materials at different speeds. Light, radio waves, infrared (heat) and ultraviolet radiation and X-rays are all examples of electromagnetic radiation. In general, electromagnetic waves are set up by electrical and magnetic vibrations that occur in atoms. These waves, which make up the ELECTROMAGNETIC SPECTRUM, can undergo REFLECTION, REFRACTION, INTERFERENCE, DIFFRACTION and POLARIZATION. Other phenomena, such as the absorption or emission of light, can be explained only by assuming the radiation to be composed of quanta of energy (PHOTONS) rather than waves.

electromagnetic spectrum Range of different types of ELECTROMAGNETIC RADIATION ordered by either frequency or wavelength. It ranges from low-frequency (high-wavelength) radio waves, through microwaves, infrared (heat) waves, light (the visible spectrum), ultraviolet waves and X-rays to very high-frequency gamma rays.

electromagnetism Branch of physics dealing with the laws and phenomena that involve the interaction or interdependence of ELECTRICITY

ELECTROMAGNETISM

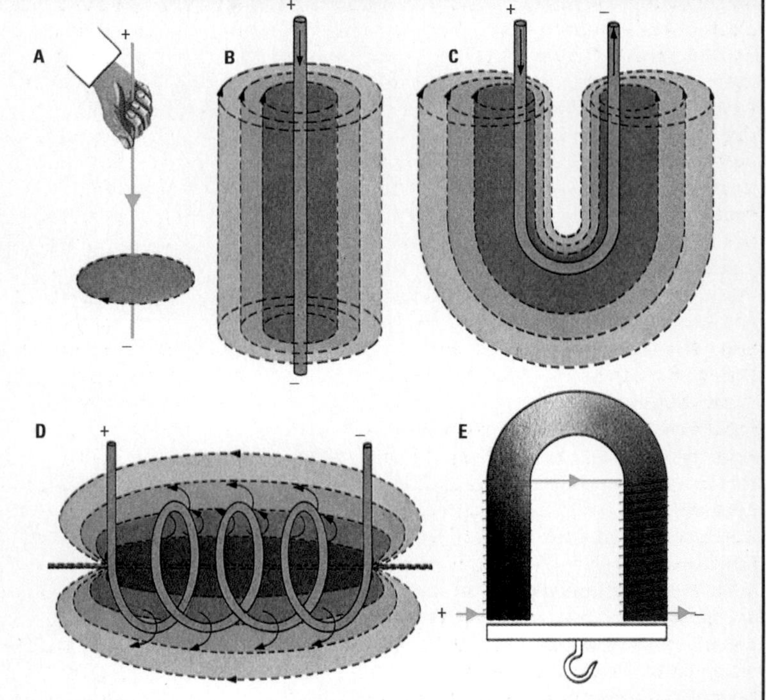

When current flows at uniform velocity along a wire, the magnetic field produced is a set of concentric rings in any plane drawn at right-angles to the wire. The number of lines of force drawn – dotted lines in B, C and D – by convention, represents the strength of the magnetic field. If the wire is bent, the field follows the change in shape. When the wire is wound (D), in a coil or solenoid, the force lines of the field all point in the same direction through the centre of the coil; overall, the field resembles that produced by a bar magnet with the north pole, in this example, on the right. If an iron bar is inserted into the core of the solenoid, the lines of force prefer to remain within it and this results in a high concentration of lines at the end of the bar. This device, called an electromagnet, can produce a very strong magnetic field which varies with the current in the wire, the number of turns in the solenoid and its cross-sectional area. In the horseshoe-shaped electromagnet (E), the strong magnetic field induced when current flows through the coil keeps the iron hook firmly in place, and it can be used to lift very heavy objects.

and MAGNETISM. The region in which the effect of an electromagnetic system can be detected is known as an **electromagnetic field**. When a magnetic field changes, an electric field can always be detected. When an electric field varies, a magnetic field can always be detected. Either type of energy field can be regarded as an electromagnetic field. A particle with an electric charge is in a magnetic field if it experiences a force only while moving; it is in an electric field if the force is experienced when stationary.

electrometallurgy Applications of ELECTROCHEMISTRY to METALLURGY. It includes the production and refining of metals by ELECTROLYSIS and such techniques as ELECTROPLATING, electrolytic machining and polishing, and ELECTROFORMING. In a wider sense, electrometallurgy includes other uses of electric current, as in alloy-steel production using an ELECTRIC FURNACE.

electrometer Instrument having an electric circuit for measuring differences of ELECTRIC POTENTIAL (voltages) without drawing appreciable current. Modern electrometers are essentially voltage AMPLIFIERS. *See also* POTENTIAL DIFFERENCE; VOLTMETER

electromotive force (emf) Sum of the POTENTIAL DIFFERENCES around an ELECTRIC CIRCUIT. When the circuit is open, so that no current flows, it is equal to the potential difference between the terminals of the power source. When a current does flow, the external potential difference decreases. The emf is equal to the energy expended in moving one unit of charge around the circuit. *See also* ELECTRICITY

electromotive series (electrochemical series) List of METALS and the gas hydrogen, the order of which indicates the relative tendency to be oxidized, or to lose electrons in chemical reactions (*see* OXIDATION-REDUCTION). The series starts with the metal that tends to lose the most electrons in reaction. A standard hydrogen ELECTRODE is arbitrarily assigned an oxidation potential of zero, so that the metals can be compared. Those that lose electrons more readily than hydrogen are termed **electropositive**; those that lose electrons less readily are called **electronegative**. The order of some common metals in the electromotive series is: lithium, potassium, calcium, sodium, magnesium, aluminum, zinc, chrominum, iron, cobalt, nickel, tin, lead, hydrogen, copper, mercury, silver, platinum, and gold.

electromyogram Record of the electrical activity of muscles by means of electromyography (EMG). Muscles produce no currents when at rest, but during activity the currents generated enable a continuous trace of action potentials to be recorded. These traces are valuable for research and in diagnosis of muscle dysfunction.

electron (symbol *e*) Stable ELEMENTARY PARTICLE with a negative charge, having a rest mass of 9.1×10^{-31} kg (1/1836 that of the PROTON). Identified in 1879 by the British physicist Joseph THOMSON, electrons move around the NUCLEUS of an ATOM in complex orbits. In a neutral atom, their total charge balances that of the protons in the NUCLEUS. Removal or addition of an atomic electron produces a charged ION. Chemical bonds are formed by the transfer or sharing of electrons between atoms. When not bound to an atom, electrons are responsible for electrical conduction. Beams of electrons are used in several electronic devices such as CATHODE-RAY TUBES, OSCILLOSCOPES and ELECTRON MICROSCOPES. In particle ACCELERATORS, electrons and POSITRONS (ANTIPARTICLES of electrons) are made to collide with other particles for the purposes of nuclear research. An electron is classified as a LEPTON. *See also* ELECTROMAGNETIC FORCE

electron diffraction Method by which the structure of gases and the surface of solids is determined. Just as light waves undergo DIFFRACTION when passing through a narrow opening, so also do electrons. The electron wavelength can be adjusted, and the way it behaves after striking the subject material is used to determine the material's CRYSTAL STRUCTURE.

electron emission Liberation of electrons from the surface of a substance. It can occur as a result of heat (THERMIONIC EMISSION), light (PHOTOELECTRIC EFFECT), high electric field (field emission), or bombardment by ions or other electrons (SECONDARY EMISSION). It may also occur from RADIOACTIVE DECAY. In all cases, the electrons must acquire energy from the outside source in excess of the WORK FUNCTION of the substance. Most ELECTRON TUBES and CATHODE-RAY TUBES depend on thermionic emission, whereas photoelectric cells rely on photoelectric emission. Field emission is important in the field-emission microscope and secondary emission is the principle used by electron multipliers and storage tubes.

electron gun In a CATHODE-RAY TUBE, such as that in a television receiver or an oscilloscope, device that emits electrons. These are focused into a beam that scans a fluorescent screen, forming an image.

electronic circuit Electronic components wired together to form a unit capable of performing a particular function. *See* ELECTRIC CIRCUIT

electronic control system Control system based on ELECTRONIC CIRCUITS. Complex control systems rely on COMPUTERS.

electronic mail (e-mail) Correspondence sent via a COMPUTER NETWORK. In a simple system, messages produced using word processing programs are transmitted over a network (which could be a small company network or the worldwide INTERNET) and stored in a computer called a **mail server**. Anyone connected to the network can contact the mail server to check whether it is holding mail for them. If it is, they can transfer the messages to their own computer, and print it if they need a permanent record. Electronic mail is transmitted rapidly, and costs are low, even for international communications.

electronic publishing Production of publications by electronic means, generally using a COMPUTER. During production, word processing and graphics programs are used to assemble "pages" on a computer screen. The publication may be printed on paper, either by conventional methods using PRINTING film generated by computer or by various computer-controlled printers (such as a LASER PRINTER). Alternatively the publication may be produced as a CD-ROM. Electronic publications may also be sent to other computer users on-line.

electronics Study and use of circuits based on the conduction of electricity through ELECTRON TUBES and semiconducting devices. John FLEMING invented the DIODE valve, which was modified in 1906 as the triode by Lee DE FOREST. These devices, with further modifications and improvements, provided the basic components for all the electronics of radio, television and radar until the end of World War 2. A major revolution occurred in 1948 when a team at Bell Telephone Laboratories, led by William SHOCKLEY, produced the first semiconducting TRANSISTOR.

ELECTROMETER

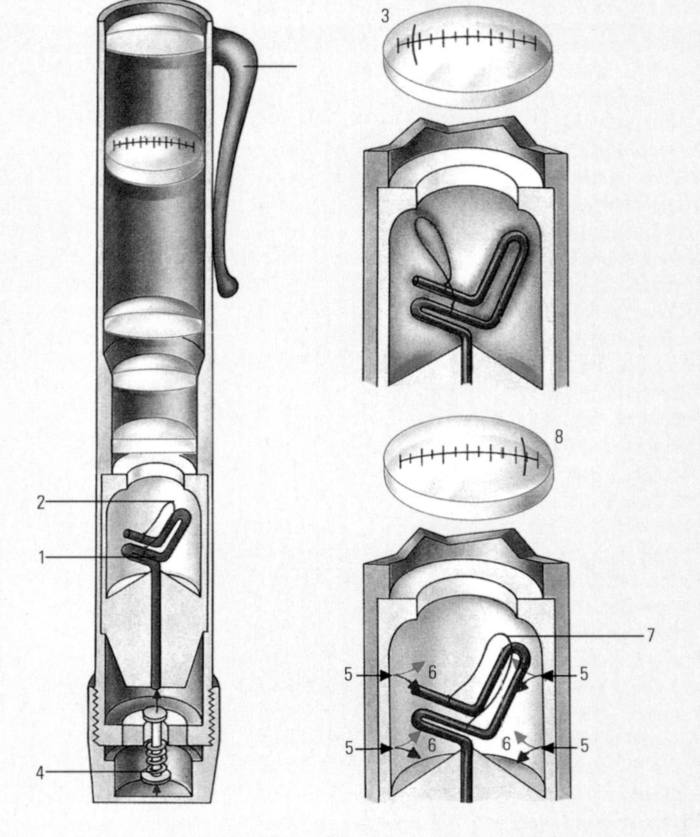

Dosimeters are a type of quartz fibre electrometer used to measure radiation doses. The quartz fibre electrometer (sometimes called the pen meter) is clipped into a pocket and monitors received radiation. It detects X-rays and gamma-rays, both of which are ionizing radiation and damage human tissue. Before the meter is given to the wearer, the inner electrode (1) is charged to between 100 and 200 volts. Electrostatic repulsion causes the flexible quartz fibre (2) to bend away from the electrode and zero reading is observed on the calibrated scale (3). The charging pin (4) springs back and the electrode is now insulated from the outside wall by air molecules in the chamber. High energy radiation entering the chamber through the walls of the meter (5) ionizes the air molecules and some of the charge on the electrode leaks away (6) to the outer wall. This reduces the voltage difference and the fibre moves back progressively to its relaxed position due to the reduced charge (7). The system of magnifying lenses allows the position of the quartz fibre to be viewed, and the dose to be read off against a calibrated scale (8).

ELECTRON MICROSCOPE

In an electron microscope, a beam of electrons (1) streams from the heated tungsten cathode (2) of an electron gun (3) and is focused by upper (4) and lower (5) electromagnetic lenses. It then passes through an aperture ring (6) and a scan coil (7) before being focused by a projector lens (8) onto the sample (9). The process takes place in a vacuum with air evacuated (10) by a pump. A computer controls the scan coil, which directs the beam across the sample. The sample is placed in an airlock (11) and manipulated into position (12). An image of the sample is created by detecting electrons dislodged (13) from the sample. These electrons correlate to the topography of the sample and are measured by a detector (14) when they hit a fluorescent target (15). The image is displayed on a computer (16).

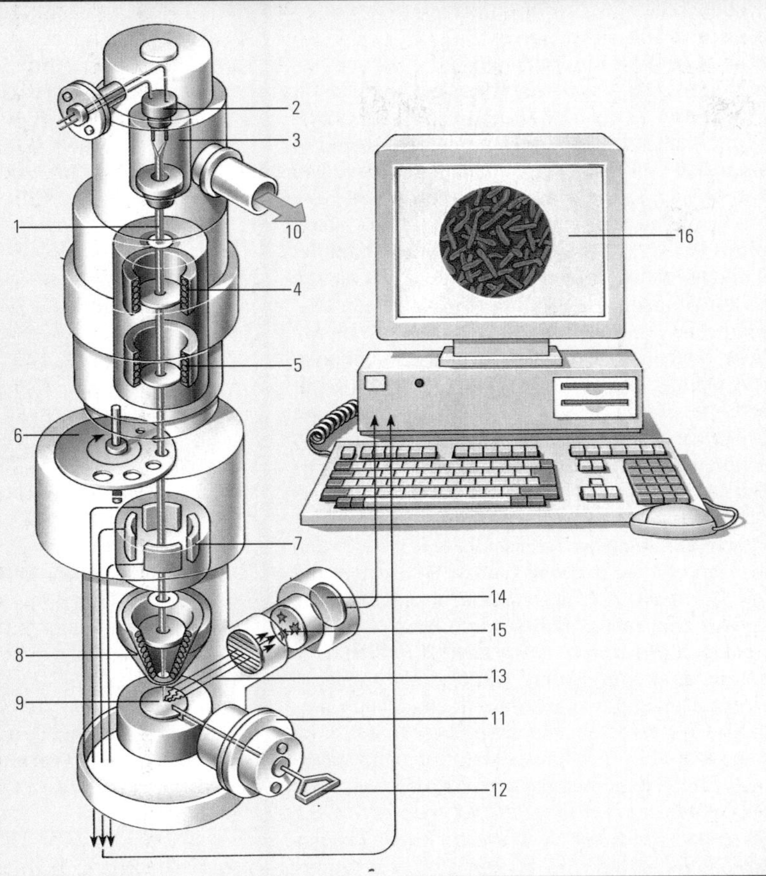

SEMICONDUCTOR devices are much lighter, smaller and more reliable than vacuum tubes. They do not require the high operating voltages of valves and can be miniaturized as an INTEGRATED CIRCUIT (IC). This has led to the production of electronic computers and automatic control devices that have changed the face of both industry and scientific research. *See also* ELECTRIC CIRCUIT; ELECTRICITY

electron microscope MICROSCOPE that "illuminates" the object with a stream of electrons. The "lenses" consist of magnets that focus the electron beam. Smaller objects can be seen, because electrons have shorter wavelengths than light and thus provide greater resolution. In a **transmission** electron microscope, the electrons pass through the target, which is in the form of a very thin slice. In a **scanning** electron microscope, a beam of electrons is swept across the target, which can be a complete object, but which generally requires to be metal-coated to reflect the electrons. The image is obtained by converting the pattern into a video display, which may be computer-processed. These microscopes can magnify millions of times.

electron spin resonance (ESR) Method of studying the structure of molecules by finding the positions of electrons within them. ESR is a spectrographic technique and is applicable only to paramagnetic substances. It locates unpaired electrons by their spin and by the way they align themselves in an external magnetic field when the molecule absorbs microwave radiation.

electron transport system (electron transport chain) Sequence of biochemical reactions that transfers electrons, through a series of carriers, in certain metabolic processes. It involves carrier substances that accept electrons and then donate them to the next carrier in the chain, while themselves undergoing a series of OXIDATION-REDUCTION reactions. It forms the last stage of aerobic RESPIRATION in cells, causing hydrogen atoms to combine with oxygen to form water and conserving energy in the form of ATP. PHOTOSYNTHESIS also involves an electron transfer system.

electron tube (valve) Electronic device consisting of electrodes arranged within an evacuated glass tube; for special purposes a gas at low pressure may be introduced into the tube. The DIODE, used for rectification, consists of a negative CATHODE, which emits electrons when heated, and a positive ANODE or plate. The TRIODE, used for amplification, has a perforated control grid between the cathode and the anode; a signal fed to the grid provides an amplified signal at the anode. Electron tubes have been largely replaced by TRANSISTORS and other semiconductor devices.

electron volt (symbol eV) Unit of energy equal to the energy acquired by an ELECTRON in falling freely through a POTENTIAL DIFFERENCE of one VOLT. It is equal to 1.602×10^{-19} joules.

electro-osmosis Flow of water caused when an electric current is passed through clay or other porous material, in a watery suspension. This phenomenon is caused by the formation of positive and negative IONS, one type of which is bound to the clay while the other (in the water) moves towards the opposite electric pole. *See also* OSMOSIS

electrophoresis Movement of electrically charged colloidal particles through a FLUID from one ELECTRODE to another when a voltage is applied between the electrodes. It is used in the analysis and separation of colloidal suspensions, especially colloidal proteins; clinical medicine uses electrophoresis to measure protein content of body fluids. It is also used as a means of depositing coatings of one material, such as paint, on another. *See also* COLLOID

electrophysiology, cardiac Study of the electrical activity of the HEART muscle. When muscles contract and relax they produce electric currents. The heart produces the principal muscular currents and these are measured for diagnostic purposes by attaching electrodes to the outside of the body and connecting them to an electrocardiograph. *See also* ELECTROCARDIOGRAM (ECG); ELECTROMYOGRAM

electroplating ELECTRODEPOSITION of a coating of metal on another by making the object to be coated the CATHODE in an electrolytic CELL. Positive IONS in the ELECTROLYTE are discharged at the cathode and deposited as metal. Electroplating is used to produce a decorative or corrosion-resistant layer, as in silverplated tableware and chromium-plated engineering parts.

electroscope Instrument for detecting the presence of an ELECTRIC CHARGE. The commonest type is the **gold-leaf** electroscope, in which two gold leaves hang from a conducting rod held in an insulated container. A charge applied to the rod causes the leaves to separate, and the amount of separation indicates the amount of charge.

electrostatic and electromagnetic systems of units In electricity and magnetism,

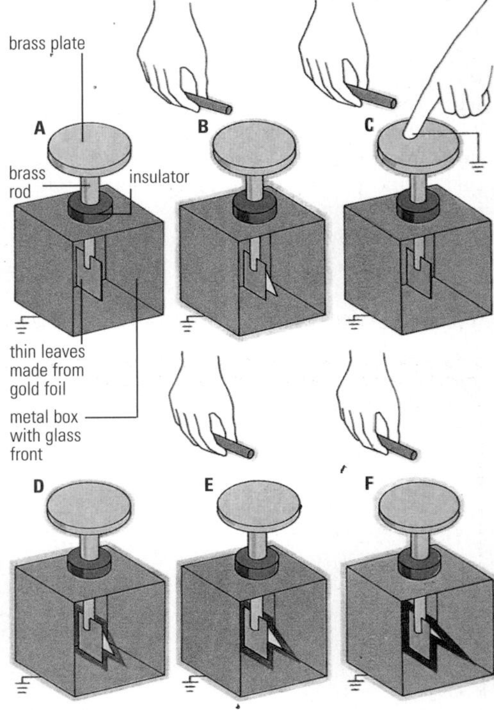

▲ **electroscope** The gold leaf electroscope is a qualitative electrostatic instrument which proved invaluable to early physicists. Among other experiments, it was used to detect the presence, and determine the polarity, of charges.
Key
A) The uncharged electroscope with leaves collapsed
B) A negatively charged rod near the plate attracts positive charges at the plate and repels negative ones to the leaves. The leaves repel each other and diverge to indicate the presence of a charge. The leaves would again diverge if instead a positively charged rod were brought near the plate. This operation is the first in a series which lead to the charging of the electroscope by induction.
C) With the charged rod still in place, the plate is earthed, and electrons flow from the leaves through the hand to earth. The leaves collapse.
D) The electroscope is positively charged when the finger, and then the rod, is removed – thus some positive charges at the plate spread the leaves and they repel each other and diverge.
E and F) The leaves diverge even further when the rod is brought nearer to the plate indicating that the influence of the charges is greater at close range.

systems of units based on different definitions of fundamental quantities. These include the centimetre-gram-second (CGS) electrostatic system, applied particularly to electrostatic problems, and the CGS electromagnetic system, for problems involving magnetic fields. These systems are combined in the Gaussian system of units. These systems of units are being superseded by the international system of SI UNITS. *See also* CGS SYSTEM; MKS UNITS

electrostatic generator *See* VAN DE GRAAFF GENERATOR

electrostatics Branch of physics that studies ELECTRIC CHARGE at rest. This is often done on a charged metal sphere (VAN DE GRAAFF GENERATOR) insulated from its surroundings, or on the insulated plates of a CAPACITOR.

electroweak theory (Weinberg-Salam theory) Theory that unifies the two FUNDAMENTAL FORCES of ELECTROMAGNETIC and WEAK NUCLEAR FORCES. First postulated in 1967 by Steven WEINBERG and Abdus SALAM, the electroweak interaction involves PHOTONS and vector BOSONS known as W BOSONS and Z BOSONS. The discovery of these particles in 1983 provided the first confirmation of the theory. The theory suggests that a yet undiscovered heavy particle, given the name HIGGS BOSON, should exist.

element Substance that cannot be split into simpler substances by chemical means. All atoms of a given element have the same ATOMIC NUMBER (at.no.) and thus the same number of ELECTRONS (the factor which determines chemical behaviour). The atoms can have different MASS NUMBERS (the number of protons and neutrons in the nucleus) and a natural sample of an element is generally a mixture of ISOTOPES. The known elements range from hydrogen (at.no. 1) to element 112; elements of the first 95 atomic numbers exist in nature, the higher numbers have been synthesized.

element 104 *See* DUBNIUM

element 105 *See* HAHNIUM

element 106 *See* RUTHERFORDIUM

element 107 to 112 Six radioactive TRANSACTINIDE ELEMENTS that have been synthesized at Darmstadt, Germany, by a team led by Peter Armbruster. In each case, only a few atoms were produced and their half-lives are extremely short.

elementary particle (subatomic particle) Although the term implies a particle that cannot be subdivided, it is used more widely than this. Elementary particles are distinguished from each other by their MASS (usually expressed in equivalent energy units) and a set of QUANTUM NUMBERS, including ELECTRIC CHARGE and SPIN. An elementary particle may be classified by its interaction as either a LEPTON, which does not experience the STRONG NUCLEAR FORCE, or as a HADRON (MESON or BARYON), which is subject to the strong interaction. Mesons consist of two QUARKS and baryons of three (baryons include the PROTON and the NEUTRON). Another method of classification divides elementary particles into BOSONS (which have an integral spin) and FERMIONS (which have non-integral spin). Many particles also undergo electromagnetic interactions. All elementary particles have an associated antiparticle that has the same mass but an opposite set of quantum numbers. Of the large number of particles known, the only stable ones are the LEPTONS, the PROTON, the NEUTRON (when in a nucleus), their antiparticles (*see* ANTIMATTER) and the PHOTON. The others DECAY after a characteristic lifetime to form stabler particles. When an elementary particle collides with its own antiparticle, mutual annihilation (destruction) occurs, with the production of RADIATION energy. Particle-antiparticle pairs can also be created in high-energy interactions *See also* FUNDAMENTAL FORCES

ELEMENTARY PARTICLE

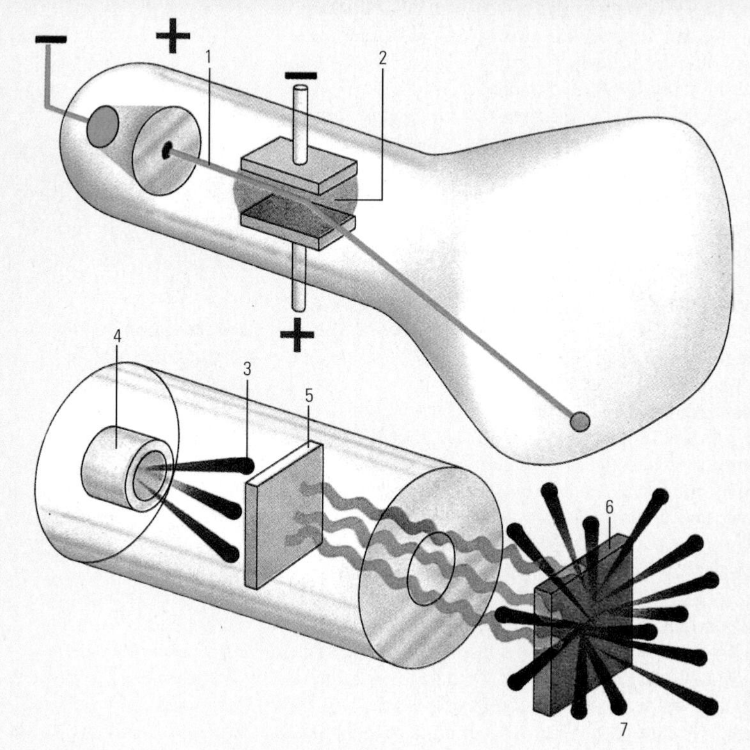

The first elementary particle to be known as such was the electron. In the late 19th century, it was known that so-called "cathode rays" (1) could be obtained using a voltage of a few thousand volts in a vacuum. That these rays were in fact electrically charged was established in the 1890s by showing that they were deflected by an electric field (2). The electrons were obviously electrically negative, being attracted by the positive field plate. In the same apparatus, but with an incomplete vacuum, positive particles were also observed, including a second kind of elementary particle, the proton. The next elementary particle was not discovered until 1932, by which time physicists had developed good techniques for measuring the energies of moving electrons, protons and alpha-particles (helium nuclei). When the alpha-particles (3) from a radioactive element, such as polonium (4), struck a piece of beryllium (5), a light metal, a radiation emerged that was capable of causing a block of wax (6) to emit protons (7) having an energy of 4.5 million electron-volts – surprisingly great. This radiation consisted of neutrons of about the same mass as protons.

EMBRYO

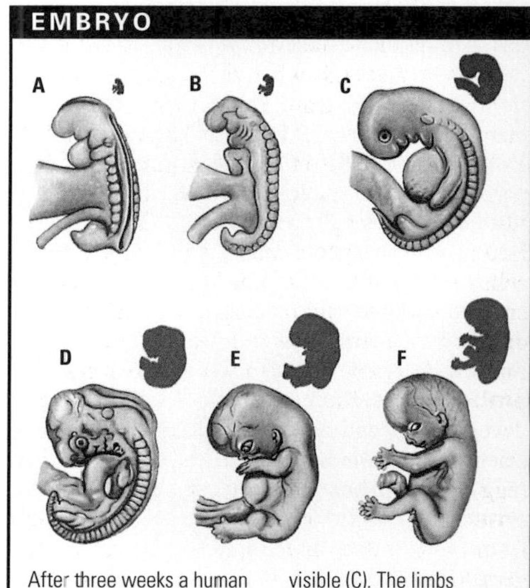

After three weeks a human embryo bears a primitive heart and head (A). By the fourth week, the heart is pumping blood around the body and into the placenta and 25 pairs of tissue blocks (somites) appear, which later give rise to bone and muscle tissue (B). After five weeks, limb buds and rudimentary eyes are visible (C). The limbs become well developed and the tail region recedes by the sixth week (D). The head grows rapidly, eyes, ears and teeth buds appear by the seventh week (E). The tail portion vanishes and almost all the organs and tissues have developed by the eighth week (F). It is now known as a fetus.

elevator *See* LIFT

ellipse CONIC section formed by cutting a right circular cone with a plane inclined at such an angle that the plane does not intersect the base of the cone. When the intersecting plane is parallel to the base, the conic section is a circle. In rectangular Cartesian coordinates (x, y), its standard equation is $x^2/a^2 + y^2/b^2 = 1$ for any non-zero values of a and b. Most planetary orbits are approximately elliptical.

elliptical galaxy Type of regular GALAXY as classified by Edwin HUBBLE.

El Niño Warm surface current that sometimes flows in the equatorial Pacific Ocean towards the South American coast. The current is believed to be closely associated with irregular variations in the global weather system and it occurs approximately every 7–11 years. It is called El Niño (Spanish for "the Christ Child") because it often begins around Christmas. The flow of warm water prevents plankton-rich cold water from the Antarctic rising to the surface off the coasts of Peru and Chile. As a result, fish do not come to feed and local fishermen make no catches. The wider consequences of El Niño can be catastrophic. The current is associated with short-term changes in worldwide climate patterns, and may cause drought in places such as Australia; flooding and severe winters in North America; and violent tropical cyclones in the Pacific Ocean. Some scientists fear that GLOBAL WARMING may be making El Niño occur more frequently.

embolism Blocking of a blood vessel by an obstruction called an **embolus** – usually a blood clot, air bubble or particle of fat. The effects depend on where the embolus lodges. For example, a **cerebral** embolism (in the brain) causes a STROKE; a **limb** embolism may cause gangrene. Treatment is with "clot-busting" drugs, anticoagulants or surgery. *See also* ARTERIOSCLEROSIS

embryo Early developing stage of an animal or plant. In animals, the embryo stage starts at FERTILIZATION, and ends when the organism emerges from the EGG or from its mother's UTERUS. In plants, the embryo is found in the seed and the embryo stage ends on GERMINATION. An embryo results when the nuclei of an OVUM and a SPERM fuse to form a single cell, called a ZYGOTE (fertilized egg). The zygote then divides into a ball of cells called a BLASTULA. The blastula and then the embryo undergo rapid changes in which the cells differentiate themselves to form features, such as limbs and organs. *See also* MEIOSIS; MITOSIS

embryology Biological study of the origin, development and activities of an EMBRYO. This science traces the sequence of events from OVUM (egg) to BIRTH, hatching or germination.

emerald Variety of BERYL, varying in colour from light to deep green and highly valued as a gemstone. The colour is due to the presence of small amounts of chromium but the stone may lose its colour if heated. Emeralds were mined in Upper Egypt in 1650 BC; now they are found mainly in Colombia.

emery Impure form of the mineral CORUNDUM (aluminium oxide, Al_2O_3) that occurs as dark granules with MAGNETITE and HAEMATITE in them. An unusually hard mineral, it is used as an ABRASIVE.

emulsion SUSPENSION of one liquid in another. For example, when a water-based paint is diluted, the paint forms an emulsion with the water. It does not dissolve in the water, but is dispersed into very fine droplets or particles which take an extremely long time to settle out. Fats can form emulsions with water. During DIGESTION, salts from the bile pass into the gut and act on the fats there, reducing their surface tension so that they can form smaller droplets – an emulsion.

enamel Decorative or protective glazed coating produced on metal surfaces, or a type of paint. Ceramic enamels are made from powdered glass and **calx** (a mixture of tin and lead). Various metal oxides are added to give colour. The mixture is melted, quenched and ground, and then applied to the surface. It is fired to give a fused vitreous coating. Enamel paints consist of zinc oxide, lithopone (brown linseed oil) and high-grade varnish. The finish is hard, glossy and highly durable. The term "enamel paint" is derived from its resemblance when dry to the finish found on ceramic-enamel products.

enamel In mammals, hard covering of the crown of the TOOTH; it is the hardest body tissue. Strongest at the biting edges, enamel varies in thickness and density over the tooth.

encephalitis Inflammation of the brain, almost always associated with a viral infection; often there is an associated MENINGITIS. The most common cause is the *Herpes simplex* virus. Symptoms include fever, headache, lassitude and intolerance of light; in severe cases there may be sensory and behavioural disturbances, paralysis, convulsions and coma. The disease is diagnosed

EL NIÑO

The combined influence of land, sea and air on weather conditions can create a global climate rhythm. In the Pacific Ocean, for example (A), trade winds normally blow from E to W (1) along the Equator, "dragging" sun-warmed surface waters into a pool off N Australia and thereby depressing the thermocline – the boundary between warm surface waters and the cooler layers beneath (2). High cumulus clouds form above these warm waters, bringing rain in the summer wet season (3). Cooler, nutrient-rich waters rise to the surface off South America (4), supporting extensive shoals of anchovies on which a vast fishing industry has developed. The weather over this cold water region is dry. Every 3–5 years a change occurs in the ocean-atmosphere interaction. The climatic pattern is reversed (B) – an event known as El Niño. The trade winds ease, or even reverse direction (5), during El Niño and the warm surface waters which have "piled up" in the W Pacific flow back to warm the waters off South America by 2–3°C (3.6–5.4°F) (6). This depresses the E thermocline (7) and dramatically affects climate. In an El Niño year, drought and bush fires occur over Australia, while floods affect Bolivia and Peru. The warm waters off South America suppress upwelling of the cold nutrient-rich waters, bringing disaster to the fishing industry.

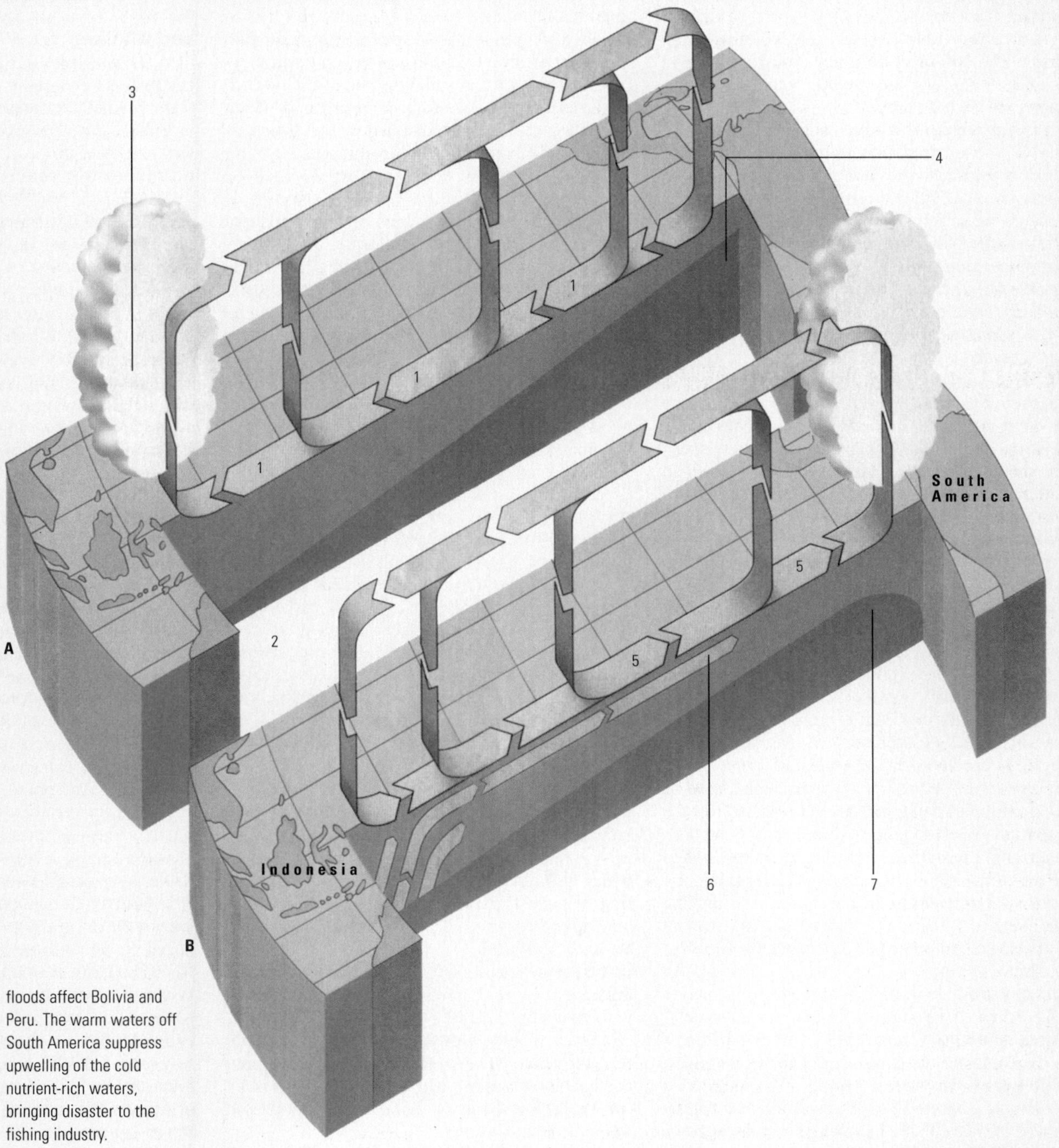

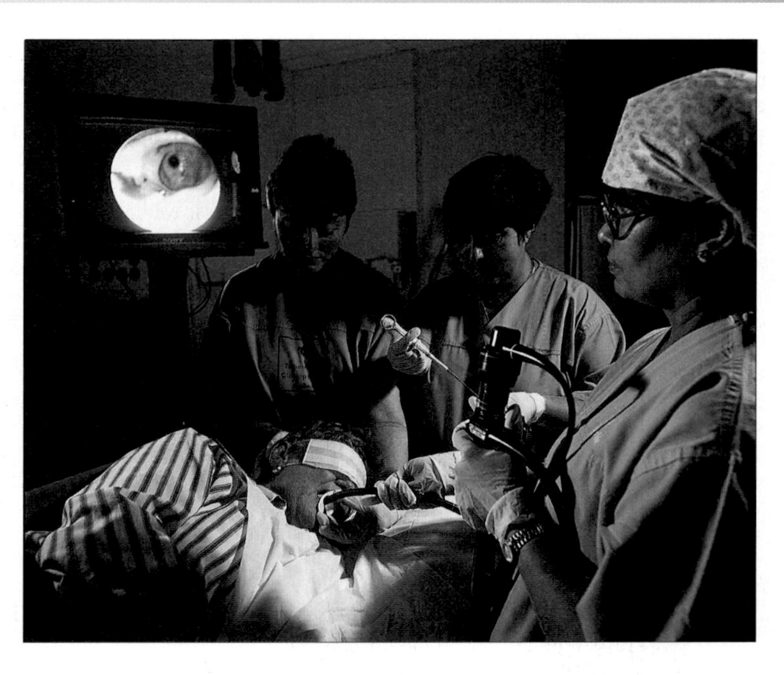

► **endoscope** This photograph shows doctors performing an endoscope examination of a woman's stomach (gastroscopy) and taking a sample of tissue (a biopsy). The endoscope has been inserted through the patient's mouth and fed down through her throat. The image obtained by the endoscope is on the screen at upper left. The doctor with the syringe-like implement is controlling a surgical instrument on the end of a fibre which has been fed through the endoscope cable. An endoscope is a flexible fibre optic cable through which internal cavities can be viewed. This technique is routinely used in the diagnosis or cancer or ulcers.

by tests on CEREBROSPINAL FLUID, which is obtained by lumbar puncture and possibly brain scans. The anti-viral drug acyclovir is effective against *H. simplex* encephalitis; otherwise treatment is mainly supportive.

endangered species Animals or plants threatened with extinction in the foreseeable future as a result of such activities as habitat destruction and pollution: nearly 1,000 different species of animals throughout the world and 20,000 plants are affected. Notable examples of animals at risk are porpoises, elephants and whales.

endemic Term to describe a disease that remains prevalent in a particular region or country. It is most often heard in connection with tropical diseases, which are difficult to eradicate.

Enders, John Franklin (1897–1985) US microbiologist. He shared the 1954 Nobel Prize for physiology or medicine with Frederick C. ROBBINS and Thomas H. WELLER for the discovery that poliomyelitis viruses can be grown in cultures of various types of tissues. This work was fundamental to the later development of the polio VACCINE.

endocrine system Body system made up of all the **endocrine glands** (ductless glands) which secrete HORMONES directly into the bloodstream to influence body processes. The endocrine system (together with the AUTONOMIC NERVOUS SYSTEM (ANS), and collectively called the **neuroendocrine system**) controls and regulates all body functions. It differs from other body systems in that its ductless glands are not structurally connected to one another. The chief endocrine glands are the PITUITARY GLAND, located at the base of the brain; the THYROID GLAND, located in the throat; the PARATHYROID GLANDS, usually embedded in the thyroid; the ADRENAL GLAND, situated on top of the kidneys; specialized cells called the **islets of Langerhans** in the PANCREAS; and the sex gland or GONAD (TESTIS in males and OVARY in females) and, in pregnancy, the placenta. *See also* EXOCRINE GLAND

endocytosis In biology, process by which a CELL takes in substances. When a cell's MEMBRANE comes into contact with food, a portion of the cytoplasm surrounds the substance and a depression forms within the cell wall. The food is eventually engulfed into a vessicle, which then moves further into the cell where it meets a LYSOSOME. Within the lysome are digestive enzymes that break down the food. There are two types of endocytosis: **pinocytosis** is the incorporation and digestion of dissolved substances, and **phagocytosis** is the engulfing and digestion of microscopic particles. Phagocytosis is the process by which many PROTOZOA obtain sustinence. In higher animals, PHAGOCYTE cells are an important part of the IMMUNE SYSTEM, helping to expel infectious microorganisms.

endoderm (entoderm) One of the three so-called GERM LAYERS of tissue formed in the early development of a fertilized OVUM (egg). It is the innermost layer of the three and forms the EPITHELIUM of the liver, pancreas, digestive tract and respiratory system. It is also the inner cell layer of a simple animal body. Other germ layers are the ECTODERM and MESODERM.

endometrium MUCOUS MEMBRANE, well supplied with blood vessels, that lines the UTERUS. It becomes thickened in the latter part of the MENSTRUAL CYCLE, but if no pregnancy takes place it is shed during menstruation.

endoparasite *See* PARASITE

endoplasmic reticulum Network of membranes and channels in the CYTOPLASM of EUKARYOTE cells (those which contain a nucleus, such as the cells of plants, animals and fungi). These help to transport material inside the cell. Parts of the endoplasmic reticulum are covered with minute granules called RIBOSOMES. These consist of protein and a form of ribonucleic acid (RNA) and are sites of protein synthesis.

endorphin Naturally occurring body PEPTIDE that has the same effects as morphine and other derivatives of OPIUM. It acts as a natural painkiller, lowering a person's perception of pain by reducing the signals traffic between neurons (nerve cells). The existence of these substances, which are also known as **endogenous opiates**, first became apparent in the course of research into opiate drugs such as heroin and morphine. *See also* ANALGESIC

endoscope Instrument used to examine internal organs or tissues. Flexible or rigid, it is equipped with an eyepiece, lenses and its own light source. Examples include a **gastroscope** (for examining the stomach) and a **cystoscope** (for examining the urinary bladder). MINIMAL ACCESS SURGERY can be performed using fine instruments passed through the endoscope.

endoskeleton Supporting system of bones and cartilage inside the body of an animal. All VERTEBRATES have an endoskeleton. It supports the body, and surrounds and protects internal organs. It also provides attachment points for muscles in order to facilitate movement. SPONGES have a type of endoskeleton consisting of a network of rigid or semi-rigid spicules. *See also* EXOSKELETON

endosperm Tissue that surrounds the developing embryo of a seed and provides food for growth. It is **triploid** (each cell has three sets of chromosomes), being derived from the FUSION of one of the male GAMETES from the germinated pollen grain and two of the HAPLOID nuclei in the embryo sac. *See also* ALTERNATION OF GENERATIONS

endosymbiosis Mutually beneficial relationship in which one organism lives inside another. For example, bacteria were engulfed by EUKARYOTE cells and formed symbiotic relationships with them, eventually becoming so interdependent that the cells behaved as a single organism; the bacteria became MITOCHONDRIA and CHLOROPLASTS. Mitochondria and chloroplasts have their own small RIBOSOMES, which resemble those of bacteria, and also their own DNA distinct from that of the host cell. *See also* SYMBIOSIS

endothelium Tissue consisting of a thin layer of cells that lines various body structures. Like EPITHELIUM, endothelium is made up of plate-like cells. It lines the inner surfaces of blood vessels and the heart; lymphatic vessels are also lined with endothelium.

endothermic reaction Chemical reaction in which heat is absorbed from the surroundings, causing a fall in temperature – as in the manufacture of water-gas from coke and steam.

energy In physics, capacity for doing work It is measured in joules. POTENTIAL ENERGY is an object's ability to do work because of a change in the object's position or shape. KINETIC ENERGY is the energy an object has because it is moving. The many forms of energy include mechanical, electrical, nuclear, thermal (heat), light and chemical. Energy undergoes transformations: thermonuclear reactions in the Sun release energy (mainly light); photosynthesis in plants stores this energy in chemical form; ingestion of the plant by animals allows muscles to transform this energy yet again into physical action. The idea that MASS is a form of energy was established in 1905 by Albert EINSTEIN, who recognized that energy (E) and mass (m) could be transformed into each other according to the relation $E = mc^2$), where c is the velocity of light. The law of conservation of energy states that energy cannot be created or destroyed. *See also* CONSERVATION, LAWS OF

energy level Fixed amount of energy possessed by a nucleus, ELECTRON, atom or molecule. For example, the energy of electrons in an atom is not continuously variable. It can have only a discrete set of values, termed energy levels. The lowest possible energy level is called the **ground state**. If the atom absorbs energy, possibly in the form of a PHOTON (a quantum of light or other ELECTROMAGNETIC RADIATION), an electron can be raised to an ORBITAL of a higher energy level, when the atom is said to be in an EXCITED STATE. When the electron in an excited atom returns to its previous energy level, the excess energy is emitted, often as a quantum of light. In a similar way, molecules have energy levels involved with rotational and vibrational motion.

energy sources Naturally occurring substances and processes from which we obtain the energy that we need. The energy in all of these sources is derived almost entirely from the Sun. The fossil fuels coal, oil and natural gas are the remains of life that was dependent for growth on SOLAR ENERGY. HYDROELECTRICITY also derives from solar energy, which maintains the Earth's HYDROLOGICAL CYCLE. Wind power also depends on the Sun, and is caused by uneven heating of the atmosphere. The movements of tides and waves depend upon solar thermal energy and on the gravitational pulls of the Sun and Moon. These movements of the oceans can be harnessed to generate electricity. We use solar energy directly for heating some domestic water supplies, and for providing electricity from PHOTOELECTRIC CELLS. GEOTHERMAL ENERGY is energy obtained from underground hot rocks. Other major energy sources are radioactive metals, such as uranium, plutonium and thorium, which provide NUCLEAR ENERGY.

engine Machine that produces useful energy of motion from some other form of energy. The term is usually restricted to combustion engines, which burn fuel. These machines include the STEAM ENGINE, DIESEL ENGINE, PETROL ENGINE, JET ENGINE and the ROCKET engine. Such engines are distinguished from ELECTRIC MOTORS which, in providing their power, do not directly alter the chemical or physical composition of a substance. Combustion engines are of two main kinds. An external combustion engine burns its fuel outside the chamber in which motion is produced. In the steam LOCOMOTIVE, for example, the fire box is separate from the cylinders. An INTERNAL COMBUSTION ENGINE burns its fuel and develops motion in the same place. In the petrol engine, for example, this all takes place in the cylinders.

engineering Application of scientific principles for practical purposes, such as the design and construction of buildings and machines. MECHANICAL ENGINEERING includes the design and testing of machinery. CIVIL ENGINEERING includes the preparation of sites and the design and building of structures, such as bridges, tunnels and harbours. Electrical engineers work on electrical systems. In ELECTRONICS, engineers are concerned with devices such as scientific instruments, radio and television equipment, radar systems and computers. Nuclear engineers install and operate nuclear reactors. CHEMICAL ENGINEERING includes the design and operation of large-scale chemical processes. There is some overlap of expertise in these many fields of engineering. For this reason, the academic training of all engineers starts with a thorough grounding in the fundamentals of science and an education in general engineering subjects. This is followed by specialized training in the student's chosen field.

engineering, genetic *See* GENETIC ENGINEERING

enstatite ($MgSiO_3$) Orthorhombic mineral of the pyroxene group, commonly found in ultrabasic IGNEOUS rocks such as norites, pyroxenites, gabbros and peridotites. It varies from colourless to yellowish grey, shading to green if iron is present. Hardness, 5.5; r.d. 3.2–3.5.

enthalpy (symbol H) Amount of thermodynamic heat energy possessed by a substance. In any system, enthalpy equals the sum of its internal energy and the product of the pressure and volume. It can be measured in terms of the change (usually uptake) of heat that accompanies a CHEMICAL REACTION that takes place at a constant pressure, and as such is equivalent to the heat of reaction (usually expressed in kilojoules per mole). For a reaction that produces heat (EXOTHERMIC REACTION), the enthalpy is taken to be negative. *See also* THERMODYNAMICS

entomophily POLLINATION of a flower by a pollen-carrying insect. Most entomophilous flowers are scented and brightly coloured, often with nectar, to attract insects. The flowers may have different structures to ensure CROSS-POLLINATION.

entropy Quantity that specifies the disorder or randomness of a physical system. In THERMODYNAMICS, it expresses the degree to which thermal energy is available for work – the less available it is, the greater the entropy. The entropy of the Universe is increasing. Energy can be extracted from a system only as it changes to a less ordered state. According to the second law of thermodynamics, an isolated system's change in entropy is either zero or positive in any process. In INFORMATION THEORY, the entropy of an information system increases with the number of random errors in transmitting the message.

E number Names of certain food additives used in the European Union. Food additives approved for use within the European Union must by law be identified in the list of ingredients of any product in which they are used. Such additives might be preservatives, flavour enhancers, stabilizers or colouring agents.

environment Physical and biological surroundings of an organism. The environment covers both non-living (ABIOTIC) factors such as temperature, soil, atmosphere and radiation, and living organisms such as plants, microorganisms and animals that make up the biotic (living) environment.

enzyme (Gk. *zymosis*, in yeast) Protein that functions as a CATALYST in biochemical reactions. The FERMENTATION properties of yeast cells have long been utilised in BREWING alcohol. In 1897 German chemist Edward BÜCHNER discovered that cell-free extracts of yeasts could ferment sugars to alcohol. In 1926 the US biochemist James B. SUMNER was the first person to isolate an enzyme in pure crystal form. He extracted urease from the jack bean and proved that enzymes are protein molecules. In the next decade, PEPSIN, TRYPSIN and CHYMOTRYPSIN were crystallized. Today, more than 1,500 catalysts have been identified, and their AMINO ACID structures revealed through X-RAY CRYSTALLOGRAPHY. In 1969 scientists first synthesized an enzyme, ribonuclease. Chemical reactions can occur several thousand or million times faster with enzymes than without them. The efficiency of an enzyme is measured in terms of its turnover rate. They operate within a narrow temperature range (usually 30°C to 40°C) and have optimal pH ranges. Many enzymes have to be bound to non-protein molecules. These include **trace elements** (such as metals) and **coenzymes** (such as vitamins). If a coenzyme is tightly attached to the protein enzyme, the unit is termed a **prosthetic group**. The lack or malfunction of enzymes can cause a variety of metabolic diseases. Enzymes are widely used in the manufacture of detergents, antibiotics, and food.

Eocene Second of the five epochs of the TERTIARY period, from c.55–38 million years ago. The name implies the "dawn" of life in which the modern families appeared. The fossil record shows members of modern plant genera, including beeches, walnuts and elms, and indicates the apparent dominance of mammals, including the ancestors of camels, horses (notably *Hyracotherium*), rodents, bats and monkeys. The world climate was warmer than at the present time.

ephedrine Drug, chemically similar to ADRENALINE. It stimulates the AUTONOMIC NERVOUS SYSTEM. Once used to treat asthma, it is still found in some cold remedies, as a nasal decongestant.

ephemeris (pl. ephemerides) Astronomical table giving the predicted positions of a celestial object, such as a planet or comet, at given intervals. The term also signifies an annual publication supplying such tables along with information concerning the Sun, eclipse and occultation data, data for certain stars, astronomical constants, and so forth.

ephemeris time (ET) System of time reckoning normally employed in the compilation of astronomical data for almanacs and ephemerides. It takes no account of irregularities in the Earth's rotation and is calculated on the basis of the tropical year. It is slightly in advance of UNIVERSAL TIME.

epicentre Spot on the Earth's surface directly above the focus of an EARTHQUAKE. Depending on the character of the focus, the epicentre may be a small circle or a line.

epidemic Outbreak of an infectious disease with rapid spread to a great many people. The study of epidemics, which includes the causes, patterns of contagion and methods of containment of disease, is known as EPIDEMIOLOGY. An epidemic that sweeps across many countries, as the Black Death did in the Middle Ages, is termed a pandemic.

epidemiology Study of the incidence and patterns of disease with a view to finding means of prevention or control. Traditionally, epidemiological methods were applied to communicable diseases – those, such as CHOLERA or MEASLES, that can be passed from one person to another. However, modern epidemiology is equally con-

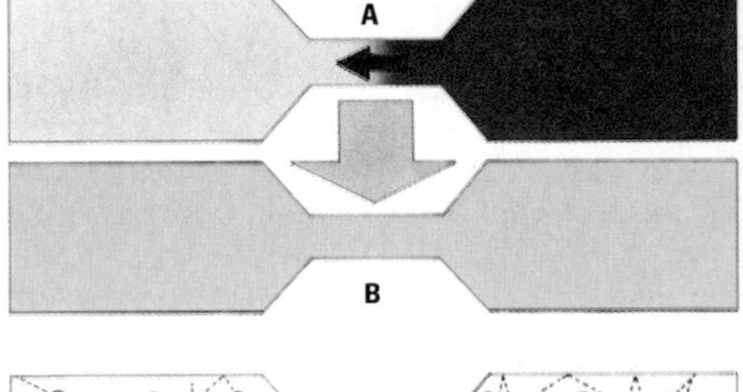

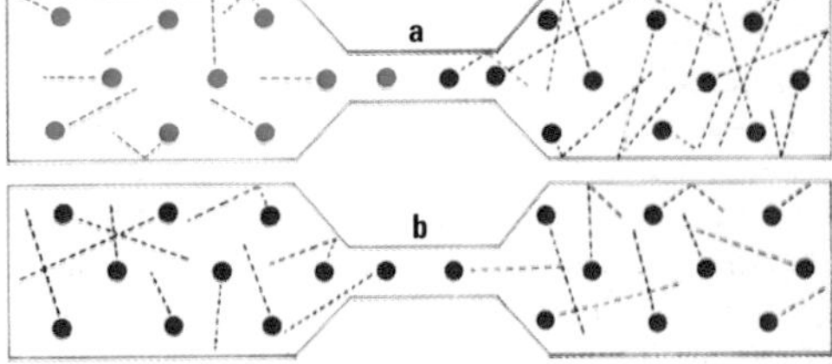

▲ **entropy** The tendency for heat to pass from a hotter to a cooler body (A), and never in the opposite direction unless externally assisted, can be stated as a quite general principle: the Second Law of Thermodynamics. The change that occurs due to the redistribution of heat – the uniform condition (B) resulting ultimately from the asymmetrical condition (A) – can be expressed as an increase in entropy: the increase in this quantity is defined as the amount of heat transferred divided by the temperature at which the heat-flow occurs. Entropy can be though of as relative disorder: B is a less organized situation than A. It can also be thought of as a measure of probability. If a given amount of heat energy – which is energy or particle-motion – is to be distributed at random throughout the object in the diagram, it is far more probably that the distribution will be a uniform mixture of speeds (b) than that the high and low speeds will be concentrated in different places (a).

cerned with the role of environmental and lifestyle factors in disease causation: for example, the link between smoking and lung CANCER or between ASBESTOS and the cancer known as mesothelioma.

epidermis In animals, outer layer of skin that contains no blood vessels; in plants, the outermost layer of a leaf or of an unthickened stem or root. In plants, the outer surface of the epidermis is usually coated in a waxy layer, the CUTICLE, which reduces water loss. The epidermis of leaves and stems may be perforated by pores, called STOMATA, which allow GAS EXCHANGE.

epididymis Network of ducts in the TESTES where sperm cells mature and are stored. *See also* VAS DEFERENS

epidote Orthosilicate mineral, hydrated calcium iron-aluminium silicate $(Ca_2Fe_3(Al_2O)(OH)(Si_2O_7)\ [Si_2O_4])$. It is found in METAMORPHIC and IGNEOUS rocks as monoclinic system prismatic crystals and fibrous or granular masses; it is typically pistachio green, glassy and brittle. Certain large crystals, 7.5–25cm (3–10in), are collector's items. Hardness 6–7; r.d. 3.4.

epidural anaesthesia Injection of a local anaesthetic into the epidural space surrounding the spinal cord. It is done to anaesthetize the spinal nerve roots and so prevent pain during surgery or childbirth.

epiglottis Small flap of CARTILAGE projecting upwards behind the root of the tongue. It closes off the LARYNX during swallowing to prevent food entering the airway.

epilepsy Disorder characterized by abnormal electrical discharges in the brain provoking seizures of varying intensity. Affecting slightly more males than females, epilepsy occurs in up to 2% of the population. It arises either spontaneously or as a result of some brain disorder. It is seen both in generalized forms, involving the whole of the CEREBRAL CORTEX, or in partial (**focal**) attacks arising in one small part of the brain. Attacks are often presaged by warning symptoms, the "aura". Seizure types vary from the momentary loss of awareness seen in *petit mal* attacks (called **absences**) to the major convulsions of *grand mal* epilepsy. They may be triggered by a number of factors, including sleep deprivation, flashing lights or excessive noise. All forms of epilepsy are controlled with anticonvulsant drugs; in some cases surgery may be recommended.

epilimnion Upper layers of water in a lake. It is warmer than the deeper water and, because light penetrates the epilimnion, it can support PHOTOSYNTHESIS and plant life. The water may be disturbed by wind and water currents. *See also* HYPOLIMNION

epinephrine *See* ADRENALINE

epiphyte (air plant) Plant that grows on another plant but is not a PARASITE. Epiphytes usually have aerial roots and produce their own food by PHOTOSYNTHESIS. They are common in tropical forests. Examples are some ferns, some orchids, Spanish moss and many bromeliads.

epithelium Layer of cells, closely packed to form a surface or a lining for a body tube or cavity. Epithelium covers not only the SKIN, but also various internal organs and surfaces such as the intestines, nasal passages and mouth. Epithelial cells may also produce protective modifications such as hair and nails, or secrete substances such as ENZYMES and MUCUS.

epoch *See* GEOLOGICAL TIME

epoxide In organic chemistry, a compound that has a three-membered ring including oxygen in its molecular make-up. An example is the gas epoxyethane (ethylene oxide, C_2H_4O), which may be regarded as a cyclic ETHER. Polymerization of epoxides with PHENOLS produces EPOXY RESINS, which are used as adhesives and electrical insulators.

epoxy resin Group of thermosetting polymers with outstandingly good mechanical and electrical properties, stability, heat and chemical resistance, and adhesion. Epoxy resins are used as adhesives, in casting and protective coatings. Adhesive epoxy resins are sold in two separate components, a viscous resin and a hardener, which are mixed just before use. *See also* THERMOSETTING RESIN

EPROM (**e**rasable **p**rogrammable **r**ead-**o**nly **m**emory) Type of read-only computer memory (ROM) in which data stored on an INTEGRATED CIRCUIT can be erased, usually by means of ultraviolet light. It can then be re-programmed using pulses of electricity. *See also* PROM; RAM

Epstein-Barr virus Organism responsible for GLANDULAR FEVER, an illness that occurs worldwide in teenagers and young adults. The symptoms include fatigue, headaches, muscular aches, sore throats and enlarged lymph nodes. Epstein-Barr virus is also implicated in the causation of a number of other diseases, in particular BURKITT'S LYMPHOMA, a common childhood cancer in Central Africa. The virus is spread by droplet infection.

equation Mathematical statement that holds for some subset of all possible values of the variables. An equation of the form $x^2 = 8 - 2x$ is true only for certain values of x ($x = 2$ and $x = -4$). These values are the SOLUTIONS of the equation.

EPOXY RESIN

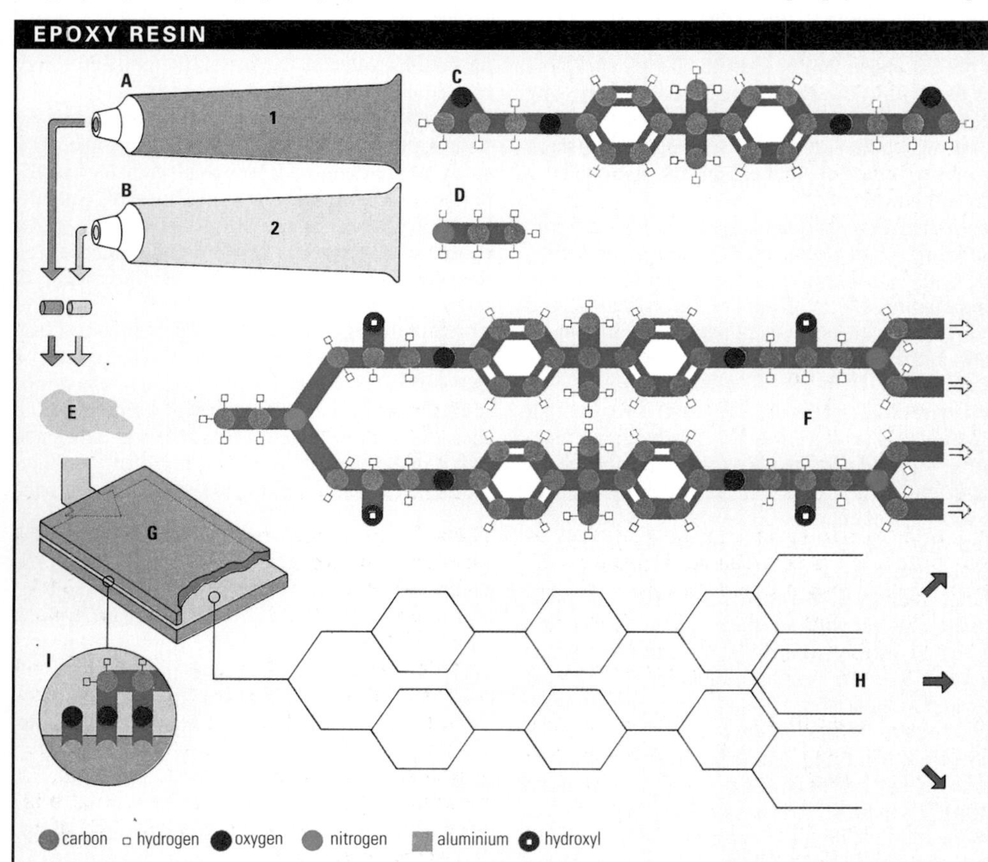

Epoxy or epoxide resins are thermosetting resins used as adhesives. The most valuable single property of the epoxy resins is their ability to transform readily from the liquid (or thermoplastic) state to tough, hard thermosetting solids. This is done by the addition of a catalyst, which cures or sets the resins. In the example shown the resin is diglycidyl ether of bisphenol A ($C_{21}H_{24}O_4$), and the catalyst is ethylamine ($C_2H_5NH_2$). Epoxy resins are known for their adhesive strength, which is due to thorough wetting and a low shrinkage rate during cure.

Key

A) Tube 1 contains the epoxy resin diglycidyl ether of bisphenol A (DGEBA). Epoxy is the term used to describe a compound in which an oxygen atom is bound to two carbon atoms, forming a three-membered ring.

B) Tube 2 contains the catalyst ethylamine. This is a primary amine as it contains one organic radical attached to the nitrogen atom. It combines with other compounds by losing the hydrogen atoms which are directly attached to the nitrogen atom.

C) The structural formula of DGEBA.

D) The structural formula of ethylamine.

E) Equal quantities of tube 1 and 2 (by volume) are mixed to a uniform paste.

F) The ethylamine attacks the epoxy group at the end of the DGEBA chain and forms a bond with the last carbon atom. Similarly with another DGEBA chain by losing another hydrogen atom. This is the curing mechanism which eventually gives the thermoset structure.

G) Section of materials to be joined by epoxy adhesive. The surfaces must be well prepared by cleaning, sanding and wire-brushing. This allows the adhesive to wet the surface more thoroughly. The material shown is aluminium.

H) As the curing of the epoxy resin proceeds, a 3-D network forms which causes the liquid to transform to a hard, tough solid. Normally the catalyst is a mixture of different amines suitably chosen to promote the network.

I) The adhesive force at the surface of the material is complex. It is a combination of mechanical, electrical and chemical binding forces. The example shown is chemical bonding to an aluminium surface. After surface preparation, some aluminium atoms in the surface-oxide layer have unsatisfied binding forces, so they interact with the adhesive at the epoxy groups.

This type of equation is contrasted with an **identity**, such as

$$(x + 2)^2 = x^2 + 4x + 4$$

which is true for all values of x. Equations are classified in various ways: for example the highest power of the variable (2 in the case above) is the degree of the equation. Equations are said to be linear, QUADRATIC, cubic, quartic, etc., according to whether their degree is 1, 2, 3, 4, etc. *See also* SIMULTANEOUS EQUATIONS

equation of state Formula that connects the pressure, volume and temperature of a system containing a given amount of substance. The most simple equation of state is the IDEAL GAS LAW.

equator Name given to two imaginary circles. The **terrestrial Equator** lies midway between the North Pole and South Pole and is the zero line from which latitude is measured. It divides the Earth into the Southern and Northern Hemispheres. The **celestial equator** lies directly above the Earth's Equator and is used as a reference to determine the position of a star using an astronomical coordinate such as the EQUATORIAL COORDINATE SYSTEM.

equatorial coordinate system Astronomical coordinate system that refers to the plane of the CELESTIAL EQUATOR. Two sets of coordinates are used: right ascension and declination, and hour angle and declination.

equatorial telescope Common TELESCOPE mounting in which the telescope rotates about two perpendicular axes, one of which, the polar axis, is parallel to the Earth's axis, and so points towards the CELESTIAL POLE. The steady movement of celestial bodies, due to the Earth's rotation, can be followed by a daily rotation of the telescope about this axis, usually by means of a clockwork drive.

equilateral triangle Planar figure with three sides all of equal length; the three interior angles are also equal and each of magnitude 60°. *See also* TRIANGLE

equilibrium In physics, a stable state in which forces acting on a particle or object negate each other, resulting in no net force. While thought of as a state of balance or rest, an object with constant velocity is also said to be in equilibrium. The term can also be ascribed to a body with a constant temperature; this is known as **thermal equilibrium**.

equilibrium constant Ratio of concentrations of products to those of reactants, characterizing the CHEMICAL EQUILIBRIUM of a particular REVERSIBLE REACTION at a specified temperature.

equilibrium sense (vestibular sense) Mammalian ability to remain upright in relation to gravity and to detect changes in position and momentum. The principal organs of equilibrium are contained in the inner EAR – the UTRICLE, which transmits orientation information, and the SEMICIRCULAR CANALS, which are concerned with acceleration and deceleration. These systems help humans to locate their bodies in space.

equinox Either of the two points at which the Sun crosses the celestial EQUATOR. The **vernal** equinox (or **spring** equinox) occurs when the Sun crosses the equator from S to N on or near 21 March each year in the Northern Hemisphere. This point is also called the First Point of Aries. The other equinox is the **autumnal** equinox, which occurs on or near 23 September in the Northern Hemisphere when the Sun crosses the equator from N to S. This point is also called the First Point of Libra. At the equinoxes, the Sun rises due E and sets due W.

EROSION

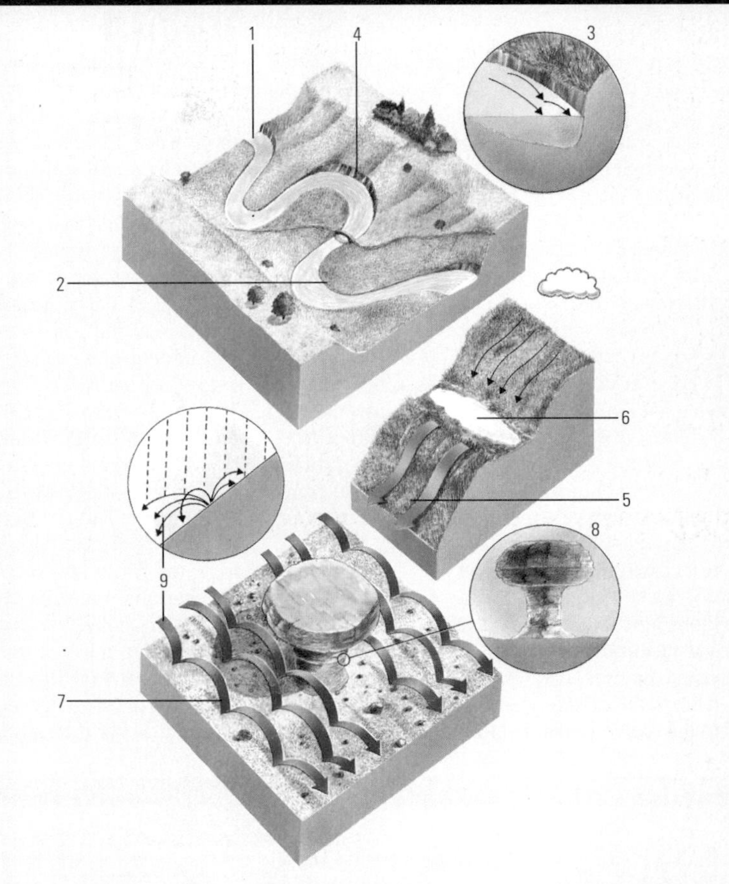

Erosion is the breakdown and transportation of rock due to the action of an outside agent. There are three main forms: river, glacial and wind. Rivers (1) erode their channels through the flow of water and the abrasion of the load they are carrying against the banks and riverbed. Erosion is most forceful at the outside of bends (2), where the banks are undercut (3) often creating cliffs or bluffs (4) down which material moves. Flood surges dramatically increase the power of the river and correspondingly magnify the erosive force.On a smaller scale, rainwater will move material down a hillside (5). Particles of soil are carried by rivulets and the impact of raindrops throws soil down slope. Vegetation reduces such erosion by binding the soil together. Where vegetation is removed, as on tracks (6), erosion is accentuated. In arid conditions wind erosion can carve distinctive features. Sand and stones blown by the wind (7) have the same effect as shot-blasting. Mushroom-shaped formations, pedestals, (8) are often the result. This is due to the maximum height at which the erosive sand is carried by the wind as it bounces across the surface (9), above which the rock is untouched.

equivalence principle Principle that observations made in a closed laboratory (from which the external Universe cannot be observed) cannot distinguish between the effects of GRAVITATION and of an ACCELERATION of the laboratory. It is a cornerstone of Einstein's general theory of RELATIVITY.

equivalent weight Weight of an element, in grams, that combines with 8g of oxygen or with an equivalent weight of some other substances. For example, the equivalent weight of hydrogen is 1 in forming water (H_2O). The equivalent weight of an acid is the number of grams that contains 1g of acidic hydrogen. A similar definition is used for bases.

Equuleus (foal) Small constellation of the N SKY. Its brightest star, Alpha, has a magnitude of 3.9.

Eratosthenes (276–194 BC) Greek scholar who first measured the Earth's circumference by using geometry. Eratosthenes administered the library of Alexandria and was renowned for his work in mathematics, geography, philosophy and literature. He invented a systematic method of identifying prime numbers in which those numbers with divisors are identified and removed from a list; the remaining numbers are prime numbers. The method is known as **Eratosthenes' sieve**, but is, unfortunately, too slow to decide whether or not a very large number is prime. *See also* NUMBER PRIME

erbium (symbol Er) Silvery, metallic element of the LANTHANIDE SERIES, first isolated in 1843. There are six naturally occurring isotopes, and the chief ores are monazite and bastnaesite. Twelve radioactive isotopes have been identified. Soft and malleable, erbium is used in some specialized alloys, and erbium oxide is used as a pink colorant for glass. Properties: at.no. 68; r.a.m. 167.26; m.p. 1,522°C (2,772°F); r.d. 9.045; b.p. 2,863°C (5,185°F); most common isotope ^{166}Er (33.41%).

erg Unit of energy in the metric CGS (centimetre-gram-second) system of units. One erg is the work done by a force of 1 DYNE acting through a distance of 1 cm. One JOULE equals 10 million ergs.

ergosterol Organic chemical compound, a STEROID, found particularly in yeast and other fungi. When irradiated with ultraviolet light, it is converted to vitamin D.

Eridanus (river) Large but inconspicuous constellation of the S sky. Achernar is its brightest star, with a magnitude of 0.46.

Eros Elongated asteroid with an irregular shaped orbit. In 1931 and 1975 it approached to within 24 million km (15 million mi) of the Earth. Longer diameter 27km (17mi); mean distance from the Sun 232 million km (144 million mi); mean sidereal period 1.76 yr.

erosion Alteration of landforms by the wearing away of rock and soil, and the removal of any debris (as opposed to WEATHERING). Erosion is carried out by the agents of wind, water, GLACIERS and living organisms. In **chemical** erosion, minerals in the rock react to other substances, such as weak acids found in rainwater, and are broken down. In **physical** erosion, powerful forces such as rivers and glaciers physically wear rock down and transport it. Erosion can have disastrous economic results, such as the removal of topsoil, the gradual destruction of buildings, and the alteration of water systems. Inland, erosion occurs most drastically by the action of rivers, and in coastal regions, by the action of waves. *See also* GEOMORPHOLOGY

erratic In geology, a rock that has been transported some distance from its source by glacial action, and is therefore of a different type to the surrounding rocks. The tracing of erratics back to their source can give important information about the movement of ice.

eruption Release of lava and gas from the Earth's interior to the Earth's surface, either on land or under the sea, and into the atmosphere. *See also* VOLCANO

erythrocyte (red blood cell) BLOOD cell that carries oxygen around the body. It contains HAEMOGLOBIN, which combines with oxygen to form OXYHAEMOGLOBIN and gives blood its red colour. In mammals erythrocytes are usually disc-shaped and have no nucleus. In other vetebrates they are more oval in shape and contain nuclei. Normal human blood contains an average of 5 million such cells per mm^3 of blood, making them the most numerous blood cell. They are manufactured in bone MARROW.

Esaki, Leo (1925–) Japanese physicist who developed the tunnel diode, a SEMICONDUCTOR that allows electrons to cross normally impassable electronic barriers. Ivar GIAEVER extended Esako's research to the field of SUPERCONDUCTIVITY, which led to the prediction of the **Josephson effect** (electrical effects of a current flowing between superconductors). For this work, they shared the 1973 Nobel Prize for physics with Brian JOSEPHSON. The tunnel diode, also called **Esaki diode**, has found extensive applications, which include its use in computer information storage.

escalator Electrically powered moving stairs, driven by chain and sprocket and held in the correct plane by two tracks. It is usually inclined at 30° and limited to a total rise of up to 18m (60ft). As the steps approach the landing, they pass through a protective comb-like device. Escalators are used to carry pedestrians between floors in busy areas such as large stores, office buildings and underground railway stations. The Otis Elevator Company exhibited the first escalator in 1900 at the Paris Exposition.

escape velocity Minimum velocity required to free a body from the gravitational field of a celestial body or stellar system. Escape velocities are, 11.2km s^{-1} (7mi s^{-1}) for the Earth and 2.4km s^{-1} (1.5mi s^{-1}) for the Moon. They can be calculated from the formula: $v = (2GM/R)^{1/2}$ where G is the gravitational constant, M the mass of the planet or system and R the distance of the rocket from the centre of mass of the system.

escarpment (scarp) Steep slope or cliff arising from a flat or gently sloping area, produced by faulting and/or differential EROSION.

ESCAPE VELOCITY

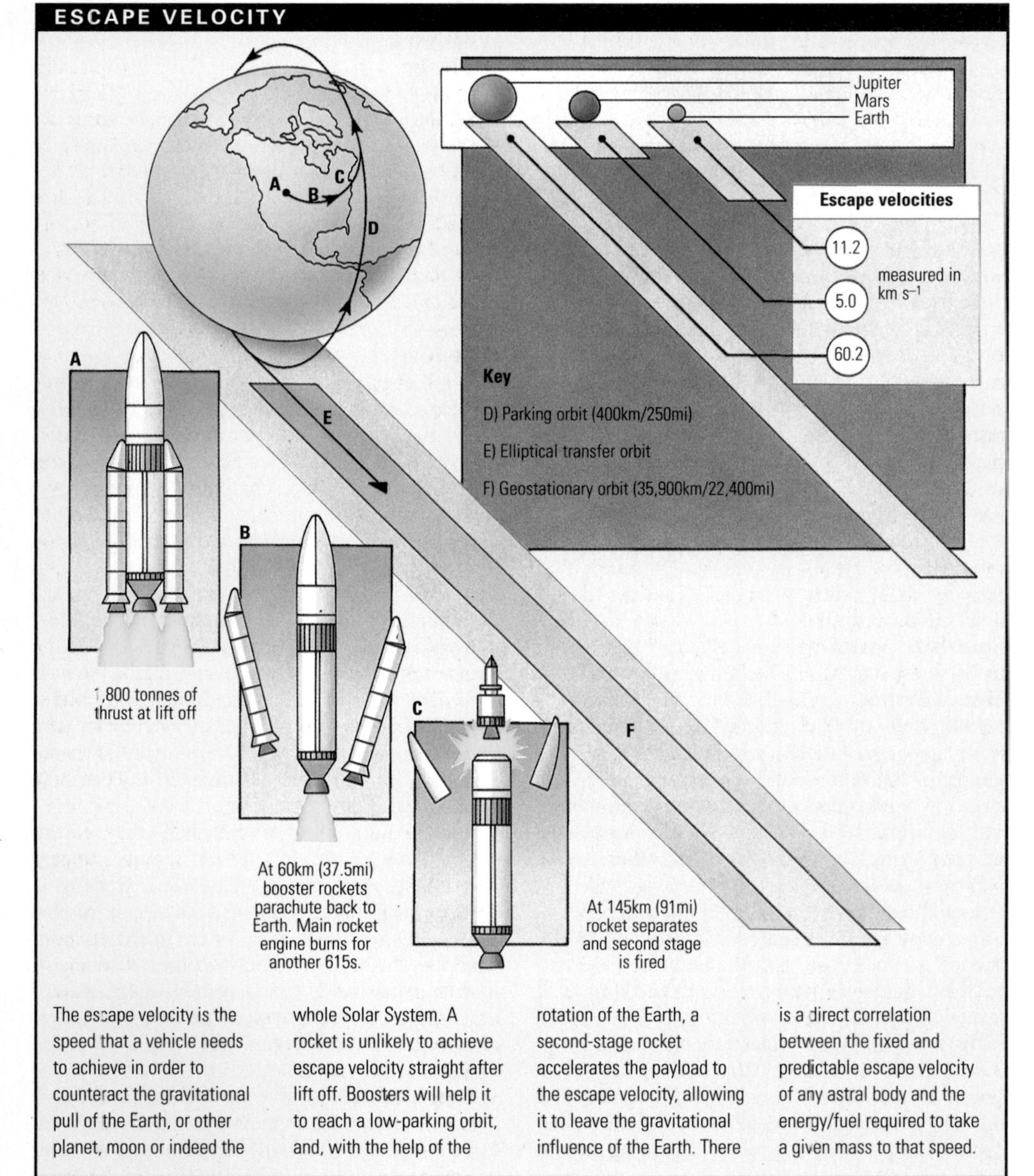

The escape velocity is the speed that a vehicle needs to achieve in order to counteract the gravitational pull of the Earth, or other planet, moon or indeed the whole Solar System. A rocket is unlikely to achieve escape velocity straight after lift off. Boosters will help it to reach a low-parking orbit, and, with the help of the rotation of the Earth, a second-stage rocket accelerates the payload to the escape velocity, allowing it to leave the gravitational influence of the Earth. There is a direct correlation between the fixed and predictable escape velocity of any astral body and the energy/fuel required to take a given mass to that speed.

esker Long, narrow ridge of rock formed by a stream flowing under a GLACIER. Subglacial streams and rivers carry gravel and sand which they deposit on their beds as they tunnel below the ice. When the ice melts, the rivers may move to other channels, leaving the eskers as long ridges 10–25m (30–80ft) high and 5–25m (15–80ft) across; they may be many kilometres long.

essential amino acid *See* AMINO ACID

essential element Element needed by a living organism so that it can grow, develop and maintain its body structure normally. All organisms require carbon, hydrogen, nitrogen and oxygen, which they take in the form of food, air and water, and which make up most of their body tissues. In addition, most organisms also need calcium, chlorine, magnesium, phosphorus, potassium, sodium and sulphur, which are termed **major elements**. Required in much smaller quantities are the so-called **trace elements**, which include cobalt, copper, chromium, iodine, iron, manganese, selenium and zinc. All of these additional essential elements – major and trace – have to be supplied by the organism's food.

essential oil Oil found in flowers, fruits or plants. It is the source of their characteristic odour and is widely used in aromatherapy, potpourri and perfumed toiletries. Many are obtained from plants that originate in dry, Mediterranean climates. These oils are usually produced by special glands.

ester Any of a class of organic compounds formed by reaction between an ALCOHOL and an ACID.

estuary Coastal region where a river mouth opens into the ocean and freshwater from the land is mixed with salt water from the sea. Many estuaries are drowned river valleys, perhaps formed after a rise in sea-level at the end of an ice age. They usually provide good harbours and are often breeding grounds for many kinds of marine life. *See also* DROWNED COAST

Eta meson (symbol η) Uncharged ELEMENTARY PARTICLE with zero SPIN. It is grouped under the HADRONS (particles that experience the STRONG NUCLEAR FORCE).

ethanal (acetaldehyde, CH_3CHO) Colourless, volatile flammable liquid manufactured by catalytic oxidation of ethene or ethanol, or catalytic hydration of ethyne (acetylene). It is used mainly to make other organic compounds. Properties: r.d. 0.788; m.p. −123.5°C (−190.3°F); b.p. 20.8°C (69.4°F).

ethane (C_2H_6) Colourless, odourless gas, the second member of the ALKANE series of HYDROCARBONS. It is a minor constituent of natural gas. *See also* SATURATED COMPOUND

ethanoate (acetate) Salt or ester of ETHANOIC ACID. A compound containing the ion CH_3COO^- or the group CH_3COO–. It is used in synthetic acetate fibres, in lacquers and in acetate film.

ethanoic acid (acetic acid, CH_3COOH) Colourless, corrosive liquid made by the oxidation of ethanol, either by catalysis or by the action of bacteria. It is the active ingredient in vinegar, and has many uses in the organic chemicals industry. Properties: r.d 1.049; m.p. 16.6°C (61.9°F); b.p. 117.9°C (244.4°F).

ethanol (ethyl alcohol, C_2H_5OH) Colourless, flammable and volatile liquid, the best-known ALCOHOL, produced by the FERMENTATION of sugars, molasses and grains, or by the catalytic HYDRATION of ethene (ethylene). Also known as grain alcohol, its many uses include alcoholic beverages (such as wine, beer, cider and spirits), cleaning solutions, antifreeze, rocket fuels, cosmetics and pharmaceuticals. Properties: r.d. 0.789; b.p. 78.5°C (173.3°F).

ethene (ethylene, $CH_2{=}CH_2$) Colourless gas derived from the CRACKING of propane and other compounds. Vast quantities of the gas are used annually for the manufacture of POLYETHENE. Ethene is also used widely for many other chemical syntheses.

ether In chemistry, colourless, volatile, flammable liquid that has the $-O-$ group in its structure. The commonest is ETHOXYETHANE (diethyl ether). It is prepared by the action of sulphuric acid on ethanol followed by DISTILLATION.

ether In physics, hypothetical medium that was supposed to fill all space and offer no resistance to motion. It was postulated as a medium to support the propagation of electromagnetic radiations, but its existence was disproved by the MICHELSON-MORLEY EXPERIMENT.

ethology Study of animal behaviour especially in the natural environment, first outlined in the 1920s by Konrad LORENZ and Nikolaas TINBERGEN. Ethologists study natural processes that range across all animal groups, such as courtship, mating and self-defence. Field observations and laboratory experiments are both used.

ethoxyethane (ether, diethyl ether, $C_2H_5OC_2H_5$) Volatile, flammable liquid with a sickly smell, once used as a general anaesthetic. It is made from ethanol (ethyl alcohol) by dehydration using concentrated sulphuric acid, or by reacting it with a chloroethane (ethyl chloride) and sodium metal (**Williamson's synthesis**). It is used mainly as a solvent. Properties: m.p. −116.2°C (−177.2°F;); b.p. 34.5°C (94.1°F).

ethyl alcohol *See* ETHANOL

ethylene *See* ETHENE

ethyne (acetylene, $CH{\equiv}CH$) Colourless, flammable gas, manufactured by the CRACKING of petroleum fractions. The simplest ALKYNE, it is explosive if mixed with air. When burned with oxygen, it produces extremely high temperatures, up to 3,480°C (6,300°F), and is therefore used in oxyacetylene torches for welding and cutting metals. It is polymerized to manufacture plastics, synthetic fibres, resins and the synthetic rubber neoprene. It is also used to produce ethanal and ethanoic acid. Properties: r.d. 0.625; m.p. −80.8°C (−113.4°F); b.p. −84°C (−119.2°F).

Eubacteria Subkingdom of the kingdom Prokaryotae, or perhaps even a separate DOMAIN. Eubacteria include all multicellular BACTERIA, as well as all those that use the energy of sunlight to make their own food (PHOTOSYNTHESIS), deriving their carbon from the air. They do not have the unique types of cell walls, RIBOSOMES and RNA of the other subkingdom, ARCHAEBACTERIA. *See also* PROKARYOTE

Euclid (*c*.300–*c*.95 BC) Ancient Greek mathematician whose ideas dominated mathematics, particularly GEOMETRY, for almost 2,000 years. He is remembered for his text books on geometry, especially *The Elements*, which was first printed in 1482 in a Latin translation from the Arabic. His other works include *Data* (on geometry) and *Phaenomena* (on astronomy). Several books have been lost. *See also* EUCLIDEAN GEOMETRY

Euclidean geometry Type of geometry that obeys the AXIOMS outlined in EUCLID's *The Elements*. Starting with a set of self-evident axioms, Euclid discovered many results using rigorous logic. The results were treated as absolute truths applicable to the physical world for almost 2,000 years. In the 19th century it was shown that his axioms are not universal and do not hold in all circumstances. Nikolai LOBACHEVSKY and Georg RIEMANN were among the mathematicians to discover types of geometry where the Euclid axioms did not hold. Their description of geometry is said to be NON-EUCLIDEAN. One striking feature of non-Euclidean geometry is that two straight lines that are parallel in one region of space can meet. In his work on general RELATIVITY Albert EINSTEIN discovered that the geometry of the Universe in which we live is non-Euclidean. Euclidean geometry, however, provides an adequate description of everything in our everyday lives.

Eudoxus of Cnidus (*c*.408–*c*.355 BC) Greek mathematician and astronomer, one of the greatest mathematicians of antiquity. Eudoxus devised the method of exhaustion, later extended by ARCHIMEDES, for calculating the areas of plane figures and volumes of solids. He also gave a precise definition of the real numbers in the framework of a general theory of proportion. In astronomy, he put forward the first system for describing the motions of the heavenly bodies. This system introduced the idea of the crystal spheres which carried the celestial bodies round the Earth, a notion subsequently taken up by ARISTOTLE. *See also* NUMBER, REAL

eugenics Study of human improvement by SELECTIVE BREEDING, founded in the 19th century by Sir Francis GALTON. It proposed the genetic "improvement" of the human species through the application of social controls on parenthood, encouraging parents who are above average in certain traits to have more children, while ensuring those who are below average have fewer. As a social movement, eugenics was discredited in the early 20th century owing to its ethical implications and its racist and class-based assumptions. Advances in GENETICS have given rise to the modern field of genetic counselling, through which people known to have defective genes that could cause physical or mental disorders in offspring are warned of the risks.

eukaryote Organism whose CELLS have a membrane-bound nucleus, with DNA contained in CHROMOSOMES. Eukaryotes include all animals, plants, fungi and PROTOCTISTA. They have a complex CYTOPLASM with an ENDOPLASMIC RETICULUM, and most of them possess MITOCHONDRIA. Most plants and algae also possess CHLOROPLASTS. Other structures specific to eukaryotic cells include microtubules, GOLGI BODIES, and membrane-bound flagella. In some classification systems eukaryotes make up the DOMAIN Eukarya. *See also* PROKARYOTE

Euler, Leonhard (1707–83) Swiss mathematician. An outstanding scholar, Euler worked in many areas of pure and applied mathematics, including analysis, differential equations, geometry, trigonometry, acoustics, fluid dynamics and celestial mechanics. His analysis of the Moon's motion laid the foundations of lunar theory, and he made important advances in studies of the tides, planetary perturbations and orbits of comets. His work on optical systems strongly influenced the technical development of telescopes and microscopes. Euler is said to have been the most prolific mathematician ever.

Euler diagram Simple diagram used in logic to illustrate syllogisms. Classes of objects are represented by circles so that, for example, a premise of the type "some *a* is *b* " can be represented by an overlap of two circles, one which represents *a* and the other representing *b*. *See also* VENN DIAGRAM

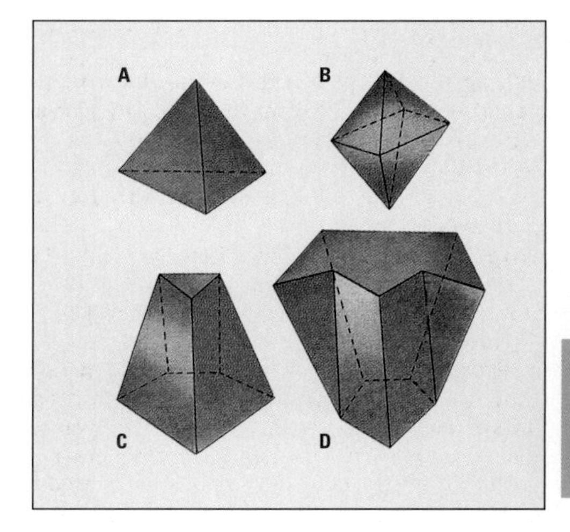

▲ **Euler's rule** All solids that do not have holes through them, and have flat faces, obey Euler's rule: $V + F - E = 2$ where V is the number of vertices (corners), F the number of faces and E the number of edges. For the tetrahedron (A) 4 + 4 − 6 = 2; for the octahedron (B), 6 + 8 −12 = 2. The shapes C and D also obey the rules, in fact the size and shape does not matter at all.

Euler's rule In solid geometry, states that for any polyhedron (many-sided solid figure) $V + F - E = 2$, where V is the number of vertices (corners), F the number of faces, and E the number of edges of the polyhedron. The rule is named after Leonhard EULER.

euphotic zone Uppermost layer of oceans. Much light penetrates this comparatively shallow zone, and many green plants and herbivores are found there, and often sea-dwelling mammals.

Europa Smallest of Jupiter's GALILEAN SATELLITES, with a diameter of 3,138km (1,950mi). It is a predominantly rocky body. Europa's smooth crust of water-ice is criss-crossed by a network of light and dark linear markings. There are very few craters, so old ones must have been removed by some form of geological activity.

European Space Agency (ESA) An organization of European nations to promote space research and technology for peaceful purposes, founded in 1975. ESA has 14 member states: Austria, Belgium, Denmark, Finland, France, Germany, Ireland, Italy, the Netherlands, Norway, Spain, Sweden, Switzerland and the United Kingdom. Its headquarters are in Paris. ESA has designed and built the Ariane series of launch rockets, and operates a launch site at Kourou in French Guiana, on the Atlantic coast of South America.

europium (symbol Eu) Silver-white, metallic element of the LANTHANIDE SERIES, being the softest and most volatile member. It was first isolated as the oxide in 1896. Its chief ores are monazite and bastnaesite. The metal is used in the manufacture of colour television screens, lasers and in control rods in NUCLEAR REACTORS. Properties: at.no. 63; r.a.m. 151.96; r.d. 5.25; m.p. 822°C (1,512°F); b.p. 1,597°C (2,907°F); most common isotope ^{153}Eu (52.18%).

Eustachian tube Small channel that connects the middle EAR to the back of the throat. It opens when swallowing, to allow the pressure in the middle ear to remain the same as the pressure of air outside the body. It is named after the Italian anatomist Bartolomeo Eustachio (1520–74).

eutectic *See* ALLOY

eutrophication Process by which a stream or lake becomes rich in inorganic nutrients by run-off from the land or by artificial means. Compounds of nitrogen, phosphorus, iron, sulphur, and potassi-

EUTROPHICATION

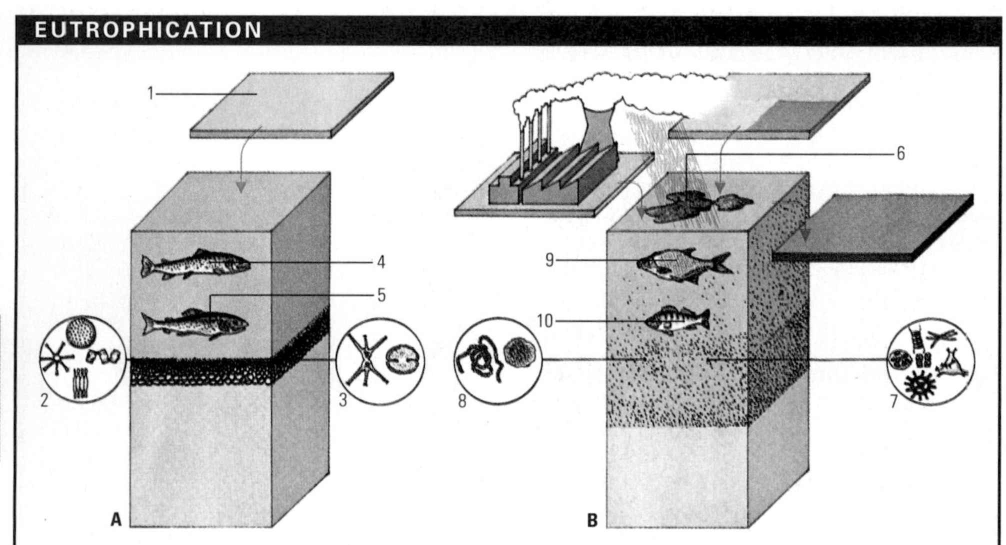

Lakes may be classified according to their fertility or productivity. Young oligotrophic lakes (A) become more fertile, or eutrophic, with age as they are efficient traps of nutrients delivered in run-off water from surrounding land (1). This natural process is hastened when the water contains agricultural fertilizers and industrial effluent. Oligotrophic lakes are characterized by clear water, a sandy or pebbly bottom and a low density of algae and plankton (mainly diatoms, 2, and desmids, 3). Typical fish are trout (4) and char (5), which thrive in deep, cold water. In eutrophic lakes (B), nitrate fertilizers and industrial organic effluent (6), which is rapidly broken down to release nitrates, phosphates and potassium, cause a great increase in algal growth, especially green algae (7) (chlorophytes) and cyanobacteria (8). The bottom of the lake becomes muddy, the water becomes turbid, and stagnation eliminates the bottom fish, which are replaced by perch (9) and bream (10). Eventually the algal bloom becomes so important that most of the dissolved oxygen is removed from the water, killing large numbers of fish. The decomposition of large amounts of organic matter will further reduce the oxygen content and a reducing condition exists in which hydrogen sulphide and methane are produced.

um are vital for plant growth in water; in excess, however, they overstimulate the growth of surface CYANOBACTERIA and microorganisms, which consume all the available dissolved oxygen.

evaporation Process by which a liquid or solid becomes a vapour. The reverse process is CONDENSATION. Solids and liquids cool when they evaporate because they give up energy (LATENT HEAT) to the escaping molecules.

evaporite Mineral deposit of precipitated salts, formed by the evaporation of saline lakes or confined volumes of salt water, usually in previous geological eras. They are important sources of GYPSUM, anhydrite, rock-salt, sylvite and small amounts of nitrates and borates.

evapotranspiration Combined processes of EVAPORATION and TRANSPIRATION in which water is transferred from the Earth's surface to the atmosphere. It is a key factor in the HYDROLOGICAL CYCLE. Water or ice is evaporated by the Sun's rays and by the wind, and plants are cooled by transpiration.

event horizon Boundary of a BLACK HOLE, from which nothing can escape. Observers outside the event horizon can therefore obtain no information about the black hole's interior. The radius of the event horizon is called the **Schwarzschild radius**.

evergreen Plant that retains its green foliage for a year or more (DECIDUOUS plants lose their leaves every autumn or dry season). Evergreens are divided into two groups: narrow-leaved (CONIFERS) and broad-leaved. Conifers include fir, spruce, pine and juniper. Among the broad-leaved evergreens are holly and rhododendron. Not all conifers are evergreens; exceptions are the deciduous larch (*Larix*) and dawn redwood (*Metasequoia*).

evolution Theory that a species undergoes gradual changes to survive and reproduce in a competitive, and often changing, environment, and that a new species is the result of development and change from the ancestral forms. Early work on evolutionary theory was initiated by Jean LAMARCK during the early 1800s, but it was not until Charles DARWIN wrote *The Origin of Species* during the mid-1800s that the theory was considered worthy of argument. Present-day evolutionary theory is largely derived from the work of Darwin and MENDEL and maintains that in any population or a GENE POOL, there is VARIATION, including random MUTATION, in genetic forms and characteristics. Most species produce greater quantities of offspring than their environment can support, so only those members best adapted to the environment survive. When new characteristics provide survival advantages those individuals that possess them pass on these characteristics to their offspring. through the process of HEREDITY (quantum of electromagnetic radiation). Since more of their offspring are likely to survive, the proportion of the population containing these new characteristics increases down the generations. *See also* NATURAL SELECTION; NEO-DARWINISM; PUNCTUATED EQUILIBRIUM

Ewing, James Alfred (1855–1935) Scottish physicist and engineer. He observed and identified hysteresis, the lag in effect when forces acting on an object are changed, and invented a number of instruments to test magnetic properties.

Ewing, William Maurice (1906–74) US geophysicist. The first person to take seismic measurements in open seas (1935), he aided understanding of marine sediments and ocean basins. He proposed that earthquakes are associated with central oceanic rifts and took the first deep-sea photographs (1939).

excited state Condition of an atom, ion or molecule, when its energy level is higher than that of the ground (lowest) state. For example, an atom can be in an excited state having absorbed a PHOTON. The increased energy causes one of the electrons to occupy an ORBITAL of higher energy; the atom may restore its former state by various emissions.

exclusion principle Basic law of QUANTUM MECHANICS, proposed by Wolfgang PAULI in 1925, stating that no two electrons in an atom can possess the same ENERGY LEVEL and SPIN. More precisely, the set of four quantum numbers characterizing certain ELEMENTARY PARTICLES called FERMIONS must be unique. In atoms, these numbers specify an electron's spin direction,

EXCRETION

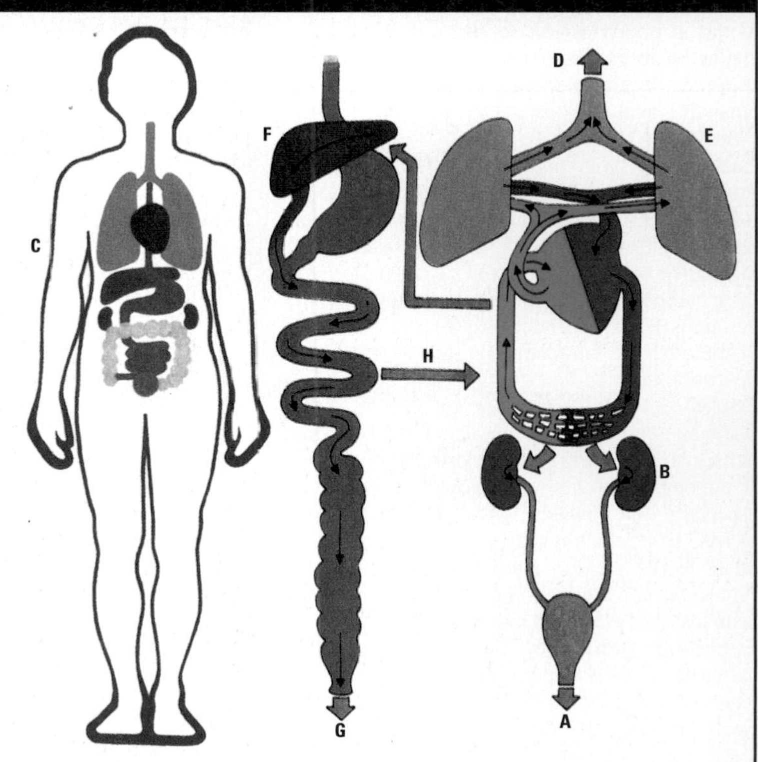

Excretion is the removal from the body of the waste products formed within it as a result of its metabolic activities. The elimination of undigested material from the gastro-intestinal tract is thus not true excretion. The main substances to be excreted from the body are carbon dioxide and water from glucose metabolism, urea from amino acid breakdown, salts, and larger molecules from the breakdown or conjunction of porphyrins and pyrimidines and other substances, such as plant constituents and drugs. Routes of excretion are in the urine (A) from the kidneys (B), in the sweat from the skin (C) (mainly urea and NaCl), and CO_2 in expired air (D) from the lungs (E). Some compounds, such as bile pigments from haemoglobin breakdown, are excreted via the liver (F) in the bile. Some of these are eliminated with the faeces (G) and some are taken into the blood from the intestines (H) and eventually excreted in the urine.

EXPANDING UNIVERSE

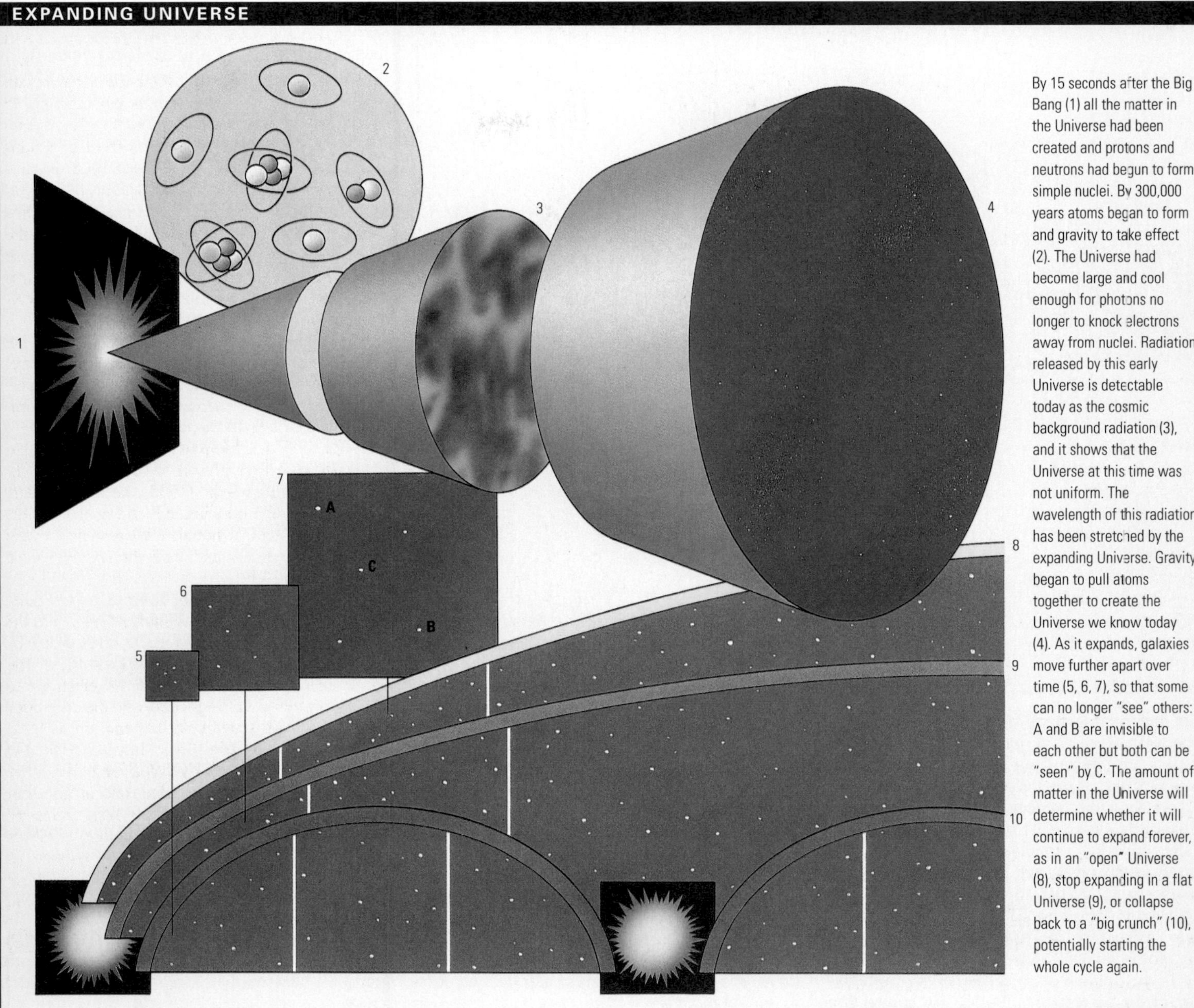

By 15 seconds after the Big Bang (1) all the matter in the Universe had been created and protons and neutrons had begun to form simple nuclei. By 300,000 years atoms began to form and gravity to take effect (2). The Universe had become large and cool enough for photons no longer to knock electrons away from nuclei. Radiation released by this early Universe is detectable today as the cosmic background radiation (3), and it shows that the Universe at this time was not uniform. The wavelength of this radiation has been stretched by the expanding Universe. Gravity began to pull atoms together to create the Universe we know today (4). As it expands, galaxies move further apart over time (5, 6, 7), so that some can no longer "see" others: A and B are invisible to each other but both can be "seen" by C. The amount of matter in the Universe will determine whether it will continue to expand forever, as in an "open" Universe (8), stop expanding in a flat Universe (9), or collapse back to a "big crunch" (10), potentially starting the whole cycle again.

ORBITAL shape and the energy level at which it resides, or would reside, in a magnetic field. For this work, Pauli was awarded the 1945 Nobel Prize for physics.

excretion Elimination of materials from the body which have been involved in METABOLISM. Such waste materials, particularly nitrogenous wastes, would be toxic if allowed to accumulate. In mammals these wastes are excreted mainly as URINE, and to some extent also by sweating. Carbon dioxide, a waste product of metabolism, is excreted mainly through the lungs during breathing. Defecation, strictly speaking, is not excretion: faeces consist mostly of material that has never been part of the body. They are eliminated through the ANUS.

exfoliation Process of weathering in which rock flakes off in small pieces. It is caused by large differences in day and night temperatures, as occur in hot deserts. During the day the rock heats up and expands; during the night it cools and contracts. Eventually the rock is weakened by the repeated expansion and contraction, and pieces flake off. The presence of dew or other water hastens the process. Because the rock flakes off like the skin of an onion, it is also known as onion WEATHERING.

exocrine gland Gland that releases its secretion onto the surface or into a body cavity. Examples of exocrine glands in the skin include SEBACEOUS GLANDS (which secrete waxy **sebum**) and SWEAT GLANDS (which secrete perspiration). MAMMARY GLANDS are temporary exocrine glands that secrete milk. Exocrine glands in the pancreas secrete digestive juices along ducts to the duodenum and intestines. *See also* ENDOCRINE SYSTEM

exoskeleton Protective skeleton or hard supporting structure forming the outside of the soft bodies of certain animals, notably ARTHROPODS and MOLLUSCS. In arthropods, it consists of a thick horny covering attached to the outside of the body and may be jointed and flexible. The exoskeleton does not grow as the animal grows; instead it is shed periodically and the animal generates a new one. *See also* ENDOSKELETON

exosphere Outer shell in the Earth's ATMOSPHERE from which light gases, including hydrogen and helium, can escape. It lies about 400km (250mi) above the surface of the Earth.

exothermic reaction CHEMICAL REACTION in which heat is evolved, causing a rise in temperature. A common example is COMBUSTION. *See also* ENDOTHERMIC REACTION

expanding Universe Theory of the origin and direction in time of the Universe. Physicists have attempted to explain the RED SHIFT phenomena of some stars as resulting from a single, huge explosion which causes these stars to be moving away from our section of the Universe. The RED SHIFT occurs when, due to the DOPPLER EFFECT, the perceived wavelengths of light from some stars are lengthened because of their outwards movement. There is also the opposite effect, a BLUE SHIFT, but this does not occur as often. Today the balance of opinion is in favour of the theory of the Universe expanding following the BIG BANG. *See also* HUBBLE CONSTANT

expansion In mathematics, a process of replacing an expression by a sum of a finite number of terms, or by an infinite series. For example, the expression $(x + 1)(x + 3)$ has the expansion $x^2 + 4x + 3$; the function $\sin x$ can be expanded into the converging series $x - x^3/3! + x^5/5! - \ldots$ and so on. *See also* BINOMIAL THEOREM

expansion (expansivity) In physics, a change in the size of an object with change in temperature. Most substances expand on heating, although there are exceptions – water expands when it cools from 4°C (39°F) to its freezing point at 0°C (32°F). A solid has three coefficients of expan-

EXTRUSION

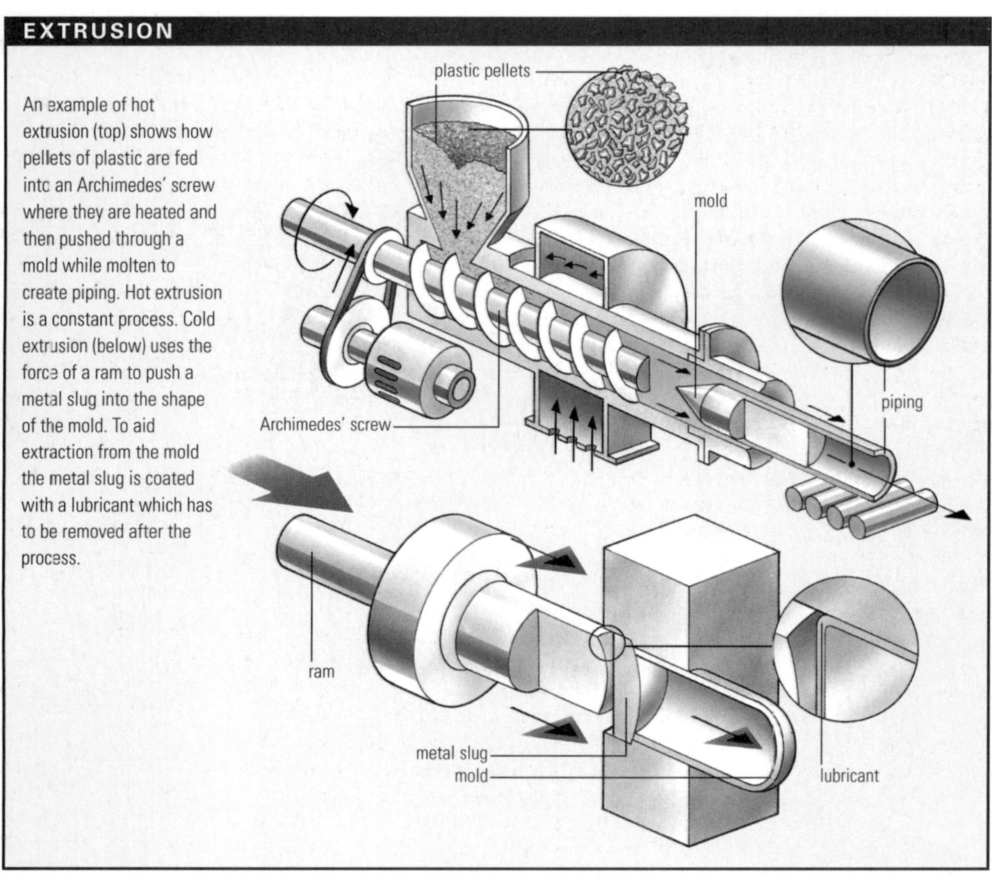

An example of hot extrusion (top) shows how pellets of plastic are fed into an Archimedes' screw where they are heated and then pushed through a mold while molten to create piping. Hot extrusion is a constant process. Cold extrusion (below) uses the force of a ram to push a metal slug into the shape of the mold. To aid extraction from the mold the metal slug is coated with a lubricant which has to be removed after the process.

sion: **linear**, **superficial** and **volume**, equal to the fractional increase in length, area or volume (respectively) per unit temperature rise. For a gas, the coefficient of expansion is the rate of change of volume with temperature (at constant pressure), or of volume with pressure (at constant temperature). For a solid, the coefficient of expansion is usually small; for a gas, it is much larger. *See also* CHARLES' LAW

expiration (exhalation) Process by which air (or water in the case of fish) is expelled from the lungs (or gills). In mammals it involves a reduction in the volume of the chest cavity by relaxation of the DIAPHRAGM muscles and contraction of the muscles between the ribs. This "squeezes" the lungs and forces gases out until the pressure in the lungs is the same as atmospheric pressure. Expiration and its opposite process, **inspiration**, are fundamental in aiding GAS EXCHANGE part of the complete process of RESPIRATION.

explosive Substances that react rapidly and violently, emitting heat, light, sound and shock waves. Chemical explosives are mostly highly nitrated compounds or mixtures that are unstable and decompose violently with the evolution of much gas. Nuclear explosives are radioactive metals, the atoms of which can undergo a CHAIN REACTION of nuclear FISSION, or light particles such as hydrogen that undergo nuclear FUSION, to release vast amounts of radiant energy and devastating shock waves. *See also* NUCLEAR WEAPON

explosive decompression Sudden drop in pressure of an aircraft or spacecraft cabin due to meteorite puncture or system failure. A pressure drop to below 47mm Hg (pressure of water vapour at 38°C) results in the boiling of body liquids. Lungs collapse completely within a few seconds, blood pressure drops and the large bowel relaxes. Death is inevitable.

exponent Superscript number placed to the right of a symbol indicating its power; for example, in a^4 (= $a \times a \times a \times a$), 4 is the exponent. Certain laws of exponents apply in mathematical operations. For example, $3^2 \times 3^5 = 3^{(2+5)} = 3^7$; $3^4/3^3 = 3^{(4-3)} = 3^1$; $(3^2)^5 = 3^{(2 \times 5)} = 3^{10}$; $3^{-5} = 1/3^5$.

exponential In general a function of x of the form a^x, where a is a constant. The exponential function e^x (where e is the base of natural logarithms, 2.7182818...) can be represented by a power series $1 + x + x^2/2! + x^3/3! + ...$

extensor MUSCLE that makes a limb extend or straighten. *See also* FLEXOR

extinction Dying out of a species or population. Extinction is part of the evolutionary process in which species of plants and animals die out, often to be replaced by others. The rate at which extinctions have occurred is very variable. Periods of the Earth's history when extinction rates have been high are called **mass extinctions**. Extinctions brought about by human impact on the environment do not necessarily involve the replacement of extinct species by others. *See also* EVOLUTION; PUNCUTATED EQUILIBRIUM

extracellular fluid Any of various body fluids that exist outside CELLS in body spaces lined with moisture-exuding membranes. Such fluids are also found in blood, in lymph, in various body tissues including muscle, and in the channels and cavities of the brain and spinal cord.

extraction In chemistry, physical separation of a liquid or solid mixture by selectively dissolving some components with a solvent. Specific ingredients for perfumes and flavourings may be extracted in this way from plant oils. Proteins are separated from carbohydrates in soya beans by controlled solvent extraction.

extrapolation In mathematics, an approximate method of finding values beyond those measured. Most of economics is extrapolation, in that statistics are compiled up to the present year, and using them as a guide, decisions are made concerning future years. *See also* INTERPOLATION

extrusion Operation of forcing copper, aluminum, magnesium, their alloys, or plastics at the optimum temperature through a die to manufacture specific shapes such as rods, tubes, and various hollow or solid sections. Plastic extrusion can produce composite sheets, and coat film and wire.

extrusive rock (volcanic rock) Type of IGNEOUS ROCK formed by MAGMA that has reached the surface of the ground. Extrusive rock generally comes from the vents of VOLCANOES or fissures producing LAVA flows. It cools quickly and is therefore fine-grained or even glassy. The term does not generally apply to PYROCLASTIC rocks.

eye Organ of SIGHT. It converts light energy to nerve impulses that are transmitted to the visual centre of the brain. Most of the mass of a human eye lies in a bony protective socket, called the **orbital cavity**, which also contains muscles and other tissues to hold and move the eye. The eyeball is spherical and composed of three layers: the **sclera** (white of the eye), which contains the transparent CORNEA; the **choroid**, which connects with the IRIS, PUPIL and LENS and contains blood vessels to provide nutrients and oxygen; and the RETINA, which contains RODS AND CONES for converting the image into nerve impulses. The AQUEOUS HUMOUR (a watery liquid between the cornea and iris) and the VITREOUS HUMOUR (a jelly-like substance behind the lens) both help to maintain the shape of the eye.

EYE

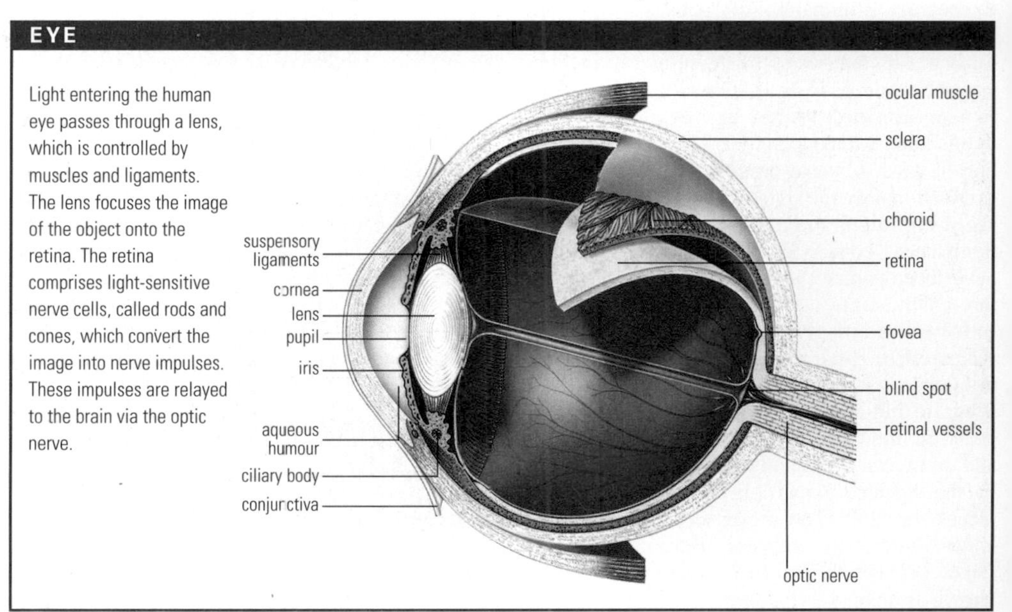

Light entering the human eye passes through a lens, which is controlled by muscles and ligaments. The lens focuses the image of the object onto the retina. The retina comprises light-sensitive nerve cells, called rods and cones, which convert the image into nerve impulses. These impulses are relayed to the brain via the optic nerve.

Fabre, Jean Henri (1823–1915) French entomologist. He made accurate and important studies of the anatomy and behaviour of INSECTS and ARACHNIDS, especially bees, ants, beetles, grasshoppers and spiders. He emphasized the importance of the theory of inherited instincts in insects. His most important work was embodied in the ten volumes of his *Souvenirs Entomologiques* (1879–1907).

Fabricius, Hieronymus (1537–1619) (Girolamo Fabrici) Italian anatomist. He was a pupil of Gabriello FALLOPIUS, whom he succeeded as professor of anatomy at Padua in 1562. His book, *On the Formed Foetus* (1600), describes his pioneering research in EMBRYOLOGY. Fabricius also gave the first complete description of the valves in veins, although he misunderstood their function.

facies In geology, all the features of a rock or stratum of rock that show the history of the rock's formation. Geologists often distinguish the age by the facies. The term is also applied to gradations of IGNEOUS ROCK.

factor In mathematics, any number that divides exactly into another given number without remainder. For example, the factors of 72 are 1, 2, 3, 4, 6, 8, 9, 12, 18, 24, 36 and 72 itself.

factor analysis In statistics and psychometrics, mathematical method for reducing a large number of measurements or tests to a smaller number of "factors" that can completely account for the results obtained on all the tests, as well as for the correlations between them.

factorial (symbol !) Number formed by the product of a given number and all the whole numbers less than it. For example, "6 factorial" is $6! = 6 \times 5 \times 4 \times 3 \times 2 \times 1 = 720$. By definition, 0 factorial is $0! = 1$.

Factor VIII (anti-haemophilic factor) Blood-clotting factor that is deficient or absent in people with HAEMOPHILIA. Factor VIII is a protein that brings about the conversion of prothrombin to thrombin, which in turn creates FIBRIN fibres round which a blood clot forms. It is extracted from blood plasma or genetically engineered and used for treating haemophilia.

faculae *See* SUNSPOT

faeces *See* EXCRETION

Fahrenheit, Gabriel Daniel (1686–1736) German physicist and instrument-maker. He invented the alcohol THERMOMETER (1709) and the first mercury thermometer (1714), for which he devised the FAHRENHEIT temperature scale. He also showed that the boiling points of liquids vary with changes in atmospheric pressure and that water can remain liquid below its freezing point.

Fahrenheit Temperature scale based on the freezing point of water (32°F) and the boiling point of water (212°F). The interval between these points is divided into 180 equal parts. Although replaced by the CELSIUS scale, the Fahrenheit scale is still sometimes used for non-scientific measurements. Degrees Fahrenheit are converted to degrees Celsius by substracting 32 and then dividing by 1.8. The system was devised by the German physicist Gabriel FAHRENHEIT.

Fallopian tube (oviduct) In mammals, either of two narrow ducts leading from the upper part of the UTERUS into the pelvic cavity and ending in finger-like projections called **fimbriae**, which almost encircle each OVARY. After ovulation, an OVUM enters and travels through the Fallopian tube where FERTILIZATION can occur. The fertilized ovum continues into the uterus where it becomes implanted. *See also* SEXUAL REPRODUCTION

Fallopius, Gabriello (1523–62) Italian anatomist who became professor of anatomy and botany at Padua in 1551. His book *Anatomical Observations* (1561) contains the first description of the FALLOPIAN TUBES and the SEMICIRCULAR CANALS of the inner EAR.

fallout (radioactive fall-out) Radioactive contamination in the atmosphere following a leakage or accident at a NUCLEAR REACTOR, or explosion of a NUCLEAR WEAPON. Large, windborne particles fall to Earth after a few hours, sometimes up to several hundred kilometres from the source (**local** fall-out); lighter particles entering the TROPOSPHERE are detected after a longer period at about the same latitude as the source (**tropospheric** fall-out). Any particles entering the STRATOSPHERE eventually fall over the Earth's surface, often many years later (**stratospheric** fall-out).

false colour image Image to which colours are added to emphasize or give a value for certain features. The colours are usually added by a computer, typically using data transmitted to Earth from satellites or space probes or from medical imaging equipment. Colours may be added to indicate such things as temperature (with, for example, red for hot and greens and blues for cool temperatures) or height (to generate a CONTOUR map).

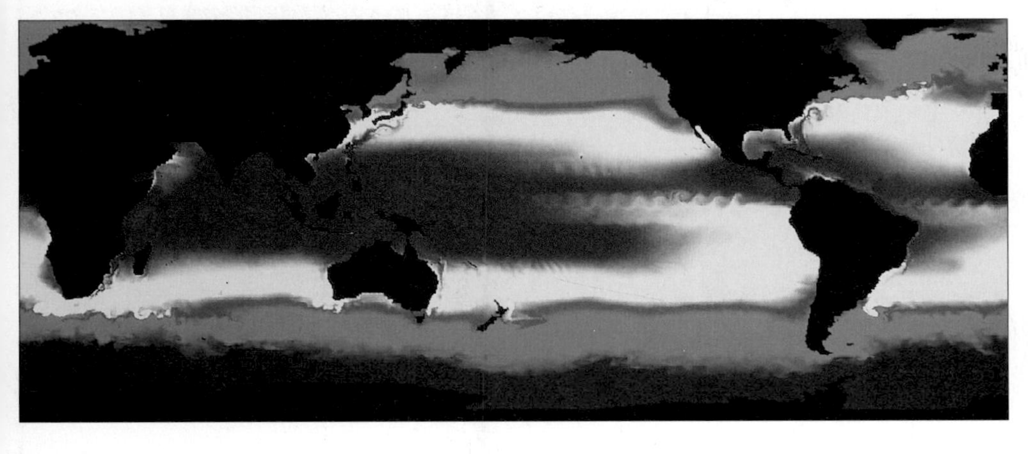

▲ **false colour image** Computer model of global sea surface temperature. Computer-generated image showing the sea surface temperature across the globe. The temperatures are colour coded from red (warmest) through green and yellow to blue (coolest). The model has predicted several real-world phenomena, such as the Gulf Stream (right of North America) and the Kuroshjo current off Japan. Large eddy streets are seen in equatorial latitudes, and large-scale waves are seen in the tropical Pacific. The model used a spatial resolution of 0.5 degrees of latitude and longitude and 20 vertical depth levels, combined with realistic ocean floor topography.

▲ **Faraday** Portrait of English chemist and physicist Michael Faraday. Faraday is considered one of the greatest experimental physicists. He was born into poverty and poorly educated. He became a laboratory assistant at the Royal Society. In 1825 he discovered benzene, but is better known for studies on electrolysis and electromagnetic phenomena. His electromagnetic research formed the background for the unified electromagnetic theory elaborated by James Maxwell. In 1832 Faraday formulated the *Laws of Electrolysis* which dictate the behaviour of a chemical solution in electric currents. He devised the first electric motor and equipment such as transformers and dynamos.

family In biology, part of the CLASSIFICATION of living organisms, ranking above GENUS and below ORDER. Family names are printed in Roman (ordinary) letters with an initial capital; for example Felidae is the family of all cats.

farad (symbol F) Unit of CAPACITANCE. It is equal to the capacitance of a capacitor that acquires a charge of 1 coulomb when a POTENTIAL DIFFERENCE of 1 volt is applied across the plates. It is a very large unit, and the microfarad (symbol μF), or one millionth of a farad, and the picofarad (symbol pF), or one million-millionth of a farad are commonly used.

Faraday, Michael (1791–1867) British physicist and chemist. Faraday worked as Sir Humphry DAVY's assistant at the Royal Institution in London, where, in 1825, he became director of the laboratories. He liquefied chlorine, discovered benzene and two compounds of carbon with chlorine, and enunciated the laws of electrolysis (FARADAY'S LAWS). Moving from chemistry to electricity, he discovered ELECTROMAGNETIC INDUCTION, made the first DYNAMO, built a primitive electric motor, and studied nonconducting materials (DIELECTRICS). The unit of CAPACITANCE (the FARAD) is named after him.

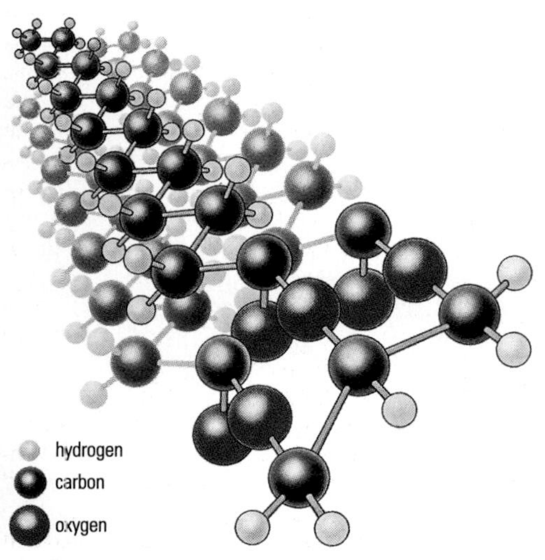

▲ **fat** Fats, found widely distributed throughout the animal and plant kingdoms, are generally esters of fatty acids and the trihydric alcohol, glycerol. Three molecules of the fatty acid combine with one molecule of glycerol, with the liberation of three molecules of water, to form the fat. The "glyceride structure" is an important common feature of fats. Those fats having one or more double bonds in the chains are known as unsaturated fats, while those only with single bonds are known as saturated fats. In both animals and plants, fats are found in the food storage areas of the organism. They are easily hydrolysed in the cells to give energy.

Faraday's laws Two laws of ELECTROLYSIS and three of ELECTROMAGNETIC INDUCTION, formulated by Michael FARADAY. In modern form, the electrolysis laws state that (1) the amount of chemical change during electrolysis is proportional to the charge passed, and (2) the amount of chemical change produced in a substance by a certain amount of electricity is proportional to the electrochemical equivalent of that substance. Faraday's laws of induction state that (1) an electromagnetic force is induced in a conductor if the magnetic field surrounding it changes, (2) the electromagnetic force is proportional to the rate of change of the field, and (3) the direction of the induced electromagnetic force opposes the change in the external field.

Farman, Henri (1874–1958) French aviation pioneer and manufacturer. In 1908 he won a prize for the first circular flight of 1km (0.6mi) and in 1909 he established a flight record of 179km (112mi). In 1912 he founded an aircraft factory which later became one of the largest in Europe and with his brother Maurice established the first air passenger service between London and Paris.

farming *See* AGRICULTURE

fast neutron High-energy NEUTRON that occurs in a nuclear FISSION reaction called **fast fission**. Their high energy is caused by undergoing fewer collisions during an initial fission reaction, and thereby unable to sustain a CHAIN REACTION. Fast fission is used in certain NUCLEAR REACTORS called **fast reactors**.

fat Semi-solid organic substance made and used by plants and animals to store energy. In animals, fats also serve to insulate the body and protect internal organs. Fats are soluble in organic solvents such as ether, carbon tetrachloride, chloroform and benzene. They are triglycerides: ESTERS of one molecule of glycerol connected to three molecules of FATTY ACIDS (carboxylic acids), such as palmitic, lauric and stearic acid, each having 12 to 18 carbon atoms. Research indicates that the consumption of high levels of animal fats can increase the risk of heart disease. Vegetable oils are similar to fats, but are viscous liquids rather than semi-solids and have a higher proportion of molecules with double carbon–carbon (C=C) bonds in the chain – that is, they are unsaturated. *See also* LIPID; SOAP

fathom Unit used in measuring depth, especially of water. One fathom equals 1.83m (6ft). It is also a quantity of material that has a cross-section 6 feet square but varies in length. Originally, a fathom was the distance spanned by a person's arms.

fatigue In general, mental or physical tiredness after activity. In physiology, an inability to function at normal levels of physical and mental activity. Muscle fatigue results from the accumulation of LACTIC ACID in the muscle tissue and the depletion of GLYCOGEN (stored carbohydrates).

fatigue, metal Weakening of the crystalline structure of a metal due to repeated rhythmic straining, bending or vibration. It can cause superficial cracks which may spread inwards, causing the metal to crumble. Components have to be designed to reduce and delay fracture due to metal fatigue.

fatty acids Organic compounds, so called because they are present widely in nature as constituents of FAT. They are CARBOXYLIC ACIDS containing a single carboxyl group (–COOH). Examples of saturated fatty acids (lacking double bonds in their hydrocarbon chain) are ethanoic acid and palmitic acid, the latter being a common fat constituent; unsaturated fatty acids (having one or more double carbon–carbon bonds – C=C) include oleic acid. Both saturated and unsaturated types have molecules shaped like a long, straight chain. *See also* LIPID

fault In geology, a crack or fracture in the Earth's CRUST along which movement has occurred. The movement will be slow and quite small, only a few centimetres, though fault movements often continue for thousands of years. In such cases, uplift or downthrow of hundreds of metres is possible. Faulting is caused by PLATE TECTONICS, when movements in the Earth's crust create stress and tension in the rocks, causing them to stretch and crack. Vertical movements cause **normal** and **reverse** faults, and horizontal movements cause **tear** faults. The vertical change of height on opposite sides of a fault is called the **throw**, and the horizontal movement is called the **heave**. Faults often occur in groups and if two or more roughly parallel faults cause a block of land to rise, it is called a HORST, or block mountain. If the land sinks between two or more parallel faults, it creates a GRABEN, or RIFT VALLEY.

fault plane Surface along which a FAULT occurs. In a normal fault, caused by tension pulling two masses of rock apart before one mass slips downwards, the fault plane is either vertical or inclined so that the mass moves downwards on the dip side of the fault plane.

fauna Collective name for all the animals that are found in a particular area, or that occurred at a particular time.

fax (facsimile transmission) Equipment by which text, photographs and drawings can be transmitted and received through a TELEPHONE system. The image, on paper, is scanned to translate it into a series of electrical pulses. Inside the fax machine, a MODEM converts the pulses into a digital form that can be transmitted through the telephone system. At the receiving end, the fax machine's modem converts the signals back into pulses, and prints these as dots to build up a copy of the original document. *See also* SCANNING

feather One of the skin appendages that makes up the plummage of birds. They are composed of the fibrous protein KERATIN, and provide insulation and enable flight. They are usually replaced at least once a year.

feedback In technology, process by which an electronic or mechanical control system regulates itself. Feedback works by returning part of the "output" of the system to its "input". In other words, the process compares how the system is performing with how it is set up to perform. If there is any difference between the two, the feedback system makes adjustments accordingly. A governor on an engine, for example, uses feedback to keep the speed within a specified limit, regardless of the load carried. The high-pitched feedback heard in electronic music systems occurs when sound from the loudspeaker is picked up by a microphone. *See also* BIOFEEDBACK

feldspar Important group of common rock-forming minerals that all contain aluminium, silicon and oxygen, but with varying proportions of potassium, sodium and calcium. They are essential constituents of IGNEOUS ROCK. ORTHOCLASE and MICROCLINE are potassium feldspars of monoclinic and triclinic system crystals, respectively. Members of the PLAGIOCLASE series (sodium and calcium feldspars) have physical properties similar to microcline, but with crystals frequently twinned. Hardness 6–6.5; r.d. 2.5–2.8.

fell Name for an upland in N England. Most fells

FAULT

The Earth's crust is subjected to enormous forces and the stress creates faults. In a tear fault (1) the stresses cause horizontal movement. The forces build up until they are released in a sudden movement (2) often causing earthquakes. In a normal fault (3) the rocks are pulled apart, causing one side to slip down along the plane of the fault. In a reverse fault (4) the rocks on either side of the fault are forced together. One side rises above the other along the fault plane. In a horst fault (5) the central section is left protruding due to compression from both sides or the sinking of the bracketing rock. A rift valley (6) has a sunken central section, formed either by compression or the outward movement of the two valley sides.

FAX

A fax machine converts text or images fed into the machine (1) into a digital code (2). The code is created by shining light on tiny strips of the document in turn (3). Sensors (4) detect the amount of light that bounces back. Where ink is present little light is reflected creating an electrical pulse of low voltage. A high voltage results when light is reflected from white paper. The digital code is converted by a modem in the fax into an analogue signal (5) and transmitted to the receiving fax machine (6) via the telephone network. A modem in the second machine converts the analogue code back into a digital code (7), a printer (8) interprets the digital code and produces the hard facsimile copy (9). Each machine has its own number which is dialled in via a keyboard (10) on the sending machine.

occur in the Lake District and the Pennines, where their moorlands provide rough grazing for sheep in summer.

femur Thigh bone, extending from the hip to the knee of four- and two-legged VERTEBRATES, including humans. It is the longest and strongest bone of the human SKELETON. Its rounded, smooth head articulates with the pelvis at the hip socket.

fen Tract of low-lying marshy land where peat is formed below the surface. The soil is only slightly acid and much drainage is needed before a fen can become arable. The term is usually applied to the swampy Wash area (The Fens) in E England where most of the land is at or below sea level.

Fermat, Pierre de (1601–65) French mathematician. With Blaise PASCAL he helped to formulate the theory of probability and, by showing that light travels along the shortest optical path (**Fermat's principle**), he laid the foundation for geometric optics. In mathematics, Fermat is best known for his work in number theory. It was also recently discovered that differential CALCULUS, thought to be an invention of NEWTON, was communicated to him by Fermat in a series of letters. Leibnitz is believed to have developed calculus independently of both men. *See also* FERMAT'S LAST THEOREM

Fermat's last theorem Hypothesis, first stated by FERMAT, that for all integers $n > 2$, there are no natural NUMBERS x, y and z that satisfy the equation $x^n + y^n = z^n$. Fermat wrote in one of his books (Diophantus' text *Arithmetica*) that he had found a "remarkable" proof of this theorem, which the margin of the book was too small to contain. He died without revealing his proof to anyone. Subsequent attempts to find a valid proof of the result baffled mathematicians for more than 350 years. In 1993 Andrew WILES of Princeton University, drawing together strands from diverse areas of mathematics, announced a proof of the theorem at a mathematical conference at the Newton Institute in Cambridge. This alleged proof was soon found to contain a gap, but further work has repaired this and the complete proof was published and widely accepted in 1995. See also NUMBER, NATURAL

fermentation Energy-yielding metabolic process by which sugar and starch molecules are broken down to carbon dioxide and ethanol in the absence of air (ANAEROBIC respiration), catalysed by ENZYMES usually in microorganisms such as yeast. Uses include bread-making, wine-making, beer-BREWING and cheese maturation. The intoxicating effect of crushed fruits stored in a warm place (where they would ferment) may have been known as early as 4000 BC.

Fermi, Enrico (1901–54) US physicist, b. Italy. He worked mainly in the fields of atomic behaviour and structure, and in QUANTUM THEORY. He discovered the element NEPTUNIUM and proved that neutron bombardment of most elements produces RADIOISOTOPES of those elements. Fermi produced the first self-sustaining CHAIN REACTION in uranium and in 1942 built the world's first nuclear reactor. He helped to develop the atom bomb for the MANHATTAN PROJECT and later worked on developing the hydrogen bomb. With Paul DIRAC, he studied quantum statistics. For his work with RADIOACTIVITY, Fermi was awarded the 1938 Nobel Prize for physics; the element FERMIUM was named after him. The US Atomic Energy Commission named their special award for outstanding work in NUCLEAR PHYSICS after Fermi, who received the first one in 1954. *See also* NUCLEAR WEAPON

fermion ELEMENTARY PARTICLE that has a half-integer SPIN. Named after the physicist Enrico FERMI, fermions are those particles that obey the EXCLUSION PRINCIPLE. This means that only one fermion can occupy a certain quantum state. Nuclear structures tend to be made up of fermions; LEPTONS (such as ELECTRONS and NEUTRINOS) and QUARKS are all fermions. *See also* BOSON

fermium (symbol Fm) Radioactive metallic TRANSURANIC ELEMENT of the ACTINIDE SERIES. It was first identified in 1952 as a decay product of uranium ^{255} after it was produced in the first large hydrogen bomb explosion. Ten isotopes have been identified, but fermium has only been produced in trace amounts. Properties: at.no. 100; most stable isotope ^{257}Fm (half-life 80 days).

fern Any member of the 10,000 or so species of the non-flowering plant phylum Filicinophyta. Many ferns grow in warm, moist areas. The best-known genus *Pteridium* (bracken) grows on moorland and in open woodland. Ferns are characterized by their ALTERNATION OF GENERATIONS: the conspicuous SPOROPHYTE, which possesses leafy fronds, stems, RHIZOMES and roots and reproduces by minute SPORES usually clustered on the leaves; and the inconspicuous GAMETOPHYTE, which resembles moss and produces gametes (sperm and ova). Fronds unroll from curled "fiddle-heads" and may be divided into leaflets. Ferns were growing in the Devonian period, some 400 million years ago.

ferric Description of compounds in which the element iron has a greater VALENCE (usually three) than in FERROUS compounds.

ferricyanide *See* HEXACYANOFERRATE

ferrite One of the crystalline forms (ALLOTROPES) of the metal iron, present together with other forms in steel, wrought iron and cast iron. It is also called α-iron, and has highly magnetic properties. The name ferrite is also given to chemical compounds such as nickel ferrite ($NiFe_2O_4$) and zinc ferrite ($ZnFe_2O_4$), which are found naturally in rocks and also synthesized as magnetic materials. *See also* FERROMAGNETISM

ferro-alloys Combinations of silicon, manganese, chromium, molybdenum, vanadium, titanium and several other elements, added to molten steel to confer such properties as greater strength and corrosion resistance. Individually they are named after their major constituents; for example, ferrochromium contains about 70% chromium, about 6% carbon and about 2% silicon as well as iron. *See also* ALLOY

ferrocene ($Fe(C_5H_5)_2$) Red, crystalline organo-iron compound, systematic name dicyclopentadienyl iron(II). Its molecules consist of an iron(II) (ferrous) ion sandwiched between two five-membered carbon rings. It is, therefore, an example of a **sandwich compound**.

ferrocyanide *See* HEXACYANOFERRATE

ferroelectrics Crystalline materials that are naturally electrically polarized. Within such a material lie regions or domains which are spontaneously polarized in specific directions, although these

FEATHER

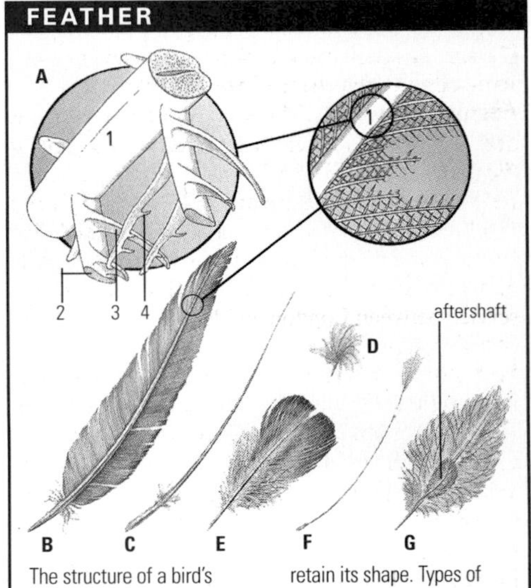

The structure of a bird's feather (A) shows how barbs (2) extend from the central midrib (1). Barbules (3) project from both sides of the barb, one side of which has tiny hooks (hamuli) (4) which catch on the next barbule. The interlocking construction adds strength and helps the feather to retain its shape. Types of feather incluce flight (B), bristle (C), down (D), contour feathers (E), which insulate, filoplumes (F), hair-like feathers that are either sensory or decorative, and body contour feathers (G), which have a smaller feather (aftershaft) growing from the base.

directions can often be changed by the application of an electric field. All ferroelectrics exhibit the PIEZOELECTRIC EFFECT; this leads to many applications of the materials in research and industry.

ferromagnetism Form of MAGNETISM exhibited by substances (such as iron, cobalt and nickel) with high magnetic PERMEABILITIES. Below a certain temperature, called the Curie temperature, a ferromagnetic material in an applied magnetic field becomes magnetized in the direction of the field. The magnetism increases with the applied field, up to a limit called the **saturation magnetization**. Some of the magnetization persists when the applied field is removed.

ferrous Description of compounds in which the element iron has a lesser VALENCE (usually two) than in FERRIC compounds.

ferry Passenger, vehicle or freight-carrying boat. Most are powered by diesel or diesel-electric engines, and some high-speed passenger ferries are HYDROFOILS. Recently there has been concern over the safety of certain ferry designs, following disasters such as the sinking of the *Estonia* in the Baltic Sea (September 1994), which resulted in over 900 fatalities.

fertility drugs Drugs taken to increase a woman's chances of conception and pregnancy. One of the many causes of female sterility results from insufficient secretion of pituitary hormones, and this malfunction is often treated with either human chorionic gonadotropin or clomiphene citrate, although use of the latter has resulted in multiple births. In cases where fertilization of the ovum does occur, but where the uterine lining is unable to support the developing fetus, the hormone progesterone may be used. There are many cases of infertility that cannot be corrected with drugs.

fertilization Key process in SEXUAL REPRODUCTION during which the nuclei of female and male GAMETES (sex cells) fuse to form a ZYGOTE. The zygote contains the genetic material (CHROMOSOMES) from both parents (*see* HEREDITY). In most animals, the female sex cell is called the OVUM and the male cell SPERM. After fertilization, the zygote begins to divide through a number of stages to form an EMBRYO. Fertilization of the female ovum by the male sperm can be external (outside the body, as in most fish, amphibians and aquatic invetebrates) or internal (inside the body, as in reptiles, birds, mammals and insects). In plants, the male gamete is found in POLLEN, and for most higher plants POLLINATION occurs before fertilization. While most animals and plants undergo **cross-fertilization**, in which the male gamete fuses with the female gamete of another animal or plant, some organisms undergo **self-fertilization**, whereby the male gamete fuses with the female gamete of the same flower or plant, or where an HERMAPHRODITE animal fertilizes itself. *See also* CELL DIVISION; DIPLOID; HAPLOID

fertilizer Natural or artificial substance added to soil, containing chemicals to improve plant growth by increasing fertility. Manure and compost were the first fertilizers. Other natural substances, such as bone meal, ashes, guano and fish, have been used for centuries. Modern chemical fertilizers, composed of nitrogen, phosphorus and potassium compounds in powdered, liquid or gaseous forms, are now widely used. Specialized fertilizers also contain essential TRACE ELEMENTS.

Fessenden, Reginald Aubrey (1866–1932) US engineer, physicist and inventor who was a pioneer in radio and echo-sounding. He is thought to have broadcast the first radio programme (using speech signals, not Morse code) in 1906 from a transmitter he constructed at Brant Rock, Massachusetts. Fessenden developed a new type of wireless system using continuous waves. Among his 500 or so patents were AMPLITUDE MODULATION (AM), the high-frequency ALTERNATOR, the electrolytic detector, the heterodyne system of radio reception and the fathometer.

fetus (foetus) EMBRYO in a mammal after the main adult features are recognizable. In humans it dates from about 8 weeks after CONCEPTION. *See also* BLASTULA

fever Elevation of the body temperature above the normal 37°C (98.6°F). It is mostly caused by bacterial or viral infection and can accompany virtually any infectious disease.

▲ **Fibonacci sequence** The leaves of a plant when seen from above grow in a spiral pattern. The angles between one leaf and the next follow a strict mathematical series known as the Fibonacci sequence. This ensures that each individual leaf on the plant stem receives the maximum amount of sunlight available.

Feynman, Richard Phillips (1918–88) US physicist, one of the most important theoretical physicists of modern times. He helped to develop the atom bomb during the MANHATTEN PROJECT, before going to Cornell University with Hans BETHE where he worked on QUANTUM ELECTRODYNAMICS (QED). His invention of **Feynman diagrams** greatly facilitated the solution of electromagnetic interactions between ELEMENTARY PARTICLES. He was professor of physics at the California Institute of Technology from 1950 until he died. He shared the 1965 Nobel Prize for physics with Julian SCHWINGER and the Shin'ichiro Tomonaga for their independent work on quantum electrodynamics. With Murray GELL-MANN, Feyman developed a theory of weak INTERACTIONS, such as those that occur in the emission of electrons from radioactive nuclei. Feyman also did research on the structure of protons and the properties of liquid helium. In 1986 he was a key member of the committee that investigated the *Challenger* space shuttle disaster.

Fibonacci, Leonardo (c.1170–c.1240) Italian mathematician. He wrote *Liber abaci* (*c*.1200), the first Western work to propose the adoption of the Arabic (originally Hindu) numerical system. He produced the mathematical sequence known as the FIBONACCI SEQUENCE.

Fibonacci sequence Mathematical SEQUENCE in which each term of the sequence is formed by the addition of the two terms preceding it. Thus, if the *n*th term of the sequence is denoted x_n, the sequence is defined by equation $x_{n+2} = x_n + x_{n+1}$, with the first two terms being $x_1 = 1$ and $x_2 = 1$. The sequence begins 1, 1, 2, 3, 5, 8, 13, 21.... the next number being 34, the sum of 13 and 21, and so on. As *n* becomes very large the ratio of successive terms becomes $1:(\sqrt{5} + 1)/2$. This ratio is called the golden ratio. The Fibonacci sequence can be seen in nature in successive spiral shell segments and sunflower petals radiating from the centre of the flower. *See also* GOLDEN MEAN

fibre Fibrous material of animal, vegetable, mineral or synthetic origin. Natural fibres can be made into yarn, textiles and other products, including carpets, rope and felt. The fibres consist of long narrow cells. **Animal** fibres are based on protein molecules and include wool, silk, mohair,

FERTILIZATION

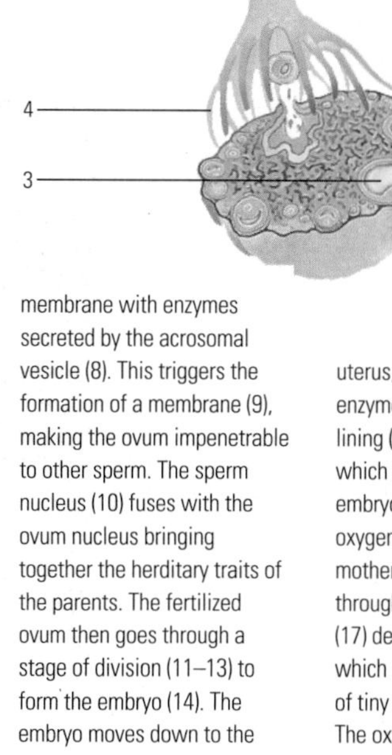

Mammalian fertilization begins with ovulation, in which an ovum (1) develops in an ovary (2) into a follicle (3). The follicle consists of the ovum, a sac of liquid and follicle cells. The pressure in the follicle increases until it bursts, releasing the ovum into the fallopian tube (4). During ovulation, the hormone oestrogen is produced by the collapsed follicle. Oestrogen causes the lining of the uterus to thicken and extend its network of blood vessels, from which the fertilized ovum will be nourished. Fertilization occurs when millions of sperm (5) are ejaculated from the male's penis during copulation. The sperms use their long tails called flagella (6), powered by mitochondria (7), to swim up the uterus. The first sperm to reach the ovum penetrates the ovum membrane with enzymes secreted by the acrosomal vesicle (8). This triggers the formation of a membrane (9), making the ovum impenetrable to other sperm. The sperm nucleus (10) fuses with the ovum nucleus bringing together the herditary traits of the parents. The fertilized ovum then goes through a stage of division (11–13) to form the embryo (14). The embryo moves down to the uterus, where it releases enzymes that break down the lining (15), creating a hole in which the embryo sits (16). The embryo indirectly obtains its oxygen and food from the mother's blood as it flows through the uterus. The embryo (17) develops the placenta which is comprised of millions of tiny appendages called villi. The oxygen and food are absorbed from the mother's blood, via capillaries in the villi, into the embryo's blood.

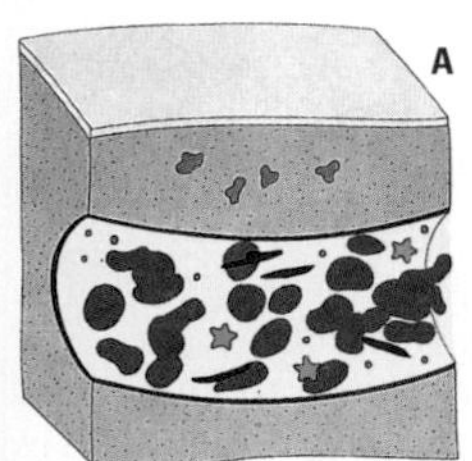

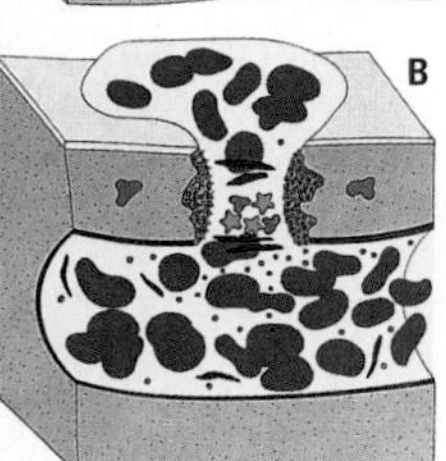

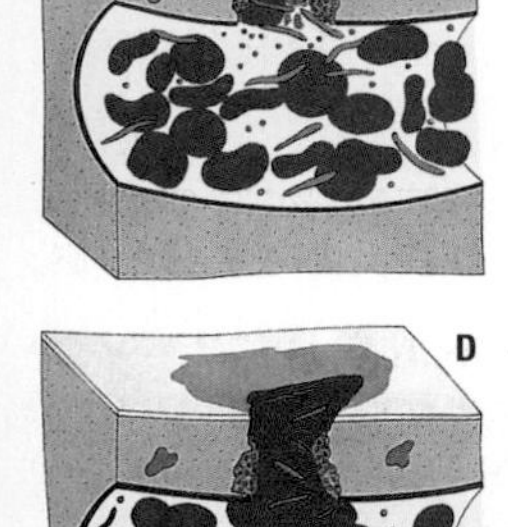

◀ **fibrin** An essential component of blood clotting, fibrin prevents excessive blood loss from a wound. Normally, circulating blood contains erythrocytes, platelets, plasma, clotting factors and fibrinogen. Tissue-clotting factors lie trapped within cells surrounding each blood vessel (A). When damage occurs, blood escapes from the broken vessel. Platelets congregate at the site and help plug the wound. Tissue-clotting factors are released (B). The reaction of the platelets with plasma and tissue-clotting factors converts the soluble fibrinogen into insoluble threads of fibrin. The fibrin forms a mesh across the break (C). Platelets and blood cells become trapped in the mesh. The jelly-like mass shrinks and serum oozes out, leaving a clot (D).

angora and horsehair. **Vegetable** fibres are based mainly on CELLULOSE and include cotton, linen, flax, jute, sisal and kapok. The mineral ASBESTOS is a natural, inorganic fibre. Regenerated fibres are manufactured from natural products, modified chemically. For example, RAYON is made from cellulose fibre obtained from cotton or wood. **Synthetic fibres** are made from a molten or dissolved plastic RESIN by forcing it through fine nozzles (spinnerets). The result is a group of filaments that are wound onto bobbins. These fibres can be used as single-strand yarn, or spun to form multi-strand yarn. The yarn may be woven into textiles. Many textiles, particularly for clothing, use both synthetic and natural fibres woven together. Some synthetic fibres are made into rope, carpets and other products. Synthetic fibres include NYLON and other polyamides, POLYESTERS and ACRYLICS. Nylon, the first synthetic fibre, was introduced in 1938. Other synthetic fibres, such as glass, carbon or metals, can be used to reinforce resins to produce extremely strong materials to be used in sports equipment, for example.

fibre glass *See* GLASS FIBRE

fibre optics Branch of OPTICS concerned with the transmission of data and images by reflecting light through very fine glass OPTICAL FIBRES.

fibrillation Rapid, uncontrolled beating of part of the heart muscle. **Atrial** fibrillation is seen in a number of conditions, including rheumatic heart disease, ATHEROSCLEROSIS and hyperthyroidism. **Ventricular** fibrillation, which occurs mostly after a heart attack, can be fatal.

fibrin Insoluble, fibrous protein that is essential to BLOOD CLOTTING. Developed in the blood from a soluble protein, FIBRINOGEN, fibrin is laid down at the site of a wound in the form of a mesh, which then dries and hardens so that the bleeding stops.

fibrinogen Soluble protein synthesized by the liver and released into the bloodstream. It is converted to FIBRIN (an insoluble protein) by THROMBIN (an enzyme) during the BLOOD CLOTTING process.

fibroblast Type of living cell found in the CONNECTIVE TISSUES of the body. Fibroblasts are mobilized to restore tissue damaged as a result of injuries. The elongated or irregularly shaped cells lay down the fibrous protein COLLAGEN and other structural materials of the connective tissues.

fibro-cartilage Form of CARTILAGE found mainly in parts of the skeleton that are subject to extreme physical tension or compression. These include INTERVERTEBRAL DISCS and the parts of TENDONS that are attached to the ends of the long bones. Fibro-cartilage consists largely of the fibrous protein COLLAGEN.

fibroid Benign (non-cancerous) tumour, consisting of fibrous tissue and muscle, in the wall of the uterus. It may grow to be enormous, causing pain and very heavy periods. If so, it may be removed in a procedure known as **myomectomy**; or HYSTERECTOMY may be recommended.

fibula Long thin outer bone of the lower leg of four- and two-legged VERTEBRATES, including humans. It articulates with the other lower leg bone, the TIBIA, just below the knee; the lower end forms the projection that may be felt on the outer side of the ankle.

field In computing, a particular item of DATA, usually part of a record, which in turn is part of a DATABASE.

field In physics, region in which an object is affected by a FORCE as a result of the presence nearby of another object or objects. There are various kinds of fields, including ELECTRIC FIELDS, MAGNETIC FIELDS and GRAVITATIONAL FIELDS.

file In computing, a block of stored data. A file may contain information (such as a group of addresses), a document or a complete program. It is usually stored on a MAGNETIC DISK, but tape or other similar media may be used. With random-access files, any item of data can be accessed immediately. With serial files, the data must be read through from the beginning until the required item is reached. *See also* DATABASE

film, photographic Sensitized strips of CELLULOSE acetate or other plastic, coated on one side with a light-sensitive emulsion, used to record photographic images. The emulsion of a black-and-white film consists of a suspension of finely divided grains of silver bromide in gelatin. After exposure the film is kept in darkness until the latent image is made visible by developing and fixing. Film is rated according to its "speed" or sensitivity to light. This is a measure of the size of the silver bromide grains contained in the emulsion. A "fast" black-and-white film (high ASA or DIN number) gives rise to a grainy, high-contrast image, whereas "slower" film captures a greater range of tones. Within the two rating systems an ASA number of 50 is equivalent to a DIN number of 18.

filter Porous device for separating solid particles from a liquid or gas. The process is known as FILTRATION. Many complex forms of filter have been devised for various uses. Most cars have a number of filters, for air, petrol and oil. These operate either by trapping solid particles in porous materials such as paper or meshes, or by circulating the material to be filtered through a maze, the pockets of which trap particles, as in the air filter.

filtration Process of removing solids from liquids or solids from gases by passage through a suitable medium such as filter paper, glass wool or sand. *See also* FILTER

fiord Alternative spelling of FJORD

fire Combustion of flammable materials, usually accompanied by flames or smoke. Chemically it is an example of rapid OXIDATION, and a supply of air or oxygen is generally necessary for a fire to continue to burn. Early humans learned to control fire for warmth, cooking, making pottery and metalworking. Although fundamental to much technology, uncontrolled it can cause great devastation.

firearm Weapon from which a projectile is fired, generally by the expansion of gases after the rapid combustion of GUNPOWDER or other explosive. Although occasionally used to describe an ARTILLERY piece, the term generally means a small arm – a weapon carried and fired by one person or a

FILTRATION

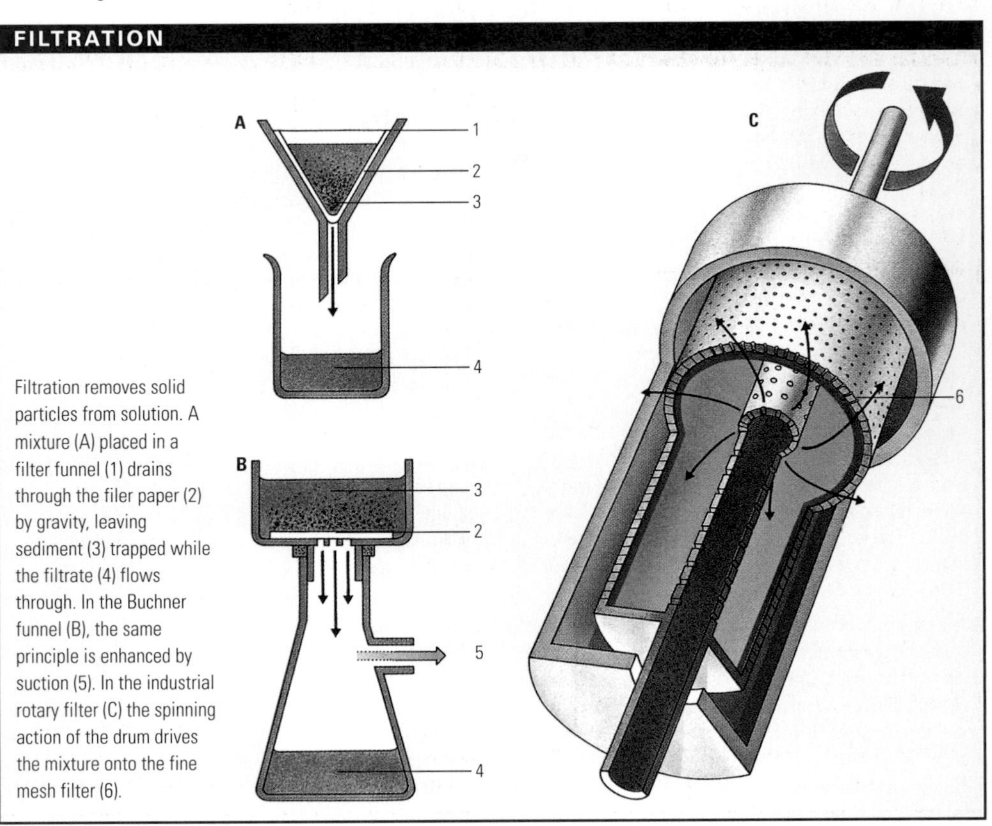

Filtration removes solid particles from solution. A mixture (A) placed in a filter funnel (1) drains through the filer paper (2) by gravity, leaving sediment (3) trapped while the filtrate (4) flows through. In the Buchner funnel (B), the same principle is enhanced by suction (5). In the industrial rotary filter (C) the spinning action of the drum drives the mixture onto the fine mesh filter (6).

small group of people. Firearms were used in Europe in the 14th century but were ineffective in close combat until *c.*1425, when a primitive trigger to bring a lighted match into contact with the gunpowder charge was invented. Such firearms, called **matchlocks**, were heavy and cumbersome, and needed a constantly lit match. The lighter flintlock (which used the spark produced by flint striking steel to ignite the powder) superseded the matchlock in the mid-17th century and became the main infantry weapon for more than 100 years. During the 19th century there were great changes. In 1805 the Rev. Alexander Forsyth (1769–1843) discovered the explosive properties of mercury fulminate; together with the percussion cap invented in 1815, it provided a surer, more efficient means of detonation, permitting the development by 1865 of both the centre-fire cartridge (which has been the type of ammunition used in firearms ever since) and breech loading, not previously practicable. Another major 19th-century advance was rifling, the cutting of spiral grooves along the inside of a barrel to make the bullet spin in flight, thus vastly increasing accuracy. During the 1830s Samuel Colt perfected the revolver, a pistol that could fire several shots without the need to reload. By the 1880s magazine rifles were also in use and were made more effective when a bolt action was incorporated after 1889. The next step was towards a weapon that could fire a continuous stream of bullets. Manually operated systems had been tried (such as the gatling gun), but the first modern machine gun was the Maxim gun, invented in the 1880s, which used the recoil energy of the fired bullet to push the next round into the breech and recock the weapon. Guns of this type dominated the trench warfare of World War 1, and by World War 2 more portable automatic weapons, light machine guns such as the Bren gun and sub-machine guns, were in use. Most 20th-century firearms operate on principles established in the 19th century. Newer developments, under the stimulus of the emergence of aerial warfare and the TANK, include gas-operated rifles, recoilless rifles, firearms with several rotating barrels and extremely high rates of fire, and small firearms that use explosive bullets.

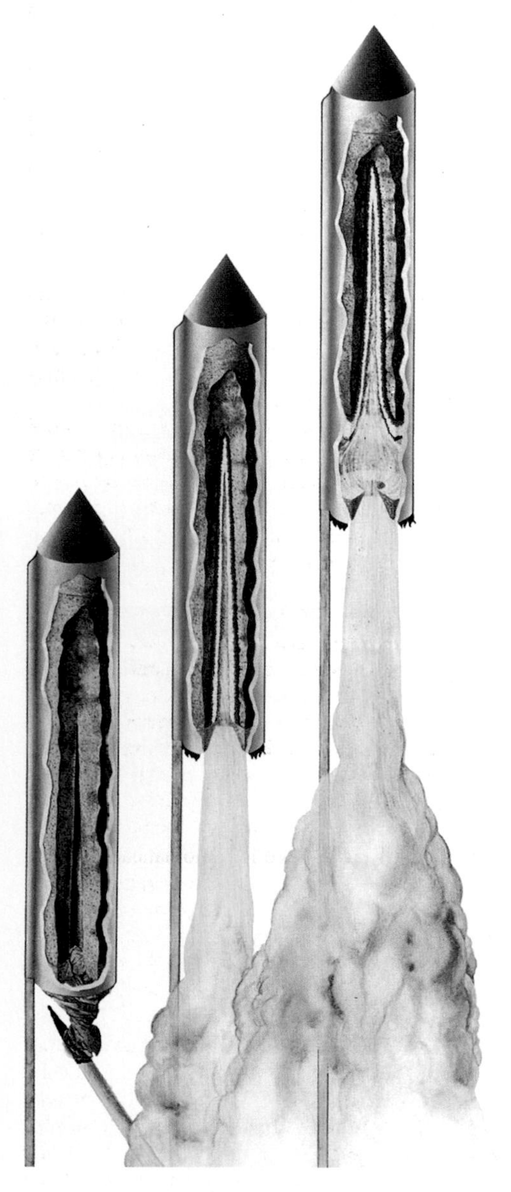

▲ **firework** The combustion of propellant in a firework produces its power. The case is wet-rolled and the thrust increased by constricting it near one end (constriction made with a cord before the case is dry). The propellant is packed into the case so that a conical cavity is formed at the burning end, allowing a large surface area of combustion to give the initial push. The cap of the rocket contains flares, or "stars", which are ejected as the propellant burns out.

fireball Exceptionally bright METEOR. Fireballs have been loosely defined as meteors brighter than the planets; with the modern estimate of the maximum brightness of Venus, this would mean that all meteors brighter than magnitude –4.7 should be classified as fireballs.

firebrick Brick formed in a variety of shapes for use in structures, such as FURNACE linings, that are exposed to high temperatures. Firebrick is composed of FIRECLAY and other nonmetallic, high melting-point minerals, particularly those rich in silica, aluminium oxide, magnesium and carbon.

fireclay Clay that can withstand high temperatures without becoming deformed. It is used for making FIREBRICK, crucibles and many refractory shapes. Fireclay approaches KAOLIN in composition; better grades contain at least 35% alumina (Al_2O_3) when fired.

firedamp In coal mining, flammable and explosive gas emitted by coal seams. It is composed mostly of methane (natural gas, CH_4) but also contains some hydrogen, oxygen and carbon dioxide, and occasionally a little ethane. Firedamp has caused many disastrous pit explosions; it is detectable with a Davy lamp (after Sir Humphry DAVY), the flame of which elongates when the gas is present.

fire retardant Chemical solution, consisting mainly of boric acid and borax, impregnated into flammable materials to slow down the process of combustion. In construction, fire-resistive buildings are made of reinforced concrete or protected steel. ASBESTOS is no longer considered to be a safe building material.

fireworks (pyrotechnics) Controlled EXPLOSIVES (and their spectacular display) that are intended for visual rather than destructive effects. They range from light and sparkle-emitting capsules and coils, to rockets that burst into miniature "galaxies of stars" after a short near-vertical flight. Fireworks began with the ancient Chinese, who first used them in the 8th or 9th centuries. They spread to Europe in the 12th and 13th centuries, along with the military uses of GUNPOWDER. Besides gunpowder, potassium chlorate mixtures came to be used as explosives and propellants; iron filings and magnesium and aluminium powders were incorporated to provide sparkle and brilliance; and fireworks were coloured brightly by the inclusion of salts of the metals sodium (yellow), barium (green), strontium (red) and copper (blue). Fireworks find serious applications in Very lights and distress rockets and flares.

firing Heating of ceramics until they are chemically changed; the process prevents a return to the plastic state. Many ceramic clays contain matter that must vitrify (fuse) to render the ware non-porous. This may require temperatures of up to 1,450°C (2,650°F). Firing is done in KILNS, or ovens.

firn (névé) Hard-packed mountain snow that has been converted into granular ice in a mountain GLACIER and, with an accumulation of broken rock materials at its base, digs out round basins called CIRQUES.

Fischer, Emil Hermann (1852–1919) German chemist. Fischer's important organic chemistry research contributed substantially to the knowledge of PURINE substances. He was the first to recognize the importance of STEREOCHEMISTRY and was fundamental in establishing ENZYME chemistry. Fischer synthesized glucose and fructose in addition to several other sugars, and for this work, and his work on purines, he was awarded the 1902 Nobel Prize for chemistry.

Fischer, Ernst Otto (1918–) German chemist. He became professor of chemistry in Munich in 1957. He shared the 1973 Nobel Prize for chemistry with Geoffrey WILKINSON for their work on organometallic chemistry.

Fischer, Hans (1881–1945) German biochemist who received the 1930 Nobel Prize for chemistry for his structural studies of CHLOROPHYLL and of the red BLOOD pigment haemin. His research indicated the close relationship between these two substances, and he was able to synthesize haemin and almost completely to synthesize one of the chlorophylls. He also studied carotene.

fish The classification of fish varies. They are usually divided into three classes: Agnatha, which are **jawless** fish, including the hagfish and lamprey; Chondrichthyes (**cartilaginous** fish), which includes shark, skate, ray and chimera; and the much more numerous Osteichthyes (**bony** fish), including subclasses of soft-rayed fish (lungfish and lobefin), and the very successful teleost fish, such as salmon and cod. There are more than 22,000 species of bony fish, and they represent about 40% of all living vertebrates. They are divided into 34 orders and 48 families. All fish are cold-blooded (POIKILOTHERMIC), aquatic, vertebrate animals characterized by fins, gills for breathing, a streamlined body almost always covered by scales or bony plates on to which a layer of mucus is secreted, and a two-chambered heart. Fish are the most ancient form of vertebrate life, with a history of about 450 million years. They reproduce sexually, and FERTILIZATION may be external or internal. The eggs develop in water or inside the female, according to species. Fish have lateral line organs, which are fluid-filled pits and channels that run under the skin of the body. Sensitive fibres link these channels to the central nervous system and detect changes of pressure in the water and changes of strength and direction in currents. About 75% of all fish live in the sea; the remainder are freshwater species that live in lakes, rivers and streams. A few fish, such as the SALMON and EEL, divide their lives between salt and freshwater habitats. *See feature on page 144*

fission In zoology, form of ASEXUAL REPRODUCTION in some single-celled organisms. In **binary** fission, the parent cell, such as a bacterium, diatom or protazoan, simply divides in two to produce two identical daughter cells. **Multiple** fission produces 4, 8, or, in the case of some protozoa, more than 1,000 daughter cells, each developing into a new organism.

fission, nuclear Type of nuclear reaction in which a heavy atomic NUCLEUS splits into two, with the release of two or three NEUTRONS and large amounts of energy. It may occur spontaneously or be made to occur by bombarding certain nuclei with low-energy, or slow NEUTRONS. The neutrons released by the initial splitting may go on to produce further fission in a nuclear CHAIN REACTION. The process is employed in NUCLEAR WEAPONS and NUCLEAR REACTORS. *See also* FUSION, NUCLEAR; NUCLEAR ENERGY

Fitzgerald, George Francis (1851–1901) Irish physicist who researched into ELECTROLYSIS and ELECTROMAGNETISM and who is noted for his theory of ELECTROMAGNETIC RADIATION. As an explanation of the MICHELSON-MORLEY EXPERIMENT to determine the Earth's movement through the ETHER, Fitzgerald suggested the theory that objects change length (the **Lorentz-Fitzgerald contraction**) due to this type of movement. This was important to Albert EINSTEIN when he devised his theory of relativity. *See also* LORENTZ TRANSFORMATION

fixer, photographic Solution in which photographic film or paper is immersed after development to remove unexposed and unreduced silver halide and render the image stable. Sodium or ammonium thiosulphate (hypo) is usually used, often acidified to avoid staining.

Fizeau, Armand Hippolyte Louis (1819–96) French physicist, the first to determine with accuracy the speed of light in both air and water. Fizeau also took, with the French physicist Jean Léon FOUCAULT, the first clear photograph of the Sun. He researched the POLARIZATION of light, the expansion of crystals and the DOPPLER EFFECT, and looked for methods of increasing the life of the early photographs called **daguerrotypes**.

fjord (fiord) Narrow, steep-sided inlet on a sea coast. These deeply cut valleys were formed by GLACIERS as they moved towards the sea, and then flooded when the ice melted and sea levels rose.

flagellate Any member of the class Mastigophora. Flagellates are single-celled organisms that possess, at some stage of their development, one or several whiplike structures (FLAGELLUM) for locomotion and sensation. They are divided into two major groups; the **phytoflagellates** resemble plants (in that they obtain their energy through photosynthesis), the **zooflagellates** resemble animals (in that they obtain their energy through feeding). Most have a single nucleus. Reproduction may be asexual (FISSION) or sexual. See also ASEXUAL REPRODUCTION; SEXUAL REPRODUCTION

flagellum (pl. flagella) Long, whip-like extension of a cell. There may be a single flagellum or a group of them. Many cells, such as SPERM cells and some BACTERIA, PROTOZOA and single-celled algae, beat their flagella as a means of locomotion – they "swim" through fluids in this way. In sperm and protozoa, the structure and movement of the flagella resembles that of CILIA, whereas in bacteria, the flagellum has a different structure and a rotary action. *See also* FLAGELLATE

flame test In chemical analysis, test for the presence of metallic elements whose main spectral emission lines give characteristic colours in a Bunsen flame. A salt of the metal is introduced into the flame by means of platinum wire first dipped in hydrochloric acid. Some metals and their colours are: lithium, red; sodium, yellow; potassium, lilac; strontium, crimson; barium, apple-green; copper, blue-green; lead and arsenic, blue.

Flamsteed, John (1646–1719) British astronomer, the first Astronomer Royal (1675). His book *British Catalogue of Stars* contained a record of his observations and listed nearly 3,000 stars; it was the most accurate of the time and became the standard work for many years.

FISSION, NUCLEAR

Most nuclear power stations use uranium-235 as fuel. When a uranium-235 nucleus is struck by a slow-moving neutron (1), it absorbs the neutron to form uranium-236. This is unstable and splits violently (2) forming two smaller nuclei, generating radiant energy (some in the form of heat), and releasing several neutrons (3). These neutrons can then start the process again (4), splitting further nuclei, which in turn release yet more neutrons (5). Such a process is known as a chain reaction and can spread at lightning speed. In a nuclear reactor, many of the neutrons are absorbed to prevent the chain reaction from running out of control and causing an excessive release of energy. Atom bombs are designed to encourage the chain reaction to spread extremely rapidly.

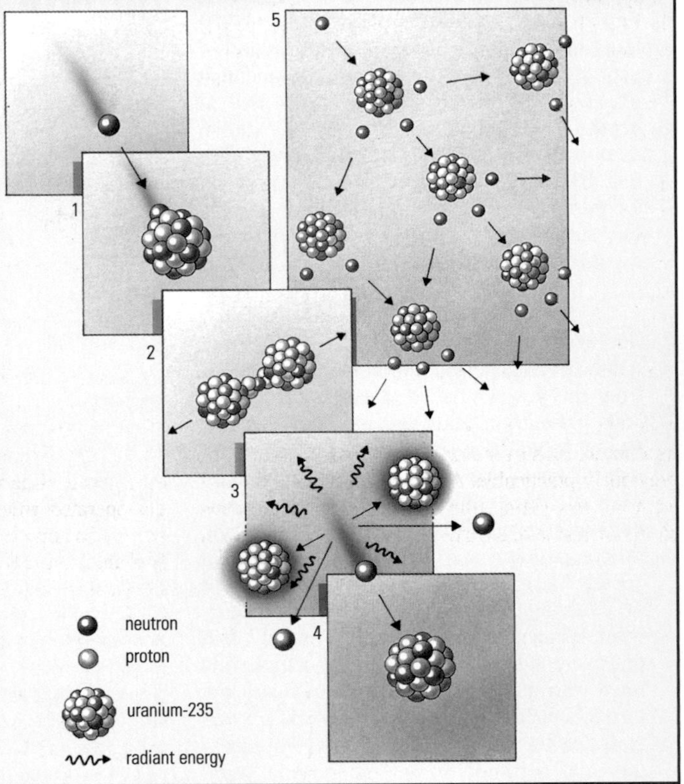

neutron
proton
uranium-235
radiant energy

flash point Lowest temperature at which a flammable liquid heated under test conditions gives off sufficient vapour to be ignited (as a flash) when a flame is applied.

F-layer (Appleton layer) Region within the IONOSPHERE of the Earth's atmosphere.

Fleming, Sir Alexander (1881–1955) Scottish bacteriologist, discoverer of PENICILLIN (the first ANTIBIOTIC). In 1922, while at St. Mary's Hospital in London, Fleming had discovered lysozyme, a natural antibacterial substance found in saliva and tears. During research on staphylococci in 1928, Fleming noticed that a mould, identified as *Penicillium notatum*, liberated a substance that inhibited the growth of some bacteria. He named it penicillin, the most powerful antibiotic known. In 1938 he became professor of bacteriology at the University of London. Howard FLOREY and Ernst CHAIN refined the drug's production, and in 1941 it was produced commercially. In 1944 Fleming was knighted, and in 1945 he shared the Nobel Prize for physiology or medicine with Florey and Chain.

Fleming, Sir John Ambrose (1849–1945) English electrical engineer who invented the thermionic ELECTRON TUBE (valve). Fleming's valve was a rectifier, or DIODE. It consisted of two electrodes in an evacuated glass envelope resembling an electric light bulb. The diode permitted current to flow in one direction only. It could detect radio signals but could not amplify them. He is also remembered for FLEMING'S RULES for motors and generators.

Fleming's rules In physics, aids for remembering the relationships between the directions of the current, field and motion in ELECTRIC MOTORS and GENERATORS. In the left-hand rule (for motors), the forefinger represents field, the second finger current, and the thumb, motion; when the digits are extended at right-angles to each other, the appropriate directions are indicated. The right-hand rule applies the same principles to generators. Sir John FLEMING devised the rules.

flexor MUSCLE that makes a limb bend. A flexor usually has two points of attachment on either side of a joint. For example, the biceps muscle attaches to the forearm and the upper arm. Flexing the biceps causes the arm to bend. Like all flexors, the biceps has an antagonistic ("opposite") muscle, called an EXTENSOR, that causes the arm to straighten in this case the triceps.

flight Movement through the air. To maintain level flight through the air, the two main requirements are thrust and lift. Thrust overcomes air resistance (DRAG) and gives forward motion. LIFT overcomes the downward pull of gravity. In an AIRCRAFT, thrust is provided by an engine-driven propeller or by the rapid emission of hot expanding gases from JET or ROCKET engines. Lift is provided by AEROFOILS, which cause the air to move faster over their more highly curved upper surfaces than under their lower surfaces. The slower-moving air exerts a higher pressure, resulting in an overall upward force. The wings of birds have a similar shape (in cross-section) to an aerofoil. Instead of a propeller, birds, insects and bats beat their wings, angling them so as to push air behind them. *See also* HELICOPTER

flight recorder Device for automatically recording data during the operation of an aircraft. Investigators analyse the data after a crash or malfunction. A small aircraft may have just a simple cockpit voice recorder (CVR). This records all cockpit sounds, and also all radio contact with air traffic control. Larger aircraft carry a separate flight data recorder (FDR). Control settings, instrument readings, and other data are recorded on magnetic wire. The FDR, often called a "black box", is usually coloured orange for ease of location at a crash site.

flight simulator Device that duplicates the instrument behaviour and physical attitude of an aircraft in flight, used for training pilots and aircrew. A more sophisticated simulator, in addition to having an exact replica of the controls of a particular aircraft, may also have a visual display of the terrain being "flown over" to contribute to realism.

flint Granular variety of QUARTZ (SiO_2) of a fine crystalline structure. It is usually smoky brown or dark grey, although the variety known as CHERT is a

There are more than 22,000 different species of fish, ranging in size from the tiny tropical species to sharks 12m (39ft) long and weighing 12 tonnes.The ranges of habitat is equally extreme; while some fish live near the ocean surface others live at depths of up to 2,000m (6,600ft). The icefish lives under the polar ice, whereas desert pupfish live in hot springs. Indeed some fish, including the lungfish and walking catfish, can survive long periods on land and are capable of breathing air.

Evolution

The evolution of fish is not straightforward. Although in general there is real progression from the jawless fishes of the oceans some 460–480 million years ago and the shark-like fish of 380 million years ago to the true bony fish (teleosts) that first appeared 175 million years ago, the evolutionary success of the earlier types ensured that they did not succumb to competition and simply die out when "newer" fish evolved. Instead they too went on evolving. The class of fish that contains most species today – the "ray-finned" fish that have a single DORSAL fin, pectoral fins lined with thin radial bones, scales that grow throughout life, a bony skeleton, and a SWIM BLADDER for flotation – derives from ancestors that appeared some 390 million years ago: they are a "modern" type of fish. The sharks, which are often thought of as relatively primitive, in fact evolved later, between 190 and 135 million years ago.

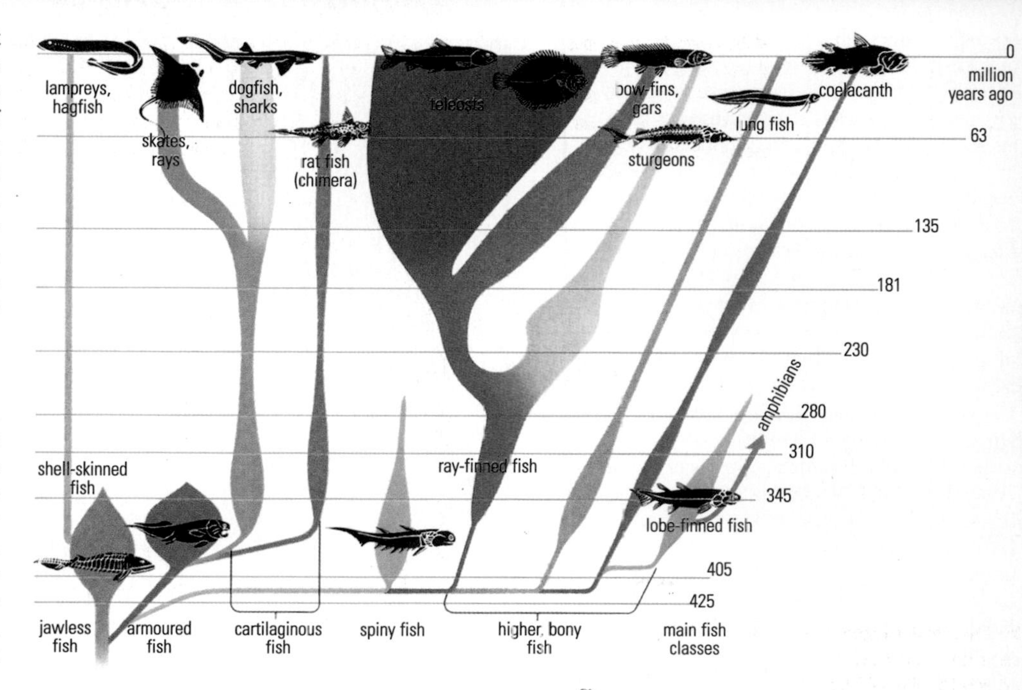

▲ **A family tree** outlines fish evolution from the most primitive armour-plated fish to modern fish that have scales, bony skeletons and paired fins – known as teleosts. The primitive coelacanth and sharks survive, but ray fins and armoured fish are extinct.

► **The sea bass**, like most modern fish, belong to the class Osteichthyes, or bony fish. They have a bony skeleton (1) with fins (2) supported by bony rays (3). Fins and powerful muscles (4) – overlapping in blocks corresponding to a pair of vertebrae – in the flexible body provide propulsion for swimming. The streamlined body, which tapers smoothly at each end, offers minimal water resistance; most fish have scales (5) – bony skin outgrowths. Gills (6), eyes (7) and nostrils (8) enable fish to breathe, see and smell underwater.

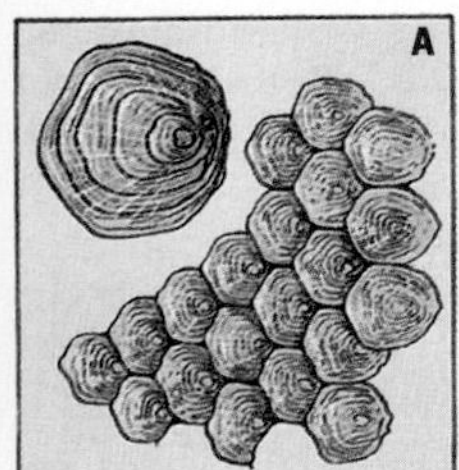

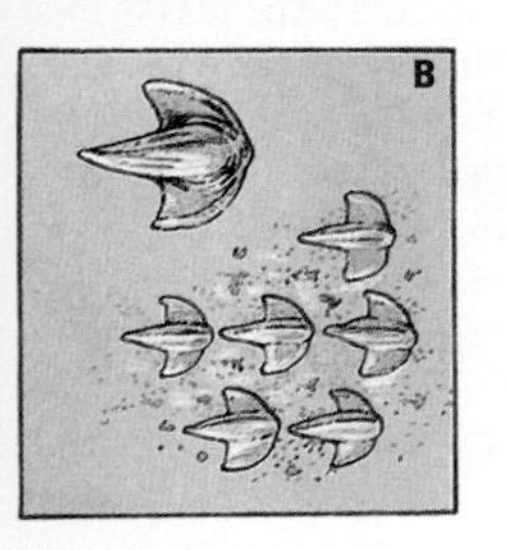

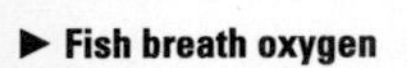

◄ **Fish** have different types of scale depending on whether they are teleosts (bony fish), such as the sea bass on the opposite page, or catilaginous, such as the mako shark shown below. Cycloid (or ctenoid) scales (A) are arranged in rows, each one having a series of tiny ring-shaped ridges (1) – growth rings that can show the fish's age. Scales are both a protection and flexible covering. Because they are translucent, the scales let the pigmentation on the skin of the fish show through from below. Placoid scales (B), found on cartilaginous exoskeletons of sharks look like tiny, closely spaced teeth. Ganoid scales are another type of scale found on primitive fish, such as the bony gar. These diamond-shaped scales contain ganoin, which gives a silvery, mirror-like look. Some fish, such as certain species of eel and freshwater catfish, have no scales at all.

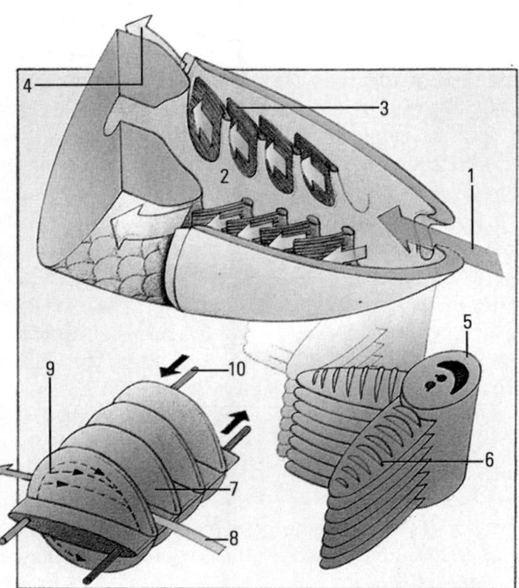

► **Fish breath oxygen** dissolved in water and extracted by gills, which can achieve 80% extraction rates, over three times the rate human lungs can extract from air. Water enters the mouth (1), passes through the gill chamber (2) over the gills (3) and exits via a flap called the operculum (4). Flow is maintained by the pumping action of the mouth, synchronized to the opening and closing of the operculum. The gills are rows of bony rods (5) to which are attached fleshy filaments (6) rich in blood capillaries to absorb oxygen. Each filament has fine secondary flaps (lamellae) (7) to maximize the gas exchange surface area, which, in active fish like mackerel, can be over 10 times the outer body area. Water (8) passes over the gills against the capillary blood flow (9); this "counter-current flow" ensures water always passes over de-oxygenated blood, maximizing oxygen absorption. Blood vessels (10) circulate the blood.

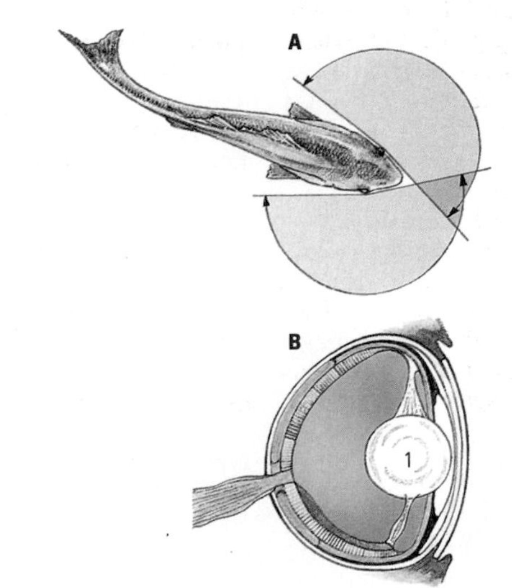

◄ **Fish eyes** (A) are adapted to see underwater. Unlike a human lens, a fish lens (1) is a perfect sphere, which may reduce image distortion. The eyes protrude somewhat to give reasonable all-round vision, but there is very little overlap between each eye's field of vision (B), hence 3-dimensional vision is poor. Fish have no eyelids – there is no need to prevent the eye from drying up – and they also lack pupils that can vary their size. Most fish have some colour vision, but sharks and rays appear to see only in black and white. Unlike those species that live in caves, fish of deep water have functioning eyes, probably used to detect luminous deep-sea creatures.

ANATOMY

There are more than 22,000 species of bony fish. Although they vary in shape and the way they swim, they share many common features. All have a tail with equal upper and lower lobe sizes, which provides neither up nor down thrust. Such fish achieve natural buoyancy by adjusting their density using the swim bladder. The fish can expand or contract the swim bladder by secreting gas into or absorbing gas out of it, so adjusting the volume and external pressure, and counteracting the tendency to sink or float to the surface.

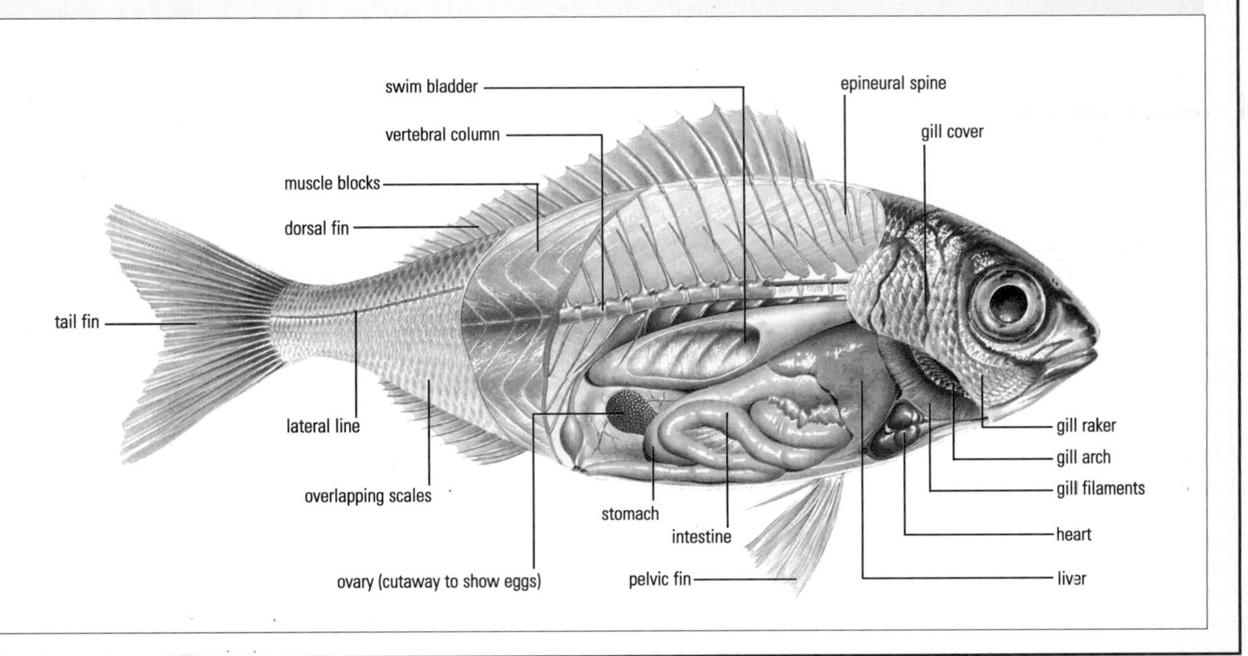

▲ **flood plain** Although they provide rich, fertile agricultural land, the major drawback of low-lying flood plains becomes apparent during the monsoon season. Here flooding on the plains of India has resulted in half-submerged field systems.

paler grey. It commonly occurs in rounded nodules and is found in chalk or other SEDIMENTARY ROCKS containing calcium carbonate. Of great importance to early humans during the Stone Age, when struck a glancing blow, flint can be flaked, leaving sharp edges appropriate for tools and weapons; two flints struck together produce a spark which can be used to make fire.

flint clay Hard, flinty FIRECLAY. It is a KAOLINITE, usually found at greater depths than most clays. It is used almost exclusively in the production of FIREBRICK and crucibles.

flip-flop (bistable circuit) Electronic circuit that can be in either of two stable states. Much used in computers, a flip-flop (or bistable circuit) changes state when it receives an electric pulse. It can be used as a one-bit storage device, and is employed in delay elements (which store a signal briefly before transferring it to an output line).

floodplain Level land alongside a river consisting of ALLUVIUM deposited by the river when in flood. Such plains usually have extremely fertile soil and are often used, as near the River Nile, for intensive cultivation.

floppy disk Portable MAGNETIC DISK for use in personal COMPUTERS. Protected by a plastic coating, floppy disks are traditionally still measured in inches, and are commonly either 3.5in (7.9cm) or, now less common, 5.25in (13.3cm) in diameter. To access the data on a disk it is inserted into a computer's disk drive. Data is stored magnetically on both sides of the disk, and is read by magnetic heads in the computer as the disk rotates at some 300rpm. Modern disks can store up to 1.4 MEGABYTES of data, which can be overwritten when it is no longer required.

flora Collective term used to describe all the plants that are found in a particular area, or that occurred at a particular time.

Florey, Sir Howard Walter (Baron Florey of Adelaide) (1898–1968) British pathologist, b. Australia. He shared, with Alexander FLEMING and Ernst CHAIN, the 1945 Nobel Prize for physiology or medicine for his part in the development of PENICILLIN. Florey isolated the antibacterial agent from the mould discovered by Alexander Fleming. This made possible the large-scale preparation of PENICILLIN.

Flory, Paul John (1910–85) US chemist whose discovery of a method of analysing POLYMERS accelerated the development of plastics and earned him the 1974 Nobel Prize for chemistry. With Wallace CAROTHERS he developed the first NYLON and a synthetic rubber, NEOPRENE. Flory also discovered similarities between the elastic properties of organic tissues and plastics.

flotation process Method of separating mineral ores from unwanted rock and other material. The ore as mined is finely ground and then mixed with water, to which chemicals called **wetting agents** are added. The mixture is aerated to make a froth. The required ore particles stick to the bubbles and float, and are skimmed off the surface; waste rock sinks to the bottom. The method works because of differences in wetting properties (SURFACE TENSION) between the various components.

flow, solid In geology, the slide, flow or creep of solid, unsuspended material downwards on a slope. The movement within the solid is achieved by rearrangement between or within its particles.

flower Reproductive structure of all ANGIOSPERMS (flowering plants). It has four sets of organs set in whorls on a short apex (RECEPTACLE). Typically the SEPALS are leaf-like structures that protect the bud; they form the CALYX. The petals, often brightly coloured, form the COROLLA; the STAMENS are stalks (filaments) tipped by ANTHERS (pollen sacs); the CARPELS form the PISTIL with an OVARY, STYLE and STIGMA. Flowers are bisexual if they contain stamens and carpels, and unisexual if only one of these is present. Reproduction occurs, following POLLINATION, when POLLEN is transferred from the anthers of one flower to the stigma of another flower of another plant (**cross-pollination**), or to the same flower or flower of the same plant (**self-pollination**). A pollen tube grows down into the ovary where FERTILIZATION occurs and a seed is produced. The ovary bearing the seed ripens into a FRUIT containing the seed, and the other parts of the flower wilt and fall.

flowering plant *See* ANGIOSPERM

fluid Any substance that is able to flow. Of the three common states of matter, GAS and LIQUID are considered fluid, while any SOLID is not. *See also* PHASE

fluid flow Behaviour of a moving fluid, determined by its velocity, pressure and density. These three quantities are related by three basic equations: the **equation of continuity**, which relates the amount of fluid flowing into a given space with the amount flowing out of that space; EULER's **equation of motion**, which shows how the velocity of the fluid changes with time at a given point in space; and the **adiabatic equation**, which describes the exchange of heat between different parts of the fluid. In incompressible fluid flow, which applies to most liquids, these equations take on a particularly simple form. Compressible flow equations are necessary for high-speed aerodynamic calculations. Often a fluid is treated as "ideal", meaning that no internal friction or viscosity is supposed. The equations for realistic fluids are so complicated that complete solutions to most problems do not exist; numerical solutions must be attempted by computer techniques.

fluidics Use of devices operated by a FLUID (gas or liquid) for controlling processes and instruments. Fluidic systems simulate electronic circuits, but use a fluid instead of electrons. Fluidic circuits were developed in the USA in the 1960s for use in rocket and aircraft guidance.

fluidization Powdering of a solid so that it can be processed as if it were a fluid. The technique not only allows solids to be conveniently transported, but also hastens gas-solid reactions – the gas is injected from below and the powder is suspended to form a fluidized bed, an ideal condition for industrial drying or roasting.

fluid mechanics Study of the behaviour of FLUIDS (liquids and gases). **Fluid statics** includes the study of pressure, density and the principles of PASCAL and ARCHIMEDES. **Fluid dynamics** includes the study of streamlines, BERNOULLI'S LAW and the propagation of waves. Engineers use fluid mechanics in the design of bridges, dams and ships. Physicists use it in studying the structure of the atomic nucleus, and astronomers use it to explain the spiral structure of some galaxies.

fluid mining Method of obtaining minerals which involves dredging or HYDRAULIC MINING. Gravel (PLACER) deposits containing ore can be broken down with high-pressure streams of water or flooded so that dredgers with digging and processing equipment can operate. Many off-shore deposits, including gold, tin, iron-bearing sands, shell and gravel, are mined by dredging.

fluorescence Emission of radiation, usually light, from a substance when its atoms have acquired excess energy from a bombarding source of radiation, usually ultraviolet light or electrons. Unlike PHOSPHORESCENCE, fluorescence ceases when the source of energy is removed. Mercury vapour is a fluorescent substance used in the fluorescent lamps of motorway lights; television tubes use fluorescent screens.

fluorescent lamp Glass tube coated on the inside with a chemical phosphor and containing, at low pressure, small quantities of mercury. When an electric current is passed through the tube, the atoms of mercury emit ultraviolet radiation, which impinges on the phosphor coating, causing it to glow brightly. *See also* DISCHARGE TUBE; FLUORESCENCE

fluoridation Addition of inorganic fluorides to the water supply to reduce tooth decay. The additive is usually sodium fluoride, at a concentration of about one part per million. Since its inception in the 1930s fluoridation has been adopted in many countries.

fluoride Any salt of hydrogen fluoride (HF); more particularly, fluoride compounds added to drinking water or toothpaste in order to build up resistance to tooth decay. Fluoride protection appears to result from the formation of a fluorophosphate complex in the outer tooth layers, which become resistant to penetration by acids made by mouth bacteria.

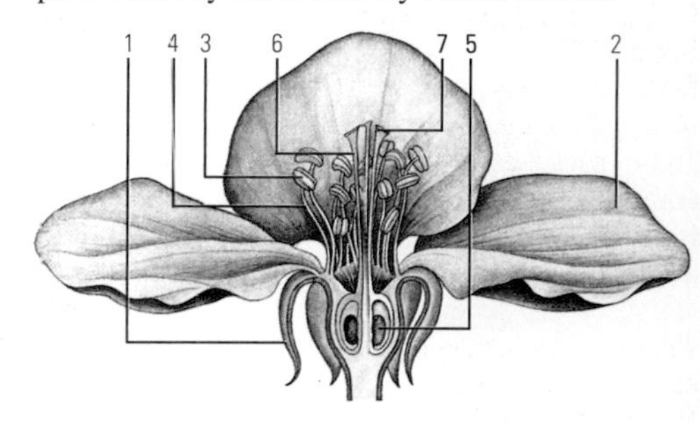

▲ **flower** A typical flower has four main parts: sepals, petals, stamens and carpels. The sepals (1) form a protective covering (the calyx) over the developing flower bud, and lie outside the showy petals (2), which collectively are called the corolla. Each male stamen is made up of an anther (3), which contains the pollen grains, borne on a filament (4). The female carpels, which together form the pistil, are found at the centre of the flower, each containing ovaries (5) which bear ovules and a style (6) which supports the stigma (7) – the structure on which pollen is deposited.

FLUIDICS

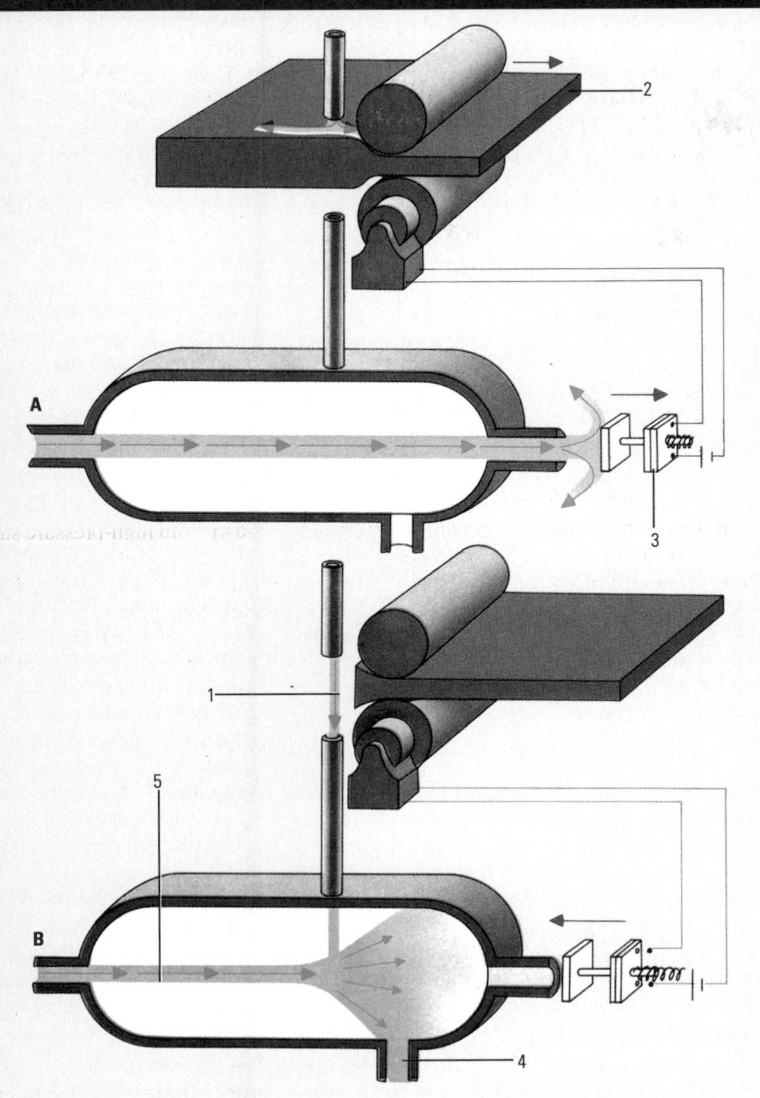

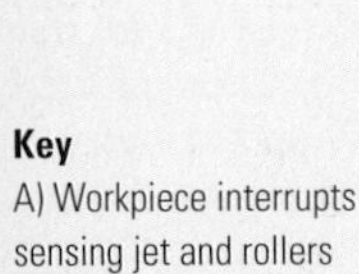

This fluidic controller is shown controlling a rolling process. So long as a workpiece is passing between the electrically driven rollers, the sensing jet is interrupted and the supply jet forces the spring contacts closed. When the trailing edge of the workpiece passes the sensor, the sensing jet impinges on the supply jet and scatters it; little or no flow reaches the contact plate which therefore springs back to open the circuit and stop the rollers. The next workpiece to pass the sensor interrupts the sensing jet and the rollers are again set in motion.

Key
A) Workpiece interrupts sensing jet and rollers continue in motion.
B) Workpiece passes through and sensing jet scatters supply jet thereby stopping rollers.
1) continuous sensing jet
2) workpiece
3) sprung contact plate
4) exit port
5) continuous supply jet

fluorine (symbol F) Gaseous, toxic element of the HALOGEN group (elements in group VII of the periodic table), first isolated in 1886. Its chief sources are fluorspar and cryolite. The pale yellow element, obtained by ELECTROLYSIS, is the most electronegative of all elements and the most reactive of non-metallic elements, attacking many compounds. It is in FLUORIDE in drinking water and used in making FLUOROCARBONS and in extracting uranium. Properties: at.no. 9; r.a.m. 19; m.p. −219.6°C (−363.3°F); b.p. −188.1°C (−306.6°F); single isotope ^{19}F.

fluorite (fluospar) Mineral, calcium fluoride (CaF_2). It has cubic system crystals with granular and fibrous masses. It is brittle and glassy and its colour varies; it can be yellow, purple or green and is frequently banded. It is found in mineral veins as a gangue mineral with metallic ores. It is used as a FLUX in STEEL production and in the ceramics and chemical industries. The deep-purple banded variety is known as Blue John. Hardness 4; r.d. 3.1.

fluorocarbon Organic compound that is produced by replacing the hydrogen atoms of HYDROCARBONS with fluorine atoms. They are inert, have low toxicity, and an ability to withstand high temperature, making them ideal for use in plastics such as PTFE (POLYTETRAFLUOROETHYLENE) or Teflon. Many of these chemicals also contain chlorine and are called CHLOROFLUOROCARBONS (CFCs), the breakdown of which by sunlight damages the OZONE LAYER.

fluorocarbon plastic Plastic made from a class of chemically inert compounds composed entirely of carbon and FLUORINE. The best known of these plastics is POLYTETRAFLUOROETHENE (PTFE), also known as Teflon. Fluorocarbons are valued for their non-flammability, low chemical activity and low resistance to friction.

fluorspar *See* FLUORITE

flux In ceramics, any substance that promotes vitrification when mixed with clay. When the ware is fired, the flux melts, filling the porous clay form. As the piece cools, it hardens, becoming glossy and non-porous. Fluxes include felsphathic rock, silica and borax. In METALLURGY, a flux is added to the charge of a smelting furnace to purge impurities from the ore and to lower the melting point of the slag.

flywheel Massive solid or heavy-rimmed wheel attached to a machine's drive shaft to minimize fluctuations in rotational velocity and to store energy. The inertia of the large wheel tends to absorb energy in the effective portion of an engine's cycle and to release energy in the least effective portion of the cycle. The flywheel was initially developed by James WATT.

FM Abbreviation of FREQUENCY MODULATION

f-number In photography, measurement used to describe the relative opening or APERTURE of a lens, often known as the f-stop or focal ratio. It is calculated by dividing the FOCAL LENGTH of a lens by the diameter of the lens opening. For example, a lens with a 50mm focal length with the aperture set at f/2 indicates an aperture diameter of 25mm, whereas f/4 indicates an aperture of 12.5mm. The higher the f-number (f/32 is usually the highest), the less light reaches the film. The f-numbers allow a photographer to gauge the amount of light to which a film is exposed.

foam Suspension of gas bubbles separated from each other by thin films (0.1–1mm thick) of liquid or solid. Foaming is not possible in a pure liquid and a promoter must be added – such as soap or proteins, which are used in such edible foams as marshmallow.

focal length Distance from the midpoint of a curved mirror or the centre of a thin LENS to the focal point of the system. For converging systems, it is given a positive value; for diverging systems, a negative value.

focal point (focus) Point on the axis of a LENS or curved mirror to which incident light rays, parallel and close to the axis, are converged (**real focus**) or from which they appear to be diverged (**virtual focus**) after REFLECTION or REFRACTION.

focus Either of two points on the major axis of an ELLIPSE such that the distance from one focus to any point on the ellipse and back to the other focus is constant. In geology, the term is also used to describe the point of maximum displacement caused by an EARTHQUAKE. In optics it is the FOCAL POINT.

fog Mass of water droplets immediately above the Earth's surface that reduces visibility to less than 1km (0.6mi). A light fog is called mist or haze. Fog is caused by water vapour condensing as a result of the air becoming cooler. This condensation takes place around particles of dust. There are four main types of fog: **advection** fog develops from air flowing over a surface of a different temperature, such as steam fog that results from cold air passing over warm water; **frontal** fog forms when warm rain

◄ **fog** Fog in the New Forest, Hampshire, UK. The type of fog shown here is radiation fog, which forms as the once warm ground, heated by daytime Sun, cools during the night..

FOLD

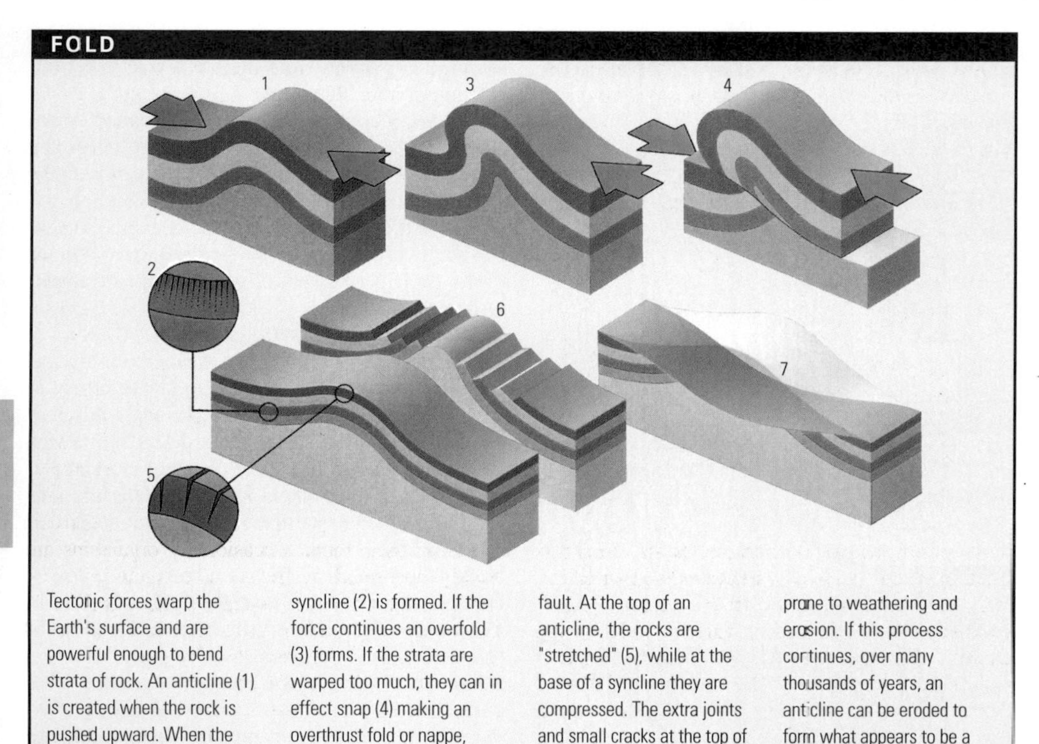

Tectonic forces warp the Earth's surface and are powerful enough to bend strata of rock. An anticline (1) is created when the rock is pushed upward. When the rock is forced down, a syncline (2) is formed. If the force continues an overfold (3) forms. If the strata are warped too much, they can in effect snap (4) making an overthrust fold or nappe, which is both a fold and a fault. At the top of an anticline, the rocks are "stretched" (5), while at the base of a syncline they are compressed. The extra joints and small cracks at the top of the anticline makes it more prone to weathering and erosion. If this process continues, over many thousands of years, an anticline can be eroded to form what appears to be a syncline (7).

falls through cold air near the ground; **radiation** fog occurs when the ground cools on a still, clear night, and is most common in valleys; **upslope** fog develops when air cools as it ascends a slope. *See also* DEW POINT

fold In geology, a bend in a layer of rock. An upfold is an ANTICLINE; a downfold, a SYNCLINE. A monocline (flexure) slopes in one direction only and usually passes into a FAULT. Folds occur as part of the process of PLATE TECTONICS, where rock strata buckle and bend under pressure. If the compression is fairly gentle and even, the resulting fold is "symmetrical". If the pressure is uneven, then asymmetrical folds will form. In many cases, the folds are pushed right over to form **recumbent folds**. Eventually the rock strata may break under the pressure, to form an overthrust or a NAPPE.

foliation Crude layering of rocks produced under compression. The layering is approximately parallel to the bisecting planes of folds, and often results in the rock splitting because of the parallel orientation of the mineral layers. Mica and slate are good examples of foliation.

folic acid Yellow crystalline derivative of glutamic acid; it forms part of the VITAMIN B complex. Found in liver and green vegetables, it is crucial for growth and is used in the treatment of ANAEMIA.

follicle In botany, fruit that splits along one side to release its seeds when it is ripe (unlike a pod, which splits along both sides). Such follicles occur in clusters called **etaerios**. They are found in plants such as delphiniums and larkspur.

follicle In zoology, group of cells forming a sac that envelops and protects a structure within it. A hair follicle surrounds and nourishes the root of a hair. A GRAAFIAN FOLLICLE encloses a developing OVUM.

follicle-stimulating hormone (FSH) HORMONE produced by the anterior PITUITARY GLAND located at the base of the brain of mammals. In females it regulates OVULATION by stimulating the GRAAFIAN FOLLICLES found in the OVARY to produce eggs (*see* OVUM). In males, FSH promotes SPERMATOGENESIS, the process by which SPERM are produced. FSH is an ingredient in most FERTILITY DRUGS.

fontanelle Space between the bony plates of an infant's SKULL, often called a "soft spot". There are in fact two fontanelles in a baby's skull, one in the centre front, one towards the rear, which close over as the skull plates grow towards each other. By the age of two they are joined, but the seams between the bones do not completely fuse until old age.

food Material taken into an organism to maintain life and growth. Important substances in food include PROTEINS, FATS, CARBOHYDRATES, MINERALS and VITAMINS. *See also* FOOD CHAIN

food chain Transfer of energy through a series of organisms, each organism consuming the previous member. In its simplest form, the main sequence of a food chain is from green plants (PRIMARY PRODUCERS) to HERBIVORES (PRIMARY CONSUMERS) and then to CARNIVORES (SECONDARY CONSUMERS). DECOMPOSITION, brought about by organisms of decay such as bacteria and fungi at each stage and at the end of the chain, breaks down waste and dead matter into forms that can be absorbed by plants, so perpetuating the chain. *See also* PHOTOSYNTHESIS

food poisoning Acute illness caused by consumption of food which is itself poisonous or which has become poisoned or contaminated with BACTERIA. Frequently implicated are SALMONELLA bacteria, found in many different strains in cattle, pigs, poultry and eggs, and LISTERIA, sometimes found in certain types of cheese. Symptoms include abdominal pain, DIARRHOEA, nausea and vomiting. Treatment includes rest, fluids to prevent dehydration and, possibly, medication to curb vomiting. *See also* BOTULISM; GASTROENTERITIS

food preservation Treatment of foodstuffs to prolong the time for which they can be kept before spoiling. Smoke-curing and heat-drying preserve meat by removing water and forming protective outer layers. Salting, pickling and FERMENTATION preserve food chemically. Chemical preservatives, such as sodium benzoate, can also be added to foods. In CANNING, meats and vegetables are sterilized by heat after being sealed into airtight cans. Cold storage at 5°C (41°F) prolongs the life of foods temporarily, while deep-freezing at −5°C (23°F) or below greatly extends the acceptable storage period. The packing of foods in sealed plastic containers is a method of temporary preservation, and in the modern technique of **freeze-drying**, the frozen foods are placed in a vacuum chamber and the water in them is removed as vapour, the foods can then be fully reconstituted at a later date. Since the early 1990s, **irradiation** (the preservation of food, especially soft fruit, by subjecting them to low levels of radiation in order to kill microorganisms) has been increasingly used.

food technology Application of scientific techniques to the generation, mass production, packaging, preparation and preservation of all types of food. Generating new and better forms of food is an entire science in itself. In the world of edible plants it corresponds to the improvement of genetic strains of grains, pulses and ordinary table vegetables in order to achieve greater yield and resistance to disease and ageing. Improving genetic strains is also important in the mass production of all forms of meat farming (including fish farming). Vegetables are grown to achieve optimum size and shape through the scientific provision of controlled environments, pest control measures and constantly metered nutrition through roots, stems and leaves. Animal stock may be afforded similar idealized conditions of growth and maturity, but can be further scientifically bioengineered through the careful introduction of hormones and other chemical substances intended to cause effects beneficial to the eventual consumer. Mechanical and electronic machinery for the mass-production of both plant and animal foodstuffs evolves year by year, sometimes radically. The presentation of food to the consumer is an aspect into which considerable technological effort has been expended in terms of packaging and preparation, both to look inviting on the shop shelf and to be appetizing on the table. The modern kitchen is full of mechanical and electric gadgets, many useful in food preparation but some aimed at FOOD PRESERVATION. *See also* BIOTECHNOLOGY

food web The often complex pattern of interrelating FOOD CHAINS. When created schematically on paper the lines that link the various animals and plants, and which make the individual food chains, form a criss-cross pattern resembling a web.

foot (symbol ft or ') Imperial unit of measurement equal to 0.3048m. There are 12 INCHES to 1ft.

foramen Cavity or aperture in a body part or organ. The foramen magnum, for example, is an aperture in the skull which allows the spinal column to pass through. In gorillas it is positioned towards the back, whereas in humans it is set well forward.

foraminifer Any member of the order Foraminiferida – amoeboid protozoan animals that live among plankton in the sea. They have multi-chambered chalky shells (**tests**) which may be spiral, straight or clustered and vary in size from microscopic to 5cm (2in) across, according to species. Many remain as fossils and are useful in geological dating. When foraminifera die, their shells sink to the oceanfloor to form large deposits, a source of chalk and limestone.

force (symbol F) Loosely speaking, a push, a pull or a turn. A force acting on an object may (1) balance an equal but opposite force or a combination of forces to maintain the object in equilibrium (so that it does not move), (2) change the state of motion of the object (in magnitude or direction), or (3) change the shape or state of the object. There are four FUNDAMENTAL FORCES in nature.

force ratio (mechanical advantage) Factor by

which a simple MACHINE multiplies an applied FORCE. It is the ratio of the load (output force) to the effort (input force). *See also* DISTANCE RATIO

forces, fundamental *See* FUNDAMENTAL FORCES

Ford, Henry (1863–1947) US industrialist. He developed a petrol-engined car in 1892 and founded Ford Motors in 1903. It achieved huge success with the economical and inexpensive Model T (1908), which from 1913 was produced on an assembly line, lowering production costs enormously. Almost as startling was Ford's introduction of an eight-hour working day and a relatively high basic wage.

forebrain One of the three primary parts of the developing BRAIN, clearly distinguishable, together with the MIDBRAIN and HINDBRAIN, in the early embryo. It becomes overlaid in the adult human by the cerebral hemispheres, which are massive developments from the forebrain.

forecasting, weather *See* WEATHER FORECASTING

forensic medicine Branch of medicine concerned with the detailed investigation of any assault, injury or sudden death that may be subject to legal action. Its practitioners include police surgeons and forensic pathologists.

Forest, Lee de *See* DE FOREST, LEE

forest Large area of land covered with a dense growth of trees and plants. Earth's first forests developed *c*.365 million years ago. In the early 1800s forests covered 60% of the Earth, today they account for *c*.30%. DEFORESTATION is a major environmental concern, since forests make such a vital contribution to Earth's atmosphere and also act as a WATERSHED. Forests have been an important source of timber, food and other resources since prehistoric times. The forest ECOSYSTEM has five basic strata; **canopy**, **understory**, **shrub layer**, **herb layer** and **forest floor**. Forests are classified into three general formations: **tropical hardwood** forests, including RAINFORESTS, are predominantly EVERGREEN. They account for *c*.7% of the Earth's landmass, but *c*.50% of the Earth's species. **Temperate** hardwood forests are mostly DECIDUOUS and are found in temperate climates. **Boreal forests**, consisting mainly of CONIFERS, lie in the far N.

forestry (silviculture) Managment of FOREST resources for human benefit, in particular the production of timber through REFORESTATION. It also includes the conservation of soil, water and wildlife. In 1907 the first US national forests were created, under the administration of the National Forest System. Today, more than 180 million acres (76 million ha) of forests are publicly owned and managed by the US Forest Service. In 1960 the service was directed to manage the national forests according to the principles of multiple use and sustained yield; to produce a continuous supply of timber while preserving the natural environment. In 1964 the Forest Service controversially adopted the commercial practice of clearcutting (removal of all trees in a certain area of the forest). The use of clearcutting came under severe criticism from conservationists. In 1974 the Forest and Rangeland Renewable Resources Planning Act was passed to establish procedures for continually reviewing the management of US forests. Education in technical forestry began in W Europe in the 19th century. Now many universities throughout the world offer forestry curricula. Modern forestry includes silviculture, dendrology, forest protection, engineering, utilization and management. Many operations are included, with emphasis on cycles of cutting and replenishment, selection and breeding, insect control, and limitation of forest fires. *See also* DEFORESTATION

forging Shaping of metal by hammering or by applying pressure against a shaped die. Blacksmiths forge horseshoes and other iron items by hammering the red-hot metal on an anvil. In mass-manufacturing processes, pressure from an hydraulic forging press or blows from a forging hammer shape metal parts by forcing them against hard-metal dies, which are carved in the shape of the required piece. Most metals are forged hot, but cold forging is also practised.

formaldehyde *See* METHANAL

formic acid *See* METHANOIC ACID

formula In chemistry, a **molecular** formula represents the elements and the number of their atoms in a compound. For example, the formula H_2O indicates that water is made up of two atoms of hydrogen to one atom of oxygen. A **structural** formula not only represents the elements and the number of atoms, but also indicates the structure of the compound and sometimes the number of bonds between molecules. For example, the structural formula of ethene (CH_2CH_2) indicates the two carbohydrate molecules are attached by a double bond.

formula In mathematics, rule or relationship expressed in mathematical symbols. An example is the formula $V = 4/3\pi r^3$ for the volume of a sphere.

Fornax (Furnace) Small constellation of the S skies. The only star above magnitude 4.0 is Alpha, 3.9.

Forssmann, Werner (1904–79) German physician and physiologist. He shared the 1956 Nobel Prize for physiology or medicine with A. F. COURNAND and D. W. RICHARDS. This was belated recognition for his development of the cardiac catheter, a tube that is inserted into a vein in the arm and advanced until it reaches the heart. It is used in the diagnosis of heart and lung diseases.

FORTRAN (Acronym for **for**mula **tran**slation) High-level COMPUTER LANGUAGE generally used for scientific applications. FORTRAN has been in general use for more than 40 years, but has developed greatly from its original form.

fossil Any direct evidence of the existence of an organism more than 10,000 years old. Fossils mostly consist of original structures, such as bones or shells or wood, often altered through MINERALIZATION or preserved as moulds and casts. Imprints such as tracks and footprints are also fossils. Leaves are often preserved as a carbonized film outlining their form. Occasionally organisms are totally preserved in frozen soil (such as mammoths), peat bogs, and asphalt lakes, or trapped in hardened resin (such as insects in AMBER). Fossil excrement, called COPROLITE, frequently contains undigested and recognizable hard parts. Very few animals and plants that die become fossilized. Since fossils reveal evolutionary changes through time, they are essential clues for geological dating.

FOSSIL

The type of fossil formed in a rock is dictated by the chemistry of both the organism and the rock itself (A). The harder parts of animals and plants contain a number of materials that do not decay, such as phosphates in bone and calcium carbonate in shells. During fossil formation, such minerals are often replaced by others that are better able to survive the rigours of the subterranean environment. For example, pyrite, haematite or quartz commonly replace minerals in a shell or a bone, molecule by molecule, so a calcium-based coral may be preserved as a hematite fossil. The organic part of the shelled creature is always the first to decay. This leaves a gap in the shell which is often filled with sediment (1). The shell is more soluble than the rock, and slowly dissolves (2), leaving a sedimentary-rock cast of the inside of the shell (3). Under the right circumstances the shell material is replaced by some other substance - often a type of silica - to give a cast of the shell's exterior (4). The "replacement" may even be the shell material, recrystallized. Sometimes, no sediment fills the empty shell (5). The solutions that permeate the strata (6) may totally dissolve a buried shell, leaving a "mould" behind in the rock (7). A replica of the original shell may later be cast, as another mineral or sediment is deposited in this mould. The tracks of living creatures can also be fossilized (B). Typically, a footprint (1) is left behind in soft mud, which partially hardens to form a cast. If the mud becomes flooded (2), sediment is laid over the mud especially quickly (3) helping to preserve the shape of the footprint. Over the course of time, the mud and sediment become compressed and turn to rock (4). The original mud-based rock forms a mould of the footprint (5) and the sediment-based rock forms a cast (6).

Oil, gas and coal are all formed from the decay of once-living organisms under heat and pressure. Over 80% of the oil and gas currently exploited formed in Mesozoic or Tertiary strata between 180 and 30 million years ago, from marine microorganisms deposited as sediment on the sea bed. The basic components of oil and gas are created when the organic remains are not completely oxidized, leaving a residual mass of carbohydrates, hydrocarbons and similar compounds. As layers of sediment bury this residue, temperatures and pressure increase and the liquid hydrocarbons are segregated into pore spaces in the rock. Coal deposits come from many epochs, but primarily the Carboniferous period (360–286 million years ago). The quantity of oil and gas formed depends on temperature and on the speed of subsidence of strata. Ideally up to 3km (2mi) of overlying strata are needed. Gas and oil migrate up – the gas and lighter oils first, with heavier bitumens staying closer to their source – until they are trapped by overlying impermeable rocks.

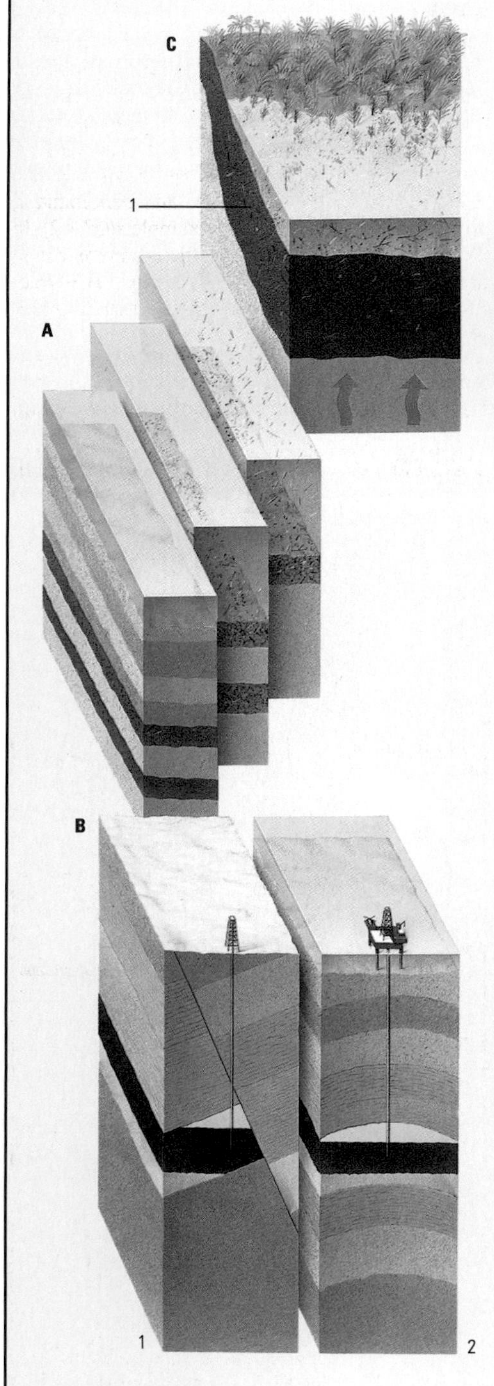

▼ **Oil and gas** are formed as layers of plankton are buried under thick piles of sediment (A). Increasing heat and pressure at depth first cause gast and oils from the bodies of the marine organisms to "link up" into a thick compound called kerogen. As temperature increases with burial, long chains made of hydrogen and carbon atoms break away from the kerogen, giving a viscous heavy oil. With even more heat, valuable light oils and natural gas are formed. The oils accumulate in "reservoir" rocks – permeable rocks such as sandstone which hold the oil like a sponge. To form an oilfield, the oil must be trapped between layers of an impermeable rock, such as shale (B). Faults, where such rock has sheared to form a seal (1), can trap oil and gas, as can convex domes (2). The making of most coal (C–G) was begun in the middle of the Carboniferous period. The Earth's equatorial regions were hot and wet, and lush tropical forests grew in extensive swamps (C). In these types of environment, thick layers of peat were laid down (1): typically they were sandwiched between layers of sediment, such as shale (2), deposited when the waters temporarily retreated. The seas receded during the Permian period and many of the tropical coastal plains turned to desert (D). Other sedimentary rocks, such as sandstones (3), were laid down over the shale and peat. With increasing temperature and pressure, the buried peat began its metamorphosis into coal. Around 150 million years ago (E) the deserts were covered over by shallow, tropical seas in which limestone (4) was deposited. Around 50 million years ago, plates of the Earth's crust collided forcing mountains up and further burying the underlying rocks (F). Metamorphism continued, converting coal into high-grade anthracite (5). Today anthracite is found in deep seams up to 30m (100ft) thick.

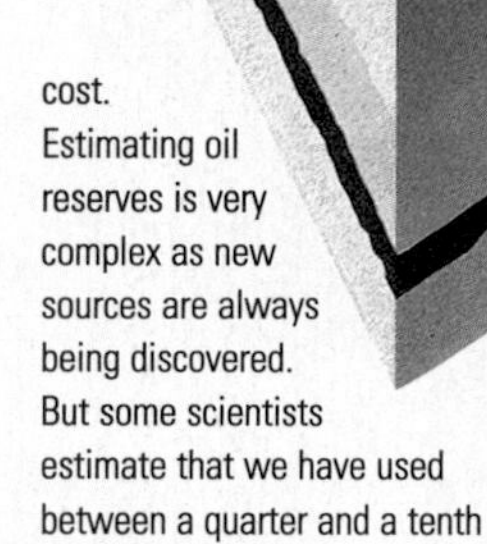

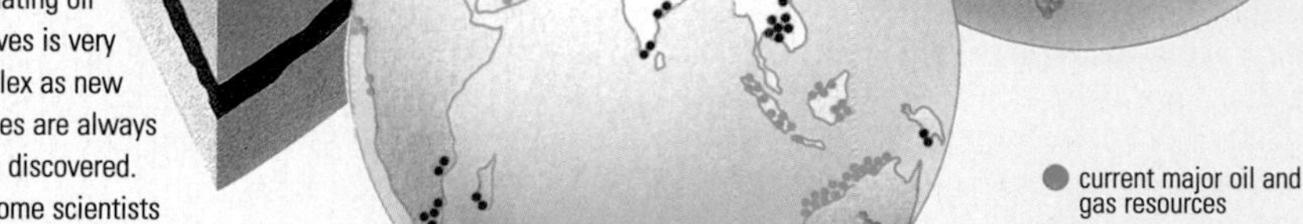

► **The predicted future life** of the world's coal reserves was once put at as low as a few decades, but the real figure is actually nearer 300 years. Some E European countries and the USA have the greatest resources. Western Europe, India, China, Brazil, South Africa and Australia also have large stocks of top-rate coal from the Carboniferous and Permian periods. Brown tertiary lignites can be found in Central Europe and the Ukraine. Antarctica also has large unused coal reserves, which could only be exploited at disastrous environmental cost.
Estimating oil reserves is very complex as new sources are always being discovered. But some scientists estimate that we have used between a quarter and a tenth of all oil reserves.

► **fractal** A computer graphics image entitled *Toesville*. The image was derived from the Julia set, a class of shapes plotted from complex number coordinates. Each point is assigned a colour depending upon its behaviour under a series of simple, but repeated, mathematical operations or mappings. The Julia set was invented and studied during World War 1 by the French mathematicians Gaston Julia and Pierre Fatou. The better-known Mandelbrot set arose as a result of attempts, made in the 1970s, to rationalise the massive variety of forms of the Julia set.

fossil fuels Term to describe COAL, OIL and NATURAL GAS – FUELS that were formed millions of years ago from the fossilized remains of plants or animals. By their very nature fossil fuels are a non-renewable energy source

Foucault, Jean Bernard Léon (1819–68) French physician and physicist who invented the GYROSCOPE. He used a pendulum (FOUCAULT'S PENDULUM) to prove that the Earth spins on its axis. He also devised a method to measure the absolute velocity of light, which he did, and in 1850 he showed it to be slower in water than in air. Foucault, with Armand FIZEAU, took the first clear photograph of the Sun. He also noted the occurrence of eddy currents, which became known as **Foucault currents**. They are electrical currents induced in a conducting medium when the conductor and a magnetic field passing through it are in relative movement.

Foucault's pendulum Heavy metal bob hung from a long, fine wire, first used by Jean FOUCAULT to demonstrate the Earth's rotation. As the Earth turns on its axis, the plane of swing of the pendulum, as observed from the Earth's surface, rotates because of the heavy bob's inertial resistance to change of absolute direction, until it has returned to its original orientation. Its period of rotation depends on its latitude; in London it rotates in 30 hours.

foundry Workshop in which metals are processed by melting and casting in moulds. Up to the 18th century, foundries were usually small, mainly casting gun barrels and bells. From 1760, however, they grew in size as the use of coke provided an efficient fuel to heat large furnaces. Foundries became increasingly sophisticated, especially where iron and steel were concerned, after the innovations of men such as Henry Cort (1740–1800) and Henry BESSEMER.

four-colour process Method of printing that reproduces COLOUR images by using only four colour printing plates. When printed one after the other the four colours recreate a complete range of colours. The colour image is "scanned" and separated into red, green and blue primary colours. Films are made representing each colour and a fourth is made representing the shades of grey and black in order to strengthen these tones in the final image. The film is then used to create the printing plates.

Fourier, Jean Baptiste Joseph (1768–1830) French mathematician and physicist, scientific adviser to Napoleon in Egypt from 1798 to 1801. His work on the mathematical analysis of heat flow led to his devising a technique, now known as **Fourier analysis**. In this technique, complex periodically varying functions are decomposed into sums of waves and a problem is analysed using these waves, which are easier to manipulate than the original function. *See also* FOURIER SERIES

Fourier series Series of sine and cosine functions used to represent other PERIODIC FUNCTIONS. Most bounded, periodic functions (on the domain $-\pi \leqslant \pi$) can be analyzed as a sum of simple harmonic components. Thus if $f(x)$ is such a function, with x taking values between $-\pi$ and π, so that

$f(x + 2\pi) = f(x)$

it may be expressed as

$f(x) = a_0/2 + (a_1\cos x + b_1\sin x) + (a_2\cos 2x + b_2\sin 2x) + ...$ In this series the nth coefficients a_n and b_n are given by $a_n = 1/\pi\int_{-\pi}^{+\pi} f(x)\cos(nx)dx$, and $b_n = 1/\pi\int_{-\pi}^{+\pi} f(x)\sin(nx)dx$.

four-stroke engine INTERNAL COMBUSTION engine in which the operation of each piston is in four stages. This kind of operation is called the **four-stroke cycle**, or the **Otto cycle**, after its inventor, Nickolaus OTTO. Most internal combustion engines of the sort found in cars, including DIESEL ENGINES, use the four-stroke principle. Each of the four stages corresponds to one movement of a reciprocating piston along a cylinder. First, the piston descends to admit the fuel-air mixture (intake stroke). Second, the piston rises, compressing the mixture (compression stroke). Third, the mixture is ignited by spark or compression, and expanding gases from the explosion force the piston downwards (ignition or power stroke). Fourth, the piston again rises to expel the spent gases from the cylinder (exhaust stroke). *See also* TWO-STROKE ENGINE

fourth dimension Time considered as an additional dimension, together with the three dimensions of space, in a full description of the motion of a particle. In RELATIVITY theory this four-dimensional framework is called SPACE-TIME.

fovea *See* BLIND SPOT; OPTIC NERVE

Fowler, William A. (Alfred) (1911–95) US physicist and astrophysicist. Fowler is best known for his contributions to the understanding of the way in which chemical elements are built up (from lighter to heavier) within stars as they evolve. He developed these ideas (published 1957) while working with three British astronomers: Fred HOYLE, and Geoffrey and Margaret Burbidge. In 1983 he shared the Nobel Prize for physics with Subrahmanyan CHANDRASEKHAR.

fps system Abbreviation of the "foot-pound-second" system of units. In the fps system, formerly used in much of the English-speaking world, the FOOT is the unit of length, the POUND the unit of mass and the SECOND the unit of time. It has been replaced by the METRIC SYSTEM or SI UNITS for all scientific and many commercial purposes.

fractal Geometrical figure in which an identical motif repeats itself on smaller and smaller scales. Such figures are said to be "self-similar". The term "fractal" was coined by Benoit MANDELBROT, and fractal geometry is closely associated with CHAOS THEORY. There are many examples of fractal-like objects in nature, from shells and cauliflowers to mountains and leaves, in addition to those produced mathematically in computer graphics.

fraction Quotient written in the form of one number divided by another, such as 3/4. In general, a fraction is a/b, where a is the numerator and b the denominator. If a and b are whole numbers, the quotient is a simple fraction. If a is smaller than b, it is a **proper** fraction; if b is smaller than a, it is an **improper** fraction. In an **algebraic** fraction, the denominator, or the numerator and denominator, are algebraic expressions, for example $x/(x^2 + 2)$. In a **composite** fraction, both the numerator and denominator are themselves fractions. Two fractions a/b and c/d can be added, subtracted, multiplied and divided according to the rules:
$a/b + c/d = (ad + bc)/bd$; $a/b - c/d = (ad - bc)/bd$; $a/b \times c/d = ac/bd$; and $a/b \div c/d = ad/bc$.

fractionation Separation of a mixture using

FRACTIONATION

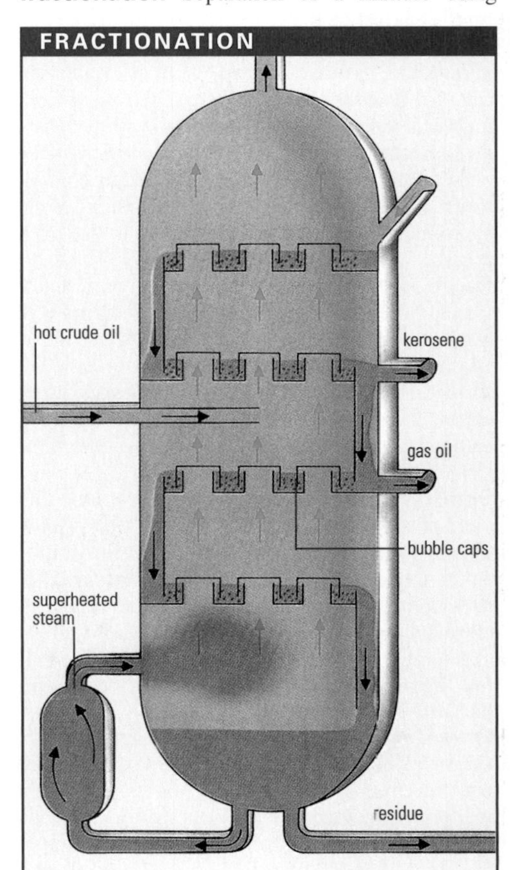

Hot crude oil is fed into the side of a tall column (up to 70m high) divided at 1 metre intervals by a series of trays. As the oil fills a tray, it overflows to the tray below, cascading slowly down the column. Hot steam entering at the base of the column rises up through the bubble caps in each tray, vapourising part of the oil. On each tray the hot vapour from the tray below vapourises some of the lower boiling material which rushes up the column. As it rises it cools, allowing part to condense back to liquid. In this way each tray separates a mixture with a slightly lower boiling point than the tray below. The mixtures required are run off from trays at different levels and passed to smaller columns (strippers) for further refining.

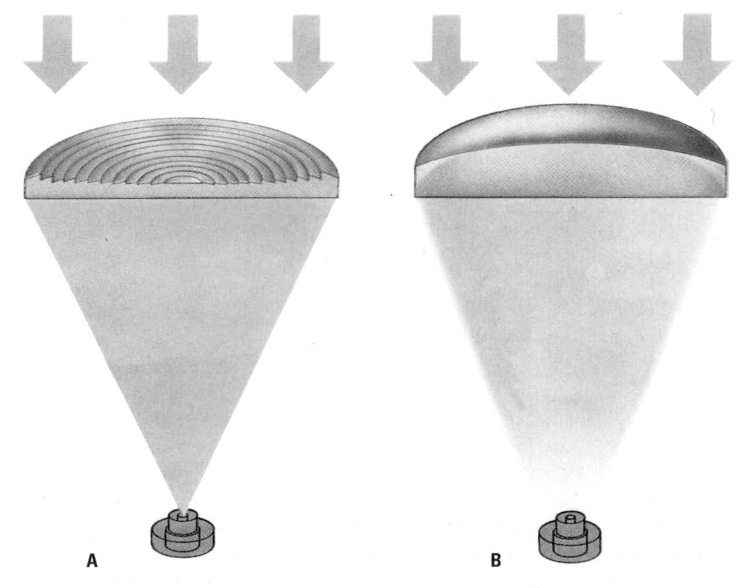

► **Fresnel lens** Once commonly used in lighthouses and spot lights, the concentric circles of the Fresnel lens (A) help to focus light more effeciently than a standard convex lens (B).

CRYSTALLIZATION (**fractional crystallization**) or DISTILLATION (**fractional distillation**). In fractional crystallization, the mixture is dissolved in a hot solvent. Then, as the solution gradually cools, the components of the mixture crystallize out one at a time – the least soluble one first. In fractional distillation, a liquid mixture is boiled and the vapour allowed to pass up a long vertical condenser, called a **fractionating column**. The lowest-boiling (most volatile) components in the mixture pass up to the top of the column, while the least volatile ones remain near the bottom. The various components – called **fractions** – can be drawn off at the appropriate levels. The technique is the key process in OIL refining.

frame of reference Mathematical coordinate system for describing events in space and time with respect to a given observer. In the theory of RELATIVITY, this frame of reference is four-dimensional, and the description of events in other frames of reference depends on the relative velocities and accelerations of those frames with respect to the frame of reference of the observer.

francium (symbol Fr) Radioactive, metallic element of group I of the periodic table; first discovered in 1939. The heaviest member of the ALKALI METALS, it occurs naturally in uranium ores and is a decay product of ACTINIUM. It is quite rare and its chemical properties are unknown, but it is thought to resemble CAESIUM. There are 21 isotopes identified. Properties: at.no. 87; most stable isotope ^{223}Fr (half-life 22 minutes).

Franck, James (1882–1964) US physicist, b. Germany. With Gustav HERTZ, he experimented with electron bombardment of gases, providing support for the theory of atomic structure proposed by Niels BOHR and information for the quantum theory of Max PLANCK. Franck and Hertz shared the 1925 Nobel Prize for physics for these studies on the changes of energy occurring when atoms collide with electrons. In 1935 he went to the USA and became a US citizen. He worked on the development of the atom bomb with the MANHATTAN PROJECT but in 1945 drew up the "Franck petition" saying the bomb should not be used against Japanese civilians.

Frankland, Sir Edward (1825–99) British chemist who formulated the theory of VALENCE. This states that each element may be assigned a whole number (1, 2, 3 or more) or valence which represents the number of hydrogen atoms with which an atom of the element will combine. He also discovered, with Sir Joseph LOCKYER, the gas helium in the Sun.

Frasch process Method of mining SULPHUR by pumping superheated water and air into the sulphur deposits, melting the mineral, and forcing it to the surface. Named after the German-born chemist Herman Frasch, the process was first put to practical use in Louisiana and made the USA independent of imported sulphur.

Fraunhofer, Joseph von (1787–1826) German physicist, the founder of astronomical SPECTROSCOPY. He studied the DIFFRACTION of light through narrow slits and developed the earliest form of DIFFRACTION GRATING. In 1814 he observed and began to map the dark lines in the Sun's spectrum, now called **Fraunhofer lines**, and was the first to appreciate their significance. Fraunhofer solved many of the scientific and technical problems of astronomical telescope-making.

free fall State of motion of an unsupported body in a gravitational field. *See* GRAVITY

free radical Short-lived (less than 1ms) molecule that has an unpaired ELECTRON and, therefore, rapidly binds with other molecules. Occurring as by-products of normal CELL chemistry, free radicals can cause extensive damage in the body, even though cells have some protective enzymes. They are thought to play a role in ageing and in a number of disease processes, including cancer and cardiovascular problems. Free radicals are used commercially to initiate POLYMERIZATION.

freeze-drying *See* FOOD PRESERVATION

freezing *See* FOOD PRESERVATION

freezing point Temperature at which a substance changes PHASE (state) from liquid to solid. The freezing point for most substances increases as pressure increases. The reverse process, from solid to liquid, is melting; melting point is the same as freezing point.

Freon Trade name for a type of CHLOROFLUOROCARBON (CFC). Freons are used as refrigerants, AEROSOL propellants, cleaning fluids and solvents. They are all clear, stable and inert liquids. The most used is freon-12 (dichlorodifluoromethane, CCl_2F_2). Implicated in the destruction of OZONE in the IONOSPHERE, the use of Freons is being phased out.

frequency Rate of occurrence. In statistics, the number of times a numerical value, event, or special property occurs. In physics, the frequency is the number of oscillations (or waves) passing a given point per second (measured in HERTZ), including those of sound, light and radio waves, a swinging PENDULUM and vibrating springs. The product of the frequency and the period of a wave (or WAVELENGTH) is constant and equals the wave velocity.

frequency, wave *See* WAVE FREQUENCY

frequency modulation (FM) Form of RADIO transmission. It is the variation of the frequency of a transmitted radio carrier wave by the signal being broadcast. The technique makes radio reception fairly free from static interference and, although restricted in range to receivers in line of sight of the transmitter, has become the most favoured transmission method as a result. *See also* AMPLITUDE MODULATION (AM)

Fresnel, Augustin Jean (1788–1827) French physicist and engineer. His pioneer work in optics was instrumental in establishing the wave theory of light. He researched into the conditions governing INTERFERENCE phenomena in POLARIZED LIGHT, studied double refraction and devised a way of producing circularly polarized light.

Fresnel lens Type of lens, used in spotlights and lighthouses, that consists of a piece of heat-resistant glass cast with concentric portions of lenses of different diameters and approximately the same focal length. The Fresnel lens is two or three times more efficient than a plano-convex lens.

friction Resistance encountered when surfaces in contact slide or roll against each other, or when a fluid (liquid or gas) flows along a surface. Friction is directly proportional to the force pressing the surfaces together and the surface roughness. Before the movement begins, it is opposed by static friction up to a maximum "limiting friction" and then slipping occurs. The streamlined design of aircraft and cars reduces air (fluid) friction. *See also* AERODYNAMICS

friction, coefficient of Number characterizing the force necessary to slide or roll one material along the surface of another. If an object has a weight N and the coefficient of FRICTION is μ, then the force (F) necessary to move it without acceleration along a level surface is $F = \mu N$. The coefficient of static friction determines the force necessary to initiate movement; the coefficient of kinetic friction determines the (lesser) force necessary to maintain movement.

Frisch, Karl von (1886–1982) Austrian zoologist. He shared with Konrad LORENZ and Nikolaas TINBERGEN the 1973 Nobel Prize for physiology or medicine for his pioneering work in

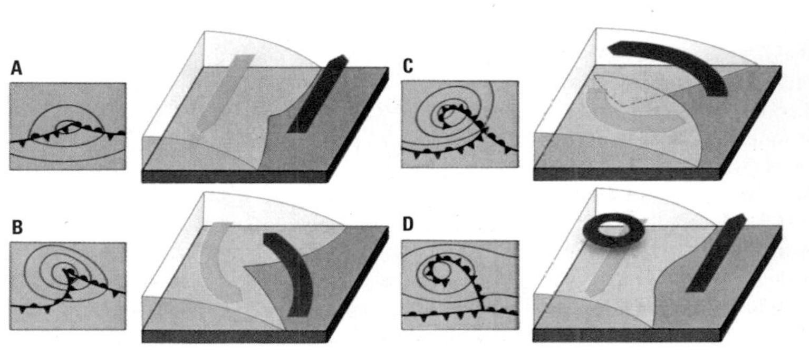

◄ **front** A front forms in temperate latitudes where a cold air mass meets a warm air mass (A). The air masses spiral round a bulge causing cold and warm fronts to develop (B). The warm air rises above the cold front and the cold air slides underneath the warm (C). Eventually, the cold air areas merge, and the warm air is lifted up or occluded (D).

► **fruit** Fruits are the ripened ovary of a flowering plant, which is enclosed by the fruit wall (the pericarp). Most fruits, such as blackcurrants (1), cherries (2), strawberries (3), oranges (4) and pepper (5) contain more than one seed. The avocado (6), however, contains only one seed, about the size of a golf ball. The largest fruit, the double coconut (7) is also single-seeded.

ETHOLOGY (the study of animal behaviour patterns). He deciphered the "language of bees" by studying their dance patterns (WAGGLE DANCE) in which one bee tells others in the hive the direction and distance of a food source. In his earlier work he showed that fish and bees see colours, fish can hear and bees can distinguish various flower scents.

front In meteorology, the boundary between two air masses of different temperatures or of different densities. **Cold** fronts occur as a relatively cold and dense air mass undercuts warmer air. The upwards movement of the warm air usually results in condensation, cloud formation and rainfall. With a **warm** front, the warmer air gradually rises over colder air. As the air rises it cools and forms cloud. Eventually it may give rise to some warm rain. An **occluded** front is a combination of a warm and a cold front in a DEPRESSION. It occurs when the cold front catches up with the warm front and undercuts it. The effects of the warm front and cold front together are to cause heavier and more prolonged rain than either front would give alone.

frost In meteorology, atmospheric temperatures at Earth's surface below 0°C (32°F). The visible result of a frost is usually a deposit of minute ice crystals formed on exposed surfaces from DEW and water vapour. In freezing weather, the "degree of frost" indicates the number of degrees below freezing point. When **white hoar-frost** is formed, water vapour changes from its gaseous state to a solid, without becoming a liquid. A frost is referred to as white or black, depending on whether or not white hoar-frost is present.

frostbite Freezing of living body tissue in subzero temperatures. Frostbite is an effect of the body's defensive response to intense cold, which is to shut down blood vessels at the extremities in order to preserve warmth at the core of the body. Consequently, it mostly occurs in the face, ears, hands and feet. In superficial frostbite the affected part turns white and cold; it can be treated by gentle thawing. If freezing continues, ice crystals form in the tissues, the flesh hardens, and there is no sensation. Deep frostbite, which causes tissue death, requires urgent medical treatment. No attempt should be made at rewarming if there is a risk of refreezing as this increases the likelihood of GANGRENE.

fructose (fruit sugar) Simple white, crystalline sugar ($C_6H_{12}O_6$) found in honey, sweet fruits and the nectar of flowers. It is a MONOSACCHARIDE that is sweeter than the DISACCARIDE sucrose (cane sugar), of which it is a component along with glucose. It is made commercially by the HYDROLYSIS of beet or cane sugar and is used in foods as a sweetener. Its derivatives play a crucial role in providing energy to living organisms.

fruit Seed-containing, mature OVARY of an ANGIOSPERM (flowering plant), in which the seeds are surrounded by the PERICARP (fruit wall). Fruits serve to disperse plants and are an important food source (they provide vitamins, acids, salts, calcium, iron and phosphates). They can be classified as simple, aggregate or multiple. **Simple** fruits, dry or fleshy, are produced by one ripened ovary of a single PISTIL (unit comprising a STIGMA, STYLE and OVARY) and include legumes (peas and beans) and nuts. **Aggregate** fruits develop from several simple pistils; examples are raspberry and blackberry. **Multiple** fruits develop from a flower cluster; each flower produces a fruit which merges into a single mass at maturity; examples are pineapples and figs. Although considered fruits in culinary terms, apples and pears are regarded botanically as "false" fruits (PSEUDOCARPS), as the edible parts are created by the RECEPTACLE and not the carpel walls. Some fruits spontaneously disperse their seeds (*see* DEHISCENT), others keep their seeds and the fruit is dispersed as a whole (*see* INDEHISCENT).

FSH Abbreviation of FOLLICLE-STIMULATING HORMONE

Fuchs, Klaus (1912–88) German physicist and Soviet spy. Interned in Britain at the start of World War 2, he was released and naturalized. He began to pass secrets to the Soviet Union in 1941. In 1943 he went to the USA to work on the atom bomb, returning to the UK in 1946 to head the theoretical physics division of the atomic research centre at Harwell. He was imprisoned in 1950 for his espionage activities, and his British citizenship was revoked. Released in 1959, he went to East Germany to work at a nuclear research centre.

fuel Substance that is burned or otherwise modified to produce energy, usually in the form of heat. Apart from FOSSIL FUELS (COAL, OIL and GAS) and firewood and charcoal, the term also applies to radioactive materials used in NUCLEAR REACTORS.

fuel cell ELECTROCHEMICAL CELL for direct conversion of the energy of oxidation of a fuel to electrical energy. Suitably designed electrodes are immersed in an ELECTROLYTE, and the fuel (such as hydrogen) is supplied to one and the oxidizer (such as oxygen) to the other. Electrode reactions occur leading to oxidation of the fuel, with production of electric current. Fuel cells are used in space vehicles.

fuel injection Method of introducing fuel into the cylinders of an INTERNAL COMBUSTION ENGINE, utilizing a pump rather than piston-created suction. It distributes fuel more evenly for greater power and less likelihood of engine knock or vapour lock.

fuel rod (fuel element) Cylindrical case containing a fissionable (able to undergo FISSION) material such as uranium-235 placed in the core of a NUCLEAR REACTOR. Fuel rods have to be able to withstand the heat generated by the nuclear reac-

FUEL CELL

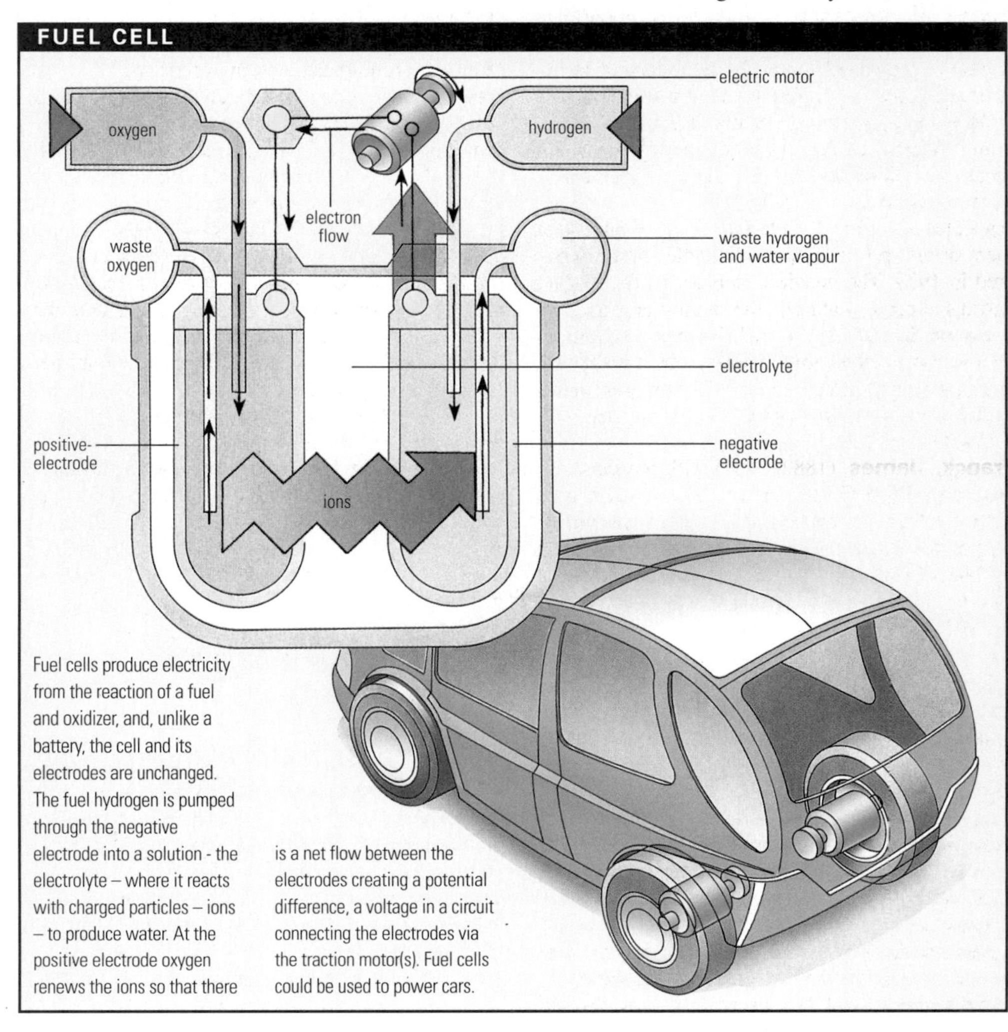

Fuel cells produce electricity from the reaction of a fuel and oxidizer, and, unlike a battery, the cell and its electrodes are unchanged. The fuel hydrogen is pumped through the negative electrode into a solution - the electrolyte – where it reacts with charged particles – ions – to produce water. At the positive electrode oxygen renews the ions so that there is a net flow between the electrodes creating a potential difference, a voltage in a circuit connecting the electrodes via the traction motor(s). Fuel cells could be used to power cars.

tion and be sufficiently strong so that there is no danger of them splitting and releasing radioactive material. Spent fuel rods are generally made safe in a reprocessing plant.

fulcrum Point about which a LEVER pivots. Its use in the PRINCIPLE OF MOMENTS was recognized by ARCHIMEDES, who is reputed to have said: "Give me a firm place on which to stand and I will. move the Earth".

fuller's earth Clay-like substance that contains more than 50% SILICA. Originally used to remove oil and grease from wool (the process of **fulling**), it is now used to bleach petroleum and in refining vegetable oils.

fulminate Explosive, a salt of fulminic acid (HONC), especially fulminate of mercury, ($Hg(ONC)_2$), which is formed by the action of nitric acid on mercury metal in the presence of alcohol. It is a highly unstable compound used in explosive DETONATORS and blasting caps.

fumarole Vent in the ground that emits gases and vapours, usually found in volcanic areas. The term also refers to a hot spring or GEYSER that emits steam. A fumarole is sometimes defined in terms of the composition of its gases, such as a chlorine fumarole.

function In mathematics, a rule that assigns to each element of a given set a uniquely specified value. The given set is the **domain** of the function, and the set of values is the **range**. Two or more elements of the domain may be assigned the same value, but a function must assign only one value to each element of the domain. We also say that a function f maps each element x of the domain to a corresponding element (or value) y in the range. Here x and y are variables, with y dependent on x through the functional relationship f. The dependent variable y is said to be a function of the independent variable x. Examples of functions are the natural logarithm, whose domain is the set of positive real numbers and whose range is the set of real numbers, and the square-root function, with domain and range the non-negative real numbers. *See also* NUMBER, REAL; TRIGONOMETRIC FUNCTION

fundamental forces Four basic forces that exist in physics. The most familiar, and the weakest, is GRAVITATION. The gravitational force between the Earth and an object accounts for the object's WEIGHT. Much stronger is the ELECTROMAGNETIC FORCE, which operates between electrically charged particles. It holds atoms together and binds them to each other chemically. The two other forces recognized both operate only on the subatomic level: the WEAK NUCLEAR FORCE, associated with the decay of particles, is intermediate in strength between the gravitational and electromagnetic force; the STRONG NUCLEAR FORCE associated with the "glue" that holds nuclei together is the strongest force known in nature.

fundamental unit One unit out of the three that are usually required to make up a system of units. For example, the CGS SYSTEM (now adapted to provide the basis of the SI UNITS system) comprises the centimetre, gram and second fundamental units.

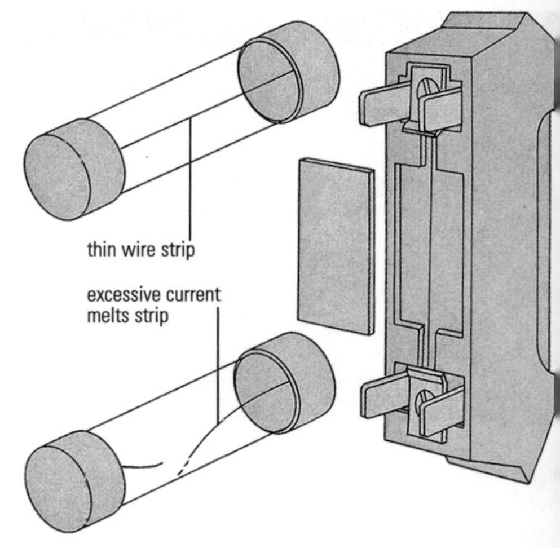

▲ **fuse** The heating effect of an electric current is used in a fuse, which consists of a thin wire that melts when excessive current passes through it, thereby cutting off the supply of electricity, and so avoiding further damage.

fungicide Chemical that kills fungi. For example, creosote is used on wood to prevent dry rot.

fungus Any of *c*.100,000 species of the kingdom Fungi, which are unable to photosynthesize and which reproduce by means of spores and never produce cells with FLAGELLA. They include mushrooms, moulds and yeasts. Fungi have relatively simple structures, with no roots, stems or leaves. Their cell walls contain the POLYSACCHARIDE CHITIN. The main body of a typical multicellular fungus consists of an inconspicuous network (MYCELIUM) of fine filaments (HYPHAE), which contain many nuclei, and which may or may not be divided into segments by cross-walls. The hypha nuclei are HAPLOID. The mycelia may develop spore-producing, often conspicuous, fruiting bodies, mushrooms and toadstools. Fungal PARASITES depend on living animals or plants: SAPROPHYTES utilize the materials of dead plants and animals, and symbionts obtain food in a mutually beneficial relationship with plants (SYMBIOSIS). Fungi feed by secreting digestive ENZYMES onto their food, then absorbing the soluble products of digestion. Many fungi cause diseases in crops, livestock and humans (athlete's foot). Moulds and yeasts are used in the production of beer and cheese; some fungi, such as *Penicillium*, are sources of ANTIBIOTICS.

Funk, Casimir (1884–1967) US biochemist who discovered VITAMINS. In 1912 he found vitamins B_1, B_2, C and D, coined the term "vitamine" (the AMINES of life) and presented a paper formulating the idea of vitamin deficiency disorders. Funk also contributed to the knowledge of cancer and of the hormones of the sex glands.

fur Soft, dense hair covering the skin of certain mammals. Such mammals include mink, fox, ermine, musquash, wolf, bear, squirrel and rabbit. Many are hunted and killed for their pelts, which, when manufactured into clothing, may command high prices. Some fur-bearing animals are now protected by law because overhunting has made extinction likely.

furlong Imperial unit of measurement, equal to 220yds (201m). There are 8 furlongs to 1 mile. The word derives from a "furrowlong"; it represented the standard length of a furrow on a square field of 10 acres in medieval times.

furnace Enclosed space that is raised to high temperatures usually by the combustion of fuels or by electric heating. Most furnaces are used in

FUNGUS

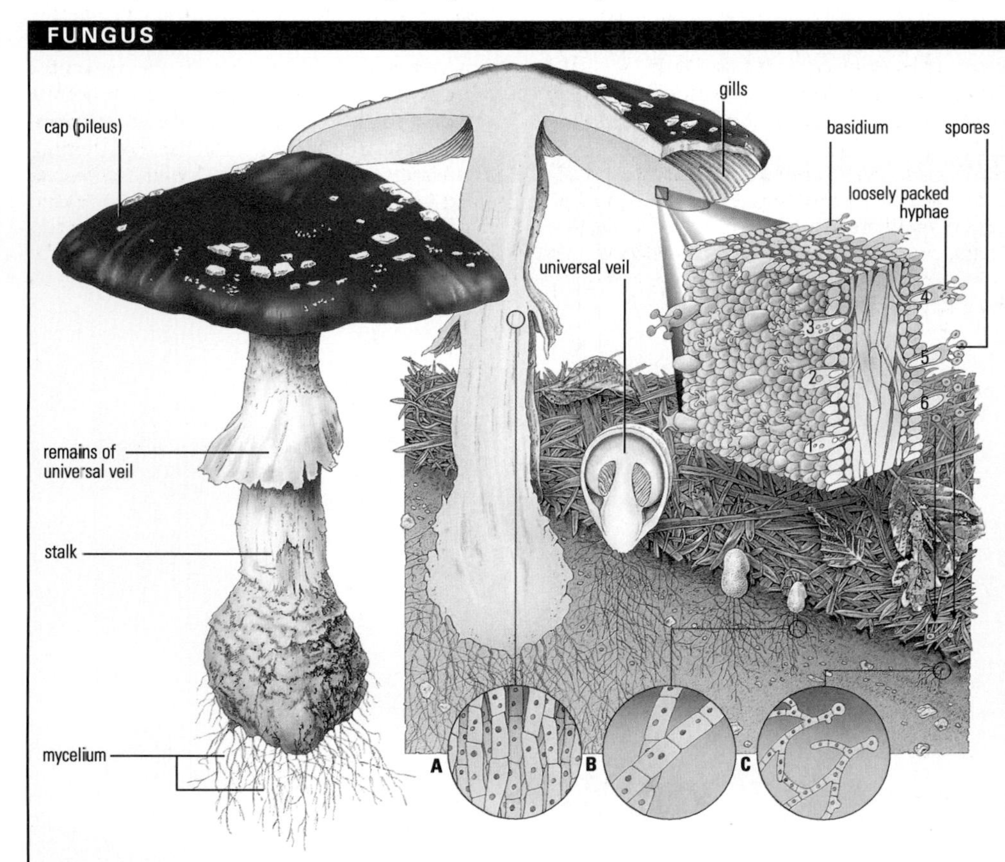

Fly Agaric's fruiting body consists of a stalk (stipe), made of closely packed hyphae (A), attached to a buried mycelium and crowned with a broad cap (pileus), which protects the delicated spore-bearing layer (hymenium) on the gills. The life cycle of this toadstool begins with the germination of spores to form a mycelium, the compartments of which contain one nucleus each (C). Hyphae of different "sex" fuse, and a secondary mycelium develops, in which the compartments contain both parental nuclei (B). The fruiting body grows from the secondary mycelium, starting life as a button enclosed in a protective universal veil, which eventually ruptures. In the hymenium, the two nuclei in each compartment (1) fuse and their chromosomes are reshuffled in meiosis (2) and the four resulting nuclei (3) migrate to the ends of club-shaped cells (basidia) (4,5), from which they bud off, and are shed as spores (6).

FUSION, NUCLEAR

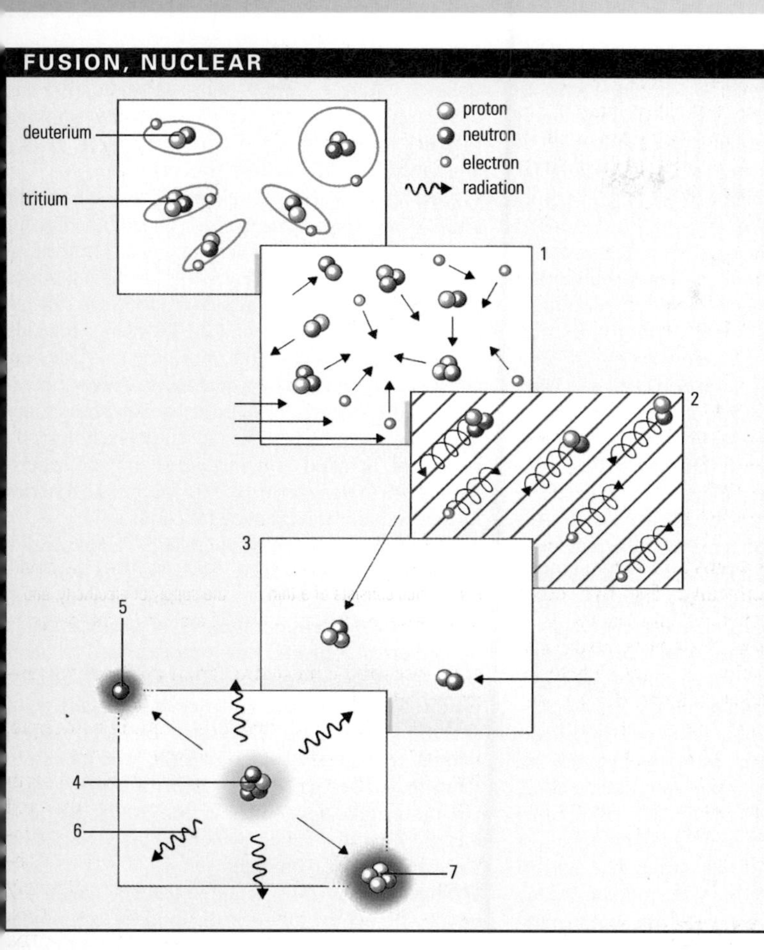

In experiments to generate power by nuclear fusion, the aim is usually to produce energy by fusing tritium and deuterium. These are isotopes of hydrogen and the process can only occur at a temperature above 100 million°C and under enormous pressure. Deuterium has one proton, one neutron and one electron, while tritium has one proton, two neutrons and an electron. In a fusion reactor, the mixture of the two isotopes is heated by intense radio emissions, ion bombardment and electrical pulses (1). The plasma which results is suspended in a magnetic field (2). The tritium and deuterium nuclei fuse (3) creating a helium nucleus (4), a loose neutron (5), radiation (6), and energy when the products hit the edge of the plasma (7).

the extraction of metals or the making of alloys. Some furnaces heat ovens, boilers and kilns. A domestic boiler, especially if it burns solid fuel, is sometimes called a furnace. There are three main types of electric furnace. An arc furnace relies on the heat generated by an electric arc (spark), often between two large carbon electrodes, which are slowly consumed. A resistance furnace is heated by passing an electric current through a heating element or directly through metallic material. An induction furnace uses ELECTROMAGNETIC INDUCTION to cause a current to flow in a metallic charge. The resulting heat is sufficient to melt the metal.

fuse In electrical engineering, a safety device to protect against overloading. Fuses are commonly strips of easily melting metal placed in series in an electric circuit such that, when overloaded, the fuse melts, breaking the circuit and preventing damage to the rest of the system.

fuselage Body of an AIRCRAFT, to which the wings, tail assembly and landing gear are attached; it houses the crew, cargo, passengers and controls. Its structural strength lies in a stress-bearing skin, internal bulkheads, lengthwise stringers and formers.

fusel oil Poisonous, clear, colourless liquid with a disagreeable smell. It consists of a mixture of amyl alcohols, obtained as a by-product of the fermentation of plant materials containing sugar and starch. Fusel oil also occurs as a dangerous impurity in badly made spirit drinks. It is used as a solvent for waxes, resins, fats and oils, and in the manufacture of explosives and pure amyl alcohols.

fusion, nuclear Type of nuclear reaction in which nuclei of light atoms (such as hydrogen) combine to form one or more heavier nuclei with the release of large amounts of energy. The process takes place in the Sun and other stars, and has been reproduced on Earth in the HYDROGEN BOMB. In a self-sustaining fusion reaction, the combining nuclei are in the form of a PLASMA. It is the maintenance of this state of matter that has proved difficult in harnessing fusion reaction as a controlled source of NUCLEAR ENERGY, though attempts have been made in TOKAMAK reactors. *See also* FISSION, NUCLEAR

fuzzy logic System of logic able to represent statements that are true or false depending on context. For example, the statement "this is warm" applied to the inside of a freezer that is not working, is true only in the context that the freezer is not literally "freezing". Fuzzy logic is based on a notion of degrees of truth rather than precise truth, which is the concern of other systems of logic. Computer-controlled devices programmed on the principles of fuzzy logic are able to put into context information they receive from external sources, thereby responding flexibly to the environment. Such ability is more associated with human-reasoning than classic machine-reasoning. For example, dishwashers have been fitted with microprocessors that program the release of powder according to how dirty the water is following an initial washing cycle.

G

g Symbol for the ACCELERATION OF FREE FALL
G Symbol for the GRAVITATIONAL CONSTANT
gabbro Coarse-grained basic IGNEOUS ROCK. It can be regarded as the plutonic equivalent of BASALT, being much coarser because of its slow crystallization. The constituents are sodium and calcium FELDSPAR and the dark minerals OLIVINE and PYROXENE.
Gabor, Dennis (1900–79) British physicist, b. Hungary. He was awarded the 1971 Nobel Prize for physics for his invention of HOLOGRAPHY. He developed the basic technique in 1947 but it was not until the invention of the laser in 1960 by Charles TOWNES that holography became commercially feasible. Gabor also researched into optics, inventing a type of colour television tube, and into communication theory. He was an environmentalist, often lecturing on the Earth's limited resources.
gadolinium (symbol Gd) Silvery white, metallic element of the LANTHANIDE SERIES, first isolated as the oxide in 1880. Chief ores are gadolinite, monazite and bastnaesite. A malleable and ductile metal, its specialized uses include neutron absorption (important in many NUCLEAR REACTORS) and the manufacture of certain alloys. Properties: at.no. 64; r.a.m. 157.25; r.d. 7.898; m.p. 1,311°C (2,392°F); b.p. 3,233°C (5,851°F); most common isotope ^{158}Gd (24.87%).
Gagarin, Yuri Alekseyevich (1934–68) Russian cosmonaut, the first man to orbit the Earth (12 April 1961). Gagarin made the single orbit in 1 hour 29 minutes. He died in a plane crash.
Gaia hypothesis Scientific theory that relates the Earth's many and varied processes – chemical, physical and biological. Popular in the 1970s, when it was proposed by UK scientist James Lovelock (1919–), the Gaia hypotheses showed the Earth as one single living organism, with its components made up of all living things. It was popular with ecological groups because it explained the delicate interrelationship between people and the environment.
Gajdusek, Daniel Carleton (1923–) US physiologist. He shared the 1976 Nobel Prize for physiology or medicine with Baruch Blumberg (1925–) for work on the neurological disease kuru. Working among the Fore tribe in Papua New Guinea, which was being decimated by this disease, he postulated its transmission by the ritual eating of human brains. Further work showing that kuru could be transmitted to animals led him to conclude that it was caused by a slow-acting agent which could remain dormant for years.
galactic cluster *See* OPEN CLUSTER
galactic coordinate system Astronomical COORDINATE SYSTEM that refers to the plane of the galaxy. The coordinates are galactic latitude and longitude. Galactic latitude is the angular distance of a celestial object above or below the galactic plane. Galactic longitude is the angular distance around the galactic equator from a defined zero point. This point is at right ascension 17hr 42.4min, declination −28°55'.
galaxy Huge assembly of stars, dust and gas, an example of which is our own Galaxy. There are three main types, as originally classified by Edwin HUBBLE in 1925. **Elliptical** galaxies (E) are round or elliptical systems, showing a gradual decrease in brightness from the centre outwards. Graded E0 to E7 according to increasing ellipticity, elliptical galaxies consist of old stars free of gas and dust. **Spiral** galaxies (S) are flattened, disc-shaped systems in which young stars, dust and gas are concentrated in spiral arms coiling out from a central bulge, the **nucleus**. They are subdivided Sa to Sc to designate tightly wound spirals (a) to loose spirals (c). **Barred spiral** galaxies (SB) are distinguished by a bright central bar from which the spiral arms emerge. They too are subdivided SBa to SBc in the same way as spirals. In addition there are **lenticular** galaxies, systems intermediate between ellipticals and spirals, having a disc and nucleus, similar to spiral galaxies, but with no apparent spiral arms. IRREGULAR GALAXIES (Ir) are systems with no symmetry. Current theories suggest that all galaxies were formed from immense clouds of gas at roughly the same time, soon after the BIG BANG. Galaxies can exist singly or in clusters (GALAXY CLUSTER), which contain anywhere from just a few to thousands of members. Our Galaxy is a star system containing the SOLAR SYSTEM. It is spiral in shape and about 100,000 light-years in diameter. Our Sun and Solar System are located at the edge of one of the spiral arms, about 30,000 light-years from the centre. The spiral arms form a disc-shaped system, with a bulging core (or nucleus) in the direction of Sagittarius. The stars of the spiral arms form the MILKY WAY, running around the sky. However, it is uncertain how many spiral arms the Galaxy has, or whether it is a normal or a barred spiral. The whole Galaxy is rotating, but the rotational rate varies with the distance from the centre. Our Sun circles the centre at about 250km (150mi) s^{-1}, taking 220 million years per orbit. The Galaxy's age is thought to be about 10,000 million years, during which time the Sun would have completed *c.*50 revolutions. The Galaxy is a member of a cluster known as the Local Group, which includes the ANDROMEDA GALAXY and the MAGELLANIC CLOUDS. *See also* DOUBLE GALAXY; RADIO GALAXY
galaxy, double *See* DOUBLE GALAXY
galaxy, radio *See* RADIO GALAXY
galaxy cluster Group of associated GALAXIES, consisting of several separate systems moving together through space. A concentration of galaxy clusters is termed a **galaxy supercluster**.
Galen (*c.*129–*c.*199) Greek physician whose work and writings provided much of the foundation for the development of medical practice. He tried to systematize all that was known of medical practice and to develop a theoretical framework for an explanation of the body and its disorders. He made numerous anatomical and physiological discoveries, including ones concerning HEART valves, secretions of the KIDNEY, RESPIRATION and nervous function. He was among the first to study physiology by means of detailed and ingenious animal experimentation. Galen's often wildly inaccurate theories influenced medical practice for centuries.
galena Grey, brittle, metallic mineral, lead sulphide (PbS). It is the major ore of lead. It occurs as granular masses, and commonly as cubic or octahedral crystals with a lead-grey streak. It is widely found in hydrothermal veins and as a replacement in limestone and dolomite rocks. Hardness 2.5–2.7; r.d. 7.5.
Galilean satellites Four chief SATELLITES of JUPITER: GANYMEDE, CALLISTO, IO and EUROPA, named collectively after GALILEO, who observed them in 1610. Apart from the Earth's Moon, they are the brightest satellites in the SOLAR SYSTEM, and all would be naked-eye objects were it not for the glare from Jupiter itself.
Galileo (1564–1642) (Galileo Galilei) Italian scientist who helped to found the modern experimental method. He was a lecturer at Pisa, later moving to Florence and then to Padua. According to legend, he observed that a hanging lamp in Pisa Cathedral took the same time to complete one oscillation however long the swing, and he suggested the pendulum could be used for timekeeping. Galileo later studied falling bodies and disproved ARISTOTLE'S view that they fall at different rates according to weight. He also discovered the parabolic flight path of projectiles. In 1609 he used one of the first astronomical TELESCOPES to discover SUNSPOTS, lunar craters, Jupiter's major satellites and the phases of

GALAXY

Hubble's classification of galaxies according to their forms has three main groups. **Elliptical**, which vary from nearly spherical (E0) to flattened ellipsoid (E7) these contain little gas and dust but many red stars. **Spiral** varying from galaxies with large, bright nuclei and tightly coiled spiral arms (Sa) to those with smaller nuclei and more prominent arms (Sc). **Barbed spiral** having a conspicuous bar intersecting the nucleus. The sequence SBa to SBc shows large bright nuclei giving way to less noticeable ones with the arms progressively less tightly coiled. (SO) type galaxies resemble the spirals in most features but lack the arms.

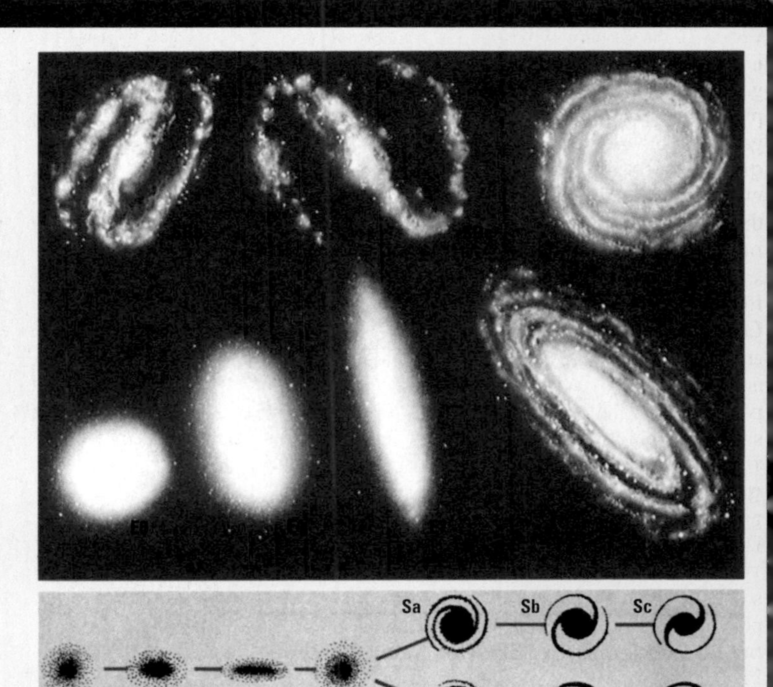

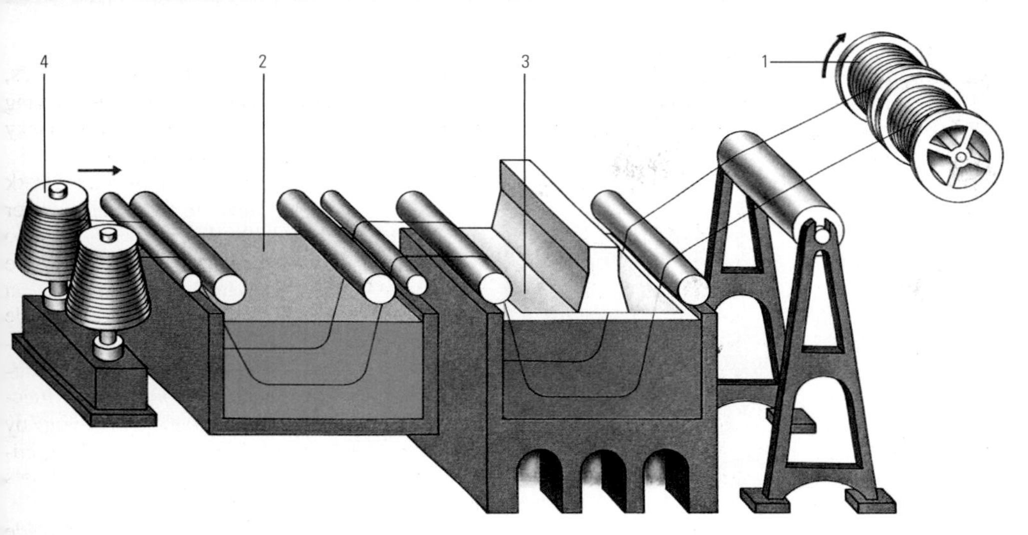

▲ **galvanizing** As a method of preventing corrosion in iron and steel, (galvanizing) dipping into a molten zinc bath was first practised in the late 18th century. The first practical process on an industrial scale was initiated in 1836. The iron was pickled in acid to clean it, then, to aid the adhesion of the zinc, it was fluxed in a bath of ammonium chloride, finally it was dipped in a tank of molten zinc. Galvanized sheets, rolls and coils soon made their appearance, but the most extensive use was in steel wiring, which could now be used in fencing, cables for suspension bridges and in similarly exposed situations. The continuous wire galvanizing plant shown dates from 1880. The mechanically driven spoolers (1) draw the wire through the acid bath (2) and zinc bath (3) from the spools (4).

Venus. In *Sidereus Nuncius* (1610) he announced his support for the Copernican view of the Universe, with the Earth moving around the Sun. He was forbidden by the Roman Catholic church to teach that this system represented physical reality, as opposed to a calculational device, but in *Dialogue on the Two Great World Systems* (1632), Galileo defied the pope by making his views even more explicit. As a result, he was brought before the Inquisition and forced to recant.

Galileo SPACE PROBE to Jupiter, launched in October 1989. Galileo flew past Venus once and the Earth twice. The probe passed the asteroids Gaspra and Ida in October 1991 and August 1993 respectively, photographing both. Galileo went into orbit around Jupiter in 1995, after dropping off a subprobe to enter the planet's atmosphere. *See also* SPACE EXPLORATION

gall Abnormal swelling or protuberance of plant tissue stimulated by an invasion of any of a wide variety of parasitic or symbiotic organisms, including bacteria, fungi, insects and nematodes. Most gall organisms stunt but do not kill the affected plants. The gall wasp larva makes a spongy "apple" gall on oak trees.

gall bladder Muscular sac, found in most VERTEBRATES, which stores BILE. In humans it lies beneath the right lobe of the LIVER and releases bile into the DUODENUM by way of the bile duct. Its release is signalled by a hormone which is secreted when food is present in the duodenum.

Galle, Johann Gottfried (1812–1910) German astronomer, the first to observe and identify Neptune as a planet (1846), although its presence had been postulated by other scientists. Galle also suggested a system for determining the scale of the SOLAR SYSTEM based on the observation of asteroids.

gallium (symbol Ga) Grey, metallic element of group III of the periodic table. It was predicted by Dmitri MENDELEYEV and discovered using SPECTROSCOPY in 1875. Chief sources are bauxite and some zinc ores. The metal, which melts at low temperature, is used in LASERS, TRANSISTOR SEMICONDUCTORS and high-temperature thermometers. Properties: at.no. 31; r.a.m. 69.72; r.d. 5.9; m.p. 29.78°C (85.60°F); b.p. 2,403°C (4,357°F); most common isotope ^{69}Ga (60.4%).

Gallo, Robert Charles (1937–) US scientist who identified the VIRUS that causes AIDS. Now called the HUMAN IMMUNODEFICIENCY VIRUS (HIV), the organism was discovered by Gallo in 1984, a year later than its independent discovery at the Pasteur Institute.

gallstone (cholelithiasis) Hard mass, usually composed of bile pigments, cholesterol and calcium salts, which forms in the GALL BLADDER. Gallstones may be present for years and give no trouble. However, they may cause severe pain, or they may become lodged in the bile duct, causing obstructive JAUNDICE or cholecystitis (inflammation). Treatment is by removal of the stones or of the gall bladder.

Galois, Évariste (1811–32) French mathematician who laid the foundation of GROUP THEORY. In 1830 he submitted several papers, some of which were lost, and one, on algebraic equations, which was dismissed by the mathematical establishment as incomprehensible and thus ignored. During the turmoil that followed the French Revolution, Galois was arrested twice. After being released from jail the second time, he was challenged to a duel by a political opponent. Reputedly, the night before the fatal duel, Galois tried to note down as many of his thoughts as possible. These notes and a few other papers were discovered 14 years later by Joseph Liouville, who recognized them as works of genius. Galois set out the theory of groups and laid down conditions for the solvability of various algebraic equations.

Galton, Sir Francis (1822–1911) British scientist and scholar. He pioneered the study of human differences and statistical analysis of data. In *Hereditary Genius* (1869) he described his theories of HEREDITY and advocated EUGENICS.

Galvani, Luigi (1737–98) Italian physician and physicist who became a lecturer in anatomy at Bologna and did pioneer studies in electrophysiology, or "animal electricity". His experiments with frogs' legs indicated a connection between muscular contraction and electricity. He believed a new type of electricity was created in the muscle and nerve, but the correct explanation was given by Alessandro VOLTA. Galvani's name was given to the galvanometer and to several processes involving electricity.

galvanizing Coating of iron or steel articles with zinc, either applied directly in a bath of molten zinc, electroplated from cold zinc sulphate solutions, or dusted on and baked. Corrugated sheets, nails, pails and wire netting may be protected from the corrosive effects of rusting by galvanizing. The metal to be protected is first derusted but if, after galvanizing, the zinc coating is pierced, protection is still provided because of the preferential attack by corrosion on zinc rather than on iron. *See also* ELECTROPLATING

galvanometer Instrument for detecting, comparing or measuring ELECTRIC CURRENTS. It is

GALVANOMETER

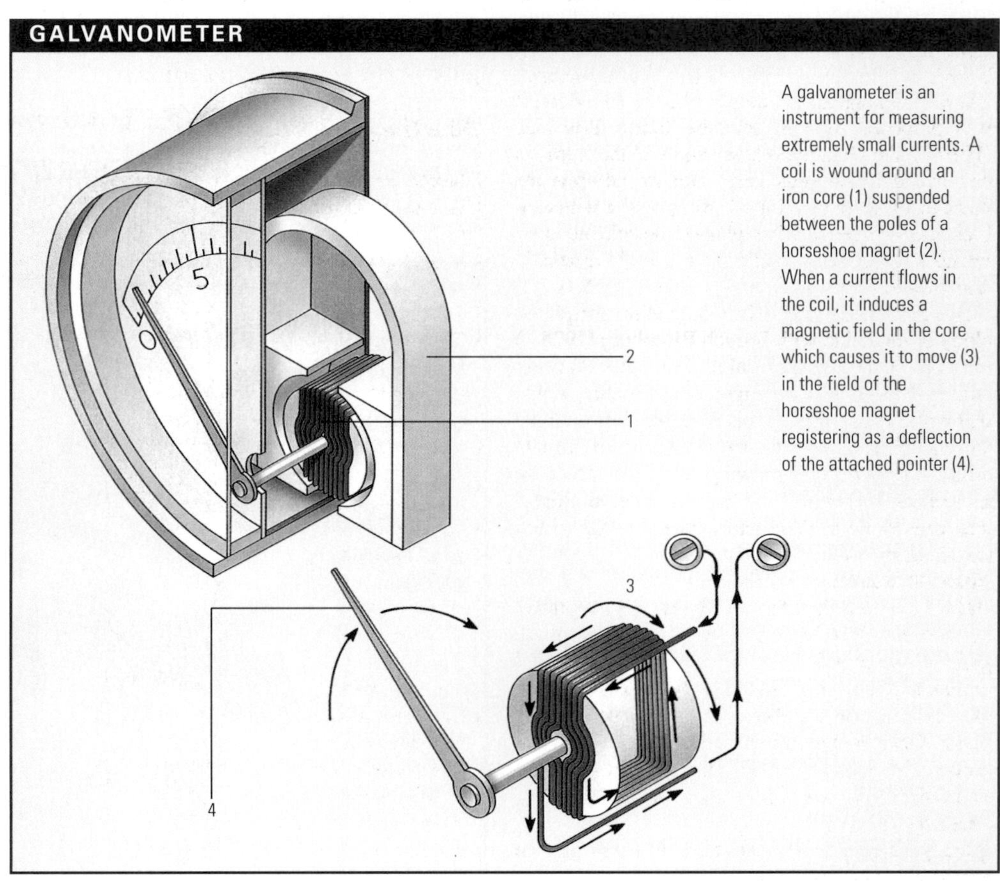

A galvanometer is an instrument for measuring extremely small currents. A coil is wound around an iron core (1) suspended between the poles of a horseshoe magnet (2). When a current flows in the coil, it induces a magnetic field in the core which causes it to move (3) in the field of the horseshoe magnet registering as a deflection of the attached pointer (4).

GAS

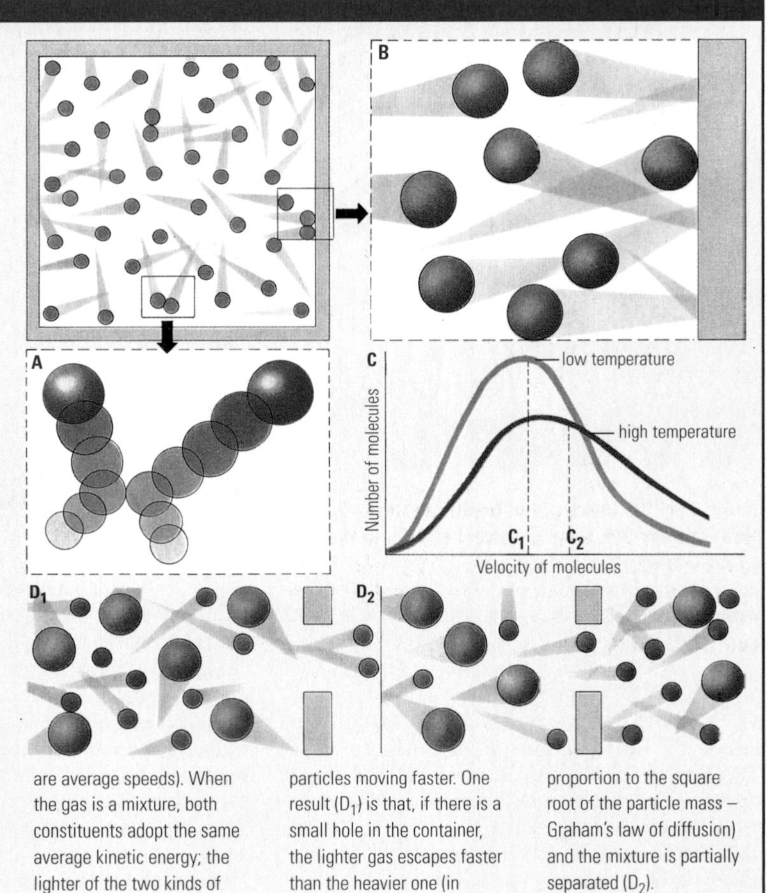

The kinetic theory of gases explains the behaviour of a gas by considering it as consisting of very small, perfectly elastic, particles (the molecules) which collide with one another (A) and with the walls of the container (B). All will have different velocities but useful statistical statements can be made about the gas as a whole. The pressure – the force exerted on a unit area of the walls due to the particles hitting it – clearly depends on the number of particles per unit volume and on their velocities and mass. In the simplest case pressure is proportional to the average kinetic energy and thus to the average of the squares of the speeds. The temperature is also proportional to the average kinetic energy and thus also to the pressure. The graph (C) shows the statistics of velocity for two different temperatures (C_1 and C_2 are average speeds). When the gas is a mixture, both constituents adopt the same average kinetic energy; the lighter of the two kinds of particles moving faster. One result (D_1) is that, if there is a small hole in the container, the lighter gas escapes faster than the heavier one (in proportion to the square root of the particle mass – Graham's law of diffusion) and the mixture is partially separated (D_2).

based on the principle that a current through a CONDUCTOR (usually a coil) creates a MAGNETIC FIELD which reacts with a magnet giving a deflection. The main types are the **moving-coil** galvanometer (the most commonly used) and the **moving-magnet instrument**.

gamete Reproductive (sex) CELL that, during FERTILIZATION, fuses with another reproductive cell of the opposite sex to form a ZYGOTE, the start of a new individual. Female gametes (*see* OVUM) are usually motionless; in many organisms, the male gametes (in this example SPERM) have a tail (FLAGELLUM) that enables them to swim to the ova. All gametes are HAPLOID – they contain only a single set of CHROMOSOMES due to MEIOSIS. On fusion, the resulting zygote is DIPLOID. *See also* SEXUAL REPRODUCTION

game theory In mathematics, analysis of problems involving conflict. Its application includes problems in business management, sociology, economics and military strategy as well as "true" games such as poker and chess. The theory was first introduced by Émile Borel and developed by John VON NEUMANN in 1928.

gametophyte Generation of PLANTS and ALGAE that bears the female and male GAMETES (reproductive cells). In flowering plants (ANGIOSPERMS) these are the germinated pollen grains (male) and the embryo sac (female) inside the OVULE. *See also* ALTERNATION OF GENERATIONS; FERN; MOSS

gamma globulin One of the protein components of the serum of mammalian blood. It contains approximately 85% of the circulating ANTIBODIES of the BLOOD. It gives temporary IMMUNITY in patients who have been exposed to certain diseases, such as measles.

gamma radiation (gamma rays) Form of very short wavelength ELECTROMAGNETIC RADIATION emitted from the nuclei of some RADIOACTIVE atoms. High-energy gamma rays have even greater powers of penetration than X-RAYS, so that sources of this radiation must be stored inside thick lead containers. These sources are used in medicine to treat cancer and in the food industry to kill microorganisms. Gamma rays can be detected with a GEIGER COUNTER. *See also* FOOD PRESERVATION; RADIOACTIVITY; RADIOTHERAPY

gamma-ray astronomy *See* ASTRONOMY

Gamow, George (1904–68) US nuclear physicist, cosmologist and author, b. Russia. His most famous work, co-authored with Ralph Alpher (1921–) and Hans BETHE, pictured a hot and dense early UNIVERSE in which the nuclei of elements were built up from protons; this was a cornerstone of the BIG BANG theory. His greatest prediction, borne out by the detection of the COSMIC MICROWAVE BACKGROUND RADIATION, was that the Universe should by now have cooled to a few degrees above ABSOLUTE ZERO. He also contributed to the deciphering of the genetic code, developed the QUANTUM THEORY of radioactivity, and proposed the liquid-drop model of atomic nuclei. With Edward TELLER, he put forward the **Gamow-Teller theory** of beta decay and a theory of the internal structure of RED GIANT stars.

ganglion Cluster of nervous tissue containing numerous cell bodies and SYNAPSES, usually enclosed in a fibrous sheath of MYELIN. In a VERTEBRATE, most ganglia occur outside the CENTRAL NERVOUS SYSTEM. *See also* NEURON

gangue Waste material in an ORE or deposit, as opposed to metal-containing or other useful minerals.

Ganymede Largest of Jupiter's GALILEAN SATELLITES, with a diameter of 5,262km (3,270mi), and the largest satellite in the SOLAR SYSTEM. Ganymede contains a high proportion of water-ice. Its surface is a mixture of dark, heavily cratered terrain and a brighter, more lightly cratered terrain covered with meandering, intersecting grooves that suggest recent geological activity. The most prominent feature is the dark Galileo Regio, some 4,000km (2,500mi) across.

garnet Two series of common orthosilicate minerals: the **pyralspite series** and the **ugradite series**. Garnet is found in METAMORPHIC ROCKS and PEGMATITES as cubic system crystals, rounded grains and granular masses. It is brittle, glassy and occurs in many hues; some varieties (particularly the red form) are important as gemstones. Hardness 6.5–7.5; r.d. 4.

garnierite Ore of nickel. It is a lustrous green SERPENTINE, also containing magnesium and silicon. It occurs in ultrabasic rocks as a decomposition product of OLIVINE.

gas State (PHASE) of MATTER in which molecules are free to move in any direction; a gas always spreads (by DIFFUSION) to fill a container of any size. Because of their low densities, most gases are poor CONDUCTORS of heat and electricity (although at high voltages a gas may be ionized and become electrically conducting). When cooled, gases become liquids. Some, such as carbon dioxide, can be liquefied by pressure alone. Others require cooling below their critical temper-

GAS EXCHANGE

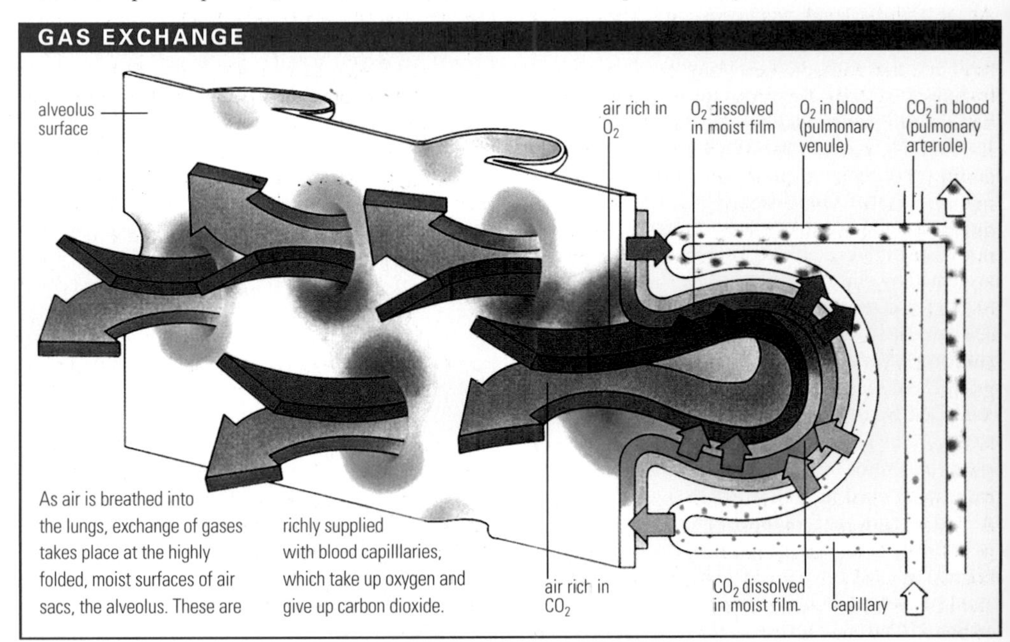

As air is breathed into the lungs, exchange of gases takes place at the highly folded, moist surfaces of air sacs, the alveolus. These are richly supplied with blood capillaries, which take up oxygen and give up carbon dioxide.

ature before they can be liquefied (*see* CRITICAL POINT). All gases follow certain laws, such as AVOGADRO's law, BOYLE'S LAW, CHARLES' LAW, GRAHAM'S LAW and the IDEAL GAS LAW. *See also* LIQUID; PLASMA; SOLID

gas, natural *See* NATURAL GAS

gas analysis Determination of the nature and quantity of constituents of gas mixtures. Three methods of analysis are available: 1) the gases can be selectively absorbed in chemical solutions and the decrease in volume measured; 2) in GAS CHROMATOGRAPHY the rates of diffusion of gases through such media as charcoal can be utilized; 3) the instrumental method such physical properties of the gases as density can be used as the basis of analysis.

gas chromatography Type of CHROMATOGRAPHY in which the carrier medium is a gas. The technique is used for analysing or separating a mixture of gases, often derived from a heated liquid mixture. A "carrier" gas (the **mobile phase**) – often hydrogen – sweeps the mixture around a very long coiled tube containing the **stationary** phase, which differentially absorbs the components of the mixture. In **gas-liquid** chromatography (GLC), the stationary phase is a liquid such as a non-volatile oil coating a solid support; in **gas-solid** chromatography (GSC), it is a solid such as diatomite (kieselguhr). The components of the mixture are detected as they leave the apparatus in turn or, if the chromatograph has been calibrated with known substances, they are identified by the time they take to pass through. This method of chromatography is used to analyse tiny amounts of a sample.

gas constant (symbol *R*) Constant in the universal gas equation (*see* IDEAL GAS LAW). *R*, also called the universal molar gas constant, has a value of $8.314510\ \mathrm{JK^{-1}\ mol^{-1}}$.

gas engine *See* ENGINE

gas exchange In biology, the uptake and output of gases, especially oxygen and carbon dioxide, by living organisms. In animals and other organisms that obtain their energy by the breakdown of food in the chemical reactions of AEROBIC RESPIRATION, gas exchange involves the uptake of oxygen and the output of carbon dioxide. In plants, algae and photosynthetic bacteria, which carry out PHOTOSYNTHESIS, the opposite may occur when photosynthesis is active, with a net carbon dioxide uptake and net oxygen output. At the cellular level, gas exchange takes place by DIFFUSION across cell MEMBRANES in solution. In multicellular animals there may be a specialized transport system for transporting these gases between the cells and the outside environment. By using up one or other of these gases and generating the other, respiration and photosynthesis maintain their own concentration gradients for diffusion. Where transport systems are involved, the gradient may be enhanced by other means. For example, in vertebrate ERYTHROCYTES (red blood cells), the affinity of the pigment HAEMOGLOBIN for oxygen changes, as the local oxygen concentration varies, enabling it to take up oxygen in the lungs and release it to respiring cells. *See also* BREATHING; CIRCULATORY SYSTEM; RESPIRATORY SYSTEM; VENTILATION

gas laws *See* IDEAL GAS LAW

gasohol Petrol to which alcohol has been added. A typical gasohol composition is 90% unleaded petrol and 10% ethanol (grain alcohol). This gasohol can be used in cars with CATALYTIC CONVERTERS and has about the same OCTANE NUMBER as unleaded petrol. The chief advantage of gasohol is that the ethanol can be produced from renewable biomass (organic matter), a relatively cheap resource. Petrol, on the other hand, is produced from expensive petroleum.

gasoline *See* PETROL

gastric juice Fluid comprising a mixture of substances, including PEPSIN and HYDROCHLORIC ACID, secreted by GLANDS of the stomach. Its principal function is to break down proteins into POLYPEPTIDES during DIGESTION.

gastroenterology Branch of medicine that deals with the diagnosis and treatment of diseases of the gastrointestinal tract and associated organs such as the pancreas, liver and gall bladder.

gastropod Member of the class Gastropoda, a class of molluscs, which includes the snail, slug, whelk, limpet, abalone and sea slug. Many possess a single spiral shell, produced by chemical precipitation from the MANTLE. Many types of gastropod live immersed in seawater, breathing through gills. Some freshwater snails, however, breathe through lungs and need to surface periodically for air. Like slugs, sea slugs are entirely without shells and are often brilliantly coloured.

gastrula Early stage in the development of the EMBRYO in animals; it follows the BLASTULA stage. The gastrula is a hollow sac with two layers (*see* GERM LAYER) of cells, the inner ENDODERM and the outer ECTODERM. The sac cavity is called the **archenteron** and its mouth the **blastopore**.

gas turbine *See* TURBINE

gate In ELECTRONICS and FLUIDICS, circuit having two or more inputs but a single output energized only under specific input conditions. Gates are used extensively in LOGIC CIRCUITS in which the sequence of operations, or their combinations, are determined by the rules of symbolic logic.

Gates, Bill (William Henry) (1955–) US businessman. In 1975 he co-founded Microsoft Corporation, which in the 1980s became the predominant computer software producer. He is noted for his innovative thinking and for his aggressive marketing and business tactics. Recently, Microsoft has been accused of improper business practices regarding software it has produced for the Internet.

gauge Term for many kinds of measuring device. For example, calipers are width gauges, dipsticks are depth gauges and thermometers are temperature gauges, although here the term is restricted usually to mechanical thermometers. Pressure gauges are of various kinds, the Bourdon gauge being the type used to measure fluctuations of air pressure in aneroid BAROMETERS and variations in gas and liquid pressure in industry. The term is also used of the perpendicular distance between railway lines, from the largest actual gauge (the "5-foot gauge") to miniature and model.

gauge boson Type of ELEMENTARY PARTICLE that mediates interactions between other particles. The four gauge bosons are: PHOTONS, for electromagnetic interactions in QUANTUM ELECTRODYNAMICS; GLUONS, for strong interactions in QUANTUM CHROMODYNAMICS; intermediate vector BOSONS designated W and Z, for WEAK NUCLEAR FORCE interactions; and GRAVITONS, for gravitational interactions.

Gauss, Karl Friedrich (1777–1855) German mathematician. He was a child prodigy from a poor family, and his education was sponsored by a wealthy aristocrat, the Duke of Brunswick, who had heard of the boy via his teacher. While still a teenager, Gauss made many discoveries, including the idea of non-EUCLIDEAN GEOMETRY (thirty or so years before Nikolai LOBACHEVSKY). He discovered the fundamental theorem in complex analysis a decade before Baron CAUCHY and discovered QUATERNIONS before William HAMILTON. Gauss chose not to publish these ideas, refusing to publish his work unless he was satisfied that it was perfect. He made contributions in many areas of mathematics, including algebra, number theory, probability theory, analysis and the theory of series. He also studied electricity and magnetism, gravitation, optics and astronomy. The unit of magnetic flux density is named after him.

▲ **Gates** US business executive and computer engineer Bill Gates made his fame and fortune in the personal computer boom of the 1980s. His company, Microsoft Corporation, produced operating systems (MS-DOS) and application programs (Windows) that became the world standard for so-called IBM-compatible computers. Microsoft Corporation is the world's leading software company, and Gates himself became the youngest billionaire when just 31 years old.

Gauss' law Total ELECTRIC FLUX over a closed surface equals the ELECTRIC CHARGE within that surface divided by the PERMITTIVITY of the medium. The law, stated by Karl GAUSS, also applies to surfaces in a magnetic field and similar statements can be made for a gravitational field.

Gay-Lussac, Joseph Louis (1778–1850) French chemist and physicist who did pioneer research into the behaviour of gases. He discovered the law of combining gas volumes (GAY-LUSSAC'S LAW) and the law of gas expansion, often also attributed to Jacques CHARLES (who discovered it earlier but did not publish his results – CHARLES' LAW). Gay-Lussac prepared (with Louis Jacques THÉNARD) the elements POTASSIUM and BORON, investigated fermentation and hydrocyanic (prussic) acid, and invented a HYDROMETER.

Gay-Lussac's law When gases react, their volumes are in a simple ratio to each other and to the volume of products, at the same temperature and pressure. It was named after Joseph GAY-LUSSAC.

gear Wheel, usually toothed, attached to a rotating shaft. The teeth of one gear engage those of another to transmit and modify rotary motion and TORQUE. The smaller member of a pair of gears is called a pinion. If the pinion is on the driving shaft, speed is reduced and turning force increased. If the larger gear is on the driving shaft, speed is increased and turning force reduced. A screw-type driving gear, called a **worm**, gives the driven gear a greatly reduced speed. Smooth wheels can be used as gears, but may slip against each other.

Geber (*c*.720–815) (Jabir or Jabir ibn-Hayyan) Arab alchemist whose works were studied by the English philosopher Roger Bacon (*c*.1214–92). The identification of Geber with Jabir is uncertain, but Geber is known to have performed scientific experiments as well as attempting to convert mercury into gold. He felt that a TRANSMUTATION of baser metals into nobler ones was possible. He prepared nitric acid and distilled vinegar (to obtain ethanoic acid).

gegenschein (counterglow) Faint luminous patch, visible at a point along the ecliptic, that is diametrically opposite to the Sun. Best seen in tropical regions, it is a phenomenon allied to the zodiacal light, and often appears as an extension of it.

Geiger, Hans Wilhelm (1882–1945) German physicist who, with Ernest RUTHERFORD, devised in 1908 a method of detecting and counting alpha particles. This device, with modifications, became the GEIGER COUNTER. Geiger and Ernest Marsden measured the deflection of alpha particles by thin metal foil in 1909, and this phenomenon was the basis of Rutherford's discovery of the atomic NUCLEUS.

Geiger counter (Geiger-Müller counter) Instrument used to detect and measure the strength of radiation by counting the numbers of ionized particles produced. It is a type of IONIZATION CHAMBER in which a high voltage is applied across a pair of electrodes. Radiation and particles entering the chamber ionize gas atoms and the resulting ions, gaining energy from the electric field between the electrodes, magnify the effect by producing many more ions. These make a current that quickly subsides, so a pulse of current is produced for each particle. Each pulse activates a counting circuit enabling up to about 10,000 particles per second to be counted.

Geissler tube Early form of DISCHARGE TUBE.

gel Homogeneous COLLOID consisting of minute particles dispersed or arranged in a liquid to form a fine network throughout the mass. Their appearance can be elastic or jellylike, as in GELATIN or fruit jelly, or quite rigid and solid, as in silica gel, a material resembling coarse white sand used as a dehumidifier and in the manufacture of silicone and rubber.

gelatin Colourless or yellowish protein obtained from COLLAGEN (a fibrous protein) in animal cartilages and bones by boiling in water. It is used in photographic film emulsions, size, capsules for medicines, as a culture medium for bacteria, and in foodstuffs such as jellies.

Gell-Mann, Murray (1929–) US theoretical physicist who was awarded the 1969 Nobel Prize for physics for his application of group theory to ELEMENTARY PARTICLES, which led to the prediction of the QUARK as the basic constituent of the BARYON and MESON. His theory also predicted the existence of a baryon called the **omega-minus particle**, subsequently discovered in 1964. Gell-Mann had also introduced in 1953 the concept of "strangeness" to explain the longevity of certain quarks.

gem Any of about 100 minerals, either opaque, transparent or translucent, valued for their beauty, rarity and durability. Transparent stones, such as DIAMOND, RUBY, EMERALD and SAPPHIRE, are the most highly valued. PEARL, AMBER and CORAL are gems of organic origin. During the Middle Ages, many gems were thought to have magical powers. The art of cutting and polishing gemstones to bring out their colour and brilliance was developed in the 15th century, particularly in Italy. Stones with a design cut into them are **intaglios**; with a design in relief, **cameos**. The study of gems is known as gemmology. *See also* SEMI-PRECIOUS STONES

Gemini (Twins) Large constellation of the N sky, lying between TAURUS and CANCER. It contains the star cluster M 35 (NGC 2168) and the planetary nebula NGC 2392. Its brightest stars are Castor (Alpha Geminorum, magnitude 1.57) and Pollux (Beta Geminorum, magnitude 1.14).

Gemini project Series of US manned experimental space flights that followed the Mercury and preceded the APOLLO PROJECT. Gemini 3 was a two-person craft first manned by Virgil Grissom and John Young on 23 March 1965; they made three Earth orbits. Subsequent flights and experiments, ending with Gemini 12 (11–15 November 1966), demonstrated space walks, space docking manoeuvres and tests of longevity including a 14-day flight by Frank Borman and James LOVELL, Jr, in December 1965.

gemma In botany and zoology, a bud that will give rise to a new individual. The term also refers to a multicellular reproductive structure found in algae, liverworts and mosses.

gene Unit by which hereditary characteristics or traits are passed on from one generation to another in plants and animals. A gene is a length of DNA that codes (*see* GENETIC CODE) for a particular protein or peptide. Genes are found along the CHROMOSOMES of the cells of plants, animals, fungi and protoctists (in bacteria the DNA is not contained in chromosomes). In most cell nuclei genes occur in pairs, one located on each of a chromosome pair. In cases where different forms of a gene (ALLELES) are present in a population, some forms may be RECESSIVE to others (DOMINANT genes) and will not be expressed unless present on both members of a chromosome pair. *See also* GENETICS; GENETIC ENGINEERING; HEREDITY

gene bank Collection of genetic material kept for possible future use. Material stored includes cultures of bacteria and moulds; seeds, spores and tubers; frozen sperm, eggs and embryos; and even live plants and animals. The material can be used in plant and animal BREEDING, GENETIC ENGINEERING and in medicine. Live species are used for restocking natural habitats in which species are extinct or in danger of EXTINCTION.

gene pool All of the GENES, and their various ALLELES, that occur in all the members of a particular species or population of species at any given moment. For a species to be vigorous and survive, the gene pool must be large enough for natural variation to take place through CROSS-FERTILIZATION. A large population does not necessarily imply a large gene pool, however. For example, there is concern about the viability of some sizeable populations of cheetahs which, despite their numbers, have insufficient variation in the gene pool.

general theory of relativity *See* RELATIVITY

generator Device for producing electrical energy. The most common is a machine that converts the mechanical energy of a TURBINE or INTERNAL COMBUSTION ENGINE into electricity by employing ELECTROMAGNETIC INDUCTION. There are two types of generators: alternating current (AC) and direct current (DC), often called an ALTERNATOR and a DYNAMO. Each has an ARMATURE (or ring) which rotates within a magnetic field creating an induced current. The direct current generator has a COMMUTATOR on the armature which divides the induced current as it changes direction, giving rise to a pulsating direct current. *See also* ELECTRIC CURRENT

gene therapy (gene replacement therapy, GRT) Treatment of disease by administering to patients healthy GENES in place of missing or defective ones. The first human being to undergo gene therapy, in the USA in 1990, was a four-year-old suffering from a rare enzyme deficiency that destroys a person's natural immunity. Capitalizing on techniques of GENETIC ENGINEERING, the gene therapist repairs or replaces a faulty gene. The healthy gene is packaged into some kind of vector (usually a suitably doctored virus) so that it can be targeted at affected cells. Originally gene therapy was intended to fight hereditary disorders, such as cystic fibrosis or sickle cell disease, but researchers are also seeking to direct it against other disorders, such as cancer, where a gene has become faulty over a period of time. Although thousands of patients have been treated, mostly in the USA, gene therapy has yet to live up to its early promise. Few of the inserted genes reach their target cells and those which do may not always work efficiently.

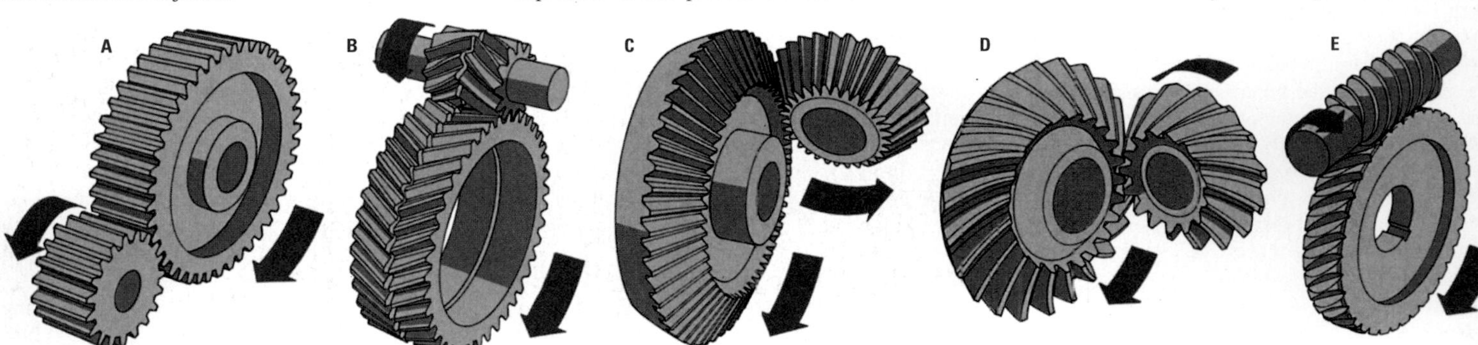

▲ **gear** For all gears, the mechanical advantage is related to the number of teeth. The teeth of spur gears (A) are cut parallel to the axis rotation, while those of helical gears (B) are "twisted" to form part of a helix and often cut double to avoid thrusts that result in wear. The teeth of bevel gears (C) are longer than those of spur gears, giving a greater area of contact that permits the transmission of much greater thrusts. This advantage applies more particularly to bevel gears with spiral teeth (D). In a worm gear (E), the worm has a single spiral thread and turns a spirally toothed gear at right-angles to its own axis.

GENE THERAPY

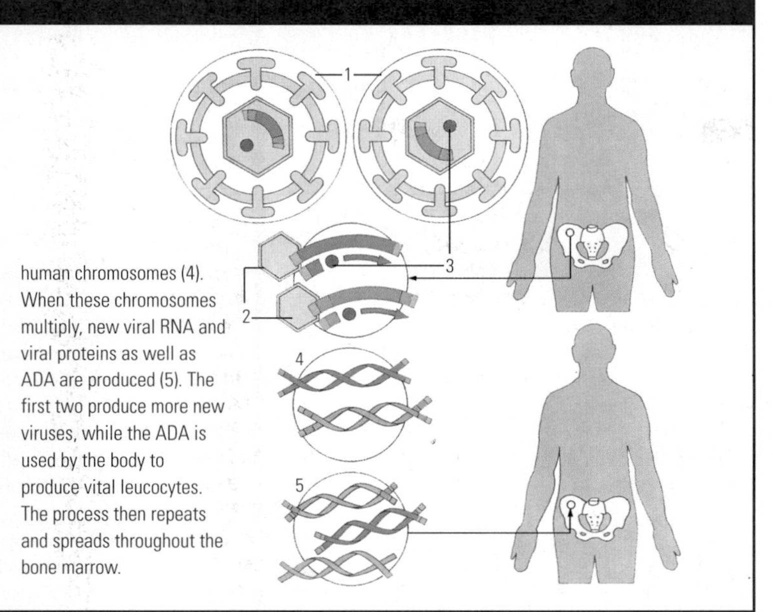

Gene therapy is used to treat severe combined immuno-deficiency (SCID), where the gene responsible for production of the enzyme adenosine deaminase (ADA) is missing. As ADA is essential for leucocyte (white blood cell) production, this renders the body open to infection. Two retroviruses (1) are introduced into the bone marrow. These have the ability to produce RNA from their DNA (2) using a reverse transcriptase enzyme (3). This DNA is then incorporated into the human chromosomes (4). When these chromosomes multiply, new viral RNA and viral proteins as well as ADA are produced (5). The first two produce more new viruses, while the ADA is used by the body to produce vital leucocytes. The process then repeats and spreads throughout the bone marrow.

Also, there are problems with the use of viruses as gene shuttles. The body treats the virus as "foreign" and some patients have suffered severe immune reactions. There is, too, a theoretical risk that the virus itself could spread and cause cancer.

genetic code Arrangement of information stored in GENES. The genetic code specifies exactly the way in which all the complex substances in living cells are made and also specifies the manner of cell reproduction. It is, therefore, the ultimate basis of HEREDITY and forms a blueprint for the entire organism. The genetic code is based on the genes that are present, which, in molecular terms, depends on the arrangement of nucleotides in the long molecules of DNA in the cell CHROMOSOMES. Each group of three nucleotides specifies or codes for an AMINO ACID or an action such as "start" or "stop". Amino acids are the basic units of which PROTEINS are made. By specifying which proteins to make and in what quantities, the genetic code not only directly controls production of structural materials, but also, by coding for the ENZYMES (also proteins) that regulate all the chemical reactions in the cell, indirectly codes for the production of other cell materials as well.

genetic engineering Construction of a DNA (deoxyribonucleic acid) molecule containing a desired GENE, which is then introduced into a bacterial, fungal (yeast), plant or mammalian cell, so that this cell then produces the desired protein. The technique has been used successfully to alter bacterial genetic material to produce substances such as human growth hormone, insulin and enzymes for biological washing powder. Many people are concerned about the ethics of genetic engineering, given the possibility that scientists may one day be able to alter the genetic structure of humans, or accidentally release genetically engineered, disease-causing bacteria into the environment.

genetic fingerprinting Forensic technique pioneered by Alec JEFFREYS using genetic material, specifically the DNA within sample body cells, to identify individuals. It is used mainly in paternity suits to detect the true father of a child, and sometimes in criminal cases to establish identity from traces of the assailant's semen, hair or skin. The chances of two individuals who are not blood relations having genetic "fingerprints" that could be confused are estimated to be at least a billion to one. The technique's first use in a court of law was in the late 1980s.

genetics Study of HEREDITY. Geneticists study how the characteristics of an individual organism depend on its GENES, how the characteristics are passed down to the next generation, and how changes may occur through MUTATION. A person's behaviour, learning ability and physiology may be explained partly by genetics, although the environment in which an individual grows up may have a considerable influence too. The deliberate modification of GENETIC CODES by scientists is called genetic manipulation, or GENETIC ENGINEERING. Scientists can modify the genes of livestock and crops in order to improve the quality or quantity of the produce. Genetics was founded by Gregor MENDEL, following experiments with successive generations of peas.

genome Entire complement of genetic material carried within the CHROMOSOMES of a single cell. In effect, a genome carries all the genetic information about an individual coded in sequence by the DNA that makes up the chromosomes. It has also been applied to the whole range of GENES in a particular species. *See also* GENETICS; HUMAN GENOME PROJECT

genotype Genetic makeup of an individual. The particular set of GENES present in each cell of an organism is distinct from the PHENOTYPE, the observable characteristics of the organism.

genus Part of the CLASSIFICATION of living organisms, ranking below FAMILY and above SPECIES. A genus is a group of closely related biological species with common characteristics. Genus names are printed in *italic* with a capital initial letter; for example the domestic cat belongs to the genus *Felis*.

geochemistry Study of the chemical composition of the Earth, particularly the abundance and distribution of the chemical elements and their isotopes, and the changes that have resulted in it from chemical and physical processes.

geochronology Dating of rocks or of Earth processes. Absolute dating techniques (RADIOMETRIC DATING) involve the measurement of radioactive decay to determine the actual date in years for a given rock. Relative dating involves the use of fossils, sediments or relationships between structures to place rock sequences and geological events in order. *See also* CARBON DATING

geode Hollow rock nodule with inner walls lined with crystals, generally quartz or calcite. It is formed by gelatinous silica and mineral-bearing water within a cavity.

geodesic dome Architectural structure of plastic, metal or even of cardboard, based in shape upon triangular or polygonal facets, which evenly distribute the tension. They were originated (1947) and perfected by R. Buckminster FULLER. A well-known example was the US Pavilion at Montreal's Expo 67. *See also* BUCKMINSTER FULLERENE

geodesic surveying Method of surveying that covers areas large enough to involve consideration of the Earth's curvature. Geodesic surveying is carried out in order to establish geographical features, such as national boundaries, and as a basis for the mapping of whole states or countries. A MAP is made using various systems for representing the Earth's curved surface as a flat drawing. These systems are the projections, such as Mercator's or polar azimuthal, which are referred to on maps of large areas.

geodesy In geophysics, the determination of the size and shape of the Earth, its gravitational field and the location of fixed points. *See also* GEODESIC SURVEYING

geography Science that studies the spatial relationships between the surface of the Earth and humankind. It includes the size and distribution of land masses, seas, resources, climatic zones and plant and animal life. Because it seeks to relate all the Earth's features to human existence, geography differs from other Earth sciences such as GEOLOGY, METEOROLOGY and OCEANOGRAPHY, which study this planet's features as specific phenomena.

geoid Theoretical surface of Earth. The roughly elliptical Earth is bulged out at the Equator and flattened at the poles. *See also* GEODESY

geological survey Study of the GEOLOGY of an area, usually with some other end in view, such as the evaluation of the area's agricultural potential, its mineral resources or its suitability as building land.

geological time Timescale of the history of the Earth, divided into periods of time millions of years long. Until recently, only methods of relative dating were possible. These include the world-wide study, comparison and correlation of sequences of

▼ **geological time** The 4,600 million years since the formation of the Earth are divided into four great eras, which are further split into periods and, in the case of the most recent era, epochs. The present era is the Cenozoic ("new life"), extending backwards through "middle life" and "ancient life" to the Precambrian. Although traces of ancient life have since been found, it was largely the proliferation of fossils from the beginning of the Palaeozoic era onwards, some 570 million years ago, which first allowed precise sub-divisions to be made.

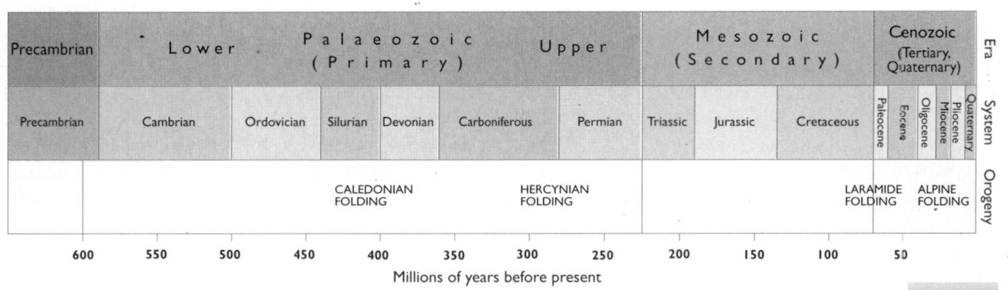

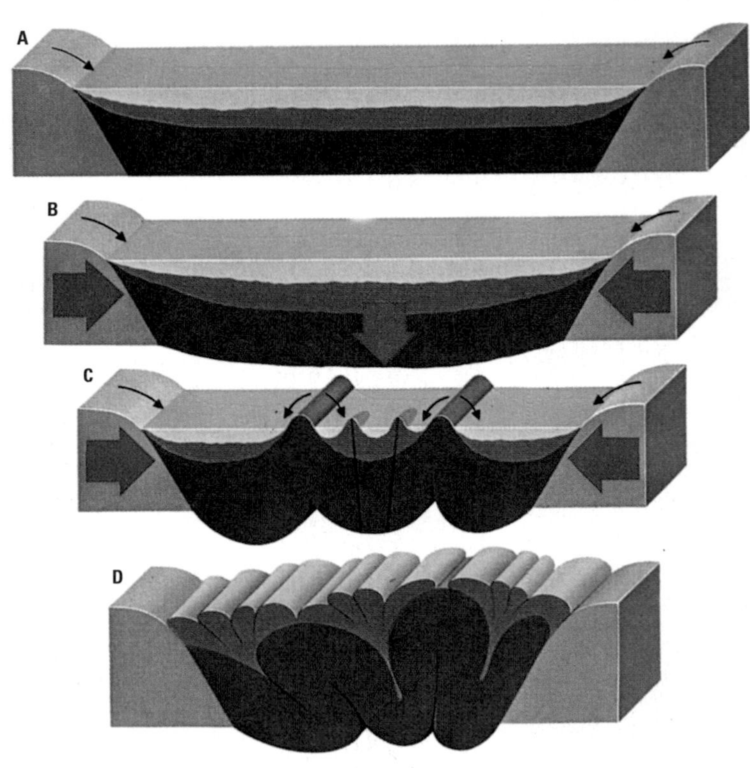

▶ geosyncline
Geosynclines are the birthplaces of mountains. They are large troughs where thick layers of sediments can accumulate (A). Where geosynclines develop between two colliding crustal plates (B) the sediments can be squeezed up as broad ridges known as geanticlines (C). Further compression creates mountain ranges (D). The whole process is usually accompanied by pressure-induced recrystallization or melting, which then forms metamorphic, plutonic and volcanic rocks; examples of each of these are gneisses, granites and rhyolites.

rock formations and the fossils they contain. The information thus gathered is used to distinguish earlier from later deposits, to estimate periods of passed time, and to reconstruct geologic and climatic events by assuming that geological processes in the past were the same as today. Geological time is divided into four eras: PRECAMBRIAN, PALAEOZOIC, MESOZOIC and CENOZOIC, which are further divided into periods. Periods, in turn, are subdivided into series or epochs. Epochs are subdivided into stages and then zones.

geology Study of the materials of the Earth, their origin, arrangement, classification, change and history. Geology is divided into several categories, the major ones being MINERALOGY (arrangement of minerals), PETROLOGY (rocks and their combination of minerals), STRATIGRAPHY (arrangement and succession of rocks in layers), PALAEONTOLOGY (study of fossilized plant and animal remains), GEOMORPHOLOGY (study of landforms), STRUCTURAL GEOLOGY (classification of rocks and the forces that produced them), and **environmental geology** (geological study applied to the best use of the environment by humans).

geomagnetism Physical properties of the Earth's MAGNETIC FIELD. Geomagnetism is thought to be caused by the metallic composition of the Earth's CORE. The gradual movements of magnetic north result from currents within the MANTLE.

geometric isomer *See* ISOMER

geometric mean Geometric mean of n numbers is the nth root of their product. An example is the square root of the product of two numbers: the geometric mean of 8 and 2 is $\sqrt{(8 \times 2)} = 4$. The geometric mean of 5, 8 and 25 is $\sqrt[3]{(5 \times 8 \times 25)} = 10$.

geometric progression Sequence of numbers in which the ratio of each term to the preceding one is a constant (called the common ratio). It has the form $a, ar, ar^2, ar^3, ...$, where r is the common ratio. The sum of these terms, $a + ar + ar^2...$, is a **geometric series**. If there are n terms, the sum equals $a(1 - r^n)/(1 - r)$. Infinite geometric series converge to $a/(1 - r)$ if r lies between -1 and $+1$.

geometry Branch of mathematics concerned broadly with studies of shape and size. To most people, geometry is the EUCLIDEAN GEOMETRY of simple plane and solid figures. It is, however, a much wider and more abstract field with many subdivisions. COORDINATE GEOMETRY (also called **analytical geometry**), introduced in 1637 by René DESCARTES, applies algebra to geometry and allows the study of more complex curves than those of Euclidean geometry. PROJECTIVE GEOMETRY is, in its simplest form, concerned with projection of shapes – operations of the type involved in drawing maps and in perspective painting – and with properties that are independent of such changes. It was introduced by Jean Victor Poncelet (1788–1867) in 1822. Even more abstraction occurred in the early 19th century with formulations of NON-EUCLIDEAN GEOMETRY by Janos Bolyai (1802–60) and N.I. LOBACHEVSKY. DIFFERENTIAL GEOMETRY – based on the application of CALCULUS and ideas such as the curvature and lengths of curves in space – was also developed during this period. Geometric reasoning was also applied to spaces with more than three dimensions. TOPOLOGY is a general form of geometry, concerned with properties that are independent of any continuous deformation of shape.

geomorphology Scientific study of features of the Earth's surface and the processes that have formed them.

geophone Seismic detection device placed on or in the ground to measure the echoes produced by volcanic or seismic (earthquake) activity.

geophysics Study of the characteristic physical properties of the Earth as a whole system. It uses parts of CHEMISTRY, GEOLOGY, ASTRONOMY, SEISMOLOGY, METEOROLOGY and many other disciplines. From the study of seismic waves, geophysicists have worked out the structure of the Earth's interior.

geostationary orbit Location of an artificial SATELLITE so that it remains above the same point on a planet's surface because it completes one orbit in the same time as it takes the planet to rotate once on its axis. Communications and REMOTE-SENSING satellites are often placed in geostationary orbits above the Earth. *See also* SYNCHRONOUS ORBIT

geostrophic current General term for an ocean or wind current whose direction is affected by the meeting of its pressure gradient with the CORIOLIS EFFECT.

geosyncline Great basin or trough in which deposits of sediments and volcanic rock, thousands of metres thick, have accumulated during slow subsidence over long geological periods. The term has generally fallen into disuse. *See also* SYNCLINE

geothermal Referring to heat from the interior of the Earth. Geothermal activity is responsible for many geological phenomena, including hot springs, GEYSERS and VOLCANOES. Some of this natural heat is being harnessed and used in the form of GEOTHERMAL ENERGY.

geothermal energy Heat contained in the Earth's rocks. It is produced by RADIOACTIVITY and by the movement of tectonic plates (*see* PLATE TECTONICS). It is released naturally in GEYSERS and VOLCANOES, and is used as a power source for generating electricity in several places, including Iceland, Italy, New Zealand and the USA.

geotropism In plant growth, the response to the stimulus of gravity. Plant stems are generally negatively geotropic and grow upwards; roots are positively geotropic and grow downwards. For example, growth curvature is caused by the accumulation of the plant hormone AUXIN in the tissue on the lower side of the stem; growth increases on that side and the stem bends upwards.

Gerhardt, Charles Frédéric (1816–1856) French chemist best known for his attempt to systematize organic chemistry under the theory of types. This based all compounds on four main types and was influential in the development of structural chemistry. He published this and other ideas in *Traité de Chimie Organique* (4 vols, 1853–56).

GERM LAYER

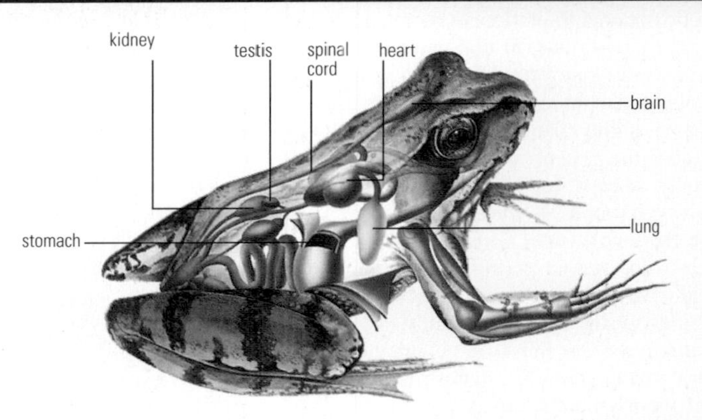

At a very early stage during the development of a frog, the embryonic cells begin to group into three distinct layers, each of which is predetermined to eventually form specific body parts. The layers, called the germ layers, are the endoderm (yellow), mesoderm (red) and ectoderm (blue and violet). The endoderm develops into the inner lining of the digestive tract and systems developed from it; the mesoderm forms the inner skin layer, skeleton and cartilage, muscles, circulatory, excretory and reproductive systems, outer layers of digestive tract and systems developed from it (e.g. the respiratory system). Finally, the ectoderm develops into the outer skin layer (epidermis) and associated structures, peripheral nervous system, sensory systems and pituitary gland; also the brain and spinal cord.

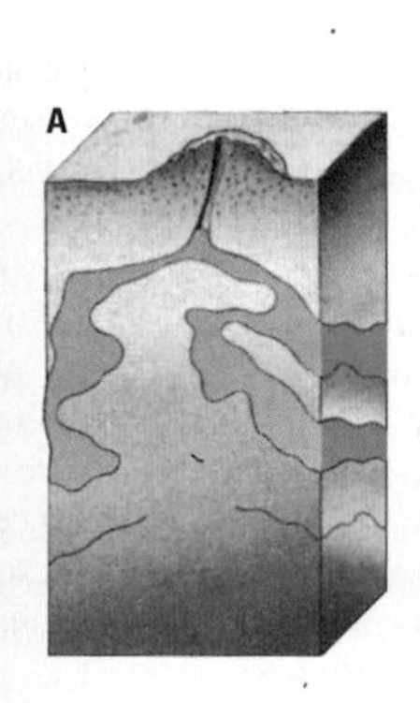

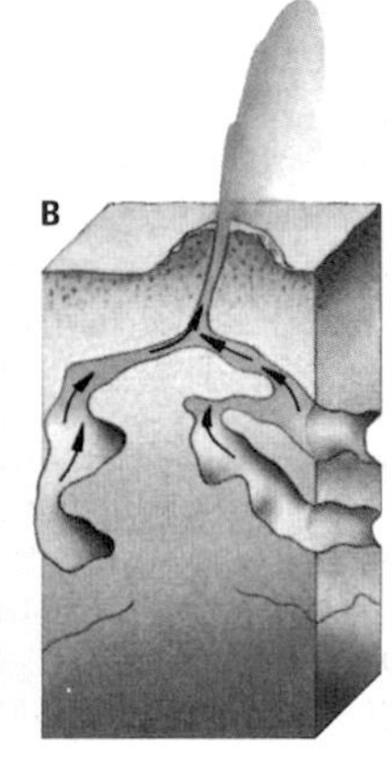

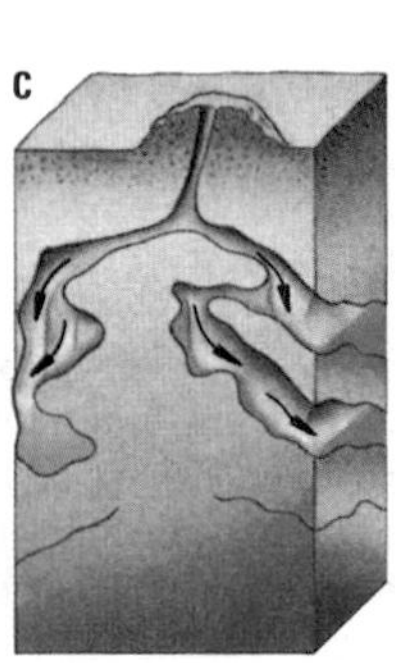

▲ **geyser** A plume of hot water and steam, a geyser is the result of the boiling of water at depth in a series of interconnecting chambers by volcanic heat (A). The expansion of steam produced drives the water and steam above it out at the surface (B), and this is followed by a period of refilling and heating making it a periodic phenomenon (C).

germ Popular term for any infectious agent. Germs can be pathogenic bacteria, fungi or viruses. The word germ is used in biology to denote a rudimentary stage in plant growth, such as an embryo in a seed, or a bud.

germanium (symbol Ge) Grey-white, metalloid element of group IV of the periodic table, predicted by the Russian chemist Dmitri MENDELEYEV and discovered in 1886. Germanium is a by-product of smelting zinc ores or may be obtained from the combustion of certain coals. It is important in transistors, rectifiers and similar semiconductor devices. Properties: at.no. 32; r.a.m. 72.59; r.d. 5.35; m.p. 937°C; (1,719°F); b.p. 2,830°C (5,126°F); most common isotope ^{74}Ge (36.54%).

germicide Any substance for destroying disease-causing microorganisms. Germicides include ANTISEPTICS, disinfectants and ANTIBIOTICS.

germination Growth of the embryo in a SEED to produce a new plant following FERTILIZATION. It may occur immediately or after a period of DORMANCY. In order to germinate, a seed or spore needs favourable conditions of temperature, light, moisture and oxygen. The process begins after the seed has taken up water. Germination is complete when a root appear outside the seed coat.

germ layer Any one of three layers of cells found in a developing EMBRYO. The three layers are the ECTODERM, MESODERM and ENDODERM, each one giving rise to particular tissues and organs.

gerontology Study of ageing. The capacity of an organism to repair tissues effectively, by the replacement of worn out cells, diminishes with age, so that "growing old" is inevitable (even when, as in many large trees, it is delayed for some thousands of years). The effect of certain types of medicine has been to relieve and treat the disease burdens of age. In itself, it has not noticeably increased life expectancy.

gestation Period of carrying a developing EMBRYO in a UTERUS between conception and birth. Gestation periods are specific to each mammalian species and range from 12 days (Virginia opossum) to 22 months (Indian elephant). In humans and many apes the period is *c*.38 weeks.

geyser Hot spring that erupts intermittently, throwing up jets of superheated water and steam to a height of *c*.60m (200ft), followed by a shaft of steam with a thunderous roar. Geysers occur in Iceland, New Zealand and the USA.

Giaever, Ivar (1929–) US physicist, b. Norway. He shared the 1973 Nobel prize for physics with Brian JOSEPHSON and Leo ESAKI for their research into tunnelling effects in SEMICONDUCTORS and superconductors. Esaki developed the tunnel diode, which enables electrons to cross normally impassable electronic barriers, and Giaever applied this to superconductors, demonstrating new effects which led to a better understanding of SUPERCONDUCTIVITY.

giant star Star belonging to a luminosity class midway between main-sequence stars and SUPERGIANTS. Giant stars lie directly above the MAIN SEQUENCE on the HERTZSPRUNG-RUSSELL DIAGRAM and are characterized by large dimensions, high luminosity and low density. They are found throughout the entire surface temperature range. Those of spectral type G–M (the RED GIANTS) contrast with corresponding DWARF STARS of the main sequence.

Giauque, William Francis (1895–1982) US chemist. Giauque is noted for discovering the isotopes of oxygen and for his studies of phenomena at extremely low temperatures. In 1926 he suggested a method for producing temperatures approaching ABSOLUTE ZERO, until then believed impossible in practice. He received the 1949 Nobel Prize for chemistry.

gibberellin Any of a group of organic compounds (HORMONES) found in plants, which stimulate cell division, stem elongation, the breaking of DORMANCY (by triggering the production of enzymes essential for germination) and response to light and temperature. Gibberellins interact with some AUXINS to promote cell enlargement and have been used to increase crop yields substantially.

Gibbs, Josiah Willard (1839–1903) US theoretical physicist and chemist. While a professor at Yale, he devoted himself to establishing the basics of physical chemistry. His application of THERMODYNAMICS to physical processes led to STATISTICAL MECHANICS. He evolved the concepts of free energy and chemical equilibrium, devised the phase rule, which specifies the number of distinct PHASES OF MATTER that can co-exist in various circumstances, and developed VECTOR ANALYSIS.

gigabyte (symbol Gb) In computing, 1,024 MEGABYTES. Often used for expressing the capacity of a COMPUTER memory and disk drives, the gigabyte is usually rounded to 1,000 megabytes.

Gilbert, Walter (1932–) US molecular biologist who determined the sequence of NUCLEOTIDES in DNA. In 1966 Gilbert identified a repressor substance, a protein that regulates gene activity. He

GEOTHERMAL ENERGY

An artificial hot spring can be used as a source of heat energy. A bore-hole is drilled several hundred metres into a natural cavity in the Earth, in which the temperature may be as high as 300°C (570°F). Water pumped down the bore is heated, turns to steam and is forced up a second bore-hole. At the surface the steam drives turbines to produce electricity.

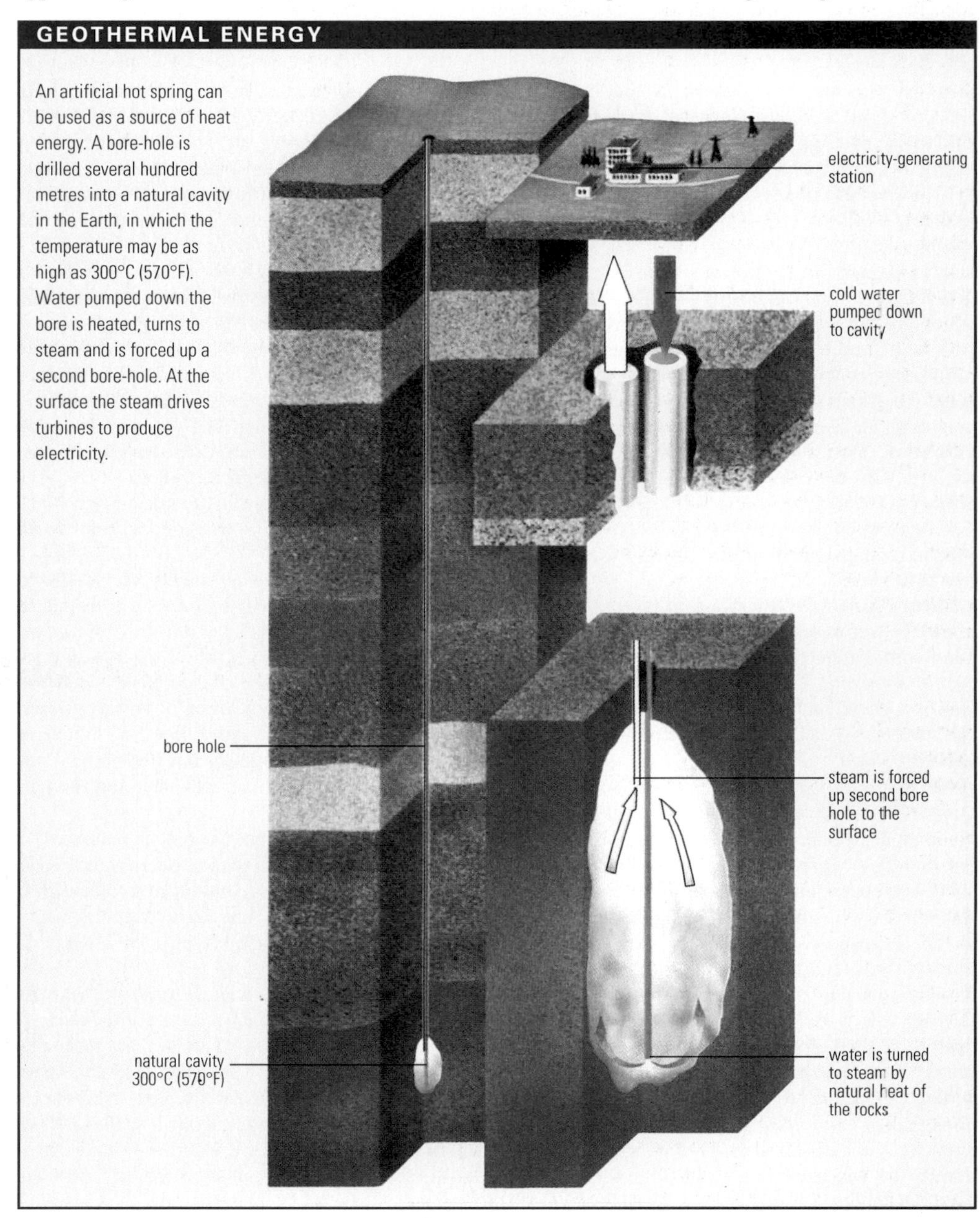

► **glacier** In spite of a return to warmer conditions, some regions of the world (namely those nearer the poles) are still covered by ice and are being greatly altered by its action. Glaciated regions have been subjected to erosion and deposition, the erosion mainly taking place in the highland areas, leaving features such as pyramidal peaks, corries, roches moutonnées, truncated spurs and hanging valleys. Most deposition has occurred on lowlands, where, after the retreat of the ice, moraines, drumlins, eskers, erratic boulders and alluvial fans remain.

Key
1 pyramidal peak
2 firn (granular snow)
3 corrie
4 tarn (corrie lake)
5 arête
6 marginal crevasse
7 lateral moraine
8 medial moraine
9 terminal moraine
10 sérac
11 subglacial moraine
12 glacial table
13 roche moutonnée
14 drumlin
15 esker
16 glacial lake
17 finger lake
18 U-shaped valley
19 erratics
20 truncated spur
21 hanging valley
22 outwash fan

then experimented with enzymes that split the DNA molecule at known places, and was gradually able to discover the sequence of nucleotides along the molecule. For devising this technique he shared the 1980 Nobel Prize for chemistry.

Gilbert, William (1544–1603) English physicist and physician to Queen Elizabeth I. His *De Magnete, Magneticisque Corporibus, et de Magno Magnete Tellure, Physiologia Nova* (1600) laid the foundation for the scientific study of MAGNETISM. He was the first to recognize terrestrial magnetism and mistakenly concluded that a type of magnetism keeps the planets in their orbits. He coined the terms magnetic pole, electric attraction and electric force.

Gilchrist, Percy Carlyle (1851–1935) British chemist who developed, with his cousin Sidney Gilchrist THOMAS, a process for smelting iron ore rich in phosphorous in a BESSEMER converter. The process was patented and extended to the open-hearth furnace.

gilding Process by which a thin layer of gold is used to cover another surface. Silver is the metal most commonly gilded, but base metals are also used extensively. The original method of gilding was by a chemical process, but in the 19th century ELECTROPLATING, a cheaper if less effective method, was introduced.

gills Organs through which most FISH, some larval amphibians, such as tadpoles, and many aquatic invertebrates obtain oxygen from water. When a fish breathes it opens its mouth, draws in water and shuts its mouth again. Water is forced over the gills, through the gill slits, and out into the surrounding water. Oxygen is absorbed into small capillary blood vessels, and at the same time, waste carbon dioxide carried by the blood diffuses into the water through the gills. The gills of young tadpoles, axolotls (a kind of salamander) and many invertebrates are on the outside of their bodies.

ginkgo (maidenhair tree) Member of the order Ginkgoales, the oldest living species of GYMNOSPERM. Gingkos flourished throughout the world during the Mesozoic era (245–74 million years ago). The only surviving species is *Ginkgo biloba*. It has fan-shaped leaves, small, foul-smelling fruits and edible, nut-like seeds. Height: to 30m (100ft).

girder In engineering, a main load-supporting beam. Steel or iron girders may be made of a single piece or of laminated strips built up of plates, latticework or bars. When the cross-section is formed in the shape of an I by riveting and welding plates or by rolling the girder, greater stiffness is achieved and larger spans are possible. Girders may also be made of reinforced or pre-stressed concrete.

glacial groove Deep, wide, usually straight furrow cut in bedrock. It is caused by the abrasive action of large rock fragments dragged along the base of a moving GLACIER. The grooves are larger and deeper than glacial striations, ranging in size from a deep scratch to a glacial valley.

glaciation Covering of a large region or a landscape by ice. The term is also used to refer to an ICE AGE.

glacier Large mass of ice, mainly recrystallized snow, which moves slowly by CREEP downslope or outward in all directions due to the stress of its own weight. The flow terminates where the rate of melting is equal to the advance of the glacier. There are three main types: the **mountain**, or **valley**, glacier, originating above the snow line in mountain regions; the PIEDMONT, which develops when valley glaciers spread out over lowland; and the ICE-SHEET and ICE-CAP.

glaciology Study of ice. No longer restricted to the study of glaciers, this science deals with ice and the action of ice in all its natural forms. Glaciology, therefore, draws upon the knowledge of many other related subjects, notably physics, chemistry, geology and meteorology.

gland Cell or tissue that manufactures and secretes special substances. The glands of animals are of two basic types. EXOCRINE GLANDS make such substances as hydrochloric acid, mucus, sweat, sebaceous fluids and ENZYMES, and secrete these usually through ducts to an external (such as the SKIN) or internal body surface. Endocrine glands (*see* ENDOCRINE SYSTEM) contain cells that secrete HORMONES directly into the bloodstream.

glandular fever Acute disease, usually of young people, caused by the EPSTEIN-BARR VIRUS. Symptoms include fever, painful enlargement of the LYMPH nodes and pronounced lassitude (which may persist for months or even years). There may also be a sore throat, skin rash and digestive disorder. It is also known as **infectious mononucleosis** due to the presence in the blood of an increased number of white blood cells called MONOCYTES.

Glaser, Donald Arthur (1926–) US physicist who in 1953 invented the BUBBLE CHAMBER and used it to devise a new method for studying ELEMENTARY PARTICLES; for this work he was awarded the 1960 Nobel Prize for physics. In 1959 Glaser joined the physics faculty at the University of California at Berkeley where he did research in applying physics to molecular biology.

Glashow, Sheldon Lee (1932–) US physicist and educator. He attended Cornell and Harvard, joining the Harvard faculty in 1966. He shared the 1979 Nobel Prize for physics with Steven WEINBERG and Abdus SALAM for their independent formulation of a unified "electroweak" theory that describes both electromagnetic interaction and the WEAK NUCLEAR FORCE of ELEMENTARY PARTICLES. He also proposed that QUARKS make up particles that interact strongly. *See also* FUNDAMENTAL FORCES

glass Brittle, transparent material. Although glass behaves in a similar way to a solid, it is not crystalline. It is really a liquid which has been cooled so quickly that the particles have not had time to organize themselves into a regular pattern (*see* SUPERCOOLING). It is made by melting together silica (sand), sodium carbonate (soda) and calcium carbonate (limestone). Glass melts slowly, can be worked only while hot and pliable, and must be cooled gradually to prevent strains or breakage. Glass appears to have been invented in the E Mediterranean, probably *c*.2500 BC in Egypt, for jewellery and small containers. There are many types of glass. Soda-lime glass is used in the manufacture of bottles and drinking vessels. Flint glass, which refracts light well, is used in lenses and prisms. Toughened glass – made by laminating with plastic – is used in car windscreens. Glass is also used in OPTICAL FIBRES, for the transmission of information such as telephone calls and in investigative medicine for the ENDOSCOPE.

glass fibre (fibre glass) Glass in the form of fine filaments. It is used widely for heat insulation (as glass wool), for fabrics, and with a plastic resin to make a construction and repair material called GRP (glass-reinforced plastic). GRP, a rigid composite material, is commonly referred to as glass fibre or fibre glass and is a popular medium for car bodies, boats, aircraft parts and containers. It resists heat, corrosion, rot and most chemicals and can be, weight for weight, stronger than steel. Short lengths of glass fibre are made by blasting air or steam into molten glass. Most forms of glass fibre are made by forcing molten glass through fine metal nozzles or spinnerets. The resulting continuous filaments are usually bundled together to form strands. These may then be chopped, twisted or woven according to the product required.

Glauber, Johann Rudolf (1604–68) German chemist and physician. He made valuable contributions to chemistry, mainly concerning the preparation of salts. He prepared hydrochloric acid, hydrated sodium sulphate (known as Glauber's salt) and tartar emetic.

glauconite Complex silicate mineral of iron, magnesium and aluminium, a member of the MICA family. It forms very small lath-shaped crystals (monoclinic), which occur as rounded grains in SEDIMENTARY ROCKS. It is bluish or dark green with a usually dull lustre; in sufficient quantity, it can colour the entire rock green, when it is known as greensand. Hardness 2.0–2.5; r.d. 2.7–2.9.

Glenn, John Herschel, Jr (1921–) First US astronaut to orbit the Earth. On 20 February 1962, Glenn made three orbits of the Earth in the spacecraft *Friendship 7*. Nearly five hours after launch from Cape Canaveral, Florida, the craft splashed down in the Pacific Ocean. Glenn later moved into politics, becoming Democrat Senator for Ohio.

glider Aircraft usually with no power source of its own. Gliders usually have long narrow wings to provide maximum lift with minimum drag. Gliders may be launched in a variety of ways. They may be pulled by a winch or an elastic shock cord, or towed by a plane, car or ground crew. Once airborne, gliders descend relative to the surrounding air. If this air is a rising "updraft" of warm air (thermal), a glider may gain altitude for a while, thus prolonging its flight.

Global Positioning System (GPS) Navigation system based on a network of Earth-orbiting SATELLITES. There are 24 satellites, each with a 24-hour orbit and its own identification signal. The satellites transmit continuous time signals. A user needs to be within range of four satellites to fix a position. One satellite sends out a reference signal and the other three give bearings indicating direction. A computer uses the minute time differences in receiving the signals from each satellite to calculate the position. Ordinary users can find their position to within 100m (330ft), although the military can pinpoint position to within 10m (33ft).

global warming Trend towards higher average temperatures on Earth's surface. During the last few million years, there have been several periods when surface temperatures have been significantly higher or lower than at present. During cold periods (ICE AGES) much of the land area has been covered by glaciers. The Earth is currently in the middle of a warm period (**inter-glacial**), which began about 10,000 years ago. Since the 1960s, some scientists have called attention to signs that the Earth is becoming unnaturally warmer as the result of an increased GREENHOUSE EFFECT caused by human activity.

Globigerina Genus of one-celled marine PROTOZOA whose empty shells are an important component of ocean floor ooze. The shell is spiralled into a lumpy sphere with needle-like extensions. One species is an indicator of sea temperature, and therefore of ocean palaeoclimatology. Shells coiling to the left indicate that the animals lived in cold water; shells coiling to the right were of animals that lived in warm water.

globular cluster Near-spherical cluster of very old stars in the halo of our GALAXY. Globular clusters contain anything from 100,000 to several million stars, concentrated so tightly near the centre that they cannot be fully separated by ground-based telescopes. Globulars are at least 10,000 million years old, as deduced from the highly evolved state of their stars. This extreme age indicates that globulars formed while our Galaxy was condensing from a huge cloud of gas. Globulars can also be seen around other galaxies. They are more plentiful around elliptical galaxies than spirals: our Galaxy has about 140 known globular clusters, whereas the largest elliptical galaxies may have thousands.

globulins Family of globular proteins that are soluble in dilute salt solutions, found in blood serum. Some globulins serve as ANTIBODIES; others are carriers of lipids, hormones and minerals.

glomerulus Mass of capillary blood vessels within a BOWMAN'S CAPSULE, the funnel-shaped end of a NEPHRON in the KIDNEY. Fluid passes from the blood in the capillaries into the capsule and then down the tubule of the nephron. It eventually passes out of the kidney along the ureter as urine as part of the process of EXCRETION.

glottis Tube less than 25mm (1in) long between the VOCAL CORDS at the lower end of the PHARYNX. It opens into the TRACHEA and its dimensions are changed by the vibration of the vocal cords.

glove-box Chamber with a viewing window into which are let, and firmly sealed, rubber gloves. It is used in laboratories for handling toxic and other harmful materials, including radioactive chemicals and metals, and cultures and toxins of pathogenic microbes.

glucagon Hormone secreted by cells in the ISLETS OF LANGERHANS in the PANCREAS. It helps to regulate blood sugar levels. *See also* INSULIN

glucose (dextrose, $C_6H_{12}O_6$) Colourless, crystalline SUGAR that occurs in fruit and honey. It requires no digestion before being absorbed into the bloodstream. The storage carbohydrate GLYCOGEN in the bodies of animals is converted to glucose before being utilized as an energy source in cellular RESPIRATION. A MONOSACCHARIDE sugar (having a single sugar unit), glucose is prepared commercially by the HYDROLYSIS of starch using hydrochloric acid. It is used in foods, especially confectionery, as a sweetener, and in tanning and pharmaceuticals.

glucoside Carbohydrate-containing compound that yields a GLUCOSE and a nonsugar component (either an alcohol or phenol) when decomposed by the process of HYDROLYSIS. The natural glucosides are important in plant metabolism, and many POLYSACCHARIDES, such as cellulose, starch and glycogen, are regarded chemically as glucosides. They may be used as drugs, colouring agents and aromatics.

glue Name for any natural ADHESIVE.

gluon Particle that can be thought of as the "glue" that holds QUARKS together. Quarks are the fundamental particles that combine to form ELEMENTARY PARTICLES such as PROTONS and NEUTRONS. The gluon is the carrier of STRONG NUCLEAR FORCES. It is analogous to the PHOTON, which is the carrier of ELECTROMAGNETIC INTERACTIONS. *See also* FUNDAMENTAL FORCES

glutamic acid ($COOH(CH_2)_2CHNH_2COOH$) Colourless AMINO ACID commonly found in proteins. One of its functions is to increase the solubility of its associated proteins by providing them with one negative charge. It also helps in the removal of poisonous ammonia from the body. Its sodium salt (sodium glutamate) is used as a food flavouring. *See also* GLUTAMINE

glutamine ($COOHCH(NH_2)(CH_2)_2CONH_2$) Colourless soluble AMINO ACID. It is formed when one of the acidic carboxyl groups of GLUTAMIC ACID couples with ammonia, a poison to the body, which is carried safely to the liver where it is converted to urea, and freely expelled as urine.

gluten Main protein substance present in wheat flour, grey in colour and elastic in texture. Not present in barley, oats or maize, gluten contributes the elasticity to dough. Being insoluble in water, it can be washed out of flour; it is then used as an additive to chocolate and coffee.

glycerol (glycerine) Syrupy, sweet liquid (1,2,3–trihydroxypropane, $HOCH_2CH(OH)CH_2OH$) obtained from animal and vegetable fats and oils, as when making soap, or from propene (propylene). It is used in the manufacture of various products including plastics, explosives, cosmetics, foods, antifreeze and paper coatings.

glyceryl trinitrate In medicine, a VASODILATOR mainly used in treating the symptoms of angina pectoris. It reduces the amount of blood returning to the heart, thereby reducing the amount of work the heart has to do. It is also used as an explosive, when it is known as NITROGLYCERINE.

glycine (NH_2CH_2COOH) Colourless, soluble crystalline AMINO ACID; the principal amino acid in sugar cane and the simplest, structurally, of the alpha-amino acids.

glycogen CARBOHYDRATE stored in the animal body, principally by the liver and muscles. Glycogen is sometimes known as animal starch, and, like starch and cellulose, it is a POLYMER of GLUCOSE. When the body needs energy, glycogen is broken down to glucose, which is further metabolized to carbon dioxide and water, providing ADENOSINE TRIPHOSPHATE (ATP), a source of chemical energy, in the process. *See also* METABOLISM; RESPIRATION

GLOBAL POSITIONING SYSTEM (GPS)

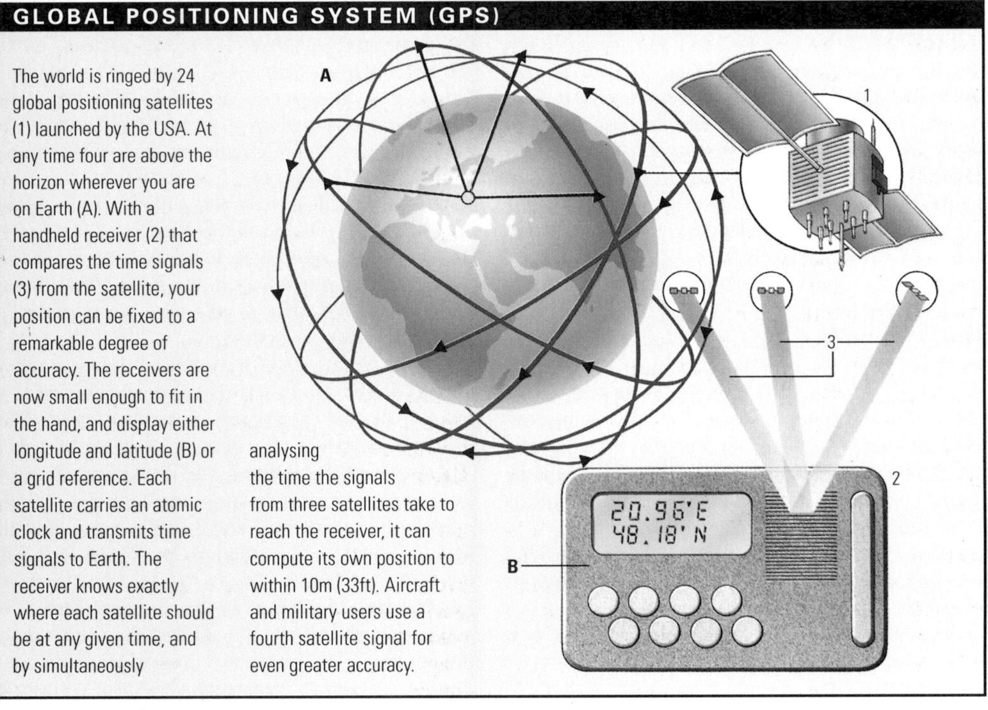

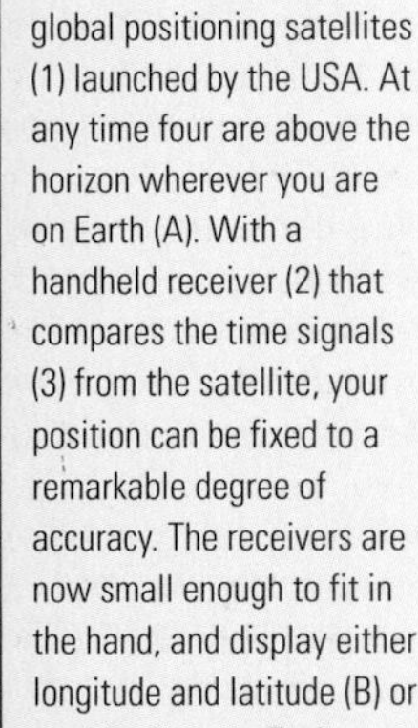

The world is ringed by 24 global positioning satellites (1) launched by the USA. At any time four are above the horizon wherever you are on Earth (A). With a handheld receiver (2) that compares the time signals (3) from the satellite, your position can be fixed to a remarkable degree of accuracy. The receivers are now small enough to fit in the hand, and display either longitude and latitude (B) or a grid reference. Each satellite carries an atomic clock and transmits time signals to Earth. The receiver knows exactly where each satellite should be at any given time, and by simultaneously analysing the time the signals from three satellites take to reach the receiver, it can compute its own position to within 10m (33ft). Aircraft and military users use a fourth satellite signal for even greater accuracy.

► **gold** This photograph shows nuggets of gold. Gold has been known since prehistoric times, as it appears in nature as a free metal, either in veins in quartz or as nuggets in alluvial gravels. Its main uses are as the basis of monetary systems and as jewellery, due to its relative rarity, attractive appearance and lack of tarnishing. Gold is also used in the electronics industry as a conductor. It has no known biological role, although compounds of gold are used in the treatment of rheumatoid arthritis.

G

glycol (ethylene glycol, ethane-1,2-diol, $(CH_2OH)_2$) Colourless, odourless liquid containing two HYDROXYL groups. It is a viscous liquid used in plastics, solvents, antifreeze and artificial fibres.

glycolysis Series of biochemical reactions in which glucose is converted to PYRUVATE. It takes place during RESPIRATION in CELLS. The nine stages of glycolysis are accompanied by the net release of two molecules of the energy-containing substance ADENOSINE TRIPHOSPHATE (ATP) per glucose molecule. During AEROBIC RESPIRATION, the pyruvate enters the KREBS CYCLE, with the ultimate yield of 12 more molecules of ATP. During ANAEROBIC RESPIRATION, the pyruvate is converted to lactic acid. *See also* ELECTRON TRANSPORT SYSTEM; PHOSPHORYLATION

GMT Abbreviation of GREENWICH MEAN TIME

gneiss METAMORPHIC ROCK with a distinctive layering or banding formed by high pressure or high temperature conditions of regional metamorphism. The darker minerals are likely to be hornblende, augite, mica or dark feldspar.

Goddard, Robert Hutchings (1882–1945) US physicist and pioneer in ROCKET development. He developed and launched (1926) the first liquid-fuelled rocket, and developed the first smokeless powder rocket and the first automatic steering for rockets.

Gödel, Kurt (1906–78) US mathematician, b. Czechoslovakia. Gödel is best known for his "undecidability" or "incompleteness" theorem, the proof of which first appeared in a German technical journal in 1931. The theorem states that an axiom based mathematical system can contain statements that can neither be proved nor disproved using results from within the system. Put simply, Gödel's theorem means that the truth of some statements in mathematics can never be decided. His work ended nearly a century of attempts to establish axioms that would provide a rigorous basis for all mathematics. In 1940 he emigrated to the USA; he was a professor, and a close friend of EINSTEIN, at the Institute for Advanced Study, Princeton, from 1953 to his death.

goethite Hydroxide mineral, iron oxyhydroxide, (FeO[OH]). It is found in secondary oxidized deposits with orthorhombic system slender plates and velvety needles. It also occurs in a massive fibrous form with uniform surfaces. It is black, brilliant and earthy. Hardness 5–5.5; r.d. 3.3–4.3.

gold (symbol Au) Naturally occurring TRANSITION ELEMENT. It is also obtained as a by-product in the electrolytic refining of copper. Gold is used in jewellery, in connectors for electronic equipment, and as a form of money for international transactions. Gold leaf can be made as thin as 0.00001mm (0.0000004in). Gold in the form of a COLLOID is sometimes used in colouring glass. It is often alloyed with copper or silver to make it harder. The gold content of an alloy is expressed as fine (parts per 1,000) or in carats (parts per 24). Gold forms two series of salts termed gold(I), or aurous, and gold(III), or auric. The isotope ^{198}Au (half-life 2.7 days) is used in RADIOTHERAPY. The metal is unreactive, being unaffected by oxygen and common acids. It dissolves in aqua regia, a mixture of nitric and hydrochloric acids. Properties: at.no. 79; r.a.m. 196.9665; r.d. 19.30; m.p. 1,063°C (1,945°F); b.p. 2,800°C (5,072°F); most common isotope ^{197}Au (100%).

golden mean (golden section or golden ratio) Classical ratio created when a line is divided into two parts in such a way that the ratio of the shorter to the longer is as the longer to the whole, that is. $a/b = b/(a + b)$, where $a + b$ is the line's length. The ratio is approximately 3:5, or exactly $1:(\sqrt{5} + 1)/2$ (*see* FIBONACCI SEQUENCE). The golden mean was cited by the Roman historian Vitruvius as the basis of proportion in Classical Greek architecture. The ratio, considered pleasing to the eye, is observed in many Renaissance paintings and buildings.

Goldmark, Peter Carl (1906–77) US scientist, b. Hungary. In 1940 he demonstrated a colour television system that he had devised, and he developed the first system to find commercial acceptance. In the late 1940s he invented the $33^1/_3$ r.p.m. long-playing gramophone record (the LP). Goldmark later developed an electronic VIDEO recorder used as an educational aid, and a system that enabled photographs to be transmitted from space to Earth.

Golgi body Collection of microscopic vesicles or packets observed near the nucleus of many living CELLS. It is a part of a cell's inner membrane structure, or ENDOPLASMIC RETICULUM, specialized for the purpose of packaging and dispatching proteins made by the cell.

Golgi, Camillo (1843–1926) Italian histologist (specializes in the structure of CELLS). In 1873 he developed a method of staining tissue with silver nitrate for microscopic study. With this he discovered the GOLGI BODY within the CELL. Golgi shared the 1906 Nobel Prize for physiology or medicine with Santiago Ramón y Cajal (1852–1934) for his work on the structure of the NERVOUS SYSTEM.

gonad Primary reproductive organ of male and female animals, in which develop the GAMETES or sex cells. Thus, the gonad in the male is a TESTIS and in the female an OVARY. Hermaphrodite animals possess both types.

gonadotrophins General term for two HORMONES secreted by the PITUITARY GLAND, which stimulate the development and function of the GONADS (ovary and testis) and similar hormones made by the PLACENTA. Some gonadotrophins are used to treat INFERTILITY.

Gondwanaland Name given to the S supercontinent which began to break away from the single land mass PANGAEA about 200 million years ago. It later became South America, Africa, India, Australia and Antarctica. The name comes from a historical region in central India. The N supercontinent, which eventually became North America and Eurasia without India, was LAURASIA. The occurrence of tillites (glacial deposits) in separate parts of India, South America and Africa provides evidence that these were once a single continent.

goniatite Any CEPHALOPOD mollusc belonging to the order Goniatitida. Goniatites have coiled shells divided into chambers by zigzag partitions. They first appeared in the upper Devonian period, about 350 million years ago. The group became extinct at the end of the Permian period. They are useful as zone fossils. Goniatites and AMMONITES belonged to the same subclass, Ammonoidea.

goniometer Instrument used mainly by mineral collectors to help in the identification of crystal forms by measuring the angles of related sets of crystal faces. These angles are characteristic for certain minerals.

Gorgas, William Crawford (1854–1920) US surgeon whose successful mosquito-control programme in Panama eradicated MALARIA and YELLOW FEVER there and made possible the building of the Panama Canal.

gorge Narrow, deep-sided valley. Gorges form only in hard rock, such as carboniferous limestone, otherwise erosion would soon break down the steep sides. A very large gorge is known as a CANYON.

Gould, Stephen Jay (1941–) US palaeontologist who proposed that EVOLUTION could occur in sudden spurts rather than gradually. Called PUNCTUATED EQUILIBRIUM, Gould's theory suggested that sudden accelerations in the evolutionary process could produce rapid changes in species over the comparatively short time of a few hundred thousand years. Gould has also written many popular science books, including *Wonderful Life* (1990) and *Bully for Brontosaurus* (1991), which often use examples taken from the modern world to explain his theories.

Graaf, Regnier de (1641–73) Dutch anatomist and doctor after whom the GRAAFIAN FOLLICLES were named. Graaf investigated the functions of the PANCREAS and the nature of the sexual organs. He was the first user of the term "OVARY".

Graafian follicle (ovarian follicle) Fluid-filled, cyst-like cavity found in the OVARIES of MAMMALS. It surrounds and protects the developing OVUM. Once the ovum is released into a FALLOPIAN TUBE, the follicle develops a yellowish mass of tissue called the CORPUS LUTEUM.

graben Elongated, trenchlike, down-dropped block of the Earth's crust bordered by two or more similarly trending normal FAULTS. The Basin and Range province in Utah and Nevada, USA, consists of grabens and HORSTS which form sedimentary basins and mountain ranges. A long graben found at the Earth's surface can also be called a RIFT VALLEY, such as the Great Rift Valley in East Africa

graft, tissue Organ or tissue transplanted to replace a part of the body which is damaged or nonfunctioning. The replacement tissue may be taken from elsewhere on the patient's body (an **autograft)**

or from a donor (**allograft**). Except in the case of an identical twin, the use of donor organs may set up rejection, the phenomenon where the IMMUNE SYSTEM attacks "foreign" tissues.

grafting In horticulture, method of plant PROPAGATION. A twig of one variety, called the **scion**, is established on the roots of a related variety, called the **stock**. Most fruit trees are propagated by a similar process called **budding**, in which the scion is a single bud. Grafting is a way of producing new plants that are genetically identical to the parent from which the scions are cut.

Graham, Thomas (1805–69) British chemist who is best remembered for the rule he formulated (GRAHAM'S LAW). He also discovered DIALYSIS.

Graham's law The DIFFUSION rate of a gas is inversely proportional to the square root of its density. This principle, formulated by Thomas GRAHAM, is used in separating isotopes by the diffusion method; it has important industrial applications.

gram (symbol g) SI UNIT of mass defined as one thousandth of a kilogram. *See also* PHYSICAL UNITS

gram-atom Quantity of an element the mass of which, in grams, is equal to its RELATIVE ATOMIC MASS (R.A.M.). It has been replaced by the SI unit, the MOLE. For example, one gram-atom of hydrogen (H, r.a.m. = 1) is 1g.

Gramme, Zénobe Théophile (1826–1901) Belgian engineer who developed the first practical industrial DYNAMO in 1869. His knowledge of electrical theory was limited, but even so he developed an ELECTRIC MOTOR from his dynamo. He was one of a group of scientists who, in 1872, succeeded in transmitting direct current electricity over long distances.

gram molecular volume *See* MOLAR VOLUME

Gram's stain In microbiology, a differential staining method named after the Danish physician Hans Christian Gram (1853–1938). It aids in the categorizing and identification of BACTERIA, which are said to be Gram-positive or Gram-negative, depending on whether or not they retain the original violet stain at the end of the process. If Gram-negative, the stain washes out to a red counterstain.

grand unified theory (GUT) Theory that would demonstrate that three of the four FUNDAMENTAL FORCES – the WEAK NUCLEAR FORCE, STRONG NUCLEAR FORCE and ELECTROMAGNETIC FORCE – are really slightly different aspects of the same fundamental force. The weak nuclear force and electromagnetic force have been incorporated as the **electroweak force** (*see* ELECTROWEAK THEORY), as demonstrated by particle ACCELERATOR experiments. In order to achieve a GUT, the electroweak force must be unified with the strong nuclear force. If GRAVITATION could be incorporated, then a UNIFIED FIELD THEORY would be produced. *See also* SUPERSTRING THEORY; SUPERSYMMETRY

granite Coarse-grained, acid, IGNEOUS ROCK from deep within the Earth, composed chiefly of feldspar and quartz, with some mica or hornblende. Its colour is usually light grey, although feldspar may redden it. Its durability makes it a valuable construction material. Granite is thought to have solidified from magma (molten rock), but the occurrence of some granite with features normally associated with rocks of metamorphic origin suggests that not all granites are igneous. It crystallizes at great depths where the pressure is high; it becomes exposed at the Earth's surface only by erosion of surface rocks or by movements in the Earth's crust.

granulite (leptites) Granular METAMORPHIC ROCK that derives largely from quartz and feldspar. Granulites often have a banded appearance.

granuloma Growth or nodule of connective tissue and capillaries usually associated with an infection such as tuberculosis or syphilis.

graph In mathematics, a diagram representing a relationship between numbers or quantities. Many graphs use the CARTESIAN COORDINATE SYSTEM. Usually axes are drawn at right-angles and marked with scales representing the variable quantities. On a temperature conversion graph, for example, the vertical axis might represent degrees Fahrenheit, with degrees Celsius along the horizontal axis. The graph itself would be a line drawn on the diagram. To convert from one reading to the other, you would have to trace from the appropriate scale to the graph, and from there to the other scale. Other forms of graph include bar charts, in which a series of figures is represented by lines of various lengths. In pie charts, the quantities are represented by sectors of a circle, which resemble the slices of a round pie.

graphical user interface (GUI) COMPUTER PROGRAM that enables a user to operate a COMPUTER using simple symbols. Early personal computers used operating systems (the programs that organize how the computer stores data and runs all other programs) that were text based. Commands were often obscure combinations of letters and numbers that made using the systems difficult for the uninitiated. A GUI replaces these commands with a screen containing symbols called icons. The user manipulates these using an on-screen pointer controlled by a MOUSE. Other important features of GUIs are menus, from which commands are selected, and windows displaying program events. The GUI makes it possible for files to be transferred and programs to be opened with a simple movement of the hand or finger. The first GUI was designed by Xerox in the 1970s. Microsoft's Windows operating system is now the most commonly used GUI, although it was predated by the Macintosh operating system used on Apple computers.

graphics pad (graphics tablet or bit pad) Computer input device consisting of a sensitive pad and a pen linked to the computer. When the pen is moved over the pad, digital signals indicating the pen's position pass to the computer. The device is generally used to input drawings and diagrams. *See also* COMPUTER GRAPHICS

graphite (plumbago) Dark grey, soft crystalline form of CARBON that occurs naturally in deposits of varying purity and which is made synthetically by heating petroleum coke. It is used in pencils (a mixture of graphite and clay is the "lead"), lubricants, electrodes, brushes of electrical machines, rocket nozzles and as a MODERATOR to slow down neutrons in NUCLEAR REACTORS. Graphite is a good CONDUCTOR of heat and electricity. It owes its lubricating properties to overlapping scale-like crystals, which tend to slide, giving it a smooth slippery feel. Properties: r.d. 2–2.25.

graptolite Any of an extinct group of colonial, marine, drifting organisms; sometimes considered

GOLDEN MEAN

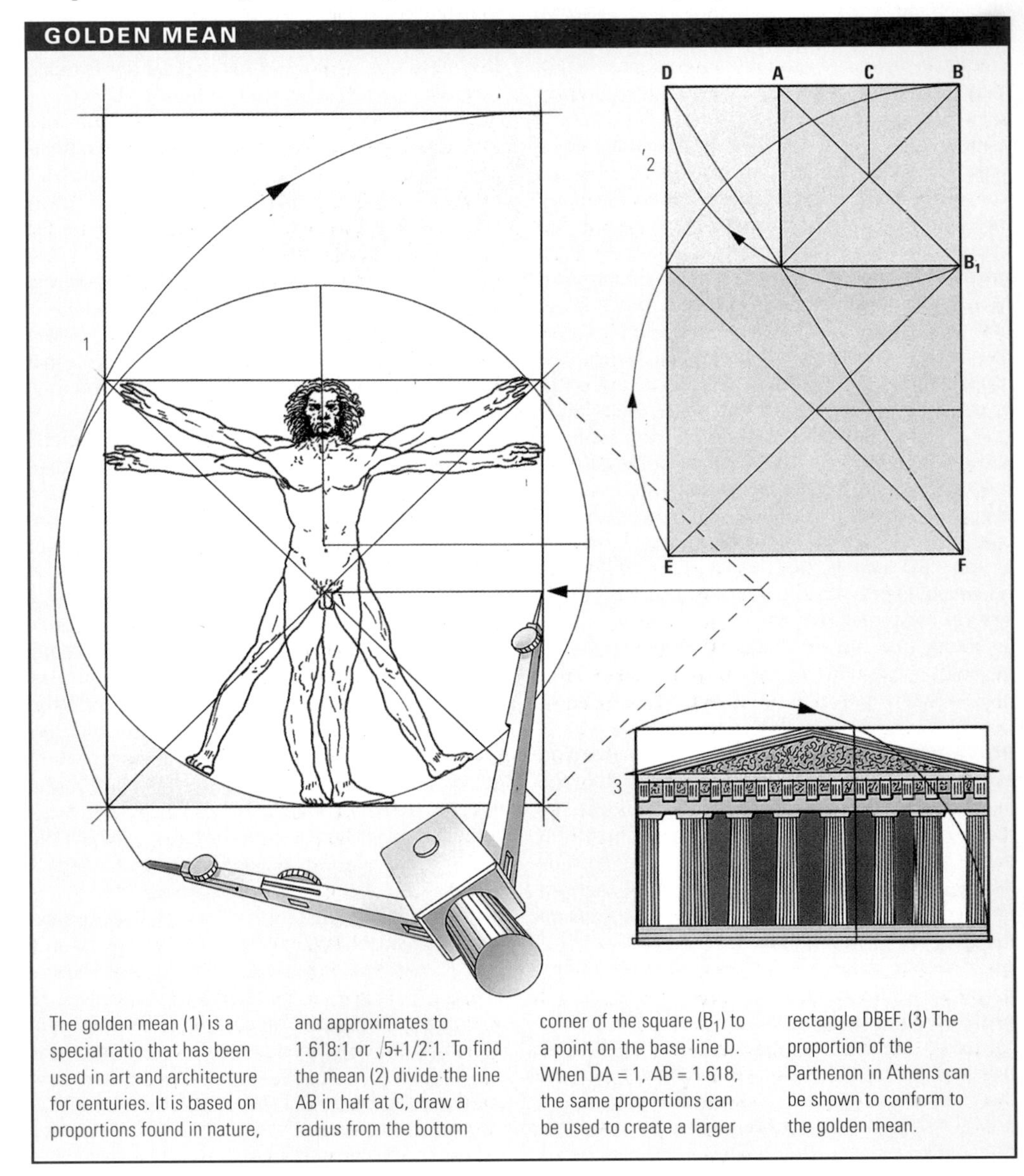

The golden mean (1) is a special ratio that has been used in art and architecture for centuries. It is based on proportions found in nature, and approximates to 1.618:1 or $\sqrt{5}$+1/2:1. To find the mean (2) divide the line AB in half at C, draw a radius from the bottom corner of the square (B_1) to a point on the base line D. When DA = 1, AB = 1.618, the same proportions can be used to create a larger rectangle DBEF. (3) The proportion of the Parthenon in Athens can be shown to conform to the golden mean.

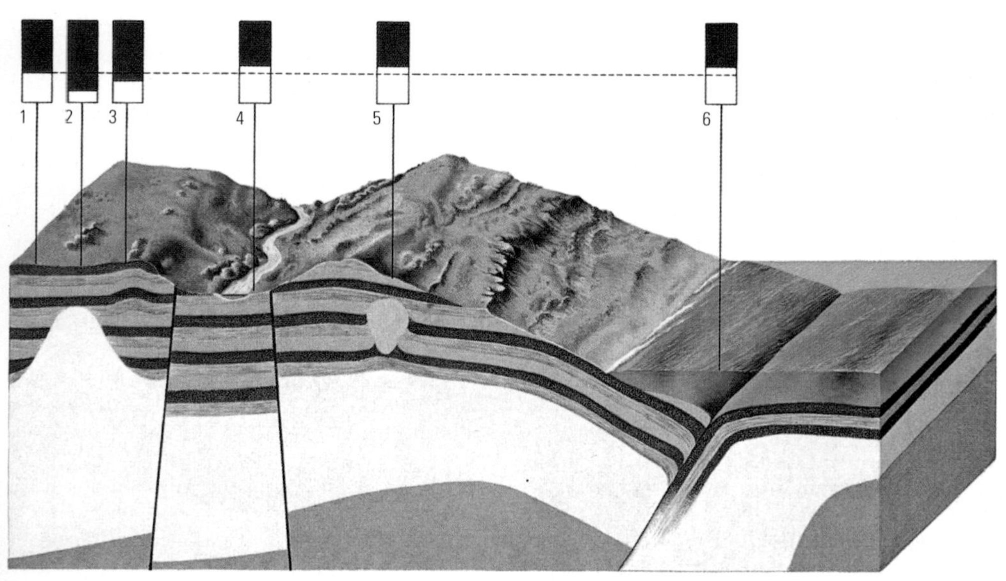

▲ **gravimetric analysis** Underground rock formations can be detected by measuring the local variations in the pull of gravity on a delicate spring balance called a gravimeter. The lighter the material the weaker the pull.

Key
1) Normal gravity reading
2) Heavy igneous material near surface gives high reading
3) Anticline gives gravity high
4) Rift valley where lighter surface material continues to a greater depth gives gravity low
5) Salt dome or upward emplacement of light material giving gravity low
6) Oceanic trough where lighter crustal material deep in mantle gives gravity low

a separate phylum but also thought to be related to the CHORDATES. They are found most frequently as flattened fibres of carbon, resembling pencil marks, in black shales of the ORDOVICIAN and SILURIAN ages. Their uncompressed skeletons etched out of limestone show that graptolites were composed of many small tubes regularly arranged along branches, which are presumed to have been attached to a common bladder-like float. Graptolites first appeared in the Middle Cambrian and the last are known from the Lower Carboniferous. They are important zone fossils, used for correlating and dating rocks of the Lower Palaeozoic.

grass Any one of c. 8,000 species of the class Monocotyledonae (*see* MONOCOTYLEDON).), being non-woody plants with fibrous roots and long, narrow leaves enclosing hollow, jointed stems; The stems may be upright or bent, lie on the ground, or grow underground. The flowers are small, without PETALS and SEPALS. The leaves grow from the base, and so removal of the tips does not inhibit growth, making grass suitable for lawns and pastures. Cereal grasses, such as rice, millet, maize and wheat, are cultivated for their edible seeds. Others are grown as food for animals and for erosion control and ornament. Family Poaceae/Gramineae.

gravel Mixed pebbles and rock fragments, 2mm to 60mm (0.1–3in) in diameter. Gravel beds are generally the remains of ancient seashores or river beds. Usually gravel is used as an aggregate in CONCRETE.

gravimeter (gravity meter) Instrument used in geophysical surveys to measure the strength of the local GRAVITATIONAL FIELD. Essentially, it is a heavy mass suspended from a spring. In the vicinity of a dense underground feature such as an ignesus deposit, the weight of the gravimeter mass increases slightly by gravitational attraction, so causing the spring to extend slightly.

gravimetric analysis In geology, study of the magnitude of the Earth's GRAVITATIONAL FIELD in a particular area. Small variations in this field can be caused by the density of the rocks beneath the surface and can provide information about structures that are otherwise inaccessible. The instrument used for such a survey is known as a GRAVIMETER. In chemistry, gravimetric analysis is a method of determining the composition of a substance by making weighings.

gravitation One of the four FUNDAMENTAL FORCES in nature. Gravitation is extremely weak compared with the others. The gravitational force F between two masses m_1 and m_2 a distance d apart was found by Isaac NEWTON to be $F = Gm_1m_2/d^2$, where G is a constant of proportionality called the universal GRAVITATIONAL CONSTANT. A more complete treatment of gravitation was developed by Albert EINSTEIN, who showed in his general theory of RELATIVITY that gravitation is a manifestation of the curvature of SPACE-TIME.

gravitational constant (symbol G) Constant in Newton's law of GRAVITATION. Its value is 6.67259×10^{-11} Nm^2 kg^{-2}.

gravitational field Region around an object that has mass in which there is an attractive force on any other object. The force divided by the mass of the second object is the gravitational field strength. A massive object, such as the Earth, has a powerful gravitational field, which we call the force of gravity. Weak gravitational forces exist between even very small particles.

gravitational red shift *See* RED SHIFT

gravitational waves Gravitational equivalent of electromagnetic waves, postulated by Einstein's general theory of RELATIVITY as being emitted from a massive accelerating body, such as an exploding or collapsing star, and travelling at the speed of light. Experimental results put forward as evidence for such waves have not yet been generally accepted.

graviton Hypothetical ELEMENTARY PARTICLE thought to be continuously exchanged between bodies of mass as the carrier of the gravitational force. The graviton is analogous to the PHOTON – the quantum of electromagnetic waves.

gravity Gravitational force of attraction at the surface of a planet or other celestial body. The Earth's gravity produces an acceleration of $c.9.8$ ms^{-2} ($c.32$fts^{-2}) for any unsupported body. If the mass M and radius R of a planet are known, the acceleration due to its gravity (g) at its surface can be determined from $g = GM/R^2$, where G is the universal constant of gravitation. The WEIGHT of a body is a measure of the force with which the Earth's gravity attracts it. Unlike MASS, which remains constant at normal speeds, weight (or the force of gravity) varies with altitude.

gravity anomaly Deviation in gravity from the expected value. Gravity measurements over deep ocean trenches are lower than average, those in mountainous regions are higher than average. Higher values are also found over deposits of dense minerals.

gravity-assisted flight Use of a close approach to a planet to obtain extra velocity from its gravitational pull. A rocket "grazing" a planet is accelerated and whipped off in a new direction with greatly increased velocity, perhaps unattainable by conventional means. All probes to the giant outer planets use this "whiplash" technique, often flying by inner planets, including the Earth, several times in order to do so.

Gray, Asa (1810–88) US botanist. He made many contributions to PLANT CLASSIFICATION, and his donation of a valuable collection of books and plants to Harvard University in 1865 led to the establishment of that school's department of botany. His *Manual of Botany for the Northern United States* (1848) is considered a classic.

Gray, Elisha (1835–1901) US inventor. He received his first patent for a self-adjusting telegraph relay in 1867. He also invented and patented the telegraphic switch, telegraphic repeater, type-printing telegraph and telautograph. He claimed priority as the inventor of the speaking telephone, but Alexander Graham BELL's patent rights were upheld by the US Supreme Court.

gray (symbol Gy) SI unit of absorbed radiation dose. One gray is equivalent to supplying 1 joule of energy per kilogram of irradiated material. It superseded the rad (1 gray = 100 rad). It was named after the British biologist L.H. Gray.

grease Semi-solid lubricant consisting of a mixture of a liquid lubricant with either a saponified FAT or a dispersion of metallic soaps. There are five types: water soluble; water-resistant; synthetic; special-purpose; and multi-purpose. Although liquid lubricants are usually preferred, greases are used in situations where a piece of machinery can-

▲ **Great Red Spot** This view of Jupiter was obtained from Voyager 1 on 25 February 1979 at a distance of 5.7 million mi. The feature at the top is the Great Red Spot, believed to be a giant storm system several times the size of the Earth, which has survived for at least the past 200 years. The colourful wavy cloud patterns below and to the left of the Spot are believed to be regions of complex atmospheric wave motion. This image was assembled from three black and white images in the Image Processing Laboratory at the Jet Propulsion Laboratory, Pasadena, California, USA.

not be relubricated or where there is a need to avoid leakage, which may occur if oil were used.

Great Bear *See* URSA MAJOR

great circle Circle on a sphere the centre of which is coincident with the centre of the sphere. On the Earth, the Equator is a great circle, as are all meridians. On the CELESTIAL SPHERE, the CELESTIAL EQUATOR is a great circle, as are the meridians (circles passing though both celestial poles).

Great Red Spot (GRS) Large oval area in the S hemisphere of JUPITER. Since it was first observed by Robert HOOKE in 1664, it has varied in size and colour, from about 11,000 to 14,000km wide by 24,000 to 40,000km long (7,000–9,000mi by 15,000–25,000mi), and from a deep red to a light pink, occasionally fading away altogether. It is an atmospheric phenomenon, a huge ANTICYCLONE, projecting above the surrounding cloud-tops. The red colour may be produced by phosphorus.

green algae *See* CHLOROPHYTE

green belt Area of open land maintained as a barrier between adjoining built-up areas. The concept of green belts was first put forward by Ebenezer Howard (1850–1928) in the UK in his plans for Garden Cities. Howard used them to distinguish residential from industrial sections. Green belts provide insulation from factories and intensive commercial areas. They are a source of recreational space and aid in the replenishment of atmospheric oxygen.

green flash Greenish or bluish hue of the upper rim of the Sun seen just as it is about to disappear at setting, or appear at rising, due to refraction, scattering and absorption of light in the atmosphere.

greenhouse effect Raised temperature at a planet's surface as a result of heat energy being trapped by gases in the atmosphere. Certain gases cause the atmosphere to act like the glass in a greenhouse. As a result, the temperature of a planet's surface may be higher than it otherwise would be – on Earth about 33°C (59°F) higher. The main gases that produce the greenhouse effect on Earth are water vapour and CARBON DIOXIDE. Scientists suspect that increased discharge of carbon dioxide from human activity (notably motor transport and industry) is contributing to GLOBAL WARMING.

Greenwich Mean Time (GMT) Local time at Greenwich, London, situated on the prime MERIDI-

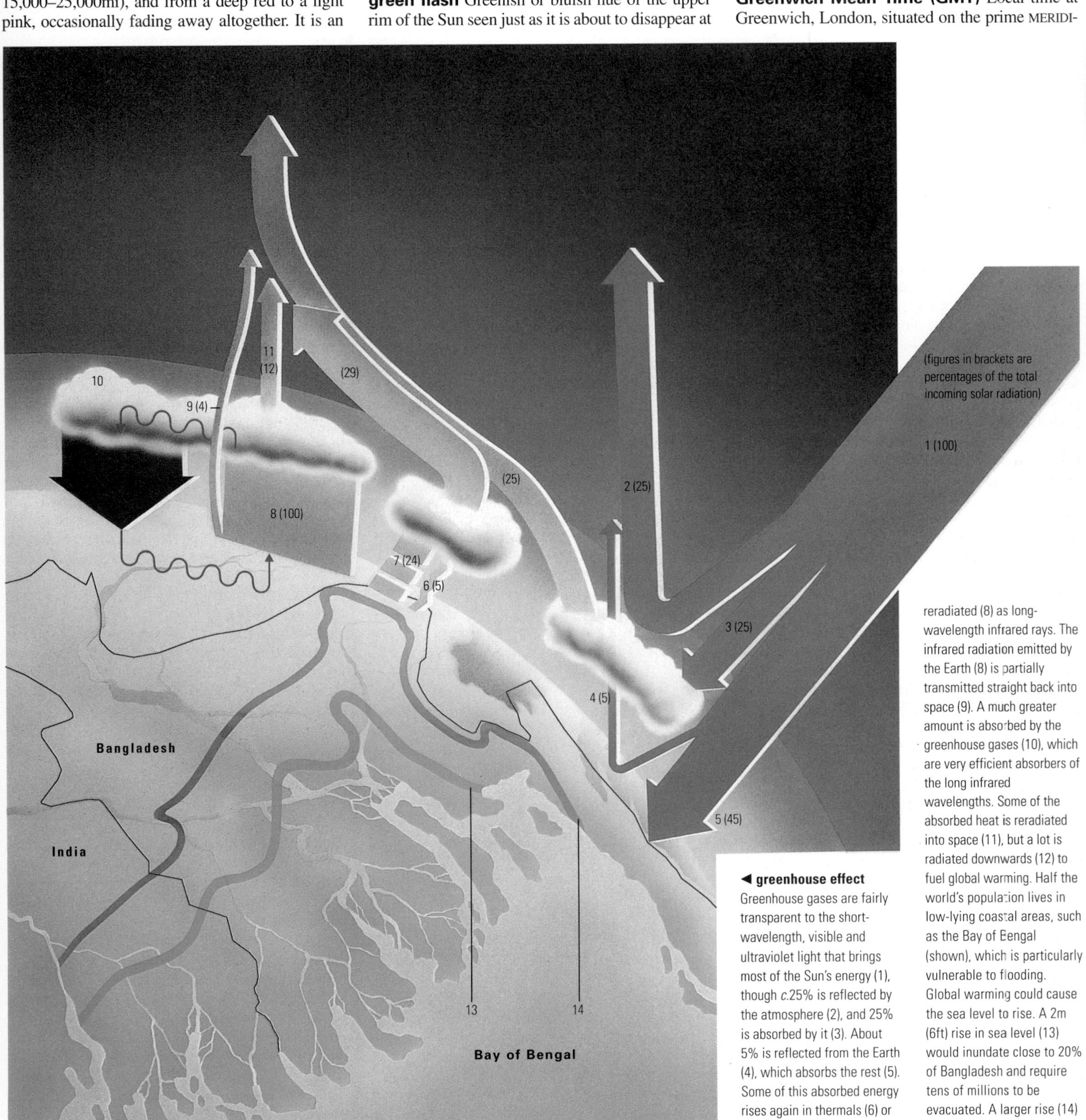

◄ **greenhouse effect** Greenhouse gases are fairly transparent to the short-wavelength, visible and ultraviolet light that brings most of the Sun's energy (1), though *c.*25% is reflected by the atmosphere (2), and 25% is absorbed by it (3). About 5% is reflected from the Earth (4), which absorbs the rest (5). Some of this absorbed energy rises again in thermals (6) or in the heat of evaporated moisture (7). The rest is reradiated (8) as long-wavelength infrared rays. The infrared radiation emitted by the Earth (8) is partially transmitted straight back into space (9). A much greater amount is absorbed by the greenhouse gases (10), which are very efficient absorbers of the long infrared wavelengths. Some of the absorbed heat is reradiated into space (11), but a lot is radiated downwards (12) to fuel global warming. Half the world's population lives in low-lying coastal areas, such as the Bay of Bengal (shown), which is particularly vulnerable to flooding. Global warming could cause the sea level to rise. A 2m (6ft) rise in sea level (13) would inundate close to 20% of Bangladesh and require tens of millions to be evacuated. A larger rise (14) of 5m (16ft) would drown close to half the country.

AN. It has been used as the basis for calculating standard time in various parts of the world since 1884. GMT corresponds with civil time in Britain during the winter months; British Summer Time (BST) is one hour ahead of GMT.

Gregg, Sir Norman McAlister (1892–1966) Australian ophthalmic surgeon. In 1941 he discovered that RUBELLA (German measles) in a pregnant woman can cause physical defects in her child. Gregg was knighted in 1953.

grey matter Darker-coloured nerve tissue which forms the CEREBRAL CORTEX of the BRAIN and is also present in the SPINAL CORD. It is distinct from the so-called white matter, which contains more nerve fibres (NEURONS) and larger quantities of the whitish insulating material called MYELIN.

greywacke Any of a variety of SANDSTONES that consist of a mixture of rock fragments, feldspar and quartz, strongly bonded together in a mud matrix. They are characterized by poor sorting of angular or sub-angular particles. They are usually interpreted as indicating short-distance, rapid-sediment depostition.

grid (control grid) Perforated electrode placed between the cathode and the anode of an ELECTRON TUBE (valve) for controlling the flow of electrons in the valve. This effect is the basis of electronic amplification. "Grid" also refers to the supply network for electricity or gas, as in the national grid.

Griess, Johann Peter (1829–88) German chemist who discovered the diazo reaction that enabled azo dyes to be prepared from a wide range of intermediate chemical compounds. *See also* AZO AND DIAZO COMPOUNDS

Grignard, François Auguste Victor (1871–1935) French chemist who shared the 1912 Nobel Prize for chemistry with Paul Sabatier (1854–1941) for his discovery and investigation of GRIGNARD REAGENTS. He also synthesized organic compounds of aluminium and mercury.

Grignard reagent Organic chemical compound containing the metal magnesium. The Grignard reagents are named after François GRIGNARD who discovered them in 1900. Examples are methylmagnesium iodide (CH_3MgI) and phenylmagnesium bromide (C_6H_5MgBr). Grignard reagents undergo many reactions to form carbon-carbon bonds and for this reason they are invaluable in organic synthesis.

grike (gryke) Enlarged vertical crack in an area of bare limestone formed by solution along JOINTS. The resulting blocks of limestone are called CLINTS and the two features together form a limestone pavement. In the UK, examples can be seen near Malham, N Yorkshire.

grit In geology, a coarse sandstone. It is usually formed in a river delta where material eroded from a landmass has been deposited rapidly before the fragments have been rounded or sorted. The Upper Carboniferous deltas left thick beds of grit in many areas of the British Isles.

gritstone Medium- to coarse-grained SEDIMENTARY ROCK with angular grains that may rub off easily. Most gritstones are formed in water. The chief mineral components are QUARTZ, FELDSPAR and MICA. Those with more than 75% quartz are called **quartz** gritstone, whereas more than 25% feldspar gives rise to **feldspathic** gritstone; the former has an orange colour and the latter is brown, sometimes tinged with pink.

grooming Behaviour pattern of self-cleaning of the body surface, usually stereotyped, practised by many animals. **Mutual** grooming, such as fur grooming among monkeys and apes, serves to cement pair-bonding and social bonds in groups.

ground state Most stable energy state of an atom, ion or molecule. It is the state of an atom when the orbiting electrons move in orbits such that the atom's energy is at a minimum.

groundwater Water that lies beneath the surface of the Earth. The water comes chiefly from rain, although some is of volcanic or sedimentary origin. Groundwater moves through porous rocks and soil and can be collected in wells. Groundwater can dissolve minerals from the rocks and leave deposits in other places, creating such structures as CAVES, STALAGMITES, STALACTITES and cavities called SINK-HOLES.

group theory Branch of mathematics applicable to sets with symmetric properties. A group is a set with elements (together with an operation) that must obey four rules: closure; associativity; every element must have an inverse; and there must be an identity element. The theory was developed by the French mathematician Évariste GALOIS. Group theory is particularly useful in QUANTUM MECHANICS, SPECTROSCOPY and theories of elementary particles, but is used widely in mathematics to describe many other natural phenomena that have symmetry.

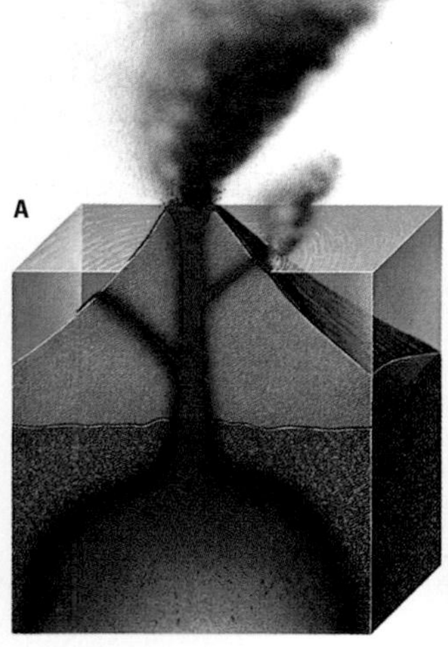

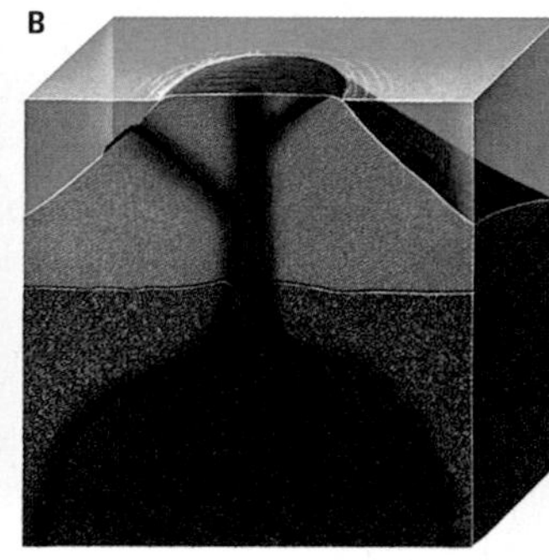

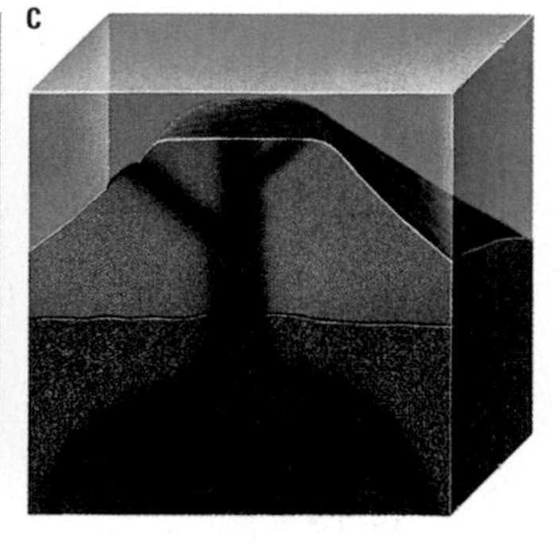

▲ **guyot** Formed originally from seamounts (submarine mountains rising at least 1,000m (3,300ft) above their surroundings), guyots have flat tops at depths down to 2,500m (8,200ft). These tops are often much too large to be explained as ancient craters filled to the rim by sediment and other material. It has been proposed that guyots were originally volcanoes above sea level (A) which after extinction were, over many years, worn flat by the action of waves (B) and which then sunk as the sea level rose or as the sea bed subsided (C). This has subsequently been confirmed by the presence of beach pebbles.

growth Process by which an organism increases in size and mass. It involves CELL DIVISION or enlargement, or both.

growth hormone (somatotrophin) Hormone produced by the anterior lobe of the PITUITARY GLAND; it is involved in general growth of the body. Over-secretion results in ACROMEGALY in the adult, and gigantism in the young; under-secretion results in dwarfism in children.

growth ring *See* ANNUAL RING

groyne Artificial dam of rocks or wooden pilings that juts out from a beach face, built in an attempt to combat LONGSHORE DRIFT. Groynes are now considered a less effective and more expensive way of maintaining beaches than a beach-nourishment programme, in which material swept along the beach is replaced by new material.

Grus (Crane) Insignificant constellation of the S skies. Its brightest star is Alpha, magnitude 1.7.

gryke *See* GRIKE

guanine One of the nitrogenous bases in DNA and RNA. Guanine is a derivative of PURINE and in DNA is always paired with the pyrimidine derivative CYTOSINE. *See* BASE PAIR

guano Dried excrement of seabirds and bats. It contains phosphorous, nitrogen and potassium, and is a natural fertilizer. It is found mainly on certain coastal islands off South America and Africa, and on some Paficic islands.

guided missile MISSILE controlled throughout its flight to target either by exterior or interior control systems. There are four types: (1) surface-to-surface, (2) surface-to-air, (3) air-to-air, (4) air-to-surface. The first guided missiles were built in Germany during World War 2. The first of these was the V1, which was powered by a pulse-jet engine and which flew at quite low speeds, making it easy to destroy in the air. The V2 was a far more sophisticated weapon – rocket-powered, with an automatic pilot and electronic guidance; it could reach a height of 100km (60 miles) and speeds of over 5,800km/h (3,600mph) and deliver a tonne of high explosive. Postwar development improved upon the V2, with missiles ranging from the huge intercontinental ballistic missiles (ICBMs) with ranges of 10,000km (6,000 miles) and nuclear warheads, to small hand-launched antitank missiles. The main strategic capability of nuclear missiles resides in the submarine systems developed by the superpowers, the first of which was Polaris. The multiple independently targeted re-entry vehicles (MIRVs) – ICBMs with many independently targetable sub-missiles – were developed in the late 1960s. The CRUISE MISSILE can evade radar by flying at low heights and its computer guidance compares satellite photographs with the terrain over which it flies, bringing the missile potentially to within 10m (33ft) of its target.

Guillaume, Charles Édouard (1861–1938) Swiss physicist who was awarded the 1920 Nobel Prize for physics for his discovery of invar, an alloy of iron and nickel with an extremely small coefficient of expansion. He also developed another alloy, platinite, which expands at about the same rate as glass.

Gulf Stream Relatively fast-moving current of the N Atlantic Ocean. It flows from the straits of Florida, USA, along the E coast of North America, then E across the Atlantic Ocean (at which point it is known as the **North Atlantic Drift**) to the NW European coast. It was long considered to be one wide mass of water, but research now indicates that it is made up of many thin streams which cause local variations in the water temperature. The current has a warming effect on the coastal climates along its course.

GYROCOMPASS

A laser gyrocompass measures rotation by comparing the wavelength of lasers (1). A current is passed from a cathode (2) to two anodes (3) creating two lasers in a gas-filled triangular chamber (4) drilled in a solid glass block (5). Part of the lasers are bled out at one end of the gyrocompass (6) and the wavelength measured. If the gyrocompass rotates to the left the path of the laser travelling to the left is reduced fractionally, thus shortening its wavelength. The opposite occurs to the other laser. A sensor (7) compares the two lasers to measure the rotation.

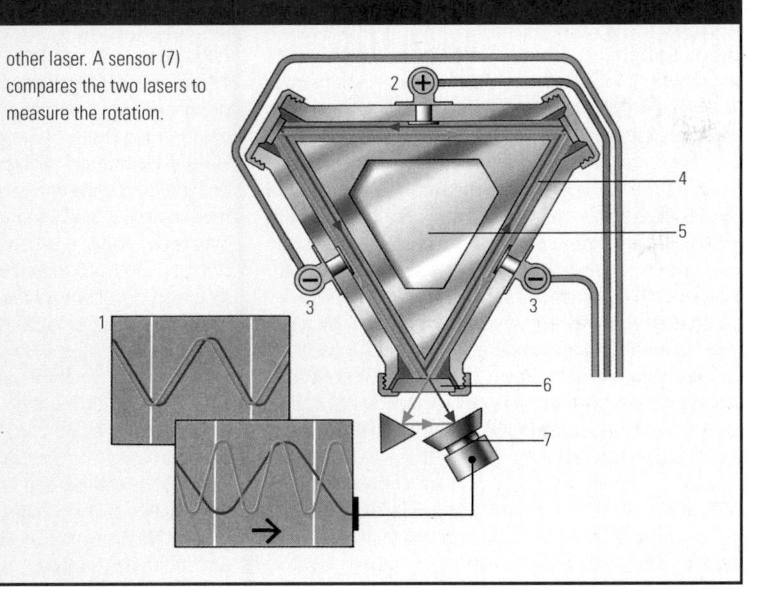

Gullstrand, Allvar (1862–1930) Swedish ophthalmologist. He received the Nobel Prize for physiology or medicine in 1911 for his contributions to dioptrics, the study of refraction of light by the eye. He invented many important ophthalmological tools, including the slit lamp, which aided the study of the structure and function of the CORNEA, and improved corrective lenses for use after surgery.

gum In botany, the secretions of some plants. Some gums are soluble in water; others absorb water and swell. Gums are chemically complex, consisting mainly of various SACCHARIDES bound to organic acids. Common examples are gum arabic, AGAR and tragacanth. Many substances of similar appearance, such as some RESINS, are classed as gums.

gums Soft tissue surrounding the base of the teeth. The gums are liable to inflammation, called GINGIVITIS, and to infection from food debris.

guncotton (cellulose nitrate, nitrocellulose) Highly nitrated and highly explosive form of CELLULOSE. It is made by soaking cotton cellulose in mixed nitric and sulphuric acids. When dry, guncotton is dangerously unstable. It is used mainly as an ingredient of explosive propellants such as CORDITE.

gunmetal Types of wear-resistant BRONZE once used to make heavy field and naval guns, but now used in gears, bearings and pump spindles. A typical gunmetal alloy contains about 88% copper, 10% zinc and 2% tin.

gunpowder Mixture of saltpetre (potassium nitrate), charcoal and sulphur. When ignited, it expands violently due to the almost instantaneous conversion of the solid ingredients into gases such as carbon dioxide, carbon monoxide, nitrogen, nitrogen oxides, sulphur oxides and steam. The sudden release of enormous volumes of these gases gives the reaction its explosive force. It was used extensively in FIREARMS until about 1900, after which it was replaced by smokeless powders such as CORDITE.

Gutenberg, Beno (1889–1960) US seismologist, noted for his analyses of EARTHQUAKE waves. He showed that 75% of earthquakes occur in the Circum-Pacific belt. He worked with Charles Richter (1900-85) to develop the RICHTER SCALE.

Gutenberg, Johann (1400–68) German goldsmith and printer credited with the invention of printing from movable metallic type. He experimented with printing in the 1430s, and his innovations included a new type of press and a typemetal alloy. He made the first printed Bible, known as the Gutenberg Bible (*c.*1455).

guyot (tablemount) Flat-topped, submarine mountains. They rise *c.*1,000m (3,300ft), with tops *c.*2.5km (1.5mi) below sea level. Before becoming submerged, they are believed to have been volcanic islands with peaks flattened by wave erosion. If so, this is evidence for the subsiding of the ocean floor.

gymnosperm Seed-bearing plant with naked seeds borne on scales, usually in cones. Most trees commonly referred to as EVERGREENS are gymnosperms. However, larch and some other CONIFERS are DECIDUOUS. All living seed-bearing plants are divided into two main groups: gymnosperms and ANGIOSPERMS (with seeds enclosed in an ovary). In the modern "Five KINGDOMS " classification system, gymnosperms no longer form a taxonomic group. Instead they comprise three distinct phyla, the Coniferophyta (such as pine, spruce and cedar); the Ginkgophyta (a single species, the GINKGO); and the Gnetophyta (strange plants such as *Welwitschia*, *Ephedra*, and *Gnetum*, a genus of woody shrubs, vines and large-leaf tropical trees). The CYCADS are no longer considered to be gymnosperms.

gynoecium Collective name for the female elements of a FLOWER. The gynoecium is composed of a flower's CARPELS, consisting of the STIGMA, STYLE and OVARY. *See also* ANDROECIUM

gypsum Most common sulphate mineral, hydrated calcium sulphate ($CaSO_4.2H_2O$). It is formed by precipitation from evaporating seawater. Huge beds of gypsum occur in SEDIMENTARY ROCKS, where it is associated with halite. It can be clear, white or tinted and it crystallizes in the monoclinic system as prismatic or bladed CRYSTALS. Varieties are ALABASTER (massive); selenite (transparent and foliated); and satinspar (silky and fibrous). It is a source of plaster of Paris. Hardness 2; r.d. 2.3.

gyrocompass Navigational aid incorporating a continuously driven GYROSCOPE. The spinning axis of the gyroscope is horizontal and its direction indicates true N, irrespective of the course or attitude of the craft. *See also* COMPASS

gyroscope Symmetrical spinning disc that can adapt to any orientation, being mounted in gimbals (a pair of rings with one swinging freely in the other). When a gyroscope is spinning, a change in the orientation of the gimbals does not change the orientation of the spinning wheel. This means that changes in direction of an aircraft or ship can be determined on board by a gyroscope without external references. A gyrostabiliser is used to stablize the roll of a ship or aircraft, and a gyrocompass is a gyroscope modified to act as a compass. When torque (twisting force) is applied to a fast-spinning gyroscope, such as leaning it out of the vertical, this produces a phenomenon known as PRECESSION: the gyroscope rotates about a fixed point with the axis of spin describing a cone around the vertical. This tendency to resist changes in the spin axis accounts partly for the stability of bicycles and of the orbits of stars and planets.

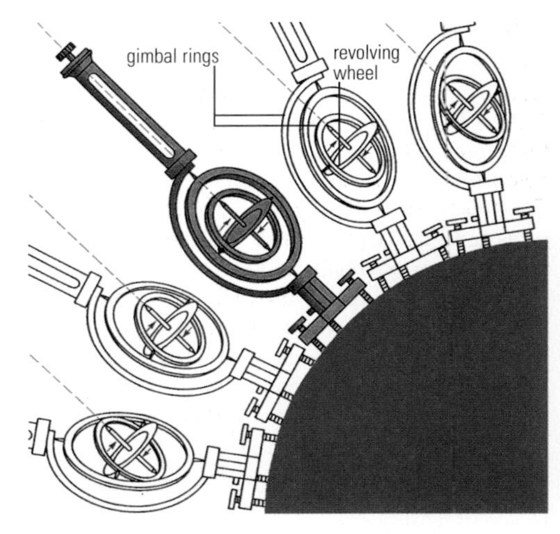

▲ **gyroscope** Stability is the vital contribution to the science of measurement made by the gyro. This instrument depends on a rapidly spinning wheel with a heavy rim which is suspended, with a minimum of friction, in a system of gimbal rings that allow it to rotate on its axis in any plane. If momentum is maintained at a given speed (by an electric motor for instance), the axis of the wheel maintains the position it took up when first spun. As the Earth revolves, the axis continues to point to a particular position in space, although the gimbals change their angle relative to it. Early applications were gun-sighting at sea and torpedo steering.

Haber, Fritz (1868–1934) German physical chemist whose early work involved ELECTROCHEMISTRY and THERMODYNAMIC gas reactions. With Carl BOSCH he invented (1908–09) the HABER PROCESS for converting atmospheric nitrogen into ammonia. He was awarded the 1918 Nobel Prize for chemistry.

Haber process Industrial process (invented by Fritz HABER and Carl BOSCH) in which nitrogen from the atmosphere is "fixed" by synthesizing ammonia. A mixture of nitrogen and hydrogen is passed over a heated catalyst at a pressure of about 1,000 atmospheres. The chemical reaction, $N_2 + 3H_2 \rightarrow 2NH_3$, is EXOTHERMIC, which in turn speeds up the reaction. *See also* NITROGEN FIXATION

habitat Place in which an organism normally lives. A habitat is defined by characteristic physical conditions and the presence of other organisms.

hacker Person who obtains unauthorized access to a computer DATABASE through a communications system. A hacker may read or even alter the information in the database. Government departments, companies and other organizations provide employees with access to their computer databases through the public telephone system. Anyone with suitable equipment and SOFTWARE can gain access from a computer if they know the right telephone numbers and are able to discover the right passwords to enter after contact is made. *See also* MODEM

Hadley cell Atmospheric circulation cell, named after the British scientist George Hadley. Hadley proposed it (1735) to explain the trade winds, in which winds rise and flow polewards from the Equator and then descend and flow Equatorwards, transferring heat by CONVECTION.

hadron Class of ELEMENTARY PARTICLE with STRONG NUCLEAR FORCE interaction. The group can be divided into BARYONS, such as the NEUTRON and PROTON, and MESONS, such as the PION and KAON. Hundreds of hadrons have been discovered, mostly since about 1950. With the exception of the proton and antiproton, they are all unstable. Unlike LEPTONS, they have a measurable size and a substructure of QUARKS. *See also* ANTIMATTER

haematite (hematite) One of the most important iron ores, containing mainly ferric oxide (Fe_2O_3). Containing 70% iron by weight, it occurs in several forms, frequently in small dome-shaped masses called **kidney-ore**. Haematite varies in colour from steel-grey to black, but sometimes red. Deposits are found on all continents. Hardness 5–6; r.d. 4.9–5.3.

haemoglobin Protein present in ERYTHROCYTES (red blood cells) of vertebrates. Haemoglobin carries oxygen to all cells in the body. It is scarlet when combined with oxygen to form OXYHAEMOGLOBIN and bluish-red when deoxygenated. Oxygen attaches to the haem– part of the protein, which contains iron; the –globin part is a globular PROTEIN. Worn-out erythrocytes are destroyed by the liver, and the iron is used again to make more haemoglobin. *See also* BLOOD

haemophilia Hereditary blood-clotting disorder in which there is prolonged external or internal bleeding, often without apparent cause. The commonest form, **haemophilia A**, is due to an inability to synthesize FACTOR VIII, a substance essential to blood clotting. Bleeding into the joints may cause joint deformities. The gene for the disorder is passed on almost exclusively from mother to son. About one in 10,000 boys are born with haemophilia. It is treated by injections of Factor VIII. A rarer form, **haemophilia B** or Christmas disease, is caused by a deficiency of blood Factor IX. It is inherited in the same way and follows the same course as haemophilia A.

hafnium (symbol Hf) Silver TRANSITION ELEMENT, first discovered in 1923. Hafnium's chief source is as a by-product in obtaining the element ZIRCONIUM. It is used as a MODERATOR in NUCLEAR REACTOR control rods. Properties: at.no. 72; r.a.m. 178.49; r.d. 13.31; m.p. 2,227°C (4,041°F); b.p. 4,602°C (8,316°F); most common isotope ^{180}Hf (35.24%).

Hahn, Otto (1879–1968) German chemist, who co-discovered nuclear FISSION in 1939 with Fritz STRASSMANN (for which Hahn won the 1944 Nobel Prize for chemistry). With Lise MEITNER, he discovered protactinium and several isomers. During World War 2 he remained in Germany and, after the war, was appointed president of the Max Planck Institute in Berlin.

Hahnemann, Christian Friedrich Samuel (1755–1843) German physician. He popularized HOMEOPATHY, a system of medical treatment based on the idea that "like cures like", or that a disease should be treated with minute doses of agents that produce the symptoms of the disease.

hahnium (symbol Ha) Synthetic, radioactive, TRANSACTINIDE ELEMENT. It has atomic number 105; six isotopes have been synthesized. It was first reported by a Soviet team at the Joint Institute for Nuclear Research at Dubna. They claimed the isotopes of mass numbers 260 and 261, as a result of bombarding AMERICIUM with neon ions. In 1970 a team at the University of California claimed the isotope 260 (half-life 1.6 seconds) obtained by bombarding CALIFORNIUM with nitrogen nuclei. The element is named after Otto HAHN.

hail PRECIPITATION from clouds in the form of balls of ice. Hailstorms are associated with atmospheric turbulence extending to great heights together with warm, moist air nearer the ground. Hailstones are usually less than 1cm (0.4in) across but some have exceeded 13cm (5in).

hair Outgrowth of mammalian SKIN; it has insulating, protective and sensory functions. In MAMMALS, a thick coating of hair is usually called FUR. Hair grows in a FOLLICLE, a tubular structure extending down through the EPIDERMIS to the upper DERMIS. New cells are continually added to the base of the hair; older hair cells become impregnated with KERATIN and die. A muscle attached to the base of the hair allows it to become erect in response to nerve signals sent to the follicle. Erecting the hairs traps a thicker layer of air close to the skin. *See also* CILIA

Haldane, John Burdon Sanderson (1892–1964) British scientist whose work formed the basis of the mathematical study of population genetics. His book *The Causes of Evolution* (1933) examined the theory of NATURAL SELECTION in the light of modern genetical research. In 1932 he was the first to estimate the mutation rate of a human gene. His book *Daedalus, or Science and the Future* (1924) was an early attempt to popularize science.

Hale, George Ellery (1868–1938) US astronomer who organized a number of observatories. He planned and secured finance for the 1-m (40-in) refracting telescope at Yerkes Observatory, completed in 1897, the 2.5-m (100-in) Hooker reflector at Mount Wilson Observatory, completed in 1917, and the 5-m (200-inch) reflector at Mount Palomar Observatory, which entered regular service in 1949. Each was in turn the largest telescope of its day. The 5-m (200-in) was named the Hale telescope in his honour. Hale was an accomplished solar observer, and he also invented the spectroheliograph.

half-cell ELECTRODE immersed in an ionic solution. A full ELECTROLYTIC CELL is formed of two connected half-cells – different solutions being separated by a membrane or conducting bridge that allows electricity to flow but prevents mixing. *See also* ELECTRODE POTENTIAL

half-life Time taken for one-half of the nuclei in a given amount of a radioactive ISOTOPE to decay (change into another element or isotope). Only the half-life is measured because the decay is never total. Half-lives remain constant under any temperature or pressure, but there is a great variety among different isotopes. Oxygen-20 has a half-life of 14 seconds and uranium-234 of 250,000 years. A radioactive isotope disintegrates by giving off alpha or beta particles, and a measurement of this rate of emission is a way of recording decay. The term

HARDNESS OF WATER

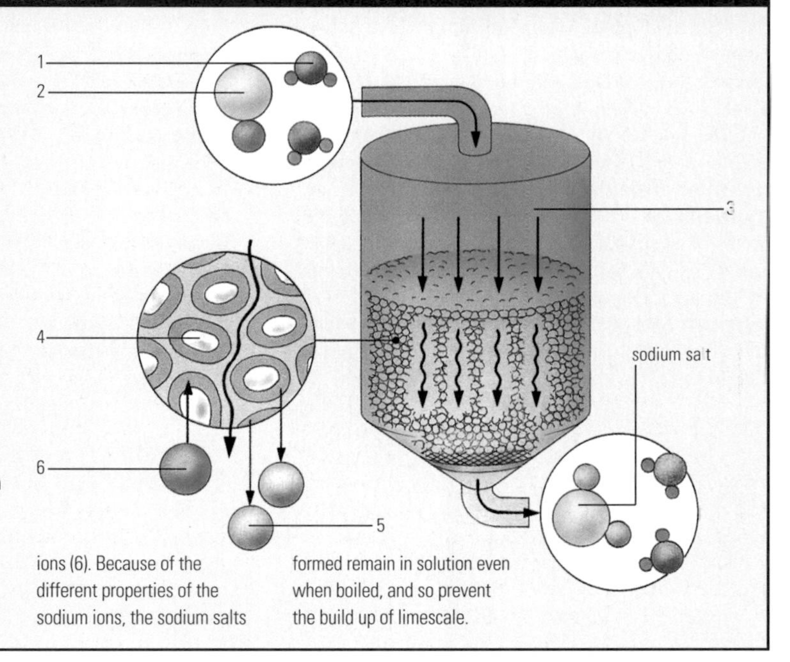

Furring in kettles and water pipes is caused by the presence in water (1) of dissolved calcium carbonate (2), usually from the chemical weathering of limestone. In hot or boiling water, calcium carbonate precipitates forming solid limescale deposits on surfaces, such as the inside of kettles. Calcium carbonate also prevents soap from lathering. In an ion exchange tank (3), the tank is filled with grains of sodium-coated material with which the water has to come into contact (4). Sodium ions (5), which are more reactive than calcium, are exchanged for calcium ions (6). Because of the different properties of the sodium ions, the sodium salts formed remain in solution even when boiled, and so prevent the build up of limescale.

"half-life" also refers to particles that spontaneously decay into new particles, such as a free neutron transforming into a proton and an electron. *See also* CARBON DATING, RADIOACTIVE DECAY

halide Salt containing one of the HALOGENS (elements in group VII of the periodic table); examples are sodium fluoride, hydrogen bromide and potassium chloride and iodide. The haloalkanes (alkyl halides) are organic compounds, such as chloromethane (methyl chloride, CH_3Cl), containing an alkane group bound to a halogen atom.

halite Mineral form of sodium chloride (NaCl), or common (rock) salt. It is found in evaporite SEDIMENTARY ROCKS, in association with gypsum and in salt domes and dried lakes. It is colourless, white, orange or grey, with a glassy lustre. It has a cubic system of interlocking cubic crystals, granules and masses. It is important as table salt and as a source of CHLORINE. Hardness 2.5; r.d. 2.2.

Hall effect Generation of an ELECTROMOTIVE FORCE across a conductor or semi-conductor when it is carrying a current perpendicular to a magnetic field around it. The voltage is at 90° to both the direction of the current and the direction of the magentic field. The effect was discovered in 1879 by the US physicist Edwin H. Hall (1855–1938) and is used to test the electrical properties of materials and in ascertaining the strength of magnetic fields.

Halley, Edmond (1656–1742) British astronomer and mathematician. His most famous achievement in astronomy was to realize that comets could be periodic, following his observation of HALLEY'S COMET in 1682. It was Halley who persuaded Isaac NEWTON to write the *Principia*, and he also financed its publication. Halley's other achievements are numerous. He founded modern GEOPHYSICS, charting variations in the Earth's magnetic field, and establishing the magnetic origin of the AURORA borealis. In meteorology he showed that ATMOSPHERIC PRESSURE decreases with altitude, and studied monsoons and trade winds. In mathematics, he showed how to use mortality statistics to cost life assurance policies.

Halley's comet Bright periodic COMET. It takes 76 years to complete an orbit that takes it from within Venus's orbit to outside Neptune's. It was observed by Edmond HALLEY in 1682; later he deduced that it was the same comet as had been seen in 1531 and 1607, and predicted it would return in 1758. There are historical records of every return since 240 BC. Several space probes were sent to Halley on its 1986 return. The closest approach was by the probe Giotto, whose cameras showed the nucleus to be an irregular object measuring 15 × 8km (9 × 5mi). The main constituent was found to be water-ice, but the surface was covered with a dark deposit. Some craters were imaged.

hallucinogen Drug that causes hallucinations, unusual perceptions without external cause. Hallucinogenic drugs, such as mescaline and LSD (lysergic acid diethylamide), have been used in primitive religious ceremonies and are illicitly taken today.

haloalkane (alkyl halide) Organic chemical compound consisting of an ALKANE in which one or more hydrogen atoms have been substituted by a HALOGEN element. Haloalkanes are used as solvents and in organic synthesis. Examples include iodomethane (methyl iodide, CH_3I), dibromomethane (methylene bromide, CH_2Br_2), trichloromethane (chloroform, $CHCl_3$) and tetrachloromethane (carbon tetrachloride, CCl_4). They are made by halogenating an alcohol or by the direct reaction between a halogen and an alkane in ultraviolet light. Chlorofluorocarbons (CFCs) are haloalkanes. *See also* HALOGENATION

halogen Element (FLUORINE, CHLORINE, BROMINE, IODINE and ASTATINE) belonging to group VII of the periodic table. They react with most other elements and with organic compounds; reactivity decreases down the group. The halogens are highly electronegative and react strongly because they require only one electron to achieve the "stable 8" inert gas configuration; they produce crystalline salts (HALIDES) containing negative ions of the type F^- and Cl^-. The name halogen means salt-producer.

halogenation Introduction of one of the HALOGEN elements – fluorine, chlorine, bromine, iodine or astatine – into an organic compound either by addition or by substitution of an atom or group of atoms. This reaction is widely used in making pharmaceuticals and dyes.

halon Any of several organic gases used in fire extinguishers. Chemically halons can be considered as simple hydrocarbons (such as METHANE or ETHANE) that have had some or all of their hydrogen atoms replaced by a HALOGEN. They are similar to CHLOROFLUOROCARBONS (CFCs) but up to ten times more destructive of the OZONE LAYER. The most commonly used halon is bromotrifluoromethane ($CBrF_3$, halon 1301).

halophyte Any plant, usually a GYMNOSPERM (seed-bearing plant), that is able to live in a salty environment. Typical examples include mangrove trees, sea lavender and rice grass.

Hamilton, William D. (1936–) New Zealand biologist. He developed the theory in genetics that sought to explain the evolution of ALTRUISM in animals in terms of Darwinian NATURAL SELECTION. Hamilton theorized that if an organism sacrificed itself to save its relatives, it was doing so to ensure that at least some of its own GENE variants, or ALLELES, were passed on to the following generation. The success with which an individual's alleles are passed on is called **inclusive fitness**. His theories were taken up by Richard DAWKINS.

Hamilton, Sir William Rowan (1805–65) Irish mathematician. In 1827 Hamilton was appointed professor of astronomy at Trinity College, Dublin, even before he had formally graduated with his first degree. He worked in the field of complex numbers, but his greatest achievement was the creation of QUATERNIONS.

hamstring *See* ACHILLES TENDON

hanging valley Tributary valley that ends high up the face of a larger main valley, possibly with a stream running through it and ending in a waterfall. Most hanging valleys result from glacial deepening of the main valley. *See also* GLACIER

haploid Term describing a CELL that has only one member of each CHROMOSOME pair. All human cells except GAMETES are DIPLOID, having 46 chromosomes. A gamete (OVUM and SPERM) is haploid, having 23 chromosomes. The body cells of many lower organisms, including many algae and single-celled organisms, and also moss plants, are also haploid. *See also* ALTERNATION OF GENERATIONS; MEIOSIS

hard disk Rigid MAGNETIC DISK for storing COMPUTER PROGRAMS and DATA. The built-in hard disk drive in a typical personal COMPUTER consists of a number of hard platters coated with a magnetic material set on a common spindle. They are housed inside a sealed container, with a motor to spin the stack of platters, a head to write (record) and read (replay) each side of each platter, and associated electronic circuits. Hard disk capacity is continually being increased: most computers are now sold with a disk of at least 500Mb (500 megabytes) capacity.

Harden, Sir Arthur (1865–1940) British biochemist. He showed that each stage in the FERMENTATION of sugar is catalyzed by a specific ENZYME and that the essential first step is the attachment of a phosphorus-containing group to the sugar. For this research he shared the 1929 Nobel Prize for chemistry with Hans von Euler-Chelpin (1873–1964). His work laid the foundation for the elucidation of the KREBS CYCLE, which describes the process of glucose METABOLISM.

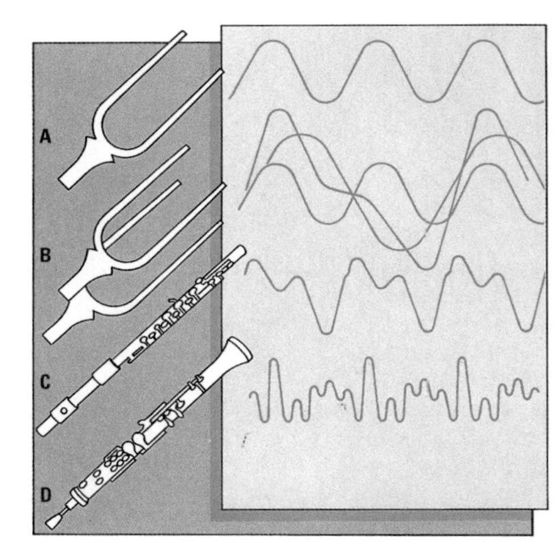

▲ **harmonics** A tuning fork vibrates, producing uniform sound of a single pitch, its fundamental frequency (A). A more complex sound pattern will arise when two forks are struck together (B). The fundamental pitch of the vibrating column of air in a flute is enhanced by other shorter wavelength sounds which are multiples of the fundamental (C). Woodwind instruments have a more reedy sound because of the interaction of all these different frequencies heard together (D).

hardness Resistance of a material to abrasion, cutting or indentation. The MOHS SCALE is a means of expressing the comparative hardness of materials.

hardness of water Reluctance of water to produce a lather with soap, due to various dissolved salts, mainly those of calcium and magnesium. These salts give rise to an insoluble precipitate, which causes "fur" or "scale" in boilers, pipes and kettles. Lather is inhibited until all the dissolved salts are precipitated as scum, which floats on the surface. Hardness may be temporary (removed by boiling), caused by calcium bicarbonate; or permanent (not affected by boiling), caused by calcium sulphate.

hardware Equipment as opposed to the programs, or software, with which a COMPUTER carries out various functions. The computer, keyboard, printer and electronic circuit boards are examples of hardware.

Hargreaves, James (1722–78) British inventor and industrialist. In 1764, while working as a weaver at Stanhill, Lancashire, England, Hargreaves invented the SPINNING JENNY. This machine greatly speeded the spinning process of cotton by producing eight threads at the same time. In 1768 local spinners destroyed his house and equipment, fearing that Hargreaves and his machines threatened their jobs. Hargreaves moved to Nottingham and, with Thomas James, built a spinning mill and became one of the first great factory owners.

harmonic progression Sequence of the form $1/a$, $1/b$, $1/c$, ..., where a, b, c, and so on, form an ARITHMETICAL PROGRESSION. The simplest is formed by the reciprocals of the positive integers: 1, 1/2, 1/3, 1/4, ... PYTHAGORAS discovered that taut strings with lengths proportional to these terms (and with identical diameter and tension) vibrate with harmonic musical tones.

harmonics (overtones) In acoustics, additional notes the frequencies of which are multiples of a basic (fundamental) note. When a violin string is

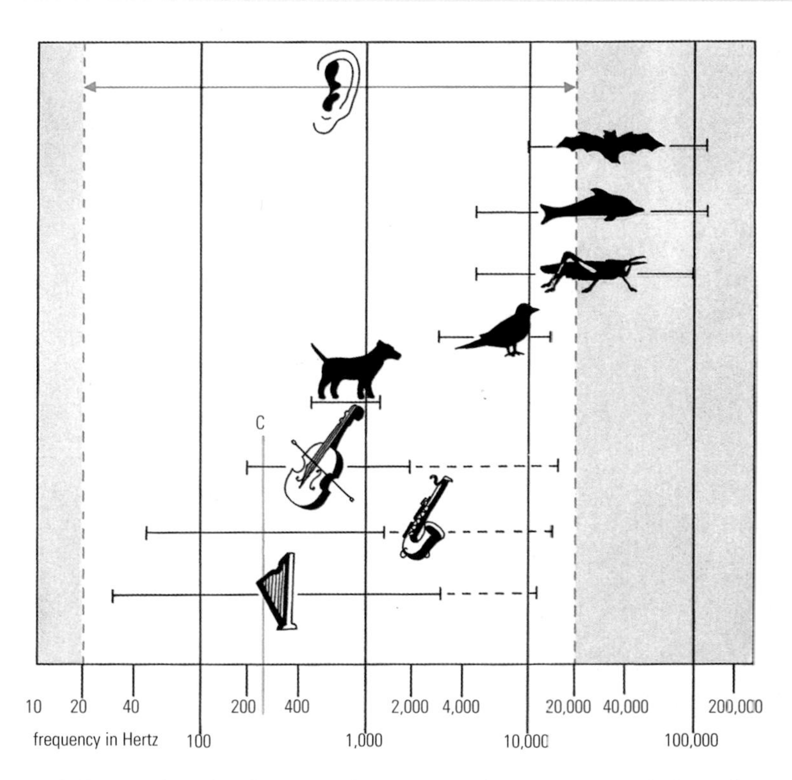

▶ **hearing** Human hearing extends from a frequency of about 20 cycles per second (Hertz, Hz) to 20,000 cycles. Some animals are able to generate sounds far beyond this range: the chart shows the frequencies generated by bats porpoises and grasshoppers, and for comparison the frequency-production ranges of birds and dogs. Musical instruments have two kinds of frequency-range: the range of notes that can be played (shown as a solid line) and the range of overtones that go to make up the characteristic sound of the instrument (broken line). The ranges shown are those of the violin, the saxophone family (from bass to soprano) and the harp. For reference, the note middle-C is marked in yellow.

plucked or a drumskin tapped, the sounds emitted correspond to vibrations of the string or skin. The loudest sound (note) corresponds to the fundamental mode of vibration. But other, weaker notes, corresponding to subsidiary vibrations, also sound at the same time. Together these notes make up a harmonic series.

Harvey, William (1578–1657) English physician and anatomist who discovered the circulation of the blood. This landmark in medical history marked the beginning of modern physiology. His findings, published (1628) in *De Motu Cordis et Sanguinis* (*On the Motions of the Heart and Blood*), were ridiculed at first and only later generally accepted. Harvey also carried out important studies in EMBRYOLOGY. *See also* CIRCULATORY SYSTEM

Hassel, Odd (1897–1981) Norwegian chemist who shared the 1969 Nobel Prize for chemistry with Derek BARTON for his work on conformational analysis (the study of the three-dimensional structures of molecules). In 1930 he began research on aromatic organic compounds and discovered the two stereoisomers of cyclohexane.

haustorium Specialized invasive sucker-like or tube-like structure in a parasitic plant or fungus. It is used to penetrate the outer tissues of a host plant in order to absorb nourishment.

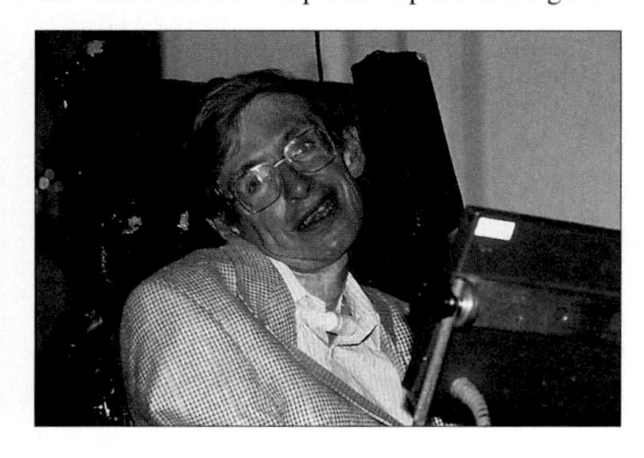

▲ **Hawking** The author of the best-selling *A Brief History of Time* (1988), Professor Stephen Hawking's work has concentrated on the nature of black holes. He was the first scientist to hypothesize that the immensely powerful gravitational field around super dense black holes can radiate matter. Since the 1960s he has suffered from a motor neuron disease, which has confined him to a wheelchair.

Hawking, Stephen William (1942–) British theoretical physicist. He studied at Oxford and later at Cambridge, where he became a professor of mathematics. During the course of his research into RELATIVITY, Hawking studied the BIG BANG origin of the Universe and did much pioneering work on the theory of BLACK HOLES. He found that small black holes should lose energy – by emitting **Hawking radiation** – and eventually "evaporate". He published a best-selling, popular account of his work in *A Brief History of Time* (1988).

Hawksbee, Francis (died *c.*1713) English experimental physicist and student of Robert BOYLE. Hawksbee was responsible for the double-cylindered air pump, with which he showed that air glows at low pressure when excited by an electric discharge. He also demonstrated that the transmission of sound through air is dependent on pressure.

Haworth, Sir Walter Norman (1883–1950) British organic chemist and Fellow of the Royal Society. Educated at Manchester and Göttingen, he held academic posts in London, St Andrews and Newcastle and was Director of chemistry at Birmingham from 1925–48. He determined the constitution of vitamin C and various carbohydrates, sharing the 1937 Nobel Prize for chemistry with Paul KARRER for this work.

hearing Process by which SOUND waves are experienced. Humans and many other mammals hear when sound waves enter the EAR canal and vibrate the eardrum. The vibrations are transmitted by three small bones (ossicles) to the COCHLEA. In the cochlea, receptors generate nerve impulses that pass via the auditory nerve to the brain to be interpreted as sound.

hearing aid Electronic sound-reproducing device to increase the sound intensity at the ear. Modern aids use a small crystal microphone, a microminiaturized battery-powered amplifier and an earpiece (often shaped to fit into the auditory canal). The device consists of two parts either connected by a thin wire (the battery and amplifier being hidden in the clothing) or of one tiny unit placed in or behind the ear or housed in spectacle frames.

heart Muscular organ that pumps BLOOD throughout the body by means of the CIRCULATORY SYSTEM. In humans, the heart is located behind the breastbone between the lower parts of the lungs. Divided longitudinally by a muscular wall, the right side contains only deoxygenated blood, the left side only oxygenated blood. Each side is divided into two chambers, an ATRIUM and a VENTRICLE, both separated by valves. The average heart-beat rate for an adult at rest is 70–80 beats per minute.

heart attack (myocardial infarction) Death of part of the heart muscle due to coronary occlusion, the blockage of a coronary ARTERY by a blood clot (thrombosis). It is accompanied by gripping chest pain, sweating and vomiting. The greatest danger is within the first few hours. Modern drug treatment treats abnormal heart rhythms and dissolves the clots in the coronary arteries. **Heart failure** occurs when the heart is unable to pump blood at the rate necessary to supply body tissues. It may be due to high BLOOD PRESSURE (HYPERTENSION) or heart disease. Symptoms include shortness of BREATH, OEDEMA and fatigue. Treatment is with a DIURETIC and heart drugs. *See also* ANGINA

heart-lung machine Apparatus used during surgery to take over the function of the heart and lungs. Its use allows the patient's own circulation to be interrupted briefly – for example, to enable open-heart procedures. It consists of a pump to circulate blood around the body and special equipment to add oxygen to the blood and remove carbon dioxide from it.

heart transplant *See* ORGAN TRANSPLANT

heat Form of energy associated with the constant vibration of atoms and molecules. It was once thought to be a material substance called caloric that was contained in all objects and could flow from one to another. The KINETIC THEORY, presently accepted, holds that the degree of hotness of a body

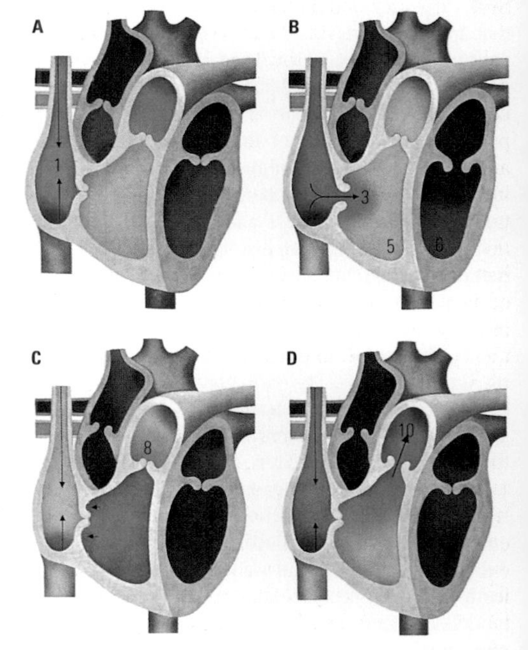

▲ **heart** The sequence of a single heartbeat takes *c.*0.9s and begins (A) as the atria (1, 2) fill with blood. Contraction (B) forces blood past the retaining tricuspid (3) and mitral (4) valves into the ventricles (5, 6). The thick muscular walls of the ventricles contract (C), snapping shut the atrio-ventricular valves and tending to cause the semi-lunar valves (7, 8) to open. With both ventricles fully contracted (D), two streams of blood are forced along the separate routes: fresh blood into the aorta (9), spent blood into the pulmonary artery (10).

HEAT EXCHANGE

The three ways in which heat moves all take place when a pan is heated (A) - conduction through the metal walls of the pan (1), convection by fluid motion (2) and radiation from the heat source to the pan (3). In theory an insulated good conductor with ice at one end and boiling water at the other varies in temperature linearly with distance along the bar (B), as in the straight-line graph. With poor insulation a curve like the dotted line results. A vacuum flask (C) has a vacuum (4) to prevent conduction and convection and silvered walls (5) to minimize heat loss by radiation.

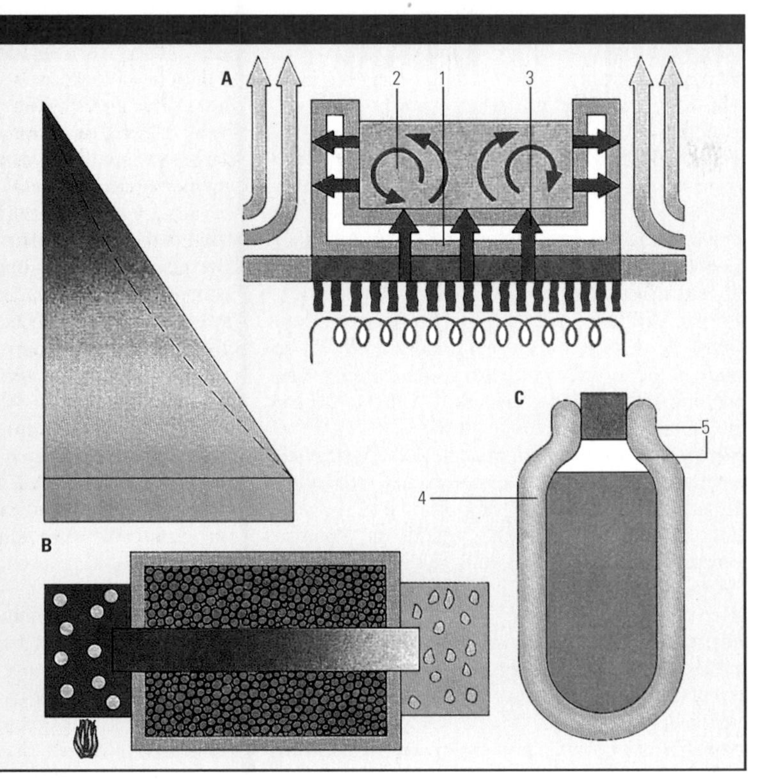

depends on the extent of vibration of its atoms. Heat is distributed in three forms: CONVECTION through FLUIDS (gases and liquids); CONDUCTION through solids; and RADIATION mainly through space. Heat and TEMPERATURE are sometimes confused; temperature is only one factor on which the heat content of an object depends. Heat and mechanical energy are interchangeable – for example hands can be warmed by rubbing them because friction produces heat.

heat balance (heat budget) Balance between the Sun's heat received by the Earth's atmosphere and that returned back into space. About two-thirds of solar radiation is absorbed by clouds, the atmosphere and the Earth, and about one-third is reflected back, mainly by clouds. The heat absorbed powers the circulation of the atmosphere, oceans and water cycle. Eventually the heat is reradiated into space, maintaining the Earth's heat equilibrium. *See also* CIRCULATION, ATMOSPHERIC; GREENHOUSE EFFECT

heat capacity (thermal capacity, symbol *C*) Ratio of the heat supplied to an object to its resultant rise in temperature. It is measured in joules/kelvin. For most purposes, the alternative SPECIFIC HEAT CAPACITY is used.

heat engine Any engine that converts heat energy, usually through the burning of a fuel, into useful mechanical energy. Thus, all INTERNAL COMBUSTION ENGINES are heat engines.

heat exchange (heat transfer) Flow of heat energy from one object to another. This flow of energy occurs at all times when two or more bodies at different temperatures are in thermal contact. Three methods of heat transfer may be distinguished: CONDUCTION, CONVECTION and RADIATION. In conduction, heat is transferred from molecule to molecule within a body, as in an iron rod stuck in a fire; in convection, heat is transferred by circulation of FLUID, as in boiling; in radiation, heat is transferred in the form of electromagnetic waves, as in sunlight. **Heat exchangers** are essential in many industrial processes where heat may be extracted from one source for use elsewhere without the two sources combining. The most simple heat exchangers use conduction to achieve this. For example, a hot fluid is run through tubes running parallel to one another (and with a high surface area). This arrangement of tubes is immersed or built around a container into, and through, which another, cold fluid flows. By conduction the heat is transferred from the hot fluid to the cold.

heat pump Device employed mainly in refrigerators and central cooling systems. In a refrigeration cycle, a liquid boiling at about −30°C, usually a FREON, is evaporated at low pressure in metal coils. This absorbs heat from the surrounding space, so producing the cooling effect. The freon vapour then passes to a pump driven by an electric motor, which delivers it at raised temperature and pressure to a condenser outside the refrigerated area. As the vapour condenses to a liquid it gives up heat to the surrounding space. The liquid, still under pressure, passes to a receiver, from which the suction intake of the pump withdraws it at lowered pressure back to the evaporator in the refrigerated area, so completing the cycle.

heat shield In aerospace technology, plate or apron attached to the leading edge of a spacecraft to absorb or dissipate heat upon re-entry into the Earth's atmosphere. It is constructed of ablative materials, which chip off or vaporize to dissipate heat and provide an insulating layer of gases to protect the spacecraft. Materials used include quartz and various plastics.

Heaviside-Kennelly layer (E-layer) Region within the IONOSPHERE of the Earth's ATMOSPHERE.

heavy metal Any metal of high density (high ATOMIC NUMBER). They are used in electron microscopy to "stain" biological specimens in order to reveal details of structure. The term may also refer to metallic pollutants in soil that restrict plant growth and to toxic metals such as cadmium and lead.

heavy water (deuterium oxide, symbol D_2O) Water in which the role of hydrogen atoms is played by DEUTERIUM, the isotope of HYDROGEN of RELATIVE ATOMIC MASS (R.A.M.) approximately 2 (normal hydrogen has a relative atomic mass of approximately 1). It occurs in small proportions in ordinary water, from which it is obtained by ELECTROLYSIS. Heavy water is used as a MODERATOR in some NUCLEAR REACTORS.

Heisenberg, Werner Karl (1901–76) German physicist and philosopher, professor at Leipzig and later in Berlin. He is best known for his discovery in 1927 of the UNCERTAINTY PRINCIPLE (1927). He was awarded the 1932 Nobel Prize for physics for his theory of matrix mechanics, a form of QUANTUM MECHANICS.

helicopter Aircraft that gains lift from power-driven rotor(s). The helicopter is capable of vertical take-off and landing (VTOL), hovering, and forward, backward and lateral flight. To manoeuvre, helicopters have a complex system of hydraulic cylinders, control rods and plates which alter the orientation of the rotor blades. *See also* ROTOR, HELICOPTER

heliosphere Region of space in which the SUN's magnetic field and the SOLAR WIND dominate the interstellar medium. It is similar to a planet's MAGNETOSPHERE, with an outer bow shock, and a boundary called the **heliopause** where the energy of the solar wind particles has fallen to the level of galactic cosmic ray particles. Measurements returned by the Voyager probes suggest that the heliosphere extends for 100–150AU.

helium (symbol He) Gaseous, nonmetallic element, a NOBLE GAS, discovered in 1868. The element was first obtained in 1895 from the mineral clevite; the chief source today is from natural gas. It is also found in some radioactive minerals and in the Earth's ATMOSPHERE (0.0005% by volume). Helium is the second-most common element in the Universe and the second lightest element (after HYDROGEN). It has the lowest melting and boiling points of any element. A colourless, odourless non-flammable gas, it is used in meteorological balloons, to make artificial "air" (mixed with oxygen) for deep-sea divers, and in welding, semiconductors, LASERS, metallurgy and other applications requiring an INERT atmosphere. Liquid helium is used in CRYOGENICS (physics dealing with low tem-

HELICOPTER

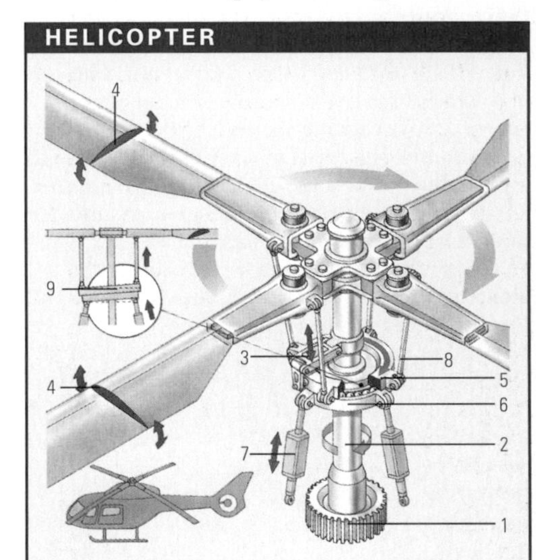

A helicopter rotor head transfers the power of the engines to the rotor blades via gears (1) and the rotor shaft (2). The swish plate controls the tilt of the rotor (3) and also the pitch of the blades (4). The upper (5) and lower (6) swish plates are controlled by hydraulic cylinders (7) attached to the lower plate. The upper plate is connected to the rotor blades by control rods (8). The pitch of the blades controls the amount of lift generated while the attitude of the whole rotor controls how the helicopter moves horizontally. If the rear of the swish plate is raised (9) the rotor dips towards the nose of the helicopter causing it to travel forward.

HELIUM

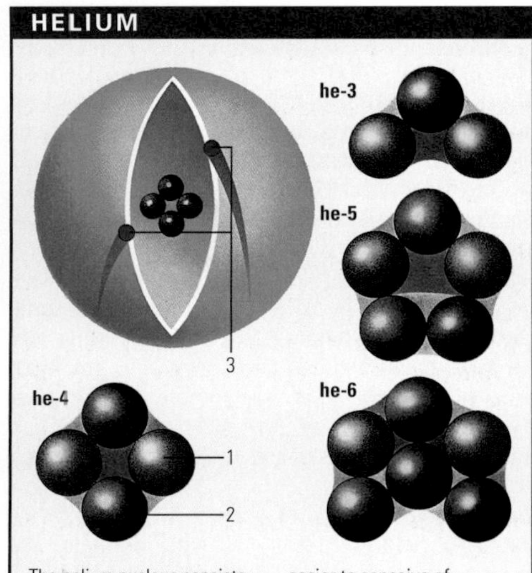

The helium nucleus consists of two protons (1) and (most usually) two neutrons (2), about which two electrons (3) revolve. The helium atom is smaller in size than the lighter hydrogen atom with its single proton and electron, because the greater charge of the helium nucleus pulls the electrons closer in. Although it is easier to conceive of electrons orbiting the nucleus in circular orbits, the shell should be considered as a "probability cloud" of locations in which the electrons are likely to be found. Isotopes of helium, although retaining 2 protons and 2 electrons, may contain from 1 to 4 neutrons.

peratures). The element forms no chemical compounds. Properties: at.no. 2; r.a.m. 4.0026; r.d. 0.178; m.p. −272.2°C (−458°F); b.p. −268.9°C (−452.02°F); single isotope ^{4}He.

helix Curve generated when a point moves over the surface of a CYLINDER in such a way that the path it traces is inclined at a constant angle to the axis of the cylinder, as in a coiled spring or the thread of a bolt.

helix, double *See* DOUBLE HELIX

Helmholtz, Hermann Ludwig Ferdinand von (1821–94) German anatomist, physicist and physiologist. He made contributions in ACOUSTICS and OPTICS, expanding the three-colour theory of vision formulated by Thomas YOUNG and inventing an OPHTHALMOSCOPE and an ophthalmometer. His experiments on the speed of nerve impulses led to a study of animal heat, which in turn led to work on the principle of conservation of energy.

Helmont, Jan Baptista van (1580–1644) Flemish chemist, physician and physiologist. He believed in the TRANSMUTATION of metals and that water was the basic substance of the Earth. He carried out quantitative experiments and coined the word "gas" (Greek *chaos*). He also discovered the gas now called carbon dioxide.

hematite *See* HAEMATITE

hemiplegia Paralysis of one side of the body. It can be caused by brain injury, and may be accompanied by spasticity.

Hemiptera Order of the class Insecta (*see* INSECT), having two pairs of wings and symmetrical mouthparts that are adapted for piercing and sucking. Within the order are the suborders **Heteroptera** (the true bugs) and **Homoptera** (which include white, black and green flies, cicadas, aphids, plant lice, leaf hoppers and scale insects).

Hench, Philip Showalter (1896–1965) US physician. He shared the 1950 Nobel Prize for physiology or medicine with Edward KENDALL and Tadeus REICHSTEIN for their pioneering work on hormones of the adrenal cortex, which led to discovery of the treatment of rheumatoid ARTHRITIS with CORTISONE. *See also* ADRENAL GLAND

Henry, Joseph (1797–1878) US physicist whose contribution to ELECTROMAGNETISM was essential for the development of the commercial telegraph. His work on INDUCTION led to the development of the TRANSFORMER. In 1850 he introduced a system of using the telegraph for sending weather reports, making possible the establishment of the US Weather Bureau. The unit of inductance is named after him.

Henry, William (1774–1836) British chemist and physician whose most important work was his study of the solubility of gases in water under varying temperatures and pressures. This led to his formulation in 1803 of HENRY'S LAW.

Henry's law Principle in physical chemistry stating that the mass of a gas dissolved in a fixed quantity of a liquid is directly proportional to the pressure of the gas on the liquid, at constant temperature. Thus the more a gas is compressed, the more it is absorbed in a liquid. It was named after William HENRY.

henry (symbol H) Unit of INDUCTANCE equal to the inductance of a closed loop that gives rise to a MAGNETIC FLUX of one weber for each ampere of current that flows.

heparin Anticoagulant substance produced in the body that inhibits the production of THROMBIN, an ENZYME involved in blood clotting. A purified extract is used after surgery to minimize the risk of thrombosis and to prevent the formation of further clots in anyone who has had a pulmonary EMBOLISM.

hepatitis Inflammation of the liver. It is usually due to a generalized infection. Early symptoms include lethargy, nausea, fever, jaundice and muscle and joint pains. Five different hepatitis viruses have come to light: A, B, C, D and E. The commonest single cause, accounting for up to 40% of cases worldwide, is the hepatitis A virus (HAV), which colonizes the digestive tract, causing an acute illness. Much more serious is infection with the hepatitis B virus (HBV). Formerly known as serum hepatitis, it can lead to chronic inflammation or complete failure of the liver and, in some cases, to liver cancer. Hepatitis C can also become chronic. HDV, common in the Mediterranean region, can only replicate in the presence of HBV. HEV is endemic in some tropical countries. Vaccines are available against HAV and HBV.

heptane Any chemical compound having the formula C_7H_{16}. Such compounds are saturated hydrocarbons (ALKANES). Normal heptane, or *n*-heptane, is used in car fuels to determine OCTANE ratings.

herb Seed-bearing plant (SPERMATOPHYTE), usually with a soft stem that withers away after one growing season. Most herbs are ANGIOSPERMS (flowering plants). The term is also applied to any plant used as a flavouring, seasoning or medicine, such as thyme, sage and mint.

herbaceous Term for any non-woody plant. Herbaceous plants can be either ANNUAL (dying completely after one season), BIENNIAL (taking two seasons to complete its lifecycle) or PERENNIAL (living for three or more seasons).

herbicide Chemical substance used to kill weeds and other unwanted plants. There are two kinds: selective herbicides kill the weeds growing with crops, leaving the crops unharmed; non-selective herbicides kill all the vegetation.

herbivore Animal that feeds solely on plants. The term is most often applied to mammals, especially UNGULATES (hoofed mammals). Herbivores are characterized by broad molars and blunt-edged teeth, which they use to pull, cut and grind their food. Their digestive systems are adapted to the assimilation of CELLULOSE. In a standard FOOD CHAIN, herbivores are PRIMARY CONSUMERS.

hereditary disorder (genetic disorder) Any disorder passed on from parents to their young in defective genes or chromosomes. Some 3,000 such disorders are known, including DOWN'S SYNDROME, CYSTIC FIBROSIS and SICKLE-CELL DISEASE.

heredity Transmission of physical and other characteristics from one generation of plants or animals to another. Characteristics such as blue eyes and red hair may be specific to the individuals involved; other features, such as erect posture and the possession of external ears, may be typical of the type of organism. The combination of characteristics that makes up an organism and makes it different from others is set out in the organism's GENETIC CODE, passed on from parent(s) to offspring. The first studies of heredity were carried out during the 19th century by Gregor MENDEL. *See also* ALLELE; CHROMOSOMES; GENE; SEXUAL REPRODUCTION

hermaphrodite Organism that has both male and female sexual organs. Most hermaphrodite animals are INVERTEBRATES, such as the earthworm and snail. They reproduce by the mating of two individuals, each of which receives SPERM from the other. Sometimes, in both plant and animal hermaphrodites, self-FERTILIZATION takes place.

hernia (rupture) Protrusion of an organ, or part of an organ, through its enclosing wall or through connective tissue. Common types of hernias are protrusion of an intestinal loop through the umbilicus (**umbilical** hernia) or inguinal canal of the groin (**inguinal** hernia), or protrusion of part of the stomach or oesophagus into the chest cavity through the opening (hiatus) of the oesophagus into the diaphragm (**hiatus** hernia).

heroin Drug derived from MORPHINE. Compared to morphine, heroin requires a smaller dose to produce similar effects more quickly, and is less likely to cause nausea and vomiting. It is prescribed for the relief of pain in terminal illness and severe injuries. Illicit heroin users soon become addicted.

Hero of Alexandria (active *c.*1st–2nd century AD) Greek engineer who built an early steam turbine, called an **aeolipile.** He also devised mechanisms to control doors automatically and to power moving statues.

Herschbach, Dudley Robert (1932–) US physical chemist who used intersecting beams of atoms or molecules to study atomic and molecular motions in chemical changes and to record the energy of the reactant and product molecules. He also invented devices to identify the products. Herschbach shared the 1986 Nobel Prize for chemistry with Yuan Tseh Lee and John Polanyi for their work in chemical reaction dynamics. They originally used their new technique for the reaction between potassium atoms and iodomethane (methyl iodide) molecules and later employed it for numerous other systems. Herschbach was professor of chemistry at Harvard and chairman of that department from 1977–80.

Herschel, Sir John Frederick William (1792–1871) English scientist and astronomer, son of Sir William HERSCHEL At first assisting his father with observations and telescope-making, he went on to extend his father's work on DOUBLE STARS and NEBULAE, discovering more than 500 more nebulae and clusters. In 1834 he took one of William's telescopes to the Cape of Good Hope and undertook a systematic survey of the S sky, discovering more than 1,200 doubles and 1,700 nebulae and clusters. He combined these and his father's observations into a *General Catalogue of Nebulae and Clusters*. He made the first good direct measurement of solar radiation, and inferred the connection between the Sun and aurorae. His *Outlines of Astronomy* (1849) was a standard textbook for many decades. He was also a pioneer of photography and its application in astronomy.

Herschel, Sir William (1738–1822) (originally Friedrich Wilhelm Herschel) English astronomer, b. Germany. He became famous in 1781 by discovering the planet Uranus, which he wanted to name after King George III; the next year he was appointed the king's private astronomer. He discovered two SATELLITES of Uranus (1787), and two of Saturn (1789). He observed and catalogued many double stars, hoping to measure their distances by detecting the parallactic movement of the nearer against the further if they were optical doubles. He observed more than 2,000 nebulae and clusters and published catalogues of them. Herschel realized that the Milky Way is the plane of a disc-shaped stellar universe, whose form he calculated by counting the numbers of stars visible in different directions, and also noted the motion of the Sun towards a point in the constellation Hercules. In 1800 he discovered and investigated the properties of INFRARED RADIATION.

Hershey, Alfred Day (1908–97) US biologist who shared the 1969 Nobel Prize for physiology or medicine for his part in discovering the way VIRUSES reproduce. Through his study of bacteriophages, Hershey confirmed that the NUCLEIC ACID RNA is solely responsible for infection and subsequent viral development. He was director of the Spring Harbour Laboratory from 1962–74.

Hertz, Gustav (1887–1975) German physicist, who shared the 1925 Nobel Prize for physics with his colleague James FRANCK for their research on the electron bombardment of gases, work that supported the QUANTUM THEORY. Hertz was Professor of Physics at Berlin and Director of the Siemens Research Laboratory until 1945. He worked as head of a research laboratory in the Soviet Union from 1945 to 1954, returning then to East Germany to become Director of the Physics Institute at Leipzig. He was the nephew of Heinrich HERTZ

Hertz, Heinrich Rudolf (1857–94) German physicist who discovered radio waves. He was assistant to Hermann HELMHOLTZ and held professorships at Karlsruhe and later at Bonn. He discovered, broadcast and received the radio waves predicted by James Clerk MAXWELL. He demonstrated that heat and light are also part of the ELECTROMAGNETIC SPECTRUM. The unit of frequency, the HERTZ (Hz), is named after him. He was the uncle of Gustav HERTZ.

hertz (symbol Hz) SI unit of wave FREQUENCY named after Heinrich HERTZ. One oscillation per second is equivalent to 1Hz. This unit replaces the cycle per second. Commonly used multiples are the kilohertz (kHz) equal to 1,000Hz and the megahertz (MHz) equal to 1,000,000Hz.

Hertzsprung, Ejnar (1873–1967) Danish astronomer who devised (1908) a diagram illustrating the relationship between the LUMINOSITIES, surface temperatures and SPECTRAL CLASSIFICATION of stars. He anticipated the work of Henry RUSSELL in the USA a few years later. The diagram is now known as the HERTZSPRUNG–RUSSELL DIAGRAM.

Hertzsprung–Russell diagram (HR diagram) Plot of the absolute MAGNITUDE of stars against their SPECTRAL CLASSIFICATION; this is equivalent to plotting their LUMINOSITY against their surface temperature or colour index. Brightness increases from bottom to top, and temperature increases from right to left. The diagram was devised by Henry RUSSELL in 1913, independently of Ejnar HERTZSPRUNG, who had had the same idea some years before. The HR diagram reveals a pattern in which most stars lie on a diagonal band, the MAIN SEQUENCE, which includes our Sun. There is another region of GIANT STARS, which lies above the main sequence. There are further groupings above the main sequence called SUPERGIANTS and subgiants. Below the main sequence are subdwarfs and WHITE DWARFS. The HR diagram is an important aid for interpreting astrophysical data. *See illustration on page 178*

Herzberg, Gerhard (1904–) Canadian scientist, b. Germany. Educated at Darmstadt, Göttingen and Bristol, he settled in Canada in 1935. He joined the National Research Council in Ottawa in 1949, and became their first Distinguished Research Scientist in 1969. He was awarded the 1971 Nobel Prize for chemistry for his work on the electronic structure and geometry of molecules, particularly FREE RADICALS.

Hess, Victor Francis (1883–1964) US physicist, b. Austria. As a result of his investigations into the ionization of air, he suggested that radiation similar to X-rays, later named COSMIC RADIATION, comes from space. He shared the 1936 Nobel Prize for physics with US physicist Carl ANDERSON, who had worked indpendently on cosmic rays. Hess then emigrated to the USA in 1938 to become Professor of Physics at Fordham University in New York City.

Hess, Walter Rudolf (1881–1973) Swiss physiologist. He shared with António de Egas MONIZ the 1949 Nobel Prize for physiology or medicine for his work on how specific, highly circumscribed areas of the BRAIN control body functions, such as blood circulation, muscle relaxation and breathing rate.

Hess's law In physical chemistry, law referring to the heat produced or absorbed during CHEMICAL REACTIONS leading to the preparation of any chemical compound. It states that the net value of this heat is the same by any chemical route.

heterocyclic compound *See* CYCLIC

heterodyne In radio, the production of one WAVE FREQUENCY as the difference of two others. It is used in radio-frequency OSCILLATORS and in receivers. The heterodyne system was invented by Reginald A. FESSENDEN in 1905.

heterostyly Condition found in some ANGIOSPERMS (flowering plants) of having STYLES of different lengths. The classic example of heterostyly is the primrose *Primula vulgaris*, in which there are two distinct flower forms, each occurring on different plants. Each form has a different-sized style and STAMEN from the other so that pollen collected by insects from the ANTHER of one type is dropped on the STIGMA of the other. Heterostyly increases the likelihood of CROSS-POLLINATION because pollen cannot be passed between plants with sex organs of the same size.

heterotroph Organism that obtains its energy from the digestion of organic matter (usually plant or animal tissue) through a process known as **heterotrophic nutrition**. All animals and fungi are heterotrophs; the digestive process breaks down the tissue to provide material with which the organism can synthesize essential nutrients, such as carbohydrates, protein, fat, vitamins and minerals in the case of humans.

heterozygote Organism possessing two contrasting forms (ALLELES) of a GENE in a CHROMOSOME pair. In cases where one of the forms is DOMINANT and one RECESSIVE only the dominant form will be expressed in the PHENOTYPE. *See also* HOMOZYGOTE

Hevesy, George Karl von (1855–1966) Hungarian-Swedish chemist, who was awarded the 1943 Nobel Prize for chemistry. Co-discoverer of the element HAFNIUM (Hf) in 1923, he was also an early researcher into the uses of radioactive ISOTOPES, including their use as "tracers" in living tissue. In the early 1940s he fled from Germany, where he had been teaching, to live in Sweden. He worked with Ernest RUTHERFORD in England and later with Niels BOHR in Copenhagen.

Hewish, Antony (1924–) British radio astronomer. He and Martin RYLE developed the technique of **aperture synthesis,** which allows an array of radio dishes to synthesize the resolution of a much larger telescope, in 1960. In 1967 his student Jocelyn Bell Burnell (1943–) obtained the first signal from a PULSAR. For subsequent work on pulsars, Hewish shared (with Ryle) the 1974 Nobel Prize for physics.

hexacyanoferrate Complex ION of iron. There are two ions: hexacyanoferrate(II) ($[Fe(CN)_6]^{4-}$), formerly called ferrocyanide, and hexacyanoferrate(III) ($[Fe(CN)_6]^{3-}$), or ferricyanide. The potassium salts are used in chemical ANALYSIS for the presence of iron. Iron(III) hexacyanoferrate(II) (ferric ferrocycanide) is the dark blue pigment Prussian blue. The pigment was once used a source of HYDROCYANIC ACID, hence its former name prussic acid.

hexagon Six-sided plane figure. Its interior angles

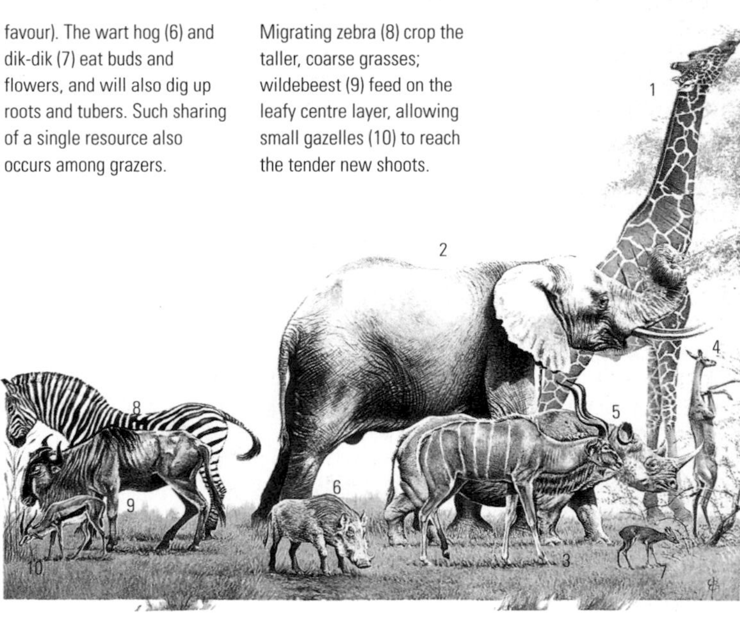

► **herbivore** Mammalian herbivores may conveniently share a habitat without competing for resources. On the African plains, giraffes (1) browse in branches up to 6m (20ft) above the ground. Elephants (2) too can browse tree canopies, using their trunks to pluck off vegetation. Eland (3) attack the middle branches with their horns, twisting twigs to break them off, while gerenuk (4) stand on their hind legs to reach higher branches. The black rhino (5) uses its hook-like upper lip to feed on bark, twigs and leaves (white rhinos have lengthened skulls and broad lips for grazing the short grasses that they favour). The wart hog (6) and dik-dik (7) eat buds and flowers, and will also dig up roots and tubers. Such sharing of a single resource also occurs among grazers. Migrating zebra (8) crop the taller, coarse grasses; wildebeest (9) feed on the leafy centre layer, allowing small gazelles (10) to reach the tender new shoots.

add up to 720°. A regular hexagon has all sides and interior angles equal; each interior angle is 120°.

Heyrovsky, Jaroslav (1890–1967) Czechoslovakian chemist. He was awarded the 1959 Nobel Prize for chemistry for the discovery of **polarography**, an electrochemical method of analysis that can be used with substances that undergo electrolytic OXIDATION (loss of electrons) or REDUCTION (gain of electrons) in solution.

hibernation Dormant (sleep-like) condition adopted by some animals to survive the harsh conditions of winter. Adaptive mechanisms to avoid starvation and extreme temperatures include reduced body temperature, a slower heartbeat, slower breathing rate and slower rate of METABOLISM.

Higgs boson ELEMENTARY PARTICLE the existence of which is necessitated by the ELECTROWEAK THEORY. Predicted by particle physicist Peter Higgs (1929–), this boson has not yet been detected. It should have a finite mass but no SPIN.

high *See* ANTICYCLONE

high-definition television (HDTV) Form of TELEVISION in which the picture is made up of 1,250 or 1,125 scanning lines instead of the previous 625 or 525. The increased number of scanning lines makes the television image much clearer and sharper (comparable with the visual clarity of movie film), and relies on digital transmission along OPTICAL FIBRES, rather than the transmission of electronic signals by RADIO waves. The use of such cables greatly increases the total number of channels available, and also allows for an interactive facility between receiver and transmitter. The screens of many HDTV sets are proportionately wider and flatter than former models.

high-fidelity (hi-fi) Sound recording and reproduction with the minimum of distortion, usually involving the use of high-quality components and circuits. The original sound source may be a COMPACT DISC (CD), gramophone record, MAGNETIC TAPE, a radio signal or a live performance. An audio amplifier is required to "magnify" the signals, and one or more loudspeakers are needed to make the source audible. Domestic high-fidelity sound systems have been produced since the 1920s, but did not become common until the advent of long-playing gramophone records in the 1950s and 1960s.

high-level language COMPUTER LANGUAGE used for programming, that is reasonably close to spoken English. The high-level language is converted to a form a computer can use by a COMPILER. Most languages in use today (such as Basic, Pascal or C) are high-level languages.

high-temperature physics Production and analysis of the effects of temperatures above 50,000°C (90,000°F). At such temperatures atoms begin to be ionized (stripped of their ELECTRONS) and a fourth state of matter, called a PLASMA, is achieved. Normal stars are plasmas; so is the matter at the heart of an exploding hydrogen bomb. To make controlled thermonuclear FUSION (joining of light nuclei and release of energy) a reality, physicists must find a way to confine a plasma at temperatures of over a million degrees. *See also* MAGNETIC BOTTLE; NUCLEAR REACTOR; NUCLEAR WEAPON

Hilbert, David (1862–1943) German mathematician, b. Russia, one of the most important figures in 20th-century mathematics. While professor of mathematics at Göttingen, Hilbert wrote *Foundations of Geometry*, which put geometry on a rigorous axiomatic basis. His *Numbers Information* systematized all the known results of algebraic number theory. He developed the concept of an infinite dimensional space, now called a **Hilbert space**, which was a tool necessary for the develop-

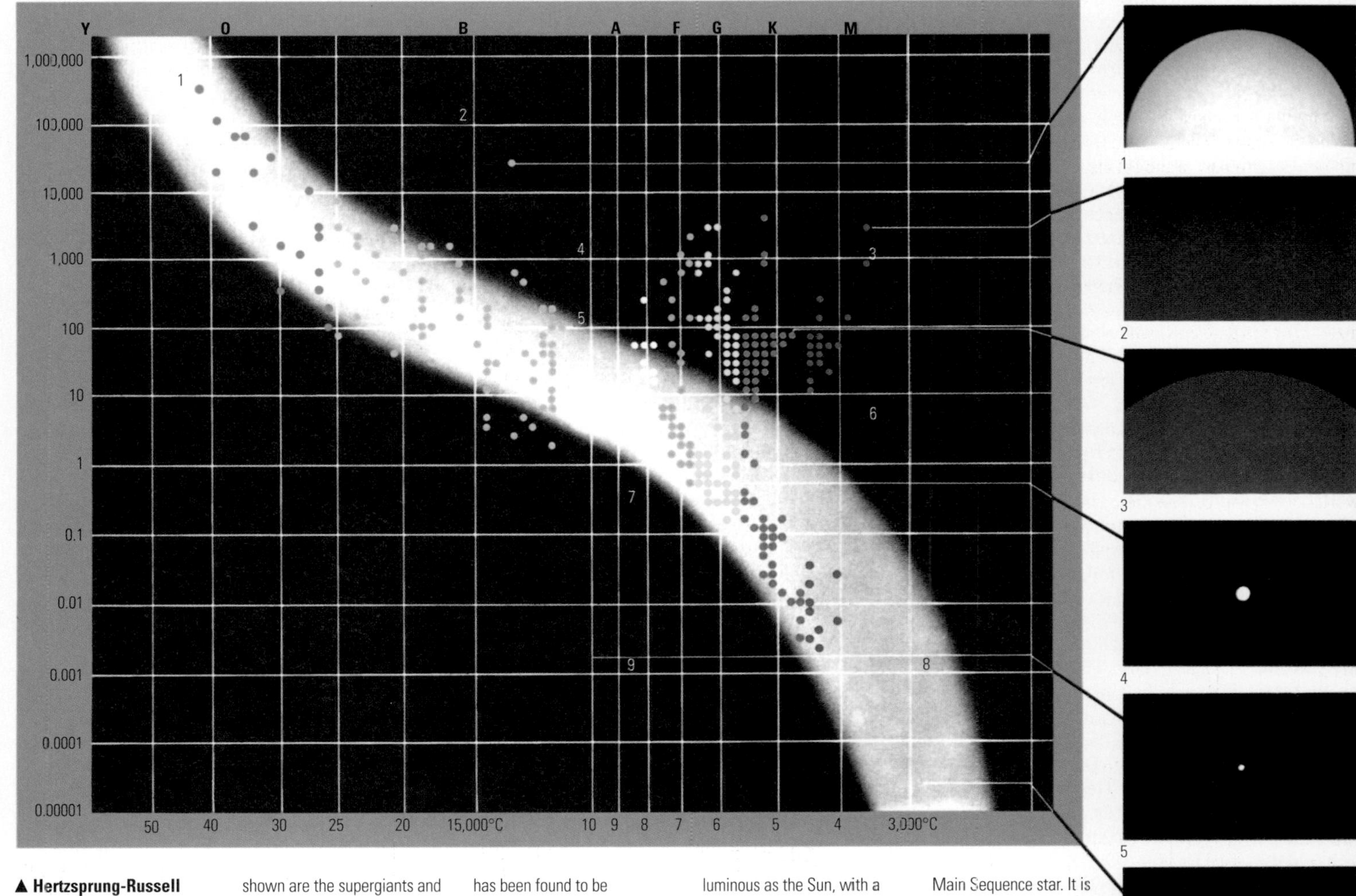

▲ **Hertzsprung-Russell diagram** In this version of a typical HR diagram, the stars are plotted according to their spectral types and surface temperatures (horizontal axis, x) and their luminosities in terms of the Sun (vertical axis, y). The Main Sequence is obvious at first glance, from the hot and powerful W and O stars (1), through to the dim red dwarfs of type M (8). Also shown are the supergiants and giants (2, 3); Cepheid variables (4); RR Lyrae variables (5); subgiants (6); subdwarfs (7); and white dwarfs (9). Originally it was believed that a star began its career as a large, cool red giant, and then heated up to join the Main Sequence; it then cooled and shrank as it passed down the Main Sequence from top left to bottom right. This theory has been found to be completely wrong. The red giants are at an advanced stage in their evolution.
1) Rigel: type B8. A very massive and luminous star, at the upper end of the Main Sequence. It is 60,000 times as luminous as the Sun, and is white, with a temperature of more than 12,000°C.
2) Betelgeuse: type M. A red supergiant, 15,000 times as luminous as the Sun, with a greater diameter than that of the Earth's orbit. It is surrounded by a very tenuous "shell" of potassium.
3) Aldebaran: type K. A giant star, orange in colour, not as large as Betelgeuse, although it is 100 times as powerful as the Sun and has a diameter believed to be at least 50 million km (30 million mi).
4) The Sun: type G2. A typical Main Sequence star. It is officially ranked as a dwarf, while Capella, also of type G (G*) is a giant. (Capella is not a single star, but a close binary).
5) Sirius B: A white dwarf which has used up all its nuclear "fuel". It has a diameter of 40,000km (25,000mi), smaller than that of Uranus, but it is very dense, and is as massive as the Sun.
6) Wolf 339: type M. A dim red dwarf, with a temperature of 3,000°C but a luminosity only 0.00002 that of the Sun. Yet its spectral type is the same as that of Betelgeuse.

ment of QUANTUM THEORY. In 1900 Hilbert put forward a list of 23 problems for the mathematicians of the new century to tackle. Most of these have led to profound new understandings in mathematics. Hilbert's dream was that it would be possible either to prove or to disprove all mathematical questions starting with only one set of well-defined rules and assumptions. In 1931 Kurt GÖDEL showed that this could not be achieved.

hindbrain One of the three primary parts of the BRAIN. In the embryo, the hindbrain develops into the CEREBELLUM, the part of the brain concerned with unconscious muscle coordination, and also into the PONS and the MEDULLA OBLONGATA. *See also* FOREBRAIN; MIDBRAIN

Hinshelwood, Sir Cyril Norman (1897–1967) British chemist, president of the Royal Society (1955–60). From 1937–64 he was Professor of Chemistry at Oxford, and Research Fellow at Imperial College, London, from 1964. He shared the 1956 Nobel Prize for chemistry with Nikolai SEMENOV for their simultaneous work on chemical reaction kinetics.

Hinton, Christopher, Baron Hinton of Bankside (1901–83) British nuclear engineer. From 1946, as deputy controller of NUCLEAR REACTOR atomic energy production, he created a new industry in Britain with the world's first full-scale atomic power station at Calder Hall, which was opened in 1956.

hip Joint on each side of the lower trunk, into which the head of each FEMUR (thigh bone) fits; the hip bones form part of the PELVIS.

Hipparchus (*c*.190–127 BC) Greek astronomer. Hipparchus estimated the distance of the Moon from the Earth and drew the first accurate star map with more than 800 stars. He divided stars into orders of magnitude based on their brightness – a system fundamentally in use today. He also developed an organization of the Universe, which, although it still had the Earth at the centre, provided for accurate prediction of the positions of the planets.

Hippocrates (460–377 BC) Greek physician, often called "the father of medicine". Although little is known about him, and the writings known as the *Hippocratic Collection* probably represent the works of several people, he exerted a tremendous influence. He freed medicine from superstition, emphasizing clinical observation and providing guidelines for surgery and for the treatment of fevers. He is credited with providing the **Hippocratic oath**, a code of professional conduct still followed by doctors today.

histamine Substance derived from the AMINO ACID HISTIDINE, occurring naturally in many plants and in animal tissues, and released on tissue injury. Histamine's several functions in the body include dilation of the capillaries. It is implicated in allergic reactions (such as hay fever), which can be treated with ANTIHISTAMINE drugs.

histidine ($C_3H_3N_2CH_2CHNH_2COOH$) Colourless, soluble, crystalline AMINO ACID, which is the percursor of HISTAMINE.

histochemistry Study of the distribution of biochemical substances and reactions in living tissue, usually through visual examination. *See also* BIOCHEMISTRY

histogram (bar chart) Chart in which bars represent the frequency (in absolute terms or as a percentage of the total) with which certain values (or ranges of value) occur within a given set of data. *See also* GRAPH

histology Biological, especially microscopic, study of TISSUES and structures in living organisms.

histone Basic protein found associated with DNA in CHROMOSOMES. Histones function during CELL DIVISION to condense CHROMATIN and coil the chromosomes. There are five kinds, each containing large amounts of the AMINO ACIDS arginine, histidine and lysine; they are soluble in water.

HIV *See* HUMAN IMMUNODEFICIENCY VIRUS

Hodgkin, Sir Alan Lloyd (1914–) British physiologist. He shared the 1963 Nobel Prize for physiology or medicine with John Eccles (1903–) and Andrew Huxley for his part in discoveries concerning the ionic mechanisms involved in the excitation and inhibition of the membranes of nerve cells (NEURONS). This work is fundamental to an understanding of the function of neurons themselves.

Hodgkin, Dorothy Mary Crowfoot (1910–94) British chemist. After teaching at Oxford and Cambridge universities, she spent some time at the University of Ghana, returning to Oxford in 1960. In 1964 she was awarded the Nobel Prize for chemistry for her determination of the structure of VITAMIN B_{12} by X-RAY CRYSTALLOGRAPHY. She also determined the structure of PENICILLIN and INSULIN.

Hodgkin's disease Rare type of cancer characterized by painless enlargement of the LYMPH NODES, lymphatic tissue and spleen, with subsequent spread to other areas. Fever is a common symptom and weight loss, anaemia, loss of appetite and night sweats may occur. It is named after the pathologist Thomas Hodgkin (1798–1866), who first described it. The condition is more common in men. Treatment varies with the stage reached by the disease but in general consists of RADIOTHERAPY, surgery, drug therapy or a combination of these. Caught in the early stages it is potentially curable.

Hoff, Jacobus Hendricus van't (1852–1911) Dutch physical chemist, professor at Amsterdam, Leipzig and then Berlin. His research involved advanced investigations of the carbon atom, the theory that gas laws are also applicable to dissolved substances, and the chemical application of THERMODYNAMICS. In 1901 he was awarded the first Nobel Prize for chemistry for his studies on the mechanism of CHEMICAL EQUILIBRIUM, reaction rates and osmotic pressure. *See also* IDEAL GAS LAW; OSMOSIS

Hofmann, August Wilhelm von (1818–92) German chemist and teacher. After serving as an assistant to Justus von LIEBIG he was appointed director of the new Royal College of Chemistry in London. Most of his work was based on the compounds of COAL TAR and their derivatives. Hofmann discovered a method, known as the **Hofman reaction**, for converting an AMIDE organic compound into an AMINE. He also discovered METHANAL (formaldehyde). One of his pupils, William Perkin (1838–1907), made the first ANILINE dye, the colour mauve.

Hofstadter, Robert (1915–) US physicist who shared the 1961 Nobel Prize for physics with Rudolf MÖSSBAUER for their investigations of nuclear structure. Hofstadter proposed that the PROTON and NEUTRON (both of which make up the NUCLEUS) have a positively charged central core surrounded by a cloud of ELEMENTARY PARTICLES (PIONS).

Holland, John Philip (1840–1914) US engineer, b. Ireland. He invented the first modern, petrol-driven SUBMARINE. He offered a design to the US Navy Department in 1875, but it was rejected as impracticable. After further development, the successful *Fenian Ram* was launched in 1881. This was not practical for extended operations, but its design was the basis for the *Holland*, which had an INTERNAL COMBUSTION ENGINE for surface cruising and an electric motor for submerged operations. The US government bought the *Holland* in 1900.

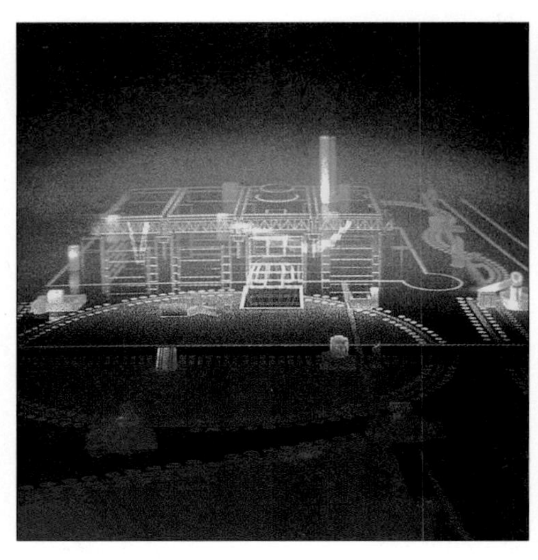

▲ **holography** Transmission hologram of the Palace of Discovery at La Villette, Paris, France. The hologram, made on film 113 x 117cm (45 x 47in) in size and illuminated by halogen lamp, was made from a model of the future Paris museum of science and technology.

Hollerith, Herman (1860–1929) US computer pioneer. In 1890 Hollerith invented a mechanical tabulating machine for use in analysing the results of the US census of that year. The machine used punched cards to record and process data. These gave rise to the HOLLERITH CODE (later employed by early electronic computers), which uses 12 bits per alphanumeric character. Hollerith's firm, the Tabulating Machine Company, later became a founder of IBM.

Hollerith code Computer code consisting of 12 levels, or bits per character, which defines the relation between an alpha-numeric character and the punched holes in an 80-column computer data card.

Holley, Robert William (1922–93) US biochemist. He shared with Har KHORANA and Marshall NIRENBERG the 1968 Nobel Prize for physiology or medicine for his work on the way GENES determine the function of CELLS. Holley described the first full sequence of sub-units in nucleic acid, the genetic material of a cell. This was an important step towards the understanding of gene action.

Holmes, Arthur (1890–1965) British geologist and geophysicist who studied the age of rocks by measuring their radioactivity. He devised the first quantitative GEOLOGICAL TIMESCALE in 1913 and estimated the Earth's age by means of temperature measurement.

holmium (symbol Ho) Metallic element of the LANTHANIDE SERIES, first identified spectroscopically in 1878. Its chief ore is monazite (a phosphate). The element has few commercial uses. Properties: at.no. 67; r.a.m. 164.930; r.d. 8.803 (25°C); m.p. 1,470°C (2,680°F); b.p. 2,720°C (4,930°F); most common isotope ^{165}Ho (100%).

Holocene (Recent or Post-Glacial Epoch) Division of the GEOLOGICAL TIME extending from roughly 10,000 years ago to the present. It includes the emergence of human beings as settled members of communities, with the first known villages dating from about 8,000 years ago.

holography Process of making a **hologram**. One or more photographs are formed on a single film or plate by means of interference between two parts of a split LASER beam. An apparently meaningless pattern is formed, but when light falls on it (ideally laser light, but possibly ordinary light), the reflected or transmitted light reconstructs a 3-D image or images. The concept of holography has been appreciated since Dennis GABOR's inventions in the

1940s, but progress was not possible until the development of a sufficiently coherent light source – the laser.

homeopathy Unorthodox medical treatment involving the administration of minute doses of a drug or remedy that causes effects or symptoms similar to those that are being treated. It is the opposite of allopathy, in which curatives are administered to create the opposite effect. Homeopathy was popularized in the 18th century by the German physician Christian HAHNEMANN.

homeostasis In biology, processes that maintain constant conditions within a cell or organism in response to either internal or external changes.

homeothermic (endothermic) Describes an animal whose body temperature does not fluctuate as the temperature of its surroundings fluctuates, often referred to as warm-blooded. Mammals and birds are warm-blooded. They maintain their body temperature through METABOLISM. *See also* POIKILOTHERMIC

hominid Any member of the family Homonidae. Members include modern humans and our earliest ancestors. *See also* HOMO

Homo Genus to which humans belong. Modern humans are classified *Homo sapiens sapiens*. *See also* HOMO ERECTUS; HOMO HABILIS; HUMAN EVOLUTION; NEANDERTHAL

Homo erectus Species of early human dating from *c.*1.5 million to 0.2 million years ago. The "Ape Man of Java" was the first early human fossil to be found, late in the 19th century. Both it and Peking man, another early discovery, represent more advanced forms of *Homo erectus* than older fossils found more recently in Africa. *See also* HUMAN EVOLUTION

homogenization Process that reduces a substance contained in a FLUID (liquid or gas) to small particles and redistributes them evenly through it. For example, the fat in milk can be broken down so thoroughly by homogenization that particles do not recombine and cream does not rise.

Homo habilis Species of early human, discovered by Louis LEAKEY in 1964 in the Olduvai Gorge, East Africa. Its fossil remains have been estimated to be between 1.8 and 1.2 million years old, being contemporary with those of AUSTRALOPITHEUS. The development of hand and skull is much more like that of modern human. *See also* HUMAN EVOLUTION

homologous Term that describes similarity in essential structure of organisms based on a common genetic heritage. It often refers to organs that now have a different superficial appearance and function in different organisms. For example, despite appearances, a human arm and a seal's flipper are homologous, having evolved from a common origin. *See also* EVOLUTION

HOMEOTHERMIC

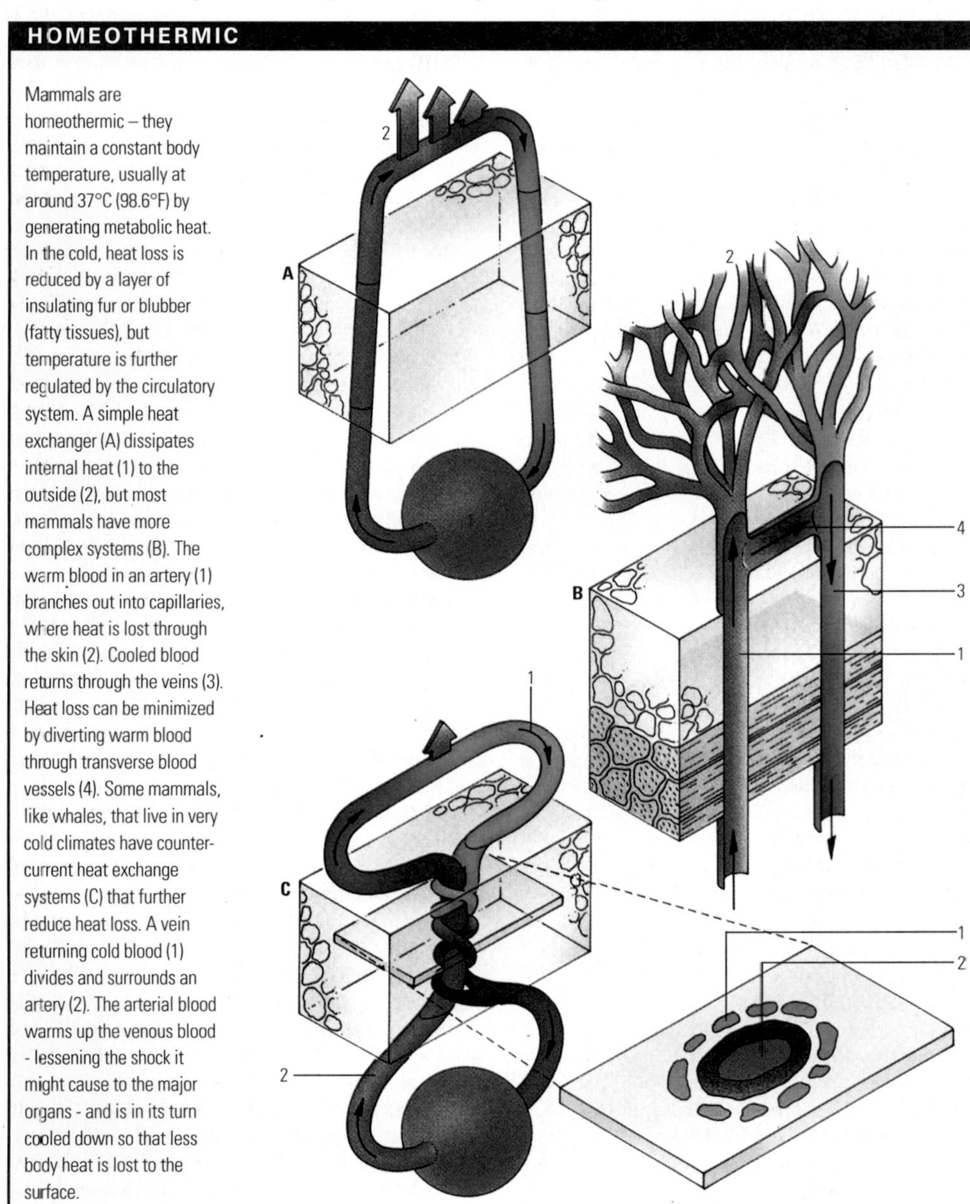

Mammals are homeothermic – they maintain a constant body temperature, usually at around 37°C (98.6°F) by generating metabolic heat. In the cold, heat loss is reduced by a layer of insulating fur or blubber (fatty tissues), but temperature is further regulated by the circulatory system. A simple heat exchanger (A) dissipates internal heat (1) to the outside (2), but most mammals have more complex systems (B). The warm blood in an artery (1) branches out into capillaries, where heat is lost through the skin (2). Cooled blood returns through the veins (3). Heat loss can be minimized by diverting warm blood through transverse blood vessels (4). Some mammals, like whales, that live in very cold climates have counter-current heat exchange systems (C) that further reduce heat loss. A vein returning cold blood (1) divides and surrounds an artery (2). The arterial blood warms up the venous blood - lessening the shock it might cause to the major organs - and is in its turn cooled down so that less body heat is lost to the surface.

homologous chromosome One of a pair of CHROMOSOMES that share the same genetic structure, but not always the same ALLELES for a given characteristic. Found in DIPLOID cells, pairs of homologous chromosomes are composed of one chromosome from the female parent, the other from the male. During MEIOSIS, when GAMETES (sex cells) are formed, homologous chromosomes undergo a complex process known as CROSSING OVER, when strands of the chromosomes (CHROMATIDS) break from the original chromosome and exchange places with the other member of the pair. This "reshuffling" gives rise to genetic variation, which together with random MUTATION is an essential part of EVOLUTION.

homologous series Series of related organic chemical compounds that share similar properties and the same general formula, differing only by a single, unchanging collection of atoms. A notable example is the ALKANE hydrocarbon group. They all have the general formula C_nH_{2n+2}. Therefore each member of the group changes by a CH_2 group – methane (CH_4), ethane (C_2H_6), propane (C_3H_8) and butane (C_4H_{10}).

homozygote Organism possessing identical forms of a GENE on a CHROMOSOME pair. It is a purebred organism and always produces the same kind of GAMETE (sex cell). *See also* HETEROZYGOTE

Hooke, Robert (1635–1703) English philosopher, experimental physicist and inventor, considered the best mechanic of his generation. Interested in astronomy, he deduced that Jupiter rotated and claimed to have stated the laws of planetary motion before Isaac NEWTON. Hooke made many improvements to astronomical instruments, watches and clocks. He also studied the elastic properties of solids, which led to HOOKE'S LAW. Among his inventions were a practical optical system and the Gregorian (reflecting) MICROSCOPE, with which he discovered plant CELLS, having coined the biological term "cell".

Hooke's law Relationship between STRESS and STRAIN in an elastic material when it is stretched. The law states that the stress (force per unit area) is proportional to the strain (a change in dimensions). The law, which holds only approximately and over a limited range, was discovered in 1676 by Robert HOOKE. *See also* ELASTICITY

Hopkins, Sir Frederick Gowland (1861–1947) British biochemist. He shared with the Dutch physician Christiaan EIJKMAN the 1929 Nobel Prize for physiology or medicine for his work on VITAMINS. Hopkins pointed out that some diseases, such as scurvy and rickets, might be caused by the deficiency in the diet of a substance necessary for proper health, the so-called **vitamin concept**. He also investigated the function of LACTIC ACID in muscular activity.

horizon In geology, a term used in three ways: 1) a continuous horizontal surface or time-plane between two STRATA that has no thickness; 2) a horizontal layer generally less than 1m (3ft) in thickness that is characterized either by a distinct fossilized FLORA or FAUNA or by a distinctive rock type; 3) a distinct layer in soil is also called a horizon. *See also* SOIL HORIZON

horizon, celestial GREAT CIRCLE on the CELESTIAL SPHERE, the plane of which contains the line through an observer's position at right angles to the vertical. It lies midway between the observer's ZENITH and NADIR and cuts the observer's meridian at the N and S points.

horizon coordinate system Astronomical COORDINATE SYSTEM referred to the plane of the observer's horizon. The coordinates are AZIMUTH and ALTITUDE.

hormone Chemical substance secreted by living CELLS which affects the METABOLISM of cells in other parts of the body. In MAMMALS, hormones are secreted by glands of the ENDOCRINE SYSTEM and released directly into the bloodstream. They exercise chemical control of physiological functions, regulating growth, development, sexual maturity and functioning, metabolism and (in part) emotional balance. Hormones circulate in the body in minute quantities and usually exert their effects at considerable distances from their sites of origin. Most are slow to take effect, exert widespread action, and are also slow to disappear from the system. The secretion and activity of the various hormones are closely inter-dependent; one can stimulate or inhibit the secretion of another, or two or more can act together to produce a certain effect. In general they maintain a delicate equilibrium which is vital to health. The HYPOTHALAMUS, which lies at the base of the brain and is adjacent to the PITUITARY GLAND, is responsible for overall coordination of the secretion of many hormones. Familiar hormones include THYROXINE, ADRENALINE, INSULIN, OESTROGEN, PROGESTERONE, TESTOSTERONE and ANTIDIURETIC HORMONE (ADH). In plants, hormones control many aspects of metabolism, including cell elongation, direction of growth, cell division, initiation of flowering, development of fruits, leaf fall, and responses to environmental factors such as light, water and gravity. The most important plant hormones include AUXIN, GIBBERELLIN, CYTOKININ and ABSCISIC ACID, a growth inhibitor that prevents buds from opening too soon in spring and keeps seeds DORMANT through adverse seasons. *See also* HOMEOSTASIS

hormone replacement therapy (HRT) Use of the female HORMONES OESTROGEN and progestogen to relieve symptoms in women who are either menopausal or who have had both ovaries removed. HRT is effective against hot flushes, vaginal dryness and the mood changes associated with the MENOPAUSE; it also gives some protection against disease of the heart and blood vessels, and OSTEOPOROSIS. However, except in people who have undergone HYSTERECTOMY, the oestrogen component causes a thickening of the lining of the uterus, which may lead to an increased risk of cancer of the ENDOMETRIUM. The progestogen component of the treatment causes a regular shedding of the lining, similar to MENSTRUATION, which may lessen this risk.

hornblende Black or green mineral found in IGNEOUS and METAMORPHIC ROCK. The commonest form of AMPHIBOLE, it contains iron and silicates of calcium, aluminium and magnesium. Hardness 5.5; r.d. 3.2.

hornfels Group of hard, fine-grained METAMORPHIC ROCKS formed originally in a high-temperature area in near contact with an IGNEOUS INTRUSION. They all contain QUARTZ and MICA, with a third component that gives them their individual names: **chiastolite** hornfels, **cordierite** hornfels, **garnet** hornfels and **pyroxine** hornfels. All are dark grey or black, although garnet hornfels has reddish patches.

horn Defensive or offensive structure that grows from the head of some mammals. A typical horn is made up of a central bony core covered by a layer of the skin protein KERATIN; in the rhinoceros, the entire horn is made of keratin. Horns are generally retained for life, although deer shed their antlers annually.

horology Science of measuring TIME, or the art of constructing CLOCKS and watches.

horse latitudes Either of two areas of high pressure found near 25°N and 25°S of the Equator. The air rises at or near the Equator because of low pressure and then circles in the atmosphere, but much falls near the tropics of Cancer and Capricorn. The falling air gives rise to fairly permanent regions of high pressure, known as the horse latitudes. Both the TRADE WINDS, which blow from the horse latitudes towards the Equator, and the westerly winds, which blow from the horse latitudes towards the poles, are caused by the high pressure.

horsepower (symbol hp) Unit indicating the rate at which work is done, adopted by James WATT in the 18th century. He defined it as the weight, 550lb (250kg), a horse could raise 1ft (0.3m) in one second – a rate of 550ft-lb/s. At the output shaft of an engine or motor, it is termed "brake horsepower" or "shaft horsepower". In large reciprocating engines it is termed "indicating horsepower" and is determined from the pressure in the cylinders. The electrical equivalent of one horsepower is 746 watts.

horsetail Any of about 30 species of flowerless, rush-like plants of the phylum Sphenophyta. They are allied to FERNS and grow in all continents except Australasia. The hollow jointed stems have a whorl of tiny leaves at each joint. Spores are produced in a cone-like structure at the top of a stem. Horsetails date from the Carboniferous period.

Horsley, Sir Victor Alexander Haden (1857–1916) British physiologist and surgeon who introduced refinements in NEUROSURGERY. He performed the first successful removal of a tumour of the spinal cord and carried out research on the functions of the thyroid gland, on rabies vaccine and on functions of various regions of the brain.

horst In geology, an elongated upthrust block bounded by parallel normal FAULTS on its long sides. *See also* GRABEN

horticulture Growing of vegetables, fruits, seeds, herbs, ornamental shrubs and flowers, on a commercial scale, in nurseries, orchards and gardens. The origins of horticulture lie in the small intensive kitchen gardens of the medieval farming system. The techniques employed include asexual, or vegetative, PROPAGATION by leaf, stem and root cuttings, and by stem and bud grafting. Fruit trees, shrubs and vines, including those bearing apples, pears, plums, cherries, citrus fruits and grapes, are usually propagated by GRAFTING the fruiting stock on to a hardier rootstock. In the case of apple trees, the rootstock is usually French crab apple. Plants propagated sexually by seed include some fruits and vegetables, including maize and many other types that produce seed abundantly. Often the seed needs to be overwintered, or stored at low temperatures and high humidity, before it will begin GERMINATION. Seed itself is a major horticultural crop. Close scientific control of POLLINATION is essential for producing crops of specific quality.

host Organism infected by a PARASITE, either **definitive**, in which the parasite reaches sexual maturity, or **intermediate**, in which it does not.

hot spot Area of high volcanic activity. Some are located on CONSTRUCTIVE MARGINS, such as Iceland; others occur within plates, for example Hawaii. Hot spots are thought to be caused by upwelling of MAGMA in the MANTLE beneath (mantle plumes). *See also* PLATE TECTONICS

Houssay, Bernardo Alberto (1887–1971) Argentinian physiologist who greatly advanced knowledge of the ENDOCRINE SYSTEM. He shared with Carl CORI the 1947 Nobel Prize for physiology or medicine for discovering the role played by the HORMONE of the anterior lobe of the PITUITARY GLAND in regulating the METABOLISM of sugar. He demonstrated the complex interlocking action of various hormones of the body – in this case, a pituitary hormone with the sugar-metabolizing hormone INSULIN.

hovercraft (air-cushion vehicle) Fast, usually amphibious craft invented by Sir Christopher COCKERELL. A horizontal fan produces a cushion of air that supports the hovercraft just above the ground or water. Vertical fans propel the craft. Most hovercraft are powered by gas turbine engines, similar to those used in aircraft, or by diesel engines. Hovercraft travel at speeds of up to 160km/h (100mph). They are invaluable for negotiating swamps, and are used as marine ferries and as military vehicles. Smaller, less speedy craft have been developed as runabouts. A **hovertrain** is a high-speed train that uses an air cushion to support it just above a special track.

Hoyle, Sir Fred (Frederick) (1915–) English astrophysicist and cosmologist. With Hermann BONDI and Thomas GOLD he developed the STEADY-STATE THEORY, in which the continuous creation of matter drives the expansion of the Universe. Although it was subsequently displaced by the BIG BANG theory (which owes its name to a disparaging remark by Hoyle), it sparked research by Hoyle and others into nucleosynthesis (the production of chemical elements from other chemical elements via nuclear reactions) in stars, which has been of lasting value. Hoyle has often attracted controversy with unorthodox ideas, such as his suggestion that life was brought to Earth by comets.

HR diagram Abbreviation of HERTZSPRUNG-RUSSELL DIAGRAM

Hubble, Edwin Powell (1889–1953) US astronomer. At Mount Wilson Observatory, Hubble used the 2.5m (100-inch) telescope for NEBULA investigations. He found that although some nebulae were clouds of gas lying within our Galaxy, many others (the so-called spiral nebulae) were resolvable as independent star systems. These results were published in 1925, along with the HUBBLE CLASSIFICATION. Also in 1925, Hubble measured the distance to the ANDROMEDA GALAXY. He attributed the RED SHIFT of the spectral lines of galaxies to the recession of galaxies, and hence to the expansion of the Universe, upon which modern cosmology is based. In 1929 he announced HUBBLE'S LAW, which claims a linear relation between the distance of galaxies from us and their velocity of recession, deduced from the red shift in their spectra. *See also* EXPANDING UNIVERSE

Hubble classification Classification of GALAXIES according to their shape or structure. The system was introduced by Edwin HUBBLE in 1925. The three major categories are elliptical (E), spiral (S) and barred spiral (SB); each category has several subdivisions depending on the observable shape or structure of the galaxy. Irregular galaxies (Ir) were not included in Hubble's original classification.

Hubble constant (symbol H_0) Rate at which the velocity of recession of galaxies (as shown by the RED SHIFT) increases with distance from us, according to HUBBLE'S LAW. The zero subscript specifies that it is the expansion rate at the present time that is meant, for the rate changes with time. The inverse of the Hubble constant is called the **Hubble time**, which gives a maximum age for the Universe on the assumption that there has been no slowing of the expansion. The constant is named after Edwin HUBBLE. *See also* EXPANDING UNIVERSE

Hubble's law Law proposed by HUBBLE claiming a linear relation between the distance of galaxies from us and their velocity of recession, deduced from the RED SHIFT in their spectra. The figure linking velocity with distance is the HUBBLE CONSTANT.

▶ **Hubble Space Telescope** One of the most ambitious and expensive pieces of astronomical equipment ever to be made, the Hubble Space Telescope initially was unable to relay useable images back to Earth due to an optical aberration of its main mirror. A shuttle mission in 1993 corrected this fault. A further mission in 1997 corrected problems with the telescope's power supply. Images now received on Earth have provided astronomers with the clearest-ever images of distant objects.

H

Hubble Space Telescope (HST) Optical TELESCOPE that was placed in Earth orbit by the SPACE SHUTTLE in 1990. It was hoped that by-passing the Earth' atmosphere, the telescope would provide scientists with the clearest images of distant objects ever seen. However, images transmitted back to Earth revealed that the telescope's main mirror was incorrectly shaped. A shuttle repair team corrected the fault in 1993, and it was again repaired in 1997. Hubble now produces accurate images of celestial bodies.

Huggins, Charles Brenton (1901–) US physician. He shared the 1966 Nobel Prize for physiology or medicine with Francis ROUS for his pioneering work in CANCER chemotherapy and his development of ways of investigating and treating cancers through the use of HORMONES.

Huggins, Sir William (1824–1910) British astronomer, the first to use the SPECTROSCOPE to determine the chemical composition of stars, comets and novae. He developed spectroscopic photography and applied the DOPPLER shift *(see* RED SHIFT) of the spectral lines of stars to the measurement of stellar motion. He distinguished between nebulae that are uniformly gaseous and those that contain stellar clusters.

Hulst, Hendrik van de (1918–) Dutch astronomer. He carried out research on interstellar matter and the Solar CORONA. In 1944 he predicted that hydrogen should emit radio waves with a wavelength of 21cm, and in 1951 he and Jan OORT used DOPPLER shifts at this wavelength to map the Galaxy.

human body Physical structure of a human. It is composed of water, PROTEIN and other organic compounds, and some inorganic material (minerals). It has a bony framework, the SKELETON, consisting of more than 200 bones, sheathed in SKELETAL MUSCLE to enable movement. A protective bony case, the SKULL, surrounds the large BRAIN. The body is fuelled by nutrients absorbed from the DIGESTIVE SYSTEM and oxygen from the LUNGS, which are pumped around the body by the CIRCULATORY SYSTEM. Metabolic wastes are eliminated mainly by the process of EXCRETION. Continuation of the species is enabled by the REPRODUCTIVE SYSTEM. Overall control is exerted by the NERVOUS SYSTEM, working closely with the ENDOCRINE SYSTEM. The body surface is covered by a protective layer of SKIN. *See also* ANATOMY

human evolution Process by which humans developed from pre-human ancestors. The FOSSIL record of human ancestors is patchy and unclear. Some scientists believe that our ancestry can be traced back to one or more species of Australopithicenes (*see* AUSTRALOPITHECUS) that flourished in S and E Africa *c.*4–1 million years ago. Other scientists believe that we are descended from some as yet undiscovered ancestor. The earliest fossils that can be identified as human are those of *Homo habilis* (handy people), which date from 2 million years ago. The next evolutionary stage was *Homo erectus* (upright people), who first appeared *c.*1.5 million years ago. The earliest fossils of our own species, *Homo sapiens* (wise people), date from *c.*250,000 years ago. An apparent side-branch, the NEANDERTHALS (*Homo sapiens neanderthalensis*), existed in Europe and W Asia *c.*130,000–30,000 years ago. Fully modern humans, *Homo sapiens sapiens*, first appeared *c.*100,000 years ago. All human species apart from *Homo sapiens sapiens* are now extinct.

Human Genome Project International scientific programme initiated in 1988 to map the human GENOME by analysing the 100,000 or so GENES making it up and charting their positions within the human chromosomes and their functions and biochemical nature. The study is expected to take 15 years and cost $3 billion. Researchers are located in more than 20 centres around the world, and their efforts are coordinated primarily in the USA and France by the Human Genome Organization (HUGO). The study is expected to increase the understanding of human evolution as well as of genetic disorders such as Alzheimer's disease and muscular dystrophy. It will also identify genetic similarities between humans and other species.

human immunodeficiency virus (HIV) Organism that causes ACQUIRED IMMUNE DEFICIENCY SYNDROME (AIDS). A RETROVIRUS, only identified in 1983, HIV attacks the IMMUNE SYSTEM, leaving the person unable to fight off infection. There are two distinct viruses: HIV-1, which has now spread worldwide; and HIV-2, which is concentrated almost entirely in West Africa. Both cause AIDS, with people infected with HIV-1 generally developing symptoms sooner than those with HIV-2. There are three main routes of transmission: from person to person by sexual contact; from mother to baby during birth; and by contact with contaminated blood or blood products (for instance, during transfusions or when drug-users share needles). People can carry the virus for many years before developing symptoms.

Humboldt, Baron Friedrich Heinrich Alexander von (1769–1859) German scientist and explorer. His special subject was mineralogy, but on his scientific trips in Europe and Central and South America he studied volcanoes, the ori-

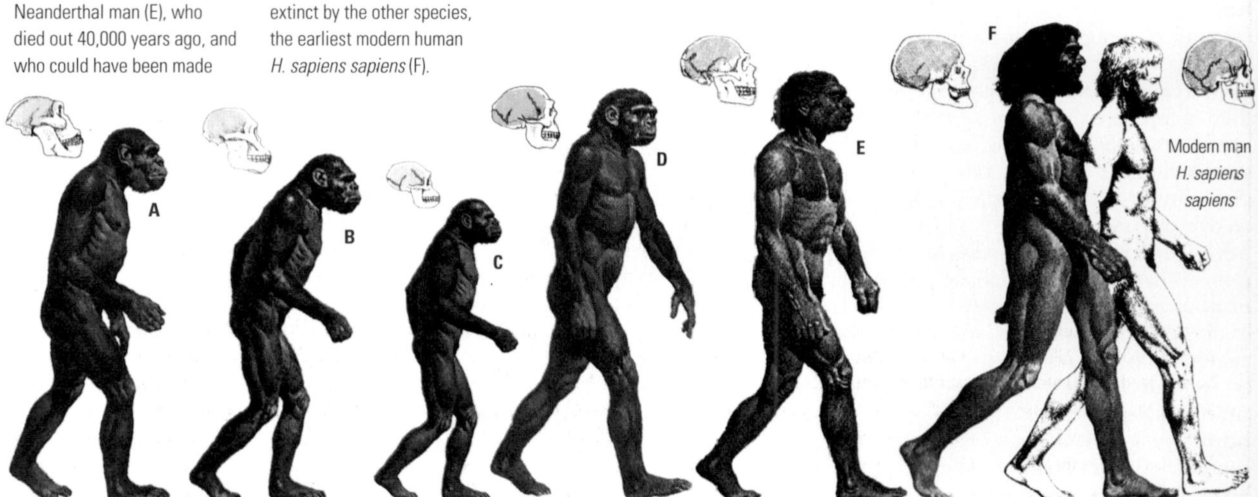

▶ **human evolution** Although the fossil record is not complete, we know that humans evolved from ape-like creatures. Our earliest ancestor, *Australopithecus afarensis* (A), lived in NE Africa some 5 million years ago. Over the next 3–4 million years *A. africanus* (B) evolved. *Homo Habilis* (C), who used primitive stone tools, appeared c.500,000 years later. *H. erectus* (D) is believed to have spread from Africa to regions all over the world 750,000 years ago. Records indicate that from *H. erectus* evolved two species, Neanderthal man (E), who died out 40,000 years ago, and who could have been made extinct by the other species, the earliest modern human *H. sapiens sapiens* (F).

The body as a whole is a compromise between mobility and rigidity. The internal organs, such as the HEART, LIVER, STOMACH, SPLEEN and the INTESTINES, are closely packed together and yet can work freely and easily. The surrounding framework of bone and muscles applies support and protection.

The SKELETON gives the upright strength to the body and in some places, such as the SKULL and THORAX, acts as a protective layer. The JOINTS give the bones mobility and the muscles strength and suppleness.

The contents of the chest and ABDOMEN are constantly moving – the beating of the heart, inhalation and exhalation of the lungs during BREATHING, and PERISTALSIS of the bowel. These structures can move without difficulty as they are surrounded by special, smooth layers of tissue known as PERICARDIUM, PLEURA and PERITONEUM.

The skin

The bodies largest organ is the SKIN. In an adult it covers approximately 2sq m (21.5sq ft), and not only envelops the whole body in a protective waterproof layer but is also part of the heat-regulating system. The liver, meanwhile, is the most complicated organ with the greatest number of functions – transforming digested food into usable materials and disposing of waste substances.

The cell

The circulation is constantly restoring and revitalizing as well as removing waste products from the basic unit of the body – the CELL. The cell is a microscopic structure of which there are billions that build up the whole body. Each cell specializes and carries out its own particular function. All the structures and organs are held together by the connective tissue, made up of cells that act as a kind of packing to protect and support the internal mechanisms.

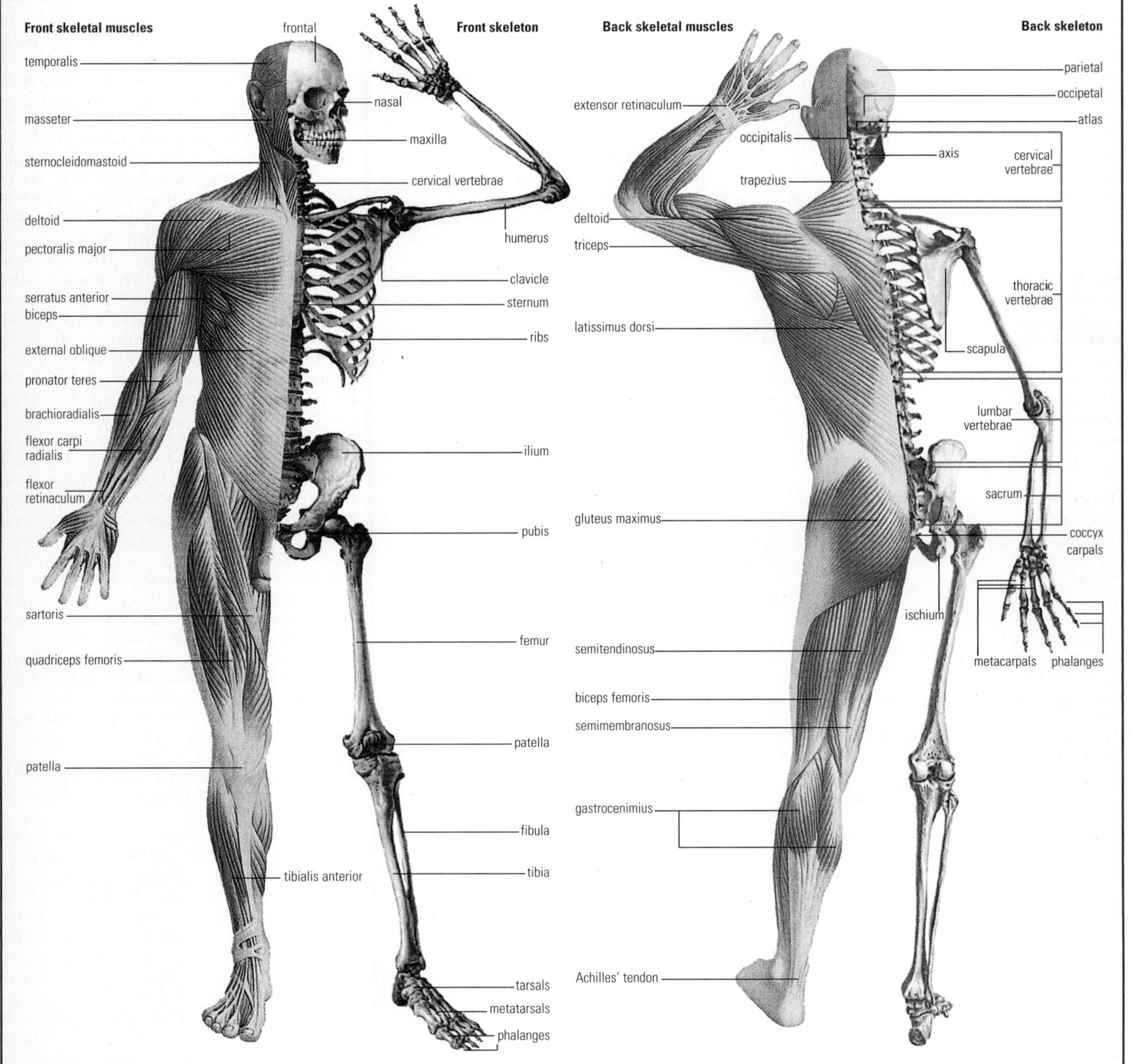

The human skeleton consists of about 206 bones, divided into two groups, the appendicular and axial skeletons. The skeleton has three functions; it provides support, protects the internal organs, and with some muscles, it provides movement. The axial skeleton (the skull, spine and rib cage) supplies the structure on which the limbs, the appendicular skeleton, are joined via the shoulder and pelvic girdles.

Muscles are made up of contractile tissue and are involved in movement. Muscles comprise 35–40% of total body weight, and there are more than 650 skeletal muscles controlled by the nervous system. Three kinds of muscle exist; smooth, cardiac and skeletal. Skeletal muscle is under voluntary control, cardiac muscle occurs only in the heart, and smooth, or involuntary, muscle controls organs such as the digestive tract.

gins of tropical storms, and the increase in magnetic intensity from the Equator towards the poles. His principal, five-volume work, *Kosmos* (1845–62), is a comprehensive description of the physical universe.

Humboldt current (Peru current) Current in the SE Pacific Ocean. An extension of the West Wind Drift, the Humboldt current flows N along the W coast of South America. A cold current, it lowers temperature along its course until it joins the warm Pacific South Equatorial current.

humerus Bone in the human upper arm, extending from the SCAPULA (or shoulder blade) to the elbow. A notch, or depression, called the **olecranon cavity**, on the posterior roughened lower end of the humerus, provides the point of articulation for the ULNA, one of the forearm bones. *See also* HUMAN BODY

humidity (relative humidity) Measure of the amount of water vapour in air. It is the ratio of the actual vapour pressure to the saturation vapour pressure at which water normally condenses, and is usually expressed as a percentage. Humidity is measured by a HYGROMETER.

humus Dark brown, organic substance, resulting from partial decay of plant and animal matter. It improves soil by retaining moisture, aerating, and increasing mineral nutrient content and bacterial activity. Often made in a compost heap or obtained naturally, types of humus include PEAT MOSS, leaf mould and soil from woods.

Hunter, John (1728–93) Scottish anatomist and surgeon who worked with his brother **William** (1718–83) on the LYMPHATIC SYSTEM, the course of the OLFACTORY NERVES and the coagulation of blood. In the 1760s, John Hunter was the first to ligate (tie off) an artery.

Huntington's disease (formerly Huntington's chorea) Rare hereditary disease of the brain characterized by jerky involuntary movements (chorea), disturbed behaviour and disintegration of personality and intellect, culminating in dementia. Fifty per cent of the children of an affected parent are at risk. The disease usually occurs in early middle life.

hurricane Wind of Force 12 or greater on the BEAUFORT WIND SCALE; intense and devastating tropical cyclone with winds ranging upwards from 118km/h (73mph), known also as a **typhoon** in the Pacific. Originating over oceans 10 to 20° from the Equator, hurricanes have a calm central hole called the **eye**, surrounded by inward spiralling winds and cumulonimbus CLOUDS. The winds and waves of hurricanes take many lives and cause extensive damage to shipping and coastal areas, but weather satellites often give adequate warning of their approach.

Hutton, James (1726–97) Scottish geologist. Hutton sought to formulate theories of the origin of the Earth and of atmospheric changes. He concluded that the Earth's history could be explained only by observing current forces at work within it, thereby laying the foundations of modern geological science.

Huxley, Sir Andrew Fielding (1917–) British physiologist. He analyzed the electrical and chemical events in NEURON (nerve cell) discharge and for this work he shared the 1963 Nobel Prize for physiology or medicine with John Eccles (1903–) and Alan HODGKIN.

Huxley, Sir Julian Sorell (1887–1975) British biologist, grandson of Thomas HUXLEY. His researches were chiefly on the behaviour of birds and other animals in relation to EVOLUTION. His writings greatly promoted public interest in the subject. He served as secretary of the Zoological Society of London (1935–48) and as the director general of the United Nations Educational, Scientific and Cultural Organization (UNESCO) (1946–48). His books include *The Individual in the Animal Kingdom* (1911) and *Evolutionary Ethics* (1943).

Huxley, Thomas Henry (1825–95) British biologist. Huxley was a champion of DARWIN's theory of EVOLUTION and of the scientific method in research. His works include *Zoological Evidences as to Man's Place in Nature* (1863), *Manual of Comparative Anatomy of Vertebrated Animals* (1871) and *Evolution and Ethics* (1893).

Huygens, Christiaan (1629–95) Dutch physicist and astronomer. In 1655 he discovered Saturn's largest satellite, Titan, and the next year explained that the planet's appearance was due to a broad ring surrounding it. He was concerned with the development of the TELESCOPE and introduced the convergent eyepiece. Huygens' many contributions to physics include the idea that light is a wave motion and the theory of the pendulum; he built the first pendulum clock.

hybrid Offspring of two parents of different GENE composition. It often refers to the offspring of different varieties of a species or of the cross between two separate species. Most inter-species hybrids, plant or animal, are unable to produce fertile offspring.

hybridization CROSS-BREEDING of plants or animals between different species to produce offspring that differ in genetically determined traits; it is commonly the result of human intervention. Changes in climate or in the environment of an organism may give rise to natural hybridization, but usually humans use specialized techniques to produce hybrids that may be hardier or more resistant to disease than the original forms, or more economical.

hybrid vigour Inordinate strength demonstrated in some HYBRID offspring, not characteristic of the parents.

Hydra (Water snake) Largest constellation in the S sky. Its brightest star is Alpha (Alphard) magnitude 2.0. There is one bright open cluster, M48.

hydration Assimilation of water by an ION or compound. A hydrate is a compound, such as copper(II) sulphate (cupric sulphate) that has a fixed amount of water weakly bonded to each crystal ($CuSO_4.5H_2O$). ETHENE can be hydrated to produce ETHANOL.

hydraulic action Erosive effect of water. The force of moving water can be an effective agent of EROSION. It is important both in rivers and on coastlines. In rivers the movement of currents causes hydraulic action, though if a load of silt or sand were added the process would be called CORRASION. Along the coast, breaking waves batter cliffs and may open out JOINTS in the rock, forming caves or undermining the cliff and leading to collapse.

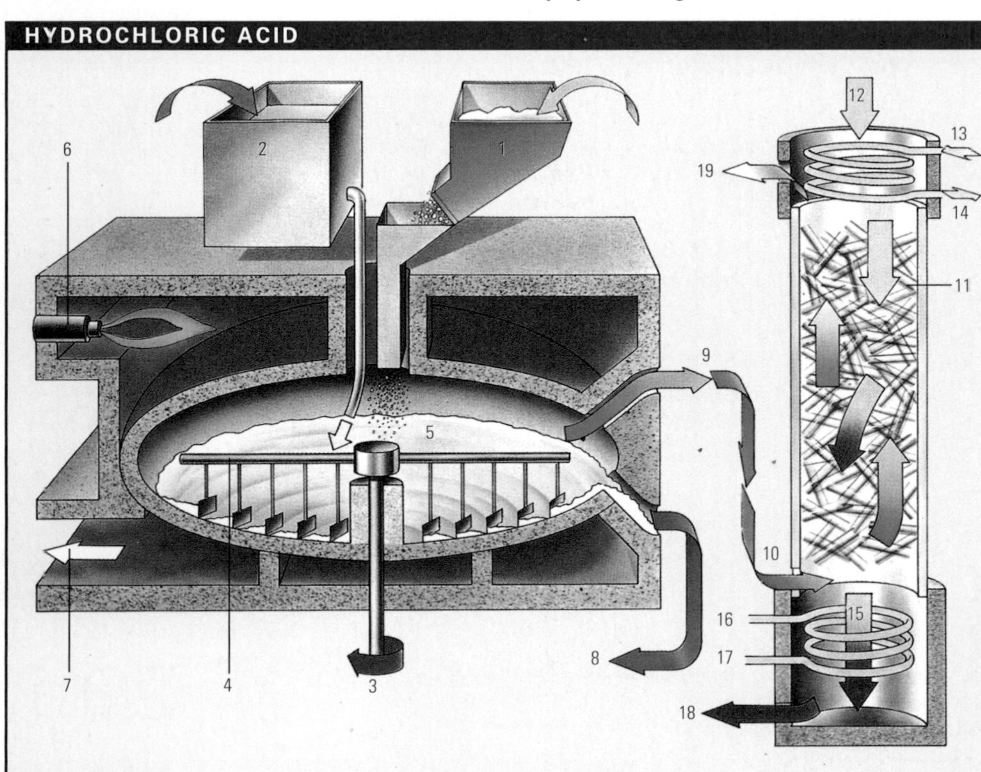

In all manufacturing processes for hydrochloric acid another useful product is obtained along with the acid. In this process (using a Manheim type furnace) sodium sulphate is produced.

Key
1) Common salt (sodium chloride) added to the furnace.
2) Sulphuric acid inlet by way of a lead-lined tank
3) Rotating shaft
4) Rotating stirrers mix reactants
5) Reaction chamber. The salt and sulphuric acid react to form sodium sulphate and hydrochloric acid, which comes off as the gas hydrogen chloride because of the high temperature.
6) Oil burner heats reaction chamber
7) Combustion gases outlet
8) Salt cake (sodium sulphate) outlet
9) Hydrogen chloride gas led off
10) Hydrogen chloride gas piped into the absorption column below the packed section
11) The absorption column is packed with Raschig rings made of glass. On the surface of these rings the hydrogen chloride combines with water, immitted at the top of the tower (12), to form hydrochloric acid. This reaction releases heat.
12) Water inlet. The water passes down the packed column and dissolves the hydrogen chloride gas.
13) Cooling water inlet
14) Cooling water outlet
15) Hot concentrated hydrochloric acid passes into the cooler at the bottom of the column
16) Cooling water inlet
17) Cooling water outlet
18) Cool hydrochloric acid led out to storage tanks
19) Spent gas vent

HYBRID

Occasionally, it is possible to interbreed two closely related species to produce a hybrid. Male donkeys (1) mated with female horses (2) produce mules (3) which are sterile because their chromosomes cannot pair properly during meiosis, the form of cell division required to produce sperm or eggs. Modern bread wheat (8) is a product of many years of selective breeding during which many varieties have been developed (7) (9). They have their origins in the hybridization of wild wheat and grass species to increase the number of chromosomes in the wheat cells from pairs of seven chromosomes, to four sets of seven (4) and then to six sets of seven (5) and (6) which could divide properly during meiosis and breed normally.

hydraulics Physical science and technology of the behaviour of FLUIDS in both static and dynamic states. It deals with practical applications of fluid in motion and devices for its utilization and control. *See also* FLUID MECHANICS; HYDRODYNAMICS

hydrazine (N_2H_4) Fuming corrosive liquid obtained by the reaction between chloramine (NH_2Cl) and ammonia (NH_3), or by the oxidation of UREA. It is used as a jet and rocket fuel, as a reducing agent and as a corrosion preventer. Properties: r.d. 1.011, m.p. 1.4°C (34.5°F); b.p. 113.5°C (236.3°F).

hydrazone In organic chemistry, a compound formed between hydrazine (N_2H_4) and an ALDEHYDE or KETONE. Hydrazones are condensation products (*see* CONDENSATION REACTION) containing the group $=C=NNH_2$. They are used in QUALITATIVE ANALYSIS to identify aldehydes and ketones. In practice, the substance phenylhydrazine is generally employed as the analytical reagent.

hydride Chemical compound of HYDROGEN with another ELEMENT, especially a more electropositive element (one that loses ELECTRONS and forms positive IONS). The hydrides of the electropositive metals, such as lithium hydride (LiH) and sodium hydride (NaH), are salt-like ionic compounds containing the negative ion H^-. They are useful reagents for HYDROGENATION reactions.

hydrocarbon Organic compound containing only CARBON and HYDROGEN. Many thousands of different hydrocarbons exist, including open-chain compounds, such as the ALKANES (paraffins), ALKENES (olefins) and ALKYNES (acetylenes). Aromatic hydrocarbons have properties similar to BENZENE; they contain at least one benzene ring, as in naphthalene and anthracene. Petroleum, natural gas and coal tar are sources of hydrocarbons.

hydrocephalus Increase in volume of CEREBROSPINAL FLUID (CSF) in the ventricles of the brain. A condition that exerts dangerous pressure on brain tissue, it can be due either to obstruction, preventing the CSF from draining away, or to a failure of natural reabsorption. In babies it is congenital, often associated with SPINA BIFIDA; in adults it may arise from brain injury or disease. It is treated by insertion of a shunting system to drain the CSF harmlessly into the abdominal cavity.

hydrochloric acid Solution of the pungent, colourless gas hydrogen chloride (HCl) in water. It is obtained by the action of sulphuric acid on common salt, as a byproduct of the chlorination of hydrocarbons, or by combination of HYDROGEN and CHLORINE. Hydrochloric acid is used to produce other chemicals and has uses in the food, oil and metallurgical industries. It is also produced in humans by cells in the stomach lining as it is needed for the enzyme PEPSIN to digest proteins. The concentrated commercial acid contains up to 35% hydrogen chloride.

hydrocyanic acid (prussic acid) Solution of hydrogen cyanide (HCN) in water; a deadly poisonous, weak acid. It is a liquid with an odour of bitter almonds, sometimes used as a fumigant to destroy rats and other pests of buildings. It also has uses in the chemical industry but is so dangerous to handle that its solid salts potassium and sodium cyanide are preferred, even though these, too, are highly poisonous.

hydrodynamics In physics, branch of MECHANICS that deals with the motion of FLUIDS (liquids and gases). It is of great importance in industry, particularly chemical, petroleum and water engineering. The properties of fluids studied include their adhesion, cohesion, viscosity and type of flow (viscous and turbulant). *See also* HYDRAULICS; FLUID MECHANICS

hydroelectricity Electricity generated from the motion of water, whether from dams, tides or waves. In nearly all installations this energy of movement, or kinetic energy, drives the blades of a water TURBINE, which in turn drives the rotor of an electric GENERATOR.

hydrofluoric acid Colourless, corrosive fuming liquid consisting of a solution of the gas hydrogen fluoride (HF) in water, prepared by distilling a mixture of sulphuric acid (H_2SO_4) and fluorspar (FLUORITE). It is used in frosting and etching glass and in FLUOROCARBONS.

HYDROELECTRICITY

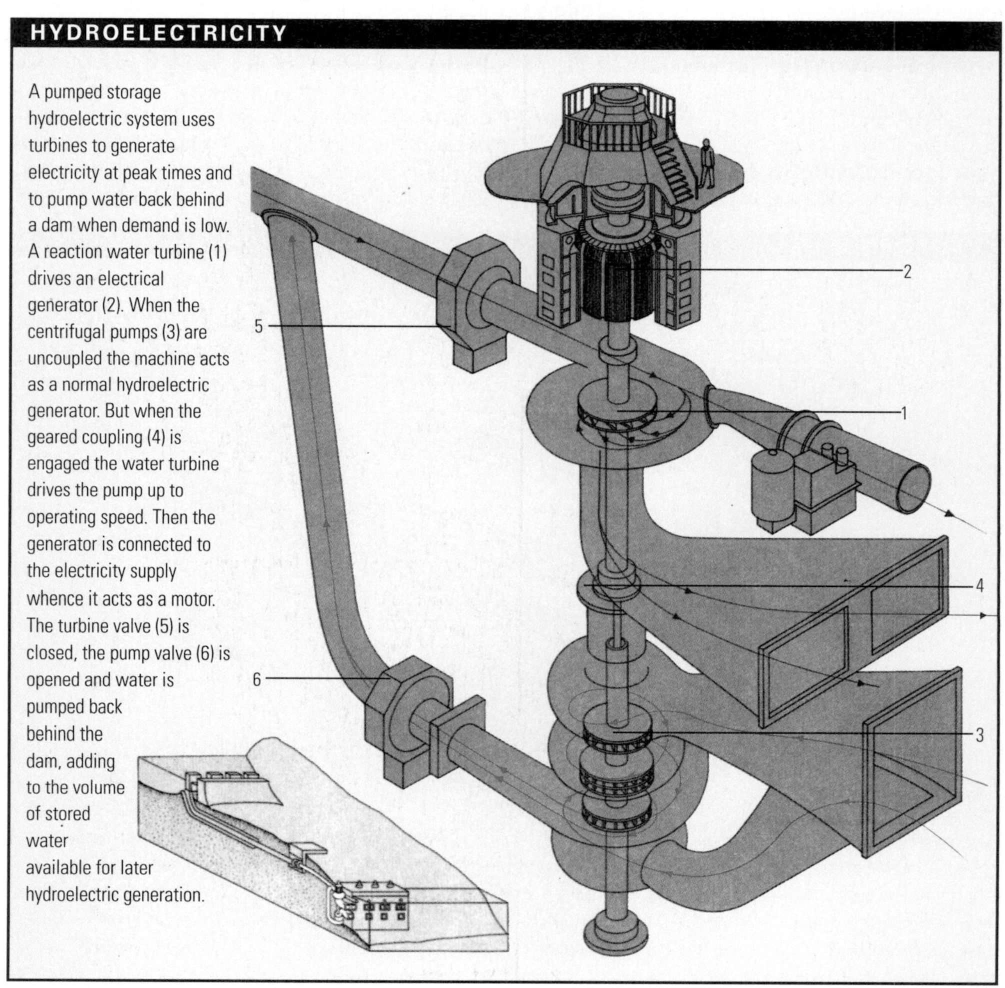

A pumped storage hydroelectric system uses turbines to generate electricity at peak times and to pump water back behind a dam when demand is low. A reaction water turbine (1) drives an electrical generator (2). When the centrifugal pumps (3) are uncoupled the machine acts as a normal hydroelectric generator. But when the geared coupling (4) is engaged the water turbine drives the pump up to operating speed. Then the generator is connected to the electricity supply whence it acts as a motor. The turbine valve (5) is closed, the pump valve (6) is opened and water is pumped back behind the dam, adding to the volume of stored water available for later hydroelectric generation.

hydrofoil Boat or ship the hull of which is lifted clear of the water, when moving at speed, by submerged wings. Hydrofoils usually have gas-turbine or diesel engines that power propellers or water jets. Speeds range from 30 to 60 knots.

hydrogen (symbol H) Gaseous, nonmetallic element, first identified as a separate element in 1766 by Henry CAVENDISH who called it "inflammable air". Colourless and odourless, it is usually classified with the ALKALI METALS in group I of the periodic table. Isotopes include DEUTERIUM and the radioactive TRITIUM. Hydrogen is the lightest and most abundant element in the Universe (76% by mass), mostly found on Earth combined with oxygen in water. It is used to manufacture AMMONIA, to harden fats and oils, and in ROCKET fuels. The hydrogen bomb (*see* NUCLEAR WEAPON) is produced by the nuclear FUSION of hydrogen isotopes. Properties: at.no. 1; r.a.m. 1.00797; r.d. 0.0899; m.p. −259.1°C (−434.4°F); b.p. −252.9°C (−423.2°F); most common isotope ^{1}H (99.985%).

hydrogenation Chemical reaction between hydrogen and an element or compound, sometimes under pressure, usually in the presence of a metal CATALYST such as nickel, platinum or palladium. An unsaturated compound, such as benzene, may accommodate the extra hydrogen, but a saturated species will break up (**destructive** hydrogenation). On an industrial scale, hydrogenation of coal produces oil, and hydrogenation of liquid vegetable oils yields solid fats (such as margarine).

hydrogen bomb *See* NUCLEAR WEAPON

hydrogen bond CHEMICAL BOND formed between certain hydrogen-containing molecules. The hydrogen atom must be bound to an electronegative (electron-withdrawing) atom; the bond is formed between the positive charge on hydrogen and the negative charge on an atom in an adjacent molecule. Hydrogen bonding occurs in water, and in many biological systems.

hydrogencarbonate (bicarbonate) Salt of CARBONIC ACID, containing the hydrogencarbonate ion (HCO_3^-). An example is SODIUM HYDROGENCARBONATE.

hydrogen chloride *See* HYDROCHLORIC ACID

hydrogen cyanide *See* HYDROCYANIC ACID

hydrogen peroxide (H_2O_2) Liquid compound of hydrogen and oxygen, usually sold in aqueous solution. It is prepared by electrolytic OXIDATION of sulphuric acid and by methods involving REDUCTION of oxygen. Hydrogen peroxide solution is used as a bleach and disinfectant. Concentrated hydrogen peroxide is dangerously explosive and is used as an oxidizer for ROCKET fuel. Properties: r.d. 1.44; m.p. −0.9°C (30.4°F); b.p. 150°C (302°F).

hydrogen sulphide (H_2S) Colourless, poisonous gas with the smell of rotten eggs. It is produced by decaying matter, found in crude oil and prepared by the action of sulphuric acid on metal sulphides. It is used in traditional QUALITATIVE ANALYSIS. Properties: m.p. −85.5°C (−121.9°F), b.p. −60.7°C (−77.3°F).

hydrography Science of charting the water-covered areas of the Earth. Navigational charts have been made since the 13th century, but these were accurate only for coastlines. Interest in the charting of oceanic areas away from coasts only developed in the 19th century. Now, detailed bathymetric (deep-sea) surveys are available for geological studies. Hydrographic offices are run by governments of maritime nations to furnish their mariners with nautical charts; the work of these offices is coordinated by the International Hydrographic Bureau, founded in 1921.

hydrological cycle Circulation of water around the Earth. Water is evaporated from the sea; most falls back into the oceans, but some is carried over land. There it falls as PRECIPITATION (rain, snow, and so on) and, either by EVAPORATION, (EVAPOTRANSPIRATION), TRANSPIRATION by plants, is returned to the ATMOSPHERE. Alternatively, it gradually finds its way back to the sea, either by surface runoff, infiltration or seepage. Less than 1% of the world's water is involved in this cycle; 97% of all water is in the oceans, and much of the remainder is held as snow or ice.

hydrology Study of the Earth's waters, their sources, circulation, distribution, uses and chemical and physical composition. The HYDROLOGICAL CYCLE is the Earth's natural water circulation system. Hydrologists are scientists and engineers involved in various aspects of hydrology, including the provision of fresh water, building dams and irrigation systems, and controlling floods and water pollution.

hydrolysis Chemical reaction in which molecules of a substance are split into smaller molecules by reaction with water, often assisted by a CATALYST. For example, in digestion, ENZYMES catalyse the hydrolysis of CARBOHYDRATES, PROTEINS and FATS into smaller, soluble molecules that the body can assimilate.

hydrometallurgy Treatment of ores by wet processes to extract their metals. It usually includes LEACHING by water and additives such as sulphuric acid, separation of the waste, purification of leach solution and precipitation of metal from solution by chemicals or electrolytic means. Most metals are extracted by this method.

hydrometer Instrument for measuring density or RELATIVE DENSITY (R.D.) of a liquid. It consists of a long-necked, sealed glass bulb that is weighted. The neck is calibrated to indicate relative density, which is read off against the surface of the liquid in which it floats. Hydrometers are used to check concentrations of liquids in storage batteries, freezing points of radiator solutions and the "proof" of alcohol.

hydrophily POLLINATION of a FLOWER through the agency of water. A rare type of pollination employed by aquatic plants, it can occur in one of two ways. Either the pollen grains themselves are carried on or under the surface of the water, or the male flowers break off and float along until they come into contact with a female flower.

hydrophone Specialized MICROPHONE designed to detect sounds and ultrasonic waves underwater. Initially used to detect submarines, hydrophones have, more recently, been used in oceanography, for example, for listening to whale song.

hydrophyte (aquatic plant) Plant that grows only in water or in damp places. Examples include water lilies, water ferns, water milfoil, water hyacinth, duckweed and various pondweeds.

hydroplane Speedboat able to skim over water by rising out of it at speed until only a small area at the stern remains immersed. This is achieved by means of a flat-bottomed hull, which is powered by a high-performance outboard motor (*see also* HYDROFOIL). The term is also used to describe a horizontal, fin-like planning surface on a submarine that controls the vessel's vertical position underwater.

hydroponics (soil-less culture or tank farming) Method of growing plants with their roots in a mineral solution or a moist inert medium (such as gravel) containing the necessary nutrients, instead of soil. Added nutrients include nitrogen, phosphorus and potassium, with trace elements such as calcium, sulphur and magnesium. The technique has been in use for more than 100 years and is employed on a commercial scale to grow such crops as tomatoes.

hydrosphere Water on the surface of the Earth, including oceans, lakes, streams and groundwater.

hydrostatics In physics, the branch of mechanics that deals with liquids at rest. Its practical applications are mainly in water engineering and in the design of such equipment as hydraulic presses, rams, lifts and vehicle braking and control systems. *See also* STATICS

hydrothermal process Geological activity produced by the action of very hot subterranean water, at temperatures between 300 and 500°C (570–930°F). Superheated water can dissolve many minerals, such as those of copper, lead and zinc, depositing them as veins in rock crevices. It can also trigger chemical changes, for example

HYDROFOIL

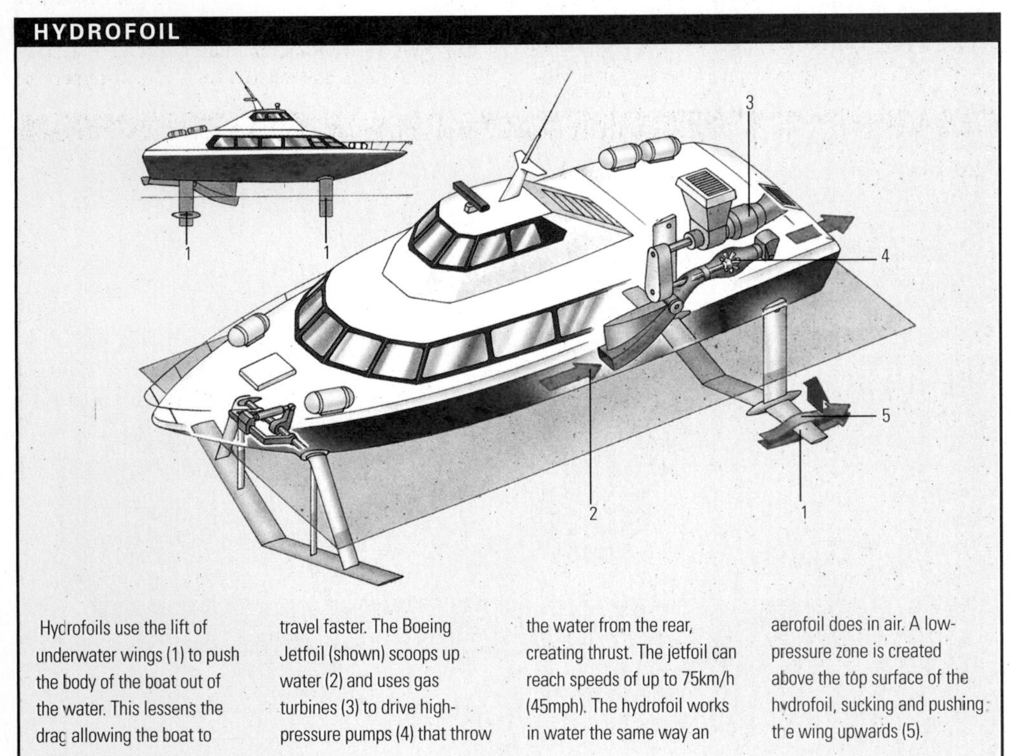

Hydrofoils use the lift of underwater wings (1) to push the body of the boat out of the water. This lessens the drag allowing the boat to travel faster. The Boeing Jetfoil (shown) scoops up water (2) and uses gas turbines (3) to drive high-pressure pumps (4) that throw the water from the rear, creating thrust. The jetfoil can reach speeds of up to 75km/h (45mph). The hydrofoil works in water the same way an aerofoil does in air. A low-pressure zone is created above the top surface of the hydrofoil, sucking and pushing the wing upwards (5).

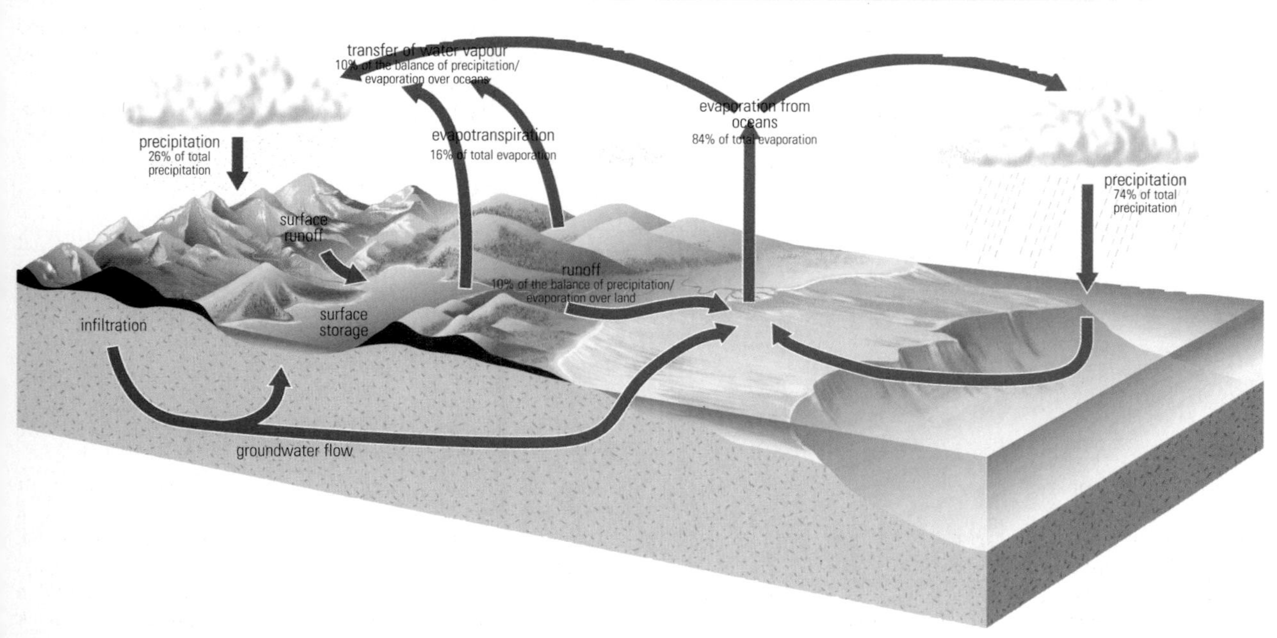

◄ **hydrological cycle** The Earth's water balance is regulated by the constant recycling of water between the oceans, the atmosphere and the land. The movement of water between these three "reservoirs" is called the hydrological cycle. The oceans play a vital role in this cycle: 74% of the total precipitation falls over the oceans and 84% of the total evaporation comes from the oceans. Water vapour in the atmosphere circulates around the planet, transporting energy as well as water itself. When the vapour cools it falls as rain or other precipitation.

turning the mineral olivine into serpentine. Much of this hydrothermal activity takes place in the deep ocean seabeds. Hot water welling from the floor of oceanic trenches is responsible for colonies of life 2.5km (1.5mi) below the surface, where tube worms, crustaceans and fish thrive without the benefit of the Sun's warmth.

hydrothermal vent (smoker) Fissure in the seafloor through which mineral-rich water escapes. Hydrothermal vents form in areas of volcanic activity. Seawater seeps down through the bottom sediments into the hot rocks beneath and dissolves various minerals. When the water erupts through a vent, it cools rapidly and the dissolved minerals are precipitated out of solution as a "smoke". Unique species of bacteria and animals have been found living in the mineral-rich water round such vents.

hydrotropism Plant growth in response to water stimulus. It is not a strong TROPISM.

hydroxide Inorganic chemical compound containing the ion OH^-, which acts as a BASE (a substance that accepts protons and reacts with an acid to yield a salt and water). The strong inorganic bases such as potassium hydroxide (KOH) dissociate in water almost completely to provide many hydroxide ions.

hydroxyl group Monovalent chemical grouping consisting of a hydrogen atom joined to an oxygen atom. Represented by the formula –OH, the hydroxyl group is an important functional group in organic compounds such as alcohols and phenols. It should not be confused with the hydroxide ion OH^-.

hygrometer Instrument used to measure the water-vapour content, or HUMIDITY, of the atmosphere. One type, the psychrometer, compares the wet and dry bulb temperatures of the air. The wet bulb is covered by a cloth, which is kept damp by water from a small reservoir. Evaporation from the cloth reduces the temperature of the wet bulb and the difference in temperature between the wet bulb and the dry bulb shows the relative humidity. Other types of hygrometer measure absorption or condensation of moisture from the air, or chemical or electrical changes caused by that moisture.

hygroscopic Term describing a substance that, on standing, reacts with or absorbs water vapour from the air. Magnesium chloride and concentrated sulphuric acid are typical examples. Some hygroscopic solids such as magnesium chloride, said to be deliquescent, absorb sufficient water to dissolve, yielding a concentrated solution.

hymen In ANATOMY, MEMBRANE that covers or partly covers the entrance to the VAGINA. Intact at birth, it normally opens spontaneously before PUBERTY or at first penetration during sexual intercourse.

hymenopteran Member of the order of insects Hymenoptera, containing sawflies, ants, wasps, hornets and bees. All have a complete life cycle – egg, larva, pupa and adult – and two pairs of membranous wings, which move together in flight. They are found worldwide, living solitarily or in social groups. The larvae (grubs or caterpillars) feed on plants or are parasitic or predaceous.

hyperbaric chamber Sealed chamber in which oxygen is used at pressures higher than normal in the treatment of certain conditions. It is most often used in the treatment of decompression sickness, carbon monoxide poisoning or gas gangrene.

hyperbola Plane curve defined relative to a fixed point (the **focus**) and a fixed line (the **directrix**) so that the distance from any point on the curve to the focus is a constant multiple of the distance to the directrix. The curve has two branches and is a CONIC section. Its standard equation in CARTESIAN COORDINATES x and y is $x^2/a^2 - y^2/b^2 = 1$, where a and b are real, non-zero numbers.

hypermetropia (hyperopia, long-sightedness) Common disorder of vision in which near objects appear blurred. It occurs when parallel light rays entering the EYE are brought to a focus behind the RETINA, often because the eyeball is too short front-to-back. It can be rectified by wearing spectacles or contact lenses with convex lenses. *See also* MYOPIA

hyperon Class of ELEMENTARY PARTICLES with an anomalously long lifetime. The group includes the lambda particle (Λ), sigma particles (Σ^+, Σ^0, Σ^-), xi particles (Ξ^0, Ξ^-), and the omega-minus particle (Ω^-). A lambda can replace a NEUTRON in a nucleus to form a **hypernucleus**. *See also* BARYON

hyperplasia Increased production of cells in an organ or tissue, causing it to enlarge in size. *See also* HYPERTROPHY

hypersensitivity Condition in which a person reacts excessively to a stimulus. Most hypersensitive reactions are synonymous with ALLERGIES, the commonest being HAY FEVER, a reaction to pollen. Some people become hypersensitive to drugs, such as penicillin, or to chemicals after initial contact.

hypertension Persistent high BLOOD PRESSURE. Hypertension may be of unknown cause (**essential** or **primary** hypertension) or may be secondary to a variety of conditions, including kidney disease. Treatment is by salt restriction and anti-hyperten-

HYGROMETER

Hygrometers are used to measure humidity, the amount of water vapour in the air. A hair hygrometer uses a bundle of human hair, or from the beard of certain varieties of goat, to measure the relative humidity. As the amount of water vapour in the air changes, the length of the bundle of hair (1) increases (2) or decreases (3) and this movement is amplified by a series of levers (4) to which the bundle is attached and recorded on a moving roll of paper (5).

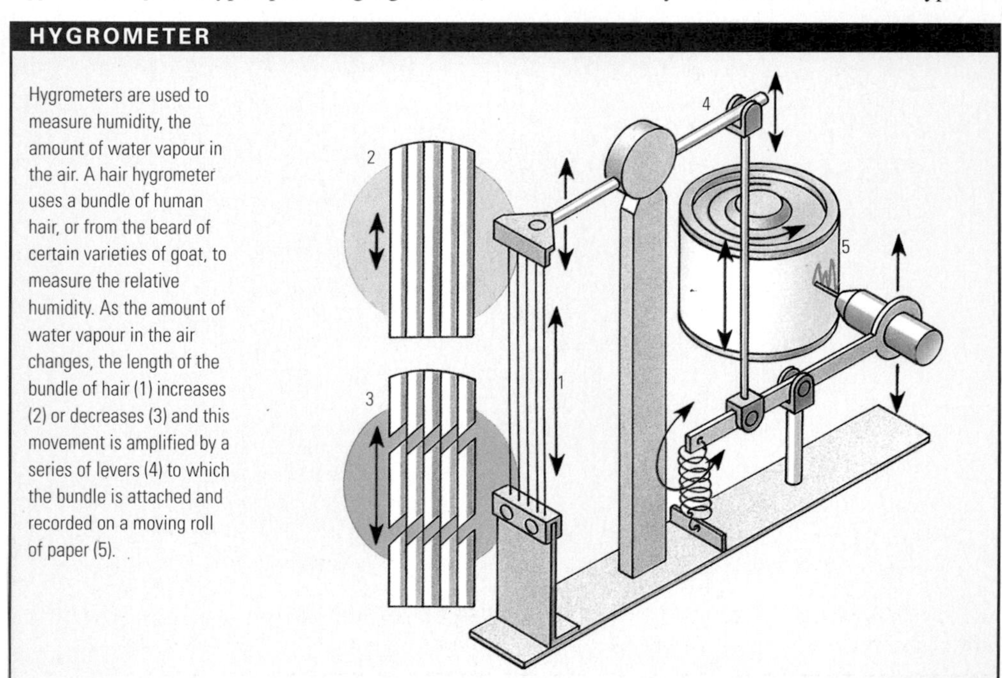

sive drugs; weight reduction is recommended where appropriate. Untreated, hypertension predisposes to heart disease, stroke and kidney failure. *See also* HYPOTENSION

hypertext Computing term that describes a method by which one piece of DATA is linked to another piece of data. Hypertext is most commonly seen on the WORLD WIDE WEB (WWW) and such things as interactive CD-ROMs. It usually manifests itself in the form of highlighted words (usually either in a different colour or underlined), which, when selected, take the viewer to associated material. This material may be many things, such as more text, a visual image or a sound clip. Hypertext can be seen as performing a similar function in MULTIMEDIA environments as cross-references do in printed encyclopedias and dictionaries.

hyperthermia Abnormally high body temperature, usually defined as being 41°C (106°F) or more. It is usually due to over-heating (as in HEAT-STROKE) or FEVER.

hypertrophy Enlargement of a tissue or organ due to an increase in size of its cells. Muscles enlarge in this way in response to increased use. *See also* HYPERPLASIA.

hyperventilation Rapid breathing that is not brought about by physical exertion. It reduces the carbon dioxide level in the blood to undesirably low levels, producing dizziness, tingling in the limbs and a feeling of tightness in the chest; if prolonged, it may cause loss of consciousness.

hypha (pl. hyphae) Very fine, theadlike projections found in FUNGI. Collectively a number will form the MYCELIUM, which is responsible for the reproduction and feeding of the fungus.

hypnotic drug (soporific) Drug that depresses the CENTRAL NERVOUS SYSTEM, inducing sleep. They include tranquillizers, BARBITURATES, chloral hydrate and some ANTIHISTAMINES.

hypodermic syringe Surgical instrument for injecting fluids beneath the skin into a muscle or blood vessel. It comprises a graduated tube containing a piston plunger, connected to a hollow needle. The whole is provided sterile – free from microorganisms – and must be resterilized after use. Many hypodermic syringes are disposable: designed to be used once and thrown away. *See also* INJECTION

hypolimnion Lower, colder layers of water in a lake, beneath the EPILIMNION. There is not enough light for PHOTOSYNTHESIS to occur and levels of dissolved oxygen are low.

hypotension Condition in which the BLOOD PRESSURE is abnormally low. It is commonly seen after heavy blood loss from any cause or excessive fluid loss due to prolonged vomiting or diarrhoea. It also occurs in many kinds of serious illness. Some people experience a drop in blood pressure when standing up from a lying down position. Temporary hypotension may cause sweating, dizziness and fainting. *See also* HYPERTENSION

hypotenuse Side opposite the right angle in a right-angled TRIANGLE. It is the longest side of the triangle.

hypothalamus Region at the base of the BRAIN, adjoining the PITUITARY GLAND, containing centres that regulate body temperature, fluid balance, hunger, thirst and sexual activity. It is also involved in emotional activity, sleep and the integration of HORMONE and nervous activity.

hypothermia Fall in body temperature to below 35°C (95°F). Most at risk are newborns (particularly if premature) and the aged living in conditions of inadequate heating. Insidious in onset, it can progress to coma and death. Hypothermia is sometimes induced during surgery to lower the body's oxygen demand. It occurs naturally in animals during HIBERNATION.

hypothyroidism Deficient functioning of the THYROID GLAND. Congenital hypothyroidism can lead to cretinism in children. In adults the condition is called **myxoedema**. More common in women, it causes physical and mental slowness, weight gain, sensitivity to cold and susceptibility to infection. It can be due to a defect of the gland itself or a lack of iodine in the diet, both of which can lead to goitre or to inadequate stimulation of the thyroid gland by the pituitary. It is treated with the HORMONE thyroxine.

hypsometer Instrument for testing a THERMOMETER'S accuracy by using the boiling point of water. The name hypsometer is also given to an instrument that measures height above sea level as a function of the boiling point of water (which decreases with increasing height).

hysterectomy Removal of the UTERUS, possibly with surrounding structures. It is performed to treat fibroids or cancer or to put an end to heavy menstrual bleeding. A woman who has undergone hysterectomy does not menstruate and is no longer fertile. *See also* MENSTRUATION

hysteresis Phenomenon occurring in the elastic behaviour of substances in which the STRAIN is greater when the STRESS is decreasing than when it is increasing because of a lag in the effect. When the stress is removed, a residual strain remains. The phenomenon also occurs in the magnetic behaviour of materials: the magnetization is greater when the magnetizing field is decreasing. This phenomenon is particularly important in FERROMAGNETISM; the magnetization of ferromagnetic materials lags behind the magnetizing force.

I

iatrogenic disease Illness or disorder caused by medical treatment. It can be applied to any untoward condition that is caused by medical staff or equipment, therapeutics, or especially drugs.

ice Water frozen to 0°C (32°F) or below, when it forms complex six-sided CRYSTALS. Ice is less dense than water and floats. When water vapour condenses below freezing point, ice crystals are formed. This occurs mainly in high, cirrus CLOUDS, but also in the grey portions of other clouds. Clusters of these ice crystals form snowflakes. It also occurs near the ground and gives rise to FROST.

ice ages Prolonged period of colder climatic conditions, during which snow and ice covered large areas of the Earth. There is evidence of about 20 ice ages having occurred throughout Earth's history, the earliest dating back to 2,300 million years ago. The most recent began about 2 million years ago and is often referred to as the "Ice Age". During this period, ice advanced for several thousand years, and then retreated for about the same period. An advance of ice is called a **glacial phase**, and the retreat is an **interglacial**. The Earth is experiencing an interglacial at present, although there are still ice-age conditions over Greenland and Antarctica. During the glacial phases, ice covered Britain as far S as the Bristol Channel and the Thames estuary. Areas farther S experienced PERMAFROST and TUNDRA conditions. When the ice melted, finally disappearing about 12,000 years ago, the climate warmed up, enabling different types of vegetation to colonize S England, and eventually spread farther N. The last ice age produced many of the landforms seen in the N continents and affected sea level on a global scale. *See also* GLACIER

iceberg Large, drifting piece of ice, broken off from a GLACIER, or polar ICE-CAP. As the ice moves slowly outwards from a snow-cap, some reaches the sea. It floats and is gradually broken off by the movement of tides and waves. Some icebergs may be several hundred metres across, but they gradually melt and shrink as they float across the oceans. Icebergs from Greenland are quite tall, whereas those from the Antarctic are flat and tabular in shape. Up to 80% or 90% of an iceberg is below the surface of the water. As they melt, icebergs sometimes turn over; this readjusts the balance. One of the major areas for icebergs is on the route of the cold Labrador Current, which brings them S along the coast of Greenland, towards Newfoundland, where they can be a hazard to shipping.

icebreaker Ship with a heavy bow and armoured sides, designed with powerful engines which enable it to make a passage through ice. For breaking thick ice, the bow is so designed that the forward motion of the ship drives it up onto the ice, and the weight of the ship breaks it. Propellers fore and aft allow great manoeuvrability. Icebreakers are used regularly to clear channels in the Great Lakes and in the Baltic Sea, and they have also been used in Arctic and Antarctic explorations. The Soviet icebreaker *Lenin*, launched in 1959, was the first nuclear-powered, non-naval vessel.

ice-cap Small ICE-SHEET, often in the shape of a flattened dome, which spreads over the mountains and valleys of polar islands. The floating ice fields surrounding the North Pole are sometimes incorrectly called an ice-cap.

Iceland spar (calcium carbonate, $CaCO_3$) Transparent form of CALCITE that has the property of **birefringerence** (or double refraction) – that is it bends light two ways so that an image seen through it appears double. It is used for polarizing PRISMS and in polarizing MICROSCOPES and other optical instruments. *See also* POLARIZED LIGHT

ice-sheet (continental GLACIER) Large expanse of snow and ice covering a land mass; for example, Antarctica. Ice-sheets can cover entire mountain ranges and can spread across lowlands and oceans. The ice may become very thick: in Greenland and Antarctica there are places where the ice measures 3,000m (9,800ft) in depth. An ICEBERG formed when ice breaks from the margins of such a sheet may be many square kilometres in area.

ichthyology Zoological study of FISH, including their classification, structure, distribution and ecology. The ancient Greeks, particularly Aristotle, are regarded as the pioneers of this science in the West.

iconoscope CATHODE-RAY TUBE used in early TELEVISION cameras, invented by Vladimir ZWORYKIN. Light entering the tube strikes a PHOTOELECTRIC PLATE, causing the plate to become positively charged. The other side of the plate is scanned by a beam of ELECTRONS from an ELECTRON GUN. Those electrons not held in the plate bounce back to be collected and transmitted as the television signal.

ideal gas law (gas laws) Law relating pressure, temperature and volume of an **ideal (perfect) gas**: $pV = nRT$, where n is the number of molecules of the gas and R is the GAS CONSTANT of proportionality. This law implies that at constant temperature (T), the product of pressure and volume (pV) is constant (BOYLE'S LAW); and at constant pressure, the volume is proportional to the temperature (CHARLES' LAW).

idiopathic In medicine, term used to describe a condition that arises spontaneously, or the cause of which is unknown.

idocrase (versuvianite) Orthosilicate mineral, consisting of hydrous calcium, iron, magnesium, aluminium silicate. Found in impure limestone, it forms tetragonal system crystals and is glassy green, brown, yellow or blue. Hardness 6.5; r.d. 3.4. Its transparent green or brown crystals are cut as gems.

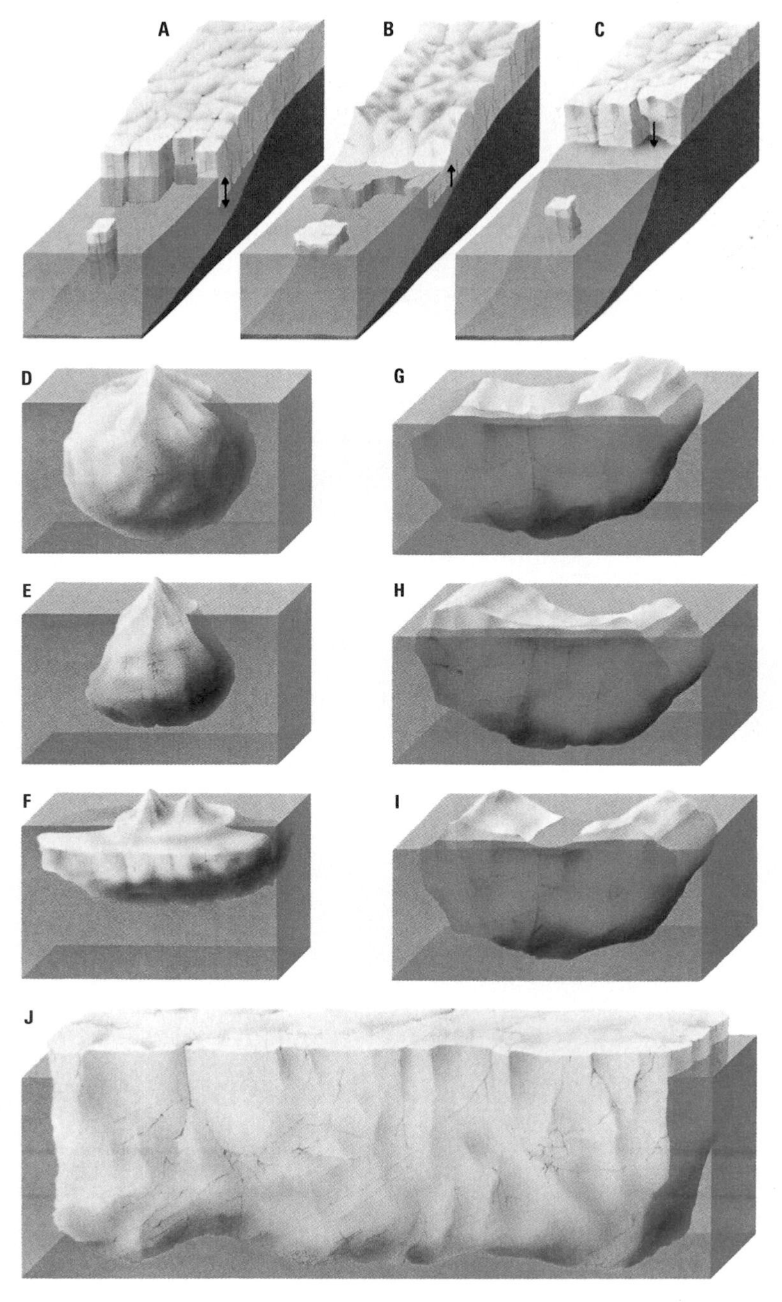

◄ **iceberg** Icebergs form in several ways. When a glacier reaches the sea it floats away from the bed. The movement of waves and tides exerts pressures on this floating ice causing lumps to break away (A). If the glacier is moving rapidly when it reaches the sea, a projecting shelf of ice forms under the water. The buoyancy of this shelf exerts an upward pressure causing pieces to break off (B). The snout of the glacier may be above the level of the sea and hence lumps may break off under the force of gravity and fall into the water (C). The forming of new icebergs is known as "calving". D–I show typical shapes of icebergs. Northern icebergs come from the Greenland ice sheet, but the largest ones originate in Antarctica (J). The largest iceberg ever seen was 336km (208mi) long and 97km (60mi) in width.

Idrisi (*c*.1099–1165) Arabian geographer. He worked at the court of Roger of Sicily after 1145 and wrote the *Kitab Rujjar* (Book of Roger, 1154), a geographical study of the world. He also made a map of the world engraved on silver.

igneous rock Term denoting rocks of volcanic origin which were formed from MAGMA. If the magma solidified beneath the Earth's surface, cooling would have been slow; if it solidified at the surface, cooling would have been more rapid. The faster the cooling process, the smaller are the crystals that make up the rock. All igneous rocks are crystalline. In addition to grain size, the other important differences found in igneous rocks are related to the chemical content of the minerals in the rock. Acidic minerals, including QUARTZ, some FELDSPARS and some MICAS, are generally light in colour. Conversely, basic minerals are generally dark in colour; they include some feldspars, biotite mica, HORNBLENDE and AUGITE. Igneous rocks formed at depth are called **plutonic**, those formed at the surface are called **volcanic**, and those formed between are **hypabyssal**. Volcanic rocks are EXTRUSIVE, and the commonest variety is BASALT. Hypabyssal rocks are found in SILLS and DYKES, and they are INTRUSIVE ROCKS. Plutonic rocks form in large bodies of magma called BATHOLITHS, stocks and "bosses" (circular intrusions, lying at a steep angle to the surface). The commonest plutonic rock is GRANITE, which is acidic in mineral content; the basic equivalent is called GABBRO. Igneous rocks do not have BEDDING PLANES, and they do not contain fossils. They are mostly quite resistant to erosion, and often form high ground; British examples include Dartmoor, Bodmin Moor, the Malvern Hills, parts of the Lake District and the Cairngorms.

ignition, engine Electrical components for igniting fuel in a petrol engine. Air and petrol vapour are usually mixed in a CARBURETTOR and delivered to the CYLINDER, where the mixture is compressed and ignited by a spark produced by the SPARK PLUG.

ileum Major part of the small intestine, *c*.4m (13ft) long. Its inner wall is lined with finger-like projections (villi), which increase the surface area available for the absorption of nutrients.

ilium Broad flat bone on either side of the PELVIS. Joined by the SACRUM, the fused bones of the base of the spine, it is fused with the ISCHIUM, and PUBIS in a triangular suture at the hip socket. The ilia form the back of the "basin" of the pelvic girdle and anchor the muscles that form the frontal wall of the abdomen. *See also* HUMAN BODY

illusion Lack of correspondence between the perception of an object and the physical reality. Illusions include such phenomena as hallucinations and can occur in any of the five senses, although optical and visual illusions are the commonest. **Optical** illusions are caused by physical properties such as REFRACTION, which explains why a rod lowered into water appears to bend. A **visual** illusion is the misinterpretation by the brain of something perceived by the eye. **Geometric** illusions occur when patterns of lines or shapes present conflicts of interpretation, such as the illusion in which two lines of equal length appear unequal because of the diagonal lines that are attached to their ends in different ways. Other visual illusions concern the perception of colour, brightness, motion and depth.

ilmenite (iron titanate, $FeTiO_3$) Black, oxide mineral, a major ore of the metal TITANIUM. It is found in basic IGNEOUS ROCKS and beach sands as tabular to fine, scaly crystals in the rhombohedral system, and as compact masses or granules. It has a metallic lustre and is magnetic. Hardness 5–6; r.d. 4.7.

image, optical Representation of an object produced by an optical instrument. A **real** image has rays of light passing through it and is formed by points to which the rays from the object converge. Rays of light do not pass through the points composing a **virtual** image: the image lies at the place from which the light rays, if traced backwards, appear to diverge. Images formed in a MIRROR are examples of virtual images. A real image can be projected onto a screen and recorded in a photographic emulsion; a virtual image, such as that produced by a plane mirror, cannot. *See also* LENS

image intensifier Electronic device in which gamma rays, X-rays and ultraviolet waves from a source can be converted into a visible image or in which the intensity of a visible image can be increased. The radiation falls onto a special surface and releases electrons that are then accelerated and focused onto a fluorescent screen, producing an enhanced visible image. *See also* ELECTROMAGNETIC SPECTRUM; PHOTOMULTIPLIER

imaginary number *See* NUMBER, IMAGINARY

imago Adult, reproductive stage of an INSECT that has undergone full METAMORPHOSIS. Imagos are the winged insects, such as butterflies and dragonflies, that emerge from PUPAS or develop from NYMPHS.

immune system System by which the body defends itself against disease. It involves many kinds of LEUCOCYTES (white blood cells) in the blood, lymph and bone marrow. Some of the cells (B-cells) make ANTIBODIES against invading microbes and other foreign substances (ANTIGENS), or neutralize TOXINS produced by PATHOGENS, whereas other antibodies encourage PHAGOCYTES and MACROPHAGES to attack and digest invaders. Another type of cell, T-CELL, provides a variety of functions in the immune system. *See also* ACQUIRED IMMUNE DEFICIENCY SYNDROME (AIDS)

immunity Resistance to attack by disease-causing microorganisms. It can be acquired naturally, as from an infection that stimulates the body to produce protective ANTIBODIES (a newborn baby carries some ANTIBODIES from its mother). Alternatively, it can be conferred by IMMUNIZATION.

immunization Practice of conferring IMMUNITY against disease by artificial means. **Passive** immunization may be conferred by the injection of an antiserum containing ANTIBODIES. **Active** immunization involves vaccination with dead or attenuated (weakened) microorganisms to stimulate production of specific antibodies and so provide lasting immunity.

immunoglobulin PROTEIN found in the bloodstream that plays a role in the body's IMMUNE SYSTEM. Immunoglobulins act as ANTIBODIES for specific ANTIGENS. They can be obtained from donor plasma and injected into patients most at risk of particular diseases.

immunology Study of IMMUNITY and all the defence mechanisms of the IMMUNE SYSTEMS mounted by the body for its protection. Immunologists are also concerned with related phenomena such as allergic responses, the rejection of transplanted organs and AUTOIMMUNE DISEASE.

immunosuppressive drug Any drug that suppresses the body's IMMUNE SYSTEM'S responses to infection or "foreign" tissue. Such drugs are used to prevent rejection of transplanted organs and to treat AUTOIMMUNE DISEASE and some cancers.

impedance (symbol *Z*) Property of a component in an ELECTRIC CIRCUIT that opposes the passage of

IGNEOUS ROCK

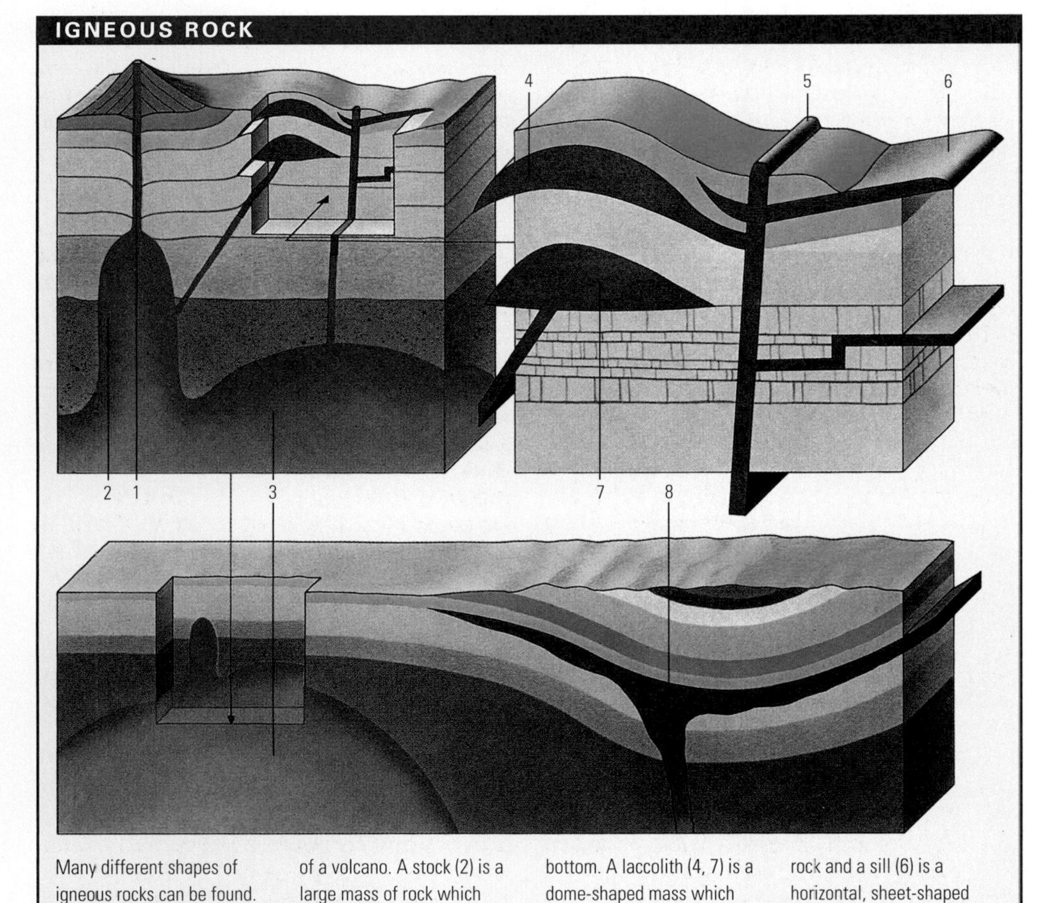

Many different shapes of igneous rocks can be found. Among the most common are the neck (1), which is a circular vertical feed channel of a volcano. A stock (2) is a large mass of rock which solidified at great depth. A batholith (3) is a large body of granite with no detectable bottom. A laccolith (4, 7) is a dome-shaped mass which has forced the rock above it to arch. A dyke (or dike) (5) is a vertical, sheet-like mass of rock and a sill (6) is a horizontal, sheet-shaped body of rock. A lopolith (8) is a saucer-shaped mass of rock.

current. In a direct current (DC) circuit, the impedance corresponds to the RESISTANCE (R). In an alternating current (AC) circuit with CAPACITANCE or INDUCTANCE, the additional property of reactance (X) has to be allowed for, as expressed in the equation $Z^2 = R^2 + X^2$. All of these quantities are measured in OHMS.

imperial system Units of measurement developed in the UK. *See* FPS SYSTEM

impermeable Term that describes rock that does not allow water to pass through it. Slate is one example. *See also* IMPERVIOUS; PERMEABILITY

impervious Term that describes rocks that contain no cracks or fissures for water to flow through, although the rock itself may absorb water. Clay is one example. *See also* IMPERMEABLE

implication Logical proposition of the type "if P then Q", connecting two simple propositions P (the **antecedent**) and Q (the **consequent**). In the form used in mathematical LOGIC the two propositions need not be connected. This is called **material** implication – an example is "if the Earth is flat then gold is a metal". A material implication is false only when the antecedent is true and the consequent false, otherwise it is true. In normal discourse **formal** implication is used, in which the simple propositions are related in meaning; for example "if it does not rain, we will go for a walk." An implication is written as $P \to Q$ (or $P \Rightarrow Q$), read "P implies Q."

imprinting Form of learning that occurs within a critical period in very young animals. A complex relationship develops between the newborn infant and the first animate object it encounters; this is usually a parent. The future emotional development of the infant depends upon this relationship. Imprinting in birds has been studied by Konrad LORENZ, who believed that it is an irreversible process.

inbreeding Mating of two closely blood-related organisms. It is the opposite of outbreeding and over successive generations causes much less variation in GENOTYPE and PHENOTYPE than is normal in a wild population. A form of GENETIC ENGINEERING, it can be used to improve breeds in domestic plants and animals. In humans, it can have harmful results, such as the persistence of HAEMOPHILIA in some European royal families.

incandescence Emission of light by a substance at a high temperature. Whether or not a hot object is seen as incandescent depends to some extent on the dark-adaptation of the human EYE, but an incandescent object is never at a temperature below *c.*400°C (750°F). An object like a FLUORESCENT LAMP can emit bright light without being incandescent.

inch (symbol in or ") Imperial unit of measurement equal to 2.54cm. There are 12in to 1 FOOT.

incisor Any of the chisel-shaped cutting TEETH between the canines in the front of the mouths of mammals. There are eight in humans, four in the centre front of each jaw. *See also* DENTITION

inclination, magnetic (magnetic dip) Angle between the direction of the Earth's magnetic field and the horizontal, measured by a free-floating magnet. At the north magnetic pole the inclination is zero; at the magnetic equator it is 90°. *See also* DECLINATION, MAGNETIC

inclined plane Simple machine that consists of a ramp. It is easier to push a load (mass m) up a slope (at an angle θ to the horizontal) than it is to lift it vertically – there is a MECHANICAL ADVANTAGE. The minimum force needed to push a load up an inclined plane is $mg \sin\theta$ (where g is the acceleration of free fall).

incomplete dominance In GENETICS, a situation in which neither GENE for a particular characteristic is DOMINANT. As a result, the organism shows the influence of both genes. For example, a plant with genes for both red flowers and white flowers may have pink flowers.

incubation In biology, process of maintaining stable, warm conditions to ensure that eggs develop and hatch. Incubation is carried out naturally by birds and by some reptiles. It is accomplished by sitting on the eggs, by making use of volcanic or solar heat or the warmth of decaying vegetation, or by covering the eggs with an insulating layer of soil or sand.

incubation period In medicine, time-lag between becoming infected with a disease and the appearance of the first symptoms. In many infectious diseases, the incubation period is quite short – anything from a few hours to a few days – although it may also be more variable. In leprosy, for example, incubation periods range from a few months to as much as 20 years.

indehiscent Describing a FRUIT or seed pod that does not open spontaneously to release its seeds when ripe. Release of the seeds takes place when the fruit rots or is eaten by an animal. Fungi that do not release their spores are also termed indehiscent. *See also* DEHISCENT

independent assortment Separation of the ALLELES of a GENE into the GAMETES (sex cells) with no regard to the way in which the alleles of other genes have separated. It takes place during MEIOSIS – the type of CELL DIVISION that gives rise to gametes. Theoretically, as a result, all possible combinations of alleles should occur with equal frequency, giving rise to maximum genetic variation. It does not always happen in practice because alleles on the same CHROMOSOME are usually inherited together.

independent variable *See* VARIABLE

index fossil Any FOSSIL of limited time distribution that clearly marks certain beds or strata of rocks. Its occurrence in rocks located kilometres apart proves that these deposits were formed at the same time. These fossils are important in mapping rock formations and in locating valuable resources.

indicator In chemistry, substance used to indicate an ACID or ALKALI. It does this usually by a change of colour. Indicators, such as the dye LITMUS, can detect a change of PH that measures a solution's acidity (litmus turns red) or alkalinity (turns blue). Universal indicator (liquid or paper) undergoes a spectral range of colour changes from pH 1 to 13.

indigo (indigotin, $C_{16}H_{10}O_2N_2$) Violet-blue dye traditionally obtained from plants of the genus *Indigofera*. It has been manufactured synthetically since the early 1900s.

indium (symbol In) Silver-white, metallic element of group III of the periodic table. Its chief source is as a by-product of zinc refining. Malleable and ductile, indium is used in certain alloys with low melting points, in semiconductors, to wet glass and coat metals, and as a mirror surface. Properties: at.no. 49; r.a.m. 114.82; r.d. 7.31; m.p. 156.6°C (313.9°F); b.p. 2,080°C (3,776°F); most common isotope ^{115}In (95.77%).

inductance Property of an electric circuit or component that produces an ELECTROMOTIVE FORCE (EMF) following a change in the current. The SI unit of inductance is the HENRY. **Self-inductance** (symbol L) occurs when the current flows through the circuit or component, and **mutual inductance** (symbol M) when current flows through two circuits or components that are linked magnetically. *See also* ELECTROMAGNETIC INDUCTION; INDUCTION

induction In medicine, initiation of LABOUR before it starts of its own accord. It involves perforating the fetal membranes and administering the hormone oxytocin to stimulate contractions of the UTERUS. It is undertaken where there is medical risk to the mother or baby in waiting for labour to commence naturally.

induction In physics, process by which magnification or electrification is produced in an object. In ELECTROMAGNETIC INDUCTION, an electric current is produced in a CONDUCTOR when placed within a varying MAGNETIC FIELD. The magnitude of the current is proportional to the rate of change of MAGNETIC FLUX. In a TRANSFORMER, the alternating current in the primary coil creates a changing magnetic field that induces a current in the secondary coil. A DYNAMO, or generator, contains a constant magnetic field, within which a conducting coil (armature) is rotated. In MAGNETIC INDUCTION, a magnetic field is created in an unmagnetized metal by another nearby magnetic field. *See also* INDUCTANCE

induction coil Type of TRANSFORMER that produces high-voltage alternating current from lower-voltage current. A common example is the induction coils used to ignite SPARK PLUGS in most petrol INTERNAL COMBUSTION ENGINES. All induction coils have two windings of wire, called the primary and secondary winding, the latter having more turns of wire. When the current in the **primary wire** is interrupted, a magnetic field, due to INDUCTION, is created around both windings, with the second winding creating greater voltage because of the greater number of turns.

inductor Device that introduces INDUCTANCE into an electric circuit. A typical inductor is a TORUS – a doughnut-shaped ring of metallic or ceramic material, around which is wound a coil carrying an electric current.

Indus (Amerindian) Small constellation of the s sky; the only star above magnitude 3.5 is Alpha.

industrial chemistry Chemical reactions and processes involved in the large-scale manufacture of chemicals in industry. While the engineering aspects of the subject belong to the study of CHEMICAL ENGINEERING, the industrial chemist is concerned with the chemical problems associated with the transfer of small-scale laboratory reactions to industrial plants.

industrial engineering Branch of ENGINEERING concerned with the application of engineering principles to industrial processes. Industrial engineers try to discover the most efficient ways for a firm to use the available workforce, equipment and materials. The engineers' work includes industrial planning, management organization, arranging technical training programmes and devising quality control methods.

inequality Mathematical statement that one expression is less than, greater than or equal to another. The symbol $>$ represents "is greater than", and $<$ stands for "is less than". An example is $2x + 4 > 12$, which is equivalent to $12 < 2x + 4$. Inequalities of this type may be handled in a somewhat similar way to equations: thus, in the case above, $x > 4$. The symbols $\geqslant$ and $\leqslant$ indicate "greater than or equal to" and "less than or equal to", respectively.

inert gas *See* NOBLE GAS

inertia Property possessed by all MATTER that is a measure of the way an object resists changes to its state of motion. Isaac NEWTON formulated the first law of motion, sometimes called the law of inertia, stating that a body will remain at rest or in a uniform motion unless acted upon by external forces.

inertial guidance In aeronautics, system using components that respond to changes in INERTIA to control missiles or spacecraft the orbits of which are largely above the Earth's atmosphere. The main

► **inflorescence** An inflorescence is a group of individual flowers borne on the same main stalk. Examples (A), (B) and (C) are inflorescences in which the main axis (red) increases in length by growth at its tip, giving rise to lateral branches (blue) which bear flowers. The flowers of *Sorbus aria*, whitebeam, form an inflorescence (A) called a corymb in which all the lateral branches are of different lengths, with the lowermost of the greater length so that all the flowers are brought to the same level. The flowers of *Bubomus umbellatus*, the water gladiolus, form an umbel (B) where all flower stalks arise from the same point. The inflorescence of *Astrantia major*, the masterwort, is a compound umbel (C) in which the lateral branches are themselves subdivided (yellow).

components include GYROSCOPES for reference and for detecting changes in orientation, motors or jets for correcting differences between planned and actual flight paths, and ACCELEROMETERS for determining velocity and position. The system can be supplemented by radar observations and control to correct gyro drift. *See also* NAVIGATION

infection Invasion of the body by disease-causing organisms that become established, multiply and give rise to symptoms. *See also* IMMUNE SYSTEM

infertility Inability to reproduce. In a woman it may be due to a failure to ovulate (release an OVUM for FERTILIZATION), obstruction of the FALLOPIAN TUBE or disease of the ENDOMETRIUM; in a man it is usually due to inadequate SPERM production. In plants, the term refers to inability to reproduce sexually. Infertility occurs in a HYBRID between different species, which are unable to produce viable GAMETES (eggs and male sex cells). Such plants cannot be grown from seed. However, many can reproduce by ASEXUAL REPRODUCTION. *See also* IN VITRO FERTILIZATION

infinity (symbol ∞) Abstract quantity that represents a number without bound. In analysis, $1/x$ approaches (positive) infinity as x approaches zero through positive values. In SET THEORY, the set of all integers is an example of an infinite set.

inflammation Reaction of body tissue to infection or injury, with resulting pain, heat, swelling and redness. It occurs when damaged cells release a substance called HISTAMINE, which causes blood vessels at the damaged site to dilate. LEUCOCYTES (white blood cells) invade the area to engulf bacteria and remove dead tissue, sometimes with the formation of pus. Inflammation can become chronic.

inflorescence FLOWER or flower cluster. Inflorescences are classified into two main types according to branching characteristics. A **racemose** inflorescence has a main axis and lateral flowering branches, with flowers opening from the bottom up or from the outer edge in; types include panicle, raceme, spike and umbel. A **cymose** inflorescence has a composite axis with the main stem ending in a flower and lateral branches bearing additional, later-flowering branches.

information storage and retrieval Branch of COMPUTER science that is concerned with the function of datasets. **Database** systems retrieval involves the searching on large computer files for specific DATA which may be organized into a variety of fields. Once a perfect match is found the information is made available. This type of system allows specific sets of data to be retrieved independently or with other sets. **Document-retrieval** systems store and retrieve entire documents, which can be searched for and retrieved using the document name or key words within the document. **Reference-retrieval** systems do not store documents themselves but references to documents. A search will make available the location of the relevant documents. This system is most commonly used in libraries where printed matter, such as books and periodicals, can be quickly located. *See also* DATABASE

information technology (IT) COMPUTER and TELECOMMUNICATIONS technologies used in processing information of any kind. Word processing, the use of a DATABASE and the sending of messages over a COMPUTER NETWORK all involve the use of information technology. Television stations use information technology to provide viewers with TELETEXT services. *See also* WORD PROCESSOR

information theory Mathematical study of the laws governing communication channels. It is primarily concerned with the measurement of information and the methods of coding, transmitting, storing and processing this information. Some of the concepts of information theory have been used in psychology to elucidate the processes involved in sensory perception and memory.

infrared astronomy *See* ASTRONOMY

infrared radiation Radiations occupying the ELECTROMAGNETIC SPECTRUM between the red end of the visible spectrum and microwaves. Often called heat radiation, it was first discovered by Sir William HERSCHEL in 1800. Its wavelength range is about 750nm to 1mm. Infrared ASTRONOMY uses infrared radiation to study celestial objects, the military uses it for missile guidance systems and night sights, and in medicine it is used for thermal imaging.

infrasound (infrasonics) SOUND waves having a frequency of 15 hertz (cycles per second) or less, below the range of human HEARING. This range is also known as the **subsonic range**, because such low frequency waves are felt as vibrations rather than heard as sounds. *See also* ULTRASONICS

inhibitor Compound that stops or substantially reduces the rate of a CHEMICAL REACTION. Inhibitors are as specific in their action as CATALYSTS and are widely used to prevent CORROSION, OXIDATION or POLYMERIZATION.

injection In medicine, use of a syringe and needle to introduce drugs or other fluids into the body to diagnose, treat or prevent disease. Most injections are either **intravenous** (into a vein), **intramuscular** (into a muscle), or **intradermal** (into the skin).

ink Coloured liquid used for writing, drawing or printing. It may be coloured by a suspended pigment or a soluble dye. The latter, often based on

INFRARED RADIATION

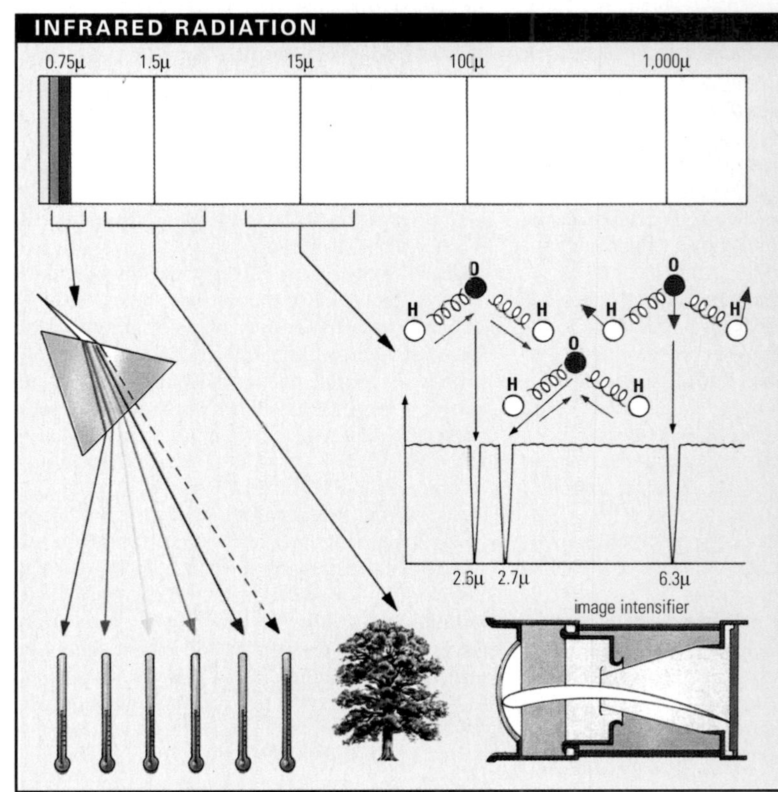

Infrared radiation has wavelength lying between 0.75–1000 microns (μ). The near infrared can be detected as heat, as discovered by Sir William Herschel. He found that a thermometer, placed beyond the red of a spectrum produced by a prism, showed a rise in temperature. The near infrared (0.75–1.5μ) has many uses, for example diagnosis of breast cancer and aerial surveying. The intermediate infrared (1.5–15μ) is used for chemical research to investigate molecular binding energies. When the molecular vibration is equal to the infrared frequency, resonance occurs. This can be detected by absorption spectroscopy. Infrared is also used in image intensifiers.

the compound ANILINE, is used in inks for ball-point pens. Printing ink usually contains finely divided carbon black suspended in a drying oil, often with added synthetic RESINS. Some inks dry by evaporation of a volatile solvent.

inlier Area of old rocks surrounded by younger ones. The old rocks may have resisted EROSION, and stand as hills above the surrounding eroded area. Alternatively, the upper and outer rocks of an ANTICLINE may have been eroded, leaving the older, once lower rocks surrounded by younger ones. *See also* OUTLIER

inner ear *See* EAR

inoculation *See* IMMUNIZATION; VACCINATION

inorganic chemistry *See* CHEMISTRY

insanity (madness) Imprecise term for severe mental illness. The condition has many manifestations, and is known by various other names which change from culture to culture and from age to age. Mental illnesses fall into one of three categories: PSYCHOSIS, which includes schizophrenia, manic depression, paronoia and functional psychoses; NEUROSIS, including states of extreme anxiety or phobia, overt obsessional or compulsive states and the hysterical or depressive neuroses; and **personality disorders**, such as the behavioural disorders of childhood, alcoholism and drug addiction.

insect Any of more than a million species of small, INVERTEBRATE animals of the order Insecta; including the beetle, bug, butterfly, ant and bee. There are more species of insects than all other species combined. Adult insects have three pairs of jointed legs, usually two pairs of wings, and a three-segmented body (**head**, **thorax** and **abdomen**) with a horny outer covering or **exoskeleton.** The head has three pairs of mouthparts, a pair of compound eyes, three pairs of simple eyes and a pair of antennae. Most insects can detect a wide range of sounds through ultra-sensitive hairs on various parts of their bodies. Some can "sing" or make sounds by rubbing together parts of their bodies. Most insects are HERBIVORES, many being serious farm and garden pests. Some prey on small animals, especially other insects, and a few are scavengers. There are two main kinds of mouthparts – chewing and sucking. Reproduction is usually sexual. Most insects go through four distinct life stages, in which complete METAMORPHOSIS is said to take place. The four stages are OVUM (egg), LARVA (caterpillar or grub), PUPA (chrysalis) and IMAGO (adult). Young grasshoppers and some other insects, called NYMPHS, resemble wingless miniatures of their parents. The nymphs develop during a series of moults (incomplete metamorphosis). Silverfish and a few other primitive, wingless insects do not undergo metamorphosis. *See feature on page 194*

insecticide Substance used to destroy or control insect pests. Insecticides may be **stomach** poisons, such as lead arsenate and sodium fluoride; **contact** poisons, such as DDT and ORGANOPHOSPHATES; or **systematic** poisons, such as octamethylpyrophosphoramide, which are toxic to insects that eat plants into which they have been absorbed. Organophosphates are preferred to chlorinated hydrocarbons (such as DDT) because they eventually break down into non-toxic substances. They are, however, hazardous to humans and must be handled with care.

insectivore Small order of carnivorous MAMMALS (Insectivora), many of which eat insects. Almost worldwide in distribution, some species live underground, some on the ground and some in streams and ponds. Most insectivores have narrow snouts, long skulls and five-clawed feet. Three families are always placed in the order: Erinaceidae, comprising moon rats, gymures and hedgehogs; Talpidae, comprising moles, shrew moles and desmans; and Soricidae, the shrews. Six other families – including tree shrews, tenrecs and solenodons – are also often included in the order.

insectivorous plant (carnivorous plant) Any of several plants that have poorly developed root sys-

▲► insectivorous plant
The hollow, jug-shaped traps of the Nepenthes pitcher plant hang at the end of its long leaves (A). Each one has a lid at the top to keep out the heavy tropical rains (1). Insects are attracted to the pitcher by its bright colours and by a sugary nectar which is produced by glands around the rim (2). The surface of this rim is waxy and very slippery, and most visiting insects rapidly lose their footing and fall in. Once inside, it is very difficult for them to escape: the upper areas of the inner surface are waxy; the lower regions are glassy smooth and covered with digestive glands; neither offer any foothold to an insect attempting to escape. They quickly tire of struggling and drown in the pool of water and digestive juices in the base of the trap (3). The digestion and absorption of the insects' bodies is very efficient in Nepenthes, taking about two days for an average fly, but only two hours for a small midge. The sticky leaves of the common European butterwort act like flypaper (B). A sticky mucilage is produced by stalked glands scattered over the leaf surface (1). When insects land on a leaf, they pull the mucilage out into strands which set and hold them fast (2). As they struggle, more glands are touched and they are held even firmer. The insects' movements also stimulate the leaf to start rolling up (3). This forms a temporary "stomach" into which digestive enzymes are released by stalkless glands (4). The enzymes are stored at the tip of the gland, in large vacuoles (5), and in the cell walls (6) of special secretory cells. Capture of prey stimulates a rush of water from the vascular system (7) through the gland, flushing out the enzymes onto the leaf surface (8) to form a pool around the insect. The products are absorbed by the leaf (9) and distributed around the plant (10). The Venus fly trap (C) is found in the bogs of North Carolina in the eastern United States of America. Ants, spiders and flies are its usual prey, though it may also capture other small animals such as snails and slugs. The leaf tips of the Venus fly trap are modified into two kidney-shaped lobes (1) hinged at the mid-rib (2). Covering the inner surface of the lobes are two kinds of glands. The alluring glands, on the green outer margins of the trap, secrete a sugary nectar that attracts insects (3). Farther in towards the midrib are the digestive absorptive glands (4), which give the lobes their red colour. Also on the inner surface of each lobe is a triangle of three tiny hairs which act as triggers (5). If two or more of these hairs are touched in quick succession by an insect, they produce a tiny electrical current. This causes a change in the water retention of the membranes of the motor cells in the midrib region. These cells quickly lose pressure and become limp, and the pressure from the cells in the outer epidermis forces the two lobes together (6). Within two-fifths of a second, the trap has closed sufficiently to prevent the escape of larger insects. As the prey struggles to escape it further stimulates the trap to close completely. The insect's soft parts are slowly broken down by acids and enzymes released from the digestive glands and then absorbed. After the insect's soft parts have been digested, the leaves open again, releasing the indigestible skeleton.

Insects are the most numerous of all living creatures, representing *c.*80% of all animal species. There are more than one million known species and probably as many again are still to be discovered – some estimates put that figure much higher. There are enough insect fossils and primitive living forms to serve as a guide to the evolution of the 29 orders into which all insects are classified. Most of the evidence dates from the advent of the Carboniferous period, when a number of winged insects inhabited the coal-forming swamps. Insects are thought to have evolved from a centipede-like ancestor. The most primitive of modern insects are possibly the wingless species, such as the North American springtail.

ADVANCED INSECTS

Of all insects, those that have a four-stage life cycle – egg, larva, pupa and adult – are the ones that are both most advanced and most successful. They account for more than two-thirds of all insect species and include such familiar groups as the moths and butterflies, beetles, bees, wasps and ants, and true flies. Because of their numbers and worldwide distribution insects are of great biological importance, both beneficial and damaging to the human health and economy. For example, pollination by insects, particulary bees, is essential for many crops, including most fruits. Other insects provide us with silk and biological pest control. However, insects also cause untold damage. Insect pests, such as locusts and the Colorado beetle, damage crops and stored products, and some, such as the tsetse fly and mosquitoes, carry diseases potentially fatal to people and livestock.

▲ **The life cycle** of the European swallowtail (*Papilio machoan*) is typical of most butterflies. The female adult (1) lays her eggs (2) on the underside of leaves in batches of 100 or more. The eggs hatch into the first stage larva or caterpillar (3). The caterpillar moults several times before it is fully grown (4). After the final moult, the caterpillar's skin hardens (4, 5) to form the case of the pupa or chrysalis (6). Within the case, the tissues of the caterpillar reorganize before the adult butterfly emerges (1).

► **The greenbottle fly** has highly adapted mouthparts. When it finds food, the greenbottle secretes enzymes from its salivary gland (1), via the salivary channel (2) in the proboscis (3), on to the food. The saliva is squirted over the food through grooves (4) in the labella (5), situated on the end of the proboscis. The grooves keep their shape because of rings of chitin (6). Once the food has been partly broken down to a liquid form by the enzymes, the greenbottle sucks up the liquid (7) through the grooves of the labella, via the food channel (8) and into the mid-gut (9), where digestion is completed.

ANATOMY

The internal anatomy of all insects, such as the honey bee, is contained and protected within the confines of a tough, flexible exoskeleton. The typical insect body contains organs of digestion, respiration, circulation, excretion and reproduction. There are muscles through which movement is effected and a nervous system that coordinates and controls insect actions on the basis of information received by the sense organs, most important of which are the large compound eyes and the feelers and antennae. All insect bodies comprise three parts: the head, thorax and abdomen.

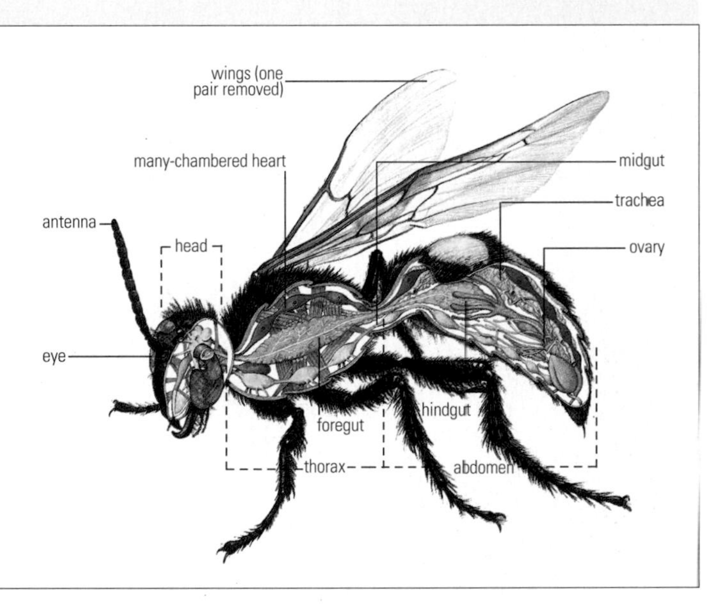

► **Highly modified** for the purposes of camouflage, stick insects, such as *Euryacantha horrida* (A), mimic the plants on which they live. Because they look like twigs, they are protected from predators. Their legs are barely noticeable at rest. Some stick insects may remain motionless for hours; others sway backwards and forwards as if moving with a breeze. Leaf insects such as *Phyllium crurufolium* (B), have legs and wings adapted to resemble leaves. In their tropical Asian habitats their predators find them difficult to spot. Some species have taken this adaptation a stage further, by laying eggs that resemble the seeds of various plants.

tems and are often found in nitrogen-deficient sandy or boggy soils. They obtain the missing nutrients by trapping, "digesting" and absorbing insects. Some, such as the Venus fly-trap (*Dionaea muscipula*), are active insect trappers; when triggered, its hinged leaves close on its prey. The sundews (*Drosera*) snare insects with a sticky substance, and then enclose them in their leaves. Bladderworts (*Utricularia*) suck insects into underwater bladders. Other plants have vase-shaped leaves, such as the pitcher plant (*Sarracenia flava*) into which insects fall.

inselberg Steep-sided, round-topped hill rising abruptly from a plain, found in semi-arid, tropical and subtropical regions. Inselbergs are probably formed by exfoliation EROSION of old mountains.

insemination, artificial Use of instruments to introduce donor semen into a female's reproductive tract in order to bring about FERTILIZATION. First developed for livestock breeding purposes, it is now routinely used to help INFERTILE couples. When SPERM from the woman's husband is used, it is known as AIH (artificial insemination by the husband); if it is acquired from an unknown donor, it is termed AID (artificial insemination by a donor).

insolation Quantity of solar radiation reaching the Earth, as measured in the SOLAR CONSTANT. It consists of a broad range of ELECTROMAGNETIC RADIATION, from infrared to X- and gamma rays.

instinct Behaviour that is innately determined, as opposed to learned. In the 19th century instincts were often cited to explain behaviour, but the term fell into disrepute with the advent of behaviourism. The term was revived in the work of such ethologists as Konrad LORENZ.

instrument landing system (ILS) Combination of three radio systems that guide an aircraft pilot to a landing when visibility is poor. A **glide slope** beam sent from the runway indicates the proper angle of descent, and the **localizer** beam indicates its direction. The **outer marker** beam, set *c*.8km (5mi) from the runway, and the **middle marker**, at *c*.0.8km (0.5mi), show distance.

insulation Technique for reducing or preventing the transfer of heat, electricity, sound or other vibrations. Wool, fibre-glass and foam plastic are good **heat** insulating materials because they contain air. This trapped air reduces the transfer of heat by CONDUCTION. Free-moving air is not an effective insulator because it circulates and transfers heat by CONVECTION. Water is also a good heat insulator. A diver's wet suit keeps the wearer warm by trapping a layer of water around the body. **Electrical** insulators are substances that provide a high resistance to an ELECTRIC CURRENT. They are used to prevent contact between CONDUCTORS in a circuit. Common insulators are rubber, polyvinyl chloride (PVC), mica, teflon, glass, asbestos, thermoplastics and porcelain. **Sound** insulating materials absorb sound. The physical structure of sound insulators determines their effectiveness at various frequencies. For example, a blanket over the head will block out mainly high frequencies, giving a muffled sound quality. **Noise** insulation, from engine vibrations in a car for example, is achieved by mounting the engine on springs or rubber fittings.

insulin HORMONE secreted by the ISLETS OF LANGHERHANS in the PANCREAS, important in the maintenance of blood-glucose levels. Insulin has the effect of lowering the blood-glucose level by helping the uptake of glucose into cells, where it is used up, and by causing the liver to convert glucose to GLYCOGEN, which is then stored in the liver. Glucose, unmetabolized because of lack of insulin, accumulates in excess amounts in the blood and urine, resulting in DIABETES.

integer Negative or positive whole NUMBER and zero, for example ... −3, −2, −1, 0, 1, 2, 3, ... There is an infinite number of such integers. The positive integers are also called the natural numbers. The existence of negative integers and zero allow any integer to be subtracted from any other integer giving an integer result.

integral (symbol $\int$) Mathematical symbol used in CALCULUS representing a summing operation. The integral of a function $f(x)$, written $\int f(x)dx$ can be interpreted as the area enclosed between the curve $y = f(x)$ and the x-axis. INTEGRATION (finding the integral of a function) is formally carried out by dividing the area into a number of rectangular strips parallel to the y-axis, and taking the limit of the sum of their areas as the number of strips increases and their width decreases. In fact, the symbol for an integration, $\int$, stands for "sum". A **definite** integral has a numerical value which represents the area bounded by the x-axis, $y = f(x)$ and the two lines $x = a$ and $x = b$. If a and b are unspecified the integral is said to be indefinite. In many cases the DERIVATIVE of the **indefinite** integral of a function gives the original function itself, so integration can be viewed as the inverse of DIFFERENTIATION.

integral calculus *See* CALCULUS

integrated circuit (IC) Complete miniature ELECTRONIC CIRCUIT incorporating SEMICONDUCTOR devices such as the TRANSISTOR and RESISTOR. The components of a **monolithic** integrated circuit occupy the surface layers of a single crystal of silicon (commonly called a SILICON CHIP). **Hybrid** integrated circuits have separate components attached to a ceramic base. Components in both types are joined by conducting film. Among their many uses, integrated circuits are found in pocket calculators and microcomputers. *See also* PRINTED CIRCUIT

Integrated Services Digital Network (ISDN) High-speed TELEPHONE lines designed to carry digital information. There are various grades of ISDN that can carry information more than a thousand times faster than conventional analogue voice lines. ISDN lines connect directly to a computer and do not need a MODEM.

integration In mathematics, name given to a set of techniques used to evaluate a given INTEGRAL. Integration, together with DIFFERENTIATION, forms the branch of mathematics known as CALCULUS.

intensive care unit (ICU) Specialized hospital unit treating patients with acute, life-threatening conditions. It is characterized by the use of continuous (electronic) monitoring and life-support technology, together with intensive nursing and the judicious use of drugs. The intention is to offer short-term support to critically ill or injured patients whose condition is deemed to be treatable.

interaction, nuclear Interaction in which ELEMENTARY PARTICLES can take part and by which they may be classified. HADRONS are subject to the STRONG NUCLEAR FORCE, which acts over a tiny range (10^{-13} cm, approximately the diameter of a PROTON). Two hadrons inside this range interact, in about 10^{-23} second, by producing other particles or being deflected. LEPTONS are subject to the WEAK NUCLEAR FORCE and a much lower probability of interaction. *See also* FUNDAMENTAL FORCE

intercom (intercommunication system) System that permits selective loudspeaker voice communication via wires between any pair of several stations, usually in the same building. The stations may be either "master stations", which may initiate calls to any in a group of stations, or "slave stations", which may initiate calls only to a master station.

interface *See* COMPUTER INTERFACE

interference In OPTICS, the interaction of two or more wave motions, such as those of light and sound, creating a disturbance pattern. The beams totally or partly reinforce or cancel each other out, producing an interference pattern. **Constructive** interference is the reinforcement of the wave motion because the component motions are in phase. **Destructive** interference occurs when two waves are out of phase and cancel each other. Interference of light radiation from a LASER is used in HOLOGRAPHY to create a hologram image.

interferometer Instrument in which a wave, especially a light wave, is split into component waves, which are made to travel unequal distances to recombine as INTERFERENCE patterns. The pat-

INTERFEROMETER

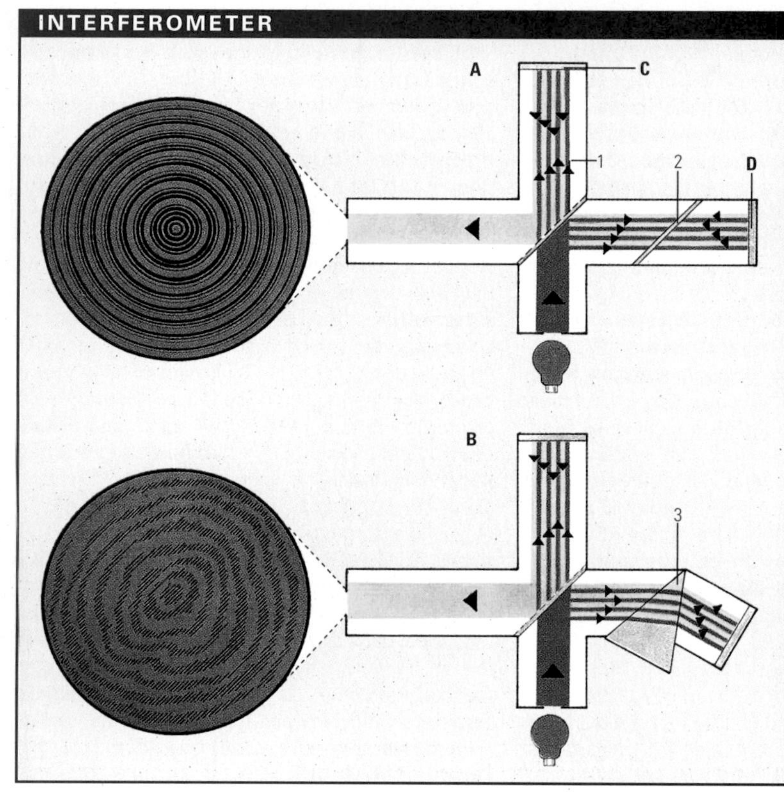

The Michelson (1890) interferometer consists of a diagonal mirror (1) which divides the beam from the lamp in half; one half is transmitted to the mirror (C) and the other reflected to the mirror (D). A glass plate (2) is inserted to equalise the path lengths. The beams then return and recombine towards the left where 'fringes' form. An observer sees both the image of C and D. If the planes of these two are parallel, circular fringes appear. Twyman (1918) modified Michelson's interferometer for looking for flaws in glass prisms and lenses (B). The prism to be tested (3) is placed in the light path. If it has imperfections the resulted image will be distorted.

INTERNAL COMBUSTION ENGINE

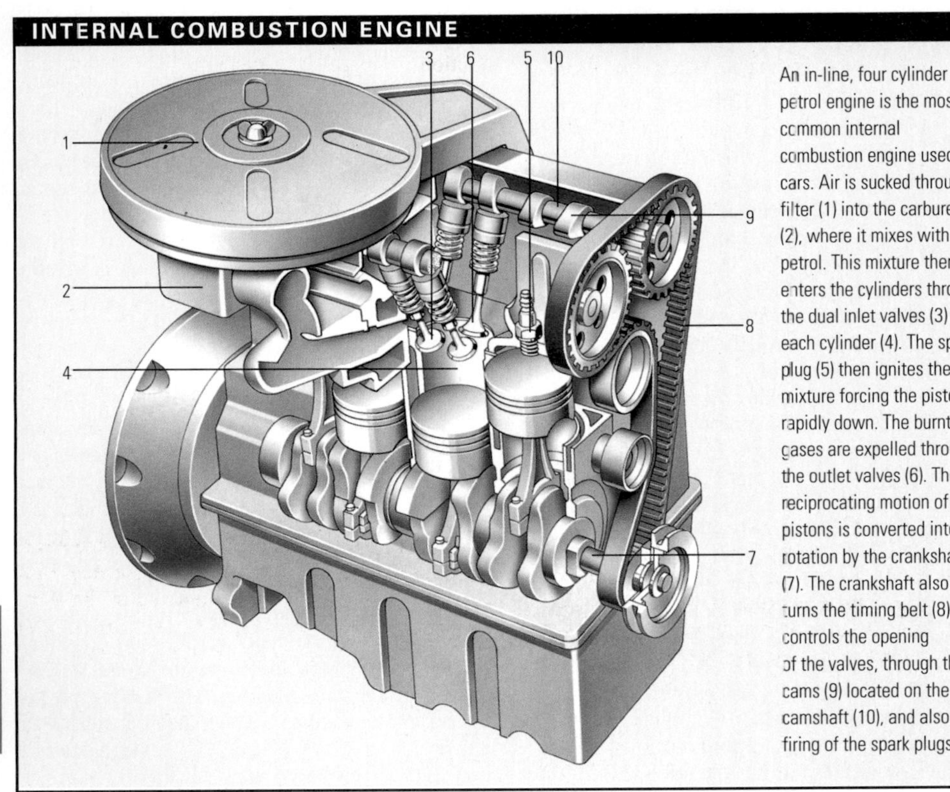

An in-line, four cylinder petrol engine is the most common internal combustion engine used in cars. Air is sucked through a filter (1) into the carburettor (2), where it mixes with the petrol. This mixture then enters the cylinders through the dual inlet valves (3) on each cylinder (4). The spark plug (5) then ignites the mixture forcing the piston rapidly down. The burnt gases are expelled through the outlet valves (6). The reciprocating motion of the pistons is converted into rotation by the crankshaft (7). The crankshaft also turns the timing belt (8) that controls the opening of the valves, through the cams (9) located on the camshaft (10), and also the firing of the spark plugs.

terns have such uses as quality control of lenses and prisms, and the measurement of wavelengths.

interferon Protein produced by body CELLS when infected with a VIRUS. Interferons can help uninfected cells to resist infection by the virus, and may also impede virus replication and PROTEIN SYNTHESIS. In some circumstances they can inhibit cell growth; human interferon is now produced by GENETIC ENGINEERING for therapeutic use, to treat some cancers, hepatitis and multiple sclerosis.

interglacial Interval between ICE AGES; the period of glacial retreat. *See also* GLACIER

internal combustion engine Engine, widely used in cars and motorbikes, in which fuel is burned inside, so that the gases formed can produce motion. An internal combustion engine may be a TWO-STROKE ENGINE or a FOUR-STROKE ENGINE. In the most common type of engine, a mixture of petrol vapour and air is ignited by a spark from a SPARK PLUG. The gases produced in the resulting explosion drive a PISTON down along a CYLINDER. A crankshaft changes the reciprocating (to-and-fro) movement of the pistons into rotary motion. In the WANKEL ROTARY ENGINE, the gases produced in the explosions drive a triangular rotor. The DIESEL ENGINE is another form of internal combustion engine.

Internet (Net) Worldwide communications system consisting of hundreds of small COMPUTER NETWORKS, interconnected by telephone systems. It is a network of networks, in which messages and data are sent and received using short local links from place to place around the world. This enables users to send a message to the other side of the world by ELECTRONIC MAIL (E-MAIL) for the cost of a local phone call. *See also* WORLD WIDE WEB (WWW)

interphase Stage following cell division (MEIOSIS or MITOSIS) in which the nucleus "rests". The nucleus is not dividing and adopts its final form in each of the daughter cells.

interplanetary matter Material in the space between the planets. It is made up of atomic particles (mainly PROTONS and ELECTRONS ejected from the Sun via the SOLAR WIND) and dust particles (mainly from COMETS but some are believed to be possibly of cosmic origin) in the plane of the ecliptic. The particles in the inner Solar System constitute the zodiacal dust cloud, scattered sunlight from which is responsible for the zodiacal light. An extremely tenuous ring of dust, over 50 million km (30 million mi) wide, orbits the Sun just outside the Earth's orbit. There is also dust in METEOR streams.

interpolation Mathematical procedure for finding intermediate unknown values of a FUNCTION lying between two or more known values. One simple method is to assume that the unknown values lie on a straight line between the known values; there are, however, many other interpolation techniques available. *See also* EXTRAPOLATION

intersection Point, or LOCUS of points, common to two or more geometrical figures. Two non-parallel lines in the same plane meet in a point; two non-parallel planes meet in a line. The intersection of two SETS is a new set consisting only of elements that are in both sets.

interstellar cloud Relatively dense concentration of INTERSTELLAR MATTER, especially dust, which usually appears in the form of a dark NEBULA. Such clouds are responsible for absorption leading to reddening or dimming of starlight, and they also may be regions in which stars are born.

interstellar matter Matter within galaxies occupying the space between stars. It includes bright NEBULAE (clouds of gas excited to luminescence by nearby stars) and dark nebulae (not excited). Farther away from stars, and filling interstellar space, is gas – mostly hydrogen – and particulate matter at exceedingly low densities, about 10–21kg per cu m. The interstellar matter in our galaxy accounts for 5% of its total mass.

interstitial cells Cells that form the connective tissue between other tissues or organized groups of cells. In cnidarians (such as jellyfish), for example, interstitial cells take the form of embryonic cells in the spaces between the columnar cells that form the body structure. In the TESTIS of vertebrates, interstitial cells between the seminiferous tubules secrete androgen male sex HORMONES.

intervertebral disc Ring of CARTILAGE that separates the VERTEBRAE – the bones of the SPINE. Intervertebral discs allow, with the exception of the axis and atlas bone *(see* HUMAN BODY), small amounts of movement, thus providing the spine with a degree of flexibility. A second function is to absorb shock, particularly for the skull and brain. The discs consist of a fibrous ring containing a soft, pulpy substance. Strain or injury to the spine may rupture or herniate a disc so that the pulpy interior protrudes and presses on a spinal nerve. This condition, commonly called a slipped disc, causes pain and immobility.

intestine Lower part of the ALIMENTARY CANAL, beyond the STOMACH, part of the digestive system. Food is moved through the intestine by the wave-like action known as PERISTALSIS. It undergoes the final stages of digestion and is absorbed into the bloodstream in the **small** intestine, a coiled tube that extends from the stomach to the large intestine. In the **large** intestine (caecum, colon and rectum) water is absorbed from undigested material, which is then passed out of the body through the anus.

intravenous drip Apparatus for delivering drugs, blood and blood products, nutrients and other fluids directly into the bloodstream. A hollow needle is inserted into an appropriate vein and then attached to a length of tubing leading from a bag containing the solution. The infusion is run into the vein at a controlled rate that may be regulated by a simple clip or a computer-operated infusion pump.

intrusion In geology, emplacement of rock material that was either forced or flowed into spaces among other rocks. An igneous intrusion, sometimes called a **pluton**, consists of MAGMA that never reached the Earth's surface but filled cracks and faults, before cooling and hardening.

intrusive rock Any IGNEOUS ROCK that forms by slow cooling under the Earth's surface (INTRUSION). In general, they are coarser grained than volcanic rocks that cooled on the surface.

invar Alloy containing about 64% iron, 36% nickel and small quantities of carbon. It has a very low thermal expansion, which leads to its use in instruments for measuring, such as surveying rods.

inverse square law Any of several laws in physics in which a quantity is inversely proportional to the square of the distance to the source of the quantity. A typical example is COULOMB'S LAW, which states that the force between two electric charges is proportional to the product of the charges and inversely proportional to the square of the distance between them. Newton's law of GRAVITATION is another example.

inversion Atmospheric condition in which a property of the air, such as moisture content or temperature, increases with altitude. In a **temperature** inversion, the air temperature rises with altitude and a cap of hot air encloses the cooler air below. With little wind or turbulence to break up the condition, POLLUTION can build up often to a dangerous extent.

invertebrate In zoology, term for an animal without a backbone. There are more than a million species of invertebrates, divided into 30 major groups. One of these is Arthropoda (ARTHROPODS), the largest animal PHYLUM in terms of number of species. Most invertebrates are INSECTS, but the term also covers CRUSTACEANS and ARACHNIDS. MOLLUSCS make up the second-largest group of invertebrates.

in vitro fertilization (IVF) Use of artificial techniques that join an OVUM with SPERM outside the woman's body to help some infertile couples to have children of their own. The basic technique of IVF involves removing ova from a woman's OVARIES, fertilizing them in the laboratory and then inserting them into the woman's UTERUS. In ZYGOTE intrafallopian transfer (ZIFT), a fertilized ovum (zygote) is returned to the FALLOPIAN TUBE, from where it makes its own way to the uterus. In GAMETE intrafallopian transfer (GIFT), the ova are removed, mixed with sperm, then both ova and sperm are inserted into a Fallopian tube to be fertilized in the natural setting. These IVF procedures result in live births in up to 20% of cases. The world's first "test-tube baby" was born in Britain in 1978.

involuntary muscle One of three types of MUSCLE in VERTEBRATES, so called because, unlike SKELETAL MUSCLE, it is not under the conscious control of the brain but is stimulated by the AUTONOMIC NERVOUS SYSTEM and by HORMONES in the bloodstream. It is of two kinds: **smooth** muscle is the muscle of the ALIMENTARY CANAL, blood vessels and bladder; **cardiac** muscle powers the HEART. *See also* HUMAN BODY

iodine (symbol I) Nonmetallic element that is the least reactive of the HALOGEN group. The black, volatile solid gives a violet vapour and has an unpleasant odour that resembles CHLORINE. Iodine, only found in combination with other elements, was discovered in 1811. Existing as iodide in seawater, seaweeds and other plants, it is also extracted from Chile saltpetre and oil-well brine. Iodine is essential for the functioning of the THYROID GLAND. It is used as a medical antiseptic and in photography (light-sensitive silver iodide). The radioactive isotope ^{131}I (half-life 8 days) is used in the treatment of thyroid gland disorders. Properties: at.no. 53; r.a.m. 126.9; r.d. 4.93; m.p. 113.5°C (236.3°F); b.p. 184.4°C (363.9°F); most stable isotope ^{127}I (100%).

iodoform *See* TRIIODOMETHANE

Io Innermost of Jupiter's four Galilean satellites, and the second-smallest, with a diameter of 3,630km (2,980mi). It has the highest density (3.53 g/cm^3) of the four, indicating a lack of water-ice in its make-up. Io is by far the most volcanically active world in the Solar System. The active regions contain volcanic vents and CALDERAS, lava flows, and volcanic plumes that send material to heights of up to 300km (200mi). A thin atmosphere of sulphur dioxide surrounds Io, and a doughnut-shaped plasma TORUS of ionized sulphur and oxygen surrounds its whole orbit and interacts with Jupiter's MAGNETOSPHERE. In 1996 the Galileo SPACE PROBE detected an iron core and magnetic field, and found the surface features to have changed significantly since the satellite was first imaged during the 1979 VOYAGER PROJECT missions.

ion ATOM or group of atoms with an electric (positive or negative) charge resulting from the loss or gain of one or more ELECTRONS. Positive ions are called **cations** and move towards the CATHODE in ELECTROLYSIS; negative ions are called **anions** and move towards the ANODE. Many crystalline inorganic solids are composed of arrays of ions of opposite charge. The process of forming ions is called IONIZATION.

ion engine Type of ROCKET engine that uses IONS, rather than hot gases, as the means of propulsion (ion propulsion). Ions are produced by projecting metal atoms (such as caesium or mercury) through an electric field; the ions are then accelerated to high speeds by an even more powerful field. A separate beam of electrons is projected behind the engine to neutralize the positively charged ions (or else the engine would acquire a high electric charge). The engine, which has been successfully tested in space, would be used to power long-distance spacecraft.

ion exchange Method of water-softening employed widely in domestic and industrial units. Permanent HARDNESS OF WATER is caused mainly by chlorides and sulphates of calcium and magnesium. In ion exchange softeners, the calcium and magnesium ions are exchanged by chemical reactions with those of sodium. Since sodium salts are much more soluble, this softens the water.

ionic bond (electrovalent bond) Type of chemical bond in which IONS of opposite charge are held together by electrostatic attraction.

ionic compound Substance formed by IONIC BONDS, chemical bonds of positively and negatively charged IONS. Salts, bases and some acids are ionic compounds. Examples include magnesium oxide (MgO) and sodium fluoride (NaF). As crystalline solids, with negative and positive ions arranged alternately in the lattice, such compounds have high melting points above 250°C (480°F) and boiling points above 500°C (930°F). They are usually soluble in water but insoluble in organic solvents. As solids, ionic compounds are nonconductors of electricity, but in solution and in their molten state, they are good CONDUCTORS.

ionic equilibrium Equilibrium state existing in solutions of ELECTROLYTES. Strong acids and bases are completely dissociated into IONS when in AQUEOUS SOLUTION, but weak acids or bases are only partly dissociated. Electrolyte, AB, when placed in solution, may partly dissociate into A^+ and B^-. $[A^+]$ $[B^-]/[AB]$, where [] indicates activities, is a constant, termed the **equilibrium constant**.

ionic radius Effective radius of a particular type of ION. An ionic solid (CRYSTAL) is composed of two or more species of ions, which may be regarded as spheres packed together in a regular pattern. By considering the CRYSTAL STRUCTURES and parameters of numbers of compounds, it is possible to obtain a consistent set of ionic radii.

ionic transport *See* TRANSPORT

ionization Process in which neutral atoms or molecules are converted into IONS. Positive ions can be formed by supplying energy to detach ELECTRONS from the atom, as by the action of X-rays, ULTRAVIOLET RADIATION or high-energy particles. The minimum energy to form an ion is the ionization energy (or potential). The opposite process – electron capture by a neutral species to yield a negative ion – is much less probable. Both types of ions can also be produced by breaking bonds, which can be induced by PHOTONS, or may occur spontaneously, as in the dissociation of acids and salts when dissolved in water.

ionization chamber Instrument for measuring the intensity of ionizing particles or radiation, such as X-rays. The gas-filled chamber contains two electrodes across which a voltage is applied. The passage of radiation through the chamber ionizes

IN VITRO FERTILIZATION

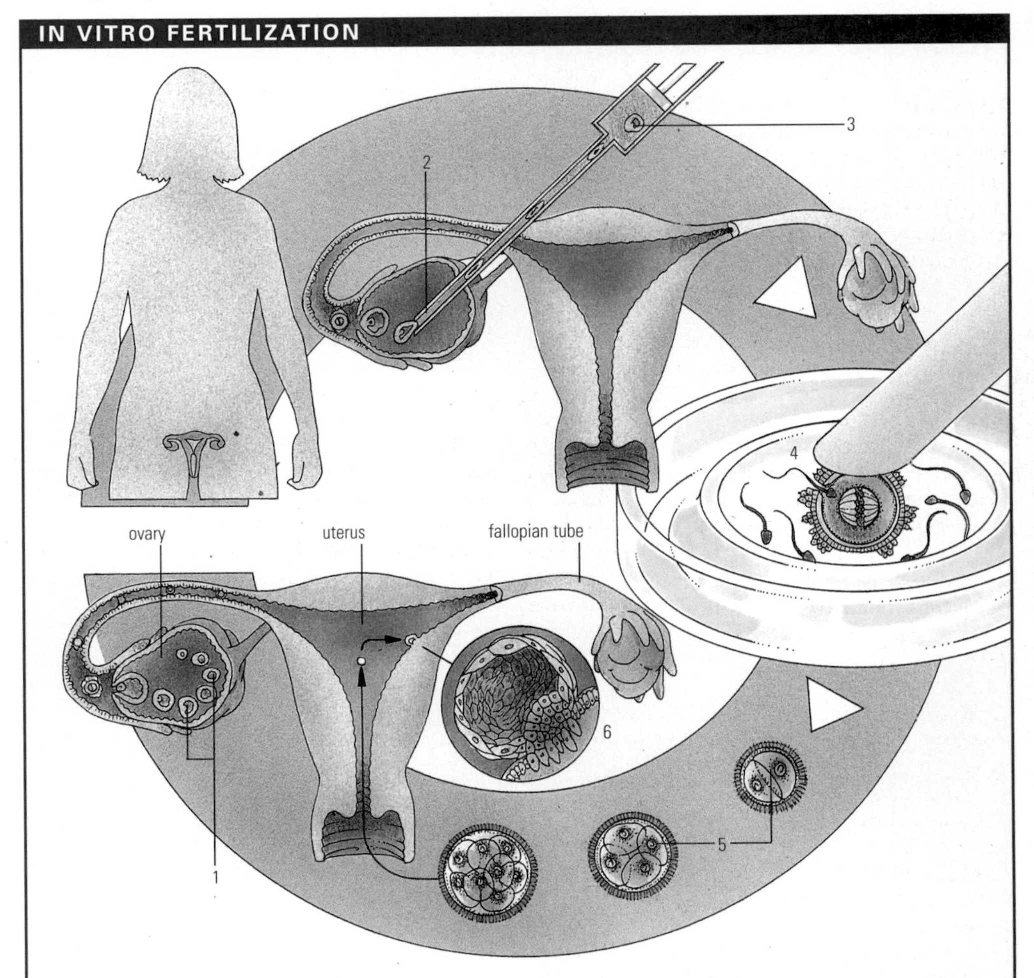

In vitro fertilization is a technique developed to help couples who are unable to conceive a baby normally. The mother, or egg donor, is given hormone treatment to cause many ovarian follicles to develop (1). While the ovary is observed through a laparoscope, a long needle (2) is inserted into each follicle, which is then sucked up (3) and transferred to a petri dish where sperm are added (4). The cells are monitored under a microscope. As the zygote divides (5) to a stage where it consists of 16 cells, it is introduced into the mother's uterus through a long plastic tube so that it can implant in the wall (6) and progress to full term in the usual way.

the gas and the ions formed move towards the charged electrodes. The current thus produced, which is amplified in an associated circuit, is proportional to the radiation intensity. A similar process is used in a GEIGER COUNTER, which detects and measures radioactivity.

ionizing radiation ELECTROMAGNETIC RADIATION having sufficient energy to react with atoms to produce IONS. There are two natural sources: radioactive elements in the Earth's crust and atmosphere, which emit beta and gamma radiation; and COSMIC RADIATION from outer space, which give rise to the IONOSPHERE. *See also* IRRADIATION

ionosphere Region around a planet in which there are free ELECTRONS and IONS produced by ULTRAVIOLET RADIATION and X-rays from the Sun. The degree of IONIZATION is greatly affected by solar activity. The Earth's ionosphere extends from *c.*80km (50mi) to 1,000km (625mi) above the surface (that is to the limits of the ATMOSPHERE in the VAN ALLEN RADIATION BELTS). It is divided into distinct layers distinguished by the concentration of electrons. The lowest layers, D and E, or Heaviside-Kennelly layers, undergo molecular ionization, while the upper layer, the F, or Appleton layer, undergoes atomic ionization. These layers reflect radio waves of long wavelengths while shorter wavelengths can pass through undisturbed. The reflecting power of these layers makes long-range radio broadcasting possible up to frequencies of about 30MHz.

ion propulsion *See* ION ENGINE

iridium (symbol Ir) Silver-white TRANSITION ELEMENT, first discovered in 1804. A platinum-type metal, iridium is hard and brittle and the most corrosion-resistant metal. It is used in making surgical instruments, scientific instruments, pen tips and electrical contacts. Properties: at.no. 77; r.a.m. 192.22; r.d. 22.42; m.p. 2,410°C (4,370°F); b.p. 4,130°C (7,466°F); most common isotope ^{193}Ir (62.6%).

iris Coloured part of the EYE. It controls the amount of light that enters the PUPIL in the centre of the eye by increasing or decreasing the size of the pupil. These changes are brought about by muscles in the iris contracting or relaxing.

iron (symbol Fe) Common TRANSITION ELEMENT, known from the earliest times. Its chief ores are HAEMATITE (Fe_2O_3), MAGNETITE (Fe_3O_4) and PYRITE (FeS_2). It is obtained in a BLAST FURNACE by reducing the oxide with carbon monoxide from coke (carbon), using limestone to form a SLAG. The pure metal – a reactive soft element – is rarely used; most iron is alloyed with carbon and other elements in the various forms of STEEL. The element has four allotropic forms, one of which is ferromagnetic. It forms two series of salts, termed iron(II), or ferrous, and iron(III), or ferric. Properties: at.no. 26; r.a.m. 55.847; r.d. 7.86; m.p. 1,535°C (2,795°F); b.p. 2,750°C (4,982°F); most common isotope ^{56}Fe (91.66%).

iron meteorite (siderite) METEORITE consisting mainly of iron (90–95%) and some nickel, with traces of other metals. Iron meteorites are classified into a number of groups according to the proportions of these other metals that they contain, and each group is thought to correspond to a different parent ASTEROID.

iron oxide One of three compounds that exist in three OXIDATION states: iron(II) oxide (ferrous oxide, FeO); iron(III) oxide (ferric oxide, Fe_2O_3), which occurs naturally as HAEMATITE; and ferrosoferric oxide (Fe_3O_4), which has iron in both oxidation states.

iron pyrites *See* PYRITE

ironstone Ancient medium- to fine-grained SEDIMENTARY ROCK containing iron-rich minerals such as HAEMATITE and SIDERITE; MAGNETITE and PYRITE may also be present. Ironstones were formed 2,000–3,000 million years ago in the Precambrian era. In **banded** ironstone, the basic rock is CHERT interspersed with layers of ferruginous minerals. **Oolitic** ironstone, formed in sediments at the bottom of ancient seas, consists of small rounded grains (OOLITHS) in which ferruginous minerals have replaced lime-rich ones. In both types, the iron-containing minerals result in a deep red colour.

irradiation Exposure to nuclear or ELECTROMAGNETIC RADIATION. Materials are often irradiated with high-energy neutrons in NUCLEAR REACTORS, to make them temporarily radioactive. More portable sources of such radiation are RADIOISOTOPES such as cobalt-60 and caesium-137, which are used in the irradiation treatment for cancer. Treatment also involves the use of particle ACCELERATORS, including proton and neutron beam machines. Radioactive sources are used industrially, such as in FOOD PRESERVATION. IONIZING RADIATION is used to destroy bacteria and microorganisms in some foodstuffs, while in other foods, such as soft fruits, it increases shelf life. The process of irradiation in the food industry is still being closely monitored. Some people fear it not only destroys harmful bacteria but also beneficial nutrients.

irrational number *See* NUMBER, IRRATIONAL

irregular galaxy Type of GALAXY characterized by an apparent absence of symmetrical or definite structure. Only about 3% of observed galaxies fall into this category, and it is possible that some represent intermediate stages in development, while others may be distorted by internal or external disturbances.

irrigation Artificial watering of land for growing crops. Irrigation enables crops to grow in regions with inadequate precipitation. The first irrigation systems date from before 3000 BC, when canals and ditches were built to carry water from rivers to nearby fields in Egypt, Asia and the Middle East. Later, simple mechanical devices, including the ARCHIMEDES' SCREW, were used to transfer water from rivers and streams into irrigation channels. Today, most water for irrigation is surface water or GROUNDWATER. Surface water comes from streams, rivers and lakes. Building a DAM across a river causes large quantities of water to accumulate in an artifi-

IONOSPHERE

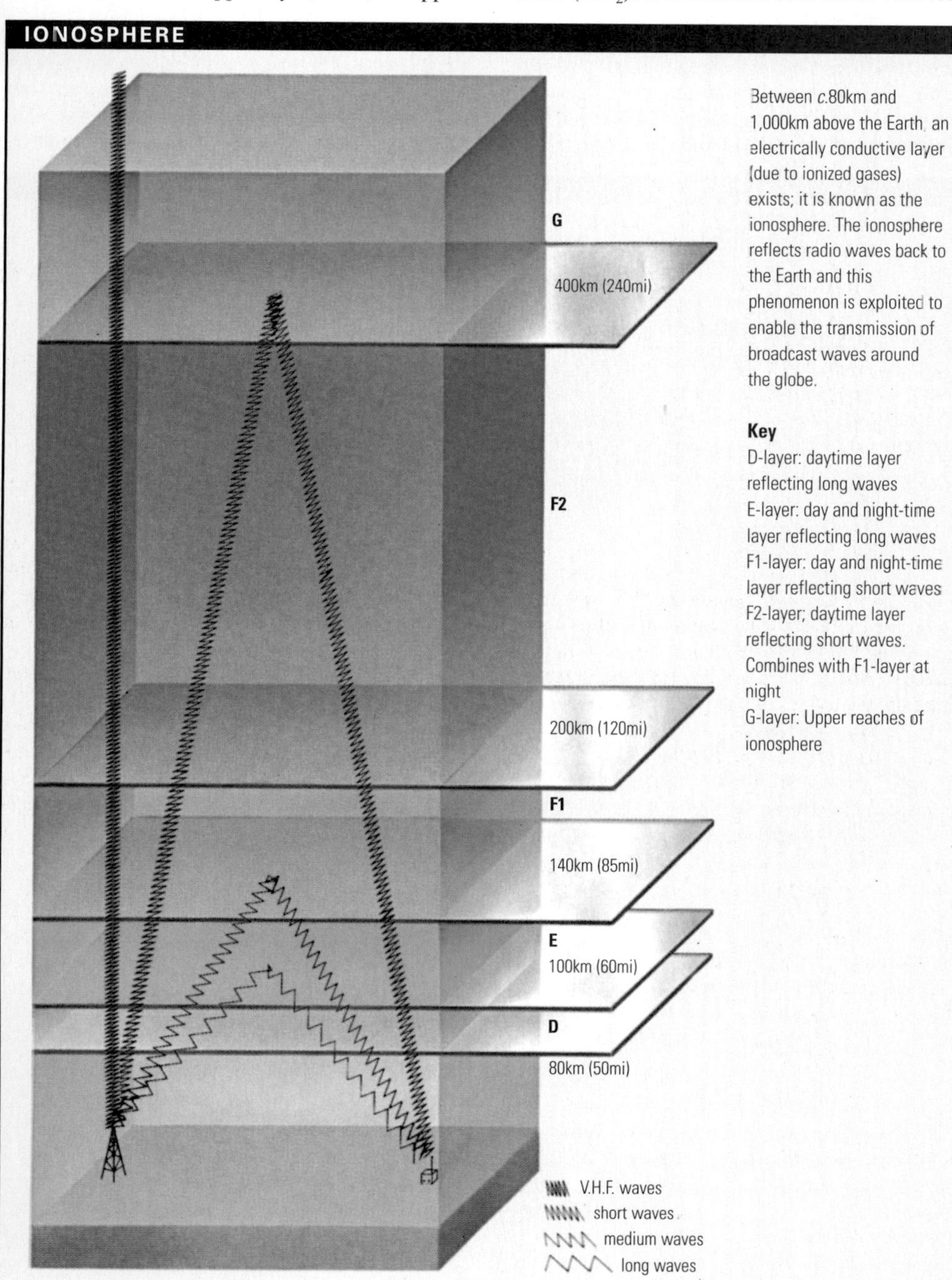

Between *c.*80km and 1,000km above the Earth, an electrically conductive layer (due to ionized gases) exists; it is known as the ionosphere. The ionosphere reflects radio waves back to the Earth and this phenomenon is exploited to enable the transmission of broadcast waves around the globe.

Key
D-layer: daytime layer reflecting long waves
E-layer: day and night-time layer reflecting long waves
F1-layer: day and night-time layer reflecting short waves
F2-layer; daytime layer reflecting short waves. Combines with F1-layer at night
G-layer: Upper reaches of ionosphere

cial lake or reservoir. This water can be used for irrigation when rainfall is scarce. Groundwater is obtained from WELLS. In some regions, fresh water for irrigation is obtained by DESALINATION. Canals, ditches, pumps and pipes convey water to fields.

ischium U-shaped bone at the base and on either side of the PELVIS. It is fused with the ILIUM and the PUBIS in the hip socket; its lower edge meets the pubis. In primates the ischium takes most of the body's weight when sitting. *See also* HUMAN BODY

ISDN Abbreviation of INTEGRATED SERVICES DIGITAL NETWORK

island arc Chain of volcanoes that occurs along one side of a deep ocean TRENCH. The volcanoes rest on the plate of LITHOSPHERE that is not moving down into the Earth. Their andesitic (*see* ANDESITE)lavas may be formed from the partially melted material of the descending plate. Northern Japan and the Aleutian Islands are examples.

islets of Langerhans Clusters of cells in the PANCREAS which produce two hormones: INSULIN to raise the level of GLUCOSE in the blood and GLUCAGON to counteract the effect of insulin when necessary.

isobar Line on a weather map connecting points of equal pressure, either at the Earth's surface or at a constant height above it. The patterns of isobars depict the variation in ATMOSPHERIC PRESSURE, showing areas of high and low pressure on the map.

isogamy In biology, state of having male and female reproductive cells that act like GAMETES (sex cells), but are similar in size and structure. Isogamy is found in algae, some protozoans, and primitive plants. It is distinct from **anisogamy,** where male and female sex cells differ in appearance.

isohyet Line on a weather map connecting points with equal rainfall. Isohyets are usually drawn for a particular period of time, to show the distribution of rainfall over a few months in the summer or winter, or perhaps for a whole year.

isoleucine $(CH_3CH_2CH(CH_3)CH(NH_2)COOH)$ One of the essential AMINO ACIDS found in protein.

isomers Chemical compounds having the same molecular formula but different properties due to the different arrangement of atoms within the molecules. **Structural** isomers have atoms connected in different ways. **Geometric** isomers, also called cis-transisomers, differ in their symmetry about a double bond. **Optical** isomers are mirror images of each other.

isomorphism In CRYSTALLOGRAPHY, resemblance between crystals of chemical compounds having similar crystal structures. In biology, isomorphism refers to the similarity of shape observed in unrelated groups due to CONVERGENT EVOLUTION.

isoprene (2-methylbuta-1,3-diene, $CH_2{=}C(CH_3)CH{=}CH_2$) Organic chemical compound. It is a synthetic product that closely resembles natural RUBBER.

isosceles triangle TRIANGLE having any two of its sides equal in length; the angles opposite these sides are also equal.

isostasy Theory describing the maintenance of an equilibrium in the total mass of the Earth's CRUST despite its crustal movement. There exists a balance between the land masses and the MANTLE of the Earth on which the continental plates float, so that the plates rise and sink on the semi-molten surface of the mantle in such a fashion that the relative mass weighing upon the Earth's crust is constant. The spread of the continental plates by the upwelling of material from deep within the Earth's crust is balanced by the submergence of the opposite edges of the plates. *See also* CONTINENTAL DRIFT; PLATE TECTONICS

isotherm Line of equal temperature on a map (usually a weather map). The patterns of isotherms depict how the temperature changes across the area of the map.

isotope Variety of a chemical element differing in the numbers of particles in the nuclei of their ATOMS. Atoms of different isotopes have the same number of electrons or protons (same ATOMIC NUMBER) but a different number of neutrons, so that both the MASS NUMBER and the mass of the nucleus are different for different isotopes. The RELATIVE ATOMIC MASS (R.A.M.) of an element is an average of the relative atomic masses of the element's naturally occurring isotopes. Isotopes are distinguished by writing the mass number by the name or symbol of the element, for example, ^{14}C or carbon-14. The isotopes of an element all have similar chemical properties, but physical properties vary slightly. Most elements have two or more naturally occurring isotopes, some of which are radioactive (radioisotopes). Many radioisotopes can be produced artificially by bombarding elements with high-energy particles such as alpha particles. Radioisotopes are used in medicine, research and industry. Isotopes are also used in radioactive DATING. *See also* RADIOACTIVITY

isotropic Term meaning the possession of physical properties that are the same regardless of the direction in which they are measured. For example, in water, heat conductivity and viscosity are isotropic but they may not be so in a very thick oil.

IVF Abbreviation of IN VITRO FERTILIZATION

IRRADIATION OF FOOD

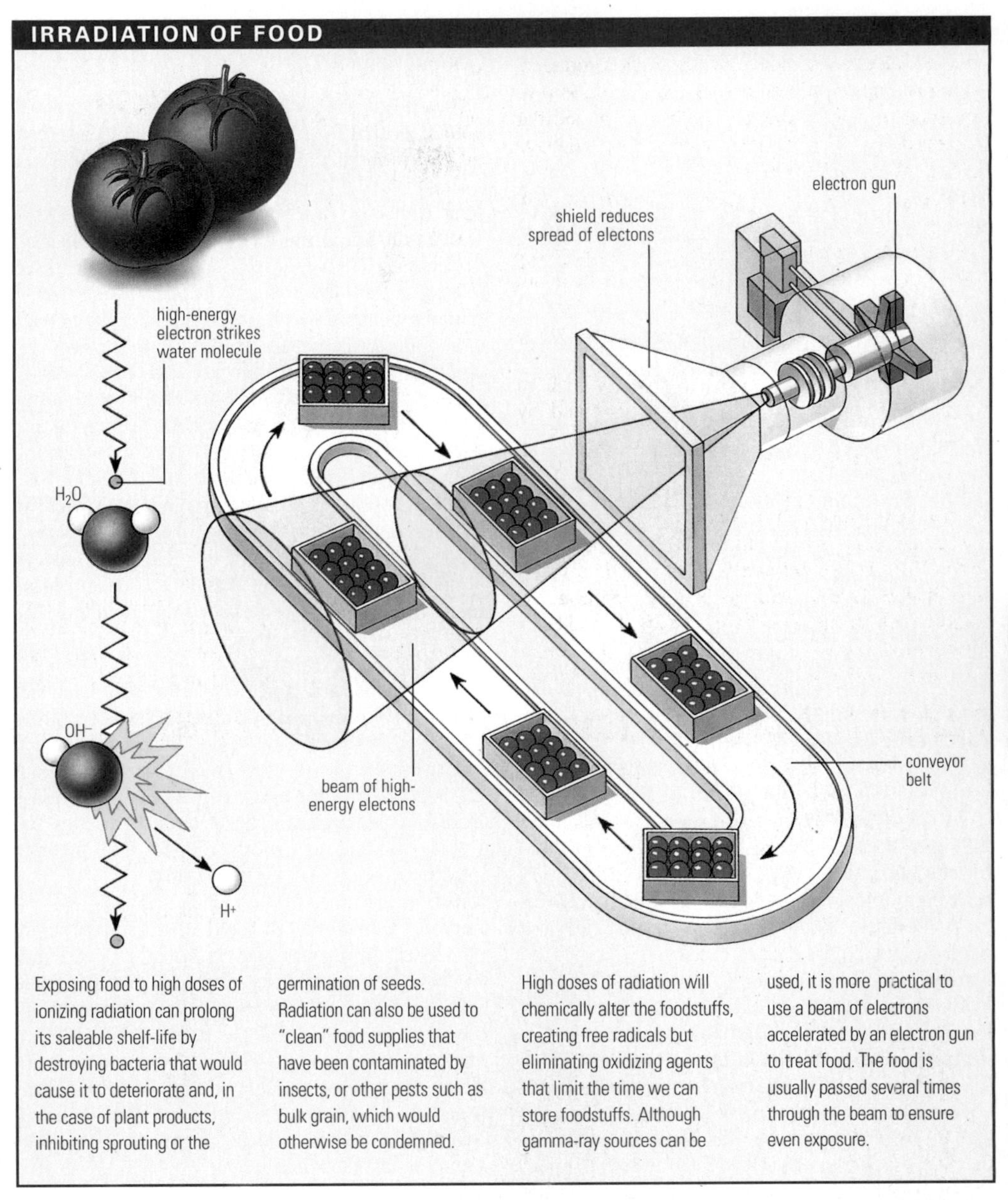

Exposing food to high doses of ionizing radiation can prolong its saleable shelf-life by destroying bacteria that would cause it to deteriorate and, in the case of plant products, inhibiting sprouting or the germination of seeds. Radiation can also be used to "clean" food supplies that have been contaminated by insects, or other pests such as bulk grain, which would otherwise be condemned. High doses of radiation will chemically alter the foodstuffs, creating free radicals but eliminating oxidizing agents that limit the time we can store foodstuffs. Although gamma-ray sources can be used, it is more practical to use a beam of electrons accelerated by an electron gun to treat food. The food is usually passed several times through the beam to ensure even exposure.

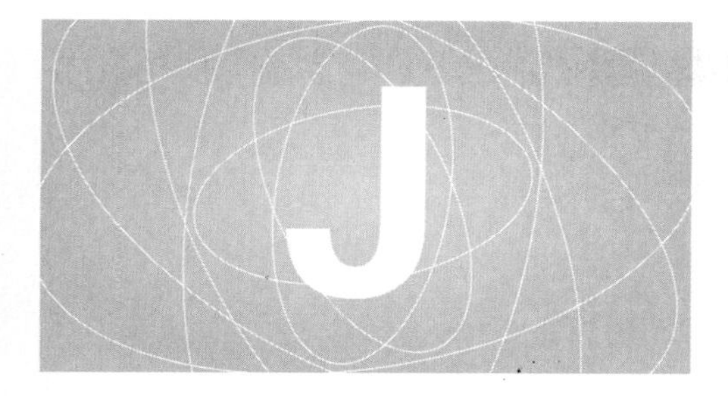

Jabir *See* GEBER

Jackson, John Hughlings (1835–1911) British neurologist. He studied speech defects as a symptom of brain disease, which he located in the left cerebral hemisphere. He also traced motor spasms to a brain disorder. Such spasms are now called Jacksonian EPILEPSY.

Jacob, François (1920–) French biologist. With Jacques MONOD he discovered that a substance that they named MESSENGER RNA (MRNA) carried hereditary information from the CELL nucleus to the sites of PROTEIN SYNTHESIS and that certain GENES, called **operator genes,** control the activity of others. They shared the 1965 Nobel Prize for physiology or medicine with another French biologist, André LWOFF. *See also* HEREDITY

Jacquard, Joseph Marie (1752–1834) French silk weaver who perfected (1801–06) a LOOM that could weave patterns automatically. By 1812 there were 11,000 Jacquard looms in use in France. The first automated machine, the loom was controlled by PUNCHED CARDS.

jade Semi-precious silicate mineral of two major types: **jadeite,** which is often translucent, and **nephrite,** which has a waxy quality. Both types are extremely hard and durable. Jade is found mainly in Burma and comes in many colours, most commonly green and white. Hardness 5–6; r.d. 3–3.4.

Jansky, Karl (1905–50) US engineer who in 1931, while investigating interference to telephone communications, discovered unidentifiable radio signals from outer space. Jansky concluded that they were stellar in origin and that the source lay in the direction of SAGITTARIUS, the direction of the centre of our Galaxy. His discovery is considered to be the beginning of RADIO ASTRONOMY. A unit measuring radio emission is named after him.

Janssen, Pierre Jules César (1824–1907) French astronomer who specialized in solar SPECTROSCOPY. He developed special instruments, including a spectrohelioscope for observing solar PROMINENCES without waiting for an ECLIPSE. He discovered the helium line in the solar spectrum, and produced a photographic atlas of the Sun. In 1893 he established an observatory on Mont Blanc.

jasper Opaque semi-precious stone, usually red to brown, sometimes yellow or grey to green. Jasper is found in Greece, Siberia, Libya and the River Nile valley. It was once believed that stomach complaints could be cured by wearing these stones. A dark form of jasper was traditionally used as a touchstone to estimate the gold content of precious alloys.

jaw Either of two bony structures that hold the TEETH and frame the mouth in humans and most vertebrates. The upper jawbone (MAXILLA) is fused to the SKULL. The lower jawbone (MANDIBLE) is movable, having two ligament hinges, one at each side of the temple near the entrance to the ear.

Jeans, Sir James Hopwood (1877–1946) British mathematician and physicist. In astronomy, he worked on stellar dynamics, the formation and evolution of stars, DOUBLE-STAR systems, and spiral GALAXIES. In 1917 he proposed that the Solar System was formed from matter pulled out of the Sun by the gravitational attraction of a passing star, a theory later developed by Harold Jeffreys (1891–1989), but since abandoned. Jeans wrote many popular books on astronomy, including *The Universe Around Us* (1929).

Jeffreys, Alec John (1950–) British geneticist. He was responsible for discovering particular elements of DNA that enabled GENETIC FINGERPRINTING to be used as an accurate method of identification.

jejunum Portion of the small INTESTINE, about 25–30cm (10–12in) long, into which food passes from the DUODENUM. The jejunum precedes the ILEUM, which at the lower end opens into the large intestine. *See also* DIGESTIVE SYSTEM

jellyfish *See* CNIDARIAN

Jenner, Edward (1749–1823) British physician, pioneer of the science of IMMUNOLOGY. Aware that cowpox, a minor disease, seemed to protect people from smallpox, in 1796 Jenner inoculated a healthy boy with cowpox from the sores of an infected dairymaid. The boy developed the disease but six weeks later, when inoculated with smallpox, was found to be immune to infection. This finding established the principle of VACCINATION as a lifesaving medical technique.

jet Dense variety of LIGNITE coal formed from wood buried on the seafloor. It is often polished and used in jewellery.

JET (Joint European Torus Research Project) Nuclear FUSION power research centre set up in 1977 at Culham, England. Sponsored by the European Union, Sweden and Switzerland, scientists hope to develop a machine that will heat PLASMA to immense temperatures within a MAGNETIC BOTTLE and thereby fuse atoms of hydrogen, with the subsequent release of nearly limitless energy. The machine uses the TOKAMAK design.

JET ENGINE

A turbofan engine is the most commonly used jet engine on civil aircraft. Fuel entering the engine (1) mixes with compressed air and burns in the combustion chamber (2). The expanding gases rotate high-speed (3) and low-speed (4) turbines. These, in turn, drive a compressor (5), which forces air into the combustion chamber, and fans (6), which push air round the combustion chamber and into the tail pipe, providing extra thrust by means of displacement. An engine of this type is able to generate up to 30,000lbs of thrust.

▲ **Jenner** Edward Jenner coined the word vaccination to describe his use of cowpox inoculation to obtain immunity to smallpox. Folk tales at the time suggested that people who contracted the mild cowpox disease did not go on to contract the deadlier smallpox. Jenner investigated this by inoculating eight-year-old James Phipps with fluid obtained from the blister of a patient with cowpox. Later, he inoculated him with smallpox. The boy was found to be immune. He first published his influential findings in 1798.

jet engine Engine that derives forward motion by the rapid discharge of a jet of fluid (gas or liquid) in the opposite direction. In a jet engine, fuel burns in oxygen from the air to produce a fast-moving stream of exhaust gases. These are ejected from the back of the jet engine and produce a forward thrust (propulsion) in accordance with NEWTON'S LAWS OF MOTION. ROCKETS work on similar principles, but have their own oxygen supply so they can work outside the Earth's atmosphere.

Jet Propulsion Laboratory (JPL) Space centre in Pasadena, California, USA, for the development and control of unmanned spacecraft. The California Institute of Technology runs JPL for the NATIONAL AERONAUTICS AND SPACE ADMINISTRATION (NASA). JPL scientists sent the Surveyor probes to the Moon in the 1960s. Other notable achievements include the GALILEO and MARINER PROJECTS, the VIKING SPACE MISSION and the VOYAGER PROJECT.

jet stream Strong current of air blowing through the atmosphere at a high altitude. The main jet streams are in the middle and subtropical latitudes. The temperate jet stream flows from W to E, though along a wave-like route, going N and S as it travels all around the world. It can blow at speeds of up to 370km/h (230mph) and so is an important factor for aircraft. If planes can fly with the jet stream, which often blows at a height of 10,000m to 13,000m (33,000–42,600ft) above the Earth's surface, it can increase the speed of the journey and save fuel costs. Jet streams influence the weather systems in the lower layers of the atmosphere, as depressions tend to follow their route.

Jex-Blake, Sophia Louisa (1840–1912) British physician who fought vigorously and successfully to get legislation through Parliament enabling women to qualify as doctors and to practise medicine and surgery. She founded a medical school for women in London in 1874 and another in Edinburgh in 1886.

joint In anatomy, place where one BONE meets another. In movable joints, such as those of the knee, elbow, spine, fingers and toes, the bones are

▲ **jet stream** Photo taken by the *Gemini 12* spacecraft, showing jet stream clouds over Egypt (foreground) and the Red Sea. The tip of the Sinai Peninsula can be seen jutting into the Red Sea, and the River Nile, with its strip of vegetation, is clearly seen as it flows through the Egyptian desert.

JOINT

At the knee, the femur (1) and tibia (2) are joined by the ligament (3). Further ligaments (4), attached to the patella (5), form the tendon for the quadriceps muscle (6). The synovial membrane forms the infrapatellar (7) and suprapatellar (8) bursae. Cartilage (9) covers the articular surfaces and there are two crescents of cartilage (10) between the femur and the tibia. Synovial membrane and fluid (11) lubricate; fatty pads (12) pack the joint. Biceps (13) and gastrocnemius (14) muscles are shown.

separated and cushioned from one another by pads of CARTILAGE. In fixed joints cartilage may be present in infancy but disappears later as the bones fuse together, as in the SKULL. In the movable joints of bony VERTEBRATES, the bones are held together by LIGAMENTS, which form a fibrous capsule. Lining the capsule is the synovial membrane, which secretes SYNOVIAL FLUID into the joint to lubricate it.

joint In geology, an upright or near-upright crack in a rock. Joints occur in SEDIMENTARY ROCKS, such as limestone, along lines of weakness caused by shrinking or weathering. Their position influences the structure of cliffs and escarpments. Joints form in IGNEOUS ROCKS, such as basalt, as magma solidifies and cooling and contraction takes place.

Joliot-Curie Name of two French physicists. In 1926 **Irène Curie** (1897–1956) married **Frédéric Joliot** (1900–58), a fellow assistant to her parents, Pierre and Marie CURIE. Irène worked on the physical aspects of RADIOACTIVITY, while Frédéric concentrated on the chemical. Their discovery of artificial RADIOISOTOPES earned them the 1935 Nobel Prize for chemistry. They were active in the French resistance and became communists. Frédéric was responsible for the construction of France's first NUCLEAR REACTOR, but later was dismissed for political reasons. *See also* ISOTOPE

Josephson, Brian David (1940–95) British physicist who shared the 1973 Nobel Prize for physics with Leo ESAKI and Ivar GIAEVER for his discovery of the **Josephson effect.** In 1962 he predicted that an electric current would flow between two superconductors separated by a thin layer of insulator (the Josephson current), in addition to the current already expected on the basis of QUANTUM THEORY. He also predicted several other related phenomena. This hypothesis was later verified experimentally, and has helped in the understanding of SUPERCONDUCTIVITY.

Joule, James Prescott (1818–89) British physicist. **Joule's law** (1841) relates the current flowing through a wire to its heat loss and laid the foundation for the law of CONSERVATION of energy. The JOULE is named after him.

joule (symbol J) SI unit of energy. One joule is the work done by a force of one NEWTON acting over a distance of one metre. It was named after James JOULE and replaced the erg.

Joule-Thomson effect (Joule-Kelvin effect) Fall in temperature that occurs when a gas expands through a small opening into an area of lower pressure. It occurs because the gas has to do work to overcome intermolecular forces so that it can expand. The effect is employed in refrigerators. It is named after its discoverers, James JOULE and William Thompson KELVIN.

jugular In anatomy, term that applies to any structure in the neck and especially to any of several VEINS. The external jugular veins receive blood from the outside of the cranium, the neck and the deep tissues of the face. Others receive blood from the back of the neck, the larynx, tissues below the lower jaw, the brain and the face.

Juno Asteroid discovered (1804) by Karl Harding. It is the tenth-largest asteroid, with a diameter of 244km (152mi).

Jupiter Fifth major planet from the Sun and the largest of the giant planets. It is one of the brightest objects in the sky. Through a telescope, Jupiter's yellowish elliptical disc is seen to be crossed by brownish-red bands, known as belts and zones. The most distinctive feature is the GREAT RED SPOT (GRS). Spots, streaks and bands are caused by Jupiter's rapid rotation and turbulent atmosphere. Eddies give rise to the spots, which are cyclones or (like the GRS) anticyclones. Hydrogen accounts for nearly 90% of Jupiter's atmosphere and helium for most of the rest. The pressure at the cloud tops is *c.*0.5 bar. At 1,000km (600mi) below the cloud tops there is an ocean of liquid molecular hydrogen. At a depth of 20,000–25,000km (12,500–15,000mi), under a pressure of 3 million bars, the hydrogen becomes so compressed that it behaves as a metal. At the centre of Jupiter there is probably a massive iron–silicate core surrounded by an ice mantle. The core temperature is estimated to be 30,000K. The deep metallic hydrogen "mantle" gives Jupiter a powerful magnetic field. It traps a large quantity of PLASMA (charged particles); high-energy plasma is funnelled into radiation belts. Its MAGNETOSPHERE is huge, several times the size of the Sun, and is the source of the planet's powerful radio emissions. Jupiter has 16 known SATELLITES, the four major ones being the GALILEAN SATELLITES. Knowledge of the planet owes much to visits by space probes: Pioneers 10 and 11, Voyagers 1 and 2, Ulysses and GALILEO.

Jurassic Central period of the MESOZOIC era; it lasted from 213 to 144 million years ago. In this period there were large saurischian DINOSAURS such as *Atlantosaurus* and *Allosaurus*; and ornithischian dinosaurs such as *Camptosaurus* and *Stegosaurus*. Plesiosaurs, pterosaurs and ARCHAEOPTERYX (believed to be the precursor to modern birds) also date from this period. Primitive mammals had begun to evolve; they were the ancestors of later MARSUPIAL and placental species.

juvenile hormone (neotenin) Hormone found in insects that prevents the development of adult characteristics when larval insects moult. It is produced in the corpora allata glands behind the brain. These organs become inactive at the last moult, and METAMORPHOSIS then begins.

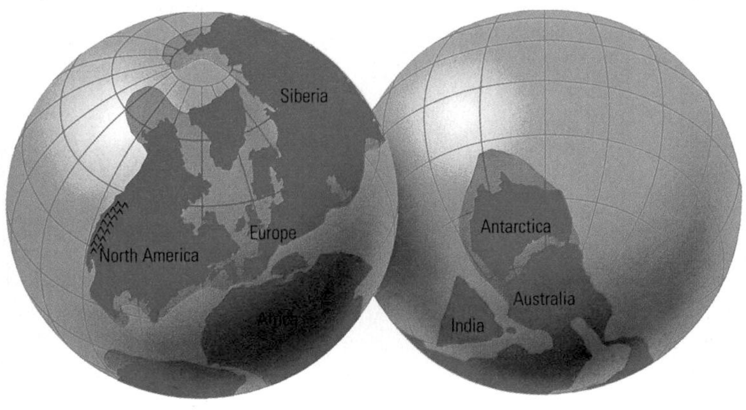

◀ **Jurassic** By the Jurassic period the Gondwanaland landmass is clearly beginning to break into separate continents.

K

kame Deposit formed by a subglacial stream near the terminal margin of a melting GLACIER. Streams flowing down mountain sides onto glaciers deposit sand and gravel at the point where the stream first reaches the ice, or where the water goes down into a crevasse. The accumulations of sand and gravel gradually fall down to the valley floor as the ice melts. Kames are generally small features, 20–30m (65–100ft) in length or breadth.

Kamerlingh-Onnes, Heike (1853–1926) Dutch physicist who was the first to liquefy HELIUM. In the late 1880s, he began studying gases at low temperatures and in 1908, using a liquid-hydrogen cooling system, he liquefied helium and found its temperature to be 4°C (7°F) above ABSOLUTE ZERO. He discovered that at this temperature some metals such as MERCURY and LEAD lose all electrical RESISTANCE and become superconductors. He was awarded the 1913 Nobel Prize for physics. *See also* SUPERCONDUCTIVITY

kaolin (china clay) Especially fine and pure clay composed chiefly of the mineral KAOLINITE, a hydrous silicate of aluminium. It is important in the manufacture of pharmaceuticals, ceramics and coated paper.

kaolinite Sheet silicate mineral of the kaolinite group, hydrous aluminium silicate ($Al_2Si_2O_5(OH)_4$). It is a product of the weathering of feldspar and has triclinic system tabular crystals, clay-like masses and particles. It is white with a dull lustre and may be tinted by impurities. Hardness 2–2.5; r.d. 2.6.

kaon (K meson) ELEMENTARY PARTICLE that is either a charged or neutral MESON with zero SPIN.

Kapitza, Peter Leonidovich (1894–1984) Russian physicist. He did extensive research on MAGNETISM (in England with Ernest RUTHERFORD) and discovered that helium-2, the stable form below −271°C (−456°F), has almost no resistance to flow – a phenomenon called SUPERFLUIDITY. He then returned to the Soviet Union and worked on satellite research and nuclear FISSION. In 1978 he was awarded the Nobel prize for physics for his work on low-temperature physics and magnetism.

Kármán, Theodore von (1881–1963) US research engineer, b. Hungary. He is best known for his work in the application of mathematics to aeronautics and space rocketry. He became a US citizen in 1936 and helped to found the NASA JET PROPULSION LABORATORY (JPL), the International Council of the Aeronautical Sciences, and the International Academy of Astronautics.

Karrer, Paul (1889–1971) Swiss chemist, b. Russia. He shared the 1937 Nobel Prize for chemistry with Sir Walter HAWORTH for research on COENZYMES, CAROTENOIDS and VITAMINS. In 1930 Karrer determined the formula of beta CAROTENE, the precursor of vitamin A, and in 1931 elucidated the structure of the vitamin itself. He also studied vitamin E and demonstrated that lactoflavin was part of the vitamin B_2 complex.

karst Limestone plateau characterized by irregular protuberant rocks, sinkholes, caves, disappearing streams and underground drainage. This type of landscape forms where solution WEATHERING wears away much of the limestone, and where rivers erode the rock. Most rivers are underground, and there are caves and large caverns. The largest caverns may collapse to form gorges, and gradually the entire area of limestone may be worn away. Such topography is named after its most typical site in the Karst region of former Yugoslavia; in Britain, the most spectacular example is above Malham, North Yorkshire.

karyotype Arrangement, structure and number of CHROMOSOMES in the nucleus of a CELL. All the body cells of an organism have the same karyotype. The number of chromosomes varies widely with species, for example 8 in fruit flies, 12 in kangaroos, 46 in humans and 200 in crayfish.

Kastler, Alfred (1902–84) French physicist, b. Germany. He was awarded the 1966 Nobel Prize for physics for the discovery and development of optical methods for studying RESONANCE in atoms. Kastler developed the technique of optical pumping, in which gas atoms absorb energy from bombardment with light and radio waves and release it a few milliseconds later, the manner of release varying according to the atomic species. This technique led to a new knowledge of atomic structure and the development of the MASER and LASER.

Katz, Sir Bernard (1911–) British biophysicist. He shared the 1970 Nobel Prize for physiology or medicine with Julius AXELROD for their work on the chemistry of NERVE transmission. Katz discovered how the NEUROTRANSMITTER acetylcholine is released by neural impulses, causing muscles to contract.

Kay, John (1704–64) British engineer. In 1733 he patented his famous flying shuttle which, by enabling a weaver to throw the shuttle automatically from side to side across the warp of a loom, doubled output. His invention, which was a major step in the development of automatic weaving, was seen as a threat to handloom weavers and he was forced to emigrate to France.

Keeler, James Edward (1857–1900) US astronomer. As head of the Allegheny Observatory (1891–98) he confirmed that SATURN's rings are not solid units but are composed of meteoric particles. While director of the Lick Observatory (1898–1900), he used the new Crossley reflecting telescope to establish that a spiral GALAXY is the most common type of observable galaxy.

Keith, Sir Arthur (1866–1955) British anatomist. He carried out work on the anatomy of the heart and applied his knowledge of general anatomy to an important study of human origins, reconstructing prehistoric man from fossil remains. His books include *The Antiquity of Man* (1915) and *A New Theory of Human Evolution* (1948).

Kekulé von Stradonitz, Friedrich August (1829–96) German chemist who discovered the

KARST

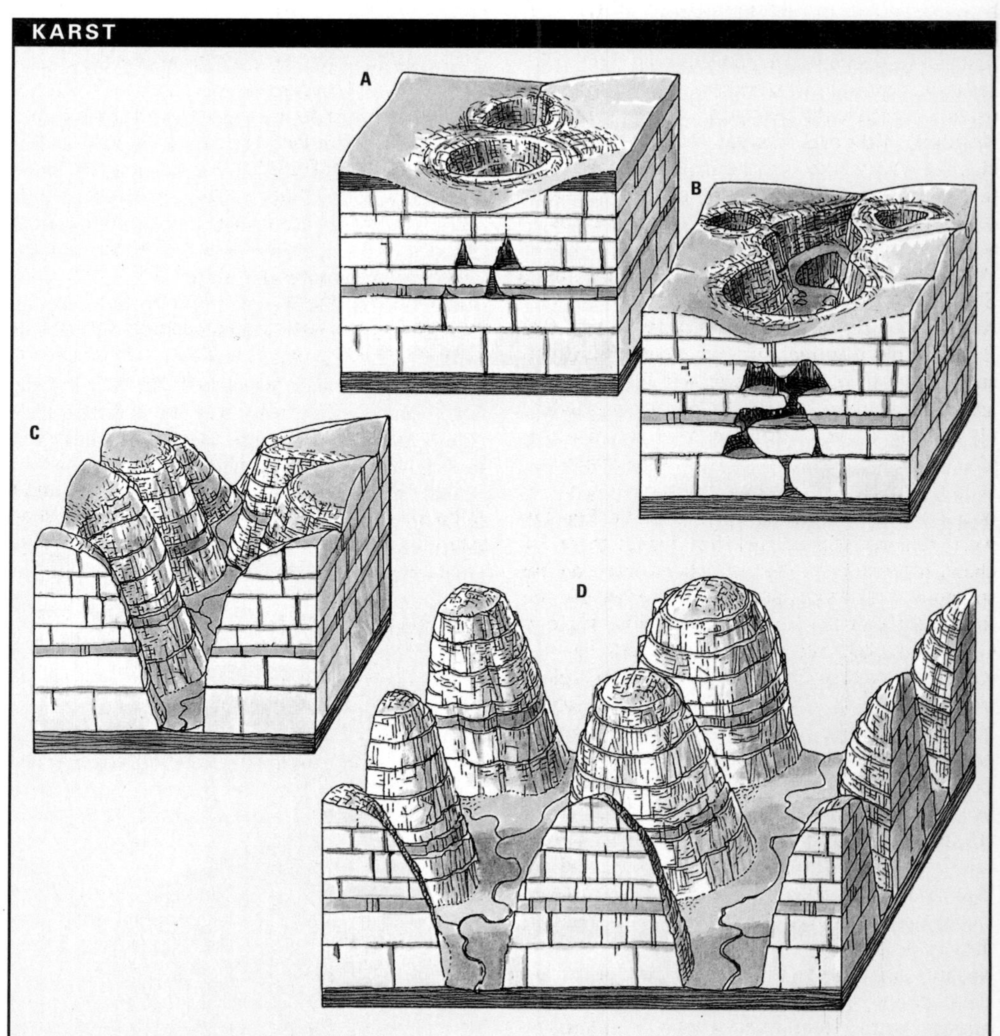

The Karst landscapes of the Dinaric Alps, Appalachian Mountains and areas of s China occur because beds of limestone rock, composed primarily of the mineral calcite (calcium carbonate) up to 200m (650ft) thick are progressively eroded by water. Carbon dioxide from the atmosphere dissolves in rain to form weak carbonic acid, which dissolves the rock, particularly along joints and bedding planes, enlarging them to produce sink holes (A), dolines (B), produced by cave rooves falling in, that can be enlarged to produce gorges (C), finally leaving the characteristic uneroded limestone cores of the Karst landscape (D).

ring structure of BENZENE. He became professor first at Ghent, and later at Bonn. He worked on the structure of organic molecules and on the concept of VALENCE, particularly with respect to benzene and other aromatic compounds.

Kelly, William (1811–88) US steelmaker. In *c.*1850 he invented a method of converting molten iron into steel by OXIDIZATION. The process, later known as the BESSEMER PROCESS, was taken up and patented by the British engineer Sir Henry Bessemer in 1856.

kelp Any of several brown SEAWEEDS commonly found on Atlantic and Pacific coasts, a type of brown ALGAE. A source of iodine and potassium compounds, kelps are now used in a number of industrial processes. Giant kelp (*Macrocystis* sp.) exceeds 45m (150ft) in length. Phylum Phaeophyta.

kelvin (symbol K) SI unit of TEMPERATURE. The Kelvin temperature scale has a zero point at ABSOLUTE ZERO and degree intervals (kelvins), the same size as degrees CELSIUS. The freezing point of water occurs at 273K (0°C or 32°F) and the boiling point at 373K (100°C or 212°F).

Kelvin, William Thomson, 1st Baron (1824–1907) British physicist and mathematician after whom the absolute scale of TEMPERATURE is named. The success of the Atlantic submarine telephone cable was due to Kelvin's research into the transmission of electric currents for which, in 1866, he was knighted. In THERMODYNAMICS, Kelvin was able to resolve conflicting interpretations of the first and second laws. He published more than 600 scientific papers. In 1892 he became the first scientist to be raised to the peerage.

Kendall, Edward Calvin (1886–1972) US chemist who worked on the biological effects of the HORMONES of the cortex of the ADRENAL GLANDS, in particular the STEROID hormone CORTISONE, which he isolated. He shared the 1950 Nobel Prize for physiology or medicine with Philip Showalter HENCH and Tadeus REICHSTEIN.

Kendrew, Sir John Cowdery (1917–97) British molecular biologist who determined the structure of MYOGLOBIN, the protein that serves as an oxygen store in muscle. Kendrew worked (1948–70) at the Cambridge Medical Research Council (MRC) Laboratory. Using X-RAY CRYSTALLOGRAPHY he was able to elucidate the arrangement of AMINO ACIDS in the myoglobin HELIX. For this work Kendrew shared the 1962 Nobel Prize for chemistry with Max PERUTZ, who worked on the structure of HAEMOGLOBIN. In 1974 he became the first director of the European Molecular Biology Organization laboratory (Embl) at Heidelberg.

Kennelly, Arthur Edwin (1861–1939) US electrical engineer. In 1902, after Guglielmo MARCONI had experimented with radio waves, Kennelly noticed that the waves could reach beyond the Earth's horizon. He proposed that they did this by bouncing off a layer of ions high in the ATMOSPHERE. The British physicist Oliver Heaviside (1850–1925) made a similar proposition, and the layer was first called the HEAVISIDE-KENNELLY LAYER (now known as the E-layer). It is found in the IONOSPHERE.

Kepler, Johannes (1571–1630) German mathematician and astronomer. He supported the Sun-centred Solar System put forward by COPERNICUS. In 1600 he went to Prague, becoming assistant to Tycho BRAHE, whom he succeeded as Imperial Mathematician to the Holy Roman Emperor, Rudolf II, with the task of completing tables of planetary motion begun by Tycho. From Tycho's accurate observations, he concluded in *Astronomia nova* (1609) that Mars moved in an elliptical orbit, and went on to establish the first of his laws (*see* KEPLER'S LAWS) of planetary motion. Erroneous reasoning led him nevertheless to the second of these laws, and his desire to match celestial and musical harmony led to the third law, published in *Harmonices mundi* (1619). The *Rudolphine Tables*, based on Tycho's observations and Kepler's laws, appeared in 1627 and remained the most accurate until the 18th century. Other works include *De stella nova*, on the supernova of 1604 , and *Dioptrice* (1611), on optics and the theory of the telescope.

Kepler's laws Three fundamental laws governing the motions of the planets around the Sun (and other celestial bodies in closed orbits), first worked out by Johannes KEPLER between 1609 and 1619, and based on observations made by Tycho BRAHE. **Law 1** deals with the shape of a planetary orbit: the orbit of each planet is an ellipse with the Sun at one of the foci. **Law 2** explains the varying speed of planetary motion such that the planet moves faster the nearer it is to the Sun: the line joining the planet to the Sun (the radius vector) sweeps out equal areas in equal times. This is known as the law of areas. **Law 3** relates the size of the orbit to the period of revolution: the square of the period of revolution (P) is directly proportional to the cube of the mean distance of the planet from the Sun (a). In practice this means that P^2/a^3 is approximately constant for all bodies orbiting the Sun. *See diagram*

keratin Fibrous PROTEIN present in large amounts in the superficial cells of the SKIN, where it serves as a protective layer. Hair and fingernails are made up of modified epidermal cells filled with keratin, which is also the basis of claws, feathers and the horns of some animals. *See also* EPIDERMIS

kerosene (paraffin) Distilled PETROLEUM product heavier than PETROL but lighter than DIESEL FUEL. Kerosene, known historically as an illuminant (paraffin oil), is now used in camping stoves, tractor fuels and gas-turbine fuels for JET ENGINES and turboprop aircraft.

ketone Organic chemical compound containing a carbonyl (C=O) group joined to two hydrocarbon groupings. Ketones are liquids or low-melting solids. Examples include propanone (acetone, CH_3COCH_3) and butanone (methyl ethyl ketone, $CH_3COC_2H_5$). Ketones can be made by oxidizing secondary alcohols. They are reactive, forming addition compounds and undergoing various condensation reactions, but they do not oxidize easily. They are used as solvents.

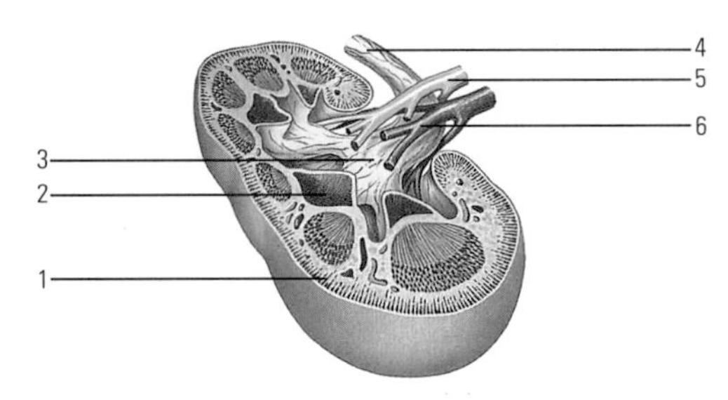

▲ **kidney** The human kidney is enclosed in a fibrous capsule, and consists of an outer cortex region (1), a medulla region (2) with pyramidal-shaped areas, and an inner pelvis region (3), which leads into the ureter (4). The renal artery (5) conducts blood into the kidney to be filtered; it is then carried away by the renal vein (6).

ketone bodies Three chemical compounds: 3-oxobutanoic acid (acetoacetic acid), 3-hydroxybutanoic acid (hydroxybutyric acid) and propanone (acetone). When present in the blood in high concentrations, they increase the blood's acidity. Known as **ketosis**, this occurs in starvation and DIABETES mellitus. The ketone bodies are products of poor carbohydrate METABOLISM, and they accumulate in the body tissue and fluids, especially the urine.

kettle hole Steep-sided basin formed when a chunk of ice left behind by a receding glacier is covered by rocks and debris previously pushed forward by the glacier. The ice melts and the rocks fall through, creating a kettle pot-shaped depression.

khamsin Hot S wind that blows across N Africa from the Sahara during spring and summer. It is very dry, with relative humidity sometimes as low as 10%. The name means "fifty" in Arabic, the number of days during which the wind is supposed to blow.

Khorana, Har Gobind (1922–) US biochemist who shared the 1968 Nobel Prize for physiology or medicine with Robert HOLLEY and Marshall NIRENBERG for discoveries about the way in which GENES determine cell function. They established that most CODONS, combinations of three of the four different bases found in DNA and RNA, eventually cause the inclusion of a specific AMINO ACID into the cell proteins.

kidney In vertebrates, one of a pair of organs responsible for regulating blood composition and the EXCRETION of waste products. The kidneys are at the back of the abdomen, one on each side of the backbone. The human kidney consists of an outer

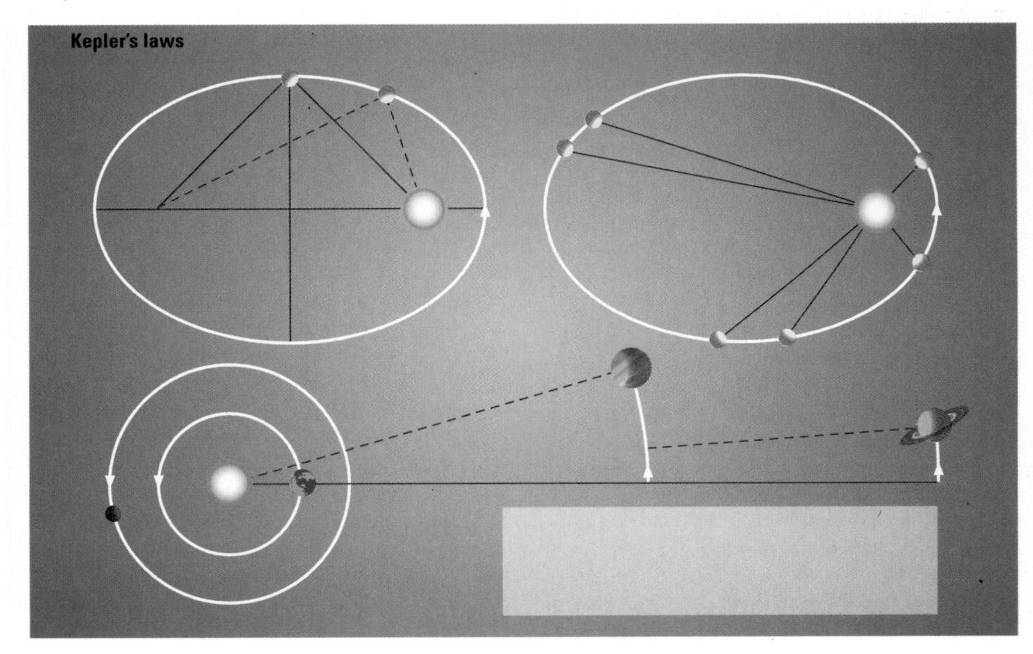

► **kinetic energy** A moving truck has kinetic energy (A). To increase its speed, more energy must be added to overcome friction and air resistance and accelerate to a higher speed. To decrease the kinetic energy of the truck requires the kinetic energy to be converted to heat in the brakes and tyres (B). The momentum of a heavily loaded truck proceeding at the same velocity will be greater because of its higher mass (C) and it will require a greater breaking force to dissipate the kinetic energy and decelerate to a stop in the same distance as the unladen truck (D).

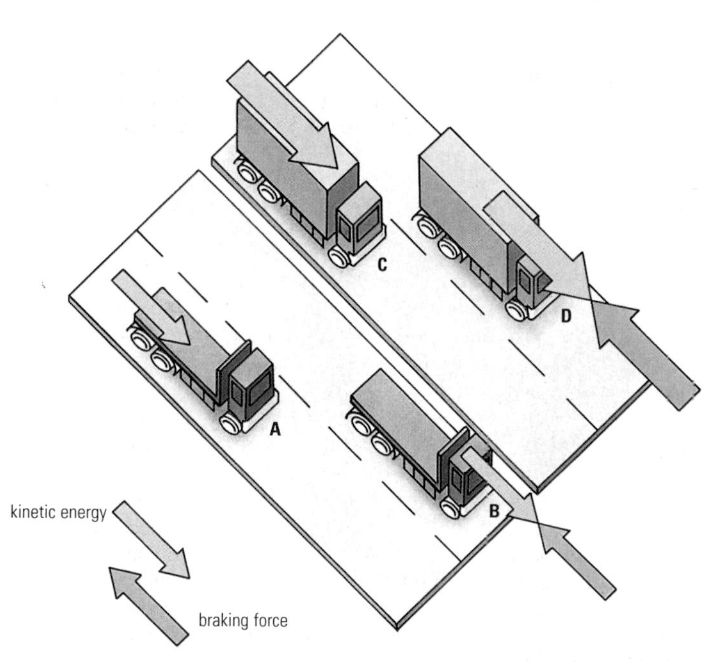

cortex and an inner medulla with about one million tubules (NEPHRONS). Nephrons contain numerous CAPILLARIES, which filter the blood entering from the renal ARTERY. Some substances, including water, are reabsorbed into the blood. URINE remains, which is passed to the URETER and on to the BLADDER. *See also* HOMEOSTASIS

kidney machine (artificial kidney) Equipment designed to perform **haemodialysis** (removing toxic wastes from the blood) in kidney failure. Plastic tubing is used to pipe blood from the body into the machine, where waste-products are filtered out by the technique of DIALYSIS. Usually two or three sessions are required a week, each lasting about four hours. Today many more patients are kept alive by the newer technique of **peritoneal dialysis**, in which fluids are passed through the peritoneal cavity to reduce blood urea.

kidney stones Small, hard, pebble-like masses formed in the KIDNEY from an accumulation of mineral substances; known medically as calculi. Passage of a stone through the URETER (tube from the kidney to the bladder) causes the excruciating pain of **renal colic**. They can be removed by surgery or, if the stones are small, they can be passed out in the urine after treatment, which includes drinking large amounts of water.

kidney transplant *See* ORGAN TRANSPLANT

kiln In ceramics, an oven for firing ware. Early kilns were holes in the ground into which the ware was placed and covered by a large fire. Later, special wood- or coalfired, oven-type kilns were built; today most kilns use gas or electricity.

kilocalorie *See* CALORIE

kilogram (symbol kg) SI unit of mass defined as the mass of the international prototype cylinder of platinum-iridium kept at the International Bureau of Weights and Measures near Paris. 1 kilogram is equal to 1,000g (2.2lb).

kilowatt hour (symbol kWh) Unit of electrical energy equal to the energy used when a power of one kilowatt (1,000 watts) is expended for one hour. It is the unit in which domestic and industrial users of electricity are charged.

kimberlite Ultrabasic rock found in diamond-bearing pipes, mainly in South Africa. It is a mica PERIDOTITE consisting chiefly of olivine and phlogopite; it weathers to the yellow and blue grounds of diamond mines.

kinaesthetic sense (proprioception) Internal sense that conveys information from the muscles and tendons of the body. Specialized receptors connect to a tract of the NERVOUS SYSTEM which provides information about the contraction and expansion of muscles. The kinaesthetic sense is the position sense, because it allows human beings to know the positions of their limbs without visual confirmation. *See also* CEREBELLUM

kinetic energy Energy that an object possesses because it is in motion. It is the energy given to an object to set it in motion; it depends on the mass (m) of the object and its velocity (v), according to the equation K.E. = $\frac{1}{2}mv^2$. On impact, it is converted into other forms of energy such as heat, sound and light. *See also* POTENTIAL ENERGY

kinetics In physics, one of the branches of DYNAMICS. In chemistry, a branch of physical chemistry that deals with the rates of chemical reactions. By studying rates at different pressures and temperatures, chemists can determine how reactions take place.

kinetic theory Theory in physics dealing with matter in terms of the forces between particles and the energies they possess. The following principles are basic to the simplest form of the kinetic theory: matter is composed of tiny particles; these are in constant motion; they do not lose energy in collision with each other or the walls of their container; there are no attractive forces between the particles or their container; and at any time the particles in a sample may not all have the same energy. More advanced forms of the theory take into account VAN DER WAALS FORCES between the particles.

kingdom In biology, the most widely adopted CLASSIFICATION for living organisms is the Five Kingdoms' system, in which the kingdom is the topmost level, or **taxon**. The five kingdoms are Animalia (ANIMAL), Plantae (PLANT), Fungi (FUNGUS), PROKARYOTAE and PROTOCTISTA. Two subkingdoms are often recognized within Prokaryotae, ARCHAEBACTERIA and EUBACTERIA, but the bacteria are so diverse that many taxonomists think they comprise more than one kingdom. Some believe that they merit the status of a new, even higher category, DOMAIN. *See also* EUKARYOTE

Kinsey, Alfred Charles (1894–1956) US zoologist, noted for his studies on human sexual behaviour. He was a professor of zoology and later director of the Institute for Sex Research, Indiana University. He is best known for *Sexual Behaviour in the Human Male* (1948) and *Sexual Behaviour in the Human Female* (1953).

Kipp's apparatus Chemical apparatus for generating a stream of any gas that can be made by reacting a liquid with a solid, for example hydrogen sulphide gas from hydrochloric acid and iron(II) (ferrous) sulphide or carbon dioxide from hydrochloric acid and calcium carbonate (marble chips). When the outlet tap is turned off, internal gas pressure separates the liquid and solid so that the reaction, and gas generation, cease.

Kirchhoff, Gustav Robert (1824–87) German physicist. He worked with Robert BUNSEN and developed the SPECTROSCOPE, with which they discovered the elements CAESIUM and RUBIDIUM. He also examined the solar spectrum and worked on BLACK BODY radiation. He is famous for two laws (KIRCHHOFF'S LAWS) which apply to multiple-loop electric circuits.

Kirchhoff's laws Two rules, based on the laws of the conservation of charge and energy, that apply to multiple-loop electric circuits. Essentially they state that (1) charge does not accumulate at one point and thin out at another, and (2) around each loop the sum of the ELECTROMOTIVE FORCES equals the sum of the POTENTIAL DIFFERENCES (voltages) across each of the resistances.

Kitasato, Shibasaburo (1852–1931) Japanese bacteriologist who isolated the bacilli that cause the diseases tetanus, anthrax (1889) and dysentery (1898). In 1890 he prepared a diphtheria antitoxin. He also discovered, independently of Alexandre YERSIN, the infectious organism that causes bubonic plague. He trained under Robert KOCH in Germany.

Kjeldahl method Rapid method for measuring the proportion of nitrogen in an organic compound, named after the Danish chemist Johan Kjeldahl (1849–1900). Nitrogen in the sample is converted into ammonium sulphate by heating it and a catalyst in concentrated sulphuric acid. Sodium hydroxide is added and the whole is boiled, liberating ammonia. This is dissolved in acid and the quantity of ammonia, and therefore the nitrogen, is determined by TITRATION.

Klaproth, Martin Heinrich (1743–1817) German chemist who began as an apothecary's apprentice. He became the first professor of chemistry at Berlin University. He was a pioneer in analytical techniques, recognizing and naming a number of elements, including uranium (1789). He also recognized and named titanium (1795), tellurium (1798) and cerium (1803), some of which had already been isolated, although left unnamed, by other scientists.

Klebs, Edwin (1834–1913) German physician and pathologist known for his work in bacteriology. In 1884 he and Friedrich LÖFFLER discovered the diphtheria bacillus. He held professorships in Europe and in the USA. His research included original observations on tuberculosis, syphilis and malaria.

klystron ELECTRON TUBE that makes use of a stream of electrons moving at controlled speeds. Klystrons are used in ultra-high-frequency (UHF) radio and microwave circuits, such as RADAR transmitters, where they operate at frequencies up to 400,000MHz.

K-meson *See* KAON

knock, engine (pre-ignition or pinking) Metallic noise produced in an INTERNAL COMBUSTION ENGINE, caused by detonation of part of the compressed fuel-air mixture before the spark from the SPARK PLUG ignites the remainder. It lowers the efficiency of the engine but can be overcome by adding an anti-knocking agent to the fuel. The traditional additive is LEAD(IV) TETRAETHYL, but because this compound releases lead into the atmosphere and thereby contributes to pollution, its use has declined with the introduction of unleaded petrol

knot Unit of measurement equal to one NAUTICAL MILE per hour – 1 knot equals 1.852km/h (1.15mph). The speeds of ships and aircraft are generally expressed in knots, as are those of winds and currents.

Kocher, Emil Theodor (1841–1917) Swiss surgeon who was awarded the 1909 Nobel Prize for physiology or medicine for his work on the physiology, pathology and surgery of the THYROID GLAND. He was the first to remove the thyroid gland in goitre cases.

Koch, Robert (1843–1910) German bacteriologist. He was awarded the 1905 Nobel Prize for physiology or medicine for his discovery of the bacillus that causes TUBERCULOSIS. This work laid the foundation for methods of determining the causative agent of a disease.

Kolbe, Adolf Wilhelm Hermann (1818–84) German chemist. In *c*.1843 he converted carbon disulphide to ethanoic acid, one of the first syntheses of an organic compound from inorganic chemicals. He also developed a theory of RADICALS and predicted the existence of secondary and tertiary alcohols.

Köppen, Wladimir Peter (1846–1940) German meteorologist and climatologist, b. Russia, who introduced a CLIMATE classification system still used today. In 1884 he produced a world map of temperature belts. In 1900 he introduced his system of climate classification, which divided climate into five major categories according to temperature and rainfall.

Kornberg, Arthur (1918–) US biochemist. He was medical director of the US Public Health Service (1951) and chairman of the department of biochemistry at Stanford University. In 1959 he shared the Nobel Prize for physiology or medicine with Severo OCHOA for work on the synthesis of RNA and DNA, an important contribution to the study of GENETICS.

Korolev, Sergei Pavlovich (1906–66) Soviet engineer. He was chief designer at the Scientific Research Institute near Moscow and directed the design and manufacture of the VOSTOK and SOYUZ manned spacecraft, including Vostok I, in which Yuri GAGARIN made the first manned space flight in 1961.

Krebs cycle (citric acid or tricarboxylic acid cycle) Biochemical pathway by which most EUKARYOTE organisms obtain much of their energy by oxidizing foodstuffs. Occuring in the MITOCHONDRIA of CELLS, the Krebs cycle comprises a number of complex chemical reactions, many of which release energy, in association with a process called the ELECTRON TRANSPORT SYSTEM, as ADENOSINE TRIPHOSPHATE (ATP) becomes ADENOSINE DIPHOSPHATE (ADP). ATP provides chemical energy for metabolic reactions. The Krebs cycle is an essential part of the process of cell RESPIRATION and METABOLISM. The cycle is named after Sir Hans KREBS. *See also* GLYCOLYSIS

Krebs, Sir Hans Adolf (1900–81) British biochemist, b. Germany. He shared with Fritz LIPMANN the 1953 Nobel Prize for physiology or medicine for his discovery of the KREBS CYCLE, the process that results in the production of energy in living organisms (RESPIRATION).

kryolite *See* CRYOLITE

krypton (symbol Kr) Gaseous, nonmetallic element that is a NOBLE GAS (inert gas). Discovered in 1898, krypton makes up about 0.0001% of the Earth's atmosphere by volume and is obtained by the fractional DISTILLATION of liquid air. It is used in FLUORESCENT LAMPS, LASERS and in electronic heart valves. The isotope ^{86}Kr was used from 1960–83 to define the standard linear metre. Properties: at.no. 36; r.a.m. 83.80; density 3.73g dm^{-3}; m.p. −156.6°C; (−249.9°F); b.p. −152.3°C; (−242.1°F); most common isotope ^{84}Kr (56.9%).

Kuhn, Richard (1900–67) German chemist who shared the 1938 Nobel Prize for chemistry with Paul KARRER. They researched CAROTENOIDS and the isolation of vitamin B_2. He was forced by the Nazis to refuse the award and did not receive it until after World War 2.

Kuiper, Gerard Peter (1905–73) US astronomer, b. The Netherlands. After World War 2 Kuiper embarked upon an observational programme to revitalize lunar and planetary astronomy. He discovered the satellites Miranda (of Uranus) in 1948, and Nereid (of Neptune) in 1949. He found methane in the atmospheres of Uranus, Neptune and Titan, and carbon dioxide in the atmosphere of Mars. He was involved with many US missions to the Moon and planets, in particular with the Ranger and Mariner probes. The Kuiper Airborne Observatory is named in his honour.

kurchatovium *See* DUBNIUM

Kuroshio current (Japan current) Western boundary CURRENT of the North Pacific gyre which flows along the E coast of Japan. It is comparable to the GULF STREAM along the E coast of North America.

Kusch, Polykarp (1911–93) US physicist, b. Germany. He shared the 1955 Nobel Prize for physics with Willis LAMB for his precise measurement of the MAGNETIC MOMENT of the ELECTRON, which he discovered had a higher value than was predicted by the theory of Paul DIRAC. This was an important step towards the theory of QUANTUM ELECTRODYNAMICS (QED).

kyanite (disthene) Mineral form of aluminium silicate (Al_2SiO_5). It occurs in SCHISTS and other METAMORPHIC ROCKS, generally as elongated, blue-bladed crystals (triclinic) or aggregates, although it may also be white, green or grey. Colour may vary within a single crystal, which is usually transparent with a vitreous lustre. Kyanite is used in making high-grade porcelain for electrical insulators; good clear crystals are used as gemstones. Hardness: 4–5 along the crystal, 6–7 across; r.d. 3.5–3.7.

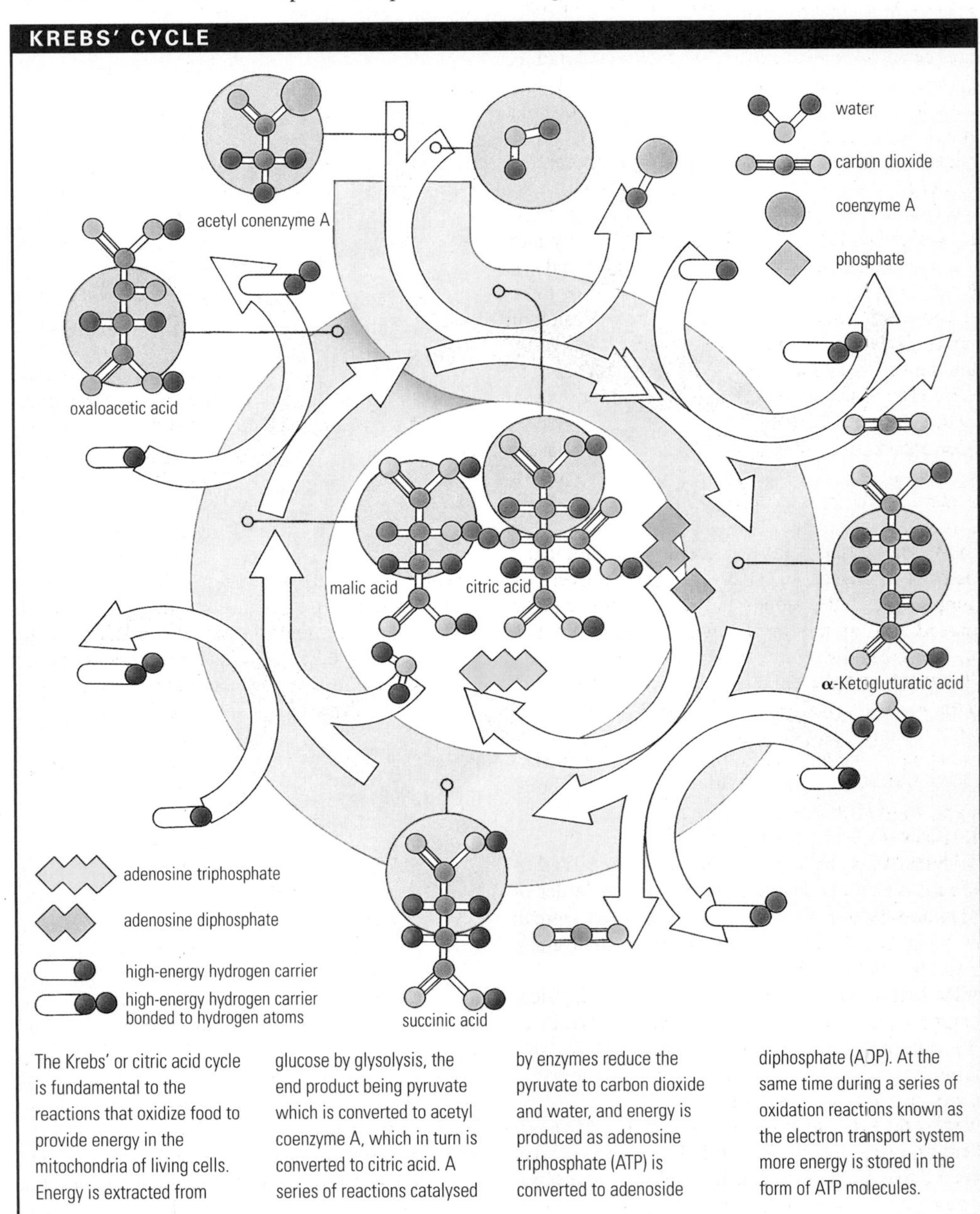

The Krebs' or citric acid cycle is fundamental to the reactions that oxidize food to provide energy in the mitochondria of living cells. Energy is extracted from glucose by glysolysis, the end product being pyruvate which is converted to acetyl coenzyme A, which in turn is converted to citric acid. A series of reactions catalysed by enzymes reduce the pyruvate to carbon dioxide and water, and energy is produced as adenosine triphosphate (ATP) is converted to adenoside diphosphate (ADP). At the same time during a series of oxidation reactions known as the electron transport system more energy is stored in the form of ATP molecules.

labelling (radioactive tracing) Technique for following the path of a biochemical or chemical reaction by replacing one of the key elements involved by one of its RADIOSOTOPES, called a RADIOSOTOPE TRACER. The method relies on the fact that the substitution of a different isotope for the normal form of an element does not affect its chemical identity or properties. But if the new form is radioactive, its presence can be traced by the radioactivity it emits (using a GEIGER COUNTER). For example, any compound containing hydrogen can be labelled by substituting tritium (a radioisotope of hydrogen of mass 3) for the normal hydrogen, or an oxygen compound can be labelled by substituting the oxygen-18 isotope for normal oxygen-16.
labium In anatomy, term meaning a lip or lip-shaped organ. In arthropods it refers to the lower lip, which may be modified in insects to form sucking mouthparts. In human anatomy, it denotes the labia majora and labia minora, skin folds of the female genitals. In botany, it refers to the lower lip of plants of the Labiatae family.
labour *See* BIRTH
Lacaille, Nicolas Louis de (1713–62) French astronomer. From 1751 to 53 he surveyed the Southern-Hemisphere skies from the Cape of Good Hope, discovering 14 new s constellations. In 1761 Lacaille made an accurate measurement of the Moon's distance, taking into account the Earth's OBLATENESS; this made possible a more accurate method of determining terrestrial longitude.
laccolith Dome of INTRUSIVE ROCK formed over the STRATA it has penetrated. The base is typically horizontal while the upper surface is convex. They are generally less than 16km (10mi) in diameter with thicknesses of 30–900m (100–3,000ft). Laccoliths are formed of MAGMA; they are smaller than BATHOLITHS.
lachrymal gland (lacrimal gland) Organ near the EYE that produces TEARS. It is located in the orbital cavity in a slight depression, and is controlled by the AUTONOMIC NERVOUS SYSTEM; it produces slightly germicidal tears that flow through ducts to the surface of the eye to lubricate it.

▲ **lagoon** An example of a lagoon on the island of Lanzarote, Canary Islands. It is called 'El Golfo', and is a green lagoon located between the sea and volcanic rocks.

lacquer Varnish used for ornamental or protective coatings, forming a film by loss of solvent through evaporation. Lacquer is usually composed of a cellulose derivative, such as CELLULOSE NITRATE, in combination with a RESIN. The solvent may be an alcohol.
lactation Secretion of milk by female MAMMALS to feed their young. In pregnant women, HORMONES induce the BREASTS to enlarge, and PROLACTIN (a HORMONE produced by the PITUITARY GLAND) stimulates breast cells to begin secreting milk. The milk "comes in" the breast immediately after the birth of the baby. Its flow is stimulated by suckling which, in turn, triggers the release of the hormone OXYTOCIN that controls the propulsion of milk out of the breast.
lactic acid (2-hydroxypropanoic acid, $CH_3CHOHCOOH$) Colourless, organic acid formed from LACTOSE in milk by the action of bacteria. It is also produced in muscles when ANAEROBIC respiration occurs due to insufficient oxygen, and causes muscle fatigue. Lactic acid is used in foods and beverages, in tanning, dyeing and adhesive manufacture. Properties: r.d. 1.206; m.p. 18°C (64.4°F); b.p. 122°C (251.6°F).
lactose (milk sugar) DISACCHARIDE present in milk, made up of a molecule of GLUCOSE linked to a molecule of galactose. It is important in cheese making, when lactic bacteria turn it into LACTIC ACID, so souring the milk in the production of cheese curd.
lagoon Stretch of shallow water partially or completely cut off from the sea. If it is completely cut off, there will be some percolation of water through the ridge or spit separating the lagoon from the sea. Lagoons are often found behind spits, bars and CORAL REEFS.
Lagrange, Joseph Louis (1736–1813) French mathematician. He became professor of mathematics in Turin at the age of 19 and later succeeded Leonhard EULER as head of the Berlin Academy of Sciences. In *Analytical Mechanics* (begun in 1753 but not finished until 1788), he introduced a new general approach to mechanics, using the idea that the motions of bodies are found by minimizing a function known as the **Lagrangian function**. **Lagrangian dynamics** is used in many areas of applied mathematics and physics.
Lagrangian points One of the points at which a celestial body can remain in a position of equilibrium with respect to two much more massive bodies orbiting each other. At these points the forces acting on the smaller body cancel out. There are five Lagrangian points in the orbital plane of two large bodies.
lake Inland body of water, generally of considerable size and too deep to have rooted vegetation completely covering the surface, occupying a hollow in the Earth's surface. Usually freshwater, lakes may have an in- or outflowing river and are not always permanent features on the landscape. The expanded part of a river and a reservoir behind a dam are also termed lakes. A particularly large lake, that is natural and saline, may be called a "sea" (for example, the Dead Sea).
Lalande, Joseph Jérôme Le Français de (1732–1807) French astronomer whose main achievement was a catalogue of more than 47,000 stars. In 1762 Lalande became professor of astronomy at the Collège de France, a post he held for 46 years. Made director of the Paris Observatory in 1768, Lalande observed Neptune in 1795, 51 years before it was located by Johann GALLE, without realizing it was a planet. In 1802 he established the Lalande Prize for achievement in astronomy.
Lamarck, Jean Baptiste Pierre Antoine de Monet, Chevalier de (1744–1829) French biologist. His theories of EVOLUTION (**Lamarckism**), according to which ACQUIRED CHARACTERISTICS are inheritable, influenced evolutionary thought throughout most of the 19th century, but were subsequently disproved by Charles DARWIN. In *Philosophie zoologique* (1809) he proposed that new biological needs of an organism promote a change in habits from which develop new physical structures. These are then transmitted to offspring as permanent characteristics. For example, he erroneously believed giraffes' long necks evolved because they reached for leaves.
Lamb, Willis Eugene, Jr (1913–) US physicist who applied new techniques to measure the lines of the hydrogen spectrum. He found that the actual positions (wavelengths) varied from the positions predicted by the theory of Paul DIRAC. For this research, which was an important step towards the theory of QUANTUM ELECTRODYNAMICS (QED), Lamb shared the 1955 Nobel Prize for physics with Polykarp KUSCH. He also developed microwave techniques for examining the helium spectrum.
laminar flow (streamline flow) Orderly flow of FLUID (liquid or gas) without turbulence. The fluid flows in layers which slide past each other. As the velocity of the layers increases, or as the VISCOSITY of the fluid decreases, a point is reached when laminar flow breaks up into turbulent flow. For a given fluid this point occurs at a certain value of the REYNOLDS NUMBER. *See also* AERODYNAMICS; FLUID MECHANICS
lamprophyre Member of a group of medium-grained IGNEOUS ROCKS that occur in DYKES and SILLS. Their composition is rich in augite, biotite and other coloured silicates, which form relatively large crystals, with little or no feldspar.
Land, Edwin Herbert (1909–91) US physicist, inventor of the "instant" camera. He worked with light POLARIZATION and by 1936 had developed Polaroid, a light-polarizing material later used in sunglasses. In 1947 Land developed a camera that produced a finished print within one minute.
Landau, Lev Davidovich (1908–68) Soviet physicist. His many contributions included the basic theories describing FERROMAGNETISM and liquid HELIUM. In 1927 he proposed a concept for energy called the **density matrix** which was later used extensively in QUANTUM MECHANICS. In the 1930s he originated the theory that underlies the superfluid behaviour of liquid helium. He received the 1962 Nobel Prize for physics for his research into condensed matter, especially helium. *See also* SUPERFLUIDITY
land bridge Temporary, natural connection between continents that can allow animal migration. One existed across the Bering Strait until comparatively recently, and a modern example is the isthmus of Panama. Before the theory of CONTINENTAL DRIFT, such bridges were proposed as a possible major influence on fossil distribution.
land reclamation Transformation of land that, for one reason or another, is unusable for building or cultivation. Although the term generally refers to reclamation of land that has been beneath the sea (that is with artificial islands or merely by extending the shore-line), it also applies to the beautifying of derelict industrial sites (with canals and recreation areas) and techniques of irrigation and fertilization.
Landsat (formerly Earth Resources Technology Satellite, ERTS) Series of scientific SATELLITES. The satellites observe the Earth's surface in different segments of the ELECTROMAGNETIC SPECTRUM, yielding valuable information concerning crops,

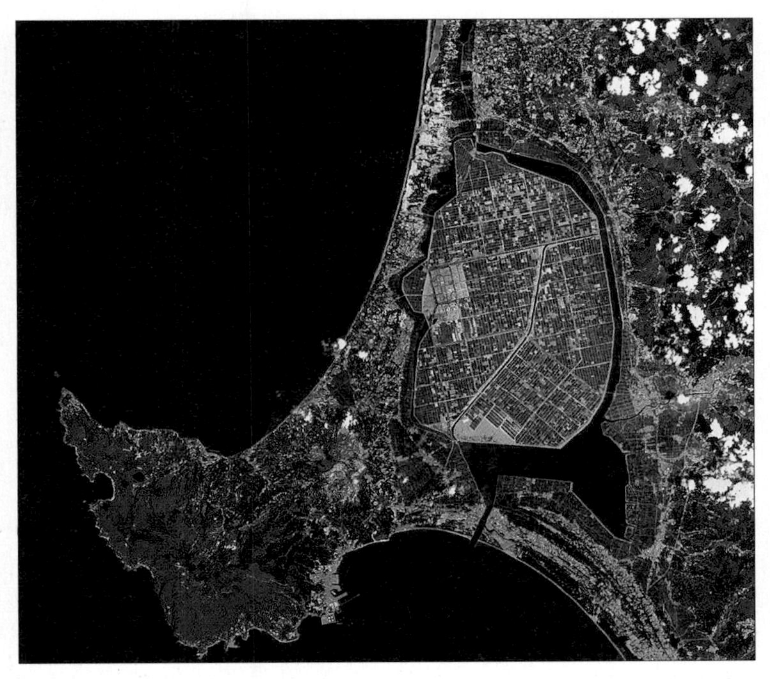

► **land reclamation** False-colour satellite view of a land reclamation project. The reclaimed land occupies most of the inlet seen just right of centre. The main drainage canal (black) runs across the land. Red patches indicate vegetation, blue indicates dry soil and dark grey indicates soil with a shallow water cover. In the surrounding areas, bright red zones of rich surface vegetation may be seen in the highlands along the right edge and in the hills of the peninsula at lower left. This reclamation project is in Akita Prefecture, on the NW coast of Japan's largest island, Honshū. The image was gathered by a Landsat satellite in June 1984.

soils, forests, population patterns, and other such matters. It is in a nearly circular polar orbit.

landslide Relatively rapid displacement of rock or soil slipping over a definite surface. It may be caused by an earthquake, but is generally the result of rain soaking the ground; the effects are exaggerated if there is an impervious layer beneath the surface. In colder climatic areas, the saturated rock may be affected by freezing and thawing of the water, and this increases the likelihood of landslides.

Landsteiner, Karl (1868–1943) US pathologist, b. Austria. He discovered the four different BLOOD GROUPS (A, B, AB and O) and demonstrated that certain blood groups are incompatible with others; clots form if blood of such groups are mixed. His findings explained why some blood transfusions in the past had been beneficial, whereas others had been fatal. He was awarded the 1930 Nobel Prize for physiology or medicine. In 1940 he helped to identify the RHESUS FACTOR.

Langevin, Paul (1872–1946) French physicist. In 1905 he was the first to interpret PARAMAGNETISM and DIAMAGNETISM in terms of the behaviour of electrons in atoms. During World War 1, he built the first submarine detector based on ULTRASONIC waves, and it was later developed into SONAR.

Langley, Samuel Pierpont (1834–1906) US astronomer who showed that mechanical flight was possible. He did this by building a large steam-powered model aircraft in 1896, which achieved the most successful flights up to that time. In 1880 he developed a BOLOMETER to measure INFRARED RADIATION.

Langmuir, Irving (1881–1957) US physical chemist who invented a gas-filled, tungsten lamp. He showed that the shortness of a lamp's life was due to the evaporation of the filament, and that this could be reduced by introducing nitrogen into the bulb instead of a high vacuum. He also devised the atomic hydrogen welding process and techniques to produce rain by cloud seeding. His researches in surface chemistry led to the general theory that CHEMICAL REACTIONS occurred between adjacent substances that were absorbed. For this work Langmuir received the 1932 Nobel Prize for chemistry.

lanolin Purified, fat-like substance derived from sheep's wool and used with water as a base for ointments and cosmetics.

lanthanide series (lanthanide elements, rare-earth metals) Series of 15 rare, metallic elements with atomic numbers from 57 to 71. They are placed in group III of the periodic table. They are, in order of increasing atomic numbers: LANTHANUM (sometimes not considered a member), CERIUM, PRASEODYMIUM, NEODYMIUM, PROMETHIUM, SAMARIUM, EUROPIUM, GADOLINIUM, TERBIUM, DYSPROSIUM, HOLMIUM, ERBIUM, THULIUM, YTTERBIUM and LUTETIUM. Their properties are similar and resemble those of lanthanum, from which the series takes its name. These shiny metals occur in MONAZITE and other rare minerals.

lanthanum (symbol La) Silver-white, metallic element of the LANTHANIDE SERIES, first identified in 1839. Its chief ores are monazite and bastnaesite. Soft, malleable and ductile, lanthanum is used as a CATALYST in CRACKING crude oil, in alloys and to manufacture optical glasses. Properties: at.no.57; r.a.m. 138.9055; r.d. 6.17; m.p. 920°C (1,688°F); b.p. 3,454°C (6,249°F); most common isotope ^{139}La (99.91%).

lanugo Soft, downy hair that covers the human fetus and those of other mammals during development in the womb. It is shed, and virtually disappears, at birth.

lapis lazuli (lazurite) Glassy, deep blue, semi-precious gemstone, a silicate mineral found in metamorphosed limestones (*see* METAMORPHIC ROCK). It occurs rarely as crystals in the cubic system, but more often as granular masses. Hardness 5–5.5; r.d. 2.4.

Laplace, Pierre Simon, Marquis de (1749–1827) French astronomer and mathematician. Laplace made significant advances in PROBABILITY theory: defining the probability of an event to be the number of ways that event can happen divided by the total number of possible outcomes. His study and application of Newton's law of GRAVITATION to the Solar System was summarized in his book *Celestial Mechanics* (1798–1827). He used Newton's law to interpret the PERTURBATIONS of planets and satellites, the shape and rotation of Saturn's rings, and the stability of the Solar System. He also did fundamental work in the study of heat, magnetism and electricity.

laptop Type of MICROCOMPUTER that is portable and carries its own power supply. Modern laptops utilize the latest developments in miniaturization of INTEGRATED CIRCUITS, giving them the ability to process information as quickly as most desktop COMPUTERS. Their independent power supply comprises rechargeable batteries, and they all have a built-in LIQUID-CRYSTAL display. The miniaturization of their components and a quality of manufacture that can withstand being carried about ensures that laptops are comparatively expensive.

larva Developmental stage in the life cycle of many INVERTEBRATES and some other animals. A common LIFE CYCLE, typified by the butterfly, is egg, larva, PUPA, IMAGO (adult) (*see* INSECT). The larva fends for itself and is mobile, but is distinctly different in form from the sexually mature adult. It has a well-developed alimentary system and often stores food until it metamorphoses (or pupates) to become an adult. Names for the larval stage in different organisms include MAGGOT, caterpillar and tadpole. *See also* METAMORPHOSIS

larynx (voice-box) Triangular cavity located between the TRACHEA and the root of the tongue. Inside it are the vocal cords, thin bands of elastic tissue that vibrate when outgoing air passes over them, setting up resonant waves that are changed into sound by the action of throat muscles, the shape of the mouth and the tongue.

LARYNX

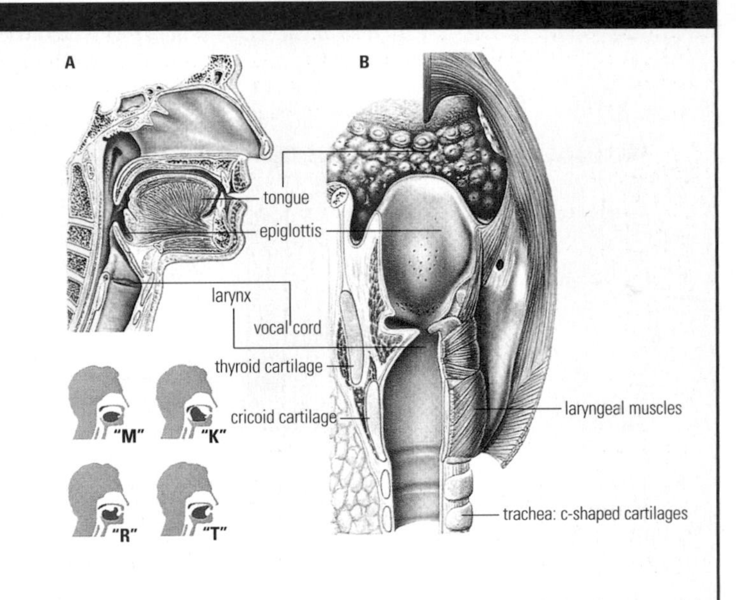

The larynx, together with the epiglottis, tongue, mouth and lips are the principal organs of speech. A side view (A) and back view (B) of these organs are shown. Air pushed out from the lungs through the larynx causes the vocal cords to vibrate, producing a continuous singing tone, the "voice". This tone can be altered in "pitch" by varying the arrangement of the cartilages of the larynx (thyroid and cricoid) by action of the associated muscles. As air passes through the mouth, the voice is modulated and broken up by changing the position and shape of the other organs to produce speech. The different vowels are produced by altering the shape of the mouth. Consonants (four shown) are formed when the stream of air is suddenly emitted or cut off.

laser (acronym for **l**ight **a**mplification by **s**timulated **e**mission of **r**adiation) Optical MASER, a source of a narrow beam of intense, coherent (same wavelength) light or ULTRAVIOLET or INFRARED RADIATION. It was first developed by Theodore MAIMAN in 1960. The source can be a solid, liquid or gas. A large number of its atoms are excited to a higher energy state. One PHOTON of radiation emitted from an excited atom then stimulates the emission of another photon, of the same frequency and direction of travel, which in turn stimulates the emission of more photons. The photon number multiplies rapidly to produce a narrow, coherent, monochromatic laser beam of very high energy content. It has applications in medicine, scientific research, engineering, telecommunications, HOLOGRAPHY and other fields.

laser printer Computer printer with a LASER diode to control image formation. The laser scans lines across an electrically charged drum. The beam flashes on and off according to whether each point is to be light or dark. Exposed areas become discharged. The charged areas attract toner powder, thus forming an image. Charged plain paper picks up the powder image from the drum. A heated roller fuses the powder onto the paper to make the image permanent. Many so-called laser printers use an array of LIGHT-EMITTING DIODES instead of a laser diode.

laser surgery Surgical treatment carried out using a LASER beam. The high energy in an extremely narrow laser beam can burn through body tissues to make a fine "cut". The heat also seals blood vessels, so there is much less bleeding than when a knife is used. A less powerful laser beam can remove coloured marks, such as tattoos, from the skin. Heat destroys the pigment because it absorbs much energy from the beam, whereas the surrounding skin is virtually unaffected. Surgeons also use laser to treat some forms of skin cancer and eye problems.

latent heat (symbol L) Heat absorbed or given out by a substance as it changes its PHASE (of matter) at constant temperature – from a solid to a liquid state or from a liquid to a gas. When ice melts, its temperature remains the same until it has been completely transformed into water; the heat necessary to do this is called the **latent heat of fusion**. Similarly the heat necessary to transform water into steam at constant temperature is called the **latent heat of vaporization.**

laterite Reddish, hard-baked soil found in tropical parts of the world. It results when hot, wet conditions wash away most of the nutrient content of the soil, leaving behind hydrated oxides of iron and aluminium. The top is baked hard by the Sun, and this factor, together with the leached nature of the soil, means that it is poor for agriculture. Some laterites, however, contain sufficient iron to be of commercial value.

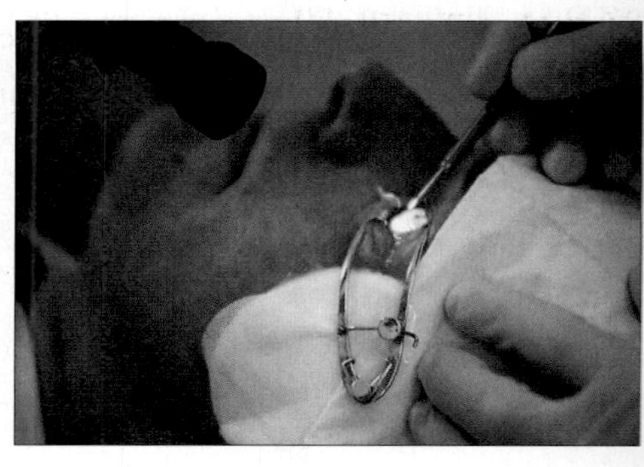

► **laser surgery** Used for a number of surgical operations today, lasers were initially restricted to surgical operations involving the eye, notably to correct detached retinas. The ultra-fine beam of light is a much more delicate instrument than a scalpel, and the energy of the laser cauterizes an incision as soon as it is made.

LASER

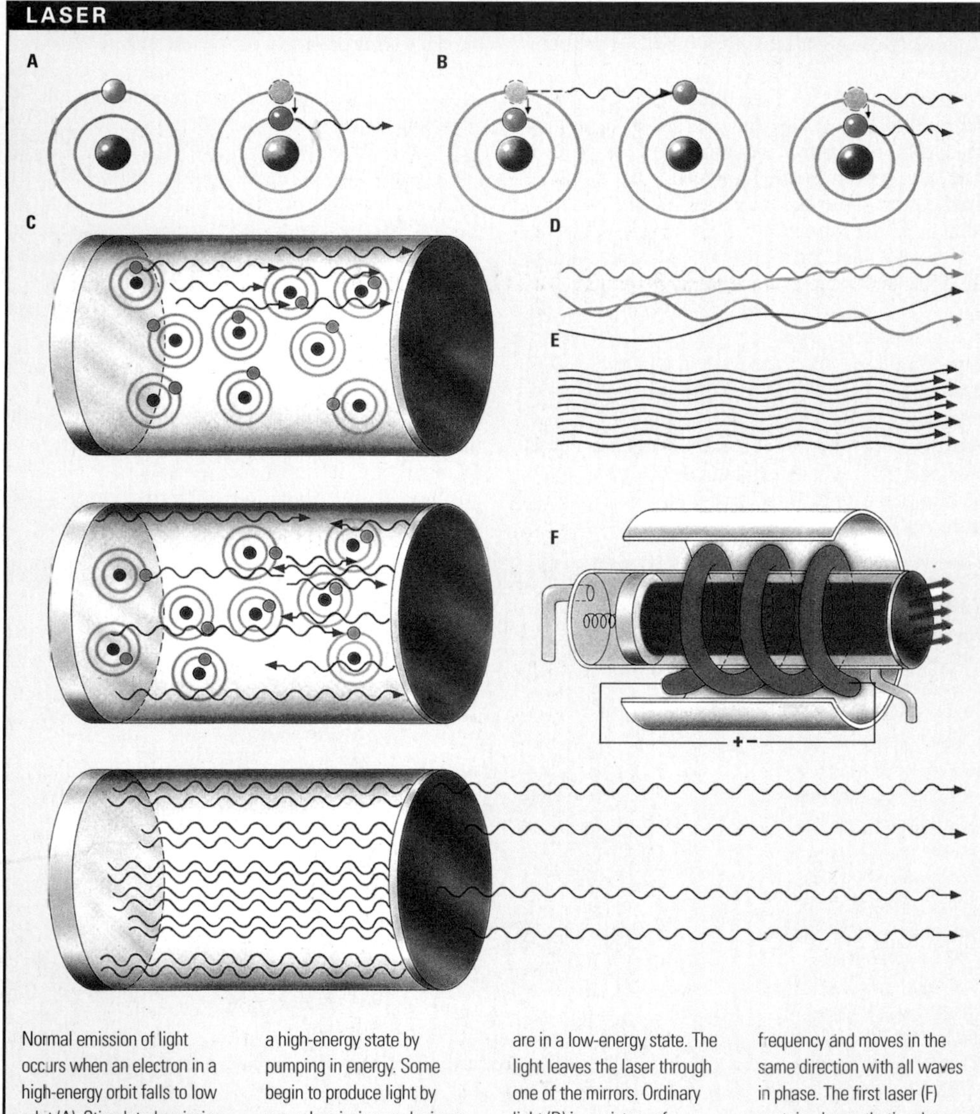

Normal emission of light occurs when an electron in a high-energy orbit falls to low orbit (A). Stimulated emission (B) is triggered by light emitted from another atom. In a laser (C), most atoms are brought to a high-energy state by pumping in energy. Some begin to produce light by normal emission, and mirrors at each end reflect the light to and fro, producing stimulated emission until all the atoms are in a low-energy state. The light leaves the laser through one of the mirrors. Ordinary light (D) is a mixture of different frequencies moving in various directions, whereas laser light (E) has a single frequency and moves in the same direction with all waves in phase. The first laser (F) contained a synthetic ruby crystal surrounded by a flash tube (to pump in light energy) and a pair of reflecting mirrors.

latex Milky fluid produced by certain species of plant, the most important being that produced by the RUBBER tree. Rubber latex is a combination of gum resins and fats in a watery medium. It is used in paints, special papers and adhesives, and to make sponge rubbers. Synthetic rubber latexes are also produced.

lathe Machine tool that turns a workpiece (either wood or metal) against a cutting tool in order to shape it.

latitude Distance N or S of the EQUATOR, measured at an angle from the Earth's centre. All lines of latitude are parallel to the Equator, which is the zero line of latitude. Each degree of latitude is *c*.111km (69mi). The tropics of CANCER and CAPRICORN are 23.5° away from the Equator, and the Arctic and Antarctic circles are at 66.5°, which is 23.5° away from the poles. In astronomy, celestial latitude is the angular distance on the CELESTIAL SPHERE of a celestial body from the ECLIPTIC; it is measured at right angles to the ecliptic, from 0° at the ecliptic to 90° at the ecliptic poles, positive to the N and negative to the S.

Laue, Max Theodor Felix von (1879–1960) German physicist. He became professor at Zurich and later director of the Institute for Theoretical Physics in Berlin. Using IONS in a crystal as a grating, Lane produced X-ray INTERFERENCE patterns, showing that X-RAYS are waves and providing a method of investigating CRYSTAL STRUCTURE. For this discovery of X-ray diffraction in crystals, he received the 1914 Nobel Prize for physics. *See also* X-RAY CRYSTALLOGRAPHY; X-RAY DIFFRACTOMETER

laughing gas *See* DINITROGEN OXIDE

launch window Interval of time – the time-slot – during which a space vehicle must be launched to achieve its required trajectory. Several launch windows may be available during a single launch opportunity.

Laurasia Ancient N continent, containing North America and Eurasia, which formed when a rift, the Tethyan trench, split PANGAEA (the Earth's single original land mass) from E to W along a line slightly N of the Equator. The name is a combination of Laurentian, a geological period in North America, and Eurasia. CONTINENTAL DRIFT and PLATE TECTONICS caused Laurasia to split up into the separate continents that exist today.

lava Molten rock or MAGMA that reaches the Earth's surface and flows out through a volcanic vent in streams or sheets. There are three main types of lava: **vesicular**, like pumice; **glassy**, like obsidian; and **even-grained**. Chemically, lavas range from acidic to ultrabasic. If it is acidic it is viscous and slow-flowing, but if it is basic it is more fluid. Basic lavas form lava flows and often erupt out of fissures, as well as volcanic craters. Basic lavas are commonest at points where plate margins are moving apart, but acidic lavas occur at colliding plate margins and are often associated with very explosive VOLCANOES. Once they reach the surface, lavas cool and solidify. If the lava is very gaseous, a rough, jagged surface will result, which is described as "scoriaceous".

Laveran, Charles Louis Alphonse (1845–1922) French physician. He contributed greatly to tropical medicine and was awarded the 1907 Nobel Prize for physiology or medicine for his work on the role played by PROTOZOA in causing diseases. Laveran discovered the parasite (*Plasmodium*) responsible for MALARIA.

Lavoisier, Antoine Laurent (1743–94) French chemist who founded modern chemistry. His careful experiments enabled him to demolish the PHLOGISTON THEORY (which said the vital essence phlogiston was lost during combusion) by demonstrating the function of OXYGEN in combustion. Lavoisier named both oxygen and hydrogen and showed how they combined to form water. In collaboration with Claude Louis Berthollet (1748–1822) and others, he published *Methods of Chemical Nomenclature* (1787), which laid down the modern method of naming substances. His *Elementary Treatise on Chemistry* (1789) established the basis of modern chemistry. He also helped to establish the METRIC SYSTEM.

Lawrence, Ernest Orlando (1901–58) US physicist. In 1930 he became professor at the University of California at Berkeley in 1930, where he built the first CYCLOTRON (a particle ACCELERATOR). Lawrence developed larger cyclotrons with higher energies, and his work in this area earned him the 1939 Nobel Prize for physics. The element LAWRENCIUM was named after him, as was the Lawrence Berkeley Radiation Laboratory.

lawrencium (symbol Lr) Radioactive, metallic element, one of the ACTINIDE SERIES. It was first made (1961) at the University of California at Berkeley by bombarding CALIFORNIUM with boron nuclei. The element has been made in only trace amounts but has been identified chemically. Properties: at.no. 103; r.a.m. 262; most stable isotope ^{256}Lr (half-life 27 seconds).

laxative Any agent used to counteract constipation. There are various kinds available, including bulk-forming drugs (mainly prescribed for people with bowel disease), stimulant laxatives, fecal softeners and saline purgatives. However, modern medical thinking is that laxatives are rarely necessary. Constipation should be kept at bay by regular exercise and a DIET high in fibre.

layering Method of plant PROPAGATION that induces or encourages root formation on a stem or branch while it is still attached to the parent plant; blackberries and loganberries multiply naturally in this way.

lazurite *See* LAPIS LAZULI

L-dopa (levodopa) Naturally occurring ISOMER of an AMINO ACID used to relieve some symptoms of PARKINSON'S DISEASE. It sometimes suppresses the trembling, unsteadiness and slowness of movement that characterize the condition. Side-effects include involuntary facial movements, mood swings, loss of appetite, nausea, vomiting and dizziness.

leaching In geology, process by which chemicals and nutrients are removed from a soil. Rainwater, especially in warm climatic regions, will dissolve anything soluble and wash it away. Once removed, these solubles can only be replaced very slowly. As a result, leached soils become coarse and are infertile. Saline soils can, however, be reclaimed for agriculture by deliberately leaching out salts. The term is also used to describe the industrial process of extracting a soluble material from a solid by washing the solid with solvents.

lead (symbol Pb) Metallic element of group IV of the periodic table. Its chief ore is GALENA (lead sulphide), from which lead is obtained by roasting. Exposure to lead from paints, pipes, petrol and other sources can lead to lead poisoning. Soft and malleable, it is used as a shield for X-rays and nuclear radiation, and in plumbing, batteries, cable sheaths and alloys such as pewter and solder. Chemically, lead is unreactive, a poor CONDUCTOR of electricity and resists corrosion. It forms two series of salts, termed lead(II), or plumbous, and lead(IV), or plumbic. Properties: at.no. 82; r.a.m. 207.19; r.d. 11.35; m.p. 327.5°C (621.5°F); b.p. 1,740°C (3,164°F); most common isotope ^{208}Pb (52.3%).

lead(II) ethanoate ($Pb[CH_3COO]_2$) (formerly lead acetate or sugar of lead) Sweet-tasting, but highly poisonous, white crystalline solid which is one of the very few lead salts soluble in water.

lead(II) oxide (lead monoxide, PbO) Yellow, crystalline compound that does not dissolve in water. It is made by heating molten lead in air, which at low temperatures produces the form called **massicot** and at high temperatures produces **litharge.** Lead(II) oxide is AMPHOTERIC; it dissolves in acids to produce lead(II) salts and in alkalis to give plumbates. It is used to make glazes for pottery, in paints and varnishes, and in the manufacture of some rubber products.

lead(II) sulphate ($PbSO_4$)White crystalline solid

LEACHING

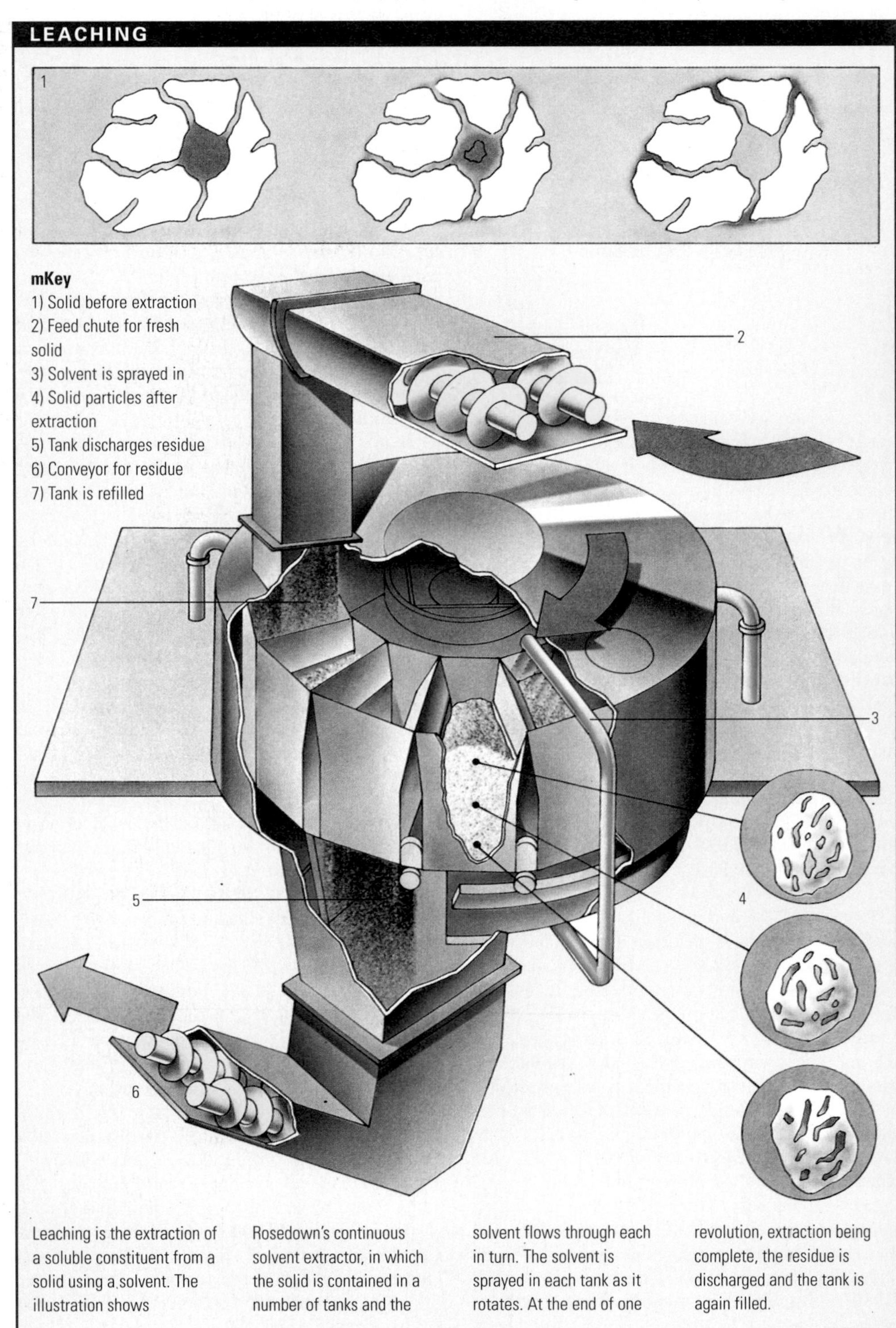

Leaching is the extraction of a soluble constituent from a solid using a solvent. The illustration shows Rosedown's continuous solvent extractor, in which the solid is contained in a number of tanks and the solvent flows through each in turn. The solvent is sprayed in each tank as it rotates. At the end of one revolution, extraction being complete, the residue is discharged and the tank is again filled.

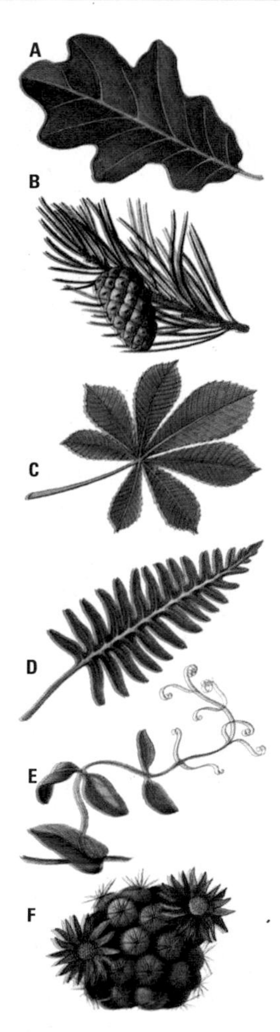

► **leaf** Leaves exhibit a wide variety of shapes. The pendunculate oak (*Quercus rober*) (A) and the Scots pine (*Pinus sylvestris*) (B) have simple leaves, with a single leaf blade, while the horse chestnut (*Aesculus hippocastrum*) (C) and ferns, such as Polypodium (D), have compound leaves. The leaflets of compound leaves either radiate from one point (palmate), as in the case of the horse chestnut, or are arranged in opposite pairs down the main stalk (pinnate) as is the case with ferns. The primary function of leaves is photosynthesis, but in addition leaflets may be modified into climbing tendrils (E), or protective spines, as in the cactus *Mammillaria zeilmannia* (F).

which is virtually insoluble in water. It occurs in nature as the mineral anglesite. With lead(II) hydroxide and water it forms the basic sulphate, which has been employed as a white paint pigment but its toxicity has resulted in a decline in this use.

lead(IV) oxide (lead monoxide, PbO_2) Brown or black amorphous solid that does not dissolve in water, but reacts slowly with concentrated acids. It is used in some kinds of safety matches and as the anode (positive plate) material in lead-acid accumulators.

lead(IV) tetraethyl (tetraethyl lead, $Pb(C_2H_5)_4$) Colourless, oily poisonous liquid. It is manufactured either by treating an alloy of lead and sodium with chloroethane (ethyl chloride) or by the action of lead chloride on a GRIGNARD REAGENT. It is used in leaded petrol as an anti-KNOCK agent. Properties: r.d. 1.65; b.p. 200°C (400°F).

lead sulphide (PbS) Toxic, black powder; a compound of LEAD. It occurs naturally as the mineral GALENA. It is insoluble in water, and is used as a SEMICONDUCTOR, and in ceramics.

leaf Part of a plant, an organ that contains the green pigment CHLOROPHYLL, and is involved in PHOTOSYNTHESIS and TRANSPIRATION. It usually consists of a blade and a stalk (**petiole**), which attaches it to a stem or twig. Most leaves are simple; some are compound, and divided into leaflets. Modifications include succulent types with fleshy tissue for water storage, tendrils that coil around supports, and needles, common in conifers.

Leakey, Louis Seymour Bazett (1903–72) Kenyan palaeoanthropologist and archaeologist. Leakey and his wife, **Mary** (1913–96), conducted important research on HUMAN EVOLUTION. Many of their most important discoveries were made in Olduvai Gorge, Tanzania. Here, they unearthed (1959) a HOMINID fossil *Zinjanthropus* or *Australopithecus boisei c.*1.75 million years old, and found (1964) the remains of *Homo habilis* – thought to be a direct ancestor of *Homo sapiens sapiens*. Leakey proposed the "Out of Africa" model of human development. After his death, Mary and their son, Richard LEAKEY, continued to make important finds.

Leakey, Richard Erskine Frere (1944–) Kenyan palaeoanthropologist and archaeologist, son of Mary and Louis LEAKEY. In 1972, at Lake Turkana, Kenya, Leakey discovered a 1.9 million year-old skull of *Homo habilis*. Other discoveries include a *Homo erectus* skeleton *c.*1.6 million years old. After serving as head of the Kenyan National Museums, Leakey became director of the Kenyan Wildlife Service (1988–94). *See also* HUMAN EVOLUTION

Leavitt, Henrietta Swan (1868–1921) US astronomer noted for her work on CEPHEID VARIABLE stars and stellar magnitudes. She discovered that the periods of Cepheids are related to their true brightness. The **Cepheid period-luminosity law** has since been used to determine the distances of stars.

Leblanc, Nicolas (1742–1806) French chemist who devised (1790) a process for producing SODA ash (sodium carbonate, Na_2CO_3) from salt (sodium chloride, NaCl) by treating it with sulphuric acid. Leblanc was physician to Louis Philippe, Duc d'Orléans, who helped him to establish a factory to produce soda ash. Leblanc's process was used extensively in the 19th century.

Le Châtelier's principle Principle announced (1888) by the French chemist Henry Louis Le Châtelier (1850–1936). It states that if a system in a state of EQUILIBRIUM is disturbed, the system tends to neutralize the disturbance and restore the equilibrium.

lecithin Substance containing FATTY ACIDS and CHOLINE and found in many animal tissues, especially in nerves, semen and the liver.

Leclanché cell Primary, electric CELL invented (1867) by French engineer Georges Leclanché (1839–82). Its ANODE was a zinc rod and its CATHODE a carbon plate surrounded by packed manganese dioxide. These electrodes were dipped into a solution of ammonium and zinc chlorides. It is the basis of the dry cell (BATTERY).

LED Abbreviation of LIGHT-EMITTING DIODE

Lederberg, Joshua (1925–) US geneticist. In 1958 he shared, with George BEADLE and Edward TATUM, the Nobel Prize for physiology or medicine for work that initiated the study of bacterial genetics. Lederberg discovered that sexual recombination of genetic materials occurs in bacteria and that genetic materials are linked in groups in bacteria as well as in other organisms.

Lee, Tsung-Dao (1926–) US physicist, b. China. He and his colleague Chen Ning YANG showed that among the WEAK NUCLEAR FORCES of ELEMENTARY PARTICLES, the law of conservation of PARITY (that nature, in effect, makes no distinction between right- and lefthandedness) does not always hold. This was subsequently verified by experiment. For their prediction, Lee and Yang were awarded the 1957 Nobel Prize for physics.

Leeuwenhoek, Anton van (1632–1723) Dutch scientist. Leeuwenhoek was a scientific amateur who built simple MICROSCOPES with a single LENS, made so accurately that they had better magnifying powers than the compound microscopes of his day. He investigated many microorganisms and their life histories, and described various microscopic structures such as spermatozoa. In 1680 Leeuwenhoek was made a Fellow of the British Royal Society.

left-hand rule *See* FLEMING'S RULES

Legendre, Adrien Marie (1752–1833) French mathematician. Legendre studied the theory of numbers and through his work on quadratic residues discovered the law of reciprocity. His most influential work was *Elements of Geometry* (1794).

legionnaire's disease Lung disease caused by infection with the bacterium *Legionella pneumophila*. It takes its name from the serious outbreak that occurred during a convention of the American Legion held in Philadelphia in 1976. Identified the following year, the bacterium thrives in water and may be found in defective heating, ventilation and air-conditioning systems. It is inhaled in fine water droplets present in the air. The disease is pneumonia-like in its effects and is treated with antibiotics.

legume Member of the pea family (Leguminosae) of ANGIOSPERMS the roots of which bear nodules that contain nitrogen-fixing bacteria. The family includes many trees, shrubs, vines and herbs. The fruit is typically a pod (legume) containing a row of seeds. Important food species include the pea, runner bean, soya bean, lentil, broad bean, kidney bean and haricot bean. *See also* NITROGEN CYCLE; NITROGEN FIXATION; ROOT NODULE

Leibniz, Gottfried Wilhelm (1646–1716) German philosopher and mathematician. Leibniz became interested in mathematics in the 1670s, when he met Christiaan HUYGENS and Robert BOYLE who described current problems in mathematics to him. While working as a lawyer, he developed CALCULUS (independently of Isaac NEWTON), combinatorial analysis and a theory of limits. He also made many practical inventions, including a calculating machine (1671). Leibniz is recognized as one of the greatest mathematicians in history. In philosophy, he created a rationalist form of metaphysics.

Leloir, Luis Federico (1906–97) Argentinian chemist, b. France. Leloir was awarded the 1970 Nobel Prize for chemistry for his research into the biochemical processes that break down complex sugars into simpler CARBOHYDRATES. In 1947 he helped to establish the Institute of Biochemical Research in Buenos Aires, where he began work on the production of LACTOSE. This research eventually led to the discovery of sugar NUCLEOTIDES, fundamental factors in the natural process of carbohydrate synthesis. Leloir also synthesized GLYCOGEN and demonstrated the necessity of certain liver enzymes in the manufacture of glycogen in the body.

Lemaître, Abbé Georges Édouard (1894–1966) Belgian astrophysicist and mathematician who first formulated the BIG BANG theory of the origin of the Universe. He saw the Universe as originally analogous to a radioactive atom, with all the energy and matter concentrated into a kernel which Lemaître called the "primeval atom". He argued that an EXPANDING UNIVERSE would have originated in the explosion of that primeval atom. His book *The Primeval Atom: An Essay on Cosmogony* (1950) describes the theory.

Lenard, Philipp Eduard Anton (1862–1947) German physicist, b. Hungary. Lenard was awarded the 1905 Nobel Prize for physics for his studies of the properties of CATHODE RAYS. His work was important in the development of ELECTRONICS and NUCLEAR PHYSICS. He also did research on ULTRAVIOLET LIGHT, the electrical conductivity of flames and PHOSPHORESCENCE.

lens Piece of transparent glass, plastic, quartz or organic matter, bounded by two surfaces, usually both spherical, that changes the direction of a light beam by REFRACTION and can produce an IMAGE. A

converging lens is convex in form (bulging at the centre) and bends light rays towards the lens axis. A **diverging** lens is concave (thinnest at the centre) and bends rays away from the axis. The optical image may be right-way-up or inverted, real or virtual, and magnified or reduced in size. Lens images suffer from various ABERRATIONS such as blurring and false colours. These characteristics also apply to the lens of the EYE.

lenticular galaxy Type of GALAXY intermediate in form between spiral and elliptical types. They resemble a convex lens in cross-section, hence their name.

Lenz's law Electromagnetic law deduced in 1834 by the Russian physicist Heinrich Lenz (1804–65). It states that an induced electric current flows in a direction that tends to oppose the charge producing it. *See also* INDUCTION

Leo (Lion) Large constellation of the N sky, lying between CANCER and VIRGO. The brightest stars are Alpha Leonis or Regulus, of the first magnitude, and Beta Leonis or Denebola, of the second magnitude.

Leonids METEOR SHOWER that occurs in November, associated with Comet Tempel–Tuttle. The shower is periodic, providing modest displays in most years, but producing magnificent meteor storms at intervals of 33 years – rates of 60,000 meteors per hour were recorded in 1966 on the night of 16–17 November.

leprosy (Hansen's disease) Chronic, progressive condition affecting the skin, and nerves, caused by infection with the microorganism *Mycobacterium leprae*. There are two main forms. **Lepromatous** leprosy is a contagious form in which raised nodules appear on the skin and there is thickening of the skin and peripheral nerves. The disease is characterized by numbness, weakness, loss of function and, finally, deformity of the affected parts. In **tuberculoid** leprosy, there is loss of sensation in parts of the skin, sometimes with loss of pigmentation and hair. The visible signs of leprosy, including deformity, paralysis and loss of digits, are the effects of nerve damage, possibly arising from injuries of which the person has been unaware (due to loss of sensation). Now confined almost entirely to the tropics, leprosy can be treated with a combination of drugs, but nerve damage is irreversible. The World Health Organization (WHO) is working towards eradication of the disease within a few years, but political instability makes this unlikely.

lepton One of a class of ELEMENTARY PARTICLES which comprises three families: the ELECTRON and the electron-NEUTRINO; the MUON and the muon-neutrino; and the TAU PARTICLE and the tau-neutrino, together with all their antiparticles (antileptons). Leptons are governed by the WEAK NUCLEAR FORCE, which is the force involved in RADIOACTIVE DECAY. *See also* ANTIMATTER

lesion Any abnormality in a body tissue caused by injury or disease. Examples are wounds of any kind, sores, blisters, ulcers and tumours.

letterpress printing Method of RELIEF PRINTING in which raised type is assembled and laid into a large frame (**chase**) by a compositor then placed either on a flat-bed press (which prints one sheet at a time) or processed into a curved plate and placed on a rotary press, which is capable of higher speeds. Ink is transferred from the raised type onto the paper. Today, this method is used only for more specialist PRINTING. *See also* LINOTYPE; LITHOGRAPHY

leucine ($[CH_3]_2$ $CHCH_2$ $CHNH_2$ $COOH$) White, soluble, essential AMINO ACID found in proteins.

leucite Grey or white FELDSPAR mineral, a potassium aluminium silicate ($KAl(SiO_3)_2$). Unstable at high pressures, it has a restricted occurrence. It can be found in potassium-rich lava flows and VOLCANIC PLUGS. Hardness 5.5–6; r.d. 2.5.

leucocyte White blood cell, a colourless structure containing a NUCLEUS and CYTOPLASM. There are two types of leucocytes, LYMPHOCYTES and PHAGOCYTES. Lymphocytes produce ANTIBODIES, and phagocytes destroy harmful organisms (by engulfing them) when the body is infected. Normal blood contains 5,000–10,000 leucocytes per mm_3 of blood. Excessive numbers of leucocytes are seen in such diseases as LEUKAEMIA. *See also* IMMUNE SYSTEM

leucotomy (prefrontal lobotomy) Surgical operation on the BRAIN. It was first performed in the 1930s to treat psychiatric disorders by severing tracts of nerve fibres leading to the frontal lobes. It may cause irreversible deterioration of personality. Drugs have almost entirely replaced this operation.

leukaemia Any of a group of CANCERS in which the bone marrow and other blood-forming tissues produce abnormal numbers of immature or defective LEUCOCYTES (white blood cells). This overproduction suppresses output of normal blood cells and PLATELETS, leaving the person vulnerable to infection, anaemia and bleeding. Leukaemias are classified into **acute** and **chronic** forms and also according to the type of leucocyte involved. **Acute lymphoblastic** leukaemia (ALL) is predominantly a disease of childhood; **acute myelogenous** leukaemia (AML) is mainly seen in older adults. Both forms are potentially curable. The outlook is less hopeful with the chronic forms, which are seen in older people. However, chronic leukaemia can often be kept under control for some years. *See also* IMMUNE SYSTEM

levée Natural embankment formed alongside a river by the deposition of silt when the river is in flood. Silt is deposited all over the FLOODPLAIN, but most is deposited near to the river banks, and so a slightly higher area is created alongside the river. Levées can help to prevent flooding, and are sometimes built up and strengthened artificially. Some levées grow so large that flood water cannot return to the main channel. When this happens, the river may change course. Levées alongside rivers in the UK are generally only a few centimetres in height, but on some larger rivers, such as the Mississippi in the USA, they may reach heights of 15m (50ft).

lever Simple MACHINE used to multiply the force applied to an object, usually to raise a heavy load. A lever consists of a rod and a point (FULCRUM) about which the rod pivots. In a crowbar, for example, the applied force (effort) and the object to be moved (load) are on opposite sides of the fulcrum, with the point of application of the effort farther from it. The lever multiplies the force applied by the ratio of the two distances. *See also* DISTANCE RATIO; FORCE RATIO

Leverrier, Urbain Jean Joseph (1811–77) French astronomer. Originally a chemist, he turned to celestial mechanics and made a study of planetary motions and PERTURBATIONS. In 1845 Leverrier predicted that a planet (which he prematurely named Vulcan) lay within the orbit of Mercury, but despite some claimed sightings it was never found. More successful was his prediction the next year, independently of John Couch ADAMS, that an unknown planet (NEPTUNE) was responsible for discrepancies between the calculated and observed orbital motion of Uranus.

Lewis, Gilbert Newton (1875–1946) US chemist. Working first in THERMODYNAMICS, he wrote a standard textbook on the subject. Later, he worked out a theory of VALENCE for non-IONIC COMPOUNDS based on shared electrons (covalency), and extended the concept of acids and bases. Lewis was the first to prepare HEAVY WATER (deuterium oxide), since used as a moderator in NUCLEAR REACTORS. *See also* COVALENT BOND

Leyden jar Earliest and simplest device for storing STATIC ELECTRICITY. It was the original electrical condenser (CAPACITOR), developed in *c.*1745 in Leiden (then Leyden), the Netherlands. It consists of a foil-lined, glass jar, partly filled with water and closed with a cork through which protrudes a knobbed brass rod or chain wired to the foil. Electricity (originally produced by friction) is passed into the jar along the rod or chain.

liana Any ground-rooting, woody vine that twines and creeps extensively over other plants for support; it is common in tropical forests. Some species may reach a diameter of 60cm (24in) and a length of 100m (330ft).

Libby, Willard Frank (1908–80) US chemist who developed the CARBON DATING technique. Libby worked (1941–45) in the war research division at Columbia University, New York, on the separation of ISOTOPES for the atom bomb. This work led to his interest in NUCLEAR PHYSICS and to his development of radioactive carbon-14 DATING, for which he was awarded the 1960 Nobel Prize for chemistry. He is the author of *Radiocarbon Dating* (1955).

Libra (Scales) Obscure constellation of the S sky, situated on the ECLIPTIC between Virgo and Scorpius. It formerly contained the First Point of Libra, the intersection of the ecliptic and the equator marking the N autumnal EQUINOX. Because of PRECESSION, this point has shifted W into Virgo. The brightest star, Beta Librae, is of magnitude 2.7.

libration Small oscillation of a celestial body about its mean position. The term is used most frequently in connection with the Moon. As a result of libration it is possible to see, at different times, 59% of the Moon's surface.

lichen Plant consisting of a FUNGUS in which microscopic (usually single-celled) ALGAE are

L

LENS

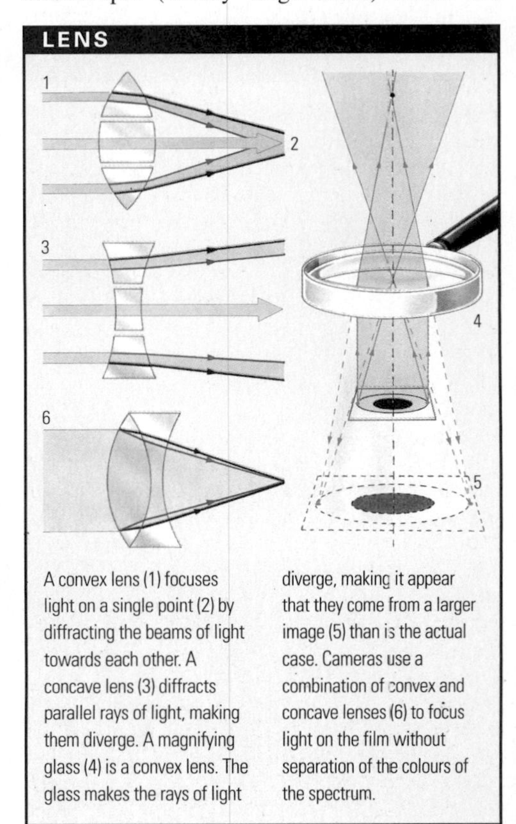

A convex lens (1) focuses light on a single point (2) by diffracting the beams of light towards each other. A concave lens (3) diffracts parallel rays of light, making them diverge. A magnifying glass (4) is a convex lens. The glass makes the rays of light diverge, making it appear that they come from a larger image (5) than is the actual case. Cameras use a combination of convex and concave lenses (6) to focus light on the film without separation of the colours of the spectrum.

embedded. The fungus and its algae form a symbiotic association in which the fungus contributes support, water and minerals, while the algae contribute food produced by PHOTOSYNTHESIS. *See also* SYMBIOSIS

Liebig, Baron Justus von (1803–73) German chemist. He was the first to realize that animals get their energy from the combustion of food (respiration) although he incorrectly thought that muscular power is a result of protein oxidation. Liebig also showed that plants derive their minerals from the soil, and introduced synthetic FERTILIZERS into agriculture.

lie detector Alternative name for a POLYGRAPH

life Widely accepted as the ability of organisms to obtain energy from the Sun or from food and to use this for growth and reproduction. All living organisms interact with their environment: they take in energy and materials from their environment and respond to changes in that environment. When they reproduce, they pass on instructions for their characteristics to the next generation in the form of DNA or RNA. Debate continues over the status of VIRUSES, which do not grow, and, although they do contain DNA or RNA, use the host cell's machinery to reproduce themselves. The current theory on life's origin is that giant molecules, similar to proteins and nucleic acids, reacted together in the watery surface environment of the young Earth that is now commonly called the **primordial soup**. In this environment the giant molecules multiplied by consuming smaller molecules. Despite much research, the evolution and development of cellular life out of this molecular pre-life remains a controversial subject and has yet to be explained fully.

life cycle Sequence of stages in the development of an organism from its formation to the creation of its offspring, followed by death. In organisms that reproduce sexually, formation corresponds to the fusion of sex cells (GAMETES) from the parents. The resulting ZYGOTE develops into an organism similar to the parents, which goes on to mature and produce new gametes that create the next generation. In organisms that reproduce asexually, formation occurs either through FISSION – simple division – BUDDING or VEGETATIVE REPRODUCTION. In plants, there is an ALTERNATION OF GENERATIONS in which the plants (GAMETOPHYTE generation) differ from the spore-forming (SPOROPHYTE) generation.

lift In AERODYNAMICS, force that acts upwards on the undersurface of an AEROFOIL, or wing. As it travels forwards, the leading edge of an aerofoil splits the airstream. Because the upper part of the airstream is forced to travel farther over the aerofoil's more highly curved upper surface before rejoining the lower part of the airstream, it travels faster and consequently has a lower pressure, in accordance with BERNOULLI'S LAW. The lift force is a result of the upward pressure underneath the aerofoil being greater than the downward pressure on the top.

lift (elevator) Machine for raising and lowering passengers and freight from one level to another inside or outside buildings. Early lifts were driven by steam or hydraulic power. The first lift was installed (1856) in a New York store by Elisha OTIS, who had invented a safety system that incorporated ratchets along the lift shaft to ensure the body of the lift could not drop accidentally. In 1889 the first electric lift was installed in a New York building.

lifting magnet Powerful ELECTROMAGNET used for lifting and transferring heavy metal objects. Such a magnet is suspended from the jib of a crane.

ligament Bands of tough fibrous, CONNECTIVE TISSUE that join bone to bone at the JOINTS. In the wrist and ankle joints, for example, they surround the bones like firm inelastic bandages. Ligaments, which contain a tough inorganic substance known as COLLAGEN, form part of the supporting tissues of the body.

ligand Ion, molecule or group of atoms linked to a metal atom or ion by coordinate chemical bonds in a metal complex. Thus chloride ions act as ligands in the ion $(CuCl_4)^{2-}$, and carbon monoxide molecules are ligands in such compounds as nickel carbonyl $(Ni(CO)_4)$. Some molecules, such as 1,2-diaminoethane (ethylene diamine, $H_2NCH_2CH_2NH_2$), can coordinate at two points in the molecule. *See also* COORDINATION COMPOUND

light Part of the total ELECTROMAGNETIC SPECTRUM that can be detected by the human eye. Visible light is in the wavelength range from about 400nm (violet) to 770nm (red). The speed (symbol *c*) at which ELECTROMAGNETIC RADIATION (including light) travels in a vacuum is equal to 299,792,458m s^{-1}. Light exhibits typical phenomena of wave motion such as REFLECTION, REFRACTION, DIFFRACTION, POLARIZATION and INTERFERENCE. The properties of light were investigated in the 17th century by Isaac NEWTON, who was the first to split white light into its component colours (SPECTRUM) with a PRISM. Newton believed in a corpuscular or particle theory of light, but the wave theory was well established by the second decade of the 19th century after the work of Thomas YOUNG and Augustin Jean FRESNEL on the diffraction and interference of light. At the beginning of the 20th century, experiments on the PHOTOELECTRIC EFFECT and the work of Max PLANCK revived the idea that light can behave like a stream of particles. This dilemma was resolved by the QUANTUM THEORY, according to which light consists of elementary particles of zero rest mass called PHOTONS. When light interacts with matter, as in the photoelectric effect, energy is exchanged in the form of photons and so light seems to consist of particles. Otherwise, it behaves as a wave. The quantum theory also explains how all electromagnetic radiations result from the acceleration of electric charges. *See also* ETHER; HOLOGRAPHY; LASER

light adaptation Shift in functional dominance from rod cells to cone cells within the RETINA of the EYE as overall illumination increases. Unlike DARK ADAPTATION, adaptation to light is swift but usually uncomfortable. Humans can discern colour and form soon after emerging from prolonged darkness.

light bulb Device consisting of a metal filament that emits light when heated to INCANDESCENCE by electricity. An early form of light bulb dates to 1860, when Joseph SWAN made a workable model. In 1879 Thomas EDISON patented an improved design in the USA. Modern light bulbs comprise a filament enclosed by a glass envelope in a vacuum containing a NOBLE GAS gas such as argon. The filament is usually made of TUNGSTEN because of its high tensile strength, low evaporation and high efficiency in converting electrical energy into light as it glows white hot. *See also* ELECTRIC LIGHT

light-emitting diode (LED) SEMICONDUCTOR device that produces light. An LED consists of a *p-n* junction DIODE that emits light when electrons and holes in the semiconductor material (typically gallium arsenide phosphide) recombine at the junction. In this way it converts electrical energy applied across the junction into light. LEDs are used for display purposes in various kinds of electronic apparatus.

lighthouse Building, often in the form of a tower, with a light at the top to guide vessels at night. Lighthouses are built on land and at sea. Some mark ports and harbours, while others warn of shallow waters or dangerous rocks. Originally, a lighthouse keeper was always in attendance, but modern lighthouses are usually unmanned and operated by remote control. The light system may produce a steady beam, a rotating beam, or a pattern of flashes, sometimes of different colours. Sailors can identify any lighthouse by the pattern of light it emits. Knowing the location of a lighthouse enables navigators to work out their own positions. In poor visibility, a siren, horn or bell may sound to warn vessels to keep clear of rocks. In addition, many lighthouses transmit radio sig-

LIGHT

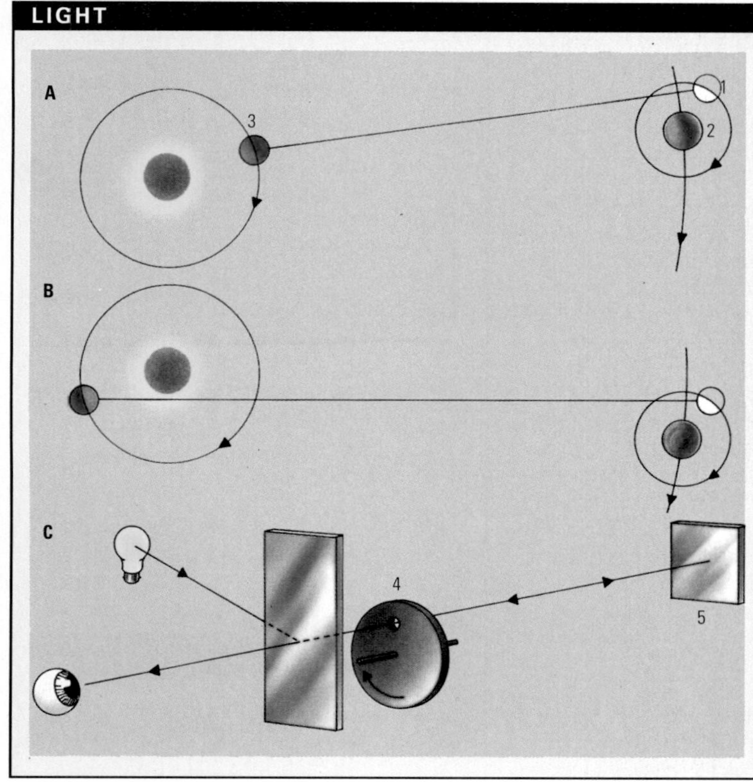

The velocity of light was first determined by Olaus Roemer in 1675 (A). He saw the eclipses of Jupiter's moons (1) by Jupiter (2). Light from the moons takes less time to reach the Earth (3) when its orbit nears Jupiter, than when it is far away towards the other side of the Sun (B). Knowing the distances and times involved, he could calculate the velocity of light. Another determination was made by Armand Fizeau in 1849 (C). Light was reflected through the teeth of a rotating wheel (4) to a mirror (5) and back through the teeth to the observer. The light was seen only when the wheel spun so fast that no teeth blocked its return journey. From the spacing of the teeth, the speed of rotation of the wheel and the distance of the mirror (8km [5mi]), the velocity of light could be accurately calculated.

► **light bulb** Heat generated by the passage of electric current is the source of light in a so-called "filament" or "incandescent" lamp. Because it is enclosed in a vacuum, or noble gas (1), the filament (2) cannot oxidize when the current passes through, causing it to become hot. It is made of a tungsten alloy, combining mechanical and thermal strength. Although it is extremely thin, it glows white hot when enough current passes through it. It is supported on two glass columns (3) through which the connecting wires pass. The whole assembly is enclosed in a thin, glass envelope (4). Only *c.*2% of the electrical energy is converted into light.

nals for use by ships' direction-finding equipment. Today, lighthouses are becoming less important as many vessels are equipped with the GLOBAL POSITIONING SYSTEM (GPS), a highly accurate, satellite-aided navigation system.

lightning Visible flash of light accompanying an electrical discharge between clouds or between clouds and the ground, most commonly produced in a THUNDERSTORM. A typical discharge consists of several lightning strokes, initiated by leaders that follow an irregular path of least resistance, the **lightning channel**. Intense heating by the discharge expands the channel rapidly up to a diameter of 13–25cm (5–10in), creating the sound waves of thunder. *See illustration on page 214*

light reaction (light-dependent reaction) One of the two distinct phases of PHOTOSYNTHESIS; the other being DARK REACTION. During this stage, the green pigment CHLOROPHYLL absorbs sunlight and uses its energy to breakdown water into oxygen gas, electrons and hydrogen ions. The energy of the electrons is used to convert ADENOSINE DIPHOSPHATE (ADP) to ADENOSINE TRIPHOSPHATE (ATP) and NADPH. Together ATP and NADPH ensure the dark reaction can take place.

light-year (symbol l.y.) Unit of astronomical distance equal to the distance travelled in free space or a VACUUM by light in one tropical year. One light-year is equal to 9.4607×10^{12} km (5.88×10^{12} mi).

lignin Complex, non-carbohydrate substance that occurs in the woody tissues (especially XYLEM) of plants, often in combination with CELLULOSE. It makes the CELL walls stronger, helping them to resist compression and tension. It is lignin that gives WOOD its strength. To obtain pure cellulose for the paper and rayon industries, the lignin has to be removed from wood. Lignin is also a major source of vanillin, a white, crystalline ALDEHYDE used as a flavouring, and in perfumes and pharmaceuticals.

lignite Soft COAL, brown-black in colour, with a carbon content higher than that of PEAT but lower than those of other coals. Sometimes the shape of its original woody structures is visible.

Lilienthal, Otto (1849–96) German engineer and pioneer of glider design. A glider designed by Sir George CAYLEY had carried a passenger in 1853, but this aircraft had no controls. In 1891 Lilienthal became the first person to control a glider in flight. He used an aircraft of his own design with cotton fabric wings. He made about 2,500 more flights before his death in a crash.

Lillie, Frank Rattray (1870–1947) US zoologist and embryologist, b. Canada. Lillie was noted for his discoveries in the field of fertilization and the role of HORMONES in SEX DETERMINATION. He spent much time helping to develop the Marine Biological Laboratory and the Oceanographic Institution at Woods Hole, Massachusetts. His book *The Development of the Chick* (1908) is a leading work in EMBRYOLOGY.

limb In astronomy, apparent edge of any celestial body presenting an observable disc, such as the Moon, the Sun or a planet.

limb Jointed extension of the VERTEBRATE body, used for locomotion and manipulation of the physical environment. The human forelimb (arm) is particularly specialized in the latter function. The form of the limbs is regarded as a criterion of evolutionary development. *See also* PENTADACTYL LIMB; HUMAN BODY

limbic system Collection of structures in the middle of the BRAIN. Looped round the HYPOTHALAMUS, the limbic system is thought to be involved in emotional responses, such as fear and aggression, and the production of mood changes, as well as the laying down of memories.

lime *See* CALCIUM HYDROXIDE; CALCIUM OXIDE

limestone SEDIMENTARY ROCK formed on the bed of a warm sea by an accumulation of dead sea creatures. At least 50% of a limestone consists of CALCITE ($CaCO_3$). There are many different types, including Carboniferous limestone, CHALK, dolomitic limestone and oolitic limestone. They all contain fossils or fragments of fossils. Because of the calcium carbonate content, limestones are soluble and permeable, and generally give rise to dry landscapes. Chalk is pure limestone and forms rounded hills and vertical cliffs. Carboniferous limestone is hard and forms high ground; oolitic hills are generally more gentle. Carboniferous limestone, like chalk and oolitic limestone, has numerous dry valleys, but also contains CAVES with STALACTITES and STALAGMITES. Limestone is used to make cement, as a source of commercial lime and as a building material.

lime water Saturated solution of CALCIUM HYDROXIDE (slaked lime, $Ca(OH)_2$). It is used to detect carbon dioxide (CO_2), which forms a milky white precipitate of calcium carbonate ($CaCO_3$) when bubbled through lime water. Excess carbon dioxide changes the calcium carbonate into calcium hydrogencarbonate (calcium bicarbonate, $Ca(HCO_3)_2$), which is soluble and so the white precipitate disappears.

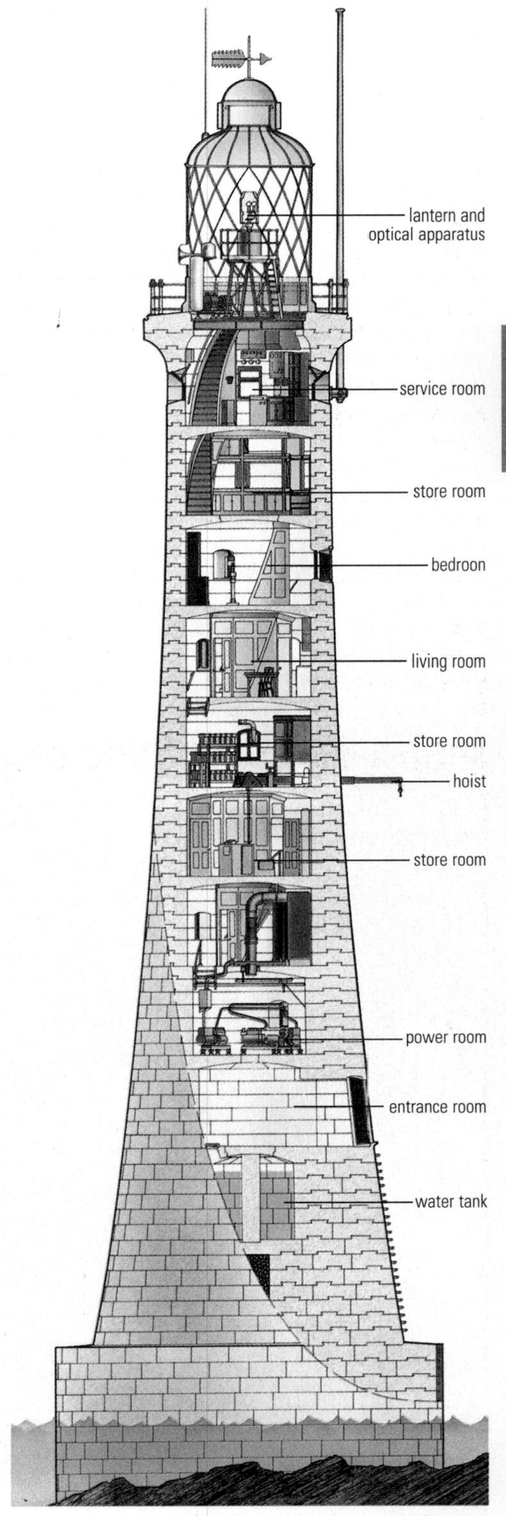

► **lighthouse** The present Eddystone lighthouse is the last of a series of four to be erected on the site and was completed on 18 May 1882. Its focal plane is 40m (133ft) above high water level and it is constructed of stones dovetailed into one another. Most modern lighthouses are unmanned, instead they are computer operated.

L

limit In mathematics, value or values approached by the terms of a sequence. The limit of a SERIES is the value approached by the sum as more and more terms are included. The concept of limits is fundamental to differential and integral CALCULUS. *See also* CONVERGENCE

limnology Science of freshwater LAKES, ponds and streams. These bodies of water are explored in terms of chemistry, physics and biology. The plants, animals and environment are quantitatively examined in light of food cycles, food chains, habitats and ZONATION of organisms. Freshwater bodies are subject to greater extremes of temperature and are therefore more fragile ecosystems and more specialized than those in marine environments. *See also* HYDROLOGY

limonite (brown iron or bog ore $FeO(OH).nH_2O$) One of the most important sources of iron. Limonite may appear lustrous black, brown to yellow and is streaky, porous and often mixed with sand or clay.

Lindbergh, Charles Augustus (1902–74) US aviator. Lindbergh became an international hero when, in *The Spirit of St Louis*, he made the first nonstop transatlantic solo flight, from New York to Paris (1927) in 33 hours 30 minutes. In 1932 his baby son was kidnapped and murdered. In 1936 he and Alexis CARREL invented an artificial heart.

Lindemann, Frederick Alexander, 1st Viscount Cherwell (1886–1957) British physicist, b. Germany. Involved in aircraft research (1914–18), he discovered how to regain control of an aircraft in an uncontrolled spin. During World War 2 he was Winston Churchill's scientific adviser. Lindemann developed the Clarendon Laboratory of Oxford University into a major research facility.

Lindley, John (1799–1865) British botanist, known for his CLASSIFICATION of plants in a natural system in which all characteristics of a plant are considered. Among Lindley's best-known books are *Theory and Practice of Horticulture* (1842) and *The Vegetable Kingdom* (1846), which included his new classification of plants.

line Continuous set of points. The term can be synonymous with curve, or can be taken to mean a straight line – the shortest distance between two points in EUCLIDEAN GEOMETRY. In a rectangular CARTESIAN COORDINATE SYSTEM, a straight line has the equation $y = mx + c$, where m is the gradient and c the intercept on the y-axis.

linear accelerator Type of particle ACCELERATOR in which charged particles travel in straight lines through a vacuum chamber. Early linear accelerators were electrostatic ACCELERATORS. In more modern types, a high-frequency, alternating electric field is used to accelerate the particles. Final energies depend on the length of the chamber, and can reach several GeV (million ELECTRON VOLTS). The linear accelerator at Stanford University is 3.2km (2mi) long and produces electrons with energies of 20GeV.

linear function Mathematical function given by a POLYNOMIAL involving no powers of its variables higher than one, for example $f(x) = 7x + 3$ is linear. The graph of a linear function, in two dimensions, is a straight line.

linear motor Type of ELECTRIC MOTOR that is being developed to power high-speed trains. Its principle is similar to that of rotary ELECTRIC MOTORS, but instead of a number of coils rotating (a rotor) within a fixed ELECTROMAGNET (a stator), the linear motor is "unfolded" so that the "rotor" windings are fixed to the locomotive's path, and the "stator" windings are incorporated into the locomotive, which is both levitated and driven forwards.

linear programming Mathematical procedure in which a multi-variable linear function is analysed to find maximum and minimum values.

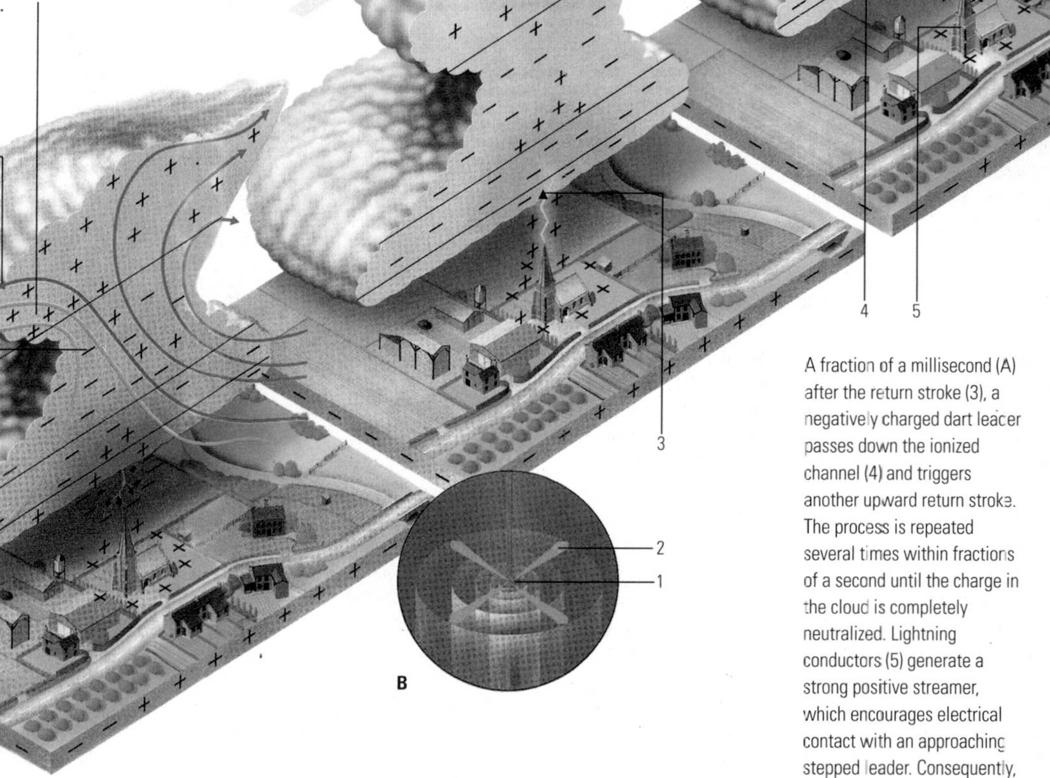

► **lightning** Once an electrical potential of a million volts/m (300,000 volts/ft) has been created in a thunder cloud, the lightning process begins (A). A stream of electrons flows down, colliding with air molecules and freeing more electrons, and in the process giving the air molecules a positive charge (ionizing the air). This intermittent, low-current discharge forms a highly branched pathway and is called the stepped leader (1). As the leading branches of the stepped leader, carrying large negative charges, near the ground they induce short upward streamers of positive electrical charges from good conducting points on the ground (2). When a branch of the stepped leader contacts an upward positive streamer, a complete channel of ionized air has been created. This allows a huge positive current, called the return stroke, to flow upward into the cloud in the form of a bright lightning stroke (3). The different strokes have been given different colours to distinguish between them. In nature, all lightning is colourless. The return stroke causes the first of the shock waves we hear as thunder (B). The flash effectively reaches the eye instantaneously, whereas the sound of thunder travels at approximately 330 m/s (1,000 ft/s). So multiply the number of seconds between the flash and the thunder by 330 to find the distance in metres.

A fraction of a millisecond (A) after the return stroke (3), a negatively charged dart leader passes down the ionized channel (4) and triggers another upward return stroke. The process is repeated several times within fractions of a second until the charge in the cloud is completely neutralized. Lightning conductors (5) generate a strong positive streamer, which encourages electrical contact with an approaching stepped leader. Consequently, potential lightning strikes within 50–100m (170–330ft) of a building are attracted to a lightning conductor. Return strokes and dart leaders are safely routed to the Earth along a wide copper strip, which has one end buried in the ground. Thunder (B) is caused by the narrow lightning stroke heating the column of air (1) surrounding it to around 30,000°C (54,000°F), expanding it explosively (2) at supersonic speeds under a force of 10–100 times normal atmospheric pressure. The immense shock wave becomes a sound wave within about a metre, producing the sound of thunder.

LINEAR MOTOR

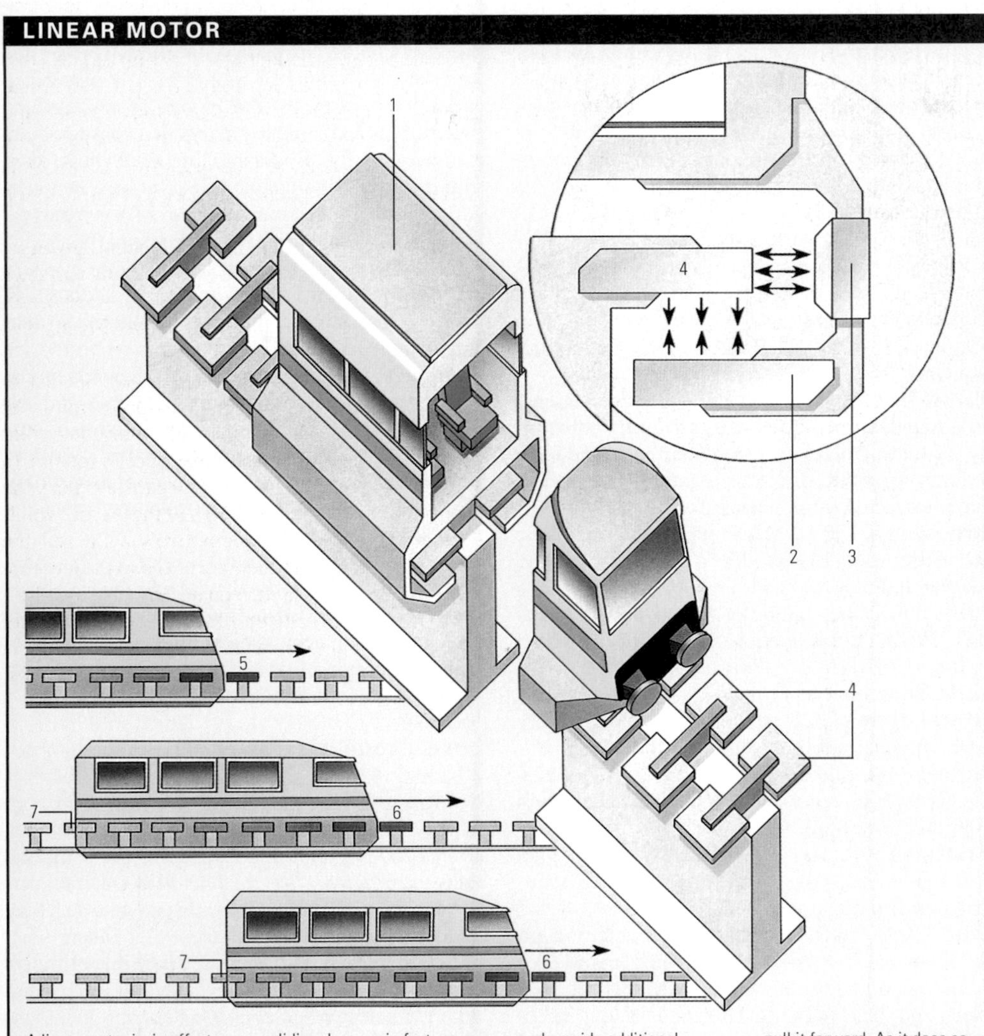

A linear motor is, in effect, a conventional motor unwound, so that its electromagnetic coils will move a flat stator (usually an aluminium plate in which magnetic fields are involved) or another set of electromagnets. Although most commonly used to open sliding doors or in factory applications, linear motors are being considered for use in the field of rapid public transport. A passenger carriage (1) carries a set of powerful electromagnets (2) to lift and propel the vehicle, and a second set to guide it and provide additional braking force (3). The track contains a series of electromagnets (4) that are energised to a polarity opposing the ones in the vehicle slightly ahead of the carriage (5). The magnetic forces lift the carriage and pull it forward. As it does so, the magnets ahead of the vehicle are switched on to continue to attract it forward, (6) while the magnets behind the vehicle have their polarity reversed (7) to help repel the magnets in the car and push it forward.

It is useful in business planning and industrial engineering to produce optimal control conditions, as in inventory control, where costs must be minimized in terms of storage time, warehouse space, customer delivery schedules, reorder times and transportation expenses.

lines of force Lines in an ELECTRIC FIELD or MAGNETIC FIELD whose direction at every point is that of the field.

line spectrum *See* SPECTRUM

linkage group Group of inherited characteristics or GENES that occur on the same CHROMOSOME and that remain connected in such a way that they are usually inherited together through successive generations.

Linnaeus, Carolus (1707–78) (Carl von Linné) Swedish botanist and taxonomist. His *Systema Naturae* (1735) laid the foundation of the modern science of TAXONOMY by including all known organisms in a single CLASSIFICATION system. Linnaeus was one of the first scientists to define clearly the differences between species, and he devised the system of BINOMIAL NOMENCLATURE, which gave standardized Latin names to every organism.

Linotype Method of setting type for printing, by the Linotype machine, invented (1884) by Ottmar MERGENTHALER in Baltimore. It was first used by the New York *Herald Tribune* in 1886, and was subsequently used by newspapers all over the world. It is operated by a keyboard like that of a typewriter, which sets letters as one complete line of hot-metal type, called a **slug**. The Linotype system has now been replaced by PHOTOSETTING.

lipid One of a large group of organic compounds in living organisms that are insoluble in water but soluble in alcohol. They include animal FATS, vegetable OILS and natural WAXES. Lipids form an important food store and energy source in plant and animal cells. Storage fat is composed chiefly of **triglycerides**, which consist of three molecules of FATTY ACID linked to GLYCEROL.

Lipmann, Fritz Albert (1899–1986) US biochemist b. Germany, who isolated and partially explained the molecular structure of COENZYME A, derived from the B vitamin pantothenic acid. For this and other work on METABOLISM, he shared the 1953 Nobel Prize for physiology or medicine with Hans KREBS (1900–81), who had discovered the KREBS CYCLE. In 1950 Lipmann demonstrated the formation of citric acid from oxalo-acetate and acetate, and found that this process required coenzyme A. During his career, Lipmann also investigated the way in which a cell acquires energy and the key role played in that process by ADENOSINE TRIPHOSPHATE (ATP).

Lippershey, Hans (1570–1619) German spectacle maker who discovered that a convex and concave LENS used in conjuction can make distant objects appear closer, thereby inventing the TELESCOPE. Its potential for astronomy was recognized by Jacques Bovedere of Paris, who reported it to GALILEO who then built his own.

Lippmann, Gabriel (1845–1921) French physicist who was awarded the Nobel Prize for physics in 1908. The first person to produce a colour photograph of the visible SPECTRUM, Lippmann also invented the direct colour process of PHOTOGRAPHY. The process used a plate consisting of a grainless emulsion in contact with a layer of liquid mercury. His work was important in the development of holograms. *See also* HOLOGRAPHY

Lipscomb, William Nunn, Jr (1919–) US chemist who made fundamental discoveries about the bonding of molecules, for which he was awarded the 1976 Nobel Prize for chemistry.

liquefaction of air Process achieved by cooling air to its critical temperature of −147°C (−232.6°F), at or below which it can be liquefied by pressure. The method involves repeated compression, followed by rapid ADIABATIC expansion, which results in progressive drops in temperature due to the JOULE-THOMSON EFFECT.

liquefied petroleum gas (LPG) Liquefied gas of light HYDROCARBONS, principally propane and butane, produced in the DISTILLATION of crude oil and the refining of NATURAL GAS. It is used as a fuel and raw material in chemical industries and as a bottled gas for domestic heating and cooking.

▲ **Linnaeus** The Swedish botanist Carl Linnaeus was a great classifier of living organisms. In 1735 he published *Systema Naturae* (*The Natural World*) in which he divided flowering plants into classes depending on their stamens and then the classes into orders according to the number of pistils. He is considered the founder of modern taxonomy. In 1749 Linnaeus introduced the so called binomial nomenclature by which each plant was given a Latin generic noun followed by a specific adjective. He opposed the evolutionary theory believing that no new species had been formed since the Creation and none had become extinct.

liquid State of MATTER intermediate between a GAS and a SOLID. A liquid substance has a relatively fixed volume but flows to take the shape of its container. The state that a substance assumes depends on the temperature and the pressure at which it is kept; a substance that is liquid, such as water, at room temperature can be changed into a vapour (its gaseous state, steam) by heating, or into a solid (ice) by cooling. Unlike the molecules in gases, those in a liquid are weakly bonded and liquids, like solids, are compressed only very slightly by pressure.

liquid crystal Substance that can exist half-way between both the liquid and solid states with its molecules partly ordered. By applying a carefully controlled electric current, liquid crystals turn dark. They are used in liquid crystal displays (LCDs) to show numbers and letters, as in pocket calculators.

Lister, Joseph, 1st Baron (1827–1912) English surgeon who introduced the principle of antisepsis, which complemented the theory of PASTEUR that bacteria cause infection. Using carbolic acid (phenol) as the ANTISEPTIC agent, and employing it in conjunction with the heat sterilization of instruments, Lister brought about a dramatic decrease in post-operative fatalities.

lithium (symbol Li) Rare, silver-coloured metallic element, one of the ALKALI METALS, first isolated in 1817. Ores include lepidolite and spodumene. Chemically it is similar to sodium. The element, which is the lightest of all metals, is used in alloys, and in glasses and glazes; its salts are used in medicine. Properties at.no. 3; r.a.m. 6.941; r.d. 0.534; m.p.180.5°C (356.9°F); b.p. 1,347°C (2,456.6°F); most stable isotope ^{7}Li (92.58%).

lithography Method of PRINTING from a flat inked surface. This technique, invented in the 1790s, is based on the fact that grease and water do not mix. In traditional lithography, the design is made on a prepared plate or stone with a greasy pencil, crayon or liquid. Water applied to the surface is absorbed where there is no design. Oil-based printing ink, rolled over the surface, sticks to the design, but not to the moist areas. Pressing paper onto the surface produces a print. Multicolour prints are made using a separate plate or stone for each colour ink used. Today, many magazines, books and newspapers are produced by a process called OFFSET PRINTING, which uses a modern form of lithography, involving photographically produced metal plates.

lithosphere Upper layer of the solid EARTH which includes the CRUST and the uppermost MANTLE. It is of variable thickness, from 60km (40mi) extending down to approximately 200km (125mi). Rigid, solid and brittle, the lithosphere is made up of a number of tectonic plates which move independently and give rise to PLATE TECTONICS.

litmus Purplish dye in neutral aqueous solutions, used to indicate acidity (by turning red) or alkalinity (by turning blue). It is most familiar in the form of litmus paper used as an acid-base indicator. The dye is extracted from lichens. *See also* PH

litre (symbol l or L) Metric unit equal to a cubic decimetre, one thousandth of a cubic metre. Another definition, used from 1901–68, was that 1 litre equalled the volume of 1kg of pure water at 4°C (39°F). A litre is equivalent to 0.22 imperial gallons or 0.2642 US gallons.

Little Bear *See* URSA MINOR

littoral zone Beach area between high and low tides. It also refers to the benthic division – between high tide and a depth of 200m (656ft). The larger zone is divided into the **eulittoral**, from high tide to a 50m (164ft) depth, and the **sublittoral**, from 50–200m (164–656ft). The lower edge of the eulittoral is the lowest limit at which abundant attached plants can grow.

liver Large organ located in the upper right abdomen of VERTEBRATES. Weighing up to 2kg (4.5lbs) in an adult human, it is divided into four lobes and has many functions. It is extremely important in the control of the body's internal environment (HOMEOSTASIS). It receives nutrients from the intestine and is a site of METABOLISM of proteins, carbohydrates and fats. It synthesizes BILE and some vitamins, regulates the blood-glucose level, produces blood-clotting factors, breaks down worn-out ERYTHROCYTES (red blood cells) and removes toxins from the blood. The many metabolic reactions that go on in the liver are the body's main source of heat, which is distributed around the body by the blood. HOMEOTHERMIC (warm-blooded) animals maintain a constant body temperature by producing too much heat, then controlling its dissipation. The liver has phenomenal powers of regeneration: if part of it is removed, it regrows within weeks. *See also* INSULIN

load In geology, sediment carried along by ice, a river or the sea. Ice-sheets and large GLACIERS can carry a greater load than a thin glacier, so that they dump larger amounts of TILL when the ice melts. The load carried by a river depends on the type and amount of material available and the speed of the current. It may be in solution or suspension, or be rolled along the river bed. A stream or slow-moving river can normally transport only mud and sand, whereas a fast-moving torrent can shift large boulders.

Lobachevsky, Nikolai Ivanovich (1792–1856) Russian mathematician. He was educated at Kazan University and appointed professor there in 1816. His outstanding achievement (announced in 1826) was the creation of one of the first comprehensive systems of NON-EUCLIDEAN GEOMETRY.

lobotomy *See* LEUCOTOMY

Local Group of galaxies Small group of about 30 GALAXIES that includes our Galaxy, the two MAGELLANIC CLOUDS, the ANDROMEDA GALAXY and the Triangulum spiral. They are distributed over a roughly ellipsoidal space, about 5 million light-years across. The members are gravitationally bound so that, unlike more distant galaxies, they are not receding from us or from each other.

lock Structure built into a stretch of inland waterway to raise or lower water levels to correspond with the surrounding countryside. Each lock consists of two sets of lock gates. A vessel enters the lock, the gates are closed, and sluices are opened to admit or release enough water to bring the vessel to the same level as the water beyond the second pair of gates.

lock Mechanism fitted to a door or case to prevent entry to people other than those with a specialized key or other device. Locks date back more than 4,000 years and were used by the Egyptians. The lever tumbler lock, an early version of the modern mortise, or dead, lock, dates back to the 18th century. This system employs a tumbler and bolt. When the right key is inserted, the tumbler, a simple lever, is raised to a height whereby the key is able to engage the bolt, turn and move the bolt back in the lock. Turning the door handle makes a cam move the latch across. The lock most commonly used today is the Yale, named after its inventor Linus YALE Jr. It is a type of cylinder lock which features a number of pins of differing lengths within the barrel. When the correct key is inserted in the lock the pins align, allowing the key to turn, and thereby moving the latch across. Other types of lock include those that require the insertion of a punched card to release the mechanism, or those that are activated by cards with magnetic strips. It is expected that in the future locks will be activated only by recognizing a particular person's unique fingerprints, or by scanning and recognizing a person's unique eye characteristics.

Lockyer, Sir Joseph Norman (1836–1920) British astronomer who discovered the element HELIUM, independently of Jules JANSSEN, in the Sun's spectrum. In 1868 he developed a technique for examining PROMINENCES at the edge of the Sun and attributed lines in the solar spectrum to a new element, which he named helium – 40 years before it was discovered on Earth.

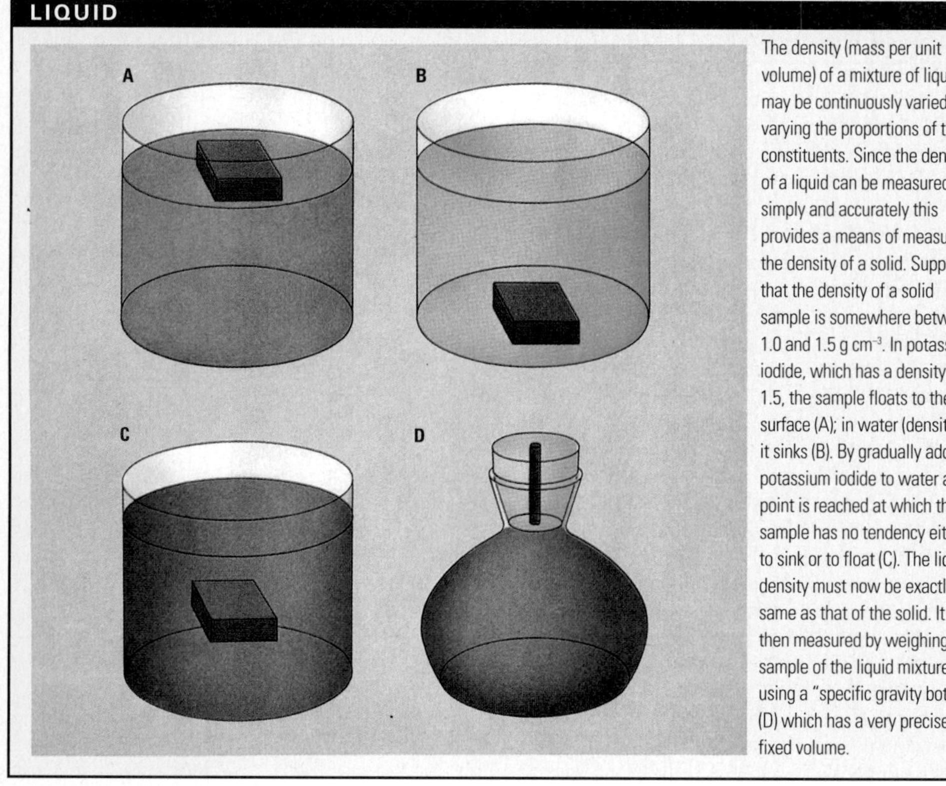

The density (mass per unit volume) of a mixture of liquids may be continuously varied by varying the proportions of the constituents. Since the density of a liquid can be measured simply and accurately this provides a means of measuring the density of a solid. Suppose that the density of a solid sample is somewhere between 1.0 and 1.5 g cm^{-3}. In potassium iodide, which has a density of 1.5, the sample floats to the surface (A); in water (density 1.0) it sinks (B). By gradually adding potassium iodide to water a point is reached at which the sample has no tendency either to sink or to float (C). The liquid's density must now be exactly the same as that of the solid. It is then measured by weighing a sample of the liquid mixture using a "specific gravity bottle" (D) which has a very precisely fixed volume.

locomotive Engine that moves under its own power, usually used on rails for pulling trucks or carriages. In 1804 British mining engineer Richard TREVITHICK built the first STEAM ENGINE locomotive for transporting heavy loads at an ironworks. His design was improved upon by George STEPHENSON, who built the *Rocket* in 1829. The first steam locomotive providing a railway service for passengers was George Stephenson's *Locomotion*, built in 1825. Electric-powered locomotives arrived in the late 1800s, and today obtain electricity from overhead cables or a rail running alongside the tracks. Diesel, diesel-electric and gas-turbine locomotives were introduced in the 1900s. Diesel locomotives burn oil to power the locomotive wheels directly. Diesel-electric locomotives use a diesel engine to generate the electricity with which the locomotive is powered. Gas-turbine locomotives use a TURBINE to create power.

locus In geometry, the path traced by a specified point when it moves according to a given formula or condition. For example, a circle is the locus of a point in a plane moving in such a way that its distance from a fixed point (the centre) is constant.

lode ORE formation consisting of a closely spaced series of veins, usually in stratified layers. Often several different minerals occur along the same lode. Lodes were probably formed as a result of liquids or gases, heated by volcanic activity, forcing their way through existing rocks, and then cooling and solidifying.

lodestone *See* MAGNETITE

Lodge, Sir Oliver Joseph (1851–1940) British physicist and author, the first principal of Birmingham University. Lodge investigated ELECTROMAGNETS and LIGHTNING, and perfected and named the **coherer**, the radio-wave detector used in the early days of wireless telegraphy. He supported the belief that the Sun might be a source of RADIO WAVES. He also wrote several books on psychical research and spiritualism.

Loeb, Jacques (1859–1924) US biologist, b. Germany, who studied the chemical processes of living organisms. He stressed the importance of physical and chemical laws in understanding life phenomena in both animals and plants. He achieved popular notoriety for experiments on artificial PARTHENOGENESIS in animals.

loess Fine-grained clay, a sedimentary material made up of rock fragments. It is earthy, porous and crumbly, usually yellowish or brown in colour. It consists of very small angular particles of mainly quartz and calcite from glaciated areas, blown by the wind and often built up into thick layers. The largest expanse is in N China in the valley of the River Huang He, where dust from the Gobi Desert has accumulated. The Huang is sometimes called the Yellow River because of all the loess it transports.

Loewi, Otto (1873–1961) German pharmacologist and medical researcher. He shared the 1936 Nobel Prize for physiology or medicine with Henry DALE for discovering that a chemical substance (later shown by Dale to be acetylcholine) mediated the transmission of NERVE IMPULSES from one NEURON to another.

Löffler, Friedrich August Johannes (1852–1915) German bacteriologist, best known for his discovery (in association with Edwin KLEBS) of the organism that causes DIPHTHERIA, *Corynebacterium diphtheriae* (Klebs-Löffler bacillus). He also made several discoveries in veterinary medicine, such as the cause of swine fever and plague, the causative organism of glanders, and the fact that foot-and-mouth disease is caused by a virus.

logarithm Aid to calculation devised by John NAPIER in 1614 and developed by the English mathematician Henry Briggs (1561–1631). The logarithm of a number (n) is the power (x) to which a base (b) must be raised to equal n, i.e. if $b^x = n$. The logarithm "of n in base b" is written $\log_b n = x$. The most widely used bases are base 10 and the so-called natural logarithm base e (2.71828...). Logarithms to the base 2 are used in COMPUTER science and information theory. A related notion is that of the antilogarithm: n may be called the antilogarithm of x in base b. Thus 100 is the antilogarithm of 2 in base 10.

LIQUID CRYSTAL

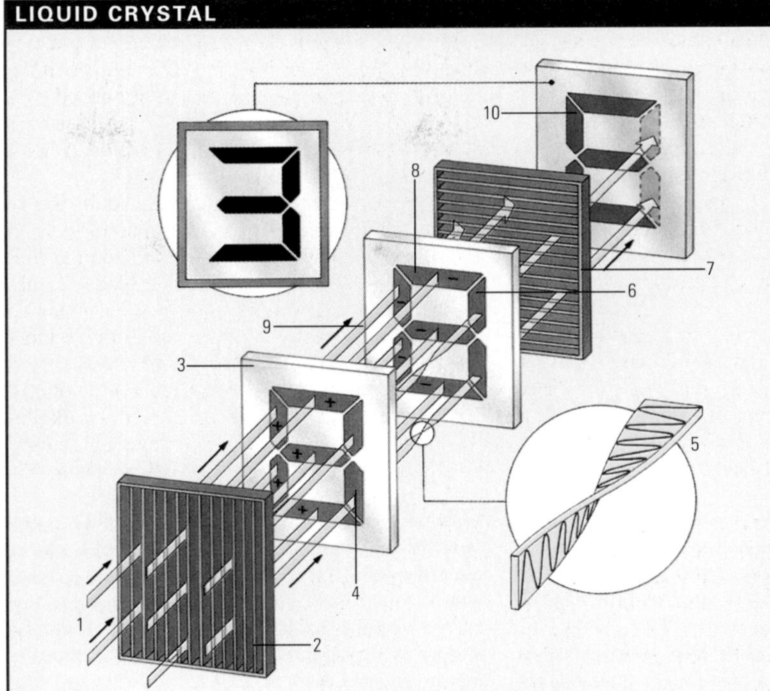

Liquid-crystal displays use the property of liquid crystals to twist the polarization of light to produce numbers or symbols. Incoming light (1) is first regimented in one plane by a polarizer (2) before it passes through the first of two plates of glass (3) on which are fixed electrodes (4). Seven electrodes are needed to represent Arabic numerals. The liquid crystal is between the two plates of glass and twists light (5) passing through uncharged electrodes (6). This light can pass through the second polarizer (7) and can be seen. The light passing through the charged electrodes (8) is not twisted (9), is blocked by the polarizer, and cannot be seen forming the components of the number (10).

logarithmic function Mathematical function the value of which depends on the LOGARITHM, to a particular base, b, of the independent variable, x, and is thus given by $f(x) = \log_b x$. For natural logarithms, where the base is exponential e, the logarithmic function is the inverse of the EXPONENTIAL function.

logic, mathematical (symbolic logic) Analytical branch of logic that attempts to give the inferences of classical logic a more precise mathematical basis. It is concerned with statements, which may be true or false, and with the relationships between statements and the operations that can be made on them, including CONJUNCTION, DISJUNCTION and negation. The statements and their relationships and operations are reduced to a symbolic notation and are manipulated according to prescribed mathematical rules. Thus inferences can be analysed and validity checked.

logic circuit Network of conductors linking modules (such as amplifiers) and components (such as transistors, diodes, resistors and capacitors), used to route and process electronic signals (such as voltages) according to the rules of symbolic logic and BOOLEAN ALGEBRA. Also, in computing and formal mathematics, a logic circuit is a network of connected logical decisions or operations called GATES.

Lomonosov, Mikhail Vasilievich (1711–65) Russian scientist and poet who helped to found the University of Moscow (1755). Lomonosov was one of the first Russian scientists to anticipate the law of CONSERVATION of mass, atomic theory and a kinetic theory of heat. He formulated Russian classical literary theory in his *Letter on the Rules of Russian Versification* (1739).

lone pair In the molecule of a compound, two unbonded electrons associated with one of the component atoms. For example, the nitrogen atom in ammonia (NH_3) and the carbon atom in carbon monoxide (CO) each have a lone pair of electrons. These electrons can form bonds in COORDINATION COMPOUNDS, in which their molecules act as LIGANDS.

Long, Crawford Williamson (1815–78) US physician. In 1842 he began using ether as an ANAESTHETIC but did not publish or publicize his work until 1849, after William Morton and others had been credited with the discovery.

long bones In human anatomy, LIMB bones characterized by a long shaft terminating at both ends in swellings (**epiphyses**) which articulate with other bones. Long bones include the RADIUS and ULNA in the forearm, the HUMERUS in the upper arm, the FEMUR in the thigh and the TIBIA and FIBULA in the lower leg. *See also* HUMAN BODY

longitude Angular measurement around the Earth, usually in degrees E or W of an imaginary N–S line through Greenwich, England. This line through Greenwich is called the prime MERIDIAN. All N–S lines are called meridians, or lines of longitude. Map makers and navigators refer to a position on the Earth by quoting its LATITUDE and longitude. In astronomy, celestial longitude is a coordinate on the CELESTIAL SPHERE or on the surface of a celestial body. It is measured E from the vernal EQUINOX to the GREAT CIRCLE passing through the pole of the ECLIPTIC and the object to be measured. It is reckoned in degrees from 0 to 360°.

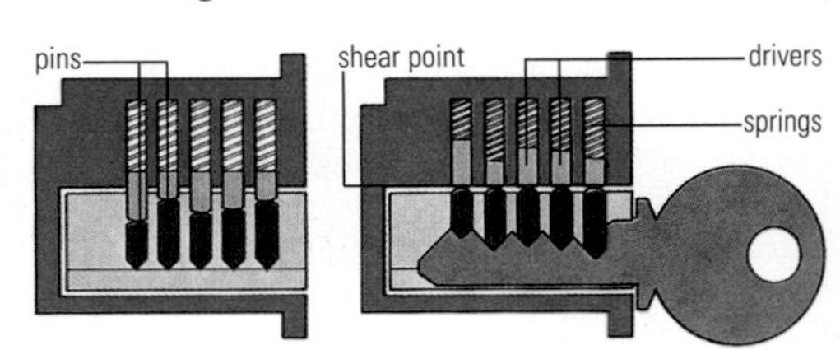

▲ **lock** When the key is inserted into the keyhold, the spring loaded drivers push the pins into the notches in the key. If the wrong key is inserted, the pins do not align correctly at the shear point and the key cannot be turned. When the right key is inserted the pins align and the key can be turned - thus unlocking the door.

► **Lorenz** Austrian zoologist Konrad Lorenz being followed by a group of ducklings. Lorenz studied medicine in Vienna before changing to zoology. His first discovery as a scientist occurred when he was given a one-day-old duckling that followed him around as if he were its mother – an instinct Lorenz called imprinting. In 1936 he met Dutch zoologist Niko Tinbergen. Together they founded a branch of animal behaviour called ethology, based on observing the instinctive behaviour of animals in the wild.

longitudinal wave Type of WAVE, such as a sound wave, in which the particles of the transmitting medium are displaced along the direction of energy propagation, that is, in the direction of wave motion.

long-playing record (LP) Disc, introduced in 1948, with a continuous groove on each side. This groove is cut as a spiral into vinyl plastic and in stereo records the walls of the groove are at right angles to each other. Each wall is a series of undulations which cause the stylus at the end of the arm of the turntable to vibrate. The vibrations are converted into two separate components of sound. Before the 33⅓ rpm discs, records were manufactured of shellac and made to revolve at 78 rpm. The LP was so named because each side took up to half an hour to play as opposed to a 78, which lasted for only about 5 minutes.

longshore drift Movement of sand and pebbles along a sea coast. The material is carried along the beach by the waves hitting the coast obliquely (swash) but is swept back at right-angles to the beach by the backwash. If the carrying power of the waves decreases the material may be deposited to form a SPIT. To combat the effect of longshore drift GROYNES are constructed. Usually wooden wall-like structures, they run perpendicular to the coast at regular intervals and so stop the EROSION of the coast. Another, increasingly used method to combat the effect is to replace the material in a process known as "beach nourishment".

long-sightedness *See* HYPERMETROPIA

loom Frame or set of frames on which threads are woven into cloth. The loom enables a set of threads, called the **weft**, to be passed over and under a set of lengthwise threads, called the **warp**. The simplest kind of loom is a single frame on which weaving is done by hand. Such looms have been used for more than 7,000 years. By the 1700s, weavers in Europe were using complex, hand-operated looms to produce many kinds of cloth. One of the most important developments came in 1785, when English clergyman Edmund CARTWRIGHT invented a loom powered by a steam engine to speed cloth production. However, it took another 30 years of development before power-operated looms became widespread. Meanwhile, French weaver Joseph JACQUARD had invented a hand loom that could weave intricate designs. These were determined by patterns of holes in PUNCHED CARDS. Changing the cards changed the pattern produced. Today's most advanced commercial looms are computer controlled and have mechanisms that thread the weft through the warp at speeds of *c.*100km/h (60mph).

loran (**lo**ng **ra**nge **n**avigation) Radio navigational system for guiding ships and aircraft. Pairs of TRANSMITTERS emit signal pulses that are picked up by an aircraft or ship's receiver. By measuring the difference in time between the signals reaching the receiver, the vessel's position can be plotted. The most widely used system is called loran C. The transmitters are about 970km (600mi) apart. The range of the system is 1,800–4,500km (1,100–2,800mi). The system has, in many cases, been superceded by the GLOBAL POSITIONING SYSTEM.

lordosis *See* CURVATURE OF THE SPINE

Lorentz, Hendrik Antoon (1853–1928) Dutch physicist and professor at Leiden. His early work was concerned with the theory of ELECTROMAGNETIC RADIATION devised by James Clerk MAXWELL. He tried to explain the fact that the speed of LIGHT appears the same to all observers by suggesting that moving objects contract in the direction of their movement. He derived equations which culminated in the LORENTZ-FITZGERALD CONTRACTION, a theory that was later relevant to Albert EINSTEIN's theory of RELATIVITY. Lorentz also worked on the ZEEMAN EFFECT, for which he and Pieter ZEEMAN were awarded the 1902 Nobel Prize for physics. *See also* FITZGERALD, GEORGE FRANCIS; LORENTZ TRANSFORMATION

Lorentz-Fitzgerald contraction (Lorentz contraction, Fitzgerald contraction) Contraction of a moving body in the direction of its motion. The theory was proposed independently by both Hendrik LORENTZ and George FITZGERALD. The theory held that any moving body is contracted in the direction of its motion by a factor $\sqrt{(1 - v^2/c^2)}$, where v is the velocity of the body and c is the velocity of light. It was proposed as an explanation for the negative result of the MICHELSON-MORLEY EXPERIMENT. Albert EINSTEIN used the theory as the mathematical basis of his special theory of RELATIVITY, but showed that it was an effect produced by differences in the position of an observer rather than a real deformation of matter.

Lorentz transformation Relation developed by Hendrik LORENTZ, connecting the space and time coordinates of an event as observed from two frames of reference, especially at relativistic velocities. It was shown by Albert EINSTEIN in 1905 to be a consequence of the special theory of RELATIVITY.

Lorenz, Konrad (1903–89) Austrian pioneer ethologist. Unlike psychologists, who studied animal behaviour in laboratories, Lorenz studied animals in their natural habitats. He observed that instinct played a major role in animal behaviour – for example in IMPRINTING, by which an animal may learn to identify its parents. Some of his views are expressed in his book *On Aggression* (1966). In 1973 he shared, with Nikolaas TINBERGEN and Karl Von FRISCH, the Nobel Prize for physiology or medicine. *See also* ETHOLOGY

loudness Magnitude of the sensation produced when the human EAR responds to a SOUND. There is no simple relationship between loudness and the intensity of the sound; the response of the ear also depends to a certain extent on the frequency.

loudspeaker Device for converting changing

LUBRICANT

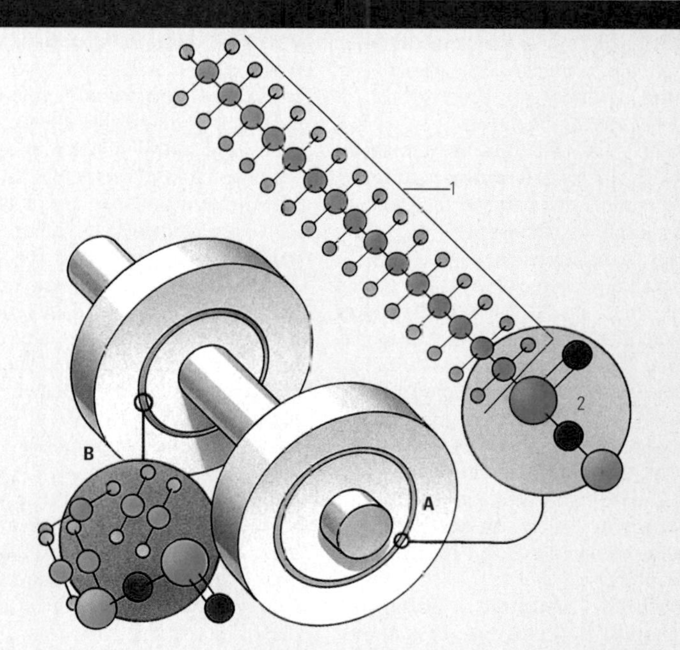

Lubricants are used to reduce wear between surfaces in rubbing contact, such as bearings. A film of oil between the bearing surfaces prevents metal contacting metal and reduces friction (A). Refined mineral oils have small quantities of fatty acids (1) added as boundary lubricants; the reactive carboxyl group at one end of the molecule (2) attracts it to the metal surface so that it stands on end. Oils for different uses usually have different viscosities, but multigrade oils contain additives which allow them to withstand the range of temperatures and pressures encountered as bearings change speed. However, with time, oil molecules degrade, break down or undergo chemical change and become less effective and should be replaced or replenished (B).

electric currents into SOUND. The most common type has a moving coil attached to a stiff paper cone and suspended in a strong magnetic field. By ELECTROMAGNETIC INDUCTION, the changing currents in the coil cause the cone to vibrate at the frequency of the currents, thus creating sound waves. Many modern loudspeakers, which are used in radio, hi-fi and television sets, give faithful sound reproduction between 80 and 20,000 HERTZ.

Lovell, James Arthur, Jr (1928–) US astronaut. His first space flight was in Gemini 7 (1965, part of the GEMINI PROJECT), which made the first successful space rendezvous (with Gemini 6). He was a member of the crew of APOLLO 8 (1968), the first manned mission around the Moon, and commander of the unsuccessful Apollo 13 mission (1970).

Lovell, Sir Alfred Charles Bernard (1913–) British astronomer. From 1951–81 he was director of the Jodrell Bank experimental station for radio ASTRONOMY near Manchester, England, and oversaw the construction there of the world's first large steerable RADIO TELESCOPE. He used the telescope to pioneer the exploration of radio emission from space.

low *See* CYCLONE

Lowell, Percival (1855–1916) US astronomer. In 1894 he built an observatory (now the Lowell Observatory) at Flagstaff, Arizona. From there he observed Mars, producing intricate maps of the so-called canals, which he ascribed to the activities of intelligent beings. He went on to study the orbits of Uranus and Neptune, and calculated that their orbital irregularities were caused by an undiscovered Planet X. His predictions led to the discovery of Pluto in 1930. It was at Lowell's urging that the US astronomer Vesto Slipher (1875–1969) carried out the observations that led to the discovery of galactic RED SHIFTS in 1916.

LSD (lysergic acid diethylamide) Hallucinogenic drug, causing changes in mental state, sensory confusion and behavioural changes, when the drug blocks the action of SEROTONIN in the brain. First synthesized in the 1940s, LSD was made illegal in the UK and USA in the mid-1960s.

lubricant Oil, grease or other substances placed between moving parts to reduce friction and dissipate heat. Most lubricants are now derived from petroleum. Solid lubricants are usually graphite or molybdenum disulphide; synthetic silicones are used where high temperatures are involved.

luciferin Proteins found in animal tissue that produce light when oxidized in the presence of the ENZYME luciferase. It is found in the light-generating organs of glow-worms and fire-flies. *See also* BIOLUMINESCENCE

Ludwig, Karl Friedrich Wilhelm (1816–95) German physiologist. He studied the cardiovascular system and invented the kymograph (1847), which records changes in arterial BLOOD PRESSURE. He also invented a device to measure arterial and venous blood flow, and investigated the functions of oxygen in the blood.

lumen (symbol lm) SI unit measuring the amount of visible light emitted in one second in a unit solid angle (one steradian) from a source of unit intensity (one CANDELA).

luminescence *See* FLUORESCENCE; PHOSPHORESCENCE

luminosity Absolute brightness of a STAR, given by the amount of energy radiated from its entire surface per second. It is expressed in watts (joules per second), or in terms of the Sun's luminosity. **Bolometric** luminosity is a measure of the star's total energy output, at all wavelengths. Absolute **magnitude** is an indication of luminosity at visual wavelengths. *See also* SPECTRAL CLASSIFICATION

LUNGS

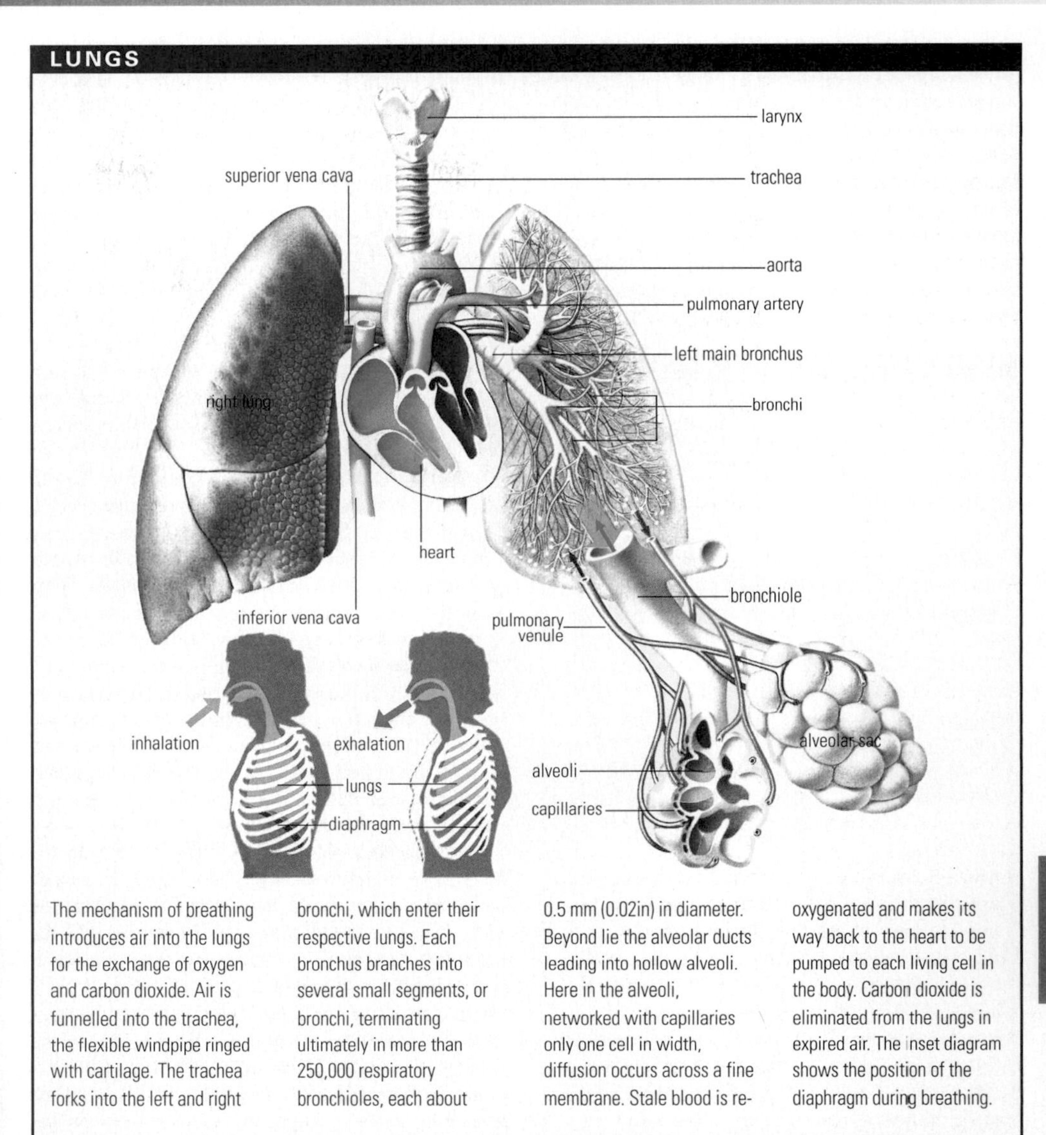

The mechanism of breathing introduces air into the lungs for the exchange of oxygen and carbon dioxide. Air is funnelled into the trachea, the flexible windpipe ringed with cartilage. The trachea forks into the left and right bronchi, which enter their respective lungs. Each bronchus branches into several small segments, or bronchi, terminating ultimately in more than 250,000 respiratory bronchioles, each about 0.5 mm (0.02in) in diameter. Beyond lie the alveolar ducts leading into hollow alveoli. Here in the alveoli, networked with capillaries only one cell in width, diffusion occurs across a fine membrane. Stale blood is re-oxygenated and makes its way back to the heart to be pumped to each living cell in the body. Carbon dioxide is eliminated from the lungs in expired air. The inset diagram shows the position of the diaphragm during breathing.

lunar eclipse *See* ECLIPSE

lunar module Part of the Saturn rocket-launched spacecraft, devised by NASA in the USA, and used in the APOLLO PROJECT, which detached itself from the command/service module when in lunar orbit and landed, with its passengers, on the Moon's surface. This happened for the first time on 21 July 1969. The upper part (ascent stage) of the lunar module later took off from the Moon using its thrust jet (with the descent stage acting as launch pad) and rejoined the command/service module.

lunar probe Unmanned, rocket-launched space vehicle aimed at either reaching the Moon's surface or going into orbit around it. The first were launched by the Soviet Union (1959) and USA (1964) and in 1966 both nations landed probes to explore the Moon's surface. An early Russian probe was the first to send photographs back to the Earth of the dark side of the Moon.

lunar roving vehicle (Moon buggy) Electrically powered, four-wheeled passenger vehicle first used in August 1971 by astronauts of the APOLLO PROJECT. It was designed to cope with the reduced gravity and rocky, uneven surface of the Moon, enabling the astronauts to explore local regions of the Moon and carry lunar rock samples back to the LUNAR MODULE for passage back to Earth.

lungs Organs of the RESPIRATORY SYSTEM of vertebrates, in which the exchange of gases between air and blood takes place. They are located in a cavity (the **pleural cavity**) within the rib cage. This cavity is lined by two sheets of TISSUE (the PLEURA), one coating the lungs and the other lining the walls of the thorax. Between the pleura is a fluid that cushions the lungs and prevents friction. Light and spongy, lung tissue is composed of tiny air sacs, called ALVEOLI, which are served by networks of fine CAPILLARIES. *See also* GAS EXCHANGE; VENTILATION

Lunokhod Name of two unmanned Russian scientific lunar surface craft. Luna 17, launched in October 1970, carried Lunokhod 1, which crawled over the surface of the Moon collecting scientific data. Lunokhod 2 was carried by Luna 21, launched in January 1973.

Luría, Salvador Edward (1912–91) US biologist, b. Italy. He shared the 1969 Nobel Prize for physiology or medicine with Max DELBRÜCK and Alfred HERSHEY for contributing to the knowledge of the growth, reproduction and mutation of VIRUSES.

luteinizing hormone (LH) Hormone produced by the anterior (front) lobe of the PITUITARY GLAND, just below the brain. In women it stimulates OVULATION by the development of the CORPUS LUTEUM and secretion of PROGESTERONE. In men it promotes the secretion of TESTOSTERONE from the TESTES.

lutetium (symbol Lu) Metallic element of the LANTHANIDE SERIES, first isolated in 1906, together with YTTERBIUM. Its chief ore is monazite. The element is used as a CATALYST, but has no other commercial uses. Properties: at.no. 71; r.a.m. 174.97; r.d. 9.842 (25°C); m.p. 1,656°C (3,013°F); b.p. 3,315°C (5,999°F); most common isotope ^{175}Lu (97.41%).

lux (symbol lx) SI unit of illumination, equal to one LUMEN per square metre.

Lwoff, André Michel (1902–94) French microbiologist. He shared the 1965 Nobel Prize for physiology or medicine with François JACOB and Jacques MONOD for work on the genetics of VIRUSES. Lwoff studied the means by which these destroy the bacteria they infect.

Lycopodophyta Phylum of about 1,000 species of VASCULAR PLANTS, which includes the CLUB MOSSES, spike mosses and quillworts. Related to FERNS, lycopods have branching underground stems (RHIZOMES) and upright shoots supported by roots. Some species are EPIPHYTES.

LYMPH NODE

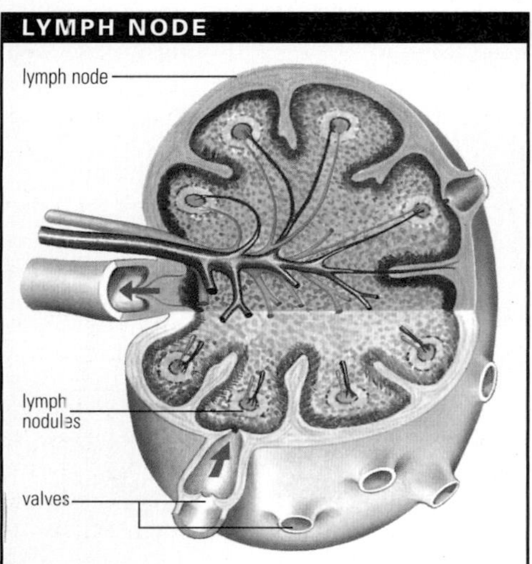

The lymph system is a network of lymphatic vessels which collects tissue fluid (the lymph) and conducts it back to the bloodstream. In the process it transports nutrients from blood to cells and cell wastes back into capillaries. Lymph drains through the system but the lymphatics possess valves to prevent backflow. Lymphatic nodes are scattered along the lymph vessels but particularly in the neck, armpits and groin. In the tissue around the nodes microorganisms are destroyed by macrophage cells, while antibody-synthesizing white blood cells, the lymphocytes, are produced by the lymph nodules.

lye Strongly alkaline solution. Originally it applied solely to POTASSIUM CARBONATE (potash), but was extended to include SODIUM HYDROXIDE and POTASSIUM HYDROXIDE. Household lye is usually sodium hydroxide.

Lyell, Sir Charles (1797–1875) British geologist. He was influential in shaping 19th-century ideas about science and wrote the popular books *Principles of Geology* (1830–33), *Elements of Geology* (1838) and *The Geological Evidence of the Antiquity of Man* (1863).

lymph Clear, slightly yellowish fluid derived from BLOOD and similar in composition to PLASMA. Circulating in the LYMPHATIC SYSTEM, it conveys LEUCOCYTES (white blood cells) and some nutrients to the tissues.

lymphatic system System of connecting vessels and organs in vertebrates that transports LYMPH through the body. Lymph flows into tiny and delicate lymph capillaries and from these into lymph vessels, or lymphatics. These extend throughout the body, joining to form larger vessels leading to LYMPH NODES that collect lymph, storing some of the LEUCOCYTES (white blood cells). Lymph nodes empty into large vessels, linking up into lymph ducts that empty back into the CIRCULATORY SYSTEM. The lymphatic system plays a major role in the body's immune system. At the lymph nodes, MACROPHAGES (large leucocytes) trapped in a network of fibres remove foreign particles, including bacteria, from the lymph. The lymph nodes also supply LYMPHOCYTES to the blood. The lymph vessels provide a route for certain nutrients to enter the bloodstream. Fat droplets that are too large to enter the capillaries of the VILLI in the small intestine pass into the lymph vessels for transport to the CIRCULATORY SYSTEM.

lymph node (lymph gland) One of many masses of tissue occurring along the major vessels of the LYMPHATIC SYSTEM. They are filters and reservoirs which collect harmful material, notably bacteria and other disease organisms, and often become swollen when the body is infected. Clusters of lymph nodes are found in the throat, neck, armpits and groin.

lymphocyte Type of LEUCOCYTE (white blood cell) found in vertebrates. Produced in the bone MARROW, it is mostly found in the LYMPH and blood and around infected sites. In human beings lymphocytes form about 25% of leucocytes and play an important role in combating disease. There are two main kinds: **B-lymphocytes**, responsible for producing antibodies, and **T-lymphocytes**, which have various roles in maintaining the IMMUNE SYSTEM.

lymphoma Any malignant growth of the LYMPH NODES. HODGKIN'S DISEASE is a lymphoma.

Lynen, Feodor Felix Konrad (1911–79) German biochemist who shared the 1964 Nobel Prize for physiology and medicine with Konrad BLOCH for research into the metabolism of FATTY ACIDS and CHOLESTEROL. He helped to ascertain the mechanism of fatty-acid breakdown and explain the pathway of the biosynthesis of several compounds, including natural rubber and cholesterol. He was also the first to isolate acetyl COENZYME A, important for the process of RESPIRATION.

Lysenko, Trofim Denisovich (1898–1976) Russian agronomist and geneticist. He expanded the theory of LAMARCK to include his own ideas of plant genetics (**Lysenkoism**). Lysenko promised the Soviet government vast increases in crop yields through the application of his theories. He enjoyed official sanction under Stalin, but in the late 1950s his influence waned.

lysergic acid diethylamide *See* LSD

lysine ($H_2N(CH_2)_4CH(NH_2)COOH$) Soluble, essential AMINO ACID found in proteins.

lysis Destruction of a CELL. A living cell may be destroyed when its outer cell wall is dissolved by LYSOSOMES or when it is digested in phagocytosis by PHAGOCYTES – a normal reaction against invading bacterial cells.

lysosome Small membranous sac that occurs within the cytoplasm of an animal CELL. It contains enzymes that control the breakdown of foreign substances entering the cell, the digestion of the contents of food VACUOLES and the breakdown of the cell itself after it dies.

maar Coneless, volcanic crater that is formed by a single explosive eruption, and not accompanied by a flow of LAVA. A crater ring surrounds the hole. *See also* VOLCANO

McAdam, John Loudon (1756–1836) Scottish engineer who is attributed with inventing the macadam road surface. He proposed that roads should be raised above the surrounding ground, with a base of large stones covered with smaller stones and bound together with fine gravel. McAdam applied these ideas when he was appointed surveyor general of Bristol's roads in 1815. They proved successful, and his views on road-making were generally adopted after a parliamentary inquiry in 1823. McAdam was made surveyor general of roads in Britain in 1827.

McClung, Clarence Erwin (1870–1946) US zoologist who discovered the mechanism of chromosomal SEX DETERMINATION. He postulated that CHROMOSOMES determine sex and that the X-chromosome is the crucial element. This stimulated further research and by 1905 Edmund WILSON had developed the theory, now generally accepted, that both X- and Y-chromosomes form the basis of sex determination.

McCormick, Cyrus Hall (1809–84) US inventor of a mechanical reaper in 1831. It was introduced into England in 1851. Large-scale manufacture of the reaper, together with McCormick's invention of the twine binder and side-rake, revolutionized the harvesting of cereal crops.

Mach, Ernst (1838–1916) Austrian physicist and philosopher. Interested in the physiology and psychology of the senses, Mach believed that physical phenomena should be explained only by data perceived by the senses. He did research on rapid air flow, and the MACH NUMBER was named after him.

machine Device that modifies or transmits a force in order to do useful work. In a basic, or simple, machine, a force (**effort**) opposes a larger force (**load**). The ratio of the load (output force) to the effort (input force) is the machine's FORCE RATIO, formerly called mechanical advantage. The ratio of the distance moved by the load to the distance moved by the effort is the DISTANCE RATIO, formerly known as the velocity ratio. The ratio of the work done by the machine to that put into it is the EFFICIENCY, usually expressed as a percentage. The three primary machines are the INCLINED PLANE (which includes the screw and the wedge), the LEVER and the wheel (which includes the PULLEY and the WHEEL AND AXLE).

machine code In computing, instructions that the CENTRAL PROCESSING UNIT (CPU) of a computer can execute directly, without the need for translation. Machine-code statements are written in a binary-coded (low-level) COMPUTER LANGUAGE. Programmers usually write COMPUTER PROGRAMS in a HIGH-LEVEL LANGUAGE (such as FORTRAN or C), which a COMPILER program then translates into machine code for execution. *See also* BINARY SYSTEM

machine tools Power-driven machines for cutting and shaping metal and other materials. Shaping may be accomplished in several ways. These include shearing, pressing, rolling, and cutting away excess material using lathes, shapers, planers, drills, milling machines, grinders and saws. Other techniques include the use of machines that use electrical or chemical processes to shape the material. Advanced machine-tool processes include cutting by means of LASER beams, high-pressure water jets, streams of PLASMA (ionized gas), and ULTRASONICS. Today, computers control many cutting and shaping processes carried out by machine tools and ROBOTS.

Mach number Ratio of the speed of a body or FLUID (gas or liquid) to the local speed of sound. Mach 1 therefore refers to the local speed of SOUND. An aircraft flying at below Mach 1 is said to be subsonic, or flying at less than the speed of sound. SUPERSONIC FLIGHT means flying at speeds above Mach 1. Mach numbers are named after Ernst MACH, who investigated supersonic speeds and shock waves.

McIndoe, Sir Archibald Hector (1900–60) New Zealand plastic surgeon. One of the pioneers in this field, he founded a small hospital for plastic surgery in East Grinstead, Surrey, which later became internationally famous. As one of the foremost plastic surgeons of his generation, McIndoe was much in demand during World War 2 and was appointed consultant in plastic surgery to the Royal Air Force.

Macintosh, Charles (1766–1843) Scottish chemist and inventor after whom the mackintosh coat is named. In 1823 he invented a method of making waterproof cloth by joining two pieces of fabric using a solution of rubber in coal-tar naphtha. However, it was not until 1839, with the advent of vulcanized rubber which withstood temperature changes, that the mackintosh became a practical garment.

Macleod, John James Rickard (1876–1935) Scottish physiologist who shared the 1923 Nobel Prize for physiology or medicine with Sir Frederick Grant BANTING for the discovery of INSULIN and studies of its use in treating DIABETES. The actual discovery was made by Banting and Charles BEST working in Macleod's laboratory at the University of Toronto. Macleod's works include *Diabetes* (1913) and *Carbohydrate Metabolism and Insulin* (1926).

McLuhan, (Herbert) Marshall (1911–80) Canadian academic and expert on communications. His view that the forms in which people receive information (such as, television, radio and computers) are more important than the messages themselves was presented in his books, *The Mechanical Bride: Folklore of Industrial Man* (1951), *Understanding Media* (1964) and *The Medium is the Message* (1967).

McMillan, Edwin Mattison (1907–91) US physicist who shared the 1951 Nobel Prize for chemistry with Glenn SEABORG for the discovery of NEPTUNIUM and other TRANSURANIC ELEMENTS. McMillan worked on the MANHATTAN PROJECT developing the atom bomb at Los Alamos, New Mexico, then on the CYCLOTRON (a type of particle ACCELERATOR) with Ernest LAWRENCE at the University of California at Berkeley. He developed the SYNCHROCYCLOTRON, an advanced type of accelerator, for which he shared the 1973 Atoms for Peace Prize with the Soviet physicist Vladimir Veksler (1907–66). *See also* NUCLEAR WEAPON

Macmillan, Kirkpatrick (1813–78) Scottish blacksmith who invented the first pedal-operated bicycle in 1839. Cranks attached to the rear axle of the machine were connected to pedals near the front. Kirkpatrick's invention became popular in 1846, when it was manufactured by a Scottish cooper called Gavin Dalzell.

macrocephaly Excessive size of the head. It is often caused by HYDROCEPHALUS and is associated with mental subnormality.

macromolecule MOLECULE up to 1,000 times greater in diameter than the molecules of most substances. Many proteins, nucleic acids, plastics, resins, rubbers and natural and synthetic fibres are made up of such giant units, each of which contains thousands of atoms.

macronutrient Element needed in relatively large amounts by growing plants. Macronutrients include carbon, hydrogen, oxygen and nitrogen, as well as calcium, magnesium, phosphorus, potassium and sulphur. *See also* MICRONUTRIENT

macrophage Large LEUCOCYTE (white blood cell) found mainly in the liver, spleen and lymph nodes. It engulfs foreign particles and microorganisms by phagocytosis. Working together with other LYMPHOCYTES, it forms part of the body's IMMUNE SYSTEM.

mad cow disease *See* BOVINE SPONGIFORM ENCEPHALOPATHY (BSE)

Magellanic clouds Two nearest GALAXIES to our own, visible to the naked eye like isolated patches of the Milky Way in the S sky. The Large Magellanic Cloud (LMC) lies in the constellations Dorado and Mensa; the Small Magellanic Cloud (SMC) lies in the constellation Tucana. Their distances from Earth are *c.*169,000 light-years and 190,000 light-years, respectively. Although usually classified as irregular galaxies, both clouds are somewhat bar-shaped with faint outer structures that can be interpreted as rudimentary spiral arms.

maggot Name commonly given to the legless LARVA of a fly. It is primarily used to describe those larvae that infest food and waste material; others are generally called grubs or caterpillars.

magic number Number of PROTONS or NEUTRONS in a very stable atomic NUCLEUS. For both particles, the magic numbers are 2, 8, 20, 28, 50 and 82 (corresponding, for protons, to the elements helium, oxygen, calcium, nickel, tin and lead). Proton number 114 is also a magic number, as are 126 and 184 for neutrons. A shell model of the nucleus (analogous to electron shells in the ATOM) has been proposed to account for the extra stability conferred by the magic numbers.

magic square Any square MATRIX divided into cells and filled with numbers or letters in ways once thought to have special magical significance. The most familiar lettered square is the SATOR square, composed of the words SATOR, AREPO, TENET, OPERA, and ROTAS. Arranged both vertically and horizontally the words read the same and form a cross through the middle that reads TENET. In arithmetical magic squares, the numbers are generally arranged so that each column, every row, and the two main diagonals add to the same constant sum.

magma Molten rock beneath the surface of the Earth which forms IGNEOUS ROCKS when it solidifies. Below the surface, cooling is slow and large crystals form as the rock solidifies. Large masses of magma are in BATHOLITHS, and thin seams of magma create SILLS and DYKES. If magma reaches the surface, it flows out as LAVA. The different igneous rocks formed by magma vary according to their chemical content as well as the depth of formation. *See also* EXTRUSIVE ROCK

magnesia *See* MAGNESIUM OXIDE

magnesite Carbonate mineral, magnesium carbonate ($MgCO_3$). It is an alteration product of CALCITE or DOLOMITE in SEDIMENTARY ROCKS and is found in microcrystalline masses and crystals like ICELAND SPAR. Magnesite is white, colourless or of light tint, with a glassy to dull lustre. Hardness 3.5–5; r.d. 3.1.

magnesium (symbol Mg) Silver-white, metallic

M

element, one of the ALKALINE-EARTH METALS. The eighth most abundant element in the Earth's crust, it was first isolated (1808) by Sir Humphry DAVY. It is necessary in human and animal diets. Magnesium is always found combined, and its chief sources are MAGNESITE and DOLOMITE. Magnesium burns in air with an intense white flame, and is used in optical instruments, flashbulbs, fireworks, flares and incendiaries. Magnesium alloys are light and used in aircraft fuselages, jet engines, missiles and rockets. Chemically, the element is similar to CALCIUM. Hydrated magnesium sulphate is called Epsom salts. Properties: at.no. 12; r.a.m. 24.312; r.d. 1.738; m.p. 648.8°C (1,200°F); b.p. 1,090°C (1,994°F); most common isotope ^{24}Mg (78.7%).

magnesium oxide (magnesia, MgO) White, neutral, stable powder formed when magnesium is burned in oxygen. It is used industrially in firebrick, and medicinally in stomach powders. Magnesium carbonate ($MgCO_3$), found as the mineral MAGNESITE and also used as an antacid, is often also called magnesia.

magnet Object that produces a MAGNETIC FIELD, an area around the magnet in which other magnetizable objects experience a force. Lodestones, which are naturally magnetic, were used as early magnets, and strong magnetic materials were later recognized as containing either iron, cobalt or nickel. A typical **permanent** magnet is a straight or horseshoe-shaped magnetized iron bar with the ends called the north and south magnetic poles. The Earth is a giant magnet, its magnetic lines of force being detectable at all latitudes. An ELECTROMAGNET is much stronger than a permanent one and is used for raising heavy steel weights and scrap. A **superconducting** magnet, the strongest of all, is made from special alloys cooled to very low temperatures.

magnetic anomaly Small variations in the Earth's MAGNETIC FIELD caused by iron objects or deposits.

magnetic bottle Configuration of MAGNETIC FIELDS used to contain the PLASMA in a fusion reactor or experimental device. It applies particularly to a linear configuration in which the ends are stoppered with magnetic mirrors, where a very high magnetic field causes the plasma to be reflected.

magnetic declination (magnetic deviation, symbol *D*) Angle between the direction of magnetic north (as indicated by a compass) and the direction of true north. In any one place, magnetic declination varies slowly with time as the position of magnetic north gradually changes. There are also small annual variations. Angles to the east of true north are taken as positive; angles to the west are negative.

MAGNETIC RECORDING

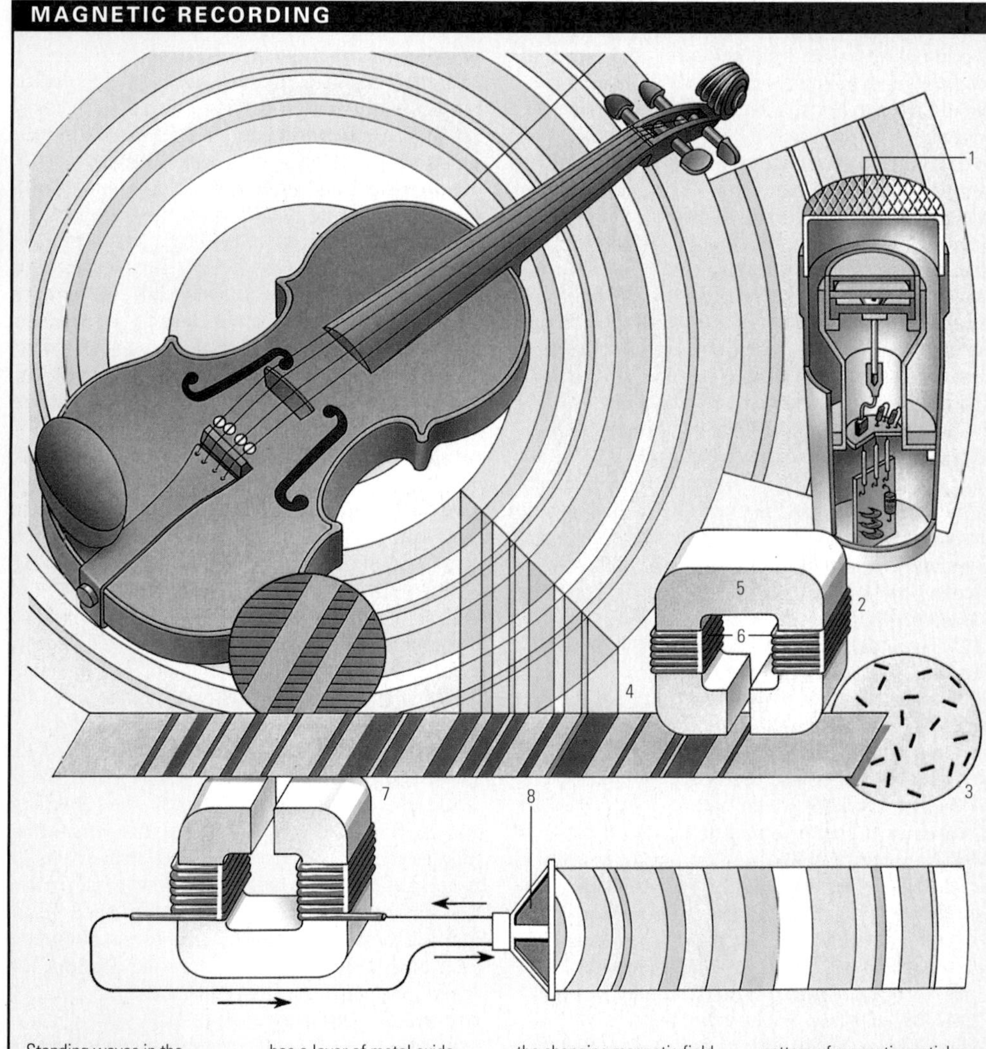

Standing waves in the molecules in the air (sounds) vibrate a metal diaphragm in the microphone (1) creating a tiny electrical signal that can be amplified and then recorded (2). Magnetic tape has a layer of metal oxide stuck to a polyester base. The metal molecules are randomly aligned (3) until forced to move into a regular pattern, corresponding to the sound being recorded (4) by the changing magnetic field in the recording head (5). This shift in polarity is caused by the amplified signal from the microphone passing through a coil of wire (6) around the iron head. The worrying pattern of magnetic particles on the tape will include a magnetic field in the playback head when the tape is played (7). The resulting electric signal is converted to sound by loudspeakers.

magnetic disk Plastic disk coated with magnetic material and used for storing COMPUTER PROGRAMS and DATA (information) as a series of magnetic spots. Most computers contain a HARD DISK unit for general storage. There is also a unit for inserting floppy disks. These lower-capacity, flimsy magnetic disks are protected by a plastic cover. Hard magnetic disks can store larger amounts of data and come in cartridges that slot into a special drive unit. *See also* CD-ROM

magnetic disturbance Change in the shape of the MAGNETOSPHERE. The most commonly observed phenomena that result from this are the polar AURORAS and interference in radio and television communications.

magnetic equator That part of the Earth's MAGNETIC FIELD where a finely balanced, magnetic needle will remain absolutely horizontal. Because of the Earth's magnetic asymmetry, the magnetic equator does not coincide with the geographical EQUATOR.

magnetic field Region surrounding a MAGNET, or a conductor through which a current is flowing, in which magnetic effects, such as the deflection of a compass needle, can be detected. A magnetic field can be represented by a set of lines of force (flux lines) spreading out from the poles of a magnet or running around a current-carrying conductor. These lines of force can be seen if iron filings are sprinkled on to a sheet of paper below which a magnet is placed. The filings line up along the lines of force, the density of the lines being greatest where the field is strongest. The direction of a magnetic field is the direction a tiny magnet takes when placed in the field. **Magnetic poles** are the field regions in which MAGNETISM appears to be concentrated. If a bar magnet is suspended to swing freely in the horizontal plane, one pole will point north; this is called the north-seeking or **north pole**. The other pole, the south-seeking or **south pole**, will point south. Unlike poles attract each other; like poles repel each other. **Earth's magnetic poles** are the ends of the huge "magnet" that is Earth.

magnetic field reversal Reversal of polarity whereby the Earth's N magnetic pole becomes the S and vice versa. Analyses of the magnetic direction of land and ocean basaltic lavas and seafloor sediments have shown that the Earth's main MAGNETIC FIELD has undergone frequent and rapid reversals. The field has changed many times in the past four million years. *See also* PALAEOMAGNETISM

magnetic flux (symbol Φ) Measure of the strength and extent of a MAGNETIC FIELD. The flux through an area A at right angles to a uniform magnetic field is $\Phi = \mu HA$, where μ is the magnetic PERMEABILITY of the medium and *H* is the magnetic field intensity. Magnetic flux density is the flux per unit area (symbol *B*), which equals μH. Change of magnetic flux through an electric circuit induces an ELECTROMOTIVE FORCE.

magnetic moment Measure of the strength of a permanent magnet or current-carrying coil. It is the maximum turning force (TORQUE) exerted on a magnet, coil or moving electric charge in a MAGNETIC FIELD divided by the strength of the field. Charged particles and atomic nuclei also have magnetic moments.

magnetic permeability *See* PERMEABILITY

magnetic pole *See* MAGNETIC FIELD

magnetic recording Formation of a record of sounds on a wire or tape by means of a pattern of magnetization. In a tape recorder, ferromagnetic tape (*see* MAGNETIC TAPE) is fed past an electromagnet that is energized by the amplified currents produced by a MICROPHONE. By ELECTROMAGNETIC INDUCTION, variations in magnetization (from the

MAGNETISM

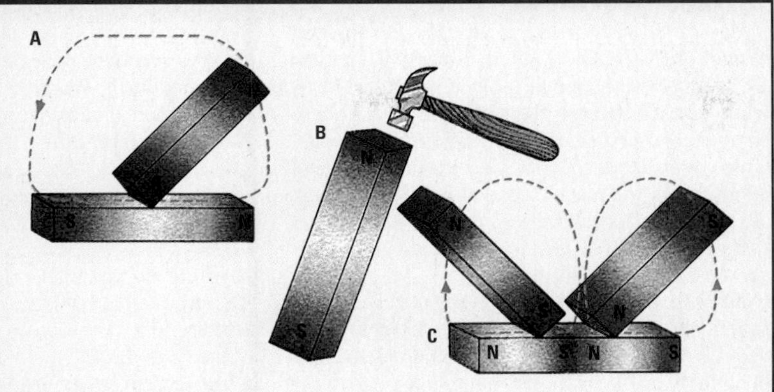

A simple way to magnetize materials, such as iron and its alloys is to stroke them with a bar magnet (A), the nearness of which, coupled with its movement, tends to align the magnetic domains within the material. In such a way they reinforce each other rather than keeping their normally random arrangements. The south-seeking ends of the domain try to follow the movement of the original magnet's north pole so that the right-hand side of the new magnet becomes a south pole. The domains lie with their south poles to the right, and so their north poles, are to the left. Another way to magnetize a bar of suitable material is simply to hit it (B). The domains receive a mechanical shock and the Earth's field tends to align them with itself. Adapting the technique shown in (A) it is possible to make a bar magnet of suitable material using two magnets (C). In this case, the right-hand side acquires a south pole, and the left-hand side a north pole.

oscillating current produced by the sound) are induced in the particles of ferromagnetic material on the tape. When played back, the tape is fed past a similar electromagnet, which converts the patterns into sound that is in turn fed to an AMPLIFIER and LOUDSPEAKER.

magnetic resonance imaging (MRI) Medical scanning system that utilizes the phenomenon of NUCLEAR MAGNETIC RESONANCE (NMR) to produce three-dimensional images of the brain, spinal cord and other soft tissues in the body. The benefit of this scanning system is that it does not expose the patient to potentially harmful X-ray radiation. The scanner's magnet causes the nuclei within the atoms of the patient's body to line themselves up in one direction. A brief radio pulse is then beamed at the nuclei, causing them to change their orientation. As they realign themselves to the magnet, they give off weak radio signals that can be recorded and converted electronically into images.

magnetic storm (geomagnetic storm) Disturbance in the Earth's MAGNETIC FIELD. Since the field encompasses all the Earth, the effects of such a storm are global: AURORAS are seen, both in areas where such displays are normal and in others as well; and radio signals are disturbed.

magnetic tape Medium for recording electrical signals, particularly audio signals in TAPE RECORDERS, video signals in video recorders, and data in COMPUTERS. It consists of a thin plastic tape coated on one side with fine particles of a ferromagnetic material, such as iron, iron oxide or chromium oxide. *See* MAGNETIC RECORDING; VIDEOTAPE

magnetism Properties of matter and of electric currents associated with a field of force (MAGNETIC FIELD) and with a north–south polarity (magnetic poles). All substances possess these properties to some degree because orbiting ELECTRONS in their atoms produce a magnetic field in the same way as an ELECTRIC CURRENT produces a magnetic field; similarly, an external magnetic field will affect the electron orbits. All substances possess weak magnetic properties and will tend to align themselves with the field, but in some cases this DIAMAGNETISM is masked by the stronger forms of magnetism: PARAMAGNETISM and FERROMAGNETISM. **Paramagnetism** is caused by electron SPIN and occurs in substances having unpaired electrons in their atoms or molecules. The most important form of magnetism, **ferromagnetism,** is shown by substances such as iron and nickel, which can be magnetized by even a weak field due to the formation of tiny regions, called domains, that behave like miniature magnets and align themselves with an external field. These domains are formed as a result of strong, interatomic forces caused by the spin of electrons in unfilled, inner electron shells of the atoms. **Permanent magnets**, which retain their magnetization after the magnetizing field has been removed, are ferromagnetic. ELECTROMAGNETS have a ferromagnetic core around which a conducting coil is wound. The passage of a current through the coil magnetizes the core.

magnetite (ferrosoferric oxide, Fe_3O_4) Oxide mineral, iron(II)-iron(III) oxide. The most magnetic mineral, it is a valuable iron ore, found in IGNEOUS and METAMORPHIC ROCKS. It displays cubic system octahedral and dodecahedral crystals, and granular masses are also common. It is black, metallic and brittle. Permanently magnetized deposits are called **lodestone**. Hardness 6; r.d. 5.2.

magneto Machine that converts mechanical energy into a high-voltage alternating electric current. Magnetos are used when batteries are impractical, such as in the ignition systems of petrol engines in small motorcycles, lawn mowers, outboard motors, and so on, or for electrically detonating explosives. A magneto has a primary winding in which a rotating magnet produces a low voltage, which in turn induces a high voltage in a surrounding secondary winding. A make-and-break device in the primary winding creates the repeated changes of magnetic flux that induce the large voltage in the secondary winding. The magneto is connected to the rotating shaft of the engine. Initially, the inner magnet is rotated by hand (often by pulling on a spring-return cord); once running, the magnet is rotated by the engine itself. *See also* INDUCTION

magnetochemistry Branch of chemistry concerned with investigating the magnetic properties of compounds. In particular, magnetic measurements made on TRANSITION-ELEMENT complexes, which are often paramagnetic due to unpaired electrons, give information on their structure and ELECTRON configuration.

magnetohydrodynamics (MHD) Study of the motion of FLUIDS, usually PLASMAS, under the influence of electric and magnetic fields. An MHD generator consists of a plasma fluid flowing in a magnetic field, thus generating an electric current. The fluid is seeded with easily ionized elements (such as potassium) to increase the power yield.

magnetometer Instrument for measuring MAGNETIC FIELD strengths. It usually consists of a short, bar magnet with a nonmagnetic pointer attached to its centre so that it is at right-angles to the axis of the magnet. The magnet is pivoted, like a compass needle, at its centre and the pointer travels over a calibrated scale. Field strengths of magnets are compared by measuring the deflections of the pointer.

magnetosphere Region of space surrounding a planet in which the planet's MAGNETIC FIELD predominates over the SOLAR WIND, and controls the behaviour of PLASMA (charged particles) trapped within it. The boundary of the magnetosphere is called the **magnetopause**, outside which is a turbulent magnetic region called the **magnetosheath**. Downwind from the planet, the solar wind draws the magnetosphere out into a long, tapering **magnetotail**. Mercury, Earth, and the giant planets have

MAGNETOSPHERE

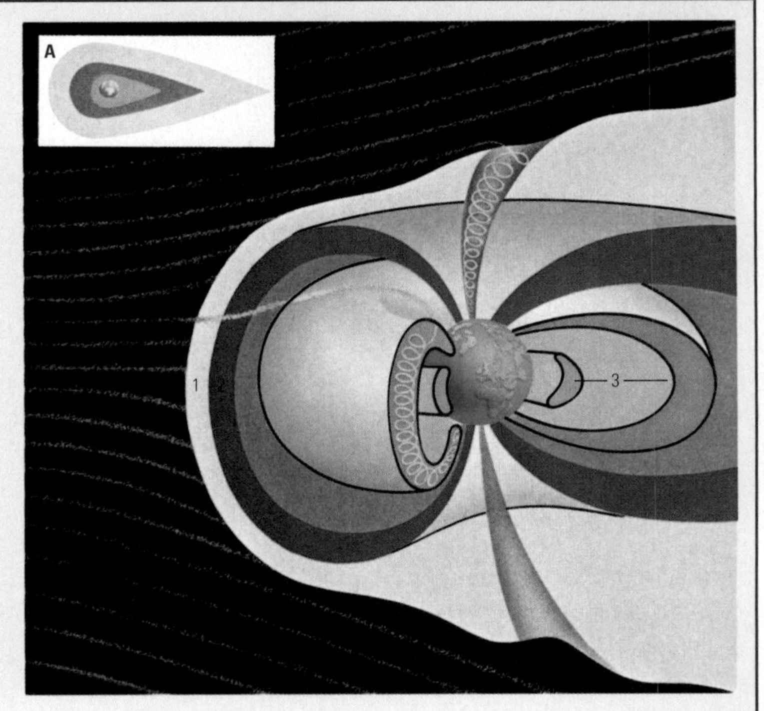

The magnetosphere is the region in which the Earth's magnetic field can be detected. It would be symmetrical were it not for electrically charged particles streaming from the Sun (A), which distort it to a teardrop shape. The particles meet the Earth's magnetic field at the shock front (1). Behind this is a region of turbulence, and inside the turbulent region is the magnetopause (2), the boundary of the magnetic field. The Van Allen belts (3) are two zones of high radiation in the magnetopause. The inner belt consists of high-energy particles produced by cosmic rays and the outer belt of solar electrons.

magnetospheres. The Earth's contains the VAN ALLEN RADIATION BELTS of charged particles.

magnetron Vacuum tube containing an ANODE and a heated CATHODE. The flow of electrons from cathode to anode is controlled by an externally applied magnetic field. When incorporated in a resonant system, a magnetron can act as an OSCILLATOR. It can generate high frequencies and high power in short bursts, and is used in radar sets and microwave ovens.

magnification Measure of the enlarging power of a MICROSCOPE or TELESCOPE. It is the size of an object's image produced by the instrument compared with the size of the object viewed with the unaided eye. In an astronomical telescope, magnification is equal to the ratio of the FOCAL LENGTH of the objective (the lens or lenses nearest the object) to the focal length of the eyepiece. With a very short eyepiece focal length increases the magnification but reduces the field of view and can produce a distorted image.

magnifying glass Convex lens that increases the apparent size of objects examined through it, usually between two and ten times. Maximum magnification is obtained when the object is situated just within the focal length of the lens. A magnifying glass generally has a handle attached to a circular clip holding the lens.

magnitude In astronomy, numerical value expressing the brightness of a celestial object on a logarithmic scale. **Apparent** magnitude is the magnitude as seen from Earth, determined either by eye, photographically or photometrically, and it ranges from positive through zero to negative values, the brightness increasing rapidly as the magnitude decreases. First magnitude is exactly 100 times brighter than sixth magnitude (just visible to the naked eye). **Absolute** magnitude indicates intrinsic luminosity and is defined as the apparent magnitude of an object at a distance of 10 parsecs (32.6 light-years) from the object.

MALARIA

The life cycle of the malaria parasite *Plasmodium* requires two hosts, the *Anopheles* mosquito and a human host, with adverse effects of infection only appearing in the human host. An infected mosquito injects thousands of *Plasmodium* organisms into the bloodstream when it bites a human (1). These penetrate liver cells, multiply and cause cell rupture (2). Released organisms may re-infect liver cells (3), but usually progress to infect red blood cells (4, 5). Male and female parasites shortly appear in the red blood cells (6). At this stage, another mosquito bites the human and takes infected blood from him or her (7). Fertilization occurs within the mosquito, the "embryo" penetrating the stomach wall (8). Within the cyst formed (9), thousands or organisms develop. The cyst ruptures, organisms released travel to the salivary glands, and from here they are injected into a second human host (10).

Maiman, Theodore Harold (1927–) US physicist who successfully built the first optical MASER, which he called the LASER. This was done in 1960 after he had first made improvements to the maser devised by Charles TOWNES. In 1962 Maiman founded the Korad Corporation to construct high-powered lasers.

mainframe Large, fast and powerful central COMPUTER and its associated storage devices. Users may be provided with small terminal units resembling personal computers; these have SOFTWARE that enables the user to access the data on the mainframe. The terminal units are linked to each other and to the mainframe by a COMPUTER NETWORK. Today's personal computers are much more powerful than early mainframe machines, so many tasks once carried out by mainframes are now done on the desktop computer, or even on a portable machine (LAPTOP) when travelling.

main sequence In astronomy, region on the HERTZSPRUNG-RUSSELL DIAGRAM where most stars lie, including the Sun. It runs diagonally from hot, bright stars at the upper left down to cool, faint stars at the bottom right. The position of a star on the main sequence depends on its mass, the most massive stars being the brightest. Stars spend most of their lives on the main sequence, producing energy from the fusion of hydrogen to helium in their cores. The zero-age main sequence is where stars lie when they first start to burn hydrogen. As stars use up their internal hydrogen they move away from the main sequence. The more massive a star, the sooner it evolves off the main sequence. All stars on the main sequence are termed DWARFS, irrespective of whether they are larger or smaller than the Sun. *See also* STELLAR EVOLUTION

malachite ($Cu_2CO_3(OH)_2$) Carbonate mineral, basic copper carbonate, found in weathered copper ore deposits. Malachite has a monoclinic system, and is green in colour; it is sometimes used as a gemstone. Hardness 3.5–4; r.d. 4.

malaria Parasitic disease resulting from infection with one of four species of *Plasmodium* PROTOZOA. Transmitted by the *Anopheles* mosquito, malaria is characterized by fever and enlargement of the spleen. The parasite multiplies in erythrocytes (red blood cells), causing their destruction. Attacks of fever, chills and sweating typify the disease and occur as new generations of parasites develop in the blood, with the frequency of attacks related to the species involved. The original antimalarial drug, QUININE, has given way to synthetics such as chloroquine, although resistant strains of the main malaria parasite, *P. falciparum*, are spreading rapidly. With 270 million people infected, malaria is one of the most widespread diseases, claiming two million lives a year, mostly in the tropics.

malic acid (HOOCCH(OH)CH_2COOH) Organic chemical compound. It has several important roles in the body's METABOLISM, notably in the KREBS CYCLE, which is concerned with the production of energy during RESPIRATION.

malleability Property of metals (or other substances) that can be permanently shaped by hammering or rolling without breaking. In some cases, it is increased by raising temperature.

mallee Scrubby Australian vegetation that consists mainly of shrubs of the genus *Eucalyptus*. They have leathery leaves which conserve moisture during dry periods.

malnutrition Condition resulting from a diet that is deficient in necessary components such as PROTEINS, FATS or CARBOHYDRATES. It can lead to DEFICIENCY DISEASES, increased vulnerability to infection, and death. Worldwide, high death rates from malnutrition are linked to poverty, famine or war.

Malpighi, Marcello (1628–94) Italian physiologist. He was a founder of microscopic anatomy, demonstrating how blood reaches the tissues through tiny vessels (CAPILLARIES), as William HARVEY had conjectured. Malpighi explained the network of capillaries on the surface of the lung. The Malpighian tubules of the kidney and the MALPIGHIAN LAYER are among the structures named after him.

Malpighian layer Innermost layer of the EPIDERMIS of the skin. A fibrous basement membrane separates the Malpighian layer from the DERMIS beneath. The layer contains polygonal cells, which continually divide by MITOSIS. As the cells grow older, they move slowly through the epidermis to replace dead surface skin cells that have been worn away. The layer is named after Marcello MALPIGHI.

maltose (malt sugar, $C_{12}H_{22}O_{11}$) Disaccharide that contains two molecules of the simple sugar GLUCOSE. It is produced by the HYDROLYSIS of STARCH by the enzyme AMYLASE and by the breakdown of starches and GLYCOGEN during digestion.

Malus, Étienne Louis (1775–1812) French physicist who, in 1809, announced his discovery of the POLARIZATION of light by reflection. In 1810 Malus also produced a theory of double REFRACTION of light in crystals. Earlier in his career, in 1798, he had taken part in the invasion of Egypt as a member of the engineer corps in the army of Napoleon I.

mammal Class (Mammalia) of VERTEBRATE animals characterized by mammary glands in the female and full, partial or vestigial hair covering. Mammals are HOMEOTHERMIC (warm-blooded). They have a four-chambered heart with circulation to the lungs separate from the rest of the body. As a group, mammals are active, alert and intelligent. They usually bear fewer young than other animals and give them longer and better parental care. Most mammals before birth grow inside the mother's body and are nourished from her by means of a PLACENTA. When born, they continue to feed on milk from the mother's mammary glands. There is a wide range of features, shapes and sizes among mammals. Mammals include 17 orders of PLACENTALS, one MARSUPIAL order – all live-bearing – and an order of egg-laying MONOTREMES. They probably evolved about 180 million years ago from a group of warm-blooded reptiles. Today, mammals range in size from shrews weighing a few grams, to the blue whale, which can weigh up to 150 tonnes. *See feature on page 226*

mammography X-ray procedure used to screen breast tissue for CANCER. It can detect a malignant tumour at an early stage of development.

man Zoological term for a human being

Mandelbrot, Benoit B. (1924–) US mathematician, b. Poland. Mandelbrot has made major contributions to CHAOS THEORY. He is best known for coining the term FRACTAL to describe irregular geometrical figures that look the same when magnified (regardless of how large the magnification). This property is called self-similarity. His book *The Fractal Geometry of Nature* (1982) contains examples of fractal-like objects seen in the world around us, such as ferns, trees and rivers. The Mandelbrot set, a well-known fractal object, is named after him.

mandible In vertebrates, the lower JAW. In insects and CRUSTACEA, one of a pair of mouthparts used for crushing and eating.

MANGANESE NODULES

manganese nodule

The commercial mining of manganese nodules lies in the future, but work is centred on two systems - hydraulic suction dredges and mechanical dredges. There are, by contrast, many ways of exploring for abyssal ore nodules, including seismic surveying, with an airgun (1) and a string of hydrophones (2), echosounders (3), grab samplers with cameras attached to them (4), oceanographic probes like the bathysonde (5), dredges (6), deep-sea cameras (7) and wire-bound samplers, such as the box corer (8).

mandrake Plant of the potato family, native to the Mediterranean region and used since ancient times as a medicine. It contains the ALKALOIDS hyoscyamine, scopolamine and mandragorine. Leaves are borne at the base of the stem, and the large greenish-yellow or purple flowers produce a many-seeded berry. The branched root resembles a distorted human figure, and for centuries the plant was believed to utter an almost human shriek when picked. Height: 40cm (16in); family Solanaceae; species *Mandragora officinarum*.

manganese (symbol Mn) Grey-white TRANSITION ELEMENT, first isolated in 1774. Its chief ores are pyrolusite, manganite and hausmannite. It is also found in MANGANESE NODULES. The metal is used in alloy steels, ferromagnetic alloys, fertilizers and paints and as a petrol additive. It forms two series of salts, termed manganese(II), or manganous, and manganese(III), or manganic. It also forms manganate(VI), containing the ion MnO_4^{2-}, and manganate(VII) (permanganate), MnO_4^-. Properties: at.no. 25; r.a.m. 54.938; r.d. 7.20; m.p. 1,244°C (2,271°F); b.p. 1,962°C (3,564°F); most common isotope ^{55}Mn (100%).

manganese nodules Concretions with diameters averaging 4cm (2in) that occur in red clay and ooze on the floor of the Pacific Ocean, with the major concentrations around the Samoan Islands. They contain high concentrations of manganese. They are thought to have formed as agglomerates from colloidal solution (*see* COLLOID) in the ocean's waters. The commercial mining of the nodules is being delayed by legal controversy over international rights to this valuable mineral resource.

manganite (MnO[OH]) Hydroxide mineral of the diaspore group, manganese oxyhydroxide. It occurs as low-temperature veins and secondary deposits and has monoclinic system striated prisms, which are often crusts of small crystals. Its colour is black or grey with a submetallic lustre. Hardness 4; r.d. 4.3.

Manhattan Project Codename given to the development of the US atom bomb (*see* NUCLEAR WEAPON) during World War 2. Work on the bomb, advocated by Albert EINSTEIN and other scientists, was carried out in great secrecy by a team headed by Enrico FERMI and including J. Robert OPPENHEIMER. This took place at several locations, notably Oak Ridge, Tennessee, and Los Alamos, New Mexico. Research was also carried out at the University of Chicago, the University of California at Berkeley and Columbia University. The first test took place on 16 July 1945, near Alamogordo, New Mexico, and in August 1945, the first bombs were dropped on the Japanese cities of Hiroshima and Nagasaki.

man-made fibre *See* SYNTHETIC FIBRE

manometer Device for measuring pressure. It consists of a U-shaped tube containing a liquid, one end open to the atmosphere and the other end attached to the vessel whose pressure is to be measured. If the gas pressure in the vessel is greater than atmospheric, it will force the liquid down on the side nearest to the vessel and up on the side open to the atmosphere. The difference in heights of the liquid in the two arms of the tube shows the difference in pressure.

Manson, Sir Patrick (1844–1922) British parasitologist known for his pioneering work in tropical medicine. He discovered that the mosquito was host to the parasite *Filaria bancrofti*, which causes the disease filariasis in humans. Manson founded the Medical School of Hong Kong and organized the London School of Tropical Medicine.

mantissa Decimal part of a LOGARITHM

mantle Layer of the EARTH between the CRUST and the CORE, down to 2,900km (1,800mi). Its boundaries are determined by major changes in seismic velocity. The mantle forms the greatest bulk of the Earth, 82% of its volume and 68% of its mass. Several distinct parts of the mantle have been recognised. The uppermost part is rigid, solid and brittle and together with the Earth's crust forms the LITHOSPHERE. The upper mantle also has a soft zone, which is called the ASTHENOSPHERE. Temperature and pressure are in delicate balance, so that much of the mantle material is near melting point, or partially melted and capable of flowing. Recent studies have suggested this layer is responsible for the production of basaltic MAGMAS. The remainder of the mantle is thought to be more solid, but still capable of creeping flow. In the lower mantle, several changes in seismic velocity can be detected, interpreted as mineral phase changes, in which the atomic packing is rearranged into denser and more compact units. The chemical constitution of the mantle is uncertain, but it is thought to be made up of iron-magnesium silicates.

map Graphic representation of part or all of the Earth's surface. The science of map making is called cartography. Maps are usually printed on a flat surface using various kinds of projections. To make a map, a cartographer uses information from land surveys, aerial photographs and other sources. A map may be formed by drawing on paper, engraving lines into plastic, or constructing an image on a computer screen.

map projection Any systematic method of drawing the Earth's meridians and parallels on a flat surface. Only on a globe can areas and shapes be represented with any fidelity. On flat maps of large areas distortions are inevitable. Projections are either geometrical derivations (cylindrical, conical or azimuthal) or networks and grids derived mathematically in the transposition from globe to flat surface.

marble Metamorphic rock composed largely of recrystallized LIMESTONES and DOLOMITES. The term is more loosely used to refer to any crystalline calcium carbonate ($CaCO_3$) rock that has a good pattern and colour when cut and polished. The colour is normally white, but when tinted by serpentine, iron oxide, or carbon can vary to shades of yellow, green, red, brown or black. It has been a favourite building and sculpting material of many ancient and modern civilizations.

marcasite Brass-yellow mineral, iron sulphide (FeS_2), found commonly in near-surface deposits. It is found as tabular or prismatic crystals in the orthorhombic system, as cockscomb aggregates and as stalactitic, globular and radiating forms. It has a metallic lustre and is brittle. Hardness 6–6.5; r.d. 4.9.

Marconi, Guglielmo (1874–1937) Italian physi-

MANOMETER

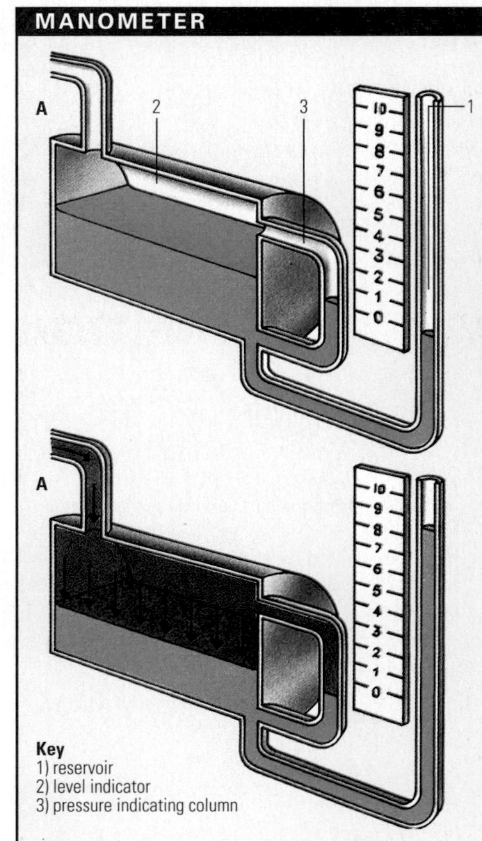

A manometer is essentially a U-shaped liquid column gauge used to measure the difference between two fluid pressures. The well type manometer is illustrated here, one column of which has a relatively small diameter; the second serves as a reservoir. The difference in cross-sectional area (sometimes as much as 1500 times) ensures that while the level of the reservoir does not change appreciably with change of pressure, the level of the small diameter column does; this makes for adequate accuracy. Small adjustments to the position of the vertical scale, with the aid of the level indicator, compensate for the little reservoir change that does occur. Readings of differential pressure may then be made directly on the vertical scale. (A) Both columns are at atmospheric pressure, so the levels are the same. (B) The reservoir is pressurised. Its level is indicated so the scale may be adjusted to the new level.

Mammals are the largest land animals to rise to prominence since the dinosaurs and are one of the most successful groups in the animal kingdom. From 4-cm (1.6-in) pygmy shrews to 100-tonne whales, the 4,500 species of mammal have explored and adapted to most of the planet's different habitats, evolving a wide range of shapes, sizes and lifestyles on the way. They owe their success to two essential factors – their warm-bloodedness (HOMEOTHERMY), which lets them, unlike reptiles, function regardless of the temperature, and their development of suckling and looking after their young.

The evolution of homeothermy gave early mammals an essential competitive edge over reptiles, because it enabled mammals to be more active at lower temperature, and so exploit more ecologicalniches. The suckling of the young on the female's milk and the investment of parental care are also highly important. Protected and fed, the young can grow rapidly and learn skills from adults: chimpanzees, for example, learn to use sticks as tools, and many carnivorous mammals learn hunting techniques.

FEEDING

A wide range of mammalian feeding habits has evolved to make full use of the Earth's food resources. Nocturnal and diurnal species coexist in the same ECOSYSTEM, allowing greater exploitation of the habitat than if all its inhabitants were competing for food. and space at the same time. Migrants, such as caribou and the vast herds of African antelopes and zebra, move around to exploit seasonal food sources. Other mammals, such as whales and fur seals, migrate to sheltered bays to breed, living on fat until they are free to return to more food-rich waters.

Many small rodents, and even a few larger animals, such as the black bear of North America, are

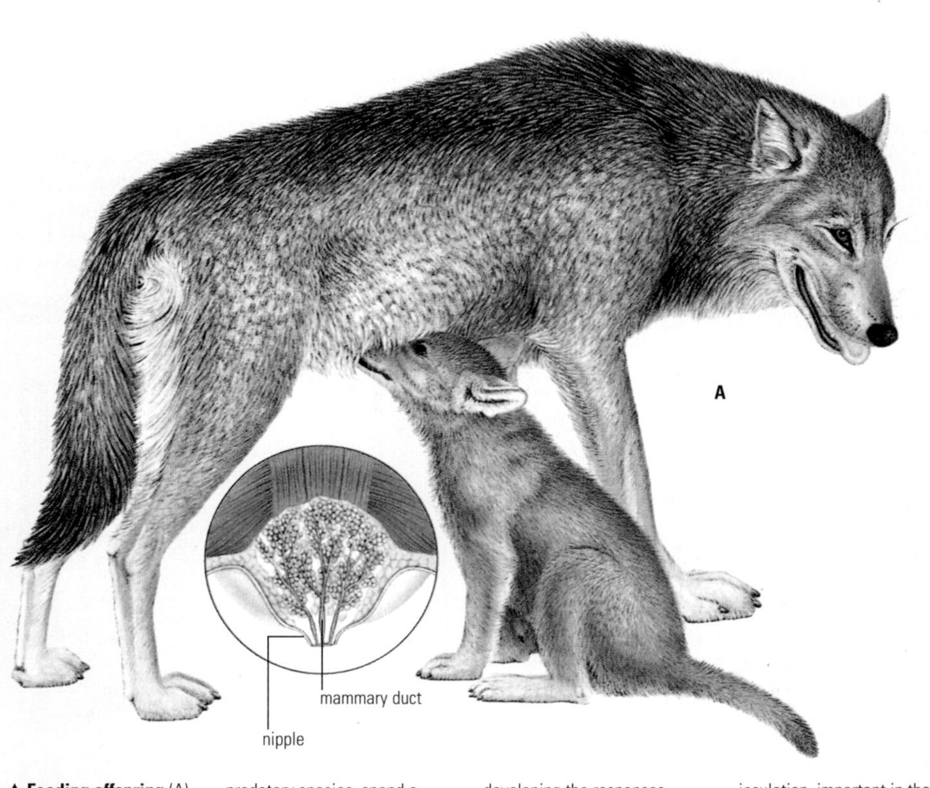

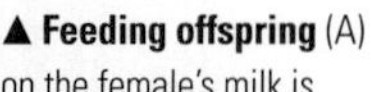

▲ **Feeding offspring** (A) on the female's milk is one of the distinctive features of mammals. In most mammals, the milk is exuded from nipples situated on the belly. The number of nipples roughly matches the largest number of young commonly produced at one time by that species. Hence pigs have many more nipples with which to feed their young than humans. Milk production is stimulated by hormones released towards the end of pregnancy. Once the young are weaned, milk production ceases, thus saving the adult energy. Milk provides a complete meal for the young mammal; fats for energy; proteins for tissue development; calcium and iron for bones and the blood. Many young mammals (B), especially predatory species, spend a long time in the care of their parents. This allows them to build up adult skills. Play is an important part of growth, improving coordination and developing the responses important for defence (and in some species also for predation). Fur is another special characteristic of mammals, providing insulation, important in the maintenance of temperature. Mutual grooming has an important social function, strengthening the bonds within family groups.

MAMMALIAN SKELETONS

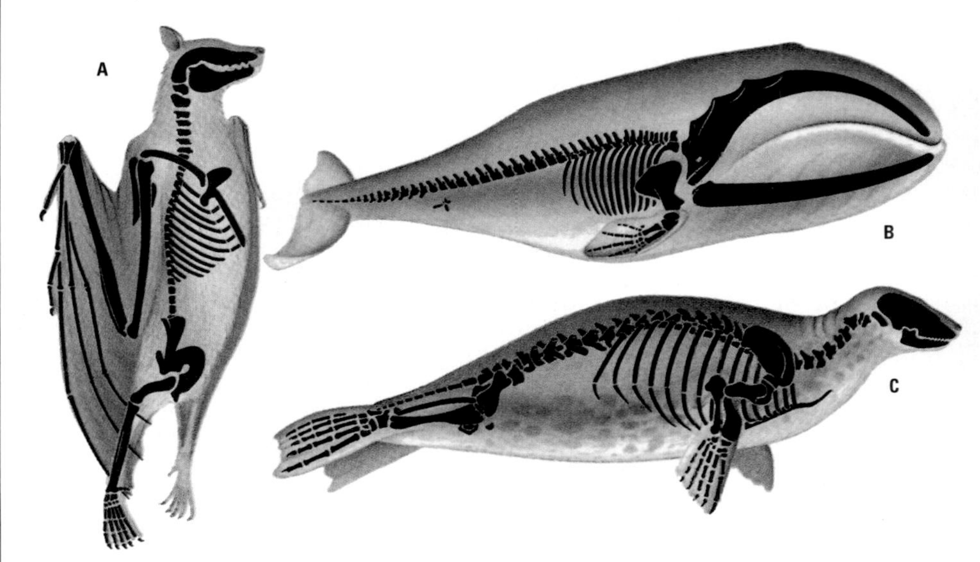

Mammalian skeletons have become modified in many different ways as animals have adapted to different habitats and lifestyles. In primitive mammals, limbs tended to be relatively short, as for instance in the opossum. But adaptations, such as running faster, led to changes, like the elongation of limbs. In bats (A), the forelimbs have become wings, while in seals (B) the limbs form paddle-like flippers. Whales' limbs have a minimal steering function; in baleen whales (C), the jawbones are greatly enlarged and curved to accommodate the animal's huge, sieve-like plates of baleen (whalebone).

able to survive severe winter conditions by hibernating. Others escape environmental fluctuations by retreating underground to burrows or to the depths of caves.

ADAPTABLE

Some mammals with terrestrial ancestors have evolved to exploit the air and the sea. The bats are highly successful, and the only vertebrates other than birds capable of sustained flight. Whales and dolphins have become so well adapted to the marine environment that they never come ashore. The baleen whales have evolved huge filters of *baleen* (whalebone) with which they sieve the vast shoals of krill that abound in the oceans, particularly in the Antarctic region.

In areas scarce of prey, predatory mammals tend to be solitary, like the polar bear, whereas where food is more plentiful, cooperative hunting techniques have evolved, where groups of predators such as hunting dogs or lions cooperate to bring down prey that may be larger than themselves. Many herbivorous mammals have evolved symbiotic relationships with cellulose-digesting bacteria so that they can take advantage of relatively indigestible plant food for which there is less competition. Bears and many primates are omnivorous, feeding on fruits, leaves and berries, but hunting other creatures when they get the opportunity. Other mammals, such as raccoons and hyenas, are mainly scavengers.

SIMILARITIES

Despite geographical isolation, many mammals have adapted in corresponding ways to similar lifestyles in different continents; examples are the anteaters and armadillos of South America, the aardvark of Africa, the pangolin of Asia, and the echidnas, or spiny anteaters of Australia and New Guinea – these creatures are all anteaters with long, tough snouts, long tongues and powerful claws for breaking into nests of ants and termites.

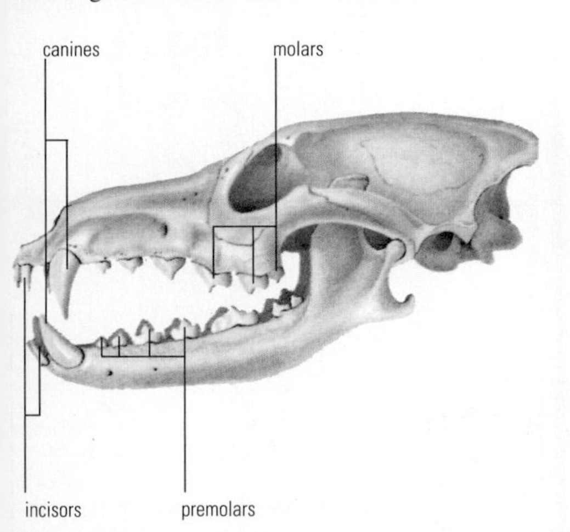

▲ Having specialized teeth of different kinds is a unique mammalian adaptation. Variety is greatest in the carnivores (meat-eaters), which have biting incisors, tearing canine teeth, and sharp molars and premolars for crushing bones. Herbivores have more uniform teeth designed for grinding vegetation such as grass.

▲ Tracing the origins of mammals depends on the availability of a good fossil record. Unfortunately for so many of the 18 orders of mammals, the record is incomplete. However, possible lines of descent have been suggested. It seems increasingly likely that mammals are all descended from lizard-like reptiles known as synapsida. The synapsida evolved many millions of years before splitting into two subclasses – the prototheria (the monotreme egg layers) and the theria (which give birth to live young). The theria themselves are divided into two further groups – the eutheria, or placental mammals, and the metatheria, the marsupials.

Key

- proposed linkage
- fossil record
- prototheria
- metatheria
- eutheria

1 Cetacea (whales)
2 Insectivora (insectivores)
3 Tubulidentata (aardavarks)
4 Chiroptera (bats)
5 Artiodactyla (even-toed ungulates)
6 Dermoptera (flying lemurs)
7 Perissodactyla (odd-toed ungulates)
8 Primates (primates)
9 Hydracoidea (hyraxes)
10 Rodentia (rodents)
11 Proboscidea (elephants)
12 Lagomorpha (rabbits)
13 Sirenia (sea cows)
14 Pholidota (pangolins)
15 Monotremata (monotremes)
16 Edentata (anteaters)
17 Carnivora (carnivores)
18 Marsupialia (marsupials)

MAMMALIAN HEART

In the mammalian heart, the right and left sides are completely separate. The left side of the heart receives oxygenated blood from the lungs via the pulmonary veins (1). This is pumped to the body's organs at high pressure via the aorta (2) and its branches. Deoxygenated blood returns to the heart via the inferior and superior venae cavae (3 and 4), and is pumped back to the lungs via the pulmonary arteries (5). This is done by the right side of the heart at a lower pressure, which is important to avoid damaging the lungs. The left side of the heart is frequently larger since it pumps at higher pressure.

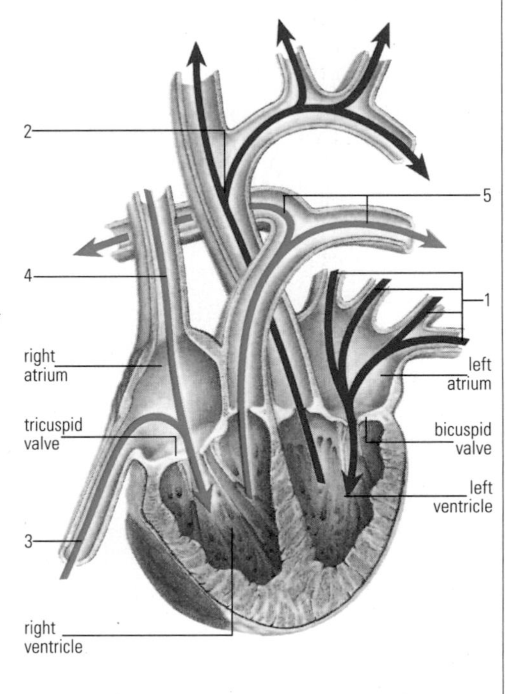

cist who developed RADIO. By 1897 Marconi was able to demonstrate radio telegraphy over a distance of 19km (12mi) and, after forming a wireless telegraph company, established radio communication between France and England in 1899. By 1901 radio transmissions were being received across the Atlantic Ocean, and in 1909 Marconi was awarded the Nobel Prize for physics. His later work on short-wave radio transmission formed the basis of nearly all modern long-distance communication.

Marcus, Rudolph (1923–) US physical chemist, b. Canada. Marcus developed what is now known as the "Marcus theory" for electron transfer (OXIDATION-REDUCTION) reactions of solvent molecules. His formula calculates the size of the energy barrier the electrons must surmount to move from one molecule to another. The theory is applicable to such processes as photosynthesis and the electrical conductivity of polymers. In 1992 Marcus received the Nobel Prize for Chemistry.

marijuana NARCOTIC drug prepared from the dried leaves of the Indian hemp plant (*Cannabis sativa*); it is different from hashish, which is prepared from resin obtained from the flowering tops of the plant. Some people smoke marijuana like tobacco. In moderate doses it can act as a relaxant and euphoriant. It has proved beneficial to some sufferers of MULTIPLE SCLEROSIS (MS). Possession of the drug is illegal in many countries. *See also* ADDICTION

marine biology Science and study of life in the sea. It covers organisms that live in the ocean water, on the sea bed and along shores. Marine biologists explore the ways in which these organisms fit into their environments, as well as their interactions with the human environment.

marine engine Power plant used to propel seagoing SHIPS, and as auxiliaries in smaller sailing vessels. In the 19th and early 20th centuries, marine engines were coal-fired, piston-driven STEAM ENGINES. Today, large steam engines are still used, but these are usually oil-fired steam TURBINE engines; steam, raised in a boiler, turns a turbine, which is geared down to turn propeller shafts. Small, medium and large diesel engines, and small petrol engines (including outboard motors) are also in general use. A few ships, notably certain ice-breakers and submarines, are fitted with nuclear engines in which steam for turbines is raised by heat from a NUCLEAR REACTOR.

Mariner project Series of US SPACE PROBES to the planets. **Mariner 2** flew past Venus in 1962, discovering that the planet was hot. **Mariner 4** flew past MARS in July 1965, photographing craters on its surface. **Mariner 5** passed Venus in October 1967, making measurements of the planet's atmosphere. **Mariners 6 and 7** obtained further photographs of Mars in 1969. **Mariner 9** became the first probe to enter planetary orbit when it went into orbit around Mars in November 1971. It made a year-long photographic reconnaissance of the planet's surface, and obtained close views of the two moons, Phobos and Deimos. **Mariner 10**, the last of the series, was the first two-planet mission, passing Venus in February 1974 and then encountering Mercury three times, in March and September 1974 and March 1975. *See also* SPACE EXPLORATION

Mariotte, Edme (1620–84) French physicist, plant physiologist and priest, noted for his independent discovery of BOYLE'S LAW. This law, which states that at constant temperature the volume of a gas is inversely proportional to its pressure, is known as **Mariotte's law** in France. Mariotte also concluded that plants synthesize materials by chemical processes, a theory verified after his death. He was a founder of the Academy of Sciences in Paris.

maritime climate CLIMATE that is strongly influenced by proximity to the sea. Britain has a maritime climate, which means that it has mild weather for its latitude. Oceans warm up more slowly than land, because the heat of the Sun is spread out through a great depth of water and because ocean currents allow the heat to move vertically as well as horizontally. However, oceans also retain heat for much longer than land. For this reason, climatic conditions near an ocean tend to be much warmer in winter and slightly cooler in summer than in inland areas at the same latitudes. In addition to the effects on temperature, the sea also influences precipitation, which is greater than in inland locations. The maritime climate of Britain is most marked in the Isles of Scilly, Cornwall, where the average temperature in winter is *c*.7°C (45°F) – about the same as the Mediterranean coast of France. In the summer, the temperature averages 16°C (61°F), and the total annual rainfall is *c*.800mm (32in). Moving eastwards into Europe, along similar latitudes, the climate gradually changes to a CONTINENTAL CLIMATE, such that in Germany the corresponding figures would be *c*.0°C (32°F), 19°C (66°F) and 550mm (22in).

marl (mudrock) Extremely fine-grained, SEDIMENTARY ROCK that is an intermediate between a clay and a limestone. It contains up to 60% of calcareous minerals such as calcite, with some quartz and particles of silt. Red marl owes its colour to the presence of iron oxide, whereas green marl contains chlorite or glauconite. They formed from deposits at the bottoms of lakes or the sea. Marls of marine origin may occur in beds with gypsum and halite.

marrow Soft tissue containing blood vessels, found in the hollow cavities of BONE. The marrow found in many adult bones, including the shafts of LONG BONES, is somewhat yellowish and functions as a store of fat. The marrow in the flattish bones, including the ribs, breast bone, skull, spinal column and the ends of the long bones, is reddish and con-

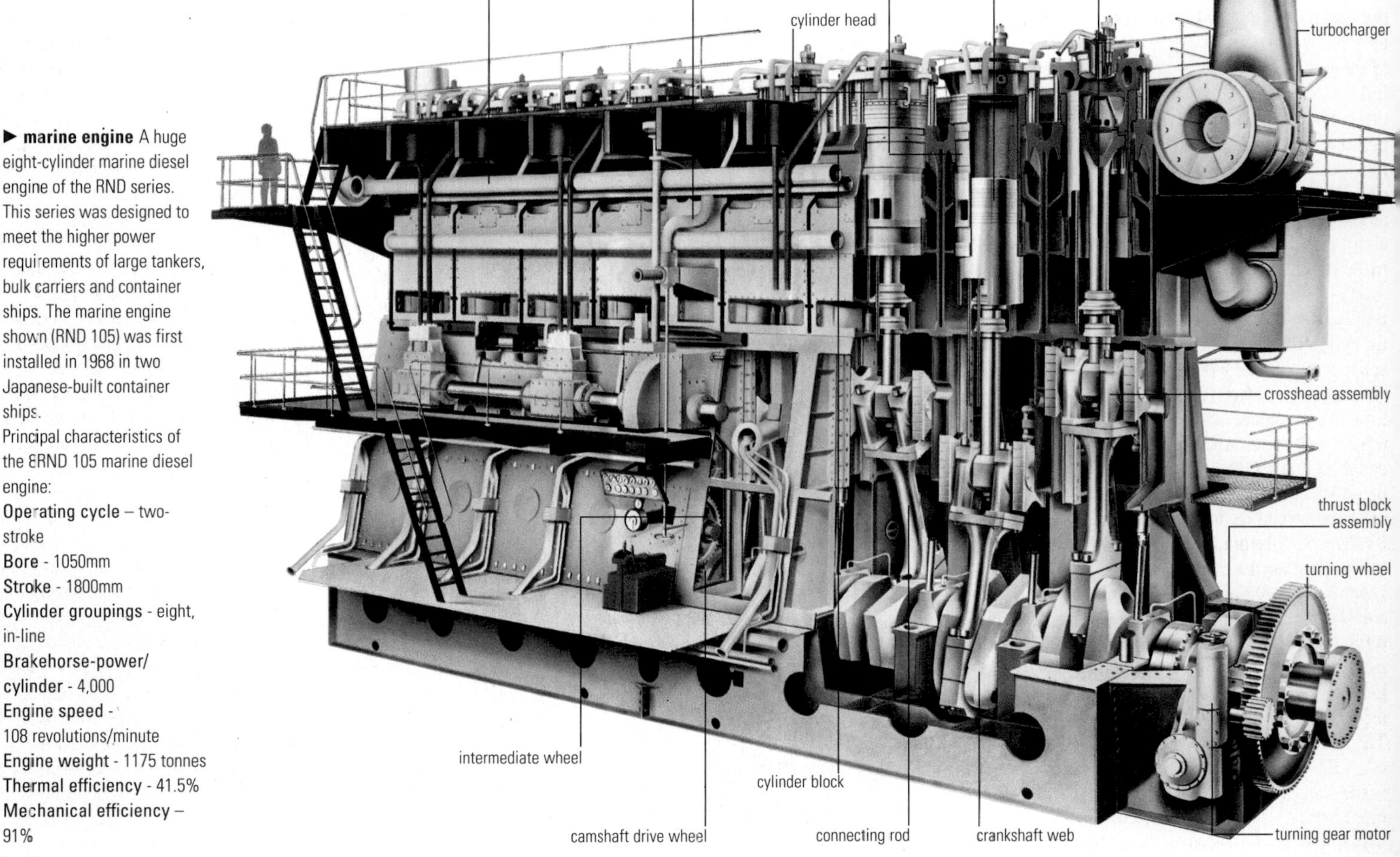

► **marine engine** A huge eight-cylinder marine diesel engine of the RND series. This series was designed to meet the higher power requirements of large tankers, bulk carriers and container ships. The marine engine shown (RND 105) was first installed in 1968 in two Japanese-built container ships.
Principal characteristics of the ERND 105 marine diesel engine:
Operating cycle – two-stroke
Bore - 1050mm
Stroke - 1800mm
Cylinder groupings - eight, in-line
Brakehorse-power/cylinder - 4,000
Engine speed - 108 revolutions/minute
Engine weight - 1175 tonnes
Thermal efficiency - 41.5%
Mechanical efficiency – 91%

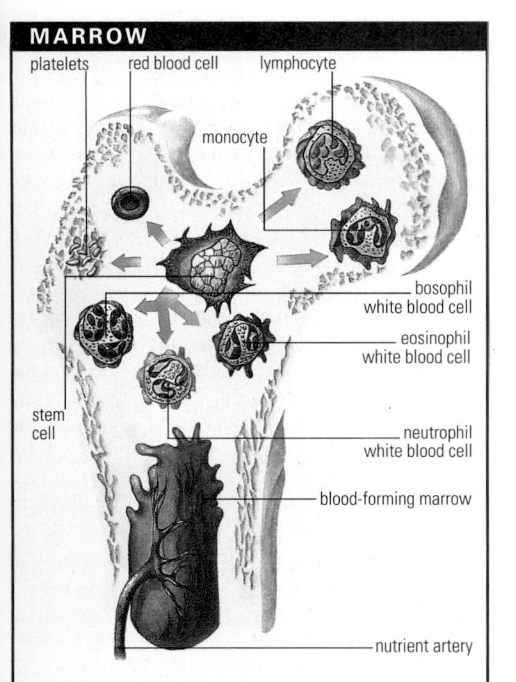

Most bones have hollow cavities that are filled with spongy bone marrow. Some of this (not, however, in the long bones) forms new red and white blood cells. Little more than 250g (0.5lb) of bone marrow is sufficient to provide the 5,000 million red cells a day need to replace old cells that have worn out after their 120-day lifetime in the body. Some white cells are also formed in the lymph nodes.

tains cells that give rise eventually to ERYTHROCYTES (red blood cells) as well as to most of the LEUCOCYTES (white blood cells), but not LYMPHOCYTES OF PLATELETS.

Mars Fourth major planet from the Sun, and the second-smallest of the inner planets. Mars appears red to the naked eye, because of the high iron content of its surface crust. When viewed through a telescope it appears as a small orange-red disc with lighter and darker markings, and white patches often visible at one pole or the other. The atmosphere consists mainly of 95% carbon dioxide, 2.5% nitrogen and 1.5% argon, with smaller quantities of oxygen, carbon monoxide and water vapour. Its axial tilt is similar to the Earth's, so it passes through a similar cycle of SEASONS, and seasonal variations in its appearance may be observed through a telescope. The surface temperature on Mars varies between extremes of 130K and 290K. Much information has been returned by various Mars probes – spacecraft that have orbited or landed on the planet. The surface of Mars reveals a long and complex history of geological activity. The major difference in terrain is between the largely smooth, lowland volcanic plains of the northern hemisphere and the heavily cratered uplands of the south. The biggest volcanic structure on Mars, and in the whole Solar System, is Olympus Mons, which is hundreds of kilometres across and 27km (17mi) high. Other geographical features, such as some giant canyons, are channels in which rivers once flowed. Perhaps all the water is now frozen into the subsurface permafrost layer; more likely, the planet's low gravity has simply not been enough to hang on to all of its air and water, which have slowly been escaping. The variable polar ice caps appear to be composed of solid carbon dioxide with underlying caps of water-ice. Mars has two tiny SATELLITES in very close orbits, Phobos and Deimos. In 1996 scientists investigating a METEORITE, thought to have originated on Mars, found fossilized microorganisms, which some believe indicates the presence of primitive life on the planet. In 1997 the US Pathfinder probe reached Mars and launched a small vehicle onto the planet's surface. It analysed the geological formations it encountered, sending information back to Earth. *See also* VIKING SPACE MISSION

marsh Flat, grassland area, devoid of peat and saturated by moisture during one or more seasons. Typical vegetation includes grasses, sedges, reeds and rushes. Marshes are sometimes the breeding sites of mosquitoes and other disease carriers, but they are valuable WETLANDS and maintain water tables in adjacent ECOSYSTEMS.

marsh gas *See* METHANE

Marsigli, Count Luigi Ferdinando (1658–1730) Italian soldier, naturalist and oceanographer. After long service in the Austrian army, during which he was captured, he was demoted in 1704. Thereafter, Marsigli engaged in scientific study, founding the Accademia delle Scienze in Bologna (1712) and becoming a member of the Royal Society (1722), to which he was presented by Isaac NEWTON.

marsupial Mammal of which the female usually has a pouch (marsupium), within which the young are suckled and protected. At birth, the young are in a very early stage of development. Most marsupials are Australasian, and include such varied types as the kangaroo, koala, wombat, tasmanian devil, bandicoot, and marsupial mole. The only marsupials to live outside Australasia are the opossums and similar species found in the Americas. *See also* MONOTREME

Martin, Archer John Porter (1910–) British biochemist who developed partition CHROMATOGRAPHY, particularly for use in amino-acid analysis. For this research, Martin shared the 1952 Nobel Prize for chemistry with Richard SYNGE. Later, he developed gas-liquid chromatography.

Martin, Pierre Émile (1824–1915) French engineer who pioneered the OPEN-HEARTH PROCESS of steel manufacture. The **Siemens-Martin** process pre-heats the air-blast by a heat-regeneration method and the excess carbon is removed by adding a calculated amount of iron ore. By applying this process to an open-hearth furnace charged with a mixture of pig-iron and wrought-iron, Martin was able to obtain the desired carbon content simply by dilution.

mascon Any of several high-density regions of the MOON's surface that produce stronger gravitational effects than the surrounding areas, causing noticeable perturbations in the orbits of spacecraft. The word is derived from the term "mass concentration" and the phenomenon seems to indicate that the lunar interior is not uniform.

maser (acronym for **m**icrowave **a**mplification by **s**timulated **e**mission of **r**adiation) Device using atoms artificially kept in states of higher energy than normal to provide amplification of radio signals. The principle of the maser was discovered by Charles TOWNES of Columbia University, for which he shared the 1964 Nobel Prize for physics with Nikolai BASOV and Alexander PROKHOROV, who also worked in this field. The first maser used electrostatic (charged) plates to separate high-energy ammonia atoms from low-energy ones. Radiation of a certain frequency stimulated the high-energy ammonium atoms to emit similar radiation and strengthen the signal. The very narrow frequency emitted made the ammonia maser one of the most accurate "atomic clocks" known. *See also* LASER

Maskelyne, Nevil (1732–1811) British astronomer. In 1765 he was appointed the fifth Astronomer Royal, a post he held for 46 years. Maskelyne introduced a method of determining longitude at sea by the measurement of lunar position, and in 1767 he published the first Nautical Almanac, an annual book for the same purpose. In 1774 he measured the mass of the Scottish mountain of Schiehallion by seeing how far it caused a plumb-line to deviate, from which he calculated the average density of the Earth to be 4.7g cm^{-3} (the true value is 5.52g cm^{-3}).

mass (symbol *m* or *M*) Measure of the quantity of matter in an object. Scientists distinguish between two types of mass: **gravitational** mass is a measure of the mutual attraction between bodies such as the Earth, as expressed in Newton's law of GRAVITATION. **Inertial** mass is a measure of a body's resistance to change in its state of MOTION, as expressed in Newton's first law of motion. Newton meters (spring balances) and platform scales measure gravitational mass; INERTIA balances measure iner-

◀ **Mars** Mosaic of images taken by the Viking 1 and 11 probes showing the globe of Mars. The image is centred on 90° longitude and 0° latitude. The equatorial band and the southern hemisphere of the planet are marked by heavily cratered areas, whereas the northern hemisphere is occupied by vast, sparsely cratered plains. The south polar cap (white bottom) is believed to consist mainly of frozen carbon dioxide; the north polar cap is formed by a mixture of carbon dioxide and water ice. The Tharsis region (upper centre left) hosts many volcanoes (round features), the largest of which is Olympus Mons (centre left) with a diameter of *c.*540km (335mi).

tial mass. EINSTEIN's general theory of RELATIVITY is based on the principle of EQUIVALENCE: that the inertial mass and the gravitational mass of an object are equivalent. *See also* WEIGHT

mass action, law of Principle that a chemical reaction rate is proportional to the product of the concentrations of each reactant.

mass defect Difference in MASS between the total rest mass of the PROTONS and NEUTRONS, when free, from which a particular NUCLEUS is formed and the slightly lower mass of the nucleus itself. The mass defect is equivalent to the energy that would be needed to separate the particles, and to indicate the stability of the nucleus. *See also* BINDING ENERGY; RELATIVITY

mass number (nucleon number, symbol *A*) Total number of NUCLEONS (NEUTRONS and PROTONS) in the NUCLEUS of an ATOM. It is usually shown as a superscript number before the chemical symbol of an element. Thus, the lightest element, hydrogen, has one proton only in its nucleus, and its nuclear notation is written ^{1}H. Heavier elements have both protons and neutrons in their nuclei: one ISOTOPE of uranium has a mass number of 238, meaning each of its nuclei contains 238 nucleons (protons plus neutrons), and is written ^{238}U. An atom also has an ATOMIC NUMBER representing the number of protons only and this is written as a subscript, for example $_{6}C$.

mass production Manufacture of goods in large quantities by standardizing parts, techniques and machinery. In 1798 US inventor Eli WHITNEY introduced mass production, when he was given the contract to make 10,000 guns for the US Army. Whitney built machinery to make identical parts for all the guns, so that all the components could be produced quickly and were guaranteed to fit. It also meant that damaged parts could be replaced with standard spares. The assembly line, with a conveyor belt carrying the work through a series of assembly areas, is another feature of some mass-production processes. In 1912 Henry FORD used this system in Detroit, USA, for the mass production of his Model T car. Many modern mass-production processes depend on computer control of machines, including ROBOTS.

mass spectrograph (mass spectrometer) Instrument for separating ions according to their masses (or more precisely, according to their charge-to-mass ratio). In the simplest types, the ions are first accelerated by an electric field and then deflected by a strong magnetic field; the lighter the ions, the greater the deflection. The mass spectrograph detects the lines of the resulting mass spectrum on a photographic plate; the mass spectrometer does this electrically. By varying the field, ions of different masses can be focused in sequence on to the photographic plate or detector and a record of charge-to-mass ratios obtained.

mastectomy In surgery, removal of all or part of the female breast. It is performed as a treatment for breast cancer. Simple mastectomy involves the breast alone; if the cancer has spread, radical mastectomy may be undertaken, removing also the lymphatic tissue from the armpit.

Masters, William Howell (1915–) US physician who, with his psychologist wife Virginia (née Johnson) (1925–), became noted for laboratory studies of the physiology and anatomy of human sexual activity. They also conducted clinical marriage counselling for people with sexual problems. Their works include *Human Sexual Response* (1966) and *Human Sexual Inadequacy* (1970).

mastoid (mastoid bone) Raised region behind the ear, containing air cells which communicate with the middle ear. In mastoiditis, the region becomes infected. This once-common disorder was frequently operated upon, but nowadays is usually cured quickly with antibiotics.

mathematical biology Modern branch of applied MATHEMATICS and theoretical BIOLOGY dealing with the growth of organisms, the development of patterns and the spread of populations and disease. Also, recent advances have been made in medicine where mathematics has been used to describe wound healing, the events leading to heart attacks and the growth of tumours within the body.

mathematical induction Method of proving that a mathematical statement is true for any positive integer *n* by proving (1) that it is true for a base value, for example 1, and (2) that if it is true for a value *k* then it is also true for $k + 1$. If (1) and (2) hold, then it follows in a finite number of steps that the statement is true for any positive integer *n*. For instance, the formula for the sum of the first *n* natural numbers is a result that can be proved using mathematical induction. The technique is a powerful way of proving that a result holds for an infinite number of values.

mathematical model Set of equations and concepts used to describe and predict a given phenomenon or behaviour. Mathematical models can be of practical or theoretical use (or both). Practically, mathematical models can be used to design new materials, to predict the weather, and to test bridges, aircraft and so on before they are built. Theoretically, mathematics is the language used to describe the properties of matter in PHYSICS and CHEMISTRY and, more recently, the formation of patterns and the behaviour of animals in MATHEMATICAL BIOLOGY.

mathematics Study concerned with the properties of numbers, space and shape and with deductions made from assumptions about abstract entities. Mathematics is often divided into **pure** mathematics, which involves purely abstract reasoning based on axioms, and **applied** mathematics, which involves the use of mathematics to model processes in engineering, physics, chemistry, biology and economics. The distinction between the two branches comes from the fact that the former does not require any observation of, or data from, the natural world. However, the two fields cannot be separated because results from pure mathematics are often useful in practical applications, and studying real problems often stimulates pure mathematics research. The main divisions of pure mathematics are GEOMETRY, ALGEBRA and ANALYSIS. *See also* ARITHMETIC; TRIGONOMETRY

matrix (pl. matrices) Rectangular array of numbers in rows and columns. The number of rows need not equal the number of columns. Matrices can be added, subtracted and multiplied according to certain rules. They are useful in the study of transformations of COORDINATE SYSTEMS and in solving sets of SIMULTANEOUS EQUATIONS. *See also* DETERMINANT

matter What things are made of. Ordinary matter is made up of ATOMS, which are combinations of ELECTRONS, PROTONS and NEUTRONS. Atoms, in turn, make up ELEMENTS, an ordered series of substances having from one proton in their nuclei (hydrogen) to a hundred or more. Many other subatomic or ELEMENTARY PARTICLES can be produced at high energies and live for short periods of time. Elements other than hydrogen and helium (such as carbon, oxygen and iron) were originally built up by nuclear FUSION in stars, including the Sun. All matter exerts an attractive force on other matter, called GRAVITATION. Charged particles exert an attractive or repulsive ELECTROMAGNETIC FORCE that accounts for nearly all everyday phenomena – the sense of touch, for instance, is dependent on the repulsion of MOLECULES at close range. The STRONG NUCLEAR FORCE is responsible for binding the protons and neutrons in an atomic NUCLEUS, and the WEAK NUCLEAR FORCE is responsible for beta decay. *See also* ANTIMATTER; MATTER, STATES OF

matter, states of Classification of matter according to its structural characteristics. Four states of matter are generally recognized: SOLID, LIQUID, GAS and PLASMA. Any one ELEMENT or compound may exist sequentially or simultaneously in two or more of these states: water, ice and water vapour can all exist at one temperature and pressure. **Solids** may be crystalline (have a regularly repeated molecular structure), as in salt and metals; or amorphous, as in tar or glass. **Liquids** have molecules that can flow past one another, but which remain almost as close as in a solid. In a **gas**, molecules are so far from one another that they travel in relatively straight lines until they collide with the walls of their container. In a **plasma**, atoms are torn apart into electrons and nuclei by high temperatures, such as those in stars.

Mauchly, John William (1907–80) US physicist and engineer who, with John Presper Eckert Jr, (1919–) invented and built (1946) the first commercial general-purpose COMPUTER – the Electronic Numerical Integrator and Computer (ENIAC). Taking up 15,000sq ft, ENIAC was bought as a military calculator by the US War Department. The two men, who met in 1943 on the faculty of the University of Pennsylvania, next built the Electronic Discrete Variable Computer (EDVAC), whose ability to store PROGRAMS helped to launch the computer revolution, and then in 1951 the Universal Automatic Computer (UNIVAC) for the US Census Bureau. In 1948 they formed the Eckert-Mauchly Computer Corporation that was sold two years later to Remington Rand.

Maudslay, Henry (1771–1831) British engineer who became a leading builder of MARINE ENGINES. Maudslay made many innovations in engineering practice, including the production of accurate plane surfaces, the use of a LATHE slide rest, an accurate bench MICROMETER and a practical screw-cutting lathe.

Maury, Matthew Fontaine (1806–73) US naval officer, a proponent of OCEANOGRAPHY. As superintendent of the US Naval Observatory and Hydrographic Office (1842–61), he produced charts showing winds and currents for the Atlantic, Pacific and Indian oceans. Maury proved the feasibility of laying a transatlantic cable, and wrote *The Physical Geography of the Sea* (1855). *See also* HYDROGRAPHY.

maxilla Upper jawbone of a vertebrate, or one of a pair of mouthparts in various arthropods. Centipedes, crustaceans, insects and millipedes have maxillae that move sideways to manipulate food. They lie close to the mouth, behind the MANDIBLES. In insects, a second pair of maxillae are united to form the LABIUM; in crustaceans, the two pairs of maxillae are separate. *See also* JAW

Maxim, Sir Hiram Stevens (1840–1916) US inventor of the Maxim gun. A mechanical genius, he took out his first patent at the age of 26 and invented numerous devices before settling in London in 1881. In 1883 Maxim built the first fully automatic machine gun, the Maxim gun. It was the first machine gun successfully to use the recoil energy of the fired bullet to eject the shell and to place another cartridge in the firing chamber. *See also* FIREARM

Maxwell, James Clerk (1831–79) Scottish mathematician and physicist who did outstanding theoretical work revealing the existence of ELECTROMAGNETIC RADIATION. In 1871 Maxwell became the first Cavendish Professor of Experimental Physics at Cambridge, England. He used the theory of the electromagnetic field to derive what are now called **Maxwell's equations**. These predict the existence of electromagnetic waves. Maxwell theoretically established (correctly) the nature of Saturn's rings, and did further work in thermodynamics and statistical mechanics. A unit of magnetic flux, the maxwell, was named after him (but is now replaced by the SI unit, the WEBER).

Maybach, Wilhelm (1846–1929) German engineer. Maybach worked with Gottlieb DAIMLER in the design and manufacture of INTERNAL COMBUSTION ENGINES and automobiles, for which he designed many improvements in fuel injection, gears, and so on. He opened his own factory in 1909, building engines for airships and, later, Maybach Cars.

Mayer, Julius Robert von (1814–78) German physicist and physician who determined the mechanical equivalent of heat (credited to James JOULE), and propounded a statement of the law of conservation of energy (credited to Hermann HELMHOLTZ). Lack of recognition led Mayer to attempt suicide and he was committed to an asylum.

Mayer, Maria Goeppert (1906–72) US physicist, b. Germany. In 1949 she proposed that protons and neutrons in an atomic NUCLEUS are arranged in orbits or shells, much as electrons are arranged around the nucleus. This provided an explanation of the apparently anomalous properties of some nuclei. Mayer shared the 1963 Nobel Prize for physics with Johannes H.D. Jensen (1907–73) and Eugene P. WIGNER.

ME (abbreviation of myalgic encephalomyelitis) Condition defined as extreme fatigue that persists for six months or more and is not relieved by rest. Also known as chronic fatigue syndrome and post-viral fatigue syndrome, it may be accompanied by many other non-specific symptoms, such as fever, headache, muscle and joint pains, gastric upset, dizziness, and mood changes. Often occurring after a VIRUS infection, it ranges in severity from chronic weariness to total physical collapse. The cause of the condition is unknown. There is no specific treatment, and recovery may take months or even years.

mean In statistics, mathematical AVERAGE of a set of values representing some data or observations. The mean of n such quantities is calculated as their sum divided by n. For instance, the mean of the numbers 4, 8, 9, 12 and 17 is 10.

meander Naturally occurring loop-like bend in the course of a river. A river will wind round any obstacle, such as hard rock. Once a meander has been created it will continue, becoming accentuated by the erosive action of the river. On the outside of a bend there will be lateral CORRASION, which will gradually work out sideways. On the inside of the bend there is likely to be some deposition, which will build up a flat flood plain. Meanders gradually move downstream. Sometimes meanders make complete loops, which, when cut off, form OXBOW LAKES. The name comes from the winding River Maeander in Asian Turkey.

mean free path Average distance a molecule moves between successive collisions. The concept is most widely used in the KINETIC THEORY of gases, for which the mean free path is usually about 10^{-7}m at normal temperature and pressure.

mean sea level Mean height of the sea calculated from measurements at regular intervals at points on the open coast. It is used as a standard or fixed geodetic point. *See also* GEODESY

measles (rubeola) Extremely infectious viral disease of children. The symptoms (fever, catarrh, skin rash and spots inside the mouth) appear about two weeks after exposure. Hypersensitivity to light is characteristic. Complications such as pneumonia and encephalitis occasionally occur, and middle-ear infection is a hazard of the disease, which was once a major cause of infant mortality in Western countries. Vaccination produces life-long IMMUNITY to measles.

measurement *See* WEIGHTS AND MEASURES

mechanical advantage *See* FORCE RATIO

mechanical engineering Field of ENGINEERING concerned with the design, construction and operation of machinery. Mechanical engineers work in many branches of industry, including transportation, power generation and tool manufacture. Achievements in mechanical engineering include the development of wind and water TURBINES; STEAM ENGINES and INTERNAL COMBUSTION ENGINES; assembly line production; and the control of machines using HYDRAULICS, FLUIDICS and COMPUTERS.

mechanics Branch of physics concerned with the behaviour of MATTER under the influence of FORCES. It may be divided into solid mechanics and fluid mechanics. Another classification is as statics – the study of matter at rest – and DYNAMICS – the study of matter in motion. In **statics**, the forces on an object are balanced and the object is said to be in equilibrium; static equilibrium may be stable, unstable or neutral. **Dynamics** may be further divided into kinematics – the description of motion without regard to cause – and KINETICS – the study of motion and force. Classical dynamics rests primarily on the three laws of MOTION formulated by Isaac NEWTON. Modern physics has shown these laws to be special cases approximating to more general laws. Relativistic mechanics deals with the behaviour of matter at high speeds, approaching that of light, whereas QUANTUM MECHANICS deals with the behaviour of matter at the level of atoms and molecules. *See also* QUANTUM THEORY; RELATIVITY

Medawar, Sir Peter Brian (1915–87) British biologist. He shared the 1960 Nobel Prize for physiology or medicine with Sir Frank Macfarlane Burnet (1899–1985) for the discovery of acquired immune tolerance. Medawar confirmed Burnet's theory that an organism can acquire the ability to recognize foreign tissue during embryonic development, and if that tissue is introduced in the embryonic stage it may be reintroduced later without inducing an immune reaction. *See also* IMMUNE SYSTEM

median In statistics, the middle item in a group found by ranking the items from smallest to largest. In the series 2, 3, 7, 9, 10, for example, the median is 7. With an even number of items, the MEAN of the two middle items is taken as the median. Thus, in the series 2, 3, 7, 9, the median is 5.

medicine Practice of the prevention, diagnosis and treatment of disease or injury; the term is also applied to any agent used in the treatment of disease. Pictographs dating from the earliest times show medical procedures and we can deduce from ancient skeletons that the practice of medicine goes back to prehistoric man. The Egyptians included magico-religious elements along with empirical therapies for many ailments, and the Hebrews placed a marked emphasis on hygiene. The Greeks had become thoroughly secular in their approach to medicine by the 6th century BC and several medical schools had been built by that time. The Romans and then the Arabs carried on and developed further the Greek medical tradition. During the Renaissance this knowledge was brought back to Europe. The dawn of modern medicine coincided with accurate anatomical and physiological observations first made in the 17th century. By the 19th century practical diagnostic criteria had been developed for many diseases; BACTERIA had been discovered and research undertaken for the production of immunizing sera in attempts to eradicate disease. The great developments of the 20th century include CHEMOTHERAPY, new surgical procedures, including organ transplants, and sophisticated diagnostic devices such as radioactive TRACERS and various scanners. *See also* ANATOMY; PHYSIOLOGY

Mediterranean climate Type of CLIMATE associated with the countries surrounding the Mediterranean Sea, but also found in S California, USA, central Chile and parts of Australia. Summers are hot and sunny, averaging 25°C (77°F) or more; winters are mild, ranging from 5°C to 15°C (41°–59°F). Summers are very dry, but winters can be quite wet, with up to 600mm (24in) of rainfall, some of which comes as thunderstorms at the end of the summer.

medulla oblongata Principal structure of the hindbrain, joining the BRAIN to the SPINAL CORD. Medulla functions are necessary to maintain life: it controls respiration and heartbeat. The medulla oblongata also contains part of the reticular formation that plays a role in arousal states, such as consciousness, wakefulness and attention. General anaesthetics, such as chloroform, probably work by depressing medulla activity.

medusa In zoology, free-swimming stage in the lifecycle of CNIDARIA. There are many types of medusa. Some are small creatures, transient stages in the life cycle of the Cnidaria, which soon change into the dominant, POLYP stage. Other medusas are larger and live longer, the polyp stage being the briefer of the two. Yet other medusas, including many jellyfish, have no polyp stage, and when they reproduce give rise directly to medusa offspring.

meerschaum (sepiolite) Clay silicate mineral, hydrated magnesium silicate, found mainly in Asia Minor. It is opaque white in colour, and has a fibrous texture. Hardness 2–2.5; r.d. 2.

▲ **Maxwell** James Clerk Maxwell applied his mathematic skills to a number of physical problems, arriving at the formulation of Maxwell's equations in 1864. With these, Maxwell unified the many varied phenomena of electricity and magnetism, binding them together in the new electromagnetic theory. He also worked out, at the same time, but independently of Austrian physicist Ludwig Boltzmann, the Maxwell-Boltzmann distribution of velocities for the molecules of a gas, discovering how the distribution depends on the temperature of the gas.

megabyte (symbol Mb) In computing, 1,024 kilobytes. Often used for expressing the capacity of computer memory or storage devices, the megabyte is sometimes rounded to stand for 1 million BYTES.

meiosis In biology, process of cell division that reduces the CHROMOSOME number from DIPLOID to HAPLOID. Meiosis involves two nuclear divisions. The first meiotic division halves the chromosome number in the two resulting cells; the second division then forms four haploid cells. In most organisms, the resulting haploid cells are the GAMETES, or sex cells, the OVA and SPERM. However, in others, which show ALTERNATION OF GENERATIONS (have two distinct generations in the life cycle, one haploid and the other diploid), these haploid cells may give rise to a new generation of plants or algae, the haploid "gametophyte" generation, which will later produce gametes by MITOSIS (normal cell division that does not involve halving of the chromosome number). *See also* ANAPHASE; CROSSING OVER; METAPHASE

Meitner, Lise (1878–1968) Austrian physicist. Working with Otto HAHN, Meitner discovered PROTACTINIUM and the fission of URANIUM, and she investigated beta decay and nuclear isomerism. She did research with Fritz STRASSMANN on the products resulting from neutron bombardment of uranium. Although her work contributed to the development of the atom bomb, Meitner refused to assist the project. She also researched the radioactive disintegration products of ACTINIUM, THORIUM and RADIUM. In 1938 Meitner fled the Nazi regime, working in Sweden until 1960, when she retired to England.

melamine ($C_3N_6H_6$) Organic chemical compound that reacts with methanal (formaldehyde, HCHO) to form a resin POLYMER, one of the thermosetting PLASTICS. Melamine resins are used as adhesives, for impregnating paper, for treating wool against shrinkage, and as constructional plastics.

melanin Dark pigment found in the skin, hair and parts of the eye. The amount of melanin determines skin colour. Absence of melanin results in an ALBINO.

melanocyte-stimulating hormone (MSH) Hormone, produced by the pituitary gland, that stimulates certain cells in the skin to synthesize MELANIN.

melanoma Type of skin CANCER. Melanomas are highly malignant tumours formed by melanocytes, cells in the skin that make the dark pigment MELANIN. They may also occur in mucous membranes and in the eye. Untreated, they may spread to the liver and lymph nodes. Excessive exposure to sunlight has been cited as a causative factor. Treatment is usually by surgery.

Melloni, Macedonio (1798–1854) Italian physicist who studied INFRARED radiation. Melloni improved the thermopile and used it to detect infrared radiation. He also showed that rock salt is transparent to infrared, and by 1850 had demonstrated that infrared undergoes reflection, refraction, polarization and interference in the same way as does visible light. Melloni laid the foundation for James Clerk MAXWELL's hypothesis of a spectrum of ELECTROMAGNETIC RADIATION beyond the visible region.

meltdown Out-of-control heating of a NUCLEAR REACTOR resulting in melting of the reactor core. The temperature of a reactor core is constantly monitored, and any dangerous rise in temperature automatically initiates a shutdown procedure. Other precautions against the possible disastrous results of a meltdown include surrounding the reactor core with a containment vessel of thick concrete, and a system of pipes that flood the core in such an emergency. Two known partial meltdowns have occurred: in 1979 at Three Mile Island, Pennsylvania, USA, and in 1986 at Chernobyl, Ukraine.

melting point Temperature at which a substance changes from SOLID to LIQUID. The melting point of the solid has the same value as the freezing point of the liquid, for example, the melting point of ice, 0°C (32°F), is the same as the freezing point of water.

meltwater Water originating from melted ice or snow. When temperatures rise, snow produces meltwater which swells rivers and may contribute to flooding. Meltwater from glaciers may add to the water in an existing river or supply streams that form a new river.

membrane In biology, boundary layer inside or around a living CELL or TISSUE. Cell membranes include the plasma membrane surrounding the cell, the network of membranes inside the cell (ENDOPLASMIC RETICULUM) and the double membrane surrounding the NUCLEUS. The multicellular membranes of the body comprise mucous membranes of the respiratory, digestive and urinogenital passages, synovial membranes of the joints, and the membranes that coat the inner walls of the abdomen, thorax and the surfaces of organs. *See also* EPITHELIUM

Mendel, Gregor Johann (1822–84) Austrian naturalist who discovered the laws of HEREDITY and in so doing laid the foundation for the modern science of GENETICS. Mendel grew pea plants and observed how various characteristics, such as tallness, flower colour and seed shape, appeared in their offspring. He proposed that the characteristics were inherited, carried by "factors" (now known to be genes) contributed by each parent. His study was published in *Experiments with Plant Hybrids* (1866), which was rediscovered in 1900.

mendelevium (symbol Md) Radioactive, metallic TRANSURANIC ELEMENT. It was first synthesized in 1955 at the University of California at Berkeley (one atom at a time) by the alpha-particle bombardment of einsteinium-253. Properties: at.no. 101; r.a.m. 258; most stable isotope ^{258}Md (half-life 2 months).

Mendeleyev, Dmitri Ivanovich (1834–1907) Russian chemist who devised the PERIODIC TABLE. Mendeleyev demonstrated that chemically similar elements appear at regular intervals if the elements are arranged in order by RELATIVE ATOMIC MASS (R.A.M.). He classified the 60 known elements and left gaps in the table, predicting the existence and properties of several unknown elements later discovered, including GALLIUM and SCANDIUM. His work was published in Russian in 1869, and was soon translated. His textbook, *The Principles of Chemistry*, was written between 1868 and 70, and translated in 1905, rapidly becoming a standard work. The element MENDELEVIUM is named after him.

Mendelism Theory of inheritance originally proposed by Gregor MENDEL. It established the science of classical GENETICS.

MENDELISM (CROSS-BREEDING OF PLANTS)

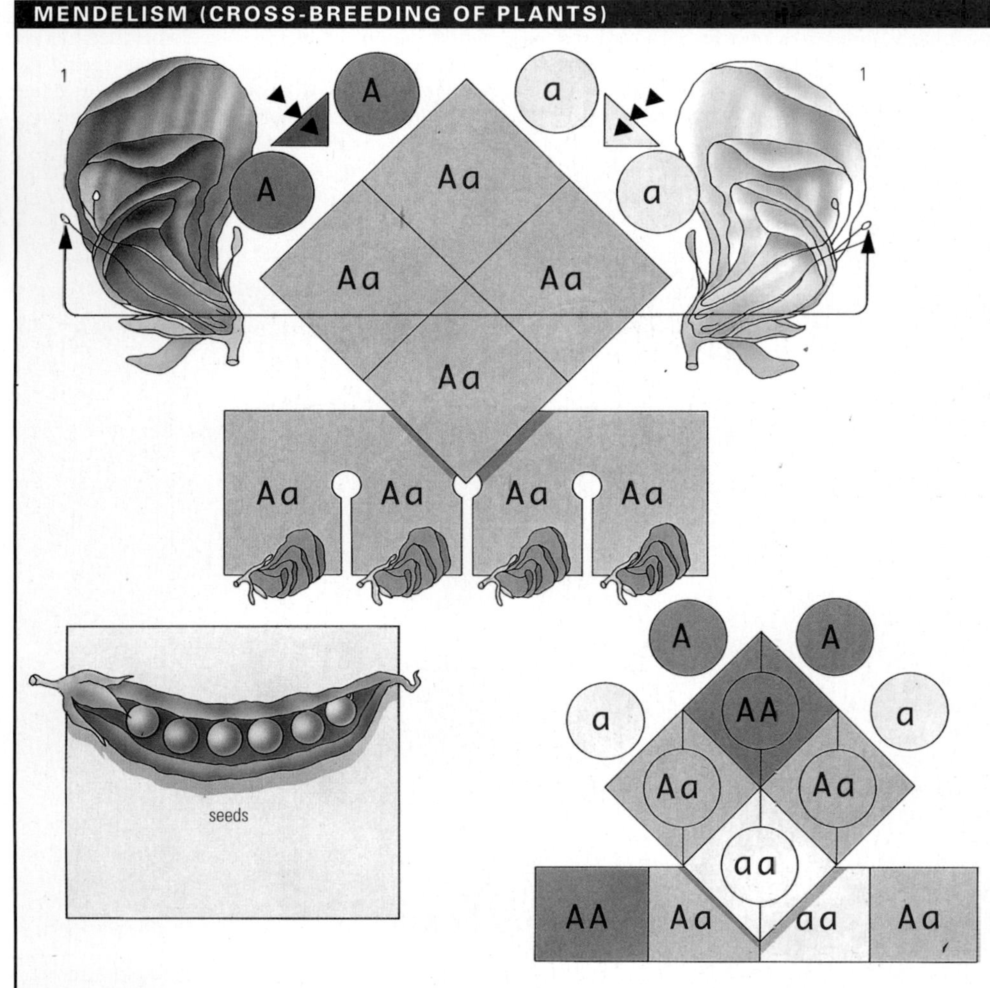

Gregor Mendel is widely accepted as making the initial discoveries of the laws of heredity and hence genetics. His discoveries were primarily based on the cross-breeding of plants. The genetic characteristics of plants and animals can be studied by cross-breeding under controlled conditions. Mendel discovered that some characteristics are determined by genes that contain a dominant allele (A) over others which contain a recessive allele (a). Flowering plant cells usually contain two alleles, but pollen or eggs (ova) contain only one. The pure-bred pea plants (1) are either purple or white. The F1 crosses produced by cross-breeding them are all purple, but carry one dominant and one recessive allele. The F2 generation will produce flowers in the ratio three purple to one white flower, as any flower with an A allele will be dominant.

meninges Three membranes that cover the BRAIN and SPINAL CORD. The outermost membrane, the **dura mater**, is a tough protective covering. Within it is the second membrane, the **arachnoid**; the innermost membrane, the **pia mater**, is a delicate layer on the surface of the brain and spinal cord containing blood vessels which supply the nervous system with nourishment. Between the arachnoid and the pia mater is the subarachnoid space, in which flows the cerebrospinal fluid.

meningitis Inflammation of the MENINGES (membranes) covering the brain and spinal cord, resulting from infection. Bacterial meningitis is more serious than the viral form. Symptoms include headache, fever, nausea and stiffness of the neck. The disease can be fatal.

meniscus Curved upper surface of a liquid in a container, due to surface tension. The surface of water in an air-water-glass system is concave, whereas in an air-mercury-glass system the mercury meniscus is convex.

menisectomy Surgical removal of damaged cartilage from the knee joint.

menopause Stage in a woman's life marking the end of the reproductive years, when the MENSTRUAL CYCLE becomes irregular and finally ceases, generally around the age of 50. Popularly known as the "change of life", it may be accompanied by unpleasant effects such as hot flushes, excessive bleeding and emotional upset. HORMONE REPLACEMENT THERAPY (HRT) is designed to relieve menopausal symptoms.

Mensa (table) Constellation of the far S sky. Although it has no star above magnitude 5, part of the Large MEGALLANIC CLOUD extends into it.

menstrual cycle Cycle in female humans and some higher primates of reproductive age, during which the body is prepared for pregnancy. In women, the average cycle is 28 days. At the beginning of the cycle, HORMONES from the PITUITARY GLAND stimulate the growth of an ovum (egg cell) contained in a follicle in one of the two OVARIES. At approximately mid-cycle, the follicle bursts, the egg is released (ovulation) and travels down the FALLOPIAN TUBE to the UTERUS. The follicle (now called the CORPUS LUTEUM) secretes two hormones, PROGESTERONE and OESTROGEN, during this secretory phase of the cycle, and the ENDOMETRIUM thickens, ready to receive the fertilized egg. Should fertilization (conception) not occur, the corpus luteum degenerates, hormone secretion ceases, the endometrium breaks down and menstruation occurs; the unfertilized egg is discharged in the blood flow from the VAGINA. In the event of conception, the corpus luteum remains and maintains the endometrium with hormones until the PLACENTA is formed. Menstruation, therefore, marks the end of the cycle. It is customary, for the purposes of CONTRACEPTION and calculating the date of an expected birth, to count the first day of bleeding as the first day of the next menstrual cycle. The onset of the menstrual cycle (menarche) is called PUBERTY (age 10–15 years); it ceases with the MENOPAUSE (around 50 years). *See also* FOLLICLE STIMULATING HORMONE (FSH); LUTEINIZING HORMONE (LH)

menthol ($C_{10}H_{19}OH$) White, waxy crystalline compound having a strong odour of peppermint. Its main source is oil of peppermint, *Mentha arvensis*. It is an ingredient of decongestant ointments and nasal sprays and is used to flavour toothpaste and cigarettes.

Mercalli scale, modified Scale of 12 points used for measuring earthquake intensity. Named after the Italian seismologist Giuseppe Mercalli (1850–1914), it is based on damage done at any point and so varies from place to place. Earthquake magnitude, on the other hand, is a function of the total energy released. *See* RICHTER SCALE

mercaptan *See* THIOL

Mercator, Gerardus (1512–94) Flemish cartographer. His huge world map of 1569 employed the system of projection now named after him, in which lines of longitude, as well as latitude, appear as straight, parallel lines. His scholarship was matched by his practical abilities, and he was respected as the greatest geographer of his time, patronized by Emperor Charles V (1519–56). Mercator also produced a globe (1541) and a huge, unfinished, atlas of the world (1595).

Mercury Innermost planet of the Solar System, and the smallest of the four inner planets. Mercury is difficult to observe because it is very close to the Sun, but through a telescope light and dark areas can be made out. The planet has no known satellite. Very little was known about Mercury's surface until the Mariner 10 probe made three close approaches to the planet in 1974 and 1975 and returned pictures of nearly half the surface. These showed a heavily cratered, lunar-like world, marked by many craters, valleys and ridges. The surface temperature at noon on the equator can reach over 800K at the perihelion, while on the night side it can fall to 90K. In 1991–92 radar mapping of Mercury's polar regions revealed what may be water-ice on the floors of craters permanently in shadow, where the temperature is estimated never to rise above 112K. There is a very tenuous atmosphere, mainly of helium and sodium. The presence of a weak magnetic field and the planet's high density for its size suggest that Mercury has a very large iron-rich core, proportionally much bigger than for any other Solar System body.

mercury (quicksilver, symbol Hg) Liquid, metallic element, known from earliest times. The chief ore is cinnabar (a sulphide), from which it is extracted by roasting. The silvery element is the only metal that is liquid at normal temperatures. It is a dangerous cumulative poison. Mercury is used in barometers, thermometers, laboratory apparatus, mercury-vapour lamps and mercury cells. It forms two series of salts, termed mecury(I), or mercurous, and mercury(II), or mercuric. Mercury compounds are used in pharmaceuticals and some dentists still fill teeth with a mercury alloy (AMALGAM), although increasingly composite resins are used. Mercury's toxic effects have always plagued workers in industries using its compounds. The industrial dumping of mercury wastes into the sea allows it to enter the human food cycle, because fish and birds ingest the mercury and these are eaten, passing on the poison to the next level in the food chain. Symptoms of mercury poisoning include loss of coordination, balance and peripheral vision, sensory disturbance and deformed offspring. Properties: at.no. 80; r.a.m. 200.59; r.d. 13.6; m.p. −38.87°C; (−37.97°F); b.p. 356.58°C (673.84°F); most common isotope ^{202}Hg (29.8%).

mercury fulminate *See* FULMINATE

mercury switch In electrical engineering, switch in which mercury is contained in a tube which, when tilted, causes the mercury to bridge two contacts. An alternative design consists of mercury cups into which movable contacts dip.

Mergenthaler, Ottmar (1854–99) US inventor, b. Germany. In 1884 Mergenthaler patented his LINOTYPE machine for setting solid lines of type, once used by most newspapers. In 1885 the machine was improved by a device that automatically justified the type (ranged it to a regular margin at both ends of the lines). The system has been superseded by PHOTOSETTING.

meridian Circle that runs through the N and S poles, at right angles to the Equator. *See also* CELESTIAL MERIDIAN; LONGITUDE

meristem In plants, a layer of cells that divides repeatedly to generate new tissues. It is present at the growing tips of shoots and roots, and at certain sites in leaves. In MONOCOTYLEDONS, the leaf meristem is at the base, explaining why grasses continue to grow when the leaf tips are removed by grazing or mowing. *See also* CAMBIUM

mesa Flat area of upland rather like a table, so called after the Spanish mesa, meaning "table". They often have horizontal strata of SEDIMENTARY ROCK.

mesentery Sheet of tissue, or double layer of the PERITONEUM, which attaches the abdominal organs to the rear wall of the abdomen.

MENSTRUAL CYCLE

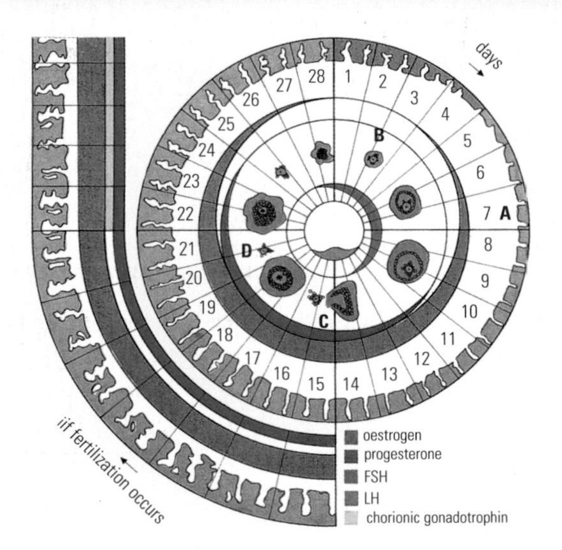

The changes occurring during the menstrual cycle are controlled by the balance of the follicle stimulating hormone (FSH) and luteinizing hormone (LH) secreted by the pituitary. The diagram shows the changing levels of these, and of oestrogen and progesterone induction from the ovarian follicle, together with changes in the structure of the uterine wall (A) and development of the follicle (B), in a circular form through a normal 28-day cycle. The sharp increase in LH at about mid-cycle causes ovulation (C) and, if fertilization does not occur, the corpus luteum (D) formed degenerates around day 26 as pituitary hormone levels fall. The consequent withdrawal of oestrogen and progesterone causes the uterine wall to shed itself in the menstrual flow. This then proliferates again under the influence of oestrogen from a new follicle. If fertilization and egg implantation do occur the placenta produces chorionic gonadotrophin, possibly as early as day 21, that allows the corpus luteum to continue to produce oestrogen and progesterone until the placenta takes over.

mesoderm One of the three so-called GERM LAYERS of tissue formed in the early development of a fertilized OVUM (egg) of higher animals. It is the middle layer and, in later development of the embryo, the mesoderm gives rise to muscles, blood and connective tissues. Other germ layers are the ECTODERM and ENDODERM.

meson ELEMENTARY PARTICLE, member of a subgroup of HADRONS, all of which have either zero or integral SPIN. They include the pions, kaons and eta mesons. There is no restriction on the number of mesons produced or destroyed in a nuclear reaction or present in a particular energy state.

mesophyll Soft tissue located between the two layers of EPIDERMIS in a plant leaf. In most plants, mesophyll cells contain chlorophyll-producing structures called CHLOROPLASTS, which are essential to PHOTOSYNTHESIS.

mesophyte Plant that grows under average moisture conditions, thriving where there is a good balance of water and evaporation. Such plants have well-developed root and leaf systems. *See also* HYDROPHYTE; XEROPHYTE

mesosphere Middle shell of gases in the Earth's ATMOSPHERE between the STRATOSPHERE and the THERMOSPHERE.

Mesozoic Third era of geologic time, extending from about 248 million to 65 million years ago. It is divided into three periods: the TRIASSIC, JURASSIC and CRETACEOUS. For most of the era, the continents are believed to have been conjoined into one huge land mass called PANGAEA. There was much volcanic activity and mountain building throughout the period, which was also characterized by the variety and size of its reptiles. For this reason it is sometimes called the "Age of Reptiles".

messenger RNA (mRNA) Type of RNA (ribonucleic acid) that carries the GENETIC CODE for PROTEIN SYNTHESIS. The mRNA transcribes the code from DNA and carries it to RIBOSOMES within the cell, where AMINO ACIDS are assembled to make polypeptides and proteins.

Messier catalogue List of 109 STAR CLUSTERS, NEBULAE and GALAXIES compiled by the French astronomer Charles Messier (1730–1817). Messier drew up the list so that he and other comet-hunters would not confuse these permanent, fuzzy-looking objects with COMETS. The first edition of the catalogue was published in 1774, with supplements in 1780 and 1781. Objects in the catalogue are given the prefix M, and are still widely known by their Messier numbers.

metabolism Chemical and physical processes and changes continuously occurring in a living organism. They include the breakdown of organic matter (CATABOLISM), resulting in energy release, and the synthesis of organic components (ANABOLISM) to store energy and build and repair TISSUES. *See also* BASAL METABOLIC RATE (BMR)

metabolite Any chemical substance that is involved in the metabolic processes of cells in organisms. These substances function in the various biological energy exchanges necessary for growth, maintenance and reproduction. *See also* METABOLISM

metacarpals In humans and other primates, bones of the palm of the hand. They articulate with the CARPALS (wrist bones) and PHALANGES (finger bones). In four-footed animals, they are the bones of the forefoot.

metal Any element that is a good conductor of heat and electricity – the atoms of which are bonded together within crystals in a unique way. Mixtures of such elements (ALLOYS) are also metals. About three-quarters of the elements are metals. Most are hard, shiny materials that form oxides (combined with oxygen). Many corrode if exposed to moist air. MERCURY is exceptional in being a liquid at room temperature. SODIUM and POTASSIUM are examples of metals that are soft and chemically very reactive: they tarnish quickly in air and are most familiar as their salts. The lightest metal is LITHIUM, which is also very reactive. The heaviest is OSMIUM, which is 22.6 times denser than water and is one of the PLATINUM group, relatively unreactive metals that include RUTHENIUM, RHODIUM, PALLADIUM and IRIDIUM. Malleability and ductility are further metallic characteristics. GOLD is the most malleable of all metals; it can be beaten so thin as to be virtually translucent. Some metals have very high melting points and various high-temperature applications: TUNGSTEN, with the highest melting point of all at 3,410°C (6,170°F), is employed as LIGHT BULB filaments. ALUMINIUM, followed by IRON, are the two most abundant and useful of metals. TITANIUM, although rarely seen as the metal, is more commonly distributed than the more familiar COPPER, ZINC and LEAD. Other metals of great economic importance, because of their radioactivity, are URANIUM and PLUTONIUM. *See also* ALKALI METALS; ALKALINE-EARTH METALS; METALLIC BOND; TRANSITION ELEMENTS

metal fatigue Progressive fracture of metals subjected to repeated cycles of STRESS. Fatigue can occur early in the metal's life, with stresses building up as it cools after the manufacturing process. The phenomenon is of particular relevance to aeronautical engineers because metals can break, tear, or otherwise deform permanently under repeated or reversed loads at stress levels much lower than for a single loading.

metallic bond In chemistry, a bond that holds atoms together in a METAL. Inside the crystals of metals, positive IONS are held by the electrostatic attraction of a cloud or "sea" of surrounding electrons which can, under various influences, move. Such electron movement under the influence of an applied voltage constitutes an electric current, so accounting for the electrical CONDUCTIVITY of metals.

metallography Study of the structure of metals and alloys, using optical and electronic microscopes and X-ray diffraction techniques. Examination by microscope reveals the size and shape of crystals and the distribution of nonmetallic inclusions. An X-RAY DIFFRACTOMETER is used to study the arrangement of the atoms in a metallic specimen.

metalloid (semimetal) Element having some properties typical of METALS, and some normally associated with NONMETALS. In moving from left to right across the PERIODIC TABLE and moving down the groups, there is a transition from metallic to nonmetallic elements. Metalloids occur as borderline cases in groups III–VI. Examples are SILICON, GERMANIUM and ARSENIC. Some metalloids are SEMICONDUCTORS.

metallurgy Science and technology concerned with METALS. Metallurgy includes the study of: methods of the extraction of metals from their ORES; physical and chemical properties of metals; ALLOY production; and the hardening, strengthening, corro-

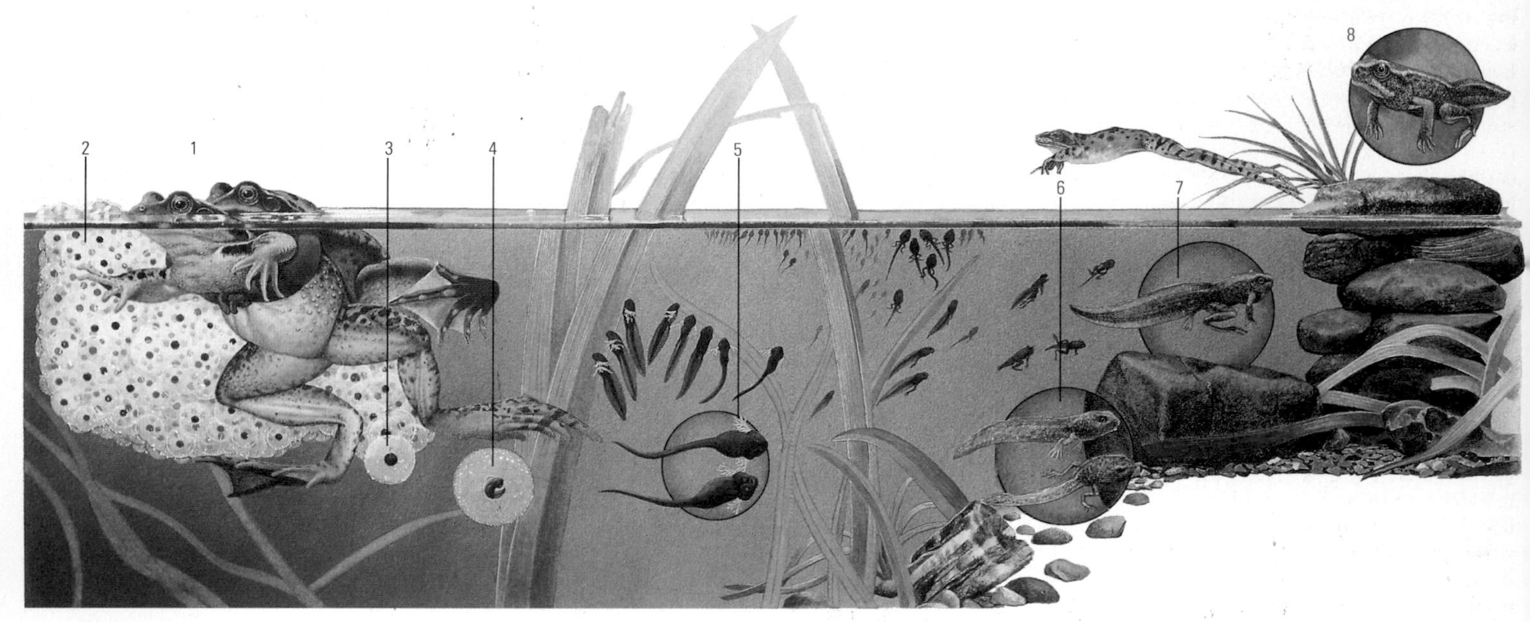

▲ **metamorphosis** When common frogs mate, fertilization and egg laying occur in water (1). Within an hour, the jelly around the egg swells to produce frogspawn (2). The eggs develop (3) and produce embryos (4) that hatch, as long-tailed tadpoles with external feathery gills, six days after fertilization (5). Mouths and eyes develop later and the tails become powerful means of propulsion. Hind legs are well formed by week eight (6); meanwhile, the tadpole has changed from a herbivore to a carnivore. Via an intermediary gill and lung stage, the tadpole changes from gill- to lung-breathing, its internal lungs growing as its external gills are absorbed; the process is complete when the gills fully disappear at month three, by which time the forelegs are well developed (7). Metamorphosis is complete when the young frog (8) loses its tail.

sion-proofing and ELECTROPLATING of metals. Analyses by chemical methods and SPECTROSCOPY are made to determine the composition of alloys during manufacture. Among the physical tests for metals is the measurement of resistance to impact and tension. Works processes include melting, SINTERING, STAMPING, FORGING, dipping and ELECTROPLATING, GALVANIZING, ANODIZING and SHERARDIZING.

metamorphic rock Broad class of rocks that have been changed by heat or pressure, or both heat and pressure, from their original nature – SEDIMENTARY, IGNEOUS, or older metamorphic. Heat is the result of volcanic activity, and pressure is the result of earth movements. Metamorphic rocks are generally hard and resistant to erosion, and are likely to form high ground. In metamorphic rocks, minerals may recrystallize or be compressed, and the new rock may look completely different from the original sedimentary or igneous rock. Thus, the metamorphic rock SLATE is made from sedimentary SHALE, the metamorphic GNEISS from igneous GRANITE.

metamorphosis Change of form or structure during the development of various organisms, such as the changing of a caterpillar into a moth, or a tadpole into a frog. Sometimes the change is gradual, as with a grasshopper, and is known as incomplete metamorphosis. Complete metamorphosis involves a change in habit or environment and usually involves the more distinct stages of LARVA, PUPA and IMAGO (adult stages).

metaphase Stage in cell division that occurs during MITOSIS and MEIOSIS. During metaphase, the nuclear membrane breaks down, the spindle forms and CHROMOSOMES become attached to the centre of the spindle. In the first metaphase of meiosis, paired chromosomes attach to the spindle. In the second metaphase of meiosis and in mitosis, individual chromosomes attach to the spindle. The division then proceeds to ANAPHASE.

metasomatism Production of mineral deposits by the movement of hot fluids from an igneous body through cracks or pores in the surrounding rock. Veins of lead and tin ores surrounding the Cornish granites were formed in this way.

metastable state Uneasy stability of a system that can easily be disturbed, making it descend to a lower energy state (a coin balanced on its edge is an example). A supersaturated solution or supercooled liquid, which remains liquid below its freezing point, is also metastable because the addition of a speck of dust or a tiny seed crystal will make it rapidly crystallize.

metastasis Spread of malignant TUMOURS beyond their original sites.

metatarsals Bones of the feet (hind feet in quadrupeds). They articulate at one end with the TARSALS (ankle bones) and at the other with the PHALANGES (toe bones).

Metchnikoff, Élie (1845–1916) Russian microbiologist. Metchnikoff shared the 1908 Nobel Prize for physiology or medicine with Paul EHRLICH for work on the mechanism of IMMUNITY, which included the discovery that LEUCOCYTES (white blood cells) are important in the body's resistance to infection and disease. He was also noted for his theories concerning longevity. He wrote *Immunity in Infectious Diseases* (1905).

meteor (shooting star) Brief streak of light in the night sky caused by a **meteoroid** (a solid particle, usually the size of a grain of dust) entering the Earth's upper atmosphere at high speed from space. They occur at altitudes of about 100km (60mi). The typical meteor lasts for a few tenths of a second to a second or two, depending on the meteoroid's impact speed, which can vary from about 11–70km s^{-1} (7–45mi s^{-1}). A few meteors per hour may be seen on any clear, moonless night at any time of year. But at certain times of the year there are METEOR SHOWERS, which occur when the Earth passes through a meteor stream – dust particles spread around the orbit of a COMET. Most of the meteors appearing during the year are sporadic meteors, not associated with cometary orbits. *See also* METEORITE

◄ **meteor shower** Time exposure showing a trail of a Perseid meteor against a starry sky at dusk. The Perseids are meteors seen annually around 12 August, and which appear to originate from the direction of the constellation Perseus. These meteors are small dust particles from an orbiting dust belt known as the meteor stream. As the Earth passes through the stream, a large number of dust particles enter the atmosphere and burn up, creating a bright trail. At its peak, the Perseid shower has about one meteor per minute. This photograph was taken in British Columbia, Canada.

meteorite That part of a large **meteoroid** (a solid particle moving in interplanetary space) that survives passage through the Earth's atmosphere and reaches the ground. Most of a meteoroid burns up in the atmosphere to produce a METEOR, but about 10% reaches the surface in the form of meteorites or micrometeorites. Meteorites generally have a pitted surface and a fused charred crust. There are three main types: IRON METEORITES (siderites); STONY METEORITES (aerolites); and STONY-IRON METEORITES. Some are tiny particles; others weigh up to 200 tonnes. Meteoroids weighing more than 100 tonnes that do not break up are not decelerated as much as lighter bodies, and produce impact craters, known as meteoric craters.

meteorology Study of weather conditions and a branch of the study of CLIMATOLOGY. Meteorologists study and analyse data from a network of weather ships, aircraft and satellites in order to compile maps showing the state of the high- and low-pressure regions in the Earth's atmosphere. They also anticipate changes in the distribution of the regions and forecast the future weather. Wind strength and direction can be predicted accurately by measuring the differences in air pressure over the surface of the Earth.

meteor shower Appearance of METEORS from the same point in the sky, the **radiant**, at around the same time each year. Nearly all showers are named after the constellation in which their radiant lies. There are a dozen or so major showers and many minor ones. During major showers there is a build-up of activity to a maximum, when 10–100 shower meteors may be visible each hour from any one location. A shower occurs when the Earth passes through a meteor stream. In an **annual** shower, the meteor rates vary little from one year to the next because the meteoroids in the stream are spread evenly around the orbit. With streams in which the meteoroids are bunched together in a swarm, meteor numbers are low except when the Earth intersects the swarm. Such **periodic** showers may produce a meteor storm (*see* LEONIDS).

meter Instrument that measures a particular quantity. For example, a gas meter measures the amount of gas that has flowed in a certain time, and a voltmeter measures the voltage between two points in an electric circuit.

methadone Synthetic opiate drug, the use of which is widespread as an institutionalized solution to HEROIN addiction. Similar to MORPHINE and heroin in chemical structure, it is an addictive drug, although the effects of withdrawal are believed to be milder than with morphine or heroin. *See also* OPIUM

methanal (formaldehyde, HCHO) Colourless, flammable, poisonous gas, with a penetrating odour. It is the simplest ALDEHYDE and is produced by the oxidation of METHANOL by air. It was discovered (1867) by August von HOFMANN. Most methanal is in the form of **formalin**, an aqueous solution of 35–40% methanal used as a preservative for biological specimens. Methanal is used in the manufacture of dyes and plastics. Synthetic resins made from methanal helped to create the first plastic objects. Chief properties: r.d. 0.82; m.p. −92°C (−133.6°F), b.p. −19 °C (−2.2°F).

methane (CH_4) Colourless, odourless HYDROCARBON, the simplest ALKANE (paraffin). It is the chief constituent of NATURAL GAS, from which it is obtained, and of FIREDAMP. Methane explodes when mixed with oxygen and ignited. It is produced naturally by decomposing organic matter, such as in marshes, which led to its original name of marsh gas. In the air, it contributes to the GREENHOUSE EFFECT and increases in global temperature. Methane is used in the form of natural gas as a fuel and, in its pure form, as a starting material for the manufacture of many chemicals. Properties: m.p. −182.5°C (−296.5°F); b.p. −164°C (−263.2°F).

methanoic acid (formic acid, HCOOH) Colourless, corrosive, pungent liquid CARBOXYLIC ACID, used to produce insecticides and for dyeing, tanning and electroplating. It occurs naturally in a wide variety of sources – stinging ants, nettles, pine needles and sweat. The simplest of the carboxylic acids, it can be produced by the action of concentrated sulphuric acid on sodium methanoate (formate). Properties: r.d. 1.22; m.p. 8.3°C (46.9°F); b.p. 100.8°C (213.4°F).

methanol (methyl alcohol, CH_3OH) Colourless, poisonous, flammable liquid that is the simplest of the ALCOHOLS. It is obtained synthetically either from carbon monoxide and hydrogen, by the oxidation of natural gas or by the destructive distillation of wood, giving it the informal name of wood alcohol. It is used as a solvent and to produce rocket fuel, and as a petrol additive. Properties: m.p. −93.9°C (−137°F); b.p. 64.9°C (148.8°F).

M

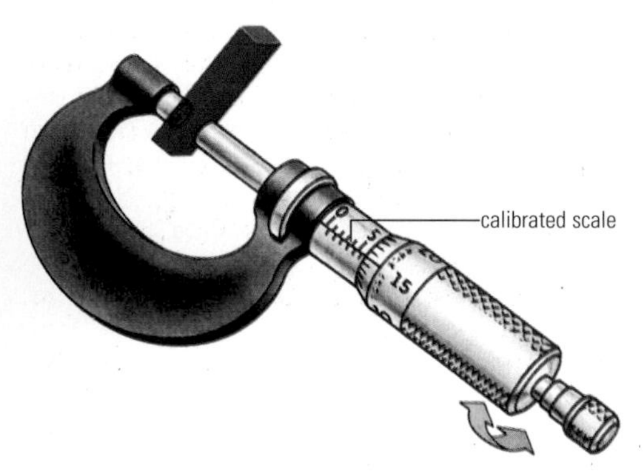

▲ **micrometer gauge** The micrometer gauge is a standard measuring instrument for dealing with precise dimensions in the engineering industry. It consists basically of a screw or spindle that can be screwed through a fixed nut. As the end of the spindle is turned in a clockwise direction, it closes towards the workpiece being measured, which is lightly held between the spindle and the anvil (an inset piece of metal opposite the spindle). The required dimension, which is related to the number of turns made by the screw, can be read off a graduated scale.

methionine ($CH_3S(CH_2)_2CH(NH_2)COOH$) Soluble, essential AMINO ACID found in proteins. It is unusual in that it contains sulphur. It is needed for growth in babies and for nitrogen equilibrium in adults.
methotrexate Chemical substance (an antimetabolite) that interferes with the synthesis of the vitamin FOLIC ACID, an essential component of cells. It is used to treat malignant TUMOURS, LEUKAEMIA, rheumatoid ARTHRITIS and psoriasis.
methyl alcohol *See* METHANOL
methylated spirit Industrial form of ETHANOL (ethyl alcohol). It contains 5% METHANOL (methyl alcohol), which is extremely poisonous, and enough pyridine to give it a foul taste. It is dyed purple and used as a solvent and fuel.
metre (symbol m) SI unit of length. Conceived as being one ten-millionth of the surface distance between the North Pole and the Equator, it was formerly defined by two marks on a platinum bar kept in Paris. It is now defined as the length of the path travelled by light in a vacuum during 1/299,792,458 of a second. One metre equals 39.3701 inches.
metric system Decimal system of WEIGHTS AND MEASURES based on a unit of length called the METRE (m) and a unit of mass called the KILOGRAM (kg). Larger and smaller metric units are related by powers of 10. Devised by the French in 1791, the metric system is used internationally by scientists (SI UNITS) and has been adopted for general use by most Western countries, although the imperial system is still commonly used in the USA and for certain measurements in Britain.
MeV (abbreviation of million ELECTRON VOLTS) Unit of energy used in particle physics to express the energy of elementary particles.
Meyer, Julius Lothar (1830–95) German chemist who worked on the relationship between ATOMIC VOLUME and atomic weight (RELATIVE ATOMIC MASS). He devised the periodic table at the same time as Dmitri MENDELEYEV, with whom he received the Davy Medal in 1882.
mica Group of common rock-forming minerals of the sheet silicate (SiO_4) type, characterized by a platy or flaky appearance. All contain aluminium, potassium and water in the form of OH^- ions; other metals such as iron and magnesium may be present. Micas have perfect basal cleavage; common members are MUSCOVITE and the BIOTITE group, which includes a number of varieties.
mica schist One of a number of METAMORPHIC ROCKS that have QUARTZ and MICA as their main constituents. Like all SCHISTS, mica schists have the bulk of their mineral components arranged in parallel, giving them a characteristic striped appearance. They often contain other metamorphic minerals such as garnet. Mica shists are found worldwide.
micelle Roughly spherical group of large molecules that come together in a COLLOID. For example, a molecule of detergent or soap has a hydrophilic ("water-loving") polar head and a long hydrophobic ("water-hating") nonpolar tail. When detergents or soaps dissolve in water, the molecules clump together to form micelles with the nonpolar tails at the centre, surrounded by a sphere of polar heads linked to water molecules. Bile salt in the intestine make the product of fat digestion form micelles, which are thereby more easily absorbed.
Michelson, Albert Abraham (1852–1931) US physicist, b. Germany. In 1887 he conducted an experiment with Edward MORLEY to determine the velocity of the Earth through the ETHER, using an INTERFEROMETER of his own design. The result of the MICHELSON-MORLEY EXPERIMENT is now seen as a key piece of evidence for the theory of RELATIVITY. Michelson also determined the speed of light and became the first US scientist to win a Nobel Prize, receiving the 1907 award for physics.
Michelson-Morley experiment Notable experiment performed in 1887 by Albert MICHELSON and Edward MORLEY to determine the motion of the Earth through the ETHER. That such a motion was not detected, discredited the ether theory and led to a crisis in physics that was resolved by EINSTEIN's theory of RELATIVITY.
Michurin, Ivan Vladimirovich (1855–1935) Russian plant breeder. His unorthodox theories of HEREDITY (Michurinism), which included belief in the theory of ACQUIRED CHARACTERISTICS, were for a time accepted as the official science of GENETICS by the Soviet Union. Through GRAFTING he managed to produce new strains of fruit, but his belief that grafting produces heritable changes has since been discredited.
microbiology Study of microorganisms, their structure, function and significance. Mainly concerned with single-cell forms, such as VIRUSES, BACTERIA, PROTOZOA and FUNGI, it has immense applications in medicine and the food industry. Microbiology began in the 17th century with the invention of the microscope, which enabled scholars to view microorganisms for the first time. Pioneers in the field include Robert HOOKE, Anton van LEEUWENHOEK and Louis PASTEUR.
microchip General term for either an INTEGRATED CIRCUIT or SILICON CHIP
microclimate *See* CLIMATE
microcline ($KAlSi_3O_8$) FELDSPAR mineral with a triclinic CRYSTAL STRUCTURE. It has a characteristic checkered twinning, seen through a microscope. It is common in granites. Hardness 6–6.5; r.d. 2.56
microcomputer Small COMPUTER with its CENTRAL PROCESSING UNIT (CPU) on an INTEGRATED CIRCUIT (SILICON CHIP) called a MICROPROCESSOR. The term microcomputer is often shortened to **micro**, and it generally has the same meaning as personal computer. It usually refers to a desktop computer, or to a portable (LAPTOP) computer that is used by only one person at a time. *See also* COMPUTER NETWORK
microelectronics Electronic systems designed and produced without wiring or other bulky components. They allow a high packing density that greatly reduces the size of component assemblies. In the years following World War 2, the application of such newly developed devices as the semiconductor TRANSISTOR and DIODE saw the beginnings of the microelectronics industry. This accelerated with the development of the PRINTED CIRCUIT, in which metal connections between miniaturized components are etched out in a single piece on a circuit board. Even further reduction in size, or **microminiaturization**, was later achieved with INTEGRATED CIRCUITS, in which complex circuits can be made smaller than a fingernail. Molecular electronics is a new development that promises to be the ultimate in size reduction, in which complex circuits of microscopic size are grown inside single crystals. The applications of microelectronics have been felt most in the COMPUTER industry, but pocket calculators, miniaturized

MICROMETER

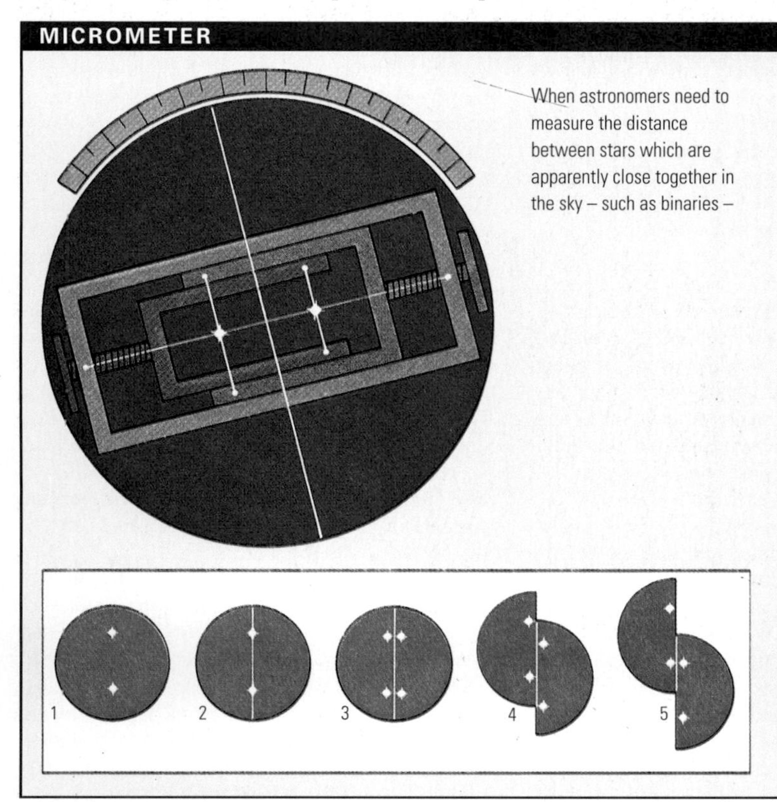

When astronomers need to measure the distance between stars which are apparently close together in the sky – such as binaries – they use a micrometer attached to the telescope. A wire micrometer uses two sets of cross-wires. One pair is used in the alignment of the instrument and a reading taken from the scale. The second set is mounted on movable carriages. The actual distance being measured by scales on the two screw adjustments. (1) shows the two stars (previously aligned) in the same field of view; (2) and (3) show how the system breaks up the field of view producing star images in each half. (4) shows what happens when two half fields are shifted optically. The total shift necessary to bring the stars into coincidence as in (5) is a measure of the distance between them.

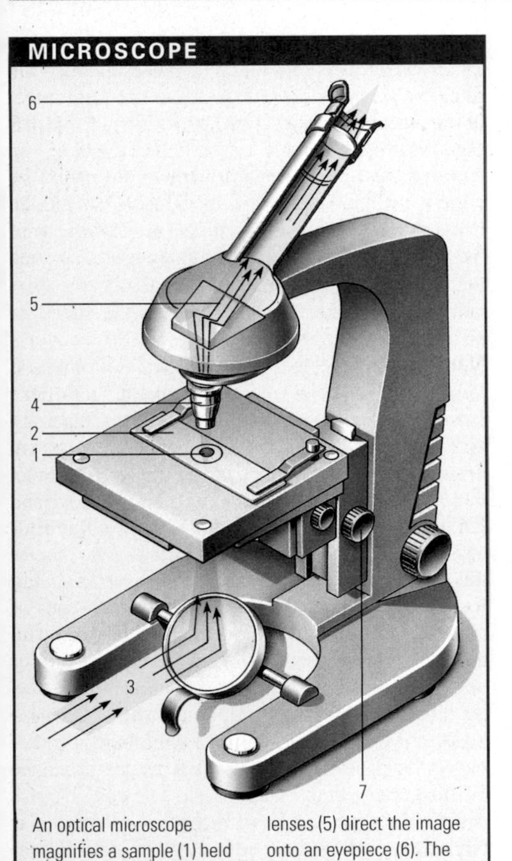

An optical microscope magnifies a sample (1) held on a slide (2). Light from below (3) illuminates the sample that is magnified by a lens (4), which can be changed. Complicated lenses (5) direct the image onto an eyepiece (6). The position of the sample can be altered physically by turning knobs (7) that bring the image into position and focus.

radios and miniaturized television sets are evidence of the importance of microelectronics to consumers. *See also* ELECTRONICS; NANOTECHNOLOGY

micrometer In astronomy, an instrument that provides an accurate measure of the apparent distance between stars. A **filar** micrometer can be fitted to the eyepiece of a TELESCOPE.

micrometer gauge Hand-held caliper instrument used in engineering machine shops to check the thickness of metal parts. The caliper of the micrometer is placed over the part to be measured and a screw is turned by the thumb, advancing a spindle until it touches the metal part (which is then just held by the caliper). The thickness is read off the uncovered part of a scale, to an accuracy of about 0.002mm (0.0001in).

micrometre (symbol μm) Unit of length equal to one millionth of a metre, formerly called the micron. In modern scientific usage, the micrometre has been largely replaced for small measurements by the nanometre (symbol nm), equal to one thousand-millionth of a metre.

microminiaturization *See* MICROELECTRONICS

micronutrient Element needed in relatively small amounts by growing plants. Micronutrients include iron, boron, zinc, copper, molybdenum and chlorine. *See also* MACRONUTRIENT

microorganism (microbe) Organism that can be seen only with the help of a MICROSCOPE. Microorganisms include BACTERIA, PROTOZOA, RICKETTSIAE, VIRUSES and some microscopic ALGAE and FUNGI. Disease-causing microorganisms are called PATHOGENS.

microphone Device for converting sound into varying electric currents with the same pattern of FREQUENCY and amplitude. Many telephones use a carbon microphone, in which the sound pressure causes a variation in the electrical resistance of carbon granules held between a diaphragm and a carbon block. Live music performers often use a moving coil microphone, in which a coil attached to a diaphragm vibrates in a stationary magnetic field. The recording industry prefers the condenser microphone, which employs a CAPACITOR. Crystal microphones use the PIEZOELECTRIC EFFECT.

microprocessor Entire CENTRAL PROCESSING UNIT (CPU) contained on an INTEGRATED CIRCUIT (SILICON CHIP) used to control the operation of a computer or other equipment, such as a washing machine.

microscope Optical device for producing an enlarged image of a minute object. The first practical microscope was made in 1668 by Anton van LEEUWENHOEK. The modern, **compound** microscope has two converging lens systems, the objective and the eyepiece, both of short focal length. The object to be examined is placed close to the objective lens and illuminated by a strong source of light. The objective produces a magnified image which is further magnified by the eyepiece to give the image seen by the observer. Most microscopes have three objective lenses on a turret, and they may be interchanged to give a choice of low, medium or high magnification. Because of the nature of visible light, an optical microscope can magnify objects only up to 2,000 times, and this is possible only by using lenses immersed in oil. Another type of instrument, the **polarizing** microscope, illuminates mineral specimens with POLARIZED LIGHT. For extremely small objects, an **optical** microscope cannot be used, because it is incapable of resolving them. The ELECTRON MICROSCOPE was invented to solve this problem. Smaller objects can be seen, because electrons have shorter wavelengths than light and thus provide greater resolution. These microscopes can magnify up to a million times.

Microscopium (microscope) Small constellation of S sky. Its brightest star is of magnitude 4.7.

microsurgery Delicate surgery performed under a binocular microscope using specialized instruments, microneedles as small as 2mm long, and sutures 20 micrometres in diameter. It is used in a number of specialized areas, including the repair of nerves and blood vessels, eye, ear and brain surgery and the reattachment of severed parts. Microsurgery was first used by the Swedish surgeon G. Holmgren in the 1920s for ear operations, and by the 1940s it was used for corneal transplants. During the 1950s, techniques were developed to suture tiny blood vessels and nerves in such a way as to retain their function, making possible the first successful limb re-implantation by microsurgery in 1962.

microswitch Electrical switch that requires only a small effort and movement to open or close it. The prefix "micro" refers to the smallness of the movement, rather than the switch. It is used for automatic, or semi-automatic, applications in electro-mechanical machines in which a moving actuator opens or closes the switch to stop or start an operation.

microtome Precision device for slicing thin sections of specimens for examination under a MICROSCOPE.

microtubule Thin, protein filament in the CYTOPLASM of a CELL, which may form pairs or bundles, that helps to maintain the shape of the cell. During cell division (MEIOSIS and MITOSIS), the microtubules from the spindle are responsible for the movement of the chromosomes. Microtubules also occur in CENTRIOLES, CILIA and FLAGELLA.

microwave Form of ELECTROMAGNETIC RADIATION having a wavelength between 1mm (0.04in) and 1m (3.3ft) and a frequency range to about 300,000MHz. This puts microwave radiation between infrared and shortwave radio waves in the ELECTROMAGNETIC SPECTRUM. Microwaves are being increasingly used today for such purposes as RADAR, RADIO and TELEVISION broadcasting, high-speed microwave heating and cellular telephones. *See also* BACKGROUND RADIATION

microwave heating Use of MICROWAVES to heat materials. Microwave ovens produce heat by causing the water molecules in food to vibrate.

Mid-Atlantic Ridge Underwater topographic feature along the margin between the American crustal plate on one side and the European and African plates on the other. It runs for 14,000km

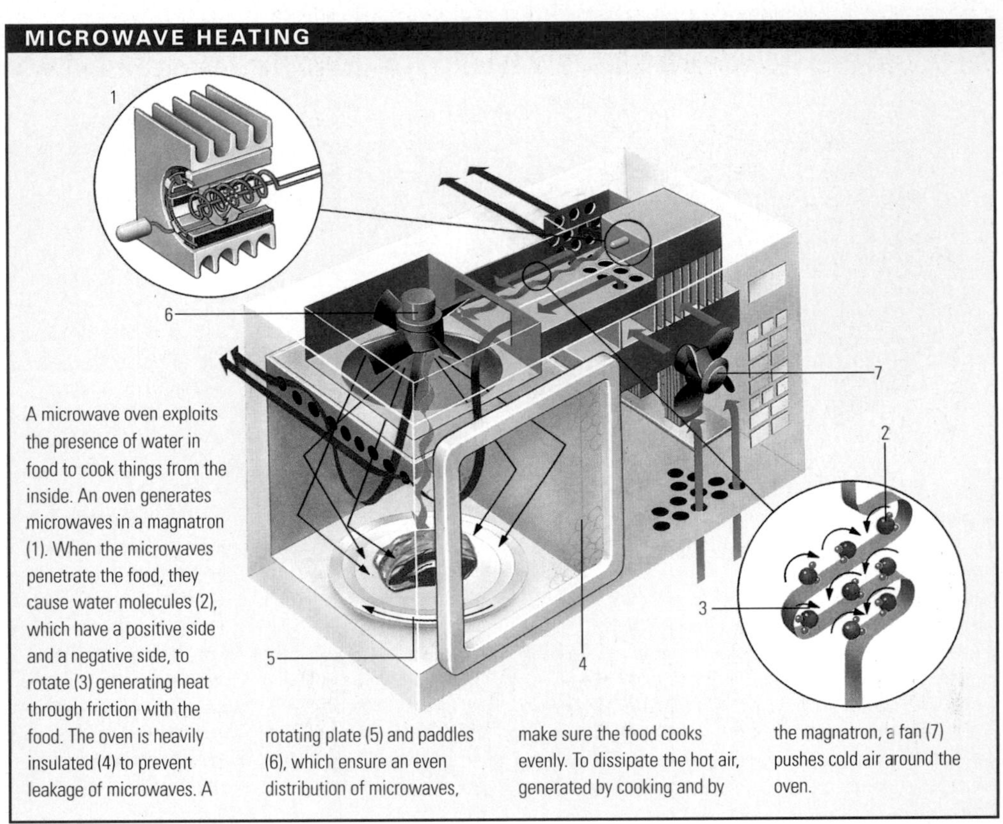

A microwave oven exploits the presence of water in food to cook things from the inside. An oven generates microwaves in a magnatron (1). When the microwaves penetrate the food, they cause water molecules (2), which have a positive side and a negative side, to rotate (3) generating heat through friction with the food. The oven is heavily insulated (4) to prevent leakage of microwaves. A rotating plate (5) and paddles (6), which ensure an even distribution of microwaves, make sure the food cooks evenly. To dissipate the hot air, generated by cooking and by the magnatron, a fan (7) pushes cold air around the oven.

(8,700mi) along the middle of the Atlantic Ocean. It is formed at a CONSTRUCTIVE MARGIN, where two plates are diverging and new magma is flowing upwards and outwards to form new ocean crust. Iceland is located on the ridge itself and was formed by the outpourings of volcanic lava. *See also* PLATE TECTONICS

midbrain (mesencephalon) One of the three divisions of the embryonic BRAIN of backboned animals. In a fully formed human being, the midbrain has become overlaid by the greatly developed cerebral lobes. It is thick-walled and is concerned particularly with sight and hearing. *See also* FOREBRAIN; HINDBRAIN

middle ear *See* EAR

mid-ocean ridge Great median ridge of the SEAFLOOR where new LITHOSPHERE is being formed. The ridges are the spreading edges of the tectonic plates that cover the Earth. They form a world-encircling system that extends, with several side branches, along the MID-ATLANTIC RIDGE, up, around and down through the Indian Ocean (the Mid-Indian Ridge) and across the Pacific (the Pacific-Antarctic Ridge). *See also* OCEANIC BASIN; PLATE TECTONICS; SEAFLOOR SPREADING

migmatite Coarse-grained, METAMORPHIC ROCK, often found in the shield of an ancient continent. It is formed at high temperatures deep within the Earth's crust, and has a granular texture consisting of dark schist or gneiss, with lighter folded bands of granitic rock.

migration Periodic movement of animals or humans, usually in groups, from one area to another, in order to find food, breeding areas or better conditions. Animal migration involves the eventual return of the migrant to its place of departure. Fish migrate between fresh and salt water or from one part of an ocean to another. Birds usually migrate along established routes. Mammals migrate usually in search of food. For thousands of years, the deserts of central Asia widened inexorably and this phenomenon resulted in the human migration of pre-historic tribes to China, the Middle East and Europe.

milk Liquid food secreted from MAMMARY GLANDS by the females of nearly all mammals to feed their young. The milk of domesticated cattle, sheep, goats, horses, camels and reindeer has been used as food by humans since prehistoric times; both directly and to make butter, cheese and fermented milks such as yogurt. Milk is a suspension of fat and protein in water, sweetened with lactose sugar. The proportions of these constituents vary with each mammal, those of cows' milk being water 87.1%, protein 3.4%, fat 5.9%. In modern dairying, cows' milk is pasteurized at about 72°C (160°F) for 16 seconds to kill all harmful microbes, then bottled under aseptic conditions. Nevertheless, milk sours after a day or so because of the action of lactic acid bacteria. *See also* PASTEURIZATION

Milky Way Faint band of light visible on clear dark nights encircling the sky along the line of the galactic equator. It is the combined light of an enormous number of stars, in places obscured by clouds of interstellar gas and dust. It is, in fact, the disc of our GALAXY, viewed from our vantage point within it.

Miller, Stanley Lloyd (1930–) US biochemist. In 1953 he carried out the first deliberate attempt to create the chemical components of life, by passing a mixture of AMMONIA, METHANE, water vapour and HYDROGEN (to simulate the Earth's early atmosphere) through water which was continuously sparked by a corona discharge. Within days, AMINO ACIDS were formed.

Millikan, Robert Andrews (1868–1953) US physicist who determined the value of the electronic charge. He also verified the photoelectric equation of Albert EINSTEIN and found a precise value for PLANCK'S CONSTANT. Millikan was awarded the 1923 Nobel Prize for physics for his measurement of the electronic charge and his work on the PHOTOELECTRIC EFFECT.

milling, ore Stage in the treatment of crude ORE, which mechanically extracts valuable MINERALS. Primary and secondary crushers break down large fragments. Fine grinding (comminution) is done with rotating cylindrical mills that contain steel balls, rods or pebbles. The fine powder is then sorted by a variety of methods.

Milne, John (1850–1913) British geologist whose invention (1880) of the SEISMOGRAPH helped to found the science of SEISMOLOGY. He helped to establish many seismological stations, particularly those for recording earthquakes in Japan. His books include *Earthquakes* (1883) and *Seismology* (1898).

Milstein, César (1927–) British molecular biologist and immunologist, b. Argentina, who shared the 1984 Nobel Prize for physiology or medicine for helping to develop ANTIBODIES that can be commercially produced for drugs and diagnostic tests. Milstein studied ENZYMES at the University of Cambridge. In 1975 he and his Cambridge colleague, the German immunochemist Georges Köhler, developed a technique for cloning MONOCLONAL ANTIBODIES (MABs) that combat diseases by targeting their sites. Milstein and Köhler shared the Nobel Prize with the immunologist Niels Jerne (1911–).

mimicry Form of animal protection through deception. The mimic, generally a harmless, edible species, imitates the warning shape or coloration of a "model", a poisonous or dangerous species. When coloration increases an animal's chances of survival, it is commonly referred to as protective coloration. Batesian mimicry, named after the

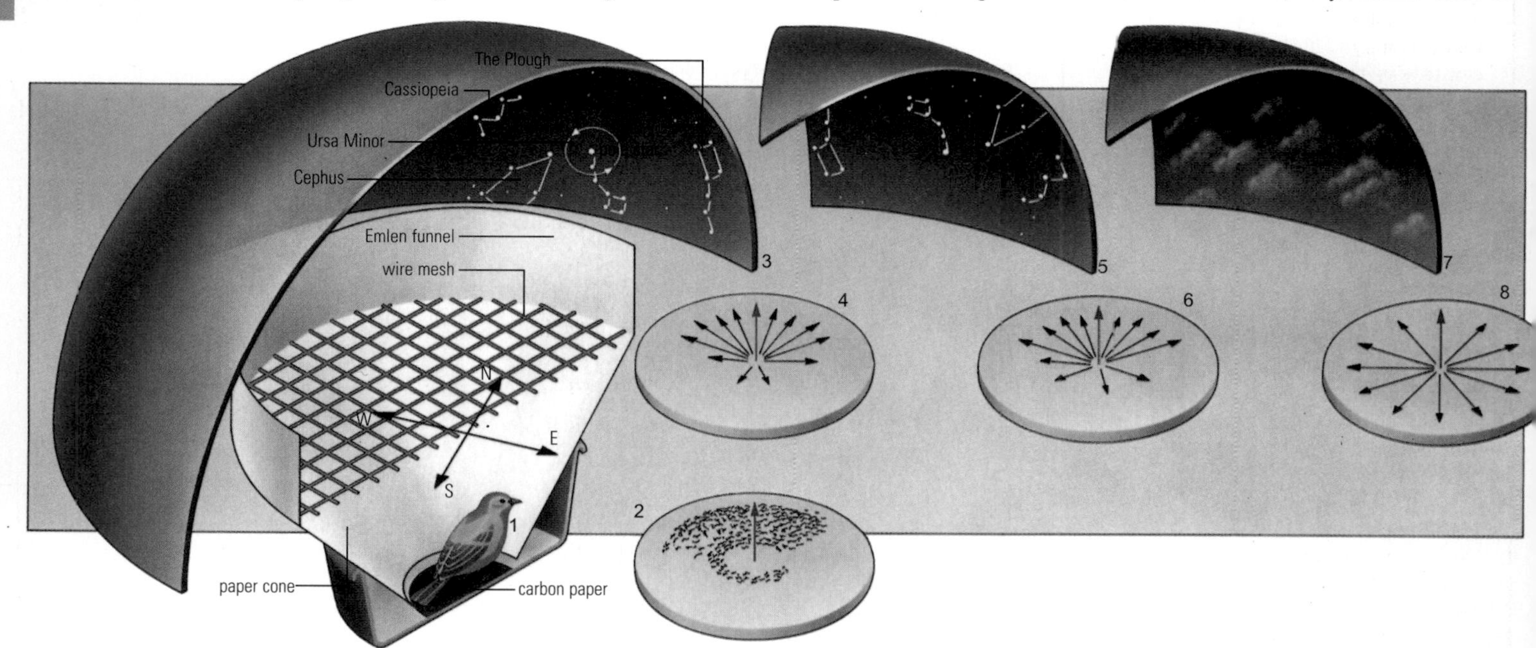

▲ **migration** Experiments with homing pigeons have shown that birds that migrate during the day have an internal body clock, which they use while referring to the position of the Sun to navigate. A night-migrating bird, however, does not use the Sun to navigate during the migrating season. A caged, night-migrating bird that can see the stars moves fretfully, orientating itself in the same direction to its normal flight – N in the spring and S in the autumn. This led to experiments to see how indigo buntings (1) were able to migrate using the stars (Moon and planet motions were ruled out as too complex) to navigate by. Did they too have an internal body clock like the homing pigeons? When caged in an Emlen funnel placed under stars projected in a planetarium, the bunting could see the sky and tried to escape, leaving carbon paper footprints on the paper funnel (2), which could then be interpreted to show the direction of orientation. The projected stars could be repositioned to change the apparent time and alter the bunting's clock. As the Earth revolves every 24 hours and its axis of revolution points N to the Pole Star, the stars appear to revolve around the Pole Star, a constant reference point from which to orientate. Under stars set at normal time (3), the bird orientated successfully N (4) (in all diagrams red arrows show the correct migratory direction and black arrows show the scatter of actual observed orientations). When the stars were shifted 12 hours forward (5), the bird still orientated in the correct direction (6), showing that the bunting made direct use of the stars, but did not compensate with an internal clock. But if the stars were eliminated and the planetarium diffusely lit to simulate stars obscured by clouds (7), the bunting's orientation attempts were entirely random (8). Further experiments confirmed that the birds navigate by observing the rotation of the stars closest to the Pole Star, which could be eliminated from the sky yet not impede correct orientation.

Many animals escape predation by looking like an inanimate object. But it is not just structural and surface appearances that have evolved, for fascinating behavioural strategies have also developed to reinforce deception. When at rest, stick caterpillars hold their bodies away from the bark of trees, rigid and unmoving like a branching twig. Mimicry can also benefit predators. A species of African jumping spider both looks and moves like an ant. This means that – often aggressive – ants will mistake it for one of their own, and relax their guard.

Instead of camouflaging themselves from predators, some animals use false signals to trick predators that they are other, more dangerous, animals that the predator would not consider attacking. A very common kind of mimicry is called Batesian mimicry, where a non-poisonous species imitates a poisonous one. Animals are not the only mimics. Some plants mimic animals in order to attract pollinators. Carrion plant flowers – with their mottled red surface and fetid smell – lure flies which come to lay their eggs on the "food". Defensive mimicry also occurs in plants. The gaily coloured *Heliconia* butterflies of South and Central American rainforests lay their eggs on passion-flower vines, so their caterpillars can feed on the vine leaves. However, to ensure an ample supply of food for her offspring, the female butterfly will not lay her eggs on stems and leaves already occupied by eggs. The passion vine produces mock eggs on its leaves to deter this invasion.

► **A stick insect** (A) shows the important defensive play of mimicking inedible objects like dry sticks. Many insects use this disguise, including caterpillars, moths and praying mantis. A stick insect has spindly legs that are barely noticeable at rest. Its head is small and the body is smooth and textured like a twig, with slight ridges and markings mimicking twig nodes. Stick insects may remain motionless for hours. Some stick insects grow up to 30cm (12in) long, yet are almost invisible until they move; even then, they move each leg very slowly, responding to any disturbance by keeping still.

► **Many different insects** (B) mimic wasps or bees that deter predators with unpleasant stings. Key recognition points of the common wasp (1) are its size, a black and yellow body, transparent wings and a tiny waist. One "impostor" is the wasp-mimic moth (2), whose transparent wings are shaped like a hornet's; the moth even mimics its role model's flight patterns. The wasp beetle (3) convinces at a distance, and also imitates the jerkiness of a wasp's flight. A hoverfly (4) has a black thorax, which gives the impression that the abdomen ends abruptly in a narrow waist, even though this is only an optical illusion.

▲ **Many coral snakes** (C) of South America have black, red and yellow warning stripes. Some have no poison, some like *Micrurus lemniscatus* (1), have lethal venom, and some, like *Oxyrhopus trigeminus* (2) are mildly poisonous: all inhabit the same areas. A mildly poisonous snake gives a predator an unpleasant taste that it remembers when it sees a similar snake. A harmless species has no poison, but is protected by mimicking the poisonous species. Lethal species also benefit - if attacked, their predators would not live to learn by experience. So both harmless and lethal species mimic moderately poisonous species, which must be the most numerous locally for a statistical probability that an attacker's experience will be unpleasant. Disputed by some scientists, Mertensian mimicry, as this is known, may be used by king snakes (3), which, although unrelated to coral snakes, have very similar colour banding but no venom.

British naturalist Henry BATES, is exemplified by hoverflies, which mimic inedible wasps. Similarly, the viceroy butterfly imitates the inedible monarch butterfly. The less common Müllerian mimicry, named after the German naturalist Fritz Müller, involves two or more unpalatable species that share a similar pattern, thus reinforcing it as one of the warnings to predators. A third form is aggressive mimicry, in which a predatory or parasitic species resembles a harmless one, thus allowing the former to remain undetected by its prey or host.

mine In civil engineering, excavation from which MINERALS (mainly coal and metal ores) are extracted. Underground mines are of two main types: shaft mines and drift mines. **Shaft** mines are sunk vertically in the Earth's crust until they reach the depth of the seams to be exploited, which are reached by tunnels or galleries. The gold mines of South Africa, the world's deepest mines, are shaft mines. **Drift** mines are generally shallower, the seams being reached by a drift, or gradually sloping shaft, which leads on to a gallery system. In OPENCAST MINING, the seams are near or on the surface and are exposed by giant dragline machines. *See also* MINING

mine In military technology, concealed or buried explosive device detonated through contact with individuals or vehicles. Underwater mines are used either to protect or to blockade coastal areas. There is now much opposition to the use of land mines as they are very difficult to remove from an area after the conflict has ceased. Each year, many thousands of civilians are killed by mines left behind in war zones.

mineral Natural, homogeneous and, with a few exceptions, solid and crystalline materials that form the Earth and make up its ROCKS. Most are formed through inorganic processes, and more than 3,000 minerals have been identified. They are classified on the basis of chemical make-up, crystal structure and physical properties such as hardness, specific gravity, cleavage, colour and lustre. Some minerals are economically important as ORES from which metals are extracted.

mineral acid Strong, inorganic acid such as HYDROCHLORIC ACID (HCl), NITRIC ACID (HNO_3) and SULPHURIC ACID (H_2SO_4).

mineralization FOSSIL-forming process whereby preserved parts of plants and animals are altered by the minerals in the soil. Circulating water dissolves certain constituents of bones and shells, which are replaced by silica, iron or other compounds. This replacement can be so exact as to preserve the finest structures. It is a type of PETRIFICATION.

mineralogy Investigation of MINERALS, naturally occurring inorganic substances found on Earth and elsewhere in the Solar System. Major subdivisions are: CRYSTALLOGRAPHY, which studies the composition and atomic arrangement of minerals; **paramagnetic** mineralogy, which deals with the associations and order of crystallization of minerals; **descriptive** mineralogy, concerned with the physical properties used in identification of minerals; and **taxonomic** mineralogy, the classification of minerals by chemical and crystal type. *See also* GEOCHEMISTRY; PETROLOGY

mineral oil (liquid paraffin, liquid petrolatum) Colourless liquid without taste or smell, distilled from PETROLEUM. It is used as a lubricant, and as a solvent in the manufacture of plastics. It is used medicinally as a laxative.

mineral water Originally natural waters valued for their mineral content, now also used to describe carbonated drinks containing salts or various flavourings. "Medicinal waters" from many famous spas are bottled and sold. They are classified according to location, use and chemical content. Synthetic mineral waters were developed to imitate natural ones.

minimal access surgery Term used to encompass operations that do not involve cutting open part of the body in the traditional way. Minimal access, or "keyhole", procedures are performed either by means of an ENDOSCOPE, an instrument for viewing the interior of the body, or by passing miniature instruments through a fine catheter, or tube, that is fed into a large blood vessel. Certain keyhole operations are said to be safer and less traumatic than conventional surgery. There is a growing list of such procedures, many of which require only local ANAESTHESIA, and can be done on a day-care basis, or at least with a much shorter hospital stay and faster recovery. The surgical LASER – the so-called bloodless scalpel – is used in many minimal access procedures.

mining Process of obtaining metallic and non-metallic materials from the Earth's crust. Included are underground, surface and underwater methods. Mining mostly involves the physical removal of rock and earth. Petroleum, gas and some sulphur are extracted by techniques which are different from those which would normally be put under the heading "mining". Any mining operation comprises four stages: prospecting, exploration, development and exploitation. Once a valuable deposit has been found and delineated, decisions are made on modes of entry, subsidiary developments and removal techniques. In underground mining, various methods are employed to obtain minerals buried deep in the ground. Access may be through oblique or vertical shafts or horizontal tunnels. Cross cuts are made at various levels and the ore body is divided into blocks by vertical "rises" that connect different levels. The ore is broken up by the use of hand tools, blasting or machinery. Coal, potash and rock salt are generally mined by the "room and pillar" method, in which haulageways open into rooms supported by tenant rock. Underwater mining is done by the use of large conveyor belts which are capable of bringing to the surface many tonnes of material each minute. The parent ships sift the material and deposit the unwanted material overboard. *See also* MINE

Minkowski, Hermann (1864–1909) Russian mathematician who worked in Germany. He contributed to the mathematical foundations of EINSTEIN's work on special RELATIVITY and formulated the concept of SPACE-TIME.

Miocene Geological epoch beginning about 25 million and ending about 5 million years ago. It falls in the middle of the TERTIARY period and is marked by an increase in grasslands over the globe at the expense of forests, and the development of most of the modern mammal groups. No British Miocene deposits are known, although some folding took place caused by the rising of the Alps to the S.

Mir Soviet SPACE STATION orbiting the Earth. The main part of Mir was launched in 1986. It weighs nearly 21 tonnes and has six docking ports for the attachment of scientific modules or cargo vessels. Over the years various modules have been added to enlarge the space station, which has been permanently manned by three-person crews who have occupied Mir for many months at a time. A series of mishaps in 1997 led to the cancellation of the project.

Mira Ceti (Omicron Ceti) RED GIANT star in the constellation of Cetus. It is a long-period VARIABLE STAR, with a mean period of 332 days, but this is subject to irregularities. Its mean range is magnitude 3.5 to 9.1, but maxima as bright as 2.0 and minima as faint as 10.1 have been observed. Mira Ceti is the most celebrated long-period variable.

mirage Type of optical illusion sometimes seen near the Earth's surface when light is refracted as it passes between cool, dense air to warmer, less dense air. Mirages are most commonly seen shimmering on hot, dry roads; the shimmer is a refracted image of the sky. *See also* REFRACTION

mirror Highly polished surface that produces an image of objects in front of it because of the laws of REFLECTION. Most mirrors are made of glass "silvered" on one side with SILVER, MERCURY or ALUMINIUM. They can be flat (plane) or curved (spherical or parabolic). **Plane** mirrors produce a virtual

MIRROR

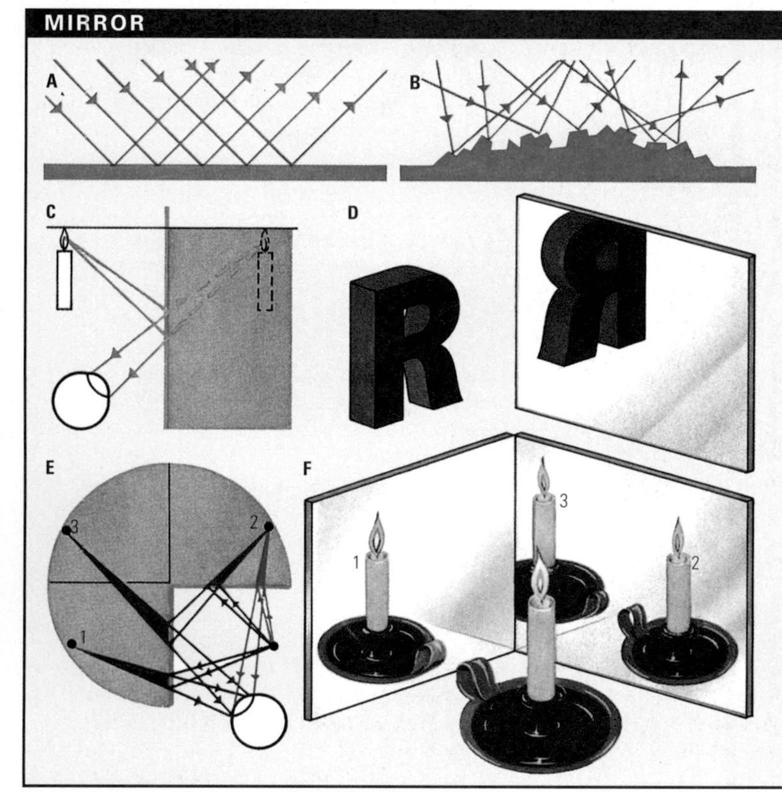

A plane mirror reflects all light rays at the same angle (A), whereas a matt surface (B) scatters light. The brain imagines that light rays reaching the eye from a mirror come to it in straight lines, and it therefore sees an image at that point where the rays would originate if their paths were not bent by the mirror (C). The image is seen laterally inverted (D), because the reflected rays reach opposite sides of the eye. An image can be seen right way round in two plane mirrors at right angles (E). Although an image reflected once appears laterally inverted (1, 2), one reflected twice (3) is seen correctly (F).

MOBILE TELEPHONE

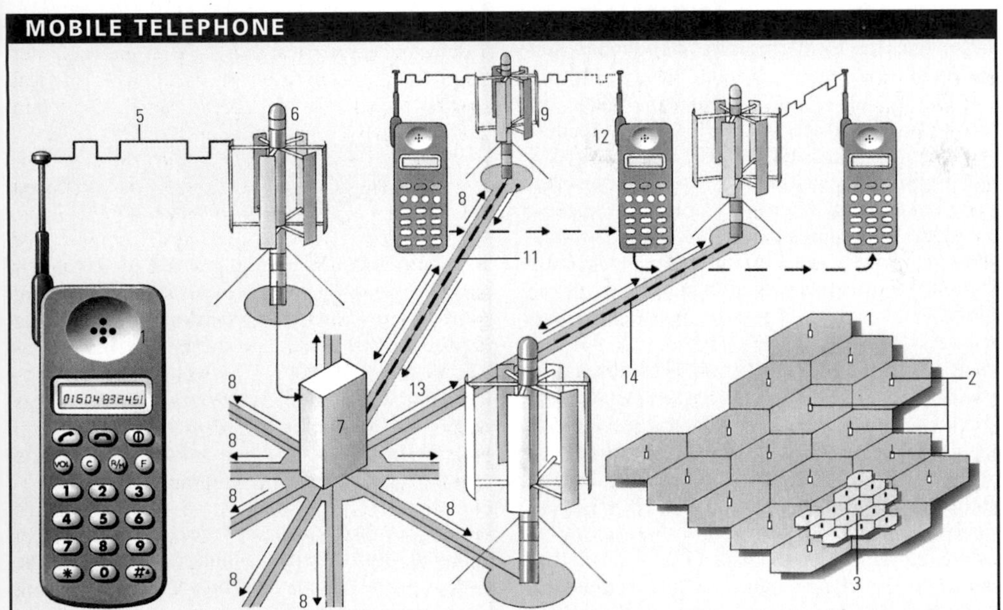

Mobile phone networks use a system of cells (1). By having a transmitter (2) in each cell, the same frequencies can be used in each cell allowing an enormous capacity for calls. Where there are many users, such as in the heart of a city (3), the cells are much smaller - further multiplying the number of frequencies available. A digital mobile phone (4) sends digital information (5) to a transmitter tower (6). Digital phones are better than analogue versions because they reduce background noise and interference, and are more difficult to bug. The transmitter passes the message to the systems' central exchange (7). If the call is for another mobile phone, the exchange sends a message (8) to the other transmitters, which in turn send out a message to locate the receiving phone. The transmitter locating the required phone (9) sends a confirmation message to the exchange (10), which then connects the conversation (11 – dotted line). When a phone moves out of range of a transmitter (12), a complicated procedure ensures the conversation can continue seamlessly. When the phone's signal to the transmitter becomes weaker, the exchange sends a message (13) to the transmitters in the surrounding cells to see which is receiving the strongest signal. It then transfers the conversation to that transmitter (14 – dotted line).

image that is the same size but inverted laterally (left to right). **Spherical** mirrors may be concave (curving inwards) or convex (bulging outwards). The image can be right-way-up or inverted, real or virtual, depending on the position of the object in relation to the focal point of the mirror; it may also be either magnified or reduced in size. A spherical mirror suffers spherical ABERRATION which is absent in a concave parabolic mirror, as used in reflecting TELESCOPES. *See also* IMAGE, OPTICAL

miscarriage Popular term for a spontaneous ABORTION, the loss of a fetus from the UTERUS before it is sufficiently developed to survive.

missile Unmanned and self-propelled flying weapons, powered by ROCKET, ram jet or turbojet. Ballistic missiles travel in the outer atmosphere and can be powered only by rockets. GUIDED MISSILES carry self-contained guidance systems or can be controlled by radio from the ground. CRUISE MISSILES, a type of guided missile, travel in the lower atmosphere and can be powered by jet engines. Unguided missiles are freeflying, with no control other than initial aim and amount of fuel.

Mississippian In North America, name given to the earlier part of the CARBONIFEROUS period.

mist Water droplets in the atmosphere at ground level that decrease visibility. By standard definition, visibility in a mist is between 2,000–1,000m (6,560–3,280ft). When visibility drops below 1,000m (3,280ft), the mist is referred to as FOG. *See also* SMOG

mistral Cold, winter, N wind that sweeps down the Rhône valley from the Massif Central of France. It is caused by cold air from Europe blowing southwards into a low-pressure area over the Mediterranean. When the mistral blows, temperatures sometimes fall as low as freezing point and strong gusts of up to 60 km/h (37mph) make conditions very unpleasant. Most farmhouses have rows of trees on their N side as protection from the wind, and many fields of fruits and vegetables shelter behind rows of cypress trees.

mitochondrion Structure (organelle) inside a CELL containing ENZYMES necessary for energy production. Mitochondria are found in the cytoplasm of most types of cell (but not in bacteria). *See also* RESPIRATION

mitosis Nuclear division of a CELL resulting in two genetically identical daughter cells with the same number of chromosomes as the parent cell. Mitosis is the normal process of TISSUE growth, and is also involved in ASEXUAL REPRODUCTION. *See also* ANAPHASE; METAPHASE; MEIOSIS

Mitscherlich, Eilhardt (1794–1863) German chemist who discovered (1819) that compounds of similar composition tend to have the same crystal structure – the phenomenon known as ISOMORPHISM. This helped in the understanding of the relationship between the molecular composition and crystalline structure of compounds. In 1823 Mitscherlich discovered the monoclinic form of sulphur and in 1827 he discovered selenic acid. Mitscherlich named BENZENE and was the first to synthesize nitrobenzene (1832); he was also among the first to recognize CATALYSIS.

mixture In chemistry, a material made up of two or more substances that retain their specific identities (such as air containing oxygen, nitrogen and other gases). The identities remain separate no matter in what proportion or how closely the components are mixed. *See also* COMPOUND; SOLUTION

mks system METRIC SYSTEM of units based on the metre, kilogram and second. *See also* WEIGHTS AND MEASURES; SI UNITS

mobile telephone Portable radio device that connects users to the public TELEPHONE system. They are also called cellular phones, because they operate within a network of radio cells. The first generation operated with analogue signals, the second generation have been designed to operate with DIGITAL SIGNALS. True mobile telephones (that operate from anywhere on the planet's surface by communicating directly with SATELLITES) became available in the 1990s.

Möbius strip Shape or figure that can be made by giving a long strip of paper a half-twist, then joining the ends together. It is of great interest in TOPOLOGY, being a one-sided surface (a line drawn along the strip of paper will appear both sides, returning to the starting-point to meet itself). It was invented by August Ferdinand Möbius (1790–1868).

mode In a statistical sample, the value of a VARIABLE that occurs most often. For example, in a survey on beer-drinking, a sample of 100 men were asked how many pints they drank in a week. The results were as follows (with numbers of men in parentheses): 0 pints (1); 10 pints (14); 12 pints (19); 15 pints (29); 20 pints (18); 25 pints (12); 30 pints (7). The mode is 15. If there are two modes, the distribution is said to be **bimodal**.

modem (**mo**dulator-**dem**odulator) Electronic device for sending and receiving COMPUTER signals through a TELEPHONE system. The streams of electrical pulses produced by computers cannot pass through a telephone system. So they are fed into a modem, which uses the pulses to modulate a continuous tone (carrier) by a process called FREQUENCY MODULATION (FM). The modulated carrier can then be sent through a telephone system. At the other end, another modem extracts the pulses (demodulation), so that they can be fed into a receiving computer. *See also* COMPUTER NETWORK; INTERNET

moderator In nuclear physics, substance consisting of light elements, such as HEAVY WATER or graphite, used in NUCLEAR REACTORS to slow down the FAST NEUTRONS produced by the FISSION reaction so that they can maintain a CHAIN REACTION.

modulation In physics, process of varying the characteristics of one wave system in accordance with those of another. It is basic to RADIO broadcasting. In AMPLITUDE MODULATION (AM), the amplitude of a high-frequency radio CARRIER WAVE is varied in accordance with the amplitude of a current generated by a sound wave. For static-free,

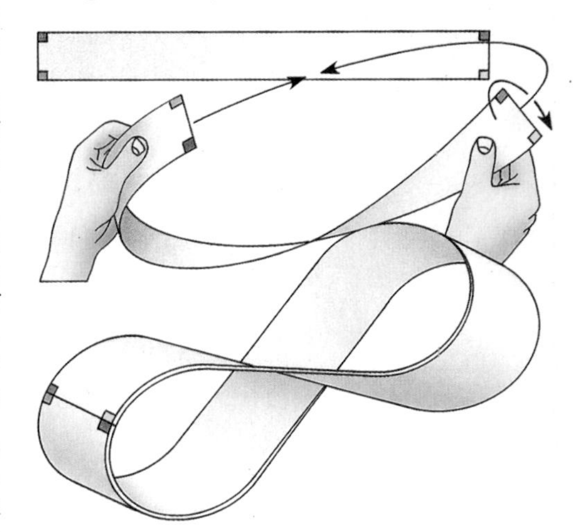

▲ **mobius strip** August Ferdinand Mobius gave his name to a strip of paper (or other flexible material) which is twisted through 180° and joined together. Join red to yellow and blue to green and the resulting topological figure possesses only one surface and one edge.

short-range broadcasting FREQUENCY MODULATION (FM) is used, in which the frequency of the carrier wave is modulated.

Mohl, Hugo von (1805–72) German botanist who did pioneering research into the anatomy and physiology of plant CELLS. He formulated the idea that the cell nucleus is surrounded by a granular, colloidal substance, which in 1846 he called PROTOPLASM, although the term had been invented by Jan Purkinje (1787–1869) in reference to the embryonic material in eggs. Von Mohl was the first to propose that new cells arise from CELL DIVISION. In 1851 he claimed that the secondary wall of a plant cell is fibrous, and this was subsequently confirmed. He also gave the first distinct account of the function of OSMOSIS.

Moho (Mohorovičić discontinuity) Boundary between the Earth's LITHOSPHERE and MANTLE. It is identified by a change in the velocity of seismic waves passing through the Earth, and is named after the Croatian geophysicist Andrija Mohorovičić (1857–1936), who first recognized it in 1909. The velocity change is explained by a change to more dense rocks in the mantle; above the Moho the waves travel at *c*.6.4km s^{-1} (4mi s^{-1}), whereas below they travel at 8.2km s^{-1} (5.1mi s^{-1}). The depth of the Moho varies from *c*.5km (3mi) to 60km (37mi) below the Earth's surface; it is usually deeper beneath landmasses than beneath the sea. Initially interpreted as a sharp boundary, it is now known to be a wider zone in some locations and its exact nature remains a matter of scientific debate. The **Gutenberg** discontinuity (discovered in 1912, by Beno GUTENBERG) separates the mantle and the core, and is *c*.2,900km (1,800mi) deep. EARTHQUAKE and other shock waves are deflected at these boundaries. *See also* MOHOLE PROJECT

Mohole, Project Attempt by US geologists and engineers, begun in the 1950s, to penetrate the Earth's crust to the deeper layers of the mantle, by drilling a borehole at least 10km (6mi) deep. It was named after the MOHO, the irregular boundary separating the crust and the mantle. Because of rising costs the Mohole was discontinued in 1966.

MOLECULAR MODEL

A	O_2	H_2O	C_6H_6
B			
C			
D	O=O	O, H H	H–C=C–H ring (H=C, C=H)

Molecules are groups of bonded atoms. They can be groups of the same atoms as in oxygen (O_2) or combinations of different elements, such as water (H_2O) and benzene (C_6H_6). Molecules can be illustrated in four ways. Line one (A), the chemical formula, lists the type and number of atoms in a molecule but does not show their structure. Line two (B) is the closest representation of the actual shape of the molecule but does not detail the bonds between the atoms. Line three (C) is less realistic but does show the bonding. Line four (D) combines the chemical formula, using the abbreviations of the periodic table, and symbolic representation of the bonding structure.

Mohr, Karl Friedrich (1806–79) German chemist. The foremost pharmacist of his time in Germany, Mohr improved analytical techniques and invented many items of laboratory equipment. He was also among the first to advance the idea of CONSERVATION of energy (in 1837). In 1867 he became professor of chemistry at the University of Bonn.

Mohs scale Range of HARDNESS used by geologists to express the comparative hardness of minerals by testing them against ten standard materials. It was devised in 1812 by the German mineralogist Friedrich Mohs (1773–1839). The hardest mineral, diamond, has a hardness of 10 on the Mohs scale. It can scratch, or mark, any mineral with a lower Mohs number, including corundum (9), topaz (8), quartz (7), orthoclase (6), apatite (5), fluorite (4), calcite (3), gypsum (2) and talc (1). A mineral hard enough to scratch material 3, but soft enough to be scratched by material 5, would be rated as having hardness 4 on the Mohs scale. Useful tools for determining hardness are the finger nail (about 2.5) and a penknife (about 5.5).

Moissan, Ferdinand Frédéric Henri (1852–1907) French chemist. He trained as an apothecary and won international recognition as the first to isolate FLUORINE (1886). He also developed an electric FURNACE that enabled him to prepare samples of several of the less common metals, such as URANIUM, long before modern metallurgy made them commercially available. He was awarded the Nobel Prize for chemistry in 1906.

molar One of the large back TEETH of MAMMALS, adapted for grinding and chewing food. In most HERBIVORES the cusps (points or ridges on the top surface) are fused to form ridges for grinding plants. In adult humans there are 12 permanent molars, three on each side of each jaw, top and bottom.

molarity *See* CONCENTRATION

molar volume Volume occupied by one MOLE of a substance. It is approximately the same for all gases at STANDARD TEMPERATURE AND PRESSURE, 22.414 litres.

mole (symbol mol) SI UNIT of amount of substance. It is the amount of substance that contains as many elementary units, such as atoms and molecules, as there are atoms in 0.012kg of carbon-12. The mass of one mole of a compound is its RELATIVE MOLECULAR MASS (molecular weight) in grams.

molecular biology Biological study of the make-up and function of MOLECULES found in living organisms. Major areas of study include the chemical and physical properties of proteins and of NUCLEIC ACIDS such as DNA. *See also* BIOCHEMISTRY

molecular formula *See* FORMULA

molecular model Geometrical structure of a molecule made by joining small coloured balls, representing atoms, using stiff metal springs or plastic rods, representing single bonds. They are invaluable aids not only in visualizing shapes of molecules but also in indicating "strain" in bonds and possible conformations that the molecule may adapt in reactions.

molecular orbital theory Explanation of how ELECTRONS are distributed in stable MOLECULES. In the simpler VALENCE theory of the CHEMICAL BOND, each ATOM of a molecule is assumed to retain its own electrons. Molecular orbital theory, however, treats each electron as associated with the molecule as a whole and it describes the configuration of electrons for each molecule. These patterns are unique for each molecule and can be considered as regions surrounding the molecule in which there is a probability of finding an electron. The calculations are very complex and only the simplest molecules can be treated exactly.

▲ **mollusc** Remarkably adept at exploring new habitats, snails originated in the sea, but gradually the *c*.22,000 species adapted to life on dry land, losing their gills and evolving air-breathing lungs. Most species of land snail, such as *Helix pomatia*, shown here, live on the ground and are dull in coloration. A few species are arboreal: these tend to be brightly coloured. Others have returned to aquatic environments and must surface periodically to breathe.

molecular weight Former term for RELATIVE MOLECULAR MASS

molecule Smallest particle of a substance (such as a compound) that exhibits the chemical properties of that substance. A molecule may consist of one ATOM, but generally consists of two or more atoms held together by CHEMICAL BONDS. For example, water molecules consist of two atoms of hydrogen bonded to one atom of oxygen (H_2O). A molecule (unlike an ION) has no electrical charge. *See also* MACROMOLECULE

mollusc Any of more than 80,000 species of INVERTEBRATE animals in the phylum Mollusca. They include the familiar snails, clams and squids, and a host of less well-known forms. Originally marine, members of the group are now found in the oceans, in freshwater and on land. Classes of mollusc include: the primitive GASTROPODS, univalves (slugs and snails), BIVALVES (clams, etc.), tusk shells and CEPHALOPODS (squids, etc.). The mollusc body is divided into three: the head, the foot and the visceral mass. Associated with the body is a fold of skin called the **mantle** which secretes the limy shell typical of most molluscs. The head is well developed only in snails and in the cephalopods, which have eyes, tentacles and a well-formed mouth. The visceral mass contains the internal organs of circulation (blood vessels and heart), respiration (gills), excretion (kidney), digestion (stomach and intestine) and reproduction (gonads). The sexes are usually separate but there are many hermaphroditic species. Cephalopods, bivalves and gastropods feature as important fossils in the geological past. *See also* HERMAPHRODITE

molybdenite Sulphide mineral, molybdenum sulphide (MoS_2). A major ore of molybdenum, molybdenite is found in PEGMATITES, IGNEOUS and METAMORPHIC ROCKS. It has hexagonal system tabular prisms, flakes and fine granules and is lead-grey in colour with a metallic lustre. Hardness 1–1.5; r.d. 4.7.

molybdenum (symbol Mo) Silver-white TRANSITION ELEMENT, first discovered in 1778. It is obtained from ores containing MOLYBDENITE (MoS_2). The concentrated mineral is roasted to yield molybdenum trioxide, which is mixed with iron in electric FURNACES to make ferromolybdenum. The pure metal is produced as a powder and converted to massive metal. Hard but malleable and ductile, molybdenum is used in alloy steels (as a hardener), X-ray tubes and missile parts;

molybdenum compounds are used as CATALYSTS, and as LUBRICANTS for bearings. It is one of the essential TRACE ELEMENTS for plant growth. Properties: at.no. 42; r.a.m. 95.94; r.d. 10.22; m.p. 2,610°C (4,730°F); b.p. 5,560°C (10,040°F); most stable isotope ^{98}Mo (23.78%).

moment of force *See* TORQUE

moment of inertia (symbol *I*) For a rotating object, the sum of the products formed by multiplying the elements of mass of the rotating object by the squares of their distances from the axis of the rotation. Finding this distribution of mass is important when determining the force needed to make the object rotate.

momentum Product of the mass and linear velocity of an object. One of the fundamental laws of physics is the principle that the total momentum of any system of objects is conserved (remains constant) at all times, even during and after collisions.

monadnock Residual hill rising above a PENEPLAIN, representing the core of the upland area that was eroded to give the plain. The plains of N Australia have many examples.

monazite Mineral containing rare-earth metals (LANTHANIDE SERIES) such as cerium, lanthanum, and thorium. It occurs widely in granitic and other rocks, and is the major source of those metals.

Mond, Ludwig (1839–1909) British chemist and industrialist, b. Germany. Mond experimented with alkalis and developed a PRODUCER GAS (a mixture of carbon monoxide and nitrogen). He also discovered nickel carbonyl, a gas formed from carbon monoxide and metallic nickel. Using nickel carbonyl, he developed a useful industrial method (the **Mond process**) for extracting pure nickel from its ore.

Monera *See* PROKARYOTAE

Moniz, António de Egas (1874–1955) Portuguese physician and medical researcher. He shared the 1949 Nobel Prize for physiology or medicine with Walter HESS for developing the surgical technique known as prefrontal LEUCOTOMY. This treatment of severely mentally ill patients, involving cutting selected brain fibres, was made popular by Moniz in the late 1940s and founded the controversial field of psychosurgery.

monoamine oxidase (MAO) ENZYME widely distributed in animals; other names are adrenaline oxidase and tryaminase. Its function in the body is the breakdown of certain biologically active AMINES, three of the most important of which are the tryptamine derivatives, the catechol amines (such as ADRENALINE and DOPAMINE) and histamine. **Monoamine oxidase inhibitors** are used as antidepressants.

Monoceros (Unicorn) Faint equatorial constellation situated S of Gemini and E of Orion. The Milky Way passes through this group, which also contains several bright and dark nebulae, including the Rosette Nebula (NGC 2237), and some star clusters, such as NGC 2244.

monochromatic Describing ELECTROMAGNETIC RADIATION, such as light, which has a single wavelength or frequency (a single colour). Pure monochromatic radiation is not possible, although the light from a LASER occupies a very narrow band of wavelengths and is virtually monochromatic, as are some spectral lines.

monocline In geology, sudden downward turn in the direction of a rock STRATUM within a fold, surrounded by horizontal rock.

monoclonal antibody ANTIBODY produced by cells that are derived by cloning a single original parent cell. Monoclonal antibodies are all identical and have unique sequences of AMINO ACIDS in their protein make-up. They are made by fusing a normal LYMPHOCYTE (an antibody-producing cell) with a cancerous cell derived from lymphatic tissue (LYMPHOMA) or bone MARROW (myeloma). The resulting "hybridoma" multiplies rapidly and produces large quantities of a single antibody. Monoclonal antibodies are used in highly specific VACCINES and in identifying ANTIGENS such as those involved with BLOOD GROUPS.

monocotyledon Any member of the Monocotyledonae subclass of flowering plants (ANGIOSPERMS) characterized by one seed leaf (COTYLEDON) in the seed embryo; the leaves are usually parallel-veined. Lilies, onions, orchids, palms and grasses are examples of monocotyledons. The larger subclass of plants is the Dicotyledonae (*see* DICOTYLEDON).

monoculture Agricultural practice of growing one crop. Monoculture has been common for a long time in many places, such as plantations or vineyards, but it has become far more widespread during the last 20 or 30 years, especially in the production of cereals. This is as a result of the availability of machinery for harvesting large quantities very quickly. There are also many fertilizers which restore fertility to the soil, so that crop rotation (whereby different crops are grown in the same fields every year or so) is unnecessary. Monoculture has created problems, however, as dependence on one crop can be serious if the price suddenly falls because of a glut, or if new pests invade the crop. Moreover, it is being realized that monoculture harms and possibly ruins the soil, in spite of the addition of artificial fertilizers. Monoculture in such areas as the prairies in Canada and the USA, the steppes in Russia, and East Anglia in the UK is now declining slightly, and there are signs of a return to crop rotation methods.

monocyte Large LEUCOCYTE (white blood cell). A monocyte is characterized by a kidney-shaped nucleus and protoplasm that stains a blue-grey colour. It is a PHAGOCYTE, actively digesting foreign particles such as bacteria and debris from dead cells. A millilitre of normal blood contains up to a million monocytes. *See also* IMMUNE SYSTEM

Monod, Jacques (1910–76) French biochemist. With François JACOB, Monod developed the idea that MESSENGER RNA carries hereditary information from the nucleus of a CELL to the cellular sites during the PROTEIN SYNTHESIS, and the concept of the operator GENE controlling the activity of other genes. Together with André LWOFF, they were awarded the 1965 Nobel Prize for physiology or medicine.

monoecious In botany, having both male and female FLOWERS on the same plant. The male flowers bear STAMENS, and the female bear one or more CARPELS.

monomer Chemical compound composed of single molecules, as opposed to a POLYMER (which is built up from repeated monomer units). For example, an AMINO ACID is a monomer of a protein, and propene (propylene) is the monomer from which polypropene (polypropylene) is made.

monorail train Passenger train that runs on a single rail; modern versions are gyroscopically stabilized and propelled by electric motors. The passenger cars are unsupported from the bottom or sides; they generally hang from wheeled axles that run along the rail. Bottom-supported cars, such as those used in zoos and amusement parks, are also referred to as monorails, as are certain types of MAGLEV (magnetic levitation) train.

monosaccharide Sweet-tasting CARBOHYDRATE that cannot be broken down by HYDROLYSIS; a simple sugar. GLUCOSE is a monosaccharide. *See also* DISACCHARIDE; POLYSACCHARIDE

monosodium glutamate (MSG) Food additive with a meat-like taste; a white crystalline powder that is the sodium salt of glutamic acid (an amino acid). It is obtained mainly from cereal glutens, and used as a flavour enhancer for meat products.

monotreme Member of the Monotremata order of primitive MAMMALS that lay eggs. The only monotremes are the platypus and two species of echidna, all native to Australasia. The eggs are temporarily transferred to a pouch beneath the female's abdomen where they eventually hatch and are nourished by rudimentary mammary glands. *See also* MARSUPIAL; PROTOTHERIAN

Monotype In printing, TYPESETTING machine that casts letters one at a time. *See also* LINOTYPE; PHOTOSETTING

◀ **monorail train** Any type of train that only uses one track on which the wheels run or by which the train is guided is strictly a monorail train. This particular monorail in Sydney, Australia does not come into contact with the track along which it runs, but floats above it. This is achieved by generating a powerful electromagnetic field between the train and the track, and its effect is to eliminate friction. The result is a quiet and efficient mode of transport.

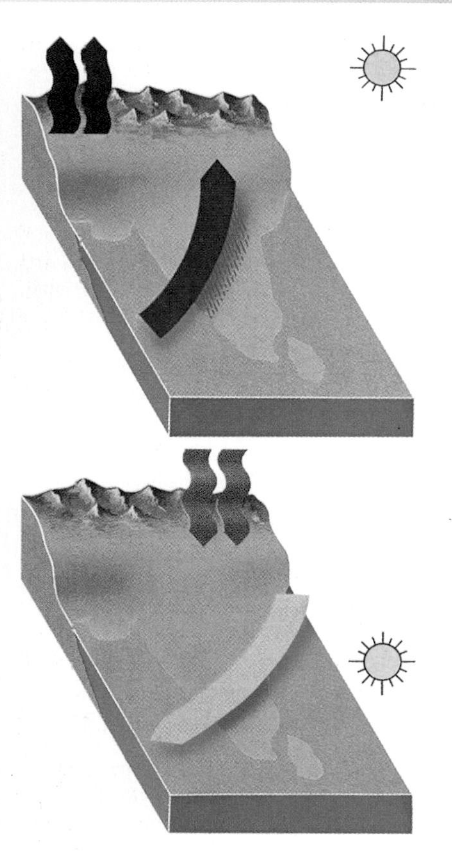

▲ **monsoon** Caused by low pressure areas over land masses in the summer, monsoons bring wet winds from the sea, and high pressure areas in the winter causing dry winds to blow from the land. During the summer (top) large areas of mainland Asia are heated by the Sun. The air over these regions expands and rises forming regions of low pressure. Wet winds from the sea then blow into these areas giving the summer monsoons. In the winter (bottom), the situation is reversed and regions of high pressure are formed over the land. Dry winter monsoon winds the blow out to sea. The paths of these winds are deflected due to the Coriolis effect.

monsoon Seasonal wind, especially in s Asia. In summer, which is the monsoon season, the winds normally blow from the sea to the land and bring RAIN, but in winter there is a complete change of direction and the winds blow out from the land, giving dry weather. Some monsoon regions are very wet. Cherrapunji in India, for example, receives over 11,000mm (433in) per annum, but others can be dry, such as the Thar Desert between India and Pakistan where the rainfall is less than 250mm (10in). The major monsoon areas are in Asia, where the seasonal reversal of wind is greatest. This is because the largest continent, Asia, is adjacent to the largest ocean, the Pacific. In the smaller continents of South America, Africa, Australia and North America, the monsoonal effects are less marked. These smaller continents do not have such wet summers or such dry winters, and are sometimes referred to as "eastern marginal" rather than true monsoon.

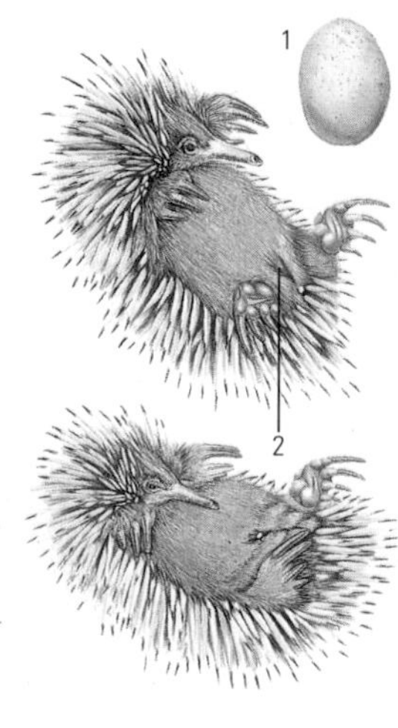

► **monotreme** An example of a primitive mammal, the echidna (spiny anteater) is classified as a monotreme. Instead of giving birth to live young like other mammals, it lays a tiny egg. The egg (1) is soft-shelled and resembles a reptile's egg. Once the egg is laid, the echidna uses its hind limbs to roll it to a special incubation groove (2). The minute hatchling is about 1.25cm (0.5in) long.

Montgolfier, Joseph Michel (1740–1810) and **Jacques Étienne** Montgolfier (1745–99) French inventors of the hot-air balloon. In 1782 the brothers experimented with paper and linen balloons filled with hot gases collected over a fire. In September 1783, one of their balloons carried some animals *c.*1.6km (1mi). In November, the brothers launched a balloon that carried Jean PILÂTRE DE ROZIER and the Marquis d'Arlandes over Paris. This was the first free flight.

montmorillonite General name for a group of clay minerals, hydrous calcium-sodium aluminium-magnesium-iron silicate. The minerals are moisture-sensitive clays weathered out of IGNEOUS ROCKS, and are a common constituent of soil. Montmorillonite is used in the paper industry for "carbon-less" copying paper. It is white, yellowish or grey. Hardness 1–2; r.d. 2.5.

moon Natural SATELLITE of a planet. The Earth's Moon, apart from the Sun, is the brightest object in the sky because of its proximity, being at a mean distance of only 384,000km (239,000mi). With an actual diameter of 3,476km (2,160mi), the Moon has 0.0123 of the Earth's mass and 0.0203 of its volume. The Earth and Moon revolve around a common centre of gravity. The gravitational attraction between the Moon and Earth produces TIDES. Although the Moon looks bright by contrast with the night sky, its surface rocks are dark. It is a cratered world, in many ways a typical SOLAR SYSTEM satellite. As the Moon orbits the Earth, it is seen to go through a sequence of PHASES as the proportion of the illuminated hemisphere visible to us changes. One complete sequence, from one new Moon, say, to the next, is called a **lunation**. The Moon's surface features may be broadly divided into the darker maria, which are low-lying volcanic plains, and the brighter highland regions (sometimes called terrae), which are found predominantly in the s part of the Moon's nearside and over the entire farside. There are impact features of all sizes. Other features include mountain peaks and ranges, valleys and elongated depressions. The origin of the Moon is uncertain. A current theory, known as the **giant impactor** or **big splash theory**, is that a Mars-sized body collided with the newly formed Earth, and debris from the impact formed the Moon. Subsequent large impacts produced the Moon's basins, and smaller impacts the craters. The chemical composition of material brought back from the Moon has been found to consist mainly of silica, iron oxide, aluminium oxide, calcium oxide, titanium dioxide and magnesium oxide. Lunar rocks are IGNEOUS ROCKS. The Moon has only the most tenuous of atmospheres. Apollo instruments detected traces of gases such as helium, neon and argon. The surface temperature variation is extreme, from 100 to 400K. In 1998 an American spacecraft, the *Lunar Prospector*, found frozen water in craters of the Moon. It is believed that there could be between 11 million and 330 million tonnes of water on the Moon's surface at the poles.

moor Tract of bleak open land, generally on a plateau or mountain. The vegetation is likely to be heather or bracken, with some coarse grasses. Many moorland areas have extensive stretches of peat bog; for example, parts of the Pennines and Dartmoor in England. Acid soils often develop, especially in areas of heavy precipitation.

Moore, Stanford (1913–82) US biochemist. He shared the 1972 Nobel Prize for chemistry with Christian ANFINSEN and William STEIN for research into the functioning and composition of the ENZYME ribonuclease. Anfinsen had discovered the composition of ribonuclease in the 1950s; subsequently Moore and Stein explained how it catalyses the digestion of food. They used CHROMATOGRAPHY to analyse AMINO ACIDS and PEPTIDES and to determine the structure of ribonuclease. By 1973, using essentially the same technique, Moore and Stein had analysed the structure of deoxyribonuclease.

moraine Accumulation of boulders and rock fragments which have been deposited by a GLACIER. Some of the rocks may have been eroded by the ice; some may be the result of freeze-thaw activity; and others are weathered blocks that have simply been transported by the ice. Morainic debris is carried by the ice and dumped at the sides or at the end, or at the bottom of the ice if it falls through crevasses. Accumulations at the side of a glacier are called **lateral moraines**; at the end of the glacier they are called **terminal moraines**. When two glaciers merge, two lateral moraines unite in the middle of the enlarged glacier to form a **medial moraine**. Beneath the ice there will be gradual accumulations of **ground moraine**, but much of it is reduced to small fragments and powder by the effect of the ice grinding it on the bedrock. Ground moraine can be an effective abrasive tool, if scraped along the valley floor by the movement of the ice.

mordant In dyeing processes, chemical that reacts with the dye or the fibre being dyed, or both, to "fix" the dye to the fibre, making it less likely to

▲ **Montgolfier** The first, manned, untethered, balloon flight took place on 21 November, 1783, when a balloon designed by the Montgolfier brothers, and carrying Jean-François Pilâtre de Rozier and the Marquis of d'Arlandes, flew across Paris for some 14km (9mi) across Paris. The flight is believed to have lasted 23min, and the balloon reached a height of 900m (3,000ft). Made of paper-lined linen and coated with alum to reduce the fire risk, the ballon was 15m (50ft) high, and weighed 785kg (1,730lb). The air inside was heated by a large mass of burning straw resting on a wire grid in the centre of the gallery.

► **moss** Mosses vary in growth and colour according to species. *Fontinalis anti pyretica* (A) is an aquatic moss, whose boat-shaped leaves have a sharp keel (1); the capsules are oblong or cylindrical (2, 3) and there is a pointed cap (4). *Polytrichum commune* (B) is extremely common and has a capsule (5) that looks like a four-sided box. It bears a long, golden brown cap (6) which is released before the spores are dispersed. *Atrichum undulatum* (C) is common on heaths and in woods, and has a capsule (7) with a long, pointed cap. *Schistostega pennata* (D), has flattened, translucent leaves.

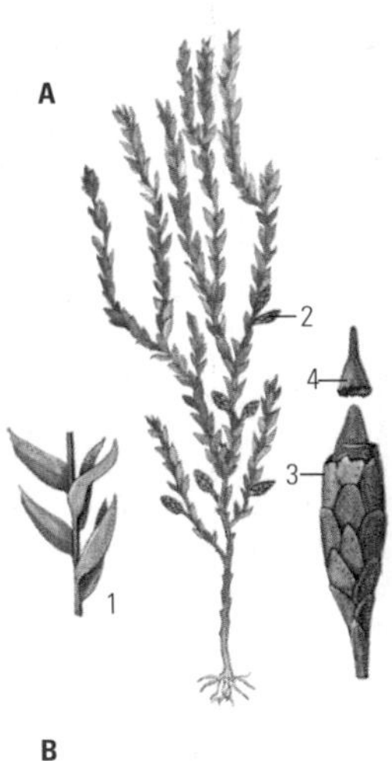

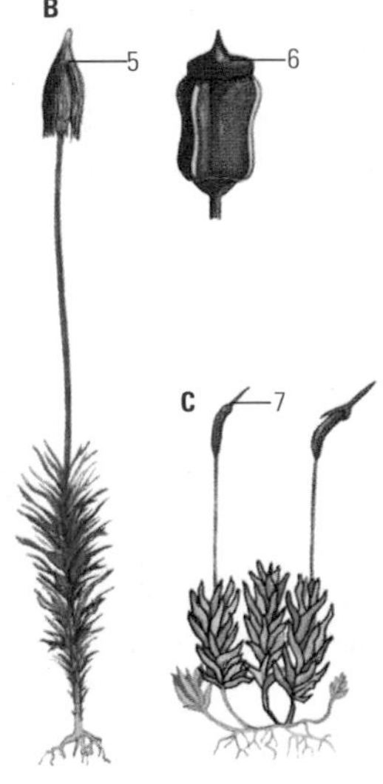

be washed out. Many mordants are metallic HYDROXIDES or SALTS (such as ALUM).

Morgan, Thomas Hunt (1866–1945) US biologist. He was awarded the 1933 Nobel Prize for physiology or medicine for the establishment of the CHROMOSOME theory of HEREDITY. His discovery of the function of chromosomes through experiments with the fruit-fly (*Drosophila*) is related in his book *The Theory of the Gene* (1926, rev. ed. 1928).

Morley, Edward Williams (1838–1923) US chemist who worked with Albert MICHELSON on the famous MICHELSON-MORLEY EXPERIMENT in 1887. Their experiment demonstrated the absence of a hypothetical substance called "ether", which had been supposed to pervade the Universe and carry light waves.

morphine White crystalline ALKALOID derived from OPIUM; first isolated in 1806. It depresses the CENTRAL NERVOUS SYSTEM and is used as an ANALGESIC for severe pain. An addictive drug, its use is associated with a number of side-effects, including nausea. *See also* HEROIN

morphology Biological study of the form and structure of living things, especially the external form. It ranges from visible characteristics to microscopic structures, and often focuses on the relation between similar features in different organisms.

Morse code Series of signals used for sending TELEGRAPH messages, either along wires or by radio telegraphy. It consists of dots and dashes created by the interruption of a continuous electric current or radio signal. The dashes are three times the length of the dots. The codes for letters of the alphabet, or numbers, vary in length from a single dot or dash to six dots or dashes or a combination thereof.

mortar In building construction, a material used to bind brick, stone, tile or concrete blocks into a structure. Modern mortar generally consists of a mixture of cement, sand and water; lime may be added to improve its spreading properties.

mortar In chemistry, a bowl of smooth, hard material in which softer substances are ground or beaten with a pestle, a club-shaped tool.

morula Early stage in the development of the EMBRYO in animals when the cells are in the process of splitting before the BLASTULA stage. The morula consists of a number of blastomeres, the cells formed from the fertilized OVUM as a result of CELL DIVISION.

Moseley, Henry Gwyn Jeffreys (1887–1915) British physicist. His initial studies involved RADIOACTIVITY; later he discovered a relationship between the X-ray spectra of the elements and their atomic numbers. He also discovered that the ATOMIC NUMBER of an element, and not its RELATIVE ATOMIC MASS, determines its major properties.

moss Any of about 14,000 species of small, simple, non-flowering green plants of the Bryophyta division, which typically grow in colonies, often forming dense carpets. They reproduce by means of SPORES produced in a capsule on a long stalk. The spores germinate into branching filaments, from which buds arise that grow into moss plants. Mosses grow on soil, rocks and tree trunks in a wide variety of land habitats, especially in shady, damp places. *See also* ALTERNATION OF GENERATIONS; BRYOPHYTE; SPHAGNUM

Mössbauer, Rudolf Ludwig (1929–) German physicist. His doctoral thesis, published in 1958, dealt with the emission of GAMMA RADIATION by radioactive nuclei within crystals (*see* MÖSSBAUER EFFECT). For this work, he shared the 1961 Nobel Prize for physics with Robert HOFSTADTER. Mössbauer's discovery has been used to test Albert EINSTEIN's theory of general RELATIVITY.

Mössbauer effect RESONANCE effect observed in certain radioactive substances emitting GAMMA RADIATION. When such a radioactive atom is part of a CRYSTAL LATTICE, the recoil momentum is taken up by the atomic nuclei, so that the gamma rays have a sharply defined energy instead of a range of energies. The rays can then be reabsorbed by a similar nucleus. This effect was first studied by Rudolf MÖSSBAUER. Motion of the source alters the frequency of the rays, by the DOPPLER EFFECT, and the Mössbauer effect may be used to study nuclear energy levels and molecular structure with great precision.

motion, laws of Three laws proposed by Isaac NEWTON in his *Principia* (1687). They form the basis of the classical study of motion and FORCE. According to the **first law**, a body resists changes in its state of motion – a body at rest tends to remain at rest unless acted on by an external force, and a body in motion tends to remain in motion at the same velocity unless acted on by an external force. This property is known as INERTIA. The **second law** states that the change in velocity of a body as a result of a force is directly proportional to the force and inversely proportional to the mass of the body; if the change in velocity, or ACCELERATION, is a, the force is F, and the mass is m, then $a = F/m$. According to the **third law**, to every action there is an equal and opposite reaction.

motor Mechanism that converts energy (such as heat or electricity) into useful work. The term is sometimes applied to the internal combustion ENGINE (which converts heat produced by burning gases into reciprocating or rotary motion), but is more often applied to the ELECTRIC MOTOR (which converts electrical energy into rotary motion). ROCKETS are motors that can leave the Earth's atmosphere because they carry both fuel and oxidiz-

MOTION, LAWS OF

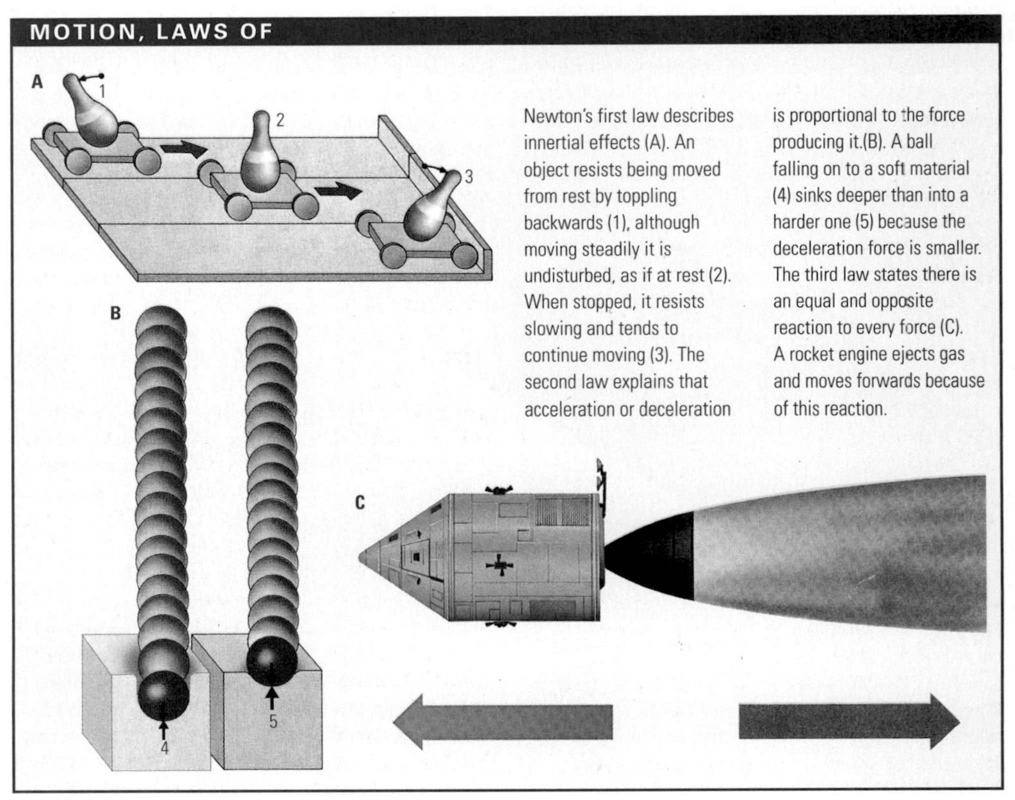

Newton's first law describes innertial effects (A). An object resists being moved from rest by toppling backwards (1), although moving steadily it is undisturbed, as if at rest (2). When stopped, it resists slowing and tends to continue moving (3). The second law explains that acceleration or deceleration is proportional to the force producing it.(B). A ball falling on to a soft material (4) sinks deeper than into a harder one (5) because the deceleration force is smaller. The third law states there is an equal and opposite reaction to every force (C). A rocket engine ejects gas and moves forwards because of this reaction.

M

er. Other types of motor presently being developed include the LINEAR MOTOR, a new motor proposed for high-speed train propulsion, and ION ENGINES, which employ ION PROPULSION to drive spacecraft.

motorcycle Powered, two-wheeled vehicle. Gottlieb DAIMLER is usually credited with building the first practical motorcycle. He mounted a four-stroke INTERNAL COMBUSTION ENGINE on the frame of a wooden bicycle. Since then the motorcycle industry has developed increasingly powerful and efficient engines, and sophisticated frames that have improved the handling of the machines. Motorcycles are classified in terms of engine capacity, generally from 50cc to 1200cc. Smaller capacity engines tend to be air-cooled and use a TWO-STROKE ENGINE, while larger engines (125cc and above) are mostly FOUR-STROKE and are usually water-cooled. Transmission of power to the rear wheel is most commonly by chain, although some motorcycles have either belt or shaft transmission. The clutch, accelerator and front brake controls are usually on the handlebars. Foot pedals control the gear change and rear brake.

motor neuron (motor neurone) NERVE that carries messages to an EFFECTOR organ (usually a muscle) from the CENTRAL NERVOUS SYSTEM (CNS), thereby causing an appropriate response. The cell bodies of some motor neurons form part of the SPINAL CORD. The AXONS (electrochemically conductive extensions) of these nerves, which are sheathed with MYELIN (an insulating substance), pass from the spinal cord to connect with the muscles. These nerves are involved in spinal REFLEX ACTIONS. Motor neurons are, however, also linked to the BRAIN by the descending spinal tracts. The cell bodies of these tracts lie in the CEREBRAL CORTEX, and their axons, which pass down the spinal cord, connect the nerves to the brain, so allowing voluntary muscular control. Some motor neurons act on glands of the ENDOCRINE SYSTEM to make them release their hormone secretions. *See also* NEURON; SENSORY NEURON

Mott, Sir Nevill Francis (1905–96) British scientist. He made important contributions in the field of QUANTUM MECHANICS, especially in the theory of atomic SCATTERING. In 1977 Mott was awarded the Nobel Prize for physics, jointly with Philip Anderson (1923–) and John VAN VLECK, for their work on the electronic structures of magnetic and disordered systems.

Mottelson, Ben Roy (1926–) Danish physicist, b. USA, who shared the 1975 Nobel Prize for physics with Aage Niels BOHR and James RAINWATER for their cooperative research into the structure of atomic nuclei. To explain why atomic nuclei were not spherical, Rainwater had suggested that nuclear particles form an inner nucleus and an outer nucleus, and these two interacting sets produce energy that deforms the symmetrical nucleus. From 1950 to 1953 Mottelson and Bohr, combining the shell and liquid drop nuclear models, provided experimental evidence to support this theory, which made nuclear FUSION a practical possibility.

mould, metal Mould for CASTING metal, usually made of sand or clay. This is packed over the face of the pattern that forms the cavity for the casting. The mould must be strong, resisting pressure of the hot, liquid metal, and permeable to allow gases to escape the cavity. It must also resist fusion with the metal, which is poured in through special channels.

moulting Process involving the shedding of the outermost layers of an organism and their replacement. Mammals moult by shedding outer skin layers and hair, often at seasonal intervals – human beings do not moult but lose dead, dry skin continuously as it is replaced from below. Birds moult their feathers, and amphibians and reptiles their skin. In all cases the process is controlled by HORMONES. It often serves, in growing animals, to replace worn out tissues of skin that have become too small. The moulting of INSECTS and other ARTHROPODS is a more elaborate affair which is fundamental to growth. The process, also called **ecdysis**, involves the resorption into the body of materials from the hard outer cuticle of the EXOSKELETON, so making the cuticle more fragile. The arthropod then swells its body and bursts free from the old cuticle, and slowly reforms a new one around its swollen body, thus increasing in size. *See also* METAMORPHOSIS

Moulton, Forest Ray (1872–1952) US astronomer who put forward the PLANETESIMAL THEORY to account for the formation of the SOLAR SYSTEM. Together with Thomas Chamberlain (1843–1928), he proposed that relatively small aggregates (planetesimals) condensed out of the original solar nebula, and that these were attracted to each other by gravity and accreted to form the planets. He also suggested that the small moons of Saturn are captured asteroids.

mountain Part of the Earth's surface that rises conspicuously higher, at least 380m (1,250ft) higher, than the surrounding area. Mountains have a restricted summit area, comparatively steep sides and considerable bare rock surface. They are formed in three main ways. First, FOLD mountains, such as the Himalayas, are formed by a squeezing-in of rock layers, caused by movements of the tectonic plates (*see* PLATE TECTONICS) of the Earth's CRUST. Second, BLOCK MOUNTAINS, such as the Sierra Nevada of North America, are formed by vertical movements between geological FAULTS, leading to the tilting of large blocks of STRATA. Third, a typical VOLCANO forms from the molten rock and ash that piles up around its original vent hole. The extent and speed of this process is illustrated by Parácutín, a Mexican volcano that rose out of a cornfield to a height of 450m (1,475ft) from its base between 1943 and 1952. Mountains may occur as single isolated masses, as ranges and in systems or chains. *See also* OROGENISIS

mouse In computing, input device that can be operated with one hand. It is designed to fit the palm of the hand, with one or more buttons that can be pressed by the fingers of the same hand. When the mouse is moved around a flat surface, it controls the movement of a cursor or pointer on the COMPUTER screen. The buttons can be "clicked" onto icons or other responsive areas on the computer display. *See also* GRAPHICAL USER INTERFACE (GUI)

mouth In animals, the anterior (front) end of the ALIMENTARY CANAL, where it opens to the outside. In humans and other higher animals it is the cavity within the JAWS, containing the TEETH and TONGUE.

MS-DOS Abbreviation of Microsoft Disk Operating System. It was the original operating system produced in 1981 for IBM-compatible personal computers.

mucopolysaccharide Class of POLYSACCHARIDE molecules composed of aminosugars that are linked into repeating units to give a linear, unbranched, polymeric compound. They are structurally similar to GLYCOGEN and STARCH. CHITIN is a mucopolysaccharide that acts like CELLULOSE as a structural polysaccharide for many phyla of lower plants and insects.

mucous membrane Sheet of TISSUE (or EPITHELIUM) lining all body channels that communicate with the air, such as the mouth and respiratory tract, the digestive and urogenital tracts and the various glands that secrete MUCUS. The "membrane" contains gland cells that secrete mucus, which serves for lubrication and protection.

mucus Slippery, viscous fluid containing mucin, produced by MUCOUS MEMBRANES of the body. It serves for lubrication and protection. Nasal mucus traps airborne particles; mucus of the stomach protects the lining from irritation by hydrochloric acid secreted during DIGESTION.

mudstone Rock made of consolidated mud which, although firmer than clay, lacks the laminated structure and tendency to cleave to SHALE and sometimes decomposes into mud when exposed to the atmosphere.

Muir, John (1838–1914) US naturalist, b. Scotland. An advocate of forest preservation, he was influential in the establishment of many national parks. Muir studied the glaciers and forests of the Sierra Nevada, and explored Alaska, where he discovered Glacier Bay and Muir Glacier. By 1867 he was urging the US government to endorse a policy of forest conservation, and his writings and lobbying influenced conservation programmes of both Presidents Grover Cleveland and Theodore Roosevelt.

Muller, Hermann Joseph (1890–1967) US geneticist. He found that he could artifically increase the rate of MUTATIONS in the fruit-fly (*Drosophila*) by the use of X-rays. He thus highlighted the human risk in exposure to RADIOACTIVITY. For this work Muller was awarded the 1946 Nobel Prize for physiology or medicine.

Müller, Johannes Peter (1801–58) German

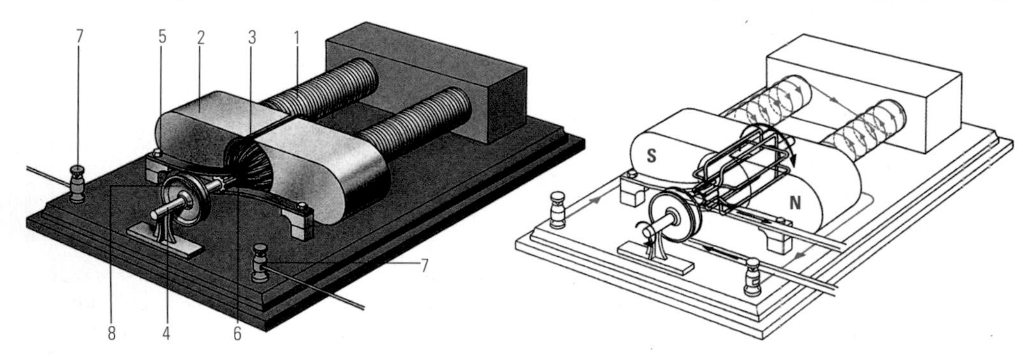

▲ **motor** Although generators made earlier in the 19th century had produced alternating current, (AC), those constructed later concentrated on the generation of direct current (DC) which at that time was considered more useful. From 1873 it had been known that the current generated by a dynamo, passed through a similar machine in a reverse direction, would rotate the armature of that machine. This principle gave rise to the electromotor. Well made dynamos always duplicate as efficient electromotors, although the reverse does not hold true. Edison's motor of 1879 featured long horizontal field electromagnets (1), with attached pole pieces (2), between which the armature revolved (3) on its shaft (4). The brushes (5) ran against a split ring commutator (6) thereby collecting the DC current and passing it to the terminals (7). A drive pulley (8) was located on the motor shaft, in order to transport the motion produced.

MOUTH

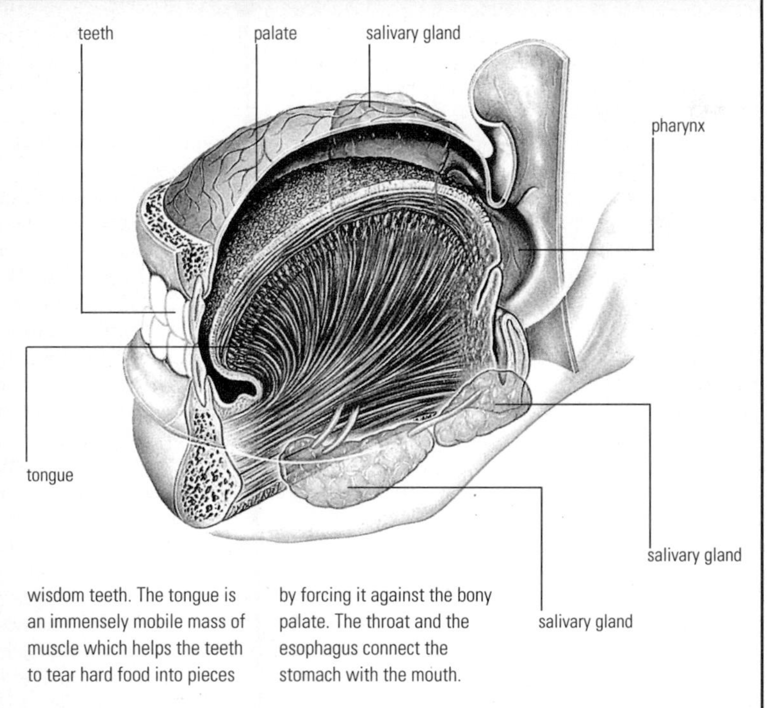

Food has to be broken down into its basic molecules before it can be absorbed in the body. In this respect, the human mouth has four main functions. It breaks up food by chewing with the teeth and tongue; it lubricates the food with saliva, to make swallowing easier; it regulates temperature by either cooling or warming the food; and it consciously initiates swallowing when the bolus is ready. There are three pairs of salivary glands. Between them they produce 1,500–2,000ml of saliva a day, containing the starch-reducing enzyme ptyalin. An adult has 32 teeth, of which there are 8 incisors, 4 canines, 8 premolars, 8 molars and 4 wisdom teeth. The tongue is an immensely mobile mass of muscle which helps the teeth to tear hard food into pieces by forcing it against the bony palate. The throat and the esophagus connect the stomach with the mouth.

physiologist and anatomist. Regarded as one of the founders of modern physiology, he conducted pioneering research into the NERVOUS SYSTEM, LYMPH and BLOOD systems, hearing and embryology.

Müller, Paul Hermann (1899–1965) Swiss chemist who was awarded the 1948 Nobel Prize for physiology or medicine for his discovery of the use of DDT as an INSECTICIDE. In 1935 he began the search for a perfect insecticide, one toxic for a wide variety of insects with few poisonous effects on other life forms. He concentrated on chlorine-containing compounds and in 1939 tested DDT. In 1944 DDT was successfully employed against a typhus epidemic in Naples, and for more than 20 years was the most widely used insecticide. In the 1970s, however, it was implicated as a hazard to animal life because it persists in FOOD CHAINS; its use has been banned in many countries.

Mulliken, Robert Sanderson (1896–1986) US chemist and physicist who was awarded the 1966 Nobel Prize for chemistry for his fundamental work on CHEMICAL BONDS and the MOLECULAR ORBITAL THEORY. He studied at the Massachusetts Institute of Technology and the University of Chicago, becoming professor at the latter in 1937. He later worked on the development of the atom bomb and devised a scale of electronegativity of the elements.

Mullis, Kary (1944–) US biochemist who discovered how to copy minute quantities of DNA millions of times for research purposes. This technique, POLYMERASE CHAIN REACTION (PCR), has been used in a variety of ways, including tests for HIV and for DNA forensic-science investigations. For his research, Mullis shared the 1993 Nobel Prize for chemistry with Michael SMITH, who worked independently on altering the genetic code.

multimedia COMPUTER system that includes text, audio (sound) and video (picture) components. Often the user can interact with the system to interrogate it or even to control or contribute to what is happening on screen. A multimedia product is usually produced on a CD-ROM, which has the necessary high-storage capacity for the audio and graphics elements. A computer to exploit it therefore has to have a CD-ROM drive, together with a high-resolution colour monitor and a sound card. It needs a minimum of 8 megabytes of RAM and at least an 80486 processor.

multiple birth Bearing of several offspring. Multiple BIRTHS are common among small MAMMALS but relatively rare in human beings, in whom the most usual form is TWINS. Identical twins develop from a single OVUM that has divided shortly after FERTILIZATION by a single SPERM; fraternal twins arise from two individually fertilized OVA. The use of FERTILITY DRUGS is associated with an increased incidence of multiple births.

multiple myeloma Malignant TUMOUR of the bone MARROW. Occurring most often in older people, it leads to anaemia, collapse of the vertebrae, blood clotting problems, and damage to the eyes and internal organs. It is treated with CHEMOTHERAPY and RADIOTHERAPY.

multiple sclerosis (MS) Incurable disease in which there is degeneration of the MYELIN SHEATH that surrounds nerves in the brain and SPINAL CORD. Striking mostly young adults (more women than men), it is mainly a disease of the world's temperate zones. Symptoms vary according to which nerves are affected, but may include unsteadiness, loss of coordination, and speech and visual disturbances. Affected people typically have relapses and remissions over many years. Its cause is unknown, but previous infection may play a part in genetically susceptible people.

multiple star Gravitationally connected group of more than two stars. An example is Castor, which has six components. In such systems there is a recognizable hierarchy or order. In Castor, the two main pairs rotate about a common centre of gravity, while a pair of cool RED DWARFS rotates about the same centre but at a larger distance.

multiplet In SPECTROSCOPY, a set of closely spaced, spectral lines arising from the splitting of a single line by, for example, electron SPIN. In NUCLEAR PHYSICS, a group of elementary particles, all HADRONS with about the same mass, identical in all other properties except electric charge, usually having two or three members. The NUCLEONS and PIONS form multiplets. In STRONG NULCEAR FORCE interactions, members of a multiplet are all equivalent. A supermultiplet is a larger, more sophisticated and symmetrical grouping of hadrons involving eight QUANTUM NUMBERS, all of whose members have identical spin.

multiplication Arithmetical operation signified by × (interpreted as repeated ADDITION). $a \times b$ is $a + a + ... + a$, in which b terms are added. In $a \times b$ ("a multiplied by b"), a is the **multiplicand**, b the **multiplier**, and the result is the **product**. *See also* ARITHMETIC

mumps Viral disease, most common in children, characterized by fever, pajn and swelling of one or both parotid salivary glands (located just in front of the ears). It has an incubation period of about 18 to 22 days and in children usually takes a relatively mild course. MENINGITIS is an occasional complication, and the infection may involve other organs such as the pancreas. The symptoms are more serious in adults and in men inflammation of the testes (**orchitis**) may occur, with the risk of sterility. Children over 18 months of age can be vaccinated against mumps, MEASLES and rubella (German measles) (MMR vaccine). One attack of mumps generally confers lifelong immunity.

muon (symbol μ^-) Negatively charged ELEMENTARY PARTICLE, originally thought to be a MESON but now classified as a LEPTON. It has SPIN $^1/_2$, a mass about 207 times that of the ELECTRON, and decays weakly into an electron, a NEUTRINO and an antineutrino. *See also* ANTIMATTER

Murchison, Sir Roderick Impey (1792–1871) Scottish geologist. His work on the rocks underlying the Old Red Sandstone in S Wales led to the definition of the strata known as the SILURIAN system (corresponding to that period of geological time). He worked with Sir Charles LYELL on the Auvergne volcanics and with Adam SEDGWICK on the structure of the Alps. He later collaborated with Sedgwick on a study of the rocks that were to become known as the DEVONIAN system. He is also responsible for founding the PERMIAN system.

Murdock, William (1754–1839) Scottish engineer and pioneer of gas lighting. In 1792 he began experiments to produce COAL GAS for lighting. He also devised ways of purifying and washing the gas. In 1802 he used coal gas to illuminate the exterior of a factory in the Soho area of Birmingham, England, where he worked for Matthew BOULTON and James WATT. Murdock also invented GEARS and valves for use on STEAM ENGINES.

Musca (fly) Small distinctive, constellation of the S sky. Its brightest star, Alpha, has magnitude 2.7.

muscle Tissue that has the ability to contract,

▲ **moulting** Because an arthropod's cuticle is dead, it is periodically shed, and a new one is secreted as the animal undergoes its next spurt of growth. Land-living vertebrates produce keratin – a hard, water-resistant protein – in the outer skin layer cells. As keratinization causes many cell components to degenerate, and eventually die, the layer of keratinized cells is shed from time to time. So snakes and other reptiles literally crawl out of their skins. Birds and mammals slough off small pieces of keratinized skin almost continually. (Much of household dust is powdered skin.) Even frogs and toads shed the surface of their skin, which they will then often eat.

M

enabling movement. There are three basic types: VOLUNTARY MUSCLE (or skeletal muscle), INVOLUNTARY MUSCLE (or smooth muscle) and cardiac muscle. **Voluntary** muscle is the largest tissue component of the human body, comprising about 40% by weight. It is attached by TENDONS to the BONES of the SKELETON and is characterized by cross-markings known as striations; it typically contains many nuclei per cell. Most voluntary muscles require conscious effort for contraction (when they move the limbs and body). A muscle whose contraction causes a limb or a part of the body to straighten (extend) is called an EXTENSOR. A muscle whose contraction causes a limb or part of the body to bend (flex) is called a FLEXOR. **Involuntary** muscle lines the digestive tract, blood vessels and many other organs. It is not striated and typically has only one nucleus per cell; it is not under conscious control, and is therefore known as involuntary muscle. **Cardiac** muscle is found only in the HEART and differs from the other types of muscle in that it beats rhythmically and does not need stimulation by a nerve impulse to contract. Cardiac muscle has some striations (but not as many as in voluntary muscle) and has only one nucleus per cell.

muscovite Sheet silicate mineral, hydrous potassium aluminium silicate ($KAl_2(Si_3Al)O_{10}(OH)_2$) the most common MICA. Found in many kinds of rocks, muscovite crystallizes in hexagonal, tabular forms in the monoclinic system. Its name derives from its use as a glass in Muscovy, Russia. It is clear or tinted with varied lustres. Hardness 2–2.5; r.d. 2.9.

M

muscular dystrophy Any of a group of hereditary disorders in which the characteristic feature is progressive weakening and atrophy of the muscles. The commonest type, **Duchenne** muscular dystrophy, affects boys, usually before the age of four. Muscle fibres degenerate, to be replaced by fatty tissue. The child develops a characteristic waddling gait and an inward curvature of the lower spine; later there is involvement of the respiratory muscles. There is no cure and Duchenne victims (one in 3,000 baby boys) usually die before the age of 20.

muskeg Area of boggy ground found in the N TUNDRA regions. Most of the ground in such regions, such as the tundra of N Canada and Russia, is PERMAFROST. But in summer the top few centimetres of soil thaw, and the MELTWATER, unable to run away, forms muskegs. The bogs support lichens and mosses, which are important food for herds of reindeer (caribou).

mutagen Anything that causes an increase in the number of MUTATIONS in a population of organisms. Mutagens work by affecting the GENES, thus producing inherited defects. They may damage chromosomes, or change the DNA and affect the DNA code. Mutagens include chemicals (such as colchicine) and ionizing radiation (such as radioactivity and X-rays). Some mutagens have been shown to cause CANCER.

mutation Sudden change in an inherited characteristic of an organism. This change occurs in the DNA of the GENES. Natural mutations during reproduction are rare, occur randomly, and usually produce an organism unable to survive in its environment. Occasionally the change results in the organism being better adapted to its environment and, through NATURAL SELECTION, the altered gene may pass on to the next generation. Natural mutation is therefore one of the key means by which organisms evolve. The mutation rate can be increased by exposing genetic material to IONIZING RADIATION, such as X-RAYS or ULTRAVIOLET RADIATION, or mutagenic chemicals. *See also* EVOLUTION

mutualism Relationship with mutual benefits for the two or more organisms involved. *See* SYMBIOSIS

mycelium Vegetative body of a FUNGUS, found underground. It is made up of a web of filaments (HYPHAE), sometimes massed like felt. The hyphae are of two types, those that feed and those responsible for reproduction.

mycology Science and study of FUNGUS.

mycoplasma Any of about 15 species of the smallest known form of cellular life; they are usually considered to be BACTERIA. Most are PARASITES on birds and mammals, some live in stagnant water, and some play a part in such diseases as pleuropneumonia in cattle.

mycorrhiza (fungus root) Association between certain fungi and the root cells of some VASCULAR PLANTS. The FUNGUS may penetrate the root cells or form a mesh around them. Water and minerals enter the roots via these threads. Sometimes the fungus digests organic material for the plant. *See also* SYMBIOSIS

myelin sheath Protective layer around the AXONS of peripheral and some central NERVE fibres. It insulates the fibre to prevent loss of electrical impulses and aids conduction. The myelin sheath is composed of specialized cells (SCHWANN CELLS) made up of proteins and fats. The disease MULTIPLE SCLEROSIS features degeneration of the myelin sheath.

mylonite Any of several laminated, fine-grained rocks formed when layers of parent rock fault, granulate or flow. It is chemically stable but partly melted and is reduced to a powder by the movement of rock along a fault line. It generally contains fragments of the parent rock.

myoglobin Protein found in animals. In vertebrates it is the pigment producing the red colour of MUSCLE tissue. Like HAEMOGLOBIN, myoglobin combines readily with oxygen for use in rapidly contracting muscles. It has been used extensively in research into the structure of PROTEINS. In 1962 John KENDREW shared the Nobel Prize for chemistry for his construction of a three-dimensional crystalline model of sperm whale myoglobin.

myopia (short-sightedness) Common disorder of vision in which near objects are seen sharply (in focus) but distant objects are hazy (out of focus). It is caused either by the eyeball being too long or the EYE's lens being too powerful, so that light rays entering the eye focus in front of the RETINA. It is corrected with concave LENSES in spectacles or contact lenses. *See also* HYPERMETROPIA

myosin Thick filamentous PROTEIN present in MUSCLE cells, associated with ACTIN in the contractile process.

myriapod Member of the Myriapoda class of ARTHROPODS with bodies made up of many similar segments. Each segment has one or more pairs of legs. Centipedes and millipedes are myriapods.

myxomatosis VIRUS disease of rabbits, sometimes spread deliberately by farmers to exterminate them, notably in Australia where early campaigns killed many millions of animals. The disease is generally fatal, although there is evidence that some rabbit populations are becoming resistant. Among rabbit populations, the disease is spread by physical contact and by insect VECTORS such as mosquitoes.

nadir Point on the CELESTIAL SPHERE vertically below the observer. It is diametrically opposite the ZENITH.

naevus Discoloured patch on the skin that has been present since birth, also termed a mole or birthmark. There are many different types, including those caused by a cluster of tiny blood vessels just below the surface of the skin, as in the "strawberry mark".

nail In anatomy, tough outgrowth from the fingers and toes of primates. Nails, like FEATHERS, are made of the same substance as HAIR, the fibrous protein KERATIN.

nanometre (symbol nm) Unit of length, equal to 10^{-9} metre. It is used in the measurement of intermolecular distances and wavelengths. It has superseded the ANGSTROM as the accepted unit for such measurements.

nanotechnology Micromechanics used to develop working devices the size of a few NANOMETRES, one billionth of a metre. US scientists have already etched from silicon an electric motor that is smaller than 0.1mm wide and have made workable gears with a diameter less than a human hair. Further developments have resulted in the manipulation of electrons and individual atoms.

napalm Gelatinous PETROLEUM used to make bombs and as fuel for flame-throwers. It was developed during World War 2 and was widely used by US forces in the Vietnam War. When napalm hits its target, it spreads out, clinging to and burning everything it touches.

naphtha Any of several volatile liquid-hydrocarbon mixtures. In the 1st century AD, "naphtha" was mentioned by Pliny the Elder. Alchemists used the word for various liquids of low boiling point. Several types of products are now called naphtha, including coal-tar naphtha and petroleum naphtha. Petroleum naphtha contains ALIPHATIC COMPOUNDS and boils at higher temperatures than petrol and at lower temperatures than kerosene.

naphthalene Important hydrocarbon ($C_{10}H_8$) composed of two BENZENE rings joined together. A white, waxy solid, naphthalene is soluble in ether and hot alcohol and is highly volatile. It is used in moth balls, dyes and synthetic resins, and is obtained from coal tar and the high-temperature CRACKING of petroleum. It crystallizes in white plates. Properties: m.p. 80°C (176°F); b.p. 218°C (424°F).

Napier, John (1550–1617) Scottish mathematician. He developed "Napier bones", a calculating apparatus that he used to invent LOGARITHMS (1614) and the present form of decimal notation.

nappe Large-scale FOLD in rocks, thrown up by mountain-building processes and transported over large distances. The Alps are a system of several such nappes.

narcotic Any DRUG that induces sleep and/or relieves pain. The term is used especially in relation to OPIUM and its derivatives, such as HEROIN and MORPHINE. These drugs have largely been replaced as sedatives because of their addictive properties, but they are still used for severe pain, notably in terminal illness. Other narcotics include ALCOHOLS (such as ethanol) and BARBITURATES.

Narlikar, Jayant Vishnu (1938–) Indian physicist who has conducted research in such areas as general relativity and gravitation, quantum theory, cosmology and astrophysics. His many publications include *Action at a Distance in Physics and Cosmology* (1974), written in collaboration with Sir Fred HOYLE.

NASA *See* NATIONAL AERONAUTICS AND SPACE ADMINISTRATION

nascent hydrogen Highly reactive form of hydrogen gas. It is generated within a chemical reaction mixture, usually by adding magnesium or zinc to dilute hydrochloric acid, where it acts as a powerful REDUCING AGENT.

Nasmyth, James (1808–90) Scottish inventor and engineer. He developed a foundry for making machine tools and steam-powered machines. In 1842 he patented a steam hammer designed to forge the drive shafts originally intended for the SS *Great Britain*.

National Aeronautics and Space Administration (NASA) US government agency that organizes civilian aeronautical and space research programmes. NASA was established in 1958. Its headquarters are in Washington, D.C., and it operates various field stations. These include the Goddard Space Flight Center at Greenbelt, Maryland, which handles space science research and Earth satellite tracking; the JET PROPULSION LABORATORY in Pasadena, California, which is mission control for NASA space probes and manages NASA's deep-space tracking stations around the world; the Ames Research Center, San Francisco, for aeronautics and planetary science; the John F. Kennedy Space Center at Cape Canaveral, Florida, the main

NANOTECHNOLOGY

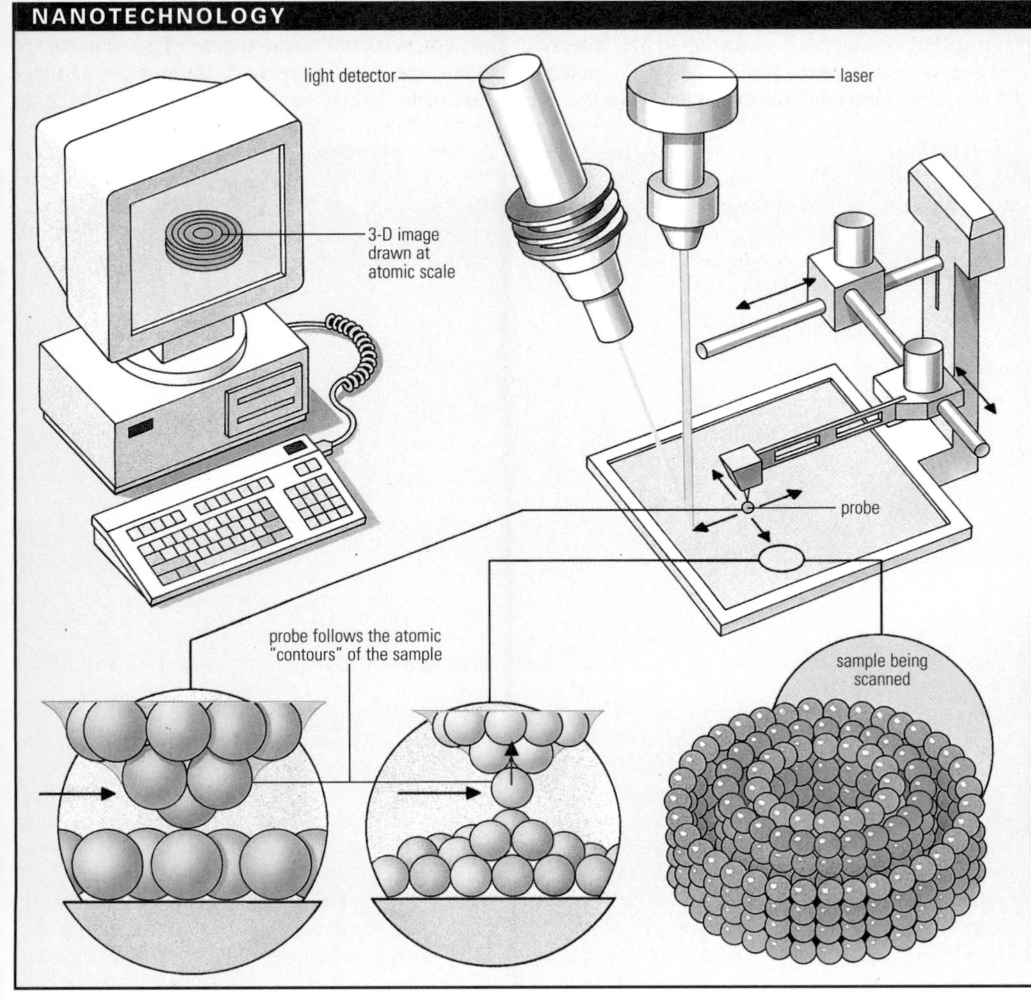

The atomic force microsope (AFM) is one of the latest instruments used in nanotechnology. It has a fine stylus-like probe that traces over the atoms of a sample. As the probe "reads" the sample, a laser beam is deflected and, via a light sensor, a 3-D image is produced on a computer screen. The AFM is also capable of manipulating individual atoms into any position required. The inset photograph shows a coloured scanning electron micrograph of the drive gear (orange) in a micromotor. This gear is smaller in diameter than a human hair and 100 times thinner than a sheet of paper. Driven by two microengines (not visible), it turns a larger gear (green, lower right) through one revolution per second. The whole device was etched into the surface of a wafer of silicon by the same techniques used to make silicon chips. Micromotors could one day be used in a range of exotic "nanotechnology" applications, such as microscopic electric drug pumps implanted in the human body. Magnification: x660 at 5x7cm size.

US launch site; and the Lyndon B. Johnson Space Center in Houston, Texas, mission control for manned space missions. NASA employs more than 20,000 scientific and technical staff.

National Physical Laboratory (NPL) Establishment responsible in the UK for national systems of measurement and for technical aspects of physical standards. Located at Teddington, England, it carries out research into many fields of physics and engineering. It was set up in 1900 under the Department of Industry and sponsored by the Royal Society.

natrolite Hydrated silicate mineral, hydrous sodium aluminium silicate ($Na_2Al_2Si_3O_{10}.2H_2O$). It has orthorhombic system, needle-like crystals, with radiating nodules or compact fibrous masses. It is colourless or white, glassy and brittle. Hardness 5–5.5; r.d. 2.2. *See also* ZEOLITE

Natta, Giulio (1903–79) Italian chemist who shared the 1963 Nobel Prize for chemistry with Karl ZIEGLER for their work on POLYMERS. In 1953 Natta began studying very large molecules and succeeded in polymerizing PROPENE (propylene) to form POLYPROPENE, a substance of great commercial importance.

natural gas Naturally occurring, combustible, gaseous mixture of HYDROCARBONS trapped in pore spaces in sedimentary rocks. It is used as a fuel and in the production of plastics, drugs, antifreeze and dyes. Natural gas is the gaseous component of PETROLEUM and is extracted from OIL WELLS. Certain wells, however, yield only natural gas. Before natural gas can be used as a fuel, the heavier hydrocarbons of BUTANE and PROPANE are extracted; in liquid form these hydrocarbons are forced into containers as bottled gas. The remaining gas, called "dry gas", is piped to consumers for use as fuel. Dry gas is composed of the light hydrocarbons METHANE and ETHANE.

natural number *See* NUMBER, NATURAL

natural selection In EVOLUTION, theory that advantageous changes in an organism will tend to be passed on to successive generations. Changes arise out of natural genetic VARIATION, especially MUTATION. Those that give an individual organism a greater capacity for survival and reproduction in a particular environment will help it to produce more offspring bearing the same beneficial characteristic or trait. Thus the proportion of individuals in the population bearing that trait will increase through successive generations. This theory was put forward by Charles DARWIN in his book *The Origin of Species* (1859). It is still regarded as the key mechanism of evolution.

nature reserve Area of land, sometimes including inland waters and estuaries, set aside for the study and CONSERVATION of wildlife, HABITAT or geological features. In Britain there are many hundreds of such reserves. National Nature Reserves are managed by the government conservation agencies for England, Wales and Scotland; other nature reserves are managed by local government authorities or non-governmental conservation organizations such as Wildlife Trusts, the Woodland Trust, the Royal Society for the Protection of Birds and Regionally Important Geological Sites schemes. Reserves are usually looked after by a warden or conservation officer. The style of management on a reserve varies according to its main interest and purpose; some may allow permission to enter only for a serious study whereas others may encourage people to visit through a nature trail.

naturopathy System of medical therapy that relies exclusively on the use of natural treatments, such as exposure to sunlight, fresh air and a healthy diet of organically grown foods. Herbal remedies are preferred to drugs.

nautical mile Unit used to measure distances at sea. It is defined as the length of one minute of arc of the Earth's circumference. The international nautical mile is equal to 1,852m (6,076.04ft), but in the UK it is defined as 6,080ft (1,853.18m). A speed of one nautical mile per hour is called a KNOT, a term used both at sea and for flying.

navigation Process for determining the position of a vehicle (usually a ship or plane) and its course. Five main techniques are used: dead reckoning, in which the distance and direction travelled are used to plot the postion; piloting (generally the navigation of ships using buoys, landmarks, and so on); celestial navigation; INERTIAL GUIDANCE (used in the navigation of missiles); and radio navigation. The latter includes the use of radio beacons, LORAN (long range navigation), RADAR navigation and satellite navigation systems, such as GPS (GLOBAL POSITIONING SYSTEM). Instruments and charts enable the navigator to determine position, expressed in terms of LATITUDE and LONGITUDE, direction in degrees of arc from true north, speed and distance travelled. *See also* COMPASS; GYROCOMPASS; SEXTANT

Neanderthal Middle PALAEOLITHIC variety of human, known from fossils in Europe and Asia. Neanderthals were discovered when a skeleton was unearthed in the Neander Valley in w Germany in 1856. The bones were thick and powerfully built and the skull had a pronounced brow ridge. Neanderthals are now considered to be a separate species of human, possibly a local adaptation during the ice ages, and are not thought to be ancestral to modern humans. Neanderthals predated modern humans in Europe, but were superseded by them about 35,000 years ago. *See also* HUMAN EVOLUTION

neap tide *See* TIDE

nebula (Lat. cloud) Region of interstellar gas and dust. There are three main types of nebula. **Emission** nebulae are bright diffuse nebulae which emit light and other radiation as a result of ionization (removal of electrons) and excitation of the gas atoms by ultraviolet radiation. The source of the ultraviolet is usually one or more hot stars. In contrast, the brightness of **reflection** nebulae results from the scattering by dust particles of light from nearby stars. **Dark** nebulae are not luminous: interstellar gas and dust absorbs light

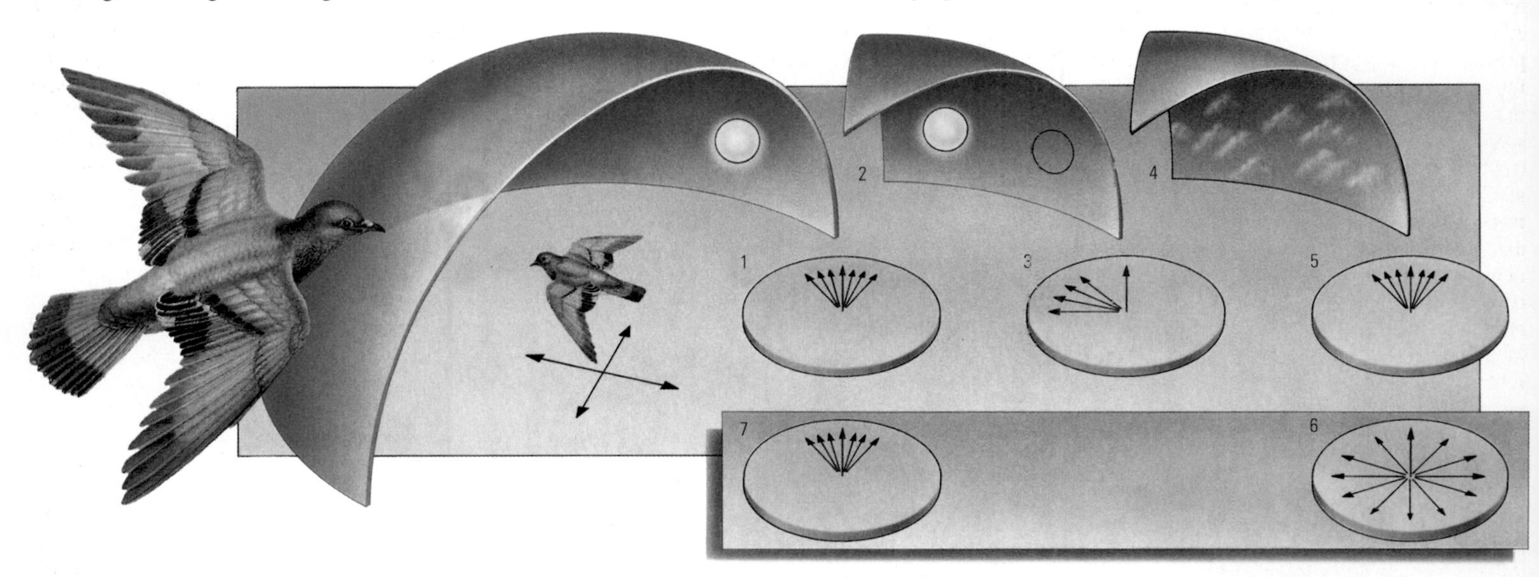

▲ **navigation** Homing pigeons (A) successfully navigate home over hundreds of kilometres from sites never previously visited; these astonishing feats have inspired many experiments, some of which are shown below. Evidence suggests visual landmarks are fairly unimportant guides, so the obvious alternative is solar navigation; but, as 16th-century sailors were only too well aware, accurate solar navigation needs an accurate clock. Bird behaviour is influenced by light and darkness, and laboratory birds can be tricked by artificial lighting to make their "body clocks" run fast or slow. On release, a homing pigeon (shown here in the Southern Hemisphere) with an unaltered clock flies in the correct direction relative to the Sun to reach its home loft when released (1) (in all diagrams red arrows show the correct direction to the pigeon loft and black arrows show the observed scatter of actual flight directions). If a pigeon with a clock "set" six hours fast is released at noon, it assumes the time is 6.00 pm and the Sun is in a 6.00 pm position (2), much farther W than in reality. The pigeon therefore flies too far E in its attempt to reach home (3). But pigeons can also navigate if the Sun is hidden by clouds (4), when birds with altered clocks navigate just as successfully as bird with normal clocks (5). In fact, they have a "back-up" navigational aid using the Earth's magnetic fields. This can be tested by attaching a small bar magnet to a pigeon's neck to distort the magnetic field around it; on an overcast day the "homing" directions flown by such a pigeon are completely random (6). Confirming the magnet's effect, a pigeon with a similar, non-magnetic, brass bar is not disorientated. On sunny days, however, a pigeon with a magnet and a normal clock reverts to solar navigation and is not disorientated (7) by its disturbed magnetic field.

from background stars, producing apparently dark patches in the sky.

neck In vertebrates, the part of the body that connects the head and the trunk. In human beings, it contains the seven cervical VERTEBRAE of the spinal column and, to the front, the PHARYNX, leading to the TRACHEA and OESOPHAGUS.

necrosis Death of plant or animal tissue. It can be caused by disease, injury or interference with the blood supply.

nectar Sweet liquid secreted by most flowering plants (ANGIOSPERMS). It consists mainly of a solution of GLUCOSE, FRUCTOSE and SUCROSE in water. The glands (nectaries) that produce it usually lie at the base of the flower petals, but may be found also in parts of the stem or at the leaf bases. Nectar attracts insects, which help with cross POLLINATION. Bees turn nectar into honey.

Néel, Louis Eugène Félix (1904–) French scientist who shared the 1970 Nobel Prize for physics with Hannes ALFVÉN for their work in SOLID-STATE PHYSICS. Néel studied the magnetic properties of solids and discovered antiferromagnetism, in which the magnetic fields of small groups of atoms are aligned in opposite directions. His discoveries have been widely used in computer memories.

nekton Active swimming organisms at the sea's surface, as opposed to the floating PLANKTON. Nekton includes the large migrating marine animals such as adult squids, fish and whales. *See also* BENTHOS; PELAGIC

nematocyst In cnidarians such as jellyfish, a tiny fluid-filled sac containing a coiled hollow thread that can be shot out at prey. It forms part of a thread cell (nematoblast) in the ECTODERM of the animal. When a "trigger" (cnidocil) on the thread cell is touched, the thread is rapidly projected to coil round the prey or inject poison. The poisonous stings on the tentacles of jellyfish derive from thread cells and their nematocycts.

nematode Worm-like invertebrates of the phylum Aschelminthes. They are identified by their unsegmented, elongated, cylindrical bodies, pointed at both ends. Covered in thick cuticle and comprising longitudinal muscles, the nematodes are found in a variety of habitats including soil, and marine and freshwater environments. Although some species are free-living, others, such as *Acaris* (roundworms), are parasitic. Reproduction is sexual. Class Nematoda.

neo-Darwinism Development of DARWINISM that incorporates the modern ideas of genetic HEREDITY (notably MENDEL's) with DARWIN's ideas of EVOLUTION through NATURAL SELECTION. It uses research into GENETICS to explain the origin of VARIATION within a species.

neodymium (symbol Nd) Silver-yellow metallic element of the LANTHANIDE SERIES. It was first isolated in the form of its oxide in 1885 and the pure metal was first obtained in 1925. Chief ores are monazite and bastnaesite. It is used to manufacture lasers, and neodymium salts are used to colour glass. Properties: at.no. 60; r.a.m. 144.24; r.d. 7.004; m.p. 1,010°C (1,850°F); b.p. 3,068°C (5,554°F); most common isotope ^{142}Nd (27.11%).

Neolithic (New Stone Age) Stage in human cultural development following the PALAEOLITHIC. The Neolithic began *c.*8000 BC in W Asia, and *c.*4000 BC in Britain. It was during this period that people first lived in settled villages, domesticated and bred animals, cultivated cereal crops and practised stone-grinding and flint mining.

neon (symbol Ne) Gaseous, nonmetallic element, a NOBLE GAS, discovered in 1898. Colourless and odourless, it is present in the atmosphere (0.0018% by volume) and is obtained by the fractional distillation of liquid air. Its main use is in discharge tubes for advertising signs (emitting a bright red glow while conducting electricity) and in gas lasers, Geiger counters and particle detectors. The element forms no compounds, except briefly and under extreme conditions. Properties: at.no. 10; r.a.m. 20.179; m.p. −248.67°C (−415.6°F); b.p. −246.05°C (−410.89°F); most common isotope ^{20}Ne (90.92%).

NEON

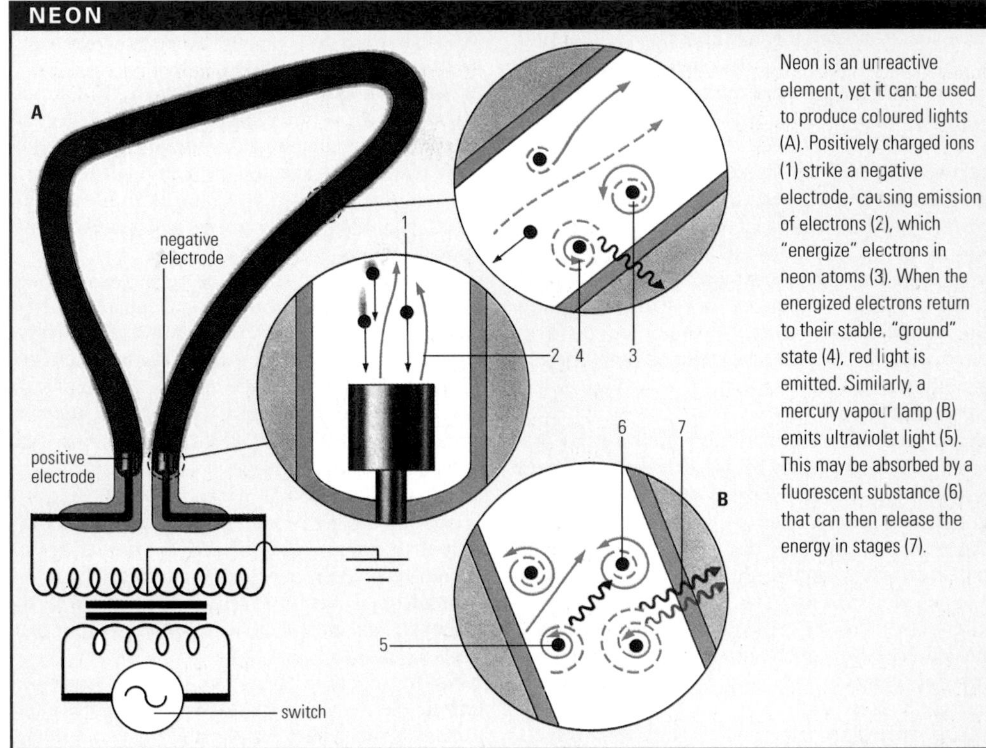

Neon is an unreactive element, yet it can be used to produce coloured lights (A). Positively charged ions (1) strike a negative electrode, causing emission of electrons (2), which "energize" electrons in neon atoms (3). When the energized electrons return to their stable, "ground" state (4), red light is emitted. Similarly, a mercury vapour lamp (B) emits ultraviolet light (5). This may be absorbed by a fluorescent substance (6) that can then release the energy in stages (7).

neoplasm Any new or abnormal growth on the body. It may be malignant (cancerous) or benign (non-cancerous).

neoprene (2-chlorobuta-1,3diene, CH_2=CCl.CH=CH_2) Synthetic RUBBER that is very resistant to corrosive chemicals and is widely used in the chemical industry. Neoprene is a POLYMER of chloroprene.

neoteny Persistence in an adult animal of larval characteristics. It includes the retention of GILLS, as in some salamanders, the best example of which is the axolotl of W USA and Mexico. An entire order of tunicates (sea squirts), the Larvacea, is permanently larval, never reaching a typical adult form.

nephritis (Bright's disease) Inflammation of the KIDNEY. It is a general term, used to describe a condition rather than any specific disease. It may be acute (due perhaps to a streptococcal infection) or chronic, often progressing to kidney failure.

nephron Basic functional unit of the mammalian KIDNEY, involved in the formation of URINE. There are more than a million nephrons in a human kidney. Each consists of a GLOMERULUS (a cluster of tiny blood capillaries) cupped in a structure called a BOWMAN'S CAPSULE, with an attached long, narrow tubule. Blood enters the kidney under pressure, and water and wastes are forced into the tubule by a process known as ultrafiltration. Some water and essential molecules are reabsorbed into the bloodstream; the remaining filtrate, urine, is passed to the BLADDER for voiding.

Neptune Eighth planet from the Sun. The mass, orbit and position of an unseen planet had been calculated by Urbain LEVERRIER and, independently, John Couch ADAMS to account for perturbations in the orbital motion of URANUS. Neptune was actually observed in 1846 by Johann GALLE and Heinrich d'Arrest. Neptune is invisible to the naked eye. Through a telescope, it appears as a small, greenish-blue disc on which very few details can be distinguished. In its size, mass, atmosphere and colour, Neptune resembles Uranus. In 1989 the fly-by of the VOYAGER 2 probe in 1989 provided most of our current knowledge of the planet. The upper atmosphere is about 85% molecular hydrogen and 15% helium. Its predominant blue colour is due to a trace of methane, which strongly absorbs red light. Several different atmospheric features were visible at the time of the Voyager encounter. There were faint bands parallel to the equator, and spots, the most prominent of which was the oval Great Dark Spot (GDS), about 12,500km (8,000mi) long and 7,500km (4,500mi) wide. It is about the same size relative to Neptune as the Great Red Spot (GRS) is to Jupiter, and like the GRS it is a giant anticyclone. White, cirrus-type clouds of methane crystals cast shadows on the main cloud deck some 50km (30mi) below. There are also the highest wind speeds recorded on any planet, over 2,000km/h (1,250mph) in places. Neptune's dynamic and changing atmosphere probably has to do with its internal heat source: it radiates more than twice as much heat as it receives from the Sun. Below the visible atmosphere, the composition and structure are less certain. Neptune has eight known SATELLITES and a ring system.

neptunium (symbol Np) Radioactive metallic element, the first of the TRANSURANIC ELEMENTS to be synthesized (1940). The silvery element is found in small amounts in uranium ores and is obtained as a by-product in nuclear reactors. Properties: at.no. 93; r.a.m. 237.0482; r.d. 20.25; m.p. 640°C (1,184°F); b.p. 3,902°C (7,056°F); most stable isotope ^{237}Np (half-life 2.2 million years).

Nereid NEPTUNE's outermost satellite, discovered by Gerard KUIPER in 1949. It has the most eccentric orbit of all known satellites, taking it from 1,4 million km (850,000mi) from the planet out to 9,7 million km (6 million mi). It is almost certainly a captured body.

Nernst, Walther Hermann (1864–1941) German chemist who was professor at Göttingen and then at Berlin. Nernst was awarded the 1920 Nobel Prize for chemistry for his discovery of the third law of THERMODYNAMICS. His other important work was concerned with chain reactions in photochemistry. He wrote *Thermodynamics* (1893) and *The New Heat Theorem* (1918).

nerve Collection of NEURONS providing a communications link between the vertebrate NERVOUS SYSTEM and other parts of the body. Afferent or sensory nerves (SENSORY NEURONS) transmit nervous impulses to the CENTRAL NERVOUS SYSTEM (CNS); efferent or motor nerves (MOTOR NEURONS) carry impulses away from the central nervous system to muscles. Some nerves contain both afferent and efferent fibres.

nerve cell *See* NEURON

nerve gas Extremely poisonous form of gas first developed as a weapon by the Germans in World War 2 but never used. Nerve gases are often organic chemical compounds containing phosphorus. Their effect is to inactivate the enzyme CHOLINESTERASE, and so prevent NERVE messages reaching the muscles. The result is death within minutes by failure of respiratory and other systems. Inhalation of even small amounts of the gas, or any physical contact with the liquid chemical, is enough to cause death.

nerve impulse Electrical signal that travels along the AXON of a nerve cell. Nerve impulses carry information throughout the NERVOUS SYSTEM. As an impulse passes, sodium and potassium ions flow into and out of the axon's membrane, creating a voltage reduction called the ACTION POTENTIAL. Nerve impulses can travel at up to 150m (500ft) per second.

nervous system Communications system consisting of interconnecting nerve cells or NEURONS that coordinate all life, growth and physical and mental activity. In simple animals, such as jellyfish and sea anemones, it consists of a **nerve net** without a centre, or brain. The vertebrate nervous system consists of the CENTRAL NERVOUS SYSTEM (CNS), comprising the BRAIN and SPINAL CORD, and, serving all parts of the body outside the CNS, the PERIPHERAL NERVOUS SYSTEM. *See also* AUTONOMIC NERVOUS SYSTEM (ANS)

network *See* COMPUTER NETWORK

Neumann, Franz Ernst (1798–1895) German scientist who, in two papers (1845, 1847), formulated the first mathematical theory of ELECTROMAGNETIC INDUCTION, the production of ELECTROMOTIVE FORCE (EMF) by changes in a magnetic field. Neumann also conducted research into optics and crystallography.

Neumann, John von *See* VON NEUMANN, JOHN

neuralgia Intense pain from a damaged nerve, possibly tracking along its course. It may be temporary or chronic. There are various forms, including trigeminal neuralgia, which features attacks of stabbing pain in the mouth area, and post-herpetic neuralgia, which follows an attack of Trigeminal neuralgia (at one time known as *tic douloureux*) usually affects people over 60, especially women. When it occurs in younger people, it is often a symptom of serious neurological disease. Most cases of neuralgia are treated with anticonvulsant drugs.

neurology Branch of medicine dealing with the diagnosis and treatment of disorders of the NERVOUS SYSTEM.

neuron (neurone, nerve cell) Basic structural unit of the NERVOUS SYSTEM which enables rapid transmission of NERVE IMPULSES between different parts of the body. It is composed of a cell body, containing a nucleus, and a number of trailing, finger-like processes. The largest of these is the AXON, which carries outgoing impulses, causing the release of NEUROTRANSMITTERS across a junction (SYNAPSE) resulting in the stimulation of another neuron; the rest are DENDRITES, which receive incoming impulses. *See also* MOTOR NEURON; SENSORY NEURON

neurosis Emotional disorder such as anxiety, depression or various phobias. It is a form of mental illness in which the main disorder is of mood, but the person does not lose contact with reality as happens in PSYCHOSIS.

neurosurgery Surgical treatment of disorders of the NERVOUS SYSTEM.

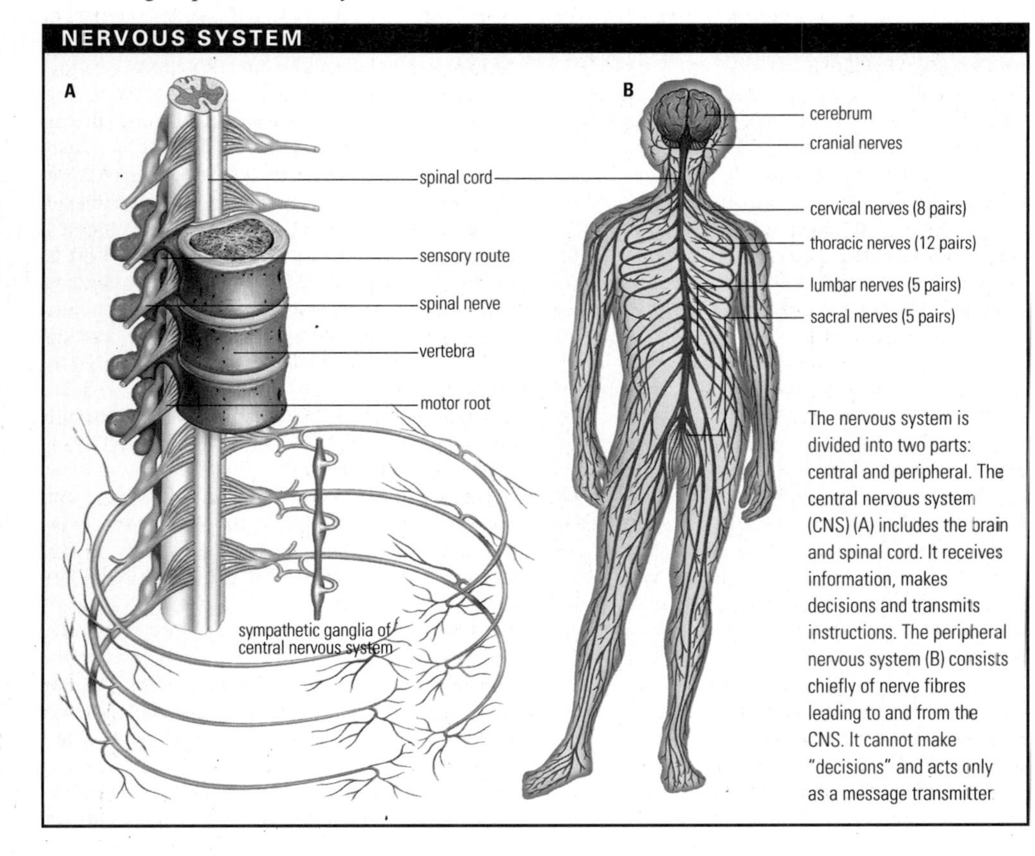

The nervous system is divided into two parts: central and peripheral. The central nervous system (CNS) (A) includes the brain and spinal cord. It receives information, makes decisions and transmits instructions. The peripheral nervous system (B) consists chiefly of nerve fibres leading to and from the CNS. It cannot make "decisions" and acts only as a message transmitter

neurotransmitter Any one of several dozen chemicals involved in communication between NEURONS (nerve cells) or between a nerve and muscle cells. When an electrical impulse arrives at a nerve ending, a neurotransmitter is released to carry the signal across the specialized junction (SYNAPSE) between the nerve cell and its neighbour. Fast-acting neurotransmitters either prompt nerves to fire or inhibit them from firing; slow-acting ones seem to be involved in modifying the activity of whole groups of nerve cells. ACETYLCHOLINE, the neurotransmitter that causes muscle to contract, is the target of NERVE GASES. Some drugs work by disrupting neurotransmission. *See also* ADRENALINE; NERVOUS SYSTEM; NORADRENALINE; SEROTONIN

neutralization In chemistry, mixing of equivalent amounts of an acid and a base in an aqueous medium until the mixture is neither acidic nor basic. It then has a PH of 7, which indicates a neutral solution. Neutralization is often accompanied by the generation of heat. *See also* TITRATION

neutrino (symbol ν) Uncharged ELEMENTARY PARTICLE with no, or very low, mass, SPIN $\frac{1}{2}$, and travelling at, or near, the speed of light. Classified as a LEPTON, it has little reaction with matter and is difficult to detect (postulated in 1930 by Wolfgang PAULI, but not directly detected until 1956). Neutrinos are created and destroyed by particle decays that involve the WEAK NUCLEAR FORCE. There are three known species. The **electron** neutrino is closely associated with the ELECTRON and is produced when, for example, PROTONS and electrons react to form NEUTRONS, as in the Sun. The electron neutrino's antiparticle is more common and occurs when a neutron decays. The **muon** neutrino is associated with the MUON, and occurs in high-energy reactions. The **tau** neutrino is associated with the TAU PARTICLE. All have their own antiparticles, called **antineutrinos**.

neutron (symbol n) Uncharged ELEMENTARY PARTICLE that occurs in the atomic nuclei of all chemical elements, except the lightest isotope of HYDROGEN. It was first identified (1932) by James CHADWICK. Outside the nucleus, it is unstable, decaying with a half-life of 11.6 minutes into a PROTON, ELECTRON and antineutrino. Its neutrality allows it to penetrate and be absorbed in nuclei and thus to induce nuclear transmutation and fission. It is a BARYON with SPIN $\frac{1}{2}$ and a mass slightly greater than that of the proton.

neutron activation analysis Highly sensitive method of identifying the chemical composition of a substance. The sample is bombarded with high-energy NEUTRONS that are then absorbed by the atoms present. The resulting radioactive nuclei emit radiation of an energy and decay rate characteristic of the original atoms, which can thus be identified. The quantity present can also be found with extreme precision.

neutron bomb *See* NUCLEAR WEAPON

neutron flux Rate at which NEUTRONS are released during the nuclear FISSION process in a controlled nuclear CHAIN REACTION. In a NUCLEAR REACTOR, MODERATORS of several kinds (especially DEUTERIUM and GRAPHITE) are used to reduce neutron speeds. *See also* FAST NEUTRON

neutron star Extremely small, dense star that consists mostly of NEUTRONS. It is the last stage of many stars. Neutron stars are formed when a massive star explodes as a SUPERNOVA, blasting off its outer layers and compressing the core so that its component PROTONS and ELECTRONS merge into neutrons. They are observed as PULSARS. They have masses comparable to that of the Sun, but diameters of only about 20km (12mi) and average densities of about 10^{15} g cm^{-3}. The maximum mass for a neutron star is about 3 solar masses, beyond which it would collapse further, into a BLACK HOLE. *See also* STELLAR EVOLUTION

névé *See* FIRN

Newcomb, Simon (1835–1909) US mathematical astronomer, b. Canada. He calculated highly accurate values for astronomical constants, such as the solar PARALLAX, and prepared extremely accurate tables of the motions of the Moon and the planets. It was Newcomb who calculated the discrepancy in the advance of Mercury's PERIHELION that was to provide a proof of Einstein's special theory of RELATIVITY.

Newcomen, Thomas (1663–1729) British engineer and inventor of the atmospheric STEAM ENGINE. Assisted by John Calley, a plumber, he spent more than 10 years perfecting the invention. In 1705 Newcomen built a model of the engine, which consisted of a piston moved by atmospheric pressure within a CYLINDER in which a partial vacuum had been created by condensing steam. A 1698 patent by Thomas SAVERY covered the principle, and Newcomen went into partnership with him. For many years, this type of engine was used to power pumps, particularly to drain mines.

New General Catalogue (NGC) In astronomy, list of stellar clusters, nebulae and galaxies compiled by Danish astronomer Johan Dreyer in 1888, based on the observations of William HERSCHEL and his son, John. Dreyer later published two Index Catalogues (IC) bringing the number of stellar objects listed in the catalogue to *c.*13,000.

Newlands, John Alexander Reina (1837–98) British chemist. In 1864 Newlands announced his law of octaves, which arranged the chemical elements in a table of eight columns according to increasing relative atomic mass. The law was ridiculed until MENDELEYEV incorporated it in his own PERIODIC TABLE five years later.

Newton, Alfred (1829–1907) British zoologist, b. Switzerland. Newton promoted the first laws for the protection of birds, and was the editor of the ornithological journals *Ibis* and *The Zoological Record*. He wrote *A Dictionary of Birds* (1893–96).

Newton, Sir Isaac (1642–1727) English scientist who proposed a theory of MECHANICS and of GRAVITATION that survived unchallenged until the 20th century. He studied at Cambridge and became professor of mathematics there (1669–1701). Many of his discoveries were made between 1664 and 1666, but not published until later. His main works were *Philosophiae Naturalis Principia Mathematica* (1687), in which he proposed the three laws of MOTION, and *Opticks* (1704). In the former, he also proposed the principle of universal GRAVITATION; in the latter, he showed that white light is made up of colours of the SPECTRUM and put forward his particle theory of light. In the 1660s Newton also created the first system of CALCULUS, but did not publish it until Gottfried LEIBNIZ had published his own system in 1684. He built a reflecting telescope *c.*1671, and studied alchemy and biblical chronology. Newton was president of the Royal Society (1703–27) and master of the mint after 1701. In 1705 he became the first man to be knighted for scientific work. Newton engaged in acrimonious scientific disputes with Leibniz, Robert HOOKE and John FLAMSTEED.

newton (symbol N) SI unit of FORCE. One newton is the force that gives a mass of one kilogram an acceleration of one metre per second per second. One kilogram weighs 9.807N; one pound weight equals 4.45N.

Newton's laws of gravitation *See* GRAVITATION

Newton's laws of motion *See* MOTION, LAWS OF

niccolite (kupfernickel) One of the chief ores of NICKEL, consisting of nickel arsenide (NiAs), often also with some cobalt, iron and sulphur. It has hexagonal system of crystals generally in columnar masses, rarely as tabular crystals. It is found in vein deposits and is copper coloured, with an easily tarnished metallic lustre. Hardness 5–5.5; r.d. 7.8.

nickel (symbol Ni) Silver-white, TRANSITION ELEMENT, first discovered in 1751. Its chief ores are nickel-iron sulphide (pentlandite) and nickel arsenide (niccolite). Refining is complex, and involves the differential separation of the various compounds to obtain concentrated nickel sulphide (or arsenide), which is reduced to the pure nickel. Hard, malleable and ductile, nickel is used in stainless steels, other special alloys, coinage, cutlery, storage batteries and as a hydrogenation catalyst. The metal is ferromagnetic. Properties: at.no. 28; r.a.m. 58.71; r.d. 8.90 (25°C); m.p. 1,453°C (2,647°F); b.p. 2,732°C (4,950°F); most common isotope ^{58}Ni (67.84%).

nickel carbonyl ($Ni(CO)_4$) Colourless, flammable, toxic liquid. It is used in the Mond process (*see* MOND, LUDWIG) for the production of pure nickel. Properties: r.d. 1.32g cm^{-3}; m.p. −25°C (−13°F); b.p. 43°C (109.4°F).

nickel silver (German silver) Alloy of copper and nickel. It may also contain small amounts of zinc. Nickel silver is hard and does not tarnish easily. It is used for making coins and cutlery, when it is often plated with silver to give E.P.N.S. (electroplated nickel silver).

Nicolle, Charles Jules Henri (1866–1936) French bacteriologist. Nicolle was awarded the 1928 Nobel Prize for physiology or medicine for his discovery (1909) that typhus is transmitted by the body louse. He later distinguished between classical typhus and murine typhus, which is passed on to humans by the rat flea. Nicolle was head of the Pasteur Institute in Tunis (1902–32), which under his direction became a distinguished centre for bacteriological research.

Nicol prism Optical device used for the production of POLARIZED LIGHT. It was invented (1828) by the British physicist William Nicol (1768–1851). A Nicol PRISM consists of a rectangular crystal of ICELAND SPAR (calcite) split along a diagonal and rejoined with Canada balsam (a type of resin). Light passing through it is separated into a reflected (ordinary) ray and a transmitted (extraordinary) ray, both polarized in different planes. It is used in the polarizing MICROSCOPE, and also in GEOLOGY to examine thin sections of rock microscopically.

nicotine Poisonous ALKALOID obtained from the leaves of tobacco, used in agriculture as a pesticide and in veterinary medicine to kill external parasites. Nicotine is the principal addictive agent in smoking tobacco.

nicotinic acid VITAMIN of the B complex, lack of which causes the disease PELLAGRA.

nielsbohrium *See* HAHNIUM

nimbostratus *See* CLOUD

nimbus CLOUD from which rain is discharged. In meteorological cloud classification it is added as a suffix to the names of clouds that typically produce rain or snow, such as cumulonimbus.

niobium (symbol Nb) Shiny, grey-white TRANSITION ELEMENT, first discovered in 1801. Its chief ore is pyrochlore. Soft and ductile, niobium is used in special stainless steels and in alloys for rockets and jet engines. Properties: at.no. 41; r.a.m. 92.9064; r.d. 8.57; m.p. 2,468°C (4,474°F); b.p. 4,742°C (8,568°F); most common isotope ^{93}Nb (100%).

▲ **Newton** Engraving of Sir Isaac Newton, English physicist, mathematician, and alchemist. As a mathematician he discovered the binomial theorem and developed differential and integral calculus. As a physicist, he devised laws of motion, formulated the general theory of gravitation and wrote on optics. As an astronomer, he invented the reflecting telescope. Newton wrote extensively on alchemy and tried in vain to discover a "philosopher's stone" which would convert common metals into gold. Newton calculated the year of biblical Creation as being 3500 BC and wrote extensively on the biblical prophesies of Daniel and the Apocalypse.

Nirenberg, Marshall Warren (1927–) US biochemist who found the key to the amino acid GENETIC CODE by deciphering different combinations of three nucleotide bases (called "codons") within long nucleotide chains in DNA and RNA. Each combination is coded to convert a different AMINO ACID to PROTEIN, a key process in transferring inherited characteristics. Nirenberg found he could decipher the unknown configurations by synthesizing a nucleic acid with a known base combination, and then recording the amino acid that it changed to protein. He was awarded the 1968 Nobel Prize for physiology or medicine, together with Robert W. HOLLEY and Har Gobind KHORANA, for his part in discovering how GENES determine cell function.

nitrate Salt of NITRIC ACID (HNO_3). Nitrate salts contain the nitrate ion (NO_3^-) and some are important naturally occurring compounds, such as saltpetre (potassium nitrate, KNO_3) and Chile saltpetre (sodium nitrate, $NaNO_3$). Nitrates are used as food preservers, fertilizers, explosives and as a source of nitric acid.

nitration Introduction of a nitro group ($-NO_2$) to the carbon atom of an organic compound, usually replacing a hydrogen atom. Strong nitric acid or a NITRATE salt with sulphuric acid is a nitrating agent. The nitro compounds formed are useful chemical intermediates, as in the production of

NITRATION

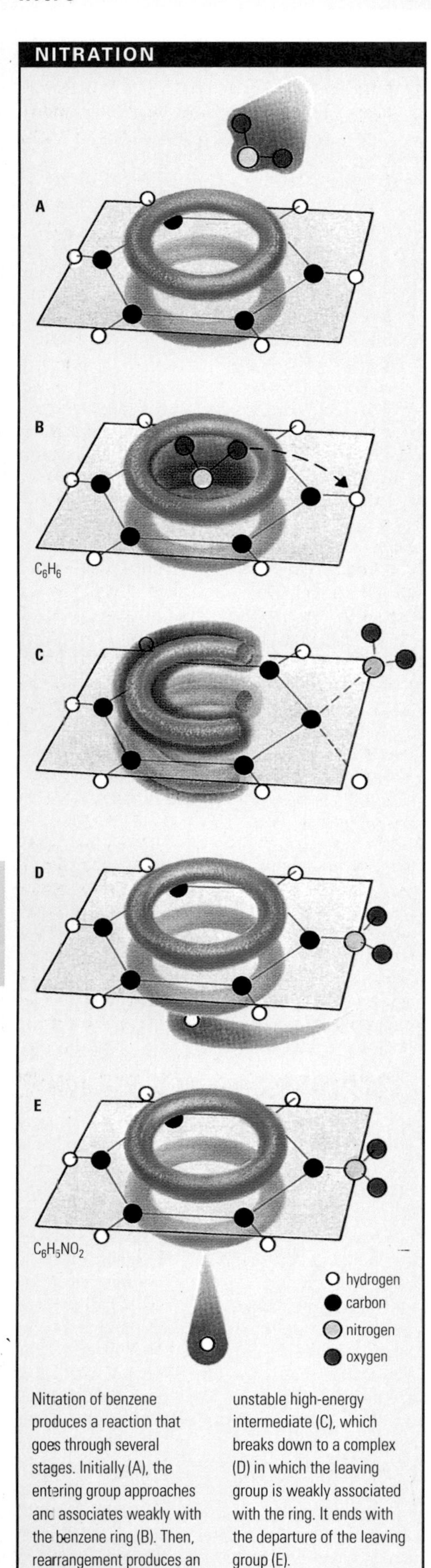

Nitration of benzene produces a reaction that goes through several stages. Initially (A), the entering group approaches and associates weakly with the benzene ring (B). Then, rearrangement produces an unstable high-energy intermediate (C), which breaks down to a complex (D) in which the leaving group is weakly associated with the ring. It ends with the departure of the leaving group (E).

ANILINE from BENZENE, and several are explosives (such as trinitrotoluene, TNT).

nitre (saltpetre) Commercial form of potassium nitrate (KNO_3), known and used since medieval times as an ingredient of GUNPOWDER.

nitric acid (HNO_3) Colourless liquid, one of the strongest mineral acids. It fumes in air and soon turns yellow as the result of decomposition into NITROGEN DIOXIDE (NO_2). Nitric acid attacks most metals, resulting in the formation of NITRATES, and is a strong oxidizing agent. It is mostly made by oxidizing AMMONIA over a catalyst, and is used for many purposes, including the manufacture of agricultural chemicals, explosives, plastics, dyes and rocket propellants.

nitric oxide *See* NITROGEN MONOXIDE

nitrile Organic chemical compound containing the cyanide group (–C≡N) connected to a HYDROCARBON group. Nitriles are used in organic synthesis, particularly for the manufacture of acrylic fabrics and synthetic RUBBER.

nitrobenzene ($C_6H_5NO_2$) Aromatic nitro compound, a pale yellow, highly toxic, oily liquid that smells of almonds. It freezes at 5.7°C (42.3°F) and boils at 211°C (411.6°F). Nitrobenzene is used in the manufacture of ANILINE and other organic chemicals.

nitrocellulose *See* CELLULOSE NITRATE

nitrogen (symbol N) Common gaseous, nonmetallic element of group V of the periodic table, discovered in 1772. Colourless and odourless, it is the major component of Earth's atmosphere (78% by volume), from which it is extracted by fractional distillation of liquid air. The NITROGEN CYCLE is an essential process for the existence of life on Earth. The main industrial use is in the HABER PROCESS, which produces ammonia for fertilizers and making nitric acid. Nitrogen compounds are used in fertilizers, explosives, dyes, foods and drugs. The element is chemically inert. Properties: at.no. 7; r.a.m. 14.0067; r.d. 1.2506; m.p. −209.86°C (−345.75°F); b.p. −195.8°C (−320.4°F); most common isotope ^{14}N (99.76%).

nitrogen cycle Circulation of nitrogen through plants and animals in the BIOSPHERE. Nitrogen-fixing BACTERIA in the soil or plant root nodules take free nitrogen from the soil and air to form the nitrogen compounds (nitrates) used by plants in ASSIMILATION (*see* NITROGEN FIXATION). HERBIVORES obtain their nitrogen from the plants, and in turn CARNIVORES obtain nitrogen by eating herbivores. SAPROPHYTES decompose the tissue of all the organisms concerned and the nitrogen is released back into the cycle.

nitrogen dioxide (NO_2) Oxide of nitrogen. It is a pungent-smelling, brown gas that readily forms the dimer (two identical molecules linked together) dinitrogen tetroxide (N_2O_4). It is made by the action of concentrated nitric acid on copper, and dissolves in water to give a mixture of nitrous and nitric acids. It is also formed by the reaction of oxygen with NITROGEN MONOXIDE. Its presence in the atmosphere from the exhaust gases of petrol engines contributes to the formation of ACID RAIN and the depletion of the OZONE LAYER.

nitrogen fixation Incorporation of atmospheric NITROGEN into chemicals for use by organisms. Nitrogen-fixing microorganisms (mainly BACTERIA and CYANOBACTERIA) absorb nitrogen gas from the air, from air spaces in the soil, or from water, and build it up into compounds of AMMONIA. Other bacteria then change these compounds into NITRATES, which can be taken up by plants. Some nitrogen-fixing bacteria live in symbiotic association with other organisms. Bacteria in the root nodules of plants, especially members of the pea family, exchange nitrogen compounds for carbohydrates. If the remains of these plants are ploughed back into the soil, they enrich it with nitrates. Cyanobacteria are important nitrogen fixers in the sea, in freshwater and in soil, and are particularly important in rice fields. *See also* NITROGEN CYCLE; SYMBIOSIS

nitrogen monoxide (nitric oxide, NO) Oxide of nitrogen, a colourless soluble gas. It is made by reducing nitric acid or by reacting sodium nitrite, sulphuric acid and iron(II) (ferrous) sulphate. It reacts with oxygen to form NITROGEN DIOXIDE.

nitroglycerine Oily liquid used in the manufacture of explosives. It is made by reacting nitrating mixture (nitric and sulphuric acids) with glycerol (glycerine). Too dangerous to handle in any quantity on its own, it is made into DYNAMITE. It is also used in medicine, known as GLYCERYL TRINITRATE, for treating angina.

nitrous oxide *See* DINITROGEN OXIDE

nivation Type of mechanical WEATHERING of rocks that results from alternate freezing and thawing. At night, when temperatures fall, water lying in cracks in the rock freezes and expands. The following morning the ice melts and contracts, only to freeze again after nightfall. The continual expansion and contraction causes stresses that eventually break the rock along the crack or other line of weakness. Accumulated fragments of rock produced in this way form SCREE on hillslopes.

NMR *See* NUCLEAR MAGNETIC RESONANCE

Nobel, Alfred Bernhard (1833–96) Swedish chemist, engineer and industrialist. In 1866 Nobel invented DYNAMITE (a mixture of NITROGLYCERINE and kieselguhr). In 1876 he patented a more powerful form of blasting gelatine, gelignite. Nobel was a pacifist and, with the immense fortune he made from the manufacture of explosives, he founded the NOBEL PRIZES, first awarded in 1901.

nobelium (symbol No) Radioactive, metallic element, one of the ACTINIDE SERIES. It was first made in 1958 by bombarding CURIUM with CARBON at the University of California at Berkeley. Seven isotopes are known. Properties: at.no. 102, most stable isotope ^{255}No (half-life 3 minutes). *See also* TRANSURANIC ELEMENT

Nobel Prize Awards given each year for outstanding contributions in the fields of physics, chemistry, physiology or medicine, literature and economics and to world peace. Established in 1901 by the will of Alfred NOBEL, the prizes are awarded annually on 10 December. The winners are selected by committees based in Sweden and Norway.

Nobili, Leopoldo (1784–1835) Italian physicist, a pioneer of ELECTROCHEMISTRY. He generated electricity using platinum ELECTRODES in an alkaline nitrate ELECTROLYTE and devised the astatic GALVANOMETER to measure the current. Nobili also invented the thermocouple and the thermopile (two types of thermometers).

noble gas (inert gas) Any of a group of colourless, odourless and unreactive gases. It includes (in order of increasing atomic number) HELIUM, NEON, ARGON, KRYPTON, XENON and RADON and forms group 0 of the periodic table. The noble gases' low reactivity is due to complete outer electron shells (two electrons for helium and eight each for the rest), thus offering no VALENCE "hooks". The noble gases were once thought to be completely inert (nonreactive), hence their alternative name, but compounds have been identified for krypton, xenon, radon and argon. They are also called the "rare gases", although argon and helium are relatively common.

noble metal Any of a group of unreactive metals found naturally in the native, metallic state and not usually as chemical compounds or ores. The group includes SILVER, GOLD, PLATINUM, OSMIUM, IRIDIUM, PALLADIUM, RHODIUM and RUTHENIUM. However, COPPER, also sometimes found as the native metal, is not included, whereas MERCURY, found both as the metal and as ores such as CINNABAR, often is, so that the noble metals are at best a rather loose assemblage. Gold and platinum are precious metals used in jewellery for their rarity, lustre, resistance to tarnishing, ductility and malleability. Osmium, the heaviest of all metals, is alloyed with iridium to make osmiridium, used to tip fountain pen nibs. Platinum and palladium are used in industry as CATALYSTS. *See also* BASE METAL

node In anatomy, a node is a thickening or enlargement of an organ or tissue, such as a lymph node (gland) or the sinoatrial node of nervous tissue in the heart which controls the heartbeat. In botany, a node is the position on the stem of a plant from which a leaf or leaves grow. In physics, nodes are points of minimum displacement (zero amplitude) on a STANDING WAVE such as the stationary points on a vibrating string.

node In astronomy, point at which one ORBIT cuts another. Specifically, a node is one of the two points in the orbit of a planet or comet at which it cuts the ECLIPTIC, or at which the orbit of a satellite cuts that of its primary. The ascending node is the node at which a celestial body passes from south to north; the descending node is the node at which a celestial body passes from north to south.

Noether, Amalie Emmy (1882–1935) German mathematician. Noether was invited by David HILBERT to lecture and conduct research at the University of Göttingen. She remained there until the Nazi purges of Jewish academics in 1933 forced her to seek refuge in the USA. Her main work was in the field of abstract ALGEBRA, where she contributed significantly to the general theory of ideals and to non-commutative algebras.

Noguchi, Hideyo (1876–1928) Japanese bacteriologist known for isolating the causative agent of syphilis in the CENTRAL NERVOUS SYSTEM. Noguchi improved the technique of the WASSERMANN reaction and devised the Wassermann skin test.

noise Any unwanted SOUND, which may come from any source, including traffic and factory machinery. Experiments carried out to determine the effects of noise on workers in manufacturing industries show that noise seriously affects only that output requiring mental concentration. Nevertheless, noise is often regarded as a form of acoustic pollution, and noise-abatement pressure groups lobby for legislation. Noise is usually measured on a DECIBEL scale: conversation is generally about 60dB and pain starts at about 120dB.

non-Euclidean geometry Self-consistent GEOMETRY that uses a different set of AXIOMS from those of EUCLIDEAN GEOMETRY, in particular a set that does not include the parallel postulate. Euclid's fifth postulate is equivalent to the statement that if a point lies outside a line, then only one line can be drawn through that point that does not cut the first line. In the early 19th century Farkas Bolyai and, independently, N.I. LOBACHEVSKY developed systems of geometry in which an infinite number of parallels could exist. This system, called hyperbolic non-Euclidean geometry, was self-consistent; that is, it had no inherent contradiction in the results obtained from the postulates. Later, Georg RIEMANN introduced a form (elliptical geometry) in which no parallel can exist. Development of these systems cast light on the fundamental nature of geometry. Non-Euclidean geometry is also used in RELATIVITY theory.

nonmetal Element that has none of the properties of a metal. Nonmetals are usually poor conductors of heat and electricity (they are thermal and electrical insulators). Examples include carbon, the halogens, oxygen, nitrogen, phosphorus and sulphur, all of which are also electronegative. Nonmetal oxides are acidic or neutral - not basic - and they usually form part of the negative ion (anion) in any compound. A few nonmetals have some metallic properties and are termed metalloids; they include elements that are semiconductors, such as silicon and germanium.

nonstoichiometric compound Chemical compound in which the atoms are not combined in simple, whole-number ratios. The composition of such compounds varies, depending on their source. For example, titanium(IV) oxide (expected formula TiO_2) obtained from the mineral rutile generally has the formula $TiO_{1.8}$.

noradrenaline (norepinephrine) Hormone secreted by nerves in the autonomic NERVOUS SYSTEM and by the ADRENAL GLANDS. Chemically, it is closely related to ADRENALINE and has similar effects on the cardiovascular system. It slows the heart rate and constricts small arteries, thus raising the blood pressure. It is used therapeutically to combat the fall in blood pressure that accompanies shock. *See also* NEUROTRANSMITTER

Norrish, Ronald George Wreyford (1897–1976) British chemist who shared the 1967 Nobel Prize for chemistry with Manfred EIGEN and George PORTER for research into very fast chemical reactions. Although primarily of theoretical importance, their work has potential applications in such areas as the hybridization of large molecules and ULTRASONIC communications.

North Atlantic Drift *See* GULF STREAM

Northern Cross (Cygnus or The Swan) Northern constellation between Pegasus and Lyra. It contains several bright stars, including DENEB (Alpha), one of the three components of the Summer Triangle, and Albireo (Beta), which is a DOUBLE STAR. Chi is a long-period VARIABLE STAR, with a period of 408 days. The MILKY WAY is particularly bright in this region and there are several STELLAR CLUSTERS.

northern lights *See* AURORA

nose In human beings, other primates and some vertebrates, the prominent structure between the eyes; it contains RECEPTORS sensitive to various chemicals (sense of SMELL) and serves as the opening to the respiratory tract, warming and moistening incoming air and trapping dust particles on the MUCOUS MEMBRANES. It has two cavities separated by a wall of CARTILAGE; the external openings are the nostrils.

notochord In CHORDATES and the early embryonic stages of vertebrates, the flexible, primitive backbone; in mature vertebrates, it is replaced by the SPINE.

nova (plural novae) Faint star that undergoes unpredictable increases in brightness by several magnitudes, and then slowly fades back to normal. A typical nova rises to maximum in a few days, and declines thereafter in brightness by a factor of about ten in 40 days. Novae occur in close binary systems, where one component is a WHITE DWARF and the other is a GIANT STAR. Matter drains from the surface of the giant, where gravity is weak, to the surface of the white dwarf, where gravity is strong. Nuclear reactions involving the infalling material cause a violent explosion on the surface of the white dwarf. The energy of a typical explosion is about the same as the Sun emits in 10,000 years. Clouds of gas are ejected during the outburst. The total amount of material lost is estimated to be about one ten-thousandth of the mass of the white dwarf.

nucellus Tissue that forms most of the OVULE in a FLOWER. It consists of nutrient tissue and an embryo sac, enclosed by a skin that has a small hole in it called the **micropyle**.

nuclear energy ENERGY released during a nuclear reaction as a result of the conversion of MASS into energy. The conversion involves the BINDING ENERGY of the NUCLEUS of an ATOM, and occurs as described by the equation $E = mc^2$ (where E is energy, m is mass and c is the speed of light), which was derived by EINSTEIN in his special theory of RELATIVITY. Nuclear energy is released in two ways: by FISSION and by FUSION. Fission is the process responsible for the atom bomb, and for the fission NUCLEAR REACTORS now contributing to energy requirements throughout the world. The fission reaction was first discovered in 1938 by Otto HAHN and Fritz STRASSMANN; and in 1942 the first sustained nuclear CHAIN REACTION was acheived by Enrico FERMI. Fusion provides the energy for the Sun and the stars and for the hydrogen bomb. It also offers the prospect of cheap energy, once a method

NITROGEN CYCLE

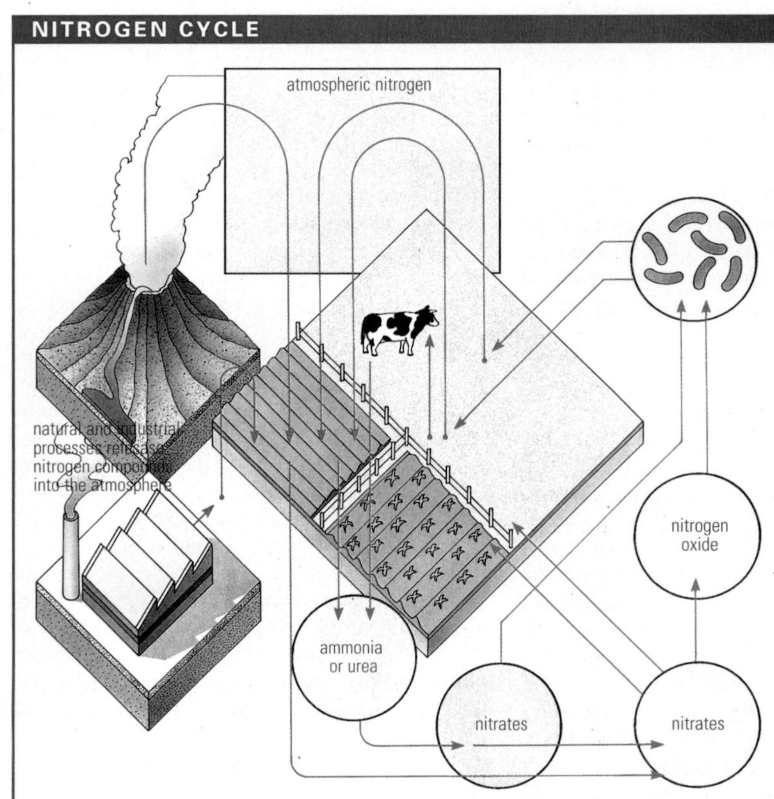

Free nitrogen (N_2) in the atmosphere cannot be absorbed directly by plants or animals. Soil-dwelling bacteria convert the nitrogen into nitrates which can then be absorbed by plants. The plants are eaten by herbivores, thus triggering the food chain. Some of the excess nitrogen in the animals' bodies is converted to ammonia or urea and returned to the soil. In addition, when the animals die, saprophytes decompose the tissue releasing any remaining nitrogen, and the cycle can begin again.

NUCLEAR REACTOR

A pressurized water reactor (PWR) is so named because the primary coolant (1), that passes through the reactor core (2), is pressurized to prevent it from boiling. The uranium-235 fuel is loaded into the reactor in pellets (3) contained by the fuel rods (4). To prevent an uncontrolled chain reaction, the fuel rods are separated by control rods of graphite (5). All the rods are loaded into the reactor from above (6). The primary coolant is heated by the fission reaction in the fuel rods and circulates into a steam generator (7), where it superheats the secondary coolant (8). The secondary coolant leaves the protective containment vessel (9) and drives turbines (10), which produce electricity through a generator (11). A third coolant loop (12) cools the secondary coolant, transferring the heat to a sea, river or lake. Reducing the temperature of the secondary coolant increases the efficiency of the transfer from the primary to the secondary coolant. The inset photograph shows the core (dark circular area) of a nuclear reactor during the "charging" period, at which time the first load of fuel is placed in the reactor core.

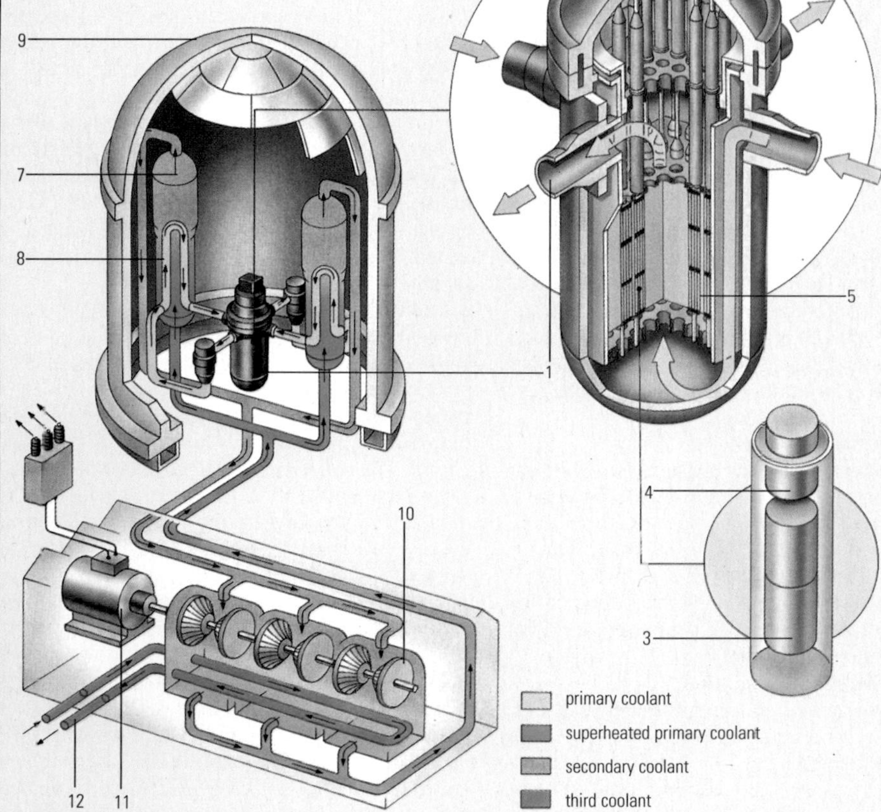

has been perfected for controlling fusion reactions. *See also* CRITICAL MASS; MANHATTAN PROJECT; MATTER OF STATES; NUCLEAR WEAPON

nuclear engineering Branch of engineering concerned with the design, construction and operation of NUCLEAR REACTORS and particle ACCELERATORS as well as coolant, shielding and moderator devices. *See also* NUCLEAR ENERGY

nuclear fission *See* FISSION, NUCLEAR

nuclear fuel Various chemical and physical forms of URANIUM and PLUTONIUM used in NUCLEAR REACTORS. Fluid fuels are required in homogeneous reactors; heterogeneous reactors use various forms of fuels – pure metals or alloys, as well as oxides or carbides. The fuel must have a high thermal conductivity, be resistant to radiation damage and be easy to fabricate.

nuclear fuel enrichment Separation of the fissionable ISOTOPE of URANIUM, uranium-235, from the more abundant uranium-238 isotope. Gaseous uranium(VI) fluoride undergoes diffusion separation using cascades of barriers with microscopically small pores. The difference in mass between the two isotopes is minimal, but sufficient so that the heavier, slower-moving uranium-238 molecules are concentrated on one side. High-speed centrifugal-force separation methods are also used.

nuclear fusion *See* FUSION, NUCLEAR

nuclear magnetic resonance (NMR) Absorption of ELECTROMAGNETIC RADIATION by certain nuclei when placed in a magnetic field. In this field the NUCLEUS, as a result of its SPIN, can have slightly different energy values. It can make transitions between these energy values, acquiring the energy by absorbing radio-frequency radiation of the appropriate wavelength. This pheneomenon is used in medicine in the form of MAGNETIC RESONANCE IMAGING (MRI), and in chemical analysis and research in nuclear physics in a method called NMR SPECTROSCOPY.

nuclear physics Branch of physics concerned with the structure and properties of the atomic NUCLEUS. The principal means of investigating the nucleus is the particle SCATTERING experiment, carried out in particle ACCELERATORS, in which a nucleus is bombarded with a beam of high-energy ELEMENTARY PARTICLES and the resultant particles analysed. Other information is obtained by studying their magnetic moments and SPIN and any FISSION and FUSION products. Study of the nucleus has led to an understanding of the processes occurring inside stars and has enabled the building of NUCLEAR REACTORS and NUCLEAR WEAPONS.

nuclear radiation Refers to all the types of emissions of waves and particles from the nuclei of atoms as the result of natural or induced RADIOACTIVITY. Natural radioactivity is the spontaneous breakdown of unstable nuclei, such as those of radium or uranium-235, with the release of ALPHA and BETA PARTICLES and GAMMA RADIATION. **Alpha** (α) particles are identical to nuclei of atoms of helium gas: they travel very fast, but have little penetrating power and so are relatively harmless. **Beta** (β) rays or particles are electrically charged, and are identical to high-energy ELECTRONS or POSITRONS. They have greater penetrating power. **Gamma** (γ) rays are very high frequency X-RAYS, and are dangerously penetrating. Other emissions from nuclei include PROTONS, NEUTRONS and NEUTRINOS. These and yet other particles, called MESONS, result from the interactions of cosmic rays with the upper atmosphere, and from the bombardment of materials with high-speed particles in particle ACCELERATORS.

nuclear reactor Device in which nuclear FISSION (and sometimes nuclear FUSION – *see* TOKAMAK) reactions are used for power generation or for the production of radioactive materials. In the reactor, the fuel is a radioactive heavy metal: URANIUM-235, uranium-233 or PLUTONIUM-239. In these metals, ATOMS break down spontaneously, undergoing a process called RADIOACTIVE DECAY. Some NEUTRONS released in this process strike the nuclei of fuel atoms, causing them to undergo fission and emit more neutrons. These in turn strike more nuclei, and in this way a CHAIN REACTION is set up. Usually, a material called a MODERATOR is used to slow down the neutrons to a speed at which the chain reaction is self-sustaining. This process occurs within the reactor **core**. The chain reaction is regulated by inserting **control rods**, which contain neutron-absorbing material such as cadmium or boron, into the core. The heat generated by the nuclear reaction is absorbed by a circulating **coolant** and transferred to a boiler to raise steam. The steam drives a TURBINE that turns a GENERATOR, that in turn produces electricity. There are a variety of nuclear reactors, named after the type of coolant they use. For example, a **boiling-water** reactor (BWR) and a **pressurized-water** reactor (PWR), presently the most common type of reactor, both use water as the coolant and the moderator. In **advanced gas-cooled** reactors (AGR), the coolant is a gas – most commonly carbon dioxide. **Fast** reactors do not use a moderator, and fission is caused by FAST NEUTRONS. This type of reactor generates greater temperature and the coolant used is a liquid metal, usually liquid sodium. Sometimes called "fast-breeder" reactors, fast reactors produce ("breed") more fissionable material than they consume. Excess neutrons from the fission of a fuel such as uranium-235, instead of being absorbed in control rods, are used to bombard atoms of relatively inactive uranium-238 which transmutes to (or "breeds") the active ISOTOPE plutonium-239. When the original fuel is spent, the plutonium can be used as a nuclear fuel in other reactors or NUCLEAR WEAPONS. *See* ELECTRICITY SOURCES

nuclear submarine Underwater vessel powered by a NUCLEAR REACTOR. The great advantage of nuclear SUBMARINES is that they use little fuel and may remain submerged for many weeks, even months. Heat produced by the reactor turns water into steam. The steam turns TURBINES, which are linked to electricity GENERATORS that charge the submarine's batteries. The vessel's propeller may be driven by an electric motor, or by linking its shaft to a turbine. The first nuclear submarine, the USS *Nautilus*, was launched by the US Navy in 1954. *Nautilus* travelled 100,000km (62,000mi) in its first two years of service, and in 1958 was the first vessel to sail under the ice cap at the North Pole. Britain's first nuclear submarine was HMS *Dreadnought* (launched 1963). Nuclear submarines carrying missiles have enormous firepower and are difficult to detect and destroy.

nuclear waste Residues, generally from NUCLEAR REACTORS, containing radioactive substances. After URANIUM, PLUTONIUM and other useful fission products have been removed, some long-lived, radioactive elements remain, such as CAESIUM-137 and STRONTIUM-90. International regulations exist for disposing of such wastes, with separate regulations applying to liquids, gases and solids.

nuclear weapon Device that derives a huge explosive force from nuclear FISSION or FUSION reactions. The first nuclear weapons were dropped by the USA on the Japanese cities of Hiroshima and Nagasaki in August 1945. These **atom bombs** consisted of two stable, sub-critical masses of URANIUM or PLUTONIUM which, when brought forcefully together, caused the CRITICAL MASS to be exceeded, thus initiating an uncontrolled nuclear fission CHAIN REACTION. In such detonations, huge amounts of energy and harmful radiation are released: the explosive force can be equivalent to 200,000 tonnes of TNT. The much more powerful **hydrogen bomb** (H-bomb or thermonuclear bomb), first tested in 1952, consists of an atom bomb that on explosion provides a temperature high enough to cause nuclear fusion in a surrounding solid layer, usually lithium deuteride. The explosive power can be that of several million tonnes (megatons) of TNT. Devastation from such bombs covers a wide area: a 15-megaton bomb will cause all flammable material within 20km (12mi) to burst into flame. A third type of weapon, the **neutron bomb**, is a small hydrogen bomb, also called an enhanced radiation weapon, that produces a small blast but a very intense burst of high-speed NEUTRONS. The lack of blast means that buildings are not heavily damaged. The neutrons, however, produce intense radiation sickness in people located within a certain range of the explosion, killing those affected within a week. *See also* NUCLEAR ENERGY

nucleic acid Giant, chemical molecules present in all living cells and in viruses. There are two types: DNA (deoxyribonucleic acid) stores the GENETIC CODE that functions as the basis of heredity; and RNA (ribonucleic acid) delivers these coded instructions to the cell's PROTEIN manufacturing sites. Chemically, nucleic acids are polymers of NUCLEOTIDES. *See also* CHROMOSOME; ENZYME; GENE; MESSENGER RNA (MRNA)

nucleon Either of the two kinds of particles found within the NUCLEUS of an atom: a NEUTRON or a PROTON.

nucleoside Subunit of NUCLEIC ACID, consisting of a sugar linked to a nitrogenous base (purine or pyrimidine).

nucleosynthesis Production of all the various chemical elements that exist in the Universe from one or two simple atomic nuclei. It is believed to have occurred by way of a huge release of NUCLEAR ENERGY during cosmogenesis and is still in progress in the Sun and other stars. Starting with hydrogen and helium, repeated nuclear FUSION reactions can account for most of the elements up to iron. Elements heavier than iron can be explained in terms of repeated reactions involving the capture of NEUTRONS by nuclei.

nucleotide Complex, naturally occurring chemical group that contains a nitrogen base linked to a sugar

NUCLEAR WEAPON

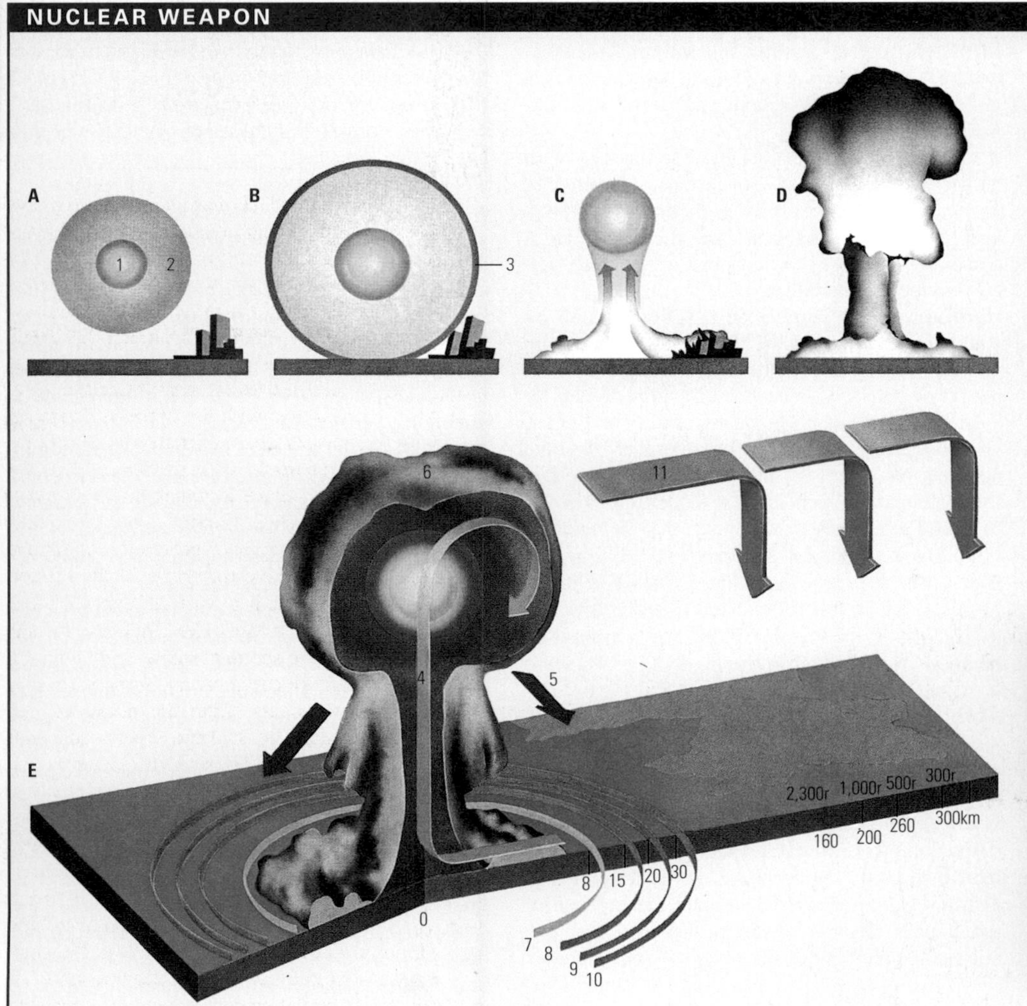

An exploding atom bomb (A) forms a fireball (1) millions of degrees centigrade and gives out radiation (2). Within a few seconds (B), it expands and creates a high-pressure shock wave (3). The fireball rises (C), sucking up dust and rubble to form a mushroom cloud (D). As the fireball expands (E), it sets up a powerful convection current (4), giving off heat radiation (5) and forming a cloud (6). With a 15-megaton bomb, blast damage to buildings is total (7) within 8km (5mi), severe (8) within 15km (9mi), and noticeable (9) at 30km (19mi). Even 20km (12mi) away (10), all inflammable material bursts into flame. Up to two days after the explosion of the bomb, fallout (11) continues at radiation doses of 300r (roentgen) at 300km (185mi) from the blast. The inset photograph shows how the detonation of a large nuclear weapon above ground creates a huge mushroom cloud of radioactive dust and debris that can reach several kilometres in height. The hazardous airborne dust is then free to be carried in any direction by the prevailing winds. The devastation covers a wide area.

and an acid phosphate. Nucleotides are the building blocks of NUCLEIC ACIDS (such as DNA and RNA). In the molecules of nucleic acids, nucleotides are linked together by bonds between the sugar and phosphate groups. Nucleotides also occur freely in a CELL as various COENZYMES, such as ADENOSINE TRIPHOSPHATE (ATP), the principal carrier of chemical energy in the metabolic pathways of the body.

nucleus In biology, membrane-bound structure that, in most CELLS, contains the CHROMOSOMES. As well as holding the genetic material, the nucleus is essential for the maintenance of cell processes. It manufactures the RNA used to build RIBOSOMES. Other RNA molecules carry the genetic code from the DNA through pores in the nuclear membrane into the CYTOPLASM, where it is used as a template for PROTEIN SYNTHESIS by the ribosomes. CELL DIVISION involves the splitting of the nucleus and CYTOPLASM. Cells without nuclei include BACTERIA and mammalian ERYTHROCYTES (red blood cells). Instead of chromosomes, bacteria have a naked, circular molecule of DNA in the cytoplasm.

nucleus In physics, central core of an ATOM. Made up of NUCLEONS (PROTONS and NEUTRONS), the nucleus accounts for almost all of an atom's mass. Because protons are positively charged and neutrons have no charge, a nucleus has an overall positive charge; this is cancelled out, however, by the negatively charged ELECTRONS that orbit the nucleus. Following the discovery (1836) of RADIOACTIVITY by Henri BECQUEREL, in 1911 Ernest RUTHERFORD proposed the existence of the nucleus after identifying ALPHA and BETA PARTICLES. The number of protons in a nucleus is called its ATOMIC NUMBER, while the number of nucleons is the MASS NUMBER. *See also* FUNDAMENTAL FORCE; QUANTUM THEORY; NUCLEAR ENERGY; RELATIVE ATOMIC MASS (R.A.M.); SPIN

nuée ardente Cloud of burning hot gas that travels quickly down the side of a VOLCANO during an eruption. Small particles of dust and ash may be carried in the cloud, which kills anyone and anything it passes over. In 1902 a nuée ardente accompanying the eruption of Mount Pelée accounted for the deaths of all but one of the inhabitants of St Pierre, on the island of Martinique.

number Symbol representing a quantity used in counting or calculation. All ancient cultures devised their own NUMBER SYSTEMS for the practical purposes of counting and measuring. The development of the idea of place-value notation and the introduction of zero into the base-10 number system culminated in the familiar DECIMAL SYSTEM that we use today. This system seems to have reached the West from India via 12th-century translations of Arabic mathematical writings. From the basic process of counting we get the natural numbers. This concept can be extended to define the INTEGERS, the rational numbers, the real numbers and the complex numbers. Other classes of number, like PRIME NUMBERS and perfect numbers, arise in a branch of mathematics called NUMBER THEORY.

number, cardinal Number expressing the content of a SET, but not the order of its members. For example, six is a cardinal number in "six books". Two sets have the same cardinal number if their members can be put in one-to-one correspondence – a concept that allows the idea of cardinal numbers of infinite sets. The infinite set of INTEGERS is said to have cardinal number χ_0 (aleph-zero). The set of all real numbers (*see* NUMBERS, REAL) cannot be put into one-to-one correspondence with aleph-zero and is a "larger" infinite set. *See also* NUMBER, ORDINAL

number, complex Number of the form $z = a + bi$, where $i = \sqrt{-1}$, and a and b are real numbers (*see* NUMBER, REAL). a is called the **real part** and b the **imaginary part**. All complex numbers can be represented on an **Argand diagram** using rectangular Cartesian coordinates, the complex number $a + bi$ being identified with the point (a, b) in the plane. Thus, real numbers are positioned on the x-axis and purely imaginary numbers are positioned on the y-axis. An equivalent description uses POLAR COORDINATES (r,θ) with $z = r\cos(\theta) + r\sin(\theta)i$, where the two descriptions are related through the equations $r = \sqrt{(a^2 + b^2)}$, $\tan(\theta) = b/a$. Complex numbers have applications in many areas of pure and applied mathematics.

number, imaginary Square root ($\sqrt{}$) of a negative quantity. The simplest, $\sqrt{-1}$, is usually represented by i. The numbers are so called because when first discovered they were widely regarded as meaningless. However, they are necessary for the solution of many QUADRATIC EQUATIONS, the roots of which can be expressed only as complex numbers (*see* NUMBER, COMPLEX), which are composed of a real part and an imaginary part. They find applications in ALTERNATING CURRENT (AC) theory.

number, irrational Number that cannot be expressed as a fraction. Examples include $\sqrt{2}$ and π (PI). Irrational numbers are, therefore, numbers with infinitely many (non-repeating) DECIMAL PLACES. (However, the reverse is not true: some rational fractions, such as ⅓, also have infinite decimal representations.) Irrational numbers, together with the rational numbers (*see* NUMBER, RATIONAL), make up the set of real numbers (*see* NUMBER, REAL).

number, natural Any of the numbers 1, 2, 3, 4. . . as used in counting; they are the simplest numbers, with no fractional, decimal or imaginary part, and of which there is an infinite number. They are all positive numbers, that is, greater than zero. Negative numbers are those less than zero. *See also* DECIMAL FRACTION; INTEGER; NUMBER, IMAGINARY

number, ordinal Number that indicates the position of a member in an ordered sequence or in a group. Thus "first", "second" and "third" are ordinal numbers.

number, perfect INTEGER equal to the sum of all its FACTORS including 1. For instance, the number 28 is a perfect number since its factors are 1, 2, 4, 7 and 14 (excluding 28 itself), and the sum of these equals 28. It is not known whether or not there exist any odd perfect numbers.

number, prime Positive INTEGER that has no FACTORS other than itself and 1. The first few primes are 2, 3, 5, 7, 11, 13 and 17. The INTEGERS 4, 6, 8, ... are not prime numbers, since they can all be divided by 2 (that is, 2 is a factor). In the 4th century BC EUCLID proved that there are an infinite number of prime numbers. But it remains an unsolved problem to find formulae that will generate the sequence of primes. In 1640 Pierre de FERMAT showed that the numbers $F_n = 2^x + 1$, where $x = 2^n$, for $n = 0, 1, 2, 3$ and 4, were prime. He conjectured that F_n would be prime for all n, but in the 18th century Leonhard EULER showed that F_5 (= 4,294,967,297) was a composite number (that is, equal to the product of two numbers) and therefore not a prime.

number, rational Number that can be written in the form a / b where a and b are INTEGERS (and b is not zero). Rational numbers include all integers. A number that cannot be so expressed is called an irrational number (*see* NUMBER, IRRATIONAL). The rational numbers are also called FRACTIONS.

number, real Number that is either a rational or an irrational number. Real numbers are expressible as decimals. They are distinguished from complex numbers, which are of the form $a + bi$ (where $i = \sqrt{-1}$). *See also* NUMBER, COMPLEX; NUMBER, IRRATIONAL; NUMBER, RATIONAL

number, transcendental Irrational number (*see* NUMBER, IRRATIONAL) that is not the solution to any algebraic equation with rational (fractional) coefficients. For example, the numbers π (PI) and e are both transcendental, but the irrational number $x = \sqrt{2}$ is not transcendental because it satisfies the algebraic equation $x^2 - 2 = 0$. *See also* ALGEBRAIC FUNCTION

number, whole *See* INTEGER

number system Counting methods that use various number BASES. The DECIMAL SYSTEM, for example, has a base of 10; the DUODECIMAL SYSTEM a base of 12. Computers do their calculations within the BINARY SYSTEM, base 2.

number theory Branch of mathematics concerned with the properties of natural numbers (*see* NUMBER, NATURAL) or special classes of natural numbers such as prime numbers (*see* NUMBER, PRIME). Typical results from number theory are that there are infinitely many prime numbers (proved by EUCLID), and that any number may be written as the sum of four square numbers (proved by Joseph LAGRANGE). Many of the most important problems in number theory are very easy to pose but have been found to be incredibly difficult to solve. FERMAT'S LAST THEOREM remained unproven for more than 350 years, until Andrew WILES found a proof in 1995. Many problems relating to the prime numbers remain completely elusive: for instance, no-one has yet found a formula that will generate all the prime numbers.

numeral Symbol used alone or in a group to denote a NUMBER. **Arabic numerals** are the 10 digits from 0 to 9. **Roman numerals**, as normally used today, consist of seven letters or marks (I =1, V = 5, X = 10, L = 50, C = 100, D = 500 and M = 1,000). The formation of numbers from numerals depends on the NUMBER SYSTEM used.

numerator In mathematics, upper number in a FRACTION. The lower number is known as the DENOMINATOR. For example, in the fraction 2/3, 2 is the numerator, 3 the denominator.

nut In botany, dry, one-seeded FRUIT with a hard woody or stony wall. It develops from a flower that has petals attached above the ovary (inferior ovary). Examples include acorns and hazelnuts.

nutation Oscillating movement (period 18.6 years) superimposed on the steady precessional movement of the Earth's axis, so that the precessional path of each celestial pole on the CELESTIAL SPHERE follows an irregular rather than a true circle. It results from the varying gravitational attraction of the Sun and Moon on the Earth, due to variations in their distances from Earth and in their relative directions. *See also* PRECESSION

nutrition All the processes by which plants and animals take in and make use of food. The science of nutrition involves identifying the kinds and amounts of nutrients necessary for growth and health. Nutrients are generally divided into PROTEINS, CARBOHYDRATES, FATS, MINERALS and VITAMINS.

nylon Any of numerous synthetic materials consisting of polyamides (with protein-like structures). Developed in the USA in the 1930s, nylon can be formed into fibres, filaments, bristles or sheets. It is characterized by elasticity and strength and is used chiefly in yarn, cordage and moulded products. Hard and tough or soft and rubbery nylon products can be made by varying their composition. *See also* SYNTHETIC FIBRE

nymph Young insect of primitive orders that do not undergo complete METAMORPHOSIS. The term is used to designate all immature stages after the OVUM. The nymph resembles the adult (IMAGO) and does so more closely with each successive moulting. Some examples are the aquatic nymphs of dragonflies, mayflies and damsel flies.

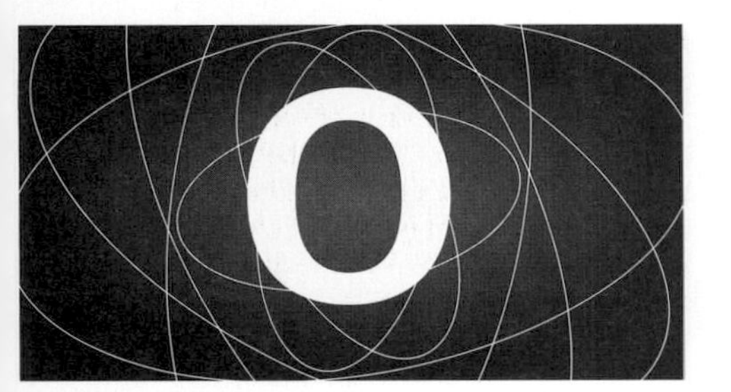

oasis Fertile location which has water in an arid landscape. Usually GROUNDWATER is brought to the surface in a well, but an oasis may occur at the point where a river flowing from a wetter region crosses the desert on its way to the sea, such as the Nile in N Africa or the Indus in S Asia. Most Saharan oases contain large numbers of date palms, and there also be many fields of cereals, vegetables and fruits. There is often a settlement at the oasis, sometimes with as many as 30,000 inhabitants.

Oberon Second-largest of the five main satellites of URANUS, discovered in 1787 by William HERSCHEL. It is 1,524km (947mi) in diameter and consists largely of a mixture of rock and water-ice. Oberon has many craters, some with rays and some with dark floors, and mountains, one of which appears to be 20km (12mi) high. The high density of craters is evidence of a world unaltered by geological activity.

Oberth, Hermann Julius (1894–1990) German physicist, b. Austro-Hungary, who made a fundamental contribution to the development of astronautics. Oberth conducted experiments to simulate and study the effects of weightlessness and designed a long-range, liquid-propellant ROCKET. During World War 2 he worked on German rocket development under Wernher VON BRAUN. Oberth later worked on the development of solid-propellant, anti-aircraft rockets.

observatory Location of TELESCOPES and other equipment for astronomical observations. Large, optical telescopes are housed in domed buildings, usually sited high up in mountain areas, well away from the lights and smog of cities. These observatories rotate to face any area of the sky, the telescope being exposed by the retraction of a panel in the dome. Radio observatories are open sites containing one or more large RADIO TELESCOPES, usually having the shape of a parabolic dish. The largest radio telescopes have been built in natural mountain hollows: the Arecibo Observatory in Puerto Rico, for example, is 300m (975ft) across. The latest observatories include telescopes carried by orbiting satellites, such as the HUBBLE SPACE TELESCOPE, entirely above the absorbing and distorting effects of Earth's atmosphere. In addition to optical and radio studies, these craft carry out measurements in the infrared, ultraviolet, X-ray and gamma-ray regions of the spectrum, which do not penetrate Earth's atmosphere and so cannot be studied from ground-based observatories.

obsidian Rare grey to black glassy volcanic rock. High in silica, it is the uncrystallized equivalent of RHYOLITE and GRANITE. It polishes well and makes an attractive semi-precious stone. Hardness 5.5; r.d. 2.4.

occultation In astronomy, temporary concealment of one celestial body by another, especially of a star by the Moon or a planet. Lunar occultation is of particular importance because it has served accurately to pinpoint a number of notable radio sources.

ocean Continuous body of salt water that surrounds the continents and fills the Earth's great depressions. Oceans cover about 71% of the Earth's surface (more than 80% of the Southern Hemisphere), and represent about 98% of all the water on the face of the Earth. There are five main oceans, the Atlantic, Pacific, Indian, Arctic and Antarctic. They may be described by distinct region (LITTORAL, BENTHOS, PELAGIC and ABYSSAL), or by depth (CONTINENTAL MARGIN, deep sea plain and deep TRENCHES). The SEAFLOOR has a varied topography. SEAWATER contains salt and other mineral deposits; the salt content, between 3.3% and 3.7%, is the result of washout from the land and interchange with the ATMOSPHERE over the ages. Light penetrates seawater to a maximum depth of *c*.300m (1,000ft), below which plant life cannot grow. The oceans are constantly moving in CURRENTS, TIDES and WAVES. They form an integral part of the Earth's HYDROLOGICAL CYCLE and CLIMATE. They are a rich source of FOSSILS, providing invaluable evidence of the evolution of life on Earth. They also provide minerals such as MANGANESE NODULES, OIL and GAS. The CONTINENTAL SHELF yields sand and gravel. Marine fauna, such as FISH and PLANKTON, are a vital part of the food chain.

Through volcanic activity the oceans have been forming over the last 200 million years. The theory of CONTINENTAL DRIFT (and associated SEAFLOOR SPREADING) has revealed that PANGAEA was surrounded by one vast ocean, Panthallasia. As Panagea began to split, a smaller and shallower ocean, the Tethys Sea, formed between the continents. By *c*.65 million years ago, the Atlantic and Indian oceans appeared. The Pacific became separated from the Atlantic and Indian oceans when the North and South American continents joined. The separation of Greenland from North America, and the widening of the North Atlantic, completed the encirclement of the Arctic Ocean. **Total area**: 360 million sq km (138 million sq mi). **Total volume**: c.1.4 billion cu km (322 million cu mi). **Average depth**: 3,500m (12,000ft). **Average temperature**: 3.9°C (39°F). *See also* DESALINATION; TIDE; WATER POLLUTION

ocean floor *See* SEAFLOOR

oceanic basin One of two major provinces of the deep ocean floor, lying at over 2km (1.2mi) in depth. The MID-OCEAN RIDGES form the other province. Together they constitute 56% of the Earth's surface. The deep ocean basin is underlain by a basaltic crust, about 7km (4.3mi) thick, and is covered in sediment and dotted by low ABYSSAL hills. *See also* TRENCH

oceanic current *See* CURRENT

oceanography Study of the OCEANS. The major sub-disciplines of OCEANOGRAPHY include marine geology (see PLATE TECTONICS), MARINE BIOLOGY, marine METEOROLOGY, and physical and chemical oceanography. The science of oceanography dates from the CHALLENGER EXPOSITION (1872–76). Jacques COUSTEAU'S invention of the SCUBA aided human exploration of the seas. In 1948 the BATHYSCAPHE, invented by August PICCARD, was first used. The DEEP SEA DRILLING PROJECT is a long-term investigation of the evolution of OCEANIC BASINS.

Ochoa, Severo (1905–93) US biochemist, b. Spain. He shared the 1959 Nobel Prize for physiology or medicine with Arthur KORNBERG for work on the synthesis of RNA, which helped greatly in the study of GENETICS.

ochre Yellowish or red pigment made by mixing a type of iron ore (LIMONITE or HAEMATITE) with fine clay and grinding the dried product to a powder.

octane number Indication of the resistance of PETROL to engine KNOCK (pre-ignition firing). It represents the percentage by volume of iso-octane in a reference fuel, consisting of a mixture of iso-octane and normal heptane, that matches the knocking properties of the fuel being tested. The higher the number, the less likely the possibility of the fuel pre-igniting. Four-star (regular) petrol has an octane number of *c*.90. Premium grade petrols have between 95 and 96.

odonatan Any member of Odonata, an order of primitive winged insects found worldwide. Damselflies, in the suborder Zygoptera, have thin bodies with wings held vertically along the body when at rest. The long, slender, aquatic nymphs have three, leaflike gills on the abdomen. Dragonflies, in the suborder Anisoptera, have heavy bodies with wings held horizontally when at rest. The stout nymphs have gills at the anal end. All prey on insects; none attack humans. Length: 18–193mm (0.7–8in).

Oersted, Hans Christian (1777–1851) Danish physicist and professor at Copenhagen University. He took the first steps in explaining the relationship between electricity and magnetism, thus founding the science of ELECTROMAGNETISM. The OERSTED unit of magnetic field strength is named after him. In 1825 he became the first scientist to isolate pure metallic ALUMINIUM.

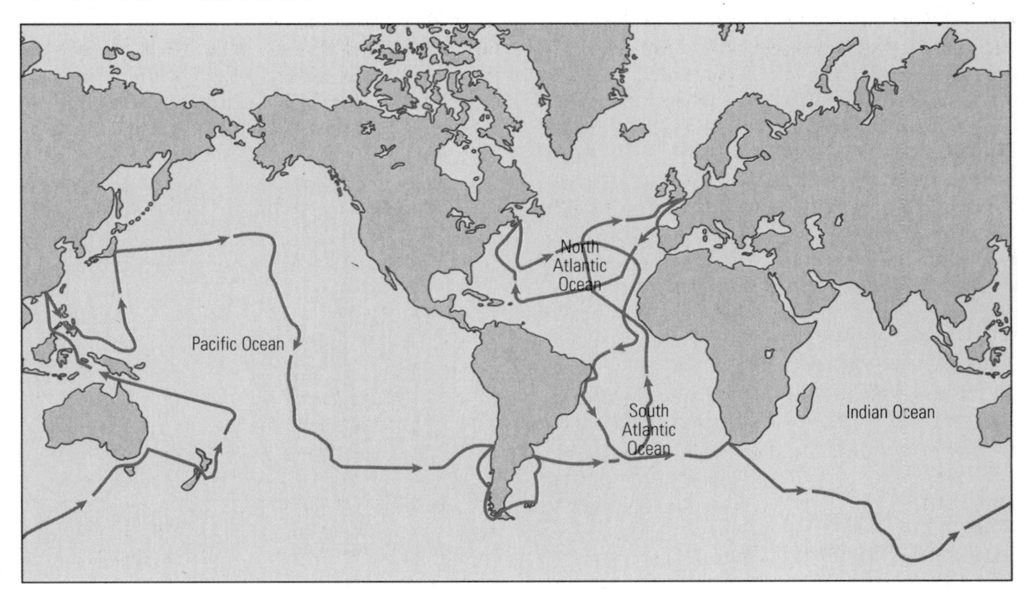

▲ **oceanography** One of the most important scientific voyages ever made, the map shows the route of the *Challenger* expedition of 1872–76. The expedition laid the foundations for the modern science of oceanography. The *Challenger* covered a distance of more than 68,900 nautical miles and established 362 observation stations in the three main oceans, where depth, temperature, surface currents were measured and samples of water, flora and fauna were taken. The ship held a cramped but very well-equipped laboratory.

oersted Unit of magnetic field strength in the CGS SYSTEM, equal to the magnetic field that would cause a unit magnetic pole to experience a force of one dyne in a vacuum. It is named after Hans OERSTED.
oesophagus (gullet) Muscular tube, part of the ALIMENTARY CANAL (or gut), that carries swallowed food from the throat to the STOMACH. Food is moved down the lubricated channel by a wave-like movement known as PERISTALSIS.
oestrogen Female SEX HORMONE. First produced by a girl at PUBERTY, oestrogen leads to the development of the secondary sexual characteristics that turn her body into a woman's: breasts, body hair and redistributed fat. It regulates the MENSTRUAL CYCLE and prepares the UTERUS for pregnancy. Oestrogen is also a constituent of the contraceptive pill (*see* CONTRACEPTION).
oestrous cycle Physiological changes occurring during the female reproductive cycle of most placental mammals. Controlled by HORMONES, it is evident among mammals other than humans. Cycles of different animals vary in frequency and length. Typically, ovulation is associated with the oestrous (heat) period. *See also* MENSTRUAL CYCLE
offset printing Method of PRINTING widely used for high-volume publications. In the printing machine, a roller applies ink to the printing plate, which is mounted on a rotating cylinder. The image is then transferred (offset) to a cylinder with a rubber covering, called the blanket. This transfers the image to the paper. Usually, the plates are made by LITHOGRAPHY, and the process is called offset lithography. Separate plates are used for each colour of ink. This book was printed using a four-colour offset-litho process.
Ohm, Georg Simon (1787–1854) German physicist. He was appointed professor at Munich following his discovery (OHM'S LAW) of the relationship linking current, electromotive force and resistance in an ELECTRIC CIRCUIT. His name is also given to the SI unit of electrical RESISTANCE.
ohm (symbol Ω) SI unit of electrical RESISTANCE, equal to the resistance between two points on a conductor when a constant potential difference of one VOLT produces a current of one AMPERE.
Ohm's law Statement that the amount of steady current through a material is proportional to the voltage across the material. Proposed in 1827 by Georg OHM, Ohm's law is expressed mathematically as $V = IR$ (where V is the voltage in volts, I is the current in amperes and R is the resistance in ohms). In alternating current (AC) circuits where deductances and capacitances are present, the formula is amended to $V = IZ$, where Z is IMPEDANCE. It holds only approximately and under limited conditions and not for all materials.
oil General term to describe a variety of substances, whose chief shared properties are viscosity at ordinary temperatures, a density less than that of water, flammability, insolubility in water and solubility in ether and alcohol. There are three main types: **mineral** oils are hydrocarbon mixtures that occur, most notably, in crude or PETROLEUM oil, and which are extracted by refining to be used primarily as fuels; **animal** oils are glycerides of FATTY ACIDS and are used as food, lubricants and as a major ingredient of soap, paints and varnishes. There are two kinds of **vegetable** (or fixed) oil: drying, such as linseed or poppyseed oil, and non-drying, such as olive and castor oil. All vegetable oils are ESTERS or mixtures of substances CALLED TERPENES. There are also ESSENTIAL OILS that are obtained from plants and which, unlike other vegetable oils, are volatile. Such oils are found in perfumes and food flavourings.
oil-drop experiment Experimental method used by Robert MILLIKAN to measure an ELECTRON'S charge by subjecting charged oil drops to a variable electric field.
oil extraction Process by which PETROLEUM (crude oil) is brought to the Earth's surface and transported to an OIL REFINERY. A borehole is made into the Earth's crust and the oil reservoir emptied with the help of NATURAL GAS pressure or pumps. The discovery of undersea deposits, as in the North Sea, has meant that many new techniques for extracting and transporting the oil have had to be developed in order to avoid WATER POLLUTION and blowouts. *See also* DRILLING RIG; OIL WELL
oil refinery Large, industrial chemical complex in which PETROLEUM (crude oil) is refined. This involves many processes whereby the oil is split into its major fractions, such as petrol, kerosene, diesel oil, lubricating oil and bitumen. These are obtained by fractional DISTILLATION, but many refineries carry out further operations on some of these fractions. These include catalytic CRACKING, alkylation – increasing the OCTANE NUMBER of petrol – HYDROGENATION and POLYMERIZATION.
oil rig *See* DRILLING RIG
oil shale Dark-coloured, soft rock containing hydrocarbon compounds which can be distilled off as SHALE oil. Oil shales yielding more than a small percentage of oil can be valuable as fuels and as sources of organic chemicals. They are found in many countries, sometimes in vast deposits; that of the Green River shales in the USA alone has been estimated to contain 960,000 million barrels of oil.
oil well Shaft through which PETROLEUM (crude oil) is brought to the surface from underground deposits. Wells are usually drilled with a rotating shaft bearing a drill bit, which, for cutting through hard rock, may be diamond-tipped. Rock particles are pumped from the hole as a fluid slurry, which also lubricates and cools the drill. Proven wells are cased with steel piping and topped above ground with an oil take-off arrangement, often of the type known as a "Christmas tree".
Olbers, Heinrich Wilhelm Matthäus (1758–1840) German astronomer and physician. He discovered five comets and the asteroids Pallas (1802) and Vesta (1807). He also devised a new method of calculating a comet's orbit. In 1826 he discussed a longstanding problem, now called **Olbers' paradox**: why is the sky dark at night? If there are an infinite number of stars, evenly distributed in space, every line of sight should end on a star, and the sky should be roughly as bright as the Sun, all over. Olbers' explanation, that interstellar dust obscures the light, was discarded in the 1970s. In fact, space if not infinite. According to the BIG BANG theory, the Universe is believed to have been born about 15,000 million years ago, and so light reaches us only from a region 15,000 million light years in radius. In addition, the light from distant galaxies is moved out of the visible spectrum by the RED SHIFT caused by the expansion of the Universe.
Oldowan Type of stone tool dating from the early PLEISTOCENE epoch (*c.*2 million years ago). The name comes from the Olduvai Gorge in Tanzania, where archaeologists found the first tools of this kind. Made from quartz or basalt stones, the edges were chipped away to form tools capable of chopping, scraping or cutting. Oldowan tools were made for about 1.5 million years.
olefin *See* ALKANE
oleic acid ($CH_3(CH_2)_7CH{=}CH(CH_2)_7COOH$) Organic chemical compound that has a long-chain molecule which is unsaturated (containing a double bond). It is the unsaturated acid found most commonly in fatty oils in the form of a glycerol ESTER.
olfaction Alternative name for the sense of SMELL
olfactory nerve Nerve of SMELL, the name of one of the 12 pairs of cranial nerves present in all vertebrate animals. Its nerve cells (NEURONS) lie in the brain and their long shafts, or AXONS, extend to their sense RECEPTORS in the upper part of the nasal cavity.
Oligocene Extent of geological time from about 38 to 25 million years ago. It is the third of five epochs of the TERTIARY period in the CENOZOIC era. During the Oligocene epoch, the climate cooled. Many modern mammals evolved, including early elephants and *Mesohippus*, an ancestor of the modern horse. Only a few archaic mammals, such as titanotheres, survived into the epoch, and they became extinct before it ended.
oligochaete Any member of the class Oligochaeta. Oligochaetes are ANNELID (segmented) worms, characterized by long, naked bodies bearing a few bristles (chaetae) on each segment. Earthworms are the best-known oligochaetes, but many are freshwater worms as small as 1mm (0.025in) long. Giant Australian earthworms grow up to several metres long.
Oliphant, Sir Mark (1901–) Australian nuclear physicist who conducted much work on the nuclear disintegration of LITHIUM. He was a member of the atom bomb team at Los Alamos (1943–45) and became Australian representative on the United Nations Atomic Energy Commission in 1946.
olivine Ferromagnesian mineral, $(Mg,Fe)_2SiO_4$,

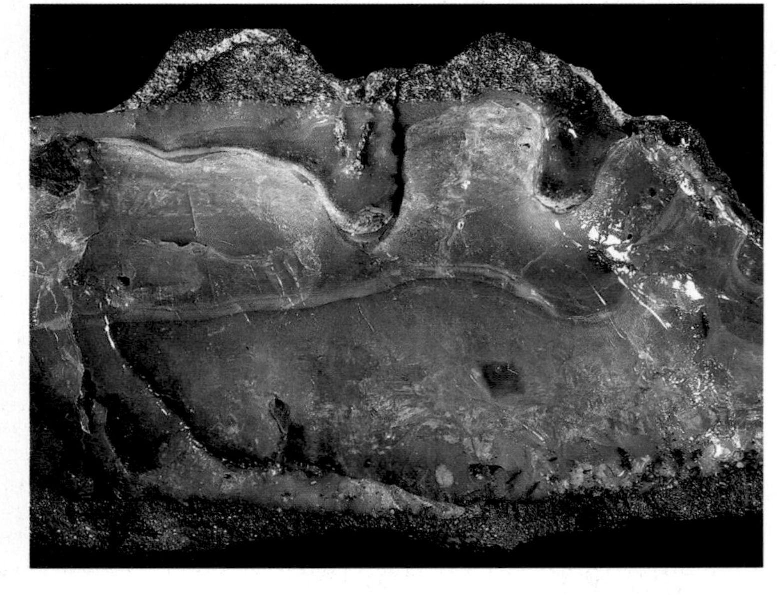

▶ **opal** An iridescent blue sample of precious opal. Opal is composed of microspheroids of hydrous silicon oxide. It is the refraction of light within these spheroids that gives the mineral its iridescent, mother-of-pearl-like sheen. Opal never occurs in a crystaline form, but as small veins, globules and crusts. Several other varieties occur; colourless, milky white, blue, red ("fire opal") and black types. There is also a type of fossilized wood known as wood opal. Black and fire opals are particularly valued as gemstones.

found in basic and ultrabasic igneous rocks. Olivine has orthorhombic system crystals, of usually granular masses. Its colour is commonly olive-green, but can be brown or grey. It is glassy and brittle with no cleavage. Hardness 6.5–7; r.d. 3.3.

omnivore Any creature that eats both animal and vegetable foods – examples are human beings and pigs. Omnivorous animals are characterized by having teeth adapted for cutting, tearing and pulping food. *See also* CARNIVORE; HERBIVORE

oncogene GENE that, by inducing a cell to divide abnormally, contributes to the development of CANCER. Proto-oncogenes are present in all normal cells and in some viruses. Oncogenes arise from the mutation of proto-oncogenes. Some of these mutations are associated with various cancers. The oncogene concept led to the development of the onco-mouse, a laboratory animal that has one of these rogue genes implanted into its cells; it is used in testing anti-cancer treatments. *See also* GENETICS; RETROVIRUS

oncology In medicine, specialty concerned with the diagnosis and treatment of CANCER.

on-line publishing Distribution of information for public access using COMPUTER NETWORKS instead of physical media, such as paper or CD-ROMs. Access to on-line publications may be free or allowed in return for payment. The content of on-line publications is usually similar to that of traditional publications, but may be enhanced by the use of such features as HYPERTEXT, search facilities and MULTIMEDIA. The majority of on-line publishing takes place over the INTERNET, particularly by means of the WORLD WIDE WEB.

Onnes, Heike Kamerlingh *See* KAMERLINGH-ONNES, HEIKE

Onsager, Lars (1903–76) US chemist, b. Norway. Onsager was awarded the 1968 Nobel Prize for chemistry for his development of a theory of irreversible chemical reactions. In 1931 he presented an explanation of the motion of IONS in solution which, although it received little attention at the time, has since been called the fourth law of THERMODYNAMICS. From this theory, Onsager developed a technique for producing uranium-235 (the basic fuel of NUCLEAR REACTORS) from natural uranium (which consists mainly of uranium-238). This technique was put into large-scale operation in 1943 with the first gaseous diffusion plant at Oak Ridge, Tennessee, and remained the primary method for obtaining uranium-235 until the 1960s.

ontogeny Total biological development of an organism. It includes the embryonic stage, birth, growth, body changes and death. *See also* BIOGENETIC LAW; PHYLOGENETICS

onyx Variety of the mineral CHALCEDONY with straight parallel bands. Black and white onyx is often used for making cameos; white and red forms are called **carnelian** onyx; white and brown, **sardonyx**. It is widely distributed, but found mostly in India and South America. It is a form of AGATE.

oocyte Female gametocyte, an animal OVUM before the first polar body is formed. *See* OOGENESIS

oogenesis Process by which a female reproductive cell develops, ready for fertilization. DIPLOID primary egg cells (**oogonia**) divide by MITOSIS to produce many prospective eggs cells (**oocytes**). Each of these oocytes divides by MEIOSIS to produce a HAPLOID secondary oocyte and a **polar body**. The secondary oocyte then divides again to produce an OVUM and a second polar body.

oolite Sedimentary rock composed of OOLITHS.

oolith Small, spherical concretion that occurs in SEDIMENTARY ROCKS. Oolithic grains, up to 2mm (0.1in) in diameter, usually consist of calcium carbonate, dolomite (magnesium calcium carbonate) or chamosite (an iron silicate mineral) and have a radiating, fibrous structure, or one consisting of concentric layers. The centre of an oolith may be a grain of shell or quartz, which suggests to geologists that they grow in oceans or streams by accreting layers of calcium carbonate as they roll about in the turbulent water. Ooliths give the rounded grainy structure to oolitic limestone.

◄ **Oppenheimer** US nuclear physicist Robert Oppenheimer, having led the Manhattan Project and successfully created the first atom bomb, was shocked by the power of the weapon he had helped to create. Oppenheimer opposed the development of the more powerful hydrogen bomb, and proposed a joint control of nuclear energy with the Soviet Union. His alleged left-wing tendencies resulted in his suspension from further nuclear research. However, his achievements in physics were finally recognized when he was awarded the Enrico Fermi award by the US Atomic Energy Commission in 1963.

Oort, Jan Hendrik (1900–92) Dutch astronomer. He carried out research on the structure and dynamics of stellar systems, especially our Galaxy, whose rotation he confirmed in 1927. Oort was also a pioneer of radio astronomy, in particular with respect to the 21cm radiation of interstellar hydrogen, collaborating with Hendrik van de HULST to map the Galaxy at this wavelength. In 1950 he proposed the existence of what has come to be called the **Oort cloud** – a spherical region of space surrounding the Solar System in which comets are thought to reside.

ooze Fine-grained, deep-ocean deposit containing materials of more than 30% organic origin. Oozes are divided into two main types. **Calcareous** ooze at depths of 2,000–3,900m (6,600–12,800ft) contains the skeletons of animals such as foraminifera and pteropods. **Siliceous** ooze at depths of more than 3,900m (12,800ft) contains skeletons of radiolarians and DIATOMS.

opal Non-crystalline variety of QUARTZ, found in recent volcanoes, deposits from hot springs and sediments. Usually colourless or white with a rainbow play of colour in gem forms, it is the most valuable of quartz gems. Hardness 5.5–6.5; r.d. 2.0.

opencast mining Stripping surface layers from the Earth's crust to obtain coal, ores or other valuable minerals. Opencast MINING is relatively cheap as it does not require the sinking of shafts, tunnelling of galleries or underground working. Large, dragline excavators strip away surface layers. Bulldozers and other mechanical shovels distribute minerals and spoil. Lorries, railway wagons and overhead skips carry minerals away for grading and processing. This form of mining is restricted in many countries, because of the damage it causes to valuable land. In many instances owners of opencast MINES are required to restore the environmental quality of the land by landscaping and replanting after mining has ceased.

open-chain compound *See* ALIPHATIC COMPOUND

open cluster (galactic cluster) Group of young stars in the spiral arms of our GALAXY, containing from a few tens of stars to a few thousand. They are usually several light-years across. One example is the Pleiades. Related to open clusters are **associations** – loose groupings of stars of common origin. Often one or more open clusters are found in the central parts of associations. *See also* STAR CLUSTER

open-hearth process Method of producing STEEL in a FURNACE heated by overhead flames. The flames come from gas or oil burners, and oxygen may be blown through the furnace to increase its temperature. Pig iron, scrap steel and limestone are heated together. Various impurities form SLAG, which is removed from the surface of the molten metal. Most of the carbon present in the iron forms carbon monoxide gas, which is removed. Various materials are added to the metal to produce steel of the required type. Use of the open-hearth process is declining because faster and cheaper processes are available, notably the BASIC OXYGEN PROCESS.

operator Mathematical object representing a specific action on a function or variable. The operator itself is meaningless unless a rule is given to explain how the operator acts on the function. For example, the DERIVATIVE operator, written $D = d/dx$, has meaning only when applied to the function $f(x)$, that is $Df(x) = df(x)/dx = f(x)$.

operon Any of a group of GENES on a CHROMOSOME comprising structural genes and an operator gene. **Structural** genes direct the synthesis of ENZYMES involved in the formation of a cell constituent or the utilization of a nutrient. The **operator** gene responds to a **repressor** molecule, and can exist open or closed. When the operator gene is open, the genes it controls are functional, producing PROTEINS. When interacting with the repressor, the operator gene is closed. REGULATORY GENES control the operons.

ophthalmology Branch of medicine that specializes in the structure, function and diseases of the eyes and surrounding structures. It includes surgical and other treatment of eye disorders and the correction of defective vision. *See also* OPTOMETRY

ophthalmoscope Instrument for examining the interior of the eye, invented by Hermann von HELMHOLTZ in 1851.

opium Drug derived from the unripe seeds of the opium poppy. Its components and derivatives have been used as NARCOTICS and ANALGESICS for many centuries. It produces drowsiness and euphoria and reduces pain, although it is addictive. MORPHINE and CODEINE are opium derivatives.

Oppenheimer, (Julius) Robert (1904–67) US theoretical physicist. After early work in nuclear physics at the University of California, Oppen-

O

heimer was appointed director (1943–45) of the Los Alamos laboratory in New Mexico, where he headed the MANHATTAN PROJECT to develop the atom bomb. In 1949, however, he strongly opposed the construction of the hydrogen bomb and in 1953 was suspended by the US Atomic Energy Commission and declared a security risk. He was subsequently reinstated and presented with the Commission's Enrico Fermi Award (1963) for his contribution to science.

opposition In astronomy, position on the ORBIT of a superior planet at which the Sun, the Earth and the planet are in alignment, with the Earth located between the planet and the Sun. Only superior planets, with larger orbits than that of the Earth, can be in opposition to the Sun, and these may be observed most clearly at this time.

optical activity Ability of some chemical compounds in solution and some crystals and transparent substances placed in an intense magnetic field to rotate the plane of POLARIZED LIGHT. Optically active crystals are used in polarimeters, instruments for measuring the degree of optical activity of chemical solutions. Optically active compounds are characterized by having molecules that are asymmetrical. Magnetic rotation of the plane of polarized light was discovered by Michael FARADAY in 1845 and is known as the Faraday effect.

optical character recognition (OCR) Technique for reading letters or numbers so that they can be inputted into a COMPUTER. The text to be read is scanned and the text "image" converted into a digital form. A special program then identifies the characters by their shapes. Early optical characters, such as those used for account numbers printed on cheque forms, had special shapes to prevent ambiguity (for example, the ordinary characters for 1 and l or 0 and O would be easy to confuse). However, modern scanners can deal with most ordinary typefaces and some can even read handwriting, although special optical characters are still retained in applications where errors could be serious.

optical disk In computing, high-density storage device consisting of a disk on which data is recorded and read by a laser. The most common type is a CD-ROM, although an audio compact disc is also a read-only device of this kind. A recording facility is provided by a WORM disk (write once, read many), on which a computer can save data once, and then read it repeatedly but not be able to change it.

optical fibre Fine strand of glass (less than 1mm (0.04in) thick) that is able to transmit digital information in the form of pulses of light. Such transmission is possible because light entering a glass optical fibre is conducted, by reflection, from one end of the fibre to the other without loss of energy. Images may be magnified, distorted or scrambled, depending on how bundles of the fibres are configured. In medicine, optical fibres have facilitated the development of the ENDOSCOPE for the internal observation of organs. They also have an increasingly widespread application for telecommunications as they are able to carry more information than conventional metal cable systems, providing high-speed data links.

optical illusion Effect in which visual information is misleading or misinterpreted. Natural optical illusions are often created by light REFRACTION, such as a MIRAGE of water on a hot, dry road.

optical rotation Angle through which a molar solution of an optically active compound rotates the plane of POLARIZATION of a beam of POLARIZED LIGHT at a given temperature. Many natural products show an ability to rotate polarized light counterclockwise; few rotate clockwise. *See also* OPTICAL ACTIVITY

optic nerve Second cranial nerve, which carries the visual stimuli from the RETINA of the EYE to the visual centre in the CORTEX of the brain. That part of the retina where the optic nerve enters the eye is known as the blind spot. About 1 million optic nerve fibres comprise the optic nerve, and these fibres are arranged in such a way that impulses from the left side of the visual field travel to the right side of the brain, and vice versa.

optics Branch of physics concerned with the study of LIGHT and its behaviour. Fundamental aspects are the physical nature of light, both as a wave phenomenon and as particles (PHOTONS), and the REFLECTION, REFRACTION and POLARIZATION of light and its transmission through various media. Optics also involves the study of MIRRORS, LENSES and lens systems (including that of the eye and those of optical instruments) and of optically active chemicals and crystals that polarize light. More generally, optics deals with the wider subject of the part of the ELECTROMAGNETIC SPECTRUM lying between short RADIO waves and soft X-RAYS, called the optical spectrum.

optometry Science that deals with the examination of eyes and the prescribing of eyewear, such as spectacles or contact lenses, in order to correct vision defects. It is distinct from OPHTHALMOLOGY, the medical and surgical treatment of the eyes.

orbit Path of a celestial body in a gravitational field. The path is usually a closed one about the focus of the system to which it belongs, as with those of the planets about the Sun, or the components of a binary system about their common centre of mass. Most celestial orbits are elliptical, although the ECCENTRICITY can vary greatly. It is rare for an orbit to be parabolic or hyperbolic. *See also* APSIS; DECLINATION, MAGNETIC; INCLINATION MAGNETIC; KEPLER'S LAWS; PERTURBATION

orbital In PARTICLE PHYSICS, region in space around an atomic NUCLEUS in which ELECTRONS can move. There is a high probability of finding an electron in such an orbital, which can accommodate one or two electrons and has a shape and energy characterized by the atom's QUANTUM NUMBERS. In molecules, the bonding electrons move in the combined electric field of all the nuclei. The atomic orbitals then become molecular orbitals – regions encompassing two nuclei, having a characteristic energy and containing two electrons. These molecular orbitals, which can be thought of as formed by the overlap of atomic orbitals, constitute CHEMICAL BONDS.

orbiter Spacecraft designed to orbit a moon or planet. The very first spacecraft, the Soviet *SPUTNIK I* (launched in 1957), was an orbiter that orbited the Earth. More Earth satellites followed before the Soviet Union and the USA sent orbiters to the Moon in 1966. Unmanned orbiters have also been sent to Mars and Venus. The term is also used to describe the main part of the SPACE SHUTTLE.

order In biology, the CLASSIFICATION of living organisms ranking below CLASS and above FAMILY. Order names are printed in Roman (ordinary) letters, and begin with a capital letter. Among animals, the names of most orders end in *-a*, such as Anura (frogs and toads) and Chiroptera (bats), although bird orders end in *-iformes*, such as Columbiformes (pigeons) and Stringiformes (owls). In plants, order names generally end in *-ales*, such as Rosales (roses).

ordinal number *See* NUMBER, ORDINAL

Ordovician Second-oldest period of the PALAEOZOIC era, 505 to 438 million years ago. All animal life was restricted to the sea. Numerous invertebrates flourished, including trilobites, brachiopods, corals, graptolites, molluscs and echinoderms. Remains of jawless fish in coastal deposits mark the first record of the vertebrates.

ore Mineral or combination of minerals from which metals and nonmetals can be extracted. Ores occur in veins, beds or seams parallel to the enclosing rock or in irregular masses. Industrial rock deposits in beds, such as gypsum and limestone, are not called ores. *See also* METALLURGY; MINING

organelle Part of a CELL, such as a MITOCHONDRION or a flagellum, with a persistent structure and a specific function. It is to a cell what an organ is to an organism.

organ In biology, group of TISSUES that form a functional and structural unit in a living organism. The major organs of the body include the brain, heart, lungs, skin, liver and kidneys. Leaves, flowers and roots are examples of plant organs.

organic chemistry *See* CHEMISTRY

organic compounds COMPOUNDS that contain the element CARBON; they are about a hundred times more numerous than inorganic compounds. Examples include HYDROCARBONS, basic structures that, when combined with atoms of other elements (such as oxygen and nitrogen), form a vast range of compounds, including those essential to life.

OPTICAL FIBRE

Optical fibres carry information as signals of light. The fibres comprise two layers of the highest quality glass. The inner, glass core (1) of a fibre optic cable is surrounded by an outer cladding of glass (2) of a different refractive index. The cladding contains the light pulses within the core. The light signal cannot leave the core because it always hits the edge of the core at too shallow an angle to escape – an effect known as "total internal reflection". A sheath (3) provides physical protection, and bundles of the sheathed cores (trunk cables) are given strength by a central steel wire (4). Narrow-core fibres (A) are increasingly used because they allow signals to be sent over greater distances without blurring. In a wide-core cable (B) more reflections can occur, causing the pulse to spread out and merge with adjacent pulses. To prevent this, more space must be left between pulses in wider cables and that limits the volume of data that can be transmitted.

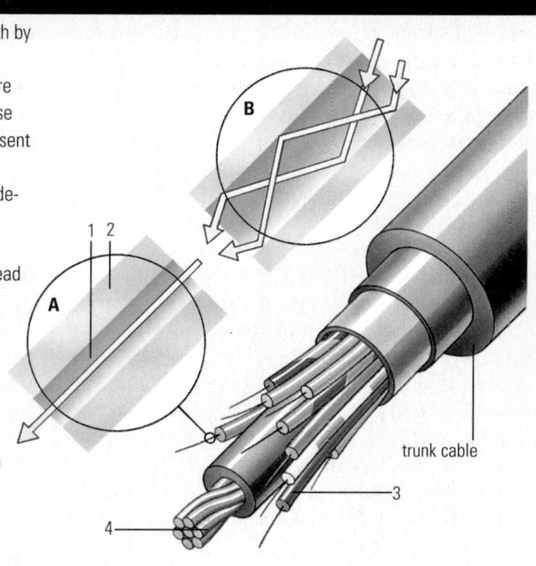

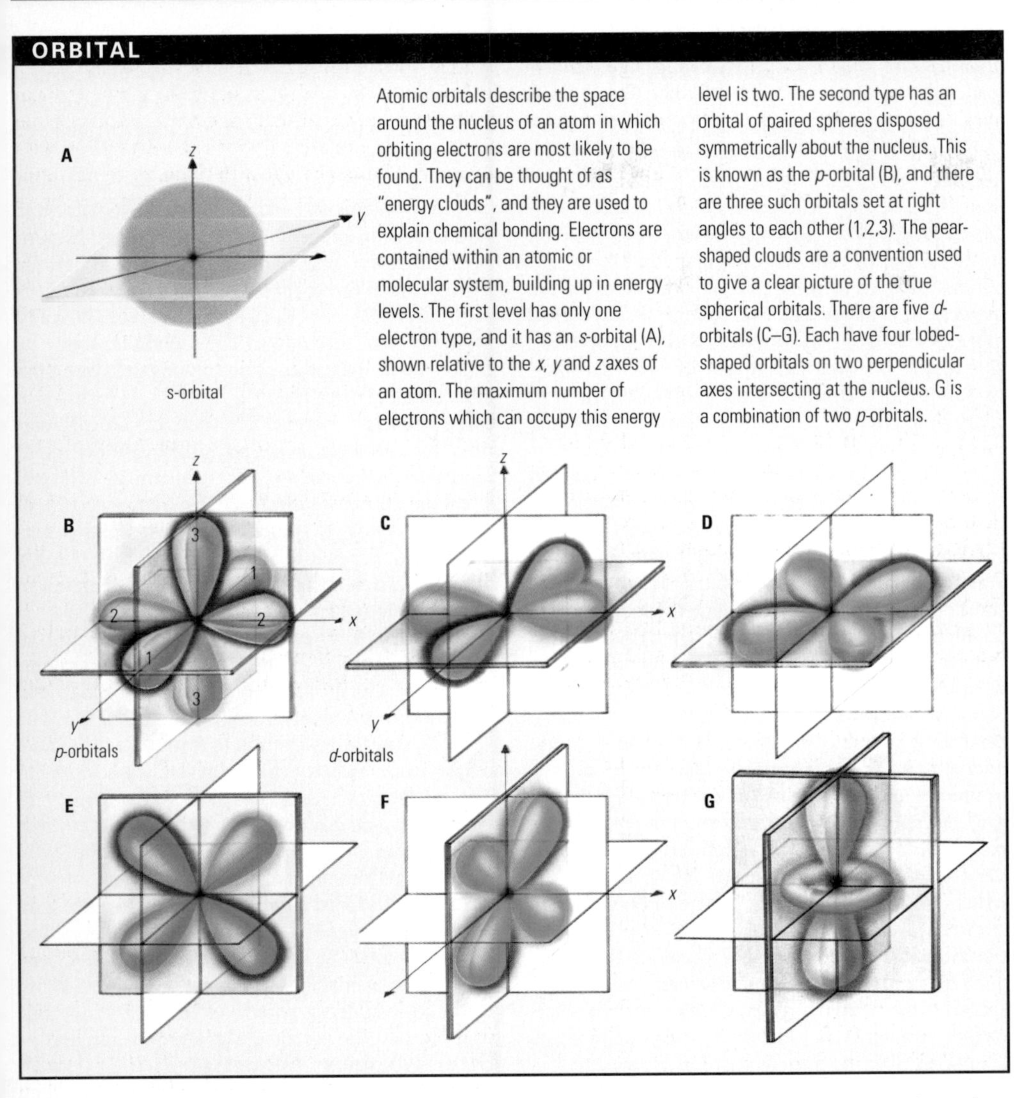

ORBITAL

Atomic orbitals describe the space around the nucleus of an atom in which orbiting electrons are most likely to be found. They can be thought of as "energy clouds", and they are used to explain chemical bonding. Electrons are contained within an atomic or molecular system, building up in energy levels. The first level has only one electron type, and it has an *s*-orbital (A), shown relative to the *x*, *y* and *z* axes of an atom. The maximum number of electrons which can occupy this energy level is two. The second type has an orbital of paired spheres disposed symmetrically about the nucleus. This is known as the *p*-orbital (B), and there are three such orbitals set at right angles to each other (1,2,3). The pear-shaped clouds are a convention used to give a clear picture of the true spherical orbitals. There are five *d*-orbitals (C–G). Each have four lobed-shaped orbitals on two perpendicular axes intersecting at the nucleus. G is a combination of two *p*-orbitals.

organic food Food grown without the use of chemical FERTILIZERS or INSECTICIDE sprays. Only natural fertilizers such as manures, bone meal, seaweed and fishmeal are used. These organic fertilizers are thought to be better for the structure of the soil than chemical or inorganic fertilizers, and plants cultivated in this manner do not contain possibly harmful chemicals.

organ of Corti Complex structure within the inner EAR of mammals, birds and reptiles, concerned with the final reception of inner ear movements resulting from sound waves striking the eardrum. It rests on a platform of membrane and bone which extends along the COCHLEA, and contains sensory hair cells that detect movements caused by sound waves and connect with nerve fibres that carry messages to the brain.

organometallic compound Chemical compound in which one or more organic groups or RADICALS are bonded to an atom of a metal. Metallic carbonates (such as sodium carbonate) and salts of common fatty acids (such as sodium ethanoate) are usually excluded from this classification. Typical examples are metallic alkyl compounds (such as LEAD(IV) TETRAETHYL and triethyl aluminium), GRIGNARD REAGENTS (such as ethylmagnesium iodide) and compounds of transition metals.

organophosphate Chemical in which the molecules contain at least one atom of carbon and one of phosphorus. The best-known organophosphates are some INSECTICIDES, certain nerve gases, essential nucleic acids and nucleotide enzymes. Organic derivatives of the phosphorus acids are important in the manufacture of FERTILIZERS. These are highly toxic.

organ transplant Surgical implantation of an organ (such as a kidney or heart) from another individual to substitute for a malfunctioning or diseased organ in the patient. Careful pre-operative preparation and matching followed by post-operative drug treatment are necessary to ensure that the transplanted organ is not rejected by the recipient's immune system. *See also* IMMUNOLOGY

Orion Prominent CONSTELLATION, named after a mythical hunter, situated S of Taurus and Gemini. Its brightest stars are BETELGEUSE and RIGEL. A row of three other stars – Zeta, Epsilon and Delta – make up Orion's belt. From the belt hangs Orion's sword, marked by the ORION NEBULA.

Orion Nebula Gaseous emission NEBULA in the ORION constellation. It is a mass of gas, lit up by a group of new-born stars called the Trapezium. Because they are very hot, about 50,000K, these stars emit most of their energy in the ultraviolet, which makes the nebula glow. The Trapezium stars have destroyed most of the volatile dust in the nebula, etching out a roughly spherical hollow that is still growing. The visible Orion Nebula is, therefore, a hole in a much larger dark nebula. Within the densest portions of the dark cloud, star formation is still under way.

ornithology Study of BIRDS. Included in general ornithological studies are classification, structure, function, evolution, distribution, migration, reproduction, ecology and behaviour.

orogenesis MOUNTAIN building, especially where the Earth's crust is compressed to produce large-scale FOLDS and FAULTS.

orrery Mechanical model of the Solar System. Orreries vary from simple ones with just the Earth, Moon and Sun, to highly complicated representations of the whole Solar System, where the planets with their satellites not only revolve in their orbits but also rotate on their axes. The name is from Charles Boyle, fourth Earl of Orrery (1676–1731), who commissioned one in 1712, although this was not the first to be made.

orthoclase Essential mineral found in igneous and metamorphic rocks containing potassium FELDSPAR. It is a potassium aluminium silicate, $KAlSi_3O_8$, with monoclinic system crystals. Its colour is generally white, but there are pink varieties. Hardness 6–6.5; r.d. 2.5–2.6.

orthopteran Any member of the order Orthoptera, an order of insects represented by more than 20,000 species found worldwide, especially in tropical regions. Orthopterans include katydids, crickets, grasshoppers and locusts in the suborders Ensifera and Caelifera. The term was formerly used to group together the orders Dictyoptera (mantids and cockroaches), Phasmida (leaf and stick insects) and Grylloblattidae (grylloblattids). Most of the species feed on plants, have chewing mouthparts and undergo incomplete METAMORPHOSIS from egg to nymph and adult. Earwigs, termites and webspinners are also considered to be Orthopteroid insects. Length: to 10cm (4in).

oscillating Universe theory Variant of the BIG BANG theory in which it is suggested that the Universe passes through successive cycles of expansion and contraction (or collapse). At the end of the collapse phase, with the Universe packed into a small volume of great density, it is possible that a "bounce" would occur. The Universe would thus oscillate between Big Bang and "Big Crunch" episodes and so be infinite in age. However, for this to happen the density of the Universe would have to be above a certain value (the critical density), which is not thought to be the case.

oscillator In the physics of SOUND, a device for producing sound waves, as in SONAR or an ultrasonic generator. In electronics, an oscillator circuit

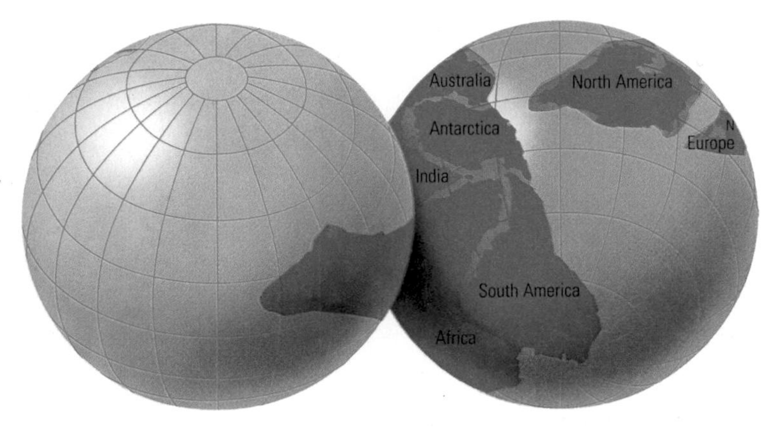

◀ **Ordovician** The most significant movement of the plates during the Ordovician is the movement of North America towards N Europe compressing the sea area between these two land masses.

converts direct current (DC) electricity into high-frequency alternating current (AC). Harmonic oscillators generate sinusoidal wave forms. *See also* ELECTRIC CURRENT

oscilloscope (cathode-ray oscilloscope) Electronic instrument in which a CATHODE-RAY TUBE (CRT) system displays how quantities, such as voltage or current, vary over a period of time. The input signal is usually fed to the vertical plates of the CRT. The electron beam that traces the pattern on the screen is moved by a time-base generator within the oscilloscope. The result is generally a curve or graph on the screen. The oscilloscope can be calibrated and used to measure the input signal.

Osler, Sir William (1849–1919) Canadian physician. Osler was the first to identify (1876) platelets in the blood. In 1888 he was appointed professor of medicine at Johns Hopkins University, Baltimore, which he developed into one of the foremost medical schools. Osler wrote *The Principles and Practice of Medicine* (1892), the leading medical text of the time and still published. Probably the most famous physician of his era, he gave his name to several medical terms, such as Osler's nodes and Osler-Vaquez disease.

osmium (symbol Os) Blue-white, metallic TRANSITION ELEMENT, discovered in 1803. The densest of the elements, osmium occurs associated with platinum; the chief source is as a by-product from smelting nickel. Like IRIDIUM, osmium is used in producing hard alloys. It is also used to make electrical contacts and pen points. Properties: at.no.76; r.a.m. 190.2; r.d. 22.57; m.p. 3,045°C (5,513°F); b.p. 5,027°C (9,081°F); most common isotope ^{192}Os (41.0%).

osmoregulation Process by which the amount of water and concentration of salts are regulated in the bodies of animals and protozoans. In saltwater environments, water tends to pass from an animal's body through the action of OSMOSIS. This effect is counteracted by the kidneys. Similarly in freshwater environments, the kidneys of higher animals, and the contractile VACUOLES of protozoans, prevent water from passing into the body.

osmosis Diffusion of a SOLVENT (such as water) through a natural or artificial, semi PERMEABLE MEMBRANE (one which only allows the passage of certain dissolved substances) into a more concentrated solution. Because the more concentrated solution contains a lower concentration of solvent molecules, the solvent flows by DIFFUSION to dilute it until concentrations of solvent are equal on both sides of the membrane. Osmosis is a vital cellular process. Plant roots absorb water by osmosis, and plant cells are kept firm as a result of the uptake of water by osmosis (turgor); membranes of all living cells use it to control the passage of required substances. *See also* DIALYSIS; REVERSE OSMOSIS; TURGOR PRESSURE

osmotic pressure (symbol Π) Pressure exerted by a dissolved substance by virtue of the motion of its molecules. In dilute solutions, it varies with the concentration and temperature as if the solute were a gas occupying the same volume. It can be measured by the pressure that must be applied to counter-balance the process of OSMOSIS into the solution. Osmotic pressure is given by $\Pi V = nRT$, where n = moles in solution, V = volume, T = temperature, and R = gas constant.

ossicle Any one of the small bones found in the middle EAR of most mammals. The ossicles are the **malleus** (hammer), **incus** (anvil) and **stapes** (stirrup). Vibrations of the eardrum (tympanum) are picked up by the malleus, amplified by the movement of the incus and passed on to the stapes, which connects with the oval window in the COCHLEA of the inner ear.

ossification (osteogenesis) Process of BONE formation in vertebrates. It occurs through the action of special cells called osteoblasts, which secrete bone-forming minerals that combine with a network of COLLAGEN fibres to form the hard bone matrix.

Osteichthyes Class of fish found in almost every environment. Their characteristics include a bony skeleton, a single flap (operculum) covering the gill openings, and erythrocytes (red blood cells) with nuclei. Most members of this class have scales. Osteichthyes first appeared during the DEVONIAN period, when they were heavily armoured and lived in freshwater.

osteomyelitis Infection of the BONE, sometimes spreading along the marrow cavity. Today, quite rare except in diabetics, it can arise from a compound fracture, where the bone breaks through the skin, or from infection elsewhere in the body. It is accompanied by fever, swelling and pain, possibly with abscess formation. The condition may be treated with immobilization, ANTIBIOTICS and possibly surgical drainage.

osteopathy System of alternative medical treatment based on the use of physical manipulation to rectify damage caused by mechanical stresses. The concept was formulated by Andrew STILL in 1874, and the first school of osteopathy was opened in Kirksville, Missouri, in 1892. Still's emphasis on treating the whole person remains an ideal of osteopathy.

Ostwald, Wilhelm (1853–1932) German chemist, b. Russia. He became professor of chemistry at Leipzig University (1887–1906), and remained in Germany for the rest of his life. His work on SOLUTIONS led him to formulate Ostwald's dilution law, which deals with the dissociation of weak ELECTROLYTES. His process for preparing nitric acid and his research on catalysis became important in industry. He was awarded the 1909 Nobel Prize for chemistry.

Otis, Elisha Graves (1811–61) US inventor who designed and manufactured lifts and escalators. His first invention (1852) was an automatic safety hoist for a lift in a factory. Otis' fame dates from 1856, when he installed a passenger lift in a New York store. In 1861 he patented a steam-powered lift.

Otto, Nikolaus August (1832–91) German engineer. In 1861 he built a gas-fired engine which won a gold medal at the 1867 Paris Exhibition. Otto later built an INTERNAL COMBUSTION ENGINE based on a four-stroke cycle (also called the **Otto cycle**). This FOUR-STROKE ENGINE was the forerunner of most of today's engines.

outcrop Exposure at the surface of the Earth of an edge of rock stratum. This phenomenon may be caused by the EROSION of soil by water, wind, ice (especially glaciers), or gravity.

outer ear *See* EAR

outlier Mass of newer rocks surrounded by older rocks. The newer rocks may have become detached from a larger formation by EROSION. *See also* INLIER

output device, computer Apparatus that transmits or records data from a COMPUTER. It may take the form of a MODEM for remote receipt of data, a hard-copy **printer** or recorder (such as a teletypewriter), a **line-printer,** or an alpha-numeric display or **monitor** (a TV-type screen on which letters or numbers are presented). Other forms include digital to analogue systems, which directly control various mechanisms and processes.

ovary In biology, part of a multicellular animal or a seed-bearing plant that produces egg cells (ova), the female reproductive cells. In vertebrates, it also produces female SEX HORMONES. In female humans, there is an ovary on each side of the UTERUS. Controlled by the PITUITARY GLAND, each ovary, as well as the ova, produces two major female sex hormones, OESTROGEN and PROGESTERONE, which in turn control development and functioning of the female reproductive system. In seed-bearing plants, the female sex cells are contained within structures called OVULES inside the ovary. After FERTILIZATION, the ovules develop into

OSCILLOSCOPE

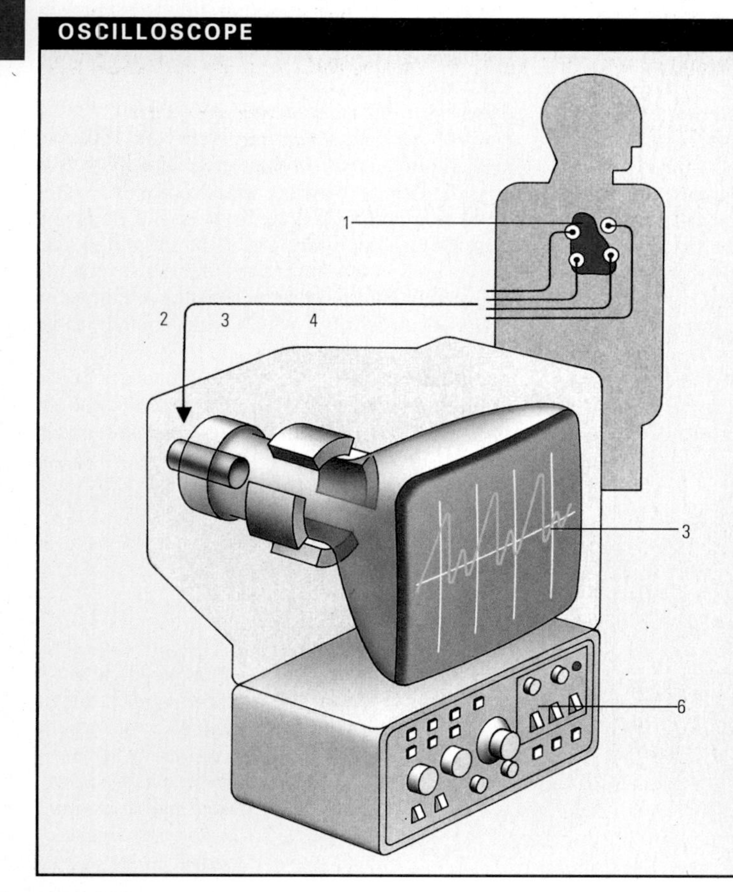

An oscilloscope displays an electronic signal on a monitor in analogue form. Usually time is represented on the x axis (horizontal), with the y axis (vertical) recording the incoming voltage - here a heartbeat (1). The signal from the object being monitored is converted into an electrical voltage (2). That voltage is shown in visible form on the screen of a cathode ray tube (3) similar to a black and white television. Deflector magnets (4) direct the stream of electrons from the electron gun (5). The magnets sweep the electron beam from left to right over a set period, while variations in the voltage of the external signal cause the wave pattern. The control box (6) allows the period of the x axis and the strength of the signal displayed to be changed.

OSMOSIS

level of sugar solution rises
level of water drops
sugar solution
pure water
1
2
3

Semi permeable membranes have small pores in them. While the pores allow tiny molecules through (1), they also block out larger molecules (2). If a semi permeable membrane is used to separate a sugar solution from a container of water (3), initially the concentration of water molecules in the solution will be lower than that in the pure water. Over time, water molecules will pass through the membrane, diffusing into the container, but as the pores are too small to allow the sugar solution to flow the other way, the level of the sugar solution rises. Osmosis, the net flow of a solvent across a membrane drives many cellular processes in organisms, like maintaining hydrostatic pressure in plant cells to keep them turgid. Man-made membranes are used in kidney dialysis machines.

seeds and the ovary develops into one or more FRUITS. *See also* HORMONE; MENSTRUAL CYCLE; OVUM

overtone Usually a HARMONIC, a constituent of a musical note with a frequency that is a whole-number multiple of that of the fundamental note. Some instruments produce nonharmonic overtones.

oviducts Tubes that connect the ovaries with the UTERUS and through which egg cells (ova) are released from the OVARY. In mammals, they are known as FALLOPIAN TUBES. These are lined with a tissue called the EPITHELIUM.

oviparity (ovipary) Type of REPRODUCTION in which fertilization of the female ova (eggs) occurs inside the body, but once achieved the eggs are laid and subsequently hatch outside her body. Unlike VIVIPARITY, the developing EMBRYO receives all its nutrition from the egg. It is the most common form of reproduction and occurs in all animals except MARSUPIALS and PLACENTAL MAMMALS. *See also* OVOVIPARITY

ovoviparity (ovoviviparity) Type of REPRODUCTION in which a female's fertilized ova (eggs) develop and then hatch inside her OVIDUCTS. Unlike VIVIPARITY, during ovoviparity development the ZYGOTES receive no nutrition from the mother as no PLACENTA is present. Ovoviparity is common in many invertebrate animals and some fish and snakes. When the young emerge, the mother appears to be giving birth to live young, although they have only just hatched. *See also* OVIPARITY

ovulation Release of a mature OVUM (egg) from the OVARY ready for FERTILIZATION. In women, one egg (ova) is released into the oviduct midway through the MENSTRUAL CYCLE, stimulated by LUTEINIZING HORMONE from the PITUITARY GLAND.

ovule In seed-bearing plants (SPERMATOPHYTES), the part of the female reproductive organ that contains an egg cell or OVUM and develops into a seed after FERTILIZATION. In ANGIOSPERMS (flowering plants), ovules develop inside an OVARY. In GYMNOSPERMS (conifers), ovules are borne on the inner surface of the female cone without any covering.

ovum Female GAMETE (reproductive cell) produced in an OVARY (in animals) or an OVULE (in plants). After FERTILIZATION by the male gamete (SPERM in animals, POLLEN in plants), it becomes a ZYGOTE that is capable of developing into a new individual.

oxalic acid (ethanedioic acid) Poisonous, colourless, crystalline organic acid ($C_2H_2O_4$) whose salts occur naturally in some plants, such as sorrel and rhubarb. It is used for metal and textile cleaning, and in tanning. Properties: r.d. 1,653; m.p. 101.5°C (214.7°F).

oxbow lake Crescent-shaped section of a river channel that no longer carries the main discharge of water. An oxbow lake forms from a MEANDER. As sediment is deposited, the meander becomes cut off from the river to create a lake. Once formed, the lake gradually shrinks as sediment fills it in; vegetation grows on the new muddy area, and the land can be reclaimed. The name derives from the shape of the lake, said to resemble an ox's collar.

oxidase In biochemistry, any enzyme that catalyses an OXIDATION REDUCTION reaction.

oxidation Chemical reaction that involves a loss of one or more ELECTRONS by an atom or molecule (always part of an OXIDATION-REDUCTION reaction in which those electrons are gained by another atom or molecule). Previously the term was more strictly applied to a reaction in which oxygen combines with another element or compound to form an oxide. Oxidation is brought about by OXIDIZING AGENTS.

oxidation number *See* OXIDATION STATE

oxidation-reduction (redox) Chemical reaction involving simultaneous OXIDATION (a loss of one or more electrons by an atom or molecule) and REDUCTION (a gain of those electrons by another atom or molecule). In general, oxidation and reduction reactions occur together and in the same quantities. For example, in the reaction between iron oxide and carbon, iron (the electron acceptor) is **reduced** by carbon (the reducing agent), and the carbon (the electron donor) is **oxidized** by iron oxide (the oxidizing agent). Oxidation-reduction reactions are important in many biochemical systems.

oxidation State in which an atom is potentially able to form a compound, depending on the number of electrons available to be transferred. It is quantified on a scale of **oxidation numbers** which indicate the degree of IONIZATION. Thus, in the compound sodium oxide, Na_2O, SODIUM (Na^+) has an oxidation number of +1, and OXYGEN (O^{2-}) an oxidation number of −2. In COVALENT and COORDINATION (complex) COMPOUNDS, the oxidation number is the electric charge that the atoms would have had if the compound was ionic; for example, in the ION $(CuCl_4)^{2-}$, regarded as formed from Cu^{2+} and $4Cl^-$, copper has an oxidation number of +2. Oxidation numbers are often used in the names of chemical compounds, as in iron(II) chloride ($FeCl_2$) and iron(III) chloride ($FeCl_3$), formerly called ferrous and ferric chloride. *See also* VALENCE

oxide Any inorganic chemical compound in which OXYGEN is combined with another element. Oxides are often formed by burning the element in

OVUM

It takes about a week for the fertilized ovum to pass down the Fallopian tube and implant itself in the uterine lining, the endometrium. Within hours of conception, mitosis begins with the development of a sphere of an increasing number of cells; the sphere starts as the blastomere (1), develops into the morula (2) of about 64 cells. At this stage, it changes into a hollow, fluid-containing ball – blastocyst (3) – with the inner cell mass at one end. It can now begin implantation (4). By the ninth day after conception, the blastocyst has sunk deep into the endometrium (5) and is already receiving nutrition from the mother.

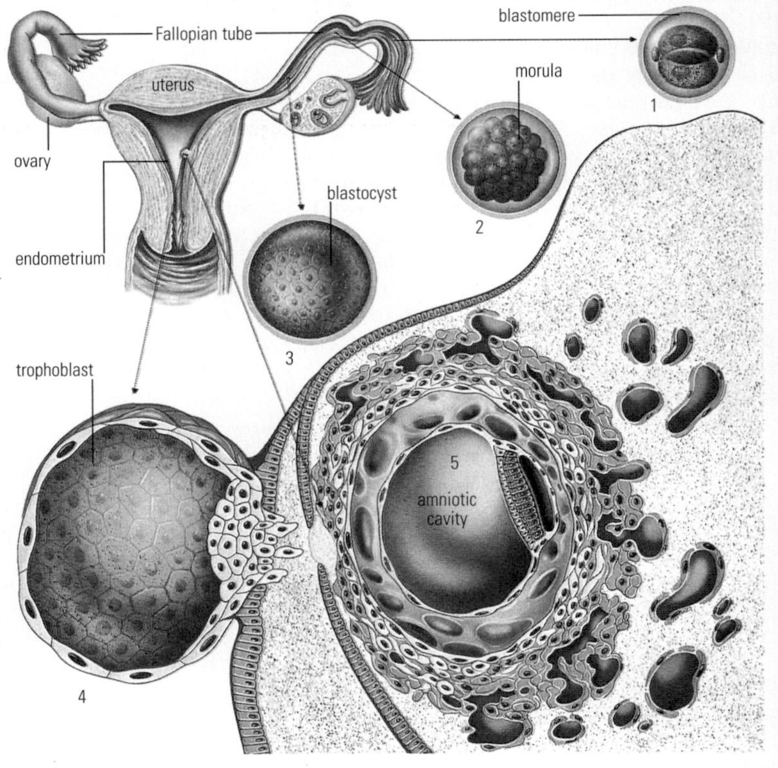

air or oxygen; thus magnesium (Mg) burns to form magnesium oxide (MgO). Oxides are acidic, basic or AMPHOTERIC, the latter capable of being acidic or basic. Other forms include PEROXIDES, such as barium peroxide, and the neutral oxides, such as carbon monoxide, which do not react with acids or bases.

oxidizing agent Substance that causes OXIDATION reactions. Thus in the oxidation of carbon to form carbon monoxide, $2C + O_2 \rightarrow 2CO$, oxygen is the oxidizing agent. Other common oxidizing agents include NITRIC ACID, HYDROGEN PEROXIDE, OZONE, potassium dichromate, iron(III) (ferric) compounds and POTASSIUM MANGANATE(VII). *See also* OXIDATION-REDUCTION

oxime In organic chemistry, a compound formed between hydroxylamine (H_2NOH) and an ALDEHYDE or KETONE. Oximes are condensation products (*see* CONDENSATION REACTION) containing the group –C=NOH, and are used in qualitative analysis to identify aldehydes and ketones. For example, ethanal (acetaldehyde) forms the oxime $CH_3CH{=}NOH$, and propanone (acetone) forms the oxime $(CH_3)_2C{=}NOH$.

oxyacetylene welding *See* WELDING

oxygen (symbol O) Common gaseous element that is necessary for the RESPIRATION of plants and animals and for combustion. It was discovered in 1774 by Joseph PRIESTLEY and independently (*c.*1772) by Karl SCHEELE. Oxygen, which is colourless and odourless, is the most abundant element in the Earth's crust (49.2% by weight) and is a constituent of water and many rocks. It is also present in the atmosphere (28% by volume), being extracted by fractional DISTILLATION of liquid air. It can also be obtained by the ELECTROLYSIS of water. Oxygen is used in steelmaking (BESSEMER PROCESS), welding, the manufacture of industrial chemicals, and in apparatus for breathing (oxygen masks) and resuscitation (oxygen tents); liquid oxygen is used in rocket fuels. It is chemically reactive, and forms compounds with nearly all other elements (especially by OXIDATION). Properties: at.no. 8; r.a.m. 15.9994; r.d. 1.429; m.p. −218.4°C (−361.1°F); b.p. −182.96°C (−297.3°F); most common isotope ^{16}O (99.759%). *See also* OXIDATION-REDUCTION; OZONE

oxygen cycle Interchange of OXYGEN among agencies such as the atmosphere, the oceans, animal and plant processes and chemical combustion. The main renewable source of the Earth's oxygen is the plant process of PHOTOSYNTHESIS, wherein oxygen is liberated. Oxygen, dissolved in water, is utilized by aquatic life-forms through RESPIRATION, a process essential to most living forms except anaerobic bacteria. *See also* CARBON CYCLE

oxygen debt Insufficient supply of oxygen in the muscles following vigorous exercise. This reduces the breakdown of food molecules that generate energy, causing the muscles to over-produce LACTIC ACID that creates a sensation of FATIGUE. The automatic rapid breathing (hyperpnoea) experienced after exercise creates extra oxygen to reduce the oxygen debt. As exercise ceases, the body uses up this extra supply of oxygen to break down the lactic acid.

oxyhaemoglobin Combination of HAEMOGLOBIN in ERYTHROCYTES (red blood cells) with OXYGEN from the lungs, in which form oxygen is transported in the blood to all cells of the body. When oxyhaemoglobin gives up its oxygen to cells, a chemical reaction is promoted which makes CARBON DIOXIDE (CO_2) from the tissues more soluble in the blood for transport back to the lungs and elimination from the body.

oxytocin Hormone produced by the posterior PITUITARY GLAND in women during the final stage of pregnancy. It stimulates the muscles of the uterus, initiating the onset of LABOUR and maintaining contractions during the BIRTH of the child. It also stimulates LACTATION.

ozone Unstable, pale-blue, gaseous allotrope of OXYGEN, being trioxygen (O_3). It has a characteristic pungent odour and decomposes into molecular oxygen. It is present in the atmosphere, mainly in the OZONE LAYER, where it is formed by the action of ultraviolet radiation on oxygen. It acts as a filter and prevents much harmful ultraviolet radiation from reaching the Earth's surface. Ozone, prepared commercially by passing a high-voltage discharge through oxygen, is used as an OXIDIZING AGENT in bleaching, air-conditioning and purifying water. *See also* ALLOTROPY

ozone layer Region of Earth's atmosphere in which OZONE (O_3) is concentrated. It is densest at altitudes of 21–26km (13–16mi). Produced by incoming sunlight, the ozone layer absorbs much of the solar ultraviolet radiation, thereby shielding the Earth's surface. Aircraft, nuclear weapons, and some aerosol sprays and refrigerants all yield chemical agents that can break down high-altitude ozone, which could lead to an increase in the amount of harmful ultraviolet radiation reaching the Earth's surface. *See also* CHLOROFLUOROCARBON (CFC).

OZONE

A naturally occurring substance, ozone acts as a sunscreen for the Earth because its molecules absorb the Sun's ultraviolet radiation (1). The presence of chlorofluorocarbon pollution (2) causes the layer to break down, allowing ultraviolet rays through. Ozone is created when ultraviolet rays split oxygen molecules. The lone oxygen atoms (3) bond with oxygen molecules (4) to make ozone (5). When, however, chlorofluorocarbons (CFCs) are present, they are split by the ultraviolet rays. The released chlorine atom (6) in turn splits ozone molecules to form a chlorine monoxide molecule (7) and oxygen. The process is continued as the chlorine monoxide absorbs the lone oxygen atoms that previously formed ozone. This frees the chlorine atom which splits another ozone molecule (8) creating another chlorine monoxide molecule and oxygen.

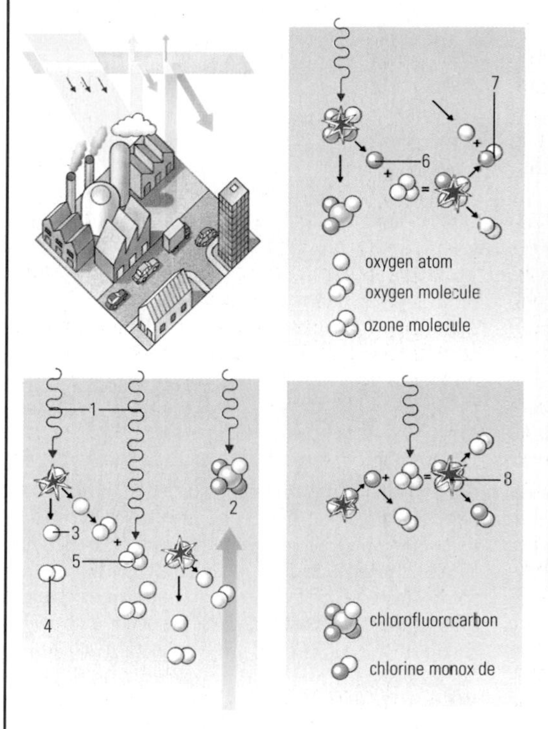

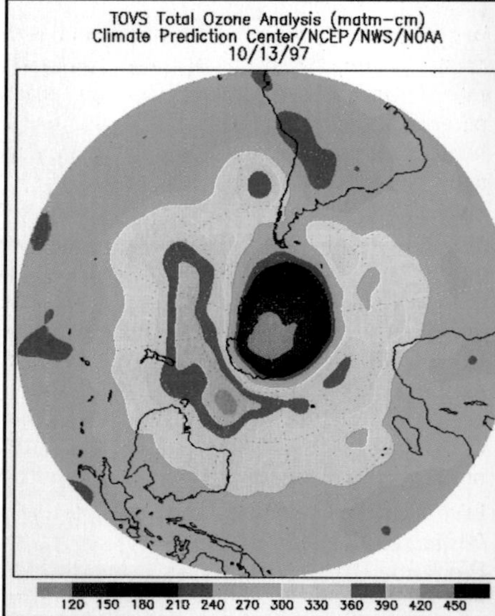

A satellite image of the hole in the ozone layer (grey and purple) over Antarctica. The scale on the bottom shows the concentration of ozone, measured in Dobson units. The continents are outlined in black.

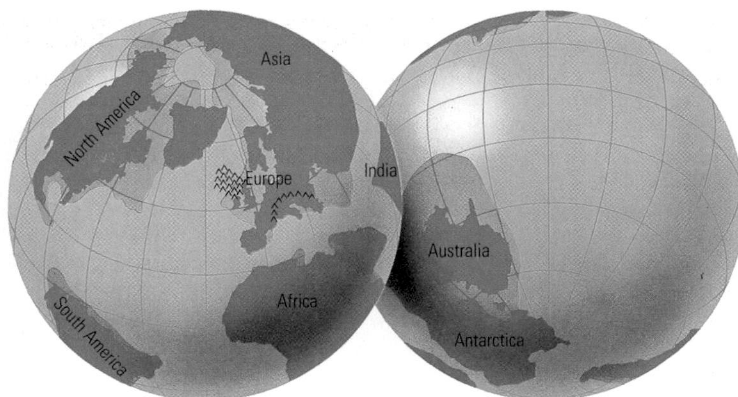

◄ **Palaeocene** The drifting apart of continents continues during this epoch, but the movement between Africa and Europe raises the Alps. A great area of volcanic activity reaches from the British Isles towards the position of Iceland.

pacemaker (sino-atrial node) Specialized group of cells in the vertebrate HEART that contract spontaneously, setting the pace for the heartbeat itself. If it fails its function can be taken over by an artificial pacemaker, an electronic unit that stimulates the heart by means of tiny electrical impulses to maintain a normal beat.

Page, Sir Frederick Handley (1885–1962) British aeronautical engineer who founded (1909) one of the first aircraft manufacturing companies (Handley Page Ltd.). The company produced civil and military aircraft, such as the Halifax bomber.

Paget, Sir James (1814–99) British surgeon, a founder of the science of PATHOLOGY. Paget discovered *Trichina spiralis*, the parasitic worm that causes trichinosis. He described the bone disease osteitis deformans (Paget's disease), and was among the first experts to recommend surgical removal of bone marrow tumours.

paint Coating applied to a surface for protective, decorative or artistic purposes. Paint is composed of PIGMENT (colour) and a liquid **vehicle** (binder or medium). The vehicle suspends the pigment, adheres to a surface and hardens when dry. Pigments are made of metallic compounds, usually oxides, or synthetic materials. Vehicles may be oils, water mixed with a binding agent, organic compounds or synthetic RESINS, which may be soluble in water or oil. Water-based paint (emulsion) mixes with water; oil-based paint (gloss) mixes with TURPENTINE or various organic solvents.

pair production Creation of an ELECTRON and a POSITRON from a PHOTON in a strong electric field of a NUCLEUS. Electrons and positrons are only formed when a photon has high quantum energy since, on the basis of the special theory of RELATIVITY ($E = mc^2$), they have a total mass of 16×10^{-14} J. *See also* PARTICLE PHYSICS

Palade, George Emil (1912–) US cell biologist, b. Romania. He shared the 1974 Nobel Prize for physiology or medicine with Albert Claude (1899–1983) and Christian de DUVE. Using sophisticated electron microscopic techniques and other methods, Palade discovered RIBOSOMES, which are the sites of PROTEIN synthesis in the cell.

palaeobotany Study of ancient plants that have been preserved by carbonization, waterlogging or freezing. Some plants have been preserved almost intact in frozen soils and in AMBER. Pollen is also very resistant and has characteristic patterns for identification. The earliest land plants grew in the CAMBRIAN period, but some water plants flourished earlier. Such plants provide important evidence of prehistoric climates and ECOSYSTEMS.

Palaeocene Geological epoch that extended from about 65 to 55 million years ago. It is the first epoch of the TERTIARY period in the CENOZOIC era, when the majority of the DINOSAURS had died out and small herbivores and ungulates were flourishing. Primates and rodents evolved towards the end of the Palaeocene epoch. *See also* GEOLOGICAL TIME

palaeoclimatology Study of prehistoric CLIMATES. Records of past climates are found in SEDIMENTARY ROCKS, in cores taken through deep layers of ice, and in fossil-bearing cores from the beds of seas and lakes. From such evidence, climatologists have discovered that the Earth is subject to alternate periods of cold, called glacials or ICE AGES, and warmth, called INTERGLACIALS. During the last two million years there have been 17 ice ages.

palaeogeography Science that studies ancient GEOGRAPHY. It seeks to determine the physical and biological conditions of the Earth throughout GEOLOGICAL TIME. Palaeogeography also seeks to determine how conditions in one period led to conditions in the next and succeeding periods. Much of the data needed for good analysis is lacking, and most studies in this area are general.

Palaeolithic (Old Stone Age) Earliest stage of human history, from *c.*2 million years ago until between 40,000 and 8,000 years ago. It was marked by the use of stone tools and covers the period of HUMAN EVOLUTION from *Homo habilis* to *Homo sapiens*. It was followed by the NEOLITHIC.

palaeomagnetism Study of changes in the direction and intensity of Earth's MAGNETIC FIELD in GEOLOGICAL TIME. It is important in the investigation of the theory of CONTINENTAL DRIFT. Since the "magnetic memory" of rocks is measurable, this determines their orientation in relation to magnetic north at the time of their solidification. The Earth's polarity has reversed at least 20 times in the past 4 to 5 million years; earlier changes cannot at present be determined. The gross displacement of large rock formations as measured by their magnetic qualities can be explained by SEAFLOOR SPREADING. *See also* MAGNETIC FIELD REVERSAL

palaeontology Study of the FOSSIL remains of plants and animals. Evidence from fossils is used in the reconstruction of ancient environments and in tracing the EVOLUTION of life on Earth.

Palaeozoic Second era of GEOLOGICAL TIME, after the PRECAMBRIAN era, lasting from 590 million to 248 million years ago. It is sub-divided into six periods: CAMBRIAN, ORDOVICIAN, SILURIAN, DEVONIAN, CARBONIFEROUS and PERMIAN. Invertebrate animals evolved hard skeletons capable of being preserved as fossils in the Cambrian, fishlike vertebrates appeared in the Ordovician, amphibians emerged in the Devonian and reptiles in the Carboniferous.

palate Roof of the mouth, comprising the bony front part known as the hard palate and the softer fleshy part at the back known as the soft palate.

palladium (symbol Pd) Silver-white TRANSITION ELEMENT, first discovered in 1803. Malleable and ductile, palladium is found in nickel ores associated with PLATINUM, to which it is chemically similar. It does not tarnish or corrode and is used in ALLOYS for ELECTROPLATING, surgical instruments, dentistry, jewellery and for CATALYTIC CONVERTERS. Properties: at.no. 46; r.a.m. 106.4; r.d. 12.02; m.p. 1,552°C (2,826°F); b.p. 3,140°C (5,684°F); most common isotope ^{106}Pd (27.3%).

palmitic acid (hexadecanoic acid) FATTY ACID

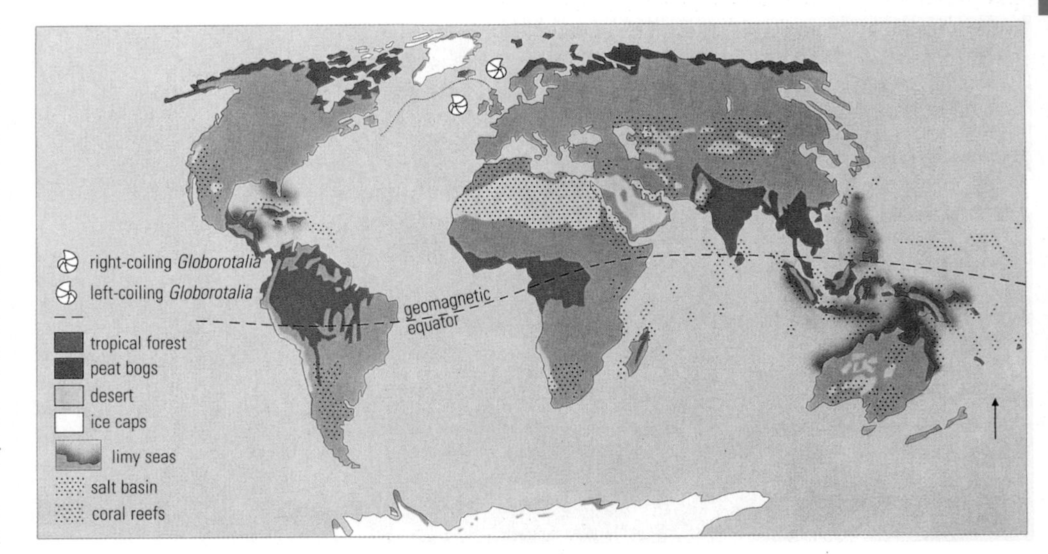

▲ **Palaeoclimatology** is the study of ancient climates through traces left in contemporary rocks. The present-day formations of such preservable climate-related features is shown. The foraminiferan *Globorotalia* is an indicator of sea temperature. It coils right in warm waters and left in cold waters. Coral reefs and major carbonate deposits are both typical of warm, shallow seas. Common to desert environments are evaporite deposits (salt basins) and reddish-hued sandstones. The lush plant life of tropical forests and swamps is the raw material from which coal is formed. Ice sheets groove and scratch the face of rocks and leave characteristic deposits of glacial till, and peat bogs are typical of the tundra environment along the fringes of the ice caps.

which is a major constituent of vegetable oils (particularly palm oil) and animal fats. Together with STEARIC ACID, it is the fatty acid most often present in soaps. Palmitic acid is a SATURATED COMPOUND, containing only single bonds, and has the formula $CH_3(CH_2)_{14}COOH$.

palynology Study of SPORES, SEEDS and POLLENS; it is a part of such disciplines as archaeology, PALAEOGEOGRAPHY and PALAEONTOLOGY. Studies of pollen in lake sediments have provided much information about the vegetation, CLIMATE and environmental conditions in past ages. Palynology has also helped to deduce the patterns of land use and settlement of primitive humans.

pancreas Elongated gland lying behind the STOMACH to the left of the mid-line. It secretes pancreatic juice into the small INTESTINE to aid digestion. Pancreatic juice contains the enzymes AMYLASE, TRYPSIN and lipase. The pancreas also contains a group of cells known as the ISLETS OF LANGERHANS, which secrete the hormones INSULIN and GLUCAGON concerned in the regulation of blood-sugar level. *See also* DIABETES

Pangaea Name for the single supercontinent that formed about 240 million years ago and which began to break up at the end of the TRIASSIC period. It was surrounded by Pantalassa, the ancestral Pacific. Using calculations based on computer data, present land masses plus their continental shelves can be fitted together into this one continent. Pangaea was first written about by Alfred WEGENER in his theory of CONTINENTAL DRIFT. *See also* GONDWANALAND; LAURASIA

pantothenic acid VITAMIN of the B complex involved in the synthesis of ACETYL COENZYME A, itself concerned in the oxidation of fats, carbohydrates and several amino acids. A common vitamin, pantothenic acid is found in cereals, yeast, liver and egg yolks.

paper Sheet or roll of compacted CELLULOSE fibres with a wide range of uses including packaging, writing, wall covering and clothing. The word "paper" derives from papyrus, the plant that the Egyptians used at least 5,500 years ago to make sheets of writing material. The papyrus reed was soaked, slit into strips, laid at right-angles and pounded and pressed into a sheet. The modern process of manufacture originated about 2,000 years ago in China and consists of reducing wood fibre, straw, rags and grasses to a pulp by the action of an ALKALI, such as SODIUM HYDROXIDE (CAUSTIC SODA). The LIGNIN and other non-cellulose material is then extracted and the residue is bleached. After washing and the addition of a filler to provide a smooth and flat surface, the pulp is allowed to settle on a wire mesh to form a thin sheet and dried. Papers of better quality are made from chemical pulps prepared in this way, but newsprint and other cheap papers are mechanical pulps made without chemical treatment and consist of wood fibres.

paper chromatography Type of CHROMATOGRAPHY in which absorbent paper is used. The mixture to be analysed is dissolved and a drop of the solution placed near one edge of sheet of paper. The sheet is suspended vertically with that edge dipping into a trough of SOLVENT. As the solvent rises up the paper by capillary action, the components in the spot move with the solvent front at different rates, depending on their composition. When the paper is dried, the components appear as a line of spots, which may need spraying with a reacting substance to colour them and reveal their presence. The distance a substance moves in a given time is characteristic of its identity.

papilla Small, conical protuberance found in mammals and plants. In mammals papillae project into the EPIDERMIS from the DERMIS; they are found in many parts of the body, including the mammalian INNER EAR and on the surface of the TONGUE.

Papin, Denis (1647–1712) French physicist and inventor who helped to develop the STEAM ENGINE. He invented (1679) a steam digester – a pressurized container which demonstrated that increased pressure raises the BOILING POINT of a LIQUID.

Pap test Alternative name for CERVICAL SMEAR

parabola Mathematical curve, a CONIC section traced by a point that moves so that its distance from a fixed point, the focus, is equal to its distance from a fixed straight line, the directrix. It may be formed by cutting a cone parallel to one side. The general equation of a parabola is $y = ax^2 + bx + c$ where a, b and c are constants. Any projectile fired from Earth, falling under the influence of GRAVITATION, follows a parabolic trajectory.

paraboloid Mathematical solid figure in which all sections parallel to the AXIS of symmetry are PARABOLAS, and sections at an angle to this axis are other CONIC sections, such as ELLIPSES, HYPERBOLAS or CIRCLES.

Paracelsus (1493–1541) Swiss physician and alchemist, b. Philippus Aureolus Theophrastus Bombastus von Hohenheim. He became professor of medicine at Basle, where his lectures discredited

PAPER PRODUCTION

Ways of making paper have changed little in 2,000 years. A suspension of cellulose fibres is made by beating the fibres in water, then separating and soaking them. The wet sheet is pressed and heated to remove the water and then further refined. The modern paper-making process is shown here in a diagrammatic and simplified flow-chart. Pre-cut logs arrive at the paper mill. They then pass to the debarker, which has cutters that penetrate the bark and force it off without damaging the timber. The wood may then go to the chipper. This machine has rotating knives that cut wood into pieces about 3mm (0.125in) thick. From there it goes to the Kamyr digester. Here treatment with boiling chemicals yields a "chemical pulp" from which chemicals are removed at the extractor. Alternatively, debarked logs may then pass to the mechanical grinder. The two streams meet in the bleacher, pass to the Hi-Lo pulper and from there to further treatment in the Jordan refiner. De-inked and bleached waste paper is pulped in the hydrapulper by means of a spinning multivane motor and joins the other pulp at the refiner. All the pulp moves on to the selectifier, which is a pressurized sieve, and into the headbox. There the pulp is adjusted for consistency and fed at a controlled rate through a sluice gate on to a fine phosphor-bronze wire-mesh which is travelling at high speed. At this point suction boxes extract most of the water and the paper then forms a web. The drained water is filtered out; both it and the recovered pulp may be recycled. The web is pressed to the required thickness in the press rolls and then dried over drying rolls. It is given its finish in the calender rolls. The paper cuttings are not wasted but are fed back into the hydrapulper. Finally the completed paper roll is ready for collection and use.

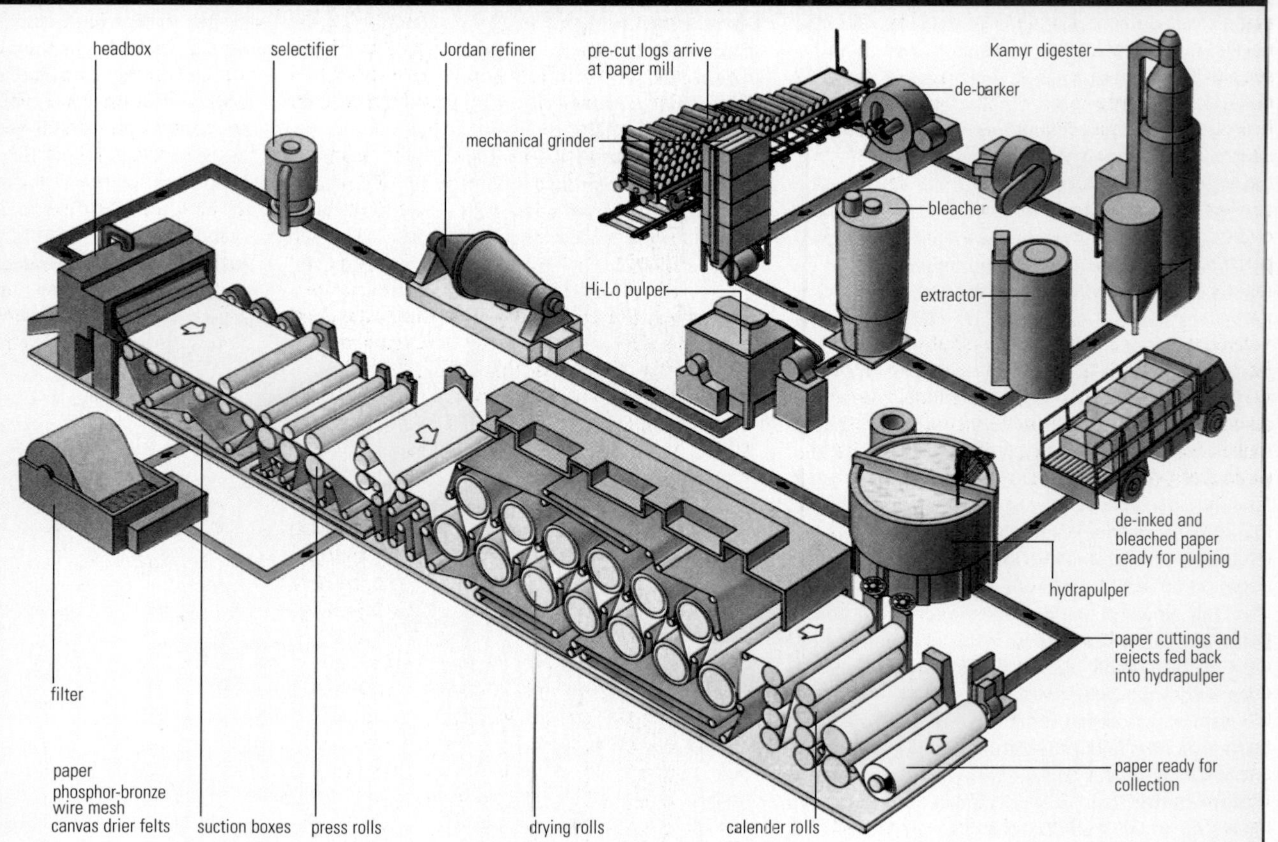

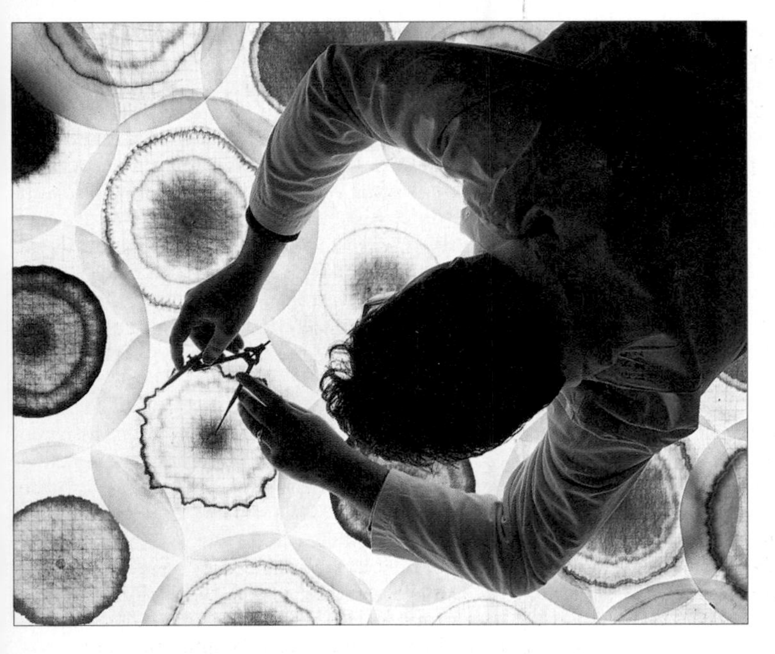

◄ **paper chromatography** A scientist studying a selection of paper chromatograms of industrial dyes. Paper chromatography involves placing a small amount of the substance under investigation on a piece of filter paper, then slowly dripping a solvent into the centre of the paper. The solvent then spreads out over the paper by capillary action, carrying with it the components of the substance at differing rates. The distance travelled along the paper by each component during the time of the experiment is a measure of this characteristic transport rate and may be used to identify each component.

past and contemporary medicine and were preceded by the burning of the works of GALEN and Avicenna (980–1037). According to Paracelsus, the human body is primarily composed of salt, sulphur and mercury, and it is the separation of these elements that causes illness. He introduced mineral baths and made opium, mercury, lead and various minerals part of the pharmacopoeia. According to his enemies, Paracelsus died in Salzburg after a drunken bout. Others say he was thrown down a hill by hirelings of jealous apothecaries.

paracetamol (acetaminophen) ANALGESIC drug that lessens pain and is also effective in reducing FEVER. It is used to treat mild to moderate pain, such as headaches, toothache and rheumatic conditions, and is particularly effective against musculoskeletal pain. Paracetamol is often used as an alternative to ASPIRIN because it is less irritating to the lining of the STOMACH. However, it lacks the anti-inflammatory action of aspirin. Paracetamol overdosage can cause LIVER damage and death.

paraffin (kerosene) Common, domestic fuel that is mainly a mixture of ALKANE hydrocarbons. It is a product of the distillation of PETROLEUM. Less volatile than petrol, paraffin is used also as a fuel for jet aircraft. Paraffin wax is a white, translucent, waxy substance consisting of a mixture of solid alkanes obtained by solvent extraction. It is used to make candles, waxed paper, polishes and cosmetics.

paraldehyde (ethanal trimer, $C_6H_{12}O_3$) Colourless liquid formed by the POLYMERIZATION of ETHANAL (acetaldehyde) by the action of SULPHURIC ACID. It regenerates ethanal on heating with dilute acids. It has a pleasant odour but disagreeable taste, and is used mainly as a SEDATIVE.

parallax Angular distance by which a celestial object appears to be displaced with respect to more distant objects, when viewed from opposite ends of a baseline. It is used as a measure of an object's distance. The parallax of a star (**annual** parallax) is the angle subtended at the star by the mean radius of the Earth's ORBIT (one ASTRONOMICAL UNIT); the smaller the angle, the more distant the star. *See also* PARSEC

parallel circuit ELECTRIC CIRCUIT in which the components are interconnected so that the CURRENT divides between them and passes through all of them at once. The total RESISTANCE, R, of a number of RESISTORS connected in parallel is given by the expression $1/R = 1/R_1 + 1/R_2 + 1/R_3 + ...$, where R_1, R_2 and R_3 are the resistances of the component resistors. The total CAPACITANCE, C, of capacitors in parallel is given by $C = C_1 + C_2 + C_3 + ...$, where C_1, C_2 and C_3 are the capacitances of the individual capacitors. *See also* SERIES CIRCUIT

parallelogram Quadrilateral (four-sided plane figure) having each pair of opposite sides parallel. Both the opposite sides and the opposite angles of a parallelogram are equal. Its area is the product of one side and its perpendicular distance from the opposite side. A parallelogram with all four sides equal is called a RHOMBUS.

parallelogram of vectors Method of calculating the sum of two VECTOR quantities. The direction and size of the sum of the vectors is ascertained by trigonometry or scle drawing. The vectors are represented by two adjacent sides of a PARALLELOGRAM and the sum is the diagonal through their point of intersection.

parallel processing In computing, a technique that allows the simultaneous execution of two or more tasks. Multitasking is performed by dedicated MICROPROCESSORS in a COMPUTER.

paramagnetism Type of MAGNETISM displayed by such metals as platinum and magnesium, which, when placed in a MAGNETIC FIELD, are magnetized parallel to the field by an extent proportional to the strength of that field. This effect is much weaker than that in ferromagnetic materials such as iron, cobalt and nickel. *See also* DIAMAGNETISM; FERROMAGNETISM

parasite Organism that lives on or in another organism (the HOST) upon which it depends for survival; this arrangement may be harmful to the host. Parasites occur in many groups of plants and in virtually all major animal groups. A parasite that lives in the host is called an **endoparasite**; a parasite that survives on the host's exterior is an **ectoparasite**. Many parasites, such as PROTOZOA, fleas and worms, carry disease or cause sores or lesions which may become infected. The European cuckoo and cowbird rely on other birds to rear their young, and are therefore considered "brood parasites". In **parasitoidism**, the relationship results in the death of the host. For example, various flying insects, such as the ichneumon flies, lay their eggs on or in a host which becomes food for the insect LARVAE. A **hyperparasite** is one that parasitizes another parasite. *See also* COMMENSALISM; MUTUALISM; SYMBIOSIS

parasympathetic nervous system One of two parts of the AUTONOMIC NERVOUS SYSTEM, the other is the SYMPATHETIC NERVOUS SYSTEM. Both are involved with the action of INVOLUNTARY MUSCLES. The parasympathetic nervous system controls muscles that prepare the body for a relaxed state, for example, by decreasing heart rate and aiding digestion by encouraging PERISTALSIS. The sympathetic nervous system does the reverse.

parathyroid glands Four small endocrine glands, usually embedded in the back of the THYROID GLAND, that secrete a HORMONE to control the level of calcium and phosphorus in the BLOOD. Over-production of parathyroid hormone causes loss of calcium from the bones to the blood; a deficiency causes tetany (involuntary muscle spasm). *See also* ENDOCRINE SYSTEM

parathyroid hormone (PTH) HORMONE secreted by the PARATHYROID GLAND. It helps to regulate calcium levels in the BLOOD by promoting the release of calcium from bones into the blood, inhibiting secretion in the kidneys and increasing reabsorption of calcium and magnesium from the intestines.

Paré, Ambroise (*c*.1517–90) French physician regarded as the founder of modern SURGERY. In 1537 Paré was employed as an army surgeon and in 1552 became surgeon to Henry II, one of four French kings he served during his lifetime. As an army surgeon, he introduced new methods of treating wounds, described in his book *The Method of Treating Wounds Made by Harquebuses and Other Guns* (1545), and revived the practice of tying ARTERIES during surgery instead of cauterizing them.

parenchyma Soft tissue made up of nonspecialized, thin-walled CELLS, either spherical or blunt-edged in shape, often with spaces between the cells. It is one of the chief tissues of plant stems, leaves and fruit pulp; it stores nutrients and water, and helps to support the plant. In animals, parenchyma is similar loose, connective, indeterminate tissue. It packs the spaces between the organs of simple animals such as worms.

parity (symbol *P*) In physics, term used to denote space-reflection symmetry. The principle of conservation of parity states that physical laws are the same in both a left- and right-handed COORDINATE SYSTEM. This was regarded as inviolable until 1956, when Chen Ning YANG and Tsung-Dao LEE showed that it was transgressed by WEAK NUCEAR FORCES between ELEMENTARY PARTICLES (*See also* SPIN). Parity is also used in INFORMATION THEORY to denote a property of sequences of digits that can be employed in message transmission to detect errors.

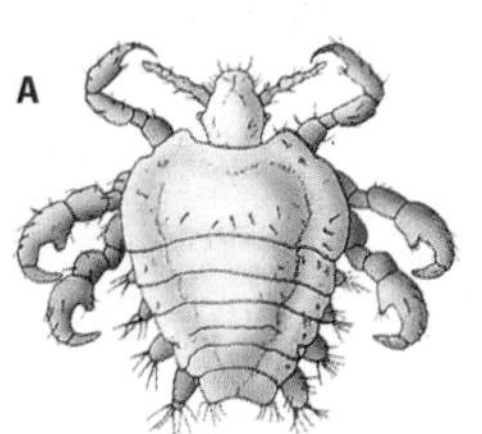

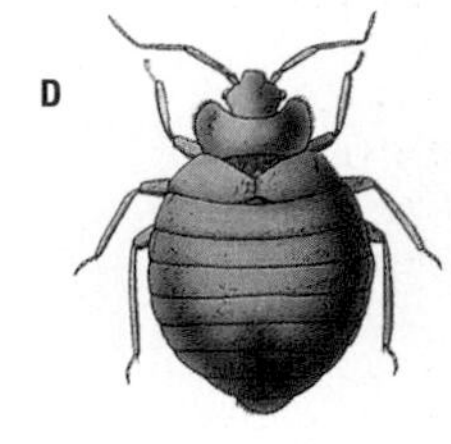

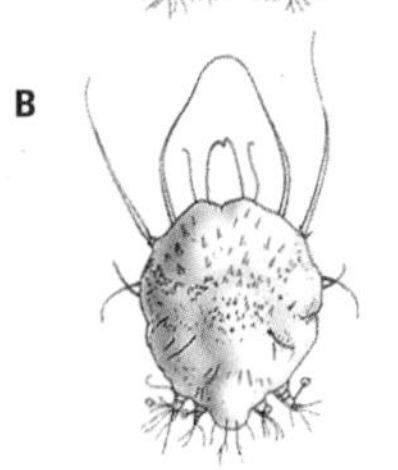

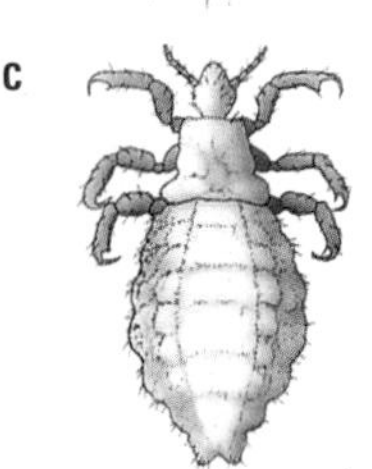

▲ **parasite** Some external parasites live either on, or nearby, humans. The crab louse (A) lives on areas of the body with widely spaced coarse hair. The acarus mite (B) is just about visible to the human eye, and is responsible for causing scabies. The body louse (C) is one of the more dangerous parasites. It is a carrier of epidemic typhus. The common bedbug (D) is found all over the world. During the day it is inactive, but at night it finds humans and sucks their blood.

P

Parkinson's disease Degenerative brain disease characterized by tremor, muscular rigidity and poverty of movement and facial expression. It arises from a lack of DOPAMINE, a NEUROTRANSMITTER that is essential for the control of voluntary movement. Slightly more common in men, it is rare before the age of 50. Foremost among the drugs used to control the disease is L-DOPA.

parotid gland One of two SALIVARY GLANDS located one on either side of the mouth just in front of the ear. With ducts opening on the inside of the cheek, it is the gland that becomes swollen during an attack of MUMPS.

parsec (*pc*) Distance at which a star would have a PARALLAX of one second of arc; equivalent to 3.2616 light-years, 206,265 astronomical units, or 30.857×10^{12} km.

parthenogenesis In biological reproduction, development of a female sex cell or GAMETE without FERTILIZATION. Since there is no involvement of a male gamete, it leads to the production of offspring that are genetically identical to the mother. This process produces little VARIATION and occurs naturally among some invertebrate animals, such as aphids and bees. The same process occurs in a few plants, such as dandelions, and is termed **parthenocarpy**.

partial differential equation Type of DIFFERENTIAL EQUATION used when a FUNCTION depends on more than one INDEPENDENT VARIABLE. For example, a wave in two dimensions has an amplitude (height) U which depends on time t and also on the two distance measurements x and y along mutually perpendicular axes. The differential equation representing the wave is

$$\partial^2 U/\partial x^2 + \partial^2 U/\partial y^2 = 1/c^2\, \partial^2 U/\partial t^2$$

where c is the wave's velocity. When solved, the solution U will give the amplitude of the wave at any point (x, y) and at any time. Symbols such as $\partial^2 U/\partial x^2$ are called **partial derivatives**. These equations are used extensively in physical science.

▲ **Pasteur** Portrait of the French chemist and microbiologist Louis Pasteur. Pasteur is considered the greatest physician of all time and "father" of the germ theory of diseases. He believed that diseases were caused by tiny organisms which were transmitted from individual to individual. In his researches he developed two vaccines: the first against anthrax, a major disease of cattle and the second against rabies. Pasteur also recognized that wine and beer, under heating, could be preserved from souring. This process is now known as pasteurization and widely applied in the food industry.

partial eclipse *See* ECLIPSE

particle accelerator *See* ACCELERATOR

particle physics Branch of physics that studies ELEMENTARY PARTICLES and interactions between them. Particle physics and NUCLEAR PHYSICS developed simultaneously, each contributing to the other. Physicists now recognize more than 300 different elementary particles, which can be grouped in various ways, although they are all composed of a few basic building blocks. The particles that are currently regarded as fundamental can be grouped into the LEPTONS, the QUARKS and the carriers of the FUNDAMENTAL FORCES, such as GLUONS and PHOTONS. Particles termed HADRONS are made up of two or more quarks. For example, PROTONS and NEUTRONS are made up of three quarks, and MESONS are made up of two quarks. Most elementary particles are unstable and decay into other particles.

Pascal, Blaise (1623–62) French mathematician and philosopher. A prodigy, he had written a book on CONIC sections by the time he was 17. Later, with Pierre de FERMAT, Pascal laid the foundations of the mathematical theory of PROBABILITY. He also contributed to CALCULUS and HYDRODYNAMICS, devising **Pascal's law** (1647). This stated that the PRESSURE applied to an enclosed FLUID is transmitted equally in all directions and to all parts of the enclosing vessel, if pressure changes due to the weight of the fluid can be neglected. In 1655 he retired from science to devote himself to religious and philosophical writing, of which *Pensées* is his best-known work. The SI unit of pressure is named after him.

pascal (symbol *Pa*) SI unit of pressure. It is equal to a pressure of 1 NEWTON per sq m, equivalent to 10 DYNES per sq cm, or 1.45×10^{-4} lb per sq in. It is named after Blaise PASCAL.

Paschen-Back effect Phenomenon of the splitting of a line of the SPECTRUM that is observed when atoms are subject to a STRONG NUCLEAR FORCE. It was discovered (1912) by Louis Paschen (1865–1947) and Ernest Back (1881–1959). *See also* ELECTRON; PRECESSION; SPIN; ZEEMAN EFFECT

Pasteur, Louis (1822–95) French chemist, one of the founders of MICROBIOLOGY. He made important contributions to chemistry, bacteriology and medicine. He discovered that microorganisms can be destroyed by heat, a technique now known as PASTEURIZATION and used to destroy harmful microorganisms in food. Pasteur also discovered that he could weaken certain disease-causing microorganisms – specifically those causing ANTHRAX in animals and rabies in man – and then use the weakened culture to vaccinate individuals against the disease.

pasteurization Controlled heat treatment of food to kill bacteria and other microorganisms, discovered by Louis PASTEUR in the 1860s. Milk is pasteurized by heating it to 72°C (161.6°F) and holding it at that temperature for 16 seconds. Ultrapasteurization is now used to produce UHT (ultra-heat-treated) milk; the milk is heated to 132°C (270°F) for one second to provide a shelf-life of several months.

patella (knee-cap) Large, flattened, roughly triangular bone just in front of the joint where the FEMUR and TIBIA are linked. It is surrounded by bursae (sacs of fluid), which cushion the joint.

pathogen Microorganism that causes disease in plants or animals. Animal pathogens are most commonly BACTERIA and VIRUSES, while common plant pathogens also include FUNGI.

pathology Study of diseases, their causes and the changes they produce in the CELLS, tissues and organs of the body. Although pathologists are not involved directly in treating patients, their work is of immense importance to the understanding and treatment of disease.

Pauli, Wolfgang (1900–58) US physicist, b. Austria. He studied QUANTUM THEORY under Niels BOHR. In 1925 Pauli formulated his EXCLUSION PRINCIPLE, which states that each FERMION (such as an ELECTRON or QUARK) has a unique quantum state. IN 1931 he correctly predicted the existence of the neutrino. Pauli received the 1945 Nobel Prize for physics for his achievements in quantum physics.

Pauling, Linus Carl (1901–94) US chemist who won both the Nobel Prize for chemistry and the Nobel Peace Prize. He applied QUANTUM THEORY to the field of chemistry. His early work on the application of WAVE MECHANICS to molecular structure, detailed in his book *The Nature of the Chemical Bond* (1939), led to the Nobel Prize for chemistry in 1954. In the 1950s Pauling worked on the structure of PROTEINS, and his suggestion that the MOLECULES of DNA were arranged in a helical structure anticipated the findings of Francis CRICK and James WATSON. He was known also for advocating massive doses of vitamin C to treat the common cold and for general health. A leading figure in the campaign for nuclear disarmament, Pauling was awarded the 1962 Nobel Peace Prize.

Pavlov, Ivan Petrovich (1849–1936) Russian neurophysiologist. His early work centred on the physiology and neurology of digestion. In 1904 Pavlov received the Nobel Prize for physiology or medicine. He is best known, however, for his classical (Pavlovian) CONDITIONING of behaviour in dogs. His major works are *Conditioned Reflexes* (1927) and *Lectures on Conditioned Reflexes* (1928).

Pavo (Peacock) Constellation of the s sky. Its brightest star, Alpha, has magnitude 1.9.

pc *See* COMPUTER

PCB Abbreviation of POLYCHLORINATED BIPHENYL

Peano, Giuseppe (1858–1932) Italian mathematician. He studied at the University of Turin, later becoming a professor there. Peano's main achievement was his AXIOMATIC METHOD for defining the natural NUMBERS, a major development in formal ARITHMETIC. Regarded as one of the founders of mathematical LOGIC, he developed a complete system of notation for logic, which was later adopted as standard. The Peano curve, an early example of a space-filling curve, is named after him.

pearl Hard, smooth, iridescent concretion of CALCIUM CARBONATE produced by certain marine and freshwater bivalve MOLLUSCS. It is composed almost entirely of nacre, or mother-of-pearl, which forms the inner layer of mollusc shells. A pearl, the only GEM of animal origin, results from an abnormal growth of nacre around minute particles of foreign matter, such as a grain of sand.

peat Soil composed of dead and decaying remains of vegetation. It is generally dark brown or black in colour, and is thought to be similar to the first stage in the formation of COAL. Peat forms in areas of high rainfall or very poor drainage, and contains a high proportion of water. In very wet conditions, plants do not decay when they die because the conditions are nearly ANAEROBIC, which means that the vegetation is not broken down by BACTERIA. Waterlogged and airless conditions create a very acidic and infertile landscape, which can become very boggy. Only specialized plants, notably SPHAGNUM MOSS, can survive in such areas. Peat is used sometimes as a fuel.

peat moss Decomposed organic matter (HUMUS) obtained from disintegrated SPHAGNUM MOSS (bog moss). The most widely obtainable source of humus, it is dug into soil and added to compost to retain moisture. Its use in horticulture is being reduced as its continuing extraction poses an ecological threat to peat bogs.

pectin Water-soluble POLYSACCHARIDE found in the

PASTEURIZATION

A 30 25 20 15 10 5 0 minutes

60 (140) 65 (150) 70 (160) 75 (170) temperature°C (°F)

B cold water · milk · hot water · 1 · 2 · 3

In the 1860s Louis Pasteur found that spoilage of wine and beer could be prevented by heating beverages to 56°C (133°F). Milk is also rendered free of bacteria by pasteurization. To do this the milk is heated for the required time. The necessary temperature and time is shown in graph A. Usually this is 62°C (144°F) for 30 minutes when pasteurised in a vat, or 73°C (163°F) for 15 seconds when pasteurized continuously. B shows pasteurization diagrammatically. Raw milk enters a pipe (lower left) in a short-time plate pasteurizer. It gains heat from adjacent pipes containing hot milk (1). It is then heated to the required temperature by pipes containing hot water (2). The pasteurized milk is partially cooled as it is pumped round pipes containing the incoming cold milk. It is finally cooled by chilled water (3).

cell walls and intercellular tissue of certain ripe fruits or vegetables. When fruit is cooked, pectin yields a gel that is the thickening agent of jellies and jams.

pectoral Relating to the chest or breast, as in the pectoral muscles, which cover the front of the chest, or the pectoral girdle, the bones of which make up the shoulder.

pedicel In botany, term for a flower stalk. The pedicel attaches an individual flower to the **peduncle** (the main floral axis), often growing out of the AXIL of a BRACT. In zoology, a pedicel is the second joint of an insect's ANTENNA or any other stalk-like appendage.

pedology Scientific study of SOILS. Pedologists divide soils according to their physical and chemical composition. These SOIL HORIZONS are determined by factors such as the presence of organic matter and the extent of drainage. Soil types are classified by such names as PODZOL and CHERNOZEM, reflecting the pioneering work of Russian pedologists.

Pegasus (Flying horse) Conspicuous constellation of the N sky. Its most famous feature, the Square of Pegasus, is made up of four stars one of which is actually part of ANDROMEDA. Beta (Scheast) is a RED GIANT irregular variable. Pegasus contains the GLOBULAR CLUSTER M15.

pegmatite In geology, any very coarse-grained IGNEOUS ROCK, generally light in colour and often of GRANITIC composition. Pegmatites are the chief sources of GEMS, MICA and FELDSPAR.

pelagic Zone of an OCEAN and the marine organisms that inhabit it. Pelagic organisms live anywhere in oceans, seas or lakes, except on the bottom (*see* BENTHOS). They are divided into NEKTON (large fish and whales) and PLANKTON (small plants and animals) on which the nekton feed.

pelecypod Alternative name for BIVALVE

Peltier effect Phenomenon of the temperature changes at a junction where an ELECTRIC CURRENT passes from one kind of metal to another. When a current passes through a THERMOCOUPLE (THERMOMETER), the temperature at one junction increases while that at the other decreases, so heat is transferred from one to the other. The effect, used in refrigeration, was discovered in 1834 by the French physicist Jean Charles Peltier (1785–1845). *See also* SEEBECK EFFECT; THERMOELECTRICITY

pelvis (pelvic girdle) Dish-shaped bony structure that supports the internal organs of the lower abdomen in vertebrates, and serves as a point of attachment for muscles that move the limbs or fins. In women, the pelvis is broader than in men, to facilitate childbirth. *See also* ILIUM; ISCHIUM; PUBIS

pendulum Any body suspended at a point so that it swings in an arc. A **simple** pendulum consists of a small heavy mass attached to a string or light rigid rod. A **compound** pendulum has a supporting rod whose mass is not negligible. A simple pendulum was first used to regulate clocks in 1673 by Christiaan HUYGENS. Huygens determined the formula for the period of such a device: $T = 2\pi\sqrt{(l/g)}$, where T is the time for one complete swing, l = length of the wire and g = acceleration of gravity. FOUCAULT'S PENDULUM, devised by Léon FOUCAULT, swings in all directions and was used to demonstrate the Earth's rotation.

peneplain Fairly flat area of land formed over millions of years due to the wearing down of ancient mountains by EROSION. Such long-term erosion, called **denudation**, often results from the action of rivers and may leave areas of more resistant rock, which stand up as monadnocks or residual hills.

Penfield, Wilder Graves (1891–1976) Canadian neurosurgeon, b. USA. In treating epileptic patients, he helped to map the sensory and motor areas of the CEREBRAL CORTEX by stimulating various areas with tiny electrical probes. His works include *Epilepsy and Functional Anatomy of the Human Brain* (1954) and *Speech and Brain Mechanisms* (1960).

penicillin ANTIBIOTIC agent derived from a mould of the genus *Penicillium*. The first antibiotic to be discovered (by Sir Alexander FLEMING in 1928), it was later developed in a soluble form and can now be produced synthetically. It is effective in combating a wide range of bacterial infections. It can produce allergic reactions, including itching and (rarely) potentially fatal shock. Because some microorganisms have become resistant to its action, other antibiotics are often used instead.

penis Male reproductive organ. It contains the URETHRA, the channel through which URINE and SEMEN pass to the exterior, and erectile tissue which, when engorged with blood, causes the penis to become erect during arousal. *See also* SEXUAL REPRODUCTION

Penney, William George, Baron (1909–91) British atomic physicist who helped to develop the first British atomic bomb (*see* NUCLEAR WEAPON). During World War 2 he was engaged in nuclear research for the British government. In 1944 Penney went to the Los Alamos Scientific Laboratory and in 1945 was official observer of the bombing of Nagasaki. In 1964 he became chairman of the UK Atomic Energy Authority.

Pennsylvanian In North America, name given to the later part of the CARBONIFEROUS period.

pentadactyl limb Limb with five digits (fingers or toes), characteristic of four-legged vertebrates. It is generally used for locomotion and is found in amphibians, reptiles, birds and mammals, but may be greatly modified (as in the flippers of seals and whales or the wings of bats). It probably evolved from the fins of primitive fish, which are the only modern vertebrates not to retain it.

pentagon Five-sided plane figure. Its interior ANGLES add up to 540°. For a regular pentagon, one whose sides and interior angles are all equal, each interior angle is 108°.

Penzias, Arno Allan (1933–) US astrophysicist, b. Germany. He and Robert WILSON discovered COSMIC MICROWAVE BACKGROUND, the residual BACKGROUND RADIATION emanating from outer space, which scientists agree supports the BIG BANG theory. In 1964, while working at Bell Laboratories, Penzias and Wilson began monitoring radio waves in the Milky Way. They detected a non-varying signal that corresponds to radiation at a temperature of *c*.3°K, or −270°C (−454°F), corresponding to thermal energy left over from the Big Bang. For this pioneer-

PENTADACTYL LIMB

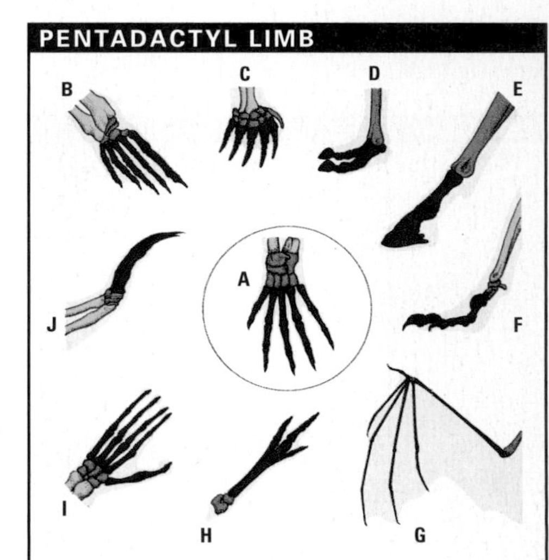

The feet of mammals have evolved in many different ways from the basic mammalian foot (A), possessed by the earliest shrew-like mammal. Seals (B) have developed evenly graduated toes for a webbed paddle. Moles (C) have truncated toes for leverage when digging. The camel's two toes (D) are padded for walking on sand. Horses have a hoof (E) instead of claws, and elongated feet for speed, as has the cheetah (F). Bats (G) have enormously elongated digits to support wings. Kangaroos' toes (H) are designed for hopping. Lemurs (I) and sloths (J) have forelimbs for grasping trees.

P

ing discovery, Penzias and Wilson shared the 1978 Nobel Prize for physics with Peter KAPITZA.

pentode ELECTRON TUBE with five electrodes: cathode, anode and three grids. The grids are made of fine wire mesh and are placed between the anode and the cathode in order to suppress the loss of ELECTRONS. This results in an output signal 1,500 times greater than the input. *See* CATHODE-RAY TUBE

pepsin Digestive ENZYME secreted by GLANDS of the STOMACH wall as part of the GASTRIC JUICE. In the presence of hydrochloric acid, pepsin catalyzes the splitting of PROTEINS in food into POLYPEPTIDES.

peptide Molecule consisting of two or more AMINO ACID molecules linked by bonds between the amino group of one acid and the carboxyl group of another. This type of linkage is called a peptide bond, and peptides containing several amino acids are called POLYPEPTIDES. PROTEINS consist of polypeptide chains having up to several hundred amino acids cross-linked to each other in various ways.

percentage Quantity or amount reckoned as part of a whole, expressed in hundredths; the rate or proportion per hundred. Thus, 75% of a quantity means 75 parts in every 100, or equivalently 3/4 of the total.

Peregrinus, Petrus (active 13th century) French scientist who described the properties of MAGNETISM. A student of Roger Bacon, he distinguished between and named the MAGNETIC POLES, and developed the existing rudimentary magnetic COMPASS into a reliable instrument for NAVIGATION.

perennial Plant with a life cycle of more than two years. It is a common term for flowering HERBACEOUS and woody plants. They include the lily, daisy and iris, and all trees. *See also* ANNUAL; BIENNIAL

perfect number *See* NUMBER, PERFECT

perfume Manufactured substance that produces a pleasing fragrance. The scents of plants such as rose, citrus, lavender and sandalwood are obtained from their ESSENTIAL OILS. These are blended and combined with a fixative of animal origin, such as musk, ambergris or civet. Fixatives add pungency and prevent the more volatile oils from evaporating too quickly. Liquid perfumes are usually alcoholic solutions containing 10–25% of the perfume concentrate; colognes and toilet waters contain about 2–6% of the concentrate.

perianth Outer region of a FLOWER. The perianth includes all the structures surrounding the reproductive organs and usually consists of an outer whorl of SEPALS (CALYX) and an inner whorl of PETALS (COROLLA).

pericardium Double membrane that surrounds the HEART, separating it from the rest of the chest cavity and protecting it from mechanical injury. The outer tough fibrous pericardial layer is separated from the inner membrane by lubricating pericardial fluid.

pericarp In SPERMATOPHYTES (seed-bearing plants), the wall of a ripened FRUIT that is derived from the ovary wall. The tissues of the pericarp may be fibrous, stony or fleshy.

perichondrium In a developing EMBRYO, membrane surrounding the CARTILAGE that eventually becomes BONE. It is well supplied with BLOOD VESSELS and is the source of the osteoblasts (bone-building cells) that invade the embryonic cartilage and lay down the hard bone matrix.

peridot Gem variety of transparent green OLIVINE, a silicate mineral. Large crystals are found on St John's Island in the Red Sea and in Burma.

peridotite Coarse-grained, heavy IGNEOUS ROCK composed of OLIVINE and PYROXENE with small flecks of MICA or HORNBLENDE. It alters readily into SERPENTINE. Rocks that consist mainly of olivine are called DUNITES.

perigee Point in the orbit about the Earth of the Moon or an artificial satellite at which the body is nearest to the Earth. *See also* APOGEE; APSIS

perihelion Point in the orbit of a planet, asteroid, comet or other body (such as a spacecraft) moving around the Sun at which the body is nearest the Sun. *See also* APHELION; APSIS

perimeter Distance around the boundary of a plane closed figure. The perimeter of a circle is its CIRCUMFERENCE; that of a polygon is the sum of the lengths of its sides.

period In astronomy, time taken by one celestial body (such as PLANET or SATELLITE) to complete one revolution around another. It can be measured in several ways, such as the SIDEREAL PERIOD and the SYNODIC PERIOD.

period, geological *See* GEOLOGICAL TIME

periodic law In chemistry, law first stated (1869) by Dmitri MENDELEYEV asserting that the properties of the ELEMENTS are a periodic function of their RELATIVE ATOMIC MASS (R.A.M.). The groupings of the elements based on this law formed the forerunner of the PERIODIC TABLE. From gaps in these groupings, Mendeleyev was able to predict the existence and properties of undiscovered elements. But his table contained anomalies, which were not resolved until Henry MOSELEY's work on X-RAYS discovered that periodicity was related to ATOMIC NUMBER (rather than relative atomic mass) and the later discovery of ISOTOPES by QUANTUM THEORY.

periodic table Arrangement of the chemical ELEMENTS in order of their ATOMIC NUMBERS in accordance with the PERIODIC LAW stated (1869) by Dmitri MENDELEYEV and later modified by Henry MOSELEY. In the modern form of the table, the elements are arranged into 18 vertical columns and seven horizontal periods. The vertical columns, containing groups, are numbered I to VII (sometimes called IA to VIIA) with a final column numbered 0. The metallic TRANSITION ELEMENTS are arranged in the middle of the table between groups II and III. ALKALI METALS are in group I and ALKALINE-EARTH METALS in group II. METALLOIDS and nonmetals are found from groups III to VII, with the HALOGENS in group VII and the NOBLE GASES (inert gases) collected into group 0. The elements in each group have the same number of VALENCE electrons and accordingly have similar chemical properties. Elements in the same horizontal period have the same number of ELECTRON shells. The elements are arranged in the periods in order of increasing atomic number from left to right.

peripheral nervous system All parts of the NERVOUS SYSTEM that lie outside the CENTRAL NERVOUS SYSTEM (CNS). It comprises the 12 pairs of cranial nerves, which principally serve the head and neck region, and 31 pairs of spinal nerves with their fibres extending to the furthermost parts of the body. It carries impulses from RECEPTORS (sensory nerves) to the CNS, and carries back responses from the CNS to EFFECTORS (motor nerves). *See also* AUTONOMIC NERVOUS SYSTEM (ANS)

periscope Optical instrument consisting of a series of MIRRORS or PRISMS that allows a person to

PERIODIC TABLE

KEY: atomic number – 43; atomic symbol – Tc; name of element – Technetium; relative atomic mass – [97] (most stable isotope in brackets)

GROUP I	II											III	IV	V	VI	VII	0
1 **H** Hydrogen 1.00794																	2 **He** Helium 4.0026
3 **Li** Lithium 6.941	4 **Be** Beryllium 9.0122											5 **B** Boron 10.81	6 **C** Carbon 12.011	7 **N** Nitrogen 14.0067	8 **O** Oxygen 15.9994	9 **F** Fluorine 18.998	10 **Ne** Neon 20.179
11 **Na** Sodium 22.9898	12 **Mg** Magnesium 24.305											13 **Al** Aluminium 26.9815	14 **Si** Silicon 28.086	15 **P** Phosphorus 30.9738	16 **S** Sulphur 32.06	17 **Cl** Chlorine 35.453	18 **Ar** Argon 39.948
19 **K** Potassium 39.098	20 **Ca** Calcium 40.06	21 **Sc** Scandium 44.956	22 **Ti** Titanium 47.90	23 **V** Vanadium 50.941	24 **Cr** Chromium 51.996	25 **Mn** Manganese 54.9380	26 **Fe** Iron 55.847	27 **Co** Cobalt 58.9332	28 **Ni** Nickel 58.70	29 **Cu** Copper 63.546	30 **Zn** Zinc 65.38	31 **Ga** Gallium 69.72	32 **Ge** Germanium 72.59	33 **As** Arsenic 74.9216	34 **Se** Selenium 78.96	35 **Br** Bromine 79.904	36 **Kr** Krypton 83.80
37 **Rb** Rubidium 85.4678	38 **Sr** Strontium 87.62	39 **Y** Yttrium 88.906	40 **Zr** Zirconium 91.22	41 **Nb** Niobium 92.906	42 **Mo** Molybdenum 95.94	43 **Tc** Technetium [97]	44 **Ru** Ruthenium 101.07	45 **Rh** Rhodium 102.905	46 **Pd** Palladium 106.4	47 **Ag** Silver 107.868	48 **Cd** Cadmium 112.40	49 **In** Indium 114.82	50 **Sn** Tin 118.69	51 **Sb** Antimony 121.75	52 **Te** Tellurium 127.75	53 **I** Iodine 126.9045	54 **Xe** Xenon 131.30
55 **Cs** Caesium 132.905	56 **Ba** Barium 137.34	57–71 Lanthanide Series	72 **Hf** Hafnium 178.49	73 **Ta** Tantalum 180.948	74 **W** Tungsten 183.85	75 **Re** Rhenium 186.207	76 **Os** Osmium 190.2	77 **Ir** Iridium 192.22	78 **Pt** Platinum 195.09	79 **Au** Gold 196.9665	80 **Hg** Mercury 200.59	81 **Tl** Thallium 204.37	82 **Pb** Lead 207.2	83 **Bi** Bismuth 208.98	84 **Po** Polonium [209]	85 **At** Astatine [210]	86 **Rn** Radon [222]
87 **Fr** Francium [223]	88 **Ra** Radium [226]	89–103 Actinide Series	104 **Db** Dubnium [261]	105 **Ha§** Hahnium [262]	106 **Rf** Rutherfordium [263]	107 **Uns** Unnilseptium [264]	108 **Uno** Unniloctium [265]	109 **Une** Unnilenium [266]	110 **Uun** Ununnilium [269]	111 **Unn** Ununnunium [272]							
LANTHANIDE SERIES (rare-earth elements)		57 **La** Lanthanum 138.9055	58 **Ce** Cerium 140.12	59 **Pr** Praseodymium 140.9077	60 **Nd** Neodymium 144.24	61 **Pm** Promethium [145]	62 **Sm** Samarium 150.36	63 **Eu** Europium 151.96	64 **Gd** Gadolinium 157.25	65 **Tb** Terbium 158.9254	66 **Dy** Dysprosium 162.50	67 **Ho** Holmium 164.9308	68 **Er** Erbium 167.26	69 **Tm** Thulium 168.9342	70 **Yb** Ytterbium 173.04	71 **Lu** Lutetium 174.97	
ACTINIDE SERIES (radioactive rare-earth elements)		89 **Ac** Actinium [227]	90 **Th** Thorium 232.0381	91 **Pa** Protactinium 231.0359	92 **U** Uranium 238.029	93 **Np** Neptunium 237.0482	94 **Pu** Plutonium [244]	95 **Am** Americium [243]	96 **Cm** Curium [247]	97 **Bk** Berkelium [247]	98 **Cf** Californium [251]	99 **Es** Einsteinium [254]	100 **Fm** Fermium [257]	101 **Md** Mendelevium [256]	102 **No** Nobelium [254]	103 **Lr** Lawrencium [255]	

§Another proposed name is unnilpentium

◄ **periodic table** The periodic table arranges chemical elements according to their atomic number. The vertical columns, called groups, contain elements with similar chemical properties. The horizontal rows, called periods, are arranged in order of increasing atomic number and all elements in a row have the same number of electron shells. Elements coloured green are non-metals; those coloured orange are metals; yellow are metalloid; blue, purple or pink are transition metals, the purple ones are the rare-earth elements, and the pink ones the transactinide elements.

P

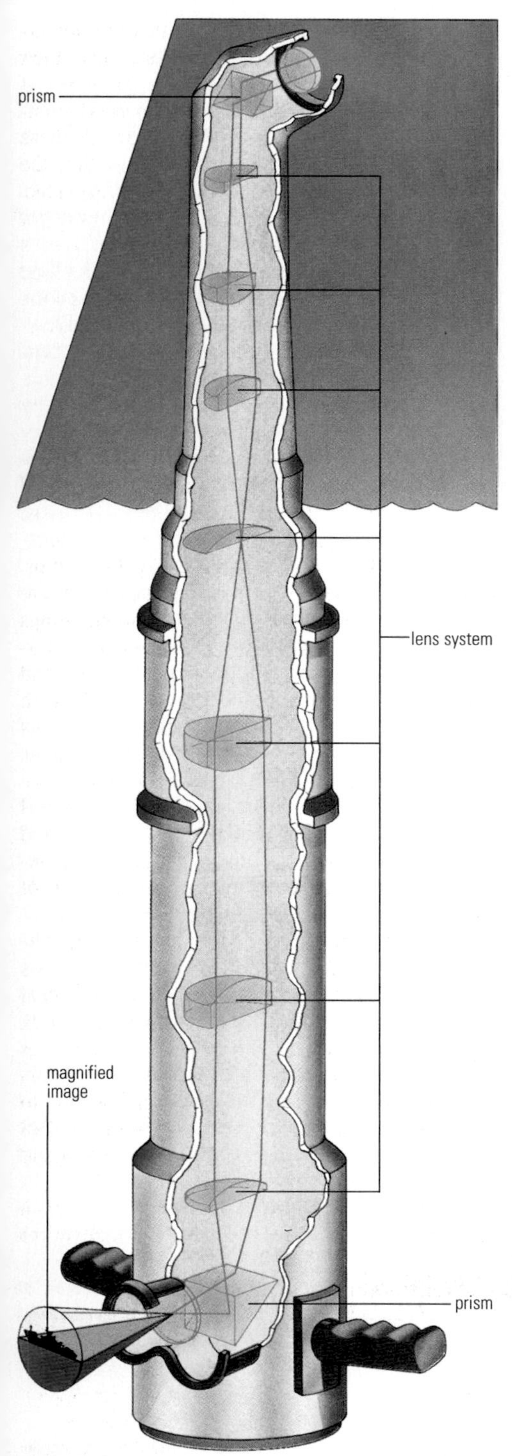

▲ **periscope** In its most simple form, a periscope consists of two mirrors angled at 45°, one above the other so that an image is reflected from the top of the instrument down to the observer at its base. A submarine periscope works on the same principle, but has prisms instead of mirrors and a system of lenses to produce a magnified image or a wide field of view. Optical adjustments can be made with the handles.

view the surroundings from a concealed position by changing the direction of the observer's line of sight. Since World War 1, the periscope has been most commonly associated with submarines.

Perissodactyla Order of mammals characterized by hoofs with an odd number of toes. All HERBIVORES, the only living members of the order are horses, tapirs and rhinoceroses; there are more than 200 extinct forms known from fossils. *See also* ARTIODACTYLA

peristalsis Series of wave-like movements that propel food through the digestive tract. It is caused by contractions of the smooth INVOLUNTARY MUSCLE of the gut wall. The reverse process, antiperistalsis, produces vomiting.

peritoneum Strong membrane of CONNECTIVE TISSUE that lines the body's abdominal wall and covers the abdominal organs. The greater omentum, a fold in the peritoneum containing layers of FAT cells, forms an apron over the INTESTINES. Inflammation of the peritoneum is known as **peritonitis**.

permafrost Land that is permanently frozen, often to a considerable depth. The top few centimetres generally thaw in summer, but the meltwater is not able to sink into the ground because of the frozen subsoil. If the landscape is fairly flat, surface water lies on the ground throughout the summer. Construction work is very difficult, and many methods have been employed in Russia, Canada and Alaska to overcome the problems. *See also* MUSKEG; TUNDRA

permeability In geology, ability of rock to transmit water; it is not directly related to POROSITY. Limestone, for example, is a permeable but nonporous rock; water percolates only through the joints and fissures.

permeability (symbol μ) In physics, ratio of the MAGNETIC FLUX density in a body to the external MAGNETIC FIELD inducing it. The permeability of free space is called the magnetic constant (symbol μ_0) and has the value $4\pi \times 10^{-7}$ henry per metre. The relative permeability of a substance (symbol μ_r) is the ratio of its permeability to the magnetic constant.

Permian Last geological period of the PALAEOZOIC era, lasting from 286 to 248 million years ago. There was widespread geologic uplift, resulting in the formation of PANAGEA. The major climatic characteristics of the period are aridity and glaciation. Many groups of marine invertebrate animals became extinct in the Permian period, but reptiles flourished.

permittivity (symbol ε) In physics, DIELECTRIC property of a material. It is the ratio of the ELECTRIC FLUX density (displacement) in a medium to the intensity of the ELECTRIC FIELD causing it. The permittivity of free space is called the ELECTRIC CONSTANT (symbol ε_0). The relative permittivity of a medium is the ratio $\varepsilon/\varepsilon_0$. It is the ratio of the CAPACITANCE of a capacitor (condenser) in which the material is the dielectric to its capacitance with a vacuum between the capacitor's plates.

peroxide Chemical COMPOUND containing two oxygen atoms bonded to each other and another element. An example is barium peroxide (BaO_2), which yields a solution of HYDROGEN PEROXIDE (H_2O_2), a popular bleach, when treated with an acid. Peroxides are powerful OXIDIZING AGENTS.

perpetual motion There are two theoretical forms of perpetual motion. In the first, a mechanism works indefinitely without an exterior supply of ENERGY. This type of machine violates the first law of THERMODYNAMICS, concerning the CONSERVATION of energy. In the second theory, a machine would take HEAT from a reservoir, such as the ocean, and convert all of it into WORK. This contravenes the second law of thermodynamics, which states that such a process must consume more energy than it can gain from the reservoir. *See also* ENTROPY

PERISTALSIS

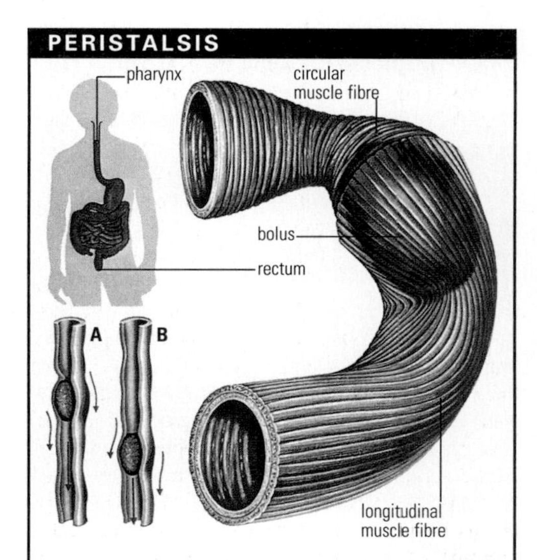

The digestive tract can be regarded as a long muscular tube extending from the pharynx to the rectum. The walls of the tube bear an inner, circular muscle fibre coat and an outer, longitudinal muscle fibre coat. As the ball of food (bolus) formed in the mouth enters the pharynx, a reflex action is initiated. This produces slow, wave-like contractions of the walls of the oesophagus and later along the whole length of the tract. These peristaltic waves involve the contraction of the circular muscle fibres behind the bolus (A) and their relaxation in front of the bolus. Longitudinal muscles provide the wave-like action. The two functions together push the ball down the tract (B). The whole process of peristalsis is an involuntary response.

Perrin, Jean Baptiste (1870–1942) French physicist and chemist. He was awarded the 1926 Nobel Prize for physics for his work on the structure of MATTER and for his investigation of CATHODE RAYS and BROWNIAN MOVEMENT of particles. Perrin wrote several books, including *The Elements of Physics* (1930).

personal computer (pc) *See* COMPUTER

Perseus Prominent N constellation. Alpha Persei is magnitude 1.8. Beta is the prototype eclipsing BINARY, **Algol**. Perseus is a rich constellation, crossed by the Milky Way.

Perspex (tradename for polymethyl methacrylate) Clourless, transparent, solid PLASTIC. It is made by POLYMERIZATION of methyl methacrylate, and is widely used as a substitute for glass.

perturbation Irregularity in the orbital motion of a celestial body, brought about by the gravitational attraction of other bodies. Attractions by PLANETS and SATELLITES cause significant displacements of bodies from where they would be in a purely ellip-

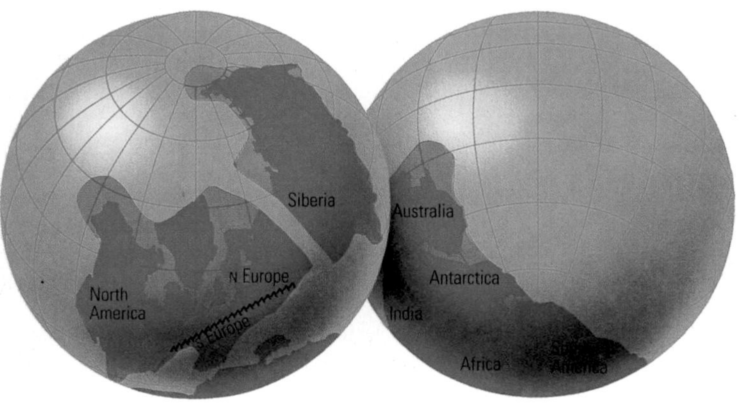

◀ **Permian** Northern Europe has now collided with S Europe, pushing up the Variscan-type fold mountains, and this combined block is moving towards the Siberian plate.

P

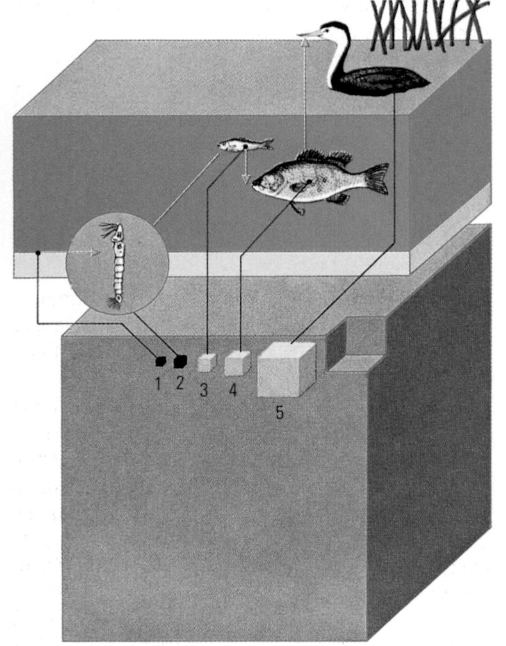

▲ **pesticide** Applied to the land, pesticides, even in small doses, are poisonous to many animals. The concentration of poison increases along the food chain, finally becoming lethal to animals at the end of the chain. A pesticide (1), such as DDT is applied to water at 0.015 parts per million (ppm) to control midge larvae, but the plankton (2) accumulates 5ppm. The fish population (3,4) builds up still higher concentrations, and finally a grebe (5), which feeds on the fish, accumulates as much as 1,600ppm of the pesticide in its body fat – enough to kill the bird.

tical ORBIT. Perturbations are responsible for dramatic changes in the orbits of COMETS, changing long-period orbits into short-period ones. *See also* KEPLER'S LAWS

Perutz, Max Ferdinand (1914–) British biochemist, b. Austria, who studied the X-ray DIFFRACTION of PROTEINS. In 1953 he discovered that adding a heavy ATOM, such as gold or mercury, to each molecule of HAEMOGLOBIN produces a slightly different diffraction pattern. By this means he demonstrated the structure of haemoglobin, for which he shared the 1962 Nobel Prize for chemistry with John C. KENDREW who, by a similar method, discovered the structure of MYOGLOBIN.

pesticide Chemical substance that is used to kill insects, rodents, weeds and other pests. Among pesticides, a HERBICIDE is used for weeds, an INSECTICIDE for insects and a FUNGICIDE for fungal pests. Pesticides are usually harmful chemicals and an important factor in their manufacture is that they should decompose after they have performed their function. Some previously common pesticides, such as DDT, have been shown to be too toxic and long-lasting, so their use is now restricted. *See also* ORGANOPHOSPHATE

petal Part of a FLOWER. The PETALS of a flower are known collectively as the COROLLA. Surrounded by SEPALS, flower petals are usually brightly coloured and often secrete NECTAR and scent to attract the insects and birds necessary for CROSS-POLLINATION. Once fertilization occurs, the petals usually drop off. The petals of wind-pollinated flowers are often small and inconspicuous, allowing the STAMENS and STIGMA to be exposed for POLLINATION.

Petit, Alexis Thérèse (1791–1820) French physicist. He and his brother-in-law, François Arago (1786–1853), worked on the REFRACTION of light in gases and eventually accepted Augustin FRESNEL's new wave theory of light. In 1818 Petit and Pierre DULONG won a prize for determining the SPECIFIC HEAT CAPACITY of solids and for devising measurements of temperature, research that led the following year to DULONG AND PETIT'S LAW.

Petri dish Shallow, flat-bottomed circular dish, often with a tight-fitting cover, used in laboratories mainly for producing cultures of microorganisms. It is named after the German bacteriologist J.R. Petri (1852–1921).

petrification Fossilizing process in which organic material (such as wood) changes into stone. Petrification is caused by mineral-rich water seeping into the empty spaces of dead, buried trees or animals, which eventually become stone. Although petrified remains can be up to 300 million years old, the stone often reproduces the original living material so clearly that the CELL structure is identifiable. The Petrified Forest National Park in Arizona, USA, has many examples of petrified wood. *See also* FOSSIL; MINERALIZATION

petrochemical Chemical substances derived from PETROLEUM or NATURAL GAS. The refining of petroleum is undertaken on a large scale not only for the fuels obtained (PETROL, PARAFFIN, fuel oil and natural gas) but also for the wide range of chemicals that can be derived. These chemicals include the common ALKANES (paraffins) and ALKENES (olefins), BENZENE, TOLUENE and NAPHTHALENE.

petrol (gasoline) Major HYDROCARBON fuel, a mixture consisting mainly of hexane, octane and HEPTANE. One of the products of OIL refining, petrol is extracted from crude oil (PETROLEUM). Frequently other fuels and substances are added to petrol to alter its properties. Most modern cars have high-compression INTERNAL COMBUSTION ENGINES, and the mixture of air and petrol vapour supplied to it can explode too quickly, pushing against a rising instead of a descending piston. This effect is known as engine KNOCK, and to eliminate it many petrol manufacturers add LEAD(IV) TETRAETHYL (an antiknock agent) to slow the rate of combustion. However, this additive has been found to be implicated in atmospheric pollution and causes brain damage, and so many manufacturers omit the additive and instead improve the OCTANE NUMBER of the petrol – a measure of its antiknock properties – by altering the mixture of hydrocarbons in it. *See also* OIL REFINERY

petrol engine Most common type of INTERNAL COMBUSTION ENGINE

petroleum (crude oil) FOSSIL FUEL that is chemically a complex mixture of HYDROCARBONS. It accumulates in underground deposits and the chemical composition of petroleum strongly suggests that it originated from the bodies of long-dead organisms, particularly marine PLANKTON. After death these organisms sank to the ocean floor, to be broken down by bacteria in oxygen-poor conditions into simpler organic materials, including hydrocarbons. Petroleum is rarely found at the original site of formation, but migrates laterally and vertically until it is trapped. Most petroleum is extracted via OIL WELLS from reservoirs in the Earth's crust sealed by upfolds of impermeable rock or by SALT DOMES which form traps. During petroleum refining, the heavier hydrocarbons, which usually have higher boiling points than lighter ones, are distilled (separated from the original crude substance) as the first stage of refining. The next stage is CRACKING, which breaks the heavy hydrocarbons down into more economically useful products, such as PETROL and PARAFFIN. Purification of the various products to remove impurities, such as SULPHUR and NITROGEN compounds, completes the refining process. The most versatile end products are ETHENE and PROPENE, which are widely used in the plastics and chemical industries. *See also* NATURAL GAS; OIL REFINERY

petrology Study of ROCKS, including their origin, chemical composition, and where they are found. Formation of the three classes of rocks – IGNEOUS (of volcanic origin); SEDIMENTARY (deposited by water); and METAMORPHIC (either of the other two changed by temperature and pressure) – are studied.

pewter Any of several silver-coloured, soft alloys that consist mainly of tin and lead. The most common form has about four parts of tin to one of lead, combined with small amounts of antimony and copper. Some food and drink can become contaminated with poisonous lead from pewter utensils, so other alloys have replaced pewter for such items.

pH (abbreviation of potential of hydrogen) Indication of the acidity or alkalinity of a solution. The **pH scale** expresses a range of pH values based on a logarithmic measure of the concentration or, more properly, activity of hydrogen ions. The scale runs from 0 to 14, and a neutral solution, such as pure water, has a pH of 7. A solution is acidic if the pH is less than 7 and alkaline if greater than 7. The pH may be measured with an INDICATOR or meter. The scale was introduced (1909) by S.P. Sørensen (1868–1939).

phaeophyte Any member of the phylum Phaeophyta, which consists of the brown ALGAE. Phaeophytes are part of the kingdom PROTOCTISTA. Classified by some biologists as plants, the simple organisms belonging to this group are mostly marine, found mainly in the intertidal zone of rocky shores. They include familiar seaweeds such as *Fucus* (wracks) and *Ascophyllum* (bladder wrack). The largest, *Macrocystis*, grows to more than 100m (320ft) at up to 0.5m (18in) per day. *Sargassum* forms vast floating masses in the Sargasso Sea in the mid-Atlantic, with their own distinctive communities of animals and microorganisms.

phagocyte Type of LEUCOCYTE (white blood cell) able to engulf other CELLS, such as bacteria. It digests what it engulfs (in a process known as **phagocytosis**) in the defence of the body against INFECTION. Phagocytes also act as scavengers by clearing the bloodstream of the remains of the cells that die as part of the body's natural processes. *See also* MACROPHAGE; MONOCYTE

phalanges Bones in the toes and fingers. In human beings there are 14 phalanges in each hand and foot, 2 in each thumb and big toe and 3 in the remaining digits. They are connected to the METACARPALS in the hand and to the METATARSALS in the foot.

phanerogam Early botanical term for seed-bearing, flowering plants, now called SPERMATOPHYTES. *See also* CRYPTOGAM

Phanerozoic Geological timescale that describes the whole of time that has elapsed since the end of the PRECAMBRIAN period. It has lasted for about 570 million years since life first evolved and thus includes the PALAEOZOIC, MESOZOIC and CENOZOIC eras. *See* GEOLOGICAL TIME

pharmacology Study of the properties of DRUGS and their effects on the body.

pharynx Cavity at the back of the nose and mouth that extends down towards the OESOPHAGUS and TRACHEA. Inflammation of the pharynx, usually caused by viral or bacterial infection, is known as pharyngitis.

phase In astronomy, proportion of the illuminated hemisphere of a body in the Solar System (in particular the Moon or an inferior planet) as seen from Earth. The phase of a body changes as the Sun and the Earth change their relative positions. All the phases of the Moon (new, crescent, half, gibbous and full) are observable with the naked eye. The inferior planets, Mercury and Venus, show phases from a slender crescent to a fully illuminated disc when observed through a telescope. Of the superior planets, Mars can show quite a marked gibbous phase, and Jupiter a very slight gibbous phase, but with Saturn and the planets beyond, no phase is ever discernible. Phase is sometimes expressed as the percentage of the visible disc that is illuminated.

phase In physics, one of the physical states of MATTER. Each homogeneous form is a distinct part within a heterogeneous system and is separated from other parts by boundary surfaces: ice is thus a phase of H_2O, water. A mixture of ice and water is called a two-phase system. Temperature and pressure determine the phase of a substance, and a PHASE DIAGRAM indicates the relationships between phases over a range of temperatures and pressures. A change in the internal ENERGY of a substance causes a phase transition (change in state), as seen in melting or boiling. *See also* BOILING POINT; MELTING POINT

phase In physics, stage in the cycle of an oscillation, such as the wave motion of light or sound waves. This is usually measured from an arbitrary starting point or compared with another motion of the same frequency. Two waves are said to be "in phase" when their maximum and minimum values happen simultaneously. If not, there is a "phase difference", resulting in INTERFERENCE.

phase-contrast microscope Type of MICROSCOPE used for examining transparent specimens. It is widely used in biology for studying CELLS and thin slices of tissue. Light passing through the specimen slows down and becomes out of PHASE with the original light. Differences in thickness or density in the specimen cause DIFFRACTION patterns. A phase-contrast plate converts these to INTERFERENCE patterns, which manifest themselves as large contrasts in brightness of different parts of the microscope image.

phase diagram Graph showing the conditions under which different equilibrium PHASES of substances exist. For example, a curve of MELTING POINT against PRESSURE of a pure solid divides the graph into two regions. Points in one represent temperatures and pressures at which the substance is solid; points in the other represent the liquid conditions. Graphs of composition against temperature are used to indicate such features as SOLUBILITIES and the ranges of stability of alloy phases.

phenacite (beryllium silicate, Be_2SiO_4) Orthosilicate mineral found in PEGMATITES and high-temperature veins. It has hexagonal system rhombohedral crystals, and is either colourless or glassy white, yellow, red or brown in colour. It is sometimes used as a gem. Hardness 7.5–8; r.d. 3.

phenol Group of AROMATIC COMPOUNDS that have one or more HYDROXYL (–OH) groups attached to a BENZENE ring. The simplest of the family is also called phenol or CARBOLIC ACID (C_6H_5OH). Phenols are colourless liquids or white solids at room temperature, with higher melting and boiling points than the parent HYDROCARBONS from which they are derived. Phenol is a major constituent of coal tar. It is used by the chemical and pharmaceutical industries for making such products as aspirin, dyes, fungicides, explosives and as a starting material for nylon and epoxy resins.

phenolphthalein ($C_{20}H_{14}O_4$) Organic compound prepared by a reaction between PHENOL and phthalic anhydride in the presence of SULPHURIC ACID. It is used as a chemical indicator (colourless in acid solutions, red-pink in alkalis), in laxatives and in dyes.

phenotype Physical characteristics of an organism resulting from HEREDITY. Phenotype is distinct from GENOTYPE, since not all aspects of GENETIC make-up manifest themselves.

phenylalanine ($C_6H_5CH_2CH(NH_2)COOH$) Crystalline, soluble AMINO ACID found in PROTEINS. It is essential to human nutrition.

phenylamine *See* ANILINE

phenylketonuria (PKU) Hereditary condition in which the amino acid PHENYLALANINE cannot be metabolized normally because the ENZYME phenylalanine hydroxylase is absent or inactive. Infants with PKU excrete phenylalanine in the urine. Special diets are used to treat the condition.

pheromone Substance secreted externally by certain animals that influences the behaviour of members of the same species. Common in mammals and insects, these substances are often sexual attractants and may be a component of body products such as URINE, or they may be secreted by specific GLANDS.

phillipsite ($KCa(Al_2Si_6O_{16}).6H_2O$) White or reddish ZEOLITE mineral. It occurs in vein formations as monoclinic CRYSTALS. Hardness 4.5–5; r.d. 2.2.

phloem Vascular tissue for distributing dissolved food materials in plants. Phloem tissue contains several types of CELLS. The most important are long, hollow cells called **sieve-tube cells**. Columns of sieve tubes are joined end to end, allowing passage of materials from cell to cell. The sieve tubes are closely associated with "**companion cells**", which have dense CYTOPLASM and many MITOCHONDRIA and are thought to produce the energy needed to transport the food substances (*see* ACTIVE TRANSPORT). Phloem may also contain FIBRES, which help to support the tissue. *See also* XYLEM

phlogiston theory Notion advanced by J.J. Becher (1635–82) and popularized in the 18th century by Georg Stahl (1660–1734). It postulated that combustible materials contained an odourless, colourless, weightless material called **phlogiston**; COMBUSTION involved loss of this phlogiston and it thereby became dephlogisticated in turning to calx (ash). The theory was disproved by Antoine LAVOISIER's discovery of the true nature of combustion.

phlogopite ($KMg_3Fe_3AlSi_3O_{10}(OH)_2$) White or brown MICA mineral. It is found in limestones as monoclinic crystals. Hardness 2.5–3; r.d. 2.8.

Phoenix Constellation of the s sky. Its brightest star, Alpha, is magnitude 2.4.

phosgene (carbonyl chloride, $COCl_2$) Colourless, toxic GAS. It was used as a poison gas in World War 1, but is now used in the manufacture of various dyestuffs and resins. Properties: b.p 8.2°C (46.8°F), m.p. −118°C (−180.5°F).

phosphate Chemical compounds derived from PHOSPHORIC(V) ACID (H_3PO_4) and other phosphorus(V) oxyacids, for example salts with the phosphate ion $(PO_4)^{3-}$.

phosphine Phosphorus hydride (PH_3), a colourless, highly poisonous GAS made by the reaction between calcium phosphide and water or dilute acid. Traces of diphosphine impurity make the gas spontaneously flammable. Pure phosphine finds application in doping SEMICONDUCTOR materials.

phospholipid Type of LIPID that contains a PHOSPHORIC ACID group or groups and an alcohol base. Phospholipids are found in egg yolk and brain tissue.

phosphor Substance capable of LUMINESCENCE (storing energy and later releasing it as light). There are two main types: the zinc phosphors, as used on CATHODE-RAY TUBES; and the oxygen type, as used on FLUORESCENT LAMPS. Zinc sulphide is often mixed with cadmium sulphide and a small quantity of metallic phosphates, silicates or fluorides to produce phosphors. *See also* PHOSPHORESCENCE

phosphorescence Form of luminescence in which a substance emits light of one wavelength without associated heat. Unlike FLUORESCENCE, it may persist for some time after the initial excitation. It is a natural as well as artificial phenomenon; in warm climates, for example, the sea often appears phosphorescent at night, as a result of the activities of millions of microscopic ALGAE, which phosphoresce when disturbed by the movements of the water.

phosphoric acid Group of ACIDS, the chief forms of which are phosphoric(V) (orthophosphoric) acid (H_3PO_4), phosphoric(VI) (metaphosphoric) acid (HPO_3) and heptaoxodiphosphoric(V) (pyrophosphoric) acid ($H_4P_2O_7$). Phosphoric acid was discovered in 1770 by Karl SCHEELE and J.G. Gahn. **Phosphoric(V)** acid is a colourless liquid

PHASE DIAGRAM

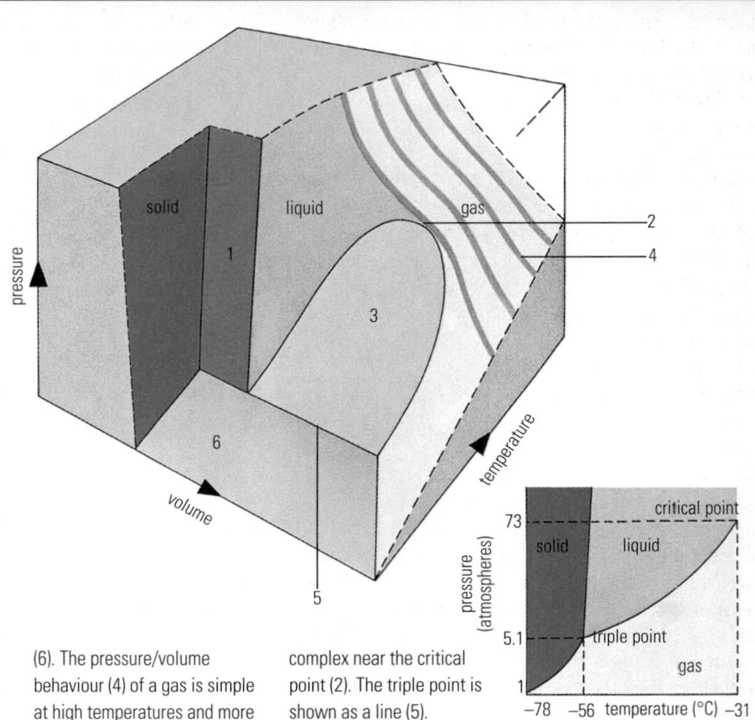

The solid, liquid and gaseous forms of a particular substance (in the small inset drawing, carbon dioxide is shown) can all exist in mutual contact at a particular temperature and pressure, called the triple point. The three lines show how the boiling and melting and sublimation change with pressure (in sublimation a solid turns to gas without melting). Above the critical point the liquid cannot exist at all; only up to that temperature can the gas be liquefied by simply compressing it. In the larger drawing, the changes of volume are included: here the solid, liquid and gas states are symbolized as surfaces, as are the conditions for melting (1), boiling (3) and sublimation (6). The pressure/volume behaviour (4) of a gas is simple at high temperatures and more complex near the critical point (2). The triple point is shown as a line (5).

P

obtained by the action of SULPHURIC ACID on phosphate rock (calcium phosphate); it is used in fertilizers, soaps and detergents. **Phosphoric(VI)** acid is obtained by heating phosphoric(V) acid; it is used as a dehydrating agent. **Heptaoxodiphosphoric(V)** acid is formed by moderately heating phosphoric(V) acid or by reacting phosphorus(V) oxide (phosphorus pentoxide, P_2O_5) with water; it is used as a CATALYST and in metallurgy.

phosphorus (symbol P) Common, nonmetallic element of group V of the periodic table, first discovered in 1669. It occurs as PHOSPHATES in minerals; APATITE is the chief source. The element is used in making PHOSPHORIC ACID for detergents and fertilizers (phosphorus is essential for plant growth). Small amounts are used in insecticides and in matches. Phosphorus exhibits ALLOTROPY; allotropes include the highly reactive white phosphorus, which ignites spontaneously on exposure to air, and the more stable form, red phosphorus. Properties: at.no. 15; r.a.m. 30.9738; r.d. 1.82 (white), 2.34 (red); m.p. 44.1°C (111.38°F) (white); b.p. 280°C (536°F) (white); most common isotope ^{31}P (100%).

phosphorylation Biochemical reaction in which a PHOSPHATE group is introduced into a MOLECULE. Controlled by the ENZYME phosphorylase, it is the initial stage in many natural biochemical processes, such as the conversion of ADENOSINE DIPHOSPHATE (ADP) to energy-rich ADENOSINE TRIPHOSPHATE (ATP). Phosphorylation is also an important reaction in the activation or deactivation of enzymes, often under hormone control.

photocell *See* PHOTOELECTRIC CELL

photochemistry Branch of chemistry concerning photochemical reactions caused by LIGHT or ULTRAVIOLET RADIATION and those reactions that produce light. Examples include PHOTOGRAPHY, PHOTOSYNTHESIS and bleaching by sunlight. Chemical effects can be produced only if light is absorbed by a system. PHOTONS of light are absorbed by reactant MOLECULES, raising them to an electronically excited state that can lead to further reactions. The quantum yield, or efficiency, of a reaction is the number of reactant molecules used, or product formed, per photon absorbed. Photolysis is the decomposition or dissociation of molecules caused by this exposure to light or ULTRAVIOLET RADIATION.

photochromic glass Form of GLASS, used in spectacle lenses and electronic devices, which darkens on exposure to ultraviolet or bright visible light. It fades to its original clear state when the light is removed. SILVER CHLORIDE or silver bromide crystals throughout the glass interact with light to cause this change.

PHOTOCOPYING

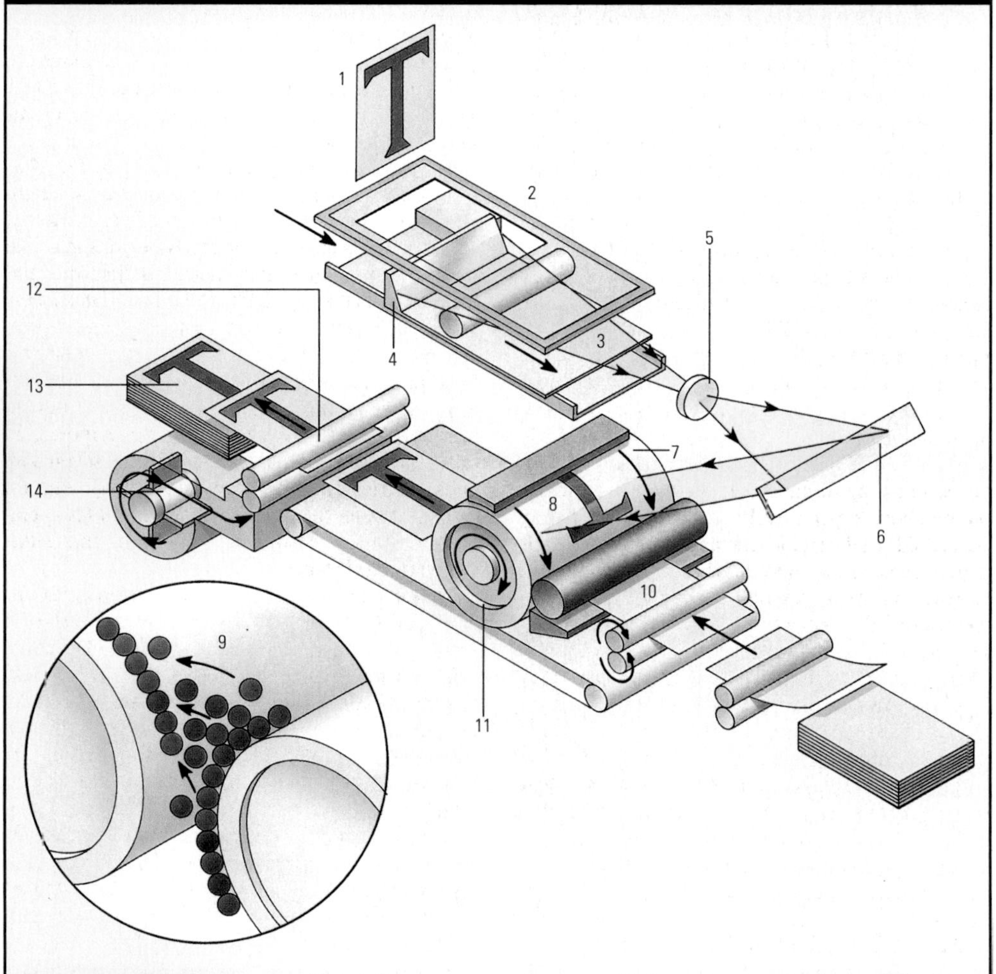

The document to be photocopied (1) is placed face down on the copier (2) and illuminated by a bright, halogen lamp (3). Light is reflected from a moving mirror (4), via a system of lenses (5) and a secondary mirror (6), onto a revolving drum (7), coated in a photosensitive polymer containing selenium. White parts of the document reflect light. Wherever light strikes the drum the overall negative charge of the drum is dissipated (8), resulting in a pattern of charged areas that will correspond to the darker parts of the original document. The revolving drum picks up the positively charged toner particles (9) only where it has a negative charge. A sheet of paper (10) is given a negative charge by a long electrode and pressed against the drum (11), where the toner is transferred to the paper. The copy passes through a set of hot rollers (12), which fuse the toner to the paper (13). Excess heat is expelled by a fan (14), blowing cold air over the rollers.

photocopying Reproduction of the written word, drawings or photographs by machine. In a photocopier, a light shines on the item to be copied, and an optical system forms an image of it. Various techniques may be used to reproduce this image on paper. In a modern, plain-paper copier using XEROGRAPHY, the image is projected onto an electrically charged **drum**, coated with a light-sensitive material such as the element SELENIUM. Light makes such a material conduct electricity, so bright areas of the drum lose their charge. The dark areas, which usually correspond to image detail, retain their charge, and this attracts particles of a fine powder called **toner**. Electrically charged paper in contact with the drum picks up the pattern of toner powder. A heated **roller** fuses the powder so that it sticks to the paper and forms a permanent image. Most machines print in one colour, normally black, but some machines can make full-colour copies.

photodiode SEMICONDUCTOR device that produces an ELECTRIC CURRENT when light shines on it. It consists of a *p-n* junction DIODE with a lens that focuses light onto the junction. It is used to measure light intensity or to detect the presence of light.

photoelectric cell (photocell) Type of electric cell whose operation depends on the amount of light to which it is exposed. It was formerly an ELECTRON TUBE with a photosensitive CATHODE, but now nearly all photoelectric cells are made using two ELECTRODES separated by light-sensitive SEMICONDUCTOR material. Photoelectric cells are used as switches (electric eyes), light detectors (burglar alarms), devices to measure light intensity (PHOTOMETERS, as for cameras) or as power sources (solar cells), most recently in space vehicles.

photoelectric effect Liberation of ELECTRONS from the surface of a material (such as silicon) when ELECTROMAGNETIC RADIATION, such as light, ultraviolet radiation, X-rays or gamma rays, falls on it. The effect can be explained only by the QUANTUM THEORY: PHOTONS in the radiation are absorbed by ATOMS in the substance and enable electrons to escape by transferring energy to them. A PHOTOELECTRIC CELL uses the photoelectric effect, producing electricity when exposed to light.

photoengraving (process-engraving) Process of preparing illustrations for letterpress printing in which the image is transferred by PHOTOGRAPHY onto metal or plastic. It involves two steps: the preparation of a photographic negative copy of the material to be reproduced; and the making of a positive printing plate. Plates are made of zinc, copper or magnesium alloy, coated with a photosensitive solution. The light passing through clear sections of the negative affects the coating, making it insoluble in water. After exposure, the plates are washed, leaving an image formed by the insoluble portions. The non-printing surface is then etched away with an acid.

photogrammetry Use of photographs for measuring distances and areas in SURVEYING. Photographs taken from aircraft or orbiting SATELLITES enable precise measurements to be taken for making maps and charts. The use of infrared film reveals details that are invisible to the naked eye or to ordinary photographic emulsions.

photography Process of obtaining a permanent image of an object, either in black and white or in colour, on treated paper or FILM. A CAMERA is used to expose a film for a set time to an image of the object to be photographed. In black and white photography, the film is covered on one side with an emulsion containing a SILVER halide (silver bromide or silver chloride). Exposure makes the silver compound easily reduced to metallic silver when treated with a **developer**. The action of the developer is to produce a black deposit of metallic silver particles on those parts of the film that were exposed to light, thus providing a "**negative**" image. After fixing in "**hypo**" (thiosulphate) and washing, the negative can be printed by placing it over a piece of sensitized paper and exposing it to light so that the silver salts in the paper are affected in the same way as those in the original film. The dark portions of the negative let through the least light, and the image on the paper is reversed back to a positive. Colour photography works on a similar, but more complex, process. In 1816 Nicéphore NIEPCE produced the first negative on paper and in 1827, working with Jacques Mandé Daguerre, he produced the first known photograph, on metal. In 1819 Sir John HERSCHEL discovered the "hypo". In 1839 Daguerre announced his invention of a direct positive image on a silver plate – the daguerreotype. At the same time, William Henry Fox TALBOT invented the calotype, the first process capable of producing an infinite number of positive prints from a single negative.

photoluminescence Type of luminescence produced by the action of PHOTONS, such as those in infrared, visible or ultraviolet light or in X-rays or gamma rays. Luminescence produced by the action of ELECTRONS is known as ELECTROLUMINESCENCE.

photometer Instrument used to measure the luminous intensity of a light source by comparing it with a light source of known intensity. This is achieved by comparing the positions of the sources when they produce equal illumination on a reference surface. Photometry is the branch of physics concerned with such measurements. Modern photometers usually use PHOTOELECTRIC CELLS. In astronomy, photoelectric photometers measure the brightness of a celestial object by generating an electrical signal that is directly proportional to the intensity of the light from that object falling on it. The unit of luminous intensity is the CANDELA (cd).

photomultiplier Electronic device that converts a light signal or light beam into an equivalent amplified electrical signal. ELECTRONS produced by the PHOTOELECTRIC EFFECT from an illuminated CATHODE, bombard a series of other electrodes and are increased in number at each bombardment. An amplified signal is obtained from the final stage.

photon Quantum of ELECTROMAGNETIC RADIATION, such as LIGHT, which can be considered to consist of streams of photons. The energy of a photon equals the frequency of the radiation multiplied by PLANCK'S CONSTANT. Absorption of photons by atoms and molecules can cause excitation or ionization. A photon is a stable ELEMENTARY PARTICLE of zero charge and spin 1, travelling at the velocity of light. Virtual photons are continuously exchanged between charged particles and are the carriers of the ELECTROMAGNETIC FORCE. *See also* ANTIMATTER; BLACK BODY; COMPTON EFFECT; PHOTOELECTRIC EFFECT; VIRTUAL PARTICLE

photoperiodism Biological mechanism that governs the timing of certain activities in an organism by reacting to the duration of its daily exposure to light and dark. For example, the start of flowering in plants and the beginning of the breeding season in animals are determined by day length. *See also* BIOLOGICAL CLOCK

photoreceptor In animals, a sensory cell (RECEPTOR) or collection of CELLS that react to the presence of light. Most photoreceptors contain a PIGMENT that undergoes a chemical change when it absorbs light, producing a NERVE IMPULSE. The RODS AND CONES in the retina of the EYE are examples of photoreceptors. In plants, photoreceptors are light-detectors containing the blue-green pigment phytochrome. They are involved in many plant processes, such as seed germination, flowering and leaf fall.

photosetting Composition of text to be printed by photographic means, developed in the 1950s as a speedier and more flexible alternative to setting type in metal. The size of type to be set is obtained from a master negative font by photographic reduction or enlargement. Computer units allied to most photosetters enable the text to be set as fast as the operator can key it in. Once programmed, these units automatically control line width, column length, spacing, justification and hyphenation. *See also* LINOTYPE; MONOTYPE; PRINTING; TYPE METAL; TYPESETTING

photosphere Visible surface of the SUN. It is a layer of highly luminous gas *c.*500km (300mi) thick and with a temperature of *c.*6,000K, falling to 4,000K at its upper level, where it meets the CHROMOSPHERE. Because this is the region of the Sun from which all its output of visible light is emitted, the photosphere is the source of the Sun's visible spectrum. The lower, hotter gases produce the continuous emission spectrum, while the higher, cooler gases absorb certain wavelengths. SUNSPOTS and other visible features of the Sun are situated in the photosphere.

photosynthesis Chemical process occurring in green plants, algae and many bacteria, by which water and CARBON DIOXIDE are converted into food and OXYGEN using energy absorbed from sunlight. There are two distinct stages of photosynthesis: LIGHT REACTION and DARK REACTION. The reactions take place in the CHLOROPLASTS. During the first part of the process, light is absorbed by CHLOROPHYLL and splits water into HYDROGEN and oxygen. The hydrogen attaches to a carrier molecule and the oxygen is set free. The hydrogen and light energy build a supply of cellular chemical energy, ADENOSINE TRIPHOSPHATE (ATP). In the dark reaction, hydrogen and ATP, by a process called the CALVIN CYCLE, convert the carbon dioxide into SUGARS, including glucose, and starch. *See also* AUTOTROPH *See feature on page 278*

phototropism (heliotropism) Growth of a plant in response to the stimulus of light, which increases CELL growth on the shaded side of the plant, resulting in curvature towards the source of light. AUXIN hormones are involved in this process. They are produced in the tip of the shoot, and when light shines on the shoot, auxin travels preferentially down the shaded side of the shoot, where it promotes cell elongation, causing the shoot to bend towards the light. Leaves and stems respond positively to light and roots respond negatively or not at all. Indoor plants lean towards windows; leaves usually grow at right-angles to light and are positioned to ensure that overlapping occurs as little as possible.

phyllite Medium-grained METAMORPHIC ROCK formed from sediments at low temperatures and only moderate pressure. It is composed of quartz, feldspar and mica, with chlorite, which imparts a green or grey colour to the shiny foliated structure. It may also contain small crystals of garnet.

phyllotaxis (phyllotaxy) Arrangement of leaves on the STEM of a plant. In nearly all LEAF-bearing plants, each leaf grows out from the stem according to a regular pattern which is characteristic of the species. Most leaves grow either in a spiral arrangement up the stem, or an alternate arrangement with one leaf on one side of the stem and the next leaf on the other side. In an opposite arrangement, two leaves originate from the same leaf NODE. In a whorled arrangement, three or more leaves grow from the same node.

phylogenetics Study of the evolutionary relationships between organisms. In molecular phylogeny, the evolutionary distances between organisms are analysed by comparing the DNA sequences of specific GENES. These analyses confirm, for example, that seals are more closely related to dogs than to whales. At the most fundamental level, molecular phylogeny has revealed that all known organisms evolved from a common ancestor and can be grouped into five KINGDOMS, whose members are more closely related to each other than to members of other kingdoms.

phylum Part of the CLASSIFICATION of living organisms, ranking above CLASS and below KINGDOM. Phylum names are written in Roman (ordinary) letters and begin with an initial capital letter; for example all animals with backbones (VERTEBRATES) are members of the phylum Chordata. Some phyla are further subdivided into subphyla (**subphylum**). In PLANT CLASSIFICATION, the analogous category is sometimes called division. *See also* TAXONOMY

physical chemistry Study of the physical changes associated with CHEMICAL REACTIONS and the relationship between physical properties and chemical composition. The main branches are THERMODYNAMICS, concerned with the changes of energy in physical systems; chemical KINETICS, concerned with rates of reaction; and molecular and atomic structure. Other topics include ELECTROCHEMISTRY, SPECTROSCOPY and some aspects of NUCLEAR PHYSICS. *See also* CHEMISTRY

physical units Units used in measuring physical quantities. In specifying a unit, it is necessary to define an instance of that physical quantity and a way in which it can be compared in making a measurement. For example, the KILOGRAM unit of mass is defined as the mass of a specified block of platinum-iridium. Other masses are measured by weighing them and comparing them, directly or indirectly, with this. Units are of two types: **base** units that, like the kilogram, have fundamental definitions, and **derived** units that are defined in terms of these base units. Various systems of units exist, founded on certain base units. They include the FPS SYSTEM (the imperial units, foot, pound, second), CGS SYSTEM (centimetre, gram, second) and MKS SYSTEM (metre, kilogram, second). For all scientific purposes, SI UNITS have been adopted, which include seven base units: METRE (length), kilogram (mass), SECOND (time), AMPERE (electric current), KELVIN (temperature), CANDELA (luminous intensity) and MOLE (amount of substance).

physics Branch of science concerned with the study of MATTER and ENERGY. Physics seeks to identify and explain their many forms and relationships. Modern physics recognizes four basic FORCES in nature: GRAVITATION, which was first adequately described by Isaac NEWTON; the ELECTROMAGNETIC FORCE, codified in the 19th century by MAXWELL'S equations; the WEAK NUCLEAR FORCE, which is responsible for the decay of some SUBATOMIC PARTICLES; and the STRONG NUCLEAR FORCE, which binds together atomic nuclei. Physics rests on six fundamental theories: Newtonian MECHANICS, THERMODYNAMICS, ELECTROMAGNETISM, STATISTICAL MECHANICS, RELATIVITY and QUANTUM MECHANICS. Branches of physics include PARTICLE PHYSICS, GEOPHYSICS, BIOPHYSICS, ASTROPHYSICS and NUCLEAR PHYSICS.

physiology Branch of biology concerned with the functions of living organisms, as opposed to ANATOMY, which is concerned with their structure. Vast in scope, physiology includes the study of single CELLS as well as multicellular organisms.

phytoplankton Free-floating, oceanic plant life as opposed to ZOOPLANKTON, the free-floating ani-

Virtually all life on Earth depends on the Sun's energy. Green plants, algae and some bacteria are able to convert this energy, by PHOTOSYNTHESIS, into chemical energy that can be used. This energy is first stored in the form of simple sugars – and every year more than 150,000 million tonnes of sugars are produced by photosynthesis. In green plants and algae, photosynthesis takes place in the CHLOROPLASTS, miniature solar converters inside the plant's CELLS. Chloroplasts contain the green pigment CHOLOROPHYLL, which absorbs light. The energy of light is transferred to the chlorophyll molecule which then passes it on through a complicated chain of reactions and biochemical processes that finally result in the formation of of simple, organic compounds. At the same time, water is split into its component atoms, producing oxygen gas: oxygen is thus a by-product of photosynthesis. Sugars are used by the plant as fuel for RESPIRATION, which generates chemical energy in MITOCHONDRIA to power biochemical reactions essential for survival and growth. Respiration also produces carbon dioxide (CO_2) as a waste product, which can then be used again for photosynthesis. The products of photosynthesis also represent the starting point for the formation of other simple organic molecules. These can then be combined into larger molecules such as PROTEINS, NUCLEIC ACIDS, POLYSACCHARIDES and LIPIDS, from which all living material is made. Plants generally store food in the form of SUCROSE, a compound of the sugars GLUCOSE and FRUCTOSE, together with STARCH.

A
waxy cuticle
epidermal cells
chloroplasts
3
3
4
4
2
1
nucleus

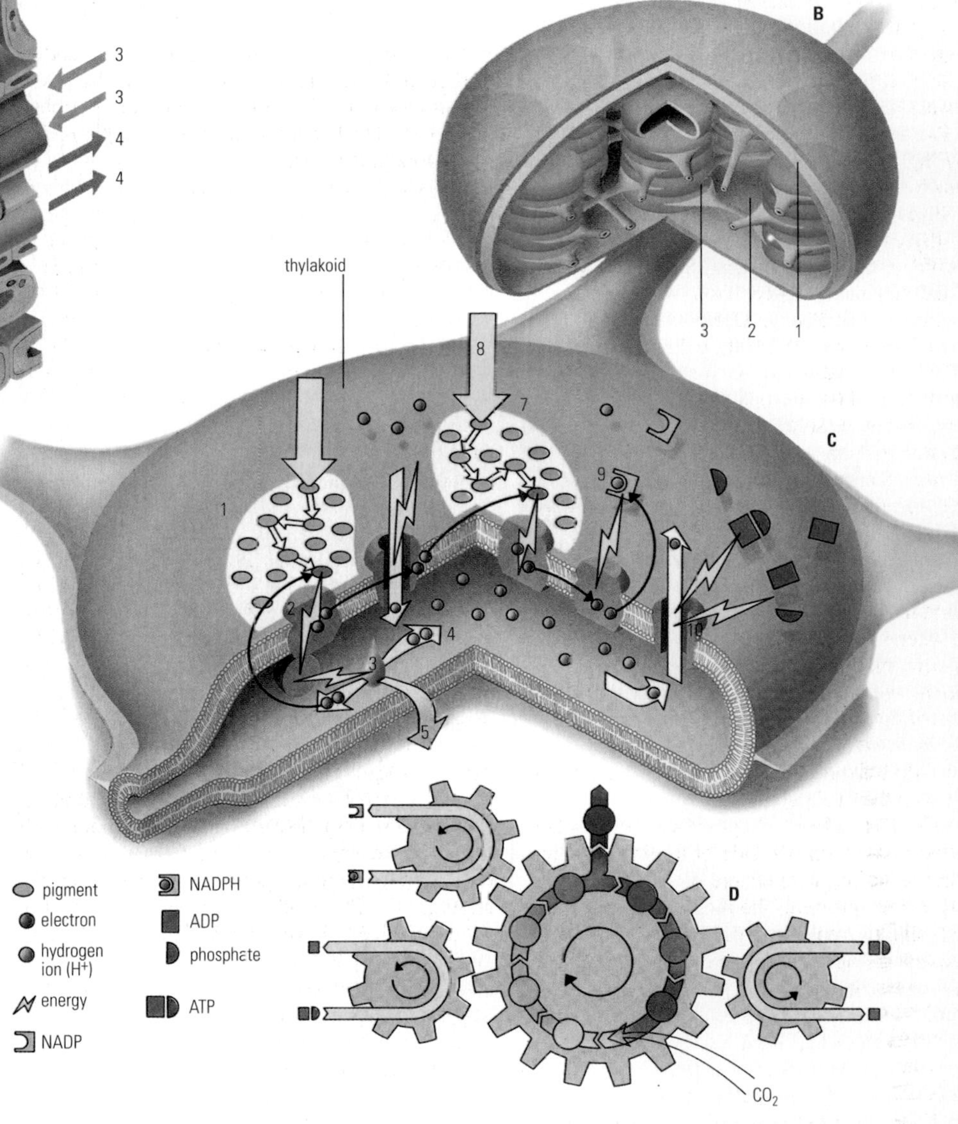

Carbon dioxide (CO_2) and water are the inorganic raw materials of photosynthesis. They arrive at a leaf's photosynthesizing cells by different routes (A): CO_2 gas simply diffuses in via pores in the leaf stomata (1) and through the air spaces between the loosely packed cells of the leaf mesophyll (2); water is drawn up from the roots through a system of woody xylem vessels (3). The products of photosynthesis – simple water-soluble sugars – are loaded into phloem sieve tubes and distributed throughout the plant (4).

Photosynthesis in a plant cell takes place within structures – organelles – called chloroplasts (B). Each chloroplast is bounded by a double membrane (1) that encloses a dense fluid known as the stroma (2). A third system of membranes within the chloroplast forms an interconnected set of flat, disc-like sacs called thylakoids (3), which are frequently stacked on top of one another to form structures called grana. The chloroplasts contain photosynthetic pigments, the most important of which is chlorophyll. This pigment absorbs light primarily in the blue, violet and red parts of the spectrum. Green light is not absorbed: it is because this light is reflected that leaves appear green. Photosynthesis involves a complex series of chemical reactions: for the sake of convenience, these are usually divided into the light-dependent reactions, which occur on the thylakoid membrane, and light-independent reactions, which take place in the stroma.

In the light-dependent reaction of photosynthesis (C), sunlight energy is trapped by chlorophyll and converted first into electrical, then into chemical energy, which is temporarily "stored" in the compounds ATP and NADPH. These compounds are later used as "fuel" to power the light-independent fixation of CO_2 into sugars. All the chemical equipment needed for the light-dependent reaction is located on the thylakoid membrane.

Light-trapping pigments, including chlorophyll, are grouped together on the outer wall of the thylakoid sac into units called photosystems (1). When light strikes a pigment molecule, one of its electrons becomes "energized" and is passed through the photosystem to an electron carrier in the membrane (2). Having lost an electron, the photosystem is left with a net positive charge: it is resupplied with electrons by the splitting of water (H_2O) (3), which also releases hydrogen ions (H^+) into the thylakoid sac (4), and liberates gaseous oxygen (O_2) (5). The energized electron is passed to another carrier in the thylakoid membrane: in this process, some of its energy is used to "pump" more H^+ into the thylakoid sac (6). The electron passes through a second photosystem (7): this absorbs more light (8), boosting the electron's energy level. The re-energized electron is now passed through other electron carriers, giving up some of its energy on the way to fuel the formation of NADPH from NADP and hydrogen ions (9).

The result is that the H^+ concentration within the thylakoid sac rises to over 1,000 times that in the stroma, generating a chemical pressure. H^+ can only "leak" back into the stroma through special membrane-spanning turbines – ATP synthetase enzymes (10). As H^+ passes through these it drives the synthesis of ATP from ADP and phosphate, as in cell mitochondria.

The "energy-rich" compounds ATP and NADPH are then used to power the formation of sugar in the light-independent reactions in the stroma (D). CO_2 is bound to a series of intermediate compounds, using energy along the way, before finally being released as a sugar (1).

P

mal life. Most of the organisms are microscopic, such as DIATOMS.

pi (π) Symbol used for the ratio of the circumference of a CIRCLE to its diameter. It is an IRRATIONAL NUMBER and as such it cannot be expressed as a FRACTION. To seven decimal places it is 3.1415926 (7d.p.). The ratio 22/7 is often used as a rougher approximation.

Piazzi, Giuseppe (1746–1826) Italian astronomer and monk. He discovered the first ASTEROID, CERES, on 1 January 1801, during a lengthy programme of taking accurate measurements of star positions (published as a catalogue in two parts, in 1803 and 1814). Piazzi proposed the name "planetoid", but "asteroid" has prevailed.

Piccard, Auguste (1884–1962) Swiss physicist who explored the STRATOSPHERE and deep seas. In 1931 a hydrogen balloon carried a pressurized aluminium sphere, containing Piccard and an assistant, to an altitude of almost 15,800m (51,800ft). This was the first ascent into the stratosphere. In 1932 Piccard set a new altitude record of 16,940m (55,577ft). From 1948 Piccard experimented with designs for a diving vessel called a bathyscaphe. In 1953 Piccard and his son, Jacques, descended in the bathyscaphe *Trieste* to a depth of *c.*3,100m (10,000ft) in the Mediterranean Sea. In 1960 Piccard and US naval officer Don Walsh used the same craft in the Pacific Ocean to reach a new record of 10,916m (35,810ft) below the surface. *See also* BATHYSPHERE AND BATHYSCAPHE

picric acid (2,4,6-trinitrophenol, $C_6H_2OH(NO_2)_3$) Yellow, crystalline solid, widely used as an EXPLOSIVE before and during World War I. It is now used in the manufacture of dyes and pigments.

Pictor (Painter's easel) Unremarkable constellation of the S sky. Its leading stae is Alpha, magnitude 3.3. *Beta Pictoris*, magnitude 3.8, is believed to be a planetary system in the process of formation.

piedmont Describing the foot of a mountain or a place where GLACIERS merge. The term is used of areas near mountains, such as the Piedmont (It. *Piemonte*) region in N Italy or the Piedmont Plateau to the E of the Appalachian Mountains in the USA. A piedmont glacier is formed when two or more glaciers flow from their valleys onto a plain and merge with their neighbours.

piezoelectric effect Appearance of positive ELECTRIC CHARGE on one side of a nonconducting CRYSTAL and a negative charge on the other when the crystal is squeezed. The pressure polarizes the crystal, slightly separating the positive and negative charge distributions in the crystal. This charge separation results in an ELECTRIC FIELD that can be detected as VOLTAGE between the opposite crystal faces. The effect has been used in gramophone pickups, crystal microphones and cigarette lighters.

pigment Coloured, insoluble substance used to impart colour to an object and added for this purpose to PAINT and INK. Unlike DYES, pigments are insoluble in the coating vehicle or medium. They generally function by absorbing and reflecting light, although some luminescent pigments emit coloured light.

pigmentation In biology, a natural chemical that gives colour to animal TISSUE. In humans, the skin, hair and the IRIS of the eye are coloured by the pigment MELANIN. HAEMOGLOBIN in erythrocytes (red blood cells) also acts as a pigment.

Pilâtre de Rozier, Jean François (1756–85) French pioneer of ballooning. In November 1783 he and the Marquis d'Arlandes were the first to fly in an untethered, hot-air BALLOON. During a flight lasting about 25 minutes, they covered a distance of 8km (5mi). In 1785 Pilâtre de Rozier was killed while attempting to cross the English Channel by balloon.

pillow lava LAVA that has occurred when MAGMA is expelled under water, or flows into water before solidifying. It commonly takes the form of a distorted globular mass, resembling the shape of a pillow. It apparently results from the rapid chilling of the outer skin, thus making a more or less spherical "balloon" that grows and flattens under its own weight.

pinchbeck Alloy of COPPER and ZINC, but with much less zinc than is contained in BRASS. Invented by the London watchmaker Christopher Pinchbeck (*c.*1670–1732), it is used to simulate gold.

Pincus, Gregory Goodwin (1903–67) US biologist who helped to develop the contraceptive pill (*see* CONTRACEPTION) using synthetic HORMONES. These hormones simulate the physiological state of women when pregnant, thus preventing FERTILIZATION. Pincus wrote several articles on hormones and a book, *The Eggs of Mammals* (1936).

pineal body Small GLAND attached to the undersurface of the vertebrate BRAIN. In human beings, it has an endocrine function, secreting the hormone melatonin, which is involved in daily rhythms. *See also* ENDOCRINE SYSTEM

pingo Cone-shaped mound that stands alone and has a core of ice. It is formed when freezing water expands and is forced upwards. Pingos are generally found in areas of PERMAFROST, where water cannot reach the surface and therefore pushes up the ground as it freezes. Pingos may be up to 90m (330ft) high and 800m (2,600ft) across.

pinna (auricle) Flap of skin and cartilage that comprises the visible, external part of the EAR. It helps to collect sound waves and direct them into the ear canal.

pinocytosis Intake and transport of fluid by living CELLS. Rather than entering and passing through the cell membrane as individual molecules, a droplet becomes bound to the membrane. A pocket then forms, which pinches off to form a vesicle in the CYTOPLASM. The vesicle may then pass across to the far side of the cell, where reverse pinocytosis occurs. It is a type of ENDOCYTOSIS.

pint Unit of liquid capacity. One pint is equal to $^1/_8$ of a gallon, $^1/_2$ of a quart, 1.201 US pint, or 0.568 litre.

pion (pi-meson) ELEMENTARY PARTICLE classified as a MESON. There are three types, forming a nuclear MULTIPLET (triplet). The **charged** pions, π^+ and π^-, have equal mass, about 280 times the electron mass, and are antiparticles of each other; the **neutral** π^0, of slightly lower mass, is its own antiparticle. All have zero SPIN. Charged pions decay into MUONS and muon NEUTRINOS; the π^0 decays into PHOTONS. Virtual pions are thought to be exchanged between NUCLEONS bound together by the STRONG NUCLEAR FORCE.

Pioneer program Series of interplanetary SPACE PROBES launched by the USA between 1958 and 1978. The first four were failed Moon probes. **Pioneers 5 to 9** were put into orbit around the Sun to study the SOLAR WIND and SOLAR FLARE activity. **Pioneer 10**, launched in March 1972, was the first probe to travel through the ASTEROID belt; it passed Jupiter in December 1973, sending back photographs. **Pioneer 11** followed it, flying past Jupiter in December 1974 and moving on to Saturn, which it reached in September 1979. Both probes are now on their way out of the Solar System. The last two in the series were the Pioneer Venus probes, launched in May and August 1978. **Pioneer Venus 1** went into orbit around Venus in December 1978, making a RADAR map of the surface and photographing the planet's clouds; it burned up in the planet's atmosphere in October 1992. **Pioneer Venus 2** dropped five probes into Venus' atmosphere in December 1978, which sent back data on the planet's atmospheric conditions. *See also* SPACE EXPLORATION

Pisces (the Fishes) Equatorial constellation situated on the ECLIPTIC between Aquarius and Aries. It is very obscure, consisting of a chain of rather faint stars south of Pegasus. Its brightest star, Eta, is magnitude 3.6.

Piscis Austrinus (Southern Fish) Constellation in the S sky. First-magnitude Formalhaut is the only star brighter than magnitude 4.

pistil Female organ located in the centre of a FLOWER. It consists of an OVARY, a slender STYLE and a STIGMA, which receives POLLEN.

pitchblende *See* URANINITE

pitch In AERODYNAMICS, angular displacement of the longitudinal relative to the axis of an aircraft relative to the horizon. The elevators control pitch by lowering (or raising) the tail relative to the nose to increase (or decrease) the angle of attack (angle at which air strikes the wing).

pitch In physics, quality of sound that determines its position in a musical scale. It is measured in terms of the FREQUENCY of sound waves (measured in hertz) – the higher the frequency, the higher the pitch. It also depends to some extent on loudness and timbre: increasing the intensity decreases the pitch of a low note and increases the pitch of a high one.

pith Central strand of parenchymatous tissue that occurs in the STEMS of most VASCULAR PLANTS. It is usually surrounded by vascular tissue and is believed to function chiefly for storage. The term is also used for the soft core at the centre of the heartwood of logs, consisting of the dried remains of the pith. *See also* PARENCHYMA

pitot tube Device for measuring the rate of flow of a fluid, either liquid or gas. It was invented by Henri Pitot (1695–1771). For **liquids**, the device used is generally a MANOMETER, with one open end facing upstream and the other open end out of the stream. The different pressures at the two ends cause a liquid to shift position within the two arms of the tube. For **gases**, a Pitot tube is generally L-shaped, with one end open and pointing towards the flow of gas and the other end connected to a pressure-measuring device. This type of pitot tube is commonly used as an air-speed indicator in aircraft.

pituitary gland Major gland of the ENDOCRINE SYSTEM located at the base of the BRAIN. In human beings it is about the size of a pea and is connected to the HYPOTHALAMUS by a stalk. It produces many HORMONES, some of which regulate the activity of other endocrine glands, while others control growth.

placenta Organ in PLACENTAL MAMMALS that connects the FETUS to the wall of its mother's UTERUS, providing a means of nutrition, GAS EXCHANGE and

▲ **pillow lava** When a lava flow enters the sea it breaks up into a number of globular structures which may come to rest piled on top of one another. This is known as a pillow lava. The same formation can occur if magma is expelled directly into the sea.

P

EXCRETION for the fetus. Part of the placenta contains tiny blood vessel branches through which oxygen and food are carried from the mother to the embryo, via the UMBILICAL CORD, and wastes are carried from the embryo to the mother's bloodstream to be excreted. The placenta secretes HORMONES that maintain pregnancy. It is discharged from the mother's body as the AFTERBIRTH, immediately after delivery.

placental mammals Mammals whose young develop to an advanced stage attached to the PLACENTA – a life-support organ inside the mother's UTERUS. All mammals except the MONOTREMES and most marsupials are placentals. *See also* VIVIPARITY

placer deposits Concentrations of heavy minerals formed by the action of GRAVITY, usually found in streams. Minerals that occur as placer deposits include gold, copper, rutile, cassiterite and magnetite.

placoderm Group of primitive jawed fishes that existed in the DEVONIAN period, characterized by armoured plates on the front part of the body. Most placoderms were quite small, but some grew to 9m (30ft).

plagioclase Type of FELDSPAR (the most abundant group of minerals on Earth). Plagioclase minerals are widely distributed and occur in many IGNEOUS and METAMORPHIC rocks. Off-white, or sometimes pink, green or brown in colour, they are composed of varying proportions of the silicates of sodium and calcium with aluminium. They show an oblique cleavage, and have triclinic system crystals. Hardness 6–6.5; r.d. 2.6.

plain Large area of flat or slightly undulating, low-lying land. Some plains, such as a PENEPLAIN, result from the wearing away of higher terrain. Most plains result from deposition of SEDIMENTS, as by rivers and lakes, which leave flat plains when the water dries up. FLOODPLAINS are formed in river valleys (*see* ALLUVIAL FAN). Glacial plains are large, level areas of TILL, which is left by retreating GLACIERS.

planarian Flatworm (*see* PLATYHELMINTH) found in marine and freshwater, identified by its triangular head with two light-sensitive eyespots. Planaria have flat, tail-like bodies, with extendable PHARYNXES for sucking in food. Reproduction is hermaphroditic or by asexual splitting, or regeneration. Length: 1in (2.54cm). Phylum Platyhelminthes; class Turbellaria; genera include *Dugesia* and *Polycelis*.

◀ **planetarium** Inside the planetarium at the Visitor Centre at Jodrell Bank radio observatory, near Manchester, England. Sophisticated computer programs can now "transport" the viewers across the Universe in order to view our Galaxy from a variety of places in space.

Planck, Max Karl Ernst Ludwig (1858–1947) German theoretical physicist whose revolutionary QUANTUM THEORY helped to establish modern physics. He was professor (1889–1928) at the University of Berlin, where he studied the characteristics of BLACK BODY radiation. In 1900 Planck came to the conclusion that the frequency distribution (SPECTRUM) of the radiation could be accounted for only if the radiation was emitted in separate "packets" called quanta, rather than continuously. PLANCK'S CONSTANT (1900) links wave and particle behaviour on the atomic scale. His equation, relating the ENERGY of a quantum to its FREQUENCY, is the basis of quantum theory. Planck was awarded the 1918 Nobel Prize for physics for his work. He remained in Germany during the Nazi era, but spoke out against the persecution of Jews. In 1944 his son, Erwin, was executed for taking part in the unsuccessful plot to assassinate Hitler. He served as president (1930–35) of the Kaiser Wilhelm Society for the Advancement of Science, Berlin, which after World War II became part of the Max Planck Institutes.

Planck's constant (symbol h) Universal constant of value 6.626×10^{-34} joule seconds, equal to the energy of a quantum of ELECTROMAGNETIC RADIATION (a PHOTON) divided by the radiation frequency. It appears in formulae describing physical quantities that can assume only certain discrete values. It is named after Max PLANCK. *See also* PARTICLE PHYSICS; QUANTUM THEORY

plane In mathematics, a flat surface such that a straight line joining any two points on it lies entirely within the surface. Its general equation in the three-dimensional CARTESIAN COORDINATE SYSTEM is: $ax + by + cz = d$, where a, b, c and d are constants. Any flat shape or surface such as a wing or craft can also be called a plane, hence aeroplane and hydroplane. In woodworking, a plane is a tool used to smooth wood by cutting away thin strips as it passes over the surface.

plane geometry Form of GEOMETRY in which all lines, angles and figures are represented in two-dimensional (PLANE) form. In plane geometry, EUCLID's axioms are valid.

planet Large, non-stellar body in orbit around a STAR, shining only by reflecting the star's light. In our SOLAR SYSTEM there are nine major planets, as opposed to the thousands of small bodies known as ASTEROIDS or minor planets. It is believed that other stars, apart from our SUN, have planets orbiting them. *See* MERCURY; VENUS; EARTH; MARS; JUPITER; SATURN; URANUS; NEPTUNE; PLUTO

planetarium Structure in which a representation of the stars and planets as visible in the night sky is **projected** on to the inside of a dome for an audience seated below. It is used to demonstrate the positions and motions of celestial bodies. In 1913 the PROJECTOR was invented by Walter Bauersfeld of the Zeiss Optical Company. In 1923 the first planetarium, built by the Zeiss company, began operating in Munich. From then until the early 1980s, planetarium projectors consisted of many individual optical projectors mounted on a complicated rotating frame. In the new generation of projectors, a computer-generated image is projected through a single, stationary fish-eye LENS.

planetesimal In theories of the origins of our Solar System (and other planetary systems), a body, between a few millimetres and a few kilometres in size, that condensed out of a cloud of gas and dust (the solar NEBULA), away from the centre where the Sun was forming. Once they had reached a few kilometres in size, planetesimals' gravitational attraction would allow them to combine with one another, by a process of ACCRETION, to form protoplanets. The term was first used for the very small planetary bodies supposed to have condensed from matter torn out of the Sun by a passing star in the "planetesimal theory" proposed by Forest Ray MOULTON and Thomas Chamberlain in the early 1900s.

plane trigonometry Branch of TRIGONOMETRY that deals with the sides and angles of PLANE triangles and their measurements and relations.

plankton All the floating or drifting life of the

▲ **placer deposits** Most offshore minerals, with the exception of oil and gas, do not occur in sufficiently high-ore grades to warrant their economic production. Nevertheless significant amounts of tin, diamonds, gold and titanium are recovered from beach and offshore placer deposits all over the world. Sulphur is extracted commercially in a number of areas, including the Gulf of Mexico, and the most important mineral resources after oil and gas are sand and gravel dredged from offshore deposits for use in the construction industry. Placer deposits occur where metal-bearing rock on land is weathered and the debris produced is washed down to the sea by rivers. There it is sorted by the currents, waves and tides so that the heavy metal particles accumulate to form deposits of mineral sand. These typically take the form of beach deposits, but where the sea level has changed they can be found well out on the continental shelf. The sands are lifted by dredgers and sifted for their metal content.

ocean, especially that near the surface. The organisms are very small or microscopic and move with the currents. There are two main kinds: PHYTOPLANKTON, which comprises the floating plants such as DIATOMS and dinoflagellates; and ZOOPLANKTON, floating animals such as radiolarians, plus the larvae and eggs of larger marine animals.

plant Any member of the kingdom Plantae, a large kingdom of multicellular organisms whose cells have CELLULOSE cell walls and contain CHLOROPLASTS or similar structures (PLASTIDS). Plants develop from DIPLOID embryos and have a regular alternation of HAPLOID and diploid generations in their life cycles. Most plants are green, contain chloroplasts, and make their own food by PHOTOSYNTHESIS. A few are colourless PARASITES or SAPROPHYTES. Simple plants reproduce by means of SPORES, whereas more advanced plants produce seeds and fruits. Plants show a wide range of biochemistry; some produce chemicals such as ALKALOIDS, NARCOTICS and even poisions such as CYANIDE; others secrete substances into the soil to prevent other plants growing near them. Many of these chemicals form the bases for the development of DRUGS. Plants are classified on the basis of their MORPHOLOGY (shape and structure). The most important phyla are the Bryophyta (BRYOPHYTES), which include the MOSSES and liverworts; LYCOPODOPHYTA, or CLUB MOSSES; Sphenophyta (HORSETAILS); Filicinophyta (FERNS); Cycadophyta (CYCADS); Ginkgophyta (GINKGO); Coniferophyta (CONIFERS); and Angiospermophyta (ANGIOSPERMS). *See also* ALTERNATION OF GENERATIONS

plant classification System devised to group PLANTS according to relationships among them. The currently accepted system of plant classificiation includes the following phyla: **Bryophyta**, **Lycopodophyta**, **Sphenophyta**, **Filicinophyta**, **Cycadophyta**, **Ginkgophyta**, **Coniferophyta** and **Angiospermophyta**. An older system split plants into four divisions: Thallophyta (ALGAE and LICHENS), Pteridophyta (CLUB MOSSES and FERNS), Spermotophyta (which included the GYMNOSPERMS and ANGIOSPERMS) and Bryophyta (LIVERWORTS and MOSSES). *See also* EICHLER, AUGUST WILHELM; TAXONOMY

plant genetics Science of HEREDITY and VARIATION in plants. Research in GENETICS since 1900 has supplied the principles of plant breeding, especially HYBRIDIZATION – the controlled crossing of plant varieties in order to reproduce new and improved plants. Breeders can select for specific characteristics, such as resistance to disease, or size and quality of yield. The development of consistently reliable and healthy first-generation crosses (**F1 hybrids**) has revolutionized the growing of food crops, ornamental annuals and bedding plants. Genetic engineers grow CELL and TISSUE CULTURES by the replication or cloning of sterile plant types. They also concentrate on isolating individual GENES with the aim of producing new colour varieties for traditional flowers, improving the flavour of food crops, breeding resistance to pests and HERBICIDES, and lengthening the shelf-life of harvested crops. *See also* CLONE; GENETIC ENGINEERING

plant hormone (phytohormone) Organic chemical produced in plant cells and functioning at various sites to effect plant growth, leaf and fruit drop, healing, cambial growth and possibly flowering. HORMONES are transported away from the STEM tip. They include abscisic acid (leaf fall), AUXINS (growth), CYTOKININS (leaf and bud growth and development) and GIBBERELLINS (growth). *See also* ABSCISSION

plant pigment Organic compound present in plant cells and tissues that colours the plant. The most common plant pigment is green CHLOROPHYLL, which exists in all higher plants. CAROTENOIDS colour plants yellow to tomato red. Located in CHLOROPLASTS and chromoplasts are more than 150 varieties of these durable pigments. Many are essential to PHOTOSYNTHESIS and are a source of vitamin A. Anthocyanins, responsible for pink, red, blue and purple, are found in the cell SAP. The shorter days and lower temperatures of autumn cause these pigments to combine with other substances and produce the brilliant foliage colours of deciduous trees. *See also* PIGMENTATION

plaque Abnormal deposit building up on a body surface, especially the film of saliva and bacteria that accumulates on TEETH. Dental plaque leads to tooth decay and gum disease.

plasma In biology, liquid portion of the BLOOD in which the CELLS are suspended. It contains an immense number of ions, inorganic and organic molecules such as IMMUNOGLOBULINS, and HORMONES and their carriers. It clots on standing.

plasma In physics, an ionized GAS. Plasma, often described as the fourth state of MATTER, occurs at enormous temperatures, as in the interiors of the Sun and other stars and in FUSION reactors. *See also* ION

plasmid Strand or loop of NUCLEIC ACID, containing GENETIC information. It can be introduced to a host CELL where it will replicate independently of the host's CHROMOSOMES. Plasmids are used in RECOMBINANT DNA RESEARCH.

plasmodium Genus of parasitic PROTOZOA that causes MALARIA. It infects the ERYTHROCYTES (red blood cells) of mammals, birds and reptiles worldwide, being transmitted by the bite of a female *Anopheles* mosquito.

plaster of Paris (gypsum cement) Powdered form of calcium sulphate hemihydrate, $2CaSO_4.H_2O$, obtained by heating GYPSUM to around 150°C (300°F). After the addition of water, it sets and hardens and is used as a PLASTER for a wide range of purposes, including the setting of broken limbs and the making of moulds.

plastic Synthetic material composed of organic molecules, often in long chains called POLYMERS, that can be shaped and then hardened. The weight and structure of the molecules determine the physical and chemical properties of a given compound. Plastics are synthesized from common materials, mostly from PETROLEUM. CELLULOSE comes from cotton or wood pulp, CASEIN from skimmed milk, others from chemicals derived from potatoes, peanuts and soya beans. There are two main types of plastics: **thermosetting** plastics, such as BAKELITE, stay hard once set; **thermoplastics**, such as POLYETHENE, can be resoftened by heat. New biodegradable plastics, more expensive to produce, are environmentally friendly because they eventually decompose (mainly due to the action of microorganisms) in landfills. The plastics industry began in 1872 with the manufacture of CELLULOID by the US inventor John Wesley Hyatt (1837–1920) and his brothers. There are three basic plastic-forming processes: calendering, casting and extrusion. **Calendering** is used to make a layered composite material from a thermosetting resin, such as phenolformaldehyde (Bakelite), and materials such as paper and cloth. CASTING uses thermosetting plastics, such as EPOXY, POLYESTER and phenolic or UREA RESINS, and enables plastic to be cast around metal parts, as in making screwdrivers. EXTRUSION forces the softened plastic through a die which shapes it into the desired form, such as a tube, rod or sheet. This is employed with THERMOPLASTIC RESINS such as polyethene, POLYSTYRENE and NYLON.

plastid Type of ORGANELLE found in the CELLS of plants and green algae. Plastids have a double membrane and contain DNA. CHLOROPLASTS and leucoplasts are two examples of plastids.

plateau Fairly flat, raised area of land. Mountains may stand up above the general level of a plateau, or it may be carved by deep river valleys or canyons to form a **dissected** plateau. An **intermontane** plateau is completely surrounded by mountains.

platelet Colourless, usually spherical structures found in mammalian BLOOD. Chemical compounds in platelets, known as factors and co-factors, are essential to the mechanism of BLOOD CLOTTING. The normal platelet count is about 300,000 per cu mm of blood.

plate tectonics Theory or model to explain the distribution, evolution and causes of the EARTH's crustal features. It proposes that the Earth's CRUST and part of the upper MANTLE (the LITHOSPHERE) are made up of several separate rigid slabs, termed **plates**, which move independently and form part of a cycle in the creation and destruction of crust. The plates collide or move apart at the CONTINENTAL MARGINS, and these produce zones of EARTHQUAKE and volcanic activity. Three types of plate bound-

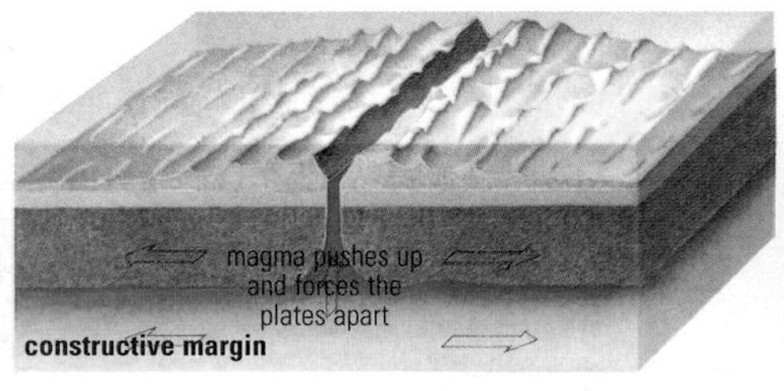

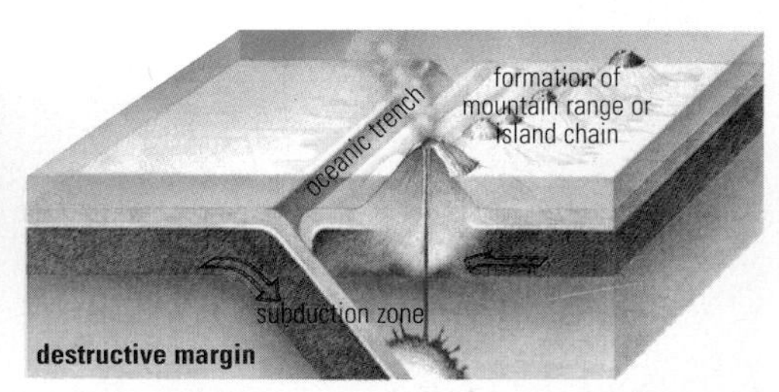

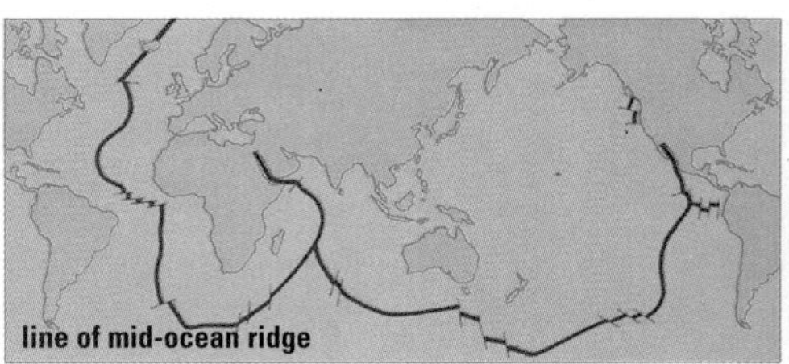

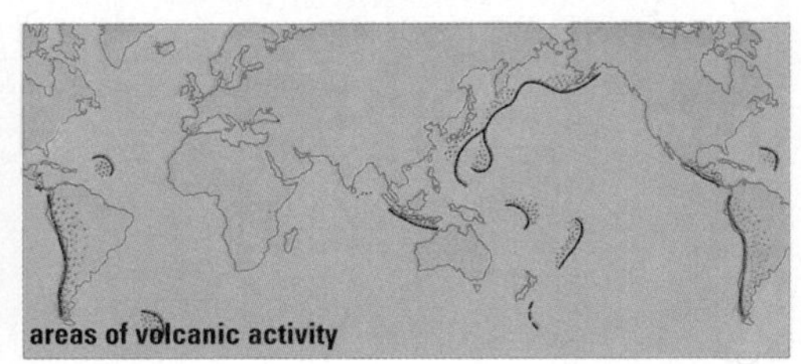

▲ **plate tectonics** The discovery that the continents are carried along on the top of slowly moving crustal plates provided the mechanism for continental drift theories to work. The plates converge and diverge along margins marked by seismic and volcanic activity. Plates diverge from mid-ocean ridges, where molten lava pushes up and forces the plates apart at a rate of up to 3.75cm (1.5in) a year; converging plates form either a trench (where the oceanic plates sink below the lighter continental rock) or mountain ranges (where two continents collide).

ary can be identified. At a CONSTRUCTIVE MARGIN (or divergent margin), new basaltic MAGMA originating in the mantle is injected into the plate which forces the crust to separate and an oceanic ridge is formed. At a DESTRUCTIVE MARGIN (or convergent margin), plates collide and one plate moves into the mantle at the site of an oceanic TRENCH. The recycling of crust by SUBDUCTION results in the melting of some crustal material and volcanic island arcs (such as the islands of Japan) are produced. Material that cannot be subducted is scraped up and fused onto the edge of plates. This can form a new CONTINENT or add to existing continents. Mountain chains are explained as the sites of former subduction or continental collision. At a **conservative margin**, plates move past each other along a transform fault. Plate movement is thought to be driven by CONVECTION CURRENTS in the mantle. *See also* SEAFLOOR SPREADING

plating *See* ELECTROPLATING

platinum (symbol Pt) Lustrous, silver-white TRANSITION ELEMENT, first discovered in 1735. It is chiefly found in certain ores of nickel. Malleable and ductile, platinum is used in jewellery, dentistry, electrical-resistance wire, magnets, thermocouples, surgical instruments, electrodes, and as a catalyst in catalytic converters for car exhausts. It is chemically unreactive and resists tarnishing and CORROSION. These traits are shared with five other elements that, with platinum, are known as the platinum metals and are often found together as alloys: IRIDIUM, OSMIUM, PALLADIUM, RHODIUM and RUTHENIUM. Properties: at.no. 78; r.a.m. 195.09; r.d. 21.45; m.p. 1,772°C (3.222°F); b.p. 3,800°C (6,872°F); most common isotope ^{195}Pt (33.8%).

platyhelminth Any member of the order Platyhelminthes, comprising the flatworms. Platyhelminths are simple, carnivorous, ribbon-like creatures which, having no circulatory system and sometimes no mouth or gut, feed by ABSORPTION through the body wall. Almost all flatworms are hermaphroditic. The fluke and TAPEWORM are both PARASITES of animals.

Pleistocene Sixth epoch of the CENOZOIC era of GEOLOGICAL TIME. The Pleistocene epoch began about 2 million years ago, during which humans and most forms of familiar mammalian life evolved. Episodes of climatic cooling in this epoch led to widespread GLACIATION in the Northern Hemisphere; the Pleistocene is the best-known glacial period or ICE AGE in the Earth's history. The present HOLOCENE epoch succeeded the Pleistocene around 8000 BC.

pleura Double membrane that lines the space between the LUNGS and the walls of the THORAX. The fluid between the pleura lubricates the two surfaces to prevent friction during breathing movements.

Pliocene Last epoch of the TERTIARY period. It lasted from 5 to 2 million years ago and preceded the PLEISTOCENE. Animal and plant life was similar to that of today. *See* GEOLOGICAL TIME.

plumbago *See* GRAPHITE

plumule In plants, embryonic shoot that develops during GERMINATION of a SEED.

Pluto Smallest and outermost planet of the SOLAR SYSTEM. Independently, William H. Pickering (1858–1938) and Percival LOWELL calculated the possible existence of Pluto, its size, ORBIT, and position in the sky. The planet was eventually located by Clyde TOMBAUGH within 5° of Lowell's predicted position, and the discovery was publicly announced in March 1930. Pluto has yet to be visited by a SPACE PROBE and most of our knowledge of it dates from the discovery (1978) of its satellite, CHARON, which allowed, among other things, Pluto's mass and unusual axial tilt to be determined. Pluto is in a much more inclined and eccentric orbit than the other planets in our Solar System, and for a 20-year period around PERIHELION (as, for example, from 1979 to 1999) it comes within the orbit of Neptune. Pluto seems to have a mottled surface with light and dark regions, and signs of polar caps. The surface is covered with icy deposits consisting of 98% nitrogen, with traces of methane, and also probably water, carbon dioxide and carbon monoxide. For a 60-year period in each orbit, when Pluto's high orbital eccentricity brings it closer to the Sun, the surface warms to above 50K, sufficient to release a thin, temporary atmosphere of nitrogen. Pluto is certainly not the massive planet whose supposed perturbations of Uranus and Neptune led to its discovery. The realization of just how small Pluto is has led some astronomers to question its status as a major planet.

plutonic Describing a rock that has formed beneath the surface of the Earth. Plutonic rocks are IGNEOUS ROCKS consisting of solidified MAGMA, which has cooled slowly, allowing large CRYSTALS to form in the rock. They are coarse-grained. The most common plutonic rock is GRANITE, which is light in colour because its mineral content is acidic. The most common of the basic plutonic rocks is GABBRO. Most plutonic rocks form in BATHOLITHS beneath the surface, but because of the EROSION of overlying rocks, they are now commonly found on the surface.

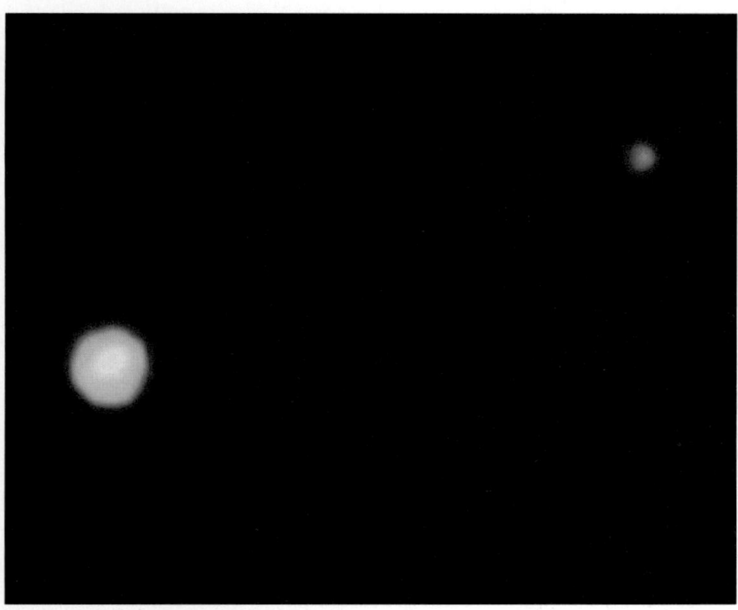

◀ **Pluto** This image, taken by the Hubble Space Telescope in 1994, represents the best view to date of the planet Pluto (left) and its satellite Charon. For the first time astronomers were able to clearly see the discs of the two bodies and then measure directly and accurately Pluto's (2,320km [1,450mi]) and Charon's (1,270km [794mi]) diameters. When the image was taken Pluto was at a distance of 4.4 billion km (2.7 billion mi) from the Earth and the separation between Pluto and Charon 19,640km. The image shows a bright (white) area parallel to the equator on Pluto; further observations will be required to confirm whether this feature is real.

plutonium (symbol Pu) Silver-white radioactive metallic element of the ACTINIDE SERIES. It was first made in 1940 at the University of California at Berkeley by the deuteron (heavy hydrogen) bombardment of URANIUM. It is found in small amounts in uranium ores but mostly produced synthetically: ^{239}Pu (half-life 24,360 years) is made in large quantities in breeder reactors. It is a fissile element (able to split into other elements) used in nuclear reactors and nuclear weapons, such as the atom bomb. The element is very toxic, a strong emitter of ALPHA PARTICLES, and absorbed by bone, making it a dangerous radiological hazard. Properties: at.no. 94; r.d. 19.84; m.p. 641°C (1,186°F); b.p. 3,232°C (5,850°F); most stable isotope ^{244}Pu (half-life 76 million years). *See also* NUCLEAR REACTOR; TRANSURANIC ELEMENTS

pneumatic Describing a device powered by compressed air normally used to produce rotary or a reciprocating (back and forth) motion to speed up operations such as sawing, grinding, digging, hammering and riveting. The pneumatic drill used in roadworks has a reciprocating, pounding motion at speeds of between 80 and 500 revs per minute.

pneumothorax Presence of air in the pleural space between the LUNGS and the chest wall. It may arise spontaneously or be caused by injury or disease. The lung is liable to collapse since it is prevented from expanding normally.

pod Fruit of any leguminous plant, such as a pea or bean. A pod is an elongated caselike structure filled with SEEDS. It develops from a single CARPEL, and when ripe splits down both sides to release the seeds. In some plants, the pods burst explosively, scattering the seeds widely. *See also* LEGUME

podiatry (chiropody) Treatment and care of the foot. Podiatrists treat such conditions as corns and bunions and devise ways to accommodate foot deformities. They can also perform certain kinds of foot surgery.

podzol Light-coloured, infertile SOIL, poor in lime and iron. The name comes from the Russian word for "ash soil". It forms under cool humid conditions and is typical of the taiga forests of North America and Eurasia. The upper layer is light grey; the lower layer is darker and contains HUMUS and minerals, such as ALUMINA, iron and lime, that have been leached out of the upper layer by rainfall. The minerals form a hard layer that is impervious to water and prevents good drainage. *See also* LEACHING

poikilothermic (ectothermic) Animal whose body temperature fluctuates with the temperature of its surroundings, often referred to as **cold-blooded**. Reptiles, amphibians, fish and invertebrates are cold-blooded. They can control their body temperature only by their behaviour – by moving in and out of the shade, or orientating themselves to absorb more or less sunlight. *See also* HOMEOTHERMIC

Poincaré, (Jules) Henri (1854–1912) French mathematician. He contributed more than 500 works in many diverse areas, including ANALYSIS, DIFFERENTIAL EQUATIONS, TOPOLOGY, PROBABILITY and NUMBER THEORY. Poincaré developed several important techniques for the study of dynamic and chaotic systems. He used the techniques to analyse the motion of planets and the nature of rotating fluids; systems that are now known to exhibit chaotic behaviour. He also made contributions to the special theory of RELATIVITY and electromagnetic theory. *See also* CHAOS THEORY

Poiseuille's equation In FLUID MECHANICS, the formula for the relation between the rate of flow, R, of an incompressible fluid of viscosity η, in a tube of radius a and length l is $R = \pi pa^4/8l\eta$, where p is the difference in pressure between the two ends of the tube. It is named after the French physiologist

► **pollen** Pollen grains are found in flowers' pollen sacs, which are themselves located in the anthers (part of the stamens). Pollen grains are safe and effective storers of the male gametes (sex cells). They come in all shapes and sizes depending on the species of plant. The selection shows mistletoe (A), venus fly trap (B), spinach (C), honeysuckle (D), touch-me-not (E), cotton (F), rice (G), dandelion (H) and hollyhock (I).

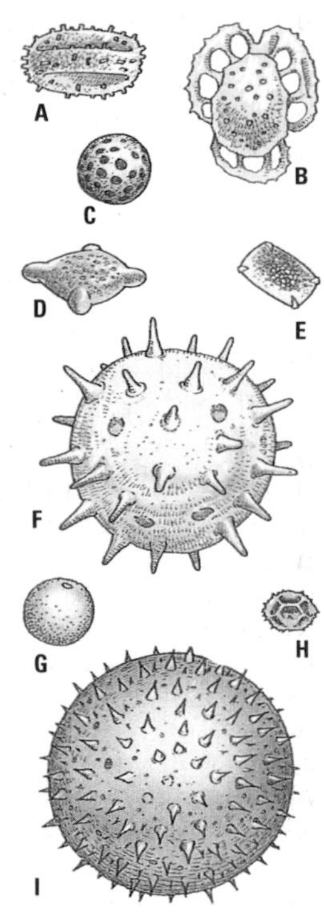

Jean Louis Marie Poiseuille (1799–1869).

Poisson, Siméon Denis (1781–1840) French scientist who applied MATHEMATICS to many areas of PHYSICS, including sound and heat propagation, ELECTROMAGNETISM and MECHANICS. Poisson extended the work of Joseph Louis LAGRANGE on CELESTIAL MECHANICS and worked on INTEGRATION and the FOURIER SERIES. In *Researches on the Probability of Opinions* (1837), he presented the **Poisson distribution**, which gives the PROBABILITY of the occurrence of a specific event among a large number of events. **Poisson's ratio**, which relates lateral contraction to longitudinal extension in stressed materials, is also named after him.

polar coordinate System of coordinates for representing the position of a point *P* on a PLANE. There are two coordinates, r and θ, where r is the point's distance from the origin and θ is an ANGLE measured anticlockwise from some reference line emanating from the origin. **Spherical** polar coordinates incorporate a second angle, ϕ, which represents the angle of inclination of a line drawn to the point from the plane surface.

polarization Suppression of certain directions of vibration of the ELECTRIC and MAGNETIC FIELDS of ELECTROMAGNETIC RADIATION, so that the wave form is no longer symmetrical about the direction of propagation. *See also* POLARIZED LIGHT

polarized light LIGHT waves that have electromagnetic vibrations in only one direction. **Ordinary** light vibrates in all directions perpendicular to the direction of propagation. Scientists distinguish between three types: plane-polarized, **circularly polarized** and **elliptically-polarized** light, each depending on the net direction of the vibrations. In **plane-polarized** light, the electric vibrations are confined to one direction only, the magnetic vibrations occurring at right angles. Plane-polarized light can be produced by REFLECTION, as from a sheet of glass or a water surface, or by passing light through certain crystals, such as quartz, tourmaline or calcite. Polarizing sunglasses use a Polaroid material to reduce glare by cutting out light polarized by reflection. *See also* LAND, EDWIN HERBERT; PRISM

polarography Method of chemical analysis that can measure minute concentrations of IONS in solution. Two ELECTRODES are inserted into the solution: one, the reference electrode, is kept at a constant VOLTAGE; the other has a variable voltage and consists of a glass capillary tube containing a head of mercury, with tiny drops of mercury on its end becoming the face of the electrode. As these drops form, a current grows to a peak between the two electrodes and then suddenly falls as the drop leaves the capillary tip. By gradually increasing the voltage, a polarogram – a graph of the current versus the voltage, which looks like a series of steps – is produced. From these, the REDUCTION potentials of the ions in the solution may be deduced and, as these are known for each metal ion, the ions can be identified.

polar solution SOLUTION in which the solvent molecules have substantial DIPOLE moments, such as water. Hydrogen bonding and solvent-solute complexes form, encouraging the association of solvent and solute molecules in preferred orientations in solution.

polar winds Prevailing WINDS that blow from the North and South Poles. They are deflected from east to west by the effects of the Earth's rotation, and are often known as the polar easterlies. They carry very cold air and their effect reaches as far as latitude 60°.

pole Generally either of the two points of intersection of the surface of a sphere and its axis of rotation. The Earth has four poles: the North and South geographic poles, where the Earth's imaginary axis meets its surface; and the north and south magnetic poles, where the Earth's MAGNETIC FIELD is most concentrated. A bar MAGNET has a north pole, where the magnetic flux leaves the magnet, and a south pole, where it enters. A pole is also one of the terminals (positive or negative) of a battery, electric machine, generator or circuit. *See also* GEOMAGNETISM; MAGNETISM

pollen Powder-like SPORES, usually yellow in colour, that give rise to the male GAMETES (sex cells) in plants. In ANGIOSPERMS (flowering plants), pollen grains are produced in the ANTHERS on the STAMEN, whereas in GYMNOSPERMS (cone-bearing plants), they are produced by the male cones. Pollen has thick resistant walls with a characteristic pattern of spines, plates or ridges, according to species. During POLLINATION, the pollen lands on the STIGMA of a flower, or a female cone, of a compatible plant. It germinates, sending a long pollen tube down through the STYLE to the OVARY. During this process, one of its nuclei divides, giving rise to two male nuclei, one of which will fuse with a female sex cell (actually a HAPLOID nucleus in the embryo sac inside the OVULE) in FERTILIZATION. The other male sex cell fuses with two more of the female nuclei to form a special tissue, the ENDOSPERM, whose cells contain three sets of CHROMOSOMES (triploid). In many species, this tissue develops into a food store for the EMBRYO in the seed.

pollination Transfer of POLLEN (containing male GAMETES) from the ANTHER to the STIGMA in ANGIOSPERMS (flowering plants), or from the male cone to the female cone in GYMNOSPERMS (cone-bearing plants), resulting in FERTILIZATION. Pollination occurs mainly by wind (ANEMOPHILY) in the cases of most trees and gymnosperms, and by flying insects (ENTOMOPHILY) for angiosperms. Other agents of pollination include grazing mammals, birds, bats and non-flying insects. **Self-pollination** occurs when the pollen of one flower (or male cone) pollinates the same flower or another flower (female cone) of the same plant. However, plants have developed intricate incompatibility mechanisms to ensure self-pollination occurs as infrequently as possible, such as chemicals that make the pollen and eggs of the same plant sterile. CROSS-POLLINATION occurs between two flowers (or male and female cones) on different plants. It results in a great variety of genetic combinations, thereby improving the chances of survival. See also SEX; SEXUAL REPRODUCTION

pollution Spoiling of the natural environment, generally by industrialized society. Pollution is usually a result of an accumulation of waste products, although excess of NOISE and heat which has adverse effects on the surrounding ECOLOGY is also included in the term. *See also* GREENHOUSE EFFECT; GLOBAL WARMING; WATER POLLUTION

polonium (symbol Po) Rare, radioactive, metallic element of group VI of the periodic table, first discovered in 1898 by Marie CURIE. It is found in trace amounts in URANIUM ores and may also be synthesized. Properties: at.no. 84; r.d. 9.40; m.p. 254°C (489°F); b.p. 962°C (1,764°F); most stable isotope ^{209}Po.

polyamide Man-made POLYMER with the generic name of NYLON, formed from the reaction between a dibasic acid (such as adipic acid) and a diamine (such as hexamethylenediamine). Nylon consists of long-chain synthetic polyamides in which the diamine and acid molecules are joined by AMIDE linkages. It was first produced in 1938 by E.I. du Pont de Nemours & Co., USA. *See also* PLASTIC

polycarbonate POLYMER plastic used for making bottles and shatterproof "glass". Polycarbonates are condensation products of carbonyl chloride (PHOSGENE) and a dihydroxy compound such as diphenylol propane. These thermosetting PLASTICS are clear, strong and tough.

polychaete Any member of the class Polychaeta, marine ANNELID worms characterized by distinctive segmentation of the body. The class includes ragworms and bristleworms. Polychaetes have a distinct head with sensory projections and most segments have outgrowths or bristles.

polychlorinated biphenyl (PCB) Several stable mixtures – liquid, resinous or crystalline – of organic compounds made by the reaction of CHLORINE with biphenyl. They are fire-resistant and are used

P

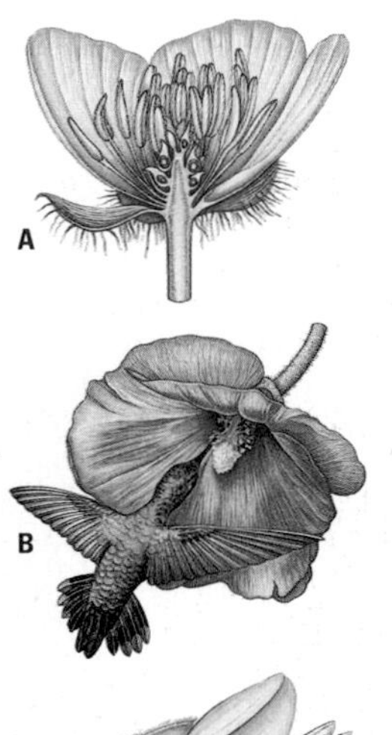

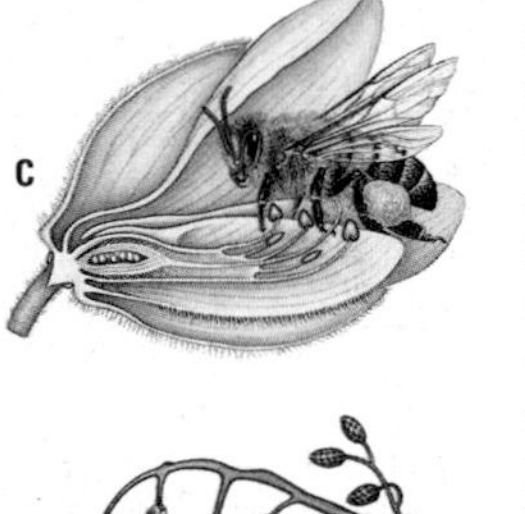

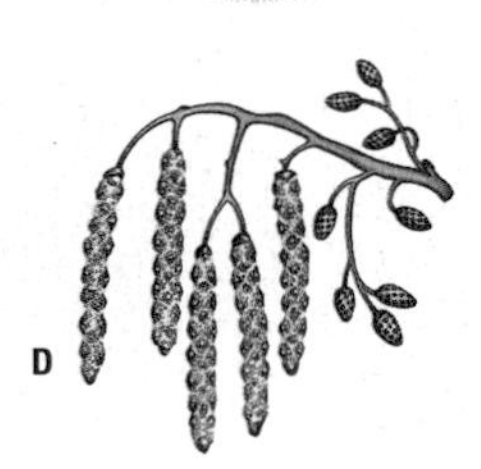

◄ **pollination** Flowers are adapted to different pollination methods. Non-specialized simple flowers, such as the buttercup (A), can be pollinated by a variety of means. Other flowers can only be pollinated by one method. There are bird-pollinated flowers, such as the hummingbird-pollinated hibiscus (B); bee-specialized flowers, including the gorse (C); and wind-pollinated flowers, such as the catkins found on the hazel (D).

as lubricants, heat-transfer fluids and fluids in transformers. They are also used to impregnate capacitors and to give flexibility to protective coatings for wood, concrete and metal. Since 1973 the use of PCBs has been restricted internationally because they are toxic, and their resistance to decomposition in streams and soils poses a threat to wildlife.

polyester Class of organic substance composed of large molecules arranged in either a chain or a network and formed from many smaller molecules through the establishment of ESTER linkages. Polyester fibres are resistant to chemicals and may be washed in alkaline solutions or dry-cleaned. They are made into woven and knitted fabrics. As a clothing material, polyester is valued for its resistance to creasing. *See also* PLASTIC

polyethene (polythene, polyethylene) POLYMER of ETHENE (ethylene), a partially crystalline, lightweight THERMOPLASTIC RESIN, with high resistance to chemicals, low moisture absorption and good insulating properties. *See also* PLASTIC

polygon PLANE geometric figure having three or more sides intersecting at three or more points (vertices). They are named according to the number of sides or vertices: TRIANGLE (three-sided), quadrilateral (four-sided), PENTAGON (five-sided) and so on. A regular polygon is equilateral (has sides equal in length) and equiangular (has equal angles).

polygraph Electronic device that may be capable of detecting lies when used by a trained examiner. The polygraph monitors such factors as heart rate, breathing rate and perspiration, all of which may be affected when a person lies.

polyhedron Three-dimensional figure or solid, the surface of which is made up of POLYGONS. The polygon faces are joined together at edges and the points at which edges meet are called vertices. *See also* EULER'S RULE

polymer Substance formed by the union of from two to several thousand simple molecules (MONOMERS) to form a large molecular structure. Some polymers, such as CELLULOSE, occur in nature; others form the basis of the PLASTICS and synthetic RESIN industry.

polymerase chain reaction (PCR) Chemical reaction that is used to make large numbers of copies of a specific sequence of DNA, or part of a GENE, starting from only one or few DNA molecules. It is speeded up by the ENZYME polymerase. The reaction enables scientists to make large enough quantities of DNA to be able to analyse it or manipulate it. The starting piece of DNA may be extracted from an organism, it may have been "designed" and synthesized in the laboratory, or it may be a forensic sample taken from the scene of a crime. PCR is extremely important in GENETIC ENGINEERING and GENETIC FINGERPRINTING.

polymerization Chemical combination of several molecules to form straight-chain molecules, cross-linked giant molecules or a combination of both, all called POLYMERS. It is the major industrial process in the manufacture of PLASTICS. In nature, large biochemical compounds (such as PROTEINS and CELLULOSE) are also formed by polymerization.

polymorphism In mineralogy, the existence of a substance in two (dimorphic) or more CRYSTAL forms. Polymorphism is caused by different temperatures or pressures or both. The separate forms are called **polymorphs**; for example, DIAMOND and GRAPHITE are polymorphs of CARBON. Polymorphism of the elements is also known as ALLOTROPY. In biology and botany, polymorphism is the existence of two or more types of individual within a single species. For example, some social insects, such as the ant and bee, have several polymorphic types of worker adapted structurally to perform different tasks within their colony. Flowers can also assume more than one form.

polynomial Sum of terms that are powers of a given VARIABLE. In general, a polynomial has the form $P_n(x) = a_n x^n + a_{n-1} x^{n-1} + a_{n-2} x^{n-2} + ... a_2 x^2 + a_1 x + a_0$, where $a_0, ... a_{n-1}, a_n$ are the COEFFICIENTS of the polynomial. The degree of a polynomial is the highest power appearing in the sum with a non-zero coefficient. For instance,

$$P_4(x) = 2x^4 - 3x^3 + x^2 + x + 5$$

is a polynomial of degree four. In this example, the values of the polynomial at $x = 0$, 1 and 2 are $P_4(0) = 5$, $P_4(1) = 6$ and $P_4(2) = 19$, respectively. Polynomials can be represented graphically by plotting values of $y = P_n(x)$ against values of x.

polyp Small, stalked growth occurring on a MUCOUS MEMBRANE, most commonly in the nose, uterus or bowel. Polyps are usually benign, but some, especially those in the intestine, may become cancerous.

polyp Sedentary stage in the life cycle of CNIDARIANS. A polyp has a mouth surrounded by extensible tentacles and a lower end that is adapted for attachment to a surface. It is distinct from the free-swimming MEDUSA. It may be solitary, as in the sea anemone, but is more often an individual of a colonial organism such as CORAL.

polypeptide Chemical consisting of a long chain of AMINO ACIDS linked by PEPTIDE bonds. PROTEINS are polypeptides.

polyploidy State of a cell or nucleus that has more than two sets of CHROMOSOMES. More likely to occur in plants than in animals, polyploidy is found in many cereals. Polyploidy can be introduced artificially using chemicals to produce HYBRID plants with desirable characteristics, such as increased size and hardiness. *See also* DIPLOID; HAPLOID

polypropene (polypropylene) POLYMER of PROPENE (propylene), resembling POLYETHENE and used chiefly in making fibres, films and moulded products. *See also* PLASTIC

polysaccharide Any of a group of complex CARBOHYDRATES made up of long chains of MONOSACCHARIDE (simple-sugar) molecules, often thousands of them. GLUCOSE is a MONOSACCHARIDE, and the polysaccharides STARCH and CELLULOSE are both POLYMERS of glucose. Higher carbohydrates are all polysaccharides and decompose by HYDROLYSIS into a large number of monosaccharide units. Polysaccharides function as both food stores (starch in plants and GLYCOGEN in animals) and as structural materials (cellulose and PECTIN in the cell walls of plants, and CHITIN in the protective skeleton of insects).

polystyrene Synthetic, organic POLYMER, composed of long chains of the aromatic compound STYRENE. It is a strong THERMOPLASTIC RESIN, acid- and alkali-resistant, non-absorbent and an excellent electrical insulator. *See also* PLASTIC

POLYP

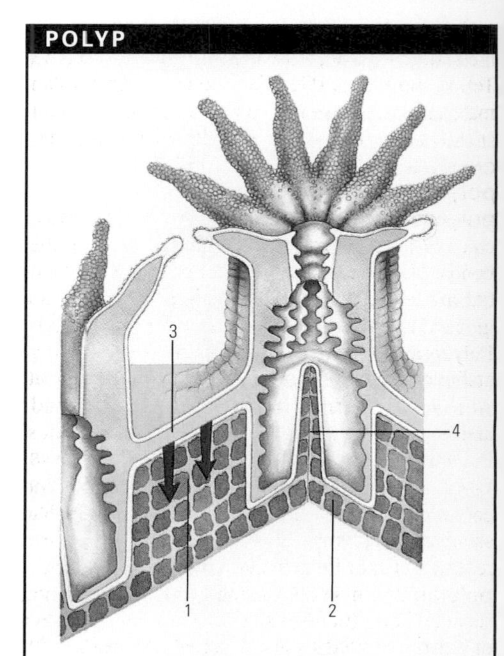

Stony coral polyps obtain support from a mineralized, cup-like theca (1), which is secreted by the animal. The theca, which is composed of minerals absorbed from surrounding seawater, anchors the polyp to the basal structure (2), and to its neighbours in the colony via connecting plates (3). All of the mineralization occurs outside the body, although mineral spikes (4) may provide additional support.

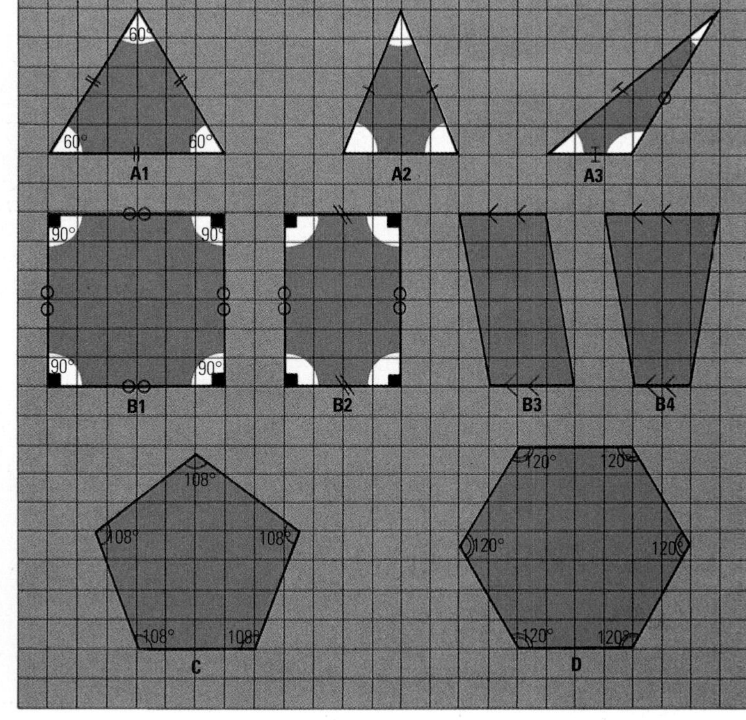

► **polygon** Polygons are many-sided plane figures, the sides and angles of which may be equal or unequal, giving symmetrical or asymmetrical shapes. The simplest figure is the equilateral triangle (A1) which has all sides and angles equal. The isosceles triangle (A2) has two sides equal while (A3), which happens to be obtuse, has sides of differing length. The simplest quadrilateral (B1) is the square with equal sides and equal internal angles. The rectangle (B2) has two pairs of equal and opposite sides with equal internal angles. The parallelogram (B3) has two pairs of equal and opposite parallel sides. The trapezium (B4) has only two parallel sides. The regular pentagon (C) has five equal sides and the regular hexagon (D), six.

polytetrafluoroethene (PTFE) Chemically inert, solid PLASTIC, known also by the trade names Teflon and Fluon. PTFE is used as a heat-resistant material for heat-shields on spacecraft, as a non-stick coating on cooking utensils, and as a LUBRICANT. It is stable up to *c*.300°C (572°F).

polythene *See* POLYETHENE

polyunsaturate Type of FAT or OIL with molecules of long CARBON chains having many double bonds. Nature's main example, considered essential in the human diet, is linoleic acid, found as a glyceryl ESTER in linseed oil and other fats and oils. Polyunsaturated fats exist in fish oils and most vegetable oils, but not those of the palm and coconut. At room temperature, unsaturated oils are liquids and SATURATED FATS, such as lard, are solids. Polyunsaturates, which have low or no CHOLESTEROL content, are widely used in margarines and cooking oils. They are considered healthier than saturated fats, which have been found by medical research to contribute to cardiovascular diseases.

polyurethane POLYMER used chiefly in making flexible foams (in upholstery), rigid foams (in cores of aeroplane wings), ELASTOMERS and RESINS (for adhesives and coatings). *See also* PLASTIC

polyvinyl acetate Colourless resin POLYMER used as an ingredient of latex paint and in lacquers, cements, adhesives and polyvinyl alcohol. It is thermoplastic, flammable, and dissolves in BENZENE. *See also* PLASTIC

polyvinyl chloride (PVC) (polychloroethene) White, tough, solid thermoplastic that is a POLYMER of CHLOROETHENE (vinyl chloride). It can be made by heating chloroethene in water with potassium persulphate or HYDROGEN PEROXIDE. PVC can be softened and made elastic with a plasticizer. Easily coloured and resistant to weather and fire, PVC is used to produce a variety of products, including fibres, bottles, shoes, windows, electrical insulation, vinyl flooring, audio discs, food containers, and coatings for raincoats. *See also* PLASTIC

pome Fleshy FRUIT formed from the flower RECEPTACLE or base. It is not developed from the CARPEL, and is therefore a PSEUDOCARP (false fruit). Familiar examples are the apple, pear and quince.

pons Upper segment of the human BRAINSTEM, which contains the nerve fibres connecting the two halves of the CEREBELLUM. The brainstem is the lower part of the BRAIN and is structurally continuous with the SPINAL CORD. Below the pons is the MEDULLA OBLONGATA, which transmits ascending and descending nerve fibres between the spinal cord and the brain.

porcelain White, glass-like, non-porous, hard, translucent ceramic material. Porcelain is widely used for tableware, decorative objects, laboratory equipment and electrical insulators. The material is often referred to as "china" because the earliest type was developed by the Chinese in the 7th or 8th century. What is referred to as china is, in fact, true or **hard-paste** porcelain, made of KAOLIN (white china clay) mixed with powdered petuntse (FELDSPAR). The mixture is fired at *c*.1,400°C (2,550°F). **Soft-paste** porcelain is composed of clay and powdered glass, fired at a comparatively low temperature, lead glazed and refired. English BONE CHINA is made from a clay paste containing bone ash, which improves the material's translucency.

poriferan Any member of the phylum Porifera – the SPONGES. Poriferans are many-celled, stationary animals that remain permanently fixed to rocks. They feed by filtering food particles from the water.

porosity In geology, degree to which rock is capable of holding GROUNDWATER. This depends on the numbers of AQUIFERS, cracks, fissures and holes present. Porosity is measured and expressed as a percentage of the rock's total volume. *See also* HYDROLOGY; PERMEABILITY

porphyrin Class of red-pigmented ORGANIC COMPOUNDS consisting of four PYRROLE residues (a ring compound, C_4H_4NH) bonded together. The porphyrins form part of the active nucleus of HAEMOGLOBIN, CHLOROPHYLLS *a* and *b* and the ENZYMES catalase and peroxidase.

porphyry Rock that contains large CRYSTALS (phenocrysts) in a fine-textured igneous MATRIX. They are found in both INTRUSIVE and EXTRUSIVE rocks. Different varieties are named after the phenocrysts.

portal vein Large vein that carries blood from the stomach and surrounding organs to the liver.

Porter, Rodney Robert Gerald (1917–85) British biochemist. Porter and Gerald EDELMAN shared the 1972 Nobel Prize for physiology or medicine for their work on the chemical structure of the GAMMA GLOBULIN molecule.

Porter, Sir George (1920–) British chemist and director (1966–85) of the Royal Institution of Great Britain. Porter shared the 1967 Nobel Prize for chemistry with Ronald NORRISH and Manfred EIGEN for their work on very fast CHEMICAL REACTIONS. He and Norrish disturbed the EQUILIBRIUM between atoms and molecules in CHLORINE gas by irradiating it with very short pulses of energy; they then observed it as it returned to equilibrium.

positron Particle that is identical to the ELECTRON, except it is positively charged, making it the ANTIPARTICLE of the electron. The theory of the electron, stated in 1928 by Paul DIRAC, required the existence of a mirror-image particle. In 1932 this particle was observed in COSMIC RADIATION by Carl ANDERSON, and it is also emitted from certain radioactive nuclei. Electron-positron pairs can be produced when GAMMA RADIATION interacts with matter. *See also* ELEMENTARY PARTICLE

positron emission tomography (PET) Medical imaging technique (used particularly on the brain) that produces three-dimensional images. RADIOISOTOPES, injected into the bloodstream prior to imaging, are taken up by tissues where they emit POSITRONS that produce detectable PHOTONS.

positronium Unstable, bound state of an ELECTRON and a POSITRON (the electron's ANTIPARTICLE) held together by electromagnetic attraction, first observed in 1951. It has a half-life of either 1.39×10^{-7} sec or 1.25×10^{-10} sec, depending on whether the SPIN axes of the two particles are parallel or antiparallel.

potash Any of several POTASSIUM compounds, especially potassium oxide (K_2O), POTASSIUM CARBONATE (K_2CO_3) or POTASSIUM HYDROXIDE (KOH). Potash is used as FERTILIZER because potassium is an ESSENTIAL ELEMENT for plant growth.

potassium (symbol K) Common metallic ele-

POSITRON EMISSION TOMOGRAPHY (PET)

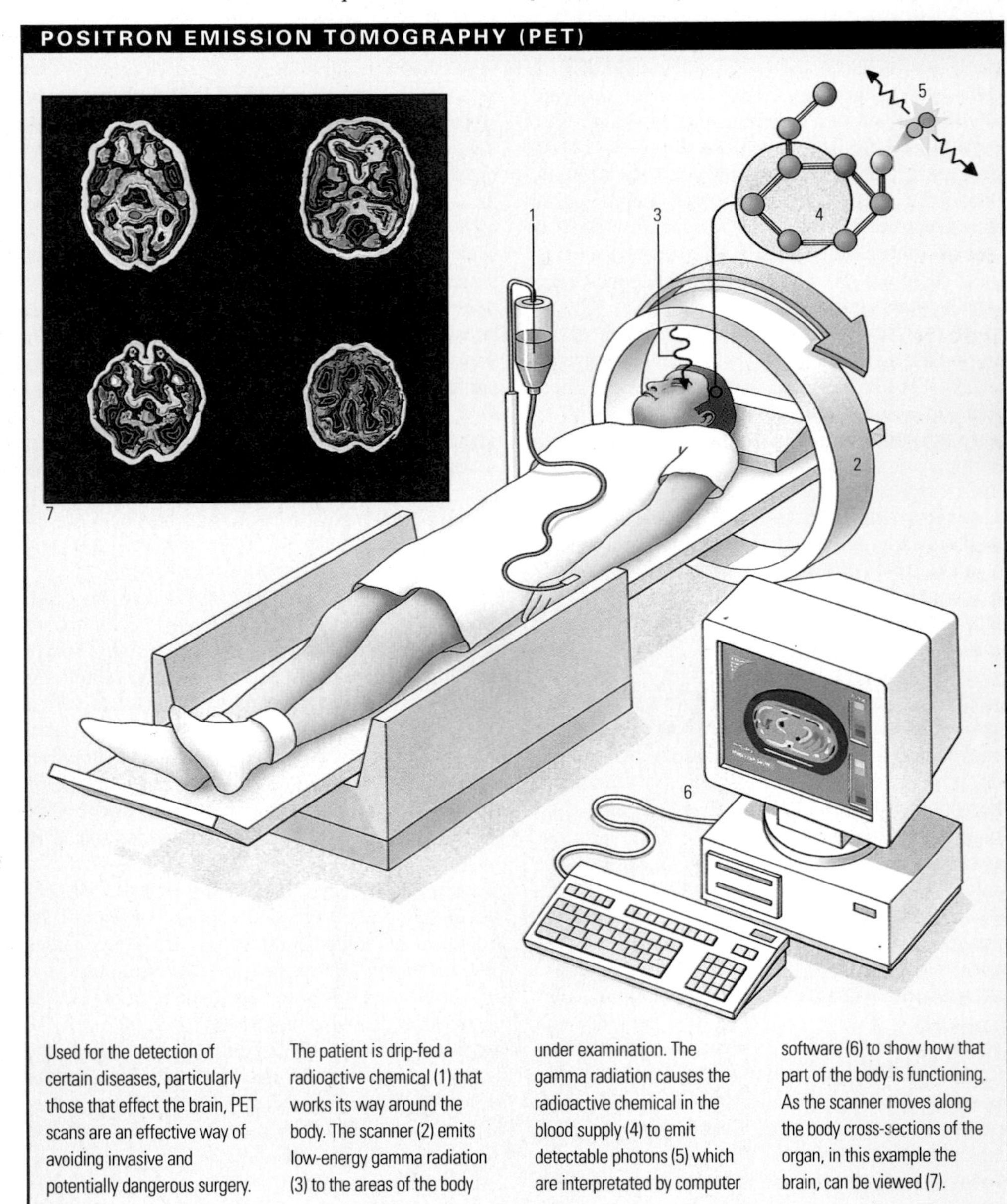

Used for the detection of certain diseases, particularly those that effect the brain, PET scans are an effective way of avoiding invasive and potentially dangerous surgery. The patient is drip-fed a radioactive chemical (1) that works its way around the body. The scanner (2) emits low-energy gamma radiation (3) to the areas of the body under examination. The gamma radiation causes the radioactive chemical in the blood supply (4) to emit detectable photons (5) which are interpretated by computer software (6) to show how that part of the body is functioning. As the scanner moves along the body cross-sections of the organ, in this example the brain, can be viewed (7).

▲ **pothole** at Mount Sodom Salt Cave on the Dead Sea in Israel. This example is in limestone rock, and the hole has formed over time by the action of rain.

ment, one of the ALKALI METALS, first isolated in 1807 by Sir Humphry DAVY. Its chief ores are sylvite (a CHLORIDE), carnallite and polyhalite. The metal is used as a heat-transfer medium in NUCLEAR REACTORS, but has few other commercial uses. Chemically, it resembles SODIUM, being rather more reactive. Potassium in the form of POTASH is used as a fertilizer, because it is an ESSENTIAL ELEMENT for plant growth. The natural element contains a radioisotope ^{40}K (half-life 1.3×10^9 yr), which is used in the radioactive DATING of rocks. Properties: at.no. 19; r.a.m. 39.102; r.d. 0.86; m.p. 63.65°C (146.6°F); b.p. 774°C (1,425°F); most common isotope ^{39} K (93.1%).

potassium-argon dating Method of assessing geological age up to about 10 million years. It is a type of radioactive DATING.

potassium carbonate (potash, K_2 CO_3) White solid, usually produced commercially by ELECTROLYSIS of POTASSIUM CHLORIDE, followed by treatment of the resulting POTASSIUM HYDROXIDE with CARBON DIOXIDE. It is used as a fertilizer, to make glass, soap and dyes, and in cleaning and ELECTROPLATING metals.

potassium chloride Colourless or white POTASSIUM salt (KCl), extracted from lake brines and from minerals such as sylvite, kainite and carnallite. It is used as a fertilizer and as a raw material in the production of POTASSIUM HYDROXIDE and POTASSIUM CARBONATE.

potassium hydroxide (caustic potash) White solid (KOH) prepared commercially by ELECTROLYSIS of POTASSIUM CHLORIDE. It is a strongly alkaline substance used for making soaps and detergents. *See also* ALKALI

potassium manganate(VII) (potassium permanganate) Purple, soluble, crystalline salt ($KMnO_4$) made from PYROLUSITE (manganese dioxide) and POTASSIUM HYDROXIDE. It is used as an oxidizing agent, disinfectant, dye, and in the preparation of other chemicals. Properties: r.d. 2.70; decomposes at 240°C (464°F).

potassium nitrate (saltpetre, nitre) Colourless solid (KNO_3) used as a fertilizer for its NITROGEN content. It is also used in the manufacture of EXPLOSIVES and as a food preservative.

potential, electric *See* ELECTRIC POTENTIAL

potential difference Difference in ELECTRIC POTENTIAL between two points in a circuit or electric field, usually expressed in VOLTS. It is equal to the work done to move a unit electric charge from one of the points to the other. *See also* ELECTROMOTIVE FORCE (EMF)

potential energy Type of ENERGY an object possesses because of its vertical position in the Earth's GRAVITATIONAL FIELD; also, the energy stored in a system like a compressed SPRING or in an oscillating system like a PENDULUM. An object on a shelf, for example, has potential energy given by mgh, where m is its mass, g the acceleration due to gravity, and h the height of the shelf. If the object falls to the ground, its potential energy is converted into KINETIC ENERGY.

potentiometer Instrument used to measure an unknown VOLTAGE or POTENTIAL DIFFERENCE by comparing it with a known one, such as that of a standard electric CELL. It consists of a length of resistance wire fixed taut on a wooden base, and connected across a BATTERY. Both the known and the unknown voltages are in turn used to balance, through a GALVANOMETER and sliding contact, the voltages in a portion of the wire, the ratio of whose lengths equals the ratio of the voltages.

pothole In geology, various formations that are pot-shaped. Most commonly it denotes a circular, bowl-shaped hollow formed in a rocky stream-bed by CORRASION, the grinding action of sand and stones whirled around by eddies or by the force of the stream. Such potholes are usually found in rapids or at the foot of a waterfall. It is also used to describe the vertical pits worn in LIMESTONE rocks by the action of water, occurring where a JOINT has been enlarged by SOLUTION; they often provide the entrances to CAVE systems.

pound Imperial unit of MASS equal to 0.454kg

powder metallurgy Manufacture of metal powders and their use in producing metal parts. Powder particles are compressed to the desired shape and then sintered (heated to just below MELTING POINT). The use of powders is more economical than molten metal in making such items as small gears. Melting may also prove impractical, such as when high melting-point metals are used, or when an ALLOY of unfusable materials is involved. Powder METALLURGY is also used to make porous metal parts.

Powell, Cecil Frank (1903–69) British physicist. During the 1930s, he developed a technique to record ELEMENTARY PARTICLES directly onto film. In 1947 Powell used this method at high altitude to investigate COSMIC RADIATION and discovered a new particle, the PION (pi MESON). This discovery supported the theory of nuclear structure proposed by Hideki YUKAWA. He subsequently discovered the antiparticle of the pion and the decay process of KAONS (K mesons). Powell was awarded the 1950 Nobel Prize for physics.

power In physics, rate of doing WORK or of producing or consuming ENERGY. It is a measure of the output of an engine or other power source. James WATT was the first to measure power; he used the unit called HORSEPOWER. The modern unit of power is the WATT. *See also* HORSEPOWER

power In mathematics, the result of multiplying a number or VARIABLE by itself a specified number of times. Thus, a^2 (= $a \times a$) is the second power of a; a^3 is the third power; a^4 the fourth, and so on. The superscript (2, 3 and 4 above) is called the EXPONENT or power.

praseodymium (symbol Pr) Silver-yellow metallic element of the LANTHANIDE SERIES, first isolated in 1885. Its chief ores are monazite and bastnaesite. Soft, malleable and ductile, praseodymium is used in carbon electrodes for arc lamps. Its green salts are used in coloured glasses, ceramics and enamels. Properties: at.no. 59; r.a.m. 140.9077; r.d. 6.77; m.p. 931°C (1,708°F); b.p. 3,512°C (6,354°F); most common isotope ^{141}Pr (100%).

Precambrian Oldest and longest era of Earth's history, lasting from the formation of the Earth about 4,600 million years ago to the beginning of a good FOSSIL record about 590 million years ago. It is often split into the subdivisions ARCHAEAN and PROTEROZOIC. Precambrian fossils are extremely rare because the earliest life forms are presumed not to have had hard parts suitable for preservation. Also, Precambrian ROCKS have been greatly changed and deformed by metamorphism. Nonetheless, primitive bacteria have been identified in deposits more than 3,000 million years old.

precession Wobble of the axis of a spinning object. This effect occurs as a result of the TORQUE on the spin axis. The Earth precesses about a line through its centre and perpendicular to the plane of the ECLIPTIC extremely slowly (a complete revolution taking 25,800 years) at an angle of 23.5°. It makes the EQUINOXES move backwards through the ZODIAC, and the positions of the stars change by a small amount each year. The motion of a GYROSCOPE is another consequence of precession, in which the entire ring containing the spinning wheel and its axle precess around the support pivot. *See also* NUTATION

precipitate Insoluble solid formed in SOLUTION either by direct reaction or by varying the solution composition to diminish the SOLUBILITY of the dissolved compound. This technique (precipitation) is used for the separation and identification of COMPOUNDS in chemical processes.

precipitation In meteorology, all forms of water particles, whether liquid or solid, that fall from the atmosphere to the ground. Distinguished from CLOUD, FOG, DEW and FROST, in that it must fall and reach the ground, precipitation includes RAIN, drizzle, SNOW (ice pellets and crystals) and HAIL. Measured by rain and snow gauges, the amount of precipitation is expressed in millimetres or inches of liquid water depth. Precipitation occurs with the CONDENSATION of water vapour in clouds into water droplets that coalesce into drops as large as 7mm (0.25in) in diameter. It also forms from melting ice crystals in the clouds. **Drizzle** consists of fine droplets, and snow of masses of many-sided ice crystals. **Sleet** is formed when raindrops freeze into small ice pellets, and hail when concentric layers of ice in cumulonimbus clouds freeze, forming lumps measuring from 0.5–10cm (0.2–4in) in diameter.

predator Animal that gets its food by hunting and killing its PREY. Many, but not all, CARNIVORES are predators (some, such as vultures and hyenas, are scavengers). Predators are SECONDARY CONSUMERS, occupying the top position in a FOOD CHAIN or web. Most predator-prey relationships naturally regulate one another's population. *See also* ECOSYSTEM

prefabrication Assembly of buildings or their components at locations other than their final position. Prefabrication methods are used for kitchen cabinets and appliances, wall and floor panels, roof trusses, window-wall elements, and whole buildings. Crystal Palace was the first building prefabricated in sections and assembled on site (1851).

pregnancy (gestation) Period of time from CONCEPTION until BIRTH; in humans, normally about

40 weeks (280 days). In humans, it is generally divided into three 3-month periods called trimesters. In the first trimester, the EMBRYO grows from a small ball of cells to a FETUS about 7.6cm (3in) in length, during which time the skeleton, brain and such vital organs as the heart and lungs develop. At the beginning of the second trimester, movements are first felt and the fetus grows to about 36cm (14in). In the third trimester, the fetus attains its full body weight. The gestation period for mammals varies greatly. For instance, a female elephant (cow) gives birth to its calf after 18 to 22 months gestation.

Prelog, Vladimir (1906–) Swiss chemist, b. Yugoslavia. He shared the 1975 Nobel Prize for chemistry with John W. Cornforth (1917–) for their work on STEREOCHEMISTRY, the branch of chemistry that deals with the arrangement of ATOMS in a MOLECULE. Prelog devised a procedure for determining whether a molecule is right- or left-handed.

premolar In the DENTITION of adult human beings and other mammals, the two crushing or cutting TEETH between the CANINES and MOLARS on both sides of the upper and lower jaws; there are usually eight in all.

presbyopia Type of HYPERMETROPIA caused by lack of ACCOMMODATION of the eye, common with ageing. As the eye lens becomes older, it loses some of its elasticity and ability to change curvature to focus on near objects. The condition is easily corrected with spectacles or contact lenses.

pressure In physics, force on an object's surface divided by the area of the surface. The SI unit is the PASCAL (symbol Pa), which is $1 Nm^{-2}$ (NEWTON per sq m). In meteorology, the millibar (symbol mb), which equals 100 pascals, is commonly used.

pressure gauge Instrument used to measure fluid PRESSURE. A liquid-column gauge, such as the mercury BAROMETER and MANOMETER, uses the shifts of liquid in U-shaped tubes to measure pressure; a PITOT TUBE measures gas pressures between two points. A mechanical gauge, like the Bourdon tube or bellows-element gauge, utilizes the elasticity of metals to measure pressure.

pressurization Reducing the volume of a substance by forcing it into an increasingly smaller space (such as compressing a gas), or by restricting expansion of a heated substance (as in pressure-cooking). Pressurization is accompanied by an increase in temperature. Pressure-reduction is accompanied by cooling – this is the principle of the process of REFRIGERATION.

prey Animal that is hunted and killed for food by a PREDATOR. Prey animals generally occupy the last place but one in a FOOD CHAIN or web. *See also* ECOSYSTEM *See feature on page 288*

Priestley, Joseph (1733–1804) British chemist and clergyman who discovered OXYGEN (which he called "dephlogisticated air") in 1774. He also discoved a number of other gases, including AMMONIA and the oxides of NITROGEN. He studied the properties of CARBON DIOXIDE (then called "fixed air") and invented carbonated drinks. Priestley was an advocate of the later discredited PHLOGISTON THEORY.

primary cell *See* CELL

primary colour *See* COLOUR

primary consumer In ecology, an animal that feeds off organisms below it in the FOOD CHAIN but does not occupy the top position. Most HERBIVORES are primary consumers, feeding on plants (below them in the food chain) but themselves providing food for SECONDARY CONSUMERS (above them in the food chain). *See also* ECOSYSTEM

primary producer In ecology, an organism at the lowest level of a FOOD CHAIN. Primary producers are AUTOTROPHS. They ultimately feed all the consumers (HETEROTROPHS) above them. Most primary producers are ALGAE or PLANTS. *See also* ECOSYSTEM

Primates Order of MAMMALS that includes monkeys, apes and human beings. Primates, native to most tropical and subtropical regions, are mostly herbivorous, diurnal (day-active), arboreal (tree-dwelling) animals. Their hands and feet, usually with flat nails instead of claws, are adapted for grasping. Most species have opposable thumbs, and all but humans have opposable big toes. They have a poor sense of smell, good hearing and acute binocular vision. The outstanding feature of primates is a large complex brain and high intelligence. Primate characteristics are less pronounced in the relatively primitive prosimians (including tree shrews, bushbabies, lorises and tarsiers) and

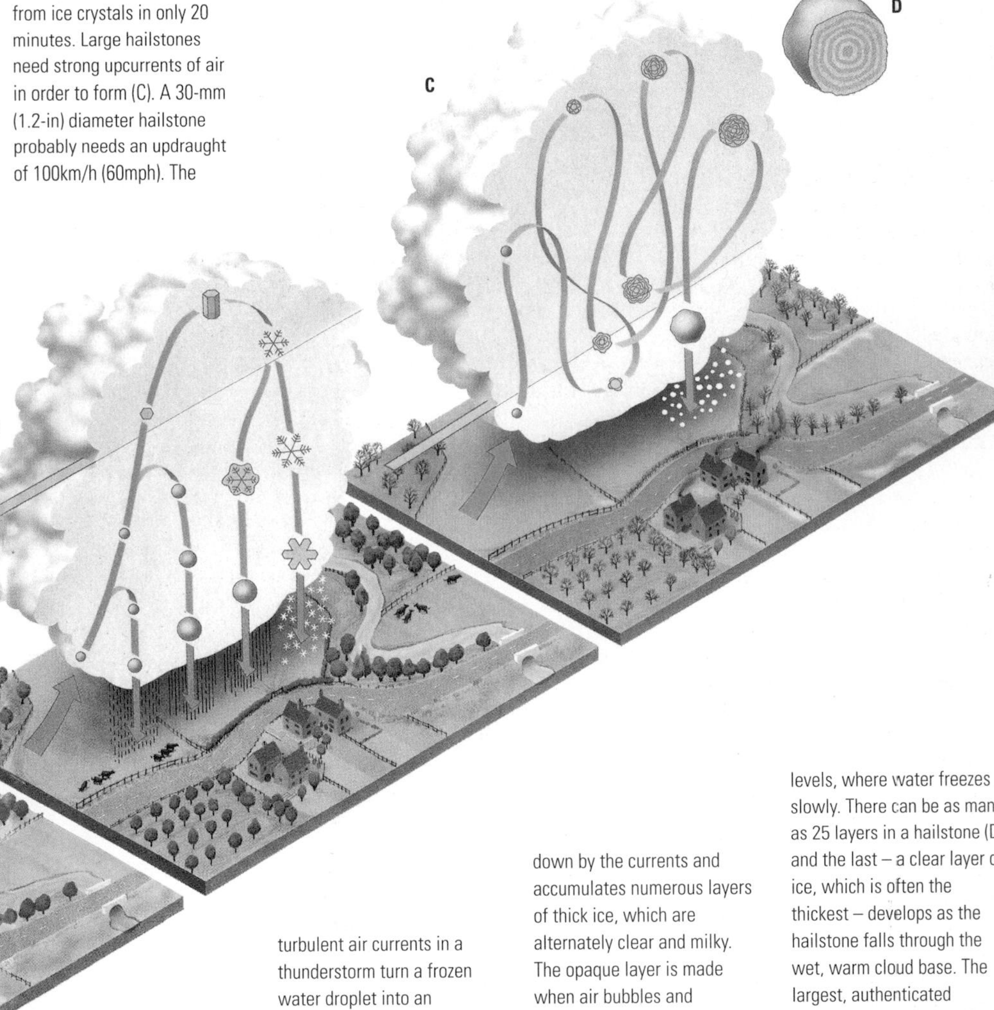

► **precipitation** Shallow clouds and those in the tropics do not reach freezing level, so ice crystals do not form (A). Instead, a larger-than-average cloud droplet may coalesce with several million other cloud droplets to reach raindrop size. Electrical charges may encourage coalescence if droplets have opposite charges. Some raindrops then break apart to produce other droplets in a chain reaction which produces an avalanche of raindrops. Most rainfall in the mid-latitudes is the result of snowflakes melting as they fall (B). It takes many millions of moisture droplets and ice crystals to make a single raindrop or snowflake heavy enough to fall from a cloud. Yet a snowflake can be grown from ice crystals in only 20 minutes. Large hailstones need strong upcurrents of air in order to form (C). A 30-mm (1.2-in) diameter hailstone probably needs an updraught of 100km/h (60mph). The turbulent air currents in a thunderstorm turn a frozen water droplet into an embryonic hailstone. The abundant supercooled moisture droplets in a storm will readily freeze on to its surface. It is swept up and down by the currents and accumulates numerous layers of thick ice, which are alternately clear and milky. The opaque layer is made when air bubbles and sometimes ice crystals are trapped during rapid freezing in the cloud's cold upper levels. The clear layers form in the cloud's warmer, lower levels, where water freezes slowly. There can be as many as 25 layers in a hailstone (D) and the last – a clear layer of ice, which is often the thickest – develops as the hailstone falls through the wet, warm cloud base. The largest, authenticated hailstone fell in Coffeyville, Kansas, on 3 September, 1970; it measured 190mm (7.5in) in diameter and weighted 766g (27oz).

When some species of sea cucumber are being chased by a predator, they expel their entrails through the mouth and anus. This has the effect of distracting the attacker so that the sea cucumber can escape to safety. The entrails grow back in a few weeks. Other animals may use startling or distracting tactics to avoid capture, although few methods are as drastic as that of the sea cucumber. Startle displays usually rely on using warning colours generally recognized throughout the animal world –reds, oranges and yellows. There are three main kinds of startle display. There is the sudden revelation of large false eyes which hopefully deceive the predator into thinking that it is attacking a much larger animal. Alternatively, flash colours (generally red, orange or yellow), unexpectedly displayed in underwing, throats or other parts of the body, will scare most predators. A sudden increase in size is also a very effective deterrent – an in the case of great horned owls, which fluff their feathers and spread their wings when in danger, or toads that puff themselves up and stand on tip-toe. Sudden noises – like the hissing of geese – are also deterrents, and can be particularly effective when combined with flashing warning colours. Animals that do not have natural armour or warning colours with which to deter their attackers usually run away or hide. These are popular defence strategies because, in the long run, they use up less energy than developing survival weapons.

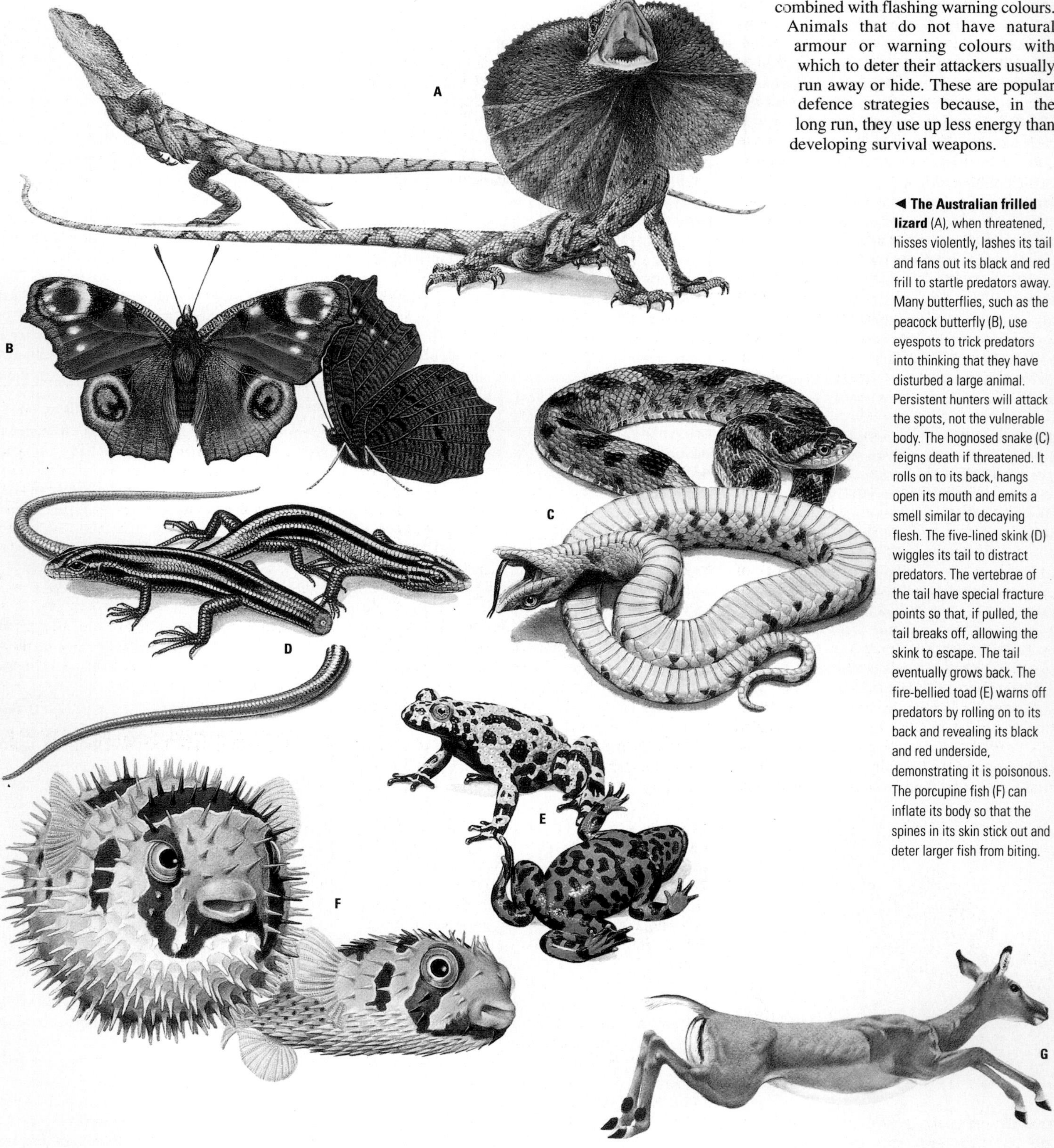

◀ **The Australian frilled lizard** (A), when threatened, hisses violently, lashes its tail and fans out its black and red frill to startle predators away. Many butterflies, such as the peacock butterfly (B), use eyespots to trick predators into thinking that they have disturbed a large animal. Persistent hunters will attack the spots, not the vulnerable body. The hognosed snake (C) feigns death if threatened. It rolls on to its back, hangs open its mouth and emits a smell similar to decaying flesh. The five-lined skink (D) wiggles its tail to distract predators. The vertebrae of the tail have special fracture points so that, if pulled, the tail breaks off, allowing the skink to escape. The tail eventually grows back. The fire-bellied toad (E) warns off predators by rolling on to its back and revealing its black and red underside, demonstrating it is poisonous. The porcupine fish (F) can inflate its body so that the spines in its skin stick out and deter larger fish from biting.

Some animals resort to flight to escape their enemies, but for some, additional physical adaptations can greatly improve the chance of survival, not just for the individual but for the whole group. Impala (G) are just one of many species of grazing antelope that have conspicuous black and white "flesh" markings on their rumps. When one member of the group senses danger and runs away, its flash markings are quickly spotted by other members of the group grazing nearby, which are then also alerted. Other species of animals, such as rabbits, have a similar survival technique.

are most pronounced in the more numerous and advanced ANTHROPOIDEA (monkeys, apes and human beings).

prime number *See* NUMBER, PRIME

principle of moments In mechanics, a law that states that the moments of two bodies balanced about a central pivot or FULCRUM are equal (the moment of a body is the product of its mass and its distance from the pivot). This principle may be observed in BALANCES and see-saws.

Pringsheim, Nathanael (1823–94) German botanist who investigated the reproductive systems of lower plants. Pringsheim discovered the occurrence of SEXUAL REPRODUCTION in ALGAE and, after further studies, concluded that NATURAL SELECTION plays only a minor part in EVOLUTION; he believed that VARIATIONS are spontaneous and tend towards greater complexity. He was the first to demonstrate **apospory**, the production of a sexual from an asexual generation without the intervention of SPORES.

printed circuit Network of electric CONDUCTORS chemically etched from a layer of copper foil coating a board of insulating material such as plastic, glass or ceramic. This interconnects components such as CAPACITORS, RESISTORS and INTEGRATED CIRCUITS (IC). The printed CIRCUIT board (PCB) represents one stage in the miniaturization of electronic circuits (*see* MICROELECTRONICS). It serves to reduce the maze of connecting wires and to compact more circuitry, such as a complete calculating complex, onto a single board or module.

PRINTING

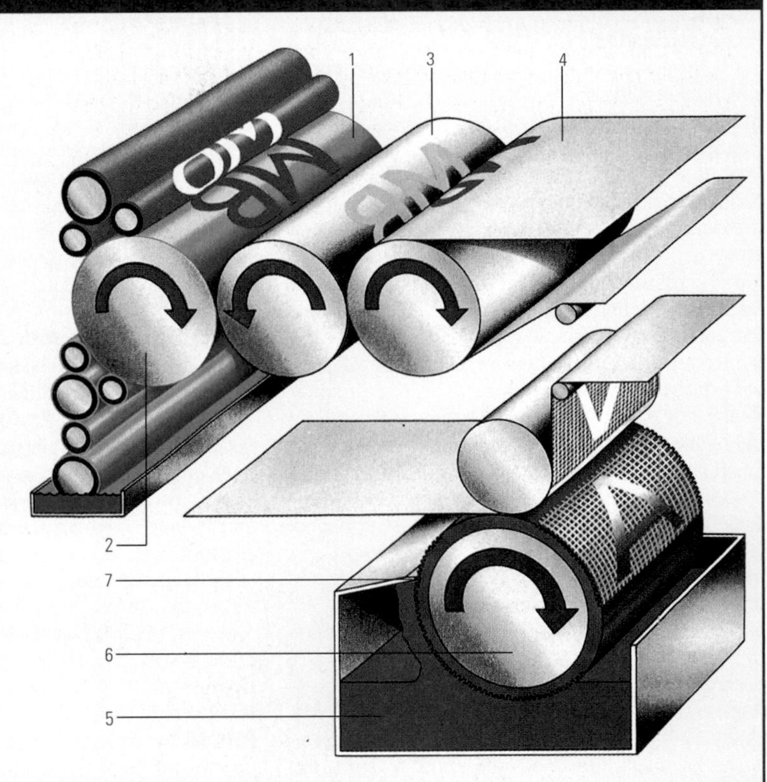

Image areas of offset lithographic printing plates (1) accept a grease-based ink and non-image areas, damped by rollers (2), reject the ink. The inked image is offset on to a rubber roller (3) and then on to the paper (4). The copper gravure printing cylinder (6) has the image etched as "cells" (small depressions beneath the surface), the volume of cell determining the quantity of ink and hence the tonal value. The cylinder is covered with ink (5) and the surface is then cleaned by a "doctor" blade (7), leaving ink only in the etched depressions. When the cylinder contacts the paper the ink in the cells is transferred to it. The gravure process is used principally for high-quality pictorial work.

printing Technique for multiple reproduction of images, such as text and pictures. There are three basic forms of printing – **relief**, **intaglio** and LITHOGRAPHY (**planographic**). The ancient Chinese and Japanese inked carved wooden blocks to print pictures. The earliest-known printed book is Chinese and dates from AD 868. Each page was printed from a solid wooden block. From the 10th century, the Chinese tried using separate pieces of type, so that each page could be printed from arrangements of standard characters. In *c.*1403 metal type made by casting first appeared in Korea. It enabled large quantities of type to be produced quickly. In the 1400s, Johann GUTENBERG and William CAXTON developed the use of LETTERPRESS PRINTING in Europe. In the 1700s and 1800s printing expanded rapidly. Lithography enabled printers to produce impressive colour prints. *The Pencil of Nature* (1844–46) by William Fox TALBOT was the first book to be illustrated with photographs. PHOTOGRAPHY, in turn aided the development of new techniques of reproduction, such as the halftone process. In 1846 Richard Hoe designed a rotary press that used STEREOTYPE plates. For text, metal printing plates were cast from the pages of type, so that the type could be used for setting other pages. In 1884 Ottmar MERGENTHALER invented the LINOTYPE machine. Other TYPESETTING machines, such as the MONOTYPE, speeded up the process of setting up pages. More recently, production speeds have greatly increased with the application of PHOTOSETTING, in which the type is set photographically on sheet film, and OFFSET PRINTING. Illustrations are often reproduced by PHOTOENGRAVING. Today, many publications are produced using a WORD PROCESSOR to enter the text. DESKTOP PUBLISHING programs allow images of the text and pictures to be arranged on screen. The computer data is used to print sheet film for each page, and the film images are transferred to printing plates.

▲ **Primates** Because Primates have large brains they can learn complex skills and pass them down through the generations. Chimpanzees, for example, demonstrate high intelligence in their use of tools. They have been observed using sticks to "fish" for termites. Young chimpanzees learn this by observing their parents – aged between two and three years, they manipulate sticks as a form of play, and by the time they are four they have mastered the use of the tool.

prion Infective agent that appears to consist simply of a PROTEIN. Prions are thought to cause diseases such as CREUTZFELDT-JAKOB DISEASE (CJD) in humans, BOVINE SPONGIFORM ENCEPHALOPATHY (BSE) in cattle, and SCRAPIE in sheep. It is not understood how prions work, or how they persist in nature: unlike VIRUSES and BACTERIA, they do not contain DNA or RNA. Infection with a prion appears to cause a change in shape of the prion protein to a form that has a harmful effect.

prism In mathematics, solid geometrical figure, the ends of which are CONGRUENT (usually triangular) and perpendicular to the other faces, which are rectangular. The volume of a prism equals the area of either end multiplied by the length of the prism. In physics, a prism is a piece of transparent material, such as glass, plastic or quartz, in which a light beam is refracted and split into its component colours (SPECTRUM) by dispersion or undergoes internal REFLECTION.

PRISM

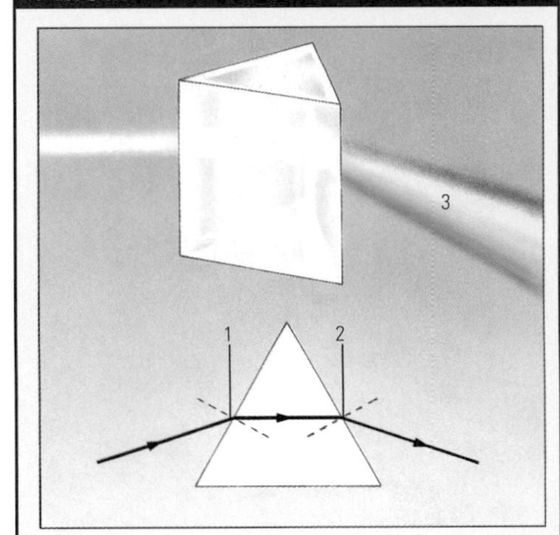

When light hits a prism it is refracted by the two surfaces it hits (1, 2). White light splits into the spectrum (3) because each of the colours of spectrum have varying wavelengths. For example, the short wavelengths of blue and indigo are refracted more than the colours further down the spectrum with longer wavelengths such as orange and red.

probability Number in the range from zero to one (inclusive) representing the likelihood of a given event. The probability of an event is defined as being equal to the number of ways that event may occur divided by the total number of possible outcomes, assuming that each possibility is equally likely. For instance, if a six-sided dice is thrown there are six possible outcomes, three of which give an even number, so the probability of throwing an even number is $^{3}/_{6}$ (or $^{1}/_{2}$). A probability of 0 represents an event that is impossible (throwing a seven on a six sided dice), while1 is the probability of an event that is certain and must occur. When a fair dice is thrown, the outcome is independent of the result of previous throws. In this case, the probabilities are said to be independent or unconditional. More generally one can consider conditional probabilities too, where the probability of an event depends on previous outcomes. The theory of prob-

P

ability was first developed (*c*.1654) by Blaise PASCAL and Pierre de FERMAT. *See also* STATISTICS

Proboscidea Order of MAMMALS that has lived on Earth from the EOCENE to the present day, but is now represented only by elephants (genera *Elephas* and *Loxodon*). They have primitive limbs but specialized trunks and teeth. Family Elephantidae.

process-engraving *See* PHOTOENGRAVING

producer gas Mixture of the flammable CARBON MONOXIDE and HYDROGEN and the nonflammable NITROGEN and CARBON DIOXIDE. It is made by the partial combustion of coal or coke in air and steam. It has a lower heating value than other fuels, but can be manufactured with simple equipment. Producer gas is often used as a fuel in large industrial FURNACES.

progesterone Steroid HORMONE secreted mainly by the CORPUS LUTEUM of the mammalian OVARY and by the PLACENTA during pregnancy. Its principal function is to prepare and maintain the inner lining (ENDOMETRIUM) of the UTERUS for pregnancy. Synthetic progesterone is one of the main components of the contraceptive pill (*see* CONTRACEPTION).

program *See* COMPUTER PROGRAM

programming *See* COMPUTER PROGRAMMING

progression *See* ARITHMETIC PROGRESSION; GEOMETRIC PROGRESSION

proinsulin Precursor of the hormone INSULIN. Proinsulin is a POLYPEPTIDE formed in the PANCREAS by the beta cells of the ISLETS OF LANGERHANS. Its long-chain molecule splits to form the insulin molecule.

projector Instrument with a LENS system, used to cast images onto a screen from an illuminated flat object. An **episcope** is a projector for opaque objects, such as a printed page; it uses light that is reflected from the object. A **diascope** is a projector for transparent objects, such as photographic slides and films; it uses light transmitted through the object. An EPIDIASCOPE can project images from either transparent or opaque objects. A motion-picture or **cine** projector produces moving images from many frames (pictures) on a transparent FILM.

prokaryote Any member of the kingdom Prokaryotae (formerly Monera), which includes BACTERIA and CYANOBACTERIA (formerly blue-green ALGAE). They have more simple CELLS than other organisms. DNA is not contained in CHROMOSOMES in the NUCLEUS, but lies in a distinct part of the CYTOPLASM, called the nucleoid. They have no distinct membrane-surrounded structures (ORGANELLES). Cell division is simple and in the rare cases where SEXUAL REPRODUCTION occurs, genetic material is simply transferred from one partner to another; there are no separate sex cells. In photosynthetic prokaryotes, PHOTOSYNTHESIS takes place on the cell membrane. At present, two subkingdoms are recognized, ARCHAEBACTERIA and EUBACTERIA. *See also* ASEXUAL REPRODUCTION; EUKARYOTE

Prokhorov, Alexsandr Mikhailovich (1916–) Russian physicist. He shared the 1964 Nobel Prize for physics with his colleague Nikolai BASOV and the US physicist Charles H. TOWNES. This was for research in quantum electronics that resulted in the development of the MASER and LASER. In 1953 Prokhorov and Basov proposed the maser principle, and Townes constructed the first operational device.

prolactin HORMONE that stimulates the production of milk after childbirth. It is secreted by the anterior PITUITARY GLAND. Prolactin also stimulates the secretion of PROGESTERONE by the CORPUS LUTEUM of the OVARY. *See also* LACTATION

proline ($C_5H_9O_2N$) White, crystalline AMINO ACID found in PROTEINS.

PROM (acronym for **pr**ogrammable **r**ead-**o**nly **m**emory) COMPUTER memory device consisting of INTEGRATED CIRCUITS, the contents of which are individually installed after manufacture (and cannot thereafter be reprogrammed). PROMS are used in control circuits for lifts and domestic appliances, such as washing machines. *See also* ROM

promethium (symbol Pm) Radioactive, metallic element of the LANTHANIDE SERIES. It was first made in 1941 by particle bombardment of NEODYMIUM and PRASEODYMIUM. Promethium occurs in URANIUM ores in minute amounts. The isotope ^{147}Pm is used in phosphorescent paints, X-ray tubes and in nuclear-powered batteries for space vehicles. Properties: at.no. 61; m.p. 1,080°C (1,976°F); b.p. 2,460°C (4,460°F); most stable isotope ^{145}Pm (half-life 17.7 years).

PROPORTION

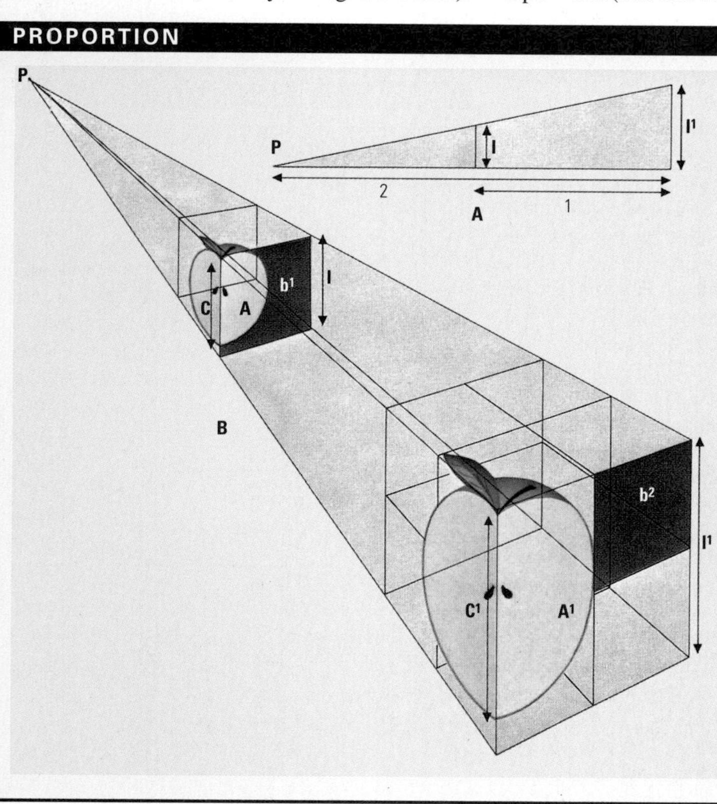

The diagram shows two cubes whose distances from the point P are in the ratio 1:2 as A shows. This measurement is taken from the nearest face of both of these cubes, thus the lengths l and l^1 are in the ratio 1:2. The areas of these faces are clearly in the ratio 1:2^2, (figure B) since area b^1 equals area b^2. The volumes however are in the ratio 1:2^3 since the larger cube is composed of 8 of the smaller ones.The apple segments show the same properties – the apple cores (C,C^1) are in the ratio 1:2. The areas A, A^1 of the cut faces are in the ratio 1:4 and the volumes of these similar solids are in the ratio 1:8.

▲ **prop root** Some species of tree, such as this example found in the cloud forests of Venezuela, grow in very shallow soil and they have little in the way of underground root systems. Instead they are supported by prop roots which are largely above ground. As well as providing support the roots, by spreading farther from the base of the tree, cover a wider area from which they can draw essential nutrients.

prominence, solar Cloud of matter extending outwards from the Sun's CHROMOSPHERE into the CORONA. Prominences are regions of higher density and lower temperature than the surrounding corona. Near the SUN's limb, they are visible as bright protrusions into the dark corona during total eclipses or with the aid of an instrument such as a coronagraph or a special filter. **Quiescent** prominences, which extend for tens of thousands of kilometres, can last for weeks or months. **Active** prominences are short-lived, high-speed, flamelike eruptions that can reach heights of as much as 700,000km (450,000mi) in just 1 hour. All prominences have spectra showing lines of neutral hydrogen, helium and ionized calcium. Their forms and behaviour are very varied, and their direction of flow is controlled by the Sun's MAGNETIC FIELD. *See also* SOLAR FLARE

promoter In chemistry, substance used in small quantities in conjunction with a CATALYST to increase the catalyst's activity. For example, in the HABER PROCESS, an iron catalyst is used to speed up the combination of hydrogen and nitrogen under pressure to form ammonia; this catalyst is usually promoted with small amounts of potassium oxide.

propagation In botany, method of producing several individual plants from one original. This may be in the form of SEXUAL REPRODUCTION, in which the plant is pollinated and the seed produced in the normal way, or VEGETATIVE REPRODUCTION (vegetative propagation), in which plants reproduce asexually by means of CORMS, BULBS, TUBERS or RUNNERS. Vegatative reproduction can be performed artificially by taking cuttings or by grafting. In zoology, ASEXUAL REPRODUCTION, for example the budding of new individuals in hydra, is also called vegetative propagation. In other areas of science, the term propagation describes a spreading out, for example of waves from a central source or of nerve impulses along the AXON of a NEURON. *See also* POLLINATION

propane (C_3H_8) Colourless, flammable GAS, the

third member of the ALKANE series of HYDROCARBONS. It occurs in NATURAL GAS, from which it is obtained; it is also acquired during petroleum REFINING. Propane is used (as bottled gas) as a fuel, as a SOLVENT and in the preparation of many chemicals. Properties: m.p. −190°C (−310°F); b.p. −42°C (−43.6°F).

propanol (propyl alcohol) Colourless ALCOHOL used as a solvent and in the manufacture of various chemicals. It exists as two ISOMERS. Normal propanol ($CH_3CH_2CH_2OH$) is a by-product of the synthesis of METHANOL (methyl alcohol). Propan-2-ol (isopropyl alcohol, $(CH_3)_2CHOH$), is a secondary alcohol which is easily oxidized into PROPANONE.

propanone (acetone) Colourless, flammable liquid (CH_3COCH_3) made by oxidizing propan-2-ol (isopropyl alcohol). It is a raw material for the manufacture of many organic chemicals and is a widely used solvent. Properties: r.d. 0.79; m.p. −94.8°C (−138.6°F); b.p. 56.2°C (133.2°F).

propellant Material that undergoes chemical, nuclear or thermoelectric reactions to propel a ROCKET. **Liquid** propellants consist of a fuel such as PARAFFIN or HYDRAZINE (N_2H_4), which reacts with an oxidizer such as liquid OXYGEN. **Solid** propellants contain both fuel and oxidizer in powder form. **Nuclear** propellants include URANIUM and PLUTONIUM. **Ion** propellants include metallic CAESIUM, which boils off ions into an electric field that accelerates them to high exhaust velocities. The pressurized gas in an AEROSOL is also called a propellant.

propeller Device for producing THRUST, usually mounted on a rotating shaft. The cross-section of an aircraft propeller reveals an AEROFOIL shape, which allows it to function as a rotating wing. It generates forward thrust by producing aerodynamic LIFT. A ship's propeller, or SCREW PROPELLER, "screws" the ship through the water.

propene (propylene) Colourless, aliphatic HYDROCARBON, $CH_3CH:CH_2$, manufactured by the thermal CRACKING of ETHENE. It is used in the manufacture of a wide range of chemicals including vinyl and acrylic RESINS. Properties: b.p. −48°C (−54.4°F); m.p. −185°C (−301°F).

proper motion Apparent motion of a star on the CELESTIAL SPHERE, as a result of its movement relative to the Sun. Most annual proper motions are smaller than 0.1 arc seconds. BARNARD'S STAR has the largest known proper motion, 10.3 arc seconds per year. Proper motion is determined by comparing the star's position on photographic plates exposed on widely separated occasions, usually many years or decades apart. Much more accurate measurements of proper motions have been obtained by the position-measuring satellite Hipparcos.

prophase Stage in cell division, the first stage of MITOSIS or MEIOSIS. During prophase, CHROMOSOMES become visible, shrink and split along their length to form CHROMATIDS.

prophylaxis Medical treatment in which the main focus is preventing a disease or other disorder. Prophylactic measures include health screening, IMMUNIZATION and, on a more general level, a healthy diet and adequate exercise. The practice of safe sex to avoid the risk of AIDS is a form of prophylaxis.

proportion Relationship between four numbers in which the ratio of the first pair (a and b) equals the ratio of the second pair (c and d), $a/b = c/d$. A continued proportion is a group of three or more quantities, each bearing the same ratio to its successor, such as the sequence 1:3:9:27:81.

proprioception Perception of internal nervous stimuli through which an animal is aware of its own movement, posture and internal condition.

prop root (stilt root) Type of ROOT that grows from the STEMS of a plant and gives it extra support. Prop roots continue to be produced as the thin, main stem grows taller, and are often necessary to prevent the plant from falling over. They occur in herbaceous plants such as maize and in some trees, such as mangroves. In other tropical trees, wide, flat-topped prop roots called **buttress roots** grow from the base of the trunk and help to support the shallow-rooted trees.

propyl alcohol *See* PROPANOL

propylene *See* PROPENE

prostaglandin Series of related FATTY ACIDS, with hormone-like action, present in semen, liver, brain and other tissues. Their biological effects include the lowering of blood pressure and the stimulation of contraction in a variety of smooth muscle tissues, such as the UTERUS.

prostate gland Gland in the male reproductive tract surrounding the URETHRA. It secretes specific chemicals that mix with SPERM cells and other secretions to make up the sperm-containing fluid, SEMEN.

prosthesis Artificial substitute for a missing organ or part of the body. Until the 17th century artificial limbs were either wooden or solid metal, but innovations in metallurgy and engineering design have enabled lighter, jointed limbs to be made. Methods of attachment, too, have greatly improved. More recent prosthetic devices indude artificial heart valves made of SILICONE materials.

protactinium (symbol Pa) Rare, radioactive, metallic element of the ACTINIDE SERIES, first identified in 1913. Its chief source is URANINITE. Properties: at.no. 91; r.a.m. 231.0359; r.d. 15.4; m.p. 1,200°C (2,192°F); b.p. 4,000°C (7,232°F); most stable isotope ^{231}Pa (half-life 3.25×10^4 yr).

protective coloration Natural CAMOUFLAGE or warning colours of organisms that serve to blend it in with the surrounding environment or to ward off PREDATORS. Tigers and some moths have permanent protective colouring. Chameleons and some flatfish can change colour to match the background. Warning colours of an animal usually mean it is poisonous, aggressive or distasteful to most predators. Predators learn to recognize and avoid these colorations, which may be mimicked by harmless species. *See also* MIMICRY; PIGMENTATION

protein Organic COMPOUND containing many AMINO ACIDS linked together by covalent, PEPTIDE bonds. Protein molecules consist of POLYPEPTIDE chains. Living CELLS use about 20 different amino acids; because proteins have thousands of amino acids in each MOLECULE, the number of possible proteins is very large. The order of amino acids in proteins is controlled by the GENES in the cell's DNA. The most important proteins are ENZYMES, which determine all the chemical reactions in the cell, and ANTIBODIES, which combat infection. **Structural** proteins include KERATIN and COLLAGEN. **Gas transport** proteins include HAEMOGLOBIN. **Nutrient** proteins include CASEIN. METABOLISM is regulated by protein HORMONES. ACTIN and MYOSIN are contractile **muscle** proteins. *See also* NUCLEIC ACID; X-RAY DIFFRACTION

protein synthesis Production within CELLS of PROTEINS from their component AMINO ACIDS. The GENETIC CODE, carried by the DNA of the CHROMOSOMES, determines which amino acids (of the 20 available) are used and in which order they are combined. The code is, in turn, encoded in MESSENGER RNA (mRNA) transcribed from the DNA in

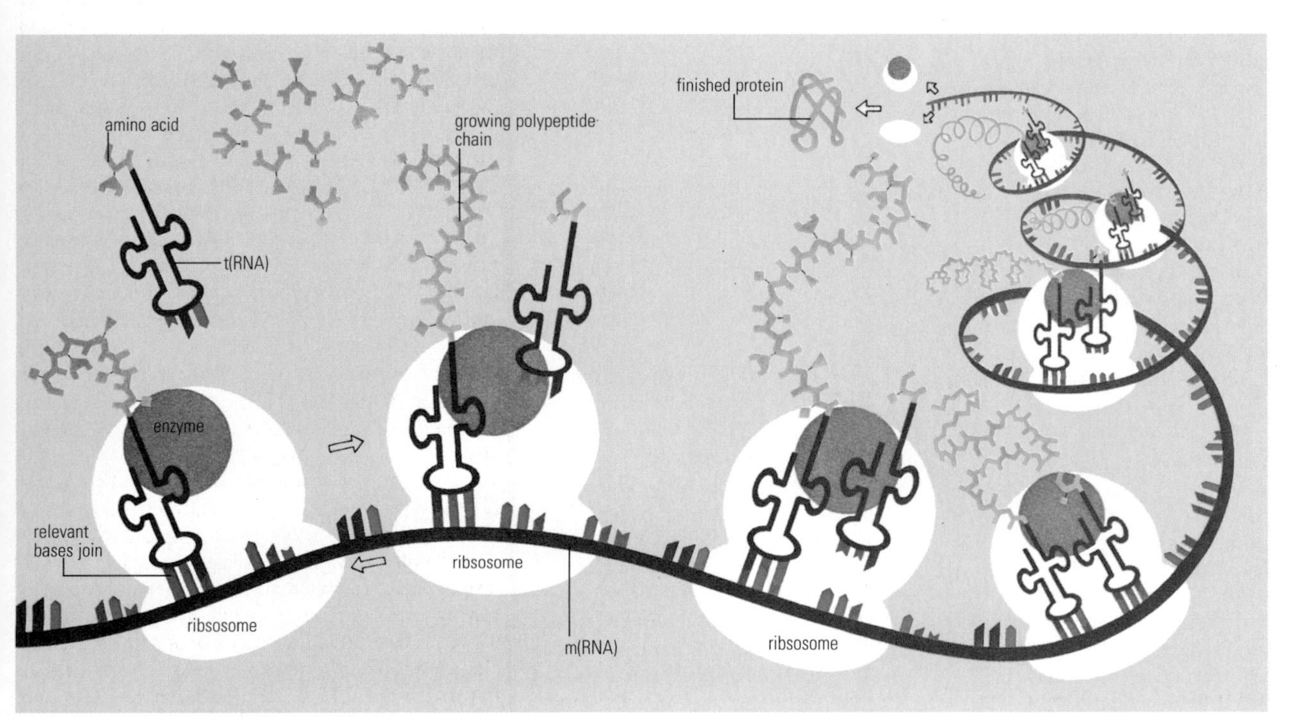

◀ **protein synthesis** occurs in ribsosomes, the sites where amino acids become linked together to form protein chains. They attach to the messenger RNA (mRNA) template, which moves relative to them. Amino acids join up in sequence as the bases of the transfer RNA (tRNA) molceules, on which they are carried, must combine with the relevant bases on (mRNA) as it passes the ribosome. Adenine (blue) always pairs with thymine (green), and cytosine (red) pairs with guanine (orange). One mRNA molecule may carry many ribosomes, with protein chains growing on each (this is called a polysome).

PULLEY

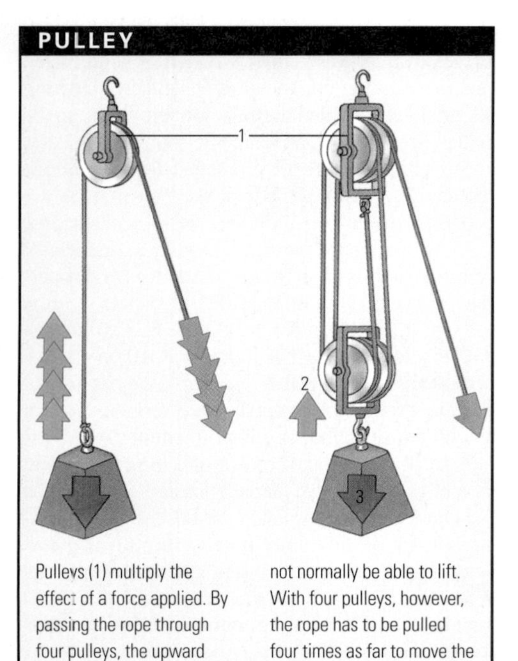

Pulleys (1) multiply the effect of a force applied. By passing the rope through four pulleys, the upward pulling force (2) is four times the applied force, which allows a person to lift a weight (3) that they would not normally be able to lift. With four pulleys, however, the rope has to be pulled four times as far to move the weight the same distance. To move the weight up 1m (3.3ft) the rope must be pulled down 4m (13ft).

the cell's NUCLEUS. The mRNA carries the information to the RIBOSOMES, where protein synthesis actually takes place. *See also* NUCLEIC ACID

Proterozoic Period of GEOLOGICAL TIME, the second or the two subdivisions of the PRECAMBRIAN era. Dating from about 2.5 billion years ago to about 590 million years ago, it saw the emergence of the very earliest life forms such as BACTERIA and ALGAE.

prothallus In some primitive plants, part that carries the sex organs. A flattened green structure, it is found in FERNS, CLUB MOSSES and HORSETAILS, and is the independent GAMETOPHYTE generation (*see* ALTERNATION OF GENERATIONS). It is anchored to damp soil by hair-like RHIZOIDS. A single prothallus may bear a male sex organ (ANTHERIDIUM) or a female sex organ (ARCHEGONIUM), or both types. The prothallus initially provides nourishment for the EMBRYO and young SPOROPHYTE after it forms in the archegonium.

prothrombin Precursor of the blood ENZYME THROMBIN. It is converted into thrombin by thromboplastin. Prothrombin plays an essential role in BLOOD CLOTTING.

protoctist Any member of the kingdom Protoctisa (formerly Protista), which includes such widely differing groups as ALGAE (including large seaweeds), AMOEBAS and other PROTOZOA, SLIME MOULDS and downy mildews.

proton (symbol p) Stable, ELEMENTARY PARTICLE with a positive charge equal in magnitude to the negative charge of the ELECTRON. It forms the NUCLEUS of the lightest isotope of HYDROGEN and, with the NEUTRON, is a constituent of the nuclei of all other elements. Also, like the neutron, it is made up of three QUARKS. The proton is a BARYON with spin $^1/_2$ and a mass 1836.12 times that of the electron. The number of protons in the nucleus of an element is equal to its ATOMIC NUMBER, also called proton number. Protons also occur in primary COSMIC RADIATION. Beams of high-velocity protons, produced by PARTICLE ACCELERATORS, are used to study nuclear reactions. In 1919, as a result of his work on the disintegration of the nucleus, Ernest Rutherford discovered the proton. Its antiparticle, the antiproton, was discovered in 1955. *See* ANTIMATTER

protonema Early stage in the development of the GAMETOPHYTE of a moss or liverwort. Most protonemata consist of green branched threads that grow on the soil surface. They carry buds that eventually develop into the adult form of the plant. *See also* ALTERNATION OF GENERATIONS

protoplasm Term referring to all the discrete structures within a plant or animal CELL, including the cell membrane. It includes both the NUCLEUS and the CYTOPLASM of cells.

protoplast (energid) In biology, term that describes the living unit of a plant CELL. The protoplast includes the cell NUCLEUS and the CYTOPLASM with its various ORGANELLES, but excludes the cell wall. *See also* PROTOPLASM

prototherian Any member of the Mammalia subclass Prototheria. Prototherians are represented in living fauna only by the MONOTREMES – echidnas and platypuses.

protozoan Any member of the phylum Protozoa, which comprises unicellular organisms found worldwide in oceans or freshwater, free-living and as PARASITES. These microscopic animals have the ability to move (by CILIA or pseudopodia) and have a NUCLEUS, CYTOPLASM and cell wall; some contain CHLOROPHYLL. Reproduction is by FISSION or encystment. Length: 0.3mm (0.1in). The 30,000 species are divided into four classes – Flagellata, Cnidospora, Ciliophora and Sporozoa.

Proust, Joseph Louis (1754–1826) French chemist who specialized in analysis, with little interest in theory. As a result of his researches, however, he concluded that COMPOUNDS were of constant and determinate composition. This discovery was very important in the establishment of the atomic theory of chemical composition.

Prozac Trade name for one of a small group of antidepressants, known as selective SEROTONIN reuptake inhibitors (SSRIs), which work by increasing levels of serotonin in the brain. Serotonin, or 5-hydroxytryptamine (5-HT), is a NEUROTRANSMITTER involved in a range of functions. Low levels of serotonin are associated with depression. Prozac is said to restore self-confidence, improve self-image and boost energy. The DRUG appears to be as effective as older antidepressants and less toxic, but may produce side-effects, such as nausea and diarrhoea.

prussic acid *See* HYDROCYANIC ACID

pseudocarp (false fruit) Fruit that is formed from other parts of a FLOWER as well as the OVARY wall. A pseudocarp may incorporate tissue from the CALYX or RECEPTACLE. In a strawberry, for example, the receptacle becomes the fleshy part while the true fruits are the "seeds" embedded in it. Other examples include apples, figs, pears and rose hips.

pseudomorphism In mineralogy, chemical or structural alteration of a mineral without change in shape. An example is **substitution**, in which the original substance is removed and replaced by another. This is exemplified by petrified wood; the wood has been gradually replaced by SILICA. *See also* PETRIFICATION

psychobiology Field, closely related to physiological psychology, that studies anatomical and biochemical structures and processes as they affect behaviour.

PULSAR

1 2 1 2

◄ **A pulsar's radiation** varies. The red on the diagram indicates the ends of the magnetic axis of a pulsar. As the pulsar rotates, the signal strength varies according to the axis. When one end faces Earth (1), the intensity is at its maximum. When the other end faces Earth (2), intensity is at its minimum.

◄ **Pulsar radiation**. It is believed that pulsars emit radiation in a narrow beam so that the rotation of the star produces regular flashes like a lighthouse.

psychogenic disorders Mental disturbances believed to have an underlying psychological origin. NEUROSIS and functional PSYCHOSIS are usually thought of as psychogenic.

psychosis Serious mental illness in which, in contrast to NEUROSIS, the patient loses contact with reality. It may feature extreme mood swings between depression and mania, delusions or hallucinations, distorted judgement and inappropriate emotional responses. The organic psychoses may spring from BRAIN DAMAGE, advanced syphilis, senile DEMENTIA or advanced EPILEPSY. The **functional** psychoses, for which there is no known organic cause, include schizophrenia and manic-depressive psychosis (bipolar disorder).

psychosomatic Descriptive of a physical complaint thought to be rooted, at least in part, in psychological factors. Extending the psychiatric concept of mind and body (psyche and soma) unity, it follows that emotional conflict can manifest itself in physical symptoms. The term has been applied to many complaints, including ASTHMA, migraine, ulcers and HYPERTENSION.

psychosurgery Surgery on the BRAIN performed with the intention of relieving psychological symptoms or altering behaviour. It may be performed on animals for experimental purposes or (rarely) upon human beings to treat severe and intractable mental illness.

pteridophyte Commonly used name for any of a group of SPORE-bearing VASCULAR PLANTS. At one time, pteridophytes were taken to include CLUB MOSSES, HORSETAILS and FERNS. These plants have similar life cycles but in other respects are quite distinct and are now classified as separate phyla. *See also* TRACHEOPHYTE

PTFE Abbreviation of POLYTETRAFLUOROETHENE

PTH Abbreviation of PARATHYROID HORMONE

Ptolemaic system Arrangement of celestial bodies and their motions, proposed by APOLLONIUS OF PERGA and formulated by PTOLEMY *c.*AD 140. In the Ptolemaic system, the Earth is fixed at the centre, with the Moon, Mercury, Venus, the Sun, Mars, Jupiter and Saturn revolving about it. Beyond these planets lies the sphere of fixed stars. Each of the bodies moves around a small circle called the **epicycle**, the centre of which in turn revolves around the Earth on a larger circle called the **deferent**. Ptolemy added to each ORBIT two more points, the **eccentric** and the **equant**, equally spaced either side of the Earth. He made the epicycle revolve around the eccentric rather than the Earth, and Earth have uniform motion with respect to the equant. The resulting model reproduced the apparent motions of the planets so well that it remained unchallenged until the revival of the heliocentric theory by Nicolas COPERNICUS in the 16th century. *See also* ECLIPTIC; ECCENTRICITY

Ptolemy (90–168) (Claudius Ptolemaeus) Greek astronomer and geographer. He worked at the library of Alexandria, Egypt, then a great centre of Greek learning. His chief astronomical work, the *Almagest*, was largely a compendium of contemporary knowledge, including a star catalogue, drawing heavily on the work of HIPPARCHUS. The PTOLEMAIC SYSTEM is based on the geocentric world system of the ancient Greeks. Ptolemy's *Geography*, which provided the basis for a world map, was a definitive text for many centuries. He was regarded as the most reliable geographical authority by Muslim and Christian scholars until the Renaissance.

ptomaine Any of several basic organic chemical COMPOUNDS, some of which are poisonous to human beings. They are derived from the DECOMPOSITION or putrefaction of animal or vegetable PROTEINS and bear a resemblance to ALKALOIDS. Ptomaine poisoning is a form of food poisoning caused by BACTERIA or TOXINS.

puberty Time in human development when sexual maturity is reached. The reproductive organs take on their adult form and SECONDARY SEXUAL CHARACTERISTICS, such as the growth of pubic hair, start to become evident. Girls develop breasts and begin their MENSTRUAL CYCLE; in boys, there is deepening of the voice and the growth of facial hair. Puberty may begin at any time from about the age of ten, usually occurring earlier in girls than in boys. The process is regulated by HORMONES.

pubis In the PELVIS (pelvic girdle), either of a pair of small bones at the front of the hip bones. They are almost U-shaped and meet at the pubic symphysis, which is closed by a pad of cartilage. Each pubis joins an ILIUM and an ISCHIUM at a triangular suture in the hip socket.

pulley Simple MACHINE used to multiply FORCE or to change the direction of its application. A simple pulley consists of a WHEEL, often with a groove, attached to a fixed structure. Compound pulleys consist of two or more such wheels, some movable, that allow a person to raise objects much heavier than he or she could lift unaided. *See also* FORCE RATIO

Pullman, George Mortimer (1831–97) US industrialist and developer of the railway sleeping carriage. A cabinet maker in his youth, he began converting old railway coaches in 1858 to make long-distance travelling more comfortable. In 1865 the first modern sleeping car went into service. Known as the Pullman car, it was designed by Pullman and Ben Field.

pulsar Celestial object that emits RADIO WAVES in pulses of great regularity. They were first noticed in 1967 by the British astronomer Jocelyn Bell (1943–), working at the Mullard Radio Astronomy Observatory, Cambridge, England. The first pulsar to be discovered pulsed every 1.337 seconds. Pulsars are believed to be rapidly rotating NEUTRON STARS, the aftermaths of SUPERNOVA explosions. A beam of radio waves emitted by the rotating pulsar sweeps past the Earth and is received in the form of pulses, in the same way as a lighthouse is seen to flash. The **Crab Pulsar** was the first to be seen flashing optically, in 1969. In some pulsars, the pulse amplitude and shape can change with time, and the pulsation can fade out temporarily. Pulsars are gradually slowing down, but some, such as the **Vela Pulsar**, occasionally increase their spin rate abruptly; such an event is called a glitch. More than 500 pulsars are now known, flashing at rates from about 4 seconds to 1 millisecond. The most common period is just under 1 second. Pulsars are concentrated towards the disc of our Galaxy, which is where supernovae most commonly occur. Most pulsars are thought to be more than a million years old.

pulsating star VARIABLE STAR, the LUMINOSITY of which is intrinsically altered as a result of periodic expansions and contractions arising from changes in the star's atmosphere. Most pulsating stars are relatively short-period variables, including the classical CEPHEID VARIABLES and the RR Lyrae variables.

pulse Regular wave of raised pressure in ARTERIES that results from the flow of BLOOD pumped into them at each beat of the HEART. The pulse is usually taken at the wrist, although it may be observed at any point where an artery runs close to the body surface. The average pulse rate is about 70 per minute in a resting adult.

pumice Rhyolitic LAVA blown when it is molten to a low density rock froth by the sudden discharge of GASES during a volcanic action. It tends to be very acidic in chemical content. When ground to a powder and pressed into cakes it is used as a light abrasive.

PUMP

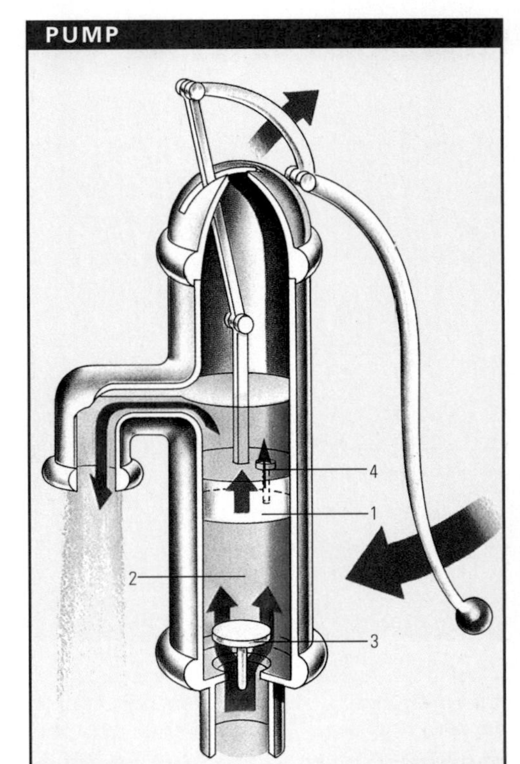

A typical lift pump, such as the ones that were commonly used to raise water from an underground supply, work by employing two one-way valves. As the handle is pumped back and forth, the action is converted into the reciprocal, up-and-down motion of the plunger (1). As the plunger moves up it creates a partial vacuum in a collecting chamber (2), into which water is sucked. As the plunger descends, a valve (3) prevents the water being forced down the bore hole, while a second valve (4) allows the water to flow through the plunger which, when it rises again, forces water from the pump outlet.

pump Device for raising, compressing, propelling or transferring fluids (liquids or gases). The lift pump, for raising water from a well, and the bicycle pump, for inflating tyres, are examples of **reciprocating** (to-and-fro) pumps. In many modern pumps, a rotating impeller (set of blades) causes the fluid to flow. **Jet** pumps move fluids by forcing a jet of liquid or gas through them.

punched card In computing, one of the earliest mediums for storing and entering data into COMPUTERS and accounting machines. The information is typed on the keyboard of a machine, which encodes it into a pattern of holes on a paper card, which the machine can interpret. This form of storage is now rarely used. An early form of punched card was developed in the early 19th century by the French weaver Joseph JACQUARD. The cards automatically changed the pattern on the material being woven. *See also* INFORMATION STORAGE AND RETRIEVAL

punctuated equilibrium Theory expounded by Stephen Jay GOULD and Niles Eldridge in 1972, strongly sceptical of the notion of "gradual" change in the EVOLUTION of the natural world, as advocated by such theorists as Charles DARWIN. Fossil records rarely document the gradual development of a new SPECIES, rather its seemingly sudden appearance. Darwin reasons that this is due to gaps in the fossil records. Punctuated equilibrium explains these "missing links" by rejecting a notion of gradual change and by invoking a different model of evolu-

▲ **Pupa** of the tortoiseshell butterfly *Aglais urticae*, attached to a stinging nettle leaf. Pupa formation takes *c.*3 hours, producing a characteristic, metallic gold sheen. As larvae, the caterpillars live socially, spinning a silken tent around a group, from behind which they feed on the nettle plant. At pupation they move off the food plant and suspend themselves, using a silk secretion, from posts, walls, sheds and tree trunks. In appearance the pupal stage is one of dormancy, but in reality substantial structural changes occur, resulting in the adult butterfly.

tion. Each species is predominantly in a steady state (equilibrium), but this is punctuated by brief but intense periods of sudden change which give rise to new species in relatively short periods of time.

pupa Non-feeding, developmental stage during which an INSECT undergoes complete METAMORPHOSIS. It generally occurs as part of a four-stage life cycle from the OVUM, through LARVA to pupa, then IMAGO (adult). Most pupae consist of a tough protective outer casing, inside which the tissues of the insect affect a drastic reorganization to form the adult body. Insects that undergo pupation include the many species of butterfly, moth and beetle and many kinds of fly. The pupa is often called a CHRYSALIS in butterflies and moths. It often acts as a resting or dormant stage, providing a resistant form in which the insect can survive severe weather or drought.

pupil In the structure of the EYE, circular APERTURE through which light falls onto the LENS; it is located in the centre of the IRIS. Its diameter changes by reflex action of the iris to control the amount of light entering the eye.

Pupin, Michael Idvorsky (1858–1935) US physicist and inventor, b. Yugoslavia. He devised a means of extending the range of long-distance TELECOMMUNICATIONS by placing loading coils at intervals along the transmitting wire, and invented a method of taking short-exposure X-RAY photographs. His autobiography *From Immigrant to Inventor* (1923) won a Pulitzer Prize.

Puppis (Argo Nevis) Constellation of the S sky. Its brightest star is Zeta (Naos), amgnitude 2.2. It has several bright, open clusters.

Purcell, Edward Mills (1912–97) US scientist who shared the 1952 Nobel Prize for physics with Felix BLOCH for their independent work in NUCLEAR MAGNETIC RESONANCE (NMR). Purcell's discovery of NMR in liquids and solids is instrumental in the measurement of the MAGNETIC MOMENTS of nuclei and molecules. He also did research in radio astronomy, astrophysics and biophysics.

purine White, crystalline, organic COMPOUND ($C_5H_4N_4$) related to URIC ACID. ADENINE and GUANINE are both purines and are two of the four bases that comprise DNA. Another derivative of purine is ADENOSINE TRIPHOSPHATE (ATP), a NUCLEOTIDE essential to the transfer of energy within living cells.

pus Yellowish fluid forming as a result of bacterial infection. It comprises blood SERUM, LEUCOCYTES (white blood cells), dead tissue and living and dead BACTERIA. An abscess is a pus-filled cavity.

putrefaction DECOMPOSITION of organic matter, especially PROTEINS, by FUNGI, BACTERIA and OXIDATION. It results in foul-smelling products. Putrefaction of meat, for example, yields HYDROGEN SULPHIDE, AMINES and THIOLS.

PVC Abbreviation of POLYVINYL CHLORIDE

pyramid In geometry, a solid figure having a POLYGON as one of its faces (the base), the other faces being TRIANGLES with a common VERTEX. The volume of a regular pyramid is one-third of the base area multiplied by the vertical height.

pyridine Heterocyclic compound (C_5H_5N) characterized by a six-membered ring structure composed of five CARBON atoms and one NITROGEN atom. It occurs in bone oil and coal tar and is synthetically produced from ETHANAL (acetaldehyde) and AMMONIA. Pyridine is used as a SOLVENT and converted to sulphapyridine (a drug used to combat bacterial and viral infection), pyribenzamine and pyrilamine (ANTIHISTAMINE).

pyridoxine VITAMIN B_6, a COENZYME in the metabolism of AMINO ACIDS in the body. Pyridoxine is found in lean meat, egg yolks, milk and fish. A deficiency can lead to ANAEMIA, and it is also taken as a vitamin supplement by those who believe it helps to ease depression.

pyrimidine Colourless liquid ($C_4H_4N_2$) characterized by a heterocyclic ring structure composed of four CARBON atoms and two NITROGEN atoms. It may be prepared from URACIL (a dihydroxy pyrimidine compound) by chemical reactions that remove two OXYGEN atoms. It is used in the manufacture of various SULPHONAMIDE DRUGS and BARBITURATES.

pyrite (iron pyrites) Most common and widespread sulphide mineral, iron sulphide (FeS_2), occurring in all types of rocks and veins. It is a brass-yellow colour and, because of this, is often called "fool's gold". It was formerly used widely to produce SULPHURIC ACID. It crystallizes as cubes and octahedra, and also as granules and globular masses. It is opaque, metallic and brittle. Hardness 6.5; r.d. 5.0.

pyroclastic Term that describes any rock fragments thrown out by volcanic activity, generally in association with some violent explosive action. Pyroclasts normally include solidified LAVA left behind by a previous eruption of the VOLCANO, as well as rocks from the crust, smaller pieces of cinders, ash and dust. The largest pyroclastic fragments, known as "volcanic bombs", weigh several tonnes. Smaller pieces are known as "lapilli". When the pyroclastic activity ceases, outpourings of lava often follow. Rocks formed by consolidation of these pyroclastic fragments, such as volcanic BRECCIA and TUFF, are also termed pyroplastic.

pyrolusite Common mineral form of manganese dioxide (MnO_2), an important ore of MANGANESE. It is formed as a PRECIPITATE in lakes and bogs at low temperatures, often in veins of quartz, but also occurs as MANGANESE NODULES on the floor of the Pacific Ocean. It is black or dark grey, usually occurring in masses that have an earthy lustre, although it may form prismatic crystals (tetragonal) and fern-like aggregates in the joints of sedimentary rocks. Hardness: massive form 2–6, crystals 6–7; r.d. 4.5–5.

pyrolysis Chemical decomposition of a complex substance into simpler ones by the action of heat alone. Waste materials and coal may be pyrolyzed to produce valuable fuels and chemicals.

pyrometer THERMOMETER for use at extremely high temperatures, well above the ranges of ordinary thermometers. An **optical** pyrometer consists essentially of a small TELESCOPE, a RHEOSTAT (a type of variable resistor) and a filament. When the telescope is aimed at a furnace or other hot object, the filament appears dark against the background. As the electric current flowing through the filament is increased slowly using the rheostat, it grows brighter until it matches the intensity of the furnace. The amount of current is a measure of the temperature.

pyrope ($Mg_3Al_2(SiO_4)_3$) Magnesium aluminium GARNET, the most common form of garnet. It may be purplish or brownish red and is found mainly in South Africa and the USA. Often used in jewellery, it has been given such misleading names as Bohemian garnet and is often mistaken for a RUBY.

pyroxene Any of an important group of rock-forming minerals. They are single-chain SILICATES. Their colours are variable but are usually dark greens, browns and blacks. Crystals are usually short PRISMS with good cleavages. Hardness 2.3–4; r.d. 5.5–6.

pyrrhotite Magnetic mineral form of iron sulphide, FeS (but variable), containing more sulphur than PYRITE. Its colour is a metallic bronze and it forms massive aggregates or hexagonal, flat crystals (monoclinic). It may contain nickel and in such cases is mined as a nickel ore. Hardness: 3.5–4.5; r.d. 4.6.

pyrrole (C_4H_5N) Heterocyclic, organic chemical COMPOUND consisting of a five-membered ring with four CARBON atoms and one NITROGEN atom. A colourless liquid, it smells like TRICHLOROMETHANE (chloroform) and is obtained from COAL TAR and bone oil. It forms part of the molecules of various natural pigments, including CHLOROPHYLL and HAEMOGLOBIN.

pyruvic acid ($CH_3COCOOH$) Colourless liquid that can be formed by the distillation of TARTARIC ACID; it has a vinegary odour. Phenylpyruvic acid, a derivative of pyruvic acid, occurs in the urine of patients suffering from PHENYLKETONURIA – a disorder in which the amino acid PHENYLALANINE is not metabolized normally.

Pythagoras (*c.*580–500 BC) Greek philosopher and founder of the Pythagorean school. He was born on the Greek island of Samos and later lived in Croton, Italy, where he founded his school. The Pythagoreans believed that numbers and their interrelationships could be used to represent everything in the Universe. Pythagoras himself is credited with many advances in MATHEMATICS, geometry, medicine and philosophy. He was probably the first to teach that the Earth is a sphere revolving about a fixed point, and is said to have been the first to discover the mathematical relationship between the length of a string and the PITCH of the sound it makes when vibrating (*see* FREQUENCY). Pythagoras is also credited with the theorem that the square of the HYPOTENUSE of a right-angled TRIANGLE equals the sum of the squares of the other two sides, but this result was actually known to the Egyptians and Babylonians many hundreds of years earlier. Profoundly at odds with contemporary religion, his brotherhood was suppressed at the end of the 6th century BC, but its doctrines were revived (**Neo-Pythagoreanism**) by the Romans *c.*500 years later.

Pyxis (Compass) Constellation of the S sky adjoining VELA and PUPPIS. Its most interesting feature is the recurrent NOVA T Pyxidis.

quadrant In plane geometry, a quarter of a CIRCLE, bounded by radii at right angles to each other and the arc of the circle. In analytic geometry, it is one of the four sections of a plane divided by an x-axis and a y-axis. A quadrant is also a device for measuring angles, based on a 90° scale.

quadrat Square frame used to mark out an area of ground for quantifying and studying the plants it encloses. The area of ground itself is sometimes also called a quadrat. The area is typically 1 or $0.5m^2$. Botanists and ecologists can, for example, count the numbers of species within a quadrat and estimate their abundance. Repeated measurements at several locations provide information about overall species distribution.

quadratic equation Polynomial EQUATION of degree two. A quadratic equation has the general form $ax^2 + bx + c = 0$, where a, b and c are constants, with a nonzero. It has two solutions (also called roots), shown by AL-KHWARIZMI to be given by the formula $x = [- b \pm \sqrt{(b^2 - 4ac)}]/2a$.

quadratic function Mathematical FUNCTION, the value of which depends on the square of the independent VARIABLE, x, and is thus given by a quadratic POLYNOMIAL, as in

$f(x) = 4x^2 + 17$ or $f(x) = x^2 + 3x + 2$

See also QUADRATIC EQUATION

qualitative analysis Identification of the chemical elements or ions in a substance or mixture. *See also* QUANTITATIVE ANALYSIS

quantitative analysis Identification of the amounts of chemical constituents in a substance or mixture. Chemical methods use reactions such as NEUTRALIZATION and OXIDATION, and determine the concentrations of substances present in terms of volume (VOLUMETRIC ANALYSIS) or weight (GRAVIMETRIC ANALYSIS). Physical methods measure qualities such as density and REFRACTIVE INDEX (how much light is refracted by a medium). *See also* QUALITATIVE ANALYSIS

quantum In physics, the smallest quantity of some physical property, such as matter, that a system can posess according to the QUANTUM THEORY.

quantum chromodynamics Study of the properties of QUARKS in which, to explain permissible combinations of quarks to form various ELEMENTARY PARTICLES, each is notionally assigned a colour. Quarks are given one of the three primary colours – red, green and blue. When three quarks combine to form BARYONS, the resulting colour is always white. Antiquarks are given one of the three complimentary colours – cyan, magenta and yellow. When a quark combines with an antiquark to form a MESON, the resulting colour is also white. *See also* QUANTUM ELECTRODYNAMICS (QED)

QUANTUM MECHANICS

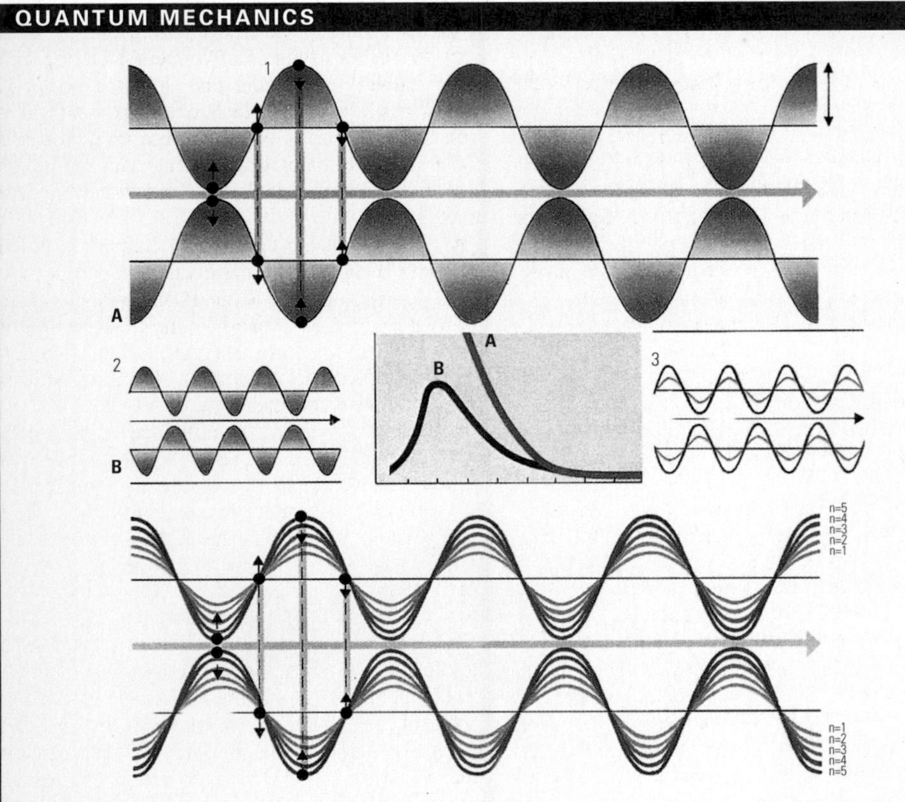

Nineteenth-century physics failed to explain certain observed phenomena – notably, the wavelength-spectrum of radiation emitted by a hot object. In the graph, curve A is the spectrum of radiation (intensities at different wavelengths) that according to theory should have been given off at a temperature of about 1,400°C (2,550°F) (the horizontal scale is in units of 10^{-4}cm). Curve B is the spectrum actually observed. Note that the theoretical spectrum is completely unrealistic, predicting infinite intensity at the shortest wavelengths – a prediction known as the "ultraviolet catastrophe". The 19th-century theory was based on the idea that the radiation was emitted by vibrating, electrically polarized entities – dipoles (1) – whose energy and frequency of vibration could in principle have any magnitude. The difficulty was resolved, around 1900, by Planck's proposal (B3) that each dipole could vibrate only at certain energies, related to its frequency (ν) by the formula $E = nh\nu$, where n is 1, 2, 3 etc. and h is a universal constant; a higher-frequency vibration (3) would thus have a higher minimum energy.

▲ **quadrant** Tycho Brahe's quadrant was one of the earliest instruments used for measuring star positions. This quadrant, used between 1576 and 1596, was mounted on a central pillar (1); a pointer (2) with sights rotated against a graduated metal circle (3). The well (4) accommodated the observer at various levels, determined by the position of the pointer. Modern work depends on fundamentals such as those established by Brahe.

quantum electrodynamics (QED) QUANTUM THEORY of ELECTROMAGNETIC RADIATION and how it interacts with charged particles. For example, the theory predicts that a collision between an ELECTRON and a PROTON should result in the production of a PHOTON of electromagnetic radiation, which is exchanged between the colliding particles.

quantum field theory Theory that embraces all ELEMENTARY PARTICLES and their interactions. In scientific terms, it applies QUANTUM MECHANICS to systems with an infinite number of degrees of freedom. Each type of particle is represented by fields that can oscillate only in quantized modes. Thus the particles become the quanta of the fields – for example, the PHOTON is the quantum of LIGHT (the electromagnetic field).

quantum mechanics In physics, use of QUANTUM THEORY to explain the behaviour of ELEMENTARY PARTICLES, such as ELECTRONS. In the quantum world, waves and particles are interchangeable concepts. In 1924 the French physicist Louis de BROGLIE suggested that particles have wave properties, the converse having been postulated in 1905 by Albert EINSTEIN. In 1926 the Austrian physicist Erwin SCHRÖDINGER used this hypothesis to predict particle behaviour on the basis of wave properties, but a year earlier the German physicist Werner HEISENBERG had produced a mathematical equivalent to Schrödinger's theory without using wave concepts at all. In 1928 the British physicist Paul DIRAC unified these approaches, while incorporating RELATIVITY into quantum mechanics. QUANTUM ELECTRODYNAMICS was derived by the US physicist Richard FEYNMAN in the 1940s. *See also* QUANTUM NUMBERS

quantum numbers Set of four numbers used to classify ELECTRONS and their atomic states. The **principal** quantum number (symbol n) gives the electron's energy level, the ORBITAL quantum number (symbol l) describes its angular momentum, the **magnetic** quantum number (symbol m) describes the energies of electrons in a magnetic field, and the SPIN quantum number (symbol m_s) gives the spin of the individual electrons. *See also* QUANTUM THEORY

quantum theory Together with the theory of RELATIVITY, the foundation of 20th-century physics. It is concerned with the relationship between MATTER and ENERGY at the elementary or subatomic level and with the behaviour of ELEMENTARY PARTICLES. According to the theory, all radiant energy is

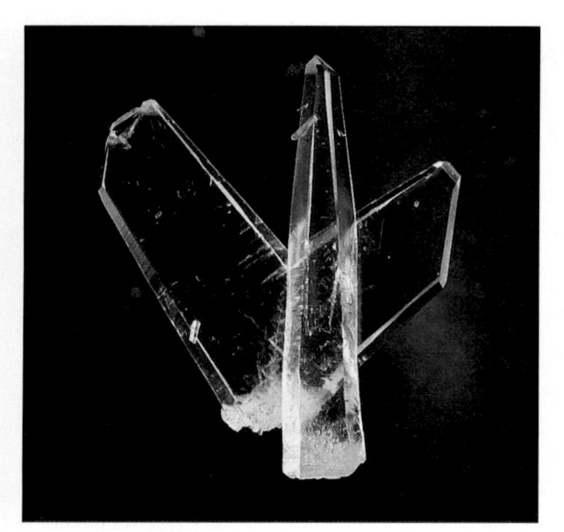

▲ **quartz** (silicon oxide) is colourless when pure, but the presence of impurities can give a range of colours including white, pink, yellow, blue, green and smoky brown. The pure, colourless variety is known as rock crystal and is highly prized by collectors. The elbow-shaped crystal is a formation known as a Japan Law twin. This specimen measures 7 x 6cm (2.8 x 2.4in) and was found in Piraja's Mine, Brumado, Brazil.

emitted and absorbed in multiples of tiny "packets" or quanta. This breaks down the traditional distinction between energy (as a wave motion) and matter (as occupying a fixed space), because quanta have properties of both. In 1900 Max PLANCK proposed the idea that energy is radiated and absorbed in quanta. He calculated that the energy E of a quantum is related to the frequency of radiation v according to the formula $E = hv$, where h (PLANCK'S CONSTANT) is a constant with the value of 6.626×10^{-34} joule seconds. Using this formula, Albert EINSTEIN quantized light radiation and in 1905 explained the PHOTOELECTRIC EFFECT. In 1913 Niels BOHR used quantum theory to explain atomic structure and the relationship between the energy levels of an atom's electrons and the frequencies of radiation emitted or absorbed by the ATOM. *See also* QUANTUM MECHANICS; QUANTUM NUMBERS

quarantine Originally a 40-day waiting period during which ships were forbidden to discharge passengers or freight to prevent transmission of plague or other diseases. Today, the term refers to any period of isolation legally imposed on people, animals, plants or goods in order to prevent the spread of contagious disease.

quark Any one of six particles (although there is increasing evidence of another two) and their antiparticles (antiquarks) that are the fundamental members of the HADRON group of ELEMENTARY PARTICLES. Quarks always exist in combination; free quarks cannot exist at the temperatures of the modern Universe, though they may have done so in the first moments after the BIG BANG. In 1964 the US physicists Murray GELL-MANN and George Zweig proposed independently that hadrons (particles that experience the STRONG NUCLEAR FORCE) are composed of quarks. A BARYON, such as a PROTON or NEUTRON, consists of three quarks; a MESON of a quark and an antiquark tightly bound together. A quark has a charge of either $-2/3$ or $-1/3$ of the charge on an ELECTRON. It can occur in one of six "flavours": up, down, top, bottom, charmed and strange. Antiquarks have similar flavours but their charge is opposite to that of their corresponding quark. Each quark and antiquark also has a "colour" (*see* QUANTUM CHROMODYNAMICS). *See also* ANTIMATTER

quartz (silica, SiO_2) Rock-forming mineral, the natural form of silicon dioxide. It is widely distributed, occurring in IGNEOUS and METAMORPHIC ROCKS (notably granite and gneiss), and in clastic sediments. It is also a common gangue mineral in mineral veins. Pure quartz is clear and colourless, but the mineral is often opaque or translucent, and may be coloured by impurities. It forms six-sided crystals. The most common varieties are colourless quartz (known as rock crystal), rose, yellow, milky and smoky. The most usual cryptocrystalline varieties, the crystals of which can be seen only under a microscope, are CHALCEDONY and FLINT. Quartz crystals exhibit the PIEZOELECTRIC EFFECT and are used in electronic clocks and watches to keep accurate time. Hardness 7; r.d. 2.65.

quartzite METAMORPHIC ROCK usually produced from sandstone, in which the quartz grains have recrystallized. Fracturing through these grains rather than between them, quartzite is a hard and massive rock. Its colour is usually white, light grey, yellow, or buff, but it can be coloured green, blue, purple, or black by various minerals.

quasar (quasi-stellar object) In astronomy, compact object that is an extremely luminous source of energy, radiated over a wide range of wavelengths from X-rays through optical to radio wavelengths. Quasars were discovered in 1963 when astronomers identified optical counterparts of certain strong radio sources. However, only a small proportion of quasars are actually radio sources. Every quasar shows a very large RED SHIFT which, if it results from the expansion of the Universe, indicates that quasars are the oldest and most distant objects in the Universe. Quasars seem to inhabit the centres of very remote galaxies. Since they vary in brightness with periods of a few days, they can be no larger than a few hundred astronomical units wide. A surrounding galaxy has been seen around the nearest quasars, almost lost in the glare from the brilliant centre. Quasar activity is thought to be caused by a BLACK HOLE accreting material at the centre of the host galaxy. The most luminous quasars would need to be powered by black holes of 100 million solar masses, swallowing stars at a rate of about one every year. Quasars are bright ultraviolet sources, and this hot radiation seems to come from a swirling ACCRETION disc. Also associated with the inner regions of quasars are emission lines of hydrogen, helium and iron that indicate motions as rapid as 5,000km s^{-1}, as would be expected near a massive black hole. Because quasars are the most luminous objects in the Universe, they can be seen at greater distances, and hence further back in time, than anything else. Quasars thus provide us with a way of examining the Universe in its youth.

Quaternary Most recent period of the CENOZOIC era, beginning about 2 million years ago and extending to the present. It is divided into the PLEISTOCENE epoch, characterized by a periodic succession of great ice ages, and the HOLOCENE epoch, which started some 10,000 years ago.

quaternions Type of abstract number, discovered by William HAMILTON. An ordinary complex number has the form $a + bi$ (where a and b are real numbers, i is the square root of -1). A quaternion has the form $a + bi + cj + dk$, in which i, j and k are defined by the equations $i^2 = j^2 = k^2 = -1$, and $-ji = k$).

quinine White, crystalline ALKALOID isolated in 1820 from the bark of the cinchona tree. It was once widely used in the treatment of MALARIA, but has been largely replaced by drugs which are less toxic and more effective.

quinone Aromatic organic compound that contains a BENZENE ring in which two hydrogen atoms have been replaced by oxygen atoms. Paraquinone (1,4-benzoquinone, $O{=}C_6H_4{=}O$) is a yellow crystalline solid obtained by the oxidation of aniline with chromic acid; it is used in the manufacture of hydroquinone for photographic developers.

QUARTZ

Many modern clocks and watches use a quartz crystal (1) to tell the time accurately. When electricity is passed through the quartz, it oscillates exactly 32,768 times each second. The oscillations are counted and on every 32,768th, a pulse of electricity is sent to a motor (2) that moves the hands (3) via gears (4). The need for a battery to power the motor can be removed if a swinging weight (5) is used to generate a current. As the watch moves, the weight rotates (6) turning a generator (7). The current produced by the generator is stored in a capacitor (8) and is smoothed before reaching the quartz crystal.

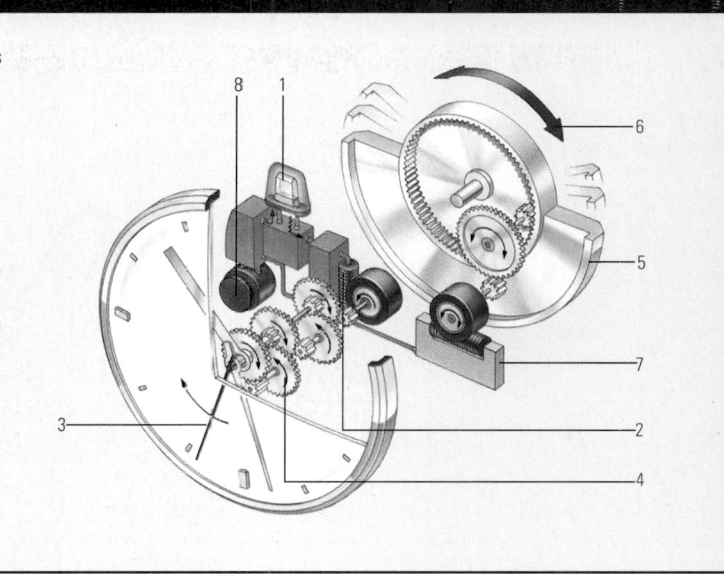

Rabi, Isidor Isaac (1898–1988) US physicist, b. Austria, who was known for his work on MAGNETISM and QUANTUM MECHANICS. Rabi was awarded the 1944 Nobel Prize for physics for his discovery and measurement of the radio-frequency spectra of atomic nuclei whose magnetic SPIN has been disturbed. This led to the development of new SPECTROSCOPIC methods.

racemization Transformation of an optically active substance into a MIXTURE of equal quantities of two mirror-image forms (**enantiomorphs**), usually by heat or the action of acids or bases. Individually, each enantiomorph rotates the plane of POLARIZED LIGHT through a characteristic angle, but an equal quantity of each will cancel each other's rotatory effect.

rack In engineering, a straight rod on which grooves are cut to mesh with a cylindrical GEAR (pinion). Rotary motion of the pinion is converted into reciprocating (to-and-fro) motion at the rack. It is used extensively for steering systems in AUTOMOBILES.

rack railway Form of mountain transport in which the locomotive powers a cogwheel that engages a cogged rail (rack) set between the track. The first rack railway in Britain came into use in 1812 between Middleton colliery and Leeds in the N of England, a distance of 5.5km (3.5mi).

radar (contraction of **ra**dio **d**etecting **a**nd **r**anging) Electronic system for determining the direction and distance of objects. First developed by the British and Germans during World War 2, it works by the transmission of pulses of RADIO WAVES to an object from the radar transmitting equipment. The object reflects the pulses, which are detected by an ANTENNA. By measuring the time it takes for the reflected waves to return, the object's distance may be calculated, and its direction ascertained from the alignment of the receiving radar antenna. Air traffic controllers use radar to indicate aircraft positions. In this case, a rotating antenna transmits the signals and picks up the reflections. The reflected signals from planes are shown as bright moving dots on a circular display called a plan position indicator. In 1946 radar beams from Earth were reflected back from the Moon, marking the beginnings of RADAR ASTRONOMY. *See also* DOPPLER EFFECT; DOPPLER RADAR

radar astronomy Branch of ASTRONOMY in which RADAR pulses, reflected back to Earth from celestial bodies inside the SOLAR SYSTEM, are studied for information concerning the distance from Earth of the bodies, their orbital motion, and large surface features. It differs from RADIO ASTRONOMY, which involves only the reception of radio waves emitted naturally by objects in the Universe. Radar echoes from the Moon were first obtained in 1946. Radar reflections from Venus were received in 1958 and from the Sun in 1959. The resulting measurements provided an accurate new idea of the scale of the Solar System. As the power of radars grew and orbiting PLANETARY PROBES were developed, techniques were developed for radar mapping of planetary surfaces, which proved particularly important for cloud-covered VENUS. Other objects in the Solar System studied by radar include ASTEROIDS, Saturn's rings, planetary SATELLITES and the nuclei of COMETS.

radial velocity In astronomy, velocity of a celestial body in the line of sight, either towards (positive) or away from (negative) an observer. The radial velocity is determined from the DOPPLER EFFECT on lines in the object's SPECTRUM. *See also* RED SHIFT

radian Angle formed by the intersection of two radii at the centre of a CIRCLE, when the length of the arc cut off by the radii is equal to one radius in length. Thus, the radian is a unit of ANGLE equal to *c*.57.296°, and there are 2π radians in 360°.

radiant heat Form of ENERGY transfer through the atmosphere or space. The Sun's energy reaches the Earth as radiant heat or INFRARED RADIATION, as well as visible light and other ELECTROMAGNETIC RADIATIONS. Domestic electric fires emit radiant HEAT, but they also heat the air by CONVECTION.

radiation Transmission of energy by SUBATOMIC PARTICLES or ELECTROMAGNETIC WAVES. All ELECTROMAGNETIC RADIATION can cross a VACUUM, which distinguishes it from such effects as CONDUCTION and CONVECTION and the transmission of sound. In a vacuum, electromagnetic waves travel at the speed of light, 299,792.5km s^{-1} (186,291mi s^{-1}), but this speed is slightly decreased as the waves pass through a medium (as when light passes through glass), an effect that explains REFRACTION. This type of radiation also exhibits some of the properties of a stream of PHOTONS. Similarly, streams of particles such as ELECTRONS exhibit properties like those of waves, which allows for electron DIFFRACTION as in an ELECTRON MICROSCOPE.

radiation, heat Energy given off from all solids, liquids or gases as a result of their temperature, however low. The energy comes from the vibrations of ATOMS in an object and is emitted as ELECTROMAGNETIC RADIATION, often in the form of INFRARED RADIATION.

radiation, nuclear Particles or ELECTROMAGNETIC RADIATION emitted spontaneously and at high ener-

RADIATION, NUCLEAR

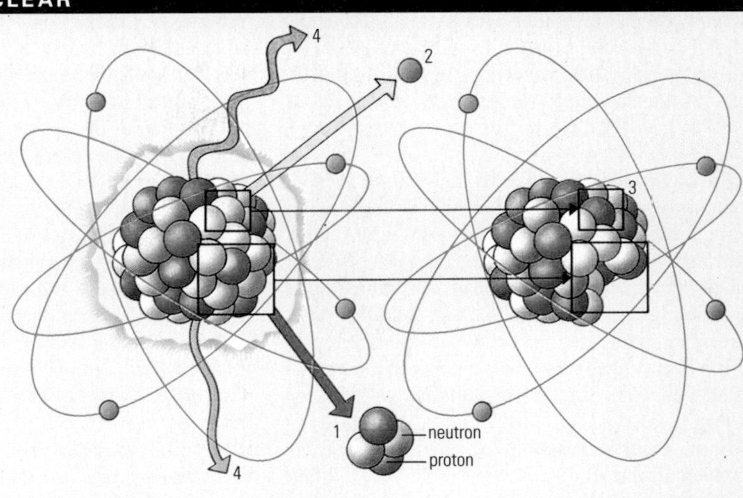

Many elements are naturally radioactive, so there is always background radiation even when there is no manmade or accidentally generated radioactivity in the vicinity. Alpha radiation occurs when an unstable atomic nucleus emits an alpha (1) particle, a helium nucleus comprising of two protons and two neutrons. Alpha radiation (α) is stopped by flesh. Beta radiation (β) occurs when an electrically neutral neutron emits a negative electron (which is the beta (2) particle) and becomes in the process a positive proton (3). Beta particles can be stopped by a thin sheet of aluminium. X-rays and gamma (γ) are both electromagnetic radiation (types of light), created when a nucleus loses energy without undergoing structural change (4). X-rays can be stopped by a sheet of lead, while high-energy gamma rays will penetrate thin sheets of lead, but may be stopped by a layer of concrete.

RADAR

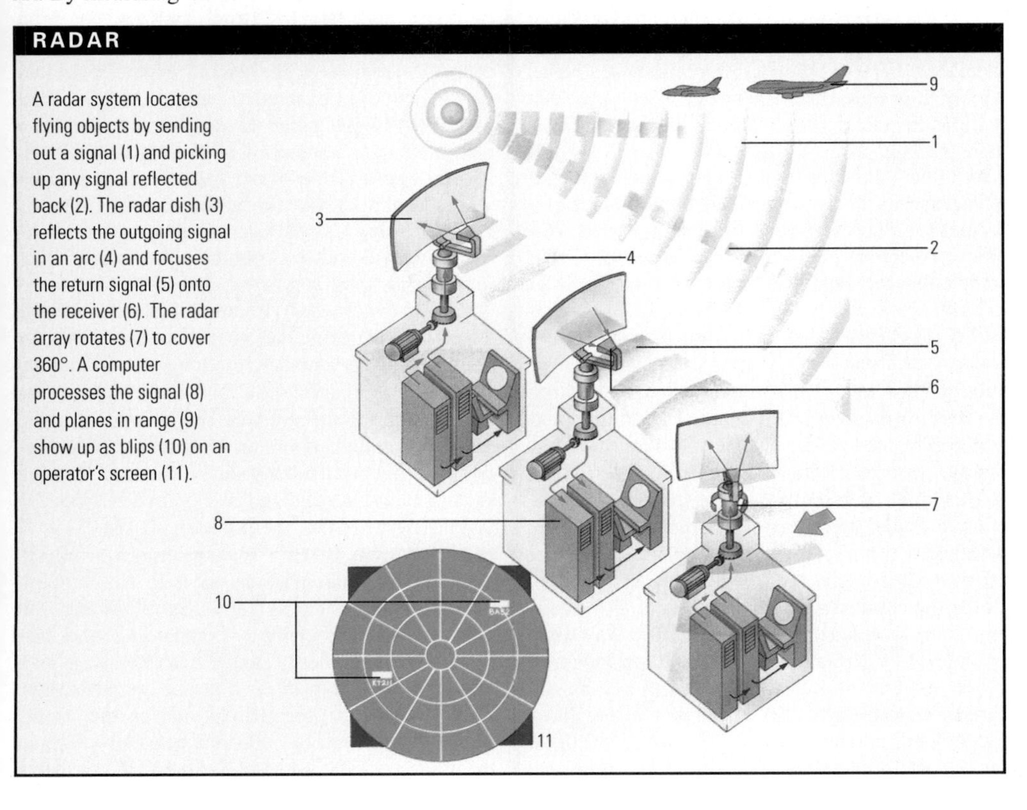

A radar system locates flying objects by sending out a signal (1) and picking up any signal reflected back (2). The radar dish (3) reflects the outgoing signal in an arc (4) and focuses the return signal (5) onto the receiver (6). The radar array rotates (7) to cover 360°. A computer processes the signal (8) and planes in range (9) show up as blips (10) on an operator's screen (11).

R

gies from atomic nuclei. Possible causes include RADIOACTIVE DECAY, which yields ALPHA PARTICLES (helium nuclei), BETA PARTICLES (electrons), GAMMA RADIATION and, more rarely, POSITRONS (antielectrons). It can also result from spontaneous FISSION of a NUCLEUS, with the ejection of NEUTRONS or gamma rays. *See also* RADIOACTIVITY

radiator Space heater in a central heating system. Commonly, it is a slim fluted steel or iron container through which hot water or steam is circulated. Electrically-powered appliances, such as convector heaters and storage heaters, can also function as radiators. In cars, lorries and other vehicles powered by an INTERNAL COMBUSTION ENGINE, a radiator is a heat exchanger, which dissipates the heat in circulating water (to the surrounding air) to keep the engine cool.

radical, atomic Single ATOM or group of atoms in which all the atomic VALENCES are not satisfied by chemical bonding. Thus, removal of a hydrogen atom from methane, CH_4, gives the methyl radical CH_3. The term is usually taken to mean a FREE RADICAL, that is, one existing free for a short time in a reaction. Free radicals are highly reactive; they can be produced by ULTRAVIOLET LIGHT, PYROLYSIS, ELECTRON impact, and other means, and play an important part in the mechanisms of many CHEMICAL REACTIONS.

radical (symbol $\sqrt{}$) Expression used to indicate the ROOT of a number. The expression $\sqrt[n]{x}$ represents the *n*th root (or index) of the number x (the radicand). The CUBE ROOT of ten in radical form is written $\sqrt[3]{10}$. Another way of expressing the *n*th root of a number x is by $x^{1/n}$.

radicle Part of a plant embryo that develops into the ROOT. When the radicle emerges from a germinating seed, its tip is protected by a root cap (**calyptra**) which enables it to break through the seed case (**testa**) and penetrate the soil. In most plants, the radicle originates the whole root system.

radio BROADCASTING or reception of ELECTROMAGNETIC RADIATION in the form of RADIO WAVES. A radio signal of fixed FREQUENCY (the carrier wave) is generated at a TRANSMITTER. The sound to be broadcast is converted by a MICROPHONE into a varying electrical signal that is combined with the carrier by means of MODULATION. FREQUENCY MODULATION (FM) minimizes interference and provides greater fidelity than AMPLITUDE MODULATION (AM). The modulated carrier wave is passed to an ANTENNA from which it is transmitted into the atmosphere. At the receiver, another aerial intercepts the signal, and it undergoes "detection", the reverse of modulation, to retrieve the sound signal. It is amplified to activate a LOUDSPEAKER that reproduces the original sound. A station is selected by tuning to a specified frequency. Radio waves travel at the speed of light and are transmitted not only by line-of-sight (ground waves), but also by reflection from the IONOSPHERE (sky waves). Sky waves enable long-range radio transmission. The ULTRA HIGH FREQUENCY (UHF) and VERY HIGH FREQUENCY (VHF) waves used in TELEVISION broadcasting penetrate the ionosphere with little reflection, and long-range broadcasting is made possible by means of artifical SATELLITES. The most efficient and commonest circuit used in radio is the SUPERHETERODYNE RECEIVER. Walkie-talkies and MOBILE TELEPHONES use radio for communications, television uses radio to send sound and vision signals and RADAR transmits pulses of radio waves. The development of radio can be traced to the mathematical studies of James Clerk MAXWELL. Heinrich HERTZ devised an apparatus for the transmission and detection of radio waves. In 1895 Guglielmo MARCONI gave a demonstration of the first wireless TELEGRAPH, and in 1901 he sent the first transatlantic message using MORSE CODE. In 1904 Sir John FLEMING invented the thermionic valve, the first vacuum ELECTRON TUBE. In 1906 Lee DE FOREST developed the audion triode valve, which was able to detect and amplify radio waves. It remained at the heart of radio and television manufacture until the invention of the TRANSISTOR in 1947.

radioactive decay Process by which a RADIOISOTOPE loses ELEMENTARY PARTICLES from its NUCLEUS and so becomes a different, more stable element. The starting isotope is called the "parent" element and the resulting isotope is termed the "daughter" element. Each radioactive element has its own constant rate of decay. This rate is measurable, and given as an element's HALF-LIFE. This property is utilized in radioactive DATING. A decaying ISOTOPE releases one, two or three of the natural types of RADIOACTIVITY: ALPHA PARTICLES, BETA PARTICLES and GAMMA RADIATION.

radioactive series Chain of ISOTOPES, each being a product of the RADIOACTIVE DECAY of its predecessor, starting with a RADIOISOTOPE and ending at one of the stable isotopes of LEAD. The three naturally occurring series are headed by uranium-238, thorium-232 and uranium-235. The neptunium-237 series does not occur in nature, because the parent isotope has a HALF-LIFE (about 2 million years) that is much less than the age of the Earth and has long decayed; this series is artificially produced. Each step in a radioactive series is called a TRANSMUTATION.

radioactive waste *See* NUCLEAR WASTE

radioactivity Disintegration of the nuclei of a RADIOISOTOPE, such as uranium-238, usually with the emission of ALPHA PARTICLES (helium nuclei) or BETA PARTICLES (ELECTRONS), often accompanied by GAMMA RADIATION. The two processes of alpha decay or beta decay cause the radioisotope to be transformed into a chemically different ATOM. **Alpha decay** results in the nucleus losing two PROTONS and two NEUTRONS; **beta decay** occurs when a neutron changes into a proton with an electron (beta particle) being emitted in the process. Thus, the ATOMIC NUMBER (number of protons in the nucleus) changes in both types of decay, and an ISOTOPE of another element is produced, which might also be radioactive. The stability of different isotopes varies widely. It is impossible to predict when a nucleus will disintegrate, but in a large collection of atoms there is a characteristic time (the HALF-LIFE) after which one-half of the total number of nuclei would have decayed. This time varies from millionths of a second to millions of years, depending on the isotope concerned. The activity of any radioactive sample decreases exponentially with time. *See also* CARBON DATING; NUCLEAR ENERGY

radio astronomy ASTRONOMY of radio emissions, ELECTROMAGNETIC RADIATION with wavelengths from about 1mm to many metres, reaching the Earth from space. Observations down to about 2cm can be made at sea level without serious interference from the atmosphere. At shorter wavelengths it is necessary to observe from high mountains to avoid atmospheric absorption, mainly by WATER VAPOUR. In 1931 radio noise from the Milky Way was discovered by Karl JANSKY, and the subject grew rapidly after World War 2. The radio emissions first detected from the Milky Way were produced by the process of SYNCHROTRON radiation – the motion of high-speed ELECTRONS in the Galaxy's MAGNETIC FIELD. A number of localized radio sources were also found. Some of these turned out to be SUPERNOVA remnants within our Galaxy, but optical identifications of sources led to the realization that most were far off in the Universe. These sources are now known as RADIO GALAXIES and QUASARS. The number of radio sources increases with distance, demonstrating that the Universe has been evolving with time. This, combined with the discovery at radio wavelengths of the COSMIC MICROWAVE BACKGROUND, is strong evidence in favour of the BIG BANG theory of the origin of the Universe. In 1967 a new class of individual radio sources was discovered in our Galaxy: the fast-flashing PULSARS, which are ultra-dense NEUTRON STARS formed when massive stars die. Individual spectral lines can also be detected at radio wavelengths, similar to optical spectral lines. In 1951 the emission from HYDROGEN at 21cm was detected, and has proved to be an especially useful means of studying the structure of our Galaxy. *See also* RADIO TELESCOPE

radio beacon Strategically located radio TRANSMITTER that emits a regular, nondirectional and easily recognizable signal by which mobile receivers can obtain a bearing. Such beacons are used particularly by marine and air navigators who collate the information from two or more such beacons to fix their positions. Directional beacons are used at airports to aid landing aircraft.

radiocarbon dating *See* CARBON DATING

radio galaxy GALAXY that is an abnormally high emitter of RADIO WAVES, around a million times stronger than the weak emissions from galaxies such as our own. Radio galaxies are usually giant ellipticals, such as M87 in the constellation Virgo. The emission from a typical radio galaxy is concentrated in two lobes well outside the galaxy, often located symmetrically either side, which appear to have been ejected from the galaxy in explosions. The extended radio lobes are almost invariably connected to the main galaxy by jets which seem to originate from the galactic nucleus. Strong, point-like radio sources exist in the nuclei of many of these objects. It is not understood why some elliptical galaxies emit such enormous amounts of radio energy. However, it does seem clear that the source of the activity lies in the galaxy's nucleus, and in this respect radio galaxies are similar to radio QUASARS and BL LACERTAE OBJECTS. A radio galaxy's enormous energy output comes from SYNCHROTRON radiation. The central "engine" produces vast quantities of energetic electrons travelling near the speed of light. When these electrons encounter a MAGNETIC FIELD, they spiral around and, in so doing, lose energy. It is this energy that we detect as radio waves. *See also* RADIO ASTRONOMY

radiography Use of X-RAYS to record the interiors of opaque bodies as photographs. Industrial X-ray photographs can show assembly faults and defects in the CRYSTAL structure of metals. In MEDICINE and DENTISTRY, radiography is invaluable for diagnosing bone damage, tooth decay and internal disease. The images show bones most clearly, as well as some tissue structure and air spaces. Using sensitive modern scanning techniques, cross-sectional outlines of the body can be obtained showing organs, blood vessels and diseased parts. *See also* COMPUTERIZED AXIAL TOMOGRAPHY (CAT)

radioisotope ISOTOPE that undergoes RADIOACTIVE DECAY. Radioisotopes occur in nature along with the stable isotopes of an ELEMENT, but they can also be made artificially by bombarding other isotopes with high-energy ELEMENTARY PARTICLES in a NUCLEAR REACTOR or a PARTICLE ACCELERATOR. Radioisotopes are formed naturally in the atmosphere as the result of collisions between ATOMS in the air and COSMIC RADIATION. Among the elements with radioactive isotopes of economic importance are COBALT, HYDROGEN, PLUTONIUM, RADIUM, STRONTIUM and URANIUM.

radioisotope tracer RADIOISOTOPE of an ELEMENT used to trace the course of that element through a biochemical or chemical system – a process called LABELLING. As the stable ISOTOPE and radioisotope behave identically from a chemical point of view, the radioisotope will follow exactly the same path as the stable isotope, but its presence in the cells and tissues of the organism can be monitored by the RADIATION it emits. Compounds labelled with radioisotopes are used in research and diagnosis.

radiology Medical specialty concerned with the use of RADIATION and radioactive materials in the diagnosis and treatment of DISEASE. *See also* RADIOGRAPHY; RADIOTHERAPY

radiometer Instrument for detecting and measuring ELECTROMAGNETIC RADIATION, especially INFRARED RADIATION. The **Crookes radiometer** was invented (1875) by the British chemist and physicist William CROOKES. It has pivoted vanes that rotate (in a partial VACUUM) when exposed to light or heat, because the vanes are painted black on one side and the sides absorb different amounts of radiant energy. More sophisticated radiometers can measure radiation in any part of the ELECTROMAGNETIC SPECTRUM. *See also* GEIGER COUNTER

radiosonde Meteorological instrument carried up into the atmosphere by a BALLOON. It monitors weather conditions and transmits the information by RADIO to scientists on the ground. When the balloon reaches a height of 30km (18mi) above the Earth it bursts, and the radiosonde returns to the ground by parachute. It is tracked by radar, in the hope that it can be recovered and reused.

radio source Discrete astronomical source of RADIO WAVES, detected by RADIO TELESCOPE. Radio sources include the Sun, SUPERNOVA remnants and PULSARS, QUASARS, and certain NEBULAE and GALAXIES, many of which have been identified optically. The strongest sources are the ANDROMEDA GALAXY, Perseus, Taurus A, Puppis A, Virgo A, Centaurus A, Cygnus A and Cassiopeia A.

radio telescope Instrument used to collect and record radio waves from space. The basic design is the large single **dish** or parabolic reflector, up to 100m (330ft) in diameter. The smoothness and accuracy of the reflecting surface determines the shortest wavelength that can be detected, since surface irregularities must be no more than a small fraction of this wavelength. However, for longer wavelengths, such as around 21cm, open wire mesh can be used for the reflector. Radio waves are reflected by the dish possibly via a secondary reflector, to a **focus**, where they are converted into electrical signals. The signals are amplified and sent to the main control room, where there is further amplification before analysis and recording. A radio telescope needs to have high **sensitivity** (the ability to detect faint signals), and high RESOLUTION (the ability to see fine detail). The largest individual dishes, such as those at Effelsberg Radio Observatory, Jodrell Bank and Parkes Observatory, are very sensitive. Better resolution can be obtained by linking two or more dishes to form a radio INTERFEROMETER, or by combining many of them to form an APERTURE synthesis array.

radiotherapy In medicine, the use of RADIATION to treat tumours or other pathological conditions.

radio waves Form of very high wavelength ELECTROMAGNETIC RADIATION. Radio waves are identified by their FREQUENCIES, expressed in kilohertz (kHz), megahertz (MHz), or gigahertz (GHz). Audio waves are lower-frequency waves. Signals broadcast in the AMPLITUDE MODULATION (AM) band range from 540kHz to 1,800kHz. FREQUENCY MODULATION (FM) broadcast frequencies range from 88MHz to 108MHz. Video, television, or digital information is broadcast at VERY HIGH FREQUENCY (VHF) or ULTRA HIGH FREQUENCY (UHF).

radium (symbol Ra) White, radioactive, metallic element of the ALKALINE-EARTH METALS, first discovered in pitchblende (1898) by Pierre and Marie CURIE; the metal, which is present in uranium ores, was isolated by Marie CURIE in 1911. Radium has been used in medical RADIOTHERAPY to treat cancer. It has 16 isotopes and they emit alpha, beta and gamma radiation, as well as heat. The gas RADON is a decay product. Properties: at.no. 88; r.a.m. 226.025; r.d. 5.0; m.p. 700°C (1,292°F); b.p. 1,140°C (2,084°F); most stable isotope ^{226}Ra (half-life 1,622 yr).

radius In anatomy, one of the two forearm bones, extending from the elbow to the wrist. The radius rotates around the ULNA, the other forearm bone, permitting the hand to rotate and be flexible. A projection just above the thumb side of the wrist marks the end of the radius.

radius In geometry, the distance or a line from the centre of a circle or sphere to any point on the CIRCUMFERENCE of the circle or surface of the sphere.

radon (symbol Rn) Radioactive, nonmetallic, gaseous element, a NOBLE GAS. It was first discovered (1899) by Ernest RUTHERFORD. The 20 known ISOTOPES, which are ALPHA PARTICLE emitters, are present in the Earth's atmosphere in trace amounts. Radon is mainly used in medical RADIOTHERAPY to treat cancer. Chemically, the element is mostly inert but does form compounds such as FLUORIDES. Properties: at.no. 86; r.d. 9.73; m.p. −71°C (−95.8°F); b.p. −61.8°C (−79.24°F); most stable isotope ^{222}Rn (half-life 3.8 days).

radula Structure found in the mouths of some MOLLUSCS with which they rasp their food. It consists of a tongue-like strip of horny material (modified EPITHELIUM), which bears many rows of chitinous or horny teeth. In slugs and snails, it is used to rasp off pieces of leafy food. In some marine molluscs, it is modified for boring or scraping.

railway (railroad) Form of transport in which carriages (wagons or bogies) run on a fixed track, usually steel rails. Railways date from the 1500s, when wagons used in mines were drawn by horses along tracks. In 1804 English mining engineer Richard TREVITHICK built the first steam LOCOMOTIVE. In 1825 George STEPHENSON's *Locomotion* became the first steam locomotive to pull a passenger train, on the Stockton and Darlington Railway. In 1830 the first full passenger-carrying railway, the

RADIO TELESCOPE

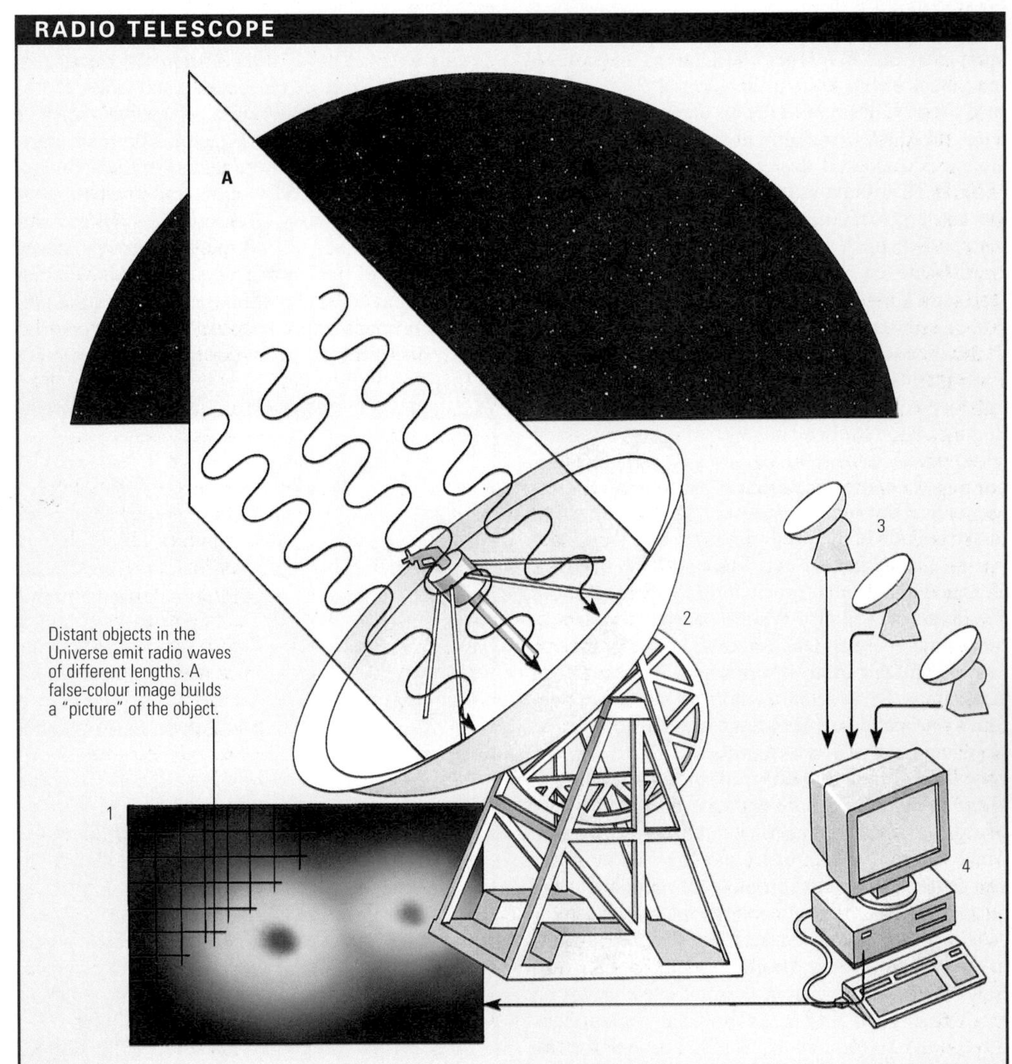

Radio telescopes are used to record radio emissions from distant stars, galaxies and other celestial bodies in the Universe that are too distant to be visible with optical telescopes. In these cases astronomers have to rely on longer wavelength radiation that can penetrate the Earth's atmosphere (A). Because of the much longer wavelengths of the radio signals, to resolve distant objects (1) large dishes that can track and follow a particular area of sky are needed (2). The largest dishes are fixed and observations need to coincide with the rotation of the Earth. To overcome this problem, it is possible to use a series of dishes in an array (called a radio interferometer) many kilometres across (3) and combine the results from each dish electronically to make a synthetic large "dish" (4).

RAILWAY

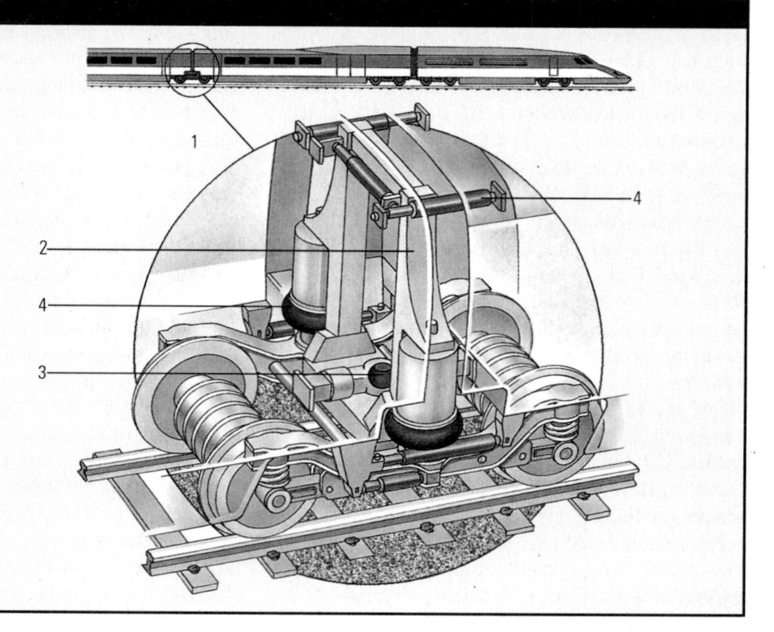

Modern railways have trains, such as the Eurostar, with bogies that use an articulated air suspension system. This has great weight savings. The bogies (1) link the carriages, which are attached to a metal frame (2) through a ball joint (3). Instead of each carriage having a bogie at either end, there is a single suspension unit between two carriages. Dampers (4) modify articulation to improve passenger comfort. The black rings between the frame and the wheel unit are the air suspension units.

Liverpool and Manchester Railway, was opened, using Stephenson's *Rocket*. The growth of rail fed the Industrial Revolution. In 1830 *Tom Thumb* became the first steam locomotive to be produced in the US. The first transcontinental railroad was completed in 1869, when the Union Pacific Railroad from Nebraska met the Central Pacific Railroad from California at Utah. The vast distances encouraged the development of sleeping cars, and George PULLMAN's design came into service in 1865. In 1851 India started to build what is now the world's largest rail network. By the 1870s, railways were a principal means of transport in Europe, North America, India, Russia and Australia, and were emerging in such countries as Japan, South Africa and New Zealand. Important technological developments included George WESTINGHOUSE's air brake in 1872. The world's first UNDERGROUND RAILWAY to carry passengers was the City and South London Railway in 1890. Many cities adopted the idea. STEAM ENGINES are still used in India, but most countries use electric, diesel, diesel-electric, or gas turbine locomotives. Modern developments include high-speed trains, such as the Japanese "Bullet" train or the French TGV (Train à Grande Vitesse), that travel at an average speed of *c.*300km/h (185mph). **Maglev** (magnetic levitation) trains use magnetic forces to hold them above a guide rail. **Air** trains hover above the track by an air cushion. Trains that do not run on wheels are usually propelled by the magnetic forces of a linear electric motor, or by jet engine.

rain Water drops that fall from the Earth's atmosphere to its surface, as opposed to FOG or DEW which drift as suspensions, and SNOW or HAIL which fall in the form of ice particles. CLOUDS are aggregates of minute droplets of moisture, and raindrops are formed when these enlarge by further CONDENSATION and by coalescence with other drops as they fall. Warm air passing over the sea absorbs water vapour and rises in thermal currents, or on reaching a mountain range. The water vapour condenses and forms clouds, which account for the usually heavier annual rainfall on windward, compared to leeward, mountain slopes. Rainfall is measured by a RAIN GAUGE. *See also* PRECIPITATION

rainbow Bright, multicoloured band, usually seen as a circular or partial arc formed opposite to the Sun or other light source. The **primary** bow is the one usually seen; in it the colours are arranged from red at the top to violet beneath. A **secondary** bow, in which the order of the colours is reversed, is sometimes seen beyond the primary bow. The colours are caused by reflection of light within spherical drops of falling RAIN, which causes white light to be dispersed into its constituent wavelengths. The colours usually seen are those of the visible SPECTRUM: red, orange, yellow, green, blue, indigo and violet.

rainforest Dense forest of tall trees that grows in hot, wet regions near the Equator. The main rainforests are in Africa, central and S America, and SE Asia. They comprise 50% of the timber growing on Earth and house 40% of the world's animal and plant species. They also, through PHOTOSYNTHESIS, supply most of the world's oxygen. This is why the present rapid destruction of the rainforests (up to 20 million hectares are destroyed annually to provide timber and land for agriculture) is a cause of great concern. Also, clearing rainforest contributes to the GREENHOUSE EFFECT and may lead to GLOBAL WARMING. There are many species of broad-leaved EVERGREEN trees in rainforests, which grow up to 60m tall. The crowns of other trees, up to 45m tall, form the upper canopy of the forest. Smaller trees form the lower canopy. Climbing vines interconnect the various levels, providing habitats for many kinds of birds, mammals and reptiles. Very little light penetrates to the forest floor, which consequently has few plants. Rainforest trees provide many kinds of food and other useful materials, such as Brazil nuts, cashews, figs and mangosteens, as well as fibrous kapok and the drugs quinine and curare.

rain gauge Instrument used for measuring rainfall and other forms of PRECIPITATION. There are many different types, including some self-recording varieties. Most rain gauges consist of a cylinder that stands up above the ground surface to prevent splashing water bouncing in. Inside the cylinder, there is a funnel to lead the water into a collecting jar. The water collected is poured into a measuring cylinder, and the total rainfall can be measured, generally in millimetres, or inches.

rainwash Process in which drops of rainwater hit rock and soil, washing away tiny particles. It contributes significantly to the EROSION of the sides of valleys, especially if they have no covering vegetation, and helps to form V-shaped valleys in areas of significant rainfall. In the absence of rainwash, as in deserts, the sides of valleys carved out by rivers remain very steep and form wadis, gorges or canyons.

Rainwater, Leo James (1917–) US physicist who proposed a new theory of atomic structure in 1950, following the discovery that atomic nuclei are not spherical. He postulated that most nuclear PARTICLES form an inner NUCLEUS while the remaining particles form an outer nucleus, the shape of each set of particles affecting the other set. Rainwater believed that if some of the outer particles

RAINBOW

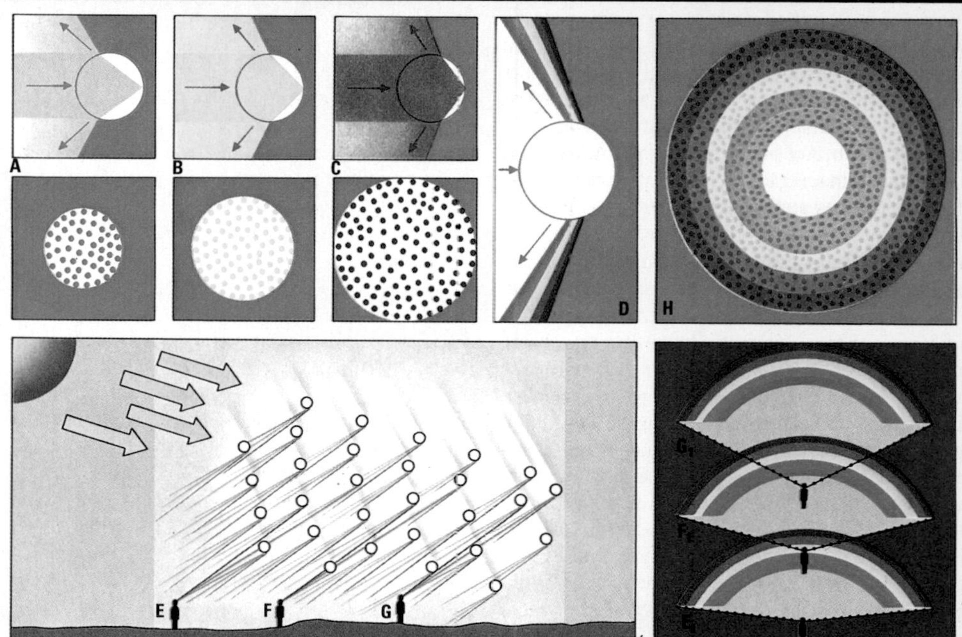

A rainbow is a natural demonstration of the mixture of wavelengths that make up white light. Drops of moisture in the atmosphere act as prisms, dispersing the light into its component colours. Violet (A) is always refracted, diffracted and finally reflected at an angle of *c.*40° parallel to the Sun's rays, yellow (B) at *c.*41° and red at *c.*42°. In this way the complete spectrum is formed (D). As each colour is formed by rays that reach the observer at a certain angle, no matter what the position of the observer (E, F and G) in relation to the Sun and the raindrops, the same spectrum is seen as E_1, F_1 and G_1. From an aircraft, however, an observer might see the complete circle (H).

▲ **rainforest** Mist rising from the forest canopy after heavy rain in the foothills of the Andes, Ecuador. This demonstrates the rapid recycling of rainwater back to the atmosphere by transpiration and evaporation (evapotranspiration) that is characteristic of rainforests.

moved in similar orbits, sufficient energy could be created from their interaction to permanently deform an otherwise symmetrical nucleus. By 1953, the Danish physicists Ben MOTTELSON and Aage Niels BOHR had provided experimental evidence to support his theory. For this research, which made NUCLEAR FUSION a practical possibility, Rainwater, Mottelson and Bohr shared the 1975 Nobel Prize for physics. *See also* NUCLEAR ENERGY; RELATIVE ATOMIC MASS (R.A.M.)

raised beach Elevated beach that is above sea level because of an uplift of the land or a retreat of the sea. A raised beach is often bounded on its inland edge by old cliffs, sometimes with caves. The uplift of land may have been the result of tectonic movements or the result of isostatic re-adjustment of the land after ice has melted. When the weight of ice has been removed by melting, the land gradually rises, relative to the sea. The same effect can be produced by the retreat of the sea, and this occurred when snow and ice increased during the glacial phases: as water becomes locked in ice and snow there is less water in the sea and it retreats from the beach.

RAM (Random Access Memory) INTEGRATED CIRCUITS (chips) that act as a temporary store for COMPUTER PROGRAMS and DATA (information). To run a program on a computer, the program is first transferred from a MAGNETIC DISK, or other storage device, to RAM. RAM also holds documents produced when the program is used. Another part of RAM stores the images to be displayed on the screen. The contents of RAM are lost when the computer is switched off.

Raman, Sir Chandrasekhara Venkata (1888–1970) Indian physicist who greatly influenced and contributed to the growth of science and research facilities in India. He was professor of physics (1917–33) at Calcutta University. He was director of the Indian Institute of Science, Bangalore, and in 1946 founded and directed the Raman Institute. Raman received the 1930 Nobel Prize for physics for his research on the DIFFUSION of light, and his discovery of the RAMAN EFFECT.

Raman effect Slight change in the frequency of monochromatic (single-wavelength) LIGHT that has been scattered. The effect, discovered by C.V. RAMAN, is seen as secondary spectral lines on each side of the primary spectral line. *See also* ELECTROMAGNETIC RADIATION; SCATTERING, LIGHT

ramjet engine Aircraft JET ENGINE, a reaction motor that relies on the speed of the aircraft to ram air (for COMBUSTION of a HYDROCARBON fuel) into the intake, thereby compressing it. The only moving part is the variable diameter intake, or diffuser, which varies the volume, and therefore the pressure, of the incoming air. The remainder of the engine is an elongated tube which forms the combustion chamber. The engine operates best at speeds above that of sound; it must first be accelerated by other means (such as solid-fuel ROCKETS) until sufficient pressure builds up.

Ramsay, Sir William (1852–1916) British chemist. Working with Lord RAYLEIGH he discovered ARGON in air. Later, he discovered HELIUM, NEON and KRYPTON. Ramsay was awarded the 1904 Nobel Prize for chemistry. He was also interested in RADIOACTIVITY and showed that helium is produced by RADIOACTIVE DECAY. He was knighted in 1902.

Ramsden, Jesse (1735–1800) British maker of astronomical instruments. He made significant improvements to the THEODOLITE, BAROMETER, MICROMETER and SEXTANT.

random access memory (RAM) *See* RAM

rangefinder Optical attachment, most commonly to guns and cameras, used for measuring distances. Optical rangefinders create images of the target as viewed from two positions; the amount of rotation or other movement needed to make the images merge indicates the range of the target. More advanced rangefinders measure the time it takes a pulse of light from a LASER to reach the target and return.

Ranger First series of US SPACE PROBES to the Moon. After several failures, in 1964 and 1965 Rangers 7, 8 and 9 transmitted thousands of photographs of the lunar surface as they approached predetermined impact points. The photographs showed details of boulders and craters as small as 1m (3ft), a much higher resolution than that attained by Earth-based TELESCOPES.

Rankine, William John Macquorn (1820–72) Scottish engineer and physicist. In 1855 he became professor of engineering at Glasgow University, where he made valuable contributions to civil and mechanical engineering and THERMODYNAMICS. Rankine wrote manuals on these subjects and also devised the absolute TEMPERATURE SCALE based on the degree FAHRENHEIT, known as the Rankine scale.

Raoult's law VAPOUR PRESSURE of a SOLUTION is the sum obtained by multiplying the vapour pressure of each component by the MOLE fraction of that component and adding these products. (The mole fraction is the number of moles of the component divided by the total number of moles present.) The law holds only approximately and over a limited range of conditions, differing for different substances. Discovered (1886) by the French physical chemist François-Marie Raoult (1830–1901), the law is used to find the RELATIVE MOLECULAR MASS of a substance and has been fundamental to many theories of solution.

rapid eye movement (REM) Phenomenon observed in sleeping people. REM has been calculated to take up to 20% of a normal night's SLEEP and has been associated with the time spent dreaming. The sleeper passes in and out of REM sleep and, although it is possible to make do with very little sleep, he or she suffers if deprived of REM sleep.

rare-earth elements *See* LANTHANIDE SERIES

rare gas *See* NOBLE GAS

ratchet Mechanism that permits leverage in one direction only. It consists of a wheel or bar with teeth sloping in one direction, which is engaged by a pawl so that the wheel or bar is unable to rotate backwards. Ratchets are often parts of MACHINE TOOLS in which non-slip movement is essential. The rear sprocket of a free-wheelable bicycle incorporates a ratchet. *See also* LEVER

ratio Measure of relative size of two quantities. Ratios can be expressed using a colon. For instance, if a classroom contains twenty girls and ten boys, the ratio of girls to boys is 20:10 or 2:1 (read two to one). Ratios can also be expressed as fractions (or percentages). In this case ⅓ of the pupils are boys and ⅔ are girls.

rational number *See* NUMBER, RATIONAL

ratite Any of a group (Ratitae) of large, usually flightless birds with flat breastbones instead of the keel-like prominences found in most flying birds. Ratites include the ostrich, rhea, cassowary, emu, kiwi and the unusual flying tinamou.

Ray, John (1627–1705) English naturalist whose work on plant and animal classification later influenced Carolus LINNAEUS and Georges CUVIER. Ray was the first to distinguish the two main types of flowering plants: MONOCOTYLEDONS and DICOTYLEDONS.

Rayleigh, John William Strutt, Lord (1842–1919) British physicist. His work was chiefly concerned with various forms of wave motion. Rayleigh was awarded the 1904 Nobel Prize for physics for his discovery (with the chemist William RAMSAY) of the noble gas ARGON and for work on gas densities.

rayon Fine, smooth, SYNTHETIC FIBRE made from solutions of CELLULOSE from soft wood pulp or cotton linters (short fibres). The first SYNTHETIC FIBRE, rayon was produced (1884) by the French chemist Hilaire de Chardonnet (1839–1924). VISCOSE rayon, the most common, is spun-dried and has a strength approaching that of NYLON. ACETATE rayon is made of filaments of cellulose ETHANOATE, while triacetate rayon has filaments of triacetate. The more costly cuprammonium rayon, known as Bemberg, resembles silk.

reactant *See* CHEMICAL REACTION

reaction propulsion Propulsion utilizing the third of Isaac NEWTON'S laws of MOTION: for every action there is an equal and opposite reaction. Both ROCKETS and JET ENGINES build up internal pressure from gases or PLASMAS that are allowed to escape in one direction, creating a propulsive force in the opposite direction.

reaction rate In chemistry, speed at which CHEMICAL REACTIONS move to completion or to EQUILIBRIUM. The rate is accelerated by one or more of the following: the presence of a CATALYST or a promoter,

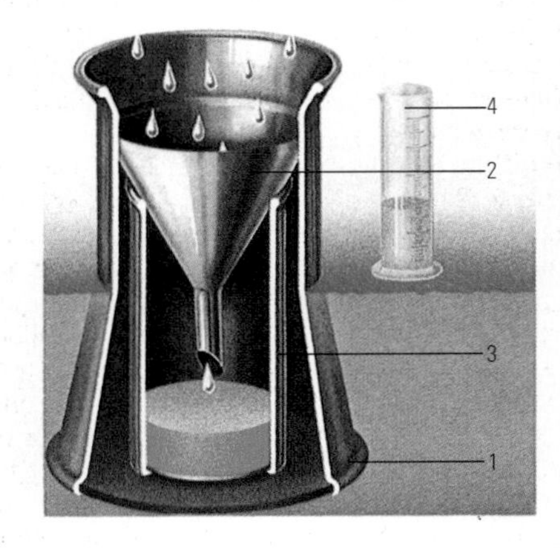

▲ **rain gauge** The example illustrated is designed for daily measurement of total rainfall. The outer case (1) is shaped to provide extra stability when the gauge is embedded in the soil. The gauge is essentially a funnel (2) of standard diameter to direct the rain to the inner can.(3). The rainwater collected is poured into a special measuring cylinder (4) graduated to read millimetres of rainfall.

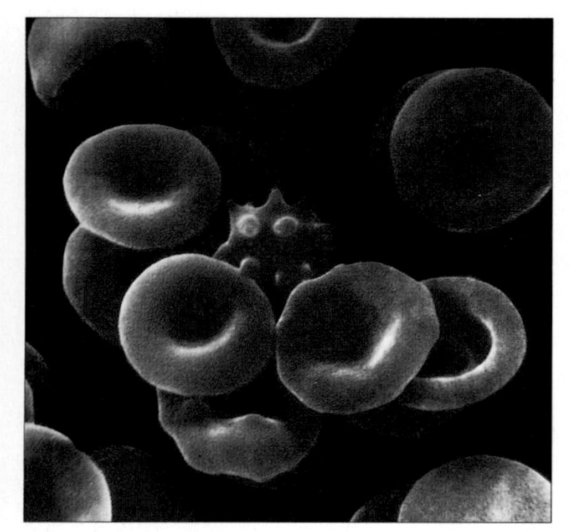

▲ **red blood cells** False-colour scanning, electron micrograph (SEM) of human red blood cells (erythrocytes) shown magnified 1,090 times. They take the form of biconcave discs, and are packed with haemoglobin (the iron-containing respiratory pigment). Their main function is the transport of oxygen and carbon dioxide around the body. The biconcave shape is an adaption for maximum surface area, over which the exchange of gases may take place. Erythocytes are also very flexible, and are able to squeeze into the smallest capillary. The shape of erythrocytes is dependant on osmotic pressure of plasma; in hypertonic plasma (higher concentration than normal) they become crenated (star-shaped – one is visible here at the centre).

heat, pressure, or REDUCTION of the cocentration of the reaction products.

reaction time Interval between the onset of a STIMULUS and the conscious execution of a response.

reactor *See* NUCLEAR REACTOR

read only memory (ROM) *See* ROM

reagent Substance that takes part in a CHEMICAL REACTION. The term is often understood to mean a common laboratory reagent, that is, a substance used in routine qualitative or quantitative analytical tests.

real number *See* NUMBER, REAL

real-time computing Description of any COMPUTER application that reacts to things as they happen. DATA is received, processed and a response given within a time set by outside events. In human terms, the reply is virtually immediate. Real-time systems are employed in air-traffic control, automatic banking, automatic pilots on aircraft, industrial robotics and many military applications.

Réaumur, René Antoine Ferchault de (1683–1757) French scientist best remembered for his THERMOMETER scale, which designates zero as the freezing point of water and 80 degrees as its boiling point. Réaumur wrote widely on natural history and conducted research in mining, metallurgy, fossils and insects.

recapitulation Biological theory which says that the embryonic development of a species repeats the evolutionary development of that species. It was called the biogenetic law by Ernst HAECKEL.

receptacle Biological structure that serves as a container for reproductive cells or organs in plants. In ANGIOSPERMS (flowering plants), the receptacle is the enlarged end of a stalk to which the FLOWER is attached. In FERNS, it is the mass of tissue that forms the sporangium (the SPORE-bearing organ). In some SEAWEEDS, it is the part that seasonally becomes swollen and carries the reproductive organs.

receptor In higher organisms, tiny organs found on both NEURONS (nerve cells) and ENDOCRINE cells. They detect physical stimuli, such as light, sound, touch and taste, and chemical changes occurring inside the body and on or near its surface. They then transfer the physical or chemical signals into electrical energy or the secretion of HORMONES. For example, the RODS AND CONES of the RETINA of the eye are receptors that transform light energy into the nerve impulses sent via SENSORY NEURONS to the brain. Others, such as those in muscles, the ear and visceral organs, tell the brain how the body is positioned and how it moves.

recessive In genetics, describes a form of GENE (ALLELE) that does not express itself when paired with a DOMINANT allele. Although it is part of the GENOTYPE (genetic make-up) of a HETEROZYGOTE (an organism containing both dominant and recessive alleles for a particular gene), it does not contribute to the PHENOTYPE (physical characteristics) of the organism. It becomes expressed only when the same recessive allele form appears on both CHROMOSOMES of a HOMOZYGOTE. Recessive alleles were discovered by Gregor MENDEL, who found that a cross between pure bred red- and white-flowering garden peas always produced red flowers in the offspring. The allele for red coloration is dominant; the allele for white is recessive. In humans, the alleles for blue eyes are recessive to those of brown eyes; alleles for HAEMOPHILIA are also recessive.

reciprocal Name given to the quantity that results when a number is divided into 1 (unity). For example, the reciprocal of 2 is 1/2, the reciprocal of 1/2 is 2, the reciprocal of 8 is 1/8 (or 0.125) and the reciprocal of 3/7 is 7/3 and so on.

reciprocating engine Any ENGINE in which a piston moves to and fro, such as a STEAM ENGINE or an INTERNAL COMBUSTION ENGINE, commonly used to power motor vehicles and aircraft. There are two basic designs of reciprocating internal combustion engine. An **in-line** engine has either one row of CYLINDERS or two rows of cylinders in a V arrangement or horizontally opposed. A **radial** engine, used only for aircraft, has cylinders arranged in a circle.

recombinant DNA research Branch of GENETIC ENGINEERING involving the transfer of a segment of DNA from a source organism into a host organism (typically a microbe). It is also known as GENE splicing because the transferred segment is spliced into the overall DNA structure of the host, thus altering the information contained in its GENETIC CODE. When the host undergoes asexual CELL DIVISION, each product cell carries a replica of the new DNA. In this way, numerous CLONES of the new cell can be made. Gene splicing can be used to design cells to produce medicines, to digest spilled oil or waste products, and for numerous other purposes.

recombination Process that rearranges GENES to increase genetic VARIATION in sexually produced offspring. Recombination takes place during MEIOSIS, the type of CELL DIVISION that leads to the formation of sex cells (GAMETES). It is achieved by CROSSING OVER of paired CHROMATIDS. Its effect is to "shuffle" genes derived from both parents and thereby create genetic variation. Some GENETIC ENGINEERING techniques can induce recombination artificially.

recording Storing of signals that represent sound or images, on a medium such as a plastic disc or MAGNETIC TAPE. In traditional SOUND RECORDING, the sounds being recorded are converted into variations in an electric current that controls the movements of a stylus. This cuts an original lacquer disc, which is then electroplated to make a master negative for pressing out plastic copies. In magnetic recording, sound is recorded on tape with a TAPE RECORDER; a VIDEO RECORDING machine records sound and vision. Optical recording of sound is used for the soundtrack of motion pictures. The sounds cause variations in thickness or brightness of a strip of film running alongside the motion picture photographs. By a reverse process this is converted back into sound when the picture is run. Modern sound recording is digital. *See also* COMPACT DISC; DIGITAL AUDIO TAPE (DAT)

rectangle Four-sided geometric figure (quadrilateral), the interior angles of which are right angles and each pair of opposite sides is of equal length and is parallel. It is a special case of a PARALLELOGRAM.

rectangular coordinate system CARTESIAN COORDINATE SYSTEM in which the axes are at right angles to each other.

rectified spirit Constant-boiling mixture of ETHANOL (alcohol) and water. It is made by distilling ethanol-water mixtures and contains 95.6% ethanol. *See also* ABSOLUTE ALCOHOL

RED SHIFT

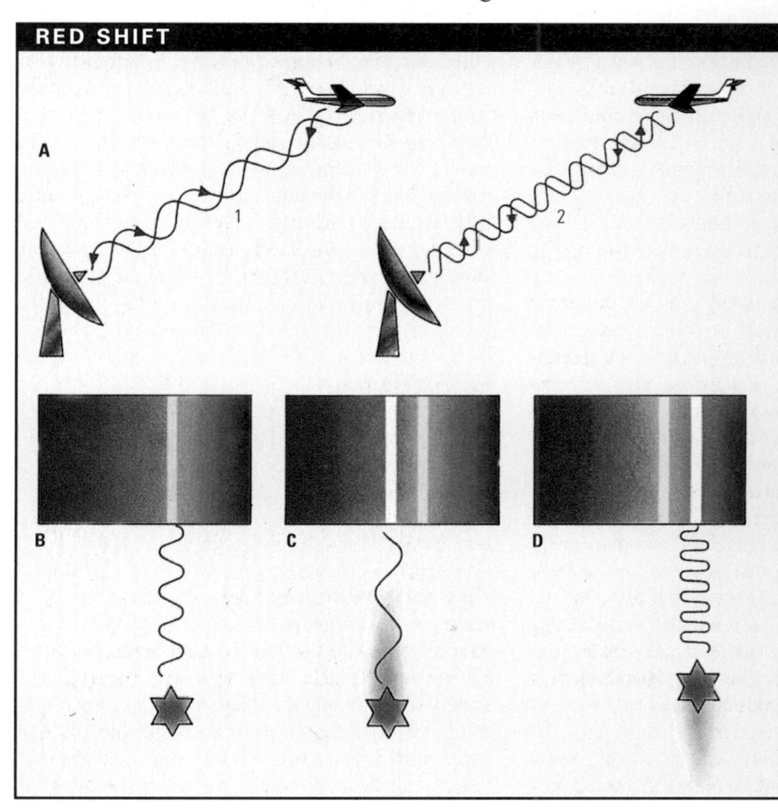

Certain red shift effects, in which light emitted from stars moves towards the longer (red) end of the spectrum, can be explained by the Doppler effect. Just as radar (A) can calculate the position of a moving object by measuring the time a transmitted signal (1) takes to return (2), in this way the movements of stars can be measured relative to Earth. The wavelength of a star that appears to be moving neither closer to Earth nor farther away (B) does not alter, and its spectrum does not change. The wavelength of a star that is moving away from Earth lengthens (C), and moves towards the red end of the spectrum. The wavelength of a star moving closer to Earth (D) shortens, and moves towards the blue end of the spectrum.

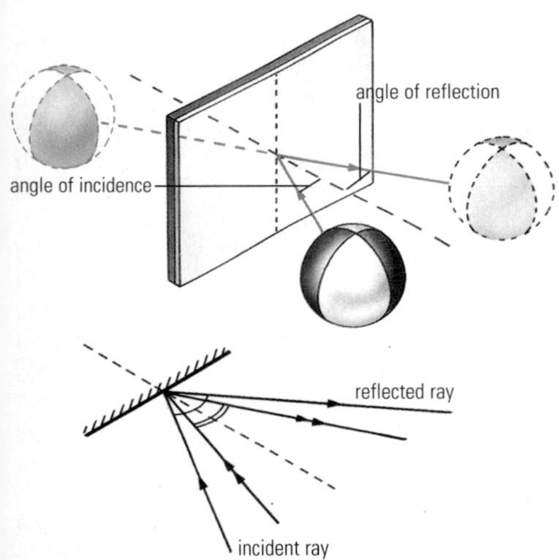

▲ **reflection** The top diagram shows the reflection of an image in a mirror. The image is reflected back at an angle of reflection equal to the angle of incidence. The image, however, appears to the eye to be behind the mirror on an extension of the angle of reflection. The bottom line drawing illustrates that the angle of reflection is always the same as the angle of incidence, whether the angle is acute or oblique.

rectifier Component of an ELECTRIC CIRCUIT that converts ALTERNATING CURRENT (AC) into DIRECT CURRENT (DC). The rectifier is usually a semiconductor DIODE that presents a high RESISTANCE to a current flowing in one direction and a low resistance to current flowing in the opposite direction. *See also* ELECTRIC CURRENT

rectum In humans and many other vertebrates, last part of the large INTESTINE, where the faeces accumulate prior to voiding.

recycling Natural and man-made processes by which substances are broken down and reconstituted. In nature, elemental cycles include the CARBON CYCLE, NITROGEN CYCLE and HYDROLOGICAL CYCLE. Natural cyclic chemical processes include the metabolic cycles, such as the KREBS CYCLE, in the bodies of living organisms. Man-made recycling includes using bacteria to break down excreta and many kitchen and factory organic wastes to harmless, or even beneficial, substances. Large quantities of inorganic waste, such as metal scrap, glass bottles and building spoil, are recycled. POLYMER wastes are often burned rather than recycled, with consequent atmospheric POLLUTION and loss of material resources. Controlled burning (PYROLYSIS) recovers useful substances from PLASTICS.

red algae *See* RHODOPHYTE

red blood cell Alternative common name for an ERYTHROCYTE

red dwarf STAR at the lower end of the MAIN SEQUENCE. Red dwarfs have masses of between 0.8 and 0.08 of a solar mass. They are of small diameter, relatively low surface temperature (between 2,500–5,000K), and low absolute magnitude. BARNARD'S STAR and Proxima Centauri are examples. *See also* HERTZSPRUNG-RUSSELL DIAGRAM

red giant Type of STAR in the late stage of its evolution. They have a diameter of between 10 to 100 times that of the Sun, and a luminosity 100 to 10,000 times greater. **Low-mass** red giants usually evolve into WHITE DWARFS, while those of a **high mass**, go through further nuclear fusion, before becoming NEUTRON STARS or BLACK HOLES. Examples include Arcturus and Aldebaran. *See also* HERTZSPRUNG-RUSSELL DIAGRAM; STELLAR EVOLUTION

redox *See* OXIDATION-REDUCTION

red shift (symbol *z*) Lengthening of the WAVELENGTH of light or other ELECTROMAGNETIC RADIATION from a source, caused either by the source moving away (the DOPPLER EFFECT) or by the expansion of the Universe (*see* EXPANDING UNIVERSE). It is defined as the change in the wavelength of a particular spectral line, divided by the unshifted, or rest, wavelength of that line. Red shifts caused by the expansion of the Universe, called **cosmological red shifts**, have nothing to do with the Doppler effect. The Doppler effect results from motion through space, whereas cosmological red shifts are caused by the expansion of space itself, which literally stretches the wavelengths of light that is travelling towards us. The longer the travel time of light, the more its wavelength is stretched, as embodied in the HUBBLE CONSTANT. GRAVITATIONAL RED SHIFT is a phenomenon predicted by Albert EINSTEIN's general theory of RELATIVITY. Light emitted by a star has to do work to overcome the star's gravitational field. As a result, there is a slight loss of energy and a consequent increase in wavelength, so that all the spectral lines are shifted towards the red.

reducing agent Substance that causes REDUCTION reactions. Thus, in the reduction of oxygen to form water, $2H_2 + O_2 \rightarrow 2H_2O$, hydrogen is the reducing agent. In biological systems, some SUGARS act as reducing agents. *See also* OXIDATION-REDUCTION

reduction Chemical reaction in which an ELECTRON is added to an ATOM or an ION. It is always part of a OXIDATION-REDUCTION reaction.

reef Rocky outcrop lying in shallow water, especially one built up by CORALS or other organisms.

refinery Any industrial plant based on processes such as DISTILLATION and ELECTROLYSIS. Spirit drinks such as whisky, foods such as SUGAR, and materials such as ores, metals or PETROLEUM are produced, purified or separated into their components in refineries. *See also* OIL REFINERY

refining Process of removing impurities from metals, sugar, petroleum and other products. For metal ORES especially, the refining process is carried out near the MINE in order to avoid unnecessary transport costs. Iron ore is between 25% to 70% pure; a low-quality ore of only 25% contains 75 tonnes of unwanted rock for every 100 tonnes extracted.

reflection Change in direction of part or all of a WAVE. This happens when a wave, such as a light or sound wave, encounters a surface separating two different media, such as air and metal, and is partly "bounced" back into the original medium. The remaining part passes into the second medium. The incident wave (striking the surface), reflected wave and the normal (line perpendicular to the surface) all lie in the same plane; the incident wave and reflected wave make equal angles with the normal.

reflex action Rapid, involuntary response to a particular STIMULUS – for example, the "knee-jerk" reflex that occurs when the bent knee is tapped. It is controlled by a **reflex arc** found in the CENTRAL NERVOUS SYSTEM. A reflex arc consists of a **sensory receptor** (which detects the stimulus), an **afferent** (or sensory) nerve, which conveys the impulse to the grey matter of the SPINAL CORD, and an **efferent** (or motor) nerve, which carries the response impulse to the EFFECTOR, in this case the muscles in the knee, causing the leg to jerk. This type of response is only detected by the brain once the reflex has taken place. This reflex is sometimes called a **spinal** reflex. A **cranial** reflex involves the brain, whereby the initial reflex can be overridden by the brain and either stopped or altered.

reforestation Planting of trees in a region where DEFORESTATION has occurred. Throughout the world, trees have been cut down for fuel or wood pulp, or to clear land for agriculture. On sloping land with no trees, soil EROSION soon takes place as rain washes away the soil. Replanting trees prevents erosion and provides a renewable resource of timber and fuel.

reforming Conversion of straight-chain ALKANES into branched-chain alkanes by CRACKING or by catalytic reaction. Reforming is used to produce more efficient PETROL, since straight-chain alkanes burh unsteadily and cause engine KNOCK.

refraction Bending of a WAVE, such as a light or sound wave, when it crosses the boundary between two transparent media, such as air and glass, and undergoes a change in speed. The incident wave (striking the surface), the refracted wave and the normal (line perpendicular to the surface) all lie in the same plane. The incident wave and refracted wave make an angle of incidence, *i*, and an angle of refraction, *r*, with the normal. The index of refraction for a transparent medium is the ratio of the speed of light in a vacuum to its speed in the medium. It is also equal to sin *i*/sin *r*. Snell's law states that this ratio is constant for a given interface. *See also* REFLECTION

refractive index Measure of the ability of a transparent medium, such as glass, to refract (bend) light. It is the ratio of the speed of light in a vacuum to its speed in the medium and is calculated by dividing the SINE of the ANGLE OF REFRACTION into the sine of the ANGLE OF INCIDENCE of the incoming light.

refractometer Instrument for measuring the REFRACTIVE INDEX of transparent materials, ie the degree to which they bend light that passes through them. Since the refractive index of a chemical solution changes according to its composition, refractometers are frequently employed in chemical analysis.

refrigeration Process by which the temperature in a REFRIGERATOR is lowered. In a domestic refrigera-

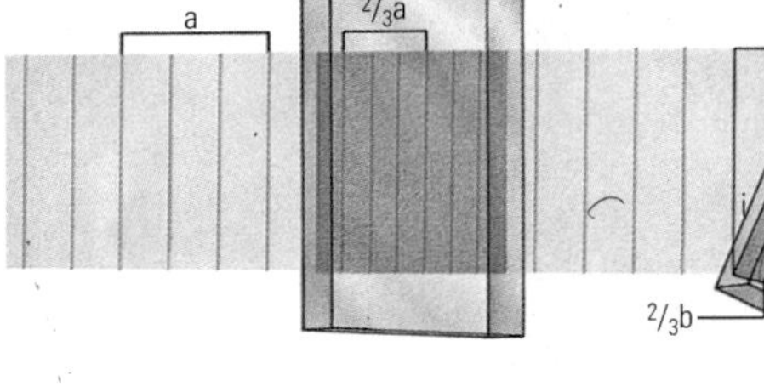
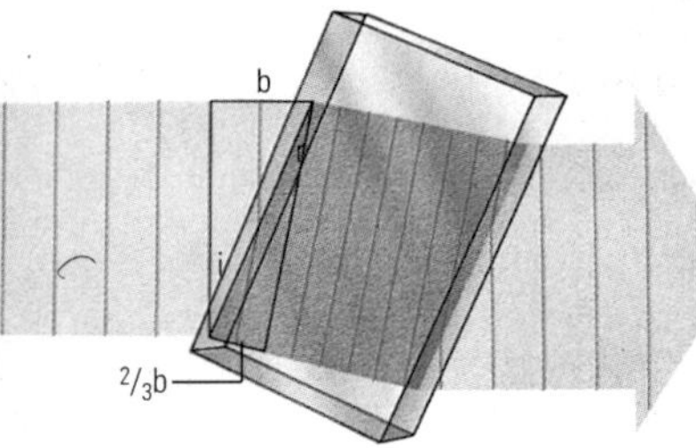

► **refraction** When light passes from one transparent substance to another, its speed changes. The bending which occurs when a beam of light crosses the boundary (between, for example, air and glass) obliquely, can be explained on the basis of the change in speed; one side of the beam is affected before the other. In the illustration, a beam passes through two glass blocks, first at right angles and then at an oblique angle.

Key
a) The length of a chosen number of wavelengths of the light in the air. Due to the reduction in speed, the corresponding length in glass is a/1.5, = 2/3a; the figure 1.5 is the refractive index of the glass.
i) The angle at which the light beam reaches the surface of the glass.
r) The angle at which the beam enters the glass.
b) The distance which the right-hand side of the beam travels in air after the left-hand side of it has entered the glass. The corresponding distance in the glass, as before, is b/1.5. From the two outlined right-angled triangles, it can be seen that (sin i)/(sin r) = the refractive index 1.5; this is the basic law of refraction.

tor, a refrigerant gas such as AMMONIA or CHLOROFLUOROCARBON (CFC) is alternately compressed and expanded. The gas is first compressed by a pump, causing it to warm up. It is then cooled in a CONDENSER where it liquefies. It is then passed into an evaporator where it expands and boils, absorbing heat from its surroundings and thus cooling the refrigerator. It is then passed through the pump again to be compressed. *See also* AIR-CONDITIONING

refrigerator Heat-insulated container, maintained at a low temperature for preserving perishable foods and other materials. The refrigerator was invented (1859) by Ferdinand Carré (1824–1900) for preserving meat during the long sea voyage from Australia to Europe. In 1913 the domestic refrigerator appeared in the USA, and was marketed in Britain in the early 1920s. In domestic refrigerators, power for the cooling process, called REFRIGERATION, is provided by electricity or gas. Some commercial refrigerators use a steam jet to power the cooling process.

regeneration Biological term for the ability of an organism to replace one of its parts if it is lost. Regeneration also refers to a form of ASEXUAL REPRODUCTION in which a new individual grows from a detached portion of a parent organism.

regolith Loose fragments of rock between the layers of soil and the BEDROCK beneath. The fragments are usually thin and have broken away from the bedrock. Eventually they break down and become mineral particles in the soil. *See also* SOIL HORIZON

regulatory gene Any GENE that controls the way in which another gene is produced. Regulatory genes activate or deactivate a group of neighbouring genes called an OPERON, which functions as a unit. Usually found in BACTERIA, operons are responsible for the formation of ENZYMES that control various processes in a metabolic pathway.

Reichstein, Tadeus (1897–1996) Swiss chemist and pharmacologist, b. Poland. After teaching chemistry (1922–38) at the Institute of Technology in Zurich, he became head of the department of pharmacy at the University of Basel. Reichstein shared the 1950 Nobel Prize for physiology or medicine with Edward KENDALL and Philip S. HENCH for his work on the HORMONES of the cortex of the ADRENAL GLANDS. He was the first to synthesize (1933) ascorbic acid (vitamin C).

Reis, Johann Phillip (1834–74) German scientist who invented (*c*.1861) an early form of the TELEPHONE. Its capacity to reproduce speech, however, was limited and the first practical telephone was patented by Alexander Graham BELL in 1876.

rejection In transplantation, the IMMUNE SYSTEM's response during which a grafted tissue or organ is destroyed for being "foreign" tissue. **Acute** rejection mostly occurs within the first few weeks after surgery and can be halted with IMMUNOSUPPRESSIVE DRUGS. **Chronic** rejection, a long-term process resulting in gradual failure of the grafted part, is irreversible.

relative atomic mass (r.a.m.) (formerly atomic weight) Mass of an ATOM of the naturally occurring form of an ELEMENT divided by 1/12 of the mass of an atom of carbon-12. The naturally occurring form may consist of two or more ISOTOPES, and the r.a.m. takes this into account in averaging.

relative density (r.d.) (formerly specific gravity) Ratio of the DENSITY of a substance to the density of water. Thus, the relative density of gold is 19.3; that is it is *c*.19 times denser than water.

relative humidity *See* HUMIDITY

relative molecular mass (formerly molecular weight) Ratio of the average mass per MOLECULE of an element or compound to 1/12th of the mass of an ATOM of carbon-12. The molecular masses of REACTANTS (elements or compounds) must be known in order to make calculations about yields in a chemical reaction.

relativity Theory, proposed by Albert EINSTEIN, based on the postulate that the motion of one body can be defined only with respect to that of a second body. This led to the concept of a four-dimensional SPACE-TIME continuum in which the three space dimensions and time are treated on an equal footing. The **special theory**, put forward in 1905, is limited to the description of events as they appear to observers in a state of uniform relative motion. The more important consequences of the theory are: (1) that the velocity of light is constant for all observers; (2) that the mass of a body increases with its velocity, although appreciably only at velocities approaching that of light; (3) that mass (m) and energy (E) are equivalent, that is, $E = mc^2$, where c is the velocity of light (this shows that when mass is converted to energy, a small mass gives rise to large energy); (4) the LORENTZ-FITZGERALD CONTRACTION, that is bodies contract as their velocity increases, again only appreciably near the velocity of light; and (5) relative to a stationary observer, time passes more slowly for a moving object, "time dilation". The **general theory**, completed in 1915, is applicable to observers not in uniform relative motion (that is accelerating). This showed the relation of space and GRAVITATION. It may be put as the idea that the presence of MATTER in space causes space to "curve", forming GRAVITATIONAL FIELDS; thus gravitation becomes a property of space itself. Light is also bent by massive gravitational fields, which may explain the existence of BLACK HOLES. *See also* SPACE-TIME

relay Electromagnetic switch that is operated (opened or closed) by variations in an ELECTRIC CURRENT in a coil of wire. Current flowing through the wire coil sets up a MAGNETIC FIELD which moves a metal contact that can open or close the CIRCUIT.

releasing hormone HORMONE secreted by the HYPOTHALAMUS in the brain which travels to the nearby PITUITARY GLAND where it stimulates the release of another hormone. Each pituitary hormone has a specific releasing hormone (also called a releasing factor). For example, the production of

RELATIVITY

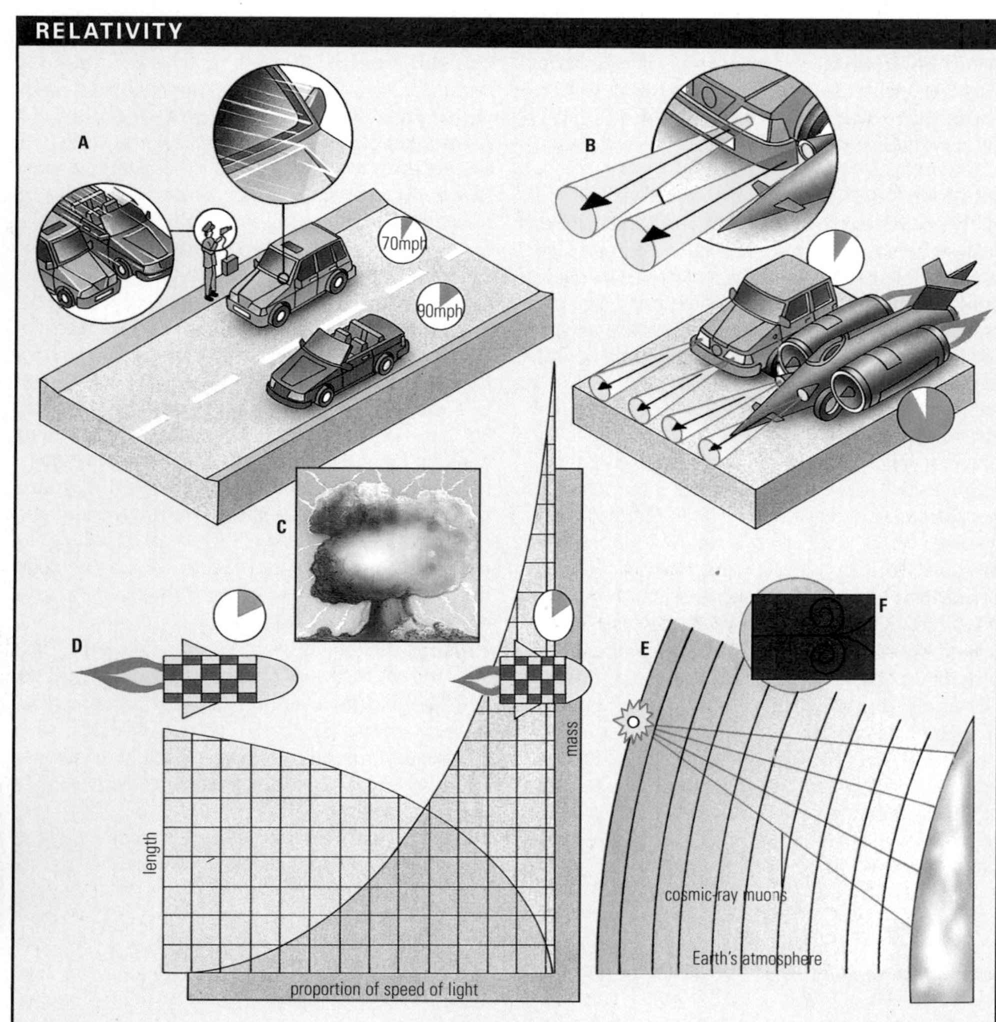

Einstein's theories of relativity are based on the postulate that the motion of one body can be defined only with respect to a second body. For example (A) shows how a stationary observer sees the red car moving at 90mph and the blue car at 70mph. To the blue car, however, the red car is moving at 20mph relative to it. Further consequences of relativity show the velocity of light is absolute. In (B), even though the supercar is travelling hundreds of mph faster, the speed of light emitted from the headlamps of both vehicles is identical, 186,000mi s^{-1}. Nuclear explosions (C) demonstrate that mass (m) and energy (E) are equivalent, that is $E = mc^2$. A small amount of mass can produce huge amounts of energy. (D) shows how, when approaching the speed of light, a body's mass increases while at the same time the body contracts. Finally E demonstrates "time-dilation". Cosmic-ray muons are unstable, and at rest decay in two nanoseconds. They should therefore only penetrate *c*.600m (2,000ft) of Earth's atmosphere before breaking down. However, muons have been detected in bubble chambers (F) 10km (6mi) below sealevel. According to relativity time runs slower for objects with speed close to that of light. To observers on Earth, therefore, it appears that a fast-moving muon lives longer than a stationary one.

REMOTE SENSING

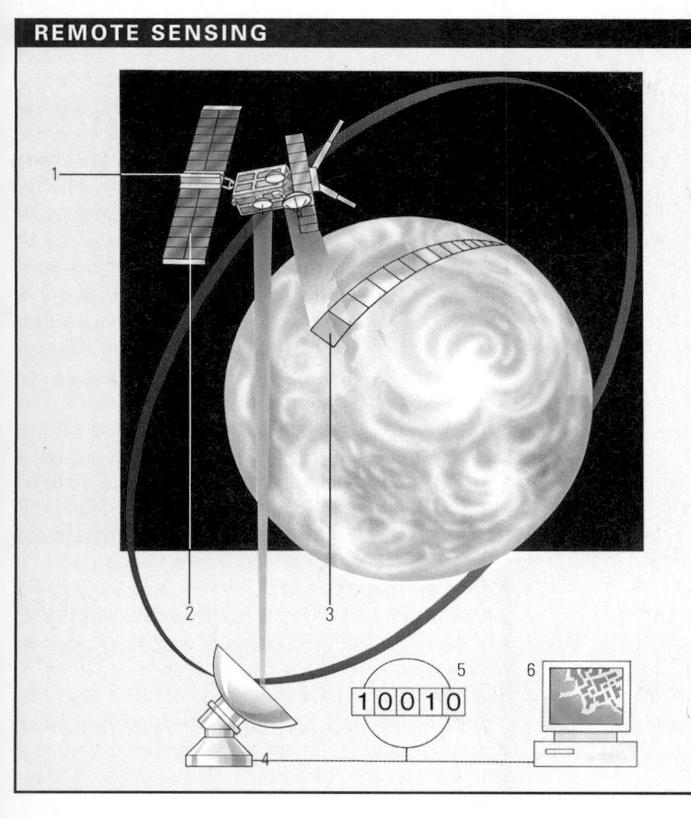

Remote sensing satellites (1) view the Earth from space using various sensors and cameras. These instruments are classed as either active and passive. Passive devices, such as optical cameras and infrared scanners, pick up reflected radiation (either light or heat). Active instruments on the other hand send out radio pulses and record the returning signal. One of the strengths of active scanning is the ability to see through cloud. Powered by a solar sail (2), the satellites use orbits which take them above the whole of the Earth during a series of days (3). Images of the Earth's surfaces are beamed down to ground stations (4) in digital form (5), and are converted into pictures by computers (6).

thyrotrophin (thyroid-stimulating hormone, TSH) by the pituitary gland is triggered by thyrotrophin-releasing hormone from the hypothalamus.

relief printing Form of PRINTING whereby a print is achieved from a raised, inked surface; the ink-bearing surface stands out above the surrounding non-printing area. *See* LETTERPRESS PRINTING

REM Acronym for RAPID EYE MOVEMENT

remote sensing Any method of obtaining and recording information from a distance. The most common sensor is the CAMERA; cameras are used in aircraft, satellites and space probes to collect information and transmit it back to Earth (often by RADIO). The resulting photographs provide a variety of information, including archaeological evidence and weather data. MICROWAVE sensors use RADAR signals that can penetrate cloud. Infrared sensors can measure temperature differences over an area. Data from sensors can be processed by computers.

renewable energy (alternative energy) ENERGY from a source that can be replenished or that replenishes itself and is generally environmentally less harmful than "traditional" energy forms such as COAL, NATURAL GAS or NUCLEAR ENERGY. SOLAR ENERGY harnesses the rays of the Sun. TIDAL POWER stations make use of the gravitational force of the Sun and Moon on the ocean. Wave power harnesses the natural movement of the sea. The power of rivers and lakes can also be tapped, usually by damming the flow and using turbines to generate HYDROELECTRICITY. WIND POWER devices have existed for centuries in the form of WINDMILLS. A less well-known renewable source is GEOTHERMAL ENERGY produced in the Earth's crust. *See also* ENERGY SOURCES

renin Polypeptide HORMONE produced by the KIDNEY in response to lowered blood pressure. It activates ANGIOTENSIN (a powerful VASOCONSTRICTOR), which reduces venous and arterial volume with a subsequent rise in blood pressure.

Rennell, James (1742–1830) English cartographer, geographer and oceanographer. He was an expert on the geography of W Asia and N Africa, and constructed the most accurate map of India of his time (1783). Rennell pioneered the scientific study of winds and ocean currents. *See also* OCEANOGRAPHY

reproduction Process by which living organisms create new organisms similar to themselves. Reproduction may be **sexual** or **asexual**, the first being the fusion of two special reproductive CELLS from different parents, and the second being the generation of new organisms from a single organism. ASEXUAL REPRODUCTION is the more limited, found mainly in PROTOZOA, some INVERTEBRATES and in many plants. By contrast, almost all living organisms have the capacity for SEXUAL REPRODUCTION. In the majority of cases, the species has two kinds of individuals – male and female – with different sex functions. Male and female sex cells (in animals, SPERM and OVA) fuse to produce a new cell, the ZYGOTE, which contains GENETIC information from both parents, and from which a new individual develops. Alternatively, organisms may be HERMAPHRODITE (like the earthworm), each individual of the species having male and female functions, so that when two of them mate each individual fertilizes the other's eggs. Usually, animals that can reproduce asexually have alternating (not always regularly) generations of asexual and sexual individuals. Examples are the aphid, jellyfish and some parasitic worms. Some plants have alternating life cycles. Sexually reproducing plants (or generations) are called GAMETOPHYTE; ones which reproduce asexually, SPOROPHYTE. *See also* ALTERNATION OF GENERATIONS; PARTHENOGENESIS

reproductive system Organs of a plant or animal necessary for SEXUAL REPRODUCTION. In plants, the reproductive system comprises the STAMENS and CARPALS. In mammals, the male system comprises the TESTES, the EPIDIDYMIS, and the PENIS, while the female system consists of the OVARIES, the UTERUS and the FALLOPIAN TUBES.

reptile Any one of about 6,000 species of VERTEBRATES distributed worldwide. Reptiles are POIKILOTHERMIC (cold blooded). Most lay yolky eggs on

REPTILE

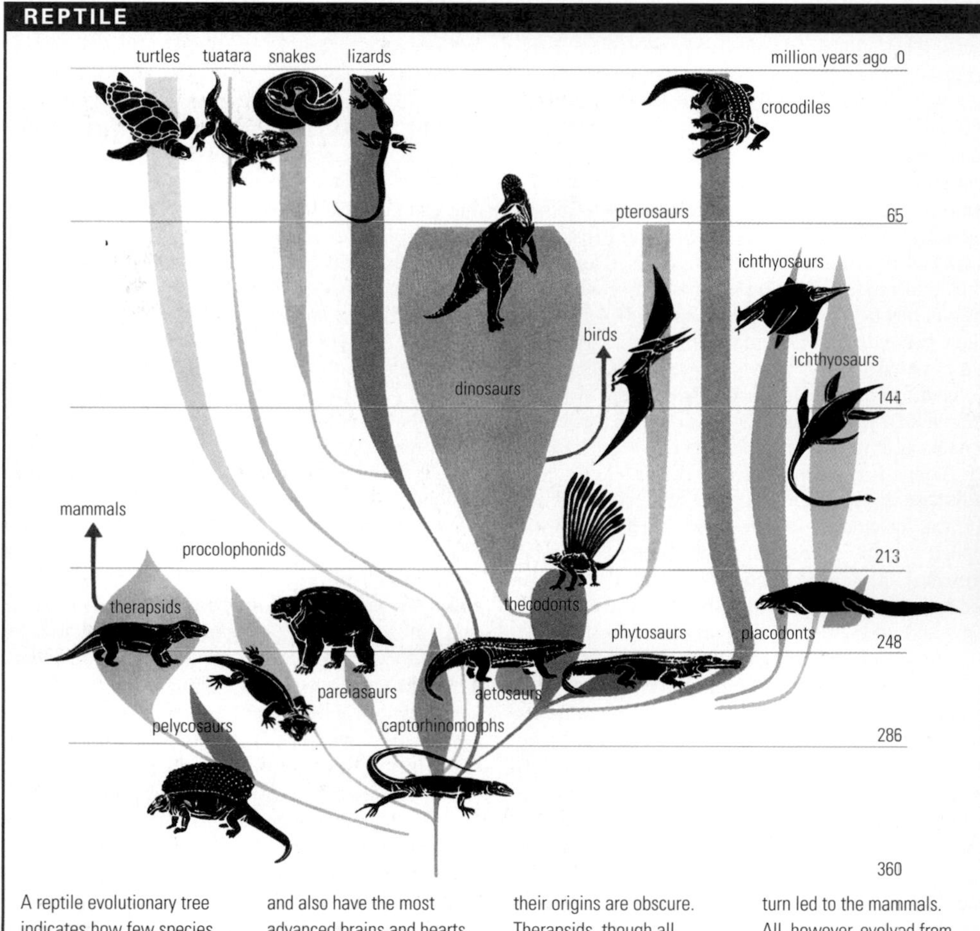

A reptile evolutionary tree indicates how few species survived from the reptiles' heyday. Crocodiles and alligators are the closest living relations to dinosaurs and also have the most advanced brains and hearts. Turtles and tortoises have changed very little over millions of years; like many areas of reptilian evolution, their origins are obscure. Therapsids, though all species are now extinct, are a vital branch of evolution, for it is from them that the synapsids evolved, which in turn led to the mammals. All, however, evolved from the earliest group of reptiles, the captorhinomorphs, otherwise known as the "stem reptiles."

RESONANCE

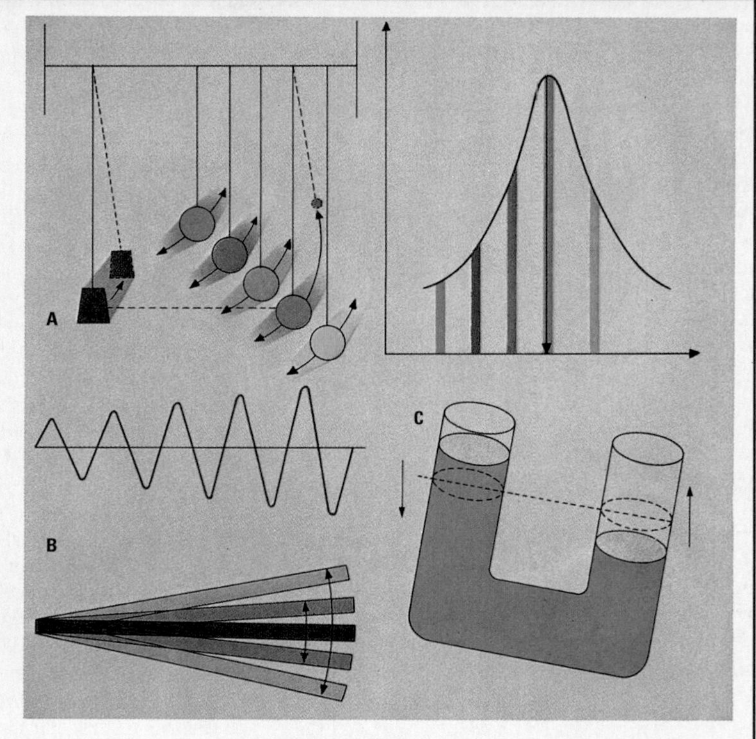

Barton's pendulum (A) provides a good illustration of mechanical resonance. A number of pendulums of varying lengths are suspended from a string. When a weighted pendulum oscillates, all the pendulums will begin to swing. The amplitude of their swing depends directly on the length of the string to which they are attached. A pendulum with the same length of string as the initial weighted pendulum, and hence the same natural frequency, will swing with the greatest amplitude.

An external force acting periodically at the resonant frequency may be used to produce resonance. A repeated thrust on one end of a diving board (B), enabling it to store up energy, will give an increasing amplitude of oscillation. A liquid in a U-shaped tube (C) can be set in oscillation by blowing in the end of the tube, but the size of oscillations are limited by friction with the walls.

land. Some species – particularly snakes – carry eggs in the body and bear live young. The skin is dry and covered with scales or embedded with bony plates. Their limbs are poorly developed or nonexistent. Those with limbs usually have five clawed toes on each foot. There are now four living orders: Chelonia (turtles); Rhynchocephalia (tuatara); Squamata (scaly reptiles such as snakes and lizards); and Crocodilia (alligators and crocodiles).

resin (rosin) Artificial or natural POLYMER which is generally viscous and sticky. **Artificial** resins, such as POLYESTERS and EPOXY RESINS, are used as ADHESIVES and binders. **Natural** resins are secreted by various plants. Resin is impermeable to water and when present in large amounts, such as in pine, makes wood resistant to rot and weather. Oleoresin, secreted by conifers, is distilled to produce TURPENTINE; ROSIN (a yellow, brown or black material) remains after the oil of turpentine has been distilled off. AMBER is fossilized resin.

resistance (symbol *R*) Property of an electric CONDUCTOR, calculated as the ratio of the VOLTAGE applied to the conductor to the current passing through it. In a conductor, this represents the opposition to the flow of ELECTRONS; the electrical energy is converted into HEAT. The SI unit of resistance is the OHM. In an ALTERNATING CURRENT circuit, resistance is the real part of the IMPEDANCE. *See also* RESISTOR

resistivity (symbol ρ) Electrical property of materials. Its value is given by $\rho = AR/l$, where A is the cross-sectional area of a CONDUCTOR, l is its length and R is its RESISTANCE in OHMS. As the temperature of the CONDUCTOR rises, the resistivity usually also rises. Resistivity is also called specific resistance and its SI UNIT is the ohm-metre.

resistor Electrical or ELECTRONIC CIRCUIT component with a specified RESISTANCE. Resistors limit the size of the current flowing in a circuit. Those for electronic circuits usually consist of finely ground CARBON particles mixed with a ceramic material and enclosed in an insulated tube. The value of the resistance is given by a coded set of coloured rings on the outside of the tube. Resistors for carrying larger currents are coils of insulated wire.

resolution In chemistry, the separation of an optically inactive MIXTURE into its optically active components (*see* OPTICAL ACTIVITY). If the components have a different CRYSTAL shape, they can be separated by hand (using a microscope). More usually, the mixture is reacted with an optically active COMPOUND to produce two new nonactive compounds that can be separated. In another technique, bacteria are used that react with one component of the mixture but do not react with the other.

resolution In electronics, ability of an imaging system, such as a COMPUTER or TELEVISION screen, to distinguish between two closely spaced objects. In computing, it is expressed as the number of dots per unit length in which an image can be formed. The usual unit is **dpi** (dots per inch) and a typical value for a colour screen is 80dpi (this compares with 150dpi for the resolution of the human eye at normal reading distance and up to 2,400dpi for printed illustrations in books). The total resolution of a computer screen is expressed as the number of pixels in each direction, typically 640×480 pixels.

resonance Increase in the amplitude of vibration of a mechanical or acoustic system when it is forced to vibrate by an external source. It occurs when the FREQUENCY of the applied FORCE is equal to the natural vibrational frequency of the system. Large vibrations can cause damage to the system.

resonance, particle Extremely short-lived ELEMENTARY PARTICLE produced in high-energy nuclear reactions occurring in particle ACCELERATORS. Such particles decay in about 10^{-24} seconds, which is characteristic of HADRONS. More than 150 have been detected, all since the 1960s. A resonance can be considered as an increase in the probability of interaction between two energetic particles, when the particles' combined energy reaches the "resonance energy".

respiration Series of chemical reactions by which complex MOLECULES are broken down to release ENERGY in living organisms. These reactions are controlled by ENZYMES and are an essential part of METABOLISM. There are two main types of respiration: AEROBIC and ANAEROBIC. In **aerobic** respiration, OXYGEN combines with the breakdown products and is necessary for the reactions to take place. **Anaerobic** respiration takes place in the absence of oxygen. Aerobic respiration results in a more complete breakdown of the original substrate, so it releases more energy than anaerobic respiration. If all the energy were released in a single burst, it would probably disrupt CELL reactions. Instead, energy is released in small amounts at different stages in respiration. In most living organisms, the energy released by respiration is used to convert ADENOSINE DIPHOSPHATE (ADP) into ADENOSINE TRIPHOSPHATE (ATP), a molecule which acts as a movable energy store to TRANSPORT energy around the cell. At the site where the energy is needed, ATP is converted back to ADP, with the aid of a special ENZYME, and energy is released. In plant and animal cells, the first stages of respiration take place in the CYTOPLASM and the later stages in the MITOCHONDRIA. *See also* COENZYME; GAS EXCHANGE; KREBS CYCLE; TRANSPIRATION

respiratory system System of the body concerned with GAS EXCHANGE. The respiratory tract of an air-breathing animal begins with the nose and mouth, through which air enters the body. The air then passes through the LARYNX and into the TRACHEA. At its lower end, the trachea branches into two bronchi and each BRONCHUS leads to a lung. Inside the LUNGS, the bronchi divide into many tiny bronchioles wich lead in turn to bunches of tiny air sacs (ALVEOLUS), where the exchange of gases between air and blood takes place. Exhaled air leaves along the same pathway.

rest mass In RELATIVITY, the MASS of an object when it is at rest. The PHOTON is described as having a rest mass of zero, but this is misleading: the photon's speed is the speed of light in all frames of reference and it can never be brought to rest. The GRAVITON, not yet observed, would have zero rest mass, and so may the NEUTRINO.

restriction enzyme In molecular biology, an ENZYME used in GENETIC ENGINEERING to cut a molecule of DNA at specific points in order to insert or remove a piece of DNA. There are many different restriction enzymes. Each cuts the DNA at a specific sequence of BASES, allowing great precision in genetic engineering. A piece of DNA may be cut out and removed for insertion into another organism. Or the geneticist may wish to insert a different piece of DNA at that point. Pieces of DNA are joined together again with enzymes called ligases. These enzymes are also used when analysing DNA to find out its sequence of bases.

reticular formation Complex mechanism in the vertebrate CENTRAL NERVOUS SYSTEM, located in the BRAINSTEM. It consists of interconnected clusters of nerve cell bodies (grey matter) and is believed to influence many physiological functions, including sleep, blood pressure and coordination.

reticulo-endothelial system Body system made up of reticular tissue, a specialized type of CONNECTIVE TISSUE that lines and supports the spleen, lymph nodes, and some other organs. Reticular tissue also contains CELLS that have MACROPHAGE action, destroying worn-out ERYTHROCYTES (red blood cells) in the spleen or invading organisms in the lymph nodes.

retina Inner layer of the EYE, composed mainly of different kinds of NEURONS (nerve cells), some of which are the visual receptors of the eye. Receptor cells (RODS AND CONES) are sensitive to light. **Cones**

respond primarily to the spectrum of visible colours; **rods** respond mainly to shades of grey and to movement. The rods and cones connect with SENSORY NEURONS that in turn connect with the OPTIC NERVE that carries the visual stimuli to the brain.

retrorocket ROCKET engine used to slow a space vehicle, separate it from another one, place it in a desired ORBIT, or allow it to make a soft landing.

retrovirus Any of a large family of VIRUSES (Retroviridae) that, unlike other living organisms, contain the genetic material RNA (ribonucleic acid) rather than the customary DNA (deoxyribonucleic acid). In order to multiply, retroviruses make use of a special ENZYME to convert their RNA into DNA, which then becomes integrated with the DNA in the CELLS of their hosts. Diseases caused by retroviruses include ACQUIRED IMMUNE DEFICIENCY SYNDROME (AIDS) and some forms of LEUKAEMIA.

reverse osmosis Movement of a liquid (the SOLVENT) through a SEMIPERMEABLE MEMBRANE from a concentrated SOLUTION to a less concentrated solution. This movement is in the opposite direction to that in normal OSMOSIS and can be achieved only by applying external pressure to the concentrated solution. It is used in DESALINATION plants to produce drinking water from seawater. The process is generally employed only on a small scale because the pressures involved are so high (up to 25 atmospheres).

reverse transcription Method of producing a double-stranded DNA copy of RNA from a VIRUS. The technique is used often in GENETIC ENGINEERING to make DNA copies of MESSENGER RNA (MRNA). It is achieved using an ENZYME called **reverse transcriptase**, which occurs in RETROVIRUSES.

reversible process Any process that can be made to go "backwards" so that parameters defining the system change in reverse order to their original values. If there were any exchanges of ENERGY or MATTER with the surroundings during the forward process, these change in direction and order when the process is reversed. In an isolated system, all processes are irreversible, because the system's ENTROPY always increases.

reversible reaction CHEMICAL REACTION in which the products can change back into the reactants. Thus nitrogen and hydrogen can be combined to give ammonia (as in the HABER PROCESS) and ammonia may be decomposed to nitrogen and hydrogen. Such processes yield an equilibrium mixture of reactants and products. *See also* CHEMICAL EQUILIBRIUM

revolution Movement of a planet or other celestial object around its ORBIT, as distinct from ROTATION of the object on its axis. A single revolution is the planet's or satellite's "year".

Reynolds number Dimensionless quantity characterizing types of FLUID FLOW. For a body of density p and linear dimension l travelling at velocity v in a fluid of viscosity ν, the Reynolds number is pvl / ν. At low values of the Reynolds number, the fluid flow is laminar, or layered, and is well understood mathematically; at higher values, the flow becomes turbulent and complicated.

rhenium (symbol Re) Silver-white TRANSITION ELEMENT, first discovered in 1925. Rhenium is found in MOLYBDENITE and PLATINUM ores from which it is obtained as a by-product. It is heavy and used in thermocouples, flashlights and filaments; it is also a useful CATALYST. Properties: at.no. 75; r.a.m. 186.2; r.d. 21.0; m.p. 3,180°C (5,756°F); b.p. 5,627°C (10,160°F); most common isotope ^{187}Re (62.93%).

rheology Study of the ways in which MATTER deforms and flows. It includes the investigation of such physical properties as VISCOSITY, ELASTICITY and plasticity (nonelastic deformation). Although fluidity is most widely recognized as a property of liquids and gases, solids also flow to some extent and are included in rheological investigation. Rheological properties are influenced by TEMPERATURE and PRESSURE. Thixotropic liquids, such as paints, decrease in viscosity as they are stirred. These and other rheological phenomena are related to the molecular structure of materials. *See also* FLUID FLOW

rheostat Variable RESISTOR for regulating an ELECTRIC CURRENT. The resistance element may be a metal wire, carbon or a conducting liquid, depending upon the application. Rheostats are used to adjust GENERATOR characteristics, to dim lights and to control the speeds of electric motors.

rhesus factor (Rh factor) Any of a group of ANTIGENS (substances that stimulate ANTIBODY production) found on the surface of ERYTHROCYTES (red blood cells), so called because they were first discovered in the rhesus monkey. Rh-negative blood lacks the Rh factor and Rh-positive blood contains it; Rh factor is present in *c*.85% of human beings. Rh incompatibility (an Rh-negative pregnant woman with an Rh-positive fetus) can give rise to ANAEMIA in newborn babies. *See also* AGGLUTINATION

rheumatism General term for a group of disorders the symptoms of which are pain, inflammation and stiffness in the bones, joints and surrounding tissues. Rheumatoid ARTHRITIS is an AUTOIMMUNE DISEASE featuring chronic inflammation of the joints, often with serious disability and disfigurement; there may also be damage to the eyes and other organs. It is more common in women.

rhizoid Fine, hairlike growth used for attachment to a solid surface by some simple organisms, such as certain fungi and mosses. The rhizoid lacks the conducting TISSUES of a ROOT.

rhizome Creeping, root-like underground stem of certain plants. It usually grows horizontally, is rich in accumulated starch, and can produce new roots and stems asexually. Examples include the iris and water lily. Rhizomes differ from roots in having nodes, buds and scale-like leaves. *See also* ASEXUAL REPRODUCTION; TUBER

rhodium (symbol Rh) Silver-white TRANSITION ELEMENT, first discovered in 1803. It occurs associated with PLATINUM and its chief source is as a by-product of nickel smelting. It resists tarnish and corrosion and is used in hard platinum alloys and to plate jewellery. Properties: at.no. 45; r.a.m. 102.906; r.d. 12.4; m.p. 1,966°C (3,571°F); b.p. 3,727°C (6,741°F); most common isotope ^{103}Rh (100%).

rhodophyte (red algae) Any member of the ALGAE phylum Rhodophyta. They are numerous in tropical and subtropical seas. Most rhodophytes are slender, branching SEAWEEDS that form shrub-like masses. Some become encrusted with CALCIUM CARBONATE and are important in REEF formation. Rhodophytes have red and purplish pigments, which help to absorb light for PHOTOSYNTHESIS. They also have CHLOROPHYLL. They have complex life cycles with two or three distinct stages, involving ALTERNATION OF GENERATIONS.

rhodopsin (visual purple) Visual PIGMENT present in the RODS eye's RETINA. It absorbs light, producing a NERVE IMPULSE that is perceived as SIGHT.

rhombus Plane figure with opposite sides parallel and all of its sides equal in length. A rhombus is a particular type of PARALLELOGRAM. The diagonals of a rhombus bisect each other at right angles.

rhyolite Fine-grained, extrusive IGNEOUS ROCK with a similar composition to GRANITE. It is an uncommon, light-coloured rock made up of alkali feldspars, quartz and sometimes mica. It forms by the rapid cooling of viscous LAVA explosively ejected from a VOLCANO and as a result is made up of tiny CRYSTALS, too small to be seen with the naked eye. It may contain small cavities (vesicles) caused by gas bubbles and areas of glassy material.

rib Any of the long, curved bones that are arranged in pairs, extending sideways from the backbone, to form the framework of the chest. There are 12 pairs of ribs in humans. The first seven pairs, known as true ribs, are joined by CARTILAGE to the BREASTBONE; the next three, called false ribs, are joined to the cartilage of the seventh pair; and the last two, called floating ribs, are not attached at the front. Adjacent ribs are joined by intercostal muscles which contract and relax to vary the capacity of the chest during BREATHING.

riboflavin VITAMIN B_2 of the B complex, lack of which impairs growth and causes skin disorders. It is a COENZYME important in transferring energy within cells. Soluble in water, riboflavin is found in milk, eggs, liver and green vegetables.

ribonucleic acid *See* RNA

ribose ($C_5H_{10}O_5$) Pentose or five-carbon sugar occurring in the structure of RNA. It is an isomeric form of arabinose. Ribose, which can be obtained

REVERSIBLE REACTION

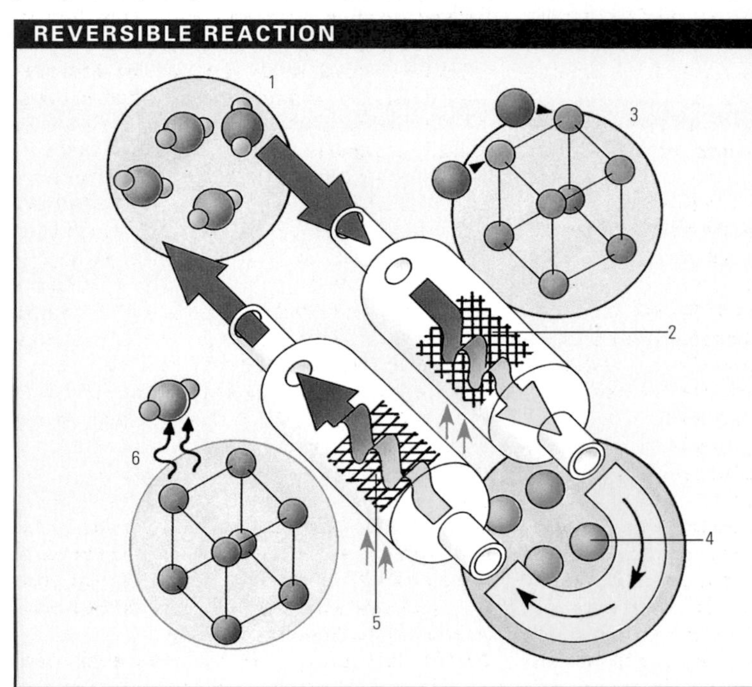

A reversible reaction is one in which the products of one reaction can react to form the chemicals that reacted to form the products. For example, if steam (1) is passed over iron that is being heated in an enclosed tube (2), a reaction will take place producing iron oxide (rust) (3) and hydrogen (4). If hydrogen is passed over rust and heated (5), water and iron are formed (6).

by the hydrolysis of RNA, is composed of a large string of nucleotides derived from nucleic acids.

ribosome Tiny structure in the CYTOPLASM of cells, involved in synthesizing PROTEIN molecules. Each protein is made up of a specific sequence of AMINO ACIDS. The genetic material DNA is made up of a long string of organic BASES of four different kinds. The order of these bases is like a code for all the instructions needed to make a living organism. It specifies the sequence of AMINO ACIDS for all the proteins the cell needs to make. Segments of DNA, called GENES, contain the instructions for individual proteins. The DNA molecule is too large to escape from the nucleus into the cytoplasm, but a "working copy" is made in the form of a molecule of MESSENGER RNA (MRNA), and this travels to the ribosomes in the cytoplasm. Ribosomes attach themselves to the messenger RNA, then assemble the amino acids in the correct sequence to form a particular protein. Each ribosome is itself made up of proteins and a special kind of RNA. *See also* GENETIC CODE

ribulose bisphosphate (RuBP) (ribulose disphosphate) Five-carbon sugar (ribulose) linked to two PHOSPHATE groups. It is involved in the CALVIN CYCLE in which it combines with CARBON DIOXIDE in the first stage of the dark reaction of PHOTOSYNTHESIS. The reaction is mediated by an ENZYME called ribulose biphosphate carboxylase, which, because of its presence in every green plant, is reputed to be the most abundant PROTEIN on Earth.

Richards, Theodore William (1868–1928) US chemist. From 1901 he was professor of chemistry at Harvard. Richards was awarded the 1914 Nobel Prize for chemistry for his accurate determinations of atomic weights (now called RELATIVE ATOMIC MASS). He also provided experimental evidence that supported Frederick SODDY's theory of the existence of ISOTOPES.

Richardson, Sir Owen Willans (1879–1959) British physicist who was awarded the 1928 Nobel Prize for physics for his work in THERMIONICS, his term for the phenomenon of the emission of electricity from hot bodies. In 1911 Richardson showed that ELECTRONS are emitted from hot metals, disproving the theory that they arose from the surrounding air. Also in 1911, he formulated a mathematical equation, known as **Richardson's law**, relating the absolute temperature of a metal to the rate of electron emission. It was of fundamental importance in the development of the thermionic valve or ELECTRON TUBE, and in the ELECTRON GUN of a CATHODE-RAY TUBE.

Richet, Charles Robert (1850–1935) French physiologist. He was awarded the 1913 Nobel Prize for physiology or medicine for his work on immune reactions, which led to research on ALLERGIES.

Richter, Burton (1931–) US physicist. Working with a very powerful particle ACCELERATOR, consisting of a pair of ELECTRON storage rings which could drive MATTER and antimatter particles together with extremely high energy, Richter discovered (1974) a new ELEMENTARY PARTICLE (which he named psi) with a relatively long lifetime; it is a type of MESON. Samuel TING, working independently of Richter and using a different method, simultaneously discovered the same particle, which he named the J-particle. For this work, Richter and Ting shared the 1976 Nobel Prize for physics.

Richter scale Classification of earthquake magnitude set up in 1935 by the US geologist Charles Richter (1900–85). The scale is logarithmic – each point on the scale increases by a factor of ten – and is based on the total energy released by an earthquake as opposed to a scale of intensity that measures the damage done at a particular place.

rickettsiae Group of tiny, BACTERIA-like PARASITES carried by fleas, lice and ticks. They cause many diseases of vertebrates, including typhus, Rocky Mountain spotted fever and Q fever, all of which can be transmitted to humans.

Riemann, Georg Friedrich Bernhard (1826–66) German mathematician. With Augustin CAUCHY, Riemann formalized the theory of INTEGRATION. He also made advances in complex analysis and NUMBER THEORY. His work on DIFFERENTIAL and NON-EUCLIDEAN GEOMETRY was used by EINSTEIN to develop his general theory of RELATIVITY.

rift valley DEPRESSION formed by the subsidence of land between two parallel FAULTS. Rift valleys are believed to be formed by THERMAL currents within the Earth's MANTLE that break up the CRUST into large slabs or blocks of rock, which then become fractured. The best example on land is the Great Rift Valley in E Africa. Rift valleys are also characteristic of MID-OCEAN RIDGES. *See also* GRABEN

Rigel (Beta Orionis) Brightest star in the constellation ORION. It is of magnitude 0.12 (the seventh-brightest star in the sky), and is a SUPERGIANT of spectral type B8. Its distance is not accurately known, but is thought to be about 1,000 LIGHT-YEARS. Rigel's LUMINOSITY is estimated to be over 100,000 times that of the Sun.

right ascension (RA) In astronomy, coordinate used to define the position of a celestial object. It is the angular distance measured eastwards from the vernal EQUINOX (also known as the First Point of Aries) to the point where the hour circle of an object meets the CELESTIAL EQUATOR. RA is usually expressed in hours, minutes, and seconds, the circumference of the celestial equator being 24 hours. *See also* CELESTIAL SPHERE

right-hand rule *See* FLEMING'S RULES

rigor mortis Stiffening of the body after death brought about by chemical changes in MUSCLE tissue. Onset is gradual from minutes to hours, and rigot mortis disappears within about 24 hours.

ring eclipse *See* ECLIPSE

rip current Narrow, swift, short-lived surface CURRENT that flows seawards at right-angles to the shoreline. It occurs at the mouth of a bay or along a coast where wind and incoming waves pile up the water until the excess rushes quickly and forcefully back into the sea.

Ritter, Johann Wilhelm (1776–1810) German physicist. In 1801 he discovered ULTRAVIOLET RADIATION beyond the violet end of the SPECTRUM. Ritter made the find while working with SILVER CHLORIDE, which decomposes in the presence of light, especially ultraviolet light. He also worked in ELECTROCHEMISTRY and in 1802 made the first DRY CELL.

river Large, natural channel containing water, which flows downhill under gravity. The water in a river is supplied by tributary streams, by direct runoff over the land, by seepage from valley side slopes, by water emerging from underground sources in springs, by water falling on the river surface, and by the melting of snow and glacier ice. A river system is a network of connecting channels. It can be divided into tributaries which collect water and SEDIMENT, the main trunk river, and the dispersing system at the river's mouth where much of the sediment is deposited. The DISCHARGE of a river is the volume of water flowing past a point in a given time. It is usually expressed as cubic metres per second (cumecs), and is calculated by multiplying the cross-sectional area of the river by the VELOCITY (speed) of the water. The velocity of a river is controlled by the slope of the river, the depth of the river and the roughness of the river bed. Rivers transport sediment as they flow, by the processes of TRACTION (rolling), saltation (jumping), SUSPENSION (carrying) and SOLUTION. A greater discharge increases the amount of sediment that can be transported. Most river sediment is transported during flood conditions, but as a river returns to normal flow it deposits sediment. A river adjusts its channel shape to be able to transport sediment most efficiently. This can result in the EROSION of a river channel or in the building up of FLOODPLAINS, sand and gravel banks. All rivers tend to flow in a twisting pattern, even if the slope is relatively steep, because water flow is naturally turbulent. The force of the water striking the river bank causes erosion and undercutting which starts a small bend in the river channel. Over time the bend grows into a large MEANDER. The current flows faster on the outside of the bend eroding the bank, while sedimentation occurs on the inside of the bend where the current is slowest. Eventually, the river channel cuts across the loop and follows a more direct course downstream and the loop is abandoned as an OXBOW LAKE. Rivers flood when their channels cannot contain the discharge. Flood risk can be reduced by straightening the channel, dredging sediment or making the channel deeper by raising the banks. *See also* DELTA; LEVÉE

RNA (ribonucleic acid) Chemical (NUCLEIC ACID) that controls the synthesis of PROTEIN in a CELL and is the genetic material in some VIRUSES. The molecules of RNA in a cell are copied from DNA and

RIFT VALLEY

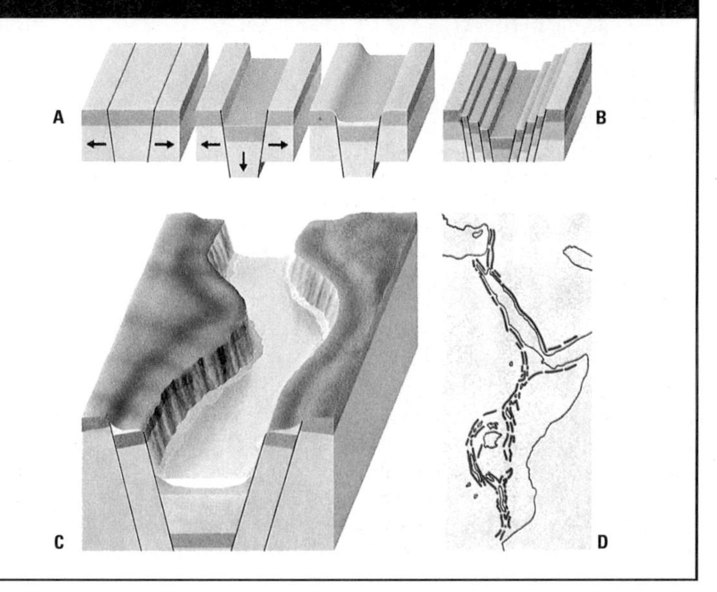

Rift valleys are formed by tension between two roughly parallel faults (A), causing downward earth movement resulting in the formation of a graben (trough of land between two faults). Sometimes a number of parallel faults results in land sinking in steps (B). A typical example of a step-faulted rift valley is shown (C). A series of block faults can occur on either side of a graben, sometimes tilting in the process of creating the block-faulted rift valley. The Great Rift Valley in E Africa (D), incorporating the Red Sea, is perhaps the world's best example of this geological formation.

RIVER

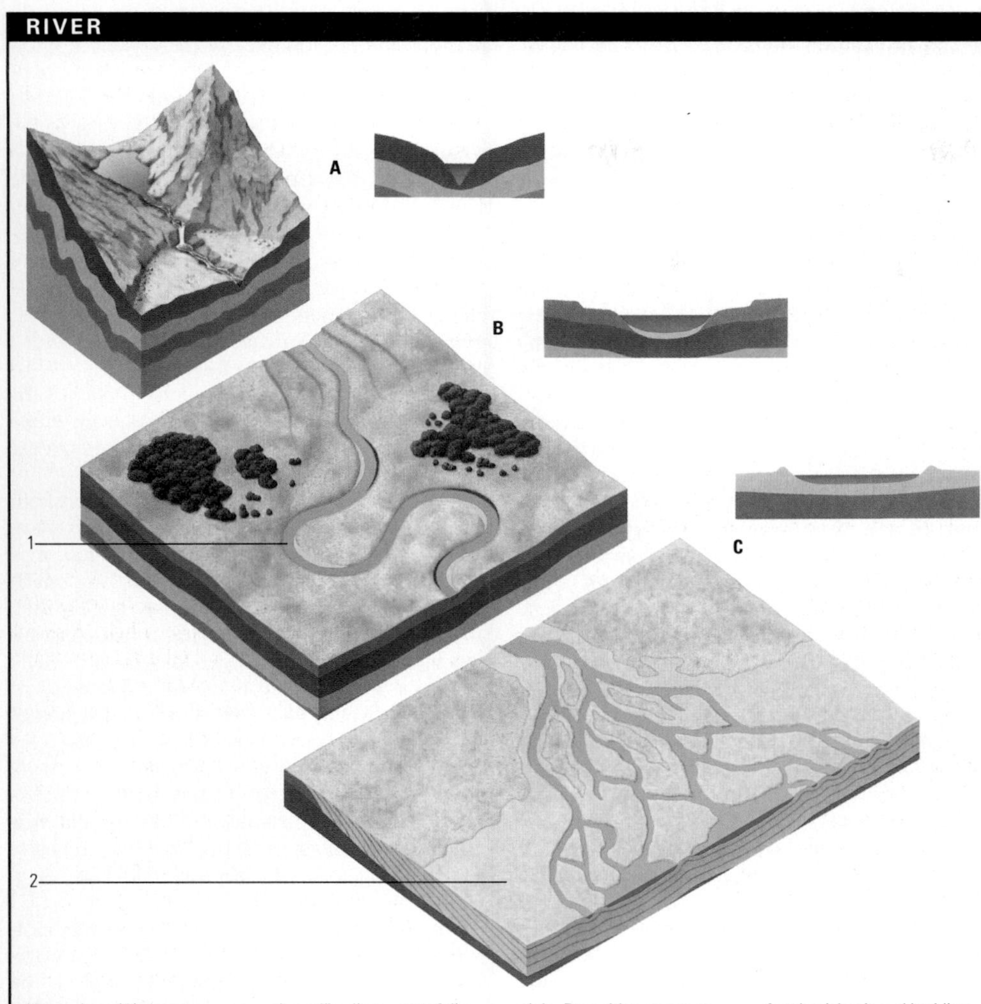

Young rivers (A) close to their source tend to be fast-flowing, high-energy environments with rapid downward and headward erosion, despite the hardness of the rock over which they may flow. Steep-sided, "V-shaped" valleys, waterfalls and rapids are characteristic features. Mature rivers (B) are lower-energy systems. Erosion takes place on the outside of bends, creating looping meanders (1) in the soft alluvium of the river plain. Deposition occurs on the inside of bends and on the river bed. At a river's mouth (C), sediment is deposited as the velocity of the river slows. As the river becomes shallower more deposition occurs, forming islands and braiding the main channel into multiple, narrower channels. As the sediment is laid down (2), the actual mouth of the river moves away from the source into the sea or lake, forming a delta.

consist of a single strand of NUCLEOTIDES, each containing the sugar RIBOSE, PHOSPHORIC ACID, and one of four BASES: ADENINE, GUANINE, CYTOSINE or URACIL. MESSENGER RNA (mRNA) carries the information for protein synthesis from DNA in the cell NUCLEUS to the RIBOSOMES in the CYTOPLASM. During a process known as TRANSLATION, TRANSFER RNA (tRNA) brings AMINO ACIDS to their correct position on the mRNA to form POLYPEPTIDE chains.

Roaring Forties Region between 40° and 50°S in the Southern Hemisphere where strong westerly winds blow. A series of DEPRESSIONS move continually from west to east, bringing mild temperatures but wet and windy conditions. The winds remain persistent and strong because there is no major land mass to slow them.

Robbins, Frederick Chapman (1916–) US physician. He shared the 1954 Nobel Prize for physiology or medicine with John F. ENDERS and Thomas H. WELLER for his part in the discovery of the ability of the poliomyelitis viruses to grow in cultures of various types of tissues. This discovery was vital to the development of a poliomyelitis VACCINE.

Robinson, Sir Robert (1886–1975) British chemist who was awarded the 1947 Nobel Prize for chemistry for research into ALKALOIDS and other plant chemicals. Robinson succeeded in synthesizing the alkaloid tropinone using three simple COMPOUNDS found in plants. This led to his theory of organic molecular structure based on electronic processes occurring during the formation and disruption of CHEMICAL BONDS. The theory proved important for the understanding of all biosynthetic mechanisms. His other research concerned plant PIGMENTS and the GENETICS of variations in flower colours. Robinson also assisted in the synthesis of PENICILLIN and female SEX HORMONES.

robot Automated MACHINE used to carry out various tasks. Robots are often COMPUTER-controlled devices, the most common type having a single arm that can move in any direction. Such robots are used to carry out various tasks in car manufacturing. A range of attachments enables the robots to weld metal parts, pick up and attach components, and carry out paint spraying. Scientists working in the field of NANOTECHNOLOGY hope to produce microscopic robots that will prolong lives by cleaning the arteries in human bodies.

Rochelle salt Colourless, crystalline compound, potassium sodium tartrate tetrahydrate ($KNaC_4H_4O_6.4H_2O$). It can be prepared from BAKING SODA and cream of tartar (crystallized potassium hydrogen bitartrate). Rochelle salt is used in baking powders and as a laxative. It is also used in electronics for the PIEZOELECTRIC EFFECT. It is named after La Rochelle, France, where it was first discovered.

roche moutonnée (Fr. sheep rock) Rock that has been scraped smooth by ice on one side, but has been plucked into a jagged shape on the other, downhill side. They are found in glacial valleys. Roches moutonnées can be only a few metres in height, but some are much bigger, reaching 30–40m (98–130ft). It is said that roches moutonnées were so named because they resemble sheep lying down.

rock Solid material that comprises the Earth's CRUST. Although solid, it is not necessarily hard – CLAY and volcanic ASH are also considered to be rocks. Rocks are classified by origin into three major groups. IGNEOUS ROCK is any rock formed by the cooling and solidification of molten material from the Earth's interior; volcanic LAVA and GRANITE are igneous rocks. Igneous rocks can also be classified as INTRUSIVE and EXTRUSIVE ROCK. SEDIMENTARY ROCK is formed from older rock that has been transported from its original position by water, glaciers or the atmosphere, and consolidated again into rock; LIMESTONE and SANDSTONE are examples. METAMORPHIC ROCK originates from igneous or sedimentary rocks, but has been changed in texture or mineral content or both by extreme pressure and heat deep within the Earth; MARBLE, derived from limestone, is a metamorphic rock.

rocket Slender, tapering MISSILE or craft powered by a rocket ENGINE. Most of its volume contains fuel; the remainder is the payload (such as an explosive, scientific instruments or a spacecraft). **Liquid-fuelled** rockets use a fuel (such as liquid hydrogen) and an oxidizer (usually liquid oxygen), which are burned together in the engine. **Solid-fuelled** rockets have both fuel and oxidizer in a solid mixture. A single-stage rocket has one fuel load, or several that are used simultaneously. A multi-stage rocket has more than one fuel load, which are ignited singly in succession as the preceding one burns out. Because rocket engines carry their own supply of both fuel and oxidizer, they can operate in outer SPACE where there is no atmosphere. They gain THRUST from the

ROBOT

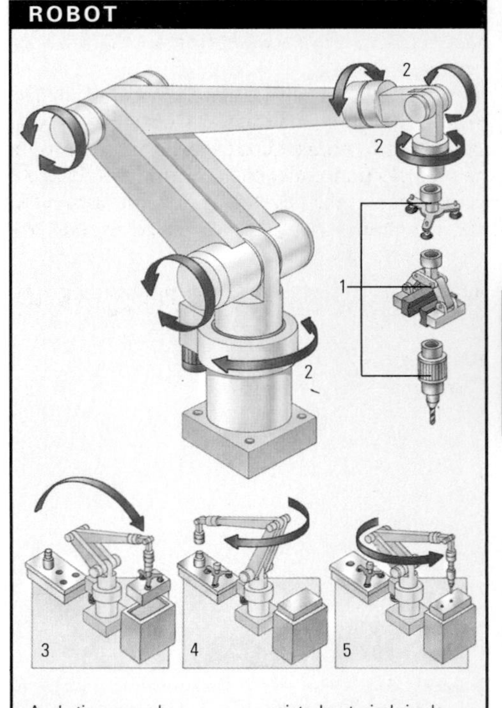

A robotic arm can be programmed to use different tools (1) to carry out a variety of tasks. The arm, head and base can all rotate (2) and can extend to reach different points. In a typical simple engineering setting, the robot first places the raw material into position (3) and then changes tools (4) before drilling the required holes (5).

reaction (referred to in Newton's third law of MOTION) produced by rapid, continuous output of exhaust gases. **Chemical** rocket engines are powered by solid or liquid PROPELLANTS that are burned in a COMBUSTION chamber and expelled at SUPERSONIC VELOCITIES from the exhaust nozzle. Nuclear engines heat fuel by RADIATION from NUCLEAR REACTOR cores. ION engines use thermoelectric power to expel IONS rather than gases. The walls of the combustion chamber and the exit nozzle must be cooled to protect them against the heat of the escaping gases, whose temperature may be as high as 3,000°C. Today, payloads are put into orbit by the SPACE SHUTTLE. In the late 19th century, Ernst MACH investigated the possibility of supersonic speed. Konstantin TSIOLKOVSKY is regarded as the "father of astronautics". He suggested that multistage rockets could escape Earth's GRAVITATIONAL FIELD. In 1926 Robert H. GODDARD launched the first liquid-fuel rocket. During World War 2 Wernher VON BRAUN and Hermann OBERTH developed the first long-range GUIDED MISSILES. After the war, von Braun helped develop the US space programme. Saturn V, the largest rocket ever built, delivered 3.4 million kg (7.5 million lb) of thrust. It was used for the US Apollo missions to the Moon. *See also* SPACE EXPLORATION

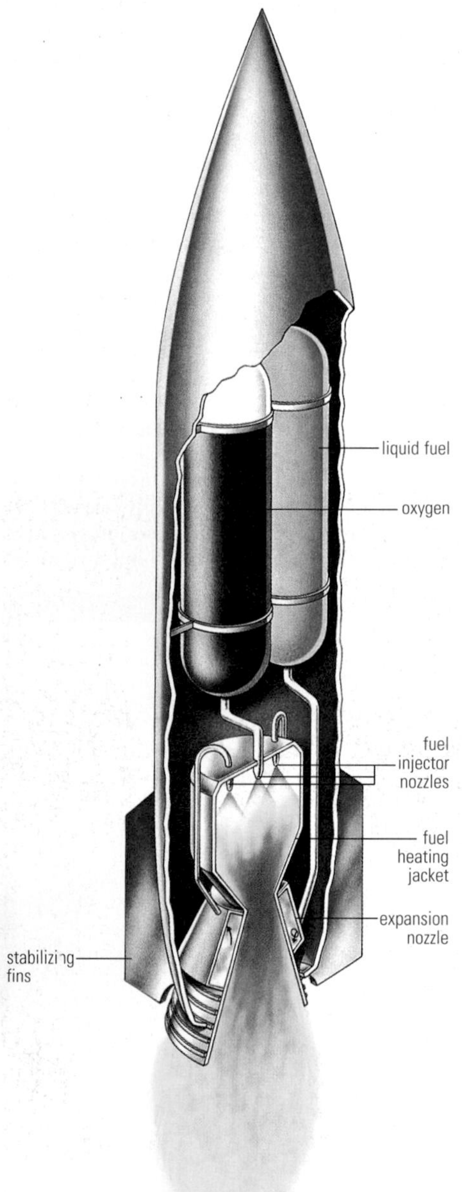

▲ **rocket** The illustration shows a very simple type of rocket engine. Liquid fuel is mixed with oxygen and burned in a combustion chamber and the exhaust gases expand through a nozzle, giving a propelling thrust in the opposite direction. The liquid fuel passes round the combustion chamber so that after starting (using an auxiliary pump) the flow is maintained by thermal expansion. The oxygen is in liquid form and boils off into the combustion chamber.

rods and cones Cells in the RETINA of the EYE that are sensitive to light. Located in the pigmented layer, the rod-shaped RHODOPSIN-secreting cells are the RECEPTORS for low-intensity light, and the cone-shaped iodopsin-secreting cells are adapted to distinguish colour. Rods detect only shades of black and white, but are particularly sensitive to movement.

Roebling, John Augustus (1806–69) US engineer and industrialist, b. Germany. He devised a method for the manufacture of wire CABLES and directed the construction of SUSPENSION BRIDGES across the Allegheny and Niagara rivers. As one of the earliest proponents of the construction of the Brooklyn Bridge, Roebling was appointed its chief engineer in 1869. Soon after, he died in an accident during a survey of the East River. His son, **Washington Augustus** Roebling (1837–1926), was also an engineer and industrialist. Washington Roebling assisted his father in the construction of the Allegheny Suspension Bridge. During the American Civil War, he served as a military engineer and balloonist. In 1869 he succeeded his father as chief engineer on the Brooklyn Bridge and supervised the work until its successful completion in 1883.

Roentgen, Wilhelm Konrad *See* RÖNTGEN, WILHELM KONRAD

roller bearing Component used in engineering to minimize FRICTION by limiting contact between rotating parts of a machine. It consists of two concentric metal rings separated by, and in contact with, cylindrical, barrel-shaped or tapered rollers made of polished hardened steel. *See also* BEARING

ROM (acronym for **r**ead-**o**nly **m**emory) INTEGRATED CIRCUITS (chips) that act as a permanent store for DATA (information) required by a COMPUTER. The contents of ordinary ROM chips are set by the manufacturer and cannot be altered by the user. The stored data is available to the computer's MICROPROCESSOR whenever the computer is switched on.

Röntgen, Wilhelm Konrad (1845–1923) German physicist. In 1895, while teaching at Würzburg University, he discovered X-RAYS, for which he was awarded the first Nobel Prize for physics in 1901. Röntgen also conducted important research on ELECTRICITY, the specific heats of gases, and the heat CONDUCTIVITY of crystals. The RÖNTGEN, the former unit to measure X-rays, is named after him.

röntgen (symbol R) Former unit used to measure the X-RAY or GAMMA RADIATION to which a body is exposed. One röntgen causes sufficient IONIZATION to produce a total electric charge of 2.58×10^{-4} COULOMBS on all the positive (or negative) IONS in one kilogram of air. The unit, named after the Wilhelm RÖNTGEN, has been replaced by the SI UNIT, the GRAY (symbol Gy).

root Underground portion of a VASCULAR PLANT that serves as an anchor and absorbs water and minerals from the soil. Some plants, such as the dandelion, have TAPROOTS with smaller lateral branches. Some of these plants, such as carrots and parsnips, have large taproots that store nutrients and are grown as food for humans and farm animals. Other plants, such as the GRASSES (including cereal crops), develop fibrous roots with lateral branches.

root In mathematics, fractional POWER of a number. The SQUARE ROOT of a number, x, is written as either $\sqrt{x}$ or $x^{\frac{1}{2}}$, and represents a number that, when multiplied by itself, gives the original number. The fourth root of x may be written in RADICAL form as $\sqrt[4]{x}$ or in power form as $x^{\frac{1}{4}}$. For example, the fourth root of 16 is 2, since $2 \times 2 \times 2 \times 2 = 16$. The term root is also used to refer to the solution of a POLYNOMIAL equation.

root-mean-square (RMS) In mathematics, the value obtained by taking the MEAN average of a set of squared values. It provides the value of a periodic quantity and was developed at the beginning of the 20th century as an accurate measure of VOLTAGE and ELECTRIC CURRENT. An RMS error of approximity is also used for equations devised to predict outcomes, such as weather forecasts.

root nodule Small swelling in the roots of various plants, such as legumes, that contain nitrogen-fixing bacteria. *See also* NITROGEN CYCLE; NITROGEN FIXATION

root pressure Means by which water travels from the soil through the roots of a plant and up its STEM. The XYLEM conducts the water, while the pressure is produced by a combination of OSMOSIS, which forces water from the soil into the CELLS of the root, and an active pumping mechanism that creates a concentration gradient of salt IONS in the xylem. Excessive root pressure may cause "bleeding" of SAP from a wound in the plant or the collection of drops of water from undamaged leaf margins.

rosin (colophony) Yellowish, amorphous RESIN obtained as a residue from the distillation of TURPENTINE; the chief constituent is abietic acid. It is used in VARNISHES, soldering FLUXES, linoleum and printing INKS. Properties: r.d. 1.08; m.p. 100–150°C (212–308°F).

Ross, Sir Ronald (1857–1932) British bacteriologist, b. India. In 1898 he demonstrated the PARASITE (*Plasmodium*) that carries MALARIA in the stomach of the *Anopheles* mosquito. Ross was awarded the 1902 Nobel Prize for physiology or medicine for his work.

rotation Turning of a celestial body about its AXIS, as distinct from its orbital REVOLUTION. In the SOLAR SYSTEM, the Sun and all the planets, with the exception of Uranus and Venus, rotate from W to E.

rotor, helicopter Engine-driven rotary wing. Each of the blades is an AEROFOIL and produces LIFT. The hub moves to permit altering of the rotors' plane of rotation in order to give sideways or forward and backward motion.

Rous, Francis Peyton (1879–1970) US pathologist who shared the 1966 Nobel Prize for physiology or medicine with Charles B. HUGGINS for the discovery of CANCER-inducing VIRUSES. In 1910 Rous began a series of experiments on SARCOMAS in hens, injecting healthy birds with a solution made from sarcomas from cancerous ones. He filtered the solution to remove all cells and bacteria, but found that the healthy hens developed cancer. This led him to postulate that cancer could be caused by viruses. The theory was rejected at the time, but later experiments proved him correct.

Royal Greenwich Observatory UK national astronomical observatory, founded at Greenwich, London, in 1675. After World War 2 the observatory moved to Herstmonceux in Sussex and, in 1990, it moved again to Cambridge. The RGO operates the UK TELESCOPES at the Roque de los Muchachos Observatory on La Palma in the Canary Islands.

royal jelly In honeybee colonies, the food given to female bee LARVAE destined to become queens. It is a glandular secretion of young worker bees who themselves are females, but have been fed on honey and POLLEN and so have not developed into queens.

Royal Society British society founded in 1660 and incorporated two years later. Its aim was to accumu-

▲ **ruminant** Impalas are found in the grassland regions of central and E Africa. Like other ruminants, they can regurgitate food in small amounts once it has been partly digested, for chewing again, reswallowing, and further digestion. This enables them to obtain a lot of food in a short time, then retreat to a safe, sheltered place to digest it. When grazing, an impala grasps vegetation between its spade-like incisors (1) and a hard upper pad (2) and pulls it up rather than biting it off. The molars (3) are ideal for chewing. The gap between incisors and molars (4) allows the tongue to mix food with saliva. The powerful masseter muscle (5) moves the jaw up and down, while other facial muscles move it side to side for grinding.

late experimental evidence on a wide range of scientific subjects, including medicine and botany as well as the physical sciences. In the 20th century, the Royal Society became an independent organization concerned with the promotion of scientific research.

rubber Elastic solid obtained from the LATEX of the rubber tree (*Hevea brasiliensis*, Family Euphorbiaceae). **Natural** rubber consists of a POLYMER of cis-ISOPRENE and is widely used for tyres and other applications, especially after VULCANIZATION. **Synthetic** rubbers are polymers tailored to excel in certain properties for specific purposes; none has the overall advantages of natural rubber. They include STYRENE-BUTADIENE rubber (SBR), NEOPRENE, NITRILE rubber and the newer stereo-rubbers based on synthetic cis-polyisoprene.

rubidium (symbol Rb) Silver-white, metallic element of the ALKALI METALS. It was discovered (1861) by Robert BUNSEN and Gustav KIRCHHOFF. The element has few commercial uses; small amounts are used in PHOTOELECTRIC CELLS. Chemically, it resembles SODIUM but is more reactive. Properties: at.no. 37; r.a.m. 85.4678; r.d. 1.53; m.p. 38.89°C (102°F); b.p. 688°C (1,270°F); most common isotope ^{85}Rb (72.15%).

ruby Gem variety of the mineral CORUNDUM (aluminium oxide), whose characteristic red colour is due to impurities of chromium and iron OXIDES. The traditional source of rubies is Burma. Today, synthetic rubies are widely used in industry, especially in LASER technology.

rudaceous Describing a coarse-grained SEDIMENTARY ROCK. Examples of rudaceous rocks include breccia, conglomerate and gravel, as well as types of stony boulder clay. The particles in the rock range in size from large grains of sand to fragments of gravel.

Rumford, Benjamin Thompson, Count (1753–1814) British scientist and administrator, b. USA. He moved to Britain during the American Revolution and was made a count of the Holy Roman Empire in 1791. In science, Rumford discovered that HEAT is a form of MOTION. He also introduced the standard candle, which became the unit of LUMINOSITY until 1940 (when it was replaced by the CANDELA). In 1799 Rumford founded the Royal Institution of Great Britain.

ruminant Cud-chewing, even-toed, hoofed mammals. They include the okapi, chevrotain, deer, giraffe, antelope, cattle, sheep and goat. All except the chevrotain have four-chambered STOMACHS. They are named for the habit of re-chewing food previously swallowed (cud) and stored in one of the chambers.

runner In botany, a long, thin STEM (STOLON) that extends along the surface of the soil from the axil of a plant's leaf. It serves to propagate the plant. At points (NODES) along its length, a runner has small leaves with buds that develop shoots and roots and turn into small independent plants as the runner dies. Runners are produced by such plants as strawberries and creeping buttercups. *See also* ASEXUAL REPRODUCTION

Russell, Henry Norris (1877–1957) US astronomer. He was director (1912–47) of the observatory at Harvard University. Russell devised a method for calculating the dimensions of ECLIPSING BINARY stars. He and Ejnar HERTZSPRUNG developed the HERTZSPRUNG-RUSSELL DIAGRAM.

rust In botany, group of FUNGI that live as PARASITES on many kinds of higher plants. The rust-red colour of their SPORES gives them their name. Rusts are of particular concern for the damage they do to cereal crops, such as wheat and barley, and to several fruits and vegetables. They have complex life cycles that involve growth on more than one host plant.

rust Product of CORROSION of iron caused by a reaction with air and water. CARBON DIOXIDE from the air dissolves in water to form an acid solution which attacks the iron to form iron(II) (ferrous) oxide. This is then oxidized by oxygen in the air to reddish-brown iron(III) (ferric) OXIDE. Rusting may be prevented by coating the iron with ZINC, a process called GALVANIZING, when the zinc is preferentially attacked by the acid solution.

ruthenium (symbol Ru) Silver-white TRANSITION ELEMENT, first discovered in 1827. It is found in PLATINUM ores. Ruthenium is used as a CATALYST and is alloyed with other platinum metals in electrical contacts and used to colour glass and ceramics. Properties: at.no. 44; r.a.m. 101.07; r.d. 12.41; m.p. 2,310°C (4,190°F); b.p. 3,900°C (7,052°F); most common isotope ^{102}Ru (31.61%).

Rutherford, Ernest, Lord (1871–1937) British physicist, b. New Zealand, who pioneered modern NUCLEAR PHYSICS. Rutherford discovered and named alpha and beta radiation, named the NUCLEUS, and proposed a theory of the radioactive transformation of ATOMS for which he received the 1908 Nobel Prize for Chemistry. He studied in Christchurch, New Zealand and in Cambridge, England. In Cambridge, under J.J. THOMSON, Rutherford discovered the URANIUM radiations. At McGill University, Canada, he formed the theory of atomic disintegration with Frederick SODDY. In 1907 Rutherford became professor at the University of Manchester. Here, he devised the nuclear theory of the atom, which was combined with QUANTUM THEORY by Niels BOHR to produce the description of the atom that is still used today. In 1919 he became director of the Cavendish Laboratory, where his research team became the first to induce a change in an atom's nucleus by bombarding it with ELEMENTARY PARTICLES (HELIUM nuclei). He predicted the existence of the NEUTRON, later discovered by his Cambridge colleague, James CHADWICK. Rutherford became a peer in 1931.

rutherfordium (symbol Rf) Synthetic, radioactive, metallic element of the TRANSACTINIDE SERIES. It has a HALF-LIFE of less than a second. It is thought to have been discovered first in 1974, but its existence was only proved in 1994, the year it was officially given the name rutherfordium. The name rutherfordium had previously been used for element 104 (now known as DUBNIUM). Properties: at.no. 106

rutile (titanium dioxide, TiO_2) Black to red-brown oxide mineral, found in IGNEOUS and METAMORPHIC rocks and quartz veins. Rutile occurs as long prismatic and needle-like CRYSTALS in the tetragonal system and as granular masses. It has a metallic lustre, is brittle and is used as a gemstone. Hardness 6–6.5; r.d. 4.2.

Ružička, Leopold Stephen (1887–1976) Swiss chemist, b. Yugoslavia, who shared the 1939 Nobel Prize for chemistry with Adolf BUTENANDT for his research into ring MOLECULES and TERPENES (hydrocarbons found in the ESSENTIAL OILS of many plants). In 1916 Ružička began investigations into the active compounds in musk and civet, finding that they were ring structures consisting of 15 and 17 CARBON atoms respectively. This discovery had a profound effect on organic CHEMISTRY, as until that time rings with more than six atoms were thought to be too unstable to exist. In the 1930s he synthesized several male SEX HORMONES, including TESTOSTERONE, having previously discovered their molecular structure.

Rydberg, Johannes Robert (1854–1919) Swedish chemist who classified optical spectra and developed (1890) empirical formulae for spectral lines, using a constant ($1.097 \times 10^7 m^{-1}$) now known as the **Rydberg constant** (symbol *R*). He also studied hydrogen line wavelengths. The **Rydberg atom** is a highly excited hydrogen atom that exhibits chaotic behaviour in a strong MAGNETIC FIELD.

Ryle, Sir Martin (1918–84) British physicist and radio astronomer. After studying RADAR during World War 2, he went to Cambridge University where he pioneered RADIO ASTRONOMY. Ryle also catalogued RADIO SOURCES, which led to his discovery of QUASARS.

▲ **Rutherford** Considered by many to be the father of nuclear physics, Ernest Rutherford he was awarded the Nobel Prize for chemistry in 1908. Earlier, in 1903, together with the British chemist Frederick Soddy, Ernest Rutherford proposed that radioactive decay occurs by successive and spontaneous disintegrations of atoms. In 1911, he elaborated an atomic model in which the positive charge of the atom (protons) was concentrated in a very small region, the nucleus. He also showed that the structure of an atom could be changed (nuclear transmutation) by bombarding it with alpha particles.

R

Sabin, Albert Bruce (1906–93) US virologist, b. Russia. He developed a **live** VIRUS oral VACCINE against **poliomyelitis**, in contrast to Jonas SALK's inactive virus vaccine, and began testing the Sabin vaccine in 1957. In 1980 Sabin worked to make the polio vaccine available in Brazil.

saccharide Organic compound based on SUGAR molecules. Monosaccharides include GLUCOSE and FRUCTOSE, found in fruit and honey. Two sugar molecules join together to make a DISACCHARIDE, such as LACTOSE (milk sugar) or SUCROSE (cane sugar). POLYSACCHARIDES have more than two sugar molecules. *See also* CARBOHYDRATE

Sachs, Julius von (1832–97) German botanist, known for his studies of the metabolism of plants, notably the role of CHLOROPHYLL. Sachs wrote *Textbook of Botany* (1868).

sacrum (sacral bone) Triangular segment of the lower backbone formed by the fusion of five vertebrae. It articulates above with the last lumbar vertebra, below with the coccyx and, to either side, with the hip bones.

safety glass Form of GLASS that is less hazardous than ordinary glass when broken. One form of safety glass consists of two sheets of 3mm (⅛in) plate glass bonded to a thinner central sheet of transparent plastic. If an impact breaks the glass, the plastic holds the fragments in place. **Bullet-proof** glass consists of several layers of glass and plastic. **Wired** glass has an embedded wire mesh to hold the fragments in place. **Toughened** glass is glass that has been treated to make it stronger than ordinary glass. When shattered, the glass forms small blunt fragments, which are much less dangerous than the sharp splinters from ordinary glass.

SALT DOME

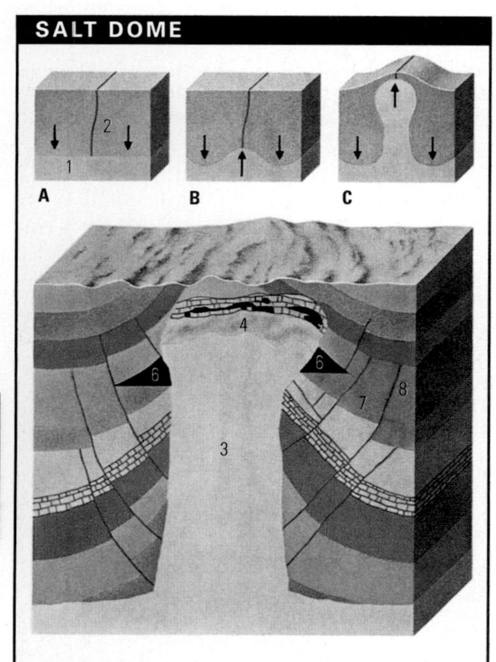

A salt dome, or diapis, is caused by a massive plug of salt that has been forced upwards through a fault in overlying sedimentary strata by subterranean pressure. Oil and gas are often trapped in rock strata associated with salt domes.

Key
1) halite beds (salt)
2) sedimentary overburden
3) salt plug
4) gypsum and anhydrite
5) porous limestone caprock
6) oil trap
7) sand
8) shale
A) sedimentary beds overlying halite beds
B) pressure of denser overlying beds causing peripheral sinking and salt to rise up fissure
C) formation of salt plug and dome

safety lamp Lamp that can be used in mines. Sir Humphry DAVY invented a miner's safety lamp in the early 1800s. Its oil flame burned inside a wire mesh cylinder, which conducted away much of the heat, so that any explosive FIREDAMP (methane) in the air would not be ignited. Others improved Davy's lamp, and its use became widespread. Today, most miners use electric lamps, but a modern version of the Davy lamp is still used to check for the presence of METHANE, which turns a low flame blue.

Sagan, Carl Edward (1934–96) US astronomer and writer. He was professor of astronomy and space science (1968–96) at Cornell University. Sagan was the greatest popularizer of science of his era. He hosted the television series *Cosmos* (1980). Sagan studied many aspects of the SOLAR SYSTEM, including the physics and chemistry of planetary atmospheres and surfaces. He also investigated the origins of extraterrestrial life and maintained the possibility of extraterrestrial life-forms. Sagan's work includes *The Dragons of Eden* (1977), *Broca's Brain* (1979) and *Contact* (1985).

Sagittarius (Archer) Southern constellation on the ecliptic between Scorpio and Capricorn. Rich in STELLAR CLUSTERS, this region of the sky also contains much interstellar matter, which obscures the central region of the MILKY WAY galaxy and is only penetrable with radio telescopes. The brightest star is Epsilon Sagittarii (*Kaus Australis*), magnitude 1.8.

St Elmo's fire (corposant) Electrical discharge illuminating the tops of tall objects. It usually occurs during a storm when the atmosphere becomes charged strongly enough to create a discharge between the air and an object. Early sailors named this phenomenon after their patron saint.

Sakharov, Andrei Dimitrievich (1921–89) Soviet physicist and social critic. His work in nuclear FUSION was instrumental in the development of the Soviet HYDROGEN BOMB. An outspoken defender of civil liberties, he created the Human Rights Committee in 1970. Sakharov received the 1975 Nobel Peace Prize. In 1989 he was elected to the Russian Congress. Sakharov's books include *Sakharov Speaks* (1974) and *My Country and My World* (1975).

Salam, Abdus (1926–96) Pakistani physicist who proposed (1967) a theory that unifies the electromagnetic and WEAK NUCLEAR FORCES within the NUCLEUS of an ATOM (*see* FUNDAMENTAL FORCES). Salam and the US physicist Steven WEINBERG worked independently on the theory (now known as the **Weinberg-Salam theory**). After the theory was proved experimentally, the pair shared the 1979 Nobel Prize for physics with US physicist Sheldon GLASHOW, who also independently came to similar conclusions. *See also* GRAND UNIFIED THEORY (GUT)

salicylic acid (1-hydroxybenzoic acid, $C_7H_6O_3$) Colourless, crystalline solid, derivatives of which are used as analgesics (including ASPIRIN, acetylsalicylic acid), antiseptics, dyes and liniments. It occurs naturally in plants, including willow bark and oil of wintergreen.

salina (playa lake) Dried lake bed found in desert or semi-arid areas. Salinas form where a stream has flowed into an area of inland drainage and then evaporated, leaving behind tiny deposits of salt. The deposits can accumulate for hundreds of years to form large salty areas.

salinity Degree of saltiness, as of the OCEANS. The average proportion of salt in the sea is about 35 parts per 1,000 (3.5%). In the Red Sea, because of high evaporation rates, the figure is over 40, but in the N parts of the Baltic Sea it is less than 10, and in the Dead Sea the figure is about 250. The salts found in the oceans include sodium chloride (common salt), which accounts for about 75%, magnesium chloride, magnesium sulphate and calcium sulphate.

saliva Fluid released into the mouth by the SALIVARY GLANDS. In vertebrates, saliva is composed of about 99% water with dissolved traces of sodium, potassium, calcium and the ENZYME amylase. Saliva softens and lubricates food to aid swallowing, and AMYLASE starts the digestion of starches.

salivary glands One of three pairs of GLANDS located on each side of the mouth that form and secrete SALIVA. The **parotid** gland, just below and in front of each ear, is the largest of the salivary glands and the one that becomes enlarged in mumps; the **submaxillary** gland is near the angle of the lower jaw; and the **sublingual** gland is under the side of the tongue.

Salk, Jonas Edward (1914–95) US medical researcher. In 1952 he developed the first vaccine against poliomyelitis, using an **inactive** polio VIRUS as an immunizing agent. Extensive tests of the vaccine began in 1954 and mass immunization programmes followed. *See also* SABIN, ALBERT

Salmonella Genus of rod-shaped BACTERIA that cause intestinal infections in human beings and animals. *Salmonella typhi* causes typhoid, but the effects of other species may result in only mild gastroenteritis. The bacteria are transmitted by carriers, particularly flies, and in food and water.

salt Ionic COMPOUND formed, along with water, when an ACID is neutralized by a BASE. The HYDROGEN of the acid is replaced by a metal or ammonium ION (NH_4^+). The most familiar salt is SODIUM CHLORIDE (NaCl). Salts are typically high-melting crystalline compounds that tend to be soluble in water. They are formed of ions held together by electrostatic forces, and in SOLUTION they conduct electricity. *See also* ELECTROLYSIS

salt dome Body of salt that has intruded into a SEDIMENTARY ROCK overlay. The flow of the relatively plastic salt into a dome may be the result of a difference in density between the salt and the overlying rock, but the process is not completely understood.

salt lake Saline lake in a desert region. The lake originates in an area of inland drainage. The salinity increases as the water evaporates, and layers of salt may crystallize out along the lake's shores. Prolonged rain may redissolve the salt and wash most of it away. The largest salt lakes are Lake Eyre in Australia and the Great Salt Lake in Utah, USA.

saltpetre *See* NITRE; POTASSIUM

samara Form of ACHENE in which the FRUIT wall dries out to form a wing-like structure. Common examples include ash and elm; the sycamore samara has two wings, and is called a schizocarp.

samarium (symbol Sm) Grey-white, metallic element of the LANTHANIDE SERIES, first identified spectroscopically in 1879. Its chief ores are MONAZITE and bastnaesite. Samarium is used in carbon-arc lamps, as a moderator in NUCLEAR REACTORS and as a catalyst. Some samarium alloys are used for making powerful permanent magnets. Properties: at.no. 62; r.a.m. 150.35; r.d. 7.52; m.p. 1,072°C (1,962°F); b.p. 1,791°C (3,256°F); most common isotope ^{152}Sm (26.72%).

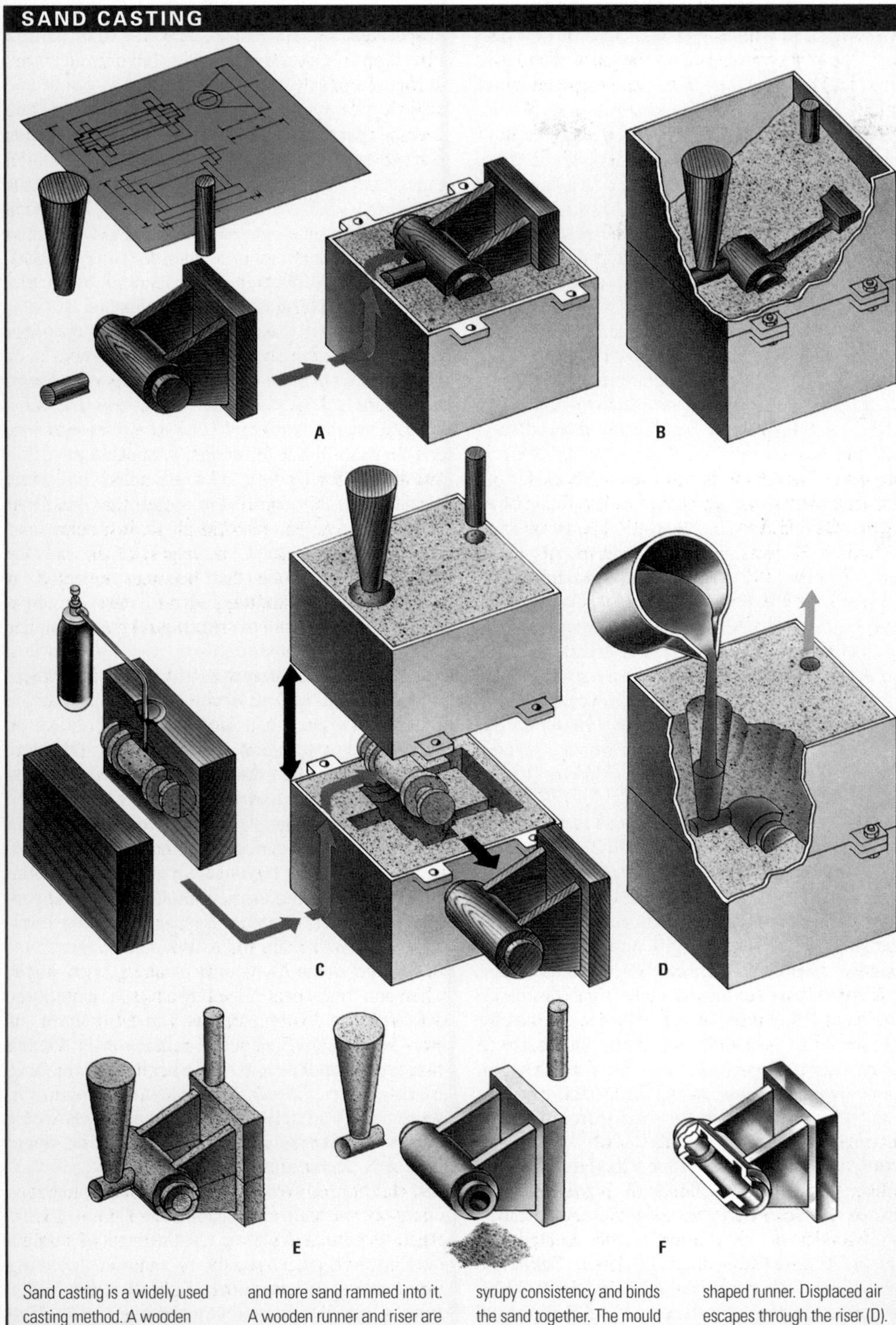

Sand casting is a widely used casting method. A wooden pattern is made up from the original design, then packed with sand up to its largest cross-sectional area in a steel box (A). The top part of the mould is assembled. It is clamped onto the bottom box and more sand rammed into it. A wooden runner and riser are fixed in position (B). The sand core for a hollow in the casting is made. The sand for this is mixed with sodium silicate that forms "silica gel" when carbon dioxide is pumped through it. This "gel" has a syrupy consistency and binds the sand together. The mould is split and the pattern is removed. The core is placed in position and the mould assembled (C). Runner and riser are removed. Molten metal is poured into the dried mould through the conical-shaped runner. Displaced air escapes through the riser (D). After cooling, the mould is split open and the casting removed (E). The runner and riser are cut off and the sand core is knocked out. The finished casting (F) shows the hollow produced by the core.

sand Mineral particles worn away from rocks by EROSION, individually large enough to be distinguished with the naked eye. Sand is composed mostly of QUARTZ, but black sand (containing volcanic rock) and coral sand also occur. It is usually classified according to grain size. *See also* DUNE; SANDSTONE

sand blasting Industrial method for cleaning the surfaces of materials such as metals and concrete. It is an efficient means of cleaning the outer walls of buildings, but its most frequent use is to remove the oxide scale from metals prior to painting, spraying, electroplating or dip-coating. Articles to be cleaned are subjected to a high pressure blast of SAND.

sand casting Method of moulding metals and alloys in which the molten metal is poured into a MOULD made of firmly packed sand. To help the sand grains cohere, they are mixed with clay, water and other binders. The method is used in the casting of steel, brass, bronze and aluminium.

sandstone SEDIMENTARY ROCK composed of SAND grains cemented in such materials as SILICA or iron oxide. Sandstones are ARENACEOUS rocks, and are very variable in character as a result of differences in grain size, chemical content and cementing materials. Most sand grains contain QUARTZ, which is hard and resistant. Other minerals in sandstones include FELDSPARS and MICAS, and iron also occurs, which tends to give sandstones a reddish or brownish colour. Most sandstones are formed by the accumulation of river sediments on the seabed. They are then compressed and uplifted to form new land. There are also a few sandstones which have been formed of wind-blown materials, especially in desert regions.

Sanger, Frederick (1918–) English biochemist who became the first person to win two Nobel Prizes for chemistry. Sanger was awarded his first prize in 1958 for finding the structure of INSULIN. His second came in 1980 (shared with the US molecular biologists Walter GILBERT and Paul Berg) after work on the chemical structure of NUCLEIC ACID. Sanger and his colleagues found the entire sequence of the more than 5,400 bases in the DNA of a VIRUS and some 17,000 bases of another DNA.

sap In plants, fluid that circulates water and nutrients through them. Water is absorbed by the roots and carried, along with minerals, through the XYLEM to the leaves. Sap from the leaves is distributed throughout the plant; it travels upwards by OSMOSIS, ROOT PRESSURE and pressure differences created by TRANSPIRATION.

saponin GLUCOSIDES present in some plants that produce a lather when agitated with water. Used as a SOAP substitute, it is found in horse chestnut, soapberry, soapnut and soapwort.

sapphire Transparent to transluscent gemstone variety of CORUNDUM. It has various colours produced by impurities of iron and titanium, the most valuable being deep blue. They are brilliant when cut and polished. *See also* RUBY

saprophyte Plant or FUNGI that obtains its food from dead or decaying plant or animal tissue. Generally it has no CHLOROPHYLL. Saprophytes include fungi, such as mushrooms and moulds, and some flowering plants.

sarcoma Cancerous growth or TUMOUR arising from muscle, fat, bone, blood or lymph vessels or connective tissue. *See also* CANCER

satellite (moon) Celestial body orbiting a PLANET

▲ **sapphire** Cut sapphires and polished star sapphire from Sri Lanka. Valued as gemstones, sapphires are gems of the mineral corundum, and can be a variety of colours such as blue and yellow. The red gem is ruby.

SATELLITE

Four artificial satellite orbits around the globe are illustrated. Two are equatorial: (1) a geostationary satellite orbits at 35,900km (22,300mi)), another (2) has a lower orbit. A polar orbit, as used by remote sensing satellites, is shown running vertically around the Earth (3) and the last is an angled elliptical orbit used by communications satellites for the high latitudes of the Earth (4). The diagrams above the orbital map show the launching of an Intelsat communication satellite by a NASA shuttle (5) and the two main types of satellites. To prevent them being knocked off course by fluctuations in the Earth's magnetic field, satellites are given centrifugal stability in one of two ways. The first is to rotate the whole satellite as is the case in "spinners" such as the Intelsat satellite (6). The other method is to have gyroscopes placed within the satellite (7).

or STAR. In the SOLAR SYSTEM, planets with natural satellites include the EARTH (1), MARS (2), JUPITER (16), SATURN (18), URANUS (17), NEPTUNE (8) and PLUTO (1). There are probably more small satellites of the giant planets awaiting discovery. They vary enormously in their size, orbit, surface features and supposed origin. One, Jupiter's GANYMEDE (the largest natural satellite in the Solar System), is bigger than Mercury, and seven are bigger than Pluto, while the smallest are tiny bodies a few kilometres across. The satellites' orbits vary just as much, from Phobos, which circles Mars' equator more than three times a day, to the outer satellites of Jupiter, which take more than two Earth years to complete their inclined, eccentric orbits. In their surface features satellites are remarkably diverse, ranging from dead, densely cratered worlds to ones which show a great variety of geological processes. Jupiter's Io is the most geologically active body known. As to origin, some, such as Jupiter's GALILEAN SATELLITES, may have condensed out of the same part of the solar NEBULA as their parent planet, while several small ones are believed to be "captured" asteroids. The MOON is currently thought to have formed out of debris flung out when a Mars-sized body collided with the Earth.

satellite, artificial Manmade object placed in orbit around the Earth or other celestial body. Satellites can perform many tasks and can send back data or pictures to the Earth. Hundreds of satellites of various types orbit the Earth. They may study the atmosphere, or photograph the surface for scientific or military purposes. **Communications** satellites relay radio, television, telephone, telegraph and data signals from one part of the Earth to another. **Navigation** satellites transmit radio signals that enable navigators to determine their positions. The GLOBAL POSITIONING SYSTEM (GPS) uses satellites in this way. **Geodetic** satellites are used to make accurate measurements of the Earth's size and shape. SPUTNIK 1, launched by the Soviet Union on 4 October 1957, was the first artificial satellite. TELSTAR, launched by the USA on 10 July 1962, was the first active communications satellite. The first satellite in geosynchronous ORBIT, Earlybird, was launched in 1965. Geosynchronous satellites circle the Earth at a height of 35,900km (22,300mi), enabling it to complete one revolution every 24 hours. Satellites that orbit the Earth at heights less than about 160km (100mi) are slowed by friction with the atmosphere and eventually spiral in and burn up. In SPACE EXPLORATION, a spacecraft may be put into orbit around a planet or a moon and thus become an artificial satellite of that body. *See also* SATELLITE, WEATHER; SYNCHRONOUS ORBIT

satellite, weather Artifical SATELLITE studying weather patterns on Earth by photographic, infrared or other means. Some satellites are in GEOSTATIONARY ORBIT, which means that they remain over the same point of the Earth, travelling at the same speed as the rotation of the Earth. Other satellites circle round the Earth in a polar direction, and with each ORBIT pass over the Equator about 30° W of the previous orbit. Photographic histories of hurricane development and discovery of large-scale coherence in cloud and other weather patterns have greatly helped WEATHER FORECASTING.

satellite television TELEVISION services transmitted to viewers via communications SATELLITES in orbit around the Earth. These satellites are in synchronous orbit above Earth's Equator. As a result, each one remains above a fixed point on the Equator (*see* GEOSTATIONARY ORBIT). Television companies beam their signals to the satellites from TRANSMITTERS on the ground. The satellites retransmit the signals back to viewers' dish-shaped receiving aerials. A fixed ANTENNA can receive many services from a single satellite. Some people use a movable antenna to receive services from more than one satellite.

saturated compound In organic chemistry, compounds in which the CARBON atoms are bonded to one another by single COVALENT BONDS only, never by the more reactive double or triple CHEMICAL BONDS. For this reason, they tend to be unreactive. A simple example is ETHANE (C_2H_6). In a molecule of ethane, each carbon atom is bonded to three hydrogen atoms and to the other carbon atom by single bonds. *See also* UNSATURATED COMPOUND

saturated fat Organic fatty compounds, the molecules of which contain only saturated FATTY ACIDS combined with GLYCEROL. These acids have long chains of carbon atoms which are bound together by single bonds only. *See also* SATURATED COMPOUNDS

saturated solution In chemistry, a SOLUTION containing so much of a dissolved compound (SOLUTE) that no more will dissolve at the same temperature or pressure, except in cases of SUPERSATURATION.

Saturn Sixth planet from the Sun and second largest in the SOLAR SYSTEM. It was the most remote planet known in the pre-telescope era. Viewed through a telescope, it appears as a flattened, golden-yellow disc encircled by white **rings**. Saturn's magnificent system of rings is its dominant feature. They were first observed by GALILEO in 1610, but Christiaan HUYGENS was the first to identify (1656) them as rings. As seen from Earth, the ring system consists of a few principal components, separated by a number of gaps. The rings are made up of particles ranging from dust to objects a few metres in size, all in individual orbits around Saturn. The main rings are only a kilometre or so thick. In the PIONEER PROGRAM, Pioneer 11 visted Saturn after Jupiter. The two VOYAGER PROJECT probes flew past Saturn, **Voyager 1** in 1980 and **Voyager 2** in 1981, and revealed the ring system to be made up of thousands of separate ringlets. The gaps were not empty, but simply regions with a lower ring density. Saturn has an internal heat source, which probably drives its weather systems. Its density is much lower than for any other planet, and it also has much more of its mass concentrated at the centre. It is assumed, therefore, to be composed predominantly of hydrogen, and to have an iron–silicate core about five times the Earth's mass, surrounded by an ice mantle of perhaps 20 Earth masses. Near the mantle, the hydrogen would be in metallic form; above this, it would exist as liquid molecular hydrogen. The upper atmosphere contains 97% hydrogen and 3% helium, with traces of other gases including methane, ethane, ammonia and phosphine. The MAGNETIC FIELD of Saturn is 500 to 1,000 times the strength of Earth's, and it generates a MAGNETOSPHERE intermediate between those of the Earth and Jupiter in its shape, extent and the intensity of its radiation belts of trapped particles. Saturn has 18 known SATELLITES, including TITAN, the only satellite in the Solar System to have a significant atmosphere.

savanna Area of tropical grassland, found between the Earth's major desert regions and the tropical forests. Savannas occur mainly between 5° and 20° N and S of the Equator, except in East Africa, where, because the land is elevated, savanna regions are situated on the Equator. The savanna climate is hot and wet in summer, when the Sun is overhead; temperatures average *c.*25°C (77°F) and the rainfall is 250mm to 1,250mm (10–50in). In winter, the tropical high-pressure system controls the climate, and so the conditions are warm but very dry. Temperatures are 15°C to 25°C (59°–77°F), and rainfall totals are less than 250mm (10in). In such a climate, grass grows very well in summer – up to 2m (6ft) in height – but it will shrivel in winter. Trees do not grow well, although baobabs and bottle trees survive by storing water in

S

their trunks. Pastoral farming is the main activity on the savanna grasslands.

Savery, Thomas (1650–1715) English engineer. In 1698 he patented a machine for pumping mines, which employed the principle of a vacuum STEAM ENGINE. Savery developed this invention with Thomas NEWCOMEN. Savery also designed an early form of paddle-driven boat.

saw Cutting tool with a toothed blade, usually operated manually, by treadle, or by motor. Alternate teeth are bent outwards in opposite directions, so that the saw cut is slightly wider than the blade thickness. This helps to prevent the blade from sticking. The blade may be straight, circular, or formed into a band. Widely spaced teeth give a rapid, but rough cut. Closely spaced teeth give a fine cut. Applying a lubricant to the blade improves its performance.

scalar Mathematical quantity that only has a magnitude, as contrasted with a VECTOR, which also has direction. Mass, energy and speed are scalars whereas weight, force and velocity are represented by vectors.

scale In biology, small hard plate that forms part of the external SKIN of an animal. It is usually a development of the skin layers. In most fish, scales are composed of bone in the dermal skin layer. The scales of reptiles and those on the legs of birds are horny growths of the epidermal skin layer, and are comprised mostly of the fibrous protein KERATIN.

scandium (symbol Sc) Silver-white, metallic element of group III of the periodic table, predicted (as ekaboron) by MENDELEYEV and discovered in 1879. It is found in thortveitite and, in small amounts, in other minerals. Scandium is a soft metal used as a radioactive tracer and in nickel alkaline storage batteries. Chemically, it resembles the rare-earth metals of the LANTHANIDE SERIES. Properties: at.no. 21; r.a.m. 44.956; r.d. 2.99; m.p. 1,539°C (2,802°F); b.p. 2,832°C (5,130°F); most common isotope ^{45}Sc (100%).

scanning In medicine, use of a non-invasive system to detect abnormalities of structure or function in the body. Detectable waves (X-RAYS, gamma rays, ultrasound) are passed through the part of the body to be investigated and the computer-analysed results are displayed as images on a viewing screen.

scanning tunelling microscope Type of ELECTRON MICROSCOPE that produces highly magnified images of the surface of a specimen. A tiny metal probe on a movable arm is held in near contact with the specimen, which has to be a conducting material, and scanned across it. An electrical signal, produced when electrons "tunnel" between the specimen and the probe, is kept constant by raising and lowering the arm. The necessary movements are analysed by a computer, which generates a "contour map" of the surface displayed on the computer screen.

scapolite (wernerite) Group of SILICATE minerals related to the FELDSPARS and of variable chemical composition. They are common in metamorphosed limestones and are grey and glassy in appearance. Hardness 5–6; r.d. 2.7.

scapula (shoulder blade) In vertebrates, either of two large, roughly triangular flat bones found one on either side of the upper back, and forming part of the PECTORAL girdle. It provides for the attachment of muscles that move the forelimb.

scarp Abbreviation of ESCARPMENT

scattering Deflection of ELECTROMAGNETIC RADIATION by particles. Where the particles are very much larger than the WAVELENGTH, scattering consists of a mixture of REFLECTION and DIFFRACTION, and the amount of scattering depends very little on wavelength. Where the particles are very much smaller than the wavelength, the amount of scattering (d) is inversely proportional to the fourth power of the wavelength (λ): $d \propto 1/\lambda^4$. Thus, blue LIGHT is scattered by small particles ten times as much as red light. Scattering of sunlight by atoms and molecules in the atmosphere is what makes the sky blue (*see* SPECTRUM; TYNDALL, JOHN). It is the main cause of **atmospheric extinction**, making the Sun appear red at sunrise and sunset by preferentially scattering blue light out of the line of sight. ELEMENTARY PARTICLES themselves can be scattered by atomic nuclei or other particles. It is the means by which the structure of ATOMS was discovered. Ernest RUTHERFORD's students, Hans GEIGER and Ernest Marsden, "fired" ALPHA PARTICLES through thin metal films and noted their scattering. From the results, Rutherford deduced the existence of the atomic NUCLEUS (positively charged and centrally placed in the atom). Most knowledge of elementary particles and the discovery of new ones has been obtained by scattering experiments carried out in particle ACCELERATORS. *See illustration on page 316.* *See also* COMPTON EFFECT; RAMAN EFFECT

Scheele, Karl Wilhelm (1742–86) Swedish chemist who discovered OXYGEN. An apothecary, his interest in chemistry led to an investigation of combustion and the discovery of oxygen in 1771, but publication of his discovery was delayed and the credit went to Joseph PRIESTLEY. Scheele made other important discoveries, including CHLORINE, GLYCEROL and a number of organic acids.

scheelite ($Ca[WO_4MoO_4]$) Mineral, calcium tungstatemolybdate, an important ore of TUNGSTEN. It is found in metamorphic deposits and in pegmatites. Scheelite has a tetragonal system and bipyramidal crystals, also occurring as massive and granular aggregates. It reveals various tints with adamantine lustre, fluoresces under ultraviolet light and is brittle. Hardness 4.5–5; r.d. 6.

Schiaparelli, Giovanni Virginio (1835–1910) Italian astronomer. He was an assiduous observer of the inner planets, MARS in particular. Schiaparelli prepared a map of the surface of Mars, introducing a nomenclature for the various features, that remained standard until the planet was mapped by space probes. His use of the term *canali* (It. canals) has caused much subsequent controversy. He also prepared maps of Venus and Mercury. Schiaparelli discovered the connection between meteor streams and comets, and studied double stars.

Schick, Bela (1877–1967) US paediatrician who devised (1913) the **Schick test** which determines susceptibility to diphtheria. A small amount of diphtheria toxin is injected into the skin and swelling indicates a level of antibody production. If the subject has adequate immunity no skin reaction occurs.

schist Large group of METAMORPHIC ROCKS that have been made cleavable, causing the rocks to split into thin plates leaving a wavy, uneven surface. Almost any type of rock will become a schist if subjected to sufficient metamorphism. Schists are named after their predominant mineral; a common example is MICA SCHIST.

Schönbein, Christian Friedrich (1799–1868) German chemist, the discoverer of OZONE. After studying at various universities, he became professor at the University of Basle (1835). Schönbein

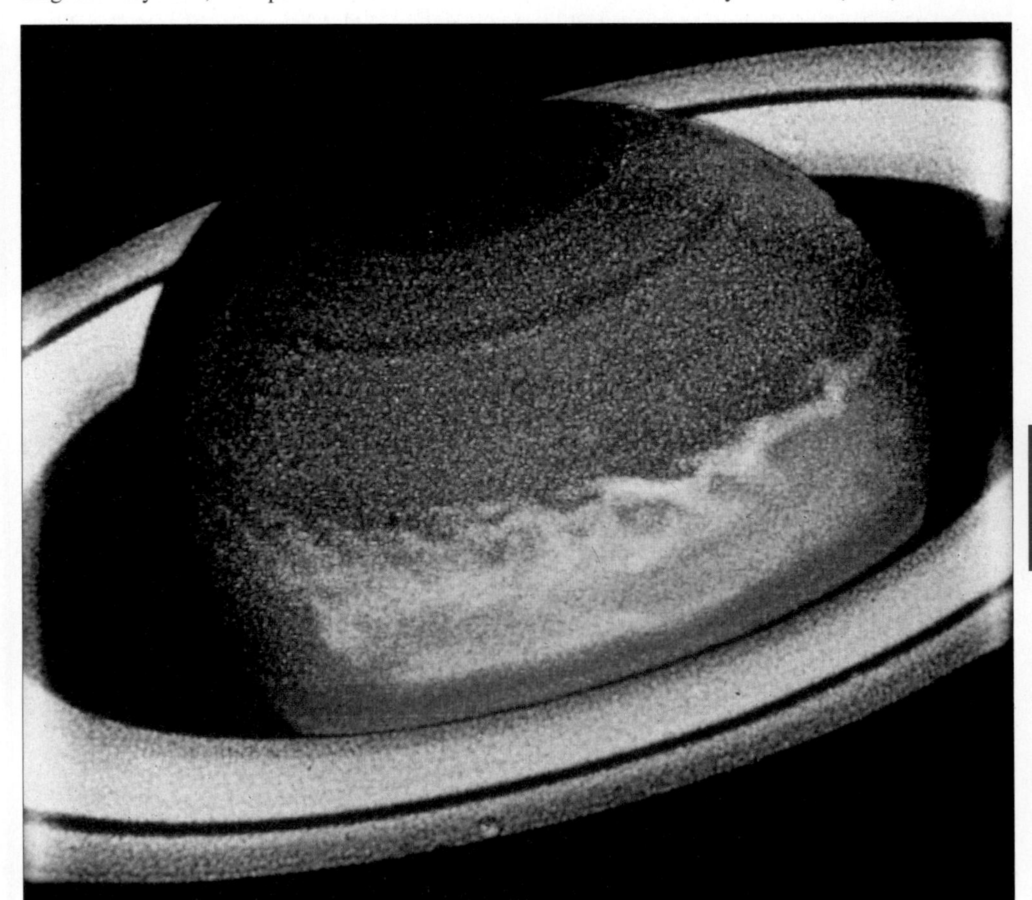

▲ **Saturn** Hubble Space Telescope image of the planet Saturn, showing an unusual long-lived storm system known as the "white spot". The white spot is the large, elongated, reddish-white region lying along the planet's equator. The spot first appeared in late September 1990, growing rapidly over the next few days and spreading along the equator as well as climbing higher in the atmosphere. This image was taken on 9 November and combines images recorded in blue and infrared light; lower parts of the cloud appear blue and the region with high clouds appears red. The clouds which form the white spot are thought to be composed of ammonia ice crystals.

SCATTERING

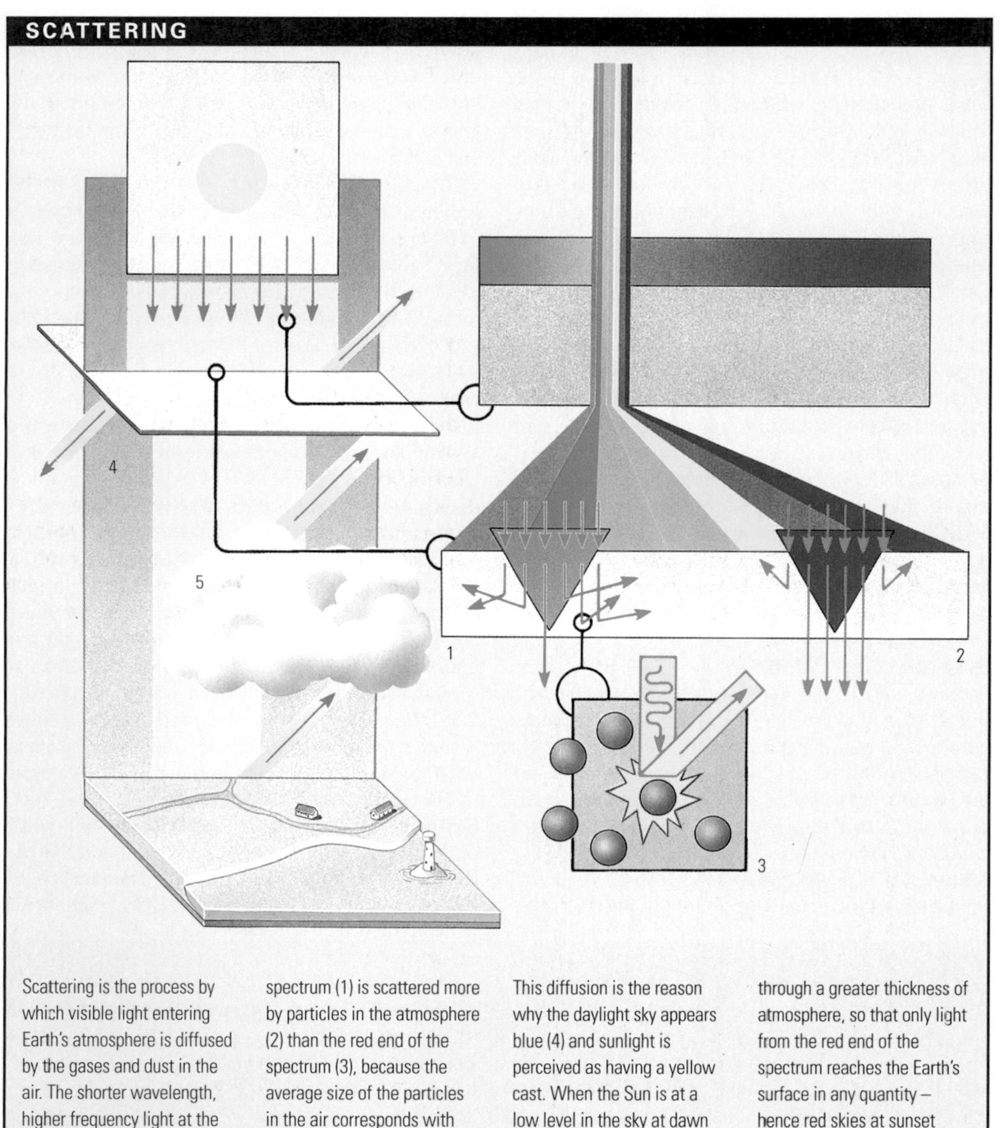

Scattering is the process by which visible light entering Earth's atmosphere is diffused by the gases and dust in the air. The shorter wavelength, higher frequency light at the blue end of the visible spectrum (1) is scattered more by particles in the atmosphere (2) than the red end of the spectrum (3), because the average size of the particles in the air corresponds with the blue light's wavelength. This diffusion is the reason why the daylight sky appears blue (4) and sunlight is perceived as having a yellow cast. When the Sun is at a low level in the sky at dawn or sunset the light has to pass through a greater thickness of atmosphere, so that only light from the red end of the spectrum reaches the Earth's surface in any quantity – hence red skies at sunset or sunrise.

also discovered GUNCOTTON, recognizing its potential as a smokeless substitute for GUNPOWDER.

Schrieffer, John Robert (1931–) US physicist. In 1957, working at the University of Illinois with John BARDEEN and Leon COOPER, Schrieffer helped to formulate a theory of SUPERCONDUCTIVITY, which is known as the **BCS theory** (after their initials). They all shared the 1972 Nobel Prize for physics.

Schrödinger, Erwin (1887–1961) Austrian physicist. He was professor at the Dublin Institute for Advanced Studies (1940–57). Schrödinger established the science of WAVE MECHANICS as part of the QUANTUM THEORY. **Schrödinger's equation** is a fundamental mathematical tool of QUANTUM MECHANICS. It formulates a relationship between WAVE FUNCTION, mass and energy:

$$\nabla^2\psi + 8\pi^2 m(E - U)\psi/h^2 = 0,$$

where ∇^2 is the LAPLACE operator, ψ is the wave function, m is a particle's mass, E its total energy, U its potential energy and h PLANK'S CONSTANT. Schrödinger shared the 1933 Nobel Prize for physics with Paul DIRAC. *See also* BROGLIE, LOUIS DE

Schultze, Max Johann Sigismund (1825–74) German biologist. He studied the nervous system of vertebrates, the electric organs of fish and the anatomy of worms and molluscs. Schultze's most important work was the identification of a CELL as an organism containing both nucleus and protoplasm and his recognition that PROTOPLASM is a basic substance of plant and animal cells.

Schuster, Sir Arthur (1851–1934) British physicist, b. Germany, whose work included electrical discharge in gases, calorimetry and seismology. Schuster was an authority in the field of SPECTROSCOPY. In 1875 he led an expedition to Siam (now Thailand) to study the solar eclipse.

Schwann, Theodor (1810–82) German biologist who laid the foundation of modern HISTOLOGY by establishing the CELL theory. In 1836 he became the first person to prepare an ENZYME from animal tissue when he isolated and named PEPSIN.

Schwann cells Cells that wrap around the AXONS of nerve fibres to form a fatty, protective MYELIN SHEATH. Between sections of myelin sheath are small gaps called **nodes of Ranvier**. The sheath serves to electrically insulate the nerve and allow the more rapid passage of nerve impulses. The cells were named after their discoverer, Theodor SCHWANN.

Schwarzschild radius Critical radius at which a very massive body under the influence of its own GRAVITATION becomes a BLACK HOLE. It is the radius of the EVENT HORIZON of a black hole, from which nothing can escape, not even light. The radius is given by the expression $R = 2GM/c^2$, where G is the GRAVITATIONAL CONSTANT, M is the MASS of the body and c is the speed of LIGHT. The Schwarzschild radius for the Sun is about 3km and for the Earth it is about 1cm. It is named after the German astronomer Karl Schwarzschild (1873–1916).

Schwinger, Julian Seymour (1918–94) US physicist who devised a theory of QUANTUM ELECTRODYNAMICS (QED) that explained mathematically the interaction of LIGHT and MATTER. For their work on QED, Schwinger shared the 1965 Nobel Prize for physics with Richard FEYNMANN and the Japanese physicst S. Tomonaga. He also conducted research on SYNCHROTRON radiation, a type of ELECTROMAGNETIC RADIATION emitted by charged particles in circular orbits.

scintillation (twinkling) In astronomy, slight variability in the brightness and colour of the stars caused by local disturbances in the Earth's atmosphere that change its REFRACTIVE INDEX. Planets do not exhibit this phenomenon because the scintillations from different points on the surface are not in PHASE, and the fluctuations are lost in the general illumination. For astronomical purposes, the inconvenience of the effects of scintillation is reduced by siting observatories at high altitudes, and it is almost entirely avoided by using telescopes carried by balloons, rockets or artificial satellites.

scintillation counter Instrument containing a crystal that emits scintillations of light when bombarded by radiation. Each light flash, corresponding to a single particle, is converted into an electric pulse by a PHOTOMULTIPLIER. The number of pulses, counted electronically, indicates the activity of the source of radiation.

sclerenchyma In botany, plant support-tissue, the cells of which have become rigid due to the formation of LIGNIN. Sclerenchyma occurs in veins, stems and midribs of leaves. As sclerenchyma cells mature, they die because the lignin in the thickened cell wall does not allow the passage of gases and water, although shallow pits (**plasmodesmata**) in

◀ **scuba diving** Diver viewing a sea fan in the Coral Sea, off Australia. Scuba diving was developed by Jacques Cousteau and Émile Gagnan. Recent developments include a system whereby the breathing rate of the diver is monitered by a microcomputer, which then releases the compressed air at a rate that corresponds to the diver's requirement. As the diver becomes more or less active, so the compressed air is released accordingly. It is thought that this system will improve the efficiency of the apparatus by up to 20%.

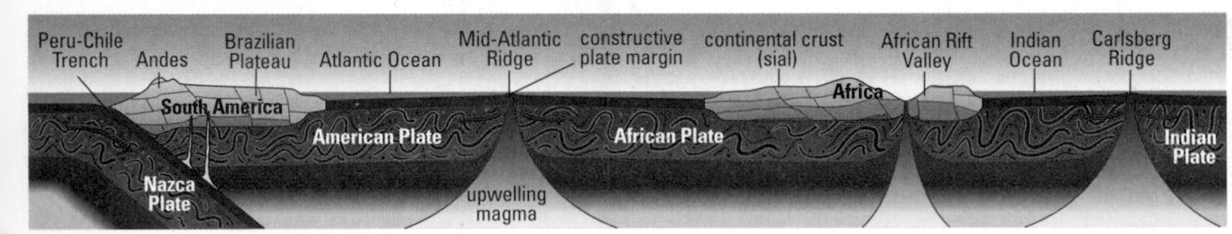

Seafloor spreading in the Atlantic Ocean

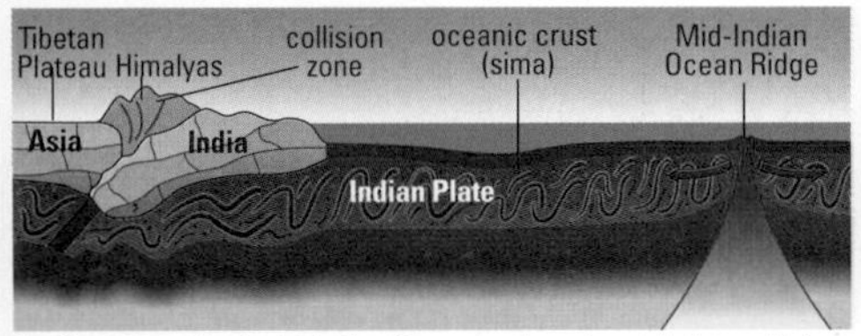

Seafloor spreading in the Indian Ocean and continental plate collision

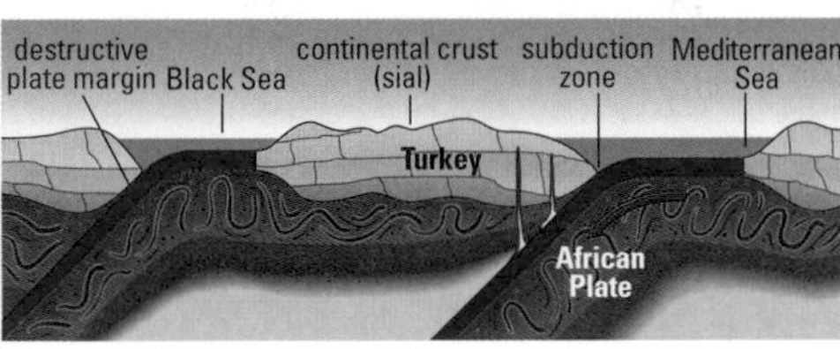

Oceanic and continental plate collision

▲ **seafloor spreading** The vast ridges that divide the Earth's crust beneath each of the world's oceans mark the boundaries between tectonic plates that are gradually moving in opposite directions. As the plates shift apart, molten magma rises from the mantle to seal the rift and the seafloor slowly spreads towards the continental landmasses. The rate of spreading has been calculated by magnetic analysis of the rock at *c*.3.75cm (1.5in) a year in the N Atlantic Ocean. Underwater volcanoes mark the line where the continental rise begins. As the plates meet, much of the denser oceanic crust dips beneath the continental plate and melts back into the magma.

the wall are free from lignin and permit some exchange of substances between adjacent cells. In zoology, the hard skeletal tissue of corals is also called sclerenchyma.

sclerosis Degenerative hardening of tissue, usually due to scarring, following inflammation or as a result of ageing. It can affect the brain and spinal cord, causing neurological symptoms, or the walls of the arteries. *See also* MULTIPLE SCLEROSIS (MS)

Scorpius (Scorpio, scorpion) Constellation of the S sky, situated on the ECLIPTIC between LIBRA and SAGITTARIUS. It contains several OPEN CLUSTERS and GLOBULAR CLUSTERS. The MILKY WAY also passes through the region. The brightest star is the 1st-magnitude Alpha Scorpii (Antares).

scrambler Electronic device used to render a transmitted message unintelligible to all but its sender and its receiver, used particularly on voice messages sent by radio or telephone. The message is distorted in a prearranged way by the sender's scrambler, for instance by oscillating its frequency or by rearranging its order; the receiver's scrambler reverses this process to make the message intelligible again.

scrapie Fatal disease of sheep and goats that affects the CENTRAL NERVOUS SYSTEM (CNS), causing staggering and itching that makes an infected animal scrape itself against various objects. It is caused by a slow-acting, virus-like, microscopic particle called a PRION. The animal usually dies within six months of exhibiting the first symptoms. The disease is thought to be related to BOVINE SPONGIFORM ENCEPHALOPATHY (BSE) and CREUTZFELDT-JAKOB DISEASE (CJD).

scree (talus) Loose fragments of rock which have accumulated on a hillside. Scree is made up of particles ranging in size from sand grains to boulders. It mainly results from WEATHERING processes, such as freeze-thaw activity. Scree forms more frequently on certain rock types, such as limestone.

screening In medicine, a test applied either to an individual or to groups of people who, although apparently healthy, are known to be at special risk of developing a particular disease. It is used to detect treatable diseases while they are still in the early stages. For example, middle-aged women may be offered mammography to rule out breast cancer, and, in most developed countries, newborns are routinely screened for phenylketonuria (PKU), an inherited metabolic defect which can lead to severe mental retardation. Screening, therefore, is an important form of preventive health care.

screw Variant of a simple MACHINE, the INCLINED PLANE. It is an inclined plane cut around a cone, usually of metal, in a helical spiral. When FORCE is exerted radially on the screw, for example by a screwdriver or the lever of a screw-jack, the screw advances to an extent determined by its pitch (the distance between crests of its thread).

screw propeller Device that converts the TORQUE of a rotating shaft into propulsion or THRUST. It is based on the principle of a helical SCREW, which advances in wood when it is rotated; the PROPELLER advances in air or water when it is rotated by an engine. **Simple** propellers have a fixed pitch so that the amount of thrust increases as the speed of rotation increases; **variable-pitch** propellers are rotated at constant speed, and thrust is varied by adjusting the angle at which the blades meet the air or water.

scrotum In males, the thin external sac of skin, divided into two compartments, which contains the TESTES and the coiled tubes in which the SPERM mature (epididymides). It is located behind the PENIS.

scuba diving Diving with the use of **s**elf-**c**ontained **u**nderwater **b**reathing **a**pparatus, or scuba. The equipment usually consists of tanks of compressed air connected to a device called a **demand regulator**, which controls the air flow through the mouth. Other equipment includes a wet suit to insulate the diver from the cold, weighted belt, face mask and flippers. Professional divers use scuba to inspect vessels below the waterline and to search underwater. Scuba diving has attracted many enthusiasts since the design of the first scuba equipment called the Aqualung by Jacques COUSTEAU and Émile Gagnan in 1943.

sea *See* OCEAN

Seaborg, Glenn Theodore (1912–) US physicist. During World War 2 he worked on the development of the atom bomb. In 1940 Seaborg and Edwin M. MCMILLAN discovered PLUTONIUM. They went on to isolate the entire ACTINIDE SERIES, Seaborg and McMillan shared the 1951 Nobel Prize for chemistry.

seafloor Floor of the OCEANS. The major features of the seafloor are the CONTINENTAL SHELF, the CONTINENTAL RISE, the ABYSSAL floor, seamounts, oceanic TRENCHES and MID-OCEAN RIDGES. The **abyssal** or deep ocean floor is *c*.3km (1.8mi) deep and is mostly made of basaltic rock covered with fine-grained (PELAGIC) sediment consisting of dust and the shells of marine organisms. Oceanic **trenches** are up to 11km (7mi) deep, typically 50–100km (30–60mi) wide and may be thousands of kilometres long. The slopes are usually asymmetrical with the steeper slope on the landward side and a more gentle slope on the side of the ocean basin. They are regarded as the site of plate SUBDUCTION. Oceanic **ridges** are long, linear volcanic structures which tend to occupy the middle of seafloors; they are the sites of crustal spreading. *See also* SEAFLOOR SPREADING

seafloor spreading Theory suggested (1960) by the US geologist Harry Hess to explain how CONTINENTAL DRIFT occurs. It is now a key part of PLATE TECTONICS. It proposes that the SEAFLOOR is moved laterally as new basalt rock is injected along MID-OCEAN RIDGES and so the seafloor spreads symmetrically, and becomes older with increasing distance from the ridge. In 1963 British geophysicists F.J. Vine and D.H. Matthews uncovered supporting evidence for the theory when they tested the patterns of FOSSIL magnetism in the rocks of the seafloor. They discovered a symmetrical pattern of stripes centred on mid-ocean ridges and by matching the stripes with known dates it became possible to calculate the rate of spreading.

seal In engineering, liquid or gas-tight union, such as that between mating pipe flanges, beneath bolted or clamped hatches and between moving parts of machines. Seals are made from rubber, plastics, fibres such as cotton and flax, or combinations of these materials.

sea level Level from which topographic heights are measured. It is usually the mean sea level and is reckoned on being the average of regular sea levels taken over a long period of time. It may also be referred to as the Ordnance Datum (OD).

seaplane AIRCRAFT that can land on and take off from water. The first practical seaplane was built in the USA by Glen CURTISS and flown in 1911. There are two main types of seaplane. **Floatplanes** have large floats that support the fusilage above the water. **Flying boats** float on their boat-shaped hulls, with small floats supporting the wings. In the 1930s flying boats carried more passengers than any other type of aircraft. Some floatplanes and fly-

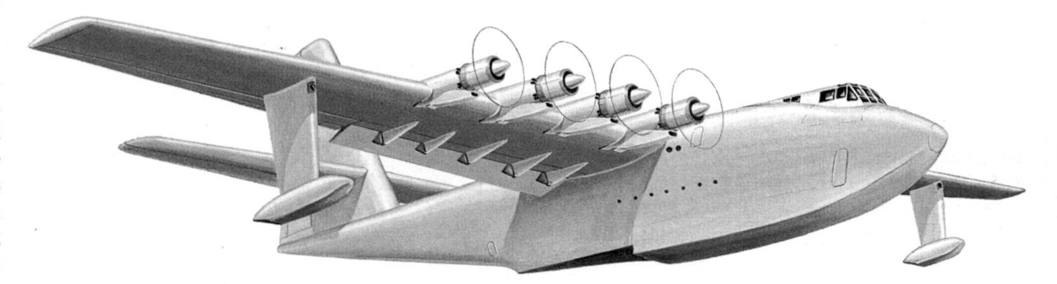
▲ **seaplane** The "Hughes Hercules" was the largest seaplane ever built. It was made of wood and took to the air in 1947. It had a wingspan of 94.49m (312ft), a gross weight of 1.8 tonnes and was powered by eight 3,000 h.p. engines.

SEASONS

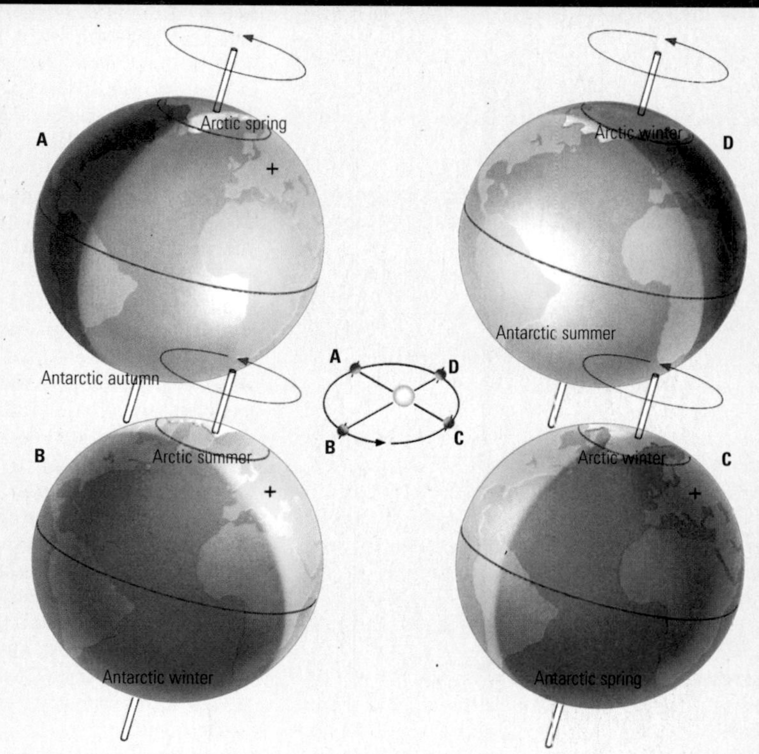

Each of the four seasons begins and ends at one of the two solstices (the shortest and longest days of the year), which are separated by six months, or one of the equinoxes (when day and night are equal), also separated by six months. The solstices mark the days when one pole is at its closest position to the Sun and the other is furthest away from it. Equinoxes mark the days when each pole is the same distance from the Sun. For example, spring in the Northern Hemisphere at the spot marked by a cross on the globe commences on the equinox of 20 March (A), when day and night are equal length, and is followed by the summer solstice on 20 or 21 June (B) (the dates vary because of leap years), when the North Pole is at its closest to the Sun and the day is the longest of the year. Autumn equinox commences on 22 or 23 September (C), when day and night are once again equal in length. Winter in the Northern Hemisphere beings on 21 or 22 December (D), when the day is shortest in the year. This seasonal sequence is reversed in the Southern Hemisphere, where, for example, winter starts on 20 or 21 June when the South Pole is furthest away from the Sun but the North Pole is closest and the Northern Hemisphere summer begins. The start and finish of each season can thus be defined in very precise astronomical terms by the Earth's position relative to the Sun.

ing boats are fitted with wheels so that they can take off and land on ground as well as water; such aircraft are called **amphibians**.

seasonal affective disorder (SAD) Mental depression apparently linked to the changing seasons of the year. Sufferers experience depressed mood during autumn and winter, and become more cheerful in spring and summer. Some research has indicated that SAD may be caused by decreased levels of light.

seasons Four astronomical and climatic periods of the year based on differential solar heating of the Earth as it makes its annual revolution of the Sun. Due to the parallelism of the Earth's axis of rotation, pointed near the Pole Star throughout the year, the Northern Hemisphere receives more solar radiation when its pole is aimed towards the Sun in summer and less in winter when it is aimed away, whereas the opposite holds for the Southern Hemisphere. The seasons are conventionally initiated at the vernal (spring) and autumnal EQUINOXES and the winter and summer SOLSTICES.

seawater Solution of several minerals in the OCEAN. It has a high concentration of microscopic life and tastes salty. The dissolved salt content varies between 3.3% and 3.7%, most of which is common salt (sodium chloride, NaCl) but nearly all elements are found in seawater, including vast reserves of magnesium, potassium and calcium. Large amounts of nitrogen, oxygen and carbon dioxide are present as dissolved gases. The freezing point of seawater of salt content 3.5% is −1.9°C (28.6°F). *See also* SALINITY

seaweed Any of numerous species of brown, green or red ALGAE, found in greatest profusion in shallow waters on rocky coasts. KELPS are the largest forms. Many species are important for the manufacture of fertilizers or food, or as a valuable source of chemicals such as iodine. Kingdom PROTOCTISTA.

sebaceous gland Gland in the skin producing the oily substance sebum and opening into a hair follicle. Sebum acts as a skin lubricant and also contains some antibacterial substances.

seborrhoea Disorder of the SEBACEOUS GLANDS characterized by over-production of sebum that results in red, scaly patches on the skin and dandruff. Common at puberty, it predisposes to ACNE.

secant In TRIGONOMETRY, ratio of the length of the hypotenuse to the length of the side adjacent to an acute angle in a right-angled TRIANGLE. The secant of angle A is usually abbreviated to sec$A$, and is equal to the reciprocal of its COSINE, that is 1/cos$A$.

Secchi, Pietro Angelo (1818–78) Italian astronomer who was the first to use PHOTOGRAPHY and SPECTROSCOPY in astronomy. In 1851 Secchi photographed a solar eclipse and between 1864 and 1868 he examined the spectra of more than 4,000 stars, classifying them into spectral types. His system forms the basis of the SPECTRAL CLASSIFICATION used today.

second (symbol s) SI unit of time defined as the time taken for 9,192,631,770 periods of vibration of the ELECTROMAGNETIC RADIATION emitted by a CAESIUM-133 atom. *See also* PHYSICAL UNITS

secondary cell *See* CELL

secondary consumer In a FOOD CHAIN, an organism that feeds on the PRIMARY CONSUMERS. The primary consumers, which themselves feed on green plants (PRIMARY PRODUCERS), are predominantly herbivores. Secondary consumers are, therefore, mainly CARNIVORES that feed on the herbivores. If a secondary consumer has no PREDATORS (that is, there are no tertiary consumers), it is known as the top carnivore. *See also* ECOSYSTEM

secondary emission In physics, emission of ELEMENTARY PARTICLES by a target, such as a gas or metal surface, that is bombarded with primary (newly generated) radiation or particles. Secondary COSMIC RAYS are emitted when primary cosmic rays bombard the atmosphere. The anode of an ELECTRON TUBE emits secondary electrons when it is bombarded by primary electrons from the cathode. This principle is employed in electron-multiplying devices, such as an image intensifer.

secondary sexual characteristics External features that identify a sexually mature animal and distinguish the sexes. Such characteristics play a part in reproductive behaviour, although they are not essential for mating. Their development is brought about by the first production of SEX HORMONES in a maturing juvenile. For example, ANDROGENS in male deer stimulate the growth of antlers (except in reindeer, in which the females also bear antlers). In human beings, secondary sexual characteristics develop during PUBERTY. They include breast development in girls, and the growth of facial hair and enlargement of the penis in boys. A boy's voice also becomes deeper. *See also* SEX

secondary thickening (secondary growth) Increase in the girth of stems and roots, especially shrubs and trees, as a result of the generation of new cells in the CAMBIUM. The cambium cells differentiate to produce additional XYLEM on the inside and PHLOEM on the outside. A further cylinder of cambium near the surface of the stem or root may produce a tough outer layer of CORK cells or BARK. *See also* VASCULAR BUNDLE

secretin Polypeptide HORMONE secreted by the small intestine. It stimulates the release of pancreatic juice and also stimulates bile secretion by the GALL BLADDER.

secretion Production and discharge of a substance, usually a fluid, by a cell or a GLAND. The substance so discharged is also known as a SECRETION. Secretions include ENZYMES, HORMONES, SALIVA and sweat.

sedative Drug used for its calming effect, to reduce anxiety and tension; at high doses it induces sleep. Sedative drugs include NARCOTICS, BARBITURATES and BENZODIAZEPINES.

sediment In geology, a general term used to describe any material (such as gravel, sand and clay) that is transported and deposited by water, ice, wind or gravity. Most sediment is eventually deposited in the sea. The term includes material such as lime that is transported in solution and later precipitated, and organic deposits such as coal and coral reefs.

sedimentary rock Type of rock formed by deposition of SEDIMENT derived from pre-existing rocks, which may have been sedimentary, IGNEOUS or METAMORPHIC. Most sediment accumulates on the bed of the sea, having been dumped there by rivers, or having accumulated there as dead sea creatures fell to the seafloor. This accumulated sediment is consolidated and compressed. Earth movements uplift the sediments, and they may be tilted, folded or faulted. The resulting rocks are sedimentary, and their type depends on their composition. The layers in sedimentary rocks, known as strata, may be a few centimetres or many metres in thickness. Sedimentaries consisting of land sediment are **clastic** rocks, and are GRAVELS, SANDS, SILTS or CLAYS, according to the size of the particles. Other types of sedimentary rock include: LIMESTONE, which

S

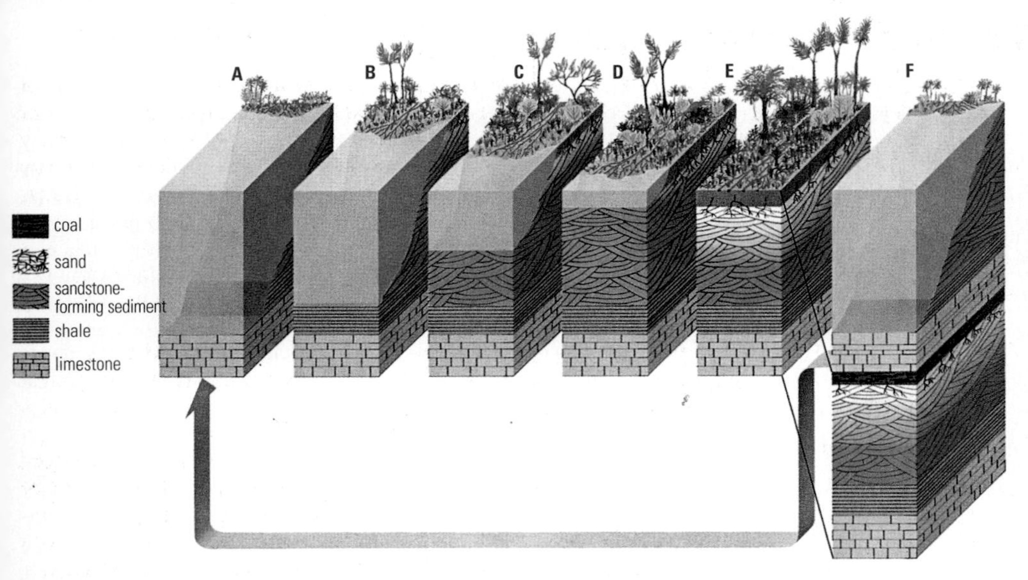

▲ **sediment** A delta's sediments are laid down in a specific order that may be endlessly repeated if the region where deposition takes place is sinking. Limestone deposits cover the seabed when the delta is too distant to be influential (A). As the delta encroaches (B), fine-grained muds that will become shale are deposited, followed by coarser, sandstone-forming sediments as the advance continues (C). As the water shallows, current bedding (D) indicates that sand is being deposited. Once the delta builds above water level (E) it can support swamp vegetation which forms coal. When the region sinks (F) the cycle restarts.

consists of fragments of dead sea creatures, sometimes mixed with land sediment; EVAPORITE, which comprises saline deposits; COAL, which is accumulated vegetation; coralline, which contains large quantities of CORAL; and CHALK, which is a pure form of limestone, with very little land sediment. *See also* FAULT; FOLD; STRATUM

sedimentation Any process or processes that deposit ROCK-forming materials. The materials deposited are continental in origin, the debris of landforms worn down and carried off by wind, water or ice. Whenever the flow of the transporting medium is interrupted or diverted, the carried material is deposited either on land (in desert, lake or river bed), or on coasts, or as marine sediments. *See also* DEPOSITION; EROSION

Seebeck, Thomas Johann (1770–1831) German physicist, b. Russia. Seebeck is celebrated chiefly for his contribution to THERMOELECTRICITY. In 1821 he discovered that if two strips of different metals were joined in a loop and the two junctions were maintained at different temperatures, an ELECTROMOTIVE FORCE (E.M.F.) was produced that was proportional to the temperature difference. He called this thermomagnetism. Today, it is called the SEEBECK EFFECT. Seebeck falsely believed that the conductors were magnetized directly by the difference in temperature. His discovery was neglected for many years, but is the basis of the modern THERMOMETER. Seebeck also studied the polarization of stressed glass.

Seebeck effect In THERMOELECTRICITY, the generation of ELECTROMOTIVE FORCE (E.M.F.) in a circuit consisting of two different metals or semiconductors joined in a loop, when the two junctions are kept at different temperatures. It was discovered by Thomas SEEBECK (1770–1831). This type of circuit is called a **thermocouple**. *See also* PELTIER EFFECT

seed Part of a flowering plant (ANGIOSPERM) that contains the EMBRYO and food store. It is formed in the OVARY by FERTILIZATION of the female GAMETE. Food may be stored in a special tissue called the ENDOSPERM, or it may be concentrated in the swollen seed leaves (COTYLEDONS). Seeds are often capable of surviving quite harsh conditions, and provide a means of survival for many plants in extreme environments. They are also the unit of dispersal of flowering plants (angiosperms) and CONIFERS. *See also* GERMINATION

Segrè, Emilio Gino (1905–89) US physicist, b. Italy. In 1937 he discovered TECHNETIUM, the first ELEMENT to be artificially produced. In 1940 he helped to discover ASTATINE and PLUTONIUM-239. Segrè is, however, chiefly remembered for his discovery of the antiproton. In 1930 Paul DIRAC had predicted the existence of antiparticles. In 1955, using a Bevatron particle ACCELERATOR at the University of California, Segrè and Owen CHAMBERLAIN bombarded copper with high-energy PROTONS to produced the antiparticle of the proton. Chamberlain and Segrè shared the 1959 Nobel Prize for physics. *See also* ANTIMATTER

seif Elongated type of sand DUNE. Seifs may be several kilometres long and often occur in groups. The narrow ridge of the dune runs parallel to the direction of the prevailing wind, and its height and width can be increased by the action of crosswinds. Seifs are found mainly in the sand deserts of N Africa.

seismic profile Continuous record of sound waves bounced off sediments on the SEAFLOOR. The sounds become SEISMIC WAVES and as such are used to determine the thickness and structure of bottom sediments. *See also* ECHO SOUNDER

seismic survey Use of SEISMIC WAVES produced by underground explosions to detect and measure the extent of minerals (such as oil and coal) under the ground. The speed of the shock waves indicates the type of material they are passing through.

seismic wave Shock wave produced by EARTHQUAKES and man-made explosions. The velocity (speed and direction) of seismic waves varies according to the material (type of rock, molten core or oil) through which they pass. Primary (P) and secondary (S) waves are transmitted by the solid Earth. **P waves** vibrate in the direction that they are advancing; **S waves** vibrate at right-angles to the direction in which they are advancing. Only P waves are transmitted through fluid (liquid or gas) zones. *See also* SEISMOLOGY

seismograph (seismometer) Instrument for measuring and recording SEISMIC WAVES caused by movement (EARTHQUAKE or explosion) in the Earth's crust. The vibrations are recorded by a pen on a revolving drum. Some seismographs are sensitive enought to pick up seismic activity thousands of kilometres away. For example, in January 1994 , the earthquake in Los Angeles, USA, was picked up by seismographs in Britain. *See also* RICHTER SCALE

seismology Study of EARTHQUAKES and the SEISMIC WAVES they produce. The movement of seismic waves is detected and recorded by SEISMOGRAPHS. The detection involves the separation of events from the ever-present background of seismic noise. Pinpointing sites of events has become very accurate since the development of precise instrumentation and the establishment of the World Wide Standard Seismograph Network (WWSSN). Seismology is very important in the exploration of the Earth's internal structure.

selection In GENETICS, the probability that one ALLELE will be passed on in favour of another. *See also* DOMINANT; NATURAL SELECTION; RECESSIVE

selective breeding Process by which stockbreeders and agriculturalists improve the strains of domesticated animals and cultivated plants; it involves selection and pairing of individuals with desirable qualities in the PHENOTYPE. Today a growing knowledge of GENETICS ensures more predictable results.

selenium (symbol Se) Grey, METALLOID element of group VI of the periodic table, discovered (1817)

terrigeneous deposits
red clay
globigerina ooze
pteropod ooze
diatom ooze
radiolarian ooze

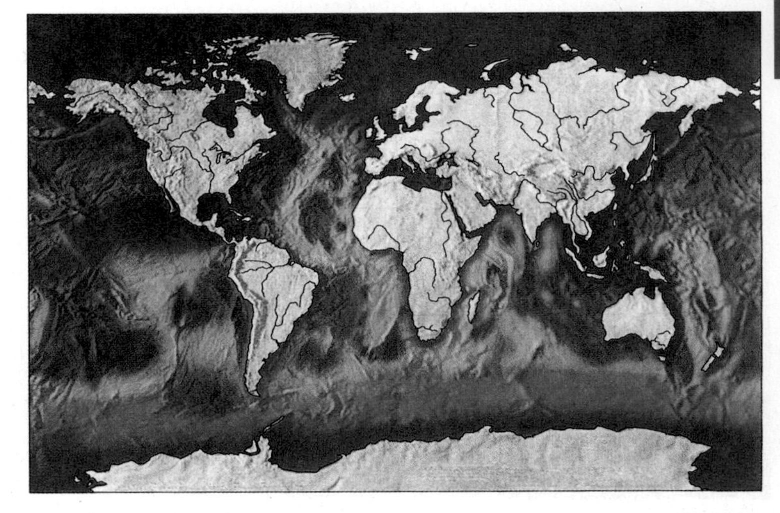

► **sedimentation** The sea floor is covered by unconsolidated sediments that are classified according to the nature of their main constituent. This constituent is determined by the distance from the landmasses, the nature of the winds and currents, the surface water temperature and by the depth. The major types are the terrigeneous deposits (debris derived from the weathering of the continents); the red (or brown) clays, which form the inorganic sediments; and the globigerina, radiolaria and diatomaceous oozes, which are formed by the accumulation of the shells of dead planktonic animals.

by Jöns BERZELIUS. Its chief source is as a by-product in the electrolytic refining of copper. The element resembles sulphur. It is used extensively in PHOTOELECTRIC CELLS, SOLAR CELLS, XEROGRAPHY and red pigments. Properties: at.no. 34; r.a.m. 78.96; r.d. 4.79; m.p. 217°C (422.6°F); b.p. 684.9°C (1,265°F); most common isotope ^{80}Se (49.82%).

selenology Study of the MOON's physical and chemical composition.

Selye, Hans (1907–82) Canadian physician, b. Austria. He is celebrated for his pioneering studies of the effects of stress on human and animal physiology. Selye outlined the "general adaptation syndrome", a series of stages the body undergoes when subjected to stress. His works include *The Stress of Life* (1956), *Physiology and Pathology of Exposure to Stress* (1960) and *Stress Without Distress* (1974).

semaphore Device or technique that communicates messages visually; a type of optical TELEGRAPH. A **railway** semaphore signal consists of a vertical post on which a single projecting arm is mounted, whose angle indicates "all clear" or "danger". In **flag** semaphore, a signaller indicates letters and numerals by the position of his outstretched arms, emphasized with flags.

semen Fluid ejaculated by the male at the climax of sexual intercourse; it contains SPERM from the TESTES and the secretions of various accessory sexual glands. In human beings, each ejaculate is normally about 3 to 6ml by volume and contains about 200 to 300 million sperm.

Semenov, Nikolai Nikolaevich (1896–1986) Russian chemist. He studied at Leningrad and later Moscow State University. Semenov shared the 1956 Nobel Prize for chemistry with Cyril HINSHELWOOD for their independent work on branched chain reactions in COMBUSTION processes.

semicircular canals Three parts of the inner EAR which function as balance organs. They are tubular ducts which project in different planes and can detect, by the movement of a liquid within them, movement in any direction in space.

semiconductor Substance with an electrical CONDUCTIVITY between that of a CONDUCTOR and an INSULATOR. The conductivity increases as temperature increases. Adding appropriate impurities also increases conductivity. A semiconductor consists of elements, such as GERMANIUM and SILICON, or compounds, such as aluminium phosphide, with a crystalline structure. At normal temperatures, some ELECTRONS break free and become carriers of the electric current. The "holes" (electron deficiencies) left by these electrons move in the opposite direction to the electrons and behave like positive charge carriers. Impurities are usually added to the semiconductor material in controlled amounts during manufacture to add more free electrons or create more holes. This process is called **doping**. There are two types of semiconductor: the ***n*-type,** in which the current carriers (electrons) are negative, and the ***p*-type**, in which the current carriers are moving, positively-charged holes. A semiconductor junction is formed when there is an abrupt change from one type of impurity to the other. Such a ***p-n* junction** acts as a very efficient RECTIFIER and is the basis of the semiconductor DIODE. Semiconductors are used in COMPUTERS, TRANSISTORS and PHOTOELECTRIC CELLS. *See also* INTEGRATED CIRCUIT (IC)

semipermeable membrane Thin, sheetlike material that permits the passage of a SOLVENT (such as water) but not larger dissolved SOLUTES (such as salt and sugar). The property of permeability depends on the molecular diameter of the dissolved substance and the nature of the membrane. Common membranes include thin palladium foil, pig's bladder, copper ferrocyanide (cyanoferrate) and the walls of plant and animal cells. *See also* OSMOSIS

semiprecious stones GEMS worn as jewellery, but not classified as precious stones. Not a strict grouping, semiprecious stones include the minerals AMETHYST, GARNET, OPAL, TOPAZ, TOURMALINE, TURQUOIS, and the non-minerals AMBER, CORAL and JET. Precious stones are generally taken to be DIAMOND, EMERALD, SAPPHIRE, PEARL and RUBY.

Semmelweiss, Ignaz Philipp (1818–65) Hungarian physician, probably the first to recognize the importance of ANTISEPTICS in preventing infection. In 1847, during an outbreak of puerperal fever, he ordered doctors at the Vienna General Hospital to wash their hands before touching a patient. His ideas were ridiculed until confirmed by Joseph LISTER.

sense receptor *See* RECEPTOR

senses Means by which animals gain information about their environment and physiological condition. The five senses (SIGHT, HEARING, TASTE, SMELL and TOUCH) all rely on specialized RECEPTORS on or near the external surface of the body. SENSORY NEURONS carry information from the sense organs to the brain. Additionally, receptors within the body detect internal physical and chemical changes.

sensory neuron NERVE that carries information from a RECEPTOR organ in any part of the body to the CENTRAL NERVOUS SYSTEM (CNS). Their nerve endings are in the sense organs. *See also* MOTOR NEURON; NEURON; SENSES

sepal Modified leaf that makes up the outermost portion of a FLOWER bud. Although usually green and inconspicuous once the flower is open, in some species, such as orchids, the sepals look like the PETALS. In other species, such as anemones, the sepals are absent.

septic tank Unit for SEWAGE disposal, usually in rural locations not connected to sewers. Watertight and airtight, septic tanks are buried in the ground. ANAEROBIC bacteria in the sewage decompose solids to form a sludge, which is pumped out regularly.

sequence Numbers or terms in an organized list. Sequences may be finite (having a limited number of terms) or infinite, such as the complete sequence of counting numbers 1, 2, 3, 4, ... Infinite sequences may or may not be convergent (tend to a single limit). The sequence 0, 0.9, 0.99, 0.999, ... for example is convergent, tending to the limit 1. Sequences are sometimes called progressions. *See also* ARITHMETIC PROGRESSION; CONVERGENCE; DIVERGENCE; GEOMETRIC PROGRESSION; HARMONIC PROGRESSION; SERIES

séracs Very irregular ice surface on a GLACIER. It may include an impenetrable system of pillars and pinnacles several metres tall. A séracs usually forms at the foot of an icefall, caused by compression of a crevassed section of the glacier. This, in turn, results from an increase in the gradient of the moving ice or because its speed is somehow reduced.

sere SUCCESSION of plant communities that eventually produces the CLIMAX COMMUNITY. Any one sere has a series of seral communities that change with time. There are various recognized types of seres, including a **hydrosere** (occurring in shallow fresh water), **lithosere** (on rocky ground), **plagiosere** (cleared ground) and **xerosere** (dry ground).

series Mathematical expression (sum) obtained by adding the terms of a SEQUENCE. Thus, the series 1 + 4 + 9 + 16 + ... is formed from the sequence 1, 4, 9, 16, ... Like a sequence, a series may involve a finite or infinite number of terms and the infinite series may or may not converge. There are many useful series for the number π, which can be used to evaluate π to several decimal places. A series formed from increasing powers of a VARIABLE is a **power series**; such power series are used to represent many functions. *See also* SUMMATION

series circuit ELECTRIC CIRCUIT in which the components are connected in sequence so that the current flows through them one after the other. The total RESISTANCE, R, of a number of resistors connected in series is given by the expression $R = R_1 + R_2 + R_3 + ...$, where R_1, R_2 and R_3 are the resistances of the individual resistors. The total CAPACITANCE, C, of capacitors in series is given by $1/C = 1/C_1 + 1/C_2 + 1/C_3 + ...$, where C_1, C_2 and C_3 are the capacitances of the component capacitors. *See also* PARALLEL CIRCUIT

serine ($HOCH_2CH(NH_2)COOH$) White, crystalline AMINO ACID found in proteins.

serotonin Chemical found in cells of the gastrointestinal tract, blood platelets and brain tissue, concentrated in the midbrain and HYPOTHALAMUS. It is a VASOCONSTRICTOR and has an important role in the functioning of the nervous system and in the stimulation of smooth muscles.

Serpens (Serpent) Extensive equatorial constellation situated between Libra and Hercules and divided into two parts, Serpens Cauda and Serpens Caput, by Ophiuchus (the Serpent-bearer). Serpens Caput contains the bright globular cluster M5 (NGC 5904), and Serpens Cauda includes M16 (NGC 6611), an open cluster shining through surrounding nebulosity. The brightest star is Alpha Serpentii (Unukalhai), magnitude 2.7.

serpentine Group of sheet silicate minerals, hydrated magnesium silicate ($Mg_3Si_2O_5(OH)_4$). Serpentine minerals come in various colours, usually green, although sometimes brownish, with a pattern of green mottling. They have monoclinic system crystals. Serpentines are secondary minerals, formed from minerals such as olivine and orthopyroxene, and they occur in igneous rocks containing these minerals. They are commonly used in decorative carving; fibrous varieties are used in asbestos cloth. Hardness 2.5–4; r.d. 2.5–2.6.

serum Clear fluid that separates out if blood is left to clot. It is essentially of the same composition as PLASMA, but without FIBRINOGEN and clotting factors.

servomechanism Device that provides remote control to activate a mechanism. This is done by converting an input signal, such as a radio impulse or mechanical movement, into a mechanical output like a lever movement or amplified hydraulic force. The device usually forms part of a control system. Some servomechanisms, such as the AUTOPILOT of an aircraft, incorporate a FEEDBACK mechanism that makes it independent of human control.

sessile In zoology, describing an animal that remains fixed in one place. Such sedentary animals are usually permanently attached to a surface, such as sea anemones, barnacles, limpets and mussels. The term sessile is also used to describe the eyes of crustaceans that lack stalks and sit directly on the animal's head. In botany, sessile describes any structure that has no stalk (in cases where one might be expected) and grows directly from a stem. Examples include the acorns and leaves of some oak trees.

sets In mathematics, a defined collection of objects. The objects are called the elements or members of the set. The number of members can be finite or infinite, or even be zero (the number of members in the null set denoted $\varnothing$). Each element in a set is counted only once. Various relations can exist between two sets. Two sets, A and B, are equal ($A = B$) if both sets contain exactly

S

the same elements. A is a **subset** of B if all members of A are also members of B. A subset of A that is not equal to B is called a **proper subset** (written $A \subset B$). **Disjoint** sets have no members in common; **overlapping** sets have one or more members in common. The **union** of two sets A and B (written $A \cup B$) is a new set that contains all the members of both A and B. The **intersection** of A and B (written $A \cap B$) contains only those elements common to both A and B. For example, the set $A = \{3, 6, 9, 12, 15\}$ is the set of positive multiples of 3 less than 16; $B = \{2, 4, 6, 8, 10, 12, 14\}$ is the set of even numbers less than 16; C = {blue, green, red} is the set of primary colours; $D = \{1, 2, 3, 4, ...\}$ is the infinite set of natural NUMBERS. Both A and B are proper subsets of D ($A \subset D$ and $B \subset D$). There are no elements common to both A and C, so $A \cap C = \varnothing$. In contrast, A intersection B contains the even multiples of 3 less than 16, that is $A \cap B = \{6, 12\}$.

set theory Branch of MATHEMATICS developed by Georg CANTOR in the late 19th century. It is based on George BOOLE's work on mathematical LOGIC, but involves manipulations of SETS of abstract or real objects rather than logical propositions. It is concerned with the properties of sets, the relations between them and operations on them, and can be applied to many other branches of mathematics.

sewage Matter that flows through sewers, including drainage water and liquid and solid waste materials (notably faeces). Sewage works remove solids and treat the WATER so that it can be discharged into rivers, lakes or the sea without causing POLLUTION. The discharged liquid is called **effluent**. At a typical sewage plant, incoming sewage is screened free from stones and other hard materials. Heavy sludge is "settled off" and removed to closed digester vessels, in which bacteria break it down to a relatively inoffensive material that can be used as FERTILIZER. Liquid sewage passes on to trickle filters or activated sludge tanks, in which it is purified by the combined digestive action of bacteria and protozoa; chemical treatment may also be used. Further settling tanks remove the inactive HUMUS materials, after which the remaining water is fully purified and can be reused. *See also* CHLORINE; EUTROPHICATION

sex Classification of an organism denoting the reproductive function of the individual. There are two divisions of the classification – male and female. In mammals, the presence of sex organs (OVARIES in the female, TESTES in the male) are primary sexual characteristics. SECONDARY SEXUAL CHARACTERISTICS, such as size, coloration and, in human beings, the development of the breasts and the growth of facial and body hair, are governed by the secretion of SEX HORMONES. In flowering plants, the female sex organs are the CARPEL (including the OVARY, STYLE and STIGMA), and the male organs the STAMENS. Male and female organs may occur in the same flower or on separate flowers or plants. In lower organisms, such as fungi, where the differences are biochemical rather than morphological, the term "mating strains" is used instead of male and female. *See also* SEXUAL REPRODUCTION

sex determination Genetic basis for the gender of an organism. Gender depends on the combination of sex CHROMOSOMES. In mammals, which have a pair of sex chromosomes, there are two types, the X-chromosome and the Y-chromosome. All ova have one **X-chromosome**, whereas sperm have either an X- or a Y-. If the sperm that fertilizes the OVUM has an X-chromosome, the resulting ZYGOTE and the organism that develops from it will be XX; or female. If the egg is fertilized with a Y-carrying sperm, the zygote will be XY, or male.

sex hormones Chemical "messengers" secreted by the gonads (TESTES and OVARIES). They regulate sexual development and reproductive activity and influence sexual behaviour. In males, they include the ANDROGEN testosterone, made by the testes; in females, the sex hormones OESTROGEN and PROGESTERONE are produced by the ovaries.

sextant Optical instrument for finding LATITUDE (angular distance N or S of the Equator). The sextant consists of a frame with a curved scale marked in degrees, a movable arm with an ordinary mirror at the pivot, a half-silvered glass and a TELESCOPE. The instrument works on the principle that the angle of a celestial body above the HORIZON depends on the observer's latitude. While looking through the telescope, the sextant arm is adjusted until an image of the Sun or a star just touches an image of the horizon. The angle of the Sun or star above the horizon is then read from the scale. A set of tables gives the corresponding latitude for various angles measured.

sexually transmitted disease (STD) Any disease that is transmitted by sexual activity involving the transfer of body fluids. It encompasses a range of conditions that are spread primarily by sexual contact, although they may also be transmitted in other ways. These include ACQUIRED IMMUNE DEFICIENCY SYNDROME (AIDS), pelvic inflammatory disease (PID), and viral HEPATITIS. One effect of the AIDS crisis may have been to divert attention away from some of the older STDs, such as SYPHILIS and GONORRHOEA, which nonetheless remain significant public health problems.

sexual reproduction Biological process of REPRODUCTION involving the combination of genetic material from two parents. It occurs in different forms throughout the plant and animal kingdoms. This process gives rise to variations of the GENOTYPE and PHENOTYPE within a species. GAMETES, HAPLOID sex cells produced by MEIOSIS, contain only half the number of CHROMOSOMES of their parent cells (which are DIPLOID). At FERTILIZATION, the gametes, generally one from each parent, fuse to form a ZYGOTE with the DIPLOID number of chromosomes. The zygote divides repeatedly and the cells differentiate to give rise to an EMBRYO and, finally, a fully formed organism. In mammalian reproduction, the male's erect PENIS is inserted into the female's VAGINA. Rhythmic, thrusting movements of the penis cause the male

SERVOMECHANISM

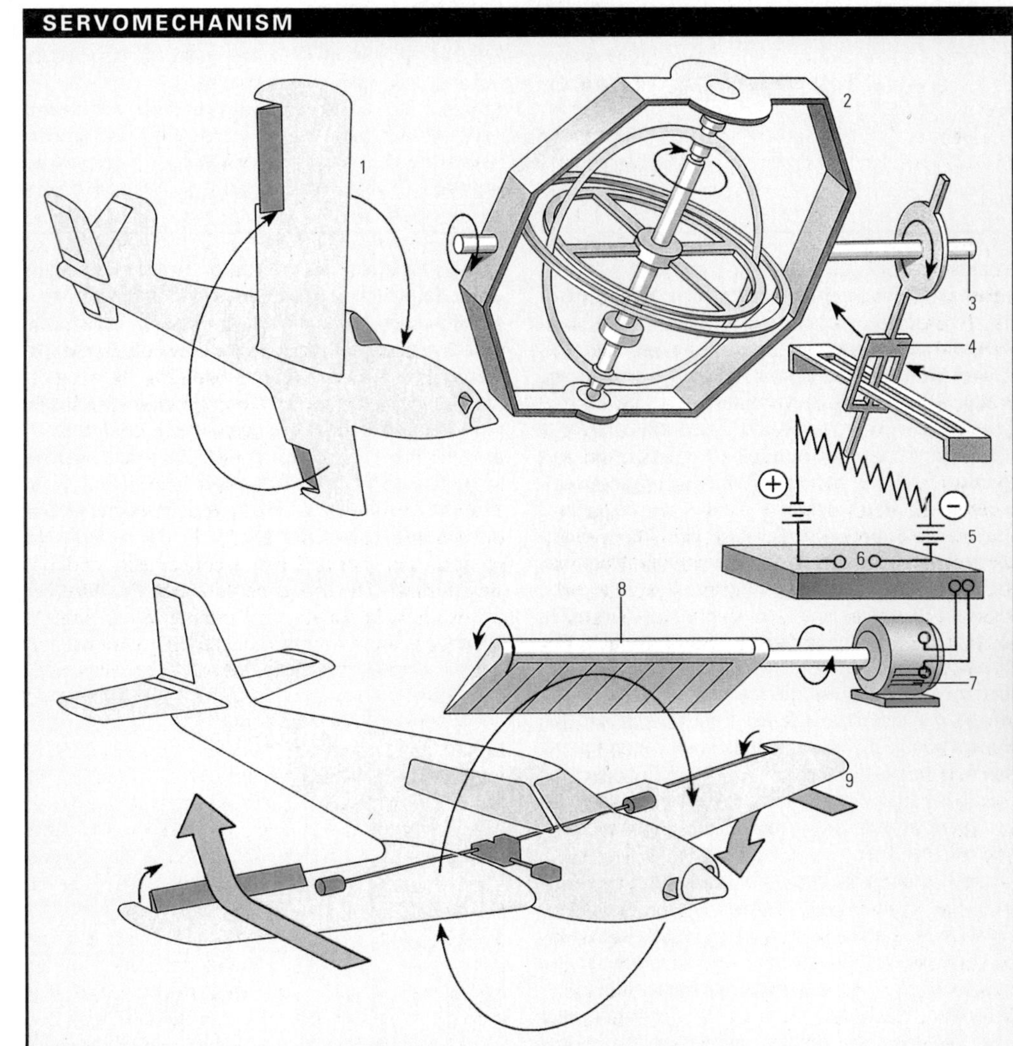

A servomechanism will automatically correct a deviation from a pre-set position of a machine. For example, the autopilots fitted to many aircraft utilize servomechanisms to manipulate the control surfaces of the plane. A change from the intended altitude of the plane – such as banking (1) will affect a gyroscope (2). This will move a linkage (3) so that a sliding contact (4) moves from a central position so that a current flows in a detector circuit (5). This current is amplified (6) and used to power a servo motor (7) – to move the ailerons (8) and correct the flight of the plane (9), returning the contact to a point where no current flows in the circuit (5). If the system was too sensitive, it could tip the plane the other way, reversing the current in the circuit and causing the plane to oscillate around the safe position – so called hunting.

S

SEXUAL REPRODUCTION

The male reproductive system is situated both inside and outside the pelvic region. Outside the body are the testes. They produce sperm in the seminiferous tubules. Once the sperm are matured they are stored in the epididymis. During intercourse they pass along the vas deferens to the urethra. The seminal vessicles, prostate and Cowper's glands all secrete fluids into the urethra, helping to make semen. The female reproductive organs lie solely within the pelvic girdle. The ovaries usually release just one mature egg each month, which is then transferred to the uterus via one or other of the Fallopian tubes. Embedded in the ovaries of the newly born female are several hundred thousand follicles that can potentially develop into eggs. Only a few hundred of them will do so, becoming mature eggs which, when they are ripe, are caught by the fimbria at the end of the tubes, near the site where fertilization usually occurs. The woman's urinary system, unlike the male's, is separate from the genitals.

to ejaculate and so introduce SEMEN (containing SPERM) into the vagina in order for fertilization of the OVUM to take place.

sexual selection Process among animals that derives from the successful finding of a mate and production of young. It is similar to NATURAL SELECTION but involves mainly the SECONDARY SEXUAL CHARACTERISTICS. For example, if a female chooses to mate with the male that has the brightest coloration and best courtship display, these features (being genetically controlled) will tend to be inherited by the male offspring. In succeeding generations the same features will be exaggerated. By the same mechanism, males may develop longer horns or louder roars to better compete with other rival males.

shadow Area screened from a light source and therefore relatively dark. It varies in size depending on the distance between the light source and the screening object. When the light source is extended, the outline of the shadow is blurred; where the source is partly visible, there is an area of mid-shadow (**penumbra**) lying outside the central, darker shadow (**umbra**).

shale Common SEDIMENTARY ROCK similar to clay. It is fine grained and consists of thin layers or sheets, with each layer probably representing a period of DEPOSITION. Shales may contain various materials such as fossils, carbonaceous matter and oil.

Shapley, Harlow (1885–1972) US astronomer who provided the first accurate model of our GALAXY. In 1914 Shapley explained how the variability of CEPHEID VARIABLES is caused by their pulsations. Later, he calibrated the period–luminosity law discovered by Henrietta LEAVITT, using Cepheid variables in GLOBULAR CLUSTERS, which enabled him to estimate of the size of our Galaxy and the position in it of the SOLAR SYSTEM. Shapley also made estimates of the sizes of ECLIPSING BINARIES, and worked on photometry and SPECTROSCOPY.

shearing force (shearing stress) Force tending to cause deformation of a material by slipping along a plane parallel to the imposed STRESS. In nature, the resulting shear is related to the downslope of Earth materials as well as to earthquakes.

shell In biology, hard protective case of various MOLLUSCS. The case is secreted by the epidermis of the mollusc and consists of a protein matrix strengthened by CALCIUM CARBONATE.

shellac Purified RESIN made from the secretions of the lac insect, *Laccifer lacca*. Shellac is soft and fluid (it flows) when heated but hard at room temperature. It is used in the production of adhesives, hair sprays, varnishes and sealants.

shell, atomic Any of the groupings of orbital ELECTRONS around the nucleus of an ATOM, named K, L, M, and so on, outward, each containing a limited number of electrons of a particular energy.

Shepard, Alan Bartlett, Jr (1923–98) US astronaut. In 1961 Shepard became the first American to be launched into space. The brief, sub-orbital flight to a height of *c.*185km (115mi) lasted just over 15

SHIP

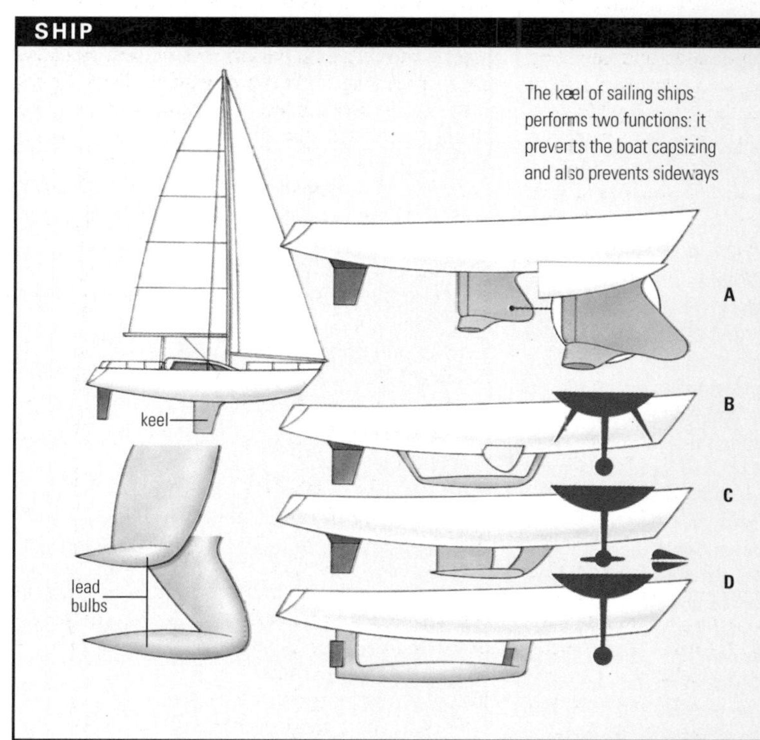

The keel of sailing ships performs two functions: it prevents the boat capsizing and also prevents sideways movement when the wind blows from the side of the boat. Large keels, however, produce drag that slows the vessel. One solution is to use lead bulbs on the end of the keel which reduce its size while still providing a counterbalance. In the America's Cup, other innovations included winglets to provide lift (A). In (B) the lead bulb providing counterbalance is suspended on thin supports with angled dagger boards preventing sideways drift. In (C) the winged bulblet is suspended on supports that are shaped to minimize drag. In (D) control tabs on the bulb supports eliminate the need for a separate rudder thereby reducing drag.

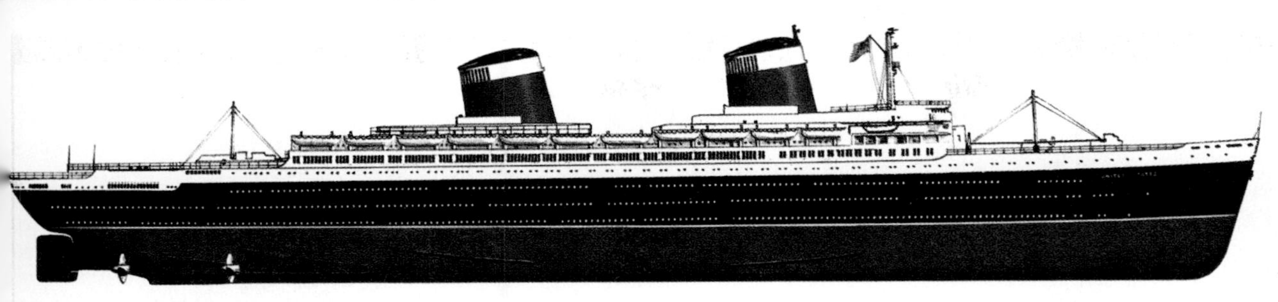

◀ **ship** The SS *United States* was launched in 1952. She had a cruising speed of 33 knots, and was the last passenger ship to hold (1952) the Blue Riband for the fastest crossing of the Atlantic (83hrs 20mins).

minutes. His *Freedom 7* capsule splashed down in the Atlantic Ocean nearly 480km (300mi) from the Cape Canaveral (Florida) launchpad. Shepard commanded *Apollo 14* on its flight to the Moon in 1971, and was the fifth man to walk on the lunar surface. Among the awards he received was the Space Medal of Honor.

sherardizing Industrial method of coating STEEL parts with ZINC for protection from CORROSION. The parts, which may be steel castings, nuts and bolts, welded tubes or sections, are tumbled in heated drums containing zinc dust. After some hours the zinc forms a very thin intermetallic surface coating the steel. *See also* GALVANIZING

Sherrington, Sir Charles Scott (1857–1952) British physiologist, pioneer in the study of how the NERVOUS SYSTEM works. His book *The Integrative Action of the Nervous System* (1906) helped to establish physiological psychology. Sherrington shared with Edgar Adrian the 1932 Nobel Prize for physiology or medicine.

shield In geology, large, low-relief, exposed mass of PRECAMBRIAN rock, commonly having a gently convex surface and surrounded by belts of younger rock. Shields contain rock that is more than 2,500 million years old, now changed by METAMORPHISM but originally composed of basaltic LAVA. Shields form the nucleus of continents; for example, the Canadian Shield, located between Hudson Bay and the Great Lakes, occupies two-thirds of Canada.

shield, radiation Material, such as loaded concrete (concrete containing lead, iron or barium) or lead, surrounding a radioactive source to absorb NEUTRONS and RADIATION. The core of a NUCLEAR REACTOR is surrounded by a massive concrete biological shield to protect personnel, and an inner iron, thermal shield to protect the concrete from high temperatures.

ship Vessel for conveying passengers and freight by water. Modern ocean-going ships developed from early **sailing ships**, such as the carracks (Mediterranean merchant ships) of the 13th century and the larger early ships called galleons. Fighting ships of the 17th and 18th centuries included frigates of various designs; later models had several rows of guns. Sailing freighters led to the development of the great **clippers** of the late 19th century, some of which had iron hulls. A century or so earlier, the first **steamships** had been built. They were powered by wood or coal-burning STEAM ENGINES that drove large paddle wheels, hence the term paddlesteamer. In 1819 the first Atlantic crossing was made by "steam-assisted sail" and this crossing soon became a regular service. By the mid-19th century steamships driven by PROPELLERS, or screws, were in general service and were voyaging on trade missions in all oceans. Marine steam TURBINES were developed at the turn of the 19th century and gradually replaced reciprocating (back-and-forth cranking) steam engines for large vessels, examples being ocean **liners** of the early 1900s with displacements (weight of water that a ship displaces when fully laden) up to 10,000 tonnes. Oil, rather than coal, was now the favourite fuel for large marine engines. Some of the newest military ships and icebreakers are fitted with nuclear engines in which heat from a NUCLEAR REACTOR raises steam in boilers to drive steam turbines. The larger modern ships are classified as warships, passenger vessels, bulk dry cargo ships, tankers for bulk liquids, freighters for mixed cargo and container ships. The largest of these are the supertankers carrying oil, with displacements of about half a million tonnes and lengths of several hundred metres. *See also* BOAT

shock Acute circulatory failure, possibly with collapse. Caused by disease, injury or emotional trauma, it is characterized by weakness, pallor, sweating, vomiting and a shallow, rapid pulse. In shock, there is a drop in blood pressure to a level below that needed to oxygenate the tissues. Treatment depends on the cause, but will involve prompt blood or fluid replacement if there has been severe loss.

shock absorber Elastic mechanism for absorbing shock in order to prevent damage to a machine. The shock absorbers of motor vehicles typically have a piston, attached to the vehicle body, moving inside a cylinder attached to a wheel. The piston moves against a liquid which escapes from and returns to the cylinder through a spring-loaded valve.

Shockley, William Bradford (1910–89) US physicist. During World War 2 he worked on anti-submarine research for the US navy. In 1947 Shockley and his colleagues, John BARDEEN and Walter BRATTAIN, invented a point-contact TRANSISTOR and a junction transistor at the Bell Telephone Laboratories. They shared the 1956 Nobel Prize for physics. As professor of electrical engineering (1958–75) at Stanford University, he caused controversy with his views on the link between race and intelligence.

shock wave In fluids (liquids or gases), region across which sharp discontinuities in pressure and density occur. Shock waves are brought about by objects moving at supersonic velocities. Because the surrounding fluid can propagate disturbances only at the local speed of sound, the moving object "piles up" the disturbances it is causing into a V-shaped "wake" attached to the object. The SONIC BOOM of supersonic aircraft is the passage of this shock wave past the eardrum. *See also* EARTHQUAKE; SEISMIC WAVES; SUPERSONIC FLIGHT

shooting star *See* METEOR

short circuit Unintended connection between two points in an ELECTRIC CIRCUIT, offering the easiest path for current to follow. A large current will flow and fire or damage may result if the voltage supply is not turned off quickly. FUSES or CIRCUIT BREAKERS are used to protect against the effects of short circuits.

short-sightedness *See* MYOPIA

shoulder blade Common name for the SCAPULA

shrub (bush) Woody, perennial plant that is smaller than a TREE. Instead of having a main STEM, it branches at or slightly above ground level into several stems. Its hard stem distinguishes it from a HERB. Shrubs such as rhododendrons and azaleas are popular ornamentals.

sial In geology, uppermost of the two main rock-classes in the Earth's CRUST. Sial rocks are so called because their main constituents are **si**licon and **al**uminium. They make up the material of the CONTINENTS and overlay the SIMA. The sial is less dense than the sima and is generally not carried beneath the surface when subduction occurs.

sickle-cell anaemia Inherited blood disorder featuring an abnormality of HAEMOGLOBIN. The haemoglobin is sensitive to a deficiency of oxygen and it distorts ERYTHROCYTES (red blood cells), causing them to become rigid and sickle-shaped. Sickle-cells are rapidly lost from the circulation, which gives rise to anaemia and jaundice. Particularly common in people of black African descent, the disease gives some protection against MALARIA.

sideband Any FREQUENCY produced by MODULATION that is added to a CARRIER WAVE. Those frequencies above the carrier wave are called **upper sidebands**; those below, lower sidebands. Sidebands contain the information that is passed from transmitter to receiver. Such FREQUENCY MODULATION (FM) minimizes interference and provides greater fidelity than AMPLITUDE MODULATION (AM). *See also* RADIO

sidereal period Orbital period of a planet or other celestial body with respect to a background star. It is the true orbital period. **Sidereal time** is local time reckoned according to the rotation of the Earth with respect to the stars. The sidereal day is 23 hours, 56 minutes and 4 seconds of mean solar time, nearly 4 minutes shorter than the mean solar day. The **sidereal year** is the time required for the Earth to complete one revolution around the Sun relative to the fixed stars. It is equal to 365.25636 mean solar days. *See also* ORBIT

siderite Green, brown or white mineral, iron(II) carbonate ($FeCO_3$), found in sedimentary iron ores

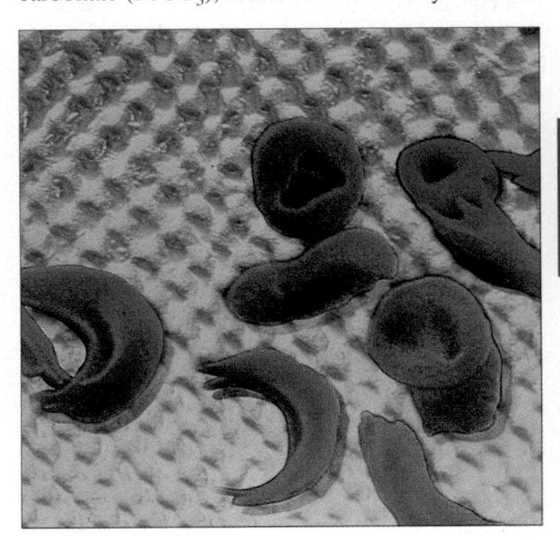

▲ **sickle cell anaemia** False-colour scanning electron micrograph of deformed erythrocytes (red blood cells) in sickle-cell anaemia, a largely hereditary blood disease. Sickle-cell anaemia is characterized by the production of abnormal haemoglobin (Hbs) in red blood cells. Hbs becomes insoluble when the blood is deprived of oxygen and precipitates, forming elongated crystals that distort the blood cell into this characteristic sickle shape. These deformed red cells are rapidly removed from the circulation, leading to anaemia.

S

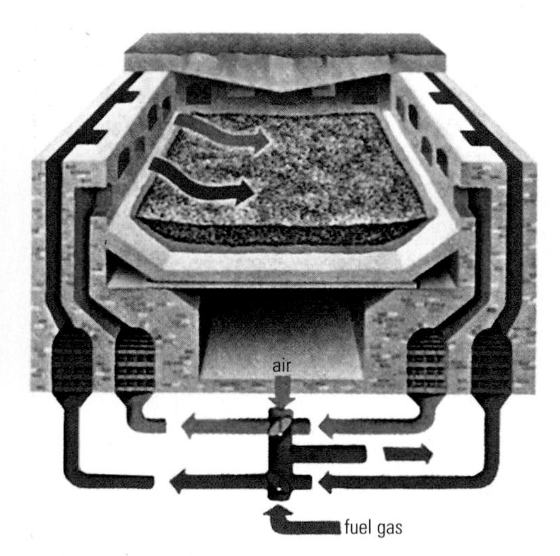

▲ **Siemens** The Siemens-Martin open-hearth process, in which streams of air and fuel gas were fed alternately onto the furnace contents, was used to make most of the steel during the 20th century, but has now been superseded by electric and basic oxygen methods. The open-hearth process used the gases from the molten charge to pre-heat the air blast and so economize on fuel. An alkaline lining was used if the ore contained phosphorus.

and as vein deposits with other ores. Its crystals are rhombohedral in the hexagonal system and it occurs as massive deposits or in granular form. 3.8. IRON METEORITES are sometimes known as siderites. Hardness 4; r.d.

Siegbahn, Karl Manne Georg (1886–1978) Swedish physicist who was awarded the 1924 Nobel Prize for physics for research into X-ray SPECTROSCOPY. Siegbahn improved techniques for measuring the WAVELENGTHS of X-RAYS and used these new methods to discover groups of X-rays both less penetrating and with longer wavelengths than had been previously thought possible. This discovery supported the SHELL THEORY of atomic structure of the Danish physicist Niels BOHR. In 1924 he also proved the similarity between X-rays and visible light by showing that X-rays could be refracted by a PRISM.

Siemens German brothers associated with the electrical engineering industry. In 1849 **Ernst Werner** von Siemens (1816–92) developed an electric TELEGRAPH system. With his brother, **Karl** (1829–1906), he set up subsidiaries of the family firm in London, Vienna and Paris. **Frederich** (1826–1904) and **Karl Wilhelm** (later Sir William) (1823–83) developed a regenerative FURNACE that led to the OPEN-HEARTH PROCESS uesd in the steel industry. In 1843 Karl Wilhelm introduced to Britain an ELECTROPLATING process he and Werner had developed.

siemens (symbol S) SI unit of electrical CONDUCTANCE. It is equal to the conductance of a circuit with a RESISTANCE of 1Ω (OHM).

sievert (symbol Sv) Derived SI unit of dose equivalent, equal to one joule per kilogram ($1Jkg^{-1}$). One sievert is equal to 100 rems. *See also* RADIATION

sight SENSE by which form, colour, size, movement and distance of objects are perceived. Essentially, it is the detection of light by the EYE, enabling the formation of visual images.

significant figures (s.f.) (significant digits) Digits of a number that express it to a desired accuracy, the last figure possibly being rounded up or down. Thus 2.871828 to six significant figures is 2.87183 (6s.f.); to three significant figures it is 2.87 (3s.f.).

Sikorsky, Igor Ivanovich (1889–1972) US aeronautical engineer and designer, b. Russia. In 1913 Sikorsky built and flew the first four-engined AIRCRAFT. In 1919 he emigrated to the USA. In 1923 he established the Sikorsky Aero Engineering Corporation, which was later merged with the United Aircraft Corporation. Sikorsky is celebrated for his devlopment of the HELICOPTER. In 1939 he designed the first commercially successful helicopter. In 1941 Sikorsky set a world endurance record for sustained flight in a helicopter.

silane (silicane, SiH_4) Colourless gas that is produced by the reduction of SILICON with lithium tetrahydridoaluminate. Silane is stable in the absence of oxygen, but is spontaneously combustible. It is insoluble in water. Silane is used for the removal of corrosion in nuclear reactor pipes. It is used often in reference to the entire class of **silicon hydride** compounds. Properties: r.d. (liquid) 0.68; m.p. −185°C (−301°F); b.p. −112°C (−170°F).

silica (SiO_2) Silicon dioxide, a compound of SILICON and oxygen. It occurs naturally as QUARTZ and CHERT (which includes flint). Silica and SILICATE minerals (silica combined with other elements) are the main constituents of 95% of all rocks and account for 59% of the Earth's crust. Silica is used in the manufacture of GLASS, CERAMICS and SILICONE.

silica gel Amorphous chemical compound. It is produced by separating out a sol of sodium silicate and then heating it to achieve partial dehydration. Because of its ability to absorb moisture silica GEL is used as a drying agent and as a CATALYST.

silicate Any of a large group of rock-forming minerals made up of silicon and oxygen in SiO_4 units bonded to various metals. The SiO_4 units may form single or double chains (as in PYROXENE and AMPHIBOLE), sheets (as in MICA), rings (as in BERYL) or ionic bonds to a metal (as in OLIVINE). Silicate minerals, such as FELDSPAR, GARNET and mica, form 95% of the material of the Earth's crust. GLASS is a mixture of silicates with small amounts of other substances. Sodium silicates are used as adhesives and in the production of detergents. QUARTZ (SiO) is also usually regarded as a silicate mineral.

silicon (symbol Si) Common, grey, nonmetallic element of group IV of the periodic table, first isolated in 1824 by J.J. BERZELIUS. Silicon is found only in combinations such as SILICA and SILICATE. After FELDSPAR it is the second most abundant element in the Earth's crust (27.7% by weight), being a common constituent of minerals, such as QUARTZ. Silicon is found in many plants and animals; it is the major constituent of the cell wall (test) of DIATOMS. It is produced commercially by heating sand and coke in a furnace. It is used as an alloy, and a purified form is used in the preparation of SILCONES. It is alo utilised in the production of GLASS. SILICON "CHIPS" are used extensively in the COMPUTER industry for TRANSISTORS and other SEMICONDUCTOR devices. Properties: at.no. 14; r.a.m. 28.086; r.d. 2.33; m.p. 1,410°C (2,570°F); b.p. 2,355°C (4,271°F); most common isotope ^{28}Si (98.21%).

silicon chip Small piece of SILICON etched to carry many tiny ELECTRIC CIRCUITS. Silicon chips are at the heart of most electronic equipment. A personal COMPUTER will contain many different types of chip, most notably the MICROPROCESSOR that is the "brain" of the computer. Chips are etched, layer by layer, onto slivers of pure silicon. Each layer is "doped" to give it particular electrical properties, and the combination of different layers form components such as TRANSISTORS, DIODES and their interconnections. Most chips are made by photographic etching processes, which impose a limit on the separation of components. New ways of etching chips using much

SILICON CHIP

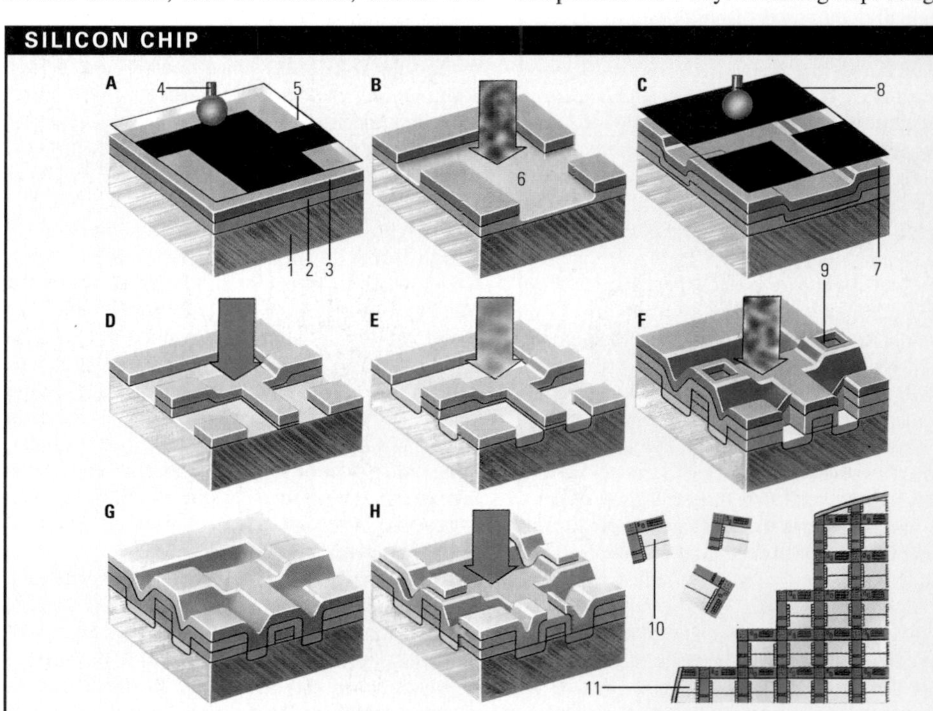

A silicon chip is manufactured by building up layers on top of a wafer of silicon (1). (A) First a layer of silicon dioxide (2), an insulator, is laid down followed by photo-sensitive photoresist (3). Photoresist hardens when hit by ultraviolet light (4). By using a mask (5) the area to be hardened can be controlled. (B) The unhardened area shielded by the mask can then be rinsed out with a solvent (6). The photoresist is then removed by hot gases. (C) The same process is used to apply a conducting polysilicon (7). Again ultraviolet light fixes the photoresist (8) in the unmasked area. (D) A solvent removes the photoresist. (E) N-type silicon, which only carries a negative charge, is then created by doping of the silicon base. (F) A third masking process creates shafts (9) to the n-type silicon. (G) An aluminium layer is then applied. (H) A fourth masking forms electrical contacts connecting the layers of silicon. Hundreds of chips are simultaneously made on a single wafer of silicon (10), before they are eventually separated (11) and mounted individually for use.

S

► **Light entering** the eye (A) is focused by the action of a lens (1) under the control of the ciliary muscles (2), which act on the suspensory ligaments (3). The image (4) formed by an object (5) on the retina is actually upside down, but the brain is able to correct this. The eyeball (A) is held in place in the orbit by muscles (6), which also allow it to move. It is covered in three layers of tissue: the sclera (7), a tough, fibrous coat; the choroid (8), which supplies nutrients, and is pigmented to reduce internal reflection; and the retina (9), where the light-sensitive cells are located. The front of the eye is protected by the transparent cornea (10) and conjunctiva (11. The aqueous (12) (behind the cornea) and vitreous humours (13) help to keep the shape of the eye and also contain blood vessels (14). The iris (15) can be dilated or contracted by muscles to control the amount of light entering the eye through the pupil (16). The optic nerve (17) carries the visual information into the brain. Where it exits the retina there are no receptors, and there is a blind spot (18). Most of the time, however, the brain can compensate. The light-sensing cells in the vertebrate retina (B) are the rods (1) and cones (2) – highly specialized nerve cells. The outer segment is comprised of membranous discs (3) containing a light sensitive pigment. The inner segment has a branched base (4) that links to nerve fibres. If sufficient light (5) is absorbed by the pigments an electrical signal is produced by the adjacent nerve fibre.

◄▼ Before reaching the brain via the optic nerve (6), messages pass through a series of neurones in the retina - horizontal cells (7), bipolar cells (8), amacrine cells (9), ganglion cells (10) - which organize the sensory information. Rods are found throughout the retina, except at the fovea, and are sensitive to different wavelengths of light. The degree to which different cones are stimulated gives the brain information about colour. Cones are highly concentrated at the fovea, giving great detail here where most images are focused. Each cone has its own connection to the brain so that very detailed information is received.

EYE DESIGNS

The simplest eyes, like those of flatworms (1), are just cups lined with a light-sensitive retina. In tube-worm eyes (2) each light receptor lies at the bottom of a pigmented tube. It receives only light from a particular angle and functions as a basic compound eye. The mirror eyes of scallops (3) form an image by reflection of incident light, in a similar way to a reflecting telescope. Shrimps and lobsters have a superimposition eye (4) in which mirrors channel light to form a single particularly bright image.

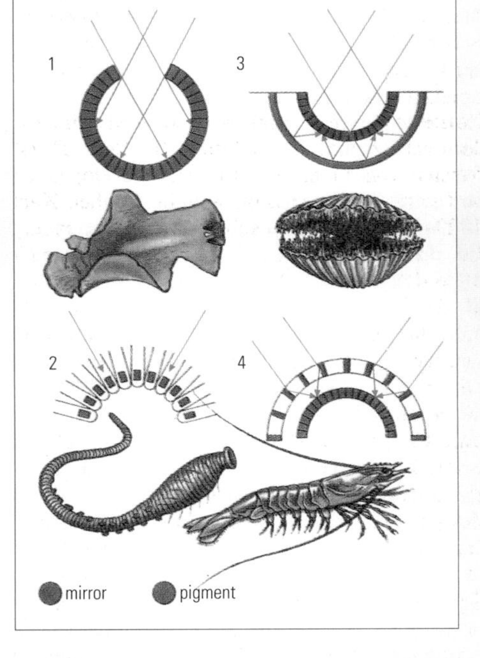

▲ **Insect compound eyes** (D and E) are made up of many individual units (from one up to 28,000) called ommatidia, which are arranged in hemispherical fashion. Each ommatidium (1) has its own cornea (2) and lens (3). A light sensitive region called the rhabdom (4) contains the visual pigment. This is surrounded by retinal cells (5) that transmit the electrical stimulus from the excited visual pigment to the brain. Cells containing screening pigment (6) prevent light entering one ommatidium infiltratin its neighbour. In the apposition eye of daytime insects (D), this means each ommatidium can only receive light from a small part of the whole field of view. Thus the whole image formed in the brain consists of the overlap of many adjacent spots of light. Nocturnal insects have a superposition eye (E), which is constructed in a very similar way to the apposition eye and acts in the same way during the day (F). In dim light, however, the pigment withdraws towards the outer surface of the eye (G), thus allowing diffracted light to reach the rhabdom from adjacent ommatidia as well. This produces an image that is brighter than it otherwise would be, though it may be less distinct.

► **The visible spectrum** (C) of deep-sea fish is limited to a little blue light. Other fish have a broad range. Many snakes can see far-red, using special pit organs, as well as ultraviolet. Birds and insects may also see into the ultraviolet. The primate range, including human beings, lies between red and blue.

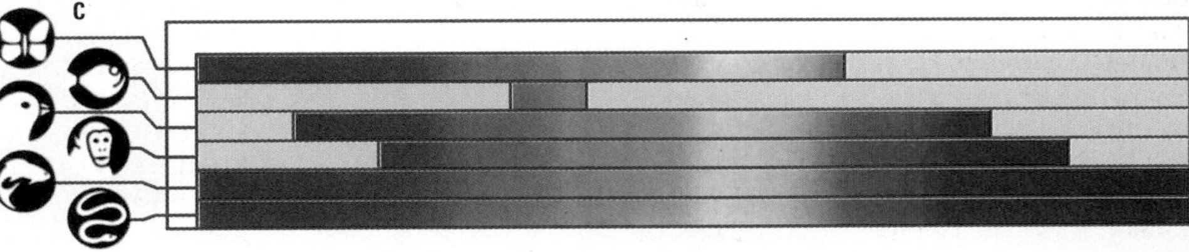

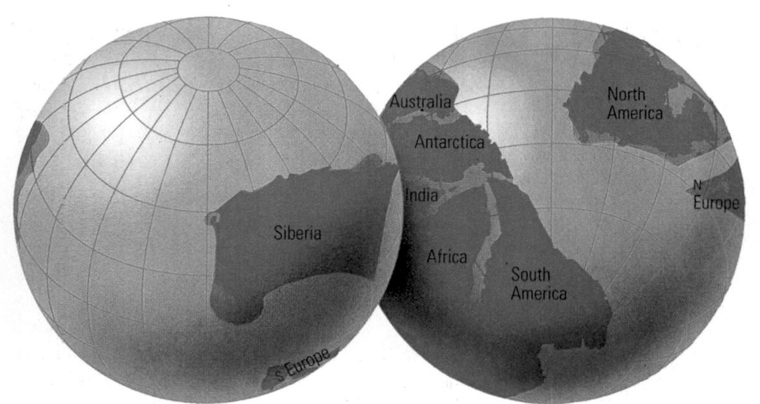

◀ **Silurian** The distance between North America and N Europe landmasses lessened considerably during the Silurian period. All the S continents had fused together forming the landmass known as Gondwanaland.

finer X-rays are being investigated. *See also* CHARGE-COUPLED DEVICE (CCD); INTEGRATED CIRCUIT (IC); PRINTED CIRCUIT; SEMICONDUCTOR

silicon dioxide *See* SILICA

silicone Odourless and colourless polymer based on SILICON. Silicones are inert and stable at high temperatures, and are used in lubricants, varnishes, adhesives, water repellents, hydraulic fluids and artificial heart valves. The modern use of silicone gel for cosmetic breast implants was drastically reduced in 1992 following health warnings.

silk Natural FIBRE produced by many creatures, notably the silkworm. Some other insects and spiders produce silk, but it is most economic to take it from the silkworm. Silk is a strong, high-quality material used to make fabric. The fibre consists mainly of fibroin (tough, elastic protein) covered with sericin (gelatinous protein). Almost all silk is obtained from silkworms reared commercially. Silkworms feed on mulberry leaves and a single COCOON can provide 600–900m (2,000–3,000ft) of filament (thread). Fibre is formed from a liquid that the caterpillar (*see* LARVA) produces from its spinning glands. It spins a silk cocoon around its body in preparation for its change into a silk moth. When the cocoons have been spun, the silk farmer heats them to kill the insects inside. The cocoons are then soaked to unstick the fibres, and the strands from several cocoons are unwound together to form a single thread of yarn. The yarn may be dyed before or after weaving into fabric. The Chinese were the first to use silk and its production was a closely guarded secret, since it could fetch its own weight in gold. Sicily was one of the first European production centres and the industry spread to Italy, Spain and France. Silk manufacturing developed in England in the 17th century. China is still the largest producer of raw silk in the world. *See also* TEXTILES

sill Sheet-like intrusion of IGNEOUS ROCK that is parallel to the bedding or other structure of the surrounding rock. Sill rock is normally medium-grained; basic sills (DOLERITES) are the commonest.

sillimanite (fibrolite) Mineral, aluminium silicate (Al_2SiO_5), found in mica SCHISTS and GNEISS. Its crystals are of the orthorhombic system, usually fibrous masses, and are satin-like or glossy white, brown, green or blue. A pale blue gem variety occurs in Sri Lanka. Hardness 6–7.5; r.d. 3.2.

silt Mineral particles, produced by the WEATHERING of rock. These particles, varying in size between grains of SAND and CLAY, are carried along in streams and RIVERS, to be deposited in the gently flowing lower reaches of rivers. When the river changes course or overflows its bank, the silt deposit forms very fertile land. Siltstone (flagstone) is a hard, durable stone which is formed from hardened silt. *See also* ALLUVIUM

Silurian Third oldest period of the PALAEOZOIC era, lasting from 438 to 408 million years ago. Marine invertebrates resembled those of ORDOVICIAN times, and fragmentary remains show that jawless fishes (agnathans) began to evolve. The earliest land plants (psilopsids) and first land animals (archaic mites and millipedes) developed. Mountains formed in NW Europe and Greenland. *See* GEOLOGICAL TIME

silver (symbol Ag) White, metallic element in the second series of TRANSITION ELEMENTS. It occurs in ARGENTITE (a sulphide) and horn silver (SILVER CHLORIDE), and is also obtained as a by-product in the refining of copper and lead. Silver ores are scattered throughout the world, Mexico being the major producer. It was one of the first metals to be crafted on by humans. Silver is an excellent conductor of heat and electricity and is used for some electrical contacts and on PRINTED CIRCUITS; other uses include jewellery, ornaments, coinage and mirrors. SILVER NITRATE is a light-sensitive material used in PHOTOGRAPHY. The metal does not oxidize in air, but tarnishes if sulphur compounds are present. Properties: at.no. 47; r.a.m. 107.868; r.d. 10.5; m.p. 961.93°C (1,763°F), b.p. 2,212°C (4,104°F); most common isotope ^{107}Ag (51.82%).

silver chloride (AgCl) White crystalline compound, insoluble in water; it occurs naturally as the mineral horn silver. It darkens on exposure to light and for this reason is, like silver bromide, used for making photographic emulsions.

silver nitrate ($AgNo_3$) Colorless, solid compound. It is the most important salt of SILVER because it is very soluble in water. Silver nitrate is used in PHOTOGRAPHY, chemical analysis, silver-plating, mirrors, inks, and dyes. It is extremely caustic. Properties: sp.gr. 4.3; m.p. 414°F.

sima In geology, undermost of the two main rock-classes that make up the Earth's CRUST, so called because its main constituents are **si**licon and **ma**gnesium. It underlies the SIAL of the CONTINENTS.

Simon, Pierre, Marquis de Laplace *See* LAPLACE, PIERRE SIMON, MARQUIS DE

simple harmonic motion (SHM) Periodic motion such as that of a PENDULUM, atomic vibrations or an oscillating electric circuit. A body has simple harmonic motion when it oscillates along a line, moving an equal distance on either side of a central point and accelerating towards that point with a speed proportional to its distance from it.

simultaneous equations Two or more equations that can be manipulated to give common solutions. In the simultaneous equations $x + 10y = 25$ and $x + y = 7$, the problem is to find values of x and y, such that these values are solutions of both the equations simultaneously. This can be done by rearranging and combining the equations to obtain each of the unknown variables in turn. In this example, one equation can be subtracted from the other to give $9y = 18$, hence $y = 2$. Substituting this value of y back into either of the equations gives $x = 5$. In general, such systems of equations may have no solutions, exactly one solution, or infinitely many solutions.

sine In TRIGONOMETRY, ratio of the length of the side opposite an acute angle to the length of the hypotenuse in a right-angled TRIANGLE. The sine of angle A is written sin$A$.

sink hole Hollow or hole in LIMESTONE formations which extends from the surface all or part of the way down to underground channels and caverns. Such holes are formed by water dissolving the limestone.

sintering Process in powder METALLURGY in which compressed particles are heated at temperatures below the melting point of the metal. This forms the particles into a coherent solid body. Glass and ceramics are nonmetals that may be sintered.

sinus Hollow space or cavity, usually in bone. Most often the term refers to the paranasal sinuses, any of the four sets of air-filled cavities in the skull

SIPHON

Bottled beverages lose their effervescence a short time after the bottle is opened, because the carbon dioxide (used to pressurize the bottle) soon bubbles off under reduced pressure. The soda water siphon provides effervescence cheaply over long periods. The siphon is filled with water and charged with compressed carbon dioxide from a small cylinder. Most of the gas dissolves, under pressure, in the water. Pressure is maintained by the bubbling out of dissolved gas. When the lever is pressed, pressure forces liquid up the siphon tube to the spout where gas bubbles out of the liquid under the reduced pressure.

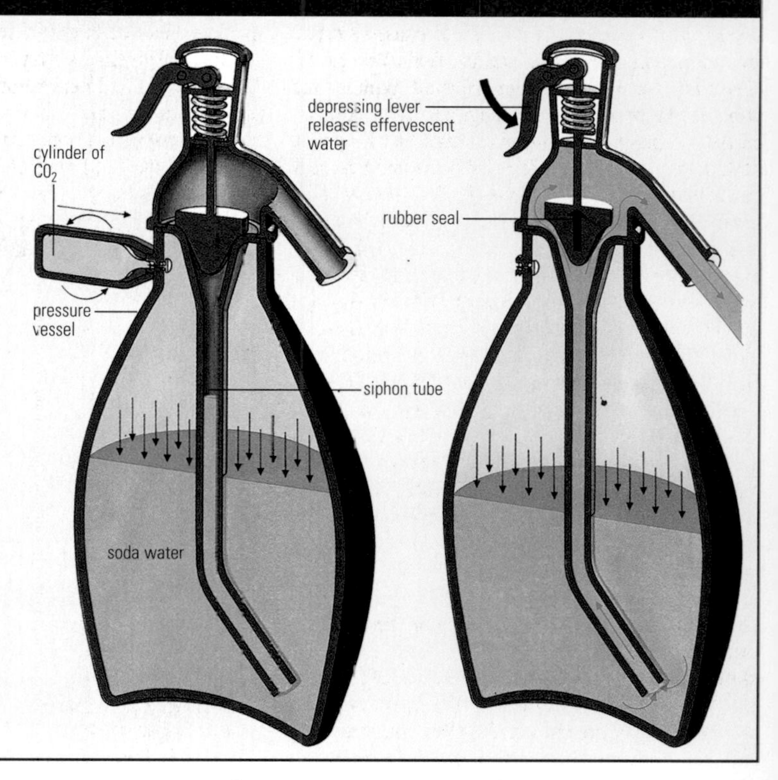

SKIN

The skin consists of epidermal and dermal layers supported by a subcutaneous layer of fat cells. The epidermis is subdivided into a cornified layer of dead, flattened cells, beneath which lies a granular layer containing living, dividing cells. Pigment-producing melanocytes, which colour the skin, lie below this third layer. The thicker dermis consists of connective tissues in which are embedded lymphatics, nerve fibres, sensory nerve endings, capillaries, sweat glands and hair follicles. A sebaceous gland and an erector muscle accompany each follicle.

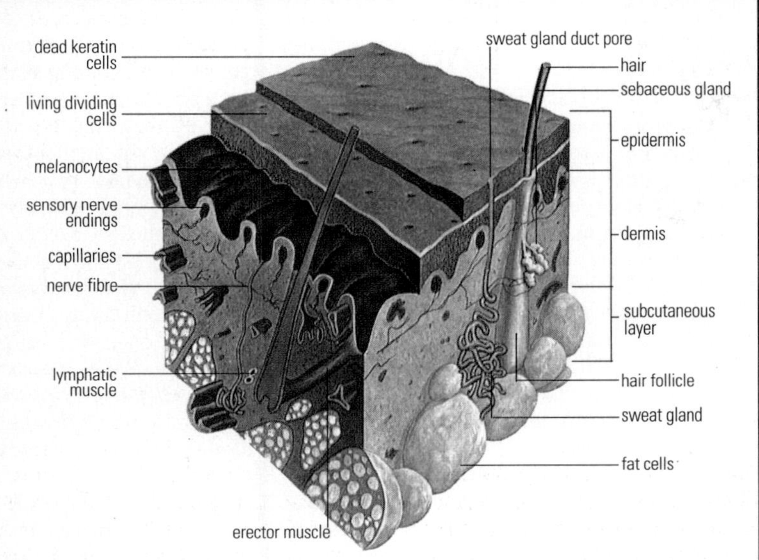

near the nose.

siphon Principle by which a liquid is raised from a container at one level and delivered to a lower level; also any inverted U-shaped tube used for this purpose. The tube is filled with the liquid, and one end of it is placed below the surface of the liquid in the container. When the other end hangs below the level of the surface (outside the container), liquid flows out due to the imbalance between the weights of the portions of liquid in the two arms of the tube.

Siphonaptera Order of small, wingless insects comprising the **fleas**. They are blood-sucking, external parasites on warm-blooded animals. Fleas have hard, laterally compressed bodies with many microscopic, backward-pointing spines. Their mouthparts are adapted for piercing their host and sucking up blood. Their powerful, bristly legs enable them to perform single leaps of more than 300mm. Flea eggs are laid in dirt or in the nest of the host; the whitish larvae feed on organic matter and the feces of adult fleas. The larvae spin cocoons as part of the METAMORPHOSIS into adults. Many species can live on more than one host species. Adults can survive for several weeks without feeding. Fleas cause irritation and some carry diseases. The rat flea (*Xenopsylla cheopsis*), for instance, can transmit typhus and bubonic plague to humans. Length: to 1cm (0.4in).

Sirius (Alpha Canis Majoris, Dog Star) Brightest STAR in the sky (magnitude −1.47), situated in the N constellation of CANIS MAJOR. Its luminosity is 23 times that of the Sun. Sirius is a main-sequence star, and is the sixth-closest star system to us. It has a binary companion, Sirius B, the existence of which was deduced by Friedrich BESSEL in 1844 from its gravitational PERTURBATIONS on Sirius. Sirius B is a WHITE DWARF, the first star to be recognized as such.

SI units (*Système International d'Unités*) Internationally agreed system of units, derived from the MKS SYSTEM (metre, kilogram and second). SI units are now used for many scientific purposes and have replaced the FPS SYSTEM (foot, pound and second) and cgs system (centimetre, gram and second). The seven basic units are: the METRE (m), KILOGRAM (kg), SECOND (s), AMPERE (A), KELVIN (K), MOLE (mol) and CANDELA (cd).

skeletal muscle Alternative name for VOLUNTARY MUSCLE

skeleton Bony framework of the body of a VERTEBRATE. It supports and protects the internal organs, provides sites of attachment for MUSCLES and a system of levers to aid locomotion. In the HUMAN BODY the skeleton consists of 206 BONES and is divided into two parts. The **axial** skeleton, or main axis of the body, includes the SKULL, the SPINAL COLUMN, the STERNUM (breastbone) and the RIBS. The **appendicular** skeleton, serving for the attachment of limbs, includes the shoulder girdle and arm bones and the pelvic, or hip, girdle and leg bones. The external skeleton of some insects is called an EXOSKELETON. *See also* ENDOSKELETON

skin Tough, elastic outer covering of the body, serving many functions. It is sometimes regarded at the largest organ of the body. The skin protects the body from injury and from the entry of some microorganisms and prevents dehydration. Nerve endings in the skin provide the sensations of TOUCH, warmth, cold and pain, each perceived at discrete points on the surface. It helps to regulate body temperature through sweating, regulates moisture loss and keeps itself smooth and pliable with an oily secretion from the SEBACEOUS GLANDS. Structurally, the skin consists of two main layers: an outer layer (EPIDERMIS) and an inner layer (DERMIS). The top layer of epidermis is made of closely packed dead cells constantly shed as microscopic scales. Below this is a layer of living cells that contain pigment and nerve fibres, and which divide to replace outer, shed layers. The dermis contains dense networks of connective tissue, blood vessels, nerves, glands and hair follicles.

Skinner, Burrhus Frederic (1904–90) US psychologist. Skinner developed the concept of operant CONDITIONING (the control of behaviour by its consequences or reinforcements). His many books include *Science and Human Behaviour* (1953), *Beyond Freedom and Dignity* (1971) and a controversial novel about social engineering, *Walden II* (1948).

skull Skeleton of the head. In mammals, it comprises the CRANIUM casing of the BRAIN and the facial and jaw bones. Although, technically speaking, the MANDIBLE is not part of the skull. There are 14 facial bones, mostly in symmetrical pairs. These are the zygomatic arches (cheekbones), the lacrimals around the eye socket, the nasals, nasal conchae and vomer (nasal septum) of the nose, the palatines (palate) of the mouth and the maxillae (upper jaw). The adult human cranium is formed of fused, skull bones with immovable joints: the frontal, occipital, temporals, parietals, ethmoid and splenoid. The occipital bone at the base of the skull forms a joint with the first (atlas) vetebra of the neck.

slag In the production of IRON and STEEL, liquid layer of ore impurities such as oxides, ash and limestone. It floats on molten steel during refining, protecting the metal from oxidation and removing some unwanted substances.

slate Grey to blue, fine-grained, homogeneous METAMORPHIC ROCK, which splits into smooth, thin layers. It is formed by the metamorphosis of SHALE, and is valuable as a roofing material.

sleep Periodic state of unconsciousness from which a person or animal can be roused. During an ordinary night's rest there are intervals of deep sleep associated with RAPID EYE MOVEMENT (REM). It is during this REM sleep that dreaming occurs. Although its importance is not known, regular sleep is essential. Studies have shown that people deprived of sleep become grossly disturbed. Sleep requirement falls sharply in old age. Difficulty in sleeping is called insomnia.

slide rule Calculating device consisting of two rules engraved with logarithmic scales of numbers, one of which slides next to the other. See also LOGARITHMS

slime mould Any of a small group of strange, basically single-celled organisms that are intermediate between the plant and animal kingdoms. During their complex life cycle they pass through several stages. These include a flagellated swimming stage, an AMOEBA-like stage, a stage consisting of a slimy mass of PROTOPLASM with many nuclei, and a flowering sporangium stage. *See* PROTOCTISTA

small intestine Part of the DIGESTIVE SYSTEM that extends – about 6m (20ft) coiled and looped – from the STOMACH to the large INTESTINE, or colon. Its

SLIME MOULD

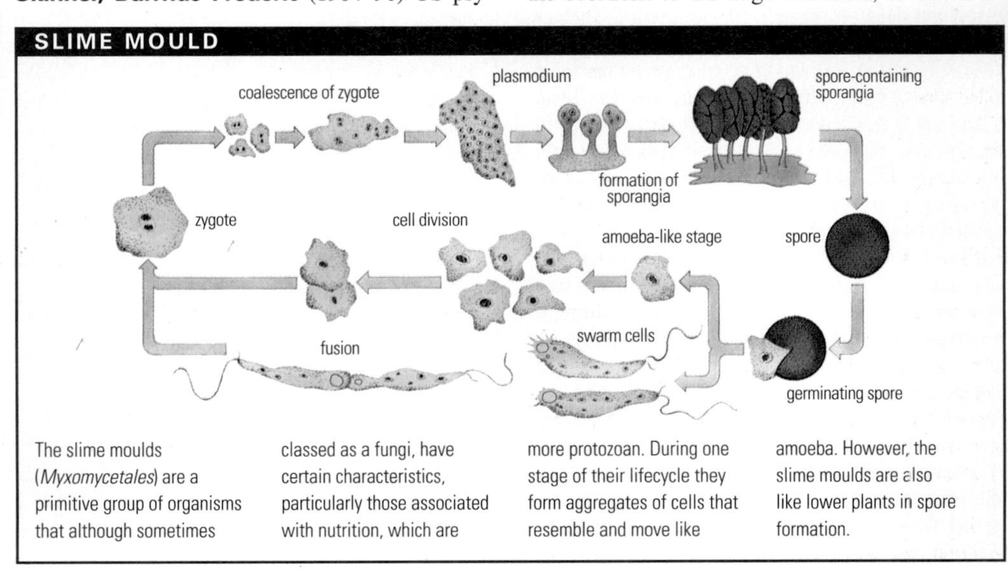

The slime moulds (*Myxomycetales*) are a primitive group of organisms that although sometimes classed as a fungi, have certain characteristics, particularly those associated with nutrition, which are more protozoan. During one stage of their lifecycle they form aggregates of cells that resemble and move like amoeba. However, the slime moulds are also like lower plants in spore formation.

S

SMOKE DETECTOR

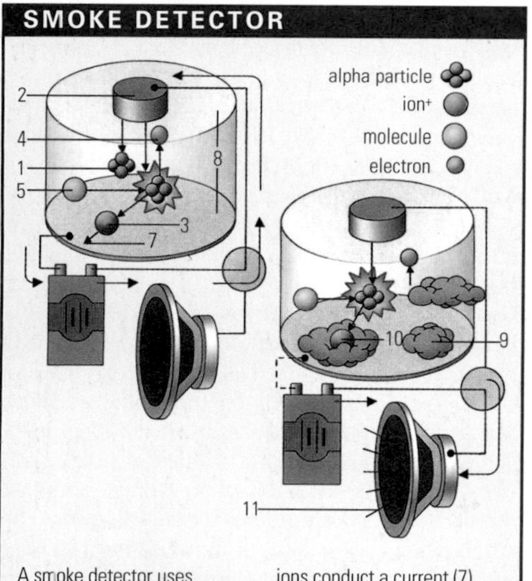

A smoke detector uses radiation to detect the presence of smoke. Alpha particles (1) are emitted from a radioactive source (2). Positively charged ions (3) and electrons (4) are created when the alpha particles react with molecules (5) in the air. The ions conduct a current (7) between two plates (8). When smoke particles (9) are present between the plates they absorb the ions (10) weakening the current. Electronic circuits register the drop in current and an alarm sounds through a loudspeaker (11).

function is the digestion and absorption of food. *See also* DUODENUM; ILEUM

smart card Plastic card that incorporates a computer MICROPROCESSOR and a small memory. Visually resembling a bank cheque card, it can hold various data, such as personal identification details and the status of a bank account. It can then be used, by means of a personal identification number (PIN), as a store card, credit card or debit card. It can have credits "stored" on it, which can then be spent through electronic tills (for example, for telephone calls or fares on public transport) and be replenished as required.

smell (olfaction) SENSE that responds to airborne molecules. The olfactory RECEPTORS can detect even a few molecules per million parts of air. There are different receptors for different chemical groups.

smelling salts (sal volatile) Mixture of ammonium carbonate crystals, alcohol and a perfume, often lavender. This sharp fragrance was formerly a common stimulant or restorative held to the nose of a person who had fainted or felt weak, dizzy or nauseous.

smelting Heat treatment for separating METALS from their ORES. The ore, often with other ingredients, is heated in a FURNACE to remove nonmetallic constituents. The metal produced is later purified. When iron is smelted in a BLAST FURNACE, for example, the ore (iron oxide) is heated with COKE and limestone. The oxide combines with carbon (from the coke) to form gaseous oxides of carbon and pig iron (an alloy of carbon, iron and other impurities).

Smith, Michael (1932–) Canadian biochemist, b. Britain. He discovered site-specific mutagenesis, a method of altering the GENETIC CODE through specific MUTATIONS instead of the previous random ones. This method has enabled the production of new proteins with a variety of functions. For this work, Smith shared the 1993 Nobel Prize for chemistry with Kary MULLIS.

Smithson, James (1765–1829) British chemist and mineralogist; the mineral SMITHSONITE was named after him. Angered at the Royal Society's rejection of a paper by him in 1826, Smithson left £105,000 to found an institution, the Smithsonian Institution in Washington, D.C., for the "increase and diffusion of knowledge among men". Today, it is one of the leading US scientific research and education centres.

smithsonite Carbonate mineral, zinc carbonate ($ZnCO_3$). It is found in ore deposits of zinc minerals, commonly associated with sphalerite, hemimorphite, galena and calcite. Its crystals, which are rare, are rhomobohedral in the trigonal system; it usually occurs as masses. Smithsonite is generally white, but may be other colours; blue specimens from New Mexico have been used as gemstones. Hardness 4–4.5; r.d. 4.4.

smog Dense, atmospheric mixture of smoke and FOG or chemical fumes, commonly occurring in urban or industrial areas. Smogs generally occur when there is radiation fog. If the air remains calm, the fog gets worse, and more industrial grime accumulates, as it is unable to escape into the atmosphere. The water droplets condense around the pollutants forming thick smog. One of the worst areas for smog is Los Angeles, California. It is a result of the exhaust emissions of the city's vast number of cars. The air pollution is exacerbated by Los Angeles' topography: it lies in a large basin, noted for inversions of temperature, which create ideal conditions for the formation of fogs and smogs.

smoke detector Device that gives a loud alarm when smoke particles enter it. There are various types of smoke detector. One type contains a PHOTOELECTRIC CELL. Smoke entering the detector cuts down on the amount of light entering the cell, thereby reducing the amount of current flowing through the cell. The drop in current sets off the alarm. In another type of detector, a steady current is sustained by IONS produced by a tiny radioactive source. Smoke particles become attached to the ions, reducing the current and setting off the alarm.

smooth muscle *See* INVOLUNTARY MUSCLE

snow Type of PRECIPITATION consisting of water vapour that has frozen into ICE crystals. Several ice crystals join together to form a **snowflake**, which will gradually fall to Earth. Snowflakes, which are symmetrical (usually hexagonal) crystalline structures, may melt as they fall, especially if the temperature is near freezing point. In such conditions, **sleet** occurs and CLOUDS yield snowfall on mountains, while producing rainfall on adjacent lowland areas. Snow is dry and powdery if it has come from a dryish area, and wetter if its source is over the sea. Interior regions, such as prairies, generally have dry snow, which is blown about in blizzards. *See also* RAIN

soap Cleansing agent made of salts of FATTY ACIDS, used to remove dirt and grease. Common soaps are produced by heating fats and oils with an alkali, such as sodium hydroxide or potassium hydroxide. Soap consists of long-chain molecules, one end of which attaches to grease while the other end dissolves in the water, causing the grease to loosen and form a floating scum. *See also* DETERGENT

soapstone (steatite) Rock with a soft soapy or greasy texture. There are many types of soapstone but all contain a large proportion of magnesium silicate, often associated with various amounts of serpentine and carbonates. Food vessels and carvings made from soapstone have been found among the remains of prehistoric human cultures. *See also* TALC

Sobrero, Ascanio (1812–88) Italian chemist who discovered (1846) the explosive NITROGLYCERINE (glyceryl trinitrate) by adding glycerine to a mixture of nitric and sulphuric acids. Horrified at the potential of his discovery, Sobrero made no attempt to develop it. *See also* EXPLOSIVE

SOIL PROFILE

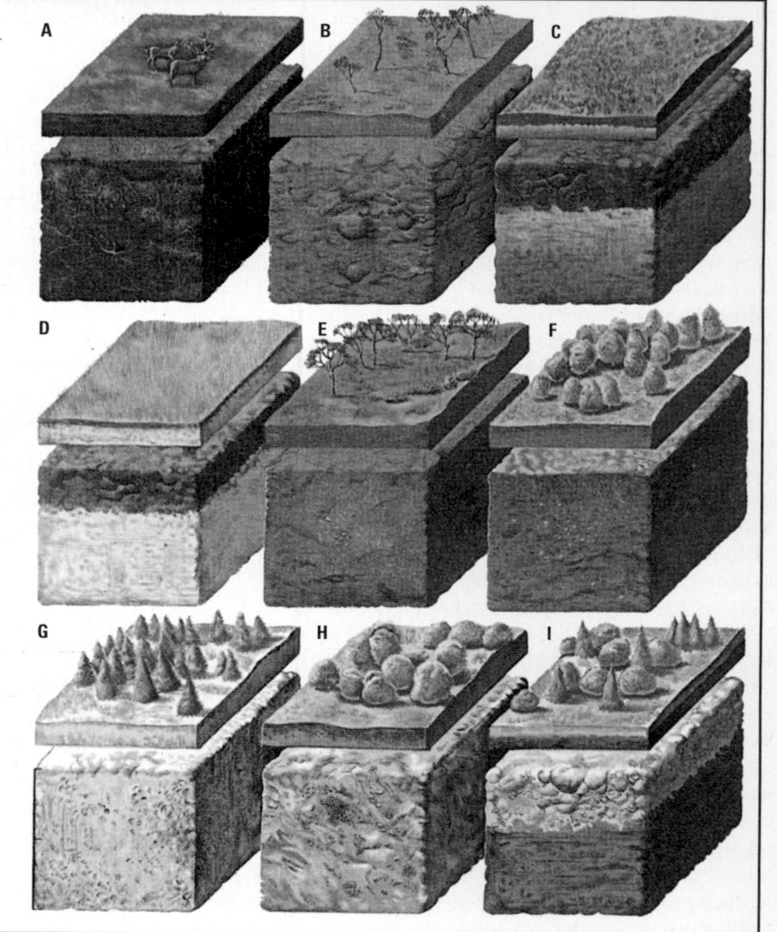

The composition and colour of a soil identifies it to a pedologist. This tundra soil (A) has a dark, peaty surface. Light-coloured, desert soil (B) is coarse and poor in organic matter. Chestnut-brown soil (C) and chernozem (D) – the Russian for "black earth" – are humus-rich grassland soils typical of the steppes and the prairies of North America. The reddish, leached latosol (E) of tropical savannas has a very thin, but rich humus layer. Podzolic soils are typical of N climates, where rainfall is heavy but evaporation is slow. They include the organically rich brown forest podzol (F), the grey-brown podzol (H) and the grey-stony podzol (I) that supports mixed growths of conifers and hardwoods. All are relatively acidic. The red-yellow podzol (G) of pine forests is quite highly leached.

soda Any of several sodium compounds, especially SODIUM CARBONATE and SODIUM HYDROGENCARBONATE.

sodalite Glassy silicate mineral, sodium aluminium silicate with some chloride, found in alkaline IGNEOUS ROCKS. It occurs as small dodecahedral crystals in the cubic system and also as masses. It may be colourless, white, blue or pink; it is sometimes used as gem. Hardness 5–6; r.d. 2.2.

Soddy, Frederick (1877–1956) British chemist. He was awarded the 1921 Nobel Prize for chemistry for his theory and definition of ISOTOPES, a term he coined. Soddy conducted research in RADIOACTIVITY with Ernest RUTHERFORD, and later Sir William RAMSAY. In 1920 he produced an explanation of the RADIOACTIVE DECAY of isotopes that proved invaluable in calculating GEOLOGICAL TIME. Soddy and Ramsay found HELIUM to be a product of URANIUM decay. His classic science books include *The Interpretation of Radium* (1909), *Matter and Energy* (1912), and *The Chemistry of the Radio-Elements* (1911–14).

sodium (symbol Na) Common, silver-white metallic element, one of the ALKALI METALS, first isolated (1807) by Sir Humphry DAVY. It occurs in the sea (as salt) and in many minerals. Its chief source is SODIUM CHLORIDE (common salt), from which it is extracted by ELECTROLYSIS. Sodium is used in the manufacture of tetraethyl lead, an anti-KNOCK compound for motor engines. It also used in arc lamps for street lights, and as a heat-transfer medium in nuclear reactors. Sodium compounds, such as SODIUM CARBONATE and SODIUM CHLORIDE, have wide industrial uses such as baking and washing sodas, fertilizers, and soap. It is an essential part of the human diet. Chemically, sodium is one of the most reactive elements. Properties: at.no. 11; r.a.m. 22.9898; r.d. 0.97; m.p. 97.81°C (208.05°F); b.p. 882°C (1,620°F).

sodium bicarbonate *See* SODIUM HYDROGENCARBONATE

sodium carbonate (soda, washing soda) White, crystalline salt (Na_2CO_3), usually manufactured from SODIUM CHLORIDE (common salt, NaCl) and AMMONIA by the SOLVAY PROCESS. The anhydrous (lacking water) form (Na_2CO_3) is known as soda ash; washing soda is hydrated sodium carbonate ($Na_2CO_3.10H_2O$)

sodium chlorate White, crystalline compound ($NaClO_3$), a strong OXIDIZING AGENT. It is used as a general herbicide and soil sterilizer and as a laboratory reagent. It is made by the ELECTROLYSIS of a solution of SODIUM CHLORIDE.

sodium chloride (NaCl) Common salt. It is the major mineral component of sea water, making up 80% of its dissolved materials. Sodium chloride is also the major ELECTROLYTE of living cells, and the loss of too much salt, through evaporation from the skin or through illness, is dangerous. It is used as a seasoning, to cure and preserve foods and, in the chemical industry, to produce SODIUM, CHLORINE, and SODIUM HYDROXIDE. Sodium chloride is an essential part of an animal's diet, and is a constituent of body fluids, such as blood, sweat and tears. Salt helps digestion by contributing to hydrchloric acid found in the stomach. Sodium chloride is combined with AMMONIA in the SOLVAY PROCESS to produce SODIUM CARBONATE.

sodium hydrogencarbonate (sodium bicarbonate, $NaHCO_3$, popularly known as bicarbonate of soda) White, crystalline salt that decomposes in acid or on heating to release carbon dioxide gas, which explains its use in baking powder. It has a slightly alkaline reaction, hence its other main application, as an ingredient of indigestion powders and tablets.

sodium hydroxide (caustic soda, NaOH) Strong ALKALI, prepared industrially by the ELECTROLYSIS of salt (sodium chloride, NaCl). It is a white solid that burns the skin, with a slippery feel because it absorbs moisture from the air. It also absorbs atmospheric carbon dioxide, forming a crust of sodium carbonate. Sodium hydroxide is used in many industries on a large scale, such as in SOAP-making to saponify fats (turn them into soap by alkali treatments) and in bauxite-processing to manufacture ALUMINIUM.

sodium thiosulphate ("hypo", $Na_2S_2O_3.5H_2O$) Chemical used as a photographic fixer. It is usually acidified by the addition of sodium metabisulphite.

softener, water *See* WATER SOFTENER

software COMPUTER PROGRAM and any associated data file. The term software is used to distinguish these coded instructions and data from computer HARDWARE, or equipment. In most cases, suppliers provide software on MAGNETIC DISK or CD-ROM for transfer to the built-in HARD DISK on the user's COMPUTER. Software can also be transferred via the INTERNET.

soil Surface layer of loose material resting on top of the rock which makes up the surface of the Earth. It consists of undissolved minerals produced by the WEATHERING and breakdown of surface rocks, organic matter, water and gases. The organic remains provide the HUMUS and the inorganic particles provide vital minerals. The inorganic fraction of the soil includes CLAY, SILT, SAND, GRAVEL and STONE. Soils are classified by structure and texture. The structure is determined by the aggregation of particles (peds). The four main textures of soil are sand, silt, clay and LOAM. Loam soils are best for cultivation, since they are able to retain more water and nutrients. Erosion and mismanagement are the chief causes of soil infertility. Fertility can be restored with the correct use of FERTILIZERS. A SOIL PROFILE reveals a number of distinct SOIL HORIZONS (layers).

soil horizon Layer of SOIL that shows in a SOIL PROFILE – a cross-section of the soil. Usually soil is divided into three horizons, designated A, B and C. A is the topsoil, B the subsoil and C the BEDROCK. Fine particles and organic HUMUS make up the **A horizon**, much more inorganic material and larger particles occur in the **B horizon**, and the **C horizon** is ROCK. There may be a layer of REGOLITH between the B and C horizons. *See also* CHERNOZEM; PODZOL

soil profile Vertical view of layers of SOIL from the surface down to the unaltered parent material. It is used in classifying soils. A layer of soil in a soil profile is known as a SOIL HORIZON.

soil testing Process of analyzing the chemical and mineral composition and porosity of SOIL. It is carried out for various reasons, including to determine the soil's ability to support plant life.

sol Colloidal SUSPENSION of particles of a solid in a liquid, such that the suspension remains liquid and does not solidify as in a GEL. *See also* COLLOID

solar cell Device that converts sunlight directly to ELECTRICITY. It normally consists of a ***p*-type** silicon crystal coated with an ***n*-type** one (*see* SEMICONDUCTOR). Light radiation causes electrons to be released and creates a POTENTIAL DIFFERENCE so current can flow between electrodes connected to the two crystals. All wavelengths shorter than one micrometre can create electrical energy. Cells convert *c*.10% of sunlight into useful energy. Solar cells are often used to power small electronic devices such as pocket calculators. Several thousand cells may be used in panels to provide power of a few hundred watts. *See also* SOLAR ENERGY

SOLAR CELL

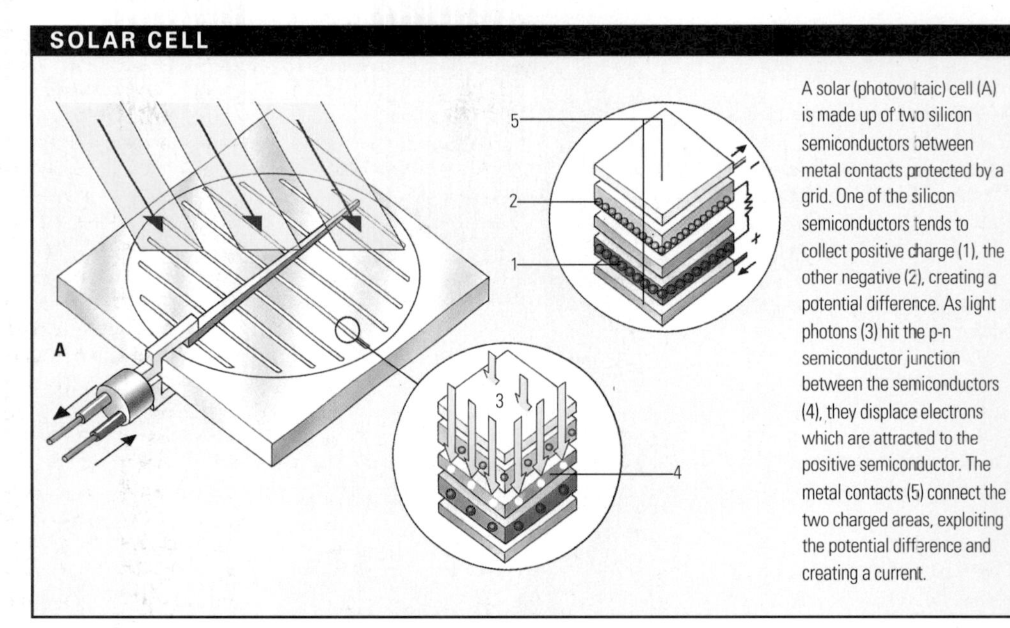

A solar (photovoltaic) cell (A) is made up of two silicon semiconductors between metal contacts protected by a grid. One of the silicon semiconductors tends to collect positive charge (1), the other negative (2), creating a potential difference. As light photons (3) hit the p-n semiconductor junction between the semiconductors (4), they displace electrons which are attracted to the positive semiconductor. The metal contacts (5) connect the two charged areas, exploiting the potential difference and creating a current.

solar constant Measure of the amount of solar energy received by a body a certain distance from the Sun. For the Earth, the solar constant is defined as the solar power received per unit area, at the top of the atmosphere, at the average Earth–Sun distance of 1AU (1 astronomical unit). The value of the solar constant is *c*.1.35kW m^{-2}. The "constant" varies from day to day with SUNSPOT activity, and also, in the longer term, with the SOLAR CYCLE.

solar cycle Periodic fluctuation in the number of SUNSPOTS and the level of other kinds of solar activity; the cycle lasts about 11 years. Over the course of a cycle, sunspots vary both in number and latitude. At solar maximum, when sunspot numbers are greatest, astronomers refer to the **active Sun**; at solar minimum, they refer to the **quiet Sun**.

solar energy Heat and light from the SUN consisting of ELECTROMAGNETIC RADIATION, including heat (infrared rays), light and radio waves. About 35% of the energy reaching the Earth is absorbed: most is spent evaporating moisture into CLOUDS, and some is converted into organic chemical energy by PHOTOSYNTHESIS in plants. All forms of energy (except NUCLEAR ENERGY) come ultimately from the Sun. SOLAR CELLS are used to power instruments on spacecraft, and experiments are being done to store solar energy in liquids from which ELECTRICITY can be generated. The effective use of solar energy is hampered by the diurnal cycle, and by seasonal and climatic variations.

solar engine Machine that converts SOLAR ENERGY into mechanical WORK, particularly for the pur-

SOLAR FLARES AND PROMINENCES

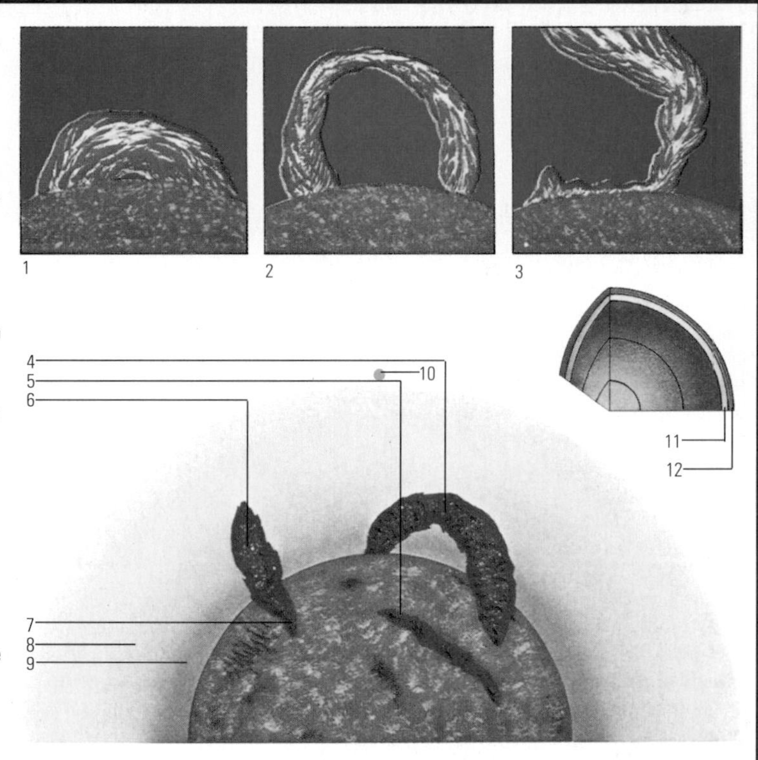

The Sun is extremely hot, 6,000°C on the surface and millions of degrees inside. There are always violent activities on the Sun, the main disturbances being the solar flares and prominences. These are great flames that leap thousands of kilometres into space. They may form thick closed loops (1), long loops (2) and open flares (3). In the main diagram, a long loop (4) is shown together with a quiescent solar flare (5) and an eruption (6). Other solar features shown include spicules (7), outer corona (8) and inner corona (9). The Earth is dwarfed in comparison (10). A section through the Sun (inset) would reveal the photosphere (11) from which prominences originate, and the chromosphere (12) through which they permeate.

pose of providing THRUST to a spacecraft. Solar engines now under development have one or more large solar collectors, which gather the energy of the Sun's rays to heat up a working fluid such as hydrogen. The heated hydrogen can be used directly to provide thrust or it can be fed to a turbogenerator.

solar flare Sudden and violent release of matter and energy from the Sun's surface, usually from the region of an active group of SUNSPOTS. In the **flash stage**, a flare builds to a maximum in a few minutes, after which it gradually fades and disappears within an hour or so. Flares emit radiation right across the ELECTROMAGNETIC SPECTRUM. Charged particles are emitted, mostly ELECTRONS and PROTONS, and smaller numbers of neutrons and atomic nuclei. A flare can cause material to be ejected in bulk, most spectacularly in the form of PROMINENCES. When energetic particles from flares reach the Earth they may cause radio interference, magnetic storms, and more intense AURORAE. Although not well understood, the origin of flares is believed to be connected with local discontinuities in the Sun's MAGNETIC FIELD.

solar heating Domestic heating system that usually employs roof-mounted heat collectors, often called solar panels. These collectors have transparent glass or plastic panels, behind which water or air circulates over black-painted metal surfaces that absorb heat energy from the Sun's rays. The warmed liquid or air passes on to a water tank or rock storage tank from which hot water or air is pumped to RADIATORS or air vents.

solar physics Study of the SUN's physical behaviour. It includes study of the solar spectrum and of the nature and causes of surface phenomena. From these may be inferred the inner thermal, electrical, magnetic, nuclear and gravitational processes.

Solar System SUN and all the celestial bodies which revolve around it: the nine PLANETS, together with their SATELLITES and ring systems, the thousands of ASTEROIDS and COMETS, meteoroids and other interplanetary material. The **inferior planets** are those planets that are closer than EARTH to the Sun; the rest are called **superior planets**. Astronomical distance is measured in ASTRONOMICAL UNITS (AU) – the mean distance between the Earth and the Sun. The boundaries of the Solar System lie beyond PLUTO, which orbits the Sun at a mean distance of 39 AU, to include the Kuiper Belt (100 AU) and the Oort Cloud of comets. The Solar System as a whole moves in a roughly circular orbit about the centre of the GALAXY, taking 2.2×10^8 years to complete one full orbit. ARISTOTLE and PTOLEMY's notion of a geocentric UNIVERSE remained fundamentally unchallenged until the 16th century. COPERNICUS developed the first Heliocentric theory of the UNIVERSE, which was championed by GALILEO. Drawing on the observations of Tycho BRAHE, Johannes KEPLER accurately described the elliptical orbit of each planet with the Sun at one of the foci. All the planets orbit the Sun in roughly the same plane (the ECLIPTIC), although Pluto's orbit is more irregular. They all move in the same direction – counterclockwise, when viewed from above. All the planets rotate about their own axes as they revolve around the Sun; the period of rotation varies (in Earth time) from under 10hrs (JUPITER) to more than 243 days (VENUS). Venus is the only planet with retrograde motion – it rotates from east to west. The equatorial plane of each planet is tilted in relation to its orbital plane: Jupiter has the smallest tilt (3°), URANUS has the largest (98°). Earth's equatorial plane is tilted at 23.5°. This tilting produces annual SEASONS. Sir Isaac NEWTON showed that all bodies in the Solar System are subject to the force of GRAVITATION. The Sun is by far the most massive component of the Solar System, accounting for 99.9% of the entire system's mass. Thus, it exerts the greatest pull. Other celestial bodies exert small variations (PERTURBATIONS) on each other's orbits. The planets are classified also by their physical properties. The inner planets (MERCURY, Venus, Earth and MARS) are called the **terrestrial planets**. They are relatively small and dense, with solid CRUSTS and molten, metallic CORES. They are composed of high-temperature condensates (mainly iron and metal silicates). The **Jovian planets** (Jupiter, SATURN, Uranus and NEPTUNE) are massive, but with relatively low density. Jupiter is heavier than all the other planets combined. Jovian planetary atmospheres are thick and gaseous, chiefly composed of hydrogen and helium. Pluto is unique and comparatively unknown. The **origin** of the Solar System is a major debate among cosmogonists. In the late 18th century, Pierre LAPLACE developed the **nebular hypothesis**. According to this theory, about 5,000 million years ago a NEBULA or INTERSTELLAR CLOUD of gas and dust slowly cooled and contracted causing gravitational collapse. The nebula began to rotate faster, throwing out rings of gaseous matter. In these rings balls of matter were formed, which shrank to form planets. At the centre of the system, mass condensed to form the Sun. In the early 20th century, Forest MOULTON and Thomas Chamberlin developed an "encounter" theory called the PLANETESIMAL theory. They proposed that a star passed close to the Sun causing huge surface tides, which ripped matter from the solar nebula and threw it into elliptical orbit. These small aggregates of matter (planetesimals) were attracted to each other by gravitation and accreted to form planets. In 1917 James JEANS proposed a similar **tidal** theory. All of these theories, however, failed to account for the distribution of angular momentum. Gerard KUIPER's **protoplanets** theory satisfied this objection. Harold UREY developed Kuiper's ideas to provide an explanation of the different physical and chemical compositions of the planets.

solar time System of time reckoning based upon the interval between successive transits of the Sun across the observer's meridian (the solar day). Because of variations in the Earth's orbital velocity and changes in the Sun's apparent position as viewed from the orbiting Earth, solar days vary in length throughout the year. A mean solar day has therefore been adopted, giving a mean solar time. As measured on the MERIDIAN at the Greenwich Observatory, mean solar time is GREENWICH MEAN TIME (GMT) or Universal time.

SOLAR HEATING

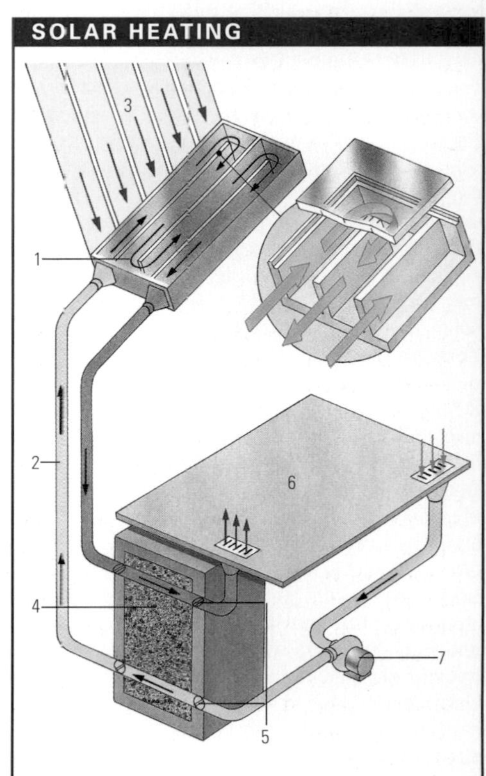

In active solar space heating, a large collector (1) is needed and air (2) is heated directly by sunlight (3). The hot air then returns to a rock storage tank, where its energy can be stored until needed (4). Valves (5) control the flow of the heat to the living areas of the house (6). The flow is controlled by a pump (7).

PLANETARY DATA

		Mercury	Venus	Earth	Mars	Jupiter	Saturn	Uranus	Neptune	Pluto
tance from	max.	69.7	109	152	249	816	1507	3004	4537	7375
ɹn, millions	mean	57.9	108.2	149.6	227.9	778	1427	2870	4497	5900
f km	min.	45.9	107.4	147	206.7	741	1347	2735	4456	4425
ital period		87.97d	224.7d	365.3d	687.0d	11.86y	29.46y	84.01y	164.8y	247.7y
odic period, ays		115.9	583.92	—	779.9	398.9	378.1	369.7	367.5	366.7
ation period quatorial)		58.646d	243.16d	23h 56m 04s	24h 37m 23s	9h 55m 30s	10h 13m 59s	17h 14m	16h 7m	6d 9h 17s
ital eccentricity		0.206	0.007	0.017	0.093	0.048	0.056	0.047	0.009	0.248
ital inclination, °		7.0	3.4	0	1.8	1.3	2.5	0.8	1.8	17.15
ial inclination, °		2	178	23.4	24.0	3.0	26.4	98	28.8	122.5
ape velocity, km/s		4.25	10.36	11.18	5.03	60.22	32.26	22.5	23.9	1.18
ss, Earth =1		0.055	0.815	1	0.11	317.9	95.2	14.6	17.2	0.002
ume, Earth = 1		0.056	0.86	1	0.15	1319	744	67	57	0.01
nsity, water =1		5.44	5.25	5.52	3.94	1.33	0.71	1.27	1.77	2.02
rface gravity, arth =1		0.38	0.90	1	0.38	2.64	1.16	1.17	1.2	0.06
rface temp., °C		+427	+480	+22	−23	−150	−180	−214	−220	−230
edo		0.06	0.76	0.36	0.16	0.43	0.61	0.35	0.35	0.4
meter, km quatorial)		4878	12,104	12,756	6794	143,884	120,536	51,118	50,538	2324
ximum magnitude		−1.9	−4.4	—	−2.8	−2.6	−0.3	+5.6	+7.7	+14

▼ The Solar System is the only part of the Universe which we can explore with the technology and design of contemporary spacecraft. Even so, we are still uncertain as to how far the Solar System extends. There may well be another planet beyond Pluto, and it is thought that comets come from a cloud of icy objects orbiting the Sun at a distance of approximately one or two light years.

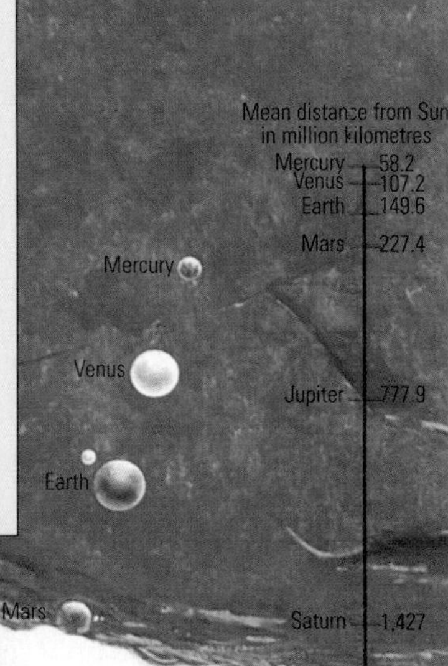

The Solar System is made up of one star (the SUN), nine planets (of which the EARTH comes third in order of distance), and various lesser bodies, such as the SATELLITES, ASTEROIDS, COMETS and METEOROIDS.

It is believed that the planets formed by ACCRETION from a cloud of material that surrounded the youthful Sun; the age of the Earth is known to be *c.*4,600 million years, and the Solar System is even older. The Solar System is divided into two. There are four small, solid planets: MERCURY, VENUS, EARTH and MARS. Then comes a gap, in which move thousands of small bodies variously known as asteroids, planetoids and minor planets. Beyond are the four giants: JUPITER, SATURN, URANUS and NEPTUNE, together with the maverick PLUTO, which is often considered too small to be classed as a planet. It seems that the four inner planets lost their light gases because of the heat of the Sun, so that they are solid and rocky; the giants, which formed in a colder region, were able to retain their lighter gases.

The Earth has one satellite: our Moon. Of the other planets, Mars has two satellites, Jupiter sixteen, Saturn eighteen, Uranus seventeen, Neptune eight and Pluto one, though only four of these are as large as our Moon.

Comets are of very low mass compared with planets. The only solid part is the nucleus, which has been discribed as a "dirty ice-ball". When a comet nears the Sun, the ice begins to evaporate, and the comet may produce a gaseous "head" with a long tail. As a comet moves along it leaves a "dusty trail" behind it. When the Earth moves through one of the trails it collects dusty particles, which burn away in the upper atmosphere and produce luminous streaks, which we call shooting stars. Larger objects, which may survive the atmosphere and reach the ground, are termed meteorites; they come from the asteroid belts, and are not associated either with comets or shooting-star meteors.

There is also a great deal of thinly spread "dust", especially in the main plane of the Solar System. Small particles of this kind catch the sunlight, and produce the glows which we call ZODIACAL LIGHT and GEGENSCHEIN.

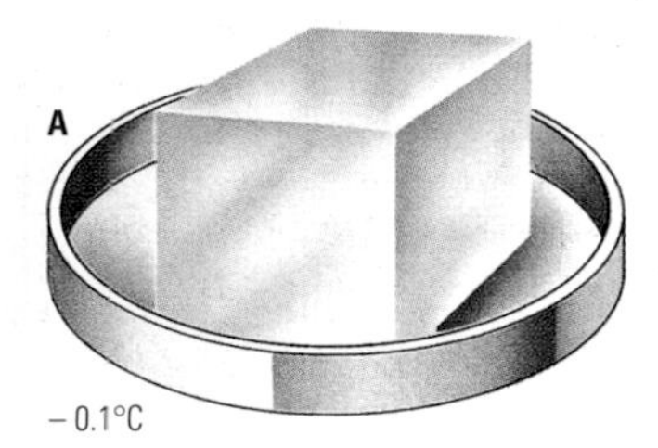

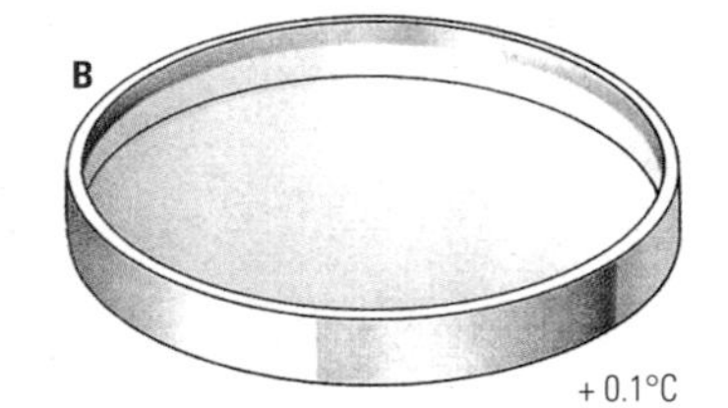

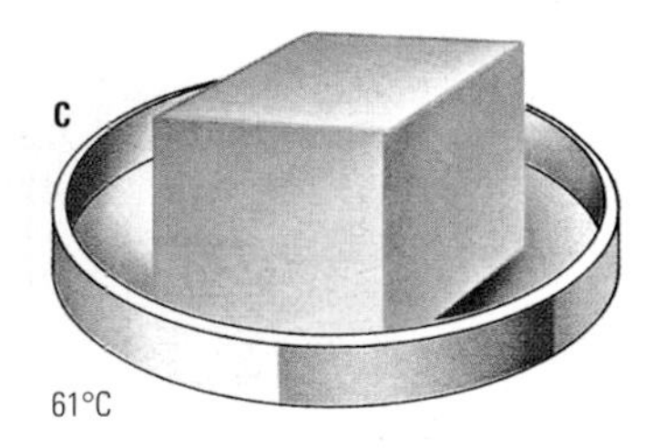

◀ **solid** Melting of solids. Ice, a crystalline solid, melts at a sharply defined temperature, but non-crystalline materials, such as beeswax, melt over a temperature range.

Key

A) Ice – water in its crystalline form
B) Water as a liquid
C) Yellow beeswax – a non-crystalline solid
D) Beeswax loses a definite shape as it is warmed
E) The wax spreads out as it becomes more liquid
F) Beeswax melts into a fluid form

solar wind Steady flow of charged particles (mainly protons and electrons) accelerated by the high temperatures of the solar CORONA to velocities great enough to allow them to escape from the SUN's gravity. The solar wind deflects the tail of the Earth's MAGNETOSPHERE and the tails of COMETS away from the Sun. Some solar wind particles get trapped in planetary magnetic fields. At the Earth, some are trapped in the outer VAN ALLEN RADIATION BELTS; others reach the Earth's upper ATMOSPHERE in the region of the magnetic poles and cause AURORAE. The solar wind carries away about 10^{-13} of the Sun's mass per year. Its intensity increases during periods of solar activity.

solder Metallic ALLOY used in a molten state to fuse together metal surfaces. **Soft solders** are usually alloys of LEAD and TIN and have melting points between 200°C and 300°C. **Hard solders** contain significant amounts of SILVER and have a much higher melting point. **Brazing solders** are alloys of COPPER and ZINC and melt at more than 800°C. *See also* BRAZING

solenoid ELECTROMAGNET in which a soft iron core moves so as to open or close an electric circuit, thus acting as a switch or RELAY.

solfatara Small CRATER or vent in the Earth's surface through which steam and gases escape. Solfataras occur mostly in volcanic regions and many of the gases contain sulphur. They probably indicate an area of declining volcanic activity. *See also* VOLCANO

solid State of MATTER in which a substance has a relatively fixed shape and size. The forces between atoms or molecules are strong enough to hold them in definite locations (about which they can vibrate) and to resist compression. *See also* CRYSTAL; GAS; LIQUID; PLASMA

solid angle Three-dimensional angle formed at the centre of a SPHERE by the VERTEX of a CONE that has its base on the surface of the sphere. Measured in **steradians**, a solid angle is defined as the ratio of the surface area occupied by the base of the cone to the square of the radius of the sphere.

solid circuit ELECTRIC CIRCUIT that is present inside a semiconductor or other solid-state device. *See also* SEMICONDUCTIVITY

solid geometry Study of figures in three dimensions. It includes the study of planes intersecting with planes, planes intersecting with solids, solids interacting with solids and the measurement of solid figures. *See also* GEOMETRY

solid-state physics Physics of SOLID materials. From the study of the structure, binding forces, electrical, magnetic and thermal properties of solids has come the development of the SEMICONDUCTOR, MASER, LASER and SOLAR CELL. **Solid-state circuits** are electronic circuits in which the individual components are formed from a single, solid SILICON CHIP.

soliton Stable, solitary wave that keeps its shape and speed without widening as it propagates. Solitons behave like particles and are important in many areas of FLUID MECHANICS and SOLID-STATE PHYSICS.

solstice Either of the two days each year when the Sun is at its greatest angular distance from the CELESTIAL EQUATOR, leading to the longest day and shortest night (**summer solstice**) in one hemisphere of the Earth, and the shortest day and longest night (**winter solstice**) in the other hemisphere. In the Northern Hemisphere, the summer solstice occurs on about 21 June, and the winter solstice on about 22 December.

solubility Mass (grams) of a SOLUTE that will saturate 100 grams of SOLVENT under given conditions to give a SATURATED SOLUTION. Solubility generally rises with temperature, but for a few solutes, such as CALCIUM SULPHATE, increasing temperature decreases solubility in water.

solute Gaseous, liquid or solid substance that is dissolved in a SOLVENT to form a SOLUTION. Ionic solids, such as common salt, and sugars and starch dissolve in water. Liquids can dissolve in liquids; for example, ethanol and water are miscible – capable of mixing – in all proportions at room temperature. Some gases, such as hydrogen chloride (HCl), are soluble in water.

solution In chemistry, a liquid (the SOLVENT) into which another substance (the SOLUTE) has dissolved. It is a liquid consisting of two or more chemically distinct compounds that are inseparable by filtering. The amount of a solute dissolved in a given volume of solvent is called the CONCENTRATION of a solution. The ability of one substance to dissolve another depends on the type of chemical bonding and the temperature. Heat can be released (an exothermic solution) or absorbed (an endothermic solution) during the formation of the solution. *See also* MIXTURE; SATURATED SOLUTION

solution In geology, a form of chemical WEATHERING. It is particularly active in limestone areas where the JOINTS can be enlarged to form GRIKES and POTHOLES. Solution can also be active in chalk areas and anywhere with rocks containing salts.

Solvay, Ernest (1838–1922) Belgian industrial chemist. He invented a technique (the SOLVAY PROCESS) for preparing SODIUM HYDROGENCARBONATE (and hence SODIUM CARBONATE) from sodium chloride (common salt) and calcium carbonate. In 1861 Solvay patented the process. By 1913 the technique was producing a large part of the world's sodium carbonate.

Solvay process (ammonia-soda process) Widely used industrial method of making SODIUM CARBONATE (soda, $[Na_2CO_3]$), invented by Ernest SOLVAY. Brine is saturated with AMMONIA and carbon dioxide gas (made by heating calcium carbonate) is bubbled into this solution in a Solvay tower. SODIUM HYDROGENCARBONATE is formed, which precipitates from the brine, and is then heated to make sodium carbonate.

solvent Liquid that dissolves a substance (the SOLUTE) without changing its composition to form a solution. Water is the most universal solvent, and many inorganic compounds dissolve in it. Ethanol, ether, propanone (acetone) and tetrachloromethane (carbon tetrachloride) are common solvents for organic substances. *See also* SOLUTION

somatotrophin Growth HORMONE made by cells of the anterior, or frontal, PITUITARY GLAND, lying under the brain. It stimulates growth in young animals including humans by increasing the amounts of PROTEIN made by the body cells and stimulating fat and carbohydrate metabolism.

Sömmering, Samuel Thomas von (1755–1828) German scientist. In 1809, using the voltaic pile invented by Alessandro VOLTA, Sömmering developed an early electric TELEGRAPH system that sent signals indicated at the receiving end by the appearance of hydrogen bubbles liberated at a gold electrode in a solution of acid.

sonar (acronym for **so**und **n**avigation **a**nd **r**anging) Underwater detection and NAVIGATION system. The system emits high-frequency sound that is reflected by underwater objects. *See also* ECHO SOUNDER

sonic boom Sudden NOISE produced by SHOCK WAVES from an AIRCRAFT flying at a SUPERSONIC VELOCITY. The shock waves are formed by the build-up of SOUND waves at the front and back of the aircraft. These waves spread out and sweep across the ground behind the aircraft, often causing a double-bang. Sometimes the shock waves are strong enough to break windows in buildings.

sound VIBRATIONS that travel through solids, liquids or gases. The human ear can respond to sounds with frequencies between about 20 and 20,000Hz. Frequencies above this audible range are called ULTRASONICS. Sound WAVES are a combination of **condensation** (close packing of molecules) and **rarefication** (spreading out of molecules). This is usually represented as a wavy line, whose crests indicate condensation and whose troughs represents rarefication. The WAVELENGTH is the distance between two crests. It is determined by dividing the velocity of sound by the FREQUENCY of vibration. A pure sound is a SINE wave of definite frequency and intensity (rate of flow of energy). Natural sounds are a mixture of sine waves characterized by PITCH and TIMBRE. The speed at which a sound wave travels through a solid depends on the ELASTICITY of the medium and its DENSITY. Solids are the best conductors of sound. If the medium is a gas, the sound wave is longitudinal and its speed depends on the gas temperature. The **speed of sound** in dry air at STANDARD TEMPERATURE AND PRESSURE (STP) is 331.4ms^{-1} (741mph). ACOUSTICS is the study of sound waves. For instance, acoustics is used in the design of a concert hall, usually to minimize the reflection of sound which can produce ECHOES and INTERFERENCE. Loudness of sound is measured in DECIBELS. *See also* DOPPLER EFFECT; HARMONICS

sound barrier Cause of an aircraft's difficulties in accelerating to a speed faster than that of SOUND (SUPERSONIC VELOCITY). When approaching the speed of sound, an aircraft experiences a sudden increase in DRAG and loss of LIFT. These effects are caused by the build-up of sound waves to form SHOCK WAVES at the front and back of the aircraft. Some aircraft designers once thought that these effects would make SUPERSONIC FLIGHT impossible. However, the problems were solved by designing aircraft with smaller surface areas, swept-back wings and more powerful propulsion systems. *See also* MACH NUMBER; SONIC BOOM

sound card (sound board) Piece of circuitry that is required for a COMPUTER to reproduce music and other SOUNDS through LOUDSPEAKERS. A sound card is necessary for various MULTIMEDIA applications using a CD-ROM drive. Eight-bit cards give poorer reproduction than the newer 16-bit sound cards, which can provide stereo. Some cards also facilitate SOUND RECORDING using a computer.

sounding Determining the depth of water. The simplest means of sounding is by dropping a measured weighted line until it reaches the bottom. Acoustic SONAR, a sonic depth finder, is often used. *See also* ECHO SOUNDER

sound recording Conversion of SOUND waves into a form that can be stored and reproduced. In 1877 Thomas EDISON invented a phonograph that recorded sound vibrations as indentations made by a stylus on a revolving cylinder wrapped in tinfoil. Emile Berliner's gramophone improved the process by using a zinc disc instead of a cylinder. The volume was amplified by the addition of acoustical horns, which were replaced before World War 1 by valve AMPLIFIERS. In 1901 moulded thermoplastic records were introduced, and subsequently improved plastics allowed finer grooving with reduced surface noise. In 1927 and 1928 patents were issued in the USA and Germany for MAGNETIC RECORDING processes. Later innovations include HIGH-FIDELITY (HI-FI), stereophonic and quadrophonic reproduction. Modern recordings on COMPACT DISC (CD) employ digital signals (rather than the analog signals used on earlier records).

Soyuz Series of Soviet manned space missions. **Soyuz 1** was launched in April 1967. In November 1968 a Soviet cosmonaut in **Soyuz 3** completed a manual approach with the unmanned **Soyuz 2**. In January 1969 **Soyuz 4** and 5 docked in space, and two cosmonauts from **Soyuz 5** completed the first first transfer in space between two spacecraft. In July 1969 a rocket explosion destroyed the Soviet's entire launch complex. In 1970 **Soyuz 9** established the record for a manned space flight – 18 days. The Soyuz program suffered another setback when **Soyuz 11** depressurized during re-entry into Earth's atmosphere, killing the three-man crew. In July 1975 APOLLO and Soyuz spacecraft docked in space, completing the first joint US-Soviet space project.

space Boundless, three-dimensional expanse in which objects are located. RELATIVITY states that space and TIME are aspects of one entity, known as SPACE-TIME. Also, space has a non-Euclidean GEOMETRY wherever there is a gravitational field. In another sense, space – sometimes referred to as outer space – is taken to mean the rest of the Universe beyond Earth's atmosphere.

space exploration Use of spacecraft to investigate outer space and heavenly bodies. On 4 October 1957, the first artificial satellite, SPUTNIK 1, was launched into Earth orbit by the Soviet Union. The first US satellite was **Explorer I** (31 January 1958). The "space race" began in earnest. Within the next decade the two superpowers launched more than 50 unmanned SPACE PROBES to the Moon. Soviet cosmonauts, and their American equivalents, astronauts, orbited the Earth soon after. Unmanned space probes crash-landed on the Moon, sending back television pictures to Earth during the descent. Then came soft landings, and probes made to orbit the Moon showed its "dark" side for the first time. By 1968 Soviet space scientists had developed techniques for returning a Moon orbiter safely to Earth. In 1969 the US APOLLO 11 mission became the first to place a man on the Moon. Meanwhile, the Soviet Union had sent probes to explore Mars and Venus. **Chronology**: First probe to land on the Moon: **Luna 2**, launched 12 September 1959. First manned spaceflight: Yuri GAGARIN, 12 April 1961. First close-up pictures of Mars: MARINER 4, received 14 July 1965. First person to walk on the Moon: Neil ARMSTRONG, 21 July 1969. First pictures from the surface of another planet: VENERA 9, received from Venus on 22 October 1975. First probes to land on Mars: VIKING 1 and 2 July 1976. Fly-bys of VOYAGER 2: Jupiter (1979), Saturn (1981), Uranus (1986), Neptune (1989). First SPACE SHUTTLE: *Columbia*, launched 12 April 1981. **Giotto probe**, launched 1985, flew within 600km (375mi) of HALLEY'S COMET, sent back photographs and data. GALILEO project, launched October 1989, photographed the asteroids Gaspra (1991) and Ida (1993), dropped a sub-probe (1995) into Jupiter's atmosphere and provided detailed data on GALILEAN SATELLITES. **Clementine probe**, launched 1994, discovered what are thought to be water-ice deposits in craters of the Moon. **Solar and Heliospheric**

SPACE EXPLORATION

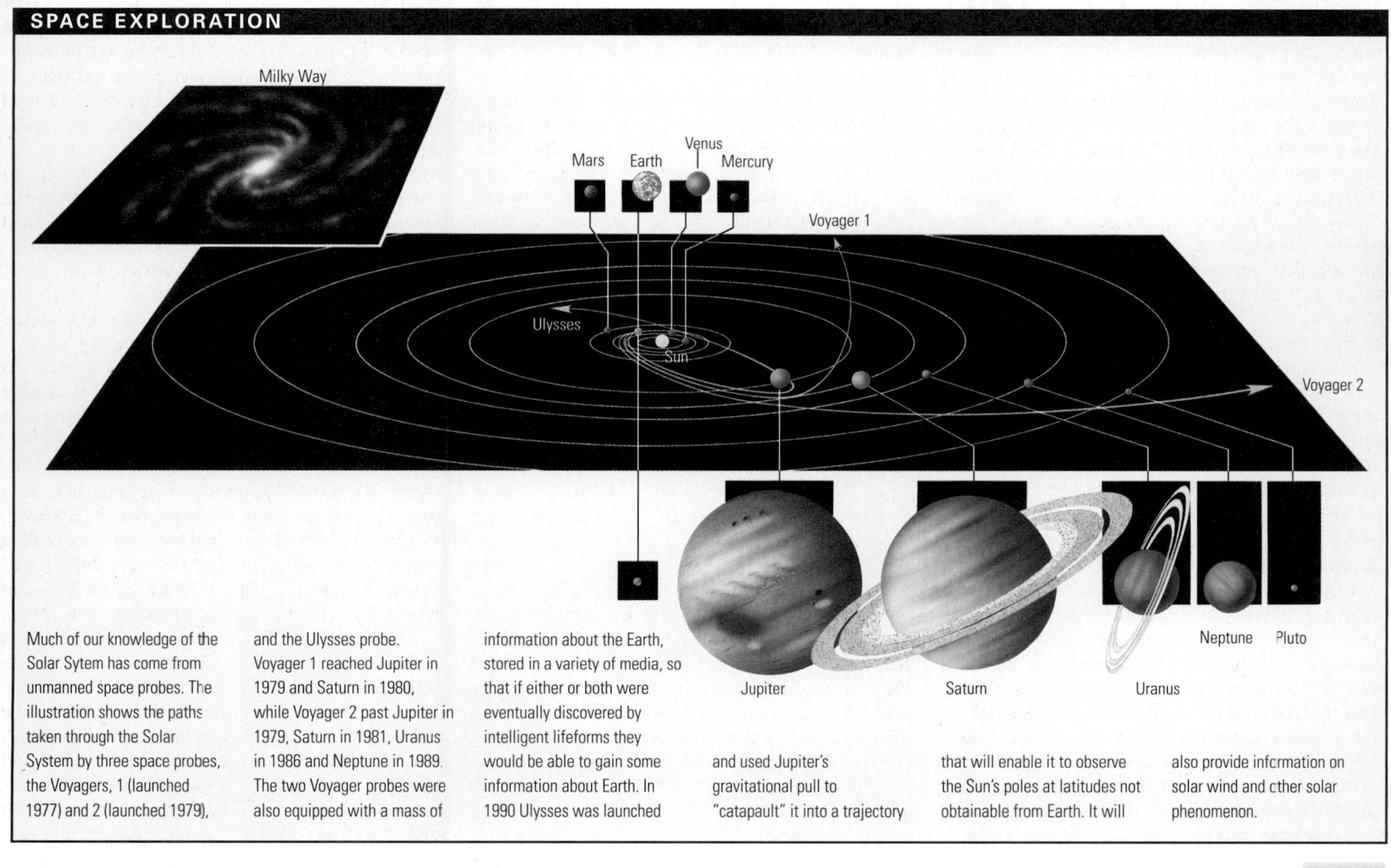

Much of our knowledge of the Solar Sytem has come from unmanned space probes. The illustration shows the paths taken through the Solar System by three space probes, the Voyagers, 1 (launched 1977) and 2 (launched 1979), and the Ulysses probe. Voyager 1 reached Jupiter in 1979 and Saturn in 1980, while Voyager 2 past Jupiter in 1979, Saturn in 1981, Uranus in 1986 and Neptune in 1989. The two Voyager probes were also equipped with a mass of information about the Earth, stored in a variety of media, so that if either or both were eventually discovered by intelligent lifeforms they would be able to gain some information about Earth. In 1990 Ulysses was launched and used Jupiter's gravitational pull to "catapault" it into a trajectory that will enable it to observe the Sun's poles at latitudes not obtainable from Earth. It will also provide information on solar wind and other solar phenomenon.

SPACE SHUTTLE

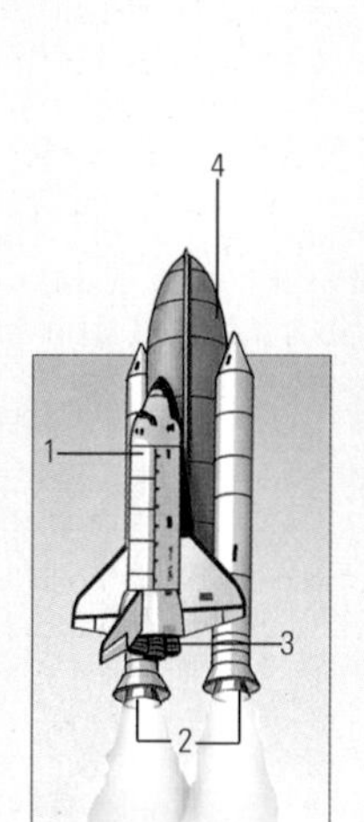

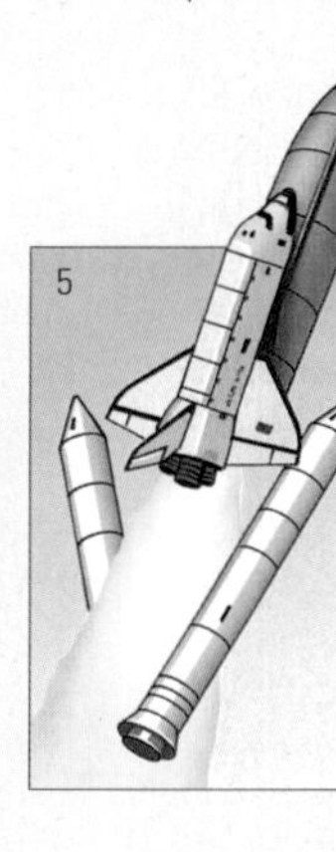

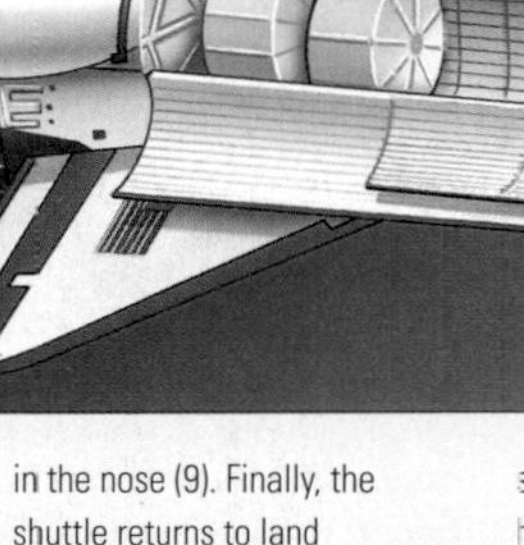

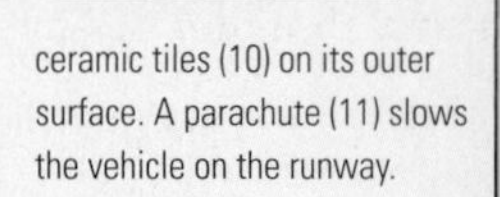

A space shuttle flight has three distinct phases. The first is reaching orbit. The orbiter (1) is propelled upward from the launch platform by two solid fuel boosters (2) and the shuttle's three engines (3), fed with liquid oxygen and liquid hydrogen fuel from the external tank (4). After the boosters have burnt out they detach (5), and float to the surface of the Earth by parachute. The external tank detaches later and burns up in the Earth's upper atmosphere. The second phase now commences. Once in orbit, the shuttle floats upside down above the Earth with its cargo bay doors (6) open to help dispel heat. Satellites (7) are launched from the cargo bay, and can be retrieved with the use of an arm (8). While in orbit, the shuttle manoeuvres using helium-fuelled thrusters in the nose (9). Finally, the shuttle returns to land unpowered. The orbiter is shielded from the enormous heat generated on re-entering the Earth's atmosphere by ceramic tiles (10) on its outer surface. A parachute (11) slows the vehicle on the runway.

Observatory (SOHO) probe, launched 1995, provided data on the Sun's interior. **Mars Pathfinder** mission, launched December 1996, landed on Mars and deployed a "microrover", *Sojourner*, to explore and collect rock samples. **Cassini** space probe, launched 1997, started on a seven-year journey towards Saturn. *See also* HUBBLE SPACE TELESCOPE; SPACE PROBE

space probe Unmanned craft sent to investigate the Moon, planets or interplanetary space. The first US space probe was **Explorer 1**, launched 31 January 1958; it discovered the inner VAN ALLEN RADIATION BELT. The first LUNAR PROBE was **Luna 2**, launched by the Soviet Union on 12 September 1959. The first planetary probe was MARINER 2, launched by the USA on 27 August 1962, which flew past Venus. PIONEERS 10 and 11 photographed Jupiter (22 October 1975). Pioneer 11 went on to Saturn. VIKING 1 landed on Mars and collected data to test the Martian surface for life (July 1976). VOYAGER 2 sent back images from its fly-bys of Jupiter (1979), Saturn (1981), Uranus (1986) and Neptune (1989). *See also* SPACE EXPLORATION

space research Scientific and technological investigations that gather knowledge through SPACE EXPLORATION. The physical makeup and behaviour of stars, planets and other cosmic objects were first determined by ground-based optical and radio TELESCOPES and later by spacecraft. Research projects have been carried into space by artificial SATELLITES, the SPACE SHUTTLE, SPACE STATIONS and SPACE PROBES.

space shuttle Re-usable, US ROCKET-powered spacecraft. The main part of the shuttle, the **orbiter** (of which four have been made, *Columbia*, *Challenger*, *Discovery* and *Atlantis*), looks like a bulky jet aircraft with swept-back wings. It ferries people and equipment between the ground and Earth ORBIT. The orbiter has a large payload bay in which it can carry SATELLITES for release into orbit. Astronauts can leave the shuttle to repair a faulty artificial satellite, and can retrieve satellites and return them to the ground. The shuttle sometimes carries Spacelab modules, developed by the European Space Agency (ESA). On such missions, scientists and astronauts carry out various experiments and observations. The orbiter takes off attached to a large fuel tank, using its own three ROCKET engines, assisted by two **booster** rockets. The boosters are jettisoned about two minutes after launch and are later recovered for re-use. Six minutes later, the orbiter's main engines cut off and the external fuel tank is dumped. Manoeuvring engines then put the craft into the required orbit. When it is time to return to Earth, these engines are used to provide reverse thrust. As a result, the craft slows down and descends into Earth's atmosphere. It glides down and lands on a runway. The first space shuttle, *Columbia*, was launched on 12 April 1981. On mission 25 in January 1986, the shuttle *Challenger* exploded soon after launch, killing all seven people on board. A leak had allowed burning gases from a booster rocket to ignite the fuel in the main tank.

space station Orbiting structure in space for use by astronauts and scientists. Space stations are more spacious than most spacecraft as the occupants may live there for several months before returning to the Earth. Space laboratories, such as the American *Skylab* (launched 1972) and the Russian MIR (launched 1986), are space stations built for scientists to carry out experiments, study the Solar System and observe distant parts of the Universe, their view undistorted by Earth's atmosphere. In 1997 *Mir* suffered a series of technical problems, and the programme was cancelled by the Russian government. In 1999 *Mir* is due to re-enter Earth's orbit. Its projected path will see the remnants of the space station land in the Pacific Ocean.

spacesuit Sealed garment enabling an astronaut to function in space. The suit is multi-layed, combining eight different materials. The outside is treated nylon to prevent perforations from tiny micrometeoroids. Four layers of aluminium material then provide insulation from extremes of temperature and protection from harmful solar radiation. These are backed by a fire- and tear-reistant layer. The astronaut is protected from the vacuum of space by a pressure suit of nylon, coated with polyeurathane. A network of tubes in a nylon chiffon undergarment carries water that maintains body temperature. Gas pressure is kept below Earth's normal atmospheric pressure; otherwise, with virtually no pressure on the outside, the suit would balloon out. At such low pressure, the only gas suitable for breathing is pure oxygen. A backpack propelled by gas jets may be worn to enable the astronaut to move through space.

space-time Central concept in the theory of RELATIVITY that unifies the three SPACE dimensions

◀ **space station** Skylab in orbit over the Earth. Launched on 14 May 1973, Skylab was the first space laboratory. Offically called Apollo Applications Program, Skylab was operational between 25 May 1973 and 8 February 1974. During this period, three-man teams of astronauts worked for a total of 84 days in space carrying out a variety of experiments. It fell out of orbit in July 1979.

(length, breadth and height) with TIME to form a **four-dimensional** frame of reference. Durations and rates of processes depend on the relative state of motion of the observer and the system observed. In 1907 Hermann MINKOWSKI, who taught Albert EINSTEIN, clarified relativity theory by describing space-time in terms of a four-dimensional geometry. An event in space-time is specified by the three coordinates of space and the time coordinate. A line drawn in this space represents a particle's path both in space and time. Einstein incorporated this viewpoint into his theory of relativity: in the **general theory**, GRAVITATION is a distortion of space-time by MATTER. A consequence of this relativistic notion of time is that two events can only occur simultaneously for one observer: another observer will not see these events as concurrent. *See also* LORENTZ CONTRACTION

spadix In some flowering plants, a spike of small flowers, generally enclosed in a sheath called a SPATHE. A familiar plant with an inflorescence of this kind is the cuckoopint (*Arum maculatum*).

spark chamber Gas-filled chamber in which ionizing radiation or charged particles can be detected by the tracks of sparks made visible after the particles cause ionization. Up to 100 parallel metal plates are stacked in the chamber, connected alternately to the positive and negative terminals of a very high-voltage supply. When a particle passes through a pair of plates, it creates a pair of IONS in its track. The gas (usually neon or helium) becomes a conductor and sparks jump between the plates. A pair of camera lenses may be used to take stereoscopic photographs of the spark tracks.

spark plug Component in an INTERNAL COMBUSTION ENGINE. It has two ELECTRODES separated by an air gap, across which electric current from the ignition discharges to form a spark. The spark plug fits into the cylinder head and the spark ignites the compressed mixture of fuel and air.

spathe Broad leaf-like organ that spreads from the base of, or enfolds, the SPADIX of certain plants, such as the cuckoopint (*Arum maculatum*).

special theory of relativity Part of EINSTEIN'S theory of RELATIVITY that applies only to observers who move with constant speed and direction relative to each other.

speciation Emergence of new species in EVOLUTION. It results from the separation of parts of a homogeneous population. Over many generations, NATURAL SELECTION operates within the separated groups to produce gradually increasing differences. New species can be said to have evolved when individuals of the separated groups are no longer capable of interbreeding.

species Part of the CLASSIFICATION of living organisms. Species are groups of physically and genetically similar individuals that can interbreed to produce fertile offspring under natural conditions. Each species has a unique two-part Latin name (BINOMIAL NOMENCLATURE), the first part being the GENUS name. This name is written in italics. For example, the tiger's species name is *Panthera tigris*, *Panthera* being the genus of big cats. So far, more than 1.5 million plant and animal species have been identified, but estimates of the total number on land and in the oceans run as high as 100 million.

specific gravity Former term for RELATIVE DENSITY (R.D.)

specific heat capacity (symbol c) Heat necessary to raise the temperature of 1kg (2.2lbs) of a substance by 1K (Kelvin). It is measured in J K^{-1} kg^{-1} (J equals JOULE). Substances with a high specific heat capacity, such as water, require more energy to raise their temperatures than do substances of low specific heat.

spectral classification Categorization of STARS based on the characteristics of their spectra. In the 1860s Angelo SECCHI made the first spectral classification of stars by dividing them into four groups according to colour and spectral lines. As SPECTROSCOPY improved, a more comprehensive system was developed at Harvard College Observatory. This was embodied in the **Henry Draper Catalogue** (1918–24) of stellar spectra. The sequence of spectral types was arranged according to the prominence or absence of certain lines in the spectra. There were seven **spectral types**, designated O, B, A, F, G, K and M (an ordering which results from revising an earlier alphabetical sequence). This sequence is a **temperature** sequence, from the hottest stars of types O and B, which appear blue-white, to type M, the coolest stars, which appear orange-red. As a result of this ordering, O, B and A stars are called **early-type stars**; K and M stars are called **late-type stars**. Three new types were added when it was found that some cool stars had strong absorption bands not usually seen in other stars of the same colour. These were classes R and N with strong bands of molecular carbon, and class S with bands of zirconium oxide. Stellar spectra can be classified into even finer divisions within these seven types, so decimal subdivisions were introduced. G5, for example, indicates a star midway in type between G0 and K0. Further refinements include the use of additional letters as suffixes to the spectral type, giving more information about the star, for example the existence of emission lines (e), metallic lines (m), broad lines due to rotation (n and nn), or a peculiar spectrum (p). However, the Harvard system could not deal with stars of different luminosities at a given temperature (that is, DWARFS, GIANTS and SUPERGIANTS). In 1943 William Morgan, Philip Keenan and Edith Kellman of Yerkes Observatory redefined the spectral types and added a classification scheme for the LUMINOSITY (absolute magnitude) of stars, which they represented using Roman numerals, ranging from I (supergiants) to VI (subdwarfs). This is now known as the **Morgan–Keenan system** (MK system), and is used universally. For example, a star classified in the MK system as O9.5 IV–V has a spectral type (and therefore a temperature) midway between that of an O9 and a B0 star, and a luminosity between that of a dwarf and a subgiant. The MK system is applicable to stars of normal chemical composition, which is to say about 95% of all stars. The various types of peculiar star are given their own special classification schemes. The Harvard R and N types are now combined into one **carbon star** class, the designations for which include a temperature type and a carbon band strength, as for example in C2,4. A similar classification is used for the S stars. WHITE DWARFS are usually classified on a Harvard-type scheme, with D preceding the type.

spectrograph SPECTROSCOPE fitted with a camera or an electronic detector such as a CHARGE-COUPLED DEVICE (CCD) used to obtain a permanent record of the SPECTRUM of ELECTROMAGNETIC RADIATION. Astronomical spectrographs, generally designed for use with a specific telescope, record the spectrum of a celestial object. Spectographs generally operate in a band of wavelengths from the near-infrared to the near-ultraviolet. There are various designs for different purposes. **High-dispersion** instruments, such as the spectroheliograph, spread the spectral lines widely so that a narrow band of wavelengths can be studied in detail.

spectrometer Type of SPECTROSCOPE equipped to measure accurately the WAVELENGTHS and intensities of spectral lines by means of an INTERFEROMETER. Spectrometers are often loosely referred to as SPECTROGRAPHS. Since spectra are now routinely recorded on charge-coupled devices (CCDs), from which information in digital form can be analysed by computer, the distinction between modern spectrometers and spectrographs lies in their application rather than their design.

spectrophotometer Type of SPECTROMETER that works in conjunction with a SPECTROGRAPH. It outputs a graph of intensity against a WAVELENGTH called a **line profile**.

spectroscope Instrument for producing and studying the SPECTRUM of light from ELECTROMAG-

SPECTROSCOPE

Prism lens combination reveals chemical spectra. The spectroscope (upper diagram) is used to analyse light. Light from a telescope is passed through a fine vertical line in a plate (1). A lens (2) concentrates the rays onto a prism (3), which splits them into a spectrum. A second prism (4) is positioned so that the spectrum is widened even further after passing through it. The wide spectrum is then focused by a lens (5) onto a screen (6). The screen may be replaced by a recording apparatus, this instrument is called a spectrograph. The lower diagram shows typical omission spectra from various chemical elements. The coloured lines become dark when superimposed on the continuous rainbow giving Fraunhofer lines.

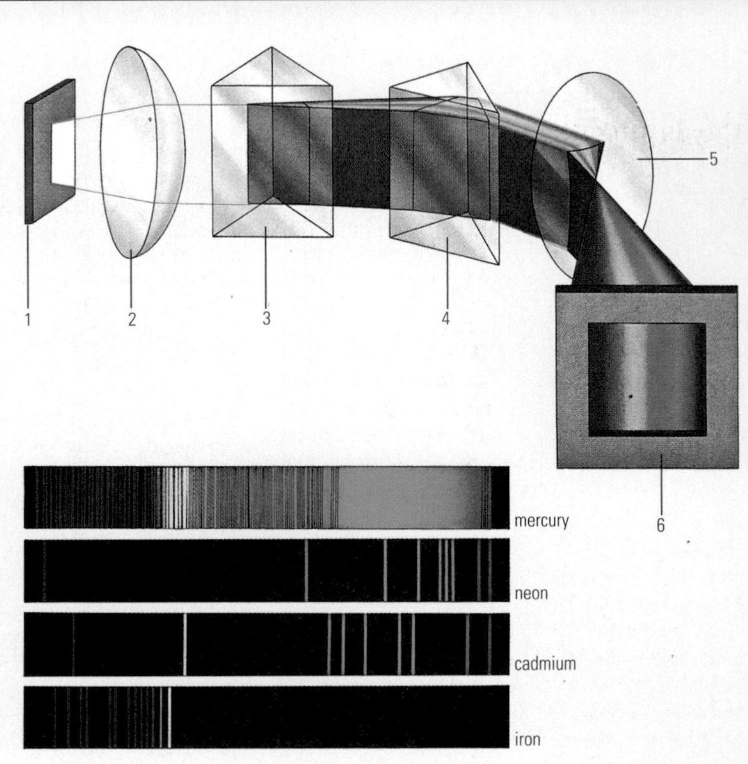

S

NETIC RADIATION (light or other wavelengths). SPECTROSCOPY is the use of such an instrument to probe the chemical composition and physical conditions of an object. Spectroscopes are used in astronomy to study the light from stars, and in chemistry to detect the presence of traces of various elements in samples that are too small to detect by other means. The light entering a spectroscope is collimated into a narrow beam by means of a slit and lens. The beam then passes through either a PRISM or DIFFRACTION GRATING so that it is dispersed into a spectrum. Combined with the grating or prism is a scale from which the spectral WAVELENGTHS may be read directly through the telescope that magnifies the spectrum. Astronomical spectroscopes are known as SPECTROGRAPHS or SPECTROMETERS. Strictly, a SPECTROGRAPH is a spectroscope equipped with a camera for recording a permanent record of a spectrum, whereas a SPECTROMETER incorporates devices for accurately measuring the wavelengths and intensities of the spectral lines.

spectroscopy Branch of OPTICS dealing with the measurement of the WAVELENGTH and intensity of lines in a SPECTRUM. The main tool in this study is the SPECTROSCOPE. An analysis of the spectrogram (record of the spectrum) can reveal the substances causing the spectrum by the position of emission and absorption lines and bands. *See also* SPECTROGRAPH; SPECTROMETER

SPECTRUM

There are two fundamental types of line spectra: an emission and an absorption spectrum. An emission (A) is the result of exciting a substance so that its electrons move to a higher energy. Photons of light are given out when the electrons fall back to their original state. Conversely, an absorption spectrum (B) is obtained when a photon of light is absorbed, raising the electrons of an atom to a higher level. A substance that emits light at a certain frequency absorbs light at the same frequency. When white light passes through a substance (C), an absorption spectrum can be seen – that is, the full spectrum (except for black lines) at the wavelengths the substance would emit if glowing alone.

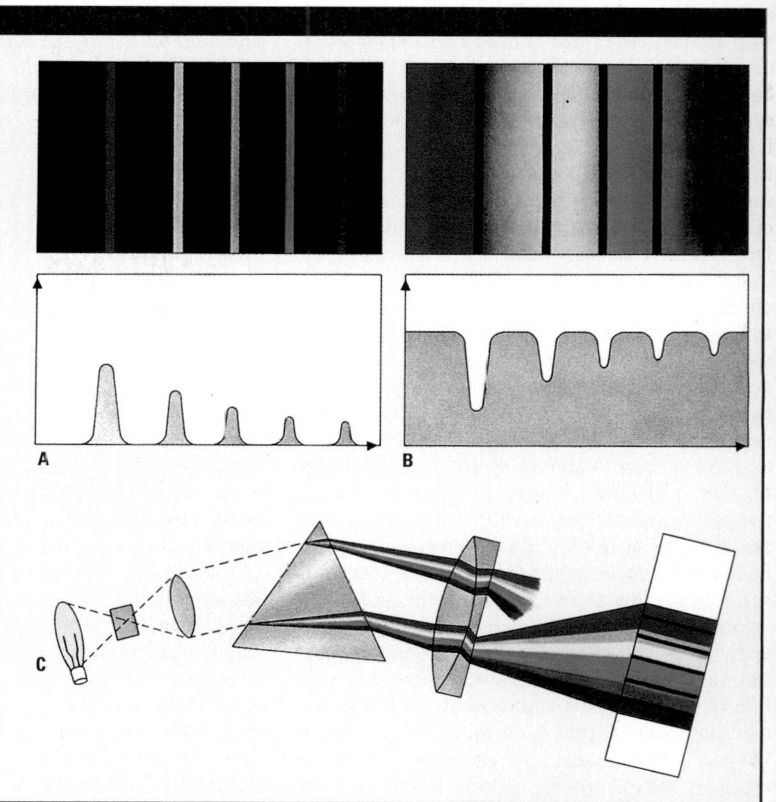

spectrum (pl. spectra) Arrangement of ELECTROMAGNETIC RADIATION ordered by WAVELENGTH or FREQUENCY. The visible light spectrum is a series of colours: red, orange, yellow, green, blue, indigo and violet. Each colour corresponds to a different wavelength of light. A spectrum is seen in a rainbow or when white light passes through a PRISM. This effect, also seen when visible light passes through a DIFFRACTION GRATING, produces a **continuous** spectrum in which all wavelengths (between certain limits) are present. Spectra consisting of bright lines or bands on a dark background are called **emission** spectra. These occur when a substance is strongly heated or bombarded by electrons. An **absorption** spectrum, consisting of dark regions on a bright background, is obtained when white light passes through a semitransparent medium that absorbs certain frequencies. A **line** spectrum is one in which only certain wavelengths or "lines" appear. **Band** spectra consist of grouped bands of lines. The emission and absorption spectra are basic characteristics of a substance and are used as identification in the science of SPECTROSCOPY. Spectra result from ELECTRONS making transitions between different energy levels in atoms or molecules of the substance, giving rise to the emission or absorption of electromagnetic radiation.

SPERMATOGENESIS

spermatogonia
mitosis
meiosis
primary spermatocyte
secondary spermatocyte
spermatid
spermatozoan

Immature sex cells in the testes of the male divide by one form of cell division, mitosis, and become primary spermatocytes. Each one then divides by meiosis, a cell division peculiar to the reproductive organs, to form two secondary spermatocytes, each containing half the full number of chromosomes. A second meiotic division splits each spermatocyte into two spermatids, which then mature into sperm.

speed Rate of movement. It is the ratio of the distance covered to the time taken by a moving object. It is a SCALAR quantity. Speed in a specified direction is VELOCITY, a VECTOR quantity.

speedometer Apparatus for recording the SPEED of a vehicle. In a traditionally designed car or lorry, it is driven by a flexible cable, connected to the TRANSMISSION, which turns at a speed proportional to the road speed. Inside the speedometer the cable rotates a magnet, which partly turns a spring-loaded drum to which the indicator needle or coloured strip is attached. In more modern designs, a sensor in the gearbox generates an electrical signal according to the speed of rotation of the transmission, and this signal is converted into a luminous display on the dashboard.

speleology Scientific study of CAVES and cave systems. Included also are the hydrological and geological studies concerned with the rate of formation of STALAGMITES and STALACTITES, and the influence of GROUNDWATER conditions on cave formation. A special aspect is the study of the animals that live in caves.

sperm (spermatozoon) Male sex cell (GAMETE) in sexually reproducing organisms. It corresponds to the female OVUM. The head of the sperm contains the genetic material of the male parent, while its tail or other motile structure provides the means of moving to the ovum to carry out FERTILIZATION. *See also* SEXUAL REPRODUCTION

spermatogenesis Process by which SPERM are formed. Sperm (spermatozoa) are produced by a series of cell divisions that take place in the seminiferous tubules of a male animal's TESTES. Initially, germ cells divide to produce **spermatogonia**, which further divide to form **spermatocytes**. These finally divide yet again to produce **spermatids**, which develop into spermatazoa.

spermatophyte Traditionally a member of the division of seed-bearing plants (Spermatophyta), including most trees, shrubs and herbaceous plants. A spermatophyte has a stem, leaves, roots and a well-developed vascular system. The dominant generation is the SPOROPHYTE. The widely accepted Five Kingdoms classification now classifies seed plants as several distinct phyla: the Angiospermophyta (ANGIOSPERMS or flowering plants), Coniferophyta (CONIFERS), Ginkgophyta (GINKGO or maidenhair tree), Cycadophyta (CYCADS) and Gnetophyta (a group of cone-bearing desert plants).

spermicide *See* CONTRACEPTIVE

Sperry, Elmer Ambrose (1860–1930) US inventor and industrialist. He is celebrated for his improvements to the GYROSCOPE. In 1910 Sperry invented the GYROCOMPASS. He also designed a number of other electrical devices, including a powerful searchlight. Sperry used gyroscopes in his development of AUTOPILOT systems for NAVIGATION. He founded the American Institute of Electrical Engineers.

sphagnum Genus of MOSS that grows on boggy soils. When decomposed under boggy conditions, sphagnum moss forms PEAT MOSS, much used in horticulture.

sphalerite (blende) Sulphide mineral composed of zinc sulphide (ZnS); it is an important source of zinc. Sphalerite has cubic system tetrahedral crystals or granular masses. It is white when pure, but more commonly yellow, black or brown with a resinous lustre. It is often found in hydrothermal veins with galena, and in limestones where it occurs by replacement. Hardness 3.5–4; r.d. 4.

Sphenophyta Phylum comprising the HORSETAILS

sphere Three-dimensional geometric figure formed by the LOCUS in space of points equidistant from a given point (the centre), or equivalently a surface generated by rotation of a CIRCLE about a diameter. If the distance from the centre to the surface is r, the volume is $4/3\pi r^3$ and the surface area is $4\pi r^2$. A **spheroid** is the name given to the figure enclosed by the sphere.

spherical triangle TRIANGLE formed by the intersection, on the surface of a SPHERE, of arcs of three great circles (circles that have the same RADIUS as the sphere). The sides of spherical triangles are measured in terms of the angles that these arcs subtend at the sphere's centre. **Spherical trigonometry**, the branch of geometry concerned with the properties of such triangles, is used in navigation.

sphincter Ring of muscle surrounding a body orifice which can open it or seal it off. Important sphincters include the pyloric sphincter in the stomach and the anal sphincter.

sphygmomanometer Instrument used to measure BLOOD PRESSURE. The device incorporates an inflatable rubber cuff connected to a column of mercury with a graduated scale. The cuff is wrapped around the upper arm and inflated to apply pressure to a major artery. When the air is slowly released, the pressure readings can be ascertained from the scale.

A B C D E

◄ **spider** When spinning its web for catching prey, the spider first casts out a thread of silk to form a horizontal strut (A). A second, drooping thread is trailed across below the bridge-line. Halfway along the second thread, the spider drops down on a vertical thread until it reaches a fixed object (B). It pulls the silk taut and anchors it, forming a "Y" shape, the centre of which forms the hub of the web. The spider then spins the framework threads and the radials, which are linked together at the hub (C). After spinning the remainder of the radials, a wide, temporary spiral of dry silk is laid down, working from the inside of the web outwards (D). This holds the web together while the spider lays down the sticky spiral. This is laid down starting from the outside, and is attached successively to each radial thread (E). The spider eats the remains of the dry spiral as it proceeds. The central dry spirals are left as a platform for the spider, which will often lie in wait there during the night. In the daytime the spider retreats to a nearby silk shelter.

spicule Spear-like column of hot gas ejected from the Sun's CHROMOSPHERE as a jet, often reaching heights of 8,000km (5,000mi) above the solar surface. Most spicules are less than 500km (300 miles) in diameter. *See also* SUN

spider Any of numerous species of terrestrial, invertebrate, ARACHNID arthropods found worldwide. Spiders have an unsegmented abdomen attached to a cephalothorax (connected head and thorax) by a slender pedicel. There are no antennae; sensory hairs are found on the appendages (four pairs of walking legs). Most species have spinnerets on the abdomen for spinning silk to make egg cases and webs.

spilite Fine-grained, extrusive IGNEOUS ROCK found near VOLCANOES. The dark-coloured rock occurs as PILLOW LAVA, which is formed where LAVA from a volcanic eruption flows into the sea; it often contains cavities formed by gas bubbles. It is a basic rock containing up to 40% silica, and more sodium than basalt, the common volcanic rock.

spin (symbol s) In QUANTUM MECHANICS, intrinsic angular momentum possessed by some ELEMENTARY PARTICLES, atoms and nuclei. This may be regarded by analogy as the spinning of the particle on its own axis. Spin is one of the quantum numbers by which a particle is specified.

spinal column *See* SPINE

spinal cord Tubular, central nerve cord, lying within the SPINE (vertebral column or backbone) and bathed in cerebrospinal fluid. It connects the PERIPHERAL NERVOUS SYSTEM to the BRAIN, with which it makes up the CENTRAL NERVOUS SYSTEM (CNS). It gives rise to the 31 pairs of spinal nerves, each of which has sensory and motor fibres. *See also* NERVOUS SYSTEM

spinal fluid *See* CEREBROSPINAL FLUID

spinal nerves Thirty-one pairs of nerves branching out from the SPINAL CORD to supply the muscles of the trunk and limbs and carry sensory information to the BRAIN. They are mixed nerves containing both sensory and motor fibres. The motor fibres arise from the anterior (frontal) side of the spinal cord and the sensory fibres from the rear.

spindle Rod-shaped structure formed from microtubules in the CYTOPLASM of cells during MITOSIS or MEIOSIS. CHROMOSOMES are attached at the bulge of the spindle (the equator). The spindle draws the chromosomes apart causing a cell to divide. *See also* METAPHASE

spine Backbone of VERTEBRATES, extending from the base of the SKULL to the tip of the tail and enclosing the SPINAL CORD. The human spine consists of 26 small vertebrae interspersed with INTERVERTEBRAL DISCS of CARTILAGE. It comprises seven cervical, 12 thoracic and five lumbar vertebrae. The five sacral and four vertebrae of the COCCYX fuse together to make two solid bones. The spine articulates with the bones of the skull, ribs and hip bones and provides points of attachment for the back muscles. *See also* SKELETON

spinel Oxide mineral, magnesium aluminium oxide ($MgAl_2O_4$), found in IGNEOUS and METAMORPHIC ROCKS. It has cubic system, frequently twinned, octahedral crystals. It is either glassy black, red, blue, brown or white in colour. Ruby spinel from Sri Lanka is a valuable gemstone. Hardness 7.5–8; r.d. 3.8.

spinning Process of making thread or yarn by twisting FIBRES together. The fibres may be of animal, vegetable or synthetic origin. Machines developed for mechanizing the spinning process include the SPINNING WHEEL (a simple machine for spinning a single strand of yarn), SPINNING FRAME, SPINNING JENNY, and SPINNING MULE. Today, large machines carry out the same basic process, but at much increased speeds. Different machines are designed to suit each kind of fibre. *See also* TEXTILES

SPIN

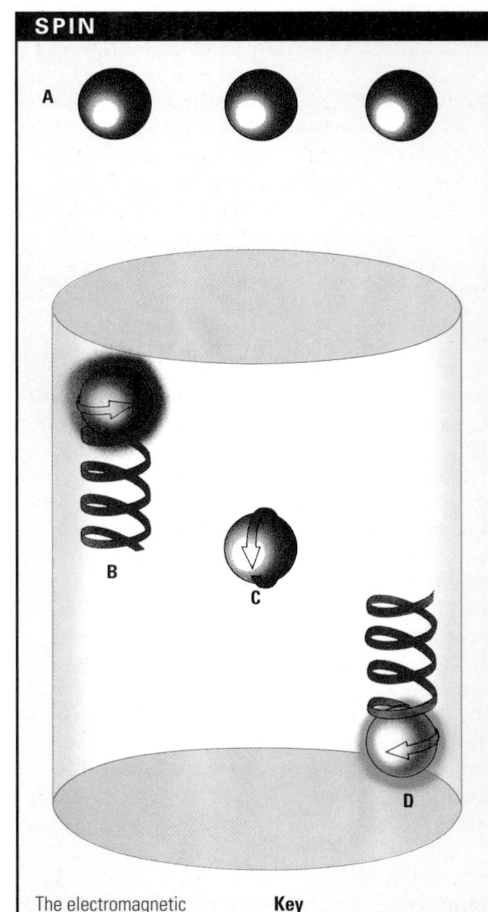

The electromagnetic interaction is one test by which different states of a particle can be recognized. An external magnetic field reacts with three, as yet indistinguishable, particles that separate into different energy states according to their respective electric charges.

Key
A) Three particles look alike in the absence of an external magnetic field
B) Positively charged particle assumes one energy state because of the interaction of its charge and an external magnetic field
C) Neutral particle
D) Negatively charged particle

spinning frame Machine for spinning yarn. Richard ARKWRIGHT invented the spinning frame in *c*.1768. He called his invention a water frame, as it was powered by a water wheel. The motion of the wheel turned several spindles, so that many threads could be spun at the same time.

spinning jenny Hand-operated spinning machine invented by James HARGREAVES in *c*.1764. It operated on the same basic principles as the SPINNING WHEEL, but spun several strands of yarn simultaneously.

spinning mule (Crompton's mule) Spinning machine invented (1779) by Samuel CROMPTON. The machine was called a mule because it was a hybrid, combining the main features of Hargreaves' SPINNING JENNY and Arkwright's SPINNING FRAME.

spinning wheel Simple machine for spinning a single strand of yarn. A hand-operated spinning wheel was used in India more than 2,500 years ago. In *c*.1500 a foot-operated wheel was invented in Germany. A spinning machine has a large wheel, which is linked by a belt to a spindle. The spindle turns rapidly, spinning the fibres into yarn.

spiracle External opening for RESPIRATION in var-

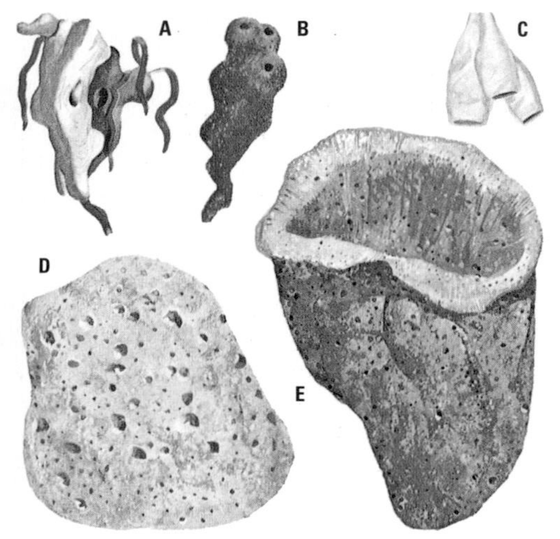

▲ **sponge** Classified according to the substance that makes up the supporting skeleton, the sponges include the breadcrumb sponge *Halichondria panica* (A), with a skeleton (B) composed of spicules of silicon. The purse sponge *Grantia compressa* (C) has a skeleton of calcium carbonate spicules and is placed in a group called calcareous sponges. Two commercial sponges, *Hippospongia equina* (D) and *Euspongia officinalis* (E), have skeletons of a horny, elastic substance called spongin. After harvesting from the seabed, they are dried, beaten and washed to remove hard debris so that the only part remaining is the skeleton.

ious animals. In insects and spiders, it is the opening to a TRACHEA. In sharks, rays and some bony fish, water passes through a pair of spiracles during GILL respiration. In whales, the nasal opening is called a spiracle.

spiral galaxy Type of regular GALAXY as classified by Edwin HUBBLE.

spirochaete General name applied to a group of protozoa-like BACTERIA that are spiral-rod shaped and capable of flexing and wriggling their bodies as they move about. SYPHILIS is the best known disease caused by a spirochete (*Treponema pallidum*).

spit Ridge of sand or shingle on a coastline, which has been built up as a result of LONGSHORE DRIFT. A spit is attached to the land at one end, and may consist of sand, mud, shingle or any combination of these. Most spits protrude across an ESTUARY, and may eventually block the estuary to form a DELTA.

spleen Dark-red organ located on the left side of the abdomen, behind and slightly below the stomach. It is important in both the BLOOD and LYMPHATIC SYSTEMS, helping to process LYMPHOCYTES, destroying worn out or damaged ERYTHROCYTES (red blood cells) and storing iron. Removal of the spleen (splenectomy) is sometimes necessary following trauma or in the treatment of some blood disorders.

spodumene Silicate mineral, lithium aluminium silicate ($LiAlSi_2O_6$). It is the chief source of LITHIUM. Spodumene displays monoclinic system prismatic crystals or masses. It is opaque or transparent in many hues, and the transparent lilac and green varieties are used as gems. Hardness 6.5–7; r.d. 3.2.

sponge Primitive, multicellular, aquatic animal. Its extremely simple structure is supported by a skeleton of lime, silica or spongin. There is no mouth, nervous system or cellular coordination, nor are there any internal organs. Sponges reproduce sexually and by asexual budding. There are *c*.5,000 species, including the simple sponge genus *Leucosolenia*. Length: 1mm–2m (0.04in–6ft). Phylum: Porifera (*see* PORIFERAN).

spontaneous combustion Outbreak of FIRE without external application of HEAT. When combustible material, such as damp hay, paper or rags, is slowly oxidized by bacteria or air, the temperature may rise to the ignition point. *See also* COMBUSTION

spontaneous generation Belief, now discredited, that living organisms arise from non-living matter. It supposedly explained the sudden appearance of maggots on decaying meat. *See* EVOLUTION

spore Small reproductive body that detaches from the parent organism to produce new offspring without having to fuse with another reproductive cell. Mostly microscopic, spores may consist of one or several cells (but do not contain an embryo) and are produced in large numbers. Some germinate rapidly, others "rest", surviving unfavourable environmental conditions. Spores are formed by ferns, horsetails, mosses, fungi, bacteria and some protozoa. Spores are formed during the SPOROPHYTE stage in the life cycle of such organisms. *See also* GAMETE

sporophyte DIPLOID (having two sets of chromosomes) stage in the life cycle of plants or algae. In most plants and algae, the sporophyte gives rise to HAPLOID (having only one set of chromosomes) spores which germinate to produce a haploid generation (the GAMETOPHYTE stage). The haploid generation produces the GAMETES. In ferns, horsetails, conifers and flowering plants, the diploid sporophyte is the dominant phase of the life cycle, the plant body we usually see. In mosses and liverworts, the main plant body is the gametophyte. *See also* ALTERNATION OF GENERATIONS

spring Point on the Earth's surface at which underground water emerges. Springs may emerge at points on dry land or in the beds of streams or ponds. They are generally located where there is a change of rock type, and the water flows out at the top of an impervious layer, such as slate, shale or clay. They are an important part of the water cycle. The mineral content of spring water varies with the surrounding soil or rocks.

spring Mechanical device designed to be elastically compressed, extended or deflected. It may be used to store energy, absorb shock, or maintain contact between two surfaces.

spring tide *See* TIDE

Sputnik 1 World's first artificial SATELLITE, launched by the Soviet Union on 4 October 1957. Weighing 83.5kg (184lb) and equipped with a radio transmitter, Sputnik 1 circled the Earth sending back signals for 21 days. **Sputnik 2** was launched in November 1957. Weighing 508kg (1,120lb), it carried the dog 'Laika'. *See also* SPACE EXPLORATION

Squamata Order of cold-blooded, scaly REPTILES, comprising the snakes (suborder Serpentes) and lizards (suborder Sauria) . They are mostly terrestrial vertebrates that first appeared at the end of the Triassic period, *c*.170 million years ago. Typically, **lizards** have cylindrical bodies with four legs, a long tail and movable eyelids. **Snakes** are legless reptiles with no external ear openings, eardrums or middle ears; sound vibrations are detected through the ground. Their eyelids are immovable. The long, forked, protractile tongue is used to detect odours. Poisonous species have hollow or grooved fangs, though which they inject venom.

square RECTANGLE with four sides of the same length. In arithmetic or algebra, a square is the result of multiplying a quantity by itself: the square of 3 is 9, and the square of x is x^2.

square root Number or quantity denoted $\sqrt{x}$ that must be multiplied by itself to give the number x. The square root of 4 is 2, i.e. $\sqrt{4} = 2$. To four decimal places, $\sqrt{2} = 1.4142$. A negative number has complex square roots. *See also* CUBE ROOT; ROOT

stabilizer In AERODYNAMICS, vertical and horizontal fins that prevent erratic rolling and pitching respectively. They are usually placed together in the form of a tailplane. Like the wings, the horizontal fins have an AEROFOIL shape and can provide LIFT at take-off. In flight, however, they act only as stabilizers and in some designs the air THRUST on them acts downwards. In marine engineering, fins or tanks are used to keep a submarine or ship steady in rough seas. *See also* PITCH

stack Pillar of rock that forms an islet at the end of a headland. A stack is the result of a series of events caused by EROSION. Sea waves striking a headland first carve a CAVE by opening up JOINTS in the rock. The cave gradually penetrates the headland until an arch is formed. Finally, the top of the arch collapses and leaves a stack separated from the rest of the headland. The process may be repeated several times, giving a series of stacks.

stained glass Coloured GLASS used for decorative effect in windows. Stained glass is most commonly found in churches. The oldest surviving examples of complete stained glass windows are in Augsburg Cathedral, Germany, and date from *c*.1100. In its purest form, stained glass is made by adding metal-oxide colouring agents during the manufacture of glass. Shapes cut from the resulting sheets are then arranged to form patterns or images. These shapes are joined and supported by flexible strips of lead that form dark, emphatic contours. Details are painted onto the glass surfaces in liquid ENAMEL and fused on by heat. Other techniques include the application of coloured surface coatings onto clear glass. Intricate designs may be scraped into the surface. In modern windows, the pieces of glass are sometimes joined using an adhesive instead of lead strips.

stainless steel Name for a group of IRON alloys that resist CORROSION. Besides carbon, contained in all STEELS, stainless steels contain from 12% to 25% CHROMIUM. This makes the steel stainless and rust-resistant by forming a thin, protective oxide coating on the surface. Most stainless steels also contain NICKEL. Various other metals and non-metals may be added to give the steel particular properties. Stainless steel is used to make kitchen utensils, cutlery, chemical process equipment, and fittings of various kinds, including parts for cars.

stalactite Icicle-shaped deposit of tiny CALCIUM CARBONATE crystals found hanging down from the roof of CAVES in Carboniferous limestone areas. Stalactites are formed by water slowly percolating through the rocks; as the water droplets are about to fall from the cave roof, a tiny layer of calcium precipitates out of the water and solidifies. Gradually, the deposits build up to form a column of calcium carbonate crystals known as CALCITE.

stalagmite Column of CALCIUM CARBONATE crystals that grows upwards from the floor of CAVES in a Carboniferous limestone areas. Stalagmites are formed as a result of water dripping from the ceiling of the cave and leaving tiny deposits of calcium carbonate on the floor. The deposits gradually build up into CALCITE crystals, and can sometimes extend upwards to meet the STALACTITE above, forming a column or pillar. Stalagmites tend to be fatter than stalactites, possibly because the water splashes out when it drops to the floor, causing a wider spread of calcium carbonate.

stamen POLLEN-producing element of a FLOWER. It consists of an ANTHER, in which pollen is produced, on the end of a stalk-like filament. The arrangement and number of stamens is important in the classification of ANGIOSPERMS (flowering plants).

stamping Method of forming metal. A flat blank of metal is stretched over a die and struck with a movable punch into the desired shape. *See also* METALLURGY

standard deviation In STATISTICS, a measure of deviation of observed data or scores from the average or MEAN. It is written as either σ or *s*. A small standard deviation indicates that observations cluster around the MEAN, while a large one indicates that the data points are spread far from the mean. The standard deviation is equal to the square root of a quantity called the variance. The variance is the mean of the sum of the squared differences of the data points from the mean.

standard form Scientific method of writing NUMBERS, especially very large or very small numbers. The first non-zero digit of the number is placed before the decimal point and all other digits are placed in order after the decimal point. This decimal is multiplied by an appropriate power of 10. For example, 1,721,448 is written as $1.721\,448 \times 10^6$ in standard form and 0.003378159 is written as $3.378\,159 \times 10^{-3}$.

standard model Model of ELEMENTARY PARTICLES and their interactions that provides the best description of current physical phenomena. Particles are divided into HADRONS (which experience the STRONG NUCLEAR FORCE and are made up of QUARKS), LEPTONS (particles such as ELECTRONS) that do not experience the strong nuclear force, and particles that "carry" interactions (such as PHOTONS, W and Z BOSONS and GLUONS; these mediate in interactions between particles in the other two categories). Physicists are striving to go beyond the standard model to achieve a GRAND UNIFIED THEORY (GUT). *See also* FUNDAMENTAL FORCES

standard electrode (reference ELECTRODE) In ELECTROCHEMISTRY, a HALF-CELL combined with other half-cells and used as a reference. The basic standard is the hydrogen electrode – a platinum electrode immersed in a molar solution of hydrogen ions and covered with an absorbent layer of platinum black, over which is bubbled molecular hydrogen. This is taken to have a ELECTRODE POTENTIAL of zero.

standard temperature and pressure (STP) (formerly normal temperature and pressure, NTP) In chemistry and physics, standard conditions for measurements, especially when comparing the volumes of gases. STP is a **temperature** of 0°C (32°F) and a **pressure** of 1 standard atmosphere (760mm of mercury or 101,325 pascals).

standing wave (stationary wave) In physics, a WAVE in which the points of maximum vibration (the **antinodes**) and the points of no vibration (the **nodes**) do not move. A standing wave is formed by the interference of TRAVELLING WAVES of equal frequency and intensity moving in opposite directions.

Stanley, Wendell Meredith (1904–71) US biochemist who shared the 1946 Nobel Prize for chemistry with John Northrop (1891–1987) and James SUMNER for his research into VIRUSES. In 1935 Stanley crystallized tobacco mosaic virus and showed that it consists of PROTEIN and NUCLEIC ACID molecules in a rod-shaped form. This enabled the exact molecular structure and means of proliferation of this and other viruses to be determined. Stanley also studied influenza viruses and developed a VACCINE against the disease.

Staphylococcus Genus of spherical bacteria that grow in grape-like clusters and are found on the skin and mucous membranes of human beings and other animals. Pathogenic staphylococci cause a range of local or generalized infections, including PNEUMONIA and SEPTICAEMIA. They are the most common cause of food poisoning. They may be destroyed by ANTIBIOTICS, although some strains have become resistant. they require oxygen for growth.

star Hot, self-luminous ball of gas, held together by its own GRAVITATION and giving off ELECTROMAGNETIC RADIATION produced by NUCLEAR ENERGY, mainly the conversion of hydrogen into helium. The stars visible to the naked eye are all in the MILKY WAY. They are classified according to their MAGNITUDE; from the brightest (first magnitude) to the dimmest (sixth magnitude). Their brightness varies because of their relative distance from Earth and their different LUMINOSITIES. Frequently, stars are grouped together in STELLAR CLUSTERS. The velocity of the movement of a star through space is measured by its PROPER MOTION and RADIAL VELOCITY. The temperatures and luminosities of stars are prescribed by their masses. The most massive stars are about 100 solar masses (a hundred times heavier than the Sun). Above this mass, a star is not stable and is likely to break up. Stars with the greatest mass are very luminous and hot (*c*.20,000°C). The hotter the star, the shorter the wavelength at which light appears most intense. Thus, the hottest stars appear blue. RED GIANTS are extremely large stars, with a diameter between 10 and 100 times that of the Sun, but are very cool. Stars with an average mass, such as the Sun, are yellow, while small stars are a dull red (*c*.3,000°C). The smallest stars (WHITE DWARFS and NEUTRON STARS) contain less than one-twentieth of a solar mass; below this the temperature does not become high enough for nuclear reactions to take place, and the object becomes a BROWN DWARF. The colours, and hence the temperatures, of stars give rise to the scheme of SPECTRAL CLASSIFICATION. Most stars are born not singly but in pairs (BINARY STARS) or groups. VARIABLE STARS (such as SUPERNOVAE) have fluctuating brightness. *See also* BLACK BODY; GALAXY; HERTZSPRUNG-RUSSELL DIAGRAM; SPECTRUM; STELLAR EVOLUTION

starch CARBOHYDRATE, stored in many plants, that provides about 70% of humankind's food in such forms as rice, potatoes and cereals. Animals and plants convert it to GLUCOSE for energy (RESPIRATION). Consisting of linked glucose units, starch exists in two forms: **amylose**, in which the glucose chains are unbranched, and **amylopectin**, in which they are branched. It is made commercially from cereals, maize, potatoes and other plants. Starch is used in the manufacture of adhesives and foods (as a thickening agent) and to stiffen laundered clothes and other fabrics.

star cluster *See* STELLAR CLUSTER

Stark, Johannes (1874–1957) German physicist. He received the 1919 Nobel Prize for physics for his discovery (1913) of the STARK EFFECT and for his work demonstrating the DOPPLER EFFECT. In the 1920s Stark rejected the theory of RELATIVITY and QUANTUM THEORY.

Stark effect In physics, splitting of spectral lines of an ATOM or MOLECULE when it is placed in a strong electric field. It was discovered (1913) by Johannes STARK. It follows from the ELECTRON theory of Hendrik LORENTZ. *See also* ZEEMAN EFFECT

Starley, James (1830–81) British inventor of sewing machines and BICYCLES. Working in Coventry from 1857, he produced his first bicycle (*The Coventry*) in 1868. His "ordinary" or "penny farthing" bicycle of 1871 had centre pivot steering. Starley introduced the *Coventry* tricycle in 1876, adding differential gears in 1877.

Starling, Ernest Henry (1866–1927) British physiologist who contributed greatly to the knowledge of body functions. Starling made major discoveries about body TISSUES, the mechanical controls of the HEART and the ENDOCRINE SYSTEM.

static electricity Study of ELECTRIC CHARGES at rest, not flowing as an ELECTRIC CURRENT. Ordinary uncharged ATOMS have equal numbers of positive and negative ELECTRONS. On the everyday (macroscopic) scale, electrically charged objects are ones that have either too many or too few electrons. COULOMB'S LAW relates the forces that charged bodies have on each other and to their charge and the distance between them. Static ELECTRICITY can be produced by friction, such as by combing hair (which acquires a negative charge) or rubbing nylon material. Electrons may then jump off as a spark, shocking anyone touching the object. Lightning is the result of a large-scale build-up of static electricity. This form of electricity is studied in the branch of physics known as ELECTROSTATICS.

statistical mechanics Branch of physics that studies large-scale properties of MATTER based on

STALACTITE/STALAGMITE

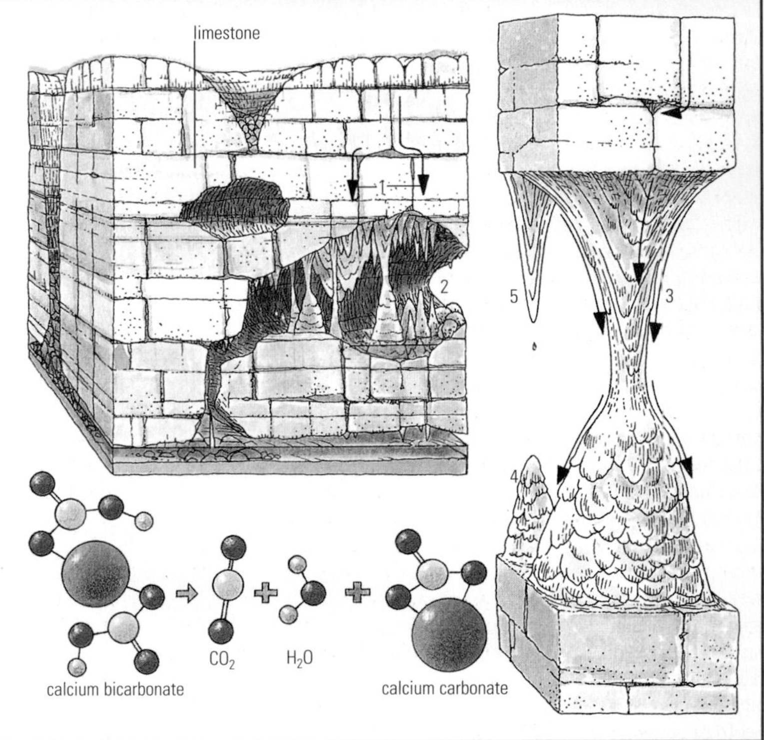

Weak carbonic acid, formed when rain absorbs carbon dioxide from the atmosphere, dissolves tiny amounts of calcium carbonate from limestone rock, as it percolates down through the rock (1). If the water enters a previously eroded cave (2), it forms drips (3). Some of the dissolved carbon dioxide escapes from the water, and the calcium bicarbonate in the water is precipitated as a calcite known as travertine, consisting of calcium carbonate. They form rising columns called stalagmites (4) and descending pillars from which water drips, called stalactites (5).

► **stealth technology** The Lockheed F-117A employs stealth technology to render itself practically invisible to radar. The aircraft's materials and design are such that its radar cross-section (RCS) – the imaginary size of a perfectly reflecting object that would reflect the same amount of energy – is reduced to within the levels of background noise. To achieve this, compromises are required in terms of the aircraft's performance and operating and maintenance costs.

the statistical laws of large numbers. The large number of molecules in such a system allows the use of statistics to predict the probability of finding the system in any state. The ENTROPY (disorder or randomness) of the system is related to the number of its possible states; a system left to itself will tend to approach the most probable distribution of energy states. *See also* THERMODYNAMICS

statistics Science of collecting and classifying numerical data. Statistics can be **descriptive** (summarizing the data obtained) or **inferential** (leading to conclusions or inferences about larger numbers of which the data obtained are a sample). Inferential statistics are used to give a greater degree of confidence to conclusions, since statistics make it possible to calculate the PROBABILITY that a conclusion is in error.

Staudinger, Hermann (1881–1965) German chemist who was awarded the 1953 Nobel Prize for chemistry for his work on the structure of PLASTICS. He showed that plastics, rather than being random aggregations of MOLECULES, are POLYMERS, similar to CELLULOSE, in which the molecules are bonded together to form a long chain. This discovery contributed greatly to the growth of the plastics industry.

staurolite Silicate mineral, iron magnesium aluminium silicate, found in SCHISTS and GNEISSES. It has orthorhombic system, prismatic and flattened crystals and is glassy brown in colour. Hardness 7–7.5; r.d. 3.6.

steady-state theory Cosmological theory put forward (1948) by the Austrian astronomers Hermann Bondi (1919–) and Thomas Gold (1920–), and further developed by Fred HOYLE and others. According to this theory, the UNIVERSE has always existed; it had no beginning and will continue forever. Although the Universe is expanding, it maintains its average density – its steady state – through the continuous creation of new MATTER. Most cosmologists now reject this theory because it cannot explain naturally the COSMIC MICROWAVE BACKGROUND or the observation that the appearance of the Universe has changed with TIME. The BIG BANG theory is the currently accepted theory. *See also* COSMOLOGY; EXPANDING UNIVERSE

stealth technology Methods used to render an aircraft nearly invisible to RADAR, heat and visual detection. To achieve "invisibility", a stealth aircraft must have sympathetic airframe design (all radar-reflecting "hard" edges smoothed away), engine exhaust dampers that mask and disperse jet efflux, and radar absorbent material that "holds" electronic emissions rather than reflecting them. This coating has to be regularly applied, since it is sensitive to water and wear. In 1983 the American Lockheed F-117A became the first "stealth" aircraft to enter frontline service; it was followed a decade later by the Northrop B-2. The F-19 stealth fighter was used in the Gulf War (1991). Few aerodynamic advances have had such a profound effect on military aviation as stealth technology.

steam Gaseous form of WATER. As commonly seen in the home, steam contains minute drops of water and is technically called **wet steam**. The pure gas, **dry steam**, is obtained by heating steam at temperatures well above the boiling point of water, 100°C (212°F). It is used widely in industry for heating, transporting heat (in heat exchangers) and to drive TURBINES.

steam engine ENGINE powered by steam. Steam, generated by heating water, is used to produce movement. In some engines, the steam forces pistons to move along cylinders. This results in a reciprocating (back-and-forth) motion. A mechanism usually changes this into rotary motion. Steam LOCOMOTIVES use RECIPROCATING ENGINES. STEAM TURBINES are engines that produce rotary motion directly by using the steam to turn sets of fan-like wheels. Steam turbines drive power station GENERATORS and ships' propellers. In any steam engine, some of the HEAT used to turn water into steam in a boiler is converted into energy of motion. The heat may be produced by burning fuel in a FURNACE, or may come from a NUCLEAR REACTOR. The first steam engine was a form of pump, used to remove water from mines. It was invented (1689) by Thomas SAVERY. In his simple machine, steam condensed to form a very small volume of water, thus producing a partial vacuum. This was used to suck water up a mine shaft. In 1712 Thomas NEWCOMEN invented a steam-operated pump with pistons. From the 1760s, James WATT improved on Newcomen's ideas and produced more efficient steam engines. This led to the use of steam engines to power machinery in factories. In 1884 British engineer Charles Parsons (1854–1931) invented the first practical steam TURBINE. His machines were so efficient that turbines soon started to replace reciprocating steam engines in power stations. The most astonishing development in steam power has been the manufacture of a completely enclosed working steam engine of microscopic size. Japanese scientists made the device using techniques developed for manufacturing INTEGRATED CIRCUITS. A small current passed through an electric heating element changes a small droplet of water into steam to move a piston. Practical applications for this device are still awaiting discovery.

steamship *See* SHIP

steam turbine STEAM ENGINE with rotating blades that are turned by steam. *See also* TURBINE

stearic acid (octadecanoic acid) Common, saturated, FATTY ACID ($CH_3(CH_2)_{16}CO_2H$) commonly found as a glyceride in animal and vegetable fats. It is used in ointments, lubicants, creams, candles and soap. Properties: r.d. 0.85; m.p. 70°C (158°F).

stearin (glyceryl tristearate, $C_{57}H_{110}O_6$) Triglyceride of STEARIC ACID ($CH_3(CH_2)_{16}CO_2H$). It is present in many animal and vegetable fats, particularly the hard ones such as tallow and cacao butter. Stearin is prepared from stearic acid and GLYCEROL, using aluminium oxide as a catalyst.

steatite *See* SOAPSTONE

steel Group of IRON alloys containing a little CARBON. The great strength of steel makes it an extremely important material in construction and manufacturing. The most common type is called **plain carbon** steel, since carbon is the main alloying material. This kind of steel usually contains less than 1% of carbon by weight, with smaller amounts

STEAM ENGINE

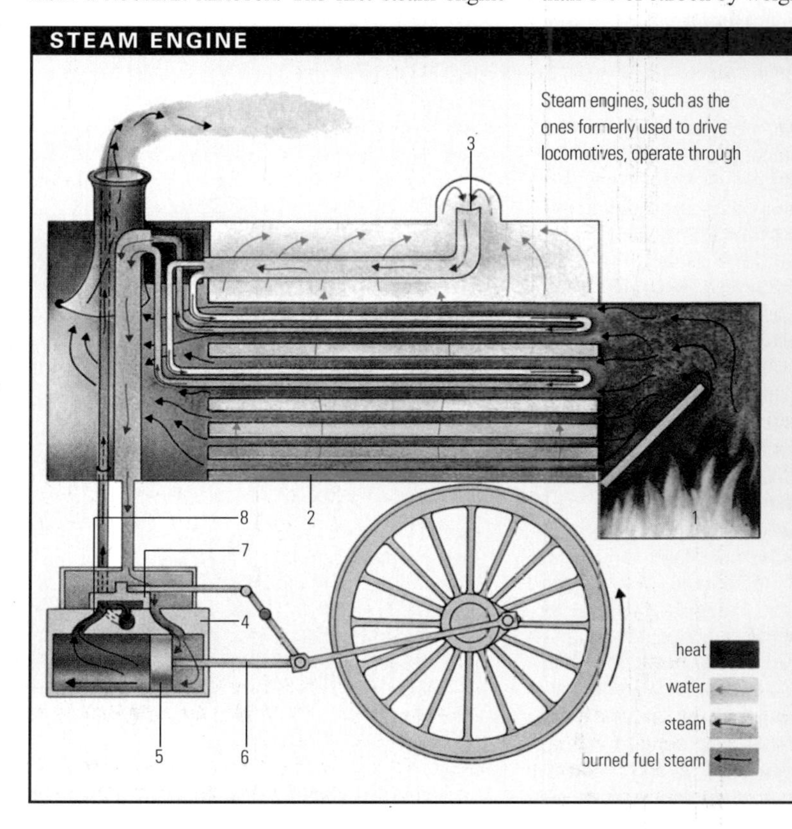

Steam engines, such as the ones formerly used to drive locomotives, operate through the production and expansion of steam as water is heated. A coal or wood-fired furnace (1) heats a water-filled boiler (2), producing steam. The steam rises and is forced, via a steam dome (3), through a series of tubes to the cylinder (4), where it drives the piston (5) in a reciprocal motion. Linked to the piston's connecting rod (6) is a slide valve (7) that at one stage allows steam to enter the cylinder (as shown), blocking off the exhaust port (8). This creates pressure that forces the piston forward, at which point the slide valve is in a position where the exhaust port is open and the steam escapes. The motion of the wheel forces the piston back and the cycle begins again.

of other alloying materials, such as manganese, silicon, sulphur and phosphorus. ALLOY steels contain some carbon, but owe their special properties to the presence of some combination of the alloying metals manganese, nickel, chromium, vanadium and molybdenum. **Low-alloy** steels, with less than 5% of alloying metals, are exceptionally strong and are used in buildings, bridges and machine parts. **High-alloy** steels contain more than 5% of alloying metals. This group includes various forms of STAINLESS STEEL. Steel has been produced for more than 2,000 years, but early methods were slow and laborious, and produced only small quantities. Large-scale steelmaking became possible in the mid-1800s, when the BESSEMER PROCESS and OPEN-HEARTH PROCESS were introduced. Today, most steel is produced by the BASIC OXYGEN PROCESS. An ELECTRIC FURNACE is used for some special steels containing materials that would become oxidized in other steel-making processes.

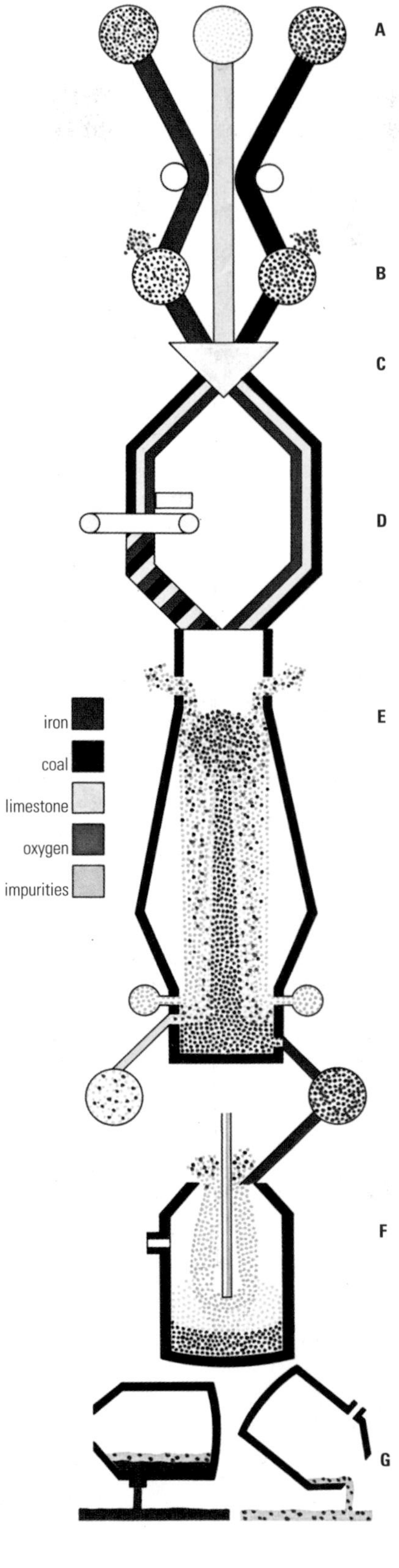

Stefan–Boltzmann law (Stefan's law) In physics, principle that the total radiated energy from a BLACK BODY per unit time per unit area is directly proportional to the fourth power of its THERMODYNAMIC temperature (T^4). The constant of proportionality (the **Stefan–Boltmann constant**) is equal to 5.6997×10^{-8} Js^{-1} m^{-2} K^{-4}. It is named after the Austrian physicist Josef Stefan (1835–93) and was derived for thermodynamic purposes by Ludwig BOLTZMANN.

Stein, William Howard (1911–80) US biochemist. He shared the 1972 Nobel Prize for chemistry with Christian ANFINSEN and Stanford MOORE for research into the molecular structure of PROTEINS. Stein developed techniques for the analysis of PEPTIDES and AMINO ACIDS obtained from proteins and applied these methods to determining the molecular structure of the ENZYME ribonuclease. This was of great importance to the understanding of processes involved in protein synthesis within CELLS.

stellar cluster (star cluster) Any of innumerable collections of gravitationally associated STARS occurring within GALAXIES. Stellar clusters are of two main types: the OPEN CLUSTER (or galactic cluster) and GLOBULAR CLUSTER. **Open clusters**, usually found in the spiral arms of galaxies, consist of up to several thousand young stars belonging to Population I; **globular clusters**, which are much more concentrated, are found in the halo surrounding the centres of galaxies and consist of old Population II stars.

stellar evolution Various phases of a STAR over its lifetime. A star "lives" for between a few thousand years to thousands of millions of years depending on its MASS. Stars form when an INTERSTELLAR CLOUD of gas and dust collapses under its own GRAVITATION. As the cloud collapses, atoms collide and generate heat. This process continues until the heat generated is sufficient to cause nuclear FUSION reactions, converting hydrogen to helium. The reactions from the core throw out considerable amounts of radiation, which stops further collapse. This stage (the MAIN SEQUENCE phase) is the longest in a star's lifetime. Eventually the mainly hydrogen core of the star is depleted, and fusion can no longer occur. With the central energy source removed, the core collapses under gravitation and heats itself further until hydrogen fusion is able to occur in a spherical shell surrounding the core. As this change takes place, the outer layers of the star expand considerably and the star becomes either a RED GIANT, or in the most massive stars, a SUPERGIANT. During this phase the core can reach temperatures of 100 million K, hot enough for fusion of helium to carbon. When this second process of fusion is complete, the core collapses once again and heats up. In **low-mass** stars the temperature will not rise sufficiently for carbon fusion to take place, and the red giant loses its outer layers leaving a WHITE DWARF. In **high-mass** stars, carbon fusion is initiated, converting the carbon into elements with atomic weights close to that of iron. At this stage no further fusion is possible, and the core collapses explosively, throwing off outer layers in a SUPERNOVA explosion. The resulting super dense core forms either a NEUTRON STAR or BLACK HOLE. *See illustration on page 342*

stem Main, upward-growing part of a plant that bears leaves, buds and flowers or other reproductive structures. In VASCULAR PLANTS, the stem contains conducting tissues (XYLEM and PHLOEM). In flowering plants, this vascular tissue is arranged in a ring (in DICOTYLEDONS) or scattered (in MONOCOTYLEDONS). Stems are usually erect, but may be climbing (vine) or prostrate (stolon). They may also be SUCCULENT (cactus) or modified into underground structures (RHIZOMES, TUBERS, CORMS, BULBS). Stems vary in size from the thread-like stalks of aquatic plants to tree-trunks. Woody stems have an outer layer of corky cells, called BARK. They increase in girth every year as specialized layers of CAMBIUM cells produce extra tissues in the cortex, and sometimes also just below the EPIDERMIS.

◄ **steel** The illustration shows how a plain carbon steel, consisting of pure iron and a trace of carbon, is obtained from iron ore in which the iron is combined with oxygen and other impurities, mainly silicon.

Key
A) The raw materials are mined: coal, limestone and iron ore.
B) After primary crushing and grading, the raw materials are concentrated. Coal is partially burned to drive off volatile material and impurities. Iron ore undergoes magnetic separation, to separate the magnetic ore from non-magnetic clays. Limestone is naturally pure
C) Raw materials are graded for use in the blast furnace.
D) While oversize material is routed for further crushing, material of the correct size goes straight to the blast furnace. Undersized iron ore is mixed with fine limestone and coke and burned to produce "sinter".
E) The limestone, iron ore and coke are subjected to a hot-air blast. Fuel oil is burned in the airblast and ignites the coke. The burning coke produces high temperatures at the centre of the furnace so that the material here is half molten. The hot coke and the gases it produces on burning remove the oxygen from the iron ore, forming carbon monoxide and carbon dioxide, a gaseous mixture which escapes through the gas off-take. The other main impurity in the iron ore is silica, which dissolves in the limestone. Liberated from oxygen, the iron and silica collect in the furnace hearth. The limestone-silica mixture is lighter than iron, and forms a "slag" layer above it, enabling easy separation via separate tap holes.
F) The molten iron absorbs excessive carbon from the coke, which must be reduced to form a usable steel. This is done by blasting pure oxygen on the surface of the molten material in the blast furnace. The carbon combines with oxygen and burns off as carbon dioxide and carbon monoxide gases. Limestone on the surface of the melt absorbs many of the impurities left over from the blast furnace.
G) The refined steel with the correct carbon content is tapped off for casting, leaving the residue in the "slag".

Stephenson, George (1781–1848) British engineer and inventor, regarded as the father of the LOCOMOTIVE. In 1814 he built his first locomotive, *Blucher*. This locomotive, the first to have flanged wheels, ran on a tramway. Stephenson's most famous locomotive, *Rocket*, was built in 1829. It reached a top speed of 47km/h (29mph). It later ran on the Liverpool to Manchester line, one of the many railway lines that he engineered.

Stephenson, Robert (1803–59) British engineer, son of George STEPHENSON. From 1827 Robert Stephenson managed his father's locomotive works. He built several railway lines and tubular bridges, including the Britannia Bridge over the Menai Strait, Wales. He served as a member of Parliament from 1847.

steppe Area of temperate grassland found in Kazakhstan, Russia, Ukraine and neigbouring parts of southern Europe. The landscape is often very flat and open and the climate is to dry to support much vegetation. When ploughed up, the steppe can become very rich farmland, especially for growing cereals. Steppes are very similar to **prairies**, and have a similar continental climate. Winters are very cold, but summers are quite warm. Rainfall is quite light, and falls mainly in the summer months.

stereochemistry Study of the chemical and physical properties of COMPOUNDS as affected by the ways in which the atoms of their molecules are arranged in space. Such arrangements can result in two or more compounds having the same numbers and kinds of atoms, but differently shaped molecules; these are known as **stereoisomers**. Stereochemistry also deals with optical isomerism, in which the configuration of one molecule is the mirror image of another. *See also* ISOMERS

stereophonic sound Method of SOUND REPRODUCTION that gives the illusion of both location and direction; at least two separate channels are required. In stereophonic LONG-PLAYING RECORDS, the groove is modulated in two planes – the lateral and the vertical (groove depth). Depth variations correspond to the difference between the left and right channels of the stereo-recording and lateral modulations correspond to their sum. Two transducers in the pick-up cartridge feed two separate AMPLIFIERS and two separated LOUDSPEAKERS. *See also* COMPACT DISC (CD); MAGNETIC TAPE

stereoscope Optical device that produces an apparently three-dimensional image by presenting two slightly different plane images, usually photographs, one to each eye. Early stereoscopes of the 1830s used mirrors and prisms, and some modern ones use POLARIZED LIGHT to project images that are viewed through polarizing filters.

stereotype In PRINTING, duplicate printing plate made from the original plate (which is reserved for making further duplicates as necessary). A traditional method of making stereotypes is to press papier mâché into the original and produce a negative mould from which a duplicate plate is electroplated or cast.

sterility Inability to reproduce. It may be due to INFERTILITY or to surgical intervention.

sterilization Surgical intervention that terminates a person's ability to reproduce. In women, the usual procedure is **tubal ligation**: sealing or tying off the FALLOPIAN TUBES so that fertilization can no longer take place. In men, a VASECTOMY is performed to block the release of sperm. The term is also applied to the practice of destroying microorganisms in order to prevent the spread of infection. Techniques include heat treatment, irradiation and the use of disinfecting agents. Sterile conditions are necessary in surgery, in the food and pharmaceutical industries, and in some laboratory experiments.

Stern, Otto (1888–1969) US physicist, b. Germany. In 1933 he resigned from the University of Hamburg and emigrated to the United States. Stern became professor of physics at the Carnegie Institute of Technology. An leading experimental physicist, Stern was awarded the 1943 Nobel Prize for physics for developing the molecular beam as a tool for studying the characteristics of the MOLECULE and for his measurement of the magnetic moment of the PROTON.

sternum (breastbone) Flat, narrow bone extending from the base of the neck to just below the diaphragm in the centre of the chest. It consists of three sections: the **manubrium**, or upper part; the **body** or gladiolus; and the **xiphoid process**, the lower and more flexible cartilaginous part. The top of the manubrium is attached by ligaments to the collarbones and the body is joined to the ribs by seven pairs of costal cartilages.

steroid Class of organic compounds characterized by a basic molecular structure of 17 carbon atoms arranged in four rings and bounded by up to 28 hydrogen atoms. Steroids are widely distributed in animals and plants, the most abundant being the sterols (steroid alcohols) such as CHOLESTEROL. Another important group are the steroid HORMONES, including the CORTICOSTEROIDS, secreted by the adrenal cortex, and the sex hormones (OESTROGEN, PROGESTERONE and TESTOSTERONE). Synthetic steroids are widely used in medicine. Athletes sometimes abuse steroids to increase their muscle mass, strength and stamina; but there are harmful side-effects. Unauthorized use of steroids is illegal in sports.

stethoscope Instrument that enables an examiner to listen to the action of various parts of the body, principally the heart and lungs, to help evaluate their condition. It consists of two earpieces attached to flexible rubber tubes which lead to either a disc or a cone.

Stevenson, Robert (1772–1850) Scottish civil engineer who designed LIGHTHOUSES. He designed and built 23 lighthouses, and systems for making the light flash an identifying signal. Stevenson also produced designs for canals and harbours.

stibnite Sulphide mineral, ANTIMONY sulphide (Sb_2S_3), found in low-temperature veins and rock impregnations. Stibnite occurs as orthorhombic system aggregates of either prismatic crystals or granular masses. It is opaque, sometimes iridescent, and grey or black in colour. Stibnite is an important ore of antimony. Hardness 2; r.d. 4.6.

stigma In botany, the free upper part of the STYLE of the female organs of a FLOWER, to which a pollen grain adheres before FERTILIZATION.

Still, Andrew Taylor (1828–1917) US founder of OSTEOPATHY. After three of his children died in an epidemic of spinal MENINGITIS, he devised a treatment based on the theory that most physical disorders were caused by "structural derangement" of the body, which could be remedied by manipulations of bones and muscles without the use of drugs.

stillbirth Delivery of a dead FETUS of more than 24 weeks' gestation. Death of a fetus before this point is termed MISCARRIAGE.

stimulant Substance that increases mental alertness and activity. There are a number of stimulants which act on the CENTRAL NERVOUS SYSTEM (CNS), notably drugs in the AMPHETAMINE group (now rarely prescribed). Many common beverages, including tea and coffee, contain small quantities of the stimulant CAFFEINE.

stimulus In psychology, any change in energy, either external or internal, that influences the RECEPTORS of the NERVOUS SYSTEM.

stoichiometric compound Chemical COMPOUND in which the atoms combine in simple whole-number ratios – that is, any compound with an exact chemical formula. *See also* NON-STOICHIOMETRIC COMPOUND

STELLAR EVOLUTION

Star formation begins with a collapsing cloud of nebular material (1). In the middle of the cloud, the temperature rises and stars begin to form (2,3). As they shine, the gas associated with them (4) is blown away and a star cluster is produced (5). This cluster is gradually disrupted and becomes a loose stellar association (6). The evolution of a star depends on its mass. A star of solar type joins the main sequence (7), and remains on it for a very long period. When its hydrogen 'fuel' begins to run low, it expands (8) and becomes a red giant (9). Eventually, the outer layers are lost and a planetary nebula is formed (10). The 'shell' of gas expands and is finally dissipated, leaving the core of the old star as a white dwarf (11). The white dwarf continues to shine feebly for an immense period before losing its heat and becoming a dead black dwarf. With a more massive star, events occur more rapidly. After its main sequence period (12), the star becomes a red supergiant (13) and may explode as a supermove (14). It may end as a neutron star or pulsar (15), although if its mass is even greater it may produce a black hole.

Stokes, Sir George Gabriel (1819–1903) British mathematician and physicist, b. Ireland. He was professor of mathematics (1849–1903) at Cambridge and served as president (1885–92) of the Royal Society. Stokes is celebrated for his law of VISCOSITY (*see* STOKES'S LAW), which describes the movement of a solid sphere in a fluid. It forms the basis of the modern theory of viscous fluids. He also developed an important theorem of VECTOR ANALYSIS, accurately described and named FLUORESCENCE and proposed the concept of an ETHER. Stokes' book, *Dynamic Theory of Diffraction* (1849) was an important work in the wave theory of light. He was also a pioneer in GEODESY.

Stokes' law In fluid mechanics, a formula that gives the terminal (final) velocity at which particles fall in a fluid (liquid or gas). For particles of radius r (less than 0.1mm) the terminal velocity is $2(d_1-d_2)gr^2/9\eta$, where d_1 and d_2 are the densities of the particle and the fluid, g is the acceleration due to gravitation, and η is the viscosity of the fluid. It is named after Sir George STOKES.

STOL aircraft (short takeoff and landing aircraft) AIRCRAFT with the ability to operate out of extremely small airfields. Present STOL aircraft are capable of taking off and climbing over a 15-m (50-ft) obstacle in less than 150m (500ft). Wings of great lifting ability and powerful engines allow for unusual performance without sacrificing load-carrying capability. *See also* VTOL AIRCRAFT

stolon Modified, horizontal, underground or aerial stem, growing from the basal node of a plant. Aerial stolons, or RUNNERS, may be slender, such as strawberry plants, or stiff and arching, such as brambles. The stolon produces a new plant at its tip, which puts out ADVENTITIOUS ROOTS (roots that arise from nodes on the stem) to anchor itself. Underground stolons are characteristic of white potato plants. *See also* TUBER; VEGETATIVE REPRODUCTION

stoma (pl. stomata) In botany, pores found mostly on the undersides of leaves through which atmospheric gases pass in and out, for RESPIRATION and PHOTOSYNTHESIS. Surrounding each stoma are two guard cells, which can swell and close the stoma to prevent excessive loss of water vapour. *See also* GAS EXCHANGE; TRANSPIRATION

stomach J-shaped organ, lying to the left and slightly below the DIAPHRAGM in human beings; one of the organs of the DIGESTIVE SYSTEM. It is connected at its upper end to the gullet (OESOPHAGUS) and at the lower end to the SMALL INTESTINE. The stomach itself is lined by three layers of muscle (longitudinal, circular and oblique) and a folded mucous layer which contains gastric GLANDS. These glands secrete HYDROCHLORIC ACID that destroys some food bacteria, and makes possible the action of PEPSIN, the ENZYME secreted by the glands, which digests PROTEINS. Gastric gland secretion is controlled by the sight, smell and taste of food, and by hormonal stimuli, chiefly the HORMONE gastrin. As the food is digested, it is churned by muscular action into a thick liquid state called chyme, at which point it passes into the small intestine.

stoneware Opaque, non-porous, high-fired clay ware, well vitrified and intermediate between PORCELAIN and EARTHENWARE. Stoneware is often salt-glazed for utility pieces and is used for vessels and tiles.

Stoney, George Johnstone (1826–1911) British physicist who coined the term "ELECTRON" for the fundamental unit of ELECTRICITY (the term was later applied to an ELEMENTARY PARTICLE). Stoney worked on the KINETIC THEORY of gases and estimated the number of molecules in a volume of gas at STANDARD TEMPERATURE AND PRESSURE (STP). His estimate of an electron's charge was incorrect because he used an incorrect value for the mass of the hydrogen atom.

stony-iron meteorite METEORITE consisting of approximately equal proportions of SILICATES (stony material) and metals, mostly nickel-iron. There are two main subtypes. **Pallasites**, which consist of olivine (a magnesium–iron silicate) mixed with nickel–iron, may have originated near the core/mantle interface of a planetary body. **Mesosiderites** are a much coarser combination of chunks of various silicates and nickel–iron, and could have been produced by impacts on a planetary surface.

stony meteorite (aerolite) METEORITE that consists mostly of SILICATES (stony material, mainly olivine, pyroxene and feldspar), with only a small amount of chemically uncombined metals (typically 5% nickel–iron). There are two subtypes: chondrites and achondrites. In **chondrites**, the rocky minerals take the form of small spherical grains (chondrules), and there is some nickel and iron. **Achondrites** are coarser-grained and have no chondrules or metal. Both types are about 4,600 million years old, dating from the formation of the Solar System.

STP Abbreviation of STANDARD TEMPERATURE AND PRESSURE

strain Change in dimensions of an object subjected to STRESS. Linear strain is the ratio of the change in length of a bar to its original length while being stretched or compressed. Shearing strain describes the change in shape of an object whose opposite faces are pushed in different directions. HOOKE'S LAW for elastic materials states that strain is proportional to stress up to the material's elastic limit.

strain gauge Device used to measure mechanical STRAIN such as that caused by the movement of a bridge or glacier. **Electrical** strain gauges use the phenomenon (first noted by Lord KELVIN in 1856) that if a wire is strained, its electrical resistance is changed. An electrical strain gauge usually consists of a grid of wire filaments printed onto a metal foil. The gauge is cemented onto the structural surface whose strain is to be measured. As the strain in the structure changes, the strain in the gauge changes too, thus producing changes in resistance that can be measured.

strangeness Property of HADRONS (ELEMENTARY PARTICLES that experience the STRONG NUCLEAR FORCE). Particles are assigned a strangeness of +1, −1 or 0. In a strong interaction, strangeness may be conserved (the total strangeness must be the same before and after the process). If strangeness is not conserved, the process can proceed only by the WEAK NUCLEAR FORCES, and hence only very slowly.

strange particles ELEMENTARY PARTICLES that have non-zero STRANGENESS. They have an anomalously long lifetime (10^{-10} to 10^{-7} s) compared to that characteristic of most other HADRONS (10^{-23} s). *See also* CONSERVATION LAW, NUCLEAR

Strassmann, Fritz (1902–80) German chemist. In 1938 he ushered in the nuclear age when he discovered, with Otto HAHN, the process of nuclear FISSION by bombarding URANIUM atoms with NEUTRONS. Strassmann shared the 1966 Fermi Award of the US Atomic Energy Commission with Otto Hahn and Lise MEITNER.

stratification Layering in ROCKS. It occurs in SEDIMENTARY ROCKS and in IGNEOUS ROCKS that formed from LAVA flows and volcanic fragmental deposits. Separations between individual layers are called stratification planes. They parallel the strata they bound, being horizontal near flat layers and exhibiting inclination on a sloping surface. *See also* STRATIGRAPHY; STRATUM

stratigraphy Branch of geology concerned with stratified or layered ROCKS. It deals with the correlation of rocks from different localities using FOSSILS and distinct rock types. *See* STRATUM

STOMACH

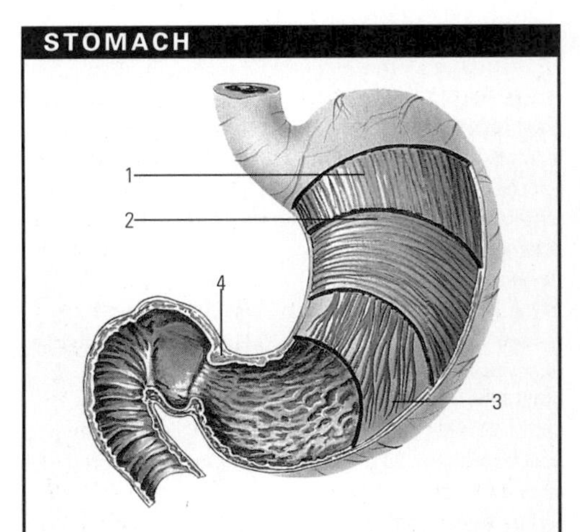

The stomach, like most of the digestive tract, is walled with involuntary or smooth muscle. The fibres of the stomach wall are built up in three layers: longitudinal (1), circular (2) and oblique (3). These muscular layers work in collaboration, contracting in turn, producing a wave-like movement (peristalsis) and forcing food through the stomach, past the circular muscle valve sphincter (4) at the base and into the adjoining duodenum.

stratosphere Section of the Earth's ATMOSPHERE between the TROPOSPHERE and the higher MESOSPHERE. It is about 40km (25mi) thick and for half this distance the temperature remains fairly constant. The stratosphere contains most of the atmosphere's OZONE LAYER.

stratum In biology, a layer of tissue, such as the *stratum basale* of the uterus.

stratum (pl. strata) In geology, a single layer or bed of SEDIMENTARY ROCK, usually one of a series of layers that have been deposited one on top of another. Each stratum indicates the conditions that prevailed at the time of its deposition. The study of the strata is called STRATIGRAPHY. The stratigraphical or geological column contains all the periods of GEOLOGICAL TIME. Rock strata may be very thin, of just a few centimetres, or they can be several metres in extent. Initially, they are usually horizontal, but they are likely to be affected by folding and faulting.

strepsiptera Order of parasitic insects, comprising about 400 species. The female is wingless and never leaves its HOST, the bee. The male has back wings and emerges from the bee as an adult.

streptococcus Genus of spherical or oval BACTERIA that grow in pairs or bead-like chains. They live mainly as parasites in the mouth, respiratory tract and intestine. Some are harmless but others are pathogenic, causing SCARLET FEVER and other infections. Treatment is with ANTIBIOTICS.

streptomycin Antibiotic drug used to treat certain bacterial diseases resistant to PENICILLIN. It may produce side-effects such as deafness and giddiness. Its discovery (1944) by Selman WAKSMAN led to the development of other related ANTIBIOTICS. Streptomycin is used, along with other drugs, to treat tuberculosis.

stress Force per unit area applied to an object. **Tensile** stress stretches an object, **compressive** stress squeezes it and **shearing** stress deforms it sideways. In a fluid, no shearing stress is possible because the fluid slips sideways, so all fluid stresses are PRESSURES (positive or negative).

string theory Recent branch of PARTICLE PHYSICS, largely developed by Ed Witten, in which ELEMENTARY PARTICLES can be represented by the vibra-

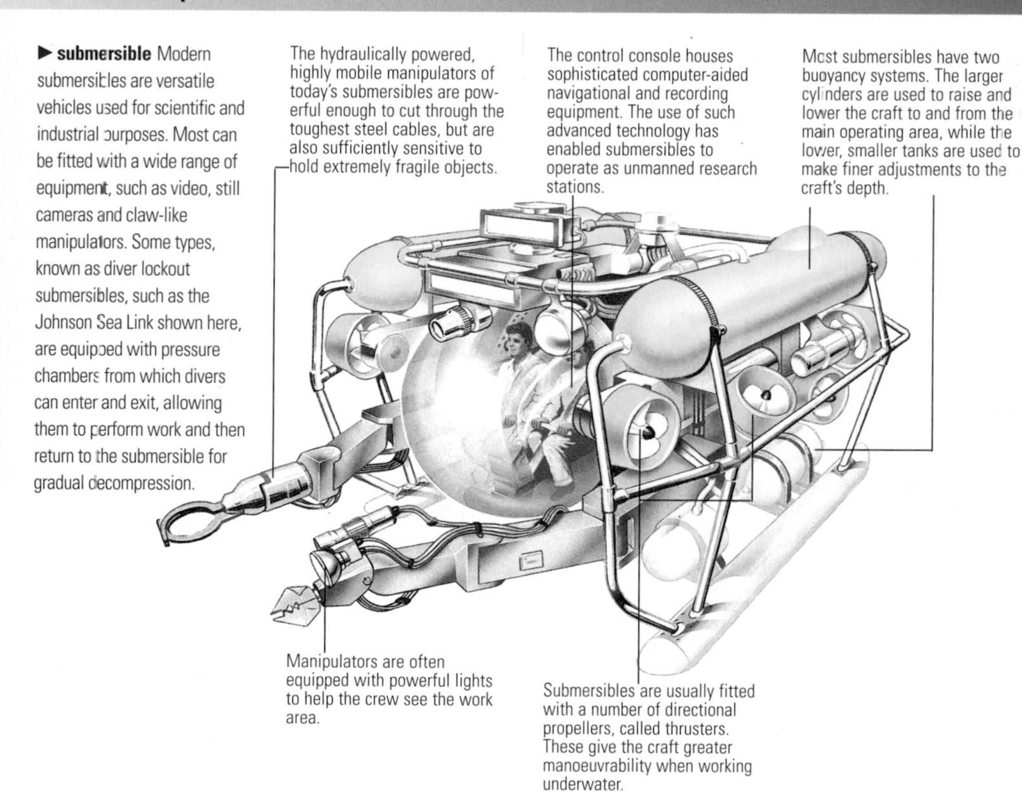

► **submersible** Modern submersibles are versatile vehicles used for scientific and industrial purposes. Most can be fitted with a wide range of equipment, such as video, still cameras and claw-like manipulators. Some types, known as diver lockout submersibles, such as the Johnson Sea Link shown here, are equipped with pressure chambers from which divers can enter and exit, allowing them to perform work and then return to the submersible for gradual decompression.

tions of a one-dimensional string. The concept of a string is completely theoretical since no string has ever been detected experimentally: the string-like nature of these objects is believed to be billions of times smaller than the scale observable in current experiments. *See also* SUPERSTRING THEORY

stroboscope (strobe) Device that emits regular flashes of light. Stroboscopes usually have a calibrated scale from which the number of flashes per minute can be read. They are used in photography to make multiple exposures of moving subjects and in engineering to apparently "slow down" moving objects for observation. A strobe is also used to find the speed of a rotating wheel by adjusting the rate of flashing until it equals the rotation speed (when the wheel seems to be stationary).

strong nuclear force One of the four FUNDAMENTAL FORCES in nature. The strongest of the four forces, it binds together protons and neutrons within the NUCLEUS of an atom. Like the WEAK NUCLEAR FORCE, it operates at very short distances (a millionth of a millionth of a centimetre) and therefore occurs only within the nucleus. It is mediated by GLUONS. *See also* GRAND UNIFIED THEORY (GUT)

strontianite Carbonate mineral, STRONTIUM carbonate ($SrCO_3$). It has orthorhombic system, hexagonal twinned crystals, and it also occurs as massive or columnar aggregates. It can be pale green, white, grey, yellow or brown. Strontianite is found in low-temperature hydrothermal veins, often in limestone. Hardness 3.5–4; r.d. 3.7.

strontium (symbol Sr) Silver-white, metallic element of the ALKALINE-EARTH METALS in group II of the periodic table. Resembling CALCIUM physically and chemically, strontium occurs naturally in STRONTIANITE and celestite and is extracted by electrolysis. Strontium salts are used to give a red colour to flares and fireworks. The isotope ^{90}Sr (half-life 28 years) is a harmful radioactive element present in fallout, from which it is absorbed into milk and bones; it is used in nuclear reactors. Properties: at.no. 38, r.a.m. 87.62, r.d. 2.554, m.p. 769°C (1,416°F), b.p. 1,384°C (2,523°F).

Strutt, John William, Lord Rayleigh *See* RAYLEIGH, JOHN WILLIAM STRUTT, LORD

Struve, Friedrich Georg Wilhelm von (1793–1864) German astronomer who catalogued more than 3,000 stars, including many previously undiscovered DOUBLE STARS. In 1839 Struve became one of the first astronomers to measure the PARALLAX of a star, Vega. His son, **Otto Wilhelm** von Struve (1819–1905), succeeded him as director (1862–99) of the Pulkovo Observatory, Russia. Otto discovered a satellite of Uranus. Friedrich's grandson, Otto Struve (1897–1963) was director (1932–47) of Yerkes Observatory and professor of astrophysics at the University of California (1950–59). He made important discoveries in the fields of stellar evolution and radial velocities.

strychnine Poisonous ALKALOID obtained from the plant *Strychnos nux-vomica*. In the past, it was believed to have therapeutic value in small doses as a tonic. Strychnine poisoning causes symptoms similar to those of tetanus, with death occurring due to spasm of the breathing muscles.

Sturgeon, William (1783–1850) British electrical engineer who demonstrated (1825) the first ELECTROMAGNET capable of useful work. It consisted of a soft, iron core insulated with varnish and wound with 18 turns of bare copper wire; it was capable of supporting 4kg (9lb) and was powered by a single cell. Sturgeon also built an electric motor, invented the COMMUTATOR and the suspended coil GALVANOMETER, and studied charges in clouds by flying kites.

style In botany, the slender tube of a flower that connects the pollen-receiving STIGMA at its tip to the OVARY at its base.

styrene ($C_6H_5CH{=}CH_2$) Colourless, liquid HYDROCARBON, important as a MONOMER of POLYSTYRENE. It is also a monomer of styrene-butadiene rubber (SBR), the artificial RUBBER most commonly used in pneumatic tyres. Styrene is a derivative of BENZENE. Properties: m.p. 145.2°C (293.4°F), b.p. −30.6°C (−23.1°F).

subatomic particle *See* ELEMENTARY PARTICLE

subduction In PLATE TECTONICS, descent of rocky material from the edge of one of the Earth's tectonic plates into the semi-molten ASTHENOSPHERE below. It occurs where two plates collide, and one rides over the other (**subduction zone**); the lower one is subducted. Often the upper plate is a continental plate and the lower one is an oceanic plate, and subduction leads to the formation of an ocean TRENCH. This type of subduction is associated with volcanic eruptions, earthquakes and other seismic activity. *See also* CONTINENT

subdwarf STAR that is less luminous by one to two MAGNITUDES than a normal DWARF STAR of the same spectral type. Subdwarfs are mainly of types F, G and K, and lie below the MAIN SEQUENCE on the HERTZSPRUNG–RUSSELL DIAGRAM. Most are Population II stars. They are placed in LUMINOSITY class VI, although an alternative designation is to prefix their spectral type with the letters "sd". *See also* SPECTRAL CLASSIFICATION

subgiant Star of smaller radius and lower LUMINOSITY than a normal GIANT STAR of the same spectral type. They are mainly of types G and K, and lie between the MAIN SEQUENCE and the giants on the HERTZSPRUNG–RUSSELL DIAGRAM. Subgiants are placed in luminosity class IV. *See also* SPECTRAL CLASSIFICATION

sublimation Direct change from SOLID to GAS, without an intervening LIQUID phase. **Dry ice** (solid carbon dioxide) sublimes from the solid phase directly to the gaseous phase. Most substances can sublimate at certain pressures, but usually not at atmospheric pressure. *See also* CONDENSATION; EVAPORATION

submarine Seagoing warship capable of travelling both on and under the water, armed principally with torpedoes. Submarines were planned in the 16th century, but they were not built until the Dutch inventor Cornelis Drebbel (1572–1633) took a greased, leather-covered boat underwater in *c.*1620. Experimental submarines were used in warfare in the late 18th and 19th centuries, to little effect. Technical advances of the late 19th century led to the general spread of underwater craft in the world's navies. Early submarines were essentially surface ships with a limited ability to remain submerged. Once underwater, they depended on battery-powered electric motors for propulsion and, with a limited air supply, were soon forced to surface. Submerging is accomplished by letting air out of internal ballast tanks; trimming underwater is done by regulating the amount of water in the ballast tanks with pumps; and surfacing is accomplished by pumping ("blowing") the water out of the tanks. In 1954 the first nuclear-powered submarine, the Nautilus, was launched. Today's, nuclear-powered submarines are propelled by steam turbines and are capable of speeds of up to 30 knots. The British *Trident* submarines have multiple-head nuclear missiles with a range of *c.*11,000km (6,750mi).

submersible Small craft for underwater exploration, research or engineering. Modern submersibles have evolved from simple devices. Diving bells were open-bottomed devices in which people were lowered into the water. A device called the BATHYSPHERE, invented in the 1930s by William BEEBE, was a spherical observation chamber. The bathyscaphe, invented in the 1940s by Auguste PICCARD, also had a spherical chamber, but this was attached to a much larger hull, which was used as a buoyancy control device. The craft also had propellers powered by electric motors. A new generation of submersibles has evolved since the late 1950s for engineering and research work. A typical craft has a spherical passenger capsule capable of withstanding water pressure down to *c.*3,600m (12,000ft). Attached to this is a structure containing batteries, an electric motor with propeller, lighting, a mechanical arm for gathering samples, and other technical equipment. A specially designed support ship launches and retrieves the submersible. Some submersibles are unmanned and operated by remote control from the surface.

subscript Small letter, number or other symbol written or printed below and to the left or right of another symbol, as in a chemical formula (such as C_2H_5) or to indicate different values, x_1, x_2, x_3, of a mathematical variable. *See also* SUPERSCRIPT

subset SET whose elements are contained in another larger set.

substitution reaction CHEMICAL REACTION in which one atom or group of atoms replaces (usually in the same structural position) another atom or group in a molecule or ion.

substrate In biochemistry, a reactant that is acted on by an ENZYME or other CATALYST. In biology, a substrate is a nutrient medium used to grow microorganisms, or the surface on which a sessile organism (such as a limpet) lives and grows. In electronics, a substrate is a single crystal or SEMICONDUCTOR used as a basis for making INTEGRATED CIRCUITS or TRANSISTORS.

subtraction Arithmetical operation signified by $-$, interpreted as the inverse process of ADDITION. The difference of two numbers, $a - b$, is the number that has to be added to b to give a. a is called the **minuend** and b the **subtrahend**. *See also* ARITHMETIC

succession In ecology, orderly change in plant and animal life in a biotic COMMUNITY over a long time period. It is the result of modifications in the community ENVIRONMENT. The process ends in establishment of a stable ECOSYSTEM, known as a CLIMAX COMMUNITY.

succulent Plant that stores water in its tissues in order to resist periods of drought. Usually PERENNIAL and evergreen, most of the succulent plant body is made up of water-storage cells, which give it a fleshy appearance. A well-developed CUTICLE and low rate of daytime TRANSPIRATION also conserve water. Swollen leaves and stems have a smaller surface area to volume ratio than thin ones, thus further reducing water loss. Succulent plants include cactus and stonecrop.

sucker In botany, an upward-growing shoot that forms at the base of the plant stem or from an ADVENTITIOUS ROOT. In zoology, a disc-like structure found on the tentacles of cephalopods used for grasping objects.

sucrase ENZYME secreted by the small intestine which breaks down SUCROSE (ordinary sugar) into GLUCOSE and FRUCTOSE so that they can be absorbed during DIGESTION.

sucrose Common, white, crystalline SUGAR, a DISACCHARIDE ($C_{12}H_{22}O_{11}$) consisting of linked GLUCOSE and FRUCTOSE molecules. It occurs in many plants, but its principal commercial sources are sugar cane and sugar beet. It is also obtained from maple trees, date palms and sorghum, and is widely used for food sweetening and in the manufacture of confectionery and preserves.

sudden infant death syndrome (SIDS) (Cot death) Unexpected death of an apparently healthy baby, usually during sleep. The peak period seems to be around two months of age, although it can occur up to a year or more after birth. Claiming more boys than girls, it is more common in winter. The cause is unknown, but a number of risk factors have been identified, including prematurity and respiratory infection.

sugar Sweet-tasting, soluble, crystalline, monoSACCHARIDE or disaccharide CARBOHYDRATE. The common sugar used in food and beverages is the disaccharide SUCROSE. This is also the main sugar transported in plant tissues. The main sugar transported around the bodies of animals to provide energy in RESPIRATION is the monosaccharide GLUCOSE. *See also* SACCHARIDE

sulpha drug *See* SULPHONAMIDE DRUG

sulphate Salt of SULPHURIC ACID (H_2SO_4). Common sulphates include copper(II) sulphate ($CuSO_4$) and iron(II) sulphate ($FeSO_4$).

sulphide Chemical compound derived from HYDROGEN SULPHIDE (H_2S) in which another element or group replaces the hydrogen. An inorganic example is lead sulphide (PbS) which occurs as the mineral GALENA; an organic example is ethane thiol (ethyl mercaptan, C_2H_5SH).

sulphite Salt or ester of sulphurous acid (H_2SO_3). Sulphurous acid is known systematically as trioxosulphuric(IV) acid and its salts, the sulphites, contain the trioxosulphate(IV) ion (SO_3^{2-}). Many sulphites are reducing agents, used as disinfectants and bleaches. Hydrogensulphates(IV) (bisulphites, HSO_3^-) also exist.

sulphonamide drug (sulpha drug) Any of a group of DRUGS derived from sulphanilamide, a red textile dye, that prevent the growth of bacteria. Discovered by Gerhard DOMAGK in the 1930s, the sulphonamides were the first commercially available antibacterials, prescribed to treat a range of infections. However, their use was associated with a number of unpleasant side-effects, and, with the arrival of the less toxic and more effective ANTIBIOTICS, they were gradually replaced. Today, their use is limited chiefly to the treatment of urinary tract infections.

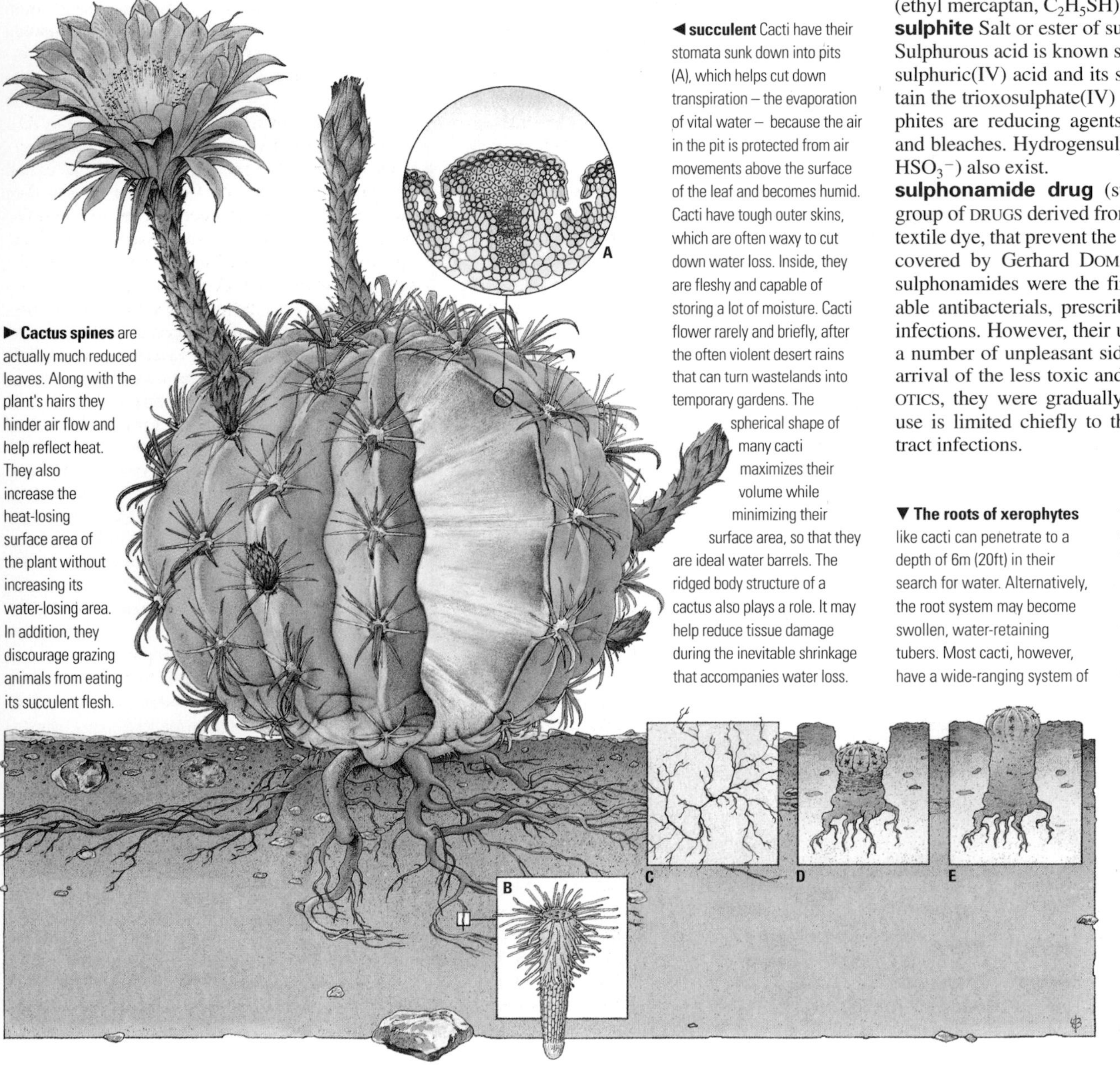

◄ **succulent** Cacti have their stomata sunk down into pits (A), which helps cut down transpiration – the evaporation of vital water – because the air in the pit is protected from air movements above the surface of the leaf and becomes humid. Cacti have tough outer skins, which are often waxy to cut down water loss. Inside, they are fleshy and capable of storing a lot of moisture. Cacti flower rarely and briefly, after the often violent desert rains that can turn wastelands into temporary gardens. The spherical shape of many cacti maximizes their volume while minimizing their surface area, so that they are ideal water barrels. The ridged body structure of a cactus also plays a role. It may help reduce tissue damage during the inevitable shrinkage that accompanies water loss.

► **Cactus spines** are actually much reduced leaves. Along with the plant's hairs they hinder air flow and help reflect heat. They also increase the heat-losing surface area of the plant without increasing its water-losing area. In addition, they discourage grazing animals from eating its succulent flesh.

▼ **The roots of xerophytes** like cacti can penetrate to a depth of 6m (20ft) in their search for water. Alternatively, the root system may become swollen, water-retaining tubers. Most cacti, however, have a wide-ranging system of fine roots, which are equipped with microscopic hairs (B). These may only penetrate a short way underground, but often cover a huge area (C) so that the cactus can quickly replenish its water supplies when water is available. In dry conditions, there are many plants that are better at absorbing water than cacti, such as small bushes with long roots.

Many cacti, such as the *Echinocereus pulchellus* (D) hide themselves underground in the dry season. Only when conditions are favourable do they extend their green tops above the surface (E). In this way, they combine drought resistance with effective drought evasion.

SULPHUR (FRASCH PROCESS)

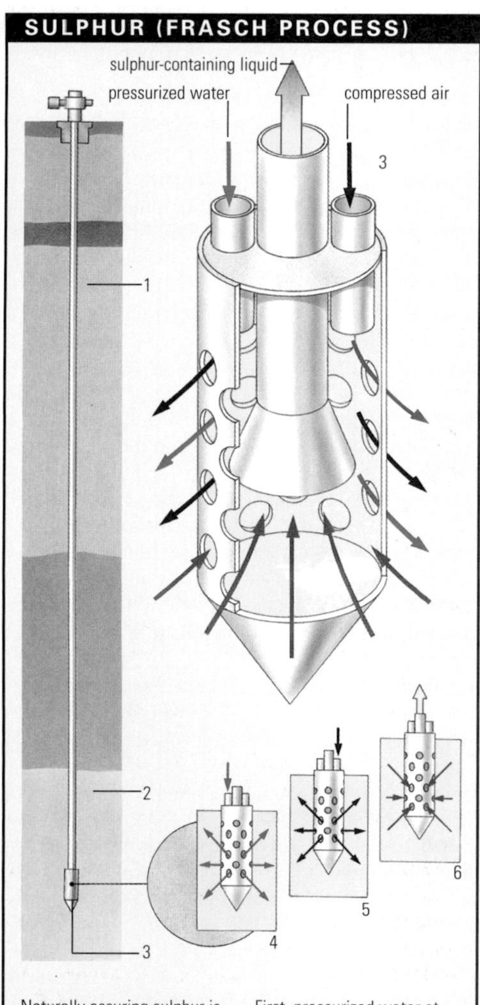

Naturally occuring sulphur is extracted from rock formations by a method known as the Frasch process. Boreholes (1) are dug to the sulphur-rich areas (2). A specialized head (3) is dropped down the borehole. First, pressurized water at 155°C (310°F) is pumped (4) into the rock melting the sulphur. The water is followed by compressed air (5), which forces the liquid containing the sulphur to the surface (6).

sulphonation Introduction of a sulphonic acid group, $-SO_3H$, into an organic molecule, widely used in the manufacture of dyes, SULPHONAMIDE DRUGS and so on. Many detergents are manufactured by sulphonation of a HYDROCARBON.

sulphonic acids Organic chemical compounds having the general formula RSO_3H, where R is a HYDROCARBON group. Examples are ethane sulphonic acid, $C_2H_5.SO_3H$, and benzene-sulphonic acid, $C_6H_5.SO_3H$. Sulphonic acids are used as chemical intermediates and their salts, the sulphonates, are widely used in DETERGENTS.

sulphur (symbol S) Nonmetallic element in group VI of the PERIODIC TABLE, known from earliest time (as brimstone). It occurs naturally as a free element and in sulphide minerals, such as GALENA and PYRITE, and sulphate minerals, such as GYPSUM. The main commercial source is native (free) sulphur, extracted by the FRASCH PROCESS. It is used in the VULCANIZATION of rubber and in the manufacture of SULPHURIC ACID, drugs, matches, dyes, fungicides, insecticides and fertilizers. Properties: at.no. 16; r.a.m. 32.064; r.d. 2.07; m.p. 112.8°C (235.0°F); b.p. 444.7°C (832.5°F). Most common isotope ^{32}S (95.1%).

sulphur cycle Continual movement of SULPHUR between living and non-living things. Most sulphur in the non-living (**abiotic**) environment occurs as SULPHUR DIOXIDE (SO_2) in the atmosphere, derived from the burning of FOSSIL FUELS, or as deposits underground and in rocks. Photochemical oxidation of sulphur dioxide in the atmosphere produces SULPHATES (SO_4^{2-}), which are carried by rain into the ground; oxidation and WEATHERING of underground deposits also produces sulphates. In the ground, sulphates are absorbed by plants in the living (**biotic**) environment to become part of sulphur-containing proteins. Animals that eat the plants assimilate the sulphur. Dead plants and animals (and their faeces) are broken down by special bacteria that return sulphur to the soil (abiotic) as HYDROGEN SULPHIDE (H_2S). Finally, other bacteria convert the hydrogen sulphide back to sulphate or native sulphur.

sulphur dioxide Colourless, poisonous gas (SO_2) with a pungent odour. It is used in the manufacture of SULPHURIC ACID and as a refrigerant, bleaching agent and preservative. Properties: m.p. −75.5°C (−103.9°F); b.p. −10.0°C (14.8°F); density 2.2 (air = 1).

sulphuric acid (formerly oil of vitriol) Colourless, odourless, poisonous liquid (H_2SO_4), one of the most corrosive acids known. It is produced by the OXIDATION of SULPHUR DIOXIDE (SO_2) in the contact process. Sulphuric acid is a major industrial chemical, the most produced chemical by tonnage, and is used in the manufacture of many acids, fertilizers, detergents, drugs, and a wide range of chemicals. Properties: r.d. 1.84; m.p. 10.3°C (50.5°F); b.p. 330°C (626°F).

sulphur trioxide (SO_3) Inorganic compound, the ANHYDRIDE of SULPHURIC ACID, that is, it reacts with water to make the acid. At room temperatures sulphur trioxide can exist either as a liquid or as a solid. It dissolves in sulphuric acid to make fuming sulphuric acid, or oleum ($H_2S_2O_7$).

summation (symbol Σ) In mathematics, finding the total of a sequence or array of numbers, or of an infinite SERIES of terms.

Sumner, James Batcheller (1887–1955) US biochemist who shared the 1946 Nobel Prize for chemistry with John Northrop (1891–1987) and Wendell STANLEY for his discovery that ENZYMES could be crystallized. In 1926 Sumner isolated and crystallized the enzyme urease; it was the first enzyme to be crystallized and this achievement finally proved the PROTEIN nature of enzymes.

Sun Star at the centre of the SOLAR SYSTEM, around which all other Solar System bodies revolve in their orbits. The apparent daily motion of the Sun across the sky and its annual motion along the ECLIPTIC are caused by the Earth rotating on its axis and revolving in its orbit. The Sun does, however, have its own motion in the Galaxy of about 20km s^{-1} (12.5mi s^{-1}) towards the point in the sky known as the APEX. The amount of sunlight normally reaching the Earth is quantified as the SOLAR CONSTANT. The Sun is a DWARF STAR, on the MAIN SEQUENCE of the HERTZSPRUNG–RUSSELL DIAGRAM, and is thus a typical, average star. It consists of about 70% hydrogen (by weight) and 28% helium, with the remainder mostly oxygen and carbon. Its temperature, pressure and density increase towards the centre, where the values are about 15×10^6 K, 10^{11} bar and 150g cm^{-3}, respectively. Like all stars, the Sun's energy is generated by nuclear FUSION reactions taking place under the extreme conditions in the core. The core is *c.*400,000km (250,000mi) across. Energy released from the core passes up through the radiative layer, which is about 300,000km (nearly 200,000mi) thick, then passes through the 200,000km (125,000mi) thick convective layer to the surface, the PHOTOSPHERE (meaning "sphere of light"), from where it is radiated into space. Most of the Sun's visible activity takes place in the 500km (300mi) thick photosphere. SUNSPOTS are darker, cooler regions of the photosphere where the local magnetic field is enhanced. Associated with sunspots are SOLAR FLARES – sudden, violent releases of energy and material. Above the photosphere lies the CHROMOSPHERE (meaning "sphere of colour", so-called because of its rosy tint when seen during a total solar ECLIPSE), which consists of hot gases and extends for thousands of kilometres. It is generally in a state of turbulence, the temperature rising extremely rapidly with height. This is the realm of SPICULES and PROMINENCES, huge eruptions of material from the Sun's limb. Extending outward from the chromosphere for millions of kilometres is the CORONA. A continuous emission from the corona of particles and radiation is known as the SOLAR WIND. The solar wind and the Sun's magnetic field dominate a region of space called the HELIOSPHERE, which extends to the boundaries of the Solar System. The Sun must never be viewed directly through any optical instrument. Temporary or permanent blindness can result from direct viewing or from the use of unsuitable filters.

sundial Instrument first used about 5,000 years ago in the Near East to measure the time of day. Traditionally a sundial is a short, flat-topped pillar on which is mounted the **gnomon**, or time pointer. This slants upwards away from the dial at an angle determined by the latitude; it points due North in N latitudes and due South in S latitudes. The position of the shadow of the gnomon indicates the time on a scale marked round the top of the sundial.

sun dog (parhelion or mock sun) Either of two bright coloured spots observed at 22° on each side of the Sun. They are caused by REFRACTION of sunlight by ice crystals in the Earth's ATMOSPHERE.

sunspot Region in the Sun's PHOTOSPHERE which is cooler than its surroundings and therefore appears darker. Sunspots consists of a dark central region, the **umbra**, and a grey outer region, the **penumbra**. They vary in size from *c.*1,000 to 50,000km (600 to 30,000mi), and are occasionally up to *c.*200,000km (125,000mi). Their duration varies from a few hours to a few weeks, or months for the very biggest. Sunspots are seen to move across the face of the Sun as it rotates. The number of spots visible depends on the stage of the 11-year SOLAR CYCLE. Sunspots occur where there is a local strengthening of the Sun's magnetic field.

superalloys Name given to ALLOYS of cobalt and nickel used as surface materials for spacecraft to withstand the rapid heating and high temperatures caused by passage through the Earth's atmosphere, particularly on re-entry from space. These alloys oxidize in air above 700°C (1,300°F) but the oxide coating initially formed prevents further oxidation.

supercluster *See* GALAXY CLUSTER

superconductivity Electrical behaviour in metals and alloys that are cooled to very low temperatures. In a superconducting circuit, an electric current flows indefinitely because there is no electrical RESISTANCE. Superconductivity was first observed in materials near ABSOLUTE ZERO. Research continues to develop superconductors that function at higher temperatures.

supercooling Lowering of the temperature of a liquid below its freezing point without the liquid solidifying. Adding particles of dust, scratching the inside surface of the container or seeding the supercooled liquid with a small crystal makes it solidify immediately and return to its freezing temperature. The term supercooling is also applied to the lowering of the temperature of a SATURATED SOLUTION without crystallization occurring. Instead, a SUPER-

SUN

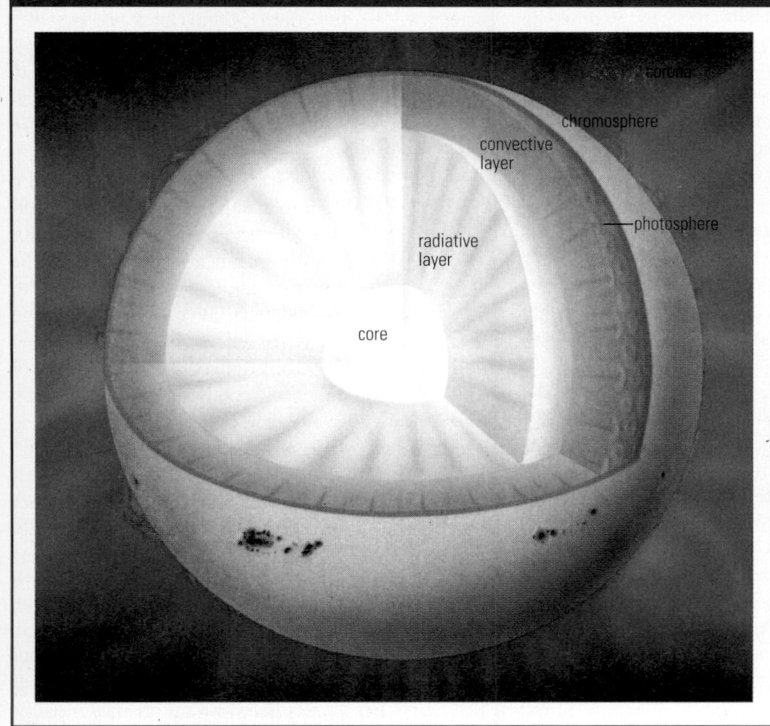

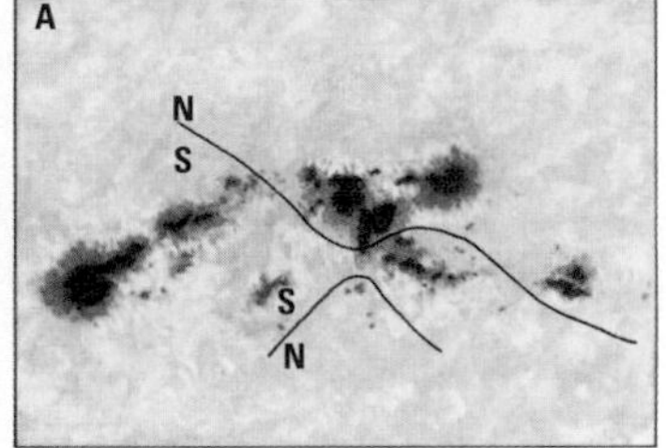

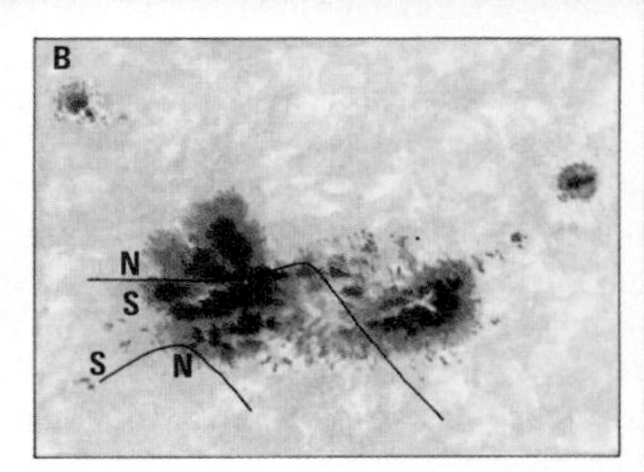

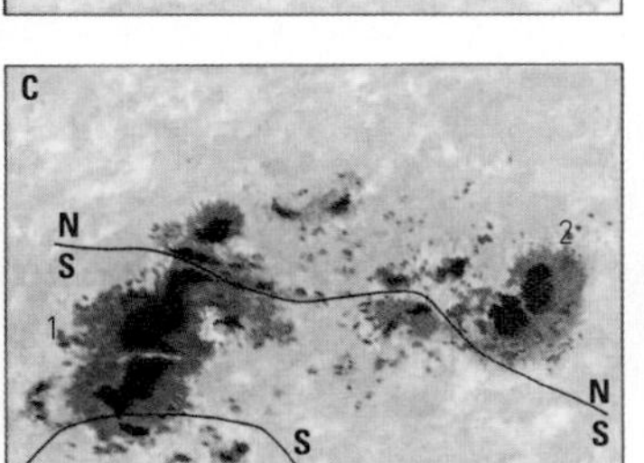

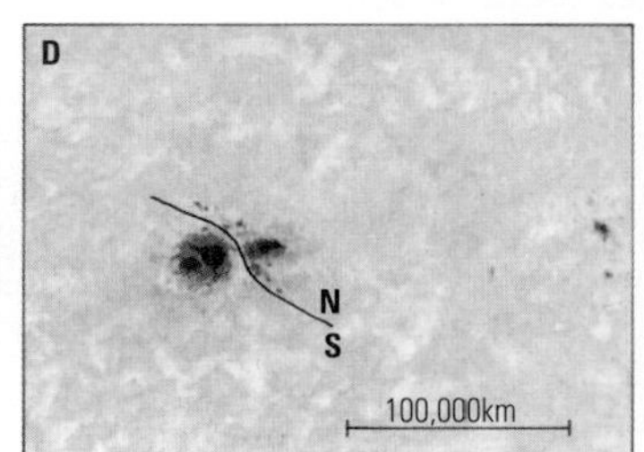

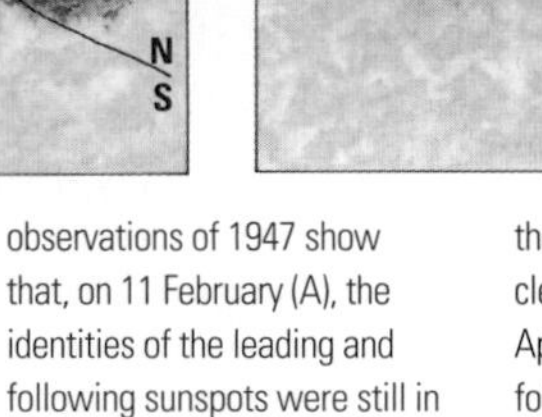

Cross-section of the Sun (left), showing the core, radiative layer, convective layer, photosphere, chromosphere and corona. The sunspot observations of 1947 show that, on 11 February (A), the identities of the leading and following sunspots were still in doubt. But as the lines indicate, the magnetic polarities were clear. From 9 March (B) to 7 April (C), the leader (1) and follower (2) are distinct. By 5 May (D), activity has ceased.

SATURATED SOLUTION is formed. Dust particles, scratching and seeding with a crystal will all make the supersaturated solution crystallize. Finally a third meaning of supercooling is the lowering of the temperature of a vapour until it is supersaturated without CONDENSATION occurring. The introduction of a particle or a disturbance leads to condensation. This is the principle of the CLOUD CHAMBER.

superfluidity Property of a liquid that has no VISCOSITY and therefore no resistance to flow. HELIUM II – liquid helium at temperatures less than 2K, or −271°C (−456°F) – was the first known superfluid. Its unusual properties were discovered independently by Peter KAPITZA in 1937 and by the British physicist John Allen in 1938. They found that helium II apparently defies gravity by flowing up a slope, and if placed in a container, it climbs up the inside walls and overflows. It also appears to contravene the laws of THERMODYNAMICS, by flowing from a cool region to a warmer one. *See also* CRYOGENICS

supergiant Extremely luminous STAR of large diameter and low density. They can be of any spectral type, from O to M. RIGEL (type B) and BETELGEUSE (type M) are examples. The LUMINOSITIES of supergiants are several magnitudes greater than those of giant stars of the same SPECTRAL CLASSIFICATION, and so they lie at the top of the HERTZSPRUNG–RUSSELL DIAGRAM. Red (M-type) supergiants, such as Betelgeuse, have the largest diameters, around 1,000 times that of the Sun. Supergiants are assigned to luminosity class I. The brightest are often given the separate class Ia, and the others placed in class Ib.

superheterodyne receiver Most common type of RADIO receiver. An oscillator produces signals of varying frequency such that the difference between them and the frequencies of the incoming modulated CARRIER WAVES – the broadcast signal – is a constant, called the intermediate-frequency (i-f). The "i-f" can then be amplified and demodulated to produce the output signal for the loudspeaker.

supernova (pl. supernovae) Stellar explosion in which virtually an entire STAR is disrupted. For a week or so, a supernova may outshine all the other stars in its galaxy. The luminosity is some 23 magnitudes (1,000 million times) brighter than the Sun, and the energy released in the explosion is the same as is released over the star's entire previous life. After a couple of years the supernova has expanded so much that it becomes thin and transparent. For hundreds or thousands of years the ejected material remains visible as a **supernova remnant**. A supernova is about 1,000 times brighter than a NOVA. There is probably about one supernova every 30 years in a galaxy like our own, but most of them are concealed by dust. Supernovae are of two main types, classified by their light curves and spectra.

superphosphate (superphosphate of lime) Mixture of dihydrogen calcium phosphate (obtained by treating calcium phosphate with sulphuric acid) and calcium sulphate. It is used as a fertilizer.

superposition, law of In geology, law that states that in undisturbed layers of sedimentary deposits, younger beds overlie older ones. *See also* STRATUM

supersaturation Term that describes the condition of a SOLUTION that contains more SOLUTE (dissolved substance) than a normal SATURATED SOLUTION. *See also* SUPERCOOLING

superscript Small letter, number or other symbol written or printed above and to the left or right of a number, letter or symbol, as indicating a mathematical exponent, such as 3^2 or x^n, or an element's mass number, such as ^{238}U. *See also* SUBSCRIPT

supersonic flight Flight at a speed greater than the local speed of sound – SUPERSONIC VELOCITY.

SUPERNOVA

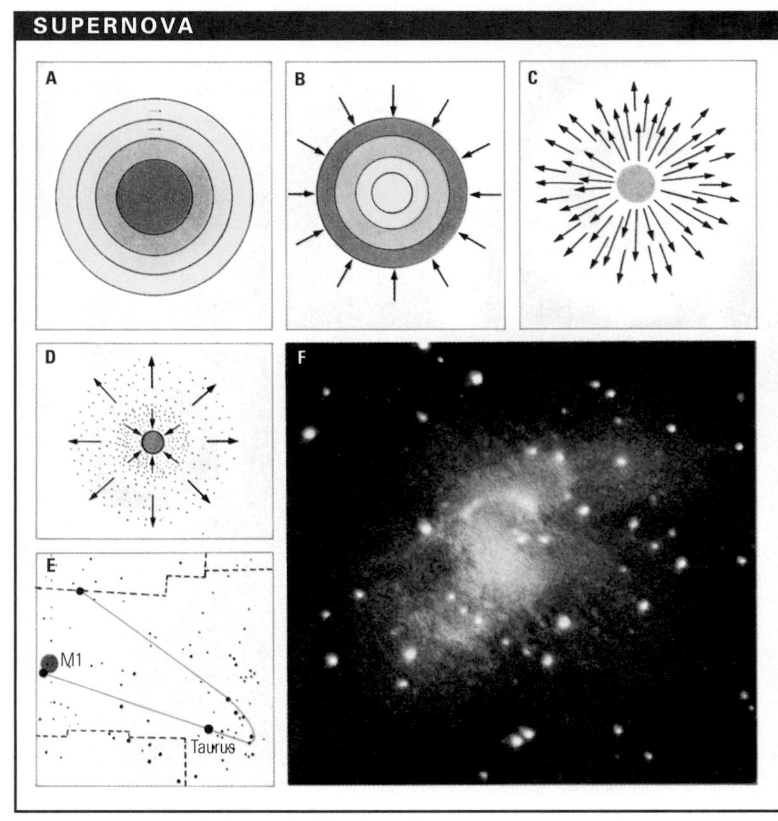

Supernovae are stars that suddenly flare up in brightness perhaps to 10,000 million times that of the Sun. They are thought to come about in stages. In (A), a massive star evolves quickly to the point where various nuclear processes exist within the stars simultaneously. In the centre, iron may be formed marking the end of nuclear energy production. In (B), the star begins to collapse. This, however, heats up the centre so that elements are broken down, and new reactions take place with explosive violence (C). This blows much of the star into space, while the central remnants collapse to obscurity probably as a very dense neutron star (D). One such supernova was seen in 1054 in Taurus (E). What remains is a cloud of gas – the Crab Nebula MI (F).

As an aircraft approaches the speed of sound, sound waves that normally move away from an aircraft build up to form SHOCK WAVES. Engineers once thought that these waves formed a SOUND BARRIER that would make supersonic flight impossible. Flight at supersonic speed is often expressed in terms of a MACH NUMBER. This indicates that an aircraft is flying at a certain number of times the local speed of sound.

supersonic velocity Velocity greater than that of the local speed of sound. In dry air at 0°C (32°F), this speed is about 330m s^{-1} (1,080ft s^{-1}) or 1,188km/h (736mph). Its magnitude is usually expressed as a MACH NUMBER. This is the ratio of the speed of a body to the speed of sound in a medium such as air. A velocity in excess of Mach 5 is said to be **hypersonic**. Any object travelling at supersonic velocity leaves behind a shock wave that a ground observer hears as a SONIC BOOM. *See also* SUPERSONIC FLIGHT

supersonic wave Wave set up in the air by any object moving at or above SUPERSONIC VELOCITY. At these speeds, sound waves created by the object's speed pile up ahead of it and are released violently behind it as a SHOCK WAVE. Observers on the ground see a supersonic aircraft pass overhead before they experience the shock wave that spreads out and reaches them as a SONIC BOOM.

superstring theory Physical theory that attempts to explain the properties of ELEMENTARY PARTICLES and the interactions between them. It combines QUANTUM THEORY and RELATIVITY, especially to explain nuclear forces and the force of gravity (*see* FUNDAMENTAL FORCES). Superstrings are hypothetical one-dimensional objects *c.*10^{-35}m long, which require a 10-dimensional universe to accommodate them. Their use in quantum calculations predicts the existence of GRAVITONS, the particles believed to be involved in gravitational interaction. *See also* UNIFIED FIELD THEORY

supersymmetry Physical theory that interrelates the basic matter particles (QUARKS and LEPTONS), both of which are classified as FERMIONS, with force-carrying BOSONS such as PHOTONS and GLUONS. Every boson is assigned a corresponding fermion, and each fermion has a corresponding boson. The bosons associated with the leptons and quarks are designated the super-particles **sleptons** and **squarks** (by adding an initial *s*). The fermions that partner the bosons, gluons and photons are called **gluinos** and **photinos**. None of these particles has yet been found, but the theory predicts an underlying superforce that would combine the electromagnetic, gravitational and strong and weak nuclear forces. *See also* FUNDAMENTAL FORCES; GRAND UNIFIED THEORY (GUT)

supplementary angles Two angles the sum of which is 180°. Each is the supplement of the other.

suprarenal glands Alternative name for the ADRENAL GLANDS.

surface tension Molecular force associated with the boundary layer of a liquid. It makes a liquid behave as if there were a "skin" on the surface. Cohesive forces in the skin tend to resist disruption, so that a needle or razor blade placed carefully on the surface "floats" even though its density is many times that of the liquid. *See also* MENISCUS

surfactant Substance that lowers the SURFACE TENSION of a SOLVENT in which it is dissolved. In water, SOAPS and DETERGENTS act as surfactants. Their molecules concentrate at the surface of oil and water molecules and act as emulsifying or foaming agents, to produce lathering and frothing.

surgery Branch of medical practice concerned with treatment by operation. Traditionally it has mainly involved open surgery: gaining access to the operative site by way of an incision. However, the practice of using ENDOSCOPES has enabled the development of MINIMAL ACCESS SURGERY. Surgery is carried out under sterile conditions, using local or general ANAESTHESIA. There is a growing number of surgical specialties, including general surgery, gynaecology, orthopaedics, cardiothoracic, neurosurgery, ophthalmology, plastic and reconstructive surgery and transplantation.

surveying Accurate measurement of the Earth's surface. It is used in establishing land boundaries, the TOPOGRAPHY of landforms, and for major construction and civil engineering work, such as dams, bridges and highways. Measurements are linear or angular, applying principles of GEOMETRY and TRIGONOMETRY. For smaller areas, the land is treated as a horizontal plane. Large areas involve considerations of the Earth's curved shape, a technique referred to as GEODESIC SURVEYING.

suspension In chemistry, liquid (or gas) medium in which small solid (or liquid) particles are uniformly dispersed. Examples are dust in air, and fine particles of sand in water. The particles are larger than those found in a COLLOID and will settle if the suspension stands undisturbed.

suspension In geology, manner in which small sedimentary particles are carried along by rivers. Turbulence and upward eddies keep particles suspended and make the water look muddy. Worldwide, the amount of material transported in this way is tremendous, estimated at 80 tonnes of solids per square kilometre of the Earth's surface every year.

suspension bridge BRIDGE that has its roadway (deck) suspended from two or more CABLES, which usually pass over towers and are anchored at the ends. The cables are of wire twisted spirally, and the deck is usually made rigid by stiffening trusses. The Humber Estuary Bridge, UK, has a span of 1,410m (4,626ft), making it one of the longest suspension bridges in the world.

Svedberg, Theodor (1884–1971) Swedish chemist who was awarded the 1926 Nobel Prize for chemistry for his development (1923) of the ultracentrifuge. Svedberg used it to study COLLOIDS and large MOLECULES and it enabled the RELATIVE MOLECULAR MASSES (molecular weights) of large PROTEINS to be determined for the first time. He also helped to develop ELECTROPHORESIS techniques for separating proteins on the basis of electrical charge.

swallow hole Large POTHOLE containing a flowing stream or river. A pothole is an enlarged JOINT, usually in limestone, formed by running water and possibly chemical WEATHERING. It may be up to 100m (330ft) deep and is termed a swallow hole if water continues to pour down it. *See also* SINK HOLE

Swammerdam, Jan (1637–80) Dutch naturalist. His researches and his book *General History of Insects* provided a system for classifying insects and laid the foundations of ENTOMOLOGY, the scientific study of insects. Swammerdam also discovered (1658) the existance of red blood corpuscles.

Swan, Sir Joseph Wilson (1828–1914) British scientist who designed the first electric LIGHT BULB. Starting in 1848, he used a carbon filament inside an evacuated glass envelope and by 1860, some 20 years earlier than Thomas EDISON, he had produced

▲ **suspension bridge** A view of the Golden Gate Bridge over Golden Gate Bay, San Francisco. The bridge was opened in 1937, and its longest span, 1,280m (4,200ft), remained the largest in the world until 1964. The span is suspended from two cables, attached to towers 227m (746ft) high. It has become one of the most famous landmarks in California.

a workable bulb, though with a short lifetime. His bulbs were used to light the House of Commons in 1881 and the British Museum in 1882.

swarm Gathering of many honey bees around a queen bee, which flies off from a hive to found another colony. This happens when a young queen supplants the old queen, and it is the latter who leads off her retinue in a swarm.

sweat gland One of many small GLANDS that open on the SKIN surface through pores and release sweat, composed mainly of water and some salt, to regulate body temperature. In humans, they are found all over the body, especially in certain places such as the armpits. In some mammals they are found only on the soles of the feet. The glands extract water, salts and urea from the blood capillaries that supply them and excrete them as sweat. The process is under the control of the AUTONOMIC NERVOUS SYSTEM, and forms an important part of the body's temperature control mechanism. *See also* HOMEOSTASIS

sweating Loss of water, salts and urea from the body surface of many mammals as a result of the action of SWEAT GLANDS. The evaporation of sweat cools the skin, and the blood passing through capillaries close to the skin surface. Sweat evaporates most rapidly when the outside air is warm and dry, and when it is windy. Excessive sweating must be compensated for by increased intake of water and salt. Sweating may also increase in response to stress or anxiety.

swim bladder (air bladder) Air-filled sac that controls the buoyancy of bony FISH. The swim bladder is located above the gut. A connection from the bladder to the gut allows air to enter and leave the bladder. Air may also travel to the bladder along capillary blood vessels. The amount of air in the bladder can be changed so that the fish's specific gravity matches that of the water surrounding it. In lungfish the swim bladder also acts as a rudimentary lung.

swing-wing aircraft Aircraft with wings the geometry of which can be changed to suit a wide range of flying speeds. A swing wing, also known as a variable sweep or variable geometry wing, can be moved forwards or backwards in flight. Moved forwards it creates a high aspect ratio (ratio of span to width) with good LIFT at low speeds for takeoff and landing. Moved backwards it forms a highly swept back wing with low DRAG, enabling high-speed flight.

Sydenham, Thomas (1624–89) English physician, often called the English Hippocrates. He initiated the cooling method of treating SMALLPOX, made a thorough study of all aspects of epidemics, and wrote descriptions of many diseases, including malaria, smallpox and GOUT.

syenite Uncommon coarse-grained, intrusive IGNEOUS ROCK. Composed mainly of alkali FELDSPAR or PLAGIOCLASE with up to 65% silica, syenite is a pink or grey rock containing also some AMPHIBOLE, which forms dark patches. A blue-grey form with large feldspar crystals, called larvikite, is used as an ornamental stone.

symbiosis Relationship between two or more different organisms that is usually mutually advantageous. Such relationships are more accurately referred to as **mutualism**. They include that of the cleaner fish that pick food from the teeth of larger fish, thus obtaining a meal while cleaning teeth; or the presence of nitrogen-fixing bacteria in the roots of many plants, where the plant gains nitrogen compounds and the bacteria are supplied with food materials such as carbohydrates. *See also* COMMENSALISM; PARASITE

SYMBIOSIS

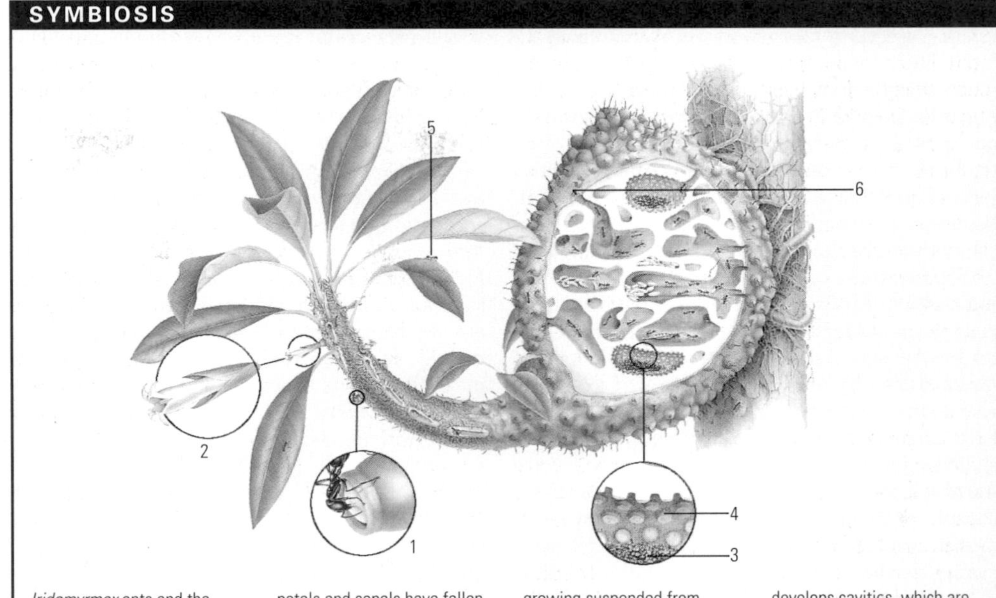

Iridomyrmex ants and the *Myrmecodia* (ant plant) benefit from a symbiotic relationship. The ants feed on the sugary nectar of the plant. This is produced in nectaries (1), which develop at the base of the flower (2) after the petals and sepals have fallen off. The plant benefits from the vital minerals in the ants' defecation and waste materials (3), which it absorbs through the warty inner surface of the chambers (4). The ant plant is epiphytic, growing suspended from trees in upland rainforests, where the soils are often lacking in nutrients. The mineral nutrients provided by the ants supplement the plant's poor diet. As the plant grows, its stem enlarges and develops cavities, which are invaded by the ants (5). These chambers do not interconnect, but have separate passages to the outside (6). A complete ant colony soon becomes established in the plant.

symmetry In biology, anatomical description of body form or geometrical pattern of a plant or animal. It can be used in the classification of living things (TAXONOMY). In mathematics, symmetry is a similarity or correspondence between parts of an object. A symmetrical figure is one that has a line or point about which certain operations produce the same figure. These operations include reflection, rotation and translation. For example, a circle has rotational symmetry about its centre point: it can be rotated any amount around this point without it's figure changing. It also has reflectional symmetry along its diameter: its form is unchanged by reflection along this line.

sympathetic nervous system One of the two parts of the AUTONOMIC NERVOUS SYSTEM, the other being the PARASYMPATHETIC NERVOUS SYSTEM. Both are involved with the action of INVOLUNTARY MUSCLES. The sympathetic nervous system controls muscles that prepare the body for action, by, for example, increasing heart rate and slowing digestion. The parasympathetic nervous system does the reverse.

synapse Connection between the nerve ending of one NEURON and the next or between a nerve cell and a muscle. It is the site at which NERVE IMPULSES are transmitted, using chemicals (NEUROTRANSMITTERS), such as acetylcholine, secreted by structures in the nerve endings.

synchrocyclotron Type of particle ACCELERATOR, a modification of the CYCLOTRON, in which the frequency of the electric field is slowly changed as the particles spiral round the device. This counteracts the increase in particle mass at relativistic velocities and prevents the particles getting out of phase with the field, the radius of the path being proportional to velocity and mass. Energies can reach 700 MeV.

synchronous orbit Path of an artificial SATELLITE with an ORBIT taking the same time as one rotation of the planet about which it moves. If a synchronous satellite is placed in a circular orbit above the equator of a planet, and moves in the same direction as the planet spins, then the satellite will keep up with the planet's movement and will always be above the same point on the equator. In the case of the Earth, this condition occurs when a satellite is orbiting at a height of about 35,900km (22,300mi) above the Equator. Such a satellite is said to be in GEOSTATIONARY ORBIT. Synchronous satellites are useful for telecommunications purposes as they can be arranged to cover specific areas. Geostationary satellites are most commonly used. Because these satellites remain in fixed positions relative to the Earth, their aerials, and those used on the ground to communicate with them, point in fixed directions. Geostationary satellites provide numerous telephone, telegraph, television, radio and data communication services around the world.

synchrotron Type of particle ACCELERATOR in which a beam of ELECTRONS or PROTONS is focused and guided around a fixed circular path by changing magnetic fields. Millions of revolutions are made. A high-frequency electric field at one point in the path accelerates the particles. Proton energies can reach hundreds of GeV, electron energies reach tens of GeV.

synchrotron radiation In physics, stream of ELECTROMAGNETIC RADIATION generated by high-energy ELECTRONS constantly accelerating around a MAGNETIC FIELD. It can take the form of X-rays, radio waves or visible light. Such radiation has uses in medicine and the electronics industry. In astronomy, the term is used to describe cosmic RADIO WAVES of a similar SPECTRUM.

syncline Downward FOLD in rocks. When rock layers fold down into a trough-like form, it is known as a syncline. (An upward arch-shaped fold is called an ANTICLINE.) The sides of the syncline are called **limbs**, and the median line between the limbs along the trough is known as the axis of the fold.

syndrome In medicine, group of symptoms that occur together, characterizing a particular disorder.

synergism Combined effect of two drugs that is greater than the sum of the effects they would produce if used separately. This synergistic effect is not necessarily beneficial and in some cases may be dangerous.

Synge, Richard Laurence Millington (1914–94) British biochemist. He shared the 1952 Nobel Prize for chemistry with Archer MARTIN for identifying the individual AMINO ACIDS in INSULIN.

synodic period Period of apparent revolution of one heavenly body about another as observed from the Earth, for example from one opposition or conjunction to the next. A **synodic month** is the period between two identical phases of the Moon. The same as the duration of one lunation, its length is 29.53059 mean solar days.

synovial fluid Viscous, colourless fluid that lubricates the movable joints between bones. It is secreted by the synovial membrane, which links the bones at a freely movable joint. Synovial fluid is also found in the bursae, membranous sacs that help to reduce friction in major joints such as the shoulder, hip and knee.

synthesizer In music, an electronic instrument capable of producing a wide variety of different sounds, pitches and timbres. Although a form of synthesizer had been conceived as early as 1929, the modern instrument was invented in 1964 by the US electronic designer Robert Moog (1934–). Computer technology is now used to control the instrument's different functions, enabling more sophisticated synthesizers to replicate the sounds of other (often non-electronic) instruments. Because they are electronic instruments, synthesizers can easily be controlled by a computer, using a program called a **sequencer**, to play rhythms and musical pieces.

synthetic fibre Materials intended to substitute for, supplement or improve on the quality of natural fibres such as cotton, wool and silk. The manufacture of synthetic fibres began early in this century with RAYON, still a familiar textile, which (like cotton) is made of the natural POLYMER cellulose. In rayon, CELLULOSE is extracted from plants, dissolved and reconstituted as a fibre by spinning it through spinnerets, devices resembling the rose of a large shower head. The first truly synthetic, or artificial-polymer, fibre was NYLON – a revolutionary material which, weight for weight, is stronger than steel. The manufacture of nylon begins in the petroleum and coal industries, from which its MONOMERS (materials such as adipic acid and hexamethylene diamine) are obtained. After polymerization, nylon can be melted and melt-spun through **spinnerets** to make fibres, which can then be further elongated by cold drawing. POLYESTERS, synthetics like nylon widely used for textiles, are also melt-spun. The equally popular ACRYLICS decompose before they melt, so are wet-spun like rayon. POLYETHENE (polythene), POLYPROPENE and VINYL PLASTICS such as POLYVINYL CHLORIDE (PVC) can also be made as fibres although they are more familiar as film or sheet. Inorganic synthetic fibres include GLASS FIBRE, spun from molten glass; CARBON FIBRE, used to reinforce ceramics and metals; and metallic fibres, used mainly for decorative purposes in clothing.

syrinx Vocal organ of a bird, consisting of thin, vibrating muscles at the base of the windpipe.

systems analysis Method of analysing a process or operation in order to improve efficiency, particularly with the aid of a COMPUTER, which has the capacity to handle the large number of interdependent activities involved in a complex operation.

systems engineering Application of such methods of analysis as operations research, CYBERNETICS and INFORMATION THEORY to the solution of problems associated with large, complex, artificial systems in which there is considerable interaction between the component parts of the system. These solutions often involve the use of COMPUTERS, feedback CONTROL SYSTEMS, and SERVOMECHANISMS. Originally developed during World War 2 for military applications, it has broadened into a variety of fields, including telephony, traffic control and social administration. *See also* FEEDBACK

systole Time in the cardiac cycle when the HEART muscle contracts, propelling blood from the heart into the circulatory system.

Szent-Györgyi, Albert von (1893–1986) US biochemist, b. Hungary. He was awarded the 1937 Nobel Prize for physiology or medicine for his work on biological oxidation processes and the isolation of vitamin C. He also studied the biochemistry of MUSCLE, discovering the muscle protein ACTIN, which is responsible for muscular contraction when combined with the muscle protein MYOSIN.

Szilard, Leo (1898–1964) US physicist, b. Hungary. His early work established the relation between information transfer and ENTROPY (disorder or randomness of a system). He devised a means of separating isotopes of artificial radioactive elements, developed the **chemostat** (a type of culture vessel in which organisms are grown), and proposed theories of aging, recall and memory. He was also involved at the University of Chicago, along with Enrico FERMI, in the creation of the first sustained nuclear CHAIN REACTION based on uranium FISSION. Szilard urged the US government to develop the atom bomb but later campaigned against nuclear weapons.

S

T

tachometer Device that measures the rate of rotation of a wheel or shaft. One type of tachometer is a simple revolution counter, with which average speeds of rotation can be measured with a stopwatch. More sophisticated instruments show instantaneous speeds of rotation. These instruments are used widely in high-performance cars and in industry to check machine efficiency. They can be made on mechanical, electrical or electronic principles.
tachycardia Increase in HEART rate beyond the normal. It may occur after exertion or due to excitement or illness, particularly during fever; or it may result from a heart condition.
tachyon Hypothetical ELEMENTARY PARTICLE that travels faster than the speed of light. It might be detectable by the CERENKOV RADIATION it emits. Such a particle would be consistent with the special theory of RELATIVITY. It would require energy to slow it down, so it could never be brought down to the speed of light.
taiga Coniferous forest extending over thousands of square kilometres in Siberia, Russia, Finland, Sweden, Norway and Canada. The climatic conditions are unsuitable for deciduous woodland, but coniferous trees can survive. Summer temperatures are around 15°C (59°F), and January averages −10°C to −20°C (14°F to −4°F). There are several months with temperatures below freezing point, and the subsoil is frozen for much of the year. The roots of the coniferous trees spread out horizontally because they cannot penetrate vertically into the ground. The main types of tree are spruce, larch, fir and pine. In the S parts of the taiga, the trees grow to 15m (50ft) in height, but further N the trees become smaller until they are little more than bushes, and TUNDRA vegetation can be seen. Wherever coniferous forests occur, the needles fall to the ground to produce the type of soil known as PODZOL, which is acid and ashy grey in colour. *See also* CONIFER; EVERGREEN
Talbot, William Henry Fox (1800–77) British scientist. Talbot improved on the work of Nicéphore Niepce (1765–1833) and Louis Daguerre (1789–1851) by inventing the first photographic process capable of producing an infinite number of positive prints from a single, original negative. *See* PHOTOGRAPHY
talc Sheet silicate mineral, hydrous magnesium silicate ($Mg_3Si_4O_{10}(OH)_2$). Talc occurs as rare tabulate crystals in a monoclinic system and as masses. It is either white, green, blue or brown in colour. It is used as base for talcum powder and in ceramics. Massive, fine-grained talc is known as SOAPSTONE. Hardness 1; r.d. 2.6.
talus *See* SCREE
Tamm, Igor Yevgenyevich (1895–1971) Russian theoretical physicist. Tamm shared the 1958 Nobel Prize for physics with Pavel CERENKOV and I.M. Frank for the discovery and interpretation (1937) of CERENKOV RADIATION, showing that light is emitted by charged particles travelling at very high speeds in a transparent medium.
tangent In TRIGONOMETRY, the RATIO between the length of the sides opposite and adjacent to an acute angle within a right-angle TRIANGLE. (The third side is the hypotenuse.) The expression denoting the tangent of angle A is commonly abbreviated to $\tan(A)$; $\tan(A) = \sin(A)/\cos(A)$. Also, the tangent at a point on a curve is the straight line whose slope matches that of the curve at that point.
tank Tracked armoured vehicle mounting a single primary weapon, usually an artillery piece, and one or more machine-guns. Modern tanks have an enclosed fully revolving turret and are heavily armoured; main battle tanks weigh from 35 to 50 tonnes and usually have a crew of four. Developed in great secrecy by the British during World War 1, tanks were first employed at the Battle of the Somme (1916).
tanker Ship designed to carry liquid cargo. The largest tankers can carry *c*.500,000 tonnes of cargo. This usually consists of PETROLEUM (crude oil) products. Spillage during loading and unloading causes considerable WATER POLLUTION. Other tanker cargoes include natural gas (liquefied by cooling) and wine. Combination carriers are tankers that can carry solid cargoes, such as grain and ore, as well as liquids.
tannin (tannic acid) Any of a group of complex organic compounds derived from tree bark, roots and galls (such as the oak), unripe fruit, tea and

TACHOMETER

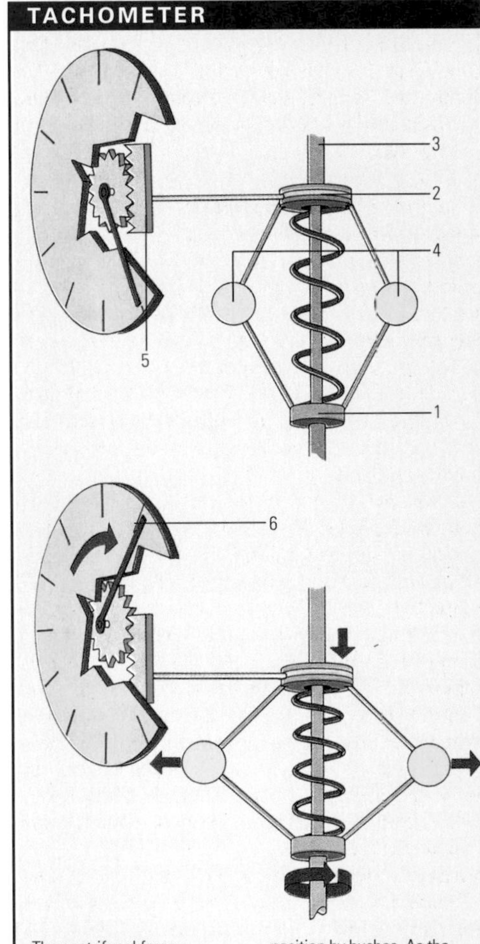

The centrifugal force tachometer measures instantaneous rate of rotation of a shaft. It is capable of accuracy to ±1% of full scale value.

Key
1) fixed bush (cylindrical sleeve)
2) moveable bush
3) rotating shaft
4) weights held in central position by bushes. As the shaft rotates, the weights move outwards due to centrifugal force. The pointer then moves indicating speed.
5) rack and pinion. When the moveable bush slides up the shaft, as the weights move outwards, the rack and pinion mechanism operates and the pointer moves.
6) pointer and graduated scale

▲▶ **tank** The cutaway view shows the main essentials of the B ritish Chieftain tank. It has a complement of four men, although all positions are not shown.

Key
1) 120mm cannon ROF main armament
2) 12.7mm ranging gun used by gunner to determine when the main armament is on target. It has the same flight trajectory as the big gun, and fires tracer for easy observance.
3) 7.62 machine gun operated by the gun loader against troops or unarmoured vehicles.
4) driver's periscope
5) foot-operated brakes and gear shift
6) track steering controls. A tank is turned by braking one track and skidding the other.
7) road wheels, each with independent suspension
8) HESH projectile stowage area on the floor of the turret
9) gun breach mechanism
10) gunner's periscope
11) commander's main periscope
12) APDS projectile stowage area
13) commander's vertically-opening hatch for overhead protection
14) fume extractor to draw powder fumes from the gun chamber and turret
15) thermal insulator wrapping to protect the barrel against extreme heat and cold effects
16) 700hp, 2400rpm diesel main engine
17) track driving sprocket

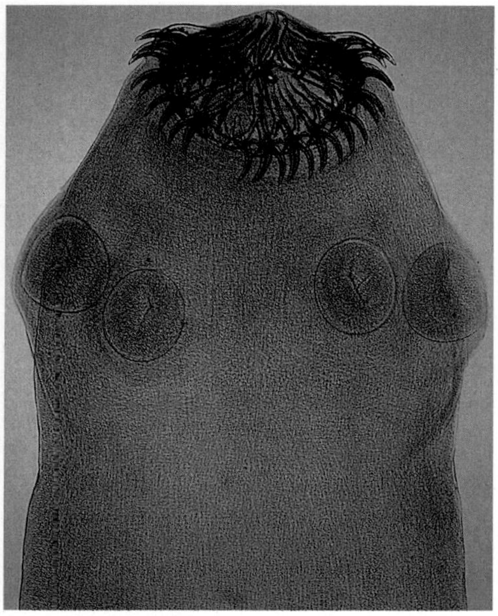

▲ **tapeworm** False-colour light micrograph of the scolex (head) of the adult pork tapeworm, *Taenia solium*, magnified 12 times. The image shows the hooks and four suckers with which the tapeworm attaches itself to the intestinal walls of its host. Humans are the principal host of this tapeworm, which annually infects some 4 million people worldwide. Tapeworms absorb their host's digested food directly through their entire surface area; they have no digestive system of their own. The eggs of the tapeworm develop into larvae in domestic pigs (hence the tapeworm's name) and other animals, their secondary hosts. Humans can be secondary as well as primary hosts.

coffee. Yellow and astringent, tannin has the property of converting the protein GELATIN into an insoluble compound that will not putrefy, and for this reason is used in TANNING to cure hides and make leather. It is also used in inks and dyes, and as an astringent in medicine.

tanning Process of converting skins and hides into leather. Traditionally, tanning liquids are based on TANNIN and, although the process takes weeks, it is still used for heavy leathers. Light leathers are now tanned in a few hours using chromium salts.

tantalum (symbol Ta) Rare, lustrous, blue-grey metallic element, first discovered in 1802. Its chief ore is columbite-tantalite. Hard but malleable, tantalum is used as a wire and in electrical components, chemical equipment and medical instruments. Properties: at.no. 73; r.a.m. 180.948; r.d. 16.6; m.p. 2,996°C (5,425°F); b.p. 5,425°C (9,797°F); most common isotope ^{181}Ta (99.988%).

tape, magnetic *See* MAGNETIC TAPE

tape recorder Device which records and plays back sound on MAGNETIC TAPE. Sound is transformed into an electric signal by a MICROPHONE and fed to a TRANSDUCER, which converts it into the magnetic variations that magnetize the particles on the treated tape. *See also* DIGITAL AUDIO TAPE (DAT); SOUND RECORDING

tapetum Reflecting structure in the EYES of some nocturnal vertebrates, which serves to improve night vision. Occurring on the retinal side of the choroid, the tapetum contains guanine crystals which reflect light back onto the RETINA. This process makes eyes shine in the dark. The eyes of some night-flying insects also contain a tapetum.

tapeworm Parasite of the genus *Taenia* which colonizes the intestines of vertebrates, including human beings. Caught from eating raw or under-cooked meat, it may cause serious disease. Class Sestoda.

taproot First ROOT of a plant that develops from the RADICLE. The taproot grows directly downwards and remains the main root of the plant, sending off lateral side roots to extend the root system. In BIENNIAL plants, whose leaves usually die down in the first winter, the root is the part of the plant that remains alive underground ready to grow new leaves the following year. In some plants (such as beets, carrots and parsnips) the taproot develops into a fleshy organ for storing STARCH. In root vegetables, it forms the edible part of the plant.

tar Black or dark brown complex liquid mixture of HYDROCARBON compounds, derived from wood, coal and other organic materials by heating them with little or no air, then condensing the tar from the distilled vapours. Tar, from petroleum, is a major source of HYDROCARBONS for the synthesis of pharmaceuticals, pesticides and plastics; cruder tar compounds such as pitch are used for road surfacing and protecting timber against rot and pests. Wood tar yields creosote and PARAFFIN. *See also* BITUMEN

Tarantula Nebula (NGC 2070 or 30 Doradus) Collection of gas and dust that is 1,000 light years across and 500,000 times the mass of the Sun. It is bigger and brighter than any NEBULA in our GALAXY or any other nearby galaxy. Faintly visible to the naked eye on the SE edge of the Large MAGELLANIC CLOUD, it has a complex filamentary structure with a cluster of stars, R136, at its centre.

tarmac Mixture of GRAVEL and TAR mainly used to cover roads, drives and paths. It is hard-wearing, impervious to water and provides an ideal surface for the rubber tyres of motor vehicles. *See also* BITUMEN; MCADAM, JOHN LOUDON

tarsals In terrestrial vertebrates, the seven bones that make up the ankle and adjoining part of the foot. Strong, compact bones, they are arranged so that the foot can be rotated (to a limited extent) in any direction. The tarsals articulate with the lower leg bones above and, below, with the METATARSALS of the foot.

tar sands Porous, often black, rocks such as limestones, sands and sandstones found on the surface. They contain deposits of BITUMEN (asphalt) in the spaces between the grains and commonly have a distinct odour of TAR. Extensive deposits are found in North America, primarily in Alberta and Texas. Estimates of the amount of bitumen contained in these deposits are extremely high, but its high viscosity makes extraction difficult.

Tarski, Alfred (1902–83) US mathematician and philosopher, b. Poland. He is celebrated for his development of a semantic method for metamathematics, a branch of mathematical LOGIC. His publications include *Introduction to Logic and the Methodology of Deductive Science* (1936) and *Logic, Semantics, Metamathematics* (1956).

Tartaglia, Niccoló Fontana (1499–1557) Italian mathematician, the first to obtain a general solution to the cubic equation of the form $x^3 + ax^2 + bx + c = 0$. His method was announced by Girolamo CARDANO. From 1534 he taught in Venice. In *Nova Scientia* (1537), Tartaglia applied mathematics to ballistics. Other publications include *Trattato di Numeri et Misure* (1556–60).

tartaric acid (dihydroxybutanedioic acid) Colourless, optically active organic compound, $(CHOH)_2(COOH)_2$, occurring naturally in fruits. It can be made from maleic anhydride and hydrogen peroxide or by ENZYME action on a succinic acid. Tartaric acid is widely used in baking powder, cream of tartar, soft drinks and in the textile industry. Properties: r.d. 1.76; m.p. 171°C (340°F).

tartrate Salt or ester of TARTARIC ACID. Tartrates, such as potassium bitartrate (cream of tartar) and sodium bitartrate, are used in baking powder, in the food industry and in tanning.

taste One of the five SENSES, responding to the chemical constituents of food. In human beings, the taste buds of the tongue differentiate four qualities: sweetness, saltiness, bitterness and sourness. The sense of taste is supplemented by the sense of smell.

Tatum, Edward Lawrie (1900–75) US geneticist and biochemist. He shared the 1958 Nobel Prize for physiology or medicine with George BEADLE and Joshua LEDERBERG for his part in the discovery that GENES act by regulating specific chemical processes, a basic principle explaining how genes determine hereditary characteristics. *See also* HEREDITY

tau particle ELEMENTARY PARTICLE that has a mass approximately 1.9 times that of the PROTON. It is a LEPTON, meaning that it does not experience the STRONG NUCLEAR FORCE. It decays via the WEAK NUCLEAR FORCE with a lifetime of *c.*3×10^{-23} s. It has a unit negative charge.

Taurus (Bull) Constellation of the N sky, on the ecliptic between Aries and Gemini; It contains the Pleiades and Hyades stellar clusters and the Crab Nebula. The brightest star is the 1st-magnitude Alpha Tauri (Aldebaran).

tautomerism Property exhibited by a COMPOUND that exists as a mixture of two ISOMERS in EQUILIBRIUM. If either isomer is removed from the mixture, the equilibrium is momentarily disturbed but is restored by the spontaneous conversion (due to the migration of an ION or FREE RADICAL) of some of the remaining isomer into the removed type.

taxis Movement of a single cell, such as a GAMETE or bacterium, in response to an external stimulus. Movement towards the stimulus is **positive** taxis and away from the stimulus is **negative** taxis. Notable stimuli include light (**phototaxis**) and chemicals (**chemotaxis**). Various types of algae, for example, exhibit positive phototaxis – they swim towards a light source in order to increase photosynthesis.

taxonomy Organization of organisms into categories based on similarities of either MORPHOLOGY and ANATOMY (**classical** taxonomy), protein and nucleaic acid structure (**biochemical** taxonomy), the behaviour and morphology of chromosomes (**cytotaxonomy**), or the analysis of numerical data (**numerical** taxonomy). Carolus LINNAEUS developed the first taxonomic system during the 1750s. *See also* BINOMIAL NOMENCLATURE; CLASSIFICATION

Taylor, Sir Geoffrey Ingram (1886–1975) British physicist and meteorologist. Specializing in FLUID MECHANICS, Taylor made significant contributions to the understanding of turbulence and diffusion in fluids.

T-cell (Thymus-cell or T-lymphocyte) Type of LYMPHOCYTE that is the key to the defence mechanism of the IMMUNE SYSTEM. There are three main kinds of T-cells designated **Th** (helper T-cells), **Ts** (suppressor T-cells) and **Tc** (cytotoxic T-cells). **Helper** T-cells recognize foreign ANTIGENS and help other immune cells to act; **suppressor** T-cells prevent specific immune reactions; and **cytotoxic** T-cells, also called effector T-cells, kill cancerous cells or cells infected with a virus. T-cells are produced in the bone marrow and then move to the THYMUS GLAND. *See also* ACQUIRED IMMUNE DEFICIENCY SYNDROME (AIDS)

tear fault (lateral fault, wrench fault) Geological FAULT in which the movement is along the horizontal. Tear faults occur frequently in areas of TECTONIC activity. The amount of movement may only be a few metres, but can be several kilometres in some cases. The San Andreas Fault in California is a well-known example of a tear fault.

tear gas Chemical compound known as a lachrymator, a gas or aerosol that causes an excessive flow of tears and blinds and incapacitates temporarily without causing permanent injury. Such gases, as used by military and civil authorities, are generally made of chlorine and bromine derivatives of acetone and acetophenone.

tears Salty fluid, secreted by the LACHRYMAL GLANDS, that moistens the surface of the eye. Having antibacterial properties, they cleanse and disinfect the surface of the eye and also bring nutrients to the CORNEA.

technetium (symbol Tc) Silver-grey, TRANSITION ELEMENT. It was first made in 1937 by bombarding MOLYBDENUM with deuterons (nuclei of atoms of DEUTERIUM), and it was the first element to be synthesized in a CYCLOTRON. Technetium is found in the fission products of uranium and is present in some stars. It is used in radioactive tracer studies. There are 16 known isotopes. Properties: at. no. 43; r.a.m. 98.9062; r.d. 11.5; m.p. 2,172°C (3,942°F); b.p. 4,877°C (8,811°F); most stable isotope ^{99}Tc (half-life 2.6×10^6 yr).

technology Systematic study of the methods and techniques employed in industry, research, agriculture and commerce. More often the term is used to describe any practical application of scientific discoveries in the production of mechanisms and in the solution of problems which confront human beings.

tectonics Deformation within the Earth's CRUST and the geological structures produced by deformation, including FOLDS, FAULTS and the development of mountain chains. PLATE TECTONICS originates from the study of the main structural features of the Earth's crust, such as MID-OCEAN RIDGES, major tear faults, ocean TRENCHES, continental blocks, earthquake belts and so on, but is now used to describe a theory that can explain the distribution and evolution of these features.

teeth Hard, bone-like structures embedded in the jaws of vertebrates, used for chewing food, defence or other purposes. The teeth of all vertebrates have a similar structure, consisting of three layers. Mammalian teeth have an outer layer of hard ENAMEL. The middle layer consists of dentine, a bone-like substance that is capable of regeneration. The core of a tooth contains pulp which is softer and has a blood supply and nerves; it provides nutrients for the dentine. *See also* DENTITION

tektite Generally dark, glassy objects, ranging in diameter from 40 micrometres to 2mm (microtektites) and larger (to 10cm), believed to be of either lunar origin or formed from splashes of liquefied rock during meteorite impact on Earth. They occur in limited areas, called strewn-fields, on continents and ocean floors.

telecommunications Technology involved in the sending of information over a distance. The information comes in a variety of forms, such as digital signals, sounds, printed words or images. The sending is achieved through TELEGRAPH, TELEPHONE and RADIO and the medium may be wires or RADIO WAVES, or a combination of the two. Telegraphy was developed in the mid-19th century and radio arrived at the end of the century. TELEVISION manufacture was perfected in the 1930s. There are two basic types of message: DIGITAL SIGNALS, in which the message to be sent is converted in simple coded pulses and then sent (as in MORSE code); and ANALOGUE SIGNALS, in which the message – for example, a voice speaking – is converted into a series of electrical pulses that are similar in waveform to the MODULATIONS of the original message. *See also* FAX; SATELLITE, ARTIFICIAL

telegraph Any communications system that transmits and receives visible or audible coded signals over a distance. The first, optical, telegraphs were forms of SEMAPHORE. Although most modern devices are electrically operated and connected by wires or cables, telegraph messages can be sent via RADIO WAVES, microwaves and artificial SATELLITES. Pioneering work was performed by many, but credit for the electric telegraph and its code is credited usually to Samuel MORSE, who inaugurated (1844) the first public line between Washington and Baltimore. In 1866 the first permanent telegraph cable was laid across the Atlantic. In 1874 Thomas EDISON invented the "quadruplex" method of simultaneously transmitting four messages over the same wire.

telemetry System of transmitting data, usually measurements, over a distance, such as to Earth from a spacecraft via RADIO WAVES. In this example, SOLAR CELLS or chemical batteries provide power for operating special ANTENNAE on the spacecraft, which beam high-frequency (but low-power) radio waves to large receiving stations on the ground.

telephone Instrument that communicates speech sounds by means of wires or MICROWAVES. In 1876 Alexander Graham BELL invented the prototype, which employed a diaphragm of soft iron which vibrated to sound waves. These VIBRATIONS caused disturbances in the MAGNETIC FIELD of a nearby bar magnet, causing an ELECTRIC CURRENT of fluctuating intensity in the thin copper wire wrapped around the magnet. This current could be transmitted along wires to a distant identical device, which reversed the process to reproduce audible sound. Later improvements separated the TRANSMITTER from the RECEIVER and replaced the bar magnet with BATTERIES. With the invention of the MOBILE TELEPHONE, which operates through a digital network, telephones have become an integral part of social and professional life, as well as a lifestyle accessory.

telephone exchange Central switching system in which incoming telephone calls are routed to their destinations, either by telephone operators or, increasingly, by automatic mechanisms that are called into operation by direct dialling codes.

telephoto lens CAMERA lens with a long FOCAL LENGTH. A true telephoto LENS has a focal length longer than the physical length of the lens, as opposed to a **long-focus** lens, in which the focal length is equal to the physical length. There is no theoretical upper limit to the focal length possible, but sheer size tends to limit the actual focal lengths

TEETH

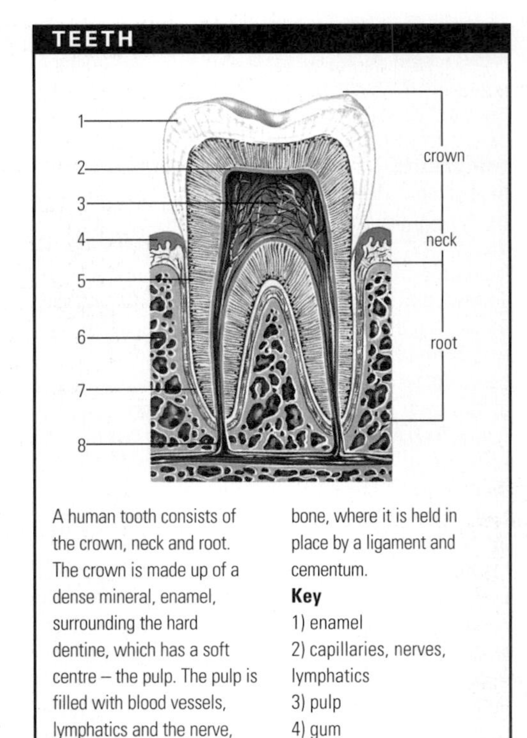

A human tooth consists of the crown, neck and root. The crown is made up of a dense mineral, enamel, surrounding the hard dentine, which has a soft centre – the pulp. The pulp is filled with blood vessels, lymphatics and the nerve, which reach the tooth through the root canal. The neck adheres to the gum, and the root penetrates the bone, where it is held in place by a ligament and cementum.

Key
1) enamel
2) capillaries, nerves, lymphatics
3) pulp
4) gum
5) dentine
6) jaw
7) cementum
8) root canal

TELEPHONE

Local telephone exchanges (1) connect local calls (2), which are analogue signals (3). Long-distance calls are routed to the long-distance exchange (4), where they are converted from analogue to digital. Digital "snapshots" are taken of the analogue signal 8,000 times a second (every 125 microseconds) – enough information to recreate the analogue signal accurately enough for the human ear. This whole process is called pulse-code modulation. Each eight-bit sample (5) is only 4 microseconds long, which leaves 121 microseconds between each one on the telephone line. To increase capacity, multiplexing combines the samples of up to 25 calls going to the same destination on the same line (6). This is done by feeding all the calls into a memory buffer (7), and then feeding them onto the long distance line in turn. At the receiving end, the process is reversed and the combined call is again fed into a memory buffer (8), separated (9), passed to the long-distance exchange (10) where it is turned back into an analogue signal (11), and sent to the local exchange (12). From there it is routed to its final destination (13). The process happens so fast that the human ear hears a continuous voice.

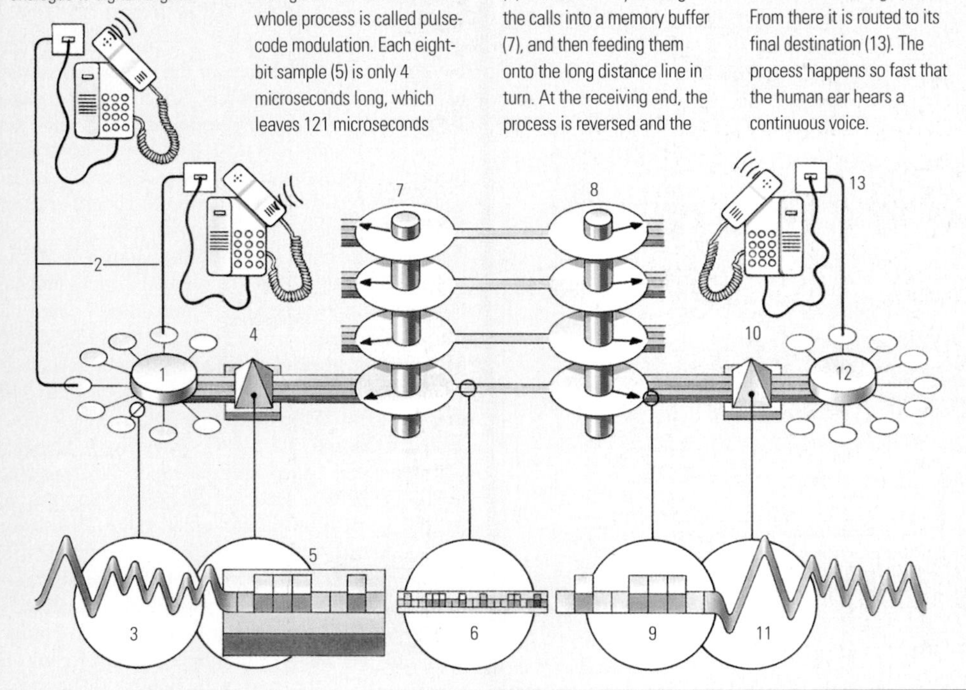

T

practicable. For a 35mm camera, any lens with a focal length of more than about 80mm may be regarded as a telephoto lens. For larger format cameras, the focal length may be as much as 1,000mm. *See also* PHOTOGRAPHY

teleprinter Machine having a keyboard similar to that of a TYPEWRITER, on which messages can be typed and telegraphed to distant RECEIVERS, and which also acts as a receiver, printing out incoming messages.

telescope Instrument for producing an enlarged image of a distant object or studying electromagnetic radiation from a distant source. Telescopes are categorized according to the area of the electromagnetic spectrum they cover. **Optical** telescopes can use lenses (**refracting** telescopes) or mirrors (**reflecting** telescopes); **catadioptric** telescopes use both in combination. The LENS or mirror is the telescope's main light-gathering part (**objective**), and its diameter is known as the APERTURE of the telescope and determines its light-gathering ability. The point at which the objective concentrates the light from the source is its FOCUS, and the distance from the focus to the objective is its FOCAL LENGTH. The light-gathering power of a telescope is proportional to the square of the aperture. Its ability to resolve (separate) two closely spaced objects into distinct images also depends on the size of the aperture: the larger the aperture, the better the resolution. Refracting telescopes were extensively used after versions were invented by Hans LIPPERSHEY (1608) and GALILEO (1609). The main disadvantage with refracting telescopes was chromatic ABERRATION, in which light of different colours is brought to a focus at different points, causing a single lens to produce a partly distorted image. This problem was solved by combining lenses so their aberrations cancel each other out. Reflecting telescopes are free of chromatic aberration – though, like lenses, they suffer from spherical aberration. In 1668 Sir Isaac NEWTON built an early astronomical reflector. Refracting telescopes also have a limit to the size of lenses: the world's largest refractor has a diameter of 1m (39in), much less than one of the larger reflectors; the Hale reflector at Mount Palomar, California, has an aperture of 508cm (200in). Earth-bound telescopes have limitations because the incoming radiation has to pass through the Earth's atmosphere. Advanced telescopes combat turbulence in the atmosphere (which makes stars twinkle) with adaptive OPTICS. Their mirrors consist of segments that are jiggled at high speed to stabilize the image. Major modern OBSERVATORIES are built on mountain peaks – for example, in Chile, the Canaries and Hawaii – in order to improve "seeing" and to observe INFRARED RADIATION from celestial bodies. Most of this infrared light is absorbed in the denser, lower part of the atmosphere. This ceases to be a problem with telescopes in Earth orbit, such as the HUBBLE SPACE TELESCOPE. Orbiting telescopes can also detect other types of electromagnetic radiation more easily, such as ULTRAVIOLET RADIATION, X-RAYS and GAMMA RADIATION. RADIO TELESCOPES are complex electronic systems that detect and analyse radio waves from beyond the Earth. Incoming RADIO WAVES are collected from a small area of the sky and focused at the centre of a large parabolic, steerable dish aerial. The signals are then amplified and analysed. The first radio telescope was built in 1937 by the US radio engineer Grote Reber.

TELESCOPE

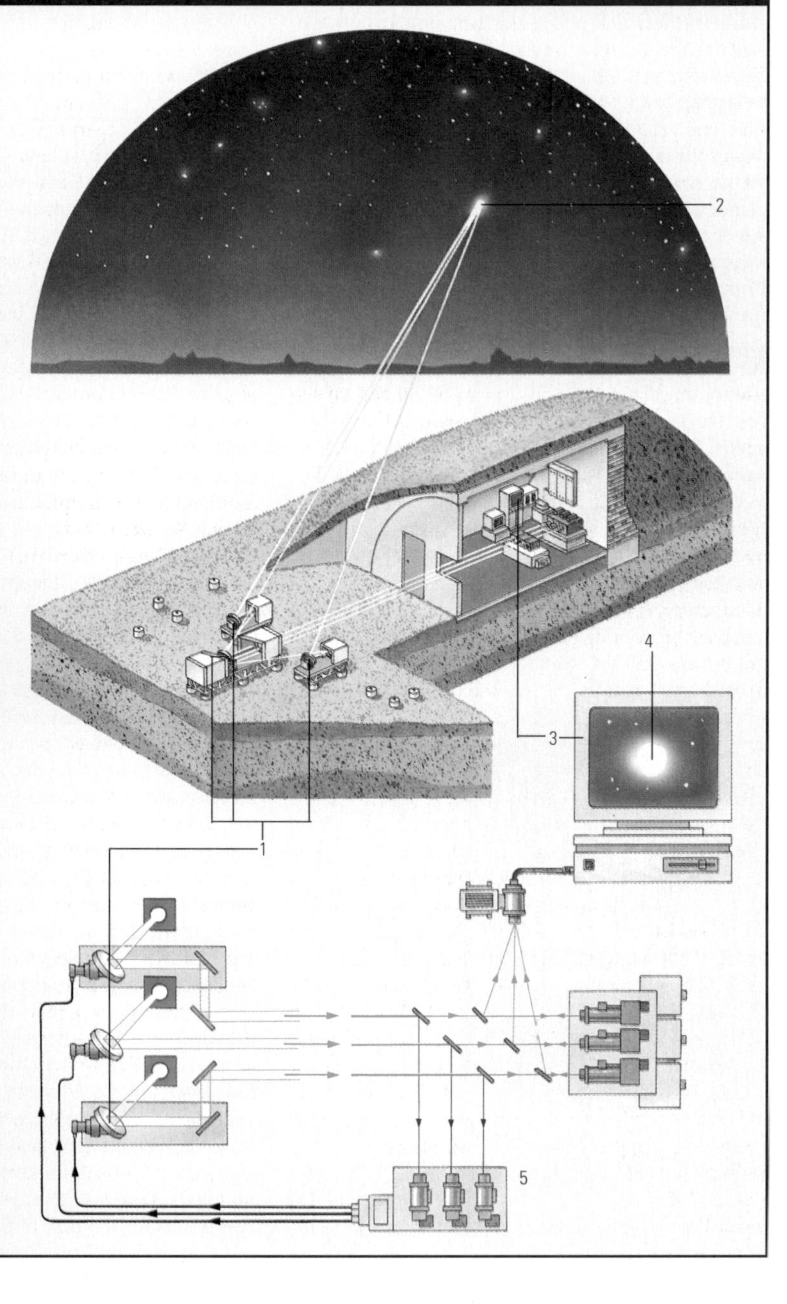

The COAST (**C**ambridge **O**ptical **A**perture **S**ynthesis **T**elescope) telescope, designed and built in Cambridge, s England, is the most powerful optical telescope ever built. Instead of a single enormous reflective surface, the Cambridge telescope uses the images collected by three small and relatively inexpensive optical telescopes, and combines them to form an extremely detailed image. The three telescopes (1) are focused on a single point (2), each one producing a fractionally different picture. As the Earth rotates, the position of the telescopes alters in relation to the target star or planet. These pictures from different angles are blended together by computer equipment (3) to provide a single, highly detailed image (4). A small portion of light reflected by the telescopes is bled off (5) and used to confirm the targeting of the star as the Earth moves.

Telescopium (the Telescope) Faint constellation of the S sky. Its brightest star is Alpha, magnitude 3.5.

television System that transmits and receives visual images by RADIO WAVES or CABLE. A television CAMERA converts the images from light rays into electrical signals. The basis of most television cameras is an **image orthicon tube**. Light rays are focused onto thin sheets of photoelectric material which emit ELECTRONS in proportion to the amount of light striking them. Behind these sheets is a positively charged target that emits additional electrons. The target has a positive charge in proportion to the amount of light in the original scene. Behind the image orthicon tube is an **electron gun** emitting a beam that scans the target. The beam is deflected back to the gun and collected by a multiplier that amplifies its intensity. The strength of this beam depends on the positive charges on the target and thus the original scene is converted into electrical signals. These are amplified and transmitted as VERY HIGH FREQUENCY (VHF) or ULTRA HIGH FREQUENCY (UHF) radio waves. Typically, a television channel has a BANDWIDTH of 5MHz (5 million cycles per second). The RECEIVER (TV set) operates in reverse to the camera. On reception, the signals are amplified and converted to light again in a CATHODE-RAY TUBE. Colour television has three synchronized image orthicon tubes in the camera, one for each of the three primary colours – red, blue and green. The tube of the receiver has three electron guns and the face of the tube is covered with a mosaic of fine phosphors in groups of three, each emitting only red, blue or green light when struck by a beam. These primary colours merge on the face of the screen to reconstitute the originally transmitted image. *See also* RADIO

telex (acronym for **tel**etype-writer **ex**change service) Telegraphic system using telephone lines through which direct current pulses representing characters typed on a typewriter keyboard are sent. It has been superseded by the FAX machine.

Telford, Thomas (1757–1834) Scottish civil engineer who built roads, bridges, canals, docks and harbours. His most notable achievements were the Caledonian Canal in Scotland, and the 177m (580ft) Menai Strait suspension bridge connecting Anglesey with mainland Wales. Telford was a founder and the first president of the Institution of Civil Engineers.

Teller, Edward (1908–) US physicist, b. Hungary, who has been called "the father of the HYDROGEN BOMB". In 1926 he received the Enrico Fermi Award. In 1935 he left Europe and settled in the USA, where he conducted research on solar energy. During World War 2, Teller contributed to atom bomb research with Enrico FERMI and was then involved in the MANHATTAN PROJECT at Los Alamos to produce the bomb. He was a central figure in developing and testing (1952) the hydrogen bomb. Teller was a major supporter of Ronald Reagan's abortive Strategic Defense Initiative ("star wars" project).

tellurium (symbol Te) Silver-white, metalloid element, discovered in 1782. It occurs naturally combined with gold in sylvanite, and its chief source is as a by-product of the electrolytic refining of copper. The lustrous, brittle element is used in semiconductor devices, as a catalyst in petroleum cracking and as an additive to increase the ductility of steel. Properties: at.no. 52; r.a.m. 127.60; r.d. 6.24; m.p. 449.5°C (841.1°F); b.p. 989.8°C (1,814°F).

telophase Stage in CELL DIVISION, following ANAPHASE. It is the final phase of MITOSIS and MEIOSIS. In **mitosis**, telophase involves the division of the CYTOPLASM to form two daughter cells with the same number of CHROMOSOMES as the original cell nucleus. The separated CHROMATIDS accumulate at the poles of the SPINDLE and a nuclear membrane forms to separate the two groups. In **meiosis**, there are two stages of telophase. In the first stage, two daughter cells are produced when a membrane forms between the separated chromatids. In the second telophase, these daughter cells divide to produce four cells, which are HAPLOID.

Telstar First active communications SATELLITE, launched by the USA on 10 July 1962. It contained a microwave radio receiver, amplifier and transmitter for relaying telephone and television signals. It operated for about 18 weeks, failed for five weeks, and then worked again for a further seven weeks before failing for good.

Temin, Howard Martin (1934–) US biologist. He shared the 1975 Nobel Prize for physiology or medicine with David Baltimore and Renato DULBECCO for his studies of the interaction between tumour viruses and the genetic material of the cell.

temperate zone Either of two regions of the Earth. The N temperate zone lies between the Arctic Circle and the Tropic of Cancer, and the S zone lies between the Tropic of Capricorn and the Antarctic Circle. A **temperate climate** is a moderate climate, as occurs in most temperate zones. The summers tend to be warm, but the winters are cool. Temperate climates have neither very high nor very low temperatures.

temperature In biology, intensity of heat. In warm-blooded (HOMEOTHERMIC) animals such as birds and mammals, body temperature is maintained within narrow limits regardless of the temperature of their surroundings. This is accomplished by muscular activity, the operation of cooling mechanisms such as vasodilation and vasoconstriction (changes in the diameter of superficial blood vessels) and sweating and metabolic activity. In humans, the normal body temperature is *c.*36.9°C (98.4°F), but this may vary with degree of activity, reaching 40°C (104°F) during exercise, and falling slightly below normal during sleep. In so-called cold-blooded (POIKILOTHERMAL) animals, such as reptiles, body temperature varies between wider limits, depending on the temperature of the surroundings (air or water).

temperature In physics, measure of the hotness or coldness of an object. Strictly, it describes the number of energy states available to a substance or system. Two objects placed in thermal contact exchange heat energy initially, but eventually arrive at thermal EQUILIBRIUM where both are said to have the same temperature – each is losing and gaining heat at equal rates so that neither has a net gain or loss of heat. At equilibrium, the most probable distribution of energy states of the atoms and molecules composing the objects has been attained. At high temperatures, the number of energy states available to the atoms and molecules of a system is large; at lower temperatures, fewer states are available (molecules become locked into position and liquids change to solids). At a sufficiently low temperature, all parts of the system are at their lowest energy levels, the ABSOLUTE ZERO of temperature. *See also* TEMPERATURE MEASUREMENT; TEMPERATURE SCALE

temperature coefficient In physics, change in any physical quantity per degree rise in temperature.

temperature inversion Anomalous increase in TEMPERATURE with height. Normally the temperature of the air decreases from ground level upwards. The average rate of decrease is 1°C (1.8°F) for every 160m (525ft). In certain meteorological conditions, this situation is reversed. On a clear, calm anticyclonic night, the cool air may roll downhill and accumulate in valleys, and the air temperature will be lower near the valley bottom than it is 100m to 200m (328–656ft) higher. Above the cold layer there will be warmer air, which is likely to form cloud or haze. Evidence of a temperature inversion can be seen if there is smoke rising from a bonfire. The smoke will rise vertically and then bend horizontally when it reaches the "inversion layer". If this situation develops on a larger scale, dust and dirt rising into the atmosphere are trapped and unable to escape, giving rise to serious pollution.

temperature measurement Method by which TEMPERATURE is specified. Several TEMPERATURE SCALES have been established, each of which uses a property (such as electrical resistance or volume expansion) of a thermometric substance (such as a metal wire or a liquid) to define a unit of temperature. The constant-volume gas THERMOMETER is the standard thermometer. Its temperatures are expressed in KELVIN degrees, but it is inconvenient for practical purposes. The International Practical Temperature Scale (IPTS), first adopted in 1968 and regularly revised since, provides scientists and technologists with a more convenient scale. The IPTS defines 0°C (32°F) as 273.15K and defines a number of reference temperatures including the boiling point of oxygen, 90.20K (−182.96°C, −297.33°F), the melting and boiling points of water and the melting points of zinc, silver and gold. It also specifies the thermometers that are used to span the range. At higher temperatures, such as in furnaces, optical pyrometers which detect variation in colour of the emitted light are used.

temperature scale Graduated scale of degrees for measuring temperature. The establishment of any temperature scale requires: a thermometric parameter which varies linearly with temperature (such as the volume of a gas at constant pressure, or the expansion of a liquid in a tube); two or more fixed points (readily reproducible reference points such as the boiling and the freezing points of water); and the assignment of arbitrary divisions (called degrees) between the fixed points. Gas, alcohol, mercury, electrical resistance and wavelength of light have been used as thermometric parameters. Common temperature scales include the FAHRENHEIT, CELSIUS (formerly the centigrade) and the KELVIN (or absolute) scales; these are abbreviated to °F, °C and K. The Fahrenheit scale originally used as fixed points the freezing point of water (taken to be 32°F) and the human body temperature (96°F, although later found to be 98.6°F). The interval between these was divided into 64 degrees; by extrapolation, the boiling point of water is 212°F. The Celsius scale uses 0°C and 100°C as the freezing and the boiling point of water, respectively; the interval is divided into 100 degrees. Zero on the Kelvin (or thermodynamic) scale (−273.15°C, −459.67°F) coincides with ABSOLUTE ZERO, the lower limit of temperature; the kelvin represents the same temperature difference as the degree Celsius.

TENDON

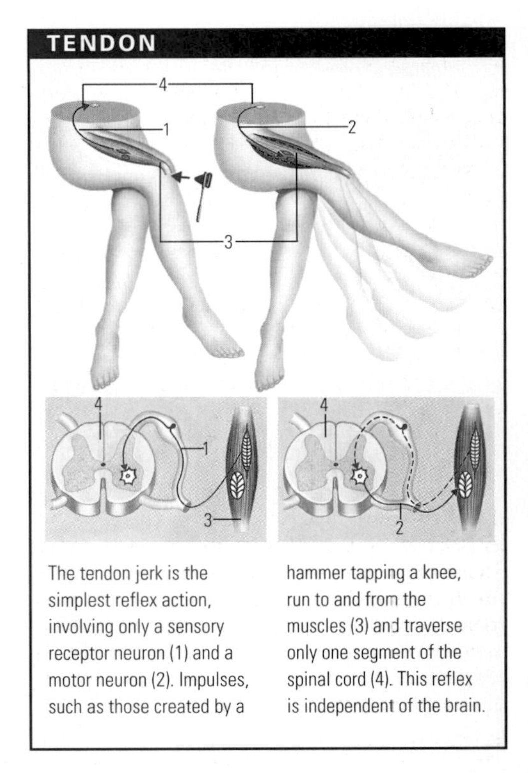

The tendon jerk is the simplest reflex action, involving only a sensory receptor neuron (1) and a motor neuron (2). Impulses, such as those created by a hammer tapping a knee, run to and from the muscles (3) and traverse only one segment of the spinal cord (4). This reflex is independent of the brain.

To convert Fahrenheit to Celsius: $C = 5(F-32)/9$; to convert Celsius to Fahrenheit, $F = (9C/5) + 32$.

tempering Heat treatment to alter the hardness of an ALLOY. The treatment alters the crystal structure of the alloy. Tempering may make the alloy harder or softer. The effect produced depends on the composition of the alloy, the temperature to which it is heated, and the rate at which it is cooled. Usually, the metal is heated slowly to a specific temperature, then cooled rapidly. Tool steel is hardened in this way.

template (templet) Mould or pattern for making repeated copies of an object. It is usually cut from thin sheets of metal, wood or paper. In architecture, a template is a supporting structure such as a beam placed over a doorway.

temporal lobes Prominent lobes of the CEREBRAL CORTEX that, in humans, lie one on either side, directly behind the temples. The temporal lobes are involved in the interpretation and generation of language. *See also* BRAIN

tendon Strong, flexible band of CONNECTIVE TISSUE that joins MUSCLE to BONE.

tendril Coiling part of stem or leaf, a slender, thread-like structure used by climbing plants for support.

Tennant, Charles (1768–1838) British industrial chemist, who invented (1799) a process for manufacturing bleaching powder from chlorine and slaked lime.

tensile strength Resistance that a material offers to tensile STRESS. It is defined as the smallest tensile stress (force divided by unit cross-sectional area) required to break the body.

tentacle Any slender, flexible organ of an animal, most notably those of the octopus and some other cephalopods, capable of feeling and grasping.

teratogen Agent that increases the likelihood of deformities developing in the fetus. Teratogens taken in by the mother during pregnancy cause malformations of the fetus. They include certain chemicals and drugs (such as ALCOHOL), some pathogenic microorganisms (such as the virus that causes rubella, or German measles) and various kinds of ionizing radiation (such as X-rays in high doses).

terbium (symbol Tb) Silver-grey, metallic element of the LANTHANIDE SERIES. Discovered in 1843, it is found in such minerals as monazite,

TERRACING

Since pre-historic times, humans have used terracing to increase the area of land available for cultivation and to help prevent soil erosion (A). Banks or walls are used to retain Earth – often using field stones for construction (B). These increase flat or near level areas and can be used for crops, including trees, or grazing. In Asia, terracing is often combined with complex systems of irrigation to raise rice in paddy fields enclosed by the terrace walls (C).

gadolinite and apatite. The soft element is used in semiconductors; sodium terbium borate is used in lasers. Properties: at.no. 65; r.a.m. 158.9254; r.d. 8.234; m.p. 1,360°C (2,480°F); b.p. 3,041°C (5,506°F). Single natural isotope ^{159}Tb.

Tereshkova, Valentina Vladimirovna (1937–) Soviet cosmonaut and the first woman in space. In June 1963 she made 45 orbits of the Earth in Vostok 6, spending nearly 71 hours in space.

terminal velocity Maximum velocity attainable by a falling body or powered aircraft. It is dependent on the shape of the body, the resistance of the air through which it is moving, and (in the case of aircraft) the THRUST of the engines.

terminator Boundary between the sunlit and dark sides of a planet or satellite. With a body lacking an atmosphere, such as the Moon, the terminator is distinct, although often broken up because of reflections from craters or mountains. Bodies with atmospheres have less well-defined terminators because atmospheric scattering causes twilight.

terpenes Group of unsaturated HYDROCARBONS related to ISOPRENE. They occur in most ESSENTIAL OILS and are colourless liquids. Examples are pinene, which is the chief ingredient of TURPENTINE, and limonene, which is found in the oils of citrus fruits.

terracing In geology, process of DEPOSITION or EROSION which produces step-like formations on a slope. Alluvial terracing is usually caused by periodic reductions in the area of the floodplain of a river.

terrigenous deposits Accumulations of sand, silt or mud that form in the sea near land as a result of EROSION.

territory In zoology, the restricted life space of an organism. It is an area selected for mating, nesting, roosting, hunting or feeding and may be occupied by one or more organisms and defended against others of the same, or a different species. The area may be defended or indicated by noise-making, chemical scent, physical displays or aggression. Many invertebrates and most vertebrates display this behaviour.

Tertiary Earlier period of the CENOZOIC era, lasting from 65 million to *c.*2 million years ago. It is divided into five epochs, starting with the PALAEOCENE, followed by the EOCENE, OLIGOCENE, MIOCENE and PLIOCENE. Early Tertiary times were marked by great mountain-building activity (Rockies, Andes, Alps and Himalayas). Both marsupial and placental mammals diversified greatly. Archaic forms of carnivores and herbivores flourished, along with early primates, bats, rodents and whales. *See also* GEOLOGICAL TIME

tertiary consumer In a FOOD CHAIN, a CARNIVORE that preys on other carnivores (SECONDARY CONSUMERS). Secondary consumers are also carnivores, but they prey on HERBIVORES (PRIMARY CONSUMERS).

tesla (symbol T) SI unit of MAGNETIC FLUX density equal to a density of one WEBER of magnetic flux per square metre.

Tesla, Nikola (1856–1943) US electrical engineer and inventor, b. Croatia, who pioneered the applications of high-voltage ELECTRICITY. Tesla developed arc lighting, the first generator of alternating current (AC) and a system of transmitting electric power without wires. He also devised the **Tesla coil**, a device for producing high-frequency currents.

testis (pl. testes) Male sex GLAND, found as a pair located in a pouch, the SCROTUM, external to the body. The testes are made up of seminiferous tubules in which SPERM are formed and mature, after which they drain into ducts and are stored in the epididymis prior to being discharged.

testosterone Steroid HORMONE secreted mainly by the mammalian TESTIS. It is responsible for the growth and development of male sex organs and male SECONDARY SEXUAL CHARACTERISTICS, such as voice change and growth of facial hair. A synthetic testosterone has been used illegally by athletes to increase their muscular development.

tetrachloromethane (carbon tetrachloride, CCl_4) Colourless, nonflammable liquid with a characteristic odour, prepared by the chlorination of methane or the catalytic reaction of carbon disulphide and chlorine. It is used as a refrigerant, insecticide, degreaser and dry-cleaning fluid. Properties: r.d. 1.59; m.p. −23°C (−9.4°F); b.p. 76.8°C (170.2°F).

tetracyclines Group of broad-spectrum antibiotics effective against a wide range of bacterial infections.

tetraethyl lead *See* LEAD(IV) TETRAETHYL

tetrahedrite Sulphide mineral, composed of varying amounts of copper, iron, zinc, silver, antimony and arsenic sulphides. It is found in medium- to low-temperature ore veins. Tetrahedrite displays cubic system, well-formed, tetrahedral crystals and also appears as masses. In colour it is metallic grey to black. It is an important ore of copper. Hardness 3–5–4; r.d. 4.9.

tetrapod Any animal that has four limbs. Tetrapods include most mammals, amphibians and reptiles (except snakes). Some authorities include also bats and birds (because their forelimbs have become adapted into wings) as well as whales and snakes (because they evolved from ancestors that once had four limbs). **Quadrupeds** use all four limbs for walking.

tetrode ELECTRON TUBE, or valve, that has a fourth electrode, a screen grid, between the anode and the control grid; the cathode is the other electrode. Tetrodes are used in power amplification and oscillator circuits.

textiles Fabrics, especially those produced by weaving yarn. The yarn is made by SPINNING natural or artificial FIBRES. Fabrics made by other processes, such as lacemaking or knitting, are now also regarded as textiles. Textiles are used to make clothing, curtains, carpets and many other products. Powered machines for spinning and weaving were introduced in the 18th century. Most textiles are now produced in factories. Twentieth-century advances in textile manufacturing have included the development of flame-resistant fabrics and "easy-care" materials that resist creasing.

thalamus One of two ovoid masses of grey matter located deep on each side of the forebrain. Sometimes called the sensory-motor receiving areas, they fulfil important relay and integration functions in respect of sensory messages reaching the BRAIN.

Thales (*c.*636–*c.*546 BC) First Greek scientist and philosopher of whom we have any knowledge. His discoveries in geometry included that the angles at the base of an isosceles TRIANGLE are equal. He predicted the ECLIPSE of the Sun that occurred in 585 BC.

thallium (symbol Tl) Shiny, metallic element of group III of the periodic table, discovered in 1861. Soft and malleable, it is obtained as a by-product of processing zinc or lead ores. It is used in electronic components, infrared detectors and optical and infrared glasses. Thallium is an extremely toxic compound, and thallium sulphide is used as a rodent and ant poison. Properties: at.no. 81; r.a.m. 204.37; r.d. 11.85; m.p. 303.5°C (578.3°F); b.p. 1,457°C (2,655°F); most common isotope ^{205}Tl (70.5%).

thallophyte Obsolete term for a subkingdom of plants that lack clearly differentiated roots, stems or leaves, and range in size from one-celled plants to 61m (200ft) SEAWEEDS. Asexual reproduction is by spores and SEXUAL REPRODUCTION is by FUSION of GAMETES. Chlorophyll-containing thallophytes are ALGAE, Euglenoids, Dinoflagellates and LICHENS. Thallophytes also included organisms formerly considered plants, lacking in chlorphyll such as BACTERIA, FUNGI and SLIME MOULDS. *See also* THALLUS

thallus Non-vascular plant body of a SEAWEED. Usually flat or ribbon-shaped, it is not differentiated into root, stem or leaves.

Thenard, Louis Jacques (1777–1857) French chemist who discovered hydrogen peroxide and the porcelain pigment now known as Thenard's blue. Thenard worked with Joseph GAY-LUSSAC and

THEODOLITE

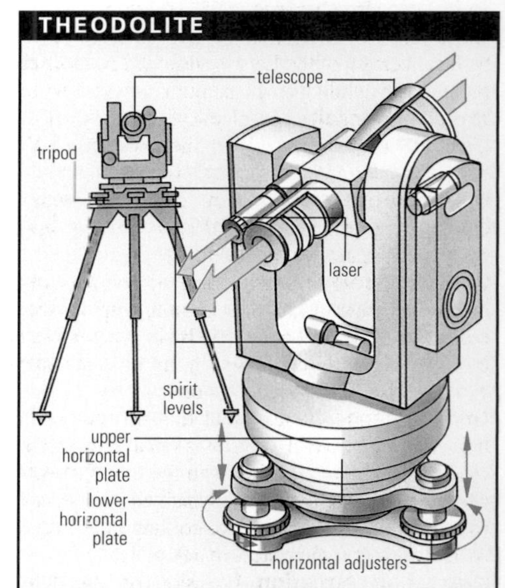

Theodolite is a surveying instrument used to measure angles between different points accurately, enabling them to be plotted in three dimensions. The telescope is used to locate measuring points on surveying poles; to align the telescope it is rotated about a horizontal or vertical axis. The precise angles of rotation in these planes are measured by using a micrometric microscope to read a scale on glass protractors. Modern computerized theodolites take these measurements electronically, and use lasers to measure distance.

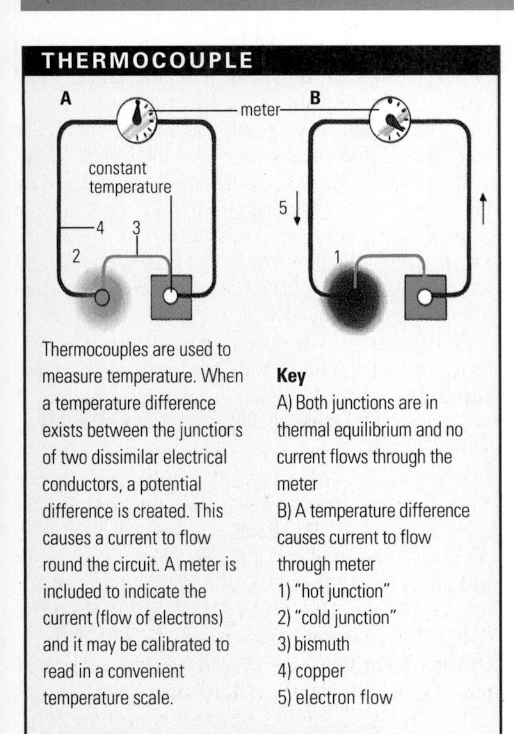

Thermocouples are used to measure temperature. When a temperature difference exists between the junctions of two dissimilar electrical conductors, a potential difference is created. This causes a current to flow round the circuit. A meter is included to indicate the current (flow of electrons) and it may be calibrated to read in a convenient temperature scale.

Key
A) Both junctions are in thermal equilibrium and no current flows through the meter
B) A temperature difference causes current to flow through meter
1) "hot junction"
2) "cold junction"
3) bismuth
4) copper
5) electron flow

wrote the standard chemical textbook of his time, *Traité de chimie élémentaire* (4 vols, 1813–16).

theodolite Surveying instrument dating back to the 16th century, used to measure horizontal and vertical angles. Its modern form consists of a telescope (with crosshairs in the lens for accurate alignment) mounted to swivel in both directions and levelled with a spirit level.

theorem Statement or proposition that is to be proved by logical reasoning from given facts and AXIOMS. *See also* FERMAT'S LAST THEOREM

therm Unit of heat used until recently, particularly for expressing the calorific value of town gas; it is now obsolete. It is equal to 100,000 British Thermal Units (BTU, also obsolete) or 25,200 kilocalories.

thermal Small-scale, rising current of air produced by local heating of the Earth's surface. Thermals are often used by gliding birds and human-built GLIDERS.

thermal capacity In physics, capacity of an object to absorb heat. It is the product of the object's mass and its SPECIFIC HEAT CAPACITY.

thermal diffusion (Soret effect) DIFFUSION caused by a temperature difference setting up a concentration gradient in a fluid.

thermal power Literally, heat power. Any heat engine, for example a STEAM ENGINE, employs thermal power. The heat generated by burning a fuel is converted into useful energy; in this case the energy drives the locomotive's pistons.

thermic lance Long, metal tube tipped with a nozzle, through which oxygen is fed at pressure. Its most important application is in the BASIC OXYGEN PROCESS of steelmaking, in which the lance, surrounded by a water jacket for cooling, jets oxygen on to a bath of molten steel in a FURNACE.

thermionic emission Emission of ELECTRONS from a substance's surface as a result of heating it.

thermionics Study of the emission of ELECTRONS or IONS from a heated CONDUCTOR. This is the principle on which ELECTRON tubes (valves) work. The heated conductor is the CATHODE and the emitted electrons are attracted to the ANODE. Such valves, known as thermionic valves, have now largely been replaced by TRANSISTORS, although electron tubes are still employed as CATHODE-RAY TUBES. A more modern aim for thermionics is the design and construction of thermionic power generators, which will convert heat directly into electricity.

thermistor Type of SEMICONDUCTOR whose resistance sharply decreases with increasing TEMPERATURE. At 20°C the resistance may be of the order of a thousand OHMS and at 100°C it may be only 10 ohms. Thermistors are used to measure temperature and to compensate for temperature changes in other parts of a circuit.

thermite process SMELTING process in which a metallic oxide ore is reduced to the metal by heating with finely divided aluminium powder. It is the basis of the **Goldschmidt process** for extracting such high melting-point metals as chromium, manganese, molybdenum and vanadium. It can also be used for welding cracks in metal, such as large castings and rails.

thermochemistry Branch of physical chemistry that deals with the HEAT effects that accompany chemical changes. Examples include the heat given off or absorbed during a chemical reaction, the dissolving of a substance, or changes of state, such as from a liquid to a gas.

thermocline Middle layers of OCEAN water, between surface and deep waters, which are defined by differing densities and temperatures. The thermocline is up to 1,000m (3,300ft) thick with a temperature only a few degrees above freezing. It is important as a stable boundary that tends to prevent interchange between layers.

thermocouple THERMOMETER made from two wires of different metals joined at one end, with the other two ends maintained at constant temperature. The junction between the two wires is placed in the substance whose temperature is to be measured. An ELECTROMOTIVE FORCE (voltage) is generated which can be measured by a millivoltmeter and which is, in turn, a measure of temperature.

thermodynamics Branch of physics that studies HEAT and how it is transformed to and from other forms of ENERGY. In general, a thermodynamic system is defined by its temperature, volume, pressure and chemical composition. The original laws of thermodynamics were conceived by observing large-scale properties of systems and with no understanding of the underlying molecular structure. The KINETIC THEORY of gases was developed in the mid-19th century. In general, the TEMPERATURE of a body is a measure of its internal energy. A body's internal energy, however, can increase if mechanical work is performed on it, and decrease if it is converted into mechanical energy. The **first law** of thermodynamics, basically a restatement of the CONSERVATION law of energy, states that the change in a system's internal energy is equal to the heat that flows into the system plus the work done on the system. Environmental sources of heat, such as oceans, are referred to as **heat reservoirs**. Any change in external conditions effects a shift in the thermodynamic system. The two main forms of change are ADIABATIC and isothermal processes. Adiabatic processes involve no change of heat, since the system is thermally insulated from the environment. The **second law** says that in an isolated system, its ENTROPY (disorder or randomness) tends to increase. The **third law**, formulated by Walter NERNST, states that a system at ABSOLUTE ZERO would effectively have an entropy of zero. *See also* CARNOT CYCLE; CLAUSIUS, RUDOLF; KELVIN, WILLIAM THOMSON, 1ST BARON

thermoelectricity Phenomena involving the conversion of HEAT into ELECTRICITY or vice versa. James JOULE described the irreversible conversion of heat into electricity. There are, however, three reversible effects. In the SEEBECK EFFECT, a current flows if different temperatures are maintained at the junctions of a circuit containing two different metals; this is the basis of the THERMOCOUPLE (thermometer). The PELTIER EFFECT acts in the reverse manner, converting electrical energy to heat energy. Lord KELVIN discovered the third thermoelectric effect, called the **Thomson effect**: if the ends of a wire conductor have different temperatures, a potential difference is created along the wire. If an electric current flows from the cooler to the hotter part of the wire, the effect is called the positive Thomson effect; the reverse is the negative Thomson effect.

thermoelectric propulsion Any of several ROCKET propulsion systems combining heat and electrical means to accelerate particles to high velocities. An arc JET ENGINE uses an electric arc to heat liquid hydrogen to 50,000°C; the resulting PLASMA of ionized hydrogen is accelerated through a conventional nozzle or, in the plasma engine, through a magnetic field for greater force. *See also* ION

thermogram Photographic or other record of the heat radiated from an object, usually obtained with an infrared camera. For example, a thermogram can show the different temperature zones of the surface of the human body, which may be useful in the diagnosis of various disorders. Or it may show isothermal contours of the Earth's surface, as recorded by an orbiting spacecraft.

thermometer Instrument for measuring TEMPERATURE. Any substance with physical properties that change with varying temperature may be used as a measure, provided the change is correctly monitored and scaled. A mercury thermometer, for example, depends on the expansion of the mercury metal that is held in a glass bulb connected to a nar-

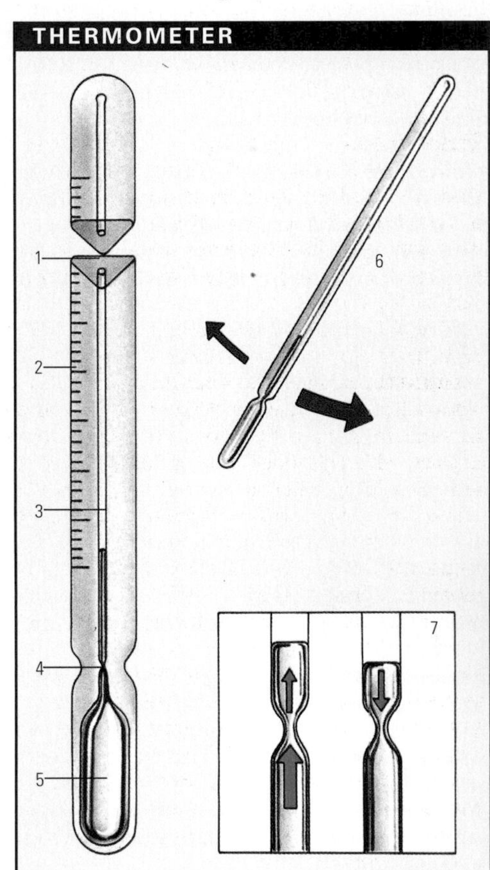

A clinical thermometer is an ordinary mercury-glass thermometer with a particularly fine capillary (3). A constriction (4) allows mercury to flow easily from the bulb (5) but, by surface tension, prevents flow back (7). A temperature reading can thus be maintained on the scale (2) until the mercury is forced back by shaking (6). For easy reading the stem is lens-shaped, as in cross-section (1), to visually magnify the mercury when the thermometer is held at the correct angle.

row, graduated tube. Temperatures are also measured by gas thermometers and resistance thermometers, which measure the RESISTANCE of a CONDUCTOR. Common scales are the CELSIUS (formerly centrigrade), FAHRENHEIT and KELVIN.
thermonuclear reaction *See* FUSION
thermopile Device used to measure radiant heat, consisting of several THERMOCOUPLES connected together in series. Alternate junctions are blackened for absorbing radiant heat, the other junctions are shielded from the RADIATION. The ELECTROMOTIVE FORCE (EMF) generated by the temperature difference between the junctions can be measured. From this the temperature of the blackened junctions can be calculated, and thus the intensity of the radiation measured.
thermoplastic resin Type of POLYMER that softens on being heated and can be repeatedly melted or softened by heat without change of properties. Examples include NYLON, POLYVINYL CHLORIDE (PVC), POLYETHENE, FLUOROCARBON PLASTIC and POLYSTYRENE. *See also* PLASTIC; THERMOSETTING RESIN
thermoreceptors In anatomy, sensory nerve endings, sited in the skin and in the deep body tissues, which detect changes in temperature and signal these to the brain. There are two types: receptors that are stimulated by cold and receptors responsive to heat.
thermosetting resin Type of POLYMER that sets hard, losing its plasticity, under heat and pressure. Examples include POLYESTER, EPOXY RESIN and SILICONE. *See also* PLASTIC; THERMOPLASTIC RESIN
thermosphere Shell of light gases between the MESOSPHERE and the EXOSPHERE, between 100km (60mi) and 400km (250mi) above the Earth's surface. The temperature steadily rises with height in the thermosphere.
thermostat Device for maintaining a constant TEMPERATURE. A common type contains a strip of two metals, one of which expands and contracts more than the other. Thus the strip bends and breaks the circuit at a set temperature. As it cools, the strip straightens, makes contact and the heating begins again once the circuit is complete.

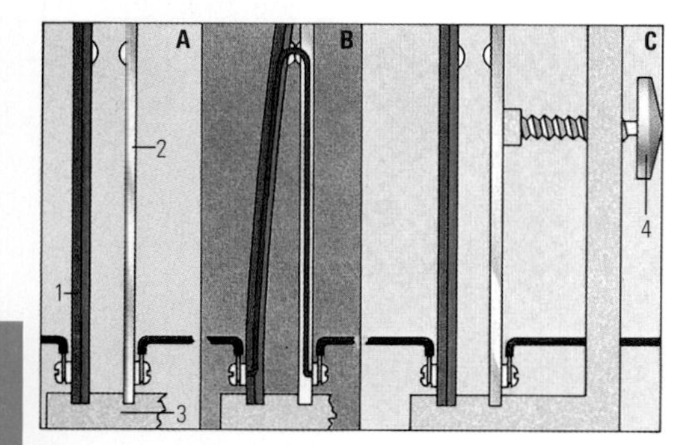

▲ **thermostat** A bimetallic strip thermostat, when included in, say, an electrical air conditioning unit, makes and breaks the circuit in response to variations in the room temperature. Two dissimilar pieces of metal, whose coefficients of thermal expansion differ appreciably, are welded together to form the strip. When the temperature rises, the unequal expansions of the two metals cause the strip to bend; this completes the circuit and switches on the cooling system. As the temperature falls, the metals contract unequally and the strip straightens and breaks the circuit.

Key
A) At the required temperature, the circuit is broken.
B) When the temperature rises, the strip bends and completes the circuit.
C) Complete thermostat with adjusting screw. By moving the fixed contact, the required temperature is adjusted.
1) bimetallic strip
2) fixed contact
3) insulator
4) temperature-adjusting screw

THIN-LAYER CHROMATOGRAPHY

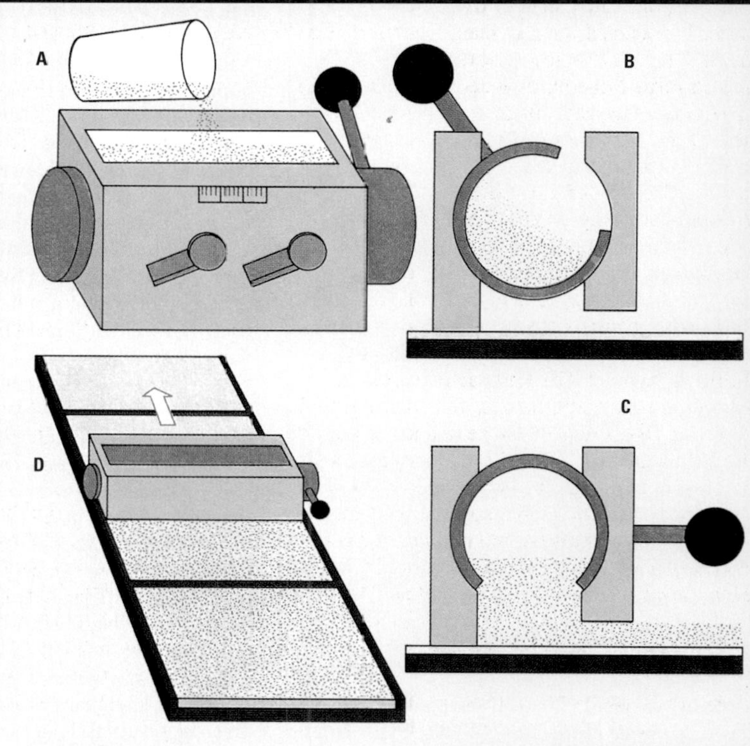

Thin-layer chromatography represents a method of quickly and accurately separating the components of a mixture. It is carried out using slurries of finely divided adsorbents, such as silica gel or kieselguhr, applied to glass plates or plastic sheets in a thin, uniform layer by a spreading procedure, usually with the aid of a commercial applicator as shown in (A). The applicator consists of a hollow metal block with exit gates on each of two opposite long faces. A section through the applicator (B) shows how the internal, rotating reservoir chamber, with a wide longitudinal slot, fits into this block. Flow of the suspension does not begin until the chamber is rotated to its open position (C). Simultaneously, the applicator is pulled across a series of plates laid out on a mounting board (D). The applicator is self-adjusting to the surface of the plates because it rides on the plate during the costing procedure.

thiamine VITAMIN B_1 of the B complex, required for carbohydrate METABOLISM. Its deficiency causes the disease beriberi. Thiamine is found in grains and seeds, nuts, liver, yeast and legumes.
Thimmonier, Barthélmy (1793–1857) French tailor who patented the first sewing machine that was put to practical use. His patent was taken out in 1830 for a wooden machine which used a single thread.
thin-layer chromatography Type of CHROMATOGRAPHY in which the components of a mixture of liquids are separated by differential absorption onto a thin layer of material supported on a glass plate. The usual absorbing material is ALUMINA, which is made into a paste with water, spread on the plate and allowed to dry. A spot of the liquid mixture is placed near one edge of the plate, which is stood vertically in a trough of SOLVENT. As the solvent rises up the plate by CAPILLARITY, the components travel with it, but at different rates depending on how readily they are absorbed by the alumina. This results in a series of spots which can be identified by a reagent or by the distance they have moved in a particular time. The technique is very similar to PAPER CHROMATOGRAPHY.
thiol (thio alcohols, mercaptans) Organic compound having a thiol (–SH) group bound to a carbon atom. Thiols are the sulphur analogues of ALCOHOLS. They have unpleasant odours resembling rotten cabbage or garlic. Ethane thiol is a typical thiol, having the formula C_2H_5SH.
Thomas, Sidney Gilchrist (1850–85) British metallurgist. In 1875 he invented a method, known as the basic OPEN HEARTH PROCESS, for removing phosphorus impurities from iron ore during conversion in the BESSEMER PROCESS.
Thompson, Benjamin, Count Rumford *See* RUMFORD, BENJAMIN THOMPSON, COUNT
Thomson, Sir Charles Wyville (1830–82) Scottish naturalist. He was appointed the director of the CHALLENGER voyage (1872–76). Thomson published *The Depths of the Sea* (1873) and an account of his expedition, *The Voyage of the Challenger* (1877).
Thomson, Elihu (1853–1937) US electrical engineer and inventor, b. Britain. With Edwin J. Houston he developed an arc lighting system that led to the founding of the American Electric Company, which later became the Thomson-Houston Electric Company. This merged in 1892 to form the General Electric Company (GEC). Thomson invented the first high-frequency DYNAMOS and TRANSFORMERS.
Thomson, Sir George Paget (1892–1975) British physicist, son of Sir Joseph John THOMSON. He shared the 1937 Nobel Prize for physics with Clinton DAVISSON for their independent work on diffracting ELECTRONS (1927). This confirmed the wave nature of particles first predicted (1923) by Louis de BROGLIE. Thomson fired fast electrons through thin gold leaf to obtain a DIFFRACTION pattern.
Thomson, Sir Joseph John (1856–1940) British physicist, father of George THOMSON. He succeeded James Clerk MAXWELL as professor of experimental physics (1884–1919) at Cambridge. Thomson's discovery (1897) of the ELECTRON is regarded as the birth of PARTICLE PHYSICS. He established that cathode rays consisted of a stream of particles. Thomson went on to prove that the electron was negatively charged and that its mass was about 2,000 times smaller than the smallest atom (hydrogen). He was awarded the 1906 Nobel Prize for physics for his investigations into the electrical conductivity of gases. Thomson and Francis ASTON produced evidence of ISOTOPES of neon. He transformed the Cavendish Laboratory into a major centre for atomic research, attracting scientists of the calibre of Ernest RUTHERFORD. Thomson served as president (1915–20) of the Royal Society.
Thomson, William, Lord Kelvin *See* KELVIN, WILLIAM THOMSON, 1ST BARON
Thomson effect Potential difference which

develops between two points on a metal conductor if the two points are maintained at different temperatures. It is named after William Thomson (Lord KELVIN). *See also* THERMOELECTRICITY

thomsonite Mineral, hydrated aluminium sodium silicate, of the ZEOLITE group. It is found in cavities of basaltic rocks. Its crystals are of the orthorhombic system and it is snow-white when pure. Hardness 5–5.5; r.d. 2.2.

thorax In anatomy, part between the neck and the abdomen. In mammals it is formed by the rib-cage and contains the lungs, heart and oesophagus. It is separated from the abdomen by the DIAPHRAGM. In arthropods it consists of several segments to which legs and other appendages are attached.

thorium (symbol Th) Radioactive, metallic element of the ACTINIDE SERIES, first discovered in 1828. The chief ore is monazite (a phosphate). The metal is used in photoelectric and thermionic emitters. One decay product is RADON-220. Thorium is sometimes used in radiotherapy, and it is increasingly used for conversion into uranium for nuclear FISSION. Chemically reactive, it burns in air but reacts slowly in water. Properties: at.no. 90; r.a.m. 232.0381; r.d. 11.72; m.p. 1,750°C (3,182°F); b.p. 4,790°C (8,654°F); most stable isotope ^{232}Th; (half-life 1.41×10^{10} yr).

three-body problem Fundamental problem in CELESTIAL MECHANICS: to determine the motions of three bodies under the influence only of their mutual gravitational attractions. While there is no exact general solution – only solutions to special cases – highly accurate approximations can be achieved with modern computers. The three-body problem was examined first by Isaac NEWTON and has been tackled by many astronomers and mathematicians. Work on the problem was stimulated by the need to understand the orbit of the Moon and, more recently, the orbits of artificial satellites.

threonine ($CH_3CH(OH)CH(NH_2)COOH$) Soluble, crystalline, essential AMINO ACID found in proteins.

throat *See* PHARYNX

thrombin Blood ENZYME that converts FIBRINOGEN to fibrin during the formation of blood clots.

thrombocyte *See* PLATELET

throttle Another name for the accelerator on any vehicle powered by an INTERNAL COMBUSTION ENGINE, so-called because such engines have a throttle valve, worked by the accelerator control, that governs the amount of fuel-air mixture entering the CARBURETTOR. Opening the throttle valve admits more fuel and makes the engine run more quickly.

thrust Driving force resulting from operation of a PROPELLER, JET ENGINE or ROCKET engine. An aircraft propeller forces air backwards, and jet and rocket engines expel gases backwards. Thrust is produced in a forward direction in accordance with the third of Newton's laws of MOTION. This states that, for every action, there is an equal and opposite reaction.

thulium (symbol Tm) Lustrous, silver-white, metallic element of the LANTHANIDE SERIES. First discovered in 1879, its chief ore is monazite. Soft, malleable and ductile, it combines with oxygen and the halogens. It is used in arc lighting; thulium-170, which emits X-rays, is used in portable X-ray units. Properties: at.no. 69; r.a.m. 168.9342; r.d. 9.31 (25°C); 1,545°C (2,813°F); b.p. 1,947deg;C (3,537°F); most stable isotope ^{169}Tm (100%).

thunderstorm Electrical storm, commonly experienced as LIGHTNING and thunder. Thunderstorms are caused by the separation of electrical charges in CLOUDS. Water drops are carried by updraughts to the top of a cloud, where they become ionized and accumulate into positive charges – the base of the cloud being negatively charged. An electrical discharge (a spark) between clouds, or a cloud and the ground, is accompanied by light (seen as a lightning stroke) and heat, which expands the air explosively and causes it to reverberate and produce sounds and echoes called thunder. Thunderstorms are usually accompanied by heavy rain; ozone and the oxides of nitrogen are produced in the air.

thymine In molecular biology, one of the four nitrogen bases in the nucleic acid DNA (the others are ADENINE, GUANINE and CYTOSINE). In RNA thymine is replaced by uracil in the base sequence.

thymus gland One of the endocrine (ductless) GLANDS, located in the upper chest in mammals; it is large in infancy and shrinks after puberty. In childhood, it controls the development of lymphoid tissue and the immune response to infection. It secretes a HORMONE, thymosin, that stimulates the activity of defensive T-CELLS. Disorder of the thymus may be associated with AUTOIMMUNE DISEASES (those caused by the body's own antibodies). *See also* ENDOCRINE SYSTEM

thyroid gland H-shaped gland of the ENDOCRINE SYSTEM lying in the base of the neck, straddling the TRACHEA (windpipe) below the Adam's apple. It secretes thyroid hormones, principally THYROXINE which is essential for growth and development and for the regulation of metabolism.

thyroid-stimulating hormone (TSH) Hormone secreted by the frontal (anterior) lobe of the PITUITARY GLAND. It stimulates the THYROID GLAND in the neck to produce the hormone thyroxine, which helps to control metabolism.

thyroxine Hormone secreted by the THYROID GLAND. It contains iodine and helps regulate the rate of METABOLISM; it is essential for normal growth and development.

tibia (shinbone) Inner and larger of the two lower leg bones. It articulates with the FEMUR, or upper leg bone, at the knee and extends to the ankle, where its lower end forms the projecting ankle bone on the inside of the leg. *See also* FIBULA

tidal bore Flow of tidal water from the sea into a funnel-shaped river mouth or estuary which, opposing out-flowing river water, builds into a surface "wall" that accelerates upstream. Notable examples occur in the rivers Amazon and Severn, with bores reaching 5m (15ft) in height and travelling at 15 to 25km/h (9–15mph). *See also* TIDE

tidal flat Extensive, nearly flat, barren land area that is alternately covered and uncovered by the action of the tide. It consists of mud and sand. A tidal marsh has a covering of salt-tolerant plants and grasses.

tidal power Energy harnessed and used by man from tidal movement of the Earth's OCEANS. This form of power is economic only where the tidal range is greater than *c*.4.6m (15ft). Modern schemes involve the use of turbo-generators driven by the passage of water through a tidal barrage. The La Rance power plant in N France has been working successfully since 1966, and in 1969 a small tidal power plant was completed by the Soviet Union on the White Sea. Special TURBINES have been developed that can be driven by water flowing in either of two directions. However, all such schemes can have very serious effects on the ecology of the ESTUARY that they enclose.

tidal wave *See* TSUNAMI

tide Periodic rise and fall of the sea caused by the

TIDAL POWER

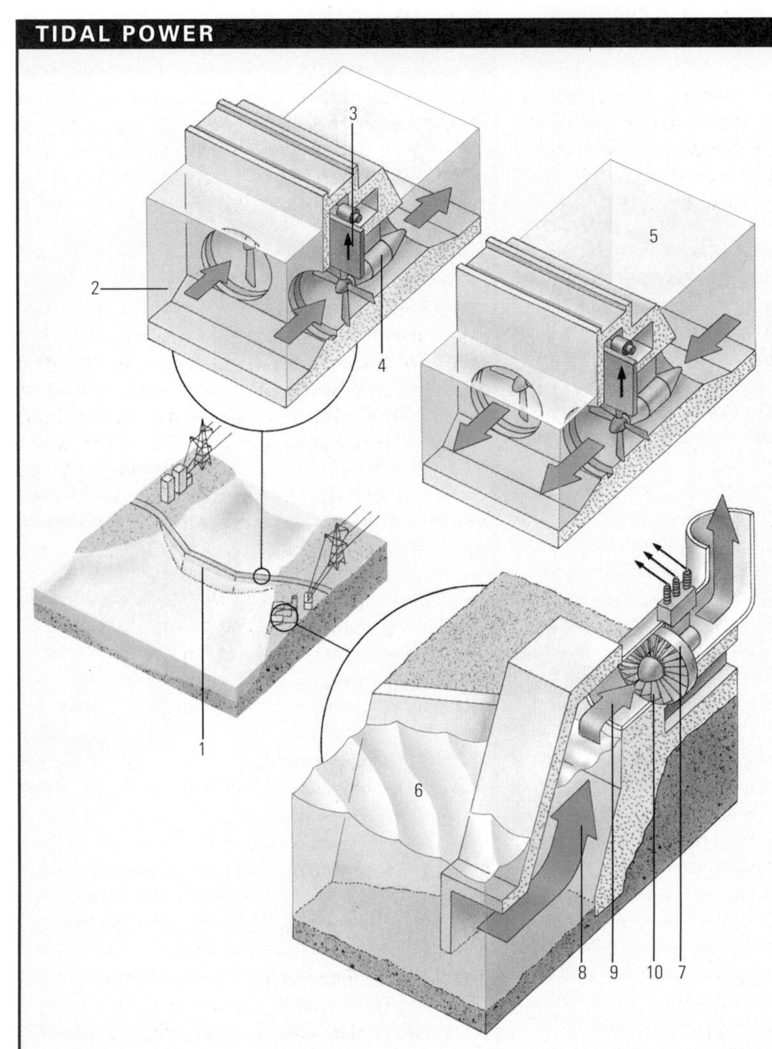

The power of the sea can be harnessed to generate electricity. **Tidal power** uses a barrage (1) across an estuary or bay. The barrage contains turbines that can spin with a flow of water in either direction. As the tide comes in, gates on the barrage remain closed until a head of water has built up on the sea side of the structure (2). The gates are then opened (3), and the incoming tide flows through the barrage driving the turbines (4). As the tide falls, the process is reversed with the gates closed until the sea has fallen below the level of water retained in the estuary (5). The second method of utilizing the power of the sea harnesses **wave power** (6). The key difference is that the turbine (7) is air driven, not turned by water. As a wave hits the shore, the force of the water (8) drives air (9) through the turbine blades (10). When the water level drops, air is sucked back down through the turbine spinning it again.

▲ **tides** The daily rise and fall of the ocean's tides are the result of the gravitational pull of the Moon as it orbits the Earth and that of the Sun, though the effect of the latter is less than half as strong as that of the Moon. The effect is greatest on the hemisphere facing the Moon and causes a tidal "bulge". When the Sun, Earth and Moon are in line, tide-raising forces are at a maximum, and spring tides occur; high tide reaches the highest values, and low tide falls to low levels. When lunar and solar forces are least coincidental, with the Sun and Moon at an angle (near the Moon's first and third quarters), neap tides occur, which have a small tidal range.

pull exerted on Earth by the Moon, and, to a lesser extent, by the Sun. In most parts of the world there are two high tides and two low tides every day. The time of each high tide is 12 hours 20–25 minutes later than the preceding tide, because the position of the Moon relative to the Earth will have changed by a small amount after 12 hours have elapsed.

till (boulder clay) In geology, SEDIMENT consisting of an unsorted mixture of clay, sand, gravel and boulders deposited directly by the ice of GLACIERS. It is not deposited in layers. **Tillite** is till that has become solid rock.

time Non-spatial order in which all events occur; also the interval between two events. Time falls within the disciplines of physics, psychology, philosophy and biology. Until the theory of RELATIVITY was devised by Albert EINSTEIN, time was perceived as absolute: a constant one-direction (past to future) flow. Since then the concept of time linked with distance in space (SPACE-TIME) has connected time with the relative velocities of those perceiving it. For clocks at velocities approaching that of light, time expands from the point of view of a stationary observer: that is, the moving clock records a smaller interval of time between two events than does the clock of the stationary observer. The passage of time (or equivalently, the rate of an accurate clock) is also slowed in a gravitational field.

timescale *See* GEOLOGICAL TIME

time-space distortion In the theory of RELATIVITY, the bending of the SPACE-TIME continuum so that its geometry is no longer Euclidean. This effect is caused by very strong gravitational fields such as that found near a BLACK HOLE, where the distortion is so great that light leaving the black hole is bent round itself and cannot escape. *See also* EUCLIDEAN GEOMETRY

time zone One of 24 divisions of the Earth's surface, each of which is 15° of LONGITUDE wide and within which the time of day is reckoned to be the same. At a conference held (1884) in Washington D.C. the meridian of Greenwich was adopted as the zero of longitude, and zones of longitude were established. Standard time in each successive zone westwards is one hour behind that in the preceding zone. Large territories such as the USA and Russia span several time zones. The adoption of time zones results in a discrepancy at longitude 180° (the International Date Line), which is resolved by omitting one day from the calendar when crossing from west to east, or repeating one day if the crossing is from east to west. *See also* GREENWICH MEAN TIME (GMT)

tin (symbol Sn) Metalloid element of group IV of the periodic table, known from ancient times. Its chief ore is CASSITERITE (an oxide). Soft, malleable and resistant to corrosion, tin is used as a protective coating for iron, steel, copper and other metals, and in such alloys as solder, pewter, bronze and type metal. It forms two series of salt, termed tin(II), or stannous, and tin(IV), or stannic. Tin compounds are used as fungicides, in glass coatings and as a tooth decay preventive (tin(II) fluoride) in toothpaste. Three allotropes exist: the common lustrous metallic form (white tin) changes slowly below 13.2°C (55.8°F) to a powder (grey tin); above 161°C, (321.8°F) a brittle form exists. Tin has been used by humans since the Bronze Age. Until the 20th century, the tin mines of Cornwall were the world's leading source. Properties: at.no. 50; r.a.m. 118.69; r.d. 7.29; m.p. 232°C (449.6°F); b.p. 2,270°C (4,118°F); most common isotope ^{118}Sn (24.03%).

Tinbergen, Nikolaas (1907–88) Dutch ethologist. He shared with Konrad LORENZ and Karl von FRISCH the 1973 Nobel Prize for physiology or medicine for his pioneering work in ETHOLOGY. Tinbergen studied how in animals certain stimuli evoke specific responses, and emphasized the importance of observing animals under natural conditions.

tin chloride Either of two chlorides: tin(II) or stannous chloride ($SnCl_2$) or tin(IV) or stannic chloride ($SnCl_4$). **Tin(II) chloride** is a white, soluble solid, which can be produced by dissolving TIN in hydrochloric acid. It is used as a reducing agent and also as a constituent of tin-plating electrolytes. **Tin(IV) chloride** is a colourless, fuming liquid. It is used in the preparation of other inorganic tin compounds and organic tin (organotin) compounds. When sprayed on to glass and fired, it provides a conductive coating.

Ting, Samuel Chao Chung (1936–) US physicist who researched ELEMENTARY PARTICLES. He used a SYNCHROTRON to fire PROTONS at a BERYLLIUM target and then observed the resultant decaying particles. From the results Ting found evidence for the existence of a heavy ELEMENTARY PARTICLE, which he named the **J particle**. At the same time but working independently, Burton RICHTER discovered the same particle, which he named **psi**. For this discovery Ting and Richter shared the 1976 Nobel Prize for physics.

tin oxide Either of two oxides: tin(II) or stannous oxide, SnO, a black powder used as a reducing agent; tin(IV) or stannic oxide, SnO_2, a white powder occuring naturally as the mineral CASSITERITE. It is used in ceramics, glass and cosmetics.

TIN

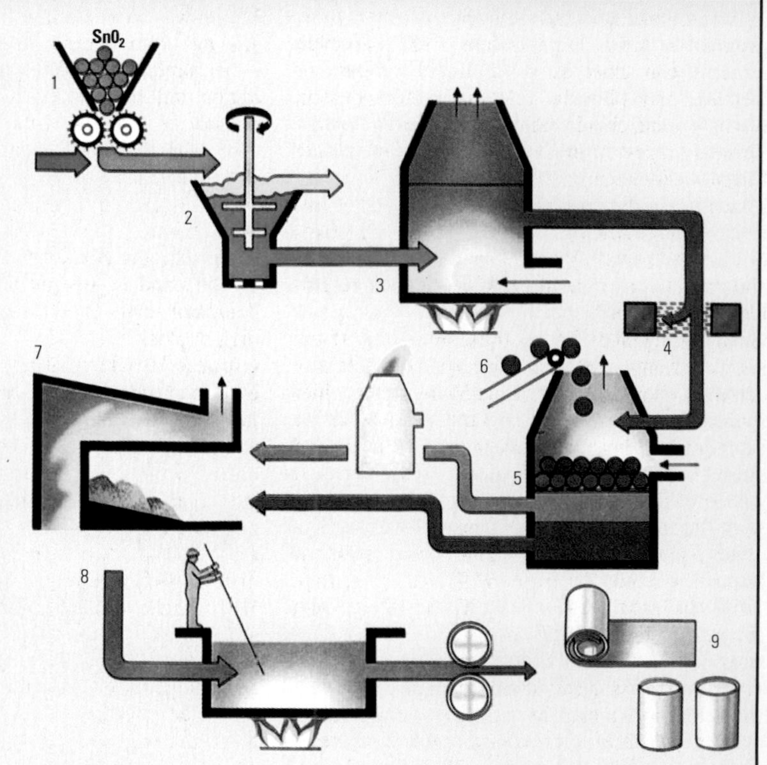

Cassiterite (SnO_2) is the sole source of commercial tin. It is extracted by the ore being crushed (1) and passed through a sink-float separation system (2). The washed ore is roasted in an oxidizing atmosphere (3) to remove arsenic and sulphur. Tungsten is removed by electromagntic separation (4). The tin oxide is then roasted in a blast furnace (5) with coke (6). The tin produced is refined in a reverbatory furnace (7). The slag produced from the blast furnace is reworked and the tin obtained is also refined in the reverbatory furnace. Further refining takes place to remove any remaining impurities (8), before the tin is pressed and rolled (9). Tin is mainly used for electroplating or in combination with other metals as an alloy. Pure tin is used principally for coating steel to prevent corrosion. The tin is applied in a layer *c.*6 x10⁻⁶mm (2.5 x 10⁻⁶in) thick by either dipping the steel into molten tin and rolling, or more commonly by electrolysis. Tin-coated steel is used extensively in the preparation of cans used for food storage because it is not poisonous. As an alloy, tin is used in solders, where its low melting point is an advantage, and in bronze, pewter and various other alloys used in machine bearings. Because of its low melting point and resistance to atmospheric corrosion, tin is used in modern glass-making processes where the molten glass is floated on a layer of molten tin in a hermostatically controlled bath and allowed to congeal. The resulting glass is so smooth that it needs no polishing or grinding.

TITANIUM

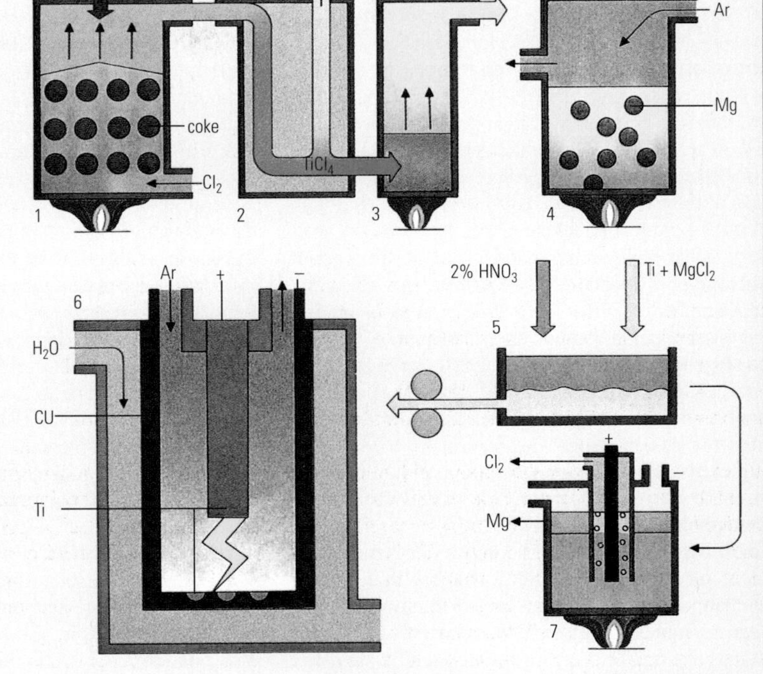

Titanium is extracted almost exclusively from rutile (TiO_2), heated with coke in an atmosphere of chlorine (Cl_2) in a brick-lined chimney (1) to produce titanium tetrachloride ($TiCl_4$). The vapour is condensed (2) and purified by distillation (3). Heating in the Kroll furnace (4) under argon (Ar) with fused magnesium (Mg) produces a titanium "sponge", Ti, and magnesium chloride ($MgCl_2$). The magnesium chloride is removed (5) by leaching with 2% nitric acid (HNO_3). The titanium is pressed into electrodes and melted under argon in a consumable-arc furnace (6) to improve the product density. The water-cooled, copper jacket of the furnace does not react with the titanium, unlike other processes in which side reactions nearly always occur. The consumable arc furnace was developed especially to cater for metals, such as titanium, which are extremely reactive with, for example, oxygen at temperatures well below their melting points. The process is repeated to ensure that the product is of uniform quality. The leached magnesium chloride is fused and electrolysed in a Dow furnace to produce the elements for recycling (7). Titanium is used to make very tough alloys with steel for the aerospace, toolmaking and automotive

Tiselius, Arne Wilhelm Kaurin (1902–71) Swedish biochemist. In 1937, while working with Theodor SVEDBERG on ELECTROPHORESIS, he observed that a current passing through a colloidal solution causes different proteins to move at different speeds and therefore separate. (Proteins are electrically charged in a colloidal solution.) Tiselius built a machine that would take advantage of the effect and for this work received the 1948 Nobel Prize for chemistry. *See also* COLLOID

tissue Material of a living body consisting of a group of similar and often interconnected CELLS, usually supporting a similar function. Tissues vary greatly in structure and complexity. In animals, they may be loosely classified according to function into epithelial, connective, skeletal, muscular, nervous and glandular tissues, although each of these categories contains more than one different type of cell.

tissue culture In biology, artificial cultivation of living TISSUE in sterile conditions. Tissue culture in laboratories is used for biological research or to help in the diagnosis of diseases. Tissue culture is also used a means of propagating plant CLONES; it avoids the mingling of genetic material that occurs with POLLINATION and normal SEXUAL REPRODUCTION, and is faster than waiting for seed set and germination. It is used in the production of genetically engineered crop plants. *See also* GENETIC ENGINEERING

Titan Largest SATELLITE of SATURN, and the second-largest satellite in the Solar System, discovered (1655) by Christiaan HUYGENS. Titan is unique among planetary satellites in having a substantial atmosphere. It is composed of rock and water-ice in roughly equal proportions. The space probe, VOYAGER 1, found no gaps in an opaque, reddish cloud layer 200km (125mi) above the surface. The atmosphere consists mostly of nitrogen, with some methane and other hydrocarbon compounds. The surface temperature is 95K, at which methane can exist as solid, liquid or gas, so methane may play the role that water does on Earth, forming clouds, rain, lakes and even snow.

Titania Largest of the five satellites of the planet URANUS, discovered (1787) by William HERSCHEL. It orbits at a distance of 438,370km (272,270mi) from Uranus and has a diameter of *c.*1,800km (1,120mi). Comprising rock and water-ice, it is almost OBERON's twin in size but its surface resembles that on Ariel.

titanium (symbol Ti) Lustrous, silver-grey TRANSITION ELEMENT, first discovered in 1791. It is a common element found in many minerals, chief sources being ILMENITE and RUTILE. Resistant to corrosion and heat, titanium is used in steels and other alloys, especially in aircraft, spacecraft and guided missiles, where strength must be combined with lightness. Titanium(IV) oxide is used as a white pigment in paints and as the gemstone titania. Properties: at.no. 22; r.a.m. 47.90; r.d. 4.54; m.p. 1,660°C (3,020°F); b.p. 3,287°C (5,949°F); most common isotope ^{48}Ti (73.94%).

titanium oxide Either of two oxides. Titanium(II) oxide (titanium monoxide, TiO) is a white powder formed by reducing the dioxide at 1,500°C (2,732°F). It has no industrial uses. Titanium(IV) oxide (titanium dioxide, TiO_2) is a white or grey powder, according to its purity. It occurs naturally as the mineral RUTILE, from which it is extracted by chlorination. It can be extracted from the mineral ILMENITE by treating it with sulphuric acid. It is used as a white pigment in ceramics, glass, cosmetics, paper and paints.

titration Method used in volumetric analysis to determine the concentration of a COMPOUND in a SOLUTION by measuring the amount needed to complete a reaction with another compound. A solution of known concentration is added in measured amounts from a burette (a graduated glass tube) to a measured volume of liquid of unknown concentration until the reaction is complete (as shown by an INDICATOR or electrochemical device). The volume added enables the unknown concentration to be calculated.

TNT (2,4,6–trinitrotoluene, $C_7H_5N_3O_6$) EXPLOSIVE organic compound made from TOLUENE by using sulphuric and nitric acids. It was discovered in 1863 and its resistance to shock (requiring a detonator to set it off) makes it one of the safest high explosives.

tobacco mosaic virus (TMV) Simple VIRUS used in experiments concerning the transference of the GENETIC CODE. It consists of a single helix of RNA containing some 6,400 NUCLEOTIDES. This is coated with some 2,100 molecules of a single PROTEIN, each molecule of which comprises a polypeptide chain of 158 AMINO ACIDS. The sequence of these acids has been determined.

Tokamak Soviet design of experimental nuclear FUSION reactor consisting of a TORUS-shaped vessel, in which a PLASMA (an extremely hot ionized gas which can reach temperatures that facilitate nuclear fusion) is contained by a magnetic field produced by coils wound over the reaction vessel. Several versions of these devices now exist outside the former Soviet Union.

toluene (methylbenzene) Aromatic HYDROCARBON ($C_6H_5CH_3$) derived from coal tar and petroleum. It is a colourless, flammable liquid widely used as an industrial solvent and in aircraft and motor fuels. Toluene is also used in the manufacture of TNT and chemicals such as PHENOL and BENZENE. Properties: r.d. 0.87; m.p. −94.5°C (−138.1°F); b.p. 110.7°C (231.3°F).

Tombaugh, Clyde William (1906–97) US astronomer. In 1930, nearly a year into a search based on predictions by PERCIVAL LOWELL, he discovered the planet PLUTO. He continued to search for other planets for more than ten years, discovering in the process star clusters, clusters of galaxies, a comet and hundreds of asteroids. After World War 2 he developed telescopic cameras for tracking rockets after launch.

tombolo Bar connecting an island with the mainland. It is usually formed by the growth of a sand spit until it reaches the island.

tomography Technique of X-RAY photography in which details of only a single slice or plane of body tissue are shown. *See also* COMPUTERIZED AXIAL TOMOGRAPHY (CAT)

Tompion, Thomas (1639–1713) British horologist. He raised the art of clockmaking to a high level, regarding the movement as more important than the exterior casing. There are examples of Tompion's art at Hampton Court and in many British museums. *See also* CLOCK

tongue Muscular organ usually rooted to the floor of the mouth. The tongue contains the taste buds, groups of cells that distinguish the four basic tastes: bitter, tasted on the back of the tongue; sweet and salty, tasted on the tip and front of the tongue; and sour, tasted mainly on the sides of the tongue. The tongue helps to move food around the mouth for chewing and swallowing; animals also use it for lapping fluids and for grooming. In human beings, the tongue is vital for the production of speech. *See also* SENSES

tonsils Two masses of lymphoid tissue located at the back of the PHARNYX. They have a pitted surface that frequently becomes infected (tonsillitis). They act as a filter against disease microorganisms. *See also* ADENOIDS

tooth *See* TEETH

topaz Transparent, glassy mineral, aluminium fluosilicate, $Al_2SiO_4(F,OH)_2$, found in pegmatites. Its crystals are orthorhombic system, columnar prisms; it occurs as granular masses. Topaz is colourless, white, blue or yellow, and some large crystals are of gem quality. Hardness 8; r.d. 3.5.

topocentric Relating to observations made from a point on the surface of the Earth. Topocentric coordinates contrast with geocentric coordinates, which are measured from the centre of the Earth.

topographic mapping Representation of the surface of the Earth in relief, using contour lines drawn through points of equal elevation. Intervals are arbitrary. Most topographic maps are made from aerial photographs. Special plotting instruments allow the cartographer to follow a "floating" dot along an elevation, producing contour lines in sequence.

topography Study of surface features such as hills, valleys, rivers, roads and lakes. It is also the representation of such features on a relief map or a plan for construction. The terrain of a region is explored using surveyors' instruments or aerial PHOTOGRAMMETRY (plotting elevations from photographs). *See also* SURVEYING

topology Branch of mathematics concerned with those properties of geometric figures that remain unchanged after a continuous deformation process such as squeezing, stretching, or twisting (but not joining, tearing or breaking). A cup with a handle is topologically equivalent to a doughnut with a hole in it; a cube, a solid cone and a solid cylinder are topologically equivalent to a sphere. These figures can be considered sets of points, each point of one set being continuously transformable into one point in another set. *See also* GEOMETRY

torbernite Complex, radioactive, phosphate mineral that contains copper and uranium. It occurs as aggregates that easily cleave to give green, brittle, mica-like flakes, which have a vitreous lustre. The platy crystals (monoclinic) are translucent or transparent. Hardness: 2.0–2.5; s.g. 3.3.

tornado (twister) Funnel-shaped, violently rotating storm extending downwards from the cumulonimbus cloud in which it forms. At the ground its diameter may be only *c.*100m (300ft). Rotational wind speeds range from 150 to 500km/h (100–300mph). The centre of a tornado is an area of extreme low pressure, which sucks up dust to give a blackish funnel rising to the sky. Around this funnel of rising air are very strong winds which destroy crops and sometimes buildings. Tornadoes occur in deep low-pressure areas, associated with FRONTS or other instabilities. They are associated with intense heating in continental areas in late summer, when land masses are at their hottest. Tornadoes are most common in the interior of the USA, but also occur in India and many other countries. When tornadoes cross over water they become WATERSPOUTS. The average life of a tornado or a waterspout is 4 minutes. The violent storms the occur at the start of the rainy season in West Africa are also called tornadoes. They bring strong winds and torrential rain, which is caused by mild air coming in from the sea and meeting the dry NE air from the Sahara Desert.

torpedo Self-propelled, underwater MISSILE used by submarines, small surface warships and aircraft to destroy enemy vessels. Modern torpedoes may be launched by rocket boosters and often have internal electronic equipment for guiding the missile to the target.

torque Turning effect of a FORCE. An example is a turbine that produces a torque on its rotating shaft to turn a generator. The output of a rotary engine, such as the familiar FOUR-STROKE ENGINE or an electric motor, is rated by the torque it can develop. The unit of measurement is Nm (newton-metre).

Torrey, John (1796–1873) US botanist and chemist. He conducted major studies of North American plant life and amassed one of the most valuable botanical libraries and herbariums of his time. His works include *Flora of the Northern and Middle Sections of the United States* (1824) and *Flora of the State of New York* (1843), which he wrote after being appointed state botanist (1836).

Torricelli, Evangelista (1608–47) Italian physicist. Assistant and secretary to GALILEO during the latter's last three months of life, he is credited with designing the first artificial VACUUM (the Torricellian vacuum). In 1643 Torricelli invented the mercury BAROMETER. He also constructed a primitive MICROSCOPE and improved the TELESCOPE. His work in geometry eventually contributed to the development of INTEGRAL CALCULUS.

TOPOLOGY

Topology explores the properties of geometric shapes that have been deformed but not torn. If A^1 and B^1 are folded so that the arrows point in the same direction, a collar A^2 and a twisted band B^2 result. A^2 has an inner and outer surface and two edges, but B^2 has only one of each – precisely because of its twist. B^2 is called a Möbius strip and, unlike A^2, a path beginning at S and traced round the band will, after one revolution, end on the other "side" of the band at T. A Möbius strip stretched as in C^1 is distorted further into one half of a Klein bottle, (C^2) which again has only one edge and one surface – compare a normal vase (C^3).

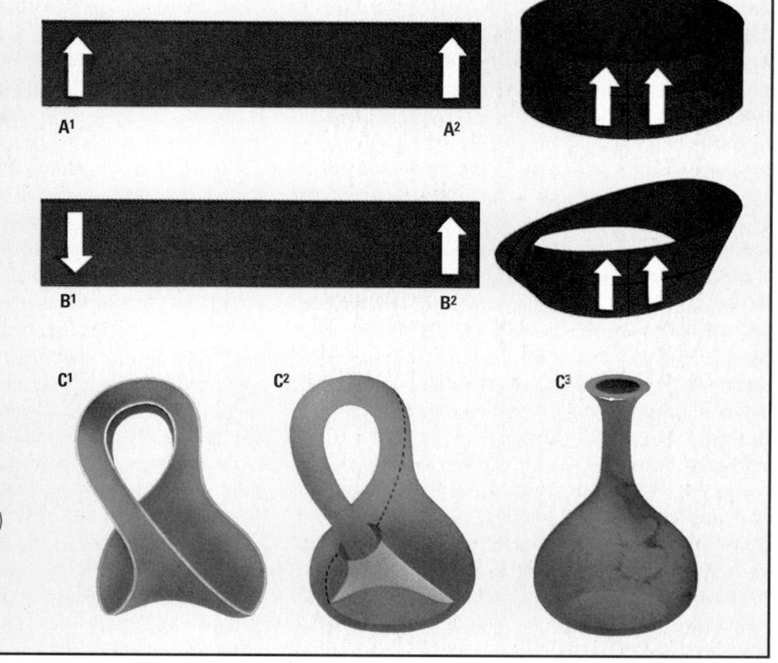

torsion In mechanics, strain in material that is subjected to a twisting FORCE. In a rod or shaft, such as an engine drive shaft, the torsion angle of **twist** is inversely proportional to the fourth power of the rod diameter multiplied by the shear modulus (a constant) of the material. Torsion bars are used in the spring mechanism of some automobile suspensions.

torsion balance Sensitive device for measuring small forces, whether gravitational, magnetic or electrical. A TORSION balance has a horizontal arm, suspended at its centre by a fine wire or fibre, which is twisted about a vertical axis by the force being measured.

torus (toroid, anchor ring) Solid figure generated by rotating a circle about an external line in its plane. It resembles a ring doughnut or the inner tube of a tyre. The volume of a torus is $2\pi^2Rr^2$ and the surface area is $4\pi^2Rr$, where R is the distance between the centre of the circle and the line, and r is the radius of the circle.

total internal reflection REFLECTION without REFRACTION of light at a boundary. When light passes from a dense medium, such as glass, to a less dense medium, such as water or air, a range of angles of incidence exists in which no light passes through the boundary; all the light is reflected within the denser medium.

touch One of the five SENSES, functioning by means of specialized nerve receptors in the SKIN.

touch screen Computer input device that allows the user to pass instructions by touching the screen of the display or monitor. The user can simply point to an icon, menu option or piece of data in order to select it. Usually the screen is covered by a pressure-sensitive, double-skinned membrane. When it is touched, fine conducting lines on the inside surfaces of the membrane come into contact and pass an electrical signal. Other types of touch screen use capacitive surface devices or a grid of intersecting light beams in front of the screen.

tourmaline Silicate mineral, sodium or calcium aluminium borosilicate, found in IGNEOUS and METAMORPHIC rocks. Its crystals are hexagonal system and glassy, either opaque or transparent. Tourmaline may be black, red, green, brown or blue. Some crystals are prized as gems. Hardness 7.5; r.d. 3.1.

Townes, Charles Hard (1915–95) US physicist. In 1953 he invented the first operational MASER, for which he shared the 1964 Nobel Prize for physics with Alexsandr PROKHOROV and Nikolai BASOV. The MICHELSON-MORLEY EXPERIMENT was accurately confirmed with the aid of masers.

toxin Poisonous substance produced by a living organism. The unpleasant symptoms of many bacterial diseases are due to the release of toxins into the body by the BACTERIA. The toxins may be secreted into the bloodstream or they may be released when the bacteria die. Many MOULDS, some larger FUNGI (such as the death cap), and seeds of some higher plants (such as LABURNUM and castor oil) produce toxins. The VENOMS of many snakes contain powerful toxins. *See also* BLOOM

trace element Chemical element that is essential to life but only in small quantities normally obtainable from the diet. They include boron, cobalt, copper, iodine, magnesium, manganese, molybdenum and zinc. They are essential to the reactions of ENZYMES and HORMONES.

tracer, radioactive Radioactive substance that is introduced into the body so that its progress can be tracked by special diagnostic equipment. Its course is followed in the body by the detection and identification of metabolic compounds "labelled" with

▶ **tornado** A tornado may form under the influence of the shearing action of strong air currents associated with an intense cold front. A strong convection current will sustain the vortex until all the potential energy is dissipated. They can cause extensive damage.

the isotope. This technique may help in the diagnosis of conditions such as thyroid disease. Radioactive tracers are used in a similar way to follow chemical reactions in complex organic industrial processes. In 1904 Franz Knoop used synthetic derivatives of fatty acids to trace the metabolic pathway. In 1923 George HEVESY became the first person to use a radioactive ISOTOPE (thorium) to investigate the metabolism.

trachea (windpipe) Airway that extends from the LARYNX to about the middle of the STERNUM (breastbone). Reinforced with rings of CARTILAGE, it is lined with hair-like CILIA that prevent dirt and other substances from entering the LUNGS. At its lower end, the trachea splits into two branches, the bronchi, which lead to the lungs.

tracheophyte In certain classification systems, any VASCULAR PLANT of the phylum Tracheophyta, defined by having some kind of vascular tissue. Within this phylum are: psilopsids (leafless, rootless primitive forms, such as whisk fern), sphenopsids (such as HORSETAIL), lycopsids (such as CLUB MOSS), pteropsids (such as FERN), GYMNOSPERMS and flowering plants. In the Five KINGDOMS classification, these groups now constitute separate phyla in their own right, and the Tracheophyta as a separate entity has been abolished.

trachoma Chronic eye infection, similar to conjunctivitis, caused by the microorganism *Chlamydia trachomatis*, characterized by inflammation of the cornea with the formation of pus. A disease of dry tropical regions, it is the major cause of blindness in the developing world.

trachyte Uncommon, fine-grained, extrusive IGNEOUS ROCK of volcanic origin. Usually light-coloured, it is rich in alkali feldspar and contains small amounts of quartz, nepheline and other minerals; its silica content is up to 60%. Trachyte is formed, along with basalt, in lava flows from island volcanoes, and it also occurs in small sills and dykes.

tracking In aerospace technology, following a moving object, usually with RADAR or radio using aerials that "lock" onto the object. The correct sweep speed of the ANTENNAE is calculated by computer from information on the object's position and velocity. Satellite tracking is of two kinds: **active** tracking uses radar to locate an object; **passive** tracking uses signals from the object itself.

traction In medicine, gradual exertion of force on a limb to overcome muscular tension and so ensure proper alignment of a fractured bone. The force is exerted gradually over a period of time, generally using weights on cords over pulleys.

traction engine Large, heavy vehicle powered by steam, once used for drawing farm wagons and other heavy loads, but now obsolete. Traction engines were also used at funfairs and circuses to drive dynamos for electricity. *See also* STEAM ENGINE

tractor Four-wheeled or tracked vehicle for moving and operating heavy implements. Tractors are used mostly in farming and construction. The first tractors were built in the 1870s. These steam-powered machines were known as TRACTION ENGINES. Modern tractors have petrol or diesel engines and can haul and power a wide range of implements, including hay balers, crop sprayers and mowing machines. Some tractors operate unmanned, having a computer to control their work and using a form of GLOBAL POSITIONING SYSTEM (GPS) to track their positions.

Tradescant, John Name of two British botanists and travellers, father and son, who became successive gardeners to Charles I. John (*c.*1570–1638) travelled to Russia and Algeria gathering plants, among them the "Algiers Apricot", and established a garden of exotic plants, at his Lambeth home. His son, also named John (1608–62), travelled to Virginia for plants for the Lambeth garden. He wrote a book about the Lambeth collection, *Musaeum Tradescantium* (1656). Their collection, transferred to Oxford after their deaths, formed the nucleus of the Ashmolean Museum.

trade winds Steady winds that blow from the tropical high-pressure zones (HORSE LATITUDES) to the equatorial low-pressure zones (DOLDRUMS). In the Northern Hemisphere, the air moving from the Tropic of CANCER towards the Equator is deflected to the right, making it a NE wind. In the Southern Hemisphere, the air moving from the Tropic of CAPRICORN to the Equator is turned to the left to become a SE wind. The trades are persistent winds – all are easterly and tropical – blowing from the deserts towards the tropical forest zone.

train *See* RAILWAY

trajectory Path of a projectile. On Earth, and in the absence of air resistance, all trajectories would be portions of an ELLIPSE with one focus at the centre of the Earth. Since this is 6,400km (4,000mi) deep, which is usually much greater than the height of the trajectory, the mathematically simpler PARABOLA provides a good approximation.

tram Passenger carriage that runs on rails which are usually sunk into the road. Horse-drawn trams first appeared in New York in 1832. Some later trams were powered by steam locomotives. Trams with electric motors, supplied by overhead cables, became popular in the early 1900s. In some places, tramways became unpopular in the 1920s. As they could not leave their rails, they tended to cause traffic jams. Motor buses and trolleybuses (which are not confined to rails) steadily became more popular, although many European cities retained tramway systems. A new generation of trams is appearing in some cities. With increasing concern over air pollution and public health, electric trams are attracting renewed interest. Unlike petrol and diesel engines, the electric motors in trams do not emit polluting exhaust gases.

tranquillizer Drug prescribed to reduce anxiety or tension and generally for its calming effects. Tranquilizers are used to control the symptoms of severe mental disturbance, such as schizophrenia or manic depression. They are also prescribed to relieve depression. Prolonged use of tranquillizers can produce a range of unwanted side-effects, including dependence.

transactinide element Element beyond LAWRENCIUM in the PERIODIC TABLE – that is, with an ATOMIC NUMBER of more than 103. Elements 104 (DUBNIUM), 105 (HAHNIUM) and 106 (RUTHERFORDIUM) have been produced in particle accelerators, and heavier elements (ELEMENTS 107 to 112) have been claimed. All are unstable and radioactive, with a very short half-life.

transatlantic cable Any submarine cable laid under the Atlantic Ocean and used to carry electrical signals. The first successful one was the telegraph cable laid in 1866.

transcendental number *See* NUMBER, TRANSCENDENTAL

transcription (DNA transcription) Stage in the synthesis of PROTEINS within a EUKARYOTE cell in which the GENETIC CODE, represented by NUCLEOTIDES on DNA, is copied into single-stranded MESSENGER RNA (MRNA). It takes place in the cell nucleus (because the DNA-containing chromosomes are located there). RNA polymerase enzymes make an RNA copy from the original unwound strand of DNA that forms the template. The section to be copied is indicated on the DNA template strand by start and termination "signals". The mRNA then migrates to the RIBOSOMES, where actual protein synthesis takes place. Different RNA polymerase ENZYMES synthesize different types of RNA. *See also* TRANSLATION

▲ **traction engine** The Garrett tractor, "Empress", was built in 1920. It was a single operator traction engine used for haulage work. The tyres are simply thick strips of rubber bolted on to the wheels

transducer Device for converting any nonelectrical signal, such as sound or light, into an electrical signal, and vice versa. Examples are microphones, loudspeakers, gramophone pickups and various measuring instruments used in ACOUSTICS.

transduction In microbiology, natural process by which genetic material is transferred between one host CELL and another by a VIRUS.

transfer orbit (transfer ellipse) Path followed by a spacecraft while changing from one orbital path to another. The tangential ELLIPSE is a tangent to the arrival and departure ORBITS and uses the least energy. Transfer ellipses intersecting the arrival and departure at higher angles use more energy but may be advantageous in other ways.

transfer RNA *See* RNA

transfinite number (symbol $\aleph$) Number denoting the size of an infinite SET, with a subscript. $\aleph_0$ represents the set of all INTEGERS and $\aleph_1$ represents the size of the set of all real NUMBERS. Although both sets are infinite, the set of real numbers is, in a sense, bigger than the set of integers (it can be shown that for every real number there is not an integer).

transformer Device for converting alternating current (AC) at one VOLTAGE to another voltage at the same frequency. It consists of two coils of wire coupled together magnetically. The input current is fed to one coil (the primary), the output being taken from the other coil (the secondary). If core losses are ignored, the ratio of the input voltage to the output voltage is equal to the ratio of the number of turns in the primary coil to the number of turns in the secondary coil. *See also* ELECTRIC CURRENT

transform fault Special class of strike-slip FAULT characteristic of MID-OCEAN RIDGES. Because of the transform faults, which are at right-angles to the ridge itself, the MID-ATLANTIC RIDGE does not run in a straight line but in offset steps. Some geologists think the four major structures of the Earth's crust are mountains, deep-sea trenches, mid-ocean ridges and strike-slip faults, and that they form continuous networks.

transgenic Describing an organism whose GENOME contains GENES artificially transferred from another species. Using GENETIC ENGINEERING techniques, transgenic organisms are created by inserting isolated DNA from another species into the fertilized egg or early embryo. Transgenic animals and plants are usually created for commercial purposes. For example, it is possible to insert the genes for extra growth hormone into the genomes of livestock in order to increase meat production.

transistor Electronic device made of SEMICONDUCTOR material that can amplify electrical signals. The material, such as SILICON or GERMANIUM, is "doped" with minute amounts of PHOSPHOROUS, ARSENIC or ANTIMONY to produce *n*-type material, in which current is carried by negative charges, ELECTRONS; or with ALUMINIUM, GALLIUM or INDIUM to give a *p*-type material. Joining together a piece of each produces a DIODE. Sandwiching one type between two of the other produces a transistor. These can thus be of two kinds: a *p-n-p* or *n-p-n* transistor. The middle region is known as the base, one of the side regions as the emitter and the other as the collector. In an *n-p-n* transistor, the signal to be amplified is fed across the collector, maintained at a constant voltage with respect to the base. The amplified signal comes out across the base and the emitter. Transistors were first developed in 1948 by John BARDEEN, Walter BRATTAIN and William SHOCKLEY, making possible many advances in technology, especially in computers, portable radios and televisions, satellites, industrial control systems and navigation.

transistor radio *See* RADIO; TRANSISTOR

transit In astronomy, passage of either of the inner planets, Mercury and Venus, across the face of the Sun. Transits of Mercury occur every 13 years, those of Venus every 100 years. The term also signifies the motion of any celestial object across the observer's celestial meridian.

transition In atomic physics, movement from one ENERGY LEVEL, or quantum state, to another. A transition from a lower to a higher energy level usually involves the absorption of a PHOTON. To move from a higher to a lower energy level, a photon is usually released.

transition elements Metallic elements that have incomplete inner electron shells. Transition elements are characterized by variable VALENCIES (combining power) and the formation of coloured ions. They include elements with atomic numbers from 21 to 112. The ACTINIDE SERIES of elements and the rare-earth metals of the LANTHANIDE SERIES are sometimes called the inner transition elements. *See also* PERIODIC TABLE

translation Part of the process of protein synthesis that takes place in living cells. MESSENGER

TRANSPIRATION

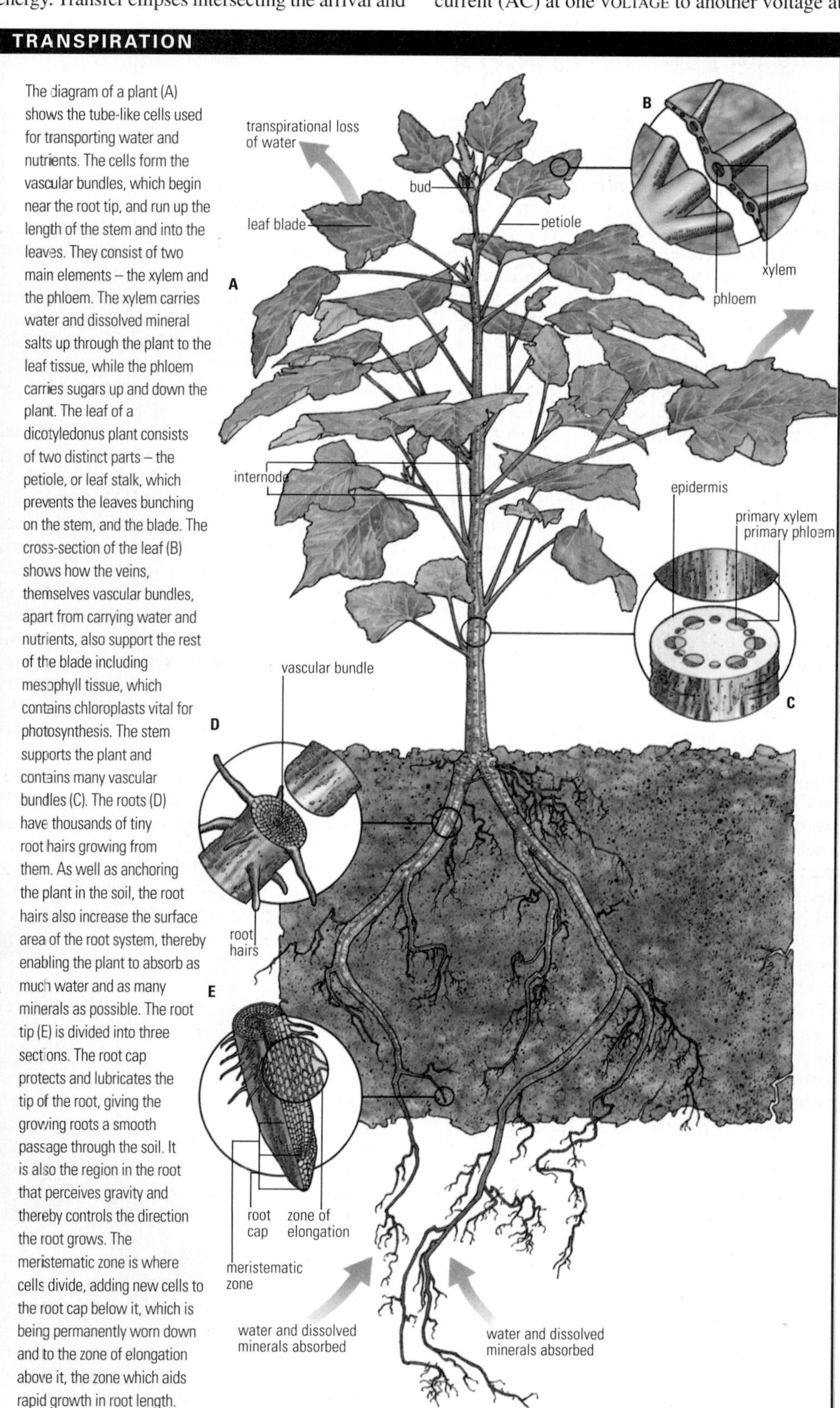

The diagram of a plant (A) shows the tube-like cells used for transporting water and nutrients. The cells form the vascular bundles, which begin near the root tip, and run up the length of the stem and into the leaves. They consist of two main elements – the xylem and the phloem. The xylem carries water and dissolved mineral salts up through the plant to the leaf tissue, while the phloem carries sugars up and down the plant. The leaf of a dicotyledonus plant consists of two distinct parts – the petiole, or leaf stalk, which prevents the leaves bunching on the stem, and the blade. The cross-section of the leaf (B) shows how the veins, themselves vascular bundles, apart from carrying water and nutrients, also support the rest of the blade including mesophyll tissue, which contains chloroplasts vital for photosynthesis. The stem supports the plant and contains many vascular bundles (C). The roots (D) have thousands of tiny root hairs growing from them. As well as anchoring the plant in the soil, the root hairs also increase the surface area of the root system, thereby enabling the plant to absorb as much water and as many minerals as possible. The root tip (E) is divided into three sections. The root cap protects and lubricates the tip of the root, giving the growing roots a smooth passage through the soil. It is also the region in the root that perceives gravity and thereby controls the direction the root grows. The meristematic zone is where cells divide, adding new cells to the root cap below it, which is being permanently worn down and to the zone of elongation above it, the zone which aids rapid growth in root length.

RNA (MRNA) carries a genetic code, in the form of a series of **codons** (triplets of NUCLEOTIDES), which ultimately determines the correct sequence for AMINO ACIDS to form protein. RIBOSOMES in the cell cytoplasm scan the mRNA for its codons, relaying the information to amino acid-bearing TRANSFER RNA (TRNA). The strands of tRNA have their own **anti-codons**, which form pairs with the codons of the mRNA. In effect, the tRNA (along with its amino acid), by pairing and attaching to the mRNA, "translates" the mRNA code to form the correct sequence of amino acids that form peptides, polypeptides and eventually proteins. As the polypeptide chain grows, the tRNA falls away.

translocation In botany, movement of food materials in solution through the tissues of VASCULAR PLANTS from one part to another. It includes the passage upwards of minerals and other inorganic salts from the soil and roots in the XYLEM tissue, and the movement of sugars and other organic compounds from the leaves to other parts of the plant in the PHLOEM tissue.

translocation In genetics, process that alters the genetic make-up of a CHROMOSOME. A section of a chromosome breaks off, leaving it with less genetic information. The fragment joins onto another chromosome, also altering its genetic make-up. Rearrangement of genes within a single chromosome is also called translocation.

translucent Describing a material that allows the passage of light (or other electromagnetic radiation) but scatters it so that no object is visible through the material. *See also* TRANSPARENT

transmission, automobile Device (a gear or gears and shaft or driving chain) for transmitting power from the engine of a vehicle to the wheels. Speed can usually be varied in discrete steps, with gears or chains providing fixed speed ratios.

transmitter Electrical apparatus that generates RADIO WAVES. It has three main parts: an **oscillator**, which generates high-frequency alternating electric current (a carrier wave); a **modulator**, which encodes the carrier wave with the pattern or sound to be broadcast; and an **amplifier**, which increases the power of the modulated carrier (signal). *See also* MODULATION

transmutation In physics, formation of one element or ISOTOPE from another by RADIOACTIVE DECAY or by bombardment with energetic particles. The term was formerly applied to alchemists' attempts to convert base metals into gold.

transparent Describing a material that allows the passage of light (or other electromagnetic radiation) with little or no scattering so that objects are clearly visible through it. Different materials may be transparent to different wavelengths of radiation. *See also* TRANSLUCENT

transpiration In plants, loss of moisture as water vapour from leaf surfaces or other plant parts. Most of the water entering plant roots is lost by transpiration. The process is speeded up in light, warm and dry conditions. The flow of water from the roots to the leaf pores (stomata) is called the **transpiration stream**. Several forces drive this flow. As moisture evaporates from the stomata in the leaves, water moves in by OSMOSIS from adjacent cells, setting up a flow of water by osmosis across the leaf from the XYLEM tubes to the stomata, pulling water out of the xylem. The force of attraction between the water molecules as they pass up the xylem tubes from the root to the leaves makes it difficult to break the water column. These tubes are also very narrow, so water rises in them by capillary action. In the roots, water being drawn into the xylem creates a lower concentration of water in adjacent root cells, thus water is drawn across the root from the root hairs by osmosis.

transplant Surgical operation to introduce organ or tissue from one person (the donor) to another (the recipient); it may also refer to the transfer of tissues from one part of the body to another, as in grafting of skin or bone. Major transplants are performed to save the lives of patients facing death from end-stage organ disease. Organs routinely transplanted include the kidneys, heart, lungs, liver and pancreas. Experimental work continues on some other procedures, including small bowel grafting. Many other tissues are commonly grafted, including heart valves, bone and bone marrow. The oldest transplant procedure (and still the most successful) is corneal grafting, undertaken to restore the sight of one or both eyes. Most transplant material is acquired from dead people, although kidneys, part of the liver, bone marrow and corneas may be taken from living donors. The two main problems in transplantation are rejection and infection. Rejection refers to the body's tendency to destroy "foreign" tissue and this cannot always be controlled by drugs given to suppress the immune system.

transport In chemistry and physics, movement of measurable entities such as molecules, ions, isotopes, electrical charges, mass, momentum or energy through or across a medium, because of the natural or applied non-uniform conditions existing within it. Transport properties include VISCOSITY, diffusivity (*see* DIFFUSION) and thermal CONDUCTIVITY. ACTIVE TRANSPORT in biochemistry is the movement of a substance against a concentration gradient.

transposon (transposable element) Section of DNA that can insert itself at many different sites on a CHROMOSOME. The presence of a transposon alters the genetic composition of the chromosome and can affect the surrounding GENES, causing some to be omitted or altered.

transuranic element (transuranium element) Any of those elements with atomic numbers higher than that of URANIUM (92), the best known of which are members of the ACTINIDE SERIES (atomic numbers 89 to 103). All transuranic elements are radioactive. Only NEPTUNIUM and PLUTONIUM occur naturally (in minute amounts) and have been synthesized in great quantity. Neptunium was the first of the transuranic elements to be produced, in 1940 by the US physicist Edwin MCMILLAN and US physical chemist Philip Abelson, who bombarded uranium with neutrons in a synchrocyclotron. The other elements have since been produced in a similar manner. The only commercially important element in the group is plutonium, which is used in NUCLEAR WEAPONS and as a fuel for NUCLEAR REACTORS.

trapezium (trapezoid) Four-sided plane figure with one pair of opposite sides parallel. The area of a trapezoid is half the sum of its parallel sides multiplied by the perpendicular distance between them.

travelling wave WAVE that continuously carries electrical energy, light energy, and so on, away from a source.

Travers, Morris William (1872–1961) British chemist who, in collaboration with Sir William RAMSAY, discovered the NOBLE GASES of the atmosphere. Those found included argon, helium, neon, krypton and xenon. These researches are well described in his book *Discovery of the Rare Gases* (1928).

travertine SEDIMENTARY ROCK consisting almost entirely of crystals of CALCITE (calcium carbonate, $CaCO_3$). It has a very light colour, unless stained yellowish by iron compounds, and is often porous. It usually forms in hot springs from water rich in dissolved calcium carbonate. *See also* TUFA

TREE

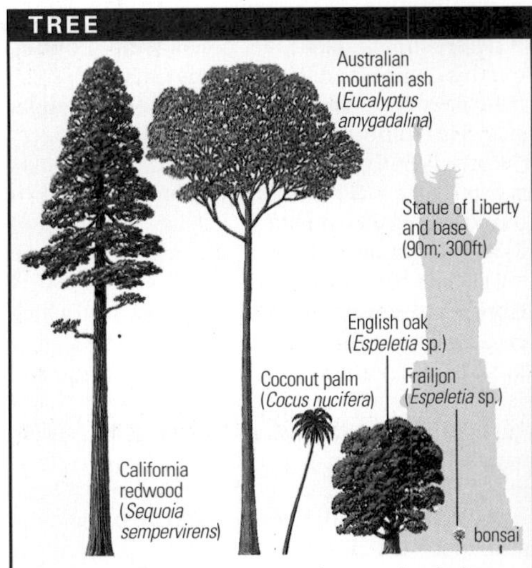

Trees grow taller than any other living thing, but can still survive in miniature form. If the roots are restricted either artificially, as in bonsai perfected in Japan, or by natural means, as when a seed germinates in very thin soil on a mountain, a fully formed tree only a few centimetres high will result. The California redwood, the world's tallest tree, is closely rivalled by a eucalyptus, such as the mountain ash of Australia. The coconut palm reaches its height of 27m (90ft) in a few years. The English oak is one of 450 species of oak that grow as trees, bushes and shrubs. It enlarges slowly – *c.*4.5m (15ft) – in ten years, but produces wood of prodigious strength. Espeletia grows on snowy ledges more than 400m (1,300ft) up in the Sierra Nevada of California.

tree Woody, PERENNIAL plant with one main stem or trunk and smaller branches. The trunk increases in diameter each year, and the leaves may be EVERGREEN or DECIDUOUS. The largest trees, redwoods, can grow more than 110m (420ft) tall; the bristlecone pine can live for over 5,000 years.

trematode Any member of the class of invertebrates Trematoda, which includes nearly 6,000 species of parasitic FLATWORMS and FLUKES. Parasitic trematodes have a thick outer cuticle and one or more suckers for attaching to the tissues of the host. They are bilaterally symmetrical, and are commonly found as endoparasites or ectoparasites in all classes of vertebrates, including human beings. Length: to 10cm (4in).

tremor Involuntary shaking of part of the body. It may be caused by fatigue, anxiety, shock, debility or any one of a range of neurological disorders, especially PARKINSON'S DISEASE.

trench Deep, V-shaped depression of the SEAFLOOR. In PLATE TECTONIC theory trenches are places where one plate is being pushed under another. They are the deepest – to 11km (7mi) – formations on Earth and are found primarily along the borders of the Pacific Ocean.

Trevithick, Richard (1771–1833) British engineer and designer of steam engines. Trevithick worked in the Cornish tin mines, where the use of steam-operated pumps to remove water led to his interest in pumps and steam engines. In 1801 Trevithick built a steam-powered road vehicle. The following year, he patented a high-pressure steam engine, his most important invention. In 1803 Trevithick built the first steam railway LOCOMOTIVE. In 1816 he went to Peru to install his steam engines in mines. On returning to England in 1827, Trevithick found that others were profiting from his engine, and he died in poverty.

triangle Plane figure bounded by three straight lines. The sum of the interior angles totals 180°. The area is measured by either (1) half the product of one of the sides and the perpendicular onto it from the opposite vertex ($^1/_2 \times$ base $\times$ height), or (2) half the product of two of the sides and the sine of the angle between them.

triangle of vectors Triangle, the sides of which represent the magnitude and direction of three VECTORS about a point that are in the same plane and are in equilibrium. A triangle of vectors is commonly used to represent forces or velocities. If the magnitude and direction of two forces (or velocities) are known, then two sides of the triangle can be drawn. Using scale drawing or trigonometry, the magnitude and direction of the third force (or velocity) can be calculated. *See also* PARALLELOGRAM OF VECTORS

triangulation In navigation and land surveying, a method of determining distance. The area under survey is divided into triangles. The baseline of one triangle is measured and the angles that it makes to its vertex are measured with a THEODOLITE. The distances from each end of the baseline to the vertex can be calculated using trigonometry.

Triangulum (Triangle) Inconspicuous N constellation situated between Andromeda and Aries. Its brightest star, Beta (magnitude 3.0), forms a well-marked triangle with Alpha (3.4) and Gamma (4.0). The main feature of interest is the spiral galaxy M33, 2.7 million light years away in our Local Group.

Triangulum Australe (Southern Triangle) Southern circumpolar constellation situated S of Norma. The three brightest stars, Alpha (magnitude 1.9), Beta and Gamma (both 2.9), form a conspicuous triangular figure. The open cluster NGC 6025 is visible with binoculars.

Triassic First period of the MESOZOIC era, lasting from 248 to 213 million years ago. Following a wave of extinctions at the close of the PERMIAN period, many new kinds of animals developed. On land lived the first DINOSAURS. MAMMAL-like reptiles were common and by the end of the period the first true mammals existed. In the seas lived the first ichthyosaurs, placodonts and nothosaurs. The first frogs, turtles, crocodilians and lizards also appeared. Plant life consisted mainly of primitive GYMNOSPERMS, with ferns and conifers predominating.

tricarboxylic acid cycle (TCA cycle) *See* KREBS CYCLE

trichloromethane (chloroform, $CHCl_3$) Colourless, volatile, sweet-smelling liquid prepared by the chlorination of methane. Formerly a major anaesthetic, trichloromethane is used in the manufacture of fluorocarbons, in cough medicines and as a solvent. Properties: r.d. 1.48; m.p. −63.5°C (−82.3°F); b.p. 61.2°C (142.2°F).

tricyclic compound Organic chemical that has three rings as its primary structure. Examples are ANTHRACENE and phenanthrene.

triglyceride *See* LIPID

trigonometric function Six ratios of the sides of a right-angled TRIANGLE containing a given acute ANGLE – they are the SINE, COSINE, TANGENT, COTANGENT, SECANT and COSECANT of the angle. These functions can be extended to cover angles of any size. Thus the sine of an angle A increases from 0 to 1 as A increases from 0° to 90°, decreases to 0 as A increases from 90° to 180°, decreases further to −1 as A increases to 270°, and increases again to 0 as A increases to 360°. The trigonometric functions can also be defined when the argument (the angle) is measured in RADIANS. Many relationships exist between different trigonometric functions, including the identities $\sin^2 A + \cos^2 A = 1$; $1 + \tan^2 A = \sec^2 A$; and $\tan A = \sin A/\cos A$.

trigonometric parallax Means of determining the distance of a star using the principle of TRIANGULATION. As the Earth revolves around the Sun, nearby stars show a small change in position with respect to more distant stars. This change in position can be measured from photographs taken six months apart, that is when the Earth is on opposite sides of the Sun. The angular displacement is a direct measure of the star's PARALLAX (π). Since the baseline (the distance from the Sun to the Earth) is known, the distance d of the star can be determined by TRIGONOMETRY. The first star to have its distance measured by this method was 61 Cygni, by Friedrich BESSEL in 1838. Parallax is measured in arc seconds. The inverse of the parallax is the distance in PARSECS. Alpha Centauri is the star with the largest known parallax, 0.752″, which corres-ponds to a distance of 1.3 parsecs. Typical errors in measuring parallax photographically are about 0.01″, but by using many plates this error can be reduced to about 0.004″. The parallax of a star at 25 parsecs is 0.04″, so the uncertainty is around 10%. The most accurate parallax determinations have been made by the Hipparcos satellite.

trigonometry Use of ratios of the sides of a right-angled TRIANGLE to calculate lengths and angles in geometrical figures. If three sides, or two sides and the included angle, or one side and two angles of a triangle are known, then all the other sides and angles may be found.

triiodomethane (iodoform, CHI_3) Yellow, crystalline chemical compound with a characteristic odour. It is used as an antiseptic.

trilobite Any of an extinct group of ARTHROPODS found as fossils in marine deposits, ranging in age from Cambrian through Permian times. The body is mostly oval, tapering towards the rear; it was covered by a skeleton made of chitin. The name refers to the division of the body into three – a central axis and two lateral lobes. Transverse divisions show segmentation, with each segment bearing a pair of jointed limbs. Most species were bottom-crawling, shallow-water forms that ranged in size from 6mm (0.25in) to 75cm (30in).

trinitrotoluene *See* TNT

triode *See* ELECTRON TUBE

triple-alpha process Nuclear reaction in which three alpha particles (helium nuclei) are transformed into carbon with the release of energy; it is thought to be the dominant energy-producing process in RED GIANTS. It takes place after all the hydrogen of a star's core has been exhausted. The core contracts and its temperature rises to over 100

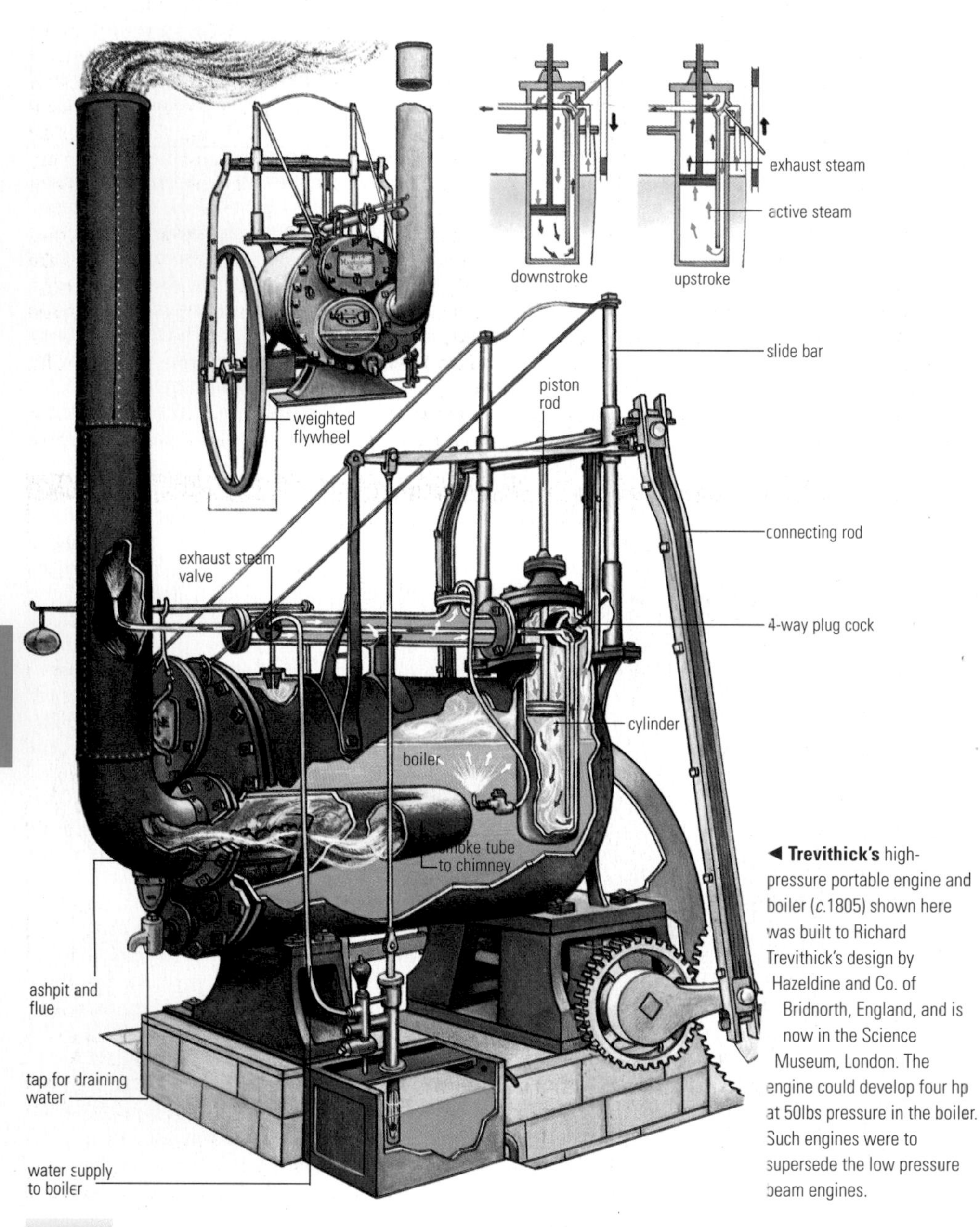

◀ **Trevithick's** high-pressure portable engine and boiler (*c.*1805) shown here was built to Richard Trevithick's design by Hazeldine and Co. of Bridnorth, England, and is now in the Science Museum, London. The engine could develop four hp at 50lbs pressure in the boiler. Such engines were to supersede the low pressure beam engines.

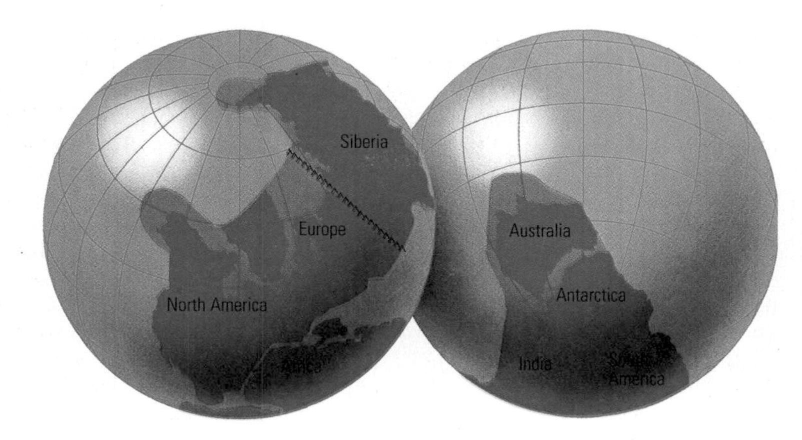

▶ **Triassic** During this period, the collision between the North America-Europe landmass and the Siberian plate pushed up the Ural Mountains.

million K, when the triple-alpha reaction begins. In each reaction two helium nuclei combine to form a beryllium nucleus, which, in turn, captures a further helium nucleus to form a carbon nucleus. Additional helium captures can produce oxygen, neon and a number of heavier elements. The process is also known as the Salpeter process after the US physicist Edwin Salpeter (1924–), who described it.

triple point Temperature and pressure at which all three PHASES (solid, liquid and gas) of a substance can coexist. For water, the triple point occurs at 273.16K and 610Pa.

trisomic Describing a DIPLOID organism that has three of one type of chromosome instead of two. Trisomy is the cause of several human genetic disorders, notably DOWN'S SYNDROME.

tritium (symbol T) Radioactive isotope of hydrogen, the nucleus of which consists of one proton and two neutrons. Only one atom in 10^{17} of natural hydrogen is tritium. Tritium compounds are used in radioactive tracing. Properties: half-life 12.3 yr.

triton Nucleus of a TRITIUM atom, consisting of one proton and two neutrons.

Triton Largest of Neptune's eight satellites, discovered by the English amateur astronomer William Lassell (1799–1880) in 1846. It has an inclined orbit in the opposite direction to Neptune's own spin (retrograde), and was almost certainly a separate planet "captured" by Neptune. Its diameter is 2,705km (1,680mi), and it probably comprises three parts rock to one part water-ice. Triton is the coldest world so far visited by probes: Voyager 2's instruments measured a surface temperature of 35K (–238°C). There is evidence of recent volcanic activity and in the S polar region eruptive plumes send fine dark material rising to altitudes of 8km (5mi).

trophic level Position that an organism occupies in a FOOD CHAIN. It is usually defined in terms of the food supply. The first trophic level is occupied by PRIMARY PRODUCERS, which are green plants that use photosynthesis to obtain energy from sunlight. Herbivores (PRIMARY CONSUMERS), which eat the plants, occupy the second level. SECONDARY CONSUMERS, on the third trophic level, are carnivores, which eat the herbivores. There may be a fourth level (TERTIARY CONSUMERS) occupied by other carnivores that eat those on the third level.

tropics *See* CANCER, TROPIC OF; CAPRICORN, TROPIC OF

tropism (tropic response) Response in growth and orientation of a plant or a part of it in relation to a directional, external stimulus. Such stimuli include light (PHOTOTROPISM), gravity (GEOTROPISM) or water (HYDROTROPISM).

tropomyosin Filamentous protein associated with ACTIN in MUSCLE fibres. When the muscle is at rest, tropomyosin separates actin from MYOSIN. When the muscle contracts, TROPONIN displaces tropomyosin, and the actin and myosin come into contact.

troponin Protein involved in MUSCLE contraction. It consists of three polypeptide chains. When a muscle contracts, troponin binds with calcium to displace the TROPOMYOSIN, thus allowing ACTIN and MYOSIN to interact.

troposphere Lowest part of the Earth's ATMOSPHERE. In the troposphere, temperature decreases with height, and it is within this region that clouds and other weather phenomena occur. The troposphere extends to about 16km (10mi) above the Earth's surface.

trough In meteorology, area of low atmospheric pressure, usually an extension to a DEPRESSION. The opposite are ridges of high pressure.

truss In engineering and building, structural member made up of straight pieces of metal or timber formed into a series of triangles lying in a vertical plane. The triangles resist distortion through stress, making the truss capable of sustaining great loads over long spans. It was probably first used almost 5,000 years ago in lake dwellings.

trypsin Digestive enzyme secreted by the PANCREAS. It is secreted in an inactive form (so it does not damage tissues en route to the intestine), which is converted into active trypsin by an enzyme in the small intestine. It breaks down peptide bonds on the amino acids lysine and arginine. *See also* ALIMENTARY CANAL; DIGESTION; DIGESTIVE SYSTEM

tryptophan AMINO ACID ($C_{13}H_{10}O_2N_2$), first isolated in 1902, which is necessary for the synthesis of the anti-pellagra vitamin NICOTINIC ACID. Tryptophan itself cannot be synthesized by animals and is therefore an essential amino acid that must be supplied in the diet. It is found in most natural proteins, but some cereals, notably maize and sorghum, contain particularly low levels.

TSH Abbreviation of THYROID STIMULATING HORMONE

Tsiolkovsky, Konstantin Eduardovich (1857–1935) Russian scientist who provided the theoretical basis for space travel. In 1898 he became the first person to stress the importance of liquid propellants in ROCKETS. He also proposed the idea of using multistage rockets to overcome the effect of GRAVITATION, and he described the principles of SPACE STATIONS.

tsunami WAVE caused by a submarine EARTHQUAKE, subsidence, or volcanic eruption. Tsunamis spread radially from their source in ever-widening circles. In mid-ocean they are shallow, 30–60cm (1–2ft) high, and are rarely detected. In shallower water they build up in force and height, occasionally reaching 30m (100ft), crashing on shore and causing enormous damage. Tsunami is a Japanese word; such waves are also called seismic sea waves, but tidal wave is an erroneous term for them, as they are never the result of the activity of the TIDES.

tuber In plants, short, swollen, sometimes edible underground stem (modified for the storage of food as in the potato) or swollen root (such as the dahlia). In stem tubers new plants develop from the buds, or eyes, growing in the axils of the scale leaves. Tubers are propagated by sections containing at least one eye. In the wild, they enable the plant to survive an adverse season (winter or dry season), providing food for the development of new shoots and roots later.

Tubulidentata Order of insectivorous mammals, found in central and S Africa. The aardvark is the only member of the order. A nocturnal, bristly-haired animal, the aardvark feeds on termites and ants, which it scoops up with its sticky, 30cm-(12in-) long tongue. Length: up to 1.5m (5ft); weight: up to 70kg (155lb).

Tucana (Toucan) Constellation of the S sky. Notable for the presence of the small MAGELLANIC

TSUNAMI

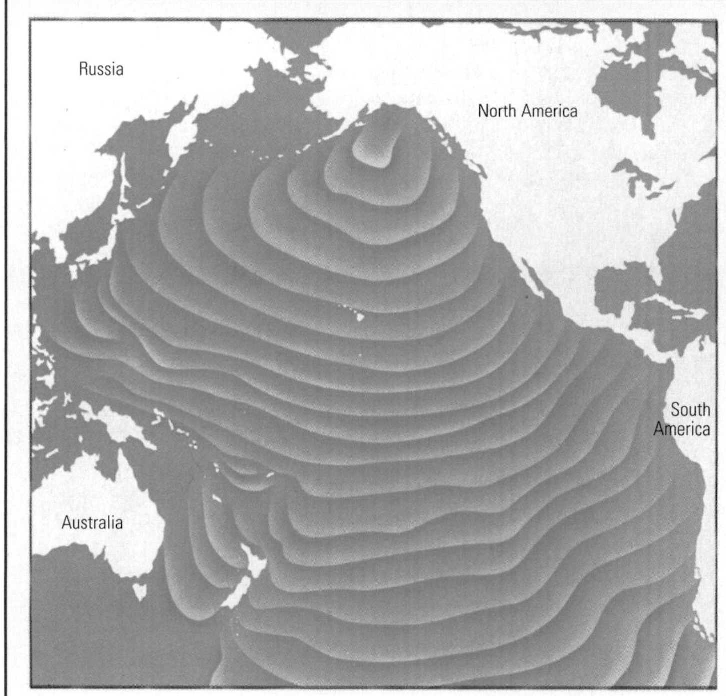

Tsunamis are dramatic waves generated when a submarine earthquake causes a sudden shift in the ocean floor along a fault line. The upheaval creates a bulge that breaks down into a series of waves travelling at speeds of up to 750km/h (450mph). The map shows the hourly position of a tsunami which originated just S of Alaska.

CLOUD and the superb globular cluster 47 Tucanae (NGC 104). It lies 15,000 light years away.

tufa SEDIMENTARY ROCK composed mainly of CALCITE (calcium carbonate, $CaCO_3$). It is a porous, non-bedded rock, similar to but less dense than TRAVERTINE. It forms when dissolved calcium carbonate precipitates from lime-rich water in limestone caves. It may rapidly encrust small growing plants or it may contain embedded sediment grains and pebbles. The presence of iron oxides gives it a yellow or red colour.

tuff SEDIMENTARY ROCK made up of particles of IGNEOUS ROCK from volcanic eruptions. The particles vary in size from fine to coarse but are generally smaller than 4mm (0.16in). They may be either stratified or heterogeneous in their arrangement. *See also* PYROCLASTIC

Tull, Jethro (1674–1741) British agriculturalist. He influenced agricultural methods through his innovations and writings. He invented a mechanical drill for sowing in 1701 and advocated the use of manure and thorough tilling during the growing period. He is considered a pioneer of the AGRICULTURAL REVOLUTION.

tumour Any uncontrolled, abnormal proliferation of cells, often leading to the formation of a lump. Tumours are classified as either benign (non-cancerous) or malignant (*see* CANCER).

tundra Treeless, level or gently undulating plain characteristic of Arctic and sub-Arctic regions. The main swathe of tundra occurs in the Northern Hemisphere, running across N North America and Eurasia. Summer days are quite long and often sunny, but average temperatures for July do not rise above 10°C (50°F). During winter, temperatures can fall to −30°C (−86°F). For six to nine months the average temperature is below freezing point, and there is a thick layer of PERMAFROST, of which only the top few centimetres thaw out in the summer. Where melting occurs, the land is wet and marshy, an ideal breeding ground for mosquitoes. The wet tundra soils are **gleys**, and waterlogging is characteristic. Generally, trees cannot grow in the tundra because of the unfavourable conditions. Mosses and lichens are common, and there are many small flowering plants in the summer.

tungsten (wolfram, symbol W) Silver-grey TRANSITION ELEMENTS, first isolated in 1783. Its chief ores are WOLFRAMITE and SCHEELITE. Tungsten has the highest melting point of all metals and is used for lamp filaments and in special alloys. TUNGSTEN CARBIDE is used in high-speed cutting tools. The sulphide is used as a lubricant. Chemically tungsten is fairly unreactive; it oxidizes only at high temperatures. Properties: at.no. 74; r.a.m. 183.85; r.d. 19.3; m.p. 3,410°C (6,170°F); b.p. 5,660°C (10,220°F); most common isotope ^{184}W (30.64%).

tungsten carbide Extremely hard, inert, grey powder used for making ABRASIVES, dyes, drill tips and armour-piercing shells. Tungsten carbide (WC) is made by heating tungsten and lamp-black (powdered carbon) in an electric furnace. Properties: r.d. 15.6; m.p. 2,780°C (5,036°F).

tunnel effect Effect in which electrons cross a potential barrier that, according to the rules of classical physics, is impossible to cross. Using QUANTUM MECHANICS, however, it can be shown that there is a probability that the electrons will cross the barrier. The effect is utilized in the **tunnel diode** (*see* ESAKI, LEO).

tunnelling Construction of passages under the ground or underwater. Tunnels in hard rock are usually made by blasting, while boring machines cut tunnels through softer material. Tunnels in soft earth have to be shored up to prevent collapse. During construction, a cylindrical steel shield is pushed through the earth, immediately followed by a permanent cast iron or concrete lining. Flooding is a danger when constructing river and sea tunnels. Usually, the air pressure inside is kept higher than the water pressure outside. This means that, if the tunnel is not quite sealed, air will escape, but no water will enter. Some underwater tunnels are formed by sinking sealed, prefabricated steel or concrete sections into the riverbed or seabed. The sections are joined together and then the structure is covered with earth.

turbellarian Any member of the class Turbellaria, comprising the PLANARIANS.

turbine Rotary device turned by a moving fluid (liquid or gas). Turbines enable the energy of moving wind, water, steam or other fluids to do useful work. The WATER WHEEL is a simple form of turbine. In early designs, a stream of water flowed over or under a vertical wheel with blades or buckets on its circumference, making it turn. The movement was used to operate simple machinery. For example, the WATER MILL could turn a stone to grind grain to make flour. The modern form of water turbine is like a multi-bladed propeller and is used to generate HYDROELECTRICITY. In power stations that burn fuels to produce electricity, the energy released by the burning is harnessed by the blades of jet engine-like steam turbines. As they spin, the turbines turn GENERATORS that produce electricity. WINDMILLS are simple turbines that use wind energy to grind grain. Modern wind generators produce electricity when the wind turns their rotors. In gas turbines, hot gases from burning fuel turn turbines that can operate generators or other machinery. Jet engines use gas turbines to drive compressors, fans or propellers.

turbocharger Device that boosts the performance of an INTERNAL COMBUSTION ENGINE. A TURBINE driven by exhaust gases compresses the fuel/air mixture before it passes through the inlet valve.

turbofan engine TURBINE engine that developed from, and is more efficient than, the TURBOJET ENGINE. It has a compressor, which forces air from the intake into the combustion chamber, and a turbine, which is driven by the exhaust gases and drives the compressor. Extra thrust is provided by large blades at the front, driven by the turbine, which push air along a bypass round the combustion chamber and into the tailpipe.

turbojet engine Aircraft engine (a type of gas TURBINE) that produces power through the reactive force of expanding gases. Air is taken in at the front through a compressor, forced into a combustion chamber, mixed with fuel and burned, producing a rush of expanding gas that propels the aircraft in a direction opposite to that of the rapid outflow. To maintain the cycle, the expanding gas also turns a turbine which drives the air compressor. *See also* TURBOFAN ENGINE

turboprop engine Aircraft engine that has a PROPELLER (airscrew) driven by a gas TURBINE through gears. The turbine compresses air which is mixed with fuel and burned in a combustion chamber, producing exhaust gases which drive the turbine, compressor and propeller. *See also* TURBOFAN ENGINE; TURBOJET ENGINE

turbulent flow In physics, type of fluid flow in which the movement of the particles is irregular. It is characteristic of a fluid with a high REYNOLDS NUMBER. *See also* LAMINAR FLOW

turgor pressure Hydrostatic pressure generated in cells of plants and bacteria as a result of the uptake of water by OSMOSIS. Water diffuses through the semi-permeable membrane of the cell, causing the cell to swell; the increase in volume is resisted by the limited elasticity of the cell wall, which generates the pressure, rather in the way that pumping air into a bicycle tyre causes it to become stiff and resist squeezing. It is turgor pressure that supports non-woody plants. When water is lost, a plant's cells collapse, and it wilts.

Turing, Alan (1912–54) British mathematician and logician who formulated basic theories about COMPUTERS. In 1937 he invented the **Turing machine**, a hypothetical machine that could modify a set of input instructions. It was the forerunner of the modern computer. Turing also used the concept of a computer to give an alternative and simpler proof of GÖDEL's incompleteness theorem. Turing played a major role in deciphering the Enigma code – a complex method of encryption used by Germany during World War 2. In 1948 he helped to construct one of the world's first computers. In 1950 he devised the **Turing test**: a proposed test of a computer's ability to "think". The test is essentially that a human would not be able to distinguish between the dialogue of the computer and that of another human. This work

TURBINE

Turbines and generators are the components that convert the rotary motion produced by steam or water power into electricity. To obtain the maximum energy from the water or steam, there are several stages of turbines – up to five in a large power station – each using water or steam at a slightly lower pressure than the previous one.

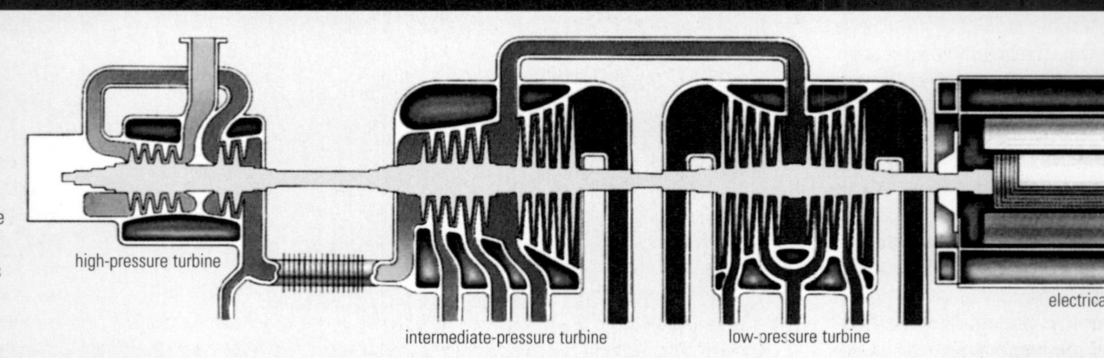

Super-heated steam at up to 600°C (1,112°F) gives up much of its energy in a high-pressure turbine. The exhaust steam from this stage is reheated and passed to an intermediate-pressure turbine and then to the low-pressure turbine. The output shaft drives a generator.

TURBOCHARGER

The diesel turbocharger uses the energy of the exhaust gases to force air into the cylinder via linked impellers (1). First air is sucked into the engine (A). The air, as it is compressed into a space 22 times smaller (B), is heated. Diesel fuel is then injected in a swirling manner to maximize mixing, accentuated by the shape of the piston head (C). Combustion occurs spontaneously, without the need of a spark plug, pushing the piston down (D). The rotation of the crankshaft pushes the piston up, expelling the waste gases (E) that turn one of the linked impellors.

paved the way for the foundation of ARTIFICIAL INTELLIGENCE (AI). Turing also worked in the area of theoretical biology. In *The Chemical Basis of Morphogenesis* (1952) he proposed a model that described the origin of pattern formation in biology. Such models have since been shown to describe and explain many patterns seen in nature. He committed suicide after being prosecuted for homosexuality.

Turner's syndrome Hereditary condition in females, in which there is only one X-chromosome instead of two. It results in infertility (due to lack of ovaries), short stature and various developmental defects, including the lack of secondary sexual characteristics.

turpentine (gum turpentine) Sticky liquid obtained from coniferous trees; it contains ROSIN and a volatile oil. Oil of turpentine, often called merely turpentine, is obtained by DISTILLATION of the gum and is used as a paint thinner, solvent and in varnishes and lacquers.

turquoise Blue mineral, hydrated basic copper aluminium phosphate, found in aluminium-rich rocks in deserts. Its crystal system is triclinic and it occurs as tiny crystals and dense masses. Its colour ranges from sky-blue and blue-green to a greenish grey, and it is a popular gemstone. Turquoise occurs as a secondary mineral in veins in association with aluminium IGNEOUS or SEDIMENTARY rocks that have undergone considerable alteration. Hardness 6; r.d. 2.7.

twins Two offspring that are born within a short time (usually minutes) of each other. In human beings, it is the most usual form of MULTIPLE BIRTH.

two-stroke engine Engine in which the operation of each piston is in two stages. This kind of operation is called the **two-stroke cycle**. Many smaller petrol engines use the two-stroke principle. In the two-stroke cycle, a piston moves up a cylinder to compress a fuel-air mixture in the top. At the same time, more of the mixture is sucked in below the piston through ports in the cylinder. A spark ignites the compressed mixture, causing an explosion. This sends the piston back down the cylinder. The piston forces the fresh fuel-air mixture out from beneath it and along a transfer port leading to the top part of the cylinder. The mixture forces the exhaust gases out from the top of the cylinder. The process then repeats. *See also* FOUR-STROKE ENGINE

tympanic membrane (tympanum) Medical term for the eardrum. It is a thin membrane that in mammals closes the inner end of the EAR canal (EUSTACHIAN TUBE), separating the outer ear from the middle ear. When sound waves vibrate the tympanic membrane, the vibrations are picked up in the middle ear by three small bones called OSSICLES, which pass them on to the COCHLEA in the inner ear. In frogs and other amphibians, the tympanic membrane is on the surface of the skin at the side of the skull.

Tyndall, John (1820–1893) Irish physicist who correctly suggested that the blue colour of the sky is due to the SCATTERING of light by particles of dust and other colloidal particles. He also studied the magnetic properties of crystals, and the scattering of light by colloidal particles in solution (now known as the **Tyndall effect**). By 1881 he had helped to disprove the theory of spontaneous generation by showing that food does not decay in germ-free air.

type metal Alloy of lead, containing from 2.5–12% of tin and 2.5–25% of antimony, used for printing type. Both tin and antimony lower the melting temperature, which is usually between 240° and 250°C (460–475°F). Antimony also hardens the alloy and enhances its moulding properties, reducing shrinkage during solidification.

typesetting Part of the PRINTING process in which metal or other type is set, or composed, into lines, paragraphs and pages. In early printing, wooden or metal type was set by hand. In hot-metal type processes, the LINOTYPE machine casts a complete line of molten type metal into a mould, and the MONOTYPE machine casts individual letters and spaces and arranges these in lines. Most type is now set using photographic or computerized processes. The text is put on computer and printed on sheet film. The film is used in making printing plates.

typhoon Name given in the NW Pacific Ocean to a HURRICANE, a violent tropical cyclonic storm.

tyre Air-filled rubber and fabric cushion that fits over the wheels of vehicles to grip the road and absorb shock. The pneumatic tyre was invented in 1845 by the Scottish engineer Robert Thomson (1822–73) but was not commonly used until the end of the century, solid tyres being more popular. The pneumatic tyre consists of a layer of fabric – either rayon, nylon or polyester – surrounded by a thick layer of rubber treated with chemicals to harden it and decrease wear and tear. The angle of the grain of the fabric to the axle defines the kind of tyre: crossply tyres have the grain at an angle of about 55°, the more modern radial-ply (radials) have the grain parallel to the axle. The latter last longer and are safer in bad weather, but they are less shock-absorbent and more expensive.

Tyrian purple Natural dye obtained from the shell of a small, purple, marine mollusc, *Murex brandaris*, which must be crushed and exposed to light. Originally made from Mediterranean rock whelks by the Phoenicians, it takes its name from the ancient city of Tyre where it was reputedly discovered.

tyrosine ($OHC_6H_4CH_2CH(NH_2)COOH$) Crystalline AMINO ACID found in proteins. It was isolated in 1849 from CASEIN.

TWO-STROKE ENGINE

In a two-stroke engine the upstroke (A) opens the inlet port (1) and compresses the mixture already in the cylinder. A spark plug ignites the fuel (B). After combustion the descending piston (C) first opens the exhaust port (2), and then the transfer port (3) letting in fresh fuel-air mixture. In this way the piston also acts as the engine's valves, and oil is added to the fuel to lubricate it. In many two-stroke engines the cylinder is air-cooled via fins attached to it.

UFO Abbreviation of UNIDENTIFIED FLYING OBJECT
UHF Abbreviation of ULTRA HIGH FREQUENCY
ulna Long bone of the inner side of the forearm. At its upper end it articulates with the HUMERUS and with the RADIUS, the slightly shorter bone of the outer side of the forearm. At its other end it articulates only with the radius, not directly with the carpal bones of the wrist.
ultrabasic Igneous rock containing very little QUARTZ or FELDSPAR. Ultrabasic rocks are usually PLUTONIC in origin.
ultrafiltration Separation of small particles from a suspension or colloidal solution using filtration under pressure. Small molecules, ions and water are forced across a semipermeable membrane or colloidal filter against a concentration gradient. The larger molecules do not cross the membrane.
ultra high frequency (UHF) Radio waves in the frequency band 300–3,000MHz, with wavelengths of about 0.1–1m (0.3–3ft). UHF waves have a range of about 80km (50mi) and are used for TELEVISION broadcasting and tracking spacecraft.
ultrasonics Study of sound waves with frequencies beyond the upper limit of human hearing (above 20,000Hz). In medicine, ultrasonics are used to locate a tumour, to scan a pregnant woman's abdomen in order to produce a "picture" of the fetus and to treat certain neurological disorders. Other applications of ultrasonics include the agitation of liquids to form emulsions, detection of flaws in metals (the ultrasonic wave passed through a metal is reflected by a hairline crack), cleaning small objects by vibrating them ultrasonically in a solvent, echo sounding in deep water (*see* ECHO SOUNDER) and soldering aluminium.
ultraviolet astronomy *See* ASTRONOMY
ultraviolet lamp Lamp that emits radiation of wavelength 220–400nm (nanometres). It is used as a germicidal lamp, for fluorescent effects and for producing an artificial "suntan". It uses a mercury vapour discharge lamp, the walls of which are made of quartz, which is much more transparent to ultraviolet light than is glass.
ultraviolet radiation Type of ELECTROMAGNETIC RADIATION of shorter wavelength and higher frequency than visible light. Typical wavelengths range from roughly 4–400nm (nanometres), and those of visible light range from 400nm (violet) to 700nm (red). Ultraviolet radiation affects living matter in a number of important ways: it kills bacteria and many other parasites (and so is used medically to sterilize equipment); it tans the skin by stimulating the formation of the pigment MELANIN; and it helps to make VITAMIN D in the body, which plays a part in the prevention of rickets. Ultraviolet light also causes certain materials to fluoresce (emit visible light), notably the brighteners added to some detergents. Sunlight contains ultraviolet rays, most of which are filtered by the OZONE LAYER. If the ozone layer is weakened, enough ultraviolet can reach the ground to harm living things. Excessive exposure to sunlight can cause sunburn and skin cancer in people with fair skins.
Ulysses European Space Agency probe launched in October 1990 to study the polar regions of the Sun. It was thrown out of the ecliptic by a fly-by of Jupiter in February 1992 and looped over the south pole of the Sun in September 1994, moving over its north pole a year later.
umbilical cord Long, thick cord that connects a developing FETUS with the PLACENTA. The umbilical cord contains two large arteries and one vein. At birth, the cord is clamped and cut from the placenta; the part of the cord remaining on the baby's abdomen dries and falls off, leaving the scar known as the navel.
uncertainty principle In physics, law that states that an object does not have a measurable position and momentum at the same time, because the act of measuring disturbs the system. The product of these two uncertainties is always greater than PLANCK'S CONSTANT. This principle was first stated by Werner HEISENBERG.
uncommitted logic array (ULA) In computing, a CIRCUIT consisting of logic gates connected in all possible ways. It consists of a single chip (INTEGRATED CIRCUIT) in which all the elements are identical. To make a circuit for a particular function, the unwanted connections are burned away. Alternatively, the chip may merely have all the elements in place and the required connections are then added to complete the required circuit.
unconformity In geology, break in the time sequence of rocks layered one above the other. The time gap may be caused by interruptions in the deposition of sediment, ancient erosion, earth movements, or other activity. An angular unconformity occurs when successive strata (*see* STRATUM) dip at different angles.
underground railway Transport system used in urban areas. The first underground railway was opened in London, England, in 1863. It was steam-powered and carried passengers between Farringdon and Paddington. Today, many cities world wide have underground railway systems for passenger transport. They are all electrically powered. Underground trains provide fast transportation for large numbers of people for short journeys that would take much longer by bus through crowded city streets. Some underground systems were once used for carrying goods. This is no longer generally convenient, although London still has a small underground railway that carries mail. In many cases, parts of underground railway systems run in cuttings, and over roads and rivers.
ungulate MAMMAL with hoofed feet. Most ungulates, including cattle, sheep, pigs and deer, are members of the order ARTIODACTYLA (with an even number of toes). The order PERISSODACTYLA (ungulates with an odd number of toes) consists of horses, tapirs and rhinoceroses. The orders PROBOSCIDEA and Hyracoidea, collectively known as sub-ungulates, contain elephants and hyraxes.
unidentified flying object (UFO) Any flying object that cannot readily be explained as either a man-made craft or a natural phenomenon. Reports of UFOs have been documented since ancient times. With the development of aeronautics and astronautics the number of sightings has increased enormously. The vast majority of supposed UFO sightings have various rational explanations, including spy planes, optical floaters (in the observer's eye), weather balloons and artificial satellites.
unified field theory Attempt to extend the general theory of RELATIVITY to give a simultaneous representation of both gravitational and electromagnetic fields. A more comprehensive theory would also include the strong and weak nuclear forces. Although success has been achieved in unifying the electromagnetic and weak nuclear forces, the general problem is still unsolved. *See also* GRAND UNIFIED THEORY (GUT)
units *See* WEIGHTS AND MEASURES
universal joint Connection between two shafts that allows them to rotate together even when they are at an angle to each other.
universal set In mathematics, a SET containing all the elements with a certain property. Also the name given to a hypothetical set which could include everything. Such an all-encompassing set contradicts the basic notion of a set.
universal time System of time reckoning based

ULTRASONICS

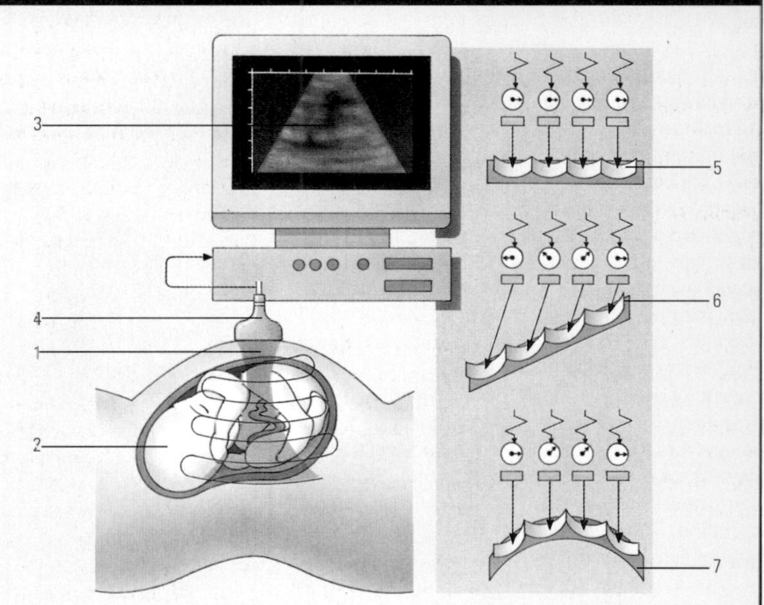

Ultrasonic, or ultrasound, scanners send out beams of sound (1) and read the returning echoes to build up a picture of structures below the surface (2). They are commonly used to view fetuses in the womb with the picture displayed on a real-time monitor (3). The sound waves used are above the range of human hearing and are created by piezoelectric crystals in the handset (4). These crystals change shape when the voltage is switched off. An ultrasonic scanner uses an oscillating voltage to make the crystals vibrate, producing sound of the right frequency. The process is reversed when an echo hits the transducer and the sound wave is converted back into a voltage, which the monitor reads to build up an image. Each handset has dozens of the piezoelectric crystals at its face and by altering the timing of the oscillations of the voltage, the beam can be steered. When the voltage hits all the crystals at the same time a flat beam is issued (5). By sending the voltage to one end of the row of crystals before the other the beam can be steered (6). The beam of sound can also be focused by hitting the end crystals before the ones on the centre (7).

on the mean solar day, the average interval between two successive transits of the Sun across the Greenwich (0°) meridian.

Universe Aggregate of all matter, energy and space, consisting of vast cold, empty regions with a distribution of high-temperature stars and other objects grouped in GALAXIES. On a large scale the Universe is considered uniform: it is identical in every part. It is believed to be expanding at a uniform rate, the galaxies all receding from each other. The origin, evolution and future characteristics of the Universe are considered in several cosmological theories. Recent developments in astronomy imply a finite Universe, as postulated in the BIG BANG theory. *See also* STEADY-STATE THEORY; COSMOLOGY

unknown Variable with values that are to be found by solving a given EQUATION or set of equations.

unnilquadium *See* DUBNIUM

unsaturated compound In organic chemistry, compound in the molecule of which two or more carbon atoms are linked, or bonded together, with double or triple bonds. Simple examples are ethene (ethylene) and ethyne (acetylene). By breakage of their double or triple bonds, unsaturated compounds can add on more atoms to their molecules – and for this reason are chemically reactive. *See also* SATURATED COMPOUND

upper atmosphere Region of the atmosphere, extending upwards from about 50km (30mi), which is free of disturbances caused by the WEATHER. It includes the MESOSPHERE, THERMOSPHERE and IONOSPHERE. At this altitude the air is rarefied, with temperatures ranging from −110°C (−170°F) in the lower regions to 250–1,500°C (500–2,700°F) higher up. The behaviour of the upper atmosphere is greatly affected by such extraterrestrial phenomena as solar and COSMIC RADIATION, which cause the gas molecules to produce the ionosphere, and atmospheric tides, which cause turbulence.

upwelling Process that brings water of greater density and lower temperature up to the surface of the ocean. It is especially characteristic of the w side of continents where winds blow parallel to the coast and the water carried away by the surface current is replaced by bottom water.

uracil One of the four bases in the NUCLEOTIDES that make up RNA (ribonucleic acid). A PYRIMIDINE derivative, uracil accounts for the major difference between RNA and DNA: in DNA uracil is replaced by thymine (the other three bases, ADENINE, CYTOSINE and GUANINE, are common to both RNA and DNA). In forming the double-stranded molecule of RNA, uracil pairs with adenine.

uraninite (pitchblende) Dense radioactive mineral form of uranium oxide, UO_2, the chief ore of uranium and the most important source for uranium and radium. The blackish and lustrous ore occurs in several varieties: crystallized varieties are called uraninite, massive varieties are called pitchblende. They usually contain small amounts of radium, lead, thorium, polonium, and sometimes the gases helium and argon. Both varieties are found around the world in veins (often quartz and silver), especially in the USA, Canada, Britain, Europe, Australia, Zaïre and South Africa. Hardness 5–6; r.d. 6.5–8.5 (pitchblende), 8–10 (uraninite)

uranium (symbol U) Radioactive metallic element, one of the ACTINIDE SERIES. It was discovered in 1789 by Martin KLAPROTH, and it is now important because of its use in nuclear reactors and bombs. Soft and silvery-white, uranium is the most abundant radioactive element in the Earth's crust, the chief ores being URANINITE, CARNOTITE, autunite and TORBERNITE. The ISOTOPE ^{238}U makes up more than 99% of natural uranium. Chemically uranium is a reactive metal: it oxidizes in air and reacts with cold water. It is the heaviest of the naturally occurring elements and is malleable and ductile. ^{235}U is fissionable and will sustain a neutron chain reaction as a fuel for reactors. ^{238}U is converted in certain nuclear reactors into the plutonium isotope ^{239}Pu, a nuclear fuel. ^{235}U and ^{239}Pu are needed for atom bombs, while ^{238}U and ^{235}U are also used to date rocks and determine the age of the Earth. Uranium is used to synthesize the TRANSURANIC ELEMENTS (atomic numbers greater than uranium). Properties: at.no. 92; r.a.m. 238.029; r.d. 19.05; m.p. 1,132°C (2,070°F); b.p. 3,818°C (6,904°F); most stable isotope ^{238}U (half-life 4.51×10^9 yr).

uranium oxide One of a series of compounds of which UO_2, U_4O_9, U_3O_8 and UO_3 are the most common, and of which U_3O_8 is the most stable. The latter is green, brown or black with an orthorhombic crystalline structure. URANINITE and PITCHBLENDE (containing UO_2 and UO_3) occur naturally and are used as a source of uranium.

uranium-thorium-lead dating Method of assessing geological age up to many millions of years. The proportions of URANIUM and its radioactive decay products, THORIUM and LEAD, show how long it is since the rock was laid down. *See also* DATING, RADIOACTIVE

Uranus Seventh planet from the Sun discovered (1781) by Sir William HERSCHEL. Uranus is visible to the naked eye under good conditions. Through a telescope it appears as a small, featureless, greenish blue disc. In size, mass, atmosphere and colour, Uranus resembles NEPTUNE. Like all the giant planets, it possesses a ring system and a retinue of SATELLITES. Uranus was the first planet to be discovered since ancient times. Although it had been observed on several occasions (and had once been catalogued as a star), its non-stellar nature was first recognized by Herschel. Like Pluto, Uranus's axis of rotation is steeply inclined, and its poles spend 42 years in sunlight followed by 42 years in darkness. Highly exaggerated seasonal variations are therefore experienced by both the planet and its satellites. The fly-by of the Voyager 2 probe in 1986 provided most of our current knowledge of the planet. The upper atmosphere is about 83% molecular hydrogen and 15% helium, with the other 2% mostly methane, whose strong absorption of red light gives Uranus its predominant hue. Other images obtained from the ground and from the Hubble Space Telescope have since revealed dark spots and bright clouds similar to those visible when Voyager flew past Neptune. The five largest satellites were known before the Voyager encounter, which led to the discovery of ten more. All 15 are regular satellites orbiting in or close to Uranus's equatorial plane. They are all darkish bodies composed of ice and rock. Two further moons were discovered by powerful telescopes in 1997. The main components of Uranus's ring system were discovered in 1977 and others were imaged by Voyager. The brightest and outermost is the Epsilon Ring; the innermost is a very diffuse sheet of material. Between the two are nine narrow, darkish rings.

urea Organic compound ($CO(NH_2)_2$), a white crystalline solid excreted in URINE. Most vertebrates excrete most of their nitrogen wastes as urea; human urine contains about 25 grammes of urea to a litre. Because it is so high in nitrogen, urea is a good fertilizer. It is also used to make urea-formaldehyde resins and barbiturates.

ureter In vertebrates, the long, narrow duct that connects the KIDNEY to the urinary BLADDER. The ureters transport URINE from the kidneys to the bladder, where it is stored until it is discharged by way of the URETHRA.

urethra Duct through which URINE is discharged from the bladder in mammals. Urine is produced in the KIDNEY, stored in the bladder until pressure in the bladder triggers specific neural responses that cause urine, under voluntary control, to be released through the urethra. In males the urethra is also the tube through which SEMEN is ejaculated.

Urey, Harold Clayton (1893–1981) US chemist. He became professor at Columbia University and in the same year (1934) was awarded the Nobel Prize for chemistry for his isolation of deuterium. Thereafter he worked on isotope separation at the universities of Chicago and California. Concerned that his work on isotopes had aided the construction of nuclear weapons, he later turned to geophysics.

uric acid Insoluble end product of PROTEIN metabolism. It is the main excretory substance in birds and reptiles, but only small quantities of it are normally produced by humans (who, like most mammals, mainly excrete soluble urea). An excessive build-up of uric acid in human blood causes gout, the deposition of urate salts around the joints.

urine Fluid filtered out from the bloodstream by the KIDNEY. It consists mainly of water, salts and waste products such as UREA. From the kidneys it passes through the URETERS to the BLADDER for discharging by way of the URETHRA. In DIABETES, urine contains glucose, and to diagnose the disease urine is tested for glucose.

urogenital system Organs comprising the body's urinary and reproductive systems. The urinary system, which provides an important route of excretion, consists of the KIDNEYS, URETERS, the BLADDER and URETHRA. In males, the reproductive system consists of paired TESTES located in an external pouch known as the scrotum; accessory glands; and the PENIS, the organ through which both urine and sperm are discharged to the exterior. In females, the reproductive system consists of paired OVARIES, one on each side of the pelvic cavity; FALLOPIAN TUBES, which provide a passage from the ovaries to the UTERUS, located in the pelvic cavity between the bladder and rectum; the CERVIX (lower part of the uterus); and the VAGINA. In females, the urethral and vaginal openings are separate but close to each other.

Ursa Major (Great Bear) Famous N constellation, the main pattern of which is known as the Plough or Big Dipper. This feature consists of seven stars. Five of the Plough stars make up a moving CLUSTER, a group of stars that share the same motion through the galaxy. There are many galaxies in Ursa Major.

Ursa Minor (Little Bear) Constellation that contains the north celestial pole. Its brightest star is Alpha (Polaris), the pole star, magnitude 2.0. The constellation's seven main stars make a pattern that gives the impression of a faint and distorted Plough, or Little Dipper.

uterus (womb) Hollow muscular organ located in the pelvis of female mammals. It protects and nourishes the growing FETUS until birth. The upper part is broad and branches out on each side into the FALLOPIAN TUBES. The lower uterus narrows into the CERVIX, which leads to the VAGINA. Its muscular walls are lined with mucous membrane (ENDOMETRIUM), to which the egg attaches itself after fertilization. *See also* MENSTRUAL CYCLE

utricle Fluid-filled chamber in the mammalian inner EAR. The three SEMICIRCULAR CANALS, the organs of balance, loop out from the utricle. Sensory cells within the utricle detect changes in the direction of movement of the ear (and hence the head). They also sense how fast the fluid in the canals is moving. All the information passes along part of the auditory nerve to the brain, which processes it to provide the sense of balance. *See also* EQUILIBRIUM SENSE

V

vaccination Injection of a VACCINE in order to produce IMMUNITY against a disease. In many countries today children are vaccinated routinely against infectious diseases such as polio, whooping cough, measles, mumps and German measles. *See also* IMMUNIZATION; JENNER, EDWARD

vaccine Agent used to give immunity against various diseases without producing symptoms. A vaccine consists of modified disease organisms, such as a live attenuated VIRUS (low in virulence), or a dead one (still able to induce the production of specific ANTIBODIES within the blood). The first vaccine to come into general use was for SMALLPOX, developed in the late 18th century. *See also* IMMUNE SYSTEM

vacuole Membrane-bounded, cavity within the CYTOPLASM of a CELL. Depending on the organism, vacuoles contain either liquid, gas or food particles. Vacuoles perform various functions. In single-celled organisms, such as *Amoeba*, **contractile** vacuoles fill with water before rapidly contracting and expelling excess water or wastes. In this way vacuoles perform OSMOREGULATION and excretion. In plants, vacuoles allow individual cells to increase in size without amassing bulk that would hinder cell METABOLISM.

vacuum Area of extremely low pressure. Interstellar space is a high vacuum, with an average density of less than 1 molecule per cubic centimetre; the best man-made vacuums contain less than 100,000 molecules per cubic centimetre. Evangelista TORRICELLI is credited with developing the first man-made vacuum in a mercury BAROMETER. In 1650 the German physicist Otto von Guericke (1602–86) invented the first vacuum pump. Vacuums are used widely in scientific research and industry, such as in the vacuum packing of foodstuffs.

vacuum deposition Method of producing a thin adherent coating, generally of a metal, on plastics, ceramics and other materials. The object to be "plated" is placed in a vacuum chamber that also contains a length of metal wire inside a heavy-duty coil. When a high-voltage electric current is passed through the coil, the metal vaporizes and coats the object. *See also* ELECTROPLATING

vacuum flask Container for keeping things (usually liquids) hot or cold. A vacuum flask is made with double, silvered glass walls separated by a near vacuum. This vessel, which has a close-fitting stopper, is held in an insulated metal or plastic case. It reduces the transfer of heat between the contents and the surroundings. The vacuum prevents heat transfer by CONDUCTION and CONVECTION, and the silvering on the glass minimizes heat transfer by radiation. The vacuum flask was invented (1892) by James DEWAR, and it is sometimes called a Dewar flask. It is also known as a Thermos flask, after the trade name of the first commercial brand.

vacuum pump Device that exhausts air from an enclosed space, which becomes a partial VACUUM. In the simplest vacuum pump, water or steam enters the chamber and it forces air out, thus creating a partial vacuum. The first vacuum pump was invented by the German physicist Otto von Guericke (1602–86).

vagina Portion of the female reproductive tract, running from the CERVIX of the UTERUS to the exterior of the body. Tube-like in shape, it receives the erect PENIS during sexual intercourse. The muscular walls of the vagina enable it to dilate massively to allow the passage of the baby during childbirth. *See also* SEXUAL REPRODUCTION

vagus nerve Tenth CRANIAL NERVE, running from the BRAIN to the abdomen. It contains motor, secretory and sensory fibres and supplies branches to the lungs, heart, stomach and other abdominal organs.

valence (valency) Measure of the "combining power" of a particular element, equal to the number of (individual) CHEMICAL BONDS one ATOM can form. The valence of an atom is determined by the number of ELECTRONS in the outermost (valence) shell. The valence of many elements is determined by their ability to combine with hydrogen or displace it in compunds (hydrogen has a valence of 1). For example, one carbon atom combines with four hydrogen atoms to make methane (CH_4) so the valence of carbon is given as 4. *See also* ATOMIC NUMBER; COVALENT BOND; IONIC BOND; OXIDATION-REDUCTION

valine ($[CH_3]_2CH(NH_2)COOH$) White, crystalline essential AMINO ACID found in proteins.

Valium Proprietary name for diazepam, a sedative drug in the BENZODIAZEPINE group. It is used in the treatment of anxiety, muscle spasms and EPILEPSY and may also feature as part of the premedication given to patients before surgery.

valley Elongated, gently sloping depression of the Earth's surface, commonly situated between mountains or hills. It often contains a stream or river that receives the drainage from the surrounding heights. A **U-shaped** valley was probably formed by a glacier, a **V-shaped** one by a stream. The term may also be applied to a broad, generally flat, area that is drained by a river, such as the Mississippi Valley.

valve, electronic *See* ELECTRON TUBE

valve In a petrol or diesel ENGINE, device that cyclicly opens and closes the intake and exhaust parts of an engine's combustion chamber or cylinder. It consists of a disc attached to a shank and held against a seat by a spring. A rotating cam pushes on the bottom of the shank, raising the valve from its seat and permitting gas to flow past it.

valves In anatomy, the structures that prevent the backflow of blood in the HEART and VEINS. Heart valves separate and connect the two atria and ventricles, the right ventricle and the pulmonary artery, and the left ventricle and the aorta.

vanadinite (lead chlorovanadate, $Pb_5Cl(VO_4)_3$) found in the upper zone of lead ore deposits. It occurs in hexagonal system prisms and needle-like

VALVE

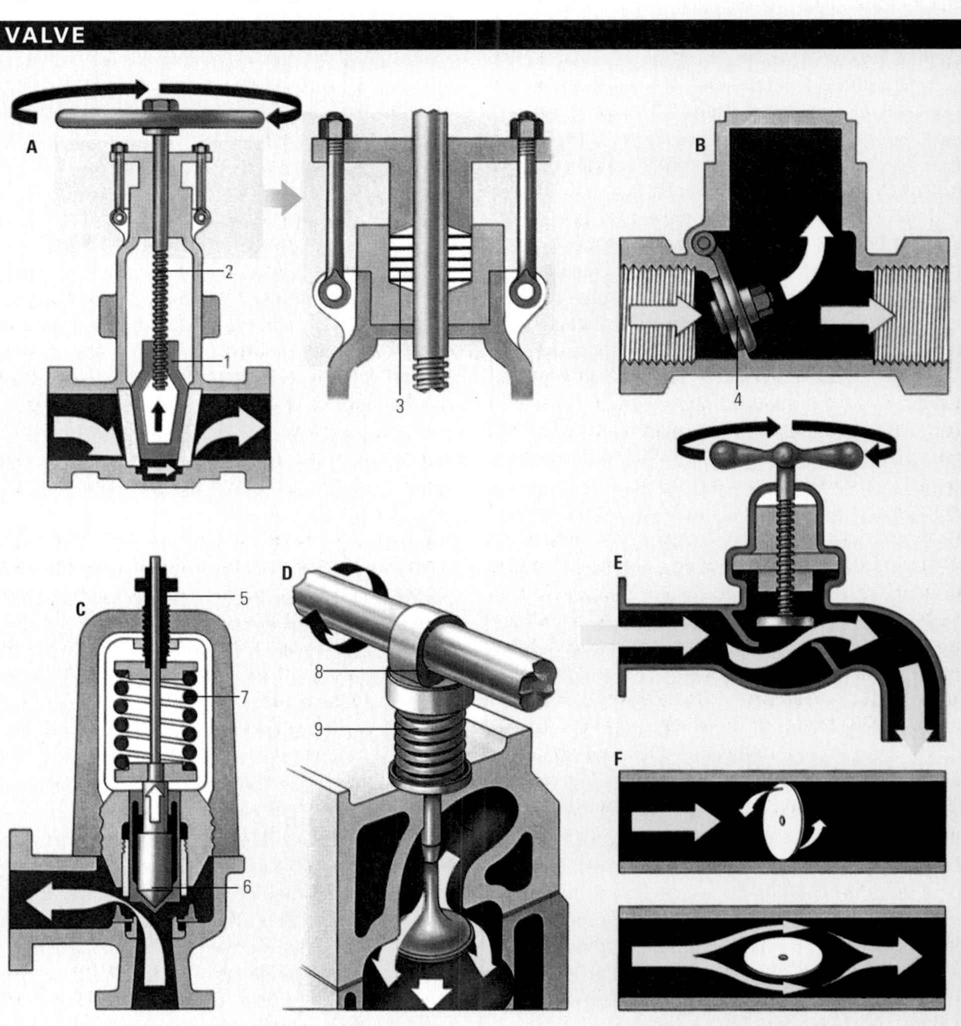

(A) is a gate valve in which the gate (1) is lifted by a screw thread on the stem (2). Gate valves are usually operated closely or wide open. The packing (3) kept under pressure forms the stuffing box (enlarged) which acts as a seal. The swing check valve (B) is an automatic device for preventing a reverse flow. During normal flow, the disc (4) is kept horizontal but any reversal causes the disc to drop. (C) is a safety valve used for boilers and steam above a pressure set by the adjusting screw (5) raises the plug (6) against the spring (7) thus relieving the pressure. The poppet valve (D) is moved by the cam (8) and returned by the spring (9). The domestic tap (E) is one of a type of pipe valve constructed with a disc that seats on to a circular aperture. (F) shows a butterfly valve that is used for throttling and has applications from hydroelectric systems to car throttles.

V

masses, resinous orange to brown. It is a commercial source of VANADIUM; hardness 2.7–3; r.d. 7.

vanadium (symbol V) Silver-white, TRANSITION ELEMENT, first discovered in 1801. A malleable and ductile metal, vanadium is found in IRON, LEAD and URANIUM ores and in coal and petroleum. It is used in steel alloys to add strength and heat resistance. Vanadium pentoxide is an important oxidation CATALYST in the chemical industry and is used in glass, ceramics and dyes. Chemically, vanadium reacts with oxygen and other nonmetals at high temperature. Properties: at.no. 23; r.a.m. 50.9414; r.d. 6.1 at 18.7°C; m.p. 1,890°C (3,434°F); b.p. 3,380 °C (6,116°F); most common isotope ^{51}V (99.76%).

Van Allen, James Alfred (1914–) US physicist. During World War 2, he helped to develop the radio proximity fuse, which used radio waves to detonate an explosive projectile when it neared a target. After the war, Van Allen supervised experiments with captured German V-2 ROCKETS, leading to research with artificial SATELLITES and his discovery of the two zones of radiation encircling the Earth, known as the VAN ALLEN RADIATION BELTS. They were discovered by a detector in the first US satellite, Explorer I, which he helped launch in 1958.

Van Allen radiation belts Two doughnut-shaped regions of RADIATION trapped by Earth's MAGNETIC FIELD in the upper ATMOSPHERE, named after James VAN ALLEN, who discovered them in 1958. The belts contain particles carrying energies of *c*.10,000 to several million electron volts. Artificial satellites are insulated against this radiation. The inner belt (of ELECTRONS and PROTONS) extends from *c*.1,000km to 4,000km (600–2,500mi) above the Equator. The outer belt (of ELECTRONS) extends from *c*.15,000 to 25,000km (9,000–15,000mi) above the Equator. It is thought that the particles come from SOLAR FLARES and are carried by the SOLAR WIND.

Van de Graaff generator Machine that generates high voltages by concentrating electrical charges on the outside of a hollow CONDUCTOR. The Cockcoft-Walton ACCELERATOR, built by John COCKCROFT and Ernest WALTON, produced high voltages through a bank of charged CAPACITORS placed in series. US physicist Robert Van de Graaff (1901–67) improved on this design by spraying positive or negative charges onto a continuously moving belt that carried them up to a large, hollow metal sphere where the voltage built up. In this manner, an applied voltage of *c*.50,000 volts generated up to 1 million electron volts. Today, the Van de Graaff generator is used primarily to "inject" particles into more powerful, linear accelerators. *See also* LAWRENCE, ERNEST

Van der Waals, Johannes Diderik (1837–1923) Dutch physicist who was awarded the 1910 Nobel Prize for physics for his work on gases and the gas equation which he derived. The Van der Waals equation takes into account intermolecular attraction and repulsion, which were ignored by the KINETIC THEORY of gases. *See also* IDEAL GAS LAW; VAN DER WAALS FORCES

Van der Waals forces Weak forces of mutual attraction that contribute towards cohesion between neighbouring ATOMS or MOLECULES. They are named after Johannes VAN DER WAALS who investigated such phenomena during the 19th century. Van der Waals forces are caused by the distribution of electrons in neighbouring atoms or molecules.

van't Hoff, Jacobus Henricus *See* HOFF, JACOBUS HENRICUS, VAN'T

Van Vleck, John Hasbrouck (1899–1980) US mathematician and physicist. A professor at Harvard University, he studied how electrons behave in non-crystalline, magnetic materials. Van Vleck was the first scientist to use QUANTUM MECHANICS to explain the phenomenon of MAGNETISM. He later developed the LIGAND field theory, which explains chemical bonding in molecules, and studied free molecules by means of their spectra. For his work on magnetism, Van Vleck shared the 1977 Nobel Prize for physics with the US physicist Philip Anderson and the British physicist Nevill MOTT.

VAN DE GRAAFF GENERATOR

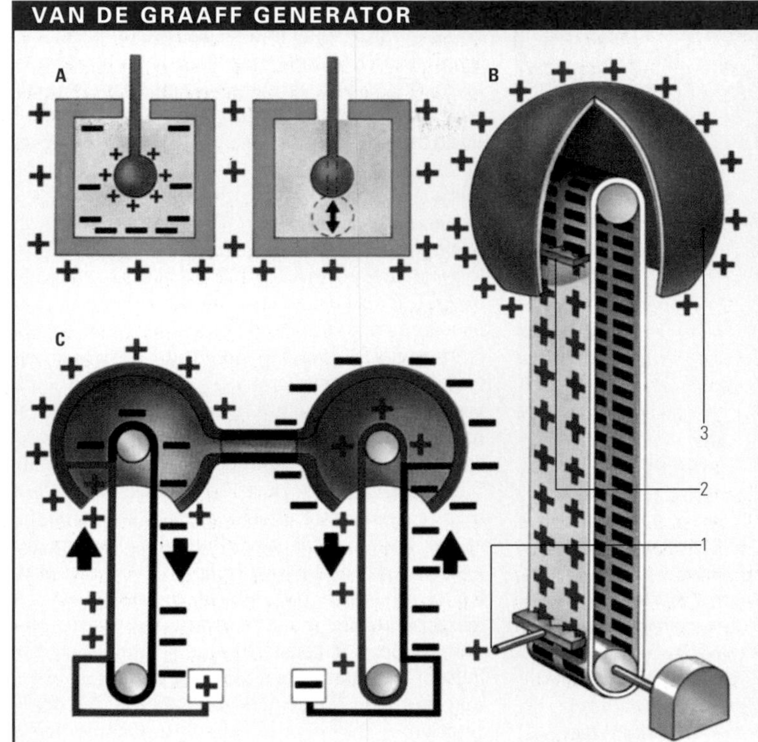

Extremely high voltages can be produced with the Van de Graaff generator (B). If a body having an excess of positive ions is placed inside a container, the inside acquires electrons (A) and the outside an equal number of positive ions. If the charged body touches the inside, all the free electrons flow into it, thus making it neutral. The outside of the container still retains its positive ions. In the Van de Graaff generator, positive ions are sprayed from a suitable source (1) onto an endless conveyor belt which carries them inside a metal sphere. The belt connects to the inside wall through a conductor in the form of a comb (2), thus permitting an electron flow to the belt. This causes positive ions to form on the sphere's outside wall (3). The effect may be enhanced by using two generators connected as shown (C).

vaporization (volatilization) Conversion of a liquid or solid into its vapour, such as water into steam. When heated, some solids (such as ammonium chloride), pass directly into the vapour state, a process known as SUBLIMATION. Vaporization of a solid or liquid can also be achieved by lowering the surrounding pressure or by EVAPORATION.

vapour Gas that is below its CRITICAL POINT and so can be liquefied by an increase in PRESSURE. It does not obey the IDEAL GAS LAW.

vapour pressure Pressure exerted by a vapour when it evaporates from a liquid or solid. In the instance where as many molecules leave to form vapour as return (in an enclosed space), this EQUILIBRIUM is termed a **saturated** vapour pressure. When a solid is dissolved in a liquid, the vapour pressure of the liquid is reduced by an amount proportional to the solid's RELATIVE MOLECULAR MASS. This is important in a chemical analysis.

Varenius, Bernhardus (1622–50) German geographer. He is celebrated for his *Geographia generalis* (1650), in which he attempted to structure geography on a scientific basis and coordinate its various branches. Varenius developed the concept of what is now known as regional geography.

variable In mathematics, symbol used to represent a quantity which may take any of several values in a range. For example, in the expression $y = x^2 + x + 1$, the quantity x may be assigned the value of any real number; here x is said to be an **independent** variable, and y is a **dependent** variable because its value depends on the value of x. Types of variables include SCALAR, VECTOR and MATRIX.

variable In statistics, item that may change or vary – unlike a constant, which maintains a given value. The mark given to each student in a class is a variable: it changes from student to student.

variable star Star whose brightness varies with time. The *c*.30,000 known variable stars fall into three broad classes according to their light curve: pulsating variables, eruptive variables and eclipsing varaibles. Most variable stars are **pulsating** variables, which have slight instabilities that cause the star to alternately contract and expand. The majority of pulsating variables are long-term variables, such as the RED GIANTS. CEPHEID VARIABLES are unusual short-term pulsating variables. Most short-term variables are found in STELLAR CLUSTERS. **Eruptive** variables, such as nova and supernova, are extremely unstable stars whose fluctuations in brightness are unpredictable. Both pulsating and eruptive variables are **intrinsic** variables – their variations are due to inherent structural features. **Eclipsing** variables are extrinsic variables: their apparent fluctuation in brightness is due to the fact that they are BINARY STARS.

variation In biology, differences between members of the same SPECIES. Variation naturally occurs due to HEREDITY and to differences in the environment during development. The difference may be in physical appearance, behaviour or fertility. *See also* ADAPTATION; EVOLUTION

variegation Term used to describe parts of plants, especially leaves, that display different colours. It occurs naturally, mostly as the result of a lack of the green pigment CHLOROPHYLL, the most common effect being yellow, cream or white patches, streaks or spots on green leaves. Variegation is exploited for ornamental purposes, but such decorative plants are usually weaker than the uniformly coloured plants. The variegated petals of some flowers can often be the result of mutation or viral infection.

variscite Glassy white to green phosphate mineral, hydrous aluminium phosphate with iron impurities, found in aluminium-rich rocks near the Earth's surface. It occurs as crystals in the orthorhombic system. It resembles TURQUOISE, but is greener, and is used in jewellery. Hardness 3.5–4.5; r.d. 2.5.

varnish Solution of a RESIN or a PLASTIC that dries to form a hard, transparent, protective and often decorative coating. Varnishes may have a matt or glossy finish. Pigments are often added to colour the varnish. Drying takes place by evaporation of the solvent or

VEIN

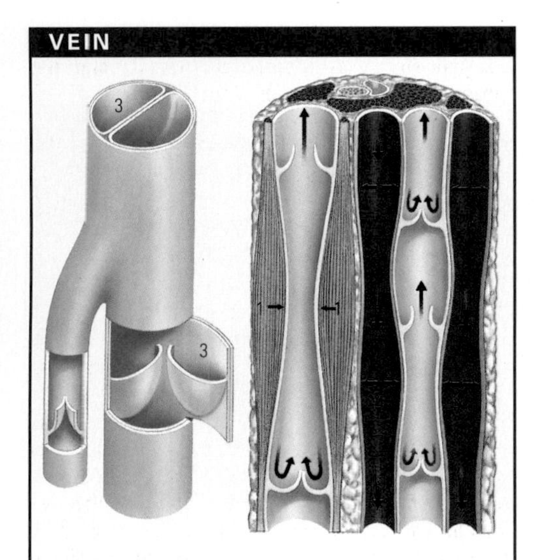

Veins carry blood to the heart. The returning venous blood moves slowly due to low pressure, and the veins can collapse or expand to accommodate variations in blood flow. Movement relies on the surrounding muscles, which contract (1) and compress the vein. Pulsation of adjacent arteries (2) has a regular pumping effect. Semilunar valves (3) are found at regular intervals throughout the larger veins and these allow the blood to move only in one direction.

by chemical action. French polish is a varnish made by dissolving SHELLAC in alcohol (ethanol). It dries when the alcohol evaporates into the air.

varve Term applied to a layer of sediment deposited in a single year in a body of still water. Specifically a varve consists of two layers of sediment, a coarse layer deposited in the summer and a fine layer deposited in the winter in glacial melt-water lakes. Varves have been used to date the age of Pleistocene glacial deposits.

vascular bundle Strand of conductive tissue that transports water and dissolved mineral salts and nutrients throughout a vascular plant. Vascular bundles are the equivalent of the blood and lymph vessels found in animals. They extend from the roots, through the stem, and out to the leaves. They consist of two types of tissue: XYLEM, which conducts water from the roots to the shoot and is located towards the centre of the bundle, and PHLOEM, which conducts salts and nutrients and forms the outer regions of the bundle. In plants that exhibit SECONDARY THICKENING, the xylem and phloem are separated by a thin layer of CAMBIUM, from which new vascular tissue is generated.

vascular plant Plant with vessels or tubes to carry water and nutrients within it. All higher plants – FERNS, CONIFERS and FLOWERING PLANTS – have a vascular system (XYLEM and PHLOEM). This system exists in roots, stems and leaves.

vascular system Network of vessels through which fluids circulate in the body of an animal or plant. They include BLOOD vessels (arteries, veins and capillaries) in the blood vascular system and lymph vessels in the LYMPHATIC SYSTEM. The network of tubes that carry fluid and nutrients in the PHLOEM and XYLEM tissues of plants is also called a vascular system or VASCULAR BUNDLE.

vas deferens Long, narrow tube that carries SPERM from the TESTES, around the bladder, and opens into the seminal vesicles, where the sperm are stored until ejaculation.

vasoconstrictor Any substance which causes constriction of BLOOD VESSELS and, therefore, decreased blood flow. Examples include NORADRENALINE, ANGIOTENSIN and the hormone VASOPRESSIN (also known as antidiuretic hormone). Vasoconstricting drugs are used for a number of reasons, such as to raise the BLOOD PRESSURE in circulatory disorders or for treating SHOCK.

vasodilator Any substance that causes widening of the blood vessels, permitting freer flow of blood. Vasodilator drugs are mostly used to treat HYPERTENSION and ANGINA.

vasopressin HORMONE secreted by the posterior lobe of the PITUITARY GLAND situated under the brain. It is also called antidiuretic hormone or ADH, its action being to prevent excessive water loss by reabsorption of water in the KIDNEY tubules during the filtration of urine.

vector In mathematics, a quantity that has both magnitude (size) and direction, as contrasted with a SCALAR, which has the property of magnitude only. For example, the VELOCITY of an object is specified by its speed and the direction in which it is moving at any instant. Similarly, a force has both magnitude and direction, hence weight (the force of gravity on a body) is a vector. In contrast, mass and temperature are scalar quantities. A vector at a point in three-dimensional space is represented by a list of three numbers which give the direction and magnitude (the value of which depends also on the position of the point in question). Geometrically, the three values specify an arrow with direction and magnitude. A vector must have two or more entries. *See also* PARALLELOGRAM OF VECTORS

vector analysis Study of the mathematical properties of VECTORS. The addition of two vectors is the diagonal of a parallelogram formed with the two vectors as adjacent sides. Multiplication of a vector by a SCALAR a gives a vector with the same direction but its magnitude multiplied by a. Two vectors can also be multiplied, and this may occur in two ways: the scalar, or dot product of vectors a and b (written $a.\ b$) is a scalar given by $ab \cos\theta$, θ being the angle between the vectors and a and b being their magnitudes; the vector, or cross product (written ab or $a \times b$), is a vector given by $ab \sin\ \theta$, directed at right-angles to the plane of a and b.

vector In disease, any agent that carries an infectious agent from one host to another. It may be an insect such as a mosquito or flea, or an inanimate object such as a cup. It also may serve as an intermediate host for the infectious agent.

vegetative reproduction (vegetative propagation) Form of ASEXUAL REPRODUCTION in higher plants. It involves an offshoot or a piece of the original plant (from leaf, stem or root) separating and giving rise to an entire new plant. It may occur naturally, as in strawberries reproducing by runners, or artificially, as in a house plant cutting yielding a new plant. The reproductive structure may include a food store, as in a BULB, CORM or TUBER.

vein In mammals, vessel that carries deoxygenated blood to the heart. An exception is the pulmonary vein, which carries oxygenated blood from the lungs to the left upper chamber of the heart. Veins have valves on their inner walls to prevent any backflow of blood. *See also* ARTERY; VENA CAVA

velocity Rate of motion of a body in a certain direction. Its symbol is v. Velocity is a VECTOR (having magnitude and direction), whereas speed, which does not specify direction, is called a SCALAR.

velocity ratio *See* DISTANCE RATIO

vena cava Main VEIN of vertebrates that supplies the HEART with deoxygenated blood, emptying into its right atrium.

veneer Extremely thin sheet of wood, or a thin sheet of a precious material such as ivory of tortoise-shell, which gives furniture or other objects the appearance of being more valuable than they are. Plain, cheaper woods or substitutes (such as plywood, chipboard or blockboard) may be completely covered in a richly coloured or grained veneer to give it the appearance of a decorative hardwood. Veneers may also be used as decorative shapes inlaid into a surface.

Venera space missions Series of SPACE PROBES sent to Venus by the Soviet Union after 1965. **Venera 3**, launched in 1965, was the first spacecraft to make an impact on another planet (Venus, 1966). Veneras 9 and 10 made soft landings on Venus in 1975 and transmitted pictures and atmospheric data for an hour. *See also* SPACE EXPLORATION

Vening Meinesz, Felix Andries (1887–1966) Dutch geophysicist and geodesist who developed (1923) a method of measuring gravity in ocean basins. He is also noted for research on the effect of solar movements on the Earth's surface and his studies of convection currents within the Earth.

Venn diagram Diagrammatical representation of the relations between mathematical SETS or logical statements, named after the British logician John Venn (1834–1923). The sets are drawn as geometrical figures, usually circles, which overlap whenever different sets share some elements.

venom, snake Toxic substance produced in the poison glands of snakes and injected into the victim through ducts in or along their fangs. Many venoms are dangerous to people and some can be lethal unless counteracted by antiserums. The effects of snake venom vary according to the species and the constituents of the poison. Blood coagulation, respiratory effects and haemorrhage are the commonest.

ventifact In geology, any stone that has been shaped and sculptured by windblown sand.

ventilation In biology, the process by which air or water is taken into and expelled from the body of an animal and passed over a surface across which GAS EXCHANGE takes place. Ventilation mechanisms include BREATHING, by which air is drawn into the LUNGS for gas exchange across the wall of the alveoli; the movements of the floor of a FISH's mouth, coupled with those of its GILL covers, which draw water across the gills; and the pumping movements of the ABDOMEN of some insects, which draw air through SPIRACLES that carry it to the tissues.

ventilator In medicine, a machine that helps an ill or injured person to breathe. A **life-support machine** is a positive-pressure ventilator. It works by forcing air along a tube into the patient's lungs; when the pressure is regularly released, the patient breathes out. An **iron lung** is a negative-pressure ventilator. A chamber encloses the patient's chest and, as the pressure in the chamber is decreased, the chest wall expands and "sucks" air into the lungs. When the pressure is regularly re-applied, the chest wall moves back and the patient breathes out.

ventral Describing that part of an organism that normally faces the ground. In plants, the ventral surface of a leaf is the under surface (which usually faces away from the light). In fish, the ventral fin is the fin pointing downwards. In four-legged animals, it is the lower or under surface. In humans and other bipedal animals, it is the forward-facing surface, more usually described as anterior. *See also* DORSAL.

ventricle Either of the two lower chambers of the HEART. The term is also used for the four fluid-filled cavities within the BRAIN.

Venturi tube Mechanism for mixing a fine spray of liquid with a gas or measuring fluid flow. In an automobile CARBURETTOR, petrol is passed through a jet into two, narrowing sections of tube joined by a slim throat. In the throat, the petrol mixes with air from the INTERNAL COMBUSTION ENGINE. The pressure differential produced by liquid flowing through the constriction gives a measure of its rate of flow. It is named after its inventor, Italian physicist G.B. Venturi (1746–1822).

Venus Second planet from the Sun and almost as large as the Earth. Visible around dawn or dusk as the so-called **morning star** or **evening star**, it is the most conspicuous celestial object, after the Sun and Moon. At its brightest it is even visible to the naked eye in the daytime, when the Sun is high in a clear sky. There is no known satellite. A telescope shows the planet's dazzling yellowish-white cloud cover, with faint markings. Venus' very high surface temperature was indicated by measurements at radio wavelengths in 1958, but SPACE PROBES were needed to reveal more about the surface. A gently undulating plain covers two-thirds of Venus. Highlands account for a further quarter, and depressions and chasms the remainder. Although there is a mountain range and some craters, most of the surface features are volcanic in origin and Venus is almost certainly still volcanically active. The atmosphere consists of 96% carbon dioxide and 3.5% nitrogen, with traces of helium, argon, neon and krypton.

verdigris Green, basic COPPER carbonate which forms on objects made of copper or bronze that have been exposed to atmospheric corrosion. The word also refers to the blue-green powder (copper ETHANOATE, or acetate) formed from metallic copper and ethanoic (acetic) acid. *See also* PATINA

vermiculite Clay mineral. Its flakes are light and are used in plaster and insulation, and as a packing material. The mineral is also used widely for conditioning soil and as a starting medium for seeds.

vernal equinox Spring EQUINOX in the N Hemisphere (September 23 in S Hemisphere).

Verneuil process Method for making gemstones artificially, particularly rubies and sapphires. Both of these gems are essentially forms of aluminium oxide (Al_2O_3) coloured with small amounts of other metal oxides. These are fused together in an oxyhydrogen flame, to form molten droplets of the coloured substance, which builds up gradually into a crystalline boule that is cut to make the gemstone.

Vernier scale On instruments for measuring length, small scale that moves alongside the main scale. Named after its French inventor Pierre Vernier (1580–1637), its function is to allow more precise measurement. On MICROMETERS the vernier scale is on the stem, at right-angles to the spiral main scale.

vertebra One of the bones making up the SPINE (backbone) or vertebral column. Each vertebra is composed of a large, solid body from the top of which wing-like processes project to either side. It has a hollow centre through which the SPINAL CORD passes. The human backbone is composed of 26 vertebrae (the five sacral and four vertebrae of the coccyx fuse together to form two solid bones) which are held together by ligaments and INTERVERTEBRAL DISCS.

vertebral column *See* SPINE

vertebrate Animal with individual discs of bone or cartilage called VERTEBRA surrounding or replacing the embryonic NOTOCHORD to form a jointed backbone that encloses the spinal column. The principal division is between the aquatic, fish-like forms and partly land-adapted forms (AMPHIBIANS), and the wholly land-adapted forms (REPTILES, BIRDS and MAMMALS), although some mammals, such as whales, have adapted for a totally aquatic existence. Birds and mammals are the only warm-blooded vertebrates with circulatory, respiratory and excretory systems that allow for constant high body temperatures. Phylum CHORDATA; subphylum Vertebrata.

vertex (pl. vertices) Point at which two sides of a triangle or other polygon intersect, or at which three or more sides of a pyramid or other polyhedron intersect. It is also the point of a cone.

very high frequency (VHF) Range or band of RADIO WAVES of frequencies between 30 and 300MHz and of wavelength between one and 10m (3–33ft). This band is used for TELEVISION and FREQUENCY MODULATION (FM) radio broadcasts to provide high-quality reception.

Vesalius, Andreas (1514–64) Flemish anatomist. His *De Corporis Humani Fabrica*, stating the results of his study and dissection of the human body, marked the beginning of modern anatomy.

vesicle Small sac within the CYTOPLASM of a CELL, usually containing containing serous fluid. *See also* GOLGI BODY

vestigial structure In biology, an organ or limb that is deformed or degenerate in appearance and no longer has any recognizable function. Human vestigial structures include the tonsils and the appendix. *See also* DEGENERATION

vesuvianite *See* IDOCRASE

VHF Abbreviation of VERY HIGH FREQUENCY

viaduct BRIDGE (usually resting on several narrow, concrete or masonry arches) with high supporting piers, which carries roads or railways. Steel bridges comprising short spans carried on high towers are also referred to as viaducts. *See also* AQUEDUCT

video Used in TELEVISION and computing to refer to electronic vision signals, and to equipment and software associated with visual displays. The picture component of a television signal is often referred to as the video. Video (or visual) display unit (VDU) is another name for a COMPUTER screen and its associated equipment. *See also* VIDEO RECORDING

video disc Vinyl disc coated with a reflective metallic surface, on one side of which a spiral of microscopic pits is etched corresponding to digital information that can be picked up by a laser scanner

VIDEO

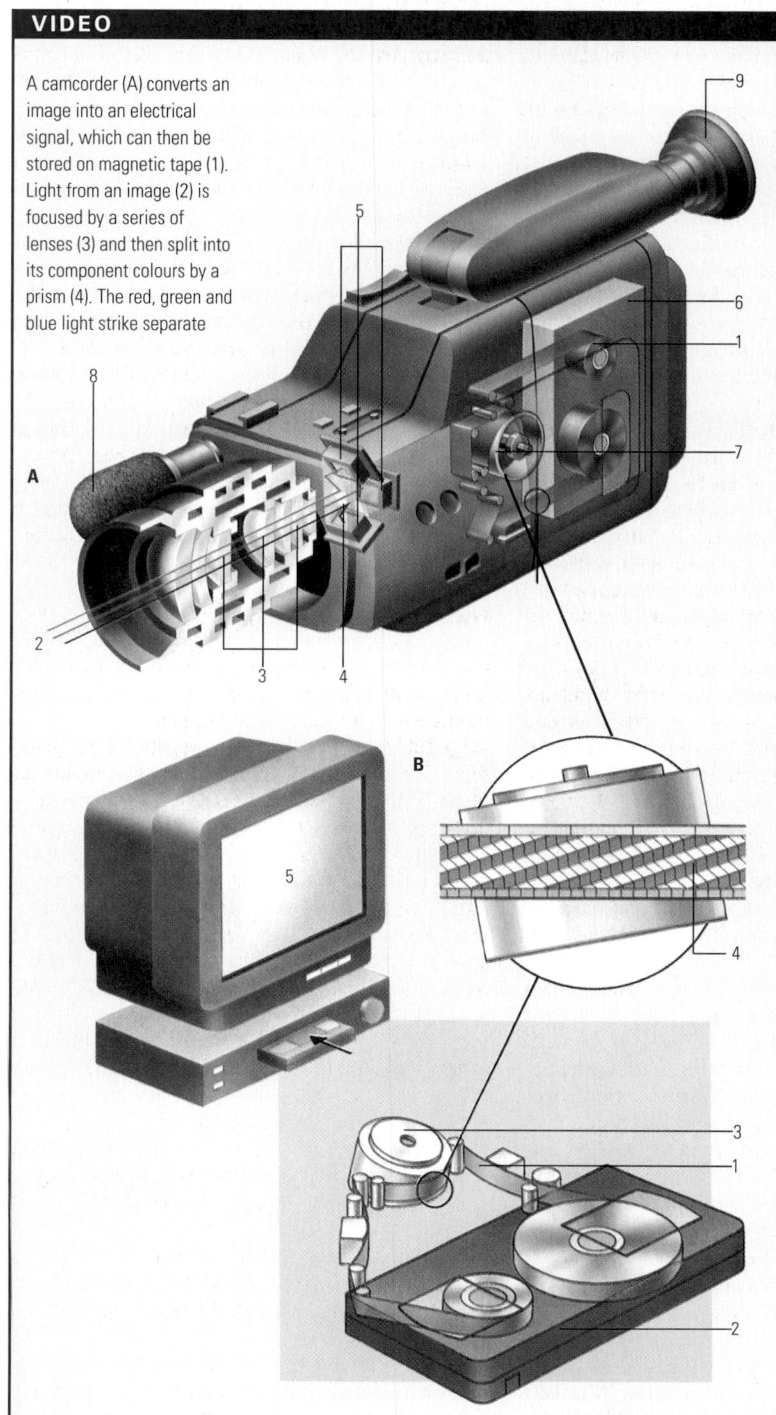

A camcorder (A) converts an image into an electrical signal, which can then be stored on magnetic tape (1). Light from an image (2) is focused by a series of lenses (3) and then split into its component colours by a prism (4). The red, green and blue light strike separate light-sensitive chips (5) that reproduce them in electronic form. The magnetic tape is housed in a protective case (6) that opens when inserted into the camcorder. The recording head (7) is angled and records information onto the tape in diagonal bands of magnetic particles. The helical scanning allows more information to be stored on a length of tape. A microphone (8) picks up sound, which is laid down in parallel to the visual information. Camcorders have a small television screen in the eyepiece (9) that allows the operator to play back and review the pictures taken. The camcorder can also be plugged directly into the television and the images played back.

A video recorder (B) reads and records onto magnetic tape (1). The tape is housed in a protective cassette (2) that opens when inserted into the recorder. The tape is pulled across an angled recording head (3). To play a tape the recording head scans the diagonal bands of magnetic particles on the tape (4) passing the information stored there to the television (5). To record the head magnetizes particles on the tape in the correct sequence. The soundtrack is laid in parallel to the visual code.

and converted electronically to video pictures and sound. Commercially marketed since 1978, most video discs are 30cm (12in) in diameter and consist of recordings of feature films or music presentations. Since the late 1980s, however, they have been almost entirely superseded by the smaller, more comprehensive type of COMPACT DISC (CD) called a CD-ROM. *See also* RECORDING

video game Game using electronically generated images displayed on a screen. Some video games test the skill of a single player, whilst other games allow two or more players to compete. High-quality graphics can produce good simulations of motor racing, football and other games. *See also* VIRTUAL REALITY

videotape Magnetic tape that records TELEVISION signals. The signals are passed through the recording head (an ELECTROMAGNET) of a videotape recorder, so that they fix, by INDUCTION, a magnetic pattern on the tape. These patterns generate the original signals when the tape passes the replay head (also an electromagnet).

videotext General term for the different methods by which information can be brought up onto a TELEVISION screen. Information that is transmitted by the broadcasting authority in parallel with the ordinary TV signals, and that may be screened simultaneously with or independently from other channels, is known as **teletext**. The system that brings information to the screen from a computer databank via a telephone landline is called **videotex** (formerly viewdata); the telephone link provides an interactive facility that allows armchair shopping and banking.

Vieta, Franciscus (1540–1603) French mathematician. He devised methods for solving algebraic equations to the fourth degree and discovered several trigonometric results. He introduced the terms COEFFICIENT and "negative".

Vigneaud, Vincent du (1901–78) US biochemist was was awarded the 1955 Nobel Prize for chemistry for determining the structure and synthesis of VASOPRESSIN and oxytocin, two polypeptide hormones. In the 1930s he discovered the metabolic processes by which a methyl group ($-CH_3$) is moved from one compound to another, called transmethylation. In 1943 Vigneaud deduced the two-ring structure of biotin. He also studied INSULIN, PENICILLIN synthesis and AMINO ACID metabolism.

Viking space mission US space project to investigate conditions on Mars, launched in 1975. Two spacecraft, Viking 1 and Viking 2, each attached to separate vehicles that orbited the planet, made the first successful landings on Mars in July and September 1976. They transmitted much information to Earth, including weather conditions and dramatic photographs of the surface. Samples of soil from the landers were analysed on board, but no organic compounds or other signs of life were detected.

Villard, Paul (1860–1934) French physicist. In 1900, while studying radiation emitted from uranium, he found that some of it was more penetrating than either alpha or beta rays and, unlike these two, was not deflected by a magnetic field. This new form of radiation was similar to X-rays and later came to be called GAMMA RADIATION.

villi In anatomy, small, finger-like projections of a mucous membrane such as that which lines the inner walls of the SMALL INTESTINE or the CHORION. They serve to increase the absorptive surface area of a particular organ. In digestion, intestinal villi absorb most of the products of food broken down in the STOMACH, DUODENUM and ILEUM. Each villus contains blood capillaries, which receive PROTEIN and CARBOHYDRATE, and a lymph vessel, which receives FAT.

vinyl chloride *See* CHLOROETHENE

vinyl plastics Any of several substances composed of large molecules linked by the POLYMERIZATION of vinyl compounds. They are used in plastic film, upholstery, floor tiles, toys, buttons and in fibres. The most common is POLYVINYL CHLORIDE (PVC).

Virchow, Rudolf (1821–1902) German pathologist. His discovery that all CELLS arise from other cells completed the formulation of the cell theory and repudiated the theories of SPONTANEOUS GENERATION. Virchow also studied the nature of disease at a cellular level and established the science of cellular pathology.

Virgo (Virgin) Equatorial constellation on the ecliptic between Leo and Libra. It lies in a region of the sky that has many GALAXIES and GALAXY CLUSTERS. The brightest star is the 1st-magnitude Alpha Virginis, or Spica.

virology Study of VIRUSES. The existence of viruses was established (1892) by the Russian botanist D. Ivanovski, who found that the causative agent of tobacco mosaic disease could pass through a porcelain filter impermeable to BACTERIA. Bacteriophages (viruses that infect bacteria) were discovered in 1915. The introduction of the electron microscope in the 1940s made it possible to view viruses.

Virtanen, Artturi Ilmari (1895–1973) Finnish biochemist who was awarded the 1945 Nobel Prize for chemistry for his work on fodder preservation. During the 1920s Virtanen discovered that fermentation in green fodder, which ruins silage stores, could be prevented by the addition of hydrochloric and sulphuric acids. He also found that this technique, known as the AIV method, had no detrimental effects on the nutritional qualities or edibility of the fodder.

virtual particle Particle that is not directly observable because it exists for an extremely short time, during an interaction between observable particles. Because of the Heisenberg UNCERTAINTY PRINCIPLE, it is not subject to the law of CONSERVATION of energy. Virtual PHOTONS, for example, are exchanged between charged particles when they interact electromagnetically.

virtual reality Use of COMPUTER GRAPHICS to simulate a three-dimensional environment so that users can explore it as if it were real. A virtual reality system can allow an architect to see what the inside of a building will look like before construction begins. The computer images are produced using the architect's drawings of the building. Some entertainment machines use virtual reality to simulate space-flight adventures and various ball games. In many cases, sensors in a special helmet and glove detect the player's movements and automatically control the screen display. Twin screens inside a special helmet show slightly different images, calculated to give a convincing three-dimensional effect.

virus Sub-microscopic infectious organism. Viruses vary in size from 10 to 300 nanometres and contain only genetic material in the form of DNA or RNA. This is enclosed in a protein coat known as the capsid. Viruses are incapable of independent existence. They can grow and reproduce only when they enter another cell, such as a bacterium or animal cell, because they lack energy-producing and protein-synthesizing functions. When they enter a cell, the host's metabolism is subverted so that viral reproduction is favoured. Pathogenesis is the result of cell death or altered metabolism as the virus multiplies. Control of viruses is difficult because harsh

VIRTUAL REALITY

A data glove (A) measures the movements of the wearer's hand and allows the user to manipulate objects in virtual reality. Fibre optic cables (1) on the glove detect the flexing of the hand. Light travels up and down the cables (2). When the cables are bent (3) they no longer reflect light back to the interface board (4). A position sensor (5) detects the movement of the glove in three dimensions. Fingertip padding (6) convinces the wearer they are touching an actual object.

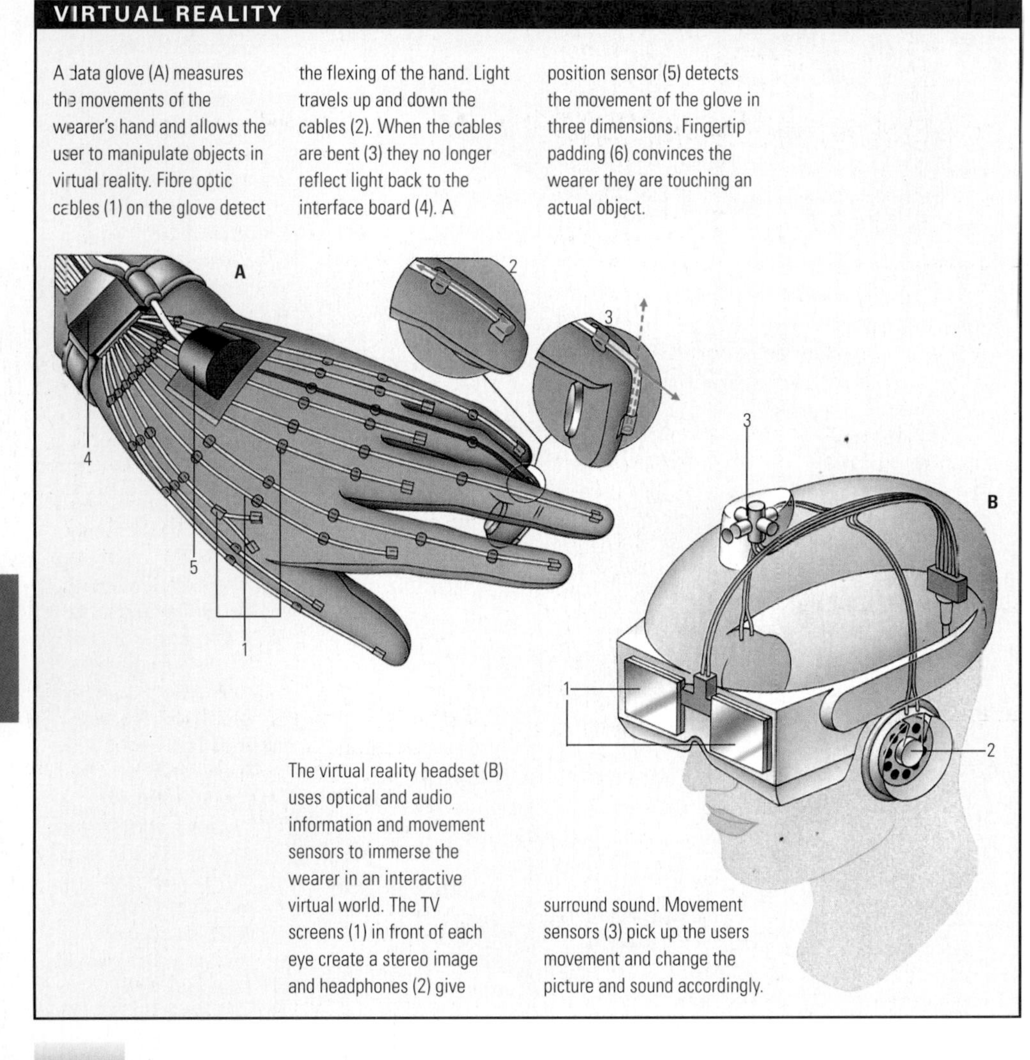

The virtual reality headset (B) uses optical and audio information and movement sensors to immerse the wearer in an interactive virtual world. The TV screens (1) in front of each eye create a stereo image and headphones (2) give surround sound. Movement sensors (3) pick up the users movement and change the picture and sound accordingly.

Viruses are unlike any other form of life. They are on a mysterious border between living and nonliving. Too small to be seen under the ordinary light microscope, they are revealed only under electron microscopes, which show them to be packets of genetic material – DNA, or the closely related RNA – surrounded by a protective coating of proteins.

Viruses are most familiar to us as the cause of human, animal and plant diseases. Smallpox (now eradicated), rabies and polio are ancient examples. Other viruses, such as the human immunodeficiency virus (HIV), the cause of AIDS, seem to have arisen in more recent times.

HIJACKING

Viruses often gain entry to cells by latching onto proteins on the cell surface – proteins that are normally used by the cells for quite different purposes. The AIDS virus, for example, attacks only a particular set of white blood cells, T-lymphocytes. These carry a special protein on their surface to which the virus sticks. Once attached to its "receptor" protein, the virus is then taken into the cell.

Upon entering a cell, a virus discards its proteins and replicates itself by copying and translating its own genetic material, then begins copying and translating the virus' nucleic acid instead. The genes encoded in viral nucleic acid direct the manufacture of more virus coating protein, and the nucleic acid itself itself is also copied many times. So the cell is tricked into making new virus particles that are eventually shed from the cell, sometimes destroying the cell in the process. Most viruses have relatively few genes, all of which are directed to replicating the virus.

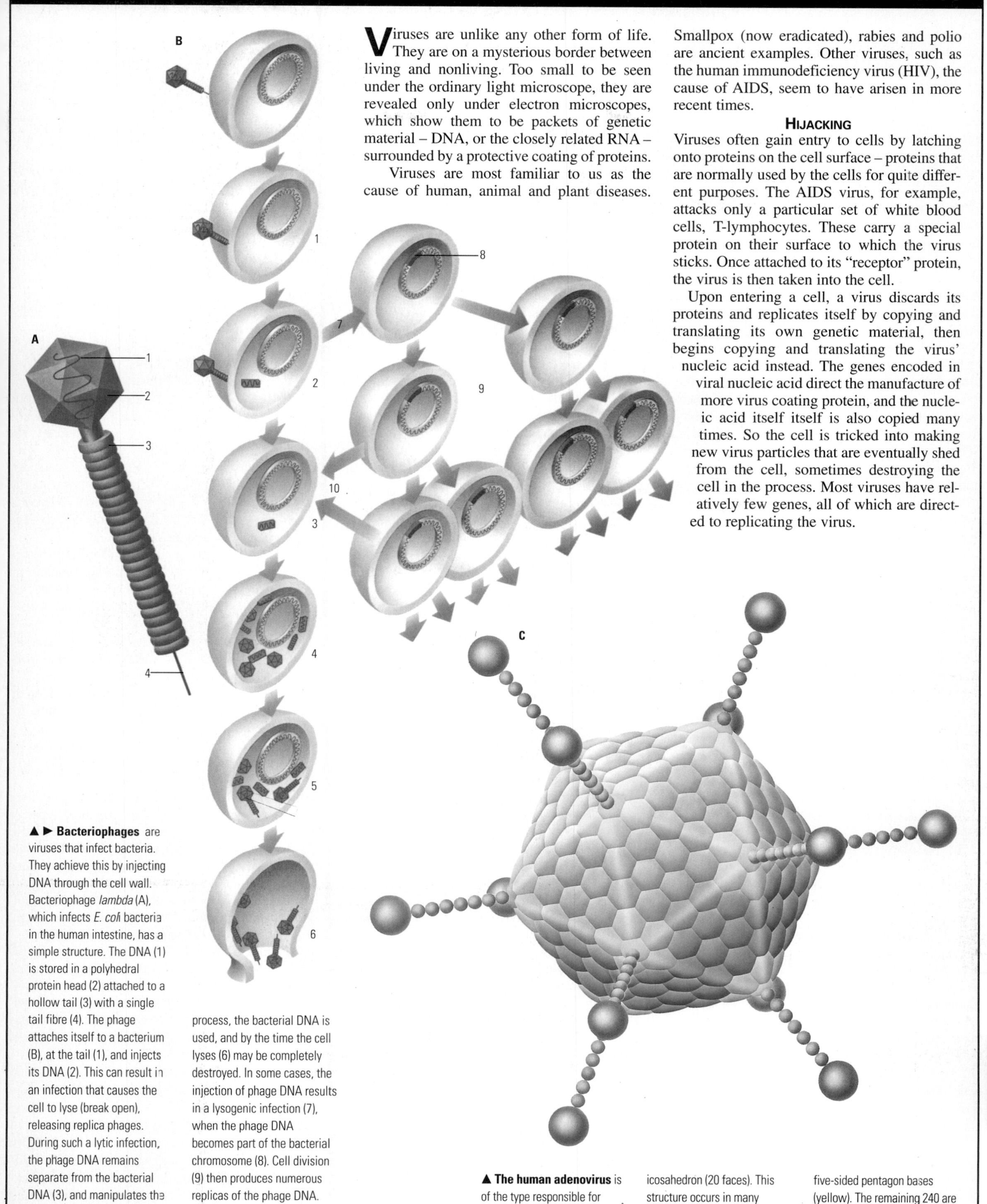

▲ ► **Bacteriophages** are viruses that infect bacteria. They achieve this by injecting DNA through the cell wall. Bacteriophage *lambda* (A), which infects *E. coli* bacteria in the human intestine, has a simple structure. The DNA (1) is stored in a polyhedral protein head (2) attached to a hollow tail (3) with a single tail fibre (4). The phage attaches itself to a bacterium (B), at the tail (1), and injects its DNA (2). This can result in an infection that causes the cell to lyse (break open), releasing replica phages. During such a lytic infection, the phage DNA remains separate from the bacterial DNA (3), and manipulates the cell's enzymes to synthesize the proteins that form the components of new phages (4). The phage DNA replicates and large numbers of phages are assembled (5). In the process, the bacterial DNA is used, and by the time the cell lyses (6) may be completely destroyed. In some cases, the injection of phage DNA results in a lysogenic infection (7), when the phage DNA becomes part of the bacterial chromosome (8). Cell division (9) then produces numerous replicas of the phage DNA. During the course of lysogenic growth, damage to the cell may result in a lytic infection (10) by causing the phage DNA to be ejected from the bacterial chromosome.

▲ **The human adenovirus** is of the type responsible for colds and sore throats. The virus is colour coded for ease of identification. The casing consists of 252 protein molecules (capsomeres) arranged into a regular icosahedron (20 faces). This structure occurs in many viruses, representing the most economical packing arrangement around the DNA inside. Twelve of the capsomeres, located at the points of the icosahedron, are five-sided pentagon bases (yellow). The remaining 240 are six-sided hexons (green). Five of these (green-yellow) adjoin each penton base, from which extends a single fibre (red) tipped with a terminal structure (blue) that begins cell entry.

measures are required to kill them. The animal body has, however, evolved some protective measures, such as production of INTERFERON and of ANTIBODIES directed against specific viruses. In most cases, one attack of a viral disease confers lifelong IMMUNITY. Where the specific agent can be isolated VACCINES can be developed, but some viruses (such as those responsible for causing influenza) change so rapidly that vaccines become ineffective. There are few antiviral drugs. *See also* PATHOGEN

virus In computing, a program that, unknown to the user, reproduces itself and can be transferred to another COMPUTER. The virus may be transferred on a floppy disk, via a modem or communicated to any other computer on a network. Viruses can corrupt data or modify programs and may even erase a memory or cause a computer to crash. Software is available to detect and counteract known viruses, but new ones are being introduced all the time. To knowingly spread a virus is a crime in some countries.

virus infection Disease caused by a VIRUS. Virus infections are among the greatest health problems facing humans, since ANTIBIOTICS have achieved some measure of control over bacterial infections. Many virus infections can be treated only with specific VACCINES and antisera. Notable among viral infections affecting human beings are YELLOW FEVER, RABIES, influenza (FLU), MEASLES and POLIOMYELITIS. Viruses are also responsible for many diseases in plants, in which they are particularly difficult to control. The role of viruses in causing CANCER has been the subject of extensive research since this effect was discovered in 1908. Living cells gain some protection by their own production of INTERFERON, which leads to the formation of another protein virus inhibitor.

viscose *See* RAYON

viscosity Resistance of a fluid to flow because of internal friction. The more viscous the fluid, the slower it flows. Viscosity is large for liquids and extremely small for gases. In most liquids, it increases with decreasing temperature.

vision *See* SIGHT

vitamin Organic compound that is essential in small amounts to the maintenance and healthy growth of all animals. Vitamins are classified as either water-soluble (B and C) or fat-soluble (A, D, E and K). They are usually taken in the DIET, but today most can be made synthetically. Some are synthesized in the body. Many vitamins act as coenzymes, helping ENZYMES in RESPIRATION and other metabolic processes such as the synthesis of cell components and the detoxification of body wastes. Lack of a particular vitamin can lead to a deficiency disease. **Vitamin E** is important in reproduction and many other biological processes. **Vitamin D** helps the body absorb phosphorus and calcium. It is essential for the normal growth of bone and teeth. Existing in human skin (activated by sunlight), vitamin D is also found in fish-liver oil, yeast and egg yolk. Deficiency causes such bone diseases as rickets, while an excess could cause kidney damage. **Vitamin C** or ascorbic acid is commonly found in many fruits and vegetables. It helps the body resist infection and stress, and is essential to normal metabolism. A deficiency causes scurvy. Vitamin C is destroyed by heating, and a lot is lost during cooking. **Vitamin B** is actually a group of 12 vitamins, known collectively as the **vitamin B complex**, important in assisting the process by which energy is produced in the body (RESPIRATION). Vitamin B_1 or THIAMINE occurs in yeast and cereals. Lack of it causes a disease called beriberi. Another B vitamin is **niacin** (nicotinic acid) found in milk, meat and green vegetables. Lack of this vitamin causes pellagra. **Vitamin B_{12}** is needed for the formation of blood cells. It is found especially in meat, liver and eggs. Deficiency causes anaemia. **Vitamin A** (retinol) is important for healthy eyes, and shortage of the vitamin caused night-blindness. It is found in fish-liver oil, but the body can convert carotene, the red pigment in carrots, mangoes and red peppers, into Vitamin A. *See also* RIBOFLAVIN

viticulture Worldwide cultivation of GRAPES. Grapes are grown throughout the warmer of temperate regions. The main centres of viticulture are southern Europe, California, South Africa, Australia and New Zealand and South America. Grapes are grown mostly for wine, though in some places they are also used in various dried forms for currants, raisins and sultanas.

vitreous humour Transparent, jelly-like medium which fills the eyeball between the lens and the retina. It constitutes the vitreous body, which serves to hold the retina in position and combines with the lens to ensure the clear passage of light to the receptor cells of the retina. *See also* AQUEOUS HUMOUR

vitrification In ceramics, process of producing a glassy phase or close crystallization by firing at a high temperature. Vitrified clay becomes non-porous and loses its plasticity.

viviparity (vivipary) Process or trait among animals of giving birth to live young. Placental mammals show the highest development of viviparity, in which the offspring develops inside the body, within the mother's UTERUS. The young receive nutrition via the mother's PLACENTA. Viviparity is also exhibited by certain arthropods, amphibians, fish and reptiles. Viviparous plants are those whose seeds germinate before separating from the mother plant or those that produce plantlets or bulbils rather than flowers.

vivisection Dissection of living bodies. Work with laboratory animals in testing drugs, vaccines and pharmaceuticals frequently involves such dissections. Smaller animals such as rats, mice and hamsters are usually killed painlessly after the survey of the test results. The ethical issue of experimenting on living animals is still a matter of controversy, because alternative methods may be available, if not as cheap or convenient to use.

vocal cords *See* LARYNX

voice In human beings, sound produced by air from the lungs passing through the vocal cords, causing them to vibrate. The cords are two wedge-shaped bands of fibrous elastic tissue attached to the cartilage of the LARYNX (voice-box), which bulges in the front of the neck as the Adam's apple. The respiratory tract above the larynx acts as a resonating chamber, lending the voice its particular timbre. The voice of a bird is produced differently by a vocal apparatus called the SYRINX, located at the base of the windpipe.

volcanic plug Hard mass of solidified LAVA in the vent of a VOLCANO. Over a period of millions of years, the volcano will have been eroded, but the hard plug will survive as a small isolated hill in the crater.

volcanism (vulcanism) Volcanic activity. The term is a general one and includes all aspects of the process: the eruption of molten and gaseous matter, the building up of cones and mountains, and the formation of lava flows, geysers and hot springs.

volcano Vent from which molten rock or LAVA, solid rock debris and gases issue. The term is also applied to the pile of rock around the vent. Volcanoes may be of the **central vent** type, where the material erupts from a single pipe, or of the **fissure type**, where material is extruded along an extensive fracture, building plains and plateaus. Volcanoes are usually classed as active, dormant or extinct.

volt (symbol V) SI unit of electric potential and ELECTROMOTIVE FORCE (EMF), or voltage. It is the difference of potential between two points on a conducting wire carrying a current of 1 ampere when the power dissipated is 1 watt. *See also* POTENTIAL DIFFERENCE

Volta, Count Alessandro Giuseppe Antonio Anastasio (1745–1827) Italian physicist who invented (1800) the first electric BATTERY, the voltaic pile. Volta also invented a charge-accumulating machine, which incorporated the principle of some modern electrical CONDENSERS. His investigations into ELECTRICITY led him to interpret correctly Luigi GALVANI's experiments with muscles, showing that the metal electrodes and not the tissue generated the current. The VOLT was named in his honour.

voltage Measure of the electrical POTENTIAL DIFFERENCE between two points in a circuit. Two points are at a potential difference of one volt if one coulomb of electricity (ELECTRIC CHARGE) does one joule of work in flowing between them. Voltage is also given by the RESISTANCE multiplied by the CURRENT (using OHM'S LAW).

voltameter (coulometer) Electrolytic CELL for measuring the magnitude of an ELECTRIC CHARGE by by determining the the total amount of decomposition on the CATHODE arising from the passage of the charge through the cell

voltmeter Instrument for measuring the voltage (POTENTIAL DIFFERENCE) between two points in an electric circuit. Voltmeters are always connected across (in parallel with) the components whose voltages are being measured. A voltmeter has a high internal RESISTANCE (compared with the resistance across which it is connected). Digital voltmeters are now common. *See also* AMMETER

volume Measure of the amount of space taken up by a body. Volume is measured in cubic units, for example cm^3 or cubic metres.

volumetric analysis In chemistry, method of finding the concentration of solutions. *See also* QUANTITATIVE ANALYSIS; TITRATION

voluntary muscle In human beings and other mammals, the most plentiful of the three types of MUSCLE, comprising the bulk of the body. It is also known as **skeletal** muscle or **striated** muscle because of its characteristic striped appearance under the microscope. Mostly attached to BONE, either directly or through TENDONS, it includes FLEXORS to bend joints and EXTENSORS to straighten them. *See also* INVOLUNTARY MUSCLE; SKELETAL MUSCLE

vomiting Act of bringing up the contents of the stomach by way of the mouth. Vomiting is a reflex mechanism that may be activated by any of a number of stimuli, including dizziness, pain, gastric irritation or shock. It may also be a symptom of serious disease. Prolonged or repeated vomiting is in itself dangerous because of the dehydration it causes.

von Braun, Wernher (1912–77) German-born engineer, known for his expertise in the development of ROCKETS for SPACE EXPLORATION. In World War 2, Von Braun worked on the development of the V2 ROCKET. In 1945 he went to the USA to work on MISSILE development, becoming a US citizen in 1955. Von Braun developed the Jupiter rocket that took the first US SATELLITE, Explorer 1, into space (1958). In 1960 he joined the NATIONAL AERONAUTICS AND SPACE ADMINISTRATION (NASA) and developed the Saturn rocket that carried astronauts to the Moon. In 1970 von Braun left NASA to join the aerospace company Fairchild Industries.

Von Frisch, Karl *See* FRISCH, KARL VON

Von Neumann, John (1903–57) US mathe-

When the small volcanic island Krakatoa in w Indonesia exploded, it blasted out $18km^3$ (4.5cu mi) of rock in under a day, creating massive waves and forming a circular depression 6km (4mi) across, and 1km (0.6mi) deep.

Volcanoes occur when the Earth's surface is breached and magma either flows out as lava or explodes into the air as TUFF. All volcanic eruptions are driven by rapidly expanding gases within the magma, so the two factors that determine the violence of an eruption are the amount of dissolved gases and how easily they can escape. It is the magma's viscosity or fluidity that controls how the gas escapes, and it is the composition of the magma that determines its viscosity.

Eruptions

When the magma has low viscosity and is mainly composed of BASALT, with a low content of dissolved gas which rises slowly through the crust, the gas escapes slowly and gradually as the magma rises. Near the surface, perhaps a few metres away, the magma may start to foam and this is what causes lava fountains.

Magma that has a high silica content with high viscosity and high dissolved gas content behaves very differently. As the magma rises the gas starts to escape from solution, but the high viscosity of the melt holds the bubbles back, so that they have an internal pressure which can be as high as several hundred atmospheres. Then, when enough bubbles form, or the external pressure decreases, the gas pressure blows the rock apart and starts off an explosive eruption. Violent eruptions that throw dust and rock fragments into the air are called PYROCLASTIC eruptions. These eruptions alternate with lava flows to produce cone-shaped, crater-topped mountains, sloping up at angles of *c*.30°, such as seen at Mount Fuji in Japan. Such mountains are called STRATOVOLCANOES.

Volcanic gases

The gases produced by volcanoes consist of water vapour – the most prevalent component – carbon dioxide, sulphur dioxide, hydrochloric acid and nitrogen. In fact the ratios of these volcanic gases are extraordinarily close to the ratios of water, carbon, chlorine, and nitrogen in the air, oceans and surface rocks of the Earth. The main difference is that volcanoes produce too much sulphur. Nevertheless, it may be that the volcanoes, expelling the Earth's gases throughout the millennia of geological time, were what created the atmosphere, oceans and rocks.

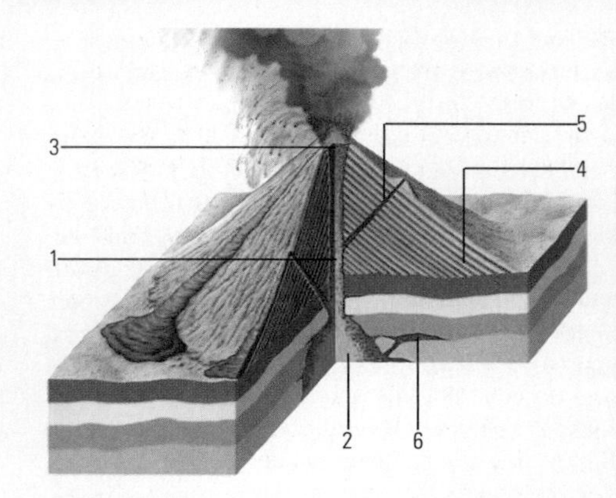

▲ **Volcanoes** are formed when molten lava (1) from a magma chamber (2) in the Earth's crust forces its way to the surface (3). The classic, cone-shaped volcano is formed of alternating layers of cooled lava and cinders (4) thrown out during an eruption. Side vents (5) can occur and when offshoots of lava are trapped below the surface, laccoliths (6) are formed.

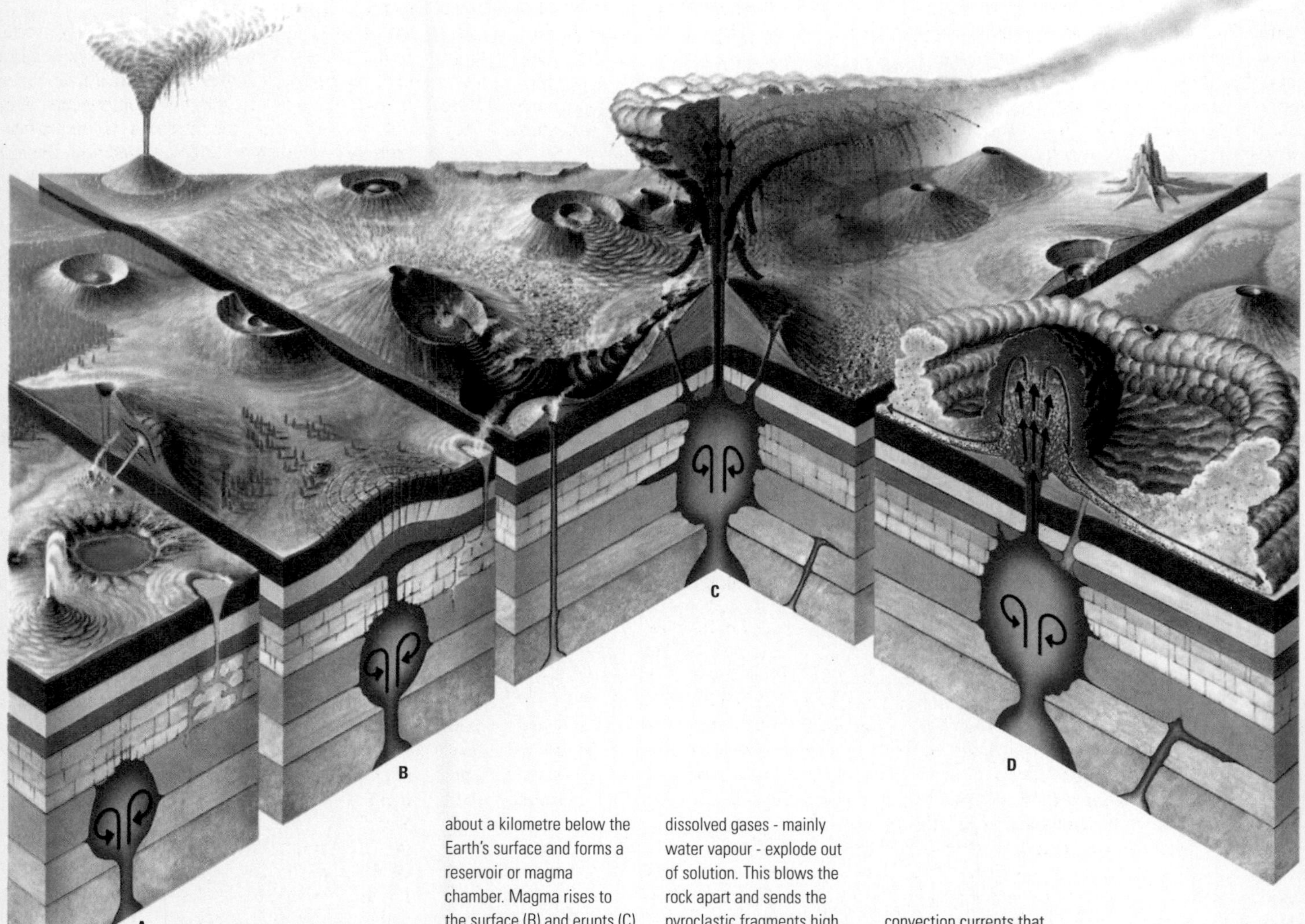

The forces that lie behind a volcano are born deep in the Earth. Mantle material upwells and becomes partly molten. The decrease in pressure as it rises causes it to melt even more and to melt the surrounding rock. The magma then ponds (A) about a kilometre below the Earth's surface and forms a reservoir or magma chamber. Magma rises to the surface (B) and erupts (C) when the pressure in the magma chamber exceeds the pressure of the surrounding rock. If the magma is viscous and the pressure drop is rapid, dissolved gases - mainly water vapour - explode out of solution. This blows the rock apart and sends the pyroclastic fragments high into the air, forming a massive eruption column composed of hot gases and incandescent pumice and ash. The particles heat up the surrounding air, causing convection currents that buoy them even higher - up to 50 km (30 miles). When the column can no longer be supported by the surrounding air (D) it collapses to create incandescent pyroclastic flows that race outwards at velocities of up to 100 m/s (more than 200 mph).

matician, b. Hungary. In 1930 he left Hungary to teach at Princeton University (1930–33) and then at the Institute for Advanced Studies. In 1937 he became a US citizen. In mathematics, Von Neumann helped found the theory of GAMES. His early contribution to QUANTUM THEORY was followed by work on the atomic bomb at Los Alamos. Von Neumann is celebrated chiefly, however, for his development of high-speed electronic COMPUTERS. One of his first designs, the maniac (**m**athematical **an**alyser **n**umerical **i**ntegrator **a**nd **c**omputer), was used to test (1952) the first hydrogen bomb.

vortex Eddy or whirlpool observed in fluid motion. Vortices cannot occur in ideal (nonviscous) fluid motion, but they are important in the study of real fluids. In particular, the vortices occurring behind aerofoils are of great interest in aerodynamic design. The study of turbulence shows that energy is dissipated most efficiently in smaller vortices.

Vostok I First of a series of manned spacecraft placed into Earth orbit by the Soviet Union on 12 April 1961. It was launched by the two-stage Vostok rocket and carried cosmonaut Yuri GAGARIN.

Voyager project SPACE EXPLORATION programme using two unmanned craft to study Jupiter, Saturn, Uranus and Neptune, launched in 1977. The two Voyager probes were launched 16 days apart from Kennedy Space Center at Cape Canaveral, Florida and beamed back stunning close-up pictures of Jupiter in 1979. The probes then passed Saturn and showed the structure of the planet's rings. While that was Voyager 1's last planetary encounter, Voyager 2 went on to study Uranus in 1986 and Neptune in 1989. Too far from the Sun to generate power from solar cells, it carried small nuclear generators. Both probes have now left the Solar System.

Vries, Hugo de *See* DE VRIES, HUGO

VTOL aircraft (**v**ertical **t**ake-**o**ff and **l**anding aircraft) Experimental AIRCRAFT designed to perform vertical take-offs and landings but to maintain flight speed and payload capabilities superior to those of a HELICOPTER. There are designs that permit rotation of the wing and engines from vertical for take-off, to horizontal for high-speed cruising. Another design allows for the diversion downward of the exhaust of fixed jet engines. A third style has two separate systems of thrust for upward and forward movement. *See also* STOL AIRCRAFT

vulcanization Chemical process, developed (1839) by Charles Goodyear, of heating sulphur or its compounds with natural or synthetic rubber to improve its durability and resilience. The process was used in the development (1845) of the first pneumatic TYRE.

vulva In females, the external genitalia. Extending downwards from the clitoris (a sensitive, erectile organ), a pair of fleshy lips (labia majora) surround the vulvar orifice. Within them two smaller folds of skin (labia minora) surround a small depression called the vestibule, within which are the urethral and vaginal openings.

Waals, Johannes Diderik van der *See* VAN DER WAALS, JOHANNES DIDERIK

wadi (arroyo) Narrow, steep-sided valley found in deserts and semi-arid regions. As rainfall is infrequent in desert areas, wadis are dry valleys for much of the year. When it does rain, however, it falls in heavy showers. As much of the land is bare with no vegetation, surface runoff is rapid. The water carries sand and small rock fragments, which cause EROSION. Vertical CORRASION forms the narrow, steep-sided wadis, which can be several hundred metres in depth and extend for long distances.

waggle dance (bee dance) Method of communication performed by worker honeybees to indicate to other workerbees the exact position and richness of sources of pollen and nectar. The dances rely on a knowledge of the position of the Sun.

Waksman, Selman Abraham (1888–1973) US microbiologist, b. Russia. He was awarded the 1952 Nobel Prize for physiology or medicine for his discovery of the antibiotic STREPTOMYCIN in 1943. Waksman developed techniques for extracting ANTIBIOTICS from various microorganisms, and he discovered several new ones, including neomycin.

Wald, George (1906–) US biologist who shared the 1967 Nobel Prize for physiology or medicine with the US physiologists Haldan Hartline (1903–83) and Ragnar Granit (1900–91) for research into the chemical processes of vision. He discovered that vitamin A is an essential component of visual pigments and later determined the chemical reactions that occur in the rods of the retina.

Wallace, Alfred Russel (1823–1913) British naturalist and evolutionist. Wallace developed a theory of NATURAL SELECTION independently of but at the same time as Charles DARWIN. He wrote *Contributions to the Theory of Natural Selection* (1870), which, with Darwin's *Origin of Species*, comprised the fundamental explanation and understanding of the theory of EVOLUTION.

Wallace's line Imaginary line dividing the islands of Borneo and Bali from Sulawesi and Lombok. It was drawn by Alfred Russel WALLACE to demarcate animal populations that live close to each other and yet are extraordinarily different. Earlier this observation had influenced Wallace to state his theory of natural selection, which was coincident with that of Charles DARWIN. He argued that homogeneous animal populations could, for various reasons, divide and diversify gradually until they had evolved into separate species.

Wallach, Otto (1847–1931) German chemist who was awarded the 1910 Nobel Prize for chemistry for his research into TERPENES. In 1884 he began investigating the ESSENTIAL OILS in plants and, by a method of repeated distillation, he isolated the components of these complex compounds.

Wallis, Sir Barnes Neville (1887–1979) British aeronautical engineer and inventor, best known for his invention of the bouncing bomb during World War 2. In the 1920s he designed the AIRSHIP R100 and after the war the first SWING-WING AIRCRAFT.

Wallis, John (1616–1703) English mathematician, who together with Robert BOYLE formed a society where scientists congregated to discuss current scientific ideas. It led to the foundation of the Royal Society in 1660. In mathematics Wallis developed various series expressions for PI and worked on problems in number theory, mechanics and algebra.

Walton, Ernest Thomas Sinton (1903–95) British physicist who shared the 1951 Nobel Prize for physics with John COCKCROFT for the development of the first nuclear particle ACCELERATOR. After two unsuccessful attempts they succeeded in 1929, using a voltage multiplier to accelerate protons. In 1931 they produced the first artificial nuclear reac-

WAGGLE DANCE

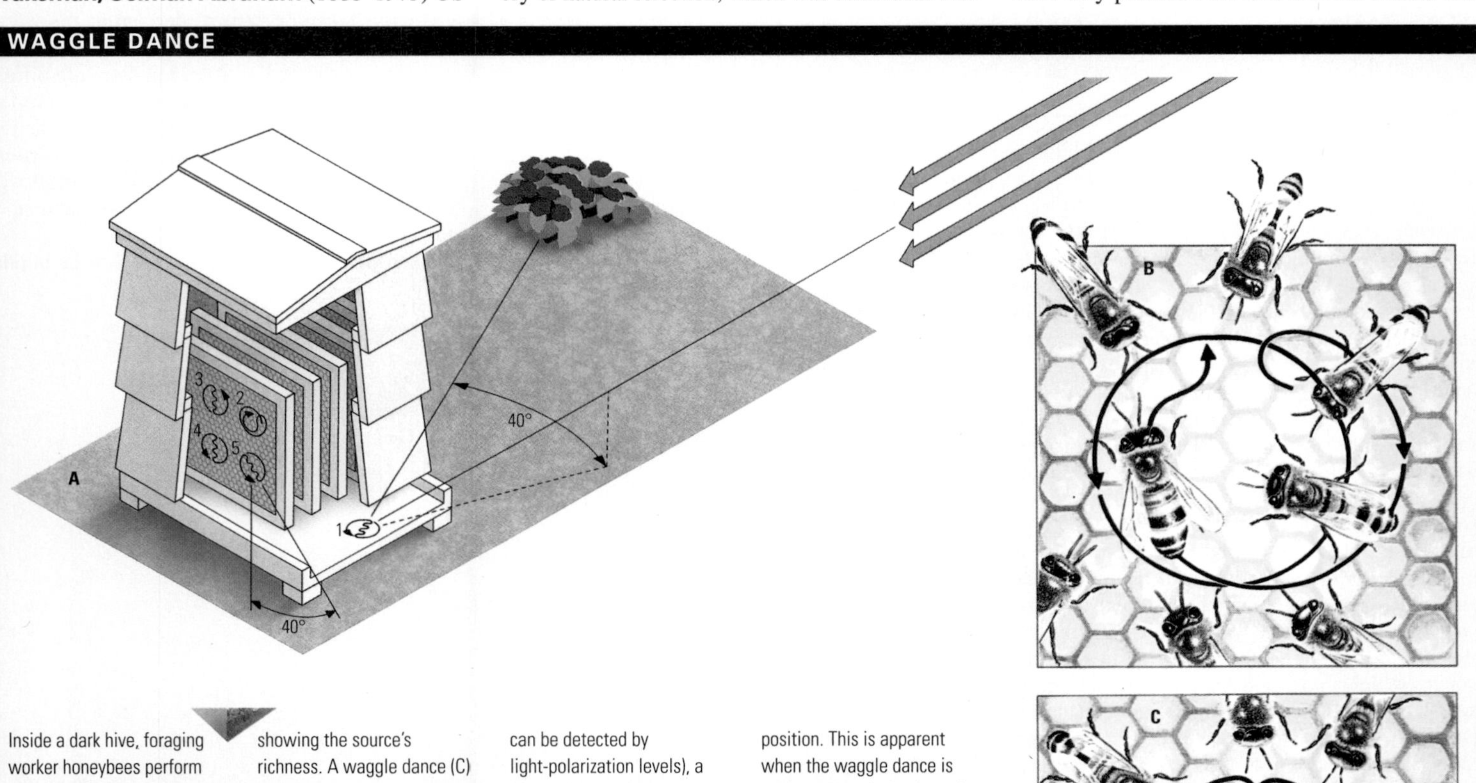

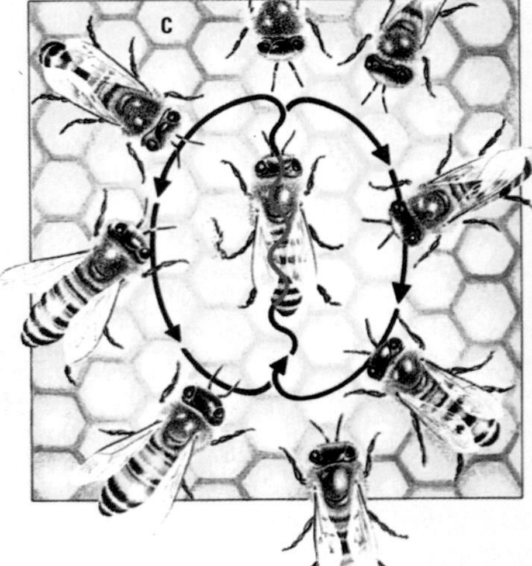

Inside a dark hive, foraging worker honeybees perform complex dances on combs, using touch (conveyed to bee "dance-followers" via their antenna), air vibrations and smell to communicate nectar and pollen locations to others. Nectar and pollen provide carbohydrates and protein – both vital to a colony. The round dance (B) indicates food *c.*50–100m (150–300ft) from the hive, the dance's length and vigour showing the source's richness. A waggle dance (C) consists of a figure of eight with a waggle in the middle and indicates more distant sources (waggle orientation shows direction and waggle frequency shows distance). The dancing bee's smell may indicate flower types. To communicate and utilize the information requires a knowledge of the Sun's position (even if hidden in a blue sky, the Sun's position can be detected by light-polarization levels), a sense of time and wind speed, the use of visual landmarks and perhaps some sort of magnetic sensory device. Bee dances (A) usually take place in the darkness of the hive on the comb sides, and are interpreted mainly by touch and smell. However, the bees are essentially communicating visual directions based on the Sun's position. This is apparent when the waggle dance is performed on the horizontal entrance board of a hive (1). The angle of the waggle indicates a nectar source 40° to the left of the Sun. Inside the hive, the vertical surfaces do not affect the round dance (2), but a downward (3) waggle indicates a direction away from the Sun, and an upward one (4) towards the Sun. The dance (5) is a vertical representation of (1).

WANKEL ROTARY ENGINE

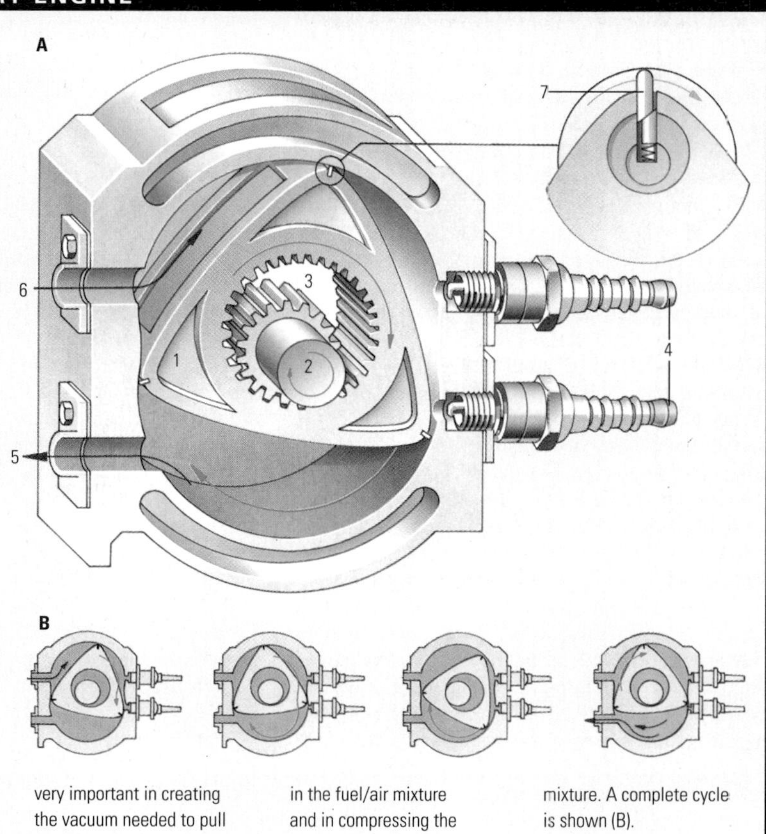

The Wankel rotary engine (A) compresses and ignites a petrol/air mixture with a spark plug like a normal combustion engine, but does so with a three-sided rotor (1), rather than an in-line piston action. The explosion of the fuel/air mixture drives a crankshaft (2) that passes through the centre of the cylinder. (3) The movement of the rotor sucks air into the cylinder and, as it continues to rotate, seals the inlet (so valves are not needed) and then compresses the mixture as it continues to turn. The spark plugs (4) then ignite the mixture, which expands, rotating the piston and driving the crankshaft. The burned fuel is expelled through an outlet (5) as the piston turns, pulling in more air to repeat the process (6). The seals (7) at the edge of the piston's faces are very important in creating the vacuum needed to pull in the fuel/air mixture and in compressing the mixture. A complete cycle is shown (B).

tion, using high-energy protons to bombard lithium nuclei. This process was of great importance for studying nuclear structure; its modern counterparts are among the most useful tools of NUCLEAR PHYSICS.

Wang, An US computer engineer, b. China. He emigrated to the USA in 1945. In 1948 he devised the core memory, which came into universal use in mainframe computers before integrated circuit chips were introduced. In 1951 he established Wang Laboratories, which grew to be one of the world's major computer companies.

Wankel rotary engine Petrol engine with rotors instead of pistons. German engineer Felix Wankel (1902–88) invented this engine in the 1950s. Each triangular rotor turns inside a close-fitting casing. Gaps between the casing and rotor form three crescent-shaped combustion chambers. Each chamber goes through a sequence of events similar to those in a FOUR-STROKE ENGINE with pistons. The movement of the rotor draws in the petrol-air mixture through an inlet port and then compresses it. A spark ignites the mixture, producing expanding gases that force the rotor around. The gases then pass out of an exhaust port, and the process repeats.

Warburg, Otto Heinrich (1883–1970) German biochemist who was awarded the 1931 Nobel Prize for physiology or medicine for his discovery of respiratory enzymes. He made significant contributions to the understanding of the mechanisms of cellular respiration and energy transfer, and of photosynthesis.

warm-blooded *See* HOMEOTHERMIC

warning coloration Markings of an animal that warn potential predators that it is dangerous. It may indicate that the animal tastes bad or is poisonous. Red, yellow and black are common warning colours, as in some venomous insects and snakes. Some non-poisonous animal species mimic the warning colours of other animals in order to trick potential predators into thinking they are not suitable prey. *See also* MIMICRY

Wassermann, August von (1866–1925) German bacteriologist. He made important contributions to immunology, the best known of which was his development of a diagnostic test for syphilis, the Wassermann test.

waste disposal, nuclear FISSION reactors and their fuel preparation plants produce waste residues containing highly radioactive substances. After URANIUM, PLUTONIUM and any other useful fission products have been removed some long-lived radioactive products remain, such as CAESIUM-137 and STRONTIUM-90. These wastes have to be disposed of so that they will not contaminate the soil, seas or atmosphere. International regulations exist for disposing of the wastes, separate regulations applying to liquids, gases and solids. In general, disposal consists of burying the materials under controlled conditions in underground tanks, in disused mines, and in the sea.

water Odourless, colourless liquid (H_2O) that covers over 70% of the Earth's surface and is the most widely used solvent. Essential to life, it makes up from 60–70% of the human body. It is a compound of hydrogen and oxygen with the two H-O links of the molecule forming an angle of 105°. This asymmetry results in polar properties and a force of attraction (hydrogen bond) between opposite ends of neighbouring water molecules. These forces maintain the substance as a liquid, in spite of its low molecular weight, and account for its unusual property of having its maximum density at 4°C (39.2°F). Properties: r.d. 1.000; m.p. 0.0°C (32°F); b.p. 100°C (212°F).

water cycle *See* HYDROLOGICAL CYCLE

waterfall Steep or vertical descent of the water of a stream or river. Where there is a change of rock type on the bed of a stream, there are different rates of EROSION. The softer rock is eroded more quickly, causing an abrupt change in the gradient. If there is a vertical or nearly vertical STRATUM of hard rock, only a small waterfall will be created. If the harder rock is horizontal, a much larger waterfall can form.

watermark Translucent impression made in oth-

► **watermill** Mills driven by water provided the motive force for many processes before the Industrial Revolution, including grinding corn. The corn (1) is hoisted to the top of the mill where it is poured, via a shoot, into the grinder (2). The resulting flour falls through the shoot and is collected (3). An old grinding wheel can be seen propped up against a wall (4).

WATERSPOUT

A waterspout occurs when a tornado passes over water, sucking up a column of water up to 100m (330ft) tall. They usually occur in tropical regions, associated with stormy weather.

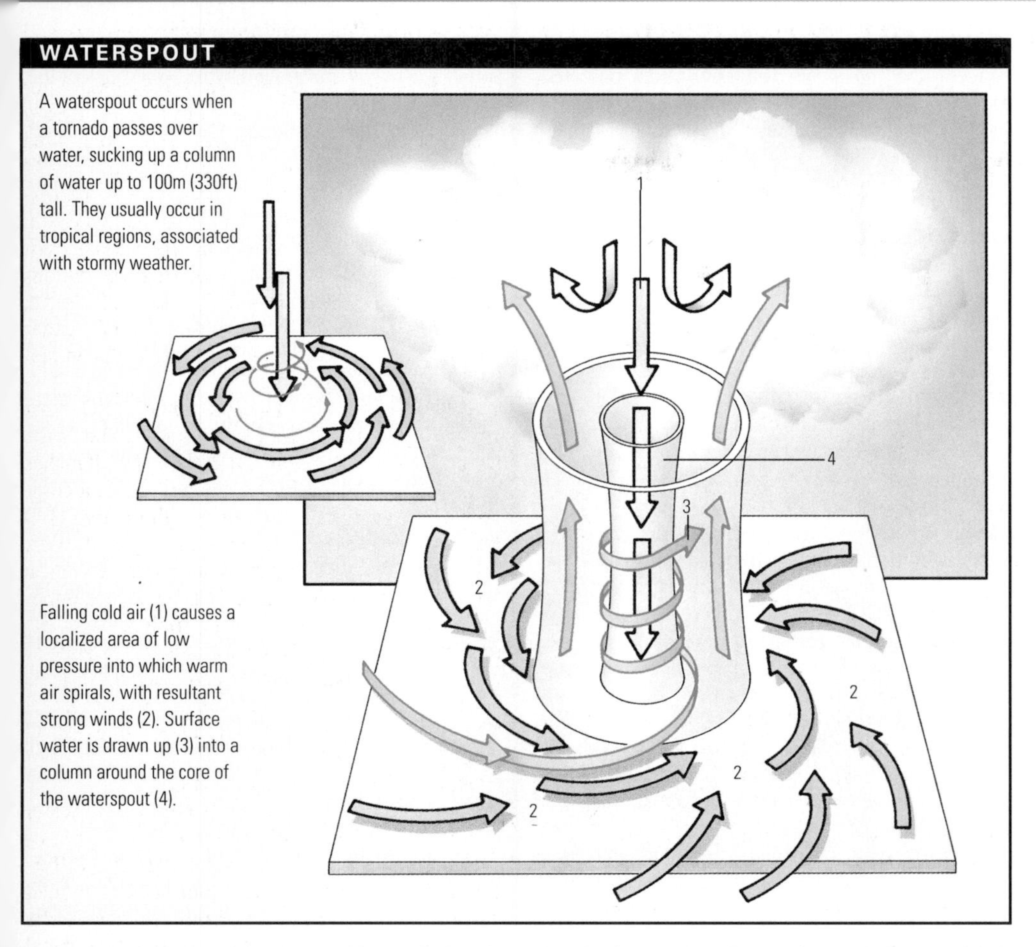

Falling cold air (1) causes a localized area of low pressure into which warm air spirals, with resultant strong winds (2). Surface water is drawn up (3) into a column around the core of the waterspout (4).

erwise opaque PAPER, either by machine or by hand. It is formed during the making of paper by a raised design on wires over which the paper is passed while still in a pulpy state. Watermarks have been used since the latter part of the 13th century as a guarantee of authenticity, as in banknotes, and to identify the makers of high-quality papers.

water mill Machinery powered by the flow of water past a WATER WHEEL. Water mills were invented about 2,000 years ago to grind grain into flour. Water also powered looms at textile mills from the early 1700s.

water of crystallization Definite molecular proportion of water that is chemically combined with certain substances in the crystalline state. Much of this water will be lost on heating to *c*.100°C (212°F) but some, the water of constitution, will be retained to a much higher temperature. Copper(II) sulphate ($CuSO_4.5H_2O$), loses four molecules of water at 100°C (212°F) becoming $CuSO_4.H_2O$.

water pollution Contamination of water by harmful wastes. The chief source of water pollution is **industrial waste**. Toxic chemicals, such as POLYCHLORINATED BIPHENYL (PCB), are discharged as effluent and cannot be disinfected with CHLORINE. The burning of fossil fuels causes ACID RAIN. Untreated or partially treated SEWAGE is another source of water pollution. Sewage treatment is unable to prevent the spread of viruses and of some phosphorus-based detergents which cause EUTROPHICATION. Agricultural chemicals and wastes, such as PESTICIDES and FERTILIZERS, are another major cause of pollution. Once pollution has affected GROUNDWATER it spreads more rapidly. Oil spills and ocean dumping are major causes of marine pollution. *See also* DDT

water potential Force that brings about OSMOSIS in a biological system such as a cell. It makes water tend to move between solutions of different concentrations. It is the difference between the chemical potential of pure water (which is zero) and the chemical potential of water in a solution (which is negative). During osmosis, water moves down the water potential gradient, which is measured in pressure units (kilopascals or bars).

waterproofing Rendering materials such as fabrics, leather and wood impervious to water by the application of substances such as rubber, synthetic resins, waxes and metallic compounds.

watershed (water divide, water parting) High ground that forms the boundary or dividing line between two river basins. On one side of the boundary line the water will drain in one direction; on the other side of the boundary line it will drain in the opposite direction. The watershed is normally a ridge or a piece of ground that is higher than the surrounding areas. It is often a very irregular line.

water softener Substance added to water to reduce its hardness – that is, its inability to form a lather with soap as a result of the presence of dissolved calcium and magnesium compounds. Some hardness, mostly due to the bicarbonates of these metals, can be removed by boiling. The remaining "permanent" hardness is mostly due to sulphates of the metals and is reduced by adding such water-softening compounds as sodium carbonate, sodium phosphate or zeolites (hydrated silicates of calcium and aluminium) to remove or sequester the metallic ions. *See also* HARDNESS OF WATER

waterspout Funnel-shaped mass of water, which is similar in formation to a TORNADO. It occurs over water when a heated patch of air rises, and whirls around in a circular or corkscrew fashion. Waterspouts are mostly seen in tropical areas, especially in late summer. When waterspouts go on to land they become tornadoes, and when tornadoes go over lakes or the sea, they become waterspouts. Waterspouts are generally less than 30m (98ft) in height, and may last for half an hour or so. Waterspouts have been known to cause great damage to shipping and coastal installations

water table In geology, level below which the rock is saturated. After rain has fallen, the water trickles downwards through the pores in the rocks. The level at which all pores are full is the water table. The height of the water table gradually changes, moving up or down depending on the recent rainfall. On average the water table lies a few metres below the surface, although in the Fens in England and the polders of the Netherlands it is never be more than a few centimetres below the surface. In upland limestone areas it can be as deep as 100m (330ft). Water located below the water table is called GROUNDWATER.

water treatment Removal of undesirable elements from a water supply. Unpleasant tastes and odours are removed by aeration. Bacteria are destroyed by the addition of chlorine (the taste of chlorine is then removed with sodium sulphite). Hardness is reduced with lime or by using zeolite as a WATER SOFTENER. Other organic and mineral matter is removed with alum. FLUORIDE is added to the water supply of many communities to reduce dental caries. *See also* SEWAGE

water vapour Water in its gaseous state. It occurs in the atmosphere and determines humidity and the formation of clouds as it condenses. Most of the

WATER TABLE

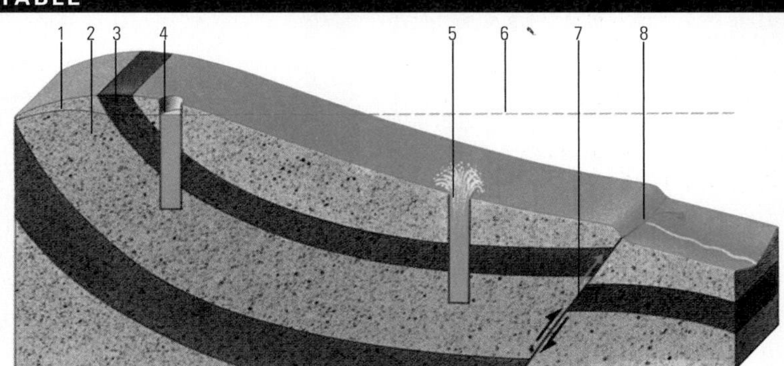

Artesian springs and wells are found where groundwater is under pressure. The water-table (1) in the confined aquifer (2) lies near the top of the dipping layers. A well (4) drilled through the top impervious layer (3) is not an artesian well because the head of the hydrostatic pressure (6) is not sufficient to force water to the surface. In such wells, the water must be pumped or drawn to the surface. The top of an artesian well (5) lies below the level of the head of hydrostatic pressure and so water gushes to the surface. Artesian springs (8) may occur along joints or faults (7), where the head of hydrostatic pressure is sufficient to force the water up along the fault. Areas with artesian wells are called artesian basins. In the London and Paris artesian basins, the water has been so heavily tapped that the water level has dropped below the level of the well heads.

atmosphere's water vapour is found in the TROPOSPHERE, mainly below an altitude of 8km (5mi). Water vapour absorbs infrared radiation and holds it in the atmosphere, because of the GREENHOUSE EFFECT, and thus plays a vital role in the transfer of energy and the Earth's heat balance.

water wheel Mechanical device used since early Roman times to supply power, mainly for grinding cereals. The first water wheels were laid horizontally in flowing water, their shafts pointing upwards and surmounted by a millstone. Later water wheels were vertical, either undershot (in which the water passed under the wheel) or overshot (in which the water was directed by a sluice on to the top of the wheel). The latter had the advantage of being operable almost independently of the level of water in the stream or river that supplied it.

Watson, James Dewey (1928–) US geneticist and biophysicist. He is known for his role in the discovery of the double-helix molecular structure of deoxyribonucleic acid (DNA) and shared the 1962 Nobel Prize for physiology or medicine with Francis CRICK and Maurice WILKINS. Watson later helped to break the GENETIC CODE of the DNA base sequences and found the ribonucleic acid messenger (MESSENGER RNA) that carries the DNA code to the cell's protein-forming structures.

Watson-Watt, Sir Robert Alexander (1892–1973) British physicist. As scientific adviser to the Air Ministry in 1940, he was a major influence in the rapid development of RADAR, which helped to detect German warplanes during World War 2. In 1941 he helped to establish the US radar system.

Watt, James (1736–1819) Scottish engineer. In 1765 Watt invented the condensing STEAM ENGINE, but he did not build a full-scale working model until some ten years later. The new engine was much more efficient than earlier designs and reduced the cost of running steam-powered machines by 75%. In 1782 he invented the double-acting engine, in which steam pressure acted alternately on each side of a piston. With Matthew BOULTON, Watt coined the term "HORSEPOWER". The unit of POWER is called the WATT in his honour.

watt (symbol W) SI unit of POWER. A machine consuming one JOULE of energy per second has a power output of one watt. One horsepower corresponds to 746 watts. An electric fire that uses, for example, 12 amps at 250 volts expends a power of 3,000 watts (or 3 kilowatts).

wattle and daub Ancient method of building, in which mud or clay (daub) is plastered on to wooden laths or branches (wattle) to form walls. The technique was used by itself, or to fill in spaces such as those of a half-timbered house.

wave In oceanography, moving disturbance travelling on or through water which does not move the water itself. Wind causes waves by frictional drag. The strength and size of the waves depends on the wind speed and the distance of open water across which the wind blows. Waves not under pressure from strong winds are called swells. Waves begin to break on shore or "feel bottom" when they reach a depth shallower than half the wave's length. When the water depth is about 1.3 times the wave height, the wave front is so steep that the top falls over and the wave breaks.

wave In physics, an oscillating disturbance in a material medium or in an electromagnetic field. In a **travelling** wave, the oscillations spread out (propagate) from the source. The velocity depends on the type of wave and on the medium. Electromagnetic waves, such as light, consist of varying magnetic and electric fields at right angles to each other and to the direction of motion; they are **transverse** waves. Sound waves are transmitted by the vibrations of the particles of the medium itself, the vibrations being in the direction of wave motion: they are **longitudinal** waves. Sound waves, unlike electromagnetic waves, cannot travel through a vacuum and cannot undergo POLARIZATION. Waves can undergo REFRACTION and REFLECTION, and give rise to INTERFERENCE phenomena. A wave is characterized by its WAVELENGTH and FREQUENCY, the VELOCITY of wave motion being the product of wavelength and frequency. *See also* ELECTROMAGNETIC RADIATION; STANDING WAVE

WATT, JAMES

Like earlier steam engines, James Watt's engine of the 1770s also condensed the steam with water, but in a separate condenser (1). This technique, and that of admitting steam to both sides of the piston (2), greatly increased the efficiency of the steam engine. Pushing the piston in both directions made it double-acting. Watt soon adapted it to produce rotary motion, used for machines other than pumps.

wave amplitude Peak value of a periodically varying quantity. This peak value may be either positive or negative, as the quantity varies either above or below its zero value (maximum or minimum).

wave dispersion Separation of the WAVELENGTHS in a wave because the REFRACTIVE INDEX of the medium is different for different wavelengths. It occurs with all electromagnetic radiation, but is most obvious at visible wavelengths, causing light to be separated into its component colours. Dispersion is seen when a beam of light passes through a refracting medium, such as a glass PRISM, and forms a SPECTRUM. Each colour has a different wavelength, and so the prism bends each colour in the light a different amount. Red (long wavelength) is bent less than violet (short wavelength). Dispersion can cause chromatic ABERRATION in lenses. *See also* REFRACTION

wave frequency Number of complete oscillations or wave cycles produced in 1 second, measured in HERTZ. It can be calculated from the wave velocity divided by wavelength. According to QUANTUM THEORY, the frequency of any ELECTROMAGNETIC RADIATION (such as light, radio waves and X-rays) is proportional to the energy of the component photons. Many of the characteristics of electromagnetic radiation depend on frequency.

wave front Contour on which at every point the medium or electromagnetic field is vibrating in step (that is, in phase). The wave front is usually at right angles to the direction of wave motion, and may be plane, spherical, or otherwise. Each point is a source of secondary wavelets, which give rise to a new wave front position a short time later.

wave function In QUANTUM MECHANICS, function that represents the probability of a quantum system being in a particular state s at time t, usually written $\psi(s)$ or $\Psi(s, t)$. The wave function is used in SCHRÖDINGER's equation.

wave interference Phenomenon in which two WAVES of the same WAVELENGTH and generally from a single source interact at a point, having travelled along different paths. If the two waves have a constant PHASE relationship – that is, if they are "in step" or "out of step" by a constant amount – they will interfere with each other. If maximum coincides with maximum, and minimum with minimum, the two waves will reinforce each other and the amplitude will be doubled; if maximum and minimum coincide, the two waves will cancel each other out, producing zero amplitude. With two light waves, an interference pattern of light and dark bands or rings can be seen on a screen in the overlap region.

wavelength (symbol λ) Distance between successive points of equal PHASE of a WAVE. For example, the wavelength of water waves could be measured as the distance from crest to crest. The wavelength of a visible light wave, which can range from 390 to 780nm (nanometres), determines its colour. Wavelength is equal to the wave VELOCITY divided by the WAVE FREQUENCY.

wave mechanics Version of QUANTUM MECHANICS developed in 1926 by Erwin SCHRÖDINGER. It

WAVE

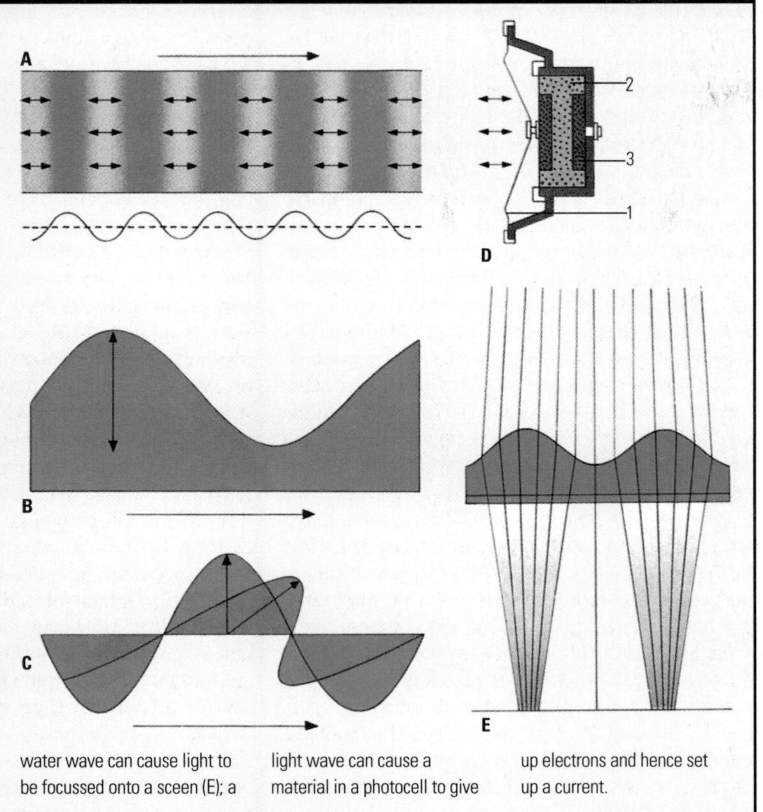

In longitudinal waves (A) the medium in which the wave is travelling moves backwards and forwards in the direction in which the wave is travelling (for example movement of air as sounds travel through it).In transverse waves (B) the motion of the medium is at right angles to the direction in which the wave travels (such as the movement of water waves). Electromagnetic waves (C) are transverse waves set up by high frequency electric and magnetic forces interlinked and acting at right angles. They can travel without a medium (such as light). To examine a wave the wave must be converted to some other form of energy: Sound waves cause (D) a microphone diaphragm (1) to vibrate, compressing the carbon granules (2), so altering the resistance between the blocks (3). A water wave can cause light to be focussed onto a sceen (E); a light wave can cause a material in a photocell to give up electrons and hence set up a current.

explains the behaviour of electrons in terms of their wave properties. Although quickly superseded by a more complex formulation by physicist Paul DIRAC, it is still widely used in calculations.

wave-particle duality Concept that combines the corpuscular theory, first outlined by Sir Isaac NEWTON, with the WAVE theory described by Christiaan HUYGENS. ELECTROMAGNETIC RADIATION, such as light, behaves like a wave when being transmitted or moved, and like particles when interacting with matter. INTERFERENCE, DIFFRACTION and POLARIZATION can be described best in terms of waves, whereas the PHOTOELECTRIC EFFECT and the COMPTON EFFECT can be described in terms of PHOTONS. *See also* PARTICLE PHYSICS; PLANCK'S CONSTANT; QUANTUM THEORY

wax Solid, insoluble substance of low melting point that is mouldable and water-repellent. There are three forms of wax: animal, vegetable and mineral. The first two are simple LIPIDS consisting of esters of fatty acids; animal waxes include beeswax, wool wax (LANOLIN) and spermaceti wax from the sperm whale; vegetable waxes include carnauba (wax palm) from the leaves of the Brazilian carnauba palm tree, and candelilla from a Mexican rush. Mineral waxes include PARAFFIN wax made from petroleum. Synthetic waxes are of diverse origins and include POLYETHENES. Waxes are used in the manufacture of lubricants, polishes, cosmetics and candles and to waterproof leather and to coat paper.

W boson (symbol W^+ or W^-) ELEMENTARY PARTICLE that is a carrier of the WEAK NUCLEAR FORCE. It carries a unit electric charge and has a mass approximately 80 times that of the PROTON. *See also* BOSON

weak nuclear force (weak interaction) One of the four FUNDAMENTAL FORCES in physics. The weak nuclear force causes radioactive decay. It can be observed only in the subatomic realm, being of very short range. It is weaker than the ELECTROMAGNETIC FORCE and the STRONG NUCLEAR FORCE (the strongest of the forces) but stronger than GRAVITATION.

weather State of the ATMOSPHERE at a given locality or over a broad area, particularly as it affects human activity. Weather refers to short-term states (days or weeks) as opposed to long-term CLIMATE conditions. Weather involves such elements as atmospheric temperature, pressure, humidity, precipitation, cloudiness, brightness, visibility and wind. *See also* CLIMATOLOGY; METEOROLOGY

weather forecasting Technique of anticipating and describing future WEATHER for a given time and place. There is a three-fold classification depending on the length of the period covered. **Short-term forecasts** cover the next 24 hours and include wind velocity, cloud cover and temperature. **Medium-range forecasts** look two to five days ahead and contain information on the dominant circulatory conditions – the paths of CYCLONES and ANTICYCLONES. **Long-range forecasts** cover periods of five days and more ahead and are more extensive medium-range forecasts. Pressures and circulatory patterns are usually given as averages. Artificial SATELLITES have increased the accuracy of weather forecasting, especially when dealing with severe turbulences such as hurricanes.

weathering In geology and physical geography, the breakdown and chemical disintegration of rocks and minerals at the Earth's surface by physical, chemical and organic processes. It is important in the formation of soil and plays a major part in shaping landscapes. In **mechanical** (physical) weathering in cold, wet climates, water seeping into cracks in the rock expands on freezing, so causing the rock to crack further and to crumble. Extreme temperature changes in drier regions, such as deserts, also cause rocks to fragment. Cracks opened by the weather can be exploited by plants, the roots of which place further stress on the rock. **Chemical** weathering can lead to a weakening of the rock structure by altering the minerals of a rock and changing their size, volume and ability to hold shape. The best-known example is the dissolution of limestone in acid rain water. Chemical weathering processes include oxidation, hydration, silication, desilication and carbonation. **Organic** weathering describes the breakdown of rock and soil by plants, such as by root action, and animals, such as worms, which are important for the breaking up of soil. Unlike EROSION, weathering does not involve the transportation of broken-down material.

weather observations Furnishing the data required for WEATHER description, analysis and WEATHER FORECASTING from thousands of weather stations and instruments around the globe, such as ships, aircraft, balloons and weather SATELLITES. Weather stations measure and record weather conditions in their vicinity and survey with radar up to 100 to 200mi (161–320km). Information about upper-air conditions is obtained from RADIOSONDE balloons, from high-flying aircraft, and from stratospheric balloons operating for long periods at heights up to 32km (20mi). Weather satellites obtain day and night observations in visible and infrared wavelengths. They provide pictures of cloud cover and storm disturbances, data on vertical temperature and moisture profiles of the atmosphere, making accurate long-term weather forecasts possible.

weather vane Device to indicate wind direction. It comprises an upright rod with cross-members

WEATHERING

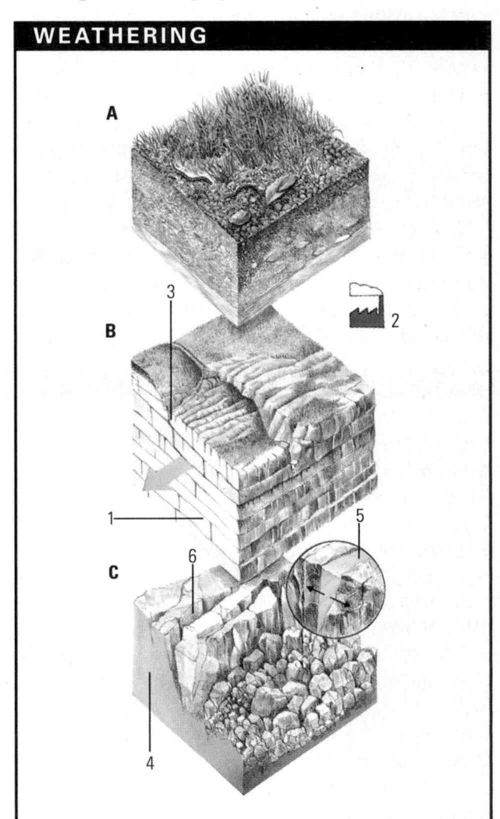

Weathering is the breakdown of rock in place. It occurs in two main ways: physical (A and C) and chemical (B). They usually occur in combination. At the surface, plant roots and animals such as worms break down rock turning it into soil (A). In chemical weathering (B), soluble rocks such as limestone (1) are dissolved by groundwater, which is a very mild solution of carbonic acid. Acid rain caused by sulphate pollution (2) also attacks the rock. The water can create cave systems deep below the surface. Both heat and cold can cause physical weathering (C). When temperatures drop below freezing, freeze-thaw weathering can split even the hardest rocks, such as granite (4). Water that settles in cracks and joints during the day, expands as it freezes at night (5). The expansion cleaves the rock along the naturally occurring joints (6). In deserts, rock expands and contracts due to the extremes of cooling and heating, resulting in layers of rock splitting off.

bearing the cardinal points of the compass and a revolving arm. The arm is arrow-shaped at one end and has a vane at the other, the large surface area of which is designed to react to wind pressure and point the arrow in the direction from which the wind is blowing. One of the oldest meteorological instruments, vanes are still widely used.

weber (symbol W) SI unit of MAGNETIC FLUX. It is defined so that an ELECTROMOTIVE FORCE (EMF) of 1 volt is induced in a circuit of a single loop when the flux changes through it at a steady rate of 1 W s^{-1}.

Weber, Ernst Heinrich (1795–1878) German physiologist who pioneered the study of sensation and perception. He laid the foundations for the branch of psychology called "psychophysics" and encouraged psychologists to be more scientific and methodical.

Weber, Wilhelm Eduard (1804–91) German physicist who, in 1846, standardized the units used in ELECTRICITY, relating them to the fundamental dimensions of mass, length, charge and time. He was the first physicist to regard electricity as composed of elementary units each carrying a charge, now known as ELECTRONS. The SI UNIT of magnetic flux is named after him.

web offset Method of PRINTING in which the press utilizes continuous rolls (webs) of paper and lithographic printing plates on rotary printing machines. Web offset is widely used in the magazine and newspaper industry, where speed is a vital production factor. *See also* LITHOGRAPHY; OFFSET PRINTING

weed Uncultivated or unwanted plant (a pest). Weeds are a threat to commercial crops because they compete for water and sunlight and harbour pests and diseases that can spread to the crop plants. HERBICIDES are used to kill weeds.

Wegener, Alfred Lothar (1880–1930) German geologist, meteorologist and explorer. In *The Origin of Continents and Oceans* (1915) he was the first to use well-presented evidence and scientific argument in support of a theory of CONTINENTAL DRIFT.

weight Force of attraction on a body due to GRAVITY. A body's weight is the product of its MASS and the gravitational field strength at that point. Mass remains constant, but weight depends on the object's position on the Earth's surface, decreasing with increasing altitude.

weightlessness Condition experienced by an object for which the effects of WEIGHT are not apparent. Weightlessness can be experienced in space and during a free fall, even though gravitational attraction on the "weightless" body is still present. Astronauts are trained in the weightless conditions that exist briefly in an aircraft flying in an arc and in water tanks. The adverse effects on the human body of prolonged weightlessness include decreased circulation of blood, less water retention in tissues and the bloodstream, loss of bone calcium and loss of muscle tone.

weights and measures Agreed units for expressing the amount of some quantity, such as capacity, length or weight. Early measurements were based on body measurements, plant grains, and so on. The French introduced the METRIC SYSTEM in 1799 in which the unit of length, the METRE, was taken as one ten-millionth of the distance from the Equator to the North Pole. The unit of mass was the GRAM, defined as the mass of a cubic centimetre of water. The metric system has developed into the system of SI UNITS.

Weinberg, Steven (1933–) US physicist who in 1967, independently of Pakistani physicist Abdus SALAM, proposed a theory, known as the ELECTROWEAK THEORY, that unifies the electromagnetic and weak interactions between subatomic particles. Later experiments proved the Salam-Weinberg hypothesis to be true. In 1979 they shared the Nobel Prize for physics with Sheldon GLASHOW, who had earlier proposed a similar theory.

Weismann, August (1834–1914) German biologist. His essay discussing the germ plasm theory, *The Continuity of the Germ Plasm* (1885), proposed the immortality of the germ line cells as opposed to body cells. It was influential in the development of modern genetic study.

welding Technique for joining metal parts, usually by controlled melting. Vehicles, domestic appliances, bridges and electronic components are just some of the many items that have parts joined by welding. Several welding processes are used. In fusion welding, the parts to be joined are heated together until the metal starts to melt. On cooling, the molten metal solidifies to form a permanent bond between the parts. Such welds are usually strengthened with filler metal from a welding rod or wire. In **arc welding**, an electric arc heats the work and filler metal. In **oxyacetylene welding**, the heat is provided by burning acetylene gas in oxygen; in **resistance** or **spot welding**, the heat is generated by passing an electric current through the joint. Spot welds are made at regular intervals to join sheet metal. In brazing and soldering, the temperature used is sufficient to melt the filler metal, but not the parts that it joins. Special techniques include the use of friction, electron beams and ultrasonics to cause bonding. Some plastics can also be heat-welded.

well Shaft sunk vertically in the Earth's CRUST through which water, oil, natural gas, brine, sulphur or other mineral substances can be extracted. Artesian wells are sunk into water-bearing rock strata, the so-called AQUIFERS, from which water rises under pressure in the wells to the surface. *See also* OIL WELL

Weller, Thomas Huckle (1915–) US bacteriologist and virologist. He shared the 1954 Nobel Prize for physiology or medicine with John F. ENDERS and Frederick C. ROBBINS for the discovery that the poliomyelitis virus can grow in cultures of various types of tissue, so making it possible to develop a VACCINE for polio.

Welsbach, Carl Auer, Freiherr von (1858–1929) Austrian chemist, inventor of the Welsbach mantle – a shaped gauze which, when heated by gas, produces a bright light. In 1885 he discovered NEODYMIUM and PRASEODYMIUM.

Werner, Abraham Gottlob (1750–1817) German geologist who was the first to classify minerals systematically. His theory of the Earth's origins, called **neptunism**, proposed that the Earth was originally a vast ocean from which solid rocks were precipitated to form land.

Werner, Alfred (1866–1919) Swiss chemist who was awarded the 1913 Nobel Prize for chemistry. After working with Marcelin BERTHELOT in Paris, he returned to a professorship at Zurich University. Werner was awarded the Nobel Prize for his coordination theory of VALENCE in which he correctly suggested that some metals can form secondary, or coordinate, bonds. *See also* CO-ORDINATION COMPOUND

wernerite *See* SCAPOLITE

Westinghouse, George (1846–1914) US engineer and inventor. In 1865 he received his first patent, for a rotary engine, but the best known of his hundreds of inventions was the air brake, which made high-speed rail travel safe. He formed the Westinghouse Electric Company in 1886.

Weston standard cell (CADMIUM cell) Primary voltaic CELL, used as a standard of ELECTROMOTIVE FORCE (EMF). It produces a constant 1.018636 volts at 20°C (68°F). The Weston cell comprises a mercury anode and a cadmium amalgam cathode. The electrolyte is a saturated solution of cadmium sulphate.

wetland Marshy ground in an intertidal zone that has prolific vegetation; coastal wetlands are said to contain a greater concentration of living matter, both flora and fauna, than any other kind of terrain.

Weyl, Herman (1885–1955) US mathematician, b. Germany. After holding professorships at Zurich and Göttingen universities he moved to Princeton University, USA, in 1933. He worked in GROUP THEORY and QUANTUM THEORY, and also on NUMBER THEORY and on GEOMETRY particularly relevant to general RELATIVITY. His findings play an important role in modern PARTICLE PHYSICS, particularly in STRING THEORY. His *Group Theory and Quantum Mechanics* (1928) is a classic text of quantum theory.

Wheatstone, Sir Charles (1802–75) British physicist and inventor. In 1843, with William COOKE, he improved a device that accurately measures electrical RESISTANCE. Widely used in laboratories, it became known as the WHEATSTONE BRIDGE. By 1837 they had patented an electric TELEGRAPH. Wheatstone also coined the term "microphone" for a sound magnifier.

Wheatstone bridge Electric circuit used for measuring RESISTANCE, named after Sir Charles WHEATSTONE. It consists of four resistances arranged to form a square with an ELECTROMOTIVE FORCE (EMF) (voltage from a battery) across one diagonal, and a GALVANOMETER across the other. When the galvanometer shows no deflection, the ratio of the values of one adjacent pair of resistances equals the ratio of the values of the other two. By adjusting the ratio of one pair (with a sliding contact along a wire which forms one of the adjacent pairs of resistances) the unknown resistance of one of the other two can be calculated.

wheel Circular structure that revolves around a central axis. Before the wheel was invented, heavy loads were sometimes moved by rolling them on logs or on rounded stones. More than 5,000 years ago sections of tree trunks were cut to form the first wheels for carts. Spoked wheels were introduced several hundred years later. Eventually, the wheel was used in simple machines, such as the water wheel and potter's wheel.

wheel and axle Machine based on the principle that a small force applied to the rim of a wheel will exert a larger force on an object attached to the axle. The FORCE RATIO, or mechanical advantage, is the ratio of the radius of the wheel to that of the axle.

Whinfield, John Rex (1901–66) British chemist who, with fellow British scientist John Dickson, invented the synthetic fibre TERYLENE in 1941. His interest was stimulated by the discovery of nylon by Wallace CAROTHERS in 1935. In 1946 terylene was made in the USA and sold as Fibre V and then as Dacron, and in 1947 Imperial Chemical Industries (ICI) began to make and market it to the rest of the world. Whinfield received little recognition for his great invention, although he was awarded the CBE in 1954.

Whipple, Fred Lawrence (1906–) US astronomer best known for his "dirty snowball" theory of comets – proposed in 1949 but not proven until 1986, when space probes were sent to HALLEY'S COMET. He discovered several comets and worked on cometary orbits. He also studied planetary nebulae, stellar evolution and the Earth's upper atmosphere.

whirlpool Circular motion of a fluid. Whirlpools in rivers occur in regions where waterfalls or sharp breaks in topographic continuity make steady flow impossible. *See also* VORTEX

White, Gilbert (1720–93) British naturalist. He held curacies at Selborne in Hampshire and devoted himself to the study of natural history around his parish. His famous *The Natural History and Antiquities of Selborne* (1789) consists of letters to his fellow naturalists.

white blood cell Alternative name for a LEUCOCYTE

white dwarf Type of star about the size of the Earth, but with a mass about that of the Sun. As a result, its density is enormously greater than that of any terrestrial material. This is because the normal atomic structure is broken down completely, with electrons and nuclei packed tightly together. A white dwarf cannot have a mass of more than about 1.4 solar masses (the CHANDRASEKHAR limit). For larger masses, gravity would always overwhelm the pressure of the electrons and the star would collapse under its own weight, forming a NEUTRON STAR or BLACK HOLE. White dwarfs are of low luminosity and gradually cool down to become cold, dark objects. They represent the final stage in the evolution of low-mass stars after they have lost their outer layers. The first white dwarf to be discovered was the companion of SIRIUS. *See also* STELLAR EVOLUTION

white hole Hypothetical localized point in space at which matter is suddenly created with explosive violence. It can be considered as being similar to a BLACK HOLE but with time reversed. The cosmological BIG BANG is an analogous phenomenon that embraced the whole Universe rather than being localized.

Whitney, Eli (1765–1825) US inventor and manufacturer. He invented the cotton gin (1793), which revolutionized cotton picking in the South and turned cotton into a profitable export. After 1798 he manufactured muskets at a factory in New Haven, Connecticut, which was one of the first to use MASS PRODUCTION methods with interchangeable parts.

Whittle, Sir Frank (1907–96) British aeronautical engineer who pioneered the development of the TURBOJET ENGINE. He entered the Royal Air Force (RAF) as an apprentice and later studied at the RAF engineering school and at Cambridge University. In 1930 he patented his first turbojet engine. In 1936 he founded a company called Power Jets and was later given special permission to carry out development work on jet propulsion. A plane powered by the Whittle engine had its first flight in May 1941 and it entered service with the RAF in 1944. He was knighted in 1948.

Whitworth, Sir Joseph (1803–87) British engineer. After studying engineering in Manchester he went to London, where he invented a technique for making perfectly flat metal surfaces. He returned to Manchester, where he produced extremely accurate machine tools. A system of threads for nuts and bolts is named after him.

whole number *See* NUMBER, NATURAL

Widmanstätten patterns Lines surrounded by dark bands found in the class of METEORITES called IRON METEORITES (composed entirely of metal). They are formed when the liquefied outer surface of a meteorite cools on entering the Earth's atmosphere, and they can be used to determine the meteorite's motions in flight. They appear when the meteorite is treated with nitric acid. Widmanstätten patterns are named after Count Aloys Joseph von Widmanstätten (1754–1849), who observed them in 1808.

Wieland, Heinrich Otto (1877–1957) German chemist who was awarded the 1927 Nobel Prize for chemistry for his research into BILE acids. In 1912 he discovered that the three bile acids then known were closely related in structure. He showed them to have a STEROID skeleton and thus found that they were also structurally related to CHOLESTEROL. Wieland also did research into oxidation reactions (*see* OXIDATION-REDUCTION) occurring in living tissues and discovered that such oxidation consisted of dehydrogenation (removal of hydrogen atoms), not the addition of oxygen.

Wien, Wilhelm (1864–1928) German physicist who was awarded the 1911 Nobel Prize for physics for his work on RADIATION. He investigated the radiation from BLACK BODIES and in 1893 calculated that the peak frequency of radiation increases with temperature. This is called **Wien's displacement law**.

Wiener, Norbert (1894–1964) US mathematician and originator of CYBERNETICS. He contributed much to the study and development of the COMPUTER and to the understanding of FEEDBACK systems that control the behaviour of humans and machines. Wiener developed a mathematical model of BROWNIAN MOVEMENT and also explained ENTROPY, which is a measure of disorder.

Wigner, Eugene Paul (1902–95) US physicist, b. Hungary, who worked on the MANHATTAN PROJECT during World War 2 to develop the atom bomb. Wigner was the first physicist to apply GROUP THEORY to QUANTUM MECHANICS. With this technique he discovered the law of conservation of PARITY. For his work on the structure of the atomic nucleus, Wigner shared the 1963 Nobel Prize for physics with the German physicist Hans Jensen and the German-born US physicist Maria Goeppert-Mayer. He also received the 1960 Atoms for Peace Award.

Wigner effect In nuclear engineering, the change in the physical properties of a MODERATOR (such as graphite) as a consequence of irradiation. In a NUCLEAR REACTOR, high-energy NEUTRONS displace atoms in the CRYSTAL LATTICE resulting in the accumulation of stored energy (**Wigner energy**). This accumulation changes the lattice and the overall size of the moderator. It was described first by Eugene WIGNER.

Wiles, Andrew (1953–) British mathematician, the number theorist who proved FERMAT'S LAST THEOREM. As a child, Wiles had been fascinated with NUMBER THEORY and, in particular, the conjectured last theorem of Fermat. After many years, working largely alone, he announced in 1993 that he had proved the theorem. Though subsequent checks revealed a gap in the proof, Wiles soon submitted a completed version, which was widely accepted in 1995. His achievement is one of the greatest in 20th-century mathematics.

Wilkins, Maurice Hugh Frederick (1916–) British biophysicist, b. New Zealand. During World War 2 he worked on the separation of uranium isotopes for the Manhattan Project. Wilkins is celebrated for his work in molecular biology, in particular the structure of NUCLEIC ACID. He extracted some fibres from a gel of DNA while working at the Biophysics Research Unit at King's College, London. Using X-ray diffraction, he noted the helical structure of the strands. Francis CRICK and James WATSON used this research to build a model of the structure of DNA and to describe its action. For this work the three men shared the 1962 Nobel Prize for Physiology or Medicine.

Wilkinson, Sir Geoffrey (1921–96) British chemist. He shared the 1973 Nobel Prize for chemistry with Ernst FISCHER for their independent work on ORGANOMETALLIC sandwich compounds. Wilkinson demonstrated the new chemical structure in which an iron atom is "sandwiched" between two rings of carbon and hydrogen (organometallic compounds have metal or metalloid atoms bound to carbon atoms or organic groups). He was appointed assistant professor at Harvard University in 1951 and became professor of inorganic chemistry at Imperial College, London, in 1955.

will-o'-the-wisp (jack-o'-lantern) Mysterious light sometimes seen at night in marshy areas. It is not well understood but is thought to be due to the spontaneous combustion of marsh gas (METHANE). Its elusive nature makes it the basis for various superstitious beliefs.

Willstätter, Richard (1872–1942) German chemist who revived the technique of CHROMATOGRAPHY, first discovered by the Russian chemist Mikhail Tsvett (1872–1920). Willstätter used it to partly deduce the structure of CHLOROPHYLL. He showed that it contained a magnesium atom surrounded by pyrrole rings and demonstrated its relationship to HAEMOGLOBIN. For this work Willstätter was awarded the 1915 Nobel Prize for chemistry.

Wilson, Charles Thomson Rees (1869–1959) British physicist who invented the Wilson CLOUD CHAMBER used to study radioactivity, X-rays and cosmic rays. It uses water droplets to track ions left by passing radiation. For this work, he shared the 1927 Nobel Prize for physics with Arthur COMPTON. Wilson also devised a way of protecting World War 2 barrage balloons from lightning.

Wilson, Edmund Beecher (1856–1939) US biologist whose research mainly concerned embryology and cytology. He traced the formation of different kinds of tissues from individual precursor cells. He also studied the relationship of CHROMOSOMES to sex determination and the hereditary function of chromosomes.

Wilson, John Tuzo (1908–) Canadian geophysicist and geologist who determined global patterns of faulting and the structure of the continents. His investigations have influenced theories of CONTINENTAL DRIFT, SEAFLOOR SPREADING and convection currents within the Earth.

Wilson, Robert Woodrow (1936–) US physicist. He shared the 1978 Nobel Prize for physics with Arno PENZIAS for their discovery, in 1964, of the residual background radiation known as COSMIC MICROWAVE BACKGROUND. This phenomena provides strong evidence for the BIG BANG theory.

wilting Drooping of plants due to shortage of water. Insufficient water causes a decline in the TURGOR PRESSURE of plant cells.

winch Drum that turns to pull or release a rope, cable or chain. A winch is used to raise, lower or pull heavy loads. It may be motor-driven or operated by hand. A winch normally has a pawl and RATCHET wheel to prevent accidental release of a load.

wind Air current that moves rapidly parallel to the

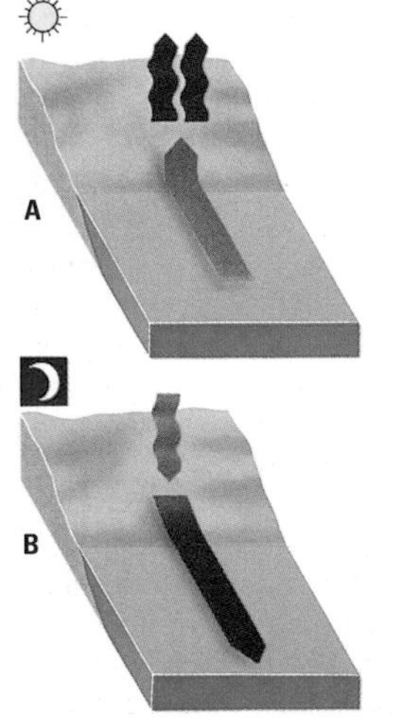

◀ **wind** Onshore winds (A) generally occur during the day. The land is heated by the Sun, causing the air over it to rise. As the warm air rises, it is replaced by cooler air overlying the sea. At night, because the land loses heat quicker than the sea, air flows down hillsides out to sea, where the air is relatively warmer, generating offshore breezes (B).

W

Earth's surface. (Air currents in vertical motion are called updrafts or downdrafts.) Wind direction is indicated by wind or WEATHER VANES, wind speed by ANEMOMETERS and wind force by the BEAUFORT WIND SCALE. Steady winds in the tropics, such as those in the DOLDRUMS, are called TRADE WINDS. MONSOONS are seasonal winds that bring predictable rains in Asia. **Foehns** (föhns) are warm, dry winds produced by adiabatic compression (compression accompanied by temperature rise) as air descends the lee of mountainous areas in the Alps; they are called **chinooks** in the Rockies. **Siroccos** are hot, dry Mediterranean winds. The MISTRAL is a cold, northerly wind that sweeps from the Massif Central of France down the Rhône valley during winter.

Windaus, Adolf (1876–1959) German chemist. In 1907 he synthesized histamine and in 1928 he was awarded the Nobel Prize for chemistry for his work on the structure of STEROIDS and the photochemistry of VITAMIN D.

windchill factor Perceived cooling effect of WIND in low-temperature conditions. The air temperature feels much colder if there is a wind blowing, and the "physiological temperature" is strongly influenced not just by the wind speed but also by its direction: an easterly wind in Britain in January, for example, will feel far colder than a south-westerly wind of similar speed. If the wind reaches speeds around 40km/h (25mph) temperatures of 5°C can feel very cold, and in sub-zero weather even moderate winds of 10km/h (6mph) can significantly reduce effective temperatures. With a wind speed of 80km/h (50mph), a zero temperature would feel like –20°C (–4°F)

windlass Hauling device operated by hand, most familiar as that used to raise buckets of water from a well. It consists of a cylinder around which the haulage rope is wound and which is turned by a hand lever or crank, using the principle of the WHEEL AND AXLE.

windmill Machine powered by the wind acting on sails or vanes. The earliest known windmills were built in the Middle East in the 7th century. The idea spread to Europe in the Middle Ages. Their use was widespread during the early years of the Industrial Revolution, but declined with the development of the steam engine in the 19th century. Today modern windmills are used to generate electricity on wind farms. *See also* RENEWABLE ENERGY; WATER MILL

window In computing, a rectangle displayed on the screen of a computer which displays what is stored in a particular part of the machine's memory or in some other storage device. A window may show text, graphics or other work in progress. The contents of a window are shown in text, often accompanied by symbols called icons.

windpipe *See* TRACHEA

wind power Technology of harnessing the WIND to provide ENERGY to drive machines and to generate electricity. Since the 1970s, advanced aerodynamic designs have been used to build wind turbines that generate electricity. The largest of these is on the island of Hawaii. It has two blades, each 50m (160ft) long, attached to a 20-storey high tower. Individual turbines are often grouped in strategic locations (wind farms) to maximize the generating potential. Wind power is a cheap form of RENEWABLE ENERGY, but cannot as yet produce sufficiently large amounts of electricity to provide a realistic alternative to fossil fuel and nuclear power stations. *See also* WINDMILL

wind sock Flexible, cone-shaped device that indicates wind direction and force. The sock is designed to swivel so that the small end of the cone points downwind. An idea of wind speed can be judged from the sock's extension. *See also* ANEMOMETER; WEATHER VANE

wind tunnel Chamber in which scale models and even full-size aircraft and road vehicles are tested in a controlled airflow. Some wind tunnels can reproduce extreme conditions of wind speed, temperature and pressure. Models of bridges and other structures are tested in wind tunnels to check that winds cannot set up destructive vibrations.

wings In biology, specialized organs for flight which are possessed by most birds, many insects and certain mammals and reptiles. The forelimbs of a bird have developed into such structures. Bats have membranous tissue supported by the digits ("fingers") of the forelimbs. Insects may have one or two pairs of veined or membranous wings.

wire Strand of metal, made by drawing a rod through progressively smaller holes in metal dies. The drawing process toughens steel, so that a rope or cable made from steel wire is much stronger than an undrawn steel rod of the same diameter. Copper and aluminium wires are used to make electric CABLES. If flexibility is important, each conductor is made of several fine strands of wire instead of one thick one.

Withering, William (1741–99) British doctor who introduced the use of digitalis (a drug obtained from the leaves of the foxglove) to treat cardiac disorders, giving details of his work in *An Account of the Foxglove* (1785). He was the first to establish a connection between dropsy and heart disease. Also a keen botanist, he published *Botanical Arrangement* (1776).

witherite Glassy white, yellow or grey mineral, barium carbonate ($BaCO_3$), found in low-temperature lead and barite ore veins. It occurs as twinned, near-hexagonal crystals in the orthorhombic system. Hardness 3–3.5; r.d. 4.5.

Wöhler, Friedrich (1800–82) German chemist who first isolated ALUMINIUM and BERYLLIUM and discovered calcium carbide. His synthesis of UREA

WIND POWER

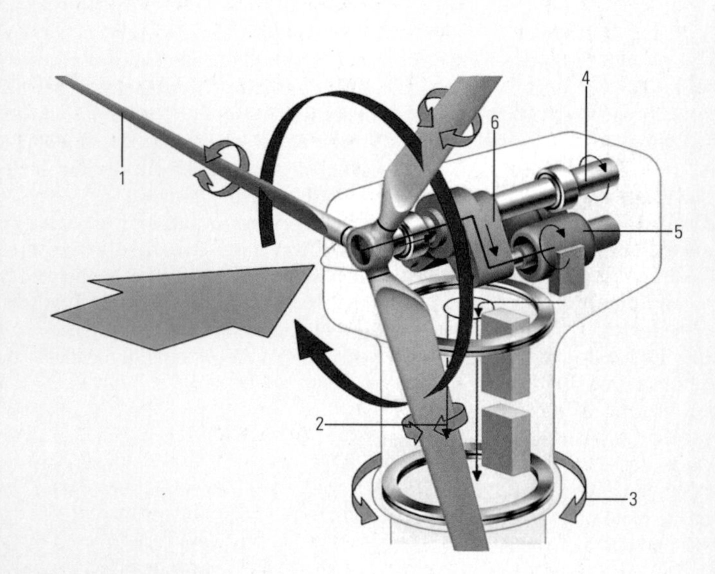

A wind generator converts the energy of the wind into electricity. Three-bladed, variable pitch designs are the most efficient. Variable pitch means the attitude of the blades (1) can be changed. By altering the pitch (2) of the blades, they can generate at maximum efficiency in varying wind conditions. The whole rotor assembly rotates (3) into the wind. The blades turn a prop shaft (4), which links to a generator (5) through gearing (6). The largest wind farms have thousands of linked turbines and can produce the same power as a fossil fuel power station. The Nordfriesland windpark (right), near Bievul, N Germany, consists of fifty wind turbines each producing 250 kilowatts of electricity. The total capacity for the park is 12.5 megawatts.

(from the inorganic substance ammonium cyanate) in 1828 was the first synthesis of an organic chemical compound from an inorganic one, and contributed to the foundation of modern organic chemistry.

Wolf, Max (Maximilian Franz Joseph Cornelius) (1863–1932) German astronomer. He was a pioneer of photographic methods in astronomy. In 1891 he made the first photographic discovery of an asteroid (Brucia, no. 323); in all he made 232 asteroid discoveries. He also detected several new nebulae.

wolfram *See* TUNGSTEN

wolframite Black to brown mineral, iron-manganese tungstate, $(Fe,Mn)WO_4$. It is the chief ore of the metal TUNGSTEN. It occurs as crystals in the monoclinic system or as granular masses. It is found in quartz veins and pegmatites associated with granitic rocks and also in high-temperature hydrothermal veins in association with other minerals. It is found extensively in Australia, China and India. Hardness 5–5.5; r.d. 7–7.5

Wolf–Rayet star (WR star, W star) Type of star, the spectrum of which contains bright emission lines rather than dark absorption ones. WR stars are divided into two kinds. In the WN type, emission lines from nitrogen dominate the spectrum. In the WC type, emission lines of carbon and oxygen predominate. Both types have strong lines of helium and a few have moderate or weak lines of hydrogen as well. WR stars are very hot, with surface temperatures between 25,000 and 50,000K, luminosities between 100,000 and 1 million times the Sun's, and masses from 10 to 50 solar masses. They have strong stellar winds that carry away 3 solar masses per million years. Many central stars of planetary nebulae are WR stars. Wolf–Rayet stars are named after their French discoverers, Charles Joseph Etienne Wolf (1827–1918) and Georges Antoine Pons Rayet (1839–1906).

Wollaston, William Hyde (1766–1828) British scientist. He developed a method of making PLATINUM malleable, discovered PALLADIUM and RHODIUM, and invented the reflecting GONIOMETER, the CAMERA LUCIDA, and a lens (named after him) for correcting spherical ABERRATION. He also did research in the fields of pharmacology, pathology, physiology and botany.

womb *See* UTERUS

wood Hard substance that forms the trunks of TREES; it is the XYLEM (the vascular tissue of a woody plant) that comprises the bulk of the stems and roots, supporting the plant. It consists of fine cellular tubes arranged vertically within the trunk, which accounts for the grain found in all wood. The relatively soft, light-coloured wood is called **sapwood**. The non-conducting, older, darker wood is called **heartwood**, and is generally filled with RESIN, gums, mineral salts and TANNIN (tannic acid). Easily worked softwood, generally from a CONIFER such as pine, is composed of simple tracheids that provide support and conduct water and food. More durable hardwood, generally from a DECIDUOUS species such as oak, derives support from woody fibres; water and food are conducted through separate vessels. Wood is still commonly used as a building material, fuel, to make some types of PAPER, and as a source of CHARCOAL, CELLULOSE, ESSENTIAL OIL, LIGNIN, tannins, DYES and SUGAR.

Woodward, Sir Arthur Smith (1864–1944) British geologist. He was keeper of Geology at the British Museum (1901–24) but is popularly remembered for his part in the Piltdown Man controversy; his belief that the remains, which appeared to be the "missing link" in the evolution of man from the apes, were human contributed to the success of the hoax. In 1953 it was proved that the jaw belonged to a modern ape, stained to appear like a prehistoric bone.

◀ **Wright brothers** Photograph of the first powered flight, made by Orville Wright on 17 December 1903 near Kill Devil Hill, Kitty Hawk, North Carolina. Wright can be seen lying on the lower wing of the 12hp, chain-driven Flyer 1. The photo was taken as the aircraft left the ground at the end of the take-off rail (visible on left). The flight lasted for about 12 seconds, covering a distance of 36.5m (120ft) at an airspeed of 48 km/h (30 mph), a groundspeed of 10.9 km/h (6.8mph) and an altitude of 2.5–3.5m (8–12ft).

Woodward, Robert Burns (1917–79) US chemist. He became professor at Harvard in 1950 and was awarded the 1965 Nobel Prize for chemistry in recognition of his synthesis of a number of complex organic substances, including quinine, cholesterol, cortisone, strychnine, lysergic acid, reserpine, chlorophyll and tetracycline.

wool Soft, generally white, brown or black animal fibre that forms the fleece (coat) of sheep. Wool is also the name of the yarns and textiles made from the fibres after spinning, dyeing and weaving. In most sheep, selective breeding has reduced or eliminated a coarse outer coat of long hairs (still present in the moufflon) leaving a wooly undercoat. Composed chiefly of KERATIN, the fibres are treated to remove a fat called LANOLIN, which is used in some ointments.

word processor COMPUTER system used for writing and printing text. The system may be designed just for this purpose, in which case it is called a **dedicated word processor**. More common is a general-purpose personal computer running a **word processing program**. Text typed on the computer keyboard is displayed on the screen. Any errors are easily corrected before a "printout" or "hard copy" is produced on a printer connected to the computer. If required, the text can be stored in code on a magnetic disk for future use.

work In physics, ENERGY transferred in moving the point of application of a FORCE. It equals the magnitude of the force multiplied by the distance moved in the direction of the force. If the force opposing movement is the object's weight mg (where m is the object's mass and g is the acceleration due to gravity), the work done in raising it a height h is mgh. This work has been transferred to the object in the form of POTENTIAL ENERGY; if the object falls a distance x, the KINETIC ENERGY that it gains equals the work done in raising it through the height x.

work function Energy required to liberate an ELECTRON from a material. It is important in the PHOTOELECTRIC EFFECT and in THERMIONICS.

World Health Organization (WHO) Intergovernmental, specialized agency of the United Nations. Founded in 1948, it collects and shares medical and scientific information and promotes the establishment of international standards for drugs and vaccines. WHO has made major contributions to the prevention of diseases such as malaria, polio, leprosy and tuberculosis. It helps to fund research into other diseases such as AIDS. Its headquarters is in Geneva, Switzerland.

World Meteorological Organization (WMO) Intergovernmental organization, founded as the International Meteorological Organization in 1873. It has been a specialized agency of the United Nations since 1950. WMO promotes international cooperation in meteorology through the establishment of a network of meteorological stations throughout the world, and by the mutual exchange of weather information. Its headquarters is in Geneva, Switzerland.

World Wide Web (WWW) Name given to a series of COMPUTER NETWORKS that can be accessed via local servers, which in turn are serviced by telephone lines. It consists of a network of sites which users can access via the INTERNET to retrieve or post data. Web documents may include text, graphics and sound, and have HYPERTEXT that the user can click on to access further related information from other Web documents at the same site or another site anywhere in the world.

Worthington, Henry Rossiter (1817–80) US engineer and inventor. He invented a direct steam pump and set up a factory for its manufacture in New York in 1859. Worthington also developed the duplex steam feed pump, still much used in oil pipelines and in waterworks.

Wright brothers Wilbur (1867–1912) and Orville (1871–1948), US aviation pioneers. The Wright brothers assembled their first aircraft in their bicycle factory. They experimented with model gliders and built a WIND TUNNEL to study the effects of using various wing shapes. In 1903 Orville made the first recorded piloted flight in a power-driven plane at Kitty Hawk, North Carolina. This flight lasted just 12 seconds, and attracted little attention. But the brothers eventually convinced manufacturers to invest in powered aircraft.

wrought iron Commercial form of smelted IRON (the other is cast iron), containing less than 0.3% carbon with 1 or 2% SLAG mixed with it. Originally it was made from ore in a forge, and later in a puddling furnace, where it never becomes molten. Wrought iron replaced bronze in Asia Minor (*c.*2000 BC) at the beginning of the IRON AGE. In the 19th century wrought iron began to be used in building construction, but was replaced by steel after the invention of the BESSEMER and OPEN-HEARTH PROCESSES. Today wrought iron is used principally for decoration, as in ornamental gates and railings.

xanthine Yellow substance found in plants, in most body tissues and fluids, and in urinary tract stones, and which can be oxidized to form URIC ACID. It is a muscle stimulant, particularly of cardiac muscle, and its synthetic derivatives are used as DIURETICS and to dilate blood vessels and bronchi.

xanthophyll Plant pigment responsible for the yellow and brown colours of leaves in autumn. Chemically it is a CAROTENOID, a group of compounds that includes CAROTENE, another pigment which gives the red colour to carrots and tomatoes. All carotenoids, including xanthophyll, also play a part in PHOTOSYNTHESIS.

X-chromosome One of the two kinds of sex CHROMOSOME in human CELL nuclei, the other being the Y-CHROMOSOME. A normal human female has two X-chromosomes in her body cells, so that each of her OVA (eggs) is HAPLOID, having only one X-chromosome. When she passes this on to one of her offspring, it influences the development of female characteristics. Her male offspring will have one X- and one Y-chromosome in his body cells. Non-sexual characteristics are also carried on the X-chromosome. Examples are the genes for colourblindness and haemophilia. X-chromosomes occur in other organisms as well, such as fruitflies. Where one sex has two identical chromosomes and the other has one of each type of sex chromosome, the former chromosomes are called the X-chromosomes. Sometimes sex is determined by the presence or absence of one particular chromosome, as in grasshoppers; this is also called the X-chromosome.

xenomorphism Formation of a mineral crystal in a restricted space so that it cannot develop crystal faces. It occurs in the late-forming minerals of IGNEOUS ROCKS.

xenon (symbol Xe) Gaseous, nonmetallic element, one of the NOBLE GASES. Discovered in 1898, xenon is present in the Earth's atmosphere (about one part in 20 million) and is obtained by FRACTIONATION of liquid air. Colourless and odourless, it is used in light bulbs, LASERS and arc lamps for cinema projection. The element, which has 9 stable isotopes, forms some compounds, mostly with FLUORINE. Properties: at.no. 54; r.a.m. 131.30; r.d. 5.88; m.p. −111.9°C (−169.42°F); b.p. −107.1°C (−160.8°F); most common isotope ^{132}Xe (26.89%).

xerography Most common process used for PHOTOCOPYING.

xerophyte Any plant that is adapted to survive in dry conditions, in areas subject to drought or in physiologically dry areas such as saltmarshes and acid bogs, where saline or acid conditions make the uptake of water difficult. SUCCULENTS, such as cacti, have thick fleshy leaves and stems for storing water. Other adaptations include the ability to reduce water loss by shedding leaves during drought, having waxy or hairy leaf coatings or reduced leaf area.

xi particle *See* HYPERON

X-ray In medicine, X-RAYS directed towards the body to permit visualization of internal structures or to destroy diseased or unwanted tissue. Passed through the body onto a photographic plate, X-rays reveal such abnormalities as fractures, tumours, foreign bodies and enlargement of organs. The capacity of X-rays to break up cells makes them useful also in the treatment of some forms of cancer.

X-ray In physics, form of ELECTROMAGNETIC RADIATION of much shorter wavelength, or higher frequency, than visible light, produced when a beam of electrons hits a solid target. X-rays were discovered in 1895 by the German physicist Wilhelm RÖNTGEN. They are normally produced for scientific use in X-RAY TUBES. Because they are able to penetrate matter that is opaque to light, X-rays are used to investigate inaccesible areas, especially of the body. *See also* RADIOGRAPHY; X-RAY ASTRONOMY

X-ray astronomy *See* ASTRONOMY

X-ray crystallography Use of X-RAYS to discover the molecular structure of CRYSTALS. It uses the phenomenon of **X-ray diffraction**, the scattering of an X-ray beam by the atomic structure of a crystal.

X-ray diffractometer Instrument used in the analysis of the atomic arrangement that determines the crystal structure of minerals. X-rays, when passing through the symmetrically arranged atoms of a crystal, are deflected in a regular pattern, with the atoms acting as a DIFFRACTION GRATING. Photos of these patterns permit deductions of interatomic dimensions, spacing, and bonding arrangements within crystals. *See also* X-RAY CRYSTALLOGRAPHY

X-RAY CRYSTALLOGRAPHY

Each substance has a different crystalline structure. X-ray crystallography allows the details of the structure to be recorded by bombarding a crystal sample (1) with a beam of X-rays (2). The X-ray is created by bombarding a tungsten anode (3) with electrons in a vacuum (4). A slit (5) focusses the X-rays on the crystal. The distance between the atomic planes in the crystal either reinforces (6) or cancels out (7) the X-rays. When the ray is reinforced the emerging X-ray (8) creates a spot (9) on photographic film. The pattern of the spots can be used by scientists to deduce the exact structure of molecules.

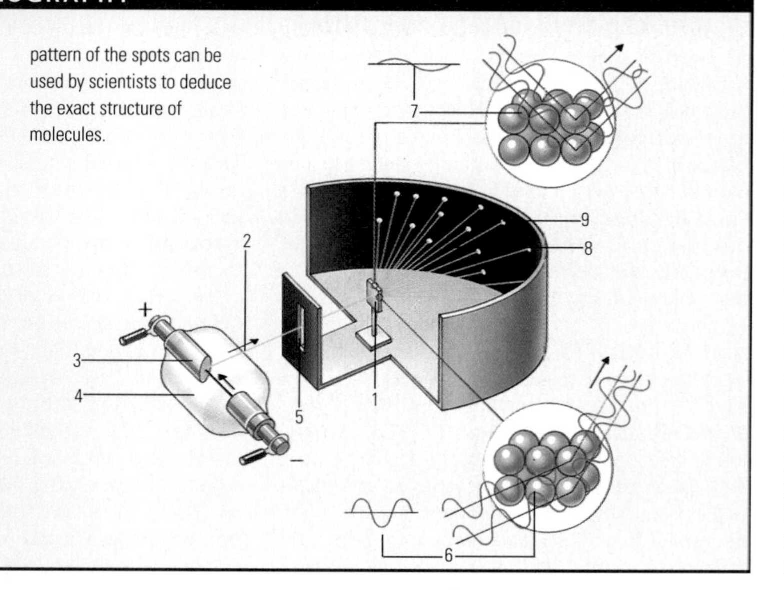

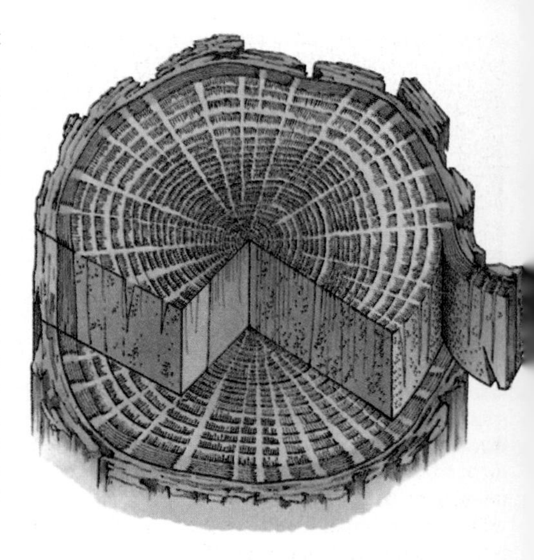

▲ **xylem** Wood is made of the old xylem of previous years' growth, the walls of which have become heavily impregnated with lignin. The cells die and are embalmed in tannins and resins, forming the strong, dark, heartwood. This is surrounded by sapwood, which is weaker and makes up most of the thickness of the trunk in young trees. A layer of new sapwood is added to the outer edge of the column of wood with each year's growth, so forming a ring visible among others in the trunk. These are most apparent in temperate regions, where patterns of growth vary from season to season.

X-ray telescope *See* TELESCOPE

X-ray tube Evacuated tube used to provide a source of X-RAYS for medical or other purposes. It consists of an electron gun producing a stream of ELECTRONS that strike an ANODE, part of which is made of a heavy metal such as TUNGSTEN. The tungsten emits X-rays when it is bombarded by the stream of high-energy electrons.

xylem Transport TISSUE of a plant, which conducts water and minerals from the roots to the rest of the plant and provides support. The most important cells are long, thin tapering cells called xylem vessels. These cells are dead and have no cross-walls; they are arranged in columns to form long tubes, up which water is drawn. As water evaporates from the leaves (TRANSPIRATION), water is drawn across the leaf by OSMOSIS to replace it, drawing water out of the xylem. This process creates TURGOR PRESSURE in the xylem vessels, which, along with rings of LIGNIN that reinforce side walls and prevent them from collapsing, provides support to the plant. Tiny holes in the walls of the xylem vessels, called **pits**, allow water to cross from one tube to another. Ferns and conifers do not have xylem vessels. Instead they have similar cells called **tracheids**, which do not lose their end walls, so water has to travel through the pits, which slows the flow. The lignin makes the walls of vessels and tracheids strong and rigid, an important support as the plant grows bigger. Xylem tissue also contains non-conducting fibres, dead cells thickened with lignin for extra support. In trees, the xylem becomes blocked with age, and new xylem forms towards the outside of the trunk to replace it. The core of dead, non-functioning xylem remains an essential part of the support system. *See also* PHLOEM; VASCULAR BUNDLE

xylene ($C_6H_4(CH_3)_2$) Organic chemical compound obtained from the distillates of coal tar and petroleum, and important as a solvent. Chemically it is dimethyl benzene which exists in three isomeric forms: ortho-, meta-, and para-xylene. The ISOMERS have different physical properties.

Yagi antenna Basic form of ANTENNA used in simple RADIO TELESCOPES. It consists of several parallel elements mounted on a straight member, and often forms the basis of cheap arrays used in **aperture synthesis** (in which arrays of radio dishes are used to synthesize the resolution of a much larger telescope); the Yagi antenna is also a familiar form of television aerial. It was developed by the Japanese engineer Hidetsugu Yagi (1886–1976).

Yale, Linus (1821–68) US inventor of various types of locks, including the modern cylinder LOCK that bears his name.

Yalow, Rosalyn (1921–) US biochemist who shared the 1977 Nobel Prize for physiology or medicine for her role in the development of a method of detecting peptide HORMONES in the blood. In the 1950s Yalow found that some people who received regular INSULIN injections developed ANTIBODIES against the hormone. She discovered that insulin, labelled with radioactive iodine, combined with the antibodies; from this she developed radio-immunological tests to detect and measure the amount of insulin present.

Yang, Chen Ning (1922–) US physicist, b. China, who shared the 1957 Nobel Prize for physics with Tsung-Dao LEE for their discovery of violations of the conservation of PARITY. They studied the decay of K MESONS, which seemed to break down in two different ways, and in 1956 they concluded that in these WEAK NUCLEAR FORCE interactions parity need not be conserved.

yard Imperial unit of length equal to 3 feet. 1 yard (yd) also equals 0.9144m precisely.

yardang Elongated ridge formed by wind erosion in arid areas. Rock is eroded into a series of ridges and furrows lying parallel to the prevailing wind direction.

Y-chromosome One of the two kinds of sex CHROMOSOME in the nuclei of DIPLOID human CELLS, the other being the X-CHROMOSOME. A normal male mammal has one X- and one Y-chromosome in his body cells, and his SPERM cells are HAPLOID, containing either an X- or a Y-chromosome. Since female OVA (eggs) always contain an X-chromosome, the resulting offspring is either XY (male) or XX (female). The Y-chromosome is smaller than the X-chromosome and contains fewer GENES.

year Length of time taken by the Earth to circle once round the Sun in its orbit. Various years, defined according to the choice of reference point, are given in the table. The civil year (the calendar year) averages 365.2425 mean solar days.

TYPES OF YEAR

anomalistic year	365.25964 mean solar days
eclipse year	346.62003 mean solar days
sidereal year	365.25636 mean solar days
tropical year	365.24219 mean solar days

yeast Any of a group of single-celled, microscopic FUNGI found in all parts of the world in the soil and in organic matter. Yeasts reproduce asexually by budding or fission. Yeasts are also produced commercially for use in baking, brewing and winemaking. They occur naturally as a bloom (white covering) on grapes and other fruit. See also ASEXUAL REPRODUCTION

yellow fever Acute infectious disease marked by sudden onset of headaches, fever, muscle and joint pain, jaundice and vomiting; the kidneys and heart may also be affected. It is caused by a VIRUS transmitted by mosquitoes in tropical and subtropical regions. It may be prevented by vaccination.

yellow-green algae Variety of flagellate ALGAE (division Xanthophyta). They differ from GREEN ALGAE by their characteristic, motile bodies and unequal FLAGELLA.

Yerkes, Robert Mearns (1876–1956) US biologist and psychologist. He was a pioneer in the comparative study of apes and the development of methods to test the abilities of lower animals and humans. His publications include *The Mind of a Gorilla* (1927) and *Chimpanzees : A Laboratory Colony (1943).*

Yersin, Alexandre Émile John (1863–1943) French bacteriologist who studied with Louis PASTEUR. In Hong Kong Yersin discovered (independently of Shibasaburo KITASATO) the plague bacillus and developed a serum against it in 1895.

yolk Rich substance found in the EGGS or OVA of most animals except those of placental MAMMALS. It consists of fats and proteins and serves as a store of food for the developing embryo.

yolk sac Membranous, saclike structure in the EGGS of most animals. It is attached directly to the ventral surface or gut of the developing embryo in the eggs of birds, reptiles, and some fish, and contains YOLK. The term also refers to an analogous, saclike membrane that develops below the mammalian embryo. It contains no yolk but is connected to the umbilical cord.

Young, Charles Augustus (1834–1908) US astronomer. He made many solar observations, including the earliest spectroscopic studies of the corona of the SUN. He also used spectral line measurements to determine the Sun's rotation period. *See also* SPECTROSCOPY

Young, Thomas (1773–1829) British physicist and physician. He revived the wave theory of LIGHT first put forward in the 17th century by Christiaan HUYGENS. He helped present the Young-Helmholtz theory of COLOUR vision and detailed the cause of ASTIGMATISM. He studied ELASTICITY, giving his name to the tensile elastic modulus (*see* YOUNG'S MODULUS). Young was also an Egyptologist who helped decipher the Rosetta stone.

Young's modulus Ratio of the STRESS exerted on a body to the longitudinal STRAIN produced.

ytterbium (symbol Yb) Silver-white, metallic element of the LANTHANIDE SERIES. First isolated in 1878, ytterbium's chief ore is monazite. The shiny, soft element is malleable and ductile is used to produce steel and other alloys. Properties: at.no. 70: r.a.m. 173.04; r.d. 6.97; m.p. 824°C; (1,515°F), b.p. 1,193°C (2,179°F); most common isotope ^{174}Yb (31.84%).

yttrium (symbol Y) Silver-grey, metallic element of group III of the PERIODIC TABLE. First isolated in 1828, it is found associated with lanthanide elements (*see* LANTHANIDE SERIES) in monazite sand, bastnaesite, and gadolinite, and resembles the lanthanides in its chemistry. Yttrium was found in lunar rock samples collected by the Apollo 11 space mission. Its compounds are used in phosphors and communications devices, such as colour TELEVISON picture tubes and superconducting ceramics. Properties: at.no. 39; r.a.m. 88.9059; r.d. 4.47; m.p. 1,523°C (2,773°F); b.p. 3,337°C (6,039°F); most common isotope ^{89}Y (100%).

Yukawa, Hideki (1907–81) Japanese physicist who was awarded the 1949 Nobel Prize for physics for his prediction of the existence of the MESON. In the 1930s he proposed that there was a nuclear force of very short range (less than 10^{-15} m), strong enough to overcome the repulsive force of protons and which diminished rapidly with distance. He predicted that this force manifested itself by the transfer of particles between neutrons and protons. In 1947 Yukawa's theory was confirmed by the discovery of the PION (pi meson) by Cecil POWELL.

Z

Z boson (symbol Z^0) ELEMENTARY PARTICLE that is a carrier of the WEAK NUCLEAR FORCE. The superscript zero (^{0}) indicates that it has no electric charge. Its mass is about 96 times that of the PROTON.

Zeeman, Pieter (1865–1943) Dutch physicist. He shared the 1902 Nobel Prize for physics with his teacher Hendrik LORENTZ for their 1896 discovery of the ZEEMAN EFFECT. This is the splitting of spectral lines when a light source is placed in a strong MAGNETIC FIELD. He also detected the magnetic fields at the surface of the SUN.

Zeeman effect In physics, effect produced by a strong MAGNETIC FIELD on the light emitted by a radiant body, observed as a splitting of its spectral lines. It was first observed in 1896 by the Dutch physicist Pieter ZEEMAN. The effect has been useful in investigating the charge/mass ratio and MAGNETIC MOMENT of an ELECTRON.

Zeiss, Carl (1816–88) German manufacturer of optical instruments. In 1846 he established a factory at Jena, Germany and made various optical components, including lenses, binoculars and microscopes.

zenith In astronomy, point on the CELESTIAL SPHERE which is directly overhead. The zenith distance of a heavenly body is the angle it makes with the zenith. In common use it means the highest point.

zeolite Group of alumino-silicates containing loosely held water that can be continuously expelled on heating. They also contain sodium, calcium or barium. Members occur as fibrous aggregates while others form robust, non-fibrous crystals. There are many zeolites varying in hardness from 3 to 5 and in specific gravity from 2 to 2.4. Zeolites include analcime ($NaAlSi_2O_6.H_2O$), stilbite ($NaCa_2(Al_5Si_{13}36.14H_2O$), and natrolite ($Na_2Al_2Si_3O_{10}.2H_2O$).

Zeppelin, Ferdinand, Count von (1838–1917) German army officer and inventor. He served in the armies of Württemburg and Prussia and while an observer with the Union army during the US Civil War (1861–65) made his first balloon ascent. In 1900 he invented the first rigid AIRSHIP, named Zeppelin after him.

Zermelo, Ernst (1871–1956) German mathematician who was one of the founders of axiomatic SET THEORY. The Zermelo-Fraenkel AXIOMS stipulate which collections of objects constitute sets.

Zernike, Frits (1888–1966) Dutch physicist who in 1935 developed the PHASE-CONTRAST MICROSCOPE, in which objects being viewed (often living-cell biological specimens) take on a different colour from their surroundings. For this work, he received the 1953 Nobel Prize for physics.

zero In mathematics, that number which, when added to any number x, leaves the number unchanged: $x + 0 = x$. Multiplication of a number by zero gives zero. Division by zero is undefined. The concept of zero was unknown to the ancient Greeks and was introduced to the West by the Arabs, who had taken it from the Hindus.

zero gravity (zero g) Condition of experiencing no apparent gravitational force. It may be attained by airplane pilots in aircraft briefly during free fall, or by astronauts in a spacecraft, when the engines have been turned off and the craft is "falling" freely. Gravitational force is still present, however. *See also* WEIGHTLESSNESS

Zeta Cancri Group of four stars in the constellation of Cancer consisting of two BINARY STAR systems. Zeta-one has a period of 59.6 years, Zeta-two is a close binary with a period of 17.6 years.

Ziegler, Karl (1898–1973) German chemist who shared the 1963 Nobel Prize for chemistry with the Italian chemist Giulio NATTA (his colleague) for research into POLYMERS. Ziegler discovered a technique of using a RESIN with metal ions attached as a CATALYST in the production of POLYETHYLENE. He also did research into AROMATIC COMPOUNDS and developed ORGANOMETALLIC COMPOUNDS (in which metal or metalloid atoms are bound to carbon atoms or organic groups) that are more reactive than GRIGNARD REAGENTS.

zinc (symbol Zn) Blue-white, metallic element of group II of the periodic table, known from early times. Chief ores are sphalerite (zinc blende), smithsonite and calamine. Zinc is a vital TRACE ELEMENT, found in erythrocytes (red blood cells). The shiny element is used in many alloys, including brass, bronze, nickel, silver and soft solder. It is corrosive-resistant and used in galvanizing iron. Zinc oxide is used in cosmetics, pharmaceuticals, paints, inks, pigments and plastics. Zinc chloride is used in dentistry and to manufacture batteries and fungicides. Properties: at.no. 30; r.a.m. 65.37; r.d. 7.133; m.p. 419.6°C (787.3°F); b.p. 907°C (1,665°F); most common isotope ^{64}Zn (48.89%).

zinc oxide (ZnO) White powder used in ointments as a mild antiseptic, astringent and protection against drying of ECZEMA and other skin conditions.

zircon Orthosilicate mineral, zirconium silicate ($ZrSiO_4$) found in IGNEOUS and METAMORPHIC ROCKS and in sand and gravel. It displays tetragonal system prismatic crystals. Its colour is variable: it is usually light or reddish brown but can be colourless, grey, yellow or green. It is used widely as a gemstone because of its hardness and high refractive index. Hardness 7.5; r.d. 4.6.

zirconium (symbol Zr) Grey-white, TRANSITION ELEMENT, first discovered in 1789. Its chief source is the gemstone ZIRCON (zirconium silicate). Lunar rocks collected during the Apollo space missions show a higher content of zirconium than Earth ones,

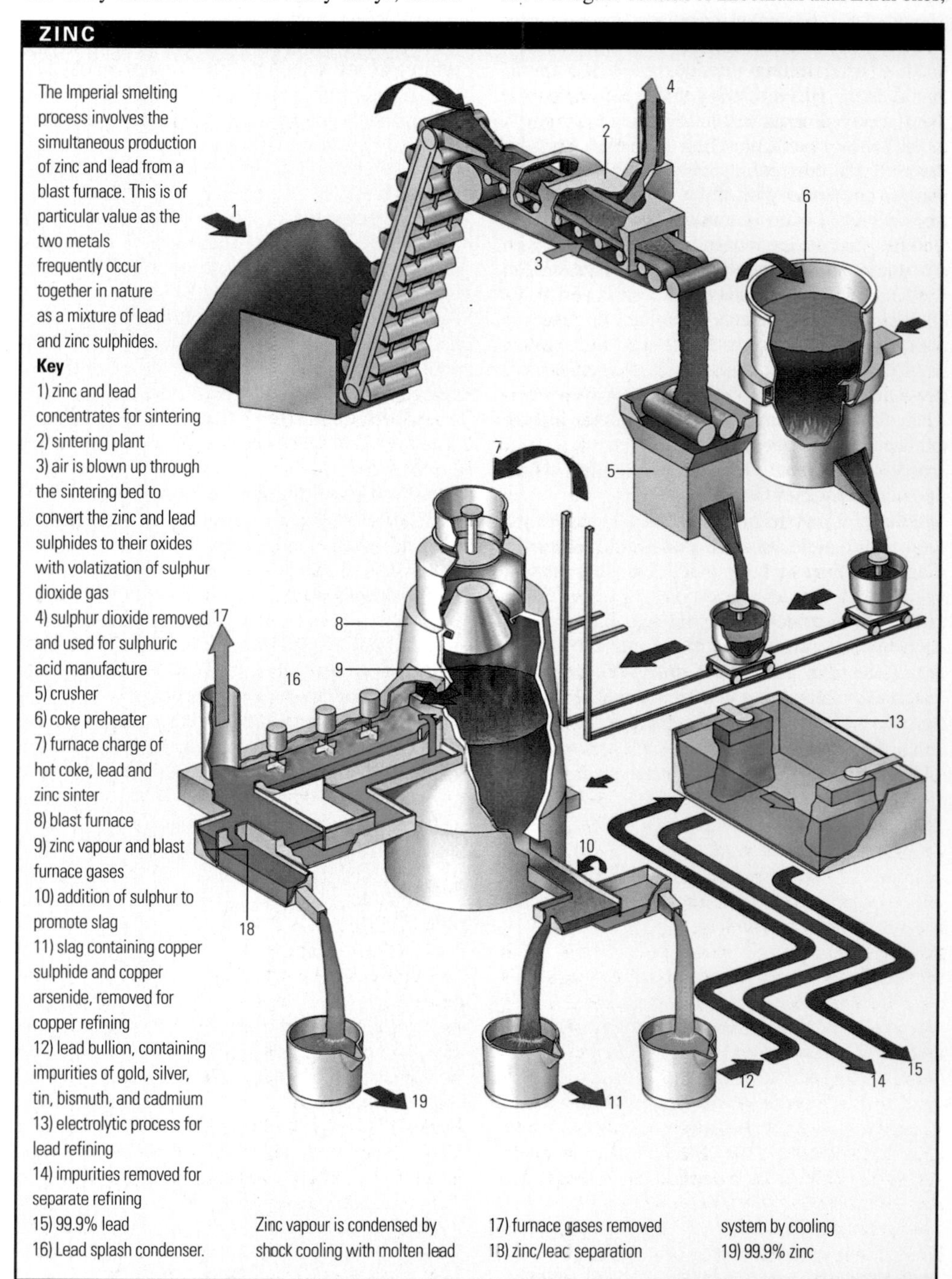

and zirconium exists in meteorites and stars, including the Sun. Chemically similar to titanium, it is used in ceramics and absorption of NEUTRONS in nuclear REACTORS. Properties: at.no. 40; r.a.m. 91.22; r.d. 6.51; m.p. 1,852°C (3,366°F); b.p. 4,377 °C (7,911°F); most common isotope Zr^{90} (51.46%).

zodiac Belt on the celestial sphere which forms the background for the motions of the Sun, Moon and planets (except Pluto). The zodiac is divided into 12 **signs**, which are named after the constellations they contained at the time of the ancient Greeks: Aries, Taurus, Gemini, Cancer, Leo, Virgo, Libra, Scorpio, Sagittarius, Capricorn, Aquarius and Pisces. The constellations now inside the zodiac do not correspond to those named by the ancients, because PRECESSION of the Earth's axis has meanwhile tilted the Earth in a somewhat different direction. Most of the zodiacal signs represent animals and the name zodiac derives from a Greek phrase meaning "circle of animals". To modern astronomers the zodiac has only historical significance. *See also* ASTROLOGY; ASTRONOMY

zodiacal light Cone of faint light, usually fainter than the Milky Way, visible at all seasons in the tropics in the absence of moonlight. It can be seen on the W horizon after evening twilight, or on the E horizon before morning twilight. The spectrum of the zodiacal light resembles the Sun's, indicating that the phenomenon results from the SCATTERING of sunlight by particles. These particles constitute the zodiacal dust cloud, and probably originate partly from matter ejected by the Sun, and partly from the decay of comets and asteroids.

zoisite Orthosilicate mineral, hydrous calcium aluminum silicate, found in METAMORPHIC ROCKS. Orthorhombic system prismatic crystals and masses. Glassy, transparent grey, white, brown, green, or pink; hardness 6–6.5; r.d. 3.2. A vivid blue variety from Tanzania, called tanzanite, is a gemstone.

zonation In ecology, occurrence of bands of characteristic vegetation in an area. Examples include the bands of different types of seaweed that occur progressively up a shore from the low-tide to high-tide levels. Species from the low-tide zone would dry out and die at the high-tide level, whereas high-tide species could not withstand prolonged immersion in seawater in the low-tide zone. The varying vegetation in turn attracts varying animal species.

zone refining Method of purifying crystals, especially for use in SEMICONDUCTOR devices. The material is placed in a long tube and passed through a furnace in which hot and cold zones alternate. As the rod moves through the furnace, impurities remain in the molten state while melting and recrystallization of the material takes place; the impurities are thus transferred to one end of the tube.

zoogeography (animal geography) Study of the geographic distribution of animals. Formerly, the approach of this science was descriptive; today it utilizes various data, including isotope DATING and ocean-bottom core sampling.

zoology Study of animals; combined with BOTANY, it comprises the science of BIOLOGY. It is concerned with the structure of the animal and the way in which animals behave, reproduce and function, their evolution and their role in interactions with humankind and their environment. There are various subdivisions of the discipline, including: TAXONOMY; ECOLOGY; PALAEONTOLOGY; ANATOMY, and ZOOGEOGRAPHY, the distribution of animals. ANTHROPOLOGY is an extension of zoology. *See also* EMBRYOLOGY; GENETICS; MORPHOLOGY

zoonosis Any infection or infestation of vertebrates that is transmissible to human beings.

zooplankton Animal portion of the PLANKTON. It consists of a wide variety of microorganisms, such as COPEPOD and larval forms of higher animals. It is a central constituent of the oceans' FOOD CHAINS. There are few levels or areas of the ocean that have no zooplankton. *See also* PHYTOPLANKTON

ZONATION

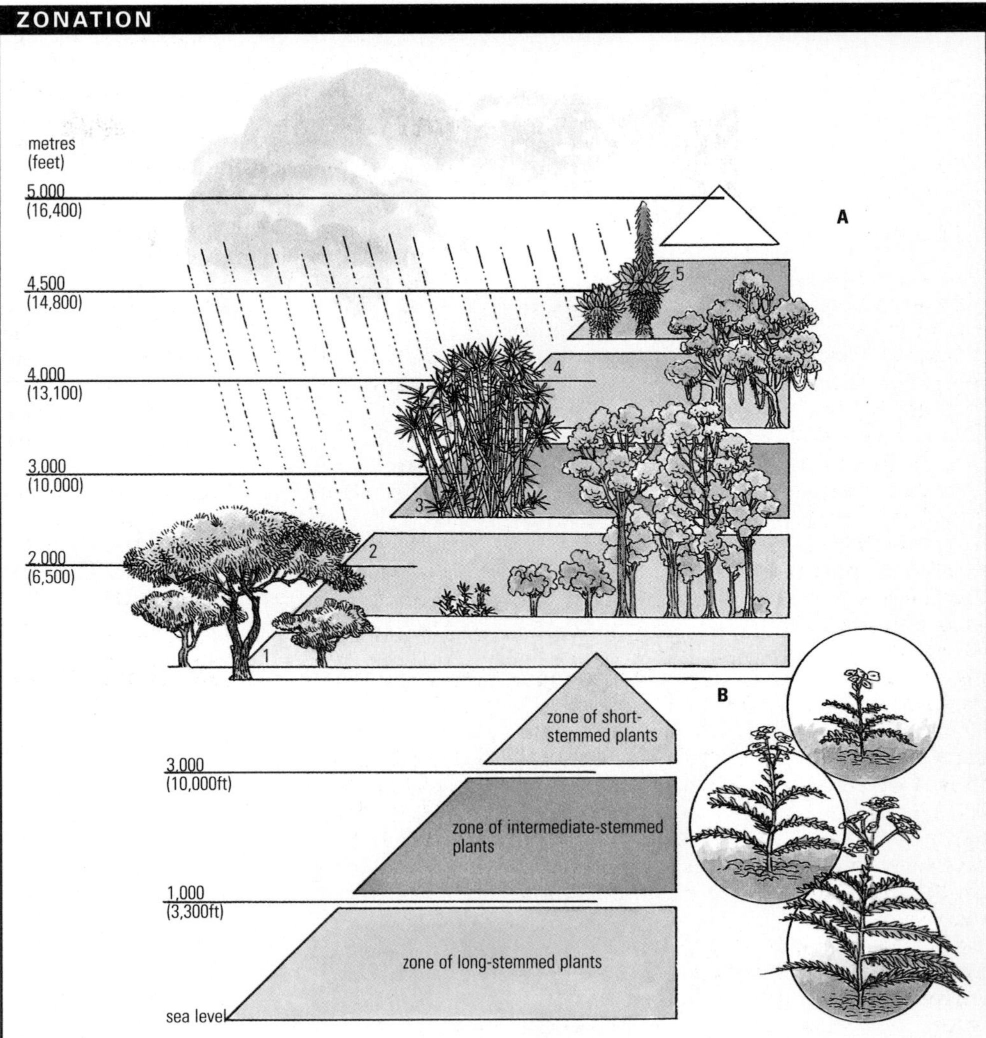

As levels of rainfall, temperature and soil type vary within an area, different groups of plants will grow producing different habitats which attract a variety of animals. This effect is most marked in areas where conditions are charged by moving from one level to another. On a mountain, for example Mount Kilimanjaro (A), zones range from dry savannah (1) through rainforest (2), bamboo (3) and cloud forest (4) to reach an area below the snow line where lobelias have evolved in isolation to giant proportion (5). These habitats arise because each level has a different amount of rain, a different temperature and will encourage a different group of animals. In (B), experiments have shown how quickly plants will adapt to different zones. California Yarrow was grown at different altitudes, the higher the zone the shorter the stem that grew. When all the plants were replanted at sea level the difference in stem height was maintained.

Zsigmondy, Richard Adolf (1865–1929) Austrian chemist who won the 1925 Nobel Prize for chemistry for his work on COLLOIDS. While employed at a glass manufacturing company (1897–1900), he discovered a water suspension of gold and proposed that the shape and size of colloids could be deduced from the way in which the particles scatter light. To aid such studies he helped to develop the **ultramicroscope**, in which the particles are suspended in a gas or liquid, through powerful light is shone. The particles scatter the light ensuring they appear to the viewer due to the **Tyndall effect**. *See also* SCATTERING; TYNDALL, JOHN

Zuckerman, Sir Solly (1904–) British science administrator, b. South Africa. He was professor of anatomy at Birmingham (1943–68), and served as chief scientific adviser to the Ministry of Defence (1960–66), as well as working on other scientific research committees, including the Royal Commission on Environmental Pollution (1970–74).

Zwicky, Fritz (1898–1974) Swiss-American astronomer, b. Bulgaria, who in 1934, with Walter BAADE suggested that what is left after a SUPERNOVA explosion is a NEUTRON STAR. This was confirmed in 1968 with the discovery of the PULSAR in the Crab Nebula. In the mid-1960s, he discovered many supernovae and began a study of CLUSTER GALAXIES.

Zworykin, Vladimir Kosma (1889–1982) US physicist and inventor, b. Russia, who was a pioneer of TELEVISION. In 1929, he joined the Radio Corporation of America (RCA), becoming its director of development and in 1947 a vice president. Zworykin developed the **iconoscope**, the forerunner of the modern television camera tube, and the **kinescope**, a CATHODE-RAY TUBE for TV sets. In 1928, he patented a colour television system. He also invented the ELECTRON MICROSCOPE. In 1967, Zworykin received the National Medal of Science for his inventions and contributions to research.

zygote In SEXUAL REPRODUCTION, a CELL formed by fusion of a male and a female GAMETE. It contains a DIPLOID number of CHROMOSOMES, half contributed by SPERM, half by the OVUM. Through successive CELL DIVISIONS, a zygote will develop into an EMBRYO.

CHRONOLOGY of SCIENCE

	ASTRONOMY AND SPACE	PHYSICS	CHEMISTRY	BIOLOGY	MEDICINE
10,000 BC				**c.10,000 BC** Dog is domesticated in Mesopotamia.	
9000 BC				**c.9000 BC** Einkorn wheat is domesticated in Palestine; sheep and goats are domesticated in Persia (for food).	
8000 BC				**c.8000** Pumpkins and squashes are domesticated in Central America. **c.7500** BC Potatoes and chile peppers are domesticated in Peru.	
7000 BC				**c.7000** BC Water buffalo and chickens are domesticated in Asia; cattle are domesticated in Asia Minor.	
6000 BC				**c.6000** BC Durum wheat is domesticated in Asia Minor.	
5000 BC				**c.5000 BC** Horse is domesticated (for food) in Russia and millet is domesticated in China.	
4500 BC				**c.4500 BC** Guinea pig is domesticated in Peru (for food); maize is domesticated in Central America.	
4000 BC				**c.4000 BC** Pigs are domesticated in southeast Asia.	
3500 BC				**c.3500 BC** Alpaca and llama are domesticated in Peru; donkeys are domesticated in Palestine.	
3000 BC				**c.3000 BC** Peanuts are domesticated in Peru; camels are domesticated in Arabia; wild boar are domesticated in Europe.	**c.3000 BC** Filling of teeth is practised in Sumeria.
2500 BC	**c.2300 BC** Earliest observations are made by Chinese and Babylonian astronomers, including comets and constellations.			**c.2500 BC** Yak is domesticated in Tibet.	**c.2500 BC** Contraception is practised by the ancient Egyptians.
2000 BC					**c.1550 BC** Central control function of the brain is discovered by physicians in Egypt.
1500 BC					
1000 BC	**c.1000 BC** Egyptians devise a calendar based on the motion of the star Sirius (Dog Star). **721 BC** Solar eclipse is first recorded in Babylon. **585 BC** Solar eclipse is (said to have been) predicted by Thales of Miletus (c.620–c.555 BC).		**c.600 BC** Three-element theory – that all matter consists of a combination of mist, earth and water – is introduced by Greek natural philosopher Thales of Miletus (c.620–c.555 BC). **c.530 BC** Fact that air is the primary substance is proposed by Greek natural philosopher Anaximenes of Miletus (d.c.500 BC).	**c.1000** BC Reindeer are domesticated in Siberia; oats are domesticated in Europe. **c.520 BC** Behaviour and structure of animals (using dissection) are studied by Greek physician Alcmaeon of Croton.	**c. 535 BC** Human dissection (for medical study) is introduced by Greek physician Alcmaeon of Croton.
450 BC	**432 BC** Nineteen-year Metonic cycle, made up of seven years containing 13 lunar months and twelve years of 12 lunar months, is discovered by Greek astronomer Melton of Athens.		**c.450 BC** Four-element theory – that all matter is made of a combination of earth, air, fire and water – is introduced by Greek natural philosopher Empedocles (d.c.430 BC).	**c.450 BC** Detailed studies of human anatomy are made by Greek physician Hippocrates of Kos.	**c.450 BC** Fact that disease has natural causes is recognized by Greek physician Hippocrates of Cos (d.c.485 BC).
400 BC	**352 BC** Supernova explosion is recorded by Chinese astronomers. **c.340 BC** Precession of the equinoxes is discovered by Babylonian astronomer Cidenas. **c.330 BC** Fact that tides are caused by the Moon is proposed by Greek mariner Pytheas of Massilia.	**c.340 BC** Fact that heavier objects fall faster than lighter ones (which is incorrect) is proposed by Greek philosopher Strato of Lampsacus (d.c.270 BC).	**c.400 BC** Atomic theory – that all matter consists of atoms – is developed by Greek natural philosopher Democritus of Abdera (c.460–c.370 BC), probably based on previous (c.450 BC) ideas of Leucippus of Miletus. **306 BC** Democritus' atomic theory gains support from Greek philosopher Epicurus (c.342–270 BC).	**c.350 BC** Animals and plants (about 500 known species) are classified into eight classes by the Greek philosopher Aristotle (384–322 BC). **c.340 BC** Arteries and veins are distinguished by Greek physician Praxagoras of Cos (who thought that arteries carry air and veins carry blood from the liver).	
300 BC	**266 BC** Observatory is built by Greek astronomer Eudoxos of Cnidus (c.408–c.353 BC), who also attempts to explain the motions of the planets. **260 BC** Sun-centered Universe (Solar		**c.270 BC** Five-element theory – that all matter is made of a combination of earth, fire, water, wood and metal – is introduced by Chinese philospher Zou Yan.	**c.300 BC** First detailed studies of plants (more than 550 species) are made by Greek natural philosopher Theophrastus of Eresus.	**c.300 BC** First Greek treatise on medicine is written by Greek physician Diocles of Carystus. **c.290 BC** Human dissection to aid understanding how the body works is

	FARMING AND FOOD	TRANSPORT	ENGINEERING AND TECHNOLOGY	COMMUNICATIONS
10,000 BC				
9000 BC				
8000 BC	**c.7500 BC** Beer is brewed in Mesopotamia.			
7000 BC		**c.7000 BC** Dugout canoes are used in Europe; reed boats are used in Egypt.		
6000 BC				
5000 BC				
4500 BC			**c.4500 BC** Copper is smelted in Egypt and central Europe.	
4000 BC	**c.4000 BC** Primitive ox-drawn plough is used in China and later (c.3500 BC) in Sumeria.			
3500 BC	**c.3500 BC** Wine is made in the Middle East.	**c.3500 BC** Single-masted square-sail ships sail on the River Nile. **c.3500 BC** Solid wheels are used in Mesopotamia.	**c.3500 BC** Solder is used to join copper and the potter's wheel is invented in Mesopotamia; wheeled vehicles are used in Sumeria.	**c.3300** Hieroglyphic writing is used in Egypt; pictographic writing is used in Sumeria. **3200 BC** Ink (made from lamp-black and egg-white) is used to write hieroglyphics on papyrus in Egypt.
3000 BC	**c.3000 BC** Barley is grown as a main crop in Sumeria.		**c.3000 BC** Glass is made in Egypt and Mesopotamia; bronze and copper axes are made in Mesopotamia. **c.2700 BC** Construction of Stonehenge, England, begins (using mainly timber). **c.2600 BC** First step pyramid is built in Egypt, followed later (c.2500 BC) by the Great Pyramid at Khufu.	**c.3000 BC** Abacus is invented independently by the Babylonians and Chinese.
2500 BC	**c.2500 BC** Bees are kept for honey in Egypt.	**c.2500 BC** Skis are used in Scandinavia.	**c.2400 BC** Bitumen is used for waterproofing in Mesopotamia.	**c.2500 BC** Positional number system is invented in Mesopotamia.
2000 BC	**c.1600 BC** Wooden ploughshare is invented in Mesopotamia; grape vines and olives are cultivated in Crete.	**c.2000 BC** Spoked wheels are used in Mesopotamia and, later (1600 BC), in Egypt. **c.2000 BC** Paved roads are constructed in Crete.	**c.2000 BC** Looms are used for weaving in Assyria. **c.1900 BC** Arches are used in building construction in Palestine.	
1500 BC	**c.1500 BC** Seed drill (a vertical tube with a funnel at the top) is invented in Sumeria. **c.1100 BC** Iron ploughshare is invented in the Middle East.	**c.1450 BC** Barges 60 metres long are built in Egypt for carrying stone obelisks. **c.1400 BC** Sea-going vessels are built in western Europe. **c.1200 BC** Wooden-hulled boats with a keel are built by the Pheonecians.	**c.1450 BC** Stonehenge is completed in its final form. **1370 BC** Iron tools and weapons are made by the Hittites.	**c.1350 BC** Decimal numbers are used in China; 22-letter alphabet is used by the Phoenicians. **c.1300 BC** Musical notation is introduced in Syria.
1000 BC		**592 BC** Anchor is described by Greek philosopher Anacharis the Scythian. **512 BC** Pontoon bridge is built across the Bosphorus by the army of Persian King Darius I; a second pontoon bridge is built there later (480 BC) for the (unsuccessful) army of Persian King Xerxes.	**800 BC** Pyramids are built in Mexico by the Olmecs. **700 BC** Pulleys are used in Assyria. **650 BC** Aqueducts are used in Assyria. **550 BC** Locks and keys are used in Egypt.	**875 BC** Symbol for zero (0) is invented in India. **c.510 BC** Map showing countries surrounding the Mediterranean Sea is drawn by Greek traveller and historian Hecataeus of Miletus.
450 BC				**c.430 BC** Optical telegraph using burning torches to indicate coded letters is used in Greece; later (150 BC) it uses a code devised by the Greek historian Polybius.
400 BC	**350 BC** Rice is cultivated in western Africa.		**400 BC** Military catapult is invented by Dionysius the Elder of Syracuse (430-367 BC). **c.360 BC** Suspended device propelled by steam issuing from a jet is invented by Archytas of Tarentum (c.420 BC-c.350 BC). **c.350 BC** Iron smelting is developed in central Africa. **312 BC** Via Appia, a road from Rome to Alba Longa, is built by Roman Emperor Appius Claudius.	
300 BC	**290 BC** Chest harness for horses is invented in China.	**c.250 BC** Collar and harness for horses is invented by the Chinese.	**c.300 BC** Cast iron is invented in China. **c.280 BC** Pharos, an 85-metre lighthouse and one of the Seven Wonders of the World, is built on an island off Alexandria by the Greek architect Sostratus of Cnidus. **c.260 BC** Simple pump using a rotating screw in an inclined cylinder (Archimedes'	

ASTRONOMY AND SPACE

System) is proposed by Greek astronomer Aristarchus of Samos (c.320–c.230 BC).

240 BC Earth's circumference is calculated by Greek astronomer Eratosthenes of Cyrene (c.276–c.194 BC).

240 BC First mention of Halley's comet is made by Chinese astronomers, who later (87 BC) again observe the comet.

c.150 BC Distance to the Moon (in terms of the Earth's diameter) is determined by Greek astronomer Hipparchus of Nicaea (c.190–c.120 BC).

c.130 BC Precession of the equinoxes is discovered by Greek astronomer Hipparchus of Nicaea (c.190–c.120 BC).

AD c.150 Earth-centered system of the Universe is formalized by Greek-Egyptian astronomer Ptolemy in his book the Almagest.

185 Supernova explosion is observed by Chinese astronomers.

365 Moons of Jupiter are discovered by Chinese astronomers (using the naked eye).

PHYSICS

CHEMISTRY

AD 250 Mica is used for making windows in China.

750 Preparation of several acids and their salts, including ethanoic (acetic) acid, is decribed by Arab alchemist Geber (Jabir ibn-Hayyan) (c.721–c.815).

BIOLOGY

c.190 BC Duodenum and prostate gland are discovered (and named) by Greek physician Herophilus of Chalcedon (c.330–260 BC).

AD c.40 Medicinal properties of 600 plants are described by Greek physician Pedanius Dioscorides of Anazarbus.

77 Encyclopedic 37-volume work Historia Naturalis (Natural History) is completed by Roman scholar Pliny the Elder (Gaius Plinius Secundus) (23–79).

c.100 Dried chrysanthemum flowers are used as an insecticide by Chinese farmers.

c.175 Principles of human anatomy and physiology are established by Greek physician Galen (c.130–201).

MEDICINE

undertaken by Greek physician Herophilus of Alexandria (c.335–c.280 BC), who identifies the brain as the centre of thought and divides the nervous system into motor and sensory systems.

AD c.30 First Latin treatise on medicine is written by Roman author Aulus Celsus; 500 years later his name was partly adopted by Swiss alchemist and physician Paracelsus, real name Theorphrastus von Hohenheim (c.1493–1541).

c.70 Use of liver from a mad dog to protect against rabies is suggested by Roman scholar Pliny the Elder (Gaius Plinius Secundus) (23–79).

116 Major work on gynaecology is published by Greek physician Soranus of Ephasus, who practised in Alexandria.

c.175 The use of the human pulse as an aid to medical diagnosis is introduced by Claudius Galenus (Galen) c.130–201).

c.640 Symptoms of diabetes are noted by Chinese physician Chen Ch'uan.

c.650 Symptoms of lead poisoning are described by Greek physician Paul of Aegina.

200 BC · 100 BC · AD1 · 50 · 100 · 150 · 200 · 250 · 300 · 350 · 400 · 450 · 500 · 550 · 600 · 650 · 700 · 750

FARMING AND FOOD

200 BC Archimedes' screw is used to pump water for irrigation.

100 BC Oysters are raised (for food) in heated seawater tanks by the Romans near Naplestea; water wheels are used to drive mills for grinding grain in Albania.

AD 80 Chain pumps are used for lifting water for irrigation in China.

90 Fan-like winnowing machine is invented in China.

110 Hand-operated multi-tube seed drill is invented in China.

530 Water-powered machine for shaking and sifting flour is invented in China.

600 Windmills are used to grind grain in Iran.

630 Cotton (imported from the east) is grown in Arabia and, later (775), Spain.

700 Tea (for making a hot drink) is grown in China.

TRANSPORT

200 BC Three-masted ships are built in Greece.

170 BC Paved roads are built in Rome.

100 BC A cart with steerable front wheels is invented in western Europe.

AD 100 Tandem harness for horses (instead of using two horses side-by-side) is invented in China.

200 Ships with several masts and a rudder are constructed by the Chinese.

240 Wheelbarrow is used in China.

475 Stirrups for riding a horse are invented in China.

550 Wind-powered land vehicles are used in China.

580 Suspension bridge with iron chains is built in China.

ENGINEERING AND TECHNOLOGY

screw) is invented, traditionally by Archimedes (c.287-212), who also investigates the functioning of levers and pulleys.

c.190 BC Astrolabe is invented in Greece, possibly by astronomer Hipparchus (c.190-125); it is later (AD 850) reinvented by the Arabs.

c.120 BC Force pump is invented by Greek engineer Ctesibus of Alexandria.

c.100 BC Piston bellows for use in metal working are invented in China.

c.100 BC Bridges are constructed of masonry in Rome.

100 BC Apartment houses in Rome are built up to five storeys high.

AD c.50 Simple steam turbine is invented by Hero of Alexandra (b.c.AD 20).

c.105 Paper is made from bark and rags in China by Tsai Lun (c.50–118).

118 Pantheon is built in Rome with a 43-metre concrete dome, under orders of Emperor Hadrian.

c.130 Crude seismograph is invented in China by Zhang (Chang) Heng (78–139).

c.350 Methane (natural gas) is used for lighting in China.

600s Cast iron is used for large ceremonial structures in China (such as a 90-metre tall column).

673 "Greek fire", probably a mixture of naphtha, quicklime and saltpetre or sulphur, is invented by Syrian architect Kallinikos (Callincus) (b.c.620).

790 Paper is made in Iraq and, later (900), Egypt.

COMMUNICATIONS

c.165 BC Parchment (vellum) is invented by King Eumenes II of Pergamum (197–159 BC).

100 BC Negative numbers are introduced by mathematicians in China.

70 BC Tironian system of shorthand is devised by Roman freedman Marcus Tullius Tiro.

AD 525 Anno Domini (AD) method of expressing dates, counting from the birth of Jesus Christ, is invented by Scythian astronomer monk Dionysus Exiguus (c.500–c.556).

c.650 Printing from engraved wooden blocks is invented in China.

767 First printed text (a million prayers) is produced in Japan by Empress Shokutu.

	ASTRONOMY AND SPACE	PHYSICS	CHEMISTRY	BIOLOGY	MEDICINE
800					
850			**c.880** Ethanol (alcohol) is distilled from wine (by alchemists) in Arabia, and later (1150) in Europe; its preparation is described (1300) by Spanish alchemist Arnau de Villanova (c.1235–1312).		
900					**c.900** Measles and smallpox are recognized as different diseases by Persian-born Arab physician Rhazes (850–923)
950	**c.990** Andromeda nebula is discovered by Arab astronomer Al Sufi.				
1000		**1010** How lenses work is described by Arab mathematician Alhazen (Abu al-Hassan ibn al Haytham) (c.965–1039).			**c.1000** The Canon of Medine, a five-volume seminal book on general medicine, is written by Arab physician Aviccena (ibn-Sina) (980–1037).
1050	**1054** Supernova explosion in the constellation Taurus (later to become the Crab nebula) is observed by Chinese astronomers.				**1067** Hospital for lepers is founded by Spanish soldier El Cid (Ruy Diaz de Vivar) (c.1043–99).
1100					
1150		**1180** Magnetic compass is first described in Europe by English scholar Alexander Neckam (1157–1217), although already known for two centuries in China.			**1150** Bloodletting as a treatment for disease is advocated by Arab physician Avenzoar (abu-Mervan ibn-Zuhr) (c.1091–1162). **1170** First European textbook on surgery, Practica chirurgiae, is written by Roger of Solerno.
1200			**c.1242** Gunpowder is introduced into Europe from the East.		
1250		**1269** Properties of magnetic poles are described by French scholar Petrus Peregrinus (b.c.1240). **1275** Scientific explanation of the rainbow (in terms of reflections within a raindrop) is given by German scientist and ecclesiastic Theodoric of Freiburg (c.1250–1310). Much later (1611), a similar explanation is offered by Italian ecclesiastic Marco de Dominis (1566–1624).	**c.1250** Semimetallic element arsenic (symbol As) is probably discovered by German scholar Albertus Magnus (Albert von Bollstädt) (1193–1280).	**c.1260** Pulmonary circulation (of the blood, through the lungs) is discovered by Arab physician Ibn al-Nafis (1200–88), although the discovery is forgotten until re-made in 1546.	
1300			**c.1300** Sulphuric acid is first described by Spanish alchemist known as the False Geber, later (1775) rediscovered by British chemist Joseph Priestley (1733–1804). **1300** Alum is discovered in Rocca, Spain, and later (1470) in Tuscany, Italy. **1315** Ammonia is described by Spanish philosopher Raymond Lully (Ramón Lull) (c.1235–1315).	**1333** Botanical garden is established in Venice, Italy.	**1320** Benefit of stitching wounds is discovered by French physician Henri de Mondeville (1260–1320). **1340** Black Death (plague) breaks out in Asia; by 1352 it reaches northern Europe.
1350					
1400					**1414** Influenza is first described (in France).
1450	**1472** Comet (later called Halley's comet) is discovered by German astronomer Regiomontanus (Johann Müller) (1436–76); it was recorded by Chinese	**1490** Capillary action is discovered by Italian scientist and artist Leonardo da Vinci (1452–1519).			**1493** Medicinal use of tobacco (by Native Americans) is recorded by Italian explorer Christopher Columbus (1451–1506).

	FARMING AND FOOD	TRANSPORT	ENGINEERING AND TECHNOLOGY	COMMUNICATIONS
800	**800** Open fields, using three-crop rotation, are used in Europe.		**c.840** Porcelain dishes are made in China.	
850		**850** Astrolable is re-invented by the Arabs, who use it for navigation.	**c.850** Windmills are in use in Europe.	**868** First printed book (a Buddhist scripture) is produced in China by Wang Chieh.
900	**900** Wheeled ploughs are used in Europe.			
950		**984** The canal lock is invented by the Chinese transport engineer Ch'iao Wei-yo.	**980** Chain drive is invented by Chang Ssu-Hsun in China.	**c.950** Abacus is introduced into Europe for mathematical calculations by French scholar Gerbert (c.940–1003), later Pope Sylvester II.
1000	**1000** Coffee is used (initially as a medicine) in Arabia.		**1035** Spinning wheels are used in China. **1044** Formula for making gunpowder is published in a Chinese military text.	**c.1000** Camera obscura is invented in Arabia, first described by Arab mathematician Alhazen (c.965–1038). **c.1040** Modern system of musical notation, using a stave, is invented by Italian monk Guido d'Arezzo (c.990–c.1050). **c.1040** Movable type for printing, made of ceramic, is invented in China, reputedly by Pi (or Bi) Sheng.
1050		**1066** Buoys are used to mark river channels in Europe.	**1050** Crossbow is in use in France. **1088** Water-powered mechanical clock is constructed by Chinese astronomer Su Song (1020-1101).	
1100	**1100s** Hunting for whales is begun in the Western Atlantic by Spanish fishermen. **1100** Wine is distilled to make brandy in Italy.	**c.1100** Magnetic compass (pioneered by Shen Kua) is in use in China.	**1126** First artesian well is sunk at Lilliers in the former French province of Artois (Roman Artesium).	**1100** Colour printing is invented in China for printing paper money. **1036** Arabic numerals are used by Italian translator Gerard of Cremona (c.1114–87) in translations of Greek astronomy texts.
1150		**1190** Sternpost rudder is first used on European vessels.	**1189** Paper mill is built in France.	
1200			**c.1200** Gunpowder rockets are used in China. **1230** Shrapnel-type bombs are used in China.	**1232** Kites are used for sending military messages in China.
1250	**1275** A whisky distillery is built in Ireland.	**1250s** Wheelbarrows are used in Europe.	**c.1250** Gunpowder is reputedly re-invented by German alchemist Berthold Schwarz; it is referred to in 1269 in writings of English monk and philosopher Roger Bacon (1220–92). **1260** Firearms (made from bamboo tubes) for shooting bullets are used by the Chinese. **1280s** First cannons are made in China. **1280s** Belt-driven spinning wheels (from China via India) are used in Europe. **1290s** Rope cable suspension bridges are built by Native Indians in South America.	**1260** Carrier pigeons are used for sending messages in Mamluke Empire (Egypt and Syria).
1300			**1300s** Spectacles are used in Europe. **1307** Guillotine is used in Ireland. **1324** Forged iron cannons are constructed at Metz, France. **c.1330** Mechanical astronomical clock is built by English astronomer Richard of Wallingford (c.1291–1336). **1340s** Wind-driven water pumps are used for draining land in the Netherlands. **1340s** Cannons are used in warfare (employed by the English army at the 1347 siege of Calais).	
1350		**1370s** Locks are built on Dutch canals.	**1380s** Weight-driven mechanical clocks are in use. **1380** Gunpowder rockets are used in warfare in Europe.	
1400	**1400** Coffee is made into a beverage in Ethiopia. **1400** Beer is imported into England from Belgium. **1410** Drift nets are used by Dutch fisherman, who also use salt to preserve the catch.	**1400s** Three-masted ships are built in Europe. **1421** Canal boat fitted with cranes is patented by Italian Filippo Brunelleschi.	**1414** Warship equipped with cannons, the Holy Ghost, is launched in England.	**1403** Metal type is used for printing in Korea by King T'ai Tsung. **1440s** Printing press is invented by German printer Johannes Gutenberg (c.1400–1468).
1450	**1495** Whiskey is distilled in Scotland by Friar John Cor.	**1450** Caravel – to become the most common sea-going vessel – is developed in Portugal.	**1450** Large (450-metre) portable astronomical observatory is built by Turkestan astronomer Ulugh Beg (1394–1449). **1450s** Hollow-post windmills, with sails that can be aligned to face the wind, are in use	**1477** Intaglio printing, using engraved metal plates, is introduced. **c.1480** Astrolabe is developed as a navi-

ASTRONOMY AND SPACE

astronomers as early as 240 BC and was also observed in 1066.

1500

1519 Magellanic clouds (two small irregular galaxies close to the Milky Way) are discovered by Portuguese navigator Ferdinand Magellan (1480–1521).

1527 Nova in the constellation Cassiopeia is discovered by Danish astronomer Tycho Brahe (1546–1601).

1538 Fact that a comet's tail always points away from the Sun is discovered by Italian astronomer Girolamo Fracastoro (c.1478–1553).

1543 Theory of the Sun-centered "Universe" (the Solar System) is published by Polish astronomer Nicolaus Copernicus (Mikolaj Kopernik) (1473–1543) – first proposed by him 30 years previously – three years after it had been revealed by his Austrian co-worker Rheticus (Georg von Lauchen) (1514–76).

1550

1550 Fact that comets are not in the Earth's atmosphere (as previously thought) is proved by Italian scholar Girolamo Cardano (Jerome Cardan) (1501–76), later (1577) confirmed by Danish astronomer Tycho Brahe (1546–1601).

1592 Nova in the constellation Cetus is discovered by astronomers in Korea.

1596 Star that varies widely in brightness (later named Mira Ceti) is discovered by German astronomer David Fabricius (1564–1617), who mistakes it for a nova.

1600

1603 Bayer letters (Greek letters assigned to stars in a constellation in order of their brightness) are introduced by German astronomer Johann Bayer (1572–1625).

1604 Supernova in the constellation Ophiuchus is discovered by German astronomer Johannes Kepler (1571–1630).

1607 Observations of Halley's comet are recorded by English scientist Thomas Harriot (1560–1621).

PHYSICS

1576 Magnetic dip is discovered by English navigator Robert Norman (b.c.1560).

1577 Principle of the siphon is discovered by Scottish mathematician William Wellwood (d.c.1622).

1586 Fact that different weights dropped from the same height fall at the same rate (in vacuum) is demonstrated by Dutch mathematician Stevinus (Simon Stevin) (1548–1620), and later (after 1610) taken up by Galileo Galilei (1564–1642).

1600 Magnetic properties of the Earth are explained by English physician William Gilbert (1544–1603).

1602 Constancy of the period of a swinging pendulum is discovered by Italian scientist Galileo Galilei (1564–1642).

c.1610 Law of falling bodies (that all objects fall to Earth at the same speed) is proposed by Italian scientist Galileo Galilei (1564–1642), although not published until 1638.

1621 Snell's law, or law of refraction (in

CHEMISTRY

1597 Preparation of hydrochloric acid is first described by German alchemist Andreas Libavius (Andreas Libau) (c.1540–1616), later (1775) rediscovered by British chemist Joseph Priestley (1733–1804).

1620 Word "gas" (from "chaos") is coined by Flemish alchemist Jan (Jean, Johannes) van Helmont (1577–1644).

c.1625 Method of making hydrochloric acid from sodium chloride (common salt) and sulphuric acid is discovered by German chemist Johann Glauber (1604–68), who also produced sodium sulphate, known as Glauber's salt.

1644 Water glass is discovered by German chemist Johann Glauber (1604–68).

BIOLOGY

c.1500 Human anatomy is studied by Italian artist Leonardo da Vinci (1452–1519).

1517 Fact that fossils are the remains of once-living organisms is proposed by Italian scientist Girolamo Fracastoro (c.1478–1553).

1517 Homologies between bones in various types of vertebrates are discovered by French naturalist Pierre Belon (1517–64).

1538 Optic nerves are discovered by Italian anatomist and surgeon Constanzo Varoli (c.1543–75).

1546 Lesser (pulmonary) circulation of the blood (from the heart to the lungs and back) is rediscovered by Italian anatomist Realdo Colombo (c.1516–59) and, independently (c.1553), Spanish biologist Michael Servetus (1511–53).

1546 Term fossil is introduced by German mineralogist Georgius Agricola (Georg Bauer) (1494–1555).

1552 Eustachian tube (connecting the middle ear to the throat) is discovered by Italian anatomist Bartolommeo Eustachio (c.1520–74), although it had earlier (1546) been described by Italian physician Giovanni Ingrassia (1510–80) and much earlier (c.520 BC) by Greek physician Alcmaeon of Crotona.

c.1555 Fallopian tubes (connecting the ovaries to the uterus) are discovered by Italian anatomist Gabriele Falloppio (1523–62).

1558 Historia Animalium, the basis of modern zoology, is completed by Swiss naturalist Konrad non Gestner (1516–65).

1573 Pons Varolii (a nerve tract in the brain) is discovered by Italian anatomist Constanzo Varolio (c.1543–75).

1574 Valves in veins are discovered by Italian anatomist Hieronymus Fabricius ab Aquapendente (Girolamo Fabrici) (1537–1619).

1580 Existence of male and female flowers (on different plants) is discovered by Italian botanist Prospero Alpini (1553–1616).

1583 Method of classifying plants by their structure is introduced by Italian botanist Andrea Cesalpino (1519–1603).

1592 First European description of the coffee plant is given by Italian botanist Propero Alpini (1553–1617).

1599 First comprehensive book on zoology is written by Italian naturalist Ulisse Aldrovani (1522–1605).

1622 Lacteal vessels are discovered (in a dog) by Italian anatmoist and physician Gasparo Aseli (Gaspar Asellius) (1581–1626).

1627 Aurochs, the ancestor of domestic cattle, becomes extinct.

1628 Circulation of the blood, with the heart as a pump, is discovered by English physician William Harvey (1578–1657).

1641 African chimpanzee is first described by Dutch anatomist Claes

MEDICINE

1493 Syphilis occurs in Spain and later (1495) Italy (following the return of Christopher Columbus and his sailors from North America); it is described and named by Italian physician Fracastorius (Girolamo Fracastoro) (1478–1553), who used mercury as a treatment.

1497 Surgical treatment of gunshot wounds is introduced by Hieronymus Brunschwygk (c.1452–1512).

1504 Iron prosthetic hand is made in Germany, for mercenary knight Götz von Berlichingen (1480–1562).

1518 Smallpox occurs in the Americas, later (1520) decimating the Aztecs.

1530 Laudanum (tincture of opium) is first used in treatment (and named) by Swiss physician and alchemist Philippus Paracelsus (Theophrastus von Hohenheim) (1493–1541).

1542 Appendicitis is first described by French physician Jean Fernel (1497–1558).

1543 Eight-volume Fabrica (in full De humani corporis fabrica), the first book on human anatomy based entirely on observation, is published by Flemish anatomist Andreas Vesalius (Andreas van Wesele) (1514–64).

1546 Fact that microbes, or germs, cause disease is proposed by Italian physician Fracastorius (Girolamo Fracastoro) (1478–1553).

1550 Ligature for stopping arterial bleeding during surgery is introduced by French surgeon Ambroise Paré (c.1510–90), much later (1674) extended to the exterior of a limb as the tourniquet by his compatriot Morel.

1597 Rhinoplasty (plastic surgery to reconstruct the nose by grafting tissue) is perfected by Italian surgeon Gaspare Taglacozzi (1545–99).

1645 Clinical description of rickets is given by English physician Daniel Whistler (1619–84) and later (1650), independently, by his compatriot anatomist Francis Glisson (1597–1677).

FARMING AND FOOD

1565 First potatoes are imported into Spain from South America.

1566 First seed drill in Europe is invented by Italian agriculturalist Camillo Torello.

1575 Commercial production of gin is introduced by Dutch distiller Lucas Bols.

1610 First tea is imported into Europe from Asia.

1635 Iron cooking range is patented by British inventor John Sibthorpe.

TRANSPORT

1460 Four-wheeled passenger coach is built in Hungary.

1480 Practical parachute is designed by Italian artist and engineer Leonardo da Vinci (1452–1519).

1492 Flying machine (an impractical wing-flapping ornithopter) is designed by Italian artist and engineer Leonardo da Vinci (1452–1519).

1500 Helicopter (impractical) is designed by Italian artist and engineer Leonardo da Vinci (1452–1519).

1506 Rocket-propelled chair (meant to fly) is built by Chinese scientist Wan Hu; it explodes and kills him.

1500s Tramways, with wooden tracks, come into use in German mines.

1522 Round-the-world voyage is completed by Spanish seaman Juan del Cano (d.1526) in the Vittoria, the only survivor of five ships that set out three years earlier under the command of Ferdinand Magellan (c.1480–1521).

1535 Glass diving bell is used in Italy.

1543 Wheeled vehicle powered by sails (a land yacht) is constructed in Germany.

1573 Ship's log, for measuring speed through the water, is invented by English instrument maker Humphray Cole (d.1580).

1580 Round-the-world voyage is completed by English sailor Francis Drake (c.1540–96) in his ship the Golden Hind (ex Pelican), the only survivor of five ships that set out in 1577.

1589 Kites are described in detail by Italian Giambattista della Porta (1535–1615).

1596 Ironclad warship is built in Korea.

1600 Land yacht ("sailing chariot") is invented by Dutch engineer Simon Stevin (Stevinus) (1548–1620).

1605 Primitive diving bell is invented by Dutch engineer Jan Leeghwater (1575–1650).

1605 Horse-hauled pithead railway, with wooden rails, is constructed by English engineer Francis Willoughby.

1621 Primitive man-powered submarine is invented by Dutch engineer Cornelis van Drebbel (1572–1634).

ENGINEERING AND TECHNOLOGY

in the Netherlands.

1487 Land mines are used in warfare in Italy.

1494 Paper mill is constructed in England.

1495 Explosive land mine is invented by Italian engineer and artist Francesco di Giorgio (1439–1502), although crude mines were used earlier (1487) by Genoese troops.

1500 Wheel lock for muzzle-loading firearms is designed by Italian artist and engineer Leonardo da Vinci (1452–1519); it is manufactured in Italy from 1515 (replacing the matchlock).

c.1505 First spring-driven watch is made by German clockmaker Peter Henlein (1480–1542).

1517 Wheel-lock muskets are produced in Nuremberg, Germany.

1520 Muzzle-loading rifles (with spirally-grooved barrel) are made by August Kotter.

1520 Fusée, for controlling the release of a clock mainspring, is invented by Jakob the Czech.

c.1540 Gear-cutting machine (for making clocks) is invented by Italian instrument-maker Juanelo Torriano (1501–75).

1550 Screwdriver is in use (at first by gunsmiths).

1551 Theodolite is invented by English mathematician Leonard Digges (c.1520–59).

1552 New glass-making process is patented in England by "Smyth"; it is the first English patent ever issued for an invention.

1561 Dredger is invented by Flemish engineer Pieter Breughel and, much later (1618), by English naval officer John Gilbert.

1570 First timber truss girder bridge is built by Italian architect Andrea Palladio (1518–80).

1585 Floating mines, detonated by a clockwork mechanism, are used by the Dutch navy against the Belgians.

1589 Knitting machine for making stockings (called a stocking frame) is invented by English clergyman William Lee (d.1610).

1589 Flush toilet is designed by English poet John Harington (1561–1612).

c.1590 Compound microscope is reputedly invented by Dutch spectacle makers Hans Janssen and his son Zacharias.

1592 Air-filled thermometer is invented by Italian scientist Galileo Galilei (1564–1642).

1598 Self-winding and regulating clock is invented by Dutch engineer Cornelis van Drebbel (1572–1634).

1600 Surveyor's chain is invented by English mathematician Edmund Gunter (1581–1626).

1608 Refracting telescope, using two lenses, is invented by Dutch optician Hans Lippershey (c.1570–c.1619), although the designed was probably anticipated (1571) by English mathematicians Leonard (c.1520–59) and Thomas (c.1546–95) Digges; its invention was also claimed (for 1586) by Dutch charlatan Zacharias Jansen (1588–c.1631).

1609 Thermostat is invented by Dutch physicist Cornelius van Drebbel (1572–1634).

1610 Modern type of compound microscope is invented by German astronomer Johannes Kepler (1571–1630).

1615 Pantograph (for copying and scaling drawings) is invented by German astronomer Christoph Scheiner (c.1579–1650).

COMMUNICATIONS

gational aid by German Martin Behaim (c.1459–1507).

1492 Graphite "lead" is used for making pencils in England.

1498 Movable type for printing music is invented by Italian printer Ottaviano dei Petrucci (1466–1539).

1550 Glass lens for a camera obscura is made by Italian mathematician and scientist Geronimo Cardano (1501–1576); the camera was reinvented at abaout that time by his fellow countryman Giambattista della Porta (1535–1615).

1554 Sealing wax is introduced in England by Gerhard Hermann.

1588 Type of modern shorthand, using hard-to-learn arbitrary characters, is invented by English physician turned priest Timothy Bright (c.1551–1615).

1600 Newspapers are published in England and Belgium.

1614 Logarithms are invented by Scottish mathematician John Napier (1550–1617) and, independently, German mathematician Joost Bürgi (1552–1632).

1615 The slide rule is invented by English mathematician William Oughtred (1575–1660).

1620 Deaf and dumb sign language is invented by Spaniard Juan Bonet.

1500

1550

1600

ASTRONOMY AND SPACE

1609 Kepler's first two laws of planetary motion are discovered by German astronomer Johannes Kepler (1571–1630). The third law was formulated by him in 1619.

1609 Drawings of the Moon's surface, made using an imported Dutch telescope, are prodcued by English scientist Thomas Harriot (1560–1621).

1610 Four of the major moons of Jupiter, the phases of Venus and mountains on the Moon are discovered by Italian scientist Galileo Galilei (1564–1642); discovery of the moons of Jupiter is also claimed (1609), by German astronomer Simon Marius (Mayr) (1570–1624), who named them Callisto, Europa, Ganymede and Io.

1610 Sunspots are discovered by German astronomer Christoph Scheiner (c.1579–1650) and later (1613), independently, by Italian scientist Galileo Galilei (1564–1642).

1612 Andromeda nebula is discovered by German astronomer Simon Marius (Mayer) (1570–1624).

1621 Aurora borealis is described and named by French philosopher Pierre Gassendi (1592–1655).

1631 Transit of Mercury (across the Sun's disc) is first observed by French philosopher Pierre Gassendi (1592–1655).

1632 Leiden Observatory, the world's first, is established.

1633 Galieleo's theory about a Sun-centered Solar System is condemned by the Inquisition.

1637 Libration of the Moon is discovered by Italian scientist Galileo Galilei (1564–1642).

1639 Transit of Venus (across the Sun's disc) is first observed by English astronomer Jeremiah Horrocks (1619–1641).

1644 Phases of Mercury are discovered by German astronomer Johannes Hevelius (Johann Hevel) (1611–87).

1647 Map of the Moon is drawn by German astronomer Johannes Hevelius (Johann Hewel) (1611–87).

PHYSICS

modern form that, during refraction of light by a transparent medium, the ratio of the sines of the angles of incidence and refraction is a constant equal to the refractive index of the medium), is discovered by Dutch scientist Willebrod van Roijen Snell (or Snellus) (1591–1626).

1635 Gradual shift in the position of the north magnetic pole is discovered by English astronomer Henry Gellibrand (1597–1636).

c.1640 Fermat's principle (that light travels the shortest path between two points, i.e. in straight lines) is discovered by French mathematician Pierre de Fermat (1601–65).

1645 Torricellian vacuum (in the space above a column of mercury) is discovered by Italian physicist Evangelista Torricelli (1608–47).

1647 Pascal's law or principle (that external pressure on a liquid is transmitted equally in all directions) is discovered by French scientist Blaise Pascal (1623–62).

CHEMISTRY

1649 Atomic theory is proposed by French philosopher Pierre Gassendi (1592–1655), after studying the works of the Greek Epicurus (c.342–270 BC) who in turn adopted the idea from Democritus.

BIOLOGY

Tulp (Nicolaes Pieterszoon) (c.1593–c.1674).

1642 Pancreatic duct, which carries digestive juices from the pancreas to the common bile duct (and thence to the duodenum) is discovered by German anatomist Georg Wirsung (1600–43).

1647 Thoracic duct (which carries lymph from the legs, abdomen and left side of the body) is discovered by French anatomist Jean Pecquet (1622–74).

1649 Capillary blood vessels linking arteries and veins are discovered by English physician Henry Power (1623–68).

MEDICINE

1650

ASTRONOMY AND SPACE

1650 First double star (Mizar, in Ursa Major) is discovered by Italian astronomer Giovanni Riccioli (1598–1671).

1659 Rings of Saturn are discovered by Dutch scientist Christiaan Huygens (1629–95).

1664 Star Gamma Arielis (in the constellation Aries) is discovered to be a double star by English scientist Robert Hooke (1635–1703).

1664 Rotation of Jupiter and its Great Red Spot are discovered by English scientist Robert Hooke (1635–1703).

1665 Titan, the first moon of Saturn, is discovered by Dutch astronomer Christiaan Huygens (1629–95).

1666 Polar ice caps on Mars are discovered by Italian astronomer Giovanni Domenico (Jean Dominique) Cassini (1625–1712).

1669 Length of the meridian (to determine a degree of longitude and subsequently a metre) is determined by French astronomer Jean Piccard (1620–82).

1671 Iapetus, a moon of Saturn, is discovered by Italian astronomer Giovanni Domenico (Jean Dominique) Cassini (1625–1712).

1672 Rhea, a moon of Jupiter, is discovered by Italian astronomer Giovanni

PHYSICS

1650 Fact that sound will not travel in a vacuum is discovered by German scientist Athanasius Kircher (1601–80), later (1705) elaborated by English physicist Francis Hawksbee (or Hauksbee) (c.1670–c.1713).

1662 Boyle's law (that the pressure of a gas is inversely proportional to its volume, at constant temperature) is formulated by Irish scientist Robert Boyle (1627–91), and later (1676) independently discovered by French physicist Edme Mariotte (1620–84) – hence the alternative name Mariotte's law.

c.1665 Law of gravity is formulated by English scientist Isaac Newton (1642–1727).

1665 Wave theory of light is proposed by English scientist Robert Hooke (1635–1703).

1665 Diffraction of light is discovered by Italian physicist Francesco Grimaldi (1618–63).

1668 Principle of conservation of momentum is discovered by English mathematician John Wallis (1616–1703).

1669 Double refraction of light (by a crystal of Iceland spar) is discovered by Danish physician Erasmus Bartholin (1625–98).

1671 Ability of a glass prism to disperse white light into a spectrum is discov-

CHEMISTRY

c.1650 Coal gas (made by distilling coal) is discovered by English scientist John Clayton.

1661 Definition of an element (as a substance that cannot be broken down into simpler substances) is given in his book The Sceptical Chemist by Irish scientist Robert Boyle (1627–1691).

1669 Nonmetallic element phosphorus (symbol P) is discovered by German alchemist Hennig Brand (c.1630–c.92), later (1680) described independently by Irish scientist Robert Boyle (1627–91).

1677 Ammonia solution is prepared by German alchemist Johann Kunckel (c.1630–1702).

1695 Epsom salts (magnesium sulphate) are discovered in natural mineral water by English physician and botanist Nehemiah Grew (1641–1712).

1697 Phlogiston theory (that all flammable substances contain phlogiston, which is released on burning) is proposed by German chemist Georg Stahl (1660–1734), developed from the ideas of his tutor Johann Becher (1635–81).

BIOLOGY

1653 Lymphatic vessels are discovered by Swedish naturalist Olof Rudbeck (1630–1702) and, independently, Danish physician Thomas Bartholin (1616–80).

1658 Erythrocites (red blood cells) are discovered by Dutch naturalist Jan Swammerdam (1637–80).

1660 Blind spot (where the optic nerve joins the retina of the eye) is discovered by French physicist Edmé Marriotte (c.1620–84).

1661 Blood capillaries linking the arterial and venous circulation in the lungs are discovered by Italian anatomist Marcello Malpighi (1628–94).

1662 Parotid duct (Stenson's duct), which carries saliva from the parotid gland below the ear into the mouth, is discovered by Danish physician Nils Stensen (1638–86).

1664 Circle of Willis (blood vessels in the brain) are discovered by English physician Thomas Willis (1621–75).

1665 Word "cell" is coined, based on his microscope studies of plants, by English scientist Robert Hooke (1635–1703).

1667 Fact that the essential feature of respiration is the modification of blood in the lungs is demonstrated by English scientist Robert Hooke (1635–1703).

MEDICINE

1658 Fact that apoplexy (stroke) is caused by a brain haemorrhage is discovered by Swiss pathologist Johann Wepfer (1620–1695).

1659 Symptoms and progess of typhoid fever are first described by English physician Thomas Willis (1621–75).

c.1660 Laudanum (tincture of opium) is rediscovered by English physician Thomas Sydenham (1624–89).

1666 Blood transfusion (between two dogs) is first demonstrated by English physician Richard Lower (1631–91); later (1667) transfusion from a sheep to a boy is made by French physician Jean Denis (or Denys) (d.1704).

1666 Use of iron to treat anaemia and chinchona bark (containing quinine) for malaria are discovered by English physician Thomas Sydenham (1624–89).

1670 Fact that the urine of a diabetic contains sugar (glucose) is discovered by English physician Thomas Willis (1621–75), although this was also known to the ancients.

1677 Peyer's patches (lymphatic glands in the small intestine) are discovered by Swiss anatomist Johann Peyer (1653–1712).

1683 Clinical description of gout is given by English physician Thomas Sydenham (1624–89).

FARMING AND FOOD

1666 Cheddar cheese is invented in England.

1668 Champagne is invented by French Benedictine monk Dom Pierre Pérignon (1638–1715).

TRANSPORT

1640 Horse-drawn taxi is invented by French coachman Nicolas Sauvage.

1662 Twin-hulled (catamaran) yacht is built by English economist and engineer William Petty (1623–87).

1663 Public passenger-carrying coaches go into service in Paris, France.

1670 Impractical airship lifted by evacuated copper spheres is designed by Jesuit monk Francesco de Lana (1631–87).

1681 Canal du Midi, across France connecting the Atlantic Ocean and the Mediterranean Sea, is opened.

1690 Paddleboat (with paddles driven by a waterwheel) is built by French physicist Denis Papin (1647–c.1712).

ENGINEERING AND TECHNOLOGY

1616 First clinical (medical) thermometer is invented by Italian physician Santorio Santorio (Sanctorius) (1561–1636).

1618 Dredger is patented by English inventor John Gilbert.

1631 Accurate measuring caliper (the Vernier gauge) is invented by French mathematician Pierre Vernier (1580–1637).

1636 Micrometer measuring gauge is invented by English astronomer William Gascoigne (c.1612–44).

1641 Sealed liquid-in-glass thermometer is invented by Italian nobleman Ferdinand II of Tuscany (1610–70).

1641 Pendulum clock is constructed by Italian scientist Vincenzio Galilei (Galileo's son).

1644 Mercury barometer is invented by Italian scientist Evangelista Torricelli (1608–47).

1650 Air (vacuum) pump is invented by German physicist Otto von Guericke (1602–86).

1650 Mansard roof is invented by French architect François Mansard (or Mansart) (1598–1666).

1656 Pendulum clock is made by Dutch scientist Christiaan Huygens (1629–95) and, later (1657), by his countryman Salomon da Coster.

1658 Hairspring for the balance wheel of a watch is invented by English scientist Robert Hooke (1635–1703) and later (1674), and probably independently, by Dutch scientist Christiaan Huygens (1629–95).

1659 Improved air (vacuum) pump is invented by English scientist Robert Boyle (1627–91).

1660 Anchor escapement for clocks is invented by English scientist Robert Hooke (1635–1703).

1661 Manometer, for measuring gas pressure, is invented by Dutch scientist Christiaan Huygens (1629–95).

1663 Earliest type of reflecting (mirror) telescope is invented by Scottish mathematician James Gregory (1638–75), later perfected by Robert Hooke.

1666 Spirit level is invented by Frenchman Jean de Thévenot (c.1620–92).

1667 Amemometer (for measuring wind speed) is invented by English scientist Robert Hooke (1635–1703).

1699 Gas thermometer that uses changes in gas pressure is invented by French physicist Guillaume Amontons (1663–1705).

1670 Improved plunger water pump (for fire engines) is invented by English diplomat (and inventor) Samuel Morland (1625–95).

1670 Wheel-cutting machine is invented by English scientist Robert Hooke (1635–1703).

1672 Leather fire hose, made with brass joints, is invented by Dutchman Jan van der Heiden. It is introduced into England in 1674 by engineer Isaac Thomson.

COMMUNICATIONS

1623 Adding machine ("calculating clock") is constructed by German academic Wilhelm Schickard (1592–1635).

1638 First printing press in America is set up by Stephen Day in Cambridge, Mass.

1641 Mechanical adding machine is invented by French scientist Blaise Pascal (1623–1662).

1642 Mezzotint process, for printing halftones, is invented by German engraver Ludwig von Siegen (1609–c.1677), who revealed the process in 1654.

1646 Magic lantern (for projecting transparent pictures) is invented by German archaeologist Athanasius Kircher (1601–1680).

1647 Semaphore flag codes, a type of optical telegraph, are adopted by the Royal Navy; later (1857) standardized for international use by Admiral Home Popham.

1650

1666 Megaphone (called a "speaking trumpet") is invented by English nobleman Samuel Morland (1625–1695).

1673 Calculating machine that can add, subtract, multiply and divide is invented by German scientist Gottfried von Liebniz (1646–1716).

1679 Binary arithmetic is invented by German mathematician Gottfried von Leibniz (1646–1716).

1683 Penny postal system is introduced in London by English merchant William Dockwray (d.1716).

ASTRONOMY AND SPACE

Domenico (Jean Dominique) Cassini (1625–1712).

1675 Dark band in the ring around Saturn (now called Cassini's division) is discovered by Italian astronomer Giovanni Domenico (Jean Dominique) Cassini (1625–1712), although it was earlier (1665) referred to by William Balle

1675 Royal Greenwich Observatory is founded in England.

1677 Transit of Mercury is first observed by English astronomer Edmond Halley (1656–1742).

1682 Halley's comet and its period (75 years) are identified and defined by English astronomer Edmond Halley (1656–1742), after calculations based on the comet's appearances that year and in 1456, 1531 and 1607.

1684 Dione and Thetys, moons of Saturn, are discovered by Italian astronomer Giovanni Domenico (Jean Dominique) Cassini (1625–1712).

1690 Fact that Jupiter rotates faster at its equator than at its poles is discovered by Italian Astronomer Giovanni Cassini (1625–1712).

1693 Cassini's laws, dealing with the rotation of the Moon, are discovered by Italian astronomer Giovanni Domenico (Jean Dominique) Cassini (1625–1712).

PHYSICS

ered by English scientist Isaac Newton (1642–1727).

1675 Corpuscular theory of light (that light consists of a stream of particles) is proposed by English scientist Isaac Newton (1642–1727).

1676 Finite speed of light is discovered by Danish astronomer Ole Römer (1644–1710).

1678 Wave theory of light is proposed by Dutch physicist Christiaan Huygens (1629–95).

1678 Hooke's law (that in a stretched elastic solid stress is proportional to strain) is discovered by English scientist Robert Hooke (1635–1703).

1684 Law of universal gravitation (that the gravitational force of attraction between any two objects is proportional to the product of their masses divided by the square of the distance between them) is finalized by English scientist Isaac Newton (1642–1727).

1685 Concept of centre of gravity (mass) – that an object behaves as if all of its mass is concentrated at its centre – is proposed by English scientist Isaac Newton (1642–1727).

1687 Principia Mathematica, summarizing many original ideas in mathematics and physics (including the laws of motion and the theory of gravitation), is published by English scientist Isaac Newton (1642–1727).

1690 Huygen's principle (that every point on a wavefront can act as a source of secondary waves) is discovered by Dutch physicist Christiaan Huygens (1629–95).

1699 Expansion of gases – specifically that all gases increase in volume by the same amount for the same rise in temperature – is discovered by French physicist Guillaume Amontons (1663–1705).

CHEMISTRY

BIOLOGY

1668 Spontaneous generation (of living organisms) is first disproved by Italian physician Francesco Redi (1626–97).

1669 Metamorphosis in insects is discovered by Dutch naturalist Jan Swammerdam (1637–80).

1672 Human ovaries are discovered by Dutch naturalist Jan Swammerdam (1637–80).

1672 Graafian follicle (surrounding the developing egg in the ovary) is discovered by Dutch anatomist Reinier de Graaf (1641–73).

1672 Studies of chick embryos by Italian anatomist Marcello Malphigi (1628–94) establishes the schience of embryology.

1677 Human sperm are discovered by Dutch scientist Anton van Leeuwenhoek (1632–1723), who called them "human larvae".

1680 Bacteria are discovered by Dutch scientist Anton van Leeuwenhoek (1632–1723), using a microscope of his own invention.

1681 Dodo becomes extinct on the island of Mauritius.

1682 Male and female parts of a flower are discovered by English botanist Nehemiah Grew (1641–1712), later (1694) confirmed by German botanist Rudolf Camerarius (1665–1721).

c.1690 Plants are divided into monocotyledons and dicotyledons by English naturalist John Ray (1627–1705).

1690 Angiosperms are defined by Paul Hermann.

1691 Haversian canals (in bone) are discovered by English physician Clopton Havers (c.1655–1702).

MEDICINE

c.1695 Bartholin's glands (greater vestibular glands in the vagina) are discovered by Danish physician Kasper Bartholin Jr (1655–1738).

1700

ASTRONOMY AND SPACE

1712 First orrery (a clockwork working model of the Solar System, invented by George Graham) is constructed by English clockmaker John Rowley, commissioned by Charles Boyle, Fourth Earl of Orrery (1676–1731).

1718 Proper motion of stars is discovered by English astronomer Edmond Halley (1656–1742).

1729 Aberration of light is discovered by English astronomer James Bradley (1693–1762).

1748 Nutation (the "wobble") of the Earth's rotational axis is discovered by English astronomer James Bradley (1693–1762).

PHYSICS

1701 Term "acoustics" is coined by French physicist Joseph Sauveur (1653–1716).

1709 Electrical discharges in low-pressure air are described by English physicist Francis Hawksbee (or Hauksbee) (c.1670–c.1713), who in the same year studied capillarity in narrow glass tubes.

1729 Difference between electrical insulators and conductors is discovered by English scientist Stephen Gray (1696–1736).

1733 Existence of two types of static electricity (positive and negative) is discovered by French physicist Charles du Fay (1698–1739).

1738 Kinetic theory of gases is proposed by Swiss mathematician Daniel Bernoulli (1700–1782).

1742 Celsius temperature scale is devised by Swedish astronomer Anders Celsius (1701–44).

1746 Leyden jar (the original electrical condenser) is discovered independently by Dutch physicist Pieter van Musschenbroek (1692–1761) and Polish cleric Ewald von Kleist.

CHEMISTRY

1701 Boric acid (boracic acid) is discovered by Dutch chemist Willem Homberg (1652–1715).

1722 Role of carbon in the hardness of steel is discovered by French physicist René Réaumur (1683–1757).

1727 Fact that some silver salts turn black on exposure to light – a phenomenon that was to become the key to photography – is discovered by German chemist Johann Schulze.

1730 Ethoxyethane (diethyl ether) is discovered by German chemist Frobenius.

1732 Gold-colored copper-zinc alloy (for making watches) is invented by English clockmaker Christopher Pinchbeck (c.1670–1732).

1736 Potassium and sodium salts are distinguished by French chemist Henri Duhamel du Monceau (1700–82).

1737 Metallic element cobalt (symbol Co) is discovered by Swedish chemist Georg Brandt (1694–1768).

c.1740 Curare is discovered by French chemist Marie de Lacondamine (1701–74).

1741 Metallic element platinum (symbol Pt) is identified as an element by British chemist William Brownrigg (1711–1800), although known 200 years previously as the native metal in Spain and described in 1557 by the Italian physician Julius Scaliger (1484–1558); it is later (1748) also described by Spanish scientist and soldier Antonio de Ulloa (1716–95).

1746 Metallic element zinc (symbol Zn), known since antiquity in the East and described by Swiss alchemist

BIOLOGY

1711 Fact that corals are animals (not plants, as previously thought) is discovered by Italian naturalist Luigi Marsigli (1658–1730).

1734 Science of entomology is established by French scientist René Réaumur (1683–1757) in his book in insects.

1735 Binomial classification of plants is introduced by Swedish botanist Carolus Linnaeus (Carl von Linné) (1707–78) in his book Systema Naturae, thus establishing the science of taxonomy.

1740 Parthenogenesis (reproduction by means of unfertilized eggs) is discovered in aphids by Swiss naturalist Charles Bonnet (1720–93).

1740 Freshwater hydra is discovered by Swiss naturalist Abraham Trembley (1710–84), who later (1744) discovers their ability to regenerate tissue.

MEDICINE

1701 Vaccination of children with smallpox (in an attempt to prevent the disease in later life) is performed by Italian physician Giacomo Pylarini (1659–1715); the practice is already carried out in Turkey (and elsewhere) and introduced into England (1714) by John Woodward and, independently, Mary Wortley (1689–1762).

1702 Cowper's glands (bulbourethral glands at the base of the penis) are discovered by English surgeon William Cowper (1666–1709).

1721 Vaccination against smallpox is introduced (in Boston, USA) by US physician Zabdiel Boylston (1679–1766).

1726 First blood pressure measurement (of a horse) is made by English scientist Stephen Hales (1677–1761).

1730 Tracheostomy (for treating diphtheria) is first performed by Scottish physician George Martine (1702–41); it is later (1825) used routinely by French physician Pierre Bretonneau (1771–1862).

1732 Sedative effect of ipecac and opium mixture (Dover's powder) is discovered by English naval officer and physician Thomas Dover (1660–1743).

1736 Scarlet fever is first described by US physician William Douglass.

1738 Caesarian section to deliver a baby is performed by Irish midwife Mary Donally (although the procedure was first decribed in 1500).

1747 Value of citrus fruits in combating scurvy is discovered by Scottish physician James Lind (1716–94) and later (1795) confined to lime juice by British physician Gilbert Blane (1749–1834).

FARMING AND FOOD

TRANSPORT

ENGINEERING AND TECHNOLOGY

1672 Improved reflecting (mirror) telescope is invented by French physician N. Cassegrain.

1674 Crystal glass is invented by English glassmaker George Ravenscroft.

1675 Simple (single-lens) microscope is invented by Dutch spectaclemaker Anton van Leeuwenhoek (1632–1723).

1675 Spring-driven clock is made by Dutch scientist Christiaan Huygens (1629–95).

1675 Watch with hair-spring balance wheel is made by English clockmaker Thomas Tompion (c.1639–1713).

1676 Universal joint is invented by English scientist Robert Hooke (1635–1703).

1679 Pressure cooker, complete with safety valve, is invented in England by French physicist Denis Papin (1647–c.1712), who called it a "digester"; it is not marketed until for domestic use until 1954 by the French Lescure Brothers.

1682 Steam engine principle is published by English inventor Samuel Morland (1625–95).

1687 Hygrometer (for measuring humidity) is invented by French scientist Guillaume Amontons (1633–1705).

1687 Repeating watch is invented by English clockmaker Daniel Quare (1648–1724).

1688 New type of reflecting telescope, using mirrors, is invented by English scientist Isaac Newton (1642–1727).

1690 Primitive atmospheric steam engine is invented by French physicist Denis Papin (1647–c.1712).

1698 Practical atmospheric steam engine (without a piston) for pumping water – which became known as the "miner's friend" – is constructed by English military engineer Thomas Savery (c.1650–1715).

COMMUNICATIONS

1700

FARMING AND FOOD

1701 Mechanical seed drill and horse-drawn hoe are invented by British agricultural engineer Jethro Tull (1674–1741).

1701 Nougat is invented at Montelimar, France.

1715 Method of producing cognac commercially is invented by Jersey-born Jean Martell.

1741 Artificial mineral water, made by aerating spring water with carbon dioxide gas, is made by British physician William Brownrigg (1711–1800); commercial production of soda water begins later (1767) by Richard Bewley, with the soda siphon invented (1813) by Charles Plinth; fruit-flavoured carbonated drinks are introduced (1807) in the USA by Townsend Speakman; tonic water is patented (1858) by Englishman Erasmus Bond.

1747 Process for extracting sugar from sugar beet is invented by German chemist Andreas Marggraf (1709–82).

TRANSPORT

1715 Practical diving suit is invented by Englishman Andrew Becker.

1716 Diving bell with an air supply is invented by English scientist Edmund Halley (1656–1742).

1718 Self-acting valve gear for steam engines is invented by Henry Brighton.

1737 Mirror sextant is invented by Scottish naval officer John Campbell (1720–90), an improvement on the octant version designed by English instrument maker John Hadley (1682–1744).

ENGINEERING AND TECHNOLOGY

1701 Metal-turning lathe is invented by French engineer Charles Plumier (1646–1704).

1703 Efficient vacuum pump is invented by English scientific instrument maker Francis Hawksbee (or Hauksbee) (c.1670–c.1713).

1706 Electrostatic generator based on a rotating glass sphere is invented by English instrument maker Francis Hawksbee (or Hauksbee) (c.1670–c.1713).

1707 Watch that runs for exactly one minute (to be used by a physician in measuring a patient's pulse rate) is invented by English physician John Floyer (1649–1734).

1708 Porcelain is reinvented by German alchemist Johann Böttger (1682–1719) (known to Chinese 200 years earlier).

1709 Iron-smelting using coke is invented by English ironfounder Abraham Darby (1678–1717).

1709 Alcohol thermometer is invented by German physicist Gabriel Fahrenheit (1686–1736).

1709 Eau-de-Colgne is invented by Italian-born perfumier Giovani (later Johann) Farina (1685–1766).

1712 Piston-operated atmospheric steam engine is invented by English engineer Thomas Newcomen (1663–1729).

1714 Mercury thermometer, and the Fahrenheit temperature scale, are invented by German physicist Gabriel Fahrenheit (1686–1736).

1718 Flint-lock mechanical machine gun is invented by English lawyer James Puckle.

1721 Mercury pendulum for clocks, which compensates for changes in length with temperature, is invented by English engineer George Graham (1673–1751).

1726 Gridiron compensating clock pendulum is invented by English clockmaker John Harrison (1693–1776).

1728 Dental drill is invented by French dentist Pierre Fauchard; later (1790) reinveneted by US dentist John Greenwood (1760–1819).

1730 Réaumur temperature scale (and an alcohol thermometer using it) is invented by French physicist René Réaumur (1683–1751).

1732 Pitot tube, for measuring the speed of fluid flow, is invented by French engineer Henri de Pitot (1695–1771).

1732 Pinchbeck (an alloy of copper and zinc, used to simulate gold) is invented by English clockmaker Christopher Pinchbeck (1670–1732).

COMMUNICATIONS

1710 Method of three-colour printing is invented by French printer Jacob le Blon (1670–1741).

1714 Typewriter is patented by British engineer Henry Mill.

1725 Stereotype printing process is invented by Scottish printer William Ged (1690–1749).

1737 Point system for measuring type sizes is invented by French printer Pierre Fournier (1712–68).

1745 Automatic loom controlled by holes punched in a rotating metal drum is invented by French engineer Jacques de Vaucanson (1709–82).

1750

ASTRONOMY AND SPACE

1755 Fact that nebulae are large star systems is observed by German philosopher Immanuel Kant (1724–1804).

1757 Masses of Venus and the Moon are calculated by French mathematician Alexis Clairaut (1713–65), using observations made by his compatriot, astronomer Nicolas de Lacaille (1713–62).

1761 Atmosphere of Venus is discovered by Russian scientist Mikhail Lomonosov (1711–65) and, independently (1768), US mathematician and astronomer David Rittenhous (1732–96).

1765 Period of free precession of the Earth's axis is discovered to be 300 days by Swiss mathematician Leonhard Euler (1707–83), later (1891) corrected to 428 days by US astronomer Seth Chandler (1846–1913); free precession became known as Chandler wobble.

1772 Fact that Algol (Beta Persei) is a variable star is discovered by German astronomer Johann Palitzsch (1723–88).

1772 "Law" for calculating the radii of planetary orbits (to become known as Bode's law) is discovered by German astronomer Johann Titius (1729–92).

1772 Biela's comet is initially discovered by French astronomer Jacques Montaigne; rediscovered (1805) by Frenchman Jean-Louis Pons (1761–1831) and (1826) by Austrian soldier and astronomer Wilhelm von Biela (1782–1856). It broke in half in 1846 and disappeared after 1852.

1774 Absolute value of the gravitational constant is discovered by British astronomer Nevil Maskelyne (1732–1811).

1778 Atmospheric markings on Venus

PHYSICS

1752 Electrical nature of lightning is discovered by US scientist and politician Benjamin Franklin (1706–90).

1758 Method of making achromatic lenses (by using a combination of crown and flint glass) is discovered by English optician John Dolland (1706–61) and his son Peter (1730–1820), although the principle had earlier (1733) been discovered independently by English scientist Chester Hall (1703–71).

1760 Lambert's laws (concerning luminance and luminous intensity) are discovered by German scientist Johann Lambert (1728–77).

1762 Compressibility of water is discovered by British physicist John Canton (1718–72).

1763 Latent heat (the heat needed to produce a change of state from a solid to a liquid or a liquid to a vapor) is discovered by Scottish physicist Joseph Black (1728–99).

1772 Latent heat of fusion (of ice) is discovered by Swedish physicist Johan Wilcke (1732–96).

1781 Concept of specific heat is proposed by German physicist Johan Wilcke (1732–96).

1784 Sulphur dioxide is liquefied (the first gas to be liquefied) by French mathematician and physicist Gaspard Monge (1746–1818).

1785 Coulomb's law (that the force of attraction or repulsion between two electric charges is proportional to the product of their magnitudes divided by the square of the distance between them) is formulated by French physicist Charles de Coulomb (1736–1806).

c.1787 Charles' law (that the volume of an ideal gas increases by 1/273 for each Celsius degree rise in tempera-

CHEMISTRY

Paracelsus (1493–1541), is rediscovered by German chemist Andreas Marggraf (1709–82).

1747 Sugar (sucrose) is discovered in beet by German chemist Andreas Marggraf (1709–82).

1751 Metallic element nickel (symbol Ni) is discovered by Swedish industrial chemist Alex Cronstedt (1722–65) and later (1775) isolated in pure form by his compatriot Torbern Bergman (1735–84).

1753 Metallic element bismuth (symbol Bi), probably discovered as its compounds by Valentine in 1450, is isolated and shown to be different from lead by French chemist Claud Geoffroy.

1755 Metallic element magnesium (symbol Mg) is identified by Scottish chemist Joseph Black (1728–99), and later (1808) isolated by British chemist Humphry Davy (1778–1829).

1758 Flame tests (for identifying metallic elements) are introduced by German chemist Andreas Marggraf (1709–82).

1766 Gaseous element hydrogen (symbol H) is discovered by British chemist Henry Cavendish (1731–1810), who called it "inflammable air".

1769 Tartaric acid is discovered by Swedish chemist Karl Scheele (1742–1786).

1770 Phosphorus in bones is discovered by Swedish chemist Karl Scheele (1742–86) and mineralogist Johan Gahn (1745–1818).

1770 Mercury fulminate, later used as a detonator for firearms, is discovered by German chemist Johann Kunckel (1630–1703).

1771 Gaseous element fluorine (symbol F) is discovered by Swedish chemist Karl Scheele (1742–86), and later (1886) isolated by French chemist Ferdinand Moissan (1852–1907).

1771 Picric acid, originally used as a yellow dye and not an explosive, is discovered by French chemist Pierre Woulfe.

BIOLOGY

1751 Reflex response (of contraction of the pupil when light is shone into the eye) is discovered by Scottish physician Robert Whytt (1714–66).

1760s New breeds of farm animals, such as Longhorn cattle and Leicestershire sheep, result from selective breeding experiments of British agriculturist Robert Bakewell (1725–95).

1765 Fact that heat and hermetical sealing (to achieve proper sterilization) prevents the growth of microbes is discovered by Italian biologist Lazzaro Spallanzani (1729–99).

1766 Fact that nerve impulses control muscle action is discovered by Swiss physiologist Albrecht von Haller (1708–77).

1770 Fact that phosphorus is an essential constituent of bone is discovered by Swedish mineralogist Johann Gahn (1745–1818).

1771 Blood-clotting protein fibrin is discovered by British physician William Henson.

1772 Labyrinth of the ear is discovered by Italian anatomist Antonio Scarpa (1747–1832).

1773 Digestive action of saliva is discovered by Italian biologist Lazzaro Spallanzani (1729–99).

1779 Plant respiration and photosynthesis are discovered by Dutch scientist Jan Ingenhousz (1730–99), later (1782) demonstrated by Swiss botanist Jean Senebier (1742–1809).

1780 Artificial insemination is discovered (using dogs) by Italian biologist Lazaro Spallanzani (1729–99).

1790 Shorthorn cattle are created by selective breeding by British farmers Robert (1749–1820) and Charles (1750–1836) Colling.

MEDICINE

1748 Clinical description of diphtheria is given by British physician John Fothergill (1712–80).

1752 Role of gastric juices in digestion is discovered by French scientist René de Réaumur (1683–1757).

1756 Method of taking a mould of a patient's mouth for making false teeth is introduced by German dentist Philipp Pfaff (1715–67).

1760 Extrauterine pregnancy is first discovered by US physician John Bard (1716–99).

1761 Technique of percussion of the chest as an aid to diagnosis (auscultation) is discovered by Austrian physician Leopold Auenbrugger (1722–1809); it is later pioneered by French physician Jean Corvisart (1755–1821).

1768 Heart condition angina pectoris is described by British physician William Heberden (1710–1801).

1768 Experimental pathology is established by Scottish surgeon John Hunter (1728–93).

1775 Fact that repeated exposure to soot can cause cancer of the scrotum in chimney sweeps is discovered by British physician Percivall Pott (1714–88).

1776 Trigeminal neuralgia (Fothergill's disease) is discovered by British physician John Fothergill (1712–80).

1785 Use of digitalis (foxglove) in treating dropsy (oedema) is discovered by British physician William Withering (1741–99); active ingredient (digitalin) is not isolated until 1904.

1793 Link between cirrhosis of the liver and excessive drinking (of alcohol) is discovered by British physician Matthew Baillie (1761–1823).

1794 Colour blindness is first described by British chemist John Dalton (1766–1844) who, like his brother Jonathan, was colour blind.

FARMING AND FOOD

1750 Incubator for chickens' eggs is invented by French physicist René Réaumur (1683–1757); later (1609) reinvented in England by Dutch engineer Cornelis van Drebbel (1572–1634).

1750 Plain chocolate bars are produced by British physician and typefounder Joseph Fry (1728–87); eating chocolate is later (1819) popularized in Switzerland by François Cailler.

1756 Mayonnaise is invented by French nobleman Louis de Richelieu (1696–1788).

1760 Soft liquorice candy (Pontefract cake) is invented by British confectioner George Dunhill.

1762 Sandwiches are invented by John Montagu, 4th Earl of Sandwich (1718–92) (who also gave his name to the Sandwich Islands, Hawaii).

1765 Preserving food using a hermetic seal is proposed by Italian biologist Lazzaro Spallanzani (1729–99).

1768 Mechanical flail for dressing (threshing) grain is invented by Scottish engineer Andrew Maikle (1719–1811), who later (1788) patented an efficient drum threshing machine.

1783 Mechanized flour mill is built by US inventor Oliver Evans (1775–1819).

1785 Cast-iron ploughshare is invented by British engineer Robert Ransome (1753–1830).

1786 Gypsum-based fertilizer is invented by US farmer John Binns (c.1761–1813).

1789 Bourbon whisky is first made (from maize) in Kentucky by clergyman Elijah Craig.

1793 Cotton gin is invented by US engineer Eli Whitney (1765–1825).

1797 First cast-iron plough in the USA is

TRANSPORT

1758 Improved sextant is invented by British instrument maker John Bird (1709–1776).

1761 Bridgewater Canal opens for carrying coal to Manchester, designed by James Brindley (1716–72); it is later (1762) extended to the River Mersey. Later still (1766) a second canal, designed by Brindley and engineer John Smeaton (1724–79), connects it to the River Severn.

1766 Fire escape (a wicker basket on a chain over a pulley) is invented by British watchmaker David Marie.

1770 Three-wheeled steam-powered gun carriage is invented by French engineer Nicolas Cugnot (1725–1804).

1775 Human-powered working submarine is invented by US engineer David Bushnell (c.1742–1824).

1775 Experimental steamboat is built by Frenchman Jean Perrier.

1775 Horse-drawn tram (running on iron rails) is invented by Englishman John Outram.

1783 Man-carrying hot-air balloon is invented by French papermakers Jacques (1745–99) and Joseph (1740–1810) Montgolfier, and flown by Pilâtre de Rozier and the Marquis d'Arlandes. De Rozier later (1785) becomes the first person to be killed in a balloon accident.

1783 Man-carrying hydrogen-filled balloon is invented and flown by French physicist Jacques Charles (1746–1823).

1783 Experimental parachute is invented and tested by French mechanic Louis Lenormand (1757–1839).

1783 Paddle steamboat (the Pyroschaphe) is invented and tested on the River Saône by French nobleman Claude de Jouffroy d'Abbans (1751–1832).

ENGINEERING AND TECHNOLOGY

1733 Flying shuttle (for looms) is invented by English clockmaker John Kay (1704–c.1764).

1734 Fire extinguisher, consisting of a glass flask filled with salt solution (for throwing into a fire), is invented by German physician M. Fuches.

1735 Chronometer (for accurately measuring time, and hence longitude, at sea) is invented by English clockmaker John Harrison (1693–1776).

1738 Caisson for building bridge piers below water level is invented by French engineer Charles de Labelye.

1738 Machine for carding wool is invented in England by French engineer Lewis Paul (d.1759); it is later (1845) reinvented by British textile engineer Samuel Lister (1815–1906).

1740 Crucible process for making steel (using coke) is invented by English clockmaker Benjamin Huntsman (1704–76).

1741 Gear-cutting machine is invented by Swedish engineer Christopher Polhem (1661–1751).

1742 Ballistic pendulum (for measuring the muzzle velocity of firearms) is invented by British mathematician Benjamin Robins (1707–51).

1742 Celsius (centigrade) temperature scale is devised by Swedish astronomer Anders Celsius (1701–44).

1742 Process for galvanizing steel (coating it with zinc) is invented by French chemist Paul Malouin.

1743 Sheffield plate (a thin layer of silver on an object made of copper) is invented, originally for making buttons, by British cutler Thomas Boulsover.

1745 First significant self-acting mechanical loom is invented by French weaver Jacques de Vaucanson (1709–82).

1746 Lead-chamber process for making sulphuric acid is invented by British engineer John Roebuck (1718–94).

1747 Electrometer (for measuring static electrical charge) is invented by French physicist Jean Nollet (1700–70).

1747 Lightning conductor is invented by US scientist and statesman Benjamin Franklin (1706–90).

1748 Bone china is invented by British potter Thomas Frye. It was perfected about 50 years later by his compatriot Josiah Spode (1754–1827).

1749 Radial ball bearings are invented by Philip Vaughan.

1750

1755 Double-pointed needle (later used in sewing machines) is invented by Englishman Charles Weisenthal.

1756 Hydraulic cement (known to the Romans), which sets underwater, is reinvented by British engineer John Smeaton (1724–92), who uses it to build the third Eddystone lighthouse.

1756 Turntable for turning steam railway locomotives is constructed by British engineer John Smeaton (1724–92).

1758 Stocking frame, a fast knitting machine for hosiery, is invented by British cotton spinner Jedediah Strutt (1726–97).

1762 Process for making wrought iron out of cast iron (using coal and an air blast) is invented by British engineer John Roebuck (1718–94).

1762 Technique of puddling canals (coating the channel with a waterproof mixture of clay and sand or gravel) is introduced by British canal-builder James Brindley (1716–72).

1763 Double-cylinder atmospheric steam engine is invented by Russian engineer Ivan Polsunov (1728–66).

1764 Spinning jenny is invented by British weaver James Hargreaves (1720–78).

1764 Steam engine with a separate condenser is invented by Scottish engineer James Watt (1736–1819).

1764 Arched stone base for road construction is invented by French engineer Pierre Trésaguet (1716–96).

1765 Cylinder-boring machine is invented by British engineer John Smeaton (1724–92).

1766 Electrometer is invented by Swiss scientist Horace-Bénédict de Saussure (1740–99).

1768 Hydrometer (for measuring liquid density) graduated with the Baumé scale is invented by French scientist Antoine Baumé (1728–1804).

1768 Machine for generating static electricity using glass plates is invented by British instrumentmaker Jesse Ramsden (1735–1800).

1768 Spinning frame is invented (patented 1769) by British inventor Richard Arkwright (1732–92), assisted by John Smalley (d.1782).

1768 Rubber tubes for use as surgical catheters are invented by Frenchc chemist Pierre Macquer (1718–84).

1769 Artificial stone for making ornaments is invented by British artist Eleanor Coade (1709–96).

1769 Venetian blinds are patented by British inventor Edward Beran.

COMMUNICATIONS

1750 Camera obscura is used as an aid to drawing by painters such as Italian Antonia Canaletto (1697–1768).

1753 Electrical telegraph using 26 wires – one for each letter – is proposed by Scotsman Charles Morrison and, later (1774) by Swiss inventor George Lesage (1724–1803).

1761 Mechanical telegraph is demonstrated by British inventor Richard Edgeworth (1744–1817).

1768 Aquatint engraving process is invented by French artist Jean le Prince (1733–81).

1780 Embossed characters (on paper) for the blind to read are invented by Frenchman Valentin Haüy (1745–1822).

1780 Hectograph, a method of copying documents written using a special ink, is patented by Scottish inventor James Watt (1736–1819).

1790 Improved method of making stereotypes is invented by French printer Firmin Didot (1764–1836).

1791 Optical (semaphore) telegraph is invented by French engineers Claude (1763–1805) and Ignace (1760–1828) Chappe.

1795 Fired graphite-and-clay pencil leads (a substitute for natural plumbago) are invented by French chemist Nicolas Conté (1755–1805).

1798 Lithography, for printing, is invented by Bavarian cartographer Aloys Senefelder (1771–1864).

ASTRONOMY AND SPACE

are first observed by German astronomer Johann Schröter (1745–1816).

1780 "Bode's law", actually discovered in 1772 by German astronomer Johann Titius (1729–92), is championed by fellow-German Johann Bode (1747–1826). The "law" enables the radii of the orbits of most planets to be calculated approximately, although it appears to be based on coincidence.

1781 Planet Uranus is discovered by British astronomer William Herschel (1738–1822).

1782 Fact that the variation of the light from the star Algol is due to an invisible orbiting companion (which periodically blocks out some of the light from the bright star – i.e. it is an eclipsing binary) is discovered by Dutch-born British astronomer John Goodricke (1764–86).

1783 Intrinsic motion of the Sun through space is discovered by British astronomer William Herschel (1738–1822).

1784 First Cepheid variable star (Delta Cephei) is discovered by Dutch-born British astronomer John Goodricke (1764–86).

1786 Comet with the shortest known period – Encke's comet – is discovered by French astronomer Pierre Méchain (1744–1804) and later (1818) recorded by French astronomer Jean Louis Pons (1761–1831); its orbit and period are then (1819) calculated by the German mathematician and astronomer Johann Encke (1791–1865).

1787 Titania and Oberon, the two major moons of Uranus, are discovered by British astronomer William Herschel (1738–1822).

1789 Enceladus, moon of Saturn, is discovered by British astronomer William Herschel (1738–1822).

1790 Planetary nebulae are discovered by British astronomer William Herschel (1738–1822).

1794 Fact that meteors are extraterrestrial is discovered by German physicist Ernst Chladni (1756–1827).

1796 Nebular hypothesis, a theory that the Solar System originated from a cloud of gas, is proposed by French mathematician and astronomer Pierre Laplace (1749–1827), who also predicts the existence of black holes.

1799 "Flattening" of the Earth (from pole to pole) is discovered by French astronomer Jean Baptiste Delambre (1749–1822).

PHYSICS

ture) is discovered by French physicist Jacques Charles (1746–1823), and is later (1802) accurately defined by his compatriot Joseph Gay-Lussac (1778–1850).

1787 Ammonia is liquefied by Dutch physicists Martinus van Marum (1750–1837) and Pars van Troostwijk (1752–1837).

1788 Blagden's law (that the decrease in freezing point of a liquid when a solute is added is proportion to the concentration of the solute) is discovered by British chemist Charles Blagden (1748–1820).

1791 Prévost's theory of exchange (that, at equilibrium, an object emits and absorbs radiant energy at equal rates) is discovered by Swiss physicist Pierre Prévost (1751–1839).

1791 Fact that heat can only travel from a hot object to a cooler object (and not the other way round) is discovered by Swiss physicist Pierre Prèvost (1751–1839).

1798 Density of the Earth is accurately determined by British scientist Henry Cavendish (1731–1810), although it had been estimated earlier (1774) by British astronomer Nevil Maskelyne (1732–1811).

1798 Relationship between mechanical work and heat is discovered by US scientist Count Rumford (Benjamin Thompson) (1753–1814).

CHEMISTRY

1772 Gaseous element oxygen (symbol O) is discovered by Swedish chemist Karl Scheele (1742–86), although his discovery was not announced until 1777. It was independently discovered in 1774 by British chemist Joseph Priestley (1733–1804) and named in 1777 by French chemist Antoine Lavoisier (1749–94). As early as 1727 it was observed in plant respiration, but not recognized as an element, by English physiologist Stephen Hales (1677–1761).

1772 Gaseous element nitrogen (symbol N) is discovered by Scottish physician Daniel Rutherford (1749–1819) and later (1790) named by French chemist Jean Chaptal (1756–1832).

1772 Fact that diamond is pure carbon is discovered by French chemist Antoine Lavoisier (1749–94), who obtains carbon dioxide by burning a diamond.

1772 Rubber is so named by British chemist Joseph Priestley (1733–1804), because it rubs out (erases) pencil marks.

1774 Gaseous element chlorine (symbol Cl) is discovered by Swedish chemist Karl Scheele (1742–86), and later (1810) shown to be an element and named by British chemist Humphry Davy (1778–1829).

1774 Ammonia gas is discovered by British chemist Joseph Priestley (1733–1804).

1774 Methanoic (formic) acid is discovered by Swedish chemist Karl Scheele (1742–86).

1774 Metallic element manganese (symbol Mn) is discovered by Swedish mineralogist Johan Gahn (1745–1818).

1775 Pure nickel is prepared by Swedish chemist Torben Bergman (1735–84).

1776 Uric acid is discovered by Swedish chemist Karl Scheele (1742–86) and, independently, by his compatriot Torbern Bergman (1735–84).

1780 Lactic acid and casein are discovered by Swedish chemist Karl Scheele (1742–1786).

1781 Water is shown to be a compound by British scientist Henry Cavendish (1731–1810).

1781 Metallic element tungsten (symbol W) is discovered by Swedish chemist Karl Scheele (later (1742–86), who called it wolfram, and later (1783) isolated by Spanish mineralogists Don Juan and Don Fausto d'Elhuyer (1755–1833).

1781 Metallic element molybdenum (symbol Mo) is isolated by Swedish chemist Peter Hjelm (1760–1813), following its discovery in 1778 by his compatriots Torbern Bergman (1735–84) and Karl Scheele (1742–86), who named it.

1782 Metallic element tellurium (symbol Te) is discovered by Austrian chemist Franz Müller (1740–1825) who sent a sample for confirmation to German chemist Martin Klaproth (1743–1817), who named it.

1782 Hydrocyanic (prussic) acid is discovered by Swedish chemist Karl Scheele (1742–86), and its composition later (1787) found by French chemist Claude Berthollet (1748–1822), who in the same year also determined the compositions of ammonia and hydrogen sulphide

1783 Fact that the atmosphere has a constant composition (at different places) is discovered by British chemist Henry Cavendish (1731–1810).

1783 Citric acid and glycerol (glycerine) are discovered by Swedish chemist Karl Scheele (1742–86).

1784 Bleaching action of chlorine is dis-

BIOLOGY

1793 Role of insects in pollination of flowers is discovered by German botanist Christian Sprengel (1750–1816).

1793 Fact that fossils are the remains of once-living organisms is proposed by French naturalist Jean Baptiste Lamarck (1744–1829).

1793 Dichogamy (the maturing of male and female parts of the same flower at different times to prevent self-fertilization) is discovered by German botanist Christian Sprengel (1750–1816).

1794 Zoonomia (proposing a Lamarckian-type theory of evolution) is published by British physician Erasmus Darwin (1731–1802).

1795 Fossil bones (Mosasaurus, found in the Netherlands in 1766) are identified as being those of an extinct reptile by French anatomist Georges Cuvier (1769–1832), who went on to found the science of palaeontology.

1798 Correlation between food supply and human population growth is discovered (and published anonymously) by British economist Thomas Malthus (1766–1834).

MEDICINE

1795 Fact that puerperal fever is contagious is discovered by Scottish physician Alexander Gordon (1752–99). This was later (1843) rediscovered by US physician Oliver Wendell Holmes (1809–94), and later still (1848) confirmed by Hungarian physician Ignaz Semmelweiss (1818–65), who died of the disease.

1796 Vaccination against smallpox is pioneered by British physician Edward Jenner (1749–1823), independently of Zabdiel Boylston (1721).

1799 Anaesthetic effect of dinitrogen oxide (nitrous oxide, or laughing gas) is discovered by British chemists Thomas Beddoes (1760–1808) and Humphry Davy (1778–1829).

FARMING AND FOOD

patented by US inventor Charles Newbold.

1799 Wool-carding machine is invented in Belgium by British engineer William Cockerill (1759–1832).

TRANSPORT

1784 First balloon ascents in Britain are made by Scottish journalist James Tytler (1747–1804) and, independently, Italian balloonist Vincenzo Lunardi (1759–1806).

1784 Working model steam-powered road carriage, with three wheels and a high-pressure steam engine, is built by British engineer William Murdock (1754–1839).

1785 "Unsinkable" lifeboat is invented by British coachbuilder Lionel Lukin (1742–1834), which is superseded later (1790) by The Original, designed by British boatbuilder Henry Greathead (1757–1816) and incorporating many ideas of British inventor William Wouldhave (1751–1821).

1785 First flight across the British Channel (Dover to Calais) is made in a hydrogen balloon by French aeronaut Jean-Pierre Blanchard (1753–1809) and his sponsor, US physician John Jeffries (1744–1819); Blanchard later (1793) makes the first balloon flight in the USA.

1787 Steam-powered boat (with an endless paddle chain resembling mechanical oars) is built by US engineer John Fitch (1743–98), but it proves unreliable. Later (from 1790) Fitch boats successfully run on the Delaware River.

1787 First iron-hulled boat (a canal barge called the Trial) is built by British ironfounder John Wilkinson (1728–1808) at Abraham Darby's Coalbrookdale works on the River Severn.

1787 Water-jet propelled boat (using a steam-driven pump) is built by US engineer James Rumsey (1743–92).

1788 Experimental steamboats are built by Scottish banker Patrick Miller (1731–1815) and engineer William Symington (1763–1831), whose vessel had twin hulls with paddle wheels between them.

1789 Steam-powered road vehicle is patented in the USA by Oliver Evans (1755–1819).

1789 Horse-drawn railways using flanged wheels on cast-iron rails are in use (in mines).

1790 Hobbyhorse-type bicycle called the célerifère, propelled by the feet on the ground, is constructed by Frenchman Count de Sivrac.

1790 Experimental steamboat is built by Samuel Morley.

1790 Practical parachute is invented and tested by jumping from a balloon by French balloonist François Blanchard (1753–1809), who died while making a practice jump; later (1797) other jumps are made from balloons by Frenchman André Garnerin (1763–1823).

ENGINEERING AND TECHNOLOGY

1772 Dial scales are invented by British engineer John Clais.

1774 Boring machine, initially for making cannon and later for boring cylinders for steam engines, is invented by British ironfounder John Wilkinson (1728–1808).

1775 Pressure anemometer for measuring high wind speeds at sea is invented by Scottish physician James Lind (1716–94).

1776 First breech-loading rifle to be adopted by the British army is invented by British army officer Patrick Ferguson (1744–80).

1776 Stopwatch is invented by Swiss clockmaker Jean-Moyse Ponzait.

1778 Flushing toilet (water closet) with ball-valve and siphon is patented by British engineer Joseph Bramah (1748–1814).

1779 Spinning mule is invented (but not patented) by Englishman Samuel Crompton (1753–1827).

1779 First cast-iron bridge is designed and built over the River Severn at Coalbrookdale by British ironfounder John Wilkinson (1728–1808) with iron produced by Abrahan Darby III (1750–91).

1779 Repeating air-powered 20-shot rifle is invented by Austrian gunsmith Bartholomew Girandoni (1744–99).

1780 Spring detent escapement for chronometers is invented by Swiss clockmaker Ferdinand Berthoud (1727–1807).

1780 Process for manufacturing caustic soda (sodium hydroxide) from sulphate wastes is invented by Scottish chemist James Keir (1735–1820).

1782 Double-acting steam engine (which admits steam alternately to each side of the piston), with sun and planet gears to produce rotary motion, is invented by Scottish engineer James Watt (1736–1819).

1782 Governor for regulating the speed of steam engines is invented by Scottish engineer James Watt (1736–1819).

1782 Pyrometer (for measuring high temepratures) is invented by British potter Josiah Wedgwood (1730–95); an improved instrument is later (1830) invented by British chemist John Daniell (1790–1845).

1783 Process for purifying iron by "puddling" and a rolling mill for metals are invented by British ironmaster Henry Cort (1740–1800).

1783 Improved hair hygrometer is invented by Swiss scientist Horace de Saussure (1740–99); another type is later (1820) invented by British chemist John Daniell (1790–1845).

1783 Accelerometer is invented by British engineer George Attwood (1746–1807).

1784 "Unpickable" safety lock is invented by British engineer Joseph Bramah (1748–1814).

1784 Bifocal spectacles are invented by US scientist and statesman Benjamin Franklin (1706–90).

1785 Steam-powered loom is invented by British engineer Edmund Cartwright (1743–1823).

1786 Nail-making machine is patented by US engineer Ezekiel Reed.

1786 Mechanical clutch is invented by Scottish engineer John Rennie (1761–1821).

1786 Steam gun is invented by British engineer William Murdock (1754–1839).

1787 Gold-leaf electrometer is invented by British physicist Abraham Bennet (1750–99).

1787 Bright oil lamp (Argand lamp) with circular wick allowing air current into the flame is invented by Swiss physicist Aimé Argand (1755–1803).

1788 Improved steam engine is patented by British engineer James Brindley (1742–72).

1789 Modern guillotine is invented by French physician Joseph Guillotin (1738–1814); it is used in France until 1939.

1790 Improved nail-making machine is invented by US engineer Jacob Perkins (1766–1849) and, independently, by British engineer Thomas Clifford.

1790 Steam-powered press for making coins is invented by British engineer Matthew Boulton (1728–1809).

1790 Industrial sewing machine is invented by Englishman Thomas Saint.

1790 Process for making soda (sodium carbonate) from common salt (sodium chloride) is invented by French chemist Nicolas Leblanc (1742–1806).

1790 Dentist's chair is invented by US dentist Josiah Flagg (1763–1816).

1791 Primitive gas turbine engine is invented by Englishman John Barber.

1791 Multitubular steam boiler is invented by US engineer Nathan Read (1759–1849); multitubular boiler for steam locomotives in patented later (1827) by French engineer Marc Séguin (1786–1875).

1792 Gas lighting using coal gas is invented by Scottish engineer William Murdock (1754–1839); first large installation is completed in 1802.

1792 Timber truss bridge (first modern use of the truss) is built in the USA over the Merrimack River by US engineer Timothy Palmer.

1795 Maximum-minimum thermometer is invented by Scottish scientist Daniel Rutherford (1749–1819).

COMMUNICATIONS

ASTRONOMY AND SPACE	PHYSICS	CHEMISTRY	BIOLOGY	MEDICINE

CHEMISTRY

covered by French chemist Claude Berthollet (1748–1822), who produces a chemical bleach called Eau de Javelle.

1785 Fact that water is a compound (of hydrogen and oxygen) is discovered by British chemist Henry Cavendish (1731–1810)..

1789 Metallic element zirconium (symbol Zr) is discovered in the mineral zirconia (its oxide) by German chemist Martin Klaproth (1743–1817), later (1824) isolated by Swedish chemist Jöns Berzelius (1779–1848).

1789 Radioactive element uranium (symbol U) is discovered (as its oxide) by German chemist Martin Klaproth (1743–1817), later (1841) isolated by French chemist Eugène Péligot (1811–90).

1791 Mineral dolomite (calcium magnesium carbonate) is discovered by French mineralogist Déodat de Dolmieu (1750–1801), who gave his name to both the mineral and the mountains where it was first found.

1791 Metallic element titanium (symbol Ti) is discovered by British amateur mineralogist (and minister) William Gregor (1761–1817), who called it "menanchinite"; it was later (1795) named titanium by German chemist Martin Klaproth (1743–1817).

1794 Metallic element yttrium (symbol Y) is identified by Finnish chemist Johan Gadolin, later (1828) isolated by German chemist Friedrich Wöhler (1800–82).

1796 Carbon disulphide is discovered by German chemist Wilhelm Lampadius (1772–1842); its composition is later (1802) determined by French scientists Charles Désormes (1777–1862) and Nicolas Clément (1779–1841).

1797 Metallic element chromium (symbol Cr) is discovered by German chemist Martin Klaproth (1743–1817) and, independently, by French chemist Louis Vauquelin (1763–1829), who named it.

1798 Liquid ammonia, the first gas to be liquefied by cooling alone, is produced by French chemist Louis Guyton de Morveau (1737–1816).

1798 Metallic element strontium (symbol Sr) is identified by German chemist Martin Klaproth (1743–1817) and, independently, Scottish chemist Thomas Hope (1766–1844); it was later (1808) isolated by British chemist Humphry Davy (1778–1829).

1798 Metallic element beryllium (symbol Be) is identified (as its oxide) by French chemist Louis Vauquelin (1763–1829), later (1828) isolated by French chemist Antoine Bussy and, independently, German chemist Friedrich Wöhler (1735–1867), who called it glucinium.

1798 Bleaching powder is discovered by British industrial chemists Charles Tennant (1761–1815) and Charles Macintosh (1766–1843).

1799 Proust's law, or law of constant composition (that chemical compounds contain elements in definite proportions), is discovered by French chemist Joseph Proust (1754–1826).

1799 Urea is discovered by French chemist Antoine de Fourcroy (1755–1809).

1799 Fructose (grape sugar) is discovered by French chemist Joseph Proust (1754–1826).

1800

ASTRONOMY AND SPACE

1800 Infrared radiation in sunlight is discovered by British astronomer William Herschel (1738–1822).

PHYSICS

1800 Infrared radiation (from the Sun) is discovered by British astronomer William Herschel (1738–1822).

CHEMISTRY

1800 Electrolysis (of water) is discovered by British chemists William Nicholson (1753–1815) and Anthony Carlisle (1768–1840).

BIOLOGY

1802 Term "biology" is coined by French naturalist Jean-Baptiste de Lamarck (1744–1829).

MEDICINE

1801 Cause of astigmatism (irregular curvature of the cornea of the eye) is discovered by British physicist and physician Thomas Young (1773–1829);

FARMING AND FOOD

1802 Wheeled threshing machine is invented by Englishman Thomas Wigful.

TRANSPORT

1801 Passenger-carrying steam-powered carriage is built by British engineer Richard Trevithick (1771–1833).

ENGINEERING AND TECHNOLOGY

1795 Hydraulic press is invented by British engineer Joseph Bramah (1748–1814).

1795 Flax-spinning machine is invented by US engineer Robert Fulton (1765–1815).

1796 Hydraulic ram, for pumping water up into a reservoir, is invented by French engineer Joseph Montgolfier (1740–1810).

1797 Portable mercury barometer is invented by French physicist Jean Fortin (1750–1831).

1798 Continuous roll paper-making machine that uses wood pulp is invented by Frenchman Nicolas Robert (1761–1828).

1799 Slide valve for steam engines is invented by Scottish engineer William Murdock (1754–1839).

1799 Mass-production method of making pulley blocks is invented by French-born engineer Marc Brunel (1769–1849).

1799 Gas lamp, using gas from heated sawdust, is invented by French chemist Phillipe Lebon (1769–1804).

1799 A solid, dry "bleaching powder" for textiles (made from chlorine and slaked lime) is invented by Scottish chemist Charles Tennant.

1800 Improved oil lamp (Carcel lamp), in which oil is pumped by clockwork into the centre of a circular wick, is invented by French clockmaker Bertrand Carcel (1750–1812).

COMMUNICATIONS

1800 Unfixable photographic images using silver salts (impregnated in paper or white leather) are produced by Englishman Thomas Wedgewood

1800

ASTRONOMY AND SPACE

1801 First asteroid, Ceres, is discovered by Italian astronomer Giuseppi Piazzi (1746–1826), and then lost because of Piazzi's illness.

1802 Second asteroid, Pallas, is discovered, and Ceres rediscovered, by German astronomer Heinrich Olbers (1758–1840).

1802 First binary stars are discovered by British astronomer William Herschel (1738–1822).

1803 Meteorites are discovered to be objects that fall from space by French physicist Jean Biot (1774–1862).

1804 Third asteroid, Vesta, is discovered by German astronomer Heinrich Olbers (1758–1840).

1804 Fourth asteroid, Juno, is discovered by German astronomer Carl Harding and later (1807) German astronomer Heinrich Olbers (1758–1840).

1814 Absorption ("dark") lines in the Sun's spectrum are discovered by German astronomer and physicist Joseph von Fraunhofer (1787–1826).

1826 Olber's paradox (that the sky at night should be bright if the Universe is uniform, infinite and unchanging) is proposed by German astronomer Heinrich Olbers (1758–1840).

1826 Short-period Biela's comet is discovered by Austrian astronomer Wilhelm von Biela (1782–1856); the comet split in two in 1846.

1831 Jupiter's Great Red Spot is discovered by German astronomer Heinrich Schwabe (1789–1875).

1836 Baily's beads (spots of light around the rim of the Moon during a total solar eclipse) are discovered by British amateur astronomer Francis Baily (1774–1844).

1838 First measurement of a distance to a star (other than the Sun) is made by German astronomer and mathematician Friedrich Bessel (1784–1846), who calculated the distance to 61 Cygni (with 50 percent accuracy).

1841 Existence of a dark star companion of Sirius is predicted by German astronomer Friedrich Bessel (1784–1846), later (1851) discovered by US astronomer Alvan Clark (1804–87).

1843 Sunspot cycle is discovered by German astronomer Heinrich Schwabe (1789–1875), who stated it lasts 10 years (not 11 as is now known).

1843 Faye's comet is discovered by French astronomer Hervé Faye (1814–1902).

1845 Position of the undiscovered planet Neptune is calculated by British astronomer John Couch Adams (1819–92), leading to its location in August 1846 by James Challis. In France, Urbain Le Verrier (1811–77) independently made a similar prediction in June 1846 and the planet was discovered in September 1846 by the German astronomer Johann Galle (1812–1910). The latter two scientists are usually credited with the discovery, although it had been observed and recorded earlier – for example in about 1800 by French astronomer Joseph de Lalande (1732–1807), who did not realize that it was a planet.

1845 Fifth asteroid, Astraea, is discovered by German astronomer Karl Hencke (1793–1866), who later (1847) discovers the sixth asteroid, Hebe.

1845 First detailed photographs of the Sun are taken by French physicists Armand Fizeau (1819–96) and Léon Foucault (1819–68).

PHYSICS

1801 Henry's law (that the equilibrium amount of gas that dissolves in a liquid at constant temperature is proportional to the partial pressure of the gas) is discovered by British chemist and physician William Henry (1775–1836).

1801 Interference of light is discovered by British physicist and physician Thomas Young (1773–1829).

1801 Ultraviolet light is discovered by German scientist Johann Ritter (1776–1810).

1802 Charles' law (that, at constant pressure, all gases increase in volume by the same amount with the same rise in temperature) – also known as Gay-Lussac's law, who independently discovered it – is formulated by French physicist Jacques Charles (1746–1823). The work of both scientists was anticipated in 1699 by Guillaume Amontons.

1807 Wave theory of light (to explain the phenomenon of interference) is proposed by British physicist and physician Thomas Young (1773–1829).

1807 Young's modulus, a measure of a material's elasticity, is discovered by British physicist and physician Thomas Young (1773–1829).

1808 Polarized light is discovered by French physicist Étienne Malus (1775–1812).

1811 Avogadro's law (hypothesis), that equal volumes of all gases – at the same temperature and pressure – contain the same number of molecules, is discovered by Italian physicist Amedeo Avogadro (1770–1856) and later (1814), independently, by French physicist André Ampère (1775–1836).

1811 Polarization of light (in quartz crystals) is discovered by French physicist Dominique Arago (1786–1853).

1812 "Two-fluid" (i.e. positive and negative) characteristics of static electricity, and the inverse square law describing attraction between unlike charges, are proposed by French physicist Siméon Poisson (1781–1840).

c.1814 Fraunhofer diffraction (in which the light source and the receiving screen are infinitely far apart) is discovered by German physicist Joseph von Fraunhofer (1787–1826).

1815 How light is refracted is discovered by French physicist Augustin Fresnel (1788–1827).

1815 Brewster's law (that the polarization of reflected light is maximized when the reflected and refracted rays are at right angles) is formulated by Scottish physicist David Brewster (1781–1868).

1815 Optical activity (the ability of certain molecules to rotate the plane of polarized light) is discovered in vegetable oils by French physicist Jean Biot (1774–1862). He later discovered the optical activity of sugar (1818) and tartaric acid (1832).

1819 Ratio of the specific heats of gases (by adiabatic expansion) is discovered by French scientists Charles Désormes (1777–1862) and his son-in-law Nicolas Clément (1779–1841).

1819 Dulong and Petit's law (that the product of an element's relative atomic mass and specific heat – its atomic heat – is approximately constant) is discovered by French scientists Pierre Dulong (1785–1838) and Alexis-Thérèse Petit (1791–1820).

1820 Electromagnetism – the production of a magnetic field by an electric current flowing in a conductor – is dis-

CHEMISTRY

1800 Dinitrogen monoxide (nitrous oxide) is discovered by British chemist Humphry Davy (1778–1829), who suggested its possible use as an anaesthetic.

1800 Quinoline synthesis is discovered by Hungarian-born Austrian chemist Zdenko Skraup (1850–1910).

1801 Carbon monoxide is discovered by French scientist Charles Désormes (1777–1862) and his son-in-law Nicolas Clément (1779–1841).

1801 Metallic element niobium (symbol Nb) is discovered by British chemist Charles Hatchett (c.1765–1847) and first isolated (1864) by Swedish chemist Christian Blomstrand (1826–97), both of whom originally called it columbium; it was named niobium by German chemist Heinrich Rose (1795–1864).

1801 Metallic element vanadium (symbol V) is discovered by Spanish-born Mexican mineralogist Andrés del Rio (1764–1849), who called it erythronium; the discovery went unacknowledged until vanadium was rediscovered (1831) and named by Swedish chemist Nils Sefström (1787–1854).

1802 Metallic element tantalum (symbol Ta) is discovered by Swedish chemist Anders Ekeberg (1767–1813), later (1820) isolated by his compatriot Jöns Berzelius (1779–1848).

1803 Atomic theory of matter is proposed by British chemist John Dalton (1766–1844), who compiled a list of atomic weights (relative atomic masses).

1803 Henry's law (that the mass of gas dissolved in a liquid is proportional to the pressure of the gas) is discovered by British chemist William Henry (1774–1836).

1803 Metallic element cerium (symbol Ce) is discovered by Swedish chemists Jöns Berzelius (1779–1848) and Wilhelm von Hisinger (1766–1852) and, independently, by German chemist Martin Klaproth (1743–1817).

1803 Metallic elements palladium (symbol Pd) and rhodium (symbol Rh) are discovered by British scientist William Wollaston (1766–1828).

1804 Metallic elements iridium (symbol Ir) and osmium (symbol Os) are discovered by British chemist Smithson Tennant (1761–1815).

1805 Morphine is discovered by German apothecary Friedrich Sertürner (1783–1841), although his discovery is not recognized until 1817.

1806 Amino acid asparagine (the first to be found) is discovered by French chemists Louis Vauquelin (1763–1829) and Pierre Robiquet (1780–1840).

1806 Role of nitrogen monoxide (nitric oxide) in the lead-chamber process for producing sulphuric acid is discovered by French scientists Charles Désormes (1777–1862) and Nicolas Clément (1779–1841).

1807 Metallic elements potassium (symbol K) and sodium (symbol Na) are isolated by British chemist Humphry Davy (1778–1829).

1808 Metallic elements barium (symbol Ba), calcium (symbol Ca), magnesium (symbol Mg) and strontium (symbol Sr) are isolated (using electrolysis) by British chemist Humphry Davy (1778–1829). He also isolates the non-metallic element boron (symbol B), which is discovered independently by French chemists Joseph Gay Lussac (1778–1850) and Louis Thénard (1777–1857).

1808 Gay Lussac's law (that gases combine in simple proportions by volume)

BIOLOGY

1804 Fact that plants require carbon dioxide (from air) and nitrogen (from the soil) for proper growth is discovered by Swiss chemist and natralist Nicholas de Saussure (1767–1845), who first used saltpetre (potassium nitrate) as a nitrogenous fertilizer.

1809 Lamarckism (an erroneous theory of evolution that acquired characteristics can be inherited) is proposed by French biologist Jean-Baptiste de Lamarck (1744–1829).

1811 Fossil bones of an ichthyiosaur, the first of their kind to be found, are discovered by British naturalist Mary Anning (1799–1847) when she was still a child; she later (1821) discovered the first complete ichthyosaur in England.

1811 Separate functions of motor and sensory nerves are discovered by Scottish anatomist Charles Bell (1774–1842).

1813 System of plant classification is introduced by Swiss botanist Augustin de Candolle (1778–1841), who also coins the term "taxonomy".

1815 Arachnids are given family status by French naturalist Jean-Baptiste de Lamarck (1744–1829).

1817 Pander layers, found originally in chick embryos, are discovered by Latvian anatomist Christian Pander (1794–1865).

1819 Alternation of generations is discovered by French poet and biologist Adelbert von Chamisso (1781–1838).

1821 Systematic classification of fungi is produced by Swedish botanist Elias Fries (1794–1878).

1825 Fossil teeth are discovered to be those of a "giant lizard" (Iguanadon dinosaur) by British palaeontologist Gideon Mantell (1790–1852).

1827 Mammalian ovum is discovered by Estonian-born German biologist Karl von Baer (1792–1876).

1828 Role of the semicircular canals (in the ear) in the sense of balance is discovered by French physician Jean Flourens (1794–1867).

1829 Gill arches and gill slits are discovered in embryo birds and mammals by German biologist Martin Rathke (1793–1860).

1830 Mechanism of pollination is discovered by Italian microscopist Giovanni Amici (1786–1863).

c.1830 Proteins are discovered by German physiologist Johannes Müller (1801–58).

1831 Nucleus of plant cells is discovered (and named) by Scottish botanist Robert Brown (1773–1858).

1832 Reflex nerve action is discovered by British physician Marshall Hall (1790–1857), although its is not correctly explained until much later (1906) by British neurophysiologist Charles Sherrington (1857–1952).

1833 Diastase, the first known enzyme, is discovered (in malt extract) by French chemists Anselme Payen (1795–1871) and Jean-François Persoz.

1834 Nitrogen fixation in plants is discovered by French chemist Jean Boussingault (1802–87).

1835 Schwann cells (which form the myelin sheath surrounding a nerve fibre) are discovered by German biologist Theodor Schwann (1810–82); the myelin sheath was later (1833) discovered independently by Polish-born German anatomist Robert Remak (1815–65).

1835 Protoplasm in bodies of unicellular

MEDICINE

treatment with corrective spectacles is later (1827) introduced by British astronomer George Airy (1831–81).

1803 Haemophilia is discovered by US physician John Otto (1774–1844) and later (1829), independently, by German physician Johann Schönlein (1793–1864).

1803 Painkilling drug morphine is discovered in opium by French chemist Charles Derosne (1780–1846) and later (1806) German chemist Friedrich Sertürner (1783–1841).

1809 Successful ovariotomy (to remove a cyst from the ovary) is performed by US surgeon Ephraim McDowell (1771–1830).

1811 Colles' fracture (of the wrist) is described by Irish surgeon Abraham Colles (1773–1843).

1817 Parkinsonism (shaking palsy, paralysis agitans or Parkinson's disease) is discovered by British surgeon and palaeontologist James Parkinson (1755–1824).

1818 Method of treating goitre with iodine is discovered by French physician Jean Dumas (1800–84).

1818 Homeopathy is founded by German physician Samuel Hahnemann (1755–1843).

1824 Carbon dioxide gas is used as an anaesthetic (on animals) by British physician Henry Hickman (1800–30).

1827 Bright's disease (non-bacterial nephritis) is discovered by British physician Richard Bright (1798–1858).

1830 Bloodletting as a treatment for disease is reintroduced by British physician Marshall Hall (1790–1857).

1831 Dupuytren's contracture (curling of the fingers caused by thickening of ligaments in the palm, usually associated with ageing) is discovered by French surgeon Guillaume Dupuytren (1777–1835).

1832 Hodgkin's disease (a progressive disorder involving enlargement of the lymph glands and spleen) is discovered by British physician Thomas Hodgkin (1798–1866).

1833 Peripheral neuritis (later found to be caused by a deficiency of vitamin B) is described by French neurologist Jules Dejerine (1849–1917).

1835 Graves' disease (thyrotoxicosis, or exophthalmic goitre) is discovered by Irish physician Robert Graves (1796–1853) and, later (1840) and independently, by German physician Karl von Basedow (1799–1854), after whom it was also called Basedow's disease.

1835 Trichina (or Trichinella), a nematode parasite in humans, is discovered in undercooked meat by British surgeon James Paget (1814–99).

1837 Correct distinction between typhus and typhoid fever is made by US physician William Gerhard (1809–72) and, later (1849) British physician William Jenner (1815–98).

1838 Stokes-Adams syndrome (loss of consciousness caused by irregular heartbeat) is discovered by Irish physician William Stokes (1804–78) and later (1842) by his compatriot Robert Adams (1791–1875).

1839 Microscopic fungus that causes favus (a kind of ringworm) is discovered by Swiss physician Johann Schönlein (1793–1864).

1840 Theory that infectious diseases are caused by parasitic organisms is proposed by German pathologist Friedrich Henle (1809–85).

FARMING AND FOOD

1802 Factory for making sugar from sugar beet is established by German chemist Karl Archard.

1806 Coffee percolator is invented (in Germany) by US-born scientist Benjamin Thompson (Count Rumford) (1753–1814).

1810 Preservation of pre-cooked food by canning (rather than by bottling, which had been used commercially since 1790) is invented by French confectioner Nicolas Appert (1752–1841).

1812 Combined heat sterilization and canning process for foods is invented by British engineer Bryan Donkin (1768–1855); tin cans are later (1818) introduced into the USA by Peter Durant.

1813 Soda siphon is invented in England by Charles Plinth.

1818 Primitive combine harvester is invented by US blacksmith John Lane; it is improved by Moore and Hascall in 1838.

1819 Cast-iron plough is invented by Stephen McCormick (with replaceable parts and a wrought-iron point) and, independently, (with a shaped mouldboard) by Jethro Wood (1774–1834).

1819 Canned sardines are manufactured by Frenchman Joseph Colin.

1820 Hygrometer for measuring moisture in the air in hothouses is invented by British physicist John Daniell (1790–1845).

1820s Nitrates (for fertilizers) are imported into Europe from South America.

1822 Method of decolourizing crude sugar using charcoal is invented by French chemist Anselme Payen (1795–1871).

1825 Canned salmon is manufactured by Scotsman John Moir.

1827 First mechanical reaper is invented by Scottish clergyman Patrick Bell (1799–1869).

1828 Cocoa (drinking chocolate) is popularized by Dutchman Coenrad van Houten, although it was originally introduced into Europe from Spanish colonies in America in 1615.

1830 Cast steel plough for bulk manufacture is invented by US industrialist John Deere (1804–86).

1830 Angostura bitters, named after the town of Angostura (since 1846 Cuidad Bolivar), Venezuela, is invented by German Johann Siegert.

1831 Improved still for manufacturing whisky is invented by Irish distiller Aeneas Coffrey (1779–1852).

1833 Reaping machine is invented by US engineer Obed Hussey (1792–1860).

1833 Plough with steel-clad mouldboard and share is invented by US blacksmith John Lane.

1834 Improved reaping and binding machine is invented by US engineer Cyrus McCormick (1809–1884).

1837 All-steel plough is invented by US engineer John Deere (1804–86).

1837 Worcester sauce is invented by British pharmacists John Lea and William Perrins.

1839 Commercial process for making superphosphate fertilizer is invented by agriculturalists Englishman John Lawes (1814–1900) and, independently, Irishman James Murray (who had invented it in 1817).

1840 Gin-based Pimms No. 1 cocktail is

TRANSPORT

1801 Man-powered submarine (the Nautilus) is constructed by US engineer Robert Fulton (1765–1815).

1802 First workable, but still experimental, stern paddle steamboat (the Charlotte Dundas) is built by Scottish engineer William Symington (1763–1831) and used to tow barges on a canal.

1804 Steam railway locomotive, running on iron rails, is built by British engineer Richard Trevithick (1771–1833).

1804 Steam-powered amphibious dredger is invented by US engineer Oliver Evans (1755–1819).

1804 Balloon ascent to a height of 5 kilometres is made to study the atmosphere by French physicist Jean Biot (1774–1862).

1804 Experimental gliders are built by British engineer George Cayley (1773–1857).

1805 Solid-fuel artillery rockets are constructed by British military engineer William Congreve (1772–1828), later (1806) used to bombard Copenhagen.

1807 First commercially successful paddle steamer (the Clermont) is constructed by US engineer Robert Fulton (1765–1815).

1807 Conveyor belt is invented by US engineer Oliver Evans (1755–1819).

1808 Circular passenger-carrying railway is built in London for demonstration purposes by British engineer Richard Trevithick (1771–1833).

1808 Glider flight (the first in a heavier-than-air machine) is made by British engineer George Cayley (1773–1857).

1811 Double-cylinder steam locomotive that runs using a toothed wheel engaging on a racked rail is invented by British engineer John Blenkinsop (1783–1831).

1812 Steamboat with 3-h.p. engine (the Comet) is built by Scottish engineer Henry Bell (1767–1830).

1812 Bogey truck for railway vehicles is invented by William Chapman.

1813 Early commercial steam locomotive (the Puffing Billy) for a mine railway is built by British engineer William Hedley (1779–1843).

1814 Improved steam locomotive (the Blücher) is built by British inventor George Stephenson (1781–1848).

1814 Vertical paddle wheel for steamboats is patented by US inventor Nicholas Roosevelt (1767–1854).

1814 Steamboats for the Mississippi River are constructed by Henry Shreve (1785–1851).

1815 Paved roads are invented by Scottish engineer John McAdam (1756–1836).

1816 A steerable bicycle (without pedals, called a draisienne or dandy horse) is invented by German engineer Karl Drais von Sauerbronn; a similar machine, called the vélocifère, is constructed by French inventor Joseph Niepce (1765–1833).

1818 Sail and steam paddle steamer Savannah crosses the Atlantic Ocean (using steam power for only about 10 percent of the four weeks it took).

1819 Modern pressurized diving suit is invented by German engineer Augustus Siebe (1788–1872).

1822 Steamship Robert Fulton is constructed and sailed from New York to Havana, Cuba, by US shipbuilder Henry Eckford (1775–1832).

ENGINEERING AND TECHNOLOGY

1800 Industrial screw-cutting lathe is invented by British engineer Henry Maudslay (1771–1831) and, independently, American David Wilkinson.

1800 Voltaic pile battery is invented by Italian physicist Alessandro Volta (1745–1827).

1800 Method of using chlorine to purify drinking water is devised by Scotsman William Cruikshank (1745–1800).

1800 Improved method of making optical glass is invented by Swiss glassmaker Pierre Guinand (c.1744–1824).

1801 Oxy-hydrogen blowpipe is invented by US chemist Robert Hare (1781–1858), who also invented Hare's apparatus for comparing or finding the densities of liquids.

1801 Method of waterproofing cloth and paper is invented by German printer Rudolph Ackermann (1764–1834).

1802 First high-pressure steam engine in the USA is invented by US engineer Oliver Evans (1755–1819).

1802 Storage battery (a rechargable cell, or accumulator) is invented by German physicist Johann Ritter (1776–1810).

1802 Industrial method of extracting sugar from beets is devised by German chemist Franz Achard (1753–1821).

1803 Practical compound steam engine (with cylinders for both high- and low-pressure steam) is constructed by British engineer Arthur Woolf (1766–1837); an earlier patent (1781) by Jonathan Hornblower (1753–1815) had been declared invalid in 1799, and the patent of James Watt (1736–1819) had expired in 1800.

1803 Spherical case shot, invented by British soldier Henry Shrapnel (1761–1842), is adopted by the British army.

1804 Steam-powered bucket dredger is built by US engineer Oliver Evans (1755–1819).

1804 Early paper-making machine is invented by British engineer Bryan Donkin (1768–1855).

1805 Percussion cap (detonator) for muzzle-loading firearms is invented by Scottish minister Alexander Forsyth (1768–1843); it is adopted by the British army in 1842.

1807 Device for saving shipwrecked persons by firing a rope from a mortar on shore is invented by Englishman George Manby (1765–1854).

1807 Rotor-spinning machine is invented by Samuel Williams.

1807 Industrial paper-making machine, which uses wood pulp to produce a continuous "web" (roll) of paper, is invented by French-born British papermakers Henry (1766–1854) and Sealy (d.1847) Fourdrinier and engineer Bryan Donkin (1765–1855).

1808 Improved electric arc lamp is invented by British scientist Humphry Davy (1778–1829).

1808 Machine for making bobbin lace is invented by British engineer John Heathcoat (1783–1861).

1810 Method of making soda (sodium carbonate) from limestone and common salt is invented by French chemist and physicist Augustin Fresnel (1788–1827).

1811 Breech-loading rifle is invented by US gunsmith John Hall (1778–1841).

1812 Practical cartridge for breech-loading firearms is invented by French gunsmith Samuel Pauly (1766–1821).

1813 Unsuccessful rotary printing press is patented by British engineer Bryan Donkin (1768–1855).

1813 Machine for making true lace is invented by British engineer John Leavers.

1815 Miner's safety lamp is invented by British scientist Humphry Davy (1778–1829).

1815 Boot-making machine and tunnelling shield (a boring machine for making tunnels) are invented by French-born engineer Marc Brunel (1769–1849).

1815 Water gas meter is invented by William Murdoch's assistant, Englishman Samuel Clegg (1781–1861).

1816 Clockwork metronome is invented by German musician Johann Mäelzel (1772–1838).

1816 Fire extinguisher, containing compressed air and water, is invented by Englishman George Manby (1765–1854).

1816 Stirling engine, a closed-cycle external combustion engine powered by the expansion of a gas such as air, is invented by Scottish scientist (and clergyman) Robert Stirling (1790–1878).

1816 Kaleidoscope is invented by Scottish physicist David Brewster (1781–1868).

1816 Pellet lock, a firing mechanism for muskets, is invented by British gunsmith Joseph Manton (1766–1835).

1817 Metal-planing machine is invented by Welsh engineer Richard Roberts (1789–1864).

1817 Stethoscope with single ear-piece is invented by French physician René Laënnec (1781–1826).

1817 Dental plate is invented by US dentist Anthony Plantson (1774–1837).

1818 Lathe that makes parts by copying a pattern is invented (for making gun stocks) by US engineer Thomas Blanchard (1788–1864).

COMMUNICATIONS

(1771–1805), and later (1802) demonstrated by British scientist Humphry Davy (1778–1829).

1800 First iron printing press is made by British scientist (and politician) Charles Stanhope (1753–1816).

1801 Automatic loom controlled by punched cards is invented by French engineer Joseph-Marie Jacquard (1751–1834).

1804 Multi-wire electrostatic telegraph, with one wire for each letter of the alphabet, is invented by Spaniard Francisco Salva.

1806 Carbon paper is invented by Englishman Ralph Wedgewood.

1806 Machine for automatically numbering banknotes as they are printed is invented by British engineer Joseph Bramah (1748–1814).

1809 Chief German shorthand system is invented by German civil servant Franz Gabelsberger (1789–1848).

1810 Steam-powered printing press is invented by German printer Friedrich König (1774–1833).

1812 Camera lucida (which enables drawings to be made of optical images in a microscope) is invented by British scientist William Wollaston (1766–1828).

1816 Primitive short-range electric telegraph is constructed by British scientist Francis Ronalds (1788–1873), who published the results in 1823.

1820 Arithometer, the first mass-produced calculating machine, is produced by Frenchman Charles de Colmar (1785–1870).

1821 System for writing Native American languages (originally Cherokee) using an 85-letter alphabet is invented by Sequoya ("George Guess") (c.1770–1843).

1822 Early typesetting machine, requiring hand justification, is patented in England by US inventor William Church (c.1778–1863).

1822 Diorama of projected scenes is invented by Frenchmen Charles Bouton (1781–1853) and Louis Daguerre (1787–1851).

1822 Abacus is reintroduced from Russia by French mathematician Jean Poncelet (1788–1867).

1826 First photograph, on metal, is taken by French physicist Nicéphore Niepce (1765–1833).

1826 Colour lithography process is invented by Bavarian cartographer Aloys Senefelder (1771–1834).

1829 Alphabet of characters consisting of raised dots, for reading by the blind, is invented by Frenchman Louis Braille (1809–52), who was himself blind.

1829 Primitive typewriter is patented by US inventor William Burt (1792–1858).

1831 Principle of the electromagnetic telegraph is put forward by US physicist and inventor Joseph Henry (1797–78).

1831 First successful electric telegraph in Britain is invented by scientists Charles Wheatstone (1802–75) and William Cooke (1806–79).

1832 Mechanical computer, termed an "analytical engine", is patented by British mathematician Charles Babbage (1792–1871), who had produced his first model of the engine ten years earlier.

1832 Five-wire electric telegraph is

ASTRONOMY AND SPACE

1845 First spiral galaxy is discovered by Irish astronomer William Rosse (1800–67).

1846 Planet Neptune is discovered by German astronomer Johann Galle (1812–1910), based on positional calculations by French astronomer Urbain Le Verrier (1811–77).

1846 Triton, the first moon of Neptune, is discovered by British astronomer William Lassell (1799–1880).

1847 New comet is discovered by US astronomer Maria Mitchell (1818–89).

1847 Asteroid Iris is discovered by British astronomer John Hind (1823–95).

1848 Hyperion, the eighth moon of Saturn, is discovered by British astronomer William Lassell (1799–1880) and, independently, US astronomer George Bond (1825–65).

1849 Roche limit (the distance within which an orbiting moon is destroyed by the gravitation forces of the parent body) is discovered by French astronomer Edouard Roche (1820–83).

PHYSICS

covered by Danish physicist Hans Oersted (1777–1851).

1820 Biot-Savart law (that the magnetic field strength near a current-carrying conductor is proportional to the current and inversely proportional to the distance from the conductor) is discovered by French physicists Jean Biot (1774–1862) and Félix Savart (1791–1841).

c.1820 Positions of dark lines in the Sun's spectrum – first observed (1802) by British scientist William Wollaston (1766–1828) – are identified by German spectroscopist Josef von Fraunhofer (1787–1826).

c.1820 Fresnel diffraction (in which the light source and receiving screen are at a finite distance from the refracting object) is discovered by French physicist Augustin Fresnel (1788–1827).

1822 Critical state of liquids (at which vapor and liquid phases become indistinguishable) is discovered by French physicist Charles Cagniard de la Tour (1777–1859).

1824 Magnetic induction is first demonstrated by French physicist Dominique Arago (1786–1853).

1824 Carnot's theorem (that all reversible heat engines working between the same pair of temperatures have the same efficiency), based on the ideal Carnot cycle, is formulated by French physicist Nicolas Carnot (1796–1832).

1827 Ohm's law (that the voltage across a current-carrying conductor is equal to the product of the current and the conductor's resistance) is discovered by German physicist Georg Ohm (1789–1854).

1827 Brownian movement (the random movement of microscopic particles suspended in a fluid) is discovered (while studying pollen grains) by British botanist Robert Brown (1773–1858).

1827 That the unit of length should be based on the wavelength of a particular colour of light is proposed by French physicist Jacques Babinet (1794–1872), a suggestion which was finally adopted in 1960.

1828 Term "potential" for electric voltage is coined by British mathamatician and physicist George Green (1793–1841).

1829 Poisson's ratio (of lateral strain to longitudinal stress in a stretched elastic object) is discovered by French physicist Siméon Poisson (1781–1840).

1829 Coriolis effect (then called the Coriolis force) is postulated and explained by French mathematician Gaspard de Coriolis (1792–1843).

c.1832 Gauss' law (that electric flux is proportional to the sum of the electric charges within a surface) is discovered by German physicist Karl Gauss (1777–1855).

1834 Peltier effect (that an electric current passing through a junction between two different metals causes a change in temperature) is discovered by French physicist Jean Peltier (1785–1845).

1831 Electromagnetic induction (the production of a voltage in a conductor in a changing magnetic field) is discovered by British scientist Michael Faraday (1791–1867).

1833 Lenz's law (that the direction of an induced electric current is such as to oppose the change producing it) is discovered by German physicist Heinrich Lenz (1804–65).

CHEMISTRY

is discovered by French chemist Joseph Gay Lussac (1778–1850).

1810 Amino acid cystine is discovered by British scientist William Wollaston (1766–1828).

1811 Glucose is prepared (by heating starch with sulphuric acid) by German-born Russian chemist Gottlieb Kirchhoff (1764–1833).

1811 Highly explosive nitrogen trichloride is discovered by French scientist Pierre Dulong (1785–1838) – who was blinded in one eye as a result.

1811 Nonmetallic element iodine (symbol I) is discovered by French chemists Bernard Courtois (1777–1838) and, independently (1813), Charles Désormes (1777–1862) and Nicolas Clément (1779–1841).

1813 Modern system of chemical symbols and formulae is devised by Swedish chemist Jöns Berzelius (1779–1848).

1815 Prout's hypothesis, that all relative atomic masses are whole-number multiples of that of hydrogen, is proposed by British chemist William Prout (1785–1850).

1815 Cyanogen is discovered by French chemist Joseph Gay-Lussac (1778–1850).

1815 Oleic acid (a constituent of fats) is discovered by French chemist Michel Chevreul (1786–1889).

1817 Chlorophyll is discovered and named by French chemists Pierre Pelletier (1788–1842) and Joseph Caventou (1795–1877).

1817 Alkaloid emetine is discovered (in ipecacuanha) by French physiologist François Magendie (1783–1855) and his compatriot chemist Joseph Pelletier (1788–1842).

1817 Metalloid element selenium (symbol Se) is discovered by Swedish chemist Jöns Berzelius (1779–1848).

1817 Metallic element cadmium (symbol Cd) is discovered by German chemist Friedrich Strohmeyer (1776–1835).

1817 Metallic element lithium (symbol Li) is discovered by Swedish chemistry student Johan Arfwedson (1792–1841) and later (1818) isolated and named by his tutor Jöns Berzelius (1779–1848).

1818 Hydrogen peroxide is discovered by French chemist Louis Thénard (1777–1857).

1818 Alkaloid strychnine is discovered by French chemists Pierre Pelletier (1788–1842) and Joseph Caventou (1795–1877).

1819 Amino acid leucine is discovered by French chemist Joseph Proust (1754–1826).

1819 Naphthalene is discovered (in coal tar) by British chemist and physician John Kidd (1775–1851).

1820 Alkaloid quinine is discovered by French chemist Joseph Caventou (1795–1877) and Joseph Pelletier (1788–1842), who went on to discover also the alkaloids brucine and cinchonine.

1820 Amino acid glycine is discovered by French naturalist Henri Braconnot (1781–1855).

1820 Isomorphism (the existence of two or more compounds with the same crystal structure) is discovered by German chemist Eilhard Mitscherlich (1794–1863).

1822 Use of animal charcoal as a decolourizing agent (originally for

BIOLOGY

animals is discovered by French zoologist Felix Dujardin (1801–60).

1836 Digestive enzyme pepsin is discovered by German physiologist Theodor Schwann (1810–82).

1836 Method of classifying plants by their fruits is introduced by US botanist Asa Gray (1810–88).

1836 Epic round-the-world voyage on HMS Beagle is completed by British naturalist Charles Darwin (1809–82).

1837 Von Baer's law of biogenesis, that the embryos of very different species are similar, is discoverd by Estonian-born German biologist Karl von Baer (1792–1876).

1838 Purkinje cells in the cortex of the brain are discovered by Czech histologist Johannes Purkinje (1787–1869).

1839 Fact that all living matter is made up of cells is proposed by German physiologist Theodor Schwann (1810–1882).

1840 Fact that plants obtain nitrogen from nitrates in the soil is discovered by French agricultural chemist Jean-Baptiste Boussingault (1802–87).

1841 Word "dinosaur" ("fearful lizard") is coined by British anti-Darwinian palaeontologist Richard Owen (1804–92).

1841 Digestive enzyme ptyalin is discovered (in saliva) by French biologist Louis Mialhe (1807–86).

1842 Alternation of generations in certain animals (such as jellyfish and some worms) is discovered by Danish zoologist Johannes Steenstrup (1813–97).

1842 Concept of cranial index (ratio of the width of a person's skull to its length), as a means of classifying human races, is introduced by Swedish anatomist Anders Retzius (1796–1860).

1842 Bowman's capsule (on a kidney nephron) is discovered by British physician William Bowman (1816–92); Bowman's glands and Bowman's membranes are also named after him.

1843 Penetration of the ovum by a spermatazoan is discovered by British physician Martin Barry (1802–55).

1843 Electrical nature of nerve impulses is discovered by German biologist Emil du Bois-Reymond (1818–96).

1844 Fact that an ovum is a cell and that embryos develop by division of this cell is discovered by Swiss anatomist Rudolf von Kölliker (1817–1905).

1845 Standard three germ layers in an embryo are discovered by Polish-born German anatomist Robery Remak (1815–65).

1846 Protoplasm is discovered (and named) by German botanist Hugo von Mohl (1805–72).

1846 Fact that protists are single cells is discovered by German biologist Karl von Siebold (1804–85).

1849 Cellular nature of nerve cells is discovered by Swiss anatomist Rudolf von Kölliker (1817–1905).

MEDICINE

1842 Bell's palsy (paralysis of the facial nerve) is discovered by Scottish surgeon Charles Bell (1774–1842).

1842 Anaesthetic effect of ethoxyethane (ether) is discovered by US physician Crawford Long (1815–78), although he did not report the fact until 1849.

1844 Anaesthesia using dinitrogen oxide (nitrous oxide, laughing gas) is discovered by US dentist Horace Wells (1815–48).

1846 Anaesthesia using ethoxyethane (ether) is demonstrated in the USA by dentist Thomas Morton (1819–68) and in Britain by physician John Snow (1813–58).

1847 Anaesthetic effect of trichloromethane (chloroform) is discovered independently by Scottish physician James Simpson (1811–70) and British physician John Snow (1813–58).

1847 Leukemia is recognized by German physician Rudolf Verchow (1821–1902) and, independently, Scottish physician John Bennet (1812–75).

1847 Silver amalgam for filling teeth is introduced by US dentist Thomas Evans (1823–97).

1848 Terms "thrombosis" and "embolus" for blood clots are coined by German pathologist Rudolph Virchow (1821–1902).

1849 Pernicious anaemia (at first called Addisonian anaemia) is discovered by British physician Thomas Addison (1793–1860); later (1872) called "pernicious" (or Biermer's) anaemia after its description by German physician Anton Biermer (1827–92).

1849 Bacterium that causes anthrax is discovered by French bacteriologist Aloys Pollender, and later (1863) by his compatriot Casimir Davaine (1812–82).

1849 Clinical description of multiple sclerosis is given by German physician Friedrich von Frerichs (1819–85).

FARMING AND FOOD

invented by British restairant-owner James Pimm.

1845 Self-raising flour is introduced by British baker Henry Jones.

1845 Fruit jelly dessert is invented by US engineer Peter Cooper (1791–1883); it is later (1897) marketed as Jell-O by manufacturer Pear Wait.

1846 Wine containing the astringent drug quinine is invented by French pharmacist Joseph Dubonnet.

1847 Ring-shaped doughnuts are invented by US baker Hanson Gregory.

1848 Chewing gum (based on spruce gum) is invented in the USA by John Curtis; gum based on chicle is not introduced commercially until 1871, although the American Mayans had been chewing chicle since AD 900.

TRANSPORT

1825 Stockton to Darlington railway opens, with locomotive Locomotion No.1 (ex Active), designed by British engineer George Stephenson (1781–1848).

1825 Experimental balloons filled with coal gas are flow by British balloonist Charles Green (1785–1870).

1825 Primitive caterpillar track is invented by British engineer George Cayley (1773–1857).

1826 Floating landing stages, which rise and fall with the tide, are invented (for Liverpool docks) by French-born British engineer Marc Brunel (1769–1849).

1828 Steam locomotive with twin boiler flues, the Lancashire Witch, is built by British engineer George Stephenson (1781–1848).

1828 Differential gears for steam traction engines are invented by French engineer Onésiphore Pecquer.

1829 Steam locomotive Rocket, built by British engineer George Stephenson (1781–1848), wins competition at Liverpool and later (1830) operates on the new Liverpool and Manchester Railway.

1829 Successful steam carriage (which ran from London to Bath at an average speed of 25 km/h) is built by British inventor Goldsworthy Gurney (1793–1875).

1829 Horse-drawn buses (with three horses) are introduced in London by British coachbuilder George Shillibeer (1797–1866).

1830 First steam locomotive in the USA (Tom Thumb, for the Baltimore and Ohio railway) is built by US engineer Peter Cooper (1791–1883).

1811 Steam ferry Juliana, driven by paddle wheels, is built by US engineer John Stevens (1749–1838).

1831 Steam-powered bus runs regularly in London, built by British engineer Walter Hancock.

1832 Early screw propeller for ships is invented by French shipbuilder Pierre Sauvage (1785–1857).

1834 "Patented Safety Cab" (the Hansom cab), a two-wheeled carriage (cabriolet) with doors at the front and an outside seat at the rear for the driver, is invented by British architect Joseph Hansom (1803–1882).

1835 First successful steam locomotive in Germany, Der Adler, is built in England by George Stephenson (1781–1848).

1835 Practical screw propeller for ships is invented independently by Swedish engineer John Ericsson (1803–89) in the USA and Francis Smith (1808–74) in England; simple propellers had previously been patented in 1827 by Robert Wilson (1803–82) in Scotland and Joseph Ressel in Austria, and in 1832 by Pierre Sauvage (1785–1857) in France. A year later (1836) another patent was granted in the USA to John Stevens (1749–1838).

1837 First steamship to regularly cross the Atlantic Ocean, the Great Western, is built by British engineer Isambard Brunel (1806–59), although the first steam-only ship to make the crossing (Queenstown, Ireland, to New York) is the British paddle steamer Sirius.

1838 Process for pressure impregnation of timber railway sleepers with wood preservative is invented in the USA by John Bethell.

1839 Treadle-driven bicycle is invented

ENGINEERING AND TECHNOLOGY

1818 Secure "detector" lock is invented by British locksmith Charles Chubb (1772–1846).

1818 Five-shot revolving flintlock pistol is patented in England by US gunsmith Elisha Collier.

1818 Method of making a combustible gas from water (water gas, a mixture of hydrogen and carbon monoxide) is patented by US inventor Samuel Morey (1762–1843).
1819 Method of security printing banknotes and postage stamps is invented by US engineer Jacob Perkins (1766–1849) – although the first regular stamps (British penny blacks) are not introduced until 1840.

1819 Siren for determining the pitch of sounds is invented by French physicist Charles Cagniard de la Tour (1777–1859).

1819 Patent leather is patented in the USA by Seth Boyden (1788–1870).

1820 Needle galvanometer (later essential for early telegraphs) is invented by German physicist Johann Schweigger (1779–1857).

1820 Fresnel lens for producing a concentrated beam of light (used in lighthouses) is invented by French chemist and physicist Augustin Fresnel (1788–1827), based on an earlier (1748) idea of French scientist George de Buffon (1707–88).

1821 Prony brake, a type of dynamomter for measuring an engine's power output, is invented by French engineer Gaspard de Prony (1755–1839).

1821 Hot-air central heating is invented by US engineer Zachariah Allen (1795–1882).

1821 Azimuthal mounting for astronomical telescopes is invented by British optician George Dollond (1774–1852).

1822 Automatic loom is invented by Welsh engineer Richard Roberts (1789–1864).

1822 Döbereiner's lamp, which uses a platinum sponge catalyst to ignite hydrogen (generated by the action of suphuric acid on zinc), is invented by German chemist Johann Döbereiner (1780–1849).

1822 Method of expressing the relative hardness of a mineral (now called Mohs scale) is devised by German mineralogist Friedrich Mohs (1773–1839).

1823 Rubberized waterproof fabric is developed by Scottish chemist Charles Macintosh (1766–1843) (patented 1835), based on a process invented earlier (1815) by his compatriot James Syme (1799–1870).

1823 Electromagnet is invented by British electrical engineer William Sturgeon (1783–1850).

1824 Portland cement (consisting of roasted powdered limestone and clay) is invented by British mason Joseph Aspidin (1779–1855).

1824 Hydrometer for measuring the alcohol content of liquors is invented by French physicist Joseph Gay-Lussac (1778–1850).

1825 Uniflow steam emgine, in which the piston is the same length as its stroke and therefore acts also as the exhaust valve, is invented by French engineer Jacques de Montgolfier and, independently (1827), US engineer Jacob Perkins (1766–1849).

1825 Actinometer (for measuring the intensity of actinic – i.e., solar – rays) is invented by British astronomer John Herschel (1792–1871), later (1856) improved by German chemist Robert Bunsen (1811–99).

1825 Drummond light (a type of limelight) is invented by Scottish engineer Thomas Drummond (1797–1840) and British engineer Goldsworthy Gurney (1793–1875).

1826 Inward-flow water turbine is designed by French engineer Jean-Victor Poncelet (1788–1867), but it is not successfully constructed until 1838.

1826 Imporoved microscope with much reduced spherical and chromatic aberrations is constructed by British biologist James Smith (d.1870), with lenses designed by microscopist Joseph Lister (1786–1869).

1827 First practical water turbine is invented by French engineer Claude Burdin, and built by Benoît Fourneyron (1802–67).

1827 Method of extracting pure aluminium is invented by German chemist Friedrich Wöhler (1800–82).

1827 Magnetometer (for measuring magnetic fields) is invented by German physicist Johann Poggendorf (1796–1877).

1827 Reflecting achromatic microscope is invented by Italian scientist Giovanni Amici (c.1786–1863).

1827 Contact lenses (to correct defective vision) are invented by British astronomer William Herschel (1738–1822).

1827 "Strike anywhere" friction matches are invented by British pharmacist John Walker (c.1781–1859); safety matches are invented later (1855) in Sweden by Johan Lundström.

1827 Date-stamping machine for railway tickets is invented by Englishman Thomas Edmondson (1792–1851).

1828 Blast furnace is invented by Scottish iron-founder James Neilson (1792–1865).

1828 Corrugated iron is invented by Richard Walker; corrugated steel is invented later (1853) by Pierre Carpentier.

1828 Ring-spinning machine is invented by US engineer John Thorp (1784–1848).

1828 Binaural stethoscope is invented by French physician Pierre Poirry.

1829 Practical electric motor is invented by US physicist Joseph Henry (1797–1878).

COMMUNICATIONS

invented in Germany by Russian diplomat Paul von Schilling-Cannstadt (1786–1837).

1833 First single-wire electric telegraph in Germany is invented by scientists Karl Gauss (1777–1855) and Wilhelm Weber (1804–91).

1833 Precursor of the modern typewriter, using bars of type, is invented by Frenchman Xavier Progin.

1834 Phonetic-based system of shorthand writing, called "Stenographic Sound-Hand", is invented by Englishman Isaac Pitman (1813–97).

1834 Method of printing art illustrations in oil colours is invented by British printmaker George Baxter (1804–67).

1835 Photographs (which he called "photogenic drawings") are taken using small cameras with glass lenses by Englishman William Fox Talbot (1800–77).

1837 Five-wire electric telegraph is invented by scientists William Cooke (1806–79) and Charles Wheatstone (1802–75) in England.

1837 Recording single-wire telegraph using punched or printed paper tape is invented by German physicist Carl Steinheil (1801–70).

1837 Chromolithography, for colour printing, is invented by Godefroy Englemann.

1838 Single-wire telegraph is invented by US artist and inventor Samuel Morse (1791–1872); with his assistant Alfred Vail (1807–59) he also invented Morse code.

1838 Stereoscope (for viewing 3-D pictures) is invented by British scientist Charles Wheatstone (1802–75).

1839 National postal system is introduced by Englishman Rowland Hill, who also invented the adhesive postage stamp known (for its price and colour) as the penny black, although adhesive stamps had earlier been proposed by British bookseller James Chalmers (1782–1853).

1839 Daguerreotype photographic process (on metal) is invented by Frenchman Louis Daguerre (1787–1851).

1839 Calotype photographic process, using paper negatives, is invented by British photography pioneer William Fox Talbot (1800–77).

1839 Gum-dichromate photographic emulsion is invented by Scottish photographer Mungo Ponton (1802–80).

1839 Microfilm is invented by British photographer John Dancer, although it is not produced commercially until 1852.

1839 Commercial telegraph system opens in England.

1840 Machine for distributing printing type after use is invented by Englishman Etienne Gaudens; it is independently patented in the USA (1843) by British inventor Frederick Rosenberg.

1843 Hand printing machine with a horizontal cylindrical platen, similar to a typewriter, is invented by US inventor Charles Thurber (1803–86).

1843 Photographic enlarger is patented by Americans John Johnson and Alexander Wolcott.

1844 Foghorn with a range of 10 kilometres is invented by British naval officer J.N. Taylor.

1845 Single-needle telegraph is invented

ASTRONOMY AND SPACE	PHYSICS	CHEMISTRY	BIOLOGY	MEDICINE

PHYSICS

1840 Joule's law (that the rate of heating by an electric current flowing in a conductor is equal to the product of the conductor's resistance and the square of the current) is discovered by British physicist James Joule (1818–89).

1842 Principle of conservation of energy is discovered by German physicist Julius von Meyer (1814–78) and, independently, by British physicist James Joule (1818–89).

1842 Doppler effect (the apparent change in frequency of a wave motion with the change in relative velocity of the observer and the source) is discovered by Austrian physicist Christian Doppler (1805–53). Applied initially only to sound waves, the effect was later (1848) applied also to light waves by French physicist Armand Fizeau (1819–96).

1843 Mechanical equivalent of heat is discovered by British physicist James Joule (1818–89).

1843 Poiseuille's formula or law (that the volume of flow of a liquid through a tube is determined by the pressure and the dimensions of the tube) is discovered by French physician Jean Poiseuille (1799–1869).

1844 Significance of the critical temperature in the liquefaction of gases is discovered by Irish chemist Thomas Andrews (1813–85).

1845 Kinetic theory of gases is proposed by Scottish physicist John Waterson (1811–83), although his work is overlooked for nearly 30 years.

1845-1847 Kirchhoff's laws (concerning electric currents in circuits and networks) are discovered by German physicist Gustav Kirchhoff (1824–87).

1846 Magnetostriction, the shortening of a bar of ferromagnetic metal when it is magnetized, is discovered by British physicist James Joule (1818–89).

1847 Principle of conservation of energy is formulated by German scientist Hermann Helmholtz (1821–94).

1847 Draper point (the temperature, about 525°C, at which all substances glow a dull red colour) is discovered by British-born US chemist John Draper (1811–82).

1848 Absolute temperature scale, and the concept of absolute zero, is proposed by British physicist William Thomson (Lord Kelvin) (1824–1907).

1849 First fairly accurate value for the speed of light is obtained by French physicist Armand Fizeau (1819–96).

1849 Stokes' law (that defines the force acting on an object as it falls through a viscous fluid) is discovered by Irish physicist George Stokes (1819–1903).

CHEMISTRY

impure sugar) is discovered by French chemist Anselme Payen (1795–1871).

1822 Potassium hexacyanoferrate(III) (potassium ferricyanide, known as Gmelin's salt) is discovered by German chemist Leopold Gmelin (1788–1853).

1822 Alkaloid caffeine is discovered (in coffee) by French chemist Joseph Caventou (1795–1877).

1822 Triiodomethane (iodoform) is discovered by French chemist Georges Serrulas (1744–1832).

1823 Fact that animal fats are esters of glycerol and a fatty acid is discovered by French chemist Michel Chevreul (1786–1889), who in the same year discovers stearin.

1823 Nonmetallic element silicon (symbol Si) is isolated by Swedish chemist Jöns Berzelius (1779–1848), having been detected earlier (1817) and named by Scottish chemist Thomas Thomson (1773–1852).

1823 Monoclinic sulphur is discovered by German chemist Eilhardt Mitscherlich (1794–1863).

1824 Metallic element aluminium (symbol Al) is discovered by Danish physicist Hans Oersted (1777–1851), later (1827) isolated by German chemist Friedrich Wöhler (1800–82); the name aluminium was suggested by British Humphry Davy (1778–1829) but afterwards changed to "aluminium" in everywhere but North America. Quantity extraction was first achieved (1855) by French chemist Henri Sainte-Claire Deville (1818–81).

1825 Benzene and benzene hexachloride (BHC) – used a century later as an insecticide – are discovered by British scientist Michael Faraday (1791–1867).

1826 Osmosis is rediscovered by French physiologist René Dutrochet (1776–1847), having been earlier (1748) described by his compatriot Jean Nollet (1700–70).

1826 Liquid non-metallic element bromine (symbol Br) is discovered by French chemist Antoine Balard (1802–76) and, independently, German chemist Carl Löwig.

1826 Phenylamine (aniline) is discovered by German chemist Otto Unverdorben (1806–73), later (1834) investigated by his compatriot Friedlieb Runge (1795–1867).

1844 Metallic element ruthenium (symbol Ru) is isolated by German chemist Karl Klaus (1796–1864).

1827 Selenic(IV) (selenous) acid is discovered by German chemist Eilhard Mitscherlich (1794–1863).

1828 Wöhler's synthesis (of urea by heating ammonium isocyante) is discovered by German chemist Friedrich Wöhler (1800–82), the first synthesis of an organic compound from an inorganic one.

1828 Radioactive element thorium (symbol Th) is discovered by Swedish chemist Jöns Berzelius (1779–1848).

1829 Graham's law (that the rate of diffusion of a gas is inversely proportional to the square root of its density) is discovered by Scottish chemist Thomas Graham (1805–69).

1829 Existence of sets of three chemically similar elements ("triads") is discovered by German chemist Johann Döbereiner (1780–1849).

1830 Manganic(VII) (permanganic) acid is discovered by German chemist Eilhard Mitscherlich (1794–1863).

FARMING AND FOOD

TRANSPORT

by Scottish blacksmith Kirkpatrick Macmillan (1813–78).

1843 Iron-hulled steamship the Great Britain, the first propeller-driven ship to cross the Atlantic Ocean, is built by British engineer Isambard Brunel (1806–59).

1844 First iron-hulled propeller-driven warship, the Princeton, is built in the USA by Swedish engineer John Ericsson (1803–89)

1844 Rocket with angled exhausts, which makes it spin in flight, is invented by British engineer William Hale.

1845 Pneumatic tyre, for horse-drawn vehicles, is patented by Scottish engineer Robert Thomson (1822–73).

1845 First clipper ship, the Rainbow, is built in the USA.

1847 British naval vessel HMS Driver becomes the first steamship to travel round the world.

1847 Fishplate for joining railway rails is invented by British engineer William Adams (1797–1872).

1848 Steam-powered, two-propeller model aeroplane is constructed by British engineer John Stringfellow.

1848 Improved practical diving bell is invented by US engineer James Eads (1820–87).

ENGINEERING AND TECHNOLOGY

1829 Practical steam fire engine is invented by British engineer John Braithwaite (1797–1870).

1830 Bimetallic strip thermostat is invented by Scottish chemist Andrew Ure (1778–1857).

1830 Standard screw threads are introduced by British engineer Joseph Whitworth (1803–87).

1830 Efficient type of cast-iron beam (Hodkinson's beam) is invented by British engineer Eaton Hodgkinson (1789–1861).

1830 Single-thread sewing machine is invented by French tailor Barthélemy Thimonnier (1793–1854).

1830 Rotary-cutter (with fixed blade) lawn mower is invented by British engineer Edwin Budding (1795–1846).

1830 Hydrometer for measuring the specific gravity of dense liquids is invented by Scottish scientist William Twaddell (d.1830).

1831 Dynamo is invented by British scientist Michael Faraday (1791–1867).

1831 Electric bell is invented by US physicist Joseph Henry (1797–1878).

1831 Polarizing prism for microscopes, made of Iceland spar, is invented by Scottish physicist and geologist William Nicol (c.1768–1851).

1831 Safety fuse for blasting explosives is invented by British leatherworker William Bickford (1774–1834).

1831 Unsuccessful steam turbine is built by US engineer Charles Avery.

1831 Contact process for making sulphuric acid is patented by British chemist Peregrine Phillips (b.c.1800), but process is not worked until much later (1875).

1831 Platform scales are invented by US engineer Thaddeus Fairbanks (1796–1886).

1832 Electric generator with a rotating field magnet (a type of magneto) is invented by French instrument maker Hippolyte Pixii (1808–35); the low-tension magneto adopted for internal combustion engines is invented later (1890) by German engineer Robert Bösch, followed (1895) by the high-tension magneto of British engineer Frederick Simms.

1832 Eye-pointed needle for sewing machines is invented in the USA by Walter Hunt (1796–1859).

1832 Method of preserving wood (kyanizing) is invented by Irish brewer John Kyan (1774–1850).

1833 Improved outward-flow water turbine is invented by French engineer Benoît Fourneyron (1802–67).

1835 Chemical apparatus for producing small quantities of common laboratory gases is invented by Dutch chemist Petrus Kipp (1808–64).

1835 Electric relay is invented by US physicist Joseph Henry (1797–1878) and, later (1836), independently by British engineer Edward Davy (1806–85).

1835 Metal-shaping machine is invented by Scottish engineer James Naysmith (1808–90).

1835 Knitting machine is invented by British engineer Joseph Whitworth (1803–87).

1836 Daniell primary cell (battery) is invented by British chemist John Daniell (1790–1845).

1836 Revolver, made from interchangeable parts, is invented by US gunsmith Samuel Colt (1814–62).

1836 Moving-coil galvanometer is invented by British physicist William Sturgeon (1783–1850).

1836 Stroboscope is invented by Belgian physicist Joseph Plateau (1801–83).

1836 Chemical method of testing for the presence of arsenic is invented by British chemist James Marsh (1794–1846).

1837 Lace-making power loom is invented by US weaver Erastus Bigelow (1814–79).

1837 Galvanized iron is patented by British inventor Henry Crawfurd.

1837 Processes for electroplating silver, nickel and chromium are devised by German-born Russian inventor Moritz von Jacobi (1801–74).

1838 Process for electroplating gold is patented by British inventor George Elkington (1801–65) who, with his cousin Henry Elkington (1810–52), pioneered the commercial electroplating industry.

1839 Steam hammer is invented by Scottish engineer James Nasmyth (1808–90).

1839 Primary cell (battery), the Grove cell, and a hydrogen-oxygen fuel cell are invented by British physicist (and judge) William Grove (1811–96).

1839 Improved bearing (a journal box lined with Babbitt metal, a new anti-friction alloy) is invented by US engineer Isaac Babbitt (1799–1862).

1839 Simple photoelectric cell is invented by French physicist Antoine Becquerel (1788–1878).

1840 First practical electric clock is constructed by Scottish engineer Alexander Bain (1810–77).

1840 Process for electroplating gold onto brass and silver is invented by Swiss physicist Auguste de la Rive (1801–73).

COMMUNICATIONS

by British scientists Charles Wheatstone (1802–75) and William Cooke (1806–79).

1845 System of raised type (on paper) for blind people to read is invented by Englishman William Moon (1818–94), who had become blind five years earlier. A similar system is invented in 1856 by US journalist Alfred Beach (1826–96).

1846 High-speed rotary printing press is invented by US printer Richard Hoe (1812–86).

1846 Printing telegraph is invented by US inventor Royal House (1814–95) and later (1850), independently, by British scientist Francis Galton (1822–1911).

1846 Dial telegraph is invented by German engineer Werner von Siemens (1816–92).

ASTRONOMY AND SPACE	PHYSICS	CHEMISTRY	BIOLOGY	MEDICINE

CHEMISTRY

1831 Trichloromethane (chloroform) is discovered by French pharmacist Eugène Soubeiran (1797–1858) and, independently, by US chemist Samuel Guthrie (1782–1848) and German chemist Justus von Liebig (1803–73).

1831 Red dye alizarin is discovered by French chemists Pierre Robiquet (1780–1840) and Jean Colin (1784–1865).

1832 Nitrobenzene is discovered by German chemist Eilhardt Mitscherlich (1794–1863).

1832 Pain-killing drug codeine is discovered by French chemist Pierre Robiquet (1780–1840).

1832 Anthracene is discovered in coal tar by French chemists Auguste Larent (1807–53) and Jean Dumas (1800–84).

1833 Ethyl radical is discovered by Irish chemist Robert Kane (1809–90) and, independently (1834), German chemist Justus von Liebig (1803–73).

1833 Alkaloid atropine is discovered by German chemist Philipp Geiger.

1833 Creosote is discovered (in coal tar) by German industrial chemist Karl von Reichenbach (1788–1869).

1834 Laws of electrolysis are discovered by British scientist Michael Faraday (1791–1867).

1834 Quinoline is discovered by German chemist Friedlieb Runge (1795–1867).

1834 Dichlorine oxide (chlorine monoxide) is discovered by French chemist Antoine Balard (1802–76).

1834 Phenol (carbolic acid) is discovered by German chemist Friedlieb Runge (1795–1867).

1834 Cellulose is extracted from wood (and named) by French chemist Anselme Payen (1795–1871).

1834 Law of substitution (that chlorine, e.g., can replace hydrogen in hydrocarbons) is discovered by French chemist Jean Baptiste Dumas (1800–84).

1835 Methanol (methyl alcohol) is discovered by French chemist Jean Dumas (1800–84).

1835 Pyruvic acid is discovered by Swedish chemist Jöns Berzelius (1779–1848).

1836 Catalysis is discovered by Swedish chemist Jöns Berzelius (1779–1848).

1836 Ethyne (acetylene) is discovered by British chemist Edmund Davy.

1838 Method of making nitric acid by oxidizing ammonia in the presence of a platinum catalyst is discovered by French chemist Charles Kuhlmann (1803–81).

1838 1-hydroxybenzoic (salicylic) acid is discovered by Italian chemist Rafaelle Piria (1815–65).

1838 Methylbenzene (toluene) is discovered by Polish chemist Philippe Walter (1810–47) and French chemist Joseph Pelletier (1788–1842).

1839 Metallic element lanthanum (symbol La) is discovered by Swedish chemist Carl Mosander (1797–1858).

1840 Ozone is discovered by German chemist Christian Schönbein (1799–1868).

c.1840 Fehling's test (for reducing sugars) is discovered by German chemist Hermann von Fehling (1812–85).

1840 Dichloroethene (dichloroethylene) is discovered by French chemist Henri Regnault (1810–78).

FARMING AND FOOD	TRANSPORT	ENGINEERING AND TECHNOLOGY	COMMUNICATIONS

1840 "Needle gun" rifle, invented by German gunsmith Johann Dreyse (1787–1867) in 1827, is adopted by the Prussian army.

1841 Machine for making wire rope (later used in suspension bridges) is invented by US engineer John Roebling (1806–69).

1841 Cold vulcanization process for rubber is invented by British chemist Alexander Parkes (1813–90).

1841 Process for softening water (using lime) is invented by British chemist Thomas Clark (1801–67).

1841 Zinc-carbon primary cell (battery) for laboratory use is invented by German chemist Robert Bunsen (1811–99).

1841 System of standard screw threads is invented by British engineer Joseph Whitworth (1803–87); they become known as BSW threads, or British Standard Whitworth.

1842 Method of making superphosphate fertilizer is patented by British agricultural chemist John Lawes (1814–1900).

1842 Machine for stitching leather is invented by US engineer George Corliss (1817–88).

1842 Street-cleaning machine is invented by British engineer Joseph Whitworth (1803–87).

1843 First tunnel under the River Thames River is constructed by French-born English engineer Marc Brunel (1769–1849); it is later (1885) used for underground trains.

1843 Pneumatic drill is invented (for tunnel construction) by Joseph Fowle; later (1861) improved by Germain Sommelier.

1844 Improved water turbine is invented by US engineer Uriah Boyden (1804–79).

1844 Improved valve gear for steam locomotives is invented by Belgian engineer Egide Walschaerts (1820–1901).

1844 Type of bridge truss (Pratt truss, or N truss), consisting of two horizontal girders joined by diagonal members, is invented by US engineer Thomas Pratt (1812–75).

1844 Aneroid barometer is invented by French scientist Lucien Vidi (1805–66).

1844 Process for vulcanizing rubber using heat and sulphur is invented by US chemist Charles Goodyear (1800–60); a similar process had been patented in England in 1843 by Thomas Hancock who later (1846) manufactured rubber tyres.

1845 Guncotton (nitrocellulose) is invented by German chemist Christian Schönbein (1799–1868).

1845 Ebonite, an insulating material made from rubber and sulphur, is invented by British engineer Thomas Hancock.

1845 Improved compound steam engine is invented by Scottish engineer William McNaught (1813–81).

1845 Elastic band made of vulcanized rubber is patented by British inventor Stephen Perry.

1845 Power loom for weaving Wilton-type carpets is invented by US weaver Erastus Bigelow (1814–79).

1845 Ball valve for cisterns is invented by Edward Chrimes.

1845 Process for making alum (much used in dyeing) from coal shale is invented by Scottish chemist Peter Spence (1806–83).

1846 Modern lock-stitch sewing machine is invented by US engineer Elias Howe (1819–67).

1846 High-speed power loom for tapestries and carpets is invented by US engineer Erastus Bigelow (1814–79).

1846 Kerosene lamp is introduced by Canadian scientist Abraham Gesner.

1846 Hydraulic crane is invented by British engineer William Armstrong (1810–1900).

1846 Spirometer, to measure the vital (maximum) capacity of the lungs, is invented by British physician John Hutchinson (1811–61).

1847 Ophthalmoscope, an instrument for examining the interior of the eye, is invented, but not announced publicly, by British scientist Charles Babbage (1792–1871); it is later (1850) invented independently by German scientist Hermann Helmholtz (1821–94).

1847 Kymograph, an instrument for continuously recording blood pressure, is invented by German physiologist Karl Ludwig (1816–95).

1847 Alarm clock is invented by French clockmaker Antoine Redier (1817–92).

1848 Pre-cast iron columns for making tall buildings are invented by US engineer James Bogardus (1800–74).

1849 Safety pin is patented by US inventor Walter Hunt (1796–1859) and, independently, Englishman Charles Rowley.

1849 Bourdon pressure gauge is patented by French watchmaker and engineer Eugène Bourdon (1808–84).

1849 Reinforced concrete is invented by French engineer Joseph Monier (1823–1906).

1849 Rifle that uses an expanding conical lead bullet (Minié rifle) is invented by French army officer and gunsmith Claude Minié (1814–79).

CHEMISTRY

1841 Nitroprussides (nitrosopentacyanoferrate(III)) are discovered by Scottish chemist Lyon Playfair (1818–98).

1843 Metallic elements erbium (symbol Er) and terbium (symbol Tb) are discovered by Swedish chemist Carl Mosander (1797–1858). In the same year he also discovered "didymium", which later turned out to be a mixture of two elements, neodymium and praseodymium.

1844 Metallic element ruthenium (symbol Ru) is discovered by German chemist Karl Claus (1796–1864).

1845 Explosive properties of cellulose nitrate (nitrocellulose, guncotton) are discovered by German chemist Christian Schönbein (1799–1868).

1845 Ethanoic (acetic) acid is synthesized (from inorganic compounds) by German chemist Hermann Kolbe (1818–84).

1846 Glyceryl trinitrate (nitroglycerine) is discovered by Italian chemist Ascanio Sobrero (1812–88).

1846 Collodion is discovered by French chemist Louyis Menard (1822–1901).

1846 Amino acid tyrosine is discovered by German chemist Justus von Liebig (1803–73).

1847 Babo's law (that the decrease in vapour pressure of a liquid when a solute is added is proportional to the amount of solute) is discovered by German chemist Lambert Babo (1818–99).

1847 Nitroglycerin is discovered by Italian chemist Ascania Sobrero (1812–88).

1848 Optical activity (of tartaric acid) is discovered by French scientist Louis Pasteur (1822–95).

1848 Accurate relative atomic mass (atomic weight) of carbon is determined by French chemist Jean Dumas (1800–84).

1849 Kolbe's method of making an alkane (by electrolytically decomposing a dissolved carboxylic acid salt) is discovered by German chemist Herman Kolbe (1818–84).

1849 Pentanol (amyl alcohol) is discovered by British chemist Edward Frankland (1825–99).

1850

ASTRONOMY AND SPACE

1850 Saturn's innermost ring (the feint "crepe" ring) is discovered using photography by US astronomer William Bond (1789–1859) and, independently, by British astronomer William Lassell (1799–1880).

1851 Sirius is discovered to be one partner of a double-star system by US astronomer Alvan Clark (1804–87).

1851 First photographs of a solar eclipse are taken by Italian astronomer Angelo Secchi (1818–78).

1851 Ariel and Umbriel, the third and fourth moons of Uranus, are discovered by British astronomer William Lassell (1799–1880).

1852 Correlation between frequency of sunspots and magnetic storms on Earth is discovered by British astronomer Edward Sabine (1778–1883).

1852 Asteroid Lutetia is discovered by German astronomer Hermann Goldschmidt.

1855 Spiral structure of galaxies is discovered by British astronomer William Parsons, Earl of Rosse (1800–67).

1857 True nature of Saturn's rings (as a vast collection of small particles) is

PHYSICS

1850 Second law of thermodynamics (that heat cannot flow by itself from one object to a hotter object) is formulated by German physicist Rudolf Clausius (1822–88) and, independently (1851), by British physicist William Thomson (Lord Kelvin) (1824–1907).

1850 Foucault's pendulum, which demonstrates the rotation of the Earth, is devised by French physicist Léon Foucault (1819–68).

1851 Weber's constant (the ratio of the electrostatic and electromagnetic units of quantity of electricity, which equals the speed of light) is discovered by German physicist Wilhelm Weber (1804–91).

1852 Joule-Thomson (Joule-Kelvin) effect (the cooling that accompanies the expansion of a gas into a region of lower pressure) is discovered by British physicists James Joule (1818–89) and William Thomson (Lord Kelvin) (1824–1907).

1852 Regnault's method (of measuring the density of a gas) is discovered by French chemist Henri Regnault (1810–78).

1853 Mechanism of formation of absorption and emission spectra is explained by Swedish physicist Anders Ångström (1814–74).

CHEMISTRY

c.1850 Barfoed's test (for reducing sugars) is discovered by Swedish physician C. Barfoed (1815–99).

1850 Williamson's synthesis (for making ethers by reacting a haloalkane with an alkoxide) is discovered by British chemist Alexander Williamson (1824–1904).

1850 Red phosphorus is discovered by Austrian chemist Anton Schrötter (1802–75), using a method later (1851) improved by British chemist Arthur Albright (1811–1900).

1851 Pyridine is discovered by Scottish chemist Thomas Anderson (1819–74).

1852 Deviations from Boyle's law by real gases is discovered by French chemist Henri Regnault (1810–78).

1852 Concept of valence is introduced by British chemist Edward Frankland (1825–99).

1852 Kerosene is discovered by Canadian geologist Abraham Gesner (1797–1864).

1853 Cannizzaro reaction (in which an aldehyde disproportionates into an alcohol and a carboxylic acid) is discovered by Italian chemist Stanislao Cannizzaro (1826–1910).

BIOLOGY

1851 Vasomotor nerves are discovered by French biologist Claude Bernard (1813–78).

1852 Parathyroid glands are discovered (in a rhinoceros) by British zoologist Richard Owen (1804–92).

1852 Meissner's corpuscles (sense organs of touch in the skin) are discovered by German physiologists George Meissner (1829–1905) and Rudolf Wagner (1805–64).

1852 Term "evolution" is coined by British philosopher Herbert Spencer (1820–1903).

1856 X- and Y-chromosomes in mammals are discovered by US biologist Edmund Wilson (1856–1939).

1856 Remains of Neanderthal man are found by French anatomist Pierre Broca (1824–80).

1856 Method of keeping animal organs alive in vitro is discovered by German physiologist Carl Ludwig (1816–95).

1857 Glycogen ("animal starch") is discovered by French physiologist Claude Berrard (1813–78) and, independently, German physiologist Victor Hensen (1835–1924); it is (1875) isolated by US chemist Russell Chittenden (1856–1943).

MEDICINE

1850 Use of the hypodermic syringe is introduced by British physician Alfred Higginson (1808–84).

1854 Fact that cholera is linked to contaminated drinking water is discovered by British physician John Snow (1813–53).

1854 Quick-drying plaster of Paris is introduced as a cast for broken bones by Dutch physician Anthonius Mathijsen (1805–78).

1859 Muscle relaxant properties of curare are discovered by French physiologist Claude Bernard (1813–78).

1855 Addison's disease (caused by excessive activity of the adrenal glands) is discovered by British physician Thomas Addison (1793–1860).

1855 Fact that cells arise only by division of other cells is discovered by German pathologist Rudolf Virchow (1821–1902).

1858 Clinical description of tabes dorsalis (locomotor ataxia), a long-term effect of syphilis, is given by French physician Guillaume Duchenne (1806–75).

1858 Fact that farsightedness is often caused by eyeballs that are too short front-to-back is discovered by Dutch

FARMING AND FOOD

1850s First sugar refineries (for sugar cane) are built in the USA by US sugar planter Valcour Aime (1798–1867).

1851 Process for the commercial production of ice cream is invented in the USA by Jacob Fussel.

1851 Meat biscuit (pemican) is invented by US food technologist Gail Borden (1801–74).

1853 Potato chips are invented by Native American chef George Crum.

1855 Self-propelled rotary cultivator is invented by Canadian engineer Robert Romaine.

1856 Condensed milk is invented by US food technologist Gail Borden (1801–74) and, independently, German-born Swiss Henri Nestlé (1814–90).

1857 Steam plough is invented by British engineer John Fowler (1826–64).

1858 Self-binding machine is invented by US engineer John Appleby (1840–1917).

1858 Can opener is invented by US Ezra Warner (previously food cans were opened using a chisel).

TRANSPORT

1850 First electric railway locomotive is made by US physicist Charles Page (1812–68).

1852 Safety passenger lift is invented by US engineer Elisha Otis (1811–61).

1852 Steerable airship (dirigible), with a 3-h.p. steam engine and a 3-bladed propeller, is invented by Frenchman Henri Giffard (1825–82).

1853 First man-carrying glider is built by British inventor George Caley (1773–1857).

1853 Vacuum, or pneumatic, tube for carrying cylinders containing messages is invented by British engineer Josiah Clark (1822–98).

1854 Cork life preserver is invented by J. Ross Ward.

1855 First propeller-driven ship in Germany is constructed by German shipbuilder Ferdinand Schichan (1814–96)

1858 Iron passenger liner with both paddles and a propeller, the Great Eastern, is built by British engineer Isambard Brunel (1806–59).

1858 First vessel built entirely of steel, the steam launch Ma Robert, is con-

ENGINEERING AND TECHNOLOGY

1850 Vortex water wheel, a type of turbine, is invented by Irish-born engineer James Thomson (1822–92).

1850 Process for mercerizing cotton cloth (making it stronger and silky) is invented by British chemist John Mercer (1791–1866).

1850 Improved type of sewing machine is patented by US inventor Allen Wilson (1824–88).

1850 Method of producing "lamp oil" (paraffin, now kerosene) from coal is invented by Scottish chemist James Young (1811–83)

1851 New type of sewing machine is patented by US inventor Isaac Singer (1811–75).

1851 Soda process for making chemical wood pulp (for paper-making) is invented by British chemists Hugh Burgess (1825–92) and Charles Watt.

1851 Simple transformer, the Ruhmkorff coil (little more than an induction coil), is invented by German physicist Heinrich Ruhmkorff (1803–77).

1851 Water meter is invented by German-born British engineer William (Wilhelm) Siemens (1823–83).

1851 Mechanical refrigerator is invented by US engineer John Gorrie (1803–55).

1851 Machine for making envelopes is invented by British scientist Warren de la Rue (1815–89).

1851 Modern denture, with porcelain teeth on a platinum plate, is invented by US dentist John Allen (1810–92).

1851 Arch-truss bridge is invented by Scottish-born US engineer Daniel McCallum (1815–78).

1852 Fink truss, for railway bridges and roofs, is invented by US engineer Albert Fink (1827–97).

COMMUNICATIONS

1850 Rotary press is invented by British publisher Thomas Nelson (Jr) (1822–1892).

1850 Typewriter using an inked ribbon is patented by US inventor Oliver Eddy; a similar machine is later (1856) marketed by his compatriot Alfred Beach (1826–96).

1850 First photographs through a microscope are taken by British-born US chemist and photographer John Draper (1811–82).

1850 Wet collodion plates for photography are invented by British artist Frederick Archer (1813–57).

1851 Submarine telegraph cable is laid across the British channel (Dover to Calais) by British engineer Thomas Crampton (1816–88).

1851 Telegraphic news service is established in London by German news gatherer Paul Reuter (pre-1844, Israel Josaphat) (1816–99).

1854 Pneumatic tube for sending written messages is patented by British engineer Josiah Clark (1822–98); it is later (1858) improved by his compatriot Cromwell Varley (1828–83), whose system employs compression and vacuum alternately.

1850

ASTRONOMY AND SPACE

proposed by Scottish physicist James Clerk Maxwell (1831–79).

1858 Donati's comet is discovered by Italian astronomer Giambattista Donati (1828–73).

1859 Explanation of dark "Fraunhofer" lines in the Sun's spectrum is given by German physicist Gustav Kirchhoff (1824–87).

1859 Fact that the outer layers of the Sun (and therefore sunspots) rotate faster at the equator than at other latitudes is discovered by British astronomer Richard Carrington (1826–75). The same effect is discovered independently by German astronomer Gustav Spörer (1822–95), and is usually known as Spörer's law.

1860 "Red flames" during an eclipse of the Sun are discovered to be solar prominences (not lunar phenomena) by British astronomer (and photographer) Warren De la Rue (1815–89).

1861 Asteroid Hesperia is discovered by Italian astronomer Giovanni Schiaparelli (1835–1910).

1861 Hydrogen is discovered in the Sun, by means of its spectrum, by Swedish physicist Anders Ångström (1814–74).

1862 Asteroid Freia is discovered by German astronomer Heinrich d'Arrest (1822–75).

1863 Fact that stars contain the same elements as those on Earth is discovered (spectroscopically) by British astronomer William Huggins (1824–1910).

1864 Tempel's comet is discovered by German astronomer Ernst Tempel (1821–89).

1866 Fact that meteors follow cometary orbits is discovered by Italian astronomer Giovanni Schiaparelli (1835–1910).

1867 Kirkwood gaps (asteroid-free regions among the orbits of asteroids) are discovered by US astronomer Daniel Kirkwood (1814–95).

1867 Three original Wolf-Reyet stars are discovered by French astronomers Charles Wolf (1827–1918) and Georges Rayet (1839–1906).

1868 Spectrographic method of studying solar prominences is discovered independently by French astronomer Pierre Janssen (1824–1907) and British astronomer Joseph Lockyer (1836–1920).

1868 Fact that comets emit light from luminescent carbon gas and the red shift in some stellar spectra (which proves that the stars emitting them are receding) are discovered by British astronomer William Huggins (1824–1910).

1869 Two different types of solar prominences (quiescent and non-quiescent) are discovered by British astronomer William Huggins (1824–1910).

1874 Asteroid Austria is discovered by Austrian astronomer Johann Palisa (1848–1925)

1877 Dark lines on Mars – termed canali but mistranslated as "canals" – are "discovered" by Italian astronomer Giovanni Schiaparelli (1835–1910).

1877 Phobos and Deimos (two moons of Mars) are discovered by US astronomer Asaph Hall (1829–1907).

1880 Existence of "Planet X" (Pluto) is predicted by US engineer George Forbes (1849–1936).

1884 Wolf's comet, with a period of seven years, is discovered by German astronomer Max Wolf (1863–1932).

PHYSICS

1853 Magnus effect, which causes a spinning ball to swerve in flight, is discovered by German physicist Heinrich Magnus (1802–70).

1856 Term "kinetic energy" is coined by British physicist William Thomson (Lord Kelvin) (1824–1907).

1857 Lissajous figures (curves produced by combining two harmonic motions) are discovered by French physicist Jules Lissajous (1822–80).

1858 Fact that electric discharges in a low-pressure gas (cathode rays) are deflected by a magnetic field is discovered by German physicist Julius Plücker (1801–68).

1860 Thomson (Kelvin) effect (the evolution of heat by an electric current flowing along a conductor whose ends are at different temperatures) is discovered by British physicist William Thomson (Lord Kelvin) (1824–1907).

c.1860 Kirchhoff's laws of radiation (that the emissivity of an object is equal to its absorptance at the same temperature) is discovered by German physicist Gustav Kirchhoff (1824–87).

1864 Velocity of electromagnetic waves is predicted to be the same as that of light by Scottish physicist James Clerk Maxwell (1831–79).

1865 Term "entropy" is coined by German mathematician and physicist Rudolf Clausius (1822–88).

1866 Boltzmann's law (concerning the equipartition of energy) is formulated by Austrian physicist Ludwig Boltzmann (1844–1906).

1869 Tyndall effect (the scattering of light by fine particles suspended in a fluid) is discovered by Irish physicist John Tyndall (1820–93), who also showed that the effect is responsible for the blue colour of the daytime sky.

1871 Method of measuring the conductivity of electrolytes (using alternating current) is discovered by German physicist Friedrich Kohlrausch (1840–1910), leading to Kohlrausch's law (that ions migrate independently during electrolysis).

1873 Electromagnetic theory of light is published by Scottish physicist James Clerk Maxwell (1831–79).

1873 Equation of state or Van der Waals' equation (relating pressure and volume of a gas and making due allowance for the force of attraction between gas molecules – the Van der Waals' force) is discovered by Dutch physicist Johannes Van der Waals (1837–1923).

1874 Term "electron" is coined by Irish physicist George Stoney (1826–1911) – later (1897) to be adopted instead of J.J. Thomson's term "corpuscle".

c.1874 Mach's principle (that the inertia of an object results from its interaction with the rest of the Universe) is discovered by Austrian physicist and philosopher Ernst Mach (1838–1916),

1875 Photoelectric properties of selenium are discovered by British physicist Willoughby Smith (1828–91).

1875 Kerr effect (the differential refraction of light by some materials when they are in a strong electric field) is discovered by Scottish physicist John Kerr (1824–1907).

1876 Cathode rays – emitted by the cathode of a discharge tube – are described and named by German physicist Eugen Goldstein (1850–1930), who went on (1886) to identify "canal rays" (positive ions emitted by the anode).

CHEMISTRY

1853 Transport number (which acknowledges that not all ions travel at the same speed during electrolysis) is discovered by German physical chemist Johann Hittorf (1824–1914).

1853 Aspirin (acetylsalicylic acid), although not its analgesic action, is discovered by French chemist Charles (Karl) Gerhardt (1816–56).

1853 Dye rosaniline is discovered by German chemist August Hofmann (1818–92).

1854 Fact that each element has its own characteristic spectrum is discovered by US physicist David Alter (1807–81).

1855 Theory of types in organic chemistry, including homologous and heterologous series, is proposed by French chemist Charles (Karl) Gerhardt (1816–56).

c.1855 Wurtz reaction (between a haloalkane and sodium to form an alkane) is discovered by French chemist Charles-Adolphe Wurtz (1817–84).

c.1855 Fact that ozone is an allotrope of oxygen is discovered by Irish physical chemist Thomas Andrews (1813–85).

1856 Mauvine, the first synthetic aniline dye, is discovered by British chemist William Perkin (1848–1907).

1857 Photochemical reaction (of hydrogen with chlorine) is discovered by German chemist Robert Bunsen (1811–99) and British chemist Henry Rosco (1833–1915).

1857 Silcon nitride is discovered by German chemist Friedrich Wöhler (1800–82) and, independently, French chemist Henri Sainte-Claire Deville (1818–81).

1857 Reversible dissociation of chemical compounds is discovered by French chemist Henri Sainte-Claire Deville (1818–81).

1857 Glycogen is discovered by French physiologist Claude Bernard (1813–78).

1858 Fact that carbon atoms are tetravalent and can combine with each other to form chains (thus introducing the concept of chemical bonds in organic compounds) is discovered by Scottish chemist Archibald Couper (1831–92), although his findings remain unpublished until long after German chemist Friedrich Kekulé von Stradonitz (1829–96) independently announces reaching the same conclusion in the same year.

1858 Distinction between atomic and molecular weights (masses) is made by Italian chemist Stanislao Cannizzaro (1826–1910).

1858 Composition of silica (silicon dioxide) is discovered by Swiss chemist Jean de Marignac (1817–94).

1859 Cocaine is prepared (from coca leaves) by German chemist Albert Niemann and, later (1860), by French chemist Albert Niemann (1834–61).

1860 Spectrum analysis for identifying elements is introduced by German physicist Gustav Kirchhoff (1824–87) and chemist Robert Bunsen (1811–99), who in that year used the technique to discover the metallic element caesium (symbol Cs).

1861 Isomerism (in organic compounds) is discovered by Russian chemist Boutlerov.

1860 Method of making synthetic ruby (a form of corundum) is discovered by French chemist Edmond Frémy (1814–94).

1861 Azo dyes are discovered by German chemist Peter Griess (1829–88).

BIOLOGY

1859 The Origin of Species by Means of Natural Selection, outlining the theory of evolution, is published by British naturalist Charles Darwin (1809–82); similar conclusions had been reached independently by his compatriot Alfred Russel Wallace (1823–1913) and communicated to Darwin in 1858. Wallace's own book, Contributions to the Theory of Natural Selection, was published in 1870.

1860 Damaging criticism of Charles Darwin's theory of natural selection (as the mechanism for evolution) is published anonymously by British palaeontologist Richard Owen (1804–92).

1861 Broca's area (the speech centre in the cortex of the brain) is discovered by French anatomist Pierre Broca (1824–80).

1861 Batesian mimicry (in which a harmless animal gains protection by mimicking the coloration of an animal that is harmful to predators) is discovered by British naturalist Henry Bates (1825–92).

1862 Spontaneous generation of life (actually bacteria from inorganic material) is finally disproved by French scientist Louis Pasteur (1822–95).

1865 Laws of heredity – the foundation of genetics – are published (but go unnoticed) by Moravian monk Gregor Mendel (1822–84).

1865 "Germ theory" of fermentation (that each type of fermentation depends on a specific microorganism) is proposed by French bacteriologist Louis Pasteur (1822–93).

1865 Chloroplasts (chlorophyll-containing structures in plant cells) are discovered by German botanist Julius von Sachs (1833–97).

1866 Lichens are discovered to be symbiotes (algae and fungi living together) by German botanist Heinrich de Barry (1831–1888).

1866 Recapitulation theory (that an animal's embryonic development mirrors its evolutionary history) is proposed by German biologist Ernst Haeckel (1834–1919), after having been first suggested (1864) by his compatriot Fritz Müller (1821–97). Haeckel also coins the term "ecology" (German Oecologie).

1866 Difference between rods and cones in the retina of the eye is discovered by German zoologist Max Schultze (1825–74).

1867 Method of gold-staining of tissue samples and the role of white blood cells in inflammation are discovered by German pathologist Julius Cohnheim (1839–84).

1868 Fossils of Cro-Magnon Man (the Earliest in Europe) are discovered in France by French palaeontologist Edouard Lartet (1801–71).

1869 Islets of Langerhans (groups of cells in the pancreas which produce insulin) are discovered by German physiologist Paul Langerhans (1847–88).

1869 Nucleic acid DNA is discovered by German biochemist Friedrich Mieschler.

1871 Enzyme invertase, which mediates the conversion of sucrose (table sugar) into glucose and fructose, is discovered by German biochemist Ernst Hoppe-Seyler (1825–95).

1873 Blood cells called platelets are discovered by Canadian physician William Osler (1849–1919).

1873 Staining technique that allows microscopic study of nerve cells is discovered by Italian histologist Camillo Golgi (1843–1926).

MEDICINE

physician Franciscus Donders (1818–89).

1859 Chemical test (on urine) for gout is discovered by French physician Alfred Garod (1819–1907).

1863 Antiseptic surgery, using phenol (carbolic acid), is introduced by British surgeon Joseph Lister (1827–1912), using the disinfectant properties of phenol as reviously advocated by British chemist Frederick Calvert (1819–73).

1863 Barbituric acid is discovered by German chemist Johann von Baeyer (1835–1917), later (1903) used to produce barbiturate drugs by Emil Fischer (1852–1919).

1864 International Red Cross is founded by Swiss banker Jean Dunant (1828–1910).

1865 Fact that tuberculosis is infectious is discovered by German physician Wilhelm Wundt (1832–1920), later (1867) confirmed by polish pathologist Julius Cohnheim (1839–84) and French physician Jean Villemin (1827–92).

1866 Trypanosome that causes filariasis (elephantiasis) is discovered by British physician Joseph Bancoft (1836–94) and, a year later, by Welsh physician Timothy Lewis (1841–86).

1867 Amyl nitrite as a treatment for angina pectoris is discovered by Scottish physician Thomas Brunton (1844–1916).

1867 Cholecystotomy, an operation to remove gallstones, is introduced by US surgeon John Bobbs (1809–70).

1867 Use of an incubator for premature babies is introduced by US gynaecologist Theodore Thomas (1831–1903).

1870 Nephrotomy, an operation to remove a kidney, is introduced by German surgeon Gustav Simon (1824–76).

1874 Osteopathy is founded by US physician Andrew Still (1828–1917).

1876 Bacterium that causes anthrax is isolated by German bacteriologist Robert Koch (1843–1910).

1877 Paget's disease (a form of osteitis causing bone thickening) is discovered by British surgeon James Paget (1814–99).

1877 Lanereuax's diabetes (diabetes mellitus with pancreatic disease) is discovered by French physician Étienne Lancereaux (1829–1910).

1877 Role of the gnat Culex in carrying the nematode worms (Filaria) that cause elephantiasis (filariasis) in humans is discovered by Scottish physician Patrick Manson (1844–1922).

1878 Cause of "the bends" in divers who have breathed compressed air (nitrogen dissolved in the blood) is discovered by French physiologist Paul Bert (1833–86).

1879 Electrocardiograph for studying the electrical activity of the brain is introduced by British physician Augustus Waller (1816–80).

1879 Bacterium (gonococcus) that causes gonorrhea is discovered by German bacteriologist Albert Niesser (1855–1916).

1880 Bacterium that causes typhoid fever is discovered by German bacteriologist Karl Eberth (1835–1926) and, independently, his compatriot Robert Koch (1843–1910) and, later (1884), by German bacteriologist Georg Gaffky (1850–1918).

1881 Fact that the mosquito is the vector of yellow fever is discovered by Cuban

FARMING AND FOOD

1858 Concentrated meat extract is invented by German chemist Justus von Liebig (1803–73).

1860 Hydroponics (the cultivation of plants without soil) is introduced by German scientist Julius von Sachs (1832–97).

1863 Pasteurization (originally for wine) is invented by French chemist Louis Pasteur (1822–95).

1863 Starch-free slimming diet ("Bantingism") is invented by British undertaker William Banting (1797–1878).

1867 Baby food mimicking mother's milk is invented by German chemist Justus von Liebig (1803–73).

1868 Refrigerated railway freight cars for transporting perishable foods are invented by US engineer William Davis (1812–68).

1868 Plough with hardened cast-iron edge to the share is invented by Scottish engineer James Oliver (1823–1908).

1869 Commercial margarine production (from tallow) is patented by French chemist Hippolyte Mergé-Mouriès; the original discovery of margarine was made 30 years earlier by Michel Chevreul (1786–1889) and the process later (1872) improved, by using also skimmed milk, by French inventor F. Boudet.

1869 Chewing gum (based on chicle) is invented by US photographer Thomas Adams (patented 1871).

1870 Differential gears for reaping machines are invented by German-born US engineer Rudolf Eickemeyer (1831–95).

1875 Refrigerated cold store for meat and dairy produce is constructed by Australian Thomas Mort.

1875 Large-scale meat processing and canning is introduced by US industrialist Philip Armour (1832–1901).

1875 Baked beans are first canned by the US Burnham and Morrill company (for the crews of their fishing boats); they became know as "Boston beans" and were later (1891) marketed in the USA canned in tomato sauce.

1876 Mass-produced canned foods are first marketed by US retailer Henry Heinz (1844–1919).

1876 Ship with refrigerated hold, the Paraguay, is used by French engineer Charles Tellier (1828–1913) to carry perishable foods across the Atlantic Ocean.

1876 Vanilla essence, the first artificial flavouring, is synthesized by German chemists William Haarman and Karl Reimer.

1877 Centrifugal cream separator is invented by Swedish engineer Carl de Laval (1845–1913).

1878 Grain binding machine is invented by US engineer John Appleby (1840–1917).

1878 Frozen mutton is sent from Argentina to France on SS Paraguay.

1880 Frozen beef is sent from Australia to England on SS Strathleven.
1881 petrol-engined farm tractor is built by US engineer John Froelich.

1882 Frozen meat is sent from New Zealand to England on SS Dunedin.

1884 Method of making evaporated milk is patented in the USA by John Mayenberg, who later (1885) manufactures it with his Helvetia Milk Condensing Company.

TRANSPORT

structed by British shipbuilder John Laird (1805–74).

1861 First iron-clad warship, HMS Warrior, is launched in England.

1861 Pedal-crank driven bicycle (the "bone-shaker") is invented by French engineers, father and son, Pierre and Ernest Michaux.

1863 Locomotive with two sets of pivoted driving wheels (like powered bogey trucks) is invented by Scottish engineer Robert Fairlie (1831–85).

1863 Submarine with compressed-air engine is launched in France.

1863 First underground railway (steam powered) opens in London, designed by British engineers Benjamin Baker (1840–1907) and John Fowler (1817–98).

1863 Roller skates are patented by US inventor James Plimpton.

1864 Pullman sleeping car for trains is invented by US cabinetmaker George Pullman (1831–97).

1865 Semaphore traffic signals, with arms lit by red or green gas lamps at night, are invented by British engineer John Knight (1828–86).

1866 Cable-hauled tram (cable-car) is constructed by German inventor Wilhelm Ritter; later (1873) re-invented by US engineer Andrew Hallidie (1836–1900).

1867 Air brake for railway vehicles is invented by US engineer George Westinghouse (1846–1914).

1867 Cog railway for steep inclines is invented by US engineer Sylvester Marsh (1803–84).

1868 Flying model triplane (three-winged aeroplane) is constructed by John Stringfellow (1799–1883).

1868 Automatic coupling for railway cars is invented by US engineer Eli Janney (1831–1912).

1869 Experimental steam-powered motorcycle is constructed by French engineers, father and son, Pierre and Ernest Michaux.

1869 Automatic electric railway signal ("Banjo" signal) is invented by US engineer Thomas Hall (1827–80).

1869 Suez Canal is opened, designed by French engineer Ferdinand de Lesseps (1805–94) and running for 165 kilometres between the Mediterranean Sea and the Red Sea.

1870 Penny-farthing bicycle is invented in France by Pierre Lallement.

1870 Elevated railway system is patented by US inventor Rufus Gilbert (1832–85).

1871 Rubber-band powered model aeroplane is constructed by Frenchman Alphonse Penaud.

1872 Steam-driven car is built by French engineer Amédée Bollé (1844–1917).

1872 Automatic block signalling system for railways is invented by Irish-born US engineer William Robinson (1840–1921).

1872 Airship powered by hydrogen-burning gas engine (the first with an internal combustion engine) is built by German engineer Paul Haenlein (1835–1905).

1873 Chain drive for bicycles is invented by Englishman J.H. Lawson.

1874 Dynamometer (for measuring power output of an engine) is invented

ENGINEERING AND TECHNOLOGY

1852 Gyroscope is invented by French physicist Jean Foucault (1819–68).

1852 Derringer pocket pistol (originally spelled Deringer) is invented by US gunsmith Henry Deringer (1786–1868).

1853 Experimental gas-burning internal combustion engine is built by Italian engineers Eugenio Barsanti and Felice Matteucci.
1854 Planimeter for measuring irregular areas is invented by Swiss mathematician Jakob Amsler (1823–1912).

1854 Compound marine steam engine is invented by Scottish engineer John Elder (1824–69).

1853 Process for making soda (sodium carbonate) out of cryolite (sodium aluminofluoride) is invented by Danish chemist Julius Thomsen (1826–1909).

1854 Artificial teeth made from kaolin (China clay) are invented by US dentist Mahlon Loomis (1826–86).

1854 Automatic (self-cocking and repeating) revolver is patented by US gunsmiths Horace Smith (1808–93) and Daniel Wesson (1825–1906).

1855 Converter for making steel by blowing air through the melt is invented by British engineer Henry Bessemer (1813–98).

1855 Breech-loading artillery piece with patented steel and wrought iron barrel is invented by British engineer William Armstrong (1810–1900).

1855 Compression refrigerator, using ether as coolant, is invented by Scottish-born Australian James Harrison (1816–93).

1855 Method of making and moulding celluloid (then called xylonite) is patented by British chemist Alexander Parkes (1813–90).

1855 Mercury vacuum pump is invented by German glassblower and physicist Heinrich Geissler (1814–79).

1855 Seismograph, for measuring earthquake intensities, is invented by Italian meteorologist Luigi Palmieri (1807–96).

1855 Electrochemical method of producing relief engraving is invented by Englishman Benjamin ("Benn") Pitman (1822–1910).

1855 Safety matches are invented by Swedish scientist J. Lundstrom.

1855 Tape primer for firearms (a roll of paper tape containing patches of fulminate caps) is invented by US dentist Edward Maynard (1813–1891).

1855 Laryngoscope, for examining the throat, is invented by French singing teacher Manuel Garcia (1805–1906).

1855 Prefabricated hospital (for use in the Crimean War) is invented by British engineer Isambard Brunel (1806–59).

1856 Regenerative smelting oven (used mainly in glass-making) is invented by German-born Freidrich Siemens (1826–1904).

1856 Improved chain-stitch sewing machine is patented by US inventor James Gibbs (1829–1902), together with James Willcox.

1857 Method of making steel by blowing air through the melt is patented by US inventor William Kelly (1811–88), independently of Henry Bessemer (1855).

1857 Method of silvering glass to make mirrors is invented by French physicist Jean Foucault (1819–68).

1858 Absorption refrigerator (using a compressor and liquid ammonia as refrigerant) is invented by Frenchman Ferdinand Carré (1824–1900).

1858 High-voltage gas discharge tube is invented by German glassblower and physicist Heinrich Geissler (1814–79).

1858 Burglar alarm is invented by US engineer Edwin Holmes.

1859 Accumulator (secondary, or storage, battery) is invented by French physicist Gaston Planté (1834–89), later improved by German physicist Karl Correns (1864–1933).

1859 Practical gas-burning internal combustion engine is invented by French engineer Étienne Lenoir (1822–1900).

1859 Battery-powered incandescent electric lamp (with a platinum filament) is invented by US electrical engineer Moses Farmer (1820–93).

1859 Direct steam pump (for pumping large volumes of water) is invented by Henry Worthington (1817–80).

1859 First oil rig to drill successfully for oil is built in Pennsylvania by US engineer Edwin Drake (1819–80).

1859 The Snider breech-loading rifle, invented by US gunsmith Jacob Snider (d.1866), is adopted by the British army.
1859 The Mitrailleuse mechanical machine gun is invented in France by Joseph Mintigny.

1860 Steam pump is invented by US engineer Lucius Knowles (1819–84).

1860 Dynamo with ring winding is invented by Italian physicist Antonio Pacinotti (1841–1912) and, later (1870), Zénobe Gramme (1826–1901).

1860 Source of a standard voltage, the Clark cell, is invented by British engineer Josiah Clark (1822–98).

1860 Improved sphygmograph, for recording a patient's pulse, is invented by French physician Étienne-Jules Marey (1830–1904).

COMMUNICATIONS

1854 Gelatin-covered dry photographic plates are invented by British photographer John Carbutt (1832–1905).

1855 Improved printing telegraph is patented by British-born US inventor David Hughes (1831–1900).

1855 Collotype printing process is invented by Frenchman Alphonse Poitevin (who rediscovered Ponton's gum-dichromate process of 1839).

1858 Punched paper tape telegraph system is invented by British scientist Charles Wheatstone (1802–75).

1858 Transatlantic submarine telegraph cable is laid by British engineer Charles Bright (1832–88) using the USS Niagara and HMS Agamemnon; the cable soon fails.

1858 Aerial photographs are taken from a balloon by French artist and balloonist Nadar (Félix Tournachon) (1820–1910); he later (1859) takes photographs underground using arc lamps.

1858 Pencil with an eraser attached at one end is patented in the USA by Hyman Lipman.

1860 Addressograph machine is invented by US engineer Christopher Sholes (1819–90).

1860 Pony Express, carrying mail from St Joseph to Sacramento, is founded by William Russel (1812–72).

1861 Single-lens reflex (plate) camera is invented by Englishman Thomas Sutton; later (1862) a steroscopic version is marketed.

1863 Propelling pencil is patented by US engineer Johann von Faber (1817–96).

1864 Carbon process (called autotype) for printing photographs is invented by British chemist Joseph Swan (1828–1914).

1865 Web-fed (i.e., using rolls of paper) rotary printing press is invented by US printer William Bullock (1813–67).

1865 Improved camera lenses are patented by German-born US inventor Joseph Zentmayer (1826–88).

1866 Fist successful transatlantic submarine telegraph cable is laid using I.K. Brunel's ship Great Eastern; a second cable was laid in 1874 using the Faraday, a ship designed by German-born British electrical engineer William (Wilhelm) Siemens (1823–83).

1866 Photoengraving process is invented by British photographer Walter Woodbury (1834–85).

1866 Indelible pencil is patented by US inventor Edson Clark.

1867 Practical and commercially successful typewriter, using the modern "QWERTY" keyboard layout, is invented by US engineer Christopher Sholes (1819–90); it was later (1873) sold to the Remington Arms Company.

1868 Searchlight and a code for flashing signals between ships at sea is invented by Scottish naval officer Philip Colomb (1831–99).

1868 Cryptograph, a machine for writing in code, is invented by British scientist Charles Wheatstone (1802–75).

1869 Ticker-tape machine is invented by US engineer Thomas Edison (1847–1931).

1869 Trichrome colour-photography process is invented by French physicist Louis Duclos du Hauron (1837–1920).
1871 Optical telegraph (used during the Siege of Paris) is invented by French physicist Jules Lissajous (1822–80).

ASTRONOMY AND SPACE

1888 New General Catalogue of Nebulae and Clusters of Stars (NGC) is compiled by Danish astronomer Johann Dreyer (1852–1926), who later (1895) also complied its supplement (Index Catalogue of Nebulae).

1889 First photographs of the Milky Way are taken by US astronomer Edward Barnard (1857–1923).

1889 Spectrographic binary stars (whose relative movements are revealed only by their spectra) are discovered by US astronomer Edward Pickering (1846–1919) and, independently (1890), German astronomer Hermann Vogel (1842–1907).

1891 Method of finding asteroids photographically is devised by German astronomer Maximillian Wolf (1862–1932), who in that year discovered Brucia (and who went on to discover more than 500 more).

1892 Amalthea, the fifth moon of Jupiter, is discovered by US astronomer Edward Barnard (1857–1923) and named by French astronomer Camille Flammarion (1842–1925).

1895 Differential speed of rotation of inner and out parts of Saturn's rings (proving that they are not solid but made up of particles) is discovered by US astronomer James Keeler (1857–1900).

1895 Theory that the Sun emits radio waves is proposed by British physicist Oliver Lodge (1851–1940), later (1942) proved to be a fact.

1898 Phoebe, the ninth moon of Saturn, is discovered by US astronomer William Pickering (1858–1938).

1898 Asteroid Eros, which can approach to within 24 million kilometres of Earth, is discovered by Carl Witt.

PHYSICS

1877 Theory of sound (as consisting of vibrations in an elastic medium) is proposed by British physicist Lord Rayleigh (John William Strutt) (1842–1919).

1877 Method of liquefying oxygen by compression and cooling is devised by Swiss physicist Raoul Pictet (1846–1929) and, independently, French physicist Louis Cailletet (1832–1913), who went on to liquefy also air, hydrogen and nitrogen.

1878 Properties of "molecular rays" (cathode rays, i.e. electrons), produced in high-voltage discharge ("Crookes") tubes, are described by British physicist William Crookes (1832–1919).

1879 Stefan's law (that the radiation emitted by a hot object – per unit area per unit time – is proportional to the fourth power of its absolute temperature) is discovered by Austrian physicist Joseph Stefan (1835–93). It was later (1884) proved theoretically by his compatriot Ludwig Boltzmann (1844–1906) and is now usually termed the Stefan-Boltzmann law.

1880 Piezoelectric effect (that pressure across a crystal of quartz produces a voltage) is discovered by French physicist Pierre Curie (1859–1906).

1883 Method of large-scale liquefaction of nitrogen is discovered by Polish physicist Zygmunt Wróblewski (1845–88).

1884 Convergent series of lines in the atomic spectrum of hydrogen (subsequently of great importance in the development of quantum theory) is discovered by Swiss physicist Johann Balmer (1825–98).

1886 Nernst effect (in which a difference in temperature between the ends of a conductor in a magnetic field results in a voltage between the opposite faces of the conductor) is discovered by German physical chemist Walter Nernst (1864–1941).

1886 Canal rays are discovered by German physicist Eugen Goldstein (1850–1931).

1887 Non-existence of the ether (a medium through which light was supposed to travel) is proved experimentally by US physicists Albert Michelson (1852–1931) and Edward Morley (1838–1923).

1887 Photoelectric effect and the existence of electromagnetic (radio) waves – first called Hertzian waves – are confirmed by German physicist Heinrich Hertz (1857–94).

1888 Liquid crystals are discovered by Austrian botanist Friedrich Reinitzer, although not utilized until much later (1964) by Scottish chemist George Gray (1926–), who made stable liquid crystals for electronic displays.

1888 Hallwachs effect (a type of photoelectric effect) is discovered by German physicist Wilhelm Hallwachs (1859–1922).

1890 Guldberg's law (relating critical temperature to boiling point) is discovered by Norwegian physicist Cato Guldberg (1836–1902), also discovered independently by P. Guye (1862–1922).

1890 Fleming's rules (relating the relative directions of the magnetic field, electric current and motion in electric machines) are proposed by British physicist John Fleming (1849–1945).

1890 Magnetic hysteresis is discovered by Scottish physicist Alfred Ewing (1855–1935) and, later (1892), by German-born electrical engineer Charles (Karl) Steinmetz (1865–1923).

1890 Rydberg formula (for calculating

CHEMISTRY

1861 Term "colloid" is coined by Scottish chemist Thomas Graham (1805–69).

1861 Metallic element thallium (symbol Tl) is discovered (spectroscopically) by British physicist William Crookes (1832–1919) and, independently, French chemist C. Lamy (1820–78).

1861 Metallic element rubidium (symbol Ru) is discovered (spectroscopically) by German chemists Robert Bunsen (1811–99) and Gustav Kirchhoff (1824–87).

1862 Chemical elements are plotted in order of atomic weights around a cylinder, creating a primitive periodic table with elements arranged in vertical groups, by French geologist Alexandre Beguyer de Chancourtois (1820–86).

1862 Crystalline haemoglobin is prepared by German biochemist Ernst Hoppe-Segler (1825–95).

1863 Law of mass action (that the rate of a chemical reaction, at a given temperature, is proportional to the product of the active masses of the reactants) is discovered by Norwegian physicist Cato Guldberg (1836–1902) and his brother-in-law Peter Waage (1833–1900) and, independently (1864), by British chemists William Harcourt (1789–1871) and W. Esson (1839–1916).

1863 Metallic element indium (symbol In) is discovered spectroscopically by German physicist Ferdinand Reich (1799–1882) and German chemist Hieronymus Richter (1824–98).

1863 Trinitrotoluene (TNT) is discovered by German chemist T. Wilbrand.

1863 Two optical isomers of lactic acid are discovered by German chemist Johannes Wislicenus (1835–1902).

1863 Law of octaves, an early attempt at a periodic classification of the elements, is formulated by British chemist John Newlands (1837–98).

1864 Existence of tertiary alcohols is predicted by Russian chemist Alexander Butlerov (1828–86).

1864 Microstructure of steel is discovered by British geologist Henry Sorby (1826–1908).

1865 First modern table of relative atomic masses (atomic weights) is drawn up by Belgian chemist Jean Stas (1813–91).

1865 Kekulé structure of benzene (with alternate single and double bonds) is proposed by German chemist Friedrich Kekulé von Stradonitz (1829–69).

1866 Method of extracting aluminium electrolytically is discovered by US chemist Charles Hall (1863–1914) and, independently, French chemist Paul Héroult (1863–1914).

1867 Methanal (formaldehyde) is discovered by German chemist August Hofmann (1818–92), who in the same year discovers methyl violet dyes.

1867 Bismark brown dye is discovered by German naturalist Karl Martinus (1794–1868).

1868 Phase rule is discovered by US chemist Josiah Gibbs (1839–1903).

1868 Composition and method of synthesizing alizarin (used to make dyes) are discovered by German chemists Karl Graebe (1841–1927) and Karl Liebermann (1842–1914), and, independently, British chemist William Perkin (1838–1907), who in the same year synthesizes coumarin.

1868 Gaseous element helium (symbol

BIOLOGY

1875 Knee-jerk reflex is discovered by German physiologists Wilhelm Erb (1840–1921) and, independently, Carl Westphal (1833–90).

1875 Method of classifying proteins is devised by German biochemist Ernst Hoppe-Seyler (1825–95).

1876 Centrosomes (structures within a cell) are discovered by Belgian cytologist Édouard van Beneden (1846–1910); the term "centrosome" was later (1888) coined by German cytologist Theodor Boveri (1862–1915).

1878 Okapi is discovered by Russian-born German explorer Wilhelm Junker (1840–92).

1878 Chromosomes are discovered by German cytologist Walther Flemming (1843–1915), who later (1879) identifies chromatin.

1878 Word "enzyme" is introduced by German physiologist Willy Kuhne (1837–1900).

1880 Fact that starch acts as an energy store in plants is discovered by German botanist Andreas Schimper (1856–1901).

c.1880 Golgi apparatus (body), in the protoplasm of cells, is discovered by Italian biologist Camillo Golgi (1843–1926).

1882 Process of cell division is discovered by German cytologist Walther Flemming (1843–1915), who named it "mitosis".

1883 Phagocytes (cells that devour foreign "invaders") are discovered by Russian-born French biologist Ilya Mechnikov (1845–1916).

1883 Term "eugenics" is coined (for the improvement of human characteristics through selective breeding) by British scientist Francis Galton (1822–1911).

1883 Quagga becomes extinct.

1884 Method of staining, and therefore classifying, bacteria is discovered by Danish bacteriologist Hans Gram (1853–1938).

1884 Fact that the energy content of foods, as made use of by the body, is exactly the same as the energy produced when the same foods are burned (the calorie content) is discovered by german physiologist Max Rubner (1854–1932).

1886 Weismannism (or "germ plasm" theory, that only the contents of sperm and ova are passed unchanged to offspring – but not acquired characteristics) is proposed by German biologist August Weismann (1834–1914).

1887 Fact that in all species the body cells, whatever their type, contain a fixed number of chromosomes is discovered by Belgian cytologist Édouard van Beneden (1846–1910).

1887 Nitrogen fixation (the conversion of nitrogen in the soil into nitrates and nitrites by bacteria in plant roots) is discovered by German agricultural chemist Hermann Hellriegel (1831–95).

1887 Centrosomes (bodies within cells) are discovered by German biologist Theodor Boveri (1862–1915) and, independently, Belgian cytologist Édouard van Beneden (1846–1910).

1888 Fact that there are different kinds of yeast (some of which are better than others for fermentation processes such as brewing and winemaking) is discovered by Danish botanist Emil Hansen (1842–1909).

1888 Word "chromosome" (for chromatin threads formed during cell division) is coined by German anatomist Heinrich von Waldeyer (1836–1921).

MEDICINE

physician Carlos Finlay (1833–1915), although his discovery attracted little attention at the time.

1881 Pneumococcus bacterium that causes pneumonia is discovered by US physician George Sternberg (1838–1915).

1882 Bacterium that causes tuberculosis is discovered by German bacteriologist Robert Koch (1843–1910).

1882 Cholecystectomy, an operation to remove the gall bladder, is introduced by German surgeon Carl Langenbuch (1846–1901).

c.1883 Down's syndrome (trisomy 21) is discovered by British physician John Haydon-Down (1828–96).

1883 Association between myxoedema and thyroid function is discovered by Swiss surgeon Emil Kocher (1841–1917).

1883 Analgesic and fever-reducing drug antipyrene is discovered by German chemist Ludwig Knorr (1859–1921).

1883 Ringer's solution (used for keeping tissues alive outside the body) is devised by British physician Sydney Ringer (1835–1910).

1883 Bacterium that causes diphtheria is discovered by German bacteriologist Edwin Klebs (1834–1913), later (1884) isolated by his compatriot Friedrich Löffler (1852–1915).

1883 Bacterium that causes cholera is discovered by German bacteriologist Robert Koch (1843–1910) who, in the same year, developed a vaccine against anthrax.

1883 Fact that the trypanasome that transmits filariasis is carried by a mosquito is discovered by Scottish physician Patrick Manson (1844–1922).

1884 Cocaine (discovered in coca leaves by Albert Niemann in 1860) is first used as a local anaesthetic (in eye surgery) by Austrian-born US physician Carl Koller (1857–1944).

1884 Operation for the removal of a brain tumour is first carried out by British surgeon Rickman Godlee (1849–1925).

1885 Vaccine against rabies is developed by French bacteriologist Louis Pasteur (1822–1895).

1885 Acromegaly (overgrowth of bones after adulthood) is discovered by French neurologist Pierre Marie (1853–1940); its cause (a defect in the pituitary gland) is later (1909) discovered by US physician Harvey Cushing (1869–1939).

1885 Method of treating goitre by removing the thyroid gland is introduced by Swiss surgeon Emil Kocher (1841–1917).

1886 Appendicitis (and appendicetomy, the surgical treatment of it) is described by US physician Reginald Fitz (1843–1913). Appendicetomy is later (1880) perfected by British surgeon Robert Tait (1845–99) and (1886) German surgeon Ulrich Krönlein (1847–1910).

1886 Use of steam for sterilizing surgical instruments is introduced by Latvian-born German brain surgeon Ernst von Bergmann (1836–1907).

1886 Muscular dystrophy is described by French neurologist Jules Dejerine (1849–1917).

1886 Bacterium that causes glanders in horses and the bacterium that causes rinderpest in cattle are discovered by German bacteriologist Friedrich Löffler (1852–1915).

FARMING AND FOOD

1885 Bordeaux mixture is introduced as a fungicide for use on grape vines by French horticulturist P.M. Millarder.

1885 Canned treacle (sold as "Golden Syrup") is marketed by Scottish sugar refiner Abram Lyle III.

1886 Coca-Cola is invented by US physician John Pemberton, later (1894) bottled commercially by Joseph Biedenham.

1887 Malted milk drink is marketed by British-born US industrialist William Horlick (1846–1936).

1889 Simple milking machine is invented by Scottish farmer William Murchland.

1889 Machines for knotting cord round bales in a binder is invented by US engineer La Verne Noyes (1849–1919).

1890 Agricultural tractor is patented by US inventor G. Edwards.

1890 Method of measuring the butterfat content of milk is invented by US agricultural chemist Stephen Babcock (1843–1931).

1890s Stationary steam engines are used for ploughing in Europe and North America.

1891 Crown cork bottle top is invented in the USA by William Painter.

1891 Process for manufacturing chewing gum is set up by US industrialist William Wrigley (1861–1932).

1892 First commercially produced petrol-driven agricultural tractor is built in the USA by John Froelich.

1893 Shredded wheat breakfast cereal is invented by US lawyer Henry Perky.

1895 Wheat flakes are invented by US physician John Kellog (1852–1943); corn flakes are invented in 1898 by his brother William (1860–1951).

1895 Health food called Postum is marketed in the USA by Charles Post (1854–1914); he later (1897) introduces a breakfast cereal called Grape Nuts.

1895 Practical "pulsating" milking machine is invented by Scottish physician Alexander Shields.

1898 Pepsi-Cola (for treating dyspepsia) is invented by US pharmacist Caleb Bradman.

1899 Oxo cubes (beef stock cubes) are manufactured in Fray Bentos, Uruguay.

TRANSPORT

by US engineer Plimmon Dudley (1843–1924).

1874 Bicycle wheel with spokes angled to each other (the tangent-spoke wheel) is invented by British engineer James Starley (1830–81).

1874 Three-wheeled battery-powered electric car is built by British electrical engineer David Salomons.

1875 Collapsible boat made of canvas with wooden stretchers (for naval use) is invented by British clergyman Edward Berthon (1813–99).

1876 Safety bicycle is invented by British engineer James Starley (1830–81).

1876 Plimsoll line (loading line for sea-going ships) is introduced by British coal merchant (and social reformer) Samuel Plimsoll (1824–98).

1876 Public service steam tram runs in Wantage, England.

1878 Electric railway is demonstrated and patented by German engineer Werner von Siemens (1816–92); he later (1881) builds an electric tramway in the streets of a Berlin suburb.

1879 Public electric railway (Volk's Electric Railway) is demonstrated in England.

1880 Electric tramway is demonstrated in New Jersey by US inventor Thomas Edison (1847–1931).

1882 Electric tricycle is built by British engineer William Ayrton (1847–1908).

1882 Electric trolley bus is built by German engineer Werner von Siemens (1816–92).

1883 Electric trams enter service in London.

1883 First motorboat (with a high-speed petrol engine) is built by German engineer (and ex-gunsmith) Gottlieb Daimler (1834–1900).

1883 First monorail railway (steam-powered) is built in Ireland by Charles Lartique; gyroscopic stabilizer for a monorail system is invented by Irishman Louis Brennan (1852–1932).

1883 Electric railway system is patented by Belgian-born US inventor Charles Van Depoele (1846–92).

1883 Electric tramway, designed by German-born British engineer Wilhelm – later William – Siemens (1823–88), opens at Portrush, Ireland.

1883 Spark plug (for gas engines) is invented by French engineer Etienne Lenoir (1822–1900).

1883 Steerable balloon is invented by French balloonist Gaston Tissandier (1843–99), who also builds an electric-powered model airship.

1884 Experimental motor tricycle is built by British engineer Edward Butler.

1884 Practical electric-powered steerable airship, La France, is built by French engineers Charles Renard (1847–1905) and Arthur Krebs (1847–1935).

1885 First successful petrol-engined motorcycles are built by German engineers Karl Benz (1844–1929) and Gottlieb Daimler (1834–1900); Benz also builds a three-wheeled motor car.

1886 Steam-powered automobile is built by US engineer Ransom Olds (1864–1950).

1886 Four-wheeled petrol-engined motor car is built by German engineer Gottlieb Daimler (1834–1900).

ENGINEERING AND TECHNOLOGY

1860 Method of making linoleum (using oxidized linseed oil) is invented by Englishman Frederick Walton.

1860 Rolled-steel armour plating for warships is invented by British engineer John Brown (1816–1896).

1860 Method of making seamless steel tubes is invented by German ironmaster Reinhard Mannesmann (1856–1922).

1861 Method of making gun barrels by shrinking wrought-iron hoops onto the breech of a cast-iron barrel is invented by US ironfounder (and soldier) Robert Parrott (1804–77).

1861 "Kettledrum" microscope condenser is invented by British chemist Joseph Reade (1801–70).

1862 Four-stroke cycle (for internal combustion engines) is patented by French engineer Alphonse Beau de Rochas (1815–91), but the principle is not applied to an actual engine.

1862 Ball-bearings are patented by French engineer Pierre Michaux, probably for use on his experimental bicycles and motorcycles.

1862 Rapid-fire mechanical machine gun is patented, independently, by US inventors Richard Gatling (1818–1903) and Wilson Agar.

1862 Industrial process for making soda (sodium carbonate) from common salt (sodium chloride) is invented by Belgian chemist Ernest Solvay (1838–1922).

1862 Universal milling machine for cutting metal is invented by British engineer Joseph Brown.

1862 Revolving gun turret for warships, as fitted to the Union's ironclad Monitor, is patented by US engineer Theodore Timby (1822–1909).

1863 Explosive rifle bullet is invented by British gunsmith William Metford (1824–99), but it is banned by the 1869 St Petersburg Convention.

1863 TNT (trinitrotoluene) explosive is invented by Swedish chemist J. Wilbrand.

1863 Improved steam pump is invented by US engineer Lucius Knowles (1819–84).

1864 Automatic cartridge feeder (magazine) for rifles is patented by US inventor John Appleby (1840–1917).

1865 Improved vacuum pump is invented by German-born British chemist Hermann Sprengel (1834–1906).

1865 Electrostatic generator is invented by German physicist Wilhelm Holtz (1836–1913).

1865 Cylinder lock is invented by US locksmith Linus Yale (1821–68).

1865 Rolling-block breech-loading rifle, with a tubular magazine, is invented by US gunsmith Philo Remington (1816–89).

1866 Accurate short clinical (medical) thermometer is invented by British physician Clifford Allbutt (1836–1925).

1866 Practical zinc-carbon primary cell (battery) is invented by French chemist Georges Leclanché (1839–82); it becomes the forerunner of dry batteries.

1866 Apparatus for measuring the speed of sound in different gases (Kundt's tube) is invented by German physicist August Kundt (1839–94).

1866 Automatic lathe is invented by British engineer C. Spencer.

1866 Rifle invented in 1833 by French gunsmith Antoine Chassepot (1833–1905) is adopted by the French army.

1866 Self-propelled torpedo (with a motor driven by compressed air) is invented by British engineer Robert Whitehead (1823–1905).

1867 Dynamite – nitroglycerin (discovered by Ascanio Sobrero in 1847) absorbed in kieselguhr and wood wool – is invented by Swedish chemist Alfred Nobel (1833–96) and, independently, US inventor Paul Oliver (1830–1912).

1867 Sulphite process for making chemical wood pulp (for paper-making) is invented by American Benjamin Tilghman (1821–1901).

1867 Water-tube boiler (with inclined tubes) for steam engines is manufactured by US engineers George Babcock (1832–93) and Stephen Wilcox (1830–93); Babcock's original patent was in 1856.

1867 Barbed wire is patented by US inventors Lucien Smith and, independently, Alphonso Dabb.

1868 Commercial method of making celluloid (the first mass-produced plastic, based on cellulose) is invented by US printer John Hyatt (1837–1920) – for making billiard balls.

1868 System of standard screw threads, invented by US engineer William Sellers (1824–1905), is adopted by the US Government.

1868 Explosive harpoon (for whaling) is invented by Norwegian Svend Foyn (1809–94).

1868 Automatic stapler, for binding magazines, is invented by Englishman Charles Gould.

1868 Magazine rifle invented by Swiss gunsmith Friedrich Vetterli (1822–82) is adopted by the Swiss army and, later (1870), by the Italian army.

1869 Metal-planing machine is invented by US engineer Francis Pratt (1827–1902), four years after he and Amos Whitney founded the Pratt & Whitney company.

COMMUNICATIONS

1871 Dry-plate photography is invented by British chemist Joseph Swan (1828–1914).

1872 Electric typewriter is patented by US inventor Thomas Edison (1847–1931).

1873 Orthochromatic photographic plates, sensitive to all colours, are invented by German chemist Hermann Vogel (1834–98).

1874 Dewey decimal system for cataloguing library books is invented by US librarian Melvil Dewey (1851–1931).

1874 Multiple telegraph is invented by Scottish-born US scientist Alexander Graham Bell (1847–1922).

1875 Photochemical engraving process (for making printing plates) is invented by German-born US David Bachrach (1845–1921) and Bohemian-born US Louis Levy (1846–1919).

1876 Telephone is invented by Scottish-born US scientist Alexander Graham Bell (1847–1922).

1877 Phonograph is invented by US engineer Thomas Edison (1847–1931).

1877 Very lights (signal flares fired from a handgun) are invented by US naval officer Edward Very (1847–1927).

1878 First telephone exchange opens in New Haven, Connecticut.

1878 Microphone is invented, independently, by British-born US engineer David Hughes (1831–1900) and German-born Emile Berliner (1851–1929).

1878 Bromide paper for making photographic prints is invented by British chemist Joseph Swan (1828–1914), who uses hypo (sodium thiosulfate, formerly hyposulfate) as a fixative, first proposed (1842) by British astronomer John Herschel (1792–1871) (who also coined the terms "photography", "negative" and "positive").

1878 Halftone engraving process for printing is invented by US photographer Frederick Ives (1856–1937); he later (1881) modifies it for three-colour working.

1878 Series of cameras is used to take consecutive photographs analyzing the movement of animals by US photographer Edweard Muybridge (1830–1904).

1878 Comparatively fast dry photographic plates are invented by Charles Bennett.

1878 Crookes tube, a type of cathode-ray tube, is invented by British physicist William Crookes (1832–1919).

1880 New kind of typewriter is patented by US inventor James Hammond (1839–1913).

1880 Half-tone engraving process is invented by American Stephen Horgan (1854–1941).

1880 Photophone, which transmits messages by means of a light beam, is patented by Scottish-born US inventor Alexander Graham Bell (1847–1922).

1881 Transmission of stereophonic sound by telephone is demonstrated by French inventor Clément Ader (1841–1925).

1881 Type of dry photographic plate is perfected by British-born US Miles Seed (1843–1913); ten years later he sells the rights to Kodak.

1882 Motion picture camera using a rotating glass plate is invented by French physician Etienne-Jules Marey (1830–1904).

ASTRONOMY AND SPACE

PHYSICS

the frequency of spectral lines, which includes a term later called the Rydberg constant) is discovered by Swedish physicist Johannes Rydberg (1854–1919).

1891 Gravitational constant is determined by British physicist John Poynting (1852–1914).

1892 Properties of cathode rays (then called Lenard rays) outside a discharge tube are investigated by German physicist Philipp von Lenard (1862–1947).

1893 Wien's law (that the wavelength of maximum energy radiated by a black body is inversely proportional to its absolute temperature) is discovered by German physicist Wilhelm Wien (1864–1928).

1894 Physical properties of radio ("Hertzian") waves are discovered by British physicist Oliver Lodge (1851–1940).

1895 X-rays (Röntgen rays) are discovered by German physicist William Röntgen (1845–1923).

1895 Explanation for the negative result of the 1887 Michelson-Morley experiment (to determine the existence of the "ether", a medium through which light was supposed to travel) is given by Irish physicist George Fitzgerald (1851–1901), who proposed that at speeds approaching the speed of light objects shorten in length. Because of a similar conclusion by Dutch physicist Hendrik Lorentz (1853–1928), the phenomenon became known as the Fitzgerald-Lorentz contraction.

1895 Curie point or temperature (at which ferromagnetic substances become paramagnetic) is discovered by French physicist Pierre Curie (1859–1906).

1895 Fact that cathode rays are negatively charged is discovered by French physicist Jean Perrin (1870–1942).

1895 Curie's law (that the magnetic susceptibility of a ferromagnetic substance is proportional to its absolute temperature) is discovered by French physicist Pierre Curie (1859–1906).

1895 Method of liquefying air in quantity is discovered by German engineer Carl von Linde (1842–1934).

1896 Alpha and beta rays are discovered by New Zealand-born British physicist Ernest Rutherford (1871–1937).

1896 Radioactivity (of uranium) is discovered by French physicist Antoine Becquerel (1852–1908).

1896 Zeeman effect (the splitting of spectral lines in a strong magnetic field) is discovered by Dutch physicist Pieter Zeeman (1865–1943); the theoretical explanation of the phenomenon is later given by Dutch physicist Hendrik Lorentz (1853–1928).

1896 Fact that X-rays are a type of electromagnetic radiation is discovered by Irish physicist George Stokes (1819–1903).

1897 Electron is discovered by British physicist J.J. (Joseph John) Thompson (1856–1940), who called it a "corpuscle".

1897 Charge on a single gas ion is determined by Irish physicist John Townsend (1868–1957).

1897 Hall effect (the generation of a voltage across a current-carrying conductor at right angles to a magnetic field) is discovered by US physicist Edwin Hall (1855–1938).

1897 Electric field emission is discovered by US physicist Robert Wood (1868–1955).

CHEMISTRY

He) is discovered in the spectrum of the Sun by French physicist Jules Janssen (1824–1907) and, independently, British astronomer Norman Lockyer (1836–1920) and British chemist Edward Frankland (1825–99). It is later (1895) discovered in mineral deposits on Earth by Swedish chenist Per Cleve (1840–1905) and in air by Scottish chemist William Ramsay (1825–1916).

1868 Periodic law (relating to properties of the elements and leading to the Periodic Table, based on relative atomic masses (atomic weights)), is proposed by Russian chemist Dmitri Mendeléev (1834–1907). A similar table was later (1870) independently drawn up by German chemist Julius Lothar Meyer (1830–95).

1870 Markovnikoff's rule (which predicts the principal product in a mixture produced by reacting an acid with an unsymmetrical alkene) is discovered by Russian chemist Vladimir Markovnikov (1837–1904).

c.1870 Fittig synthesis (for making aromatic hydrocarbons from their halogen compounds, using sodium) is discovered by German chemist Rudolph Fittig (1835–1910).

1871 Phenolphthalein (used as a dye and in medicine) is discovered by German chemist Adolf von Baeyer (1835–1917).

1872 Nitromethane is discovered by German chemist Hermann Kölbe (1818–84).

1874 Stereochemistry is established by Dutch chemist Jacobus van't Hoff (1852–1911), who discovers the tetrahedral arrangement of carbon's valences.

1874 Fact that optical activity in organic compounds is due to the presence of an asymmetrical carbon atom is discovered by French chemist Joseph Le Bel (1847–1930).

1874 Nitrosyl chloride is discovered by British chemist William Tilden (1842–1926).

1874 DDT is discovered by German chemist Othmar Zeidler, who does not recognize its insecticidal properties.

1875 Metallic element gallium (symbol Ga) is discovered by French chemist Paul Lecoq Boisbaudran (c.1838–c.1912).

1875 Eosin scarlet dye is discovered by German chemist Adolf von Baeyer (1835–1917).

1875 Phenylhydrazine (an important chemical in organic analysis) is discovered by German chemist Emil Fischer (1852–1919)

1876 Idea of chemical potential (extending thermodynamics into chemistry) is proposed by US chemist Josiah Gibbs (1839–1903).

1876 Tiemann-Reimer reaction (for synthesizing hydroxy-aldehydes) is discovered by German chemists Johann Tiemann (1848–99) and, independently, C. Reimer (1856–1921).

1877 Friedel-Crafts reaction (in which an alkyl or acyl group is substituted into a benzene ring) is discovered by French chemist Charles Friedel (1832–99) and US chemist James Crafts (1839–1917).

1878 Metallic element ytterbium (symbol Yb) is discovered by Swiss chemist Jean de Marignac (1817–94), although his "element" is later (1907) found by French chemist Georges Urbain (1872–1938) to be a mixture of two elements, which he called neoytterbium and lutetium; in the event, the name ytterbium was retained for neoytterbium.

BIOLOGY

1888 Fixation of atmospheric nitrogen by nodules of the roots of leguminous plants is discovered by German chemist Hermann Hellriegel (1831–95).

1891 Mechanism of the knee-jerk reflex is discovered by British physiologist Charles Sherrington (1857–1952), also described by British chemist Stephen Hales (1677–1761) and Scottish neurologist Robert Whytt (1714–1766).

1891 Fossil bones of Pithecanthropus (later Homo) erectus ("Java man", a link in human evolution) are discovered by Dutch palaeontologist Marie Eugène Dubois (1858–1940).

1893 Dollo's law (of irreversibility in evolution) is formulated by French palaeontologist Louis Dollo (1857–1931).

1894 Hormone adrenaline (epinephrine) is discovered by British endocrinologist Edward Sharpey-Schafer (1850–1935) and physiologist George Oliver (1841–1915), later (1897) isolated by US biochemist John Abel (1857–1938) and crystallized in pure form (1901) by Japanese-born US chemist Jokichi Takamine (1854–1922).

1895 Fact that gas exchange in plants occurs through "pores" (stomata) in the leaves is discovered by British plant physiologist Frederick Blackman (1866–1947).

1896 Medelian ratio of 3:1 in the first generation in plant-breeding experiments (first discovered by Gregor Mendel in 1866) is rediscovered by Dutch physiologist Hugo de Vries (1848–1935) and, independently, Karl Correns (1864–1933) and E. von Tschermak (1871–1962).

1897 Plant viruses are discovered by Dutch botanist Martinus Beijerinck (1851–1931).

1898 Golgi body (apparatus) in cells is discovered by Italian histologist Camillo Golgi (1844–1926).

1899 Natural pacemaker of the heart (the sino-atrial node) is discovered by Scottish anatomist and anthropologist Arthur Keith (1866–1955) and Englishman Martin Flack (1882–1931).

1899 "Lock-and-key" mechanism of enzyme action is proposed by German chemist Emil Fischer (1852–1919).

MEDICINE

1886 Cause of Weil's disease (infectious jaundice, later called leptospirosis) in sewer workers is shown to be due to contact with rat's urine by German physician Adolf Weil (1848–1916).

1887 Bacterium that causes undulant fever (Malta fever or brucellosis) is discovered by Scottish bacteriologist David Bruce (1855–1931).

1888 Bacterium that causes salmonella (food poisoning) is discovered by German bacteriologist August Gärtner (1848–1934), later named by US veterinary surgeon Daniel Salmon (1850–1914).

1895 Tse-tse fly is identified as the carrier of sleeping sickness by Australian-born British bacteriologist David Bruce (1855–1931).

1899 Vaccine against foot-and-mouth disease is developed by German bacteriologist Friedrich Löffler (1852–1915).

1890 Bacterium that causes phthisis (pulmonary tuberculosis) is discovered by German bacteriologist Robert Koch (1843–1910).

1890 Antitoxins (for diphtheria and tetanus) are discovered by German bacteriologist Emil von Behring (1854–1914).

1890 Antituberculosis serum is first used on humans by French physiologist Charles Richet (1850–1935).

1891 Injection of thyroid extract as a treatment for myxoedema is discovered by German physiologist Moritz Schiff (1823–96).

1891 Technique of lumbar puncture is introduced by German physician Heinrich Quincke (1834–1924).

1892 Guarnieri bodies, cell inclusions that are diagnostic of smallpox and cowpox (vaccinia), are discovered by Italian pathologist Giuseppi Guarnieri (1856–1918).

1892 Vaccine against typhoid fever is discovered by British bacteriologist Almoth Wright (1861–1947).

1892 Clostridium bacterium which causes gas gangrene is discovered by US pathologists William Welch (1850–1934) and G. Nuttall (1862–1937).

1893 Fact that Texas cattle fever is caused by a protozoan parasite and spread by cattle ticks is discovered by US microbiologist Theobald Smith (1859–1934).

1893 Inoculation against cholera is developed by Russian bacteriologist Waldemar Haffkine (1860–1930).

1894 Bacterium that causes bubonic plague is discovered by Swiss-born French bacteriologist Alexandre Yersin (1863–1943) and, independently, Japanese bacteriologist Shibasaburo Kitasato (1856–1931).

1894 Bacteriolysis – the destruction of bacteria by vaccine-induced antibodies – is discovered by Polish bacteriologist Richard Pfeiffer (1858–1945).

1894 Anaphylaxis – a fatal allergic reaction – is discovered by French physiologist Charles Richet (1850–1935).

1894 Reaction for detecting cholera is discovered by German bacteriologist Richard Pfeiffer (1879–1945).

1895 Parasite that causes malaria is discovered by British physician Ronald Ross (1857–1932).

1896 Diagnostic X-ray photographs are first taken by Hungarian-born US physicist Michael Pupin (1858–1935).

FARMING AND FOOD

TRANSPORT

1886 Motorcycle with a two-cylinder water-cooled petrol engine is built by British engineer Edward Butler.

1887 Pneumatic tyre, for bicycles, is invented by Scotsman John Dunlop (1840–1921).

1887 Modern electric tram system (for Richmond, Virginia) is invented by US engineer Frank Sprague (1857–1934).

1887 Electric railway locomotive is built by German-born US inventor Rudolf Eickenmeyer (1831–1895) and US engineer Stephen Field (1846–1913).

1887 Electric lift is constructed by German Siemens & Halske company.

1888 First successful French submarine, the Gymnote, invented by Gustave Zédé (1825–91), is adopted by the French navy.

1888 Petrol-engined trams, built by the Daimler company, run in Germany.

1889 Articulated steam locomotive is invented by Swiss engineer Anatole Mallet (1837–1919).

1890 First electric underground railway opens in London.

1890 Short (50-metre) but uncontrolled steam-powered aeroplane flight is made by French engineer Clément Ader (1841–1925), who crashes on landing.

1891 Tubular steel car chassis with an internal combustion engine mounted on it is built by French engineers René Panhard (1841–1908) and Emile Levassor (1843–97).

1891 Man-carrying gliders are flown by German inventor Otto Lilienthal (1848–96).

1891 Boat built of aluminium is constructed by the Swiss Escher Wyss company.

1892 Escalator is patented by US engineer Jesse Reno (1861–1947).

1892 Motor car with pneumatic tyres is built by French engineer Emile Levassor (1843–97).

1893 Four-wheeled motor car is built by German engineer Karl Benz (1844–1929).

1893 Improved float-feed carburetor for petrol engines is invented by German engineer Wilhelm Maybach (1847–1929).

1893 Toughened glass, later used in cars, is invented by Frenchman Leon Appert.

1893 Pneumatic tyres for cars are first manufactured in quantity by French engineers André (1853–1931) and Edouard (1859–1940) Michelin.

1893 Box kite, capable of lifting heavy loads, is invented in Australia by British aviation pioneer Lawrence Hargrave (1850–1915); later (1894) four of his box kites lifted him from the ground.

1893 Corinth Canal, 5.6 kilometres long across Greece, is opened.

1894 Man-carrying kite is built by British soldier Robert Baden-Powell (1857–1941); it was later (1901) improved for the British military by US-born British pioneer aviator Samuel Cody (1862–1913).

1895 Part-controlled gliding flight is achieved by German aviation pioneer Otto Lilienthal (1848–96) and, independently, Scottish inventor Percy Pilcher (1866–99) in his man-carrying glider called the Bat; both gliders were launched downhill. Lilienthal is later (1896) killed in a gliding accident.

ENGINEERING AND TECHNOLOGY

1869 Patent for making celluloid is granted in the USA to US printer John Hyatt (1837–1920).

1869 Hydroelectric generator is invented by French papermaker Aristide Bergès (1833–1904).

1869 Improved tunelling shield is invented by South African-born British engineer James Greathead (1844–96), who later (1884) made further improvements to the shield by incorporating hydraulic jacks.

1869 Weather balloon is invented by US meteorologist Cleveland Abbe (1838–1916).

1869 Towed torpedo is invented by Captain Harvey of the Royal Navy.

1871 Rock drill is invented by US engineer Simon Ingersoll (1818–94).

1871 Sand-blasting machine is invented by US engineer Benjamin Tilghman (1821–1901).

1871 Engraving machine for ruling extremely fine lines on diffraction gratings (for spectroscopes) is invented by US astronomer Lewis Rutherfurd (1816–92).

1871 Steam-powered vacuum cleaner is invented in the USA by Ives McGaffey.

1871 Martini-Henry rifle, with a falling-breech action invented by Hungarian engineer Frêdéric Martini (1832–97) and barrel invented by US gunsmith Benjamin Henry, is adopted by the British army.

1871 Breech-loading Mauser rifle, invented by German gunsmiths Wilhelm (1834–82) and Peter (1838–1914) Mauser, is adopted by the Prussian army.

1872 Industrial dynamo is invented by Belgian engineer Zénobe Gramme (1826–1901).

1872 Revolving-barrel machine gun is invented by American Benjamin Hotchkiss (1826–85).

1873 Electric motor that can also act as a dynamo is invented by French engineer Hippolyte Fontaine (1833–1917), who uses it for the first electricity supply system in France.

1873 "Giant powder", an improved blasting explosive, is invented by US chemist Egbert Judson (1812–93).

1873 Machine for mass-producing barbed wire is invented by US engineer Joseph Glidden (1813–1906).

1873 Apparatus for commercial production of water gas (carbon monoxide and hydrogen) is invented by US balloonist and inventor Thaddeus Lowe (1832–1913).

1873 Water closet for railway carriages is invented by US engineer Lewis Latimer (1848–1928).

1874 Gras fusil modèle rifle, invented by French soldier Basile Gras (1836–1901), is adopted by the French army.

1875 Magazine rifle, invented by US gunsmith Benjamin Hotchkiss (1826–85), is adopted by armies in Britain, France and the USA.

1875 Contact process for making sulphuric acid (originally patented by Perigrine Phillips in 1831) is re-invented by German chemist Rudolph Messel (1848–1920) and put into production in England.

1875 Electric searchlight, invented by British engineer Henry Wilde (1833–1919), is adopted by the British navy.

1875 Electric dental drill is invented by US dentist Georg Green.

1875 Process for removing phosphorus impurity from iron in steelmaking is invented by British metallurgists Sidney Thomas (1850–85) and Percy Gilchrist (1851–1935).

1875 Process for making ultrapure nickel is developed by US metallurgist Joseph Wharton (1826–1909).

1876 Four-stroke internal combustion engine, running on gas, is invented and patented by German engineer Nikolaus Otto (1832–91), employing ideas pioneered (1867) by his compatriot Eugen Langen (1833–95) and previously patented by French engineer Alphonse Beau de Rochas (1815–93).

1876 Surface-type carburetor is invented by German engineer Gotlieb Daimler (1834–1900).

1876 Compound steam engine for railway locomotives is invented by Swiss engineer Anatole Mallet (1837–1919).

1876 Low-current arc lamp is invented by Russian-born engineer Pavel Jablochkoff (1847–94).

1876 Mechanical carpet sweeper is invented by US shop-keeper Melville Bissell (1843–89), whose patent supersedes an earlier one (1811) granted to James Hume.

1877 Thermocouple for measuring high temperatures is invented by French chemist Henri Le Châtelier (1850–1936).

1877 Arc welder is invented by British-born US electrical engineer Elihu Thomson (1853–1937).

1877 Improved two-stroke internal combustion (gas) engine, working on the so-called Clerk cycle, is invented by Scottish engineer Dugald Clerk (1854–1932).

1878 Long-life arc lamp for industrial and street lighting is invented by US scientist Charles Brush (1849–1929).

1878 Spinthariscope, which produces scintillations (tiny flashes of light) to reveal the presence of alpha particles, is invented by British physicist William Crookes (1832–1919).

COMMUNICATIONS

1883 "Zoopraxiscope", which presents moving pictures of animals in motion, is invented by British-born American photographer Eadweard Muybridge (1830–1904).

1884 Linotype machine is invented by German-born US printer Ottmar Mergenthaler (1854–99).

1884 Rotating-disc optical scanning system is invented by German engineer Paul Nipkow (1860–1940).

1884 Fountain pen is patented by US inventor Lewis Waterman (1837–1901).

1884 High-speed stock ticker machine is invented by Stephen Field.

1885 Dictaphone is invented by US engineer Charles Tainter (1854–1940).

1885 Comptometer, a key-operated adding machine, is marketed by US inventor Dorr Felt (1862–1930); he later (1889) adds a printer to the machine.

1887 Gramophone is invented by German-born US inventor Emil Berliner (1851–1929).

1887 Synthetic "universal" language (Esperanto) is invented by Polish philologist Ludwig Zamenhof (1859–1917).

1887 Mechanical typesetting machine is invented by US printer Tolbert Lanston (1844–1913); in 1897 it becomes known as Monotype.

1887 Duplicating machine in which writing is inscribed with a stylus on waxed paper (based on Thomas Edison's patent of 1875) is marketed (as the Mimeograph) by American Albert Dick.

1888 Duplicating machine using wax typewriter stencils (called the Cyclostyle) is invented in England by Hungarian-born Englishman David Gestetner.

1888 Key-set mechanical adding machines are invented by US William Burroughs (1857–98).

1888 Movie camera using a ribbon of paper "film" is invented by French physician Etienne-Jules Marey (1830–1904).

1888 Discs, instead of cylinders, for phonograph records are invented by US inventor Emile Berliner (1851–1929).

1888 System of shorthand is invented by Irish-born John Gregg (1867–1948).

1889 Integral printer is fitted to Dorr Felt's Comptometer of 1885.

1889 Movie camera, using paper "film", is invented in England by William Friese-Greene (1855–1921); he later (1890) uses celluloid film.

1889 35-mm film is invented by US engineer Thomas Edison (1847–1931).

1889 Jukebox, a coin-operated cylinder gramophone, is invented in the USA by Louis Glass.

1890 Electrically-driven punched card system, for dealing with statistics in the US census, is invented by US statistician Herman Hollerith (1860–1929), who later (1896) founds the Tabulating Machine Company, which in 1924 becomes International Business Machines (IBM).

1890 Telephoto camera lens is invented by New Zealand geologist Alexander McKay and, independently in 1891, British optician Thomas Dallmayer.

1890 Submarine telephone cable is laid across the English Channel (Dover to Sangatte).

ASTRONOMY AND SPACE

PHYSICS

1897 Larmor progression (of charged particles in a magnetic field) is discovered by Joseph Lamor (1857–1942).

1898 Method of producing liquid hydrogen in quantity is discovered by Scottish physicist James Dewar (1842–1923).

1898 Sabine's law of acoustics (that, for a given room, the product of the reverberation time and the absorptivity equals the volume) is discovered by US physicist Wallace Sabine (1868–1919).

1899 Identity between a beam of electrons and cathode rays is established by Dutch physicist Hendrik Lorentz (1853–1928), who also coined the word electron.

CHEMISTRY

1879 Saccharin is discovered by US chemists Ira Remsen (1846–1927) and Constantin Fahlberg (1850–1910).

1879 Nucleic acids are discovered by German biochemist Albrecht Kossel (1853–1927), later (1929) to be redicovered by Russian-born US chemist Phoebus Levene (1869–1940).

1879 Four-carbon ring compounds are discovered by Russian chemist chemist Vladimir Markovnikov (1837–1904).

1879 Metallic elements holmium (symbol Ho), detected spectroscopically a year earlier by Swiss chemist J.L. Soret, and thulium (symbol Tm) are isolated by Swedish chemist Per Cleve (1840–1905).

1879 Metallic element samarium (symbol Sm) is discovered by French chemist Paul Lecoq Boisbaudran (c.1838–1912).

1879 Metallic element scandium (symbol Sc) is discovered (as its oxide) by Swedish physicist Lars Nilson (1840–99).

1880 Wallach rearrangement (of organic azo compounds) is discovered by German chemist Otto Wallach (1847–1931).

1880 Étard reaction – the oxidation of methylbenzene (toluene) to benzenecarbaldehyde (benzaldehyde) by chromium oxychloride (chromyl chloride) – is discovered by French chemist Alexandre Étard (1852–1910).

1880 Method of synthesizing indigo is discovered by German chemist Adolf von Baeyer (1835–1917), who later (1883) works outs its structure.

1880 Metallic element gadolinium (symbol Gd) is discovered by Swiss chemist Jean de Marignac (1817–94), later (1886) rediscovered by Paul Lecoq Boisbaudran (1838–1912).

1882 Raoult's law (that, at a given temperature, the relative lowering of vapour pressure of a solution is proportional to the concentration of solute) is discovered by French chemist François Racult (1830–1901).

1882 Oximes are discovered by German chemist Viktor Meyer (1848–97).

1883 Thiophene is discovered by German chemist Viktor Meyer (1848–97).

1883 Method of analyzing the amount of nitrogen in organic compounds is discovered by Danish chemist Johan Kjeldahl (1849–1900).

c.1883 Ostwald's dilution law (that the degree of dissociation of a weak electrolyte is proportional to the square root of the dilution) is discovered by Latvian-born German chemist Friedrich Ostwald (1853–1932).

1884 Amino acid cysteine is discovered by German chemist Eugen Baumann (1846–96).

1884 Sandmeyer reaction (for making aromatic halides using diazonium compounds) is discovered by German chemist Traugott Sandmeyer (1854–1922).

1884 Dissociation of ionic compounds in aqueous solution (to form an electrically-conducting electrolyte) is discovered by Swedish chemist Svante Arrhenius (1859–1927), later (1893) reaffirmed by German physical chemist Hermann Nernst (1864–1941).

1884 Van't Hoff factor (the ratio of the number of particles in an electrolyte to the number of undissociated particles) is discovered by Dutch chemist Jacobus van't Hoff (1852–1911).

BIOLOGY

MEDICINE

1896 Bacterium that causes Bang's disease (infectious abortion in cattle) is discovered by Danish veterinary surgeon Bernhard Bang (1848–1932).

1896 Vaccine against rinderpest in cattle is developed by German bacteriologist Robert Koch (1843–1910).

1896 Fact that the disease beriberi is caused by the lack of some factor in food is discovered by Dutch physician Christiaan Eijkman (1858–1930).

1897 Inoculation against plague is developed by Russian bacteriologist Waldemar Haffkine (1860–1930).

1897 Use of ultraviolet light as a treatment for skin disorders, such as lupus vulgaris, is discovered by Danish physician Niels Finsen (1860–1904).

1897 Barium meal, which can be swallowed to make the stomach and intestines opaque to X-rays (and therefore show up on an X-ray photograph), is introduced by US physician Walter Cannon (1871–1945).

1898 Fact that mosquitoes transmit malaria to humans by biting is discovered by Italian parisitologists Giovanni Grassi (1854–1925), Arnico Bignami (1862–1929) and Giuseppi Bastianelli (1862–1959).

1898 Tobacco mosaic virus (the first known virus) is discovered by Dutch botanist Martinus Beijerinck (1851–1931).

1898 Fact that hoof-and-mouth disease is caused by a virus is discovered by German bacteriologists Friedrich Löffler (1852–1915) and Paul Frosch.

1898 Causative agent of one type of dysentery is discovered by Japanese bacteriologist Shibasaburo Kitasato (1856–1931).

1898 Chiropractic is established in the USA by Canadian osteopath Daniel Palmer (1845–1939).

1899 Toxin produced by the diphtheria bacterium is discovered by French bacteriologist Pierre Roux (1853–1933) and Swiss-born French bacteriologist Alexandre Yerson (1863–1943).

1899 Aspirin (acetylsalicylic acid) is marketed (as a prescription drug, in powder form) by the German Bayer company, later (1915) as tablets.

FARMING AND FOOD

TRANSPORT

1895 Petrol-engined motorcycle is built by French engineers Albert de Dion and George Bouton.

1895 Petrol-engined bus is built by German engineer Karl Benz (1844–1929).

1895 First US patent for a petrol-engined car is granted to US lawyer George Selden (1846–1922).

1895 Control system for multiple-unit trains is invented by US engineer Frank Sprague (1857–1934).

1896 First car to be built in the USA is marketed by Charles (1862–1938) and J. Frank Duryea; Henry Ford (1863–1947) builds his first car.

1896 Detachable solid rubber tyres (originally for carriages) are pioneered by US industrialist Harvey Firestone (1868–1938).

1896 Gliding flight is achieved in the Hawk, launched by tow line, by Scottish inventor Percy Pilcher (1866–99); later (1899) Pilcher is killed flying the Hawk.

1896 Scientific study of gliders is begun by French-born US engineer Octave Chanute (1832–1910).

1896 Steam-powered pilotless aeroplane is built by US physicist Samuel Langley (1834–1906); it crashes in the Potomac River.

1896 Quadrilateral tailless kite (the Eddy kite), for carrying a camera to photograph the Earth, is invented by US meteorologist William Eddy (1858–1909).

1897 Unsubstantiated claim of a flight by a steam-powered aeroplane, Avion III, is made by French engineer Clément Ader (1841–1925)

1897 Rigid airship, made of aluminium, is built by German engineer David Schwarz.

1897 Commercially successful steam car is invented by US engineer Francis Stanley (1849–1918).

1897 Steam-turbine powered boat is built by British inventor Charles Parsons (1854–1931).

1897 Hydrofoil boat is constructed by French engineer Comte de Lambert.

1897 First submarine to work in the open sea (the Argonaut) is built by US engineer Simon Lake (1866–1945).

1897 Dynamometer car (for measuring performance of railway locomotives) is invented by John Beckinridge.

1897 Electric taxi cabs run in London, England.

1897 Radiator for cooling petrol engines is invented by German engineer Wilhelm Maybach (1847–1929).

1898 Pressurized dirigible (steerable airship) is invented by Brazilian aviator Alberto Santos-Dumont (1873–1932).

1898 First completely modern submarine (with electric motors for underwater and a petrol engine for surface running) is built by Irish-born US inventor John Holland (1840–1914).

1898 Radio-controlled model boat is patented by Croatian-born US inventor Nokola Tesla (1857–1943).

1898 Armoured car is invented, independently, by Americans E.J. Pennington and R.P. Davidson, Englishman Frederic Simms, and Russian W. Lutski; production cars are built later by Charron (France 1904), Daimler (England 1904), Fiat (Italy 1912) and Rolls-Royce (England 1914).

ENGINEERING AND TECHNOLOGY

1879 Carbon-filament incandescent lamp is invented independently by American Thomas Edison (1854–1932) and Englishman Joseph Swan (1828–1914).
1879 Electric arc furnace is invented by German-born British engineer William (Wilhelm) Siemens (1823–83).

1879 Gabardine cloth is developed by British draper Thomas Burberry.

1879 Unsuccessful cash register is invented by US saloon keeper James Ritty.

1879 Practical steam turbine is invented by French engineer Carl de Laval (1845–1913).

1879 Apparatus for finding the flash point of petrol is invented by British chemist Frederick Abel (1827–1902).

1879 Nordenfelt machine gun is invented by Helge Palmkrantz.

1879 Magazine for holding rifle cartridges is invented by Scottish-born US engineer James Lee (1831–1904).

1880 Chain drive (originally for textile machinery) is invented by German engineer Hans Renold.

1880 Bolometer (originally for measuring the heat of the Sun) is invented by US astronomer Samuel Langley (1834–1906).

1880 Jacketed bullet (lead sheathed with copper) is invented by Swiss army officer Eduard Rubin (1846–1920).

1881 Air-cooled petrol engine is invented by French engineer Fernand Forest (1851–1914).

1881 System of arc lighting is patented by US electrical engineers Edwin Houston (1847–1914) and Elihu Thomson (1853–1937).

1881 Metal detector (for finding bullets in the human body) is invented by US engineer Alexander Graham Bell (1847–1922).

1882 Alternator for the commercial production of electricity is invented by British engineer Sebastian Ferranti (1864–1930).

1882 Three-phase electricity distribution is invented by British electric engineer John Hopkinson (1849–98).

1882 Method of transmitting electric power along wires over long distances (by increasing the voltage) is invented by French engineer Marcel Deprez (1843–1918).

1882 Electrically-propelled torpedo is invented by US engineer Winfield Sims (1844–1918).

1882 Electric iron is patented by US inventor Henry Seely.

1882 Manganese steel (for toolmaking) is invented by British metallurgist Robert Hadfield (1859–1940).

1883 Process for recovering sulphur from wastes from the Leblanc process for making soda (sodium carbonate) is invented by British chemist Alexander Chance (1844–1917).

1883 Automatic machine gun, actuated by recoil energy, is invented by US engineer Hiram Maxim (1840–1916).

1883 Spark plug (for gas engines) is invented by French engineer Étienne Lenoir (1822–1900), later (1889) patented in England by Thomas Parker.

1883 Coin-operated vending machine is patented by British inventor Percival Everitt.

1884 Multistage reaction steam turbine is invented by British engineer Charles Parsons (1854–1931), who in the same year patents a turbo-generator (a steam turbine driving a dynamo, for producing electricity).

1884 Gear-cutting machine is invented by US engineer Francis Pratt (1827–1902).

1884 Type of artificial silk made from cotton (cellulose), now called rayon, is invented by French chemist Hilaire de Chardonnet (1839–1924).

1884 Heat-resistant borosilicate glass is invented by German scientist Carl Zeiss (1816–88).

1884 Direct current electric motor for an electric railway locomotive is invented by US engineer Frank Sprague (1857–1934).

1884 Instrument for dividing a line into any number of equal parts is invented by British electrical engineer Hertha Ayrton (1854–1923).

1885 Modern type of transformer, suitable for large-scale electricity supply networks, is invented by US engineer William Stanley (1858–1916).

1885 Gas mantle is invented by Austro-Hungarian chemist Carl Auer von Welsbach (1858–1929).
1885 Improved internal combustion engine that burns petrol is invented by German engineers Karl Benz (1844–1929) and, independently, Gottlieb Daimler (1834–1900).

1885 Cartridge clip for rapid reloading of magazine rifles is invented by Austrian engineer Ferdinand Mannlicher (1848–1904).

1886 Lebel rifle, a small-calibre weapon using smokeless powder and invented by French army officer Nicolas Lebel (1838–91), is adopted by the French army.

1886 Apochromatic lenses are invented by German physicist Ernst Abbe (1840–1905).

1886 Flush toilet (water closet) is invented by British engineer Thomas Crapper (1837–1910).

1886 Electrolytic process for extracting aluminium from its ore (alumina) is invented by

COMMUNICATIONS

1890 Submarine telephone cable is laid across the River Plate between Montevideo and Buenos Aires.

1890 Coherer (later used as a detector in the first radios) is invented by French physicist Edouard Branly (1844–1940).

1891 New colour photography process, using interference of light, is invented by French physicist Gabriel Lippman (1845–1921).

1891 Kinetoscope movie camera/projector is invented by Americans Thomas Edison (1854–1932) and Edwin Porter (1870–1941); it was first exhibited in 1893.

1891 Sound movie camera is patented in the USA by William Dickson.

1892 Commercially successful adding machine, with a printer, is patented by US inventor William Burroughs (1857–98).

1892 Automatic telephone exchange opens in La Porte, Indiana, designed by undertaker Almon Strowger.

1892 Typewriter on which the user can see the words as they are typed is invented by US engineer Thomas Oliver.

1893 Improved typewriter is patented by Franz Wagner, and later (1895) marketed by US industrialist John Underwood (c.1857–1935).

1893 A four-function calculator (with addition, subtraction, multiplication and division) is marketed in the USA.

1893 Etched glass screen (for making halftone engravings) is patented by US invetors Louis (1846–1919) and Max (1857–1926) Levy.

1893 Stereoscopic (3D) photography, using superimposed red and green images, is invented by French physicist Louis Duclos du Hauron (1837–1920).

1893 First film studio (to make movies for peep shows) is opened in New Jersey, USA, by Thomas Edison (1847–1931).

1893 Magnetic sound recorder, called the Telegraphone and using steel wire not tape, is invented by Danish engineer Valdemar Poulsen (1869–1942).

1894 Improved coherer, for detecting radio waves, is invented by British physicist Oliver Lodge (1851–1940) and, independently, Russian radio pioneer Alexandr Popov (1859–1905).

1894 Radio communications are developed by Italian inventor Guglielmo Marconi (1874–1937) and later (1896), independently, by Russian physicist Aleksandr Popov (1859–1905).

1895 Motion pictures are is demonstrated by French chemists Auguste (1862–1954) and Louis (1864–1948) Lumière.

1895 Motion-picture projector with intermittent motion is invented in the USA by Charles Jenkins (1867–1934).

1895 Photogravure printing process is invented by Czech-born Austrian Karl (Karel) Klic (1841–1926).

1896 Early type of motion-picture projector (called the theatrograph or animatograph) is invented by British engineer Robert Paul (1869–1943).

1896 Synchronized sound for movies, using discs and a Berliner gramophone, is attempted by French film pioneer Charles Pathé (1863–1957).

1896 Optical soundtrack for movie film is patented by US inventor Lee De Forest (1873–1961).

ASTRONOMY AND SPACE	PHYSICS	CHEMISTRY	BIOLOGY	MEDICINE

1885 Mélinite explosive (ammonium picrate) is discovered by French chemist Eugène Turpin.

1885 Metallic elements neodymium (symbol Nd) and praseodymium (symbol Pr) are discovered by Austrian chemist Carl von Welsbach (1856–1929).

1886 Beckman rearrangement (of oximes of ketones into amides) is discovered by German chemist Ernst Beckmann (1853–1923).

1886 Metalloid element germanium (symbol Ge) is discovered by German chemist Clemens Winkler (1838–1904).

1886 Metallic element dysprosium (symbol Dy) is discovered by French chemist Paul Lecoq Boisbaudran (1838–1912).

1887 Hydrazine is discovered by German chemist Theodor Curtius (1875–1928).

1887 Fructose is synthesized by German chemist Emil Fischer (1852–1919), who founds the science of biochemistry.

1887 Rhodamine dyes are discovered by German chemist Adolf von Baeyer (1835–1917).

1887 Gabriel's synthesis (of primary amines from potassium phthalimide and haloalkanes) is discovered by German chemist Siegmund Gabriel (1851–1924).

1888 Le Chatelier's principle, that if a constraint is imposed on a system in equilibrium the system adjusts itself to minimize the effects of the constraint, is discovered by French chemist Henry le Chatelier (1850–1936).

1889 Arrhenius equation (for the rate of a chemical reaction) is discovered by Swedish chemist Svante Arrhenius (1859–1927).

1889 Seven-carbon ring compounds are discovered by Russian chemist Vladimir Markovnikov (1837–1904).

1889 Metallic element europium (symbol Eu) is discovered spectroscopically by British scientist William Crookes (1832–1919) and, later (1896), isolated by French chemist Eugène Demarçay (1852–1903).

1889 Method of making pure nickel by heating nickel carbonyl is discovered by German-born British chemist Ludwig Mond (1839–1909).

1890 Tautomerism (in which two isomers are in equilibrium) is discovered by German chemist Ludwig Claisen (1851–1930), who called it pseudomerism, and, independently, by his compatriot Johannes Wislicenus (1835–1902).

1890 Gattermann reaction (for making aromatic halides using diazonium compounds) is discovered by German chemist Ludwig Gattermann (1860–1920).

1890 Babcock test for the amount of fat in milk is discovered by US agricultural chemist Stephen Babcock (1843–1931).

1890 Guldberg's law (relating critical temperature to boiling point) is discovered by Swedish chemist Cato Guldberg (1836–1902) and, independently, P. Guye (1862–1922).

1890 Dye base acridine is discovered by German chemists Karl Graebe (1841–1927) and Heinrich Caro (1834–1910).

c.1890 Claisen condensation (in which two molecules of an ester combine to form a keto ester) is discovered by German chemist Ludwig Claisen (1851–1930).

FARMING AND FOOD

TRANSPORT

1889 Steam car is constructed by French engineer Armand Peugeot (1849–1915).

ENGINEERING AND TECHNOLOGY

US chemist Charles Hall (1863–1914) and, independently, French chemist Paul Héroult (1863–1914); an altenative chemical method is invented by US chemist Hamilton Castner (1859–99).

1887 Float-feed spray carburetor is invented by British engineer Edward Butler.

1887 Smokeless powder (Poudre B) for firearms is invented by French engineer Paul Vieille (1854–1934) and, independently, US chemist Charles Munroe (1849–1938).

1897 Modeling clay (trade name Plasticine) is invented by Wlliam Harbutt.

1888 Ballistite smokeless explosive is invented by Swedish chemist Alfred Nobel (1833–96).

1888 Alternating current (AC) motor is invented by Serbian-born US engineer Nikola Tesla (1856–1943).

1888 Carbon brushes (to carry electric current into or out of commutators on dynamos and electric motors) are invented by Belgian-born US engineers Charles Van de Poele (1846–92) and, independently, George Forbes (1849–1936).

1888 Photometer is invented by Irish physicist John Joly (1857–1933).

1888 Steam-powered blowing engine for forcing air into blast furnaces is invented by US engineer Edwin Reynolds (1831–1909).

1889 Propellant explosive nitrocellulose (guncotton) is invented by British scientists Frederick Abel (1826–1902) and James Dewar (1842–1923).

1888 Lee-Metford rifle, developed by Scottish-born American James Lee (1831–1904) and Englishman William Metford (1824–99), is adopted by British army.

1889 Spectroheliograph (an instrument for taking photographs of the Sun at specified wavelengths of light) is invented by Americam astronomer George Hale (1868–1938).

1889 Automatic dishwasher is invented in the USA by Mrs W. A. Cockran.

1889 System of electric lighting for trains is invented by US engineer Harry Leonard (1861–1915), who later (1892) patented a control system for electric lifts.

1890 Improved reaction steam turbine is invented by Swedish engineer Carl de Laval (1845–1913).

1890 System of machine lubrication using galleries to channel oil to moving parts in contact is invented by engineer Albert Pain (1856–1929).

1890 High-frequency generator is invented by British-born US electrical engineer Elihu Thomson (1853–1937).

1890 Surgical rubber gloves are invented by US surgeon William Halsted (1852–1922).

1890 Electric ovens are marketed by the US Carpenter company.

1890 Electric chair, using alternating current, is first used to execute criminals in the USA.

1890 Clockwork model train set is marketed by the German Märklin company.

1891 Machine for mining coal is invented by Belgian-born US Charles Van Depoele (1846–92).

1891 Synthetic rubber is made by British chemist William Tilden (1842–1926).

1891 Method of making silicon carbide (Carborundum) is invented by American scientist Edward Acheson (1856–1931).

1892 Compression-ignition (Diesel) engine is invented by French-born German engineer Rudolf Diesel (1858–1913).

1892 Viscose process for making rayon (artificial silk) is invented by British chemists Edward Bevan (1856–1921) and Charles Cross (1855–1935); it is later (1904) put into commercial production by British industrialist Samuel Courtauld (1793–1881).

1892 Vacuum bottle with internal silvering of all glass surfaces is invented by Scottish physicist James Dewar (1842–1943); an earlier (1882) version with no silvering was invented by French physicist Jules Violle (1841–1923).

1892 Simple electrolytic method of extracting bromine from sea water is invented by Canadian-born US chemist Henry Dow (1866–1930).

1892 Electric furnace is invented by French chemist Henri Moissan (1852–1907).

1892 Mercury-vapour lamp is invented by German engineer Leon Arons (1860–1919) and, independently (1901), Peter Cooper-Hewitt (1861–1921).

1892 Acetylene lamp is invented by French chemist Henri Moissan (1852–1907).

1892 Primus stove (fuelled by pressurized kerosene) is invented by Swedish engineer F. Lindquist.

1892 Ferris wheel for fairgrounds is invented by US engineer George Ferris (1859–96).

1897 Wind tunnel for testing aircraft design is invented by Russian scientist Konstantin Tsiolkovsky (1857–1935).

1892 Pre-stressed concrete is invented by French engineer François Hennebique (1842–1921), later (1904) improved by French civil engineer Eugène Freyssinet (1879–1962).

1892 Successful cash register is invented by US engineer William Burroughs (1857–98).

1892 Zip fastener, called a clasp locker, is invented by US Whitcomb Judson (patented 1893). It was later (1913) improved, and a machine for manufacturing it developed, by Swedish engineer Gideon Sundback.

COMMUNICATIONS

1897 Cathode-ray tube, a modified Crookes tube, is invented by German physicist Karl Ferdinand Braun (1850–1918).

1898 High-speed focal-plane camera shutter is patented by US inventor William Folmer (1862–1936), who also invented the Graflex camera.

1899 Focusing camera lens is invented by British optician Thomas Dallmeyer (1859–1906).

1899 Repeater (amplifier) for submarine telegraph cables is invented by British electrical engineer Sidney Brown (1873–1948).

ASTRONOMY AND SPACE

1900

1904 Himalia, the sixth moon of Jupiter, is discovered by US astronomer Charles Perrine (1867–1951), who later (1905) discovers Elara, Jupiter's seventh moon.

1906 Asteroid Achilles (the first of the Trojan asteroids, which occupy Jupiter's orbit) is discovered by German astronomer Maximillian Wolf (1863–1932).

PHYSICS

1900 Quantum theory is proposed by German physicist Max Planck (1858–1947).

1900 Gamma rays are discovered by French physicist Paul Villard (1860–1934).

1901 Fact that light exerts a pressure is discovered, and the pressure mea-

CHEMISTRY

1891 Silicon carbide (Carborundum) is discovered by US chemist Edward Acheson (1856–1931).

1891 Nature of coordination compounds (complexes) is discovered by German-born Swiss chemist Alfred Werner (1866–1919).

1892 Polymerization of isoprene to form a synthetic rubber is discovered by British chemist William Tilden (1842–1926).

1892 Crum Brown rule (for aromatic substitution) is discovered by Scottish chemist Alexander Crum Brown (1838–1922).

1893 Dissociation of ionic compounds in aqueous solution (to form an electrically-conducting electrolyte) is discovered by German physical chemist Hermann Nernst (1864–1941).

1893 Formula of camphor is discovered by German chemist Julius Bredt (1855–1937), later (1903) synthesized by G. Komppa (1867–1949).

1894 Gaseous element argon (symbol Ar) is discovered by British scientist Lord Rayleigh (1842–1919) and Scottish chemist William Ramsay (1852–1916).

1896 Iron-nickel alloy invar (which has a very low coefficient of expansion) is discovered by Swiss-born French physicist Charles Guillaume (1861–1938).

1896 Amino acid histidine is discovered by German biochemist Albrecht Kossel (1853–1927) and Swede Sven Hedin (1865–1952).

1896 Cell-free fermentation of sugar (i.e., using ground-up yeast cells) is discovered by German chemist Eduard Buchner (1860–1917).

1896 Walden inversion (the interconversion of two optical isomers) is discovered by Russian chemist Paul Walden (1863–1957).

1897 Ability of finely divided nickel to catalyze the hydrogenation of unsaturated hydrocarbons (an important reaction in organic synthesis) is discovered by French chemists Paul Sabatier (1854–1941) and Jean-Baptiste Senderens (1856–1937).

1897 Method of manufacturing pure aspirin (acetylsalicylic acid) is discovered by German chemist Felix Hoffman, later (1899) marketed by the Bayer company.

1898 Purine is discovered by German chemist Emil Fischer (1852–1919).

1898 Radioactive elements polonium (symbol Po) and radium (symbol Ra) are discovered (in pitchblende) by Polish-born French chemists Marie (1867–1934) and Pierre (1859–1906) Curie.

1898 Gaseous elements krypton (symbol Kr), neon (symbol Ne) and xenon (symbol Xe) are discovered by Scottish chemists William Ramsay (1852–1916) and Morris Travers (1872–1961).

1899 Radioactive element actinium (symbol Ac) is discovered by French chemist André Debierne (1874–1949).

1899 Radioactive gas "thoron" (actually an isotope of radon) is discovered by New Zealand-born British physicist Ernest Rutherford (1871–1937).

1900 Grignard reagents (organo-metallic compounds of magnesium used in organic synthesis) are discovered by French chemist François Grignard (1871–1935).

1900 Fulvene (a red hydrocarbon) is discovered by German chemist Friedrich Thiele (1865–1918).

BIOLOGY

1900 Mendel's laws of inheritance are rediscovered by German botanist Karl Correns (1864–1933) and, independently, Dutch botanist Hugo de Vries (1848–1935).

1902 Hormone secretin (produced in the walls of the small intestine to stimulate the production of digestive juices by the pancreas and liver) is discov-

MEDICINE

1900

1900 A, B and O blood groups are discovered by Austrian pathologist Karl Landsteiner (1868–1943).

1900 Bacterium that causes one kind of dysentery is discovered by US microbiologist Simon Flexner (1863–1946); another type is isolated later (1915) by US biochemist Edward Kendall (1886–1972).

FARMING AND FOOD

TRANSPORT

ENGINEERING AND TECHNOLOGY

1892 Smokeless powder (propellant explosive) is invented by US gunsmith Hudson Maxim (1853–1927).

1893 Process for synthesizing potassium cyanide (used to extract gold from its ores) is invented by Scottish chemist George Beilby (1850–1924).

1894 Thermit process for producing high-melting point metals and for welding is invented by German chemist Johann (Hans) Goldschmidt (1861–1923).

1894 Safety fuse box is invented by US electrical engineer James Packard (1863–1928).

1895 Practical photoelectric cell is developed by German physicists Julius Elster (1854–1920) and Hans Geitel (1855–1923).

1895 Safety razor is patented by US inventor King Gillette (1855–1932); it goes on sale in 1902.

1895 First portable electric hand drill is invented by German engineer Wilhelm Fein.

1895 Large-scale process for producing liquid air is invented by German engineer Carl von Linde (1842–1934).

1896 Sphygmomanometer, for measuring blood pressure, is invented by Italian physician Scipione Riva-Rocci (1863–1937).

1896 Impulse steam turbine is invented by US patent lawyer Charles Curtis (1860–1953).

1896 Lubricatnts based on colloidal graphite are devised by US chemist Edward Acheson (1856–1931).

1897 Steam turbine to power boats is invented by British engineer Charles Parsons (1854–1931).

1897 Improved oscillograph is invented by British engineer William Duddell (1872–1917).

1897 Process for making caustic soda (sodium hydroxide) from brine (sodium chloride) is invented independently by US chemist Hamilton Castner (1859–1899) and German chemist Karl Kellner (1851–1905).

1897 Mauser magazine rifle, invented by Peter Mauser (1838–1914), is adopted by the German army.

1898 Tapered roller bearings (for carriages) are invented by German-born US engineer Henry Timken (1831–1909).

1898 Osmium filament for electric lamps is invented by Austro-Hungarian chemist Carl Auer von Welsbach (1858–1929).

1899 Rotary engine for aeroplanes is invented in Australia by British engineer Lawrence Hargrave (1850–1915); a much improved engine is developed later (1910) by French engineer Pierre Clerget (1875–1943).

1899 Method for the commercial synthesis of aspirin (from salicylic acid) is devised by Felix Hoffman and patented by C. Witthauer, both of the German Bayer company.

1899 Machine for measuring the hardness of metals is invented by Swedish engineer Johann Brinell (1849–1925).

1899 Process for heat-treating tool steel to increase its hardness is invented by US engineers Frederick Taylor (1856–1915) and Maunsel White (1856–1912).

1899 Process for making synthetic graphite for lubricants is invented by US chemist Edward Acheson (1856–1931).

COMMUNICATIONS

1900

FARMING AND FOOD

1900 Hamburger is invented in the USA by Louis Lassen.

1901 Commercially successful petrol-engined tractor is produced in the USA by Charles Hart and Charles Parr.

1902 Tea-making machine (with an alarm clock) is invented by British gunsmith Frank Smith.

TRANSPORT

1900 The first Zeppelin, the Deutschland, flies (with a rigid hull filled with hydrogen), designed by German engineer Ferdinand Graf Zeppelin (1838–1917); it is wrecked on landing.

1900 Hydrofoil twin-hulled boat is invented by Italian engineer Enrico Forlanini.

ENGINEERING AND TECHNOLOGY

1900 Nickel-iron alkali accumulator is invented by US engineer Thomas Edison (1854–1932).

1900 Magnetic clutch is invented by US electrical engineer Bion Arnold (1861–1942).

1900 Brinell test, for measuring the hardness of substances, is invented by Swedish metallurgist Johann Brinell (1849–1925).

1900 Domestic gas heater with fireclay burners is invented by British engineer John Wright.

COMMUNICATIONS

1900 Loudspeaker is invented by Horace Short.

1900 Wide-screen cinema (with a circular screen 100 metres across) is invented by Frenchman Raoul Grimoin-Sanson.

1901 Transatlantic radio telegraphy is introduced by Italian inventor Guglielmo Marconi (1874–1937).

ASTRONOMY AND SPACE

1907 First useful photograph of Mars is taken by US astronomer Percival Lowell (1855–1916).

1908 Classification of stars into giants and dwarfs – the key to the understanding of stellar evolution – is discovered by Danish astronomer Ejnar Hertzsprung (1872–1967).

1908 Strong magnetic fields associated with sunspots are discovered by US astronomer George Hale (1868–1938).

1908 Morehouse's comet is discovered by US astronomer Davis Morehouse (1876–1941).

1909 Halley's comet is rediscovered on its return towards Earth by German astronomer Maximillian Wolf (1863–1932).

PHYSICS

sured, by Russian physicist Pyotr Lebedev (1866–1912).

1902 Heaviside-Kennelly layer (of ionized gas in the upper atmosphere that reflects radio signals) is discovered independently by British physicist Oliver Heaviside (1850–1925) and US electrical engineer Arthur Kennelly (1861–1939).

1902 Photoelectric effect (the emission of electrons from the surface of a metal exposed to light or other electromagnetic radiation) is quantified by German physicist Philipp von Lenard (1862–1947).

1902 Claude process (for the bulk liquefaction of air) is discovered by French scientist George Claude (1870–1960).

1904 Model of the atom as a "pudding" of positive charges containing negatively charged electrons (the "plums" in the pudding) is proposed by British physicist J.J. Thomson (1856–1940).

1905 Special Theory of Relativity is proposed by German-born US physicist Albert Einstein (1879–1955).

1905 Third law of thermodynamics (that entropy change tends to zero at absolute zero) is formulated by German physical chemist Hermann Nernst (1864–1941).

1905 Explanation of paramagnetism (as being caused by electron charges within an atom) is given by French physicist Paul Langevin (1872–1946).

1906 X-rays characteristic of the element that scatters them – related to the element's atomic number – are discovered by British physicist Charles Barkla (1877–1944).

1906 Lines in the ultraviolet region of the hydrogen spectrum (the Lyman series) are discovered by US physicist Theodore Lyman (1874–1954).

1907 Theory that explains ferromagnetism in terms of magnetic domains within the material is proposed by French physicist Pierre Weiss (1865–1940).

1908 Method of liquefying helium is discovered by Dutch physicist Heike Kamelingh Onnes (1853–1926).

1909 Avogadro's number (the number of molecules in 1 cubic centimetre of gas at normal temperature and pressure) is determined by French physicist Jean Perrin (1870–1942).

CHEMISTRY

1900 Radioactive gaseous element radon (symbol Rn) is discovered by German chemist Ernst Dorn (1848–1916), and later (1908) extracted by Scottish chemist William Ramsay (1852–1916) and British chemists Frederick Soddy (1877–1956) and Robert Whytlaw-Gray (1877–1958), who initially named it niton.

1900 Basic process for cracking crude oil to make petrol is discovered by US chemist Charles Palmer (1858–1939).

1900 Absorption chromatography (for separating petroleum mixtures) is discovered by US chemist David Day (1859–1925).

1900 Triphenylmethyl, the first free radical to be isolated, is discovered by Russian-born US chemist Moses Gomberg (1866–1947).

1901 Amino acid proline is discovered by German chemist Emil Fischer (1852–1919).

1901 First essential amino acid, tryptophan, is discovered by British chemists Frederick Gowland Hopkins (1861–1947) and S. Cole.

1902 Amino acid hydroxyproline is discovered by German chemist Emil Fischer (1852–1919).

1902 Optically active inorganic compounds (previously all such compounds were organic) are discovered by British chemist William Pope (1870–1939).

1902 Method of making artificial corundum (aluminium oxide) is discovered by French chemist Auguste Verneuil (1856–1913).

1903 Buffer solutions, for stabilizing pH (acidity), are discovered by German physical chemist Hermann Nernst (1864–1941).

1903 Amino acid isoleucine is discovered by German bacteriologist Paul Ehrlich (1854–1915).

1904 Abegg's "rule of eight" (that an outer atomic shell containing eight electrons confers extra stability on an atom or ion) is discovered by German chemist Richard Abegg (1869–1910).

1904 Highly toxic chemical divinylchloroarsine, later (World War I) used as the war gas Lewisite, is discovered by Belgian-born US chemist Julius Nieuwland (1878–1936).

1905 Fact that radioactive metals all eventually decay to (an isotope of) lead is discovered by US chemist Bertram Boltwood (1870–1927), who later (1907) discovered "ionium", thought to be a new element but now known to be a radioactive isotope of thorium.

1905 Cyclooctatetraene (an 8-carbon cyclic compound) is discovered by German chemist Richard Willstätter (1872–1942).

1906 Tricarbon dioxide (carbon suboxide) is discovered by German chemist Otto Diels (1876–1954).

1906 Metallic element lutetium (symbol Lu) is discovered by French chemist George Urbain (1872–1938) and, independently (1907), Austrian chemist Carl von Welsbach (1858–1929).

1906 Isotope thorium-230 is discovered by US chemist Bertram Boltwood (1870–1927), who calls it "ionium".

1908 Food additive monosodium glutamate (MSG) is discovered by Japanese food technologist Ikeda Kikunae (1864–1936).

1909 Concept of pH (as a measure of acidity/alkalinity) is introduced by

BIOLOGY

ered by British physiologists Ernest Starling (1866–1927) and William Bayliss (1860–1924).

1903 Enzyme zymase is discovered by German biologist Eduard Buchner (1860–1917)

1904 Conditioned reflexes are discovered by Russian biologist Ivan Pavlov (1849–1936).

1904 Breeding ground of the European eel is found to be in the western Atlantic by Danish biologist Johannes Schmidt (1877–1933).

1904 Coenzymes are discovered by British biochemist Arthur Harden (1865–1940).

c. 1905 Nissl bodies (particles in nerve cells) are discovered by German neurologist Franz Nissl (1860–1919).

1905 Mammalian sex chromosomes (XX for females and XY for males) are discovered by US biologist Clarence McClung.

1905 Technique for growing tissue cells in vitro (outside the body) is discovered by US biologist Ross Harrison (1870–1959).

1907 Fact that proteins are composed of amino acids is discovered by German chemist Emil Fischer (1852–1919).

1908 Phagocytes (digestive white blood cells) are discovered by Russian biologist Ilya Metchnikoff (1845–1916).

1909 Pituitary hormone oxytocin (which stimulates contractions of the womb during childbirth) is discovered by British physiologist Henry Dale (1875–1968).

1909 Word "gene" is coined by Danish botanist Wilhelm Johannsen (1857–1927).

MEDICINE

1900 Microorganism (a protozoan) that causes kala-azar (leishmaniasis) is discovered by Scottish physician William Leishman (1865–1926).

1901 Role of white blood cells in combating infection is discovered by Russian biologist Ilya Metchnikoff (1845–1916).

1901 Fact that yellow fever is caused by a virus is discovered by US surgeon Walter Reed (1851–1902).

1902 Anaphylaxis (a life-threatening allergic reaction) is discovered by French physician Charles Richet (1850–1935).

1903 X-ray treatment for cancerous tumours is discovered by German physician Georg Perthes (1869–1927).

1904 Local anaesthetic novocaine is discovered by US physician J. Leonard Corring.

1905 Successful direct blood transfusion between humans is performed by US physician George Coile (1864–1943).

1905 Bacterium (a spirochete) that causes syphilis is discovered by German bacteriologists Erich Hoffman (1868–1959) and Fritz Schaudinn (1871–1906).

1906 Allergies are discovered by Austrian physician Clement von Pirquet (1874–1929).

1906 Bacterium that causes whooping cough (pertusis) is discovered by Belgian physician Jules Bordet (1870–1961)

1906 Fact that Rocky Mountain spotted fever is spread by cattle ticks is discovered by US pathologist Howard Ricketts (1871–1910), who went on to discover the microorganisms (rickettsia) that cause the disease.

1906 Corneal transplant operation is introduced by French surgeon Edouard Zirm (1863–1944).

1906 Wassermann test (for syphilis) is discovered by German bacteriologist August von Wassermann (1866–1925).

1906 Atoxyl, the first synthetic drug (for treating sleeping sickness), is discovered by German physician Paul Ehrlich (1854–1915).

1907 Role of protozoans in causing various tropical diseases is discovered by French physician Charles Laveran (1845–1922).

1907 Serum for treating cerebrospinal menigitis is discovered by US microbiologist Simon Flexner (1863–1946).

1907 Full clinical description of presenile dementia (Alzheimer's disease) is given by German neuropathologist Alois Alztheimer (1864–1915).

1907 AB blood group is discovered by Czech physician Jan Jansky (1873–1921) and, independently, later (1910) by US physician William Moss (1876–).

1908 Vaccine against tuberculosis is developed by French physicians Albert Calmette (1863–1933) and Charles Guérin, later (1923) used as BCG (bacille Calmette Guérin) vaccine.

1909 First cancer-causing virus (Rous chicken sarcoma) is discovered by US physician Francis Rous (1879–1970).

FARMING AND FOOD

1903 Agricultural tractor is marketed by the British Petter company.

1903 Processed cheese is marketed by US manufacturer James Kraft.

1903 Decaffeinated coffee, trade-marked Sanka, is introduced by German coffee dealer Ludwig Roselius.

1904 Ovaltine (then called Ovamaltine) is patented in Switzerland by George Wander.

1906 Freeze-drying to preserve food is invented by French biophysicists Jacques Arsène d'Arsonval (1851–1940) and Georges Bordas.

1908 Caterpiller farm tractor is manufactured by US engineer Benjamin Holt, using track designs purchased from the British Hornsby company.

1909 Electric toaster is marketed by US General Electric Company.

TRANSPORT

1901 First commercial ship powered by steam turbines, the King Edward, is launched in Scotland.

1900 Paris underground (the Metro) opens.

1901 Front-wheel drive car is constructed (in France).

1901 Non-rigid airship with a large gas-bag (and suspended gondola) is built by German aeronautical engineer August von Parseval (1861–1942).

1901 Acetylene gas lamps are introduced for cars.

1902 Steam superheater for railway locomotives is invented by German engineer Wilhelm Schmidt.

1902 Underground railway (U-Bahn) opens in Berlin.

1902 First diesel-engined vessel, the canal boat Petit-Pierre, is built in France.

1902 First motor scooter, called l'Autofauteuil (motor-armchair), is built by French engineer G. Gauthier, although it was preceded (1901) by a motor-assisted bicycle marketed by Michel and Eugene Werner.

1902 Three-wheeled petrol-engined farm tractor is invented by British engineer Dan Albone (1860–1906).

1902 Drum brakes for cars are invented by French engineer Louis Renault (1877–1944); and disc brakes for cars are invented by British engineer Frederick Lanchester (1868–1946).

1903 Short powered flight is made by New Zealand inventor William Pearce (1877–1953).

1903 First sustained flight in an aeroplane is made by US inventors Orville (1871–1948) and Wilbur (1867–1912) Wright, using an engine of their own design.

1903 Car seat belt is patented by Frenchman Gustave Liebau.

1903 Four-wheel drive car is manufactured in the Netherlands.

1903 Theory of multi-stage rocket propulsion is propounded by Russian engineer Konstantin Tsiolkovsky (1857–1935).

1904 Practical steam-powered caterpillar tractor is invented by US engineer Benjamin Holt; he built petrol-engined tractors after 1906, in the year that British engineer David Roberts also, and independently, patented a caterpillar track (sold to Holt in 1912).

1904 First section of the underground railway opens in New York City.

1904 Rubber car bumpers are invented by British engineer Frederic Simms.

1905 Non-shatter safety glass for car windscreens is invented by British solicitor John Wood.

1906 First German U-boat, the U-1 built by Nordenfeldt, is launched.

1906 Heavier-than-air flight is a achieved by Brazilian aviator Alberto Santos-Dumont (1873–1932).

1906 First monoplane is built by Romanian-born French aircraft engineer Trajan Vuia.

1906 Car rear-view mirror is patented by French inventor Alfred Faucher.

1907 Primitive helicopter, the four-rotor Gyroplane, is built by French engineer Louis Bréguet (1881–1955).

1907 Vertical takeoff helicopter is built in

ENGINEERING AND TECHNOLOGY

1900 Practical photoelectric cell is invented by German physicists Johan Elster (1854–1920) and Hans Geitel (1855–1923).

1900 High-power Browning automatic pistol is produced by US gunsmith John Browning (1855–1926).

1900 Paper-clip is invented by Norwegian Johaan Vaaler (and patented in Germany).

1901 Mass production using an assembly line (for cars) is introduced in the USA by US manufacturer Ransome Olds (1864–1950); the system was later (1908) adopted by Henry Ford (1863–1947) to produce his Model T, and in 1913 he added a moving conveyor belt to the line.

1901 Electric lamp using rare-earth oxide filament is patented by German chemist and physicist Walter Nernst (1864–1941).

1901 Electric typewriter is invented by US engineer Thaddeus Cahill (1867–1934).

1901 Practical vacuum cleaner (powered by a petrol engine and mounted on a horse-drawn wagon) is invented by British bridge engineer Hubert Booth (1871–1955).

1901 Press-stud is invented by George Abraham.

1902 Spark plug for petrol engines is invented by German engineer Robert Bosch (1861–1942).

1902 Process using superheated water for extracting sulphur from underground deposits is invented by German-born US chemist Herman Frasch (c.1851–1914).

1902 Process for making nitric acid out of ammonia is invented by German chemist Wilhelm Ostwald (1853–1932).

1902 Improved lawn mower is invented by James Ransome.

1902 Air conditioning unit (for controlling humidity in a printing works) is invented by US engineer Willis Carrier; it is later (1906) improved by Stuart Cramer.

1902 Commercial process for liquefying air (and thus producing liquid oxygen and nitrogen cheaply) is invented by French scientist Georges Claude (1870–1960).

1902 Electrocardiograph is invented by Dutch physician Willem Einthoven (1860–1927).

1903 Ultramicroscope is invented by German scientists Richard Zsigmondy (1865–1929) and Henry Siedentopf (b.1872).

1903 Oxyacetylene torch is invented by Frenchman Edmond Fouché.

1903 No. 3 Lee-Enfield rifle, developed in 1893 by Scottish-born US engineer James Lee (1831–1904), is adopted by the British army.

1904 Diode (rectifying) valve is invented by British engineer John Fleming (1849–1945).

1904 Practical photocell is invented by German physicists Julius Elster (1854–1920) and Hans Geitel (1855–1923).

1904 Pneumatic hammer is introduced (for coal mines in Germany).

1904 Electric cash register is invented by US engineer Charles Kettering (1876–1958).

1904 Machine for making bottles (in moulds) is patented by US inventors Michael Owens (1859–1923) and Edward Libbey (1854–1925).

1904 Invar, a nickel-steel alloy that does not expand on heating, is invented by French physicist Charles-Edouard Guillaume (1861–1938).

1904 Wrist watch (for women) is invented by French jeweller Louis Cartier.

1905 Bakelite is invented by Belgian-born US chemist Leo Baekeland (1863–1944).

1905 Automatic pistol is patented by US gunsmith Arthur Savage (1857–1938).

1905 Contraceptive intrauterine device (IUD coil) is invented by R. Richter.

1905 Ammonium picrate high explosive (Dunnite or Explosive D) is invented for armour-piercing shells by US army officer B. W. Dunn (1860–1936).

1906 Triode (amplifying) valve is invented by US engineer Lee De Forest (1873–1961).

1906 Nichrome, a high-melting nickel-iron alloy used for making electric heater elements, is invented by US metallurgist Albert Marsh.

1906 Outboard motor for boats is invented by Norwegian-borm US engineer Ole Evinrude.

1907 Meccano metal constructional toy (first devised in 1900) is marketed by British toymaker Frank Hornby, sold later (1913) in the USA as Erector Set and in Germany as Märklin.

1908 Process for synthesizing ammonia is invented by German chemist Fritz Haber (1868–1934).

1908 Reliable tungsten-filament electric light bulb is invented by US chemist William Coolidge (1873–1975).

1908 Alloy Duralumin is invented by German metallurgist Alfred Wilm (1869–1937).

1908 Portable electric vacuum cleaner, invented in 1907 by US James Sprangler, is marketed in the USA by leather-maker William Hoover (1849–1932).

1908 Cellophane is invented by British chemist Charles Cross (1855–1935) and, independently, Swiss chemist Jacques Brandenberger.

1908 Simple radiation counter is invented by German physicist Hans Geiger (1882–1945).

COMMUNICATIONS

1901 25-centimetre (10-inch) shellac gramophone records are introduced in Britain.

1901 Electric hearing aid is patented by US engineer Miller Hutchinson (1875–).

1901 Method of synchronizing a motion-picture projector with a gramophone is developed by French inventor Léon Gaumont (1864–1946).

1902 Commercial electric typewriter is marketed by US inventor George Blickensderfer; an earlier (1901) design by Thaddeus Cahill was not a commercial success.

1902 Radio telephony is invented by US physicist and engineer Reginald Fessenden (1866–1932); it was first used in that year on two US ships.

1903 Arc transmitter for radio telegraphy is invented by Danish engineer Valdemar Poulsen (1869–1942).

1903 Commercially successful three-colour photographic process is invented by the French Lumière brothers, Auguste (1862–1954) and Louis (1864–1948).

1904 Method of sending photographs by telegraph using a photocell scanner (a primitive facsimile machine) is invented by German physicist Arthur Korn (1870–1945).

1904 Answerphone is introduced in the USA.

1905 Dial telephone is invented by US undertaker Almon Strowger.

1906 Silicon crystal detector (for radio receivers) is patented by US electrical engineer Greenleaf Pickard (1877–1956).

1906 Commercial AM radio broadcasting of speech and music is introduced in the USA by US inventor Reginald Fessenden (1866–1932).

1906 Crude television pictures are produced by Russian engineer Boris Rosing, using a Nipkow disc and a cathode-ray tube.

1906 Single-emulsion (transparency) colour film for movies, called Kinemacolor, is invented by British photographer George Smith (1864–1959).

1908 Electronic scanning circuit for capturing images is invented by Scottish electrical engineer Alan Campbell Swinton.

1908 The Morse letters SOS (... ---...)are introduced as an international distress signal for ships at sea.

ASTRONOMY AND SPACE

1910

1912 Rotational periods of the planets Venus, Mars, Jupiter, Saturn and Uranus are determined (using the Doppler effect) by US astronomer Vesto Slipher (1875–1969).

1914 Fact that spiral nebulae are galaxies is discovered by British astronomer Arthur Eddington (1882–1944).

1915 First white dwarf star, Sirius B, is discovered by US astronomer Walter Adams (1876–1956).

1915 Proxima Centauri, the nearest star to Earth after the Sun, is discovered by British astronomer Robert Innes (1861–1933).

1916 Star with the largest proper motion, later called Barnard's star, is discovered by US astronomer Edward Barnard (1857–1923).

1916 Schwarzschild radius (below which a star becomes a black hole, from inside which no radiation can escape) is discovered by German astronomer Karl Schwarzchild (1873–1916).

1918 Milky Way is discovered to be ten times larger than previously thought by US astrophysicist Harlow Shapley (1885–1972).

1919 23-year cycle of polarity reversal of the magnetic field of sunspots is discovered by US astronomer George Hale (1868–1936).

1919 Explanation accounting for the evolution of stars is published by British astronomer James Jeans (1877–1946).

1919 Cephid variable stars are discovered in the Andromeda Nebula by US astronomer Edwin Hubble (1889–1953).

PHYSICS

1911 Atomic nucleus is discovered by New Zealand-born British physicist Ernest Rutherford (1871–1937).

1911 Cosmic rays (although not named as such until 1925 by US physicist Robert Millikan) are discovered by Austrian physicist Victor Hess (1883–1964), who also showed that they come from outer space, not from the Sun.

1911 Superconductivity at very low temperatures is discovered by Dutch physicist Kamerlingh Onnes (1853–1926).

1912 X-ray diffraction by crystals is discovered by French physicist Max von Laue (1879–1960).

1912 Bragg's law (relating the diffraction of X-rays by crystals to the spacing of the atoms in the crystal) is formulated by British physicist William Henry Bragg (1862–1942), assisted by his son William Lawrence Bragg (1890–1971).

1912 Electric charge on an electron is determined by US physicist Robert Millikan (1868–1953).

1912 Born-Haber cycle (for calculating the lattice energy of an ionic crystal) is discovered by German-born British physicist Max Born (1882–1970) and German chemist Fritz Haber (1868–1934).

1912 Paschen-Back effect (the splitting of spectral lines in a strong magnetic field) is discovered by German physicists Louis Paschen (1865–1947) and Ernest Back (1881–1959).

1913 Stark effect (the splitting of spectral lines in a strong electric field) is discovered by German physicist Johannes Stark (1874–1957).

CHEMISTRY

Danish biochemist Sören Sörensen (1868–1939).

1909 Identity of the sugar component (D-ribose) in the nucleic acid RNA is discovered by Russian-born US biochemist Phoebus Levene (1869–1940), who went on (1929) to identify the sugar (deoxyribose) in DNA.

1910 Amino acid tryptophan is discovered by British chemist Frederick Gowland Hopkins (1861–1947).

1910 Histamine is synthesized by British physiologists George Barker (1878–1939) and Henry Dale (1875–1968).

1911 Method of making propanone (acetone) by the bacterial fermentation of grain is discovered by Russian-born Israeli chemist Chaim Weizmann (1874–1952).

1911 Metallic element hafnium (symbol Hf) is discovered by French chemist Georges Urbain (1872–1932), and first extracted (1923) by Dutch physicist Dirk Coster (1889–1950) and Hungarian-born Swedish chemist Georg von (or George de) Hevesy (1886–1966).

1912 Existence of two forms of uranium (now known to be isotopes) is discovered by German physicist Hans Geiger (1882–1945) and British physicist J. Nuttall (1890–1958).

1912 Use of radioactive tracers in chemical analysis is established by Austrian chemist Friedrich Paneth (1887–1958) and Hungarian-born Swedish chemist Georg von (or George de) Hevesy (1886–1966).

1913 Fact that lead from different natural sources can have different relative atomic masses (now known to be due to their having different combinations of isotopes) is discovered by US chemist Theodore Richards (1868–1928).

1913 Isotopes are discovered by British chemist Frederick Soddy (1877–1956).

BIOLOGY

1910 Role of chromosomes in inheritance (and occurrence of genes along chromosomes) is discovered by US geneticist Thomas Hunt Morgan (1866–1945).

1911 Histamine (which can cause allergies) is discovered by British physiologist Henry Dale (1875–1968).

1911 Gradient theory of regeneration, that the dominant part of a regenerating organ develops first, is proposed by US biologist Charles Child (1869–1954).

1912 Fossil remains of Piltdown man are "discovered" by British naturalist Charles Dawson (1864–1916), later (1953) shown to fraudulent.

1913 Chromosome mapping is devised by US geneticist Alfred Sturtevant (1891–1970).

1914 Role of ATP (adenosine triphosphate) in cell metabolism is discovered by Russian-born US biochemist Fritz Lipmann (1899–1986).

1914 Amino acids in blood are discovered by US physiologist John Abel (1857–1938).

1915 Bacteriophages (viruses that attack bacteria) are discovered by British bacteriologist Frederick Twort (1877–1950) and later (1966), independently, by Canadian bacteriologist Felix D'Herelle (1873–1949), who named them.

1915 Thyroid hormone thyroxine is discovered by US biochemist Edward Kendall (1886–1972).

1917 Role of calories (in food) as an energy source is discovered by US

MEDICINE

1910 Bacterium that causes typhus is discovered by US pathologist Howard Ricketts (1871–1910), who also showed, independently of French physician Charles Nicolle (1866–1936), that the disease is transmitted by body lice.

1910 Chagas'disease, the South American form of trypanosomiasis, is discovered by Brazilian bacteriologist Carlos Chagas (1879–1934).

1910 Antiseptic and disinfectant properties of iodine are discovered by US physician F. Woodbury.

1910 Antitoxin against botulism is discovered by German physician L. Leuchs.

1910 Sickle-cell anaemia is discovered by US physician James Herrick (1861–1954) and, later (1917), his compatriot V. Emmel (1878–1928).

1910 Antisyphilis drug Salvarsan (arsphenamine) is discovered by German physician Paul Ehrlich (1854–1915); its synthesis is aided by US dermatologist Jay Shamberg (1870–1934).

1911 Plasma transfusion to treat haemophilia is discovered by US physician Thomas Addis (1881–1949).

1911 Vitamin B1 (thiamine) – the first vitamin to be found – is discovered by Polish-born biochemist Casimir Funk (1884–1967) and its deficiency shown to be the cause of beriberi.

1912 Nicotinic acid (niacin), a B vitamin, is discovered in rice polishings by Polish-born biochemist Casimir Funk (1884–1967); in the same year he coined the word "vitamine", later shortened to "vitamin".

FARMING AND FOOD

TRANSPORT

France by engineer Paul Cornu (1881–1944), but it proves to be unreliable.

1908 First aeroplane flight in Britain is made by US-born British aviator Samuel Cody (1862–1913) in a plane of his own design.

1908 First aeroplane flight around a 1-kilometre circuit is made by French aviation pioneer Henri Farman (1874–1958) who, with his brother Maurice (1878–1964), also invented ailerons for controlling aeroplanes in flight.

1908 Gyrocompass is invented by German inventor Hermann Anschütz-Kämpfe (1872–1931); later (1910) improved by US engineer Elmer Sperry (1860–1930).

1908 Hand-operated car windscreen wipers are invented by Englishman Gladstone Adams (1880–1966); automatic wipers, using the vacuum from the car's exhaust system, followed later (1921), invented by Englishman W.M. Folberth.

1908 Model T is marketed by the US Ford Company.

1909 First aeroplane flight across the English Channel (Baraques to Dover) is made by French airman Louis Blériot (1872–1936).

1909 Hydrofoil ship is built by Italian inventor Enrico Forlanini.

1909 Gyro-stabilized monorail railway system is invented by Irish engineer Louis Brennan (1852–1932).

1909 Air-speed indicator for aircraft is invented by British engineer Alec Ogilvie.

1909 Distributor, for supplying high-voltage current to a car's spark plugs, is invented by US engineers Edward Deeds and Charles Kettering (1876–1958).

ENGINEERING AND TECHNOLOGY

1908 Cadmium cell (battery), invented by US scientist Edward Weston (1850–1936), is adopted as the standard of electrical voltage.

1909 Electric toaster is marketed by the US General Electric company.

COMMUNICATIONS

0 — 1910

FARMING AND FOOD

1910 Petrol-driven combine harvester is invented in the USA.

1910 Coffee-making machine is invented by Englishman Alfred Cohn.

1912 Process for making sugar from wood is invented by German chemist Friedrich Bergius (1884–1949).

1913 Vacuum milking machine is invented by Swede Carl de Laval (1845–1913).

1917 Uncooked quick-frozen foods are invented by US food technologist Clarence Birdseye (1886–1956).

1918 Domestic mechanical refrigerator is marketed in the USA by the Kelvinator Company.

1919 Paper tea bags are invented and mass-produced in the USA by Joseph Krieger; silk tea bags were made earlier (1904) by tea importer Thomas Sullivan.

1919 First supermarket, called Piggly-Wiggly, is opened in Memphis, Tennessee, by US retailer Clarence Saunders (1881–1953).

1920 Steam-powered rotary hoe (called a Rotovator) is invented by Australian blacksmiths Albert and Cliff Howard.

1921 Aeroplanes are first used for dusting crops with pesticides in the USA.

1925 First successful hydroponics experiments (growing plants without soil) are carried out in the USA.

1926 Artificial PKN (phosphorus, potash and nitrogen) fertilizers are first produced in Britain.

TRANSPORT

1910 Automatic gearbox (a torque converter) for cars is invented by German engineer Hermann Föttinger.

1910 Rotary petrol engine for aeroplanes is invented by French engineer Pierre Clerget (1875–1943).

1910 Anti-rolling mechanism for ships is invented by H. Frahm; later (1913) true gyro-stabilizer is patented by US engineer Elmer Sperry (1860–1930).

1910 First take-off of an aeroplane from a ship is made by US airman Eugène Ely from USS Birmingham; he later (1911) makes the first on-ship landing on USS Pennsylvania.

1910 Successful seaplane, a rotary-engined biplane, is constructed and flown by French engineer Henri Fabre; the first commercial seaplanes were built later (1911), independently, by US engineers Glenn Curtis (1879–1930) and (1912) Glenn Martin (1886–1955).

1911 Quadruplane (four-winged aeroplane) is patented by US inventor Matthew Sellers (1869–1932).

1911 Electric self-starter for cars is invented by US engineer Charles Kettering (1876–1958).

1912 Diesel railway locomotive is built by the Swiss Sulzer company.

1912 Non-stop flight from Paris to London is made by French airman Henri Seimet.

1912 Parachute jump is made from an aeroplane by American Albert Berry.

1912 Kiel Canal is opened (as the Kaiser Wilhelm Canal), running 99 kilometres across Germany to connect the Baltic Sea with the North Sea.

ENGINEERING AND TECHNOLOGY

1910 Neon lighting is invented by French scientist Georges Claude (1870–1960).

1910 High-pressure steam turbine is invented by B. Ljundstrom.

1910 Electrostatic precipitator for extracting harmful dust from gases in factory chimneys is invented by US chemist Frederick Cottrell (1877–1948).

1910 Spectroheliograph (for photographing the Sun's spectrum) is invented by US astronomer George Hale (1877–1945).

1910 Electric washing machine is patented by US engineer Alva Fisher (1862–1947).

1911 Cloud chamber is invented by Scottish physicist Charles Wilson (1869–1959).

1911 High output carbon-arc lamp for movie studios is invented by German-born US brothers John (1869–1959) and Anton (1872–1927) Klieg.

1911 Automatic light machine gun is invented by US army officer Isaac Lewis (1858–1931), based on a design by his compatriot Samuel MacLean.

1912 Thermal cracking technique (for oil refining) is invented by US chemist William Burton (1865–1954).

1912 Low-pressure water turbine is invented by Austrian engineer Viktor Kaplan (1876–1934).

1912 Ultraviolet microscope is invented by British microscopist Joseph Barnard (1870–1949).

1912 Method of making oil by hydrogenating coal is invented by German chemist Friedrich Bergius (1884–1949).

1912 Stainless steel (containing chromium and nickel) is invented by British metallurgist Harold Brearley (1871–1948), and later (1919) patented as non-rusting steel by US inventor Elwood Haynes (1857–1925).

1912 "Sun" valve (responsive to sunlight) for controlling actylene gas supply to unmanned lighthouses, buoys and railway signals is invented by Swedish engineer Nils Dalén (1869–1937).

1912 Portable domestic electric room heater is marketed in England by the Belling company.

1913 Hot-cathode X-ray tube is invented by US chemist William Coolidge (1873–1975).

1913 Industrial version of the Haber process for making ammonia is invented by German chemist Karl Bosch (1874–1940).

COMMUNICATIONS

1911 Cathode-ray tube image scanning (later used in television) is invented by German Karl Braun (1850–1918) and Scotsman Alan Campbell-Swinton.

1911 Regular cinema newsreels are inaugurated by French photographer Charles Pathé (1873–1932).

1912 Heterodyne radio receiver, using positive feedback, is invented by US engineer Reginald Fessenden (1866–1932) and, independently, Austrian radio engineer Alexander Meissner (1883–1958).

1912 Audion (triode) valve for radio amplifiers is developed by US electronics engineer Lee De Forest (1873–1961).

1912 Teleprinter machine is perfected by Canadian inventor Frederick Creed (1871–1957), whose system develops into Telex.

1912 Colour photography process that employs three dye-coupled emulsions on one plate is invented by R. Fischer.

1912 Type of colour motion-picture film is introduced by French inventor Léon Gaumont (1864–1946).

1916 Type of sonar for detecting icebergs and submarines is invented by French physicist Paul Langevin (1872–1946).

1916 Radio direction finder is invented by Frederick Kolster, giving rise to navigational radio beacons.

1917 Condenser microphone is invented by E. C. Wente.

1917 Method of transmitting radio signals underwater is invented by US engineer James Rogers (1856–1929).

ASTRONOMY AND SPACE

1919 Bending of starlight by the Sun's gravity (during a solar eclipse), thus proving Einstein's theory of gravitation, is discovered by British astronomers Frank Dyson (1868–1939) and Arthur Eddington (1882–1944); similar confirmation is obtained later (1922) by Swiss-born US astronomer Robert Trumpler (1886–1956).

1920

1920 Red shift in the spectra of galaxies (implying that they are receding at speed) is discovered by US astronomer Vesto Slipher (1875–1969).

1920 Structure of the Milky Way is discovered by German astronomer Maximillian Wolf (1863–1932).

1920 Saha's equation, which relates the

PHYSICS

1913 Atom is conceived as having a central nucleus surrounded by orbiting electrons by Danish physicist Niels Bohr (1885–1962).

1913 Atomic number is equated with the positive charge on the atomic nucleus by British physicist Henry Moseley (1887–1915).

1914 Relationship between the atomic number of an element and the frequency of the lines in its X-ray spectrum (Moseley's law) is discovered by British physicist Henry Moseley (1887–1915).

1915 Elliptical (rather than circular) orbits of electrons in atoms is proposed by German physicist Arnold Sommerfield (1868–1951) and, independently, W. Wilson (1875–1965).

1915 Rectifying effect (ability to convert alternating current into direct current) of germanium is discovered by Swedish physicist Carl Benedicks (1875–).

1916 General Theory of Relativity is proposed by German-born US physicist Albert Einstein (1879–1955).

1919 Proton is discovered by New Zealand-born British physicist Ernest Rutherford (1871–1937).

1919 Atomic fission – splitting of an atom – (by alpha-ray bombardment of nitrogen to convert it into oxygen) is achieved by New Zealand-born British physicist Ernest Rutherford (1871–1937).

1919 First separation of isotopes (of neon) is achieved by British physicist Francis Aston (1877–1945), using his mass spectroscope.

1919 Barkhausen effect (the discontinuity of magnetization of a ferromagnetic substance) is discovered by German physicist Heinrich Barkausen (1881–1956).

1920 Existence of the neutron is postulated by New Zealand-born British physicist Ernest Rutherford (1871–1937).

1923 Compton effect – the scattering of X-rays, with an increase in wavelength (reduction in energy), by collisions with matter – is discovered by US physicist Arthur Compton (1892–1962).

CHEMISTRY

1913 Presence of ozone in the upper atmosphere is discovered by French physicist Charles Fabry (1867–1945).

1913 Composition of chlorophyll is discovered by German chemist Richard Willstätter (1872–1942).

1914 Phenarsazine chloride (Adamsite), a potential war gas that causes sneezing, is discovered by US chemist Roger Adams (1889–1971).

1914 Xylyl bromide (Cyclite, T-Stoff) is developed as a toxic war gas in World War I by German chemist von Tappen.

1915 Carbonyl chloride (phosgene, Collognite or D-Stoff) is developed as a toxic war gas by Germany, and later used by other combatants in World War I.

1916 Explanation that a covalent bond involves the sharing of electrons between two atoms is given by US chemist Gilbert Lewis (1875–1946).

1916 Bromacetone (Martonite, BA or B-Stoff) is developed as a tear gas by Germany, and later used by other combatants in World War I.

1916 Chloropicrin (nitrochloroform) is developed as a tear gas by Russia, and later used (sometimes mixed with chlorine as Yellow Star gas) by other combatants in World War I.

1916 Powerfully magnetic cobalt-tungsten steel is discovered by Japanese metallurgist Kotaro Honda (1870–1954).

1916 Dichloroethyl sulphide (mustard gas, a toxic and blinding gas) is developed by Germany.

1917 Toxic war gases diphenylchlorarsine (Clark I, a sneezing gas), diphenylcyanarsine (Clark II, a lethal gas) and phenyldichlorarsine (Sternite, a tear gas) – the so-called Blue Cross gases – are developed by Germany.

1917 Isotope of the radioactive element protactinium (Pa-234) is discovered and named by German physicists Lise Meitner (1878–1968) and Otto Hahn (1879–1968), although another isotope had been found earlier (1913) by Polish-born physical chemist Kasimir Fajans (1887–1975).

1918 Chemical chain reactions are discovered by German physical chemist Hermann Nernst (1864–1941).

1918 Lewisite poison gas (for warfare) is developed by US chemist Winford Lewis (1878–1943).

1918 Toxic gas bromobenzoyl cyanide (BBC, Camite) is developed in France as a tear gas for warfare.

1919 Whole-number rule (that isotopes have integral atomic masses) is discovered by British physicist Francis Aston (1877–1945).

1920 Electronic nature of organic chemical reactions is discovered by Scottish chemist Arthur Lapworth (1872–1941).

1921 Glutathione, a tripeptide (combination of three amino acids) involved in cell metabolism, is discovered by British biochemist Frederick Gowland Hopkins (1861–1947).

BIOLOGY

physiologist Graham Lusk (1866–1932) and British bacteriologist Frederick Twort (1877–1950).

1918 Fact that humans have 48 chromosomes in all body cells is discovered by US embryologist Herbert Evans (1882–1971).

1919 Acetylcholine is discovered by British physiologist Henry Dale (1875–1968), although its role as a neurotransmitter is not established until later (1921 by German physiologist Otto Loewi (1873–1961).

1919 Communication among honeybees by means of body movements (the "bee's dance") is discovered by Austrian-born German biologist Karl von Frisch (1886–1982), who later (1947) discovers that the bees use the polarization of light for orientation.

1921 Insulin is discovered by Canadian physiologists Frederick Banting (1891–1941) and Charles Best (1899–1978) and Scotsman John Macleod (1876–1935).

1921 Chemical mechanism for the transmission of nerve impulses is discovered by German-born US pharmacologist Otto Loewi (1873–1961).

MEDICINE

1913 Schick test (for immunity to diphtheria) is discovered by Hungarian-born US physician Béla Schick (1877–1967).

1913 Vaccine against diphtheria is developed by German bacteriologist Emil von Behring (1854–1917).

1913 Cancer is produced artificially in experimental animals by Danish physician Johannes Fibiger (1867–1928), first using nematodes to produce tumours in rats and later (1920) using coal tar as a carcinogen.

1913 Vitamin A (retinol), soluble in fats, is discovered by US biochemist Elmer McCollum (1879–1967), who distiguished it from water-soluble compounds, which he classified as B vitamins. Vitamin A is simultaneously, and independently, discovered by his biochemist compatriots Thomas Osborne (1859–1929) and Lafayette Mendel (1872–1935).

1914 Successful open-heart surgery is first performed (on a dog) by French surgeon Alexis Carrel (1873–1944).

1914 Anticoagulent properties of sodium citrate (used in stored blood for transfusions) is discovered by US physician Richard Lewisohn (1875–).

1914 Dakin's solution (0.5% sodium hypochlorite), an antiseptic for treating war wounds, is developed by British chemist Henry Dakin (1880–1952) and French biologist Alexis Carrel (1873–1944).

1914 Fact that Brill's disease (named after US physician Nathan Brill (1860–1925)) is actually typhus is discovered by US bacteriologist Harry Plotz (1890–1947), who went on to produce a vaccine against it.

1915 Fact that pellagra is a dietary deficiency disease is discovered by Austrian-born US physician Joseph Goldberger (1874–1929); the vitamin concerned (nicotinic acid, or niacin) is later (1937) used in the treatment of pellagra by US biochemist Conrad Elvehjem (1901–62).

1915 Streptomyces bacterium is discovered by Russian-born US microbiologist Selman Waksman (1888–1973).

1917 Vaccine against Rocky Mountain spotted fever is developed by US bacteriologist Ralph Parker (1888–1949).

1917 Bacteriophages (viruses that attack bacteria) are discovered by French bacteriologist Félix d'Hérelle (1873–1949).

1917 Treatment for general paralysis by deliberately infecting the patient with malaria – a precursor of shock therapy – is introduced by Austrian neurologist Julius von Wagner-Jauregg (1857–1940).

1918 Anticoagulant drug heparin is discovered by US physiologists William Howell (1860–1945) and Luther Holt (1855–1924).

1919 Fact that there are two strains of botulism (each requiring a different antitoxin) is discovered by US physician Georgiana Burke.

1919 Drug tryparsamide is introduced as a treatment for sleeping sickness by US physician Louise Pearce (1885–1959).

1920 Ergotamine treatment for migraine is discovered by German chemist Karl Spiro (1867–1932).

1920 Band-Aid sterile first-aid dressings are marketed by the US Johnson & Johnson company; its British equivalent, called Elasoplast, did not appear until 1928.

FARMING AND FOOD

1927 Improved machine for picking cotton is invented in the USA by the brothers John (1892–1954) and Mack (1900–66) Rust.

1928 Wrapped sliced bread is invented by US jeweller Otto Rohwedder.

TRANSPORT

1913 Four-engined aeroplane, the Grand, is built and flown in Russia by Russian-born US inventor Igor Sikorsky (1889–1972).

1913 Diesel-electric railcar enters service in Sweden.

1914 All-metal aeroplanes are built by German engineer Claude Dornier (1884–1969).

1914 Automatic traffic lights, first installed in Detroit in 1919, are invented by Alfred Benesch.

1914 Panama Canal (begun in 1880) is opened, running for 82 kilometres across the isthmus of Panama to connect the Atlantic and Pacific oceans.

1915 All-metal cantilever-wing aircraft, the J-1, is built by German engineer Hugo Junkers (1859–1935).

1915 Military tank is invented by British army officer Ernest Swinton (1868–1951).

1916 All-steel bodied cars are built in the USA.

1917 Cantilever wing for monoplane airplanes is invented by US engineer Clyde Cessna (1874–1954).

1917 Radio-controlled unmanned aeroplane (for bombing an enemy) is invented by British air force officer Archibald Low.

1917 Trans-Australian railway is completed.

1918 High-speed hydrofoil boat is built by Scottish-born US inventor Alexander Bell (1847–1922).

1918 Experimental helicopter is built by British engineer Peter Cooper-Hewitt (1861–1921).

1919 Hydraulic brakes for cars are introduced by the US Lockheed Company.

1919 Non-stop aeroplane flight across the Atlantic Ocean (Newfoundland to Ireland) is made by British airmen John Alcock (1892–1919) and Arthur Whitten Brown (1886–1948).

ENGINEERING AND TECHNOLOGY

1913 Heatproof plastic laminate, made using formaldehyde (methanal) resins, is marketed by the US Formica Insulation company.

1913 First skyscraper, the Woolworth Building, is built in New York City.

1913 Bomb sight is invented by US engineer Glenn Martin (1886–1955).

1914 First successful submachine gun designed to fire pistol ammunition, the Villar Perosa, is invented by Italian gunsmith B.A. Revelli.

1914 Brassiere is invented by Mary Jacob.

1914 Mercury-vapour vacuum pump is invented by US chemist Irving Langmuir (1881–1957).

1915 Gas-filled tungsten filament lamp is invented by US chemist Irving Langmuir (1881–1957).

1915 Borosilicate (Pyrex) glass is marketed by the US Corning company.

1915 High explosive amatol (a mixture of TNT and ammonium nitrate) is invented in England.

1915 Mills bomb (hand grenade) is invented by British metallurgist William Mills (1856–1932).

1915 Interrupter gear (to allow a machine gun to fire through an aeroplane's propeller) is introduced in Germany by the Fokker company.

1915 Remote-controlled tank for the German army is invented by German engineer Anton Flettner (1855–1961).

1915 Adaptable trench mortar is invented by British engineer Frederick Stokes (1860–1927); it is adopted by the British army in 1917.

1916 Alnico magnetic cobalt alloy is invented by Japanese metallurgist Kotaro Honda (1870–1954).

1916 Schneider-Creusot military tank is adopted by the French army.

1916 Agitator-type washing machine is invented by US engineer John Fisher.

1917 Bergmann Machinen Pistole (submachine gun) is invented by German engineer Hugo Schmeisser.

1917 Commercial detergents are developed in Germany.

1917 Lightweight portable electric drill is marketed by US inventors Duncan Black and Alonso Decker.

1918 Diamond-tipped rock-boring drill is developed by the Shell company.

1918 Self-loading (automatic) rifle is invented by US gunsmith John Browning (1855–1926).

1918 High-intensity arc searchlight for the USA military is invented by US engineer Elmer Sperry (1860–1930).

1918 Lewisite poison gas (for warfare) is invented by US chemist Winford Lewis (1878–1943).

1919 Mass spectrograph is invented by British physicist Francis Aston (1877–1945).

COMMUNICATIONS

1918 System of talking motion pictures (with sound on film), called Tri-Ergon, is invented by German engineer Hans Vogt (1890–); it has its first public showing in Berlin in 1922. A similar system (Phonofilm) is developed in the USA by US electronics engineer Lee De Forest (1873–1961).

1919 Superheterodyne radio receiver is invented by US electrical engineer Edwin Armstrong (1890–1954).

1919 Experimental radio broadcasts using gramophone records are made by US engineer Frank Conrad (1874–1941).

1919 Dial telephones are introduced in the USA by the AT & T company.

1919 Flash camera and aerial camera are invented by US engineer Sherman Fairchild (1896–1971).

1920

TRANSPORT

1920 The first all-welded ship is built in Britain.

1921 Hydraulic brakes for cars are introduced in France and the USA.

1921 Manufacture of the "Baby" Austin 7 car, British equivalent of Ford's Model T, is begun by British industrialist Herbet Austin (1866–1941).

ENGINEERING AND TECHNOLOGY

1920 Polymer plastics are first made by German chemist Hermann Staudinger (1881–1965).

1920 0.45-in. calibre submachine gun (Tommy gun) is invented by US soldier John Thompson (1860–1940); it is first used in military action by the US Marines in 1925.

1920 Invar (low-expansion alloy of iron and nickel) is invented by Swiss physicist Charles Guillaume (1861–1938).

1922 First all-metal aeroplane in the USA is built by US engineer William Stout (1880–1956).

COMMUNICATIONS

1920 Regular radio broadcasts begin in the USA (from station KDKA, Pittsburgh).

1921 Magnetron valve, for generating microwaves, is invented by US physicist Albert Hull (1880–1966).

1922 Echo-sounder for detecting submarines is invented by US engineer Reginald Fessenden (1866–1932).

ASTRONOMY AND SPACE

ionization of gases to temperature in the atmospheres of stars and is important in understanding stellar evolution, is discovered by Indian astrophysicist Meghnad Saha (1894–1956).

1922 Plaskett's star, a binary star consisting of two massive supergiants, is discovered by Canadian astronomer John Plaskett (1865–1941).

1923 Fact that galaxies are not part of the Milky Way is discovered (through the location of the Andromeda galaxy) by US astronomer Edwin Hubble (1889–1953).

1924 Fact that the luminosity of a star is an approximate function of its mass is discovered by British astronomer Arthur Eddington (1882–1944).

1925 Cosmic radiation is discovered in the upper atmosphere by US physicist Robert Millikan (1868–1953).

1925 Classification scheme for galaxies is introduced by US astronomer Edwin Hubble (1889–1953).

1926 Fact that stars consist mainly pf hydrogen and helium is disceovered by Celia Payne-Gaposchkin (1900–79).

1927 Big Bang theory of the Universe is proposed by Belgian astrophysicist George Lemaitre (1894–1966), revised (1946) by Russian-born US physicist George Gamov (1904–88) et al.

1928 Highly ionized forms of oxygen and nitrogen atoms are discovered in emission nebulae (and shown to be the cause of their colours) by US astronomer Ira Bowen (1898–1973).

1928 Abundances of elements in the Sun's atmosphere are discovered (spectrographically) by US astronomer Henry Russell (1877–1957).

1929 Hubble's law (which relates the velocity of recession of stars to their distances from our Galaxy) is discovered by US astronomer Edwin Hubble (1889–1953).

PHYSICS

1924 Existence of the ionosphere is proved by British physicist Edward Appleton (1892–1965).

1924 Exclusion principle (that no two subatomic particles can have the same set of quantum numbers) is discovered by Austrian physicist Wolfgang Pauli (1900–58).

1924 Fact that moving subatomic particles have an associated wavelength (i.e., that they can behave as both particles and waves) is discovered by French physicist Louis de Broglie (1892–1987).

1924 Method of refracting X-rays (and thereby measuring their wavelength) is discovered by Swedish physicist Karl Siegbahn (1886–1978).

1925 Fact that electrons spin is discovered by Dutch-born US physicists Samuel Goudsmit (1901–78) and fellow research student George Ulenbeck (1900–).

1925 Auger effect (the emission of an electron from an atom with accompanying X-rays or gamma rays) is discovered by French physicist Pierre Auger (1899–).

1926 Wave equation for the hydrogen atom (which describes the distribution of electrons) is discovered by Austrian physicist Erwin Schrödinger (1887–1961), thus founding wave mechanics.

1926 Method of attaining temperatures within a thousandth of a degree of absolute zero is discovered by US chemist William Giauque (1895–1982).

1926 Term "photon" for a quantum of light is coined by US chemist Gilbert Lewis (1875–1946).

1927 Fact that parity is conserved in nuclear reactions is discovered by Hungarian-born US physicist Eugene Wigner (1901–95).

1927 Method of diffracting a beam of electrons (thus proving their wave nature) using a nickel crystal is discovered by US physicists Clinton Davisson (1881–1958) and Lester Germer (1896–1971) and, independently, using gold foil by British physicist George Thomson (1892–1975).

1927 First X-ray crystallographic analysis of proteins (animal fibres) is carried out by British physicist William Astbury (1898–1961).

1928 Existence of the positron is postulated by US physicist Julius Oppenheimer (1904–67) and, independently, British physicist Paul Dirac (1902–84).

1928 Raman effect or scattering (the scattering of electromagnetic radiation by the molecules of a medium through which it passes) is discovered by Indian physicist Chandrasekhara Raman (1888–1970).

1929 Unified Field Theory, combining electromagnetism and gravitation, is proposed by German-born US physicist Albert Einstein (1879–1955).

CHEMISTRY

1922 Natural polymers (in rubber) are discovered by German chemist Hermann Staudinger (1881–1965); he later (1926) identified the polymeric nature of all plastics.

1923 Unified theory of acids and bases is proposed by Danish chemist Johannes Bronsted (1879–1947) and, independently, British chemist Thomas Lowry (1874–1936).

1924 Parachor (the molecular volume of a substance with unit surface tension, a useful quantity in organic chemistry) is discovered by British chemist Samuel Sugden (1892–1950).

1925 Structure of morphine is discovered by British chemist Robert Robinson (1886–1975).

1925 Metallic element rhenium (symbol Re) is discovered by German chemists Walter Noddack (1893–1960) and Ida Tacke (later Noddack) (1896–).

1927 Modern theory of valence – that valence depends on the number of electrons in an atom's outer shells – is proposed by British chemist Nevil Sidgwick (1873–1952).

1927 Quantum theory of chemical bonding is developed by German physicists Fritz London (1900–54) and Walter Heitler.

1927 Zyklon-B, a solid fumigant that gives off toxic hydrogen cyanide gas when exposed to air, is developed by German chemist Bruno Tesch (1946–); it is later (World War II) used in Nazi extermination camps.

1928 Diels-Alder reaction (in which two double-bonded organic molecules join to form a ring compound) is discovered by German chemists Otto Diels (1876–1954) and Kurt Alder (1902–58).

1929 Short-life free radicals (e.g. methyl radical) are discovered by Austrian chemist Friedrich Paneth (1887–1958).

1929 Polymerization of ethyne (acetylene) is discovered by Belgian-born US chemist Julius Nieuwland (1878–1936).

1929 Fact that oxygen consists of three different isotopes (O-16, O-17 and O-18) is discovered by US chemist William Giauque (1895–1982).

1929 Goldschmidt's law, relating crystal structure to the properties of the component ions, is discovered by Swiss chemist Victor Goldschmidt (1888–1947).

BIOLOGY

1922 Lysozyme (an enzyme that destroys bacteria) is discovered by Scottish bacteriologist Alexander Fleming (1881–1955).

1923 Method of studying tissue respiration is discovered by German botanist Otto Warburg (1859–1933).

1924 Electric currents in the brain are discovered by German physiologist Hans Berger (1873–1941).

1924 Role of ultraviolet light in increasing vitamin D from food is discovered by US biochemist Harry Steenbock (1886–1967).

1924 Fossils of Australopithecus (an apelike link in human evolution) are discovered by Australian-born South African anthropologist Raymond Dart (1893–1988).

1925 Fact that iron is an important component of erythrocytes (red blood cells) is discovered by US biologist George Whipple (1878–1976).

1925 Cytochrome, an enzyme that acts as a respiratory catalyst in cells, is discovered by Russian-born Polish (later British) chemist David Keilin (1887–1963).

1926 Crystalline urease (the first enzyme to be crystallized) is produced by US chemist James Sumner (1877–1955), who also shows that it is a protein.

1926 Crystalline insulin is produced by US biochemist John Abel (1857–1938), who also shows that it is a protein.

1927 Mutations in fruit flies (Drosophila) are produced using X-rays by US geneticist Hermann Muller (1890–1967).

1929 Female sex hormone oestrogen is discoveed by German chemist Adolf Butenandt (1903–95) and, independently, US biologist Edward Doisy (1893–1986).

1929 Structure of haem (the non-protein part of haemoglobin) is discovered by German chemist Hans Fischer (1881–1945), who also synthisized it.

MEDICINE

1922 Vitamin D (calciferol) is discovered by US biochemist Elmer McCollum (1879–1967).

1922 Vitamin E (tocopherol) is discovered by US embryologist Herbert Evans (1882–1971).

1922 Cod liver oil or sunlight as a treatment for rickets is discovered by Austrian physician H. Chick, later shown to be effective because of the formation of vitamin D in the body.

1922 Insulin is first used successfully to treat a diabetic patient (in Toronto, Canada); it is later (1925) used on children in England.

1923 Dick test, for determining a person's susceptibility to scarlet fever, is discovered by US bacteriologists Gladys (1881–1963) and George (1881–1967) Dick.

1924 Use of liver in the diet as a treatment for pernicious anaemia is discovered by US physicians George Minot (1885–1950) and William Murphy (1892–1987).

1924 Bacterium that causes scarlet fever is discovered by US bacteriologists Gladys (1881–1963) and George (1881–1967) Dick.

1925 Carcinogenetic effect of ultraviolet radiation is discovered by US biologist Ernest Just (1883–1941).

1925 Use of parathyroid gland extract as a treatment for pernicious anaemia is discovered by Canadian biochemist James Collip (1892–1965).

1928 Antibiotic penicillin is discovered by Scottish bacteriologist Alexander Fleming (1881–1955), later (1935) developed by Australian-born British pathologist Howard Florey (1898–1968) and German-born British bacteriologist Ernst Chain (1906–79).

1928 Vitamin C (ascorbic acid) is discovered by Hungarian biochemist Albert Szent-Györgyi (1893–1986) and later (1932), independently, by US biochemist Glen King (1896–1988).

1929 Technique of diagnosing heart conditions by passing a catheter into the heart (via an artery in the arm) is introduced by German surgeon Werner Forssmann (1904–1979), and developed by French physician André Cournard (1895–1988) and US physician Dickinson Richards (1895–1973).

1930

ASTRONOMY AND SPACE

1930 "Planet X" (Pluto) is discovered by US astronomer Clyde Tombaugh (1906–).

1931 Chandrasekhar limit (1.4 times the mass of the Sun, which cannot be exceeded by a white dwarf star) is discovered by Indian-born US astrophysicist Subrahmanyan Chandrasekhar (1910–95).

1932 Radio waves from outer space are discovered by US radio engineer Karl Jansky (1905–50), so establishing the science of radio astronomy.

PHYSICS

1930 Fact that cosmic rays consist of protons (and other positively charged particles) is discovered by Italian-born US physicist Bruno Rossi (1905–94).

1930 Value of Van der Waals', or London, forces between molecules is discovered by German physicist Fritz London (1900–54).

c.1930 Néel temperature (above which an antiferromagnetic material becomes paramagnetic) is discovered by French physicist Louis Néel (1904–).

CHEMISTRY

1930 Refrigerant gas Freon (dichlorodifluoromethane, the first of the CFCs) is discovered by US chemist Thomas Midgley (1889–1944).

1930 Electrophoresis (as a method of separating proteins) is developed by Swedish chemist Arne Tiselius (1902–1971).

1930 Sedative drug sodium pentothal (Nembutal) is discovered by US pharmacologists Ernest Volwiler (1893–1992) and Donalee Tabern (1900–74).

BIOLOGY

1930 Enzyme pepsin is crystallized (and shown to be a protein) by US biochemist John Northrop (1891–1987), who later (1932) also crystallizes trypsin.

1931 Male sex hormone androsterone is discovered by German chemist Adolf Butenandt (1903–95), and later (1934) synthesized by Croatian-born Swiss chemist Leopold Ruzicka (1887–1976).

1932 Muscle protein myoglobin is crystallized by Swedish biochemist Hugo Theorell (1903–82); its structure is later (1957) determined by British biochemist John Kendrew (1917–).

MEDICINE

1930 Vaccine against typhus is developed by US bacteriologist Hans Zinsser (1878–1940).

1930 Vaccine against yellow fever is developed by South African bacteriologist Max Theiler (1899–1972).

1931 Method of culturing viruses (for making vaccines) in chick embryos is discovered by US pathologist Ernest Goodpasture (1886–1960).

1931 Structure of vitamin A (retinol) is determined by German chemist Paul Karrer (1889–1971).

FARMING AND FOOD

TRANSPORT

1923 Rotating-wing autogyro is invented and flown by Spanish engineer Juan de la Cierva (1896–1936).

1923 Diesel-engined lorries are introduced by the Benz Company.

1923 Steam bulldozer with caterpillar tracks is marketed by US La Plante Choate company.

1924 Variable-pitch aeroplane propeller is invented by British engineer Henry Hele-Shaw (1854–1941).

1926 Successful liquid-fuel rocket (using petrol and liquid oxygen) is launched in the USA by Robert Goddard (1882–1945).

1927 Solo non-stop flight across the Atlantic Ocean (New York to Paris) is made by US airman Charles Lindbergh (1902–74); the first east-west flight is not made until later (1932) by Scottish aviator James Mollison (1905–59).

1927 Power steering for cars is invented by Francis Davis.

1928 Rocket-propelled car (Opel-Rak I) is constructed by German engineer Fritz von Opel (1899–1971).

1929 German airship Graf Zeppelin is flown around the world by German aeronaut Hugo Eckner (1868–1954).

1929 Metal-clad airship is invented by US engineer Ralph Upson (1888–1968).

1929 Synchromesh gearboxes for cars are introduced in the USA.

1929 Decompression chamber for divers and escape apparatus for submarines are invented by British engineer Robert Davis.

ENGINEERING AND TECHNOLOGY

1922 Polygraph machine ("lie detector") is invented by Canadian-born American John Larson (1892–1965).

1924 Smooth-faced fibreboard (marketed as Masonite) is produced by US inventor William Mason.

1925 High-power sleeve-valve petrol engine for aircraft is developed by British engineer Roy Fedden (1885–1973).

1925 Synthetic petrol is produced by German chemists Franz Fischer (1877–1947) and H. Tropsch.

1925 Masking tape is invented by American Richard Drew (1899–1980) of the 3M (Minnesota Mining and Manufacturing) company.

1926 Aerosol spray is patented by Norwegian inventor Erik Rotheim (1898–1938), and later (1939) made into disposable cans by US inventor Julius Khan.

1927 Automatic electric toaster is invented by US engineer Charles Strite.

1927 Synthetic rubber is first made on a commercial scale in the USA.

1927 Precision bomb sight is invented by Dutch-born US engineer Carl Norden (1880–1965).

1928 Quartz clock is invented by US electrical engineers Joseph Horton (1899–) and Warren Marrison (1896–1980).

1928 "Iron lung" respirator is invented by US public health engineer Philip Drinker (1894–).

1928 Electric shaver is patented by US inventor Joseph Schick.

1928 Modern form of Geiger counter is invented by German physicists Hans Geiger (1882–1945) and W. Müller.

1928 First working robot is invented by A. Refell et al.

1929 Flight simulator (the Link trainer) is invented by US engineer Edwin Link (1904–81).

1929 Rotary internal combustion engine is invented by German engineer Felix Wankel (1902–88), although it was not developed until the 1950s.

1929 Bofers light anti-aircraft gun is invented by the Swedish Bofers company.

COMMUNICATIONS

1922 Technicolor for full-colour cinema films is invented by H. Kalmus; later perfected in the USA by Leonard Troland (1889–1932).

1922 Method of recording sound as an optical sound track on film is invented by US engineer Lee De Forest (1873–1961).

1922 BBC begins radio broadcasts in England (from station 2LO, London).

1923 Iconoscope television camera tube is invented by Russian-born US engineer Vladimir Zworykin (1889–1982).

1923 Amateur format 16-mm ciné film is marketed in the USA by the Eastman Kodak company; cameras and projectors for the new film are made available by the Bell & Howell company.

1923 "Wire photo" (a facsimile machine using radio telephony) is demonstrated by German physicist Arthur Korn (1870–1945).

1924 High-intensity discharge lamp for use in radio facsimile (wire-photo) machines is invented by US engineer Daniel Moore (1869–1926).

1925 High-power long-wave radio station is built by British radio engineer Peter Eckersley (1892–).

1925 Mechanical-scanning television (using a Nipkow disc) is developed by Scottish inventor John Logie Baird (1888–1946).

1925 Anamorphic lens, which compresses images sideways, is invented by French astronomer Henri Chrétien (1870–1956).

1925 Leica camera is invented by German engineer Oskar Barnack (1879–1936).

1926 Transatlantic telephone service, between London and New York, is inaugurated (using radio telephony).

1927 Audio amplifier using negative feedback is invented by Harold Block.

1928 Colour television system is invented by Russian-born US engineer Vladimir Zworikin (1889–1982) and, independently, Scottish inventor John Logie Baird (1888–1946).

1928 Plastic magnetic tape, for sound recording, is patented by German engineer Fritz Pfleumer; a tape recorder using it is later (1930) marketed by the German I.G. Farben company.

1929 Experimental television broadcasts are made by the British Broadcasting Company, using Baird's system.

1930

FARMING AND FOOD

1935 Canned beer is first marketed in the USA by the Krueger Brewing company.

1937 Instant coffee (Nescafé, produced by freeze drying) is marketed by the Swiss Nestlé company.

1939 Precooked frozen foods are marketed by the US Birds Eye company.

TRANSPORT

1930 Six-cylinder diesel engine for lorries is designed by British engineer Bernard Dicksee (1888–1981).

1930 Bathysphere submersible for deep-sea exploration is constructed by US engineer Otis Barton and naturalist Charles Beebe (1877–1962).

1931 First liquid-fuelled rockets in Europe are launched by German engineers Johannes Winkler and, later (1932), Walter Dornberger (1895–1980).

1931 Gyroscopic stabilizer for ships is invented by US engineers Elmer

ENGINEERING AND TECHNOLOGY

1930 Geodetic construction is invented by British engineer Barnes Wallis (1887–1979).

1930 Gas turbine (jet) engine is patented by British inventor Frank Whittle (1907–96); the prototype is not produced until 1937.

1930 PVC (polyvinyl chloride) plastic is developed by US chemist Waldo Semon (1898–).

1930 Plexiglass (also trademarked Lucite and Perspex) is invented in Canada by student chemist William Chalmers, later (1933) manufactured by the German Röhm & Haas company.

1930 Scotch tape (Sellotape) is invented by American Richard Drew (1899–1980) of the 3M (Minnesota Mining and Manufacturing) company.

1930 Astronomical telescope for taking wide-angle photographs of the heavens is invented by Estonian astronomer Bernhard Schmidt (1879–1935).

COMMUNICATIONS

1930 Complete television system is invented by Swedish-born US electrical engineer Ernst Alexanderson (1878–1975).

1930 Analog computer for solving differential equations is built by US electrical engineer Vannevar Bush (1890–1974) et al.

1931 Electronic flash, for photography, is invented by US engineer Harold Edgerton (1903–90).

1932 Television receiver that uses a cathode-ray tube is invented in the USA.

ASTRONOMY AND SPACE

1932 Asteroid Apollo, whose orbit crosses that of Earth, is discovered by German astronomer Karl Reinmuth.

1934 Theory that stars of mass greater than the Chandrasekhar limit (1.4 solar masses) will collapse into a neutron star is proposed by Swiss astrophysicist Fritz Zwicky (1898–1974) and German-born US astronomer Walter Baade (1893–1960).

1938 Asteroid Hermes is discovered by German astronomer Karl Reinmuth; it was immediately lost and has not been rediscovered.

1938 Ionized hydrogen in interstellar space is discovered by Russian-born US astronomer Otto Struve (1897–1963).

1938 Nuclear fusion is proposed as the mechanism of energy production in stars by German physicist Hans Bethe (1906–) and, independently, his compatriot Carl von Weizsäcker (1912–).

1939 Strömgren sphere (a volume of luminous gas surrounding a supergiant star) is discovered by Danish astronomer Bengt Strömgren (1820–1947).

PHYSICS

1931 Existence of neutrino is postulated by Hungarian-born US physicist Wolfgang Pauli (1900–58).

1932 Neutron is discovered by British physicist James Chadwick (1891–1974); its existence had been predicted earlier (1920) by US physicist William Harkins (1873–1951).

1932 Positron is discovered (in cosmic rays) by US physicist Carl Anderson (1905–91).

1932 Atoms (of lithium and boron) are split by fast protons accelerated in a machine devised by British physicist John Cockroft (1897–1967) and Irish physicist Ernest Walton (1903–95).

1933 Meissner effect (the absence of magnetism in a superconductor cooled below its critical temperature in a magnetic field) is discovered by Walther Meissner and R. Ochsenfeld.

1934 Radioisotopes are produced artificially (by alpha-particle bombardment) by French physicists Irène (1897–1956) and Jean-Frédéric Joliot (1900–58) and, independently, (by neutron bombardment) by Italian-born US physicist Enrico Fermi (1901–54). In the same year US physicist Ernest Lawrence (1901–58) used a cyclotron to produce useful amounts of radioisotopes, which were later used to treat cancer.

1934 Cherenkov radiation (produced by charged particles moving through a medium faster than the speed of light through that medium) is discovered by Soviet physicist Pavel Cherenkov (1904–90), later (1937) explained by his compatriots Ilya Frank (1908–90) and Igor Tamm (1895–1971).

1934 Dislocations in crystal structures, a common cause of failure in metals, are discovered by British physicist Geoffrey Taylor (1886–1975).

1934 First general theory of superconductivity is proposed by Dutch physicist Hendrik Casimir (1909–).

1935 Interference effect of sunspot activity on radio communications is discovered by US physicist John Dellinger (1886–1962).

1935 London equations, which explain superconductivity, are discovered by German-born US physicists Fritz (1900–54) and Heinz (1907–70) London.

1936 Wigner effect (the storage in a crystal of energy from irradiation) is discovered by Hungarian-born US physicist Eugene Wigner (1902–95).

1937 Muon (mu-meson) is discovered in cosmic rays by US physicist Carl Anderson (1905–91), and by US physicists Jabez Street (1906–89) and Edward Stevenson.

1938 Nuclear fission (by means of neutron bombardment) is achieved unknowingly by Austrian-born Swedish physicist Lise Meitner (1878–1968) and German physical chemists Otto Hahn (1879–1968) and Fritz Strassmann (1902–80).

1939 Fact that the Earth's magnetic field is produced by eddy currents in its rotating liquid metallic core is proposed by German-born US physicist Walter Elsasser (1904–).

CHEMISTRY

1930 Method of catalytic cracking of crude oil is introduced by French-born US engineer Eugene Houdry (1892–1962).

1931 Synthetic rubber neoprene is developed by Belgian-born US chemist Julius Nieuwland (1878–1936).

1932 Deuterium (heavy hydrogen, isotope of mass 2) is discovered by US chemist Harold Urey (1893–1981) and, independently (1933), by his compatriot Gilbert Lewis (1875–1946).

1933 Molecular orbital theory, which explains chemical bonding, is developed by British mathematician Charles Coulson (1910–1974) and, independently, German theoretical chemist Erich Hückel (1896–1980).

1933 Vitamin C is synthesized by Polish-born Swiss chemist Tadeusz Reichstein (1897–) and later (1934), independently, by British chemist Walter Haworth (1883–1950), who names it ascorbic (meaning "anti-scurvy") acid.

1933 Thyroid hormone thyroxine is synthesized by British chemist Charles Harington (1897–1972).

1934 Tritium (hydrogen isotope of mass 3) is discovered by Australian physicist Marcus Oliphant (1901–).

1934 Male sex hormone androsterone is synthesized by Swiss chemist Leopold Ruzicka (1887–1976), who later (1954) also synthesizes oxytocin.

1935 Amino acid threonine, the last essential amino acid to be found, is discovered by US biochemist William Rose (1887–1984).

1935 Physostygmine (an alkaloid drug used to treat the eye disorder glaucoma) is synthesized by US chemist Percy Julian (1899–1975).

1935 Fissionable uranium-235 isotope is discovered by US physicist Arthur Dempster (1886–1950).

1936 Tabun, the first nerve gas, is discovered by German chemist Gerhard Schracher.

1937 Radioactive element technetium (symbol Tc) is discovered by Italian physicists Emilio Segrè (1905–89) and Carlo Perrier, who first called it masurium.

1939 Radioactive metallic element francium (symbol Fr) is discovered by French chemist Marguérite Perey (1909–75).

1939 Insecticidal properties of DDT are discovered by Swiss chemist Paul Müller (1899–1965), who also synthesized it; DDT is first used on a large scale in Italy in 1944.

BIOLOGY

1932 Concept of homeostasis (the automatic self-regulation of the body's internal environment in such a way that biochemical process proceed in the best way) is proposed by US physiologist Walter Cannon (1871–1945).

1933 Role of rhodopsin in the reina of the eye, and its relation to vitamin A, is discovered by US biochemist George Wald (1906–).

1933 Tasmanian wolf probably becomes extinct.

1934 Female sex hormone progesterone is discovered and isolated by German chemist Adolf Butenandt (1903–95).

1935 Adrenal hormone cortisone is discovered by US biochemist Edward Kendall (1886–1972) and later (1936) isolated by Polish-born Swiss chemist Tadeusz Reichstein (1897–), who determined its structure.

1935 Hormone-like prostaglandin is discovered by Swedish physiologist Ulf von Euler (1905–83), later isolated by his compatriots Sune Bergström (1916–) and Bengt Samuelsson (1934–).

1935 Imprinting in young animals is discovered by Austrian biologist Konrad Lorenz (1903–89), who establishes the science of ethology.

1935 Term "ecosystem" is coined by British botanist and ecologist Arthur Tansley (1871–1955).

1935 Tobacco mosaic virus is produced in crystalline form (the first virus to be crystallized) by US biochemist Wendell Stanley (1904–71) and later (1936), independently, by British biochemists Frederick Bawden (1908–72) and Norman Pirie (1907–), who also showed that it contains RNA, establishing this nucleic acid as a basic component of life.

1936 Fact that a virus can cause cancer (in mice) is established by US biologist John Bittner (1904–61).

1937 Essential amino acids (which cannot be made in the body and must therefore be supplied by foods) are discovered by US biochemist William Rose (1887–1984).

1937 Electrical nature of nerve transmission is discovered by US physiologist Joseph Erlanger (1874–1965).

1937 Mutations produced by polyploidy (multiplication of chromosomes) are discovered by US botanist Albert Blakeslee (1874–1954).

1938 First bacteriophage (virus that attacks bacteria) is isolated by US biochemist John Northrop (1891–1987).

1938 Coelacanth (a "living fossil" fish) is discovered and identified by South African biologist J.-L.-B. Smith; the fish is later (1953) found to be relatively common near the Comoro Islands.

1939 Plant growth hormone gibberellin is discovered by Japanese biologist Teijiro Yabuta (1888–).

MEDICINE

1931 Vitamin D2 (cholecalciferol) is crystallized by German chemist Adolf Windaus (1876–1959).

1932 First sulpha drug, Prontosil, is discovered by German biochemist Gerhard Domagk (1895–1964); it is later (1935) used to treat streptococcal infections.

1933 Natural childbirth, without the use of anaesthetics, is advocated by British gynaecologist Grantly Dick-Read (1890–1959).

1934 Vitamin K is discovered by Danish biochemist Carl Dam (1895–1976) and, independently, US biochemist Edward Doisy (1893–1986) who later (1939) determines its structure and the fact that it exists in two forms.

1934 Method of extracting vitamin B1 (thiamine, deficiency of which causes beriberi) from rice polishings is discovered by US chemist Robert Williams (1886–1965), who later (1936) works out its structure and a method of synthesizing it.

1935 Structure and synthesis of vitamin B2 (riboflavin) are discovered by German chemist Paul Karrer (1889–1971) and, independently, Austrian-born German chemist Richard Kuhn (1900–67).

1935 Prefrontal lobotomy (leucotomy), as a treatment of mental illness, is introduced by Portuguese surgeon António Egaz Moniz (1874–1955).

1936 Crystalline tobacco mosaic virus (the first virus to be crystallized) is produced by British biochemists Norman Pirie (1907–) and Frederick Bawden (1908–72).

1937 First antihistamine, pyrilamine (mepyramine), is discovered by Swiss physiologist Daniel Bovet (1907–92).

1938 Vitamin B6 (pyridoxine) is isolated by Austrian-born German biochemist Richard Kuhn (1900–67), who later (1939) determines its structure.

1938 Vitamin E (tocopherol) is synthesized by German chemist Paul Karrer (1889–1971).

1939 Antibiotic tyrothricin (gramicidin), the first to be produced commercially, is discovered by French-born US microbiologist René Dubos (1901–82).

1939 Sulphapyridine (M & B 693 or sulphadiazine, a sulpha drug) is discovered by British chemist Arthur Ewins (1882–1957) and, independently, US chemist Richard Roblin (1907–).

1940

1942 Radio maps of the Universe are made by US radio astronomer Grote Reber (1911–).

1942 Accurate value for the Earth-Sun distance (1 astronomical unit) is determined by British astronomer Harold Jones (1890–1960).

1943 Seyfert galaxies are discovered by US astronomer Carl Seyfert (1911–60).

1940 Decay of a cosmic-ray meson into an electron is discovered by Welsh physicist Evan Williams (1903–45).

1940 Fact that it is the uranium-235 isotope (not the much more common uranium-238) that undergoes nuclear fission, as carried out earlier (1938) by Lise Meitner (1878–1968), is discovered by US physicist John Dunning (1907–75).

1940 Radioactive carbon-14 isotope is discovered by Canadian-born US biochemist Martin Kamen (1913–).

1940 Radioactive element plutonium (symbol Pu) is discovered by US chemist Glenn Seaborg (1912–) et al.

1940 Radioactive element astatine (symbol At) is discovered by Italian-born US physicist Emilio Segrè (1905–89) et al.

1940 Krebs cycle, also called tricarboxylic acid (TCA) or citric acid cycle, is discovered by German-born British physiologist Hans Krebs (1900–81).

1940 Rhesus factor (in blood) is discovered by Austrian-born US biologist Karl Landsteiner (1868–1943).

1940 Role of iodine in thyroid function is discovered by US anatomist Herbert Evans (1882–1971).

1940 Vitamin H (biotin) is discovered by Hungarian biochemist Albert Szent-Györgyi (1893–1986) and US biochemist Vincent Du Vigneaud (1901–78), who later (1942) determines its structure and then (1943) synthesizes it.

1940 Structure of the B vitamin pantothenic acid is determined by US chemist Robert Williams (1886–1965).

194

FARMING AND FOOD

TRANSPORT

Sperry (1860–1930) and Carl Norden (1880–1965).

1932 First flight into the stratosphere (to a height of 17 kilometres) is made in a balloon by Swiss scientists, and twin brothers, Auguste and Jean Piccard (1884–1962).

1933 Petrol/liquid oxygen fuelled rocket is launched by Sergei Korolev (1906–66) and Mikhail Tikhanravov in the Soviet Union.

1933 Solo round-the-world flight is made by US airman Wiley Post (1900–35); in his plane Winnie Mae he made ten refuelling stops during the journey, which took nearly 8 days.

1933 Tracked amphibious vehicle (the Alligator) is invented by Donald Roebling.

1934 Cars with a monocoque construction (combined body and chassis) are manufactured in France.

1934 Front-wheel drive car, the 7A or Traction-Avant, is mass-produced by the French Citroën company.

1934 Bathysphere for deep-sea exploration is invented by US naturalist Charles Beebe (1877–1962).

1934 Cat's eyes (reflectors in the road surface) are invented by Englishman Percy Shaw.

1935 Parking meter is invented by US journalist Carlton Magee.

1936 Practical helicopter, the Fa-61, is built in Germany by German engineer Heinrich Focke.

1937 First fully pressurized aircraft, the Lockheed XC-35, enters service in the USA.

1937 Automatic transmission is introduced for cars.

1938 Breathalyzer ("Drunkometer") is invented in the USA by S. R. Harger.

1938 British steam locomotive Mallard, designed by Scottish engineer Nigel Gresley (1876–1941), establishes world speed record (for steam) of 203 kilometres per hour.

1939 First mass-produced helicopter is built by Russian-born US inventor Igor Sikorsky (1889–1972).

1939 First jet aeroplane, the Heinkel He-178, is flown in Germany by Erich Warsitz; it is designed by Hans Pabst von Ohain (1911–).

ENGINEERING AND TECHNOLOGY

1930 Coronagraph, for seeing the Sun's corona when it is not eclipsed, is invented by French astronomer Bernard Lyot (1897–1952).

1931 Nylon is invented by US chemist Wallace Carothers (1896–1937); it is not patented until 1937.

1931 Electrostatic generator capable of supplying millions of volts is invented by US physicist Robert Van de Graaff (1901–67).

1932 Cyclotron is invented by US physicists Ernest Lawrence (1901–58) and Milton Livingston (1905–86).

1932 Cockroft-Walton generator of high voltages is invented by British physicist John Cockroft (1897–1967) and Irish physicist Ernest Walton (1903–).

1932 Phase-contrast microscope is invented by Dutch physicist Frits Zernike (1888–1966).

1934 Electric tone-wheel organ is invented by US engineer Laurens Hammond (1895–1973).

1935 Fluorescent lighting is introduced by the General Electric company, and sodium vapour lamps are developed for street lighting.

1936 Field-emission microscope is invented by German-born US physicist Erwin Mueller (1911–77).

1936 Semi-automatic rifle is invented by US marine and gunsmith Melvin Johnson (1884–1937), who a year later patented a light machine gun.

1936 M1 semiautomatic rifle, invented by US gunsmith John Garand (1888–), is adopted as the standard shoulder arm by the US army.

1936 Commercial process for making high-octane petrol (by catalytic hydrogenation of lignite) is invented by French-born US chemist Eugene Houdry (1892–1962).

1938 Bren gun (light machine gun ZB 26) is developed at Brno Armoury, Czechoslovakia.

1938 Electron microscope is developed independently by German engineer Ernst Ruṣka (1906–88) and R. Ruedenberg.

1939 Polyethylene plastic goes into commercial production in Britain.

1939 Anti-glare coating for glass (later applied to camera lenses) is invented by US physicist Katharine Blodgett (1898–).

COMMUNICATIONS

1932 Electronic television camera is patented in England by the EMI (Electrical and Musical Industries) company.

1932 Telex service is introduced by the British Post Office.

1933 Frequency modulation (FM) radio transmission is invented by US engineer Edwin Armstrong (1890–1954).

1933 Stereophonic sound recording is invented by British engineer Alan Blumlein.

1935 Radar is invented independently by Robert Watson-Watt (1892–1973) in Britain and Rudolf Kühnold of the Germany navy.

1935 Kodachrome (transparency) colour film is invented in the USA.

1936 Electronic television system using a 405-line picture is developed by British engineer Alan Blumlein (1903–42).

1936 Regular television broadcasts are begun by the BBC in London.

1936 Mathematical theory of computing is devised by British mathematician Alan Turing (1912–54).

1936 Primitive digital computer, using electric relays, is built by German engineer Konrad Zuse (1910–95).

1936 Kine Exacta, the first single-lens reflex 35mm camera, is marketed by the Ihagee company.

1937 Xerography (dry photocopying) is invented by US engineer Chester Carlson (1906–68).

1937 Cinerama wide-screen motion pictures (using three projectors) are invented by Fred Waller (1886–1954) of the US Paramount company (although not introduced commercially until 1952).

1938 Klystron radar (microwave) transmitting valve is invented by US physicists Russel (1898–1959) and Sigurd (1901–61) Varian.

1937 Binary calculator is constructed by US engineer George Stibitz (1904–95), followed later (1938) by a machine (the Z1) using Boolean algebra, by German engineer Konrad Zuse (1910–95).

1938 Reliable ballpoint pen is invented by Hungarian-born Argentinian Laszlo (later Ladislao) Biró (1899–1985), but not patented until 1943.

1939 Electronic television receivers using cathode-ray tubes are produced by US engineer Allen Du Mont (1901–65).

1939 Television service is inaugurated in the USA, set up by David Sarnoff (1891–1971) of the RCA company.

1939 Electronic computer for solving linear equations is built by American physicist John Atanasoff (1903–).

1939 Phototypesetting machine is invented by US William Heubner.

1939 The cavity magnetron, an improved microwave and radar transmitter valve, is invented by British engineers John Randall (1905–84) and Henry Boot (1917–83).

1940

1946 Espresso coffee machine is produced by Italian inventor Achille Gaggia.

1941 Jet aeroplane, the Gloster Meteor, flies in England with an engine designed by Frank Whittle (1907–); in Italy the Caproni-Campini CC2 jet plane also flies.

1942 Ejection seat, operated by compressed air, is invented by the German Heinkel company; later (1943) explosive-powered seats were made by the Swedish Saab company.

1940 Practical electron microscope is constructed by Candian scientist James Hillier (1915–).

1940 Portable bridge for the military is invented by British engineer Donald Bailey (1901–85)

1940 Gas-operated automatic rifle, invented by US gunsmith Jonathan Browning (1859–1939), is patented.

1941 Dacron (Terylene) is invented by British chemists John Whinfield (1901–66) and J. Dickson.

1940 The Plan Position Indicator (PPI) radar screen is invented in Britain.

1940 Experimental colour television broadcasts begin in the USA.

1941 Computer using electric relays and punched paper tape (the Z2) is built by German engineer Konrad Zuse (1910–95).

1942 The ABC (Atansoff-Berry

ASTRONOMY AND SPACE

1943 Fact that Saturn's moon Titan has an atmosphere is discovered by Dutch-born US astronomer Gerard Kuiper (1905–73).

1944 Classification of stars into Population I (young) and Population II (old) is introduced by German-born US astronomer Walter Baade (1893–1960).

1944 Fact that interstellar hydrogen should emit microwave radio waves is predicted by Dutch astronomer Hendrik van de Hulst (1918–); such waves are later (1951) detected by US physicist Edward Purcell (1912–).

1944 Planetesimal theory of the origin of the Solar System is proposed by German astronomer Carl von Weizsäcker (1912–).

1946 Fact that sunspots emit radio waves is discovered by astronomers Edward Apello and Donald Hay.

1946 Cygnus A, the first known radio galaxy, is discovered by British astronomer Martin Ryle (1918–84).

1948 Miranda, the fifth moon of Uranus, is discovered by Dutch-born US astronomer Gerard Kuiper (1905–73).

1948 Steady State theory of the Universe is proposed by British astronomers Hermann Bondi (1919–), Thomas Gold (1920–) and Fred Hoyle (1915–).

1948 Theory to account for the abundances of the elements in terms of the fusion processes following the Big Bang (the alpha, beta, gamma theory) is proposed by US physicist Ralph Alpher (1921–), German physicist Hams Bethe (1906–) and Russian-born US physicist George Gamov (1904–88).

1948 Magnetic field of the Sun is detected by US astronomers Horace (1912–) and Harold (1882–1968) Babcock.

1949 Nereid, the second moon of Neptune, is discovered by Dutch-born US astronomer Gerard Kuiper (1905–73).

1949 Asteroid Icarus (which has the smallest known orbit) is discovered by German-born US astronomer Walter Baade (1893–1960).

1949 Fact that a comet's nucleus consists mainly of ice and dust is proposed by US astronomer Fred Whipple (1906–).

1950

1950 Source of long-period comets, the Oort cloud, is proposed by Dutch astronomer Jan Oort (1900–92).

1951 Source of short-period (less than 200 years) comets, the Kuiper belt, is

PHYSICS

1940 Method of separating the two uranium isotopes (U-235 and U-238), by thermal diffusion of gaseous uranium hexafluoride, is proposed by US physical chemist Philip Abelson (1913–).

1940 Fact that beryllium will slow down fast neutrons (and therefore act as a moderator in a nuclear reactor) is discovered by Austrian-born US physicist Maurice Goldhaber (1911–).

1941 Spontaneous fission of uranium is discovered by Soviet physicist Georgii Flerov (1913–).

1942 First nuclear reactor ("atomic pile") is completed in the USA under the direction of Italian-born US physicist Enrico Fermi (1901–54).

1942 Existence of magnetohydrodynamic waves (Alfvén waves) in plasmas is predicted by Swedish physicist Hannes Alfvén (1908–95).

1946 Proton linear accelerator is developed by US physicist Luis Alvarez (1911–88).

1946 First fast nuclear reactor (called Clementine) is built at Los Alamos, New Mexico, USA.

1946 Nuclear magnetic resonance (NMR) spectroscopy is developed by Swiss-born US physicist Felix Bloch (1905–83) and, independently, US physicist Edward Purcell (1912–97).

1947 Pion (pi-meson) is discovered in cosmic rays by British physicist Cecil Powell (1903–69).

1947 Lamb shift (between two energy levels in the hydrogen spectrum) is discovered by US physicist Willis Lamb (1913–).

1948 Basic concept of holography is discovered by Hungarian-born British physicist Dennis Gabor (1900–79).

1948 "Shell" structure for the protons and neutrons in the atomic nucleus is proposed by German-born US physicist Marie Goeppert-Mayer (1906–72) and, independently, German physicist Hans Jensen (1907–73).

1948 Quantum electrodynamics, which deals with the interactions of charged subatomic particle, is formulated by US physicists Richard Feynman (1918–88) and Julian Schwinger (1918–94), and, independently, Japanese physicist Sin-Itiro Tomonaga (1906–79).

1950 "Shell" and "liquid drop" models of the atomic nucleus are combined in a single theory by US physicist Leo Rainwater (1917–86) with Aage Bohr (1922–) and B. Mottelson (1926–).

CHEMISTRY

1940 Radioactive element neptunium (symbol Np) is discovered by US physical chemists Edwin McMillan (1907–91) and Philip Abelson (1913–).

1943 Silicones are first manufactured by the US Dow Corning company.

1944 Quinine is synthesized by US chemists Robert Woodward (1917–79) and William Doering.

1945 Radioactive elements americium (symbol Am) and curium (symbol Cm) are discovered by Glenn Seaborg (1912–) et al.

1945 Structure of penicillin is discovered by British chemist Dorothy Hodgkin (1910–94).

1946 Radiocarbon dating (based on the amount of the isotope carbon-14 in an organic sample) is discovered by US chemist Willard Libby (1908–80).

1946 Structure of strychnine is discovered by British chemist Robert Robinson (1886–1975); it is later (1954) synthesized by US chemist Robert Woodward (1917–79).

1947 ADP (adenosine diphospahte) and ATP (adenosine triphosphate), important energy-containing chemicals involved in cell metabolism, are synthesized by Scottish chemist Alexander Todd (1907–).

1947 Radioactive element promethium (symbol Pm) is discovered by J. Marinsky et al.

1949 Method of studying ultrafast chemical reactions is discovered by British physical chemists George Porter (1920–) and Ronald Norrish (1897–1978).

1949 Radioactive element berkelium (symbol Bk) is discovered by US chemist Glenn Seaborg (1912–) et al.

1950 Element Californium (Cf) is discovered by US chemist Glenn Seaborg (1912–)

1951 Cholesterol and cortisone are synthesized by US chemist Robert Woodward (1917–79).

BIOLOGY

1940 Paedomorphism, the fact that adult animals retain some ancestral juvenile features, is proposed by British zoologist Gavin de Beer (1899–1972), thus refuting Haeckel's recapitulation theory of 1866.

1941 Role of ATP (adenosine triphosphate) in cellular energy release is discovered by German-born US biochemist Fritz Lipmann (1899–).

1941 Role of genes in controlling chemical reactions in cells is discovered by US biologists George Beadle (1903–89) and Edward Tatum (1909–75).

1942 Electron microscope is first used in biological research by Belgian-US biologist Albert Claude (1898–1983).

1943 Formation of enzyme-substrate complexes (the key to how enzymes work) is discovered by US biochemist Britton Chance (1913–).

1944 Structure of the bile pigment bilirubin is discovered by German chemist Hans Fischer (1881–1945), who also synthisized it.

1946 Fact that viruses can combine to form new viruses is discovered by German-born US biologist Max Delbrück (1906–81) and US biologist Alfred Hershey (1908–97).

1946 Noradrenaline (norepiniphrine) is discovered by Swedish physiologist Ulf von Euler (1905–83).

1947 Coenzyme A is discovered by German-born US biochemist Fritz Lipmann (1899–1986).

1947 Nerve growth factor (in embryos) is discovered by Italian neurophysiologist Rita Levi-Montalcini (1909–).

1948 Fossils of Proconsul africanus, a 20 million-year-old ape, are found in Kenya by British archaeologist Mary Leakey (1913–).

1949 Barr bodies (condensed X-chromosomes in nondividing nuclei of cells of female animals) are discovered by Canadian geneticist Murray Barr (1908–).

1950 Fact that a single organism has many different kinds of RNA (but only one kind of DNA) is discovered by Czech-born US biochemist Erwin Chargaff (1905–).

MEDICINE

1940 Fact that blood plasma can be used instead of whole blood for transfusion and a method of storing plasma in blood banks are discovered by US physician Charles Drew (1904–50).

1941 Use of female sex hormones to treat prostate cancer is discovered by US physician Charles Huggins (1901–).

1941 Connection between birth defects and German measles (rubella) during pregnancy is discovered by Australian physician Norman Gregg (1892–1966).

1941 Term "antibiotic" is coined by Russian-born US microbiologist Selman Waksman (1888–1973).

1943 Antibiotic streptomycin (effective against tuberculosis) is discovered in soil by Russian-born US microbiologist Selman Waksman (1888–1973).

1943 Hallucinogenic drug LSD (lysergic acid diethylamide, in German Lyserg-Saure-Diathylamid) is discovered by German chemist Albert Hofmann.

1943 Pap smear test (for uterine cancer) is discovered by Greek-born US physician George Papanicolaou (1883–1962).

1944 Antibiotic Aureomycin (chlortetracycline, the first of the tetracyclines) is discovered by US botanist Benjamin Dugger (1872–1956).

1944 Surgical treatment of the heart defect in newborn "blue babies" is developed by US physicians Helen Taussig (1899–1986) and Alfred Blalock (1899–1964).

1944 Method of making viruses visible under the electron microscope (by "shadowing" them with a thin layer of metal) is discovered by US biophysicists Robley Williams (1908–) and Ralph Wyckoff (1897–).

1947 Powerful antibiotic chloramphenicol is discovered in a Streptomyces bacillus in a sample of soil from Venezuela.

1948 Operation to enlarge the mitral valve in the heart (which can malfunction because of narrowing called mitral stenosis) is introduced by US physician Dwight Harken (1910–) and, independently, British surgeon Russell Brock (1903–80).

1948 Cortisone treatment for rheumatoid arthritis is discovered by US physician Philip Hench (1896–1965) and biochemist Edward Kendall (1886–1972).

1948 Use of ultrasound to scan the fetus in pregnant women is pioneered by Scottish physician Ian Donald (1910–87).

1948 Structure of antibiotic streptomycin is determined by US biochemists Karl Folkers (1906–) et al. of the US Merck company. In the same year they isolated vitamin B12 (cyanocobalamin), whose structure was later (1956) determined by the British biochemist Dorothy Hodgkin (1910–94).

1949 Role of the immune system in tissue rejection (in skin grafts) is discovered by Australian biologist Frank Burnet (1899–1985).

1949 Cause of sickle-cell anaemia is discovered by US biochemist Linus Pauling (1901–94).

1949 Method of growing poliomyelitis virus in a tissue culture (treated with antibiotic) is discovered by US microbiologist John Enders (1897–1985).

1950 Statistical correlation between cigarette smoking and the incidence of lung cancer is discovered by British physicians William Doll (1912–) and Austin Hill (1897–1991).

19

FARMING AND FOOD

TRANSPORT

1941 Jeep, a lightweight four-wheel drive vehicle for the US Army, is manufactured by the Willys-Overland company, following previous (1940) prototypes designed by Karl Pabst of the Bantam Car Company.

1941 "Human torpedo" – a large steerable torpedo, with a detachable warhead, carrying two frogmen as crew – is introduced by the Italian navy.

1941 Gas-turbine locomotive runs in Switzerland.

1942 First jet fighter plane, the Messerschmidt Me-262, is flown in Germany.

1942 First surface-to-surface missile, the autopilot-controlled V-1 "flying bomb" powered by a ram jet, is launched in Germany; the V-2 (or AS-4) rocket is also tested, designed by German rocket engineer Werner von Braun (1912–77) et al.

1943 Air-to-air missile is designed by German inventor Herbert Wagner.

1944 Rocket-powered interceptor fighter, the Messchersmitt Komet, flies in Germany.

1947 Bathyscape for deep-sea exploration is invented by Swiss physicist Auguste Piccard (1884–1962).

1947 US rocket-powered Bell X-1 aircraft, piloted by Chuck Yeager (1923–), becomes the first plane to fly faster than the speed of sound.

1947 Unmanned Skymaster aeroplane flies from the USA to England under automatic radio control.

1947 Tubeless tyres for cars are introduced in the USA by the B.F. Goodrich company.

1947 Two-stage rocket is built by Holgar Toftoy.

1948 Radial-ply tyres are introduced.

1948 First turbo-prop aeroplane, the Vickers Viscount, flies in Britain.

1949 First jet airliner, the De Havilland Comet designed by British engineer Ronald Bishop (1903–89), enters service in Britain.

1949 Non-stop round-the-world flight is made by US Air Force officer James Gallagher in a Boeing B50 Superfortress, Lucky Lady II, which was refueled in flight four times.

1949 First two-stage rocket is launched in the USA (to a height of 400 kilometres).

ENGINEERING AND TECHNOLOGY

1941 Polyurethane plastics are first produced by the IG Farbenindustrie company.

1942 Turbo-prop aircraft engine is invented by Max Muller.

1942 Bazooka anti-tank rocket is adopted by the US army.

1942 Aqualung is invented by French diver Jacques Cousteau (1910–97) and Emile Gagnan.

1942 The incendiary material napalm is invented by Louis Fieser.

1943 Kidney dialysis machine is invented by Dutch-born US physician Willem Kolff (1911–).

1943 Plastics based on silicones, discovered in 1899 by British chemist Frederick Kipping (1863–1949), are produced by the Dow Corning company.

1943 The V-1 (Vergeltungswaffe, or "vengence weapon") flying bomb, propelled by a pulse jet, is produced in Germany and used to bombard England in World War II.

1944 Paper chromatography (method of chemical analysis) is invented by British chemists Archer Martin (1910–) and Richard Synge (1914–94).

1944 The V-2 rocket bomb (a ballistic missile), propelled by a liquid fuel, is produced in Germany and used to bombard England in World War II.

1944 Tupperware plastic food containers are invented by Earl Tupper.

1945 First atomic bomb is exploded in the USA; within the same year, two atomic bombs are dropped on Japan.

1945 Afterburner, for giving additional thrust to a jet engine, is developed by the British Rolls-Royce company.

1945 The synchrocyclotron particle accelerator is invented by Soviet physicist Vladimir Veksler (1907–66).

1945 Microwave oven is patented by US radar engineer Percy Spencer.

1946 Procedure for producing falling snow (by dropping dry ice – solid carbon dioxide – into clouds) is invented by US chemist Vincent Schaefer (1906–93). It was later adapted, using silver iodide crystals, by US physicist Bernard Vonnegut (1914–).

1946 Nuclear magnetic resonace (NMR) spectrometer is invented by Swiss-born US physiciat Felix Bloch (1905–83).

1947 Point-contact transistor is invented by US physicists John Bardeen (1908–91), Walter Brattain (1902–87) and William Shockley (1910–89).

1947 Geodesic dome construction is invented by US architect and engineer Buckminster Fuller (1895–1983).

1948 Junction transistor is invented by US physicist William Shockley (1910–89).

1948 Velcro fastener is invented by Swiss engineer Georges de Mestral (1907–90), but not introduced commercially until 1956.

1948 Atomic clock, using ammonia atoms, is constructed in the USA.

COMMUNICATIONS

Computer), a punched-card version of Atanasoff's machine of 1939, is built but proves unreliable.

1942 Kodacolor colour print film is invented in the USA.

1943 First all electronic, valve-operated calculating machine (the Colossus) is built (for cracking enemy military codes) by British mathematician Alan Turing (1912–54) et al.

1943 Decca radio navigation system is invented in Britain by US electronics engineer Harvey Schwarz (1905–88).

1944 World's second electronic computer (the Automatic Sequence Controlled Calculator, or Harvard University Mark I), which uses punched tape and valves, is built by US engineer Howard Aiken (1900–73) et al.

1945 Pilot ACE (Automatic Computing Engine), an electronic stored-program computer, is constructed by British mathematician Alan Turing (1912–54) et al.

1946 ENIAC (Electronic Numerical Integrator and Computer), the first true stored-program computer, is developed by Americans John Eckert (1919–95) and John Mauchly (1907–80).

1946 Assembler computer programming language is devised by Englishman Maurice Wilkes (1913–).

1947 Polaroid camera is patented by US inventor Edwin Land (1909–).

1948 Long-playing gramophone record is invented by Hungarian-born American Peter Goldmark (1906–77).

1948 Commercial solid-body electric guitar (the Broadcaster) is produced by US instrument maker Leo Fender (1909–91), a few years after earlier prototypes by Les Paul (Lester Polfuss) (1915–), whose Gibson was not marketed until 1952

1948 Hologram is invented by Hungarian-born British engineer Dennis Gabor (1900–79).

1948 The Mark I prototype stored-program electronic computer, designed by British electrical engineers Tom Kilburn (1921–) and Frederic Williams (1911–77), begins operation at Manchester University.

1948 Phototypestting machine, Rotophoto, is developed by British engineer George Westover, followed later (1949) by the Lumitype 200, made by the US Photon company.

1949 Cambridge University's EDSAC (Electronic Delay Storage Automatic Calculator), a stored-program computer designed by British scientist Maurice Wilkes (1913–) et al, begins operation.

1950

1953 Edible synthetic protein made from soya beans is patented by US scientist Robert Boyer (who first made it when looking for a substitute for natural leather).

1950 Gas-turbine car is built in Britain by the Rover Company.

1950 Power-assisted steering for cars is introduced in the USA.

1951 Field-ion microscope, capable of revealing individual atoms, is invented by German-born US physicist Erwin Mueller (1911–77).

1952 Implantable artificial heart valve is invented by US surgeon Charles Hufnagel.

1953 The heart-lung machine is invented by US surgeon John Gibbon (1903–74).

1950 The Videcon television camera tube is invented in the USA.

1950 Commercial Xerox photocopier is marketed by the US Haloid company.

ASTRONOMY AND SPACE

proposed by Dutch-born US astronomer Gerard Kuiper (1905–73).

1951 "Earth-grazing" asteroid Geographos is discovered by German-born US astronomer Rudolph Minkowski (1895–).

1951 Fact that the Milky Way galaxy has spiral arms is discovered by US astronomer William Morgan (1906–94).

1951 First visible objects that are also sources of radio waves from space are discovered by German-born US astronomer Walter Baade (1893–1960).

1953 Superclusters of galaxies are discovered by Polish-born US astronomer Jerzy Neymann (1899–1981).

1955 Radio emissions from Jupiter are discovered by US radio astronomer Kenneth Franklin (1923).

1956 Fact that solar flares emit X-rays is discovered by US astronomer Herbert Friedman (1916–).

1957 First orbiting Earth satellite, Sputnik 1, is launched by the Soviet Union; Sputnik 2, carrying a dog, is launched a month later.

1957 Mechanism by which elements build up (from lighter to heavier) within stars during stellar evolution is discovered by US astrophysicist William Fowler (1911–95).

1958 Solar wind is discovered by US physicist Eugene Parker (1927–).

1958 Van Allen radiation belts around the Earth are discovered by US physicist James Van Allen (1914–) from data collected by the first US Earth satellite, Explorer 1.

1958 Evidence of volcanic activity on the Moon is observed by Soviet astronomer Nikolai Kozyrev (1908–).

1958 US government agency the National Aeronautics and Space Administration (NASA) is founded.

1959 Soviet Union launches three space probes to the Moon; Lunik I flies past, Lunik II crashes on the Moon, and Lunik III bypasses and sends back photographs of the far side of the Moon.

PHYSICS

1955 Antiproton is discovered by US physicists Emilio Segrè (1905–89) and Owen Chamberlain (1920–).

1956 Neutrino is discovered by US physicists Clyde Cowan (1919–) and Frederick Reines (1918–), who later (1965) found neutrinos in cosmic rays.

1956 Continuous emission maser is developed by Dutch-born US physicist Nicolaas Bloembergen (1920–).

1956 Cooper pairs (bound pairs of electrons that carry electric current in superconductors) are discovered by US physicist Leon Cooper (1930–).

1956 Use of colliding-beam storage-rings for increasing the energy of particle accelerators is proposed by US physicist Gerard O'Neill (1927–92).

1957 Mössbauer effect (the recoil-less emission of gamma rays by the nuclei of atoms in a crystal) is discovered by German physicist Rudolf Mössbauer (1929–).

1957 Tunnel effect (in which electrons penetrate a narrow potential barrier) is discovered by Japanese physicist Leo Esaki (1925–).

1959 Xi-zero subatomic particles is discovered by US physicist Luis Alvarez (1911–88) et al.

CHEMISTRY

1952 Radioactive elements einsteinium (symbol Es) and fermium (symbol Fm) are discovered by US chemists Glenn Seaborg (1912–), Albert Ghiorso et al.

1953 Ziegler process (for making high-density polyethene) is discovered by German chemist Karl Ziegler (1898–1973), later (1954) improved by Italian chemist Giulio Natta (1903–79).

1954 Method of studying the progress of very rapid chemical reactions is discovered by German chemist Manfred Eigen (1927–).

1954 Series of reactions by which the body synthesizes cholesterol are discovered by German-born US chemist Konrad Bloch (1912–).

1955 Method of making synthetic diamonds is discovered by US physicist Percy Bridgman (1882–1961).

1955 Radioactive element mendelevium (symbol Md) is discovered by US chemist Albert Ghiorso et al.

1956 Structure of vitamin B12 is discovered by British chemist Dorothy Hodgkin (1910–94).

1958 Method of identifying the amino acids and their sequence in proteins and nucleic acids is developed by US biochemists Stanford Moore (1913–82) and William Stein (1911–80).

1958 Radioactive element nobelium (symbol No) is discovered by US chemists Albert Ghiorso and Glenn Seaborg (1912–) and, independently, scientists in the Soviet Union.

BIOLOGY

1950 Alpha-helix structure of proteins is discovered by US biochemist Linus Pauling (1901–94).

1950 Human population is divided into 13 different races on the basis of blood group by US biochemist William Boyd (1903–.)

1950 Parasexual cyle in fungi is discovered by Italian geneticist Guido Pontecorvo (1907–93).

1951 "Jumping genes" are discovered by US geneticist Barbara McClintock (1902–92).

1952 Fact that DNA carries genetic information is discovered by US biologists Alfred Hershey (1908–) and M. Chase.

1952 "Sodium pump" mechanism of nerve transmission is discovered by British physiologists Alan Hodgkin (1914–) and Andrew Huxley (1917–), and, independently, Australian physiologist John Eccles (1903–).

1952 Plasmids, mobile rings of DNA found in bacteria, are discovered by US biologist and geneticist Joshua Lederberg (1925–).

1952 Rapid eye movement (REM) sleep, which occurs when a person is dreaming, is discovered by Russian-born US physiologist Nathaniel Kleitman (1895–).

1953 Structure of DNA is discovered by British biophysicist Francis Crick (1916–) and US biologist James Watson (1928–), using measurements made by New Zealand-born British physicist Maurice Wilkins (1916–).

1953 Structures of vasopressin and oxytocin (two hormones from the pituitary gland) are discovered by US biochemist Vincent Du Vigneaud (1901–78), who also synthesizes them.

1954 Fact that triplets of nucleotides (in nucleic acids) act as the genetic code in enzyme formation is discovered by Russian-born US physicist George Gamov (1909–68).

1955 Lysosomes (structures within cells) are discovered by British-born Belgian biochemist Christian de Duve (1917–78).

1955 Method of synthesizing RNA is discovered by Spanish-born US molecular biologist Severo Ochoa (1905–93).

1955 Amino-acid sequence of insulin is discovered by British biochemist Frederick Sanger (1918–).

1956 Ribosomes, and the fact that they are mostly RNA, are discovered by Romanian-born US physiologist George Palade (1912–).

1956 Transfer RNA (tRNA) is discovered by Mahlon Haoglanc and Paul Zamecnick.

1956 Enzyme that catalyzes DNA synthesis is discovered by US biochemist Arthur Kornberg (1918–).

1957 Interferons (which attack viruses) are discovered by British biologists Alick Isaacs and Jean Lindermann.

1957 Mechanism by which glycogen is synthesized in the body is discovered by French biochemist Luis Leloir (1906–).

1957 Details of how the DNA double helix carries genetic information are discovered by US molecular biologists Matthew Meselsor (1930–) and Franklin Stahl (1929–).

1957 Crystalline prostaglandins are isolated by Swedish biochemist Sune Bergström (1916–).

1958 All-female species of lizard that

MEDICINE

1950 Reserpine treatment for hypertension is discovered by US physician Robert Wilkins (1906–), who later (1952) also discovers the sedative effect of the drug, which becomes the first tranquilizer.

1952 Apgar score test for newborns is introduced by US physician Virginia Apgar (1909–74).

1952 Sex-change operation is first performed by Danish surgeon Karl Hamburger, when US soldier George Jorgensen becomes Christine Jorgensen.

1953 Salk oral vaccine against poliomyelitis is developed by US microbiologist Jonas Salk (1914–95).

1953 Carcinogenic properties of tars from tobacco are discovered by US surgeon Evarts Graham (1883–1957).

1954 Human kidney transplant (between identical twins) is first performed by US surgeon Joseph Murray (1919–).

1954 Contraceptive pill is developed by Chinese biologist Min-Chueh Chang (1909–91), US physiologist Gregory Pincus (1903–67) and Polish-born US chemist Frank Colton (1923–), who patented the first pill Enovid.

1955 Fact that some viruses can be split into a protein and a nucleic acid (the infective part) is discovered by German-born US biochemist Heinz Fraenkel-Conrat (1910–).

1955 Structure of vitamin B12 (cyanocobalamin) is determined by British biochemist Dorothy Hodgkin (1910–94).

1956 Succesful bone marrow transplant is carried out by US physician Edward Thomas (1920–).

1957 Sabin vaccine against poliomyelitis is developed by Polish-born US microbiologist Albert Sabin (1906–93).

1957 Interferon is discovered by Scottish virologist Alick Isaacs (1921–67).

1958 Vaccine against measles is developed by US bacteriologist John Enders (1897–1985), and later (1962) put into quantity production.

1958 Sedative drug thalidomide is introduced in West Germany; later (1961) it is withdrawn because it causes foetal deformities.

1959 Abnormal chromosome responsible for Down's syndrome (trisomy 21) is discovered by French geneticist Jérôme Lejeune (1926–94).

1959 Cause of Burkitt's lymphoma (a cancer of the lymphatic system) is discovered by Irish physician Denis Burkitt (1911–93).

1959 External heart massage (as a first-aid technique) is introduced by US engineer William Kouwenhoven (1886–1975).

FARMING AND FOOD

1954 Non-stick cooking pans are invented by French research engineer Marc Grégoire.

1959 Tab-opening aluminium drinks can is invented in the USA by Ermal Fraze (patented in 1963).

TRANSPORT

1953 Hovercraft (air-cushion vehicle) is invented by British engineer Christopher Cockerell (1910–).

1953 Vertical take-off aircraft, the Rolls-Royce "Flying Bedstead", is tested in Britain.

1954 Swing-wing aeroplane is developed in the USA.

1954 Fuel injection is introduced for cars in Germany.

1955 Nuclear-powered submarine, Nautilus, is built in the USA; it later (1958) travels under the Arctic ice cap.

1956 Snowmobile is patented by Canadian inventor Joseph Bombardier.

1957 First turboprop airliner (Bristol Britannia) enters service.

1958 Folding bicycle, with small wheels and no crossbar, is invented by British engineer Alex Moulton.

1959 First commercial hovercraft (the SRN1) is built in Britain.

1959 First nuclear-powered surface ships are built in the Soviet Union (ice-breaker Lenin) and the USA (the merchant ship Savannah) and aircraft carrier USS Enterprise (1960); later vessels include the Japanese Mutsu (1967) and West German Otto Hahn (1969).

ENGINEERING AND TECHNOLOGY

1953 Maser is invented by US physicist Charles Townes (1915–).

1954 First silicon transistors are produced by Gordon Teal.

1954 Robot arm is invented by George Devol.

1954 Solar-powered battery is invented by Gerald Pearson (1905–) et al at the Bell Telephone company.

1955 Caesium atomic clock is constructed in Britain, and later (1960) in the USA by US physicist Norman Ramsey Jr (1915–).

1955 Artificial diamonds are produced by the General Electric company.

1955 Lego building blocks constructional toy is invented by Danish cabinet-maker Ole Christiansen.

1957 Semiconductor tunnel diode is invented by Japanese physicist Leo Esaki (1925–) of the Sony company.

1957 First intercontinental ballistic missile (ICBM) is constructed in the Soviet Union.

1957 Implantable artificial heart pacemaker is constructed by US electronics engineer Earl Bakken and, later (1959), Swedish physician Rune Elmqvist (1935–).

1958 Integrated circuits (first proposed in 1952 by British physicist G. Dunner) are produced by US electronics engineers Jack Kilby and Robert Noyce (1927–90), who later (1986) helped to found the Intel company.

1958 Float-glass process (for making plate glass) is invented by Englishman Alistair Pilkington (1920–95).

1958 "Superglue" (cyanoacrylate adhesive) is invented by US chemists Harry Cooper and Fred Joyner working for the Kodak company, which first marketed it as Eastman 910.

1959 First industrial robot (Unimate) is constructed in the USA.

1959 Practical high-power hydrogen-oxygen fuel cell is demonstrated by British physicist Francis Bacon (1904–92).

1959 Metallic glass is invented by Belgian-born US scientist Pol Duwez (1907–).

COMMUNICATIONS

1950 Floppy computer disk is patented by Japanese inventor Yoshiro Nakamata.

1951 UNIVAC I, the first commercially available electronic computer, which uses magnetic tape storage, is built by Americans John Eckert (1919–) and John Mauchly (1907–80).

1951 33-rpm long-playing gramophone record is introduced by the Deutsche Grammophon company.

1951 Stereoscopic (3D) cinema films, using polarized glasses, are produced.

1952 EDVAC (Electronic Discrete Variable Computer) is built in the USA by John Von Neumann et al.

1953 CinemaScope wide-screen motion pictures are introduced by the US Twentieth Century Fox company (using a single anamorphic lens invented by French optician Henri Crétien in 1925).

1953 Regular colour television transmissions begin in the USA.

1953 Magnetic core computer memory is invented by Chinese-born US computer engineer An Wang (1920–90).

1955 First transistor computer (TRADIC) is built in the USA.

1955 Fibre optics are demonstrated in the USA by Narinder Kapany (1927–), and in Japan.

1955 Electronic music synthesizer is invented by US radio engineer Harry Olson (1901–82).

1956 The videophone is invented in the USA.

1956 Transatlantic telephone cable (Scotland to Newfoundland) comes into commission.

1956 Videotape recorder is invented by Russian-born US engineer Alexander Poniatoff (1892–80) of the US Ampex company.

1956 FORTRAN, the first computer programming language, is devised by John Backus et al at the US IBM company.

1958 Stereo recordings become available in the USA and Britain.

1958 Modem, allowing binary data to be sent over telephone lines, is introduced by the US Bell company.

1958 Audio cassette tape recorder is developed in the USA by the RCA Victor company.

1959 Portable television using transistors is marketed in Japan.

ASTRONOMY AND SPACE

1960

1960 Planet is discovered orbiting the star Lelande 21185 by Dutch-born US astronomer Peter Van de Kamp (1901–), who later (1963) discovers a second non-Solar System planet orbiting Barnard's star.

1960 Quasars are predicted by US astronomers Thomas Matthews and Alexander Sandage (1926–).

1960 US Pioneer 5 space probe investigates the solar wind between the planets.

1961 Soviet space probe Venera 1 bypasses Venus, but contact is lost.

1962 US Ranger 4 space probe hits the Moon; Mariner 2 bypasses Venus.

1963 First quasar (3C 273) is discovered (because of its huge red shift) by Dutch-born US astronomer Maarten Schmidt (1929–).

1963 Strong source of X-rays in the constellation Scorpio is discovered by US astronomers Bruno Rossi (1905–94), Herbert Friedman (1916–) et al (using a detector carried by a space rocket), later (1966) identified as Scorpio X-1 by US astronomer Allan Sandage (1926–).

1964 Existence of microwave cosmic background radiation (evidence of the Big Bang) is predicted by US physicist Robert Dicke (1916–); it is later (1965) discovered by US radio astronomers Arno Penzias (1933–) and Robert Wilson (1936–).

1964 US Ranger 7 space probe sends back more than 4,000 photographs before crash-landing on the Moon.

1965 US Ranger 8 space probe sends back television pictures from the surface of the Moon; Mariner 4 bypasses Mars and sends back close-up photographs; Soviet Zond 3 bypasses the Moon and sends back photographs of the rear side.

1965 US Pioneer 6 space probe becomes the first of four consecutive Pioneers to monitor the Sun's activity.

1966 Soviet Luna 9 probe soft lands on the Moon; Luna 10 and 11 orbit the Moon, Luna 13 soft lands; Venera 3 crash-lands on Venus; US Lunar Orbiters 1 and 2 orbit the Moon, photographing possible landing sites; Surveyors 5 and 6 make soft landings on the Moon.

1966 Janus, a small moon of Saturn, is discovered by French astronomer Andouin Dollfus.

1966 Galaxy Cygnus A is discovered to be a powerful source of X-rays by US astrophysicist Herbert Friedman (1916–).

1967 First pulsar is discovered by British astronomers Jocelyn Bell (now Burnell) (1943–) and Antony Hewish (1924–).

1967 Ultraviolet-emitting Markarian galaxies are discovered by Soviet astronomer Benjamin Markarian.

1967 US Surveyors 3, 5 and 6 soft land on the Moon and Lunar Orbiters 3 and 4 photograph the surface; Mariner 5 makes close bypass of Venus; Soviet Venera 4 drops instruments into the atmosphere of Venus.

PHYSICS

1962 Muon neutrino is discovered by US physicists Leon Lederman (1922–), Melvin Schwartz (1932–) and Jack Steinberger (1921–).

1974 Leda, the thirteenth moon of Jupiter, is discovered by US astronomer Charles Kowal (1940–).

1964 Existence of the quark is proposed by US physicists Murray Gell-Mann (1929–) and George Zweig.

1964 Existence of an elementary particle (the Higgs boson) that accounts for the mass of other particles is proposed by British physicists Peter Higgs (1929–) and Thomas Kibble (1932–).

1964 Non-conservation of parity in certain reactions involving subatomic particles is discovered by US nuclear physicists James Cronin (1931–) and Val Fitch (1923–).

1964 Test to detect an interconnection between two widely separated subatomic particles that were once connected is devised by British physicist John Bell (1928–90).

1967 Fact that the electromagnetic force and the weak nuclear force are variations of a single force (the electroweak force) is proposed by US physicists Steven Weinberg (1933–) and Sheldon Glashow (1932–) and, independently, Pakistani physicist Abdus Salam (1926–).

CHEMISTRY

1960 Chlorophyll is synthesized by US chemist Robert Woodward (1917–79).

1961 Radioactive element lawrencium (symbol Lr, formerly Lw) is discovered by US chemist Albert Ghiorso et al.

1962 First compound of a rare gas, xenon heptafluoroplatinate, is made by British chemist Neil Bartlett (1932–).

1962 Solid phase method of synthesizing peptides and proteins (from amino acids) is discovered by US chemist Bruce Merrifield (1921–).

1962 Structure of human immunoglobulin is discovered by British biochemist Rodney Porter (1917–85).

1962 Crown ethers and cryptate metal complexes are discovered by Korean-born US chemist Charles Pedersen (1904–89).

1969 Structure of insulin is discovered by British chemist Dorothy Hodgkin (1910–94).

1969 Woodward-Hoffmann rules (concerning the behaviour of molecular orbitals during some organic reactions) are discovered by US chemist Robert Woodward (1917–79) and Polish-born US Roald Hoffmann (1937–)

BIOLOGY

reproduces parthenogenically (without the intervention of a male) is discovered in Armenia by Soviet biologist J. Darevsky.

1958 Human histocompatability system is discovered by French biologist J. Dausset.

1959 Method of staining and sorting human chromosomes is discovered by British biologist C. Ford.

1960 Messenger RNA (mRNA) is discovered by Sydney Brenner (1927–) and François Jacob (1920–).

1960 Sequence of amino acids (124 of them) in ribonuclease is discovered by biochemists S. Moore and W. Stein.

1960 Structure of haemoglobin is discovered by British biochemist John Kendrew (1917–1997).

1960 Chlorophyll is synthesized by US chemist Robert Woodward (1917–).

1960 Hormone parathormone (from the parathyroid gland) is isolated by Lyman Craig (1906–).

1960 Operons (genes that regulate other genes) are discovered by French bacteriologists Jacques Monod (1910–76) and François Jacob (1920).

1961 Genetic code of DNA is discovered by British biophysicist Francis Crick (1916–) and South-African born molecular biologist Sydney Brenner (1927–).

1961 Role of ATP in energy transfer in chloroplasts (in plant cells) and mitochondria (in animal cells) is discovered by British biochemist Peter Mitchell (1920–92).

1962 Role of the thymus gland in establishing the immune system in young animals is discovered by French-born Australian physician Jacques Miller (1931–).

1962 Fact that natural selection applies to both K type and R type animal species is discovered by Canadian ecologist Robert Macarthur (1930–71).

1965 Nucleotide sequence of transfer-RNA is discovered by US biochemist Robert Holley (1922–93).

1965 Insulin is first synthesized

1966 Human growth hormone is discovered by Chinese-born US biochemist Choh Hao Li (1913–); he later (1970) synthesizes it.

1967 Method of synthesizing biologically active DNA is discovered by US biochemist Arthur Kornberg (1918–).

1968 Enzyme (produced by bacteria) that selectively cuts viral DNA is discovered by Swiss microbiologist Werner Arber (1929–).

1969 Amino acid sequence of immunoglobulin is discovered by US Biochemist Gerald Edelman (1929–).

MEDICINE

1960 Metal-and-plastic artificial hip joints are fitted by British orthopaedic surgeon John Charnley (1911–82), who later (1963) settled on a combination of polished stainless steel and high molecular weight polyethylene (HMWP).

1962 Killed-virus vaccine against German measles (rubella) is developed by US biologist Thomas Weller (1915–).

1962 Successful kidney transplant from an unrelated donor is first performed by US surgeon Joseph Murray (1919–) and between nonidentical twins by French physician Jean Hamberger (1909–92), who later (1963) pioneered immunosuppressive therapy and the use of donor kidneys from cadavers.

1963 Liver transplant is first performed by US surgeons Thomas Starzl (1926–) and Francis Moore (1913–).

1964 Beta-blocker drug propranalol, used to treat heart disorders, is discovered by British pharmacologist James Black (1924–).

1964 First medical tests in space are carried out in orbiting Soviet spacecraft Voskod 1 by Soviet physician Boris Yegorov (1937–94).

1966 Live-virus vaccine against German measles (rubella) is developed by US bacteriologists Harry Meyer Jr (1928–) and Paul Parkman (1932–).

1967 Successful human heart transplant is carried out by South African surgeon Christiaan Barnard (1922–) and, later (1968), US surgeon Norman Shumway (1923–).

1967 Coronary bypass operation is introduced by US cardiovascular surgeon Rene Favaloro (1923–).

FARMING AND FOOD

1963 Process for preserving food using radiation is developed in Britain.

1967 Domestic microwave oven is marketed by the Raytheon Company.

TRANSPORT

1960 US nuclear submarine Triton travels round the world underwater.

1961 Orbital flight round the Earth is made by Soviet cosmonaut Yuri Gagarin (1934–1968).

1963 Wankel-engined car is produced in Japan.

1963 First diesel-engined passenger car is produced in England by the Rolls-Royce Company.

1964 Container cargo ships are introduced.

1964 "Bullet" trains, running at 210 kilometres per hour, are intoduced by Japan National Railways.

1968 First supersonic airliner, the Soviet TU-144 designed by Andrei Tupolev (1888–1972), flies in the Soviet Union. The Anglo-French Concorde has its maiden flight two months later in early 1969.

1968 First supertankers go into service.

ENGINEERING AND TECHNOLOGY

1960 Ruby laser is constructed by US physicist Theodore Maiman (1927–), using principles worked out by his compatriot Arthur Schawlow (1921–) and his brother-in-law Charles Townes (1915–).

1961 Silicon chip is patented in the USA.

1961 Field-effect transistor is invented by Steven Hofstein.

1963 CAT (computerized axial tomography) X-ray scanner is invented by South African-born US physicist Allan Macleod (1924–) and later (1971), independently, by British engineer Godfrey Hounsfield (1919–).

1963 Friction welding is invented in the Soviet Union.

1964 Carbon fibre is developed in the USA and Britain.

1966 First tidal power station is commissioned on a barrage across the River Rance, France.

1967 Cash dispensing machine is invented by Englishman John Shepherd-Barron.

1967 First quartz wrist watch is produced in Japan.

1968 "Hydromassage" Whirlpool bath is invented by Italian engineer Candido Jacuzzi (1903–86).

COMMUNICATIONS

1960 Passive communications satellite Echo is launched into Earth orbit.

1960 Fibre-tipped pen (Pentel) is marketed by the Japan Stationery company.

1961 Golfball typewriter (the Selectric) is marketed in the USA by the IBM company.

1962 Active communications satellite, Telstar, allows live transatlantic television transmissions.

1963 Audio cassette tapes are first marketed.

1963 Electronic calculator is produced by the Bell Punch company in the USA.

1963 First minicomputer is made by Digital Equipment (DEC) in the USA.

1963 Instamatic camera, using film cartridges, is launched in the USA by the Kodak company.

1963 Cibachrome, a photographic paper for making colour prints directly from transparencies, is marketed by the Swiss Ciba-Geigy company.

1964 First word processor is introduced in the USA by the IBM company.

1964 ASCII character codes are adopted by the US Standard Association and later (1966) by the International Standard Organization.

1964 Moog synthesizer (of music) is invented by US electronics engineer Robert Moog (1934–).

1965 BASIC computer language is invented by John Kemeny and Thomas Kurtz (1928–).

1965 Commercial (Intelsat) geostationary communications satellite, Early Bird, is placed in Earth orbit.

1965 Computer typesetting is introduced in Germany.

1966 Fibre-optic telephone cable is invented by British engineers Charles Kao and George Hockham of Standard Telecommunications Laboratories.

1966 Laser radar is invented in Japan.

1967 Noise-reduction system for sound recording is invented by US engineer Ray Dolby (1933–).

1968 Bubble memories, which retain data even when the computer is switched off, are invented.

1968 Computer "mouse" input device is invented by US computer engineer Douglas Engelbart (1925–).

1969 Live television pictures are sent to Earth by US astronauts on the Moon.

ASTRONOMY AND SPACE

1968 US Surveyor 7 soft lands on the Moon and surveys the surface. Soviet Zond 5 and 6 orbit the Moon and return to Earth.

1968 US Orbiting Astronomical Observatory (OAO-2), an ultravioilet telescope, is launched, followed (1972) by OAO-3 (Copernicus).

1969 Soviet Venera 5 and 6 land on Venus. US astronauts from Apollos 11, and 12 land on the Moon; Mariners 6 and 7 fly by Mars, photographing the surface.

1970

1970 Soviet Luna 16 with Lunokhod 1 robot spacecraft soft lands on the Moon; Venera 7 soft lands on Venus.

1971 US Apollo 14 and 15 astronauts land on the Moon.

1971 Soviet Mars 2 and 3, and US Mariner 9, all reach Mars; Soviet Venera 8 soft lands on Venus.

1971 First orbital space station, Salyut 1, is put into orbit by the Soviet Union; later (1973) the USA launches the space station Skylab I into orbit, which returns to Earth in 1979.

1972 US Apollo 16 and 17 Moon shots land astronauts on the Moon, who set up an astronomical observatory and return with rock samples; Soviet Veneras 9 and 10 probes soft land on Venus.

1973 Soviet Luna 21 soft lands on the Moon and deploys the Lunokhod II Moon rover; US Pioneer 10 flies close by Jupiter before later (1983) leaving the Solar System.

1974 US Mariner 10 sends back close-up pictures of Venus and Mercury; Pioneer 11 passes close to Jupiter on its way to Saturn; Soviet Mars 4, 5, 6 and 7 all each Mars.

1974 Leda, the thirteenth moon of Jupiter, is discovered by US astronomer Charles Kowal (1940–).

1975 Soviet Veneras 9 and 10 soft land on Venus and send back pictures.

1975 US Apollo 18 and Soviet Soyuz 19 spacecraft link up in Earth orbit.

1975 US COS B gamma-ray satellite is launched into Earth orbit.

1975 The proper motion of the Milky Way is discovered by US astronomers V. Rubin and W. Kent.

1976 US space probes Viking 1 and 2 land on Mars and send back pictures.

1977 Asteroid or cometary nucleus Chiron, which has the largest known orbit of any object in the Solar System, is discovered by US astronomer Charles Kowal (1940–).

1978 Charon, the moon of Pluto, is discovered by US astronomers James Christy and Robert Harrington.

1978 US space probe Pioneer-Venus 1 and 2 land on Venus.

1978 Joint US-European Space Agency International Ultraviolet Explorer (IUE) satellite is launched into Earth orbit.

1979 US space probe Pioneer 11 sends back close-up pictures of Saturn and discovers an eleventh moon, before later (1990) leaving the Solar System; Voyagers 1 and 2 photograph the moons of Jupiter and discover a faint ring system.

PHYSICS

1974 J-psi particle (a type of meson) is discovered by US physicists Samuel Ting (1936–) and, independently, Burton Richter (1931–).

1977 Upsilon particle is discovered by US physicist Leon Lederman (1922–).

1977 Mechanism of the quantum Hall effect is discovered by German physicist Klaus von Klitzing (1943–).

CHEMISTRY

BIOLOGY

1970 Hormone LRF (luteinizing releasing factor) from the hypothalamus is discovered by French physiologist Henri Guillemin and, independently, Polish scientist Andrew Schally.

1970 Enzyme reverse transcriptase (which transcribes RNA into DNA in some viruses) is discovered by US virologists Howard Temin (1934–94) and David Baltimore (1938–).

1970 Restriction enzymes are discovered by US molecular biologist Hamilton Smith (1931–).

1971 Theory of punctuated equilibrium (that evolution takes place in short "bursts") is proposed by US palaeontologist Stephen Gould (1941–).

1972 Hybrid DNA is made by splicing bacterial and viral DNA by US molecular biologist Paul Berg (1926–).

1973 Technique of using restriction enzymes to "cut and splice" DNA (i.e. using recombinant DNA in genetic engineering) is discovered by US biochemists Stanley Cohen (1935–) and Herbert Boyer (1936–).

1974 Fossils of Australopithcus afarensis, the oldest humanoid fossil yet found, are discovered in Ethiopia by US paleoanthropologist Donald Johanson (1943–).

1976 Endorphins (pain-relieving chemicals in the brain) are discovered by French physiologist Roger Guillemin (1924–).

1976 First functional artificial gene is made by Indian-born US chemist Har Khorana (1922–) et al.

1977 Full sequences of bases in DNA is discovered by British biochemist Frederick Sanger (1918–).

1977 Introns (DNA sequences that do not code for proteins) are discovered by British biochemists Alec Jeffreys (1950–) and R. Flavell (1945–).

1979 Gaia hypothesis, which regards the whole Earth as a single organism, is proposed by British scientist James Lovelock (1919–).

MEDICINE

1972 Drug cimetridine, which blocks acid-producing sites in the stomach and thereby aids healing of ulcers, is discovered by British pharmacologist James Black (1924–).

1975 Technique for large-scale production of monoclonal antibodies is discovered by Argentinian-born British molecular biologist César Milstein (1927–) and Georges Köhler (1926–95).

1976 Legionnaire's disease is recognized for the first time, in Philadelphia, USA.

1976 Mechanism by which dormant oncogenes initiate the devlopment of cancer is discovered by US virologists John Bishop (1936–) and Harold Varmus (1939–).

1977 Smallpox virus becomes extinct in the wild; World Health Organization later (1979) announces that smallpox has been eliminated worldwide.

1977 Balloon angioplasty (for dilating and thereby unblocking a constricted artery) is introduced by German surgeon Andreas Grüntzig (1939–85).

1978 World's first test-tube baby, Louise Brown, is born in England.

1979 AIDS (acquired immune deficiency syndrome) is first diagnosed, in the USA.

1970

FARMING AND FOOD

1972 Perrier mineral water, from Vergèze, France, is marketed internationally.

TRANSPORT

1970 The Boeing 747 jumbo jet enters service.

1972 Self-sealing tyres are introduced by the Dunlop company.

1972 First plastic-hulled warship, the Royal Navy's minehunter HMS Wilton, is launched.

1972 European A300 Airbus enters service.

1974 Catalytic converter for car exhausts is developed by US General Motors company.

1975 Hang glider is perfected in the USA by Francis Rogallo.

1976 Anglo-French supersonic Concorde airliner enters service.

1977 Human-powered aircraft Gossamer Condor, designed and piloted by American Paul MacCready (1925–), makes its first flight.

1978 Transatlantic crossing by balloon, the helium-filled Double Eagle II, is made by a US team.

1979 Maglev test vehicle of Japan National Railways reaches a speed of 517 kilometres per hour.

1979 Human-powered aircraft Gossamer Albatross, piloted by American Bryen Allen, makes a crossing of the English Channel.

ENGINEERING AND TECHNOLOGY

1970 Commercial LED (light-emitting diode) is patented in Switzerland by F. Hoffman.

1972 Orbital two-stroke reciprocating piston engine is invented by Australian engineer Ralph Sarich (1939–).

1972 Appartus for converting solar energy into power is invented (in his hundredth year) by US astrophysicist Charles Abbot (1872–1973).

1974 Holographic electron microscope is invented in the USA.

1978 Scanning ion microscope is invented by Robert Seliger et al of the Hughes Research Laboratory.

COMMUNICATIONS

1970 Floppy disks (originally 20 centimetres across and holding 240 kilobytes) are introduced for data storage by the American IBM company.

1970 Daisy wheel printer is introduced.

1970 Videodisc systems are produced by AEG-Telefunken in Germandy and Decca in Britain.

1971 Pocket calculator is marketed in the USA by Texas Instruments.

1971 First computer microprocessor, the Intel 4004, is invented by US engineer Ted Hoff (1937–), although covered by previous 1968 patent of Gilbert Hyatt.

1971 Dot matrix printer is introduced by the US Centron company.

1971 PASCAL computer language is invented by Niklaus Wirth.

1972 Dedicated word processers are introduced.

1972 Laserdisk recording system, called Laservision, is developed by the Dutch Philips company.

1973 Teletext is introduced in Britain by BBC Television (Ceefax) and Independent Televison (Oracle).

1973 Image intensifier, for use as a television camera tube in poor light, is invented in Germany.

1973 Home computer video game (Pong or Tele-Tennis) is invented by US engineer Nolan Bushnell (1943–) of the US Atari company.

1974 Programmable pocket calculator is introduced in the USA.

1974 Intel 8080 microprocessor is launched; a general-use operating system for it (CP/M) is designed by US software engineer Gary Kildall (1942–).

1974 Bar codes and bar-code readers are introduced into supermarkets in the USA.

1975 Altair 8800, the first personal computer (PC), is marketed in kit form in the USA.

1975 Prestel, the world's first viewdata system on television, is launched by the British Post Office.

1975 Laser printer is introduced by the US IBM company.

1975 Betamax domestic videocassette system is introduced by the Sony company.

1975 Liquid-crystal digital displays are marketed in Britain.

1976 Ink-jet printer is introduced by the US IBM company.

1976 First camera with a microprocessor-controlled exposure system, the Canon AE-1, is launched.

1976 Videocassette recorder is invented by Japanese electronics engineer Shizuo Takano (1923–92).

1976 VHS (video home system) format

ASTRONOMY AND SPACE

1980

1980 Inflationary Universe theory (an extension of the Big Bang theory) is proposed by A. Guth.

1980 US space probe Voyager 1 flies close to Saturn, discovering new moons; Soviet probes Venera 11 and 12 land on Venus.

1981 R 136a, a star 2,500 times as massive as the Sun, is discovered by US astronomers J. Cassinelli et al.

1981 US space probe Voyager 2 sends back close-up pictures of Saturn.

1981 US space shuttle Columbia makes its first flight.

1982 Soviet Veneras 13 and 14 make soft landings on Venus and send pack colour photographs.

1983 Joint US-British-Dutch InfraRed Astronomy Satellite (IRAS) is launched into Earth orbit.

1983 Soviet Venera 15 orbits Venus, making radar maps of the surface; Venera 16 repeats the mission in 1984.

1984 Two more rings of Saturn are discovered by Indian astronomers J. Bhattacharya et al.

1985 Soviet space probes Vega 1 and 2 bypass Venus en route to Halley's comet.

1986 US Voyager 2 bypasses Uranus, discovering several new moons.

1986 European Space Agency probe Giotto sends back close-up pictures of Halley's comet; Soviet probes Vega 1 and Vega 2 pass through the comet's coma; Japanese probes Susisei and Sakigake fly by the comet.

1987 Pisces-Cetus Supercluster Complex is discovered by B. Tally.

1987 Supernova SN1987A flares up, the first to be visible to the naked eye since 1604.

1989 US space probe Voyager 2 passes close by Neptune and discovers six new moons (including Proteus) and a ring system.

1989 US space probe Magellan is launched towards Venus from the space shuttle Atlantis; a few months later Galileo is launched towards Jupiter, also from Atlantis.

1989 European Space Agency astrometry satellite Hipparcos is launched into Earth orbit.

PHYSICS

1983 W and Z particles ("weakons") are discovered by Italian physicist Carlo Rubbia (1934–) and Dutch physicist Simon van der Meer (1925–).

1986 Superconductors that work at relatively high temperatures (-243°C) are discovered by German physicist Georg Bednorz (1950–) and Swiss physicist Karl Müller (1927–); later (1987) a higher temperature (-196°C) is achieved by Chinese physicist Ching-Wu Chu (1941–).

1986 Brief controlled production of energy by laser-induced nuclear fusion is achieved by the Nova device at the Lawrence Livermore National Laboratory.

1989 Large Electron Positron Collider (LEP) particle accelerator comes into operation at the CERN laboratories near Geneva, Switzeralnd, and proves the existence of the Z particle.

1989 Production of energy by cold nuclear fusion is announced by US phsyicists Martin Fleischmann (1927–) and Stanley Pons (1943–), but the results were not substantiated and were refuted by Austrian-born US physicist Harold Furth (1930–).

CHEMISTRY

1985 Buckminsterfullerenes, allotropes of carbon whose molecules consist of spherical structures of 60 or more carbon atoms, are discovered by Harold Kroto and David Walton.

BIOLOGY

1981 Genes are transplanted between different organisms by geneticists at Ohio University.

1982 Bacteria that live in very hot (105°C) sea water are discovered by German biologist Karl Setter.

1984 Fossil skeleton of *Homo erectus* is found in Africa by British archaeologist Richard Leaky (1944–).

1984 Clone of a lamb is produced.

1984 Human Genone Project is proposed by Robert Sinsheimer of the University of California; it is launched in 1989.

1985 Genetic fingerprinting technique is devised by British biochemist Alec Jeffreys (1950–).

1985 Retinoblastoma, the first human cancer gene, is isolated in the United States.

1987 Fossilized dinosaur eggs containing embryos are discovered by Canadian palaeontologist Kevin Aulenback.

MEDICINE

1980 Device to break up kidney stones using ultrasound is introduced by the German Dornier Medical Systems company.

1982 Prions (virus-like infective agents consisting simply of proteins) are discovered by US molecular biologist Stanley Prusiner.

1982 Genetically-engineered human insulin (for treating diabetes) is marketed by the US Eli Lilly company.

1983 Human immunodeficiency virus (HIV), responsible for AIDS, is discovered by French virologist Luc Montagnier (1932–) and, independently (1984), US Robert Gallo (1937–).

1984 Vaccine against leprosy becomes available.

1986 Defective gene responsible for Duchenne muscular dystrophy is discovered by US geneticist Louis Kunkel (1949–) et al.